2B (12)	3A (13)	4A (14)	5A (15)	6A (16)	7A (17)	8A (18)
						Helium 2 He 4.0026
	Boron 5 B 10.811	Carbon 6 C 12.011	Nitrogen 7 N 14.0067	Oxygen 8 O 15.9994	Fluorine 9 F 18.9984	Neon 10 Ne 20.1797
	Aluminum 13 Al 26.9815	Silicon 14 Si 28.0855	Phosphorus 15 P 30.9738	Sulfur 16 S 32.066	Chlorine 17 Cl 35.4527	Argon 18 Ar 39.948
Zinc 30 Zn 65.38	Gallium 31 Ga 69.723	Germanium 32 Ge 72.61	Arsenic 33 As 74.9216	Selenium 34 Se 78.96	Bromine 35 Br 79.904	Krypton 36 Kr 83.80
Cadmium 48 Cd 112.411	Indium 49 In 114.818	Tin 50 Sn 118.710	Antimony 51 Sb 121.760	Tellurium 52 Te 127.60	Iodine 53 I 126.9045	Xenon 54 Xe 131.29
Mercury 80 Hg 200.59	Thallium 81 Tl 204.3833	Lead 82 Pb 207.2	Bismuth 83 Bi 208.9804	Polonium 84 Po (208.98)	Astatine 85 At (209.99)	Radon 86 Rn (222.02)
— 112 — Discovered 1996	— 113 — Discovered 2004	— 114 — Discovered 1999	— 115 — Discovered 2004	— 116 — Discovered 1999		— 118 — Discovered 2002

Terbium 65 Tb 158.9254	Dysprosium 66 Dy 162.50	Holmium 67 Ho 164.9303	Erbium 68 Er 167.26	Thulium 69 Tm 168.9342	Ytterbium 70 Yb 173.054	Lutetium 71 Lu 174.9668
Berkelium 97 Bk (247.07)	Californium 98 Cf (251.08)	Einsteinium 99 Es (252.08)	Fermium 100 Fm (257.10)	Mendelevium 101 Md (258.10)	Nobelium 102 No (259.10)	Lawrencium 103 Lr (262.11)

SEVENTH EDITION

CHEMISTRY
& Chemical Reactivity

John C. Kotz
SUNY Distinguished Teaching Professor
State University of New York
College of Oneonta

Paul M. Treichel
Professor of Chemistry
University of Wisconsin–Madison

John R. Townsend
Associate Professor
West Chester University of Pennsylvania

THOMSON
BROOKS/COLE

Australia • Canada • Mexico • Singapore • Spain • United Kingdom • United States

THOMSON

BROOKS/COLE

Senior Acquisitions Editor: Lisa Lockwood
Senior Development Editor: Peter McGahey
Assistant Editor: Ashley Summers
Editorial Assistant: Liz Woods
Technology Project Manager: Lisa Weber
Marketing Manager: Amee Mosley
Marketing Assistant: Elizabeth Wong
Marketing Communications Manager: Talia Wise
Project Manager, Editorial Production: Teresa L. Trego
Creative Director: Rob Hugel
Art Director: John Walker
Print Buyer: Rebecca Cross
Permissions Editor: Mari Masalin-Cooper
Production Service: Graphic World Inc.
Text Designer: Brian Salisbury
Photo Researcher: Marcy Lunetta
Copy Editor: Graphic World Inc.
Illustrators: Patrick A. Harman and Graphic World Inc.
OWL Producers: Stephen Battisti, Cindy Stein, David Hart (Center for Educational Software Development, University of Massachusetts, Amherst)
Cover Designer: John Walker
Cover Image: Felice Frankel, Harvard University
Compositor: Graphic World Inc.

Thomson Higher Education
10 Davis Drive
Belmont, CA 94002-3098
USA

Library of Congress Control Number: 2007940546

Student Edition
ISBN-13: 978-0-495-38703-9
ISBN-10: 0-495-38703-7

Volume 1
ISBN-13: 978-0-495-38711-4
ISBN-10: 0-495-38711-8

Volume 2
ISBN-13: 978-0-495-38712-1
ISBN-10: 0-495-38712-6

Printed in Canada
1 2 3 4 5 6 7 11 10 09 08

CREDITS

This page constitutes an extension of the copyright page. We have made every effort to trace the ownership of all copyrighted material and to secure permission from copyright holders. In the event of any question arising as to the use of any material, we will be pleased to make the necessary corrections in future printings. Thanks are due to the following authors, publishers, and agents for permission to use the material indicated.

264: Based on L. Schlarbach and A. Zuttle: Nature, Vol. 414, pp. 353-358, 2001;

667: Reprinted with permission of Dr. Klaus Hermann of the Fritz Haber Institution;

961: From www.acs.org. Copyright © American Chemical Society. Reprinted with permission from the American Chemical Society.

Brief Contents

Contents

This text is available in these student versions:
- Complete text ISBN 978-0-495-38703-9 • Volume 1 (Chapters 1–11) ISBN 978-0-495-38711-4
- Volume 2 (Chapters 11–23) ISBN 978-0-495-38712-1 • Two-Volume set ISBN 978-0-495-63323-5

5 Principles of Chemical Reactivity: Energy and Chemical Reactions | 208

INTERCHAPTER

The Chemistry of Fuels and Energy Resources | 254

PART 2
ATOMS AND MOLECULES

6 The Structure of Atoms | 268

23 Nuclear Chemistry | 1060

A Appendices | A-1

Index/Glossary | I-1

Go Chemistry Modules

The new Go Chemistry modules are mini video lectures included in ChemistryNow that are designed for portable use on video iPods, iPhones, MP3 players, and iTunes. Modules are referenced in the text and may include animations, problems, or e-Flashcards for quick review of key concepts. Modules may also be purchased at **www.ichapters.com.**

Chapter 1	Basic Concepts of Chemistry	Module 1	Exploring the Periodic Table
Chapter 2	Atoms, Molecules, and Ions	Module 2	Ion Charges
		Module 3	Naming: Names to Formulas of Ionic Compounds
		Module 4	The Mole
Chapter 3	Chemical Reactions	Module 5	Solubility of Ionic Compounds
		Module 6	Net Ionic Equations
Chapter 4	Stoichiometry: Quantitative Information About Chemical Reactions	Module 7	Simple Stoichiometry
		Module 8a	Limiting Reactants – part 1
		Module 8b	Limiting Reactants – part 2
		Module 9a	pH – part 1
		Module 9b	pH – part 2
Chapter 5	Principles of Chemical Reactivity: Energy and Chemical Reactions	Module 10	Using Hess's Law
Chapter 7	The Structure of Atoms and Periodic Trends	Module 11	Periodic Trends
Chapter 8	Bonding and Molecular Structure	Module 12	Lewis Electron Dot Structures
		Module 13	Molecular Polarity
Chapter 9	Bonding and Molecular Structure: Orbital Hybridization and Molecular Orbitals	Module 14	Hybrid Orbitals
Chapter 10	Carbon: More Than Just Another Element	Module 15	Naming Organic Compounds
Chapter 11	Gases and Their Properties	Module 16	The Gas Laws and Kinetic Molecular Theory
Chapter 12	Intermolecular Forces and Liquids	Module 17	Identifying Intermolecular Forces
Chapter 13	The Chemistry of Solids	Module 18	Unit Cells and Compound Formulas
Chapter 14	Solutions and Their Behavior	Module 19	Colligative Properties
Chapter 15	Chemical Kinetics: The Rates of Chemical Reactions	Module 20	Half-Life and the Integrated First Order Equation
Chapter 16	Principles of Reactivity: Chemical Equilibria	Module 21	Solving an Equilibrium Problem
Chapter 17	The Chemistry of Acids and Bases	Module 22	Equilibrium – pH of a Weak Acid
Chapter 18	Principles of Reactivity: Other Aspects of Aqueous Equilibria	Module 23	Understanding Buffers
Chapter 19	Principles of Reactivity: Entropy and Free Energy	Module 24	Free Energy and Equilibrium
Chapter 20	Principles of Reactivity: Electron Transfer Reactions	Module 25	Balancing Redox Equations

Preface

The authors of this book have more than 100 years of experience teaching general chemistry and other areas of chemistry at the college level. Although we have been at different institutions during our careers, we share several goals in common. One is to provide a broad overview of the principles of chemistry, the reactivity of the chemical elements and their compounds, and the applications of chemistry. To reach that goal with our students, we have tried to show the close relation between the observations chemists make of chemical and physical changes in the laboratory and in nature and the way these changes are viewed at the atomic and molecular level.

Another of our goals has been to convey a sense of chemistry as a field that not only has a lively history but also one that is currently dynamic, with important new developments occurring every year. Furthermore, we want to provide some insight into the chemical aspects of the world around us. Indeed, a major objective of this book is to provide the tools needed for you to function as a chemically literate citizen. Learning something of the chemical world is just as important as understanding some basic mathematics and biology and as important as having an appreciation for history, music, and literature. For example, you should know what materials are important to our economy, some of the reactions in plants and animals and in our environment, and the role that chemists play in protecting the environment.

These goals and our approach have been translated into *Chemistry & Chemical Reactivity*, a book that has been used by more than 1 million students in its first six editions. We are clearly gratified by this success. But, at the same time, we know that the details of our presentation and organization can always be improved. In addition, there are significant advances in the technology of communicating information, and we want to take advantage of those new approaches. These have been the impetus behind the preparation of this new edition, which incorporates a new organization of material, new ways to describe contemporary uses of chemistry, new technologies, and improved integration with existing technologies.

Emerging Developments in Content Usage and Delivery: OWL, the e-Book, and Go Chemistry™

The use of media, presentation tools, and homework management tools has expanded significantly in the last 3 years. More than 10 years ago we incorporated electronic media into this text with the first edition of our interactive CD-ROM, a learning tool used by thousands of students worldwide.

Multimedia technology has evolved over the past 10 years, and so have our students. Our challenge as authors and educators is to use our students' focus on assessment as a way to help them reach a higher level of conceptual understanding. In light of this we have made major changes in our integrated media program. We have redesigned the media so that students now have the opportunity to interact with media based on clearly stated chapter goals that are correlated to end-of-chapter questions. This has been achieved through ***OWL (Online Web-based Learning)***, a system developed at the University of Massachusetts and in use by general chemistry students for more than 10 years. In the past few years the system has been used successfully by over 100,000 students.

In addition, as outlined in *What's New in this Edition*, the ***electronic book (e-book)*** has been enhanced for this edition, and we have developed new ***Go Chemistry*** modules that consist of mini-lectures of the most important aspect in each chapter.

Audience for *Chemistry & Chemical Reactivity* and OWL

The textbook and OWL are designed for introductory courses in chemistry for students interested in further study in science, whether that science is chemistry, biology, engineering, geology, physics, or related subjects. Our assumption is that students beginning this course have had some preparation in algebra and in general

What's New in This Edition

1. ***New chapter introductions*** on topics such as altitude sickness (page 514) and the contribution of ethanol to environmental goals (page 860). Each of these chapter-opening topics has a question or two that is answered in Appendix Q.

2. One or more ***Case Studies*** are presented in each chapter. These cover practical chemistry and pose questions that can be answered using the concepts of that chapter. Case Studies cover such topics as silver in washing machines (page 148), using isotopes to catch cheaters (page 58), aquarium chemistry (page 992), why garlic stinks (page 541), what is in those French fries (page 96), why Beethoven died at an early age (page 989), and many others.

3. New and completely revised ***Interchapters.*** John Emsley, a noted science writer, revised the interchapter on the environment (page 949) and wrote a new interchapter on the history of chemistry (page 338).
4. ***Reorganization/addition of material:***
   - The first four chapters in particular have been revised and condensed.
   - The "moles of reaction" concept is used in thermodynamics.
   - The material on intermolecular forces (Chapter 12) has been separated from solids (Chapter 13).
   - The chapter on entropy and free energy (Chapter 19) has been thoroughly revised.
   - A brief discussion of modern organometallic chemistry has been added to Chapter 22.
   - Additional challenging questions have been added to each chapter.
   - Additional *Chemical Perspectives* and *Case Studies* boxes have been authored by Jeffrey Kaeffaber (University of Florida) and Eric Scerri (UCLA).
5. The ***OWL*** (Online Web-based Learning) system has been used by over 100,000 students. The contents of OWL are the contents and organization of *Chemistry & Chemical Reactivity*. For the sixth edition, about 20 end-of-chapter questions were assignable in OWL. That number has been approximately doubled for the seventh edition. In addition, the assets of ChemistryNow—Exercises, Tutorials, and Simulations that allow students to practice chemistry—are now fully incorporated in OWL.
6. The new ***e-Book in OWL*** is a complete electronic version of the text, fully assignable and linked to OWL homework. The e-book can be purchased with the printed book or as an independent text replacement.
7. ***Go Chemistry*** modules. There are 27 mini-lectures that can be played on an iPod or other MP3 player or on a computer. The modules feature narrated examples of the most important material from each chapter and focus on areas in which we know from experience that students may need extra help.

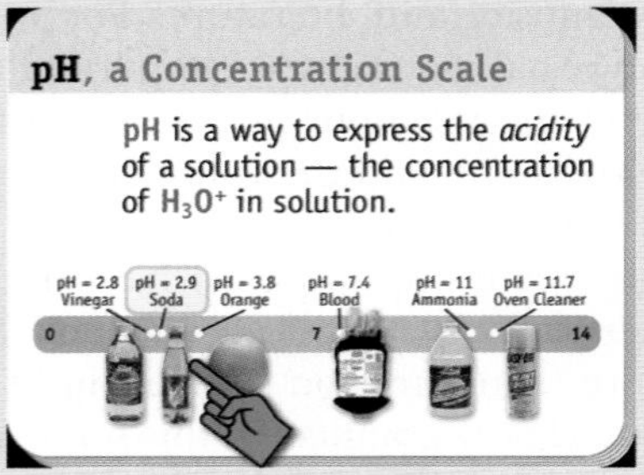

9. ***How Do I Solve It?*** modules in OWL help students learn how to approach the types of questions asked in each chapter.
8. ***In the Laboratory*** end-of-chapter Study Questions. These questions pertain directly to situations that the student may confront in a typical laboratory experiment.

science. Although undeniably helpful, a previous exposure to chemistry is neither assumed nor required.

Philosophy and Approach of the Chemistry & Chemical Reactivity Program

We have had several major, but not independent, objectives since the first edition of the book. The first was to write a book that students would enjoy reading and that would offer, at a reasonable level of rigor, chemistry and chemical principles in a format and organization typical of college and university courses today. Second, we wanted to convey the utility and importance of chemistry by introducing the properties of the elements, their compounds, and their reactions as early as possible and by focusing the discussion as much as possible on these subjects. Finally, with the new ***Go Chemistry*** modules and even more complete integration of ***OWL***, we wanted to give students new and proven tools to bring them to a higher level of conceptual understanding.

The American Chemical Society has been urging educators to put "chemistry" back into introductory chemistry courses. We agree wholeheartedly. Therefore, we have tried to describe the elements, their compounds, and their reactions as early and as often as possible by:

- Using numerous **color photographs** of reactions occurring, of the elements and common compounds, and of common laboratory operations and industrial processes.
- Bringing **material on the properties of elements and compounds** as early as possible into the Exercises and Study Questions and to introduce new principles using realistic chemical situations.
- Introducing each chapter with a **problem in practical chemistry**—a short discussion of the color of an aurora borealis or ethanol in gasoline—that is relevant to the chapter.
- Introducing ***Case Studies*** on practical chemistry.

General Organization of the Book and Its Features

Chemistry & Chemical Reactivity has two overarching themes: *Chemical Reactivity* and *Bonding and Molecular Structure.* The chapters on *Principles of Reactivity* introduce the factors that lead chemical reactions to be successful in converting reactants to products. Thus, under this topic there is a discussion of common types of reactions, the energy involved in reactions, and the factors that affect the speed of a reaction. One reason for the enormous advances in chemistry and molecular biology in the last several decades has been an understanding of molecular structure. Therefore, sections of the book on *Principles of Bonding and Molecular Structure* lay the groundwork for understanding these developments. Particular attention is paid to an understanding of the structural aspects of such biologically important molecules as DNA.

Flexibility of Chapter Organization

A glance at the introductory chemistry texts currently available shows that there is a generally common order of topics used by educators. With a few minor variations, we have followed that order as well. That is not to say that the chapters in our book cannot be used in some other order. We have written it to be as flexible as possible. The most important example is the chapter on the behavior of gases (Chapter 11), which is placed with chapters on liquids, solids, and solutions (Chapters 12–14) because it logically fits with these topics. It can easily be read and understood, however, after covering only the first four or five chapters of the book.

Similarly, chapters on atomic and molecular structure (Chapters 6–9) could be used before the chapters on stoichiometry and common reactions (Chapters 3 and 4). Also, the chapters on chemical equilibria (Chapters 16–18) can be covered before those on solutions and kinetics (Chapters 14 and 15).

Organic chemistry (Chapter 10) is often left to one of the final chapters in chemistry textbooks. However, we believe the importance of organic compounds in biochemistry and in consumer products means we should present that material earlier in the sequence of chapters. Therefore, it follows the chapters on structure and bonding because organic chemistry nicely illustrates the application of models of chemical bonding and molecular structure. However, one can use the remainder of the book without including this chapter.

The order of topics in the text was also devised to introduce as early as possible the background required for the laboratory experiments usually performed in introductory chemistry courses. For this reason, chapters on chemical and physical properties, common reaction types, and stoichiometry begin the book. In addition, because an understanding of energy is so important in the study of chemistry, thermochemistry is introduced in Chapter 5.

Interchapters

In addition to the regular chapters, uses and applications of chemistry are described in more detail in supplemental chapters on *The Chemistry of Fuels and Energy Sources; Milestones in the Development of Chemistry and the Modern View of Atoms and Molecules; The Chemistry of Life: Biochemistry; The Chemistry of Modern Materials;* and *The Chemistry of the Environment.*

Other Book Sections

As in the sixth edition, we continue with boxed sections titled *Chemical Perspectives, Historical Perspectives, A Closer Look* (for a more in-depth look at relevant material), and *Problem Solving Tips.* As described in "What's New . . ." we have now introduced one or more *Case Studies* in each chapter.

Organization and Purposes of the Sections of the Book

Part 1: The Basic Tools of Chemistry

There are basic ideas and methods that are the basis of all chemistry, and these are introduced in Part 1. Chapter 1 defines important terms, and the accompanying *Let's Review* section reviews units and mathematical methods. Chapter 2 introduces basic ideas of atoms, molecules, and ions, and the most important organizational device in chemistry, the periodic table. In Chapters 3 and 4 we begin to discuss the principles of chemical reactivity and to introduce the numerical methods used by chemists to extract quantitative information from chemical reactions. Chapter 5 is an introduction to the energy involved in chemical processes. The supplemental chapter *The Chemistry of Fuels and Energy Sources* follows Chapter 5 and uses many of the concepts developed in the preceding chapters.

Part 2: The Structure of Atoms and Molecules

The goal of this section is to outline the current theories of the arrangement of electrons in atoms (Chapters 6 and 7). This discussion is tied closely to the arrangement of elements in the periodic table so that these properties can be recalled and predictions made. In Chapter 8 we discuss for the first time how the electrons of atoms in a molecule lead to chemical bonding and the properties of these bonds. In addition, we show how to derive the three-dimensional structure of simple molecules. Finally, Chapter 9 considers the major theories of chemical bonding in more detail.

This part of the book is completed with a discussion of organic chemistry (Chapter 10), primarily from a structural point of view.

This section includes the interchapter on *Milestones in the Development ...*, and *The Chemistry of Life: Biochemistry* provides an overview of some of the most important aspects of biochemistry.

Part 3: States of Matter

The behavior of the three states of matter—gases, liquids, and solids—is described in that order in Chapters 11–14. The discussion of liquids and solids is tied to gases through the description of intermolecular forces in Chapter 12, with particular attention given to liquid and solid water. In Chapter 14 we describe the properties of solutions, intimate mixtures of gases, liquids, and solids.

The supplemental chapter on *The Chemistry of Modern Materials* is placed after Chapter 14, following coverage of the solid state. Designing and making new materials with useful properties is one of the most exciting areas of modern chemistry.

Part 4: The Control of Chemical Reactions

This section is wholly concerned with the *Principles of Reactivity.* Chapter 15 examines the important question of the rates of chemical processes and the factors controlling these rates. With this in mind, we move to Chapters 16-18, chapters that describe chemical reactions at equilibrium. After an introduction to equilibrium in Chapter 16, we highlight the reactions involving acids and bases in water (Chapters 17 and 18) and reactions leading to slightly soluble salts (Chapter 18). To tie together the discussion of chemical equilibria, we again explore thermodynamics in Chapter 19. As a final topic in this section we describe in Chapter 20 a major class of chemical reactions, those involving the transfer of electrons, and the use of these reactions in electrochemical cells.

The Chemistry of the Environment supplemental chapter is at the end of Part 4. This chapter uses ideas from kinetics and chemical equilibria, in particular, as well as principles described in earlier chapters in the book.

Part 5: The Chemistry of the Elements and Their Compounds

Although the chemistry of the various elements has been described throughout the book to this point, Part 5 considers this topic in a more systematic way. Chapter 21 is devoted to the chemistry of the representative elements, whereas Chapter 22—which has been expanded to include an introduction to organometallic chemistry—is a discussion of the transition elements and their compounds. Finally, Chapter 23 is a brief discussion of nuclear chemistry.

Supporting Materials for the Instructor

Supporting instructor materials are available to qualified adopters. Please consult your local Cengage Learning, Brooks/Cole representative for details. Visit **academic.cengage.com/kotz** to:

- See samples of materials
- Request a desk copy
- Locate your local representative
- Download electronic files of the *Instructor's Manual*, the *Test Bank*, and other helpful materials for instructors and students

Instructor's Resource Manual

by Susan Young, Hartwick College

ISBN-10: 0-495-38705-3; ISBN-13: 978-0-495-38705-3

Contains worked-out solutions to *all* end-of-chapter Study Questions and features ideas for instructors on how to fully utilize resources and technology in their courses. The *Manual* provides questions for electronic response systems, suggests classroom demonstrations, and emphasizes good and innovative teaching practices. Electronic files of the *Instructor's Resource Manual* are available for download on the PowerLecture CD-ROM and on the instructor's companion site at **academic.cengage.com/kotz**.

OWL: Online Web-based Learning

by Roberta Day and Beatrice Botch of the University of Massachusetts, Amherst, and William Vining of the State University of New York at Oneonta

OWL Instant Access (2 Semesters) ISBN-10: 0-495-05099-7; ISBN-13: 978-0-495-05099-5

e-Book in OWL Instant Access (2 Semesters) ISBN-10: 0-495-55499-5; ISBN-13: 978-0-495-55499-8

Used by more than 300 institutions and proven reliable for tens of thousands of students, OWL offers an online homework and quizzing system with unsurpassed ease of use, reliability, and dedicated training and service. OWL makes homework management a breeze and helps students improve their problem-solving skills and visualize concepts, providing instant analysis and feedback on a variety of homework problems, including tutors, simulations, and chemically and/or numerically parameterized short-answer questions. OWL is the only system specifically designed to support mastery learning, where students work as long as they need to master each chemical concept and skill. To view an OWL demo and for more information, visit **academic.cengage.com/owl** or contact your Brooks/Cole representative.

New to OWL!

For the seventh edition, approximately 20 new end-of-chapter questions (marked in the text with ■) can be assigned in OWL for a total of approximately 40 end-of-chapter Study Questions for each chapter available in OWL.

The **e-Book in OWL** is a complete electronic version of the text, fully assignable and linked to OWL homework. This exclusive option is available to students with instructor permission. Instructors can consult their Brooks/Cole representative for details and to determine the best option: access to the e-book can be bundled with the text and/or ordered as a text replacement.

Learning Resources allow students to quickly access valuable help to master each homework question with integrated e-book readings, tutors, simulations, and exercises that accompany each question. Learning Resources are configurable by instructors.

More new OWL features:

- New student Learning Resources and Toolbars
- New Answer Input tool for easy subscript and superscript formatting
- Enhanced reports that give instant snapshots of your class progress
- Easier grading access for quick report downloads
- New Survey and Authoring features for creating your own content
- Enhanced security to help you comply with FERPA regulations

A fee-based access code is required for OWL. OWL is available only to North American adopters.

Instructor's PowerLecture CD-ROM with ExamView® and JoinIn™ for *Chemistry & Chemical Reactivity*

ISBN-10: 0-495-38706-1; ISBN-13: 978-0-495-38706-0

PowerLecture is a dual platform, one-stop digital library and presentation tool that includes:

- Prepared Microsoft® PowerPoint® Lecture Slides covering all key points from the text in a convenient format that you can enhance with your own materials or with additional interactive video and animations on the CD-ROM for personalized, media-enhanced lectures.
- Image Libraries in PowerPoint and in JPEG format that contain electronic files for all text art, most photographs, and all numbered tables in the text. These files can be used to print transparencies or to create your own PowerPoint lectures.
- Electronic files for the complete *Instructor's Resource Manual* and *Test Bank*.
- Sample chapters from the *Student Solutions Manual* and *Study Guide*.
- ExamView testing software, with all the test items from the printed *Test Bank* in electronic format, enables you to create customized tests of up to 250 items in print or online.
- JoinIn STUDENT RESPONSE SYSTEM JoinIn "clicker" questions written specifically for the use of *Chemistry & Chemical Reactivity* with the classroom response system of your choice that allows you to seamlessly display student answers.

Test Bank

by David Treichel, Nebraska Wesleyan University

ISBN-10: 0-495-38709-6; ISBN-13: 978-0-495-38709-1

A printed test bank of more than 1250 questions in a range of difficulty and variety are correlated directly to the chapter sections found in the main text. Numerical, open-ended, or conceptual problems are written in multiple choice, fill-in-the-blank, or short-answer formats. Both single- and multiple-step problems are presented for each chapter. Electronic files of the *Test Bank* are included on the PowerLecture CD-ROM. WebCT and Blackboard versions of the test bank are available on the instructor's companion site at **academic.cengage.com/kotz.**

Transparencies

ISBN-10: 0-495-38714-2; ISBN-13: 978-0-495-38714-5

A collection of 150 full-color transparencies of key images selected from the text by the authors. The Power Lecture CD-ROM includes all text art and many photos to aid in preparing transparencies for material not present in this set.

Supporting Materials for the Student

Visit the student companion website at **academic.cengage.com/kotz** to see samples of selected student supplements. Students can purchase any Brooks/Cole products at your local college store or at our preferred online store **www.ichapters.com**.

Student Solutions Manual

by Alton J. Banks, North Carolina State University

ISBN-10: 0-495-38707-X; ISBN-13: 978-0-495-38707-7

This manual contains detailed solutions to the text's blue-numbered end-of-chapter Study Questions that match the problem-solving strategies from the text. Sample chapters are available for review on the PowerLecture CD and on the student companion website at **academic.cengage.com/kotz.**

Study Guide

by John R. Townsend and Michael J. Moran, West Chester University of Pennsylvania

ISBN-10: 0-495-38708-8; ISBN-13: 978-0-495-38708-4

This study guide contains chapter overviews, key terms with definitions, and sample tests explicitly linked to the goals introduced in each chapter. Emphasis is placed on the text's chapter goals by means of further commentary, study tips, worked examples, and direct references back to the text. Sample chapters are available for review on the student companion website at **academic.cengage.com/kotz.**

ChemistryNow Chemistry Now™

ChemistryNow's online self-assessment tools give you the choices and resources you need to study smarter. You can explore a variety of tutorials, exercises, and simulations (cross-referenced throughout the text by margin annotations), view Active Figure interactive versions of key pieces of art from the text, or take chapter-specific Pre-Tests and get a Personalized Study Plan that directs you to specific interactive materials that can help you master the areas in which you need additional work. Includes access to one-on-one tutoring and Go Chemistry mini video lectures. Access to ChemistryNow for two semesters may be included with each new textbook or can be purchased at **www.ichapters.com** using ISBN 0-495-39431-9.

Go Chemistry for General Chemistry

27-Module Set ISBN-10: 0-495-38228-0; ISBN-13: 978-0-495-38228-7

These new mini video lectures, playable on video iPods, iPhones, and MP3 players as well as on iTunes, include animations and problems for a quick summary of key concepts. In selected Go Chemistry modules, e-Flashcards briefly introduce a key concept and then test student understanding of the basics with a series of questions. Modules are also available separately. Go Chemistry is included in ChemistryNow. To purchase, enter ISBN 0-495-38228-0 at **www.ichapters.com.**

OWL for General Chemistry

See the above description in the instructor support materials section.

***Essential Math for Chemistry Students,* Second Edition** by David W. Ball, Cleveland State University

ISBN-10: 0-495-01327-7; ISBN-13: 978-0-495-01327-3

This short book is intended for students who lack confidence and/or competency in the essential mathematical skills necessary to survive in general chemistry. Each chapter focuses on a specific type of skill and has worked-out examples to show how these skills translate to chemical problem solving.

***Survival Guide for General Chemistry with Math Review,* Second Edition** by Charles H. Atwood, University of Georgia

ISBN-10: 0-495-38751-7; ISBN-13: 978-0-495-38751-0

Intended to help you practice for exams, this "survival guide" shows you how to solve difficult problems by dissecting them into manageable chunks. The guide includes three levels of proficiency questions—A, B, and minimal—to quickly build confidence as you master the knowledge you need to succeed in your course.

For the Laboratory

Brooks/Cole Lab Manuals

Brooks/Cole offers a variety of printed manuals to meet all general chemistry laboratory needs. Visit the chemistry site at **academic.cengage.com/chemistry** for a full listing and description of these laboratory manuals and laboratory notebooks. All Brooks/Cole lab manuals can be customized for your specific needs.

Signature Labs . . . for the customized laboratory

Signature Labs (**www.signaturelabs.com**) combines the resources of Brooks/Cole, CER, and OuterNet Publishing to provide you unparalleled service in creating your ideal customized lab program. Select the experiments and artwork you need from our collection of content and imagery to find the perfect labs to match your course. Visit **www.signaturelabs.com** or contact your Brooks/Cole representative for more information.

Acknowledgments

Because significant changes have been made from the sixth edition, preparing this new edition of *Chemistry & Chemical Reactivity* took almost 3 years of continuous effort. However, as in our work on the first six editions, we have had the support and encouragement of our families and of some wonderful friends, colleagues, and students.

CENGAGE LEARNING Brooks/Cole

The sixth edition of this book was published by Thomson Brooks/Cole. As often happens in the modern publishing industry, that company was recently acquired by another group and the new name is Cengage Learning Brooks/Cole. In spite of these changes in ownership, we continue with the same excellent team we have had in place for the previous several years.

The sixth edition of the book was very successful, in large part owing to the work of David Harris, our publisher. David again saw us through much of the development of this new edition, but Lisa Lockwood recently assumed his duties as our acquisitions editor; she has considerable experience in textbook publishing and was also responsible for the success of the sixth edition. We will miss David but are looking forward to a close association with Lisa.

Peter McGahey has been our the Development Editor for the fifth and sixth editions and again for this edition. Peter is blessed with energy, creativity, enthusiasm, intelligence, and good humor. He is a trusted friend and confidant and cheerfully answers our many questions during almost-daily phone calls.

No book can be successful without proper marketing. Amee Mosley was a great help in marketing the sixth edition and she is back in that role for this edition. She is knowledgeable about the market and has worked tirelessly to bring the book to everyone's attention.

Our team at Brooks/Cole is completed with Teresa Trego, Production Manager, and Lisa Weber, Technology Project Manager. Schedules are very demanding in textbook publishing, and Teresa has helped to keep us on schedule. We certainly appreciate her organizational skills. Lisa Weber has directed the development of the *Go Chemistry* modules and our expanded use of OWL.

People outside of publishing often do not realize the number of people involved in producing a textbook.

Anne Williams of Graphic World Inc. guided the book through its almost year-long production. Marcy Lunetta was the photo researcher for the book and was successful in filling our sometimes offbeat requests for a particular photo.

Photography, Art, and Design

Most of the color photographs for this edition were again beautifully created by Charles D. Winters. He produced several dozen new images for this book, always with a creative eye. Charlie's work gets better and better with each edition. We have worked with Charlie for more than 20 years and have become close friends. We listen to his jokes, both new and old—and always forget them. When we finish the book, we look forward to a kayaking trip.

When the fifth edition was being planned, we brought in Patrick Harman as a member of the team. Pat designed the first edition of the *General ChemistryNow* CD-ROM, and we believe its success is in no small way connected to his design skill. For the fifth edition of the book, Pat went over almost every figure, and almost every word, to bring a fresh perspective to ways to communicate chemistry and he did the same for the sixth edition. Once again he has worked on designing and producing new illustrations for the seventh edition, and his creativity is obvious in their clarity and beauty. Finally, Pat also designed and produced the Go Chemistry modules. As we have worked together so closely for so many years, Pat has become a good friend, as well, and we share interests not only in beautiful books but in interesting music.

Other Collaborators

We have been fortunate to have a number of other colleagues who have played valuable roles in this project.

- Bill Vining (State University of New York, Oneonta), was the lead author of the *General ChemistryNow* CD-ROM and of the media assets in OWL. He has been a friend for many years and recently took the place of one of the authors at SUNY-Oneonta. Bill has again applied his considerable energy and creativity in preparing many more OWL questions with tutorials.
- Susan Young (Hartwick College) has been a good friend and collaborator through five editions and has again prepared the *Instructor's Resource Manual.* She has always been helpful in proofreading, in answering questions on content, and in giving us good advice.
- Alton Banks (North Carolina State University) has also been involved for a number of editions preparing the *Student Solutions Manual.* Both Susan and Alton have been very helpful in ensuring the accuracy of the Study Question answers in the book, as well as in their respective manuals.
- Michael Moran (West Chester University) has updated and revised the *Study Guide* that was written by John Townsend for the sixth edition. This book has had a history of excellent study guides, and this manual follows that tradition.
- We also wish to acknowledge the support of George Purvis and Fujitsu for use of the CAChe Scientific software for molecular modeling. All the molecular models and the electrostatic potential surfaces in the book were prepared using CAChe software.
- Jay Freedman once again did a masterful job compiling the index/glossary for this edition.

A major task is proofreading the book after it has been set in type. The book is read in its entirety by the authors and accuracy reviewers. After making corrections, the book is read a second time. Any errors remaining at this point are certainly the responsibility of the authors, and students and instructors should contact the authors by email to offer their suggestions. If this is done in a timely manner, corrections can be made when the book is reprinted.

We want to thank the following accuracy reviewers for their invaluable assistance. The book is immeasurably improved by their work.

- William Broderick, Montana State University
- Stephen Z. Goldberg, Adelphi University
- Jeffrey Alan Mack, California State University, Sacramento
- Clyde Metz, College of Charleston
- David Shinn, University of Hawaii, Manoa
- Scott R. White, Southern Arkansas University, Magnolia

Reviewers for the Seventh Edition

- Gerald M. Korenowski, Rensselaer Polytechnic Institute
- Robert L. LaDuca, Michigan State University
- Jeffrey Alan Mack, California State University, Sacramento
- Armando M. Rivera-Figueroa, East Los Angeles College
- Daniel J. Williams, Kennesaw State University
- Steven G. Wood, Brigham Young University
- Roger A. Hinrichs, Weill Cornell Medical College in Qatar (reviewed the Energy interchapter)
- Leonard Fine, Columbia University (reviewed the Materials interchapter)

Advisory Board for the Seventh Edition

As the new edition was being planned, this board listened to some of our ideas and made other suggestions. We hope to continue our association with these energetic and creative chemical educators.

- Donnie Byers, Johnson County Community College
- Sharon Fetzer Gislason, University of Illinois, Chicago
- Adrian George, University of Nebraska
- George Grant, Tidewater Community College, Virginia Beach Campus
- Michael Hampton, University of Central Florida
- Milton Johnston, University of South Florida
- Jeffrey Alan Mack, California State University, Sacramento
- William Broderick, Montana State University
- Shane Street, University of Alabama
- Martin Valla, University of Florida

About the Authors

JOHN C. KOTZ, a State University of New York Distinguished Teaching Professor, Emeritus, at the College at Oneonta, was educated at Washington and Lee University and Cornell University. He held National Institutes of Health postdoctoral appointments at the University of Manchester Institute for Science and Technology in England and at Indiana University.

He has coauthored three textbooks in several editions (*Inorganic Chemistry, Chemistry & Chemical Reactivity,* and *The Chemical World*) and the *General ChemistryNow CD-ROM.* His research in inorganic chemistry and electrochemistry also has been published.

He was a Fulbright Lecturer and Research Scholar in Portugal in 1979 and a Visiting Professor there in 1992. He was also a Visiting Professor at the Institute for Chemical Education (University of Wisconsin, 1991-1992), at Auckland University in New Zealand (1999), and at Potchefstroom University in South Africa in 2006. He has been an invited speaker on chemical education at conferences in South Africa, New Zealand, and Brazil. He also served 3 years as a mentor for the U.S. National Chemistry Olympiad Team.

He has received several awards, among them a State University of New York Chancellor's Award (1979), a National Catalyst Award for Excellence in Teaching (1992), the Estee Lecturership at the University of South Dakota (1998), the Visiting Scientist Award from the Western Connecticut Section of the American Chemical Society (1999), the Distinguished Education Award from the Binghamton (NY) Section of the American Chemical Society (2001), the SUNY Award for Research and Scholarship (2005), and the Squibb Lectureship in Chemistry at the University of North Carolina-Asheville (2007). He may be contacted by email at kotzjc@oneonta.edu.

Left to right: Paul Treichel, John Townsend, and John Kotz.

PAUL M. TREICHEL received his B.S. degree from the University of Wisconsin in 1958 and a Ph.D. from Harvard University in 1962. After a year of postdoctoral study in London, he assumed a faculty position at the University of Wisconsin-Madison. He served as department chair from 1986 through 1995 and was awarded a Helfaer Professorship in 1996. He has held visiting faculty positions in South Africa (1975) and in Japan (1995). Retiring after 44 years as a faculty member in 2007, he is currently Emeritus Professor of Chemistry. During his faculty career he taught courses in general chemistry, inorganic chemistry, organometallic chemistry, and scientific ethics. Professor Treichel's research in organometallic and metal cluster chemistry and in mass spectrometry, aided by 75 graduate and undergraduate students, has led to more than 170 papers in scientific journals. He may be contacted by email at treichel@chem.wisc.edu.

JOHN R. TOWNSEND, Associate Professor of Chemistry at West Chester University of Pennsylvania, completed his B.A. in Chemistry as well as the Approved Program for Teacher Certification in Chemistry at the University of Delaware. After a career teaching high school science and mathematics, he earned his M.S. and Ph.D. in biophysical chemistry at Cornell University. At Cornell he also performed experiments in the origins of life field and received the DuPont Teaching Award. After teaching at Bloomsburg University, Dr. Townsend joined the faculty at West Chester University, where he coordinates the chemistry education program for prospective high school teachers and the general chemistry lecture program for science majors. His research interests are in the fields of chemical education and biochemistry. He may be contacted by email at jtownsend@wcupa.edu.

Contributors

When we designed this edition, we decided to seek chemists outside of our team to author some of the supplemental chapters and other materials.

John Emsley, University of Cambridge

Milestones in the Development of Chemistry and the Modern View of Atoms and Molecules and *The Chemistry of the Environment*

After 22 years as a chemistry lecturer at King's College London, John Emsley became a full-time science writer in 1990. As the Science Writer in Residence at Imperial College London from 1990 to 1997, he wrote the "Molecule of the Month" column for *The Independent* newspaper. Emsley's main activity is writing popular science books that feature chemistry and its role in everyday life. Recent publications include *The Consumer's Good Chemical Guide,* which won the Science Book Prize of 1995; *Molecules at an Exhibition; Was it Something You Ate?; Nature's Building Blocks; The Shocking History of Phosphorus; Vanity, Vitality & Virility;* and *The Elements of Murder.* His most recent book, published in 2007, is *Better Looking, Better Living, Better Loving.*

Jeffrey J. Keaffaber, University of Florida

Case Study: A Healthy Aquarium and the Nitrogen Cycle

Jeffrey J. Keaffaber received his B.S. in biology and chemistry at Manchester College, Indiana, and his Ph.D. in physical organic chemistry at the University of Florida. After finishing his doctoral work, Keaffaber joined the environmental research and development arm of Walt Disney Imagineering. He has worked as a marine environmental consultant and has taught chemistry and oceanography in the California Community College system. His research is in the fields of marine environmental chemistry and engineering, and his contributions have included the design of nitrate reduction and ozone disinfection processes for several large aquarium projects.

Eric Scerri, University of California, Los Angeles

Historical Perspectives: The Story of the Periodic Table

Eric Scerri is a continuing lecturer in the Department of Chemistry and Biochemistry at University of California, Los Angeles. After obtaining an undergraduate degree from the University of London, and a master's degree from the University of Southampton, he obtained his Ph. D. in the history and philosophy of science from King's College, London, focusing on the question of the reduction of chemistry to quantum mechanics. Scerri is the founder and editor of the international journal *Foundations of Chemistry* and recently authored *The Periodic Table: Its Story and Its Significance* (Oxford University Press, 2007), which has been described as the definitive book on the periodic table. He is also the author of more than 100 articles on the history and philosophy of chemistry, as well as chemical education. At UCLA, Scerri regularly teaches general chemistry classes of 350 students and smaller classes in the history and philosophy of science.

Felice Frankel, Harvard University

Cover photograph

As a senior research fellow, Felice Frankel heads the Envisioning Science program at Harvard University's Initiative in Innovative Computing (IIC). Frankel's images have appeared in more than 300 articles and covers in journals and general audience publications. Her awards include the 2007 Lennart Nilsson Award for Scientific Photography, a Guggenheim Fellowship, and grants from the National Science Foundation, the National Endowment for the Arts, the Alfred P. Sloan Foundation, the Graham Foundation for the Advanced Studies in the Fine Arts, and the Camille and Henry Dreyfus Foundation. Frankel's books include *On the Surface of Things, Images of the Extraordinary in Science,* and *Envisioning Science: The Design and Craft of the Science Image,* and she has a regularly appearing column, "Sightings," in *American Scientist* magazine.

About the Cover

© narcisa - floricica buzlea/istockphoto.com

The lotus is the national flower of India and Vietnam. The flowers, seeds, young leaves, and rhizomes of the plant are edible and have been used for centuries in Asia and India. Hindus associate the lotus blossom with the story of creation, and Buddhists believe it represents purity of body, speech, and mind. These ideas have come in part from the fact that the lotus flower grows in a muddy, watery environment, but, when the flower and leaves open, the mud and water are completely shed to leave a clean surface. Chemists recently discovered the underlying reasons for this phenomenon. First, the surface of the leaves is not smooth; it is covered with micro- and nanostructured wax crystals, and these tiny bumps allow only minimal contact between the leaf surface and the water droplet. Thus, only about 2% to 3% of the droplet's surface is actually in contact with the leaf. Second, the surface of the leaf itself is hydrophobic, that is, the forces of attraction between water molecules and the surface of the leaf are relatively weak. Because of strong hydrogen bonding, water molecules within a droplet are strongly attracted to one another instead of the leaf's surface and so form spherical droplets. As these droplets roll off of the surface, any dirt on the surface is swept away. On less hydrophobic surfaces, water molecules interact more strongly with the surface and drops glide off rather than roll off. This self-cleaning property of lotus leaves has been called the "lotus effect," an effect beautifully illustrated by the photograph on the cover of this book. Chemists are now trying to mimic this in new materials that can be incorporated into consumer products such as self-cleaning textiles, paint, and roofing tiles.

1 Basic Concepts of Chemistry

Charles D. Winters

Sports Drinks

Sports drinks are popular among athletes and nonathletes alike. The original sports drink, Gatorade, contained sucrose, glucose-fructose syrup, citric acid, sodium chloride, sodium citrate, potassium dihydrogen phosphate, and flavoring and coloring agents. This is not unlike the usual soft drink, but the carbohydrates in the sports drink provide only about half of the calories in fruit juice or in an ordinary soda or soft drink. To understand the chemistry behind sports drinks, we have to know the names and composition of the various organic and inorganic compounds in the drink and understand such important areas of chemistry as thermodynamics and colligative properties. As you study chemistry you will learn about these compounds, their structures, and their functions, and you will come to understand more about the ingredients in sports drinks, among other things. For now, think how you would describe a sports drink.

Questions:

1. What are its physical properties? Is it a homogeneous or heterogeneous mixture? Is the density of a sports drink more or less than that of water?
2. What is the volume of a bottle of a typical sports drink in milliliters? In liters? In deciliters?

Answers to these questions are in Appendix Q.

Chapter Goals

See Chapter Goals Revisited (page 20) for Study Questions keyed to these goals and assignable in OWL.

- Understand the differences between hypotheses, laws, and theories.
- Apply the kinetic-molecular theory to the properties of matter.
- Classify matter.
- Recognize elements, atoms, compounds, and molecules.
- Identify physical and chemical properties and changes.

Chapter Outline

In February 2006 an athlete from the U.S. was sent home from the Winter Olympics Games in Italy because he had used a common treatment for his baldness. The reason was, as the Olympic committee stated, that the remedy can also be used to mask other, illegal drugs. Similarly, athletes were banned from the Summer Olympics Games in Greece in 2004 and the winner of the 2006 Tour de France was stripped of his title for using banned steroids. How can these drugs be detected or identified?

On June 13, 2003, a colorless liquid arrived at the Olympic Analytical Laboratory (OAL) in Los Angeles, California. This laboratory annually tests about 25,000 samples for the presence of illegal drugs. Among its clients are the U.S. Olympic Committee, the National Collegiate Athletic Association, and the National Football League.

About the time of the U.S. Outdoor Track and Field Championships in the summer of 2003, a coach in Colorado tipped off the U.S. Anti-Doping Agency (USADA) that several athletes were using a new steroid. The coach had found a syringe containing an unknown substance and had sent it to the USADA. The USADA chemists dissolved the contents of the syringe in a few milliliters of an alcohol and sent the solution to the OAL. That submission initiated weeks of intense work that led to the identification of a previously unknown steroid that was presumably being used by athletes (Figure 1.1).

Chemistry Now™

Throughout the text this icon introduces an opportunity for self-study or to explore interactive tutorials by signing in at **www.thomsonedu.com/login**.

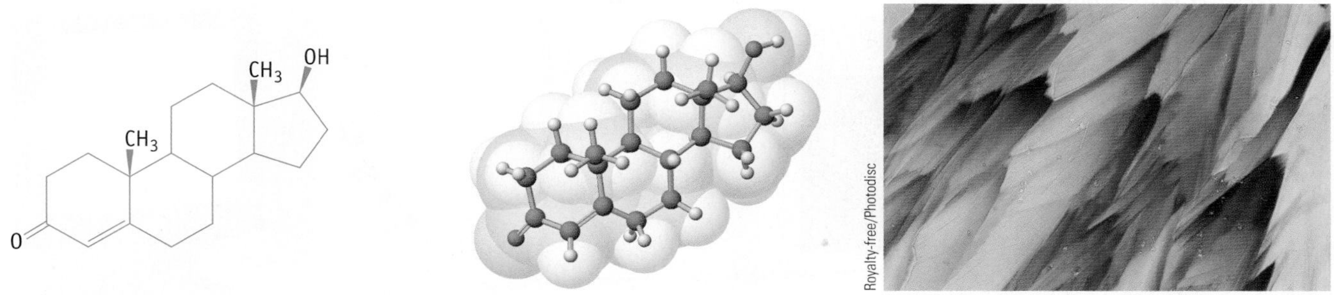

The steroid testosterone. All steroids, including cholesterol, have the same four-ring structure.

A molecular model of testosterone.

A photo of crystals of the steroid cholesterol taken with a microscope using polarized light.

FIGURE 1.1 The steroid testosterone. The unknown compound discussed in the text is a steroid closely related to testosterone.

© Jeff Minton, Courtesy of the UCLA Olympic Laboratory

Dr. Donald Catlin, the director of the Olympic Analytical Laboratory in Los Angeles, California.

To identify the unknown substance, chemists at the Olympic Analytical Laboratory used a GC-MS, an instrument widely employed in forensic science work (Figure 1.2). They first passed the sample through a gas chromatograph (GC), an instrument that can separate different chemical compounds in a mixture. A GC has a small-diameter, coiled tube (a typical inside diameter is 0.025 mm) in which the inside surface has been specially treated so that chemicals are attracted to the surface. This tube is placed in an oven and heated to temperatures of 200 °C or higher. The substances in a sample are swept along the tube with a stream of helium gas. Because each component in the sample binds differently to the material on the inside surface of the tube, each component moves through the column at a different rate and exits from the end of the column at a different time. Thus, separation of the components in the mixture is achieved.

After exiting the GC, each compound is routed directly into a mass spectrometer (MS). In a mass spectrometer, the compounds are bombarded with high-energy electrons and each compound is turned into ions, a chemical species with an electric charge. These ions are then passed through a strong magnetic field, causing the ions to be deflected. The path an ion takes in the magnetic field (the extent of deflection) is related to its mass. The masses of the ions are a key piece of information that helps to identify the compound.

Such a straightforward process: separate the compounds in a GC and identify them in an MS. What can go wrong? In fact, many things can go wrong that require ingenuity to overcome. In this case, the unknown steroid did not survive the high temperatures of the GC. It broke apart into pieces, and so it was only possible to study the pieces of the original molecule. However, this gave enough evidence to convince scientists that the compound was indeed a steroid. But what steroid? Dr. Catlin, the director of OAL, said that one hypothesis was that "the new steroid was made by people who knew it was not going to be detectable," that the molecule had been designed in a way that would guarantee that it would not be detected by the standard GC-MS procedure.

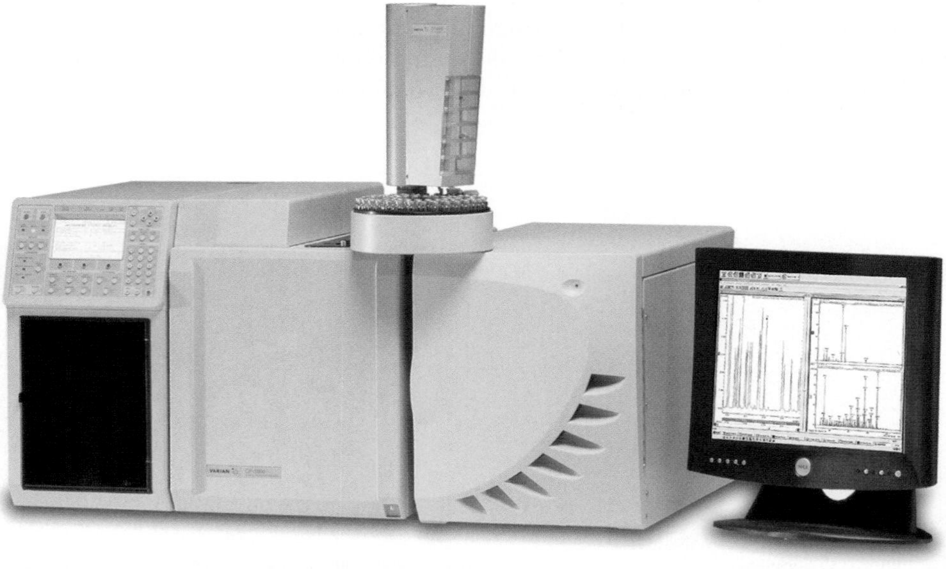

Courtesy of Varian, Inc.

FIGURE 1.2 A GC-MS (gas chromatograph-mass spectrometer). A GC-MS is one of the major tools used in forensic chemistry. The GC portion of the instrument separates the components in a mixture of volatile compounds, and the MS portion then analyzes and identifies them. The GC-MS pictured here has an automated sample changer (carousel, center). An operator will load dozens of samples into the carousel, and the instrument will then process the samples automatically, with the data recorded and stored in a computer.

So, Catlin and his colleagues set out to identify the steroid. The first thing they did was to make the molecule stable during the analysis. This was done by attaching new atoms to the molecule to make what chemists call a *derivative.* A number of approaches to making derivatives were tested, and, within a few weeks, they believed they knew the identity of the unknown steroid.

The final step in solving the mystery was to try to make a sample of the compound in the laboratory and then to use the gas chromatograph and mass spectrometer on this sample. If the material behaved the same way as the unknown sample, then they could be as certain as possible they knew the identify of what they had received from the track coach. These experiments worked, confirming the identity of the compound. It was an entirely new steroid, never seen before in nature or in the laboratory. Its formula is $C_{21}H_{28}O_2$, and its name is tetrahydrogestrinone or THG. It resembled two well-known steroids: gestrinone, used to treat gynecological problems, and trenbolone, a steroid used by ranchers to beef up cattle.

There are two sequels to the story. First, a scientific problem is not solved until it has been verified in another laboratory. Not only was this done, but a test was soon devised to find THG in urine samples. Second, with new analytical procedures, the USADA asked OAL to retest 550 urine samples—and THG was found in several.

Throughout this mystery, chemistry played a critical role, from using the analytical technique of mass spectrometry, to using chemical intuition to determine possible arrangements of the molecular fragments, and finally to synthesizing the derivative and the proposed compound. A knowledge of chemistry is crucial to solving problems not only like this one but many others as well.

■ **Athletes and Steroids** What is the problem with athletes taking steroids? THG is one of a class of steroids called anabolic steroids. They elevate the body's natural testosterone levels and increase body mass, muscle strength, and muscle definition. They can also improve an athlete's capacity to train and compete at the highest levels. Aside from giving steroid users an illegal competitive advantage, the potential side effects of steroids are liver damage, heart disease, anxiety, and rage.

A check of the internet shows that there are hundreds of sources of steroids for athletes. The known performance-enhancing drugs can be detected and their users banned from competitive sports. But what about as-yet unknown steroids? The director of the OAL, Dr. Donald Catlin, believes there are other steroids out in the market, made by secret labs without safety standards, a problem he calls horrifying.

1.1 Chemistry and Its Methods

Chemistry is about change. It was once only about changing one natural substance into another—wood and oil burn; grape juice turns into wine; and cinnabar (Figure 1.3), a red mineral from the earth, changes ultimately into shiny quicksilver (mercury). Chemistry is still about change, but now chemists focus on the change of one pure substance, whether natural or synthetic, into another (Figure 1.4).

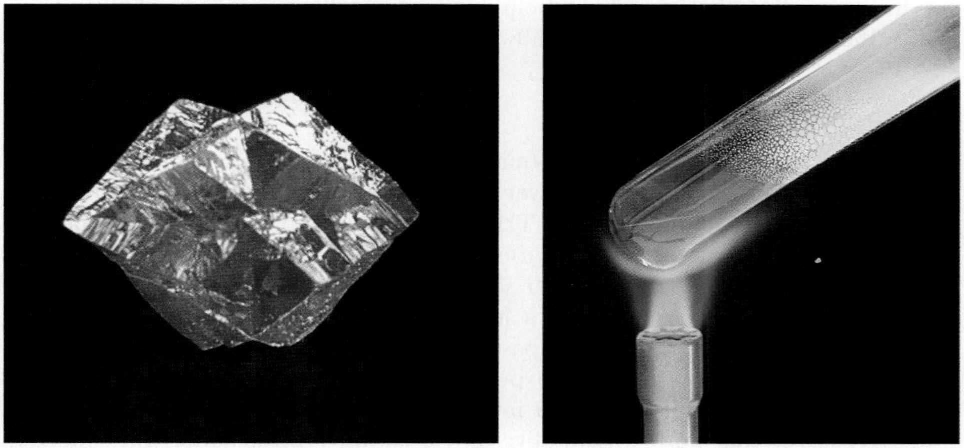

Charles D. Winters

(a) (b)

FIGURE 1.3 Cinnabar and mercury. (a) The red crystals of cinnabar are the chemical compound mercury(II) sulfide. (b) It is heated in air to change it into orange mercury oxide, which, on further heating, is decomposed to the elements oxygen and mercury metal. (The droplets you see on the inside of the test tube wall are mercury.)

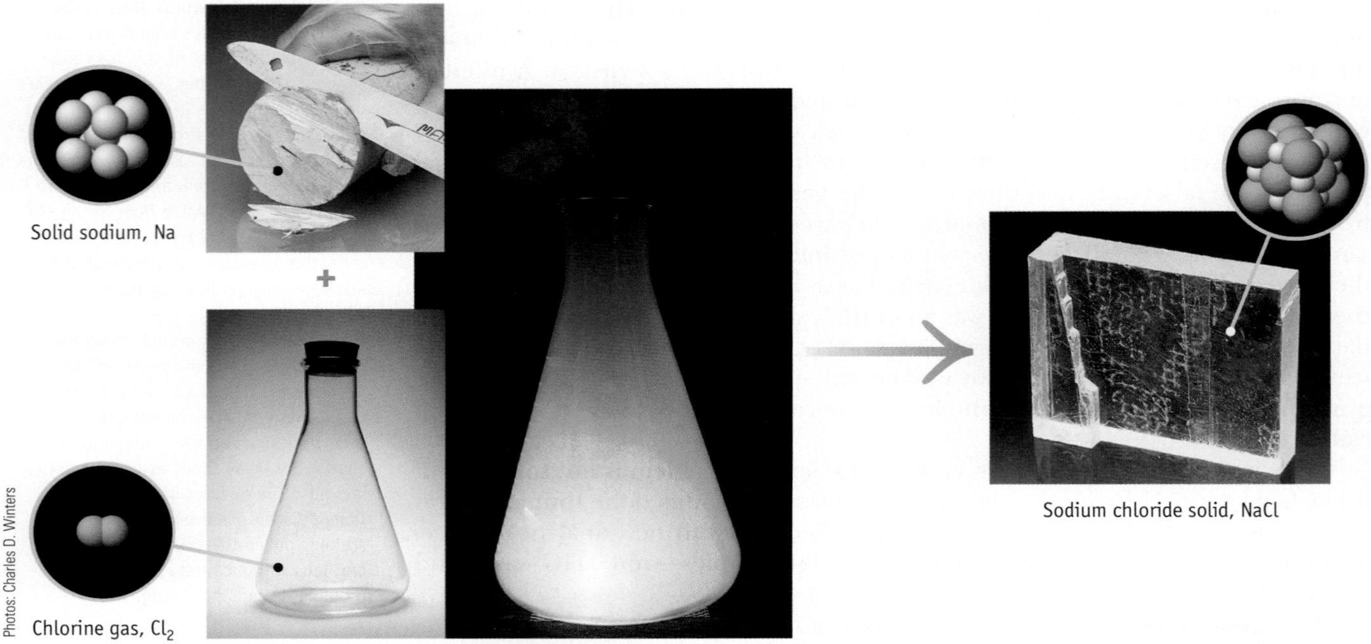

FIGURE 1.4 Forming a chemical compound. Sodium chloride, table salt, can be made by combining sodium metal (Na) and yellow chlorine gas (Cl_2). The result is a crystalline solid, common salt. (The tiny spheres show how the atoms are arranged in the substances. In the case of the salt crystal, the spheres represent electrically charged sodium and chlorine ions.)

Although chemistry is endlessly fascinating—at least to chemists—why should you study chemistry? Each person probably has a different answer, but many students take a chemistry course because someone else has decided it is an important part of preparing for a particular career. Chemistry is especially useful because it is central to our understanding of disciplines as diverse as biology, geology, materials science, medicine, physics, and many branches of engineering. In addition, chemistry plays a major role in the economy of developed nations, and chemistry and chemicals affect our daily lives in a wide variety of ways. Furthermore, a course in chemistry can help you see how a scientist thinks about the world and how to solve problems. The knowledge and skills developed in such a course will benefit you in many career paths and will help you become a better-informed citizen in a world that is becoming technologically more complex—and more interesting.

Hypotheses, Laws, and Theories

As scientists, we study questions of our own choosing or ones that someone else poses in the hope of finding an answer or of discovering some useful information. In the story of the banned steroid, THG, the chemists at the Olympic Analytical Laboratory were handed a problem to solve, and they followed the usual methods of science to get to the answer. After some preliminary tests they recognized that the mystery substance was probably a steroid. That is, they formed a **hypothesis**, a tentative explanation or prediction based on experimental observations.

After formulating one or more hypotheses, scientists perform experiments designed to give results that confirm or invalidate these hypotheses. In chemistry this usually requires that both quantitative and qualitative information be collected.

Charles D. Winters

FIGURE 1.5 Qualitative and quantitative observations. A new substance is formed by mixing two known substances in solution. We can make several observations about the substances involved. *Qualitative* observations: The solutions before mixing are colorless and yellow; a yellow, fluffy solid is formed on mixing. *Quantitative* observations: Mixing measured volumes of the solutions produces a measureable mass of solid.

Quantitative information is numerical data, such as the temperature at which a chemical substance melts or its mass (Figure 1.5). **Qualitative** information, in contrast, consists of nonnumerical observations, such as the color of a substance or its physical appearance.

The chemists at the OAL assembled a great deal of qualitative and quantitative information on various drugs. Based on their experience, and on published reports of experiments done in the past by other chemists who study steroids, they became more certain that they knew the identity of the substance. Their preliminary experiments led them to perform still more experiments, such as looking for a way to stabilize the molecule so it would not decompose, and they looked for a way to make the molecule in the laboratory. Final confirmation came when their work was reproduced by scientists in other laboratories.

After scientists have done a number of experiments and the results have been checked to ensure they are reproducible, a pattern of behavior or results may emerge. At this point it may be possible to summarize the observations in the form of a general rule or conclusion. After making a number of experimental observations, the chemists at OAL could conclude, for example, that the unknown substance was a steroid because it had properties characteristic of many other steroids they had observed.

Finally, after numerous experiments by many scientists over an extended period of time, the original hypothesis may become a **law**—a concise verbal or mathematical statement of a behavior or a relation that is consistently observed in nature without contradiction. An example might be the *law of mass conservation* in chemical reactions.

We base much of what we do in science on laws because they help us predict what may occur under a new set of circumstances. For example, we know from experience that if the chemical element sodium comes in contact with water, a violent reaction occurs and new substances are formed (Figure 1.6), and we know that the mass of the substances produced in the reaction is exactly the same as the mass of sodium and water used in the reaction. That is, *mass is always conserved in chemical reactions.* But the result of an experiment might be different from what is expected based on a general rule. When that happens, chemists get excited because experiments that do not follow our expectations are often the most interesting. We know that understanding the exceptions almost invariably gives new insights.

Charles D. Winters

FIGURE 1.6 The metallic element sodium reacts with water.

Once enough reproducible experiments have been conducted, and experimental results have been generalized as a law or general rule, it may be possible to conceive a theory to explain the observations. A **theory** is a well-tested, unifying principle that explains a body of facts and the laws based on them. It is capable of suggesting new hypotheses that can be tested experimentally.

Sometimes nonscientists use the word "theory" to imply that someone has made a guess and that an idea is not yet substantiated. But, to scientists, a theory is based on carefully determined and reproducible evidence. Theories are the cornerstone of our understanding of the natural world at any given time. Remember, though, that theories are inventions of the human mind. Theories can and do change as new facts are uncovered.

Goals of Science

The sciences, including chemistry, have several goals. Two of these are prediction and control. We do experiments and seek generalities because we want to be able to predict what may occur under a given set of circumstances. We also want to know how we might control the outcome of a chemical reaction or process.

A third goal is explanation and understanding. We know, for example, that certain elements such as sodium will react vigorously with water. But why should this be true? To explain and understand this, we turn to theories such as those developed in Chapters 6 and 7.

Dilemmas and Integrity in Science

You may think research in science is straightforward: Do an experiment; draw a conclusion. But, research is seldom that easy. Frustrations and disappointments are common enough, and results can be inconclusive. Experiments sometimes contain some level of uncertainty, and spurious or contradictory data can be collected. For example, suppose you do an experiment expecting to find a direct relation between two experimental quantities. You collect six data sets. When plotted, four of the sets lie on a straight line, but two others lie far away from the line. Should you ignore the last two points? Or should you do more experiments when you know the time they take will mean someone else could publish their results first and thus get the credit for a new scientific principle? Or should you consider that the two points not on the line might indicate that your original hypothesis is wrong, and that you will have to abandon a favorite idea you have worked on for a year? Scientists have a responsibility to remain objective in these situations, but it is sometimes hard to do.

It is important to remember that scientists are human and therefore subject to the same moral pressures and dilemmas as any other person. To help ensure integrity in science, some simple principles have emerged over time that guide scientific practice:

- Experimental results should be reproducible. Furthermore, these results should be reported in the scientific literature in sufficient detail that they can be used or reproduced by others.
- Conclusions should be reasonable and unbiased.
- Credit should be given where it is due.

■ **The Stem Cell Scandal of 2005** In 2004–2005 researchers at Seoul National University in South Korea published several papers in which they claimed to have cloned DNA from human embryonic stem cells. Because it was such an important discovery, scientists around the world examined the data closely. Within months, however, it was discovered that many of the reported results were based on fabricated data, and the results were retracted. Misrepresentation of information does great harm because it misleads scientists into spending time, energy, and funds in trying to replicate the information and using that information in other experiments. The stem cell scandal illustrates, however, the ways in which the scientific community ensures that scientific research is correct and accurate.

Chemical Perspectives — Moral Issues in Science

Moral and ethical issues frequently arise in science. One of these concerns the use of the pesticide DDT. This is a classic case of the law of "unintended consequences." The pesticide was developed during World War II and promoted as effective in controlling pests but harmless to people. In fact, it was thought to be so effective that it was used in larger and larger quantities around the world. It was especially effective in controlling mosquitoes carrying malaria, although it was soon evident there were consequences. In Borneo, the World Health Organization (WHO) used large quantities of DDT to kill mosquitoes. The mosquito population did indeed decline and so did malaria. Soon, however, the thatch roofs of people's houses fell down because parasitic wasps that ate thatch-eating caterpillars were wiped out by the DDT.

The problem did not end there, however. Small lizards also normally ate the caterpillars and helped to keep that population in check, and the lizard population was controlled by cats. However, DDT was passed up the food chain from caterpillars to lizards to cats, and cats began to die. When the cat population declined, this led to an infestation of rats. Unintended consequences, indeed.

DDT use has been banned in many parts of the world because of its very real, but unforeseen, environmental consequences. The DDT ban began in the United States in 1972 because evidence accumulated that the pesticide affected the reproduction of birds such as the bald eagle.

As expected, the ban on DDT affected the control of malaria-carrying insects. Several million people, primarily children in sub-Saharan Africa, die every year from malaria. The chairman of the Malaria Foundation International has said that "the malaria epidemic is like loading up seven Boeing 747 airliners each day and crashing them into Mt. Kilimanjaro." Consequently, there has been a movement to return DDT to the arsenal of weapons in fighting the spread of malaria, and in 2006 the WHO approved indoor spraying of DDT in some areas in Africa.

There are many, many moral and ethical issues for chemists. Chemistry has extended and improved lives for millions of people. But just as clearly, chemicals can cause harm, particularly when misused. It is incumbent on all of us to understand enough science to ask pertinent questions and to evaluate sources of information sufficiently to reach reasonable conclusions regarding the health and safety of ourselves and our communities.

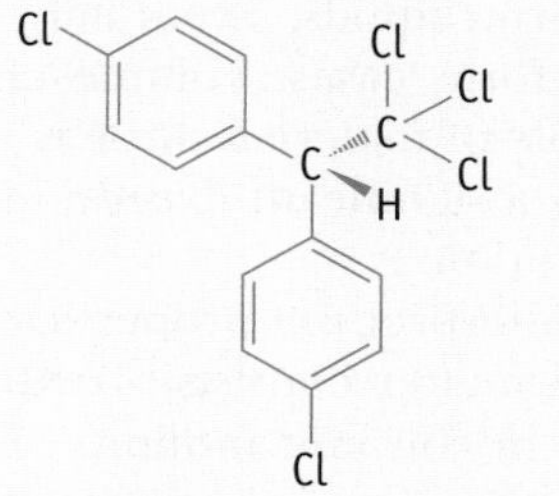

(a) The molecular structure of DDT.

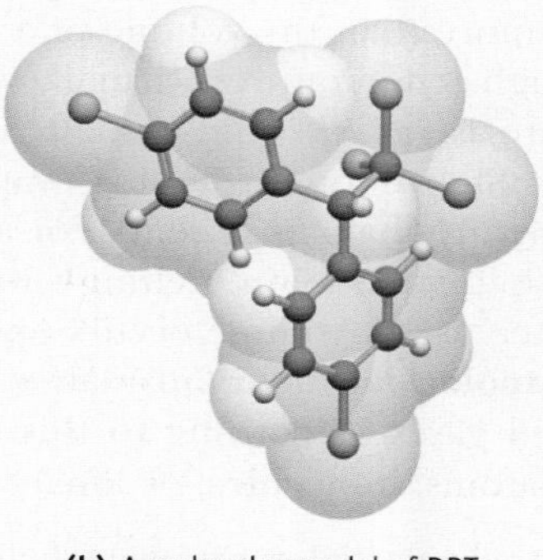

(b) A molecular model of DDT.

(c) DDT can be used to control malaria-carrying insects such as mosquitos.

1.2 Classifying Matter

This chapter begins our discussion of how chemists think about science in general and about matter in particular. After looking at a way to classify matter, we will turn to some basic ideas about elements, atoms, compounds, and molecules and describe how chemists characterize these building blocks of matter.

States of Matter and Kinetic-Molecular Theory

An easily observed property of matter is its **state**—that is, whether a substance is a solid, liquid, or gas (Figure 1.7). You recognize a material as a solid because it has a rigid shape and a fixed volume that changes little as temperature and pressure

Active Figure 1.7 **States of matter—solid, liquid, and gas.** Elemental bromine exists in all three states near room temperature. The tiny spheres represent bromine (Br) atoms. In elemental bromine, two Br atoms join to form a Br_2 molecule. (See Section 1.3 and Chapter 2.)

Chemistry Now™ Sign in at www.thomsonedu.com/login and go to the Chapter Contents menu to explore an interactive version of this figure accompanied by an exercise.

Bromine solid and liquid

Bromine gas and liquid

■ **Water—Changes in Volume on Freezing** Water is an exception to the general statement that a given mass of a substance has a smaller volume as a solid than as a liquid. Water is almost unique in that, for a given mass, the volume *increases* on changing from a liquid to a solid. (That is, its density decreases. See page 15.)

■ **Gases, Liquids, and Solids** Gases and kinetic-molecular theory are discussed in detail in Chapter 11, liquids in Chapter 12, and solids in Chapter 13.

change. Like solids, liquids have a fixed volume, but a liquid is fluid—it takes on the shape of its container and has no definite shape of its own. Gases are fluid as well, but the volume of a gas is determined by the size of its container. The volume of a gas varies more than the volume of a liquid with temperature and pressure.

At low enough temperatures, virtually all matter is found in the solid state. As the temperature is raised, solids usually melt to form liquids. Eventually, if the temperature is high enough, liquids evaporate to form gases. Volume changes typically accompany changes in state. For a given mass of material, there is usually a small increase in volume on melting—water being a significant exception—and then a large increase in volume occurs upon evaporation.

The **kinetic-molecular theory of matter** helps us interpret the properties of solids, liquids, and gases. According to this theory, all matter consists of extremely tiny particles (atoms, molecules, or ions), which are in constant motion.

- In solids these particles are packed closely together, usually in a regular array. The particles vibrate back and forth about their average positions, but seldom does a particle in a solid squeeze past its immediate neighbors to come into contact with a new set of particles.
- The atoms or molecules of liquids are arranged randomly rather than in the regular patterns found in solids. Liquids and gases are fluid because the particles are not confined to specific locations and can move past one another.
- Under normal conditions, the particles in a gas are far apart. Gas molecules move extremely rapidly because they are not constrained by their neighbors. The molecules of a gas fly about, colliding with one another and with the container walls. This random motion allows gas molecules to fill their container, so the volume of the gas sample is the volume of the container.

An important aspect of the kinetic-molecular theory is that the higher the temperature, the faster the particles move. The energy of motion of the particles (their **kinetic energy**) acts to overcome the forces of attraction between particles. A solid melts to form a liquid when the temperature of the solid is raised to the point at which the particles vibrate fast enough and far enough to push one another out of the way and move out of their regularly spaced positions. As the temperature increases even more, the particles move even faster until finally they can escape the clutches of their comrades and enter the gaseous state. *Increasing temperature* *corresponds to faster and faster motions of atoms and molecules,* a general rule you will find useful in many future discussions.

Matter at the Macroscopic and Particulate Levels

The characteristic properties of gases, liquids, and solids are observed by the unaided human senses. They are determined using samples of matter large enough to be seen, measured, and handled. Using such samples, we can also determine, for example, what the color of a substance is, whether it dissolves in water, or whether it conducts electricity or reacts with oxygen. Observations such as these generally take place in the **macroscopic** world of chemistry (Figure 1.8). This is the world of experiments and observations.

Now let us move to the level of atoms, molecules, and ions—a world of chemistry we cannot see. Take a macroscopic sample of material and divide it, again and again, past the point where the amount of sample can be seen by the naked eye, past the point where it can be seen using an optical microscope. Eventually you reach the level of individual particles that make up all matter, a level that chemists refer to as the **submicroscopic** or **particulate** world of atoms and molecules (Figures 1.7 and 1.8).

Chemists are interested in the structure of matter at the particulate level. Atoms, molecules, and ions cannot be "seen" in the same way that one views the macroscopic world, but they are no less real. Chemists imagine what atoms must look like and how they might fit together to form molecules. They create models to represent atoms and molecules (Figures 1.7 and 1.8)—where tiny spheres are used to represent atoms—and then use these models to think about chemistry and to explain the observations they have made about the macroscopic world.

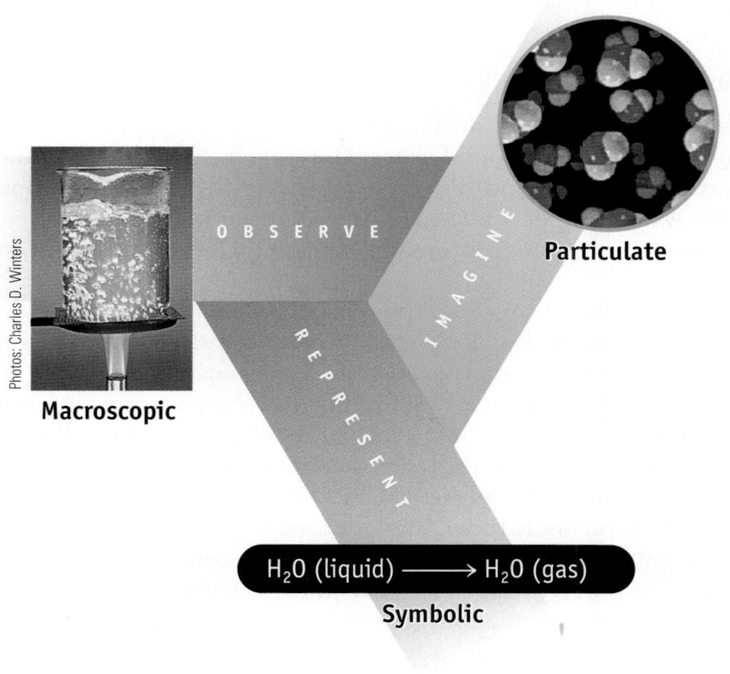

Active Figure 1.8 Levels of matter. We observe chemical and physical processes at the macroscopic level. To understand or illustrate these processes, scientists often try to imagine what has occurred at the particulate atomic and molecular levels and write symbols to represent these observations. A beaker of boiling water can be visualized at the particulate level as rapidly moving H_2O molecules. The process is symbolized by the chemical equation H_2O(liquid) ⟶ H_2O(gas).

ChemistryNow™ Sign in at www.thomsonedu.com/login and go to the Chapter Contents menu to explore an interactive version of this figure accompanied by an exercise.

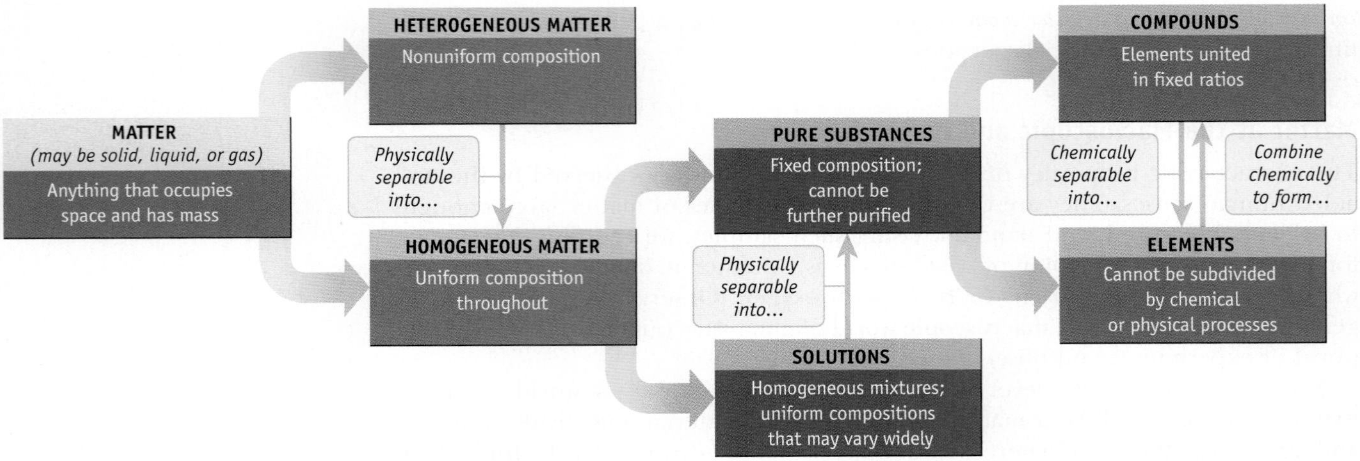

Active Figure 1.9 **Classifying Matter.**

Chemistry Now™ Sign in at www.thomsonedu.com/login and go to the Chapter Contents menu to explore an interactive version of this figure accompanied by an exercise.

It has been said that chemists carry out experiments at the macroscopic level, but they think about chemistry at the particulate level. They then write down their observations as "symbols," the letters (such as H_2O for water or Br_2 for bromine molecules) and drawings that signify the elements and compounds involved. This is a useful perspective that will help you as you study chemistry. Indeed, one of our goals is to help you make the connections in your own mind among the symbolic, particulate, and macroscopic worlds of chemistry.

Pure Substances

A chemist looks at a glass of drinking water and sees a liquid. This liquid could be the pure chemical compound water. More likely, though, the liquid is a *homogeneous* mixture of water and dissolved substances—that is, a **solution.** It is also possible the water sample is a *heterogeneous* mixture, with solids suspended in the liquid. These descriptions represent some of the ways we can classify matter (Figure 1.9).

Every substance has a set of unique properties by which it can be recognized. Pure water, for example, is colorless and odorless and certainly does not contain suspended solids. If you wanted to identify a substance conclusively as water, you would have to examine its properties carefully and compare them against the known properties of pure water. Melting point and boiling point serve the purpose well here. If you could show that the substance melts at 0 °C and boils at 100 °C at atmospheric pressure, you can be certain it is water. No other known substance melts and boils at precisely these temperatures.

A second feature of a pure substance is that it generally cannot be separated into two or more different species by any physical technique at ordinary temperatures. If it could be separated, our sample would be classified as a mixture.

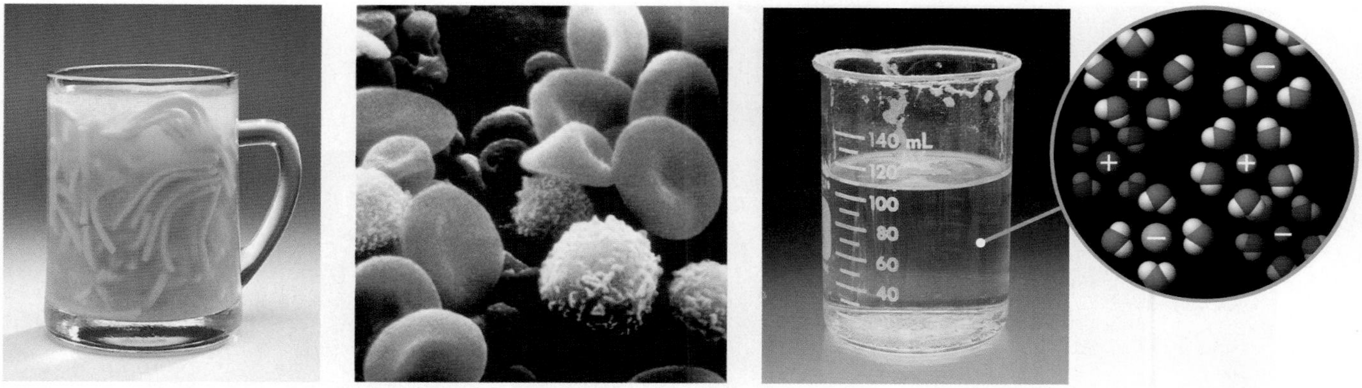

a and c, Charles D. Winters; b, Kenneth Eward/BioGrafx/ Photo Researchers, Inc.

(a) (b) (c)

FIGURE 1.10 Mixtures. (a) A cup of noodle soup is a heterogeneous mixture. (b) A sample of blood may look homogeneous, but examination with an optical microscope shows it is, in fact, a heterogeneous mixture of liquids and suspended particles (blood cells). (c) A homogeneous mixture, here consisting of salt in water. The model shows that salt in water consists of separate, electrically charged particles (ions), but the particles cannot be seen with an optical microscope.

Mixtures: Homogeneous and Heterogeneous

A cup of noodle soup is obviously a mixture of solids and liquids (Figure 1.10a). A mixture in which the uneven texture of the material can be detected is called a **heterogeneous** mixture. Heterogeneous mixtures such as blood may appear completely uniform but on closer examination are not (Figure 1.10b). Milk, for example, appears smooth in texture to the unaided eye, but magnification would reveal fat and protein globules within the liquid. In a heterogeneous mixture the properties in one region are different from those in another region.

A **homogeneous** mixture consists of two or more substances in the same phase (Figure 1.10c). No amount of optical magnification will reveal a homogeneous mixture to have different properties in different regions. Homogeneous mixtures are often called **solutions.** Common examples include air (mostly a mixture of nitrogen and oxygen gases), gasoline (a mixture of carbon- and hydrogen-containing compounds called *hydrocarbons*), and an unopened soft drink.

When a mixture is separated into its pure components, the components are said to be **purified.** Efforts at separation are often not complete in a single step, however, and repetition almost always gives an increasingly pure substance. For example, soil particles can be separated from water by filtration (Figure 1.11). When the mixture is passed through a filter, many of the particles are removed. Repeated filtrations will give water a higher and higher state of purity. This purification process uses a property of the mixture, its clarity, to measure the extent of purification. When a perfectly clear sample of water is obtained, all of the soil particles are assumed to have been removed.

Chemistry Now™

Sign in at **www.thomsonedu.com/login** and go to Chapter 1 Contents to see:
- Screen 1.5 for an exercise on **identifying pure substances and types of mixtures**
- Screen 1.6 to watch a video on **heterogeneous mixtures**

a, Charles D. Winters; b, Littleton, Massachusetts, Spectacle Pond Iron and Manganese Treatment Facility

(a) (b)

FIGURE 1.11 Purifying water by filtration. (a) A laboratory setup. A beaker full of muddy water is passed through a paper filter, and the mud and dirt are removed. (b) A water treatment plant uses filtration to remove suspended particles from the water.

Module 1

1.3 Elements and Atoms

Passing an electric current through water can decompose it to gaseous hydrogen and oxygen (Figure 1.12a). Substances like hydrogen and oxygen that are composed of only one type of atom are classified as **elements.** Currently, 117 elements are known. Of these, only about 90—some of which are illustrated in Figure 1.12—are found in nature. The remainder have been created by scientists. The name and symbol for each element are listed in the tables at the front and back of this book. Carbon (C), sulfur (S), iron (Fe), copper (Cu), silver (Ag), tin (Sn), gold (Au), mercury (Hg), and lead (Pb) were known to the early Greeks and Romans and to the alchemists of ancient China, the Arab world, and medieval Europe. However,

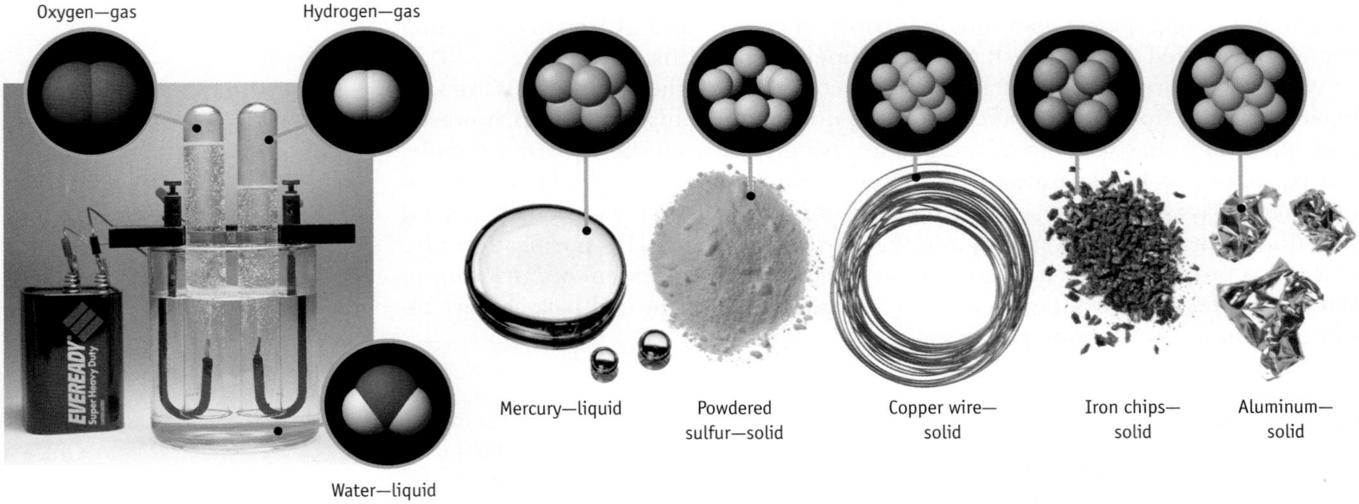

Photos: Charles D. Winters

(a) (b)

FIGURE 1.12 Elements. (a) Passing an electric current through water produces the elements hydrogen (test tube on the right) and oxygen (test tube on the left). (b) Chemical elements can often be distinguished by their color and their state at room temperature.

Sign in at **www.thomsonedu.com/login** to download the Go Chemistry module for this section or go to **www.ichapters.com** to purchase modules.

many other elements—such as aluminum (Al), silicon (Si), iodine (I), and helium (He)—were not discovered until the 18th and 19th centuries. Finally, scientists in the 20th and 21st centuries have made elements that do not exist in nature, such as technetium (Tc), plutonium (Pu), and americium (Am).

The table inside the front cover of this book, in which the symbol and other information for the elements are enclosed in a box, is called the **periodic table.** We will describe this important tool of chemistry in more detail beginning in Chapter 2.

An **atom** is the smallest particle of an element that retains the characteristic chemical properties of that element. Modern chemistry is based on an understanding and exploration of nature at the atomic level (► Chapters 6 and 7).

Chemistry Now™

Sign in at **www.thomsonedu.com/login** and go to Chapter 1 Contents to see Screen 1.7 for a self-study module on **Elements and Atoms,** and the Periodic Table tool on this screen or in the Toolbox.

EXERCISE 1.1 Elements

Using the periodic table inside the front cover of this book:

(a) Find the names of the elements having the symbols Na, Cl, and Cr.

(b) Find the symbols for the elements zinc, nickel, and potassium.

1.4 Compounds and Molecules

A pure substance like sugar, salt, or water, which is composed of two or more different elements held together by **chemical bonds,** is referred to as a **chemical compound.** Even though only 117 elements are known, there appears to be no limit to the number of compounds that can be made from those elements. More than 20 million compounds are now known, with about a half million added to the list each year.

When elements become part of a compound, their original properties, such as their color, hardness, and melting point, are replaced by the characteristic properties of the compound. Consider common table salt (sodium chloride), which is composed of two elements (see Figure 1.4):

- Sodium is a shiny metal that reacts violently with water. Its solid state structure has sodium atoms tightly packed together.
- Chlorine is a light yellow gas that has a distinctive, suffocating odor and is a powerful irritant to lungs and other tissues. The element is composed of Cl_2 units in which two chlorine atoms are tightly bound together.
- Sodium chloride, or common salt, is a colorless, crystalline solid composed of sodium and chloride ions bound tightly together (NaCl). Its properties are completely unlike those of the two elements from which it is made.

It is important to distinguish between a mixture of elements and a chemical compound of two or more elements. Pure metallic iron and yellow, powdered sulfur (Figure 1.13a) can be mixed in varying proportions. In the chemical compound iron pyrite (Figure 1.13b), however, there is no variation in composition. Not only does iron pyrite exhibit properties peculiar to itself and different from those of either iron or sulfur, or a mixture of these two elements, but it also has a definite percentage composition by mass (46.55% Fe and 53.45% S). Thus, two

■ **Writing Element Symbols** Notice that only the first letter of an element's symbol is capitalized. For example, cobalt is Co, not CO. The notation CO represents the chemical compound carbon monoxide. Also note that the element name is not capitalized, except at the beginning of a sentence.

■ **Origin of Element Names and Symbols** Many elements have names and symbols with Latin or Greek origins. Examples include helium (He), from the Greek word *helios* meaning "sun," and lead, whose symbol, Pb, comes from the Latin word for "heavy," *plumbum*. More recently discovered elements have been named for their place of discovery or for a person or place of significance. Examples include americium (Am), californium (Cf), and curium (Cm; for Marie Curie).

■ **Periodic Tables Online** Sign in at **www.thomsonedu.com/login** and go to Chapter 1 Contents to see Screen 1.7 or the Toolbox. See also the extensive information on the periodic table and the elements at the American Chemical Society website:

- acswebcontent.acs.org/games/pt.html

FIGURE 1.13 Mixtures and compounds. (a) The material in the dish is a mixture of iron chips and sulfur. The iron can be removed easily by using a magnet. (b) Iron pyrite is a chemical compound composed of iron and sulfur. It is often found in nature as perfect, golden cubes.

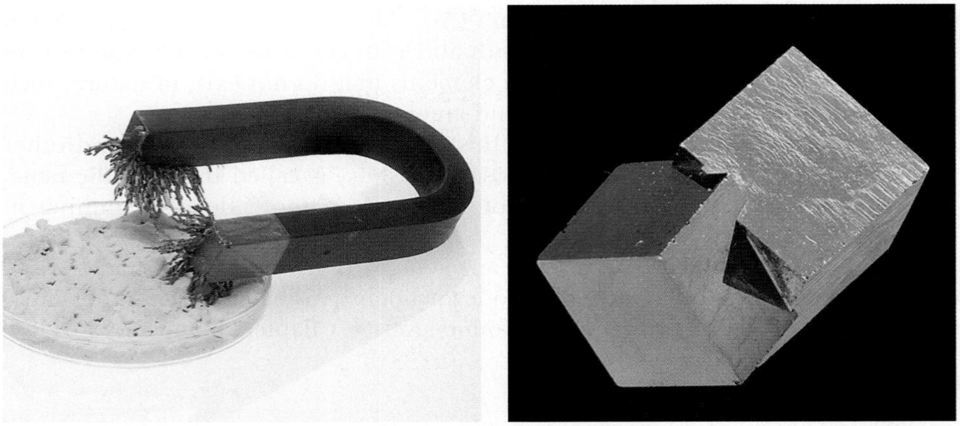

Photos: Charles D. Winters

(a) (b)

major differences exist between mixtures and pure compounds: Compounds have distinctly different characteristics from their parent elements, and they have a definite percentage composition (by mass) of their combining elements.

Some compounds—such as table salt, NaCl—are composed of **ions,** which are electrically charged atoms or groups of atoms [▶ Chapter 2]. Other compounds—such as water and sugar—consist of **molecules,** the smallest discrete units that retain the composition and chemical characteristics of the compound.

The composition of any compound is represented by its **chemical formula.** In the formula for water, H_2O, for example, the symbol for hydrogen, H, is followed by a subscript "2" indicating that two atoms of hydrogen occur in a single water molecule. The symbol for oxygen appears without a subscript, indicating that one oxygen atom occurs in the molecule.

As you shall see throughout this book, molecules can be represented with models that depict their composition and structure. Figure 1.14 illustrates the names, formulas, and models of the structures of a few common molecules.

1.5 Physical Properties

You recognize your friends by their physical appearance: their height and weight and the color of their eyes and hair. The same is true of chemical substances. You can tell the difference between an ice cube and a cube of lead of the same size not only because of their appearance (one is clear and colorless, and the other is a lustrous metal) (Figure 1.15), but also because one is more dense (lead) than the other (ice). Properties such as these, which can be observed and measured without changing the composition of a substance, are called **physical properties.** The chemical elements in Figure 1.12, for example, clearly differ in terms of their color,

FIGURE 1.14 Names, formulas, and models of some common molecules. Models of molecules appear throughout this book. In such models, C atoms are gray, H atoms are white, N atoms are blue, and O atoms are red.

NAME	Water	Methane	Ammonia	Carbon dioxide
FORMULA	H_2O	CH_4	NH_3	CO_2

MODEL

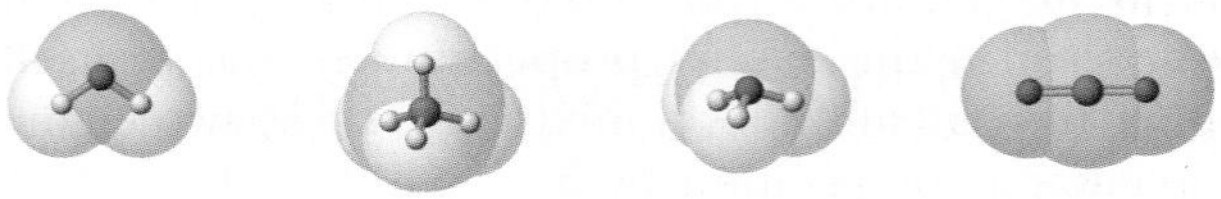

TABLE 1.1 Some Physical Properties

Property	Using the Property to Distinguish Substances
Color	Is the substance colored or colorless? What is the color, and what is its intensity?
State of matter	Is it a solid, liquid, or gas? If it is a solid, what is the shape of the particles?
Melting point	At what temperature does a solid melt?
Boiling point	At what temperature does a liquid boil?
Density	What is the substance's density (mass per unit volume)?
Solubility	What mass of substance can dissolve in a given volume of water or other solvent?
Electric conductivity	Does the substance conduct electricity?
Malleability	How easily can a solid be deformed?
Ductility	How easily can a solid be drawn into a wire?
Viscosity	How easily will a liquid flow?

Charles D. Winters

FIGURE 1.15 Physical properties. An ice cube and a piece of lead can be differentiated easily by their physical properties (such as density, color, and melting point).

appearance, and state (solid, liquid, or gas). Physical properties allow us to classify and identify substances. Table 1.1 lists a few physical properties of matter that chemists commonly use.

EXERCISE 1.2 Physical Properties

Identify as many physical properties in Table 1.1 as you can for the following common substances: (a) iron, (b) water, (c) table salt (chemical name is sodium chloride), and (d) oxygen.

Density, the ratio of the mass of an object to its volume, is a physical property useful for identifying substances.

$$\text{Density} = \frac{\text{mass}}{\text{volume}} \qquad \textbf{(1.1)}$$

For example, you can readily tell the difference between an ice cube and a cube of lead of identical size (Figure 1.15). Lead has a high density, 11.35 g/cm^3 (11.35 grams per cubic centimeter), whereas the density of ice is slightly less than 0.917 g/cm^3. An ice cube with a volume of 16.0 cm^3 has a mass of 14.7 g, whereas a cube of lead with the same volume has a mass of 182 g.

The **temperature** of a sample of matter often affects the numerical values of its properties. Density is a particularly important example. Although the change in water density with temperature seems small (Table 1.2), it affects our environment profoundly. For example, as the water in a lake cools, the density of the water increases, and the denser water sinks (Figure 1.16a). This continues until the water temperature reaches 3.98 °C, the point at which water has its maximum density (0.999973 g/cm^3). If the water temperature drops further, the density decreases slightly, and the colder water floats on top of water at 3.98 °C. If water is cooled below about 0 °C, solid ice forms. Water is unique among substances in the universe: Ice is less dense than water, so the solid ice floats on liquid water.

Because the density of liquids changes with temperature, the volume of a given mass of liquid also changes with temperature. This is the reason laboratory glassware used to measure precise volumes of solutions always specifies the temperature at which it was calibrated (Figure 1.16b).

■ **Units of Density** As described on page 25, the SI unit of mass is the kilogram and the SI unit of length is the meter. Therefore, the SI unit of density is kg/m^3. In chemistry, the more commonly used unit is g/cm^3. To convert from kg/m^3 to g/cm^3, divide by 1000.

■ **Calculations Involving Density and Mathematics Review** See *Let's Review* beginning on page 24 for a review of some of the mathematics used in introductory chemistry.

■ **Temperature Scales** Scientists use the Celsius (°C) and Kelvin scales (K) for temperature. See page 26.

TABLE 1.2 Temperature Dependence of Water Density

Temperature (°C)	Density of Water (g/cm^3)
0 (ice)	0.917
0 (liq water)	0.99984
2	0.99994
4	0.99997
10	0.99970
25	0.99707
100	0.95836

Chemical Perspectives — Thermophilic Bacteria

When you think of chemistry, you probably think of colored liquids bubbling in flasks and maybe a fire or even an explosion. That is not what we usually see in a university laboratory, but pay a visit to Yellowstone National Park in Wyoming (or to areas of the North Island of New Zealand) and you will see just that: bubbling, steaming hot water springs with colorful substances in a natural "laboratory."

Yellowstone Park is unique in having one of the highest concentrations of geysers, hot springs, steam vents, and mudpots on the planet. The reason rests in the geology of the area—the earth's crust is thinner here (about 64 km) compared with the crust covering the rest of the earth (144 km). Hot magma lies not far below Yellowstone's surface (6–16 km), and it heats the rocks above and the reservoirs of water closer to the surface (a physical process). The superheated water dissolves some of the minerals (another physical process), and it is forced upwards through fissures in the rocks and sometimes explodes through the surface as geysers and hot springs.

The hot water shooting to the surface carries with it dissolved minerals such as limestone (calcium carbonate), and they are deposited around the geysers and hot springs as limestone and travertine.

Once the hot water reaches the surface, it can harbor thermophilic or "heat-loving" bacteria. These can grow in enormous colonies with brilliant colors. Different bacteria grow at different temperatures, usually in the range of 50 °C to 70 °C, and the colors of the pools and streams can change with temperature. In the Grand Prismatic Spring shown in the photo, you can see that bacteria grow in the slightly cooler water around the edge of the spring, but the deep blue, very hot water in the center is devoid of living organisms. The bacteria growing around the hot springs are single-cell organisms ranging in size from 0.2 to 50 μm in diameter. They are often highly colored, owing to pigments such as carotenoids and chlorophylls, and different organisms with different colors grow at different temperatures. Some are anaerobic bacteria and use sulfur instead of oxygen for respiration (a chemical process).

A study of the hot springs of Yellowstone National Park is a good example of the intersection of chemistry, geology, and biology and of all of their subdisciplines (biochemistry, geochemistry, bacteriology, and mineralogy, among others). Our goal in this book is to introduce the chemistry background you will need to study more chemistry or to carve out a career in another field of science.

© James Cowlin

A travertine formation in Yellowstone National Park. The formation consists largely of limestone (calcium carbonate) mixed with silica.

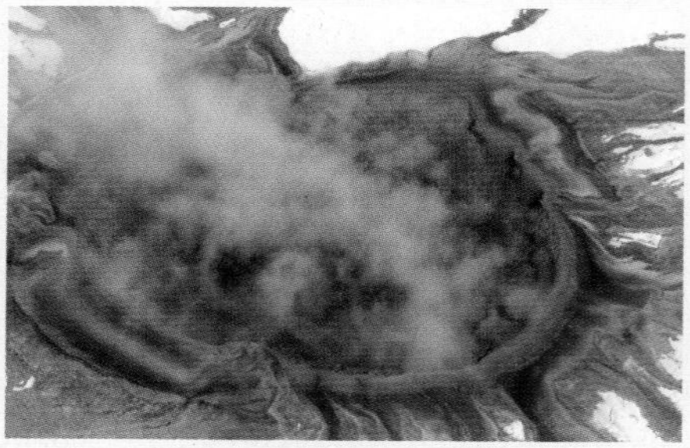

Stephen Hoerold

Grand Prismatic Spring, Yellowstone National Park.

Extensive and Intensive Properties

Extensive properties depend on the amount of a substance present. The mass and volume of the samples of elements in Figure 1.12, or the amount of heat obtained from burning gasoline, are extensive properties, for example. In contrast, **intensive properties** do *not* depend on the amount of substance. A sample of ice will melt at 0 °C, no matter whether you have an ice cube or an iceberg. Density is also an intensive property. The density of gold, for example, is the same (19.3 g/cm^3 at 20 °C) whether you have a flake of pure gold or a solid gold ring. Intensive properties are often useful in identifying a material. For example, the temperature at which a material melts (its melting point) is often so characteristic that it can be used to identify the solid (Figure 1.17).

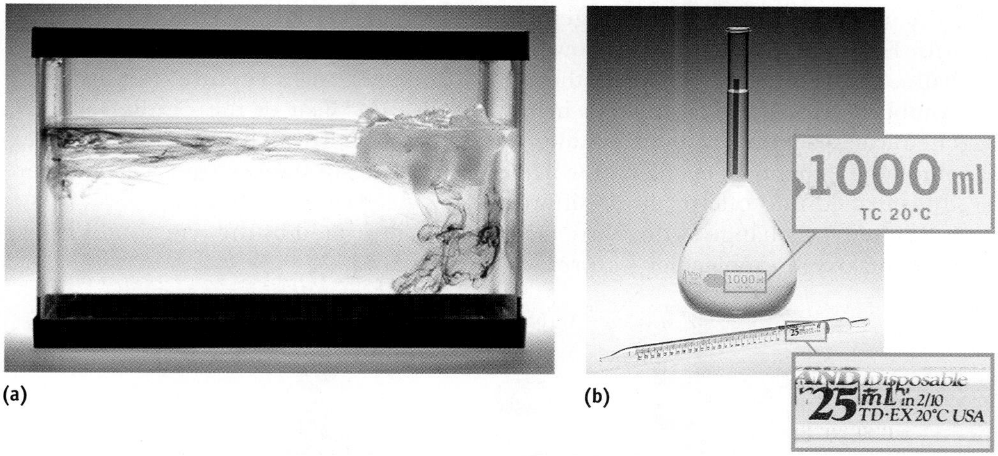

Photos: Charles D. Winters

FIGURE 1.16 Temperature dependence of physical properties. (a) Change in density with temperature. Ice cubes were placed in the right side of the tank and blue dye in the left side. The water beneath the ice is cooler and denser than the surrounding water, so it sinks. The convection current created by this movement of water is traced by the dye movement as the denser, cooler water sinks. (b) Temperature and calibration. Laboratory glassware is calibrated for specific temperatures. The pipet will deliver and the volumetric flask will contain the specified volume at the indicated temperature.

1.6 Physical and Chemical Changes

Changes in physical properties are called **physical changes.** In a physical change, the identity of a substance is preserved even though it may have changed its physical state or the gross size and shape of its pieces. A physical change does not result in a new chemical substance being produced. The substances (atoms, molecules, or ions) present before and after the change are the same. An example of a physical change is the melting of a solid (Figure 1.17). In the case of ice melting, the molecules present both before and after the change are H_2O molecules. Their chemical identity has not changed; they are now simply able to flow past one another in the liquid state instead of being locked in position in the solid.

Photos: Charles D. Winters

FIGURE 1.17 A physical property used to distinguish compounds. Aspirin and naphthalene are both white solids at 25 °C. You can tell them apart by, among other things, a difference in physical properties. At the temperature of boiling water, 100 °C, naphthalene is a liquid (left), whereas aspirin is a solid (right).

A physical property of hydrogen gas (H_2) is its low density, so a balloon filled with H_2 floats in air. Suppose, however, that a lighted candle is brought up to the balloon. When the heat causes the skin of the balloon to rupture, the hydrogen combines with the oxygen (O_2) in the air, and the heat of the candle sets off a chemical reaction, producing water, H_2O (Figure 1.18). This reaction is an example of a **chemical change,** in which one or more substances (the **reactants**) are transformed into one or more different substances (the **products**).

A chemical change at the particulate level is illustrated by the reaction of hydrogen and oxygen molecules to form water molecules.

$$2\ H_2(\text{gas}) + O_2(\text{gas}) \longrightarrow 2\ H_2O(\text{gas})$$

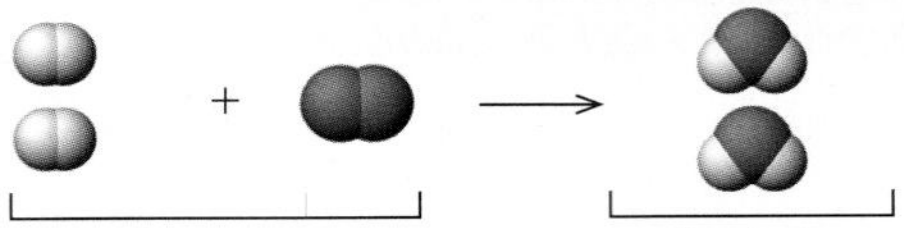

Case Study — Ancient and Modern Hair Coloring

Humankind has always been interested in pigments for artistic uses and to decorate their bodies. One of the oldest pigments is galena, PbS (Figure 1), which was first brought from Asia to Egypt thousands of years ago. When powdered, it is black and has been widely used as a cosmetic, particularly for dyeing hair. Some evidence for this is that small piles of galena were found next to the skeleton of a young Egyptian woman whose remains were buried in 3080 BC ± 110 years. Analysis by chemical archeologists found that the galena in this tomb was not from Asia, though. Instead, using modern forensic techniques, they found it came from nearby cities on the Red Sea.

According to recent research, the same effect on hair that galena has could be achieved by applying a mixture of the chemical compounds PbO and $Ca(OH)_2$. This was originally described by Claudius Galen, a Roman physician who lived between about 130 and 200 AD. Similar formulations have been used through the centuries for dyeing wool, and the present-day hair-coloring product Grecian Formula™ still uses this technique.

In recent research, chemists in France found that only a few hours after applying the PbO and $Ca(OH)_2$ mixture to hair (Figure 2a), the hair was blackened (Figure 2b), and that the blackening came from tiny particles of PbS (about 5 nanometers in diameter).

But where did the sulfur in PbS come from? The French research found it came from the amino acids in hair. (The amino acids cysteine and methionine both have sulfur as part of their structure.)

(a) (b)

FIGURE 1 Lead sulfide, galena. (a) Small crystal of the mineral galena, PbS. (b) The particulate view of PbS (where the gray spheres are Pb and the yellow spheres are S).

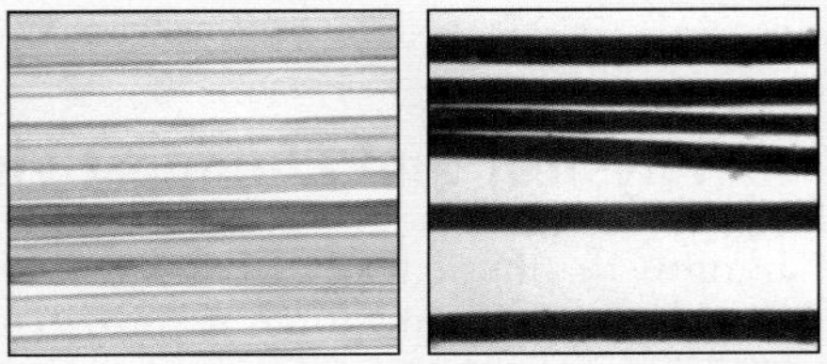

(a) (b)

FIGURE 2 Photographs of strands of hair before (a) and after (b) treating with a mixture of PbO and $Ca(OH)_2$. From *C & EN*, Vol. 84, No. 37, p. 12, 2006, "Still Dyeing After 2,000 Years," Dr. B. Halford. Copyright © 2006 American Chemical Society. Used with permission.

Questions:

1. *What is the name of the element Pb? Of Ca?*
2. *What is the density of Pb?*
3. *What is the symbol for the element sulfur?*
4. *Can you find (on the World Wide Web, for example) the common name for the compound $Ca(OH)_2$?*
5. *A particulate view of galena is illustrated in Figure 1b. Briefly describe the shape of this tiny piece of PbS.*
6. *Compare the particulate view of galena with that of NaCl on page 4. Are there similarities? Any differences?*

Answers to these questions are in Appendix Q.

References: Nano Letters, 2006, p. 2215; J. L. Lambert, *Traces of the Past,* Addison-Wesley, 1997, p. 80.

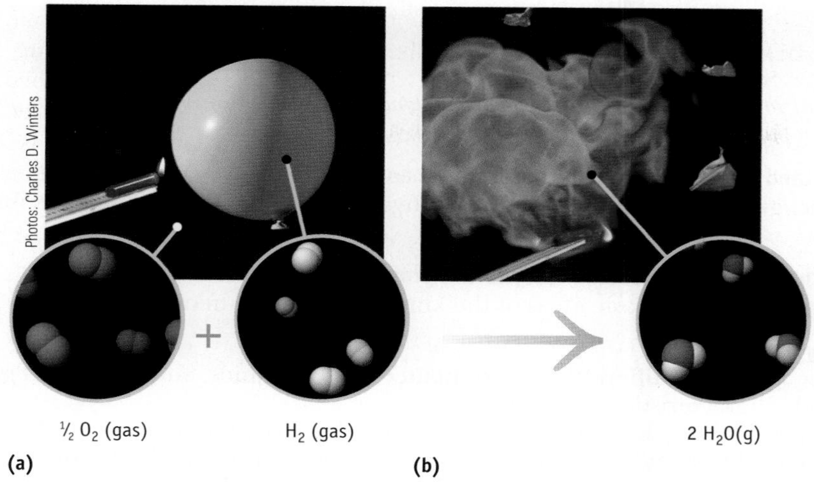

FIGURE 1.18 A chemical change—the reaction of hydrogen and oxygen. (a) A balloon filled with molecules of hydrogen gas and surrounded by molecules of oxygen in the air. (The balloon floats in air because gaseous hydrogen is less dense than air.) (b) When ignited with a burning candle, H_2 and O_2 react to form water, H_2O.

Chemistry Now™ Sign in at **www.thomsonedu.com/login** and go to Chapter 1 Contents to see Screen 1.11 Chemical Change, for a video of this reaction.

The representation of the change using chemical formulas is called a **chemical equation.** It shows that the substances on the left (the reactants) produce the substances on the right (the products). As this equation shows, there are four atoms of H and two atoms of O before *and* after the reaction, but the molecules before the reaction are different from those after the reaction.

A **chemical property** indicates whether and sometimes how readily a material undergoes a chemical change with another material. For example, a chemical property of hydrogen gas is that it reacts vigorously with oxygen gas.

Chemical and physical changes. A pot of water has been put on a campfire. What chemical and physical changes are occurring here (Exercise 1.3)?

Chemistry Now™

Sign in at **www.thomsonedu.com/login** and go to Chapter 1 Contents to see:

- Screen 1.12 for an exercise on **identifying physical and chemical changes**
- Screen 1.13 to watch a video and view an animation of the **molecular changes when chlorine gas and solid phosphorus react**

EXERCISE 1.3 Chemical Reactions and Physical Changes

When camping in the mountains, you boil a pot of water on a campfire. What physical and chemical changes take place in this process?

Chapter Goals Revisited

Chemistry.♁.Now™ Sign in at **www. thomsonedu.com/login** to:

- Assess your understanding with Study Questions in OWL keyed to each goal in the Goals and Homework menu for this chapter
- For quick review, download Go Chemistry mini-lecture flashcard modules (or purchase them at **www.ichapters.com**)
- Check your readiness for an exam by taking the Pre-Test and exploring the modules recommended in your Personalized Study plan.

Access **How Do I Solve It?** tutorials on how to approach problem solving using concepts in this chapter.

Now that you have studied this chapter, you should ask whether you have met the chapter goals. In particular, you should be able to:

Understand the nature of hypotheses, laws, and theories

a. Recognize the difference between a hypothesis and a theory and describe how laws are established.

Apply the kinetic-molecular theory to the properties of matter

a. Understand the basic ideas of the kinetic-molecular theory (Section 1.2).

Classify matter

a. Recognize the different states of matter (solids, liquids, and gases) and give their characteristics (Section 1.2).

b. Appreciate the difference between pure substances and mixtures and the difference between homogeneous and heterogeneous mixtures (Section 1.2).

c. Recognize the importance of representing matter at the macroscopic level and at the particulate level (Section 1.2).

Recognize elements, atoms, compounds, and molecules

a. Identify the name or symbol for an element, given its symbol or name (Section 1.3). **Study Question(s) assignable in OWL: 2, 4; Go Chemistry Module 1.**

b. Use the terms atom, element, molecule, and compound correctly (Sections 1.3 and 1.4).

Identify physical and chemical properties and changes

a. List commonly used physical properties of matter (Section 1.5).

b. Identify several physical and chemical properties of common substances (Sections 1.5 and 1.6). **Study Question(s) assignable in OWL: 8, 10, 13, 14, 31, 37.**

c. Relate density to the volume and mass of a substance (Section 1.5). **Study Question(s) assignable in OWL: 18, 19, 21, 23, 29, 30, 34.**

d. Explain the difference between chemical and physical changes (Section 1.6). **Study Question(s) assignable in OWL: 8, 13, 33.**

e. Understand the difference between extensive and intensive properties and give examples of them (Section 1.5). **Study Question(s) assignable in OWL: 11.**

KEY EQUATIONS

Equation 1.1 (page 15) Density. In chemistry the common unit of density is g/cm^3, whereas kg/m^3 is common in geology and oceanography.

$$\text{Density} = \frac{\text{mass}}{\text{volume}}$$

STUDY QUESTIONS

Online homework for this chapter may be assigned in OWL.

▲ denotes challenging questions.

■ denotes questions assignable in OWL.

Blue-numbered questions have answers in Appendix O and fully-worked solutions in the *Student Solutions Manual*.

Practicing Skills

Matter: Elements and Atoms, Compounds, and Molecules
(See Exercise 1.1)

1. Give the name of each of the following elements:
   (a) C (c) Cl (e) Mg
   (b) K (d) P (f) Ni

2. ■ Give the name of each of the following elements:
   (a) Mn (c) Na (e) Xe
   (a) Cu (d) Br (f) Fe

3. Give the symbol for each of the following elements:
(a) barium (c) chromium (e) arsenic
(b) titanium (d) lead (f) zinc

4. ■ Give the symbol for each of the following elements:
(a) silver (c) plutonium (e) technetium
(b) aluminum (d) tin (f) krypton

5. In each of the following pairs, decide which is an element and which is a compound.
(a) Na and NaCl
(b) Sugar and carbon
(c) Gold and gold chloride

6. ■ In each of the following pairs, decide which is an element and which is a compound.
(a) $Pt(NH_3)_2Cl_2$ and Pt
(b) Copper or copper(II) oxide
(c) Silicon or sand

Physical and Chemical Properties

(See Exercises 1.2 and 1.3)

7. In each case, decide if the underlined property is a physical or chemical property.
(a) The color of elemental bromine is orange-red.
(b) Iron turns to rust in the presence of air and water.
(c) Hydrogen can explode when ignited in air (Figure 1.18).
(d) The density of titanium metal is 4.5 g/cm^3.
(e) Tin metal melts at 505 K.
(f) Chlorophyll, a plant pigment, is green.

8. ■ In each case, decide if the change is a chemical or physical change.
(a) A cup of household bleach changes the color of your favorite T-shirt from purple to pink.
(a) Water vapor in your exhaled breath condenses in the air on a cold day.
(b) Plants use carbon dioxide from the air to make sugar.
(c) Butter melts when placed in the sun.

9. Which part of the description of a compound or element refers to its physical properties and which to its chemical properties?
(a) The colorless liquid ethanol burns in air.
(b) The shiny metal aluminum reacts readily with orange-red bromine.

10. ■ Which part of the description of a compound or element refers to its physical properties and which to its chemical properties?
(a) Calcium carbonate is a white solid with a density of 2.71 g/cm^3. It reacts readily with an acid to produce gaseous carbon dioxide.
(b) Gray, powdered zinc metal reacts with purple iodine to give a white compound.

General Questions

These questions are not designated as to type or location in the chapter. They may combine several concepts.

11. ■ A piece of turquoise is a blue-green solid; it has a density of 2.65 g/cm^3 and a mass of 2.5 g.
(a) Which of these observations are qualitative and which are quantitative?
(b) Which of the observations are extensive and which are intensive?
(c) What is the volume of the piece of turquoise?

12. Give a physical property and a chemical property for the elements hydrogen, oxygen, iron, and sodium. (The elements listed are selected from examples given in Chapter 1.)

13. ■ Eight observations are listed below. What observations identify chemical properties?
(a) Sugar is soluble in water.
(b) Water boils at 100 °C.
(c) Ultraviolet light converts O_3 (ozone) to O_2 (oxygen).
(d) Ice is less dense than water.
(e) Sodium metal reacts violently with water.
(f) CO_2 does not support combustion.
(g) Chlorine is a green gas.
(i) Heat is required to melt ice.

14. ■ Azurite, a blue, crystalline mineral, is composed of copper, carbon, and oxygen.

Azurite is a deep blue crystalline mineral. It is surrounded by copper pellets and powdered carbon (in the dish).

(a) What are the symbols of the three elements that combine to make the mineral azurite?
(b) Based on the photo, describe some of the physical properties of the elements and the mineral. Are any the same? Are any properties different?

15. The mineral fluorite contains the elements calcium and fluorine and has colors that range from blue, to violet, to green and yellow.

The mineral fluorite, calcium fluoride.

What are the symbols of these elements? How would you describe the shape of the fluorite crystals in the photo? What can this tell us about the arrangement of the particles (ions) inside the crystal?

16. Small chips of iron are mixed with sand (see the following photo). Is this a homogeneous or heterogeneous mixture? Suggest a way to separate the iron from the sand.

Charles D. Winters

Chips of iron mixed with sand.

17. In Figure 1.4 you see a piece of salt and a representation of its internal structure. Which is the macroscopic view and which is the particulate view? How are the macroscopic and particulate views related?

18. ■ The following photo shows copper balls, immersed in water, floating on top of mercury. What are the liquids and solids in this photo? Which substance is most dense? Which is least dense?

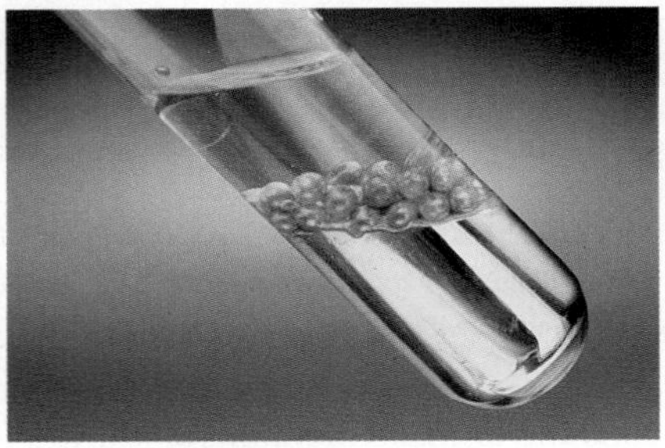

Charles D. Winters

Water, copper, and mercury.

19. ■ Carbon tetrachloride, CCl_4, a common liquid compound, has a density of 1.58 g/cm^3. If you place a piece of a plastic soda bottle (d = 1.37 g/cm^3) and a piece of aluminum (d = 2.70 g/cm^3) in liquid CCl_4, will the plastic and aluminum float or sink?

20. ▲ You have a sample of a white crystalline substance from your kitchen. You know that it is either salt or sugar. Although you could decide by taste, suggest another property that you could use to decide. (Hint: You may use the World Wide Web or a handbook of chemistry in the library to find some information.)

21. ■ Hexane (C_6H_{14}, density = 0.766 g/cm^3), perfluorohexane (C_6F_{14}, density = 1.669 g/cm^3), and water are immiscible liquids; that is, they do not dissolve in one another. You place 10 mL of each in a graduated cylinder, along with pieces of high-density polyethylene (HDPE, density 0.97 g/cm^3), polyvinyl chloride (PVC, density = 1.36 g/cm^3), and Teflon (density = 2.3 g/cm^3). None of these common plastics dissolves in these liquids. Describe what you expect to see.

22. Milk in a glass bottle was placed in the freezing compartment of a refrigerator overnight. By morning, a column of frozen milk emerged from the bottle. Explain this observation.

Charles D. Winters

Frozen milk in a glass bottle.

23. ■ You can figure out whether a substance floats or sinks if you know its density and the density of the liquid. In which of the liquids listed below will high-density polyethylene (HDPE) float. (HDPE, a common plastic, has a density of 0.97 g/cm^3. It does not dissolve in any of these liquids.)

Substance	Density (g/cm^3)	Properties, Uses
Ethylene glycol	1.1088	Toxic; the major component of automobile antifreeze
Water	0.9997	
Ethanol	0.7893	The alcohol in alcoholic beverages
Methanol	0.7914	Toxic; gasoline additive to prevent gas line freezing
Acetic acid	1.0492	Component of vinegar
Glycerol	1.2613	Solvent used in home care products

24. Describe an experimental method that can be used to determine the density of an irregularly shaped piece of metal.

25. ▲ Make a drawing, based on the kinetic-molecular theory and the ideas about atoms and molecules presented in this chapter, of the arrangement of particles in each of the cases listed here. For each case, draw 10 particles of each substance. It is acceptable for your diagram to be two dimensional. Represent each atom as a circle, and distinguish each different kind of atom by shading.
(a) A sample of solid iron (which consists of iron atoms)
(b) A sample of *liquid* water (which consists of H_2O molecules)
(c) A sample of water *vapor*

▲ more challenging ■ in OWL Blue-numbered questions answered in Appendix O

26. ▲ Make a drawing, based on the kinetic-molecular theory and the ideas about atoms and molecules presented in this chapter, of the arrangement of particles in each of the cases listed here. For each case, draw 10 particles of each substance. It is acceptable for your diagram to be two dimensional. Represent each atom as a circle, and distinguish each different kind of atom by shading.
(a) A homogeneous mixture of water vapor and helium gas (which consists of helium atoms)
(b) A heterogeneous mixture consisting of liquid water and solid aluminum; show a region of the sample that includes both substances
(c) A sample of brass (which is a homogeneous solid mixture of copper and zinc)

27. You are given a sample of a silvery metal. What information would you seek to prove that the metal is silver?

28. Suggest a way to determine if the colorless liquid in a beaker is water. If it is water, does it contain dissolved salt? How could you discover if there is salt dissolved in the water?

29. ■ Diabetes can alter the density of urine, and so urine density can be used as a diagnostic tool. Diabetics can excrete too much sugar or excrete too much water. What do you predict will happen to the density of urine under each of these conditions? (*Hint*: Water containing dissolved sugar is more dense than pure water.)

30. ■ Three liquids of different densities are mixed. Because they are not miscible (do not form a homogeneous solution with one another), they form discrete layers, one on top of the other. Sketch the result of mixing carbon tetrachloride (CCl_4, $d = 1.58$ g/cm^3), mercury ($d = 13.546$ g/cm^3), and water ($d = 1.00$ g/cm^3).

31. ■ The following photo shows the element potassium reacting with water to form the element hydrogen, a gas, and a solution of the compound potassium hydroxide.

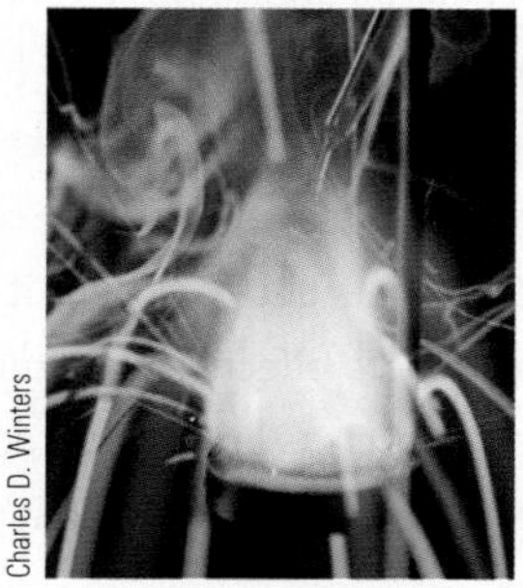

Potassium reacting with water to produce hydrogen gas and potassium hydroxide.

(a) What states of matter are involved in the reaction?
(b) Is the observed change chemical or physical?
(c) What are the reactants in this reaction, and what are the products?
(d) What qualitative observations can be made concerning this reaction?

32. A copper-colored metal is found to conduct an electric current. Can you say with certainty that it is copper? Why or why not? Suggest additional information that could provide unequivocal confirmation that the metal is copper.

33. ■ What experiment can you use to:
(a) Separate salt from water?
(b) Separate iron filings from small pieces of lead?
(c) Separate elemental sulfur from sugar?

34. ■ Four balloons are each filled with a different gas of varying density:

Helium, $d = 0.164$ g/L

Neon, $d = 0.825$ g/L

Argon, $d = 1.633$ g/L

Krypton, $d = 4.425$ g/L

If the density of dry air is 1.12 g/L, which balloon or balloons float in air?

35. Many foods are fortified with vitamins and minerals. Some breakfast cereals have elemental iron added. Iron chips are used instead of iron compounds because the compounds can be converted by the oxygen in air to a form of iron that is not biochemically useful. Iron chips, on the other hand, are converted to useful iron compounds in the gut, and the iron can then be absorbed. Outline a method by which you could remove the iron (as iron chips) from a box of cereal and determine the mass of iron in a given mass of cereal. (*See ChemistryNow Screens 1.1 and 1.18, Chemical Puzzler.*)

36. Study the animation of the conversion of P_4 and Cl_2 molecules to PCl_3 molecules in *ChemistryNow, Screen 1.12 (Chemical Change on the Molecular Scale)*.
(a) What are the reactants in this chemical change? What are the products?
(b) Describe how the structures of the reactant molecules differ from the structures of the product molecules.

37. ■ The photo below shows elemental iodine dissolving in ethanol to give a solution. Is this a physical or chemical change?

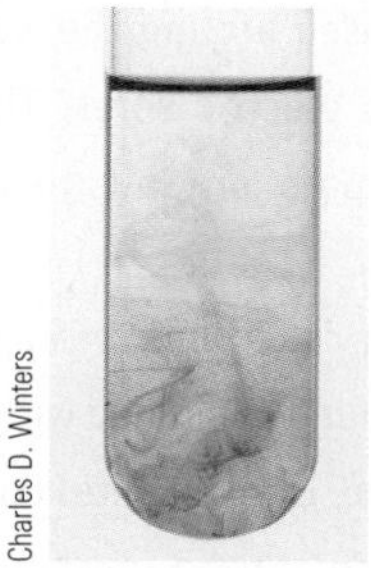

Elemental iodine dissolving in ethanol.

(See also the ChemistryNow Screen 1.12, Exercise, Physical Properties of Matter)

Let's Review

The Tools of Quantitative Chemistry

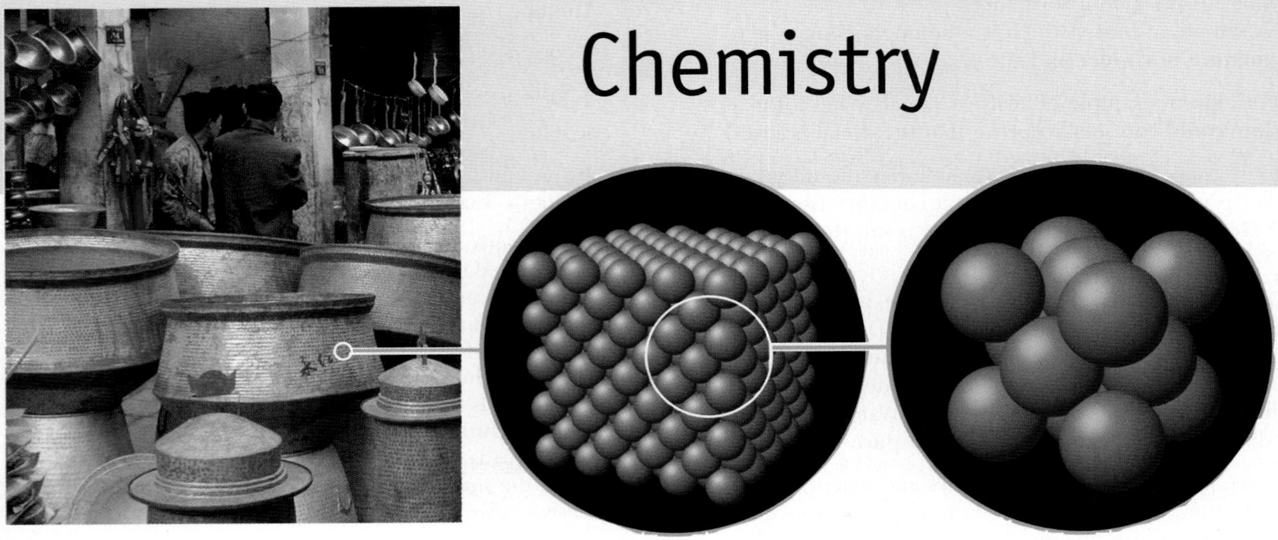

John Kotz

Copper

Copper (Cu) is the 26th element in abundance in the Earth's crust (not too different from its near neighbors nickel and zinc in the periodic table), but it and its minerals are widely distributed, and it is relatively easy to obtain the metal from its ores. As a result, elemental copper is used around the world for many useful items, from cooking pots to electric wires. The photo at the left above shows large copper pots on sale in a market in southwestern China.

Pure copper (often called *native copper*) is found in nature, but more commonly it is found combined with other elements in minerals such as cuprite, azurite, or malachite. Copper metal is relatively soft but, when combined in a ratio of about 2 to 1 with tin, it forms bronze. Bronze was important in early civilizations and gave its name to an epoch of human development, the Bronze Age, which started around 3000 BC and lasted until about 1000 BC. The development of bronze was significant because bronze is stronger than copper and can be shaped into a sharper edge. This improved the cutting edges of plows and weapons, thus giving cultures that possessed bronze advantages over those that did not.

Copper is now used in wiring because it conducts electricity well, and it is used in cooking pots because it conducts heat well. It is also described as one of the "coinage metals" (along with silver and gold) because it has been used in coins for centuries.

Compounds of copper are common, and copper is one of the eight essential metals in our bodies, where it is needed for some enzymes to use oxygen more effectively. Fortunately, it is found in common foods (in meats such as lamb, duck, pork, and beef, and in almonds and walnuts). The average person has about 72 mg of copper in his or her body.

The figure above also shows what happens as we zoom into copper at the particulate level. We begin to see atoms arranged in a regular array, or *lattice,* as chemists call it. Zooming in even closer, we see the smallest repeating unit of the crystal.

You can learn more about copper and its properties by answering the Study Questions 56 and 57 at the end of this *Let's Review* section.

At its core, chemistry is a quantitative science. Chemists make measurements of, among other things, size, mass, volume, time, and temperature. Scientists then manipulate that information to search for relationships among properties and to provide insight into the molecular basis of matter.

This section reviews the units used in chemistry, briefly describes the proper treatment of numerical data, and reviews some mathematical skills you will need in chemical calculations. After studying this section you should be able to:

- use the common units for measurements in chemistry and make unit conversions (such as liters to milliliters).
- express and use numbers in exponential or scientific notation.
- express quantitative information in an algebraic expression and solve that expression.
- read information from graphs.
- prepare a graph of numerical information. If the graph produces a straight line, find the slope and equation of the line.
- recognize and express uncertainties in measurements.

1 Units of Measurement

Doing chemistry requires observing chemical reactions and physical changes. We make qualitative observations—such as changes in color or the evolution of heat—and quantitative measurements of temperature, time, volume, mass, and length or size. To record and report measurements, the scientific community has chosen a modified version of the **metric system.** This decimal system, used internationally in science, is called the Systême International d'Unités (International System of Units), abbreviated **SI.**

All SI units are derived from base units, some of which are listed in Table 1. Larger and smaller quantities are expressed by using appropriate prefixes with the base unit (Table 2). The nanometer (nm), for example, is 1 billionth of a meter. That is, it is equivalent to 1×10^{-9} m (meter). Dimensions on the nanometer scale are common in chemistry and biology because a typical molecule is about 1 nm across and a bacterium is about 1000 nm in length. The prefix *nano-* is also used in the name for a new area of science, *nanotechnology* (▶ *Materials Chemistry,* pages 656–669) which involves the synthesis and study of materials around this tiny size.

TABLE 1 Some SI Base Units

Measured Property	Name of Unit	Abbreviation
Mass	kilogram	kg
Length	meter	m
Time	second	s
Temperature	kelvin	K
Amount of substance	mole	mol
Electric current	ampere	A

■ **Common Conversion Factors**
1000 g = 1 kg
1×10^9 nm = 1 m
10 mm = 1 cm
100 cm = 10 dm = 1 m
1000 m = 1 km
Conversion factors for SI units are given in Appendix C and inside the back cover of this book.

TABLE 2 Selected Prefixes Used in the Metric System

Prefix	Abbreviation	Meaning	Example
giga-	G	10^9 (billion)	1 gigahertz = 1×10^9 Hz
mega-	M	10^6 (million)	1 megaton = 1×10^6 tons
kilo-	k	10^3 (thousand)	1 kilogram (kg) = 1×10^3 g
deci-	d	10^{-1} (tenth)	1 decimeter (dm) = 1×10^{-1} m
centi-	c	10^{-2} (one hundredth)	1 centimeter (cm) = 1×10^{-2} m
milli-	m	10^{-3} (one thousandth)	1 millimeter (mm) = 1×10^{-3} m
micro-	μ	10^{-6} (one millionth)	1 micrometer (μm) = 1×10^{-6} m
nano-	n	10^{-9} (one billionth)	1 nanometer (nm) = 1×10^{-9} m
pico-	p	10^{-12}	1 picometer (pm) = 1×10^{-12} m
femto-	f	10^{-15}	1 femtometer (fm) = 1×10^{-15} m

Temperature Scales

Two temperature scales are commonly used in scientific work: Celsius and Kelvin (Figure 1). The Celsius scale is generally used worldwide for measurements in the laboratory. When calculations incorporate temperature data, however, the Kelvin scale must be used.

The Celsius Temperature Scale

The size of the Celsius degree is defined by assigning zero as the freezing point of pure water (0 °C) and 100 as its boiling point (100 °C). You may recognize that a comfortable room temperature is around 20 °C and your normal body temperature is 37 °C. And we find that the warmest water we can stand to immerse a finger in is about 60 °C.

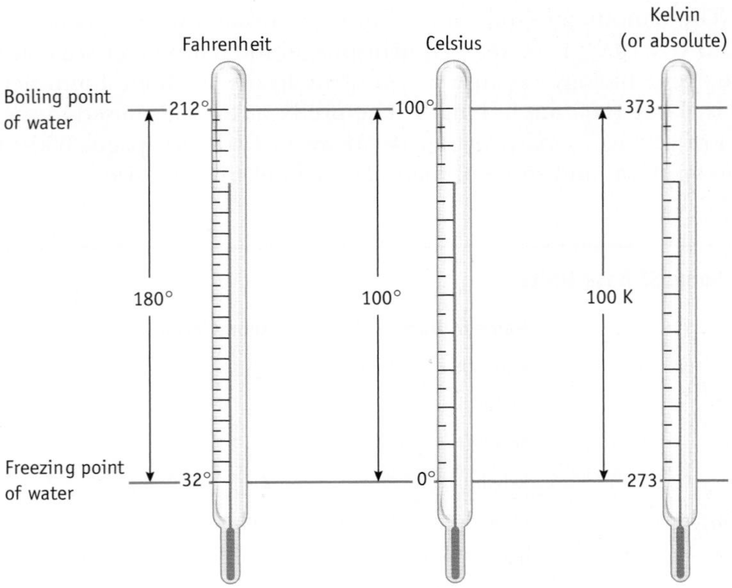

Active Figure 1 A comparison of Fahrenheit, Celsius, and Kelvin scales. The reference, or starting point, for the Kelvin scale is absolute zero (0 K = −273.15 °C), which has been shown theoretically and experimentally to be the lowest possible temperature.

Chemistry•Now™ Sign in at www.thomsonedu.com/login and go to the Chapter Contents menu to explore an interactive version of this figure accompanied by an exercise.

The Kelvin Temperature Scale

William Thomson, known as Lord Kelvin (1824–1907), first suggested the temperature scale that now bears his name. The Kelvin scale uses the same size unit as the Celsius scale, but it assigns zero as the lowest temperature that can be achieved, a point called **absolute zero.** Many experiments have found that this limiting temperature is −273.15 °C (−459.67 °F). *Kelvin units and Celsius degrees are the same size.* Thus, the freezing point of water is reached at 273.15 K; that is, 0 °C = 273.15 K. The boiling point of pure water is 373.15 K. Temperatures in Celsius degrees are readily converted to kelvins, and vice versa, using the relation

$$T\ (\mathrm{K}) = \frac{1\ \mathrm{K}}{1\ ^\circ\mathrm{C}}\ (T\ ^\circ\mathrm{C} + 273.15\ ^\circ\mathrm{C}) \qquad \textbf{(1)}$$

Thus, a common room temperature of 23.5 °C is

$$T\ (\mathrm{K}) = \frac{1\ \mathrm{K}}{1\ \cancel{^\circ\mathrm{C}}}\ (23.5\ \cancel{^\circ\mathrm{C}} + 273.15\ \cancel{^\circ\mathrm{C}}) = 296.7\ \mathrm{K}$$

Finally, notice that the degree symbol (°) is not used with Kelvin temperatures. The name of the unit on this scale is the *kelvin* (not capitalized), and such temperatures are designated with a capital K.

E. F. Smith Collection/Van Pelt Library/University of Pennsylvania

■ **Lord Kelvin** William Thomson (1824–1907), known as Lord Kelvin, was a professor of natural philosophy at the University in Glasgow, Scotland, from 1846 to 1899. He was best known for his work on heat and work, from which came the concept of the absolute temperature scale.

■ **Temperature Conversions** When converting 23.5 °C to kelvins, adding 273.15 gives 296.65. However, the rules of "significant figures" (page 35) tell us that the sum or difference of two numbers can have no more decimal places than the number with the fewest decimal places. Thus, we round the answer to 296.7 K, a number with one decimal place.

EXERCISE 1 Temperature Scales

Liquid nitrogen boils at 77 K. What is this temperature in Celsius degrees?

Length, Volume, and Mass

The meter is the standard unit of *length*, but objects observed in chemistry are frequently smaller than 1 meter. Measurements are often reported in units of centimeters (cm), millimeters (mm), or micrometers (μm) (Figure 2), and objects on the atomic and molecular scale have dimensions of nanometers (nm; 1 nm = 1×10^{-9} m) or picometers (pm; 1 pm = 1×10^{-12} m) (Figure 3).

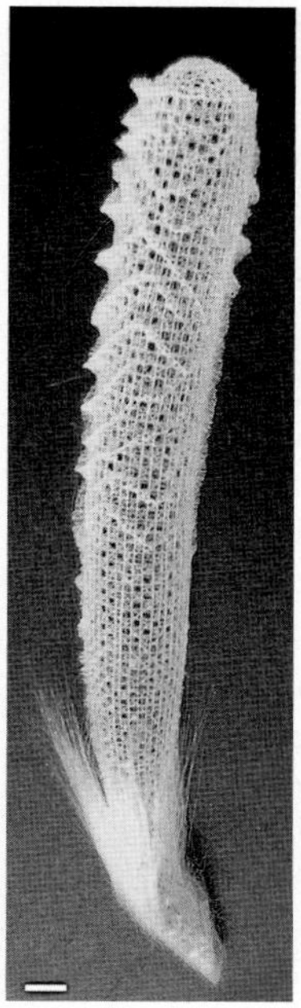

(a)

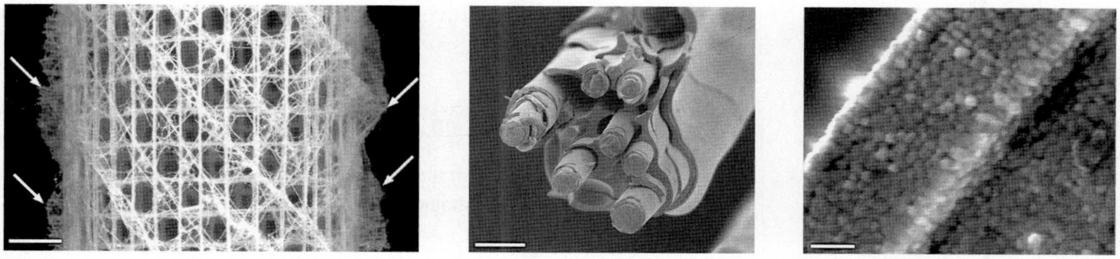

(b) (c) (d)

Photos courtesy of Joanna Aizenberg, Bell Laboratories. Reference: J. Aizenberg, et al., *Science*, Vol. 309, pages 275-278, 2005.

FIGURE 2 Dimensions in chemistry and biology. (a) Photograph of the glassy skeleton of a sea sponge, *Euplectella*. Scale bar = 5 cm. (b) Fragment of the structure showing the square grid of the lattice with diagonal supports. Scale bar = 1 mm. (c) Scanning electron microscope (SEM) image of a single strand showing its ceramic-composite structure. Scale bar = 20 μm. (d) SEM image of the surface of a strand showing that is it composed of nanoscale spheres of hydrated silica. Scale bar = 500 nm.

FIGURE 3 Dimensions in the molecular world. Dimensions on the molecular scale are often given in terms of nanometers (1 nm = 1×10^{-9} m) or picometers (1 pm = 1×10^{-12} m). Here, the distance between C atoms in diamond is 0.154 nm or 154 pm. An older, but often-used non-SI unit is the Ångstrom unit (Å), where 1 Å = 1.0×10^{-10} m. The C–C distance in diamond would be 1.54 A.

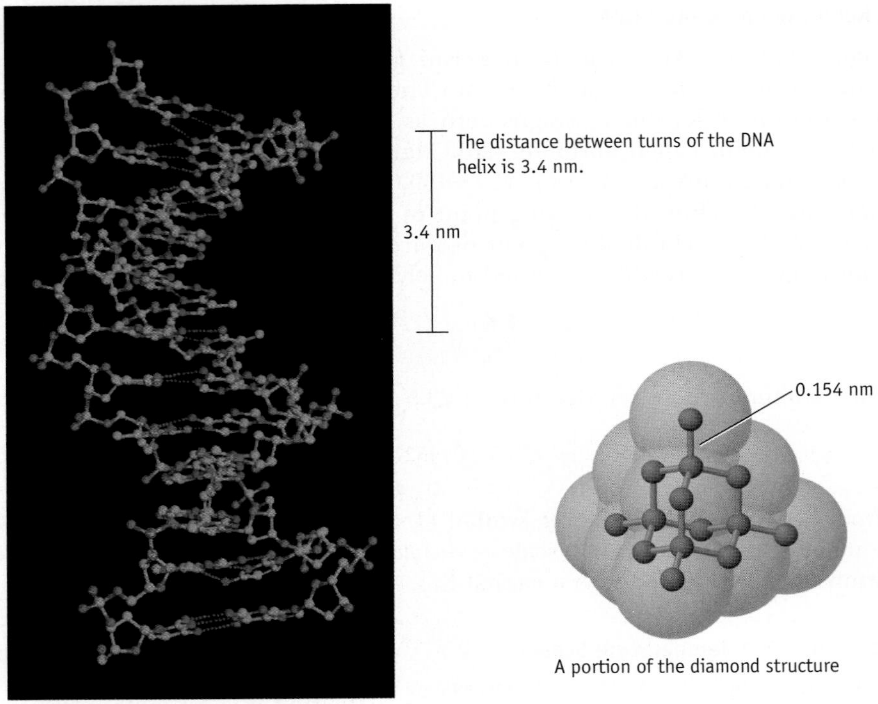

To illustrate the range of dimensions used in science, let us look at a recent study of the glassy skeleton of a sea sponge. The sea sponge in Figure 2a is about 20 cm long and a few centimeters in diameter. A closer look (Figure 2b) shows more detail of the lattice-like structure. Scientists at Bell Laboratories found that each strand of the lattice is a ceramic-fiber composite of silica (SiO_2) and protein less than 100 μm in diameter (Figure 2c). These strands are composed of "spicules," which, at the nanoscale level, consist of silica nanoparticles just a few nanometers in diameter (Figure 2d).

EXAMPLE 1 Distances on the Molecular Scale

Problem The distance between an O atom and an H atom in a water molecule is 95.8 pm. What is this distance in meters (m)? In nanometers (nm)?

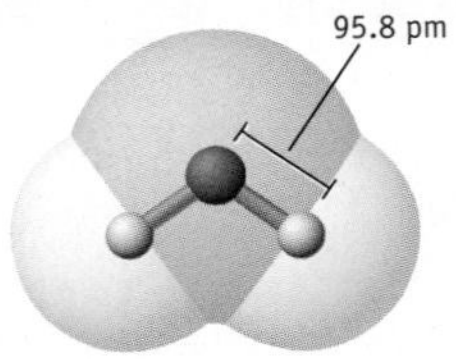

Strategy You can solve this problem by knowing the relationship or conversion factor between the units in the information you are given (picometers) and the desired units (meters or nanometers). (For more about conversion factors and their use in problem solving, see page 38.) There is no conversion factor given in Table 2 to change nanometers to picometers directly, but relationships are listed between meters and picometers and between meters and nanometers. Therefore, we first convert picometers to meters, and then we convert meters to nanometers.

$$\text{Picometers} \xrightarrow{\times\ \text{m/pm}} \text{Meters} \xrightarrow{\times\ \text{nm/m}} \text{Nanometers}$$

Solution Using the appropriate conversion factors (1 pm = 1×10^{-12} m and 1 nm = 1×10^{-9} m), we have

$$95.8 \text{ pm} \times \frac{1 \times 10^{-12} \text{ m}}{1 \text{ pm}} = 9.58 \times 10^{-11} \text{ m}$$

$$9.58 \times 10^{-11} \text{ m} \times \frac{1 \text{ nm}}{1 \times 10^{-9} \text{ m}} = 9.58 \times 10^{-2} \text{ nm} \text{ or } 0.0958 \text{ nm}$$

Comment Notice how the units cancel to leave an answer whose unit is that of the numerator of the conversion factor. The process of using units to guide a calculation is called *dimensional analysis*. It is explored further on pages 38–39.

EXERCISE 2 Using Units of Length

A platinum sheet is 2.50 cm square and has a thickness of 0.25 mm. What is the volume of the platinum sheet (in cm^3)?

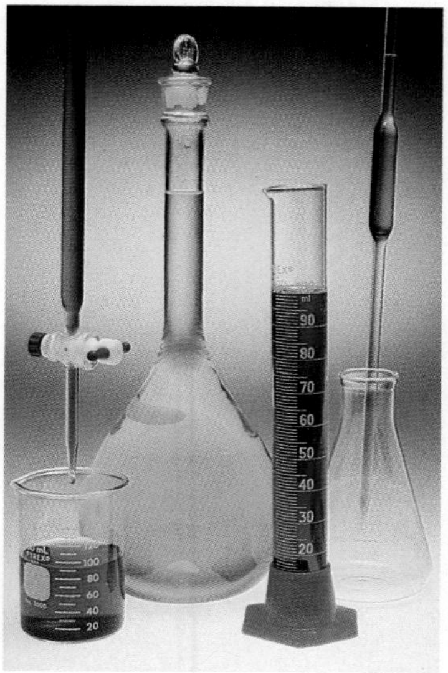

Charles D. Winters

FIGURE 4 Some common laboratory glassware. Volumes are marked in units of milliliters (mL). Remember that 1 mL is equivalent to 1 cm^3.

Chemists often use glassware such as beakers, flasks, pipets, graduated cylinders, and burets, which are marked in volume units (Figure 4). The SI unit of volume is the cubic meter (m^3), which is too large for everyday laboratory use. Chemists usually use the liter, symbolized by L, for volume measurements. One liter is equivalent to the volume of a cube with sides equal to 10 cm [= $(0.1 \text{ m})^3$ = 0.001 m^3].

$$1 \text{ liter (L)} = 1000 \text{ cm}^3 = 1000 \text{ mL} = 0.001 \text{ m}^3$$

The liter is a convenient unit to use in the laboratory, as is the milliliter (mL). Because there are exactly 1000 mL (= 1000 cm^3) in a liter, this means that

$$1 \text{ mL} = 0.001 \text{ L} = 1 \text{ cm}^3$$

The units *milliliter and cubic centimeter* (or "cc") *are interchangeable.* Therefore, a flask that contains exactly 125 mL has a volume of 125 cm^3.

Although not widely used in the United States, the cubic decimeter (dm^3) is a common unit in the rest of the world. A length of 10 cm is called a decimeter (dm). Because a cube 10 cm on a side defines a volume of 1 liter, *a liter is equivalent to a cubic decimeter.* 1 L = 1 dm^3. Products in Europe, Africa, and other parts of the world are often sold by the cubic decimeter.

The *deciliter, dL,* which is exactly equivalent to 0.100 L or 100 mL, is widely used in medicine. For example, standards for concentrations of environmental contaminants are often set as a certain mass per deciliter. The state of Massachusetts recommends that children with more than 10 micrograms (10×10^{-6} g) of lead per deciliter of blood undergo further testing for lead poisoning.

EXERCISE 3 Volume

(a) A standard wine bottle has a volume of 750 mL. What volume, in liters, does this represent? How many deciliters?

(b) One U.S. gallon is equivalent to 3.7865 L. What is the volume in liters of a 2.0-quart carton of milk? (There are 4 quarts in a gallon.) How many cubic decimeters?

Finally, when chemists prepare chemicals for reactions, they often take given quantities or masses of materials. The *mass* of a body is the fundamental measure of the quantity of matter, and the SI unit of mass is the kilogram (kg). Smaller masses are expressed in grams (g) or milligrams (mg).

$$1 \text{ kg} = 1000 \text{ g } \textit{and } 1 \text{ g} = 1000 \text{ mg}$$

> **EXERCISE 4 Mass**
>
> **(a)** A new U.S. quarter has a mass of 5.59 g. Express this mass in kilograms and milligrams.
>
> **(b)** An environmental study of a river found a pesticide present to the extent of 0.02 microgram per liter of water. Express this amount in grams per liter.

2 Making Measurements: Precision, Accuracy, Experimental Error, and Standard Deviation

The **precision** of a measurement indicates how well several determinations of the same quantity agree. This is illustrated by the results of throwing darts at a target. In Figure 5a, the dart thrower was apparently not skillful, and the precision of the dart's placement on the target is low. In Figures 5b and 5c, the darts are clustered together, indicating much better consistency on the part of the thrower—that is, greater precision.

Accuracy is the agreement of a measurement with the accepted value of the quantity. Figure 5c shows that our thrower was accurate as well as precise—the average of all shots is close to the targeted position, the bull's eye.

Figure 5b shows it is possible to be precise without being accurate—the thrower has consistently missed the bull's eye, although all the darts are clustered precisely around one point on the target. This is analogous to an experiment with some flaw (either in design or in a measuring device) that causes all results to differ from the correct value by the same amount.

The accuracy of a result in the laboratory is often expressed in terms of percent error, whereas the precision is expressed as a standard deviation.

■ **Accuracy and NIST** The National Institute for Standards and Technology (NIST) is an important resource for the standards used in science. Comparison with NIST data is a test of the accuracy of the measurement. See www.nist.gov.

Experimental Error

If you measure a quantity in the laboratory, you may be required to report the error in the result, the difference between your result and the accepted value,

$$\text{Error} = \text{experimentally determined value} - \text{accepted value}$$

Charles D. Winters

(a) Poor precision and poor accuracy **(b)** Good precision and poor accuracy **(c)** Good precision and good accuracy

FIGURE 5 Precision and accuracy.

or the **percent error.**

$$\text{Percent error} = \frac{\text{error in measurement}}{\text{accepted value}} \times 100\%$$

■ **Percent Error** Percent error can be positive or negative, indicating whether the experimental value is too high or too low compared to the accepted value. In Example 2, Student B's error is −0.2%, indicating it is 0.2% lower than the accepted value.

■ **EXAMPLE 2 Precision, Accuracy, and Error**

Problem A coin has an "accepted" diameter of 28.054 mm. In an experiment, two students measure this diameter. Student A makes four measurements of the diameter of the coin using a precision tool called a micrometer. Student B measures the same coin using a simple plastic ruler. The two students report the following results:

Student A	**Student B**
28.246 mm	27.9 mm
28.244	28.0
28.246	27.8
28.248	28.1

What is the average diameter and percent error obtained in each case? Which student's data are more accurate?

Strategy For each set of values, we calculate the average of the results and then compare this average with 28.054 mm.

Solution The average for each set of data is obtained by summing the four values and dividing by 4.

Average value for Student A = 28.246 mm
Average value for Student B = 28.0 mm

Although Student A has four results very close to one another (and so of high precision), student A's result is less accurate than that of Student B. The average diameter for Student A differs from the "accepted" value by 0.192 mm and has a percent error of 0.684%:

$$\text{Percent error} = \frac{28.246 \text{ mm} - 28.054 \text{ mm}}{28.054 \text{ mm}} \times 100\% = 0.684\%$$

Student B's measurement has a percent error of only about −0.2%.

Comment We noted that Student A had precise results; the standard deviation calculated as described below is 2×10^{-3}. In contrast, Student B had less precise results (standard deviation = 0.14). Possible reasons for the error in Students A's result are incorrect use of the micrometer or a flaw in the instrument.

Standard Deviation

Laboratory measurements can be in error for two basic reasons. First, there may be "determinate" errors caused by faulty instruments or human errors such as incorrect record keeping. So-called "indeterminate" errors arise from uncertainties in a measurement where the cause is not known and cannot be controlled by the lab worker. One way to judge the indeterminate error in a result is to calculate the standard deviation.

The **standard deviation** of a series of measurements is equal to the square root of the sum of the squares of the deviations for each measurement from the average divided by one less than the number of measurements. It has a precise statistical significance: assuming a large number of measurements is used to calculate the average, 68% of the values collected are expected to be within one standard deviation of the value determined, and 95% are within two standard deviations.

Suppose you carefully measured the mass of water delivered by a 10-mL pipet. (A pipet containing a green solution is shown in Figure 4.) For five attempts at the measurement (shown in column 2 of the table below), the standard deviation is found as follows: First, the average of the measurements is calculated (here, 9.984). Next, the deviation of each individual measurement from this value is determined (column 3). These values are squared, giving the values in column 4, and the sum of these values is determined. The standard deviation is then calculated by dividing this sum by 4 (the number of determinations minus 1) and taking the square root of the result.

Determination	Measured Mass (g)	Difference between Average and Measurement (g)	Square of Difference
1	9.990	−0.006	4×10^{-5}
2	9.993	−0.009	8×10^{-5}
3	9.973	0.011	12×10^{-5}
4	9.980	0.004	2×10^{-5}
5	9.982	0.002	0.4×10^{-5}

$$\text{Average mass} = 9.984 \text{ g}$$

$$\text{Sum of squares of differences} = 26 \times 10^{-5}$$

$$\text{Standard deviation} = \sqrt{\frac{26 \times 10^{-5}}{4}} = \pm 0.008$$

Based on this calculation, it would be appropriate to represent the measured mass as 9.984 ± 0.008 g. This would tell a reader that if this experiment were repeated, a majority of the values would fall in the range of 9.976 g to 9.992 g.

EXERCISE 5 Accuracy, Error, and Standard Deviation

Two students measured the freezing point of an unknown liquid. Student A used an ordinary laboratory thermometer calibrated in 0.1 °C units. Student B used a thermometer certified by NIST (National Institute for Standards and Technology) and calibrated in 0.01 °C units. Their results were as follows:

Student A: −0.3 °C; 0.2 °C; 0.0 °C; and −0.3 °C

Student B: −0.02 °C, +0.02 °C, 0.00 °C, and +0.04 °C

Calculate the average value, and, knowing that the liquid was water, calculate the percent error and standard deviation for each student. Which student has the more precise values? Which has the smaller error?

Charles D. Winters

FIGURE 6 Lake Otsego. This lake, with a surface area of 2.33×10^7 m^2, is located in northern New York State. Cooperstown is a village at the base of the lake where the Susquehanna River originates. To learn more about the environmental biology and chemistry of the lake, go to **www.oneonta.edu/academics/biofld**

3 Mathematics of Chemistry

Exponential or Scientific Notation

Lake Otsego in northern New York is also called *Glimmerglass*, a name suggested by James Fenimore Cooper (1789–1851), the great American author and an early resident of the village now known as Cooperstown. Extensive environmental studies have been done along this lake (Figure 6), and some quantitative information useful to chemists, biologists, and geologists is given in the following table:

Lake Otsego Characteristics	Quantitative Information
Area	2.33×10^7 m^2
Maximum depth	505 m
Dissolved solids in lake water	2×10^2 mg/L
Average rainfall in the lake basin	1.02×10^2 cm/year
Average snowfall in the lake basin	198 cm/year

W. Keel, U. Alabama/NASA

FIGURE 7 Exponential numbers in astronomy. The spiral galaxy M-83 is 3.0×10^6 parsecs away from Earth and has a diameter of 9.0×10^3 parsecs. The unit used in astronomy, the parsec (pc), is equivalent to 206265 AU (astronomical units) where 1 AU is 1.496×10^8 km. What is the distance between Earth and M-83 in km?

All of the data collected are in metric units. However, some data are expressed in **fixed notation** (505 m, 198 cm/year), whereas other data are expressed in **exponential,** or **scientific, notation** (2.33×10^7 m^2). Scientific notation is a way of presenting very large or very small numbers in a compact and consistent form that simplifies calculations. Because of its convenience, scientific notation is widely used in sciences such as chemistry, physics, engineering, and astronomy (Figure 7).

In scientific notation a number is expressed as a product of two numbers: $N \times 10^n$. N is the digit term and is a number between 1 and 9.9999. . . . The second number, 10^n, the exponential term, is some integer power of 10. For example, 1234 is written in scientific notation as 1.234×10^3, or 1.234 multiplied by 10 three times:

$$1234 = 1.234 \times 10^1 \times 10^1 \times 10^1 = 1.234 \times 10^3$$

Conversely, a number less than 1, such as 0.01234, is written as 1.234×10^{-2}. This notation tells us that 1.234 should be divided twice by 10 to obtain 0.01234:

$$0.01234 = \frac{1.234}{10^1 \times 10^1} = 1.234 \times 10^{-1} \times 10^{-1} = 1.234 \times 10^{-2}$$

When converting a number to scientific notation, notice that the exponent n is positive if the number is greater than 1 and negative if the number is less than 1. The value of n is the number of places by which the decimal is shifted to obtain the number in scientific notation:

$$1\,2\,3\,4\,5. = 1.2345 \times 10^4$$

(a) Decimal shifted four places to the left. Therefore, n is positive and equal to 4.

$$0.0\,0\,1\,2 = 1.2 \times 10^{-3}$$

(b) Decimal shifted three places to the right. Therefore, n is negative and equal to 3.

If you wish to convert a number in scientific notation to one using fixed notation (that is, not using powers of 10), the procedure is reversed:

$$6\,.\,2\,7\,3 \times 10^2 = 627.3$$

(a) Decimal point moved two places to the right because n is positive and equal to 2.

$$0\;0\;6.273 \times 10^{-3} = 0.006273$$

(b) Decimal point shifted three places to the left because n is negative and equal to 3.

Two final points should be made concerning scientific notation. First, be aware that calculators and computers often express a number such as 1.23×10^3 as 1.23E3

Problem Solving Tip 1 Using Your Calculator

You will be performing a number of calculations in general chemistry, most of them using a calculator. Many different types of calculators are available, but this problem-solving tip describes several of the kinds of operations you will need to perform on a typical calculator. Be sure to consult your calculator manual for specific instructions to enter scientific notation and to find powers and roots of numbers.

1. Scientific Notation

When entering a number such as 1.23×10^{-4} into your calculator, you first enter 1.23 and then press a key marked EE or EXP (or something similar). This enters the "$\times$ 10" portion of the notation for you. You then complete the entry by keying in the exponent of the number, −4. (To change the exponent from +4 to −4, press the "+/−" key.)

A common error made by students is to enter 1.23, press the multiply key ($\times$), and then key in 10 before finishing by pressing EE or EXP followed by −4. This gives you an entry that is 10 times too large.

2. Powers of Numbers

Electronic calculators often offer two methods of raising a number to a power. To square a number, enter the number and then press the x^2 key. To raise a number to any power, use the y^x (or similar key such as ^). For example, to raise 1.42×10^2 to the fourth power:

1. Enter 1.42×10^2.
2. Press y^x.
3. Enter 4 (this should appear on the display).
4. Press =, and 4.0659×10^8 appears on the display.

3. Roots of Numbers

A general procedure for finding any root is to use the y^x key. For a square root, x is 0.5 (or 1/2), whereas it is 0.3333 (or 1/3) for a cube root, 0.25 (or 1/4) for a fourth root, and so on. For example, to find the fourth root of 5.6×10^{-10}:

1. Enter the number.
2. Press the y^x key.
3. Enter the desired root. Because we want the fourth root, enter 0.25.
4. Press =. The answer here is 4.9×10^{-3}.

To make sure you are using your calculator correctly, try these sample calculations:

1. $(6.02 \times 10^{23})(2.26 \times 10^{-5})/367$
   (Answer = 3.71×10^{16})
2. $(4.32 \times 10^{-3})^3$
   (Answer = 8.06×10^{-8})
3. $(4.32 \times 10^{-3})^{1/3}$
   (Answer = 0.163)

or 6.45×10^{-5} as 6.45E-5. Second, some electronic calculators can readily convert numbers in fixed notation to scientific notation. If you have such a calculator, you may be able to do this by pressing the EE or EXP key and then the "=" key (but check your calculator manual to learn how your device operates).

In chemistry, you will often have to use numbers in exponential notation in mathematical operations. The following five operations are important:

■ **Comparing the Earth and a Plant Cell—Powers of Ten**

Earth = 12,760,000 meters wide
= 12.76 million meters
= 1.276×10^7 meters

Plant cell = 0.00001276 meter wide
= 12.76 millionths of a meter
= 1.276×10^{-5} meters

- *Adding and Subtracting Numbers Expressed in Scientific Notation*
  When adding or subtracting two numbers, first convert them to the same powers of 10. The digit terms are then added or subtracted as appropriate:

$$(1.234 \times 10^{-3}) + (5.623 \times 10^{-2}) = (0.1234 \times 10^{-2}) + (5.623 \times 10^{-2}) = 5.746 \times 10^{-2}$$

- *Multiplication of Numbers Expressed in Scientific Notation*
  The digit terms are multiplied in the usual manner, and the exponents are added algebraically. The result is expressed with a digit term with only one nonzero digit to the left of the decimal:

$$(6.0 \times 10^{23}) \times (2.0 \times 10^{-2}) = (6.0)(2.0 \times 10^{23-2}) = 12 \times 10^{21} = 1.2 \times 10^{22}$$

- *Division of Numbers Expressed in Scientific Notation*
  The digit terms are divided in the usual manner, and the exponents are subtracted algebraically. The quotient is written with one nonzero digit to the left of the decimal in the digit term:

$$\frac{7.60 \times 10^3}{1.23 \times 10^2} = \frac{7.60}{1.23} \times 10^{3-2} = 6.18 \times 10^1$$

- *Powers of Numbers Expressed in Scientific Notation*
When raising a number in exponential notation to a power, treat the digit term in the usual manner. The exponent is then multiplied by the number indicating the power:

$$(5.28 \times 10^3)^2 = (5.28)^2 \times 10^{3\times2} = 27.9 \times 10^6 = 2.79 \times 10^7$$

- *Roots of Numbers Expressed in Scientific Notation*
Unless you use an electronic calculator, the number must first be put into a form in which the exponent is exactly divisible by the root. For example, for a square root, the exponent should be divisible by 2. The root of the digit term is found in the usual way, and the exponent is divided by the desired root:

$$\sqrt{3.6 \times 10^7} = \sqrt{36 \times 10^6} = \sqrt{36} \times \sqrt{10^6} = 6.0 \times 10^3$$

Significant Figures

In most experiments, several kinds of measurements must be made, and some can be made more precisely than others. It is common sense that a result calculated from experimental data can be no more precise than the least precise piece of information that went into the calculation. This is where the rules for significant figures come in. **Significant figures** are the digits in a measured quantity that were observed with the measuring device.

Determining Significant Figures

Suppose we place a U.S. dime on the pan of a standard laboratory balance such as the one pictured in Figure 8 and observe a mass of 2.265 g. This number has four significant figures or digits because all four numbers are observed. However, you will learn from experience that the final digit (5) is somewhat uncertain because you may notice the balance readings can change slightly and give masses of 2.264, 2.265, and 2.266, with the mass of 2.265 observed most of the time. Thus, of the four significant digits (2.265) the last (5) is uncertain. In general, *in a number representing a scientific measurement, the last digit to the right is taken to be inexact.* Unless stated otherwise, it is common practice to assign an uncertainty of ±1 to the last significant digit.

Suppose you want to calculate the density of a piece of metal (Figure 9). The mass and dimensions were determined by standard laboratory techniques. Most of these data have two digits to the right of the decimal, but they have different numbers of significant figures.

Measurement	Data Collected	Significant Figures
Mass of metal	13.56 g	4
Length	6.45 cm	3
Width	2.50 cm	3
Thickness	3.1 mm = 0.31 cm	2

The quantity 0.31 cm has two significant figures. That is, the 3 in 0.31 is exactly known, but the 1 is uncertain. This means the thickness of the metal piece may have been as small as 0.30 cm or as large as 0.32 cm, but it is most likely 0.31 cm.

Charles D. Winters

FIGURE 8 Standard laboratory balance and significant figures. Such balances can determine the mass of an object to the nearest milligram. Thus, an object may have a mass of 13.456 g (13456 mg, five significant figures), 0.123 g (123 mg, three significant figures), or 0.072 g (72 mg, two significant figures).

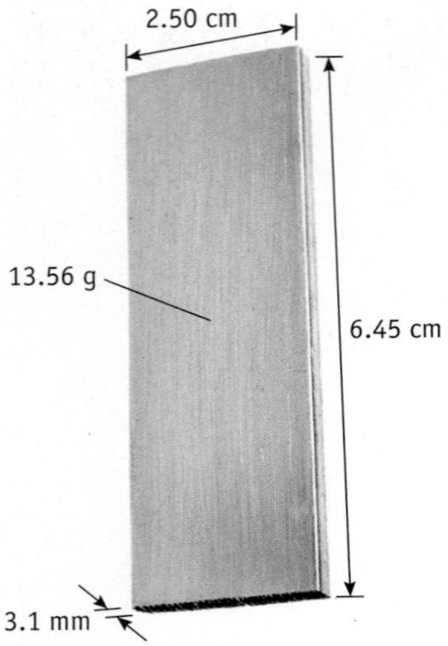

FIGURE 9 Data used to determine the density of a metal.

■ **Zeroes and Common Laboratory Mistakes** We often see students find the mass of a chemical on a balance and fail to write down trailing zeroes. For example, if you find the mass is 2.340 g, the final zero is significant and must be reported as part of the measured value. The number 2.34 g has only three significant figures and implies the 4 is uncertain, when in fact the balance reading indicated the 4 is certain.

In the case of the width of the piece, you found it to be 2.50 cm, where 2.5 is known with certainty, but the final 0 is uncertain. There are three significant figures in 2.50.

When you first read a number in a problem, or collect data in the laboratory, how do you determine how many significant figures it contains?

First, is the number an exact number or a measured quantity? If it is an exact number, you don't have to worry about the number of significant figures. For example, there are exactly 100 cm in 1 m. We could add as many zeros after the decimal place, and the expression would still be true. Using this number in a calculation will not affect how many significant figures you can report in your answer.

If, however, the number is a measured value, you must take into account significant figures. The number of significant figures in our data above is clear, with the possible exception of 0.31 and 2.50. Are the zeroes significant?

1. *Zeroes between two other significant digits are significant.* For example, the zero in 103 is significant.
2. *Zeroes to the right of a nonzero number and also to the right of a decimal place are significant.* For example, in the number 2.50 cm, the zero is significant.
3. *Zeroes that are placeholders are not significant.* There are two types of numbers that fall under this rule.
   a) The first are decimal numbers with zeroes that occur *before* the first nonzero digit. For example, in 0.0013, only the 1 and the 3 are significant; the zeroes are not. This number has two significant figures.
   b) The second are *numbers with trailing zeroes* that must be there to indicate the magnitude of the number. For example, the zeroes in the number 13,000 may or may not be significant; it depends on whether they were measured or not. To avoid confusion with regard to such numbers, we *shall assume in this book that trailing zeroes are significant when there is a decimal point to the right of the last zero.* Thus, we would say that 13,000 has only two significant figures but that 13,000. has five. We suggest that the best way to be unambiguous when writing numbers with trailing zeroes is to use scientific notation. For example 1.300×10^4 clearly indicates four significant figures, whereas 1.3×10^4 indicates two and 1.3000×10^4 indicates five.

Using Significant Figures in Calculations

When doing calculations using measured quantities, we follow some basic rules so that the results reflect the precision of all the measurements that go into the calculations. *The rules used for significant figures in this book are as follows:*

Rule 1. When adding or subtracting numbers, the number of decimal places in the answer is equal to the number of decimal places in the number with the fewest digits after the decimal.

0.12	2 decimal places	2 significant figures
+ 1.9	1 decimal place	2 significant figures
+10.925	3 decimal places	5 significant figures
12.945	3 decimal places	

The sum should be reported as 12.9, a number with one decimal place, because 1.9 has only one decimal place.

Rule 2. In multiplication or division, the number of significant figures in the answer is determined by the quantity with the fewest significant figures.

$$\frac{0.01208}{0.0236} = 0.512\text{, or in scientific notation, } 5.12 \times 10^{-1}$$

Because 0.0236 has only three significant digits, while 0.01208 has four, the answer should have three significant digits.

Rule 3. When a number is rounded off, the last digit to be retained is increased by one only if the following digit is 5 or greater.

Full Number	Number Rounded to Three Significant Digits
12.696	12.7
16.349	16.3
18.35	18.4
18.351	18.4

Now let us apply these rules to calculate the density of the piece of metal in Figure 9.

$$\text{Length} \times \text{width} \times \text{thickness} = \text{volume}$$

$$6.45 \text{ cm} \times 2.50 \text{ cm} \times 0.31 \text{ cm} = 5.0 \text{ cm}^3$$

$$\text{Density} = \frac{\text{mass (g)}}{\text{volume (cm}^3)} = \frac{13.56 \text{ g}}{5.0 \text{ cm}^3} = 2.7 \text{ g/cm}^3$$

The calculated density has two significant figures because *a calculated result can be no more precise than the least precise data used,* and here the thickness has only two significant figures.

One last word on significant figures and calculations: When working problems, you should do the calculation with all the digits allowed by your calculator and round off only at the end of the calculation. *Rounding off in the middle of a calculation can introduce errors.*

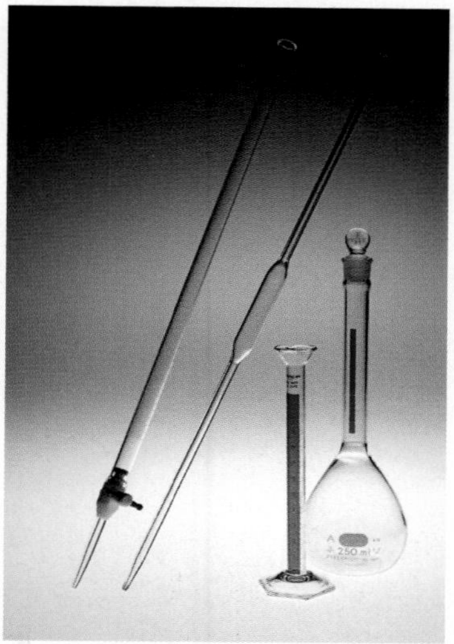

Charles D. Winters

Glassware and significant figures. The 10-mL graduated cylinder is marked in 0.1-mL increments. Graduated cylinders are not considered precision glassware, so, at best, you can expect no more than two significant figures when reading a volume. Conversely, a 50-mL buret is marked in 0.10-mL increments, so it can be read to the nearest 0.01 mL. A volumetric flask is meant to be filled to the mark on the neck. When you have this volume, it is known to the nearest 0.01 mL, so a 250-mL volumetric flask contains 250.00 mL when full to the mark (or five significant figures). Finally, a pipet is like a volumetric flask in that the volume is known to the nearest 0.01 mL.

Chemistry Now™

Sign in at **www.thomsonedu.com/login** and go to Screen 1.17 for a self-study module on **using numerical information.**

EXAMPLE 3 Using Significant Figures

Problem An example of a calculation you will do later in the book (Chapter 11) is

$$\text{Volume of gas (L)} = \frac{(0.120)(0.08206)(273.15 + 23)}{(230/760.0)}$$

Calculate the final answer to the correct number of significant figures.

Strategy Let us first decide on the number of significant figures represented by each number and then apply Rules 1–3.

■ **To Multiply or to Add?** Take the number 4.68.
(a) Take the sum of 4.68 + 4.68 + 4.68. The answer is 14.04, a number with four significant figures.
(b) Multiply 4.68 times 3. The answer can have only three significant figures (14.0). You should recognize that different outcomes are possible, depending on the type of mathematical operation.

Solution

Number	Number of Significant Figures	Comments
0.120	3	The trailing 0 is significant.
0.08206	4	The first 0 to the immediate right of the decimal is not significant.
273.15 + 23 = 296	3	23 has no decimal places, so the sum can have none.
230/760.0 = 0.30	2	230 has two significant figures because the last zero is not significant. In contrast, there is a decimal point in 760.0, so there are four significant digits. The quotient may have only two significant digits.

Analysis shows that one of the pieces of information is known to only two significant figures. Therefore, the volume of gas is 9.6 L, a number with two significant figures.

EXERCISE 6 Using Significant Figures

(a) How many significant figures are indicated by 2.33×10^7, by 50.5, and by 200?

(b) What are the sum and the product of 10.26 and 0.063?

(c) What is the result of the following calculation?

$$x = \frac{(110.7 - 64)}{(0.056)(0.00216)}$$

Problem Solving by Dimensional Analysis

Figure 9 illustrated the data that were collected to determine the density of a piece of metal. The thickness was measured in millimeters, whereas the length and width were measured in centimeters. To find the volume of the sample in cubic centimeters, we first had to have the length, width, and thickness in the same units and so converted the thickness to centimeters.

$$3.1\ \cancel{\text{mm}} \times \frac{1\ \text{cm}}{10\ \cancel{\text{mm}}} = 0.31\ \text{cm}$$

Here, we multiplied the number we wished to convert (3.1 mm) by a *conversion factor* (1 cm/10 mm) to produce the result in the desired unit (0.31 cm). Notice that units are treated like numbers. Because the unit "mm" was in both the numerator and the denominator, dividing one by the other leaves a quotient of 1. The units are said to "cancel out." Here, this leaves the answer in centimeters, the desired unit.

This approach to problem solving is often called **dimensional analysis** (or sometimes the **factor-label method**). It is a general problem-solving approach that uses the dimensions or units of each value to guide us through calculations. And, it is often the case that conversion factors are used to change measured quantities to chemically useful information.

A **conversion factor** expresses the equivalence of a measurement in two different units (1 cm ≡ 10 mm; 1 g ≡ 1000 mg; 12 eggs ≡ 1 dozen; 12 inches ≡ 1 foot). Because the numerator and the denominator describe the same quantity, the conversion factor is equivalent to the number 1. Therefore, multiplication by this factor does not change the measured quantity, only its units. A conversion factor is always written so that it has the form "new units divided by units of original number."

$$\text{Number in original unit} \left[\frac{\text{new unit}}{\text{original unit}}\right] = \text{new number in new unit}$$

↑ Quantity to express in new units — Conversion factor — ↑ Quantity now expressed in new units

Chemistry Now™

Sign in at **www.thomsonedu.com/login** and go to Screen 1.17 for a self-study module on **dimensional analysis and using numerical information.**

■ **Using Conversion Factors and Doing Calculations** As you work problems in this book and read Example problems, notice that proceeding from given information to an answer very often involves a series of multiplications. That is, we multiply the given data by a conversion factor, multiply that answer of that step by another factor, and so on to the answer.

EXAMPLE 4 Using Conversion Factors and Dimensional Analysis

Problem Oceanographers often express the density of sea water in units of kilograms per cubic meter. If the density of sea water is 1.025 g/cm^3 at 15 °C, what is its density in kilograms per cubic meter?

Strategy To simplify this problem, break it into three steps. First, change the mass in grams to kilograms. Next, convert the volume in cubic centimeters to cubic meters. Finally, calculate the density by dividing the mass in kilograms by the volume in cubic meters.

Solution First convert the mass in grams to a mass in kilograms.

$$1.025\ \not{g} \times \frac{1\ \text{kg}}{1000\ \not{g}} = 1.025 \times 10^{-3}\ \text{kg}$$

No conversion factor is available in one of our tables to directly change units of cubic centimeters to cubic meters. You can find one, however, by cubing (raising to the third power) the relation between the meter and the centimeter:

$$1\ \text{cm}^3 \times \left(\frac{1\ \text{m}}{100\ \text{cm}}\right)^3 = 1\ \not{\text{cm}^3} \times \left(\frac{1\ \text{m}^3}{1 \times 10^6\ \not{\text{cm}^3}}\right) = 1 \times 10^{-6}\ \text{m}^3$$

Therefore, the density of sea water is

$$\text{Density} = \frac{1.025 \times 10^{-3}\ \text{kg}}{1 \times 10^{-6}\ \text{m}^3} = 1.025 \times 10^3\ \text{kg/m}^3$$

EXERCISE 7 Using Dimensional Analysis

(a) The annual snowfall at Lake Otsego is 198 cm each year. What is this depth in meters? In feet (where 1 foot = 30.48 cm)?

(b) The area of Lake Otsego is 2.33 × 10^7 m^2. What is this area in square kilometers?

(c) The density of gold is 19,320 kg/m^3. What is this density in g/cm^3?

(d) See Figure 7. Show that 9.0 × 10^3 pc is 2.8 × 10^{17} km.

■ **Who Is Right—You or the Book?** If your answer to a problem in this book does not quite agree with the answers in Appendix N through Q, the discrepancy may be the result of rounding the answer after each step and then using that rounded answer in the next step. This book follows these conventions:
(a) Final answers to numerical problems in this book result from retaining four or more digits past the decimal place throughout the calculation and rounding only at the end.
(b) In Example problems, the answer to each step is given to the correct number of significant figures for that step, but a number of digits are carried to the next step. The number of significant figures in the final answer is dictated by the number of significant figures in the original data.

Graphing

In a number of instances in this text, graphs are used when analyzing experimental data with a goal of obtaining a mathematical equation that may help us predict new results. The procedure used will often result in a straight line, which has the equation

$$y = mx + b$$

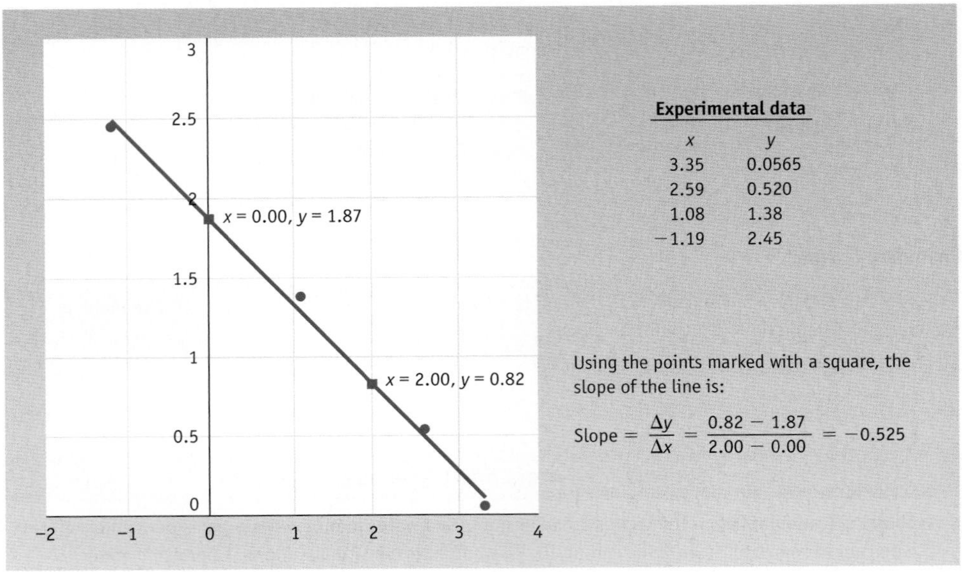

FIGURE 10 Plotting data. Data for the variable x are plotted on the horizontal axis (abscissa), and data for y are plotted on the vertical axis (ordinate). The slope of the line, m in the equation $y = mx + b$, is given by $\Delta y/\Delta x$. The intercept of the line with the y-axis (when $x = 0$) is b in the equation.

Using Microsoft Excel with these data, and doing a linear regression analysis, we find $y = -0.525x + 1.87$.

■ **Determining the Slope with a Computer Program—Least-Squares Analysis** Generally, the easiest method of determining the slope and intercept of a straight line (and thus the line's equation) is to use a program such as Microsoft Excel. These programs perform a "least squares" or "linear regression" analysis and give the best straight line based on the data. (This line is referred to in Excel as a trendline.)

In this equation, y is usually referred to as the dependent variable; its value is determined from (that is, is dependent on) the values of x, m, and b. In this equation, x is called the independent variable, and m is the slope of the line. The parameter b is the y-intercept—that is, the value of y when $x = 0$. Let us use an example to investigate two things: (a) how to construct a graph from a set of data points, and (b) how to derive an equation for the line generated by the data.

A set of data points to be graphed is presented in Figure 10. We first mark off each axis in increments of the values of x and y. Here, our x-data are within the range from -2 to 4, so the x-axis is marked off in increments of 1 unit. The y-data falls within the range from 0 to 2.5, so we mark off the y-axis in increments of 0.5. Each data point is marked as a circle on the graph.

After plotting the points on the graph (round circles), we draw a straight line that comes as close as possible to representing the trend in the data. (Do not connect the dots!) Because there is always some inaccuracy in experimental data, this line will not pass exactly through every point.

To identify the specific equation corresponding to our data, we must determine the y-intercept (b) and slope (m) for the equation $y = mx + b$. The y-intercept is the point at which $x = 0$. (In Figure 10, $y = 1.87$ when $x = 0$). The slope is determined by selecting two points *on the line* (marked with squares in Figure 10) and calculating the difference in values of y ($\Delta y = y_2 - y_1$) and x ($\Delta x = x_2 - x_1$). The slope of the line is then the ratio of these differences, $m = \Delta y/\Delta x$. Here, the slope has the value -0.525. With the slope and intercept now known, we can write the equation for the line

$$y = -0.525x + 1.87$$

and we can use this equation to calculate y-values for points that are not part of our original set of x–y data. For example, when $x = 1.50$, we find $y = 1.08$.

EXERCISE 8 Graphing

To find the mass of 50 jelly beans, we weighed several samples of beans.

Number of Beans	Mass (g)
5	12.82
11	27.14
16	39.30
24	59.04

Plot these data with the number of beans on the horizontal or *x*-axis, and the mass of beans on the vertical or *y*-axis. What is the slope of the line? Use your equation of a straight line to calculate the mass of exactly 50 jelly beans.

Case Study Out of Gas!

On July 23, 1983, a new Boeing 767 jet aircraft was flying at 26,000 ft from Montreal to Edmonton as Air Canada Flight 143. Warning buzzers sounded in the cockpit. One of the world's largest planes was now a glider—the plane had run out of fuel!

How did this modern airplane, having the latest technology, run out of fuel? A simple mistake had been made in calculating the amount of fuel required for the flight!

Like all Boeing 767s, this plane had a sophisticated fuel gauge, but it was not working properly. The plane was still allowed to fly, however, because there is an alternative method of determining the quantity of fuel in the tanks. Mechanics can use a stick, much like the oil dipstick in an automobile engine, to measure the fuel level in each of the three tanks. The mechanics in Montreal read the dipsticks, which were calibrated in centimeters, and translated those readings to a volume in liters. According to this, the plane had a total of 7682 L of fuel.

Pilots always calculate fuel quantities in units of mass because they need to know the total mass of the plane before take-off. Air Canada pilots had always calculated the quantity of fuel in pounds, but the new 767's fuel consumption was given in kilograms. The pilots knew that 22,300 kg of fuel was required for the trip. If 7682 L of fuel remained in the tanks, how much had to be added? This involved using the fuel's density to convert 7682 L to a mass in kilograms. The mass of fuel to be added could then be calculated, and that mass converted to a volume of fuel to be added.

The First Officer of the plane asked a mechanic for the conversion factor to do the volume-to-mass conversion, and the mechanic replied "1.77." Using that number, the First Officer and the mechanics calculated that 4917 L of fuel should be added. But later calculations showed that this is only about one fourth of the required amount of fuel! Why? Because no one thought about the units of the number 1.77. They realized later that 1.77 has units of pounds per liter and not kilograms per liter.

Out of fuel, the plane could not make it to Winnipeg, so controllers directed them to the town of Gimli and to a small airport abandoned by the Royal Canadian Air Force. After gliding for almost 30 minutes, the plane approached the Gimli runway. The runway, however, had been converted to a race course for cars, and a race was underway. Furthermore, a steel barrier had been erected across the runway. Nonetheless, the pilot managed to touch down very near the end of the runway. The plane sped down the concrete strip; the nose wheel collapsed; several tires blew—and the plane skidded safely to a stop just before the barrier. The Gimli glider had made it! And somewhere an aircraft mechanic is paying more attention to units on numbers.

Question:

1. *What is the fuel density in units of kg/L?*
2. *What mass and what volume of fuel should have been loaded? (1 lb = 453.6 g) (See Study Question 58, page 48.)*

Answers to these questions are in Appendix Q.

The Gimli glider. After running out of fuel, Air Canada Flight 143 glided 29 minutes before landing on an abandoned airstrip at Gimli, Manitoba, near Winnipeg.

Problem Solving and Chemical Arithmetic

Problem-Solving Strategy

Some of the calculations in chemistry can be complex. Students frequently find it is helpful to follow a definite plan of attack as illustrated in examples throughout this book.

Step 1: Problem. State the problem. Read it carefully—and then read it again.

Step 2: Strategy. What key principles are involved? What information is known or not known? What information might be there just to place the question in the context of chemistry? Organize the information to see what is required and to discover the relationships among the data given. Try writing the information down in table form. If it is numerical information, be sure to include units.

One of the greatest difficulties for a student in introductory chemistry is picturing what is being asked for. Try sketching a picture of the situation involved. For example, we sketched a picture of the piece of metal whose density we wanted to calculate, and put the dimensions on the drawing (page 36).

Develop a plan. Have you done a problem of this type before? If not, perhaps the problem is really just a combination of several simpler ones you have seen before. Break it down into those simpler components. Try reasoning backward from the units of the answer. What data do you need to find an answer in those units?

Step 3: Solution. Execute the plan. Carefully write down each step of the problem, being sure to keep track of the units on numbers. (Do the units cancel to give you the answer in the desired units?) Don't skip steps. Don't do anything except the simplest steps in your head. Students often say they got a problem wrong because they "made a stupid mistake." Your instructor—and book authors—make them, too, and it is usually because they don't take the time to write down the steps of the problem clearly.

Step 4: Comment and Check Answer. As a final check, ask yourself whether the answer is reasonable.

EXAMPLE 5 Problem Solving

Problem A mineral oil has a density of 0.875 g/cm^3. Suppose you spread 0.75 g of this oil over the surface of water in a large dish with an inner diameter of 21.6 cm. How thick is the oil layer? Express the thickness in centimeters.

Strategy It is often useful to begin solving such problems by sketching a picture of the situation.

This helps recognize that the solution to the problem is to find the volume of the oil on the water. If we know the volume, then we can find the thickness because

$$\text{Volume of oil layer} = (\text{thickness of layer}) \times (\text{area of oil layer})$$

So, we need two things: (a) the volume of the oil layer and (b) the area of the layer.

Solution First, calculate the volume of oil. The mass of the oil layer is known, so combining the mass of oil with its density gives the volume of the oil used:

$$0.75\ \not{g} \times \frac{1\ \text{cm}^3}{0.875\ \not{g}} = 0.86\ \text{cm}^3$$

Next, calculate the area of the oil layer. The oil is spread over a circular surface, whose area is given by

$$\text{Area} = \pi \times (\text{radius})^2$$

The radius of the oil layer is half its diameter (= 21.6 cm) or 10.8 cm, so

$$\text{Area of oil layer} = (3.142)(10.8\ \text{cm})^2 = 366\ \text{cm}^2$$

With the volume and the area of the oil layer known, the thickness can be calculated.

$$\text{Thickness} = \frac{\text{Volume}}{\text{Area}} = \frac{0.86\ \text{cm}^3}{366\ \text{cm}^2} = 0.0023\ \text{cm}$$

Comment In the volume calculation, the calculator shows 0.857143. . . . The quotient should have two significant figures because 0.75 has two significant figures, so the result of this step is 0.86 cm^3. In the area calculation, the calculator shows 366.435. . . . The answer to this step should have three significant figures because 10.8 has three. When these interim results are combined in calculating thickness, however, the final result can have only two significant figures. Premature rounding can lead to errors.

EXERCISE 9 Problem Solving

A particular paint has a density of 0.914 g/cm^3. You need to cover a wall that is 7.6 m long and 2.74 m high with a paint layer 0.13 mm thick. What volume of paint (in liters) is required? What is the mass (in grams) of the paint layer?

STUDY QUESTIONS

Online homework for this chapter may be assigned in OWL.

▲ denotes challenging questions.

■ denotes questions assignable in OWL.

Blue-numbered questions have answers in Appendix O and fully-worked solutions in the *Student Solutions Manual.*

Practicing Skills

Temperature Scales

(Exercise 1)

1. Many laboratories use 25 °C as a standard temperature. What is this temperature in kelvins?

2. The temperature on the surface of the sun is 5.5×10^3 °C. What is this temperature in kelvins?

3. ■ Make the following temperature conversions:

°C	K
(a) 16	_____
(b) _____	370
(c) 40	_____

4. Make the following temperature conversions:

°C	K
(a) _____	77
(b) 63	_____
(c) _____	1450

Length, Volume, Mass, and Density

(See Example 1 and Exercises 2–4)

5. A marathon distance race covers a distance of 42.195 km. What is this distance in meters? In miles?

6. ■ The average lead pencil, new and unused, is 19 cm long. What is its length in millimeters? In meters?

7. A standard U.S. postage stamp is 2.5 cm long and 2.1 cm wide. What is the area of the stamp in square centimeters? In square meters?

8. ■ A compact disc has a diameter of 11.8 cm. What is the surface area of the disc in square centimeters? In square meters? [Area of a circle = $(\pi)(\text{radius})^2$.]

9. A typical laboratory beaker has a volume of 250. mL. What is its volume in cubic centimeters? In liters? In cubic meters? In cubic decimeters?

10. ■ Some soft drinks are sold in bottles with a volume of 1.5 L. What is this volume in milliliters? In cubic centimeters? In cubic decimeters?

11. A book has a mass of 2.52 kg. What is this mass in grams?

12. A new U.S. dime has a mass of 2.265 g. What is its mass in kilograms? In milligrams?

13. ■ Ethylene glycol, $C_2H_6O_2$, is an ingredient of automobile antifreeze. Its density is 1.11 g/cm^3 at 20 °C. If you need 500. mL of this liquid, what mass of the compound, in grams, is required?

14. ■ A piece of silver metal has a mass of 2.365 g. If the density of silver is 10.5 g/cm^3, what is the volume of the silver?

15. ■ You can identify a metal by carefully determining its density (d). An unknown piece of metal, with a mass of 2.361 g, is 2.35 cm long, 1.34 cm wide, and 1.05 mm thick. Which of the following is the element?
(a) Nickel, d = 8.91 g/cm^3
(b) Titanium, d = 4.50 g/cm^3
(c) Zinc, d = 7.14 g/cm^3
(d) Tin, d = 7.23 g/cm^3

16. ■ Which occupies a larger volume, 600 g of water (with a density of 0.995 g/cm^3) or 600 g of lead (with a density of 11.35 g/cm^3)?

Accuracy, Precision, Error, and Standard Deviation

(See Example 2 and Exercise 5)

17. You and your lab partner are asked to determine the density of an aluminum bar. The mass is known accurately (to four significant figures). You use a simple metric ruler to measure its dimensions and find the results in A. Your partner uses a precision micrometer, and obtains the results in B.

Method A (g/cm^3)	Method B (g/cm^3)
2.2	2.703
2.3	2.701
2.7	2.705
2.4	5.811

The accepted density of aluminum is 2.702 g/cm^3.
(a) Calculate the average density for each method. Should all the experimental results be included in your calculations? If not, justify any omissions.
(b) Calculate the percent error for each method's average value.
(c) Calculate the standard deviation for each set of data.
(d) Which method's average value is more precise? Which method is more accurate?

18. ■ The accepted value of the melting point of pure aspirin is 135 °C. Trying to verify that value, you obtain 134 °C, 136 °C, 133 °C, and 138 °C in four separate trials. Your partner finds 138 °C, 137 °C, 138 °C, and 138 °C.
(a) Calculate the average value and percent error for you and your partner.
(b) Which of you is more precise? More accurate?

Exponential Notation and Significant Figures

(See Example 3)

19. ■ Express the following numbers in exponential or scientific notation, and give the number of significant figures in each.
(a) 0.054 g
(b) 5462 g
(c) 0.000792 g
(d) 1600 mL

20. ■ Express the following numbers in fixed notation (e.g., $1.23 \times 10^2 = 123$), and give the number of significant figures in each.
(a) 1.623×10^3
(b) 2.57×10^{-4}
(c) 6.32×10^{-2}
(d) 3.404×10^3

21. ■ Carry out the following operations. Provide the answer with the correct number of significant figures.
(a) $(1.52)(6.21 \times 10^{-3})$
(b) $(6.217 \times 10^3) - (5.23 \times 10^2)$
(c) $(6.217 \times 10^3) \div (5.23 \times 10^2)$
(d) $(0.0546)(16.0000)\left[\dfrac{7.779}{55.85}\right]$

22. Carry out the following operations. Provide the answer with the correct number of significant figures.
(a) $(6.25 \times 10^2)^3$
(b) $\sqrt{2.35 \times 10^{-3}}$
(c) $(2.35 \times 10^{-3})^{1/3}$
(d) $(1.68)\left[\dfrac{23.56 - 2.3}{1.248 \times 10^3}\right]$

Graphing

(See Exercise 8)

23. To determine the average mass of a popcorn kernel, you collect the following data:

Number of kernels	Mass (g)
5	0.836
12	2.162
35	5.801

Plot the data with number of kernels on the x-axis and mass on the y-axis. Draw the best straight line using the points on the graph (or do a least-squares or linear regression analysis using a computer program), and then write the equation for the resulting straight line. What is the slope of the line? What does the slope of the line signify about the mass of a popcorn kernel? What is the mass of 20 popcorn kernels? How many kernels are there in a handful of popcorn with a mass of 20.88 g?

24. Using the graph below:
(a) What is the value of x when $y = 4.0$?
(b) What is the value of y when $x = 0.30$?
(c) ■ What are the slope and the y-intercept of the line?
(d) What is the value of y when $x = 1.0$?

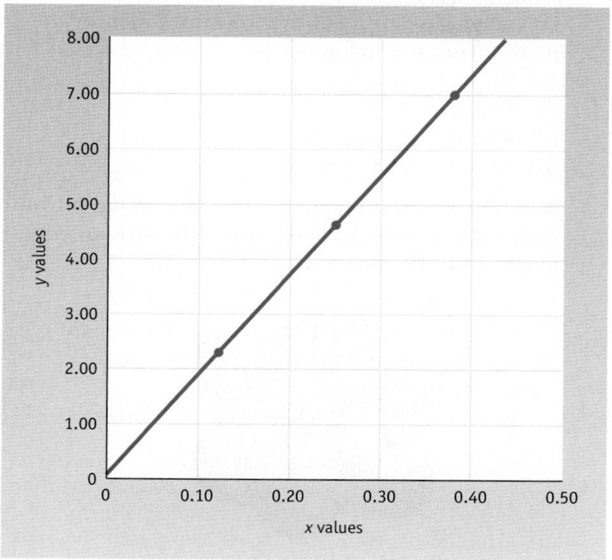

25. ■ Use the graph below to answer the following questions.
(a) Derive the equation for the straight line, $y = mx + b$.
(b) What is the value of y when $x = 6.0$?

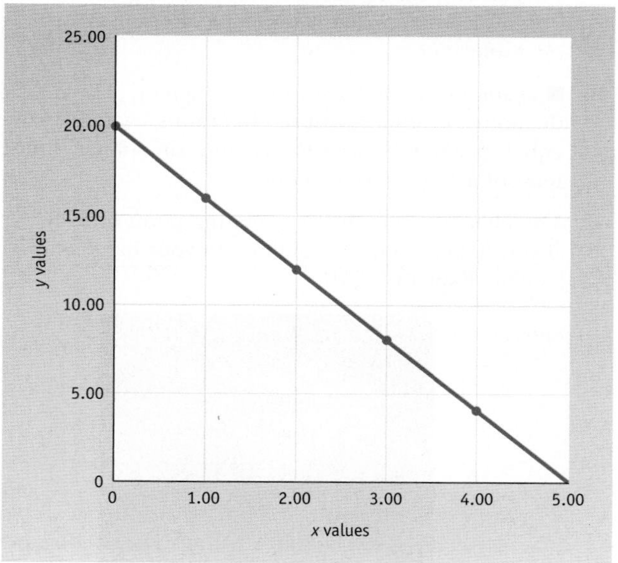

26. The following data were collected in an experiment to determine how an enzyme works in a biochemical reaction.

Amount of H_2O_2	Reaction Speed (amount/second)
1.96	4.75×10^{-5}
1.31	4.03×10^{-5}
0.98	3.51×10^{-5}
0.65	2.52×10^{-5}
0.33	1.44×10^{-5}
0.16	0.585×10^{-5}

(a) Plot these data as 1/amount on the y-axis and 1/speed on the x-axis.
(b) Determine the equation for the data, and give the values of the y-intercept of the slope. *(Note: in biochemistry this is known as a Lineweaver-Burk plot, and the y-intercept is related to the maximum speed of the reaction.)*

Solving Equations

27. Solve the following equation for the unknown value, C.

$$(0.502)(123) = (750.)C$$

28. Solve the following equation for the unknown value, n.

$$(2.34)(15.6) = n(0.0821)(273)$$

29. Solve the following equation for the unknown value, T.

$$(4.184)(244)(T - 292.0) + (0.449)(88.5)(T - 369.0) = 0$$

30. Solve the following equation for the unknown value, n.

$$-246.0 = 1312\left[\frac{1}{2^2} - \frac{1}{n^2}\right]$$

General Questions

These questions are not designated as to type or location in the chapter. They may combine several concepts.

31. Molecular distances are usually given in nanometers ($1\ \text{nm} = 1 \times 10^{-9}$ m) or in picometers ($1\ \text{pm} = 1 \times 10^{-12}$ m). However, the angstrom (Å) unit is sometimes used, where $1\ \text{Å} = 1 \times 10^{-10}$ m. (The angstrom unit is not an SI unit.) If the distance between the Pt atom and the N atom in the cancer chemotherapy drug cisplatin is 1.97 Å, what is this distance in nanometers? In picometers?

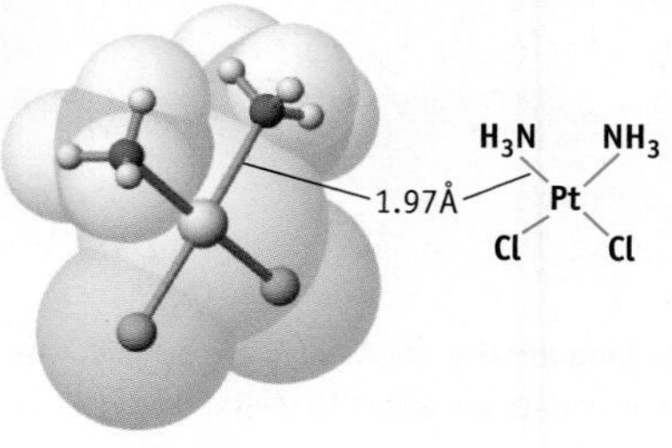

Cisplatin.

32. ■ The separation between carbon atoms in diamond is 0.154 nm. What is their separation in meters? In picometers (pm)? In Angstroms (Å)? (See Study Question 31.)

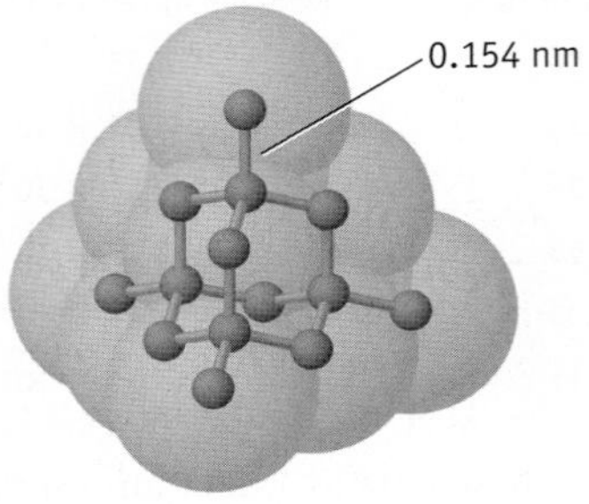

A portion of the diamond structure.

33. ■ A red blood cell has a diameter of 7.5 μm (micrometers). What is this dimension in (a) meters, (b) nanometers, and (c) picometers?

34. ■ The platinum-containing cancer drug *cisplatin* (Study Question 31) contains 65.0 mass-percent of the metal. If you have 1.53 g of the compound, what mass of platinum (in grams) is contained in this sample?

35. ■ The anesthetic procaine hydrochloride is often used to deaden pain during dental surgery. The compound is packaged as a 10.% solution (by mass; d = 1.0 g/mL) in water. If your dentist injects 0.50 mL of the solution, what mass of procaine hydrochloride (in milligrams) is injected?

36. ■ You need a cube of aluminum with a mass of 7.6 g. What must be the length of the cube's edge (in cm)? (The density of aluminum is 2.698 g/cm^3.)

37. ■ You have a 250.0-mL graduated cylinder containing some water. You drop 3 marbles with a total mass of 95.2 g into the water. What is the average density of a marble?

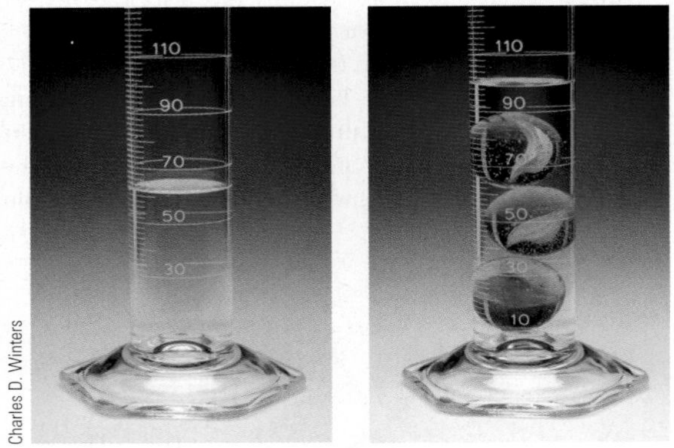

(a) (b)

Determining density. (a) A graduated cylinder with 61 mL of water. (b) Three marbles are added to the cylinder.

38. ■ You have a white crystalline solid, known to be one of the potassium compounds listed below. To determine which, you measure its density. You measure out 18.82 g and transfer it to a graduated cylinder containing kerosene (in which salts will not dissolve). The level of liquid kerosene rises from 8.5 mL to 15.3 mL. Calculate the density of the solid, and identify the compound from the following list.
(a) KF, d = 2.48 g/cm^3
(b) KCl, d = 1.98 g/cm^3
(c) KBr, d = 2.75 g/cm^3
(d) KI, d = 3.13 g/cm^3

39. ■ ▲ The smallest repeating unit of a crystal of common salt is a cube (called a unit cell) with an edge length of 0.563 nm.

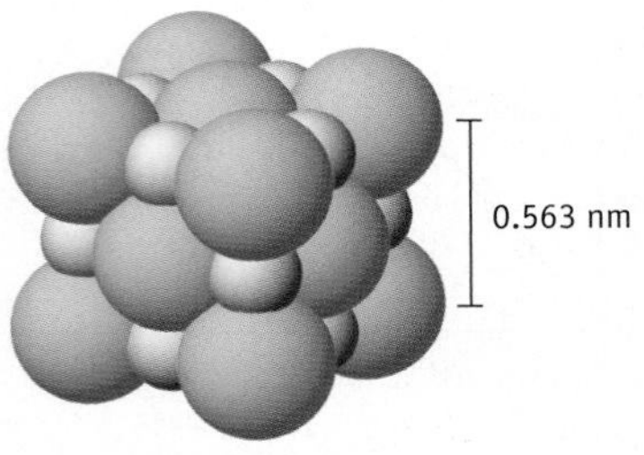

Sodium chloride, NaCl.

(a) What is the volume of this cube in cubic nanometers? In cubic centimeters?
(b) The density of NaCl is 2.17 g/cm^3. What is the mass of this smallest repeating unit ("unit cell")?
(c) Each repeating unit is composed of four NaCl "molecules." What is the mass of one NaCl molecule?

40. ■ Diamond has a density of 3.513 g/cm^3. The mass of diamonds is often measured in "carats," where 1 carat equals 0.200 g. What is the volume (in cubic centimeters) of a 1.50-carat diamond?

41. The element gallium has a melting point of 29.8 °C. If you held a sample of gallium in your hand, should it melt? Explain briefly.

Gallium metal.

42. ■ ▲ The density of pure water is given at various temperatures.

t (°C)	*d* (g/cm³)
4	0.99997
15	0.99913
25	0.99707
35	0.99406

Suppose your laboratory partner tells you the density of water at 20 °C is 0.99910 g/cm³. Is this a reasonable number? Why or why not?

43. When you heat popcorn, it pops because it loses water explosively. Assume a kernel of corn, with a mass of 0.125 g, has a mass of only 0.106 g after popping.
(a) What percentage of its mass did the kernel lose on popping?
(b) ■ Popcorn is sold by the pound in the United States. Using 0.125 g as the average mass of a popcorn kernel, how many kernels are there in a pound of popcorn? (1 lb = 453.6 g)

44. ■ ▲ The aluminum in a package containing 75 ft² of kitchen foil weighs approximately 12 ounces. Aluminum has a density of 2.70 g/cm³. What is the approximate thickness of the aluminum foil in millimeters? (1 oz = 28.4 g)

45. ■ ▲ The fluoridation of city water supplies has been practiced in the United States for several decades. It is done by continuously adding sodium fluoride to water as it comes from a reservoir. Assume you live in a medium-sized city of 150,000 people and that 660 L (170 gal) of water is consumed per person per day. What mass of sodium fluoride (in kilograms) must be added to the water supply each year (365 days) to have the required fluoride concentration of 1 ppm (part per million)—that is, 1 kilogram of fluoride per 1 million kilograms of water? (Sodium fluoride is 45.0% fluoride, and water has a density of 1.00 g/cm³.)

46. ■ ▲ About two centuries ago, Benjamin Franklin showed that 1 teaspoon of oil would cover about 0.5 acre of still water. If you know that 1.0×10^4 m² = 2.47 acres, and that there is approximately 5 cm³ in a teaspoon, what is the thickness of the layer of oil? How might this thickness be related to the sizes of molecules?

47. ■ ▲ Automobile batteries are filled with an aqueous solution of sulfuric acid. What is the mass of the acid (in grams) in 500. mL of the battery acid solution if the density of the solution is 1.285 g/cm³ and if the solution is 38.08% sulfuric acid by mass?

48. ■ A 26-meter-tall statue of Buddha in Tibet is covered with 279 kg of gold. If the gold was applied to a thickness of 0.0015 mm, what surface area is covered (in square meters)? (Gold density = 19.3 g/cm³)

49. At 25 °C, the density of water is 0.997 g/cm³, whereas the density of ice at −10 °C is 0.917 g/cm³.
(a) If a soft-drink can (volume = 250. mL) is filled completely with pure water at 25 °C and then frozen at −10 °C, what volume does the solid occupy?
(b) Can the ice be contained within the can?

50. ■ Suppose your bedroom is 18 ft long, 15 ft wide, and the distance from floor to ceiling is 8 ft, 6 in. You need to know the volume of the room in metric units for some scientific calculations.
(a) What is the room's volume in cubic meters? In liters?
(b) What is the mass of air in the room in kilograms? In pounds? (Assume the density of air is 1.2 g/L and that the room is empty of furniture.)

51. ■ A spherical steel ball has a mass of 3.475 g and a diameter of 9.40 mm. What is the density of the steel? [The volume of a sphere = $(4/3)\pi r^3$ where r = radius.]

52. ■ ▲ The substances listed below are clear liquids. You are asked to identify an unknown liquid that is known to be one of these liquids. You pipet a 3.50-mL sample into a beaker. The empty beaker had a mass of 12.20 g, and the beaker plus the liquid weighed 16.08 g.

Substance	Density at 25 °C (g/cm³)
Ethylene glycol	1.1088 (the major component of antifreeze)
Water	0.9971
Ethanol	0.7893 (the alcohol in alcoholic beverages)
Acetic acid	1.0492 (the active component of vinegar)
Glycerol	1.2613 (a solvent, used in home care products)

(a) Calculate the density and identify the unknown.
(b) If you were able to measure the volume to only two significant figures (that is, 3.5 mL, not 3.50 mL), will the results be sufficiently accurate to identify the unknown? Explain.

53. ■ ▲ You have an irregularly shaped piece of an unknown metal. To identify it, you determine its density and then compare this value with known values that you look up in the chemistry library. The mass of the metal is 74.122 g. Because of the irregular shape, you measure the volume by submerging the metal in water in a graduated cylinder. When you do this, the water level in the cylinder rises from 28.2 mL to 36.7 mL.
(a) What is the density of the metal? (Use the correct number of significant figures in your answer.)
(b) The unknown is one of the seven metals listed below. Is it possible to identify the metal based on the density you have calculated? Explain.

Metal	Density (g/cm³)	Metal	Density (g/cm³)
zinc	7.13	nickel	8.90
iron	7.87	copper	8.96
cadmium	8.65	silver	10.50
cobalt	8.90		

54. ■ ▲ There are 5 hydrocarbon compounds (compounds of C and H) that have the formula C_6H_{14}. (These are isomers; they differ in the way that C and H atoms are attached. See Chapters 8 and 10.) All are liquids at room temperature but have slightly different densities.

Hydrocarbon	Density (g/mL)
hexane	0.6600
2,3-dimethylbutane	0.6616
1-methylpentane	0.6532
2,2-dimethylbutane	0.6485
2-methylpentane	0.6645

(a) You have a pure sample of one of these hydrocarbons, and to identify it you decide to measure its density. You determine that a 5.0-mL sample (measured in a graduated cylinder) has a mass of 3.2745 g (measured on an analytical balance.) Assume that the accuracy of the values for mass and volume is expressed by the number of significant figures, that is, plus or minus one (± 1) in the last significant figure. What is the density of the liquid?

(b) Express the estimated uncertainty of your value in two other ways:

i) The value you have calculated for the density is uncertain to ____ g/mL.

ii) The value calculated for density is between x g/mL and y g/mL.

(c) Can you identify the unknown hydrocarbon based on your experiment?

(d) Can you eliminate any of the five possibilities based on the data? If so, which one(s)?

(e) You need a more accurate volume measurement to solve this problem, and you redetermine the volume to be 4.93 mL. Based on these new data, what is the unknown compound?

55. ■ ▲ Suppose you have a cylindrical glass tube with a thin capillary opening, and you wish to determine the diameter of the capillary. You can do this experimentally by weighing a piece of the tubing before and after filling a portion of the capillary with mercury. Using the following information, calculate the diameter of the capillary.

Mass of tube before adding mercury = 3.263 g

Mass of tube after adding mercury = 3.416 g

Length of capillary filled with mercury = 16.75 mm

Density of mercury = 13.546 g/cm^3

Volume of cylindrical capillary filled with mercury = $(\pi)(\text{radius})^2(\text{length})$

56. ■ **COPPER**: Copper has a density of 8.96 g/cm^3. An ingot of copper with a mass of 57 kg (126 lb) is drawn into wire with a diameter of 9.50 mm. What length of wire (in meters) can be produced? [Volume of wire = $(\pi)(\text{radius})^2(\text{length})$]

57. ▲ COPPER: See the illustration of the copper lattice on page 24.

(a) Suppose you have a cube of copper metal that is 0.236 cm on a side with a mass of 0.1206 g. If you know that each copper atom (radius = 128 pm) has a mass of 1.055×10^{-22} g (you will learn in Chapter 2 how to find the mass of one atom), how many atoms are there in this cube? What fraction of the cube is filled with atoms? (Or conversely, how much of the lattice is empty space?) Why is there "empty" space in the lattice?

(b) Now look at the smallest, repeating unit of the crystal lattice of copper. Knowing that an edge of this cube is 361.47 pm and the density of copper is 8.960 g/cm^3, estimate the number of copper atoms in this smallest, repeating unit.

58. ■ ▲ CASE STUDY: In July 1983, an Air Canada Boeing 767 ran out of fuel over central Canada on a trip from Montreal to Edmonton. (The plane glided safely to a landing at an abandoned airstrip.) The pilots knew that 22,300 kg of fuel were required for the trip, and they knew that 7682 L of fuel were already in the tank. The ground crew added 4916 L of fuel, which was only about one fifth of what was required. The crew members used a factor of 1.77 for the fuel density—the problem is that 1.77 has units of pounds per liter and not kilograms per liter! What is the fuel density in units of kg/L? What mass and what volume of fuel should have been loaded? (1 lb = 453.6 g)

In the Laboratory

59. ■ A sample of unknown metal is placed in a graduated cylinder containing water. The mass of the sample is 37.5 g, and the water levels before and after adding the sample to the cylinder are as shown in the figure. Which metal in the following list is most likely the sample? (d is the density of the metal.)

(a) Mg, d = 1.74 g/cm^3 (d) Al, d = 2.70 g/cm^3
(b) Fe, d = 7.87 g/cm^3 (e) Cu, d = 8.96 g/cm^3
(c) Ag, d = 10.5 g/cm^3 (f) Pb, d = 11.3 g/cm^3

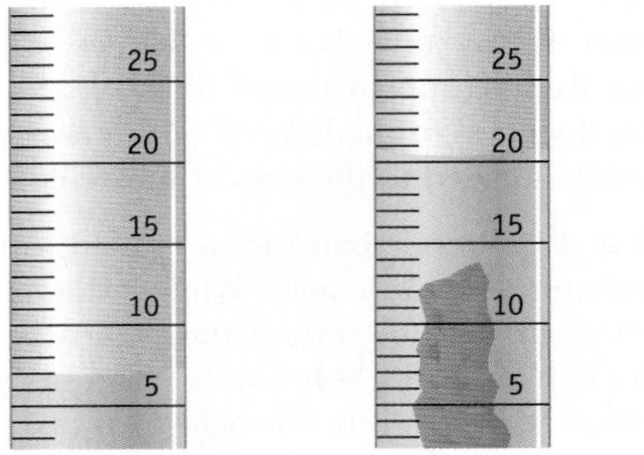

Graduated cylinders with unknown metal *(right)*.

60. ■ Iron pyrite is often called "fool's gold" because it looks like gold (see page 14). Suppose you have a solid that looks like gold, but you believe it to be fool's gold. The sample has a mass of 23.5 g. When the sample is lowered into the water in a graduated cylinder (see Study Question 37), the water level rises from 47.5 mL to 52.2 mL. Is the sample fool's gold (d = 5.00 g/cm^3) or "real" gold (d = 19.3 g/cm^3)?

61. You can analyze for a copper compound in water using an instrument called a spectrophotometer. In this technique, the light passing through an aqueous solution of a compound can be absorbed, and the amount of light absorbed (at a given wavelength of light) depends directly on the amount of compound per liter of solution. To calibrate the spectrophotometer, you collect the following data:

Absorbance (A)	Concentration of Copper Compound (g/L)
0.000	0.000
0.257	1.029×10^{-3}
0.518	2.058×10^{-3}
0.771	3.087×10^{-3}
1.021	4.116×10^{-3}

Plot the absorbance (A) against the mass of copper compound per liter (g/L), and find the slope (m) and intercept (b) (assuming that A is y and the amount in solution is x in the equation for a straight line, $y = mx + b$). What is the amount of copper compound in the solution in g/L and mg/mL when the absorbance is 0.635?

62. A gas chromatograph (page 2) is calibrated for the analysis of isooctane (a major gasoline component) using the following data:

Percent Isooctane (x-data)	Instrument Response (y-data)
0.352	1.09
0.803	1.78
1.08	2.60
1.38	3.03
1.75	4.01

If the instrument response is 2.75, what percentage of isooctane is present? (Data are taken from *Analytical Chemistry, An Introduction,* by D.A. Skoog, D.M. West, F. J. Holler, and S. R. Crouch, Thomson-Brooks/Cole, Belmont, CA, 7th Edition, 2000.)

2 Atoms, Molecules, and Ions

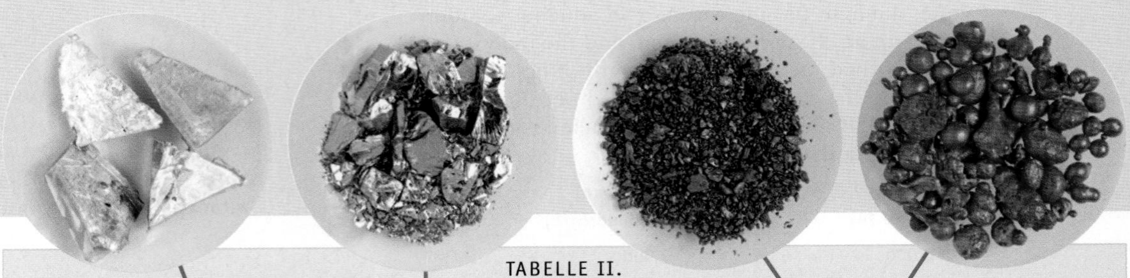

The Periodic Table, the Central Icon of Chemistry

TABELLE II.

REIHEN	GRUPPE I. — R^2O	GRUPPE II. — RO	GRUPPE III. — R^2O^3	GRUPPE IV. RH^4 RO^2	GRUPPE V. RH^3 R^2O^5	GRUPPE VI. RH^2 RO^3	GRUPPE VII. RH R^2O^7	GRUPPE VIII. — RO^4
1	H=1							
2	Li=7	Be=9,4	B=11	C=12	N=14	O=16	F=19	
3	Na=23	Mg=24	Al=27,3	Si=28	P=31	S=32	Cl=35,5	
4	K=39	Ca=40	—=44	Ti=48	V=51	Cr=52	Mn=55	Fe=56, Co=59, Ni=59, Cu=63.
5	(Cu=63)	Zn=65	—=68	—=72	As=75	Se=78	Br=80	
6	Rb=85	Sr=87	?Yt=88	Zr=90	Nb=94	Mo=96	—=100	Ru=104, Rh=104, Pd=106, Ag=108.
7	(Ag=108)	Cd=112	In=113	Sn=118	Sb=122	Te=125	J=127	
8	Cs=133	Ba=137	?Di=138	?Ce=140	—	—	—	— — — —
9	(—)	—	—	—	—	—	—	
10	—	—	?Er=178	?La=180	Ta=182	W=184	—	Os=195, Ir=197, Pt=198, Au=199.
11	(Au=199)	Hg=200	Tl=204	Pb=207	Bi=208	—	—	
12	—	—	—	Th=231	—	U=240	—	— — — —

Charles D. Winters

Nineteenth-century chemists such as Newlands, Chancourtois, Mayer, and others devised ways to organize the chemistry of the elements with varying degrees of success. However, it was Dmitri Mendeleev in 1870 who first truly recognized the periodicity of the chemistry of the elements, who proposed the first periodic table, and who used this to predict the existence of yet-unknown elements.

Mendeleev placed the elements in a table in order of increasing atomic weight. In doing so Li, Be, B, C, N, O, and F became the first row of the table. The next element then known, sodium (Na), had properties quite similar to those of lithium (Li), so Na began the next row of the table. As additional elements were added in order of increasing atomic weight, elements with similar properties fell in columns or groups.

If you compare the periodic table published by Mendeleev in 1871 (shown here) with the table in the front of this book, you will see that many elements are missing in the 1871 table. Mendeleev's genius was that he recognized there must be yet-undiscovered elements, and so he left a place for them in the table (marking the empty places with a —). For example, Mendeleev concluded that "Gruppe IV" was missing an element between silicon (Si) and tin (Sn) and marked its position as "— = 72." He called the missing element *eka-silicon* and predicted the element would have, for example, an atomic weight of 72 and a density of 5.5 g/cm^3. Based on this and other predictions, chemists knew what to look for in mineral samples, and soon many of the missing elements were discovered.

Questions:

1. What is eka-silicon, and how close were Mendeleev's predictions to the actual values for this element?
2. How many of the missing elements can you identify?

Answers to these questions are in Appendix Q.

Chapter Goals

See Chapter Goals Revisited (page 72) for Study Questions keyed to these goals and assignable in OWL.

- Describe atomic structure, and define atomic number and mass number.
- Understand the nature of isotopes, and calculate atomic weights from the isotopic masses and abundances.
- Know the terminology of the periodic table.
- Interpret, predict, and write formulas for ionic and molecular compounds.
- Name ionic and molecular compounds.
- Understand some properties of ionic compounds.
- Explain the concept of the mole, and use molar mass in calculations.
- Calculate percent composition for a compound and derive formulas from experimental data.

Chapter Outline

The chemical elements are forged in stars, and from these elements molecules such as water and ammonia are made in outer space. These simple molecules and much more complex ones such as DNA and hemoglobin are found on earth. To comprehend the burgeoning fields of molecular biology, as well as all modern chemistry, we have to understand the nature of the chemical elements and the properties and structures of molecules. This chapter begins our exploration of the chemistry of the elements, the building blocks of chemistry, and of the compounds they form.

Chemistry.Now™

Throughout the text this icon introduces an opportunity for self-study or to explore interactive tutorials by signing in at **www.thomsonedu.com/login**.

2.1 Atomic Structure—Protons, Electrons, and Neutrons

Around 1900, a series of experiments done by scientists such as Sir John Joseph Thomson (1856–1940) and Ernest Rutherford (1871–1937) in England established a model of the atom that is still the basis of modern atomic theory. Three subatomic particles make up all atoms: electrically positive protons, electrically neutral neutrons, and electrically negative electrons. The model places the more massive protons and neutrons in a very small nucleus (Figure 2.1), which contains all the positive charge and almost all the mass of an atom. Electrons, with a much smaller mass than protons or neutrons, surround the nucleus and occupy most of the volume.

The chemical properties of elements and molecules depend largely on the electrons of the atoms involved. We shall look more carefully at their arrangement and how they influence the properties of atoms in Chapters 6 and 7. In this chapter, however, we first want to describe how the composition of the atom relates to its mass and then to the mass of molecules. This is crucial information when we consider the quantitative aspects of chemical reactions in later chapters.

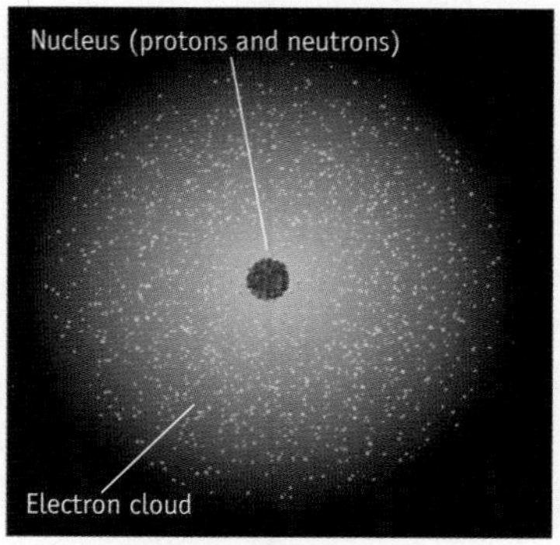

FIGURE 2.1 The structure of the atom. All atoms contain a nucleus with one or more protons (positive electric charge) and, except for H atoms, neutrons (no charge). Electrons (negative electric charge) are found in space as a "cloud" around the nucleus. In an electrically neutral atom, the number of electrons equals the number of protons. Note that this figure is not drawn to scale. If the nucleus were really the size depicted here, the electron cloud would extend over 200 m. The atom is mostly empty space!

2.2 Atomic Number and Atomic Mass

Atomic Number

All atoms of a given element have the same number of protons in the nucleus. Hydrogen is the simplest element, with one nuclear proton. All helium atoms have two protons, all lithium atoms have three protons, and all beryllium atoms have four protons.

■ **How Small Is an Atom?** The radius of the typical atom is between 30 and 300 pm (3×10^{-11} m to 3×10^{-10} m). To get a feeling for the incredible smallness of an atom, consider that one teaspoon of water (about 1 cm^3) contains about three times as many atoms as the Atlantic Ocean contains teaspoons of water.

■ **Periodic Table Entry for Copper**

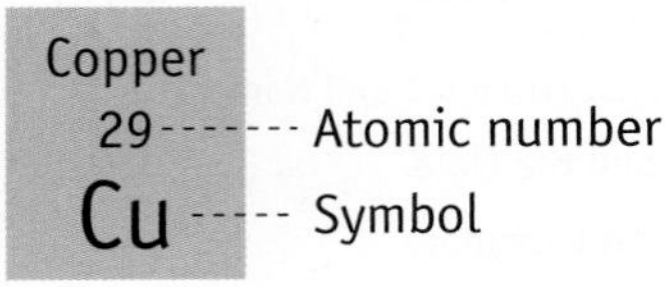

■ **Historical Perspective on the Development of Our Understanding of Atomic Structure** A brief history of important experiments and the scientists involved in developing the modern view of the atom is on pages 338–347. See also ChemistryNow Screens 2.3–2.10.

The number of protons in the nucleus of an element is its **atomic number**, which is generally given the symbol **Z**.

Currently known elements are listed in the periodic table inside the front cover of this book and on the list inside the back cover. The integer number at the top of the box for each element in the periodic table is its atomic number. A sodium atom (Na), for example, has an atomic number of 11, so its nucleus contains 11 protons. A uranium atom (U) has 92 nuclear protons and $Z = 92$.

Atomic Weight and the Atomic Mass Unit

With the quantitative work of the great French chemist Antoine Laurent Lavoisier (1743–1794), chemistry began to change from medieval alchemy to a modern field of study. As 18th- and 19th-century chemists tried to understand how the elements combined, they carried out increasingly quantitative studies aimed at learning, for example, how much of one element would combine with another. Based on this work, they learned that the substances they produced had a constant composition, and so they could define the relative masses of elements that would combine to produce a new substance. At the beginning of the 19th century, John Dalton (1766–1844) suggested that the combinations of elements involve atoms, and so he proposed a relative scale of atom masses. Apparently for simplicity, Dalton chose a mass of 1 for hydrogen on which to base his scale.

The atomic weight scale has changed since 1800, but like the 19th-century chemists, we still use *relative* masses. Our standard today, however, is carbon-12. A carbon atom having six protons and six neutrons in the nucleus is assigned a mass value of exactly 12. From chemical experiments and physical measurements, we know an oxygen atom having eight protons and eight neutrons has 1.3329 times the mass of carbon, so it has a relative mass of 15.9949. Masses of atoms of other elements have been assigned in a similar manner.

Masses of fundamental atomic particles are often expressed in **atomic mass units (u)**. *One atomic mass unit, 1 u, is one twelfth of the mass of an atom of carbon with six protons and six neutrons.* Thus, such a carbon atom has a mass of 12.000 u. The atomic mass unit can be related to other units of mass using the conversion factor $1 \text{ u} = 1.66054 \times 10^{-24}$ g.

Mass Number

Protons and neutrons have masses very close to 1 u (Table 2.1). The mass of an electron, in contrast, is only about 1/2000 of this value. Because proton and neutron masses are so close to 1 u, the approximate mass of an atom can be estimated if the

TABLE 2.1 Properties of Subatomic Particles*

	Mass			
Particle	***Grams***	***Atomic Mass Units***	**Charge**	**Symbol**
Electron	9.109383×10^{-28}	0.0005485799	1−	$_{-1}^{0}e$ or e^-
Proton	1.672622×10^{-24}	1.007276	1+	$_{1}^{1}p$ or p^+
Neutron	1.674927×10^{-24}	1.008665	0	$_{0}^{1}n$ or n

* These values and others in the book are taken from the National Institute of Standards and Technology website at http://physics.nist.gov/cuu/Constants/index.html

number of neutrons and protons is known. The sum of the number of protons and neutrons for an atom is called its **mass number** and is given the symbol ***A***.

$$A = \text{mass number} = \text{number of protons} + \text{number of neutrons}$$

For example, a sodium atom, which has 11 protons and 12 neutrons in its nucleus, has a mass number of $A = 23$. The most common atom of uranium has 92 protons and 146 neutrons, and a mass number of $A = 238$. Using this information, we often symbolize atoms with the notation

$$\begin{matrix}\text{Mass number} \rightarrow \\ \text{Atomic number} \rightarrow\end{matrix} {}^{A}_{Z}\text{X} \leftarrow \text{Element symbol}$$

The subscript Z is optional because the element's symbol tells us what the atomic number must be. For example, the atoms described previously have the symbols ${}^{23}_{11}\text{Na}$ or ${}^{238}_{92}\text{U}$, or just ${}^{23}\text{Na}$ or ${}^{238}\text{U}$. In words, we say "sodium-23" or "uranium-238."

Chemistry•**Now**™

Sign in at **www.thomsonedu.com/login** and go to Chapter 2 Contents to see Screen 2.11 for a tutorial on **the notation for symbolizing atoms.**

EXAMPLE 2.1 Atomic Composition

Problem What is the composition of an atom of phosphorus with 16 neutrons? What is its mass number? What is the symbol for such an atom? If the atom has an actual mass of 30.9738 u, what is its mass in grams? Finally, what is the mass of this phosphorus atom relative to the mass of a carbon atom with a mass number of 12?

Strategy All P atoms have the same number of protons, 15, which is given by the atomic number. The mass number is the sum of the number of protons and neutrons. The mass of the atom in grams can be obtained from the mass in atomic mass units using the conversion factor $1 \text{ u} = 1.66054 \times 10^{-24}$ g.

Solution A phosphorus atom has 15 protons and, because it is electrically neutral, also has 15 electrons. A P atom with 16 neutrons has a mass number of 31.

$$\text{Mass number} = \text{number of protons} + \text{number of neutrons} = 15 + 16 = 31$$

The atom's complete symbol is ${}^{31}_{15}\text{P}$.

$$\text{Mass of one } {}^{31}\text{P atom} = (30.9738 \text{ u}) \times (1.66054 \times 10^{-24} \text{ g/u}) = 5.14332 \times 10^{-23} \text{ g}$$

An atom of ${}^{31}\text{P}$ is 2.58115 times heavier than an atom of ${}^{12}\text{C}$: $30.9738/12.0000 = 2.58115$

EXERCISE 2.1 Atomic Composition

(a) What is the mass number of an iron atom with 30 neutrons?

(b) A nickel atom with 32 neutrons has a mass of 59.930788 u. What is its mass in grams?

(c) How many protons, neutrons, and electrons are in a ${}^{64}\text{Zn}$ atom?

(d) What is the mass of ${}^{64}\text{Zn}$ (63.929 u) relative to ${}^{12}\text{C}$?

2.3 Isotopes

In only a few instances (for example, aluminum, fluorine, and phosphorus) do all atoms in a naturally occurring sample of a given element have the same mass. Most elements consist of atoms having several different mass numbers. For example, there are two kinds of boron atoms, one with a mass of about 10 u (${}^{10}\text{B}$) and a second with a mass of about 11 u (${}^{11}\text{B}$). Atoms of tin can have any of 10 different masses. Atoms with the same atomic number but different mass numbers are called **isotopes**.

Atomic Masses of Some Isotopes

Atom	Relative Mass
${}^{4}\text{He}$	4.0092603
${}^{13}\text{C}$	13.003355
${}^{16}\text{O}$	15.994915
${}^{58}\text{Ni}$	57.935346
${}^{60}\text{Ni}$	59.930788
${}^{79}\text{Br}$	78.918336
${}^{81}\text{Br}$	80.916289
${}^{197}\text{Au}$	196.966543
${}^{238}\text{U}$	238.050784

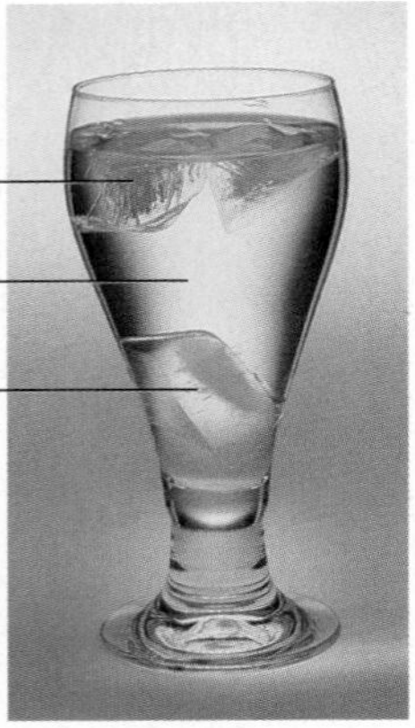

FIGURE 2.2 Ice made from "heavy water." Water containing ordinary hydrogen (^1_1H, protium) forms a solid that is less dense (d = 0.917 g/cm³ at 0 °C) than liquid H_2O (d = 0.997 g/cm³ at 25 °C) and so floats in the liquid. (Water is unique in this regard. The solid phase of virtually all other substances sinks in the liquid phase of that substance.) Similarly, "heavy ice" (D_2O, deuterium oxide) floats in "heavy water." D_2O-ice is denser than liquid H_2O, however, so cubes made of D_2O sink in liquid H_2O.

All atoms of an element have the same number of protons—five in the case of boron. To have different masses, isotopes must have different numbers of neutrons. The nucleus of a ^{10}B atom (Z = 5) contains five protons and five neutrons, whereas the nucleus of a ^{11}B atom contains five protons and six neutrons.

Scientists often refer to a particular isotope by giving its mass number (for example, uranium-238, ^{238}U), but the isotopes of hydrogen are so important that they have special names and symbols. All hydrogen atoms have one proton. When that is the only nuclear particle, the isotope is called *protium,* or just "hydrogen." The isotope of hydrogen with one neutron, ^2_1H, is called *deuterium,* or "heavy hydrogen" (symbol = D). The nucleus of radioactive hydrogen-3, ^3_1H, or *tritium* (symbol = T), contains one proton and two neutrons.

The substitution of one isotope of an element for another isotope of the same element in a compound sometimes can have an interesting effect (Figure 2.2). This is especially true when deuterium is substituted for hydrogen because the mass of deuterium is double that of hydrogen.

Isotope Abundance

A sample of water from a stream or lake will consist almost entirely of H_2O where the H atoms are the ^1H isotope. A few molecules, however, will have deuterium (^2H) substituted for ^1H. We can predict this outcome because we know that 99.985% of all hydrogen atoms on earth are ^1H atoms. That is, the **percent abundance** of ^1H atoms is 99.985%.

$$\text{Percent abundance} = \frac{\text{number of atoms of a given isotope}}{\text{total number of atoms of all isotopes of that element}} \times 100\% \quad \textbf{(2.1)}$$

The remainder of naturally occurring hydrogen is deuterium, whose abundance is only 0.015% of the total hydrogen atoms. Tritium, the radioactive ^3H isotope, occurs naturally in only trace amounts.

Consider again the two isotopes of boron. The boron-10 isotope has an abundance of 19.91%; the abundance of boron-11 is 80.09%. Thus, if you could count out 10,000 boron atoms from an "average" natural sample, 1991 of them would be boron-10 atoms, and 8009 of them would be boron-11 atoms.

Chemistry Now™

Sign in at **www.thomsonedu.com/login** and go to Chapter 2 Contents to see Screen 2.12, **Isotopes.**

EXERCISE 2.2 Isotopes

Silver has two isotopes, one with 60 neutrons (percent abundance = 51.839%) and the other with 62 neutrons. What is the mass number and symbol of the isotope with 62 neutrons, and what is its percent abundance?

Determining Atomic Mass and Isotope Abundance

The masses of isotopes and their percent abundances are determined experimentally using a mass spectrometer (Figure 2.3). A gaseous sample of an element is introduced into the evacuated chamber of the spectrometer, and the atoms or

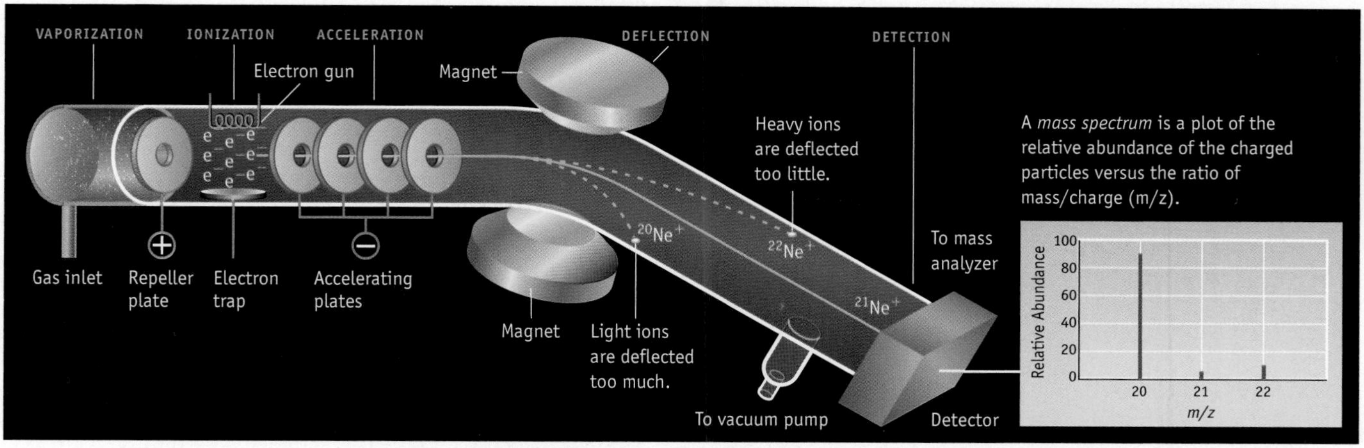

1. A sample is introduced as a vapor into the ionization chamber.

There it is bombarded with high-energy electrons that strip electrons from the atoms or molecules of the sample.

2. The resulting positive particles are accelerated by a series of negatively charged accelerator plates into an analyzing chamber.

3. This chamber is in a magnetic field, which is perpendicular to the direction of the beam of charged particles.

The magnetic field causes the beam to curve. The radius of curvature depends on the mass and charge of the particles (as well as the accelerating voltage and strength of the magnetic field).

4. Here, particles of $^{21}Ne^+$ are focused on the detector, whereas beams of ions of $^{20}Ne^+$ and $^{22}Ne^+$ (of lighter or heavier mass) experience greater and lesser curvature, respectively, and so fail to be detected.

By changing the magnetic field, charged particles of different masses can be focused on the detector to generate the observed spectrum.

Active Figure 2.3 **Mass spectrometer.**

Chemistry Now™ Sign in at www.thomsonedu.com/login and go to the Chapter Contents menu to explore an interactive version of this figure accompanied by an exercise.

molecules of the sample are converted to positively charged particles (called *ions*). A beam of these ions is injected into a magnetic field, which causes the paths of the ions to be deflected. The extent of deflection depends on particle mass: The less massive ions are deflected more, and the more massive ions are deflected less. The ions, now separated by mass, are detected at the end of the chamber. Chemists using modern instruments (Figure 1.2) can measure isotopic masses to as many as nine significant figures.

Except for carbon-12, whose mass is defined to be exactly 12 u, isotopic masses do not have integer values. However, the isotopic masses are always very close to the mass numbers for the isotope. For example, the mass of an atom of boron-11 (^{11}B, 5 protons and 6 neutrons) is 11.0093 u, and the mass of an atom of iron-58 (^{58}Fe, 26 protons and 32 neutrons) is 57.9333 u.

■ **Isotopic Masses and the Mass Defect** Actual masses of atoms are always less than the sum of the masses of subatomic particles composing that atom. This is called the *mass defect,* and the reason for it is discussed in Chapter 23.

2.4 Atomic Weight

Because every sample of boron has some atoms with a mass of 10.0129 u and others with a mass of 11.0093 u, the average atomic mass must be somewhere between these values. The **atomic weight** of an element is the average mass of a representa-

TABLE 2.2 Isotope Abundance and Atomic Weight

Element	Symbol	Atomic Weight	Mass Number	Isotopic Mass	Natural Abundance (%)
Hydrogen	H	1.00794	1	1.0078	99.985
	D*		2	2.0141	0.015
	T†		3	3.0161	0
Boron	B	10.811	10	10.0129	19.91
			11	11.0093	80.09
Neon	Ne	20.1797	20	19.9924	90.48
			21	20.9938	0.27
			22	21.9914	9.25
Magnesium	Mg	24.3050	24	23.9850	78.99
			25	24.9858	10.00
			26	25.9826	11.01

*D = deuterium; †T = tritium, radioactive.

■ **Atomic Mass, Relative Atomic Mass, and Atomic Weight** The atomic mass is the mass of an atom at rest. The relative atomic mass, also known as the atomic weight or the average atomic weight, is the average of the atomic masses of all of the element's isotopes. The term "atomic weight" is slowly being phased out in favor of "relative atomic mass."

tive sample of atoms. For boron, for example, the atomic weight is 10.811. If isotope masses and abundances are known, the atomic mass of an element can be calculated using Equation 2.2.

$$\text{Atomic weight} = \left(\frac{\%\text{ abundance isotope 1}}{100}\right)(\text{mass of isotope 1}) + \left(\frac{\%\text{ abundance isotope 2}}{100}\right)(\text{mass of isotope 2}) + \ldots \quad \textbf{(2.2)}$$

For boron with two isotopes (^{10}B, 19.91% abundant; ^{11}B, 80.09% abundant), we find

$$\text{Atomic weight} = \left(\frac{19.91}{100}\right) \times 10.0129 + \left(\frac{80.09}{100}\right) \times 11.0093 = 10.81$$

Equation 2.2 gives an average mass, weighted in terms of the abundance of each isotope for the element. As illustrated by the data in Table 2.2, *the atomic mass of an element is usually closer to the mass of the most abundant isotope or isotopes.*

The atomic weight of each stable element has been determined experimentally, and these numbers appear in the periodic table inside the front cover of this book. In the periodic table, each element's box contains the atomic number, the element symbol, and the atomic weight. For unstable (radioactive) elements, the atomic weight or mass number of the most stable isotope is given in parentheses.

Chemistry Now™

Sign in at **www.thomsonedu.com/login** and go to Chapter 2 Contents to see Screen 2.13 for an exercise and a tutorial on **mass spectrometers and on calculating atomic mass.**

EXAMPLE 2.2 Calculating Atomic Weight from Isotope Abundance

Problem Bromine has two naturally occurring isotopes. One has a mass of 78.918338 and an abundance of 50.69%. The other isotope, of mass 80.916291, has an abundance of 49.31%. Calculate the atomic weight of bromine.

Strategy The atomic weight of any element is the weighted average of the masses of the isotopes in a representative sample. To calculate the atomic weight, multiply the mass of each isotope by its percent abundance divided by 100 (Equation 2.2).

Solution

$$\text{Atomic weight of bromine} = (50.69/100)(78.918338) + (49.31/100)(80.916291)$$
$$= 79.90$$

Charles D. Winters

Elemental bromine. Bromine is a deep orange, volatile liquid at room temperature. It consists of Br_2 molecules in which two bromine atoms are chemically bonded together. There are two, stable, naturally-occurring isotopes of bromine atoms: ^{79}Br (50.69% abundant) and ^{81}Br (49.31% abundant).

EXAMPLE 2.3 Calculating Isotopic Abundances

Problem Antimony, Sb, has two stable isotopes: ^{121}Sb, 120.904 u, and ^{123}Sb, 122.904 u. What are the relative abundances of these isotopes?

Strategy The atomic mass of antimony is 121.760 u (see the periodic table). Before we do the calculation we can infer that the lighter isotope (^{121}Sb) must be the more abundant because the atomic weight is closer to 121 than to 123. Next, to calculate the abundances we recognize there are two unknown but related quantities, and we can write the following expression (where the fractional abundance of an isotope is the percent abundance of the isotope divided by 100):

$$\text{Atomic weight} = 121.760$$
$$= (\text{fractional abundance of } ^{121}\text{Sb})(120.904) + (\text{fractional abundance of } ^{123}\text{Sb})(122.904)$$

or

$$121.760 = x(120.904) + y(122.904)$$

where x = fractional abundance of ^{121}Sb and y = fractional abundance of ^{123}Sb. Because we know that the sum of the fractional abundances of the isotopes must equal 1, $x + y = 1$, and we can solve the simultaneous equations for x and y.

Solution

Because y = fractional abundance of $^{123}Sb = 1 - x$, we can make a substitution for y.

$$121.760 = x(120.904) + (1 - x)(122.904)$$

Expanding this equation, we have

$$121.760 = 120.904x + 122.904 - 122.904x$$

Finally, solving for x, we find

$$121.760 - 122.904 = (120.904 - 122.904)x$$
$$x = 0.5720$$

The fractional abundance of ^{121}Sb is 0.5720, and its percent abundance is 57.20%. This means that the percent abundance of ^{123}Sb must be 42.80%. The result confirms our initial inference that the lighter isotope is the more abundant of the two.

EXERCISE 2.3 Calculating Atomic Weight

Verify that the atomic weight of chlorine is 35.45, given the following information:

^{35}Cl mass = 34.96885; percent abundance = 75.77%

^{37}Cl mass = 36.96590; percent abundance = 24.23%

Case Study — Catching Cheaters with Isotopes

The U.S. Anti-Doping Agency is responsible for testing for performance-enhancing drugs such as synthetic testosterone that has been used by athletes (page 1). But, if an athlete were to take this synthetic steroid, how can chemists tell the difference between that and the testosterone normally occurring in a male athlete's body?

Testosterone (T) and epitestosterone (E) are closely related steroids, and both are produced naturally. (The latter is an *isomer* of testosterone, a compound with an identical formula to testosterone but with a difference in the way the molecule fills space.) In most adult men the compounds are found in about equal amounts, although the natural T/E ratio can be as high as 4/1. To allow for elevated natural testosterone levels, the Anti-Doping Agency considers that a ratio of 6/1 is an indication an athlete must be using synthetic testosterone. This situation arose when the winner of the 2006 Tour de France bicycle race, Floyd Landis, was found to have a ratio of T/E = 11/1 based on a urine test after one stage of the race.

When the race was over, and the drug-testing results were announced, Landis strongly denied taking testosterone. He argued that his body may have produced more testosterone than normal because of the rigors of the race. This prompted the doping-control laboratory for the Tour to seek additional evidence. They used a technique developed by the U.S. Anti-Doping Agency called "isotope ratio mass spectrometry," which measures the ratio of carbon isotopes, ^{12}C and ^{13}C, in compounds.

© Robert Houser/Index Stock Imagery

Carbon-13 is a naturally occurring isotope of carbon. When most plants grow using CO_2 from the atmosphere, about 1% of the C atoms incorporated in the plant are ^{13}C. The ^{13}C is ingested either directly by humans when eating plants or indirectly when eating meat from grazing animals, and it is then incorporated into the carbon-containing molecules, including testosterone, that our bodies build.

Synthetic testosterone is made from wild yams and soy, plants that are so-called warm climate "C3 plants" that take up atmospheric CO_2 differently than temperate-zone "C4 plants." One important result of this difference is that C3 plants have a lower $^{13}C/^{12}C$ ratio than C4 plants. Because diets in most industrialized countries derive from a mixture of C3 and C4 plants, the natural testosterone in male athletes in most countries will have a different $^{13}C/^{12}C$ ratio than synthetic testosterone. A skilled scientist with a mass spectrometer (Figure 2.3), can relatively easily detect the difference. In the Tour de France case, the $^{13}C/^{12}C$ ratio added further evidence to the case that illegal steroid use had occurred.

Questions:

1. *How many neutrons are there in atoms of ^{13}C?*
2. *^{14}C is a radioactive isotope of carbon that occurs in trace amounts in all living materials. How many neutrons are in a ^{14}C atom?*
3. *Use your library or the World Wide Web to find the source of ^{14}C in living materials.*

Answers to these questions are in Appendix Q.

Module 1

2.5 The Periodic Table

The periodic table of elements is one of the most useful tools in chemistry. Not only does it contain a wealth of information, but it can also be used to organize many of the ideas of chemistry. It is important to become familiar with its main features and terminology.

Developing the Periodic Table

Although the arrangement of elements in the periodic table is now understood on the basis of atomic structure (► Chapters 6 and 7), the table was originally developed from many experimental observations of the chemical and physical properties of elements and is the result of the ideas of a number of chemists in the 18th and 19th centuries.

In 1869, at the University of St. Petersburg in Russia, Dmitri Ivanovitch Mendeleev (1834–1907) was pondering the properties of the elements as he wrote a textbook on chemistry. On studying the chemical and physical properties of the elements, he realized that, if the elements were arranged in order of increasing atomic weight, elements with similar properties appeared in a regular pattern. That is, he saw a **periodicity** or periodic repetition of the properties of elements. Mendeleev organized the known elements into a table by lining them up in a horizontal row in

■ **About the Periodic Table**

For more information on the periodic table, we recommend the following:

- The American Chemical Society has a description of every element on its website (http://pubs.acs.org/cen/80th/elements.html).
- J. Emsley: *Nature's Building Blocks—An A–Z Guide to the Elements*, New York, Oxford University Press, 2001.
- O. Sacks: *Uncle Tungsten—Memories of a Chemical Boyhood*, New York, Alfred A. Knopf, 2001.

Historical Perspectives — The Story of the Periodic Table

by Eric R. Scerri, UCLA

Dimitri Mendeleev was probably the greatest scientist produced by Russia. The youngest of 14 children, he was taken by his mother on a long journey, on foot, in order to enroll him into a university. However, several attempts initially proved futile because, as a Siberian, Mendeleev was barred from attending certain institutions. His mother did succeed in enrolling him in a teacher training college, thus giving Mendeleev a lasting interest in science education, which contributed to his eventual discovery of the periodic system that essentially simplified the subject of inorganic chemistry.

After completing a doctorate, Mendeleev headed to Germany for a postdoctoral fellowship and then returned to Russia, where he set about writing a book aimed at summarizing all of inorganic chemistry. It was while writing this book that he identified the organizing principle with which he is now invariably connected—the periodic system of the elements.

More correctly, though, the periodic system was developed by Mendeleev, as well as five other scientists, over a period of about 10 years, after the Italian chemist Cannizzaro had published a consistent set of atomic weights in 1860. It appears that Mendeleev was unaware of the work of several of his co-discoverers, however.

In essence, the periodic table groups together sets of elements with similar properties into vertical columns. The underlying idea is that if the elements are arranged in order of increasing atomic weights, there are approximate repetitions in their chemical properties after certain intervals. As a result of the existence of the periodic table, students and even professors of chemistry were no longer obliged to learn the properties of all the elements in a disorganized fashion. Instead, they could concentrate on the properties of representative members of the eight columns or groups in the early short-form periodic table, from which they could predict properties of other group members.

Mendeleev is justly regarded as the leading discoverer of the periodic table since he continued to champion the finding and drew out its consequences to a far greater extent than any of his contemporaries. First, he accommodated the 65 or so elements that were known at the time into a coherent scheme based on ascending order of atomic weight while also reflecting chemical and physical similarities. Next, he noticed gaps in his system, which he reasoned would eventually be filled by elements that had not yet been discovered. In addition, by judicious interpolation between the properties of known elements, Mendeleev predicted the nature of a number of completely new elements. Within a period of about 20 years, three of these elements—subsequently called gallium, scandium, and germanium—were isolated and found to have almost the exact properties that Mendeleev had predicted.

John Kotz

Statue of Dmitri Mendeleev and a periodic table mural. This statue and mural are at the Institute for Metrology in St. Petersburg, Russia.

What is not well known is that about half of the elements that Mendeleev predicted were never found. But given the dramatic success of his early predictions, these later lapses have largely been forgotten.

Eric Scerri, *The Periodic Table: Its Story and Its Significance*, Oxford University Press, New York, 2007.

order of increasing atomic weight (page 50). Every time he came to an element with properties similar to one already in the row, he started a new row. For example, the elements Li, Be, B, C, N, O, and F were in a row. Sodium was the next element then known; because its properties closely resembled those of Li, Mendeleev started a new row. As more and more elements were added to the table, new rows were added, and elements with similar properties (such as Li, Na, and K) were found in the same vertical column.

An important feature of Mendeleev's table—and a mark of his genius—was that he left an empty space in a column when an element was not known but should exist and have properties similar to the element above it in his table. He deduced that these spaces would be filled by undiscovered elements. For example, he left a space between Si (silicon) and Sn (tin) in Group 4A for an element he called *eka-silicon.* Based on the progression of properties in this group, Mendeleev was able to predict the properties of this missing element. With the discovery of germanium (Ge) in 1886, Mendeleev's prediction was confirmed.

■ **Mendeleev and Atomic Numbers** Mendeleev developed the periodic table based on atomic weights because the concept of atomic numbers was not known until after the development of the structure of the atom in the early 20th century.

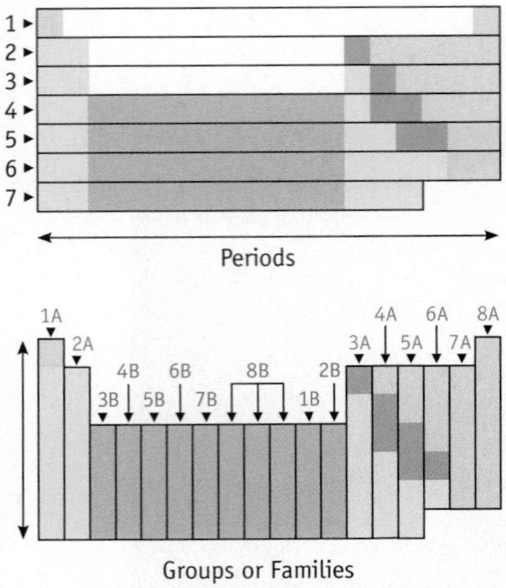

Periods and groups in the periodic table. One way to designate periodic groups is to number them 1 through 18 from left to right. This method is generally used outside the United States. The system predominant in the United States labels main group elements as Groups 1A–8A and transition elements as Groups 1B–8B. This book uses the A/B system.

In Mendeleev's table, the elements were ordered by increasing mass. A glance at a modern table, however, shows that, if some elements (such as Ni and Co, Ar and K, and Te and I) were ordered by mass and not chemical and physical properties, they would be reversed in their order of appearance. Mendeleev recognized these discrepancies and simply assumed the atomic weights known at that time were inaccurate—not a bad assumption based on the analytical methods then in use. In fact, his order is correct, and what was wrong was his assumption that element properties were a function of their mass.

In 1913, H. G. J. Moseley (1887–1915), a young English scientist working with Ernest Rutherford (1871–1937), corrected Mendeleev's assumption. Moseley was doing experiments in which he bombarded many different metals with electrons in a cathode-ray tube (page 343) and examined the x-rays emitted in the process. In seeking some order in his data, he realized that the wavelength of the x-rays emitted by a given element was related in a precise manner to the atomic number of the element. Indeed, once the concept of an atomic number was recognized early in the 20th century, chemists realized that organizing the elements in a table by increasing atomic number corrected the inconsistencies in the Mendeleev table. The **law of chemical periodicity** is now stated as *the properties of the elements are periodic functions of atomic number.*

Features of the Periodic Table

The main organizational features of the periodic table are the following:

- Elements are arranged so those with similar chemical and physical properties lie in vertical columns called **groups** or **families**. The periodic table commonly used in the United States has groups numbered 1 through 8, with each number followed by a letter: A or B. The A groups are often called the **main group elements,** and the B groups are the **transition elements**.
- The horizontal rows of the table are called **periods**, and they are numbered beginning with 1 for the period containing only H and He. For example, sodium, Na, in Group 1A, is the first element in the third period. Mercury, Hg, in Group 2B, is in the sixth period (or sixth row).

The periodic table can be divided into several regions according to the properties of the elements. On the table inside the front cover of this book, elements that behave as *metals* are indicated in purple, those that are *nonmetals* are indicated in yellow, and elements called *metalloids* appear in green. Elements gradually become less metallic as one moves from left to right across a period, and the metalloids lie along the metal–nonmetal boundary. Some elements are shown in Figure 2.4.

You are probably familiar with many properties of **metals** from everyday experience (Figure 2.5a). Metals are solids (except for mercury), can conduct electricity, are usually ductile (can be drawn into wires) and malleable (can be rolled into sheets), and can form alloys (solutions of one or more metals in another metal). Iron (Fe) and aluminum (Al) are used in automobile parts because of their ductility, malleability, and low cost relative to other metals. Copper (Cu) is used in electric wiring because it conducts electricity better than most other metals.

The **nonmetals** lie to the right of a diagonal line that stretches from B to Te in the periodic table and have a wide variety of properties. Some are solids (carbon, sulfur, phosphorus, and iodine). Five elements are gases at room temperature (hydrogen, oxygen, nitrogen, fluorine, and chlorine). One nonmetal, bromine, is a liquid at room temperature (Figure 2.5b). With the exception of carbon in the form of graphite, nonmetals do not conduct electricity, which is one of the main features that distinguishes them from metals.

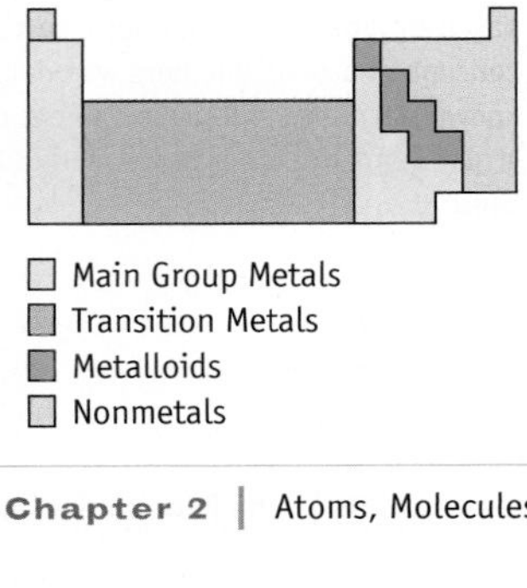

Active Figure 2.4 Some of the known 117 elements.

Chemistry Now™ Sign in at www.thomsonedu.com/login and go to the Chapter Contents menu to explore an interactive version of this figure accompanied by an exercise.

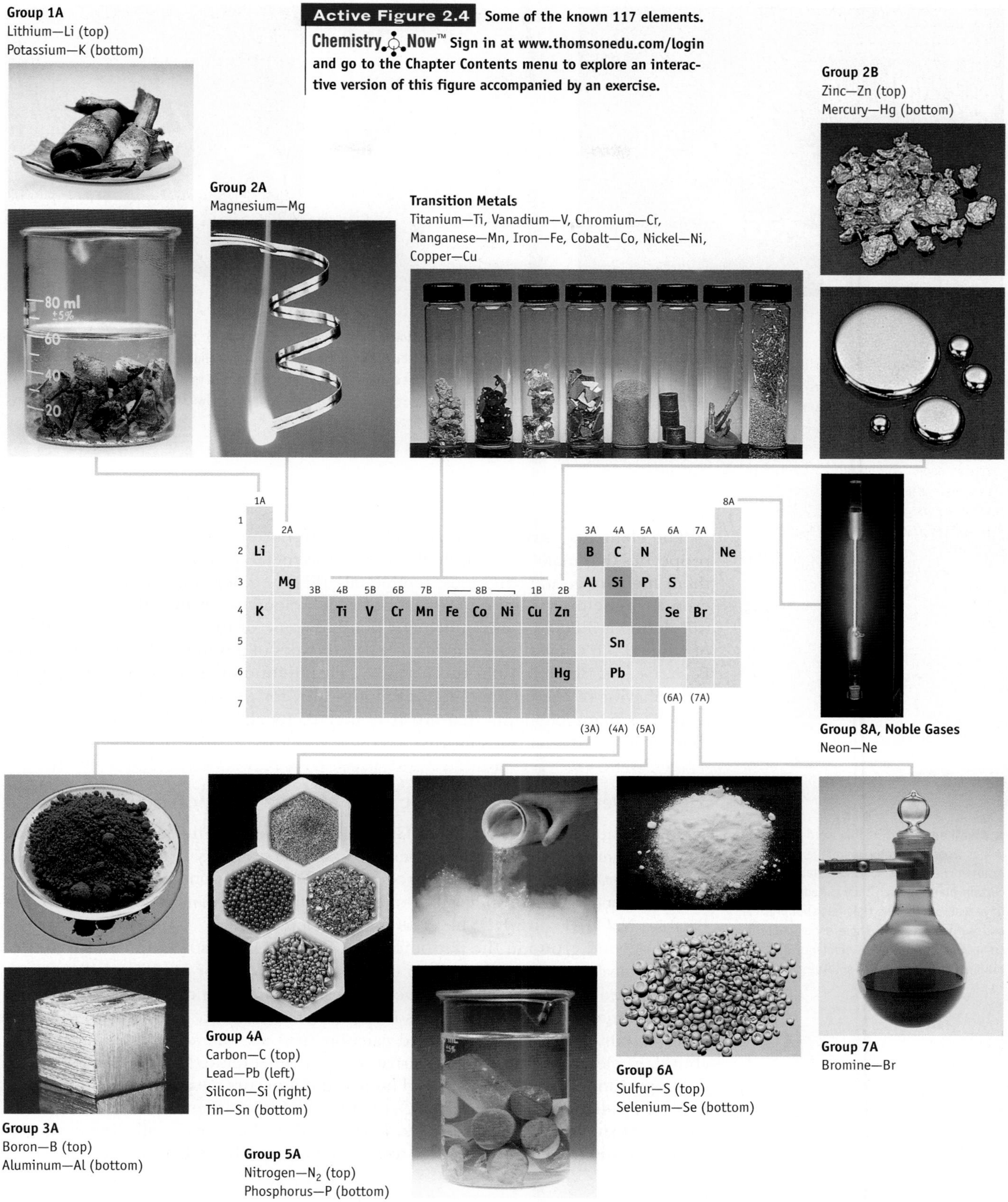

Charles D. Winters

(a) Metals **(b)** Nonmetals **(c)** Metalloids

FIGURE 2.5 Metals, nonmetals, and metalloids. (a) Molybdenum (Mo, wire), bismuth (Bi, center object), and copper (Cu) are metals. Metals can generally be drawn into wires, and they conduct electricity. (b) Only 15 or so elements can be classified as nonmetals. Here are orange liquid bromine and purple solid iodine. (c) Only six elements are generally classified as metalloids or semimetals. This photograph shows solid silicon in various forms, including a wafer that holds printed electronic circuits.

The elements next to the diagonal line from boron (B) to tellurium (Te) have properties that make them difficult to classify as metals or nonmetals. Chemists call them **metalloids** or, sometimes, **semimetals** (Figure 2.5c). You should know, however, that chemists often disagree about which elements fit into this category. We will define a metalloid as an element that has some of the physical characteristics of a metal but some of the chemical characteristics of a nonmetal; we include only B, Si, Ge, As, Sb, and Te in this category. This definition reflects the ambiguity in the behavior of these elements. Antimony (Sb), for example, conducts electricity as well as many elements that are true metals. Its chemistry, however, resembles that of the nonmetal phosphorus.

A Brief Overview of the Periodic Table and the Chemical Elements

■ **Alkali and Alkaline** The word "alkali" comes from the Arabic language; ancient Arabian chemists discovered that ashes of certain plants, which they called al-qali, gave water solutions that felt slippery and burned the skin. These ashes contain compounds of Group 1A elements that produce alkaline (basic) solutions.

The metals in the leftmost column, **Group 1A**, are known as the **alkali metals**. All are metals and are solids at room temperature. They are all very reactive. For example, they react with water to produce hydrogen and alkaline solutions (Figure 2.6). Because of their reactivity, these metals are only found in nature combined in compounds (such as NaCl) (Figure 1.4), never as the free element.

The second group in the periodic table, **Group 2A**, is also composed entirely of metals that occur naturally only in compounds. Except for beryllium (Be), these elements react with water to produce alkaline solutions, and most of their oxides (such as lime, CaO) form alkaline solutions; hence, they are known as the **alkaline earth metals**. Magnesium (Mg) and calcium (Ca) are the seventh and fifth most abundant elements in the earth's crust, respectively (Table 2.3). Calcium is one of the important elements in teeth and bones, and it occurs naturally in vast limestone deposits. Calcium carbonate ($CaCO_3$) is the chief constituent of limestone and of corals, sea shells, marble, and chalk. Radium (Ra), the heaviest alkaline earth element, is radioactive and is used to treat some cancers by radiation.

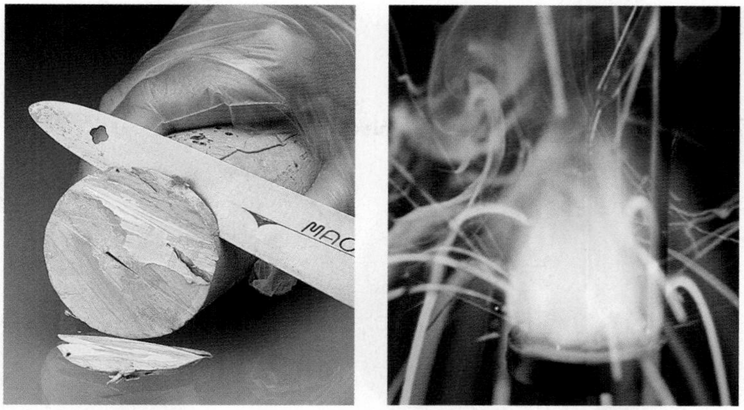

(a) Cutting sodium. **(b)** Potassium reacts with water.

FIGURE 2.6 Alkali metals. (a) Cutting a bar of sodium with a knife is about like cutting a stick of cold butter. (b) When an alkali metal such as potassium is treated with water, a vigorous reaction occurs, giving an alkaline solution and hydrogen gas, which burns in air.

TABLE 2.3 The 10 Most Abundant Elements in the Earth's Crust

Rank	Element	Abundance (ppm)*
1	Oxygen	474,000
2	Silicon	277,000
3	Aluminum	82,000
4	Iron	41,000
5	Calcium	41,000
6	Sodium	23,000
7	Magnesium	23,000
8	Potassium	21,000
9	Titanium	5,600
10	Hydrogen	1,520

*ppm = g per 1000 kg.

Group 3A contains one element of great importance, aluminum (see Figure 2.4). This element and three others (gallium, indium, and thallium) are metals, whereas boron (B) is a metalloid. Aluminum (Al) is the most abundant metal in the earth's crust at 8.2% by mass. It is exceeded in abundance only by the nonmetal oxygen and metalloid silicon. These three elements are found combined in clays and other common minerals. Boron occurs in the mineral borax, a compound used as a cleaning agent, antiseptic, and flux for metal work.

As a metalloid, boron has a different chemistry than the other elements of the group, all of which are metals. Nonetheless, all form compounds with analogous formulas such as BCl_3 and $AlCl_3$, and this similarity marks them as members of the same periodic group.

Thus far, all the elements we have described, except boron, have been metals. Beginning with **Group 4A**, however, the groups contain more and more nonmetals. In Group 4A, there are a nonmetal, carbon (C), two metalloids, silicon (Si) and germanium (Ge), and two metals, tin (Sn) and lead (Pb) (Figure 2.4). Because of the change from nonmetallic to metallic character, more variation occurs in the properties of the elements of this group than in most others. Nonetheless, there are similarities. For example, these elements form compounds with analogous formulas such as CO_2, SiO_2, GeO_2, and PbO_2.

Carbon is the basis for the great variety of chemical compounds that make up living things. It is found in Earth's atmosphere as CO_2, on the surface of the earth in carbonates like limestone and coral (calcium carbonate, $CaCO_3$), and in coal, petroleum, and natural gas—the fossil fuels.

One interesting aspect of the chemistry of the nonmetals is that a particular element can often exist in several different and distinct forms, called **allotropes**, each having its own properties. Carbon has a number of allotropes, the best known of which are graphite and diamond. Graphite consists of flat sheets in which each carbon atom is connected to three others (Figure 2.7a). Because the sheets of carbon atoms cling only weakly to one another, one layer can slip easily over another. This explains why graphite is soft, is a good lubricant, and is used in pencil lead. (Pencil "lead" is not the element lead (Pb) but a composite of clay and graphite that leaves a trail of graphite on the page as you write.)

Liquid gallium. Bromine and mercury are the only elements that are liquids under ambient conditions. Gallium (29.8 °C) and cesium (28.4 °C) melt slightly above room temperature. Here gallium melts when held in the hand.

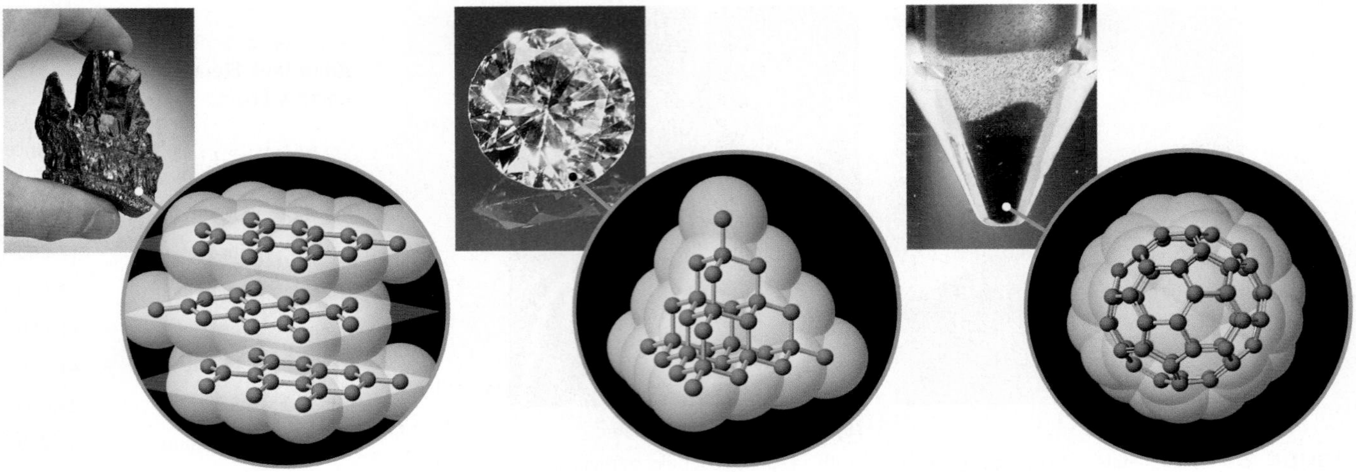

Charles D. Winters

(a) Graphite (b) Diamond (c) Buckyballs

FIGURE 2.7 The allotropes of carbon. (a) Graphite consists of layers of carbon atoms. Each carbon atom is linked to three others to form a sheet of six-member, hexagonal rings. (b) In diamond, the carbon atoms are also arranged in six-member rings, but the rings are not planar because each C atom is connected tetrahedrally by four other C atoms. (c) Buckyballs. A member of the family called buckminsterfullerenes, C_{60} is an allotrope of carbon. Sixty carbon atoms are arranged in a spherical cage that resembles a hollow soccer ball. Notice that each six-member ring shares an edge with three other six-member rings and three five-member rings. Chemists call this molecule a "buckyball." C_{60} is a black powder; it is shown here in the tip of a pointed glass tube.

■ **Special Group Names** Some groups have common and widely used names. (See Figure 2.4)

Group 1A: Alkali metals
Group 2A: Alkaline earth metals
Group 7A: Halogens
Group 8A: Noble gases

In diamond, each carbon atom is connected to four others at the corners of a tetrahedron, and this extends throughout the solid (see Figure 2.7b). This structure causes diamonds to be extremely hard, denser than graphite (d = 3.51 g/cm^3 for diamond and d = 2.22 g/cm^3 for graphite), and chemically less reactive. Because diamonds are not only hard but are excellent conductors of heat, they are used on the tips of metal- and rock-cutting tools.

In the late 1980s, another form of carbon was identified as a component of black soot, the stuff that collects when carbon-containing materials are burned in a deficiency of oxygen. This substance is made up of molecules with 60 carbon atoms arranged as a spherical "cage" (Figure 2.7c). You may recognize that the surface is made up of five- and six-member rings and resembles a hollow soccer ball. The shape also reminded its discoverers of an architectural dome conceived several decades ago by the American philosopher and engineer, R. Buckminster Fuller. This led to the official name of the allotrope, buckminsterfullerene, although chemists often simply call these C_{60} molecules "buckyballs."

Oxides of silicon are the basis of many minerals such as clay, quartz, and beautiful gemstones like amethyst (Figure 2.8). Tin and lead have been known for centuries because they are easily obtained from their ores. Tin alloyed with copper makes bronze, which was used in ancient times in utensils and weapons. Lead has been used in water pipes and paint, even though the element is toxic to humans.

Nitrogen in **Group 5A** occurs naturally in the form of the diatomic molecule N_2 (Figure 2.9) and makes up about three fourths of Earth's atmosphere. It is also incorporated in biochemically important substances such as chlorophyll, proteins, and DNA. Therefore, scientists have long sought ways to make compounds from atmospheric nitrogen, a process referred to as "nitrogen fixation." Nature accomplishes this easily in some prokaryotic organisms, but severe conditions (high temperatures, for example) must be used in the laboratory and in industry to cause N_2 to react with other elements (such as H_2 to make ammonia, NH_3, which is widely used as a fertilizer).

Charles D. Winters

FIGURE 2.8 Compounds containing silicon. Ordinary clay, sand, and many gemstones are based on compounds of silicon and oxygen. Here, clear, colorless quartz and dark purple amethyst lie in a bed of sand. All are silicon dioxide, SiO_2. The different colors are due to impurities.

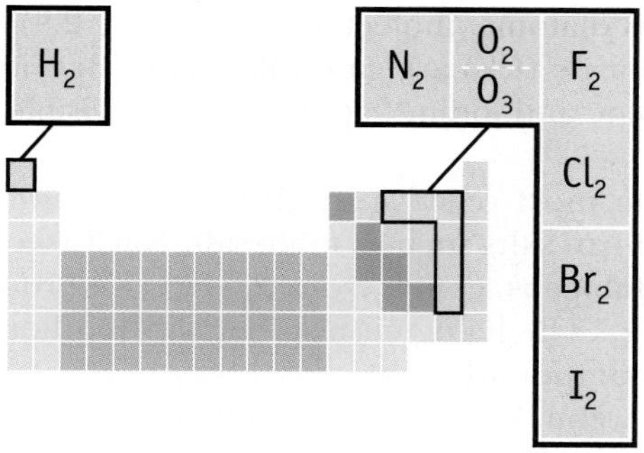

FIGURE 2.9 Elements that exist as diatomic or triatomic molecules. Seven of the elements in the periodic table exist as diatomic, or two-atom, molecules. Oxygen has an additional allotrope, ozone, with three O atoms in each molecule. See also ChemistryNow Screen 2.16, Elements that Exist as Molecules.

■ **Placing H in the Periodic Table** Where to place H? Tables often show it in Group 1A even though it is clearly not an alkali metal. However, in its reactions it forms a 1+ ion just like the alkali metals. For this reason, H is often placed in Group 1A.

Phosphorus is also essential to life. It is an important constituent in bones, teeth, and DNA. The element glows in the dark if it is exposed to air (owing to its reaction with O_2), and its name, based on Greek words meaning "light-bearing," reflects this. This element also has several allotropes, the most important being white (Figure 2.4) and red phosphorus. Both forms of phosphorus are used commercially. White phosphorus ignites spontaneously in air and so is normally stored under water. When it does react with air, it forms P_4O_{10}, which can react with water to form phosphoric acid (H_3PO_4), a compound used in food products such as soft drinks. Red phosphorus is used in the striking strips on match books. When a match is struck, potassium chlorate in the match head mixes with some red phosphorus on the striking strip, and the friction is enough to ignite this mixture.

As with Group 4A, we again see nonmetals (N and P), metalloids (As and Sb), and a metal (Bi, Figure 2.5a) in Group 5A. In spite of these variations, they also form analogous compounds such as the oxides N_2O_5, P_2O_5 and As_2O_5.

Oxygen, which constitutes about 20% of Earth's atmosphere and which combines readily with most other elements, is at the top of **Group 6A**. Most of the energy that powers life on Earth is derived from reactions in which oxygen combines with other substances.

Sulfur has been known in elemental form since ancient times as brimstone or "burning stone" (Figure 2.10). Sulfur, selenium, and tellurium are often referred to collectively as **chalcogens** (from the Greek word, *khalkos*, for copper) because most copper ores contain these elements. Their compounds can be foul smelling and poisonous; nevertheless, sulfur and selenium are essential components of the human diet. By far the most important compound of sulfur is sulfuric acid (H_2SO_4), which is manufactured in larger amounts than any other compound.

As in Group 5A, the second- and third-period elements of Group 6A have different structures. Like nitrogen, oxygen is also a diatomic molecule (see Figure 2.9). Unlike nitrogen, however, oxygen has an allotrope, the triatomic molecule ozone, O_3. Sulfur, which can be found in nature as a yellow solid, has many allotropes. The most common allotrope consists of eight-member, crown-shaped rings of sulfur atoms (see Figure 2.10).

Polonium, a radioactive element in Group 6A, was isolated in 1898 by Marie and Pierre Curie, who separated it from tons of a uranium-containing ore and named it for Madame Curie's native country, Poland.

In Group 6A, we once again observe a variation of properties. Oxygen, sulfur, and selenium are nonmetals; tellurium is a metalloid; and polonium is a metal. Nonetheless, there is a family resemblance in their chemistries. All form oxygen-containing compounds such as SO_2, SeO_2, and TeO_2 and sodium-containing compounds (Na_2O, Na_2S, Na_2Se, and Na_2Te).

At the far right of the periodic table are two groups composed entirely of nonmetals. The **Group 7A** elements—fluorine, chlorine, bromine, iodine, and radioac-

FIGURE 2.10 Sulfur. The most common allotrope of sulfur consists of S atoms arranged in eight-member, crown-shaped rings.

tive astatine—are nonmetals, and all exist as diatomic molecules (see Figure 2.9). At room temperature, fluorine (F_2) and chlorine (Cl_2) are gases. Bromine (Br_2) is a liquid, and iodine (I_2) is a solid, but bromine and iodine vapor are clearly visible over the liquid or solid (see Figure 2.5b).

The Group 7A elements are among the most reactive of all elements, and all combine violently with alkali metals to form salts such as table salt, NaCl (see Figure 1.4). The name for this group, the **halogens**, comes from the Greek words *hals*, meaning "salt," and *genes*, for "forming." The halogens also react with other metals and with most nonmetals to form compounds.

The **Group 8A** elements—helium, neon, argon, krypton, xenon, and radioactive radon—are the least reactive elements (Figure 2.11). All are gases, and none is abundant on Earth or in Earth's atmosphere. Because of this, they were not discovered until the end of the 19th century. Helium, the second most abundant element in the universe after hydrogen, was detected in the sun in 1868 by analysis of the solar spectrum. (The name of the element comes from the Greek word for the sun, *helios.*) It was not found on Earth until 1895, however. Until 1962, when a compound of xenon was first prepared, it was believed that none of these elements would combine chemically with any other element. The name **noble gases** for this group, a term meant to denote their general lack of reactivity, derives from this fact. For the same reason, they are sometimes called the *inert gases* or, because of their low abundance, the *rare gases.*

Stretching between Groups 2A and 3A is a series of elements called the **transition elements.** These fill the B-groups (1B through 8B) in the fourth through the seventh periods in the center of the periodic table. All are metals (see Figure 2.4), and 13 of them are in the top 30 elements in terms of abundance in the earth's crust. Some, like iron (Fe), are abundant in nature (Table 2.3). Most occur naturally in combination with other elements, but a few—copper (Cu), silver (Ag), gold (Au), and platinum (Pt)—are much less reactive and so can be found in nature as pure elements.

Virtually all of the transition elements have commercial uses. They are used as structural materials (iron, titanium, chromium, copper); in paints (titanium, chromium); in the catalytic converters in automobile exhaust systems (platinum and rhodium); in coins (copper, nickel, zinc); and in batteries (manganese, nickel, cadmium, mercury).

A number of the transition elements play important biological roles. For example, iron, a relatively abundant element (see Table 2.3), is the central element in the chemistry of hemoglobin, the oxygen-carrying component of blood.

Charles D. Winters

FIGURE 2.11 The noble gases. This kit is sold for detecting the presence of radioactive radon in the home. Neon gas is used in advertising signs, and xenon-containing headlights are increasingly popular on automobiles.

Two rows at the bottom of the table accommodate the **lanthanides** [the series of elements between the elements lanthanum (Z = 57) and hafnium (Z = 72)] and the **actinides** [the series of elements between actinium (Z = 89) and rutherfordium (Z = 104)]. Some lanthanide compounds are used in color television picture tubes; uranium (Z = 92) is the fuel for atomic power plants, and americium (Z = 95) is used in smoke detectors.

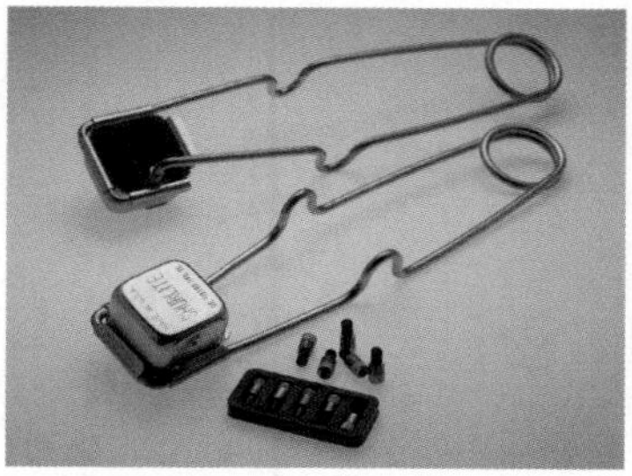

Charles D. Winters

Lanthanides. If you use a Bunsen burner in the lab, you may light it with a "flint" lighter. The flints are composed of iron and "mischmetal," a mixture of lanthanide elements, chiefly Ce, La, Pr, and Nd with traces of other lanthanides. (The word "mischmetal" comes from the German for "mixed metals.") It is produced by the electrolysis of a mixture of lanthanide oxides.

Chemistry Now™

Sign in at **www.thomsonedu.com/login** and go to Chapter 2 Contents to see:
- Screen 2.14 for an exercise on **periodic table organization**
- Screen 2.15 for an exercise on **chemical periodicity**

EXERCISE 2.4 The Periodic Table

How many elements are in the third period of the periodic table? Give the name and symbol of each. Tell whether each element in the period is a metal, metalloid, or nonmetal.

2.6 Molecules, Compounds, and Formulas

A molecule is the smallest identifiable unit into which a pure substance like sugar or water can be divided and still retain the composition and chemical properties of the substance. Such substances are composed of identical molecules consisting of atoms of two or more elements bound firmly together. For example, atoms of the element aluminum, Al, combine with molecules of the element bromine, Br_2, to produce the compound aluminum bromide, Al_2Br_6 (Figure 2.12).

$$2\ Al(s) + 3\ Br_2(\ell) \rightarrow Al_2Br_6(s)$$
$$\text{aluminum} + \text{bromine} \rightarrow \text{aluminum bromide}$$

Photos: Charles D. Winters

(a) (b) (c)

Active Figure 2.12 **Reaction of the elements aluminum and bromine.** (a) Solid aluminum and (in the beaker) liquid bromine. (b) When the aluminum is added to the bromine, a vigorous chemical reaction produces white, solid aluminum bromide, Al_2Br_6 (c).

Chemistry Now™ Sign in at www.thomsonedu.com/login and go to the Chapter Contents menu to explore an interactive version of this figure accompanied by an exercise.

To describe this chemical change (or chemical reaction) on paper, the composition of each element and compound is represented by a symbol or formula. Here, one molecule of Al_2Br_6 is composed of two Al atoms and six Br atoms.

How do compounds differ from elements? When a compound is produced from its elements, the characteristics of the constituent elements are lost. Solid, metallic aluminum and red-orange liquid bromine, for example, react to form Al_2Br_6, a white solid.

Formulas

For molecules more complicated than water, there is often more than one way to write the formula. For example, the formula of ethanol (also called ethyl alcohol) can be represented as C_2H_6O (Figure 2.13). This **molecular formula** describes the composition of ethanol molecules—two carbon atoms, six hydrogen atoms, and one atom of oxygen occur per molecule—but it gives us no structural information. Structural information—how the atoms are connected and how the molecule fills space—is important, however, because it helps us understand how a molecule can interact with other molecules, which is the essence of chemistry.

■ **Writing Formulas** When writing molecular formulas of organic compounds (compounds with C, H, and other elements), the convention is to write C first, then H, and finally other elements in alphabetical order. For example, acrylonitrile, a compound used to make consumer plastics, has the condensed formula CH_2CHCN. Its molecular formula is C_3H_3N.

To provide some structural information, it is useful to write a **condensed formula**, which indicates how certain atoms are grouped together. For example, the condensed formula of ethanol, CH_3CH_2OH (see Figure 2.13), informs us that the molecule consists of three "groups": a CH_3 group, a CH_2 group, and an OH group. Writing the formula as CH_3CH_2OH also shows that the compound is not dimethyl ether, CH_3OCH_3, a compound with the same molecular formula but with a different structure and distinctly different properties.

That ethanol and dimethyl ether are different molecules is further apparent from their **structural formulas** (Figure 2.13). This type of formula gives us an even higher level of structural detail, showing how all of the atoms are attached within a molecule. The lines between atoms represent the chemical bonds that hold atoms together in this molecule (► Chapter 8).

Chemistry Now™

Sign in at **www.thomsonedu.com/login** and go to Chapter 2 Contents to see Screen 2.17 for an exercise and tutorial on **representations of molecules.**

■ **Ethanol and Dimethyl Ether Are Isomers** Compounds having the same molecular formula but different structures are called isomers. (See Chapter 10, and sign in to ChemistryNow and see Screen 2.17, Representing Compounds.)

NAME	MOLECULAR FORMULA	CONDENSED FORMULA	STRUCTURAL FORMULA	MOLECULAR MODEL
Ethanol	C_2H_6O	CH_3CH_2OH	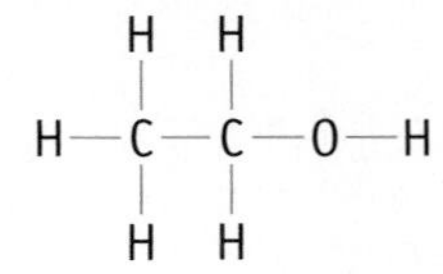	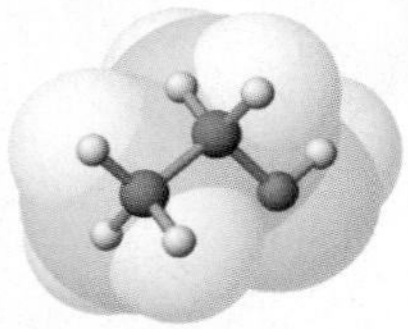
Dimethyl ether	C_2H_6O	CH_3OCH_3	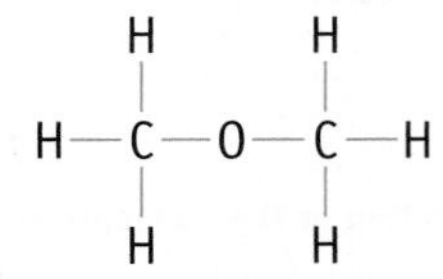	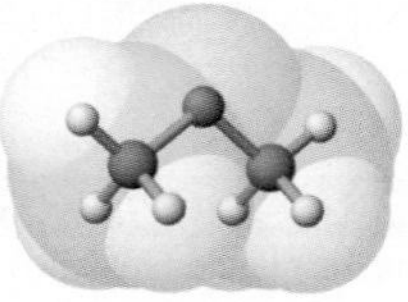

FIGURE 2.13 Four approaches to showing molecular formulas. Here, the two molecules have the same molecular formula. Condensed or structural formulas, and a molecular model, show that these molecules are different.

EXERCISE 2.5 Molecular Formulas

Cysteine, whose molecular model and structural formula are illustrated here, is an important amino acid and a constituent of many living things. What is its molecular formula?

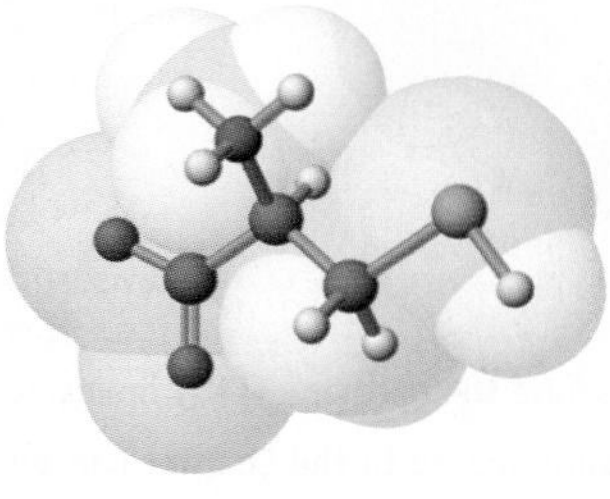

Molecular model

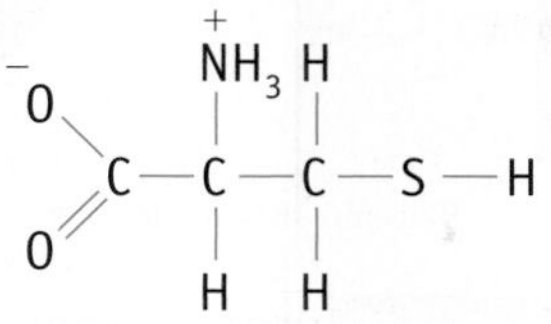

Structural formula

■ **Standard Colors for Atoms in Molecular Models** The colors listed here are used for molecular models in this book and are generally used by chemists. Note that Cl and S atoms have the same color, but it is usually apparent what atom is being designated.

carbon atoms

hydrogen atoms

oxygen atoms

nitrogen atoms

chlorine atoms

Molecular Models

Molecular structures are often beautiful in the same sense that art is beautiful, and there is something intrinsically beautiful about the pattern created by water molecules assembled in ice (Figure 2.14).

More important, however, is the fact that the physical and chemical properties of a molecular compound are often closely related to its structure. For example, two well-known features of ice are related to its structure. The first is the shape of ice crystals: The sixfold symmetry of macroscopic ice crystals also appears at the particulate level in the form of six-sided rings of hydrogen and oxygen atoms. The second is water's unique property of being less dense when solid than it is when liquid. The lower density of ice, which has enormous consequences for Earth's climate, results from the fact that molecules of water are not packed together tightly.

Because molecules are three-dimensional, it is often difficult to represent their shapes on paper. Certain conventions have been developed, however, that help represent three-dimensional structures on two-dimensional surfaces. Simple perspective drawings are often used (Figure 2.15).

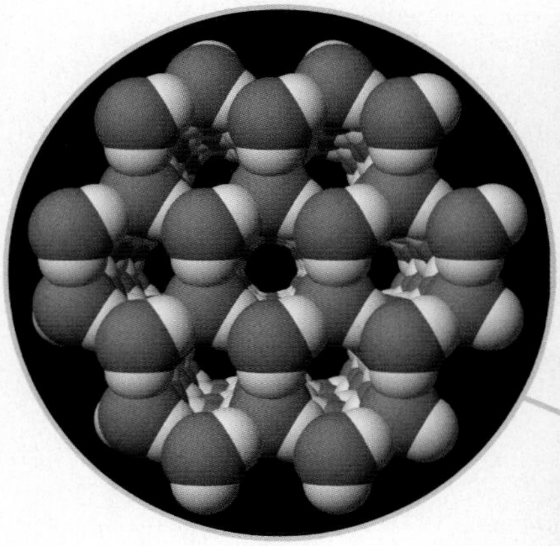

Mehau Kulyk/Science Photo Library/Photo Researchers, Inc.; model by S.M. Young

FIGURE 2.14 Ice. Snowflakes are six-sided structures, reflecting the underlying structure of ice. Ice consists of six-sided rings formed by water molecules, in which each side of a ring consists of two O atoms and an H atom.

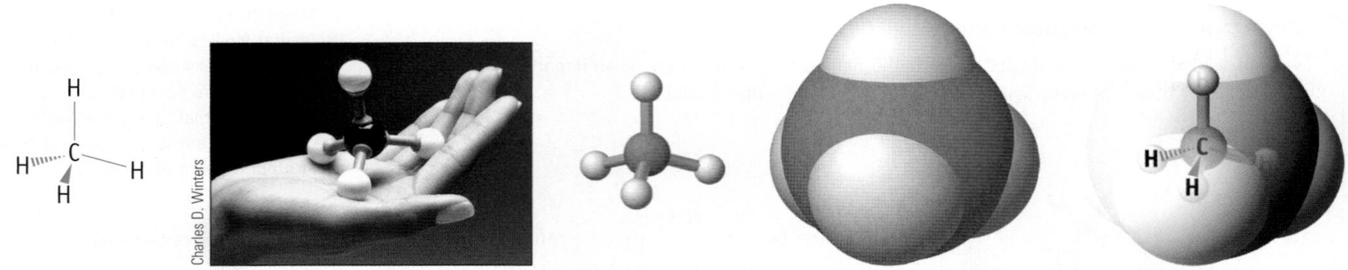

Simple perspective drawing

Plastic model

Ball-and-stick model

Space-filling model

All visualizing techniques represent the same molecule.

Active Figure 2.15 **Ways of depicting a molecule, here the methane (CH_4) molecule.**

Chemistry Now™ Sign in at www.thomsonedu.com/login and go to the Chapter Contents menu to explore an interactive version of this figure accompanied by an exercise.

Several kinds of molecular models exist. In the **ball-and-stick model**, spheres, usually in different colors, represent the atoms, and sticks represent the bonds holding them together. These models make it easy to see how atoms are attached to one another. Molecules can also be represented using **space-filling models**. These models are more realistic because they offer a better representation of relative sizes of atoms and their proximity to each other when in a molecule. A disadvantage of pictures of space-filling models is that atoms can often be hidden from view.

Module 2
Module 3

2.7 Ionic Compounds: Formulas, Names, and Properties

The compounds you have encountered so far in this chapter are **molecular compounds**—that is, compounds that consist of discrete molecules at the particulate level. **Ionic compounds** constitute another major class of compounds. They consist of **ions**, atoms or groups of atoms that bear a positive or negative electric charge. Many familiar compounds are composed of ions (Figure 2.16). Table salt, or sodium chloride (NaCl), and lime (CaO) are just two. To recognize ionic compounds, and to be able to write formulas for these compounds, it is important to know the formulas and charges of common ions. You also need to know the names of ions and be able to name the compounds they form.

FIGURE 2.16 **Some common ionic compounds.**

Common Name	Name	Formula	Ions Involved
Calcite	Calcium carbonate	$CaCO_3$	Ca^{2+}, CO_3^{2-}
Fluorite	Calcium fluoride	CaF_2	Ca^{2+}, F^-
Gypsum	Calcium sulfate dihydrate	$CaSO_4 \cdot 2\ H_2O$	Ca^{2+}, SO_4^{2-}
Hematite	Iron(III) oxide	Fe_2O_3	Fe^{3+}, O^{2-}
Orpiment	Arsenic sulfide	As_2S_3	As^{3+}, S^{2-}

Ions

Atoms of many elements can gain or lose electrons in the course of a chemical reaction. To be able to predict the outcome of chemical reactions (► Sections 3.1–3.9), you need to know whether an element will likely gain or lose electrons and, if so, how many.

Cations

If an atom loses an electron (which is transferred to an atom of anot her element in the course of a reaction), the atom now has one fewer negative electrons than it has positive protons in the nucleus. The result is a positively charged ion called a **cation** (Figure 2.17). (The name is pronounced "cat'-ion.") Because it has an excess of one positive charge, we write the cation's symbol as, for example, Li^+:

$$\underset{\text{(3 protons and 3 electrons)}}{\text{Li atom}} \rightarrow \text{e} + \underset{\text{(3 protons and 2 electrons)}}{\text{Li}^+\text{ cation}}$$

Anions

Conversely, if an atom gains one or more electrons, there is now a greater number of negatively charged electrons than protons. The result is an **anion** (Figure 2.17). (The name is pronounced "ann'-ion.")

$$\underset{\text{(8 protons and 8 electrons)}}{\text{O atom}} + 2\text{ e}^- \rightarrow \underset{\text{(8 protons and 10 electrons)}}{\text{O}^{2-}\text{ anion}}$$

Here, the O atom has gained two electrons, so we write the anion's symbol as O^{2-}.

How do you know whether an atom is likely to form a cation or an anion? It depends on whether the element is a metal or a nonmetal.

- Metals generally lose electrons in the course of their reactions to form cations.
- Nonmetals frequently gain one or more electrons to form anions in the course of their reactions.

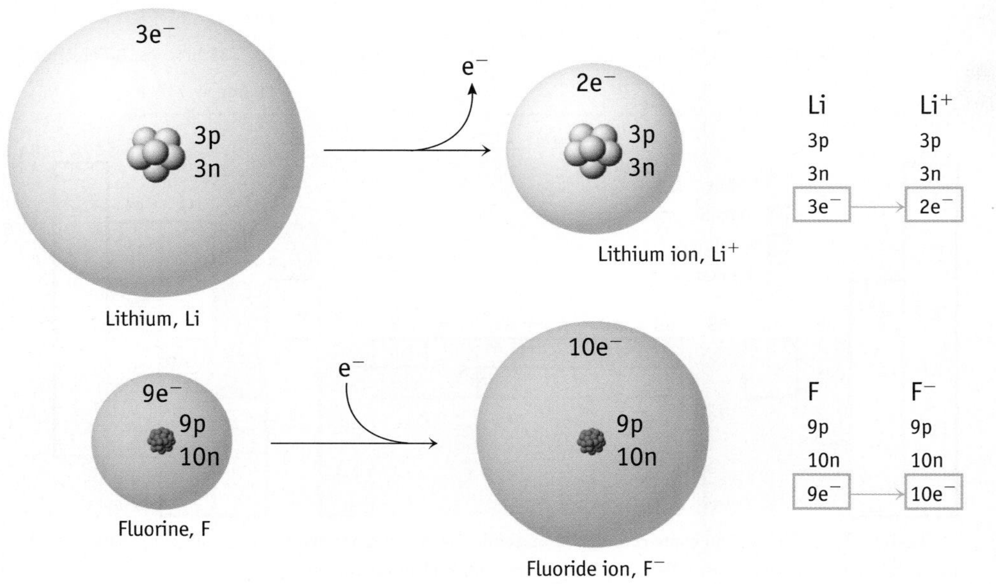

Active Figure 2.17 Ions. A lithium-6 atom is electrically neutral because the number of positive charges (three protons) and negative charges (three electrons) are the same. When it loses one electron, it has one more positive charge than negative charge, so it has a net charge of 1+. We symbolize the resulting lithium cation as Li^+. A fluorine atom is also electrically neutral, having nine protons and nine electrons. A fluorine atom can acquire an electron to produce an F^- anion. This anion has one more electron than it has protons, so it has a net charge of 1−.

Chemistry Now™ Sign in at www.thomsonedu.com/login and go to the Chapter Contents menu to explore an interactive version of this figure accompanied by an exercise.

Monatomic Ions

Monatomic ions are single atoms that have lost or gained electrons. As indicated in Figure 2.18, metals typically lose electrons to form monatomic cations, and nonmetals typically gain electrons to form monatomic anions.

How can you predict the number of electrons gained or lost? Like lithium in Figure 2.17, *metals of Groups 1A–3A form positive ions having a charge equal to the group number of the metal.*

■ **Writing Ion Formulas** When writing the formula of an ion, the charge on the ion must be included.

Group	Metal Atom	Electron Change		Resulting Metal Cation
1A	Na (11 protons, 11 electrons)	−1	→	Na^+ (11 protons, 10 electrons)
2A	Ca (20 protons, 20 electrons)	−2	→	Ca^{2+} (20 protons, 18 electrons)
3A	Al (13 protons, 13 electrons)	−3	→	Al^{3+} (13 protons, 10 electrons)

Transition metals (B-group elements) also form cations. Unlike the A-group metals, however, no easily predictable pattern of behavior occurs for transition metal cations. In addition, transition metals often form several different ions. An iron-containing compound, for example, may contain either Fe^{2+} or Fe^{3+} ions. Indeed, 2+ and 3+ ions are typical of many transition metals (see Figure 2.18).

Group	Metal Atom	Electron Change		Resulting Metal Cation
7B	Mn (25 protons, 25 electrons)	−2	→	Mn^{2+} (25 protons, 23 electrons)
8B	Fe (26 protons, 26 electrons)	−2	→	Fe^{2+} (26 protons, 24 electrons)
8B	Fe (26 protons, 26 electrons)	−3	→	Fe^{3+} (26 protons, 23 electrons)

Nonmetals often form ions having a negative charge equal to the group number of the element minus 8. For example, nitrogen is in Group 5A, so it forms an ion having a 3− charge because a nitrogen atom can gain three electrons.

Group	Nonmetal Atom	Electron Change		Resulting Nonmetal Anion
5A	N (7 protons, 7 electrons)	+3	→	N^{3-} (7 protons, 10 electrons)
				Charge = 5 − 8
6A	S (16 protons, 16 electrons)	+2	→	S^{2-} (16 protons, 18 electrons)
				Charge = 6 − 8
7A	Br (35 protons, 35 electrons)	+1	→	Br^- (35 protons, 36 electrons)
				Charge = 7 − 8

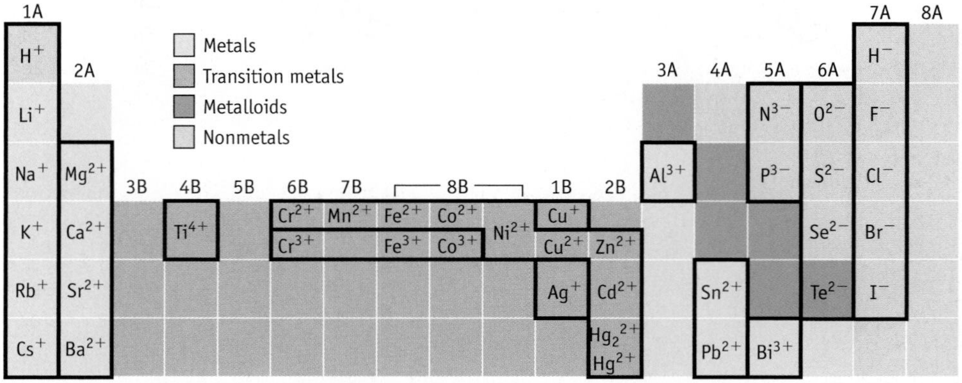

FIGURE 2.18 Charges on some common monatomic cations and anions. Metals usually form cations, and nonmetals usually form anions. (The boxed areas show ions of identical charge.)

Notice that hydrogen appears at two locations in Figure 2.18. The H atom can either lose or gain electrons, depending on the other atoms it encounters.

Electron lost: H (1 proton, 1 electron) $\rightarrow$ H^+ (1 proton, 0 electrons) + e^-
Electron gained: H (1 proton, 1 electron) + e^- $\rightarrow$ H^- (1 proton, 2 electrons)

Finally, the noble gases *very* rarely form monatomic cations or anions in chemical reactions.

Ion Charges and the Periodic Table

The metals of Groups 1A, 2A, and 3A form ions having 1+, 2+, and 3+ charges (Figure 2.18); that is, their atoms lose one, two, or three electrons, respectively. *For Group 1A and 2A metals and aluminum, the number of electrons remaining on the cation is the same as the number of electrons in an atom of the noble gas that precedes it in the periodic table.* For example, Mg^{2+} has 10 electrons, the same number as in an atom of the noble gas neon (atomic number 10).

An atom of a nonmetal near the right side of the periodic table would have to lose a great many electrons to achieve the same number as a noble gas atom of lower atomic number. (For instance, Cl, whose atomic number is 17, would have to lose seven electrons to have the same number of electrons as Ne.) If a nonmetal atom were to gain just a few electrons, however, it would have the same number as a noble gas atom of higher atomic number. For example, an oxygen atom has eight electrons. By gaining two electrons per atom, it forms O^{2-}, which has ten electrons, the same number as neon. *Anions having the same number of electrons as the noble gas atom succeeding it in the periodic table are commonly observed in chemical compounds.*

Cation charges and the periodic table

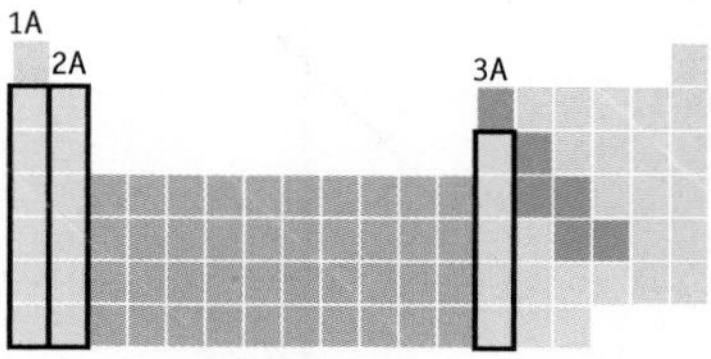

Group 1A, 2A, 3A metals form M^{n+} cations where n = group number.

Chemistry Now™

Sign in at **www.thomsonedu.com/login** and go to Chapter 2 Contents to see Screen 2.18 for an exercise and tutorial on **ion charge.**

EXERCISE 2.6 Predicting Ion Charges

Predict formulas for monatomic ions formed from **(a)** K, **(b)** Se, **(c)** Ba, and **(d)** Cs. In each case, indicate the number of electrons gained or lost by an atom of the element in forming the anion or cation, respectively. For each ion, indicate the noble gas atom having the same total number of electrons.

Polyatomic Ions

Polyatomic ions are made up of two or more atoms, and the collection has an electric charge (Figure 2.19 and Table 2.4). For example, carbonate ion, CO_3^{2-}, a common polyatomic anion, consists of one C atom and three O atoms. The ion has two units of negative charge because there are two more electrons (a total of 32) in the ion than there are protons (a total of 30) in the nuclei of one C atom and three O atoms.

The ammonium ion, NH_4^+, is a common polyatomic cation. In this case, four H atoms surround an N atom, and the ion has a 1+ electric charge. This ion has 10 electrons, but there are 11 positively charged protons in the nuclei of the N and H atoms (seven and one each, respectively).

Chemistry Now™

Sign in at **www.thomsonedu.com/login** and go to Chapter 2 Contents to see Screen 2.19 for a tutorial on **polyatomic ions.**

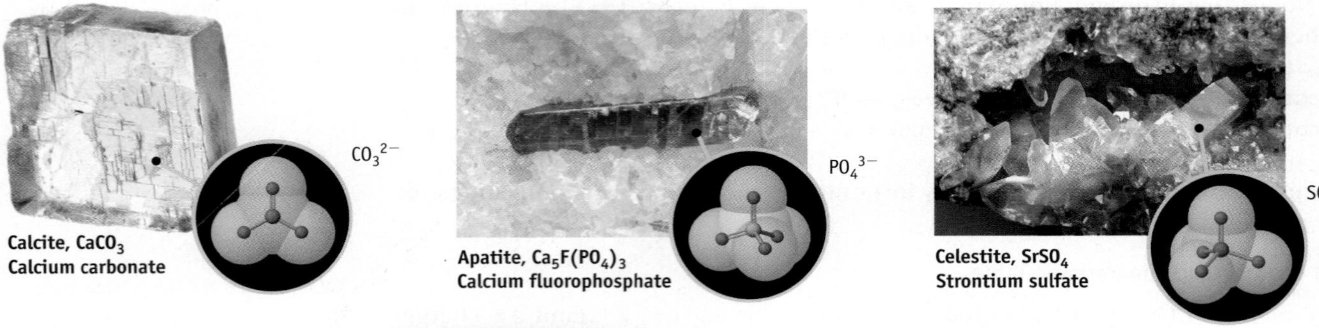

Photos: Charles D. Winters

Calcite, $CaCO_3$
Calcium carbonate

Apatite, $Ca_5F(PO_4)_3$
Calcium fluorophosphate

Celestite, $SrSO_4$
Strontium sulfate

Active Figure 2.19 **Common ionic compounds based on polyatomic ions.**

Chemistry Now™ Sign in at www.thomsonedu.com/login and go to the Chapter Contents menu to explore an interactive version of this figure accompanied by an exercise.

Formulas of Ionic Compounds

Compounds are electrically neutral; that is, they have no net electric charge. Thus, in an ionic compound, the numbers of positive and negative ions must be such that the positive and negative charges balance. In sodium chloride, the sodium ion has a 1+ charge (Na^+), and the chloride ion has a 1− charge (Cl^-). These ions must be present in a 1 : 1 ratio, and the formula is NaCl.

TABLE 2.4 Formulas and Names of Some Common Polyatomic Ions

Formula	Name	Formula	Name
CATION: Positive Ion			
NH_4^+	ammonium ion		
ANIONS: Negative Ions			
Based on a Group 4A element		**Based on a Group 7A element**	
CN^-	cyanide ion	ClO^-	hypochlorite ion
$CH_3CO_2^-$	acetate ion	ClO_2^-	chlorite ion
CO_3^{2-}	carbonate ion	ClO_3^-	chlorate ion
HCO_3^-	hydrogen carbonate ion (or bicarbonate ion)	ClO_4^-	perchlorate ion
Based on a Group 5A element		**Based on a transition metal**	
NO_2^-	nitrite ion	CrO_4^{2-}	chromate ion
NO_3^-	nitrate ion	$Cr_2O_7^{2-}$	dichromate ion
PO_4^{3-}	phosphate ion	MnO_4^-	permanganate ion
HPO_4^{2-}	hydrogen phosphate ion		
$H_2PO_4^-$	dihydrogen phosphate ion		
Based on a Group 6A element			
OH^-	hydroxide ion		
SO_3^{2-}	sulfite ion		
SO_4^{2-}	sulfate ion		
HSO_4^-	hydrogen sulfate ion (or bisulfate ion)		

The gemstone ruby is largely the compound formed from aluminum ions (Al^{3+}) and oxide ions (O^{2-}) (but the color comes from a trace of Cr^{3+} ions.) Here, the ions have positive and negative charges that are of different absolute value. To have a compound with the same number of positive and negative charges, two Al^{3+} ions [total charge = 2 × (3+) = 6+] must combine with three O^{2-} ions [total charge = 3 × (2−) = 6 −] to give a formula of Al_2O_3.

Calcium is a Group 2A metal, and it forms a cation having a 2+ charge. It can combine with a variety of anions to form ionic compounds such as those in the following table:

Compound	Ion Combination	Overall Charge on Compound
$CaCl_2$	$Ca^{2+} + 2\ Cl^-$	(2+) + 2 × (1−) = 0
$CaCO_3$	$Ca^{2+} + CO_3^{2-}$	(2+) + (2−) = 0
$Ca_3(PO_4)_2$	$3\ Ca^{2+} + 2\ PO_4^{3-}$	3 × (2+) + 2 × (3−) = 0

In writing formulas of ionic compounds, the convention is that *the symbol of the cation is given first, followed by the anion symbol.* Also notice the use of parentheses when more than one of a given polyatomic ion is present.

■ Balancing Ion Charges in Formulas Aluminum, a metal in Group 3A, loses three electrons to form the Al^{3+} cation. Oxygen, a nonmetal in Group 6A, gains two electrons to form an O^{2-} anion. Notice that in the compound formed from these ions, the charge on the cation is the subscript on the anion, and vice versa.

$$2\ Al^{3+} + 3\ O^{2-} \rightarrow Al_2O_3$$

This often works well, but there are exceptions. For example, the formula of titanium (IV) oxide is TiO_2, the simplest ratio, and not Ti_2O_4.

$$Ti^{4+} + 2\ O^{2-} \rightarrow TiO_2$$

Chemistry Now™

Sign in at **www.thomsonedu.com/login** and go to Chapter 2 Contents to see Screen 2.20 for a video of the **sodium and chlorine reaction** and for a simulation on the **relationship between cations and anions in ionic compounds.**

EXAMPLE 2.4 Ionic Compound Formulas

Problem For each of the following ionic compounds, write the symbols for the ions present, and give the number of each: (a) Li_2CO_3, and (b) $Fe_2(SO_4)_3$.

Strategy Divide the formula of the compound into the cation and the anion. To accomplish this, you will have to recognize, and remember, the composition and charges of common ions.

Solution

(a) Li_2CO_3 is composed of two lithium ions, Li^+, and one carbonate ion, CO_3^{2-}. Li is a Group 1A element and always has a 1+ charge in its compounds. Because the two 1+ charges balance the negative charge of the carbonate ion, the latter must be 2−.

(b) $Fe_2(SO_4)_3$ contains two iron ions, Fe^{3+}, and three sulfate ions, SO_4^{2-}. The way to recognize this is to recall that sulfate has a 2− charge. Because three sulfate ions are present (with a total charge of 6−), the two iron cations must have a total charge of 6+. This is possible only if each iron cation has a charge of 3+.

Comment Remember that the formula for an ion must include its composition and its charge. Formulas for ionic compounds are always written with the cation first and then the anion, but ion charges are not included.

EXAMPLE 2.5 Ionic Compound Formulas

Problem Write formulas for ionic compounds composed of aluminum cations and each of the following anions: (a) fluoride ion, (b) sulfide ion, and (c) nitrate ion.

Strategy First decide on the formula of the Al cation and the formula of each anion. Combine the Al cation with each type of anion to form an electrically neutral compound.

Solution An aluminum cation is predicted to have a charge of 3+ because Al is a metal in Group 3A.

(a) Fluorine is a Group 7A element. The charge of the fluoride ion is predicted to be 1− (from 7 − 8 = −1). Therefore, we need 3 F^- ions to combine with one Al^{3+}. The formula of the compound is AlF_3.

(b) Sulfur is a nonmetal in Group 6A, so it forms a 2− anion. Thus, we need to combine two Al^{3+} ions [total charge is 6+ = 2 × (3+)] with three S^{2-} ions [total charge is 6− = 3 × (2−)]. The compound has the formula Al_2S_3.

(c) The nitrate ion has the formula NO_3^- (see Table 2.4). The answer here is therefore similar to the AlF_3 case, and the compound has the formula $Al(NO_3)_3$. Here, we place parentheses around NO_3 to show that three polyatomic NO_3^- ions are involved.

Comment The most common error students make is not knowing the correct charge on an ion.

EXERCISE 2.7 Formulas of Ionic Compounds

(a) Give the number and identity of the constituent ions in each of the following ionic compounds: NaF, $Cu(NO_3)_2$, and $NaCH_3CO_2$.

(b) Iron, a transition metal, forms ions having at least two different charges. Write the formulas of the compounds formed between chloride ions and the two different iron cations.

(c) Write the formulas of all neutral ionic compounds that can be formed by combining the cations Na^+ and Ba^{2+} with the anions S^{2-} and PO_4^{3-}.

Names of Ions

Naming Positive Ions (Cations)

With a few exceptions (such as NH_4^+), the positive ions described in this text are metal ions. Positive ions are named by the following rules:

1. For a monatomic positive ion (that is, a metal cation) the name is that of the metal plus the word "cation." For example, we have already referred to Al^{3+} as the aluminum cation.
2. Some cases occur, especially in the transition series, in which a metal can form more than one type of positive ion. In these cases, the charge of the ion is indicated by a Roman numeral in parentheses immediately following the ion name. For example, Co^{2+} is the cobalt(II) cation, and Co^{3+} is the cobalt(III) cation.

■ **"-ous" and "-ic" Endings** An older naming system for metal ions uses the ending -ous for the ion of lower charge and -ic for the ion of higher charge. For example, there are cobaltous (Co^{2+}) and cobaltic (Co^{3+}) ions, and ferrous (Fe^{2+}) and ferric (Fe^{3+}) ions. We do not use this system in this book, but some chemical manufacturers continue to use it.

Finally, you will encounter the ammonium cation, NH_4^+, many times in this book and in the laboratory. Do not confuse the ammonium cation with the ammonia molecule, NH_3, which has no electric charge and one less H atom.

Naming Negative Ions (Anions)

There are two types of negative ions: those having only one atom (*monatomic*) and those having several atoms (*polyatomic*).

1. A monatomic negative ion is named by adding *-ide* to the stem of the name of the nonmetal element from which the ion is derived (Figure 2.20). The anions of the Group 7A elements, the halogens, are known as the fluoride, chloride, bromide, and iodide ions and as a group are called **halide ions**.
2. Polyatomic negative ions are common, especially those containing oxygen (called **oxoanions**). The names of some of the most common oxoanions are given in Table 2.4. Although most of these names must simply be learned, some guidelines can help. For example, consider the following pairs of ions:

   NO_3^- is the nitrate ion, whereas NO_2^- is the nitrite ion
   SO_4^{2-} is the sulfate ion, whereas SO_3^{2-} is the sulfite ion

The oxoanion having the greater number of oxygen atoms is given the suffix -ate, and the oxoanion having the smaller number of oxygen atoms has the suffix -ite. For a series of oxoanions having more than two members, the ion with the largest number of oxygen atoms has the prefix *per-* and the suffix *-ate*. The ion

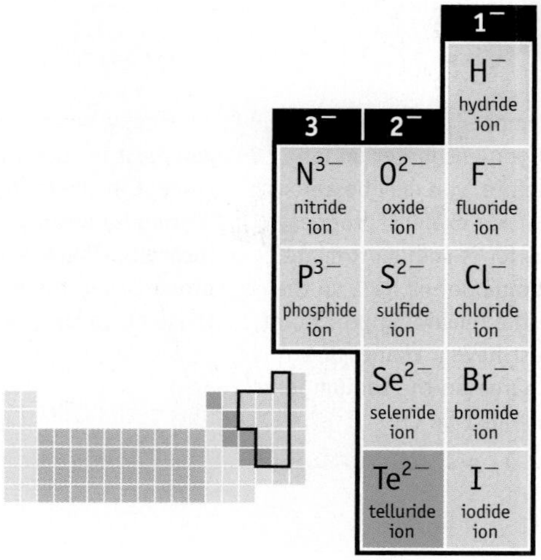

FIGURE 2.20 Names and charges of some common monatomic anions.

having the smallest number of oxygen atoms has the prefix *hypo-* and the suffix *-ite*. The chlorine oxoanions are the most commonly encountered example.

ClO_4^-	*per*chlor*ate* ion
ClO_3^-	chlor*ate* ion
ClO_2^-	chlor*ite* ion
ClO^-	*hypo*chlor*ite* ion

■ **Naming Oxoanions**

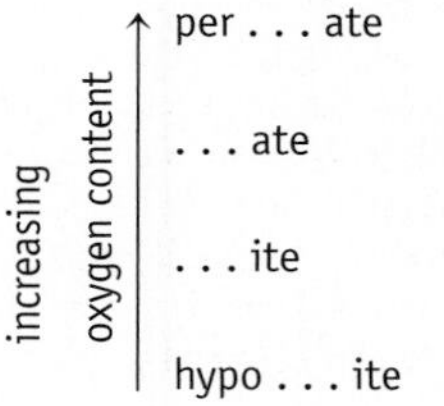

Oxoanions that contain hydrogen are named by adding the word "hydrogen" before the name of the oxoanion. If two hydrogens are in the anion, we say "dihydrogen." Many hydrogen-containing oxoanions have common names that are used as well. For example, the hydrogen carbonate ion, HCO_3^-, is called the bicarbonate ion.

Ion	Systematic Name	Common Name
HPO_4^{2-}	hydrogen phosphate ion	
$H_2PO_4^-$	dihydrogen phosphate ion	
HCO_3^-	hydrogen carbonate ion	bicarbonate ion
HSO_4^-	hydrogen sulfate ion	bisulfate ion
HSO_3^-	hydrogen sulfite ion	bisulfite ion

Names of Ionic Compounds

The name of an ionic compound is built from the names of the positive and negative ions in the compound. The name of the positive cation is given first, followed by the name of the negative anion. Examples of ionic compound names are given below.

Ionic Compound	Ions Involved	Name
$CaBr_2$	Ca^{2+} and 2 Br^-	calcium bromide
$NaHSO_4$	Na^+ and HSO_4^-	sodium hydrogen sulfate
$(NH_4)_2CO_3$	2 NH_4^+ and CO_3^{2-}	ammonium carbonate
$Mg(OH)_2$	Mg^{2+} and 2 OH^-	magnesium hydroxide
$TiCl_2$	Ti^{2+} and 2 Cl^-	titanium(II) chloride
Co_2O_3	2 Co^{3+} and 3 O^{2-}	cobalt(III) oxide

■ **Names of Compounds Containing Transition Metal Cations** Be sure to notice that the charge on a transition metal cation is indicated by a Roman numeral and is included in the name.

Problem Solving Tip 2.1 Formulas for Ions and Ionic Compounds

Writing formulas for ionic compounds takes practice, and it requires that you know the formulas and charges of the most common ions. The charges on monatomic ions are often evident from the position of the element in the periodic table, but you simply have to remember the formula and charges of polyatomic ions—especially the most common ones such as nitrate, sulfate, carbonate, phosphate, and acetate.

If you cannot remember the formula of a polyatomic ion, or if you encounter an ion you have not seen before, you may be able to figure out its formula. For example, suppose you are told that $NaCHO_2$ is sodium formate. You know that the sodium ion is Na^+, so the formate ion must be the remaining portion of the compound; it must have a charge of 1− to balance the 1+ charge on the sodium ion. Thus, the formate ion must be CHO_2^-.

Finally, when writing the formulas of ions, you *must* include the charge on the ion (except in the formula of an ionic compound). Writing Na when you mean sodium ion is incorrect. There is a vast difference in the properties of the element sodium (Na) and those of its ion (Na^+).

Chemistry Now™

Sign in at **www.thomsonedu.com/login** and go to Chapter 2 Contents to see Screens 2.20 and 2.21 for a tutorial, an exercise, and a simulation on **ionic compounds.**

EXERCISE 2.8 Names of Ionic Compounds

1. Give the formula for each of the following ionic compounds. Use Table 2.4 and Figure 2.20.

(a) ammonium nitrate
(b) cobalt(II) sulfate
(c) nickel(II) cyanide
(d) vanadium(III) oxide
(e) barium acetate
(f) calcium hypochlorite

2. Name the following ionic compounds:

(a) $MgBr_2$
(b) Li_2CO_3
(c) $KHSO_3$
(d) $KMnO_4$
(e) $(NH_4)_2S$
(f) CuCl and $CuCl_2$

Properties of Ionic Compounds

When a substance having a negative electric charge is brought near a substance having a positive electric charge, there is a force of attraction between them (Figure 2.21). In contrast, there is a repulsive force when two substances with the same charge—both positive or both negative—are brought together. These forces are called **electrostatic** forces, and the force of attraction between ions is given by **Coulomb's law** (Equation 2.3)

$$\text{Force of attraction} = k\,\frac{(n^+e)(n^-e)}{d^2} \qquad (2.3)$$

charge on + and − ions (n^+, n^-); charge on electron (e); proportionality constant (k); distance between ions (d)

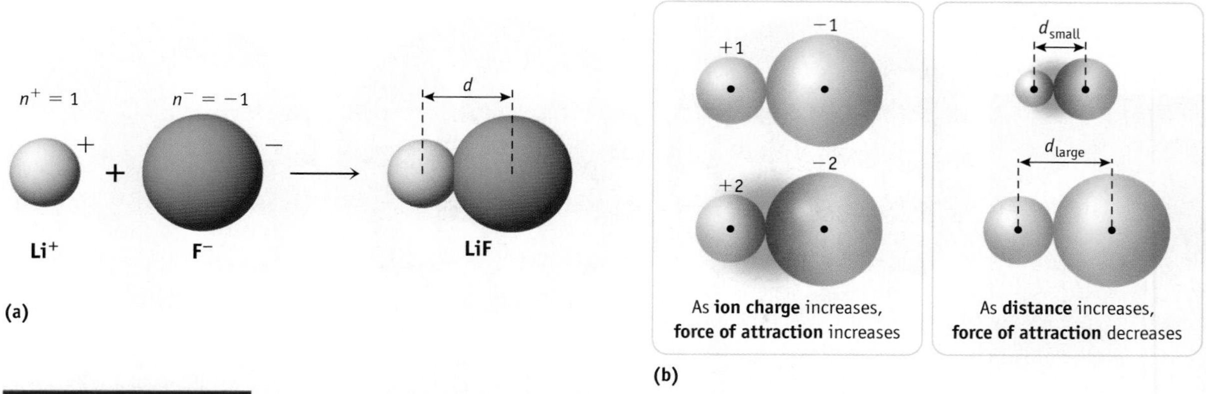

Active Figure 2.21 Coulomb's law and electrostatic forces. (a) Ions such as Li^+ and F^- are held together by an electrostatic force. Here, a lithium ion is attracted to a fluoride ion, and the distance between the nuclei of the two ions is *d*. (b) Forces of attraction between ions of opposite charge increase with increasing ion charge and decrease with increasing distance (d).

ChemistryNow™ Sign in at www.thomsonedu.com/login and go to the Chapter Contents menu to explore an interactive version of this figure accompanied by an exercise.

where, for example, n^+ is 3 for Al^{3+} and n^- is −2 for O^{2-}. Based on Coulomb's law, the force of attraction between oppositely charged ions increases:

- As the ion charges (n^+ and n^-) increase. Thus, the attraction between ions having charges of 2+ and 2− is greater than that between ions having 1+ and 1− charges (Figure 2.21).
- As the distance between the ions becomes smaller (Figure 2.21).

Ionic compounds do not consist of simple pairs or small groups of positive and negative ions. The simplest ratio of cations to anions in an ionic compound is represented by its formula, but an ionic solid consists of millions upon millions of ions arranged in an extended three-dimensional network called a **crystal lattice.** A portion of the lattice for NaCl, illustrated in Figure 2.22, illustrates a common way of arranging ions for compounds that have a 1:1 ratio of cations to anions.

Ionic compounds have characteristic properties that can be understood in terms of the charges of the ions and their arrangement in the lattice. Because each ion is surrounded by oppositely charged nearest neighbors, it is held tightly in its allotted location. At room temperature, each ion can move just a bit around its average position, but considerable energy must be added before an ion can escape the attraction of its neighboring ions. Only if enough energy is added will the lattice structure collapse and the substance melt. Greater attractive forces mean that ever more energy—higher and higher temperatures—is required to cause melting. Thus, Al_2O_3, a solid composed of Al^{3+} and O^{2-} ions, melts at a much higher temperature (2072 °C) than NaCl (801 °C), a solid composed of Na^+ and Cl^- ions.

Most ionic compounds are "hard" solids. That is, the solids are not pliable or soft. The reason for this characteristic is again related to the lattice of ions. The nearest neighbors of a cation in a lattice are anions, and the force of attraction makes the lattice rigid. However, a blow with a hammer can cause the lattice to break cleanly along a sharp boundary. The hammer blow displaces layers of ions just enough to cause ions of like charge to become nearest neighbors, and the repulsion between these like-charged ions forces the lattice apart (Figure 2.23).

Photo: Charles D. Winters; model, S. M. Young

FIGURE 2.22 Sodium chloride. A crystal of NaCl consists of an extended lattice of sodium ions and chloride ions in a 1 : 1 ratio. (Sign in to ChemistryNow, and see Screen 2.20, Ionic Compounds, to view an animation on the formation of a sodium chloride crystal lattice.)

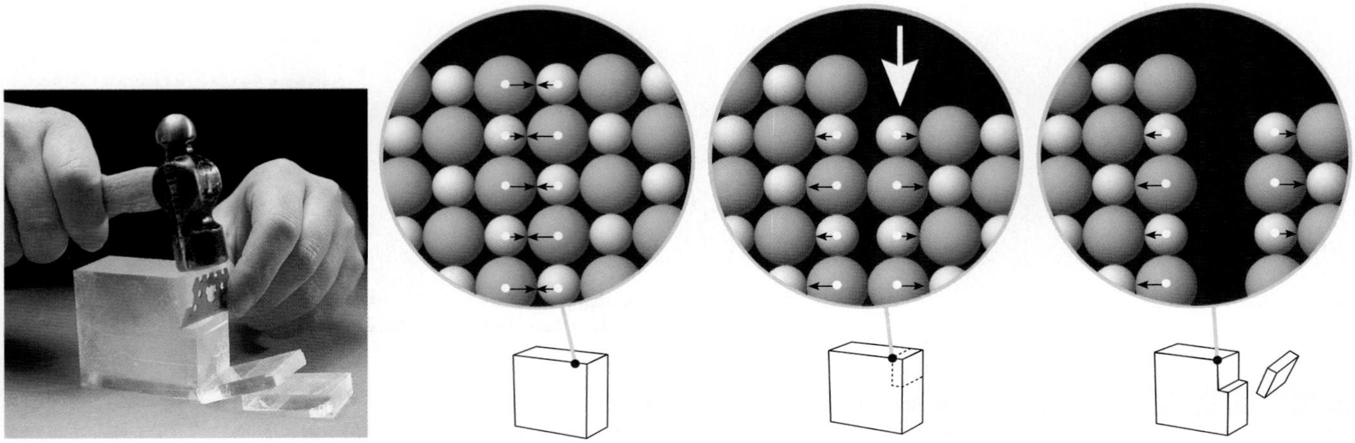

Charles D. Winters

(a)

(b)

FIGURE 2.23 Ionic solids. (a) An ionic solid is normally rigid, owing to the forces of attraction between oppositely charged ions. When struck sharply, however, the crystal can cleave cleanly. (b) When a crystal is struck, layers of ions move slightly, and ions of like charge become nearest neighbors. Repulsions between ions of similar charge cause the crystal to cleave. (Sign in to ChemistryNow, and see Screen 2.23, Properties of Ionic Compounds, to watch a video of cleaving a crystal.)

Chemistry Now™

Sign in at **www.thomsonedu.com/login** and go to Chapter 2 Contents to see:

- Screen 2.22 for a simulation on **Coulomb's Law**
- Screen 2.23 for an exercise and a simulation on **properties of ionic compounds**

EXERCISE 2.9 Coulomb's Law

Explain why the melting point of MgO (2830 °C) is much higher than the melting point of NaCl (801 °C).

2.8 Molecular Compounds: Formulas and Names

Many familiar compounds are not ionic; they are molecular: the water you drink, the sugar in your coffee or tea, or the aspirin you take for a headache.

Problem Solving Tip 2.2 Is a Compound Ionic?

Students often ask how to know whether a compound is ionic. Here are some useful guidelines.

1. Most metal-containing compounds are ionic. So, if a metal atom appears in the formula of a compound, a good first guess is that it is ionic. (There are interesting exceptions, but few come up in introductory chemistry.) It is helpful in this regard to recall trends in metallic behavior: All elements to the left of a diagonal line running from boron to tellurium in the periodic table are metallic.
2. If there is no metal in the formula, it is likely that the compound is not ionic. The exceptions here are compounds composed of polyatomic ions based on nonmetals (e.g., NH_4Cl or NH_4NO_3).
3. Learn to recognize the formulas of polyatomic ions (see Table 2.4). Chemists write the formula of ammonium nitrate as NH_4NO_3 (not as $N_2H_4O_3$) to alert others to the fact that it is an ionic compound composed of the common polyatomic ions NH_4^+ and NO_3^-.

As an example of these guidelines, you can be sure that $MgBr_2$ (Mg^{2+} with Br^-) and K_2S (K^+ with S^{2-}) are ionic compounds. On the other hand, the compound CCl_4, formed from two nonmetals, C and Cl, is not ionic.

FIGURE 2.24 Molecular compounds. Ionic compounds are generally solids at room temperature. In contrast, molecular compounds can be gases, liquids, or solids. The molecular models are of caffeine (in coffee), water, and citric acid (in lemons).

Ionic compounds are generally solids, whereas molecular compounds can range from gases to liquids to solids at ordinary temperatures (see Figure 2.24). As size and molecular complexity increase, compounds generally exist as solids. We will explore some of the underlying causes of these general observations in Chapter 12.

Some molecular compounds have complicated formulas that you cannot, at this stage, predict or even decide if they are correct. However, there are many simple compounds you will encounter often, and you should understand how to name them and, in many cases, know their formulas.

Let us look first at molecules formed from combinations of two nonmetals. These "two-element" compounds of nonmetals, often called **binary compounds**, can be named in a systematic way.

Hydrogen forms binary compounds with all of the nonmetals except the noble gases. For compounds of oxygen, sulfur, and the halogens, the H atom is generally written first in the formula and is named first. The other nonmetal is named as if it were a negative ion.

Compound	Name
HF	hydrogen fluoride
HCl	hydrogen chloride
H_2S	hydrogen sulfide

■ **Formulas of Binary Nonmetal Compounds Containing Hydrogen** Simple hydrocarbons (compounds of C and H) such as methane (CH_4) and ethane (C_2H_6) have formulas written with H following C, and the formulas of ammonia and hydrazine have H following N. Water and the hydrogen halides, however, have the H atom preceding O or the halogen atom. Tradition is the only explanation for such irregularities in writing formulas.

Although there are exceptions, *most binary molecular compounds are a combination of nonmetallic elements from Groups 4A–7A with one another or with hydrogen.* The formula is generally written by putting the elements in order of increasing group number. When naming the compound, the number of atoms of a given type in the compound is designated with a prefix, such as "di-," "tri-," "tetra-," "penta-," and so on.

Compound	Systematic Name
NF_3	nitrogen trifluoride
NO	nitrogen monoxide
NO_2	nitrogen dioxide
N_2O	dinitrogen monoxide
N_2O_4	dinitrogen tetraoxide
PCl_3	phosphorus trichloride
PCl_5	phosphorus pentachloride
SF_6	sulfur hexafluoride
S_2F_{10}	disulfur decafluoride

■ **Hydrocarbons** Compounds such as methane, ethane, propane, and butane belong to a class of hydrocarbons called alkanes. (Sign in to ChemistryNow, and see Screen 2.24, Alkanes.).

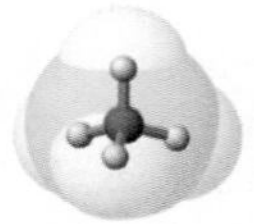

methane, CH_4

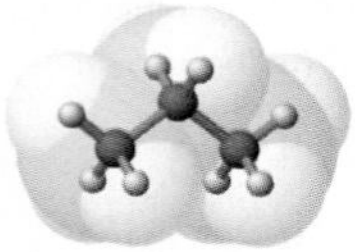

propane, C_3H_8

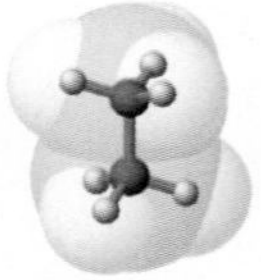

ethane, C_2H_6

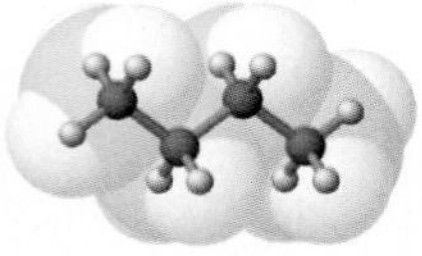

butane, C_4H_{10}

Finally, many of the binary compounds of nonmetals were discovered years ago and have common names.

Compound	Common Name	Compound	Common Name
CH_4	methane	N_2H_4	hydrazine
C_2H_6	ethane	PH_3	phosphine
C_3H_8	propane	NO	nitric oxide
C_4H_{10}	butane	N_2O	nitrous oxide ("laughing gas")
NH_3	ammonia	H_2O	water

ChemistryNow™

Sign in at **www.thomsonedu.com/login** and go to Chapter 2 Contents to see:
- Screen 2.25 for a tutorial on **naming compounds of the nonmetals**
- Screen 2.26 for an exercise on **naming alkanes**

EXERCISE 2.10 Naming Compounds of the Nonmetals

1. Give the formula for each of the following binary, nonmetal compounds:

(a) carbon dioxide
(b) phosphorus triiodide
(c) sulfur dichloride
(d) boron trifluoride
(e) dioxygen difluoride
(f) xenon trioxide

2. Name the following binary, nonmetal compounds:

(a) N_2F_4
(b) HBr
(c) SF_4
(d) BCl_3
(e) P_4O_{10}
(f) ClF_3

Module 4

2.9 Atoms, Molecules, and the Mole

One of the most exciting aspects of chemical research is the discovery of some new substance, and part of this process of discovery involves quantitative experiments. When two chemicals react with each other, we want to know how many atoms or molecules of each are used so that formulas can be established for the reaction products. To do so, we need some method of counting atoms and molecules. That is, we must discover a way of connecting the macroscopic world, the world we can see, with the particulate world of atoms, molecules, and ions. The solution to this problem is to define a unit of matter that contains a known number of particles. That chemical unit is the mole.

The **mole** (abbreviated mol) is the SI base unit for measuring an *amount of a substance* and is defined as follows:

■ **An Important Difference between the Terms "Amount" and "Quantity"** The terms "amount" and "quantity" are used in a specific sense by chemists. The amount of a substance is the number of moles of that substance. Quantity refers, for example, to the mass or volume of the substance. (See W. G. Davies and J. W. Moore. *Journal of Chemical Education*, Vol. 57, p. 303, 1980.)

A mole is the amount of a substance that contains as many elementary entities (atoms, molecules, or other particles) as there are atoms in exactly 12 g of the carbon-12 isotope.

The key to understanding the concept of the mole is recognizing that *one mole always contains the same number of particles, no matter what the substance.* One mole of sodium contains the same number of atoms as one mole of iron or as the number of mol-

Sign in at **www.thomsonedu.com/login** to download the Go Chemistry module for this section or go to **www.ichapters.com** to purchase modules.

Historical Perspectives — Amedeo Avogadro and His Number

Amedeo Avogadro, conte di Quaregna, (1776–1856) was an Italian nobleman and a lawyer. In about 1800, he turned to science and was the first professor of mathematical physics in Italy.

Avogadro did not himself propose the notion of a fixed number of particles in a chemical unit. Rather, the number was named in his honor because he had performed experiments in the 19th century that laid the groundwork for the concept.

Just how large is Avogadro's number? One mol of unpopped popcorn kernels would cover the continental United States to a depth of about 9 miles. One mole of pennies divided equally among every man, woman, and child in the U.S. would allow each person to pay off the national debt (\$9.1 trillion or 9.1×10^{12} dollars), and still have trillions of dollars left over!

Is the number a unique value like π? No. It is fixed by the definition of the mole as exactly 12 g of carbon-12. If one mole of carbon were defined to have some other mass, then Avogadro's number would have a different value.

Photo: E. F. Smith Collection/Van Pelt/Library/University of Pennsylvania

---

ecules in one mole of water. How many particles? Many, many experiments over the years have established that number as

$$\textbf{1 mole} = \textbf{6.0221415} \times \textbf{10}^{23} \textbf{ particles}$$

This value is known as **Avogadro's number** in honor of Amedeo Avogadro, an Italian lawyer and physicist (1776–1856) who conceived the basic idea (but never determined the number).

■ **The "Mole"** The term "mole" was introduced about 1895 by Wilhelm Ostwald (1853–1932), who derived the term from the Latin word *moles*, meaning a "heap" or a "pile."

Atoms and Molar Mass

The mass in grams of one mole of any element (6.0221415×10^{23} atoms of that element) is the **molar mass** of that element. Molar mass is conventionally abbreviated with a capital italicized *M*, and it has units of grams per mole (g/mol). *An element's molar mass is the quantity in grams numerically equal to its atomic weight.* Using sodium and lead as examples,

Molar mass of sodium (Na) = mass of 1.000 mol of Na atoms
= 22.99 g/mol
= mass of 6.022×10^{23} Na atoms

Molar mass of lead (Pb) = mass of 1.000 mol of Pb atoms
= 207.2 g/mol
= mass of 6.022×10^{23} Pb atoms

Figure 2.25 shows the relative sizes of a mole of some common elements. Although each of these "piles of atoms" has a different volume and different mass, each contains 6.022×10^{23} atoms.

The mole concept is the cornerstone of quantitative chemistry. It is essential to be able to convert from moles to mass and from mass to moles. Dimensional analysis, which is described in *Let's Review,* page 38, shows that this can be done in the following way:

MASS ⟷ MOLES CONVERSION

Moles to Mass

$$\text{Moles} \times \frac{\text{grams}}{\text{1 mol}} = \text{grams}$$

↑ molar mass (the factor grams/1 mol)

Mass to Moles

$$\text{Grams} \times \frac{\text{1 mol}}{\text{grams}} = \text{moles}$$

↑ 1/molar mass (the factor 1 mol/grams)

FIGURE 2.25 **One-mole of common elements.** *(left to right)* Sulfur powder, magnesium chips, tin, and silicon. *(Above)* Copper beads.

For example, what mass, in grams, is represented by 0.35 mol of aluminum? Using the molar mass of aluminum (27.0 g/mol), you can determine that 0.35 mol of Al has a mass of 9.5 g.

$$0.35 \text{ mol Al} \times \frac{27.0 \text{ g Al}}{1 \text{ mol Al}} = 9.5 \text{ g Al}$$

Molar masses are generally known to at least four significant figures. The convention followed in calculations in this book is to use a value of the molar mass with one more significant figure than in any other number in the problem. For example, if you weigh out 16.5 g of carbon, you use 12.01 g/mol for the molar mass of C to find the amount of carbon present.

$$16.5 \text{ g C} \times \frac{1 \text{ mol C}}{12.01 \text{ g C}} = 1.37 \text{ mol C}$$

↑ Note that four significant figures are used in the molar mass, but there are three in the sample mass.

Using one more significant figure for the molar mass means the precision of this value will not affect the precision of the result.

Chemistry Now™

Sign in at **www.thomsonedu.com/login** and go to Chapter 2 Contents to see:

- Screen 2.25 for a tutorial on **moles and atoms conversion**
- Screen 2.26 for two tutorials on **molar mass conversion**

EXAMPLE 2.6 Mass, Moles, and Atoms

Problem Consider two elements in the same vertical column of the periodic table, lead and tin.

(a) What mass of lead, in grams, is equivalent to 2.50 mol of lead (Pb, atomic number = 82)?

(b) What amount of tin, in moles, is represented by 36.6 g of tin (Sn, atomic number = 50)? How many atoms of tin are in the sample?

Strategy The molar masses of lead (207.2 g/mol) and tin (118.7 g/mol) are required and can be found in the periodic table inside the front cover of this book. Avogadro's number is needed to convert the amount of each element to number of atoms.

Solution

(a) Convert the amount of lead in moles to mass in grams.

$$2.50 \text{ mol Pb} \times \frac{207.2 \text{ g}}{1 \text{ mol Pb}} = \boxed{518 \text{ g Pb}}$$

(b) First convert the mass of tin to the amount in moles.

$$36.6 \text{ g Sn} \times \frac{1 \text{ mol Sn}}{118.7 \text{ g Sn}} = \boxed{0.308 \text{ mol Sn}}$$

Finally, use Avogadro's number to find the number of atoms in the sample.

$$0.308 \text{ mol Sn} \times \frac{6.022 \times 10^{23} \text{ atoms Sn}}{1 \text{ mol Sn}} = \boxed{1.85 \times 10^{23} \text{ atoms Sn}}$$

Charles D. Winters

Lead. A 150-mL beaker containing 2.50 mol or 518 g of lead.

Charles D. Winters

Tin. A sample of tin having a mass of 36.6 g (or 1.85×10^{23} atoms).

EXERCISE 2.11 Mass/Mole Conversions

(a) What is the mass, in grams, of 1.5 mol of silicon?

(b) What amount (moles) of sulfur is represented by 454 g? How many atoms?

EXERCISE 2.12 Atoms

The density of gold, Au, is 19.32 g/cm^3. What is the volume (in cubic centimeters) of a piece of gold that contains 2.6×10^{24} atoms? If the piece of metal is a square with a thickness of 0.10 cm, what is the length (in centimeters) of one side of the piece?

Molecules, Compounds, and Molar Mass

The formula of a compound tells you the type of atoms or ions in the compound and the relative number of each. For example, one molecule of methane, CH_4, is made up of one atom of C and four atoms of H. But suppose you have Avogadro's number of C atoms (6.022×10^{23}) combined with the proper number of H atoms. The compound's formula tells us that four times as many H atoms are required ($4 \times 6.022 \times 10^{23}$ H atoms). What masses of atoms are combined, and what is the mass of this many CH_4 molecules?

C	+	4 H	→	CH_4
6.022×10^{23} C atoms		$4 \times 6.022 \times 10^{23}$ H atoms		6.022×10^{23} CH_4 molecules
= 1.000 mol of C		= 4.000 mol of H atoms		= 1.000 mol of CH_4 molecules
= 12.01 g of C atoms		= 4.032 g of H atoms		= 16.04 g of CH_4 molecules

■ **Molar Mass or Molecular Weight** Although chemists often use the term "molecular weight," the more correct term is molar mass. The SI unit of molar mass is kg/mol, but chemists worldwide usually express it in units of g/mol. See "NIST Guide to SI Units" at www.NIST.gov

Because we know the number of moles of C and H atoms, we know the masses of carbon and hydrogen that combine to form CH_4. It follows that the mass of CH_4 is the sum of these masses. That is, 1 mol of CH_4 has a mass equivalent to the mass of 1 mol of C atoms (12.01 g) plus 4 mol of H atoms (4.032 g). Thus, the *molar mass, M*, of CH_4 is 16.04 g/mol. The molar masses of some substances are:

Molar and Molecular Masses

Element or Compound	Molar Mass, *M* (g/mol)	Average Mass of One Molecule (g/molecule)
O_2	32.00	5.314×10^{-23}
NH_3	17.03	2.828×10^{-23}
H_2O	18.02	2.992×10^{-23}
$NH_2CH_2CO_2H$ (glycine)	75.07	1.247×10^{-22}

Ionic compounds such as NaCl do not exist as individual molecules. Thus, we write the simplest formula that shows the relative number of each kind of atom in a "formula unit" of the compound, and the molar mass is calculated from this formula (M for NaCl = 58.44 g/mol). To differentiate substances like NaCl that do not contain molecules, chemists sometimes refer to their *formula weight* instead of their molecular weight.

Figure 2.26 illustrates 1-mole quantities of several common compounds. To find the molar mass of any compound, you need only to add up the atomic masses for each element in one formula unit. As an example, let us find the molar mass of aspirin, $C_9H_8O_4$. In 1 mole of aspirin, there are 9 mol of carbon atoms, 8 mol of

FIGURE 2.26 One-mole quantities of some compounds. Notice the molar mass for $CuCl_2 \cdot 2H_2O$. This is called a *hydrated compound* because water is associated with the $CuCl_2$ (see page 96). Thus, one "formula unit" consists of one Cu^{2+} ion, two Cl^- ions, and two water molecules. The molar masses are the sum of the mass of 1 mol of Cu, 2 mol of Cl, and 2 mol of H_2O.

hydrogen atoms, and 4 mol of oxygen atoms, which add up to 180.2 g/mol of aspirin:

$$\text{Mass of C in 1 mol } C_9H_8O_4 = 9 \text{ mol C} \times \frac{12.01 \text{ g C}}{1 \text{ mol C}} = 108.1 \text{ g C}$$

$$\text{Mass of H in 1 mol } C_9H_8O_4 = 8 \text{ mol H} \times \frac{1.008 \text{ g H}}{1 \text{ mol H}} = 8.064 \text{ g H}$$

$$\text{Mass of O in 1 mol } C_9H_8O_4 = 4 \text{ mol O} \times \frac{16.00 \text{ g O}}{1 \text{ mol O}} = 64.00 \text{ g O}$$

$$\text{Total mass of 1 mol of } C_9H_8O_4 = \text{molar mass of } C_9H_8O_4 = 180.2 \text{ g}$$

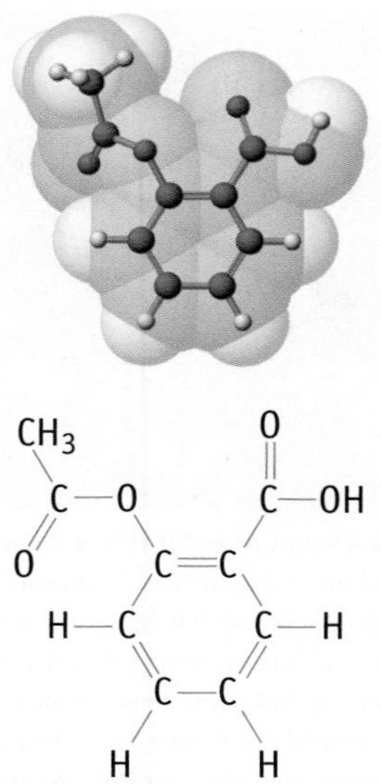

Aspirin formula. Aspirin has the molecular formula $C_9H_8O_4$ and a molar mass of 180.2 g/mol. Aspirin is the common name of the compound acetylsalicylic acid.

As was the case with elements, it is important to be able to convert between amounts (moles) and mass (grams). For example, if you take 325 mg (0.325 g) of aspirin in one tablet, what amount of the compound have you ingested? Based on a molar mass of 180.2 g/mol, there are 0.00180 mol of aspirin per tablet.

$$0.325 \text{ g aspirin} \times \frac{1 \text{ mol aspirin}}{180.2 \text{ g aspirin}} = 0.00180 \text{ mol aspirin}$$

Using the molar mass of a compound, it is possible to determine the number of molecules in any sample from the sample mass and to determine the mass of one molecule. For example, the number of aspirin molecules in one tablet is

$$0.00180 \text{ mol aspirin} \times \frac{6.022 \times 10^{23} \text{ molecules}}{1 \text{ mol aspirin}} = 1.08 \times 10^{21} \text{ molecules}$$

and the mass of one molecule is

$$\frac{180.2 \text{ g aspirin}}{1 \text{ mol aspirin}} \times \frac{1 \text{ mol aspirin}}{6.022 \times 10^{23} \text{ molecules}} = 2.99 \times 10^{-22} \text{ g/molecule}$$

Chemistry Now™

Sign in at **www.thomsonedu.com/login** and go to Chapter 2 Contents to see:
- Screen 2.27 for a simulation on **compounds and moles** and a tutorial on **determining molar mass**
- Screen 2.28 for tutorials on **using molar mass**

EXAMPLE 2.7 Molar Mass and Moles

Problem You have 16.5 g of oxalic acid, $H_2C_2O_4$.

(a) What amount (moles) is represented by 16.5 g of oxalic acid?

(b) How many molecules of oxalic acid are in 16.5 g?

(c) How many atoms of carbon are in 16.5 g of oxalic acid?

(d) What is the mass of one molecule of oxalic acid?

Strategy The first step in any problem involving the conversion of mass and moles is to find the molar mass of the compound in question. Then you can perform the other calculations as outlined by the scheme shown here to find the number of molecules from the amount of substance and the number of atoms of a particular kind:

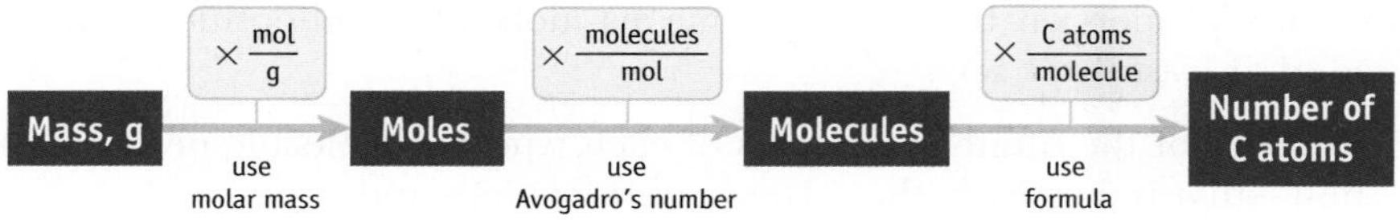

Kenneth G. Libbrecht

What amount of water is in a snowflake? According to K. G. Libbrecht, there are about a billion billion water molecules in a snowflake. Given that this is 1×10^{18} molecules, how many moles of water are in a snowflake and what mass of water? Libbrecht also calculates that "each of us on Earth has contributed by exhalation and evaporation about 1,000 of the molecules in each snowflake." (D. Overbye, *New York Times*, December 23, 2003, page F3. See also K. G. Libbrecht, *American Scientist*, Vol. 95, pages 52–59, January-February 2007.)

Solution

(a) *Moles represented by 16.5 g*

Let us first calculate the molar mass of oxalic acid:

$$2\ \cancel{\text{mol C}}\ \text{per mol}\ H_2C_2O_4 \times \frac{12.01\ \text{g C}}{1\ \cancel{\text{mol C}}} = 24.02\ \text{g C per mol}\ H_2C_2O_4$$

$$2\ \cancel{\text{mol H}}\ \text{per mol}\ H_2C_2O_4 \times \frac{1.008\ \text{g H}}{1\ \cancel{\text{mol H}}} = 2.016\ \text{g H per mol}\ H_2C_2O_4$$

$$4\ \cancel{\text{mol O}}\ \text{per mol}\ H_2C_2O_4 \times \frac{16.00\ \text{g O}}{1\ \cancel{\text{mol O}}} = 64.00\ \text{g O per mol}\ H_2C_2O_4$$

$$\text{Molar mass of}\ H_2C_2O_4 = 90.04\ \text{g per mol}\ H_2C_2O_4$$

Now calculate the amount in moles. The molar mass (expressed in units of 1 mol/90.04 g) is the conversion factor in all mass-to-mole conversions.

$$16.5\ \cancel{\text{g}\ H_2C_2O_4} \times \frac{1\ \text{mol}}{90.04\ \cancel{\text{g}\ H_2C_2O_4}} = 0.183\ \text{mol}\ H_2C_2O_4$$

(b) *Number of molecules*

Use Avogadro's number to find the number of oxalic acid molecules in 0.183 mol of $H_2C_2O_4$.

$$0.183\ \cancel{\text{mol}} \times \frac{6.022 \times 10^{23}\ \text{molecules}}{1\ \cancel{\text{mol}}} = 1.10 \times 10^{23}\ \text{molecules}$$

(c) *Number of C atoms*

Because each molecule contains two carbon atoms, the number of carbon atoms in 16.5 g of the acid is

$$1.10 \times 10^{23}\ \cancel{\text{molecules}} \times \frac{2\ \text{C atoms}}{1\ \cancel{\text{molecule}}} = 2.20 \times 10^{23}\ \text{C atoms}$$

(d) *Mass of one molecule*

Use the molar mass and Avogadro's number to carry out this calculation.

$$\frac{90.04\ \text{g}}{1\ \cancel{\text{mol}}} \times \frac{1\ \cancel{\text{mol}}}{6.0221 \times 10^{23}\ \text{molecules}} = 1.495 \times 10^{-22}\ \text{g/molecule}$$

EXERCISE 2.13 Molar Mass and Moles-to-Mass Conversions

(a) Calculate the molar masses of citric acid ($H_3C_6H_5O_7$) and $MgCO_3$.

(b) If you have 454 g of citric acid, what amount (moles) does this represent?

(c) To have 0.125 mol of $MgCO_3$, what mass (g) must you have?

2.10 Describing Compound Formulas

Given a sample of an unknown compound, how can its formula be determined? The answer lies in *chemical analysis*, a major branch of chemistry that deals with the determination of formulas and structures.

Percent Composition

Any sample of a pure compound always consists of the same elements combined in the same proportion by mass. This means molecular composition can be expressed in at least three ways:

- In terms of the number of atoms of each type per molecule or per formula unit—that is, by giving the formula of the compound

- In terms of the mass of each element per mole of compound
- In terms of the mass of each element in the compound relative to the total mass of the compound—that is, as a mass percent

Suppose you have 1.0000 mol of NH_3 or 17.031 g. This mass of NH_3 is composed of 14.007 g of N (1.0000 mol) and 3.0237 g of H (3.0000 mol). If you compare the mass of N to the total mass of compound, 82.244% of the total mass is N (and 17.755% is H).

$$\text{Mass of N per mole of } NH_3 = \frac{1 \text{ mol N}}{1 \text{ mol } NH_3} \times \frac{14.007 \text{ g N}}{1 \text{ mol N}} = 14.007 \text{ g N/1 mol } NH_3$$

$$\text{Mass percent N in } NH_3 = \frac{\text{mass of N in 1 mol } NH_3}{\text{mass of 1 mol } NH_3}$$

$$= \frac{14.007 \text{ g N}}{17.031 \text{ g } NH_3} \times 100\%$$

$$= 82.244\% \text{ (or 82.244 g N in 100.000 g } NH_3\text{)}$$

$$\text{Mass of H per mole of } NH_3 = \frac{3 \text{ mol H}}{1 \text{ mol } NH_3} \times \frac{1.0079 \text{ g H}}{1 \text{ mol H}}$$

$$= 3.0237 \text{ g H/1 mol } NH_3$$

$$\text{Mass percent H in } NH_3 = \frac{\text{mass of H in 1 mol } NH_3}{\text{mass of 1 mol } NH_3} \times 100\%$$

$$= \frac{3.0237 \text{ g H}}{17.031 \text{ g } NH_3} \times 100\%$$

$$= 17.755\% \text{ (or 17.755 g H in 100.000 g } NH_3\text{)}$$

These values represent the mass percent of each element, or percent composition by mass. They tell you that in a 100.000-g sample there are 82.244 g of N and 17.755 g of H.

■ **Molecular Composition** Molecular composition can be expressed as a percent (mass of an element in a 100-g sample). For example, NH_3 is 82.244% N. Therefore, it has 82.244 g of N in 100.000 g of compound.

82.244% of NH_3 mass is **nitrogen.**

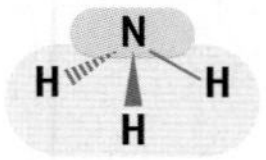

17.755% of NH_3 mass is **hydrogen.**

Chemistry Now™

Sign in at **www.thomsonedu.com/login** and go to Chapter 2 Contents to see Screen 2.29 for a tutorial on **using percent composition.**

EXAMPLE 2.8 Using Percent Composition

Problem What is the mass percent of each element in propane, C_3H_8? What mass of carbon is contained in 454 g of propane?

Strategy First, find the molar mass of C_3H_8, and then calculate the mass percent of C and H per mole of C_3H_8. Using the mass percent of C, calculate the mass of carbon in 454 g of C_3H_8.

Solution

(a) The molar mass of C_3H_8 is 44.097 g/mol (= 3 mol C + 8 mol H = 36.03 g + 8.064 g).

(b) Mass percent of C and H in C_3H_8:

$$\frac{3 \text{ mol C}}{1 \text{ mol } C_3H_8} \times \frac{12.01 \text{ g C}}{1 \text{ mol C}} = 36.03 \text{ g C/1 mol } C_3H_8$$

$$\text{Mass percent of C in } C_3H_8 = \frac{36.03 \text{ g C}}{44.097 \text{ g } C_3H_8} \times 100\% = 81.71\% \text{ C}$$

$$\frac{8 \text{ mol H}}{1 \text{ mol } C_3H_8} \times \frac{1.008 \text{ g H}}{1 \text{ mol H}} = 8.064 \text{ g H/1 mol } C_3H_8$$

$$\text{Mass percent of H in } C_3H_8 = \frac{8.064 \text{ g H}}{44.097 \text{ g } C_3H_8} \times 100\% = 18.29\% \text{ H}$$

(c) Mass of C in 454 g of C_3H_8:

$$454 \text{ g } C_3H_8 \times \frac{81.71 \text{ g C}}{100.0 \text{ g } C_3H_8} = 371 \text{ g C}$$

EXERCISE 2.14 Percent Composition

(a) Express the composition of ammonium carbonate, $(NH_4)_2CO_3$, in terms of the mass of each element in 1.00 mol of compound and the mass percent of each element.

(b) What is the mass of carbon in 454 g of octane, C_8H_{18}?

Empirical and Molecular Formulas from Percent Composition

Now let us consider the reverse of the procedure just described: using relative mass or percent composition data to find a molecular formula. Suppose you know the identity of the elements in a sample and have determined the mass of each element in a given mass of compound by chemical analysis (► Section 4.4). You can then calculate the relative amount (moles) of each element and from this the relative number of atoms of each element in the compound. For example, for a compound composed of atoms of A and B, the steps from percent composition to a formula are

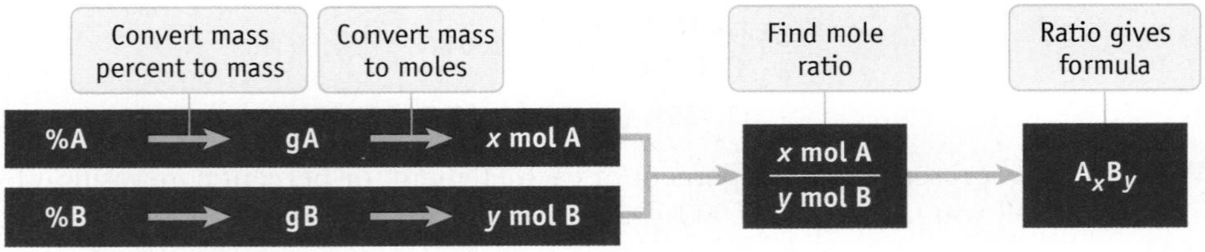

■ **Deriving a Formula** Percent composition gives the mass of an element in 100 g of sample. However, in deriving a formula, any amount of sample is appropriate if you know the mass of each element in that sample mass.

Let us derive the formula for hydrazine, a compound used to remove oxygen from water in heating and cooling systems. It is composed of 87.42% N and 12.58% H and is a close relative of ammonia.

Step 1: *Convert mass percent to mass.* The mass percentages of N and H in hydrazine tell us there are 87.42 g of N and 12.58 g of H in a 100.00 g sample.

Step 2: *Convert the mass of each element to moles.* The amount of each element in the 100.00-g sample is

$$87.42 \text{ g N} \times \frac{1 \text{ mol N}}{14.007 \text{ g N}} = 6.241 \text{ mol N}$$

$$12.58 \text{ g H} \times \frac{1 \text{ mol H}}{1.0079 \text{ g H}} = 12.48 \text{ mol H}$$

■ **Deriving a Formula—Mole Ratios** When finding the ratio of moles of one element relative to another, *always* divide the larger number by the smaller one.

Step 3: *Find the mole ratio of elements.* Use the amount (moles) of each element in the 100.00 g of sample to find the amount of one element relative to the other. For hydrazine, this ratio is 2 mol of H to 1 mol of N,

$$\frac{12.48 \text{ mol H}}{6.241 \text{ mol N}} = \frac{2.00 \text{ mol H}}{1.00 \text{ mol N}} \longrightarrow NH_2$$

showing that there are 2 mol of H atoms for every 1 mol of N atoms in hydrazine. Thus, in one molecule, two atoms of H occur for every atom of N; that is, the formula is NH_2. This simplest, whole-number ratio of atoms in a formula is called the **empirical formula**.

Percent composition data allow us to calculate the atom ratios in a compound. A *molecular formula*, however, must convey two pieces of information: (1) the relative numbers of atoms of each element in a molecule (the atom ratios) and (2) the total number of atoms in the molecule. For hydrazine, there are twice as many H atoms as N atoms, so the molecular formula could be NH_2. Recognize, however, that percent composition data give only the simplest possible ratio of atoms in a molecule. The empirical formula of hydrazine is NH_2, but the true molecular formula could be NH_2, N_2H_4, N_3H_6, N_4H_8, or any other formula having a 1:2 ratio of N to H.

To determine the molecular formula from the empirical formula, the molar mass must be obtained from experiment. For example, experiments show that the molar mass of hydrazine is 32.0 g/mol, twice the formula mass of NH_2, which is 16.0 g/mol. Thus, the molecular formula of hydrazine is two times the empirical formula of NH_2, that is, N_2H_4.

Chemistry Now™

Sign in at **www.thomsonedu.com/login** and go to Chapter 2 Contents to see:

- Screen 2.30 for a tutorial on **determining empirical formulas**
- Screen 2.31 for a tutorial on **determining molecular formulas**

EXAMPLE 2.9 Calculating a Formula from Percent Composition

Problem Eugenol is the major component in oil of cloves. It has a molar mass of 164.2 g/mol and is 73.14% C and 7.37% H; the remainder is oxygen. What are the empirical and molecular formulas of eugenol?

Strategy To derive a formula, we need to know the mass percent of each element. Because the mass percents of all elements must add up to 100.0%, we find the mass percent of O from the difference between 100.0% and the mass percents of C and H. Next, we assume that the mass percent of each element is equivalent to its mass in grams, and convert each mass to moles. Finally, the ratio of moles gives the empirical formula. The mass of a mole of compound having the calculated empirical formula is compared with the actual, experimental molar mass to find the true molecular formula.

Problem Solving Tip 2.3 Finding Empirical and Molecular Formulas

- The experimental data available to find a formula may be in the form of percent composition or the masses of elements combined in some mass of compound. No matter what the starting point, the first step is always to convert masses of elements to moles.
- Be sure to use *at least* three significant figures when calculating empirical formulas. Using fewer significant figures can give a misleading result.
- When finding mole ratios, always divide the larger number of moles by the smaller one.
- Empirical and molecular formulas can differ for molecular compounds. In contrast, the formula of an ionic compound is generally the same as its empirical formula.
- Determining the molecular formula of a compound after calculating the empirical formula requires knowing the molar mass.
- When *both* the percent composition and the molar mass are known for a compound, the alternative method mentioned in the comment to Example 2.9 could be used.

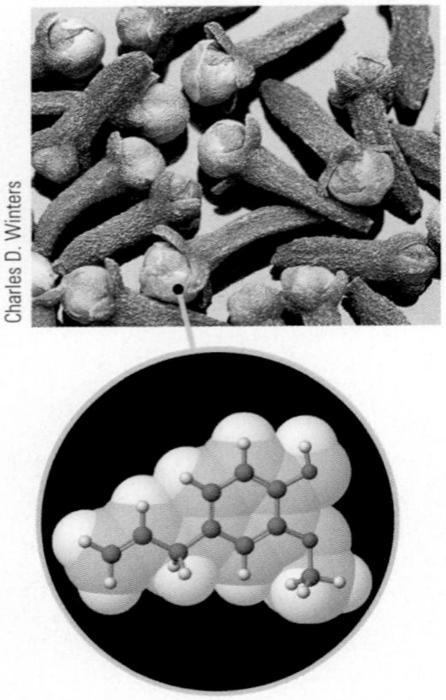

Eugenol, $C_{10}H_{12}O_2$, is an important component in oil of cloves.

Solution The mass of oxygen in a 100.00 g sample of eugenol is

$$100.00 \text{ g} = 73.14 \text{ g C} + 7.37 \text{ g H} + \text{mass of O}$$
$$\text{Mass of O} = 19.49 \text{ g}$$

The amount of each element is

$$73.14 \text{ g C} \times \frac{1 \text{ mol C}}{12.011 \text{ g C}} = 6.089 \text{ mol C}$$

$$7.37 \text{ g H} \times \frac{1 \text{ mol H}}{1.008 \text{ g H}} = 7.31 \text{ mol H}$$

$$19.49 \text{ g O} \times \frac{1 \text{ mol O}}{15.999 \text{ g O}} = 1.218 \text{ mol O}$$

To find the mole ratio, the best approach is to base the ratios on the smallest number of moles present—in this case, oxygen.

$$\frac{\text{mol C}}{\text{mol O}} = \frac{6.089 \text{ mol C}}{1.218 \text{ mol O}} = \frac{4.999 \text{ mol C}}{1.000 \text{ mol O}} = \frac{5 \text{ mol C}}{1 \text{ mol O}}$$

$$\frac{\text{mol H}}{\text{mol O}} = \frac{7.31 \text{ mol H}}{1.218 \text{ mol O}} = \frac{6.00 \text{ mol H}}{1.000 \text{ mol O}} = \frac{6 \text{ mol H}}{1 \text{ mol O}}$$

Now we know there are 5 mol of C and 6 mol of H per 1 mol of O. Thus, the empirical formula is C_5H_6O.

The experimentally determined molar mass of eugenol is 164.2 g/mol. This is twice the molar mass of C_5H_6O (82.1 g/mol).

$$\frac{164.2 \text{ g/mol of eugenol}}{82.10 \text{ g/mol of } C_5H_6O} = 2.000 \text{ mol } C_5H_6O \text{ per mol of eugenol}$$

The molecular formula is $C_{10}H_{12}O_2$.

Comment There is another approach to finding the molecular formula here. Knowing the percent composition of eugenol and its molar mass, we could calculate that in 164.2 g of eugenol there are 120.1 g of C (10 mol of C), 12.1 g of H (12 mol of H), and 32.00 g of O (2 mol of O). This gives us a molecular formula of $C_{10}H_{12}O_2$. However, you must recognize that *this approach can only be used when you know both the percent composition and the molar mass.*

EXERCISE 2.15 Empirical and Molecular Formulas

(a) What is the empirical formula of naphthalene, $C_{10}H_8$?

(b) The empirical formula of acetic acid is CH_2O. If its molar mass is 60.05 g/mol, what is the molecular formula of acetic acid?

EXERCISE 2.16 Calculating a Formula from Percent Composition

Isoprene is a liquid compound that can be polymerized to form synthetic rubber. It is composed of 88.17% carbon and 11.83% hydrogen. Its molar mass is 68.11 g/mol. What are its empirical and molecular formulas?

EXERCISE 2.17 Calculating a Formula from Percent Composition

Camphor is found in "camphor wood," much prized for its wonderful odor. It is composed of 78.90% carbon and 10.59% hydrogen. The remainder is oxygen. What is its empirical formula?

Determining a Formula from Mass Data

The composition of a compound in terms of mass percent gives us the mass of each element in a 100.0-g sample. In the laboratory, we often collect information on the composition of compounds slightly differently. We can:

1. Combine known masses of elements to give a sample of the compound of known mass. Element masses can be converted to moles, and the ratio of moles gives the combining ratio of atoms—that is, the empirical formula. This approach is described in Example 2.10.
2. Decompose a known mass of an unknown compound into "pieces" of known composition. If the masses of the "pieces" can be determined, the ratio of moles of the "pieces" gives the formula. An example is a decomposition such as

$$Ni(CO)_4(\ell) \rightarrow Ni(s) + 4\ CO(g)$$

The masses of Ni and CO can be converted to moles, whose 1 : 4 ratio would reveal the formula of the compound. We will describe this approach in Chapter 4 (► Section 4.4).

EXAMPLE 2.10 Formula of a Compound from Combining Masses

Problem Gallium oxide, Ga_xO_y, forms when gallium is combined with oxygen. Suppose you allow 1.25 g of gallium (Ga) to react with oxygen and obtain 1.68 g of Ga_xO_y. What is the formula of the product?

Strategy Calculate the mass of oxygen in 1.68 g of product (which you already know contains 1.25 g of Ga). Next, calculate the amounts of Ga and O (in moles), and find their ratio.

Solution The masses of Ga and O combined in 1.68 g of product are

$$1.68 \text{ g product} - 1.25 \text{ g Ga} = 0.43 \text{ g O}$$

Next, calculate the amount of each reactant:

$$1.25 \text{ g Ga} \times \frac{1 \text{ mol Ga}}{69.72 \text{ g Ga}} = 0.0179 \text{ mol Ga}$$

$$0.43 \text{ g O} \times \frac{1 \text{ mol O}}{16.0 \text{ g O}} = 0.027 \text{ mol O}$$

Find the ratio of moles of O to moles of Ga:

$$\text{Mole ratio} = \frac{0.027 \text{ mol O}}{0.0179 \text{ mol Ga}} = \frac{1.5 \text{ mol O}}{1.0 \text{ mol Ga}}$$

It is 1.5 mol O/1.0 mol Ga, or 3 mol O to 2 mol Ga. Thus, the product is gallium oxide, Ga_2O_3.

EXAMPLE 2.11 Determining a Formula from Mass Data

Problem Tin metal (Sn) and purple iodine (I_2) combine to form orange, solid tin iodide with an unknown formula.

$$\text{Sn metal} + \text{solid } I_2 \rightarrow \text{solid } Sn_xI_y$$

Weighed quantities of Sn and I_2 are combined, where the quantity of Sn is more than is needed to react with all of the iodine. After Sn_xI_y has been formed, it is isolated by filtration. The mass of excess tin is also determined. The following data were collected:

Mass of tin (Sn) in the original mixture	1.056 g
Mass of iodine (I_2) in the original mixture	1.947 g
Mass of tin (Sn) recovered after reaction	0.601 g

Strategy The first step is to find the masses of Sn and I that are combined in Sn_xI_y. The masses are then converted to moles, and the ratio of moles reveals the compound's empirical formula.

(a) Weighed samples of tin (left) and iodine (right).

(b) The tin and iodine are heated in a solvent.

(c) The hot reaction mixture is filtered to recover unreacted tin.

(d) When the solvent cools, solid, orange tin oxide forms and is isolated.

Charles D. Winters

The formula of a compound of tin and iodine can be found by determining the mass of iodine that combines with a given mass of tin.

Solution First, let us find the mass of tin that combined with iodine

Mass of Sn in original mixture	1.056 g
Mass of Sn recovered	− 0.601 g
Mass of Sn combined with 1.947 g I_2	0.455 g

Now convert the mass of tin to the amount of tin.

$$0.455 \text{ g Sn} \times \frac{1 \text{ mol Sn}}{118.7 \text{ g Sn}} = 0.00383 \text{ mol Sn}$$

No I_2 was recovered; it all reacted with Sn. Therefore, 0.00383 mol of Sn combined with 1.947 g of I_2. Because we want to know the amount of I that combined with 0.00383 mol of Sn, we calculate the amount of I from the mass of I_2.

$$1.947 \text{ g } I_2 \times \frac{1 \text{ mol } I_2}{253.81 \text{ g } I_2} \times \frac{2 \text{ mol I}}{1 \text{ mol } I_2} = 0.01534 \text{ mol I}$$

Finally, we find the ratio of moles.

$$\frac{\text{mol I}}{\text{mol Sn}} = \frac{0.01534 \text{ mol I}}{0.00383 \text{ mol Sn}} = \frac{4.01 \text{ mol I}}{1.00 \text{ mol Sn}} = \frac{4 \text{ mol I}}{1 \text{ mol Sn}}$$

There are four times as many moles of I as moles of Sn in the sample. Therefore, there are four times as many atoms of I as atoms of Sn per formula unit. The empirical formula is SnI_4.

EXERCISE 2.18 Determining a Formula from Combining Masses

Analysis shows that 0.586 g of potassium metal combines with 0.480 g of O_2 gas to give a white solid having a formula of K_xO_y. What is the empirical formula of the compound?

Determining a Formula by Mass Spectrometry

We have described chemical methods of determining a molecular formula, but there are many instrumental methods as well. One of them is *mass spectrometry* (Figure 2.27). We introduced this technique where it was used to describe the existence of isotopes and to measure their relative abundance (Figure 2.3). If a compound can be turned into a vapor, the vapor can be passed through an electron beam in a mass spectrometer where high energy electrons collide with the gas phase molecules. These high energy collisions cause the molecule to lose electrons and turn the molecules into positive ions. These ions usually break apart or fragment into smaller pieces. As illustrated in Figure 2.27, the cation created from

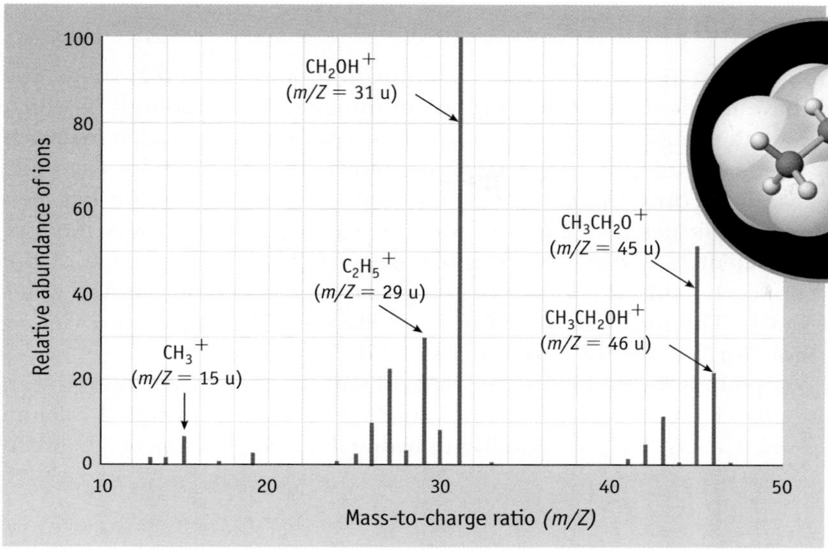

FIGURE 2.27 Mass spectrum of ethanol, CH_3CH_2OH. A prominent peak or line in the spectrum is the "parent" ion ($CH_3CH_2OH^+$) at mass 46. (The "parent" ion is the heaviest ion observed.) The mass designated by the peak for the "parent" ion confirms the formula of the molecule. Other peaks are for "fragment" ions. This pattern of lines can provide further, unambiguous evidence of the formula of the compound. (The horizontal axis is the mass-to-charge ratio of a given ion. Because almost all observed ions have a charge of $Z = +1$, the value observed is the mass of the ion.) (See *A Closer Look: Mass Spectrometry, Molar Mass, and Isotopes.*)

ethanol ($CH_3CH_2OH^+$) fragments (losing an H atom) to give another cation ($CH_3CH_2O^+$), which further fragments. A mass spectrometer detects and records the masses of the different particles. Analysis of the spectrum can help identify a compound and can give an accurate molar mass.

A Closer Look — Mass Spectrometry, Molar Mass, and Isotopes

Bromobenzene, C_6H_5Br, has a molar mass of 157.010 g/mol. Why, then, are there two prominent lines at a mass-to-charge ratio (m/Z) 156 and 158 in the mass spectrum of the compound (when $Z = 1$)? The answer shows us the influence of isotopes on molar mass.

Bromine has two naturally occurring isotopes, ^{79}Br and ^{81}Br. They are 50.7% and 49.3% abundant, respectively. What is the mass of C_6H_5Br based on each isotope? If we use the most abundant isotopes of C and H (^{12}C and ^{1}H), the mass of the molecule having only the ^{79}Br isotope, $C_6H_5{}^{79}Br$, is 156. The mass of the molecule containing only the ^{81}Br isotope, $C_6H_5{}^{81}Br$, is 158.

The calculated molar mass of bromobenzene is 157.010, a value derived from the atomic masses of the elements. Atomic masses reflect the abundances of all of the isotopes. In contrast, the mass spectrum has a line for each possible combination of isotopes. This explains why there are small lines at the mass-to-charge ratios of 157 and 159. They arise from various combinations of ^{1}H, ^{12}C, ^{13}C, ^{79}Br, and ^{81}Br atoms. In fact, careful analysis of such patterns can identify a molecule unambiguously.

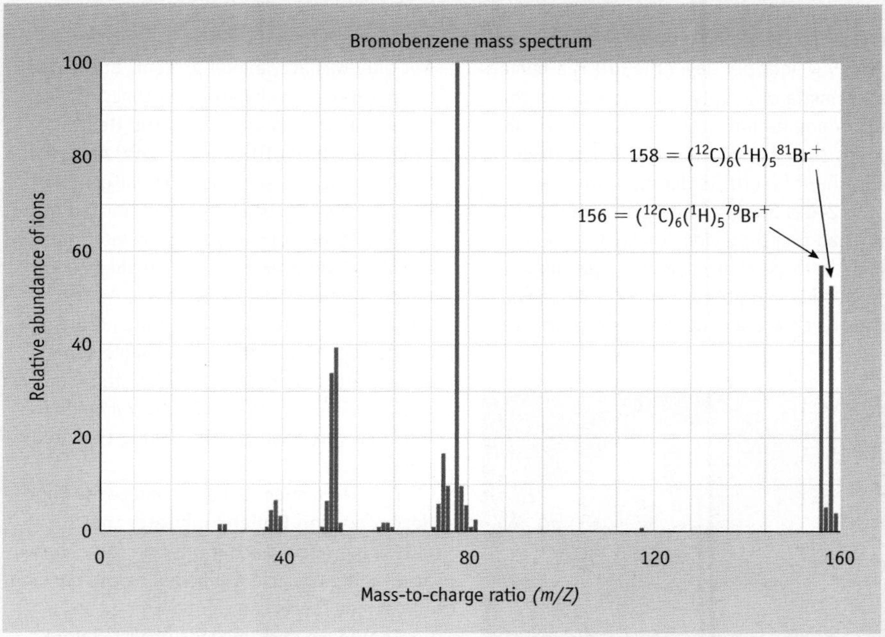

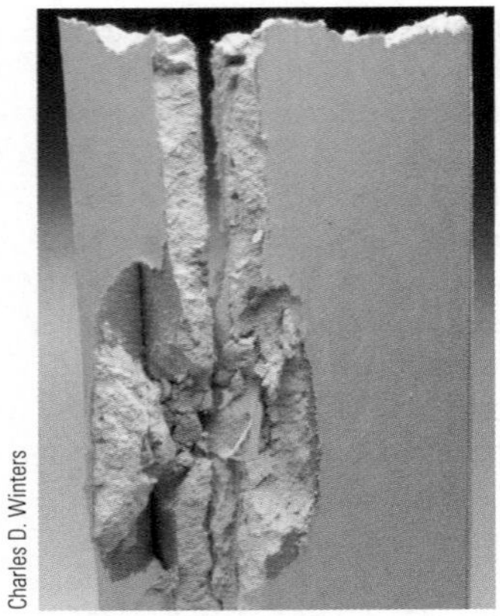

Charles D. Winters

FIGURE 2.28 Gypsum wallboard. Gypsum is hydrated calcium sulfate, $CaSO_4 \cdot 2H_2O$.

2.11 Hydrated Compounds

If ionic compounds are prepared in water solution and then isolated as solids, the crystals often have molecules of water trapped in the lattice. Compounds in which molecules of water are associated with the ions of the compound are called **hydrated compounds.** The beautiful blue copper(II) compound in Figure 2.26, for example, has a formula that is conventionally written as $CuCl_2 \cdot 2\ H_2O$. The dot between $CuCl_2$ and $2\ H_2O$ indicates that 2 mol of water are associated with every mole of $CuCl_2$; it is equivalent to writing the formula as $CuCl_2(H_2O)_2$. The name of the compound, copper(II) chloride dihydrate, reflects the presence of 2 mol of water per mole of $CuCl_2$. The molar mass of $CuCl_2 \cdot 2\ H_2O$ is 134.5 g/mol (for $CuCl_2$) plus 36.0 g/mol (for $2\ H_2O$) for a total mass of 170.5 g/mol.

Hydrated compounds are common. The walls of your home may be covered with wallboard, or "plaster board" (Figure 2.28) These sheets contain hydrated calcium sulfate, or gypsum ($CaSO_4 \cdot 2\ H_2O$), as well as unhydrated $CaSO_4$, sandwiched between

Case Study

What's in Those French Fries?

The U.S. Environmental Protection Agency (EPA) maintains a database of toxicities of chemicals. One compound on that list is acrylamide, $CH_2{=}CHCONH_2$, which is listed as a possible human carcinogen (cancer-causing substance). Although not confirmed by human data, carcinogenicity has been observed in rats. Based on the animal studies, the EPA suggests that the "Reference Dose" (RfD) of acrylamide should be 0.0002 mg per kilogram of body weight per day. (The RfD is a numerical estimate of a daily oral exposure to the human population, including sensitive subgroups such as children, that is not likely to cause harmful effects during a lifetime.)

In 2002, Swedish chemists announced that they had found previously undetected acrylamide in foods that many find appealing: french fries and potato chips (or chips and crisps as they are called in other countries).

Charles D. Winters

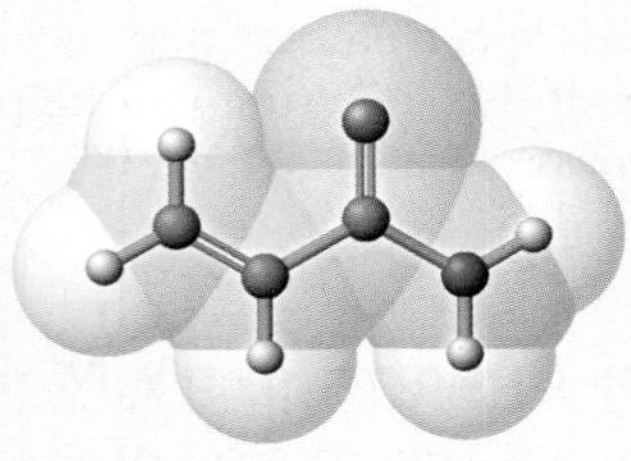

Acrylamide

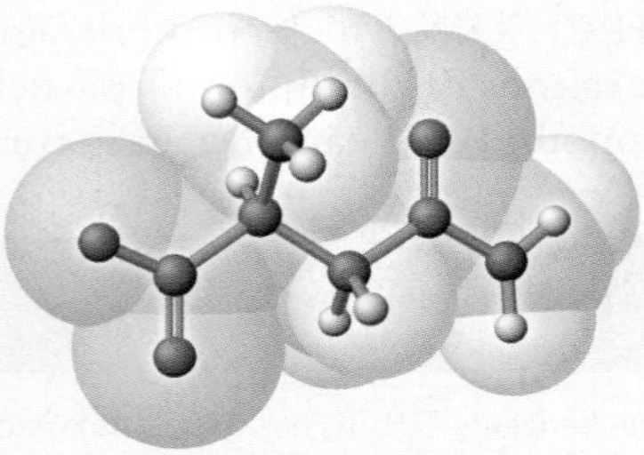

Asparagine, an amino acid

Not only was acrylamide present, but it was in concentrations hundreds of times higher than what the EPA and the World Health Organization (WHO) consider safe. And soon thereafter it was also found in coffee, pastries, cookies, cereals, rolls, and toasted bread.

Where does the acrylamide come from, and can the amount be reduced? Chemists soon understood that the likely source was an interaction between the naturally occurring amino acid asparagine and a simple sugar such as fructose or glucose when the food was cooked. However, the level of acrylamide in food can vary widely with cooking time and temperature.

Can acrylamide levels be reduced in foods? Recent work in England indicates that if 0.39% by weight each of glycine (an amino acid) and citric acid are added before cooking, acrylamide levels can be reduced by 40%.

Should we give up french fries? Before acting precipitously, a closer look is called for. Even the report from the Swedish government's National Food Administration counseled that "...there is not sufficient data to warrant changing the current dietary recommendation." Other scientists who study carcinogenic compounds point out that acrylamide is not a proven human carcinogen and that the dose of acrylamide from fried foods is 700 times less that the dose that causes cancer in rodents. Furthermore, many common foods that we eat regularly—among them cantaloupe, carrots, cauliflower, cherries, chocolate, and coffee—have substances that have been proven to cause cancer in rodents.

Nonetheless, the warning is there. Clearly, more information is needed, and chemists are the ones with the background to do such studies.

Questions:

1. *Which has the higher mass percent of nitrogen, acrylamide or asparagine?*
2. *If you weigh 150 pounds (1 pound = 453.6 g), how many molecules of acrylamide are you consuming per day if you consume 0.0002 mg per kilogram?*

Answers to these questions are in Appendix Q.

Charles D. Winters

Active Figure 2.29
Dehydrating hydrated cobalt(II) chloride, $CoCl_2 \cdot 6\ H_2O$. *(left)* Cobalt (II) chloride hexahydrate [$CoCl_2 \cdot 6\ H_2O$] is a deep red compound. *(left and center)* When it is heated, the compound loses the water of hydration and forms the deep blue compound $CoCl_2$.

Chemistry Now™ Sign in at www.thomsonedu.com/login and go to the Chapter Contents menu to explore an interactive version of this figure accompanied by an exercise.

paper. Gypsum is a mineral that can be mined. Now, however, it is usually obtained as a byproduct in the manufacture of hydrofluoric acid and phosphoric acid.

If gypsum is heated between 120 and 180 °C, the water is partly driven off to give $CaSO_4 \cdot \frac{1}{2}\ H_2O$, a compound commonly called "plaster of Paris." If you have ever broken an arm or leg and had to have a cast, the cast may have been made of this compound. It is an effective casting material because, when added to water, it forms a thick slurry that can be poured into a mold or spread out over a part of the body. As it takes on more water, the material increases in volume and forms a hard, inflexible solid. These properties also make plaster of Paris a useful material for artists, because the expanding compound fills a mold completely and makes a high-quality reproduction.

Hydrated cobalt(II) chloride is the red solid in Figure 2.29. When heated, it turns purple and then deep blue as it loses water to form anhydrous $CoCl_2$; "anhydrous" means a substance without water. On exposure to moist air, anhydrous $CoCl_2$ takes up water and is converted back into the red hydrated compound. It is this property that allows crystals of the blue compound to be used as a humidity indicator. You may have seen them in a small bag packed with a piece of electronic equipment.

There is no simple way to predict how much water will be present in a hydrated compound, so it must be determined experimentally. Such an experiment may involve heating the hydrated material so that all the water is released from the solid and evaporated. Only the anhydrous compound is left. The formula of hydrated copper(II) sulfate, commonly known as "blue vitriol," is determined in this manner in Example 2.12.

■ **Invisible Ink** $CoCl_2 \cdot 6\ H_2O$ also makes a good "invisible ink." A solution of cobalt(II) chloride in water is red, but if you write on paper with the solution, it cannot be seen. When the paper is warmed, however, the cobalt compound dehydrates to give the deep blue anhydrous compound, and the writing becomes visible.

EXAMPLE 2.12 Determining the Formula of a Hydrated Compound

Problem You want to know the value of *x* in blue, hydrated copper(II) sulfate, $CuSO_4 \cdot x\ H_2O$; that is, the number of water molecules for each unit of $CuSO_4$. In the laboratory, you weigh out 1.023 g of the solid. After heating the solid thoroughly in a porcelain crucible (see Figure), 0.654 g of nearly white, anhydrous copper(II) sulfate, $CuSO_4$, remains.

$$1.023\ \text{g}\ CuSO_4 \cdot x\ H_2O + \text{heat} \longrightarrow 0.654\ \text{g}\ CuSO_4 + ?\ \text{g}\ H_2O$$

Strategy To find *x*, we need to know the amount of H_2O per mole $CuSO_4$. Therefore, first find the mass of water lost by the sample from the difference between the mass of hydrated compound and the anhydrous form. Finally, find the ratio of the amount of water lost (moles) to the amount of anhydrous $CuSO_4$.

Solution Find the mass of water.

Mass of hydrated compound	1.023 g
− Mass of anhydrous compound, $CuSO_4$	−0.654
Mass of water	0.369 g

Charles D. Winters

Heating a hydrated compound. The formula of a hydrated compound can be determined by heating a weighed sample enough to cause the compound to release its water of hydration. Knowing the mass of the hydrated compound before heating and the mass of the anhydrous compound after heating, we can determine the mass of water in the original sample.

Next, convert the masses of $CuSO_4$ and H_2O to moles.

$$0.369 \text{ g } H_2O \times \frac{1 \text{ mol } H_2O}{18.02 \text{ g } H_2O} = 0.0205 \text{ mol } H_2O$$

$$0.654 \text{ g } CuSO_4 \times \frac{1 \text{ mol } CuSO_4}{159.6 \text{ g } CuSO_4} = 0.00410 \text{ mol } CuSO_4$$

The value of *x* is determined from the mole ratio.

$$\frac{0.0205 \text{ mol } H_2O}{0.00410 \text{ mol } CuSO_4} = \frac{5.00 \text{ mol } H_2O}{1.00 \text{ mol } CuSO_4}$$

The water-to-$CuSO_4$ ratio is 5-to-1, so the formula of the hydrated compound is $CuSO_4 \cdot 5\ H_2O$. Its name is copper(II) sulfate pentahydrate.

EXERCISE 2.19 Determining the Formula of a Hydrated Compound

Hydrated nickel(II) chloride is a beautiful, green, crystalline compound. When heated strongly, the compound is dehydrated. If 0.235 g of $NiCl_2 \cdot x\ H_2O$ gives 0.128 g of $NiCl_2$ on heating, what is the value of *x*?

Chapter Goals Revisited

Chemistry.Now™ Sign in at **www.thomsonedu.com/login** to:

- Assess your understanding with Study Questions in OWL keyed to each goal in the Goals and Homework menu for this chapter
- For quick review, download Go Chemistry mini-lecture flashcard modules (or purchase them at **www.ichapters.com**)
- Check your readiness for an exam by taking the Pre-Test and exploring the modules recommended in your Personalized Study plan.

Access **How Do I Solve It?** tutorials on how to approach problem solving using concepts in this chapter.

Now that you have studied this chapter, you should ask whether you have met the chapter goals. In particular, you should be able to:

Describe atomic structure, and define atomic number and mass number

a. Describe electrons, protons, and neutrons, and the general structure of the atom (Section 2.1). **Study Question(s) assignable in OWL: 2.**

b. Understand the relative atomic weight scale and the atomic mass unit (Section 2.2).

Understand the nature of isotopes, and calculate atomic weight from isotopic abundances and isotopic masses

a. Define isotope and give the mass number and number of neutrons for a specific isotope (Sections 2.2 and 2.3). **Study Question(s) assignable in OWL: 4, 5, 8.**

b. Do calculations that relate the atomic weight of an element and isotopic abundances and masses (Section 2.4). **Study Question(s) assignable in OWL: 10, 12, 15, 86, 88.**

Know the terminology of the periodic table

a. Identify the periodic table locations of groups, periods, metals, metalloids, nonmetals, alkali metals, alkaline earth metals, halogens, noble gases, and the transition elements (Section 2.5). **Study Question(s) assignable in OWL: 20, 21, 23, 92; Go Chemistry Module 1.**

b. Recognize similarities and differences in properties of some of the common elements of a group.

Interpret, predict, and write formulas for ionic and molecular compounds

a. Recognize and interpret molecular formulas, condensed formulas, and structural formulas (Section 2.6).

b. Recognize that metal atoms commonly lose one or more electrons to form positive ions, called cations, and nonmetal atoms often gain electrons to form negative ions, called anions (see Figure 2.7).

c. Recognize that the charge on a metal cation in Groups 1A, 2A, and 3A is equal to the group number in which the element is found in the periodic table (M^{n+}, n = Group number) (Section 2.7). Charges on transition metal cations are often 2+ or 3+, but other charges are observed. **Study Question(s) assignable in OWL: 29, 33, 35; Go Chemistry Module 2.**

d. Recognize that the negative charge on a single-atom or monatomic anion, X^{n-}, is given by n = Group number − 8 (Section 2.7).

e. Write formulas for ionic compounds by combining ions in the proper ratio to give no overall charge (Section 2.7).

Name ionic and molecular compounds

a. Give the names or formulas of polyatomic ions, knowing their formulas or names, respectively (Table 2.4 and Section 2.7).

b. Name ionic compounds and simple binary compounds of the nonmetals (Sections 2.7 and 2.8). **Study Question(s) assignable in OWL: 41, 43, 49, 51; Go Chemistry Module 3.**

Understand some properties of ionic compounds

a. Understand the importance of Coulomb's law (Equation 2.3), which describes the electrostatic forces of attraction and repulsion of ions. Coulomb's law states that the force of attraction between oppositely charged species increases with electric charge and with decreasing distance between the species (Section 2.7). **Study Question(s) assignable in OWL: 48.**

Explain the concept of the mole, and use molar mass in calculations

a. Understand that the molar mass of an element is the mass in grams of Avogadro's number of atoms of that element (Section 2.9). **Study Question(s) assignable in OWL: 53, 55, 57, 89, 93, 96.**

b. Know how to use the molar mass of an element and Avogadro's number in calculations (Section 2.9). **Study Question(s) assignable in OWL: 55, 57, 93, 98.**

c. Understand that the molar mass of a compound (often called the molecular weight) is the mass in grams of Avogadro's number of molecules (or formula units) of a compound (Section 2.9). For ionic compounds, which do not consist of individual molecules, the sum of atomic masses is often called the formula mass (or formula weight).

d. Calculate the molar mass of a compound from its formula and a table of atomic masses (Section 2.9). **Study Question(s) assignable in OWL: 59, 61, 105.**

e. Calculate the number of moles of a compound that is represented by a given mass, and vice versa (Section 2.9). **Study Question(s) assignable in OWL: 63; Go Chemistry Module 4.**

Derive compound formulas from experimental data

a. Express the composition of a compound in terms of percent composition (Section 2.10). **Study Question(s) assignable in OWL: 67, 69.**

b. Use percent composition or other experimental data to determine the empirical formula of a compound (Section 2.10). **Study Question(s) assignable in OWL: 71, 76, 77, 79, 81, 120.**

c. Understand how mass spectrometry can be used to find a molar mass (Section 2.10).

d. Use experimental data to find the number of water molecules in a hydrated compound (Section 2.11) **Study Question(s) assignable in OWL: 141.**

KEY EQUATIONS

Equation 2.1 (page 54) Percent abundance of an isotope

$$\text{Percent abundance} = \frac{\text{number of atoms of a given isotope}}{\text{total number of atoms of all isotopes of that element}} \times 100\%$$

Equation 2.2 (page 56) Calculate the average atomic mass (atomic weight) from isotope abundances and the exact atomic mass of each isotope of an element.

$$\text{Atomic weight} = \left(\frac{\%\text{ abundance isotope 1}}{100}\right)(\text{mass of isotope 1}) + \left(\frac{\%\text{ abundance isotope 2}}{100}\right)(\text{mass of isotope 2}) + \ldots$$

Equation 2.3 (page 78) **Coulomb's Law,** the force of attraction between oppositely charged ions.

$$\text{Force of attraction} = k\,\frac{(n^{+}\text{e})(n^{-}\text{e})}{d^{2}}$$

charge on + and − ions (n^+, n^-); charge on electron (e); proportionality constant (k); distance between ions (d)

STUDY QUESTIONS

Online homework for this chapter may be assigned in OWL.

▲ denotes challenging questions.

■ denotes questions assignable in OWL.

Blue-numbered questions have answers in Appendix O and fully-worked solutions in the *Student Solutions Manual.*

Practicing Skills

Atoms: Their Composition and Structure

(See ChemistryNow Screen 2.11.)

1. What are the three fundamental particles from which atoms are built? What are their electric charges? Which of these particles constitute the nucleus of an atom? Which is the least massive particle of the three?

2. ■ If a gold atom has a radius of 145 pm and you could string gold atoms like beads on a thread, how many atoms would you need to have a necklace 36 cm long?

3. Give the complete symbol ($^{A}_{Z}X$), including atomic number and mass number, for each of the following atoms: (a) magnesium with 15 neutrons, (b) titanium with 26 neutrons, and (c) zinc with 32 neutrons.

4. ■ Give the complete symbol ($^{A}_{Z}X$), including atomic number and mass number, of (a) a nickel atom with 31 neutrons, (b) a plutonium atom with 150 neutrons, and (c) a tungsten atom with 110 neutrons.

5. ■ How many electrons, protons, and neutrons are there in each of the following atoms?
(a) magnesium-24, ^{24}Mg
(b) tin-119, ^{119}Sn
(c) thorium-232, ^{232}Th
(d) carbon-13, ^{13}C;
(e) copper-63, ^{63}Cu
(f) bismuth-205, ^{205}Bi

6. Atomic Structure
   (a) The synthetic radioactive element technetium is used in many medical studies. Give the number of electrons, protons, and neutrons in an atom of technetium-99.
   (b) Radioactive americium-241 is used in household smoke detectors and in bone mineral analysis. Give the number of electrons, protons, and neutrons in an atom of americium-241.

Isotopes

(See ChemistryNow Screen 2.12.)

7. Cobalt has three radioactive isotopes used in medical studies. Atoms of these isotopes have 30, 31, and 33 neutrons, respectively. Give the symbol for each of these isotopes.

8. ■ Which of the following are isotopes of element X, the atomic number for which is 9: $^{19}_{9}X$, $^{20}_{9}X$, $_{18}^{9}X$ and $^{21}_{9}X$?

Isotope Abundance and Atomic Weight

(See Examples 2.2 and 2.3, Exercises 2.2 and 2.3, and ChemistryNow Screens 2.12 and 2.13.)

9. Thallium has two stable isotopes, ^{203}Tl and ^{205}Tl. Knowing that the atomic weight of thallium is 204.4, which isotope is the more abundant of the two?

10. ■ Strontium has four stable isotopes. Strontium-84 has a very low natural abundance, but ^{86}Sr, ^{87}Sr, and ^{88}Sr are all reasonably abundant. Knowing that the atomic weight of strontium is 87.62, which of the more abundant isotopes predominates?

11. Verify that the atomic weight of lithium is 6.94, given the following information:
    ^{6}Li, mass = 6.015121 u; percent abundance = 7.50%
    ^{7}Li, mass = 7.016003 u; percent abundance = 92.50%

12. ■ Verify that the atomic weight of magnesium is 24.31, given the following information:
    ^{24}Mg, mass = 23.985042 u; percent abundance = 78.99%
    ^{25}Mg, mass = 24.985837 u; percent abundance = 10.00%
    ^{26}Mg, mass = 25.982593 u; percent abundance = 11.01%

13. Silver (Ag) has two stable isotopes, ^{107}Ag and ^{109}Ag. The isotopic weight of ^{107}Ag is 106.9051, and the isotopic mass of ^{109}Ag is 108.9047. The atomic weight of Ag, from the periodic table, is 107.868. Estimate the percent of ^{107}Ag in a sample of the element.
    (a) 0% (b) 25% (c) 50% (d) 75%

14. Copper exists as two isotopes: ^{63}Cu (62.9298 u) and ^{65}Cu (64.9278 u). What is the approximate percent of ^{63}Cu in samples of this element?
    (a) 10% (c) 50% (e) 90%
    (b) 30% (d) 70%

15. ■ Gallium has two naturally occurring isotopes, ^{69}Ga and ^{71}Ga, with masses of 68.9257 u and 70.9249 u, respectively. Calculate the percent abundances of these isotopes of gallium.

16. Europium has two stable isotopes, ^{151}Eu and ^{153}Eu, with masses of 150.9197 u and 152.9212 u, respectively. Calculate the percent abundances of these isotopes of europium.

The Periodic Table

(See Section 2.5 and Exercise 2.4. See also the Periodic Table Tool on the ChemistryNow website.)

17. Titanium and thallium have symbols that are easily confused with each other. Give the symbol, atomic number, atomic weight, and group and period number of each element. Are they metals, metalloids, or nonmetals?

18. In Groups 4A–6A, there are several elements whose symbols begin with S. Name these elements, and for each one give its symbol, atomic number, Group number, and period. Describe each as a metal, metalloid, or nonmetal.

19. How many periods of the periodic table have 8 elements; how many have 18 elements, and how many have 32 elements?

20. ■ How many elements occur in the seventh period? What is the name given to the majority of these elements, and what well-known property characterizes them?

21. ■ Select answers to the questions listed below from the following list of elements whose symbols start with the letter C: C, Ca, Cr, Co, Cd, Cl, Cs, Ce, Cm, Cu, and Cf. (You should expect to use some symbols more than once.)
    (a) Which are nonmetals?
    (b) Which are main group elements?
    (c) Which are lanthanides?
    (d) Which are transition elements?
    (e) Which are actinides?
    (f) Which are gases?

22. Give the name and chemical symbol for the following.
    (a) a nonmetal in the second period
    (b) an alkali metal in the fifth period
    (c) the third-period halogen
    (d) an element that is a gas at 20°C and 1 atmosphere pressure

23. ■ Classify the following elements as metals, metalloids, or nonmetals: N, Na, Ni, Ne, and Np.

24. Here are symbols for five of the seven elements whose names begin with the letter B: B, Ba, Bk, Bi, and Br. Match each symbol with one of the descriptions below.
    (a) a radioactive element
    (b) a liquid at room temperature
    (c) a metalloid
    (d) an alkaline earth element
    (e) a Group 5A element

Molecular Formulas and Models

(See Exercise 2.5.)

25. A model of sulfuric acid is illustrated here. Write the molecular formula for sulfuric acid, and draw the structural formula. Describe the structure of the molecule. Is it flat? That is, are all the atoms in the plane of the paper? (Color code: sulfur atoms are yellow; oxygen atoms are red; and hydrogen atoms are white.)

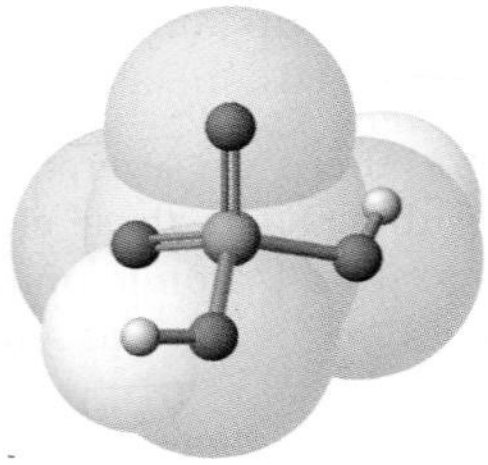

26. A model of the platinum-based chemotherapy agent cisplatin is given here. Write the molecular formula for the compound, and draw its structural formula.

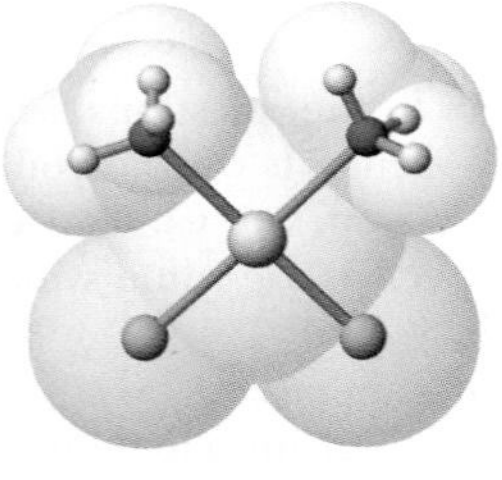

Ions and Ion Charges

(See Exercise 2.6, Figure 2.18, Table 2.4, and ChemistryNow Screens 2.18 and 2.19.)

27. What charges are most commonly observed for monatomic ions of the following elements?
    (a) magnesium (c) nickel
    (b) zinc (d) gallium

28. What charges are most commonly observed for monatomic ions of the following elements?
    (a) selenium (c) iron
    (b) fluorine (d) nitrogen

29. ■ Give the symbol, including the correct charge, for each of the following ions:
    (a) barium ion (e) sulfide ion
    (b) titanium(IV) ion (f) perchlorate ion
    (c) phosphate ion (g) cobalt(II) ion
    (d) hydrogen carbonate ion (h) sulfate ion

30. Give the symbol, including the correct charge, for each of the following ions:
    (a) permanganate ion (d) ammonium ion
    (b) nitrite ion (e) phosphate ion
    (c) dihydrogen phosphate ion (f) sulfite ion

31. When a potassium atom becomes a monatomic ion, how many electrons does it lose or gain? What noble gas atom has the same number of electrons as a potassium ion?

32. When oxygen and sulfur atoms become monatomic ions, how many electrons does each lose or gain? Which noble gas atom has the same number of electrons as an oxide ion? Which noble gas atom has the same number of electrons as a sulfide ion?

Ionic Compounds

(See Examples 2.4 and 2.5 and ChemistryNow Screen 2.20.)

33. ■ Predict the charges of the ions in an ionic compound containing the elements barium and bromine. Write the formula for the compound.

34. What are the charges of the ions in an ionic compound containing cobalt(III) and fluoride ions? Write the formula for the compound.

35. ■ For each of the following compounds, give the formula, charge, and the number of each ion that makes up the compound:
    (a) K_2S (d) $(NH_4)_3PO_4$
    (b) $CoSO_4$ (e) $Ca(ClO)_2$
    (c) $KMnO_4$ (f) $NaCH_3CO_2$

36. For each of the following compounds, give the formula, charge, and the number of each ion that makes up the compound:
    (a) $Mg(CH_3CO_2)_2$ (d) $Ti(SO_4)_2$
    (b) $Al(OH)_3$ (e) KH_2PO_4
    (c) $CuCO_3$ (f) $CaHPO_4$

37. Cobalt forms Co^{2+} and Co^{3+} ions. Write the formulas for the two cobalt oxides formed by these transition metal ions.

38. Platinum is a transition element and forms Pt^{2+} and Pt^{4+} ions. Write the formulas for the compounds of each of these ions with (a) chloride ions and (b) sulfide ions.

39. Which of the following are correct formulas for ionic compounds? For those that are not, give the correct formula.
    (a) $AlCl_2$ (c) Ga_2O_3
    (b) KF_2 (d) MgS

40. Which of the following are correct formulas for ionic compounds? For those that are not, give the correct formula.
    (a) Ca_2O (c) Fe_2O_5
    (b) $SrBr_2$ (d) Li_2O

▲ more challenging ■ in OWL Blue-numbered questions answered in Appendix O

Naming Ionic Compounds

(See Exercise 2.8 and ChemistryNow Screen 2.21.)

41. ■ Name each of the following ionic compounds:
(a) K_2S (c) $(NH_4)_3PO_4$
(b) $CoSO_4$ (d) $Ca(ClO)_2$

42. Name each of the following ionic compounds:
(a) $Ca(CH_3CO_2)_2$ (c) $Al(OH)_3$
(b) $Ni_3(PO_4)_2$ (d) KH_2PO_4

43. ■ Give the formula for each of the following ionic compounds:
(a) ammonium carbonate (d) aluminum phosphate
(b) calcium iodide (e) silver(I) acetate
(c) copper(II) bromide

44. Give the formula for each of the following ionic compounds:
(a) calcium hydrogen carbonate
(b) potassium permanganate
(c) magnesium perchlorate
(d) potassium hydrogen phosphate
(e) sodium sulfite

45. Write the formulas for the four ionic compounds that can be made by combining each of the cations Na^+ and Ba^{2+} with the anions CO_3^{2-} and I^-. Name each of the compounds.

46. Write the formulas for the four ionic compounds that can be made by combining the cations Mg^{2+} and Fe^{3+} with the anions PO_4^{3-} and NO_3^-. Name each compound formed.

Coulomb's Law

(See Equation 2.3, Figure 2.21, and ChemistryNow Screen 2.22.)

47. Sodium ions, Na^+, form ionic compounds with fluoride ions, F^-, and iodide ions, I^-. The radii of these ions are as follows: $Na^+ = 116$ pm; $F^- = 119$ pm; and $I^- = 206$ pm. In which ionic compound, NaF or NaI, are the forces of attraction between cation and anion stronger? Explain your answer.

48. ■ Consider the two ionic compounds NaCl and CaO. In which compound are the cation–anion attractive forces stronger? Explain your answer.

Naming Binary, Nonmetal Compounds

(See Exercise 2.10 and ChemistryNow Screens 2.24 and 2.25.)

49. ■ Name each of the following binary, nonionic compounds:
(a) NF_3 (b) HI (c) BI_3 (d) PF_5

50. Name each of the following binary, nonionic compounds:
(a) N_2O_5 (b) P_4S_3 (c) OF_2 (d) XeF_4

51. ■ Give the formula for each of the following compounds:
(a) sulfur dichloride
(b) dinitrogen pentaoxide
(c) silicon tetrachloride
(d) diboron trioxide (commonly called boric oxide)

52. Give the formula for each of the following compounds:
(a) bromine trifluoride
(b) xenon difluoride
(c) hydrazine
(d) diphosphorus tetrafluoride
(e) butane

Atoms and the Mole

(See Example 2.6, Exercises 2.11 and 2.12, and ChemistryNow Screens 2.25 and 2.26.)

53. ■ Calculate the mass, in grams, of each the following:
(a) 2.5 mol of aluminum (c) 0.015 mol of calcium
(b) 1.25×10^{-3} mol of iron (d) 653 mol of neon

54. Calculate the mass, in grams, of each the following:
(a) 4.24 mol of gold (c) 0.063 mol of platinum
(b) 15.6 mol of He (d) 3.63×10^{-4} mol of Pu

55. ■ Calculate the amount (moles) represented by each of the following:
(a) 127.08 g of Cu (c) 5.0 mg of americium
(b) 0.012 g of lithium (d) 6.75 g of Al

56. Calculate the amount (moles) represented by each of the following:
(a) 16.0 g of Na (c) 0.0034 g of platinum
(b) 0.876 g of tin (d) 0.983 g of Xe

57. ■ You are given 1.0 g samples of He, Fe, Li, Si, and C. Which sample contains the largest number of atoms? Which contains the smallest?

58. A semiconducting material is composed of 52 g of Ga, 9.5 g of Al, and 112 g of As. Which element has the largest number of atoms in the final mixture?

Molecules, Compounds, and the Mole

(See Example 2.7 and ChemistryNow Screens 2.27 and 2.28.)

59. ■ Calculate the molar mass of each of the following compounds:
(a) Fe_2O_3, iron(III) oxide
(b) BCl_3, boron trichloride
(c) $C_6H_8O_6$, ascorbic acid (vitamin C)

60. Calculate the molar mass of each of the following compounds:
(a) $Fe(C_6H_{11}O_7)_2$, iron(II) gluconate, a dietary supplement
(b) $CH_3CH_2CH_2CH_2SH$, butanethiol, has a skunk-like odor
(c) $C_{20}H_{24}N_2O_2$, quinine, used as an antimalarial drug

61. ■ Calculate the molar mass of each hydrated compound. Note that the water of hydration is included in the molar mass. (See Section 2.11.)
(a) $Ni(NO_3)_2 \cdot 6\ H_2O$
(b) $CuSO_4 \cdot 5\ H_2O$

62. Calculate the molar mass of each hydrated compound. Note that the water of hydration is included in the molar mass. (See Section 2.11.)
(a) $H_2C_2O_4 \cdot 2\ H_2O$
(b) $MgSO_4 \cdot 7\ H_2O$, Epsom salts

63. ■ What mass is represented by 0.0255 mol of each of the following compounds?
(a) C_3H_7OH, propanol, rubbing alcohol
(b) $C_{11}H_{16}O_2$, an antioxidant in foods, also known as BHA (butylated hydroxyanisole)
(c) $C_9H_8O_4$, aspirin
(d) $(CH_3)_2CO$, acetone, an important industrial solvent

64. Assume you have 0.123 mol of each of the following compounds. What mass of each is present?
(a) $C_{14}H_{10}O_4$, benzoyl peroxide, used in acne medications
(b) Dimethylglyoxime, used in the laboratory to test for nickel(II) ions

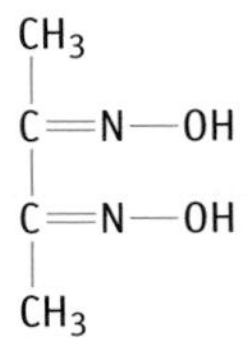

(c) The compound below is responsible for the "skunky" taste in poorly made beer.

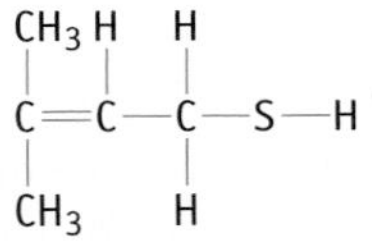

(d) DEET, a mosquito repellent

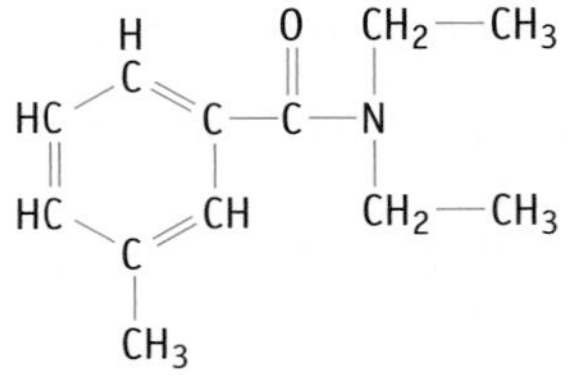

65. Sulfur trioxide, SO_3, is made industrially in enormous quantities by combining oxygen and sulfur dioxide, SO_2. What amount (moles) of SO_3 is represented by 1.00 kg of sulfur trioxide? How many molecules? How many sulfur atoms? How many oxygen atoms?

66. An Alka-Seltzer tablet contains 324 mg of aspirin ($C_9H_8O_4$), 1904 mg of $NaHCO_3$, and 1000. mg of citric acid ($H_3C_6H_5O_7$). (The last two compounds react with each other to provide the "fizz," bubbles of CO_2, when the tablet is put into water.)
(a) Calculate the amount (moles) of each substance in the tablet.
(b) If you take one tablet, how many molecules of aspirin are you consuming?

Percent Composition

(See Example 2.8 and ChemistryNow Screen 2.29.)

67. ■ Calculate the mass percent of each element in the following compounds:
(a) PbS, lead(II) sulfide, galena
(b) C_3H_8, propane
(c) $C_{10}H_{14}O$, carvone, found in caraway seed oil

68. Calculate the mass percent of each element in the following compounds:
(a) $C_8H_{10}N_2O_2$, caffeine
(b) $C_{10}H_{20}O$, menthol
(c) $CoCl_2 \cdot 6\ H_2O$

69. ■ Calculate the mass percent of copper in CuS, copper(II) sulfide. If you wish to obtain 10.0 g of copper metal from copper(II) sulfide, what mass of CuS (in grams) must you use?

70. Calculate the mass percent of titanium in the mineral ilmenite, $FeTiO_3$. What mass of ilmenite (in grams) is required if you wish to obtain 750 g of titanium?

Empirical and Molecular Formulas

(See Example 2.9 and ChemistryNow Screens 2.31 and 2.32.)

71. ■ Succinic acid occurs in fungi and lichens. Its empirical formula is $C_2H_3O_2$, and its molar mass is 118.1 g/mol. What is its molecular formula?

72. An organic compound has the empirical formula C_2H_4NO. If its molar mass is 116.1 g/mol, what is the molecular formula of the compound?

73. Complete the following table:

Empirical Formula	Molar Mass (g/mol)	Molecular Formula
(a) CH	26.0	______
(b) CHO	116.1	______
(c) ______	______	C_8H_{16}

74. Complete the following table:

Empirical Formula	Molar Mass (g/mol)	Molecular Formula
(a) $C_2H_3O_3$	150.0	______
(b) C_3H_8	44.1	______
(c) ______	______	B_4H_{10}

▲ more challenging ■ in OWL Blue-numbered questions answered in Appendix O

75. Acetylene is a colorless gas used as a fuel in welding torches, among other things. It is 92.26% C and 7.74% H. Its molar mass is 26.02 g/mol. What are the empirical and molecular formulas of acetylene?

76. ■ A large family of boron-hydrogen compounds has the general formula B_xH_y. One member of this family contains 88.5% B; the remainder is hydrogen. What is its empirical formula?

77. ■ Cumene is a hydrocarbon, a compound composed only of C and H. It is 89.94% carbon, and its molar mass is 120.2 g/mol. What are the empirical and molecular formulas of cumene?

78. In 2006, a Russian team discovered an interesting molecule they called "sulflower" because of its shape and because it was based on sulfur. It is composed of 57.17% S and 42.83% C and has a molar mass of 448.70 g/mol. Determine the empirical and molecular formulas of "sulflower."

79. ■ Mandelic acid is an organic acid composed of carbon (63.15%), hydrogen (5.30%), and oxygen (31.55%). Its molar mass is 152.14 g/mol. Determine the empirical and molecular formulas of the acid.

80. Nicotine, a poisonous compound found in tobacco leaves, is 74.0% C, 8.65% H, and 17.35% N. Its molar mass is 162 g/mol. What are the empirical and molecular formulas of nicotine?

Determining Formulas from Mass Data

(See Examples 2.10 and 2.11 and ChemistryNow Screens 2.30 and 2.31.)

81. ■ A new compound containing xenon and fluorine was isolated by shining sunlight on a mixture of Xe (0.526 g) and excess F_2 gas. If you isolate 0.678 g of the new compound, what is its empirical formula?

82. Elemental sulfur (1.256 g) is combined with fluorine, F_2, to give a compound with the formula SF_x, a very stable, colorless gas. If you have isolated 5.722 g of SF_x, what is the value of x?

83. Zinc metal (2.50 g) combines with 9.70 g of iodine to produce zinc iodide, Zn_xI_y. What is the formula of this ionic compound?

84. You combine 1.25 g of germanium, Ge, with excess chlorine, Cl_2. The mass of product, Ge_xCl_y, is 3.69 g. What is the formula of the product, Ge_xCl_y?

General Questions

These questions are not designed as to type or location in the chapter. They may combine several concepts.

85. Fill in the blanks in the table (one column per element).

Symbol	^{58}Ni	^{33}S	______	______
Number of protons	______	______	10	______
Number of neutrons	______	______	10	30
Number of electrons in the neutral atom	______	______	______	25
Name of element	______	______	______	______

86. ■ Potassium has three naturally occurring isotopes (^{39}K, ^{40}K, and ^{41}K), but ^{40}K has a very low natural abundance. Which of the other two isotopes is the more abundant? Briefly explain your answer.

87. Crossword Puzzle: In the 2 × 2 box shown here, each answer must be correct four ways: horizontally, vertically, diagonally, and by itself. Instead of words, use symbols of elements. When the puzzle is complete, the four spaces will contain the overlapping symbols of 10 elements. There is only one correct solution.

1	2
3	4

Horizontal
1–2: Two-letter symbol for a metal used in ancient times
3–4: Two-letter symbol for a metal that burns in air and is found in Group 5A

Vertical
1–3: Two-letter symbol for a metalloid
2–4: Two-letter symbol for a metal used in U.S. coins

Single squares: All one-letter symbols
1: A colorful nonmetal
2: Colorless, gaseous nonmetal
3: An element that makes fireworks green
4: An element that has medicinal uses

Diagonal
1–4: Two-letter symbol for an element used in electronics
2–3: Two-letter symbol for a metal used with Zr to make wires for superconducting magnets

This puzzle first appeared in *Chemical & Engineering News*, p. 86, December 14, 1987 (submitted by S. J. Cyvin) and in *Chem Matters*, October 1988.

88. ■ *The abundance of the elements in the solar system from H to Zn.* The chart shows a general decline in abundance with increasing mass among the first 30 elements. The decline continues beyond zinc. (Notice that the scale on the vertical axis is logarithmic, that is, it progresses in powers of 10. The abundance of nitrogen, for example, is 1/10,000 ($1/10^4$) of the abundance of hydrogen. All abundances are plotted as the number of atoms per 10^{12} atoms of H. (The fact that the abundances of Li, Be, and B, as well as those of the elements near Fe, do not follow the general decline is a consequence of the way that elements are synthesized in stars.)

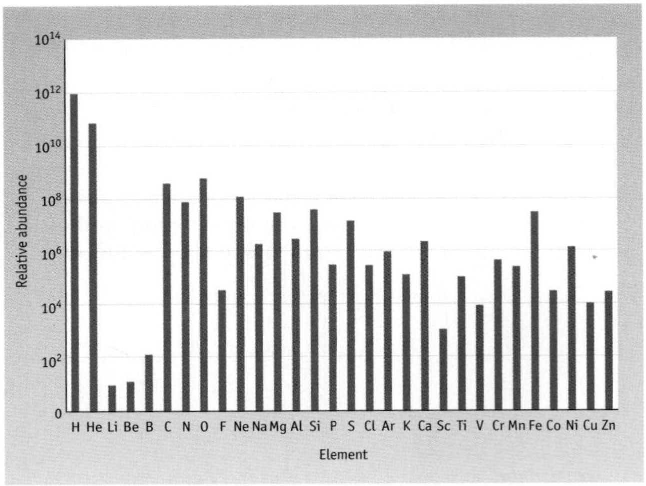

(a) What is the most abundant main group metal?
(b) What is the most abundant nonmetal?
(c) What is the most abundant metalloid?
(d) Which of the transition elements is most abundant?
(e) Which halogens are included on this plot, and which is the most abundant?

89. Copper atoms
(a) ■ What is the average mass of one copper atom?
(b) Students in a college computer science class once sued the college because they were asked to calculate the cost of one atom and could not do it. But you are in a chemistry course, and you can do this. (See E. Felsenthal, *Wall Street Journal,* May 9, 1995.) If the cost of 2.0 mm diameter copper wire (99.999% pure) is currently $41.70 for 7.0 g, what is the cost of one copper atom?

90. Which of the following is impossible?
(a) silver foil that is 1.2×10^{-4} m thick
(b) a sample of potassium that contains 1.784×10^{24} atoms
(c) a gold coin of mass 1.23×10^{-3} kg
(d) 3.43×10^{-27} mol of S_8 molecules

91. Reviewing the periodic table.
(a) Name the element in Group 2A and the fifth period.
(b) Name the element in the fifth period and Group 4B.
(c) Which element is in the second period in Group 4A?
(d) Which element is in the fourth period in Group 5A?
(e) Which halogen is in the fifth period?
(f) Which alkaline earth element is in the third period?
(g) Which noble gas element is in the fourth period?
(h) Name the nonmetal in Group 6A and the third period.
(i) Name a metalloid in the fourth period.

92. ■ Give two examples of nonmetallic elements that have allotropes. Name those elements, and describe the allotropes of each.

93. ■ In each case, decide which represents more mass:
(a) 0.5 mol of Na, 0.5 mol of Si, or 0.25 mol of U
(b) 9.0 g of Na, 0.50 mol of Na, or 1.2×10^{22} atoms of Na
(c) 10 atoms of Fe or 10 atoms of K

94. The recommended daily allowance (RDA) of iron in your diet is 15 mg. How many moles is this? How many atoms?

95. Put the following elements in order from smallest to largest mass:
(a) 3.79×10^{24} atoms Fe
(b) 19.921 mol H_2
(c) 8.576 mol C
(d) 7.4 mol Si
(e) 9.221 mol Na
(f) 4.07×10^{24} atoms Al
(g) 9.2 mol Cl_2

96. ■ ▲ When a sample of phosphorus burns in air, the compound P_4O_{10} forms. One experiment showed that 0.744 g of phosphorus formed 1.704 g of P_4O_{10}. Use this information to determine the ratio of the atomic weights of phosphorus and oxygen (mass P/mass O). If the atomic weight of oxygen is assumed to be 16.000 u, calculate the atomic weight of phosphorus.

97. ▲ Although carbon-12 is now used as the standard for atomic weights, this has not always been the case. Early attempts at classification used hydrogen as the standard, with the weight of hydrogen being set equal to 1.0000 u. Later attempts defined atomic weights using oxygen (with a weight of 16.0000). In each instance, the atomic weights of the other elements were defined relative to these masses. (To answer this question, you need more precise data on current atomic weights: H, 1.00794 u; O, 15.9994 u.)
(a) If H = 1.0000 u was used as a standard for atomic weights, what would the atomic weight of oxygen be? What would be the value of Avogadro's number under these circumstances?
(b) Assuming the standard is O = 16.0000, determine the value for the atomic weight of hydrogen and the value of Avogadro's number.

98. ■ A reagent occasionally used in chemical synthesis is sodium–potassium alloy. (Alloys are mixtures of metals, and Na-K has the interesting property that it is a liquid.) One formulation of the alloy (the one that melts at the lowest temperature) contains 68 atom percent K; that is, out of every 100 atoms, 68 are K and 32 are Na. What is the mass percent of potassium in sodium–potassium alloy?

99. Write formulas for all of the compounds that can be made by combining the cations NH_4^+ and Ni^{2+} with the anions CO_3^{2-} and SO_4^{2-}.

100. How many electrons are in a strontium atom (Sr)? Does an atom of Sr gain or lose electrons when forming an ion? How many electrons are gained or lost by the atom? When Sr forms an ion, the ion has the same number of electrons as which one of the noble gases?

101. Which of the following compounds has the highest mass percent of chlorine?
(a) BCl_3
(b) $AsCl_3$
(c) $GaCl_3$
(d) $AlCl_3$
(e) PCl_3

102. Which of the following samples has the largest number of ions?
(a) 1.0 g of $BeCl_2$
(b) 1.0 g of $MgCl_2$
(c) 1.0 g of CaS
(d) 1.0 g of $SrCO_3$
(e) 1.0 g of $BaSO_4$

103. The structure of one of the bases in DNA, adenine, is shown here. Which represents the greater mass: 40.0 g of adenine or 3.0×10^{23} molecules of the compound?

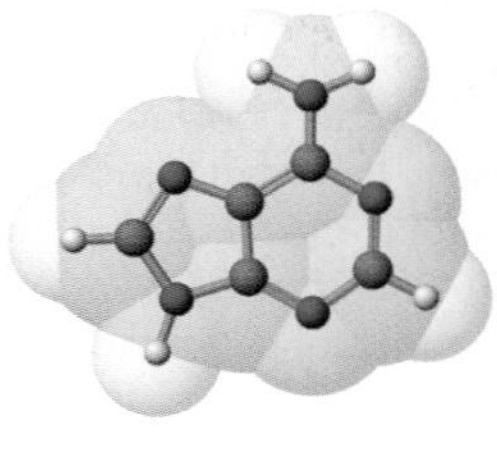

104. Ionic and molecular compounds of the halogens.
(a) What are the names of BaF_2, $SiCl_4$, and $NiBr_2$?
(b) Which of the compounds in part (a) are ionic, and which are molecular?
(c) Which has the larger mass, 0.50 mol of BaF_2, 0.50 mol of $SiCl_4$, or 1.0 mol of $NiBr_2$?

105. ■ A drop of water has a volume of about 0.050 mL. How many molecules of water are in a drop of water? (Assume water has a density of 1.00 g/cm³.)

106. Capsaicin, the compound that gives the hot taste to chili peppers, has the formula $C_{18}H_{27}NO_3$.
(a) Calculate its molar mass.
(b) If you eat 55 mg of capsaicin, what amount (moles) have you consumed?
(c) Calculate the mass percent of each element in the compound.
(d) What mass of carbon (in milligrams) is there in 55 mg of capsaicin?

107. Calculate the molar mass and the mass percent of each element in the blue solid compound $Cu(NH_3)_4SO_4 \cdot H_2O$. What is the mass of copper and the mass of water in 10.5 g of the compound?

108. Write the molecular formula, and calculate the molar mass for each of the molecules shown here. Which has the larger percentage of carbon? Of oxygen?
(a) Ethylene glycol (used in antifreeze)

(b) Dihydroxyacetone (used in artificial tanning lotions)

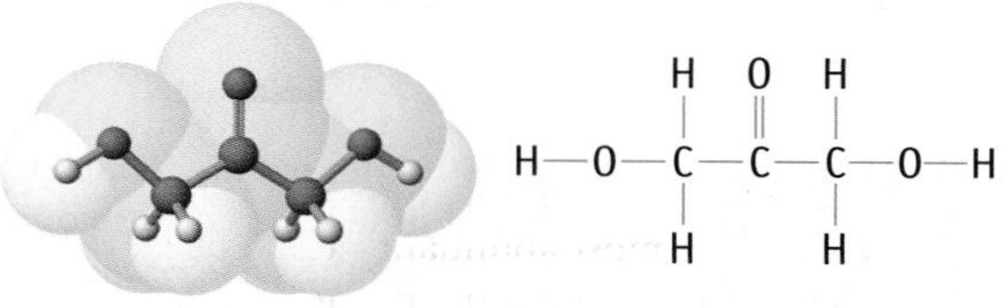

(c) Ascorbic acid, commonly known as vitamin C

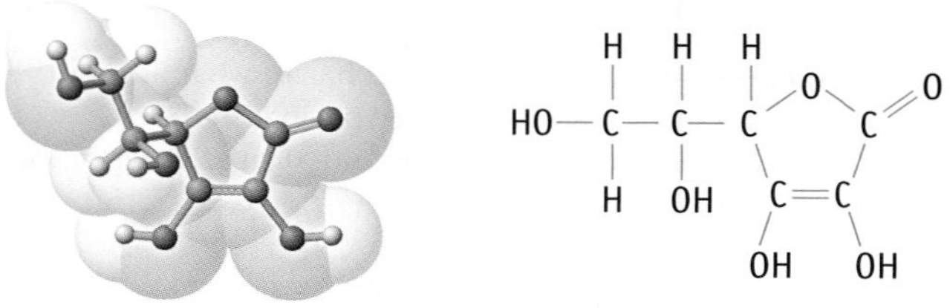

109. Malic acid, an organic acid found in apples, contains C, H, and O in the following ratios: $C_1H_{1.50}O_{1.25}$. What is the empirical formula of malic acid?

110. Your doctor has diagnosed you as being anemic—that is, as having too little iron in your blood. At the drugstore, you find two iron-containing dietary supplements: one with iron(II) sulfate, $FeSO_4$, and the other with iron(II) gluconate, $Fe(C_6H_{11}O_7)_2$. If you take 100. mg of each compound, which will deliver more atoms of iron?

111. A compound composed of iron and carbon monoxide, $Fe_x(CO)_y$, is 30.70% iron. What is the empirical formula for the compound?

112. Ma huang, an extract from the ephedra species of plants, contains ephedrine. The Chinese have used this herb for more than 5000 years to treat asthma. More recently, the substance has been used in diet pills that can be purchased over the counter in herbal medicine shops. However, very serious concerns have been raised regarding these pills following reports that their use led to serious heart problems.
(a) Write the molecular formula for ephedrine, and calculate its molar mass.
(b) What is the weight percent of carbon in ephedrine?
(c) Calculate the amount (moles) of ephedrine in a 0.125 g sample.
(d) How many molecules of ephedrine are there in 0.125 g? How many C atoms?

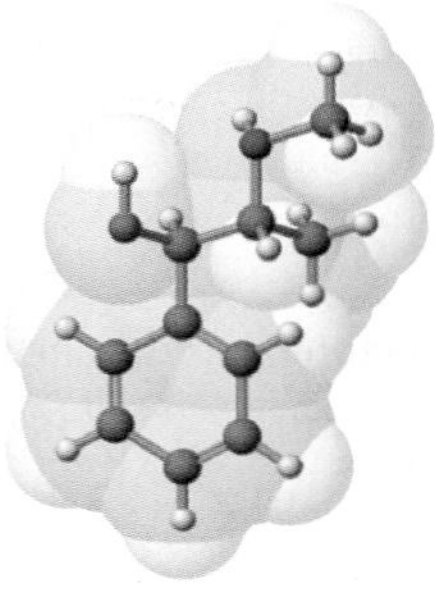

113. Saccharin is more than 300 times sweeter than sugar. It was first made in 1897, a time when it was common practice for chemists to record the taste of any new substances they synthesized.
(a) Write the molecular formula for the compound, and draw its structural formula. (S atoms are yellow.)
(b) If you ingest 125 mg of saccharin, what amount (moles) of saccharin have you ingested?
(c) What mass of sulfur is contained in 125 mg of saccharin?

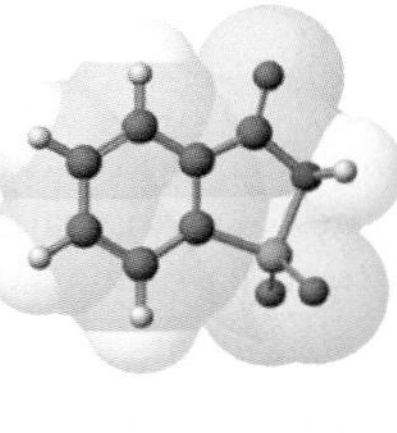

114. Name each of the following compounds, and tell which ones are best described as ionic:
(a) ClF_3
(b) NCl_3
(c) $SrSO_4$
(d) $Ca(NO_3)_2$
(e) XeF_4
(f) OF_2
(g) KI
(h) Al_2S_3
(i) PCl_3
(j) K_3PO_4

115. Write the formula for each of the following compounds, and tell which ones are best described as ionic:
(a) sodium hypochlorite
(b) boron triiodide
(c) aluminum perchlorate
(d) calcium acetate
(e) potassium permanganate
(f) ammonium sulfite
(g) potassium dihydrogen phosphate
(h) disulfur dichloride
(i) chlorine trifluoride
(j) phosphorus trifluoride

116. Complete the table by placing symbols, formulas, and names in the blanks.

Cation	Anion	Name	Formula
______	______	ammonium bromide	______
Ba^{2+}	______	______________	BaS
______	Cl^-	iron(II) chloride	______
______	F^-	______________	PbF_2
Al^{3+}	CO_3^{2-}	______________	______
______	______	iron(III) oxide	______

117. Empirical and molecular formulas.
(a) Fluorocarbonyl hypofluorite is composed of 14.6% C, 39.0% O, and 46.3% F. If the molar mass of the compound is 82 g/mol, determine the empirical and molecular formulas of the compound.
(b) Azulene, a beautiful blue hydrocarbon, is 93.71% C and has a molar mass of 128.16 g/mol. What are the empirical and molecular formulas of azulene?

118. Cacodyl, a compound containing arsenic, was reported in 1842 by the German chemist Robert Wilhelm Bunsen. It has an almost intolerable garlic-like odor. Its molar mass is 210 g/mol, and it is 22.88% C, 5.76% H, and 71.36% As. Determine its empirical and molecular formulas.

119. The action of bacteria on meat and fish produces a compound called cadaverine. As its name and origin imply, it stinks! (It is also present in bad breath and adds to the odor of urine.) It is 58.77% C, 13.81% H, and 27.40% N. Its molar mass is 102.2 g/mol. Determine the molecular formula of cadaverine.

120. ■ ▲ Transition metals can combine with carbon monoxide (CO) to form compounds such as $Fe_x(CO)_y$ (Study Question 2.111). Assume that you combine 0.125 g of nickel with CO and isolate 0.364 g of $Ni(CO)_x$. What is the value of x?

121. ▲ A major oil company has used a gasoline additive called MMT to boost the octane rating of its gasoline. What is the empirical formula of MMT if it is 49.5% C, 3.2% H, 22.0% O, and 25.2% Mn?

▲ more challenging ■ in OWL Blue-numbered questions answered in Appendix O

122. ▲ Elemental phosphorus is made by heating calcium phosphate with carbon and sand in an electric furnace. What is the mass percent of phosphorus in calcium phosphate? Use this value to calculate the mass of calcium phosphate (in kilograms) that must be used to produce 15.0 kg of phosphorus.

123. ▲ Chromium is obtained by heating chromium(III) oxide with carbon. Calculate the mass percent of chromium in the oxide, and then use this value to calculate the quantity of Cr_2O_3 required to produce 850 kg of chromium metal.

124. ▲ Stibnite, Sb_2S_3, is a dark gray mineral from which antimony metal is obtained. What is the mass percent of antimony in the sulfide? If you have 1.00 kg of an ore that contains 10.6% antimony, what mass of Sb_2S_3 (in grams) is in the ore?

125. ▲ Direct reaction of iodine (I_2) and chlorine (Cl_2) produces an iodine chloride, I_xCl_y, a bright yellow solid. If you completely consume 0.678 g of I_2 (when reacted with excess Cl_2) and produce 1.246 g of I_xCl_y, what is the empirical formula of the compound? A later experiment showed that the molar mass of I_xCl_y was 467 g/mol. What is the molecular formula of the compound?

126. ▲ In a reaction, 2.04 g of vanadium combined with 1.93 g of sulfur to give a pure compound. What is the empirical formula of the product?

127. ▲ Iron pyrite, often called "fool's gold," has the formula FeS_2. If you could convert 15.8 kg of iron pyrite to iron metal, what mass of the metal would you obtain?

128. Which of the following statements about 57.1 g of octane, C_8H_{18}, is (are) *not* true?
(a) 57.1 g is 0.500 mol of octane.
(b) The compound is 84.1% C by weight.
(c) The empirical formula of the compound is C_4H_9.
(d) 57.1 g of octane contains 28.0 g of hydrogen atoms.

129. The formula of barium molybdate is $BaMoO_4$. Which of the following is the formula of sodium molybdate?
(a) Na_4MoO (c) Na_2MoO_3 (e) Na_4MoO_4
(b) $NaMoO$ (d) Na_2MoO_4

130. ▲ A metal M forms a compound with the formula MCl_4. If the compound is 74.75% chlorine, what is the identity of M?

131. Pepto-Bismol, which helps provide soothing relief for an upset stomach, contains 300. mg of bismuth subsalicylate, $C_{21}H_{15}Bi_3O_{12}$, per tablet. If you take two tablets for your stomach distress, what amount (in moles) of the "active ingredient" are you taking? What mass of Bi are you consuming in two tablets?

132. ▲ The weight percent of oxygen in an oxide that has the formula MO_2 is 15.2%. What is the molar mass of this compound? What element or elements are possible for M?

133. The mass of 2.50 mol of a compound with the formula ECl_4, in which E is a nonmetallic element, is 385 g. What is the molar mass of ECl_4? What is the identity of E?

134. ▲ The elements A and Z combine to produce two different compounds: A_2Z_3 and AZ_2. If 0.15 mol of A_2Z_3 has a mass of 15.9 g and 0.15 mol of AZ_2 has a mass of 9.3 g, what are the atomic masses of A and Z?

135. ▲ Polystyrene can be prepared by heating styrene with tribromobenzoyl peroxide in the absence of air. A sample prepared by this method has the empirical formula $Br_3C_6H_3(C_8H_8)_n$, where the value of n can vary from sample to sample. If one sample has 10.46% Br, what is the value of n?

136. A sample of hemoglobin is found to be 0.335% iron. If hemoglobin contains one iron atom per molecule, what is the molar mass of hemoglobin? What is the molar mass if there are four iron atoms per molecule?

137. ▲ Consider an atom of ^{64}Zn.
(a) Calculate the density of the nucleus in grams per cubic centimeter, knowing that the nuclear radius is 4.8×10^{-6} nm and the mass of the ^{64}Zn atom is 1.06×10^{-22} g. (Recall that the volume of a sphere is $[4/3]\pi r^3$.)
(b) Calculate the density of the space occupied by the electrons in the zinc atom, given that the atomic radius is 0.125 nm and the electron mass is 9.11×10^{-28} g.
(c) Having calculated these densities, what statement can you make about the relative densities of the parts of the atom?

138. ▲ Estimating the radius of a lead atom.
(a) You are given a cube of lead that is 1.000 cm on each side. The density of lead is 11.35 g/cm^3. How many atoms of lead are in the sample?
(b) Atoms are spherical; therefore, the lead atoms in this sample cannot fill all the available space. As an approximation, assume that 60% of the space of the cube is filled with spherical lead atoms. Calculate the volume of one lead atom from this information. From the calculated volume (V) and the formula $(4/3)\pi r^3$ for the volume of a sphere, estimate the radius (r) of a lead atom.

139. A piece of nickel foil, 0.550 mm thick and 1.25 cm square, is allowed to react with fluorine, F_2, to give a nickel fluoride.
(a) How many moles of nickel foil were used? (The density of nickel is 8.902 g/cm^3.)
(b) If you isolate 1.261 g of the nickel fluoride, what is its formula?
(c) What is its complete name?

140. ▲ Uranium is used as a fuel, primarily in the form of uranium(IV) oxide, in nuclear power plants. This question considers some uranium chemistry.

(a) A small sample of uranium metal (0.169 g) is heated to between 800 and 900°C in air to give 0.199 g of a dark green oxide, U_xO_y. How many moles of uranium metal were used? What is the empirical formula of the oxide, U_xO_y? What is the name of the oxide? How many moles of U_xO_y must have been obtained?

(b) The naturally occurring isotopes of uranium are ^{234}U, ^{235}U, and ^{238}U. Knowing that uranium's atomic weight is 238.02 g/mol, which isotope must be the most abundant?

(c) If the hydrated compound $UO_2(NO_3)_2 \cdot z\ H_2O$ is heated gently, the water of hydration is lost. If you have 0.865 g of the hydrated compound and obtain 0.679 g of $UO_2(NO_3)_2$ on heating, how many waters of hydration are in each formula unit of the original compound? (The oxide U_xO_y is obtained if the hydrate is heated to temperatures over 800°C in the air.)

In the Laboratory

141. ■ If Epsom salt, $MgSO_4 \cdot x\ H_2O$, is heated to 250°C, all the water of hydration is lost. On heating a 1.687-g sample of the hydrate, 0.824 g of $MgSO_4$ remains. How many molecules of water occur per formula unit of $MgSO_4$?

142. The "alum" used in cooking is potassium aluminum sulfate hydrate, $KAl(SO_4)_2 \cdot x\ H_2O$. To find the value of x, you can heat a sample of the compound to drive off all of the water and leave only $KAl(SO_4)_2$. Assume you heat 4.74 g of the hydrated compound and that the sample loses 2.16 g of water. What is the value of x?

143. ■ In an experiment, you need 0.125 mol of sodium metal. Sodium can be cut easily with a knife (Figure 2.6), so if you cut out a block of sodium, what should the volume of the block be in cubic centimeters? If you cut a perfect cube, what is the length of the edge of the cube? (The density of sodium is 0.97 g/cm^3.)

144. Mass spectrometric analysis showed that there are four isotopes of an unknown element having the following masses and abundances:

Isotope	Mass Number	Isotope Mass	Abundance (%)
1	136	135.9090	0.193
2	138	137.9057	0.250
3	140	139.9053	88.48
4	142	141.9090	11.07

Three elements in the periodic table that have atomic weights near these values are lanthanum (La), atomic number 57, atomic weight 139.9055; cerium (Ce), atomic number 58, atomic weight 140.115; and praeseodymium (Pr), atomic number 59, atomic weight 140.9076. Using the data above, calculate the atomic weight, and identify the element if possible.

145. ▲ Most standard analytical balances can measure accurately to the nearest 0.0001 g. Assume you have weighed out a 2.0000-g sample of carbon. How many atoms are in this sample? Assuming the indicated accuracy of the measurement, what is the largest number of atoms that can be present in the sample?

146. ▲ When analyzed, an unknown compound gave these experimental results: C, 54.0%; H, 6.00%; and O, 40.0%. Four different students used these values to calculate the empirical formulas shown here. Which answer is correct? Why did some students not get the correct answer?

(a) $C_4H_5O_2$
(b) $C_5H_7O_3$
(c) $C_7H_{10}O_4$
(d) $C_9H_{12}O_5$

147. ▲ Two general chemistry students working together in the lab weigh out 0.832 g of $CaCl_2 \cdot 2\ H_2O$ into a crucible. After heating the sample for a short time and allowing the crucible to cool, the students determine that the sample has a mass of 0.739 g. They then do a quick calculation. On the basis of this calculation, what should they do next?

(a) Congratulate themselves on a job well done.
(b) Assume the bottle of $CaCl_2 \cdot 2\ H_2O$ was mislabeled; it actually contained something different.
(c) Heat the crucible again, and then reweigh it.

148. The mass spectrum of CH_3Cl is illustrated here. You know that carbon has two stable isotopes, ^{12}C and ^{13}C with relative abundances of 98.9% and 1.1%, respectively, and chlorine has two isotopes, ^{35}Cl and ^{37}Cl with abundances of 75.77% and 24.23%, respectively.

(a) What molecular species gives rise to the lines at m/Z of 50 and 52? Why is the line at 52 about 1/3 the height of the line at 50?
(b) What species might be responsible for the line at $m/Z = 51$?

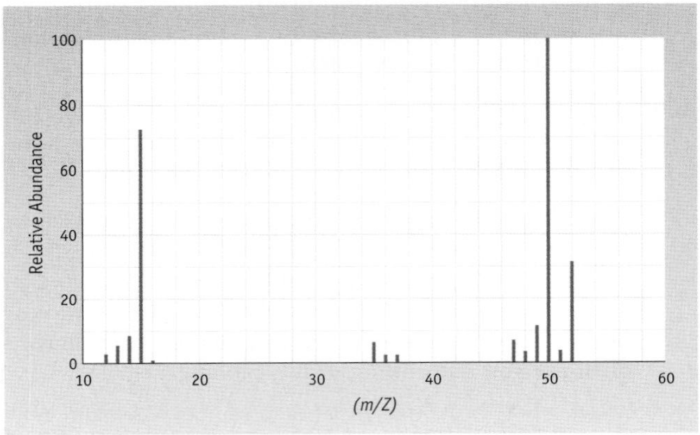

Summary and Conceptual Questions

The following questions may use concepts from this and the previous chapter.

149. ▲ Identify, from the list below, the information needed to calculate the number of atoms in 1.00 cm^3 of iron. Outline the procedure used in this calculation.
   (a) the structure of solid iron
   (b) the molar mass of iron
   (c) Avogadro's number
   (d) the density of iron
   (e) the temperature
   (f) iron's atomic number
   (g) the number of iron isotopes

150. Consider the plot of relative element abundances on page 106. Is there a relationship between abundance and atomic number? Is there any difference between the relative abundance of an element of even atomic number and the relative abundance of an element of odd atomic number?

151. The photo here depicts what happens when a coil of magnesium ribbon and a few calcium chips are placed in water.
   (a) Based on their relative reactivities, what might you expect to see when barium, another Group 2A element, is placed in water?
   (b) Give the period in which each element (Mg, Ca, and Ba) is found; what correlation do you think you might find between the reactivity of these elements and their positions in the periodic table?

Charles D. Winters

Magnesium *(left)* and calcium *(right)* in water.

152. A jar contains some number of jelly beans. To find out precisely how many are in the jar, you could dump them out and count them. How could you estimate their number without counting each one? (Chemists need to do just this kind of "bean counting" when they work with atoms and molecules. They are too small to count one by one, so they have worked out other methods to "count atoms.")

Charles D. Winters

How many jelly beans are in the jar?

153. Cobalt(II) chloride hexahydrate, dissolves readily in water to give a red solution. If we use this solution as an "ink," we can write secret messages on paper. The writing is not visible when the water evaporates from the paper. When the paper is heated, however, the message can be read. Explain the chemistry behind this observation.

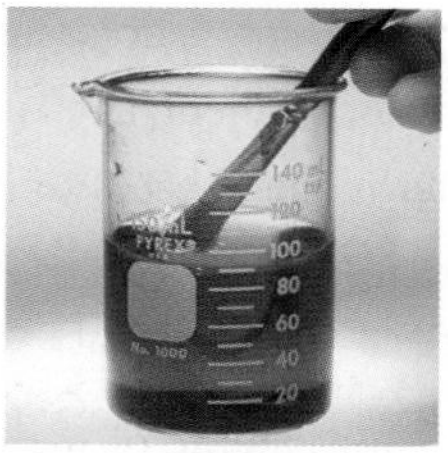

Charles D. Winters

A solution of $CoCl_2 \cdot 6\ H_2O$.

Charles D. Winters

Using the secret ink to write on paper.

Charles D. Winters

Heating the paper reveals the writing.

3 Chemical Reactions

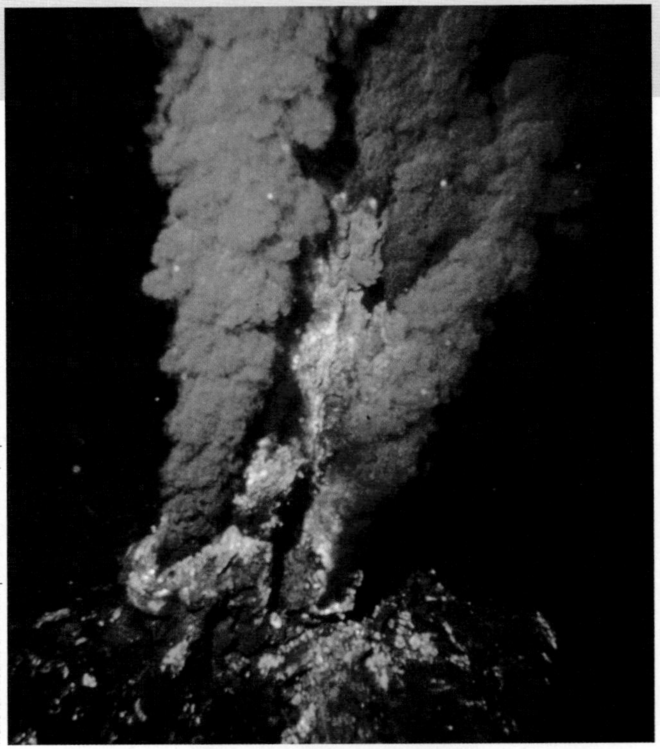

National Oceanic and Atmospheric Adminstration/Department of Commerce

A "black smoker" deep in the Pacific Ocean along the East Pacific Rise.

Black Smokers

In 1977, scientists were exploring the junction of two of the tectonic plates that form the floor of the Pacific Ocean. There they found thermal springs gushing a hot, black soup of minerals. Seawater seeps into cracks in the ocean floor, and, as it sinks deeper into the earth's crust, the water is superheated to between 300 and 400 °C by the magma of the earth's core. This superhot water dissolves minerals in the crust and is pushed back to the surface. When this hot water, now laden with dissolved metal cations and rich in anions such as sulfide and sulfate, gushes through the surface, it cools, and metal sulfates, such as calcium sulfate, and sulfides—such as those of copper, manganese, iron, zinc, and nickel—precipitate. Many metal sulfides are black, and the plume of material coming from the sea bottom looks like black "smoke"; thus, the vents have been called "black smokers." The solid sulfides and other minerals settle around the edges of the vent on the sea floor and eventually form a "chimney" of precipitated minerals.

Question:

1. Write balanced, net ionic equations for the reactions of Fe^{2+} and Bi^{3+} with H_2S and for Ca^{2+} with sulfate ions.

Answer to this question is in Appendix Q.

Chapter Goals

See Chapter Goals Revisited (page 151) for Study Questions keyed to these goals and assignable in OWL.

- Balance equations for simple chemical reactions.
- Understand the nature and characteristics of chemical equilibria.
- Understand the nature of ionic substances dissolved in water.
- Recognize common acids and bases, and understand their behavior in aqueous solution.
- Recognize the common types of reactions in aqueous solution.
- Write chemical equations for the common types of reactions in aqueous solution.
- Recognize common oxidizing and reducing agents, and identify oxidation-reduction reactions.

Chapter Outline

Chemical reactions are the heart of chemistry. We begin a chemical reaction with one set of materials and end up with different materials. Just reading this sentence involves an untold number of chemical reactions in your body. Indeed, every activity of living things depends on carefully regulated chemical reactions. Our objective in this chapter is to introduce you to the symbolism used to represent chemical reactions and to describe various types of common chemical reactions.

Chemistry Now™

Throughout the text this icon introduces an opportunity for self-study or to explore interactive tutorials by signing in at **www.thomsonedu.com/login**.

3.1 Introduction to Chemical Equations

When a stream of chlorine gas, Cl_2, is directed onto solid phosphorus, P_4, the mixture bursts into flame, and a chemical reaction produces liquid phosphorus trichloride, PCl_3 (Figure 3.1). We can depict this reaction using a **balanced chemical equation**È

$$\underbrace{P_4(s) + 6\,Cl_2(g)}_{\text{Reactants}} \longrightarrow \underbrace{4\,PCl_3(\ell)}_{\text{Product}}$$

$$\underbrace{P_4(s) + 6\,Cl_2(g)}_{\text{REACTANTS}} \longrightarrow \underbrace{4\,PCl_3(\ell)}_{\text{PRODUCT}}$$

FIGURE 3.1 Reaction of solid white phosphorus with chlorine gas. The product is liquid phosphorus trichloride.

Historical Perspectives — Antoine Laurent Lavoisier, 1743–1794

On Monday, August 7, 1774, the Englishman Joseph Priestley (1733–1804) isolated oxygen. (The Swedish chemist Carl Scheele [1742–1786] also discovered the element, perhaps in 1773 or earlier.) Priestley heated solid mercury(II) oxide, HgO, causing the oxide to decompose to mercury and oxygen.

$$2\ HgO(s) \longrightarrow 2\ Hg(\ell) + O_2(g)$$

He did not immediately understand the significance of the discovery, but he mentioned it to the French chemist Antoine Lavoisier in October, 1774. One of Lavoisier's contributions to science was his recognition of the importance of exact scientific measurements and of carefully planned experiments, and he applied these methods to the study of oxygen. From this work, Lavoisier proposed that oxygen was an element, that it was one of the constituents of the compound water, and that burning involved a reaction with oxygen. He also mistakenly came to believe Priestley's gas was present in all acids, and so he named it "oxygen," from the Greek words meaning "to form an acid."

In other experiments, Lavoisier observed that the heat produced by a guinea pig when exhaling a given amount of carbon dioxide is similar to the quantity of heat produced by burning carbon to give the same amount of carbon dioxide. From these and other experiments he concluded that, "Respiration is a combustion, slow it is true, but otherwise perfectly similar to that of charcoal." Although he did not understand the details of the process, this was an important step in the development of biochemistry.

Lavoisier was a prodigious scientist, and the principles of naming chemical substances that he introduced are still in use today. Further, he wrote a textbook in which he applied the principles of the conservation of matter to chemistry, and he used the idea to write early versions of chemical equations.

Because Lavoisier was an aristocrat, he came under suspicion during the Reign of Terror of the French Revolution. He was an investor in the Ferme Générale, the infamous tax-collecting organization in 18th-century France. Tobacco was a monopoly product of the Ferme Générale, and it was common to cheat the purchaser by adding water to the tobacco, a practice that Lavoisier opposed. Nonetheless, because of his involvement with the Ferme, his career was cut short by the guillotine on May 8, 1794, on the charge of "adding water to the people's tobacco."

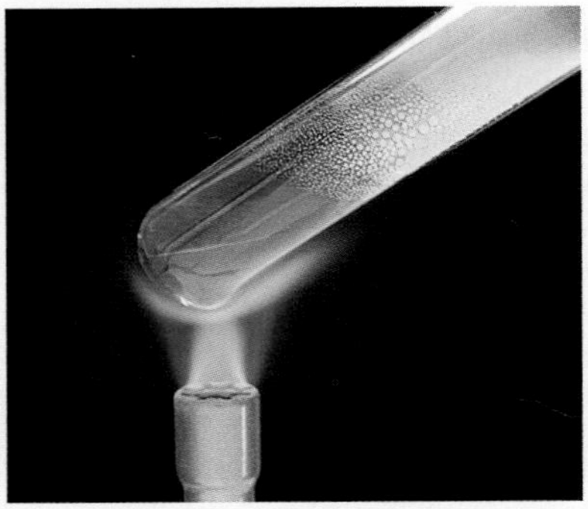
Charles D. Winters

The decomposition of red mercury(II) oxide. The decomposition reaction gives mercury metal and oxygen gas. The mercury is seen as a film on the surface of the test tube.

Lavoisier and his wife, as painted in 1788 by Jacques-Louis David. Lavoisier was then 45, and his wife, Marie Anne Pierrette Paulze, was 30. (The Metropolitan Museum of Art, Purchase, Mr. and Mrs. Charles Wrightsman gift, in honor of Everett Fahy, 1997. Photograph © 1989 The Metropolitan Museum of Art.)

■ **Information from Chemical Equations** The same number of atoms must exist after a reaction as before it takes place. However, these atoms are arranged differently. In the phosphorus/chlorine reaction, for example, the P atoms were in the form of P_4 molecules before reaction but appear in the PCl_3 molecules after reaction.

In a chemical equation, the formulas for the **reactants** (the substances combined in the reaction) are written to the left of the arrow, and the formulas of the **products** (the substances produced) are written to the right of the arrow. The physical states of reactants and products can also be indicated. The symbol (s) indicates a solid, (g) a gas, and (ℓ) a liquid. A substance dissolved in water, that is, an *aqueous* solution of a substance, is indicated by (aq).

In the 18th century, the French scientist Antoine Lavoisier (1743–1794) introduced the **law of conservation of matter**, which states that *matter can neither be created nor destroyed.* This means that if the total mass of reactants is 10 g, and if the reaction completely converts reactants to products, you must end up with 10 g of products. This also means that if 1000 atoms of a particular element are contained in the reactants, then those 1000 atoms must appear in the products in some fashion.

When applied to the reaction of phosphorus and chlorine, the law of conservation of matter tells us that 1 molecule of phosphorus, P_4 (with 4 phosphorus atoms) and 6 diatomic molecules of Cl_2 (with 12 atoms of Cl) are required to produce

$$2\ Fe(s) + 3\ Cl_2(g) \longrightarrow 2\ FeCl_3(s)$$

REACTANTS — PRODUCT

FIGURE 3.2 The reaction of iron and chlorine. Here, hot iron gauze is inserted into a flask containing chlorine gas. The heat from the reaction causes the iron gauze to glow, and brown iron(III) chloride forms.

four molecules of PCl_3. Because each PCl_3 molecule contains 1 P atom and 3 Cl atoms, the four PCl_3 molecules are needed to account for 4 P atoms and 12 Cl atoms in the product.

$6 \times 2 =$ 12 Cl atoms — $4 \times 3 =$ 12 Cl atoms

$$P_4(s) + 6\ Cl_2(g) \longrightarrow 4\ PCl_3(\ell)$$

4 P atoms — 4 P atoms

Next, consider the balanced equation for the reaction of iron and and chlorine (Figure 3.2). In this case, there are two iron atoms and six chlorine atoms on both sides of the equation.

$$2\ Fe(s) + 3\ Cl_2(g) \longrightarrow 2\ FeCl_3(s)$$

stoichiometric coefficients

The numbers in front of formulas in balanced chemical equations are required by the law of conservation of matter. They can be read as a number of atoms (2 atoms of Fe), molecules (3 molecules of Cl_2), or formula units (2 formula units of the ionic compound $FeCl_3$). They can refer equally well to amounts of reactants and products: 2 moles of solid iron combine with 3 moles of chlorine gas to produce 2 moles of solid $FeCl_3$. The relationship between the quantities of chemical reactants and products is called **stoichiometry** (pronounced "stoy-key-AHM-uh-tree") (► Chapter 4), and the coefficients in a balanced equation are the **stoichiometric coefficients**.

Chemistry Now™

Sign in at **www.thomsonedu.com/login** and go to Chapter 3 Contents to see Screens 3.2 and 3.3 for exercises on **the conservation of mass in reactions.**

The Importance of Balanced Chemical Equations Balanced chemical equations are fundamentally important for understanding the quantitative basis of chemistry. *You must always begin with a balanced equation before carrying out a quantitative study of a chemical reaction.*

EXERCISE 3.1 Chemical Reactions

The reaction of aluminum with bromine is shown on page 67. The equation for the reaction is

$$2\ Al(s) + 3\ Br_2(\ell) \rightarrow Al_2Br_6(s)$$

(a) What are the stoichiometric coefficients in this equation?

(b) If you were to use 8000 atoms of Al, how many molecules of Br_2 are required to consume the Al completely?

3.2 Balancing Chemical Equations

Balancing a chemical equation ensures that the same number of atoms of each element appears on both sides of the equation. Many chemical equations can be balanced by trial and error, although some will involve more trial than others.

One general class of chemical reactions is the reaction of metals or nonmetals with oxygen to give oxides of the general formula M_xO_y. For example, iron reacts with oxygen to give iron(III) oxide (Figure 3.3a).

$$4\ Fe(s) + 3\ O_2(g) \rightarrow 2\ Fe_2O_3(s)$$

The nonmetals sulfur and oxygen react to form sulfur dioxide (Figure 3.3b),

$$S(s) + O_2(g) \rightarrow SO_2(g)$$

and phosphorus, P_4, reacts vigorously with oxygen to give tetraphosphorus decaoxide, P_4O_{10} (Figure 3.3c).

$$P_4(s) + 5\ O_2(g) \rightarrow P_4O_{10}(s)$$

The equations written above are balanced. The same number of iron, sulfur, or phosphorus atoms and oxygen atoms occurs on each side of these equations.

The **combustion**, or burning, of a fuel in oxygen is accompanied by the evolution of energy. You are familiar with combustion reactions such as the burning of octane, C_8H_{18}, a component of gasoline, in an automobile engine:

$$2\ C_8H_{18}(\ell) + 25\ O_2(g) \rightarrow 16\ CO_2(g) + 18\ H_2O(g)$$

FIGURE 3.3 Reactions of a metal and two nonmetals with oxygen. (See ChemistryNow, Screen 3.4, Balancing Chemical Equations, for a video of the phosphorus and oxygen reaction.)

Charles D. Winters

(a) Reaction of iron and oxygen to give iron(III) oxide, Fe_2O_3.

(b) Reaction of sulfur (in the spoon) with oxygen.

(c) Reaction of phosphorus and oxygen to give tetraphosphorus decaoxide, P_4O_{10}.

In all combustion reactions, some or all the elements in the reactants end up as oxides, compounds containing oxygen. When the reactant is a hydrocarbon (a compound such as gasoline, natural gas, or propane that contains only C and H), the products of complete combustion are always just carbon dioxide and water.

When balancing chemical equations, there are two important things to remember.

- Formulas for reactants and products must be correct, or the equation is meaningless.
- Subscripts in the formulas of reactants and products cannot be changed to balance equations. Changing the subscripts changes the identity of the substance. For example, you cannot change CO_2 to CO to balance an equation; carbon monoxide, CO, and carbon dioxide, CO_2, are different compounds.

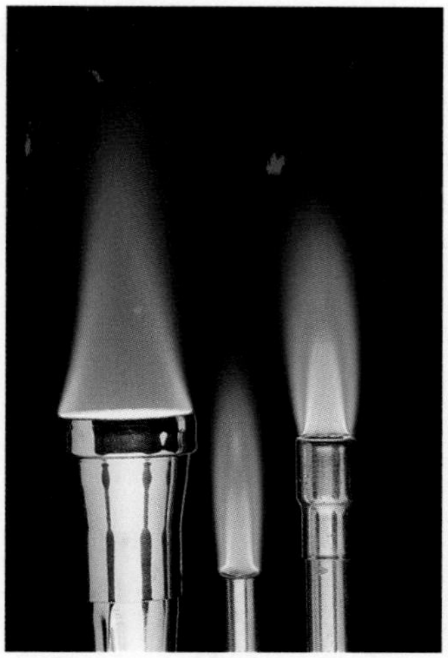
Charles D. Winters

A combustion reaction. Here, propane, C_3H_8, burns to give CO_2 and H_2O. These simple oxides are always the products of the complete combustion of a hydrocarbon.

As an example of equation balancing, let us write the balanced equation for the complete combustion of propane, C_3H_8.

Step 1. *Write correct formulas for the reactants and products.*

$$C_3H_8(g) + O_2(g) \xrightarrow{\text{unbalanced equation}} CO_2(g) + H_2O(\ell)$$

Here, propane and oxygen are the reactants, and carbon dioxide and water are the products.

Step 2. *Balance the C atoms.* In combustion reactions such as this, it is usually best to balance the carbon atoms first and leave the oxygen atoms until the end (because the oxygen atoms are often found in more than one product). In this case, three carbon atoms are in the reactants, so three must occur in the products. Three CO_2 molecules are therefore required on the right side:

$$C_3H_8(g) + O_2(g) \xrightarrow{\text{unbalanced equation}} 3\ CO_2(g) + H_2O(\ell)$$

Step 3. *Balance the H atoms.* Propane, the reactant, contains 8 H atoms. Each molecule of water has two hydrogen atoms, so four molecules of water account for the required eight hydrogen atoms on the right side:

$$C_3H_8(g) + O_2(g) \xrightarrow{\text{unbalanced equation}} 3\ CO_2(g) + 4\ H_2O(\ell)$$

Step 4. *Balance the O atoms.* Ten oxygen atoms are on the right side ($3 \times 2 = 6$ in CO_2 plus $4 \times 1 = 4$ in H_2O). Therefore, five O_2 molecules are needed to supply the required 10 oxygen atoms:

$$C_3H_8(g) + 5\ O_2(g) \rightarrow 3\ CO_2(g) + 4\ H_2O(\ell)$$

Step 5. *Verify that the number of atoms of each element is balanced.* The equation shows three carbon atoms, eight hydrogen atoms, and ten oxygen atoms on each side.

Chemistry Now™

Sign in at **www.thomsonedu.com/login** and go to Chapter 3 Contents to see Screen 3.4 for an exercise and a tutorial on **balancing the chemical equations for a series of combustion reactions.**

EXAMPLE 3.1 Balancing an Equation for a Combustion Reaction

Problem Write the balanced equation for the combustion of ammonia ($NH_3 + O_2$) to give NO and H_2O.

Strategy First, write the unbalanced equation. Next, balance the N atoms, then the H atoms, and finally, balance the O atoms.

Solution

Step 1. Write correct formulas for the reactants and products. The unbalanced equation for the combustion is

$$NH_3(g) + O_2(g) \xrightarrow{\text{unbalanced equation}} NO(g) + H_2O(\ell)$$

Step 2. *Balance the N atoms.* There is one N atom on each side of the equation. The N atoms are in balance, at least for the moment.

$$NH_3(g) + O_2(g) \xrightarrow{\text{unbalanced equation}} NO(g) + H_2O(\ell)$$

Step 3. *Balance the H atoms.* There are three H atoms on the left and two on the right. To have the same number on each side, let us use two molecules of NH_3 on the left and three molecules of H_2O on the right (which gives us six H atoms on each side).

$$2\ NH_3(g) + O_2(g) \xrightarrow{\text{unbalanced equation}} NO(g) + 3\ H_2O(\ell)$$

Notice that when we balance the H atoms, the N atoms are no longer balanced. To bring them into balance, let us use 2 NO molecules on the right.

$$2\ NH_3(g) + O_2(g) \xrightarrow{\text{unbalanced equation}} 2\ NO(g) + 3\ H_2O(\ell)$$

Step 4. *Balance the O atoms.* After Step 3, there are two O atoms on the left side and five on the right. That is, there are an even number of O atoms on the left and an odd number on the right. Because there cannot be an odd number of O atoms on the left (O atoms are paired in O_2 molecules), multiply each coefficient on both sides of the equation by 2 so that an even number of oxygen atoms (10) can now occur on the right side:

$$4\ NH_3(g) + O_2(g) \xrightarrow{\text{unbalanced equation}} 4\ NO(g) + 6\ H_2O(\ell)$$

Now the oxygen atoms can be balanced by having five O_2 molecules on the left side of the equation:

$$4\ NH_3(g) + 5\ O_2(g) \xrightarrow{\text{balanced equation}} 4\ NO(g) + 6\ H_2O(\ell)$$

Step 5. *Verify the result.* Four N atoms, 12 H atoms, and 10 O atoms occur on each side of the equation.

Comment An alternative way to write this equation is

$$2\ NH_3(g) + {}^5/_2\ O_2(g) \rightarrow 2\ NO(g) + 3\ H_2O(\ell)$$

where a fractional coefficient has been used. This equation is correctly balanced and will be useful under some circumstances. In general, however, we balance equations with whole-number coefficients.

EXERCISE 3.2 Balancing the Equation for a Combustion Reaction

(a) Butane gas, C_4H_{10}, can burn completely in air [use $O_2(g)$ as the other reactant] to give carbon dioxide gas and water vapor. Write a balanced equation for this combustion reaction.

(b) Write a balanced chemical equation for the complete combustion of liquid tetraethyllead, $Pb(C_2H_5)_4$ (which was used until the 1970s as a gasoline additive). The products of combustion are $PbO(s)$, $H_2O(\ell)$, and $CO_2(g)$.

3.3 Introduction to Chemical Equilibrium

To this point, we have treated chemical reactions as proceeding in one direction only, with reactants being converted completely to products. Nature, however, is more complex than this. Chemical reactions are reversible, and many reactions lead to incomplete conversion of reactants to products.

A good example of a reversible reaction that does not proceed completely to products is the reaction of nitrogen with hydrogen to form ammonia gas, a

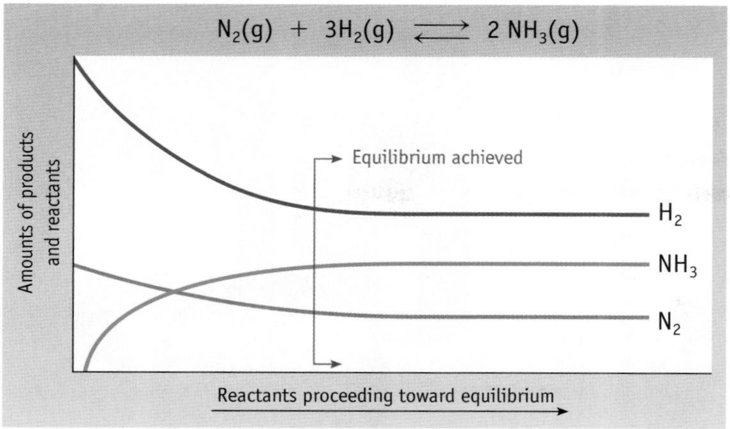

FIGURE 3.4 The reaction of N_2 and H_2 to produce NH_3. N_2 and H_2 in a 1:3 mixture react to produce some NH_3. As the reaction proceeds, the rate or speed of NH_3 production slows, as does the rate of consumption of N_2 and H_2. Eventually, the amounts of N_2 and H_2, and NH_3 no longer change. At this point, the reaction has reached equilibrium. Nonetheless, the forward reaction to produce NH_3 continues, as does the reverse reaction (the decomposition of NH_3).

compound used extensively both as a fertilizer and in the production of other fertilizers.

$$N_2(g) + 3\,H_2(g) \rightarrow 2\,NH_3(g)$$

Nitrogen and hydrogen react to form ammonia, but, under the conditions of the reaction, the product ammonia also breaks down into nitrogen and hydrogen in the reverse reaction.

$$2\,NH_3(g) \rightarrow N_2(g) + 3\,H_2(g)$$

Let us consider what would happen if we mixed nitrogen and hydrogen in a closed container under the proper conditions for the reaction to occur. At first, N_2 and H_2 react to produce some ammonia. As the ammonia is produced, however, some NH_3 molecules decompose to re-form nitrogen and hydrogen in the reverse reaction. At the beginning of the process, the forward reaction to give NH_3 predominates, but, as the reactants are consumed, the rate of the forward reaction is progressively slower. At the same time, the reverse reaction speeds up as the amount of ammonia increases. Eventually, the rate or speed of the forward reaction will equal the rate of the reverse reaction. Once this occurs, no further *macroscopic* change is observed; the amounts of nitrogen, hydrogen, and ammonia in the container stop changing (Figure 3.4). We say the system has reached **chemical equilibrium**. The reaction vessel will contain all three substances: nitrogen, hydrogen, and ammonia. Because both the forward and reverse processes are still occurring (but at equal rates), we refer to this state as a **dynamic equilibrium**. We represent a system at dynamic equilibrium by writing a double arrow symbol ($\rightleftharpoons$) connecting the reactants and products.

$$N_2(g) + 3\,H_2(g) \rightleftharpoons 2\,NH_3(g)$$

The formation of stalactites and stalagmites in a limestone cave is another example of a system that depends on the reversibility of a chemical reaction (Figure 3.5). Stalactites and stalagmites are made chiefly of calcium carbonate, a mineral found in underground deposits in the form of limestone, a leftover from ancient oceans. If water seeping through the limestone contains dissolved CO_2, a reaction occurs in which the mineral dissolves, giving an aqueous solution of $Ca(HCO_3)_2$.

$$CaCO_3(s) + CO_2(aq) + H_2O(\ell) \rightarrow Ca(HCO_3)_2(aq)$$

■ **Progression Toward Equilibrium** Reactions always proceed spontaneously toward equilibrium. A reaction will never proceed on its own in a direction that takes a system further from equilibrium.

Dr. Arthur N. Palmer

FIGURE 3.5 Cave chemistry. Calcium carbonate stalactites cling to the roof of a cave, and stalagmites grow up from the cave floor. The chemistry producing these formations is a good example of the reversibility of chemical reactions.

Equilibrium Equation: $$Ca^{2+}(aq) + 2\ HCO_3^-(aq) \rightleftharpoons CaCO_3(s) + CO_2(g) + H_2O(\ell)$$

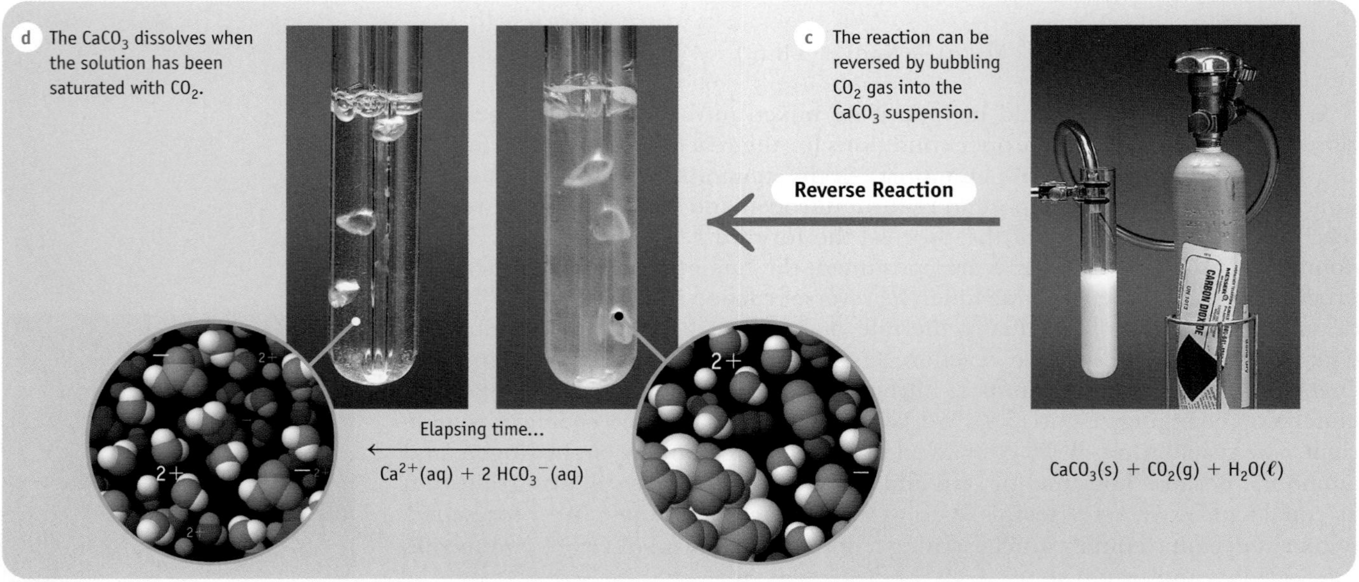

Photos: Charles D. Winters

FIGURE 3.6 The nature of chemical equilibrium. The experiments here demonstrate the reversibility of chemical reactions. (*top*) Solutions of $CaCl_2$ (a source of Ca^{2+} ions) and $NaHCO_3$ (a source of HCO_3^- ions) are mixed (a) and produce a precipitate of $CaCO_3$ and CO_2 gas (b). (*bottom*) If CO_2 gas is bubbled into a suspension of $CaCO_3$ (c), the reverse of the reaction displayed in the top panel occurs. That is, solid $CaCO_3$ and gaseous CO_2 produce Ca^{2+} and HCO_3^- ions (d).

When the mineral-laden water reaches a cave, the reverse reaction occurs, with CO_2 being evolved into the cave and solid $CaCO_3$ being deposited as stalagmites and stalactites.

$$\mathrm{Ca(HCO_3)_2(aq) \longrightarrow CaCO_3(s) + CO_2(g) + H_2O(\ell)}$$

Cave chemistry can be done in a laboratory (Figure 3.6) using reactions that further demonstrate the reversiblity of the reactions involved.

A key question that arises is, "When a reaction reaches equilibrium, will the reactants be converted largely to products, or will most of the reactants still be present?" The answer will depend on the nature of the compounds involved, the temperature, and other factors, and that is the subject of later chapters (► Chapters 16–18). For the present, though, it is useful to define **product-favored reactions** as *reactions in which reactants are completely or largely converted to products at equilibrium.* The combustion reactions we have been studying are examples of reactions that are product-favored at equilibrium, in contrast to the N_2/H_2 reaction in Figure 3.4. In fact, most of the reactions that we shall study in the rest of this chapter are product-favored reactions at equilibrium. We usually write the equations for reactions that are very product-favored using only the single arrows we have been using up to this point.

The opposite of a product-favored reaction is one that is **reactant-favored** at equilibrium. Such reactions lead to the conversion of only a small amount of the reactants to products. An example of such a reaction is the ionization of acetic acid in water, in which only a tiny fraction of the acid reacts to produce ions.

$$\mathrm{CH_3CO_2H(aq) + H_2O(\ell) \rightleftharpoons H_3O^+(aq) + CH_3CO_2^-(aq)}$$

■ **Quantitative Description of Chemical Equilibrium** As you shall see in Chapters 16–18, the extent to which a reaction is product-favored can be described by a simple mathematical expression, called the equilibrium constant expression. Each chemical reaction has a numerical value for the equilibrium constant, symbolized by *K*. Product-favored reactions have large values of *K*; small *K* values indicate reactant-favored reactions. For the ionization of acetic acid in water, $K = 1.8 \times 10^{-5}$.

■ **Acetic Acid, a Weak Acid** Acetic acid is an example of a large number of acids called "weak acids" because only a few percent of the molecules ionize to form ionic products.

3.4 Chemical Reactions in Aqueous Solution

Many of the reactions you will study in your chemistry course and the reactions that occur in living systems are carried out in aqueous solution. Because reactions in aqueous solution are so important, the remainder of this chapter is an introduction to the behavior of compounds in solution and to some of the types of reactions you will observe.

A **solution** is a homogeneous mixture of two or more substances. One substance is generally considered the **solvent**, the medium in which another substance—the **solute**—is dissolved. In the human body, the solvent for chemical reactions is usually water. Water assists in transporting nutrients and waste products in and out of cells and is necessary for digestive, absorption, circulatory, and excretory functions. In fact, the human body is two-thirds water. Water is an excellent solvent to use for biochemical reactions and also for many other chemical reactions. For the next several sections of this chapter, we shall study chemical reactions that occur in **aqueous solutions** where water is the solvent.

So that you are familiar with types of reactions as you work through the book, we also want to introduce you to four major categories of reactions in aqueous solution: precipitation, acid-base, gas-forming, and oxidation-reduction reactions. As you learn about these reactions, it will be useful to look for patterns that allow you to predict the reaction products. You will notice that many of the reactions are **exchange reactions** in which *the ions of the reactants change partners.*

$$\mathrm{A^+B^- + C^+D^- \longrightarrow A^+D^- + C^+B^-}$$

FIGURE 3.7 Precipitation of silver chloride. (a) Mixing aqueous solutions of silver nitrate and potassium chloride produces white, insoluble silver chloride, AgCl. In (b) through (d), you see a model of the process at the molecular and ionic level.

Photo, a, Charles D. Winters; b–d, model from an animation by Roy Tasker, University of Western Sydney, Australia

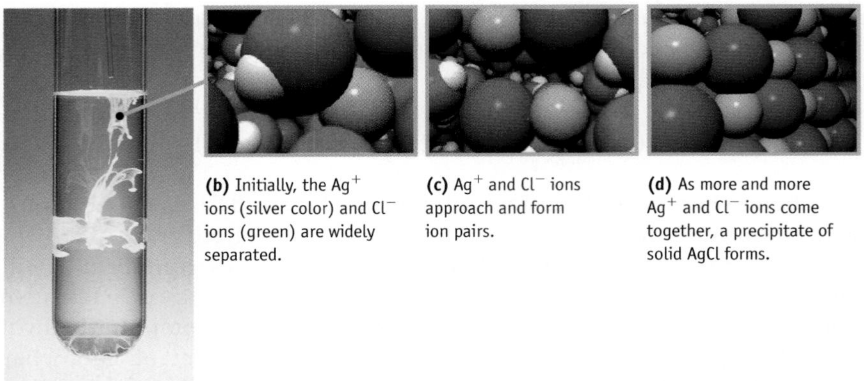

(a)

(b) Initially, the Ag^+ ions (silver color) and Cl^- ions (green) are widely separated.

(c) Ag^+ and Cl^- ions approach and form ion pairs.

(d) As more and more Ag^+ and Cl^- ions come together, a precipitate of solid AgCl forms.

For example, aqueous solutions of silver nitrate and potassium chloride react to produce solid silver chloride and aqueous potassium nitrate. (Figure 3.7a)

$$AgNO_3(aq) + KCl(aq) \rightarrow AgCl(s) + KNO_3(aq)$$

Recognizing that cations exchange anions in many chemical reactions gives us a good way to predict the products of precipitation, acid-base, and many gas-forming reactions.

Module 5

3.5 Ions and Molecules in Aqueous Solution

To understand reactions occurring in aqueous solution, it is important first to understand something about the behavior of compounds in water. The water you drink every day, the oceans, and the aqueous solutions in your body contain many ions, most of which result from dissolving solid materials present in the environment (Table 3.1).

TABLE 3.1 Concentrations of Some Cations and Anions in the Environment and in Living Cells

Element	Dissolved Species	Sea Water	*Valonia*†	Red Blood Cells	Blood Plasma
Chlorine	Cl^-	550	50	50	100
Sodium	Na^+	460	80	11	160
Magnesium	Mg^{2+}	52	50	2.5	2
Calcium	Ca^{2+}	10	1.5	10^{-4}	2
Potassium	K^+	10	400	92	10
Carbon	HCO_3^-, CO_3^{2-}	30	<10	<10	30
Phosphorus	$H_2PO_4^-$, HPO_4^{2-}	<1	5	3	<3

*Data are taken from J. J. R. Fraústo da Silva and R. J. P. Williams: *The Biological Chemistry of the Elements,* Oxford, England, Clarendon Press, 1991. Concentrations are given in millimoles per liter. (A millimole is 1/1000 of a mole.)

†*Valonia* are single-celled algae that live in sea water.

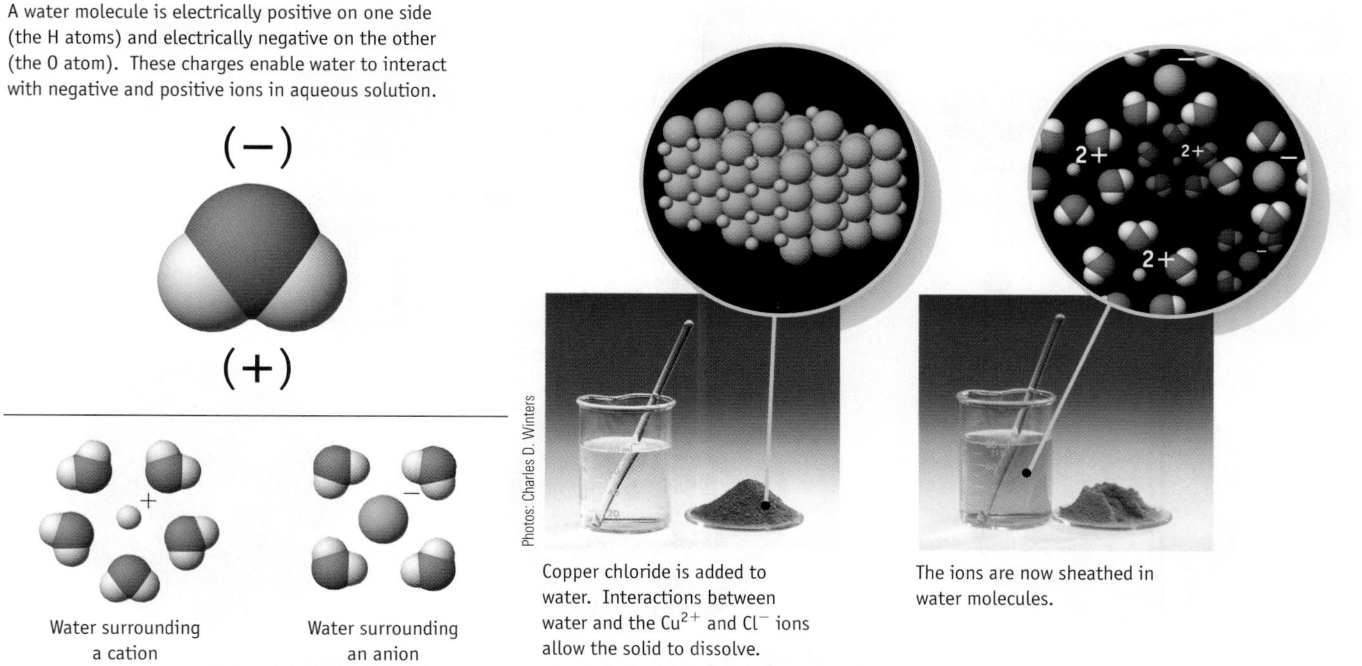

FIGURE 3.8 Water as a solvent for ionic substances. (a) Water molecules are attracted to both positive cations and negative anions in aqueous solution. (b) When an ionic substance dissolves in water, each ion is surrounded by water molecules. (The number of water molecules around an ion is often 6.)

Dissolving an ionic solid requires separating each ion from the oppositely charged ions that surround it in the solid state. Water is especially good at dissolving ionic compounds because each water molecule has a positively charged end and a negatively charged end (Figure 3.8). When an ionic compound dissolves in water, each negative ion becomes surrounded by water molecules with the positive end of water molecules pointing toward it, and each positive ion becomes surrounded by the negative ends of several water molecules.

The water-encased ions produced by dissolving an ionic compound are free to move about in solution. Under normal conditions, the movement of ions is random, and the cations and anions from a dissolved ionic compound are dispersed uniformly throughout the solution. However, if two **electrodes** (conductors of electricity such as copper wire) are placed in the solution and connected to a battery, ion movement is no longer random. Positive cations move through the solution to the negative electrode, and negative anions move to the positive electrode (Figure 3.9). If a light bulb is inserted into the circuit, the bulb lights, showing that ions are available to conduct charge in the solution just as electrons conduct charge in the wire part of the circuit. Compounds whose aqueous solutions conduct electricity are called **electrolytes.** *All ionic compounds that are soluble in water are electrolytes.*

For every mole of NaCl that dissolves, 1 mol of Na^+ and 1 mol of Cl^- ions enter the solution.

$$\text{NaCl(s)} \rightarrow \text{Na}^+\text{(aq)} + \text{Cl}^-\text{(aq)}$$

$$100\%\ \text{Dissociation} \equiv \text{strong electrolyte}$$

Because the solute has dissociated (broken apart) completely into ions, the solution will be a good conductor of electricity. Substances whose solutions are good

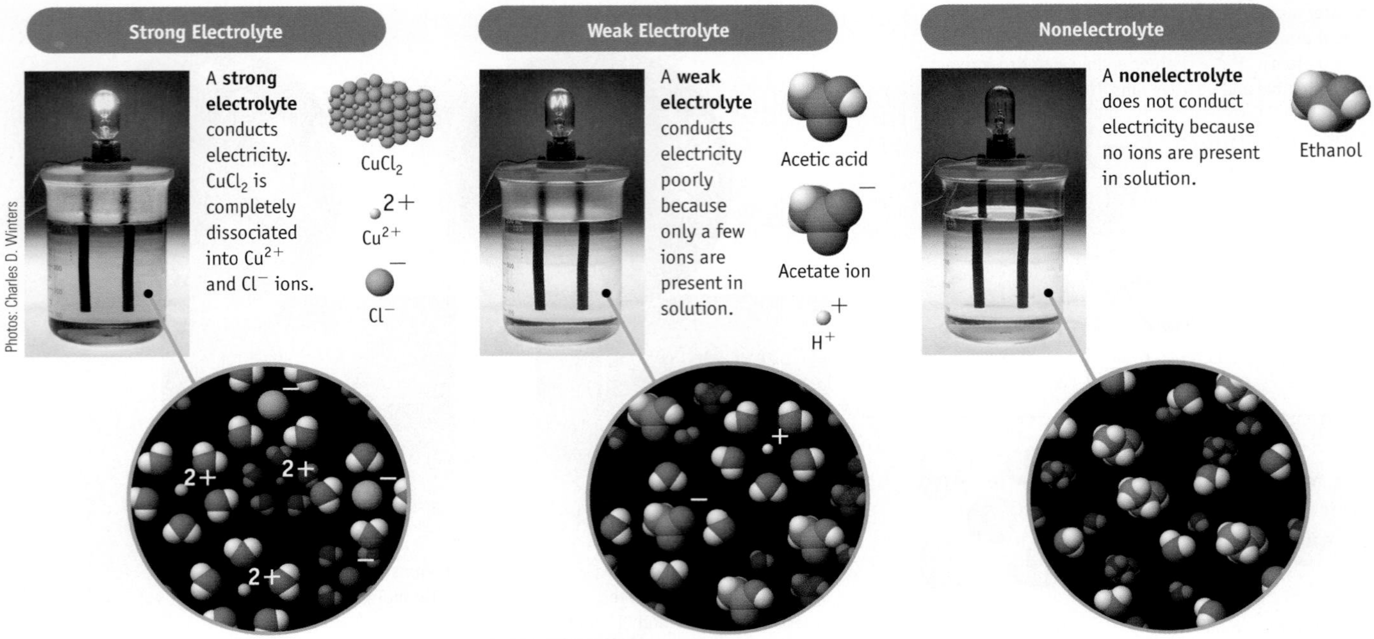

Active Figure 3.9 **Classifying solutions by their ability to conduct electricity.**

Chemistry Now™ Sign in at www.thomsonedu.com/login and go to the Chapter Contents menu to explore an interactive version of this figure accompanied by an exercise.

electrical conductors owing to the presence of ions are **strong electrolytes** (see Figure 3.9). The ions into which an ionic compound will dissociate are given by the compound's name, and the relative amounts of these ions are given by its formula. For example, as we have seen, sodium chloride yields sodium ions (Na^+) and chloride ions (Cl^-) in solution in a 1:1 ratio. The ionic compound barium chloride, $BaCl_2$, is also a strong electrolyte. In this case, there are two chloride ions for each barium ion in solution.

■ **Dissolving Halides** When an ionic compound with halide ions dissolves in water, the halide ions are released into aqueous solution. Thus, $BaCl_2$ produces two Cl^- ions for each Ba^{2+} ion (and not Cl_2 or Cl_2^{2-} ions).

$$BaCl_2(s) \rightarrow Ba^{2+}(aq) + 2\ Cl^-(aq)$$

Notice also that the chloride ions do not stay together as one unit but separate from each other into two separate chloride ions. In yet another example, the ionic compound barium nitrate yields barium ions and nitrate ions in solution. For each Ba^{2+} ion in solution, there are two NO_3^- ions.

$$Ba(NO_3)_2(s) \rightarrow Ba^{2+}(aq) + 2\ NO_3^-(aq)$$

Notice that the polyatomic ion stays together as one unit, NO_3^-, and that the two nitrate ions separate from each other.

Compounds whose aqueous solutions do not conduct electricity are called **nonelectrolytes**. The solute particles present in these aqueous solutions are molecules, not ions. *Most molecular compounds that dissolve in water are nonelectrolytes.* For example, when the molecular compound ethanol (C_2H_5OH) dissolves in water, each molecule of ethanol stays together as a single unit. We do not get ions in the solution.

$$C_2H_5OH(\ell) \rightarrow C_2H_5OH(aq)$$

Other examples of nonelectrolytes are sucrose ($C_{12}H_{22}O_{11}$) and antifreeze (ethylene glycol, $HOCH_2CH_2OH$).

Some molecular compounds (strong acids, weak acids, and weak bases) (► Section 3.7), however, react with water to form ions and are thus electrolytes. Hydrogen chloride is a molecular compound, but it reacts with water to form ions, and the solution is referred to as *hydrochloric acid.*

$$HCl(g) + H_2O(\ell) \rightarrow H_3O^+(aq) + Cl^-(aq)$$

This reaction is very product-favored. Each molecule of HCl produces ions in solution so hydrochloric acid is a strong electrolyte.

A **weak electrolyte** is a molecular substance in whose aqueous solutions some of the molecules react with water to form ions but where some of the molecules (usually most) remain as molecules. Their aqueous solutions are poor conductors of electricity (see Figure 3.9). As described on page 135, the interaction of acetic acid with water is very reactant-favored. In vinegar, an aqueous solution of acetic acid, fewer than 100 molecules in every 10,000 molecules of acetic acid are ionized to form acetate and hydronium ions. Thus, aqueous acetic acid is a weak electrolyte.

$$CH_3CO_2H(aq) + H_2O(\ell) \rightleftharpoons CH_3CO_2^-(aq) + H_3O^+(aq)$$

acetic acid < 1% ionized = weak electrolyte | water | acetate ion | hydronium ion

■ **H^+ Ions in Water** As illustrated for acetic acid in Figure 3.9, the H^+ ions from the acid are surrounded by water molecules. When writing an equation for acid ionization, we symbolize this with the H_3O^+ or hydronium ion. For more on this, see page 134.

Chemistry Now™

Sign in at **www.thomsonedu.com/login** and go to Chapter 3 Contents to see:

- Screen 3.5 for exercises on **the dissolving of an ionic compound**
- Screen 3.6 for information on **the types of electrolytes**

EXERCISE 3.3 Electrolytes

Epsom salt, $MgSO_4 \cdot 7\ H_2O$, is sold in drugstores and, as a solution in water, is used for various medical purposes. Methanol, CH_3OH, is dissolved in gasoline in the winter in colder climates to prevent the formation of ice in automobile fuel lines. Which of these compounds is an electrolyte, and which is a nonelectrolyte?

Solubility of Ionic Compounds in Water

Many ionic compounds dissolve completely in water, but some dissolve only to a small extent, and still others are essentially insoluble. Fortunately, we can make some general statements about which ionic compounds are water soluble. In this chapter, we consider solubility as an "either–or" question, referring to those materials that are soluble beyond a certain extent as "soluble" and to those that do not dissolve to that extent as "insoluble." To get a better idea of the amounts that will actually dissolve in a given quantity of water, we could do an experiment or perform a calculation that uses the concept of equilibrium (► Chapter 18).

■ **Solubility Guidelines** Observations such as those shown in Figure 3.10 were used to create the solubility guidelines. Note, however, that these are general guidelines and not rules followed under all circumstances. There are exceptions, but the guidelines are a good place to begin. See B. Blake, *Journal of Chemical Education*, Vol. 80, pp. 1348–1350, 2003.

Figure 3.10 lists broad guidelines that help predict whether a particular ionic compound is soluble in water. For example, sodium nitrate, $NaNO_3$, contains both an alkali metal cation, Na^+, and the nitrate anion, NO_3^-. The presence of either of these ions ensures that the compound is soluble in water. By contrast, calcium hydroxide is poorly soluble in water. If a spoonful of solid $Ca(OH)_2$ is added to 100 mL of water, only 0.17 g, or 0.0023 mol, will dissolve at 10 °C. Nearly all of the $Ca(OH)_2$ remains as a solid (Figure 3.10c).

Chemistry Now™

Sign in at **www.thomsonedu.com/login** and go to Chapter 3 Contents to see Screen 3.7 for a tutorial and simulation on **the solubility of ionic compounds in water.**

SILVER COMPOUNDS

$AgNO_3$ AgCl AgOH

(a) Nitrates are generally soluble, as are chlorides (except AgCl). Hydroxides are generally not soluble.

SULFIDES

$(NH_4)_2S$ CdS Sb_2S_3 PbS

(b) Sulfides are generally not soluble (exceptions include salts with NH_4^+ and Na^+).

HYDROXIDES

NaOH $Ca(OH)_2$ $Fe(OH)_3$ $Ni(OH)_2$

Photos: Charles D. Winters

(c) Hydroxides are generally not soluble, except when the cation is a Group 1A metal.

SOLUBLE COMPOUNDS	EXCEPTIONS
Almost all salts of Na^+, K^+, NH_4^+	
Salts of nitrate, NO_3^- chlorate, ClO_3^- perchlorate, ClO_4^- acetate, $CH_3CO_2^-$	
Almost all salts of Cl^-, Br^-, I^-	Halides of Ag^+, Hg_2^{2+}, Pb^{2+}
Salts containing F^-	Fluorides of Mg^{2+}, Ca^{2+}, Sr^{2+}, Ba^{2+}, Pb^{2+}
Salts of sulfate, SO_4^{2-}	Sulfates of Ca^{2+}, Sr^{2+}, Ba^{2+}, Pb^{2+}

INSOLUBLE COMPOUNDS	EXCEPTIONS
Most salts of carbonate, CO_3^{2-} phosphate, PO_4^{3-} oxalate, $C_2O_4^{2-}$ chromate, CrO_4^{2-} sulfide, S^{2-}	Salts of NH_4^+ and the alkali metal cations
Most metal hydroxides and oxides	Alkali metal hydroxides and $Ba(OH)_2$

Active Figure 3.10 Guidelines to predict the solubility of ionic compounds. If a compound contains one of the ions in the column on the left in the top chart, it is predicted to be at least moderately soluble in water. There are exceptions, which are noted at the right. Most ionic compounds formed by the anions listed at the bottom of the chart are poorly soluble (with the exception on compounds with NH_4^+ and the alkali metal cations).

Chemistry Now™ Sign in at www.thomsonedu.com/login and go to the Chapter Contents menu to explore an interactive version of this figure accompanied by an exercise.

EXAMPLE 3.2 Solubility Guidelines

Problem Predict whether the following ionic compounds are likely to be water-soluble. List the ions present in solution for soluble compounds.

(a) KCl

(b) $MgCO_3$

(c) Fe_2O_3

(d) $Cu(NO_3)_2$

Strategy You must first recognize the cation and anion involved and then decide the probable water solubility based on the guidelines outlined in Figure 3.10.

Solution

(a) KCl is composed of K^+ and Cl^- ions. The presence of *either* of these ions means that the compound is likely to be soluble in water. The solution contains K^+ and Cl^- ions dissolved in water.

$$KCl(s) \rightarrow K^+(aq) + Cl^-(aq)$$

(The solubility of KCl is about 35 g in 100 mL of water at 20 °C.)

(b) Magnesium carbonate is composed of Mg^{2+} and CO_3^{2-} ions. Salts containing the carbonate ion are usually insoluble, unless combined with an ion like Na^+ or NH_4^+. Therefore, $MgCO_3$ is predicted to be insoluble in water. (The solubility of $MgCO_3$ is less than 0.2 g/100 mL of water.)

(c) Iron(III) oxide is composed of Fe^{3+} and O^{2-} ions. Oxides are soluble only when O^{2-} is combined with an alkali metal ion; Fe^{3+} is a transition metal ion, so Fe_2O_3 is insoluble.

(d) Copper(II) nitrate is composed of $Cu^{2+}(aq)$ and $NO_3^-(aq)$ ions. Nitrate salts are soluble, so this compound dissolves in water, giving ions in solution as shown in the equation below

$$Cu(NO_3)_2(s) \rightarrow Cu^{2+}(aq) + 2\ NO_3^-(aq)$$

EXERCISE 3.4 Solubility of Ionic Compounds

Predict whether each of the following ionic compounds is likely to be soluble in water. If it is soluble, write the formulas of the ions present in aqueous solution.

(a) $LiNO_3$ **(b)** $CaCl_2$ **(c)** CuO **(d)** $NaCH_3CO_2$

3.6 Precipitation Reactions

Module 6

A **precipitation reaction** produces a water-insoluble solid product, known as a **precipitate**. The reactants in such reactions are generally water-soluble ionic compounds. When these substances dissolve in water, they dissociate to give the appropriate cations and anions. If the cation from one compound can form an insoluble compound with the anion from the other compound in the solution, precipitation occurs. As described earlier, both silver nitrate and potassium chloride are water-soluble ionic compounds. When combined in water, they undergo an *exchange reaction* to produce insoluble silver chloride and soluble potassium nitrate (Figure 3.7).

$$AgNO_3(aq) + KCl(aq) \rightarrow AgCl(s) + KNO_3(aq)$$

Reactants	Products
$Ag^+(aq) + NO_3^-(aq)$	Insoluble $AgCl(s)$
$K^+(aq) + Cl^-(aq)$	$K^+(aq) + NO_3^-(aq)$

Predicting the Outcome of a Precipitation Reaction

Many combinations of positive and negative ions give insoluble substances (see Figure 3.10). For example, the solubility guidelines indicate that most compounds containing the chromate ion are not soluble (alkali metal chromates and ammonium chromate are exceptions). Thus, we can predict that yellow, solid lead(II)

FIGURE 3.11 Precipitation reactions. Many ionic compounds are insoluble in water. Guidelines for predicting the solubilities of ionic compounds are given in Figure 3.10.

Charles D. Winters

(a) $Pb(NO_3)_2$ and K_2CrO_4 produce yellow, insoluble $PbCrO_4$ and soluble KNO_3.

(b) $Pb(NO_3)_2$ and $(NH_4)_2S$ produce black, insoluble PbS and soluble NH_4NO_3.

(c) $FeCl_3$ and NaOH produce red, insoluble $Fe(OH)_3$ and soluble NaCl.

(d) $AgNO_3$ and K_2CrO_4 produce red, insoluble Ag_2CrO_4 and soluble KNO_3. See Example 3.3.

chromate will precipitate when a water-soluble lead(II) compound is combined with a water-soluble chromate compound (Figure 3.11a).

$$Pb(NO_3)_2(aq) + K_2CrO_4(aq) \rightarrow PbCrO_4(s) + 2\ KNO_3(aq)$$

Reactants	Products
$Pb^{2+}(aq) + 2\ NO_3^-(aq)$	Insoluble $PbCrO_4(s)$
$2\ K^+(aq) + CrO_4^{2-}(aq)$	$2\ K^+(aq) + 2\ NO_3^-(aq)$

Similarly, we know from the solubility guidelines that almost all metal sulfides are insoluble in water (Figure 3.11b). If a solution of a soluble metal compound comes in contact with a source of sulfide ions, the metal sulfide precipitates.

$$Pb(NO_3)_2(aq) + (NH_4)_2S(aq) \rightarrow PbS(s) + 2\ NH_4NO_3(aq)$$

Reactants	Products
$Pb^{2+}(aq) + 2\ NO_3^-(aq)$	Insoluble PbS(s)
$2\ NH_4^+(aq) + S^{2-}(aq)$	$2\ NH_4^+(aq) + 2\ NO_3^-(aq)$

Charles D. Winters

Black tongue. Pepto-Bismol™ has antidiarrheal, antibacterial, and antacid effects in the digestive tract, and has been used for over 100 years as an effective remedy. However, some people find their tongues blackened after taking this over-the-counter medicine. The active ingredient in Pepto-Bismol is bismuth subsalicylate. (It also contains pepsin, zinc salts, oil of wintergreen, and salol, a compound related to aspirin.) The tongue blackening comes from the reaction of bismuth ions with traces of sulfide ions found in saliva to form black Bi_2S_3. The discoloration is harmless and lasts only a few days.

In still another example, the solubility guidelines indicate that with the exception of the alkali metal cations (and Ba^{2+}), all metal cations form insoluble hydroxides. Thus, water-soluble iron(III) chloride and sodium hydroxide react to give insoluble iron(III) hydroxide (Figures 3.10c and 3.11c).

$$FeCl_3(aq) + 3\ NaOH(aq) \rightarrow Fe(OH)_3(s) + 3\ NaCl(aq)$$

Reactants	Products
$Fe^{3+}(aq) + 3\ Cl^-(aq)$	Insoluble $Fe(OH)_3(s)$
$3\ Na^+(aq) + 3\ OH^-(aq)$	$3\ Na^+(aq) + 3\ Cl^-(aq)$

Chemistry Now™

Sign in at **www.thomsonedu.com/login** and go to Chapter 3 Contents to see:

- Screen 3.8 for a self-study module on **types of reactions in aqueous solutions**
- Screen 3.9 for a tutorial and simulation on **precipitation reactions**

EXAMPLE 3.3 Writing the Equation for a Precipitation Reaction

Problem Is an insoluble product formed when aqueous solutions of potassium chromate and silver nitrate are mixed? If so, write the balanced equation.

Strategy First, decide what ions are formed in solution when the reactants dissolve. Then use information in Figure 3.10 to determine whether a cation from one reactant will combine with an anion from the other reactant to form an insoluble compound.

Solution Both reactants—$AgNO_3$ and K_2CrO_4—are water-soluble. The ions Ag^+, NO_3^-, K^+, and CrO_4^{2-} are released into solution when the compounds are dissolved.

$$AgNO_3(s) \rightarrow Ag^+(aq) + NO_3^-(aq)$$

$$K_2CrO_4(s) \rightarrow 2\ K^+(aq) + CrO_4^{2-}(aq)$$

Here, Ag^+ could combine with CrO_4^{2-}, and K^+ could combine with NO_3^-. Based on the solubility guidelines, we know that the former combination, Ag_2CrO_4, is an insoluble compound, whereas KNO_3 is soluble in water. Thus, the balanced equation for the reaction of silver nitrate and potassium chromate is

$$2\ AgNO_3(aq) + K_2CrO_4(aq) \rightarrow Ag_2CrO_4(s) + 2\ KNO_3(aq)$$

Comment This reaction is illustrated in Figure 3.11d.

■ **Hair Coloring and Black Smokers** The hair-darkening reaction described on page 18 is a precipitation reaction and is much like that in "black smokers" and in the formation of Bi_2S_3 with Pepto-Bismol.

EXERCISE 3.5 Precipitation Reactions

In each of the following cases, does a precipitation reaction occur when solutions of the two water-soluble reactants are mixed? Give the formula of any precipitate that forms, and write a balanced chemical equation for the precipitation reactions that occur.

(a) Sodium carbonate and copper(II) chloride

(b) Potassium carbonate and sodium nitrate

(c) Nickel(II) chloride and potassium hydroxide

Net Ionic Equations

We have seen that when aqueous solutions of silver nitrate and potassium chloride are mixed, insoluble silver chloride forms, leaving potassium nitrate in solution (see Figure 3.7). The balanced chemical equation for this process is

$$AgNO_3(aq) + KCl(aq) \rightarrow AgCl(s) + KNO_3(aq)$$

We can represent this reaction in another way by writing an equation in which we show that the soluble ionic compounds are present in solution as dissociated ions. An aqueous solution of silver nitrate contains Ag^+ and NO_3^- ions, and an aqueous solution of potassium chloride contains K^+ and Cl^- ions. In the products, the potassium nitrate is present in solution as K^+ and NO_3^- ions. The silver chloride, however, is insoluble and thus is not present in the solution as dissociated ions. It is shown in the equation by its entire formula, AgCl.

$$\underbrace{Ag^+(aq) + NO_3^-(aq) + K^+(aq) + Cl^-(aq)}_{\text{before reaction}} \rightarrow \underbrace{AgCl(s) + K^+(aq) + NO_3^-(aq)}_{\text{after reaction}}$$

This type of equation is called a **complete ionic equation**.

The K^+ and NO_3^- ions are present in solution before and after reaction and so appear on both the reactant and product sides of the complete ionic equation. Such ions are often called **spectator ions** because they do not participate in the net reaction; they only "look on" from the sidelines. Little chemical information is lost if the equation is written without them, and so we can simplify the equation to

$$Ag^+(aq) + Cl^-(aq) \rightarrow AgCl(s)$$

Problem Solving Tip 3.1 Writing Net Ionic Equations

Net ionic equations are commonly written for chemical reactions in aqueous solution because they describe the actual chemical species involved in a reaction. To write net ionic equations, we must know which compounds exist as ions in solution.

1. Strong acids, strong bases, and soluble salts exist as ions in solution. Examples include the acids HCl and HNO_3, a base such as NaOH, and salts such as NaCl and $CuCl_2$.
2. All other species should be represented by their complete formulas. Weak acids such as acetic acid (CH_3CO_2H) exist in solutions primarily as molecules. (See Section 3.7.) Insoluble salts such as $CaCO_3$(s) or insoluble bases such as $Mg(OH)_2$(s) should not be written in ionic form, even though they are ionic compounds.

The best way to approach writing net ionic equations is to follow precisely a set of steps:

1. Write a complete, balanced equation. Indicate the state of each substance (aq, s, ℓ, g).
2. Next, rewrite the whole equation, writing all strong acids, strong bases, and soluble salts as ions. (Consider only species labeled with an "(aq)" suffix in this step.)
3. Some ions may remain unchanged in the reaction (the ions that appear in the equation both as reactants or products). These "spectator ions" are not part of the chemistry that is going on. You can cancel them from each side of the equation.
4. Like molecular equations, net ionic equations must be balanced. The same number of atoms must appear on each side of the arrow, and the sum of the ion charges on the two sides must also be equal.

■ **Net Ionic Equations** All chemical equations, including net ionic equations, must be balanced. The same number of atoms of each kind must appear on both the product and reactant sides. In addition, the sum of positive and negative charges must be the same on both sides of the equation.

The balanced equation that results from leaving out the spectator ions is the **net ionic equation** for the reaction. *Only the aqueous ions, insoluble compounds, and weak- or nonelectrolytes* (which can be soluble molecular compounds such as sugar, weak acids, weak bases, or gases) *that participate in a chemical reaction are included in the net ionic equation.*

Leaving out the spectator ions does not imply that K^+ and NO_3^- ions are unimportant in the $AgNO_3$ + KCl reaction. Indeed, Ag^+ ions cannot exist alone in solution; a negative ion, in this case NO_3^-, must be present to balance the positive charge of Ag^+. Any anion will do, however, as long as it forms a water-soluble compound with Ag^+. Thus, we could have used $AgClO_4$ instead of $AgNO_3$. Similarly, there must be a positive ion present to balance the negative charge of Cl^-. In this case, the positive ion present is K^+ in KCl, but we could have used NaCl instead of KCl. The net ionic equation would have been the same.

Finally, notice that there must always be a *charge balance* as well as a mass balance in a balanced chemical equation. Thus, in the $Ag^+ + Cl^-$ net ionic equation, the cation and anion charges on the left add together to give a net charge of zero, the same as the zero charge on AgCl(s) on the right.

Charles D. Winters

Precipitation reaction. The reaction of barium chloride and sodium sulfate produces insoluble barium sulfate and water-soluble sodium chloride.

Chemistry Now™

Sign in at **www.thomsonedu.com/login** and go to Chapter 3 Contents to see Screen 3.10 for a tutorial on **writing net ionic equations.**

EXAMPLE 3.4 Writing and Balancing Net Ionic Equations

Problem Write a balanced, net ionic equation for the reaction of aqueous solutions of $BaCl_2$ and Na_2SO_4.

Strategy Follow the strategy outlined in Problem Solving Tip 3.1.

Solution

Step 1. First, notice that this is an *exchange reaction*. That is, the Ba^{2+} and Na^+ cations exchange anions (Cl^- and SO_4^{2-}) to give $BaSO_4$ and NaCl. Now that the reactants and products are known, we can write an equation for the reaction. To balance the equation, we place a 2 in front of the NaCl.

$$BaCl_2 + Na_2SO_4 \rightarrow BaSO_4 + 2\ NaCl$$

Step 2. Decide on the solubility of each compound (Figure 3.10). Compounds containing sodium ions are always water-soluble, and those containing chloride ions are almost always soluble. Sulfate salts are also usually soluble, one important exception being $BaSO_4$. We can therefore write

$$BaCl_2(aq) + Na_2SO_4(aq) \rightarrow BaSO_4(s) + 2\ NaCl(aq)$$

Step 3. Identify the ions in solution. All soluble ionic compounds dissociate to form ions in aqueous solution.

$$BaCl_2(s) \rightarrow Ba^{2+}(aq) + 2\ Cl^-(aq)$$
$$Na_2SO_4(s) \rightarrow 2\ Na^+(aq) + SO_4^{2-}(aq)$$
$$NaCl(s) \rightarrow Na^+(aq) + Cl^-(aq)$$

This results in the following complete ionic equation:

$$Ba^{2+}(aq) + 2\ Cl^-(aq) + 2\ Na^+(aq) + SO_4^{2-}(aq) \rightarrow BaSO_4(s) + 2\ Na^+(aq) + 2\ Cl^-(aq)$$

Step 4. Identify and eliminate the spectator ions (Na^+ and Cl^-) to give the net ionic equation.

$$Ba^{2+}(aq) + SO_4^{2-}(aq) \rightarrow BaSO_4(s)$$

Comment Notice that the sum of ion charges is the same on both sides of the equation. On the left, 2+ and 2− give zero; on the right, the charge on $BaSO_4$ is also zero.

EXERCISE 3.6 Net Ionic Equations

Write a balanced net ionic equation for each of the following reactions:

(a) $AlCl_3 + Na_3PO_4 \rightarrow AlPO_4 + NaCl$ (not balanced)

(b) Solutions of iron(III) chloride and potassium hydroxide give iron(III) hydroxide and potassium chloride when combined. See Figure 3.10c.

(c) Solutions of lead(II) nitrate and potassium chloride give lead(II) chloride and potassium nitrate when combined.

3.7 Acids and Bases

Acids and bases are two important classes of compounds. You may already be familiar with some common properties of acids. They produce bubbles of CO_2 gas when added to a metal carbonate such as $CaCO_3$ (Figure 3.12a), and they react with many metals to produce hydrogen gas (H_2) (Figure 3.12b). Although tasting substances is *never* done in a chemistry laboratory, you have probably experienced the sour taste of acids such as acetic acid in vinegar and citric acid (commonly found in fruits and added to candies and soft drinks). Acids and bases have some related properties. Solutions of acids or bases, for example, can change the colors of vegetable pigments (Figure 3.12c). You may have seen acids change the color of litmus, a dye derived from certain lichens, from blue to red. If an acid has made blue litmus paper turn red, adding a base reverses the effect, making the litmus blue again. Thus, acids and bases seem to be opposites. A base can neutralize the effect of an acid, and an acid can neutralize the effect of a base. See Table 3.2 for a list of common acids and bases.

Over the years, chemists have examined the properties, chemical structures, and reactions of acids and bases and have proposed different definitions of the terms "acid" and "base." In this section, we shall examine the two most commonly used definitions, one proposed by Svante Arrhenius (1859–1927) and another proposed by Johannes N. Brønsted (1879–1947) and Thomas M. Lowry (1874–1936).

Charles D. Winters

(a) A piece of coral (mostly $CaCO_3$) dissolves in acid to give CO_2 gas.

(b) Zinc reacts with hydrochloric acid to produce zinc chloride and hydrogen gas.

(c) The juice of a red cabbage is normally blue-purple. On adding acid, the juice becomes more red. Adding base produces a yellow color.

FIGURE 3.12 Some properties of acids and bases. (a) Acids react readily with coral ($CaCO_3$) and other metal carbonates to produce gaseous CO_2 (and a salt). (b) Acids react with many metals to produce hydrogen gas (and a metal salt). (c) The colors of natural dyes, such as the juice from a red cabbage, are affected by acids and bases.

Acids and Bases: The Arrhenius Definition

The Swedish chemist Svante Arrhenius made a number of important contributions to chemistry, but he is perhaps best known for his studies of the properties of solutions of salts, acids, and bases. In the late 1800s, Arrhenius proposed that these compounds dissolve in water and ultimately form ions. This theory of electrolytes

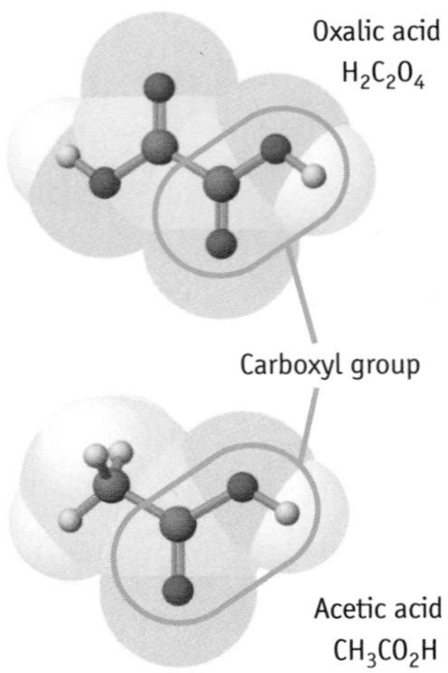

■ **Weak Acids** Common acids and bases are listed in Table 3.2. There are numerous other weak acids and bases, and many of these are natural substances. Oxalic and acetic acid are among them. Many of these natural acids contain CO_2H groups. (The H of this group is lost as H^+.)

TABLE 3.2 Common Acids and Bases

Strong Acids (Strong Electrolytes)		Soluble Strong Bases	
HCl (aq)	Hydrochloric acid	LiOH	Lithium hydroxide
HBr (aq)	Hydrobromic acid	NaOH	Sodium hydroxide
HI (aq)	Hydroiodic acid	KOH	Potassium hydroxide
HNO_3	Nitric acid	$Ba(OH)_2$	Barium hydroxide
$HClO_4$	Perchloric acid		
H_2SO_4	Sulfuric acid		
Weak Acids (Weak Electrolytes)*		**Weak Base (Weak Electrolyte)**	
H_3PO_4	Phosphoric acid	NH_3	Ammonia
H_2CO_3	Carbonic acid		
CH_3CO_2H	Acetic acid		
$H_2C_2O_4$	Oxalic acid		
$H_2C_4H_4O_6$	Tartaric acid		
$H_3C_6H_5O_7$	Citric acid		
$HC_9H_8O_4$	Aspirin		

* These are representative of hundreds of weak acids.

predated any knowledge of the composition and structure of atoms and was not well accepted initially. With a knowledge of atomic structure, however, we now take it for granted.

The Arrhenius definitions for acids and bases derives from his theory of electrolytes and focuses on formation of H^+ and OH^- ions in aqueous solutions.

- An acid is a substance that, when dissolved in water, increases the concentration of hydrogen ions, H^+ in solution.

$$HCl(g) \rightarrow H^+(aq) + Cl^-(aq)$$

- A base is a substance that, when dissolved in water, increases the concentration of hydroxide ions, OH^-, in the solution.

$$NaOH(s) \rightarrow Na^+(aq) + OH^-(aq)$$

- The reaction of an acid and a base produces a salt and water. Because the characteristic properties of an acid are lost when a base is added, and vice versa, acid–base reactions were logically described as resulting from the combination of H^+ and OH^- to form water.

$$HCl(aq) + NaOH(aq) \rightarrow NaCl(aq) + H_2O(\ell)$$

Arrhenius further proposed that acid strength was related to the extent to which the acid ionized. Some acids such as hydrochloric acid (HCl) and nitric acid (HNO_3) ionize completely in water; they are strong electrolytes, and we now call them **strong acids**. Other acids such as acetic acid and hydrofluoric acid are incompletely ionized; they are weak electrolytes and are **weak acids**. Weak acids exist in solution primarily as acid molecules, and only a fraction of these molecules ionize to produce $H^+(aq)$ ions along with the appropriate anion in solution.

Water-soluble compounds that contain hydroxide ions, such as sodium hydroxide (NaOH) or potassium hydroxide (KOH), are strong electrolytes and strong bases.

Aqueous ammonia, $NH_3(aq)$, is a weak electrolyte. Even though it does not have an OH^- ion as part of its formula, it does produce ammonium ions and hydroxide ions from its reaction with water and so is a base (Figure 3.13). The fact that this is a weak electrolyte indicates that only a fraction of ammonia molecules react with water to form ions; most of the base remains in solution in molecular form.

Although the Arrhenius theory is still used to some extent and is interesting in an historical context, modern concepts of acid–base chemistry such as the Brønsted–Lowry theory have gained preference among chemists.

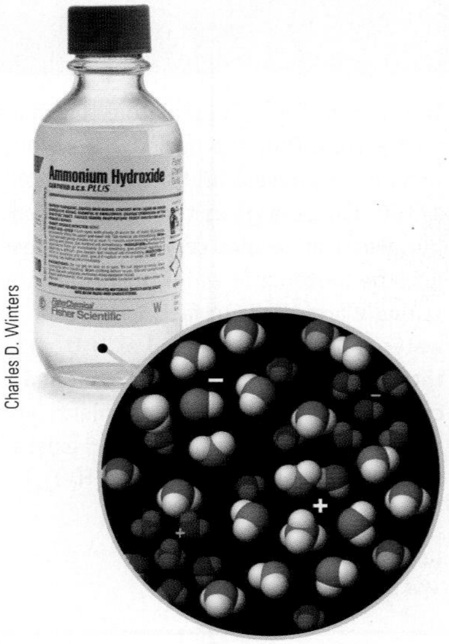

FIGURE 3.13 Ammonia, a weak electrolyte. Ammonia, NH_3, interacts with water to produce a very small number of NH_4^+ and OH^- ions per mole of ammonia molecules. (The name on the bottle, ammonium hydroxide, is misleading. The solution consists almost entirely of NH_3 molecules dissolved in water. It is better referred to as "aqueous ammonia.")

EXERCISE 3.7 Acids and Bases

(a) What ions are produced when nitric acid dissolves in water?

(b) Barium hydroxide is moderately soluble in water. What ions are produced when it dissolves in water?

Acids and Bases: The Brønsted-Lowry Definition

In 1923, Brønsted in Copenhagen (Denmark) and Lowry in Cambridge (England) independently suggested a new concept of acid and base behavior. They viewed acids and bases in terms of the transfer of a proton (H^+) from one species to an-

A Closer Look

The Hydronium Ion—The H^+ Ion in Water

The H^+ ion is a hydrogen atom that has lost its electron. Only the nucleus, a proton, remains. Because a proton is only about 1/100,000 as large as the average atom or ion, water molecules can approach closely, and the proton and water molecules are strongly attracted. In fact, the H^+ ion in water is better represented as H_3O^+, called the **hydronium ion.** This ion is formed by combining H^+ and H_2O. Experiments also show that other forms of the ion exist in water, one example being $[H_3O(H_2O)_3]^+$.

There will be instances when, for simplicity, we will use $H^+(aq)$. However, we will usually use the H_3O^+ symbol to represent the hydrogen ion in water in this book. Thus, hydrochloric acid is better represented as a solution of H_3O^+ and Cl^-.

hydronium ion $H_3O^+(aq)$ chloride ion $Cl^-(aq)$

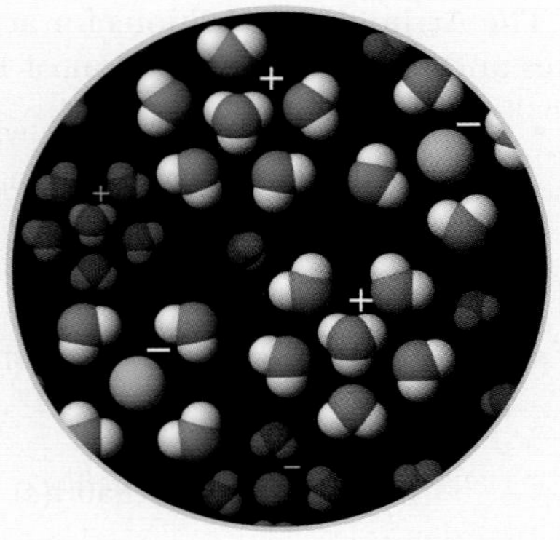

When HCl ionizes in aqueous solution, it produces the hydronium ion, H_3O^+, and the chloride ion, Cl^-.

other, and they described all acid–base reactions in terms of equilibria. The Brønsted–Lowry theory expanded the scope of the definition of acids and bases and helped chemists make predictions of product or reactant favorability based on acid and base strength. We will describe this theory here qualitatively; a more complete discussion will be given in Chapter 17.

The main concepts of the Brønsted-Lowry theory are the following:

- *An acid is a proton donor.* This is similar to the Arrhenius definition.
- *A base is a proton acceptor.* This definition includes the OH^- ion, but it also broadens the number and type of bases.
- *An acid–base reaction involves the transfer of a proton from an acid to a base to form a new acid and a new base. The reaction is written as an equilibrium reaction, and the equilibrium favors the weaker acid and base.* This allows the prediction of product- or reactant-favored reactions based on acid and base strength.

From the point of view of the Brønsted–Lowry theory, the behavior of acids such as HCl or CH_3CO_2H in water is seen to involve an acid–base reaction. Both species (both Brønsted acids) donate a proton to water (a Brønsted base), forming $H_3O^+(aq)$. Hydrochloric acid, HCl(aq), is a strong electrolyte because it ionizes completely in aqueous solution; it is thus classified as a strong acid.

Hydrogen chloride, a strong acid. 100% ionized. Equilibrium strongly favors products.

$$HCl(aq) + H_2O(\ell) \rightleftharpoons H_3O^+(aq) + Cl^-(aq)$$

hydrochloric acid strong electrolyte = 100% ionized   water   hydronium ion   chloride ion

Chemical Perspectives — Sulfuric Acid

For many years, sulfuric acid has been the chemical produced in the largest quantity in the United States (and in many other industrialized countries). About 40–50 billion kilograms (40–50 million metric tons) are made annually in the United States. The acid is so important to the economy of industrialized nations that some economists have said sulfuric acid production is a measure of a nation's industrial strength.

Sulfuric acid is a colorless, syrupy liquid with a density of 1.84 g/mL and a boiling point of 337 °C. It has several desirable properties that have led to its widespread use: it is generally less expensive to produce than other acids, is a strong acid, can be handled in steel containers, reacts readily with many organic compounds to produce useful products, and reacts readily with lime (CaO), the least expensive and most readily available base, to give calcium sulfate, a compound used to make wall board for the construction industry.

Farrel Grehan/Photo Researchers, Inc.

Sulfur. Much of the sulfur used in the U.S. is produced by the Frasch process. This works by injecting superheated water into pockets of the element deep in the earth. The sulfur is forced to the surface in the molten state by compressed air.

Fleck Chemical, United Kingdom

A sulfuric acid plant.

The first step in the industrial preparation of sulfuric acid is combustion of sulfur in air to give sulfur dioxide.

$$S(s) + O_2(g) \rightarrow SO_2(g)$$

This gas is then combined with more oxygen, in the presence of a catalyst (a substance that speeds up a reaction), to give sulfur trioxide,

$$2\ SO_2(g) + O_2(g) \rightarrow 2\ SO_3(g)$$

which can give sulfuric acid when absorbed in water.

$$SO_3(g) + H_2O(\ell) \rightarrow H_2SO_4(aq)$$

Currently, over two thirds of the production is used in the fertilizer industry. The remainder is used to make pigments, explosives, alcohol, pulp and paper, and detergents, and is employed as a component in storage batteries.

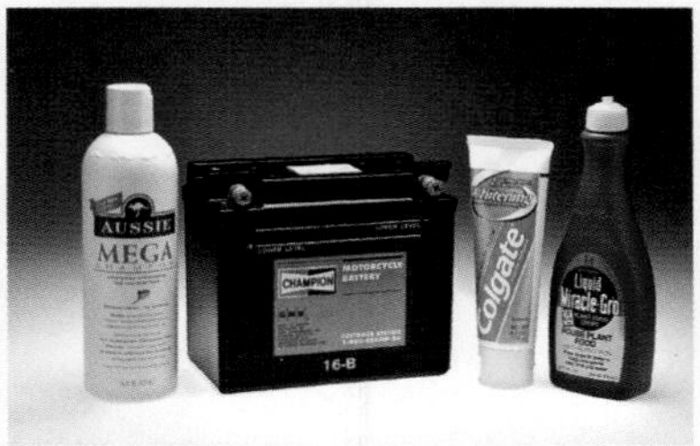

Charles D. Winters

Some products that require sulfuric acid for their manufacture or use.

In contrast, CH_3CO_2H is a weak electrolyte, evidence that it is ionized to a small extent in water and therefore is a weak acid.

Acetic acid, a weak acid, << 100% ionized. Equilibrium favors reactants.

$$\underset{\text{Weak Brønsted acid}}{CH_3CO_2H(aq)} + H_2O(\ell) \rightleftharpoons H_3O^+(aq) + \underset{\text{Brønsted base}}{CH_3CO_2^-(aq)}$$

The different extent of ionization for these two acids relates to their acid strengths. Hydrochloric acid is a strong acid, and the equilibrium strongly favors the products. In contrast, acetic acid is a weak acid, and equilibrium for its reaction with water is reactant-favored.

Sulfuric acid, a *diprotic acid* (an acid capable of transferring two H^+ ions), reacts with water in two steps. The first step strongly favors products, whereas the second step is reactant-favored.

Strong Acid: $$\underset{\text{sulfuric acid, 100\% ionized}}{H_2SO_4(aq)} + H_2O(\ell) \rightleftharpoons \underset{\text{hydronium ion}}{H_3O^+(aq)} + \underset{\text{hydrogen sulfate ion}}{HSO_4^-(aq)}$$

Weak Acid: $$\underset{\text{hydrogen sulfate ion, <100\% ionized}}{HSO_4^-(aq)} + H_2O(\ell) \rightleftharpoons \underset{\text{hydronium ion}}{H_3O^+(aq)} + \underset{\text{sulfate ion}}{SO_4^{2-}(aq)}$$

■ **Acetic Acid** Acetic acid, CH_3CO_2H, is the substance that gives the taste and odor to vinegar. Fermentation of carbohydrates such as sugar produces ethanol (CH_3CH_2OH), and the action of bacteria on the alcohol results in acetic acid. Even a trace of acetic acid will ruin the taste of wine. This is the source of the name vinegar, which comes from the French *vin egar* meaning "sour wine." In addition to its use in food products such as salad dressings, mayonnaise, and pickles, acetic acid is used in hair-coloring products and in the manufacture of cellulose acetate, a commonly used synthetic fiber.

Ammonia, a weak base, reacts with water to produce OH^-(aq) ions. The reaction is reactant-favored at equilibrium.

Ammonia, a weak base, < 100% ionized. Equilibrium favors reactants.

$$NH_3(aq) + H_2O(\ell) \rightleftarrows NH_4^+(aq) + OH^-(aq)$$

ammonia, base weak electrolyte < 100% ionized | water | ammonium ion | hydroxide ion

Some species are described as **amphiprotic**, that is, they can function either as acids or as bases depending on the reaction. In the examples above, water functions as a base in reactions with acids (it accepts a proton) and as an acid in its reaction with ammonia (where it donates a proton to ammonia forming the ammonium ion.)

Chemistry Now™

Sign in at **www.thomsonedu.com/login** and go to Chapter 3 Contents to see:

- Screen 3.11 for a simulation on **acid ionization**
- Screen 3.12 for a self-study module on **weak and strong bases**

EXERCISE 3.8 Brønsted Acids and Bases

(a) Write a balanced equation for the reaction that occurs when H_3PO_4, phosphoric acid, donates a proton to water to form the dihydrogen phosphate ion.

(b) Write a net ionic equation showing the dihydrogen phosphate ion acting as a Brønsted acid in a reaction with water. Write another net ionic equation showing the dihydrogen phosphate ion acting as a Brønsted base in a reaction with water. What term is used to describe a species such as dihydrogen phosphate that can act either as an acid or as a base?

(c) Write a balanced net ionic equation for the reaction that occurs when the cyanide ion, CN^-, accepts a proton from water to form HCN. Is CN^- a Brønsted acid or base?

Reactions of Acids and Bases

Acids and bases in aqueous solution react to produce a salt and water. For example (Figure 3.14),

$$\underset{\text{hydrochloric acid}}{HCl(aq)} + \underset{\text{sodium hydroxide}}{NaOH(aq)} \longrightarrow \underset{\text{water}}{H_2O(\ell)} + \underset{\text{sodium chloride}}{NaCl(aq)}$$

The word "salt" has come into the language of chemistry to describe any ionic compound whose cation comes from a base (here Na^+ from NaOH) and whose anion comes from an acid (here Cl^- from HCl). The reaction of any of the acids listed in Table 3.2 with any of the hydroxide-containing bases listed there produces a salt and water.

Hydrochloric acid and sodium hydroxide are strong electrolytes in water (see Figure 3.14 and Table 3.2), so the complete ionic equation for the reaction of HCl(aq) and NaOH(aq) should be written as

$$\underbrace{H_3O^+(aq) + Cl^-(aq)}_{\text{from HCl(aq)}} + \underbrace{Na^+(aq) + OH^-(aq)}_{\text{from NaOH(aq)}} \longrightarrow \underset{\text{water}}{2\,H_2O(\ell)} + \underbrace{Na^+(aq) + Cl^-(aq)}_{\text{from salt}}$$

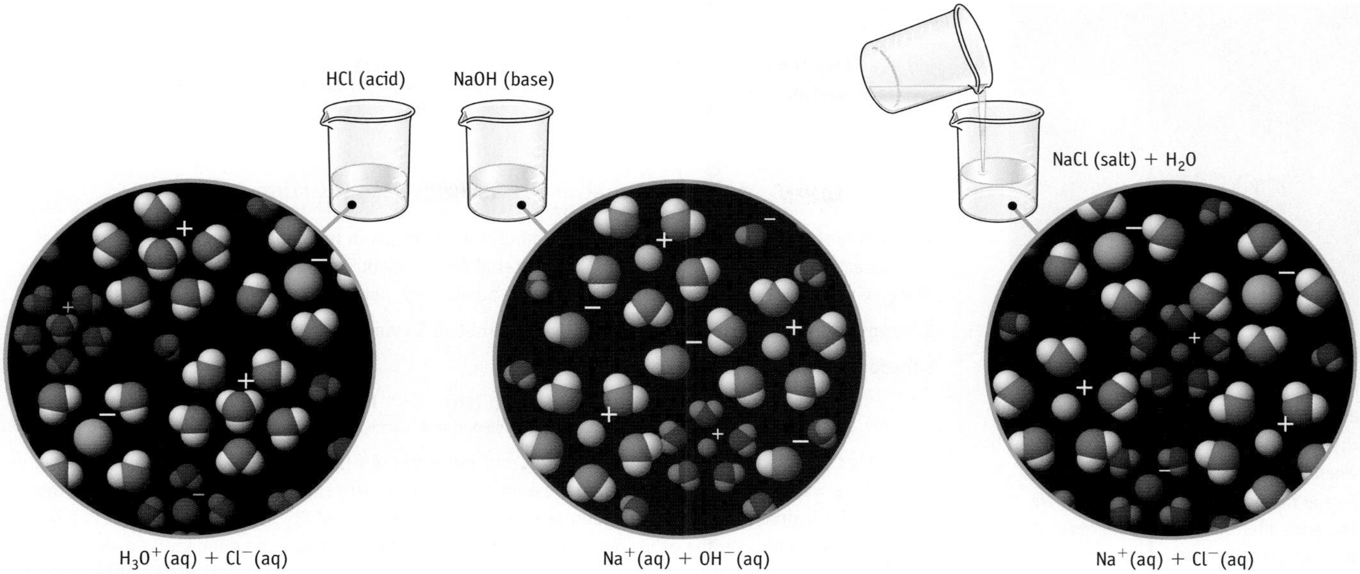

Active Figure 3.14 **An acid–base reaction, HCl and NaOH.** On mixing, the H_3O^+ and OH^- ions combine to produce H_2O, whereas the ions Na^+ and Cl^- remain in solution.

Chemistry Now™ Sign in at www.thomsonedu.com/login and go to the Chapter Contents menu to explore an interactive version of this figure accompanied by an exercise.

Because Na^+ and Cl^- ions appear on both sides of the equation, the *net ionic equation* is just the combination of the ions H_3O^+ and OH^- to give water.

$$H_3O^+(aq) + OH^-(aq) \rightarrow 2\ H_2O(\ell)$$

This is always the net ionic equation when a strong acid reacts with a strong base.

Reactions between *strong acids* and *strong bases* are called **neutralization reactions** because, on completion of the reaction, the solution is neutral if exactly the same amounts (number of moles) of the acid and base are mixed; that is, it is neither acidic nor basic. The other ions (the cation of the base and the anion of the acid) remain unchanged. If the water is evaporated, however, the cation and anion form a solid salt. In the example above, NaCl can be obtained. If nitric acid, HNO_3, and NaOH were allowed to react, the salt sodium nitrate, $NaNO_3$ (and water) would be obtained.

$$HNO_3(aq) + NaOH(aq) \rightarrow H_2O(\ell) + NaNO_3(aq)$$

If acetic acid and sodium hydroxide are mixed, the following reaction will take place.

$$CH_3CO_2H(aq) + NaOH(aq) \rightarrow H_2O(\ell) + NaCH_3CO_2(aq)$$

Because acetic acid is a weak acid and ionizes to such a small extent (Figure 3.9), the molecular species is the predominant form in aqueous solutions. In ionic equations, therefore, acetic acid is shown as molecular $CH_3CO_2H(aq)$. The *complete ionic equation* for this reaction is

$$CH_3CO_2H(aq) + Na^+(aq) + OH^-(aq) \rightarrow H_2O(\ell) + Na^+(aq) + CH_3CO_2^-(aq)$$

The only spectator ions in this equation are the sodium ions, so the *net ionic equation* is

$$CH_3CO_2H(aq) + OH^-(aq) \rightarrow H_2O(\ell) + CH_3CO_2^-(aq)$$

Charles D. Winters

Reaction of gaseous HCl and NH_3. Open dishes of aqueous ammonia and hydrochloric acid were placed side by side. When gas molecules of NH_3 and HCl escape from solution to the atmosphere and encounter one another, we observe a cloud of solid ammonium chloride, NH_4Cl.

Chemistry Now™

Sign in at **www.thomsonedu.com/login** and go to Chapter 3 Contents to see Screen 3.13 for an exercise on **acid–base reactions.**

EXAMPLE 3.5 Net Ionic Equation for an Acid–Base Reaction

Problem Ammonia, NH_3, is one of the most important chemicals in industrial economies. Not only is it used directly as a fertilizer but it is the raw material for the manufacture of nitric acid. As a base, it reacts with acids such as hydrochloric acid. Write a balanced, net ionic equation for this reaction.

Strategy Follow the general strategy outlined in Problem Solving Tip 3.1.

Solution The complete balanced equation is

$$\underset{\text{ammonia}}{NH_3(aq)} + \underset{\text{hydrochloric acid}}{HCl\,(aq)} \rightarrow \underset{\text{ammonium chloride}}{NH_4Cl(aq)}$$

Notice that the reaction produces a salt, NH_4Cl, but not water. An H^+ ion from the acid transfers to ammonia, a weak Brønsted base, to give the ammonium ion. To write the net ionic equation, start with the fact that hydrochloric acid is a strong acid and produces H_3O^+ and Cl^- ions and that NH_4Cl is a soluble, ionic compound. On the other hand, ammonia is a weak base and so is predominantly present in the solution as the molecular species, NH_3.

$$NH_3(aq) + H_3O^+(aq) + Cl^-(aq) \longrightarrow NH_4^+(aq) + Cl^-(aq) + H_2O(\ell)$$

Eliminating the spectator ion, Cl^-, we have

$$NH_3(aq) + H_3O^+(aq) \longrightarrow NH_4^+(aq) + H_2O(\ell)$$

Comment The net ionic equation shows that the important aspect of the reaction between the weak base ammonia and the strong acid HCl is the transfer of an H^+ ion from the acid to the NH_3. Any strong acid could be used here (HBr, HNO_3, $HClO_4$, H_2SO_4) and the net ionic equation would be the same.

EXERCISE 3.9 Acid–Base Reactions

Write the balanced, overall equation and the net ionic equation for the reaction of magnesium hydroxide with hydrochloric acid. *(Hint: Think about the solubility guidelines.)*

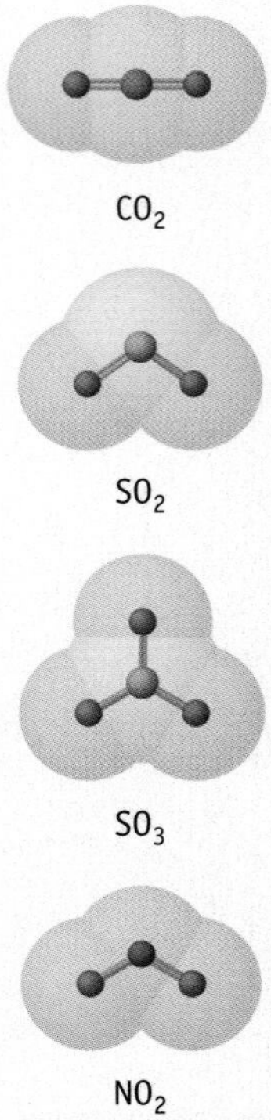

Some common nonmetal oxides that form acids in water.

Oxides of Nonmetals and Metals

Each acid shown in Table 3.2 has one or more H atoms in the molecular formula that dissociate in water to form H_3O^+ ions. There are, however, less obvious compounds that form acidic solutions. Oxides of nonmetals, such as carbon dioxide and sulfur trioxide, have no H atoms but react with water to produce H_3O^+ ions. Carbon dioxide, for example, dissolves in water to a small extent, and some of the dissolved molecules react with water to form the weak acid, carbonic acid. This acid then ionizes to a small extent to form the hydronium ion, H_3O^+, and the hydrogen carbonate (bicarbonate) ion, HCO_3^-.

$$CO_2(g) + H_2O(\ell) \rightleftharpoons H_2CO_3(aq)$$

$$H_2CO_3(aq) + H_2O(\ell) \rightleftharpoons HCO_3^-(aq) + H_3O^+(aq)$$

The HCO_3^- ion can also function as an acid, ionizing to produce H_3O^+ and the carbonate ion, CO_3^{2-}.

$$HCO_3^-(aq) + H_2O(\ell) \rightleftharpoons CO_3^{2-}(aq) + H_3O^+(aq)$$

These reactions are important in our environment and in the human body. Carbon dioxide is normally found in small amounts in the atmosphere, so rainwater is always slightly acidic. In the human body, carbon dioxide is dissolved in body fluids, where the HCO_3^- and CO_3^{2-} ions perform an important "buffering" action (► Chapter 18).

Oxides like CO_2 that can react with water to produce H_3O^+ ions are known as **acidic oxides.** Other acidic oxides include those of sulfur and nitrogen. For example, sulfur dioxide, SO_2, from human and natural sources, can react with oxygen to give sulfur trioxide, SO_3, which then reacts with water to form sulfuric acid.

$$2\ SO_2(g) + O_2(g) \rightarrow 2\ SO_3(g)$$

$$SO_3(g) + H_2O(\ell) \rightarrow H_2SO_4(aq)$$

Nitrogen dioxide, NO_2, reacts with water to give nitric and nitrous acids.

$$2\ NO_2(g) + H_2O(\ell) \rightarrow \underset{\text{nitric acid}}{HNO_3(aq)} + \underset{\text{nitrous acid}}{HNO_2(aq)}$$

Oxides of metals are called **basic oxides** because they give basic solutions if they dissolve appreciably in water. Perhaps the best example is calcium oxide, CaO, often called *lime*, or *quicklime*. Almost 20 billion kg of lime is produced annually in the United States for use in the metals and construction industries, in sewage and pollution control, in water treatment, and in agriculture. This metal oxide reacts with water to give calcium hydroxide, commonly called *slaked lime*. Although only slightly soluble in water (about 0.2 g/100 g H_2O at 10 °C), $Ca(OH)_2$ is widely used in industry as a base because it is inexpensive.

$$\underset{\text{lime}}{CaO(s)} + H_2O(\ell) \rightarrow \underset{\text{slaked lime}}{Ca(OH)_2(s)}$$

EXERCISE 3.10 Acidic and Basic Oxides

For each of the following, indicate whether you expect an acidic or basic solution when the compound dissolves in water. Remember that compounds based on elements in the same group usually behave similarly.

(a) SeO_2 **(b)** BaO **(c)** P_4O_{10}

■ **Acid Rain** Oxides of sulfur and nitrogen are the major source of the acid in what is called *acid rain*. These acidic oxides arise from the burning of fossil fuels such as coal and gasoline. The gaseous oxides mix with water and other chemicals in the troposphere, and the rain that falls is more acidic than if it contained only dissolved CO_2. When the rain falls on areas that cannot easily tolerate this greater-than-normal acidity, serious environmental problems can occur.

3.8 Gas-Forming Reactions

Several different chemical reactions lead to gas formation (Table 3.3), but the most common are those leading to CO_2 formation. All metal carbonates (and bicarbonates) react with acids to produce carbonic acid, H_2CO_3, which in turn decomposes

Charles D. Winters

FIGURE 3.15 Dissolving limestone (calcium carbonate, $CaCO_3$) in vinegar. Notice the bubbles of CO_2 rising from the surface of the limestone. This reaction shows why vinegar can be used as a household cleaning agent. It can be used, for example, to clean the calcium carbonate deposited from hard water.

TABLE 3.3 Gas-Forming Reactions

Metal carbonate or bicarbonate + acid → metal salt + $CO_2(g)$ + $H_2O(\ell)$
$Na_2CO_3(aq) + 2\ HCl(aq) \rightarrow 2\ NaCl(aq) + CO_2(g) + H_2O(\ell)$
$NaHCO_3(aq) + HCl(aq) \rightarrow NaCl(aq) + CO_2(g) + H_2O(\ell)$
Metal sulfide + acid → metal salt + $H_2S(g)$
$Na_2S(aq) + 2\ HCl(aq) \rightarrow 2\ NaCl(aq) + H_2S(g)$
Metal sulfite + acid → metal salt + $SO_2(g)$ + $H_2O(\ell)$
$Na_2SO_3(aq) + 2\ HCl(aq) \rightarrow 2\ NaCl(aq) + SO_2(g) + H_2O(\ell)$
Ammonium salt + strong base → metal salt + $NH_3(g)$ + $H_2O(\ell)$
$NH_4Cl(aq) + NaOH(aq) \rightarrow NaCl(aq) + NH_3(g) + H_2O(\ell)$

rapidly to carbon dioxide and water. For example, the reaction of calcium carbonate and hydrochloric acid is:

$$CaCO_3(s) + 2\ HCl(aq) \rightarrow CaCl_2(aq) + H_2CO_3(aq)$$

$$H_2CO_3(aq) \rightarrow H_2O(\ell) + CO_2(g)$$

$$\text{Overall reaction:}\quad CaCO_3(s) + 2\ HCl(aq) \rightarrow CaCl_2(aq) + H_2O(\ell) + CO_2(g)$$

If the reaction is done in an open beaker, most of the CO_2 gas bubbles out of the solution.

Calcium carbonate is a common residue from hard water in home heating systems and cooking utensils. Washing with vinegar is a good way to clean the system or utensils because the insoluble calcium carbonate is turned into water-soluble calcium acetate in the following gas-forming reaction (Figure 3.15).

$$2\ CH_3CO_2H(aq) + CaCO_3(s) \rightarrow Ca(CH_3CO_2)_2(aq) + H_2O(\ell) + CO_2(g)$$

What is the net ionic equation for this reaction? Acetic acid is a weak acid. Calcium carbonate is insoluble in water. Therefore, the reactants are simply $CH_3CO_2H(aq)$ and $CaCO_3(s)$. On the products side, calcium acetate is water-soluble and forms aqueous calcium and acetate ions. Water and carbon dioxide are molecular compounds, so the net ionic equation is

$$2\ CH_3CO_2H(aq) + CaCO_3(s) \rightarrow Ca^{2+}(aq) + 2\ CH_3CO_2^-(aq) + H_2O(\ell) + CO_2(g)$$

There are no spectator ions in this reaction.

Have you ever made biscuits or muffins? As you bake the dough, it rises in the oven (Figure 3.16). But what makes it rise? A gas-forming reaction occurs between an acid and baking soda, sodium hydrogen carbonate (bicarbonate of soda, $NaHCO_3$). One acid used for this purpose is tartaric acid, a weak acid found in many foods. The net ionic equation for a typical reaction would be

$H_2C_4H_4O_6(aq)$	+	$HCO_3^-(aq)$	→	$HC_4H_4O_6^-(aq)$	+	$H_2O(\ell)$	+	$CO_2(g)$
tartaric acid		hydrogen carbonate ion		hydrogen tartrate ion				

Charles D. Winters

FIGURE 3.16 Muffins rise because of a gas-forming reaction. The acid and sodium bicarbonate in baking powder produce carbon dioxide gas. The acid used in many baking powders is $CaHPO_4$, but tartaric acid and $NaAl(SO_4)_2$ are also common. (Aqueous solutions containing the aluminum ion are acidic.)

In dry baking powder, the acid and $NaHCO_3$ are kept apart by using starch as a filler. When mixed into the moist batter, however, the acid and sodium hydrogen carbonate dissolve and come into contact. Now they can react to produce CO_2, causing the dough to rise.

Chemistry Now™

Sign in at **www.thomsonedu.com/login** and go to Chapter 3 Contents to see Screen 3.14 for a tutorial identifying **the type of reaction that will result from the mixing of solutions** and to watch videos about **four of the most important gases produced in reactions.**

EXAMPLE 3.6 Gas-Forming Reactions

Problem Write a balanced equation for the reaction that occurs when nickel(II) carbonate is treated with sulfuric acid.

Strategy First, identify the reactants and write their formulas (here $NiCO_3$ and H_2SO_4). Next, recognize this as a typical gas-forming reaction (Table 3.3) between a metal carbonate and an acid. The products are water, CO_2, and a salt. The anion of the salt is the anion from the acid (SO_4^{2-}), and the cation is from the metal carbonate (Ni^{2+}).

Solution The complete, balanced equation is

$$NiCO_3(s) + H_2SO_4(aq) \longrightarrow NiSO_4(aq) + H_2O(\ell) + CO_2(g)$$

EXERCISE 3.11 Gas-Forming Reactions

(a) Barium carbonate, $BaCO_3$, is used in the brick, ceramic, glass, and chemical manufacturing industries. Write a balanced equation that shows what happens when barium carbonate is treated with nitric acid. Give the name of each of the reaction products.

(b) Write a balanced equation for the reaction of ammonium sulfate with sodium hydroxide.

3.9 Oxidation-Reduction Reactions

The terms "oxidation" and "reduction" come from reactions that have been known for centuries. Ancient civilizations learned how to change metal oxides and sulfides into the metal, that is, how to "reduce" ore to the metal. A modern example is the reduction of iron(III) oxide with carbon monoxide to give iron metal (Figure 3.17a).

Fe_2O_3 loses oxygen and is reduced.

$$Fe_2O_3(s) + 3\ CO(g) \longrightarrow 2\ Fe(s) + 3\ CO_2(g)$$

CO is the reducing agent. It gains oxygen and is oxidized.

In this reaction, carbon monoxide is the agent that brings about the reduction of iron ore to iron metal, so carbon monoxide is called the **reducing agent.**

When Fe_2O_3 is reduced by carbon monoxide, oxygen is removed from the iron ore and added to the carbon monoxide. The carbon monoxide, therefore, is "oxidized" by the addition of oxygen to give carbon dioxide. *Any process in which oxygen is added to another substance is an oxidation.* In the reaction of oxygen with magnesium,

FIGURE 3.17 Oxidation-reduction. (a) Iron ore, which is largely Fe_2O_3, is reduced to metallic iron with carbon or carbon monoxide in a blast furnace, a process done on a massive scale. (b) Burning magnesium metal in air produces magnesium oxide.

a, Jan Halaska/Photo Researchers, Inc.; b, Charles D. Winters

(a) **(b)**

for example (see Figure 3.17b), oxygen is the **oxidizing agent** because it is the agent responsible for the oxidation of magnesium.

Mg combines with oxygen and is oxidized.

$$2\ Mg(s) + O_2(g) \longrightarrow 2\ MgO(s)$$

O_2 is the oxidizing agent

Oxidation-Reduction Reactions and Electron Transfer

Not all redox reactions involve oxygen, but *all oxidation and reduction reactions can be accounted for by considering them to occur by means of a transfer of electrons between substances.* When a substance accepts electrons, it is said to be **reduced** because there is a reduction in the numerical value of the charge on an atom of the substance. In the net ionic equation for the reaction of a silver salt with copper metal, positively charged Ag^+ ions accept electrons from copper metal and are reduced to uncharged silver atoms (Figure 3.18).

Ag^+ ions accept electrons from Cu and are reduced to Ag. Ag^+ is the oxidizing agent.
$Ag^+(aq) + e^- \longrightarrow Ag(s)$

$$2\ Ag^+(aq) + Cu(s) \longrightarrow 2\ Ag(s) + Cu^{2+}(aq)$$

Cu donates electrons to Ag^+ and is oxidized to Cu^{2+}.
Cu is the reducing agent.
$Cu(s) \longrightarrow Cu^{2+}(aq) + 2\ e^-$

Because copper metal supplies the electrons and causes Ag^+ ions to be reduced, Cu is the reducing agent.

When a substance *loses electrons,* the numerical value of the charge on an atom of the substance increases. The substance is said to have been **oxidized**. In our example, copper metal releases electrons on going to Cu^{2+}, so the metal is oxidized. For this to happen, something must be available to accept the electrons from cop-

Charles D. Winters

FIGURE 3.18 The oxidation of copper metal by silver ions. A clean piece of copper wire is placed in a solution of silver nitrate, $AgNO_3$. Over time, the copper reduces Ag^+ ions, forming silver crystals, and the copper metal is oxidized to copper ions, Cu^{2+}. The blue color of the solution is due to the presence of aqueous copper(II) ions. (Sign in to ChemistryNow and Screen 3.15, Redox Reactions and Electron Transfer, to watch a video of the reaction.)

per. In this case, Ag^+ is the electron acceptor, and its charge is reduced to zero in silver metal. Therefore, Ag^+ is the "agent" that causes Cu metal to be oxidized; that is, Ag^+ is the *oxidizing agent.*

In every oxidation-reduction reaction, one reactant is reduced (and is therefore the oxidizing agent) and one reactant is oxidized (and is therefore the reducing agent). We can show this by dividing the general redox reaction $X + Y \rightarrow X^{n+} + Y^{n-}$ into two parts or *half-reactions*:

Half Reaction	Electron Transfer	Result
$X \rightarrow X^{n+} + n\,e^-$	X transfers electrons to Y	X is oxidized to X^{n+}. X is the reducing agent
$Y + n\,e^- \rightarrow Y^{n-}$	Y accepts electron from X	Y is reduced to Y^{n-}. Y is the oxidizing agent

■ **Balancing Equations for Redox Reactions** The notion that a redox reaction can be divided into an oxidizing portion and a reducing portion will lead us to a method of balancing more complex equations for redox reactions described in Chapter 20.

In the reaction of magnesium and oxygen, O_2 is reduced because it gains electrons (four electrons per molecule) on going to two oxide ions. Thus, O_2 is the oxidizing agent.

Mg releases 2 e^- per atom. Mg is oxidized to Mg^{2+} and is the reducing agent.

$$2\,Mg(s) + O_2(g) \longrightarrow 2\,MgO(s)$$

O_2 gains 4 e^- per molecule to form 2 O^{2-}. O_2 is reduced and is the oxidizing agent.

In the same reaction, magnesium is the reducing agent because it releases two electrons per atom on being oxidized to the Mg^{2+} ion (and so two Mg atoms are required to supply the four electrons required by one O_2 molecule). All redox reactions can be analyzed in a similar manner.

■ **Chemical Safety and Redox Reactions** A strong oxidizing agent and a strong reducing agent may react violently. For this reason, it would not be a good idea to store a strong oxidizing agent next to a strong reducing agent. Certain chemicals are often stored separately from the bulk of the other chemicals. Examples of chemicals that might be stored in their own separate areas are strong oxidizing agents, acids, water-reactive chemicals, and highly flammable materials.

The observations outlined so far lead to several important conclusions:

- If one substance is oxidized, another substance in the same reaction must be reduced. For this reason, such reactions are called oxidation-reduction reactions, or **redox reactions** for short.
- The reducing agent is itself oxidized, and the oxidizing agent is reduced.
- Oxidation is the opposite of reduction. For example, the removal of oxygen is reduction, and the addition of oxygen is oxidation. The gain of electrons is reduction, and the loss of electrons is oxidation.

Oxidation Numbers

How can you tell an oxidation-reduction reaction when you see one? How can you tell which substance has gained or lost electrons and so decide which substance is the oxidizing (or reducing) agent? Sometimes it is obvious. For example, if an uncombined element becomes part of a compound (Mg becomes part of MgO, for example), the reaction is definitely a redox process. If it's not obvious, then the answer is to *look for a change in the oxidation number of an element in the course of the reaction.* The **oxidation number** of an atom in a molecule or ion is defined as the charge an atom has, *or appears to have,* as determined by the following guidelines for assigning oxidation numbers.

■ **Writing Charges on Ions** Conventionally, charges on ions are written as (number, sign), whereas oxidation numbers are written as (sign, number). For example, the oxidation number of the Cu^{2+} ion is +2 and its charge is 2+.

■ **Peroxides** In peroxides, the oxidation number of oxygen is −1. For example, in hydrogen peroxide (H_2O_2), each hydrogen atom has an oxidation number of +1. To balance this, each oxygen must have an oxidation number of −1. A 3% aqueous solution of H_2O_2 is sometimes used as an antiseptic.

1. **Each atom in a pure element has an oxidation number of zero.** The oxidation number of Cu in metallic copper is 0, and it is 0 for each atom in I_2 and S_8.
2. **For monatomic ions, the oxidation number is equal to the charge on the ion.** You know that magnesium forms ions with a 2+ charge (Mg^{2+}); the oxidation number of magnesium in this ion is therefore +2.
3. **When combined with another element, fluorine always has an oxidation number of −1.**
4. **The oxidation number of O is −2 in most compounds.** The exceptions to this rule occur
   a) when oxygen is combined with fluorine (where oxygen takes on a positive oxidation number)
   b) in compounds called peroxides (such as Na_2O_2) and superoxides (such as KO_2) in which oxygen has an oxidation number of −1 and −1/2, respectively
5. **Cl, Br, and I have oxidation numbers of −1 in compounds, except when combined with oxygen and fluorine.** This means that Cl has an oxidation

A Closer Look — Are Oxidation Numbers "Real"?

Do oxidation numbers reflect the actual electric charge on an atom in a molecule or ion? With the exception of monatomic ions such as Cl^- or Na^+, the answer is no.

Oxidation numbers assume that the atoms in a molecule are positive or negative ions, which is not true. For example, in H_2O, the H atoms are not H^+ ions, and the O atoms are not O^{2-} ions. This is not to say, however, that atoms in molecules do not bear an electric charge of any kind. Calculations on water indicate the O atom has a charge of about −0.4 (or 40% of the electron charge) and the H atoms are each about +0.2.

So why use oxidation numbers? Oxidation numbers provide a way of dividing up the electrons among the atoms in a molecule or polyatomic ion. Because the distribution of electrons changes in a redox reaction, we use this method as a way to decide whether a redox reaction has occurred and to distinguish the oxidizing and reducing agents.

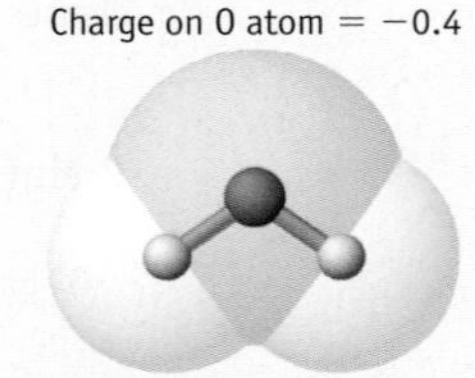

number of −1 in NaCl (in which Na is +1, as predicted by the fact that it is an element of Group 1A). In the ion ClO^-, however, the Cl atom has an oxidation number of +1 (and O has an oxidation number of −2; see guideline 4).

6. **The oxidation number of H is +1 in most compounds.** The key exception to this guideline occurs when H forms a binary compound with a metal. In such cases, the metal forms a positive ion, and H becomes a hydride ion, H^-. Thus, in CaH_2 the oxidation number of Ca is +2 (equal to the group number), and that of H is −1.
7. **The algebraic sum of the oxidation numbers for the atoms in a neutral compound must be zero; in a polyatomic ion, the sum must be equal to the ion charge.** For example, in $HClO_4$ the H atom is assigned +1, and each O atom is assigned −2. This means the Cl atom must be +7.

■ **Why Use Oxidation Numbers?** The reason for learning about oxidation numbers at this point is to be able to identify which reactions are oxidation-reduction processes and to identify the oxidizing agent and the reducing agent in a reaction. We return to a more detailed discussion of redox reactions in Chapter 20.

Chemistry•Now™

Sign in at **www.thomsonedu.com/login** and go to Chapter 3 Contents to see Screen 3.16 for exercises and a tutorial on **oxidation numbers.**

EXAMPLE 3.7 Determining Oxidation Numbers

Problem Determine the oxidation number of the indicated element in each of the following compounds or ions:

(a) aluminum in aluminum oxide, Al_2O_3

(b) phosphorus in phosphoric acid, H_3PO_4

(c) sulfur in the sulfate ion, SO_4^{2-}

(d) each Cr atom in the dichromate ion, $Cr_2O_7^{2-}$

Strategy Follow the guidelines in the text, paying particular attention to guidelines 4, 6, and 7.

Solution

(a) Al_2O_3 is a neutral compound. Assuming that O has its usual oxidation number of −2, we can solve the following algebraic equation for the oxidation number of aluminum.

Net charge on Al_2O_3 = sum of oxidation numbers for two Al atoms + three O atoms

$$0 = 2(x) + 3(-2)$$

$$0 = 2x + (-6)$$

$$+6 = 2x$$

$$x = +3$$

The oxidation number of Al must be +3, in agreement with its position in the periodic table.

(b) H_3PO_4 has an overall charge of 0. If each of the oxygen atoms has an oxidation number of −2 and each of the H atoms is +1, then we can determine the oxidation number of phosphorus as follows:

Net charge on H_3PO_4 = sum of oxidation numbers for three H atoms + one P atom + four O atoms

$$0 = 3(+1) + (x) + 4(-2)$$

$$x = +5$$

The oxidation number of phosphorus in this compound is therefore +5.

(c) The sulfate ion, SO_4^{2-}, has an overall charge of 2−. Oxygen is assigned its usual oxidation number of −2.

Net charge on SO_4^{2-} = sum of oxidation numbers of one S atom + 4 O atoms

$$2- = (x) + 4(-2)$$

$$x = +6$$

The sulfur in this ion has an oxidation number of +6.

(d) The net charge on the $Cr_2O_7^{2-}$ ion is 2-. Oxygen is assigned its usual oxidation number of -2.

Net charge on $Cr_2O_7^{2-}$ = sum of oxidation numbers for two Cr atoms + seven O atoms

$2- = 2(x) + 7(-2)$

$2- = 2x + (-14)$

$12 = 2x$

$x = +6$

The oxidation number of each chromium in this polyatomic ion is +6.

EXERCISE 3.12 Determining Oxidation Numbers

Assign an oxidation number to the underlined atom in each ion or molecule.

(a) $\underline{Fe}_2O_3$ **(b)** $H_2\underline{S}O_4$ **(c)** $\underline{C}O_3^{2-}$ **(d)** $\underline{N}O_2^{+}$

Recognizing Oxidation-Reduction Reactions

You can always tell if a reaction involves oxidation and reduction by assessing the oxidation number of each element and noting whether any of these numbers change in the course of the reaction. In many cases, however, this will not be necessary. For example, it will be obvious that a redox reaction has occurred if an uncombined element is converted to a compound or if a well-known oxidizing or reducing agent is involved (Table 3.4).

Like oxygen (O_2), the halogens (F_2, Cl_2, Br_2, and I_2) are oxidizing agents in their reactions with metals and nonmetals. An example is the reaction of chlorine with sodium metal (see Figure 1.4).

Na releases 1 e^- per atom.
Oxidation number increases.
Na is oxidized to Na^+ and is the reducing agent.

$$2\,Na(s) + Cl_2(g) \longrightarrow 2\,NaCl(s)$$

Cl_2 gains 2 e^- per molecule.
Oxidation number decreases by 1 per Cl.
Cl_2 is reduced to Cl^- and is the oxidizing agent.

NO$_2$ gas

Charles D. Winters

Copper metal oxidized to green $Cu(NO_3)_2$

Active Figure 3.19 The reaction of copper with nitric acid. Copper (a reducing agent) reacts vigorously with concentrated nitric acid (an oxidizing agent) to give the brown gas NO_2 and a deep green solution of copper(II) nitrate.

Chemistry Now™ Sign in at www.thomsonedu.com/login and go to the Chapter Contents menu to explore an interactive version of this figure accompanied by an exercise.

TABLE 3.4 Common Oxidizing and Reducing Agents

Oxidizing Agent	Reaction Product	Reducing Agent	Reaction Product
O_2, oxygen	O^{2-}, oxide ion or O combined in H_2O	H_2, hydrogen	$H^+(aq)$, hydrogen ion or H combined in H_2O or other molecule
Halogen, F_2, Cl_2, Br_2, or I_2	Halide ion, F^-, Cl^-, Br^-, or I^-	M, metals such as Na, K, Fe, and Al	M^{n+}, metal ions such as Na^+, K^+, Fe^{2+} or Fe^{3+}, and Al^{3+}
HNO_3, nitric acid	Nitrogen oxides* such as NO and NO_2	C, carbon (used to reduce metal oxides)	CO and CO_2
$Cr_2O_7^{2-}$, dichromate ion	Cr^{3+}, chromium(III) ion (in acid solution)		
MnO_4^-, permanganate ion	Mn^{2+}, manganese(II) ion (in acid solution)		

* NO is produced with dilute HNO_3, whereas NO_2 is a product of concentrated acid.

A chlorine molecule ends up as two Cl^- ions, having acquired two electrons (from two Na atoms). Thus, the oxidation number of each Cl atom has decreased from 0 to −1. This means Cl_2 has been reduced, and therefore it is the oxidizing agent.

Figure 3.19 illustrates the chemistry of another excellent oxidizing agent, nitric acid, HNO_3. Here, copper metal is oxidized to give copper(II) nitrate, and the nitrate ion is reduced to the brown gas NO_2. The net ionic equation for the reaction is

Oxidation number of Cu changes from 0 to +2. Cu is oxidized to Cu^{2+} and is the reducing agent.

$$Cu(s) + 2\ NO_3^-(aq) + 4\ H_3O^+(aq) \longrightarrow Cu^{2+}(aq) + 2\ NO_2(g) + 6\ H_2O(\ell)$$

N in NO_3^- changes from +5 to +4 in NO_2. NO_3^- is reduced to NO_2 and is the oxidizing agent.

Nitrogen has been reduced from +5 (in the NO_3^- ion) to +4 (in NO_2); therefore, the nitrate ion in acid solution is the oxidizing agent. Copper metal is the reducing agent; each metal atom has given up two electrons to produce the Cu^{2+} ion.

In the reactions of sodium with chlorine and copper with nitric acid, the metals are oxidized. This is typical of metals. In yet another example of this, aluminum metal, a good reducing agent, is capable of reducing iron(III) oxide to iron metal in a reaction called the *thermite reaction* (Figure 3.20).

$$\underset{\text{oxidizing agent}}{Fe_2O_3(s)} + \underset{\text{reducing agent}}{2\ Al(s)} \rightarrow 2\ Fe(\ell) + Al_2O_3(s)$$

Such a large quantity of energy is evolved as heat in the reaction that the iron is produced in the molten state.

Tables 3.4 and 3.5 may help you organize your thinking as you look for oxidation-reduction reactions and use their terminology.

Charles D. Winters

FIGURE 3.20 Thermite reaction. Here, Fe_2O_3 is reduced by aluminum metal to produce iron metal and aluminum oxide.

Chemistry Now™

Sign in at **www.thomsonedu.com/login** and go to Chapter 3 Contents to see Screen 3.17 for an exercise on **redox reaction.**

EXAMPLE 3.8 Oxidation-Reduction Reaction

Problem For the reaction of iron(II) ion with permanganate ion in aqueous acid,

$$5\ Fe^{2+}(aq) + MnO_4^-(aq) + 8\ H_3O^+(aq) \rightarrow 5\ Fe^{3+}(aq) + Mn^{2+}(aq) + 12\ H_2O(\ell)$$

decide which atoms are undergoing a change in oxidation number, and identify the oxidizing and reducing agents.

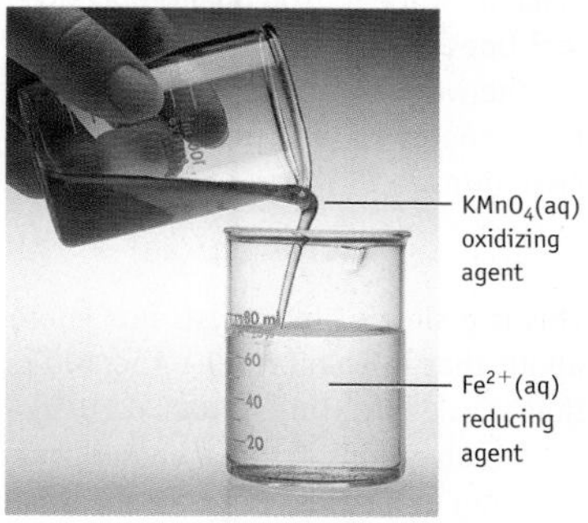

Charles D. Winters

The reaction of iron(II) ion and permanganate ion. The reaction of purple permanganate ion (MnO_4^-, the oxidizing agent) with the iron(II) ion (Fe^{2+}, the reducing agent) in acidified aqueous solution gives the nearly colorless manganese(II) ion (Mn^{2+}) and the iron(III) ion (Fe^{3+}).

TABLE 3.5 Recognizing Oxidation-Reduction Reactions

	Oxidation	Reduction
In terms of oxidation number	Increase in oxidation number of an atom	Decrease in oxidation number of an atom
In terms of electrons	Loss of electrons by an atom	Gain of electrons by an atom
In terms of oxygen	Gain of one or more O atoms	Loss of one or more O atoms

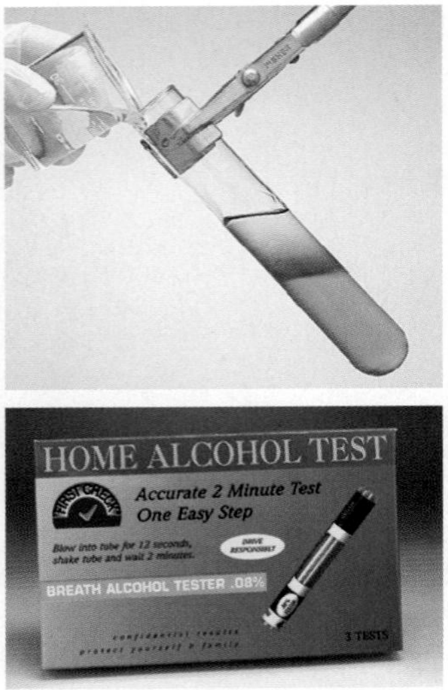

Charles D. Winters

FIGURE 3.21 The redox reaction of ethanol and dichromate ion is the basis of the test used in a Breathalyzer. When ethanol, an alcohol, is poured into a solution of orange-red dichromate ion, it reduces the dichromate ion to green chromium(III) ion. The bottom photo is a breath tester that can be purchased in grocery or drug stores. See Exercise 3.13.

Strategy Determine the oxidation numbers of the atoms in each ion or molecule involved in the reaction. Decide which atoms have increased in oxidation number (oxidation) and which have decreased in oxidation number (reduction).

Solution The Mn oxidation number in MnO_4^- is +7, and it decreases to +2 in the product, the Mn^{2+} ion. Thus, the MnO_4^- ion has been reduced and is the oxidizing agent (see Table 3.4).

$$\underset{+2}{5\ Fe^{2+}(aq)} + \underset{+7,\ -2}{MnO_4^-(aq)} + \underset{+1,\ -2}{8\ H_3O^+(aq)} \rightarrow \underset{+3}{5\ Fe^{3+}(aq)} + \underset{+2}{Mn^{2+}(aq)} + \underset{+1,\ -2}{12\ H_2O(\ell)}$$

The oxidation number of iron has increased from +2 to +3, so each Fe^{2+} ion has lost one electron upon being oxidized to Fe^{3+} (see Table 3.5). This means the Fe^{2+} ion is the reducing agent.

Comment If one of the reactants in a redox reaction is a simple substance (here Fe^{2+}), it is usually obvious whether its oxidation number has increased or decreased. Once a species has been established as having been reduced (or oxidized), you know another species has been oxidized (or reduced). It is also helpful to recognize common oxidizing and reducing agents (Table 3.4).

EXERCISE 3.13 Oxidation-Reduction Reactions

The following reaction occurs in a device for testing the breath for the presence of ethanol. Identify the oxidizing and reducing agents, the substance oxidized, and the substance reduced (Figure 3.21).

$$\underset{\text{ethanol}}{3\ CH_3CH_2OH(aq)} + \underset{\text{dichromate ion; orange-red}}{2\ Cr_2O_7^{2-}(aq)} + 16\ H_3O^+(aq) \longrightarrow$$

$$\underset{\text{acetic acid}}{3\ CH_3CO_2H(aq)} + \underset{\text{chromium(III) ion; green}}{4\ Cr^{3+}(aq)} + 27\ H_2O(\ell)$$

Case Study — Killing Bacteria with Silver

We recently read about a new washing machine that injects silver ions into the wash water, the purpose being to kill bacteria in the wash water. The advertisement told us that 100 quadrillion silver ions are injected. How does this machine work? Is 100 quadrillion silver ions a lot? Does the silver kill bacteria?

The washing machine works by using electrical energy to oxidize silver metal to give silver ions.

$$Ag(s) \rightarrow Ag^+(aq) + e^-$$

This is a simple electrolysis procedure. (More about that in Chapter 20.) And 100 quadrillion silver ions? This is 100×10^{15} ions.

Do silver ions act as a bacteriocide? There is plenty of medical evidence for this property. In fact, when you were born, the physician or nurse may have put drops of a very dilute silver nitrate solution in your eyes to treat neonatal conjunctivitis. And severely burned patients are treated with silver sulfadiazine ($C_{10}H_9AgN_4O_2S$) to prevent bacterial or fungal infections.

Charles D. Winters

Silver ions as a bacteriocide in dental floss.

The use of silver to prevent infections has a long history. Phoenicians kept wine, water, and vinegar in silver vessels. Early settlers in America put silver coins into water barrels. And you might have been born with a "silver spoon in your mouth." Babies in wealthier homes, who were fed from silver spoons and used a silver pacifier, were found to be healthier.

The historical uses of silver carry over to modern society. One can buy many different kinds of silver-containing water purifiers for the home, and dental floss coated with silver nitrate is available.

Although silver does have health benefits, beware of fraudulent claims. For example, consuming large amounts of "colloidal silver" (nanosized particles of silver suspended in water) is claimed to have health-giving properties. One person who tried this, Stan Jones, ran for the U.S. Congress in 2002 and 2006. From consuming silver, he acquired argyria, a medically irreversible condition in which the skin turns a gray-blue color.

Questions:

1. *How many moles of silver are used in a wash cycle?*
2. *What mass of silver is used?*

Answers to these questions are in Appendix Q.

EXERCISE 3.14 Recognizing Oxidation-Reduction Reactions

Decide which of the following reactions are oxidation-reduction reactions. In each case, explain your choice, and identify the oxidizing and reducing agents.

(a) $NaOH(aq) + HNO_3(aq) \rightarrow NaNO_3(aq) + H_2O(\ell)$

(b) $Cu(s) + Cl_2(g) \rightarrow CuCl_2(s)$

(c) $Na_2CO_3(aq) + 2\ HClO_4(aq) \rightarrow CO_2(g) + H_2O(\ell) + 2\ NaClO_4(aq)$

(d) $2\ S_2O_3^{2-}(aq) + I_2(aq) \rightarrow S_4O_6^{2-}(aq) + 2\ I^-(aq)$

Charles D. Winters

FIGURE 3.22 A gas-forming reaction. An Alka-Seltzer tablet contains an acid (citric acid) and sodium hydrogen carbonate ($NaHCO_3$), the reactants in a gas-forming reaction.

3.10 Classifying Reactions in Aqueous Solution

One goal of this chapter has been to explore the most common types of reactions that can occur in aqueous solution. This helps you decide, for example, that a gas-forming reaction occurs when an Alka-Seltzer tablet (containing citric acid and $NaHCO_3$) is dropped into water (Figure 3.22).

$$\underset{\text{citric acid}}{H_3C_6H_5O_7(aq)} + \underset{\text{hydrogen carbonate ion}}{HCO_3^-(aq)} \longrightarrow \underset{\text{dihydrogen citrate ion}}{H_2C_6H_5O_7^-(aq)} + H_2O(\ell) + CO_2(g)$$

We have examined four types of reactions in aqueous solution: precipitation reactions, acid-base reactions, gas-forming reactions, and oxidation-reduction reactions. Three of these four (precipitation, acid-base, and gas-forming) fall into the category of exchange reactions.

Precipitation Reactions (see Figure 3.11): Ions combine in solution to form an insoluble reaction product.

Overall Equation

$$Pb(NO_3)_2(aq) + 2\ KI(aq) \longrightarrow PbI_2(s) + 2\ KNO_3(aq)$$

Net Ionic Equation

$$Pb^{2+}(aq) + 2\ I^-(aq) \longrightarrow PbI_2(s)$$

Acid–Base Reactions (see Figures 3.12 and 3.14): Water is a product of many acid–base reactions, and the cation of the base and the anion of the acid form a salt.

Overall Equation for the Reaction of a Strong Acid and a Strong Base

$$HNO_3(aq) + KOH(aq) \longrightarrow HOH(\ell) + KNO_3(aq)$$

Net Ionic Equation for the Reaction of a Strong Acid and a Strong Base

$$H_3O^+(aq) + OH^-(aq) \rightarrow 2\ H_2O(\ell)$$

Overall Equation for the Reaction of a Weak Acid and a Strong Base

$$CH_3CO_2H(aq) + NaOH(aq) \rightarrow NaCH_3CO_2(aq) + HOH(\ell)$$

Net Ionic Equation for the Reaction of a Weak Acid and a Strong Base

$$CH_3CO_2H(aq) + OH^-(aq) \rightarrow CH_3CO_2^-(aq) + H_2O(\ell)$$

Gas-Forming Reactions (see Figures 3.15 and 3.22): The most common examples involve metal carbonates and acids, but others exist (see Table 3.3). One product with a metal carbonate is always carbonic acid, H_2CO_3, most of which decomposes to H_2O and CO_2. Carbon dioxide is the gas in the bubbles you see during these reactions.
Overall Equation:

$$CuCO_3(s) + 2\ HNO_3(aq) \rightarrow Cu(NO_3)_2(aq) + CO_2(g) + H_2O(\ell)$$

Net Ionic Equation

$$CuCO_3(s) + 2\ H_3O^+(aq) \rightarrow Cu^{2+}(aq) + CO_2(g) + 3\ H_2O(\ell)$$

Oxidation-Reduction Reactions (see Figure 3.18): These reactions are *not* ion exchange reactions. Rather, electrons are transferred from one material to another.
Overall Equation

$$Cu(s) + 2\ AgNO_3(aq) \rightarrow Cu(NO_3)_2(aq) + 2\ Ag(s)$$

Net Ionic Equation

$$Cu(s) + 2\ Ag^+(aq) \rightarrow Cu^{2+}(aq) + 2\ Ag(s)$$

These four types of reactions are usually easy to recognize, but keep in mind that a reaction may fall into more than one category. For example, barium hydroxide reacts readily with sulfuric acid to give barium sulfate and water, a reaction that is both a precipitation and an acid–base reaction.

$$Ba(OH)_2(aq) + H_2SO_4(aq) \rightarrow BaSO_4(s) + 2\ H_2O(\ell)$$

Chemistry Now™

Sign in at **www.thomsonedu.com/login** and go to Chapter 3 Contents to see Screen 3.8 for a self-study module on the **four reaction types.**

EXAMPLE 3.9 Types of Reactions

Problem Classify each of the following reactions as precipitation, acid–base, gas-forming, or oxidation-reduction.

(a) $2\ HNO_3(aq) + Ca(OH)_2(s) \rightarrow Ca(NO_3)_2(aq) + 2\ H_2O(\ell)$

(b) $2\ MnO_4^-(aq) + 5\ HSO_3^-(aq) + H_3O^+(aq) \rightarrow 2\ Mn^{2+}(aq) + 5\ SO_4^{2-}(aq) + 4\ H_2O(\ell)$

Strategy An acid–base reaction is usually easy to identify. Next, check the oxidation numbers of each element. If they change, then it is a redox reaction. If there is no change, then check to see if it is a simple precipitation or gas-forming process.

Solution Reaction (a) involves a common acid (nitric acid, HNO_3) and a common base [calcium hydroxide, $Ca(OH)_2$]; it produces a salt, calcium nitrate, and water. It is an acid–base reaction. Reaction (b) is a redox reaction because the oxidation numbers of S and Mn change.

$$2\ MnO_4^-(aq) + 5\ HSO_3^-(aq) + H_3O^+(aq) \rightarrow 2\ Mn^{2+}(aq) + 5\ SO_4^{2-}(aq) + 4\ H_2O(\ell)$$

+7, −2   +1, +4, −2   +1, −2   +2   +6, −2   +1, −2

The oxidation number of S changes from +4 to +6, and that of Mn changes from +7 to +2. Therefore, permanganate ion, MnO_4^-, has been reduced (and is the oxidizing agent), and HSO_3^- has been oxidized (and is the reducing agent).

Comment Note that no changes occur in the oxidation numbers of the elements in reaction (a).

$HNO_3(aq)$	+	$Ca(OH)_2(s)$	$\rightarrow$	$Ca(NO_3)_2(aq)$	+	$2\ H_2O(\ell)$
+1, +5, −2		+2, −2, +1		+2, +5, −2		+1, −2

EXERCISE 3.15 Classifying and Predicting Reactions

Classify each of the following reactions as a precipitation, acid–base, gas-forming reaction, or oxidation-reduction reaction. Predict the products of the reaction, and then balance the completed equation. Write the net ionic equation for each.

(a) $CuCO_3(s) + H_2SO_4(aq) \rightarrow$

(b) $Ga(s) + O_2(g) \rightarrow$

(c) $Ba(OH)_2(s) + HNO_3(aq) \rightarrow$

(d) $CuCl_2(aq) + (NH_4)_2S(aq) \rightarrow$

Chapter Goals Revisited

Now that you have studied this chapter, you should ask if you have met the chapter goals. In particular, you should be able to:

Balance equations for simple chemical reactions.

a. Understand the information conveyed by a balanced chemical equation (Section 3.1).

b. Balance simple chemical equations (Section 3.2). Study Question(s) assignable in OWL: 2, 4, 47, 54.

Understand the nature and characteristics of chemical equilibria.

a. Recognize that chemical reactions are reversible (Section 3.3).

b. Describe what is meant by the term dynamic equilibrium.

c. Recognize the difference between reactant-favored and product-favored reactions.

Understand the nature of ionic substances dissolved in water.

a. Explain the difference between electrolytes and nonelectrolytes, and recognize examples of each (Section 3.5 and Figure 3.9).

b. Predict the solubility of ionic compounds in water (Section 3.5 and Figure 3.10). Study Question(s) assignable in OWL: 9, 11, 13, 15, 55, 57, 60, 65, 67; Go Chemistry Module 5.

c. Recognize what ions are formed when an ionic compound or acid or base dissolves in water (Sections 3.5–3.7). Study Question(s) assignable in OWL: 13, 15, 60.

Recognize common acids and bases, and understand their behavior in aqueous solution.

a. Know the names and formulas of common acids and bases (Section 3.7 and Table 3.2). Study Question(s) assignable in OWL: 21, 26, 27.

b. Categorize acids and bases as strong or weak.

c. Define and use the Arrhenius concept of acids and bases.

d. Define and use the Brønsted–Lowry concept of acids and bases.

e. Appreciate when a substance can be amphiprotic.

f. Recognize the Brønsted acid and base in a reaction. Study Question(s) assignable in OWL: 21, 27, 28.

Chemistry Now™ Sign in at **www.thomsonedu.com/login** to:

- Assess your understanding with Study Questions in OWL keyed to each goal in the Goals and Homework menu for this chapter
- For quick review, download Go Chemistry mini-lecture flashcard modules (or purchase them at **www.ichapters.com**)
- Check your readiness for an exam by taking the Pre-Test and exploring the modules recommended in your Personalized Study plan.

Access **How Do I Solve It?** tutorials on how to approach problem solving using concepts in this chapter.

Recognize the common types of reactions in aqueous solution.

a. Recognize the key characteristics of four types of reactions in aqueous solution.

Reaction Type	Key Characteristic
Precipitation	Formation of an insoluble compound
Acid–strong base	Formation of a salt and water; H^+ ion transfer
Gas-forming	Evolution of a water-insoluble gas such as CO_2
Oxidation-reduction	Transfer of electrons (with changes in oxidation numbers)

Study Question(s) assignable in OWL: 47, 48, 51, 52, 70, 76–78.

b. Predict the products of precipitation reactions (Section 3.6), acid–base reactions (Section 3.7), and gas-forming reactions (Section 3.8). These are all examples of exchange reactions, which involve the exchange of anions between the cations involved in the reaction. Study Question(s) assignable in OWL: 17, 19, 21, 27, 37–39, 71.

Write chemical equations for the common types of reactions in aqueous solution.

a. Write overall balanced equations for precipitation, acid–base, and gas-forming reactions.

b. Write net ionic equations (Sections 3.6–3.8). Study Question(s) assignable in OWL: 17, 19, 33, 63, 67, 68; Go Chemistry Module 6.

c. Understand that the net ionic equation for the reaction of a strong acid with a strong base is $H_3O^+(aq) + OH^-(aq) \rightarrow 2\ H_2O(\ell)$ (Section 3.7).

Recognize common oxidizing and reducing agents, and identify oxidation-reduction reactions.

a. Determine oxidation numbers of elements in a compound and understand that these numbers represent the charge an atom has, or appears to have, when the electrons of the compound are counted according to a set of guidelines (Section 3.9). Study Question(s) assignable in OWL: 41, 42, 69.

b. Identify oxidation-reduction reactions (often called redox reactions), and identify the oxidizing and reducing agents and substances oxidized and reduced in the reaction (Section 3.9 and Tables 3.4 and 3.5). Study Question(s) assignable in OWL: 44–46, 61, 73.

STUDY QUESTIONS

Online homework for this chapter may be assigned in OWL.

▲ denotes challenging questions.

■ denotes questions assignable in OWL.

Blue-numbered questions have answers in Appendix O and fully-worked solutions in the *Student Solutions Manual*.

Practicing Skills

Balancing Equations

(See Example 3.1 and ChemistryNow Screen 3.4.)

1. Write a balanced chemical equation for the combustion of liquid pentane, C_5H_{12}.

2. ■ Write balanced chemical equations for the following reactions:
   (a) production of ammonia, $NH_3(g)$, by combining $N_2(g)$ and $H_2(g)$
   (b) production of methanol, $CH_3OH(\ell)$ by combining $H_2(g)$ and $CO(g)$
   (c) production of sulfuric acid by combining sulfur, oxygen, and water

3. Balance the following equations:
   (a) $Cr(s) + O_2(g) \rightarrow Cr_2O_3(s)$
   (b) $Cu_2S(s) + O_2(g) \rightarrow Cu(s) + SO_2(g)$
   (c) $C_6H_5CH_3(\ell) + O_2(g) \rightarrow H_2O(\ell) + CO_2(g)$

4. ■ Balance the following equations:
   (a) $Cr(s) + Cl_2(g) \rightarrow CrCl_3(s)$
   (b) $SiO_2(s) + C(s) \rightarrow Si(s) + CO(g)$
   (c) $Fe(s) + H_2O(g) \rightarrow Fe_3O_4(s) + H_2(g)$

5. Balance the following equations, and name each reactant and product:
(a) $Fe_2O_3(s) + Mg(s) \rightarrow MgO(s) + Fe(s)$
(b) $AlCl_3(s) + NaOH(aq) \rightarrow Al(OH)_3(s) + NaCl(aq)$
(c) $NaNO_3(s) + H_2SO_4(\ell) \rightarrow Na_2SO_4(s) + HNO_3(\ell)$
(d) $NiCO_3(s) + HNO_3(aq) \rightarrow Ni(NO_3)_2(aq) + CO_2(g) + H_2O(\ell)$

6. Balance the following equations, and name each reactant and product:
(a) $SF_4(g) + H_2O(\ell) \rightarrow SO_2(g) + HF(\ell)$
(b) $NH_3(aq) + O_2(aq) \rightarrow NO(g) + H_2O(\ell)$
(c) $BF_3(g) + H_2O(\ell) \rightarrow HF(aq) + H_3BO_3(aq)$

Chemical Equilibrium

(See Sections 3.3 and 3.5.)

7. Equal amounts of two acids—HCl and HCO_2H (formic acid)—are placed in solution. When equilibrium has been achieved, the HCl solution has a much greater electrical conductivity than the HCO_2H solution. Which reaction is more product favored at equilibrium?

$$HCl(aq) + H_2O(\ell) \rightleftharpoons H_3O^+(aq) + Cl^-(aq)$$

$$HCO_2H(aq) + H_2O(\ell) \rightleftharpoons H_3O^+(aq) + HCO_2^-(aq)$$

8. Equal amounts of two compounds, AgBr and H_3PO_4, are placed in solution. When equilibrium has been achieved, the H_3PO_4 solution has a greater electrical conductivity than the AgBr solution (which is almost a nonelectrolyte). Which reaction is more product favored at equilibrium?

$$AgBr(s) \rightleftharpoons Ag^+(aq) + Br^-(aq)$$

$$H_3PO_4(aq) + H_2O(\ell) \rightleftharpoons H_3O^+(aq) + H_2PO_4^-(aq)$$

Ions and Molecules in Aqueous Solution

(See Exercise 3.3, Example 3.2, and ChemistryNow Screens 3.5–3.7.)

9. ■ What is an electrolyte? How can you differentiate experimentally between a weak electrolyte and a strong electrolyte? Give an example of each.

10. Name two acids that are strong electrolytes and one acid that is a weak electrolyte. Name two bases that are strong electrolytes and one base that is a weak electrolyte.

11. ■ Which compound or compounds in each of the following groups is (are) expected to be soluble in water?
(a) CuO, $CuCl_2$, $FeCO_3$
(b) AgI, Ag_3PO_4, $AgNO_3$
(c) K_2CO_3, KI, $KMnO_4$

12. Which compound or compounds in each of the following groups is (are) expected to be soluble in water?
(a) $BaSO_4$, $Ba(NO_3)_2$, $BaCO_3$
(b) Na_2SO_4, $NaClO_4$, $NaCH_3CO_2$
(c) AgBr, KBr, Al_2Br_6

13. ■ The following compounds are water-soluble. What ions are produced by each compound in aqueous solution?
(a) KOH (c) $LiNO_3$
(b) K_2SO_4 (d) $(NH_4)_2SO_4$

14. The following compounds are water-soluble. What ions are produced by each compound in aqueous solution?
(a) KI (c) K_2HPO_4
(b) $Mg(CH_3CO_2)_2$ (d) NaCN

15. ■ Decide whether each of the following is water-soluble. If soluble, tell what ions are produced.
(a) Na_2CO_3 (c) NiS
(b) $CuSO_4$ (d) $BaBr_2$

16. Decide whether each of the following is water-soluble. If soluble, tell what ions are produced.
(a) $NiCl_2$ (c) $Pb(NO_3)_2$
(b) $Cr(NO_3)_3$ (d) $BaSO_4$

Precipitation Reactions and Net Ionic Equations

(See Examples 3.3 and 3.4 and ChemistryNow Screens 3.9 and 3.10.)

17. ■ Balance the equation for the following precipitation reaction, and then write the net ionic equation. Indicate the state of each species (s, ℓ, aq, or g).

$$CdCl_2 + NaOH \rightarrow Cd(OH)_2 + NaCl$$

18. Balance the equation for the following precipitation reaction, and then write the net ionic equation. Indicate the state of each species (s, ℓ, aq, or g).

$$Ni(NO_3)_2 + Na_2CO_3 \rightarrow NiCO_3 + NaNO_3$$

19. ■ Predict the products of each precipitation reaction. Balance the completed equation, and then write the net ionic equation.
(a) $NiCl_2(aq) + (NH_4)_2S(aq) \rightarrow ?$
(b) $Mn(NO_3)_2(aq) + Na_3PO_4(aq) \rightarrow ?$

20. Predict the products of each precipitation reaction. Balance the completed equation, and then write the net ionic equation.
(a) $Pb(NO_3)_2(aq) + KBr(aq) \rightarrow ?$
(b) $Ca(NO_3)_2(aq) + KF(aq) \rightarrow ?$
(c) $Ca(NO_3)_2(aq) + Na_2C_2O_4(aq) \rightarrow ?$

Acids and Bases and Their Reactions

(See Exercise 3.8, Example 3.5, Exercise 3.10, and ChemistryNow Screens 3.11–3.13.)

21. ■ Write a balanced equation for the ionization of nitric acid in water.

22. Write a balanced equation for the ionization of perchloric acid in water.

23. Oxalic acid, $H_2C_2O_4$, which is found in certain plants, can provide two hydronium ions in water. Write balanced equations (like those for sulfuric acid on page 135) to show how oxalic acid can supply one and then a second H_3O^+ ion.

24. Phosphoric acid can supply one, two, or three H_3O^+ ions in aqueous solution. Write balanced equations (like those for sulfuric acid on page 135) to show this successive loss of hydrogen ions.

25. Write a balanced equation for reaction of the basic oxide, magnesium oxide, with water.

26. ■ Write a balanced equation for the reaction of sulfur trioxide with water.

27. ■ Complete and balance the following acid–base equations. Name the reactants and products. Decide which is the Brønsted acid and which is the Brønsted base.
(a) $CH_3CO_2H(aq) + Mg(OH)_2(s) \rightarrow$
(b) $HClO_4(aq) + NH_3(aq) \rightarrow$

28. ■ Complete and balance the following acid–base equations. Name the reactants and products. Decide which is the Brønsted acid and which is the Brønsted base.
(a) $H_3PO_4(aq) + KOH(aq) \rightarrow$
(b) $H_2C_2O_4(aq) + Ca(OH)_2(s) \rightarrow$
($H_2C_2O_4$ is oxalic acid, an acid capable of donating two H^+ ions.)

29. Write a balanced equation for the reaction of barium hydroxide with nitric acid.

30. Write a balanced equation for the reaction of aluminum hydroxide with sulfuric acid.

31. Name two strong Brønsted acids and one strong Brønsted base.

32. Name three weak Brønsted acids and one weak Brønsted base.

Writing Net Ionic Equations
(See Examples 3.4 and 3.5 and ChemistryNow Screen 3.10.)

33. ■ Balance the following equations, and then write the net ionic equation.
(a) $(NH_4)_2CO_3(aq) + Cu(NO_3)_2(aq) \rightarrow CuCO_3(s) + NH_4NO_3(aq)$
(b) $Pb(OH)_2(s) + HCl(aq) \rightarrow PbCl_2(s) + H_2O(\ell)$
(c) $BaCO_3(s) + HCl(aq) \rightarrow BaCl_2(aq) + H_2O(\ell) + CO_2(g)$
(d) $CH_3CO_2H(aq) + Ni(OH)_2(s) \rightarrow Ni(CH_3CO_2)_2(aq) + H_2O(\ell)$

34. ■ Balance the following equations, and then write the net ionic equation:
(a) $Zn(s) + HCl(aq) \rightarrow H_2(g) + ZnCl_2(aq)$
(b) $Mg(OH)_2(s) + HCl(aq) \rightarrow MgCl_2(aq) + H_2O(\ell)$
(c) $HNO_3(aq) + CaCO_3(s) \rightarrow Ca(NO_3)_2(aq) + H_2O(\ell) + CO_2(g)$
(d) $(NH_4)_2S(aq) + FeCl_3(aq) \rightarrow NH_4Cl(aq) + Fe_2S_3(s)$

35. Balance the following equations, and then write the net ionic equation. Show states for all reactants and products (s, ℓ, g, aq).
(a) the reaction of silver nitrate and potassium iodide to give silver iodide and potassium nitrate
(b) the reaction of barium hydroxide and nitric acid to give barium nitrate and water
(c) the reaction of sodium phosphate and nickel(II) nitrate to give nickel(II) phosphate and sodium nitrate

36. Balance each of the following equations, and then write the net ionic equation. Show states for all reactants and products (s, ℓ, g, aq).
(a) the reaction of sodium hydroxide and iron(II) chloride to give iron(II) hydroxide and sodium chloride
(b) the reaction of barium chloride with sodium carbonate to give barium carbonate and sodium chloride
(c) the reaction of ammonia with phosphoric acid

Gas-Forming Reactions
(See Example 3.6 and ChemistryNow Screen 3.14.)

37. ■ Siderite is a mineral consisting largely of iron(II) carbonate. Write an overall, balanced equation for its reaction with nitric acid, and name the products.

38. ■ The beautiful red mineral rhodochrosite is manganese(II) carbonate. Write an overall, balanced equation for the reaction of the mineral with hydrochloric acid, and name the products.

Charles D. Winters

Rhodochrosite, a mineral consisting largely of $MnCO_3$.

39. ■ Write an overall, balanced equation for the reaction of $(NH_4)_2S$ with HBr, and name the reactants and products.

40. Write an overall, balanced equation for the reaction of Na_2SO_3 with CH_3CO_2H, and name the reactants and products.

Oxidation Numbers
(See Example 3.7 and ChemistryNow Screens 3.15 and 3.16.)

41. ■ Determine the oxidation number of each element in the following ions or compounds.
(a) BrO_3^- (d) CaH_2
(b) $C_2O_4^{2-}$ (e) H_4SiO_4
(c) F^- (f) HSO_4^-

42. ■ Determine the oxidation number of each element in the following ions or compounds.
(a) PF_6^- (d) N_2O_5
(b) $H_2AsO_4^-$ (e) $POCl_3$
(c) UO^{2+} (f) XeO_4^{2-}

Oxidation–Reduction Reactions
(See Example 3.8 and ChemistryNow Screen 3.17.)

43. Which two of the following reactions are oxidation–reduction reactions? Explain your answer in each case. Classify the remaining reaction.
(a) $Zn(s) + 2\ NO_3^-(aq) + 4\ H_3O^+(aq) \rightarrow Zn^{2+}(aq) + 2\ NO_2(g) + 6\ H_2O(\ell)$
(b) $Zn(OH)_2(s) + H_2SO_4(aq) \rightarrow ZnSO_4(aq) + 2\ H_2O(\ell)$
(c) $Ca(s) + 2\ H_2O(\ell) \rightarrow Ca(OH)_2(s) + H_2(g)$

44. ■ Which two of the following reactions are oxidation–reduction reactions? Explain your answer briefly. Classify the remaining reaction.
(a) $CdCl_2(aq) + Na_2S(aq) \rightarrow CdS(s) + 2\ NaCl(aq)$
(b) $2\ Ca(s) + O_2(g) \rightarrow 2\ CaO(s)$
(c) $4\ Fe(OH)_2(s) + 2\ H_2O(\ell) + O_2(g) \rightarrow 4\ Fe(OH)_3(aq)$

45. ■ In the following reactions, decide which reactant is oxidized and which is reduced. Designate the oxidizing agent and the reducing agent.
(a) $C_2H_4(g) + 3\ O_2(g) \rightarrow 2\ CO_2(g) + 2\ H_2O(\ell)$
(b) $Si(s) + 2\ Cl_2(g) \rightarrow SiCl_4(\ell)$

▲ more challenging ■ in OWL Blue-numbered questions answered in Appendix O

46. ■ In the following reactions, decide which reactant is oxidized and which is reduced. Designate the oxidizing agent and the reducing agent.
(a) $Cr_2O_7^{2-}(aq) + 3\ Sn^{2+}(aq) + 14\ H_3O^+(aq) \longrightarrow 2\ Cr^{3+}(aq) + 3\ Sn^{4+}(aq) + 21\ H_2O(\ell)$
(b) $FeS(s) + 3\ NO_3^-(aq) + 4\ H_3O^+(aq) \longrightarrow 3\ NO(g) + SO_4^{2-}(aq) + Fe^{3+}(aq) + 6\ H_2O(\ell)$

Types of Reactions in Aqueous Solution

(See Example 3.9 and ChemistryNow Screen 3.8.)

47. ■ Balance the following equations, and then classify each as a precipitation, an acid–base, or a gas-forming reaction.
(a) $Ba(OH)_2(aq) + HCl(aq) \longrightarrow BaCl_2(aq) + H_2O(\ell)$
(b) $HNO_3(aq) + CoCO_3(s) \longrightarrow Co(NO_3)_2(aq) + H_2O(\ell) + CO_2(g)$
(c) $Na_3PO_4(aq) + Cu(NO_3)_2(aq) \longrightarrow Cu_3(PO_4)_2(s) + NaNO_3(aq)$

48. ■ Balance the following equations, and then classify each as a precipitation, an acid–base, or a gas-forming reaction.
(a) $K_2CO_3(aq) + Cu(NO_3)_2(aq) \longrightarrow CuCO_3(s) + KNO_3(aq)$
(b) $Pb(NO_3)_2(aq) + HCl(aq) \longrightarrow PbCl_2(s) + HNO_3(aq)$
(c) $MgCO_3(s) + HCl(aq) \longrightarrow MgCl_2(aq) + H_2O(\ell) + CO_2(g)$

49. Balance the following equations, and then classify each as a precipitation, an acid–base, or a gas-forming reaction. Show states for the products (s, ℓ, g, aq), and then balance the completed equation. Write the net ionic equation.
(a) $MnCl_2(aq) + Na_2S(aq) \longrightarrow MnS + NaCl$
(b) $K_2CO_3(aq) + ZnCl_2(aq) \longrightarrow ZnCO_3 + KCl$

50. Balance the following equations, and then classify each as a precipitation, an acid–base, or a gas-forming reaction. Write the net ionic equation.
(a) $Fe(OH)_3(s) + HNO_3(aq) \longrightarrow Fe(NO_3)_3 + H_2O$
(b) $FeCO_3(s) + HNO_3(aq) \longrightarrow Fe(NO_3)_2 + CO_2 + H_2O$

51. ■ Balance each of the following equations, and classify them as precipitation, acid–base, gas-forming, or oxidation-reduction reactions.
(a) $CuCl_2 + H_2S \longrightarrow CuS + HCl$
(b) $H_3PO_4 + KOH \longrightarrow H_2O + K_3PO_4$
(c) $Ca + HBr \longrightarrow H_2 + CaBr_2$
(d) $MgCl_2 + H_2O \longrightarrow Mg(OH)_2 + HCl$

52. ■ ▲ Complete and balance the equations below, and classify them as precipitation, acid–base, gas-forming, or oxidation-reduction reactions.
(a) $NiCO_3 + H_2SO_4 \longrightarrow ?$
(b) $Co(OH)_2 + HBr \longrightarrow ?$
(c) $AgCH_3CO_2 + NaCl \longrightarrow ?$
(d) $NiO + CO \longrightarrow ?$

General Questions

These questions are not designated as to type or location in the chapter. They may combine concepts from several chapters.

53. Balance the following equations:
(a) the synthesis of urea, a common fertilizer

$$CO_2(g) + NH_3(g) \longrightarrow NH_2CONH_2(s) + H_2O(\ell)$$

(b) reactions used to make uranium(VI) fluoride for the enrichment of natural uranium
$UO_2(s) + HF(aq) \longrightarrow UF_4(s) + H_2O(\ell)$
$UF_4(s) + F_2(g) \longrightarrow UF_6(s)$
(c) the reaction to make titanium(IV) chloride, which is then converted to titanium metal
$TiO_2(s) + Cl_2(g) + C(s) \longrightarrow TiCl_4(\ell) + CO(g)$
$TiCl_4(\ell) + Mg(s) \longrightarrow Ti(s) + MgCl_2(s)$

54. ■ Balance the following equations:
(a) reaction to produce "superphosphate" fertilizer

$$Ca_3(PO_4)_2(s) + H_2SO_4(aq) \longrightarrow Ca(H_2PO_4)_2(aq) + CaSO_4(s)$$

(b) reaction to produce diborane, B_2H_6

$$NaBH_4(s) + H_2SO_4(aq) \longrightarrow B_2H_6(g) + H_2(g) + Na_2SO_4(aq)$$

(c) reaction to produce tungsten metal from tungsten(VI) oxide

$$WO_3(s) + H_2(g) \longrightarrow W(s) + H_2O(\ell)$$

(d) decomposition of ammonium dichromate

$$(NH_4)_2Cr_2O_7(s) \longrightarrow N_2(g) + H_2O(\ell) + Cr_2O_3(s)$$

55. ■ Give a formula for each of the following:
(a) a soluble compound containing the bromide ion
(b) an insoluble hydroxide
(c) an insoluble carbonate
(d) a soluble nitrate-containing compound
(e) a weak Brønsted acid

56. Give the formula for the following compounds:
(a) a soluble compound containing the acetate ion
(b) an insoluble sulfide
(c) a soluble hydroxide
(d) an insoluble chloride
(e) a strong Brønsted base

57. ■ Which of the following copper(II) salts are soluble in water and which are insoluble: $Cu(NO_3)_2$, $CuCO_3$, $Cu_3(PO_4)_2$, $CuCl_2$?

58. Name two anions that combine with Al^{3+} ion to produce water-soluble compounds.

59. Identify the spectator ion or ions in the reaction of nitric acid and magnesium hydroxide, and write the net ionic equation. What type of reaction is this?

$$2\ H_3O^+(aq) + 2\ NO_3^-(aq) + Mg(OH)_2(s) \longrightarrow 4\ H_2O(\ell) + Mg^{2+}(aq) + 2\ NO_3^-(aq)$$

60. ■ Identify and name the water-insoluble product in each reaction and write the net ionic equation:
(a) $CuCl_2(aq) + H_2S(aq) \longrightarrow CuS + 2\ HCl$
(b) $CaCl_2(aq) + K_2CO_3(aq) \longrightarrow 2\ KCl + CaCO_3$
(c) $AgNO_3(aq) + NaI(aq) \longrightarrow AgI + NaNO_3$

61. ■ Bromine is obtained from sea water by the following reaction:

$$Cl_2(g) + 2\ NaBr(aq) \longrightarrow 2\ NaCl(aq) + Br_2(\ell)$$

(a) What has been oxidized? What has been reduced?
(b) Identify the oxidizing and reducing agents.

62. Identify each of the following substances as a likely oxidizing or reducing agent: HNO_3, Na, Cl_2, O_2, $KMnO_4$.

63. ■ The mineral dolomite contains magnesium carbonate. This reacts with hydrochloric acid.

$$MgCO_3(s) + 2\ HCl(aq) \longrightarrow CO_2(g) + MgCl_2(aq) + H_2O(\ell)$$

(a) Write the net ionic equation for the reaction of magnesium carbonate and hydrochloric acid, and name the spectator ions.
(b) What type of reaction is this?

64. Ammonium sulfide, $(NH_4)_2S$, reacts with $Hg(NO_3)_2$ to produce HgS and NH_4NO_3.
(a) Write the overall balanced equation for the reaction. Indicate the state (s, aq) for each compound.
(b) Name each compound.
(c) What type of reaction is this?

65. ■ What species (atoms, molecules, or ions) are present in an aqueous solution of each of the following compounds? Decide which are Brønsted acids or bases and whether they are strong or weak.
(a) NH_3 (b) CH_3CO_2H (c) NaOH (d) HBr

66. (a) Name two water-soluble compounds containing the Cu^{2+} ion. Name two water-insoluble compounds based on the Cu^{2+} ion.
(b) Name two water-soluble compounds containing the Ba^{2+} ion. Name two water-insoluble compounds based on the Ba^{2+} ion.

67. ■ Balance equations for these reactions that occur in aqueous solution, and then classify each one as a precipitation, acid–base, or gas-forming reaction. Show states for the products (s, ℓ, g, aq), give their names, and write the net ionic equation.
(a) $K_2CO_3 + HClO_4 \longrightarrow KClO_4 + CO_2 + H_2O$
(b) $FeCl_2 + (NH_4)_2S \longrightarrow FeS + NH_4Cl$
(c) $Fe(NO_3)_2 + Na_2CO_3(aq) \longrightarrow FeCO_3 + NaNO_3$
(d) $NaOH + FeCl_3 \longrightarrow NaCl + Fe(OH)_3$

68. ■ For each reaction, write an overall, balanced equation and the net ionic equation.
(a) the reaction of aqueous lead(II) nitrate and aqueous potassium hydroxide
(b) the reaction of aqueous copper(II) nitrate and aqueous sodium carbonate

In the Laboratory

69. ■ The following reaction can be used to prepare iodine in the laboratory.

$$2\ NaI(s) + 2\ H_2SO_4(aq) + MnO_2(s) \longrightarrow Na_2SO_4(aq) + MnSO_4(aq) + I_2(g) + 2\ H_2O(\ell)$$

(a) Determine the oxidation number of each atom in the equation.
(b) What is the oxidizing agent, and what has been oxidized? What is the reducing agent, and what has been reduced?
(c) Is the reaction observed product-favored or reactant-favored?
(d) Name the reactants and products.

Charles D. Winters

Preparation of iodine. A mixture of NaI and MnO_2 was placed in a flask in a hood *(left)*. On adding concentrated H_2SO_4 *(right)*, brown gaseous I_2 was evolved.

70. ■ ▲ If you have "silverware" in your home, you know it tarnishes easily. Tarnish is from the oxidation of silver in the presence of sulfur-containing compounds (in the atmosphere or in your food) to give black Ag_2S. To remove the tarnish, you can warm the tarnished object with some aluminum foil (in water with a small amount of baking soda). The silver sulfide reacts with aluminum to produce silver as well as aluminum oxide and hydrogen sulfide.

$$3\ Ag_2S(s) + 2\ Al(s) + 3\ H_2O(\ell) \longrightarrow 6\ Ag(s) + Al_2O_3(s) + 3\ H_2S(aq)$$

Hydrogen sulfide is foul smelling, but it is removed by reaction with the baking soda.

$$NaHCO_3(aq) + H_2S(aq) \longrightarrow NaHS(aq) + H_2O(\ell) + CO_2(g)$$

Classify the two reactions, and identify any acids, bases, oxidizing agents, or reducing agents.

Charles D. Winters

(a) (b)

Removing silver tarnish. A badly tarnished piece of silver (a) is placed in a dish with aluminum foil and aqueous sodium hydrogen carbonate. The portion of the silver in contact with the solution is now free of tarnish (b).

71. ■ ▲ Suppose you wish to prepare a sample of magnesium chloride. One way to do this is to use an acid–base reaction, the reaction of magnesium hydroxide with hydrochloric acid.

$$Mg(OH)_2(s) + 2\ HCl(aq) \longrightarrow MgCl_2(aq) + 2\ H_2O(\ell)$$

When the reaction is complete, evaporating the water will give solid magnesium chloride. Can you suggest at least one other way to prepare $MgCl_2$?

72. ▲ Suggest a laboratory method for preparing barium phosphate. (See Question 71 for a way to approach this question.)

73. ■ One way to test for the presence of sugars (say in a urine sample) is to treat the sample with silver ions in aqueous ammonia. (This is called the Tollen's test.) Using glucose, $C_6H_{12}O_6$ to illustrate this test, the oxidation-reduction reaction occurring is

$$C_6H_{12}O_6\ (aq) + 2\ Ag^+(aq) + 2\ OH^-(aq) \longrightarrow C_6H_{12}O_7(aq) + 2\ Ag(s) + H_2O(\ell)$$

What has been oxidized, and what has been reduced? What is the oxidizing agent, and what is the reducing agent?

Charles D. Winters

(a) (b)

Tollen's test. The reaction of silver ions with a sugar such as glucose produces metallic silver. (a) The set up for the reaction. (b) The silvered testtube

Summary and Conceptual Questions

The following questions may use concepts from this and previous chapters.

74. There are many ionic compounds that dissolve in water to a very small extent. One example is lead(II) chloride. Suppose you stir some solid $PbCl_2$ into water. Explain how you would prove that the compound dissolves but to a small extent? Is the dissolving process product-favored or reactant-favored?

$$PbCl_2(s) \rightleftharpoons Pb^{2+}(aq) + 2\ Cl^-(aq)$$

75. ▲ Most naturally occurring acids are weak acids. Lactic acid is one example.

$$CH_3CH(OH)CO_2H(s) + H_2O(\ell) \rightleftharpoons H_3O^+(aq) + CH_3CH(OH)CO_2^-(aq)$$

If you place some lactic acid in water, it will ionize to a small extent, and an equilibrium will be established eventually. Suggest some experiments to prove that this is a weak acid and that the establishment of equilibrium is a reversible process.

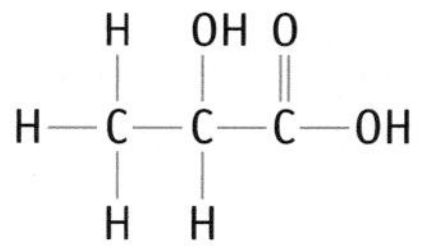

Lactic acid

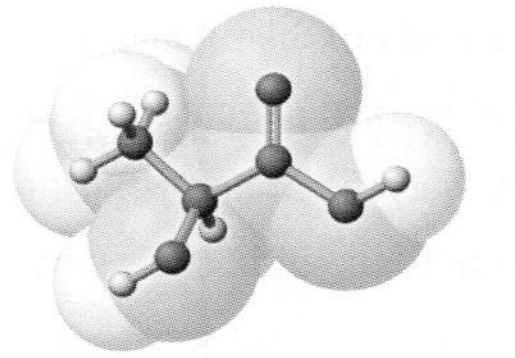

76. ■ ▲ You want to prepare barium chloride, $BaCl_2$, using an exchange reaction of some type. To do so, you have the following reagents from which to select the reactants: $BaSO_4$, $BaBr_2$, $BaCO_3$, $Ba(OH)_2$, HCl, $HgSO_4$, $AgNO_3$, and HNO_3. Write a complete, balanced equation for the reaction chosen. *(Note: There are several possibilities.)*

77. ■ Describe how to prepare $BaSO_4$, barium sulfate, by (a) a precipitation reaction and (b) a gas-forming reaction. To do so, you have the following reagents from which to select the reactants: $BaCl_2$, $BaCO_3$, $Ba(OH)_2$, H_2SO_4, and Na_2SO_4. Write complete, balanced equations for the reactions chosen. (See page 130 for an illustration of the preparation of a compound.)

78. ■ Describe how to prepare zinc chloride by (a) an acid–base reaction, (b) a gas-forming reaction, and (c) an oxidation-reduction reaction. The available starting materials are $ZnCO_3$, HCl, Cl_2, HNO_3, $Zn(OH)_2$, NaCl, $Zn(NO_3)_2$, and Zn. Write complete, balanced equations for the reactions chosen.

4 Stoichiometry: Quantitative Information About Chemical Reactions

Charles D. Winters

The Chemistry of a Sparkler

This "sparkler," like many forms of fireworks, depends on some very straightforward chemistry. It consists of a mixture of finely powdered metals, such as Al and Fe, other substances such as $KClO_3$, KNO_3, and a binder that holds the mixture onto a wire handle. When ignited, the powdered metal reacts with oxygen in the air (or extracted from $KClO_3$), and the sparks fly!

Questions:

1. What kind of chemical reaction is occurring here?
2. What are the likely products from the reaction of aluminum metal or iron with oxygen in the air or with $KClO_3$?
3. If the sparkler contains 1.0 g of Al, what is the mass of the product of the reaction of aluminum with oxygen?

Answers to these questions are in Appendix Q.

Chapter Goals

See Chapter Goals Revisited (page 194) for Study Questions keyed to these goals and assignable in OWL.

- Perform stoichiometry calculations using balanced chemical equations.
- Understand the meaning of a limiting reactant in a chemical reaction.
- Calculate the theoretical and percent yields of a chemical reaction.
- Use stoichiometry to analyze a mixture of compounds or to determine the formula of a compound.
- Define and use concentration in solution stoichiometry.

Chapter Outline

The objective of this chapter is to introduce the quantitative study of chemical reactions. Quantitative studies are needed to determine, for example, how much oxygen is required for the complete combustion of a given quantity of gasoline and what masses of carbon dioxide and water can be obtained. This part of chemistry is fundamental to much of what chemists, chemical engineers, biochemists, molecular biologists, geochemists, and many others do.

Chemistry Now™
Throughout the text this icon introduces an opportunity for self-study or to explore interactive tutorials by signing in at **www.thomsonedu.com/login**.

4.1 Mass Relationships in Chemical Reactions: Stoichiometry

Module 7

A balanced chemical equation shows the quantitative relationship between reactants and products in a chemical reaction. Let us apply this concept to the reaction of phosphorus and chlorine (◄ Figure 3.1).

$$P_4(s) + 6\ Cl_2(g) \longrightarrow 4\ PCl_3(\ell)$$

Suppose you use 1.00 mol of phosphorus (P_4, 124 g/mol) in this reaction. The balanced equation shows that 6.00 mol (= 425 g) of Cl_2 must be used for complete reaction with 1.00 mol of P_4 and that 4.00 mol (= 549 g) of PCl_3 can be produced. The mole and mass relationships of reactants and products in a reaction can be summarized in an *amounts table.* You will find such tables helpful in identifying the amounts of reactants and products and the changes that occur upon reaction.

Equation	$P_4(s)$ +	$6\ Cl_2(g)$ →	$4\ PCl_3(\ell)$
Initial amount (mol)	1.00 mol (124 g)	6.00 mol (425 g)	0 mol (0 g)
Change in amount upon reaction (mol)	−1.00 mol	−6.00 mol	+4.00 mol
Amount after complete reaction (mol)	0 mol (0 g)	0 mol (0 g)	4.00 mol [549 g = 124 g + 425 g]

The balanced equation for a reaction tells us the correct *mole ratios* of reactants and products. Here, 6 mol of Cl_2 should be used per mole of P_4. Now, what if only 0.0100 mol of P_4 (1.24 g) is available? Six times as many moles of Cl_2 are still required (0.0600 mol of Cl_2; 4.25 g), and 0.0400 mol of PCl_3 (5.49 g) will be formed. The only requirement is that there should be a ratio of 6 to 1 for the amount of Cl_2 relative to the amount of P_4.

■ **Amounts Tables** Amounts tables are useful not only here but will also be used extensively when you study chemical equilibria more thoroughly in Chapters 16–18.

Stoichiometry and Reaction Completeness *Stoichiometry calculations assume reactions are complete.* That is, *all* of at least one of the reactants has been converted to the product or products. Recall from the discussion of chemical equilibrium in Chapter 3 that some reactions do not reach completion, however.

Amount The amount of a substance is the number of moles of that substance.

Following this line of reasoning, let us determine (a) what mass of Cl_2 is required to react completely with 1.45 g of phosphorus and (b) what mass of PCl_3 can be produced.

Part (a): Mass of Cl_2 Required

Step 1. *Write the balanced equation* (using correct formulas for reactants and products). This is always the first step when dealing with chemical reactions.

$$P_4(s) + 6\ Cl_2(g) \longrightarrow 4\ PCl_3(\ell)$$

Step 2. *Calculate amount (moles) from mass (grams).* From the mass of P_4, calculate the amount of P_4 available.

$$1.45\ \text{g}\ P_4 \times \frac{1\ \text{mol}\ P_4}{123.9\ \text{g}\ P_4} = 0.0117\ \text{mol}\ P_4$$

1/molar mass of P_4

Step 3. *Use a stoichiometric factor.* The amount of Cl_2 required is related by the balanced equation to the amount of the other reactant (P_4) available.

$$0.0117\ \text{mol}\ P_4 \times \frac{6\ \text{mol}\ Cl_2\ \text{required}}{1\ \text{mol}\ P_4\ \text{required}} = 0.0702\ \text{mol}\ Cl_2\ \text{required}$$

stoichiometric factor from balanced equation

To perform this calculation the amount of phosphorus available has been multiplied by a **stoichiometric factor**, a mole ratio based on the coefficients for the two chemicals in the balanced equation. Here, the balanced equation specifies that 6 mol of Cl_2 is required for each mole of P_4, so the stoichiometric factor is (6 mol Cl_2/1 mol P_4). Calculation shows that 0.0702 mol of Cl_2 is required to react with all the available phosphorus (1.45 g, 0.0117 mol).

Problem Solving Tip 4.1 Stoichiometry Calculations

You are asked to determine what mass of product can be formed from a given mass of reactant. It is not possible to calculate the mass of product in a single step. Instead, you must follow a route such as that illustrated here for the reaction of a reactant A to give the product B according to an equation such as $x\ A \longrightarrow y\ B$.

- The mass (g) of reactant A is converted to the amount (moles) of A.
- Next, using the stoichiometric factor, find the amount (moles) of B.
- Finally, the mass (g) of B is obtained by multiplying amount of B by its molar mass.

When solving a stoichiometry problem, remember that you will always use a stoichiometric factor at some point.

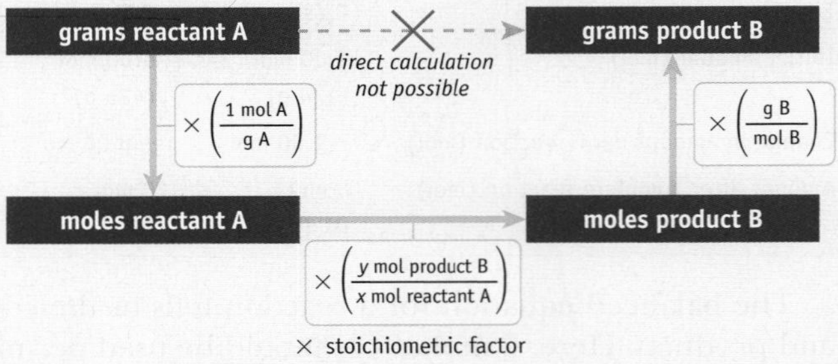

Step 4. *Calculate mass from amount.* Convert the amount (moles) of Cl_2 calculated in Step 3 to mass (in grams) of Cl_2 required.

$$0.0702\ \cancel{\text{mol Cl}_2} \times \frac{70.91\ \text{g Cl}_2}{1\ \cancel{\text{mol Cl}_2}} = 4.98\ \text{g Cl}_2$$

■ **Mass Balance and Moles of Reactants and Products** Mass is always conserved in chemical reactions. The total mass of the reactants is always the same as the products. This does not mean, however, that the total amount (moles) of reactants is the same as that of the products. Atoms are rearranged into different "units" (molecules) in the course of a reaction. In the $P_4 + Cl_2$ reaction, 7 mol of reactants gives 4 mol of product.

Part (b) Mass of PCl_3 Produced from P_4 and Cl_2

What mass of PCl_3 can be produced from the reaction of 1.45 g of phosphorus with 4.98 g of Cl_2? From part (a), we know that these masses are the correct quantities needed for complete reaction. Because matter is conserved, the answer can be obtained by adding the masses of P_4 and Cl_2 used (giving 1.45 g + 4.98 g = 6.43 g of PCl_3 produced). Alternatively, Steps 3 and 4 can be repeated, but with the appropriate stoichiometric factor and molar mass.

Step 3b. *Use a stoichiometric factor. Convert the amount of available P_4 to the amount of PCl_3 produced.* Here, the balanced equation specifies that 4 mol PCl_3 is produced for each mole of P_4 used, so the stoichiometric factor is (4 mol PCl_3/1 mol P_4)

$$0.0117\ \text{mol P}_4 \times \frac{4\ \text{mol PCl}_3\ \text{produced}}{1\ \text{mol P}_4\ \text{available}} = 0.0468\ \text{mol PCl}_3\ \text{produced}$$

↑

stoichiometric factor from balanced equation

Step 4b. *Calculate the mass of product from its amount.* Convert the amount of PCl_3 produced to a mass in grams.

$$0.0468\ \cancel{\text{mol PCl}_3} \times \frac{137.3\ \text{g PCl}_3}{1\ \cancel{\text{mol PCl}_3}} = 6.43\ \text{g PCl}_3$$

Chemistry Now™

Sign in at **www.thomsonedu.com/login** and go to Chapter 4 Contents to see:

- Screen 4.2 for (a) a video and animation of **the phosphorus and chlorine reaction discussed in this section,** and (b) an exercise that examines **the reaction between chlorine and elemental phosphorus**
- Screen 4.3 for a tutorial on **yield**

EXAMPLE 4.1 Mass Relations in Chemical Reactions

Problem Glucose, $C_6H_{12}O_6$, reacts with oxygen to give CO_2 and H_2O. What mass of oxygen (in grams) is required for complete reaction of 25.0 g of glucose? What masses of carbon dioxide and water (in grams) are formed?

Strategy After writing a balanced equation, you can perform the stoichiometric calculations using the scheme in Problem Solving Tip 4.1.

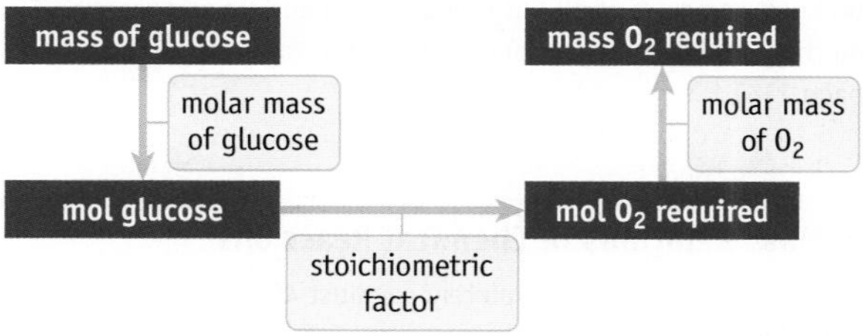

First, find the amount of glucose available; then relate it to the amount of O_2 required using the stoichiometric factor based on the coefficients in the balanced equation. Finally, find the mass of O_2 required from the amount of O_2. Follow the same procedure to calculate the masses of carbon dioxide and water.

Solution

Step 1. Write a balanced equation.

$$C_6H_{12}O_6(s) + 6\ O_2(g) \longrightarrow 6\ CO_2(g) + 6\ H_2O(\ell)$$

Step 2. Convert the mass of glucose to amount.

$$25.0\ \cancel{g}\ \text{glucose} \times \frac{1\ \text{mol}}{180.2\ \cancel{g}} = 0.139\ \text{mol glucose}$$

Step 3. Use the stoichiometric factor. Here we calculate the amount of O_2 required.

$$0.139\ \cancel{\text{mol glucose}} \times \frac{6\ \text{mol}\ O_2}{1\ \cancel{\text{mol glucose}}} = 0.832\ \text{mol}\ O_2$$

Step 4. Calculate mass from amount. Convert the required amount of O_2 to a mass in grams.

$$0.832\ \cancel{\text{mol}\ O_2} \times \frac{32.00\ \text{g}}{1\ \cancel{\text{mol}\ O_2}} = 26.6\ \text{g}\ O_2$$

Repeat Steps 3 and 4 to find the mass of CO_2 produced in the combustion. First, relate the amount (moles) of glucose available to the amount of CO_2 produced using a stoichiometric factor. Then convert the amount of CO_2 to its mass in grams.

$$0.139\ \cancel{\text{mol glucose}} \times \frac{6\ \cancel{\text{mol}\ CO_2}}{1\ \cancel{\text{mol glucose}}} \times \frac{44.01\ \text{g}\ CO_2}{1\ \cancel{\text{mol}\ CO_2}} = \boxed{36.6\ \text{g}\ CO_2}$$

Now, how can you find the mass of H_2O produced? You could go through Steps 3 and 4 again. However, recognize that the total mass of reactants

$$25.0\ \text{g}\ C_6H_{12}O_6 + 26.6\ \text{g}\ O_2 = 51.6\ \text{g reactants}$$

must be the same as the total mass of products. The mass of water that can be produced is therefore

$$\text{Total mass of products} = 51.6\ \text{g} = 36.6\ \text{g}\ CO_2\ \text{produced} + ?\ \text{g}\ H_2O$$

$$\text{Mass of}\ H_2O\ \text{produced} = \boxed{15.0\ \text{g}}$$

The results of this calculation can be summarized in an amounts table.

Equation	**$C_6H_{12}O_6(s)$**	**+ $6\ O_2(g)$**	**$\rightarrow 6\ CO_2(g)$**	**+ $6\ H_2O(\ell)$**
Initial amount (mol)	0.139 mol	6(0.139 mol) = 0.832 mol	0	0
Change (mol)	−0.139 mol	−0.832 mol	+0.832 mol	+0.832 mol
Amount after reaction (mol)	0	0	0.832 mol	0.832 mol

Comment When you know the mass of all but one of the chemicals in a reaction, you can find the unknown mass using the principle of mass conservation (the total mass of reactants must equal the total mass of products; page 114).

EXERCISE 4.1 Mass Relations in Chemical Reactions

What mass of oxygen, O_2, is required to completely combust 454 g of propane, C_3H_8? What masses of CO_2 and H_2O are produced?

4.2 Reactions in Which One Reactant Is Present in Limited Supply

Reactions are often carried out with an excess of one reactant over that required by stoichiometry. This is usually done to ensure that one of the reactants in the reaction is consumed completely, even though some of another reactant remains unused.

Suppose you burn a toy "sparkler," a wire coated with a mixture of aluminum or iron powder and potassium nitrate or chlorate (Figure 4.1 and page 158). The aluminum or iron burns, consuming oxygen from the air or from the potassium salt and producing a metal oxide.

$$4\ Al(s) + 3\ O_2(g) \longrightarrow 2\ Al_2O_3(s)$$

The sparkler burns until the metal powder is consumed completely. What about the oxygen? Four moles of aluminum require three moles of oxygen, but there is much, much more O_2 available in the air than is needed to consume the metal in a sparkler. How much metal oxide is produced? That depends on the quantity of metal powder in the sparkler, not on the quantity of O_2 in the atmosphere. The metal powder in this example is called the **limiting reactant** because its amount determines, or limits, the amount of product formed.

Let us look at an example of a limiting reactant situation using the reaction of oxygen and carbon monoxide to give carbon dioxide. The balanced equation for the reaction is

$$2\ CO(g) + O_2(g) \longrightarrow 2\ CO_2(g)$$

Suppose you have a mixture of four CO molecules and three O_2 molecules. The four CO molecules require only two O_2 molecules (and produce four CO_2 molecules). This means that one O_2 molecule remains after reaction is complete.

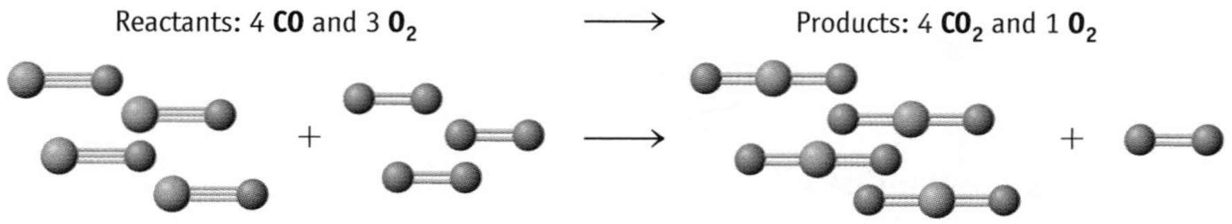

Because more O_2 molecules are available than are required, the number of CO_2 molecules produced is determined by the number of CO molecules available. Carbon monoxide, CO, is therefore the limiting reactant in this case.

Charles D. Winters

FIGURE 4.1 Burning aluminum and iron powder. A toy sparkler contains a metal powder such as Al or Fe and other chemicals such as KNO_3 or $KClO_3$. When ignited, the metal burns with a brilliant white light.

A Stoichiometry Calculation with a Limiting Reactant

The first step in the manufacture of nitric acid is the oxidation of ammonia to NO over a platinum-wire gauze (Figure 4.2).

$$4\ NH_3(g) + 5\ O_2(g) \longrightarrow 4\ NO(g) + 6\ H_2O(\ell)$$

Suppose that equal masses of NH_3 and O_2 are mixed (750. g of each). Are these reactants mixed in the correct stoichiometric ratio, or is one of them in short supply? That is, will one of them limit the quantity of NO that can be produced? How much NO can be formed if the reaction using this reactant mixture goes to completion? And how much of the excess reactant is left over when the maximum amount of NO has been formed?

■ **Comparing Reactant Ratios** For the CO/O_2 reaction, the stoichiometric ratio of reactants should be (2 mol CO/1 mol O_2). However, the ratio of amounts of reactants available in the text example is (4 mol CO/3 mol O_2) or (1.33 mol CO/1 mol O_2). The fact that the CO/O_2 ratio is not large enough tells us that there is not enough CO to react with all of the available O_2. Carbon monoxide is the limiting reactant, and some O_2 will be left over when all of the CO is consumed.

(a)

(b)

Active Figure 4.2 **Oxidation of ammonia.** (a) Burning ammonia on the surface of a platinum wire produces so much heat that the wire glows bright red. (b) Billions of kilograms of HNO_3 are made annually starting with the oxidation of ammonia over a wire gauze containing platinum.

Chemistry Now™ Sign in at www.thomsonedu.com/login and go to the Chapter Contents menu to explore an interactive version of this figure accompanied by an exercise.

Step 1. Find the amount of each reactant.

$$750.\ \text{g NH}_3 \times \frac{1\ \text{mol NH}_3}{17.03\ \text{g NH}_3} = 44.0\ \text{mol NH}_3\ \text{available}$$

$$750.\ \text{g O}_2 \times \frac{1\ \text{mol O}_2}{32.00\ \text{g O}_2} = 23.4\ \text{mol O}_2\ \text{available}$$

Step 2. What is the limiting reactant? Examine the ratio of amounts of reactants. Are the reactants present in the correct stoichiometric ratio as given by the balanced equation?

$$\text{Stoichiometric ratio of reactants required by balanced equation} = \frac{5\ \text{mol O}_2}{4\ \text{mol NH}_3} = \frac{1.25\ \text{mol O}_2}{1\ \text{mol NH}_3}$$

$$\text{Ratio of reactants } \textit{actually available} = \frac{23.4\ \text{mol O}_2}{44.0\ \text{mol NH}_3} = \frac{0.532\ \text{mol O}_2}{1\ \text{mol NH}_3}$$

Dividing moles of O_2 available by moles of NH_3 available shows that the ratio of available reactants is much smaller than the 5 mol O_2/4 mol NH_3 ratio required by the balanced equation. Thus, there is not sufficient O_2 available to react with all of the NH_3. In this case, oxygen, O_2, is the limiting reactant. That is, 1 mol of NH_3 requires 1.25 mol of O_2, but we have only 0.532 mol of O_2 available.

Step 3. Calculate the mass of product. We can now calculate the expected mass of product, NO, based on the amount of the limiting reactant, O_2.

$$23.4\ \text{mol O}_2 \times \frac{4\ \text{mol NO}}{5\ \text{mol O}_2} \times \frac{30.01\ \text{g NO}}{1\ \text{mol NO}} = 562\ \text{g NO}$$

Step 4. Calculate the mass of excess reactant. Ammonia is the "excess reactant" in this NH_3/O_2 reaction because more than enough NH_3 is available to react with 23.4 mol of O_2. Let us calculate the quantity of NH_3 remaining after all the O_2 has been used. To do so, we first need to know the amount of NH_3 required to consume all the limiting reactant, O_2.

$$23.4 \text{ mol } O_2 \text{ available} \times \frac{4 \text{ mol } NH_3 \text{ required}}{5 \text{ mol } O_2} = 18.8 \text{ mol } NH_3 \text{ required}$$

Because 44.0 mol of NH_3 is available, the amount of excess NH_3 can be calculated,

$$\begin{aligned}\text{Excess } NH_3 &= 44.0 \text{ mol } NH_3 \text{ available} - 18.8 \text{ mol } NH_3 \text{ required}\\ &= 25.2 \text{ mol } NH_3 \text{ remaining}\end{aligned}$$

and then converted to a mass.

$$25.2 \text{ mol } NH_3 \times \frac{17.03 \text{ g } NH_3}{1 \text{ mol } NH_3} = 429 \text{ g } NH_3 \text{ in excess of that required}$$

Finally, because 429 g of NH_3 is left over, this means that 321 g of NH_3 has been consumed (= 750. g − 429 g).

In limiting reactant problems, it is helpful to summarize your results in an amounts table.

Equation	$4\ NH_3(g)$	+	$5\ O_2(g)$	$\longrightarrow$	$4\ NO(g)$	+	$6\ H_2O(\ell)$
Initial amount (mol)	44.0		23.4		0		0
Change in amount (mol)	−(4/5)(23.4) = −18.8		−23.4		+(4/5)(23.4) = +18.8		+(6/5)(23.4) = +28.1
After complete reaction (mol)	25.2		0		18.8		28.1

All of the limiting reactant, O_2, is consumed. Of the original 44.0 mol of NH_3, 18.8 mol is consumed and 25.2 mol remains. The balanced equation indicates that the amount of NO produced is equal to the amount of NH_3 consumed, so 18.8 mol of NO is produced from 18.8 mol of NH_3. In addition, 28.1 mol of H_2O is produced.

■ **Conservation of Mass** Mass is conserved in the $NH_3 + O_2$ reaction. The total mass present before reaction (1500. g) is the same as the total mass produced in the reaction plus the mass of NH_3 remaining. That is, 562 g of NO (18.8 mol) and 506 g of H_2O (28.1 mol) are produced. Because 429 g of NH_3 (25.2 mol) remains, the total mass after reaction (562 g + 506 g + 429 g) is the same as the total mass before reaction.

Chemistry Now™

Sign in at **www.thomsonedu.com/login** and go to Chapter 4 Contents menu to see:

- Screen 4.4 for a video and animation of **the limiting reactant in the methanol and oxygen reaction**
- Screen 4.5 for (a) an exercise on **zinc and hydrochloric acid in aqueous solution** and (b) a simulation **using limiting reactants**

EXAMPLE 4.2 A Reaction with a Limiting Reactant

Problem Methanol, CH_3OH, which is used as a fuel, can be made by the reaction of carbon monoxide and hydrogen.

$$CO(g) + 2\ H_2(g) \longrightarrow \underset{\text{methanol}}{CH_3OH(\ell)}$$

Suppose 356 g of CO and 65.0 g of H_2 are mixed and allowed to react.

(a) Which is the limiting reactant?

(b) What mass of methanol can be produced?

(c) What mass of the excess reactant remains after the limiting reactant has been consumed?

Reuters/Corbis

A car that uses methanol as a fuel. In this car, methanol is converted to hydrogen, which is then combined with oxygen in a fuel cell. The fuel cell generates electric energy to run the car (see Chapter 20). See Example 4.2.

Strategy There are usually two steps to a limiting reactant problem:

(a) After calculating the amount of each reactant, compare the ratio of reactant amounts to the required stoichiometric ratio, here 2 mol H_2/1 mol CO.

- If [mol H_2 available/mol CO available] > 2/1, then CO is the limiting reactant.
- If [mol H_2 available/mol CO available] < 2/1, then H_2 is the limiting reactant.

(b) Use the amount of limiting reactant to find the masses of product and excess reactant.

Solution

(a) *What is the limiting reactant?* The amount of each reactant is

$$\text{Amount of CO} = 356 \text{ g CO} \times \frac{1 \text{ mol CO}}{28.01 \text{ g CO}} = 12.7 \text{ mol CO}$$

$$\text{Amount of } H_2 = 65.0 \text{ g } H_2 \times \frac{1 \text{ mol } H_2}{2.016 \text{ g } H_2} = 32.2 \text{ mol } H_2$$

Are these reactants present in a perfect stoichiometric ratio?

$$\frac{\text{Mol } H_2 \text{ available}}{\text{Mol CO available}} = \frac{32.2 \text{ mol } H_2}{12.7 \text{ mol CO}} = \frac{2.54 \text{ mol } H_2}{1.00 \text{ mol CO}}$$

The required mole ratio is 2 mol of H_2 to 1 mol of CO. Here, we see that more hydrogen is available than is required to consume all the CO. It follows that not enough CO is present to use up all of the hydrogen. CO is the limiting reactant.

(b) *What is the maximum mass of CH_3OH that can be formed?* This calculation must be based on the amount of limiting reactant.

$$12.7 \text{ mol CO} \times \frac{1 \text{ mol } CH_3OH \text{ formed}}{1 \text{ mol CO available}} \times \frac{32.04 \text{ g } CH_3OH}{1 \text{ mol } CH_3OH} = 407 \text{ g } CH_3OH$$

(c) *What mass of H_2 remains when all the CO has been converted to product?* First, we must find the amount of H_2 required to react with all the CO, then calculate the mass from the amount.

$$12.7 \text{ mol CO} \times \frac{2 \text{ mol } H_2}{1 \text{ mol CO}} = 25.4 \text{ mol } H_2 \text{ required}$$

Because 32.2 mol of H_2 is available, but only 25.4 mol is required by the limiting reactant, 32.2 mol − 25.4 mol = 6.8 mol of H_2 is in excess. This is equivalent to 14 g of H_2.

Comment The amounts table for this reaction is

Equation	CO(g)	+	2 H_2(g)	→	$CH_3OH(\ell)$
Initial amount (mol)	12.7		32.2		0
Change (mol)	−12.7		−2(12.7)		+12.7
After complete reaction (mol)	0		6.8		12.7

The mass of product formed plus the mass of H_2 remaining after reaction (407 g CH_3OH produced + 14 g H_2 remaining = 421 g) is equal to the mass of reactants present before reaction (356 g CO + 65.0 g H_2 = 421 g).

Charles D. Winters

Thermite reaction. Iron(III) oxide reacts with aluminum metal to produce aluminum oxide and iron metal. The reaction produces so much heat that the iron melts and spews out of the reaction vessel. See Exercise 4.2.

EXERCISE 4.2 A Reaction with a Limiting Reactant

The thermite reaction produces iron metal and aluminum oxide from a mixture of powdered aluminum metal and iron(III) oxide.

$$Fe_2O_3(s) + 2\ Al(s) \longrightarrow 2\ Fe(\ell) + Al_2O_3(s)$$

A mixture of 50.0 g each of Fe_2O_3 and Al is used.

(a) Which is the limiting reactant?

(b) What mass of iron metal can be produced?

Problem Solving Tip 4.2 Moles of Reaction and Limiting Reactants

There is another method of solving stoichiometry problems that applies especially well to limiting reactant problems. This involves the useful concept of "moles of reaction."

One "mole of reaction" is said to have occurred when the reaction has taken place according to the number of moles given by the coefficients in the equation. For example, for the reaction of CO and O_2,

$$2\ CO(g) + O_2(g) \longrightarrow 2\ CO_2(g)$$

one mol of reaction occurs when 2 mol of CO and 1 mol of O_2 produce 2 mol of CO_2. If the reaction mixture consists of only 1 mol of CO and 0.5 mol of O_2, then only 1 mol of CO_2 is produced, and 0.5 mol of reaction has occurred according to this balanced equation.

To pursue this example further, suppose 9.5 g of CO and excess O_2 are combined. What amount of CO_2 (moles) can be produced?

$$9.5\text{ g CO} \times \frac{1\text{ mol CO}}{28.0\text{ g CO}} \times \frac{1\text{ mol-rxn}}{2\text{ mol CO}} = 0.34\text{ mol-rxn}$$

$$0.34\text{ mol-rxn} \times \frac{2\text{ mol CO}_2}{1\text{ mol-rxn}} = 0.68\text{ mol CO}_2$$

You can see in this example that the number of moles of reaction that occurred is calculated by multiplying the amount (moles) of the reactant CO by the factor, 1 mol-rxn/2 mol CO (which amounts to dividing the amount of CO by its stoichiometric coefficient).

All reactants and products involved in a chemical reaction undergo the same number of moles of reaction because the reaction can only occur a certain number of times before the reactants are consumed and the reaction reaches completion.

If one of the reactants is in short supply, the actual number of times a reaction can be carried out—the number of "moles of reaction"—will be determined by the limiting reactant. To use this approach, we first calculate the amount of each reactant initially present and then calculate the **moles of reaction** that could occur with each amount of reactant. (This is equivalent to dividing amount [moles] of each reactant by its stoichiometric coefficient.) The reactant producing the smallest number of moles of reaction is the limiting reactant. Once the limiting reactant is known, we proceed as before.

Consider again the NH_3/O_2 reaction on page 163:

1. *Calculate the moles of reaction predicted for each reactant, and decide on the limiting reactant.*

   In the case of the NH_3/O_2 reaction,

$$4\ NH_3(g) + 5\ O_2(g) \longrightarrow 4\ NO(g) + 6\ H_2O(\ell)$$

1 "mole of reaction" uses 4 mol of NH_3 and 5 mol of O_2 and produces 4 mol of NO and 6 mol of H_2O. In the example on page 163, we started with 44.0 mol of NH_3, so 11.0 mol of reaction can result.

$$44.0\ \cancel{\text{mol NH}_3} \times \frac{1\text{ mol-rxn}}{4\ \cancel{\text{mol NH}_3}} = 11.0\text{ mol-rxn}$$

Based on the amount of O_2 available, 4.68 mol of reaction can occur.

$$23.4\ \cancel{\text{mol O}_2} \times \frac{1\text{ mol-rxn}}{5\ \cancel{\text{mol O}_2}} = 4.68\text{ mol-rxn}$$

Fewer moles of reaction can occur with the amount of O_2 available, so O_2 is the limiting reactant.

2. *Calculate the change in amount and the amount upon completion of the reaction, for each reactant and product.* The number of moles of reaction predicted by the limiting reactant corresponds to the number of moles of reaction that can *actually* occur. Each reactant and product will undergo this number of moles of reaction, 4.68 mol-rxn in this case. To calculate the *change* in amount for a given reactant or product, multiply this number of moles of reaction by the stoichiometric coefficient of the reactant or product. To illustrate this step, for NH_3 this corresponds to the following calculation:

$$4.68\text{ mol-rxn} \times \left(\frac{4\text{ mole NH}_3}{1\text{ mol-rxn}}\right) = 18.8\text{ mol NH}_3.$$

The amount of each reactant and product after reaction is calculated as usual.

Equation	**$4\ NH_3(g)$**	**$+\ 5\ O_2(g)$**	**$\longrightarrow 4\ NO(g)$**	**$+\ 6\ H_2O(g)$**
Initial amount (mol)	44.0	23.4	0	0
Moles of reaction based on limiting reactant (mol)	4.68	4.68	4.68	4.68
Change in amount (mol)	−4.68(4) = −18.8	−4.68(5) = −23.4	+4.68(4) = +18.8	+4.68(6) = +28.1
Amount remaining after complete reaction (mol)	25.2	0	18.8	28.1

Finally, from the amounts present after completion, we can calculate the masses of the products and of any reactant remaining.

You may find this approach easier to use particularly when there are more than two reactants, each present initially in some designated quantity.

A final note: the concept of "moles of reaction" will be applied in this text in the discussion of thermochemistry in Chapters 5 and 19.

4.3 Percent Yield

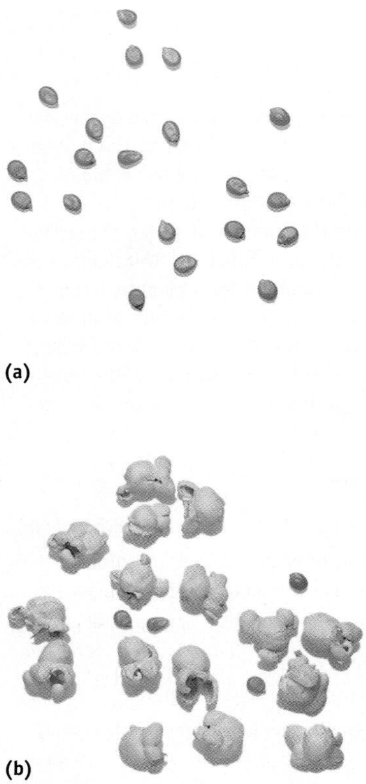

(a)

(b)

FIGURE 4.3 Percent yield. Although not a chemical reaction, popping corn is a good analogy to the difference between a theoretical yield and an actual yield. Here, we began with 20 popcorn kernels and found that only 16 of them popped. The percent yield from our "reaction" was (16/20) x 100%, or 80%.

The maximum mass of product that can be obtained from a chemical reaction is the **theoretical yield**. Frequently, however, the **actual yield** of the product—the mass of material that is actually obtained in the laboratory or a chemical plant—is less than the theoretical yield. Loss of product often occurs during the isolation and purification steps. In addition, some reactions do not go completely to products, and reactions are sometimes complicated by giving more than one set of products. For all these reasons, the actual yield is almost always less than the theoretical yield (Figure 4.3).

To provide information to other chemists who might want to carry out a reaction, it is customary to report a percent yield. **Percent yield**, which specifies how much of the theoretical yield was obtained, is defined as

$$\text{Percent yield} = \frac{\text{actual yield}}{\text{theoretical yield}} \times 100\% \qquad \textbf{(4.1)}$$

Suppose you made aspirin in the laboratory by the following reaction:

$$C_7H_6O_3(s) + C_4H_6O_3(\ell) \longrightarrow C_9H_8O_4(s) + CH_3CO_2H(\ell)$$

salicylic acid | acetic anhydride | aspirin | acetic acid

and that you began with 14.4 g of salicylic acid and an excess of acetic anhydride. That is, salicylic acid is the limiting reactant. If you obtain 6.26 g of aspirin, what is the percent yield of this product? The first step is to find the amount of the limiting reactant, salicylic acid ($C_6H_4(OH)CO_2H$).

$$14.4 \text{ g } C_6H_4(OH)CO_2H \times \frac{1 \text{ mol } C_6H_4(OH)CO_2H}{138.1 \text{ g } C_6H_4(OH)CO_2H} = 0.104 \text{ mol } C_6H_4(OH)CO_2H$$

Next, use the stoichiometric factor from the balanced equation to find the amount of aspirin expected based on the limiting reactant, $C_6H_4(OH)CO_2H$.

$$0.104 \text{ mol } C_6H_4(OH)CO_2H \times \frac{1 \text{ mol aspirin}}{1 \text{ mol } C_6H_4(OH)CO_2H} = 0.104 \text{ mol aspirin}$$

The maximum amount of aspirin that can be produced—the theoretical yield—is 0.104 mol. Because the quantity you measure in the laboratory is the mass of the product, it is customary to express the theoretical yield as a mass in grams.

$$0.104 \text{ mol aspirin} \times \frac{180.2 \text{ g aspirin}}{1 \text{ mol aspirin}} = 18.7 \text{ g aspirin}$$

Finally, with the actual yield known to be only 6.26 g, the percent yield of aspirin can be calculated.

$$\text{Percent yield} = \frac{6.26 \text{ g aspirin obtained (actual yield)}}{18.7 \text{ g aspirin expected (theoretical yield)}} \times 100\% = 33.5\% \text{ yield}$$

Chemistry Now™

Sign in at **www.thomsonedu.com/login** and go to Chapter 4 Contents to see Screen 4.6 for tutorials on (a) **determining the theoretical yield of a reaction** and on (b) **determining the percent yield of a reaction.**

EXERCISE 4.3 Percent Yield

Aluminum carbide, Al_4C_3, reacts with water to produce methane.

$$Al_4C_3(s) + 12\ H_2O(\ell) \longrightarrow 4\ Al(OH)_3(s) + 3\ CH_4(g)$$

If 125 g of aluminum carbide is decomposed, what is the theoretical yield of methane? If only 13.6 g of methane is obtained, what is the percent yield of this gas?

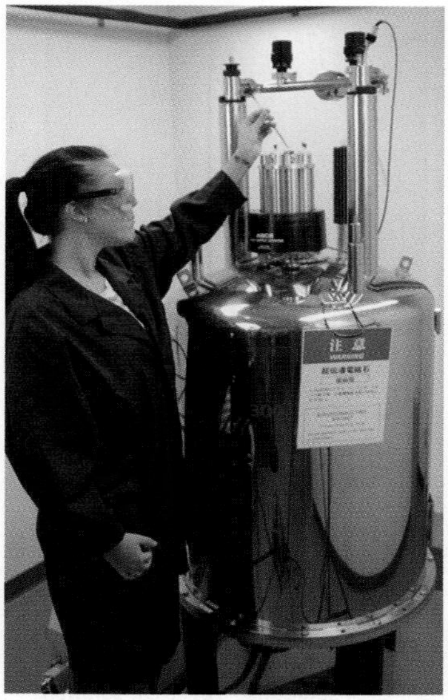

Charles D. Winters

FIGURE 4.4 A modern analytical instrument. This nuclear magnetic resonance (NMR) spectrometer is closely related to a magnetic resonance imaging (MRI) instrument found in a hospital. NMR is used to analyze compounds and to decipher their structure. (The instrument is controlled by a computer and console not seen in this photo.)

4.4 Chemical Equations and Chemical Analysis

Analytical chemists use a variety of approaches to identify substances as well as to measure the quantities of components of mixtures. Analytical chemistry is often done now using instrumental methods (Figure 4.4), but classical chemical reactions and stoichiometry still play a central role.

Quantitative Analysis of a Mixture

Quantitative chemical analysis generally depends on one of the following basic ideas:

- A substance, present in unknown amount, can be allowed to react with a known quantity of another substance. If the stoichiometric ratio for their reaction is known, the unknown amount can be determined.
- A material of unknown composition can be converted to one or more substances of known composition. Those substances can be identified, their amounts determined, and these amounts related to the amount of the original, unknown substance.

An example of the first type of analysis is the analysis of a sample of vinegar containing an unknown amount of acetic acid, the ingredient that makes vinegar acidic. The acid reacts readily and completely with sodium hydroxide.

$$\underset{\text{acetic acid}}{CH_3CO_2H(aq)} + NaOH(aq) \longrightarrow CH_3CO_2Na(aq) + H_2O(\ell)$$

If the exact amount of sodium hydroxide used in the reaction can be measured, the amount of acetic acid present can be calculated. This type of analysis is the subject of a later section in this chapter (► Section 4.7).

The second type of analysis is exemplified by the analysis of a sample of a mineral, thenardite, which is largely sodium sulfate, Na_2SO_4 (Figure 4.5). Sodium sulfate is soluble in water. Therefore, to find the quantity of Na_2SO_4 in an impure mineral sample, we would crush the rock and then wash the powdered sample thoroughly with water to dissolve the sodium sulfate. Next, we would treat this solution of sodium sulfate with barium chloride to precipitate the water-insoluble compound barium sulfate. The barium sulfate is collected on a filter and weighed (Figure 4.6).

Charles D. Winters

FIGURE 4.5 Thenardite. The mineral thenardite is sodium sulfate, Na_2SO_4. It is named after the French chemist Louis Thenard (1777–1857), a co-discoverer (with J. L. Gay-Lussac and Humphry Davy) of boron. Sodium sulfate is used in making detergents, glass, and paper.

$$Na_2SO_4(aq) + BaCl_2(aq) \longrightarrow BaSO_4(s) + 2\ NaCl(aq)$$

Charles D. Winters

(a)

$Na_2SO_4(aq)$, clear solution

$BaCl_2(aq)$, clear solution

(b)

$BaSO_4$, white solid

NaCl(aq), clear solution

(c)

NaCl(aq), clear solution

$BaSO_4$, white solid caught in filter

(d)

Mass of dry $BaSO_4$ determined

Active Figure 4.6 **Analysis for the sulfate content of a sample.** The sulfate ions in a solution of Na_2SO_4 react with barium ions (Ba^{2+}) to form $BaSO_4$. The white, solid precipitate, barium sulfate ($BaSO_4$), is collected on a filter and weighed. The amount of $BaSO_4$ obtained can be related to the amount of Na_2SO_4 in the sample.

Chemistry Now™ Sign in at www.thomsonedu.com/login and go to the Chapter Contents menu to explore an interactive version of this figure accompanied by an exercise.

■ **Analysis and 100% Yield** Quantitative analysis requires reactions in which the yield is 100%.

We can then find the amount of sodium sulfate in the mineral sample because it is directly related to the amount of $BaSO_4$.

$$1 \text{ mol } Na_2SO_4(aq) \longrightarrow 1 \text{ mol } BaSO_4(s)$$

Example 4.3 illustrates another instance of the analysis of a mineral in this way.

Chemistry Now™

Sign in at **www.thomsonedu.com/login** and go to Chapter 4 Contents to see Screen 4.7 for a tutorial on **chemical analysis.**

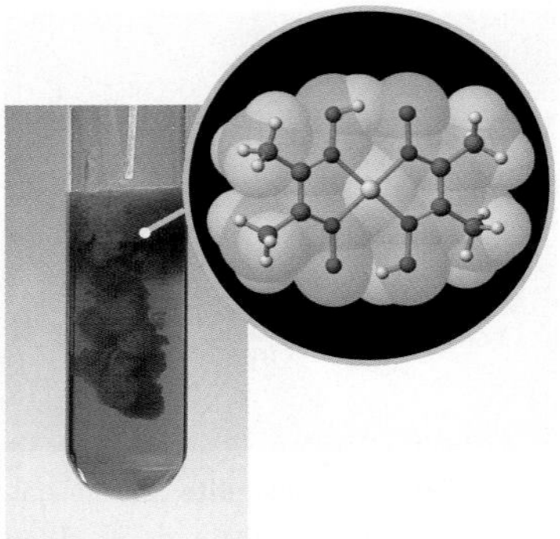

Photo: Charles D. Winters

A precipitate of nickel with dimethylglyoxime. Red, insoluble $Ni(C_4H_7N_2O_2)_2$ precipitates when dimethylglyoxime ($C_4H_8N_2O_2$) is added to an aqueous solution of nickel(II) ions. (See Example 4.3.)

EXAMPLE 4.3 Mineral Analysis

Problem Nickel(II) sulfide, NiS, occurs naturally as the relatively rare mineral millerite. One of its occurrences is in meteorites. To analyze a mineral sample for the quantity of NiS, the sample is dissolved in nitric acid to form a solution of $Ni(NO_3)_2$.

$$NiS(s) + 4\ HNO_3(aq) \longrightarrow Ni(NO_3)_2(aq) + S(s) + 2\ NO_2(g) + 2\ H_2O(\ell)$$

The aqueous solution of $Ni(NO_3)_2$ is then treated with the organic compound dimethylglyoxime ($C_4H_8N_2O_2$, DMG) to give the red solid $Ni(C_4H_7N_2O_2)_2$.

$$Ni(NO_3)_2(aq) + 2\ C_4H_8N_2O_2(aq) \longrightarrow Ni(C_4H_7N_2O_2)_2(s) + 2\ HNO_3(aq)$$

Suppose a 0.468-g sample containing millerite produces 0.206 g of red, solid $Ni(C_4H_7N_2O_2)_2$. What is the mass percent of NiS in the sample?

Strategy The balanced equations for the reactions show the following "road map":

$$1 \text{ mol NiS} \longrightarrow 1 \text{ mol } Ni(NO_3)_2 \longrightarrow 1 \text{ mol } Ni(C_4H_7N_2O_2)_2$$

If we know the mass of $Ni(C_4H_7N_2O_2)_2$, we can calculate its amount and thus the amount of NiS. The amount of NiS allows us to calculate the mass and mass percent of NiS in the sample.

Solution The molar mass of $Ni(C_4H_7N_2O_2)_2$ is 288.9 g/mol. The amount of this red solid is

$$0.206 \text{ g } Ni(C_4H_7N_2O_2)_2 \times \frac{1 \text{ mol } Ni(C_4H_7N_2O_2)_2}{288.9 \text{ g } Ni(C_4H_7N_2O_2)_2} = 7.13 \times 10^{-4} \text{ mol } Ni(C_4H_7N_2O_2)_2$$

Because 1 mol of $Ni(C_4H_7N_2O_2)_2$ is ultimately produced from 1 mol of NiS, the amount of NiS in the sample must have been 7.13×10^{-4} mol.

With the amount of NiS known, we calculate the mass of NiS.

$$7.13 \times 10^{-4} \text{ mol NiS} \times \frac{90.76 \text{ g NiS}}{1 \text{ mol NiS}} = 0.0647 \text{ g NiS}$$

Finally, the mass percent of NiS in the 0.468-g sample is

$$\text{Mass percent NiS} = \frac{0.0647 \text{ g NiS}}{0.468 \text{ g sample}} \times 100\% = 13.8\% \text{ NiS}$$

EXERCISE 4.4 Analysis of a Mixture

One method for determining the purity of a sample of titanium(IV) oxide, TiO_2, an important industrial chemical, is to react the sample with bromine trifluoride.

$$3\ TiO_2(s) + 4\ BrF_3(\ell) \longrightarrow 3\ TiF_4(s) + 2\ Br_2(\ell) + 3\ O_2(g)$$

This reaction is known to occur completely and quantitatively. That is, all of the oxygen in TiO_2 is evolved as O_2. Suppose 2.367 g of a TiO_2-containing sample evolves 0.143 g of O_2. What is the mass percent of TiO_2 in the sample?

Determining the Formula of a Compound by Combustion

The empirical formula of a compound can be determined if the percent composition of the compound is known (◀ Section 2.10). But where do the percent composition data come from? One chemical method that works well for compounds that burn in oxygen is analysis by combustion. In this technique, each element in the compound combines with oxygen to produce the appropriate oxide.

Consider an analysis of the hydrocarbon methane, CH_4. A balanced equation for the combustion of methane shows that every atom of C in the original compound appears as CO_2 and every atom of H appears in the form of water. In other words, for every mole of CO_2 observed, there must have been one mole of carbon in the unknown compound. Similarly, for every mole of H_2O observed from combustion, there must have been two moles of H atoms in the unknown carbon-hydrogen compound.

$$CH_4(g) + 2\ O_2(g) \longrightarrow CO_2(g) + 2\ H_2O(\ell)$$

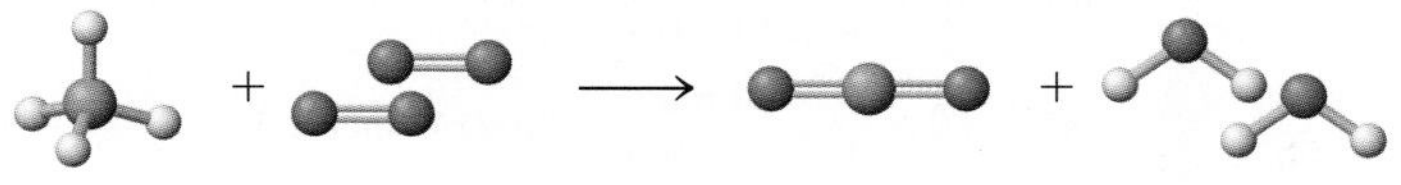

In the combustion experiment, gaseous carbon dioxide and water are separated (as illustrated in Figure 4.7) and their masses determined. From these masses, it is possible to calculate the amounts of C and H in CO_2 and H_2O, respectively, and

■ **Finding an Empirical Formula by Chemical Analysis** Finding the empirical formula of a compound by chemical analysis always uses the following procedure:

1. The unknown but pure compound is converted in a chemical reaction into known products.
2. The reaction products are isolated, and the amount of each is determined.
3. The amount of each product is related to the amount of each element in the original compound.
4. The empirical formula is determined from the relative amounts of elements in the original compound.

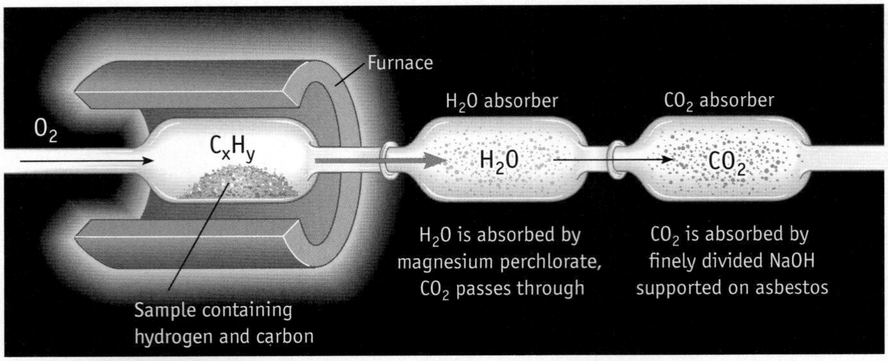

Active Figure 4.7 **Combustion analysis of a hydrocarbon.** If a compound containing C and H is burned in oxygen, CO_2 and H_2O are formed, and the mass of each can be determined. The H_2O is absorbed by magnesium perchlorate, and the CO_2 is absorbed by finely divided NaOH supported on asbestos. The mass of each absorbent before and after combustion gives the masses of CO_2 and H_2O. Only a few milligrams of a combustible compound are needed for analysis.

Chemistry Now™ Sign in at www.thomsonedu.com/login and go to the Chapter Contents menu to explore an interactive version of this figure accompanied by an exercise.

the ratio of amounts of C and H in a sample of the original compound can then be found. This ratio gives the empirical formula.

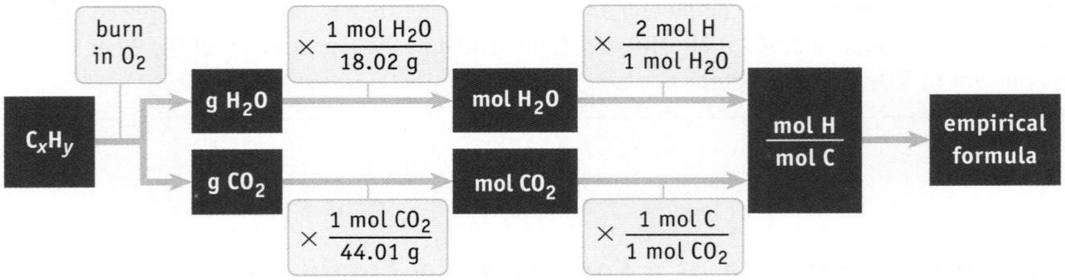

EXAMPLE 4.4 Using Combustion Analysis to Determine the Formula of a Hydrocarbon

Problem When 1.125 g of a liquid hydrocarbon, C_xH_y, was burned in an apparatus like that shown in Figure 4.7, 3.447 g of CO_2 and 1.647 g of H_2O were produced. The molar mass of the compound was found to be 86.2 g/mol in a separate experiment. Determine the empirical and molecular formulas for the unknown hydrocarbon, C_xH_y.

Strategy As outlined in the preceding diagram, we first calculate the amounts of CO_2 and H_2O. These are then converted to amounts of C and H. The ratio (mol H/mol C) is used to determine the empirical formula of the compound. The molar mass of the compound and the molar mass of the empirical formula are then used to determine the molecular formula.

Solution The amounts of CO_2 and H_2O isolated from the combustion are

$$3.447 \text{ g } \cancel{CO_2} \times \frac{1 \text{ mol } CO_2}{44.010 \text{ g } \cancel{CO_2}} = 0.07832 \text{ mol } CO_2$$

$$1.647 \text{ g } \cancel{H_2O} \times \frac{1 \text{ mol } H_2O}{18.015 \text{ g } \cancel{H_2O}} = 0.09142 \text{ mol } H_2O$$

For every mole of CO_2 isolated, 1 mol of C must have been present in the unknown compound.

$$0.07832 \text{ } \cancel{\text{mol } CO_2} \times \frac{1 \text{ mol C in unknown}}{1 \text{ } \cancel{\text{mol } CO_2}} = 0.07832 \text{ mol C}$$

For every mole of H_2O isolated, 2 mol of H must have been present in the unknown.

$$0.09142 \text{ mol } H_2O \times \frac{2 \text{ mol H in unknown}}{1 \text{ mol } H_2O} = 0.1828 \text{ mol H}$$

The original 1.125 g sample of compound therefore contained 0.07832 mol of C and 0.1828 mol of H. To determine the empirical formula of the unknown, we find the ratio of moles of H to moles of C (◀ Section 2.10).

$$\frac{0.1828 \text{ mol H}}{0.07832 \text{ mol C}} = \frac{2.335 \text{ mol H}}{1.000 \text{ mol C}}$$

Atoms combine to form molecules in whole-number ratios. The translation of this ratio (2.335/1) to a whole-number ratio can usually be done quickly by trial and error. Multiplying the numerator and denominator by 3 gives 7/3. So, we know the ratio is 7 mol H to 3 mol C, which means the empirical formula of the hydrocarbon is C_3H_7.

Comparing the experimental molar mass with the molar mass calculated for the empirical formula,

$$\frac{\text{Experimental molar mass}}{\text{Molar mass of } C_3H_7} = \frac{86.2 \text{ g/mol}}{43.1 \text{ g/mol}} = \frac{2}{1}$$

we find that the molecular formula is twice the empirical formula. That is, the molecular formula is $(C_3H_7)_2$, or C_6H_{14}.

Comment As noted in Problem Solving Tip 2.3 (page 91), for problems of this type be sure to use data with enough significant figures to give accurate atom ratios. Finally, note that the determination of the molecular formula does not end the problem for a chemist. In this case, the formula C_6H_{14} is appropriate for several distinctly different compounds. Two of the five compounds having this formula are shown here:

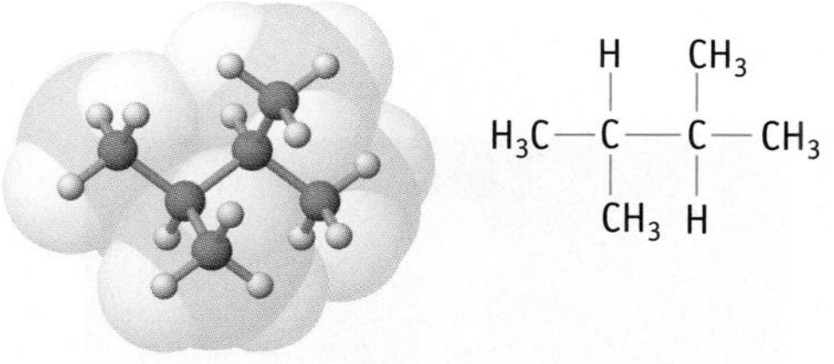

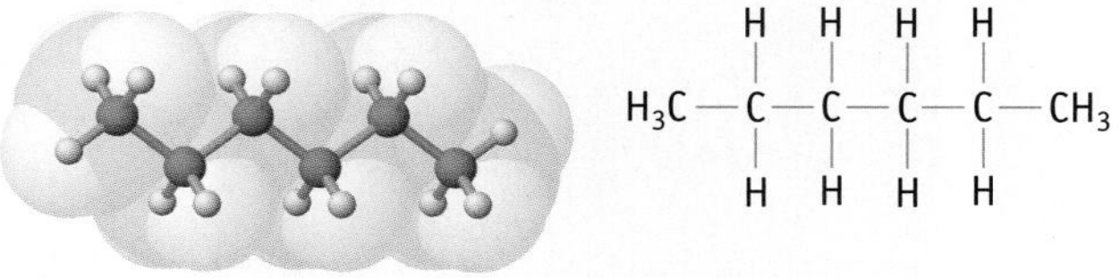

To determine the identity of the unknown compound, more laboratory experiments have to be done. One option is to use an NMR spectrometer such as is pictured in Figure 4.4 or to compare the properties of the unknown with values listed in the chemical literature.

EXERCISE 4.5 Determining the Empirical and Molecular Formulas for a Hydrocarbon

A 0.523-g sample of the unknown compound C_xH_y was burned in air to give 1.612 g of CO_2 and 0.7425 g of H_2O. A separate experiment gave a molar mass for C_xH_y of 114 g/mol. Determine the empirical and molecular formulas for the hydrocarbon.

EXERCISE 4.6 Determining the Empirical and Molecular Formulas for a Compound Containing C, H, and O

A 0.1342-g sample of a compound with C, H, and O ($C_xH_yO_z$) was burned in oxygen, and 0.240 g of CO_2 and 0.0982 g of H_2O was isolated. What is the empirical formula of the compound? If the experimentally determined molar mass was 74.1 g/mol, what is the molecular formula of the compound? (Hint: The carbon atoms in the compound are converted to CO_2, and the hydrogen atoms are converted to H_2O. The O atoms are found both in CO_2 and H_2O. To find the mass of O in the original sample, use the masses of CO_2 and H_2O to find the masses of C and H in the 0.1342-g sample. Whatever of the 0.1342-g sample is not C and H is the mass of O.)

4.5 Measuring Concentrations of Compounds in Solution

■ **Molar and Molarity** Chemists use "molar" as an adjective to describe a solution. We use "molarity" as a noun. For example, we refer to a 0.1 molar solution or say the solution has a molarity of 0.1 mole per liter.

Most chemical studies require quantitative measurements, including experiments involving aqueous solutions. When doing such experiments, we continue to use balanced equations and moles, but we measure volumes of solution rather than masses of solids, liquids, or gases. Solution concentration expressed as molarity relates the volume of solution in liters to the amount of substance in moles.

Solution Concentration: Molarity

The concept of concentration is useful in many contexts. For example, about 5,500,000 people live in Wisconsin, and the state has a land area of roughly 56,000 square miles; therefore, the average concentration of people is about (5.5×10^6 people/5.6×10^4 square miles) or 98 people per square mile. In chemistry, the amount of solute dissolved in a given volume of solution, the concentration of the

FIGURE 4.8 Volume of solution versus volume of solvent. To make a 0.100 M solution of $CuSO_4$, 25.0 g or 0.100 mol of $CuSO_4 \cdot 5\ H_2O$ (the blue crystalline solid) was placed in a 1.00-L volumetric flask.

For this photo, we measured out exactly 1.00 L of water, which was slowly added to the volumetric flask containing $CuSO_4 \cdot 5\ H_2O$. When enough water had been added so that the solution volume was exactly 1.00 L, approximately 8 mL (the quantity in the small graduated cylinder) was left over from the original 1.00 L of water. This emphasizes that molar concentrations are defined as moles of solute per liter of solution and not per liter of water or other solvent.

solution, can be found in the same way. A useful unit of solute concentration, c, is **molarity**, which is defined as amount of solute per liter of solution.

$$\text{Molarity of } x\ (c_x) = \frac{\text{amount of solute } x \text{ (mol)}}{\text{volume of solution (L)}} \qquad \textbf{(4.2)}$$

For example, if 58.4 g (1.00 mol) of NaCl is dissolved in enough water to give a total solution volume of 1.00 L, the concentration, c, is 1.00 mol/L. This is often abbreviated as 1.00 M, where the capital "M" stands for "moles per liter." Another common notation is to place the formula of the compound in square brackets (for example, [NaCl]); this implies that the concentration of the solute in moles of compound per liter of solution is being specified.

$$c_{NaCl} = [NaCl] = 1.00 \text{ mol/L} = 1.00 \text{ M}$$

It is important to notice that molarity refers to the amount of solute per liter of solution and not per liter of solvent. If one liter of water is added to one mole of a solid compound, the final volume will not be exactly one liter, and the final concentration will not be exactly one mol/L (Figure 4.8). When making solutions of a given molarity, it is always the case that we dissolve the solute in a volume of solvent smaller than the desired volume of solution, then add solvent until the final solution volume is reached.

Potassium permanganate, $KMnO_4$, which was used at one time as a germicide in the treatment of burns, is a shiny, purple-black solid that dissolves readily in water to give a deep purple solution. Suppose 0.435 g of $KMnO_4$ has been dissolved in enough water to give 250. mL of solution (Figure 4.9). What is the concentration

■ **Volumetric Flask** A volumetric flask is a special flask with a line marked on its neck (see Figures 4.8 and 4.9). If the flask is filled with a solution to this line (at a given temperature), it contains precisely the volume of solution specified.

■ **NIST and Solution Concentration** The guidelines from NIST specify that the term "molarity" with its symbol M are obsolete and should no longer be used. Instead, the preferred name is "amount of substance concentration of X" or "amount concentration of X." The numerical value should be followed by the units mol/L. Thus, a solution of salt would be described as having a concentration of c_{NaCl} = 1.00 mol/L. Nonetheless, the use of the symbol M, of square brackets, and of the term molarity is so widespread that we shall continue to use them in this edition of the text. See http://physics.nist.gov/Pubs/SP811/sec08.html

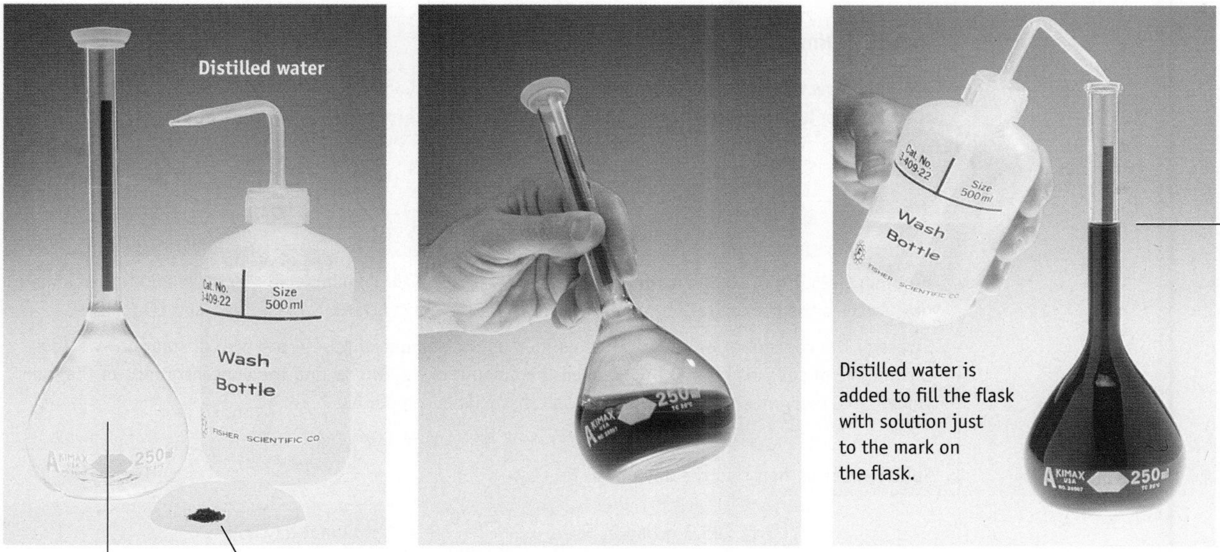

Charles D. Winters

The $KMnO_4$ is first dissolved in a small amount of water.

A mark on the neck of a volumetric flask indicates a volume of exactly 250. mL at 25 °C.

Active Figure 4.9 **Making a solution.** A 0.0110 M solution of $KMnO_4$ is made by adding enough water to 0.435 g of $KMnO_4$ to make 0.250 L of solution.

Chemistry Now™ Sign in at www.thomsonedu.com/login and go to the Chapter Contents menu to explore an interactive version of this figure accompanied by an exercise.

of $KMnO_4$? The first step is to convert the mass of $KMnO_4$ to an amount (moles) of solute.

$$0.435 \text{ g } KMnO_4 \times \frac{1 \text{ mol } KMnO_4}{158.0 \text{ g } KMnO_4} = 0.00275 \text{ mol } KMnO_4$$

Now that the amount of $KMnO_4$ is known, this information can be combined with the volume of solution—which must be in liters—to give the concentration. Because 250. mL is equivalent to 0.250 L,

$$\text{Concentration of } KMnO_4 = c_{KMnO_4} = [KMnO_4] = \frac{0.00275 \text{ mol } KMnO_4}{0.250 \text{ L}} = 0.0110 \text{ M}$$

Photo: Charles D. Winters

Ion concentrations for a soluble ionic compound. Here, 1 mol of $CuCl_2$ dissociates to 1 mol of Cu^{2+} ions and 2 mol of Cl^- ions. Therefore, the Cl^- concentration is twice the concentration calculated for $CuCl_2$.

The $KMnO_4$ concentration is 0.0110 mol/L, or 0.0110 M. This is useful information, but it is often equally useful to know the concentration of each type of ion in a solution. Like all soluble ionic compounds, $KMnO_4$ dissociates completely into its ions, K^+ and MnO_4^-, when dissolved in water.

$$KMnO_4(aq) \longrightarrow K^+(aq) + MnO_4^-(aq)$$
100% dissociation

One mole of $KMnO_4$ provides 1 mol of K^+ ions and 1 mol of MnO_4^- ions. Accordingly, 0.0110 M $KMnO_4$ gives a concentration of K^+ in the solution of 0.0110 M; similarly, the concentration of MnO_4^- is also 0.0110 M.

Another example of ion concentrations is provided by the dissociation of $CuCl_2$.

$$CuCl_2(aq) \longrightarrow Cu^{2+}(aq) + 2\ Cl^-(aq)$$
100% dissociation

If 0.10 mol of $CuCl_2$ is dissolved in enough water to make 1.0 L of solution, the concentration of the copper(II) ion is $[Cu^{2+}] = 0.10$ M. However, the concentration of chloride ions, $[Cl^-]$, is 0.20 M because the compound dissociates in water to provide 2 mol of Cl^- ions for each mole of $CuCl_2$.

Chemistry Now™

Sign in at **www.thomsonedu.com/login** and go to Chapter 4 Contents to see Screen 4.9 for a tutorial on determining **solution concentration** and for a tutorial on **determining ion concentration.**

EXAMPLE 4.5 Concentration

Problem If 25.3 g of sodium carbonate, Na_2CO_3, is dissolved in enough water to make 250. mL of solution, what is the concentration of Na_2CO_3? What are the concentrations of the Na^+ and CO_3^{2-} ions?

Strategy The concentration of Na_2CO_3 is defined as the amount of Na_2CO_3 per liter of solution. We know the volume of solution (0.250 L). We need the amount of Na_2CO_3. To find the concentrations of the individual ions, recognize that the dissolved salt dissociates completely.

$$Na_2CO_3(s) \longrightarrow 2\ Na^+(aq) + CO_3^{2-}(aq)$$

Solution Let us first find the amount of Na_2CO_3.

$$25.3 \text{ g } Na_2CO_3 \times \frac{1 \text{ mol } Na_2CO_3}{106.0 \text{ g } Na_2CO_3} = 0.239 \text{ mol } Na_2CO_3$$

and then the concentration of Na_2CO_3,

$$\text{Concentration of } Na_2CO_3 = \frac{0.239 \text{ mol } Na_2CO_3}{0.250 \text{ L}} = 0.955 \text{ mol/L}$$

The ion concentrations follow from the concentration of Na_2CO_3 and the knowledge that each mole of Na_2CO_3 produces 2 mol of Na^+ ions and 1 mol of CO_3^{2-} ions.

$$0.955 \text{ M } Na_2CO_3(aq) \equiv 2 \times 0.955 \text{ M } Na^+(aq) + 0.955 \text{ M } CO_3^{2-}(aq)$$

That is, $[Na^+] = 1.91$ M and $[CO_3^{2-}] = 0.955$ M.

EXERCISE 4.7 Concentration

Sodium bicarbonate, $NaHCO_3$, is used in baking powder formulations and in the manufacture of plastics and ceramics, among other things. If 26.3 g of the compound is dissolved in enough water to make 200. mL of solution, what is the concentration of $NaHCO_3$? What are the concentrations of the ions in solution?

Preparing Solutions of Known Concentration

Chemists often have to prepare a given volume of solution of known concentration. There are two common ways to do this.

Combining a Weighed Solute with the Solvent

Suppose you wish to prepare 2.00 L of a 1.50 M solution of Na_2CO_3. You have some solid Na_2CO_3 and distilled water. You also have a 2.00-L volumetric flask (see Figures 4.8 and 4.9). To make the solution, you must weigh the necessary quantity of Na_2CO_3 as accurately as possible, carefully place all the solid in the volumetric flask, and then add some water to dissolve the solid. After the solid has dissolved completely, more water is added to bring the solution volume to 2.00 L. The solution then has the desired concentration and the volume specified.

But what mass of Na_2CO_3 is required to make 2.00 L of 1.50 M Na_2CO_3? First, calculate the amount of Na_2CO_3 required,

$$2.00 \text{ L} \times \frac{1.50 \text{ mol } Na_2CO_3}{1.00 \text{ L solution}} = 3.00 \text{ mol } Na_2CO_3 \text{ required}$$

and then the mass in grams.

$$3.00 \text{ mol } Na_2CO_3 \times \frac{106.0 \text{ g } Na_2CO_3}{1 \text{ mol } Na_2CO_3} = 318 \text{ g } Na_2CO_3$$

Thus, to prepare the desired solution, you should dissolve 318 g of Na_2CO_3 in enough water to make 2.00 L of solution.

EXERCISE 4.8 Preparing Solutions of Known Concentration

An experiment in your laboratory requires 250. mL of a 0.0200 M solution of $AgNO_3$. You are given solid $AgNO_3$, distilled water, and a 250.-mL volumetric flask. Describe how to make up the required solution.

Diluting a More Concentrated Solution

Another method of making a solution of a given concentration is to begin with a concentrated solution and add water until the desired, lower concentration is reached (Figure 4.10). Many of the solutions prepared for your laboratory course are probably made by this dilution method. It is more efficient to store a small volume of a concentrated solution and then, when needed, add water to make a much larger volume of a dilute solution.

Suppose you need 500. mL of 0.0010 M potassium dichromate, $K_2Cr_2O_7$, for use in chemical analysis. You have some 0.100 M $K_2Cr_2O_7$ solution available. To make

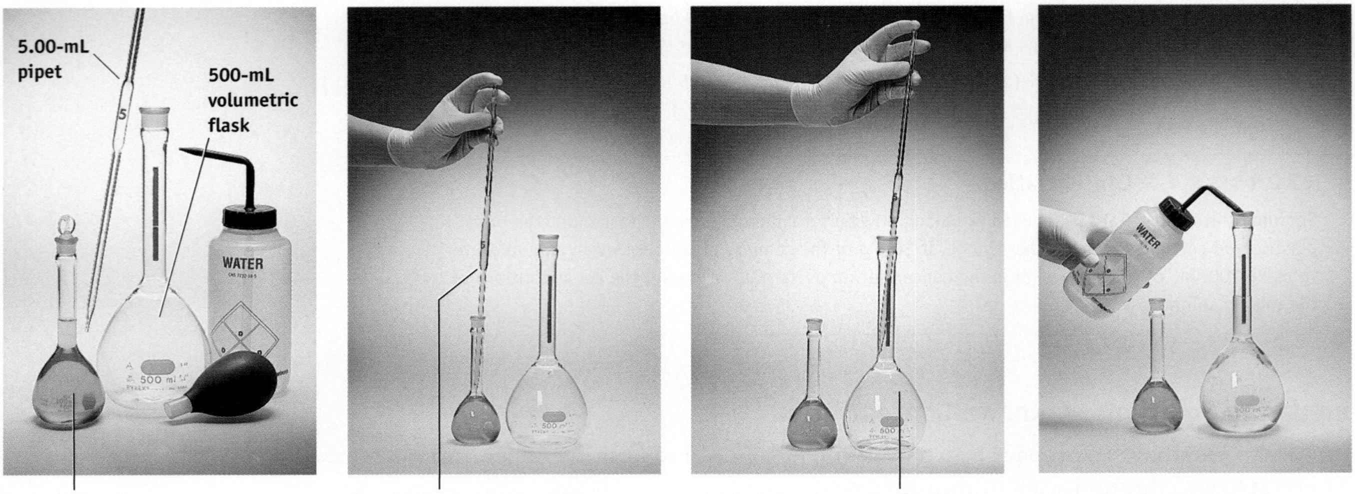

$0.100\ M\ K_2Cr_2O_7$

Use a 5.00-mL pipet to withdraw 5.00 mL of 0.100 M $K_2Cr_2O_7$ solution.

Add the 5.00-mL sample of 0.100 M $K_2Cr_2O_7$ solution to a 500-mL volumetric flask.

Fill the flask to the mark with distilled water to give 0.00100 M $K_2Cr_2O_7$ solution.

FIGURE 4.10 Making a solution by dilution. Here, 5.00 mL of a $K_2Cr_2O_7$ solution is diluted to 500. mL. This means the solution is diluted by a factor of 100, from 0.100 M to 0.00100 M.

the required 0.0010 M solution, place a measured volume of the more concentrated $K_2Cr_2O_7$ solution in a flask, and then add water until the $K_2Cr_2O_7$ is contained in the appropriate larger volume of water (Figure 4.10).

What volume of a 0.100 M $K_2Cr_2O_7$ solution must be diluted to make the 0.0010 M solution? If the volume and concentration of a solution are known, the amount of solute is also known. Therefore, the amount of $K_2Cr_2O_7$ that must be in the final dilute solution is

$$\text{Amount of } K_2Cr_2O_7 \text{ in dilute solution} = c_{K_2Cr_2O_7} \times V_{K_2Cr_2O_7} = \left(\frac{0.0010 \text{ mol}}{\cancel{L}}\right) \times (0.500\ \cancel{L})$$
$$= 0.00050 \text{ mol } K_2Cr_2O_7$$

■ **Diluting Concentrated Sulfuric Acid** The instruction that one prepares a solution by adding water to a more concentrated solution is correct, except for sulfuric acid solutions. When mixing water and sulfuric acid, the resulting solution becomes quite warm. If water is added to concentrated sulfuric acid, so much heat is evolved that the solution may boil over or splash and burn someone nearby. To avoid this problem, chemists always add concentrated sulfuric acid to water to make a dilute solution.

A more concentrated solution containing this amount of $K_2Cr_2O_7$ must be placed in a 500.-mL flask and then be diluted to the final volume. The volume of 0.100 M $K_2Cr_2O_7$ that must be transferred and diluted is 5.0 mL.

$$0.00050\ \cancel{\text{mol } K_2Cr_2O_7} \times \frac{1.00 \text{ L}}{0.100\ \cancel{\text{mol } K_2Cr_2O_7}} = 0.0050 \text{ L or } 5.0 \text{ mL}$$

Thus, to prepare 500. mL of 0.0010 M $K_2Cr_2O_7$, place 5.0 mL of 0.100 M $K_2Cr_2O_7$ in a 500.-mL flask and add water until a volume of 500. mL is reached (see Figure 4.10).

Chemistry Now™

Sign in at **www.thomsonedu.com/login** and go to Chapter 4 Contents to see Screen 4.11 for an exercise and a tutorial on **the direct addition method of preparing a solution** and for an exercise and tutorial on **the dilution method of preparing a solution.**

Problem Solving Tip 4.3 Preparing a Solution by Dilution

There is a straightforward method to use for problems involving dilution. The central idea is that the amount of solute in the final, dilute solution has to be equal to the amount of solute in the more concentrated solution. If c is the concentration (molarity) and V is the volume (and the subscripts d and c identify the dilute and concentrated solutions, respectively), then the amount of solute in either solution (in the case of the $K_2Cr_2O_7$ example in the text) can be calculated as follows:

Amount of $K_2Cr_2O_7$ in the final dilute solution is

$$c_dV_d = 0.00050 \text{ mol}$$

Amount of $K_2Cr_2O_7$ taken from the more concentrated solution is

$$c_cV_c = 0.00050 \text{ mol}$$

Because both cV products are equal to the same amount of solute, we can use the following equation:

Amount of reagent in concentrated solution = Amount of reagent in dilute solution

$$c_cV_c = c_dV_d$$

This equation is valid for all cases in which a more concentrated solution is used to make a more dilute one.

EXAMPLE 4.6 Preparing a Solution by Dilution

Problem What is the concentration of iron(III) ion in a solution prepared by diluting 1.00 mL of a 0.236 M solution of iron(III) nitrate to a volume of 100.0 mL?

Strategy First, calculate the amount of iron(III) ion in the 1.00-mL sample. The concentration of the ion in the final, dilute solution is equal to this amount of iron(III) divided by the new volume.

Solution The amount of iron(III) ion in the 1.00 mL sample is

$$\text{Amount of Fe}^{3+} = c_{Fe^{3+}}V_{Fe^{3+}} = \frac{0.236 \text{ mol Fe}^{3+}}{\cancel{L}} \times 1.00 \times 10^{-3}\ \cancel{L} = 2.36 \times 10^{-4} \text{ mol Fe}^{3+}$$

This amount of iron(III) ion is distributed in the new volume of 100.0 mL, so the final concentration of the diluted solution is

$$c_{Fe^{3+}} = [Fe^{3+}] = \frac{2.36 \times 10^{-4} \text{ mol Fe}^{3+}}{0.100 \text{ L}} = 2.36 \times 10^{-3} \text{ M}$$

EXERCISE 4.9 Preparing a Solution by Dilution

An experiment calls for you to use 250. mL of 1.00 M NaOH, but you are given a large bottle of 2.00 M NaOH. Describe how to make desired volume of 1.00 M NaOH.

4.6 pH, a Concentration Scale for Acids and Bases

Module 9

Vinegar, which contains the weak acid, acetic acid, has a hydronium ion concentration of only 1.2×10^{-3} M, and "pure" rainwater has $[H_3O^+] = 2.5 \times 10^{-6}$ M. These small values can be expressed using scientific notation, but a more convenient way to express such numbers is the logarithmic pH scale.

The pH of a solution is the negative of the base-10 logarithm of the hydronium ion concentration.

$$\text{pH} = -\log[H_3O^+] \tag{4.3}$$

Taking vinegar, pure water, blood, and ammonia as examples,

$$\begin{aligned}
\text{pH of vinegar} &= -\log(1.2 \times 10^{-3} \text{ M}) = -(-2.92) = 2.92 \\
\text{pH of pure water (at 25 °C)} &= -\log(1.0 \times 10^{-7} \text{ M}) = -(-7.00) = 7.00 \\
\text{pH of blood} &= -\log(4.0 \times 10^{-8} \text{ M}) = -(-7.40) = 7.40 \\
\text{pH of household ammonia} &= -\log(4.3 \times 10^{-12} \text{ M}) = -(-11.37) = 11.37
\end{aligned}$$

you see that something you recognize as acidic has a relatively low pH, whereas ammonia, a common base, has a very low hydronium ion concentration and a high

■ **pH of Pure Water** Highly purified water, which is said to be "neutral," has a pH of exactly 7 at 25 °C. This is the "dividing line" between acidic substances ($pH < 7$) and basic substances ($pH > 7$) at 25 °C.

A Closer Look — Serial Dilutions

We often find in the laboratory that a solution is too concentrated for the analytical technique we want to use. You might want to analyze a seawater sample for its chloride ion content, for instance. To obtain a solution with a chloride concentration of the proper magnitude for analysis by the Mohr method (Case Study, page 186), for example, you might want to dilute the sample, not once but several times.

Suppose you have 100.0 mL of a seawater sample that has a NaCl concentration of 0.550 mol/L. You transfer 10.0 mL of that sample to a 100.0-mL volumetric flask and fill to the mark with distilled water. You then transfer 5.00 mL of that diluted sample to another 100.0 mL flask and fill to the mark with distilled water. What is the NaCl concentration in the final 100.0-mL sample?

The original solution contains 0.550 mol/L of NaCl. If you remove 10.00 mL, you have removed

$$0.01000 \text{ L} \times 0.550 \text{ mol/L} = 5.50 \times 10^{-3} \text{ mol NaCl}$$

and the concentration in 100.0 mL of the diluted solution is

$$c_{NaCl} = 5.50 \times 10^{-3} \text{ mol}/0.100 \text{ L} = 5.50 \times 10^{-2} \text{ M}$$

or 1/10 of the concentration of the original solution (because we diluted the sample by a factor of 10).

Now we take 5.0 mL of the diluted solution and dilute that once again to 100.0 mL. The final concentration is

$$0.00500 \text{ L} \times 5.50 \times 10^{-2} \text{ mol/L} = 2.75 \times 10^{-4} \text{ mol NaCl}$$

$$c_{NaCl} = 2.75 \times 10^{-4} \text{ mol}/0.1000 \text{ L} = 2.75 \times 10^{-3} \text{ M}$$

This is 1/200 of the concentration of the original solution.

A fair question at this point is why we did not just take 1 mL of the original solution and dilute to 200 mL. The answer is that there is less error in using larger pipets such as 5.00- or 10.00-mL pipets rather than a 1.00-mL pipet. And then there is a limitation in available glassware. A 200.00-mL volumetric flask is not often available.

Question: *You have a 100.0-mL sample of a blue dye having a concentration of 0.36 M. You dilute a 10.0-mL sample of this to 100.0 mL and then a 2.00-mL sample of that solution to 100.0 mL. What is the final dye concentration? (Answer: 7.2 × 10^{-4} M)*

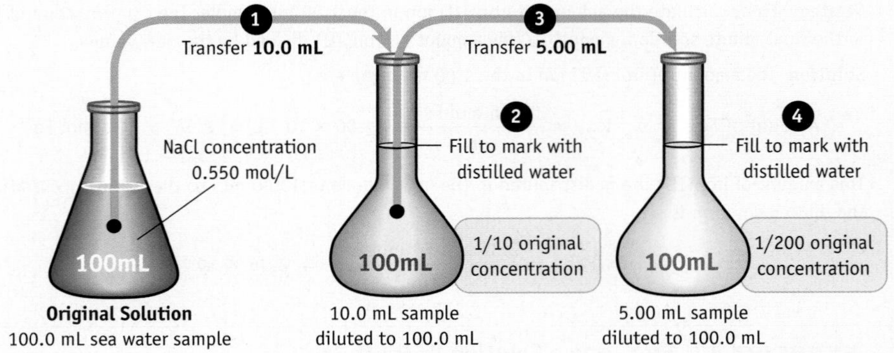

pH. Blood, which your common sense tells you is likely to be neither acidic nor basic, has a pH near 7. Indeed, for aqueous solutions at 25 °C, we can say that acids will have pH values less than 7, bases will have values greater than 7, and a pH of 7 represents a neutral solution (Figure 4.11).

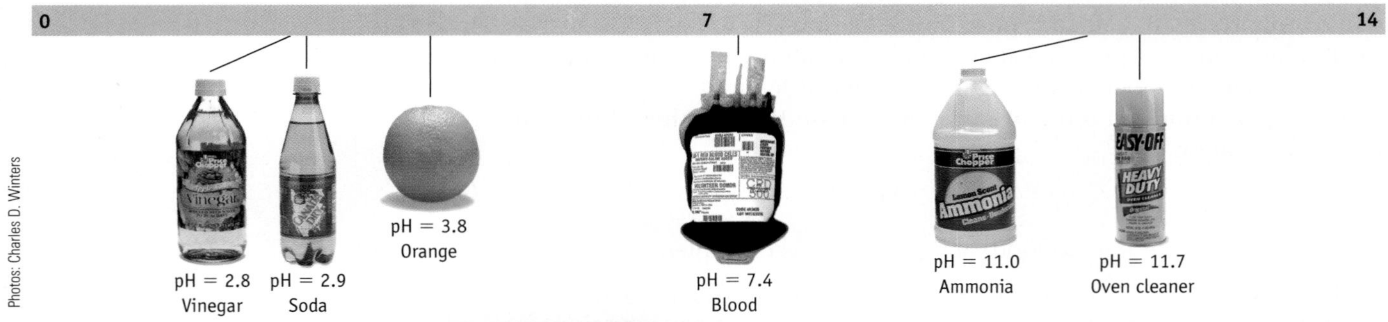

Active Figure 4.11 **pH values of some common substances.** Here, the "bar" is colored red at one end and blue at the other. These are the colors of litmus paper, commonly used in the laboratory to decide whether a solution is acidic (litmus is red) or basic (litmus is blue).

Chemistry Now™ Sign in at www.thomsonedu.com/login and go to the Chapter Contents menu to explore an interactive version of this figure accompanied by an exercise.

Charles D. Winters

(a) (b)

FIGURE 4.12 Determining pH. (a) Some household products. Each solution contains a few drops of a universal indicator, a mixture of several acid–base indicators. A color of yellow or red indicates a pH less than 7. A green to purple color indicates a pH greater than 7. (b) The pH of a soda is measured with a modern pH meter. Soft drinks are often quite acidic, owing to the dissolved CO_2 and other ingredients.

Suppose you know the pH of a solution. To find the hydronium ion concentration, you take the antilog of the pH. That is,

$$[H_3O^+] = 10^{-pH} \tag{4.4}$$

■ **Logarithms** Numbers less than 1 have negative logs. Defining pH as $-\log[H^+]$ produces a positive number. See Appendix A for a discussion of logs.

For example, the pH of a diet soda is 3.12, and the hydronium ion concentration of the solution is

$$[H_3O^+] = 10^{-3.12} = 7.6 \times 10^{-4}\text{ M}$$

The approximate pH of a solution may be determined using any of a variety of dyes. Litmus paper contains a dye extracted from a type of lichen, but many other dyes are also available (Figure 4.12a). A more accurate measurement of pH is done with a pH meter such as that shown in Figure 4.12b. Here, a pH electrode is immersed in the solution to be tested, and the pH is read from the instrument.

■ **Logs and Your Calculator** All scientific calculators have a key marked "log." To find an antilog, use the key marked "10^x" or the inverse log. In determining $[H_3O^+]$ from a pH, when you enter the value of x for 10^x, make sure it has a negative sign.

Chemistry Now™

Sign in at **www.thomsonedu.com/login** and go to Chapter 4 Contents to see Screen 4.11 for a tutorial on **determining the pH of a solution.**

■ **pH-Indicating Dyes** Many natural substances change color in solution as pH changes. See the extract of red cabbage in Figure 3.12. Tea changes color when acidic lemon juice is added.

EXAMPLE 4.7 pH of Solutions

Problem

(a) Lemon juice has $[H_3O^+] = 0.0032$ M. What is its pH?

(b) Sea water has a pH of 8.30. What is the hydronium ion concentration of this solution?

(c) A solution of nitric acid has a concentration of 0.0056 mol/L. What is the pH of this solution?

Strategy Use Equation 4.3 to calculate pH from the H_3O^+ concentration. Use Equation 4.4 to find $[H_3O^+]$ from the pH.

Solution

(a) Lemon juice: Because the hydronium ion concentration is known, the pH is found using Equation 4.3.

$$pH = -\log[H_3O^+] = -\log(3.2 \times 10^{-3}) = -(-2.49) = 2.49$$

(b) Sea water: Here, pH = 8.30. Therefore,

$$[H_3O^+] = 10^{-pH} = 10^{-8.30} = 5.0 \times 10^{-9}\text{ M}$$

■ **Weak and Strong Acids and Hydronium Ion Concentration** Because a weak acid (e.g., acetic acid) does not ionize completely in water, the hydronium ion concentration in an aqueous solution of a weak acid does not equal the concentration of the acid (as is the case for strong acids) but must be calculated using the principles of chemical equilibrium (see Chapter 17).

(c) Nitric acid: Nitric acid, a strong acid (Table 3.2, page 132), is completely ionized in aqueous solution. Because the concentration of HNO_3 is 0.0056 mol/L, the ion concentrations are

$$[H_3O^+] = [NO_3^-] = 0.0056 \text{ M}$$

$$\text{pH} = -\log[H_3O^+] = -\log(0.0056 \text{ M}) = 2.25$$

Comment A comment on logarithms and significant figures (Appendix A) is useful. The number to the left of the decimal point in a logarithm is called the *characteristic,* and the number to the right is the *mantissa*. The mantissa has as many significant figures as the number whose log was found. For example, the logarithm of 3.2×10^{-3} (two significant figures) is 2.49 (two numbers to the right of the decimal point).

EXERCISE 4.10 pH of Solutions

(a) What is the pH of a solution of HCl in which $[HCl] = 2.6 \times 10^{-2}$ M?

(b) What is the hydronium ion concentration in orange juice with a pH of 3.80?

4.7 Stoichiometry of Reactions in Aqueous Solution

Solution Stoichiometry

Suppose we want to know what mass of $CaCO_3$ is required to react completely with 25 mL of 0.750 M HCl. The first step in finding the answer is to write a balanced equation. In this case, we have a gas-forming exchange reaction involving a metal carbonate and an aqueous acid (Figure 4.13).

$$CaCO_3(s) + 2\,HCl(aq) \longrightarrow CaCl_2(aq) + H_2O(\ell) + CO_2(g)$$

metal carbonate + acid → salt + water + carbon dioxide

Charles D. Winters

FIGURE 4.13 A commercial remedy for excess stomach acid. The tablet contains calcium carbonate, which reacts with hydrochloric acid, the acid present in the digestive system. The most obvious product is CO_2 gas.

This problem can be solved in the same way as all the stoichiometry problems you have seen so far, except that the quantity of one reactant is given as a volume of a solution of known concentration instead of as a mass in grams. The first step is to find the amount of HCl.

$$\text{Amount of HCl} = c_{HCl}V_{HCl} = \frac{0.750 \text{ mol HCl}}{1 \text{ L HCl}} \times 0.025 \text{ L HCl} = 0.019 \text{ mol HCl}$$

This is then related to the amount of $CaCO_3$ required.

$$0.019 \text{ mol HCl} \times \frac{1 \text{ mol } CaCO_3}{2 \text{ mol HCl}} = 0.0094 \text{ mol } CaCO_3$$

Finally, the amount of $CaCO_3$ is converted to a mass in grams.

$$0.0094 \text{ mol } CaCO_3 \times \frac{100.\text{ g } CaCO_3}{1 \text{ mol } CaCO_3} = 0.94 \text{ g } CaCO_3$$

Chemists are likely to do such calculations many times in the course of their work. If you follow the general scheme outlined in Problem Solving Tip 4.4 and pay attention to the units on the numbers, you can successfully carry out any kind of stoichiometry calculations involving concentrations.

Chemistry Now™

Sign in at **www.thomsonedu.com/login** and go to Chapter 4 Contents to see Screen 4.12 for an exercise on **solution stoichiometry,** for a tutorial on **determining the mass of a product,** and for a tutorial on **determining the volume of a reactant.**

Problem Solving Tip 4.4 Stoichiometry Calculations Involving Solutions

In Problem Solving Tip 4.1, you learned about a general approach to stoichiometry problems. We can now modify that scheme for a reaction involving solutions such as x A(aq) + y B(aq) ⟶ products.

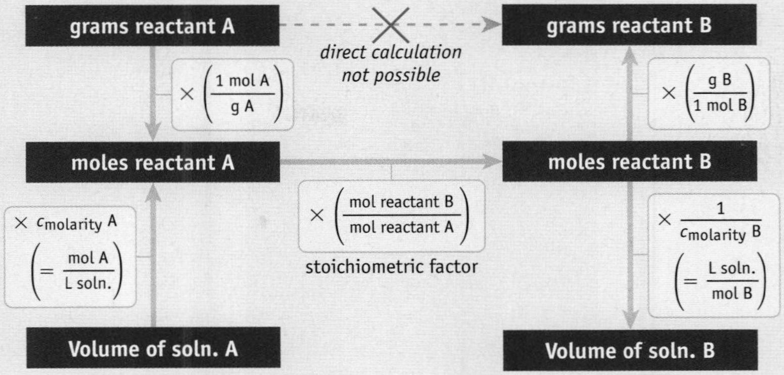

EXAMPLE 4.8 Stoichiometry of a Reaction in Solution

Problem Metallic zinc reacts with aqueous HCl.

$$Zn(s) + 2\ HCl(aq) \longrightarrow ZnCl_2(aq) + H_2(g)$$

What volume of 2.50 M HCl, in milliliters, is required to convert 11.8 g of Zn completely to products?

Strategy Here, the mass of zinc is known, so you first calculate the amount of zinc. Next, use a stoichiometric factor (= 2 mol HCl/1 mol Zn) to relate amount of HCl required to amount of Zn available. Finally, calculate the volume of HCl from the amount of HCl and its concentration.

Solution Begin by calculating the amount of Zn.

$$11.8\ \cancel{g\ Zn} \times \frac{1\ mol\ Zn}{65.39\ \cancel{g\ Zn}} = 0.180\ mol\ Zn$$

Use the stoichiometric factor to calculate the amount of HCl required.

$$0.180\ \cancel{mol\ Zn} \times \frac{2\ mol\ HCl}{1\ \cancel{mol\ Zn}} = 0.360\ mol\ HCl$$

Use the amount of HCl and the solution concentration to calculate the volume.

$$0.360\ \cancel{mol\ HCl} \times \frac{1.00\ L\ solution}{2.50\ \cancel{mol\ HCl}} = 0.144\ L\ HCl$$

The answer is requested in units of milliliters, so we convert the volume to milliliters and find that 144 mL of 2.50 M HCl is required to convert 11.8 g of Zn completely to products.

EXERCISE 4.11 Solution Stoichiometry

If you combine 75.0 mL of 0.350 M HCl and an excess of Na_2CO_3, what mass of CO_2, in grams, is produced?

$$Na_2CO_3(s) + 2\ HCl(aq) \longrightarrow 2\ NaCl(aq) + H_2O(\ell) + CO_2(g)$$

Titration: A Method of Chemical Analysis

Oxalic acid, $H_2C_2O_4$, is a naturally occurring acid. Suppose you are asked to determine the mass of this acid in an impure sample. Because the compound is an acid, it reacts with a base such as sodium hydroxide.

$$H_2C_2O_4(aq) + 2\ NaOH(aq) \longrightarrow Na_2C_2O_4(aq) + 2\ H_2O(\ell)$$

■ **Titrations** Acid–base titrations are discussed in more detail in Chapter 18.

(a)
Buret containing aqueous NaOH of accurately known concentration.

(b)
A solution of NaOH is added slowly to the sample being analyzed.

(c)
When the amount of NaOH added from the buret equals the amount of H_3O^+ supplied by the acid being analyzed, the dye (indicator) changes color.

Active Figure 4.14 **Titration of an acid in aqueous solution with a base.** (a) A buret, a volumetric measuring device calibrated in divisions of 0.1 mL, is filled with an aqueous solution of a base of known concentration. (b) Base is added slowly from the buret to the solution containing the acid being analyzed and an indicator. (c) A change in the color of the indicator signals the equivalence point. (The indicator used here is phenolphthalein.)

Chemistry.Now™ Sign in at www.thomsonedu.com/login and go to the Chapter Contents menu to explore an interactive version of this figure accompanied by an exercise.

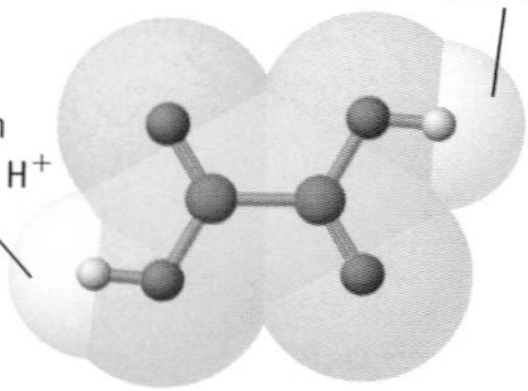

Oxalic acid $H_2C_2O_4$

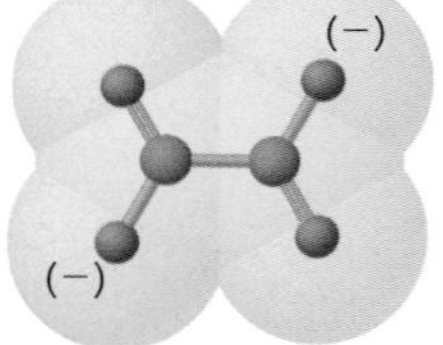

Oxalate anion $C_2O_4^{2-}$

Oxalic acid. Oxalic acid has two groups that can supply an H^+ ion to solution. Hence, 1 mol of the acid requires 2 mol of NaOH for complete reaction.

You can use this reaction to determine the quantity of oxalic acid present in a given mass of sample if the following conditions are met:

- You can determine when the amount of sodium hydroxide added is just enough to react with all the oxalic acid present in solution.
- You know the concentration of the sodium hydroxide solution and the volume that has been added at the point of complete reaction.

These conditions are fulfilled in a titration, a procedure illustrated in Figure 4.14. The solution containing oxalic acid is placed in a flask along with an acid–base indicator, a dye that changes color when the pH of the reaction solution reaches a certain value. Aqueous sodium hydroxide of accurately known concentration is placed in a buret. The sodium hydroxide in the buret is added slowly to the acid solution in the flask. As long as some acid is present in solution, all the base supplied from the buret is consumed, the solution remains acidic, and the indicator color is unchanged. At some point, however, the amount of OH^- added exactly equals the amount of H_3O^+

that can be supplied by the acid. This is called the **equivalence point**. As soon as the slightest excess of base has been added beyond the equivalence point, the solution becomes basic, and the indicator changes color (see Figure 4.14). The example that follows shows how to use the equivalence point and the other information to determine the percentage of oxalic acid in a mixture.

Chemistry Now™

Sign in at **www.thomsonedu.com/login** and go to Chapter 4 Contents to see Screen 4.13 for a tutorial on the **volume of titrant used,** for a tutorial on **determining the concentration of acid solution,** and for a tutorial on **determining the concentration of an unknown acid.**

EXAMPLE 4.9 Acid–Base Titration

Problem A 1.034-g sample of impure oxalic acid is dissolved in water and an acid–base indicator added. The sample requires 34.47 mL of 0.485 M NaOH to reach the equivalence point. What is the mass of oxalic acid, and what is its mass percent in the sample?

Strategy The balanced equation for the reaction of NaOH and $H_2C_2O_4$ is

$$H_2C_2O_4(aq) + 2\ NaOH(aq) \longrightarrow Na_2C_2O_4(aq) + 2\ H_2O(\ell)$$

The concentration and volume of NaOH delivered in the titration are used to determine the amount of NaOH. A stoichiometric factor is used to relate the amount of NaOH to the amount of $H_2C_2O_4$, and the amount of $H_2C_2O_4$ is converted to a mass. The mass percent of acid in the sample is then calculated. See Problem Solving Tip 4.4.

Solution The amount of NaOH is given by

$$\text{Amount of NaOH} = c_{NaOH} \times V_{NaOH} = \frac{0.485\ \text{mol NaOH}}{\cancel{\text{L}}} \times 0.03447\ \cancel{\text{L}} = 0.0167\ \text{mol NaOH}$$

The balanced equation for the reaction shows that 1 mol of oxalic acid requires 2 mol of sodium hydroxide. This is the required stoichiometric factor to obtain the amount of oxalic acid present.

$$0.0167\ \cancel{\text{mol NaOH}} \times \frac{1\ \text{mol}\ H_2C_2O_4}{2\ \cancel{\text{mol NaOH}}} = 0.00836\ \text{mol}\ H_2C_2O_4$$

The mass of oxalic acid is found from the amount of the acid.

$$0.00836\ \cancel{\text{mol}\ H_2C_2O_4} \times \frac{90.04\ \text{g}\ H_2C_2O_4}{1\ \cancel{\text{mol}\ H_2C_2O_4}} = \boxed{0.753\ \text{g}\ H_2C_2O_4}$$

This mass of oxalic acid represents 72.8% of the total sample mass.

$$\frac{0.753\ \text{g}\ H_2C_2O_4}{1.034\ \text{g sample}} \times 100\% = \boxed{72.8\%\ H_2C_2O_4}$$

EXERCISE 4.12 Acid–Base Titration

A 25.0-mL sample of vinegar (which contains the weak acid, acetic acid, CH_3CO_2H) requires 28.33 mL of a 0.953 M solution of NaOH for titration to the equivalence point. What mass of acetic acid, in grams, is in the vinegar sample, and what is the concentration of acetic acid in the vinegar?

$$CH_3CO_2H(aq) + NaOH(aq) \longrightarrow NaCH_3CO_2(aq) + H_2O(\ell)$$

Case Study — How Much Salt Is There in Seawater?

There is a French legend about a princess who told her father, the king, that she loved him as much as she loved salt. Thinking that this was not a great measure of love, he banished her from the kingdom. Only later did he realize how much he needed, and valued, salt.

Salt has played a key role in history. The earliest written record of salt production dates from around 800 BC, but the sea has always been a source of salt, and there is evidence of the Chinese harvesting salt from seawater by 6000 BC.

The average human body contains about 50 g of salt. Because we continually lose salt in urine, sweat, and other excretions, salt must be a part of our diet. Early humans recognized that salt deficiency causes headaches, cramps, loss of appetite, and, in extreme cases, death. Consuming meat provides salt, but consuming vegetables does not. This is the reason herbivorous animals seek out salt.

Saltiness is one of the basic taste sensations, and a taste of seawater quickly reveals it is salty. How did the oceans become salty? And why is chloride ion the most abundant ion?

A result of the interaction of atmospheric CO_2 and water is hydronium ions and bicarbonate ions.

$$CO_2(g) + H_2O(\ell) \longrightarrow H_2CO_3(aq)$$

$$H_2CO_3(aq) + H_2O(\ell) \rightleftarrows H_3O^+(aq) + HCO_3^-(aq)$$

Indeed, this is the reason rain is normally acidic, and this slightly acidic rainwater can then cause substances such as limestone or corals to dissolve, producing calcium ions and more bicarbonate ions.

$$CaCO_3(s) + CO_2(g) + H_2O(\ell) \longrightarrow Ca^{2+}(aq) + 2\ HCO_3^-(aq)$$

John Kotz

Seawater contains many dissolved salts. Among the ions in seawater are halide anions, alkali metal cations, and anions such as carbonate and hydrogen phosphate. See Table 3.1, page 122.

Sodium ions arrive in the oceans by a similar reaction with sodium-bearing minerals such as albite, $NaAlSi_3O_6$. Acidic rain falling on the land extracts sodium ions that are then carried by rivers to the ocean.

The average chloride content of rocks in the earth's crust is only 0.01%, so only a minute proportion of the chloride ion in the oceans can come from the weathering of rocks and minerals. What then is the origin of the chloride ions in seawater? The answer is volcanoes. Hydrogen chloride gas, HCl, is a constituent of volcanic gases. Early in Earth's history, the planet was much hotter, and volcanoes were much more widespread. The HCl gas emitted from these volcanoes is very soluble in water and is quickly dissolved to give a dilute solution of hydrochloric acid. The chloride ions from dissolved HCl gas and sodium ions from weathered rocks are the source of the salt in the sea.

Suppose you are an oceanographer, and you want to determine the concentration of chloride ions in a sample of seawater. How can you do this? And what results might you find?

There are several ways to analyze a solution for its chloride ion content, among them the classic "Mohr method." Here, a solution containing chloride ions is titrated with standardized silver nitrate. You know that the following reaction should occur,

$$Ag^+(aq) + Cl^-(aq) \longrightarrow AgCl(s)$$

and will continue until the chloride ions have been precipitated completely. To detect the equivalence point of the titration of Cl^- with Ag^+, the Mohr method specifies the addition of a few drops of a solution of potassium chromate. This "indicator" works because silver chromate is slightly more soluble than AgCl, so the red Ag_2CrO_4 precipitates only after all of the AgCl is precipitated.

$$2\ Ag^+(aq) + CrO_4^{2-}(aq) \longrightarrow Ag_2CrO_4(s)$$

The appearance of the red color of Ag_2CrO_4 (see Figure 3.11d) signals the equivalence point.

Question:

1. *Using the following information, calculate the chloride ion concentration in a sample of seawater.*
   a. *Volume of original seawater sample = 100.0 mL.*
   b. *A 10.00 mL sample of the seawater was diluted to 100.0 mL with distilled water.*
   c. *10.00 mL of the diluted sample was again diluted to 100.0 mL.*
   d. *A Mohr titration was done on 50.00 mL of the diluted sample (from step 3) and required 26.25 mL of 0.100 M $AgNO_3$. What was the chloride ion concentration in the original seawater sample?*

Answer to this question is in Appendix Q.

Standardizing an Acid or Base

In Example 4.9, the concentration of the base used in the titration was given. In actual practice, this usually has to be found by a prior measurement. The procedure by which the concentration of an analytical reagent is determined accurately is called **standardization**, and there are two general approaches.

One approach is to weigh accurately a sample of a pure, solid acid or base (known as a *primary standard*) and then titrate this sample with a solution of the base or acid to be standardized (Example 4.10). An alternative approach to standardizing a solution is to titrate it with another solution that is already standardized (Exercise 4.13). This is often done using standard solutions purchased from chemical supply companies.

EXAMPLE 4.10 Standardizing an Acid by Titration

Problem Sodium carbonate, Na_2CO_3, is a base, and an accurately weighed sample can be used to standardize an acid. A sample of sodium carbonate (0.263 g) requires 28.35 mL of aqueous HCl for titration to the equivalence point. What is the concentration of the HCl?

Strategy The balanced equation for the reaction is written first.

$$Na_2CO_3(aq) + 2\ HCl(aq) \longrightarrow 2\ NaCl(aq) + H_2O(\ell) + CO_2(g)$$

The amount of Na_2CO_3 can be calculated from its mass, and then, using the stoichiometric factor, the amount of HCl in 28.35 mL can be calculated. The amount of HCl divided by the volume of solution (in liters) gives its concentration (mol/L).

Solution Convert the mass of Na_2CO_3 used as the standard to amount.

$$0.263\ \text{g}\ Na_2CO_3 \times \frac{1\ \text{mol}\ Na_2CO_3}{106.0\ \text{g}\ Na_2CO_3} = 0.00248\ \text{mol}\ Na_2CO_3$$

Use the stoichiometric factor to calculate the amount of HCl in 28.35 mL.

$$0.00248\ \text{mol}\ Na_2CO_3 \times \frac{2\ \text{mol HCl required}}{1\ \text{mol}\ Na_2CO_3\ \text{available}} = 0.00496\ \text{mol HCl}$$

The 28.35-mL (0.02835-L) sample of aqueous HCl contains 0.00496 mol of HCl, so the concentration of the HCl solution is 0.175 M.

$$[HCl] = \frac{0.00496\ \text{mol HCl}}{0.02835\ \text{L}} = 0.175\ \text{M}$$

Comment In this example, Na_2CO_3 is a primary standard. Sodium carbonate can be obtained in pure form, can be weighed accurately, and reacts completely with a strong acid.

EXERCISE 4.13 Standardization of a Base

Hydrochloric acid, HCl, can be purchased from chemical supply houses with a concentration of 0.100 M, and this solution can be used to standardize the solution of a base. If titrating 25.00 mL of a sodium hydroxide solution to the equivalence point requires 29.67 mL of 0.100 M HCl, what is the concentration of the base?

Determining Molar Mass by Titration

In Chapter 2 and this chapter, we used analytical data to determine the empirical formula of a compound. The molecular formula could then be derived if the molar mass were known. If the unknown substance is an acid or a base, it is possible to determine the molar mass by titration.

EXAMPLE 4.11 Determining the Molar Mass of an Acid by Titration

Problem To determine the molar mass of an organic acid, HA, we titrate 1.056 g of HA with standardized NaOH. Calculate the molar mass of HA assuming the acid reacts with 33.78 mL of 0.256 M NaOH according to the equation

$$HA(aq) + OH^-(aq) \longrightarrow A^-(aq) + H_2O(\ell)$$

Strategy The key to this problem is to recognize that the molar mass of a substance is the ratio of the mass of a sample (g) to the amount of substance (mol) in the sample. Here, molar mass of HA = 1.056 g HA/*x* mol HA. Because 1 mol of HA reacts with 1 mol of NaOH in this case, the amount of acid (*x* mol) is equal to the amount of NaOH used in the titration, which is determined by its concentration and volume.

Solution Let us first calculate the amount of NaOH used in the titration.

$$\text{Amount of NaOH} = c_{NaOH}V_{NaOH} = \frac{0.256\ \text{mol}}{\text{L}} \times 0.03378\ \text{L} = 8.65 \times 10^{-3}\ \text{mol NaOH}$$

Next, recognize that the amount of NaOH used in the titration is the same as the amount of acid titrated. That is,

$$8.65 \times 10^{-3}\ \text{mol NaOH} \times \frac{1\ \text{mol HA}}{1\ \text{mol NaOH}} = 8.65 \times 10^{-3}\ \text{mol HA}$$

Finally, calculate the molar mass of HA.

$$\text{Molar mass of acid} = \frac{1.056\ \text{g HA}}{8.65 \times 10^{-3}\ \text{mol HA}} = 122\ \text{g/mol}$$

EXERCISE 4.14 Determining the Molar Mass of an Acid by Titration

An acid reacts with NaOH according to the net ionic equation

$$HA(aq) + OH^-(aq) \longrightarrow A^-(aq) + H_2O(\ell)$$

Calculate the molar mass of HA if 0.856 g of the acid requires 30.08 mL of 0.323 M NaOH.

Titrations Using Oxidation-Reduction Reactions

Analysis by titration is not limited to acid–base chemistry. Many oxidation-reduction reactions go rapidly to completion in aqueous solution, and methods exist to determine their equivalence point.

EXAMPLE 4.12 Using an Oxidation-Reduction Reaction in a Titration

Problem The iron in a sample of an iron ore can be converted quantitatively to the iron(II) ion, Fe^{2+}, in aqueous solution, and this solution can then be titrated with aqueous potassium permanganate, $KMnO_4$. The balanced, net ionic equation for the reaction occurring in the course of this titration is

$$\underset{\text{purple}}{MnO_4^-(aq)} + \underset{\text{colorless}}{5\ Fe^{2+}(aq)} + 8\ H_3O^+(aq) \longrightarrow \underset{\text{colorless}}{Mn^{2+}(aq)} + \underset{\text{pale yellow}}{5\ Fe^{3+}(aq)} + 12\ H_2O(\ell)$$

Case Study — Forensic Chemistry: Titrations and Food Tampering

The U.S. Food and Drug Administration (FDA) has recently discovered cases of product tampering involving the addition of bleach to products such as soup, infant formula, and soft drinks. Household bleach is a dilute solution of sodium hypochlorite (NaClO), a compound that is an oxidizing agent and is dangerous if swallowed.

One method of detecting bleach uses starch-iodide paper. The bleach oxidizes the iodide ion to iodine in an acid solution,

$$2\ I^-(aq) + HClO(aq) + H_3O^+(aq) \longrightarrow I_2(aq) + 2\ H_2O(\ell) + Cl^-(aq)$$

and the I_2 is then detected by a deep blue color in the presence of starch.

This reaction is also used in the quantitative analysis of solutions containing bleach. Excess iodide ion (in the form of KI) is added to the sample. The bleach in the sample (which forms HClO in acid solution) oxidizes I^- in a ratio of 1 mol HClO to 2 mol I^-. The iodine formed in the reaction is then titrated with sodium thiosulfate, $Na_2S_2O_3$ in another oxidation-reduction reaction (as in Exercise 4.15).

$$I_2(aq) + 2\ S_2O_3^{2-}(aq) \longrightarrow 2\ I^-(aq) + S_4O_6^{2-}(aq)$$

The amount of $Na_2S_2O_3$ used in the titration can then be used to determine the amount of NaClO in the sample.

Question:

Excess KI is added to a 100.0 mL sample of a soft drink that had been contaminated with bleach, NaClO. The iodine (I_2) generated in the solution was then titrated with 0.0425 M $Na_2S_2O_3$ and required 25.3 mL to reach the equivalence point. What mass of NaClO was contained in the 100.0-mL sample of adulterated soft drink?

Answer to this question is in Appendix Q.

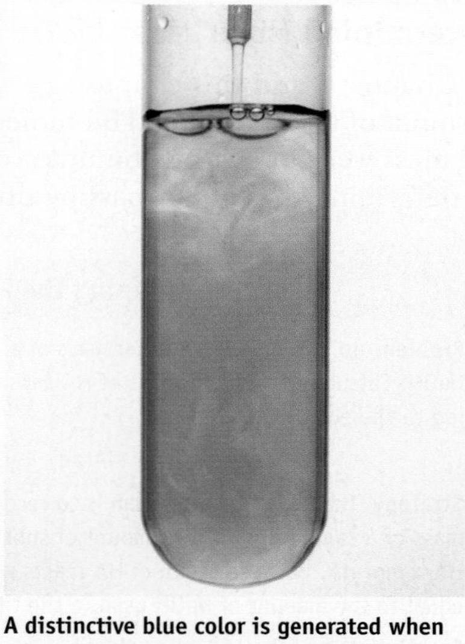

Charles D. Winters

A distinctive blue color is generated when iodine reacts with water-soluble starch.

A 1.026-g sample of iron-containing ore requires 24.35 mL of 0.0195 M $KMnO_4$ to reach the equivalence point. What is the mass percent of iron in the ore?

Strategy Because the volume and concentration of the $KMnO_4$ solution are known, the amount of $KMnO_4$ used in the titration can be calculated. Using the stoichiometric factor, the amount of $KMnO_4$ is related to the amount of iron(II) ion. The amount of iron(II) is converted to its mass, and the mass percent of iron in the sample is determined.

Solution First, calculate the amount of $KMnO_4$.

$$\text{Amount of } KMnO_4 = c_{KMnO_4} \times V_{KMnO_4} = \frac{0.0195 \text{ mol } KMnO_4}{\cancel{L}} \times 0.02435 \cancel{L} = 0.000475 \text{ mol}$$

Use the stoichiometric factor to calculate the amount of iron(II) ion.

$$0.000475 \cancel{\text{mol } KMnO_4} \times \frac{5 \text{ mol } Fe^{2+}}{1 \cancel{\text{mol } KMnO_4}} = 0.00237 \text{ mol } Fe^{2+}$$

The mass of iron can now be calculated,

$$0.00237 \cancel{\text{mol } Fe^{2+}} \times \frac{55.85 \text{ g } Fe^{2+}}{1 \cancel{\text{mol } Fe^{2+}}} = 0.133 \text{ g } Fe^{2+}$$

Finally, the mass percent can be determined.

$$\frac{0.133 \text{ g } Fe^{2+}}{1.026 \text{ g sample}} \times 100\% = 12.9\% \text{ iron}$$

Comment This is a useful analytical reaction because it is easy to detect when all the iron(II) ion has reacted. The MnO_4^- ion is a deep purple color, but when it reacts with Fe^{2+}, the color disappears because the reaction product, Mn^{2+}, is colorless. Therefore, $KMnO_4$ solution is added from a buret until the initially colorless, Fe^{2+}-containing solution just turns a faint purple color (due to unreacted $KMnO_4$), the signal that the equivalence point has been reached.

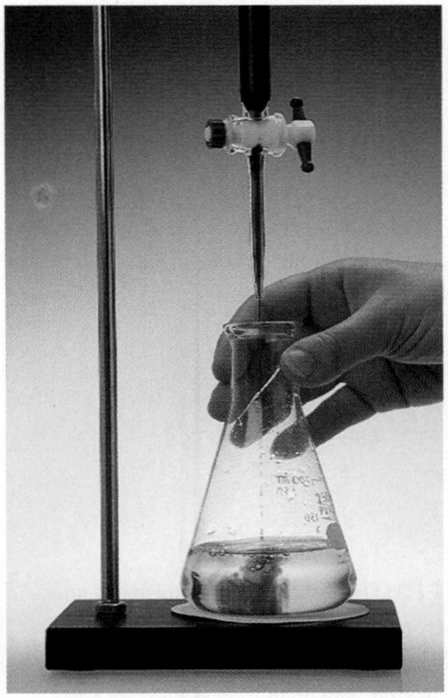

Charles D. Winters

Using an oxidation-reduction reaction for analysis by titration. Purple, aqueous $KMnO_4$ is added to a solution containing Fe^{2+}. As $KMnO_4$ drops into the solution, colorless Mn^{2+} and pale yellow Fe^{3+} form. Here, an area of the solution containing unreacted $KMnO_4$ is seen. As the solution is mixed, this disappears until the equivalence point is reached.

EXERCISE 4.15 Using an Oxidation-Reduction Reaction in a Titration

Vitamin C, ascorbic acid ($C_6H_8O_6$), is a reducing agent. One way to determine the ascorbic acid content of a sample is to mix the acid with an excess of iodine,

$$C_6H_8O_6(aq) + I_2(aq) + 2\ H_2O(\ell) \longrightarrow C_6H_6O_6(aq) + 2\ H_3O^+(aq) + 2\ I^-(aq)$$

and then titrate the iodine that did not react with the ascorbic acid with sodium thiosulfate. The balanced, net ionic equation for the reaction occurring in this titration is

$$I_2(aq) + 2\ S_2O_3^{2-}(aq) \longrightarrow 2\ I^-(aq) + S_4O_6^{2-}(aq)$$

Suppose 50.00 mL of 0.0520 M I_2 was added to the sample containing ascorbic acid. After the ascorbic acid/I_2 reaction was complete, the I_2 not used in this reaction required 20.30 mL of 0.196 M $Na_2S_2O_3$ for titration to the equivalence point. Calculate the mass of ascorbic acid in the unknown sample.

4.8 Spectrophotometry, Another Method of Analysis

Solutions of many compounds are colored, a consequence of the absorption of light (Figure 4.15). It is possible to measure, quantitatively, the extent of light absorption and to relate this to the concentration of the dissolved solute. This kind of experiment, called **spectrophotometry**, is an important analytical method.

Every substance absorbs or transmits certain wavelengths of radiant energy but not others (Figures 4.15 and 4.16). For example, nickel(II) ions (and chlorophyll) absorb red and blue/violet light, while transmitting or reflecting green light. Your eyes "see" the transmitted or reflected wavelengths, those not absorbed, as the color

Charles D. Winters

FIGURE 4.15 Light absorption and color. A beam of white light shines on a solution of nickel(II) ions in water, and the light that emerges is green. The color of a solution is due to the color of the light *not* absorbed by the solution. Here, red and blue/violet light was absorbed, and green light is transmitted.

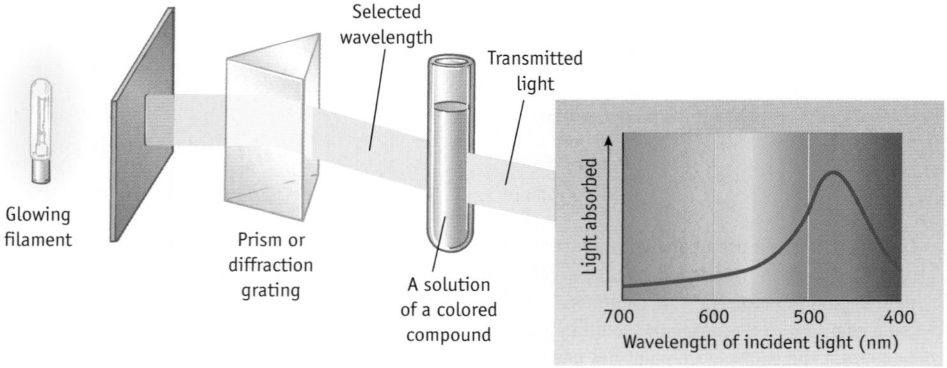

FIGURE 4.16 An absorption spectrophotometer. A beam of white light passes through a prism or diffraction grating, which splits the light into its component wavelengths. After passing through the sample, the light reaches a detector. The spectrophotometer "scans" all wavelengths of light and determines the amount of light absorbed at each wavelength. The output is a *spectrum*, a plot of the amount of light absorbed as a function of the wavelength or frequency of the incoming or incident light. Here, the sample absorbs light in the green-blue part of the spectrum and transmits light in the remaining wavelengths. The sample would appear red to orange to your eye.

green. Furthermore, the specific wavelengths absorbed and transmitted are characteristic for a substance, and so a spectrum serves as a "fingerprint" of the substance that can help identify an unknown.

Now suppose you look at two solutions of the same substance, one a deeper color than the other. Your common sense tells you that the intensely colored one is the more concentrated (Figure 4.17a). This is true, and the intensity of the color is a measure of the concentration of the material in the solution.

In recent years, spectrophotometry has become one of the most frequently used methods of quantitative analysis. It is applicable to many industrial and clinical problems involving the quantitative determination of compounds that are colored or that react to form a colored product.

Transmittance, Absorbance, and the Beer–Lambert Law

To understand the exact relationship of light absorption and solution concentration, we need to define several terms. **Transmittance** (T) is the ratio of the amount of light transmitted by or passing through the sample relative to the amount of light that initially fell on the sample (the incident light).

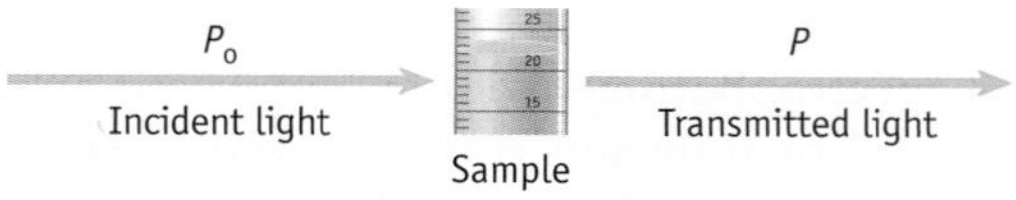

$$\text{Transmittance } (T) = \frac{P}{P_0} = \frac{\text{intensity of transmitted light}}{\text{intensity of incident light}}$$

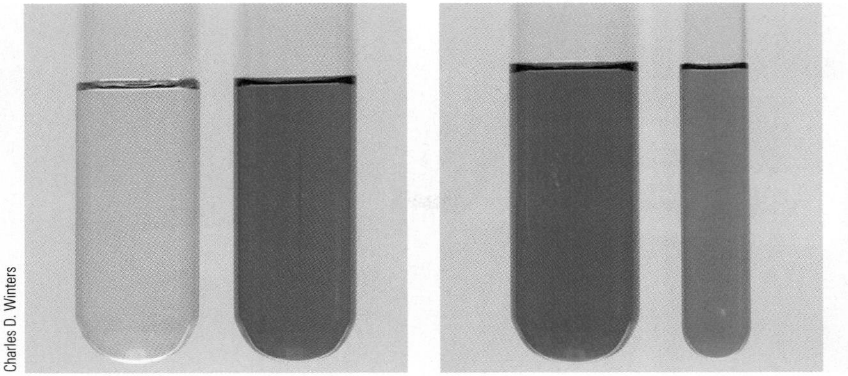

(a) (b)

FIGURE 4.17 Light absorption, concentration, and path length.
(a) The test tube on the left has a solution of copper(II) sulfate with a concentration of 0.05 M. On the right, the concentration is 1.0 M in copper(II) sulfate. More light is absorbed by the more concentrated sample, and it appears more blue.
(b) The amount of light absorbed by a solution depends on the path length. Here, both solutions have the same concentration, but the distance the light travels is longer in one than the other.

Absorbance is defined as the negative logarithm of the transmittance, and you will note that absorbance and transmittance bear an inverse relationship. That is, as the absorbance of a solution increases, the transmittance decreases

$$\text{Absorbance} = -\log T = -\log P/P_0$$

Going back to our example of an aqueous solution of copper(II) ions in Figure 4.17, if you have two colored solutions, you may deduce that the bluer solution appears more blue because it absorbs more of the light falling on it. That is, the *absorbance, A, of a sample increases as the concentration increases.*

Next, suppose that there are two test tubes, both containing the same solution at the same concentration. The only difference is that one of the test tubes has a smaller diameter than the other (Figure 4.17b). We shine light of the same intensity (P_0) on both test tubes. In the first case, the light has to travel only a short distance through the sample, whereas in the second case it has to pass through more of the sample. In the second case more of the light will be absorbed because the path length is longer. In other words, *absorbance increases as path length increases.*

■ **Beer–Lambert Law** The Beer–Lambert law applies strictly to relatively dilute solutions. At higher solute concentrations, the dependence of absorbance on concentration may not be linear.

The two observations described above constitute the **Beer–Lambert law**.

$$\text{Absorbance } (A) \propto \text{path length } (\ell) \times \text{concentration } (c)$$
$$A = \varepsilon \times \ell \times c \qquad (4.5)$$

where

- A, the absorbance of the sample, is a dimensionless number.
- ε, proportionality constant, is called the *molar absorptivity.* It is a constant for a given substance, provided the temperature and wavelength are constant. It has units of L/mol·cm.
- ℓ and c have the units of length (cm) and concentration (mol/L), respectively.

The Beer–Lambert law shows that *there is a linear relationship between a sample's absorbance and its concentration for a given path length.*

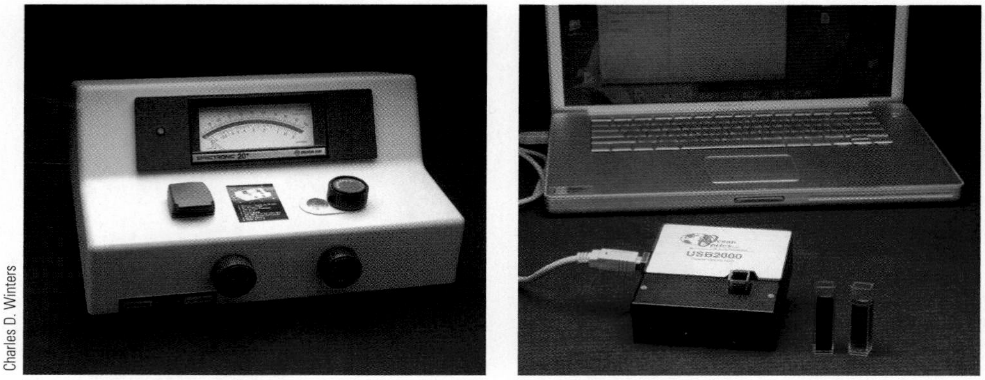

FIGURE 4.18 Spectrophotometers. The instruments illustrated here are often found in introductory chemistry laboratories. *(left)* Spectronic 20 from Spectronic Instruments. *(right)* Ocean Optics spectrometer (where the digital data are collected by a computer).

Spectrophotometric Analysis

There are usually four steps in carrying out a spectrophotometric analysis.

- **Record the absorption spectrum of the substance to be analyzed.** In introductory chemistry laboratories, this is often done using an instrument such as the ones shown in Figure 4.18. The result is a spectrum such as that for aqueous permanganate ions (MnO_4^-) in Figure 4.19. The spectrum is a plot of the absorbance of the sample versus the wavelength of incident light. Here, the maximum in absorbance is at about 525 nm.
- **Choose the wavelength for the measurement.** According to the Beer–Lambert Law, the absorbance at each wavelength is proportional to concentration. Therefore, in theory we could choose any wavelength for quantitative estimations of concentration. However, the magnitude of the absorbance is important, especially when you are trying to detect very small amounts of material. In the spectra of permanganate ions in Figure 4.19, note that the difference in absorbance between curves 1 and 2 is at a maximum at about 525 nm, and at this wavelength the change in absorbance is greatest for a given change in concentration. That is, the measurement of concentration as a function of concentration is most sensitive at this wavelength. For this reason, *we generally select the wavelength of maximum absorbance for our measurements.*
- **Prepare a calibration plot.** Once we have chosen the wavelength, the next step is to construct a **calibration curve** or **calibration plot**. This consists of a plot of absorbance versus concentration for a series of standard solutions whose concentrations are accurately known. Because of the linear relation between concentration and absorbance (at a given wavelength and pathlength), this plot is a straight line with a positive slope. Once the plot has been made, and the equation for the line is known, you can find the concentration of an unknown sample from its absorbance.

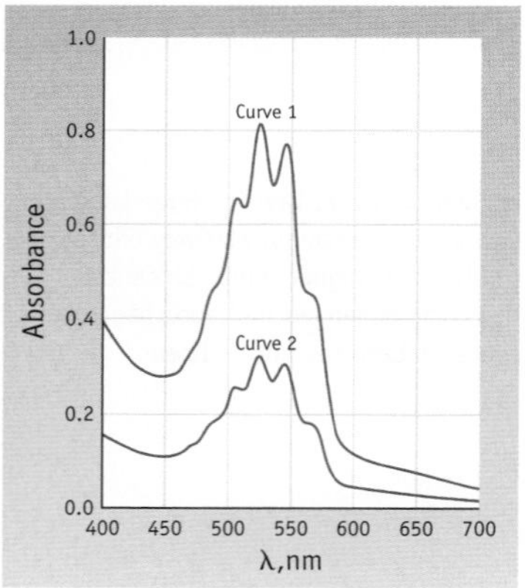

FIGURE 4.19 The absorption spectrum of solutions of potassium permanganate ($KMnO_4$) at different concentrations. The solution for curve 1 has a higher concentration than that for curve 2.

Example 4.13 illustrates the preparation of a calibration curve and its use in determining the concentration of a species in solution.

EXAMPLE 4.13 Using Spectrophotometry in Chemical Analysis

Problem A solution of $KMnO_4$ has an absorbance of 0.539 when measured at 540 nm in a 1.0-cm cell. What is the concentration of the $KMnO_4$?

Prior to determining the absorbance for the unknown solution, the following calibration data were collected for the spectrophotometer.

Concentration of $KMnO_4$ (M)	Absorbance
0.0300	0.162
0.0600	0.330
0.0900	0.499
0.120	0.670
0.150	0.840

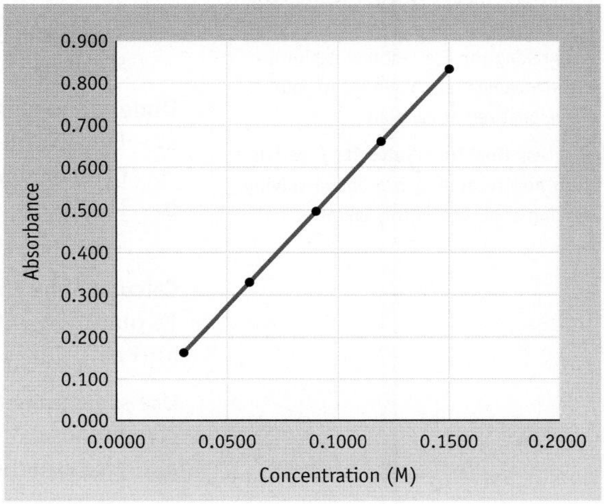

Strategy The first step is to prepare a calibration plot from the data above. You can then use the plot to estimate the unknown concentration from the measured absorbance or, better, find the equation for the straight line in the calibration plot (see pages 39 and 40) and calculate the unknown concentration. We shall do the latter.

Solution Using Excel or a calculator, prepare a calibration plot from the experimental data. The equation for the straight line (as determined using Excel) is

$$y = 5.633x - 0.009$$

$$\text{Absorbance} = 5.633\ (\text{Conc}) - 0.009$$

If we put in the absorbance for the unknown solution,

$$0.539 = 5.633\ (\text{Conc}) - 0.009$$

$$\text{Unknown concentration} = 0.0973$$

EXERCISE 4.16 Analysis Using Spectrophotometry

Using the following data, calculate the concentration of copper(II) ions in the unknown solution. (The cell pathlength is 1.00 cm in all cases, and the wavelength used in the determination was 645 nm.)

Calibration data

Concentration of Cu^{2+} (M)	Absorbance
0.0562	0.720
0.0337	0.434
0.0281	0.332
0.0169	0.219

Absorbance of unknown solution containing Cu^{2+} ions = 0.418

Chapter Goals Revisited

Chemistry Now™ Sign in at **www.thomsonedu.com/login** to:

- Assess your understanding with Study Questions in OWL keyed to each goal in the Goals and Homework menu for this chapter
- For quick review, download Go Chemistry mini-lecture flashcard modules (or purchase them at **www.ichapters.com**)
- Check your readiness for an exam by taking the Pre-Test and exploring the modules recommended in your Personalized Study plan.

Access **How Do I Solve It?** tutorials on how to approach problem solving using concepts in this chapter.

Now that you have studied this chapter, you should ask whether you have met the chapter goals. In particular, you should be able to:

Perform stoichiometry calculations using balanced chemical equations

a. Understand the principle of the conservation of matter, which forms the basis of chemical stoichiometry.

b. Calculate the mass of one reactant or product from the mass of another reactant or product by using the balanced chemical equation (Section 4.1). Study Question(s) assignable in OWL: 2, 5, 8, 77, 81, 93, 95, 97, 99, 100; Go Chemistry Module 7.

c. Use amounts tables to organize stoichiometric information. Study Question(s) assignable in OWL: 8.

Understand the meaning of a limiting reactant in a chemical reaction

a. Determine which of two reactants is the limiting reactant (Section 4.2). Study Question(s) assignable in OWL: 12, 14, 96, 132; Go Chemistry Module 8.

b. Determine the yield of a product based on the limiting reactant. Study Question(s) assignable in OWL: 12, 14, 16, 18.

Calculate the theoretical and percent yields of a chemical reaction

Explain the differences among actual yield, theoretical yield, and percent yield, and calculate percent yield (Section 4.3). Study Question(s) assignable in OWL: 19.

Use stoichiometry to analyze a mixture of compounds or to determine the formula of a compound

a. Use stoichiometry principles to analyze a mixture (Section 4.4). Study Question(s) assignable in OWL: 23, 123, 125, 127.

b. Find the empirical formula of an unknown compound using chemical stoichiometry (Section 4.4). Study Question(s) assignable in OWL: 29, 34.

Define and use concentrations in solution stoichiometry

a. Calculate the concentration of a solute in a solution in units of moles per liter (molarity), and use concentrations in calculations (Section 4.5). Study Question(s) assignable in OWL: 37, 39, 41.

b. Describe how to prepare a solution of a given concentration from the solute and a solvent or by dilution from a more concentrated solution (Section 4.5). Study Question(s) assignable in OWL: 46, 47, 51.

c. Calculate the pH of a solution from the concentration of hydronium ion in the solution. Calculate the hydronium ion concentration of a solution from the pH (Section 4.6). Study Question(s) assignable in OWL: 54, 55; Go Chemistry Module 9.

d. Solve stoichiometry problems using solution concentrations (Section 4.7). Study Question(s) assignable in OWL: 59, 62, 106, 107.

e. Explain how a titration is carried out, explain the procedure of standardization, and calculate concentrations or amounts of reactants from titration data (Section 4.7). Study Question(s) assignable in OWL: 67, 71.

f. Understand and use the principles of spectrophotometry to determine the concentration of a species in solution. (Secton 4.8). Study Question(s) assignable in OWL: 75.

KEY EQUATIONS

Equation 4.1 (page 168) Percent yield

$$\text{Percent yield} = \frac{\text{actual yield}}{\text{theoretical yield}} \times 100\%$$

Equation 4.2 (page 175) Definition of molarity, a measure of the concentration of a solute in a solution.

$$\text{Molarity of } x\ (c_x) = \frac{\text{amount of solute} \times \text{(mol)}}{\text{volume of solution (L)}}$$

A useful form of this equation is

$$\text{Amount of solute } x \text{ (mol)} = c_x \text{ (mol/L)} \times \text{volume of solution (L)}$$

Dilution Equation (page 179) This is a shortcut to find, for example, the concentration of a solution (c_d) after diluting some volume (V_c) of a more concentrated solution (c_c) to a new volume (V_d).

$$c_c \times V_c = c_d \times V_d$$

Equation 4.3 (page 179) pH. The pH of a solution is the negative logarithm of the hydronium ion concentration.

$$\text{pH} = -\log[\text{H}_3\text{O}^+]$$

Equation 4.4 (page 181) Calculating $[H_3O^+]$ from pH. The equation for calculating the hydronium ion concentration of a solution from the pH of the solution.

$$[\text{H}_3\text{O}^+] = 10^{-\text{pH}}$$

Equation 4.5 (page 192) Beer–Lambert Law. The absorbance of light (A) by a substance in solution is equal to the molar absorptivity of the substance (ε), the pathlength of the cell (ℓ), and the concentration of the solute (c).

$$\text{Absorbance } (A) \propto \text{path length } (\ell) \times \text{concentration } (c)$$
$$A = \varepsilon \times \ell \times c$$

STUDY QUESTIONS

Online homework for this chapter may be assigned in OWL.

▲ denotes challenging questions.

■ denotes questions assignable in OWL.

Blue-numbered questions have answers in Appendix O and fully-worked solutions in the *Student Solutions Manual*.

Practicing Skills

Mass Relationships in Chemical Reactions: Basic Stoichiometry
(See Example 4.1 and ChemistryNow Screens 4.2 and 4.3.)

1. Aluminum reacts with oxygen to give aluminum oxide.

$$4\ \text{Al(s)} + 3\ \text{O}_2\text{(g)} \longrightarrow 2\ \text{Al}_2\text{O}_3\text{(s)}$$

What amount of O_2, in moles, is needed for complete reaction with 6.0 mol of Al? What mass of Al_2O_3, in grams, can be produced?

2. ■ What mass of HCl, in grams, is required to react with 0.750 g of $Al(OH)_3$? What mass of water, in grams, is produced?

$$\text{Al(OH)}_3\text{(s)} + 3\ \text{HCl(aq)} \longrightarrow \text{AlCl}_3\text{(aq)} + 3\ \text{H}_2\text{O}(\ell)$$

3. Like many metals, aluminum reacts with a halogen to give a metal halide (see Figure 2.12).

$$2\ Al(s) + 3\ Br_2(\ell) \longrightarrow Al_2Br_6(s)$$

What mass of Br_2, in grams, is required for complete reaction with 2.56 g of Al? What mass of white, solid Al_2Br_6 is expected?

4. The balanced equation for a reaction in the process of the reduction of iron ore to the metal is

$$Fe_2O_3(s) + 3\ CO(g) \longrightarrow 2\ Fe(s) + 3\ CO_2(g)$$

(a) What is the maximum mass of iron, in grams, that can be obtained from 454 g (1.00 lb) of iron(III) oxide?
(b) What mass of CO is required to react with 454 g of Fe_2O_3?

5. ■ Methane, CH_4, burns in oxygen.
(a) What are the products of the reaction?
(b) Write the balanced equation for the reaction.
(c) What mass of O_2, in grams, is required for complete combustion of 25.5 g of methane?
(d) What is the total mass of products expected from the combustion of 25.5 g of methane?

6. The formation of water-insoluble silver chloride is useful in the analysis of chloride-containing substances. Consider the following *unbalanced* equation:

$$BaCl_2(aq) + AgNO_3(aq) \longrightarrow AgCl(s) + Ba(NO_3)_2(aq)$$

(a) Write the balanced equation.
(b) What mass of $AgNO_3$, in grams, is required for complete reaction with 0.156 g of $BaCl_2$? What mass of AgCl is produced?

Amounts Tables and Chemical Stoichiometry

For each question below, set up an amounts table that lists the initial amount or amounts of reactants, the changes in amounts of reactants and products, and the amounts of reactants and products after reaction. See page 159 and Example 4.1.

7. A major source of air pollution years ago was the metals industry. One common process involved "roasting" metal sulfides in the air:

$$2\ PbS(s) + 3\ O_2(g) \longrightarrow 2\ PbO(s) + 2\ SO_2(g)$$

If you heat 2.50 mol of PbS in the air, what amount of O_2 is required for complete reaction? What amounts of PbO and SO_2 are expected?

8. ■ Iron ore is converted to iron metal in a reaction with carbon.

$$2\ Fe_2O_3(s) + 3\ C(s) \longrightarrow 4\ Fe(s) + 3\ CO_2(g)$$

If 6.2 mol of $Fe_2O_3(s)$ is used, what amount of C(s) is needed, and what amounts of Fe and CO_2 are produced?

9. Chromium metal reacts with oxygen to give chromium(III) oxide, Cr_2O_3.
(a) Write a balanced equation for the reaction.
(b) If a piece of chromium has a mass of 0.175 g, what mass (in grams) of Cr_2O_3 is produced if the metal is converted completely to the oxide?
(c) What mass of O_2 (in grams) is required for the reaction?

10. Ethane, C_2H_6, burns in oxygen.
(a) What are the products of the reaction?
(b) Write the balanced equation for the reaction.
(c) What mass of O_2, in grams, is required for complete combustion of 13.6 of ethane?
(d) What is the total mass of products expected from the combustion of 13.6 g of ethane?

Limiting Reactants

(See Example 4.2 and Exercise 4.2. See also ChemistryNow Screens 4.4 and 4.5.)

11. Sodium sulfide, Na_2S, is used in the leather industry to remove hair from hides. The Na_2S is made by the reaction

$$Na_2SO_4(s) + 4\ C(s) \longrightarrow Na_2S(s) + 4\ CO(g)$$

Suppose you mix 15 g of Na_2SO_4 and 7.5 g of C. Which is the limiting reactant? What mass of Na_2S is produced?

12. ■ Ammonia gas can be prepared by the reaction of a metal oxide such as calcium oxide with ammonium chloride.

$$CaO(s) + 2\ NH_4Cl(s) \longrightarrow 2\ NH_3(g) + H_2O(g) + CaCl_2(s)$$

If 112 g of CaO and 224 g of NH_4Cl are mixed, what is the limiting reactant, and what mass of NH_3 can be produced?

13. The compound SF_6 is made by burning sulfur in an atmosphere of fluorine. The balanced equation is

$$S_8(s) + 24\ F_2(g) \longrightarrow 8\ SF_6(g)$$

If you begin with 1.6 mol of sulfur, S_8, and 35 mol of F_2, which is the limiting reagent?

14. ■ Disulfur dichloride, S_2Cl_2, is used to vulcanize rubber. It can be made by treating molten sulfur with gaseous chlorine:

$$S_8(\ell) + 4\ Cl_2(g) \longrightarrow 4\ S_2Cl_2(\ell)$$

Starting with a mixture of 32.0 g of sulfur and 71.0 g of Cl_2,
(a) Which is the limiting reactant?
(b) What is the theoretical yield of S_2Cl_2?
(c) What mass of the excess reactant remains when the reaction is completed?

15. The reaction of methane and water is one way to prepare hydrogen for use as a fuel:

$$CH_4(g) + H_2O(g) \rightarrow CO(g) + 3\ H_2(g)$$

If you begin with 995 g of CH_4 and 2510 g of water,
(a) Which reactant is the limiting reactant?
(b) What is the maximum mass of H_2 that can be prepared?
(c) What mass of the excess reactant remains when the reaction is completed?

16. ■ Aluminum chloride, $AlCl_3$, is made by treating scrap aluminum with chlorine.

$$2\ Al(s) + 3\ Cl_2(g) \rightarrow 2\ AlCl_3(s)$$

If you begin with 2.70 g of Al and 4.05 g of Cl_2,
(a) Which reactant is limiting?
(b) What mass of $AlCl_3$ can be produced?
(c) What mass of the excess reactant remains when the reaction is completed?
(d) Set up an amounts table for this problem.

17. Hexane (C_6H_{14}) burns in air (O_2) to give CO_2 and H_2O.
(a) Write a balanced equation for the reaction.
(b) If 215 g of C_6H_{14} is mixed with 215 g of O_2, what masses of CO_2 and H_2O are produced in the reaction?
(c) What mass of the excess reactant remains after the hexane has been burned?
(d) Set up an amounts table for this problem.

18. ■ Aspirin, $C_6H_4(OCOCH_3)CO_2H$, is produced by the reaction of salicylic acid, $C_6H_4(OH)CO_2H$, and acetic anhydride, $(CH_3CO)_2O$ (page 168).

$$C_6H_4(OH)CO_2H(s) + (CH_3CO)_2O(\ell) \rightarrow C_6H_4(OCOCH_3)CO_2H(s) + CH_3CO_2H(\ell)$$

If you mix 100. g of each of the reactants, what is the maximum mass of aspirin that can be obtained?

Percent Yield

(See Exercise 4.3 and ChemistryNow Screen 4.3.)

19. ■ In Example 4.2, you found that a particular mixture of CO and H_2 could produce 407 g CH_3OH.

$$CO(g) + 2\ H_2(g) \rightarrow CH_3OH(\ell)$$

If only 332 g of CH_3OH is actually produced, what is the percent yield of the compound?

20. Ammonia gas can be prepared by the following reaction:

$$CaO(s) + 2\ NH_4Cl(s) \rightarrow 2\ NH_3(g) + H_2O(g) + CaCl_2(s)$$

If 112 g of CaO and 224 g of NH_4Cl are mixed, the theoretical yield of NH_3 is 68.0 g (Study Question 12). If only 16.3 g of NH_3 is actually obtained, what is its percent yield?

21. The deep blue compound $Cu(NH_3)_4SO_4$ is made by the reaction of copper(II) sulfate and ammonia.

$$CuSO_4(aq) + 4\ NH_3(aq) \rightarrow Cu(NH_3)_4SO_4(aq)$$

(a) If you use 10.0 g of $CuSO_4$ and excess NH_3, what is the theoretical yield of $Cu(NH_3)_4SO_4$?
(b) If you isolate 12.6 g of $Cu(NH_3)_4SO_4$, what is the percent yield of $Cu(NH_3)_4SO_4$?

22. Black smokers are found in the depths of the oceans (page 112). Thinking that the conditions in these smokers might be conducive to the formation of organic compounds, two chemists in Germany found the following reaction could occur in similar conditions.

$$2\ CH_3SH + CO \rightarrow CH_3COSCH_3 + H_2S$$

If you begin with 10.0 g of CH_3SH and excess CO,
(a) What is the theoretical yield of CH_3COSCH_3?
(b) If 8.65 g of CH_3COSCH_3 is isolated, what is its percent yield?

Analysis of Mixtures

(See Example 4.3 and ChemistryNow Screen 4.7.)

23. ■ A mixture of $CuSO_4$ and $CuSO_4 \cdot 5\ H_2O$ has a mass of 1.245 g. After heating to drive off all the water, the mass is only 0.832 g. What is the mass percent of $CuSO_4 \cdot 5\ H_2O$ in the mixture? (See page 97.)

24. A 2.634-g sample containing impure $CuCl_2 \cdot 2H_2O$ was heated. The sample mass after heating to drive off the water was 2.125 g. What was the mass percent of $CuCl_2 \cdot 2\ H_2O$ in the original sample?

25. A sample of limestone and other soil materials was heated, and the limestone decomposed to give calcium oxide and carbon dioxide.

$$CaCO_3(s) \rightarrow CaO(s) + CO_2(g)$$

A 1.506-g sample of limestone-containing material gave 0.558 g of CO_2, in addition to CaO, after being heated at a high temperature. What is the mass percent of $CaCO_3$ in the original sample?

26. At higher temperatures, $NaHCO_3$ is converted quantitatively to Na_2CO_3.

$$2\ NaHCO_3(s) \rightarrow Na_2CO_3(s) + CO_2(g) + H_2O(g)$$

Heating a 1.7184-g sample of impure $NaHCO_3$ gives 0.196 g of CO_2. What was the mass percent of $NaHCO_3$ in the original 1.7184-g sample?

27. A pesticide contains thallium(I) sulfate, Tl_2SO_4. Dissolving a 10.20-g sample of impure pesticide in water and adding sodium iodide precipitates 0.1964 g of thallium(I) iodide, TlI.

$$Tl_2SO_4(aq) + 2\ NaI(aq) \rightarrow 2\ TlI(s) + Na_2SO_4(aq)$$

What is the mass percent of Tl_2SO_4 in the original 10.20-g sample?

28. ▲ The aluminum in a 0.764-g sample of an unknown material was precipitated as aluminum hydroxide, $Al(OH)_3$, which was then converted to Al_2O_3 by heating strongly. If 0.127 g of Al_2O_3 is obtained from the 0.764-g sample, what is the mass percent of aluminum in the sample?

Using Stoichiometry to Determine Empirical and Molecular Formulas

(See Example 4.4, Exercise 4.6, and ChemistryNow Screen 4.8.)

29. ■ Styrene, the building block of polystyrene, consists of only C and H. If 0.438 g of styrene is burned in oxygen and produces 1.481 g of CO_2 and 0.303 g of H_2O, what is the empirical formula of styrene?

30. Mesitylene is a liquid hydrocarbon. Burning 0.115 g of the compound in oxygen gives 0.379 g of CO_2 and 0.1035 g of H_2O. What is the empirical formula of mesitylene?

31. Cyclopentane is a simple hydrocarbon. If 0.0956 g of the compound is burned in oxygen, 0.300 g of CO_2 and 0.123 g of H_2O are isolated.
(a) What is the empirical formula of cyclopentane?
(b) If a separate experiment gave 70.1 g/mol as the molar mass of the compound, what is its molecular formula?

32. Azulene is a beautiful blue hydrocarbon. If 0.106 g of the compound is burned in oxygen, 0.364 g of CO_2 and 0.0596 g of H_2O are isolated.
(a) What is the empirical formula of azulene?
(b) If a separate experiment gave 128.2 g/mol as the molar mass of the compound, what is its molecular formula?

33. An unknown compound has the formula $C_xH_yO_z$. You burn 0.0956 g of the compound and isolate 0.1356 g of CO_2 and 0.0833 g of H_2O. What is the empirical formula of the compound? If the molar mass is 62.1 g/mol, what is the molecular formula? (See Exercise 4.6.)

34. ■ An unknown compound has the formula $C_xH_yO_z$. You burn 0.1523 g of the compound and isolate 0.3718 g of CO_2 and 0.1522 g of H_2O. What is the empirical formula of the compound? If the molar mass is 72.1 g/mol, what is the molecular formula? (See Exercise 4.6.)

35. Nickel forms a compound with carbon monoxide, $Ni_x(CO)_y$. To determine its formula, you carefully heat a 0.0973-g sample in air to convert the nickel to 0.0426 g of NiO and the CO to 0.100 g of CO_2. What is the empirical formula of $Ni_x(CO)_y$?

36. To find the formula of a compound composed of iron and carbon monoxide, $Fe_x(CO)_y$, the compound is burned in pure oxygen to give Fe_2O_3 and CO_2. If you burn 1.959 g of $Fe_x(CO)_y$ and obtain 0.799 g of Fe_2O_3 and 2.200 g of CO_2, what is the empirical formula of $Fe_x(CO)_y$?

Solution Concentration

(See Example 4.5 and ChemistryNow Screen 4.9.)

37. ■ If 6.73 g of Na_2CO_3 is dissolved in enough water to make 250. mL of solution, what is the molar concentration of the sodium carbonate? What are the molar concentrations of the Na^+ and CO_3^{2-} ions?

38. Some potassium dichromate ($K_2Cr_2O_7$), 2.335 g, is dissolved in enough water to make exactly 500. mL of solution. What is the molar concentration of the potassium dichromate? What are the molar concentrations of the K^+ and $Cr_2O_7^{2-}$ ions?

39. ■ What is the mass of solute, in grams, in 250. mL of a 0.0125 M solution of $KMnO_4$?

40. What is the mass of solute, in grams, in 125 mL of a 1.023×10^{-3} M solution of Na_3PO_4? What is the molar concentration of the Na^+ and PO_4^{3-} ion?

41. ■ What volume of 0.123 M NaOH, in milliliters, contains 25.0 g of NaOH?

42. What volume of 2.06 M $KMnO_4$, in liters, contains 322 g of solute?

43. For each solution, identify the ions that exist in aqueous solution, and specify the concentration of each ion.
(a) 0.25 M $(NH_4)_2SO_4$
(b) 0.123 M Na_2CO_3
(c) 0.056 M HNO_3

44. For each solution, identify the ions that exist in aqueous solution, and specify the concentration of each ion.
(a) 0.12 M $BaCl_2$
(b) 0.0125 M $CuSO_4$
(c) 0.500 M $K_2Cr_2O_7$

Preparing Solutions

(See Exercises 4.7–4.9, Example 4.6, and ChemistryNow Screen 4.10.)

45. An experiment in your laboratory requires 500. mL of a 0.0200 M solution of Na_2CO_3. You are given solid Na_2CO_3, distilled water, and a 500.-mL volumetric flask. Describe how to prepare the required solution.

46. ■ What mass of oxalic acid, $H_2C_2O_4$, is required to prepare 250. mL of a solution that has a concentration of 0.15 M $H_2C_2O_4$?

47. ■ If you dilute 25.0 mL of 1.50 M hydrochloric acid to 500. mL, what is the molar concentration of the dilute acid?

48. If 4.00 mL of 0.0250 M $CuSO_4$ is diluted to 10.0 mL with pure water, what is the molar concentration of copper(II) sulfate in the diluted solution?

49. Which of the following methods would you use to prepare 1.00 L of 0.125 M H_2SO_4?
(a) Dilute 20.8 mL of 6.00 M H_2SO_4 to a volume of 1.00 L.
(b) Add 950. mL of water to 50.0 mL of 3.00 M H_2SO_4.

▲ more challenging ■ in OWL Blue-numbered questions answered in Appendix O

50. Which of the following methods would you use to prepare 300. mL of 0.500 M $K_2Cr_2O_7$?
(a) Add 30.0 mL of 1.50 M $K_2Cr_2O_7$ to 270. mL of water.
(b) Dilute 250. mL of 0.600 M $K_2Cr_2O_7$ to a volume of 300. mL.

Serial Dilutions

(See A Closer Look: Serial Dilutions, page 180.)

51. ■ You have 250. mL of 0.136 M HCl. Using a volumetric pipet, you take 25.00 mL of that solution and dilute it to 100.00 mL in a volumetric flask. Now you take 10.00 mL of that solution, using a volumetric pipet, and dilute it to 100.00 mL in a volumetric flask. What is the concentration of hydrochloric acid in the final solution?

52. ▲ Suppose you have 100.00 mL a solution of a dye and transfer 2.00 mL of the solution to a 100.00-mL volumetric flask. After adding water to the 100.00 mL mark, you take 5.00 mL of that solution and again dilute to 100.00 mL. If you find the dye concentration in the final diluted sample is 0.000158 M, what was the dye concentration in the original solution?

Calculating and Using pH

(See Example 4.7 and ChemistryNow Screen 4.11.)

53. A table wine has a pH of 3.40. What is the hydronium ion concentration of the wine? Is it acidic or basic?

54. ■ A saturated solution of milk of magnesia, $Mg(OH)_2$, has a pH of 10.5. What is the hydronium ion concentration of the solution? Is the solution acidic or basic?

55. ■ What is the hydronium ion concentration of a 0.0013 M solution of HNO_3? What is its pH?

56. What is the hydronium ion concentration of a 1.2×10^{-4} M solution of $HClO_4$? What is its pH?

57. Make the following conversions. In each case, tell whether the solution is acidic or basic.

pH	$[H_3O^+]$
(a) 1.00	_____
(b) 10.50	_____
(c) _____	1.3×10^{-5} M
(d) _____	2.3×10^{-8} M

58. Make the following conversions. In each case, tell whether the solution is acidic or basic.

pH	$[H_3O^+]$
(a) _____	6.7×10^{-10} M
(b) _____	2.2×10^{-6} M
(c) 5.25	_____
(d) _____	2.5×10^{-2} M

Stoichiometry of Reactions in Solution

(See Example 4.8 and ChemistryNow Screen 4.12.)

59. ■ What volume of 0.109 M HNO_3, in milliliters, is required to react completely with 2.50 g of $Ba(OH)_2$?

$$2\ HNO_3(aq) + Ba(OH)_2(s) \longrightarrow 2\ H_2O(\ell) + Ba(NO_3)_2(aq)$$

60. What mass of Na_2CO_3, in grams, is required for complete reaction with 50.0 mL of 0.125 M HNO_3?

$$Na_2CO_3(aq) + 2\ HNO_3(aq) \longrightarrow 2\ NaNO_3(aq) + CO_2(g) + H_2O(\ell)$$

61. When an electric current is passed through an aqueous solution of NaCl, the valuable industrial chemicals $H_2(g)$, $Cl_2(g)$, and NaOH are produced.

$$2\ NaCl(aq) + 2\ H_2O(\ell) \longrightarrow H_2(g) + Cl_2(g) + 2\ NaOH(aq)$$

What mass of NaOH can be formed from 15.0 L of 0.35 M NaCl? What mass of chlorine is obtained?

62. ■ Hydrazine, N_2H_4, a base-like ammonia, can react with sulfuric acid.

$$2\ N_2H_4(aq) + H_2SO_4(aq) \longrightarrow 2\ N_2H_5^+(aq) + SO_4^{2-}(aq)$$

What mass of hydrazine reacts with 250. mL of 0.146 M H_2SO_4?

63. In the photographic developing process, silver bromide is dissolved by adding sodium thiosulfate.

$$AgBr(s) + 2\ Na_2S_2O_3(aq) \longrightarrow Na_3Ag(S_2O_3)_2(aq) + NaBr(aq)$$

If you want to dissolve 0.225 g of AgBr, what volume of 0.0138 M $Na_2S_2O_3$, in milliliters, should be used?

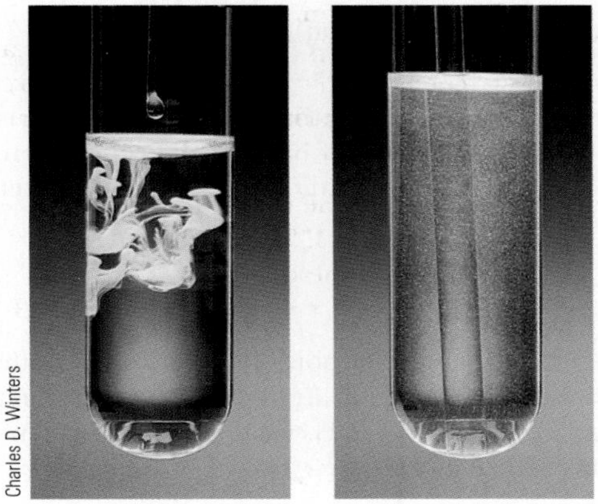

(a) (b)

Silver chemistry. (a) A precipitate of AgBr formed by adding $AgNO_3(aq)$ to KBr(aq). (b) On adding $Na_2S_2O_3(aq)$, sodium thiosulfate, the solid AgBr dissolves.

64. You can dissolve an aluminum soft-drink can in an aqueous base such as potassium hydroxide.

$$2\ Al(s) + 2\ KOH(aq) + 6\ H_2O(\ell) \rightarrow 2\ KAl(OH)_4(aq) + 3\ H_2(g)$$

If you place 2.05 g of aluminum in a beaker with 185 mL of 1.35 M KOH, will any aluminum remain? What mass of $KAl(OH)_4$ is produced?

65. What volume of 0.750 M $Pb(NO_3)_2$, in milliliters, is required to react completely with 1.00 L of 2.25 M NaCl solution? The balanced equation is

$$Pb(NO_3)_2(aq) + 2\ NaCl(aq) \rightarrow PbCl_2(s) + 2\ NaNO_3(aq)$$

66. What volume of 0.125 M oxalic acid, $H_2C_2O_4$ is required to react with 35.2 mL of 0.546 M NaOH?

$$H_2C_2O_4(aq) + 2\ NaOH(aq) \rightarrow Na_2C_2O_4(aq) + 2\ H_2O(\ell)$$

Titrations

(See Examples 4.9–4.12 and ChemistryNow Screen 4.13.)

67. ■ What volume of 0.812 M HCl, in milliliters, is required to titrate 1.45 g of NaOH to the equivalence point?

$$NaOH(aq) + HCl(aq) \rightarrow H_2O(\ell) + NaCl(aq)$$

68. What volume of 0.955 M HCl, in milliliters, is required to titrate 2.152 g of Na_2CO_3 to the equivalence point?

$$Na_2CO_3(aq) + 2\ HCl(aq) \rightarrow H_2O(\ell) + CO_2(g) + 2\ NaCl(aq)$$

69. If 38.55 mL of HCl is required to titrate 2.150 g of Na_2CO_3 according to the following equation, what is the concentration (mol/L) of the HCl solution?

$$Na_2CO_3(aq) + 2\ HCl(aq) \rightarrow 2\ NaCl(aq) + CO_2(g) + H_2O(\ell)$$

70. Potassium hydrogen phthalate, $KHC_8H_4O_4$, is used to standardize solutions of bases. The acidic anion reacts with strong bases according to the following net ionic equation:

$$HC_8H_4O_4^-(aq) + OH^-(aq) \rightarrow C_8H_4O_4^{2-}(aq) + H_2O(\ell)$$

If a 0.902-g sample of potassium hydrogen phthalate is dissolved in water and titrated to the equivalence point with 26.45 mL of NaOH, what is the molar concentration of the NaOH?

71. ■ You have 0.954 g of an unknown acid, H_2A, which reacts with NaOH according to the balanced equation

$$H_2A(aq) + 2\ NaOH(aq) \rightarrow Na_2A(aq) + 2\ H_2O(\ell)$$

If 36.04 mL of 0.509 M NaOH is required to titrate the acid to the second equivalence point, what is the molar mass of the acid?

72. An unknown solid acid is either citric acid or tartaric acid. To determine which acid you have, you titrate a sample of the solid with aqueous NaOH and from this determine the molar mass of the unknown acid. The appropriate equations are as follows:

Citric acid:

$$H_3C_6H_5O_7(aq) + 3\ NaOH(aq) \rightarrow 3\ H_2O(\ell) + Na_3C_6H_5O_7(aq)$$

Tartaric acid:

$$H_2C_4H_4O_6(aq) + 2\ NaOH(aq) \rightarrow 2\ H_2O(\ell) + Na_2C_4H_4O_6(aq)$$

A 0.956-g sample requires 29.1 mL of 0.513 M NaOH to consume the acid completely. What is the unknown acid?

73. To analyze an iron-containing compound, you convert all the iron to Fe^{2+} in aqueous solution and then titrate the solution with standardized $KMnO_4$. The balanced, net ionic equation is

$$MnO_4^-(aq) + 5\ Fe^{2+}(aq) + 8\ H_3O^+(aq) \rightarrow Mn^{2+}(aq) + 5\ Fe^{3+}(aq) + 12\ H_2O(\ell)$$

A 0.598-g sample of the iron-containing compound requires 22.25 mL of 0.0123 M $KMnO_4$ for titration to the equivalence point. What is the mass percent of iron in the sample?

74. Vitamin C has the formula $C_6H_8O_6$. Besides being an acid, it is a reducing agent. One method for determining the amount of vitamin C in a sample is therefore to titrate it with a solution of bromine, Br_2, an oxidizing agent.

$$C_6H_8O_6(aq) + Br_2(aq) \rightarrow 2\ HBr(aq) + C_6H_6O_6(aq)$$

A 1.00-g "chewable" vitamin C tablet requires 27.85 mL of 0.102 M Br_2 for titration to the equivalence point. What is the mass of vitamin C in the tablet?

Spectrophotometry

(See Section 4.8. The problems below are adapted from Fundamentals of Analytical Chemistry, 8th ed., by D. A. Skoog, D. M. West, F. J. Holler, and S. R. Crouch, Thomson/Brooks-Cole, Belmont, CA 2004.)

75. ■ A solution of a dye was analyzed by spectrophotometry, and the following calibration data were collected.

Dye Concentration	Absorbance at 475 nm
0.50×10^{-6} M	0.24
1.5×10^{-6} M	0.36
2.5×10^{-6} M	0.44
3.5×10^{-6} M	0.59
4.5×10^{-6} M	0.70

(a) Construct a calibration plot, and determine the slope and intercept.
(b) What is the dye concentration in a solution with $A = 0.52$?

▲ more challenging ■ in OWL Blue-numbered questions answered in Appendix O

76. The nitrite ion is involved in the biochemical nitrogen cycle. You can analyze for the nitrite ion content of a sample using spectrophotometry by first using several organic compounds to create a colored compound from the ion. The following data were collected.

NO_2^- Ion Concentration	Absorbance at 550 nm of Nitrite-Ion Containing Solution
2.00×10^{-6} M	0.065
6.00×10^{-6} M	0.205
10.00×10^{-6} M	0.338
14.00×10^{-6} M	0.474
18.00×10^{-6} M	0.598
Unknown solution	0.402

(a) Construct a calibration plot, and determine the slope and intercept.
(b) What is the nitrite ion concentration in the unknown solution?

General Questions on Stoichiometry

These questions are not designated as to type or location in the chapter. They may combine several concepts from the chapter.

77. ■ Suppose 16.04 g of benzene, C_6H_6, is burned in oxygen.
(a) What are the products of the reaction?
(b) What is the balanced equation for the reaction?
(c) What mass of O_2, in grams, is required for complete combustion of benzene?
(d) What is the total mass of products expected from 16.04 g of benzene?

78. The metabolic disorder diabetes causes a buildup of acetone, CH_3COCH_3, in the blood. Acetone, a volatile compound, is exhaled, giving the breath of untreated diabetics a distinctive odor. The acetone is produced by a breakdown of fats in a series of reactions. The equation for the last step, the breakdown of acetoacetic acid to give acetone and CO_2, is

$$CH_3COCH_2CO_2H \longrightarrow CH_3COCH_3 + CO_2$$

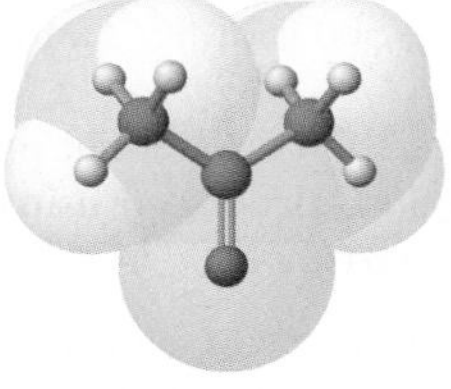

acetone, CH_3COCH_3

What mass of acetone can be produced from 125 mg of acetoacetic acid?

79. Your body deals with excess nitrogen by excreting it in the form of urea, NH_2CONH_2. The reaction producing it is the combination of arginine ($C_6H_{14}N_4O_2$) with water to give urea and ornithine ($C_5H_{12}N_2O_2$).

$$\underset{\text{Arginine}}{C_6H_{14}N_4O_2} + H_2O \longrightarrow \underset{\text{Urea}}{NH_2CONH_2} + \underset{\text{Ornithine}}{C_5H_{12}N_2O_2}$$

If you excrete 95 mg of urea, what mass of arginine must have been used? What mass of ornithine must have been produced?

80. The reaction of iron metal and chlorine gas to give iron(III) chloride is illustrated in Figure 3.2.
(a) Write the balanced chemical equation for the reaction.
(b) Beginning with 10.0 g of iron, what mass of Cl_2, in grams, is required for complete reaction? What mass of $FeCl_3$ can be produced?
(c) If only 18.5 g of $FeCl_3$ is obtained from 10.0 g of iron and excess Cl_2, what is the percent yield?
(d) If 10.0 g each of iron and chlorine are combined, what is the theoretical yield of iron(III) chloride?

81. ■ Some metal halides react with water to produce the metal oxide and the appropriate hydrogen halide (see photo). For example,

$$TiCl_4(\ell) + 2\ H_2O(\ell) \longrightarrow TiO_2(s) + 4\ HCl(g)$$

Charles D. Winters

The reaction of $TiCl_4$ with the water in moist air.

(a) Name the four compounds involved in this reaction.
(b) If you begin with 14.0 mL of $TiCl_4$ (d = 1.73 g/mL), what mass of water, in grams, is required for complete reaction?
(c) What mass of each product is expected?

82. The reaction of 750. g each of NH_3 and O_2 was found to produce 562 g of NO (see pages 163–165).

$$4\ NH_3(g) + 5\ O_2(g) \longrightarrow 4\ NO(g) + 6\ H_2O(\ell)$$

(a) What mass of water is produced by this reaction?
(b) What mass of O_2 is required to consume 750. g of NH_3?

83. Sodium azide, the explosive chemical used in automobile airbags, is made by the following reaction:

$$NaNO_3 + 3\ NaNH_2 \longrightarrow NaN_3 + 3\ NaOH + NH_3$$

If you combine 15.0 g of $NaNO_3$ (85.0 g/mol) with 15.0 g of $NaNH_2$, what mass of NaN_3 is produced?

84. Iodine is made by the following reaction

$$2\ NaIO_3(aq) + 5\ NaHSO_3(aq) \longrightarrow 3\ NaHSO_4(aq) + 2\ Na_2SO_4(aq) + H_2O(\ell) + I_2(aq)$$

(a) Name the two reactants.
(b) If you wish to prepare 1.00 kg of I_2, what mass of $NaIO_3$ is required? What mass of $NaHSO_3$?
(c) What is the theoretical yield of I_2 if you mixed 15.0 g of $NaIO_3$ with 125 mL of 0.853 M $NaHSO_3$?

85. Saccharin, an artificial sweetener, has the formula $C_7H_5NO_3S$. Suppose you have a sample of a saccharin-containing sweetener with a mass of 0.2140 g. After decomposition to free the sulfur and convert it to the SO_4^{2-} ion, the sulfate ion is trapped as water-insoluble $BaSO_4$ (see Figure 4.6). The quantity of $BaSO_4$ obtained is 0.2070 g. What is the mass percent of saccharin in the sample of sweetener?

86. ■▲ Boron forms an extensive series of compounds with hydrogen, all with the general formula B_xH_y.

$$B_xH_y(s) + \text{excess } O_2(g) \longrightarrow \frac{x}{2}\ B_2O_3(s) + \frac{y}{2}\ H_2O(g)$$

If 0.148 g of B_xH_y gives 0.422 g of B_2O_3 when burned in excess O_2, what is the empirical formula of B_xH_y?

87. ▲ Silicon and hydrogen form a series of compounds with the general formula Si_xH_y. To find the formula of one of them, a 6.22-g sample of the compound is burned in oxygen. All of the Si is converted to 11.64 g of SiO_2, and all of the H is converted to 6.980 g of H_2O. What is the empirical formula of the silicon compound?

88. ▲ Menthol, from oil of mint, has a characteristic odor. The compound contains only C, H, and O. If 95.6 mg of menthol burns completely in O_2, and gives 269 mg of CO_2 and 110 mg of H_2O, what is the empirical formula of menthol?

89. ▲ Quinone, a chemical used in the dye industry and in photography, is an organic compound containing only C, H, and O. What is the empirical formula of the compound if 0.105 g of the compound gives 0.257 g of CO_2 and 0.0350 g of H_2O when burned completely in oxygen?

90. ▲ Iron(II) chloride and sodium sulfide react to form iron(II) sulfide and sodium chloride (ChemistryNow Screen 4.8.)
(a) Write the balanced equation for the reaction.
(b) If you combine 40 g each of Na_2S and $FeCl_2$, what is the limiting reactant?
(c) What mass of FeS is produced?
(d) What mass of Na_2S or $FeCl_2$ remains after the reaction?
(e) What mass of $FeCl_2$ is required to react completely with 40 g of Na_2S?

91. Sulfuric acid can be prepared starting with the sulfide ore, cuprite (Cu_2S). If each S atom in Cu_2S leads to one molecule of H_2SO_4, what is the theoretical yield of H_2SO_4 from 3.00 kg of Cu_2S?

92. ▲ In an experiment, 1.056 g of a metal carbonate, containing an unknown metal M, is heated to give the metal oxide and 0.376 g CO_2.

$$MCO_3(s) + \text{heat} \longrightarrow MO(s) + CO_2(g)$$

What is the identity of the metal M?
(a) M = Ni (c) M = Zn
(b) M = Cu (d) M = Ba

93. ■▲ An unknown metal reacts with oxygen to give the metal oxide, MO_2. Identify the metal based on the following information:

Mass of metal = 0.356 g

Mass of sample after converting metal completely to oxide = 0.452 g

94. ▲ Titanium(IV) oxide, TiO_2, is heated in hydrogen gas to give water and a new titanium oxide, Ti_xO_y. If 1.598 g of TiO_2 produces 1.438 g of Ti_xO_y, what is the empirical formula of the new oxide?

95. ■▲ Potassium perchlorate is prepared by the following sequence of reactions:

$$Cl_2(g) + 2\ KOH(aq) \longrightarrow KCl(aq) + KClO(aq) + H_2O(\ell)$$

$$3\ KClO(aq) \longrightarrow 2\ KCl(aq) + KClO_3(aq)$$

$$4\ KClO_3(aq) \longrightarrow 3\ KClO_4(aq) + KCl(aq)$$

What mass of $Cl_2(g)$ is required to produce 234 kg of $KClO_4$?

96. ■▲ Commercial sodium "hydrosulfite" is 90.1% $Na_2S_2O_4$. The sequence of reactions used to prepare the compound is

$$Zn(s) + 2\ SO_2(g) \longrightarrow ZnS_2O_4(s)$$

$$ZnS_2O_4(s) + Na_2CO_3(aq) \longrightarrow ZnCO_3(s) + Na_2S_2O_4(aq)$$

(a) What mass of pure $Na_2S_2O_4$ can be prepared from 125 kg of Zn, 500 g of SO_2, and an excess of Na_2CO_3?
(b) What mass of the commercial product would contain the $Na_2S_2O_4$ produced using the amounts of reactants in part (a)?

97. ■ What mass of lime, CaO, can be obtained by heating 125 kg of limestone that is 95.0% by mass $CaCO_3$?

$$CaCO_3(s) \longrightarrow CaO(s) + CO_2(g)$$

98. ▲ The elements silver, molybdenum, and sulfur combine to form Ag_2MoS_4. What is the maximum mass of Ag_2MoS_4 that can be obtained if 8.63 g of silver, 3.36 g of molybdenum, and 4.81 g of sulfur are combined?

99. ■▲ A mixture of butene, C_4H_8, and butane, C_4H_{10}, is burned in air to give CO_2 and water. Suppose you burn 2.86 g of the mixture and obtain 8.80 g of CO_2 and 4.14 g of H_2O. What are the mass percentages of butene and butane in the mixture?

100. ■▲ Cloth can be waterproofed by coating it with a silicone layer. This is done by exposing the cloth to $(CH_3)_2SiCl_2$ vapor. The silicon compound reacts with OH groups on the cloth to form a waterproofing film (density = 1.0 g/cm³) of $[(CH_3)_2SiO]_n$, where n is a large integer number.

$$n\ (CH_3)_2SiCl_2 + 2n\ OH^- \longrightarrow 2n\ Cl^- + n\ H_2O + [(CH_3)_2SiO]_n$$

The coating is added layer by layer, each layer of $[(CH_3)_2SiO]_n$ being 0.60 nm thick. Suppose you want to waterproof a piece of cloth that is 3.00 m square, and you want 250 layers of waterproofing compound on the cloth. What mass of $(CH_3)_2SiCl_2$ do you need?

101. ■▲ Copper metal can be prepared by roasting copper ore, which can contain cuprite (Cu_2S) and copper(II) sulfide.

$$Cu_2S(s) + O_2(g) \longrightarrow 2\ Cu(s) + SO_2(g)$$

$$CuS(s) + O_2(g) \longrightarrow Cu(s) + SO_2(g)$$

Suppose an ore sample contains 11.0% impurity in addition to a mixture of CuS and Cu_2S. Heating 100.0 g of the mixture produces 75.4 g of copper metal with a purity of 89.5%. What is the weight percent of CuS in the ore? The weight percent of Cu_2S?

102. Which has the larger concentration of hydronium ions, 0.015 M HCl or aqueous HCl with a pH of 1.2?

103. The mineral dolomite contains magnesium carbonate.

$$MgCO_3(s) + 2\ HCl(aq) \longrightarrow CO_2(g) + MgCl_2(aq) + H_2O(\ell)$$

(a) Write the net ionic equation for the reaction of $MgCO_3$ and HCl(aq).
(b) What type of reaction is this?
(c) What mass of $MgCO_3$ will react with 125 mL of HCl(aq) with a pH of 1.56?

104. An Alka-Seltzer tablet contains exactly 100. mg of citric acid, $H_3C_6H_5O_7$, plus some sodium bicarbonate. If the following reaction occurs, what mass of sodium bicarbonate must the tablet also contain if citric acid is completely consumed by the following reaction?

$$H_3C_6H_5O_7(aq) + 3\ NaHCO_3(aq) \longrightarrow 3\ H_2O(\ell) + 3\ CO_2(g) + Na_3C_6H_5O_7(aq)$$

105. ▲ Sodium bicarbonate and acetic acid react according to the equation

$$NaHCO_3(aq) + CH_3CO_2H(aq) \longrightarrow NaCH_3CO_2(aq) + CO_2(g) + H_2O(\ell)$$

What mass of sodium acetate can be obtained from mixing 15.0 g of $NaHCO_3$ with 125 mL of 0.15 M acetic acid?

106. ■ A noncarbonated soft drink contains an unknown amount of citric acid, $H_3C_6H_5O_7$. If 100. mL of the soft drink requires 33.51 mL of 0.0102 M NaOH to neutralize the citric acid completely, what mass of citric acid does the soft drink contain per 100. mL? The reaction of citric acid and NaOH is

$$H_3C_6H_5O_7(aq) + 3\ NaOH(aq) \longrightarrow Na_3C_6H_5O_7(aq) + 3\ H_2O(\ell)$$

107. Sodium thiosulfate, $Na_2S_2O_3$, is used as a "fixer" in black-and-white photography. Suppose you have a bottle of sodium thiosulfate and want to determine its purity. The thiosulfate ion can be oxidized with I_2 according to the balanced, net ionic equation

$$I_2(aq) + 2\ S_2O_3^{2-}(aq) \longrightarrow 2\ I^-(aq) + S_4O_6^{2-}(aq)$$

If you use 40.21 mL of 0.246 M I_2 in a titration, what is the weight percent of $Na_2S_2O_3$ in a 3.232-g sample of impure material?

108. You have a mixture of oxalic acid, $H_2C_2O_4$, and another solid that does not react with sodium hydroxide. If 29.58 mL of 0.550 M NaOH is required to titrate the oxalic acid in the 4.554-g sample to the second equivalence point, what is the mass percent of oxalic acid in the mixture? Oxalic acid and NaOH react according to the equation

$$H_2C_2O_4(aq) + 2\ NaOH(aq) \longrightarrow Na_2C_2O_4(aq) + 2\ H_2O(\ell)$$

109. (a) What is the pH of a 0.105 M HCl solution?
(b) What is the hydronium ion concentration in a solution with a pH of 2.56? Is the solution acidic or basic?
(c) A solution has a pH of 9.67. What is the hydronium ion concentration in the solution? Is the solution acidic or basic?
(d) A 10.0-mL sample of 2.56 M HCl is diluted with water to 250. mL. What is the pH of the dilute solution?

110. A solution of hydrochloric acid has a volume of 125 mL and a pH of 2.56. What mass of $NaHCO_3$ must be added to completely consume the HCl?

111. ▲ One half liter (500. mL) of 2.50 M HCl is mixed with 250. mL of 3.75 M HCl. Assuming the total solution volume after mixing is 750. mL, what is the concentration of hydrochloric acid in the resulting solution? What is its pH?

112. A solution of hydrochloric acid has a volume of 250. mL and a pH of 1.92. Exactly 250. mL of 0.0105 M NaOH is added. What is the pH of the resulting solution?

113. ▲ You place 2.56 g of $CaCO_3$ in a beaker containing 250. mL of 0.125 M HCl. When the reaction has ceased, does any calcium carbonate remain? What mass of $CaCl_2$ can be produced?

$$CaCO_3(s) + 2\ HCl(aq) \longrightarrow CaCl_2(aq) + CO_2(g) + H_2O(\ell)$$

▲ more challenging ■ in OWL Blue-numbered questions answered in Appendix O

114. The cancer chemotherapy drug cisplatin, $Pt(NH_3)_2Cl_2$, can be made by reacting $(NH_4)_2PtCl_4$ with ammonia in aqueous solution. Besides cisplatin, the other product is NH_4Cl.
(a) Write a balanced equation for this reaction.
(b) To obtain 12.50 g of cisplatin, what mass of $(NH_4)_2PtCl_4$ is required? What volume of 0.125 M NH_3 is required?
(c) ▲ Cisplatin can react with the organic compound pyridine, C_5H_5N, to form a new compound.

$$Pt(NH_3)_2Cl_2(aq) + x\ C_5H_5N(aq) \longrightarrow Pt(NH_3)_2Cl_2(C_5H_5N)_x(s)$$

Suppose you treat 0.150 g of cisplatin with what you believe is an excess of liquid pyridine (1.50 mL; d = 0.979 g/mL). When the reaction is complete, you can find out how much pyridine was not used by titrating the solution with standardized HCl. If 37.0 mL of 0.475 M HCl is required to titrate the excess pyridine,

$$C_5H_5N(aq) + HCl(aq) \longrightarrow C_5H_5NH^+(aq) + Cl^-(aq)$$

what is the formula of the unknown compound $Pt(NH_3)_2Cl_2(C_5H_5N)_x$?

115. ▲ You need to know the volume of water in a small swimming pool, but, owing to the pool's irregular shape, it is not a simple matter to determine its dimensions and calculate the volume. To solve the problem, you stir in a solution of a dye (1.0 g of methylene blue, $C_{16}H_{18}ClN_3S$, in 50.0 mL of water). After the dye has mixed with the water in the pool, you take a sample of the water. Using a spectrophotometer, you determine that the concentration of the dye in the pool is 4.1×10^{-8} M. What is the volume of water in the pool?

116. ▲ Calcium and magnesium carbonates occur together in the mineral dolomite. Suppose you heat a sample of the mineral to obtain the oxides, CaO and MgO, and then treat the oxide sample with hydrochloric acid. If 7.695 g of the oxide sample requires 125 mL of 2.55 M HCl,

$$CaO(s) + 2\ HCl(aq) \longrightarrow CaCl_2(aq) + H_2O(\ell)$$

$$MgO(s) + 2\ HCl(aq) \longrightarrow MgCl_2(aq) + H_2O(\ell)$$

What is the weight percent of each oxide (CaO and MgO) in the sample?

117. ■ Gold can be dissolved from gold-bearing rock by treating the rock with sodium cyanide in the presence of oxygen.

$$4\ Au(s) + 8\ NaCN(aq) + O_2(g) + 2\ H_2O(\ell) \longrightarrow 4\ NaAu(CN)_2(aq) + 4\ NaOH(aq)$$

(a) Name the oxidizing and reducing agents in this reaction. What has been oxidized, and what has been reduced?
(b) If you have exactly one metric ton (1 metric ton = 1000 kg) of gold-bearing rock, what volume of 0.075 M NaCN, in liters, do you need to extract the gold if the rock is 0.019% gold?

118. ▲ You mix 25.0 mL of 0.234 M $FeCl_3$ with 42.5 mL of 0.453 M NaOH.
(a) What mass of $Fe(OH)_3$ (in grams) will precipitate from this reaction mixture?
(b) One of the reactants ($FeCl_3$ or NaOH) is present in a stoichiometric excess. What is the molar concentration of the excess reactant remaining in solution after $Fe(OH)_3$ has been precipitated?

In the Laboratory

119. ■ Suppose you dilute 25.0 mL of a 0.110 M solution of Na_2CO_3 to exactly 100.0 mL. You then take exactly 10.0 mL of this diluted solution and add it to a 250-mL volumetric flask. After filling the volumetric flask to the mark with distilled water (indicating the volume of the new solution is 250. mL), what is the concentration of the diluted Na_2CO_3 solution?

120. ▲ In some laboratory analyses, the preferred technique is to dissolve a sample in an excess of acid or base and then "back-titrate" the excess with a standard base or acid. This technique is used to assess the purity of a sample of $(NH_4)_2SO_4$. Suppose you dissolve a 0.475-g sample of impure $(NH_4)_2SO_4$ in aqueous KOH.

$$(NH_4)_2SO_4(aq) + 2\ KOH(aq) \longrightarrow 2\ NH_3(aq) + 2\ K_2SO_4(aq) + 2\ H_2O(\ell)$$

The NH_3 liberated in the reaction is distilled from the solution into a flask containing 50.0 mL of 0.100 M HCl. The ammonia reacts with the acid to produce NH_4Cl, but not all of the HCl is used in this reaction. The amount of excess acid is determined by titrating the solution with standardized NaOH. This titration consumes 11.1 mL of 0.121 M NaOH. What is the weight percent of $(NH_4)_2SO_4$ in the 0.475-g sample?

121. You wish to determine the weight percent of copper in a copper-containing alloy. After dissolving a 0.251-g sample of the alloy in acid, an excess of KI is added, and the Cu^{2+} and I^- ions undergo the reaction

$$2\ Cu^{2+}(aq) + 5\ I^-(aq) \longrightarrow 2\ CuI(s) + I_3^-(aq)$$

The liberated I_3^- is titrated with sodium thiosulfate according to the equation

$$I_3^-(aq) + 2\ S_2O_3^{2-}(aq) \longrightarrow S_4O_6^{2-}(aq) + 3\ I^-(aq)$$

(a) Designate the oxidizing and reducing agents in the two reactions above.
(b) If 26.32 mL of 0.101 M $Na_2S_2O_3$ is required for titration to the equivalence point, what is the weight percent of Cu in the alloy?

122. ▲ A compound has been isolated that can have either of two possible formulas: (a) $K[Fe(C_2O_4)_2(H_2O)_2]$ or (b) $K_3[Fe(C_2O_4)_3]$. To find which is correct, you dissolve a weighed sample of the compound in acid and then titrate the oxalate ion ($C_2O_4^{2-}$, which in acid be-

comes $H_2C_2O_4$) with potassium permanganate, $KMnO_4$ (the source of the MnO_4^- ion). The balanced, net ionic equation for the titration is

$$5\ H_2C_2O_4(aq) + 2\ MnO_4^-(aq) + 6\ H_3O^+(aq) \longrightarrow 2\ Mn^{2+}(aq) + 10\ CO_2(g) + 14\ H_2O(\ell)$$

Titration of 1.356 g of the compound requires 34.50 mL of 0.108 M $KMnO_4$. Which is the correct formula of the iron-containing compound: (a) or (b)?

123. ▲ Chromium(III) ion forms many compounds with ammonia. To find the formula of one of these compounds, you titrate the NH_3 in the compound with standardized acid.

$$Cr(NH_3)_xCl_3(aq) + x\ HCl(aq) \longrightarrow x\ NH_4^+(aq) + Cr^{3+}(aq) + (x + 3)\ Cl^-(aq)$$

Assume that 24.26 mL of 1.500 M HCl is used to titrate 1.580 g of $Cr(NH_3)_xCl_3$. What is the value of x?

124. ▲ Thioridazine, $C_{21}H_{26}N_2S_2$, is a pharmaceutical agent used to regulate dopamine. (Dopamine, a neurotransmitter, affects brain processes that control movement, emotional response, and ability to experience pleasure and pain.) A chemist can analyze a sample of the pharmaceutical for the thioridazine content by decomposing it to convert the sulfur in the compound to sulfate ion. This is then "trapped" as water-insoluble barium sulfate (see Figure 4.6).

$$SO_4^{2-}(aq, \text{from thioridazine}) + BaCl_2(aq) \longrightarrow BaSO_4(s) + 2\ Cl^-(aq)$$

Suppose a 12-tablet sample of the drug yielded 0.301 g of $BaSO_4$. What is the thioridazine content, in milligrams, of each tablet?

125. ■▲ A herbicide contains 2,4-D (2,4-dichlorophenoxyacetic acid), $C_8H_6Cl_2O_3$. A 1.236-g sample of the herbicide was decomposed to liberate the chlorine as Cl^- ion. This was precipitated as AgCl, with a mass of 0.1840 g. What is the mass percent of 2,4-D in the sample?

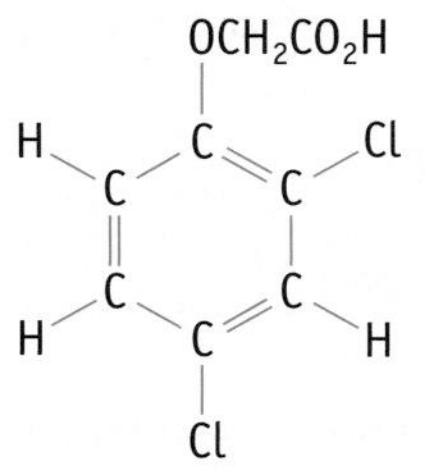

126. ▲ Sulfuric acid is listed in a catalog with a concentration of 95–98%. A bottle of the acid in the stockroom states that 1.00 L has a mass of 1.84 kg. To determine the concentration of sulfuric acid in the stockroom bottle, a student dilutes 5.00 mL to 500. mL. She then takes four 10.00-mL samples and titrates each with standardized sodium hydroxide (c = 0.1760 mol/L).

Sample	1	2	3	4
Volume NaOH (mL)	20.15	21.30	20.40	20.35

(a) What is the average concentration of the diluted sulfuric acid sample?
(b) What is the mass percent of H_2SO_4 in the original bottle of the acid?

127. ▲ Anhydrous calcium chloride is a good drying agent as it will rapidly pick up water. Suppose you have stored some carefully dried $CaCl_2$ in a dessicator. Unfortunately, someone did not close the top of the dessicator tightly, and the $CaCl_2$ became partially hydrated. A 150-g sample of this partially hydrated material was dissolved in 80 g of hot water. When the solution was cooled to 20 °C, 74.9 g of $CaCl_2 \cdot 6\ H_2O$ precipitated. Knowing the solubility of calcium chloride in water at 20 °C is 74.5 g $CaCl_2$/100 g water, determine the water content of the 150-g sample of partially hydrated calcium chloride (in moles of water per mole of $CaCl_2$).

128. ▲ A sample consisting of a mixture of iron and iron(III) oxide was dissolved completely in acid (which converted the iron to iron(III) ions.) After adding a reducing agent to ensure that all of the iron was in the form of iron(II) ions, the solution was titrated with the standardized $KMnO_4$ (0.04240 M); 37.50 mL of the $KMnO_4$ solution was required. Calculate the mass percent of Fe and Fe_2O_3 in the 0.5510-g sample. (See Example 4.12 for the reaction of iron(II) and $KMnO_4$.)

129. ▲ Phosphate in urine can be determined by spectrophotometry. After removing protein from the sample, it is treated with a molybdenum compound to give, ultimately, a deep blue polymolybdate. The absorbance of the blue polymolybdate can be measured at 650 nm and is directly related to the urine phosphate concentration. A 24-hour urine sample was collected from a patient; the volume of urine was 1122 mL. The phosphate in a 1.00 mL portion of the urine sample was converted to the blue polymolybdate (P) and diluted to 50.00 mL. A calibration curve was prepared using phosphate-containing solutions.

Solution (mass P/L)	**Absorbance at 650 nm in a 1.0 cm cell**
1.00×10^{-6} g	0.230
2.00×10^{-6} g	0.436
3.00×10^{-6} g	0.638
4.00×10^{-6} g	0.848
Urine sample	0.518

(a) What are the slope and intercept of the calibration curve?
(b) What is the mass of phosphorus per liter of urine?
(c) What mass of phosphate did the patient excrete per day?

130. ▲ A 4.000-g sample containing KCl and $KClO_4$ was dissolved in sufficient water to give 250.00 mL of solution. A 50.00-mL portion of the solution required 41.00 mL of 0.0750 M $AgNO_3$ in a Mohr titration (page 186). Next, a 25.00-mL portion of the original solution was treated with $V_2(SO_4)_3$ to reduce the perchlorate ion to chloride,

$$8\ V^{3+}(aq) + ClO_4^-(aq) + 12\ H_2O(\ell) \longrightarrow Cl^-(aq) + 8\ VO^{2+}(aq) + 8\ H_3O^+(aq)$$

and the resulting solution was titrated with $AgNO_3$. This titration required 38.12 mL of 0.0750 M $AgNO_3$. What is the mass percent of KCl and $KClO_4$ in the mixture?

Summary and Conceptual Questions

The following questions may use concepts from this and preceding chapters.

131. Two beakers sit on a balance; the total mass is 167.170 g. One beaker contains a solution of KI; the other contains a solution of $Pb(NO_3)_2$. When the solution in one beaker is poured completely into the other, the following reaction occurs:

$$2\ KI(aq) + Pb(NO_3)_2(aq) \longrightarrow 2\ KNO_3(aq) + PbI_2(s)$$

Charles D. Winters

Solutions of KI and $Pb(NO_3)_2$ before reaction.

Charles D. Winters

Solutions after reaction.

What is the total mass of the beakers and solutions after reaction? Explain completely.

132. ▲ A weighed sample of iron (Fe) is added to liquid bromine (Br_2) and allowed to react completely. The reaction produces a single product, which can be isolated and weighed. The experiment was repeated a number of times with different masses of iron but with the same mass of bromine. (See the graph below.)

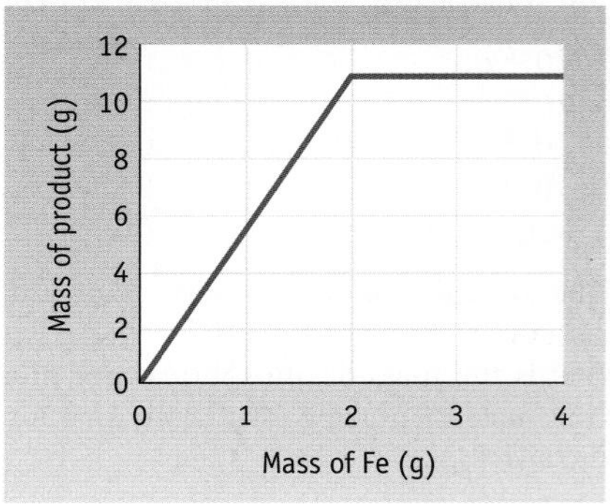

(a) What mass of Br_2 is used when the reaction consumes 2.0 g of Fe?
(b) What is the mole ratio of Br_2 to Fe in the reaction?
(c) What is the empirical formula of the product?
(d) Write the balanced chemical equation for the reaction of iron and bromine.
(e) What is the name of the reaction product?
(f) Which statement or statements best describe the experiments summarized by the graph?
   (i) When 1.00 g of Fe is added to the Br_2, Fe is the limiting reagent.
   (ii) When 3.50 g of Fe is added to the Br_2, there is an excess of Br_2.
   (iii) When 2.50 g of Fe is added to the Br_2, both reactants are used up completely.
   (iv) When 2.00 g of Fe is added to the Br_2, 10.0 g of product is formed. The percent yield must therefore be 20.0%.

133. Let us explore a reaction with a limiting reactant. (See ChemistryNow Screens 4.4 and 4.5.) Here, zinc metal is added to a flask containing aqueous HCl, and H_2 gas is a product.

$$Zn(s) + 2\ HCl(aq) \longrightarrow ZnCl_2(aq) + H_2(g)$$

The three flasks each contain 0.100 mol of HCl. Zinc is added to each flask in the following quantities.

Flask 1: 7.00 g Zn
Flask 2: 3.27 g Zn
Flask 3: 1.31 g Zn

Charles D. Winters

When the reactants are combined, the H_2 inflates the balloon attached to the flask. The results are as follows:

Flask 1: Balloon inflates completely, but some Zn remains when inflation ceases.
Flask 2: Balloon inflates completely. No Zn remains.
Flask 3: Balloon does not inflate completely. No Zn remains.

Explain these results. Perform calculations that support your explanation.

134. The reaction of aluminum and bromine is pictured in Figure 2.12 and below. The white solid on the lip of the beaker at the end of the reaction is Al_2Br_6. In the reaction pictured below, which was the limiting reactant, Al or Br_2? (See ChemistryNow Screen 4.2.)

Charles D. Winters

Before reaction.

Charles D. Winters

After reaction.

135. ▲ Two students titrate different samples of the same solution of HCl using 0.100 M NaOH solution and phenolphthalein indicator (see Figure 4.14). The first student pipets 20.0 mL of the HCl solution into a flask, adds 20 mL of distilled water and a few drops of phenolphthalein solution, and titrates until a lasting pink color appears. The second student pipets 20.0 mL of the HCl solution into a flask, adds 60 mL of distilled water and a few drops of phenolphthalein solution, and titrates to the first lasting pink color. Each student correctly calculates the molarity of an HCl solution. What will the second student's result be?
(a) four times less than the first student's result
(b) four times greater than the first student's result
(c) two times less than the first student's result
(d) two times greater than the first student's result
(e) the same as the first student's result

136. A video on Screen 4.12 of ChemistryNow shows the reaction of Fe^{2+} with MnO_4^- in aqueous solution.
(a) What is the balanced equation for the reaction that occurred?
(b) What is the oxidizing agent, and what is the reducing agent?
(c) Equal volumes of Fe^{2+}-containing solution and MnO_4^- -containing solution were mixed. The amount of Fe^{2+} was just sufficient to consume all of the MnO_4^-. Which ion (Fe^{2+} or MnO_4^-) was initially present in larger concentration?

137. In some states, a person will receive a "driving while intoxicated" (DWI) ticket if the blood alcohol level (BAL) is 100 mg per deciliter (dL) of blood or higher. Suppose a person is found to have a BAL of 0.033 mol of ethanol (C_2H_5OH) per liter of blood. Will the person receive a DWI ticket?

5 Principles of Chemical Reactivity: Energy and Chemical Reactions

©Nicolas Raymond

A Hot Air Balloon

These colorful balloons usually consist of a gas bag or envelope of nylon with a basket suspended below for passengers. A propane burner sits on top of the basket and below the gas envelope. When the air inside the envelope is heated by burning propane, the balloon can ascend (because the density of heated air in the bag is less than that of the cooler surrounding air). Under normal conditions, about 3 m^3 of envelope volume is required to lift 1 kg of mass. Thus, to carry one person and the needed equipment, most balloons have a volume of about 1000 m^3.

Question:

You have a balloon with a volume of 1100 m^3 and want to heat the air inside of the envelope from 22 °C to 110 °C. What mass of propane must you burn to accomplish this? (The specific heat capacity of air is 1.01 J/g · K, and the density of dry air [at sea level] is about 1.2 kg/m^3. Other information you need is in this chapter or Appendix L.)

Answer to this question is in Appendix Q.

Chapter Goals

See Chapter Goals Revisited (page 241) for Study Questions keyed to these goals and assignable in OWL.

- Assess the transfer of energy as heat associated with changes in temperature and changes of state.
- Understand and apply the first law of thermodynamics.
- Define and understand state functions (enthalpy, internal energy).
- Learn how energy changes are measured.
- Calculate the energy evolved or required for physical changes and chemical reactions using tables of thermodynamic data.

Chapter Outline

The importance of energy is evident in our daily lives—in heating and cooling our homes, in powering our appliances, and in propelling our vehicles, among other things. Most of the energy we use for these purposes is obtained by carrying out chemical reactions, mostly by burning fossil fuels. We use natural gas for heating, coal and natural gas to generate most of our electric power, and fuels derived from petroleum for automobiles and for heat. In addition, energy is required for all life processes. Chemical reactions in our bodies provide the energy for all body functions, for movement, and to maintain body temperature. It is not surprising that the topic of energy is a prominent part of our discussion of chemistry.

To scientists, however, energy has significance that goes well beyond these many practical uses. In this chapter, we will begin the discussion of **thermodynamics**, the science of heat and work. This subject will provide important insights on the following questions:

- How do we measure and calculate the energy changes that are associated with physical changes and chemical reactions?
- What is the relationship between energy changes, heat, and work?
- How can we determine whether a chemical reaction is product-favored or reactant-favored at equilibrium?
- How can we determine whether a chemical reaction or physical process will occur spontaneously, that is, without outside intervention?

We will concentrate attention on the first two questions in this chapter and address the last two questions in Chapter 19.

Chemistry•Now™

Throughout the text this icon introduces an opportunity for self-study or to explore interactive tutorials by signing in at **www.thomsonedu.com/login**.

■ **World Energy Consumption** Burning fossil fuels provides about 85% of the total energy used by people on our planet. Nuclear and hydroelectric power each contribute about 6%. The remaining 3% is provided from biomass, solar, wind, and geothermal sources.

5.1 Energy: Some Basic Principles

Energy is defined as the capacity to do work. You do work against the force of gravity when carrying yourself and hiking equipment up a mountain. The energy to do this is provided by the food you have eaten. Food is a source of chemical energy—energy stored in chemical compounds and released when the compounds undergo the chemical reactions of metabolism in your body.

James Cowlin

Royalty-Free/Corbis

William James Warren/Corbis

(a) Gravitational energy

(b) Chemical potential energy

(c) Electrostatic energy

Active Figure 5.1 **Energy and its conversion.** (a) Water at the top of a waterfall represents stored, or potential, energy. As water falls, its potential energy is converted to mechanical energy. (b) Chemical potential energy of the fuel and oxygen is converted to thermal and mechanical energy. (c) Lightning converts electrostatic energy into radiant and thermal energy.

Chemistry Now™ **Sign in at www.thomsonedu.com/login and go to the Chapter Contents menu to explore an interactive version of this figure accompanied by an exercise.**

Energy can be classified as kinetic or potential. **Kinetic energy** is energy associated with motion, such as:

- The motion of atoms, molecules, or ions at the submicroscopic (particulate) level (*thermal energy*). All matter has thermal energy.
- The motion of macroscopic objects like a moving tennis ball or automobile (*mechanical energy*).
- The movement of electrons through a conductor (*electrical energy*).
- The compression and expansion of the spaces between molecules in the transmission of sound (*acoustic energy*).

Potential energy results from an object's position and includes:

- Energy possessed by a ball held above the floor and by water at the top of a waterfall (*gravitational energy*) (Figure 5.1a).
- Energy stored in fuels (*chemical energy*) (Figure 5.1b). All chemical reactions involve a change in chemical energy.
- The energy associated with the separation of two electrical charges (*electrostatic energy*) (Figure 5.1c).

Potential energy and kinetic energy can be interconverted. For example, as water falls over a waterfall, its potential energy is converted into kinetic energy. Similarly, kinetic energy can be converted into potential energy: The kinetic energy of falling water can turn a turbine to produce electricity, which can then be used to convert water into H_2 and O_2 by electrolysis. Hydrogen gas contains stored chemical potential energy because it can be burned to produce heat and light or electricity.

Chemistry Now™

Sign in at **www.thomsonedu.com/login** and go to Chapter 5 Contents to see Screen 5.3 for an exercise that examines examples of **energy transfer.**

Conservation of Energy

Standing on a diving board, you have considerable potential energy because of your position above the water. Once you jump off the board, some of that potential energy is converted into kinetic energy (Figure 5.2). During the dive, the force of gravity accelerates your body so that it moves faster and faster. Your kinetic energy increases, and your potential energy decreases. At the moment you hit the water, your velocity is abruptly reduced, and much of your kinetic energy is transferred to the water as your body moves it aside. Eventually, you float to the surface, and the water becomes still again. If you could see them, however, you would find that the water molecules are moving a little faster in the vicinity of your entry into the water; that is, the kinetic energy of the water molecules is slightly higher.

This series of energy conversions illustrates the **law of conservation of energy**, which states that *energy can neither be created nor destroyed.* Or, to state this law differently, *the total energy of the universe is constant.* The law of conservation of energy summarizes the results of a great many experiments in which the amounts of energy transferred have been measured and in which the total energy content has been found to be the same before and after an event.

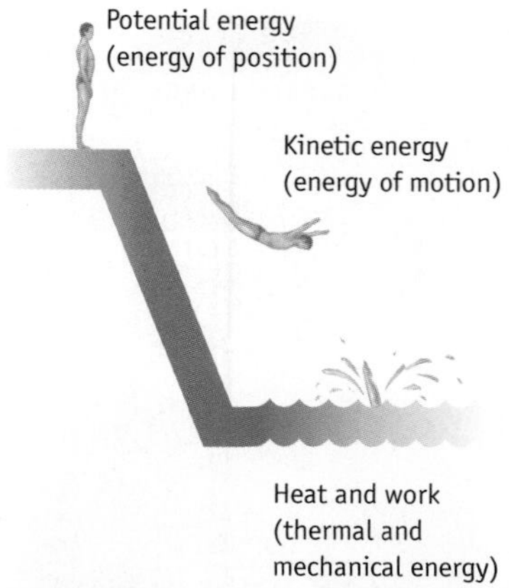

FIGURE 5.2 The law of energy conservation. The diver's potential energy is converted to kinetic energy, and this is then transferred to the water, illustrating the law of conservation of energy.
See ChemistryNow Screen 5.2 Energy, to view an animation of this figure.

Temperature and Heat

The temperature of an object is a measure of its ability to transfer energy as heat. When two objects at different temperatures are brought into contact, energy will be transferred as heat from the one at the higher temperature to the one at the lower temperature. One way to measure temperature is with a thermometer containing mercury or some other liquid (Figure 5.3). When the thermometer is placed

Photos: Charles D. Winters

FIGURE 5.3 Measuring temperature. The volume of liquid mercury in a thermometer increases slightly when immersed in warm water. The volume increase causes the mercury to rise in the thermometer, which is calibrated to give the temperature.

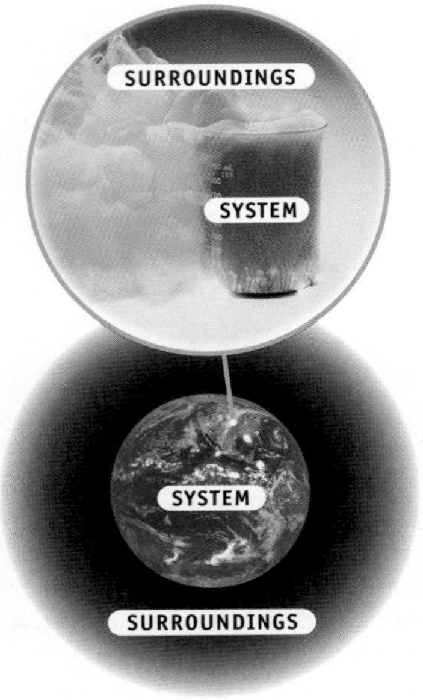

Photos: (Top) Charles D. Winters; (Bottom) NASA

FIGURE 5.4 Systems and their surroundings. Earth can be considered a thermodynamic system, with the rest of the universe as its surroundings. A chemical reaction occurring in a laboratory is also a system, with the laboratory its surroundings.

in hot water, thermal energy is transferred from the water to the thermometer (heating the thermometer and cooling the water). This causes the atoms of liquid mercury to move more rapidly (increasing their kinetic energy) and the space between them to increase slightly. The resulting increase in volume causes the column of liquid to rise higher in the thermometer tube.

Several important aspects of thermal energy and temperature should be recognized:

- Temperature determines the direction of thermal energy transfer.
- The higher the temperature of a given object, the greater the thermal energy (energy associated with molecular motion) of its atoms, ions, or molecules.
- Heating and cooling are processes by which energy is transferred as heat from an object at a higher temperature to one at a lower temperature. Heat is not a substance. (See *A Closer Look: What Is Heat?*)

Systems and Surroundings

In thermodynamics, the terms "system" and "surroundings" have precise and important scientific meanings. A **system** is defined as an object, or collection of objects, being studied (Figure 5.4). The **surroundings** include everything outside the system that can exchange energy and/or matter with the system. In the discussion that follows, we will need to define systems precisely. If we are studying the energy evolved in a chemical reaction carried out in solution, for example, the system might be defined as the reactants, products, and solvent. The surroundings would be the reaction vessel and the air in the room and anything else in contact with the vessel that might exchange energy or matter. At the atomic level, the system could be a single atom or molecule, and the surroundings would be the atoms or molecules in its vicinity. How we choose to define the system and its surroundings for each situation depends on the information we are trying to obtain or convey.

This concept of a system and its surroundings applies to nonchemical situations as well. If we want to study the energy balance on this planet, we might choose to define Earth as the system and outer space as the surroundings. On a cosmic level, the solar system might be defined as the system being studied, and the rest of the galaxy would be the surroundings.

Directionality and Extent of Transfer of Heat: Thermal Equilibrium

Energy is transferred as heat if two objects at different temperatures are brought into contact. In Figure 5.5, for example, the beaker of water and the piece of metal being heated in a Bunsen burner flame have different temperatures. When the hot

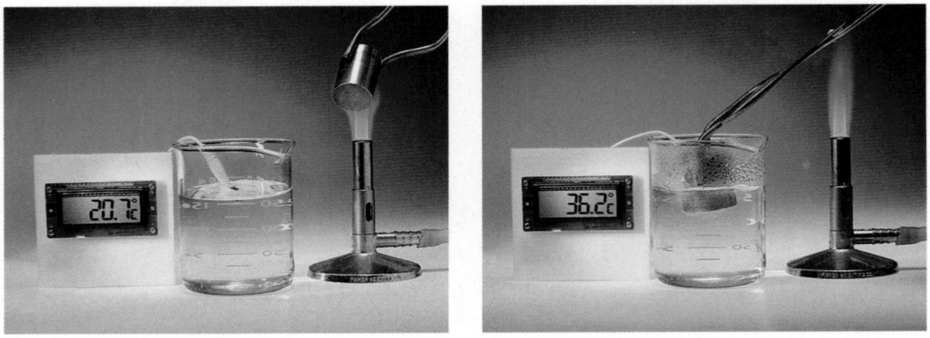

Charles D. Winters

FIGURE 5.5 Energy transfer. Energy transfer as heat occurs from the hotter metal bar to the cooler water. Eventually, the water and metal reach the same temperature and are said to be in thermal equilibrium.

See ChemistryNow Screen 5.4, Energy Transfer Between Substances, for a simulation and tutorial.

A Closer Look — What Is Heat?

Two hundred years ago, scientists characterized heat as a real substance called a caloric fluid. The caloric hypothesis supposed that when a fuel burned and a pot of water was heated, for example, caloric fluid was transferred from the fuel to the water. Burning the fuel released caloric fluid, and the temperature of the water increased as the caloric fluid was absorbed.

Over the next 50 years, however, the caloric hypothesis lost favor, and we now know it is incorrect. Experiments by James Joule (1818–1889) and Benjamin Thompson (1753–1814) that showed the interrelationship between heat and other forms of energy such as mechanical energy provided the key to dispelling this idea. Even so, some of our everyday language retains the influence of this early theory. For example, we often speak of heat flowing as if it were a fluid.

From our discussion so far, we know one thing that "heat" is not—but what is it? Heat is said to be a "process quantity" as opposed to a "state quantity." That is, heating is a *process* that changes the internal energy of a system. It is the process by which energy is transferred across the boundary of a system owing to a difference in temperature between the two sides of the boundary. In this process, the energy of one object increases, and the energy of another object decreases.

Heating is not the only way to transfer energy. Work is another process by which energy can be transferred between objects.

The idea of energy transfer by the processes of heat and work is embodied in the definition of thermodynamics: the science of heat and work.

Work and heat. A classic experiment that showed the relationship between work and heat was performed by Benjamin Thompson (also known as Count Rumford) (1753–1814) using an apparatus similar to that shown here. Thompson measured the rise in temperature of water (in the vessel mostly hidden at the back of the apparatus) that resulted from the energy expended to turn the crank.

metal is plunged into the cold water, energy is transferred as heat from the metal to the water. The thermal energy (molecular motion) of the water molecules increases; the thermal energy of the metal atoms decreases. Eventually, the two objects reach the same temperature. At that point, the system has reached **thermal equilibrium**. The distinguishing feature of thermal equilibrium is that, on the macroscopic scale, no further temperature change occurs; both the metal and water are at the same temperature.

■ **Thermal Equilibrium** Although no change is evident at the macroscopic level when thermal equilibrium is reached, on the molecular level transfer of energy between individual molecules will continue to occur. A general feature of systems at equilibrium is that there is no change on a macroscopic level but that processes still occur at the particulate level. (See Section 3.3, page 118.)

Putting a hot metal bar into a beaker of water and following the temperature change may seem like a rather simple experiment with an obvious outcome. Illustrated in the experiment, however, are three principles that are important in our further discussion:

- Energy transfer as heat will occur spontaneously from an object at a higher temperature to an object at a lower temperature.
- Transfer of energy as heat continues until both objects are at the same temperature and thermal equilibrium is achieved.
- After thermal equilibrium is attained, the object whose temperature increased has gained thermal energy, and the object whose temperature decreased has lost thermal energy.

For the specific case where energy is transferred as heat within an isolated system (that is, a system that cannot transfer either energy or matter with its surroundings), we can also say that the quantity of thermal energy lost by a hotter object and the quantity of thermal energy gained by a cooler object are numerically equal. (This is required by the law of conservation of energy.) When energy is transferred as

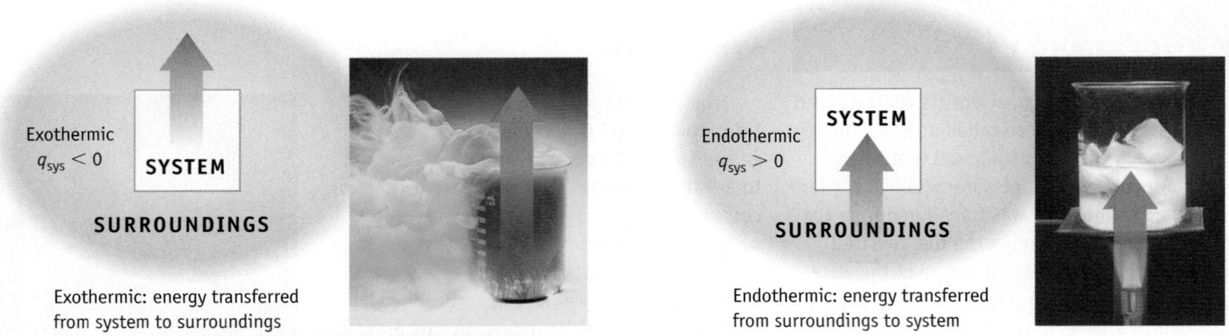

Photos: Charles D. Winters

Active Figure 5.6 **Exothermic and endothermic processes.** The symbol q represents the energy transfered as heat, and the subscript sys refers to the system.

Chemistry Now™ Sign in at www.thomsonedu.com/login and go to the Chapter Contents menu to explore an interactive version of this figure accompanied by an exercise.

■ **Exothermic and Endothermic** The terms "endothermic" and "exothermic" apply specifically to energy transfer as heat. The more general terms "endoergic" and "exoergic" are sometimes used, encompassing any type of energy transfer between system and surroundings.

heat between a system and its surroundings, we describe the directionality of this transfer as exothermic or endothermic (Figure 5.6).

- In an **exothermic process**, energy is transferred as heat from a system to its surroundings. The energy of the system decreases, and the energy of the surroundings increases.
- An **endothermic process** is the opposite of an exothermic process. Energy is transferred as heat from the surroundings to the system, increasing the energy of the system, decreasing the energy of the surroundings.

Energy Units

■ **James Joule** The joule is named for James P. Joule (1818–1889), the son of a wealthy brewer in Manchester, England. The family wealth and a workshop in the brewery gave Joule the opportunity to pursue scientific studies. Among the topics that Joule studied was the issue of whether heat was a massless fluid. Scientists at that time referred to this idea as the caloric hypothesis. Joule's careful experiments showed that heat and mechanical work are related, providing evidence that heat is not a fluid. (See *A Closer Look: What Is Heat?*)

Oesper Collection in the History of Chemistry/University of Cincinnati

When expressing energy quantities, most chemists (and much of the world outside the United States) use the **joule** (J), the SI unit. The joule is related directly to the units used for mechanical energy: 1 J equals 1 kg · m²/s². Because the joule is inconveniently small for most uses in chemistry, the kilojoule (kJ), equivalent to 1000 joules, is often the unit of choice.

To give you some feeling for joules, suppose you drop a six-pack of soft-drink cans, each full of liquid, on your foot. Although you probably will not take time to calculate the kinetic energy at the moment of impact, it is between 4 J and 10 J.

The calorie (cal) is an older energy unit. It is defined as the energy transferred as heat that is required to raise the temperature of 1.00 g of pure liquid water from 14.5 °C to 15.5 °C. A kilocalorie (kcal) is equivalent to 1000 calories. The conversion factor relating joules and calories is

1 calorie (cal) = 4.184 joules (J)

The dietary Calorie (with a capital C) is often used in the United States to represent the energy content of foods. The dietary Calorie (Cal) is equivalent to the kilocalorie or 1000 calories. Thus, a breakfast cereal that gives you 100.0 Calories of nutritional energy per serving provides 100.0 kcal or 418.4 kJ.

Chemistry Now™

Sign in at **www.thomsonedu.com/login** and go to Chapter 5 Contents to see Screen 5.4 to view an animation on **endothermic and exothermic systems** and Screen 5.5 for a tutorial on **converting between different energy units.**

Chemical Perspectives — Food and Calories

The U.S. Food and Drug Administration (FDA) mandates that nutritional data, including energy content, be included on almost all packaged food. The Nutrition Labeling and Education Act of 1990 requires that the total energy from protein, carbohydrates, fat, and alcohol be specified. How is this determined? Initially, the method used was calorimetry. In this method, which is described in Section 5.6, a food product is burned, and the energy transferred as heat in the combustion is measured. Now, however, energy contents are estimated using the Atwater system. This specifies the following average values for energy sources in foods:

1 g protein = 4 kcal (17 kJ)
1 g carbohydrate = 4 kcal (17 kJ)
1 g fat = 9 kcal (38 kJ)
1 g alcohol = 7 kcal (29 kJ)

Because carbohydrates may contain some indigestible fiber, the mass of fiber is subtracted from the mass of carbohydrate when calculating the energy from carbohydrates.

As an example, one serving of cashew nuts (about 28 g) has

14 g fat = 126 kcal
6 g protein = 24 kcal
7 g carbohydrates − 1 g fiber = 24 kcal
Total = 174 kcal (728 kJ)

A value of 170 kcal is reported on the package.

You can find data on more than 6000 foods at the Nutrient Data Laboratory website (www.ars.usda.gov/ba/bhnrc/ndl).

Charles D. Winters

Energy and food labels. All packaged foods must have labels specifying nutritional values, with energy given in Calories (where 1 Cal = 1 kilocalorie).

EXERCISE 5.1 Energy Units

(a) In an old textbook, you read that the burning 1.00 g of hydrogen to form liquid water produces 3800 calories. What is this energy in units of joules?

(b) The label on a cereal box indicates that one serving (with skim milk) provides 250 Cal. What is this energy in kilojoules (kJ)?

■ **Kinetic Energy** Kinetic energy is calculated by the equation KE = 1/2 mv^2. One joule is the kinetic energy of a 2.0 kg mass (m) moving at 1.0 m/s (v).
KE = (1/2)(2.0 kg)(1.0 m/s)^2
= 1.0 kg · m^2/s^2 = 1.0 J

5.2 Specific Heat Capacity: Heating and Cooling

When an object is heated or cooled, the quantity of energy transferred depends on three things:

- The quantity of material
- The magnitude of the temperature change
- The identity of the material gaining or losing energy

Specific heat capacity (*C*) is defined as *the energy transferred as heat that is required to raise the temperature of 1 gram of a substance by one kelvin.* It has units of joules per gram per kelvin (J/g · K). A few specific heat capacities are listed in Figure 5.7, and a longer list of specific heat capacities is given in Appendix D (Table 11).

The energy gained or lost as heat when a given mass of a substance is warmed or cooled can be calculated using Equation 5.1.

$$q = C \times m \times \Delta T \qquad (5.1)$$

Here, q is the energy gained or lost as heat by a given mass of substance (m); C is the specific heat capacity, and ΔT is the change in temperature. The change in temperature, ΔT, is calculated as the final temperature minus the initial temperature.

$$\Delta T = T_{final} - T_{initial} \qquad (5.2)$$

■ **Change in Temperature, ΔT**

Sign of ΔT	Meaning
Positive	$T_{final} > T_{initial}$, so T has increased, and q will be positive. Energy has been transferred to the object under study.
Negative	$T_{initial} > T_{final}$, so T has decreased, and q will be negative. Energy has been transferred out of the object under study.

Specific Heat Capacities of Some Elements, Compounds, and Substances

Substances	Specific Heat Capacity (J/g · K)	Molar Heat Capacity (J/mol · K)
Al, aluminum	0.897	24.2
Fe, iron	0.449	25.1
Cu, copper	0.385	24.5
Water (liquid)	4.184	75.4
Water (ice)	2.06	37.1
$HOCH_2CH_2OH(\ell)$, ethylene glycol (antifreeze)	2.39	14.8
Wood	1.8	—
Glass	0.8	—

Charles D. Winters

FIGURE 5.7 Specific heat capacity. Metals have different values of specific heat capacity on a per-gram basis. However, their molar heat capacities are all in the range of 25 J/mol · K. Among common substances, liquid water has the highest specific heat capacity on a per-gram or per-mole basis (except for liquid ammonia), a fact that plays a significant role in Earth's weather and climate.

■ **Molar Heat Capacity** Heat capacities can be expressed on a per-mole basis. The amount of energy that is transferred as heat in raising the temperature of one mole of a substance by one Kelvin is the molar heat capacity. For water, the molar heat capacity is 75.4 J/mol · K. The molar heat capacity of metals at room temperature is always near 25 kJ/mol · K.

Calculating a change in temperature using Equation 5.2 will give a result with an algebraic sign that indicates the direction of energy transfer. For example, we can use the specific heat capacity of copper, 0.385 J/g · K, to calculate the energy that must be transferred as heat to a 10.0-g sample of copper if its temperature is raised from 298 K (25 °C) to 598 K (325 °C).

$$q = \left(0.385\ \frac{\text{J}}{\text{g} \cdot \text{K}}\right)(10.0\ \text{g})(\underset{\substack{\uparrow \\ T_{\text{final}} \\ \text{Final temp.}}}{598\ \text{K}} - \underset{\substack{\uparrow \\ T_{\text{initial}} \\ \text{Initial temp.}}}{298\ \text{K}}) = +1160\ \text{J}$$

Notice that the answer has a positive sign. This indicates that the thermal energy of the sample of copper has increased by 1160 J, which is in accord with energy being transferred as heat to the copper from the water.

The relationship between energy, mass, and specific heat capacity has numerous implications. The high specific heat capacity of liquid water, 4.184 J/g · K, is a major reason why large bodies of water have a profound influence on weather. In spring, lakes warm up more slowly than the air. In autumn, the energy given off by a large lake as it cools moderates the drop in air temperature. The relevance of specific heat capacity is also illustrated when bread is wrapped in aluminum foil (specific heat capacity 0.897 J/g · K) and heated in an oven. You can remove the foil with your fingers after taking the bread from the oven. The bread and the aluminum foil are very hot, but the small mass of aluminum foil used and its low specific heat capacity result in only a small quantity of energy being transferred to your fingers (which have a larger mass and a higher specific heat capacity) when you touch the hot foil. This is also the reason why a chain of fast-food restaurants warns you that the filling of an apple pie can be much warmer than the paper wrapper or the pie crust. Although the wrapper, pie crust, and filling are at the same temperature, the quantity of energy transferred to your fingers (or your mouth!) from the filling is greater than that transferred from the wrapper and crust.

Charles D. Winters

A practical example of specific heat capacity. The filling of the apple pie has a higher specific heat (and higher mass) than the pie crust and wrapper. Notice the warning on the wrapper.

Chemistry Now™

Sign in at **www.thomsonedu.com/login** and go to Chapter 5 Contents to see Screen 5.7 for a simulation and exercise on **energy transfer as heat.**

EXAMPLE 5.1 Specific Heat Capacity

Problem How much energy must be transferred to raise the temperature of a cup of coffee (250 mL) from 20.5 °C (293.7 K) to 95.6 °C (368.8 K)? Assume that water and coffee have the same density (1.00 g/mL), and specific heat capacity (4.184 J/g · K).

Strategy Use Equation 5.1. For the calculation, you will need the specific heat capacity for H_2O, the mass of the coffee (calculated from its density and volume), and the change in temperature ($T_{final} - T_{initial}$).

Solution

$$\text{Mass of coffee} = (250\ \text{mL})(1.00\ \text{g/mL}) = 250\ \text{g}$$

$$\Delta T = T_{final} - T_{initial} = 368.8\ \text{K} - 293.7\ \text{K} = 75.1\ \text{K}$$

$$q = C \times m \times \Delta T$$

$$q = (4.184\ \text{J/g} \cdot \text{K})(250\ \text{g})(75.1\ \text{K})$$

$$q = 79{,}000\ \text{J (or 79 kJ)}$$

Comment The positive sign in the answer indicates that thermal energy has been transferred to the coffee as heat. The thermal energy of the coffee is now higher.

EXERCISE 5.2 Specific Heat Capacity

In an experiment, it was determined that 59.8 J was required to raise the temperature of 25.0 g of ethylene glycol (a compound used as antifreeze in automobile engines) by 1.00 K. Calculate the specific heat capacity of ethylene glycol from these data.

Quantitative Aspects of Energy Transferred as Heat

Like melting point, boiling point, and density, specific heat capacity is a characteristic intensive property of a pure substance. The specific heat capacity of a substance can be determined experimentally by accurately measuring temperature changes that occur when energy is transferred as heat from the substance to a known quantity of water (whose specific heat capacity is known).

Suppose a 55.0-g piece of metal is heated in boiling water to 99.8 °C and then dropped into cool water in an insulated beaker (Figure 5.8). Assume the beaker contains 225 g of water and its initial temperature (before the metal was dropped in) was 21.0 °C. The final temperature of the metal and water is 23.1 °C. What is the specific heat capacity of the metal? Here are the important aspects of this experiment.

- Let us define the metal and the water as the system and the beaker and environment as the surroundings. We will assume that energy is transferred only within the system and not between the system and the surroundings. (This assumption is good, but not perfect; for a more accurate result, we would also want to account for any energy transfer to the surroundings.)
- The water and the metal bar end up at the same temperature. (T_{final} is the same for both.)
- We will also assume energy is transferred *only as heat* within the system.
- The energy transferred as heat from the metal to the water, q_{metal}, has a negative value because the temperature of the metal drops. Conversely, q_{water} has a positive value because its temperature increases.
- The values of q_{water} and q_{metal} are numerically equal but of opposite sign.

Because of the law of conservation of energy, *in an isolated system the sum of the energy changes within the system must be zero.* If energy is transferred only as heat, then

$$q_1 + q_2 + q_3 + \ldots = 0 \qquad (5.3)$$

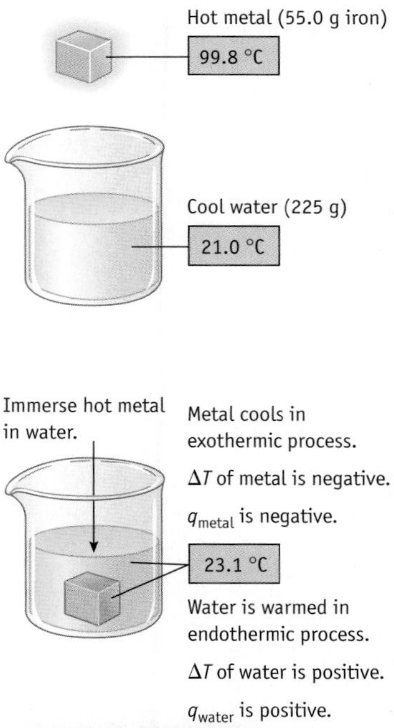

Active Figure 5.8 Transfer of energy as heat. When energy is transferred as heat from a hot metal to cool water, the thermal energy of the metal decreases, and that of the water increases. The value of q_{metal} is thus negative, and that of q_{water} is positive.

Chemistry Now™ Sign in at www.thomsonedu.com/login and go to the Chapter Contents menu to explore an interactive version of this figure accompanied by an exercise.

Problem Solving Tip 5.1 Calculating ΔT

Specific heat capacity values are given in units of joules per gram per kelvin (J/g · K). Virtually all calculations that involve temperature in chemistry are expressed in kelvins. In calculating ΔT, however, we can use Celsius temperatures because a kelvin and a Celsius degree are the same size. That is, the difference between two temperatures is the same on both scales. For example, the difference between the boiling and freezing points of water is

$$\Delta T\text{, Celsius} = 100\ °C - 0\ °C = 100\ °C$$

$$\Delta T\text{, kelvin} = 373\ K - 273\ K = 100\ K$$

where the quantities q_1, q_2, and so on represent the energies transferred as heat for the individual parts of the system. For this specific problem, there are thermal energy changes associated with water and metal, q_{water} and q_{metal}, the two components of the system; thus

$$q_{water} + q_{metal} = 0$$

Each of these quantities is related individually to specific heat capacities, mass, and change of temperature, as defined by Equation 5.1. Thus

$$[C_{water} \times m_{water} \times (T_{final} - T_{initial,\ water})] + [C_{metal} \times m_{metal} \times (T_{final} - T_{initial,\ metal})] = 0$$

The specific heat capacity of the metal, C_{metal}, is the unknown in this problem. Using the specific heat capacity of water (4.184 J/g · K) and converting Celsius to kelvin temperature gives

$$[(4.184\ J/g \cdot K)(225\ g)(296.3\ K - 294.2\ K)] + [(C_{metal})(55.0\ g)(296.3\ K - 373.0\ K)] = 0$$

$$C_{metal} = 0.469\ J/g \cdot K$$

Chemistry Now™

Sign in at **www.thomsonedu.com/login** and go to Chapter 5 Contents to see Screens 5.8 and 5.10 for exercises, tutorials, and simulations on **energy transfers as heat between substances** and **calculating energy transfer.**

EXAMPLE 5.2 Using Specific Heat Capacity

Problem An 88.5-g piece of iron whose temperature is 78.8 °C (352.0 K) is placed in a beaker containing 244 g of water at 18.8 °C (292.0 K). When thermal equilibrium is reached, what is the final temperature? (Assume no energy is lost to warm the beaker and its surroundings.)

Strategy First, define the system as consisting of the iron and water. Within the system, energy is transferred as heat from the metal to the water; the metal thus loses thermal energy, and the water gains thermal energy. The sum of these energy changes must equal zero. Each of these energy changes is related to the specific heat capacity, mass, and temperature change of the substance (Equation 5.1). The specific heat capacities of iron and water are given in Appendix D, and the final temperature is unknown. The change in temperature, ΔT, may be in °C or K (see Problem Solving Tip 5.1).

Solution

$$q_{metal} + q_{water} = 0$$

$$[C_{water} \times m_{water} \times (T_{final} - T_{initial,\ water})] + [C_{Fe} \times m_{Fe} \times (T_{final} - T_{initial,\ Fe})] = 0$$

$$[(4.184\ J/g \cdot K)(244\ g)(T_{final} - 292.0\ K)] + [(0.449\ J/g \cdot K)(88.5\ g)(T_{final} - 352.0\ K)] = 0$$

$$T_{final} = 295\ K\ (22\ °C)$$

Comment Be sure to notice that $T_{initial}$ for the metal and $T_{initial}$ for the water in this problem have different values. Also, the low specific heat capacity and smaller quantity of iron result in the temperature of iron being reduced by about 60 degrees; in contrast, the temperature of the water has been raised by only a few degrees.

EXERCISE 5.3 Using Specific Heat Capacity

A 15.5-g piece of chromium, heated to 100.0 °C, is dropped into 55.5 g of water at 16.5 °C. The final temperature of the metal and the water is 18.9 °C. What is the specific heat capacity of chromium? (Assume no energy as heat is lost to the container or to the surrounding air.)

5.3 Energy and Changes of State

A *change of state* is a change, for example, between solid and liquid or between liquid and gas. When a solid melts, its atoms, molecules, or ions move about vigorously enough to break free of the attractive forces holding them in rigid positions in the solid lattice. When a liquid boils, particles move much farther apart from one another, to distances at which attractive forces are minimal. In both cases, energy must be furnished to overcome attractive forces among the particles.

The energy transferred as heat that is required to convert a substance from a solid at its melting point to a liquid is called the **heat of fusion.** The energy transferred as heat to convert a liquid at its boiling point to a vapor is called the **heat of vaporization.** Heats of fusion and vaporization for many substances are provided along with other physical properties in reference books. Values for a few common substances are given in Appendix D (Table 12).

■ **Heats of Fusion and Vaporization for H_2O at the Normal Melting and Boiling Points**

Heat of fusion = 333 J/g
= 6.00 kJ/mol

Heat of vaporization = 2256 J/g
= 40.65 kJ/mol

It is important to recognize that *temperature is constant throughout a change of state* (Figure 5.9). During a change of state, the added energy is used to overcome the forces holding one molecule to another, not to increase the temperature (Figures 5.9 and 5.10).

For water, the heat of fusion at 0 °C is 333 J/g, and the heat of vaporization at 100 °C is 2256 J/g. These values are used to calculate the heat required for a given mass of water to melt or boil, respectively. For example, the energy required to convert 500. g of water from the liquid to gaseous state at 100 °C is

$$(2256 \text{ J/g})(500.\text{ g}) = 1.13 \times 10^6 \text{ J } (= 1130 \text{ kJ})$$

In contrast, to melt the same mass of ice to form liquid water at 0 °C requires only 167 kJ.

$$(333 \text{ J/g})(500.\text{ g}) = 1.67 \times 10^5 \text{ J } (= 167 \text{ kJ})$$

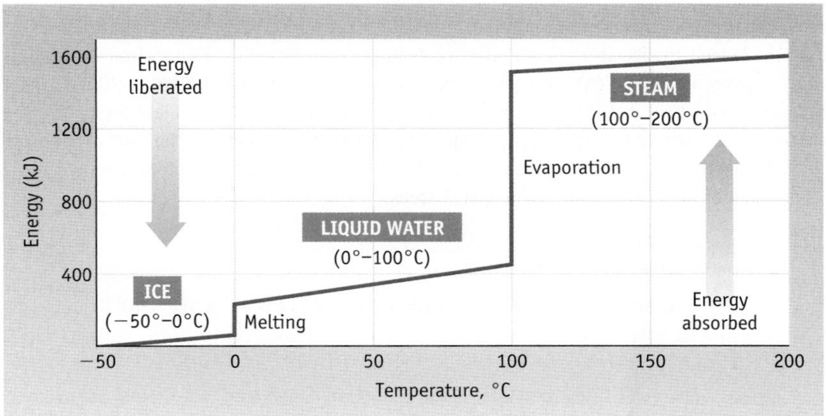

FIGURE 5.9 Energy transfer as heat and the temperature change for water. This graph shows the energy transferred as heat to 500. g of water and the consequent temperature change as the water warms from − 50 °C to 200 °C (at 1 atm).

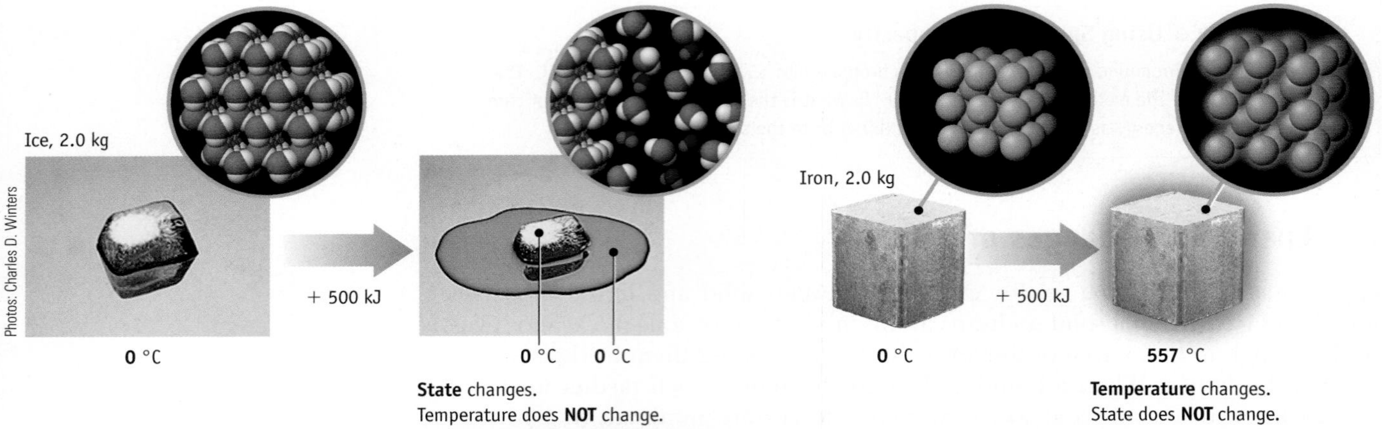

Active Figure 5.10 **Changes of state.** (*left*) Transferring 500 kJ of energy as heat to 2.0 kg of ice at 0 °C will cause 1.5 kg of ice to melt to water at 0 °C (and 0.5 kg of ice will remain). No temperature change occurs. (*right*) In contrast, transferring 500 kJ of energy as heat to 2.0 kg of iron at 0 °C will cause the temperature to increase to 557 °C (and the metal to expand slightly).

Chemistry Now™ Sign in at www.thomsonedu.com/login and go to the Chapter Contents menu to explore an interactive version of this figure accompanied by an exercise.

Figure 5.9 gives a profile of a process in which 500. g of ice at − 50 °C is converted to water vapor at 200 °C . This process involves a series of steps: (1) warming ice to 0 °C, (2) conversion to liquid water at 0 °C, (3) warming liquid water to 100 °C, (4) evaporation at 100 °C, and (5) warming the water vapor to 200 °C. Each step requires the input of additional energy. The energy transferred as heat to raise the temperature of solid, liquid, and vapor can be calculated with Equation 5.1, using the specific heat capacities of ice, liquid water, and water vapor (which are different), and the energies transferred as heat for the changes of state can be calculated using heats of fusion and vaporization. These calculations are carried out in Example 5.3.

EXAMPLE 5.3 Energy and Changes of State

Problem Calculate the energy that is transferred as heat to convert 500. g of ice at −50.0 °C to steam at 200.0 °C. (The temperature change occurring in each step is illustrated in Figure 5.9.) The heat of fusion of water is 333 J/g, and the heat of vaporization is 2256 J/g. The specific heat capacities of ice, liquid water, and water vapor are given in Appendix D.

Strategy The problem is broken down into a series of steps as noted above: (1) warm the ice from −50 °C to 0 °C; (2) melt the ice at 0 °C; (3) raise the temperature of the liquid water from 0 °C to 100 °C; (4) boil the water at 100 °C; (5) raise the temperature of the steam from 100 °C to 200 °C. Use Equation 5.1 and the specific heat capacities of solid, liquid, and gaseous water to calculate the energy transferred as heat associated with the temperature changes. Use the heat of fusion and the heat of vaporization to calculate the energy transferred as heat associated with changes of state. The total energy transferred as heat is the sum of the energies of the individual steps.

Solution

Step 1. (to warm ice from −50.0 °C to 0.0 °C)

$$q_1 = (2.06 \text{ J/g} \cdot \text{K})(500.\text{ g})(273.2 \text{ K} - 223.2 \text{ K}) = 5.15 \times 10^4 \text{ J}$$

Step 2. (to melt ice at 0.0 °C)

$$q_2 = (500.\ \text{g})(333\ \text{J/g}) = 1.67 \times 10^5\ \text{J}$$

Step 3. (to raise temperature of liquid water from 0.0 °C to 100.0 °C)

$$q_3 = (4.184\ \text{J/g} \cdot \text{K})(500.\ \text{g})(373.2\ \text{K} - 273.2\ \text{K}) = 2.09 \times 10^5\ \text{J}$$

Step 4. (to evaporate water at 100.0 °C)

$$q_4 = (2256\ \text{J/g})(500.\ \text{g}) = 1.13 \times 10^6\ \text{J}$$

Step 5. (to raise temperature of water vapor from 100.0 °C to 200.0 °C)

$$q_5 = (1.86\ \text{J/g} \cdot \text{K})(500.\ \text{g})\ (473.2\ \text{K} - 373.2\ \text{K}) = 9.30 \times 10^4\ \text{J}$$

The total energy transferred as heat is the sum of the energies of the individual steps.

$$q_{\text{total}} = q_1 + q_2 + q_3 + q_4 + q_5$$

$$q_{\text{total}} = 1.65 \times 10^6\ \text{J (or 1650 kJ)}$$

Comment The conversion of liquid water to steam is the largest increment of energy added by a considerable margin. (You may have noticed that it does not take much time to heat water to boiling on a stove, but to boil off the water takes a much greater time.)

EXAMPLE 5.4 Change of State

Problem What is the minimum amount of ice at 0 °C that must be added to the contents of a can of diet cola (340. mL) to cool the cola from 20.5 °C to 0.0 °C? Assume that the specific heat capacity and density of diet cola are the same as for water.

Strategy It is easiest to define the system as the ice and cola; the calculation will then involve energy transfers between the two components in the system. We need to assume that, within the system, energy transfers only as heat and that there is no transfer of energy between the surroundings and the system. The law of conservation of energy then dictates that $q_{\text{ice}} + q_{\text{cola}} = 0$. The value of q_{cola} can be calculated using the specific heat capacity of the cola and Equation 5.1, and the energy transferred as heat required to melt ice can be calculated using the heat of fusion for water.

Solution The mass of cola is 340. g [(340. mL)(1.00 g/mL) = 340. g], and its temperature changes from 293.7 K to 273.2 K. The heat of fusion of water is 333 J/g, and the mass of ice is the unknown.

$$q_{\text{cola}} + q_{\text{ice}} = 0$$

$$C_{\text{cola}} \times m \times (T_{\text{final}} - T_{\text{initial}}) + q_{\text{ice}} = 0$$

$$[(4.184\ \text{J/g} \cdot \text{K})(340.\ \text{g})(273.2\ \text{K} - 293.7\ \text{K})] + [(333\ \text{J/g})\ (m_{\text{ice}})] = 0$$

$$m_{\text{ice}} = 87.6\ \text{g}$$

Comment This quantity of ice is just sufficient to cool the cola to 0 °C. If more than 87.6 g of ice is added, then the final temperature will still be 0 °C when thermal equilibrium is reached, and some ice will remain (see Exercise 5.4). If less than 87.6 g of ice is added, the final temperature will be greater than 0 °C. In this case, all the ice will melt, and the liquid water formed by melting the ice will absorb additional energy to warm up to the final temperature (an example is given in Study Question 71, page 248).

EXERCISE 5.4 Changes of State

To make a glass of iced tea, you pour 250 mL of tea, whose temperature is 18.2 °C, into a glass containing five ice cubes. Each cube has a mass of 15 g. What quantity of ice will melt, and how much ice will remain to float at the surface in this beverage? Assume that iced tea has a density of 1.0 g/mL and a specific heat capacity of 4.2 J/g · K, that energy is transferred only as heat within the system, that ice is at 0.0 °C, and that no energy is transferred between system and surroundings.

Case Study — Abba's Refrigerator

If you put a pot of water on a kitchen stove or a campfire, or if you put the pot in the sun, the water will evaporate. You must supply energy in some form because evaporation requires the input of energy. This well-known principle was applied in a novel way by a young African teacher, Mohammed Bah Abba in Nigeria, to improve the life of his people.

Life is hard in northern Nigerian communities. In this rural semi-desert area, most people eke out a living by subsistence farming. Without modern refrigeration, food spoilage is a major problem. Using a simple thermodynamic principle, Abba developed a refrigerator that cost about 30 cents to make and does not use electricity.

Abba's refrigerator is made of two earthen pots, one inside the other, separated by a layer of sand. The pots are covered with damp cloth and placed in a well-ventilated area. Water seeps through the pot's outer wall and rapidly evaporates in the dry desert air. The water remaining in the pot and its contents drop in temperature, so much so that food in the inner pot can stay cool for days and not spoil.

In the 1990s, at his own expense, Abba made and distributed almost 10,000 pots in the villages of northern Nigeria. He estimates that about 75% of the families in this area are now using his refrigerator. The impact of this simple device has implications not only for the health of his people but also for their economy and their social structure. Prior to the development of the pot-in-pot device for food storage, it was necessary to sell produce immediately upon harvesting. The young girls in the family who sold food on the street daily could now be released from this chore to attend school and improve their lives.

Every two years, the Rolex Company, the Swiss maker of timepieces, gives a series of awards for enterprise. For his pot-in-pot refrigerator, Abba was one of the five recipients of a Rolex Award in 2000.

Questions:

1. *What quantity of energy must be transferred as heat to evaporate 95 g of water at 25 °C? (The heat of vaporization at 25 °C is 44.0 kJ/mol.)*
2. *If this quantity of energy is transferred as heat out of 750 g of water, what is the temperature change of the water?*

Answers to these questions are in Appendix Q.

The Rolex Awards for Enterprise/Tomas Bertelsen/Scientific American, Nov. 2000, p. 26

Mohammed Bah Abba. Abba earned a college degree in business, and while still in his 20s, became an instructor in a college of business in Jigwa, Nigeria, and a consultant to the United Nations Development Program. That brought him into close contact with rural communities in Northern Nigeria and made him aware of the hardships of the families there.

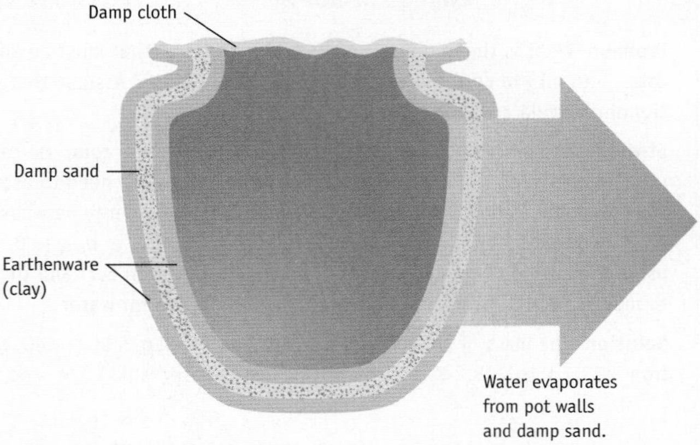

The pot-in-pot refrigerator. Water seeps through the outer pot from the damp sand layer separating the pots, or from food stored in the inner pot. As the water evaporates from the surface of the outer pot, the food inside the pot is cooled.

5.4 The First Law of Thermodynamics

To this point, we have only considered energy transfers as heat. Now we need to broaden the discussion. Recall the definition of thermodynamics as *the science of heat and work.* Work is done whenever a mass is moved against an opposing force; some form of energy is required for work to be done.

As described earlier, energy transferred as heat between a system and its surroundings changes the energy of the system. Work done by a system or on a system will also affect the energy in the system. If a system does work on its surroundings, energy must be expended by the system, and the system's energy will decrease. Conversely, if work is done by the surroundings on a system, the energy of the system increases.

A system doing work on its surroundings is illustrated in Figure 5.11. A small quantity of dry ice, solid CO_2, is sealed inside a plastic bag, and a weight (a book) is placed on top of the bag. When energy is transferred as heat from the surround-

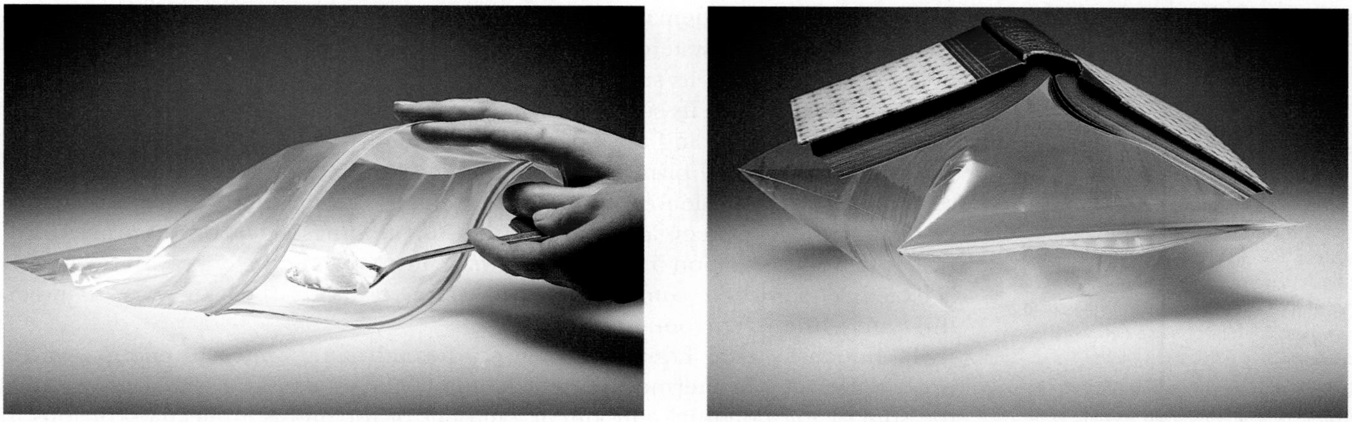

Charles D. Winters

(a) Pieces of dry ice [CO_2(s), −78 °C] are placed in a plastic bag. The dry ice will sublime (change directly from a solid to a gas) upon the input of energy.

(b) Energy is absorbed by CO_2(s) when it sublimes, and the system (the contents of the bag) does work on its surroundings by lifting the book against the force of gravity.

Active Figure 5.11 **Energy changes in a physical process.**

ChemistryNow™ Sign in at www.thomsonedu.com/login and go to the Chapter Contents menu to explore an interactive version of this figure accompanied by an exercise.

ings to the dry ice, the dry ice changes directly from solid to gas at −78 °C, in a process called **sublimation:**

$$CO_2(s, -78\ °C) \longrightarrow CO_2(g, -78\ °C)$$

As sublimation proceeds, gaseous CO_2 expands within the plastic bag, lifting the book against the force of gravity. The system (the CO_2 inside the bag) is expending energy to do this work.

Even if the book had not been on top of the plastic bag, work would have been done by the expanding gas because the gas must push back the atmosphere when it expands. Instead of raising a book, the expanding gas moves a part of the atmosphere.

Now let us restate this example in terms of thermodynamics. First, we must identify the system and the surroundings. The system is the CO_2, initially a solid and later a gas. The surroundings consist of the objects that exchange energy with the system. This includes the plastic bag, the book, the table-top, and the surrounding air. Thermodynamics focuses on energy transfer. Sublimation of CO_2 requires energy, which is transferred as heat to the system (the CO_2) from the surroundings. At the same time, the system does work on the surroundings by lifting the book. An energy balance for the system will include both quantities, heat and work.

This example can be generalized. For any system, we can identify energy transfers both as heat and as work between system and surroundings. We can express the change of energy in a system explicitly as an equation:

Change in energy content ↓ — Energy transferred as work to or from the system ↓

$$\Delta U = q + w \qquad (5.4)$$

↑ Energy transferred as heat to or from the system

More on Heat and Work To better understand the nature of heat and work, consider this passage from a classic book on chemical thermodynamics:

"Heat and work are birds of passage—never found as such in residence. Quantities of heat and work are perfectly determinate as they are transferred across the boundary that delimits the system. But after completion of the transfers we cannot speak of the system (or of its surroundings) as having some new heat content or some new work content."

L. K. Nash, *Elements of Chemical Thermodynamics*, Addison-Wesley, 1962.

Equation 5.4 is a mathematical statement of the **first law of thermodynamics:** The energy change for a system (ΔU) is the sum of the energy transferred as heat between the system and its surroundings (q) and the energy transferred as work between the system and its surroundings (w).

The equation defining the first law of thermodynamics can be thought of as a version of the general principle of conservation of energy. Because energy is conserved, we must be able to account for any change in the energy of the system. All energy transfers between a system and the surroundings occur by the processes of heat and work. Equation 5.4 thus states that the change in the energy of the system is exactly equal to the sum of all of the energy transfers (heat or work) between the system and its surroundings.

The quantity U in Equation 5.4 has a formal name—**internal energy**—and a precise meaning in thermodynamics. The internal energy in a chemical system is the sum of the potential and kinetic energies of the atoms, molecules, or ions in the system. The potential energy here is the energy associated with the attractive and repulsive forces between all the nuclei and electrons in the system. It includes the energy associated with bonds in molecules, forces between ions, and forces between molecules. The kinetic energy is the energy of motion of the atoms, ions, and molecules in the system. Actual values of internal energy are rarely determined or needed. In most instances, we are interested in the *change* in internal energy, and this is a measurable quantity. In fact, Equation 5.4 tells us how to determine ΔU: *Measure the energy transferred as heat and work to or from the system.*

The sign conventions for Equation 5.4 are important. The following table summarizes how the internal energy of a system is affected by energy transferred as heat and work.

Heat and Work For an in-depth look at heat and work in thermodynamics, see E. A. Gislason and N. C. Craig, *Journal of Chemical Education*, Vol. 64, No. 8, pages 660–668, 1987.

Sign Conventions for q and w of the System

Change	Sign Convention	Effect on U_{system}
Energy transferred as heat to the system (endothermic)	$q > 0$ (+)	U increases
Energy transferred as heat from the system (exothermic)	$q < 0$ (−)	U decreases
Energy transferred as work done on system	$w > 0$ (+)	U increases
Energy transferred as work done by system	$w > 0$ (−)	U decreases

The work in the example involving the sublimation of CO_2 (Figure 5.11) is of a specific type, called P–V (pressure–volume) work. It is the work (w) associated with a change in volume (ΔV) that occurs against a resisting external pressure (P). For a system in which the external pressure is constant, the value of P–V work can be calculated using Equation 5.5:

Work (at constant pressure) → w; Change in volume → ΔV; Pressure → P

$$w = -P \times \Delta V \quad (5.5)$$

ChemistryNow™

Sign in at **www.thomsonedu.com/login** and go to Chapter 5 Contents to see Screen 5.11 for a self-study module on **energy changes in a physical process.**

Enthalpy

Most experiments in a chemical laboratory are carried out in beakers or flasks open to the atmosphere, where the external pressure is constant. Similarly, chemical processes that occur in living systems are open to the atmosphere. Because many processes in chemistry and biology are carried out under conditions of constant pressure, it is useful to have a specific measure of the energy transferred as heat under these conditions.

Let us first examine ΔU under conditions of constant pressure:

$$\Delta U = q_p + w_p$$

where the subscript p indicates conditions of constant pressure. If the only type of work that occurs is P–V work, then

$$\Delta U = q_p - P\Delta V$$

Rearranging this gives

$$q_p = \Delta U + P\Delta V$$

We now introduce a new thermodynamic function called the **enthalpy, *H*,** which is defined as

$$H = U + PV$$

■ **Energy Transfer Under Conditions of Constant Volume** Under conditions of constant volume, $\Delta V = 0$. If energy is transferred as heat under these conditions and if the only type of work possible is P–V work, the equation for the first law of thermodynamics simplifies to $\Delta U = q_v$. The subscript v indicates conditions of constant volume. In this case, the energy transferred as heat is equal to ΔU.

A Closer Look — *P–V* Work

Work is done when an object of some mass is moved against an external resisting force. We know this from common experience, such as when we use a pump to blow up a bicycle tire.

To evaluate the work done when a gas is compressed, we can use, for example, a cylinder with a movable piston, as would occur in a bicycle pump (see figure). The drawing on the left shows the initial position of the piston, and the one on the right shows its final position. To depress the piston, we would have to expend some energy (the energy of this process comes from the energy obtained by food metabolism in our body). The work required to depress the piston is calculated from a law of physics, $w = F \times d$, or work equals the magnitude of the force (F) applied times the distance (d) over which the force is applied.

Pressure is defined as a force divided by the area over which the force is applied: $P = F/A$. In this example, the force is being applied to a piston with an area A. Substituting $P \times A$ for F in the equation gives $w = (P \times A) \times d$. The product of $A \times d$ is equivalent to the change in the volume of the gas in the pump, and, because $\Delta V = V_{final} - V_{initial}$, this change in volume is negative. Finally, because work done on a system is defined as positive, this means that $w = -P\Delta V$.

Pushing down on the piston means we have done work on the system, the gas contained within the cylinder. The gas is now compressed to a smaller volume and has attained a higher energy as a consequence. The additional energy is equal to $-P\Delta V$.

Notice how energy has been converted from one form to another—from chemical energy in food to mechanical energy used to depress the piston, to potential energy stored in a system of a gas at a higher pressure. In each step, energy was conserved, and the total energy of the universe remained constant.

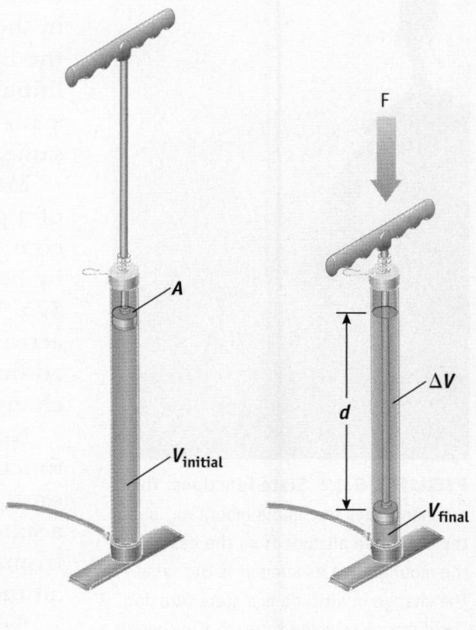

■ **Enthalpy and Internal Energy Differences** The difference between ΔH and ΔU will be quite small unless a large volume change occurs. For water at 1 atm pressure and 273 K, for example, the difference between ΔH and ΔU ($= P\Delta V$) is 0.142 J/mol for the conversion of ice to liquid water at 273 K, whereas it is 3, 100 J/mol for the conversion of liquid water to water vapor at 373 K.

Changes in enthalpy for a system at constant pressure would be calculated from the following equation:

$$\Delta H = \Delta U + P\Delta V$$

Thus,

$$\Delta H = q_p$$

Now we see that, for a system where the only type of work possible is *P–V* work, the change in enthalpy, ΔH, is equal to the energy transferred as heat at constant pressure, often symbolized by q_p.

Under conditions of constant pressure and where the only type of work possible is *P–V* work, ΔU ($= q_p - P\Delta V$) and ΔH ($= q_p$) differ by $P\Delta V$ (the energy transferred to or from the system as work). We observe that in many processes—such as the melting of ice—the volume change, ΔV, is small, and hence the amount of work is small. Under these circumstances, ΔU and ΔH have almost the same value. The amount of work will be significant, however, in processes in which the volume change is large. This usually occurs when gases are formed or consumed. In the evaporation or condensation of water, the sublimation of CO_2, and chemical reactions in which the number of moles of gas changes, ΔU and ΔH have significantly different values.

Similar sign and symbol conventions apply to both ΔU and ΔH.

- Negative values of ΔH specify that energy is transferred as heat from the system to the surroundings.
- Positive values of ΔH specify that energy is transferred as heat from the surroundings to the system.

State Functions

© Ross Woodhall/Taxi/Getty Images

FIGURE 5.12 State functions. There are many ways to climb a mountain, but the change in altitude from the base of the mountain to its summit is the same. The change in altitude is a state function. The distance traveled to reach the summit is not.

Internal energy and enthalpy share a significant characteristic—namely, changes in these quantities that accompany chemical or physical processes depend only on the initial and final states. They do not depend on the path taken to go from the initial state to the final state. No matter how you go from reactants to products in a reaction, for example, the value of ΔH and ΔU for the reaction is always the same. A quantity that has this property is called a **state function**.

Many commonly measured quantities, such as the pressure of a gas, the volume of a gas or liquid, the temperature of a substance, and the size of your bank account, are state functions. You could have arrived at a current bank balance of $25 by having deposited $25 or you could have deposited $100 and then withdrawn $75. You can blow up a balloon to a large volume and then let some air out to arrive at the desired volume. Alternatively, you can blow up the balloon in stages, adding tiny amounts of air at each stage. The change in your bank balance or change in volume of the balloon does not depend on how you got there.

Not all quantities are state functions. For instance, distance traveled is not a state function (Figure 5.12). The travel distance from New York City to Denver depends on the route taken. Nor is the elapsed time of travel between these two locations a state function. In contrast, the altitude above sea level is a state function; in going from New York City (at sea level) to Denver (1600 m above sea level), there is an altitude change of 1600 m, regardless of the route followed.

Significantly, neither the energy transferred as heat nor the energy transferred as work individually is a state function but their sum, the change in internal energy,

ΔU, is. The value of ΔU is fixed by $U_{initial}$ and U_{final}. A transition between the initial and final states can be accomplished by different routes having different values of q and w, but the sum of q and w for each path must always give the same ΔU.

Enthalpy is also a state function. The enthalpy change occurring when 1.0 g of water is heated from 20 °C to 50 °C is independent of how the process is carried out.

5.5 Enthalpy Changes for Chemical Reactions

Enthalpy changes accompany chemical reactions. In this book, we shall follow the conventions used by physical chemists and report the **standard reaction enthalpy,** $\Delta_r H°$ for reactions. For example, for the decomposition of water vapor to hydrogen and oxygen, with the reactant and products all in their standard states at 25 °C, the standard reaction enthalpy is +241.8 kJ/mol-rxn.

$$H_2O(g) \longrightarrow H_2(g) + \tfrac{1}{2}\ O_2(g) \qquad \Delta_r H° = +241.8 \text{ kJ/mol-rxn}$$

The positive sign of $\Delta_r H°$ in this case indicates that the decomposition is an endothermic process.

There are several important things to know about $\Delta_r H°$.

- The designation of $\Delta_r H°$ as a "standard enthalpy change" means that the pure, unmixed reactants in their standard states have formed pure, unmixed products in their standard states (where the superscript ° indicates standard conditions). The **standard state** of an element or a compound is defined as the most stable form of the substance in the physical state that exists at a pressure of 1 bar and at a specified temperature. [Most sources report standard reaction enthalpies at 25 °C (298 K).]
- The "per mol-rxn" designation in the units for $\Delta_r H°$ means this is the enthalpy change for a "mole of reaction" (where *rxn* is an abbreviation for reaction). For example, for the reaction $H_2O(g) \longrightarrow H_2(g) + 1/2\ O_2(g)$, a mole of reaction has occurred when 1 mol of water vapor has been converted completely to 1 mol of H_2 and 1/2 mol of O_2.

Now consider the opposite reaction, the combination of hydrogen and oxygen to form 1 mol of water. The magnitude of the enthalpy change for this reaction is the same as that for the decomposition reaction, but the sign of $\Delta_r H°$ is reversed. The exothermic formation of 1 mol of water vapor from 1 mol of H_2 and 1/2 mol of O_2 transfers 241.8 kJ to the surroundings (Figure 5.13).

$$H_2(g) + \tfrac{1}{2}\ O_2(g) \longrightarrow H_2O(g) \qquad \Delta_r H° = -241.8 \text{ kJ/mol-rxn}$$

The value of $\Delta_r H°$ depends on the chemical equation used. For example, $\Delta_r H°$ for 1 mole of the reaction

$$2\ H_2(g) + O_2(g) \longrightarrow 2\ H_2O(g) \qquad \Delta_r H° = -483.6 \text{ kJ/mol-rxn}$$

will be twice that of $\Delta_r H°$ for the reaction

$$H_2(g) + \tfrac{1}{2}\ O_2(g) \longrightarrow H_2O(g) \qquad \Delta_r H° = -241.8 \text{ kJ/mol-rxn}$$

This happens because 1 mole of reaction for the first equation uses twice the amount of reactants and produces twice the amount of product as the second equation.

■ **Notation for Thermodynamic Parameters** NIST and IUPAC (International Union of Pure and Applied Chemistry) specify that parameters such as ΔH should have a subscript, between Δ and the thermodynamic parameter, that specifies the type of process. Among the subscripts you will see are: a lower case r for "reaction," f for "formation," c for "combustion," fus for "fusion," and vap for "vaporization."

■ **Moles of Reaction, Mol-rxn** One "mole of reaction" is said to have occurred when the reaction has occurred according to the number of moles given by the coefficients in the balanced equation. (This concept was described as a way to solve limiting reactant problems on page 167.)

■ **Fractional Stoichiometric Coefficients** When writing balanced equations to define thermodynamic quantities, chemists often use fractional stoichiometric coefficients. For example, to define $\Delta_r H$ for the decomposition or formation of 1 mol of H_2O, the coefficient for O_2 must be 1/2.

(a) A lighted candle is brought up to a balloon filled with hydrogen gas.

(b) When the balloon breaks, the candle flame ignites the hydrogen.

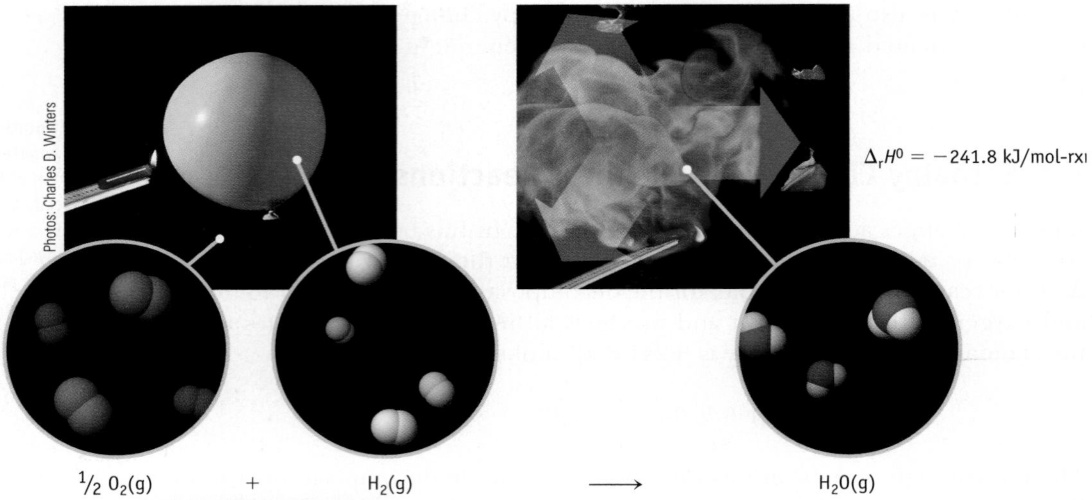

Active Figure 5.13 **The exothermic combustion of hydrogen in air.** The reaction transfers energy to the surroundings in the form of heat, work, and light.

Chemistry Now™ Sign in at www.thomsonedu.com/login and go to the Chapter Contents menu to explore an interactive version of this figure accompanied by an exercise.

■ **Standard Conditions** The superscript ° indicates standard conditions. It is applied to any type of thermodynamic data, such as enthalpy of fusion and vaporization ($\Delta_{fus}H°$ and $\Delta_{vap}H°$) and enthalpy of a reaction ($\Delta_r H°$). Standard conditions refers to reactants and products in their standard states at a pressure of 1 bar. One bar is approximately one atmosphere (1 atm = 1.013 bar; see Appendix B).

It is important to identify the states of reactants and products in a reaction because the magnitude of $\Delta_r H°$ depends on whether they are solids, liquids, or gases. For the formation of 1 mol of *liquid water* from the elements, the enthalpy change is −285.8 kJ.

$$H_2(g) + \tfrac{1}{2}\ O_2(g) \longrightarrow H_2O(\ell) \qquad \Delta_r H° = -285.8 \text{ kJ/mol-rxn}$$

Notice that this value is not the same as $\Delta_r H°$ for the formation of *water vapor* from hydrogen and oxygen. The difference between the two values is equal to the enthalpy change for the condensation of 1 mol of water vapor to 1 mol of liquid water.

These examples illustrate several general features of the enthalpy changes for chemical reactions.

- Enthalpy changes are specific to the reaction being carried out. The identities of reactants and products and their states (s, ℓ, g) are important, as are the amounts of reactants and products.
- The enthalpy change depends on the number of moles of reaction; that is, the number of times the reaction *as written* is carried out.
- $\Delta_r H°$ has a negative value for an exothermic reaction. It has a positive value for an endothermic reaction.
- Values of $\Delta_r H°$ are numerically the same, but opposite in sign, for chemical reactions that are the reverse of each other.

■ **Enthalpy of Fusion and Enthalpy of Vaporization** Previously, we called $\Delta_{fus}H°$ and $\Delta_{vap}H°$ the heat of fusion and heat of vaporization, respectively. You can see now that, based on the way that the process is carried out (at constant pressure) and by the use of *H* in their symbols, these are more properly referred to as the enthalpy of fusion and the enthalpy of vaporization. From this point on, we will refer to them by these designations.

Standard reaction enthalpies can be used to calculate the quantity of energy transferred as heat under conditions of constant pressure by any given mass of a reactant or product. Suppose you want to know the energy transferred to the surroundings as heat if 454 g of propane, C_3H_8, is burned (at constant pressure), given the equation for the exothermic combustion and the enthalpy change for the reaction.

$$C_3H_8(g) + 5\ O_2(g) \longrightarrow 3\ CO_2\ (g) + 4\ H_2O(\ell) \qquad \Delta_r H° = -2220 \text{ kJ/mol-rxn}$$

Two steps are needed. First, find the amount of propane present in the sample:

$$454 \text{ g } C_3H_8 \left(\frac{1 \text{ mol } C_3H_8}{44.10 \text{ g } C_3H_8}\right) = 10.3 \text{ mol } C_3H_8$$

Second, multiply $\Delta_r H°$ by the amount of propane:

$$\Delta_r H° = 10.3 \text{ mol } C_3H_8 \left(\frac{1 \text{ mol-rxn}}{1 \text{ mol } C_3H_8}\right)\left(\frac{-2220 \text{ kJ}}{1 \text{ mol-rxn}}\right) = -22{,}900 \text{ kJ}$$

Chemistry Now™

Sign in at **www.thomsonedu.com/login** and go to Chapter 5 Contents to see Screens 5.12 and 5.13 for self-study modules and a tutorial on **enthalpy changes.**

EXAMPLE 5.5 Calculating the Enthalpy Change for a Reaction

Problem Sucrose (sugar, $C_{12}H_{22}O_{11}$) can be oxidized to CO_2 and H_2O and the enthalpy change for the reaction can be measured (under conditions of constant pressure).

$$C_{12}H_{22}O_{11}(s) + 12\ O_2(g) \longrightarrow 12\ CO_2(g) + 11\ H_2O(\ell) \qquad \Delta_r H° = -5645 \text{ kJ/mol-rxn}$$

What is the energy transferred as heat by burning 5.00 g of sugar?

Strategy We will first determine the amount (mol) of sucrose in 5.00 g, then use this with the value given for the enthalpy change for the oxidation of 1 mol of sucrose.

Solution

$$5.00 \text{ g sucrose} \times \frac{1 \text{ mol sucrose}}{342.3 \text{ g sucrose}} = 1.46 \times 10^{-2} \text{ mol sucrose}$$

$$\Delta_r H° = 1.46 \times 10^{-2} \text{ mol sucrose} \left(\frac{1 \text{ mol-rxn}}{1 \text{ mol sucrose}}\right)\left(\frac{-5645 \text{ kJ}}{1 \text{ mol-rxn}}\right)$$

$$\Delta_r H° = -82.5 \text{ kJ}$$

Comment A person on a diet might note that a (level) teaspoonful of sugar (about 3.5 g) supplies about 15 Calories (dietary Calories; the conversion is 4.184 kJ = 1 Cal). As diets go, a single spoonful of sugar doesn't have a large caloric content. But will you use just one level teaspoonful?

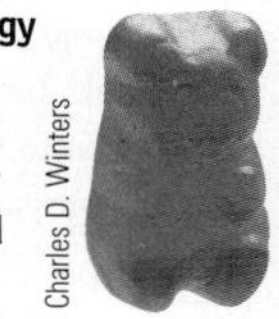

Charles D. Winters

■ Chemical Potential Energy Gummi Bears are mostly sugar. See Example 5.5 for a calculation of the energy in a spoonful of sugar, and see ChemistryNow Screen 5.2 for a video of a Gummi bear consumed by an oxidizing agent.

EXERCISE 5.5 Enthalpy Calculation

The combustion of ethane, C_2H_6, has an enthalpy change of −2857.3 kJ for the reaction as written below. Calculate the value of energy transferred as heat when 15.0 g of C_2H_6 is burned.

$$2\ C_2H_6(g) + 7\ O_2(g) \longrightarrow 4\ CO_2(g) + 6\ H_2O(g) \qquad \Delta_r H° = -2857.3 \text{ kJ/mol-rxn}$$

5.6 Calorimetry

The energy evolved or required as heat in a chemical or physical process can be measured by **calorimetry**. The apparatus used in this kind of experiment is a *calorimeter.*

Constant Pressure Calorimetry, Measuring ΔH

A constant pressure calorimeter can be used to measure the energy change for a chemical reaction as energy is transferred as heat under constant pressure conditions, that is, it measures the enthalpy change.

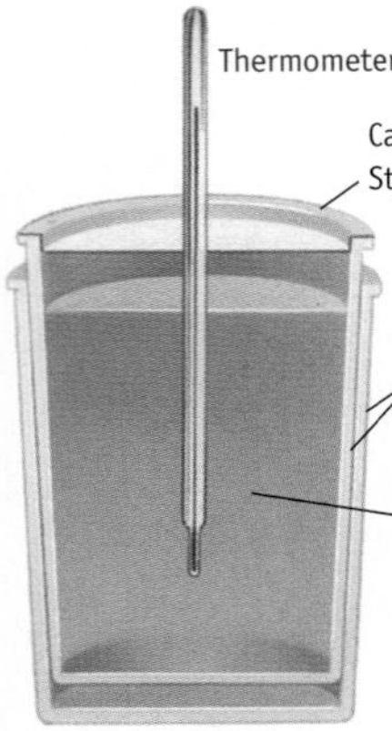

FIGURE 5.14 A coffee-cup calorimeter. A chemical reaction produces a change in temperature of the solution in the calorimeter. The Styrofoam container is fairly effective in preventing the transfer of energy as heat between the solution and its surroundings. Because the cup is open to the atmosphere, this is a constant pressure measurement.

In general chemistry laboratories, a "coffee-cup calorimeter" is often used to estimate enthalpy changes for chemical reactions. This inexpensive device consists of two nested Styrofoam coffee cups with a loose-fitting lid and a temperature-measuring device such as a thermometer (Figure 5.14) or thermocouple. Styrofoam, a fairly good insulator, minimizes energy transfer as heat between the system and the surroundings. The reaction is carried out in solution in the cup. If the reaction is exothermic, it releases energy as heat to the solution, and the temperature of the solution rises. If the reaction is endothermic, energy is absorbed as heat from the solution, and a decrease in the temperature of the solution will be seen. The change in temperature of the solution is measured. Knowing the mass and specific heat capacity of the solution and the temperature change, the enthalpy change for the reaction can be calculated.

In this calorimetry experiment, it will be convenient to define the chemicals and the solution as the system. The surroundings are the cup and everything beyond the cup. As noted above, we assume that there is no energy transfer to the cup or beyond and that energy is transferred only as heat within the system. Two energy changes occur within the system. One is the change that takes place as the chemical reaction occurs, either releasing the potential energy stored in the reactants or absorbing energy and converting it to potential energy stored in the products. We label this energy as q_r. The other energy change is the energy gained or lost as heat by the solution ($q_{solution}$). Based on the law of conservation of energy,

$$q_r + q_{solution} = 0$$

The value of $q_{solution}$ can be calculated from the specific heat capacity, mass, and change in temperature of the solution. The quantity of energy evolved or absorbed as heat for the reaction (q_r) is the unknown in the equation.

The accuracy of a calorimetry experiment depends on the accuracy of the measured quantities (temperature, mass, specific heat capacity). In addition, it depends on how closely the assumption is followed that there is no energy transfer beyond the solution. A coffee-cup calorimeter is an unsophisticated apparatus, and the results obtained with it are not highly accurate, largely because this assumption is poorly met. In research laboratories, calorimeters are used that more effectively limit the energy transfer between system and surroundings. In addition, it is also possible to estimate and correct for the minimal energy transfer that occurs between the system and the surroundings.

EXAMPLE 5.6 Using a Coffee-Cup Calorimeter

Problem Suppose you place 0.0500 g of magnesium chips in a coffee-cup calorimeter and then add 100.0 mL of 1.00 M HCl. The reaction that occurs is

$$Mg(s) + 2\ HCl(aq) \longrightarrow H_2(g) + MgCl_2(aq)$$

The temperature of the solution increases from 22.21 °C (295.36 K) to 24.46 °C (297.61 K). What is the enthalpy change for the reaction per mole of Mg? Assume that the specific heat capacity of the solution is 4.20 J/g · K and the density of the HCl solution is 1.00 g/mL.

Strategy The energy evolved in the reaction is absorbed by the solution. Solving the problem has three steps. First, calculate $q_{solution}$ from the values of the mass, specific heat capacity, and ΔT using Equation 5.1. Second, calculate q_r, assuming no energy transfer as heat occurs beyond the solution, that is, $q_r + q_{solution} = 0$. Third, use the value of q_r and the amount of Mg to calculate the enthalpy change per mole of Mg.

Solution

Step 1. Calculate $q_{solution}$. The mass of the solution is the mass of the 100.0 mL of HCl plus the mass of magnesium.

$$q_{solution} = (100.0 \text{ g HCl solution} + 0.0500 \text{ g Mg})(4.20 \text{ J/g K})(297.61 \text{ K} - 295.36 \text{ K})$$
$$= 9.45 \times 10^2 \text{ J}$$

Step 2. Calculate q_r.

$$q_r + q_{solution} = 0$$
$$q_r + 9.45 \times 10^2 \text{ J} = 0$$
$$q_r = -9.45 \times 10^2 \text{ J}$$

Step 3. Calculate the value of ΔH per mole of Mg. Note that q_r found in Step 2 resulted from the reaction of 0.0500 g of Mg. The enthalpy change per mole of Mg is therefore

$$\Delta_r H = (-9.45 \times 10^2 \text{ J}/0.0500 \text{ g Mg})(24.31 \text{ g Mg}/1 \text{ mol Mg})$$
$$= -4.60 \times 10^5 \text{ J/mol Mg } (= -460. \text{ kJ/mol-rxn Mg})$$

Comment The calculation gives the correct sign of q_r and $\Delta_r H$. The negative sign indicates that this is an exothermic reaction.

EXERCISE 5.6 Using a Coffee-Cup Calorimeter

Assume 200. mL of 0.400 M HCl are mixed with 200. mL of 0.400 M NaOH in a coffee-cup calorimeter. The temperature of the solutions before mixing was 25.10 °C; after mixing and allowing the reaction to occur, the temperature is 27.78 °C. What is the enthalpy change when one mole of acid is neutralized? (Assume that the densities of all solutions are 1.00 g/mL and their specific heat capacities are 4.20 J/g · K.)

Constant Volume Calorimetry: Measuring ΔU

Constant volume calorimetry is often used to evaluate heats of combustion of fuels and the caloric value of foods. A weighed sample of a combustible solid or liquid is placed inside a "bomb," often a cylinder about the size of a large fruit juice can with thick steel walls and ends (Figure 5.15). The bomb is placed in a water-filled container with well-insulated walls. After filling the bomb with pure oxygen, the sample is ignited, usually by an electric spark. The heat generated by the combustion reaction warms the bomb and the water around it. The bomb, its contents, and the water are defined as the system. Assessment of energy transfers as heat within the system shows that

$$q_r + q_{bomb} + q_{water} = 0$$

where q_r is the energy produced by the reaction, q_{bomb} is the energy involved in heating the calorimeter bomb, and q_{water} is the energy involved in heating the water in the calorimeter. Because the volume does not change in a constant volume calorimeter, energy transfer as work cannot occur. Therefore, the energy transferred as heat at constant volume (q_v) is the change in internal energy, ΔU.

■ **Calorimetry, ΔU, and ΔH** The two types of calorimetry (constant volume and constant pressure) highlight the differences between enthalpy and internal energy. The energy transferred as heat at constant pressure, q_p, is, by definition, ΔH, whereas the energy transferred as heat at constant volume, q_v, is ΔU.

Chemistry Now™

Sign in at **www.thomsonedu.com/login** and go to Chapter 5 Contents to see Screen 5.14 for a simulation and exercise exploring **reactions in a constant volume calorimeter** and for a tutorial on **calculating the enthalpy change for a reaction from a calorimetry experiment.**

Active Figure 5.15 Constant volume calorimeter. A combustible sample is burned in pure oxygen in a sealed metal container or "bomb." Energy released as heat warms the bomb and the water surrounding it. By measuring the increase in temperature, the energy evolved as heat in the reaction can be determined.

Chemistry Now™ Sign in at www.thomsonedu.com/login and go to the Chapter Contents menu to explore an interactive version of this figure accompanied by an exercise.

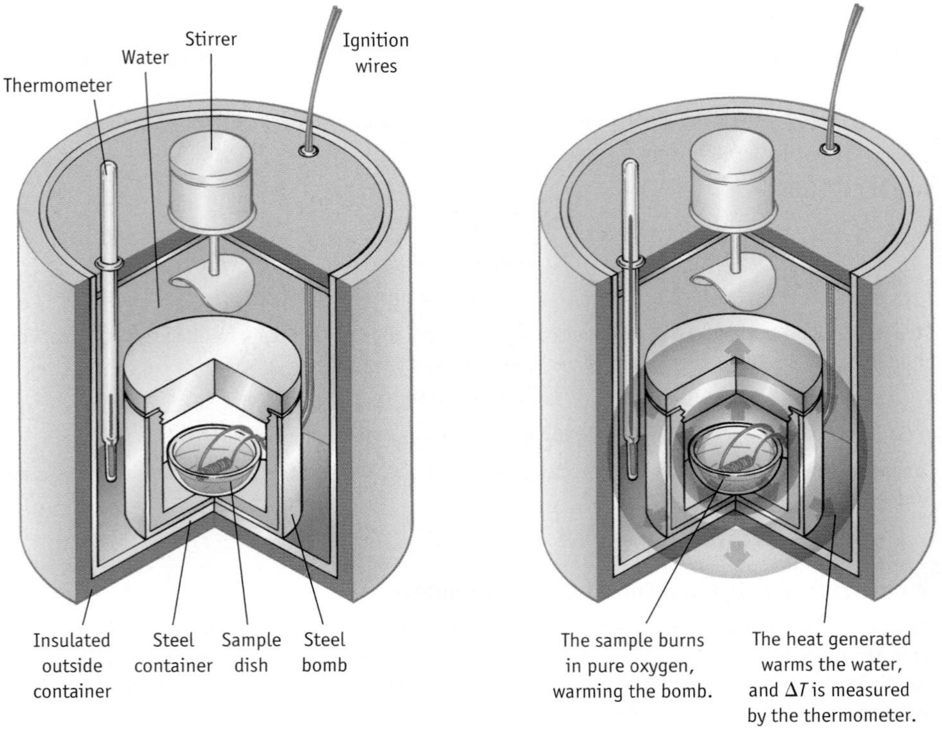

EXAMPLE 5.7 Constant Volume Calorimetry

Problem Octane, C_8H_{18}, a primary constituent of gasoline, burns in air:

$$C_8H_{18}(\ell) + 25/2\ O_2(g) \longrightarrow 8\ CO_2(g) + 9\ H_2O(\ell)$$

A 1.00-g sample of octane is burned in a constant volume calorimeter similar to that shown in Figure 5.15. The calorimeter is in an insulated container with 1.20 kg of water. The temperature of the water and the bomb rises from 25.00 °C (298.15 K) to 33.20 °C (306.35 K). The heat capacity of the bomb, C_{bomb}, is 837 J/K. (a) What is the heat of combustion per gram of octane? (b) What is the heat of combustion per mole of octane?

Strategy (a) The sum of all the energies transferred as heat in the system will be zero; that is, $q_r + q_{bomb} + q_{water} = 0$. The first term, q_r, is the unknown. The second and third terms in the equation can be calculated from the data given: q_{bomb} is calculated from the bomb's heat capacity and ΔT, and q_{water} is determined from the specific heat capacity, mass, and ΔT for water. (b) The value of q_r calculated in part (a) is the energy evolved in the combustion of 1.00 g of octane. Use this and the molar mass of octane (114.2 g/mol) to calculate the energy evolved as heat per mole of octane.

Solution

(a)

$$q_{water} = C_{water} \times m_{water} \times \Delta T$$

$$= (4.184\ J/g \cdot K)(1.20 \times 10^3\ g)(306.35\ K - 298.15\ K) = +41.2 \times 10^3\ J$$

$$q_{bomb} = (C_{bomb})\ (\Delta T) = (837\ J/K)(306.35\ K - 298.15\ K) = 6.86 \times 10^3\ J$$

$$q_r + q_{water} + q_{bomb} = 0$$

$$q_r + 41.2 \times 10^3\ J + 6.86 \times 10^3\ J = 0$$

$$q_r = -48.1 \times 10^3\ J\ (\text{or } -48.1\ kJ)$$

Heat of combustion per gram = −48.1 kJ

(b) Heat of combustion per mol of octane = (-48.1 kJ/g)(114.2 g/mol) = -5.49×10^3 kJ/mol

Comment Because the volume does not change, no energy transfer in the form of work occurs. The change of internal energy, ΔU, for the combustion of $C_8H_{18}(\ell)$ is -5.49×10^3 kJ/mol. Also note that C_{bomb} has no mass units. It is the heat required to warm the whole object by 1 kelvin.

EXERCISE 5.7 Constant Volume Calorimetry

A 1.00-g sample of ordinary table sugar (sucrose, $C_{12}H_{22}O_{11}$) is burned in a bomb calorimeter. The temperature of 1.50×10^3 g of water in the calorimeter rises from 25.00 °C to 27.32 °C. The heat capacity of the bomb is 837 J/K, and the specific heat capacity of the water is 4.20 J/g · K. Calculate (a) the heat evolved per gram of sucrose and (b) the heat evolved per mole of sucrose.

5.7 Enthalpy Calculations

Module 10

Enthalpy changes for an enormous number of chemical and physical processes are available on the World Wide Web and in reference books. These data have been collected by scientists over a number of years from many experiments and are used to calculate enthalpy changes for chemical processes. Now we want to discuss how to use such data.

Hess's Law

The enthalpy change can be measured by calorimetry for many, but not all, chemical processes. Consider, for example, the oxidation of carbon to form carbon monoxide.

$$\text{C(graphite)} + \tfrac{1}{2}\,\text{O}_2\text{(g)} \longrightarrow \text{CO(g)}$$

Even if a deficiency of oxygen is used, the primary product of the reaction of carbon and oxygen is CO_2. As soon as CO is formed, it reacts with O_2 to form CO_2. Because the reaction cannot be carried out in a way that allows CO to be the sole product, it is not possible to measure the change in enthalpy for this reaction by calorimetry.

The enthalpy change for the reaction forming CO(g) from C(s) and O_2(g) can be determined indirectly, however, from enthalpy changes for other reactions that can be measured. The calculation is based on **Hess's law**, which states that if a reaction is the sum of two or more other reactions, $\Delta_r H°$ for the overall process is the sum of the $\Delta_r H°$ values of those reactions.

The oxidation of C(s) to CO_2(g) can be viewed as occurring in two steps: first the oxidation of C(s) to CO(g) (Equation 1), and then the oxidation of CO(g) to CO_2(g) (Equation 2). Adding these two equations gives the equation for the oxidation of C(s) to CO_2(g) (Equation 3).

Equation 1:	$\text{C(graphite)} + \tfrac{1}{2}\,\text{O}_2\text{(g)} \longrightarrow \text{CO(g)}$	$\Delta_r H_1^\circ = ?$
Equation 2:	$\text{CO(g)} + \tfrac{1}{2}\,\text{O}_2\text{(g)} \longrightarrow \text{CO}_2\text{(g)}$	$\Delta_r H_2^\circ = -283.0$ kJ/mol-rxn
Equation 3:	$\text{C(graphite)} + \text{O}_2\text{(g)} \longrightarrow \text{CO}_2\text{(g)}$	$\Delta_r H_3^\circ = -393.5$ kJ/mol-rxn

Hess's law tells us that the enthalpy change for the overall reaction ($\Delta_r H_3^\circ$) will equal the sum of the enthalpy changes for reactions 1 and 2 ($\Delta_r H_1^\circ + \Delta_r H_2^\circ$). Both $\Delta_r H_2^\circ$ and $\Delta_r H_3^\circ$ can be measured, and these values are then used to calculate the enthalpy change for reaction 1.

$$\Delta_r H_3^\circ = \Delta_r H_1^\circ + \Delta_r H_2^\circ$$
$$-393.5 \text{ kJ/mol-rxn} = \Delta_r H_1^\circ + (-283.0 \text{ kJ/mol-rxn})$$
$$\Delta_r H_1^\circ = -110.5 \text{ kJ/mol-rxn}$$

Hess's law also applies to physical processes. The enthalpy change for the reaction of H_2(g) and O_2(g) to form 1 mol of H_2O vapor is different from the enthalpy change to form 1 mol of liquid H_2O. The difference is the negative of the *enthalpy of vaporization* of water, $\Delta_r H^\circ_2$ ($= -\Delta_{vap}H^\circ$) as shown in the following analysis

Equation 1:	$H_2(g) + \frac{1}{2}\, O_2(g) \longrightarrow H_2O(g)$	$\Delta_r H^\circ_1 = -241.8$ kJ/mol-rxn
Equation 2:	$H_2O(g) \longrightarrow H_2O(\ell)$	$\Delta_r H^\circ_2 = -44.0$ kJ/mol-rxn
Equation 3:	$H_2(g) + \frac{1}{2}\, O_2(g) \longrightarrow H_2O(\ell)$	$\Delta_r H^\circ_3 = -285.8$ kJ/mol-rxn

Energy Level Diagrams

When using Hess's law, it is often helpful to represent enthalpy data schematically in an energy level diagram. In such drawings, the various substances being studied—the reactants and products in a chemical reaction, for example—are placed on an arbitrary energy scale. The relative enthalpy of each substance is given by its position on the vertical axis, and numerical differences in enthalpy between them are shown by the vertical arrows. Such diagrams provide a visual perspective on the magnitude and direction of enthalpy changes and show how enthalpy changes of the substances are related.

Energy level diagrams that summarize the two examples of Hess's law discussed earlier are shown in Figure 5.16. In Figure 5.16a, the elements, C(s) and $O_2(g)$ are at the highest enthalpy. The reaction of carbon and oxygen to form $CO_2(g)$ lowers the enthalpy by 393.5 kJ. This can occur either in a single step, shown on the left in Figure 5.16a, or in two steps via initial formation of CO(g), as shown on the right. Similarly, in Figure 5.16b, the mixture of $H_2(g)$ and $O_2(g)$ is at the highest enthalpy. Both liquid and gaseous water have lower enthalpies, with the difference between the two being the enthalpy of vaporization.

Chemistry.Now™

Sign in at **www.thomsonedu.com/login** and go to Chapter 5 Contents to see Screen 5.15 for a simulation and exercise on **Hess's law.**

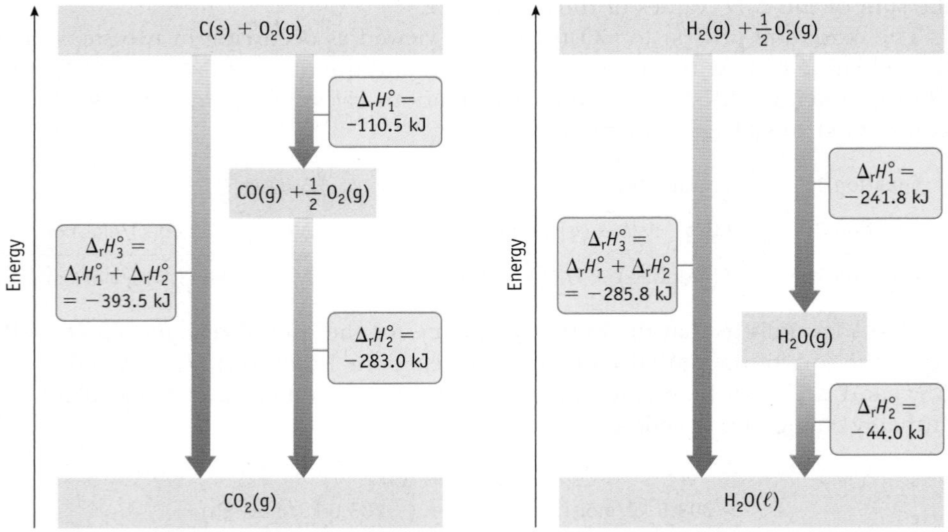

(a) The formation of CO_2 can occur in a single step or in a succession of steps. $\Delta_r H^\circ$ for the overall process is −393.5 kJ, no matter which path is followed.

(b) The formation of $H_2O(\ell)$ can occur in a single step or in a succession of steps. $\Delta_r H^\circ$ for the overall process is −285.8 kJ, no matter which path is followed.

Active Figure 5.16 Energy level diagrams. (a) Relating enthalpy changes in the formation of $CO_2(g)$. (b) Relating enthalpy changes in the formation of $H_2O(\ell)$. Enthalpy changes associated with changes between energy levels are given alongside the vertical arrows.

Chemistry.Now™ Sign in at www.thomsonedu.com/login and go to the Chapter Contents menu to explore an interactive version of this figure accompanied by an exercise.

EXAMPLE 5.8 Using Hess's Law

Problem Suppose you want to know the enthalpy change for the formation of methane, CH_4, from solid carbon (as graphite) and hydrogen gas:

$$C(s) + 2\ H_2(g) \longrightarrow CH_4(g) \qquad \Delta_r H° = ?$$

The enthalpy change for this reaction cannot be measured in the laboratory because the reaction is very slow. We can, however, measure enthalpy changes for the combustion of carbon, hydrogen, and methane.

Equation 1:	$C(s) + O_2(g) \longrightarrow CO_2(g)$	$\Delta_r H_1° = -393.5$ kJ/mol-rxn
Equation 2:	$H_2(g) + \frac{1}{2}\ O_2(g) \longrightarrow H_2O(\ell)$	$\Delta_r H_2° = -285.8$ kJ/mol-rxn
Equation 3:	$CH_4(g) + 2\ O_2(g) \longrightarrow CO_2(g) + 2\ H_2O(\ell)$	$\Delta_r H_3° = -890.3$ kJ/mol-rxn

Use this information to calculate $\Delta_r H°$ for the formation of methane from its elements.

Strategy The three reactions (1, 2, and 3), as they are written, cannot be added together to obtain the equation for the formation of CH_4 from its elements. Methane, CH_4, is a product in the reaction for which we wish to calculate $\Delta_r H°$, but it is a reactant in Equation 3. Water appears in two of these equations although it is not a component of the reaction forming CH_4 from carbon and hydrogen. To use Hess's law to solve this problem, we will first have to manipulate the equations and adjust the $\Delta_r H°$ values accordingly before adding equations together. Recall, from Section 5.5, that writing an equation in the reverse direction changes the sign of $\Delta_r H°$ and that doubling the amount of reactants and products doubles the value of $\Delta_r H°$. Adjustments to Equations 2 and 3 will produce new equations that, along with Equation 1, can be combined to give the desired net reaction.

Solution To have CH_4 appear as a product in the overall reaction, we reverse Equation 3, which changes the sign of $\Delta_r H°$.

Equation 3′: $CO_2(g) + 2\ H_2O(\ell) \longrightarrow CH_4(g) + 2\ O_2(g)$

$$\Delta_r H_3°{}' = -\Delta_r H_3° = +890.3 \text{ kJ/mol-rxn}$$

Next, we see that 2 mol of $H_2(g)$ is on the reactant side in our desired equation. Equation 2 is written for only 1 mol of $H_2(g)$ as a reactant. Therefore we multiply the stoichiometric coefficients in Equation 2 by 2 and multiply the value of $\Delta_r H°$ by 2.

Equation 2′: $2\ H_2(g) + O_2(g) \longrightarrow 2\ H_2O(\ell)$

$$\Delta_r H_2°{}' = 2\ \Delta_r H_2° = 2\ (-285.8 \text{ kJ/mol-rxn}) = -571.6 \text{ kJ/mol-rxn}$$

We now have three equations that, when added together, will give the targeted equation for the formation of methane from carbon and hydrogen. In this summation process, $O_2(g)$, $H_2O(\ell)$, and $CO_2(g)$ all cancel.

Equation 1:	$C(s) + O_2(g) \longrightarrow CO_2(g)$	$\Delta_r H_1° = -393.5$ kJ/mol-rxn
Equation 2′:	$2\ H_2(g) + O_2(g) \longrightarrow 2\ H_2O(\ell)$	$\Delta_r H_2°{}' = 2\ \Delta_r H_2° = -571.6$ kJ/mol-rxn
Equation 3′:	$CO_2(g) + 2\ H_2O(\ell) \longrightarrow CH_4(g) + 2\ O_2\ (g)$	$\Delta_r H_3°{}' = -\Delta_r H_3° = +890.3$ kJ/mol-rxn
Net Equation:	$C(s) + 2\ H_2(g) \longrightarrow CH_4(g)$	$\Delta_r H°_{net} = \Delta_r H_1° + 2\ \Delta_r H_2° + (-\Delta_r H_3°)$

$$\Delta_r H°_{net} = (-393.5 \text{ kJ/mol-rxn}) + (-571.6 \text{ kJ/mol-rxn}) + (+890.3 \text{ kJ/mol-rxn})$$
$$= -74.8 \text{ kJ/mol-rxn}$$

Thus, for the formation of 1 mol of $CH_4(g)$ from the elements, we find $\Delta_r H° = -74.8$ kJ/mol-rxn.

EXERCISE 5.8 Using Hess's Law

Use Hess's law to calculate the enthalpy change for the formation of $CS_2(\ell)$ from C(s) and S(s) [$C(s) + 2\ S(s) \longrightarrow CS_2(\ell)$] from the following enthalpy values.

$C(s) + O_2(g) \longrightarrow CO_2(g)$	$\Delta_r H° = -393.5$ kJ/mol-rxn
$S(s) + O_2(g) \longrightarrow SO_2(g)$	$\Delta_r H° = -295.8$ kJ/mol-rxn
$CS_2(\ell) + 3\ O_2(g) \longrightarrow CO_2(g) + 2\ SO_2(g)$	$\Delta_r H° = -1103.9$ kJ/mol-rxn

Problem Solving Tip 5.2 Using Hess's Law

How did we know how the three equations should be adjusted in Example 5.8? Here is a general strategy for solving this type of problem.

Step 1. Inspect the equation whose $\Delta_r H°$ you wish to calculate, identifying the reactants and products, and locate those substances in the equations available to be added. In Example 5.8, the reactants, C(s) and H_2(g), are reactants in Equations 1 and 2, and the product, CH_4(g), is a reactant in Equation 3. Equation 3 was reversed to get CH_4 on the product side.

Step 2. Get the correct amount of the substances on each side. In Example 5.8, only one adjustment was needed. There was 1 mol of H_2 on the left (reactant side) in Equation 2. We needed 2 mol of H_2 in the overall equation; this required doubling the quantities in Equation 2.

Step 3. Make sure other substances in the equations cancel when the equations are added. In Example 5.8, equal amounts of O_2 and H_2O appeared on the left and right sides in the three equations, and so they canceled when the equations were added together.

Standard Enthalpies of Formation

Calorimetry and the application of Hess's law have made available a great many $\Delta_r H°$ values for chemical reactions. Often, these values are assembled into tables. The table in Appendix L, for example, lists **standard molar enthalpies of formation,** $\boldsymbol{\Delta_f H°}$. *The standard molar enthalpy of formation is the enthalpy change for the formation of 1 mol of a compound directly from its component elements in their standard states.*

■ **$\Delta_f H°$ Values** Consult the National Institute for Standards and Technology website (webbook.nist.gov/chemistry) for an extensive compilation of enthalpies of formation.

Several examples of standard molar enthalpies of formation will be helpful to illustrate this definition.

$\Delta_f H°$ for NaCl(s): At 25 °C and a pressure of 1 bar, Na is a solid, and Cl_2 is a gas. The standard enthalpy of formation of NaCl(s) is defined as the enthalpy change that occurs when 1 mol of NaCl(s) is formed from 1 mol of Na(s) and ½ mol of Cl_2(g).

$$Na(s) + \tfrac{1}{2}\,Cl_2(g) \longrightarrow NaCl(s) \qquad \Delta_f H° = -411.12 \text{ kJ/mol}$$

$\Delta_f H°$ for NaCl(aq): The enthalpy of formation for an aqueous solution of a compound refers to the enthalpy change for the formation of a 1 mol/L solution of the compound starting with the elements making up the compound. It is thus the enthalpy of formation of the compound plus the enthalpy change that occurs when the substance dissolves in water.

$$Na(s) + \tfrac{1}{2}\,Cl_2(g) \longrightarrow NaCl(aq) \qquad \Delta_f H° = -407.27 \text{ kJ/mol}$$

■ **Units for Enthalpy of Formation** The units for values of $\Delta_f H°$ are usually given simply as kJ/mol where the denominator is really mol-rxn. However, because an enthalpy of formation is defined as the change in enthalpy for the formation of 1 mol of compound, it is understood that "per mol" means "per mol of compound," which, in this case, is the same thing as "per mol-rxn."

$\Delta_f H°$ for $C_2H_5OH(\ell)$: At 25 °C and 1 bar, the standard states of the elements are C(s, graphite), H_2(g), and O_2(g). The standard enthalpy of formation of $C_2H_5OH(\ell)$ is defined as the enthalpy change that occurs when 1 mol of $C_2H_5OH(\ell)$ is formed from 2 mol of C(s), 3 mol of H_2(g), and 1/2 mol of O_2(g).

$$2\,C(s) + 3\,H_2(g) + \tfrac{1}{2}\,O_2(g) \longrightarrow C_2H_5OH(\ell) \qquad \Delta_f H° = -277.0 \text{ kJ/mol}$$

Notice that the reaction defining the enthalpy of formation for liquid ethanol is not a reaction that a chemist can carry out in the laboratory. This illustrates an important point: *the enthalpy of formation of a compound does not necessarily correspond to a reaction that can be carried out.*

Appendix L lists values of $\Delta_f H°$ for some common substances, and a review of these values leads to some important observations.

- *The standard enthalpy of formation for an element in its standard state is zero.*
- Most $\Delta_f H°$ values are negative, indicating that formation of most compounds from the elements is exothermic. A very few values are positive, and these

represent compounds that are unstable with respect to decomposition to the elements. (One example is NO(g) with $\Delta_f H^\circ = +90.29$ kJ/mol.)

- Values of $\Delta_f H^\circ$ can often be used to compare the stabilities of related compounds. Consider the values of $\Delta_f H^\circ$ for the hydrogen halides. Hydrogen fluoride is the most stable of these compounds with respect to decomposition to the elements, whereas HI is the least stable (as indicated by $\Delta_f H^\circ$ of HF being the most negative value and that of HI being the most positive).

■ **$\Delta_f H^\circ$ Values of Hydrogen Halides**

Compound	$\Delta_f H^\circ$ (kJ/mol)
HF(g)	−273.3
HCl(g)	−92.31
HBr(g)	−35.29
HI(g)	+25.36

EXERCISE 5.9 Standard Enthalpies of Formation

Write equations for the reactions that define the standard enthalpy of formation of $FeCl_3$(s) and of solid sucrose (sugar, $C_{12}H_{22}O_{11}$).

Enthalpy Change for a Reaction

Using standard molar enthalpies of formation and Equation 5.6, it is possible to calculate the enthalpy change for a reaction under standard conditions.

$$\Delta_r H^\circ = \Sigma\Delta_f H^\circ(\text{products}) - \Sigma\Delta_f H^\circ(\text{reactants}) \qquad (5.6)$$

■ **Δ = Final − Initial** Equation 5.6 is another example of the convention that a change (Δ) is always calculated by subtracting the value for the initial state (the reactants) from the value for the final state (the products).

In this equation, the symbol Σ (the Greek capital letter sigma) means "take the sum." To find $\Delta_r H^\circ$, add up the molar enthalpies of formation of the products, each multiplied by its stoichiometric coefficient, and subtract from this the sum of the molar enthalpies of formation of the reactants, each multiplied by its stoichiometric coefficient. This equation is a logical consequence of the definition of $\Delta_f H^\circ$ and Hess's law (see *A Closer Look: Hess's Law and Equation 5.6*).

Suppose you want to know how much heat is required to decompose 1 mol of calcium carbonate (limestone) to calcium oxide (lime) and carbon dioxide under standard conditions:

$$CaCO_3(s) \longrightarrow CaO(s) + CO_2(g) \qquad \Delta_r H^\circ = ?$$

You would use the following enthalpies of formation (from Appendix L):

Compound	$\Delta_f H^\circ$ (kJ/mol)
$CaCO_3$(s)	−1207.6
CaO(s)	−635.1
CO_2(g)	−393.5

and then use Equation 5.6 to find the standard enthalpy change for the reaction, $\Delta_r H^\circ$.

$$\Delta_r H^\circ = \left[\left(\frac{1\ \cancel{\text{mol CaO}}}{1\ \text{mol-rxn}}\right)\left(\frac{-635.1\ \text{kJ}}{\cancel{\text{mol CaO}}}\right) + \left(\frac{1\ \cancel{\text{mol CO}_2}}{1\ \text{mol-rxn}}\right)\left(\frac{-393.5\ \text{kJ}}{1\ \cancel{\text{mol CO}_2}}\right)\right] - \left[\left(\frac{1\ \cancel{\text{mol CaCO}_3}}{1\ \text{mol-rxn}}\right)\left(\frac{-1207.6\ \text{kJ}}{1\ \cancel{\text{mol CaCO}_3}}\right)\right]$$

$$= +179.0\ \text{kJ/mol-rxn}$$

The decomposition of limestone to lime and CO_2 is endothermic. That is, energy (179.0 kJ) must be supplied to decompose 1 mol of $CaCO_3$(s) to CaO(s) and CO_2(g).

Chemistry Now™

Sign in at **www.thomsonedu.com/login** and go to Chapter 5 Contents to see Screen 5.16 for a tutorial on **calculating the standard enthalpy change for a reaction.**

A Closer Look — Hess's Law and Equation 5.6

Equation 5.6 is an application of Hess's law. To illustrate this, let us look further at the decomposition of calcium carbonate.

$$CaCO_3(s) \longrightarrow CaO(s) + CO_2(g) \qquad \Delta_r H° = ?$$

Because enthalpy is a state function, the change in enthalpy for this reaction is independent of the route from reactants to products. We can imagine an alternate route from reactant to products that involves first converting the reactant to elements in their standard states, then recombining these elements to give the reaction products. Notice that the enthalpy changes for these processes are the enthalpies of formation of the reactants and products in the equation above:

$CaCO_3(s) \longrightarrow Ca(s) + C(s) + 3/2\ O_2(g)$	$-\Delta_f H°[CaCO_3(s)] = \Delta_r H°_1$
$C(s) + O_2(g) \longrightarrow CO_2(g)$	$\Delta_f H°[CO_2(g)] = \Delta_r H°_2$
$Ca(s) + \frac{1}{2}\ O_2(g) \longrightarrow CaO(s)$	$\Delta_f H°[CaO(s)] = \Delta_r H°_3$
$CaCO_3(s) \longrightarrow CaO(s) + CO_2(g)$	$\Delta_r H°_{net}$

$$\Delta_r H°_{net} = \Delta_r H°_1 + \Delta_r H°_2 + \Delta_r H°_3$$

$$\Delta_r H° = \Delta_f H°[CaO(s)] + \Delta_f H°[CO_2(g)] - \Delta_f H°[CaCO_3(s)]$$

That is, the change in enthalpy for the reaction is equal to the enthalpies of formation of products (CO_2 and CaO) minus the enthalpy of formation of the reactant ($CaCO_3$), which is, of course, what one does when using Equation 5.6 for this calculation. The relationship among these enthalpy quantities is illustrated in the energy-level diagram.

Energy level diagram for the decomposition of $CaCO_3(s)$

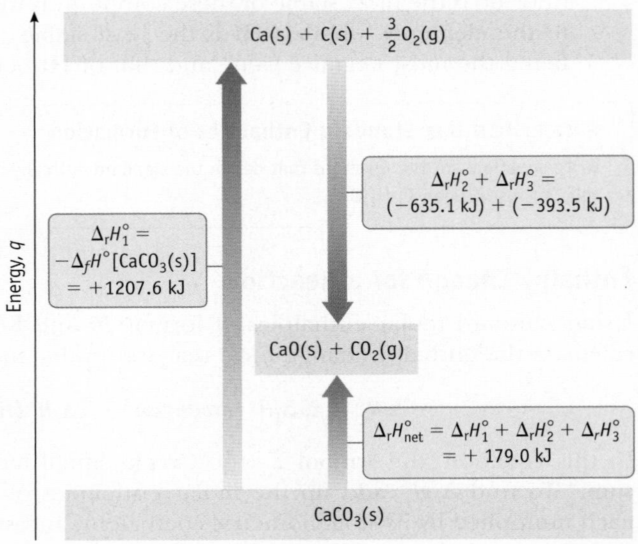

EXAMPLE 5.9 Using Enthalpies of Formation

Problem Nitroglycerin is a powerful explosive that forms four different gases when detonated:

$$2\ C_3H_5(NO_3)_3(\ell) \longrightarrow 3\ N_2(g) + \tfrac{1}{2}\ O_2(g) + 6\ CO_2(g) + 5\ H_2O(g)$$

Calculate the enthalpy change that occurs when 10.0 g of nitroglycerin is detonated. The standard enthalpy of formation of nitroglycerin, $\Delta_f H°$, is -364 kJ/mol. Use Appendix L to find other $\Delta_f H°$ values that are needed.

Strategy Use values of $\Delta_f H°$ for the reactants and products in Equation 5.6 to calculate the enthalpy change produced by one mole of reaction ($\Delta_r H°$). From Appendix L, $\Delta_f H°[CO_2(g)] = -393.5$ kJ/mol, $\Delta_f H°[H_2O(g)] = -241.8$ kJ/mol, and $\Delta_f H° = 0$ for $N_2(g)$ and $O_2(g)$. Determine the amount (mol) represented by 10.0 g of nitroglycerin; then use this value with $\Delta_r H°$ and the balanced chemical equation to obtain the answer.

Solution Using Equation 5.6, we find the enthalpy change for the explosion of 2 mol of nitroglycerin is

$$\Delta_r H° = \left(\frac{6\text{ mol } CO_2}{1\text{ mol-rxn}}\right)\Delta_f H°[CO_2(g)] + \left(\frac{5\text{ mol } H_2O}{1\text{ mol-rxn}}\right)\Delta_f H°[H_2O(g)] - \left(\frac{2\text{ mol } C_3H_5(NO_3)_3}{1\text{ mol-rxn}}\right)\Delta_f H°[C_3H_5(NO_3)_3(\ell)]$$

$$\Delta_r H° = \left(\frac{6\text{ mol } CO_2}{1\text{ mol-rxn}}\right)\left(\frac{-393.5\text{ kJ}}{1\text{ mol } CO_2}\right) + \left(\frac{5\text{ mol } H_2O}{1\text{ mol-rxn}}\right)\left(\frac{-241.8\text{ kJ}}{1\text{ mol } H_2O}\right) - \left(\frac{2\text{ mol } C_3H_5(NO_3)_3}{1\text{ mol-rxn}}\right)\left(\frac{-364\text{ kJ}}{1\text{ mol } C_3H_5(NO_3)_3}\right) = -2842\text{ kJ/mol-rxn}$$

The problem asks for the enthalpy change using 10.0 g of nitroglycerin. We next need to determine the amount of nitroglycerin in 10.0 g.

$$10.0 \text{ g nitroglycerin} \left(\frac{1 \text{ mol nitroglycerin}}{227.1 \text{ g nitroglycerin}}\right) = 0.0440 \text{ mol nitroglycerin}$$

The enthalpy change for the detonation of 0.0440 mol of nitroglycerin is

$$\Delta H° = 0.0440 \text{ mol nitroglycerin} \left(\frac{1 \text{ mol-rxn}}{2 \text{ mol nitroglycerin}}\right)\left(\frac{-2842 \text{ kJ}}{1 \text{ mol-rxn}}\right)$$

$$= -62.6 \text{ kJ}$$

Comment The large value of $\Delta H°$ is in accord with the fact that this reaction is highly energetic.

EXERCISE 5.10 Using Enthalpies of Formation

Calculate the standard enthalpy of combustion for benzene, C_6H_6.

$C_6H_6(\ell) + 15/2\ O_2(g) \longrightarrow 6\ CO_2(g) + 3\ H_2O(\ell) \qquad \Delta_rH° = ?$

$\Delta_fH°[C_6H_6(\ell)] = +49.0$ kJ/mol.

Other values needed can be found in Appendix L.

5.8 Product- or Reactant-Favored Reactions and Thermodynamics

At the beginning of this chapter, we noted that thermodynamics would provide answers to four questions. How much energy is evolved or required in physical changes and in chemical reactions, and the relationship of heat and work have been the primary topics of this chapter. The first two questions were addressed in this chapter, but there are two other important questions: How can we determine whether a reaction is product-favored or reactant-favored at equilibrium? And what determines whether a chemical reaction will occur spontaneously; that is, without outside intervention?

In Chapter 3, we learned that chemical reactions proceed toward equilibrium, and spontaneous changes occur in a way that allows a system to approach equilibrium. Reactions in which reactants are largely converted to products when equilibrium is reached are said to be *product-favored.* Reactions in which only a small amount of products are present at equilibrium are called *reactant-favored* (◀ page 121).

Let us look back at the many chemical reactions that we have seen. For example, all combustion reactions are exothermic, and the oxidation of iron (Figure 5.17) is clearly exothermic.

$$4\ Fe(s) + 3\ O_2(g) \longrightarrow 2\ Fe_2O_3(s)$$

$$\Delta_rH° = 2\ \Delta_fH°[Fe_2O_3(s)] = \left(\frac{2 \text{ mol } Fe_2O_3}{1 \text{ mol-rxn}}\right)\left(\frac{-825.5 \text{ kJ}}{1 \text{ mol } Fe_2O_3}\right) = -1651.0 \text{ kJ/mol-rxn}$$

The reaction has a negative value for $\Delta_rH°$, and it is also spontaneous and product-favored.

Conversely, the decomposition of calcium carbonate is endothermic.

$$CaCO_3(s) \longrightarrow CaO(s) + CO_2(g) \qquad \Delta_rH° = +179.0 \text{ kJ/mol-rxn}$$

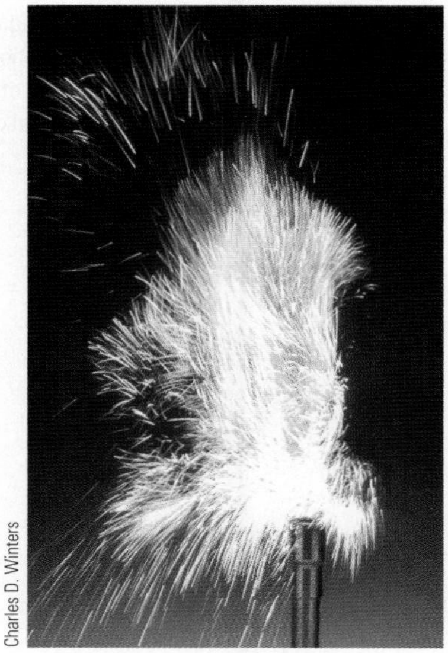

FIGURE 5.17 The product-favored oxidation of iron. Iron powder, sprayed into a bunsen burner flame, is rapidly oxidized. The reaction is exothermic and is product-favored.

Case Study — The Fuel Controversy: Alcohol and Gasoline

It is clear that supplies of fossil fuels are declining, and their price is increasing, just as the nations of the earth have ever greater energy needs. We will have more to say about this in the Interchapter (Energy) that follows. Here, however, let's analyze the debate about replacing gasoline with ethanol (C_2H_5OH).

As Matthew Wald says in the article "Is Ethanol in for the Long Haul?" (*Scientific American*, January 2007), "The U.S. has gone on an ethanol binge." In 2005, the U.S. Congress passed an energy bill stating that ethanol production should be 7.5 billion gallons a year by 2012, up from about 5 billion gallons a year presently. The goal is to at least partially replace gasoline with ethanol.

Is the goal of replacing gasoline with ethanol reasonable? This is a lofty goal, given that present gasoline consumption in the U.S. is about 140 billion gallons annually. Again, according to Matthew Wald, "Even if 100 percent of the U.S. corn supply was distilled into ethanol, it would supply only a small fraction of the fuel consumed by the nation's vehicles." Wald's thesis in his article, which is supported by numerous scientific studies, is that if ethanol is to be pursued as an alternative to gasoline, more emphasis should be placed on deriving ethanol from sources other than corn, such as cellulose from cornstalks and various grasses.

© Stephen Lunetta Photography, 2007

Ethanol available at a service station. E85 fuel is a blend of 85% ethanol and 15% gasoline. Be aware that you can only use E85 in vehicles designed for the fuel. In an ordinary vehicle, the ethanol leads to deterioration of seals in the engine and fuel system.

Beyond this, there are other problems associated with ethanol. One is that it cannot be distributed through a pipeline system as gasoline can. Any water in the pipeline is miscible with ethanol, which causes the fuel value to decline. Instead, ethanol must be trucked to service stations.

Finally, E85 fuel—a blend of 85% ethanol and 15% gasoline—cannot be used in most current vehicles because relatively few vehicles as yet have engines designed for fuels with a high ethanol content (so-called "flexible fuel" engines). The number of these vehicles would need to be increased in order for E85 to have a significant effect on our gasoline usage.

For more information, see the references in Wald's *Scientific American* article.

Questions:

For the purposes of this analysis, let us use octane (C_8H_{18}) as a substitute for the complex mixture of hydrocarbons in gasoline. Data you will need for this question (in addition to Appendix L) are:

$\Delta_f H°\ [C_8H_{18}(\ell)] = -250.1$ kJ/mol
Density of ethanol = 0.785 g/mL
Density of octane = 0.699 g/mL

1. *Calculate $\Delta_r H°$ for the combustion of ethanol and octane, and compare the values per mol and per gram. Which provides more energy per gram?*
2. *Compare the energy produced per liter of the two fuels. Which produces more energy for a given volume (something useful to know when filling your gas tank)?*
3. *What mass of CO_2, a greenhouse gas, is produced per liter of fuel (assuming complete combustion)?*
4. *Now compare the fuels on an energy-equivalent basis. What volume of ethanol would have to be burned to get the same energy as 1.00 L of octane? When you burn enough ethanol to have the same energy as a liter of octane, which fuel produces more CO_2?*
5. *On the basis of this analysis and assuming the same price per liter, which fuel will propel your car further? Which will produce less greenhouse gas?*

Answers to these questions are in Appendix Q.

The decomposition of $CaCO_3$ proceeds spontaneously to an equilibrium that favors the reactants; that is, it is reactant-favored.

Are all exothermic reactions product-favored and all endothermic reactions reactant-favored? From these examples, we might formulate that idea as a hypothesis that can be tested by experiment and by examination of other examples. We would find that *in most cases, product-favored reactions have negative values of $\Delta_r H°$, and reactant-favored reactions have positive values of $\Delta_r H°$.* But this is not *always* true; there are exceptions.

Clearly, a further discussion of thermodynamics must be tied to the concept of equilibrium. This relationship, and the complete discussion of the third and fourth questions, will be presented in Chapter 19.

■ **Reactant-Favored or Product-Favored?** In most—but not all—cases exothermic reactions are product-favored at equilibrium and endothermic reactions are reactant-favored at equilibrium.

Chemistry Now™

Sign in at **www.thomsonedu.com/login** and go to Chapter 5 Contents to see Screen 5.17 Product-Favored Systems, for an exercise on **the reaction when a Gummi Bear is placed in molten potassium chlorate.**

Chapter Goals Revisited

Now that you have studied this chapter, you should ask whether you have met the chapter goals. In particular, you should be able to:

Assess the transfer of energy as heat associated with changes in temperature and changes of state

a. Describe various forms of energy and the nature of energy transfers as heat (Section 5.1).

b. Use the most common energy unit, the joule, and convert between other energy units and joules (Section 5.1). **Study Question(s) assignable in OWL: 5.**

c. Recognize and use the language of thermodynamics: the system and its surroundings; exothermic and endothermic reactions (Section 5.1). **Study Question(s) assignable in OWL: 61, 92.**

d. Use specific heat capacity in calculations of energy transfer as heat and of temperature changes (Section 5.2). **Study Question(s) assignable in OWL: 8, 10, 12, 13, 16, 18, 83.**

e. Understand the sign conventions in thermodynamics.

f. Use enthalpy (heat) of fusion and enthalpy (heat) of vaporization to find the quantity of energy transferred as heat that is involved in changes of state (Section 5.3). **Study Question(s) assignable in OWL: 20, 22, 23, 24, 28, 68, 70, 71, 88, 93, 97.**

Understand and apply the first law of thermodynamics

a. Understand the basis of the first law of thermodynamics (Section 5.4).

b. Recognize how energy transferred as heat and work done on or by a system contribute to changes in the internal energy of a system (Section 5.4).

Define and understand state functions (enthalpy, internal energy)

a. Recognize state functions whose values are determined only by the state of the system and not by the pathway by which that state was achieved (Section 5.4).

Learn how energy changes are measured

a. Recognize that when a process is carried out under constant pressure conditions, the energy transferred as heat is the enthalpy change, ΔH (Section 5.5). **Study Question(s) assignable in OWL: 28, 29, 30, 52, 54.**

b. Describe how to measure the quantity of energy transferred as heat in a reaction by calorimetry (Section 5.6). **Study Question(s) assignable in OWL: 32, 33, 34, 36, 38, 40, 42.**

Calculate the energy evolved or required for physical changes and chemical reactions using tables of thermodynamic data.

a. Apply Hess's law to find the enthalpy change for a reaction (Section 5.7). **Study Question(s) assignable in OWL: 44, 73, 74, 79; Go Chemistry Module 10.**

b. Know how to draw and interpret energy level diagrams (Section 5.7).

c. Use standard molar enthalpies of formation, $\Delta_f H^\circ$, to calculate the enthalpy change for a reaction $\Delta_r H^\circ$ (Section 5.7). **Study Question(s) assignable in OWL: 49, 53, 58.**

Chemistry•Now™ Sign in at **www.thomsonedu.com/login** to:

- Assess your understanding with Study Questions in OWL keyed to each goal in the Goals and Homework menu for this chapter
- For quick review, download Go Chemistry mini-lecture flashcard modules (or purchase them at **www.ichapters.com**)
- Check your readiness for an exam by taking the Pre-Test and exploring the modules recommended in your Personalized Study plan.

Access **How Do I Solve It?** tutorials on how to approach problem solving using concepts in this chapter.

KEY EQUATIONS

Equation 5.1 (page 215) The energy transferred as heat when the temperature of a substance changes. Calculated from the specific heat capacity (C), mass (m), and change in temperature (ΔT).

$$q(\text{J}) = C(\text{J/g} \cdot \text{K}) \times m(\text{g}) \times \Delta T(\text{K})$$

Equation 5.2 (page 215) Temperature changes are always calculated as final temperature minus initial temperature.

$$\Delta T = T_{\text{final}} - T_{\text{initial}}$$

Equation 5.3 (page 217) If no energy is transferred between a system and its surroundings and if energy is transferred within the system only as heat, the sum of the thermal energy changes within the system equals zero.

$$q_1 + q_2 + q_3 + \ldots = 0$$

Equation 5.4 (page 223) The first law of thermodynamics: The change in internal energy (ΔU) in a system is the sum of the energy transferred as heat (q) and the energy transferred as work (w).

$$\Delta U = q + w$$

Equation 5.5 (page 224) Work (w) at constant pressure is the product of pressure (P) and change in volume (ΔV)

$$w = -P(\Delta V)$$

Equation 5.6 (page 237) This equation is used to calculate the standard enthalpy change of a reaction ($\Delta_r H°$) when the enthalpies of formation ($\Delta_f H°$) of all of the reactants and products are known.

$$\Delta_r H° = \Sigma \Delta_f H°(\text{products}) - \Sigma \Delta_f H°(\text{reactants})$$

STUDY QUESTIONS

Online homework for this chapter may be assigned in OWL.

▲ denotes challenging questions.

■ denotes questions assignable in OWL.

Blue-numbered questions have answers in Appendix O and fully-worked solutions in the *Student Solutions Manual.*

Practicing Skills

Energy

(See Section 5.1 and ChemistryNow Screen 5.2.)

1. The flashlight in the photo does not use batteries. Instead, you move a lever, which turns a geared mechanism and results finally in light from the bulb. What type of energy is used to move the lever? What type or types of energy are produced?

Charles D. Winters

A hand-operated flashlight.

2. A solar panel is pictured in the photo. When light shines on the panel, it generates an electric current that is used by a small electric motor to propel the car. What types of energy are involved in this setup?

Charles D. Winters

A solar panel operates a toy car.

Energy Units

(See Exercise 5.1 and ChemistryNow Screen 5.5.)

3. You are on a diet that calls for eating no more than 1200 Cal/day. What is this energy in joules?

4. A 2-in. piece of chocolate cake with frosting provides 1670 kJ of energy. What is this in dietary Calories (Cal)?

5. ■ One food product has an energy content of 170 kcal per serving, and another has 280 kJ per serving. Which food provides the greater energy per serving?

6. Which provides the greater energy per serving, a raw apple or a raw apricot? Go to the USDA Nutrient Database on the World Wide Web for the information: http://www.ars.usda.gov/main/site_main.htm?modecode=12354500. Report the energy content of the fruit in kcal and kJ.

Specific Heat Capacity

(See Examples 5.1 and 5.2 and ChemistryNow Screens 5.6–5.10.)

7. The molar heat capacity of mercury is 28.1 J/mol · K. What is the specific heat capacity of this metal in J/g · K?

8. ■ The specific heat capacity of benzene (C_6H_6) is 1.74 J/g · K. What is its molar heat capacity (in J/mol · K)?

9. The specific heat capacity of copper is 0.385 J/g · K. How much energy is required to heat 168 g of copper from −12.2 °C to +25.6 °C?

10. ■ How much energy is required to raise the temperature of 50.00 mL of water from 25.52 °C to 28.75 °C? (The density of water at this temperature is 0.997 g/mL.)

11. The initial temperature of a 344-g sample of iron is 18.2 °C. If the sample absorbs 2.25 kJ of energy as heat, what is its final temperature?

12. ■ After absorbing 1.850 kJ of energy as heat, the temperature of a 0.500-kg block of copper is 37 °C. What was its initial temperature?

13. ■ A 45.5-g sample of copper at 99.8 °C is dropped into a beaker containing 152 g of water at 18.5 °C. What is the final temperature when thermal equilibrium is reached?

14. A 182-g sample of gold at some temperature is added to 22.1 g of water. The initial water temperature is 25.0 °C, and the final temperature is 27.5 °C. If the specific heat capacity of gold is 0.128 J/g · K, what was the initial temperature of the gold?

15. One beaker contains 156 g of water at 22 °C, and a second beaker contains 85.2 g of water at 95 °C. The water in the two beakers is mixed. What is the final water temperature?

16. ■ When 108 g of water at a temperature of 22.5 °C is mixed with 65.1 g of water at an unknown temperature, the final temperature of the resulting mixture is 47.9 °C. What was the initial temperature of the second sample of water?

17. A 13.8-g piece of zinc was heated to 98.8 °C in boiling water and then dropped into a beaker containing 45.0 g of water at 25.0 °C. When the water and metal come to thermal equilibrium, the temperature is 27.1 °C. What is the specific heat capacity of zinc?

18. ■ A 237-g piece of molybdenum, initially at 100.0 °C, is dropped into 244 g of water at 10.0 °C. When the system comes to thermal equilibrium, the temperature is 15.3 °C. What is the specific heat capacity of molybdenum?

Changes of State

(See Examples 5.3 and 5.4 and ChemistryNow Screen 5.8.)

19. How much energy is evolved when 1.0 L of water at 0 °C solidifies to ice? (The heat of fusion of water is 333 J/g.)

20. ■ The energy required to melt 1.00 g of ice at 0 °C is 333 J. If one ice cube has a mass of 62.0 g and a tray contains 16 ice cubes, what quantity of energy is required to melt a tray of ice cubes to form liquid water at 0 °C?

21. ■ How much energy is required to vaporize 125 g of benzene, C_6H_6, at its boiling point, 80.1 °C? (The heat of vaporization of benzene is 30.8 kJ/mol.)

22. ■ Chloromethane, CH_3Cl, arises from microbial fermentation and is found throughout the environment. It is also produced industrially and is used in the manufacture of various chemicals and has been used as a topical anesthetic. How much energy is required to convert 92.5 g of liquid to a vapor at its boiling point, −24.09 °C? (The heat of vaporization of CH_3Cl is 21.40 kJ/mol.)

23. The freezing point of mercury is −38.8 °C. What quantity of energy, in joules, is released to the surroundings if 1.00 mL of mercury is cooled from 23.0 °C to −38.8 °C and then frozen to a solid? (The density of liquid mercury is 13.6 g/cm^3. Its specific heat capacity is 0.140 J/g · K and its heat of fusion is 11.4 J/g.)

24. ■ What quantity of energy, in joules, is required to raise the temperature of 454 g of tin from room temperature, 25.0 °C, to its melting point, 231.9 °C, and then melt the tin at that temperature? (The specific heat capacity of tin is 0.227 J/g · K, and the heat of fusion of this metal is 59.2 J/g.)

25. Ethanol, C_2H_5OH, boils at 78.29 °C. How much energy, in joules, is required to raise the temperature of 1.00 kg of ethanol from 20.0 °C to the boiling point and then to change the liquid to vapor at that temperature? (The specific heat capacity of liquid ethanol is 2.44 J/g · K, and its enthalpy of vaporization is 855 J/g.)

26. ■ A 25.0-mL sample of benzene at 19.9 °C was cooled to its melting point, 5.5 °C, and then frozen. How much energy as heat was given off in this process? (The density of benzene is 0.80 g/mL; its specific heat capacity is 1.74 J/g · K, and its heat of fusion is 127 J/g.)

Enthalpy Changes

(See Example 5.5 and ChemistryNow Screens 5.12 and 5.13.)

27. Nitrogen monoxide, a gas recently found to be involved in a wide range of biological processes, reacts with oxygen to give brown NO_2 gas.

$$2\ NO(g) + O_2(g) \rightarrow 2\ NO_2(g) \qquad \Delta_r H° = -114.1\ kJ/mol\text{-}rxn$$

Is this reaction endothermic or exothermic? What is the enthalpy change if 1.25 g of NO is converted completely to NO_2?

28. ■ Calcium carbide, CaC_2, is manufactured by the reaction of CaO with carbon at a high temperature. (Calcium carbide is then used to make acetylene.)

$$CaO(s) + 3\ C(s) \rightarrow CaC_2(s) + CO(g) \qquad \Delta_r H° = +464.8\ kJ/mol\text{-}rxn$$

Is this reaction endothermic or exothermic? What is the enthalpy change if 10.0 g of CaO is allowed to react with an excess of carbon?

29. ■ Isooctane (2,2,4-trimethylpentane), one of the many hydrocarbons that make up gasoline, burns in air to give water and carbon dioxide.

$$2\ C_8H_{18}(\ell) + 25\ O_2(g) \rightarrow 16\ CO_2(g) + 18\ H_2O(\ell) \qquad \Delta_r H° = -10{,}922\ kJ/mol\text{-}rxn$$

What is the enthalpy change if you burn 1.00 L of isooctane (density = 0.69 g/mL)?

30. ■ Acetic acid, CH_3CO_2H, is made industrially by the reaction of methanol and carbon monoxide.

$$CH_3OH(\ell) + CO(g) \rightarrow CH_3CO_2H(\ell) \qquad \Delta_r H° = -355.9\ kJ/mol\text{-}rxn$$

If you produce 1.00 L of acetic acid (d = 1.044 g/mL) by this reaction, how much energy as heat is evolved?

Calorimetry

(See Examples 5.6 and 5.7 and ChemistryNow Screens 5.8, 5.9, and 5.14.)

31. Assume you mix 100.0 mL of 0.200 M CsOH with 50.0 mL of 0.400 M HCl in a coffee-cup calorimeter. The following reaction occurs:

$$CsOH(aq) + HCl(aq) \rightarrow CsCl(aq) + H_2O(\ell)$$

The temperature of both solutions before mixing was 22.50 °C, and it rises to 24.28 °C after the acid–base reaction. What is the enthalpy change for the reaction per mole of CsOH? Assume the densities of the solutions are all 1.00 g/mL and the specific heat capacities of the solutions are 4.2 J/g · K.

32. ■ You mix 125 mL of 0.250 M CsOH with 50.0 mL of 0.625 M HF in a coffee-cup calorimeter, and the temperature of both solutions rises from 21.50 °C before mixing to 24.40 °C after the reaction.

$$CsOH(aq) + HF(aq) \rightarrow CsF(aq) + H_2O(\ell)$$

What is the enthalpy of reaction per mole of CsOH? Assume the densities of the solutions are all 1.00 g/mL and the specific heats of the solutions are 4.2 J/g · K.

33. ■ A piece of titanium metal with a mass of 20.8 g is heated in boiling water to 99.5 °C and then dropped into a coffee-cup calorimeter containing 75.0 g of water at 21.7 °C. When thermal equilibrium is reached, the final temperature is 24.3 °C. Calculate the specific heat capacity of titanium.

34. ■ A piece of chromium metal with a mass of 24.26 g is heated in boiling water to 98.3 °C and then dropped into a coffee-cup calorimeter containing 82.3 g of water at 23.3 °C. When thermal equilibrium is reached, the final temperature is 25.6 °C. Calculate the specific heat capacity of chromium.

35. Adding 5.44 g of $NH_4NO_3(s)$ to 150.0 g of water in a coffee-cup calorimeter (with stirring to dissolve the salt) resulted in a decrease in temperature from 18.6 °C to 16.2 °C. Calculate the enthalpy change for dissolving $NH_4NO_3(s)$ in water, in kJ/mol. Assume that the solution (whose mass is 155.4 g) has a specific heat capacity of 4.2 J/g · K. (Cold packs take advantage of the fact that dissolving ammonium nitrate in water is an endothermic process.)

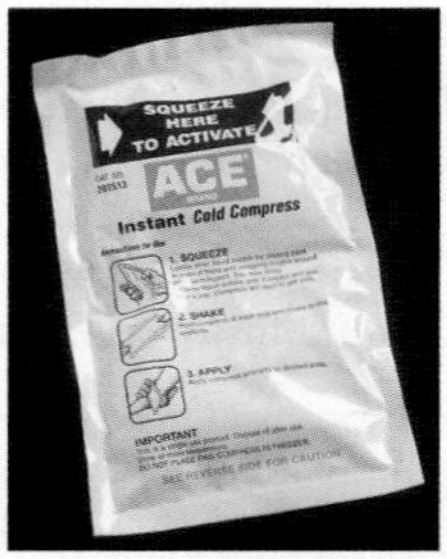

Charles D. Winters

A cold pack uses the endothermic enthalpy of solution of ammonium nitrate.

36. ■ You should use care when dissolving H_2SO_4 in water because the process is highly exothermic. To measure the enthalpy change, 5.2 g $H_2SO_4(\ell)$ was added (with stirring) to 135 g of water in a coffee-cup calorimeter. This resulted in an increase in temperature from 20.2 °C to 28.8 °C. Calculate the enthalpy change for the process $H_2SO_4(\ell) \rightarrow H_2SO_4(aq)$, in kJ/mol.

37. Sulfur (2.56 g) is burned in a constant volume calorimeter with excess $O_2(g)$. The temperature increases from 21.25 °C to 26.72 °C. The bomb has a heat capacity of 923 J/K, and the calorimeter contains 815 g of water. Calculate ΔU per mole of SO_2 formed, for the reaction

$$S_8(s) + 8\ O_2(g) \longrightarrow 8\ SO_2(g)$$

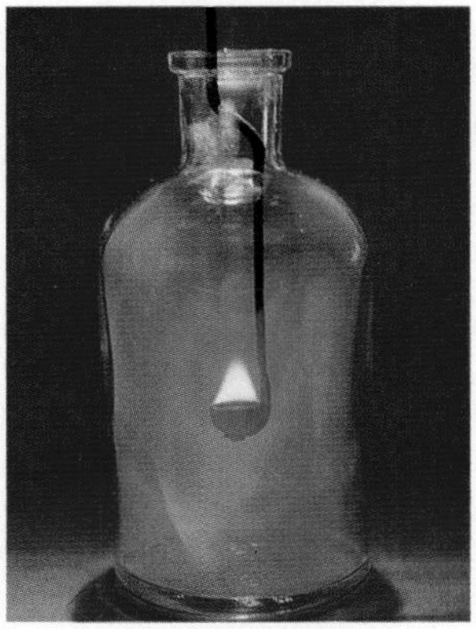

Charles D. Winters

Sulfur burns in oxygen with a bright blue flame to give $SO_2(g)$.

38. ■ Suppose you burn 0.300 g of C(graphite) in an excess of $O_2(g)$ in a constant volume calorimeter to give $CO_2(g)$.

$$C(graphite) + O_2(g) \longrightarrow CO_2(g)$$

The temperature of the calorimeter, which contains 775 g of water, increases from 25.00 °C to 27.38 °C. The heat capacity of the bomb is 893 J/K. Calculate ΔU per mole of carbon.

39. Suppose you burn 1.500 g of benzoic acid, $C_6H_5CO_2H$, in a constant volume calorimeter and find that the temperature increases from 22.50 °C to 31.69 °C. The calorimeter contains 775 g of water, and the bomb has a heat capacity of 893 J/K. Calculate ΔU per mole of benzoic acid.

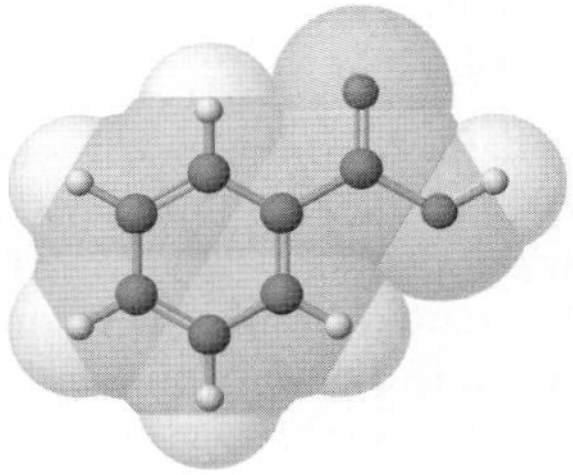

Benzoic acid, $C_6H_5CO_2H$, occurs naturally in many berries. Its heat of combustion is well known, so it is used as a standard to calibrate calorimeters.

40. ■ A 0.692-g sample of glucose, $C_6H_{12}O_6$, is burned in a constant volume calorimeter. The temperature rises from 21.70 °C to 25.22 °C. The calorimeter contains 575 g of water, and the bomb has a heat capacity of 650 J/K. What is ΔU per mole of glucose?

41. An "ice calorimeter" can be used to determine the specific heat capacity of a metal. A piece of hot metal is dropped onto a weighed quantity of ice. The energy transferred from the metal to the ice can be determined from the amount of ice melted. Suppose you heat a 50.0-g piece of silver to 99.8 °C and then drop it onto ice. When the metal's temperature has dropped to 0.0 °C, it is found that 3.54 g of ice has melted. What is the specific heat capacity of silver?

42. ■ A 9.36-g piece of platinum is heated to 98.6 °C in a boiling water bath and then dropped onto ice. (See Study Question 41.) When the metal's temperature has dropped to 0.0 °C, it is found that 0.37 g of ice has melted. What is the specific heat capacity of platinum?

Hess's Law

(See Example 5.8 and ChemistryNow Screen 5.15.)

43. The enthalpy changes for the following reactions can be measured:

$$CH_4(g) + 2\ O_2(g) \longrightarrow CO_2(g) + 2\ H_2O(g)$$
$$\Delta_rH° = -802.4\ kJ/mol\text{-}rxn$$

$$CH_3OH(g) + \tfrac{3}{2}\ O_2(g) \longrightarrow CO_2(g) + 2\ H_2O(g)$$
$$\Delta_rH° = -676\ kJ/mol\text{-}rxn$$

(a) Use these values and Hess's law to determine the enthalpy change for the reaction

$$CH_4(g) + \tfrac{1}{2}\ O_2(g) \longrightarrow CH_3OH(g)$$

(b) Draw an energy-level diagram that shows the relationship between the energy quantities involved in this problem.

44. ■ The enthalpy changes of the following reactions can be measured:

$$C_2H_4(g) + 3\ O_2(g) \longrightarrow 2\ CO_2(g) + 2\ H_2O(\ell)$$
$$\Delta_rH° = -1411.1\ kJ/mol\text{-}rxn$$

$$C_2H_5OH(\ell) + 3\ O_2(g) \longrightarrow 2\ CO_2(g) + 3\ H_2O(\ell)$$
$$\Delta_rH° = -1367.5\ kJ/mol\text{-}rxn$$

(a) ■ Use these values and Hess's law to determine the enthalpy change for the reaction

$$C_2H_4(g) + H_2O(\ell) \longrightarrow C_2H_5OH(\ell)$$

(b) Draw an energy-level diagram that shows the relationship between the energy quantities involved in this problem.

45. Enthalpy changes for the following reactions can be determined experimentally:

$N_2(g) + 3\ H_2(g) \longrightarrow 2\ NH_3(g)$ $\quad \Delta_rH° = -91.8$ kJ/mol-rxn

$4\ NH_3(g) + 5\ O_2(g) \longrightarrow 4\ NO(g) + 6\ H_2O(g)$ $\quad \Delta_rH° = -906.2$ kJ/mol-rxn

$H_2(g) + \frac{1}{2}\ O_2(g) \longrightarrow H_2O(g)$ $\quad \Delta_rH° = -241.8$ kJ/mol-rxn

Use these values to determine the enthalpy change for the formation of NO(g) from the elements (an enthalpy change that cannot be measured directly because the reaction is reactant-favored).

$$\frac{1}{2}\ N_2(g) + \frac{1}{2}\ O_2(g) \longrightarrow NO(g) \qquad \Delta_rH° = ?$$

46. You wish to know the enthalpy change for the formation of liquid PCl_3 from the elements.

$$P_4(s) + 6\ Cl_2(g) \longrightarrow 4\ PCl_3(\ell) \qquad \Delta_rH° = ?$$

The enthalpy change for the formation of PCl_5 from the elements can be determined experimentally, as can the enthalpy change for the reaction of $PCl_3(\ell)$ with more chlorine to give $PCl_5(s)$:

$P_4(s) + 10\ Cl_2(g) \longrightarrow 4\ PCl_5(s)$ $\quad \Delta_rH° = -1774.0$ kJ/mol-rxn

$PCl_3(\ell) + Cl_2(g) \longrightarrow PCl_5(s)$ $\quad \Delta_rH° = -123.8$ kJ/mol-rxn

Use these data to calculate the enthalpy change for the formation of 1.00 mol of $PCl_3(\ell)$ from phosphorus and chlorine.

Standard Enthalpies of Formation

(See Example 5.9 and ChemistryNow Screen 5.16.)

47. Write a balanced chemical equation for the formation of $CH_3OH(\ell)$ from the elements in their standard states. Find the value for $\Delta_fH°$ for $CH_3OH(\ell)$ in Appendix L.

48. Write a balanced chemical equation for the formation of $CaCO_3(s)$ from the elements in their standard states. Find the value for $\Delta_fH°$ for $CaCO_3(s)$ in Appendix L.

49. (a) Write a balanced chemical equation for the formation of 1 mol of $Cr_2O_3(s)$ from Cr and O_2 in their standard states. Find the value for $\Delta_fH°$ for $Cr_2O_3(s)$ in Appendix L.
(b) ■ What is the standard enthalpy change if 2.4 g of chromium is oxidized to $Cr_2O_3(s)$?

50. (a) Write a balanced chemical equation for the formation of 1 mol of MgO(s) from the elements in their standard states. Find the value for $\Delta_fH°$ for MgO(s) in Appendix L.
(b) What is the standard enthalpy change for the reaction of 2.5 mol of Mg with oxygen?

51. Use standard enthalpies of formation in Appendix L to calculate enthalpy changes for the following:
(a) 1.0 g of white phosphorus burns, forming $P_4O_{10}(s)$
(b) 0.20 mol of NO(g) decomposes to $N_2(g)$ and $O_2(g)$
(c) 2.40 g of NaCl(s) is formed from Na(s) and excess $Cl_2(g)$
(d) 250 g of iron is oxidized with oxygen to $Fe_2O_3(s)$

52. ■ Use standard enthalpies of formation in Appendix L to calculate enthalpy changes for the following:
(a) 0.054 g of sulfur burns, forming $SO_2(g)$
(b) 0.20 mol of HgO(s) decomposes to $Hg(\ell)$ and $O_2(g)$
(c) 2.40 g of $NH_3(g)$ is formed from $N_2(g)$ and excess $H_2(g)$
(d) 1.05×10^{-2} mol of carbon is oxidized to $CO_2(g)$

53. The first step in the production of nitric acid from ammonia involves the oxidation of NH_3.

$$4\ NH_3(g) + 5\ O_2(g) \longrightarrow 4\ NO(g) + 6\ H_2O(g)$$

(a) Use standard enthalpies of formation to calculate the standard enthalpy change for this reaction.
(b) ■ How much energy as heat is evolved or absorbed in the oxidation of 10.0 g of NH_3?

54. ■ The Romans used calcium oxide, CaO, to produce a strong mortar to build stone structures. The CaO was mixed with water to give $Ca(OH)_2$, which reacted slowly with CO_2 in the air to give $CaCO_3$.

$$Ca(OH)_2(s) + CO_2(g) \longrightarrow CaCO_3(s) + H_2O(g)$$

(a) Calculate the standard enthalpy change for this reaction.
(b) How much energy as heat is evolved or absorbed if 1.00 kg of $Ca(OH)_2$ reacts with a stoichiometric amount of CO_2?

55. The standard enthalpy of formation of solid barium oxide, BaO, is −553.5 kJ/mol, and the standard enthalpy of formation of barium peroxide, BaO_2, is −634.3 kJ/mol.
(a) Calculate the standard enthalpy change for the following reaction. Is the reaction exothermic or endothermic?

$$2\ BaO_2(s) \longrightarrow 2\ BaO(s) + O_2(g)$$

(b) Draw an energy-level diagram that shows the relationship between the enthalpy change of the decomposition of BaO_2 to BaO and O_2 and the enthalpies of formation of BaO(s) and $BaO_2(s)$.

▲ more challenging ■ in OWL Blue-numbered questions answered in Appendix O

56. An important step in the production of sulfuric acid is the oxidation of SO_2 to SO_3.

$$SO_2(g) + \frac{1}{2} O_2(g) \rightarrow SO_3(g)$$

Formation of SO_3 from the air pollutant SO_2 is also a key step in the formation of acid rain.
(a) Use standard enthalpies of formation to calculate the enthalpy change for the reaction. Is the reaction exothermic or endothermic?
(b) Draw an energy-level diagram that shows the relationship between the enthalpy change for the oxidation of SO_2 to SO_3 and the enthalpies of formation of $SO_2(g)$ and $SO_3(g)$.

57. The enthalpy change for the oxidation of naphthalene, $C_{10}H_8$, is measured by calorimetry.

$$C_{10}H_8(s) + 12\ O_2(g) \rightarrow 10\ CO_2(g) + 4\ H_2O(\ell) \qquad \Delta_r H° = -5156.1 \text{ kJ/mol-rxn}$$

Use this value, along with the standard enthalpies of formation of $CO_2(g)$ and $H_2O(\ell)$, to calculate the enthalpy of formation of naphthalene, in kJ/mol.

58. ■ The enthalpy change for the oxidation of styrene, C_8H_8, is measured by calorimetry.

$$C_8H_8(\ell) + 10\ O_2(g) \rightarrow 8\ CO_2(g) + 4\ H_2O(\ell) \qquad \Delta_r H° = -4395.0 \text{ kJ/mol-rxn}$$

Use this value, along with the standard enthalpies of formation of $CO_2(g)$ and $H_2O(\ell)$, to calculate the enthalpy of formation of styrene, in kJ/mol.

General Questions on Thermochemistry

These questions are not designated as to type or location in the chapter. They may combine several concepts.

59. The following terms are used extensively in thermodynamics. Define each and give an example.
(a) exothermic and endothermic
(b) system and surroundings
(c) specific heat capacity
(d) state function
(e) standard state
(f) enthalpy change, ΔH
(g) standard enthalpy of formation

60. For each of the following, tell whether the process is exothermic or endothermic. (No calculations are required.)
(a) $H_2O(\ell) \rightarrow H_2O(s)$
(b) $2\ H_2(g) + O_2(g) \rightarrow 2\ H_2O(g)$
(c) $H_2O(\ell, 25\ °C) \rightarrow H_2O(\ell, 15\ °C)$
(d) $H_2O(\ell) \rightarrow H_2O(g)$

61. ■ For each of the following, define a system and its surroundings, and give the direction of energy transfer between system and surroundings.
(a) Methane is burning in a gas furnace in your home.
(b) Water drops, sitting on your skin after a dip in a swimming pool, evaporate.
(c) Water, at 25 °C, is placed in the freezing compartment of a refrigerator, where it cools and eventually solidifies.
(d) Aluminum and $Fe_2O_3(s)$ are mixed in a flask sitting on a laboratory bench. A reaction occurs, and a large quantity of energy is evolved as heat.

62. What does the term "standard state" mean? What are the standard states of the following substances at 298 K: H_2O, NaCl, Hg, CH_4?

63. Use Appendix L to find the standard enthalpies of formation of oxygen atoms, oxygen molecules (O_2), and ozone (O_3). What is the standard state of oxygen? Is the formation of oxygen atoms from O_2 exothermic? What is the enthalpy change for the formation of 1 mol of O_3 from O_2?

64. See the ChemistryNow website, Screen 5.9, Heat Transfer Between Substances. Use the Simulation section of this screen to do the following experiment: Add 10.0 g of Al at 80 °C to 10.0 g of water at 20 °C. What is the final temperature when equilibrium is achieved? Use this value to estimate the specific heat capacity of aluminum.

65. See the ChemistryNow website, Screen 5.15, Hess's Law. Use the Simulation section of this screen to find the value of $\Delta_r H°$ for

$$SnBr_2(s) + TiCl_4(\ell) \rightarrow SnCl_4(\ell) + TiBr_2(s)$$

66. Which gives up more energy on cooling from 50 °C to 10 °C, 50.0 g of water or 100. g of ethanol (specific heat capacity of ethanol = 2.46 J/g · K)?

67. You determine that 187 J of heat is required to raise the temperature of 93.45 g of silver from 18.5 °C to 27.0 °C. What is the specific heat capacity of silver?

68. ■ Calculate the quantity of energy required to convert 60.1 g of $H_2O(s)$ at 0.0 °C to $H_2O(g)$ at 100.0 °C. The heat of fusion of ice at 0 °C is 333 J/g; the heat of vaporization of liquid water at 100 °C is 2260 J/g.

69. ■ You add 100.0 g of water at 60.0 °C to 100.0 g of ice at 0.00 °C. Some of the ice melts and cools the water to 0.00 °C. When the ice and water mixture has come to a uniform temperature of 0 °C, how much ice has melted?

70. ▲ ■ Three 45-g ice cubes at 0 °C are dropped into 5.00×10^2 mL of tea to make iced tea. The tea was initially at 20.0 °C; when thermal equilibrium was reached, the final temperature was 0 °C. How much of the ice melted, and how much remained floating in the beverage? Assume the specific heat capacity of tea is the same as that of pure water.

71. ▲ ■ Suppose that only two 45-g ice cubes had been added to your glass containing 5.00×10^2 mL of tea (see Study Question 70). When thermal equilibrium is reached, all of the ice will have melted, and the temperature of the mixture will be somewhere between 20.0 °C and 0 °C. Calculate the final temperature of the beverage. (Note: The 90 g of water formed when the ice melts must be warmed from 0 °C to the final temperature.)

72. You take a diet cola from the refrigerator and pour 240 mL of it into a glass. The temperature of the beverage is 10.5 °C. You then add one ice cube (45 g). Which of the following describes the system when thermal equilibrium is reached?
(a) The temperature is 0 °C, and some ice remains.
(b) The temperature is 0 °C, and no ice remains.
(c) The temperature is higher than 0 °C, and no ice remains.

Determine the final temperature and the amount of ice remaining, if any.

73. ▲ ■ The standard molar enthalpy of formation of diborane, $B_2H_6(g)$, cannot be determined directly because the compound cannot be prepared by the reaction of boron and hydrogen. It can be calculated from other enthalpy changes, however. The following enthalpy changes can be measured.

$$4\ B(s) + 3\ O_2(g) \rightarrow 2\ B_2O_3(s) \quad \Delta_r H^\circ = -2543.8\ kJ/mol\text{-}rxn$$

$$H_2(g) + \tfrac{1}{2}\ O_2(g) \rightarrow H_2O(g) \quad \Delta_r H^\circ = -241.8\ kJ/mol\text{-}rxn$$

$$B_2H_6(g) + 3\ O_2(g) \rightarrow B_2O_3(s) + 3\ H_2O(g) \quad \Delta_r H^\circ = -2032.9\ kJ/mol\text{-}rxn$$

(a) Show how these equations can be added together to give the equation for the formation of $B_2H_6(g)$ from B(s) and $H_2(g)$ in their standard states. Assign enthalpy changes to each reaction.
(b) Calculate $\Delta_f H^\circ$ for $B_2H_6(g)$.
(c) Draw an energy-level diagram that shows how the various enthalpies in this problem are related.
(d) Is the formation of $B_2H_6(g)$ from its elements product-favored or reactant-favored?

74. Chloromethane, CH_3Cl, a compound found ubiquitously in the environment, is formed in the reaction of chlorine atoms with methane.

$$CH_4(g) + 2\ Cl(g) \rightarrow CH_3Cl(g) + HCl(g)$$

(a) ■ Calculate the enthalpy change for the reaction of $CH_4(g)$ and Cl atoms to give $CH_3Cl(g)$ and HCl(g). Is the reaction product-favored or reactant-favored?
(b) Draw an energy-level diagram that shows how the various enthalpies in this problem are related.

75. When heated to a high temperature, coke (mainly carbon, obtained by heating coal in the absence of air) and steam produce a mixture called water gas, which can be used as a fuel or as a chemical feedstock for other reactions. The equation for the production of water gas is

$$C(s) + H_2O(g) \rightarrow CO(g) + H_2(g)$$

(a) Use standard enthalpies of formation to determine the enthalpy change for this reaction.
(b) Is the reaction product-favored or reactant-favored?
(c) What is the enthalpy change if 1.0 metric ton (1000.0 kg) of carbon is converted to water gas?

76. Camping stoves are fueled by propane (C_3H_8), butane [$C_4H_{10}(g)$, $\Delta_f H^\circ = -127.1$ kJ/mol], gasoline, or ethanol (C_2H_5OH). Calculate the enthalpy of combustion per gram of each of these fuels. [Assume that gasoline is represented by isooctane, $C_8H_{18}(\ell)$, with $\Delta_f H^\circ = -259.2$ kJ/mol.] Do you notice any great differences among these fuels? Are these differences related to their composition?

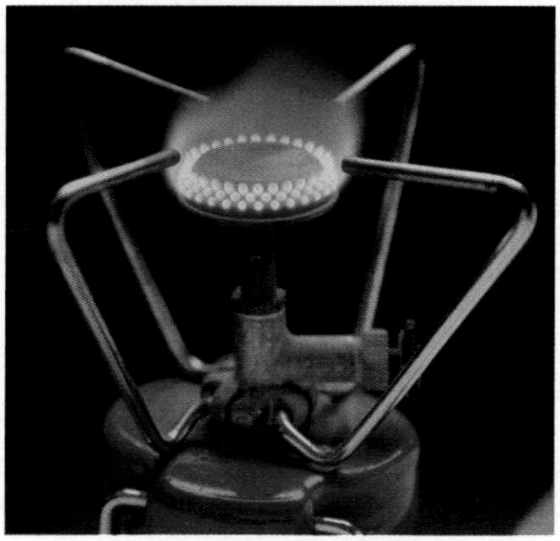
Charles D. Winters

A camping stove that uses butane as a fuel.

77. Methanol, CH_3OH, a compound that can be made relatively inexpensively from coal, is a promising substitute for gasoline. The alcohol has a smaller energy content than gasoline, but, with its higher octane rating, it burns more efficiently than gasoline in combustion engines. (It has the added advantage of contributing to a lesser degree to some air pollutants.) Compare the enthalpy of combustion per gram of CH_3OH and C_8H_{18} (isooctane), the latter being representative of the compounds in gasoline. ($\Delta_f H^\circ = -259.2$ kJ/mol for isooctane.)

78. Hydrazine and 1,1-dimethylhydrazine both react spontaneously with O_2 and can be used as rocket fuels.

$$N_2H_4(\ell) + O_2(g) \longrightarrow N_2(g) + 2\ H_2O(g)$$
hydrazine

$$N_2H_2(CH_3)_2(\ell) + 4\ O_2(g) \longrightarrow 2\ CO_2(g) + 4\ H_2O(g) + N_2(g)$$
1,1-dimethylhydrazine

The molar enthalpy of formation of $N_2H_4(\ell)$ is +50.6 kJ/mol, and that of $N_2H_2(CH_3)_2(\ell)$ is +48.9 kJ/mol. Use these values, with other $\Delta_f H°$ values, to decide whether the reaction of hydrazine or 1,1-dimethylhydrazine with oxygen provides more energy per gram.

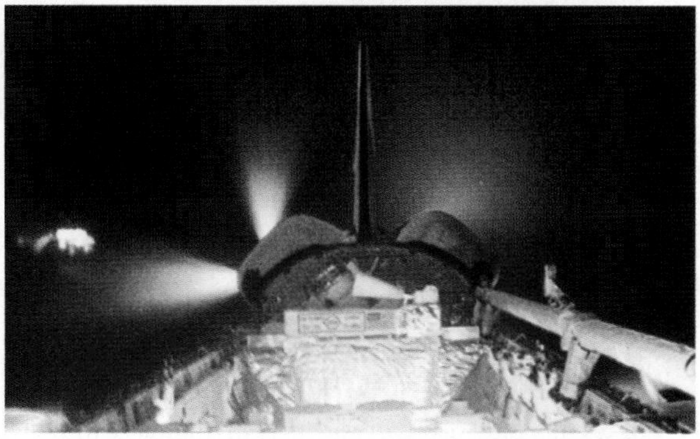

NASA

A control rocket in the Space Shuttle uses hydrazine as the fuel.

79. ■ (a) Calculate the enthalpy change, $\Delta_r H°$, for the formation of 1.00 mol of strontium carbonate (the material that gives the red color in fireworks) from its elements.

$$Sr(s) + C(graphite) + \tfrac{3}{2}\ O_2(g) \longrightarrow SrCO_3(s)$$

The experimental information available is

$$Sr(s) + \tfrac{1}{2}\ O_2(g) \longrightarrow SrO(s) \quad \Delta_f H° = -592\ kJ/mol\text{-}rxn$$

$$SrO(s) + CO_2(g) \longrightarrow SrCO_3(s) \quad \Delta_r H° = -234\ kJ/mol\text{-}rxn$$

$$C(graphite) + O_2(g) \longrightarrow CO_2(g) \quad \Delta_f H° = -394\ kJ/mol\text{-}rxn$$

(b) Draw an energy-level diagram relating the energy quantities in this problem.

80. You drink 350 mL of diet soda that is at a temperature of 5 °C.

(a) How much energy will your body expend to raise the temperature of this liquid to body temperature (37 °C)? Assume that the density and specific heat capacity of diet soda are the same as for water.

(b) Compare the value in part (a) with the caloric content of the beverage. (The label says that it has a caloric content of 1 Calorie.) What is the net energy change in your body resulting from drinking this beverage?

(c) Carry out a comparison similar to that in part (b) for a nondiet beverage whose label indicates a caloric content of 240 Calories.

81. ▲ Chloroform, $CHCl_3$, is formed from methane and chlorine in the following reaction.

$$CH_4(g) + 3\ Cl_2(g) \longrightarrow 3\ HCl(g) + CHCl_3(g)$$

Calculate $\Delta_r H°$, the enthalpy change for this reaction, using the enthalpies of formation of $CO_2(g)$, $H_2O(\ell)$, and $CHCl_3(g)$ ($\Delta_f H° = -103.1$ kJ/mol), and the enthalpy changes for the following reactions:

$$CH_4(g) + 2\ O_2(g) \longrightarrow 2\ H_2O(\ell) + CO_2(g) \quad \Delta_r H° = -890.4\ kJ/mol\text{-}rxn$$

$$2\ HCl(g) \longrightarrow H_2(g) + Cl_2(g) \quad \Delta_r H° = +184.6\ kJ/mol\text{-}rxn$$

82. Water gas, a mixture of carbon monoxide and hydrogen, is produced by treating carbon (in the form of coke or coal) with steam at high temperatures. (See Question 75.)

$$C(s) + H_2O(g) \longrightarrow CO(g) + H_2(g)$$

Not all of the carbon available is converted to water gas since some is burned to provide the heat for the endothermic reaction of carbon and water. What mass of carbon must be burned (to CO_2 gas) to provide the energy to convert 1.00 kg of carbon to water gas?

In the Laboratory

83. ■ A piece of lead with a mass of 27.3 g was heated to 98.90 °C and then dropped into 15.0 g of water at 22.50 °C. The final temperature was 26.32 °C. Calculate the specific heat capacity of lead from these data.

84. A 192-g piece of copper is heated to 100.0 °C in a boiling water bath and then dropped into a beaker containing 751 g of water (density = 1.00 g/cm^3) at 4.0 °C. What is the final temperature of the copper and water after thermal equilibrium is reached? (The specific heat capacity of copper is 0.385 J/g · K.)

85. Insoluble AgCl(s) precipitates when solutions of $AgNO_3(aq)$ and NaCl(aq) are mixed.

$$AgNO_3(aq) + NaCl(aq) \longrightarrow AgCl(s) + NaNO_3(aq) \quad \Delta_r H° = ?$$

To measure the energy evolved in this reaction, 250. mL of 0.16 M $AgNO_3(aq)$ and 125 mL of 0.32 M NaCl(aq) are mixed in a coffee-cup calorimeter. The temperature of the mixture rises from 21.15 °C to 22.90 °C. Calculate the enthalpy change for the precipitation of AgCl(s), in kJ/mol. (Assume the density of the solution is 1.0 g/mL and its specific heat capacity is 4.2 J/g · K.)

86. Insoluble $PbBr_2(s)$ precipitates when solutions of $Pb(NO_3)_2(aq)$ and $NaBr(aq)$ are mixed.

$$Pb(NO_3)_2(aq) + 2\ NaBr(aq) \rightarrow PbBr_2(s) + 2\ NaNO_3(aq) \quad \Delta_r H° = ?$$

To measure the enthalpy change, 200. mL of 0.75 M $Pb(NO_3)_2(aq)$ and 200. mL of 1.5 M $NaBr(aq)$ are mixed in a coffee-cup calorimeter. The temperature of the mixture rises by 2.44 °C. Calculate the enthalpy change for the precipitation of $PbBr_2(s)$, in kJ/mol. (Assume the density of the solution is 1.0 g/mL and its specific heat capacity is 4.2 J/g · K.)

87. The value of ΔU in the decomposition of 7.647 g of ammonium nitrate can be measured in a bomb calorimeter. The reaction that occurs is

$$NH_4NO_3(s) \rightarrow N_2O(g) + 2\ H_2O(g)$$

The temperature of the calorimeter, which contains 415 g of water, increases from 18.90 °C to 20.72 °C. The heat capacity of the bomb is 155 J/K. What is the value of ΔU for this reaction, in kJ/mol?

Charles D. Winters

The decomposition of ammonium nitrate is clearly exothermic.

88. ■ A bomb calorimetric experiment was run to determine the heat of combustion of ethanol (a common fuel additive). The reaction is

$$C_2H_5OH(\ell) + 3\ O_2(g) \rightarrow 2\ CO_2(g) + 3\ H_2O(\ell)$$

The bomb had a heat capacity of 550 J/K, and the calorimeter contained 650 g of water. Burning 4.20 g of ethanol, $C_2H_5OH(\ell)$ resulted in a rise in temperature from 18.5 °C to 22.3 °C. Calculate the enthalpy of combustion of ethanol, in kJ/mol.

89. The meals-ready-to-eat (MREs) in the military can be heated on a flameless heater. You can purchase a similar product called "Heater Meals." Just pour water into the heater unit, wait a few minutes, and you have a hot meal. The source of energy in the heater is

$$Mg(s) + 2\ H_2O(\ell) \rightarrow Mg(OH)_2(s) + H_2(g)$$

Charles D. Winters

The "heater meal" uses the reaction of magnesium with water as a source of energy as heat.

Calculate the enthalpy change under standard conditions, in joules, for this reaction. What quantity of magnesium is needed to supply the energy required to warm 25 mL of water (d = 1.00 g/mL) from 25 °C to 85 °C? (See W. Jensen: *Journal of Chemical Education*, Vol. 77, pp. 713–717, 2000.)

90. On a cold day, you can warm your hands with a "heat pad," a device that uses the oxidation of iron to produce energy as heat.

$$4\ Fe(s) + 3\ O_2(g) \rightarrow 2\ Fe_2O_3(s)$$

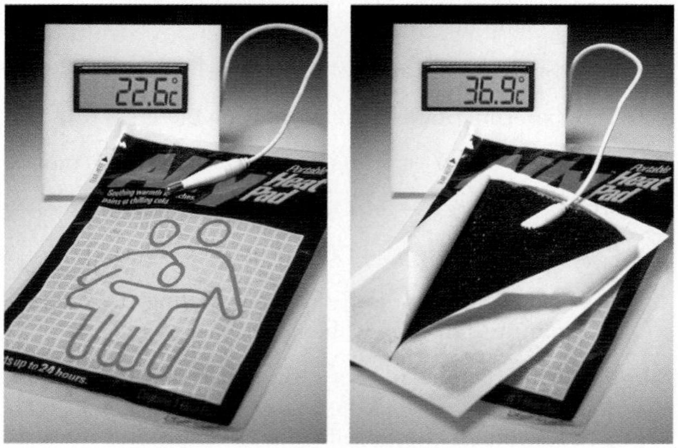

Charles D. Winters

A hand warmer uses the oxidation of iron as a source of thermal energy.

What mass of iron is needed to supply the energy required to warm 15 mL of water (d = 1.00 g/mL) from 23 °C to 37 °C?

Summary and Conceptual Questions

The following questions may use concepts from this and previous chapters.

91. Without doing calculations, decide whether each of the following is product-favored or reactant-favored.
    (a) the combustion of natural gas
    (b) the decomposition of glucose, $C_6H_{12}O_6$, to carbon and water

92. ■ Which of the following are state functions?
    (a) the volume of a balloon
    (b) the time it takes to drive from your home to your college or university
    (c) the temperature of the water in a coffee cup
    (d) the potential energy of a ball held in your hand

93. ▲ ■ You want to determine the value for the enthalpy of formation of $CaSO_4(s)$.

$$Ca(s) + S(s) + 2\ O_2(g) \longrightarrow CaSO_4(s)$$

This reaction cannot be done directly. You know, however, that both calcium and sulfur react with oxygen to produce oxides in reactions that can be studied calorimetrically. You also know that the basic oxide CaO reacts with the acidic oxide $SO_3(g)$ to produce $CaSO_4(s)$ with $\Delta_r H° = -402.7$ kJ. Outline a method for determining $\Delta_f H°$ for $CaSO_4(s)$, and identify the information that must be collected by experiment. Using information in Appendix L, confirm that $\Delta_f H°$ for $CaSO_4(s) = -1433.5$ kJ/mol.

94. Prepare a graph of specific heat capacities for metals versus their atomic weights. Combine the data in Figure 5.7 and the values in the following table. What is the relationship between specific heat capacity and atomic weight? Use this relationship to predict the specific heat capacity of platinum. The specific heat capacity for platinum is given in the literature as 0.133 J/g · K. How good is the agreement between the predicted and actual values?

Metal	Specific Heat Capacity (J/g · K)
Chromium	0.450
Lead	0.127
Silver	0.236
Tin	0.227
Titanium	0.522

95. Observe the molar heat capacity values for the metals in Figure 5.7. What observation can you make about these values—specifically, are they widely different or very similar? Using this information, estimate the specific heat capacity for silver. Compare this estimate with the correct value for silver, 0.236 J/g · K.

96. ▲ Suppose you are attending summer school and are living in a very old dormitory. The day is oppressively hot. There is no air-conditioner, and you can't open the windows of your room because they are stuck shut from layers of paint. There is a refrigerator in the room, however. In a stroke of genius, you open the door of the refrigerator, and cool air cascades out. The relief does not last long, though. Soon the refrigerator motor and condenser begin to run, and not long thereafter the room is hotter than it was before. Why did the room warm up?

97. ■ You want to heat the air in your house with natural gas (CH_4). Assume your house has 275 m^2 (about 2800 ft^2) of floor area and that the ceilings are 2.50 m from the floors. The air in the house has a molar heat capacity of 29.1 J/mol · K. (The number of moles of air in the house can be found by assuming that the average molar mass of air is 28.9 g/mol and that the density of air at these temperatures is 1.22 g/L.) What mass of methane do you have to burn to heat the air from 15.0 °C to 22.0 °C?

98. Water can be decomposed to its elements, H_2 and O_2, using electrical energy or in a series of chemical reactions. The following sequence of reactions is one possibility:

$$CaBr_2(s) + H_2O(g) \longrightarrow CaO(s) + 2\ HBr(g)$$

$$Hg(\ell) + 2\ HBr(g) \longrightarrow HgBr_2(s) + H_2(g)$$

$$HgBr_2(s) + CaO(s) \longrightarrow HgO(s) + CaBr_2(s)$$

$$HgO(s) \longrightarrow Hg(\ell) + \tfrac{1}{2}\ O_2(g)$$

   (a) Show that the net result of this series of reactions is the decomposition of water to its elements.
   (b) If you use 1000. kg of water, what mass of H_2 can be produced?
   (c) Calculate the value of $\Delta_r H°$ for each step in the series. Are the reactions predicted to be product-favored or reactant-favored?

$$\Delta_f H°\ [CaBr_2(s)] = -683.2\ \text{kJ/mol}$$

$$\Delta_f H°\ [HgBr_2(s)] = -169.5\ \text{kJ/mol}$$

   (e) Comment on the commercial feasibility of using this series of reactions to produce $H_2(g)$ from water.

99. Suppose that an inch of rain falls over a square mile of ground. (Density of water is 1.0 g/cm^3.) The enthalpy of vaporization of water at 25 °C is 44.0 kJ/mol. How much energy as heat is transferred to the surroundings from the condensation of water vapor in forming this quantity of liquid water? (The huge number tells you how much energy is "stored" in water vapor and why we think of storms as such great forces of energy in nature. It is interesting to compare this result with the energy given off, 4.2×10^6 kJ, when a ton of dynamite explodes.)

100. ▲ Peanuts and peanut oil are organic materials and burn in air. How many burning peanuts does it take to provide the energy to boil a cup of water (250 mL of water)? To solve this problem, we assume each peanut, with an average mass of 0.73 g, is 49% peanut oil and 21% starch; the remainder is noncombustible. We further assume peanut oil is palmitic acid, $C_{16}H_{32}O_2$, with an enthalpy of formation of −848.4 kJ/mol. Starch is a long chain of $C_6H_{10}O_5$ units, each unit having an enthalpy of formation of −960 kJ. *(See ChemistryNow Screens 5.1 and 5.19: Chemical Puzzler.)*

Charles D. Winters

How many burning peanuts are required to provide the energy to boil 250 mL of water?

101. ▲ Isomers are molecules with the same elemental composition but a different atomic arrangement. Three isomers with the formula C_4H_8 are shown in the models below. The enthalpy of combustion ($\Delta_cH°$) of each isomer, determined using a calorimeter, is:

Compound	$\Delta_{com}H°$ **(kJ/mol-rxn)**
cis-2 butene	−2687.5
trans-2-butene	−2684.2
1-butene	−2696.7

(a) Draw an energy level diagram relating the energy content of the three isomers to the energy content of the combustion products, CO_2(g) and H_2O(g).
(b) Use the $\Delta_cH°$ data in part (a), along with the enthalpies of formation of CO_2(g) and H_2O(g) from Appendix L, to calculate the enthalpy of formation for each of the isomers.
(c) Draw an energy level diagram that relates the enthalpies of formation of the three isomers to the energy of the elements in their standard states.
(d) What is the enthalpy change for the conversion of *cis*-2-butene to *trans*-2-butene?

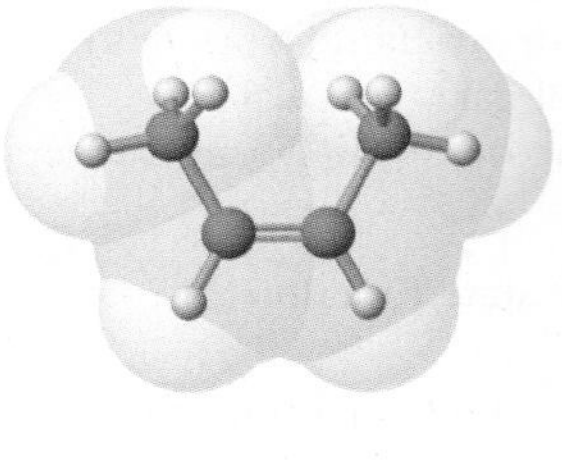

cis-2-butene

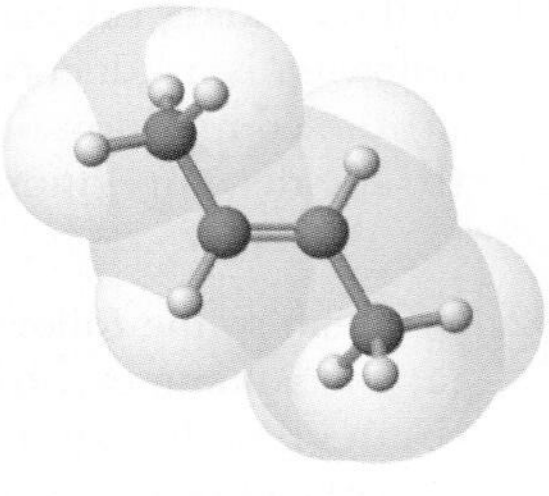

trans-2-butene

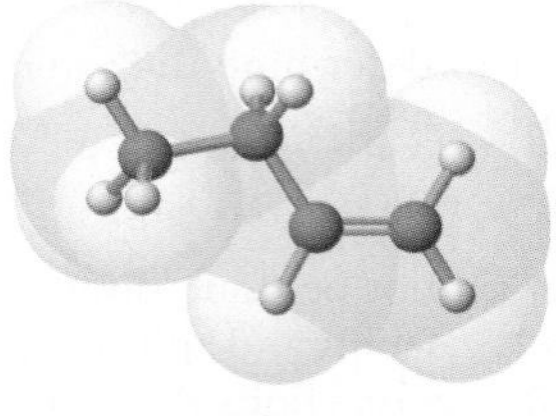

1-butene

102. Several standard enthalpies of formation (from Appendix L) are given below. Use these data to calculate:
(a) The standard enthalpy of vaporization of bromine.
(b) The energy required for the reaction $Br_2(g) \rightarrow 2\ Br(g)$. (This is the Br—Br bond energy.)

Species	$\Delta_fH°$ **(kJ/mol)**
Br(g)	111.9
$Br_2(\ell)$	0
Br_2(g)	30.9

103. When 0.850 g of Mg is burned in oxygen in a constant volume calorimeter, 25.4 kJ of energy as heat is evolved. The calorimeter is in an insulated container with 750. g of water at an initial temperature of 18.6 °C. The heat capacity of the calorimeter is 820. J/K.
(a) Calculate ΔU for the oxidation of Mg (in kJ/mol Mg).
(b) What will be the final temperature of the water and the bomb calorimeter in this experiment?

104. A piece of gold (10.0 g, C = 0.129 J/g · K) is heated to 100.0 °C. A piece of copper (also 10.0 g, C = 0.385 J/g · K) is chilled in an ice bath to 0 °C. Both pieces of metal are placed in a beaker containing 150. g H_2O at 20 °C. Will the temperature of the water be greater than or less than 20 °C when thermal equilibrium is reached? Calculate the final temperature.

105. Methane, CH_4, can be converted to methanol which, like ethanol, can be used as a fuel. The energy level diagram shown here presents relationships between energies of the fuels and their oxidation products. Use the information in the diagram to answer the following questions. (The energy terms are per mol-rxn.)

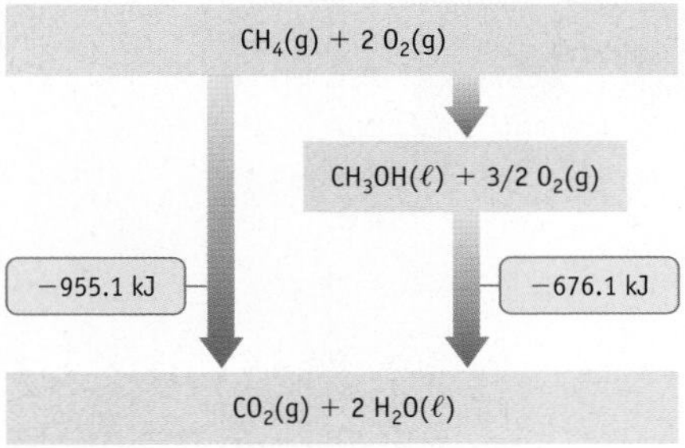

(a) Which fuel, methanol or methane, yields the most energy per mole when burned?
(b) Which fuel yields the most energy per gram when burned?
(c) What is the enthalpy change for the conversion of methane to methanol?
(d) Each arrow on the diagram represents a chemical reaction. Write the equation for the reaction that converts methane to methanol.

106. Calculate $\Delta_r H°$ for the reaction

$2\ C(s) + 3\ H_2(g) + \frac{1}{2}\ O_2(g) \rightarrow C_2H_5OH(\ell)$

given the information below.

$C(s) + O_2(g) \rightarrow CO_2(g) \quad \Delta_r H° = -393.5$ kJ/mol-rxn

$2\ H_2(g) + O_2(g) \rightarrow 2\ H_2O(\ell) \quad \Delta_r H° = -571.6$ kJ/mol-rxn

$C_2H_5OH(\ell) + 3\ O_2(g) \rightarrow 2\ CO_2(g) + 3\ H_2O(\ell) \quad \Delta_r H° = -1367.5$ kJ/mol-rxn

107. You have the six pieces of metal listed below, plus a beaker of water containing 3.00×10^2 g of water. The water temperature is 21.00 °C.

Metals	Specific Heat (J/g K)	Mass (g)
1. Al	0.9002	100.0
2. Al	0.9002	50.0
3. Au	0.1289	100.0
4. Au	0.1289	50.0
5. Zn	0.3860	100.0
6. Zn	0.3860	50.0

(a) In your first experiment you select one piece of metal and heat it to 100 °C, and then select a second piece of metal and cool it to −10 °C. Both pieces of metal are then placed in the beaker of water and the temperatures equilibrated. You want to select two pieces of metal to use, such that the final temperature of the water is as high as possible. What piece of metal will you heat? What piece of metal will you cool? What is the final temperature of the water?
(b) The second experiment is done in the same way as the first. However, your goal now is to cause the temperature to change the least, that is, the final temperature should be be as near to 21.00 °C as possible. What piece of metal will you heat? What piece of metal will you cool? What is the final temperature of the water?

108. In lab, you plan to carry out a calorimetry experiment to determine the $\Delta_r H$ for the exothermic reaction of $Ca(OH)_2(s)$ and HCl(aq). Predict how each of the following will affect the calculated value of $\Delta_r H$. (The value calculated for $\Delta_r H$ for this reaction is a negative value so choose your answer from the following: $\Delta_r H$ will be too low [that is, a larger negative value], $\Delta_r H$ will be unaffected, $\Delta_r H$ will be too high [that is, a smaller negative value.])
(a) You spill a little bit of the $Ca(OH)_2$ on the benchtop before adding it to the calorimeter.
(b) Because of a miscalculation, you add an excess of HCl to the measured amount of $Ca(OH)_2$ in the calorimeter.
(b) $Ca(OH)_2$ readily absorbs water from the air. The $Ca(OH)_2$ sample you weighed had been exposed to the air prior to weighing and had absorbed some water.
(c) After weighing out $Ca(OH)_2$, the sample sat in an open beaker and absorbed water.
(d) You delay too long in recording the final temperature.
(e) The insulation in your coffee cup calorimeter was poor and so some energy as heat was lost to the surroundings during the experiment.
(e) You have ignored the fact that energy as heat also raised the temperature of the stirrer and the thermometer in your system.

The Chemistry of Fuels and Energy Resources

with contributions from
Roger Hinrichs Weill Cornell Medical College in Quatar

Energy is necessary for everything we do. Look around you—energy is involved in anything that is moving or is emitting light, sound, or heat. Heating and lighting your home, propelling your automobile, powering your iPod—all are commonplace examples in which energy is used, and all are, at their origin, primarily based on chemical processes. In this interchapter, we want to examine how chemistry is fundamental to understanding and addressing current energy issues.

Supply and Demand: The Balance Sheet on Energy

We take for granted that energy is available and that it will always be there to use. But will it? Recently, a chemist and Nobel Prize winner, the late Richard Smalley, stated that among the top 10 problems humanity will face over the next 50 years, the energy supply ranks as number one. What is the source of this dire prediction? Information such as the following is often quoted in the popular press:

- Global demand for energy has tripled in the past 50 years and may triple again in the next 50 years. Most of the demand comes from industrialized nations, but most of the increase is coming from developing countries.
- Fossil fuels account for 85% of the total energy used by humans on our planet. (Of this total, petroleum accounts for 37%, coal 26%, and natural gas 22%.) Nuclear and hydroelectric power each contribute about 6% of the total energy budget. The remaining 3% derives from biomass, solar, wind, and geothermal energy–generating facilities.
- With only 4.6% of the world's population, the United States consumes 23% of all the energy used in the world. This usage is equivalent to the consumption of 7 gallons of oil or 70 pounds of coal per person per day.
- China and India, growing economic powerhouses, are seeing their use of energy grow by about 8% per year. In 2007, China passed the United States as the number one emitter of greenhouse gases in the world.

Two basic issues, *energy usage* and *energy resources*, instantly leap out from these statistics and form the basis for this discussion of energy.

Energy Usage

Data indicate that energy usage is related to the degree to which a country has industrialized. The more industrialized a country, the more energy is used on a per capita basis. Another way to say this is that energy usage per capita correlates with gross domestic product per capita. As a higher degree of industrialization occurs in developing nations, energy usage worldwide will increase proportionally. The rapid growth in energy usage over the last two decades is strong evidence in support of predictions of similar growth in the next half-century (Figure 1).

One way to alter energy consumption is through conservation. Energy conservation can mean consciously using less energy (such as driving less, turning off lights when not in use, and turning the thermostat down [for heating] or up [for cooling]). It can also mean using energy more

Charles D. Winters

• **Methane hydrate, a potential fuel source.** Methane, CH_4, can be trapped in a lattice of water molecules. The methane is released when the pressure is reduced or temperature is raised. See Figure 5 on page 260.

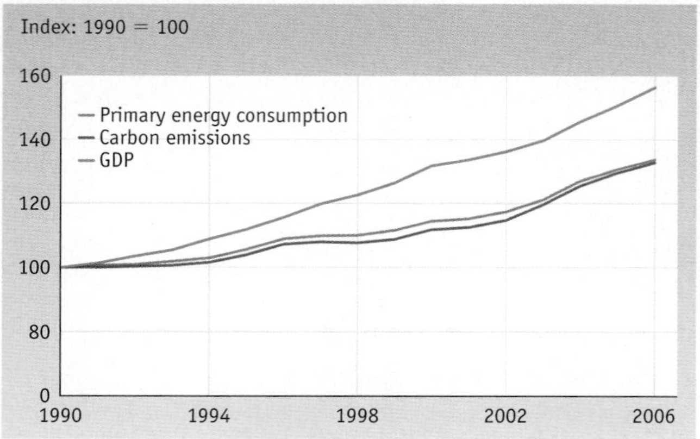

Figure 1 **World energy usage, 1990–2006.** Gross domestic product (GDP) is rising faster than energy use, indicating increased energy efficiency. The link between carbon emissions and energy use continues to show a strong correlation.

efficiently. Some examples (Figure 2) of this latter approach are:

- Aluminum is recycled because recovering aluminum requires only one third of the energy needed to produce the metal from its ore.
- Light-emitting diodes (LEDs) are being used in streetlights, and compact fluorescent lights are finding wider use in the home. Both use a fraction of the energy required by incandescent bulbs (in which only 5% of the energy used is delivered in the form of light; the remaining 95% is wasted as heat).
- Hybrid cars offer up to twice the gas mileage available with conventional cars.
- Many appliances (from refrigerators to air conditioners) are equipped to use less energy per delivered output.

One of the exciting areas of current research in chemistry relating to energy conservation focuses on superconductivity. Superconductors are materials that, at temperatures of 30–150 K, offer virtually no resistance to electrical conductivity (see "The Chemistry of Modern Materials," page 657). When an electric current passes through a typical conductor such as a copper wire, some of the energy is lost as heat. As a result, there is substantial energy loss in power transmission lines. Substituting a superconducting wire for copper has the potential to greatly decrease this loss, so the search is on for materials that act as superconductors at moderate temperatures.

Charles D. Winters

Figure 2 **Energy-conserving devices.** Energy-efficient home appliances, hybrid automobiles, and compact fluorescent bulbs all provide alternatives that consume less energy than their conventional counterparts to do the same task.

Energy Resources

On the other side of the energy balance sheet are energy resources, of which many exist. The data cited earlier make it obvious that we are hugely dependent on fossil fuels as a source of energy. We rely almost entirely on gasoline and diesel fuel in transportation. Fuel oil and natural gas are the standards for heating, and approximately 70% of the electricity in the United States is generated using fossil fuels, mostly coal (Table 1).

Why is there such a dominance of fossil fuels on the resource side of the equation? An obvious reason is that fossil fuels are cheap raw materials compared to other energy sources. In addition, societies have made an immense investment in the infrastructure needed to distribute and use this energy. Power plants using coal or natural gas cannot be converted readily to accommodate another fuel. The infrastructure for distribution of energy—gas pipelines, gasoline dispensing for cars, and the grid distributing electricity to users—is already in place. Much of this infrastructure may have to change if the source of energy changes. Some countries already have energy distribution systems that do not depend nearly as much as the U.S. system on fossil fuels. For example, countries in Europe (such as France) make much greater use of nuclear power, and certain regions on the planet (such as Iceland and New Zealand) are able to exploit geothermal power as an energy source. Germany and Spain plan on meeting 25% of their electrical energy needs with wind by the year 2020.

TABLE 1 Producing Electricity in the United States (2006)

Coal	50%
Nuclear	19%
Natural gas	19%
Hydroelectric	7%
Petroleum	3%
Other renewables	2%

We have become accustomed to an energy system based on fossil fuels. The internal combustion engine is the result of years of engineering. It is now well understood and can be produced in large quantities quickly and for a relatively low cost. The electric grid is well established to supply our buildings and roads. Natural gas supply to our homes is nearly invisible. The system works well.

Why do we worry about using fossil fuels? One major problem is that fossil fuels are *nonrenewable energy sources.* Nonrenewable resources are those in which the energy source is used and not concurrently replenished. Fossil fuels are the obvious example. Nuclear energy is also in this category. (This does not include nuclear fusion, which combines hydrogen nuclei to produce energy, as in the stars. But this technology is a long way from practical use.) Conversely, energy sources that involve using the sun's energy are examples of *renewable resources.* These include solar energy and energy derived from wind, biomass, and moving water. Likewise, geothermal energy is a renewable resource.

There is a limited supply of fossil fuels. No more sources are being created. As a consequence, we must ask how long our fossil fuels will last. Regrettably, there is not an exact answer to this question. One current estimate suggests that at current consumption rates the world's oil reserves will be depleted in 30–80 years. Natural gas and coal supplies are projected to last longer. It is estimated that natural gas reserves will last about 60–200 years, whereas coal reserves are projected to last from 150 to several hundred years. These numbers are highly uncertain, however—in part because the estimates are based on guesses regarding fuel reserves not yet discovered; in part because assumptions must be made about the rate of consumption in future years. If the use of a commodity (such as oil) continues to rise by a fixed percentage every year, then we say that we are experiencing "exponential growth" for that usage. Even though the amount of oil consumed every year might rise by only 3%, this still is a rapid growth in the total used if we look forward many years. A global growth rate of 4% per year for oil will reduce the estimate of petroleum resources lasting 80 years to only 36 years. A growth rate of 2% per year changes this estimate to 48 years. Estimates of how long these resources will last do not mean anything unless assumed growth rates are accurate.

Despite our current state of comfort with our energy system, we cannot ignore the fact that a change away from fossil fuels must occur someday. As supply diminishes and demand increases, expansion to other fuel types will inevitably occur. Increased cost of energy based on fossil fuels will encourage these changes. The technologies to facilitate change, and the answers regarding which alternative fuel types will be the most efficient and cost-effective, can be aided by research in chemistry.

Fossil Fuels

Fossil fuels originate from organic matter that was trapped under the earth's surface for many millennia. Due to the particular combination of temperature, pressure, and available oxygen, the decomposition process from the compounds that constitute organic matter resulted in the hydrocarbons we extract and use today: coal, crude oil, and natural gas—the solid, liquid, and gaseous forms of fossil fuels, respectively. These hydrocarbons have varying ratios of carbon to hydrogen.

Fossil fuels are simple to use and relatively inexpensive to extract, compared with the current cost requirements of other sources for the equivalent amount of energy. To use the energy stored in fossil fuels, these materials are burned. The combustion process, when it goes to completion, converts fossil fuels to CO_2 and H_2O (Section 3.2). The energy evolved as heat is then converted to mechanical and then electrical energy (Chapter 5).

The energy output from burning fossil fuels (Table 2) is related to the carbon-to-hydrogen ratio. We can analyze this relationship by considering data on enthalpies of formation and by looking at an example that is 100% carbon and another that is 100% hydrogen. The oxidation of 1.0 mol (12.01 g) of pure carbon produces 393.5 kJ of energy or 32.8 kJ per gram.

$$\mathrm{C(s)} + \mathrm{O_2(g)} \rightarrow \mathrm{CO_2(g)}$$

$$\Delta_r H° = -393.5 \text{ kJ/mol-rxn or } -32.8 \text{ kJ/g C}$$

Burning hydrogen to form water is much more exothermic on a per-gram basis, with about 120 kJ evolved per gram of hydrogen consumed.

$$\mathrm{H_2(g)} + \tfrac{1}{2}\,\mathrm{O_2(g)} \longrightarrow \mathrm{H_2O(g)}$$

$$\Delta_r H° = -241.8 \text{ kJ/mol-rxn or } -119.9 \text{ kJ/g } \mathrm{H_2}$$

Coal is mostly carbon, so its heat output is similar to that of pure carbon. In contrast, methane is 25% hydrogen (by mass), and the higher–molecular-weight hydrocarbons in petroleum and products refined from petroleum average 16–17% hydrogen content. Therefore, their heat output on a per-gram basis is greater than that of pure carbon, but less than that of hydrogen itself.

TABLE 2 Energy Released by Combustion of Fossil Fuels

Substance	Energy Released (kJ/g)
Coal	29–37
Crude petroleum	43
Gasoline (refined petroleum)	47
Natural gas (methane)	50

While the basic chemical principles for extracting energy from fossil fuels are simple, complications arise in practice. Let us look at each of these fuels in turn.

Coal

The solid rock-like substance that we call coal began to form almost 290 million years ago. Decomposition of plant matter occurred to a sufficient extent that the primary component of coal is carbon. Describing coal simply as carbon is a simplification, however. Samples of coal vary considerably in their composition and characteristics. Carbon content may range from 60% to 95%, with variable amounts of hydrogen, oxygen, sulfur, and nitrogen present in the coal in various forms.

Sulfur is a common constituent in some coals. The element was incorporated into the mixture partly from decaying plants and partly from hydrogen sulfide, H_2S, which is the waste product from certain bacteria. In addition, coal is likely to contain traces of many other elements, including some that are hazardous (such as arsenic, mercury, cadmium, and lead) and some that are not (such as iron).

When coal is burned, some of the impurities are dispersed into the air, and some end up in the ash that is formed. In the United States, coal-fired power plants are responsible for 60% of the emissions of SO_2 and 33% of mercury emissions into the environment. (U.S. plants emit about 50 tons of mercury per year in the U.S.; worldwide, about 5500 tons are emitted.) Sulfur dioxide reacts with water and oxygen in the atmosphere to form sulfuric acid, which contributes (along with nitric acid) to the phenomenon known as acid rain.

$$2\ SO_2(g) + O_2(g) \rightarrow 2\ SO_3(g)$$

$$SO_3(g) + H_2O(\ell) \rightarrow H_2SO_4(aq)$$

Because these acids are harmful to the environment, legislation limits the extent of sulfur oxide emissions from coal-fired plants. Chemical scrubbers have been developed that can be attached to the smokestacks of power plants to reduce sulfur-based emissions. Simply, the combustion gases are passed through a water spray with chemicals such as limestone (calcium carbonate) to form solids that can be removed:

$$2\ SO_2(g) + 2\ CaCO_3(s) + O_2(g) \rightarrow 2\ CaSO_4(s) + 2\ CO_2(g)$$

However, these devices are expensive and can increase the cost of the energy produced from these facilities.

Coal is classified into three categories (Table 3). Anthracite, or hard coal, is the highest-quality coal. Among the forms of coal, *anthracite coal* releases the largest amount of heat per gram and has a low sulfur content. Unfortunately, anthracite coal is fairly uncommon, with only 2% of the U.S. coal reserves occurring in this form. *Bituminous coal*, also referred to as *soft coal*, accounts for about 45% of the U.S. coal reserves and is the coal most widely used in electric power generation (Figure 3). Soft coal typically has the highest sulfur content. *Lignite*, also called brown coal because of its paler color, is geologically the "youngest" form of coal. It releases a smaller amount of heat per gram than the other forms of coal, often contains a significant amount of water, and is the least popular as a fuel.

TABLE 3 Types of Coal

Type	Consistency	Sulfur Content	Heat Content (kJ/g)
Lignite	Very soft	Very low	28–30
Bituminous coal	Soft	High	29–37
Anthracite	Hard	Low	36–37

Courtesy of BLM Wyoming

Figure 3 **Bituminous coal being extracted from a strip mine in Montana.**

Coal can be converted to coke by heating in the absence of air. Coke is almost pure carbon and an excellent fuel. In the process of coke formation, a variety of organic compounds are driven off. These compounds are used as raw materials in the chemical industry for the production of polymers, pharmaceuticals, synthetic fabrics, waxes, tar, and numerous other products.

Technology to convert coal into gaseous fuels (coal gasification) (Figure 4) or liquid fuels (liquefaction) is under development, but hampered by cost. These processes provide fuels that burn more cleanly than coal, except that 30–40% of the available energy is lost in the process. As petroleum and natural gas reserves dwindle and the costs of these fuels increase, liquid and gaseous fuels derived from coal are likely to become more important.

Natural Gas

Natural gas is found deep under the earth's surface, where it was formed by bacteria decomposing organic matter in an anaerobic environment (in which no O_2 is present). The major component of natural gas (70–95%) is methane (CH_4). Lesser quantities of other gases such as ethane (C_2H_6), propane (C_3H_8), and butane (C_4H_{10}) are also present, along with other gases including N_2, He, CO_2, and H_2S. The impurities and higher–molecular-weight components of

Figure 4 **Coal gasification plant.** Advanced coal-fired power plants, such as this 2544-ton-per-day coal gasification demonstration pilot plant, will have energy conversion efficiencies 20% to 35% higher than those of conventional pulverized-coal steam power plants.

natural gas are separated out during the refining process, so that the gas piped into our homes is primarily methane.

Natural gas is an increasingly popular choice as a fuel. It burns more cleanly than the other fossil fuels, emits fewer pollutants, and produces relatively more energy than the other fossil fuels. Natural gas can be transported by pipelines over land and piped into buildings such as your home, where it is used to heat ambient air, and to heat water for washing, bathing, or for cooking. It is also a popular choice for new electrical power plants, which have high efficiencies due to new gas turbines and recovery of waste heat.

Petroleum

Petroleum is often found in porous rock formations that are bounded by impermeable rock. Petroleum is a complicated mixture of hydrocarbons, whose molar masses range from low to very high (▶ page 495). The hydrocarbons may have anywhere from one to twenty or more carbon atoms in their structures, and compounds containing sulfur, nitrogen, and oxygen may also be present in small amounts.

Petroleum goes through extensive processing at refineries to separate the various components and convert less valuable compounds into more valuable ones. Nearly 85% of the crude petroleum pumped from the ground ends up being used as a fuel, either for transportation (gasoline and diesel fuel) or for heating (fuel oils).

Other Fossil Fuel Sources

When natural gas pipelines were laid across the United States and Canada, pipeline operators soon found that, unless water was carefully kept out of the line, chunks of methane hydrate would form and clog the pipes. Methane hydrate was a completely unexpected substance because it is made up of methane and water, two chemicals that would appear to have little affinity for each other. In methane hydrate, methane becomes trapped in cavities in the crystal structure of ice (Figure 5). Methane hydrate is stable only at temperatures below the freezing point of water. If a sample of methane hydrate is warmed above 0° C, it melts, and methane is released. The volume of gas released (at normal pressure and temperature) is about 165 times larger than the volume of the hydrate.

If methane hydrate forms in a pipeline, is it found in nature as well? In May 1970, oceanographers drilling into the seabed off the coast of South Carolina pulled up samples of a whitish solid that fizzed and oozed when it was removed from the drill casing. They quickly realized it was methane hydrate. Since this original discovery, methane hydrate has been found in many parts of the oceans as well as under permafrost in the Arctic. It is estimated that 1.5×10^{13} tons of methane hydrate are buried under the sea floor around the world. In fact, the energy available from this source may surpass that of all the other known fossil fuel reserves by as much as a factor of 2! Clearly, this is a potential source of an important fuel in the future. Today, however, the technology to extract methane from these hydrate deposits is very expensive, especially in comparison to the well-developed technologies used to extract crude oil, coal, and gaseous methane.

There are other sources of methane in our environment. For example, methane is generated in swamps, where it is called *swamp gas* or *marsh gas*. Here, methane is formed by bacteria working on organic matter in an anaerobic environment—namely, in sedimentary layers of coastal waters and in marshes. The process of formation is similar to the processes occurring eons ago that generated the natural gas deposits that we currently use for fuel. In a marsh, the gas can escape if the sediment layer is thin. You see it as bubbles rising to the surface. Unfortunately, because of the relatively small amounts generated, it is impractical to collect and use this gas as a fuel.

In a striking analogy to what occurs in nature, methane also forms in human-made landfill sites. A great deal of organic matter is buried in landfills. It remains out of contact with oxygen in the air, and methane is formed as the organic matter is degraded by bacteria. In the past, landfill gases have been deemed a nuisance. Today, it is possible to collect this methane and use it as a fuel. You might have seen plastic pipes in the ground in a landfill that vent the methane to a holding tank.

Another source of fossil fuels, and one that is being used right now, is oil from tar sands. Tar sands (also called oil sands) contain a very viscous organic liquid called "bitumen." This is chemically similar to the highest–molecular-weight fraction obtained by distillation of crude oil. What makes this source so enticing is the huge quantity of oil that could be obtained from such sites. The largest resource of tar sands

(a) Methane hydrate burns as methane gas escapes from the solid hydrate.

(b) Methane hydrate consists of a lattice of water molecules with methane molecules trapped in the cavity.

(c) A colony of worms on an outcropping of methane hydrate in the Gulf of Mexico.

Figure 5 **Methane hydrate.** (a) This interesting substance is found in huge deposits hundreds of feet down on the floor of the ocean. When a sample is brought to the surface, the methane oozes out of the solid, and the gas readily burns. (b) The structure of the solid hydrate consists of methane molecules trapped within a lattice of water molecules. Each point of the lattice shown here is an oxygen atom of a water molecule. The edges are O—H—O bonds. Such structures are often called "clathrates." (c) An outcropping of methane hydrate on the floor of the Gulf of Mexico. See E. Suess, G. Bohrmann, J. Greinert, and E. Lausch: *Scientific American*, pp. 76–83, November 1999.

in the world is found in Alberta, Canada (the *Athabasca Sands*). This is followed closely by those in Venezuela. Resources approaching 3.5 trillion barrels of oil are estimated in these two locations—twice the world's known reserves of petroleum. The U.S. imports more oil from Canada than any other country (0.8 million barrels per day), and most of this is from the Athabasca Tar Sands!

Extracting the oil from tar sands is quite costly. Essentially, the sands must be mined and then mixed with hot water or steam to extract the bitumen. In order to avoid an environmental catastrophe, the mined land must be restored (reclaimed). This adds to the cost of the process. Also, most of the Canadian tar sands are located in dry areas, so obtaining an adequate water supply for extraction might pose a constraint on increased production.

Environmental Impacts of Fossil Fuel Use

As mentioned earlier, about 85% of the energy used in the world today comes from fossil fuels. We are a carbon-based society. While this percentage is relatively stable, the amount of gaseous emissions of carbon compounds into our environment continues to rise. These include mainly CO_2 but also CH_4, CO, and chlorofluorocarbons (CFCs). The correlation is quite distinct—rising energy use correlates well with rising carbon emissions (Figure 1).

The "greenhouse effect" is a name given to the trapping of energy in the earth's atmosphere by a process very similar to that used in greenhouses (Figure 6). The atmosphere, like window glass, is transparent to incoming solar radiation. This is absorbed by the earth and re-emitted as infrared radiation. Gases in the atmosphere, like window glass, trap some of these longer infrared rays, keeping the earth warmer than it would otherwise be. In the last century, there has been an increase in concentrations in the atmosphere of carbon dioxide and other so-called greenhouse gases (methane, nitrogen oxides) due to increases in fossil fuel burning. There has also been a corresponding increase in global average temperatures that most scientists attribute to increases in these greenhouse gas concentrations (Figure 7). This correlates very well with increased concentrations of CO_2 in the atmosphere. For the next two decades, a warming of about 0.2 °C per decade is projected by some models. Such temperature changes will affect the earth's climate in many ways, such as more intense storms, precipitation changes, and sea level rise. Health issues will also be a factor.

Global warming—the increase in average global temperatures, which is probably owing to human activities increasing the greenhouse effect—has become one of the biggest issues facing us worldwide. Indeed, many of the steps made in the last decade to put increased

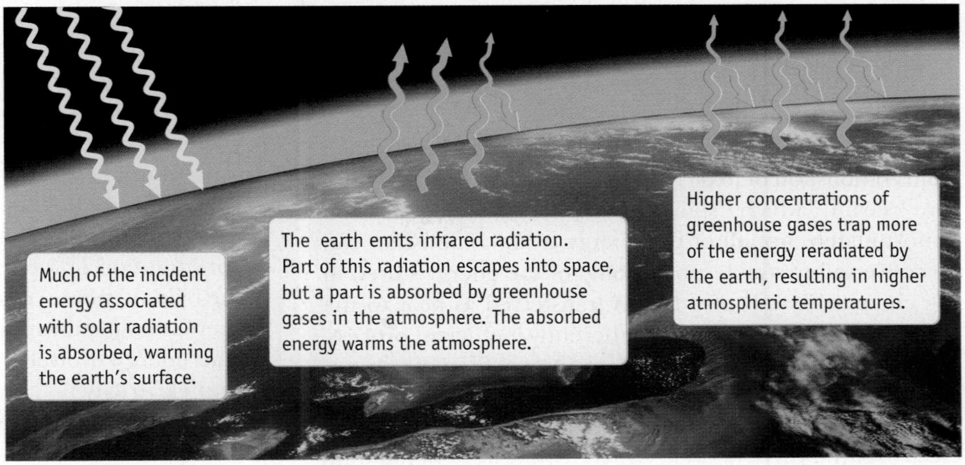

Figure 6 **The greenhouse effect.**

emphasis on renewable energies is due to the concern for the earth's climate. (For more on the greenhouse effect see "The Chemistry of the Environment," page 949.)

Another problem due to increased burning of fossil fuels is local and international air pollution. The high temperature and pressure used in the combustion process in automobile engines have the unfortunate consequence of also causing a reaction between atmospheric nitrogen and oxygen that results in some NO formation. The NO can then react further with oxygen to produce nitrogen dioxide. This poisonous, brown gas is further oxidized to form nitric acid, HNO_3, in the presence of water.

$$N_2(g) + O_2(g) \rightarrow 2\ NO(g) \qquad \Delta_rH° = 180.58\ kJ/mol\text{-}rxn$$

$$2\ NO(g) + O_2(g) \rightarrow 2\ NO_2(g) \qquad \Delta_rH° = -114.4\ kJ/mol\text{-}rxn$$

$$3\ NO_2(g) + H_2O(\ell) \rightarrow 2\ HNO_3(aq) + NO(g) \qquad \Delta_rH° = -71.4\ kJ/mol\text{-}rxn$$

To some extent, the amounts of pollutants released can be limited by use of automobile catalytic converters. Catalytic converters are high–surface-area metal grids that are coated with platinum or palladium. These very expensive metals can catalyze a complete combustion reaction, helping to combine oxygen in the air with unburned hydrocarbons or other by-products in the vehicle exhaust. As a result, the combustion products can be converted to water and carbon dioxide (or other oxides), provided they land on the grid of the catalytic converter before exiting the vehicle's tailpipe. However, some nitric acid and NO_2 inevitably remain in automobile exhaust, and they are major contributors to environmental pollution in the form of acid rain and smog. The brown, acidic atmospheres in highly congested cities such as Beijing, Los Angeles, Mexico City, and Houston result largely from the emissions from automobiles (Figure 8). Such pollution problems have led to stricter emission standards for automobiles, and a high priority in the automobile industry (motivated by impending emission standards of such states as California) is the development of low-emission or emission-free vehicles.

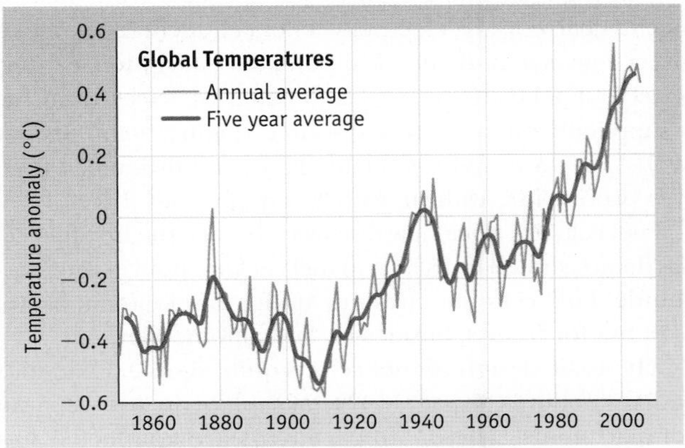

Figure 7 **Variation in global mean surfaces temperatures for 1850 to 2006.** These are relative to the period 1961–1990.

Figure 8 **Smog.** The brown cloud that hangs over Santiago, Chile contains nitrogen oxides emitted by millions of automobiles in that city. Other substances are also present, such as ozone (O_3), nitric oxide (NO_2), carbon monoxide (CO), and water.

Energy in the Future: Choices and Alternatives

Fuel Cells

In the generation of electricity, the energy derived as heat from combustion of fossil fuels is used to produce high-pressure steam, which spins a turbine in a generator. Unfortunately, not all of the energy from combustion can be converted to usable work. Some of the energy stored in the chemical bonds of a fuel is transferred as heat to the surroundings, making this an inherently inefficient process. The efficiency is about 35–40% for a coal-fired steam turbine (and 50–55% for the newer natural gas turbines). A much more efficient process would be possible if mobile electrons, the carriers of electricity, could be generated directly from the chemical bonds themselves, rather than going through an energy conversion process from heat to mechanical work to electricity. Fuel cell technology makes direct conversion of chemical potential energy to electricity possible. Fuel cells are similar to batteries, except that fuel is supplied from an external source (Figure 9 and Section 20.3). They are more efficient than combustion-based energy production, with up to 60% energy conversion.

Fuel cells are not a new discovery. In fact, the first fuel cell was demonstrated in 1839, and fuel cells are used in the Space Shuttle. Fuel cells are currently under development as well as in use for homes, businesses, and automobiles.

The basic design of fuel cells is quite simple. Oxidation and reduction (◀ page 141) take place in two separate compartments. These compartments are connected in a way that allows electrons to flow from the oxidation compartment to the reduction compartment through a conductor such as a wire. In one compartment, a fuel is oxidized, producing electrons. The electrons move through the conductor to the other compartment, where they react with an oxidizing agent, typically O_2. The spontaneous flow of electrons in the electrical circuit constitutes the electric current. While electrons flow through the external circuit, ions move between the two compartments so that the charges in each compartment remain in balance.

The net reaction is the oxidation of the fuel and the consumption of the oxidizing agent. Because the fuel and the oxidant never come directly in contact with each other, there is no combustion and minimal loss of energy as heat. The energy evolved in the reaction is converted directly into electricity.

Hydrogen is the fuel employed in the fuel cells on board the Space Shuttle. The overall reaction in these fuel cells involves the combination of hydrogen and oxygen to form water (Figure 9). Hydrocarbon-based fuels such as methane (CH_4) and methanol (CH_3OH) are also candidates for use as the fuel in fuel cells; for these compounds, the reaction products are CO_2 and H_2O. When methanol is used in fuel cells, for example, the net reaction in the cell is

$$CH_3OH(\ell) + \tfrac{3}{2}\,O_2(g) \rightarrow CO_2(g) + 2\,H_2O(\ell)$$

$$\Delta_r H° = -727 \text{ kJ/mol-rxn or } -23 \text{ kJ/g } CH_3OH$$

Using enthalpies of formation data (Section 5.8), we can calculate that the energy generated is 727 kJ/mol (or 23 kJ/g) of liquid methanol. That is equivalent to 200 watt-hours (W-h) of energy per mol of methanol (1 W = 1 J/s), or 5.0 kW-h per liter of methanol.

Prototypes of phones and laptop computers powered by fuel cells have been developed recently. The small methanol cartridges used to fuel them are no bigger than a standard AA battery, yet they are longer lasting.

Many automobile manufacturers are actively exploring the use of fuel cells that use hydrogen or methanol. Honda's FCX (Figure 10) uses hydrogen (stored in high-pressure tanks) and has a range of 350 miles. The hydro-

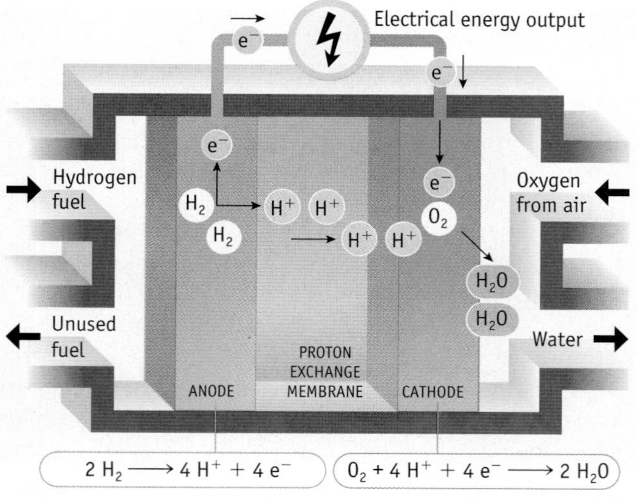

Figure 9 **Hydrogen-oxygen fuel cell.** The cell uses hydrogen gas, which is converted to hydrogen ions and electrons. The electrons flow through the external circuit and are consumed by the oxygen, which, along with H^+ ions, produces water. (H_2 is oxidized to H^+ and is the reducing agent. O_2 is reduced and is the oxidizing agent.)

Figure 10 **A hydrogen fuel cell passenger car from Honda.** The car is powered by a fuel cell using hydrogen and oxygen. The hydrogen is stored in a 171-L, high-pressure (350 atm) tank. It is scheduled for limited sale in Japan and the U. S. in 2008.

gen can be produced at home using a natural gas reformer (see next section).

A Hydrogen Economy

Predictions about the diminished supply of fossil fuels have led some to speculate about alternative fuels. In particular, hydrogen, H_2, has been suggested as a possible choice. The term *hydrogen economy* has been coined to describe the combined processes of producing, storing, and using hydrogen as a fuel. As is the case with fuel cells, the hydrogen economy does not rely on a new energy resource; it merely provides a different scheme for the use of existing resources.

There are reasons to consider hydrogen an attractive option. Oxidation of hydrogen yields almost three times as much energy per gram as the oxidation of fossil fuels. Comparing the combustion of hydrogen with that of propane, a fuel used in some cars, we find that H_2 produces about 2.6 times more energy as heat per gram than propane.

$$H_2(g) + \tfrac{1}{2}\,O_2(g) \rightarrow H_2O(g)$$

$$\Delta_r H° = -241.83 \text{ kJ/mol-rxn or } -119.95 \text{ kJ/g } H_2$$

$$C_3H_8(g) + 5\,O_2(g) \rightarrow 3\,CO_2(g) + 4\,H_2O(g)$$

$$\Delta_r H° = -2043.15 \text{ kJ/mol-rxn or } -46.37 \text{ kJ/g } C_3H_8$$

Another advantage of using hydrogen instead of a hydrocarbon fuel is that the only product of H_2 oxidation is H_2O, which is environmentally benign.

Some have proposed that hydrogen might be able to replace gasoline in automobiles and natural gas in heating homes and even that it could be used as a fuel to generate electricity or to run industrial processes. Before this can occur, however, there are many practical problems to be solved, including the following as-yet-unmet needs:

- An inexpensive method of producing hydrogen
- A practical means of storing hydrogen
- A distribution system (hydrogen refueling stations)

Perhaps the most serious problem in the hydrogen economy is the task of producing hydrogen. Hydrogen is abundant on Earth, but not as the free element. Thus, elemental hydrogen has to be obtained from its compounds. Currently, most hydrogen is produced industrially from the reaction of natural gas and water by steam reforming at high temperature (Figure 11).

Steam reforming $\quad CH_4(g) + H_2O(g) \rightarrow 3\,H_2(g) + CO(g)$

$$\Delta_r H° = +206.2 \text{ kJ/mol-rxn}$$

Hydrogen can also be obtained from the reaction of coal and water at high temperature (the so-called *water–gas reaction*).

Water–gas reaction $\quad C(s) + H_2O(g) \rightarrow H_2(g) + CO(g)$

$$\Delta_r H° = +131.3 \text{ kJ/mol-rxn}$$

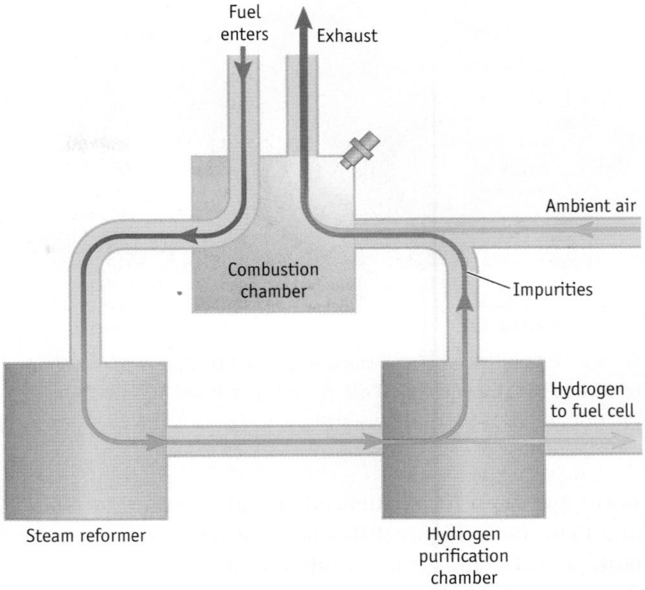

Figure 11 **Steam reforming.** A fuel such as methanol (CH_3OH) or natural gas and water is heated and then passed into a steam reformer chamber. There, a catalyst promotes the decomposition to hydrogen and other compounds such as CO. The hydrogen gas passes out to a fuel cell, and the CO and unused carbon-based compounds are burned in a combustion chamber. A small unit may be suitable for a car or light truck. However, at this time the known technology to carry this out requires temperatures of 700–1000 °C to run the reformer, and the CO can be a poison to the fuel cell.

Both steam reforming and the water–gas reaction are highly endothermic, and both rely on use of a fossil fuel as a raw material. This, of course, makes no sense if the overriding goal is to replace fossil fuels. If the hydrogen economy is ever to take hold, the logical source of hydrogen is water.

$$H_2O(\ell) \rightarrow H_2(g) + \tfrac{1}{2}\,O_2(g)$$

$$\Delta_r H° = +285.83 \text{ kJ/mol-rxn}$$

The electrolysis of water provides hydrogen (◀ page 12) but also requires considerable energy. The first law of thermodynamics tells us that we can get no more energy from the oxidation of hydrogen than we expended to obtain H_2 from H_2O. In fact, we cannot even reach this break-even point because some of the energy produced will inevitably be dispersed (Chapter 19). Hence, the only way to obtain hydrogen from water in the amounts that would be needed and in an economically favorable way is to use a cheap and abundant source of energy to drive this process. A logical candidate is solar energy. Unfortunately, the technology to use solar energy in this way has yet to become practical. This is another problem for chemists and engineers of the future to solve.

Hydrogen storage is another problem to be solved (Figure 12). A number of ways to accomplish this in a vehicle, in your home, or at a distribution point have been proposed. An experimental passenger car from Honda

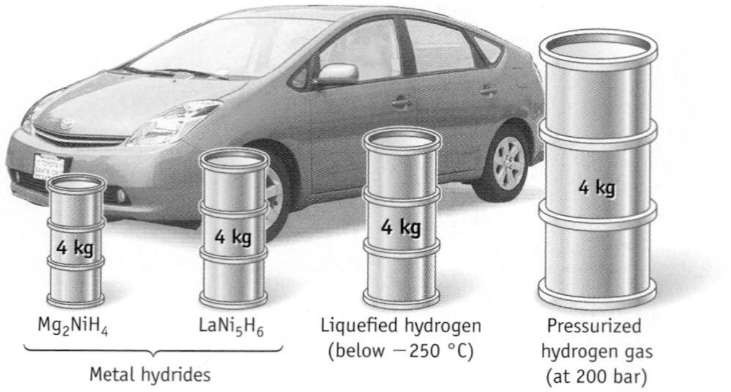

Figure 12 **Comparison of the volumes required to store 4 kg of hydrogen relative to the size of a typical car.** (L. Schlapbach and A. Züttel: *Nature*, Vol. 414, pp. 353–358, 2001.)

stores hydrogen for its fuel cell at high pressure (350 atm) in a 171-L tank (Figure 10). This is larger than the gasoline tanks found in most automobiles, so other storage methods that have smaller volumes and yet are safe are sought.

One possibility for hydrogen storage relies on the fact that certain metals will absorb hydrogen reversibly (Figure 13). When a metal absorbs hydrogen, H atoms fill the holes, called interstices, between metal atoms in a metallic crystal lattice. Palladium, for example, will absorb up to 935 times its volume of hydrogen. This hydrogen can be released upon heating, and the process of absorption and release can be repeated.

No matter how hydrogen is used, it has to be delivered to vehicles and homes in a safe and practical manner. Work has also been done in this area, but many problems remain to be solved. European researchers have found that a tanker truck that can deliver 2400 kg of compressed natural gas (mostly methane) can deliver only 288 kg of H_2 at the same pressure. Although hydrogen oxidation delivers about 2.4 times more energy per gram (119.95 kJ/g) than methane,

$$CH_4(g) + 2\ O_2(g) \rightarrow CO_2(g) + 2\ H_2O(g)$$

$$\Delta_r H° = -802.30 \text{ kJ/mol-rxn or } -50.14 \text{ kJ/g } CH_4$$

the tanker can carry about eight times more methane than H_2. That is, it will take more tanker trucks to deliver the hydrogen needed to power the same number of cars or homes running on hydrogen than those running on methane.

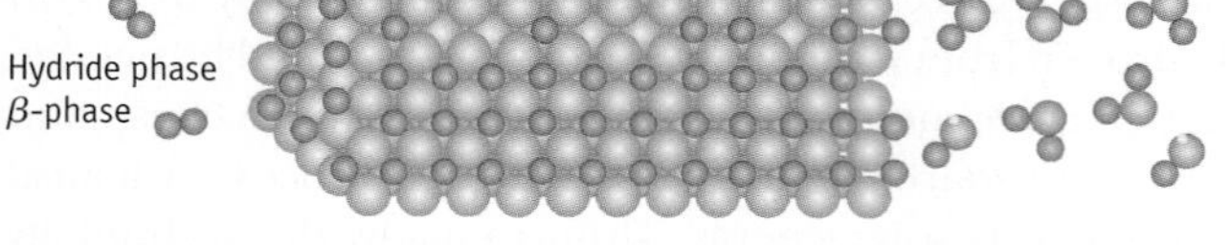

Figure 13 **Hydrogen absorbed by a metal or metal alloy.** Many metals and metal alloys reversibly absorb large quantities of hydrogen. On the left side of the diagram, H_2 molecules are adsorbed onto the surface of a metal. The H_2 molecules can dissociate into H atoms, which form a solid solution with the metal (α-phase). Under higher hydrogen pressures, a true hydride forms in which H atoms become H^- ions (β-phase). On the right side, H atoms can also be adsorbed from solution if the metal is used as an electrode in an electrochemical device.

Figure 14 **Iceland, a "carbon-free," hydrogen-based economy.** A geothermal field in Iceland. The country plans to use such renewable resources to produce hydrogen from water and then to use the hydrogen to produce electricity in fuel cells.

There is an interesting example in which the hydrogen economy has gained a real toehold. Iceland has announced that the country will become a "carbon-free economy." Icelanders plan to rely on hydrogen-powered electric fuel cells to run vehicles and fishing boats. Iceland is fortunate in that two thirds of its energy and all of its electricity already come from renewable sources—hydroelectric and geothermal energy (Figure 14). The country has decided to use the electricity produced by geothermal heat or hydroelectric power to separate water into hydrogen and oxygen. The hydrogen will then be used in fuel cells or combined with CO_2 to make methanol, CH_3OH, a liquid fuel that can either be burned or be used in different types of fuel cells.

Biosources of Energy

Biofuels now supply about 1% of the fuel used worldwide for transportation, but some project that it may contribute 30% to U.S. transportation needs by 2030. Gasoline sold today often contains ethanol, C_2H_5OH. In addition to being a fuel, ethanol improves the burning characteristics of gasoline. Every state in the U.S. now has available a blend of at least 10% ethanol and 90% gasoline. (See "Case Study: The Fuel Controversy: Alcohol and Gasoline," page 240, and the questions on ethanol on page 860.)

Ethanol is readily made by fermentation of glucose from renewable resources such as corn, sugar cane, or agricultural residues. While it may not emerge as the sole fuel of the future, this material is likely to contribute to the phasing-out process of fossil fuels and may be one of the fuel sources in the future. While the U.S. is stepping up its program to produce more ethanol from corn or other plant matter, Brazil has made the production of ethanol from sugar cane a top priority. About 40% of its motor fuel is ethanol. Most new cars sold in Brazil are "flex-fuel" cars that run on either gasoline or ethanol. The most common fuel used in such cars consists of 85% ethanol and 15% petroleum-based fuels and is labeled E85. The U.S. and Brazil produce 70% of the world's ethanol, with the U.S. having moved into the top position in 2006.

While ethanol is currently the predominant biofuel, biodiesel makes up almost 80% of Europe's total biofuel production. This comes from sunflower seeds, rapeseed, soybeans, or used cooking oil. *Biodiesel* is a mixture of methyl esters of organic acids, formed from various plant-derived oils (▶ page 479).

There are several points to be made about the use of ethanol as a fuel. Green plants use the sun's energy to make biomass from CO_2 and H_2O by photosynthesis. The sun is a renewable resource, as, in principle, is the ethanol derived from biomass. In addition, the process recycles CO_2. Plants use CO_2 to create biomass, which is in turn used to make ethanol. In the final step in this cycle, oxidation of ethanol returns CO_2 to the atmosphere. One serious issue concerning the use of corn-derived ethanol is the net energy balance. One has to consider the energy expended in the fuel to run the tractors and trucks, harvest the corn, make the fertilizer, among other things, versus the energy available in the ethanol produced as the end product. Recent analyses and improvements in corn-to-ethanol preparation seem to indicate more energy is available than is used in production, but not by much.

While production of ethanol from corn has been increasing at 20–25% per year, energy analysts believe that non-food plants that can grow on marginal lands with a minimal input of fertilizers are the best hope for biofuels. To re-engineer such cellulose plants as grasses or trees will require a lot of chemical and biological research.

Recent research on ethanol has taken this topic in a new direction. Namely, ethanol can be used as a source of hydrogen. It has been possible to create hydrogen gas from ethanol by using a steam reforming process like the methane-related process. The recently developed method involves the partial oxidation of ethanol mixed with water in a small fuel injector, like those used in cars to deliver gasoline, along with rhodium and cerium catalysts to create hydrogen gas exothermically (Figure 15). The net reaction is

$$C_2H_5OH(g) + 2\ H_2O(g) + \tfrac{1}{2}\ O_2(g) \rightarrow 2\ CO_2(g) + 5\ H_2(g)$$

The standard enthalpy of this reaction is approximately −70 kJ/mol-rxn (or about 1.5 kJ/g of ethanol).

For further insight into this process, let us analyze the overall energy cycle, starting with the photosynthetic combination of CO_2 and water to generate glucose (Figure 16). The sun provides the initial 2540 kJ input of energy for this cycle to produce 1 mol of glucose ($C_6H_{12}O_6$). The sugar is then converted to 2 mol of ethanol per 1 mol of glucose. This conversion process requires a small energy input, 20 kJ. At this point, hydrogen can be generated exothermically using the catalytic fuel-injector method described earlier. Once the hydrogen is generated, it can be used in a hydrogen fuel cell to produce electricity and water.

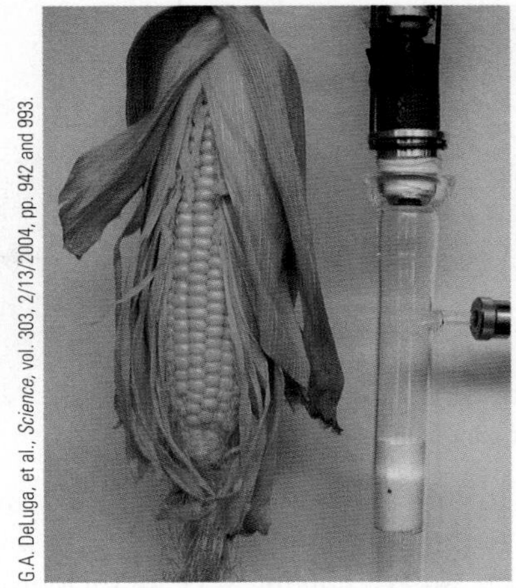

Figure 15 Hydrogen from ethanol. Ethanol can be obtained by fermentation of corn. In a prototype reactor (right), ethanol, water, and oxygen are converted by a catalyst (glowing white solid) to hydrogen (and CO_2).

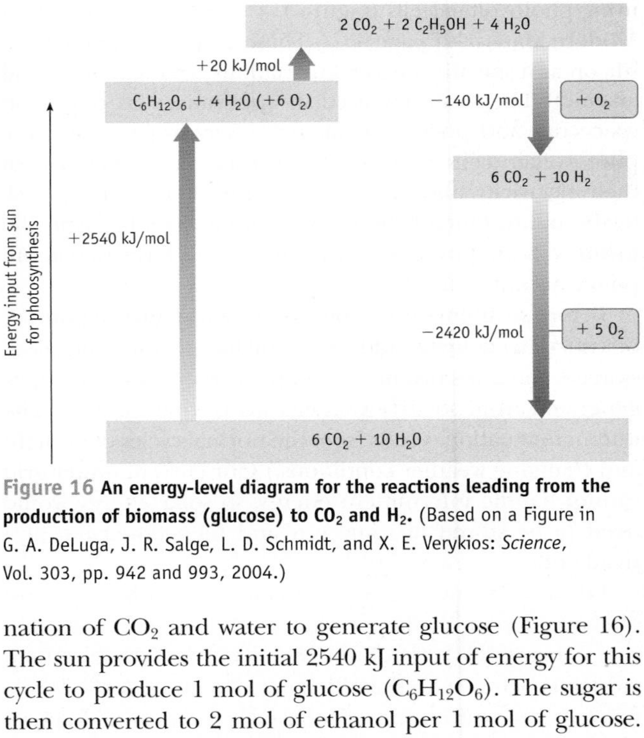

Figure 16 An energy-level diagram for the reactions leading from the production of biomass (glucose) to CO_2 and H_2. (Based on a Figure in G. A. DeLuga, J. R. Salge, L. D. Schmidt, and X. E. Verykios: *Science*, Vol. 303, pp. 942 and 993, 2004.)

Solar Energy

Every year, the earth's surface receives about 10 times as much energy from sunlight as is contained in all the known reserves of coal, oil, natural gas, and uranium combined! The amount of solar energy incident on the earth's surface

Figure 17 **Solar panels on a home.**

is equivalent to about 15,000 times the world's annual consumption of energy. Although solar energy is a renewable resource, today we are making very inefficient use of the sun's energy. Less than 3% of the electricity produced in the United States is generated using solar energy.

How might the sun's energy be better exploited? One strategy is the direct conversion of solar energy to electricity using photovoltaic cells (Figure 17) (see "The Chemistry of Modern Materials," page 657). These devices are made from silicon and specific metal combinations (often gallium and arsenic). They are now used in applications as diverse as spacecraft and pocket calculators. Many homes today use photovoltaic cells to provide a substantial percentage of their electricity, and what they don't use they can sell back to the utility. One of the largest photovoltaic farms in the world is located in southern Germany and has a maximum power output of 5 MW.

Before solar energy can be a viable alternative, a number of issues need to be addressed, including the collection, storage, and transmission of energy. Furthermore, electricity generated from solar power stations is intermittent. (The output fluctuation results from the normal cycles of daylight and changing weather conditions.) Our current power grid cannot handle intermittent energy, so solar energy would need to be stored in some way and then doled out at a steady rate.

Likewise, we need to find ways to make solar cells cost effective. Research has produced photovoltaic cells that can convert 20–30% of the energy that falls on them (which is better than the efficiency of a green leaf). While the cost per watt output of solar cells has declined appreciably over the last few decades, currently, 1 kW-h of energy generated from solar cells costs about 35 cents, compared to about 4–5 cents per kW-h generated from fossil fuels.

WHAT DOES THE FUTURE HOLD FOR ENERGY?

Our society is at an energy crossroads. The modern world is increasingly reliant on energy, but we have built an energy infrastructure that depends primarily on a type of fuel that is not renewable. Fossil fuels provide an inexpensive and simple approach for providing energy, but it is increasingly evident that their use also has serious drawbacks, among them climate change and atmospheric contamination. Alternative fuels, especially from renewable sources, and new ways of converting energy do exist. A great deal more research and resources must be put into them to make them affordable and reliable, however. This is where the study of chemistry fits into the picture. Chemists will have a great deal of work to do in coming years to develop new means of generating, storing, and delivering energy. Meanwhile, numerous ways exist to conserve the resources we have. Ultimately, it will be necessary to bear in mind the various benefits and drawbacks of each technology so that they can be combined in the most rational ways, rather than remaining in a system that is dependent on a single type of nonrenewable energy resource.

SUGGESTED READINGS

1. R. A. Hinrichs and M. Kleinbach: *Energy—Its Use and the Environment*, 3rd ed. Orlando, Harcourt, 2002.
2. M. L. Wald: "Questions About a Hydrogen Economy," *Scientific American*, pp. 67–73, May 2004.
3. U.S. Department of Energy: Energy Efficiency and Renewable Energy, www1.eere.energy.gov/hydrogenandfuelcells. Accessed November 2007.
4. G. T. Miller: *Living in the Environment*, 12th ed. Philadelphia, Brooks/Cole, 2001.
5. L. D. Burns, J. B. McCormick, and C. E. Borroni-Bird: "Vehicle of Change," *Scientific American*, pp. 64–73, October 2002.
6. M. S. Dresselhaus and I. L. Thomas: "Alternative Energy Technologies," *Nature*, Vol. 414, pp. 332–337, November 15, 2001.
7. M. R. Simmons: *Twilight in the Desert: the Coming Saudi Oil Shock and the World Economy*, New York, Wiley & Sons, 2005.
8. F. Keppler and T. Rockmann: "Methane, Plants and Climate Change," *Scientific American*, pp. 52–57, February 2007.
9. S. Ashley: "Diesels Come Clean," *Scientific American*, pp. 66–71, March 2007.
10. Special Issue: "Energy's Future Beyond Carbon," *Scientific American*, September 2006.
11. V. Smil: *Energy at the Crossroads*, Cambridge, MIT Press, 2003.
12. P. Hoffman: *Tomorrow's Energy: Hydrogen, Fuel Cells, and the Prospects for a Cleaner Planet*, Cambridge, MIT Press, 2002.
13. Worldwatch Institute: Biofuels for Transport: *Global Potential and Implications for Sustainable Agriculture and Energy in the 21st Century*, New York, 2007.

STUDY QUESTIONS

Blue-numbered questions have answers in Appendix P and fully-worked solutions in the *Student Solutions Manual*.

1. Hydrogen can be produced using the reaction of steam (H_2O) with various hydrocarbons. Compare the mass of H_2 expected from the reaction of steam with 100. g each

of methane, petroleum, and coal. (Assume complete reaction in each case. Use CH_2 and C as the representative formulas for petroleum and coal, respectively.)

2. Use the value for "energy released" in kilojoules per gram from gasoline listed in Table 2 to estimate the percentage of carbon, by mass in gasoline. (Hint: Compare the value for gasoline to the $\Delta_r H°$ values for burning pure C and H_2.)

3. Per capita energy consumption in the United States was equated to the energy obtained by burning 70. lb of coal per day. Use enthalpy of formation data to calculate the energy evolved, in kilojoules, when 70. lb of coal is burned. (Assume the enthalpy of combustion of coal is 33 kJ/g.)

4. Some gasoline contains 10% (by volume) ethanol. Using enthalpy of formation data in Appendix L, calculate the enthalpy change for the combustion of 1.00 g of ethanol to CO_2(g) and H_2O(g). Compare this value to the enthalpy change for the combustion of 1.00 g of ethane to the same products. Why should you expect the energy evolved in the combustion of ethanol to be less than the energy evolved in the combustion of ethane?

5. Energy consumption in the United States amounts to the equivalent of the energy obtained by burning 7.0 gal of oil or 70. lb of coal per day per person. Using data in Table 2, carry out calculations to show that the energy evolved from these quantities of oil and coal is approximately equivalent. The density of fuel oil is approximately 0.8 g/mL.

6. The energy required to recover aluminum is one third of the energy required to prepare aluminum from Al_2O_3 (bauxite). How much energy is saved by recycling 1.0 lb (= 454 g) of aluminum?

7. The enthalpy of combustion of isooctane (C_8H_{18}) is 5.45×10^3 kJ/mol. Calculate the enthalpy change per gram of isooctane and per liter of isooctane (d = 0.688 g/mL). (Isooctane is one of the many hydrocarbons in gasoline, and its enthalpy of combustion will approximate the energy obtained when gasoline burns.)

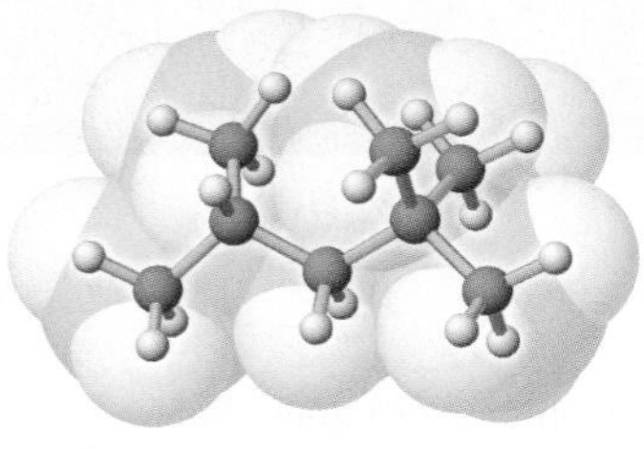

Isooctane
C_8H_{18}

8. Calculate the energy used, in kilojoules, to power a 100-W light bulb continuously over a 24-h period. How much coal would have to be burned to provide this quantity of energy, assuming that the enthalpy of combustion of coal is 33 kJ/g and the power plant has an efficiency of 35%? [Electrical energy for home use is measured in kilowatt hours (kW-h). One watt is defined as 1 J/s, and 1 kW-h is the quantity of energy involved when 1000 W is dispensed over a 1.0-h period.]

9. Major home appliances purchased in the United States are now labeled (with bright yellow "Energy Guide" tags) showing anticipated energy consumption. The tag on a recently purchased washing machine indicated the anticipated energy use would be 940 kW-h per year. Calculate the anticipated annual energy use in kilojoules. (See Question 8 for a definition of kilowatt-hour.) At 8 cents/kW-h, how much would this cost per month to operate?

10. Define the terms *renewable* and *nonrenewable* as applied to energy resources. Which of the following energy resources are renewable: solar energy, coal, natural gas, geothermal energy, wind power?

11. Confirm the statement in the text that oxidation of 1.0 L of methanol to form CO_2(g) and $H_2O(\ell)$ in a fuel cell will provide at least 5.0 kW-h of energy. (The density of methanol is 0.787 g/mL.)

12. List the following substances in order of energy released per gram: C_8H_{18}, H_2, C(s), CH_4. (See Question 7 for the enthalpy of combustion of C_8H_{18}.)

13. A parking lot in Los Angeles receives an average of 2.6×10^7 J/m^2 of solar energy per day in the summer. If the parking lot is 325 m long and 50.0 m wide, what is the total quantity of energy striking the area per day?

14. Your home loses energy in the winter through doors, windows, and any poorly insulated walls. A sliding glass door (6 ft × 6.5 ft with 0.5 in. of insulating glass) allows 1.0×10^6 J/h to pass through the glass if the inside temperature is 22 °C (72 °F) and the outside temperature is 0 °C (32 °F). What quantity of energy, expressed in kilojoules, is lost per day? Assume that your house is heated by electricity. How many kilowatt-hours of energy are lost per day through the door? (See Question 8.)

15. Palladium metal can absorb up to 935 times its volume in hydrogen, H_2. Assuming that 1.0 cm^3 of Pd metal can absorb 0.084 g of the gas, and assuming that the palladium and hydrogen actually formed a compound, what would be the approximate formula of the resulting hydride? (The α-form of hydrogen-saturated palladium has about the same density as palladium metal, 12.0 g/cm^3.)

16. Microwave ovens are highly efficient, compared to other means of cooking. A 1100-watt microwave oven, running at full power for 90 s, will raise the temperature of 1 cup of water (225 mL) from 20 °C to 67 °C. As a rough measure of the efficiency of the microwave oven, compare its energy consumption with the energy required to raise the water temperature.

17. Some fuel-efficient hybrid cars are rated at 55.0 miles per gallon of gasoline. Calculate the energy used to drive 1.00 mile if gasoline produces 48.0 kJ/g and the density of gasoline is 0.737 g/cm^3.

6 The Structure of Atoms

©Arctic-Images/Corbis

Aurora Borealis

The beautiful display of "northern lights" can light up the night sky in the Northern Hemisphere. Colors can range from white to red, green, orange, and others. The display comes from the collision of electrons in the solar wind with atoms and molecules in the upper atmosphere near Earth's poles. This excites the atoms or molecules energetically, and they emit light.

Questions:

This photo shows an aurora with green light.

1. Which has the longer wavelength, red light or green light?
2. Which has the greater energy?
3. How do the colors of light emitted by excited atoms contribute to our understanding of electronic structure?

Answer to these questions are in Appendix Q.

Chapter Goals

See Chapter Goals Revisited (page 295) for Study Questions keyed to these goals and assignable in OWL.

- Describe the properties of electromagnetic radiation.
- Understand the origin of light emitted by excited atoms and its relationship to atomic structure.
- Describe the experimental evidence for particle–wave duality.
- Describe the basic ideas of quantum mechanics.
- Define the four quantum numbers (n, ℓ, m_ℓ, and m_s) and recognize their relationship to electronic structure.

Chapter Outline

The work of the Curies, Rutherford, and other scientists early in the 20th century (Chapter 2 and pages 338–347) led to a model for an atom with a small nucleus of neutrons and protons containing most of the mass and with electrons surrounding the nucleus and filling most of the volume. This is still the basic model of the atom. But, at the beginning of the 21st century, is there a more useful model? Will a more complete model help us understand why atoms of different elements have different properties and help us predict properties of an element? To answer these and other questions, we want to probe more deeply into atomic structure in this and the next chapter.

Much of our understanding of atomic structure comes from a knowledge of how atoms interact with light and how excited atoms emit light. The first three sections of this chapter, therefore, describe radiation and its relation to our modern view of the atom.

Chemistry Now™

Throughout the text this icon introduces an opportunity for self-study or to explore interactive tutorials by signing in at **www.thomsonedu.com/login**.

6.1 Electromagnetic Radiation

In 1864, James Clerk Maxwell (1831–1879) developed a mathematical theory to describe all forms of radiation in terms of oscillating, or wave-like, electric and magnetic fields (Figure 6.1). Thus, light, microwaves, television and radio signals, x-rays, and other forms of radiation are now called **electromagnetic radiation**.

Electromagnetic radiation is characterized by its wavelength and frequency.

- **Wavelength**, symbolized by the Greek letter *lambda* (λ), is defined as the distance between successive crests or high points of a wave (or between successive troughs or low points). This distance can be given in meters, nanometers, or whatever unit of length is convenient.
- **Frequency**, symbolized by the Greek letter *nu* (ν), refers to the number of waves that pass a given point in some unit of time, usually per second. The unit for frequency, written either as s^{-1} or 1/s and standing for 1 per second, is now called a **hertz**.

Wavelength and frequency are related to the speed (c) at which a wave is propagated (Equation 6.1).

$$c \text{ (m/s)} = \lambda \text{ (m)} \times \nu \text{ (1/s)} \qquad (6.1)$$

■ **Heinrich Hertz** Heinrich Hertz (1857–1894) was the first person to send and receive radio waves. He showed that they could be reflected and refracted the same as light, confirming that different forms of radiation such as radio and light waves are related. Scientists now use "hertz" as the unit of frequency.

Active Figure 6.1 Electromagnetic radiation is characterized by its wavelength and frequency. In the 1860s, James Clerk Maxwell developed the currently accepted theory that all forms of radiation are propagated through space as vibrating electric and magnetic fields at right angles to one another. Each field is described by a sine wave (the mathematical function describing the wave). Such oscillating fields emanate from vibrating charges in a source such as a light bulb or radio antenna.

Chemistry Now™ Sign in at www.thomsonedu.com/login and go to the Chapter Contents menu to explore an interactive version of this figure accompanied by an exercise.

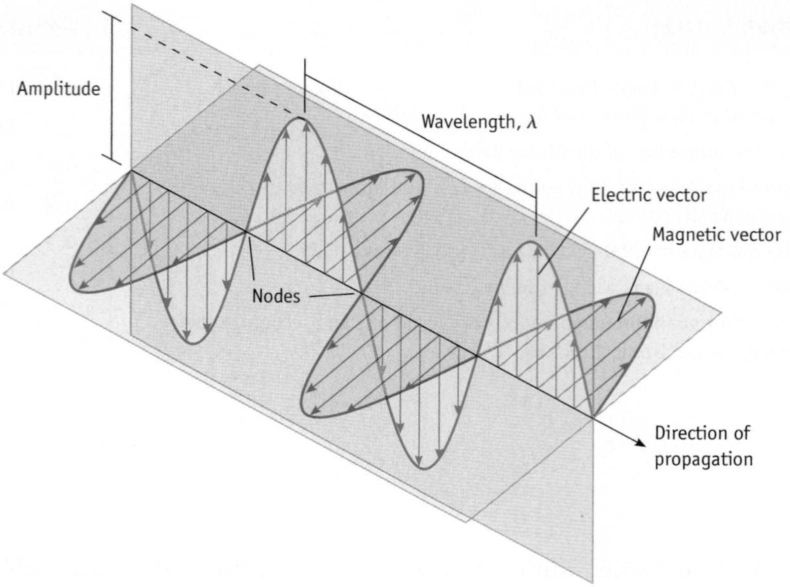

The speed of visible light and all other forms of electromagnetic radiation in a vacuum is a constant, c (= 2.99792458×10^8 m/s; approximately 186,000 miles/s or 1.079×10^9 km/h). For calculations, we will generally use the value of c with four or fewer significant figures.

Different types of electromagnetic radiation are related, as shown in Figure 6.2. Notice that visible light is only a very small portion of the total spectrum of electromagnetic radiation. Ultraviolet (UV) radiation, the radiation that can lead to sunburn, has wavelengths shorter than those of visible light. X-rays and γ-rays, the latter emitted in the process of radioactive disintegration of some atoms, have even shorter wavelengths. At wavelengths longer than those of visible light, we first encounter infrared radiation (IR). At even longer wavelengths is the radiation used in microwave ovens and in television and radio transmissions.

■ **Speed of Light** The speed of light passing through a substance (air, glass, water, for example) depends on the chemical constitution of the substance. It is also dependent on the wavelength of the light. This is the basis for using a glass prism to disperse light and is the explanation for rainbows.

Chemistry Now™

Sign in at **www.thomsonedu.com/login** and go to Chapter 6 Contents to see:

- Screen 6.3 for a tutorial on **calculating the frequency of ultraviolet light** and for a tutorial on **calculating the wavelength of visible light**
- Screen 6.4 for a simulation exploring **the wavelength and frequency of visible light**

EXAMPLE 6.1 Wavelength–Frequency Conversions

Problem The frequency of the radiation used in microwave ovens sold in the United States is 2.45 GHz. (GHz stands for "gigahertz"; 1 GHz = 10^9 1/s.) What is the wavelength of this radiation in meters?

Strategy Rearrange Equation 6.1 to solve for λ, and then substitute the appropriate values into this equation.

Solution

$$\lambda = \frac{c}{\nu} = \frac{2.998 \times 10^8 \text{ m/s}}{2.45 \times 10^9 \text{ 1/s}} = 0.122 \text{ m}$$

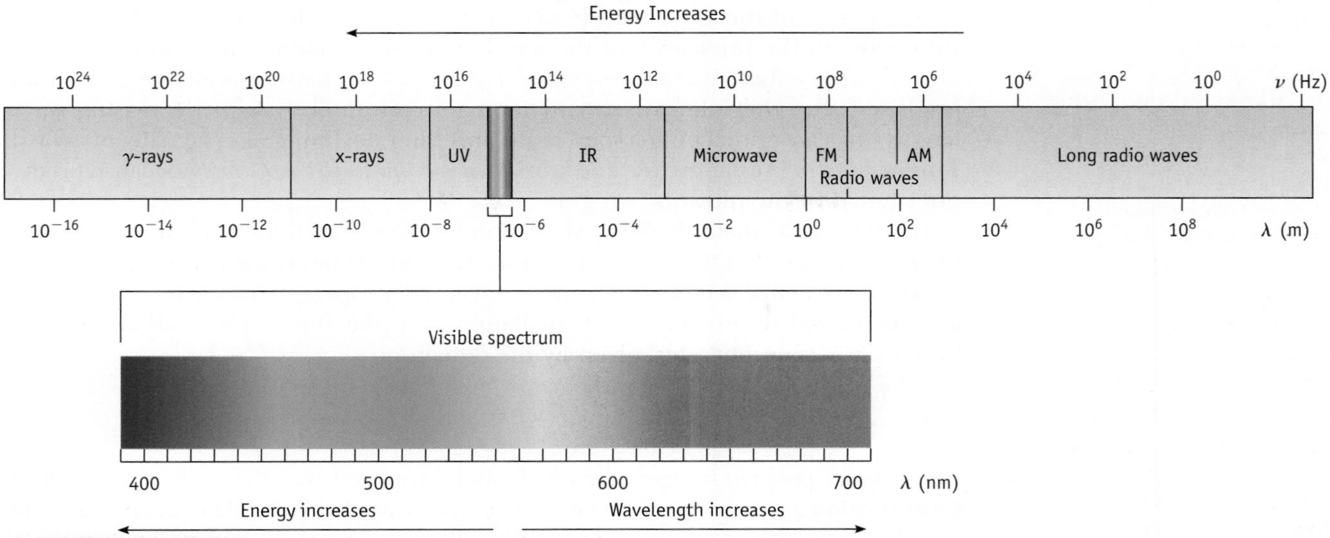

Active Figure 6.2 **The electromagnetic spectrum.** Visible light (enlarged portion) is a very small part of the entire spectrum. The radiation's energy increases from the radio-wave end of the spectrum (low frequency, ν, and long wavelength, λ) to the γ-ray end (high frequency and short wavelength).

ChemistryNow™ Sign in at www.thomsonedu.com/login and go to the Chapter Contents menu to explore an interactive version of this figure accompanied by an exercise.

EXERCISE 6.1 Radiation, Wavelength, and Frequency

(a) Which color in the visible spectrum has the highest frequency? Which has the lowest frequency?

(b) Is the frequency of the radiation used in a microwave oven higher or lower than that from your favorite FM radio station (for example, 91.7 MHz), where 1 MHz (megahertz) = 10^6 1/s?

(c) Is the wavelength of x-rays longer or shorter than that of ultraviolet light?

6.2 Quantization: Planck, Einstein, Energy, and Photons

Planck's Equation

If you heat a piece of metal to a high temperature, electromagnetic radiation is emitted with wavelengths that depend on temperature. At lower temperatures, the color is a dull red (Figure 6.3a). As the temperature increases, the red color brightens, and at even higher temperatures a brilliant white light is emitted.

Your eyes detect the radiation that occurs in the visible region of the electromagnetic spectrum. However, radiation with wavelengths both shorter (in the ultraviolet region) and longer (in the infrared region) than those of visible light is also given off by the hot metal (Figure 6.3b). In addition, it is observed that the wavelength of the most intense radiation is related to temperature: as the temperature of the metal is raised, the maximum intensity shifts toward shorter wavelengths (Figure 6.3b). This corresponds to the change in color observed as the temperature is raised.

At the end of the 19th century, scientists were not able to explain the relationship between the intensity and the wavelength for radiation given off by a heated object (often called *blackbody radiation,* Figure 6.3c). Theories available at the time predicted that the intensity should increase continuously with decreasing wavelength, instead of reaching a maximum and then declining as is actually observed. This perplexing situation became known as the *ultraviolet catastrophe* because predictions failed in the ultraviolet region.

In 1900, a German physicist, Max Planck (1858–1947), offered an explanation. Planck assumed that the electromagnetic radiation emitted was caused by vibrating atoms (called *oscillators*) in the heated object. He proposed that each oscillator had a fundamental frequency (ν) of oscillation and that the emitted radiation could have only certain energies, given by the equation

$$E = nh\nu$$

where n is a positive integer. That is, Planck proposed that the energy is *quantized.* **Quantization** means that only certain energies are allowed. The proportionality constant h in the equation is now called **Planck's constant** and its experimental

Photos: Charles D. Winters

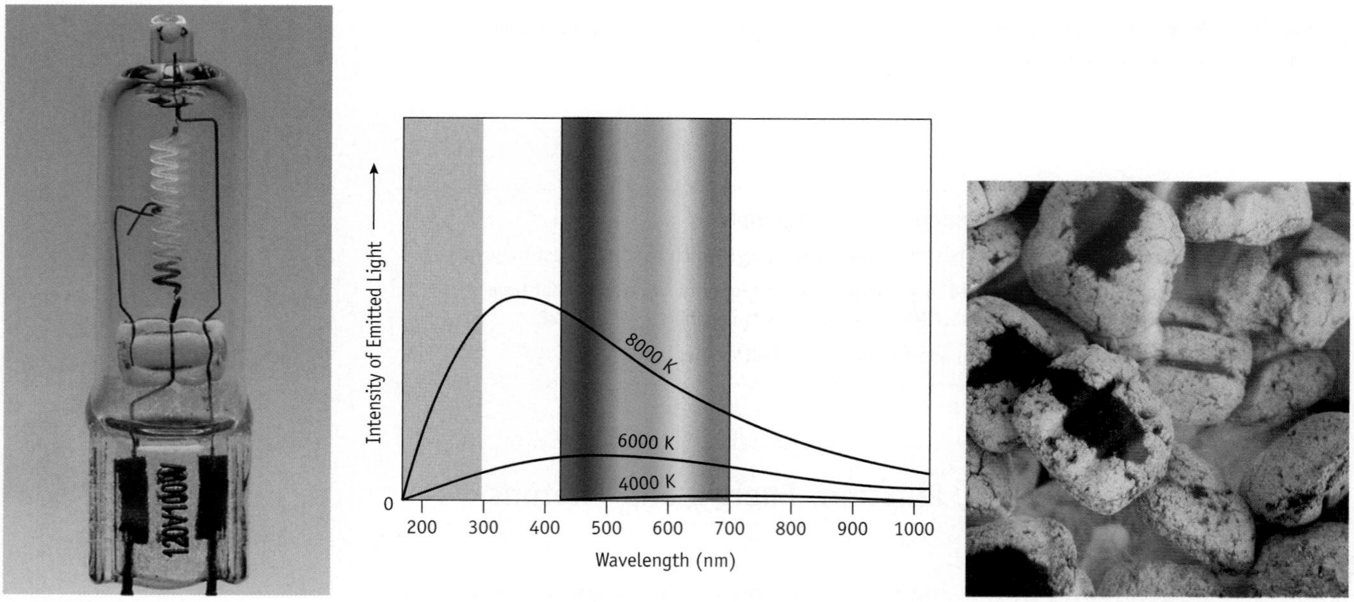

(a) Light emitted by a heated metal.

(b) The spectrum of light emitted by a heated metal at different temperatures.

(c) Blackbody radiation from burning charcoal.

FIGURE 6.3 The radiation given off by a heated body. (a) The heated filament of an incandescent bulb emits radiation at the long wavelength or red end of the visible spectrum. (b) When an object is heated, it emits radiation covering a spectrum of wavelengths. For a given temperature, some of the radiation is emitted at long wavelengths and some at short wavelengths. Most, however, is emitted at some intermediate wavelength, the maximum in the curve. As the temperature of the object increases, the maximum moves from the red end of the spectrum to the violet end. At still higher temperatures, intense light is emitted at all wavelengths in the visible region, and the maximum in the curve is in the ultraviolet region. The object is described as "white hot." (Stars are often referred to as "red giants" or "white dwarfs," a reference to their temperatures and relative sizes.) (c) In physics a blackbody is a theoretical concept in which a body absorbs all radiation that falls on it. However, it will emit energy with a temperature-dependent wavelength. The light emanating from the spaces between the burning charcoal briquets in this photo is a close approximation to blackbody radiation. The color of the light depends on the temperature of the briquets.

value is $6.6260693 \times 10^{-34}$ J · s. The unit of frequency is 1/s, so the energy calculated using this equation is in joules (J). If an oscillator changes from a higher energy to a lower one, energy is emitted as electromagnetic radiation, where the difference in energy between the higher and lower energy states is

$$\Delta E = E_{\text{higher } n} - E_{\text{lower } n} = \Delta nh\nu$$

If the value of Δn is 1, which corresponds to changing from one energy level to the next lower one for that oscillator, then the energy change for the oscillator and the electromagnetic radiation emitted would have an energy equal to

$$E = h\nu \qquad \textbf{(6.2)}$$

This equation is called **Planck's equation.**

Now, assume as Planck did that there must be a *distribution* of vibrations of atoms in an object—some atoms are vibrating at a high frequency; some are vibrating at a low frequency, but most have some intermediate frequency. The few atoms with high-frequency vibrations are responsible for some of the light, as are those few with low-frequency vibrations. However, most of the light must come from the majority of the atoms that have intermediate vibrational frequencies. That is, a spectrum of light is emitted with a maximum intensity at some intermediate wavelength, in accord with experiment. The intensity should not become greater and greater on approaching the ultraviolet region. With this realization, the ultraviolet catastrophe was solved.

Einstein and the Photoelectric Effect

As often happens, the explanation of one fundamental phenomenon leads to other important discoveries. A few years after Planck's work, Albert Einstein (1879–1955) incorporated Planck's ideas into an explanation of the photoelectric effect and in doing so changed the model that described electromagnetic radiation.

■ **The Photoelectric Effect** The photoelectric effect is put to use in photoelectric cells, light-operated switches that are commonly used in automatic door openers in stores and elevators.

In the photoelectric effect, electrons are ejected when light strikes the surface of a metal (Figure 6.4), but only if the frequency of the light is high enough. If light with a lower frequency is used, no electrons are ejected, regardless of the light's intensity (its brightness). If the frequency is at or above a minimum, critical frequency, increasing the light intensity causes more electrons to be ejected.

Einstein decided the experimental observations could be explained by combining Planck's equation ($E = h\nu$) with a new idea, that light has particle-like properties. Einstein characterized these massless particles, now called **photons,** as packets of energy, and stated that the energy of each photon is proportional to the frequency of the radiation as defined by Planck's equation. In the photoelectric effect, photons striking atoms on a metal surface will cause electrons to be ejected only if the photons have high enough energy. The greater the number of photons that strike the surface at or above the threshold energy, the greater the number of electrons dislodged. The metal atoms will not lose electrons, however, if no individual photon has enough energy to dislodge an electron from an atom.

■ **The Relationship of Energy, Wavelength, and Frequency**

As frequency (ν) increases, energy (E) increases

$$E = h\nu = \frac{hc}{\lambda}$$

As wavelength (λ) decreases, energy (E) increases

Energy and Chemistry: Using Planck's Equation

Compact disc players use lasers that emit red light with a wavelength of 685 nm. What is the energy of one photon of this light? What is the energy of 1 mole of photons of red light? To answer these questions, first convert the wavelength to the frequency of the radiation, and then use the frequency to calculate the energy per

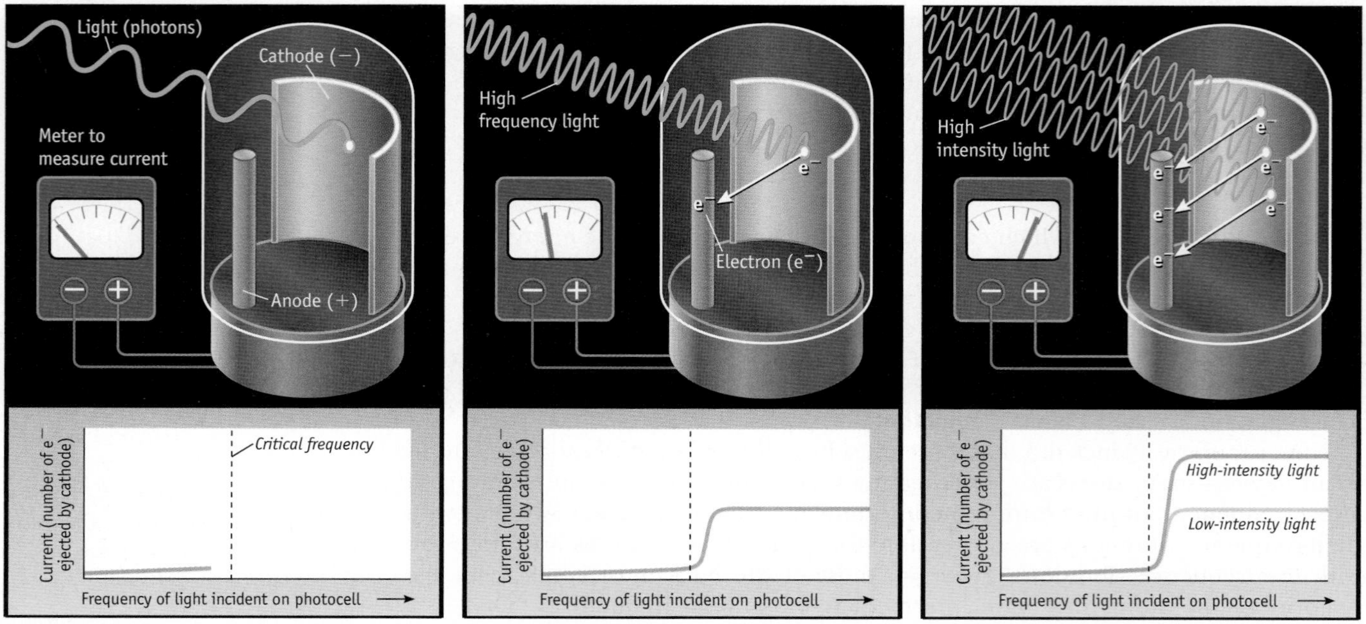

(a) A photocell operates by the photoelectric effect. The main part of the cell is a light-sensitive cathode. This is a material, usually a metal, that ejects electrons if struck by photons of light of sufficient energy. No current is observed until the critical frequency is reached.

(b) When light of higher frequency than the minimum is used, the excess energy of the photon allows the electron to escape the atom with greater velocity. The ejected electrons move to the anode, and a current flows in the cell. Such a device can be used as a switch in electric circuits.

(c) If higher intensity light is used, the only effect is to cause more electrons to be released from the surface. The onset of current is observed at the same frequency as with lower intensity light, but more current flows.

FIGURE 6.4 A photoelectric cell.

photon. Finally, calculate the energy of a mole of photons by multiplying the energy per photon by Avogadro's number:

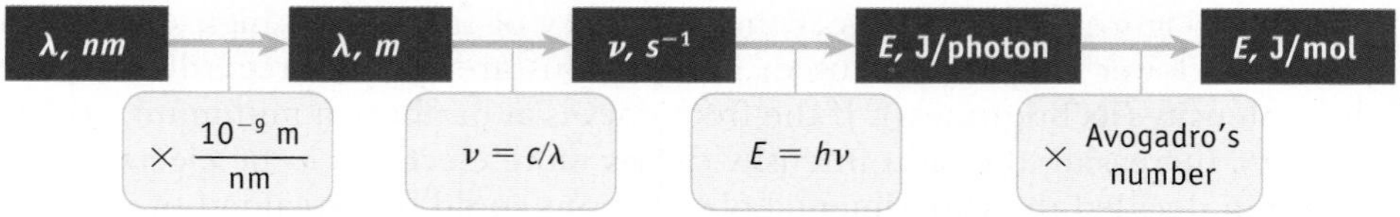

$$
\begin{aligned}
\text{For } \lambda = 685 \text{ nm}, \nu &= 4.38 \times 10^{14} \text{ 1/s (Calculated using Equation 6.1)}\\
E \text{ per photon} = h\nu &= (6.626 \times 10^{-34} \text{ J}\cdot\text{s/photon}) \times (4.38 \times 10^{14} \text{ 1/s})\\
&= 2.90 \times 10^{-19} \text{ J/photon}\\
E \text{ per mole} &= (2.90 \times 10^{-19} \text{ J/photon}) \times (6.022 \times 10^{23} \text{ photons/mol})\\
&= 1.75 \times 10^{5} \text{ J/mol (or 175 kJ/mol)}
\end{aligned}
$$

The energy of red light photons with a wavelength of 685 nm is 175 kJ/mol, whereas the energy of blue light photons (λ = 400 nm) is about 300 kJ/mol. The energy of the blue light photons, and of ultraviolet light photons in particular, is in the range of the energies necessary to break the chemical bonds in proteins. This is what happens if you spend too much time unprotected in the sun (Figure 6.5). In contrast, the energy of light at the red end of the spectrum and infrared radiation has a lower energy and, although it is generally not energetic enough to break chemical bonds, it can affect the vibrations of molecules. We sense infrared radiation as heat, such as the heat given off by a glowing burner on an electric stove.

Chemistry Now™

Sign in at **www.thomsonedu.com/login** and go to Chapter 6 Contents to see Screen 6.5 for a simulation exploring **the relationship between wavelength, frequency, and photon energy,** and for an exercise on **using Planck's equation to calculate wavelength.**

EXERCISE 6.2 Photon Energies

Compare the energy of a mole of photons of orange light (625 nm) with the energy of a mole of photons of microwave radiation having a frequency of 2.45 GHz (1 GHz = $10^9\ s^{-1}$). Which has the greater energy?

Lowell Georgia/Corbis

FIGURE 6.5 Damage from radiation. Various manufacturers have developed mixtures of compounds that protect skin from UVA and UVB radiation. These sunscreens are given "sun protection factor" (SPF) labels that indicate how long the user can stay in the sun without burning. Sunscreens produced by Coppertone, for example, contain the organic compounds 2-ethylhexyl-p-methoxycinnamate and oxybenzene. These molecules absorb UV radiation, preventing it from reaching your skin.

6.3 Atomic Line Spectra and Niels Bohr

If a high voltage is applied to atoms of an element in the gas phase at low pressure, the atoms absorb energy and are said to be "excited." The excited atoms can then emit light, and a familiar example is the colored light from neon advertising signs.

The light falling on Earth from the Sun, or the light emitted by a very hot object, consists of a continuous spectrum of wavelengths (Figures 6.2 and 6.3b). In contrast, the light from excited atoms consists of only a few different wavelengths of light (Figure 6.6). We can demonstrate this by passing a beam of light from excited neon or hydrogen through a prism; only a few colored lines are seen. The spectrum obtained in this manner, such as that for excited H atoms (Figure 6.6), is called a **line emission spectrum.**

The line emission spectra of hydrogen, mercury, and neon are shown in Figure 6.7, and you can see that every element has a unique spectrum. Indeed, the characteristic lines in the emission spectrum of an element can be used in chemical analysis, both to identify the element and to determine how much of it is present.

Charles D. Winters

Neon sign. If a high voltage is applied to a tube containing a gas like neon, light is emitted. Different colors result if different gases are used.

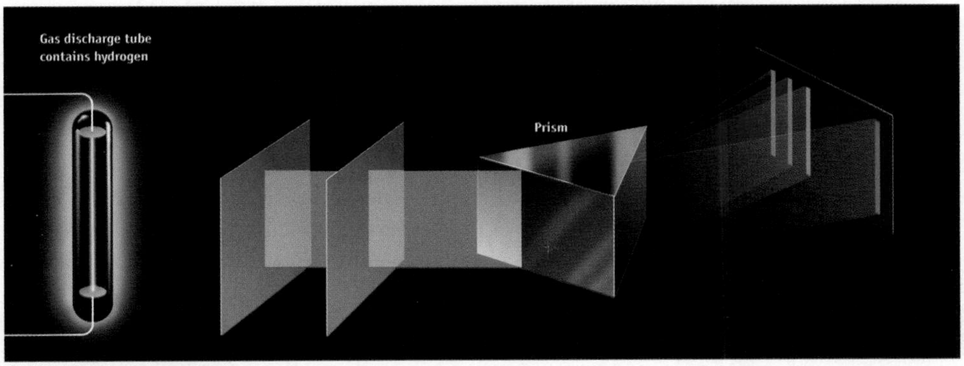

Active Figure 6.6 The line emission spectrum of hydrogen. The emitted light is passed through a series of slits to create a narrow beam of light, which is then separated into its component wavelengths by a prism. A photographic plate or photocell can be used to detect the separate wavelengths as individual lines. Hence, the name "line spectrum" for the light emitted by a glowing gas.

Chemistry Now™ Sign in at www.thomsonedu.com/login and go to the Chapter Contents menu to explore an interactive version of this figure accompanied by an exercise.

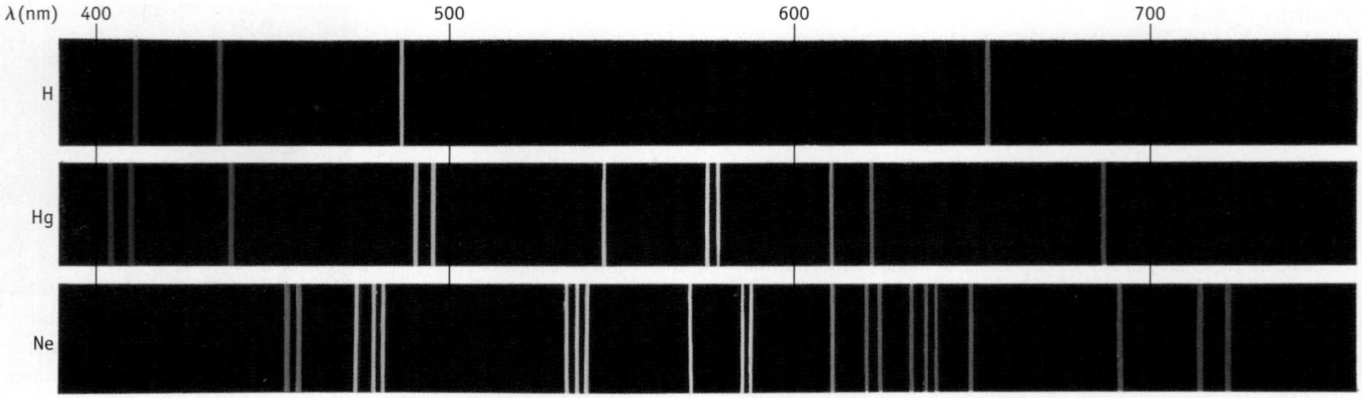

FIGURE 6.7 Line emission spectra of hydrogen, mercury, and neon. Excited gaseous elements produce characteristic spectra that can be used to identify the elements as well as to determine how much of each element is present in a sample.

A goal of scientists in the late 19th century was to explain why excited gaseous atoms emitted light of only certain frequencies. One approach was to look for a mathematical relationship among the observed frequencies because a regular pattern of information implies a logical explanation. The first steps in this direction were taken by Johann Balmer (1825–1898) and later by Johannes Rydberg (1854–1919). From these studies, an equation—now called the **Balmer equation** (Equation 6.3)—was found that could be used to calculate the wavelength of the red, green, and blue lines in the visible emission spectrum of hydrogen (Figure 6.7).

$$\frac{1}{\lambda} = R\left(\frac{1}{2^2} - \frac{1}{n^2}\right) \quad \text{when } n > 2 \tag{6.3}$$

In this equation n is an integer, and R, now called the **Rydberg constant,** has the value 1.0974×10^7 m^{-1}. If $n = 3$, the wavelength of the red line in the hydrogen spectrum is obtained (6.563×10^{-7} m, or 656.3 nm). If $n = 4$, the wavelength for the green line is calculated. Using $n = 5$ and $n = 6$ in the equation gives the wavelengths of the blue lines. The four visible lines in the spectrum of hydrogen atoms are now known as the **Balmer series**.

The Bohr Model of the Hydrogen Atom

Early in the 20th century, the Danish physicist Niels Bohr (1885–1962) proposed a model for the electronic structure of atoms and with it an explanation for the emission spectra of excited atoms. Bohr proposed a planetary structure for the hydrogen atom in which the electron moved in a circular orbit around the nucleus, similar to a planet revolving about the sun. In proposing this model, however, he had to contradict the laws of classical physics. According to classical theories, a charged electron moving in the positive electric field of the nucleus should lose energy, and, eventually, the electron should crash into the nucleus. This is clearly not the case; if it were so, matter would eventually self-destruct. To solve this contradiction, Bohr postulated that there are certain orbits corresponding to particular energy levels where this would not occur. As long as an electron is in one of these energy levels, the system is stable. That is, Bohr introduced *quantization* into the description of electronic structure. By combining this quantization postulate with the laws of motion from classical physics, Bohr derived an equation for the energy possessed by the single electron in the nth orbit (energy level) of the H atom.

Planck's constant
Rydberg constant ↘ ↓ ↙ Speed of light

$$\text{Potential energy of electron in the } n\text{th level} = E_n = -\frac{Rhc}{n^2} \tag{6.4}$$

↑
Principal quantum number

Here, E_n is the energy of the electron (in J/atom); and R, h, and c are constants (the Rydberg constant, Planck's constant, and the speed of light, respectively). The symbol n is a positive, unitless integer called the **principal quantum number**. It can have integral values of 1, 2, 3, and so on.

Equation 6.4 has several important features (which are illustrated in Figure 6.8).

- The quantum number n defines the energies of the allowed orbits in the H atom.
- The energy of an electron in an orbit has a negative value. (Because the negative electron is attracted to the positive nucleus, the energy of attraction is a negative value).
- An atom with its electrons in the lowest possible energy levels is said to be in its **ground state**; for the hydrogen atom, this is the level defined by the quantum number $n = 1$. States for the H atom with higher energies (and $n > 1$) are called **excited states**, and, as the value of n increases, states have less negative energy values.

Bohr also showed that, as the value of n increases, the distance of the electron from the nucleus increases. An electron in the $n = 1$ orbit is closest to the nucleus and has the lowest (most negative) energy. For higher integer values of n, the electron is further from the nucleus and has a higher (less negative) energy.

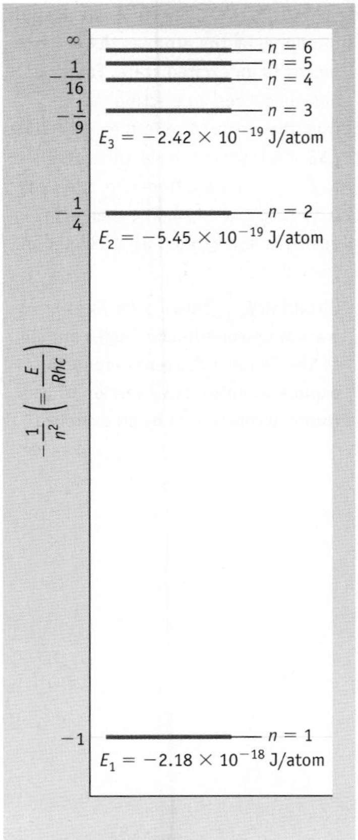

Active Figure 6.8 Energy levels for the H atom in the Bohr model. The energies of the electron in the hydrogen atom depend on the value of the principal quantum number n ($E_n = -Rhc/n^2$). The larger the value of n, the larger the Bohr radius and the less negative the value of the energy. Energies are given in joules per atom (J/atom). Notice that the difference between successive energy levels becomes smaller as n becomes larger.

Chemistry Now™ Sign in at www.thomsonedu.com/login and go to the Chapter Contents menu to explore an interactive version of this figure accompanied by an exercise.

EXAMPLE 6.2 Energies of the Ground and Excited States of the H Atom

Problem Calculate the energies of the $n = 1$ and $n = 2$ states of the hydrogen atom in joules per atom and in kilojoules per mole. What is the difference in energy of these two states in kJ/mol?

Strategy Use Equation 6.4 with the following constants: $R = 1.097 \times 10^7 \text{ m}^{-1}$, $h = 6.626 \times 10^{-34} \text{ J} \cdot \text{s}$, and $c = 2.998 \times 10^8$ m/s.

Solution When $n = 1$, the energy of an electron in a single H atom is

$$E_1 = -Rhc$$
$$E_1 = -(1.097 \times 10^7 \text{ m}^{-1})(6.626 \times 10^{-34} \text{ J} \cdot \text{s})(2.998 \times 10^8 \text{ m/s})$$
$$= -2.179 \times 10^{-18} \text{ J/atom}$$

In units of kJ/mol,

$$E_1 = \frac{-2.179 \times 10^{-18} \text{ J}}{\text{atom}} \times \frac{6.022 \times 10^{23} \text{ atoms}}{\text{mol}} \times \frac{1 \text{ kJ}}{1000 \text{ J}}$$
$$= -1312 \text{ kJ/mol}$$

When $n = 2$, the energy is

$$E_2 = -\frac{Rhc}{2^2} = -\frac{E_1}{4} = -\frac{2.179 \times 10^{-18} \text{ J/atom}}{4}$$
$$= -5.448 \times 10^{-19} \text{ J/atom}$$

Active Figure 6.9 Absorption of energy by the atom as the electron moves to an excited state. Energy is absorbed when an electron moves from the $n = 1$ state to the $n = 2$ state ($\Delta E > 0$). When the electron returns to the $n = 1$ state from $n = 2$, energy is evolved ($\Delta E < 0$). The change in energy is 984 kJ/mol, as calculated in Example 6.2.

Chemistry Now™ Sign in at www.thomsonedu.com/login and go to the Chapter Contents menu to explore an interactive version of this figure accompanied by an exercise.

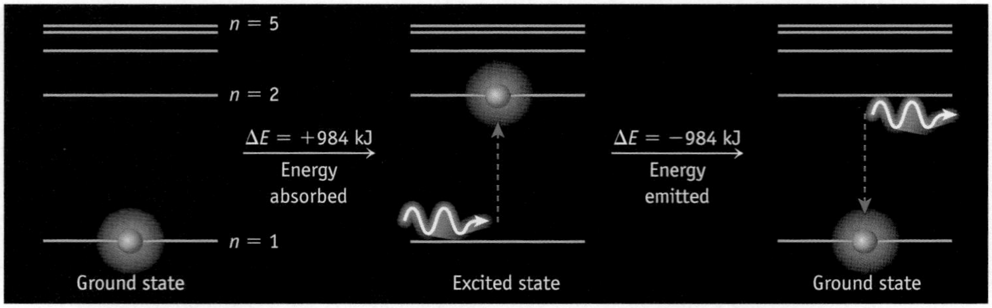

In units of kJ/mol,

$$E_1 = \frac{-5.448 \times 10^{-19}\ \text{J}}{\text{atom}} \times \frac{6.022 \times 10^{23}\ \text{atoms}}{\text{mol}} \times \frac{1\ \text{kJ}}{1000\ \text{J}}$$
$$= -328.1\ \text{kJ/mol}$$

The difference in energy, ΔE, between the first two energy states of the H atom is

$$\Delta E = E_2 - E_1 = (-328.1\ \text{kJ/mol}) - (-1312\ \text{kJ/mol}) = 984\ \text{kJ/mol}$$

Comment The calculated energies are negative, with E_1 more negative than E_2. The $n = 2$ state is higher in energy than the $n = 1$ state by 984 kJ/mol. Also, be sure to notice that 1312 kJ/mol is the value of Rhc multiplied by Avogadro's number N_A (i.e., N_ARhc). This will be useful in future calculations.

EXERCISE 6.3 Electron Energies

Calculate the energy of the $n = 3$ state of the H atom in **(a)** joules per atom and **(b)** kilojoules per mole.

The Bohr Theory and the Spectra of Excited Atoms

Bohr's theory describes electrons as having only specific orbits and energies. If an electron moves from one energy level to another, then energy must be absorbed or evolved. This idea allowed Bohr to relate energies of electrons and the emission spectra of hydrogen atoms.

To move an electron from the $n = 1$ state to an excited state, such as the $n = 2$ state, the atom must absorb energy. When E_{final} has $n = 2$ and $E_{initial}$ has $n = 1$, then 984 kJ of energy must be absorbed (Figure 6.9). This is the difference in energy between final and initial states:

$$\Delta E = E_{\text{final state}} - E_{\text{initial state}} = (-N_ARhc/2^2) - (-N_ARhc/1^2) = (0.75)N_ARhc = 984\ \text{kJ/mol}$$

(where $N_ARhc/1^2$ is the energy in kJ/mol calculated in Example 6.2 for an electron in the $n = 1$ energy level in the H atom). Moving an electron from the first to the second energy state requires input of 984 kJ/mol of atoms—no more and no less. If $0.7N_ARhc$ or $0.8N_ARhc$ is provided, a transition between states is not possible. Requiring a specific and precise amount of energy is a consequence of quantization.

Moving an electron from a state of low n to one of higher n requires that energy be absorbed. The opposite process, in which an electron "falls" from a level of higher n to one of lower n, leads to emission of energy (Figure 6.9). For example, for a transition from the $n = 2$ level to $n = 1$ level,

$$\Delta E = E_{\text{final state}} - E_{\text{initial state}} = -984\ \text{kJ/mol}$$

The negative sign indicates energy is evolved; 984 kJ must be *emitted* per mole of H atoms.

We can now visualize the mechanism by which the characteristic line emission spectrum of hydrogen originates according to the Bohr model. Energy is provided to the atoms from an electric discharge or by heating. Depending on how much energy is added, some atoms have their electrons excited from the $n = 1$ state to the $n = 2$, 3, or even higher states. After absorbing energy, these electrons can return to any lower level (either directly or in a series of steps), releasing energy. We observe this released energy as photons of electromagnetic radiation, and, because only certain energy levels are possible, only photons with particular energies and wavelengths are observed. A line spectrum is thus predicted for the emission spectrum of H atoms, which is exactly what is observed.

The energy of any emission line (in kJ/mol) for excited hydrogen atoms can be calculated using Equation 6.5.

$$\Delta E = E_{\text{final}} - E_{\text{initial}} = -N_A Rhc \left(\frac{1}{n^2_{\text{final}}} - \frac{1}{n^2_{\text{initial}}} \right) \quad \textbf{(6.5)}$$

For hydrogen, a series of emission lines having energies in the ultraviolet region (called the **Lyman series**; Figure 6.10) arises from electrons moving from states with $n > 1$ to the $n = 1$ state. The series of lines that have energies in the visible region—the **Balmer series**—arises from electrons moving from states with $n > 2$ to the $n = 2$ state. There are also series of lines in the infrared spectral region, arising from transitions from higher levels to the $n = 3$, 4 or 5 levels.

Bohr's model, introducing quantization into a description of the atom, tied the unseen (the structure of the atom) to the seen (the observable lines in the hydrogen spectrum). Agreement between theory and experiment is taken as evidence

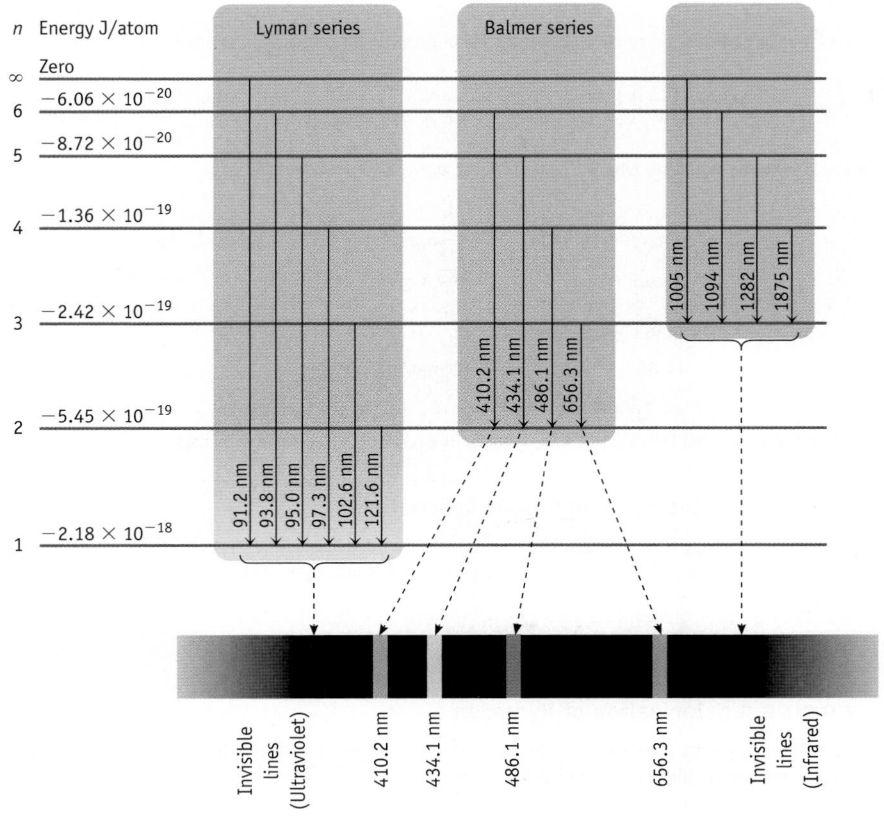

Active Figure 6.10 Some of the electronic transitions that can occur in an excited H atom. The Lyman series of lines in the ultraviolet region results from transitions to the $n = 1$ level. Transitions from levels with values of n greater than 2 to $n = 2$ occur in the visible region (Balmer series; see Figures 6.6 and 6.7). Lines in the infrared region result from transitions from levels with n greater than 3 or 4 to the $n = 3$ or 4 levels. (Only the series ending at $n = 3$ is illustrated.)

Chemistry Now™ Sign in at www.thomsonedu.com/login and go to the Chapter Contents menu to explore an interactive version of this figure accompanied by an exercise.

that the theoretical model is valid. It was apparent, however, that Bohr's theory was inadequate. This model of the atom explained only the spectrum of hydrogen atoms and of other systems having one electron (such as He^+), but failed for all other systems. A better model of electronic structure was needed.

Chemistry Now™

Sign in at **www.thomsonedu.com/login** and go to Chapter 6 Contents to see Screen 6.6 for a simulation and exercise exploring **the radiation emitted when electrons of excited hydrogen atoms return to the ground state,** and for a tutorial on **calculating the wavelength of radiation emitted when electrons change energy levels.**

EXAMPLE 6.3 Energies of Emission Lines for Excited Atoms

Problem Calculate the wavelength of the green line in the visible spectrum of excited H atoms.

Strategy First, locate the green line in Figure 6.10 and determine $n_{initial}$ and n_{final}. Then use Equation 6.5 to calculate the difference in energy, ΔE, between these states. Finally, calculate the wavelength from the value of ΔE.

Solution The green line arises from electrons moving from $n = 4$ to $n = 2$. Using Equation 6.5 where $n_{final} = 2$ and $n_{initial} = 4$, and the value of N_ARhc (1312 kJ/mol), we have

$$\Delta E = E_{final} - E_{initial} = \left(-\frac{N_ARhc}{2^2}\right) - \left(-\frac{N_ARhc}{4^2}\right)$$

$$\Delta E = -N_ARhc\left(\frac{1}{4} - \frac{1}{16}\right) = -N_ARhc(0.1875)$$

$$\Delta E = -(1312 \text{ kJ/mol})(0.1875) = -246.0 \text{ kJ/mol}$$

The wavelength can now be calculated. First, the photon energy, E_{photon}, is expressed as J/photon.

$$E_{photon} = \left(246.0 \frac{\text{kJ}}{\text{mol}}\right)\left(1 \times 10^3 \frac{\text{J}}{\text{kJ}}\right)\left(\frac{1 \text{ mol}}{6.022 \times 10^{23} \text{ photons}}\right) = 4.085 \times 10^{-19} \frac{\text{J}}{\text{photon}}$$

Now apply Planck's equation where $E_{photon} = h\nu = hc/\lambda$, and so $\lambda = hc/E_{photon}$.

$$\lambda = \frac{hc}{E_{photon}} = \frac{\left(6.626 \times 10^{-34} \frac{\text{J}\cdot\text{s}}{\text{photon}}\right)(2.998 \times 10^8 \text{ m}\cdot\text{s}^{-1})}{4.085 \times 10^{-19} \text{ J/photon}}$$

$$= 4.863 \times 10^{-7} \text{ m}$$

$$= (4.863 \times 10^{-7} \text{ m})(1 \times 10^9 \text{ nm/m})$$

$$= 486.3 \text{ nm}$$

Comment The experimental value of 486.1 nm is in excellent agreement with this.

EXERCISE 6.4 Energy of an Atomic Spectral Line

The Lyman series of spectral lines for the H atom, in the ultraviolet region, arises from transitions from higher levels to $n = 1$. Calculate the frequency and wavelength of the least energetic line in this series.

EXERCISE 6.5 Ionization of Hydrogen

Calculate the energy per mole for the process in which hydrogen is ionized [$H(g) \longrightarrow H^+(g) + e^-(g)$]; that is, the energy for the transition from $n = 1$ to $n = \infty$.

Case Study **What Makes the Colors in Fireworks?**

The colors of the beautiful fireworks displays you see on holidays such as July 4 in the U.S. are just the emission spectra of excited atoms. But what is their chemistry? What elements are involved?

Typical fireworks have several important chemical components. For example, there must be an oxidizer. Today, this is usually potassium perchlorate ($KClO_4$), potassium chlorate ($KClO_3$), or potassium nitrate (KNO_3). Potassium salts are used instead of sodium salts because the latter have two important drawbacks. They are hygroscopic—they absorb water from the air—and so do not remain dry on storage. Also, when heated, sodium salts give off an intense, yellow light that is so bright it can mask other colors.

The parts of any fireworks display we remember best are the vivid colors and brilliant flashes. White light can be produced by oxidizing magnesium or aluminum metal at high temperatures. The flashes you see at rock concerts or similar events, for example, are typically $Mg/KClO_4$ mixtures.

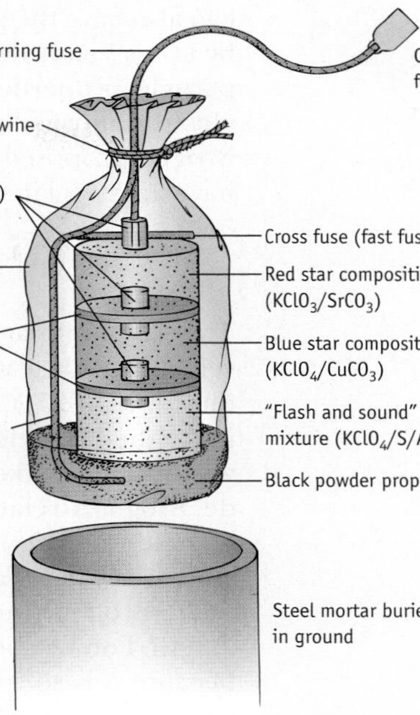

The design of an aerial rocket for a fireworks display. When the fuse is ignited, it burns quickly to the delay fuses at the top of the red star mixture as well as to the black powder propellant at the bottom. The propellant ignites, sending the shell into the air. Meanwhile, the delay fuses burn. If the timing is correct, the shell bursts high in the sky into a red star. This is followed by a blue burst and then a flash and sound.

Emission of light by excited atoms. Flame tests are often used to identify elements in a chemical sample. Shown here are the colors produced in a flame (burning methanol) by NaCl (yellow), $SrCl_2$ (red), and boric acid (green).
(See ChemistryNow Screen 6.1, Chemical Puzzler, for a description of colors in fireworks.)

Yellow light is easiest to produce because sodium salts give an intense light with a wavelength of 589 nm. Fireworks mixtures usually contain sodium in the form of nonhygroscopic compounds such as cryolite, Na_3AlF_6. Strontium salts are most often used to produce a red light, and green is produced by barium salts such as $Ba(NO_3)_2$.

The next time you see a fireworks display, watch for the ones that are blue. Blue has always been the most difficult color to produce. Recently, however, fireworks designers have learned that the best way to get a really good "blue" is to decompose copper(I) chloride at low temperatures. To achieve this effect, CuCl is mixed with $KClO_4$, copper powder, and the organic chlorine-containing compound hexachloroethane, C_2Cl_6.

Questions:

1. *The main lines in the emission spectrum of sodium are at wavelengths (nm) of 313.5, 589, 590, 818, and 819. Which one or ones are most responsible for the characteristic yellow color of excited sodium atoms?*
2. *Does the main emission line for $SrCl_2$ (in the photo) have a higher or lower wavelength than that of the yellow line from NaCl?*
3. *Mg is oxidized by $KClO_4$ to make white flashes. One product of the reaction is KCl. Write a balanced equation for the reaction.*

Answers to these questions are in Appendix Q.

6.4 Particle–Wave Duality: Prelude to Quantum Mechanics

The photoelectric effect demonstrated that light, usually considered to be a wave, can also have the properties of particles, albeit without mass. This fact was pondered by Louis Victor de Broglie (1892–1987), who asked if light can have both wave and particle properties, would matter behave similarly? Could an object such as an electron, normally considered a particle, also exhibit wave properties? In 1925, de Broglie proposed that a free electron of mass m moving with a velocity v should have an associated wavelength λ, calculated by the equation

$$\lambda = \frac{h}{mv} \tag{6.6}$$

This revolutionary idea linked the particle properties of the electron (mass and velocity) with a wave property (wavelength). Experimental proof was soon produced. In 1927, C. J. Davisson (1881–1958) and L. H. Germer (1896–1971), working at the Bell Telephone Laboratories in New Jersey, found that a beam of electrons was diffracted like light waves by the atoms of a thin sheet of metal foil and that de Broglie's relation was followed quantitatively. Because diffraction is best explained based on the wave properties of radiation (Figure 6.11), it follows that electrons can be described as having wave properties in certain situations.

De Broglie's equation suggests that any moving particle has an associated wavelength. For λ to be measurable, however, the product of m and v must be very small because h is so small. A 114-g baseball, traveling at 110 mph, for example, has a large mv product (5.6 kg · m/s) and therefore an incredibly small wavelength, 1.2×10^{-34} m! Such a small value cannot be measured with any instrument now available, nor is such a value meaningful. As a consequence, wave properties are never assigned to a baseball or any other massive object. It is possible to observe wave-like properties only for particles of extremely small mass, such as protons, electrons, and neutrons.

■ **Cathode Rays** Experiments with cathode rays led to the discovery and understanding of electrons. See the interchapter on *Milestones in the Development of the Modern View of Atoms and Molecules*.

Cathode ray tubes, such as were found in television sets before the advent of LCD and plasma TVs, generate a beam of electrons. When the electrons impact the screen, the beam gives rise to tiny flashes of colored light. In contrast to this effect, best explained by assuming electrons are particles, diffraction experiments suggest that electrons are waves. But, how can an electron be both a particle and a wave? In part, we are facing limitations in language; the words "particle" and "wave" accurately describe things encountered on a macroscopic scale. However, they apply less well on the submicroscopic scale associated with subatomic particles.

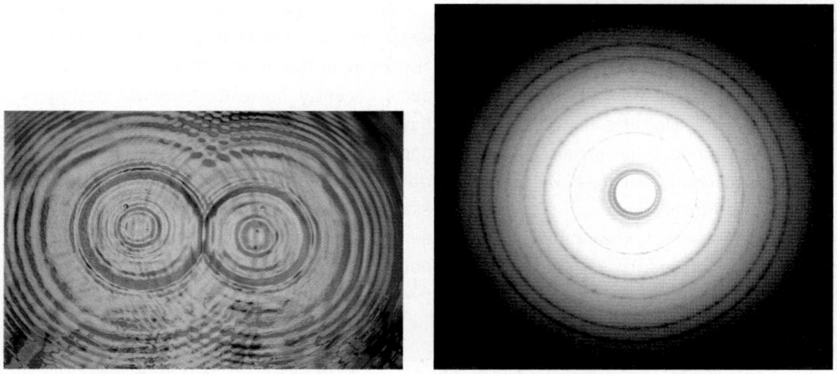

Charles D. Winters

R. K. Bohn, Department of Chemistry, University of Connecticut

(a) Constructive/destructive interference in water waves.

(b) Diffraction of an electron beam by a thin film of magnesium oxide.

FIGURE 6.11 Diffraction. (a) When two water waves come together, constructive and destructive interference occurs. Similar interference patterns are observed when electrons, which have wave properties, encounter atoms in the gas or solid phase. (b) The pattern observed when a thin film of magnesium oxide diffracts a beam of electrons.

In some experiments, electrons behave like particles. In other experiments, we find that they behave like waves. No single experiment can be done to show the electron behaving *simultaneously* as a wave and a particle. Scientists now accept this **wave–particle duality**—that is, the idea that the electron has the properties of both a wave and a particle. Which is observed depends on the experiment.

Chemistry Now™

Sign in at **www.thomsonedu.com/login** and go to Chapter 6 Contents to see Screen 6.8 for a tutorial on **calculating the wavelength of a moving electron.**

EXAMPLE 6.4 Using de Broglie's Equation

Problem Calculate the wavelength associated with an electron of mass $m = 9.109 \times 10^{-28}$ g that travels at 40.0% of the speed of light.

Strategy First, consider the units involved. Wavelength is calculated from h/mv, where h is Planck's constant expressed in units of joule seconds (J · s). As discussed in Chapter 5 (page 214), 1 J = 1 kg · m²/s². Therefore, the mass must be in kilograms and speed in meters per second.

Solution

Electron mass = 9.109×10^{-31} kg

Electron speed (40.0% of light speed) = $(0.400)(2.998 \times 10^8 \text{ m} \cdot \text{s}^{-1}) = 1.20 \times 10^8 \text{ m} \cdot \text{s}^{-1}$

Substituting these values into de Broglie's equation, we have

$$\lambda = \frac{h}{mv} = \frac{6.626 \times 10^{-34}\ (\text{kg} \cdot \text{m}^2/\text{s}^2)(\text{s})}{(9.109 \times 10^{-31}\ \text{kg})(1.20 \times 10^8\ \text{m/s})} = 6.07 \times 10^{-12}\ \text{m}$$

In nanometers, the wavelength is

$$\lambda = (6.07 \times 10^{-12}\ \text{m})(1.00 \times 10^9\ \text{nm/m}) = 6.07 \times 10^{-3}\ \text{nm}$$

Comment The calculated wavelength is about 1/12 of the diameter of the H atom.

EXERCISE 6.6 De Broglie's Equation

Calculate the wavelength associated with a neutron having a mass of 1.675×10^{-24} g and a kinetic energy of 6.21×10^{-21} J. (Recall that the kinetic energy of a moving particle is $E = \frac{1}{2}mv^2$.)

6.5 The Modern View of Electronic Structure: Wave or Quantum Mechanics

How does wave–particle duality affect our model of the arrangement of electrons in atoms? Following World War I, German scientists Erwin Schrödinger (1887–1961), Werner Heisenberg (1901–1976), and Max Born (1882–1970) provided the answer.

In Bohr's model of the atom, both the energy and location (the orbit) for the electron in the hydrogen atom can be described accurately. However, Heisenberg determined that, for a tiny object such as an electron in an atom, it is impossible to determine accurately *both* its position and its energy. That is, any attempt to determine accurately either the location or the energy will leave the other uncertain. This is now known as Heisenberg's **uncertainty principle.**

Born proposed the following application of Heisenberg's idea to understand the arrangement of electrons in atoms: *If we choose to know the energy of an electron in an atom with only a small uncertainty, then we must accept a correspondingly large uncertainty*

■ **History of the Modern View of Structure** For the historical background to efforts to understand atomic structure see the interchapter *Milestones in the Development of the Modern View of Atoms and Molecules* (pages 338–347).

in its position. The importance of this idea is that we can assess only the likelihood, or *probability,* of finding an electron with a given energy within a given region of space. Because electron energy is the key to understanding the chemistry of an atom, chemists accept the notion of knowing only the approximate location of the electron.

Erwin Schrödinger (1887–1961) worked on a comprehensive theory of the behavior of electrons in atoms. Starting with de Broglie's hypothesis that an electron could be described as a wave, Schrödinger developed a model for electrons in atoms that has come to be called **quantum mechanics** or **wave mechanics.** This model used the mathematical equations of wave motion to generate a series of equations called wave equations or **wave functions,** which are designated by the Greek letter ψ (psi).

■ **Wave Functions and Energy** In Bohr's theory, the electron energy for the H atom is given by $E_n = -Rhc/n^2$. Schrödinger's electron wave model gives the same result.

Unlike Bohr's model, Schrödinger's model can be difficult to visualize, and the mathematical approach is complex. Nonetheless, the consequences of the model are important, and understanding its implications is essential to understanding the modern view of the atom. The following points summarize the important issues concerning wave mechanics:

1. An electron in the atom is described as a **standing wave.** If you tie down a string at both ends, as you would the string of a guitar, and then pluck it, the string vibrates as a standing wave (Figure 6.12). A two-dimensional standing wave such as a vibrating string must have two or more points of zero amplitude (called **nodes**), and only certain vibrations are possible. These allowed vibrations have wavelengths of $n(\lambda/2)$, where n is an integer ($n = 1$, 2, 3, ...). In the first vibration illustrated in Figure 6.12, the distance between the ends of the string is half a wavelength, or $\lambda/2$. In the second, the string's length equals one complete wavelength, or $2(\lambda/2)$. In the third vibration, the string's length is $3(\lambda/2)$. That is, for standing waves, vibrations are quantized, and the integer n is a **quantum number.**
2. By defining the electron as a standing wave, quantization is introduced into the description of electronic structure. The mathematics describing a one-dimensional vibrating string requires one "quantum number" (n). Schrödinger's equations for an electron in three-dimensional space requires three quantum numbers: n, ℓ, and m_ℓ, all integers. Only certain combinations of their values are possible as outlined below.

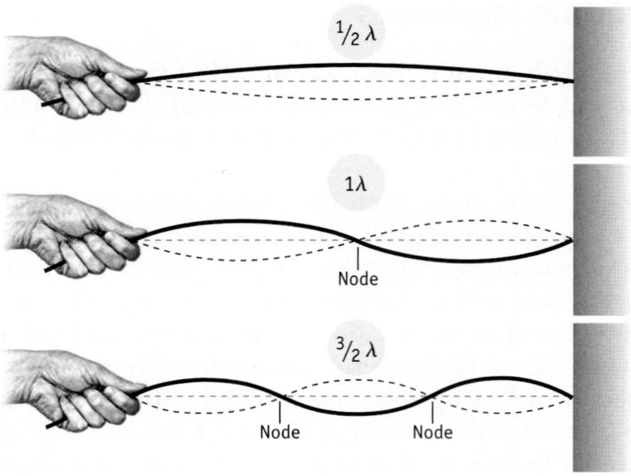

FIGURE 6.12 Standing waves. In the first wave, the end-to-end distance is $(1/2)\lambda$; in the second wave, it is λ, and in the third wave, it is $(3/2)\ \lambda$. (In addition to the nodes marked, there are nodes at the ends of the string.)

3. Each wave function is associated with an allowed energy value. That is, the energy is quantized because only certain values of energy are possible for the electron.
4. The value of the wave function ψ at a given point in space is the amplitude (height) of the wave. This value has both a magnitude and a sign that can be either positive or negative. (Visualize a vibrating string in a guitar or piano, for example. Points of positive amplitude are above the axis of propagation, and points of negative amplitude are below it.)
5. At any point in space, the square of the value of the wave function (ψ^2) defines the *probability* of finding the electron. Scientists refer to this probability as the **electron density.**
6. Schrödinger's theory defines the energy of the electron precisely. The uncertainty principle, however, tells us there must be uncertainty in the electron's position. Thus, we describe the **probability** of the electron being within a certain region in space when in a given energy state. **Orbitals** define the region of space within which an electron of a given energy is most likely to be located.

Quantum Numbers and Orbitals

Quantum numbers are used to identify the energy states and orbitals available to electrons. We will first describe the quantum numbers and the information they provide and then turn to the connection between quantum numbers and the energies and shapes of atomic orbitals.

n, the Principal Quantum Number ($n = 1, 2, 3, \ldots$)

The principal quantum number n can have any integer value from 1 to infinity. The value of n is the primary factor in determining the *energy* of an orbital. It also defines the *size* of an orbital: for a given atom, the greater the value of n, the greater the size of the orbital.

In atoms having more than one electron, two or more electrons may have the same n value. These electrons are then said to be in the same **electron shell.**

■ **Electron Energy and Quantum Numbers** The electron energy in the H atom depends only on the value of n. In atoms with more electrons, the energy depends on both n and ℓ.

ℓ, the Azimuthal Quantum Number ($\ell = 0, 1, 2, 3, \ldots, n - 1$)

Orbitals of a given shell can be grouped into **subshells**, where each subshell is characterized by a different value of the quantum number ℓ. The quantum number ℓ can have any integer value from 0 to a maximum of $n - 1$. This quantum number defines the *characteristic shape of an orbital*; different ℓ values correspond to different orbital shapes.

Because ℓ can be no larger than $n - 1$, the value of n limits the number of subshells possible for each shell. For the shell with $n = 1$, ℓ must equal 0; thus, only one subshell is possible. When $n = 2$, ℓ can be either 0 or 1. Because two values of ℓ are now possible, there are two subshells in the $n = 2$ electron shell.

Subshells are usually identified by letters. For example, an $\ell = 1$ subshell is called a "*p* subshell," and an orbital in that subshell is called a "*p* orbital."

■ **Orbital Symbols** Early studies of the emission spectra of elements classified lines into four groups on the basis of their appearance. These groups were labeled sharp, principal, diffuse, and fundamental. From these names came the labels we now apply to orbitals: *s, p, d,* and *f.*

Subshell labels

Value of ℓ	Subshell Label
0	*s*
1	*p*
2	*d*
3	*f*

■ **Subshells and Orbitals**

Subshell	Number of Orbitals in Subshell ($= 2\ell + 1$)
s	1
p	3
d	5
f	7

m_ℓ, the Magnetic Quantum Number ($m_\ell = 0, \pm1, \pm2, \pm3, \ldots, \pm\ell$)

The magnetic quantum number, m_ℓ, is related to the *orientation in space of the orbitals within a subshell.* Orbitals in a given subshell differ in their orientation in space, not in their energy.

The value of m_ℓ can range from $+\ell$ to $-\ell$, with 0 included. For example, when $\ell = 2$, m_ℓ can have five values: +2, +1, 0, −1, and −2. The number of values of m_ℓ for a given subshell ($= 2\ell + 1$) specifies the number of orbitals in the subshell.

Shells and Subshells

Allowed values of the three quantum numbers are summarized in Table 6.1. By analyzing the sets of quantum numbers in this table, you will discover the following:

- n = the number of subshells in a shell.
- $2\ell + 1$ = the number of orbitals in a subshell = the number of values of m_ℓ
- n^2 = the number of orbitals in a shell.

The First Electron Shell, $n = 1$

When $n = 1$, the value of ℓ can only be 0, and so m_ℓ must also have a value of 0. This means that, in the shell closest to the nucleus, only one subshell exists, and that subshell consists of only a single orbital, the 1*s* orbital.

The Second Electron Shell, $n = 2$

When $n = 2$, ℓ can have two values (0 and 1), so there are two subshells in the second shell. One of these is the 2*s* subshell ($n = 2$ and $\ell = 0$), and the other is the 2*p* subshell ($n = 2$ and $\ell = 1$). Because the values of m_ℓ can be −1, 0, and +1 when $\ell = 1$, three 2*p* orbitals exist. All three orbitals have the same shape. However, because each has a different m_ℓ value, the three orbitals differ in their orientation in space.

TABLE 6.1 Summary of the Quantum Numbers, Their Interrelationships, and the Orbital Information Conveyed

Principal Quantum Number	Azimuthal Quantum Number	Magnetic Quantum Number	Number and Type of Orbitals in the Subshell
Symbol = n Values = 1, 2, 3, . . . n = number of subshells	Symbol = ℓ Values = 0 . . . $n - 1$	Symbol = m_ℓ Values = $+\ell$ . . . 0 . . . $-\ell$	Number of orbitals in shell = n^2 and number of orbitals in subshell = $2\ell + 1$
1	0	0	one 1*s* orbital (one orbital of one type in the $n = 1$ shell)
2	0 1	0 +1, 0, −1	one 2*s* orbital three 2*p* orbitals (four orbitals of two types in the $n = 2$ shell)
3	0 1 2	0 +1, 0, −1 +2, +1, 0, −1, −2	one 3*s* orbital three 3*p* orbitals five 3*d* orbitals (nine orbitals of three types in the $n = 3$ shell)
4	0 1 2 3	0 +1, 0, −1 +2, +1, 0, −1, −2 +3, +2, +1, 0, −1, −2, −3	one 4*s* orbital three 4*p* orbitals five 4*d* orbitals seven 4*f* orbitals (16 orbitals of four types in the $n = 4$ shell)

The Third Electron Shell, $n = 3$

When $n = 3$, three subshells are possible for an electron because ℓ has the values 0, 1, and 2. The first two subshells within the $n = 3$ shell are the $3s$ ($\ell = 0$, one orbital) and $3p$ ($\ell = 1$, three orbitals) subshells. The third subshell is labeled $3d$ ($n = 3$, $\ell = 2$). Because m_ℓ can have five values (-2, -1, 0, $+1$, and $+2$) for $\ell = 2$, there are five d orbitals in this d subshell.

The Fourth Electron Shell, $n = 4$, and Beyond

There are four subshells in the $n = 4$ shell. In addition to $4s$, $4p$, and $4d$ subshells, there is the $4f$ subshell for which $\ell = 3$. Seven such orbitals exist because there are seven values of m_ℓ when $\ell = 3$ (-3, -2, -1, 0, $+1$, $+2$, and $+3$).

■ **Shells, Subshells, and Orbitals—A Summary** Electrons in atoms are arranged in shells. Within each shell, there can be one or more electron subshells, each comprised of one or more orbitals.

Electron Location	Quantum Number
Shell	n
Subshell	ℓ
Orbital	m_ℓ

Chemistry Now™

Sign in at **www.thomsonedu.com/login** and go to Chapter 6 Contents to see:
- Screen 6.9 to view an animation on **the quantum mechanical view of the atom**
- Screen 6.12 for a tutorial on **determining values for the quantum numbers for an orbital**

EXERCISE 6.7 Using Quantum Numbers

Complete the following statements:

(a) When $n = 2$, the values of ℓ can be ______ and ______.

(b) When $\ell = 1$, the values of m_ℓ can be ______, ______, and ______, and the subshell has the letter label ______.

(c) The subshell with $\ell = 2$ is called a ______ subshell.

(d) When a subshell is labeled s, the value of ℓ is ______, and m_ℓ has the value ______.

(e) There are ______________ orbitals in the p subshell.

(f) When a subshell is labeled f, there are ______ values of m_ℓ, corresponding to ______ orbitals.

6.6 The Shapes of Atomic Orbitals

We often say the electron is assigned to, or "occupies," an orbital. But what does this mean? What is an orbital? What does it look like? To answer these questions, we must examine the wave functions for the orbitals.

s Orbitals

A $1s$ orbital is associated with the quantum numbers $n = 1$ and $\ell = 0$. If we could photograph a $1s$ electron at one-second intervals for a few thousand seconds, the composite picture would look like the drawing in Figure 6.13a. This resembles a cloud of dots, and chemists often refer to such representations of electron orbitals as *electron cloud pictures*. In Figure 6.13a, the density of dots is greater close to the nucleus, that is, the electron cloud is denser close to the nucleus. This indicates that the $1s$ electron is most likely to be found near the nucleus. However, the density of dots declines on moving away from the nucleus and so, therefore, does the probability of finding the electron.

■ **Atomic Orbitals for the H Atom** Orbitals and their shapes discussed here are for the H atom, that is, for a one-electron atom. Orbitals for multielectron atoms are approximated by assuming they are hydrogen-like. For most purposes this is a reasonable assumption.

The thinning of the electron cloud at increasing distance is illustrated in a different way in Figure 6.13b. Here we have plotted the square of the wave function for the electron in a $1s$ orbital (ψ^2), times 4π and the distance squared ($4\pi r^2$), as a function of the distance of the electron from the nucleus. This plot represents

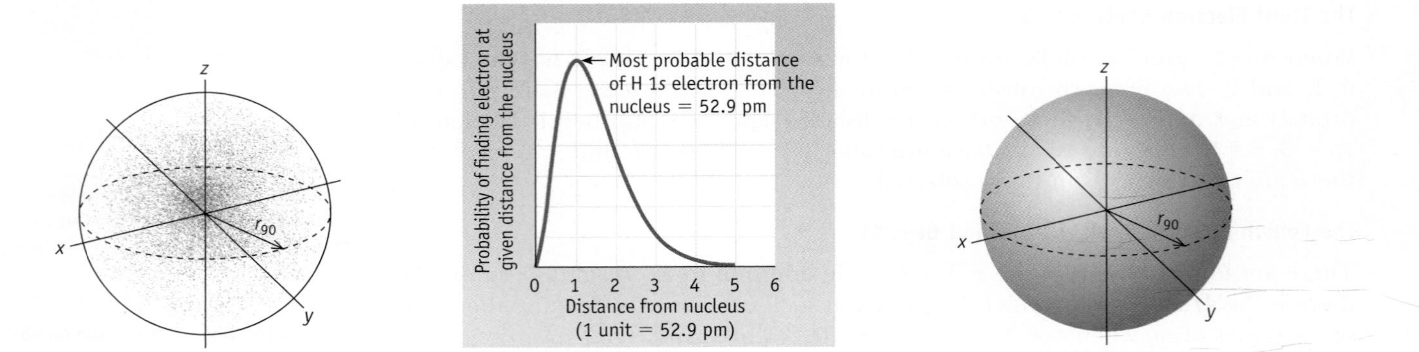

(a) Dot picture of an electron in a 1*s* orbital. Each dot represents the position of the electron at a different instant in time. Note that the dots cluster closest to the nucleus. r_{90} is the radius of a sphere within which the electron is found 90% of the time.

(b) A plot of the surface density ($4\pi r^2\psi^2$) as a function of distance for a hydrogen atom 1*s* orbital. This gives the probability of finding the electron at a given distance from the nucleus.

(c) The surface of the sphere within which the electron is found 90% of the time for a 1*s* orbital. This surface is often called a "boundary surface." (A 90% surface was chosen arbitrarily. If the choice was the surface within which the electron is found 50% of the time, the sphere would be considerably smaller.)

Active Figure 6.13 Different views of a 1*s* ($n = 1$, $\ell = 0$) orbital.

Chemistry Now™ Sign in at www.thomsonedu.com/login and go to the Chapter Contents menu to explore an interactive version of this figure accompanied by an exercise.

■ **Surface Density Plot for 1*s*** The maximum value of the radial distribution plot for a 1*s* electron in a hydrogen atom occurs at 52.9 pm. It is interesting to note that this maximum is at exactly the same distance from the nucleus that Niels Bohr calculated for the radius of the orbit occupied by the $n = 1$ electron.

the probability of finding the electron in a thin spherical shell at a distance *r* from the nucleus. For this reason, the plot of $4\pi r^2\psi^2$ vs. r is sometimes called a **surface density plot** or a **radial distribution plot.** For the 1*s* orbital, $4\pi r^2\psi^2$ is zero at the nucleus—there is no probability the electron will be exactly at the nucleus—but the probability rises rapidly on moving away from the nucleus, reaches a maximum a short distance from the nucleus (at 52.9 pm), and then decreases rapidly as the distance from the nucleus increases. Notice that the probability of finding the electron approaches but never quite reaches zero, even at very large distances.

Figure 6.13a shows that, for the 1*s* orbital, the probability of finding an electron is the same at a given distance from the nucleus, no matter in which direction you proceed from the nucleus. Consequently, the *1s orbital is spherical in shape.*

Because the probability of finding the electron approaches but never quite reaches zero, there is no sharp boundary beyond which the electron is never found (although the probability can be incalculably small). Nonetheless, the *s* orbital (and other types of orbitals as well) is often depicted as having a **boundary surface** (Figure 6.13c), largely because it is easier to draw such pictures. To create Figure 6.13c, we drew a sphere about the nucleus in such a way that the probability of finding the electron somewhere inside the sphere is 90%. The choice of 90% is arbitrary—we could have chosen a different value—and if we do, the shape would be the same, but the size of the sphere would be different.

There are misconceptions about pictures of orbitals. First, there is not an impenetrable surface within which the electron is "contained." Second, the probability of finding the electron is not the same throughout the volume enclosed by the surface. (An electron in the 1*s* orbital of a H atom has a greater probability of being 52.9 pm from the nucleus than of being closer or farther away.) Third, the terms "electron cloud" and "electron distribution" imply that the electron is a particle, but the basic premise in quantum mechanics is that the electron is treated as a wave, not a particle.

A Closer Look

H Atom Orbital Shapes—Wave Functions and Nodes

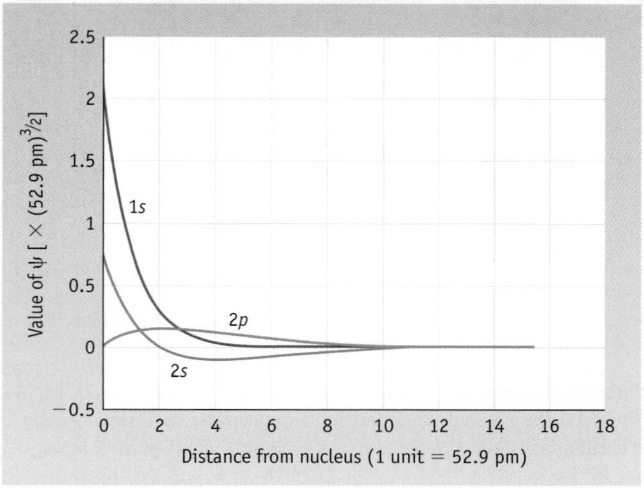

FIGURE A Plot of the wave functions for 1*s*, 2*s*, and 2*p* orbitals versus distance from the nucleus.

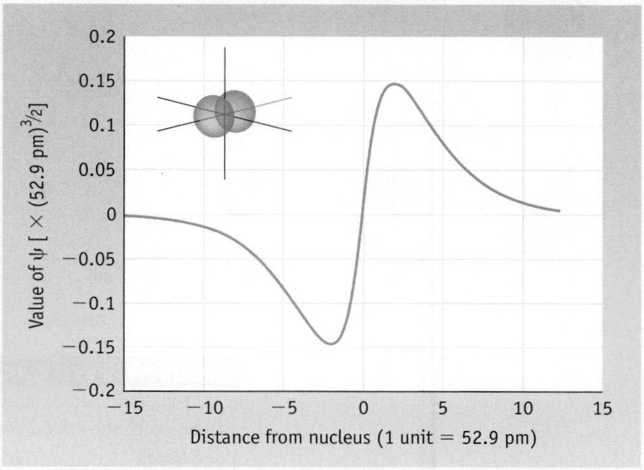

FIGURE B Wave functions for a 2*p* orbital. The sign of ψ for a 2*p* orbital is positive on one side of the nucleus and negative on the other (but it has a 0 value at the nucleus). A nodal plane separates the two lobes of this "dumbbell-shaped" orbital. (The vertical axis is the value of ψ, and the horizontal axis is the distance from the nucleus, where 1 unit = 52.9 pm.)

Waves have crests, troughs, and nodes, and these terms can be applied to the description of an electron as a wave. For a 1*s* orbital of the H atom, the wave function ψ approaches a maximum at the nucleus, but the wave's amplitude declines rapidly at points farther removed from the nucleus (Figure A). The sign of ψ is positive at all points in space.

For a 2*s* orbital, there is a different profile; the sign of ψ is positive near the nucleus, drops to zero (there is a node at 2 × 52.9 pm), and then becomes negative before approaching zero at greater distances.

For the 2*p* orbital, the value of ψ is zero at the nucleus because there is a nodal surface passing through the nucleus. Moving away from the nucleus in one direction, say along the *x*-axis, we see the value of ψ rises to a maximum around 106 pm before falling off at greater distances. Moving away along the $-x$ direction, the value of ψ is the same but opposite in sign (Figure B). The 2*p* electron is a wave with a node at the nucleus. (In drawing orbitals, we indicate this with + or − signs or with two different colors as in Figure 6.14 or Figure B.)

For the 2*s* orbital, there is a node at 105.8 pm when plotting the wave function. However, because ψ has the same value in all directions, this means there is a *spherical nodal surface* surrounding the nucleus as illustrated in Figure C.

As noted on page 290, the number of nodal surfaces passing through the nucleus for any orbital is equal to ℓ. The number of spherical nodes is $n - \ell - 1$. Thus, for a 2*s* orbital, $2 - 0 - 1 = 1$ spherical nodes, as we have seen. In general, the consequence of this is that electron density in all *s* orbitals (except 1*s*) occurs as a series of nested shells. A possible analogy to this picture is that an *s* orbital resembles an onion with layers of electron density, the number of layers increasing with *n*.

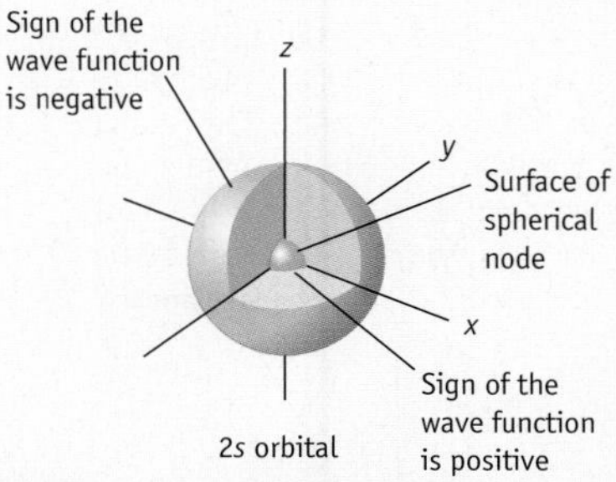

FIGURE C Wave functions for a 2*s* orbital. A 2*s* orbital for the H atom showing the spherical node (at 105.8 pm) around the nucleus.

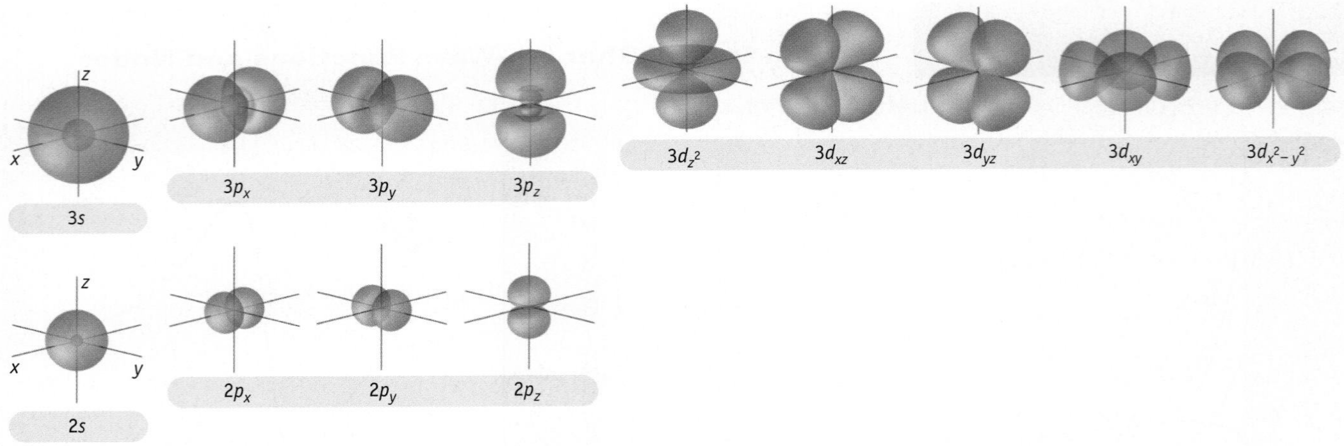

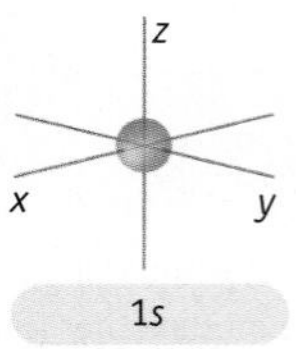

Active Figure 6.14 **Atomic orbitals.** Boundary surface diagrams for electron densities of 1*s*, 2*s*, 2*p*, 3*s*, 3*p*, and 3*d* orbitals for the hydrogen atom. For the *p* orbitals, the subscript letter indicates the cartesian axis along which the orbital lies. For more about orbitals, see *A Closer Look: H Atom Orbital Shapes—Wave Functions and Nodes.*

Chemistry Now™ Sign in at www.thomsonedu.com/login and go to the Chapter Contents menu to explore an interactive version of this figure accompanied by an exercise.

All *s* orbitals (1*s*, 2*s*, 3*s* ...) are spherical in shape. However, for any atom, the size of *s* orbitals increases as *n* increases (Figure 6.14). For a given atom, the 1*s* orbital is more compact than the 2*s* orbital, which is in turn more compact than the 3*s* orbital.

p Orbitals

All atomic orbitals for which $\ell = 1$ (*p* orbitals) have the same basic shape. If you enclose 90% of the electron density in a *p* orbital within a surface, the electron cloud has a shape that resembles a weight lifter's "dumbbell," and chemists describe *p* orbitals as having dumbbell shapes (Figures 6.14 and 6.15) because there is a **nodal surface**—a surface on which the electron has no probability—that passes through the nucleus. (The nodal surface is a consequence of the wave function for *p* orbitals,

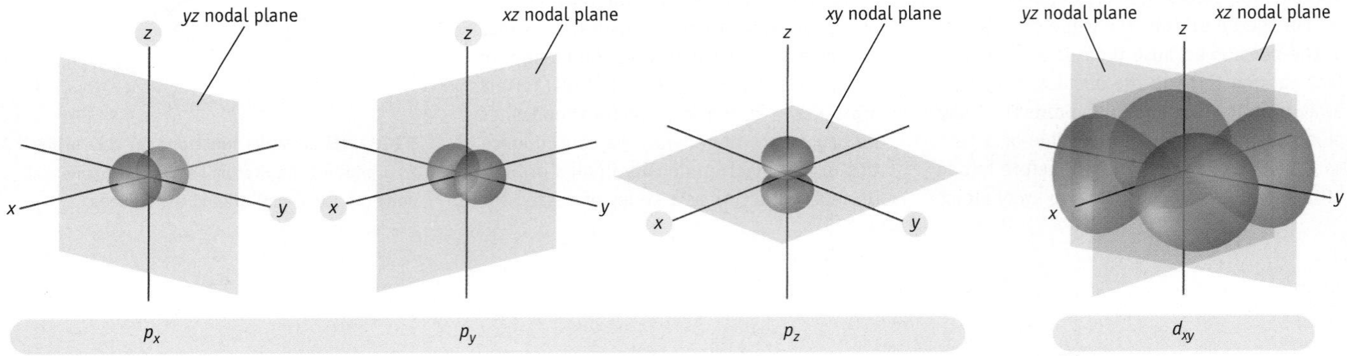

(a) (b)

FIGURE 6.15 Nodal surfaces of *p* and *d* orbitals. A plane passing through the nucleus (perpendicular to the axis) is called a nodal surface. (a) The three *p* orbitals each have one nodal surface ($\ell = 1$). (b) The d_{xy} orbital. All five *d* orbitals have two nodal surfaces ($\ell = 2$) through the nucleus. Here, the nodal surfaces are the *xz*- and *yz*-planes, so the regions of electron density lie in the *xy*-plane and between the *x*- and *y*-axes.

which has no value at the nucleus but which rises rapidly in value on moving way from the nucleus. See *A Closer Look: H Atom Orbital Shapes—Wave Functions and Nodes.*)

There are three *p* orbitals in a subshell, and all have the same basic shape with one planar node through the nucleus. Usually, *p* orbitals are drawn along the *x*-, *y*-, and *z*-axes and labeled according to the axis along which they lie (p_x, p_y, or p_z).

d Orbitals

Orbitals with $\ell = 0$, *s* orbitals, have no nodal surfaces through the nucleus, and *p* orbitals, for which $\ell = 1$, have one nodal surface through the nucleus. *The value of ℓ is equal to the number of nodal surfaces slicing through the nucleus.* It follows that the five *d* orbitals, for which $\ell = 2$, have two nodal surfaces through the nucleus, resulting in four regions of electron density. The d_{xy} orbital, for example, lies in the *xy*-plane, and the two nodal surfaces are the *xz*- and *yz*-planes (Figure 6.15). Two other orbitals, d_{xz} and d_{yz}, lie in planes defined by the *xz*- and *yz*-axes, respectively; they also have two, mutually perpendicular nodal surfaces (Figure 6.14).

Of the two remaining *d* orbitals, the $d_{x^2-y^2}$ orbital is easier to visualize. In the $d_{x^2-y^2}$ orbital, the nodal planes bisect the *x*- and *y*-axes, so the regions of electron density lie along the *x*- and *y*-axes. The d_{z^2} orbital has two main regions of electron density along the *z*-axis, and a "doughnut" of electron density also occurs in the *xy*-plane. This orbital has two cone-shaped nodal surfaces.

f Orbitals

Seven *f* orbitals arise with $\ell = 3$. Three nodal surfaces through the nucleus cause the electron density to lie in eight regions of space. One of the *f* orbitals is illustrated in Figure 6.16.

Chemistry.Now™

Sign in at **www.thomsonedu.com/login** and go to Chapter 6 Contents to see Screen 6.13 for exercises on **orbital shapes, quantum numbers, and nodes.**

EXERCISE 6.8 Orbital Shapes

(a) What are the *n* and ℓ values for each of the following orbitals: 6*s*, 4*p*, 5*d*, and 4*f*?

(b) How many nodal planes exist for a 4*p* orbital? For a 6*d* orbital?

Charles D. Winters

FIGURE 6.16 One of the seven possible *f* orbitals. Notice the presence of three nodal planes as required by an orbital with $\ell = 3$.

■ **Nodal Surfaces** Nodal surfaces through the nucleus occur for all *p*, *d*, and *f* orbitals. These surfaces are usually flat, so they are referred to as nodal planes. In some cases (for example, d_{z^2}), however, the "plane" is not flat and so is better referred to as a "surface."

■ **ℓ and Nodal Surfaces** The number of nodal surfaces passing through the nucleus for an orbital = ℓ.

Orbital	ℓ	**Number of Nodal Surfaces through the Nucleus**
s	0	0
p	1	1
d	2	2
f	3	3

6.7 One More Electron Property: Electron Spin

There is one more property of the electron that plays an important role in the arrangement of electrons in atoms and gives rise to properties of elements you observe every day: electron spin.

The Electron Spin Quantum Number, m_s

In 1921 Otto Stern and Walther Gerlach performed an experiment that probed the magnetic behavior of atoms by passing a beam of silver atoms in the gas phase through a magnetic field. Although the results were complex, they were best interpreted by imagining the electron has a spin and behaves as a tiny magnet that can be attracted or repelled by another magnet (Figure 6.17). If atoms with a single unpaired electron are placed in a magnetic field, the Stern-Gerlach ex-

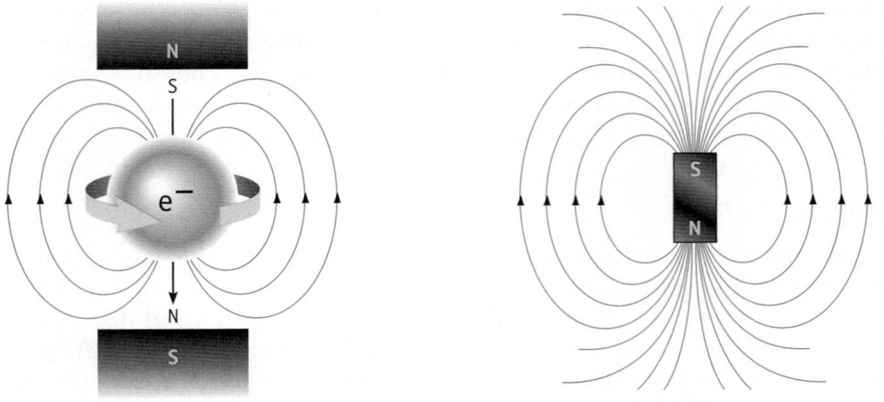

(a) Electron spin

(b) A bar magnet

FIGURE 6.17 Magnetic fields—a bar magnet and an electron. The electron, with its spin and negative electric charge, can be thought of as a small bar magnet. Relative to a magnetic field, only two spin directions are possible for the electron, clockwise or counterclockwise. The north pole of the spinning electron can therefore be either aligned with an external magnetic field or opposed to that field.

A Closer Look

Paramagnetism and Ferromagnetism

Magnetic materials are relatively common, and many are important in our economy. For example, a large magnet is at the heart of the magnetic resonance imaging (MRI) used in medicine, and tiny magnets are found in stereo speakers and in telephone handsets. Magnetic oxides are used in recording tapes and computer disks.

The magnetic materials we use are **ferromagnetic.** The magnetic effect of ferromagnetic materials is much larger than that of paramagnetic ones. Ferromagnetism occurs when the spins of unpaired electrons in a cluster of atoms (called a *domain*) in the solid align themselves in the same direction. Only the metals of the iron, cobalt, and nickel subgroups, as well as a few other metals such as neodymium, exhibit this property. They are also unique in that, once the domains are aligned in a magnetic field, the metal is permanently magnetized.

Many alloys exhibit greater ferromagnetism than do the pure metals themselves. One example of such a material is Alnico, which is composed of aluminum, nickel, and cobalt as well as copper and iron.

Audio and video tapes are plastics coated with crystals of ferromagnetic oxides such as Fe_2O_3 or CrO_2. The recording head uses an electromagnetic field to create a varying magnetic field based on signals from a microphone. This magnetizes the tape as it passes through the head, with the strength and direction of magnetization varying with the frequency of the sound to be recorded. When the tape is played back, the magnetic field of the moving tape induces a current, which is amplified and sent to the speakers.

Charles D. Winters

Magnets. Many common consumer products such as loud speakers contain permanent magnets.

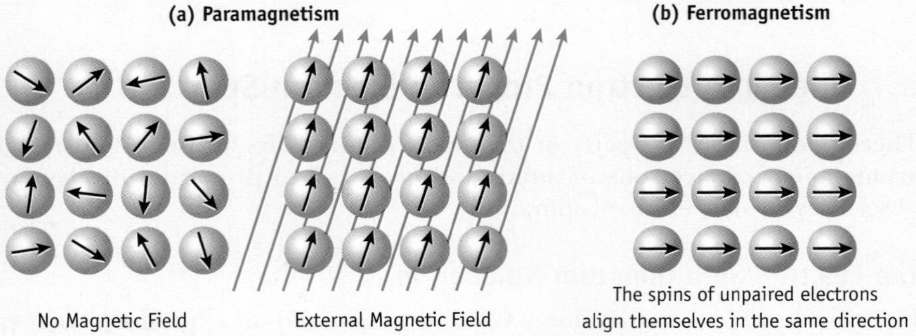

Magnetism. (a) Paramagnetism: In the absence of an external magnetic field, the unpaired electrons in the atoms or ions of the substance are randomly oriented. If a magnetic field is imposed, however, these spins will tend to become aligned with the field. (b) Ferromagnetism: The spins of the unpaired electrons in a cluster of atoms or ions align themselves in the same direction in the absence of a magnetic field.

periment showed there are two orientations for the atoms: with the electron spin aligned with the field or opposed to the field. That is, *the electron spin is quantized,* which introduces another quantum number, the **electron spin quantum number,** m_s. One orientation is associated with a value of m_s of $+\frac{1}{2}$ and the other with m_s of $-\frac{1}{2}$.

When it was recognized that electron spin is quantized, scientists realized that a complete description of an electron in any atom requires four quantum numbers, n, ℓ, m_ℓ, and m_s. The important consequences of this fact are explored in Chapter 7.

Diamagnetism and Paramagnetism

A hydrogen atom has a single electron. If a hydrogen atom is placed in a magnetic field, the magnetic field of the single electron will tend to align with the external field like the needle of a compass. There is an attractive force. Helium atoms, each with two electrons, are not attracted to a magnet, however. In fact, they are slightly repelled by the magnet. To account for this observation, we assume the two electrons of helium have opposite spin orientations. We say their spins are *paired,* and the result is that the magnetic field of one electron can be canceled out by the magnetic field of the second electron with opposite spin. To account for this, the two electrons are assigned different values of m_s.

It is important to understand the relationship between electron spin and magnetism. *Elements and compounds that have unpaired electrons are attracted to a magnet.* These species are referred to as **paramagnetic.** The effect can be quite weak, but, by placing a sample of an element or compound in a magnetic field, it can be observed (Figure 6.18). For example, the oxygen you breathe, is paramagnetic. You can observe this experimentally because liquid oxygen sticks to a magnet of the kind you may have in the speakers of a music player (Figure 6.18b).

Substances in which all electrons are paired (with the two electrons of each pair having opposite spins) experience a slight repulsion when subjected to a magnetic

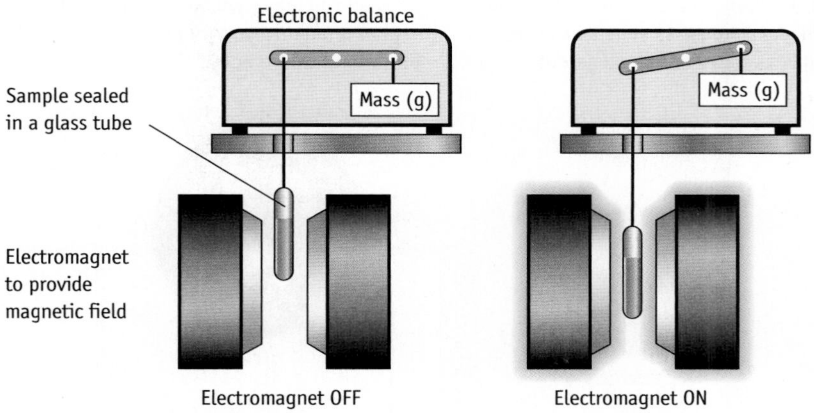

(a) Electromagnetic balance

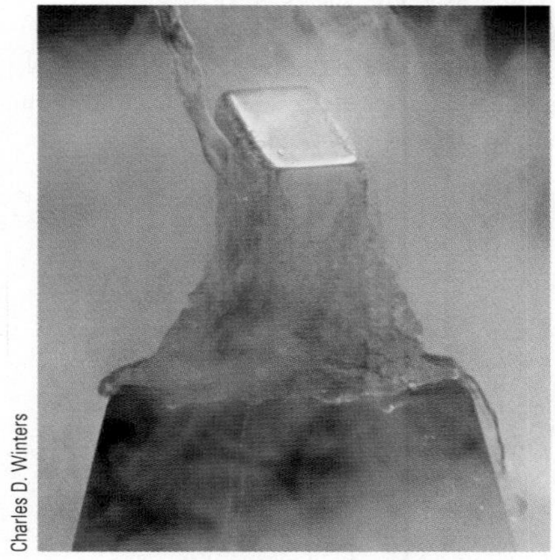

(b) Liquid oxygen clings to a magnet.

Active Figure 6.18 Observing and measuring paramagnetism. (a) A magnetic balance is used to measure the magnetism of a sample. The sample is first weighed with the electromagnet turned off. The magnet is then turned on and the sample reweighed. If the substance is paramagnetic, the sample is drawn into the magnetic field, and the apparent weight increases. (b) Liquid oxygen (boiling point 90.2 K) clings to a strong magnet. Elemental oxygen is paramagnetic because it has unpaired electrons. (See Chapter 9.)

ChemistryNow™ Sign in at www.thomsonedu.com/login and go to the Chapter Contents menu to explore an interactive version of this figure accompanied by an exercise.

Chemical Perspectives — Quantized Spins and MRI

Just as electrons have a spin, so do atomic nuclei. In the hydrogen atom, the single proton of the nucleus spins on its axis. For most heavier atoms, the atomic nucleus includes both protons and neutrons, and the entire entity has a spin. This property is important, because nuclear spin allows scientists to detect these atoms in molecules and to learn something about their chemical environments.

The technique used to detect the spins of atomic nuclei is *nuclear magnetic resonance* (NMR). It is one of the most powerful methods currently available to determine molecular structures. About 20 years ago, it was adapted as a diagnostic technique in medicine, where it is known as *magnetic resonance imaging* (MRI).

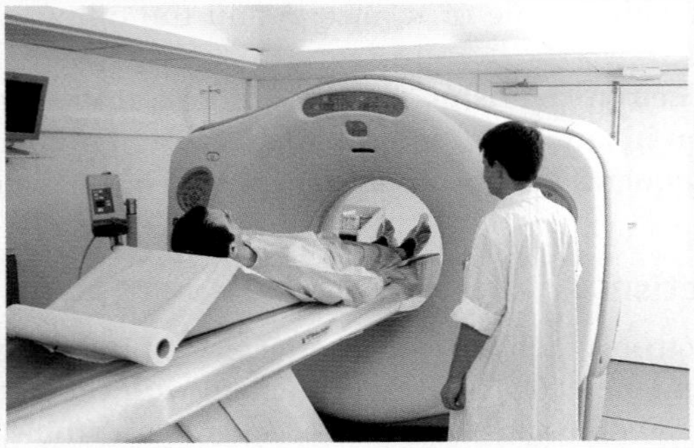

Aj/Photo Researchers, Inc.

(a)

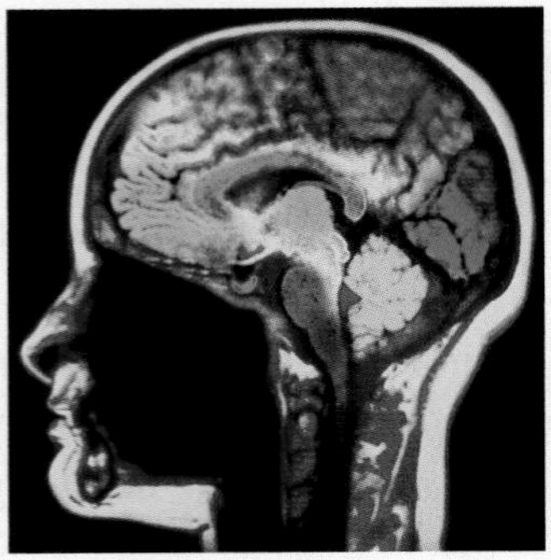

Scott Camazine & Sue Trainor/Photo Researchers, Inc.

(b)

Magnetic resonance imaging. (a) MRI instrument. The patient is placed inside a large magnet, and the tissues to be examined are irradiated with radio-frequency radiation. (b) An MRI image of the human brain.

Just as electron spin is quantized, so too is nuclear spin. The H atom nucleus can spin in either of two directions. If the H atom is placed in a strong, external magnetic field, however, the spinning nuclear magnet can align itself with or against the external field. If a sample of ethanol (CH_3CH_2OH), for example, is placed in a strong magnetic field, a slight excess of the H atom nuclei (and ^{13}C atom nuclei) is aligned with the lines of force of the field.

The nuclei aligned with the field have a slightly lower energy than when aligned against the field. The NMR and MRI technologies depend on the fact that energy in the radio-frequency region can be absorbed by the sample and can cause the nuclear spins switch alignments—that is, to move to a higher energy state. This absorption of energy is detected by the instrument.

The most important aspect of the magnetic resonance technique is that the difference in energy between two different spin states depends on the electronic environment of atoms in the molecule. In the case of ethanol, the three CH_3 protons are different from the two CH_2 protons, and both sets are different from the OH proton. These three different sets of H atoms absorb radiation of slightly different energies. The instrument measures the frequencies absorbed, and a scientist familiar with the technique can quickly distinguish the three different environments in the molecule.

The MRI technique closely resembles the NMR method. Hydrogen is abundant in the human body as water and in numerous organic molecules. In the MRI device, the patient is placed in a strong magnetic field, and the tissues being examined are irradiated with pulses of radio-frequency radiation.

The MRI image is produced by detecting how fast the excited nuclei "relax"; that is, how fast they return to the lower energy state from the higher energy state. The "relaxation time" depends on the type of tissue. When the tissue is scanned, the H atoms in different regions of the body show different relaxation times, and an accurate "image" is built up.

MRI gives information on soft tissue—muscle, cartilage, and internal organs—which is unavailable from x-ray scans. This technology is also noninvasive, and the magnetic fields and radio-frequency radiation used are not harmful to the body.

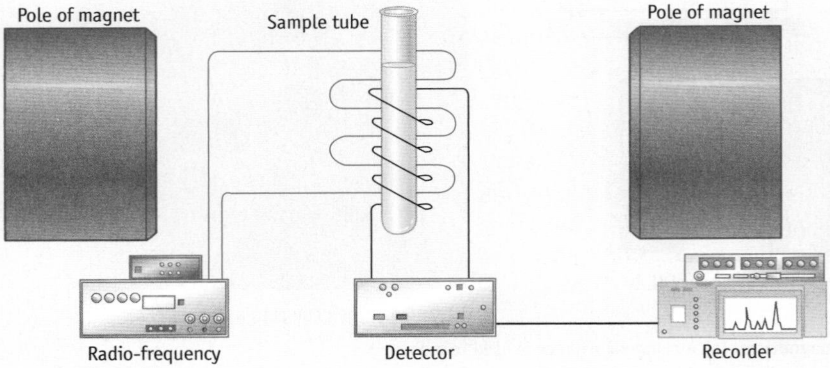

(a) An nmr spectrometer (see Figure 4.4, page 169.)

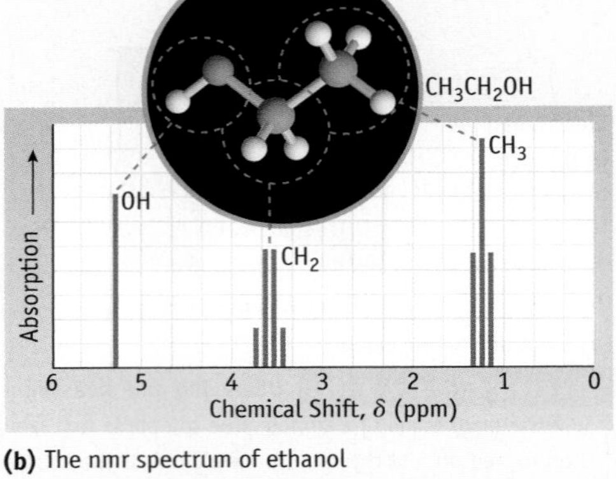

(b) The nmr spectrum of ethanol

Nuclear magnetic resonance. (a) A schematic diagram of an NMR spectrometer. (b) The NMR spectrum of ethanol, showing that the three different types of protons appear in distinctly different regions of the spectrum. The pattern observed for the CH_2 and CH_3 protons is characteristic of these groups of atoms and signals the chemist that they are present in the molecule.

field; they are called **diamagnetic**. Therefore, by determining the magnetic behavior of a substance (see Figure 16.8) we can gain information on the electronic structure.

In summary, *paramagnetism is the attraction to a magnetic field of substances in which the constituent ions or atoms contain unpaired electrons.* Substances in which all electrons are paired with partners of opposite spin are diamagnetic. This explanation opens the way to understanding the arrangement of electrons in atoms with more than one electron.

Chemistry Now™

Sign in at **www.thomsonedu.com/login** and go to Chapter 6 Contents to see:
- Screen 6.15 for a self-study module on **electron spin**
- Screen 6.16 for an exercise on **spinning electrons and magnetism**

Chapter Goals Revisited

Now that you have studied this chapter, you should ask whether you have met the chapter goals. In particular, you should be able to:

Describe the properties of electromagnetic radiation

a. Use the terms wavelength, frequency, amplitude, and node (Section 6.1). Study Question(s) assignable in OWL: 3.

b. Use Equation 6.1 ($c = \lambda\nu$), relating wavelength (λ) and frequency (ν) of electromagnetic radiation and the speed of light (c).

c. Recognize the relative wavelength (or frequency) of the various types of electromagnetic radiation (Figure 6.2). Study Question(s) assignable in OWL: 1.

d. Understand that the energy of a photon, a massless particle of radiation, is proportional to its frequency (Planck's equation, Equation 6.2) (Section 6.2). Study Question(s) assignable in OWL: 5, 8, 9, 12, 14, 56, 57, 58, 63, 64, 66, 72, 73, 78, 83.

Understand the origin of light from excited atoms and its relationship to atomic structure

a. Describe the Bohr model of the atom, its ability to account for the emission line spectra of excited hydrogen atoms, and the limitations of the model (Section 6.3).

b. Understand that, in the Bohr model of the H atom, the electron can occupy only certain energy states, each with an energy proportional to $1/n^2$ ($E_n = -Rhc/n^2$), where n is the principal quantum number (Equation 6.4, Section 6.3). If an electron moves from one energy state to another, the amount of energy absorbed or emitted in the process is equal to the difference in energy between the two states (Equation 6.5, Section 6.3). Study Question(s) assignable in OWL: 16, 18, 22, 60.

Describe the experimental evidence for particle-wave duality

a. Understand that in the modern view of the atom, electrons can be described either as particles or as waves (Section 6.4). The wavelength of an electron or any subatomic particle is given by de Broglie's equation (Equation 6.6). Study Question(s) assignable in OWL: 24, 26, 47.

Chemistry Now™ Sign in at **www.thomsonedu.com/login** to:
- Assess your understanding with Study Questions in OWL keyed to each goal in the Goals and Homework menu for this chapter
- For quick review, download Go Chemistry mini-lecture flashcard modules (or purchase them at **www.ichapters.com**)
- Check your readiness for an exam by taking the Pre-Test and exploring the modules recommended in your Personalized Study plan.

Access **How Do I Solve It?** tutorials on how to approach problem solving using concepts in this chapter.

Describe the basic ideas of quantum mechanics

a. Recognize the significance of quantum mechanics in describing atomic structure (Section 6.5).

b. Understand that an orbital for an electron in an atom corresponds to the allowed energy of that electron.

c. Understand that the position of the electron is not known with certainty; only the probability of the electron being at a given point of space can be calculated. This is a consequence of the Heisenberg uncertainty principle.

Define the four quantum numbers (n, ℓ, m_ℓ, and m_s), and recognize their relationship to electronic structure

a. Describe the allowed energy states of the orbitals in an atom using three quantum numbers n, ℓ, and m_ℓ (Section 6.5). Study Question(s) assignable in OWL: 28, 30, 32, 34, 36, 37, 38, 40, 43, 44, 80, 83.

b. Describe the shapes of the orbitals (Section 6.6). Study Question(s) assignable in OWL: 46, 51, 53, 67f.

c. Recognize the spin quantum number, m_s, which has values of $\pm\frac{1}{2}$. Classify substances as paramagnetic (attracted to a magnetic field; characterized by unpaired electron spins) or diamagnetic (repelled by a magnetic field, all electrons paired) (Section 6.7).

KEY EQUATIONS

Equation 6.1 (page 269) The product of the wavelength (λ) and frequency (ν) of electromagnetic radiation is equal to the speed of light (c).

$$c = \lambda \times \nu$$

Equation 6.2 (page 273) Planck's equation: the energy of a photon, a massless particle of radiation, is proportional to its frequency (ν). The proportionality constant, h, is called Planck's constant (6.626×10^{-34} J · s).

$$E = h\nu$$

Equation 6.4 (page 277) In Bohr's theory, the potential energy of the electron, E_n, in the nth quantum level of the H atom is proportional to $1/n^2$, where n is a positive integer (the principal quantum number and $Rhc = 2.179 \times 10^{-18}$ J/atom or $N_ARhc = 1312$ kJ/mol).

$$E_n = -\frac{Rhc}{n^2}$$

Equation 6.5 (page 279) The energy change for an electron moving between two quantum levels (n_{final} and $n_{initial}$) in the H atom.

$$\Delta E = E_{final} - E_{initial} = -Rhc\left(\frac{1}{n^2_{final}} - \frac{1}{n^2_{initial}}\right)$$

Equation 6.6 (page 282) De Broglie's equation: the wavelength of a particle (λ) is related to its mass (m) and speed (v) and to Planck's constant (h).

$$\lambda = \frac{h}{mv}$$

STUDY QUESTIONS

Online homework for this chapter may be assigned in OWL.

▲ denotes challenging questions.

■ denotes questions assignable in OWL.

Blue-numbered questions have answers in Appendix O and fully-worked solutions in the *Student Solutions Manual.*

Practicing Skills

Electromagnetic Radiation

(See Example 6.1, Exercise 6.1, Figure 6.2, and ChemistryNow Screen 6.3.)

1. ■ Answer the following questions based on Figure 6.2:
   (a) Which type of radiation involves less energy, x-rays or microwaves?
   (b) Which radiation has the higher frequency, radar or red light?
   (c) Which radiation has the longer wavelength, ultraviolet or infrared light?

2. Consider the colors of the visible spectrum.
   (a) Which colors of light involve less energy than green light?
   (b) Which color of light has photons of greater energy, yellow or blue?
   (c) Which color of light has the greater frequency, blue or green?

3. ■ Traffic signals are often now made of LEDs (light-emitting diodes). Amber and green ones are pictured here.
   (a) The light from an amber signal has a wavelength of 595 nm, and that from a green signal has wavelength of 500 nm. Which has the higher frequency?
   (b) Calculate the frequency of amber light.

Mike Condren/UW/MRSEC

(a) (b)

4. Suppose you are standing 225 m from a radio transmitter. What is your distance from the transmitter in terms of the number of wavelengths if:
   (a) the station is broadcasting at 1150 kHz (on the AM radio band)? (1 kHz = 1×10^3 Hz.)
   (b) the station is broadcasting at 98.1 MHz (on the FM radio band)? (1 MHz = 10^6 Hz.)

Electromagnetic Radiation and Planck's Equation

(See page 274, Exercise 6.2, and ChemistryNow Screens 6.4 and 6.5.)

5. ■ Green light has a wavelength of 5.0×10^2 nm. What is the energy, in joules, of one photon of green light? What is the energy, in joules, of 1.0 mol of photons of green light?

6. Violet light has a wavelength of about 410 nm. What is its frequency? Calculate the energy of one photon of violet light. What is the energy of 1.0 mol of violet photons? Compare the energy of photons of violet light with those of red light. Which is more energetic?

7. The most prominent line in the spectrum of aluminum is at 396.15 nm. What is the frequency of this line? What is the energy of one photon with this wavelength? Of 1.00 mol of these photons?

8. ■ The most prominent line in the spectrum of magnesium is 285.2 nm. Other lines are found at 383.8 and 518.4 nm. In what region of the electromagnetic spectrum are these lines found? Which is the most energetic line? What is the energy of 1.00 mol of photons with the wavelength of the most energetic line?

9. ■ Place the following types of radiation in order of increasing energy per photon:
   (a) yellow light from a sodium lamp
   (b) x-rays from an instrument in a dentist's office
   (c) microwaves in a microwave oven
   (d) your favorite FM music station at 91.7 MHz

10. Place the following types of radiation in order of increasing energy per photon:
   (a) radiation within a microwave oven
   (b) your favorite radio station
   (c) gamma rays from a nuclear reaction
   (d) red light from a neon sign
   (e) ultraviolet radiation from a sun lamp

Photoelectric Effect

(See page 274 and Figure 6.4.)

11. An energy of 2.0×10^2 kJ/mol is required to cause a cesium atom on a metal surface to lose an electron. Calculate the longest possible wavelength of light that can ionize a cesium atom. In what region of the electromagnetic spectrum is this radiation found?

12. ■ You are an engineer designing a switch that works by the photoelectric effect. The metal you wish to use in your device requires 6.7×10^{-19} J/atom to remove an electron. Will the switch work if the light falling on the metal has a wavelength of 540 nm or greater? Why or why not?

Atomic Spectra and the Bohr Atom

(See Examples 6.2 and 6.3, Figures 6.6–6.10, and ChemistryNow Screens 6.6 and 6.7.)

13. The most prominent line in the spectrum of mercury is at 253.652 nm. Other lines are located at 365.015 nm, 404.656 nm, 435.833 nm, and 1013.975 nm.
    (a) Which of these lines represents the most energetic light?
    (b) What is the frequency of the most prominent line? What is the energy of one photon with this wavelength?
    (c) Are any of these lines found in the spectrum of mercury shown in Figure 6.7? What color or colors are these lines?

14. ■ The most prominent line in the spectrum of neon is found at 865.438 nm. Other lines are located at 837.761 nm, 878.062 nm, 878.375 nm, and 1885.387 nm.
    (a) In what region of the electromagnetic spectrum are these lines found?
    (b) Are any of these lines found in the spectrum of neon shown in Figure 6.7?
    (c) Which of these lines represents the most energetic radiation?
    (d) What is the frequency of the most prominent line? What is the energy of one photon with this wavelength?

15. A line in the Balmer series of emission lines of excited H atoms has a wavelength of 410.2 nm (Figure 6.10). What color is the light emitted in this transition? What quantum levels are involved in this emission line? That is, what are the values of $n_{initial}$ and n_{final}?

16. ■ What are the wavelength and frequency of the radiation involved in the least energetic emission line in the Lyman series? What are the values of $n_{initial}$ and n_{final}?

17. Consider only transitions involving the $n = 1$ through $n = 5$ energy levels for the H atom (See Figures 6.8 and 6.10).
    (a) How many emission lines are possible, considering only the five quantum levels?
    (b) Photons of the highest frequency are emitted in a transition from the level with $n =$ ____ to a level with $n =$ ____.
    (c) The emission line having the longest wavelength corresponds to a transition from the level with $n =$ ____ to the level with $n =$ ____.

18. ■ Consider only transitions involving the $n = 1$ through $n = 4$ energy levels for the hydrogen atom (See Figures 6.8 and 6.10).
    (a) How many emission lines are possible, considering only the four quantum levels?
    (b) Photons of the lowest energy are emitted in a transition from the level with $n =$ ____ to a level with $n =$ ____.
    (c) The emission line having the shortest wavelength corresponds to a transition from the level with $n =$ ____ to the level with $n =$ ____.

19. The energy emitted when an electron moves from a higher energy state to a lower energy state in any atom can be observed as electromagnetic radiation.
    (a) Which involves the emission of less energy in the H atom, an electron moving from $n = 4$ to $n = 2$ or an electron moving from $n = 3$ to $n = 2$?
    (b) Which involves the emission of more energy in the H atom, an electron moving from $n = 4$ to $n = 1$ or an electron moving from $n = 5$ to $n = 2$? Explain fully.

20. If energy is absorbed by a hydrogen atom in its ground state, the atom is excited to a higher energy state. For example, the excitation of an electron from the level with $n = 1$ to the level with $n = 3$ requires radiation with a wavelength of 102.6 nm. Which of the following transitions would require radiation of *longer wavelength* than this?
    (a) $n = 2$ to $n = 4$
    (b) $n = 1$ to $n = 4$
    (c) $n = 1$ to $n = 5$
    (d) $n = 3$ to $n = 5$

21. Calculate the wavelength and frequency of light emitted when an electron changes from $n = 3$ to $n = 1$ in the H atom. In what region of the spectrum is this radiation found?

22. ■ Calculate the wavelength and frequency of light emitted when an electron changes from $n = 4$ to $n = 3$ in the H atom. In what region of the spectrum is this radiation found?

DeBroglie and Matter Waves

(See Example 6.4 and ChemistryNow Screen 6.8.)

23. An electron moves with a velocity of 2.5×10^8 cm/s. What is its wavelength?

24. ■ A beam of electrons ($m = 9.11 \times 10^{-31}$ kg/electron) has an average speed of 1.3×10^8 m/s. What is the wavelength of electrons having this average speed?

25. Calculate the wavelength, in nanometers, associated with a 1.0×10^2-g golf ball moving at 30. m/s (about 67 mph). How fast must the ball travel to have a wavelength of 5.6×10^{-3} nm?

26. ■ A rifle bullet (mass = 1.50 g) has a velocity of 7.00×10^2 mph (miles per hour). What is the wavelength associated with this bullet?

Quantum Mechanics

(See Sections 6.5–6.7 and ChemistryNow Screens 6.9–6.14.)

27. (a) When $n = 4$, what are the possible values of ℓ?
    (b) When ℓ is 2, what are the possible values of m_ℓ?
    (c) For a 4*s* orbital, what are the possible values of n, ℓ, and m_ℓ?
    (d) For a 4*f* orbital, what are the possible values of n, ℓ, and m_ℓ?

28. ■ (a) When $n = 4$, $\ell = 2$, and $m_\ell = -1$, to what orbital type does this refer? (Give the orbital label, such as 1*s*.)
(b) How many orbitals occur in the $n = 5$ electron shell? How many subshells? What are the letter labels of the subshells?
(c) If a subshell is labeled *f*, how many orbitals occur in the subshell? What are the values of m_ℓ?

29. A possible excited state of the H atom has the electron in a 4*p* orbital. List all possible sets of quantum numbers n, ℓ, and m_ℓ for this electron.

30. ■ A possible excited state for the H atom has an electron in a 5*d* orbital. List all possible sets of quantum numbers n, ℓ, and m_ℓ for this electron.

31. How many subshells occur in the electron shell with the principal quantum number $n = 4$?

32. ■ How many subshells occur in the electron shell with the principal quantum number $n = 5$?

33. Explain briefly why each of the following is not a possible set of quantum numbers for an electron in an atom.
(a) $n = 2$, $\ell = 2$, $m_\ell = 0$
(b) $n = 3$, $\ell = 0$, $m_\ell = -2$
(c) $n = 6$, $\ell = 0$, $m_\ell = 1$

34. ■ Which of the following represent valid sets of quantum numbers? For a set that is invalid, explain briefly why it is not correct.
(a) $n = 3$, $\ell = 3$, $m_\ell = 0$
(b) $n = 2$, $\ell = 1$, $m_\ell = 0$
(c) $n = 6$, $\ell = 5$, $m_\ell = -1$
(d) $n = 4$, $\ell = 3$, $m_\ell = -4$

35. ■ What is the maximum number of orbitals that can be identified by each of the following sets of quantum numbers? When "none" is the correct answer, explain your reasoning.
(a) $n = 3$, $\ell = 0$, $m_\ell = +1$
(b) $n = 5$, $\ell = 1$
(c) $n = 7$, $\ell = 5$
(d) $n = 4$, $\ell = 2$, $m_\ell = -2$

36. ■ What is the maximum number of orbitals that can be identified by each of the following sets of quantum numbers? When "none" is the correct answer, explain your reasoning.
(a) $n = 4$, $\ell = 3$
(b) $n = 5$
(c) $n = 2$, $\ell = 2$
(d) $n = 3$, $\ell = 1$, $m_\ell = -1$

37. ■ Explain briefly why each of the following is not a possible set of quantum numbers for an electron in an atom. In each case, change the incorrect value (or values) to make the set valid.
(a) $n = 4$, $\ell = 2$, $m_\ell = 0$, $m_s = 0$
(b) $n = 3$, $\ell = 1$, $m_\ell = -3$, $m_s = -\frac{1}{2}$
(c) $n = 3$, $\ell = 3$, $m_\ell = -1$, $m_s = +\frac{1}{2}$

38. Explain briefly why each of the following is not a possible set of quantum numbers for an electron in an atom. In each case, change the incorrect value (or values) to make the set valid.
(a) $n = 2$, $\ell = 2$, $m_\ell = 0$, $m_s = +\frac{1}{2}$
(b) $n = 2$, $\ell = 1$, $m_\ell = -1$, $m_s = 0$
(c) $n = 3$, $\ell = 1$, $m_\ell = +2$, $m_s = +\frac{1}{2}$

39. State which of the following orbitals cannot exist according to the quantum theory: 2*s*, 2*d*, 3*p*, 3*f*, 4*f*, and 5*s*. Briefly explain your answers.

40. ■ State which of the following are incorrect designations for orbitals according to the quantum theory: 3*p*, 4*s*, 2*f*, and 1*p*. Briefly explain your answers.

41. Write a complete set of quantum numbers (n, ℓ, and m_ℓ) that quantum theory allows for each of the following orbitals: (a) 2*p*, (b) 3*d*, and (c) 4*f*.

42. ■ Write a complete set of quantum numbers (n, ℓ, and m_ℓ) for each of the following orbitals: (a) 5*f*, (b) 4*d*, and (c) 2*s*.

43. ■ A particular orbital has $n = 4$ and $\ell = 2$. What must this orbital be: (a) 3*p*, (b) 4*p*, (c) 5*d*, or (d) 4*d*?

44. ■ A given orbital has a magnetic quantum number of $m_\ell = -1$. This could *not* be a (an)
(a) *f* orbital
(b) *d* orbital
(c) *p* orbital
(d) *s* orbital

45. How many planar nodes are associated with each of the following orbitals?
(a) 2*s*
(b) 5*d*
(c) 5*f*

46. ■ How many planar nodes are associated with each of the following atomic orbitals?
(a) 4*f*
(b) 2*p*
(c) 6*s*

General Questions on Atomic Structure

These questions are not designated as to type or location in the chapter. They may combine several concepts.

47. ■ Which of the following are applicable when explaining the photoelectric effect? Correct any statements that are wrong.
(a) Light is electromagnetic radiation.
(b) The intensity of a light beam is related to its frequency.
(c) Light can be thought of as consisting of massless particles whose energy is given by Planck's equation, $E = h\nu$

48. In what region of the electromagnetic spectrum for hydrogen is the Lyman series of lines found? The Balmer series?

49. Give the number of nodal surfaces through the nucleus (planar nodes) for each orbital type: *s, p, d,* and *f.*

50. What is the maximum number of *s* orbitals found in a given electron shell? The maximum number of *p* orbitals? Of *d* orbitals? Of *f* orbitals?

51. ■ Match the values of ℓ shown in the table with orbital type (*s, p, d,* or *f*).

ℓ Value	Orbital Type
3	______
0	______
1	______
2	______

52. Sketch a picture of the 90% boundary surface of an *s* orbital and the p_x orbital. Be sure the latter drawing shows why the *p* orbital is labeled p_x and not p_y, for example.

53. ■ Complete the following table.

Orbital Type	Number of Orbitals in a Given Subshell	Number of Nodal Surfaces through the Nucleus
s	______	______
p	______	______
d	______	______
f	______	______

54. Excited H atoms have many emission lines. One series of lines, called the Pfund series, occurs in the infrared region. It results when an electron changes from higher energy levels to a level with $n = 5$. Calculate the wavelength and frequency of the lowest energy line of this series.

55. An advertising sign gives off red light and green light.
(a) Which light has the higher-energy photons?
(b) One of the colors has a wavelength of 680 nm, and the other has a wavelength of 500 nm. Which color has which wavelength?
(c) Which light has the higher frequency?

56. ■ Radiation in the ultraviolet region of the electromagnetic spectrum is quite energetic. It is this radiation that causes dyes to fade and your skin to develop a sunburn. If you are bombarded with 1.00 mol of photons with a wavelength of 375 nm, what amount of energy, in kilojoules per mole of photons, are you being subjected to?

57. ■ A cell phone sends signals at about 850 MHz ($1 \text{ MHz} = 1 \times 10^6$ Hz or cycles per second).
(a) What is the wavelength of this radiation?
(b) What is the energy of 1.0 mol of photons with a frequency of 850 MHz?
(c) Compare the energy in part (b) with the energy of a mole of photons of blue light (420 nm).
(d) Comment on the difference in energy between 850 MHz radiation and blue light.

58. ■ Assume your eyes receive a signal consisting of blue light, $\lambda = 470$ nm. The energy of the signal is 2.50×10^{-14} J. How many photons reach your eyes?

59. If sufficient energy is absorbed by an atom, an electron can be lost by the atom and a positive ion formed. The amount of energy required is called the ionization energy. In the H atom, the ionization energy is that required to change the electron from $n = 1$ to $n =$ infinity. (See Exercise 6.5, page 280.) Calculate the ionization energy for the He^+ ion. Is the ionization energy of the He^+ more or less than that of H? (Bohr's theory applies to He^+ because it, like the H atom, has a single electron. The electron energy, however, is now given by $E = -Z^2Rhc/n^2$, where Z is the atomic number of helium.)

60. ■ Suppose hydrogen atoms absorb energy so that electrons are excited to the $n = 7$ energy level. Electrons then undergo these transitions, among others: (a) $n = 7 \longrightarrow n = 1$; (b) $n = 7 \longrightarrow n = 6$; and (c) $n = 2 \longrightarrow n = 1$. Which transition produces a photon with (i) the smallest energy, (ii) the highest frequency, and (iii) the shortest wavelength?

61. Rank the following orbitals in the H atom in order of increasing energy: 3*s*, 2*s*, 2*p*, 4*s*, 3*p*, 1*s*, and 3*d*.

62. ■ How many orbitals correspond to each of the following designations?

(a) 3*p*	(d) 6*d*	(g) $n = 5$
(b) 4*p*	(e) 5*d*	(h) 7*s*
(c) $4p_x$	(f) 5*f*	

63. ■ Cobalt-60 is a radioactive isotope used in medicine for the treatment of certain cancers. It produces β particles and γ rays, the latter having energies of 1.173 and 1.332 MeV. ($1 \text{ MeV} = 10^6$ electron-volts and $1 \text{ eV} = 9.6485 \times 10^4$ J/mol.) What are the wavelength and frequency of a γ-ray photon with an energy of 1.173 MeV?

64. ▲ ■ Exposure to high doses of microwaves can cause tissue damage. Estimate how many photons, with $\lambda = 12$ cm, must be absorbed to raise the temperature of your eye by 3.0 °C. Assume the mass of an eye is 11 g and its specific heat capacity is 4.0 J/g · K.

65. When the Sojourner spacecraft landed on Mars in 1997, the planet was approximately 7.8×10^7 km from Earth. How long did it take for the television picture signal to reach Earth from Mars?

66. ■ The most prominent line in the emission spectrum of chromium is found at 425.4 nm. Other lines in the chromium spectrum are found at 357.9 nm, 359.3 nm, 360.5 nm, 427.5 nm, 429.0 nm, and 520.8 nm.
(a) Which of these lines represents the most energetic light?
(b) What color is light of wavelength 425.4 nm?

67. Answer the following questions as a summary quiz on the chapter.
(a) The quantum number n describes the ______ of an atomic orbital.
(b) The shape of an atomic orbital is given by the quantum number ______.
(c) A photon of green light has ______ (less or more) energy than a photon of orange light.
(d) The maximum number of orbitals that may be associated with the set of quantum numbers $n = 4$ and $\ell = 3$ is ______.
(e) The maximum number of orbitals that may be associated with the quantum number set $n = 3$, $\ell = 2$, and $m_\ell = -2$ is ______.
(f) ■ Label each of the following orbital pictures with the appropriate letter:

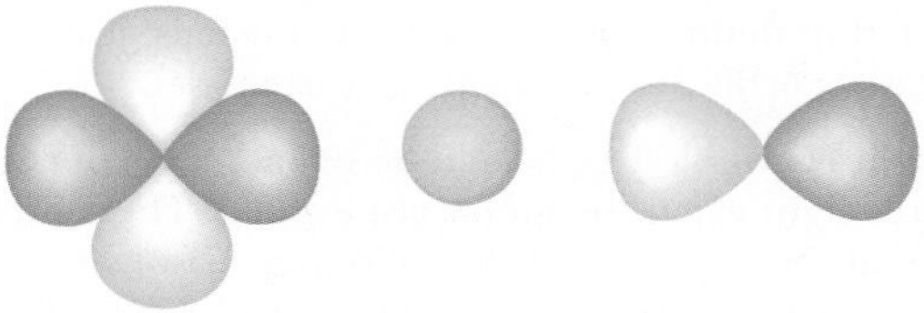

(g) When $n = 5$, the possible values of ℓ are ______.
(h) The number of orbitals in the $n = 4$ shell is ______.
(i) A Co^{2+} ion has three unpaired electrons. A sample of $CoCl_2$ is (paramagnetic) (diamagnetic).

68. Answer the following questions as a summary quiz on this chapter.
(a) The quantum number n describes the ______ of an atomic orbital, and the quantum number ℓ describes its ______.
(b) When $n = 3$, the possible values of ℓ are ______.
(c) What type of orbital corresponds to $\ell = 3$? _____
(d) For a $4d$ orbital, the value of n is ______, the value of ℓ is ______, and a possible value of m_ℓ is ______.
(e) Each of the following drawings represents a type of atomic orbital. Give the letter designation for the orbital; give its value of ℓ, and specify the number of nodal surfaces.

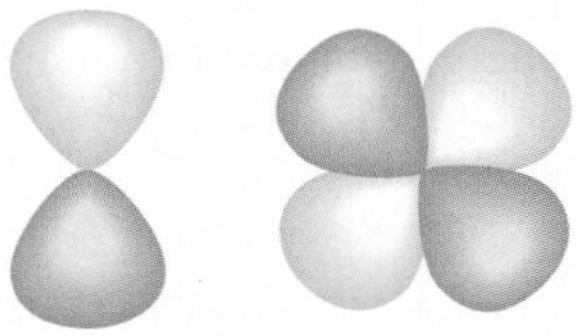

Letter = ______ ______
ℓ value = ______ ______
Planar nodes = ______ ______

(f) An atomic orbital with three nodal surfaces through the nucleus is ____.
(g) Which of the following orbitals cannot exist according to modern quantum theory: $2s$, $3p$, $2d$, $3f$, $5p$, $6p$?
(h) Which of the following is not a valid set of quantum numbers?

n	ℓ	m_ℓ	m_s
3	2	1	$-\frac{1}{2}$
2	1	2	$+\frac{1}{2}$
4	3	0	0

(i) What is the maximum number of orbitals that can be associated with each of the following sets of quantum numbers? (One possible answer is "none.")
(i) $n = 2$ and $\ell = 1$
(ii) $n = 3$
(iii) $n = 3$ and $\ell = 3$
(iv) $n = 2$, $\ell = 1$, and $m_\ell = 0$
(j) A Cu^{2+} ion has one unpaired electron. Is a sample of $CuBr_2$ paramagnetic or diamagnetic?

69. The diagrams below represent a small section of a solid. Each circle represents an atom, and an arrow represents an electron.

(a) (b) (c)

(a) Which represents a diamagnetic solid, which a paramagnetic solid, and which a ferromagnetic solid?
(b) Which is most strongly attracted to a magnetic field? Which is least strongly attracted?

In the Laboratory

70. A solution of $KMnO_4$ absorbs light at 540 nm (page 192). What is the frequency of the light absorbed? What is the energy of one mole of photons with $\lambda = 540$ nm?

71. A large pickle is attached to two electrodes, which are then attached to a 110-V power supply (see the problem on Screen 6.7 of ChemistryNow). As the voltage is increased across the pickle, it begins to glow with a yellow color. Knowing that pickles are made by soaking the vegetable in a concentrated salt solution, describe why the pickle might emit light when electrical energy is added.

Charles D. Winters

The "electric pickle."

72. ■ The spectrum shown here is for aspirin. The vertical axis is the amount of light absorbed, and the horizontal axis is the wavelength of incident light (in nm).

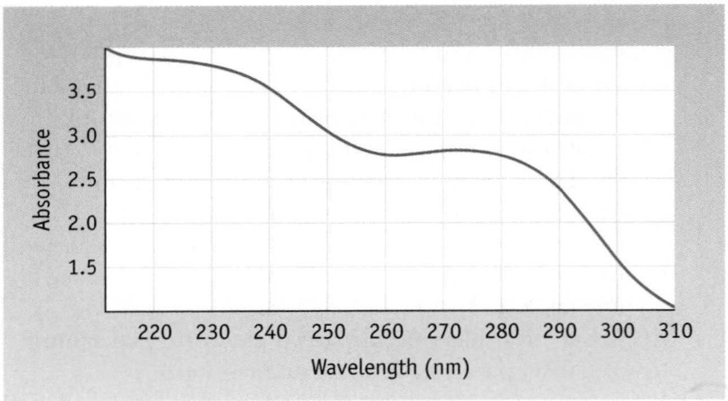

What is the frequency of light with a wavelength of 278 nm? What is the energy of one mole of photons with $\lambda = 278$ nm? What region of the electromagnetic spectrum is covered by the spectrum above? Knowing that aspirin only absorbs light in the region depicted by this spectrum, what is the color of aspirin?

73. ■ The infrared spectrum for methanol, CH_3OH, is illustrated below. It shows the amount of light in the infrared region that methanol transmits as a function of wavelength. The vertical axis is the amount of light transmitted. At points near the top of the graph, most of the incident light is being transmitted by the sample (or, conversely, little light is absorbed.) Therefore, the "peaks" or "bands" that descend from the top indicate light absorbed; the longer the band, the more light is being absorbed (or, conversely, the less is being transmitted). The horizontal scale is in units of "wavenumbers," abbreviated cm^{-1}. The energy of light is given by Planck's law as $E = hc/\lambda$; that is, E is proportional to $1/\lambda$. Therefore, the horizontal scale is in units of $1/\lambda$ and reflects the energy of the light incident on the sample.

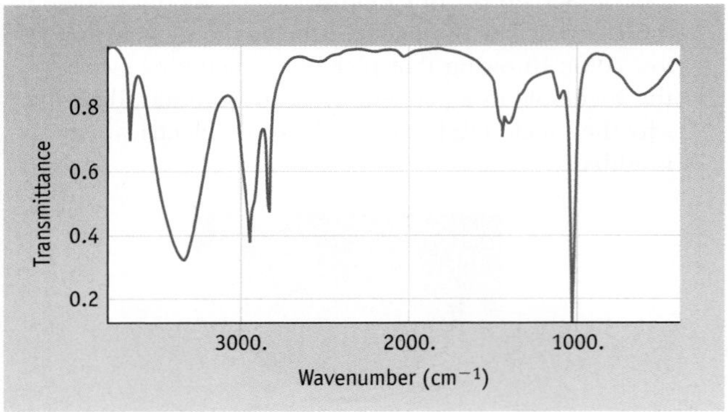

(a) One point on the horizontal axis is marked as 2000 cm^{-1}. What is the wavelength of light at this point?
(b) Which is the low energy end of this spectrum (left or right), and which is the high energy end?
(c) The broad absorption at about 3300–3400 cm^{-1} indicates that infrared radiation is interacting with the OH group of the methanol molecule. The narrower absorptions around 2800–3000 cm^{-1} are for interactions with C—H bonds. Which interaction requires more energy, with O—H or with C—H?

Summary and Conceptual Questions

The following questions use concepts from this and previous chapters.

74. Bohr pictured the electrons of the atom as being located in definite orbits about the nucleus, just as the planets orbit the sun. Criticize this model.

75. Light is given off by a sodium- or mercury-containing streetlight when the atoms are excited. The light you see arises for which of the following reasons?
(a) Electrons are moving from a given energy level to one of higher energy.
(b) Electrons are being removed from the atom, thereby creating a metal cation.
(c) Electrons are moving from a given energy level to one of lower energy.

76. How do we interpret the physical meaning of the square of the wave function? What are the units of $4\pi r^2\psi^2$?

77. What does "wave–particle duality" mean? What are its implications in our modern view of atomic structure?

78. ■ Which of these are observable?
(a) position of an electron in an H atom
(b) frequency of radiation emitted by H atoms
(c) path of an electron in an H atom
(d) wave motion of electrons
(e) diffraction patterns produced by electrons
(f) diffraction patterns produced by light
(g) energy required to remove electrons from H atoms
(h) an atom
(i) a molecule
(j) a water wave

79. In principle, which of the following can be determined?
(a) the energy of an electron in the H atom with high precision and accuracy
(b) the position of a high-speed electron with high precision and accuracy
(c) at the same time, both the position and the energy of a high-speed electron with high precision and accuracy

80. ▲ ■ Suppose you live in a different universe where a different set of quantum numbers is required to describe the atoms of that universe. These quantum numbers have the following rules:

N, principal	1, 2, 3, . . . , ∞
L, orbital	= *N*
M, magnetic	−1, 0, +1

How many orbitals are there altogether in the first three electron shells?

81. A photon with a wavelength of 93.8 nm strikes a hydrogen atom, and light is emitted by the atom. How many emission lines would be observed? At what wavelengths? Explain briefly (see Figure 6.10).

82. Explain why you could or could not measure the wavelength of a golf ball in flight.

83. ■ The radioactive element technetium is not found naturally on earth; it must be synthesized in the laboratory. It is a valuable element, however, because it has medical uses. For example, the element in the form of sodium pertechnetate ($NaTcO_4$) is used in imaging studies of the brain, thyroid, and salivary glands and in renal blood flow studies, among other things.

(a) In what group and period of the periodic table is the element found?

(b) The valence electrons of technetium are found in the 5*s* and 4*d* subshells. What is a set of quantum numbers (n, ℓ, and m_ℓ) for one of the electrons of the 5*s* subshell?

(c) ■ Technetium emits a γ-ray with an energy of 0.141 MeV. (1 MeV = 10^6 electron-volts, where 1 eV = 9.6485×10^4 J/mol.) What are the wavelength and frequency of a γ-ray photon with an energy of 0.141 MeV?

(d) To make $NaTcO_4$, the metal is dissolved in nitric acid.

$$7\ HNO_3(aq) + Tc(s) \longrightarrow HTcO_4(aq) + 7\ NO_2(g) + 3\ H_2O(\ell)$$

and the product, $HTcO_4$, is treated with NaOH to make $NaTcO_4$.

(i) Write a balanced equation for the reaction of $HTcO_4$ with NaOH.

(ii) If you begin with 4.5 mg of Tc metal, how much $NaTcO_4$ can be made? What mass of NaOH, in grams, is required to convert all of the $HTcO_4$ into $NaTcO_4$?

(e) If you synthesize 1.5 micromoles of $NaTcO_4$, what mass of compound do you have? If the compound is dissolved in 10.0 mL of solution, what is the concentration?

84. See ChemistryNow Screen 6.1, Chemical Puzzler. This screen shows that light of different colors can come from a "neon" sign or from certain salts when they are placed in a burning organic liquid. ("Neon" signs are glass tubes filled with neon, argon, and other gases, and the gases are excited by an electric current.) What do these two sources of light have in common? How is the light generated in each case?

85. See ChemistryNow Screen 6.7, Bohr's Model of the Hydrogen Atom, Simulation. A photon with a wavelength of 97.3 nm is fired at a hydrogen atom and leads to the emission of light. How many emission lines are emitted? Explain why more than one line is emitted.

7 The Structure of Atoms and Periodic Trends

Charles D. Winters

The Chromium-Bearing Mineral Crocoite, $PbCrO_4$

Minerals containing the chromate ion are extremely rare. The best example is crocoite, lead(II) chromate, which is found almost exclusively in Tasmania. Although it is nearly insoluble in water, traces will nonetheless dissolve and contaminate groundwater with Pb^{2+} and CrO_4^{2-} ions.

Chromium compounds in groundwater can be a problem. In the United States, 56% of the population relies on groundwater for drinking water, and the shallow aquifers from which the water is often obtained are susceptible to contamination. One such contaminant is chromium in its various ionic forms (Cr^{3+}, CrO_4^{2-}, and $Cr_2O_7^{2-}$). The source of such ions can be mineral deposits, but more significant sources are industries involved in leather tanning [which uses $Cr(OH)SO_4$] or electroplating of chromium for corrosion protection. In a case depicted in the movie *Erin Brockovich*, a gas and electric utility was found to have contaminated groundwater in southern California with chromates that had been used in water-cooling towers to prevent rust.

Chromium compounds can also be biochemically active, although they are not implicated in as many important processes as another element in Group 6B, molybdenum. There is some evidence that chromium is one of the essential elements, and it is implicated in insulin regulation. In fact, the recommended daily dose of chromium is 5–200 μg. A compound called chromium(III) picolinate is marketed as a "nutritional" supplement and is widely used as a weight-loss aid. The majority of research has found, however, that it is neither helpful nor beneficial in weight-loss programs.

Questions:

1. What is the electron configuration for the Cr atom, for the Cr^{3+} ion, and for the chromium in the CrO_4^{2-} ion?
2. Is chromium in any of the ionic forms paramagnetic?
3. What is the electron configuration for the lead ions in $PbCrO_4$?

Answers to these questions are in Appendix Q.

Chapter Goals

See Chapter Goals Revisited (page 331) for Study Questions keyed to these goals and assignable in OWL.

- Recognize the relationship of the four quantum numbers (n, ℓ, m_ℓ, and m_s) to atomic structure.
- Write the electron configuration for elements and monatomic ions.
- Rationalize trends in atom and ion sizes, ionization energy, and electron affinity.

Chapter Outline

The wave mechanical model of the atom has spinning electrons assigned to orbitals that are best described as matter waves. The orbitals are arranged in subshells that are in turn part of electron shells. One objective of this chapter is to apply this model to the electronic structure of all of the elements.

A second objective is to explore some of the physical properties of elements, among them the ease with which atoms lose or gain electrons to form ions and the sizes of atoms and ions. These properties are directly related to the arrangement of electrons in atoms and thus to the chemistry of the elements and their compounds.

Chemistry Now™

Throughout the text this icon introduces an opportunity for self-study or to explore interactive tutorials by signing in at **www.thomsonedu.com/login**.

7.1 The Pauli Exclusion Principle

To make the quantum theory consistent with experiment, the Austrian physicist Wolfgang Pauli (1900–1958) stated in 1925 his **exclusion principle:** no two electrons in an atom can have the same set of four quantum numbers (n, ℓ, m_ℓ, and m_s). The consequence of this is that *no atomic orbital can be assigned more than two electrons,* and the two electrons assigned to an orbital must have different values of m_s.

An electron assigned to the 1s orbital of the H atom may have the set of quantum numbers $n = 1$, $\ell = 0$, and $m_\ell = 0$, and $m_s = +\frac{1}{2}$. Let us represent an orbital by a box and the electron spin by an arrow (↑ or ↓). A representation of the hydrogen atom is then:

Electrons in 1s orbital: [↑] Quantum number set
1s $n = 1,\ \ell = 0,\ m_\ell = 0,\ m_s = +\frac{1}{2}$

■ **Orbitals Are Not Boxes** Orbitals are not boxes in which electrons are placed. Thus, it is not conceptually correct to talk about electrons being in orbitals or occupying orbitals, although this is commonly done for the sake of simplicity.

The choice of m_s (either $+\frac{1}{2}$ or $-\frac{1}{2}$) and the direction of the electron spin arrow are arbitrary; that is, we could choose either value, and the arrow may point in either direction. Diagrams such as these are called **orbital box diagrams.**

A *helium* atom has two electrons, both assigned to the 1s orbital. The Pauli exclusion principle requires that each electron must have a different set of quantum numbers, so the orbital box diagram now is:

1s
Two electrons in 1s orbital: [↑↓] ← This electron has $n = 1,\ \ell = 0,\ m_\ell = 0,\ m_s = -\frac{1}{2}$
↑ This electron has $n = 1,\ \ell = 0,\ m_\ell = 0,\ m_s = +\frac{1}{2}$

By having opposite spins, the two electrons in the 1s orbital of an He atom have different sets of the four quantum numbers.

TABLE 7.1 Number of Electrons Accommodated in Electron Shells and Subshells with $n = 1$ to 6

Electron Shell (n)	Subshells Available	Orbitals Available ($2\ell + 1$)	Number of Electrons Possible in Subshell [$2(2\ell + 1)$]	Maximum Electrons Possible for nth Shell ($2n^2$)
1	*s*	1	2	2
2	*s*	1	2	8
	p	3	6	
3	*s*	1	2	18
	p	3	6	
	d	5	10	
4	*s*	1	2	32
	p	3	6	
	d	5	10	
	f	7	14	
5	*s*	1	2	50
	p	3	6	
	d	5	10	
	f	7	14	
	*g**	9	18	
6	*s*	1	2	72
	p	3	6	
	d	5	10	
	*f**	7	14	
	*g**	9	18	
	*h**	11	22	

*These orbitals are not occupied in the ground state of any known element.

Spin Quantum Number and Arrows In this book, we arbitrarily use an arrow pointing up (↑) to represent $m_s = +1/2$ and an arrow pointing down (↓) to represent $m_s = -1/2$. We will usually designate the first electron assigned to an orbital as having $m_s = +1/2$ though it could just as readily have $m_s = -1/2$.

Our understanding of orbitals and the knowledge that an orbital can accommodate no more than two electrons tell us the maximum number of electrons that can occupy each electron shell or subshell. For example, because each of the three orbitals in a *p* subshell can hold two electrons, *p* subshells can hold a maximum of six electrons. By the same reasoning, the five orbitals of a *d* subshell can accommodate a total of 10 electrons, and the seven *f* orbitals can accommodate 14 electrons. Recall that there are *n* subshells in the *n*th shell, and that there are n^2 orbitals in that shell (◀ Table 6.1, page 286). Thus, *the maximum number of electrons in any shell is* $2n^2$. The relationship among the quantum numbers and the numbers of electrons is shown in Table 7.1.

7.2 Atomic Subshell Energies and Electron Assignments

Our goal in this section is to understand and predict the orbital distribution of electrons in atoms with many electrons. The procedure by which electrons are assigned to orbitals is known as the *aufbau* principle (aufbau means "building up").

Electrons in an atom are assigned to shells (defined by the quantum number n) and subshells (defined by the quantum number ℓ) in order of increasingly higher energy. In this way, the total energy of the atom is as low as possible.

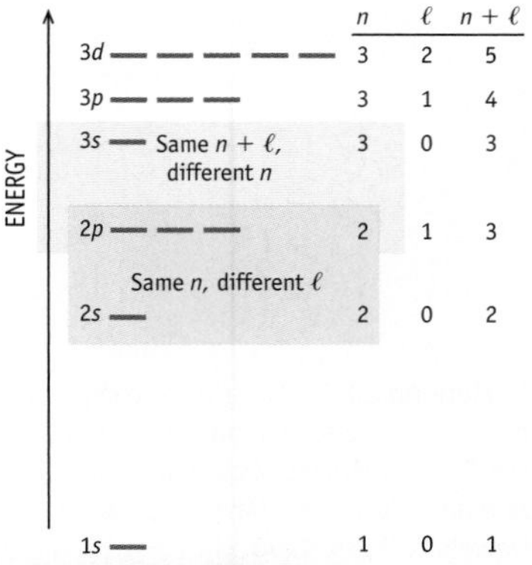

Active Figure 7.1 Order of subshell energies. Energies of electron shells increase with increasing n, and, within a shell, subshell energies increase with increasing ℓ. (The energy axis is not to scale.) The energy gaps between subshells of a given shell become smaller as n increases.

Chemistry•Now™ Sign in at www.thomsonedu.com/login and go to the Chapter Contents menu to explore an interactive version of this figure accompanied by an exercise.

Order of Subshell Energies and Assignments

Quantum theory and the Bohr model state that the energy of the H atom, with a single electron, depends only on the value of n ($E_n = -Rhc/n^2$). For atoms with more than one electron, however, the situation is more complex. The order of subshell energies for $n = 1$, 2, and 3 in Figure 7.1 shows that subshell energies in multielectron atoms depend on *both n and ℓ*.

Based on theoretical and experimental studies of orbital electron distributions in atoms, chemists have found that there are two general rules that help predict these arrangements:

- Electrons are assigned to subshells in order of increasing "$n + \ell$" value.
- For two subshells with the same value of "$n + \ell$," electrons are assigned first to the subshell of lower n.

The following are examples of these rules:

- Electrons are assigned to the $2s$ subshell ($n + \ell = 2 + 0 = 2$) before the $2p$ subshell ($n + \ell = 2 + 1 = 3$).
- Electrons are assigned to $2p$ orbitals ($n + \ell = 2 + 1 = 3$) before the $3s$ subshell ($n + \ell = 3 + 0 = 3$) because n for the $2p$ electrons is less than for the $3s$ electrons.
- Electrons are assigned to $4s$ orbitals ($n + \ell = 4 + 0 = 4$) before the $3d$ subshell ($n + \ell = 3 + 2 = 5$) because $n + \ell$ is less for $4s$ than for $3d$.

Figure 7.2 summarizes the assignment of electrons according to increasing $n + \ell$ values, and the discussion that follows explores the underlying causes and their consequences and connects atomic electron configurations to the periodic table.

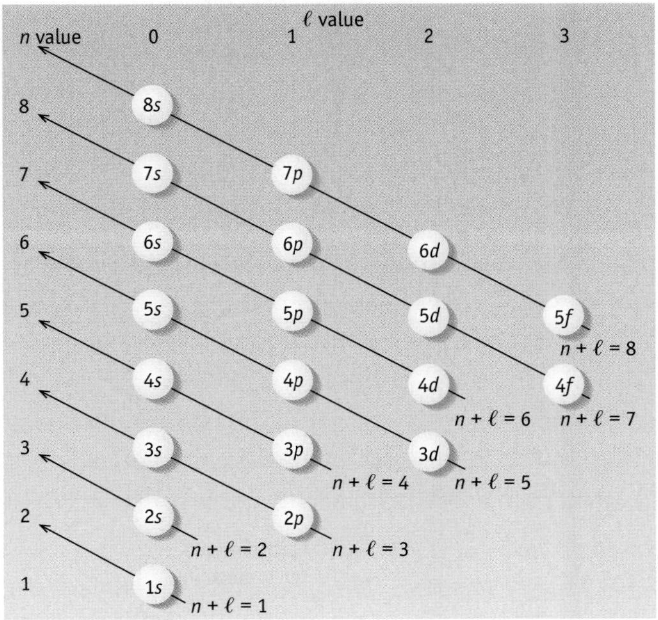

FIGURE 7.2 Subshell filling order. Subshells in atoms are filled in order of increasing $n + \ell$. When two subshells have the same $n + \ell$ value, the subshell of lower n is filled first. To use the diagram, begin at $1s$ and follow the arrows of increasing $n + \ell$. (Thus, the order of filling is $1s \Rightarrow 2s \Rightarrow 2p \Rightarrow 3s \Rightarrow 3p \Rightarrow 4s \Rightarrow 3d$ and so on.)

EXERCISE 7.1 Order of Subshell Assignments

To which of the following subshells should an electron be assigned first?

(a) 4*s* or 4*p* **(b)** 5*d* or 6*s* **(c)** 4*f* or 5*s*

Effective Nuclear Charge, Z^*

■ **More About Z^*** For a more complete discussion of effective nuclear charge, see D. M. P. Mingos: *Essential Trends in Inorganic Chemistry*, New York, Oxford University Press, 1998.

The order in which electrons are assigned to subshells in an atom, and many atomic properties, can be rationalized by introducing the concept of **effective nuclear charge** ***(Z*)***. This is the net charge experienced by a particular electron in a multielectron atom resulting from the nucleus and the other electrons. Knowing Z^* provides a convenient way to assess the attractive and repulsive forces on that electron by the nucleus and the other electrons and to assess the energy of that electron.

The surface density plot ($4\pi r^2\psi^2$) for a 2*s* electron for lithium in plotted in Figure 7.3. (Lithium has three protons in the nucleus, two 1*s* electrons in the first shell, and a 2*s* electron in the second shell.) The probability of finding the 2*s* electron (recorded on the vertical axis) changes as one moves away from the nucleus (horizontal axis). Lightly shaded on this figure is the region in which the two 1*s* electrons have their highest probability. Observe that the 2*s* electron wave occurs partly within the region of space occupied by 1*s* electrons. Chemists say that the 2*s* orbital *penetrates* the region defining the 1*s* orbital.

At a large distance from the nucleus, the lithium 2*s* electron will experience a +1 charge, the net effect of the two 1*s* electrons (total charge = −2) and the nucleus (+3 charge.) The 1*s* electrons are said to *screen* the 2*s* electron from experiencing the full nuclear charge. However, this screening of the nuclear charge varies with the distance of the 2*s* electron from the nucleus. As the 2*s* electron wave penetrates the 1*s* electron region, it experiences an increasingly higher net positive charge. Very near the nucleus, the 1*s* electrons do not effectively screen the electron from the nucleus, and the 2*s* electron experiences a charge close to +3. Figure 7.3 shows that a 2*s* electron has some probability of being both inside and outside the

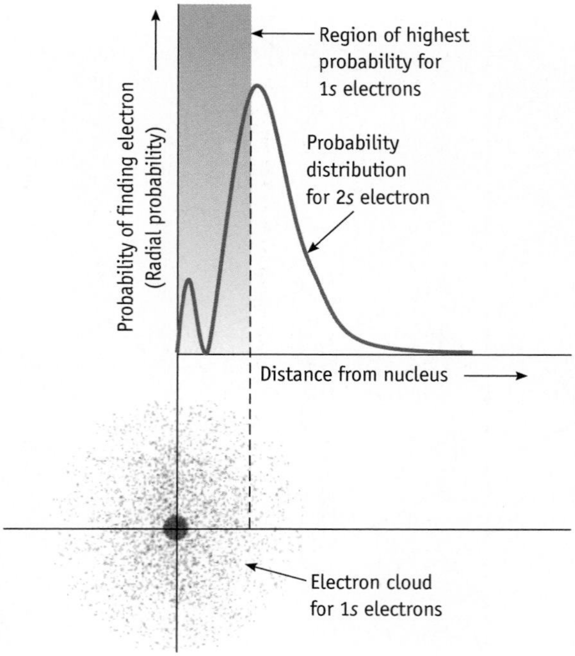

FIGURE 7.3 Effective nuclear charge, Z^*. The two 1*s* electrons of lithium have their highest probability in the shaded region, but this region is penetrated by the 2*s* electron (whose approximate surface density plot is shown here). As the 2*s* electron penetrates the 1*s* region, however, the 2*s* electron experiences a larger and larger positive charge, to a maximum of +3. On average, the 2*s* electron experiences a charge, called the effective nuclear charge *(Z^*)* that is smaller than +3 but greater than +1.

region occupied by the 1*s* electrons. Thus, on average, a 2*s* electron experiences a positive charge greater than +1 but much smaller than +3. The *average* charge experienced by the electron is called the *effective nuclear charge* (Z^*).

In the hydrogen atom, with only one electron, the 2*s* and 2*p* subshells have the same energy. However, in atoms with two or more electrons, the energies of the 2*s* and 2*p* subshells are different. Why should this be true? It is observed that the relative extent to which an outer electron penetrates inner orbitals occurs in the order $s > p > d > f$. Thus, the effective nuclear charge experienced by electrons in a multielectron atom is in the order $ns > np > nd > nf$. The values of Z^* for *s* and *p* electrons for the second-period elements (Table 7.2) illustrate this. In each case, Z^* is greater for *s* electrons than for *p* electrons. In a given shell, *s* electrons always have a lower energy than *p* electrons; *p* electrons have a lower energy than *d* electrons, and *d* electrons have a lower energy than *f* electrons. A consequence of this is that subshells within an electron shell are filled in the order *ns* before *np* before *nd* before *nf*.

Table 7.2 also shows that for the second-period elements the value of Z^* for the higher energy electrons increases across the period. As you will see in Section 7.5, this effect is important in understanding trends in properties of elements across a period.

What emerges from this analysis is the order of shell and subshell energies for any given atom and the filling order in Figure 7.2. With this as background, we turn to the periodic table and use it as a guide to electron arrangements in atoms.

■ **Z^* for *s* and *p* Subshells** Z^* is greater for *s* electrons than for *p* electrons in the same shell. This difference becomes larger as *n* becomes larger. For example, compare the Group 4A elements.

Atom	$Z^*(ns)$	$Z^*(np)$	Value of *n*
C	3.22	3.14	2
Si	4.90	4.29	3
Ge	8.04	6.78	4

7.3 Electron Configurations of Atoms

Arrangements of electrons in the elements up to 109—their **electron configurations**—are given in Table 7.3. Specifically, these are the ground state electron configurations, where electrons are found in the shells, subshells, and orbitals that result in the lowest energy for the atom. In general, electrons are assigned to orbitals in order of increasing $n + \ell$. The emphasis here, however, will be to connect the configurations of the elements with their positions in the periodic table (Figure 7.4).

Electron Configurations of the Main Group Elements

Hydrogen, the first element in the periodic table, has one electron in a 1*s* orbital. One way to depict its electron configuration is with the orbital box diagram used earlier, but an alternative and more frequently used method is the *spdf* notation. Using this method, the electron configuration of H is $1s^1$, read "one *s* one." This indicates that there is one electron (indicated by the superscript) in the 1*s* orbital.

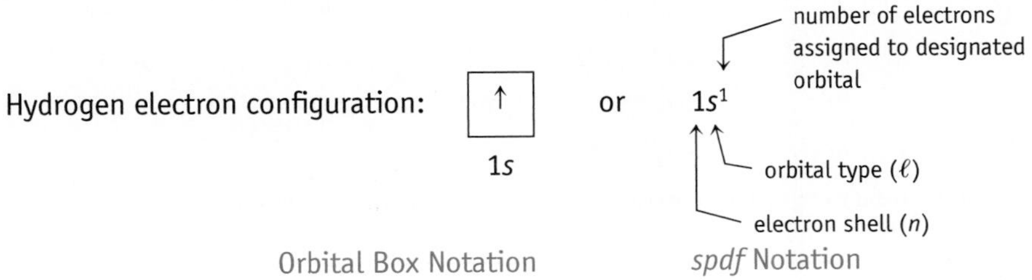

TABLE 7.2 Effective Nuclear Charges, Z^*, for $n = 2$ Elements

Atom	$Z^*(2s)$	$Z^*(2p)$
Li	1.28	
B	2.58	2.42
C	3.22	3.14
N	3.85	3.83
O	4.49	4.45
F	5.13	5.10

TABLE 7.3 Ground State Electron Configurations

Z	Element	Configuration	Z	Element	Configuration	Z	Element	Configuration
1	H	$1s^{1}$	37	Rb	$[Kr]5s^{1}$	74	W	$[Xe]4f^{14}5d^{4}6s^{2}$
2	He	$1s^{2}$	38	Sr	$[Kr]5s^{2}$	75	Re	$[Xe]4f^{14}5d^{5}6s^{2}$
3	Li	$[He]2s^{1}$	39	Y	$[Kr]4d^{1}5s^{2}$	76	Os	$[Xe]4f^{14}5d^{6}6s^{2}$
4	Be	$[He]2s^{2}$	40	Zr	$[Kr]4d^{2}5s^{2}$	77	Ir	$[Xe]4f^{14}5d^{7}6s^{2}$
5	B	$[He]2s^{2}2p^{1}$	41	Nb	$[Kr]4d^{4}5s^{1}$	78	Pt	$[Xe]4f^{14}5d^{9}6s^{1}$
6	C	$[He]2s^{2}2p^{2}$	42	Mo	$[Kr]4d^{5}5s^{1}$	79	Au	$[Xe]4f^{14}5d^{10}6s^{1}$
7	N	$[He]2s^{2}2p^{3}$	43	Tc	$[Kr]4d^{5}5s^{2}$	80	Hg	$[Xe]4f^{14}5d^{10}6s^{2}$
8	O	$[He]2s^{2}2p^{4}$	44	Ru	$[Kr]4d^{7}5s^{1}$	81	Tl	$[Xe]4f^{14}5d^{10}6s^{2}6p^{1}$
9	F	$[He]2s^{2}2p^{5}$	45	Rh	$[Kr]4d^{8}5s^{1}$	82	Pb	$[Xe]4f^{14}5d^{10}6s^{2}6p^{2}$
10	Ne	$[He]2s^{2}2p^{6}$	46	Pd	$[Kr]4d^{10}$	83	Bi	$[Xe]4f^{14}5d^{10}6s^{2}6p^{3}$
11	Na	$[Ne]3s^{1}$	47	Ag	$[Kr]4d^{10}5s^{1}$	84	Po	$[Xe]4f^{14}5d^{10}6s^{2}6p^{4}$
12	Mg	$[Ne]3s^{2}$	48	Cd	$[Kr]4d^{10}5s^{2}$	85	At	$[Xe]4f^{14}5d^{10}6s^{2}6p^{5}$
13	Al	$[Ne]3s^{2}3p^{1}$	49	In	$[Kr]4d^{10}5s^{2}5p^{1}$	86	Rn	$[Xe]4f^{14}5d^{10}6s^{2}6p^{6}$
14	Si	$[Ne]3s^{2}3p^{2}$	50	Sn	$[Kr]4d^{10}5s^{2}5p^{2}$	87	Fr	$[Rn]7s^{1}$
15	P	$[Ne]3s^{2}3p^{3}$	51	Sb	$[Kr]4d^{10}5s^{2}5p^{3}$	88	Ra	$[Rn]7s^{2}$
16	S	$[Ne]3s^{2}3p^{4}$	52	Te	$[Kr]4d^{10}5s^{2}5p^{4}$	89	Ac	$[Rn]6d^{1}7s^{2}$
17	Cl	$[Ne]3s^{2}3p^{5}$	53	I	$[Kr]4d^{10}5s^{2}5p^{5}$	90	Th	$[Rn]6d^{2}7s^{2}$
18	Ar	$[Ne]3s^{2}3p^{6}$	54	Xe	$[Kr]4d^{10}5s^{2}5p^{6}$	91	Pa	$[Rn]5f^{2}6d^{1}7s^{2}$
19	K	$[Ar]4s^{1}$	55	Cs	$[Xe]6s^{1}$	92	U	$[Rn]5f^{3}6d^{1}7s^{2}$
20	Ca	$[Ar]4s^{2}$	56	Ba	$[Xe]6s^{2}$	93	Np	$[Rn]5f^{4}6d^{1}7s^{2}$
21	Sc	$[Ar]3d^{1}4s^{2}$	57	La	$[Xe]5d^{1}6s^{2}$	94	Pu	$[Rn]5f^{6}7s^{2}$
22	Ti	$[Ar]3d^{2}4s^{2}$	58	Ce	$[Xe]4f^{1}5d^{1}6s^{2}$	95	Am	$[Rn]5f^{7}7s^{2}$
23	V	$[Ar]3d^{3}4s^{2}$	59	Pr	$[Xe]4f^{3}6s^{2}$	96	Cm	$[Rn]5f^{7}6d^{1}7s^{2}$
24	Cr	$[Ar]3d^{5}4s^{1}$	60	Nd	$[Xe]4f^{4}6s^{2}$	97	Bk	$[Rn]5f^{9}7s^{2}$
25	Mn	$[Ar]3d^{5}4s^{2}$	61	Pm	$[Xe]4f^{5}6s^{2}$	98	Cf	$[Rn]5f^{10}7s^{2}$
26	Fe	$[Ar]3d^{6}4s^{2}$	62	Sm	$[Xe]4f^{6}6s^{2}$	99	Es	$[Rn]5f^{11}7s^{2}$
27	Co	$[Ar]3d^{7}4s^{2}$	63	Eu	$[Xe]4f^{7}6s^{2}$	100	Fm	$[Rn]5f^{12}7s^{2}$
28	Ni	$[Ar]3d^{8}4s^{2}$	64	Gd	$[Xe]4f^{7}5d^{1}6s^{2}$	101	Md	$[Rn]5f^{13}7s^{2}$
29	Cu	$[Ar]3d^{10}4s^{1}$	65	Tb	$[Xe]4f^{9}6s^{2}$	102	No	$[Rn]5f^{14}7s^{2}$
30	Zn	$[Ar]3d^{10}4s^{2}$	66	Dy	$[Xe]4f^{10}6s^{2}$	103	Lr	$[Rn]5f^{14}6d^{1}7s^{2}$
31	Ga	$[Ar]3d^{10}4s^{2}4p^{1}$	67	Ho	$[Xe]4f^{11}6s^{2}$	104	Rf	$[Rn]5f^{14}6d^{2}7s^{2}$
32	Ge	$[Ar]3d^{10}4s^{2}4p^{2}$	68	Er	$[Xe]4f^{12}6s^{2}$	105	Db	$[Rn]5f^{14}6d^{3}7s^{2}$
33	As	$[Ar]3d^{10}4s^{2}4p^{3}$	69	Tm	$[Xe]4f^{13}6s^{2}$	106	Sg	$[Rn]5f^{14}6d^{4}7s^{2}$
34	Se	$[Ar]3d^{10}4s^{2}4p^{4}$	70	Yb	$[Xe]4f^{14}6s^{2}$	107	Bh	$[Rn]5f^{14}6d^{5}7s^{2}$
35	Br	$[Ar]3d^{10}4s^{2}4p^{5}$	71	Lu	$[Xe]4f^{14}5d^{1}6s^{2}$	108	Hs	$[Rn]5f^{14}6d^{6}7s^{2}$
36	Kr	$[Ar]3d^{10}4s^{2}4p^{6}$	72	Hf	$[Xe]4f^{14}5d^{2}6s^{2}$	109	Mt	$[Rn]5f^{14}6d^{7}7s^{2}$
			73	Ta	$[Xe]4f^{14}5d^{3}6s^{2}$			

*This table follows the general convention of writing the orbitals in order of increasing n when writing electron configurations. For a given n, the subshells are listed in order of increasing ℓ.

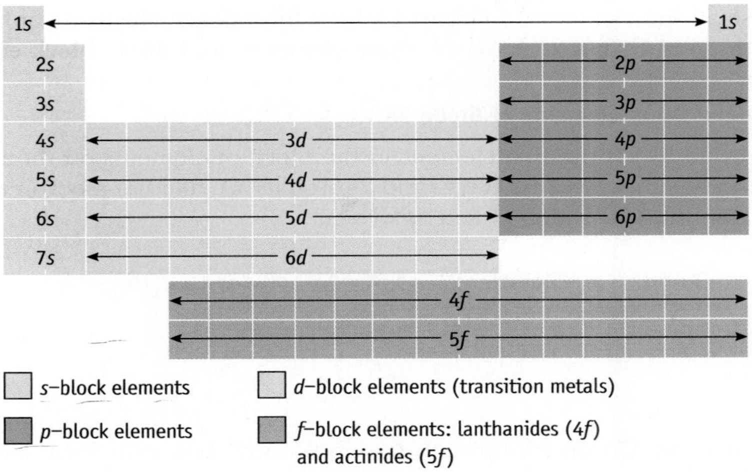

Active Figure 7.4 Electron configurations and the periodic table. The periodic table can serve as a guide in determining the order of filling of atomic orbitals. As one moves from left to right in a period, electrons are assigned to the indicated orbitals. See Table 7.3.

Chemistry Now™ Sign in at www.thomsonedu.com/login and go to the Chapter Contents menu to explore an interactive version of this figure accompanied by an exercise.

Lithium (Li) and Other Elements of Group 1A

Lithium, with three electrons, is the first element in the second period of the periodic table. The first two electrons are in the 1*s* subshell, and the third electron must be in the 2*s* subshell of the $n = 2$ shell. The *spdf* notation, $1s^22s^1$, is read "one *s* two, two *s* one."

Lithium: *spdf* notation $1s^22s^1$

Box notation [↑↓] 1*s* [↑] 2*s* [ ][ ][ ] 2*p*

Electron configurations are often written in abbreviated form by writing in brackets the symbol for the noble gas preceding the element (called the **noble gas notation**) and then indicating any electrons beyond those in the noble gas by using *spdf* or orbital box notation. In lithium, the arrangement preceding the 2*s* electron is the electron configuration of the noble gas helium, so, instead of writing out $1s^22s^1$, using this shorthand lithium's configuration would be written as $[He]2s^1$.

The electrons included in the noble gas notation are often referred to as the **core electrons** of the atom. The core electrons can generally be ignored when considering the chemistry of an element. The electrons beyond the core electrons—the $2s^1$ electron in the case of lithium—are called **valence electrons**; these are the electrons that determine the chemical properties of an element.

All the elements of Group 1A have one electron assigned to an *s* orbital of the *n*th shell, for which *n* is the number of the period in which the element is found (Figure 7.4). For example, potassium is the first element in the $n = 4$ row (the fourth period), so potassium has the electron configuration of the element preceding it in the table (Ar) plus a final electron assigned to the 4*s* orbital: $[Ar]4s^1$.

Beryllium (Be) and Other Elements of Group 2A

All elements of Group 2A have electron configurations of [electrons of preceding noble gas]ns^2, where *n* is the period in which the element is found in the periodic table. Beryllium, for example, has two electrons in the 1*s* orbital plus two additional electrons.

Beryllium: *spdf* notation $1s^22s^2$ or $[He]2s^2$

Box notation [↑↓] 1*s* [↑↓] 2*s* [ ][ ][ ] 2*p*

Because all the elements of Group 1A have the valence electron configuration ns^1, and those in Group 2A have ns^2, these elements are called ***s*-block elements**.

Boron (B) and Other Elements of Group 3A

Boron (Group 3A) is the first element in the block of elements on the right side of the periodic table. Because the 1*s* and 2*s* orbitals are filled in a boron atom, the fifth electron must be assigned to a 2*p* orbital.

Boron:	*spdf* notation	$1s^22s^22p^1$	or	$[He]2s^22p^1$
	Box notation	[↑↓] [↑↓] [↑		]		
		1*s* 2*s* 2*p*		

Elements from Group 3A through Group 8A are often called the ***p*-block elements**. All have the outer shell configuration ns^2np^x, where *x* varies from 1 to 6. The elements in Group 3A, for example, have two *s* electrons and one *p* electron (ns^2np^1) in their outer shells.

Carbon (C) and Other Elements of Group 4A

Carbon (Group 4A) is the second element in the *p*-block, with two electrons assigned to the 2*p* orbitals. You can write the electron configuration of carbon by referring to the periodic table: Starting at H and moving from left to right across the successive periods, you write $1s^2$ to reach the end of period 1 and then $2s^2$ and finally $2p^2$ to bring the electron count to six. For carbon to be in its lowest energy (ground) state, these electrons must be assigned to different *p* orbitals, and both will have the same spin direction.

Carbon:	*spdf* notation	$1s^22s^22p^2$	or	$[He]2s^22p^2$
	Box notation	[↑↓] [↑↓] [↑	↑	]		
		1*s* 2*s* 2*p*		

When assigning electrons to *p, d,* or *f* orbitals, each successive electron is assigned to a different orbital of the subshell, and each electron has the same spin as the previous one, until the subshell is half full. Additional electrons must then be assigned to half-filled orbitals. This procedure follows **Hund's rule,** which states that *the most stable arrangement of electrons is that with the maximum number of unpaired electrons, all with the same spin direction.*

All elements in Group 4A have similar outer shell configurations, ns^2np^2, where *n* is the period in which the element is located in the periodic table.

Nitrogen (N) and Oxygen (O) and Elements of Groups 5A and 6A

Nitrogen (Group 5A) has five valence electrons. Besides the two 2*s* electrons, it has three electrons, all with the same spin, in three different 2*p* orbitals.

Nitrogen:	*spdf* notation	$1s^22s^22p^3$	or	$[He]2s^22p^3$
	Box notation	[↑↓] [↑↓] [↑	↑	↑]		
		1*s* 2*s* 2*p*		

Oxygen (Group 6A) has six valence electrons. Two of these six electrons are assigned to the 2*s* orbital, and the other four electrons are assigned to 2*p* orbitals.

Oxygen:	*spdf* notation	$1s^22s^22p^4$	or	$[He]2s^22p^4$
	Box notation	[↑↓] [↑↓] [↑↓|↑|↑]		
		1*s* 2*s* 2*p*		

The fourth 2*p* electron must pair up with one already present. It makes no difference to which orbital this electron is assigned (the 2*p* orbitals all have the same energy), but it must have a spin opposite to the other electron already assigned to that orbital so that each electron has a different set of quantum numbers.

All elements in Group 5A have an outer shell configuration of ns^2np^3, and all elements in Group 6A have an outer shell configuration of ns^2np^4, where *n* is the period in which the element is located in the periodic table.

Fluorine (F) and Neon (Ne) and Elements of Groups 7A and 8A

Fluorine (Group 7A) has seven electrons in the $n = 2$ shell. Two of these electrons occupy the 2*s* subshell, and the remaining five electrons occupy the 2*p* subshell.

Fluorine:	*spdf* notation	$1s^22s^22p^5$	or	$[He]2s^22p^5$
	Box notation	[↑↓] [↑↓] [↑↓|↑↓|↑]		
		1*s* 2*s* 2*p*		

All halogen atoms have similar outer shell configurations, ns^2np^5, where *n* is the period in which the element is located.

Like all the elements in Group 8A, neon is a noble gas. The Group 8A elements (except helium) have eight electrons in the shell of highest *n* value, so all have the outer shell configuration ns^2np^6, where *n* is the period in which the element is found. That is, all the noble gases have filled *ns* and *np* subshells. The nearly complete chemical inertness of the noble gases is associated with this electron configuration.

Neon:	*spdf* notation	$1s^22s^22p^6$	or	$[He]2s^22p^6$
	Box notation	[↑↓] [↑↓] [↑↓|↑↓|↑↓]		
		1*s* 2*s* 2*p*		

Elements of Period 3

The elements of the third period have valence electron configurations similar to those of the second period, except that the preceding noble gas is neon and the valence shell is the third energy level. For example, silicon has four electrons and a neon core. Because it is the second element in the *p* block, it has two electrons in 3*p* orbitals. Thus, its electron configuration is

Silicon:	*spdf* notation	$1s^22s^22p^63s^23p^2$	or	$[Ne]3s^23p^2$
	Box notation	[↑↓] [↑↓] [↑↓|↑↓|↑↓] [↑↓] [↑|↑|]		
		1*s* 2*s* 2*p* 3*s* 3*p*		

Chemistry Now™

Sign in at **www.thomsonedu.com/login** and go to Chapter 7 Contents to see Screen 7.5 for a simulation exploring **the relationship between an element's electron configuration and its position in the periodic table** and for tutorials on **determining an element's box and spdf notations.**

EXAMPLE 7.1 Electron Configurations

Problem Give the electron configuration of sulfur, using the *spdf*, noble gas, and orbital box notations.

Strategy Sulfur, atomic number 16, is the sixth element in the third period ($n = 3$) and is in the *p*-block. The last six electrons assigned to the atom, therefore, have the configuration $3s^23p^4$. These are preceded by the completed shells $n = 1$ and $n = 2$, the electron arrangement for Ne.

Solution The electron configuration of sulfur is

Complete *spdf* notation:	$1s^22s^22p^63s^23p^4$
spdf with noble gas notation:	$[Ne]3s^23p^4$
Orbital box notation:	[Ne] [↑↓] [↑↓][↑][↑] 3s 3p

EXAMPLE 7.2 Electron Configurations and Quantum Numbers

Problem Write the electron configuration for Al using the noble gas notation, and give a set of quantum numbers for each of the electrons with $n = 3$ (the valence electrons).

Strategy Aluminum is the third element in the third period. It therefore has three electrons with $n = 3$. Because Al is in the *p*-block of elements, two of the electrons are assigned to 3*s*, and the remaining electron is assigned to 3*p*.

Solution The element is preceded by the noble gas neon, so the electron configuration is $[Ne]3s^23p^1$. Using box notation, the configuration is

Aluminum configuration:	[Ne] [↑↓] [↑][ ][ ] 3s 3p

The possible sets of quantum numbers for the two 3*s* electrons are

	n	ℓ	m_ℓ	m_s
For ↑	3	0	0	$1\frac{1}{2}$
For ↓	3	0	0	$-\frac{1}{2}$

For the single 3*p* electron, one of six possible sets is $n = 3$, $\ell = 1$, $m_\ell = 11$, and $m_s = 1\frac{1}{2}$.

EXERCISE 7.2 *spdf* Notation, Orbital Box Diagrams, and Quantum Numbers

(a) Which element has the configuration $1s^22s^22p^63s^23p^5$?

(b) Using *spdf* notation and a box diagram, show the electron configuration of phosphorus.

(c) Write one possible set of quantum numbers for the valence electrons of calcium.

Electron Configurations of the Transition Elements

The elements of the fourth through the seventh periods use *d* and *f* subshells, in addition to *s* and *p* subshells, to accommodate electrons (see Figure 7.4 and Tables 7.3 and 7.4). Elements whose atoms are filling *d* subshells are called **transition elements.** Those elements for which atoms are filling *f* subshells are sometimes called the **inner transition elements** or, more usually, the **lanthanides** (filling 4*f* orbitals) and **actinides** (filling 5*f* orbitals).

In a given period in the periodic table, the transition elements are always preceded by two *s*-block elements. After filling the *ns* orbital in the period, we begin filling the $(n-1)d$ orbitals. Scandium, the first transition element, has the configuration [Ar]$3d^14s^2$, and titanium follows with [Ar]$3d^24s^2$ (Table 7.4).

The general procedure for assigning electrons would suggest that the configuration of a chromium atom is [Ar]$3d^44s^2$. The actual configuration, however, has one electron assigned to each of the six available 3*d* and 4*s* orbitals: [Ar]$3d^54s^1$. This apparently anomalous configuration is explained by assuming that the 4*s* and 3*d* orbitals have approximately the same energy in Cr, and each of the six valence electrons of chromium is assigned to one of these orbitals.

Following chromium, atoms of manganese, iron, and nickel have the configurations that would be expected from the order of orbital filling in Figure 7.2. Copper ([Ar]$3d^{10}4s^1$) is the second exception in this series; it has a single electron in the 4*s* orbital, and the remaining 10 electrons beyond the argon core are assigned to the 3*d* orbitals. Zinc, with the configuration [Ar]$3d^{10}4s^2$, ends the first transition series.

The fifth period transition elements follow the pattern of the fourth period with minor variations.

■ **Writing Configurations for Transition Metals** We follow the convention of writing configurations with shells listed in order of increasing *n* and, within a given shell, writing subshells in order of increasing ℓ. Many educators write them as, for example, [Ar]$4s^23d^2$ to reflect the order of orbital filling. In fact, either notation is correct.

TABLE 7.4 Orbital Box Diagrams for the Elements Ca Through Zn

		3*d*	4*s*
Ca	[Ar]$4s^2$	[ ][ ][ ][ ][ ]	[↑↓]
Sc	[Ar]$3d^14s^2$	[↑][ ][ ][ ][ ]	[↑↓]
Ti	[Ar]$3d^24s^2$	[↑][↑][ ][ ][ ]	[↑↓]
V	[Ar]$3d^34s^2$	[↑][↑][↑][ ][ ]	[↑↓]
Cr*	[Ar]$3d^54s^1$	[↑][↑][↑][↑][↑]	[↑]
Mn	[Ar]$3d^54s^2$	[↑][↑][↑][↑][↑]	[↑↓]
Fe	[Ar]$3d^64s^2$	[↑↓][↑][↑][↑][↑]	[↑↓]
Co	[Ar]$3d^74s^2$	[↑↓][↑↓][↑][↑][↑]	[↑↓]
Ni	[Ar]$3d^84s^2$	[↑↓][↑↓][↑↓][↑][↑]	[↑↓]
Cu*	[Ar]$3d^{10}4s^1$	[↑↓][↑↓][↑↓][↑↓][↑↓]	[↑]
Zn	[Ar]$3d^{10}4s^2$	[↑↓][↑↓][↑↓][↑↓][↑↓]	[↑↓]

*These configurations do not follow the "$n + \ell$" rule.

Lanthanides and Actinides

The sixth period includes the lanthanide series beginning with lanthanum, La. As the first element in the *d*-block, lanthanum has the configuration $[Xe]5d^16s^2$. The next element, cerium (Ce), is set out in a separate row at the bottom of the periodic table, and it is with the elements in this row (Ce through Lu) that electrons are first assigned to *f* orbitals. Thus, the configuration of cerium is $[Xe]4f^15d^16s^2$. Moving across the lanthanide series, the pattern continues (although with occasional variations in occupancy of the 5*d* and 4*f* orbitals). The lanthanide series ends with 14 electrons being assigned to the seven 4*f* orbitals in the last element, lutetium (Lu, $[Xe]4f^{14}5d^16s^2$) (see Table 7.3).

The seventh period also includes an extended series of elements utilizing *f* orbitals, the actinides, which begin with actinium (Ac, $[Rn]6d^17s^2$). The next element is thorium (Th), which is followed by protactinium (Pa) and uranium (U). The electron configuration of uranium is $[Rn]5f^36d^17s^2$.

EXAMPLE 7.3 Electron Configurations of the Transition Elements

Problem Using the *spdf* and noble gas notations, give electron configurations for (a) technetium, Tc, and (b) osmium, Os.

Strategy Base your answer on the positions of the elements in the periodic table. That is, for each element, find the preceding noble gas, and then note the number of *s, p, d,* and *f* electrons that lead from the noble gas to the element.

Solution

(a) Technetium, Tc: The noble gas that precedes Tc is krypton, Kr, at the end of the $n = 4$ row. After the 36 electrons of Kr are assigned to the Kr core as [Kr], seven electrons remain. Two of these electrons are in the 5*s* orbital, and the remaining five are in 4*d* orbitals. Therefore, the technetium configuration is $[Kr]4d^55s^2$.

(b) Osmium, Os: Osmium is a sixth-period element and the twenty-second element following the noble gas xenon. Of the 22 electrons to be added after the Xe core, two are assigned to the 6*s* orbital and 14 to 4*f* orbitals. The remaining six are assigned to 5*d* orbitals. Thus, the osmium configuration is $[Xe]4f^{14}5d^66s^2$.

EXERCISE 7.3 Electron Configurations

Using the periodic table and without looking at Table 7.3, write electron configurations for the following elements:

(a) P	**(c)** Zr	**(e)** Pb
(b) Zn	**(d)** In	**(f)** U

Use the *spdf* and noble gas notations. When you have finished, check your answers with Table 7.4.

7.4 Electron Configurations of Ions

We can also determine electron configurations for ions. To form a cation from a neutral atom, one or more of the valence electrons is removed. Electrons are always removed from the electron shell of highest *n*. If several subshells are present within the *n*th shell, the electron or electrons of maximum ℓ are removed. Thus, a sodium ion is formed by removing the $3s^1$ electron from the Na atom,

$$Na: [1s^22s^22p^63s^1] \longrightarrow Na^+: [1s^22s^22p^6] + e^-$$

A Closer Look — Questions About Transition Element Electron Configurations

Why don't all of the $n = 3$ subshells fill before beginning to fill the $n = 4$ subshells? Why is scandium's configuration $[Ar]3d^1 4s^2$ and not $[Ar]3d^3$?

Theoretical chemists have calculated that for the atoms from scandium to zinc the energies of the 3*d* orbitals are always lower than the energy of the 4*s* orbital, so for scandium the configuration $[Ar]3d^3$ would seem to be preferred. One way to understand why it is not is to consider the effect of electron–electron repulsion in 3*d* and 4*s* orbitals. The ground state configuration will be the one that most effectively minimizes electron–electron repulsions and that leads to the lowest total energy.

If we prepare plots for 3*d* and 4*s* orbitals such as that in Figure 7.3, we find that the most probable distance of a 3*d* electron from the nucleus is less than that for a 4*s* electron. Being closer to the nucleus, the 3*d* orbitals are more compact than the 4*s* orbital. This means the 3*d* electrons are closer together, and so two 3*d* electrons would repel each other more strongly than two 4*s* electrons, for example. A consequence is that placing electrons in the slightly higher energy 4*s* orbital lessens the effect of electron–electron repulsions and lowers the overall energy of the atom.

For more on this question, and for insight into an interesting scientific debate, see a series of papers in the *Journal of Chemical Education*, and *The Periodic Table* by E. Scerri, Oxford, 2007. For example,

a) F. L. Pilar, "4*s* is Always Above 3*d*," *Journal of Chemical Education*, Vol. 55, pages 1–6, 1978.
b) E. R. Scerri, "Transition Metal Configurations and Limitations of the Orbital Approximation," *Journal of Chemical Education*, Vol. 66, pages 481–483, 1989.
c) L. G. Vanquickenborne, K. Pierloot, and D. Devoghel, "Transition Metals and the Aufbau Principle," *Journal of Chemical Education*, Vol. 71, pages 469–471, 1994.
d) M. P. Melrose and E. R. Scerri, "Why the 4*s* Orbital is Occupied before the 3*d*," *Journal of Chemical Education*, Vol. 73, pages 498–503, 1996.

and Ge^{2+} is formed by removing two 4*p* electrons from a germanium atom,

$$\text{Ge: } [Ar]3d^{10}4s^2 4p^2 \longrightarrow \text{Ge}^{2+}\text{: } [Ar]3d^{10}4s^2 + 2\ e^-$$

The same general rule applies to transition metal atoms. This means, for example, that the titanium(II) cation has the configuration $[Ar]3d^2$

$$\text{Ti: } [Ar]3d^2 4s^2 \longrightarrow \text{Ti}^{2+}\text{: } [Ar]3d^2 + 2\ e^-$$

Iron(II) and iron(III) cations have the configurations $[Ar]3d^6$ and $[Ar]3d^5$, respectively:

$$\text{Fe: } [Ar]3d^6 4s^2 \longrightarrow \text{Fe}^{2+}\text{: } [Ar]3d^6 + 2\ e^-$$

$$\text{Fe}^{2+}\text{: } [Ar]3d^6 \longrightarrow \text{Fe}^{3+}\text{: } [Ar]3d^5 + e^-$$

Note that in the ionization of transition metals the *ns* electrons are lost before $(n - 1)d$ electrons. All the common transition metals lose their *ns* electrons first, and the cations formed have electron configurations of the general type [noble gas core]$(n - 1)d^x$. This point is important to remember because the magnetic properties of transition metal cations are determined by the number of unpaired electrons in *d* orbitals. For example, the Fe^{3+} ion is paramagnetic to the extent of five unpaired electrons (Figures 6.18, 7.5, and 7.6 and *A Closer Look: Paramagnetism*, page 292). If three 3*d* electrons had been removed instead of two *s* electrons and one *d* electron, the Fe^{3+} ion would still be paramagnetic but only to the extent of three unpaired electrons.

Chemistry Now™

Sign in at **www.thomsonedu.com/login** and go to Chapter 7 Contents to see Screen 7.6 for a simulation of the **changes to an element's electron configuration when it ionizes,** a tutorial on **determining an ion's box notation,** and a tutorial on **determining whether an element is diamagnetic or paramagnetic.**

Charles D. Winters

FIGURE 7.5 Formation of iron(III) chloride. Iron reacts with chlorine (Cl_2) to produce $FeCl_3$. The paramagnetic Fe^{3+} ion has the configuration $[Ar]3d^5$.

FIGURE 7.6 Paramagnetism of Transition Metals and Their Compounds. (a) A sample of iron(III) oxide is packed into a plastic tube and suspended from a thin nylon filament. (b) When a powerful magnet is brought near, the paramagnetic iron(III) ions in Fe_2O_3 cause the sample to be attracted to the magnet. (The magnet is made of neodymium, iron, and boron [$Nd_2Fe_{14}B$]. These powerful magnets are used in acoustic speakers.)

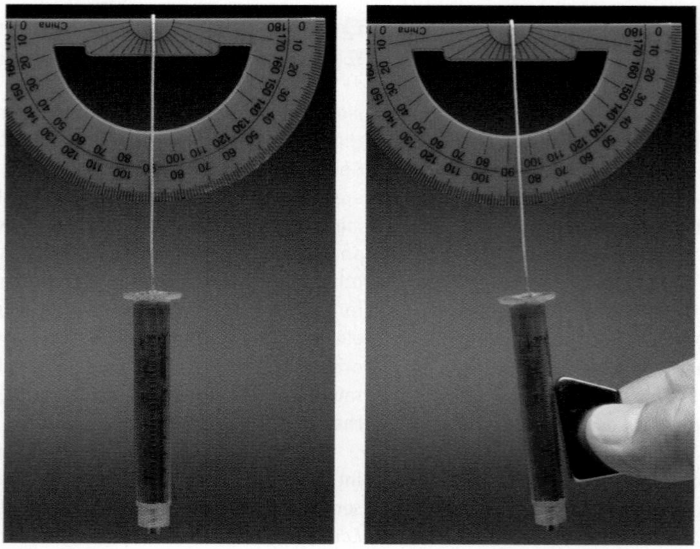

Charles D. Winters

(a) (b)

EXAMPLE 7.4 Configurations of Transition Metal Ions

Problem Give the electron configurations for Cu, Cu^+ and Cu^{2+}. Are either of the ions paramagnetic? How many unpaired electrons does each have?

Strategy Observe the configuration of copper in Table 7.4. Recall that *s* and then *d* electrons are removed to form a transition metal ion.

Solution Copper has only one electron in the 4*s* orbital and ten electrons in 3*d* orbitals:

Cu: [Ar]$3d^{10}4s^1$ [↿⇂][↿⇂][↿⇂][↿⇂][↿⇂] (3*d*) [↑] (4*s*)

When copper is oxidized to Cu^+, the 4*s* electron is lost.

Cu^+: [Ar]$3d^{10}$ [↿⇂][↿⇂][↿⇂][↿⇂][↿⇂] (3*d*) [ ] (4*s*)

The copper(II) ion is formed from copper(I) by removal of one of the 3*d* electrons.

Cu^{2+}: [Ar]$3d^9$ [↿⇂][↿⇂][↿⇂][↿⇂][↑] (3*d*) [ ] (4*s*)

A copper(II) ion (Cu^{2+}) has one unpaired electron, so it is paramagnetic. In contrast, Cu^+ is diamagnetic.

EXERCISE 7.4 Metal Ion Configurations

Depict the electron configurations for V^{2+}, V^{3+}, and Co^{3+}. Use orbital box diagrams and the noble gas notation. Are any of the ions paramagnetic? If so, give the number of unpaired electrons.

7.5 Atomic Properties and Periodic Trends

Once electron configurations were understood, chemists realized that *similarities in properties of the elements are the result of similar valence shell electron configurations.* An objective of this section is to describe how atomic electron configurations are related to some of the physical and chemical properties of the elements and why those properties change in a reasonably predictable manner when moving down groups and across periods. This background should make the periodic table an even more useful tool in your study of chemistry. With an understanding of electron configurations and their relation to properties, you should be able to organize and predict many chemical and physical properties of the elements and their compounds.

Atomic Size

An orbital has no sharp boundary (◀ Figure 6.13a), so how can we define the size of an atom? There are actually several ways, and they can give slightly different results.

One of the simplest and most useful ways to define atomic size is to relate it to the distance between atoms in a sample of the element. Let us consider a diatomic molecule such as Cl_2 (Figure 7.7a). The radius of a Cl atom is assumed to be half the experimentally determined distance between the centers of the two atoms (198 pm), so the radius of one Cl atom is 99 pm. Similarly, the C—C distance in diamond is 154 pm, so a radius of 77 pm can be assigned to carbon. To test these estimates, we can add them together to estimate the C—Cl distance in CCl_4. The predicted distance of 176 pm agrees with the experimentally measured C—Cl distance of 176 pm. (Radii determined this way are often called "covalent radii.")

This approach to determining atomic radii applies only if molecular compounds of the element exist (and so it is limited to nonmetals and metalloids). For metals, atomic radii are sometimes estimated from measurements of the atom-to-atom distance in a crystal of the element (Figure 7.7b).

Some interesting periodic trends are seen immediately on looking at a table of radii (Figure 7.8). *For the main group elements, atomic radii generally increase going down*

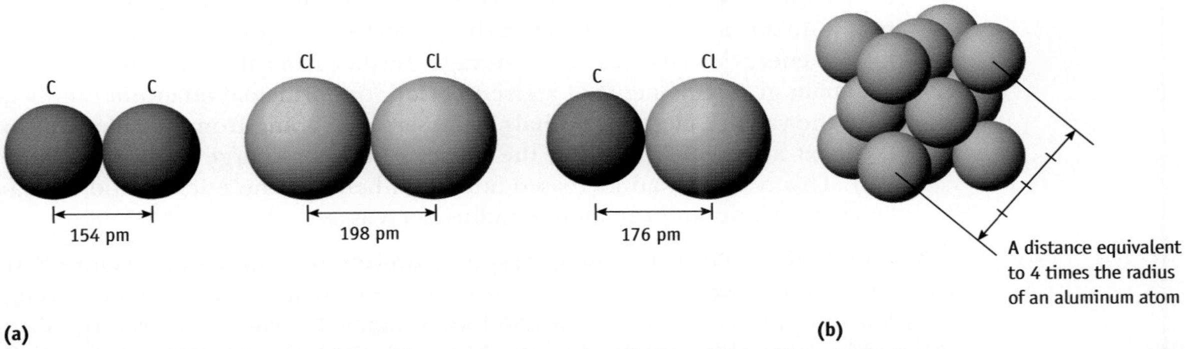

FIGURE 7.7 Determining atomic radii. (a) The sum of the atomic radii of C and Cl provides a good estimate of the C–Cl distance in a molecule having such a bond. (b) Pictured here is a tiny piece of an aluminum crystal. Each sphere represents an aluminum atom. Measuring the distance shown allows a scientist to estimate the radius of an aluminum atom.

Active Figure 7.8 Atomic radii in picometers for main group elements. 1 pm = 1×10^{-12} m = 1×10^{-3} nm. *(Data taken from J. Emsley:* The Elements, *Clarendon Press, Oxford, 1998, 3rd ed.)*

Chemistry Now™ Sign in at www.thomsonedu.com/login and go to the Chapter Contents menu to explore an interactive version of this figure accompanied by an exercise.

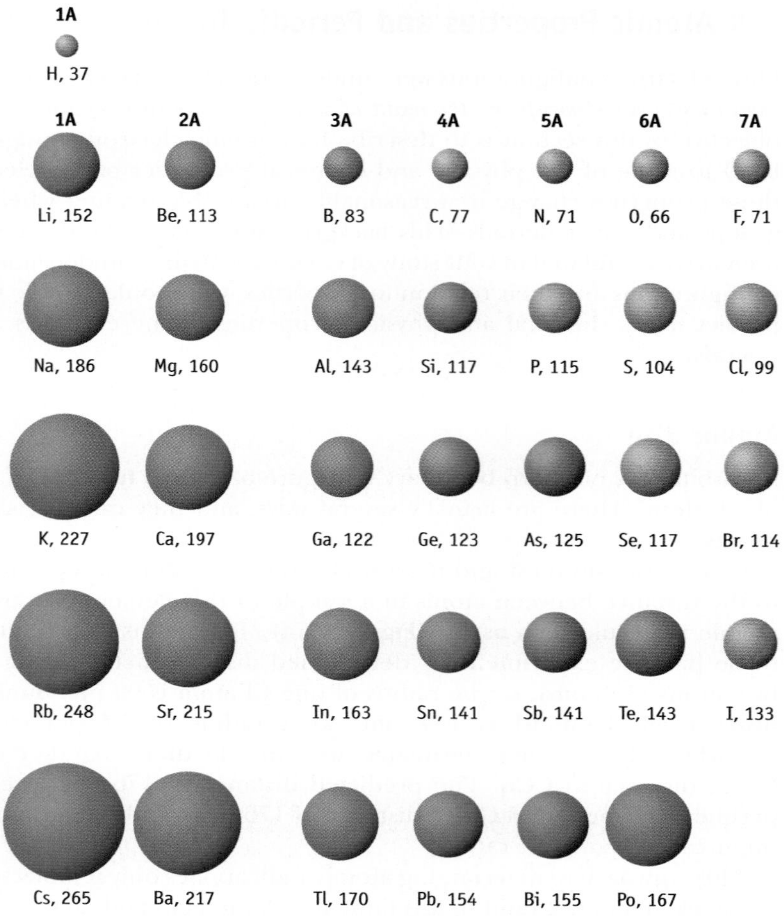

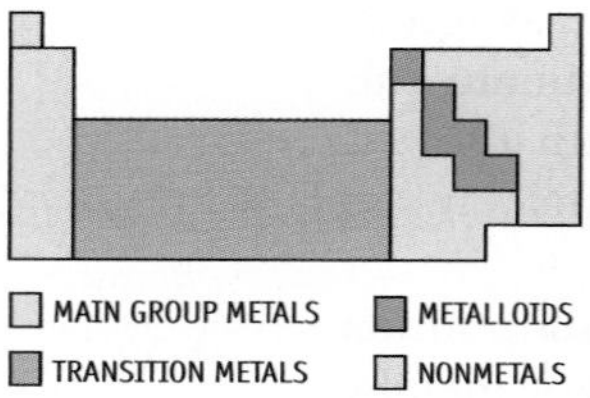

a group in the periodic table and decrease going across a period. These trends reflect two important effects:

- The size of an atom is determined by the outermost electrons. In going from the top to the bottom of a group in the periodic table, the outermost electrons are assigned to orbitals with increasingly higher values of the principal quantum number, n. Because the underlying electrons require some space, these higher energy electrons are, on average, further from the nucleus.
- For main group elements of a given period, the principal quantum number, n, of the valence electron orbitals is the same. In going from one element to the next across a period, Z^*, the effective nuclear charge increases (Table 7.2). This results in an increased attraction between the nucleus and the valence electrons, and the atomic radius decreases.

■ **Atomic Radii—Caution** Numerous tabulations of atomic and covalent radii exist, and the values quoted in them often differ somewhat. The variation comes about because several methods are used to determine the radii of atoms.

The periodic trend in the atomic radii of transition metal atoms (Figure 7.9) across a period is somewhat different from that for main group elements. Going from left to right in a given period, the radii initially decrease. However, the sizes of the elements in the middle of a transition series change very little, and a small increase in size occurs at the end of the series. The size of transition metal atoms is determined largely by electrons in the outermost shell—that is, by the electrons of the ns subshell—but electrons are being added to the $(n-1)d$ orbitals across the series. The increased nuclear charge on the atoms as one moves from left to

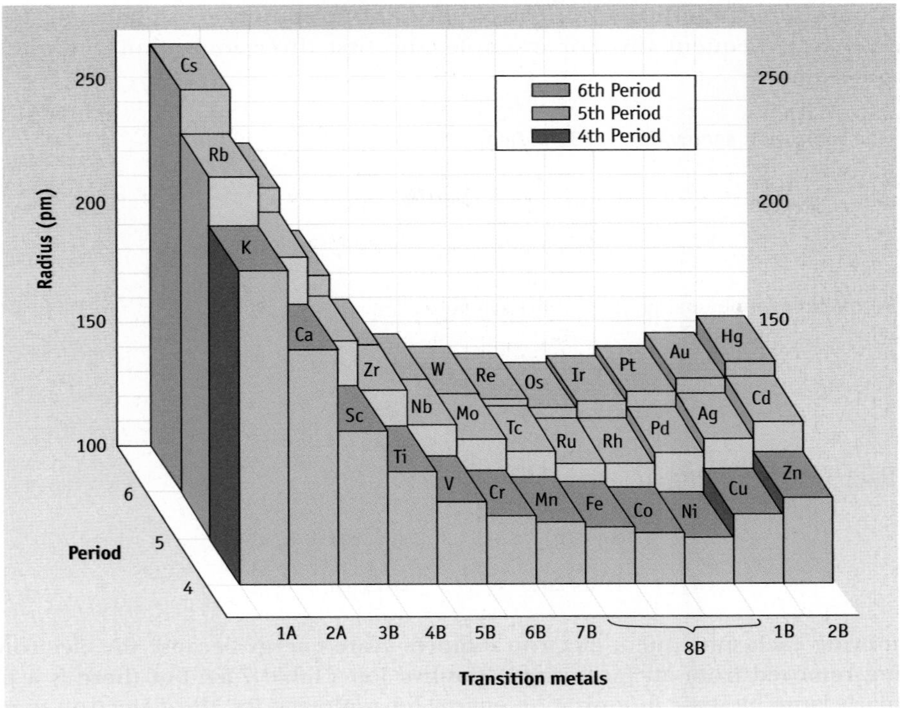

FIGURE 7.9 Trends in atomic radii for the transition elements. Atomic radii of the Group 1A and 2A metals and the transition metals of the fourth, fifth, and sixth periods.

right should cause the radius to decrease. This effect, however, is mostly canceled out by increased electron–electron repulsion. On reaching the Group 1B and 2B elements at the end of the series, the size increases slightly because the *d* subshell is filled and electron–electron repulsions dominate.

■ **Trends in Atomic Radii** General trends in atomic radii of *s*- and *p*-block elements with position in the periodic table.

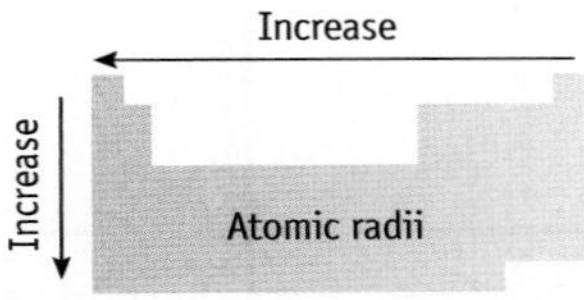

Chemistry Now™

Sign in at **www.thomsonedu.com/login** and go to Chapter 7 Contents to see:

- Screen 7.8 for a simulation exploring **the trends in atomic size moving across and down the periodic table**
- Screen 7.9 for a simulation exploring **energy levels of orbitals and the ability to retain electrons**

EXERCISE 7.5 Periodic Trends in Atomic Radii

Place the three elements Al, C, and Si in order of increasing atomic radius.

Ionization Energy

Ionization energy *(IE)* is the energy required to remove an electron from an atom in the gas phase.

$$\text{Atom in ground state(g)} \longrightarrow \text{Atom}^+(\text{g}) + \text{e}^-$$

$$\Delta U = \text{ionization energy, } IE$$

To separate an electron from an atom, energy must be supplied to overcome the attraction of the nuclear charge. Because energy must be supplied, ionization energies always have positive values.

Atoms other than hydrogen have a series of ionization energies as electrons are removed sequentially. For example, the first three ionization energies of magnesium are

■ **Valence and Core Electrons** Removal of core electrons requires much more energy than removal of a valence electron. Core electrons are not lost in chemical reactions.

■ **Measuring Ionization Energy** Ionization energy values can be measured accurately as compared with the estimations that must be made when measuring atomic radii.

First ionization energy, IE_1 = 738 kJ/mol

$$Mg(g) \rightarrow Mg^+(g) + e^-$$
$$1s^2 2s^2 2p^6 3s^2 \qquad 1s^2 2s^2 2p^6 3s^1$$

Second ionization energy, IE_2 = 1451 kJ/mol

$$Mg^+(g) \rightarrow Mg^{2+}(g) + e^-$$
$$1s^2 2s^2 2p^6 3s^1 \qquad 1s^2 2s^2 2p^6$$

Third ionization energy, IE_3 = 7732 kJ/mol

$$Mg^{2+}(g) \rightarrow Mg^{3+}(g) + e^-$$
$$1s^2 2s^2 2p^6 \qquad 1s^2 2s^2 2p^5$$

Removing each subsequent electron requires more energy because the electron is being removed from an increasingly positive ion (Table 7.5), but there is a particularly large increase in ionization energy for removing the third electron to give Mg^{3+}. The first two ionization steps are for the removal of electrons from the outermost or valence shell of electrons. The third electron, however, must come from the $2p$ subshell, which has a much lower energy than the $3s$ subshell. *This large increase is experimental evidence for the electron shell structure of atoms.*

For main group (s- and p-block) elements, *first ionization energies generally increase across a period and decrease down a group* (Figure 7.10, Table 7.5, and Appendix F). The trend across a period corresponds to the increase in effective nuclear charge, Z^*, with increasing atomic number. As Z^* increases, the energy required to remove an electron increases. The general decrease in ionization energy down a group

TABLE 7.5 First, Second, and Third Ionization Energies for the Main Group Elements in Periods 2–4 (kJ/mol)

2nd Period	Li	Be	B	C	N	O	F	Ne
1st	513	899	801	1086	1402	1314	1681	2080
2nd	7298	1757	2427	2352	2856	3388	3374	3952
3rd	11815	14848	3660	4620	4578	5300	6050	6122
3rd Period	**Na**	**Mg**	**Al**	**Si**	**P**	**S**	**Cl**	**Ar**
1st	496	738	577	787	1012	1000	1251	1520
2nd	4562	1451	1817	1577	1903	2251	2297	2665
3rd	6912	7732	2745	3231	2912	3361	3826	3928
4th Period	**K**	**Ca**	**Ga**	**Ge**	**As**	**Se**	**Br**	**Kr**
1st	419	590	579	762	947	941	1140	1351
2nd	3051	1145	1979	1537	1798	2044	2104	2350
3rd	4411	4910	2963	3302	2735	2974	3500	3565

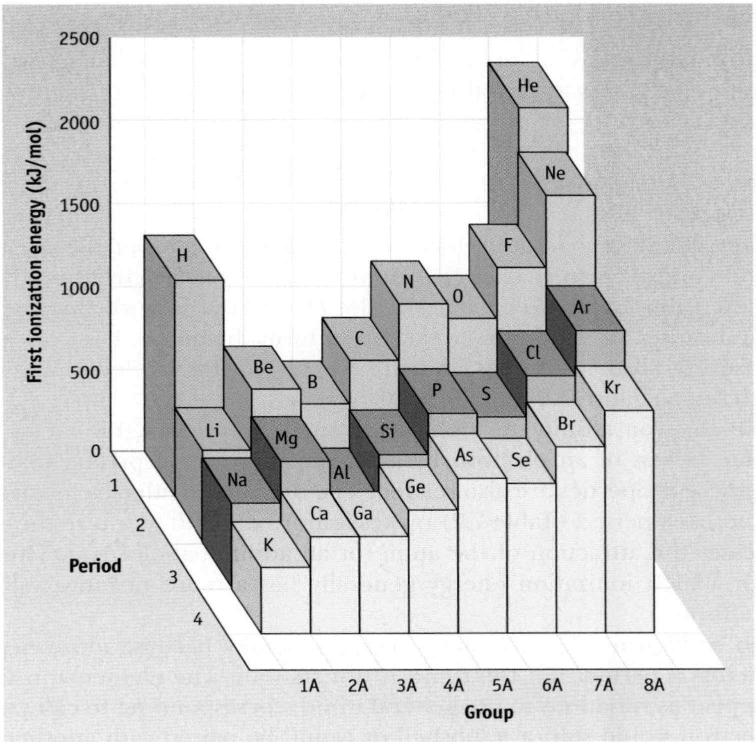

Active Figure 7.10 First ionization energies of the main group elements these first four periods. (For data on these elements see Table 7.5 and Appendix F.)

Chemistry Now™ Sign in at www.thomsonedu.com/login and go to the Chapter Contents menu to explore an interactive version of this figure accompanied by an exercise.

occurs because the electron removed is increasingly farther from the nucleus and thus held less strongly.

Notice that atomic radius and ionization energy are both linked to Z^*. They are inversely related: *as the atomic radius decreases, the ionization energy increases.*

A closer look at ionization energies reveals that there are exceptions to the general trend in a period. One exception occurs on going from *s*-block to *p*-block elements—from beryllium to boron, for example. The $2p$ electrons are slightly higher in energy than the $2s$ electrons so the ionization energy for boron is slightly less than that for beryllium. Another dip to lower ionization energy occurs on going from nitrogen to oxygen. No change occurs in either n or ℓ, but electron–electron repulsions increase for the following reason. In Groups 3A–5A, electrons are assigned to separate p orbitals (p_x, p_y, and p_z). Beginning in Group 6A, however, two electrons are assigned to the same p orbital. The fourth p electron shares an orbital with another electron and thus experiences greater repulsion than it would if it had been assigned to an orbital of its own.

$$\text{O (oxygen atom)} \xrightarrow{+1314\ \text{kJ/mol}} \text{O}^+ \text{ (oxygen cation)} + e^-$$

[Ne] [↑↓] [↑↓][↑][↑] → [Ne] [↑↓] [↑][↑][↑]

$2s$ $2p$ $2s$ $2p$

The greater repulsion experienced by the fourth $2p$ electron makes it easier to remove. The usual trend resumes on going from oxygen to fluorine to neon, however, reflecting the increase in Z^*.

■ **Trends in Ionization Energy** General trends in first ionization energies of *s*- and *p*-block elements with position in the periodic table.

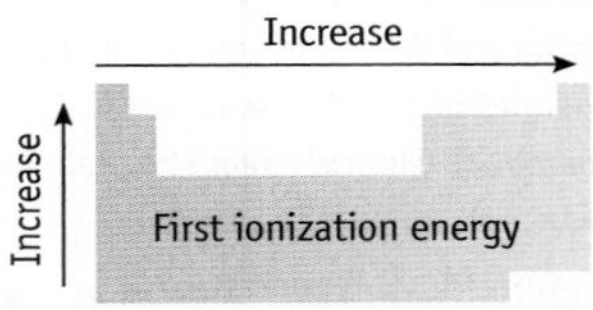

Electron Affinity

■ **Trends in *EA*** General trends in electron affinities of A-group elements. Exceptions occur at Groups 2A and 5A and in moving from period 2 to period 3 in the *p*-block.

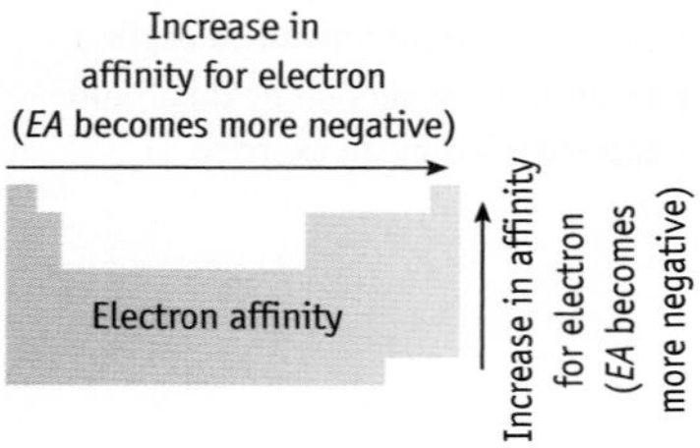

The electron affinity, *EA*, of an atom is defined as the energy change for a process in which an electron is acquired by the atom in the gas phase (Figure 7.11 and Appendix F).

$$A(g) + e^-(g) \longrightarrow A^-(g) \qquad \Delta U \equiv \text{electron affinity, } EA$$

The greater the affinity an atom has for an electron, the lower the energy of the ion will be compared to that of the atom and the free electron, and the more negative the value of *EA*. For example, the electron affinity of fluorine is −328 kJ/mol, indicating an exothermic reaction to form the anion, F^-, from a fluorine atom and an electron. Boron has a much lower affinity for an electron, as indicated by a much less negative *EA* value of −26.7 kJ/mol.

Because electron affinity and ionization energy represent the energy involved in the gain or loss of an electron by an atom, it is not surprising that periodic trends in these properties are also related. The increase in effective nuclear charge of atoms across a period (Table 7.2) makes it more difficult to ionize the atom and also increases the attraction of the atom for an additional electron. Thus, an element with a high ionization energy generally has a more negative value for its electron affinity.

■ ***EA* and Nonmetals** Nonmetals generally have much more negative values of *EA* than metals. This observation agrees with chemical experience; nonmetals form anions, and metals generally do not.

As seen in Figure 7.11, the values of *EA* generally become more negative on moving across a period, but the trend is not smooth. The elements in Group 2A and 5A appear as variations to the general trend, corresponding to cases where the added electron would start a *p* subshell or would be paired with another electron in the *p* subshell, respectively.

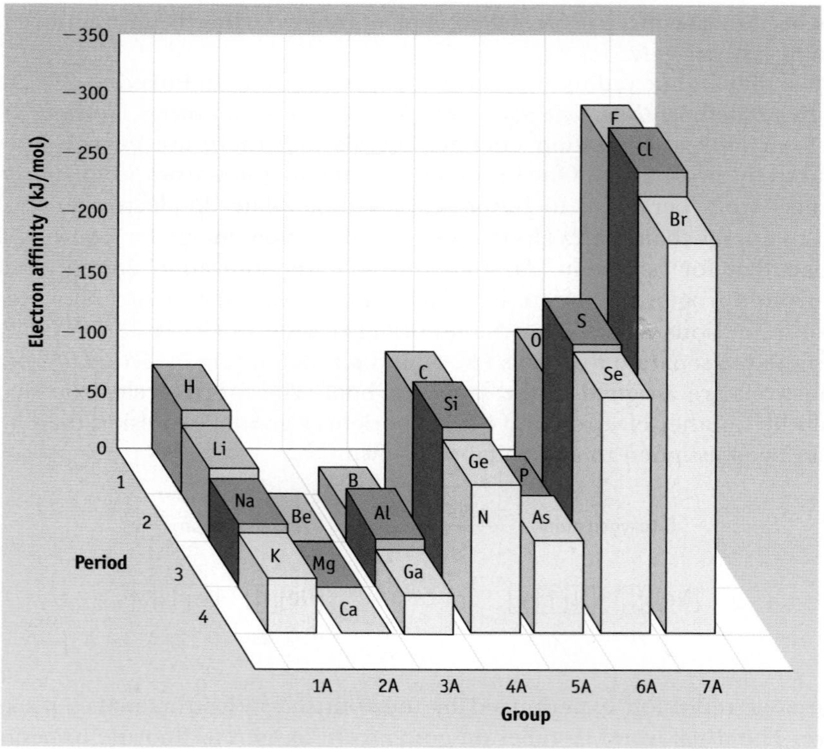

Active Figure 7.11 Electron affinity. The larger the affinity *(EA)* of an atom for an electron, the more negative the value. For numerical values, see Appendix F. (Data were taken from H. Hotop and W. C. Lineberger: "Binding energies of atomic negative ions," *Journal of Physical Chemistry*, Reference Data, Vol. 14, p. 731, 1985.)

Chemistry Now™ Sign in at www.thomsonedu.com/login and go to the Chapter Contents menu to explore an interactive version of this figure accompanied by an exercise.

The value of electron affinity usually becomes less negative on descending a group of the periodic table. Electrons are added increasingly farther from the nucleus, so the attractive force between the nucleus and electrons decreases. This general trend does not apply to second period elements, however. For example, the value of the electron affinity of fluorine is higher (less negative) than the *EA* value for chlorine. The same phenomenon is observed in Groups 3A through 6A. One explanation is that significant electron–electron repulsions occur in small anions such as F^-. That is, adding an electron to the seven electrons already present in the $n = 2$ shell of the small F atom leads to considerable repulsion between electrons. Chlorine has a larger atomic volume than fluorine, so adding an electron does not result in such significant electron–electron repulsions.

A few elements, such as nitrogen and the Group 2A elements, have no affinity for electrons and are listed as having an *EA* value of zero. The noble gases are generally not listed in electron affinity tables. They have no affinity for electrons, because any additional electron must be added to the next higher electron shell.

No atom has a negative electron affinity for a second electron. So what accounts for the existence of ions such as O^{2-} that occur in many compounds? The answer is that doubly charged anions can be stabilized in crystalline environments by electrostatic attraction to neighboring positive ions (see Chapters 8 and 13).

■ **Electron Affinity and Sign Conventions** Changes in sign conventions for electron affinities over the years have caused confusion. For a useful discussion of electron affinity, see J. C. Wheeler: "Electron affinities of the alkaline earth metals and the sign convention for electron affinity," *Journal of Chemical Education*, Vol. 74, pp. 123–127, 1997.

Chemistry Now™

Sign in at **www.thomsonedu.com/login** and go to Chapter 7 Contents to see Screens 7.8–7.10 for simulations exploring the **periodic trends in these properties.**

■ EXAMPLE 7.5 Periodic Trends

Problem Compare the three elements C, O, and Si.

(a) Place them in order of increasing atomic radius.

(b) Which has the largest ionization energy?

(c) Which has the more negative electron affinity, O or C?

Strategy Review the trends in atomic properties in Figures 7.8–7.11, Table 7.5, and Appendix F.

Solution

(a) Atomic size: Atomic radius declines on moving across a period, so oxygen must have a smaller radius than carbon. However, the radius increases on moving down a periodic group. Because C and Si are in the same group (Group 4A), Si must be larger than C. The trend is O < C < Si.

(b) Ionization energy: Ionization energies generally increase across a period and decrease down a group. Thus, the trend in ionization energies is Si (787 kJ/mol) < C (1086 kJ/mol) < O (1314 kJ/mol).

(c) Electron affinity: Electron affinity values generally become less negative down a group (except for the second period elements) and more negative across a period. Therefore, the *EA* for O (= −141.0 kJ/mol) has a more negative *EA* than C (= −121.9 kJ/mol).

EXERCISE 7.6 Periodic Trends

Compare the three elements B, Al, and C.

(a) Place the three elements in order of increasing atomic radius.

(b) Rank the elements in order of increasing ionization energy. (Try to do this without looking at Table 7.5; then compare your prediction with the table.)

(c) Which element, B or C, is expected to have the more negative electron affinity value?

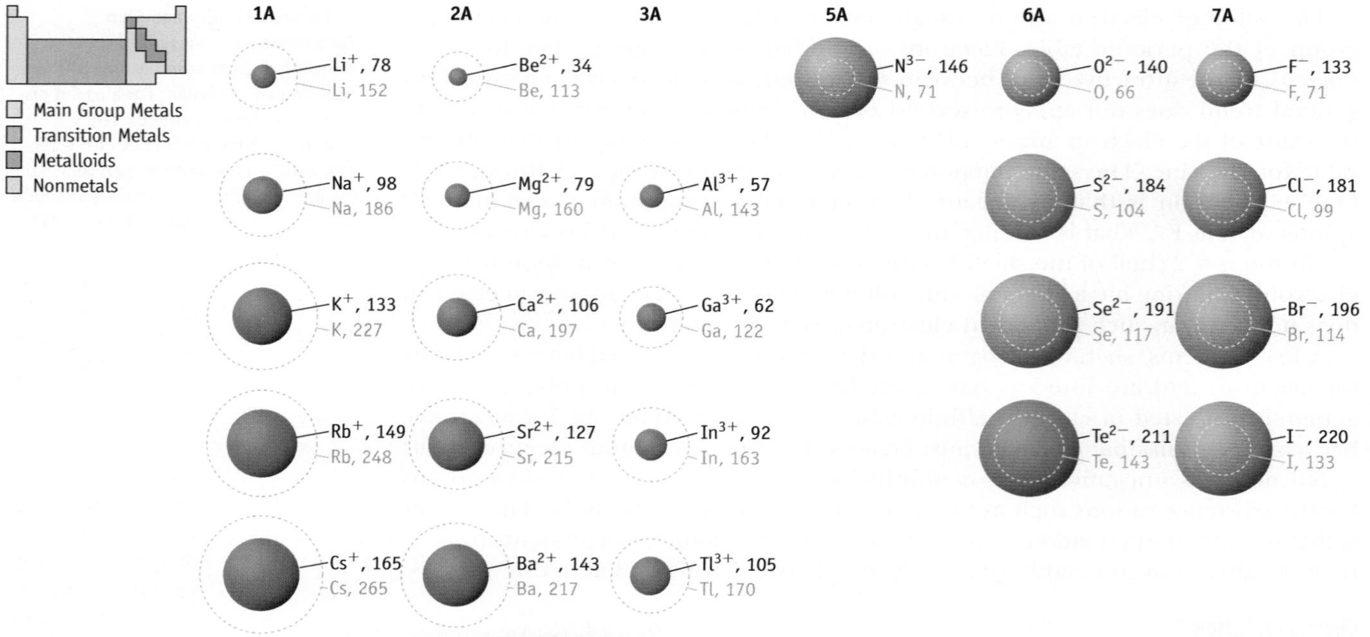

Active Figure 7.12 **Relative sizes of some common ions.** Radii are given in picometers (1 pm = 1×10^{-12} m). (Data taken from J. Emsley, *The Elements*, Clarendon Press, Oxford, 1998, 3rd edition.)

Chemistry Now™ Sign in at www.thomsonedu.com/login and go to the Chapter Contents menu to explore an interactive version of this figure accompanied by an exercise.

Trends in Ion Sizes

The trend in the sizes of ions down a periodic group are the same as those for neutral atoms: Positive and negative ions increase in size when descending the group (Figure 7.12). Pause for a moment, however, and compare the ionic radii with the atomic radii, as illustrated in Figure 7.12. When an electron is removed from an atom to form a cation, the size shrinks considerably. The radius of a cation is always smaller than that of the atom from which it is derived. For example, the radius of Li is 152 pm, whereas the radius of Li^+ is only 78 pm. When an electron is removed from an Li atom, the attractive force of three protons is now exerted on only two electrons, so the remaining electrons are drawn closer to the nucleus. The decrease in ion size is especially great when the last electron of a particular shell is removed, as is the case for Li. The loss of the 2*s* electron from Li leaves Li^+ with no electrons in the $n = 2$ shell.

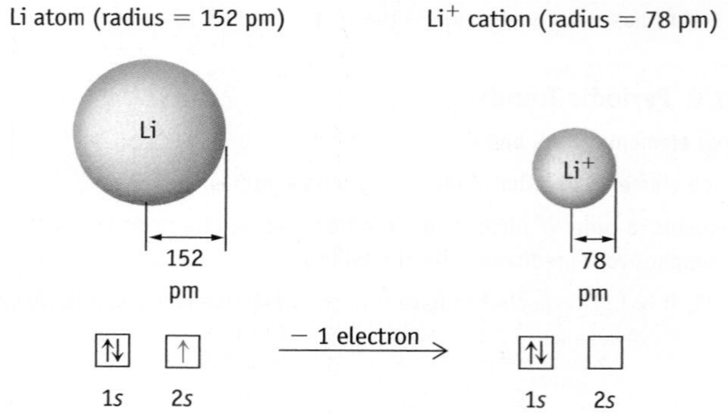

Case Study: Metals in Biochemistry and Medicine

Many main group and transition metals play an important role in biochemistry and in medicine. Your body has low levels of the following metals in the form of various compounds: Ca, 1.5%; Na, 0.1%; Mg, 0.05%, and the metals iron, cobalt zinc, and copper, all less than about 0.05%. (Levels are percentages by mass.)

Much of the 3–4 g of iron in your body is found in hemoglobin, the substance responsible for carrying oxygen to cells. Iron deficiency is marked by fatigue, infections, and mouth inflammations.

Iron in your diet can come from eggs, and brewer's yeast has a very high iron content. In addition, foods such as many breakfast cereals are "fortified" with metallic iron [made by the decomposition of $Fe(CO)_5$]. (In an interesting experiment you can do at home, you can remove the iron by stirring the cereal with a strong magnet.) Vitamin pills often contain iron(II) compounds with anions such as sulfate and succinate ($C_4H_4O_4^{2-}$).

The average person has about 75 mg of copper, about one third of which is found in the muscles. Copper is involved in many biological functions, and a deficiency shows up in many ways: anemia, degeneration of the nervous system, and impaired immunity. Wilson's disease, a genetic disorder, leads to the overaccumulation of copper in the body and results in hepatic and neurological damage.

Like silver ions (page 148), copper ions can also act as a bacteriocide. Scientists from Britain and India recently investigated a long-held belief among people in India that storing water in brass pitchers can ward off illness. (Brass is an alloy of copper and zinc.) They filled brass pitchers with sterile water inoculated with *E. coli* bacteria and filled other brass pitchers with contaminated river water from India. In both cases, they found that fecal bacteria counts dropped from as high as 1,000,000 bacteria per milliliter to zero in two days. In contrast, bacteria levels stayed high in plastic or earthenware pots. Apparently, just enough copper ions are released by the brass to kill the bacteria but not enough to affect humans.

© Jeremy Horner/Corbis

Filling a brass water jug for drinking water in India. Copper ions released in tiny amounts from the brass kill bacteria in contaminated water.

Questions:

1. *Give the electron configurations for iron and the iron(II) and iron(III) ions.*
2. *In hemoglobin, iron can be in the iron(II) or iron(III) state. Are either of these iron ions paramagnetic?*
3. *Give the electron configurations for copper and the copper(I) and copper(II) ions. Is copper in any of these forms paramagnetic?*
4. *Why are copper atoms (radius = 128 pm) slightly larger than iron atoms (radius = 124 pm)?*
5. *In hemoglobin, the iron is enclosed by the porphyrin group, a flat grouping of carbon, hydrogen, and nitrogen atoms. (This is in turn encased in a protein.) When iron is in the form of the Fe^{3+} ion, it just fits into the space within the four N atoms, and the arrangement is flat. Speculate on what occurs to the structure when iron is reduced to the Fe^{2+} ion.*

Answers to these questions are in Appendix Q.

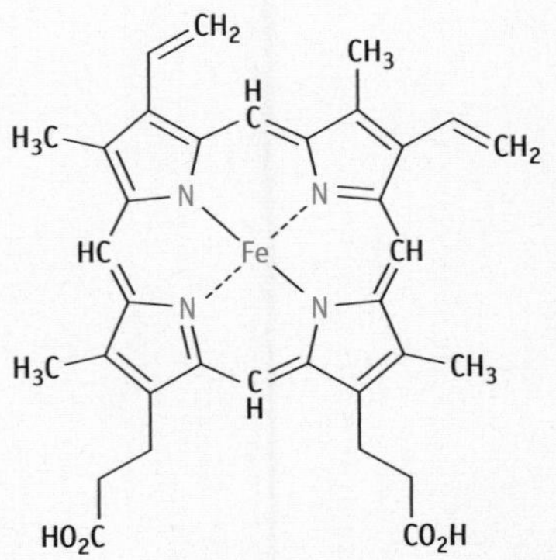

A large decrease in size is also expected if two or more electrons are removed. For example, an aluminum ion, Al^{3+}, has a radius of 57 pm; in contrast, the atomic radius of an aluminum atom is 143 pm.

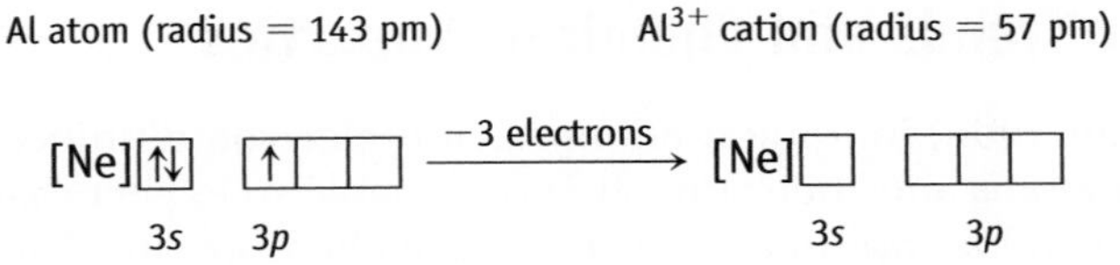

You can also see by comparing Figures 7.8 and 7.12 that anions are always larger than the atoms from which they are derived. Here, the argument is the opposite of that used to explain positive ion radii. The F atom, for example, has nine protons and nine electrons. On forming the anion, the nuclear charge is still +9, but the anion has ten electrons. The F^- ion is larger than the F atom because of increased electron–electron repulsions.

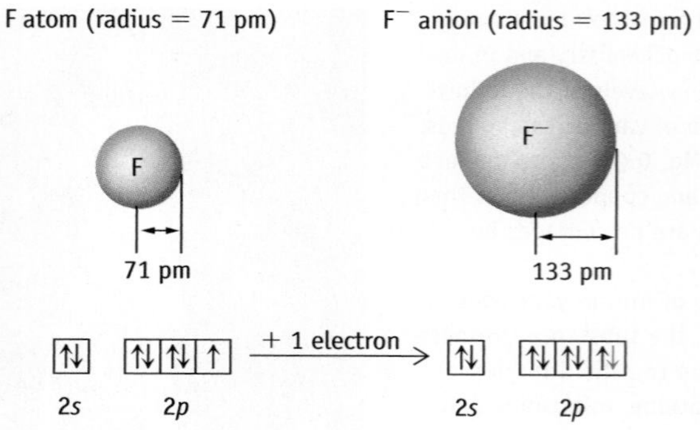

Finally, it is useful to compare the sizes of isoelectronic ions across the periodic table. **Isoelectronic** ions have the same number of electrons (but a different number of protons). One such series of ions is N^{3-}, O^{2-}, F^-, Na^+, and Mg^{2+}:

Ion	N^{3-}	O^{2-}	F^-	Na^+	Mg^{2+}
Number of electrons	10	10	10	10	10
Number of nuclear protons	7	8	9	11	12
Ionic radius (pm)	146	140	133	98	79

All these ions have 10 electrons but they differ in the number of protons. As the number of protons increases in a series of isoelectronic ions, the balance between electron–proton attraction and electron–electron repulsion shifts in favor of attraction, and the radius decreases.

Chemistry Now™

Sign in at **www.thomsonedu.com/login** and go to Chapter 7 Contents to see Screen 7.12 for simulations on **the relationship between ion formation and orbital energies in main group elements** and on the **relationship between orbital energies and electron configurations on the size of the main group element ions.**

EXERCISE 7.7 Ion Sizes

What is the trend in sizes of the ions K^+, S^{2-}, and Cl^-? Briefly explain why this trend exists.

7.6 Periodic Trends and Chemical Properties

Atomic and ionic radii, ionization energies, and electron affinities are properties associated with atoms and their ions. It is reasonable to expect that knowledge of these properties will be useful as we explore the chemistry involving formation of ionic compounds.

The periodic table was created by grouping together elements having similar chemical properties (Figure 7.13). Alkali metals, for example, characteristically form compounds containing a 1+ ion, such as Li^+, Na^+, or K^+. Thus, the reaction

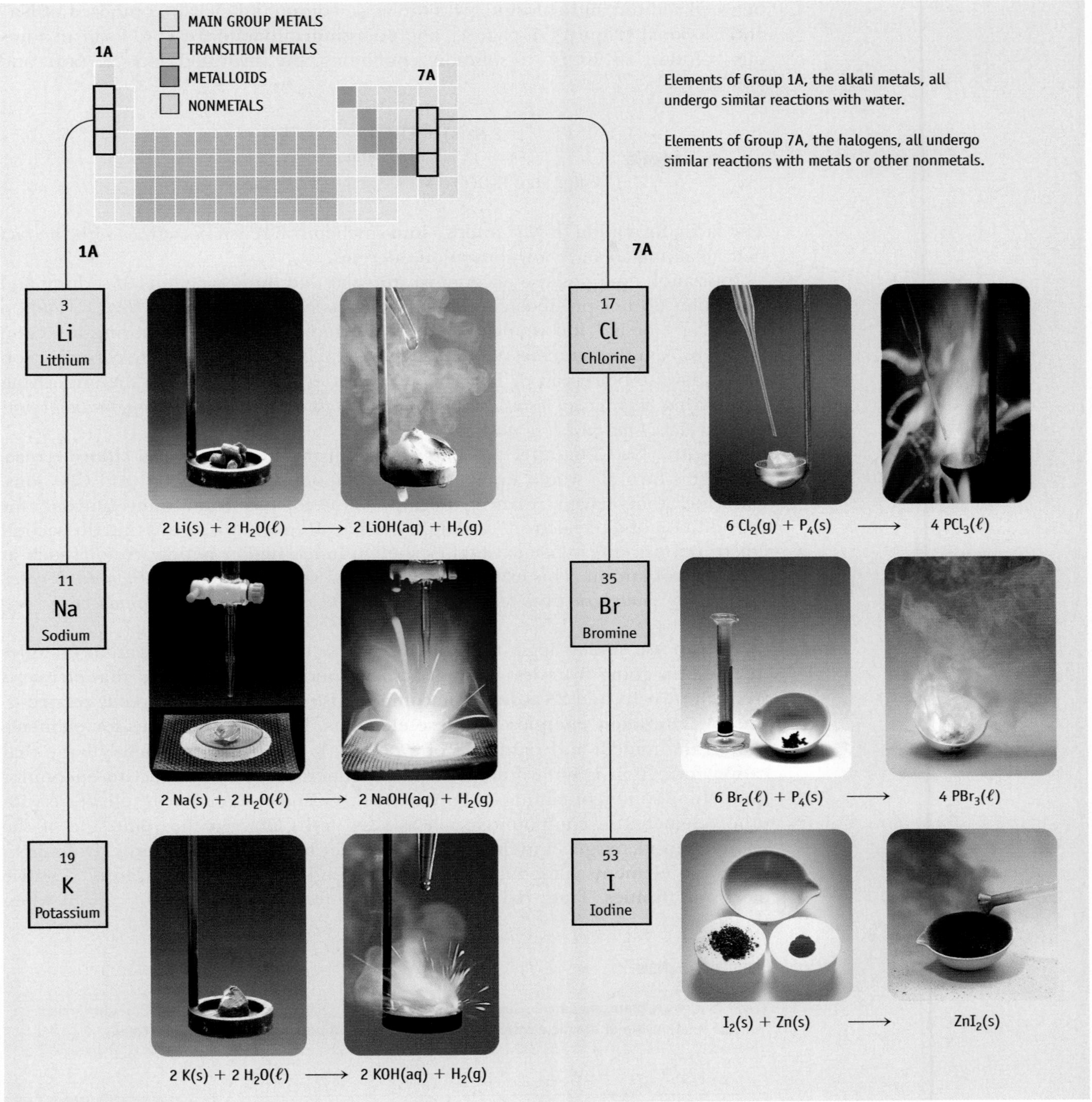

Photos: Charles D. Winters

Active Figure 7.13 Examples of the periodicity of Group 1A and Group 7A elements. Dimitri Mendeleev developed the first periodic table by listing elements in order of increasing atomic weight. Every so often, an element had properties similar to those of a lighter element, and these were placed in vertical columns or groups. We now recognize that the elements should be listed in order of increasing atomic number and that the periodic occurrence of similar properties is related to the electron configurations of the elements.

Chemistry Now™ Sign in at www.thomsonedu.com/login and go to the Chapter Contents menu to explore an interactive version of this figure accompanied by an exercise.

between sodium and chlorine gives the ionic compound, NaCl (composed of Na^+ and Cl^- ions) [Figure 1.4, page 4], and potassium and water react to form an aqueous solution of KOH, a solution containing the hydrated ions $K^+(aq)$ and $OH^-(aq)$.

$$2\ Na(s) + Cl_2(g) \longrightarrow 2\ NaCl(s)$$

$$2\ K(s) + 2\ H_2O(\ell) \longrightarrow 2\ K^+(aq) + 2\ OH^-(aq) + H_2(g)$$

The facile formation of Na^+ and K^+ ions in chemical reactions agrees with the fact that alkali metals have low ionization energies.

Ionization energies also account for the fact that these reactions of sodium and potassium do not produce compounds such as $NaCl_2$ or $K(OH)_2$. The formation of an Na^{2+} or K^{2+} ion would be a very unfavorable process. Removing a second electron from these metals requires a great deal of energy because a core electron would have to be removed. The energetic barrier to this process is the underlying reason that *main group metals generally form cations with an electron configuration equivalent to that of the preceding noble gas.*

Why isn't Na_2Cl another possible product from the sodium and chlorine reaction? This formula would imply that the compound contains Na^+ and Cl^{2-} ions. Chlorine atoms have a relatively negative value for electron affinity, but only for the addition of one electron. Adding two electrons per atom means that the second electron must enter the next higher shell at much higher energy. Anions such as Cl^{2-} are not known. This example leads us to a general statement: *nonmetals generally acquire enough electrons to form an anion with the electron configuration of the next noble gas.*

We can use similar logic to rationalize other observations. Ionization energies increase on going from left to right across a period. We have seen that elements from Groups 1A and 2A form ionic compounds, an observation directly related to the low ionization energies for these elements. Ionization energies for elements toward the middle and right side of a period, however, are sufficiently large that cation formation is unfavorable. Thus, we generally do not expect to encounter ionic compounds containing carbon; instead, we find carbon sharing electrons with other elements in compounds such as CO_2 and CCl_4. On the right side of the second period, oxygen and fluorine much prefer taking on electrons to giving them up; these elements have high ionization energies and relatively large, negative electron affinities. Thus, oxygen and fluorine form anions and not cations when they react.

Chemistry Now™

Sign in at **www.thomsonedu.com/login** and go to Chapter 7 Contents to see Screen 7.13 to watch videos on **the relationship of atomic electron configurations and orbital energies on periodic trends.**

EXERCISE 7.8 Energies and Compound Formation

Give a plausible explanation for the observation that magnesium and chlorine react to form $MgCl_2$ and not $MgCl_3$.

Chapter Goals Revisited

Now that you have studied this chapter, you should ask whether you have met the chapter goals. In particular, you should be able to:

Recognize the relationship of the four quantum numbers (n, ℓ, m_ℓ, and m_s) to atomic structure

a. Recognize that each electron in an atom has a different set of the four quantum numbers, n, ℓ, m_ℓ, and m_s (Sections 6.5–6.7, 7.1, and 7.3). Study Questions assignable in OWL: 11, 13, 35, 37, 52.

b. Understand that the Pauli exclusion principle leads to the conclusion that no atomic orbital can be assigned more than two electrons and that the two electrons in an orbital must have opposite spins (different values of m_s) (Section 7.1).

Write the electron configuration for atoms and monatomic ions

a. Recognize that electrons are assigned to the subshells of an atom in order of increasing subshell energy (Aufbau principle, Section 7.2). In the H atom, the subshell energies increase with increasing n, but, in a many-electron atom, the energies depend on both n and ℓ (see Figure 7.2).

b. Understand effective nuclear charge, Z^*, and its ability to explain why different subshells in the same shell of multielectron atoms have different energies. Also, understand the role of Z^* in determining the properties of atoms (Section 7.2).

c. Using the periodic table as a guide, depict electron configurations of neutral atoms (Section 7.3) and monatomic ions (Section 7.4) using the orbital box or *spdf* notation. In both cases, configurations can be abbreviated with the noble gas notation. Study Question(s) assignable in OWL: 2, 3, 6, 10, 13, 15, 18, 20, 21, 33, 34, 35, 36, 39, 44, 52, 59, 71.

d. When assigning electrons to atomic orbitals, apply the Pauli exclusion principle and Hund's rule (Sections 7.3 and 7.4).

e. Understand the role magnetism plays in revealing atomic structure (Section 7.4). Study Question(s) assignable in OWL: 20, 21, 33, 34, 39, 52.

Rationalize trends in atom and ion sizes, ionization energy, and electron affinity

a. Predict how properties of atoms—size, ionization energy (*IE*), and electron affinity (*EA*)—change on moving down a group or across a period of the periodic table (Section 7.5). The general periodic trends for these properties are as follows: Study Question(s) assignable in OWL: 24, 26, 28, 30, 32, 40–43, 45–50, 52, 56–58, 64.

(i) Atomic size decreases across a period and increases down a group.

(ii) *IE* increases across a period and decreases down a group.

(iii) The value of *EA* becomes more negative across a period and becomes less negative down a group.

b. Recognize the role that ionization energy and electron affinity play in forming ionic compounds (Section 7.6). Study Question(s) assignable in OWL: 72.

Chemistry Now™ Sign in at **www.thomsonedu.com/login** to:

- Assess your understanding with Study Questions in OWL keyed to each goal in the Goals and Homework menu for this chapter
- For quick review, download Go Chemistry mini-lecture flashcard modules (or purchase them at **www.ichapters.com**)
- Check your readiness for an exam by taking the Pre-Test and exploring the modules recommended in your Personalized Study plan.

Access **How Do I Solve It?** tutorials on how to approach problem solving using concepts in this chapter.

STUDY QUESTIONS

Online homework for this chapter may be assigned in OWL.

▲ denotes challenging questions.

■ denotes questions assignable in OWL.

Blue-numbered questions have answers in Appendix O and fully-worked solutions in the *Student Solutions Manual*.

Practicing Skills

Writing Electron Configurations of Atoms

(See Examples 7.1–7.3; Tables 7.1, 7.3, and 7.4; and Screen 7.5 and the Toolbox in ChemistryNow.)

1. Write the electron configurations for P and Cl using both *spdf* notation and orbital box diagrams. Describe the relationship between each atom's electron configuration and its position in the periodic table.

2. ■ Write the electron configurations for Mg and Ar using both *spdf* notation and orbital box diagrams. Describe the relationship of the atom's electron configuration to its position in the periodic table.

3. ■ Using *spdf* notation, write the electron configurations for atoms of chromium and iron, two of the major components of stainless steel.

4. Using *spdf* notation, give the electron configuration of vanadium, V, an element found in some brown and red algae and some toadstools.

5. Depict the electron configuration for each of the following atoms using *spdf* and noble gas notations.
   (a) Arsenic, As. A deficiency of As can impair growth in animals even though larger amounts are poisonous.
   (b) Krypton, Kr. It ranks seventh in abundance of the gases in Earth's atmosphere.

6. ■ Using *spdf* and noble gas notations, write electron configurations for atoms of the following elements, and then check your answers with Table 7.3.
   (a) Strontium, Sr. This element is named for a town in Scotland.
   (b) Zirconium, Zr. The metal is exceptionally resistant to corrosion and so has important industrial applications. Moon rocks show a surprisingly high zirconium content compared with rocks on Earth.
   (c) Rhodium, Rh. This metal is used in jewelry and in catalysts in industry.
   (d) Tin, Sn. The metal was used in the ancient world. Alloys of tin (solder, bronze, and pewter) are important.

7. Use noble gas and *spdf* notations to depict electron configurations for the following metals of the third transition series.
   (a) Tantalum, Ta. The metal and its alloys resist corrosion and are often used in surgical and dental tools.
   (b) Platinum, Pt. This metal was used by pre-Columbian Indians in jewelry. It is used now in jewelry and for anticancer drugs and catalysts (such as those in automobile exhaust systems).

8. The lanthanides, once called the rare earth elements, are really only "medium rare." Using noble gas and *spdf* notations, depict reasonable electron configurations for the following elements.
   (a) Samarium, Sm. This lanthanide is used in magnetic materials.
   (b) Ytterbium, Yb. This element was named for the village of Ytterby in Sweden, where a mineral source of the element was found.

9. The actinide americium, Am, is a radioactive element that has found use in home smoke detectors. Depict its electron configuration using noble gas and *spdf* notations.

10. ■ Predict reasonable electron configurations for the following elements of the actinide series of elements. Use noble gas and *spdf* notations.
    (a) Plutonium, Pu. The element is best known as a by-product of nuclear power plant operations.
    (b) Curium, Cm. This actinide was named for Marie Curie (page 342).

Quantum Numbers and Electron Configurations

(See Example 7.2 and ChemistryNow Screens 6.12 and 7.2–7.5.)

11. ■ What is the maximum number of electrons that can be identified with each of the following sets of quantum numbers? In one case, the answer is "none." Explain why this is true.
    (a) $n = 4, \ell = 3, m_\ell = 1\frac{1}{2}$
    (b) $n = 6, \ell = 1, m_\ell = -1, m_s = -\frac{1}{2}$
    (c) $n = 3, \ell = 3, m_\ell = -3$

12. What is the maximum number of electrons that can be identified with each of the following sets of quantum numbers? In some cases, the answer may be "none." In such cases, explain why "none" is the correct answer.
    (a) $n = 3$
    (b) $n = 3$ and $\ell = 2$
    (c) $n = 4, \ell = 1, m_\ell = -1$, and $m_s = +\frac{1}{2}$
    (d) $n = 5, \ell = 0, m_\ell = -1, m_s = +\frac{1}{2}$

13. ■ Depict the electron configuration for magnesium using an orbital box diagram and noble gas notation. Give a complete set of four quantum numbers for each of the electrons beyond those of the preceding noble gas.

14. Depict the electron configuration for phosphorus using an orbital box diagram and noble gas notation. Give one possible set of four quantum numbers for each of the electrons beyond those of the preceding noble gas.

15. ■ Using an orbital box diagram and noble gas notation, show the electron configuration of gallium, Ga. Give a set of quantum numbers for the highest-energy electron.

16. Using an orbital box diagram and noble gas notation, show the electron configuration of titanium. Give one possible set of four quantum numbers for each of the electrons beyond those of the preceding noble gas.

Electron Configurations of Atoms and Ions and Magnetic Behavior

(See Example 7.4, Section 6.7, and ChemistryNow Screens 6.16, 7.5, and 7.6.)

17. Using orbital box diagrams, depict an electron configuration for each of the following ions: (a) Mg^{2+}, (b) K^+, (c) Cl^-, and (d) O^{2-}.

18. ■ Using orbital box diagrams, depict an electron configuration for each of the following ions: (a) Na^+, (b) Al^{3+}, (c) Ge^{2+}, and (d) F^-.

19. Using orbital box diagrams and noble gas notation, depict the electron configurations of (a) V, (b) V^{2+}, and (c) V^{5+}. Are any of the ions paramagnetic?

20. ■ Using orbital box diagrams and noble gas notation, depict the electron configurations of (a) Ti, (b) Ti^{2+}, and (c) Ti^{4+}. Are any of the ions paramagnetic?

21. ■ Manganese is found as MnO_2 in deep ocean deposits.
    (a) Depict the electron configuration of this element using the noble gas notation and an orbital box diagram.
    (b) Using an orbital box diagram, show the electrons beyond those of the preceding noble gas for the 4+ ion.
    (c) Is the 4+ ion paramagnetic?
    (d) How many unpaired electrons does the Mn^{4+} ion have?

22. One compound found in alkaline batteries is NiOOH, a compound containing Ni^{3+} ions. When the battery is discharged, the Ni^{3+} is reduced to Ni^{2+} ions [as in $Ni(OH)_2$]. Using orbital box diagrams and the noble gas notation, show electron configurations of these ions. Are either of these ions paramagnetic?

Periodic Properties

(See Section 7.5, Example 7.5, and ChemistryNow Screens 7.7–7.13.)

23. Arrange the following elements in order of increasing size: Al, B, C, K, and Na. (Try doing it without looking at Figure 7.8, and then check yourself by looking up the necessary atomic radii.)

24. ■ Arrange the following elements in order of increasing size: Ca, Rb, P, Ge, and Sr. (Try doing it without looking at Figure 7.8; then check yourself by looking up the necessary atomic radii.)

25. Select the atom or ion in each pair that has the larger radius.
    (a) Cl or Cl^- (b) Al or O (c) In or I

26. ■ Select the atom or ion in each pair that has the larger radius.
    (a) Cs or Rb (b) O^{2-} or O (c) Br or As

27. Which of the following groups of elements is arranged correctly in order of increasing ionization energy?
    (a) C $<$ Si $<$ Li $<$ Ne (c) Li $<$ Si $<$ C $<$ Ne
    (b) Ne $<$ Si $<$ C $<$ Li (d) Ne $<$ C $<$ Si $<$ Li

28. ■ Arrange the following atoms in order of increasing ionization energy: Li, K, C, and N.

29. Compare the elements Na, Mg, O, and P.
    (a) Which has the largest atomic radius?
    (b) Which has the most negative electron affinity?
    (c) Place the elements in order of increasing ionization energy.

30. ■ Compare the elements B, Al, C, and Si.
    (a) Which has the most metallic character?
    (b) Which has the largest atomic radius?
    (c) Which has the most negative electron affinity?
    (d) Place the three elements B, Al, and C in order of increasing first ionization energy.

31. Explain each answer briefly.
    (a) Place the following elements in order of increasing ionization energy: F, O, and S.
    (b) Which has the largest ionization energy: O, S, or Se?
    (c) Which has the most negative electron affinity: Se, Cl, or Br?
    (d) Which has the largest radius: O^{2-}, F^-, or F?

32. ■ Explain each answer briefly.
    (a) Rank the following in order of increasing atomic radius: O, S, and F.
    (b) Which has the largest ionization energy: P, Si, S, or Se?
    (c) Place the following in order of increasing radius: O^{2-}, N^{3-}, and F^-.
    (d) Place the following in order of increasing ionization energy: Cs, Sr, and Ba.

General Questions

These questions are not designated as to type or location in the chapter. They may combine several concepts.

33. ■ Using an orbital box diagram and noble gas notation, show the electron configurations of uranium and of the uranium(IV) ion. Is either of these paramagnetic?

34. ■ The rare earth elements, or lanthanides, commonly exist as 3+ ions. Using an orbital box diagram and noble gas notation, show the electron configurations of the following elements and ions.
    (a) Ce and Ce^{3+} (cerium)
    (b) Ho and Ho^{3+} (holmium)

35. ■ A neutral atom has two electrons with $n = 1$, eight electrons with $n = 2$, eight electrons with $n = 3$, and two electrons with $n = 4$. Assuming this element is in its ground state, supply the following information:
    (a) atomic number
    (b) total number of *s* electrons
    (c) total number of *p* electrons
    (d) total number of *d* electrons
    (e) Is the element a metal, metalloid, or nonmetal?

36. ■ Element 109, now named meitnerium (in honor of the Austrian–Swedish physicist, Lise Meitner [1878–1968]), was produced in August 1982 by a team at Germany's Institute for Heavy Ion Research. Depict its electron configuration using *spdf* and noble gas notations. Name another element found in the same group as meitnerium.

Lise Meitner (1878–1968) and Otto Hahn (1879–1968). Element 109 (Mt) was named after Meitner. She earned her Ph.D. in physics under Ludwig Boltzmann at the University of Vienna, the first woman to do so at that university.

37. Which of the following is *not* an allowable set of quantum numbers? Explain your answer briefly. For those sets that are valid, identify an element in which an outermost valence electron could have that set of quantum numbers.

	n	ℓ	m_ℓ	m_s
(a)	2	0	0	$-\frac{1}{2}$
(b)	1	1	0	$+\frac{1}{2}$
(c)	2	1	-1	$-\frac{1}{2}$
(d)	4	2	$+2$	$-\frac{1}{2}$

38. A possible excited state for the H atom has an electron in a $4p$ orbital. List all possible sets of quantum numbers (n, ℓ, m_ℓ, m_s) for this electron.

39. ■ The magnet in the photo is made from neodymium, iron, and boron.

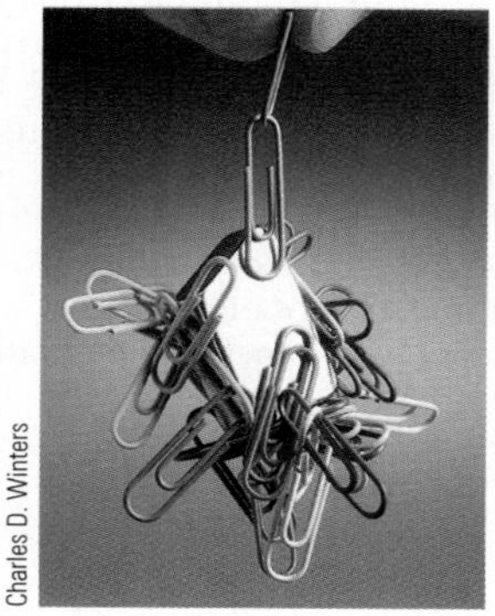

A magnet made of an alloy containing the elements Nd, Fe, and B.

(a) Write the electron configuration of each of these elements using an orbital box diagram and noble gas notation.
(b) Are these elements paramagnetic or diamagnetic?
(c) Write the electron configurations of Nd^{3+} and Fe^{3+} using orbital box diagrams and noble gas notation. Are these ions paramagnetic or diamagnetic?

40. ■ Name the element corresponding to each characteristic below.
(a) the element with the electron configuration $1s^22s^22p^63s^23p^3$
(b) the alkaline earth element with the smallest atomic radius
(c) the element with the largest ionization energy in Group 5A
(d) the element whose 2+ ion has the configuration $[Kr]4d^5$
(e) the element with the most negative electron affinity in Group 7A
(f) the element whose electron configuration is $[Ar]3d^{10}4s^2$

41. ■ Arrange the following atoms in the order of increasing ionization energy: Si, K, P, and Ca.

42. ■ Rank the following in order of increasing ionization energy: Cl, Ca^{2+}, and Cl^-. Briefly explain your answer.

43. ■ Answer the questions below about the elements A and B, which have the electron configurations shown.

$$A = [Kr]5s^1 \quad B = [Ar]3d^{10}4s^24p^4$$

(a) Is element A a metal, nonmetal, or metalloid?
(b) Which element has the greater ionization energy?
(c) Which element has the less negative electron affinity?
(d) Which element has the larger atomic radius?
(e) What is the formula for a compound formed between A and B?

44. ■ Answer the following questions about the elements with the electron configurations shown here:

$$A = [Ar]4s^2 \quad B = [Ar]3d^{10}4s^24p^5$$

(a) Is element A a metal, metalloid, or nonmetal?
(b) Is element B a metal, metalloid, or nonmetal?
(c) Which element is expected to have the larger ionization energy?
(d) Which element has the smaller atomic radius?

45. ■ Which of the following ions are unlikely to be found in a chemical compound: Cs^+, In^{4+}, Fe^{6+}, Te^{2-}, Sn^{5+}, and I^-? Explain briefly.

46. ■ Place the following ions in order of decreasing size: K^+, Cl^-, S^{2-}, and Ca^{2+}.

47. ■ Answer each of the following questions:
(a) Of the elements S, Se, and Cl, which has the largest atomic radius?
(b) Which has the larger radius, Br or Br^-?
(c) Which should have the largest difference between the first and second ionization energy: Si, Na, P, or Mg?
(d) Which has the largest ionization energy: N, P, or As?
(e) Which of the following has the largest radius: O^{2-}, N^{3-}, or F^-?

48. ■ The following are isoelectronic species: Cl^-, K^+, and Ca^{2+}. Rank them in order of increasing (a) size, (b) ionization energy, and (c) electron affinity.

49. ■ Compare the elements Na, B, Al, and C with regard to the following properties:
(a) Which has the largest atomic radius?
(b) Which has the most negative electron affinity?
(c) Place the elements in order of increasing ionization energy.

50. ■ ▲ Two elements in the second transition series (Y through Cd) have four unpaired electrons in their 3+ ions. What elements fit this description?

51. The configuration for an element is given here.

[Ar] [↑↓][↑↓][↑][↑][↑] [↑↓]
3d 4s

(a) What is the identity of the element with this configuration?
(b) Is a sample of the element paramagnetic or diamagnetic?
(c) How many unpaired electrons does a 3+ ion of this element have?

52. ■ The configuration of an element is given here.

[Ar] [↑][↑][↑][ ][ ] [↑↓]
3d 4s

(a) What is the identity of the element?
(b) In what group and period is the element found?
(c) Is the element a nonmetal, a main group element, a transition metal, a lanthanide, or an actinide?
(d) Is the element diamagnetic or paramagnetic? If paramagnetic, how many unpaired electrons are there?
(e) Write a complete set of quantum numbers (n, ℓ, m_ℓ, m_s) for each of the valence electrons.
(f) What is the configuration of the 2+ ion formed from this element? Is the ion diamagnetic or paramagnetic?

In the Laboratory

53. Nickel(II) formate [$Ni(HCO_2)_2$] is widely used as catalyst precursor and to make metallic nickel. It can be prepared in the general chemistry laboratory by treating readily available nickel(II) acetate with formic acid (HCO_2H).

$$Ni(CH_3CO_2)_2(aq) + 2\ HCO_2H(aq) \longrightarrow Ni(HCO_2)_2(aq) + 2\ CH_3CO_2H(aq)$$

Green crystalline $Ni(HCO_2)_2$ is precipitated after adding ethanol to the solution.
(a) What is the theoretical yield of nickel(II) formate from 0.500 g of nickel(II) acetate and excess formic acid?
(b) Is nickel(II) formate paramagnetic or diamagnetic? If it is paramagnetic, how many unpaired electrons would you expect?
(c) If nickel(II) formate is heated to 300 °C in the absence of air for 30 minutes, the salt decomposes to form pure nickel powder. What mass of nickel powder should be produced by heating 253 mg of nickel(II) formate? Are nickel atoms paramagnetic?

54. ▲ Spinels are solids with the general formula $M^{2+}(M'^{3+})_2O_4$ (where M^{2+} and M'^{3+} are metal cations of the same or different metals). The best-known example is common magnetite, Fe_3O_4 [which you can formulate as $(Fe^{2+})(Fe^{3+})_2O_4$].

Charles D. Winters

A crystal of a spinel.

(a) Given its name, it is evident that magnetite is ferromagnetic. How many unpaired electrons are there in iron(II) and in iron(III) ions?
(b) Two other spinels are $CoAl_2O_4$ and $SnCo_2O_4$. What metal ions are involved in each? What are their electron configurations? Are the metal ions also paramagnetic, and if so how many unpaired electrons are involved?

Summary and Conceptual Questions

The following questions use concepts from this and previous chapters.

55. Why is the radius of Li^+ so much smaller than the radius of Li? Why is the radius of F^- so much larger than the radius of F?

56. ■ Which ions in the following list are not likely to be found in chemical compounds: K^{2+}, Cs^+, Al^{4+}, F^{2-}, and Se^{2-}? Explain briefly.

57. ■ ▲ Two elements have the following first through fourth ionization energies. Deduce the group in the periodic table to which they probably belong. Explain briefly.

Ionization Energy (kJ/mol)	Element 1	Element 2
1st *IE*	1086.2	577.4
2nd *IE*	2352	1816.6
3rd *IE*	4620	2744.6
4th *IE*	6222	11575

58. ■ ▲ The ionization of the hydrogen atom can be calculated from Bohr's equation for the electron energy.

$$E = -(N_ARhc)(Z^2/n^2)$$

where $N_ARhc = 1312$ kJ/mol and Z is the atomic number. Let us use this approach to calculate a possible ionization energy for helium. First, assume the electrons of the He experience the full 2+ nuclear charge. This gives us the upper limit for the ionization energy. Next, assume one electron of He completely screens the nuclear charge from the other electrons, so $Z = 1$. This gives us a lower limit to the ionization energy. Compare these calculated values for the upper and lower limits to the experimental value of 2372.3 kJ/mol. What does this tell us about the ability of one electron to screen the nuclear charge?

59. ■ Compare the configurations below with two electrons located in *p* orbitals. Which would be the most stable (have the lowest energy)? Which would be the least stable? Explain your answers.

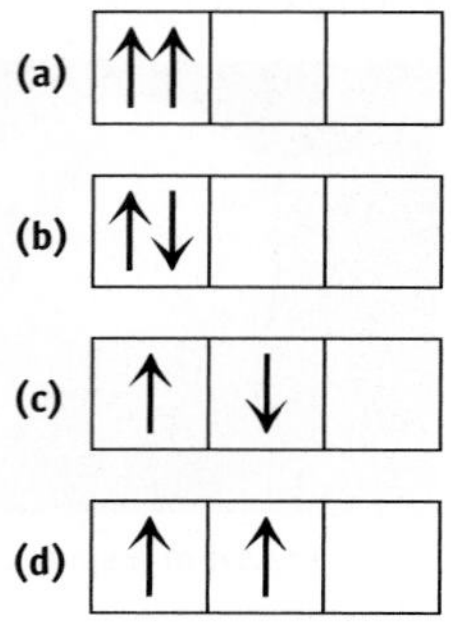

60. The bond lengths in Cl_2, Br_2, and I_2 are 200, 228, and 266 pm, respectively. Knowing that the tin radius is 141 pm, estimate the bond distances in Sn—Cl, Sn—Br, and Sn—I. Compare the estimated values with the experimental values of 233, 250, and 270 pm, respectively.

61. Write electron configurations to show the first two ionization processes for potassium. Explain why the second ionization energy is much greater than the first.

62. Explain how the ionization energy of atoms changes and why the change occurs when proceeding down a group of the periodic table.

63. (a) Explain why the sizes of atoms change when proceeding across a period of the periodic table.
(b) Explain why the sizes of transition metal atoms change very little across a period.

64. ■ Which of the following elements has the greatest difference between its first and second ionization energies: C, Li, N, Be? Explain your answer.

65. ▲ What arguments would you use to convince another student in general chemistry that MgO consists of the ions Mg^{2+} and O^{2-} and not the ions Mg^{+} and O^{-}? What experiments could be done to provide some evidence that the correct formulation of magnesium oxide is $Mg^{2+}O^{2-}$?

66. Explain why the first ionization energy of Ca is greater than that of K, whereas the second ionization energy of Ca is lower than the second ionization energy of K.

67. The energies of the orbitals in many elements have been determined. For the first two periods they have the following values:

Element	1*s* (kJ/mol)	2*s* (kJ/mol)	2*p* (kJ/mol)
H	−1313		
He	−2373		
Li		−520.0	
Be		−899.3	
B		−1356	−800.8
C		−1875	−1029
N		−2466	−1272
O		−3124	−1526
F		−3876	−1799
Ne		−4677	−2083

(a) ▲ Why do the orbital energies generally become more negative on proceeding across the second period?
(b) How are these values related to the ionization energy and electron affinity of the elements?
(c) Use these energy values to explain the observation that the ionization energies of the first four second-period elements are in the order Li < Be > B < C.

Note that these energy values are the basis for the discussion in the Simulation on ChemistryNow Screen 7.8. (Data from J. B. Mann, T. L. Meek, and L. C. Allen: *Journal of the American Chemical Society*, Vol. 122, p. 2780, 2000.)

68. ▲ The ionization energies for the removal of the first electron in Si, P, S, and Cl are as listed in the table below. Briefly rationalize this trend.

Element	First Ionization Energy (kJ/mol)
Si	780
P	1060
S	1005
Cl	1255

69. Using your knowledge of the trends in element sizes on going across the periodic table, explain briefly why the density of the elements increases from K through V.

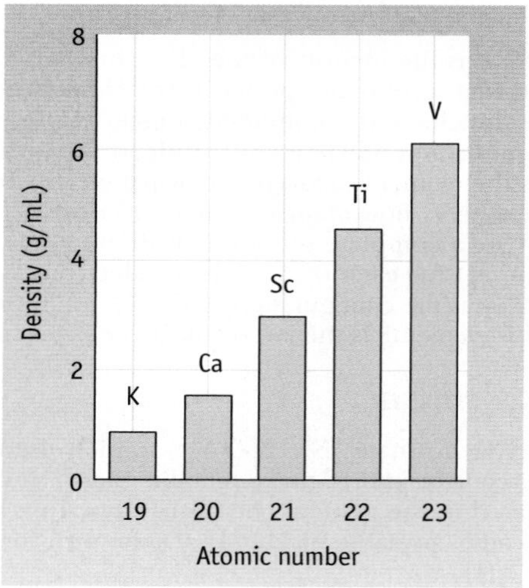

70. The densities (in g/cm^3) of elements in Groups 6B, 8B, and 1B are given in the table below.

Period 4	Cr, 7.19	Co, 8.90	Cu, 8.96
Period 5	Mo, 10.22	Rh, 12.41	Ag, 10.50
Period 6	W, 19.30	Ir, 22.56	Au, 19.32

Transition metals in the sixth period all have much greater densities than the elements in the same groups in the fourth and fifth periods. Refer to Figure 7.9, and explain this observation.

71. ■ The discovery of two new elements (atomic numbers 113 and 115) was announced in February 2004.

Courtesy of Lawrence Livermore National Laboratory

Some members of the team that discovered elements 113 and 115 at the Lawrence Livermore National Laboratory (left to right): Jerry Landrum, Dawn Shaughnessy, Joshua Patin, Philip Wilk, and Kenton Moody.

(a) Use *spdf* and noble gas notations to give the electron configurations of these two elements.
(b) Name an element in the same periodic group as the two elements.
(c) Element 113 was made by firing a light atom at a heavy americium atom. The two combine to give a nucleus with 113 protons. What light atom was used as a projectile?

72. ■ Explain why the reaction of calcium and fluorine does not form CaF_3.

73. ▲ Thionyl chloride, $SOCl_2$, is an important chlorinating and oxidizing agent in organic chemistry. It is prepared industrially by oxygen atom transfer from SO_3 to SCl_2.

$$SO_3(g) + SCl_2(g) \longrightarrow SO_2(g) + SOCl_2(g)$$

(a) Give the electron configuration for an atom of sulfur using an orbital box diagram. Do not use the noble gas notation.
(b) Using the configuration given in part (a), write a set of quantum numbers for the highest-energy electron in a sulfur atom.
(c) What element involved in this reaction (O, S, Cl) should have the smallest ionization energy? The smallest radius?
(d) Which should be smaller: the sulfide ion, S^{2-}, or a sulfur atom, S?
(e) If you want to make 675 g of $SOCl_2$, what mass of SCl_2 is required?
(f) If you use 10.0 g of SO_3 and 10.0 g of SCl_2, what is the theoretical yield of $SOCl_2$?
(g) $\Delta_r H°$ for the reaction of SO_3 and SCl_2 is −96.0 kJ/mol $SOCl_2$ produced. Using data in Appendix L, calculate the standard molar enthalpy of formation of SCl_2.

74. Sodium metal reacts readily with chlorine gas to give sodium chloride. *(See ChemistryNow Screen 7.17 Chemical Puzzler.)*

$$Na(s) + \tfrac{1}{2}\,Cl_2(g) \longrightarrow NaCl(s)$$

(a) What is the reducing agent in this reaction? What property of the element contributes to its ability to act as a reducing agent?
(b) What is the oxidizing agent in this reaction? What property of the element contributes to its ability to act as an oxidizing agent?
(c) Why does the reaction produce NaCl and not a compound such as Na_2Cl or $NaCl_2$?

75. ▲ Slater's rules are a simple way to estimate the effective nuclear charge experienced by an electron. In this approach, the "shielding constant," σ, is calculated. The effective nuclear charge is then the difference between σ and the atomic number, Z. (Note that the results in Table 7.2 were calculated in a slightly different way.)

$$Z^* = Z - \sigma$$

The shielding constant, σ, is calculated using the following rules:

1. The electrons of an atom are grouped as follows: (1*s*) (2*s*, 2*p*) (3*s*, 3*p*) (3*d*) (4*s*, 4*p*) (4*d*), and so on.
2. Electrons in higher groups (to the right) do not shield those in the lower groups.
3. For *ns* and *np* valence electrons
   a) Electrons in the same *ns*, *np* group contribute 0.35 (for 1*s* 0.30 works better).
   b) Electrons in the $n - 1$ group contribute 0.85.
   c) Electrons in the $n - 2$ group (and lower) contribute 1.00.
4. For *nd* and *nf* electrons, electrons in the same *nd* or *nf* group contribute 0.35, and those in groups to the left contribute 1.00.

As an example, let us calculate Z^* for the outermost electron of oxygen:

$$\sigma = (2 \times 0.85) + (5 \times 0.35) = 3.45$$
$$Z^* = 8 - 3.45 = 4.55$$

(a) Calculate Z^* for F and Ne. Relate the Z^* values for O, F, and Ne to their relative atomic radii and ionization energies.
(b) Calculate Z^* for one of the 3*d* electrons of Mn, and compare this with Z^* for one of the 4*s* electrons of the element. Do the Z^* values give us some insight into the ionization of Mn to give the cation?

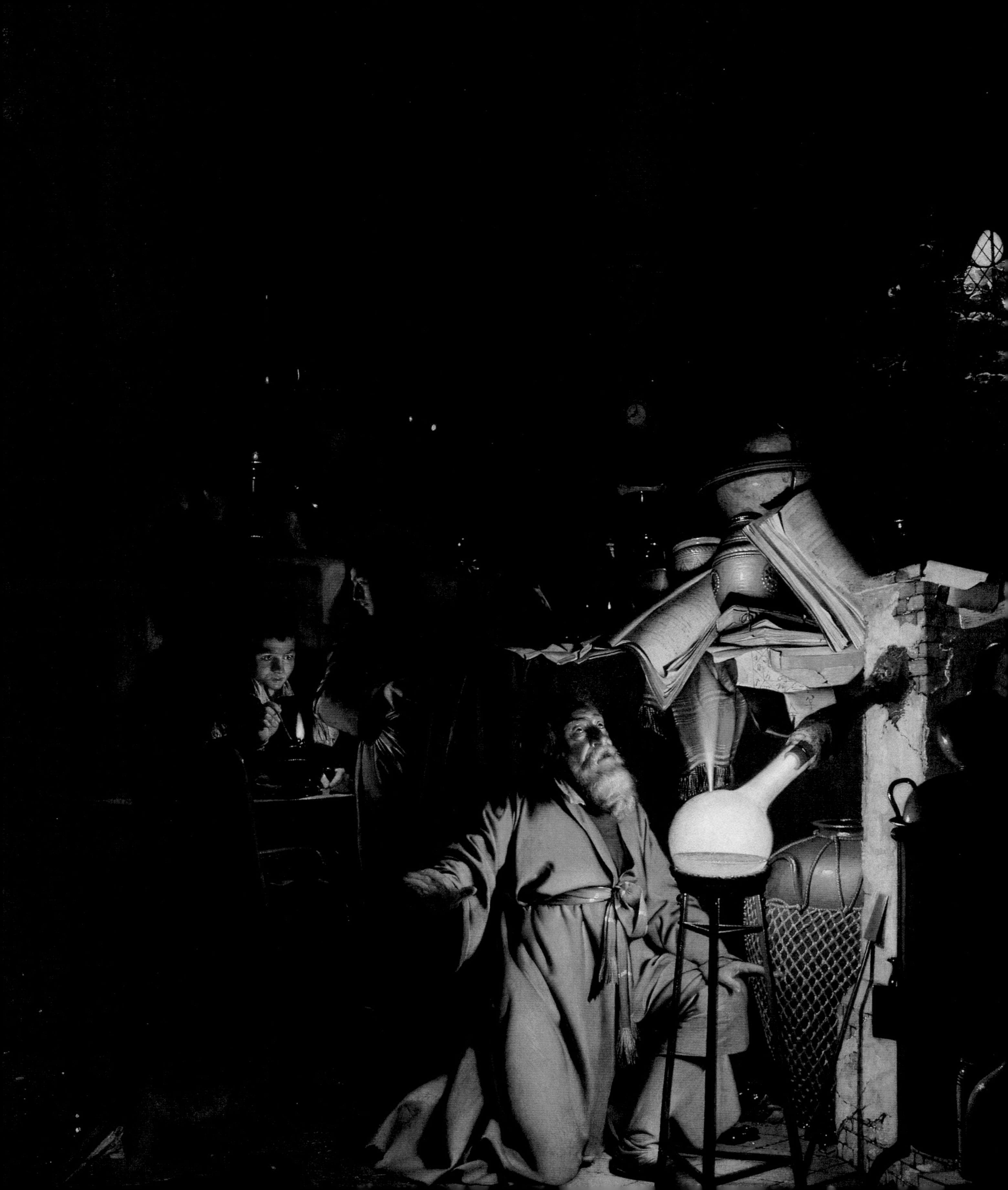

Milestones in the Development of Chemistry and the Modern View of Atoms and Molecules

John Emsley
University of Cambridge

The journey to our understanding of atoms and molecules began 2500 years ago in ancient Greece and continues today with developments in our understanding of chemical bonding and molecular structure. The journey may have started long ago, but it was only when chemistry developed in the 1700s that real progress was made. Today, we have instruments like the scanning tunneling microscope to help us visualize these tiny objects.

But let us start at the beginning and pay a short visit to the Greeks of 450 BC. Their civilization valued learning, and this led to various schools of philosophy. Their teachers did not engage in scientific research as we know it, so their theories were never more than thought experiments. Nevertheless, they correctly deduced that the world was made up of a few basic elements and that these existed as atoms, but they had no way of knowing which elements existed, or how small atoms were. Today, we know of 117 elements—at the last count—and that atoms are so tiny that the dot at the end of this sentence contains billions of them.

Greek Philosophers and Medieval Alchemists

The early Greek philosophers thought that there would be just one element and debated what it might be. Some favored air, some fire, some water, and some said earth. Eventually, Empedocles (who lived from around 490 to 430 BC) argued that all four were elements. This theory was believed for 2000 years—and yet it was wrong.

What form did these elements take? The first person to give an answer was Leucippus, around 450 BC, who said they must exist as atoms. This idea was developed by his pupil Democritus (460–370 BC), who was the first to use the word "atom," which means un-cut-able or indivisible. Epicurus (341–270 BC) said atoms were spherical, varied in size, and were constantly in motion. This theory remained also unchanged for 2000 years—but it was right.

Alchemy can trace its roots to ancient Egypt. The most famous Egyptian alchemist was Zosimos, who lived around 300 AD. He described such chemical processes as distillation and sublimation, crediting a woman alchemist, Maria the Jewess, with their invention. She lived about 100 AD and experimented with mercury and sulfur, but her best-known invention was the bain-Marie, which is still used in cooking.

Early Muslim rulers encouraged learning, and alchemy flourished. The best-known Arab alchemists were Geber (Jabir ibn Hayyan, 721–815 AD) and Rhazes (Abu Bakr Mohammad ibn Zakariyya al-Razi, 865–925 AD). Geber knew that when mercury and sulfur were combined, the product was a red compound, which we know as mercury(II) sulfide, but he believed that if the recipe were exactly right, then gold would be formed. Rhazes thought that all metals were made from mercury and sulfur, and his influential book, *Secret of Secrets*, contained a long list of chemicals, minerals, and apparatus that a modern chemist would recognize.

In the early Middle Ages, alchemists were at work in Europe, but only a little progress was made. A Spaniard, who also called himself Geber, discovered how to make nitric acid and knew that a mixture of nitric and hydrochloric acids *(aqua regia)* would dissolve gold. The European center for alchemy was Prague, where many alchemists practiced their art while some of their number merely

• **The Alchymist in Search of the Philosophere's Stone, Discovers Phosphorus.** Painted by J. Wright of Derby (1734–1797).

practiced deception, often convincing onlookers that they could turn base metals into gold.

The 1600s saw the gradual emergence of chemistry from alchemy, and in this period we find several men who are now recognized as true scientists but who were also secret alchemists, such as Robert Boyle (1626–1691) and even the great Isaac Newton (1642–1727). Today, Boyle is considered one of the founding fathers of chemistry. His book, *The Sceptical Chymist*, is regarded as the seminal work that broke the link between chemistry and alchemy.

The Philosopher's Stone, a legendary substance that can supposedly turn ordinary substances into gold, was the ultimate prize sought by alchemists. So, when the alchemist Hennig Brandt of Hamburg discovered phosphorus in 1669, he believed it would lead him to the Philosopher's Stone because of the almost miraculous ability of phosphorus to shine in the dark and burst into flames. Some of this new wonder material was shown to Boyle, who eventually was able to make it himself. (It was formed by heating evaporated urine residues to red heat.) What Boyle did next distinguished him as a true chemist: he researched the properties of phosphorus and its reactions with other materials and published his findings, not in the secret language of the alchemists but in plain English, and in a manner that would allow even a modern chemist to repeat what he had done. Phosphorus was a new element although it was not recognized as such for another century.

Boyle too thought about the nature of matter, and he came up with a definition of an element. He wrote:

> "I mean by Elements, as those Chymists that speak plainest do by the Principles, certain Primitive and Simple, or perfectly unmingled bodies; which not being made of any other bodies, or of one another, are the ingredients of which all those call'd perfectly mixt Bodies are immediately compounded, and into which they are ultimately resolved."

We can see that Boyle was struggling with the idea of elements and how these can be compounded together.

Chemists of the 18th–19th Centuries

The theory of four elements was dealt its first body blow by a shy but extremely wealthy Englishman, Henry Cavendish (1731–1810), who had his own laboratory near London. There, he investigated gases. In 1784, he discovered hydrogen and observed that when it burned, water was formed. The French chemist Antoine Lavoisier (1743–1794) (◄ page 114) also studied gases, but he went one step further and noted that when a *mixture* of hydrogen

Edgar Fahs Smith Collection/University of Pennsylvania Library

Edgar Fahs Smith Collection/University of Pennsylvania Library

Henry Cavendish (1731–1810, left) and John Dalton (1766–1844, right).

and another newly discovered gas, oxygen, was sparked, it formed only water. Obviously, water was not an element, but hydrogen and oxygen were.

In 1789, Lavoisier wrote his influential book *Elements of Chemistry*, in which he defined a chemical element as something that could not be further broken down, and he listed 33 of them. Most of these are still considered elements today, but some—such as light, heat, and the earths (later shown to be oxides)—are not. His list is given in Table 1. Lavoisier was the founder of modern chemistry, but, as described on page 114, he fell out of favor with the leaders of the French Revolution and was guillotined.

The next major advance in chemical understanding came from a modest school teacher, John Dalton (1766–1844), who lived in Manchester, England. He knew of the *Law of Fixed Proportions*, which said elements combined in definite ratios by weight, and said it could only be so if they were composed of atoms. The idea of atoms had been revived by scientists like Newton in the 1600s, who said elements would cluster together, but nothing had come of such speculations because chemistry was still little more than mystical alchemy.

In 1803, Dalton gave a talk to the Manchester (England) Literary and Philosophical Society on the way gases dis-

TABLE 1 The Elements According to Lavoisier (1789)

Gases	Nonmetals	Metals		Earths
Light	Sulfur	Antimony	Mercury	Lime
Heat	Phosphorus	Arsenic	Molybdenum	Magnesia
Oxygen	Carbon	Bismuth	Nickel	Barytes
Nitrogen	Chloride	Cobalt	Platinum	Alumina
Hydrogen	Fluoride	Copper	Silver	Silica
	Borate	Gold	Tin	
		Iron	Tungsten	
		Lead	Zinc	
		Manganese		

solved in water, and when his talk was published in the Society's Proceedings he included a table of relative atomic weights, which were based on hydrogen having a value of 1. He listed 20 elements with their weights and said they combined to form "compound atoms," his name for what we now call molecules. At a stroke, Dalton not only revived the idea of atoms, but gave them weight. In his book of 1808, *New System of Chemical Philosophy*, he went further, and he said an atom was a "solid, massy, hard, impenetrable, moveable particle." This description was eventually proved wrong on many counts, but he was right that they existed.

Chemists in the early 1800s were puzzled by the fact that most of the new atomic weights were whole numbers. An explanation was suggested in 1815 by William Prout (1785–1850), who went further than Dalton and reasoned that atoms were not indivisible but were composed of hydrogen. If the atomic weight of hydrogen was taken as 1, then it explained why all the other elements had weights that were whole numbers, or nearly so. In fact, most elements have atomic weights that fall within the limits of ± 0.1 of a whole number, with very few having fractional numbers. Nevertheless, it was the few exceptions that seemed to disprove Prout's theory, which could not explain the atomic weight of chlorine, which was 35.5, or copper, which was 63.5. (The explanation, of course, lies with their isotopic composition, a concept that lay 100 years in the future.)

Prout was almost right. Hydrogen, or at least 99.99% of it, consists of a single proton surrounded by a single electron. The proton is the nucleus, and that accounts for virtually all of the mass. We now classify elements based on how many protons their nuclei contain, with each element on the periodic table differing from the one immediately preceding it by one proton. In some ways, then, they differ according to the nucleus of hydrogen. When the proton was finally identified in 1919, and its importance realized, it took its name from the Greek word *protos* meaning first, although Ernest Rutherford, who made the discovery, also said the name was chosen partly in honor of Prout.

Back in the early 1800s, chemists preferred atomic weights that were calculated relative to that of oxygen because oxygen forms compounds with almost all other elements. (Today, atomic weights are based on the carbon isotope carbon-12, which is taken as exactly 12.) Like Dalton, they believed water consisted of one oxygen and one hydrogen atom, so naturally their scale of atomic weights was of little use.

Inaccurate atomic weights, however, did not hinder the discovery of more and more elements. The 1860s were a particularly fruitful decade with the introduction of the atomic spectroscope, which revealed that each element had a characteristic "fingerprint" pattern of lines in its visible spectrum. As a result, the discoveries of rubidium, cesium, thallium, and indium were announced in the years 1860–1863. The total of known elements was now 65, and chemists were beginning to ask whether there was a limit to their number.

Meanwhile, the Italian chemist Stanislao Cannizzaro (1826–1910) had published a correct list of atomic weights, which he circulated at the First International Chemical Congress, held in Karlsruhe, Germany, in 1860. Dimitri Mendeleev attended the conference and took a copy of Cannizzaro's atomic weights back to St. Petersburg in Russia. What he did with it was to revolutionize chemistry, a story that is told in Chapter 2.

Dalton had suggested ways in which atoms might combine to form larger units. It was clear that the world was composed primarily not of single atoms but of molecules, and chemistry was the science of studying them. The word *molecule* was first given a chemical meaning in 1811; before then, it had simply been a French word for something extremely small. In 1873, its chemical meaning was spelled out by James Clerk Maxwell, who wrote a milestone paper in the journal *Nature* in which he defined a molecule as "the smallest possible portion of a particular substance" beyond which it would no longer have the properties associated with that substance. This is the meaning it still has.

In the 1800s, chemical analysis became quite sophisticated, and the elements in a chemical compound could be identified and expressed as numbers. For example, alcohol was C_2H_6O, but what was it really? And why did dimethyl ether, which was a different substance, have exactly the same formula? The concept of valency could explain how elements combined. Hydrogen had a valency of 1, oxygen of 2 as in H_2O, nitrogen of 3 as in NH_3, and carbon of 4 as in CH_4, but this in itself was not enough to explain what molecules really were. Turning valences into actual molecular arrangements was the next step, and two people were instrumental in doing this: 29-year-old August Kekulé (1829–1896) in 1858, and 22-year-old Jacobus van't Hoff (1852–1911) in 1878.

Edgar Fahs Smith Collection/University of Pennsylvania Library

August Kekulé (1829–1896).

Kekulé was one of the great chemists of the second half of the 1800s. He is best known for his theory of molecule structures based on valence and especially of organic molecules in which carbon is four-valent and bonds in various ways to other atoms including carbon atoms. He published his paper only weeks before one by Archibald Scott Couper (1831–1892), a young Scottish chemist, who was studying in Paris. In fact, Couper had written his paper before Kekulé, but his supervisor took rather a long time to read it, so he lost out. In some ways, Scott Couper's paper was even more advanced than Kekulé's because he drew lines between atoms to indicate actual chemical bonds. (Scott Couper's sad life was to end in an insane asylum.)

Kekulé also claimed some undeserved fame for having deduced that benzene, a molecule whose formula, C_6H_6, seemed to violate the laws of valency, consisted of a ring of six carbon atoms each with hydrogen attached. Late in life, he said it had come to him in a dream while he was working in London in the mid-1850s and fell asleep on the bus taking him home one evening. What he had conveniently forgotten was that Johann Loschmidt (1821–1895), a modest high school teacher from Vienna, had deduced the structure as many as 4 years earlier and published it in an essay that Kekulé had read.

Van't Hoff focused on the problem of how two compounds with exactly the same formula and physical properties could differ in two respects: their crystal shapes could vary but were mirror images of each other and the way they rotated a beam of polarized light, one clockwise, the other counterclockwise. He put forward his theory, in 1874, that this could be explained if carbon formed four bonds arranged tetrahedrally. The idea was ridiculed by older chemists as "fantastic foolishness" and the "shallow speculations" of a youth, yet it could not be ignored because it explained why molecules could be left and right handed (► page 446). In any event, he had the laugh on them because he was awarded the first ever Nobel Prize in Chemistry in 1901.

Emilio Segre Visual Archives

Marie Curie (1867–1934) and Pierre Curie (1859–1906). Marie Curie is one of very few people and the only woman to have ever received two Nobel Prizes. She was born in Poland but studied and carried out her research in Paris. In 1903, she shared the Nobel Prize in physics with H. Becquerel and her husband Pierre for their discovery of radioactivity. She received a second Nobel Prize in 1911 for chemistry, for the discovery of two new chemical elements, radium and polonium (the latter named from her homeland, Poland). A unit of radioactivity (curie, Ci) and an element (curium, Cm) are named in her honor. Pierre, who died in an accident in 1906, was also well known for his research on magnetism. One of their daughters, Irène, married Frédèric Joliot, and they shared in the 1935 Nobel Prize in chemistry for their discovery of artificial radioactivity.

Atomic Structure—Remarkable Discoveries—1890s and Beyond

Like the 1860s, the 1890s was another decade of remarkable chemical discovery, the most surprising being that atoms could spontaneously disintegrate. It began in 1896 when Henri Becquerel (1852–1908) started to investigate the mineral potassium uranyl sulfate [$K_2SO_4 \cdot UO_2(SO_4)_2 \cdot 2\,H_2O$]. He found by chance that it was emitting invisible rays that caused a photographic plate to produce an image. Other uranium compounds also gave off these rays. What was equally intriguing was the observation that the common uranium ore pitchblende contained something that gave off more of this invisible radiation than could be explained by the uranium it contained. The husband and wife team of Pierre Curie (1859–1906) and Marie Curie (1867–1934) worked for weeks in an old shed in Paris to separate this impurity. Eventually, they succeeded, and in 1898 they announced the discovery of two new, intensely radioactive elements: polonium and radium. Radioactivity was the word they invented to describe the new phenomenon of invisible rays (Figure 1), and they called one of the new elements radium because of its intense rays, and the other polonium after Marie's native country Poland.

In 1897, J. J. Thomson (1856–1940) reported his studies of another type of ray, cathode rays. Cathode ray tubes were glass vacuum tubes containing two metal electrodes. When a high voltage is applied to the electrodes, electricity flows from the negative electrode (cathode) to the positive electrode (anode) even though there is nothing there to conduct it. Thomson showed that there was, in fact, a stream of charged particles moving from the cathode to the anode and that these could be deflected by electric and magnetic fields, which showed they were negatively charged (Figure 2). He deduced they were two thousand times lighter than even the lightest element, hydrogen. They became known as *electrons*, a term already invented to describe the smallest particle of electricity.

Oesper Collection in the History of Chemistry/University of Cincinnati

Sir Joseph John Thomson (1856–1940). Cavendish Professor of Experimental Physics at Cambridge University in England. In 1896, he gave a series of lectures at Princeton University in the U.S. on the discharge of electricity in gases. It was this work on cathode rays that led to his discovery of the electron, which he announced at a lecture on the evening of Friday, April 30, 1897. He later published a number of books on the electron and was awarded the Nobel Prize in physics in 1906.

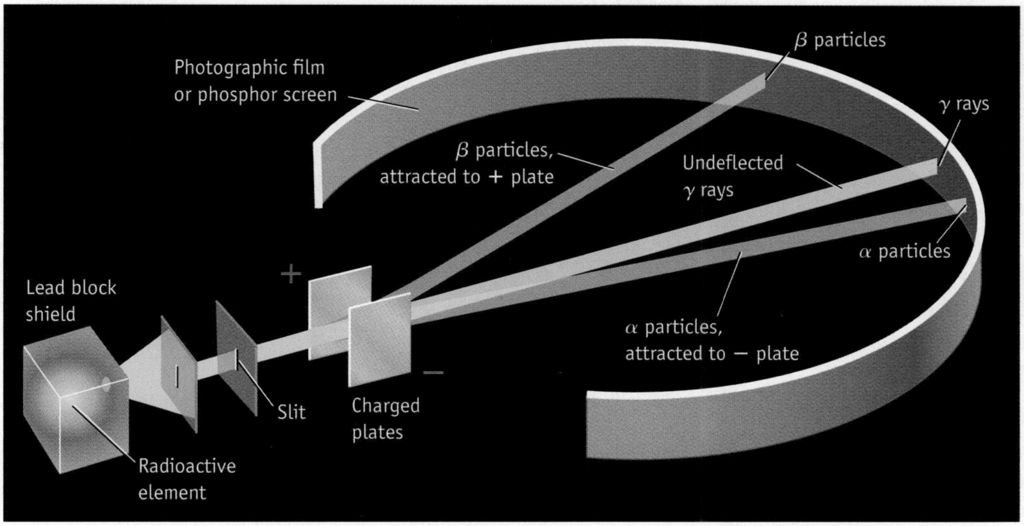

Figure 1 Radioactivity. Alpha (α), beta (β), and gamma (γ) rays from a radioactive element are separated by passing them between electrically charged plates. Positively charged α particles are attracted to the negative plate, and negatively charged β particles are attracted to the positive plate. (Note that the heavier α particles are deflected less than the lighter β particles.) Gamma rays have no electric charge and pass undeflected between the charged plates. (See ChemistryNow Screen 2.5 for an interactive version of this figure.)

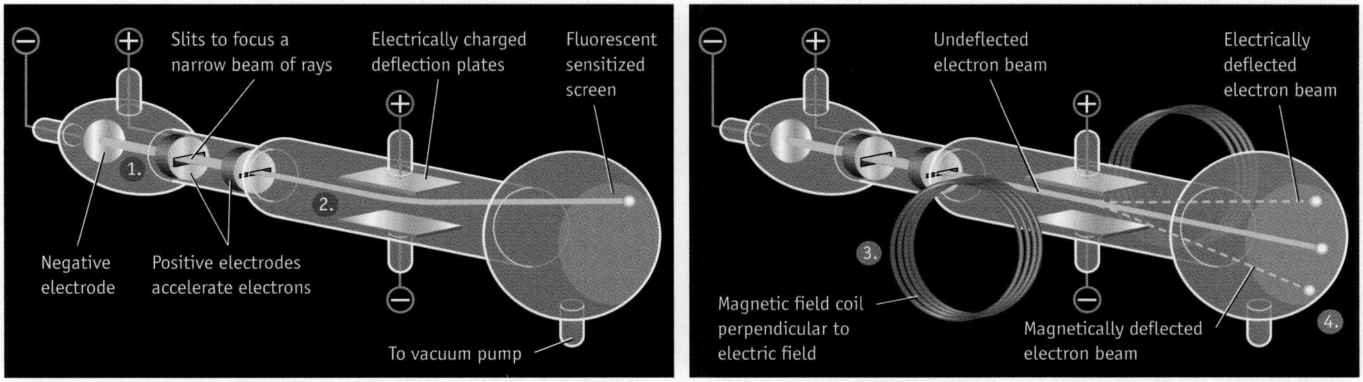

1. A beam of electrons (cathode rays) is accelerated through two focusing slits.
2. When passing through an electric field, the beam of electrons is deflected.
3. The experiment is arranged so that the electric field causes the beam of electrons to be deflected in one direction. The magnetic field deflects the beam in the opposite direction.
4. By balancing the effects of the electrical and magnetic fields, the charge-to-mass ratio of the electron can be determined.

Figure 2 Thomson's experiment to measure the electron's charge-to-mass ratio. This experiment was done by J. J. Thomson in 1896–1897. (See ChemistryNow Screen 2.6 for an interactive version of this figure.)

Thomson reasoned that electrons must originate from the atoms of the cathode, and he suggested that an atom was a uniform sphere of positively charged matter in which negative electrons were embedded. That view of an atom was not to persist for long.

In 1886, a few years before Thomson's report, Eugene Goldstein (1850–1930) had also explored cathode rays and had noticed something rather unexpected. Although negatively charged particles were streaming from the cathode to the anode, there were also positively charged particles moving in the opposite direction, and these could be observed if tiny holes were drilled into the cathode plate. These rays, which he called *Kanalstrahlen* from the German word meaning channel rays, became known as canal rays or anode rays (Figure 3). Goldstein thought he had discovered a basic type of atomic particle, the opposite of the electron, but his positive particles were just traces of residual gas ions in the cathode-ray tube.

The man most associated with discovering the true nature of the atom was Ernest Rutherford (1871–1937), better known as Lord Rutherford. He contributed to the story of atoms in three important ways: he identified the rays that radioactive atoms emitted; he proved that an atom has a tiny nucleus of positively charged protons; and he split the atom; in other words, he converted one element into another. Rutherford, who was born near Nelson on the South Island of New Zealand, went to Cambridge University in England, where he studied under Thomson. He concerned himself with radioactive phenomena and in the years 1898–1900 he identified the various types of radiation: alpha (α), beta (β), and gamma (γ) rays. Alpha and beta rays were particles, the former with an electric charge (+2) twice as

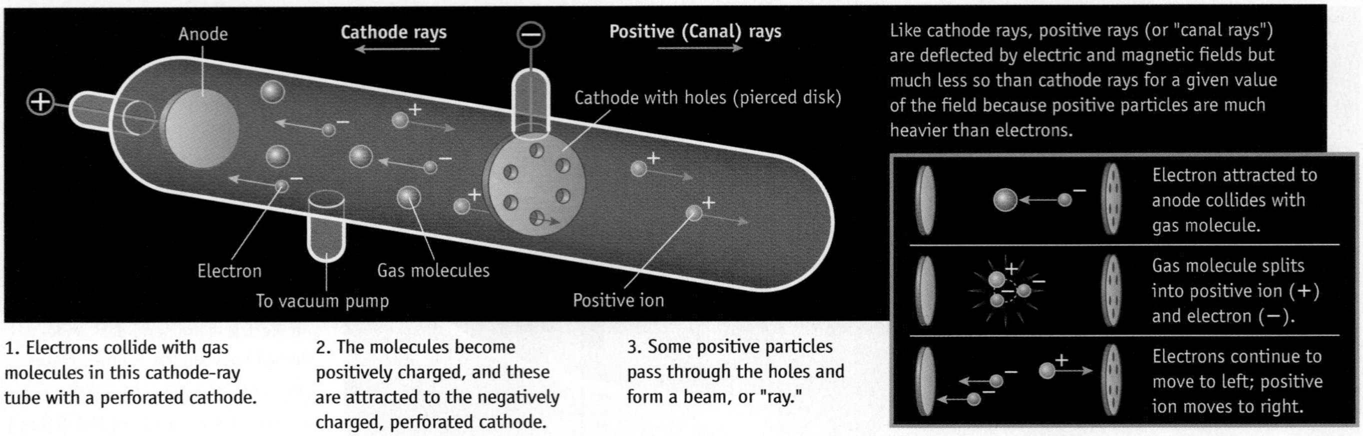

Figure 3 **Canal rays.** In 1886, Eugene Goldstein detected a stream of particles traveling in the direction opposite to that of the negatively charged cathode rays. We now know that these particles are positively charged ions formed by collisions of electrons with gaseous molecules in the cathode-ray tube. (See ChemistryNow Screen 2.8 for an animation of this experiment.)

large as that of the latter, which was negatively charged (−1). We now know that α particles are helium nuclei and that β particles are electrons. Gamma rays are like light rays but with much shorter wavelengths.

Rutherford moved to McGill University in Montreal, Canada, where he collaborated with Frederick Soddy (1877–1956), and together they were able to show that another element, thorium, was radioactive and that it decayed via a series of elements, finally ending up as stable lead. Together, they published their theory of radioactivity in 1908. That same year, Rutherford returned to England to become Professor of Physics at Manchester, where his research was to reveal even more spectacular discoveries about atoms.

Soddy had already returned to England in 1903 and was now working in University College London, where he was able to prove that as radium decayed it formed helium gas. He then moved to Glasgow, Scotland, and there his research showed that atoms of the same element could have more than one atomic mass, and the idea of isotopes was born. These made it possible to explain why an element could have an atomic weight which was not a whole number. For example, chlorine's was 35.5 because it consisted of 76% the isotope chlorine-35 plus 24% of chlorine-37.

Meanwhile at Manchester, two of Rutherford's students—Hans Geiger (1882–1945) and Ernest Marsden (1889–1970)—bombarded thin gold foil with α particles to test whether Thomson's model of a solid atom with embedded electrons was correct (Figure 4). Almost all the particles passed straight through the gold foil as if there was nothing there. However, they were surprised to find that a few were deflected sideways; some even bounced right back. This experiment proved that an atom of gold is mostly empty space with a tiny nucleus at its center. It was the electrons that accounted for most of its volume. Rutherford calculated that the central nucleus of an atom occupied only 1/10,000th of its volume. He also calculated that a gold nucleus had a positive charge of around 100 units and a radius of about 10^{-12} cm. (The currently accepted values are +79 for atomic charge and 10^{-13} cm for the radius.) Just as Dalton had done more than a century before, Rutherford announced his findings at a meeting of the Manchester Literary and Philosophical Society. The date was March 7, 1911.

In 1908, the American physicist Robert Millikan (1868–1953), based at the California Institute of Technology (Caltech), measured the charge on the electron as 1.592×10^{-19} coulombs, not far from today's accepted value of 1.602×10^{-19} C (Figure 5). Millikan rightly assumed this was the fundamental unit of charge. Knowing this, and the charge-to-mass ratio determined by Thomson, enabled the mass of an electron to be calculated as 9.109×10^{-28} g.

In 1913, Henry G. J. Moseley (1887–1915) realized it was not its atomic weight that defined an element but its atomic number, which he deduced from a study of the wavelengths of lines in x-ray spectra. Moseley was then able to put the periodic table of elements on a more secure footing. (Sadly, his life came to an end when a bullet from a sniper killed him in World War I.) Mendeleev had been

Oesper Collection in the History of Chemistry/University of Cincinnati

Ernest Rutherford (1871–1937). Rutherford was born in New Zealand in 1871 but went to Cambridge University in England to pursue his Ph.D. in physics in 1895. There, he worked with J. J. Thomson, and it was at Cambridge that he discovered α and β radiation. At McGill University in Canada in 1899, Rutherford did further experiments to prove that α radiation is composed of helium nuclei and that β radiation consists of electrons. He received the Nobel Prize in chemistry for his work in 1908. His research on the structure of the atom was done after he moved to Manchester University in England. In 1919, he returned to Cambridge University, where he took up the position formerly held by Thomson. In his career, Rutherford guided the work of 10 future recipients of the Nobel Prize. Element 104 has been named *rutherfordium* in his honor.

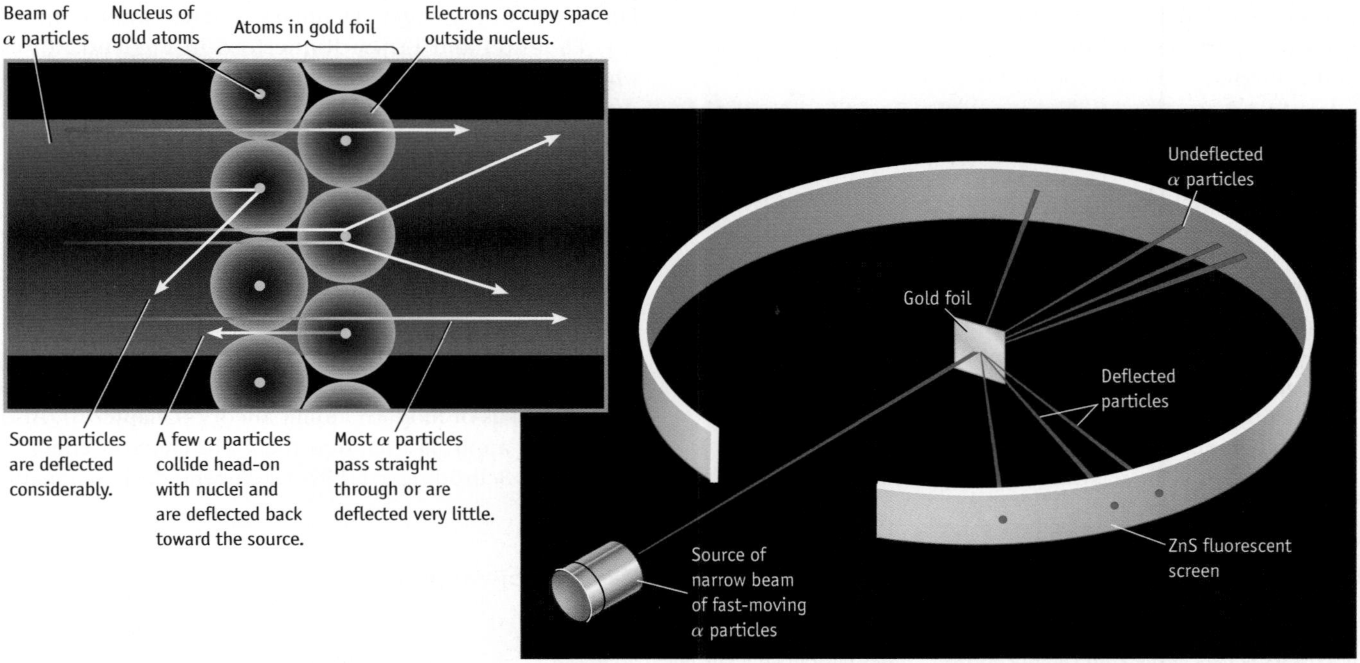

Figure 4 **Rutherford's experiment to determine the structure of the atom.** A beam of positively charged α particles was directed at a thin gold foil. A fluorescent screen coated with zinc sulfide (ZnS) was used to detect particles passing through or deflected by the foil. (A flash of light is seen when a particle strikes the screen.) Most of the particles passed through the foil, but some were deflected from their path. A few were even deflected backward. (See ChemistryNow Screen 2.10 to explore an interactive version of this figure accompanied by an exercise.)

perplexed by some elements when he arranged them in order of increasing atomic weight. For example, tellurium has an atomic weight of 127.6, and iodine has an atomic weight of 126.9. The problem was that their properties indicated that tellurium should be placed before iodine in the periodic table. Arranging the elements in order of increasing atomic number, however, removed this problem. Tellurium has an atomic number of 52, and iodine has an atomic number of 53, thus justifying placing tellurium before iodine in the table.

Although he was not to know it, Moseley's atomic number corresponded to the number of protons in the nucleus. In 1919, Rutherford proved there were such things as protons, and these were the positive charges located at the center of an atom. Atoms were at last correctly seen as consisting of positive protons balanced by the same number of negative electrons.

In 1919, Rutherford "split the atom" according to the newspapers of the day. What he had performed was the first-ever successful experiment deliberately designed to

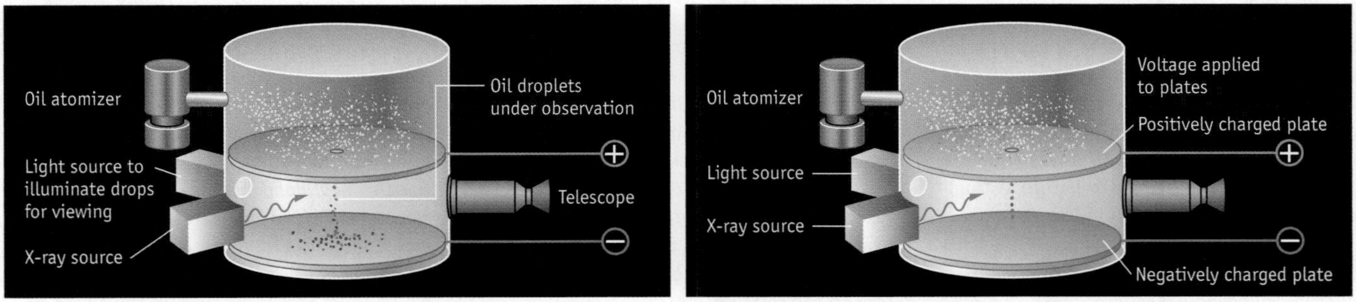

1. A fine mist of oil drops is introduced into one chamber.

2. The droplets fall one by one into the lower chamber under the force of gravity.

3. Gas molecules in the bottom chamber are ionized (split into electrons and a positive fragment) by a beam of x-rays. The electrons adhere to the oil drops, some droplets having one electron, some two, and so on.

These negatively charged droplets continue to fall due to gravity.

4. By carefully adjusting the voltage on the plates, the force of gravity on the droplet is exactly counterbalanced by the attraction of the negative droplet to the upper, positively charged plate.

Analysis of these forces led to a value for the charge on the electron.

Figure 5 **Millikan's experiment to determine the electron charge.** The experiment was done by R. A. Millikan in 1909. (See ChemistryNow Screen 2.7 for an interactive exercise on this experiment.)

convert one element into another. He bombarded nitrogen gas with α particles, and this led to its conversion to oxygen and hydrogen. We can express this as an equation:

$$^{14}\text{N (7 protons)} + \alpha \text{ particle } (^{4}\text{He, 2 protons}) \rightarrow \, ^{17}\text{O (8 protons)} + \, ^{1}\text{H (1 proton)}$$

By about 1920, the image of an atom was that it consisted of a tight nucleus, where all the protons were located, surrounded by a fuzzy cloud of negatively charged electrons, and that these were circling the nucleus rather like planets orbiting the sun. And just as the planets don't move in a random way, so electrons were confined to particular orbits, some nearer the nucleus, some farther away.

The physical nature of atoms was known, but how could the arrangement of electrons be explained? Niels Bohr (1885–1962) was the scientist who helped solve that puzzle. He lived in Copenhagen, Denmark, and had trained under J. J. Thomson and Ernest Rutherford in England. It was while studying atomic spectra in England that he began to develop his theory of electrons circulating in orbits around the nucleus and with specific quantized energies (◀ page 276). Bohr postulated that the electrons were confined to specific energy levels called orbits. He could then understand the atomic emission spectrum of hydrogen by postulating that the lines it displayed corresponded to discrete quantities of energy (quanta) that an electron emitted as it jumped from one orbit to another.

Bohr's idea of specific energy levels is still retained, but his idea of orbits at specific distances has been revised to be the orbitals of modern atomic theory (Chapter 6). Also involved in applying quantum theory to electron energies were Erwin Schrödinger (1887–1961), who devised a math-

Historical Perspectives

20th-Century Giants of Science

Many of the advances in science occurred during the early part of the 20th century, as the result of theoretical studies by some of the greatest minds in the history of science.

Max Planck

Edgar Fahs Smith Collection/University of Pennsylvania Library

Max Karl Ernst Ludwig Planck (1858–1947) was raised in Germany, where his father was a professor at a university. While still in his teens, Planck decided to become a physicist, against the advice of the head of the physics department at Munich, who told him, "The important discoveries [in physics] have been made. It is hardly worth entering physics anymore." Fortunately, Planck did not take this advice and went on to study thermodynamics. This interest led him eventually to consider the ultraviolet catastrophe in explanations of blackbody radiation and to develop his revolutionary hypothesis of quantized energy, which was announced 2 weeks before Christmas in 1900. He was awarded the Nobel Prize in physics in 1918 for this work. Einstein later said it was a longing to find harmony and order in nature, a "hunger in his soul," that spurred Planck on.

Erwin Schrödinger

Edgar Fahs Smith Collection/University of Pennsylvania Library

Erwin Schrödinger (1887–1961) was born in Vienna, Austria. Following his service as an artillery officer in World War I, he became a professor of physics. In 1928, he succeeded Planck as professor of physics at the University of Berlin. He shared the Nobel Prize in physics in 1933.

Niels Bohr

Edgar Fahs Smith Collection/University of Pennsylvania Library

Niels Bohr (1885–1962) was born in Copenhagen, Denmark. He earned a Ph.D. in physics in Copenhagen in 1911 and then went to work first with J. J. Thomson and later with Ernest Rutherford in England. It was there that he began to develop his theory of atomic structure and his explanation of atomic spectra. (He received the Nobel Prize in physics in 1922 for this work.) Bohr returned to Copenhagen, where he eventually became director of the Institute for Theoretical Physics. Many young physicists worked with him at the Institute, seven of whom eventually received Nobel Prizes in chemistry and physics. Among these scientists were Werner Heisenberg, Wolfgang Pauli, and Linus Pauling. Element 107 was recently named bohrium in Bohr's honor.

Werner Heisenberg

Emilio Segre Visual Archives

Werner Heisenberg (1901–1976) studied with Max Born and later with Bohr. He received the Nobel Prize in physics in 1932. The recent play *Copenhagen*, which has been staged in London and New York, centers on the relationship between Bohr and Heisenberg and their involvement in the development of atomic weapons in World War II.

Gilbert Newton Lewis (1875–1946) introduced the theory of the shared electron-pair chemical bond in a paper published in the *Journal of the American Chemical Society* in 1916. Lewis also made major contributions in acid–base chemistry, thermodynamics, and on the interaction of light with substances.

Gilbert Newton Lewis

Oesper Collection in the History of Chemistry/University of Cincinnati

Lewis was born in Massachusetts but raised in Nebraska. After earning his B. A. and Ph.D. at Harvard, he began his career in 1912 at the University of California at Berkeley. He was not only a productive researcher, but was also an influential teacher. Among his ideas was the use of problem sets in teaching, an idea still in use today.

Linus Pauling

© Tom Hollyman/Photo Researchers, Inc.

Linus Pauling (1901–1994) was born in Portland, Oregon, earned a B.Sc. degree in chemical engineering from Oregon State College in 1922, and completed his Ph.D. in chemistry at the California Institute of Technology in 1925. In chemistry, he is best known for his book *The Nature of the Chemical Bond*. He also studied protein structure, and, in the words of Francis Crick, was "one of the founders of molecular biology." It was this work and his study of chemical bonding that were cited in the award of the Nobel Prize in chemistry in 1954. Although chemistry was the focus of his life, at the urging of his wife, Ava Helen, he was also involved in nuclear disarmament issues, and he received the Nobel Peace Prize in 1962 for the role he played in advocating for the nuclear test ban treaty.

ematical equation that described orbitals, and Werner Heisenberg (1901–1976), whose *Uncertainty Principle* said that we cannot ever know exactly both the position and the energy of an electron.

The nucleus of an atom still presented a problem in 1930. How could a large number of positively charged protons co-exist in a nucleus without their repelling one another so much so that the atom falls apart? Lead, for example, had 82 protons. There had to be something else in the nucleus, and it had to be a heavy particle to account for the atomic weight of an element, which was more than double its atomic number and in the case of lead was 207. In 1932, the British physicist James Chadwick (1891–1974) found the missing particles. He directed the very powerful α rays that were released from radioactive polonium towards a beryllium target. The secondary emanations emitted by the latter metal were strange in that they carried no charge but were massive enough to knock protons out of the nuclei of other atoms. These new particles, now known as *neutrons,* had no electric charge and a mass of 1.674927×10^{-24} g, slightly greater than the mass of a proton. (At the same time, Hans Falkenhagen in Germany discovered neutrons also, but he did not publish his results.) Chadwick had found the missing particle. It completed the chemists' picture of an atom. It also made it possible to produce elements heavier than uranium—as well as to create atomic bombs.

The Nature of the Chemical Bond

Molecules presented a more complex problem: how did atoms join together to form them? The first person to provide an answer based on the new view of the atom was the American chemist Gilbert Newton Lewis (1875–1946). In his chemistry lectures at the University of California–Berkeley in the early 1900s, he used dots to symbolize electrons, and he developed this idea so that it became more than just a teaching aid. Lewis said in 1916 that a single chemical bond was the sharing of a pair of electrons between two atoms; a double bond was the sharing of two pairs; and a triple bond the sharing of three pairs. This simple concept explained valency and structure and had enormous influence because it made so much of the chemistry of atoms and molecules understandable.

More sophisticated concepts of bonding were developed based on Max Planck's (1858–1947) theory that energy was quantized. Neils Bohr (1885–1962) and Robert Mulliken (1896–1986) saw that this implied there were only certain energy levels within an atom that its electrons could inhabit. Mulliken proposed a theory of chemical bonding based on combining atomic orbitals into molecular orbitals and showed how the energies of these related to the way the atomic orbitals overlapped. He also developed the theory of electronegativity, which is based on the relative abilities of atoms in molecules to attract electrons, again a concept useful in explaining chemical behavior.

For chemists, the man whose name is most famously linked to bonding was Linus Pauling (1901–1994). He wrote his first paper on the subject when he was only 27 years old, in 1928, and followed it with several more. He brought all his thoughts together in his seminal work *The Nature of the Chemical Bond* in 1939.

Of course, there are still many things to be discovered about atoms and molecules, but as far as chemistry was concerned, the age-old questions of what elements, atoms, and molecules really were had been answered by the mid-twentieth century. The world of the nucleus and of subatomic particles could be left to the physicists to investigate.

SUGGESTED READINGS

1. Eric Scerri, *The Periodic Table: Its Story and Its Significance,* Oxford University Press, New York, 2007.
2. John Emsley, *The 13th Element: The Sordid Tale of Murder, Fire, and Phosphorus,* John Wiley and Sons, New York, 2000.
3. John Emsley, *Nature's Building Blocks,* Oxford University Press, 2002.
4. Arthur Greenberg, *A Chemical History Tour: Picturing Chemistry from Alchemy to Modern Molecular Science,* Wiley-Interscience, New York, 2000.
5. Aaron Ihde, *The Development of Modern Chemistry,* Dover Publications, New York, 1984.
6. L. K. James, ed., *Nobel Laureates in Chemistry, 1901–1992,* American Chemical Society and Chemical Heritage Foundation, 1993.

STUDY QUESTIONS

Blue-numbered questions have answers in Appendix P and fully-worked solutions in the *Student Solutions Manual.*

1. Dalton proposed that an atom was a "solid, massy, hard, impenetrable, moveable particle." Critique this description. How does this description misrepresent atomic structure.

2. Dalton's hypotheses on the structure of atoms was based in part on the observation of a "Law of definite proportions," which said that atoms combined in a definite ratio by weight. Using the formula for water and atomic weights from the current atomic mass scale, calculate the ratio of the mass of hydrogen to the mass of oxygen in water.

3. From cathode ray experiments, J. J. Thomson estimated that the mass of an electron was "about a thousandth" of the mass of a proton. How accurate is that estimate? Calculate the ratio of the mass of an electron to the mass of a hydrogen atom.

4. Goldstein observed positively charged particles moving in the opposite direction to electrons in a cathode ray tube. From their mass, he concluded that these particles were formed from residual gas in the tube. For example, if the cathode ray tube contained helium, the cathode rays consisted of He^+ ions. Describe the process that forms these ions.

8 Bonding and Molecular Structure

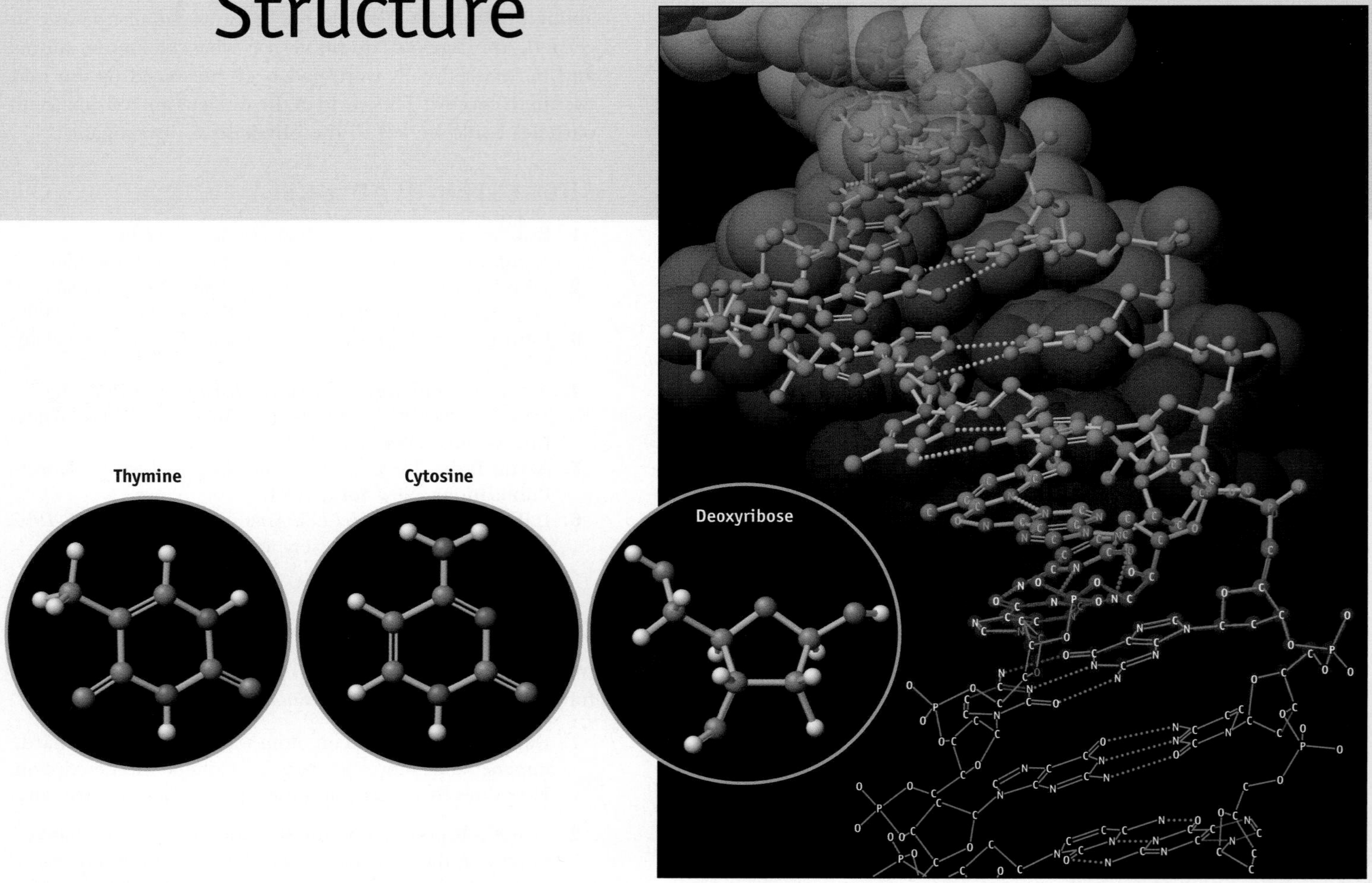

Chemical Bonding in DNA

The theme of this chapter and the next is molecular bonding and structure, and the subject is well-illustrated by the structure of DNA. This molecule is a helical coil of two chains of tetrahedral phosphate groups and deoxyribose groups. Organic bases (such as thymine and cytosine) on one chain interact with complementary bases on the other chain.

Questions:

Among the many questions you can answer from studying this chapter are the following:

1. Why are there four bonds to carbon and phosphorus?
2. Why are the C atoms and P atoms in the backbone, and the C atoms in deoxyribose, surrounded by other atoms at an angle of 109°?
3. What are the angles in the six-member rings of the bases thymine and cytosine? Why are the six-member rings flat?
4. Are thymine and cytosine polar molecules?

Answers to these questions are in Appendix Q.

Chapter Goals

See Chapter Goals Revisited (page 393) for Study Questions keyed to these goals and assignable in OWL.

- Understand the difference between ionic and covalent bonds.
- Draw Lewis electron dot structures for small molecules and ions.
- Use the valence shell electron-pair repulsion theory (VSEPR) to predict the shapes of simple molecules and ions and to understand the structures of more complex molecules.
- Use electronegativity and formal charge to predict the charge distribution in molecules and ions, to define the polarity of bonds, and to predict the polarity of molecules.
- Understand the properties of covalent bonds and their influence on molecular structure.

Chapter Outline

Scientists have long known that the key to interpreting the properties of a chemical substance is first to recognize and understand its structure and bonding. **Structure** refers to the way atoms are arranged in space, and **bonding** describes the forces that hold adjacent atoms together.

Our discussion of structure and bonding begins with small molecules and then progresses to larger molecules. From compound to compound, atoms of the same element participate in bonding and structure in a predictable way. This consistency allows us to develop a group of principles that apply to many different chemical compounds, including such complex molecules as DNA.

Chemistry Now™

Throughout the text this icon introduces an opportunity for self-study or to explore interactive tutorials by signing in at **www.thomsonedu.com/login**.

8.1 Chemical Bond Formation

When a chemical reaction occurs between two atoms, their valence electrons are reorganized so that a net attractive force—a **chemical bond**—occurs between atoms. There are two general types of bonds, ionic and covalent, and their formation can be depicted using Lewis symbols.

An **ionic bond** forms when *one or more valence electrons is transferred from one atom to another,* creating positive and negative ions. When sodium and chlorine react (Figure 8.1a), an electron is transferred from a sodium atom to a chlorine atom to form Na^+ and Cl^-.

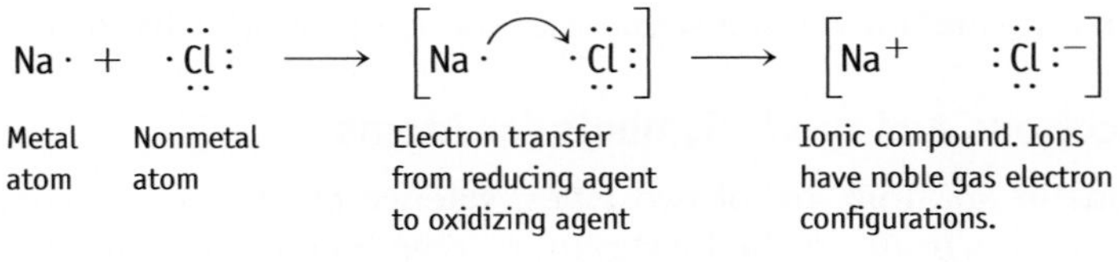

The "bond" is the attractive force between the positive and negative ions.

Covalent bonding, in contrast, *involves sharing of valence electrons between atoms.* Two chlorine atoms, for example, share a pair of electrons, one electron from each atom, to form a covalent bond.

$$:\ddot{\underset{..}{Cl}}\cdot + \cdot\ddot{\underset{..}{Cl}}: \longrightarrow :\ddot{\underset{..}{Cl}}:\ddot{\underset{..}{Cl}}:$$

■ **Valence Electron Configurations and Ionic Compound Formation** For the formation of NaCl:

Na changes from $1s^22s^22p^63s^1$ to Na^+ with $1s^22s^22p^6$, equivalent to the Ne configuration.

Cl changes from $[Ne]3s^23p^5$ to Cl^- with $[Ne]3s^23p^6$, equivalent to the Ar configuration.

FIGURE 8.1 Formation of ionic compounds. Both reactions shown here are quite exothermic, as reflected by the very negative molar enthalpies of formation for the reaction products. (See ChemistryNow Screen 8.3 for a video of the formation of sodium chloride from the elements.)

(a) The reaction of elemental sodium and chlorine to give sodium chloride. $\Delta_f H°$ [NaCl(s)] = −411.12 kJ/mol

(b) The reaction of elemental calcium and oxygen to give calcium oxide. $\Delta_f H°$ [CaO(s)] = −635.09 kJ/mol

As bonding is described in greater detail, you will discover that the two types of bonding—complete electron transfer and the equal sharing of electrons—are extreme cases. In most chemical compounds, electrons are shared unequally, with the extent of sharing varying widely from very little sharing (largely ionic) to considerable sharing (largely covalent).

Ionic bonding will be described in more detail in Chapter 13, while the present chapter focuses on bonding in covalent compounds.

Module 12

8.2 Covalent Bonding and Lewis Structures

There are many examples of compounds having covalent bonds, including the gases in our atmosphere (O_2, N_2, H_2O, and CO_2), common fuels (CH_4), and most of the compounds in your body. Covalent bonding is also responsible for the atom-to-atom connections in common ions such as CO_3^{2-}, CN^-, NH_4^+, NO_3^-, and PO_4^{3-}. We will develop the basic principles of structure and bonding using these and other small molecules and ions, but the same principles apply to larger molecules from aspirin to proteins and DNA with thousands of atoms.

The molecules and ions just mentioned are composed entirely of *nonmetal* atoms. A point that needs special emphasis is that, in molecules or ions made up *only* of nonmetal atoms, the atoms are attached by covalent bonds. Conversely, the presence of a metal in a formula is often a signal that the compound is likely to be ionic.

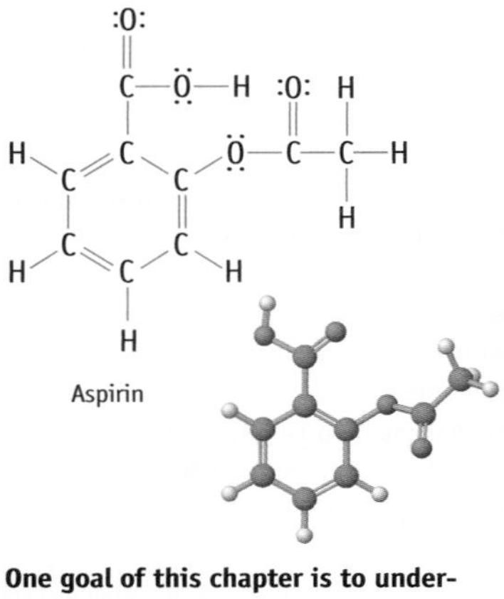

One goal of this chapter is to understand why a molecule such as aspirin has the shape that it exhibits.

Valence Electrons and Lewis Symbols for Atoms

The electrons in an atom are of two types: **valence electrons** and **core electrons.** Chemical reactions result in the loss, gain, or rearrangement of valence electrons. The core electrons are not involved in bonding or in chemical reactions.

For main group elements (elements of the A groups in the periodic table), the valence electrons are the *s* and *p* electrons in the outermost shell (Table 8.1). All electrons in inner shells are core electrons. A useful guideline for *main group elements* is that *the number of valence electrons is equal to the group number.* The fact that all elements in a periodic group have the same number of valence electrons accounts for the similarity of chemical properties among members of the group.

TABLE 8.1 Core and Valence Electrons for Several Common Elements

Element	Periodic Group	Core Electrons	Valence Electrons	Total Configuration
Main Group Elements				
Na	1A	$1s^22s^22p^6$ = [Ne]	$3s^1$	[Ne]$3s^1$
Si	4A	$1s^22s^22p^6$ = [Ne]	$3s^23p^2$	[Ne]$3s^23p^2$
As	5A	$1s^22s^22p^63s^23p^63d^{10}$ = [Ar]$3d^{10}$	$4s^24p^3$	[Ar]$3d^{10}4s^24p^3$
Transition Elements				
Ti	4B	$1s^22s^22p^63s^23p^6$ = [Ar]	$3d^24s^2$	[Ar]$3d^24s^2$
Co	8B	[Ar]	$3d^74s^2$	[Ar]$3d^74s^2$
Mo	6B	[Kr]	$4d^55s^1$	[Kr]$4d^55s^1$

Valence electrons for transition elements include the electrons in the ns and $(n-1)d$ orbitals (see Table 8.1). The remaining electrons are core electrons. As with main group elements, the valence electrons for transition metals determine the chemical properties of these elements.

The American chemist Gilbert Newton Lewis (1875–1946) introduced a useful way to represent electrons in the valence shell of an atom. The element's symbol represents the atomic nucleus together with the core electrons. Up to four valence electrons, represented by dots, are placed one at a time around the symbol; then, if any valence electrons remain, they are paired with ones already there. Chemists now refer to these pictures as **Lewis electron dot symbols.** Lewis dot symbols for main group elements of the second and third periods are shown in Table 8.2.

Arranging the valence electrons of a main group element around an atom in four groups suggests that the valence shell can accommodate four pairs of electrons. Because this represents eight electrons in all, this is referred to as an **octet** of electrons. An octet of electrons surrounding an atom is regarded as a stable configuration. The noble gases, with the exception of helium, have eight valence electrons and demonstrate a notable lack of reactivity. (Helium, neon, and argon do not undergo any chemical reactions, and the other noble gases have very limited chemical reactivity.) Because chemical reactions involve changes in the valence electron shell, the limited reactivity of the noble gases is taken as evidence of the stability of their noble gas (ns^2np^6) electron configuration. Hydrogen, which in its compounds has two electrons in its valence shell, obeys the spirit of this rule by matching the electron configuration of He.

Chemistry Now™

Sign in at **www.thomsonedu.com/login** and go to Chapter 8 Contents to see Screen 8.2 for more on **the correlation of the periodic table and valence electrons.**

TABLE 8.2 Lewis Dot Symbols for Main Group Atoms

1A ns^1	2A ns^2	3A ns^2np^1	4A ns^2np^2	5A ns^2np^3	6A ns^2np^4	7A ns^2np^5	8A ns^2np^6
Li ·	· Be ·	· B ·	· C ·	· N ·	: O ·	: F ·	: Ne :
Na ·	· Mg ·	· Al ·	· Si ·	· P ·	: S ·	: Cl ·	: Ar :

■ **Gilbert Newton Lewis (1875–1946)** Lewis introduced the theory of the shared electron-pair chemical bonds in a paper published in the *Journal of the American Chemical Society* in 1916. Lewis also made major contributions in acid–base chemistry, thermodynamics, and on the interaction of light with substances.

Lewis was born in Massachusetts but raised in Nebraska. After earning his B. A. and Ph.D. at Harvard, he began his career in 1912 at the University of California at Berkeley. He was not only a productive researcher, but was also an influential teacher. Among his ideas was the use of problem sets in teaching, an idea still in use today.

Oesper Collection in the History of Chemistry/University of Cincinnati

Gilbert Newton Lewis

Lewis Electron Dot Structures and the Octet Rule

In a simple description of covalent bonding, a bond results when one or more electron pairs are shared between two atoms. The electron pair bond between the two atoms of an H_2 molecule is represented by a pair of dots or, alternatively, a line.

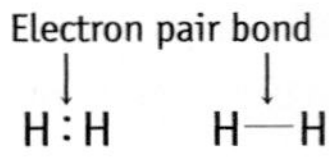

The representation of a molecule in this fashion is called a **Lewis electron dot structure** or just a **Lewis structure** in honor of G. N. Lewis.

Simple Lewis structures, such as that for F_2, can be drawn starting with Lewis dot symbols for atoms and arranging the valence electrons to form bonds. Fluorine, an element in Group 7A, has seven valence electrons. The Lewis symbol shows that an F atom has a single unpaired electron along with three electron pairs. In F_2, the single electrons, one on each F atom, pair up in the covalent bond.

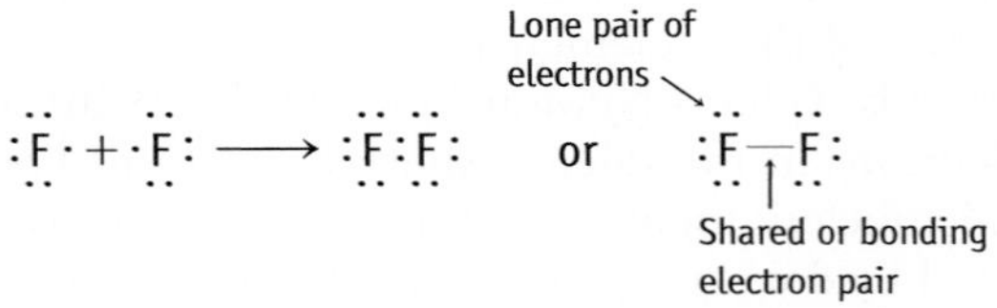

In the Lewis structure for F_2 the pair of electrons in the F—F bond is the bonding pair, or **bond pair.** The other six pairs reside on single atoms and are called **lone pairs.** Because they are not involved in bonding, they are also called **nonbonding electrons.**

Carbon dioxide, CO_2, and dinitrogen, N_2, are examples of molecules in which two atoms are multiply bonded; that is, they share more than one electron pair.

■ **H Atoms and Electron Octets** Hydrogen atoms cannot be surrounded by an octet of electrons. An atom of H, which has only a 1*s* valence electron orbital, can accommodate only a pair of electrons.

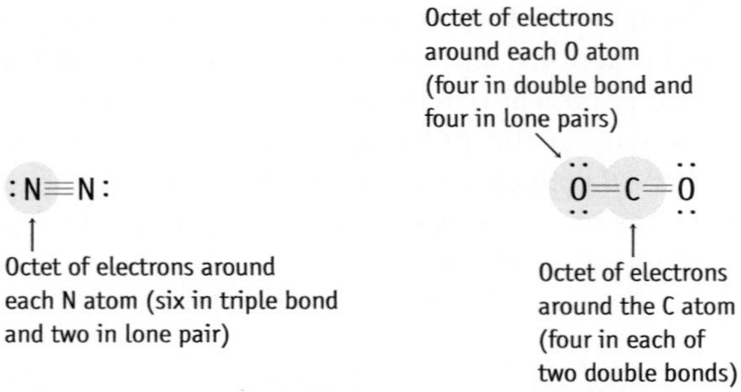

■ **Importance of Lone Pairs** Lone pairs can be important in a structure. Since they are in the same valence electron shell as the bonding electrons, they can influence molecular shape. See Section 8.6.

In carbon dioxide, the carbon atom shares two pairs of electrons with each oxygen and so is linked to each O atom by a double bond. The valence shell of each oxygen atom in CO_2 has two bonding pairs and two lone pairs. In dinitrogen, the two nitrogen atoms share three pairs of electrons, so they are linked by a triple bond. In addition, each N atom has a single lone pair.

An important observation can be made about the molecules you have seen so far: each atom (except H) has a share in four pairs of electrons, so each has achieved a *noble gas configuration. Each atom is surrounded by an octet of eight electrons.* (Hydrogen typically forms a bond to only one other atom, resulting in two electrons in its valence shell.) *The tendency of molecules and polyatomic ions to have structures in which eight electrons surround each atom* is known as the **octet rule.** As an example, a triple bond is necessary in dinitrogen in order to have an octet around each nitrogen atom. The carbon atom and both oxygen atoms in CO_2 achieve the octet configuration by forming double bonds.

The octet rule is extremely useful, but keep in mind that it is more a *guideline* than a rule. Particularly for the second period elements C, N, O, and F, a Lewis structure in which each atom achieves an octet is likely to be correct. Although there are a few exceptions, if an atom such as C, N, O, or F in a Lewis structure does not follow the octet rule, you should question the structure's validity. If a structure obeying the octet rule cannot be written, then it is possible an incorrect formula has been assigned to the compound or the atoms have been assembled in an incorrect way.

■ **Exceptions to the Octet Rule** Although the octet rule is widely applicable, there are exceptions. Fortunately, many will be obvious, such as when there are more than four bonds to an element or when an odd number of electrons occur. See Section 8.5.

Drawing Lewis Electron Dot Structures

There is a systematic approach to constructing Lewis structures of molecules and ions. Let us take formaldehyde, CH_2O, as an example.

■ **Choosing the Central Atom**
1. The electronegativities of atoms can also be used to choose the central atom. Electronegativity is discussed in Section 8.7.
2. For simple compounds, the first atom in a formula is often the central atom (e.g., SO_2, NH_4^+, NO_3^-). This is not always a reliable predictor, however. Notable exceptions include water (H_2O) and most common acids (HNO_3, H_2SO_4), in which the acidic hydrogen is usually written first but where another atom (such as N or S) is the central atom.

1. *Determine the arrangement of atoms within a molecule.* The central atom is *usually* the one with the lowest electron affinity. In CH_2O, the central atom is C. You will come to recognize that certain elements often appear as the central atom, among them C, N, P, and S. Halogens are often terminal atoms forming a single bond to one other atom, but they can be the central atom when combined with O in oxoacids (such as $HClO_4$). Oxygen is the central atom in water, but in conjunction with nitrogen, phosphorus, and the halogens it is usually a terminal atom. Hydrogen is a terminal atom because it typically bonds to only one other atom.

2. *Determine the total number of valence electrons in the molecule or ion.* In a neutral molecule, this number will be the sum of the valence electrons for each atom. For an *anion, add* the number of electrons equal to the negative charge; for a *cation, subtract* the number of electrons equal to the positive charge. The number of valence electron pairs will be half the total number of valence electrons. For CH_2O,

$$\text{Valence electrons} = 12 \text{ electrons (or 6 electron pairs)}$$
$$= 4 \text{ for C} + (2 \times 1 \text{ for two H atoms}) + 6 \text{ for O}$$

3. *Place one pair of electrons between each pair of bonded atoms to form a single bond.*

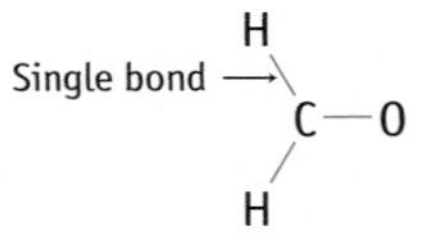

Here, three electron pairs are used to make three single bonds (which are represented by single lines). Three pairs of electrons remain to be used.

4. *Use any remaining pairs as lone pairs around each terminal atom (except H) so that each terminal atom is surrounded by eight electrons.* If, after this is done, there are electrons left over, assign them to the central atom. (If the central atom is an element in the third or higher period, it can have more than eight electrons. See page 364.)

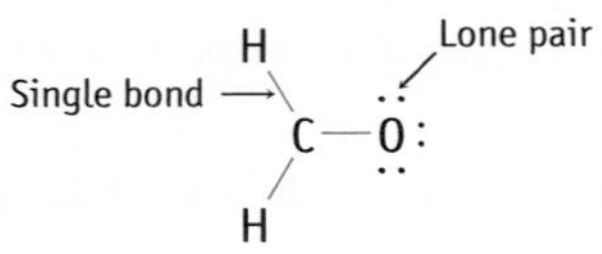

Here, all six pairs have been assigned, but notice that the C atom has a share in only three pairs.

5. *If the central atom has fewer than eight electrons at this point, change one or more of the lone pairs on the terminal atoms into a bonding pair between the central and terminal atom to form a multiple bond.*

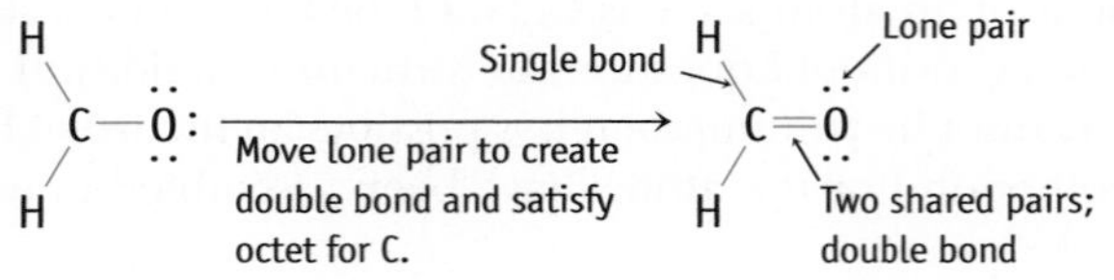

■ **O as a Terminal Atom** Oxygen atoms are usually terminal atoms when combined with many other atoms such as B, C, Si, S, N, P, and the halogens.

As a general rule, double or triple bonds are most often encountered when *both* atoms are from the following list: C, N, or O. That is, bonds such as C=C, C=N, and C=O will be encountered frequently.

Chemistry Now™

Sign in at **www.thomsonedu.com/login** and go to Chapter 8 Contents to see:

- Screen 8.4 for an animation of the **factors influencing bond formation**
- Screen 8.5 for a tutorial on **Lewis structures**
- Screen 8.6 for a tutorial and an exercise on **drawing Lewis structures**

EXAMPLE 8.1 Drawing Lewis Structures

Problem Draw Lewis structures for the chlorate ion (ClO_3^-) and the nitronium ion (NO_2^+).

Strategy Follow the five steps outlined for CH_2O in the preceding text.

Solution for chlorate ion

1. Cl is the central atom, and the O atoms are terminal atoms.
2. Valence electrons = 26 (13 pairs)

   = 7 (for Cl) + 18 (six for each O) + 1 (for the negative charge)
3. Three electron pairs form single bonds from Cl to the O terminal atoms.

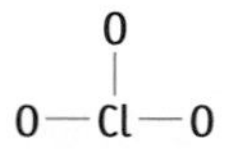

4. Distribute three lone pairs on each of the terminal O atoms to complete the octet of electrons around each of these atoms.

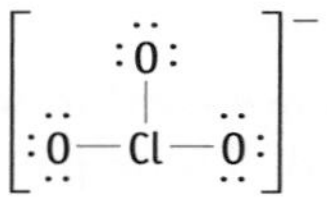

5. One pair of electrons remains, and it is placed on the central Cl atom to complete its octet.

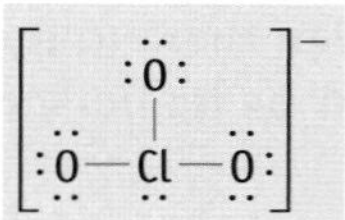

Each atom now has a share in four pairs of electrons, and the Lewis structure is complete.

Solution for nitronium ion

1. Nitrogen is the central atom, because its electron affinity is lower than that of oxygen.

2. Valence electrons = 16 (8 valence pairs)

$= 5$ (for N) + 12 (six for each O) − 1 (for the positive charge)

3. Two electron pairs form single bonds from the nitrogen to each oxygen:

O—N—O

4. Distribute the remaining six pairs of electrons on the terminal O atoms:

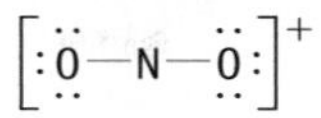

5. The central nitrogen atom is two electron pairs short of an octet. Thus, a lone pair of electrons on each oxygen atom is converted to a bonding electron pair to give two N=O double bonds. Each atom in the ion now has four electron pairs. Nitrogen has four bonding pairs, and each oxygen atom has two lone pairs and shares two bond pairs.

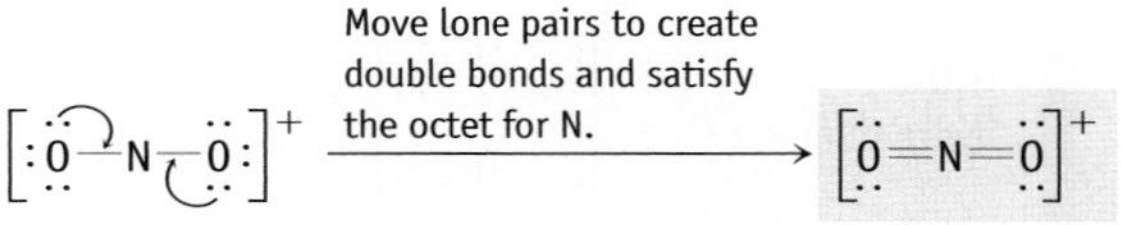

Comment Why don't we take two lone pairs from one side and none from the other? We shall discuss that after describing charge distribution in molecules and ions (page 377).

> **EXERCISE 8.1 Drawing Lewis Structures**
>
> Draw Lewis structures for NH_4^+, CO, NO^+, and SO_4^{2-}.

Predicting Lewis Structures

Lewis structures are useful in gaining a perspective on the structure and chemistry of a molecule or ion. The guidelines for drawing Lewis structures are helpful, but chemists also rely on patterns of bonding in related molecules.

Hydrogen Compounds

Some common compounds and ions formed from second-period nonmetal elements and hydrogen are shown in Table 8.3. Their Lewis structures illustrate the fact that the Lewis symbol for an element is a useful guide in determining the number of bonds formed by the element. For example, if there is no charge, nitrogen has five valence electrons. Two electrons occur as a lone pair; the other three occur as unpaired electrons. To reach an octet, it is necessary to pair each of the unpaired electrons with an electron from another atom. Thus, N is predicted

Problem Solving Tip 8.1 Useful Ideas to Consider When Drawing Lewis Electron Dot Structures

- The octet rule is a useful guideline when drawing Lewis structures.
- Carbon forms four bonds (four single bonds; two single bonds and one double bond; two double bonds; or one single bond and one triple bond). In uncharged species, nitrogen forms three bonds and oxygen forms two bonds. Hydrogen typically forms only one bond to another atom.
- When multiple bonds are formed, both of the atoms involved are usually one of the following: C, N, and O. Oxygen has the ability to form multiple bonds with a variety of elements. Carbon forms many compounds having multiple bonds to another carbon or to N or O.
- Nonmetals may form single, double, and triple bonds but never quadruple bonds.
- Always account for single bonds and lone pairs before determining whether multiple bonds are present.
- Be alert for the possibility the molecule or ion you are working on is isoelectronic (page 358) with a species you have seen before.

TABLE 8.3 Lewis Structures of Common Hydrogen-Containing Molecules and Ions of Second-Period Elements

Group 4A		Group 5A		Group 6A		Group 7A	
CH_4 methane	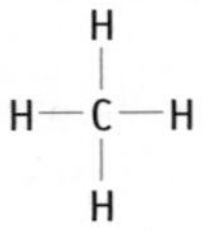	NH_3 ammonia	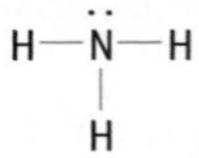	H_2O water	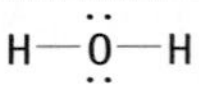	HF hydrogen fluoride	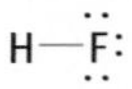
C_2H_6 ethane	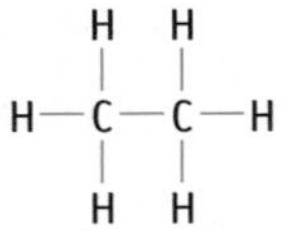	N_2H_4 hydrazine	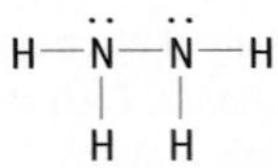	H_2O_2 hydrogen peroxide	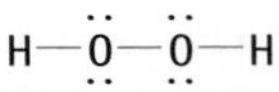		
C_2H_4 ethylene	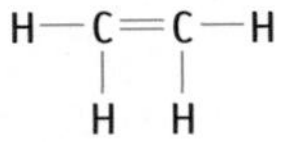	NH_4^+ ammonium ion	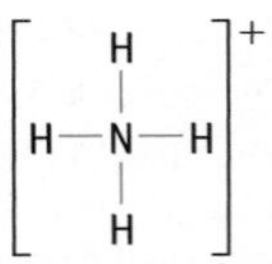	H_3O^+ hydronium ion	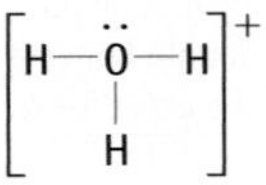		
C_2H_2 acetylene		NH_2^- amide ion	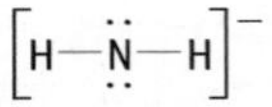	OH^- hydroxide ion	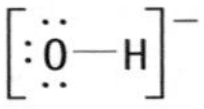		

to form three bonds in uncharged molecules, and this is indeed the case. Similarly, carbon is expected to form four bonds, oxygen two, and fluorine one.

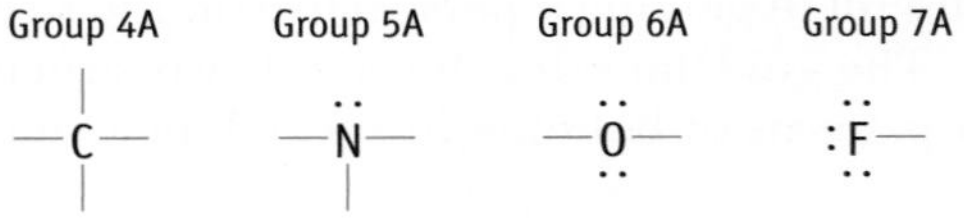

EXAMPLE 8.2 Predicting Lewis Structures

Problem Draw Lewis electron dot structures for CCl_4 and NF_3.

Strategy One way to answer this is to recognize that CCl_4 and NF_3 are similar to CH_4 and NH_3 (in Table 8.3), respectively, except that H atoms have been replaced by halogen atoms.

Solution Recall that carbon is expected to form four bonds and nitrogen three bonds to give an octet of electrons. In addition, halogen atoms have seven valence electrons, so both Cl and F can attain an octet by forming one covalent bond, just as hydrogen does.

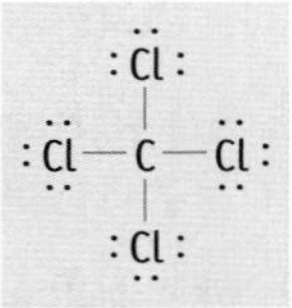

carbon tetrachloride

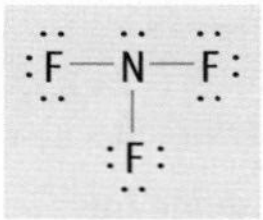

nitrogen trifluoride

As a check, count the number of valence electrons for each molecule, and verify that all are present.

CCl_4: Valence electrons = 4 for C + 4 × 7 (for Cl) = 32 electrons (16 pairs)

The structure shows eight electrons in single bonds and 24 electrons as lone pair electrons, for a total of 32 electrons. The structure is correct.

NF_3: Valence electrons = 5 for N + 3 × 7 (for F) = 26 electrons (13 pairs)

The structure shows six electrons in single bonds and 20 electrons as lone pair electrons, for a total of 26 electrons. The structure is correct.

> **EXERCISE 8.2 Predicting Lewis Structures**
>
> Predict Lewis structures for methanol, H_3COH, and hydroxylamine, H_2NOH. (Hint: The formulas of these compounds are written to guide you in choosing the correct arrangement of atoms.)

■ **Lewis Structures for Anions of Oxoacids** Stuctures for oxoanions such as PO_4^{3-}, SO_4^{2-}, and ClO_4^- are sometimes drawn with multiple bonds between the central atom and oxygen. Theory suggests that this does not accurately represent the bonding in these species, and that structures obeying the octet rule are more appropriate. See L. Suidan, J. K. Badenhoop, E. D. Glendening, and F. Weinhold, *Journal of Chemical Education*, Vol. 72, pages 583–585, 1995.

Oxoacids and Their Anions

Lewis structures of common acids and their anions are illustrated in Table 8.4. In the absence of water, these acids are covalently bonded molecular compounds, a conclusion that we should draw because all elements in the formula are nonmetals. (Nitric acid, for example, has properties that we associate with a covalent compound: it is a colorless liquid with a boiling point of 83 °C.) In aqueous solution, however, HNO_3, H_2SO_4, and $HClO_4$ are ionized to give a hydronium ion and the appropriate anion. A Lewis structure for the nitrate ion, for example, can be created using the guidelines on page 353, and the result is a structure with two N—O single bonds and one N═O double bond. To form nitric acid from the nitrate ion, a hydrogen ion is attached to one of the O atoms that has a single bond to the central N.

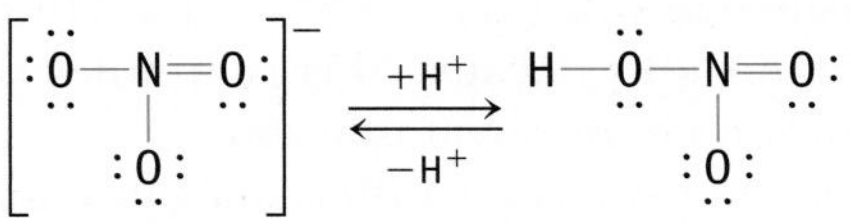

nitrate ion nitric acid

A characteristic property of acids in aqueous solution is their ability to donate a hydrogen ion (H^+, which combines with water to give the hydronium ion). The NO_3^- anion is formed when the acid, HNO_3, loses a hydrogen ion. The H^+ ion

TABLE 8.4 Lewis Structures of Common Oxoacids and Their Anions

HNO_3 nitric acid	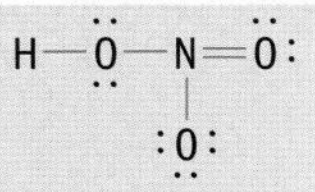
NO_3^- nitrate ion	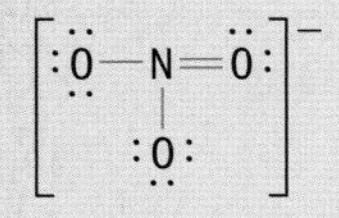
H_3PO_4 phosphoric acid	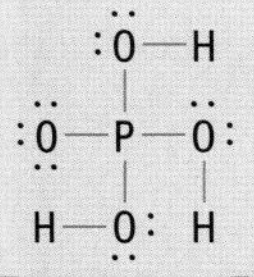
PO_4^{3-} phosphate ion	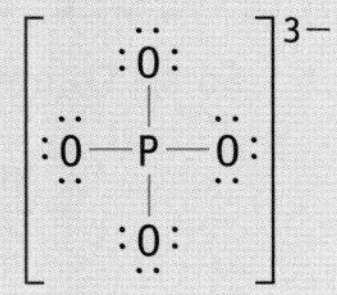
H_2SO_4 sulfuric acid	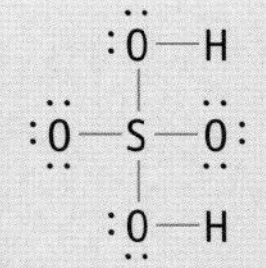
HSO_4^- hydrogen sulfate ion	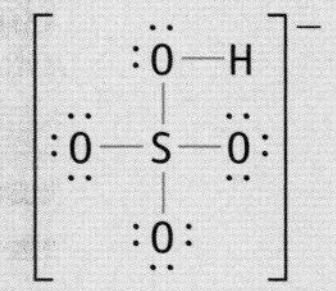
$HClO_4$ perchloric acid	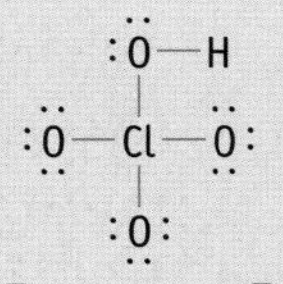
ClO_4^- perchlorate ion	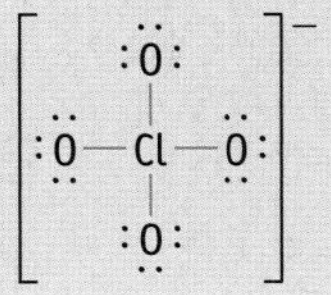
HOCl hypochlorous acid	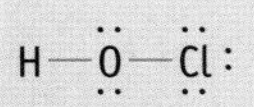
OCl^- hypochlorite ion	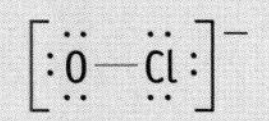
SO_4^{2-} sulfate ion	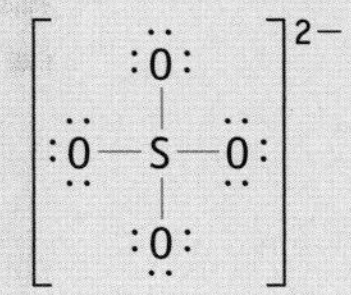

separates from the acid by breaking the H—O bond, the electrons of the bond staying with the O atom. As a result, HNO_3 and NO_3^- have the same number of electrons, 24, and their structures are closely related.

> **EXERCISE 8.3 Lewis Structures of Acids and Their Anions**
>
> Draw a Lewis structure for the anion $H_2PO_4^-$, derived from phosphoric acid.

■ **Isoelectronic and Isostructural** The term **isostructural** is often used in conjunction with isoelectronic species. Species that are isostructural have the same structure. For example, the PO_4^{3-}, SO_4^{2-}, and ClO_4^- ions in Table 8.5 all have four oxygens bonded to the central atom. In addition, they are isoelectronic in that all have 32 valence electrons.

Isoelectronic Species

The species NO^+, N_2, CO, and CN^- are similar in that they each have two atoms and the same total number of valence electrons, 10, which leads to the same Lewis structure for each molecule or ion. The two atoms in each are linked with a triple bond. With three bonding pairs and one lone pair, each atom thus has an octet of electrons.

$$[:N{\equiv}O:]^+ \qquad :N{\equiv}N: \qquad :C{\equiv}O: \qquad [:C{\equiv}N:]^-$$

Molecules and ions having the *same number of valence electrons and the same Lewis structures* are said to be **isoelectronic** (Table 8.5). You will find it helpful to recognize isoelectronic molecules and ions because this is another way to see relationships in bonding among common chemical substances.

There are similarities and important differences in chemical properties of isoelectronic species. For example, both carbon monoxide, CO, and cyanide ion, CN^-, are very toxic, which results from the fact that they can bind to the iron of hemoglobin in blood and block the uptake of oxygen. They are different, though, in their acid–base chemistry. In aqueous solution, cyanide ion readily adds H^+ to form hydrogen cyanide, whereas CO does not protonate in water.

> **EXERCISE 8.4 Identifying Isoelectronic Species**
>
> **(a)** Is the acetylide ion, C_2^{2-}, isoelectronic with N_2?
>
> **(b)** Identify a common molecular (uncharged) species that is isoelectronic with nitrite ion, NO_2^-. Identify a common ion that is isoelectronic with HF.

TABLE 8.5 Some Common Isoelectronic Molecules and Ions

Formulas	Representative Lewis Structure	Formulas	Representative Lewis Structure
BH_4^-, CH_4, NH_4^+	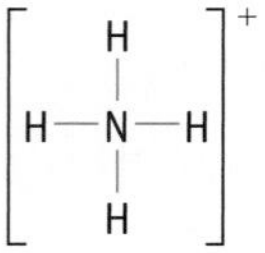	CO_3^{2-}, NO_3^-	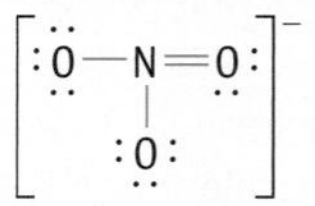
NH_3, H_3O^+	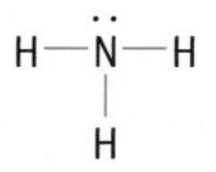	PO_4^{3-}, SO_4^{2-}, ClO_4^-	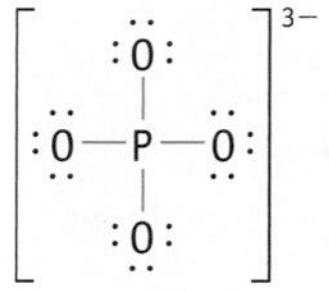
CO_2, OCN^-, SCN^-, N_2O, NO_2^+, OCS, CS_2	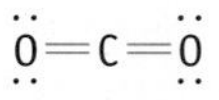		

8.3 Atom Formal Charges in Covalent Molecules and Ions

You have seen that Lewis structures show how electron pairs are placed in a covalently bonded species, whether it is a neutral molecule or a polyatomic ion. Now we turn to one of the consequences of the placement of electron pairs in this way: individual atoms can be negatively or positively charged or have no electric charge. The location of a positive or negative charge in a molecule or ion will influence, among other things, the atom at which a reaction occurs. For example, does a positive H^+ ion attach itself to the Cl or the O of ClO^-? Is the product HClO or HOCl? It is reasonable to expect H^+ to attach to the more negatively charged atom. We can predict this by evaluating atom formal charges in molecules and ions.

The **formal charge** is the charge on an atom in a molecule or polyatomic ion, and the sum of the formal charges for the atoms in a species equals the overall charge on the ion or is zero (for an uncharged molecule). The formal charge for an atom in a molecule or ion is calculated based on the Lewis structure of the molecule or ion, using Equation 8.1.

$$\text{Formal charge of an atom in a molecule or ion} = \text{group number of the atom} - [\text{LPE} + \tfrac{1}{2}(\text{BE})] \quad \textbf{(8.1)}$$

In this equation:

- The group number gives the number of valence electrons brought by a particular atom to the molecule or ion.
- LPE = number of lone pair electrons on an atom.
- BE = number of bonding electrons around an atom.

The term in square brackets is the number of electrons assigned by the Lewis structure to an atom in a molecule or ion. The difference between this term and the group number is the formal charge. An atom in a molecule or ion will be positive if it "contributes" more electrons to bonding than it "gets back." The atom's formal charge will be negative if the reverse is true.

There are two important assumptions in Equation 8.1. First, lone pairs are assumed to belong to the atom on which they reside in the Lewis structure. Second, bond pairs are assumed to be divided equally between the bonded atoms. (The factor of ½ divides the bonding electrons equally between the atoms linked by the bond.)

The sum of the formal charges on the atoms in a molecule or ion always equals the net charge on the molecule or ion. Consider the hypochlorite ion. Oxygen is in Group 6A and so has six valence electrons. However, oxygen can lay claim to seven electrons (six lone pair electrons and one bonding electron), and so the atom has a formal charge of −1. The O atom has "formally" gained an electron as part of the ion.

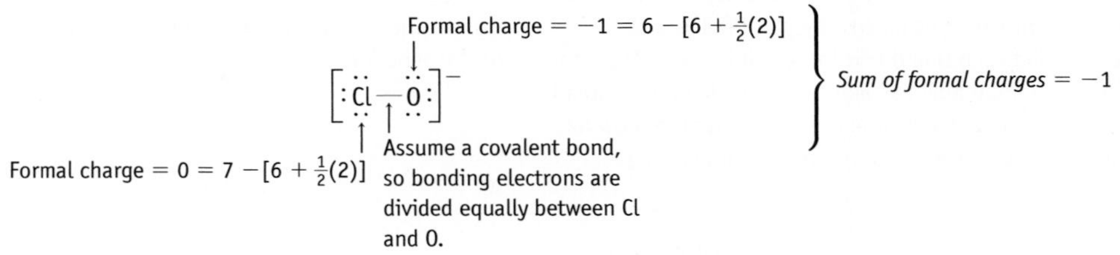

The formal charge on the Cl atom in ClO^- is zero. So we have −1 for oxygen and 0 for chlorine, and the sum of these equals the net charge of −1 for the ion. An important conclusion we can draw from the formal charges in ClO^- is that, if an

H^+ ion approaches the ion, it should attach itself to the negatively charged O atom to give hypochlorous acid, HOCl.

Chemistry Now™

Sign in at **www.thomsonedu.com/login** and go to Chapter 8 Contents to see Screen 8.11 for practice **determining formal charge.**

■ **$HClO_x$ Acids and Formal Charge** Both ClO^- and ClO_3^- ions attract a proton to give the corresponding acid, HClO and $HClO_3$. In all of the $HClO_x$ acids, the H^+ ion is attached to an O atom, owing to the negative formal charge on that atom. See Table 8.4.

EXAMPLE 8.3 Calculating Formal Charges

Problem Calculate formal charges for the atoms of the ClO_3^- ion.

Strategy The first step is always to write the Lewis structure for the molecule or ion. (The Lewis structure for the ClO_3^- ion is in Example 8.1.) Then Equation 8.1 can be used to calculate the formal charges.

Solution

Formal charge $= -1 = 6 - [6 + \frac{1}{2}(2)]$

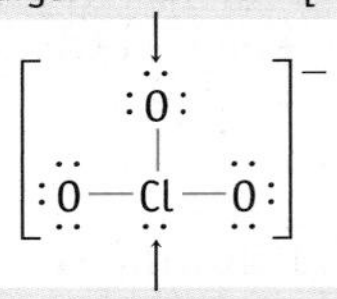

Formal charge $= +2 = 7 - [2 + \frac{1}{2}(6)]$

The formal charge on each O atom is -1, whereas for the Cl atom it is $+2$. The sum of the atom's formal charges is the charge on the ion, which is -1 for ClO_3^-.

EXERCISE 8.5 Calculating Formal Charges

Calculate formal charges on each atom in **(a)** CN^- and **(b)** SO_3^{2-}.

A Closer Look: Comparing Formal Charge and Oxidation Number

In Chapter 3, you learned to calculate the oxidation number of an atom as a way to tell if a reaction involves oxidation and reduction. Are an atom's oxidation number and its formal charge related? To answer this question, let us look at the hydroxide ion. The formal charges are -1 on the O atom and 0 on the H atom. Recall that these formal charges are calculated assuming the O—H bond electrons are shared equally in an O—H covalent bond.

Formal charge $= -1 = 6 - [6 + \frac{1}{2}(2)]$

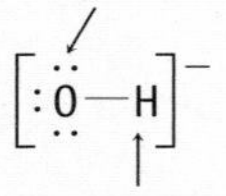

Sum of formal charges $= -1$

Formal charge $= 0 = 1 - [0 + \frac{1}{2}(2)]$

In contrast, in Chapter 3 (page 144), you learned that O has an oxidation number of -2 and H has a number of $+1$. Oxidation numbers are determined by assuming that the bond between a pair of atoms is ionic, not covalent. For OH^- this means the pair of electrons between O and H is located fully on O. Thus, the O atom now has eight valence electrons instead of six and a charge of -2. The H atom now has no valence electrons and a charge of $+1$.

Oxidation number $= -2$

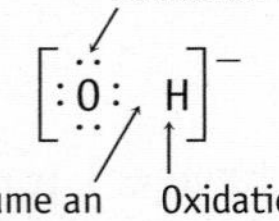

Sum of oxidation numbers $= -1$

Assume an ionic bond

Oxidation number $= +1$

Formal charges and oxidation numbers are calculated using different assumptions. Both are useful, but for different purposes. Oxidation numbers allow us to follow changes in redox reactions. Formal charges provide insight into atom charges in molecules and polyatomic ions.

8.4 Resonance

Ozone, O_3, an unstable, blue, diamagnetic gas with a characteristic pungent odor, protects the earth and its inhabitants from intense ultraviolet radiation from the sun. An important feature of its structure is that the two oxygen–oxygen bonds are the same length, which suggests that the two bonds are equivalent. That is, equal O—O bond lengths imply an equal number of bond pairs in each O—O bond. Using the guidelines for drawing Lewis structures, however, you might come to a different conclusion. There are two possible ways of writing the Lewis structure for the molecule:

Alternative Ways of Drawing the Ozone Structure

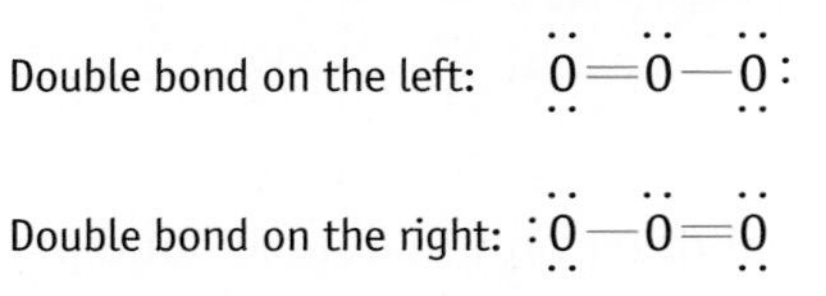

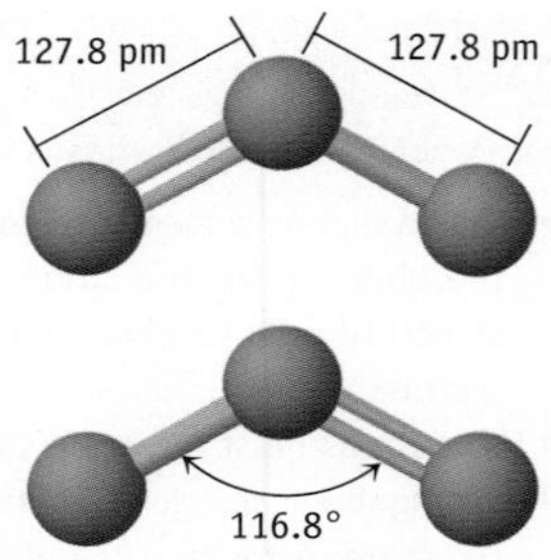

Ozone, O_3, is a bent molecule with oxygen–oxygen bonds of the same length.

These structures are equivalent in that each has a double bond on one side of the central oxygen atom and a single bond on the other side. If either were the actual structure of ozone, one bond (O═O) should be shorter than the other (O—O). The actual structure of ozone shows this is not the case. The inescapable conclusion is that these Lewis structures do not correctly represent the bonding in ozone.

Linus Pauling (1901–1994) proposed the theory of **resonance** to solve the problem. *Resonance structures are used to represent bonding in a molecule or ion when a single Lewis structure fails to describe accurately the actual electronic structure.* The alternative structures shown for ozone are called **resonance structures.** They have identical patterns of bonding and equal energy. The actual structure of this molecule is a *composite,* or **resonance hybrid,** of the equivalent resonance structures. In this hybrid, the bonds between the oxygens are between a single bond and a double bond in length, in this case corresponding to one and a half bonds. This is a reasonable conclusion because we see that the O—O bonds both have a length of 127.8 pm, intermediate between the average length of an O═O double bond (121 pm) and an O—O single bond (132 pm). Because we cannot accurately draw fractions of a bond, chemists draw the resonance structures and connect them with double-headed arrows (⟷) to indicate that the true structure is somewhere in between these extremes.

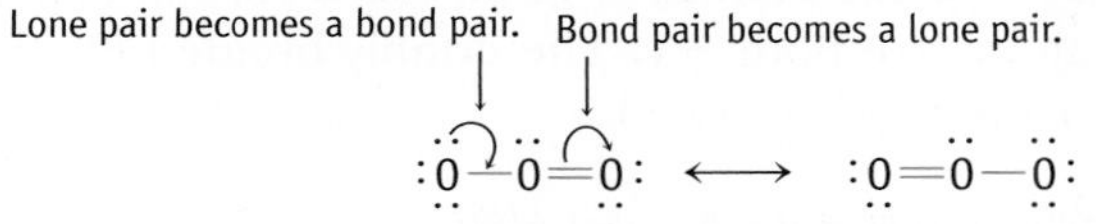

■ **Depicting Resonance Structures** The use of an arrow (⟷) as a symbol to link resonance structures and the name "resonance" are somewhat unfortunate. An arrow might seem to imply that a change is occurring, and the term resonance has the connotation of vibrating or alternating back and forth between different forms. Neither view is correct. Electron pairs are not actually moving from one place to another.

Benzene is the classic example of the use of resonance to represent a structure. The benzene molecule is a six-member ring of carbon atoms with six equivalent carbon–carbon bonds (and a hydrogen atom attached to each carbon atom). The carbon–carbon bonds are 139 pm long, intermediate between the average length of a C═C double bond (134 pm) and a C—C single bond (154 pm).

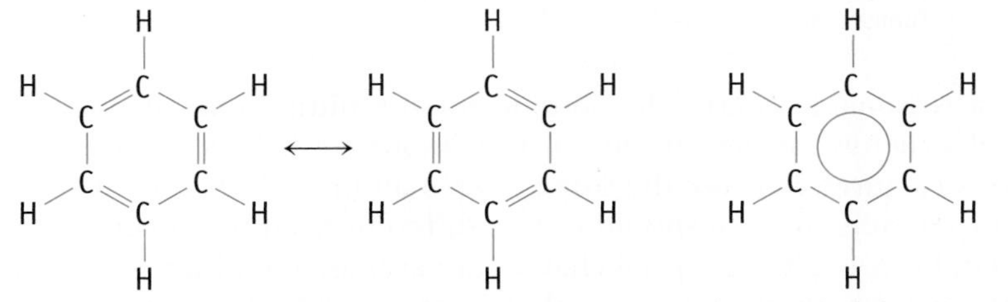

Resonance structures of benzene, C_6H_6

Abbreviated representation of resonance structures

Problem Solving Tip 8.2 Resonance Structures

- Resonance is a means of representing the bonding when a single Lewis structure fails to give an accurate picture.
- The atoms must have the same arrangement in each resonance structure. Attaching the atoms in a different fashion creates a different compound.
- Resonance structures differ only in the assignment of electron-pair positions, never atom positions.
- Resonance structures differ in the number of bond pairs between a given pair of atoms.
- Even though the formal process of converting one resonance structure to another seems to move electrons about, resonance is not meant to indicate the motion of electrons.
- The actual structure of a molecule is a composite or hybrid of the resonance structures.
- There will always be at least one multiple bond (double or triple) in each resonance structure.

Two resonance structures that differ only in double bond placement can be written for the molecule. A hybrid of these two structures, however, will lead to a molecule with six equivalent carbon–carbon bonds.

Let us apply the concepts of resonance to describe bonding in the carbonate ion, CO_3^{2-}, an anion with 24 valence electrons (12 pairs).

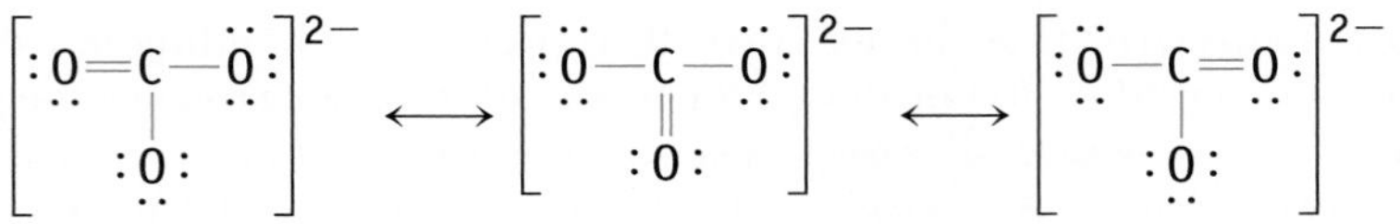

Three equivalent structures can be drawn for this ion, differing only in the location of the C=O double bond. This fits the classical situation for resonance, so it is appropriate to conclude that no single structure correctly describes this ion. Instead, the actual structure is the composite of the three structures, in good agreement with experimental results. In the CO_3^{2-} ion, all three carbon–oxygen bond distances are 129 pm, intermediate between C—O single bond (143 pm) and C=O double bond (122 pm) distances.

Formal charges can be calculated for each atom in the resonance structure for a molecule or ion. For example, using one of the resonance structures for the nitrate ion, we find that the central N atom has a formal charge of +1, and the singly bonded O atoms are both −1. The doubly bonded O atom has no charge. The net charge for the ion is thus −1.

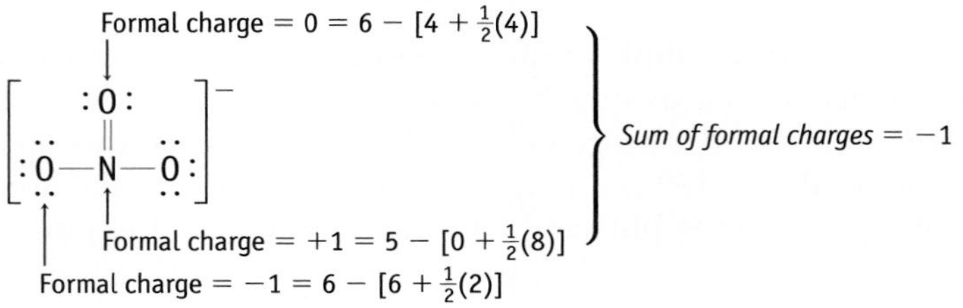

Is this a reasonable charge distribution for the nitrate ion? The answer is no. The actual structure of the nitrate ion is a resonance hybrid of three equivalent resonance structures. Because the three oxygen atoms in NO_3^- are equivalent, the charge on one oxygen atom should not be different from the other two. This can be resolved, however, if the formal charges are averaged to give a formal charge of $-\frac{2}{3}$ on the oxygen atoms. Summing the charges on the three oxygen atoms and the +1 charge on the nitrogen atom then gives −1, the charge on the ion.

In the resonance structures for O_3, CO_3^{2-}, and NO_3^-, for example, all the possible resonance structures are equally likely; they are "equivalent" structures. The molecule or ion therefore has a symmetrical distribution of electrons over all the atoms involved—that is, its electronic structure consists of an equal "mixture," or "hybrid," of the resonance structures.

Chemistry Now™

Sign in at **www.thomsonedu.com/login** and go to Chapter 8 Contents to see Screen 8.7 for a tutorial on **drawing resonance structures.**

EXAMPLE 8.4 Drawing Resonance Structures

Problem Draw resonance structures for the nitrite ion, NO_2^-. Are the N—O bonds single, double, or intermediate in value? What are the formal charges on the N and O atoms?

Strategy Draw the Lewis structure in the usual manner. If multiple bonds are required, resonance structures may exist. This will be the case if the octet of an atom can be completed by using an electron pair from more than one terminal atom to form a multiple bond. Bonds to the central atom cannot then be "pure" single or double bonds but rather are somewhere between the two.

Solution Nitrogen is the central atom in the nitrite ion, which has a total of 18 valence electrons (nine pairs).

Valence electrons = 5 (for the N atom) + 12 (6 for each O atom) + 1 (for negative charge)

After forming N—O single bonds, and distributing lone pairs on the terminal O atoms, a pair remains, which is placed on the central N atom.

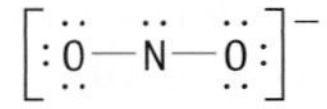

To complete the octet of electrons about the N atom, form an N=O double bond.

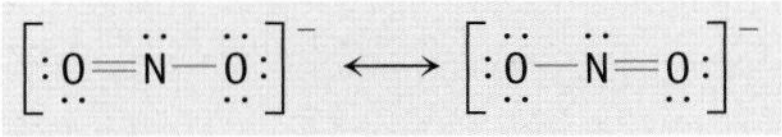

Because there are two ways to do this, two equivalent structures can be drawn, and the actual structure must be a resonance hybrid of these two structures. The nitrogen–oxygen bonds are neither single nor double bonds but have an intermediate value.

Taking one of the resonance structures, we find the formal charge for the N atom is 0. The charge on one O atom is 0 and −1 for the other O atom. Because the two resonance structures are of equal importance, however, the net formal charge on each O atom is −½.

Formal charge = $0 = 6 - [4 + \frac{1}{2}(4)]$

Formal charge = $-1 = 6 - [6 + \frac{1}{2}(2)]$

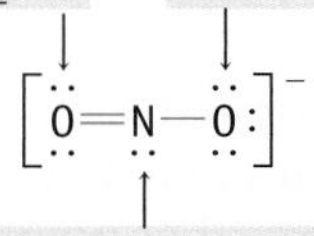

Formal charge = $0 = 5 - [2 + \frac{1}{2}(6)]$

EXERCISE 8.6 Drawing Resonance Structures

Draw resonance structures for the bicarbonate ion, HCO_3^-.

(a) Does HCO_3^- have the same number of resonance structures as the CO_3^{2-} ion?

(b) What are the formal charges on the O and C atoms in HCO_3^-? What is the average formal charge on the O atoms? Compare this with the O atoms in CO_3^{2-}.

(c) What do formal charges predict about the point of attachment of the H atom in HCO_3^-?

8.5 Exceptions to the Octet Rule

Although the vast majority of molecular compounds and ions obey the octet rule, there are exceptions. These include molecules and ions that have fewer than four pairs of electrons on a central atom, those that have more than four pairs, and those that have an odd number of electrons.

Compounds in Which an Atom Has Fewer Than Eight Valence Electrons

Boron, a metalloid in Group 3A, has three valence electrons and so is expected to form three covalent bonds with other nonmetallic elements. This results in a valence shell for boron in its compounds with only six electrons, two short of an octet. Many boron compounds of this type are known, including such common compounds as boric acid ($B(OH)_3$), borax ($Na_2B_4O_5(OH)_4 \cdot 8\ H_2O$) (Figure 8.2), and the boron trihalides (BF_3, BCl_3, BBr_3, and BI_3).

Charles D. Winters

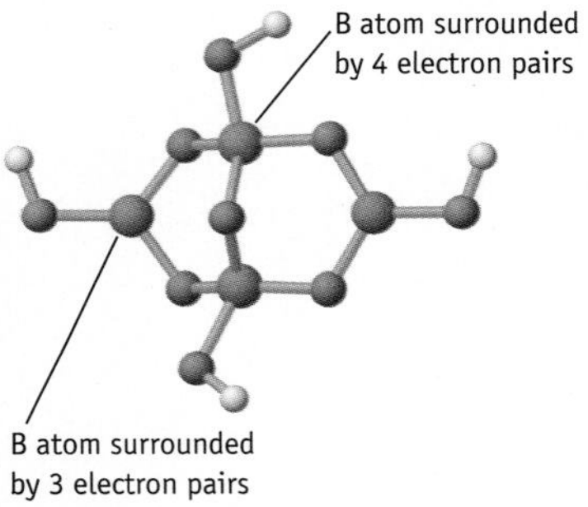

FIGURE 8.2 The anion in Borax. Borax is a common mineral, which is used in soaps and contains an interesting anion, $B_4O_5(OH)_4^{2-}$. The anion has two B atoms surrounded by four electron pairs, and two B atoms surrounded by only three pairs.

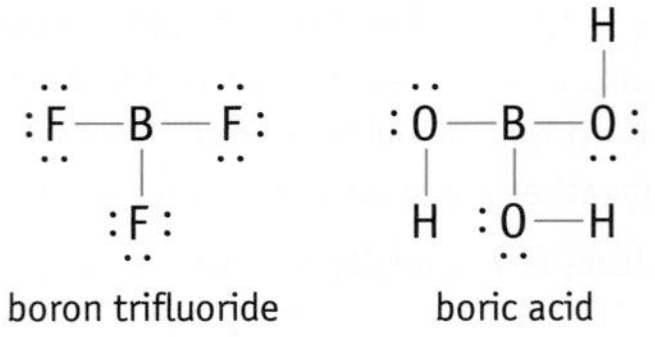

Boron compounds such as BF_3 that are two electrons short of an octet are quite reactive. The boron atom can accommodate a fourth electron pair when that pair is provided by another atom, and molecules or ions with lone pairs can fulfill this role. Ammonia, for example, reacts with BF_3 to form $H_3N{\rightarrow}BF_3$.

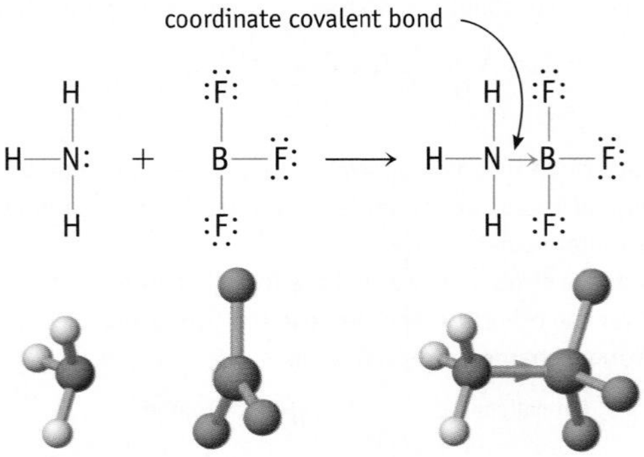

If a bonding pair of electrons originates on one of the bonded atoms, the bond is called a **coordinate covalent bond.** In Lewis structures, a coordinate covalent bond is often designated by an arrow that points away from the atom donating the electron pair.

Compounds in Which an Atom Has More Than Eight Valence Electrons

Elements in the third or higher periods often form compounds and ions in which the central element is surrounded by more than four valence electron pairs (Table 8.6). With most compounds and ions in this category, the central atom is bonded to fluorine, chlorine, or oxygen.

It is often obvious from the formula of a compound that an octet around an atom has been exceeded. As an example, consider sulfur hexafluoride, SF_6, a gas

TABLE 8.6 Lewis Structures in Which the Central Atom Exceeds an Octet

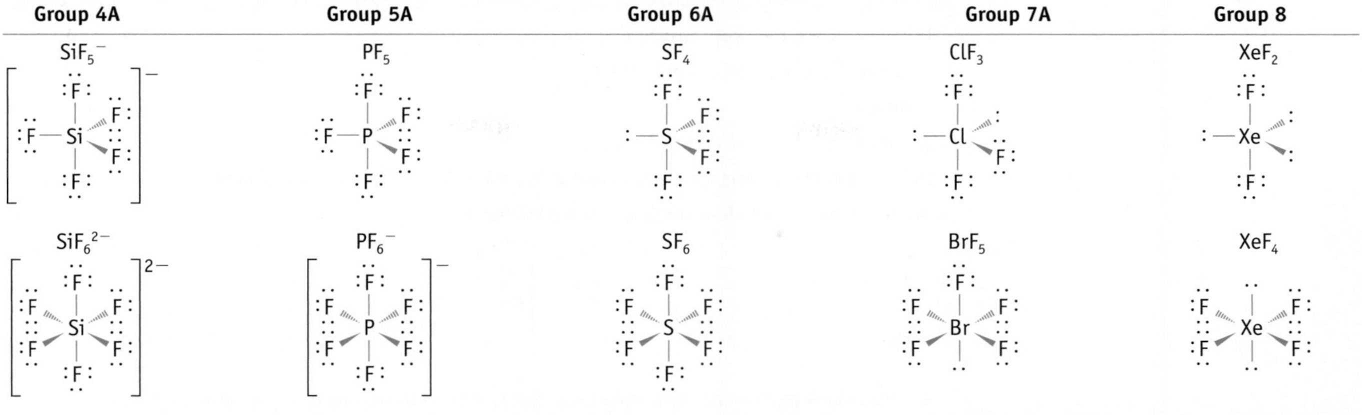

formed by the reaction of sulfur and excess fluorine. Sulfur is the central atom in this compound, and fluorine typically bonds to only one other atom with a single electron pair bond (as in HF and CF_4). Six S—F bonds are required in SF_6, meaning there will be six electron pairs in the valence shell of the sulfur atom.

If there are more than four groups bonded to a central atom, this is a reliable signal that there are more than eight electrons around a central atom. But be careful—the central atom octet can also be exceeded with four or fewer atoms bonded to the central atom. Consider three examples from Table 8.6: the central atom in SF_4, ClF_3, and XeF_2 has five electron pairs in its valence shell.

A useful observation is that *only elements of the third and higher periods in the periodic table may form compounds and ions in which an octet is exceeded.* Second-period elements (B, C, N, O, F) are restricted to a maximum of eight electrons in their compounds. For example, nitrogen forms compounds and ions such as NH_3, NH_4^+, and NF_3, but NF_5 is unknown. Phosphorus, the third-period element just below nitrogen in the periodic table, forms many compounds similar to nitrogen (PH_3, PH_4^+, PF_3), but it also readily accommodates five or six valence electron pairs in compounds such as PF_5 or in ions such as PF_6^-. Arsenic, antimony, and bismuth, the elements below phosphorus in Group 5A, resemble phosphorus in their behavior.

The usual explanation for the contrasting behavior of second- and third-period elements centers on the number of orbitals in the valence shell of an atom. Second-period elements have four valence orbitals (one 2*s* and three 2*p* orbitals). Two electrons per orbital result in a total of eight electrons being accommodated around an atom. For elements in the third and higher periods, the *d* orbitals in the outer shell are traditionally included among valence orbitals for the elements. Thus, for phosphorus, the 3*d* orbitals are included with the 3*s* and 3*p* orbitals as valence orbitals. The extra orbitals provide the element with an opportunity to accommodate up to 12 electrons.

Chemistry Now™

Sign in at **www.thomsonedu.com/login** and go to Chapter 8 Contents to see Screen 8.8 for a tutorial on **identifying electron-deficient compounds.**

■ **Xenon Compounds** Compounds of xenon are among the more interesting entries in Table 8.6 because noble gas compounds were not discovered until the early 1960s. One of the more intriguing compounds is XeF_2, in part because of the simplicity of its synthesis. Xenon difluoride can be made by placing a flask containing xenon gas and fluorine gas in the sunlight. After several weeks, crystals of colorless XeF_2 are found in the flask (see page 404).

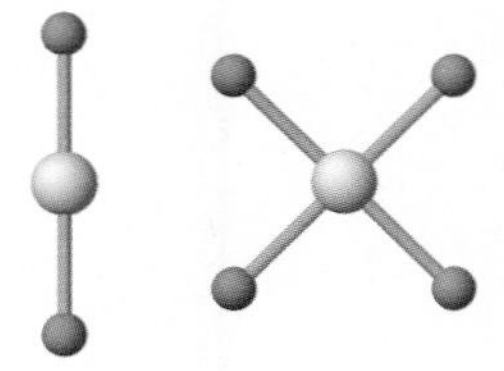

XeF_2 XeF_4

EXAMPLE 8.5 Lewis Structures in Which the Central Atom Has More Than Eight Electrons

Problem Sketch the Lewis structure of the $[ClF_4]^-$ ion.

Strategy Use the guidelines on page 353.

Solution

1. The Cl atom is the central atom.
2. This ion has 36 valence electrons [= 7 (for Cl) + 4 × 7 (for F) + 1 (for ion charge)] or 18 pairs.
3. Draw the ion with four single covalent Cl–F bonds.

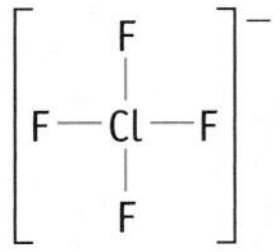

4. Place lone pairs on the terminal atoms. Because two electron pairs remain after placing lone pairs on the four F atoms, and because we know that Cl can accommodate more than four pairs, these two pairs are placed on the central Cl atom.

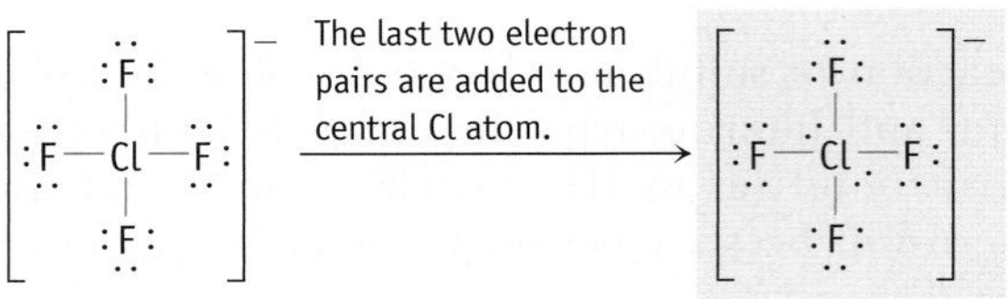

EXERCISE 8.7 Lewis Structures in Which the Central Atom Has More Than Eight Electrons

Sketch the Lewis structures for $[ClF_2]^+$ and $[ClF_2]^-$. How many lone pairs and bond pairs surround the Cl atom in each ion?

Molecules with an Odd Number of Electrons

Two nitrogen oxides—NO, with 11 valence electrons, and NO_2, with 17 valence electrons—are among a very small group of stable molecules with an odd number of electrons. Because they have an odd number of electrons, it is impossible to draw a structure obeying the octet rule; at least one electron must be unpaired.

Even though NO_2 does not obey the octet rule, an electron dot structure can be written that approximates the bonding in the molecule. This Lewis structure places the unpaired electron on nitrogen. Two resonance structures show that the nitrogen–oxygen bonds are expected to be equivalent.

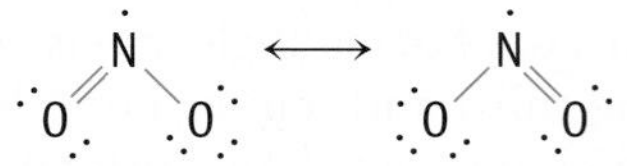

Experimental evidence for NO indicates that the bonding between N and O is intermediate between a double and a triple bond. It is not possible to write a Lewis structure for NO that is in accord with the properties of this substance, so a different theory is needed to understand bonding in this molecule. We shall return to compounds of this type when molecular orbital theory is introduced in Section 9.3.

The two nitrogen oxides, NO and NO_2, are members of a class of chemical substances called free radicals. **Free radicals** are chemical species—both atomic and molecular—with an unpaired electron. Free radicals are generally quite reac-

Case Study — The Importance of an Odd-Electron Molecule, NO

Small molecules such as H_2, O_2, H_2O, CO, and CO_2 are among the most important molecules commercially, environmentally, and biologically. Imagine the surprise of chemists and biologists when it was discovered a few years ago that nitrogen monoxide (nitric oxide, NO), which was widely considered toxic, also has an important biological role.

Nitric oxide is a colorless, paramagnetic gas that is moderately soluble in water. In the laboratory, it can be synthesized by the reduction of nitrite ion with iodide ion:

$$KNO_2(aq) + KI(aq) + H_2SO_4(aq) \rightarrow NO(g) + K_2SO_4(aq) + H_2O(\ell) + \tfrac{1}{2} I_2(aq)$$

The formation of NO from the elements is an unfavorable, energetically uphill reaction ($\Delta_f H° = 90.2$ kJ/mol). Nevertheless, small quantities of this compound form from nitrogen and oxygen at high temperatures. For example, conditions in an internal combustion engine are favorable for this to happen.

Nitric oxide reacts rapidly with O_2 to form the reddish-brown gas NO_2.

$$2\,NO(\text{colorless, g}) + O_2(g) \rightarrow 2\,NO_2(\text{brown, g})$$

The result is that compounds such as NO_2 and HNO_3 arising from reactions of NO with O_2 and H_2O are among the air pollutants produced by automobiles.

A few years ago, chemists learned that NO is synthesized in a biological process by animals as diverse as barnacles, fruit flies, horseshoe crabs, chickens, trout, and humans. Even more recently, chemists have found that NO is important in an astonishing range of physiological processes in humans and other animals. These include a role in neurotransmission, blood clotting, and blood pressure control as well as in the immune system's ability to kill tumor cells and intracellular parasites.

Questions: *Oxygen is needed by many living organisms, but some reactions with oxygen can lead to oxidative damage. One species that can produce damage in an organism is the superoxide ion, O_2^-. Fortunately, this ion is removed extremely rapidly by reaction with NO to produce the peroxynitrite ion, $ONOO^-$.*

1. Draw the Lewis structure for the ion.

2. Are there any multiple bonds in the ion?

3. Are there any resonance structures needed?

Answers to these questions are in Appendix Q.

Charles D. Winters

The colorless gas NO is bubbled into water from a high-pressure tank. When the gas emerges into the air, the NO reacts rapidly with O_2 to give brown NO_2 gas.

tive. Free atoms such as H and Cl, for example, are free radicals and readily combine with each other to give molecules such as H_2, Cl_2, and HCl.

Free radicals are involved in many reactions in the environment. For example, small amounts of NO are released from vehicle exhausts. The NO rapidly forms NO_2, which is even more harmful to human health and to plants. Exposure to NO_2 at concentrations of 50–100 parts per million can lead to significant inflammation of lung tissue. Nitrogen dioxide is also generated by natural processes. For example, when hay, which has a high level of nitrates, is stored in silos on farms, NO_2 can be generated as the hay ferments, and there have been reports of farm workers dying from exposure to this gas in the silo.

The two nitrogen oxides, NO and NO_2, are unique in that they can be isolated and neither has the extreme reactivity of most free radicals. When cooled, however, two NO_2 molecules join or "dimerize" to form colorless N_2O_4; the unpaired electrons combine to form an N—N bond in N_2O_4 (Figure 8.3).

8.6 Molecular Shapes

One reason for drawing Lewis electron dot structures is to be able to predict the three-dimensional geometry of molecules and ions. Because the physical and chemical properties of compounds are tied to their structures, the importance of this subject cannot be overstated.

FIGURE 8.3 Free radical chemistry. When cooled, the brown gas NO_2, a free radical, forms colorless N_2O_4, a molecule with an N—N single bond. The coupling of two free radicals is a common type of chemical reaction. Because identical free radicals come together, the product is called a dimer, and the process is called a dimerization. (Sign in to ChemistryNow, Screen 8.9, to see a video of this reaction.)

A flask of brown NO_2 gas in warm water

A flask of NO_2 gas in ice water

The **valence shell electron-pair repulsion (VSEPR)** model is a reliable method for predicting the shapes of covalent molecules and polyatomic ions. This model is based on the idea that *bond and lone electron pairs in the valence shell of an element repel each other and seek to be as far apart as possible.* The positions assumed by the valence electrons of an atom thus define the angles between bonds to surrounding atoms. VSEPR is remarkably successful in predicting structures of molecules and ions of main group elements. However, it is less effective (and seldom used) to predict structures of compounds containing transition metals.

■ **VSEPR Theory** The VSEPR theory was devised by Ronald J. Gillespie (1924–) and Ronald S. Nyholm (1917–1971).

To have a sense of how valence shell electron pairs repel and determine structure, blow up several balloons to a similar size. Imagine that each balloon represents an electron cloud. When two, three, four, five, or six balloons are tied together at a central point (representing the nucleus and core electrons of a central atom), the balloons naturally form the shapes shown in Figure 8.4. These geometric arrangements minimize interactions between the balloons.

Central Atoms Surrounded Only by Single-Bond Pairs

The simplest application of VSEPR theory is to molecules and ions in which all the electron pairs around the central atom are involved in single covalent bonds. Figure 8.5 illustrates the geometries predicted for molecules or ions with the general formulas AX_n, where A is the central atom and n is the number of X groups bonded to it.

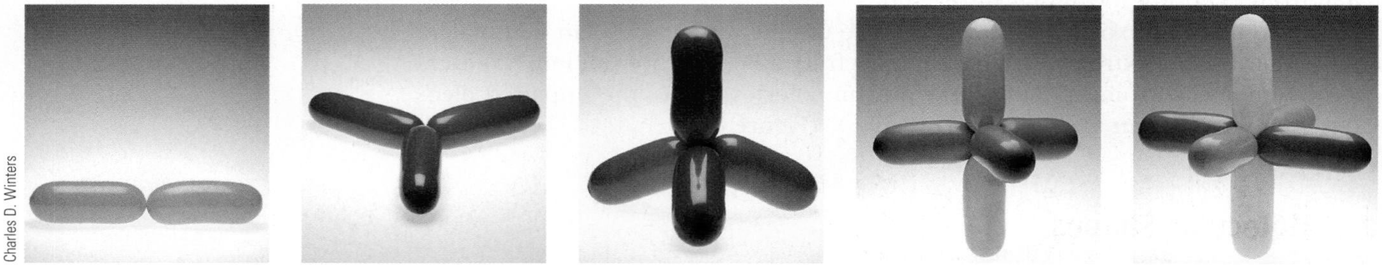

Linear | Trigonal planar | Tetrahedral | Trigonal bipyramidal | Octahedral

FIGURE 8.4 Balloon models of electron-pair geometries for two to six electron pairs. If two to six balloons of similar size and shape are tied together, they will naturally assume the arrangements shown. These pictures illustrate the predictions of VSEPR.

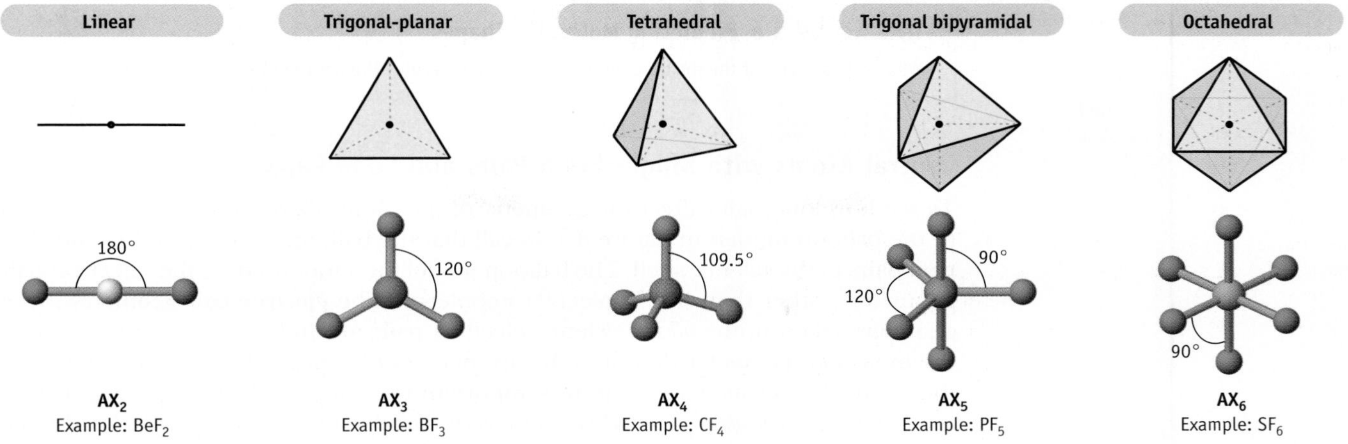

Active Figure 8.5 **Various geometries predicted by VSEPR.** Geometries predicted by VSEPR for molecules that contain only single covalent bonds around the central atom.

Chemistry Now™ Sign in at www.thomsonedu.com/login and go to the Chapter Contents menu to explore an interactive version of this figure accompanied by an exercise.

The linear geometry for two bond pairs and the trigonal-planar geometry for three bond pairs involve a central atom that does not have an octet of electrons (see Section 8.5). The central atom in a tetrahedral molecule obeys the octet rule with four bond pairs. The central atoms in trigonal-bipyramidal and octahedral molecules have five and six bonding pairs, respectively, and are expected only when the central atom is an element in Period 3 or higher of the periodic table (► page 372).

Chemistry Now™

Sign in at **www.thomsonedu.com/login** and go to Chapter 8 Contents to see:

- Screen 8.13 for an animation of **the electron-pair geometries** and on **identifying geometries**
- Screen 8.14 for practice **predicting molecular geometry**

EXAMPLE 8.6 Predicting Molecular Shapes

Problem Predict the shape of silicon tetrachloride, $SiCl_4$.

Strategy The first step is to draw the Lewis structure. The Lewis structure does not need to be drawn in any particular way because its purpose is only to describe the number of bonds around an atom and to determine if there are any lone pairs. The number of bond and lone pairs of electrons around the central atom determines the molecular shape (Figure 8.5).

Solution The Lewis structure of $SiCl_4$ has four electron pairs, all of them bond pairs, around the central Si atom. Therefore, a tetrahedral structure is predicted for the $SiCl_4$ molecule, with Cl—Si—Cl bond angles of 109.5°. This agrees with the actual structure for $SiCl_4$.

Lewis structure   Molecular geometry

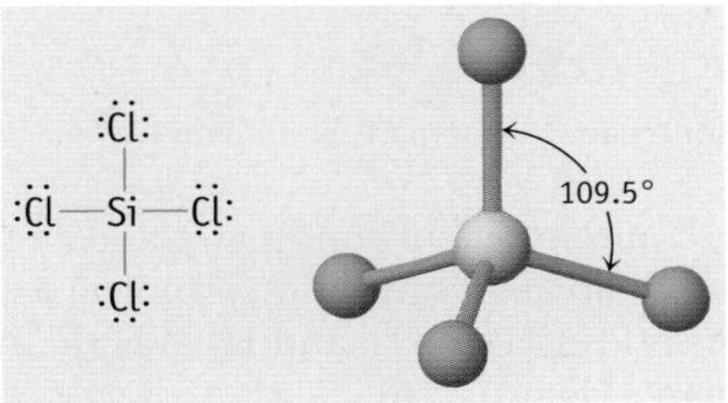

■ **Lewis Structures and Molecular Shapes** Drawing the Lewis structure is the first step in determining the shape of a molecule or ion.

EXERCISE 8.8 Predicting Molecular Shapes

What is the shape of the dichloromethane (CH_2Cl_2) molecule? Predict the Cl—C—Cl bond angle.

Central Atoms with Single-Bond Pairs and Lone Pairs

To see how lone pairs affect the geometry of the molecule or polyatomic ion, return to the balloon models in Figure 8.4. Recall that the balloons represented *all* the electron pairs in the valence shell. The balloon model therefore predicts the "electron-pair geometry" rather than the "molecular geometry." The **electron-pair geometry** is the geometry taken up by *all* the valence electron pairs around a central atom, whereas the **molecular geometry** describes the arrangement in space of the central atom and the atoms directly attached to it. It is important to recognize that *lone pairs of electrons on the central atom occupy spatial positions, even though their location is not included in the verbal description of the shape of the molecule or ion.*

Let us use the VSEPR model to predict the molecular geometry and bond angles in the NH_3 molecule. On drawing the Lewis structure, we see there are four pairs of electrons in the nitrogen valence shell, three bond pairs, and one lone pair. Thus, the predicted *electron-pair geometry* is tetrahedral. The *molecular geometry*, however, is said to be *trigonal pyramidal* because that describes the location of the atoms. The nitrogen atom is at the apex of the pyramid, and the three hydrogen atoms form the trigonal base.

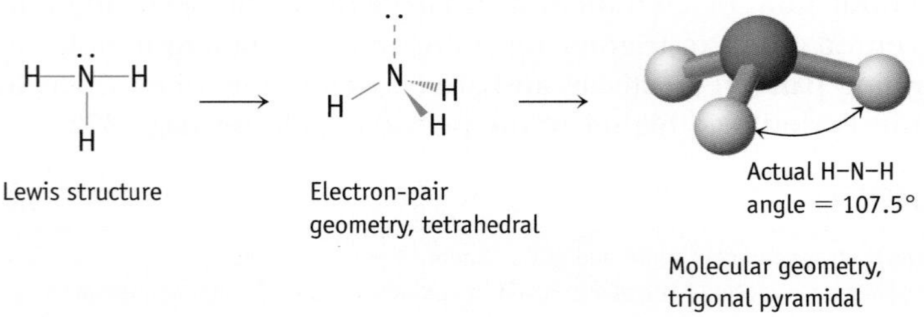

Effect of Lone Pairs on Bond Angles

Because the electron-pair geometry in NH_3 is tetrahedral, we would expect the H—N—H bond angle to be 109.5°. However, the experimentally determined bond angles in NH_3 are 107.5°, and the H—O—H angle in water is smaller still (104.5°) (Figure 8.6). These angles are close to the tetrahedral angle but not exactly that value. This highlights the fact that VSEPR is not an accurate model; it can only predict the approximate geometry. Small variations in geometry (e.g., bond angles a few degrees different from predicted) are quite common and often arise because there is a difference between the spatial requirements of lone pairs and bond pairs. Lone pairs of electrons seem to occupy a larger volume than bonding pairs, and the increased volume of lone pairs causes bond pairs to squeeze closer together. In general, the relative strengths of repulsions are in the order

Lone pair–lone pair > lone pair–bond pair > bond pair–bond pair

The different spatial requirements of lone pairs and bond pairs can be used to predict variations in the bond angles in series of molecules. For example, the bond angles decrease in the series CH_4, NH_3, and H_2O as the number of lone pairs on the central atom increases (Figure 8.6).

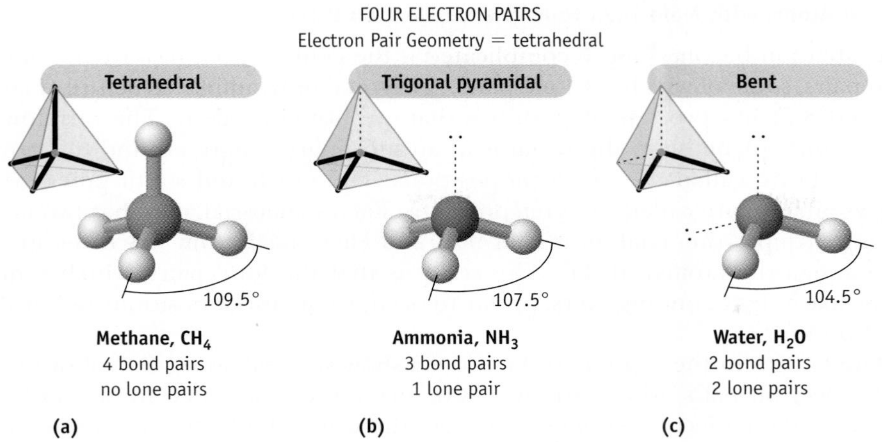

FIGURE 8.6 The molecular geometries of methane, ammonia, and water. All have four electron pairs around the central atom, so all have a tetrahedral electron-pair geometry. (a) Methane has four bond pairs and so has a tetrahedral molecular shape. (b) Ammonia has three bond pairs and one lone pair, so it has a trigonal-pyramidal molecular shape. (c) Water has two bond pairs and two lone pairs, so it has a bent, or angular, molecular shape. The decrease in bond angles in the series can be explained by the fact that the lone pairs have a larger spatial requirement than the bond pairs.

EXAMPLE 8.7 Finding the Shapes of Molecules

Problem What are the shapes of the ions H_3O^+ and ClF_2^+?

Strategy Draw the Lewis structures for each ion. Count the number of lone and bond pairs around the central atom. Use Figure 8.5 to decide on the electron-pair geometry. Finally, the location of the atoms in the ion—which is determined by the bond and lone pairs—gives the geometry of the ion.

Solution

(a) The Lewis structure of the hydronium ion, H_3O^+, shows that the oxygen atom is surrounded by four electron pairs, so the electron-pair geometry is tetrahedral.

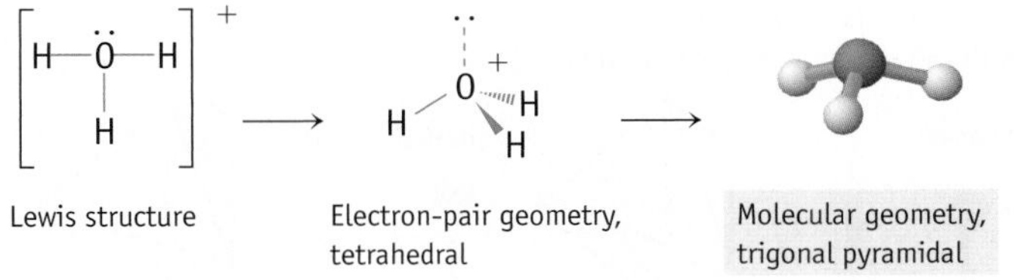

Because three of the four pairs are used to bond terminal atoms, the central O atom and the three H atoms form a trigonal-pyramidal molecular shape like that of NH_3.

(b) Chlorine is the central atom in ClF_2^+. It is surrounded by four electron pairs, so the electron-pair geometry around chlorine is tetrahedral. Because only two of the four pairs are bonding pairs, the ion has a bent geometry.

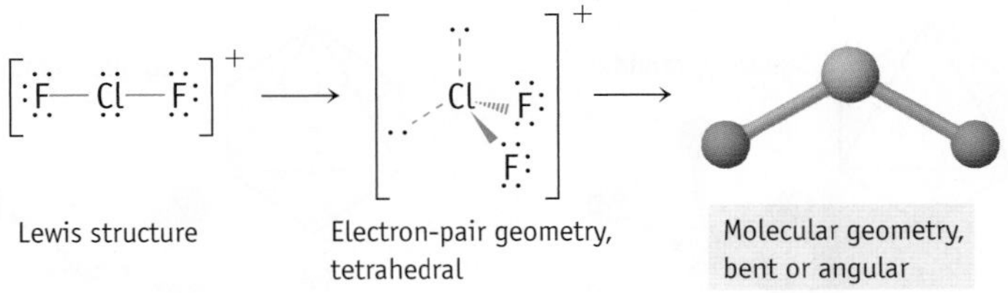

EXERCISE 8.9 VSEPR and Molecular Shape

Give the electron-pair geometry and molecular shape for BF_3 and BF_4^-. What is the effect on the molecular geometry of adding an F^- ion to BF_3 to give BF_4^-?

■ **"Energized Water" with a Bond Angle of 114°!** There are many dubious products sold over the internet, and one of them claims that water is "energized" by increasing its bond angle. One advertisement said that in the past water had a bond angle of a healthy 110°, but now it is "wimpy" and unhealthy with an angle of only 104°. Further, it is claimed that distilled water only has a bond angle of 101° and is biologically dead. To cure this problem, you can buy a costly machine that "energizes" water and causes a bond angle increase to as much as 114°. Now, it is also claimed, this water has enough energy to destroy pathogens. The old circus master, P. T. Barnum, once said there is a sucker born every minute.

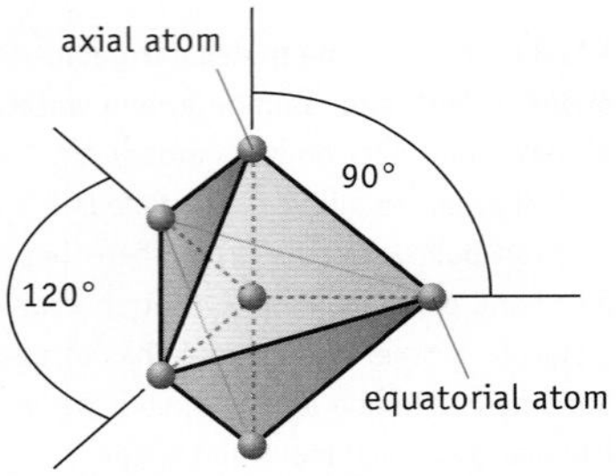

FIGURE 8.7 The trigonal bipyramid showing the axial and equatorial atoms. The angles between atoms in the equator are 120°. The angles between equatorial and axial atoms are 90°.

Central Atoms with More Than Four Valence Electron Pairs

The situation becomes more complicated if the central atom has five or six electron pairs, some of which are lone pairs. A trigonal-bipyramidal structure (Figures 8.5 and 8.7) has two sets of positions that are not equivalent. The positions in the trigonal plane lie in the equator of an imaginary sphere around the central atom and are called the *equatorial* positions. The north and south poles in this representation are called the *axial* positions. Each equatorial atom has two neighboring groups (the axial atoms) at 90°, and each axial atom has three groups (the equatorial atoms) at 90°. The result is that the lone pairs, which require more space than bonding pairs, prefer to occupy equatorial positions rather than axial positions.

The entries in the top line of Figure 8.8 show species having a total of five valence electron pairs, with zero, one, two, and three lone pairs. In SF_4, with one lone pair, the molecule assumes a "seesaw" shape with the lone pair in one of the equatorial positions. The ClF_3 molecule has three bond pairs and two lone pairs. The two lone pairs in ClF_3 are in equatorial positions; two bond pairs are axial, and the third is in the equatorial plane, so the molecular geometry is T-shaped. The third molecule shown is XeF_2. Here, all three equatorial positions are occupied by lone pairs so the molecular geometry is linear.

The geometry assumed by six electron pairs is octahedral (see Figure 8.8), and all the angles at adjacent positions are 90°. Unlike the trigonal bipyramid, the octahedron has no distinct axial and equatorial positions; all positions are the same. Therefore, if the molecule has one lone pair, as in BrF_5, it makes no difference which position it occupies. The lone pair is often drawn in the top or bottom posi-

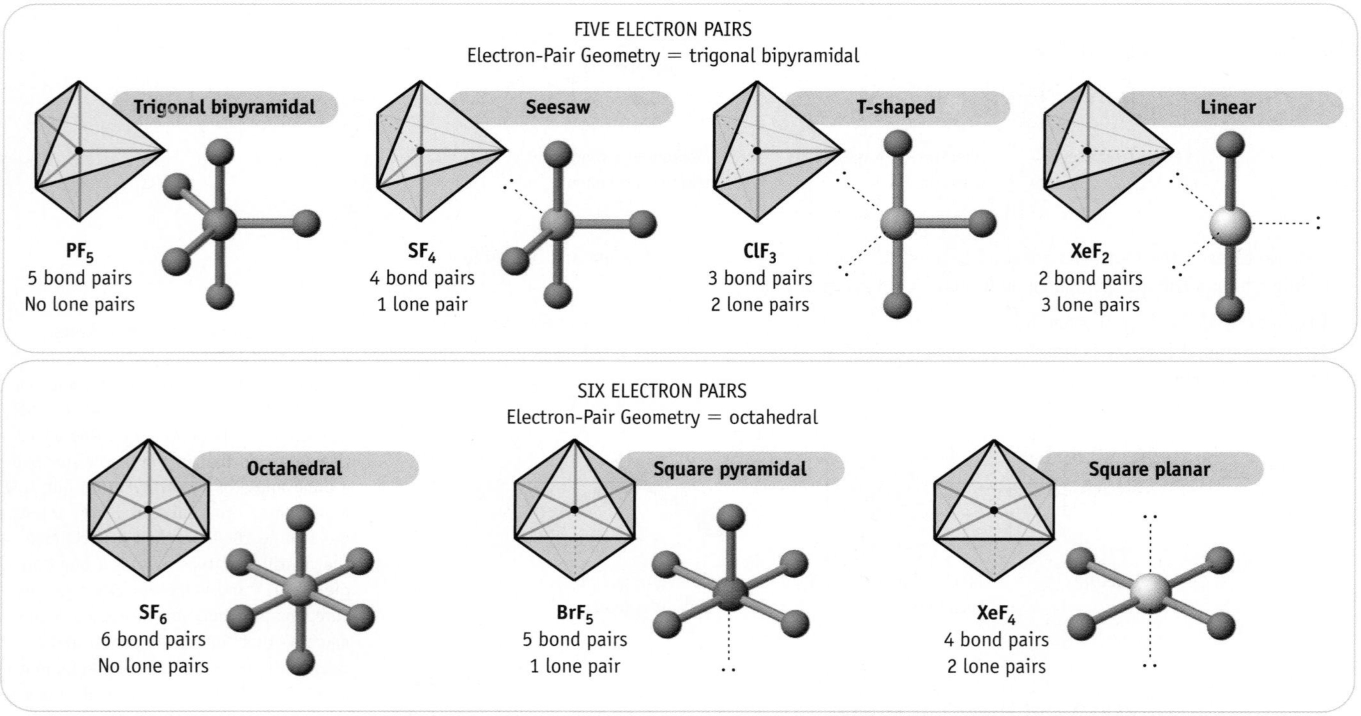

FIGURE 8.8 Electron-pair geometries and molecular shapes for molecules and ions with five or six electron pairs around the central atom.

tion to make it easier to visualize the molecular geometry, which in this case is square-pyramidal. If two pairs of electrons in an octahedral arrangement are lone pairs, they seek to be as far apart as possible. The result is a square-planar molecule, as illustrated by XeF_4.

EXAMPLE 8.8 Predicting Molecular Shape

Problem What is the shape of the ICl_4^- ion?

Strategy Draw the Lewis structure, and then decide on the electron-pair geometry. The position of the atoms gives the molecular geometry of the ion. (See Example 8.7 and Figure 8.8.)

Solution A Lewis structure for the ICl_4^- ion shows that the central iodine atom has six electron pairs in its valence shell. Two of these are lone pairs. Placing the lone pairs on opposite sides leaves the four chlorine atoms in a square-planar geometry.

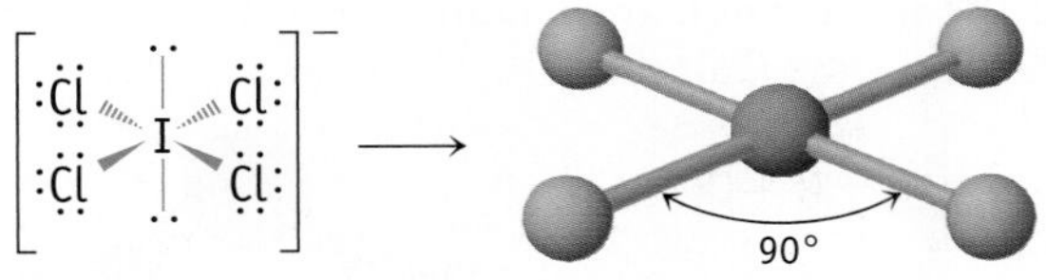

Electron-pair geometry, octahedral

Molecular geometry, square planar

EXERCISE 8.10 Predicting Molecular Shape

Draw the Lewis structure for ICl_2^-, and then decide on the geometry of the ion.

Multiple Bonds and Molecular Geometry

Double and triple bonds involve more electron pairs than single bonds, but this has little effect on the overall molecular shape. All of the electron pairs in a multiple bond are shared between the same two nuclei and therefore occupy the same region of space. Because they must remain in that region, two electron pairs in a double bond (or three pairs in a triple bond) have the same effect on the structure as one electron pair in a single bond. That is, all electron pairs in a multiple bond count as one bond and contribute to molecular geometry the same as a single bond does. For example, the carbon atom in CO_2 has no lone pairs and participates in two double bonds. Each double bond counts as one for the purpose of predicting geometry, so the structure of CO_2 is linear.

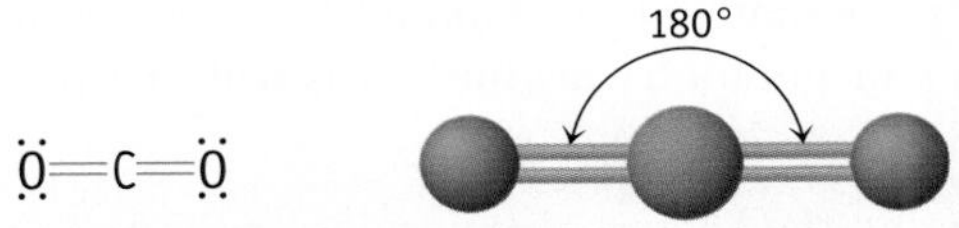

Lewis structure, electron-pair geometry = linear

Molecular structure, linear

When resonance structures are possible, the geometry can be predicted from any of the Lewis resonance structures or from the resonance hybrid structure. For example, the geometry of the CO_3^{2-} ion is predicted to be trigonal planar because the carbon atom has three sets of bonds and no lone pairs.

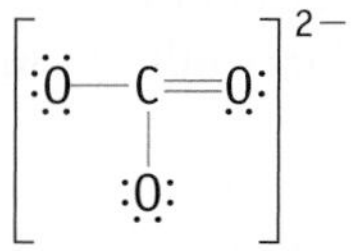

Lewis structure, one resonance structure, electron-pair geometry = trigonal planar

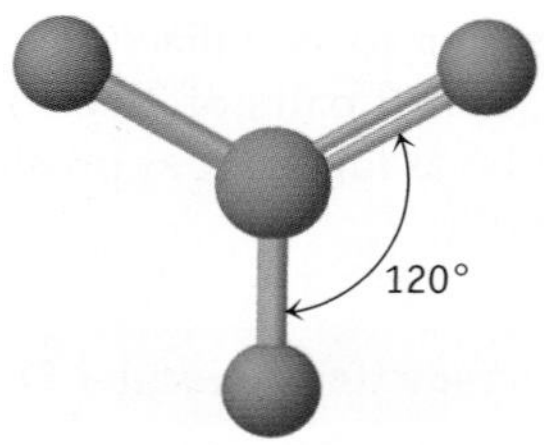

Molecular structure, trigonal planar

The NO_2^- ion also has a trigonal-planar electron-pair geometry. Because there is a lone pair on the central nitrogen atom, and bonds in the other two positions, the geometry of the ion is angular or bent.

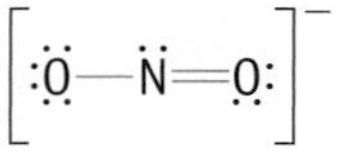

Lewis structure, one resonance structure, electron-pair geometry = trigonal planar

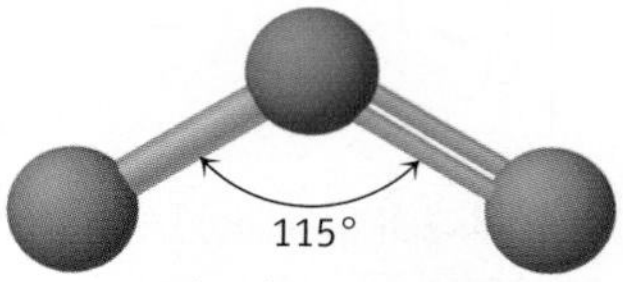

Molecular structure, angular or bent

The techniques just outlined can be used to find the geometries around the atoms in more complicated molecules. Consider, for example, cysteine, one of the natural amino acids.

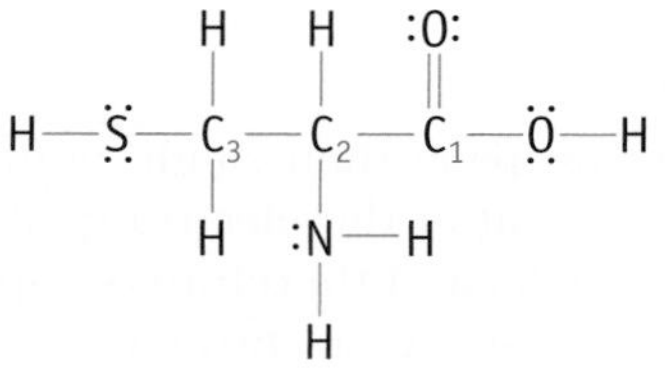

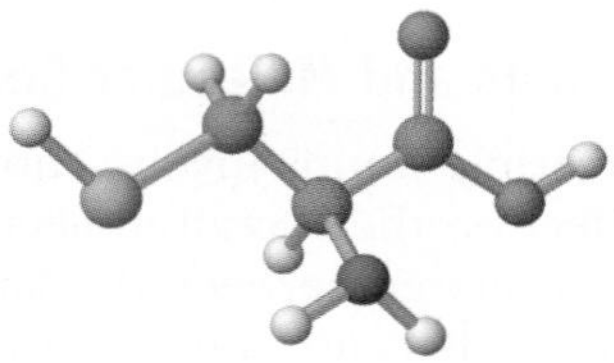

Cysteine, $HSCH_2CH(NH_2)CO_2H$

Four pairs of electrons occur around the S, N, C_2, and C_3 atoms, so the electron-pair geometry around each is tetrahedral. Thus, the S—C—H and H—N—H angles are predicted to be approximately 109°. The O atom in the grouping C—O—H and the S atom in the grouping H—S—C are also surrounded by four pairs, and so these angles are likewise approximately 109°. Finally, the angle made by O—C_1—O is 120° because the electron-pair geometry around C_1 is trigonal planar.

EXAMPLE 8.9 Finding the Shapes of Molecules and Ions

Problem What are the shapes of the nitrate ion, NO_3^-, and $XeOF_4$?

Strategy Draw the Lewis structure, and then decide on the electron pair geometry. The position of the atoms gives the molecular geometry of the ion. Follow the procedure used in Examples 8.6–8.8.

Solution

(a) The NO_3^- and CO_3^{2-} ions are isoelectronic. Thus, like the carbonate ion described in the text above, the electron-pair geometry and molecular shape of NO_3^- are trigonal planar.

(b) The $XeOF_4$ molecule has a Lewis structure with a total of six electron pairs about the central Xe atom, one of which is a lone pair. It has a square-pyramidal molecular structure. Two structures are possible, based on the position occupied by the oxygen, but there is no way to predict which is correct. The actual structure is the one shown, with the oxygen in the apex of the square pyramid.

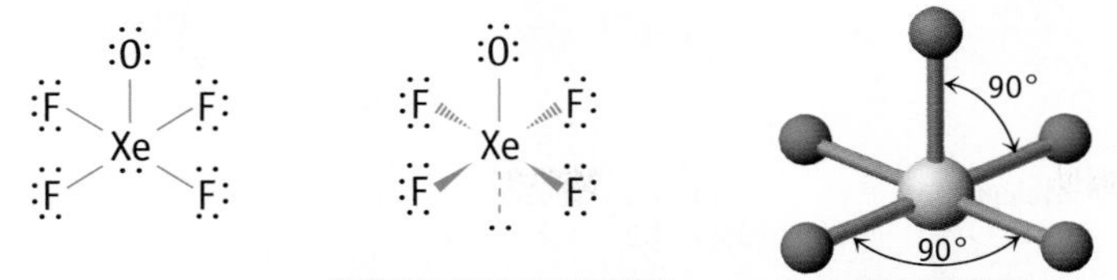

Lewis structure | Electron-pair geometry, octahedral | Molecular geometry, square pyramidal

EXERCISE 8.11 Determining Molecular Shapes

Use Lewis structures and the VSEPR model to determine the electron-pair and molecular geometries for **(a)** the phosphate ion, PO_4^{3-}; **(b)** the sulfite ion, SO_3^{2-}; and **(c)** IF_5.

8.7 Bond Polarity and Electronegativity

The models used to represent covalent and ionic bonding are the extreme situations in bonding. Pure covalent bonding, in which atoms share an electron pair equally, occurs *only* when two identical atoms are bonded. When two dissimilar atoms form a covalent bond, the electron pair will be unequally shared. The result is a **polar covalent bond,** a bond in which the two atoms have residual or partial charges (Figure 8.9).

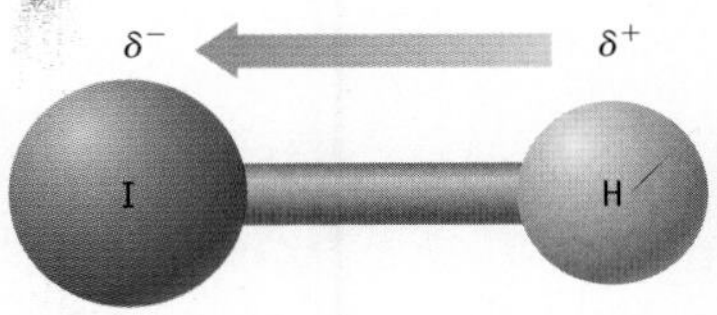

FIGURE 8.9 A polar covalent bond. Iodine has a larger share of the bonding electrons, and hydrogen has a smaller share. The result is that I has a partial negative charge ($\delta-$), and H has a partial positive charge ($\delta+$).

Bonds are polar because not all atoms hold onto their valence electrons with the same force, nor do atoms take on additional electrons with equal ease. Recall from the discussion of atom properties that different elements have different values of ionization energy and electron affinity (Section 7.5). These differences in behavior for free atoms carry over to atoms in molecules.

If a bond pair is not equally shared between atoms, the bonding electrons are on average nearer to one of the atoms. The atom toward which the pair is displaced has a larger share of the electron pair and thus acquires a partial negative charge. At the same time, the atom at the other end of the bond is depleted in electrons and acquires a partial positive charge. The bond between the two atoms has a positive end and a negative end; that is, it has negative and positive poles. The bond is called a **polar bond.**

In ionic compounds, displacement of the bonding pair to one of the two atoms is essentially complete, and + and − symbols are written alongside the atom symbols in the Lewis drawings. For a polar covalent bond, the polarity is indicated by writing the symbols $\delta+$ and $\delta-$ alongside the atom symbols, where δ (the Greek letter "delta") stands for a *partial* charge. Hydrogen fluoride, water, and ammonia are three simple molecules having polar, covalent bonds.

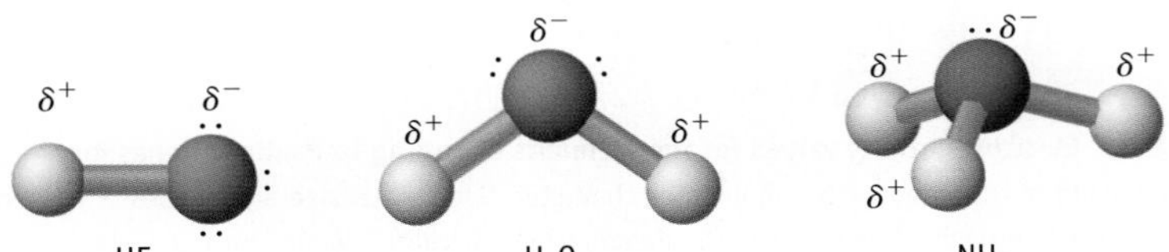

HF | H_2O | NH_3

Three simple molecules with polar covalent bonds. In each case, F, O, and N are more electronegative than H.

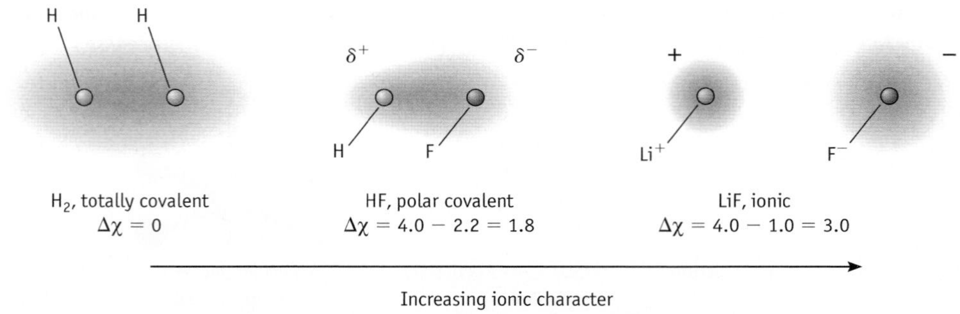

FIGURE 8.10 Covalent to ionic bonding. As the electronegativity difference increases between the atoms of a bond, the bond becomes increasingly ionic.

With so many atoms to use in covalent bond formation, it is not surprising that bonds between atoms can fall anywhere in a continuum from pure covalent to pure ionic (Figure 8.10). There is no sharp dividing line between an ionic bond and a covalent bond.

In the 1930s, Linus Pauling proposed a parameter called atom electronegativity that allows us to decide if a bond is polar, which atom of the bond is negative and which is positive, and if one bond is more polar than another. The **electronegativity, χ,** of an atom is defined as a measure of *the ability of an atom in a molecule to attract electrons to itself.*

Values of electronegativity are given in Figure 8.11. Several features and periodic trends are apparent. The element with the largest electronegativity is fluorine; it is assigned a value of $\chi = 4.0$. The element with the smallest value is the alkali metal cesium. Electronegativities generally increase from left to right across a period and decrease down a group. This is the opposite of the trend observed for metallic character. Metals typically have low values of electronegativity, ranging from slightly less than 1 to about 2. Electronegativity values for the metalloids are around 2, whereas nonmetals have values greater than 2.

1A	2A	3B	4B	5B	6B	7B	8B	8B	8B	1B	2B	3A	4A	5A	6A	7A
								H 2.2								
Li 1.0	Be 1.6											B 2.0	C 2.5	N 3.0	O 3.5	F 4.0
Na 0.9	Mg 1.3											Al 1.6	Si 1.9	P 2.2	S 2.6	Cl 3.2
K 0.8	Ca 1.0	Sc 1.4	Ti 1.5	V 1.6	Cr 1.7	Mn 1.5	Fe 1.8	Co 1.9	Ni 1.9	Cu 1.9	Zn 1.6	Ga 1.8	Ge 2.0	As 2.2	Se 2.6	Br 3.0
Rb 0.8	Sr 1.0	Y 1.2	Zr 1.3	Nb 1.6	Mo 2.2	Tc 1.9	Ru 2.2	Rh 2.3	Pd 2.2	Ag 1.9	Cd 1.7	In 1.8	Sn 2.0	Sb 1.9	Te 2.1	I 2.7
Cs 0.8	Ba 0.9	La 1.1	Hf 1.3	Ta 1.5	W 2.4	Re 1.9	Os 2.2	Ir 2.2	Pt 2.3	Au 2.5	Hg 2.0	Tl 1.6	Pb 2.3	Bi 2.0	Po 2.0	At 2.2

<1.0   1.5–1.9   2.5–2.9
1.0–1.4   2.0–2.4   3.0–4.0

FIGURE 8.11 Electronegativity values for the elements according to Pauling. Trends for electronegativities are the opposite of the trends defining metallic character. Nonmetals have high values of electronegativity; the metalloids have intermediate values, and the metals have low values. Values for these elements as well as for the noble gases and for the lanthanides and actinides are available in the following handbook: Emsley, J., *The Elements*, 3rd edition, Clarendon Press, Oxford, 1998.

There is a large *difference* in electronegativity for atoms from the left- and right-hand sides of the periodic table. For cesium fluoride, for example, the difference in electronegativity values, $\Delta\chi$, is 3.2 [= 4.0 (for F) − 0.8 (for Cs)]. The bond is decidedly ionic in CsF, therefore, with Cs the cation (Cs^+) and F the anion (F^-). In contrast, the electronegativity difference between H and F in HF is only 1.8 [= 4.0 (for F) − 2.2 (for H)]. We conclude that bonding in HF must be more covalent, as expected for a compound formed from two nonmetals. The H—F bond is polar, however, with hydrogen being the positive end of the molecule and fluorine the negative end ($H^{\delta+}$—$F^{\delta-}$).

Chemistry Now™

Sign in at **www.thomsonedu.com/login** and go to Chapter 8 Contents to see Screen 8.15 for **relative electronegativity values.**

©Ted Streshinsky/Corbis

Linus Pauling (1901–1994). Linus Pauling was born in Portland, Oregon, earned a B.Sc. degree in chemical engineering from Oregon State College in 1922, and completed his Ph.D. in chemistry at the California Institute of Technology in 1925. In chemistry, he is well known for his book *The Nature of the Chemical Bond.* He also studied protein structure and, in the words of Francis Crick, was "one of the founders of molecular biology." It was this work and his study of chemical bonding that were cited in the award of the Nobel Prize in chemistry in 1954. Although chemistry was the focus of his life, at the urging of his wife, Ava Helen, he was also involved in nuclear disarmament issues, and he received the Nobel Peace Prize in 1962 for the role he played in advocating for the nuclear test ban treaty.

EXAMPLE 8.10 Estimating Bond Polarities

Problem For each of the following bond pairs, decide which is the more polar and indicate the negative and positive poles.

(a) B–F and B–Cl

(b) Si–O and P–P

Strategy Locate the elements in the periodic table. Recall that electronegativity generally increases across a period and up a group.

Solution

(a) B and F lie relatively far apart in the periodic table. B is a metalloid, and F is a nonmetal. Here, χ for B = 2.0, and χ for F = 4.0. Similarly, B and Cl are relatively far apart in the periodic table, but Cl is below F in the periodic table (χ for Cl = 3.2) and is therefore less electronegative than F. The difference in electronegativity for B—F is 2.0, and for B—Cl it is 1.2. Both bonds are expected to be polar, with B positive and the halide atom negative, but a B—F bond is more polar than a B—Cl bond.

(b) Because the bond is between two atoms of the same kind, the P—P bond is nonpolar. Silicon is in Group 4A and the third period, whereas O is in Group 6A and the second period. Consequently, O has a greater electronegativity (3.5) than Si (1.9), so the bond is highly polar ($\Delta\chi$ = 1.6), with O the more negative atom.

EXERCISE 8.12 Bond Polarity

For each of the following pairs of bonds, decide which is the more polar. For each polar bond, indicate the positive and negative poles. First, make your prediction from the relative atom positions in the periodic table; then check your prediction by calculating $\Delta\chi$.

(a) H—F and H—I **(b)** B—C and B—F **(c)** C—Si and C—S

Charge Distribution: Combining Formal Charge and Electronegativity

The way electrons are distributed in a molecule or ion is called its **charge distribution.** The charge distribution can profoundly affect the properties of a molecule. Examples include its physical properties, such as its melting and boiling points, and its chemical properties, such as its susceptibility to attack by an anion or cation or whether it is an acid or a base.

We saw earlier (◀ page 359) that formal charge calculations can locate the site of a charge in a molecule or an ion. However, this can sometimes lead to results that are incorrect because formal charge calculations assume that there is equal sharing of electrons in all bonds. The ion BF_4^- illustrates this point. Boron has a formal charge of −1 in this ion, whereas the formal charge calculated for the fluorine atoms is 0. This is not logical: fluorine is the more electronegative atom so the negative charge should reside on F and not on B.

A Closer Look — Electronegativity

Electronegativity is a useful, if somewhat vague, concept. It is, however, related to the ionic character of bonds. Chemists have found, as illustrated in the figure, that a correlation exists between the difference in electronegativity of bonded atoms and the degree of ionicity expressed as "% ionic character."

As the difference in electronegativity increases, ionic character increases. Does this trend allow us to say that one compound is ionic and another is covalent? No, we can say only that one bond is more ionic or more covalent than another.

Electron affinity was introduced in Section 7.5. At first glance, it may appear that electronegativity and electron affinity measure the same property, but they do not. Electronegativity is a parameter that applies only to atoms in molecules, whereas electron affinity is a measurable energy quantity for atoms in the gas phase.

Although electron affinity was introduced earlier as a criterion with which to predict the central atom in a molecule, experience indicates that electronegativity is a better choice. That is, *the central atom is generally the atom of lowest electronegativity.*

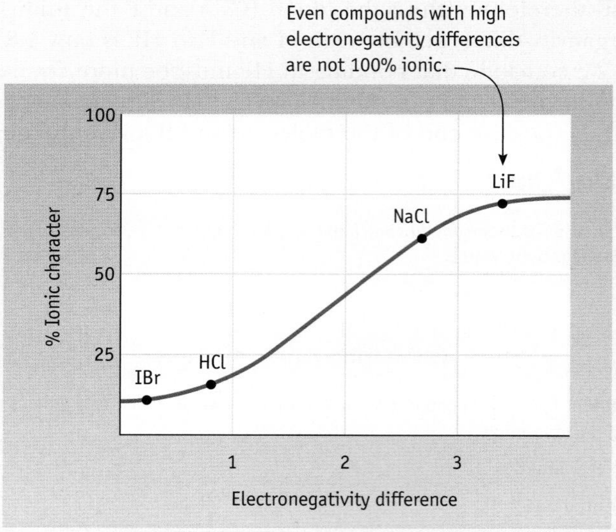

The way to resolve this dilemma is to consider electronegativity in conjunction with formal charge. Based on the electronegativity difference between fluorine and boron ($\Delta\chi = 2.0$), the B—F bonds are expected to be polar, with fluorine being the negative end of the bond, $B^{\delta+}—F^{\delta-}$. So, in this instance, predictions based on electronegativity and formal charge work in opposite directions. The formal charge calculation places the negative charge on boron, but the electronegativity difference leads us to say the negative charge on boron is distributed onto the fluorine atoms, effectively spreading it out over the molecule.

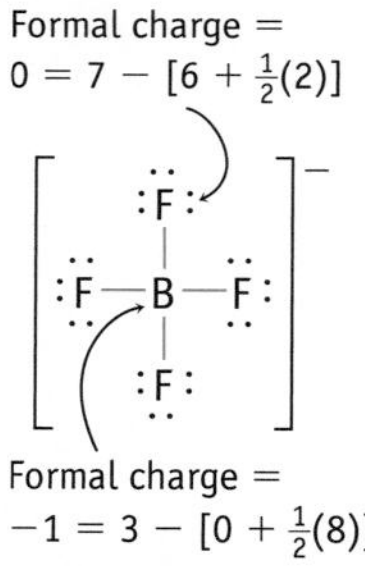

Formal charges for the B and F atoms of the BF_4^- anion.

Linus Pauling pointed out two basic guidelines to use when describing charge distributions in molecules and ions. The first is the **electroneutrality principle.** This declares that electrons will be distributed in such a way that the charges on all atoms are as close to zero as possible. Second, he noted that if a negative charge is present, it should reside on the most electronegative atoms. Similarly, positive charges are expected on the least electronegative atoms. The effect of these principles is clearly seen in the case of BF_4^-, where the negative charge is distributed over the four fluorine atoms rather than residing on boron.

Considering the concepts of electronegativity and formal charge together can help to decide which of several resonance structures is the more important. For example, Lewis structure A for CO_2 is the logical one to draw. But what is wrong with B, in which each atom also has an octet of electrons?

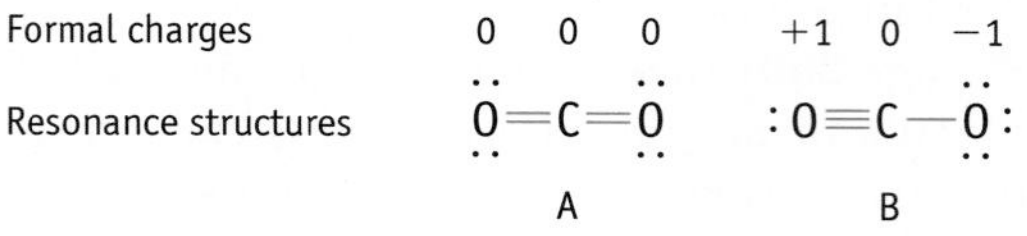

For structure A, each atom has a formal charge of 0, a favorable situation. In B, however, one oxygen atom has a formal charge of +1, and the other has −1. This is contrary to the principle of electroneutrality. In addition, B places a positive charge on the more electronegative O atom. Thus, we can conclude that structure B is a much less satisfactory structure than A.

Now use the logic applied to CO_2 to decide which of the three possible resonance structures for the OCN^- ion is the most reasonable. Formal charges for each atom are given above the element's symbol.

Formal charges

Resonance structures

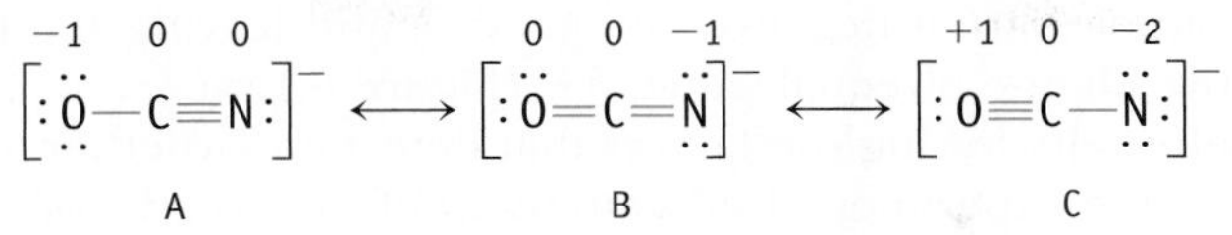

■ Formal Charges in OCN^-

Example of formal charge calculation: For resonance form *C* for OCN^-, we have

$O = 6 - [2 + (1/2)(6)] = +1$

$C = 4 - [0 + (1/2)(8)] = 0$

$N = 5 - [6 + (1/2)(2)] = -2$

Sum of formal charges = −1 = charge on the ion.

Structure C will not contribute significantly to the overall electronic structure of the ion. It has a −2 formal charge on the N atom and a +1 formal charge on the O atom. Not only is the charge on the N atom high, but O is more electronegative than N and would be expected to take on a negative charge. Structure A is more significant than structure B because the negative charge in A is placed on the most electronegative atom (O). We predict, therefore, that structure A is the best representation for this ion and that the carbon–nitrogen bond will resemble a triple bond. The result for OCN^- also allows us to predict that protonation of the ion will lead to HOCN and not HNCO. That is, an H^+ ion will add to the more negative oxygen atom.

EXAMPLE 8.11 Calculating Formal Charges

Problem Boron-containing compounds often have a boron atom with only three bonds (and no lone pairs). Why not form a double bond with a terminal atom to complete the boron octet? To answer this, consider possible resonance structures of BF_3, and calculate the atoms' formal charges. Are the bonds polar in BF_3? If so, which is the more negative atom?

Strategy Calculate the formal charges on each atom in the resonance structures. The preferred structure will have atoms with low formal charges. Negative formal charges should be on the most electronegative atoms.

Solution The two possible structures for BF_3 are illustrated here with the calculated formal charges on the B and F atoms.

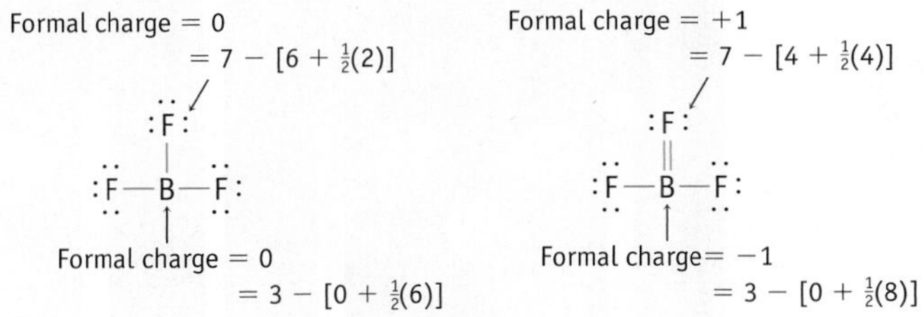

The structure on the left is strongly preferred because all atoms have a zero formal charge and the very electronegative F atom does not have a charge of +1.

F ($\chi = 4.0$) is more electronegative than B ($\chi = 2.0$), so the B—F bond is highly polar, the F atom being partially negative and the B atom partially positive.

EXERCISE 8.13 Formal Charge, Bond Polarity, and Electronegativity

Consider all possible resonance structures for SO_2. What are the formal charges on each atom in each resonance structure? What are the bond polarities? Do they agree with the formal charges?

Module 13

8.8 Bond and Molecular Polarity

The term "polar" was used in Section 8.7 to describe a bond in which one atom has a partial positive charge and the other a partial negative charge. Because most molecules have polar bonds, molecules as a whole can also be polar. In a polar molecule, electron density accumulates toward one side of the molecule, giving that side a partial negative charge, $\delta-$, and leaving the other side with a partial positive charge of equal value, $\delta+$ (Figure 8.12a).

Before describing the factors that determine whether a molecule is polar, let us look at the experimental measurement of the polarity of a molecule. When placed in an electric field, polar molecules experience a force that tends to align them with the field (Figure 8.12). When the electric field is created by a pair of oppositely charged plates, the positive end of each molecule is attracted to the negative plate, and the negative end is attracted to the positive plate (Figure 8.12b). The extent to which the molecules line up with the field depends on their **dipole moment, μ,** which is defined as the product of the magnitude of the partial charges ($\delta+$ and $\delta-$) on the molecule and the distance by which they are separated. The SI unit of the dipole moment is the coulomb-meter, but dipole moments have traditionally been given using a derived unit called the *debye* (D; 1 D = 3.34×10^{-30} C · m). Experimental values of some dipole moments are listed in Table 8.7.

■ **Dipole–Dipole Forces** The force of attraction between the negative end of one polar molecule and the positive end of another (called a dipole–dipole force and discussed in Section 12.2) affects the properties of polar compounds. Intermolecular forces (forces between molecules) influence the temperature at which a liquid freezes or boils, for example.

To predict if a molecule is polar, we need to consider if the molecule has polar bonds and how these bonds are positioned relative to one another. Diatomic molecules composed of two atoms with different electronegativities are always polar (see Table 8.7); there is one bond, and the molecule has a positive and a negative end. But what happens with a molecule with three or more atoms, in which there are two or more polar bonds?

Consider first a linear triatomic molecule such as carbon dioxide, CO_2 (Figure 8.13). Here, each C═O bond is polar, with the oxygen atom the negative end of the bond dipole. The terminal atoms are at the same distance from the C atom; they both have the same $\delta-$ charge, and they are symmetrically arranged around the central C atom. Therefore, CO_2 has no molecular dipole, even though each

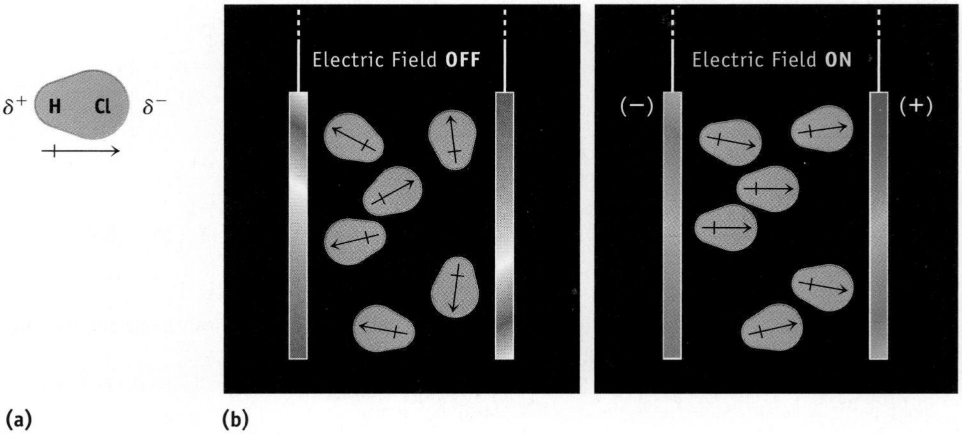

FIGURE 8.12 Polar molecules in an electric field. (a) A representation of a polar molecule. To indicate the direction of molecular polarity, an arrow is drawn with the head pointing to the negative side and a plus sign placed at the positive end. (b) When placed in an electric field (between charged plates), polar molecules experience a force that tends to align them with the field. The negative end of the molecules is drawn to the positive plate, and vice versa. The orientation of the polar molecule affects the electrical capacitance of the plates (their ability to hold a charge), and this provides a way to measure experimentally the magnitude of the dipole moment.

TABLE 8.7 Dipole Moments of Selected Molecules

Molecule (AX)	Moment (μ, D)	Geometry	Molecule (AX_2)	Moment (μ, D)	Geometry
HF	1.78	linear	H_2O	1.85	bent
HCl	1.07	linear	H_2S	0.95	bent
HBr	0.79	linear	SO_2	1.62	bent
HI	0.38	linear	CO_2	0	linear
H_2	0	linear			
Molecule (AX_3)	**Moment (μ, D)**	**Geometry**	**Molecule (AX_4)**	**Moment (μ, D)**	**Geometry**
NH_3	1.47	trigonal pyramidal	CH_4	0	tetrahedral
NF_3	0.23	trigonal pyramidal	CH_3Cl	1.92	tetrahedral
BF_3	0	trigonal planar	CH_2Cl_2	1.60	tetrahedral
			$CHCl_3$	1.04	tetrahedral
			CCl_4	0	tetrahedral

bond is polar. This is analogous to a tug-of-war in which the people at opposite ends of the rope are pulling with equal force.

In contrast, water is a bent triatomic molecule. Because O has a larger electronegativity ($\chi = 3.5$) than H ($\chi = 2.2$), each of the O—H bonds is polar, with the H atoms having the same $\delta+$ charge and oxygen having a negative charge ($\delta-$) (Figure 8.13). Electron density accumulates on the O side of the molecule, making the molecule electrically "lopsided" and therefore polar ($\mu = 1.85$ D).

In trigonal-planar BF_3, the B—F bonds are highly polar because F is much more electronegative than B (χ of B = 2.0 and χ of F = 4.0) (Figure 8.14). The molecule is nonpolar, however, because the three terminal F atoms have the same $\delta-$ charge, are the same distance from the boron atom, and are arranged symmetrically and in the same plane as the central boron atom. In contrast, the trigonal-planar molecule phosgene is polar (Cl_2CO, $\mu = 1.17$ D) (Figure 8.14). Here, the angles are all about 120°, so the O and Cl atoms are symmetrically arranged around the C atom. The

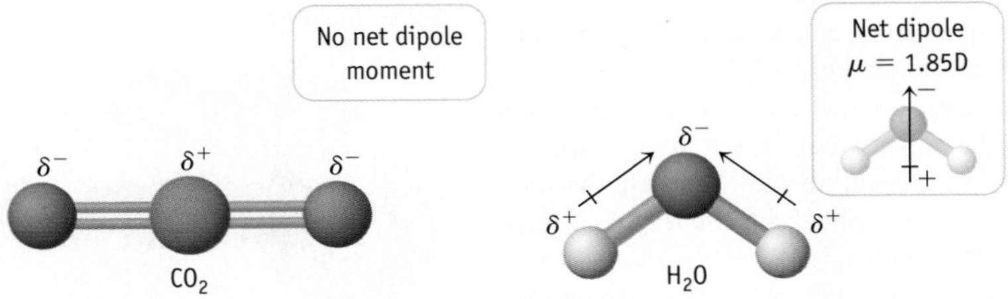

(a) (b)

Active Figure 8.13 Polarity of triatomic molecules, AX_2. For CO_2, the CO bonds are polar, but the electron density is distributed evenly over the molecule, and the charges of δ^- lie 180° apart. (Charges calculated using advanced molecular modeling software: C = +0.42 and O = −0.21.) The molecule has no net dipole. In the water molecule, the O atom is negative, and the H atoms are positive. (Calculated charges: H = +0.19 and O = −0.38.) However, the positively charged H atoms lie on one side of the molecule, and the negatively charged O atom is on the other side. The molecule is polar. (The calculated dipole of 1.86 D is in good agreement with experiment.)

ChemistryNow™ Sign in at www.thomsonedu.com/login and go to the Chapter Contents menu to explore an interactive version of this figure accompanied by an exercise.

■ **Peter Debye and Dipoles** The commonly used unit of dipole moments is named in honor of Peter Debye (1884–1966). He was born in The Netherlands, but attended university in Germany and later studied for his Ph.D. in physics in Munich. He developed a theory on the diffraction of x-rays by solids, a new concept for magnetic cooling, and (with E. Hückel) a model for interionic attractions in aqueous solution. As his interests turned more to chemistry, he worked on methods of determining the shapes of polar molecules. Debye received the Nobel Prize in chemistry in 1936.

Rare Book & Manuscript Collections/Carl A. Kroch Library/Cornell University

Peter Debye (1884–1966)

A Closer Look

Visualizing Charge Distributions and Molecular Polarity—Electrostatic Potential Surfaces and Partial Charge

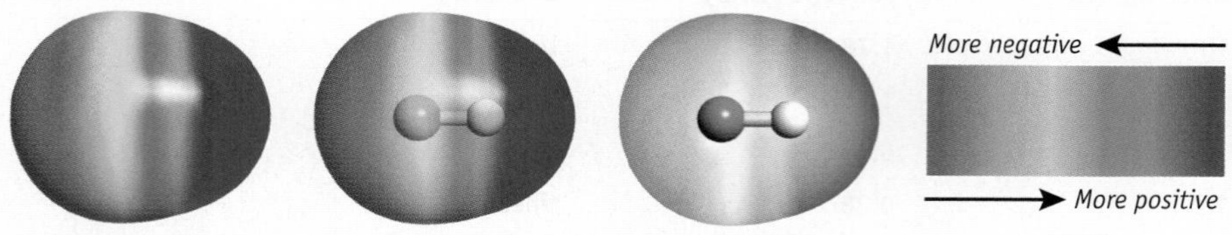

FIGURE A Three views of the electrostatic potential surface for HF.

(*left*) The electron density surface around HF. The F atom is at the left. The surface is made up of all of the points in space around the HF molecule where the electron density is 0.002 $e^-/Å^3$ (where 1 Å = 0.1 nm).

(*middle*) The surface is made more transparent, so you can see the HF atom nuclei inside the surface.

(*right*) The front of the electron density surface has been "peeled away" for a view of the HF molecule inside.

Color scheme: The colors on the electron density surface reflect the charge in the different regions of the molecule 384. Colors to the blue end of the spectrum indicate a positive charge, whereas colors to the red end of the spectrum indicate a negative charge.

In Chapter 6, you saw atomic orbitals, regions of space within which an electron is most probably found. The boundary surface of these orbitals was created in such a way that the electron wave amplitude at all points of the surface was the same value (◀ page 288). Using advanced molecular modeling software, we can generate the same type of pictures for molecules, and in Figure A you see a surface defining the electron density in the HF molecule. The electron density surface, calculated using software from CAChe, is made up of all of the points in space around the HF molecule where the electron density is at least 0.002 $e^-/Å^3$ (where 1 Å = 0.1 nm). You can see that the surface bulges toward the F end of the molecule, an indication of the larger size of the F atom. The larger size of the F atom here is mainly related to the fact that it has more valence electrons than H, and to a lesser extent to the fact that H—F bond is polar and electron density in that bond is shifted toward the F atom.

We can add another layer of information. The electron density surface can be colored according to the *electrostatic potential*. (Hence, this figure is called an *electrostatic potential surface*.) The computer program calculates the electrostatic potential that would be observed by a proton (H^+) on the surface. This is the sum of the attractive and repulsive forces on that proton due to the nuclei and the electrons in the molecule. Regions of the molecule in which there is an attractive potential are colored red. That is, this is a region of negative charge on the molecule. Repulsive potentials occur in regions where the molecule is positively charged; these regions are colored blue. As might be expected, the net electrostatic potential will change continuously as one moves from a negative portion of a molecule to a positive portion, and this is indicated by a progression of colors from blue to red (from positive to negative).

The electrostatic potential surface for HF shows the H atom is positive (the H atom end of the molecule is blue), and the F atom is negative (the F atom end is red). This is, of course, what we would predict based on electronegativity.

Our program also calculated that the F atom has a charge of −0.29 and H has a charge of +0.29. Finally, the calculated dipole moment for the molecule is 1.74 D, in good agreement with the experimental value in Table 8.7.

Other examples of electrostatic potential surfaces illustrate the polarity of water and methylamine, CH_3NH_2.

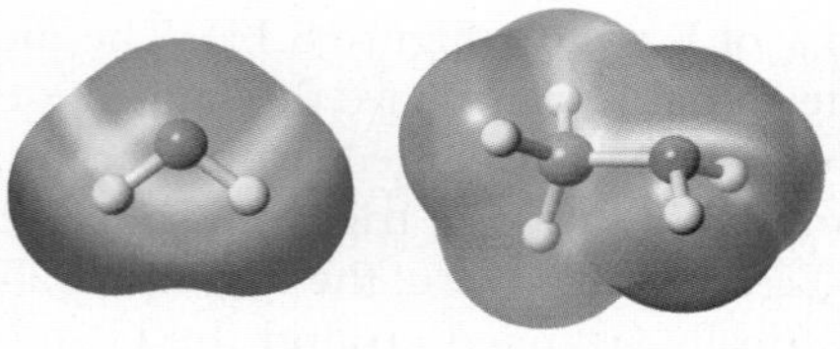

Water Methylamine

The surface shows the O atom of the water molecule bears a partial negative charge and the H atoms are positive. The surface for the amine clearly shows the molecule is polar and that the region around the N atom is also negative. Indeed, we know from experiment that an H^+ ion will attack the N atom to give the cation $CH_3NH_3^+$.

Electrostatic potential surfaces are becoming more widely used, particularly in organic chemistry and biochemistry, to probe the reactive sites of more and more complex molecules. One example is the dipeptide glycylglycine.

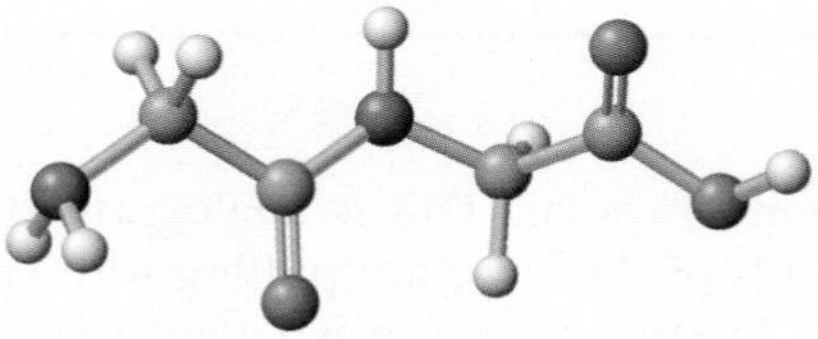

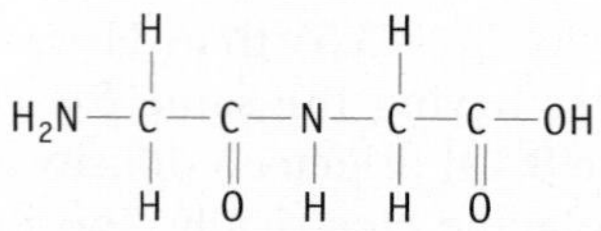

The electrostatic surface for the molecule shows that the O atoms of the C=O groups have a partial negative charge as does the N of the NH_2 group. Positive regions of the molecule include the H atom of the C(O)—NH grouping (the amide grouping) and the H atom of the OH group. Such pictures can help you see quickly the regions of a molecule that may be a proton donor (an acid) or a proton acceptor (a base).

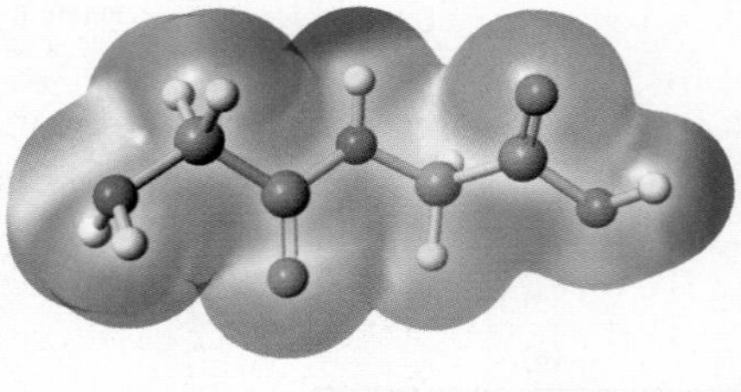

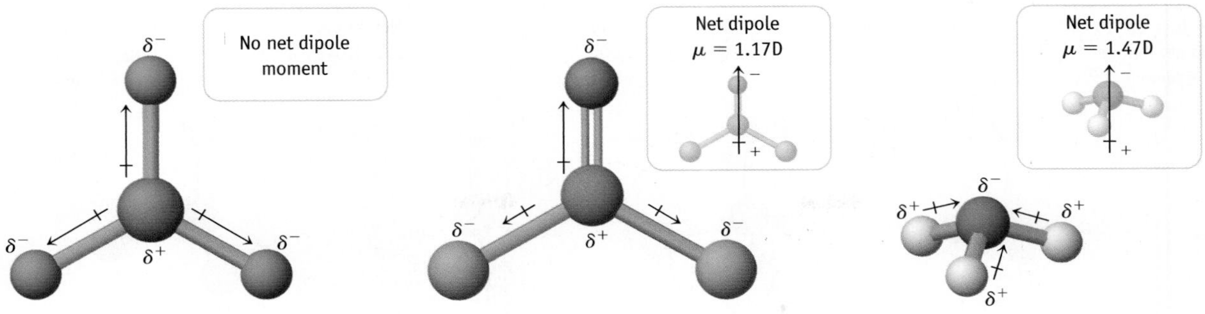

BF$_3$ Cl$_2$CO NH$_3$

Active Figure 8.14 **Polar and nonpolar molecules of the type AX_3.** In BF_3, the negative charge on the F atoms is distributed symmetrically, so the molecular dipole is zero. In contrast, in Cl_2CO and NH_3, the negative charge in the molecules is shifted to one side and the positive charge to the other side.

Molecule	Calculated Partial Charges	Calculated dipole
BF_3	B = 0.44, F = −0.15	0
Cl_2CO	O = −0.21, C = 0.23, Cl = −0.01	1.25
NH_3	N = −0.40, H = 0.13	1.58

The calculated dipoles are in reasonable agreement with experimentally measured dipoles. (Calculations were done with molecular modeling software from CAChe.)

Chemistry Now™ Sign in at www.thomsonedu.com/login and go to the Chapter Contents menu to explore an interactive version of this figure accompanied by an exercise.

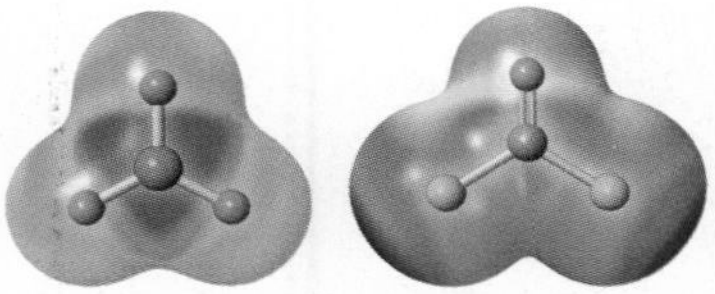

■ **Electrostatic Potential Surfaces for BF_3 and Cl_2CO** Notice that the charge distribution for BF_3 (left) is symmetrical whereas that for Cl_2CO (right) has a partial negative charge on the O atom and much less negative charges on the Cl atoms.

electronegativities of the three atoms in the molecule differ, however: $\chi(O) > \chi(Cl) > \chi(C)$. There is therefore a net displacement of electron density away from the center of the molecule, more toward the O atom than the Cl atoms.

Ammonia, like BF_3, has AX_3 stoichiometry and polar bonds. In contrast to BF_3, however, NH_3 is a trigonal-pyramidal molecule. The positive H atoms are located in the base of the pyramid, and the negative N atom is on the apex of the pyramid. As a consequence, NH_3 is polar (Figure 8.14). Indeed, trigonal-pyramidal molecules are generally polar.

Molecules like carbon tetrachloride, CCl_4, and methane, CH_4, are nonpolar, owing to their symmetrical, tetrahedral structures. The four atoms bonded to C have the same partial charge and are the same distance from the C atom. Tetrahedral molecules with both Cl and H atoms ($CHCl_3$, CH_2Cl_2, and CH_3Cl) are polar, however (Figure 8.15). The electronegativity for H atoms (2.2) is less than that of Cl atoms (3.2), and the carbon–hydrogen distance is different from the carbon–chlorine distances. Because Cl is more electronegative than H, the Cl atoms are on the more negative side of the molecule. This means the positive end of the molecular dipole is toward the H atom.

To summarize this discussion of molecular polarity, look again at Figure 8.5 (page 369). These are sketches of molecules of the type AX_n where A is the central atom and X is a terminal atom. You can predict that a molecule AX_n will *not* be polar, regardless of whether the A—X bonds are polar, if

- All the terminal atoms (or groups), X, are identical, and
- All the X atoms (or groups) are arranged symmetrically around the central atom, A.

On the other hand, if one of the X atoms (or groups) is different in the structures in Figure 8.5 (as in Figures 8.14 and 8.15), or if one of the X positions is occupied by a lone pair, the molecule will be polar.

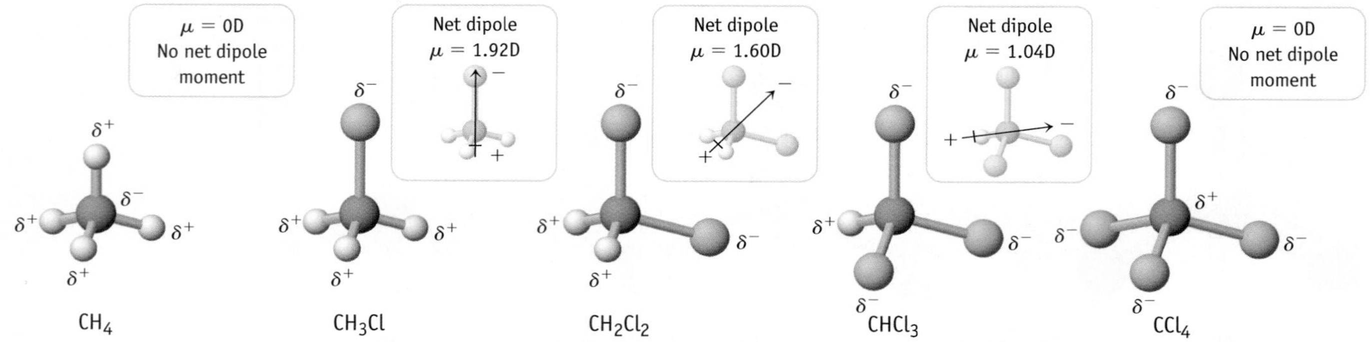

FIGURE 8.15 Polarity of tetrahedral molecules. The electronegativities of the atoms involved are in the order Cl (3.2) > C(2.5) > H (2.2). This means the C—H and C—Cl bonds are polar with a net displacement of electron density away from the H atoms and toward the Cl atoms [$H^{\delta+}$–$C^{\delta-}$ and $C^{\delta+}$–$Cl^{\delta-}$]. Although the electron-pair geometry around the C atom in each molecule is tetrahedral, only in CH_4 and CCl_4 are the polar bonds totally symmetrical in their arrangement. Therefore, CH_3Cl, CH_2Cl_2, and $CHCl_3$ are polar molecules, with the negative end toward the Cl atoms and the positive end toward the H atoms.

Chemistry Now™

Sign in at **www.thomsonedu.com/login** and go to Chapter 8 Contents to see Screen 8.16 for practice **determining polarity.**

EXAMPLE 8.12 Molecular Polarity

Problem Are nitrogen trifluoride (NF_3) and sulfur tetrafluoride (SF_4) polar or nonpolar? If polar, indicate the negative and positive sides of the molecule.

Strategy You cannot decide if a molecule is polar without determining its structure. Therefore, start with the Lewis structure, decide on the electron-pair geometry, and then decide on the molecular geometry. If the molecular geometry is one of the highly symmetrical geometries in Figure 8.5, the molecule is not polar. If it does not fit one of these categories, it will be polar.

Solution

(a) NF_3 has the same trigonal-pyramidal structure as NH_3. Because F is more electronegative than N, each bond is polar, the more negative end being the F atom. Because this molecule contains polar bonds and because the geometry is not symmetrical but has instead three positions of the tetrahedron occupied by bonding groups and one by a lone pair, the NF_3 molecule as a whole is expected to be polar.

You will notice, however, that the dipole moment for NF_3 is quite small (0.23 D in Table 8.7), much smaller than that of NH_3. This illustrates that *lone pairs have an effect on polarity*. For NH_3, the N-atom lone pair adds to the overall polarity of the molecule. (The lone electron pair extends into space beyond the N atom and increases the charge separation in NH_3; this enhances the dipole.) For NF_3, however, the effect of the lone pair on the nitrogen atom of the molecule is counterbalanced by the highly polar N—F bonds on the other side and the magnitude of the dipole pointing toward the side with the F atoms is reduced.

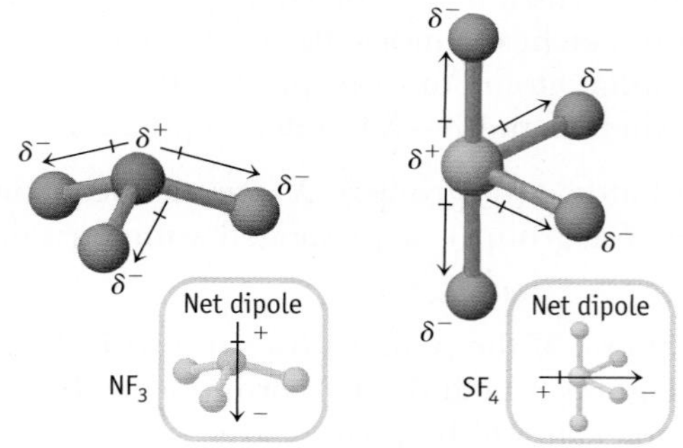

(b) The S—F bonds in sulfur tetrafluoride, SF_4, are highly polar, the bond dipole having F as the negative end (χ for S is 2.6 and χ for F is 4.0). The molecule has an electron-pair geometry of a trigonal bipyramid (see Figure 8.8). Because the lone pair occupies one of the positions, the S—F bonds are not arranged symmetrically. The axial S—F bond dipoles cancel each other because they point in opposite directions. The equatorial S—F bonds, however, both point to one side of the molecule.

Comment In general, we do not consider lone pair effects on molecular dipoles. Nonetheless, they do have an effect, as seen when comparing the dipole moment for NF_3 (0.23 D) with that for NH_3 (1.47 D).

This is a case in which electrostatic potential surfaces and calculated atom charges are useful in showing the difference between these molecules. The N atom in NH_3 is decidedly negative (−0.40), and the H atoms are positive (+0.13). In contrast, in NF_3 the N atom is positively charged, and the F atoms are negatively charged (N = +0.3 and F = −0.1). The difference in charge between N and F is not as great as between N and H in NH_3.

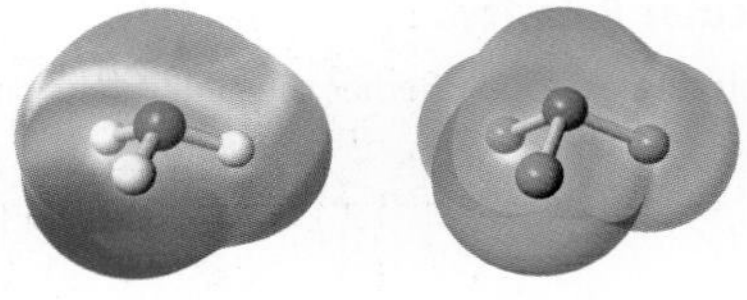

Electrostatic potential maps.

EXAMPLE 8.13 Molecular Polarity

Problem 1,2-Dichloroethylene can exist in two forms. Is either of these planar molecules polar?

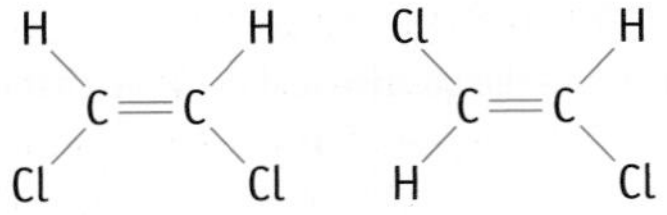

A B

Strategy To decide if a molecule is polar, we first sketch the structure and then, using electronegativity values, decide on the bond polarity. Finally, we decide if the electron density in the bonds is distributed symmetrically or if it is shifted to one side of the molecule.

Solution Here, the H and Cl atoms are arranged around the C=C double bonds with all bond angles 120° (and all the atoms lie in one plane). The electronegativities of the atoms involved are in the order Cl(3.2) > C(2.5) > H (2.2). This means the C—H and C—Cl bonds are polar with a net displacement of electron density away from the H atoms and toward the Cl atoms [$H^{\delta+}$—$C^{\delta-}$ and $C^{\delta+}$—$Cl^{\delta-}$]. In structure A, the Cl atoms are located on one side of the molecule, so electrons in the H—C and C—Cl bonds are displaced toward the side of the molecule with Cl atoms and away from the side with the H atoms. Molecule A is polar. In molecule B, the displacement of electron density toward the Cl atom on one end of the molecule is counterbalanced by an opposing displacement on the other end. Molecule B is not polar.

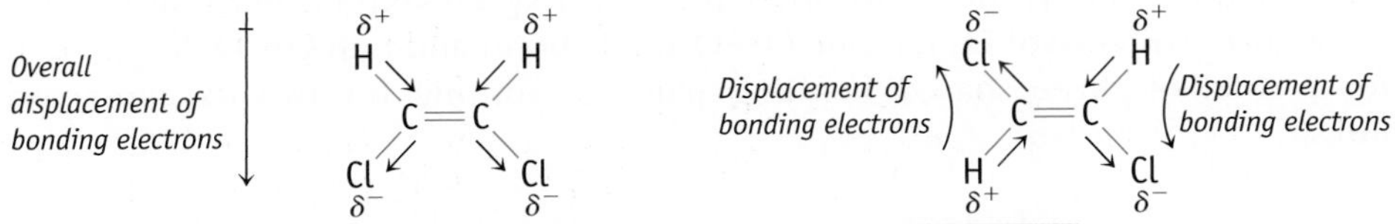

A, polar, diplacement of bonding electrons to one side of the molecule

B, not polar, no net displacement of bonding electrons to one side of the molecule

Comment The electrostatic potential surfaces reflect the fact that molecule A is polar because the electron density is shifted to one side of the molecule. Molecule B is not polar because the electron density is distributed symmetrically.

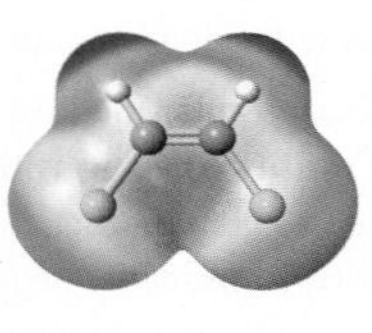

Molecule A

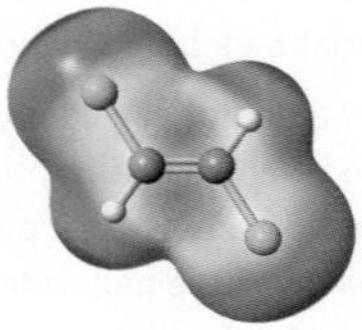

Molecule B

EXERCISE 8.14 Molecular Polarity

For each of the following molecules, decide whether the molecule is polar and which side is positive and which negative: $BFCl_2$, NH_2Cl, and SCl_2.

EXERCISE 8.15 Molecular Polarity

The electrostatic potential surface for $OSCl_2$ is pictured here.

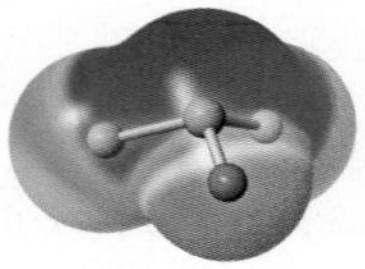

(a) Draw a Lewis electron dot picture for the molecule, and give the formal charge of each atom.

(b) What is the molecular geometry of $OSCl_2$?

(c) Is the molecule polar? If so, locate the positive and negative charges and the direction of the dipole.

8.9 Bond Properties: Order, Length, Energy

Bond Order

The **order of a bond** is the number of bonding electron pairs shared by two atoms in a molecule (Figure 8.16). You will encounter bond orders of 1, 2, and 3, as well as fractional bond orders.

When the bond order is 1, there is only a single covalent bond between a pair of atoms. Examples are the bonds in molecules such as H_2, NH_3, and CH_4. The bond order is 2 when two electron pairs are shared between atoms, such as the C=O bonds in CO_2 and the C=C bond in ethylene, $H_2C{=}CH_2$. The bond order is 3 when two atoms are connected by three bonds. Examples include the carbon–oxygen bond in carbon monoxide, CO and the nitrogen–nitrogen bond in N_2.

Fractional bond orders occur in molecules and ions having resonance structures. For example, what is the bond order for each oxygen–oxygen bond in O_3? Each resonance structure of O_3 has one O—O single bond and one O=O double bond, for a total of three shared bonding pairs accounting for two oxygen–oxygen links.

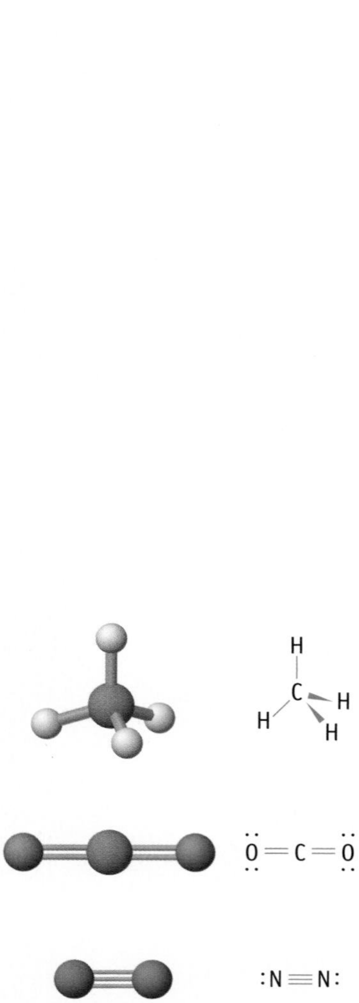

FIGURE 8.16 Bond order. The four C—H bonds in methane each have a bond order of 1. The two C=O bonds of CO_2 each have a bond order of two, whereas the nitrogen–nitrogen bond in N_2 has an order of 3.

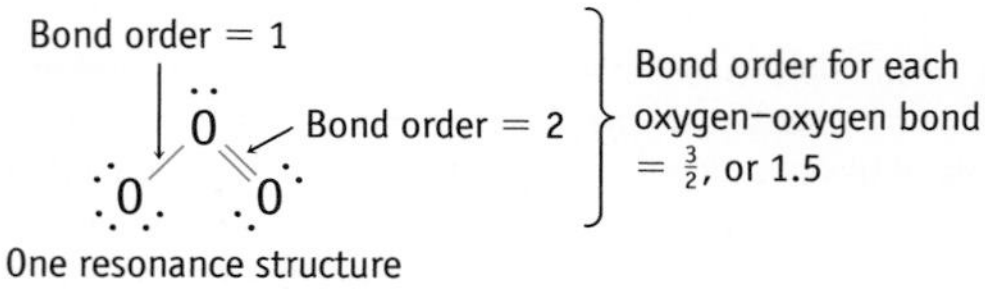

We can define the bond order between any bonded pair of atoms X and Y as

$$\text{Bond order} = \frac{\text{number of shared pairs in all X—Y bonds}}{\text{number of X—Y links in the molecule or ion}} \quad (8.2)$$

For ozone, there are three bond pairs involved in two oxygen-oxygen links, so the bond order for each oxygen–oxygen bond is $\frac{3}{2}$, or 1.5.

Chemistry Now™

Sign in at **www.thomsonedu.com/login** and go to Chapter 8 Contents to see Screen 8.17 to see **how bond order, bond length, and bond energy are related.**

Bond Length

Bond length is the distance between the nuclei of two bonded atoms. Bond lengths are therefore related to the sizes of the atoms (Section 7.5), but, for a given pair of atoms, the order of the bond also plays a role.

Table 8.8 lists average bond lengths for a number of common chemical bonds. It is important to recognize that these are *average* values. Neighboring parts of a molecule can affect the length of a particular bond. For example, Table 8.8 specifies that the average C—H bond has a length of 110 pm. In methane, CH_4, the measured bond length is 109.4 pm, whereas the C—H bond is only 105.9 pm long in acetylene, H—C≡C—H. Variations as great as 10% from the average values listed in Table 8.8 are possible.

TABLE 8.8 Some Average Single- and Multiple-Bond Lengths in Picometers (pm)*

Single Bond Lengths

	Group										
	1A	4A	5A	6A	7A	4A	5A	6A	7A	7A	7A
	H	C	N	O	F	Si	P	S	Cl	Br	I
H	74	110	98	94	92	145	138	132	127	142	161
C		154	147	143	141	194	187	181	176	191	210
N			140	136	134	187	180	174	169	184	203
O				132	130	183	176	170	165	180	199
F					128	181	174	168	163	178	197
Si						234	227	221	216	231	250
P							220	214	209	224	243
S								208	203	218	237
Cl									200	213	232
Br										228	247
I											266

Multiple Bond Lengths

C=C	134	C≡C	121
C=N	127	C≡N	115
C=O	122	C≡O	113
N=O	115	N≡O	108

*1 pm = 10^{-12} m.

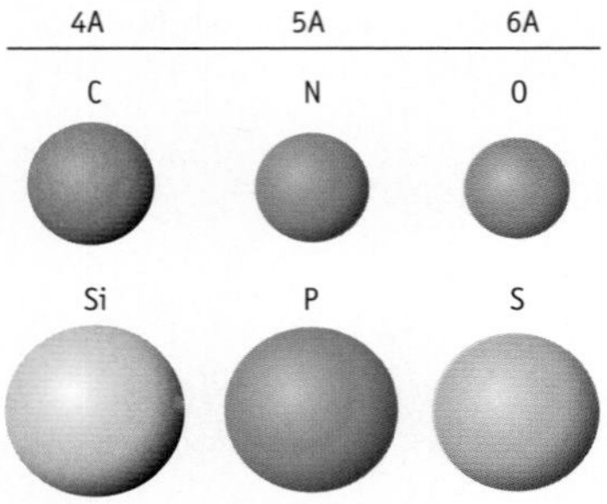

Relative sizes of some atoms of Groups 4A, 5A, and 6A.

Bond lengths are related to atom sizes.

C—H	**N—H**	**O—H**
110	98	94 pm
Si—H	**P—H**	**S—H**
145	138	132 pm

Because atom sizes vary in a regular fashion with the position of the element in the periodic table (Figure 7.8), predictions of trends in bond length can be made quickly. For example, the H—X distance in the hydrogen halides increases in the order predicted by the relative sizes of the halogens: H—F < H—Cl < H—Br < H—I. Likewise, bonds between carbon and another element in a given period decrease going from left to right, in a predictable fashion; for example, C—C > C—N > C—O > C—F. Trends for multiple bonds are similar. A C=O bond is shorter than a C=S bond, and a C=N bond is shorter than a C=C bond.

The effect of bond order is evident when bonds between the same two atoms are compared. For example, the bonds become shorter as the bond order increases in the series C—O, C=O, and C≡O:

Bond	**C—O**	**C=O**	**C≡O**
Bond Order	1	2	3
Bond Length (pm)	143	122	113

Double bonds are shorter than single bonds between the same set of atoms, and triple bonds between those same atoms are shorter still.

The carbonate ion, CO_3^{2-}, has three equivalent resonance structures. Each CO bond has a bond order of 1.33 (or 4/3) because four electron pairs are used to form three carbon–oxygen links. The CO bond distance (129 pm) is intermediate between a C—O single bond (143 pm) and a C=O double bond (122 pm).

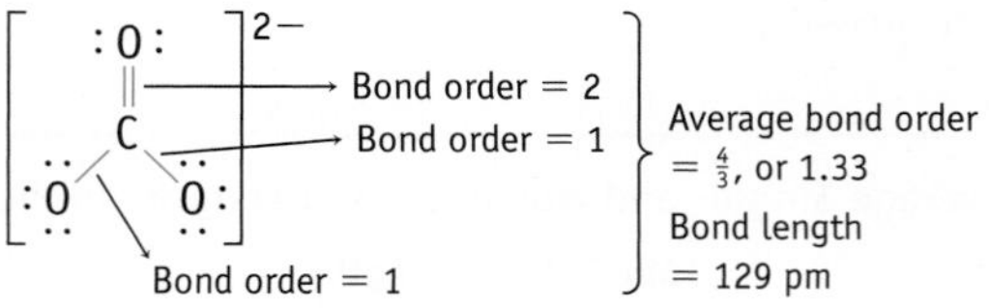

> **EXERCISE 8.16 Bond Order and Bond Length**
>
> **(a)** Give the bond order of each of the following bonds, and arrange them in order of decreasing bond distance: C=N, C≡N, and C—N.
>
> **(b)** Draw resonance structures for NO_2^-. What is the NO bond order in this ion? Consult Table 8.8 for N—O and N=O bond lengths. Compare these with the NO bond length in NO_2^- (124 pm). Account for any differences you observe.

Bond Dissociation Enthalpy

The **bond dissociation enthalpy** is the enthalpy change for breaking a bond in a molecule with the reactants and products in the gas phase.

$$\text{Molecule (g)} \underset{\text{Energy released} = \Delta H < 0}{\overset{\text{Energy supplied} = \Delta H > 0}{\rightleftharpoons}} \text{Molecular fragments (g)}$$

Suppose you wish to break the carbon–carbon bonds in ethane ($H_3C—CH_3$), ethylene ($H_2C=CH_2$), and acetylene (HC≡CH). The carbon–carbon bond orders in these molecules are 1, 2, and 3, respectively, and these bond orders are reflected in the bond dissociation enthalpies. Carbon–carbon bond breaking in ethane requires the least energy in this group, and acetylene requires the most energy.

$$H_3C{-}CH_3(g) \rightarrow H_3C(g) + CH_3(g) \qquad \Delta_rH = +368 \text{ kJ/mol-rxn}$$

$$H_2C{=}CH_2(g) \rightarrow H_2C(g) + CH_2(g) \qquad \Delta_rH = +682 \text{ kJ/mol-rxn}$$

$$HC{\equiv}CH(g) \rightarrow HC(g) + CH(g) \qquad \Delta_rH = +962 \text{ kJ/mol-rxn}$$

Because ΔH represents the energy transferred to the molecule from its surroundings, ΔH has a positive value; that is, *the process of breaking bonds in a molecule is always endothermic.*

The energy supplied to break carbon–carbon bonds must be the same as the energy released when the same bonds form. *The formation of bonds from atoms or radicals in the gas phase is always exothermic.* This means, for example, that Δ_rH for the formation of $H_3C{-}CH_3$ from two $CH_3(g)$ radicals is -368 kJ/mol-rxn.

$$H_3C\cdot(g) + \cdot CH_3(g) \rightarrow H_3C{-}CH_3(g) \qquad \Delta_rH = -368 \text{ kJ/mol-rxn}$$

Generally, the bond energy for a given type of bond (a C—C bond, for example) varies somewhat, depending on the compound, just as bond lengths vary from one molecule to another. They are sufficiently similar, however, so it is possible to create a table of *average bond dissociation enthalpies* (Table 8.9). The values in such tables may be used to *estimate* the enthalpy change for a reaction, as described below.

Variability in Bond Dissociation Enthalpies The values of Δ_rH for ethane, ethylene, and acetylene in the text are for those molecules in particular. The bond dissociation enthalpies in Table 8.9 are average values for a range of molecules containing the indicated bond.

TABLE 8.9 Some Average Bond Dissociation Enthalpies (kJ/mol)*

Single Bonds

	H	C	N	O	F	Si	P	S	Cl	Br	I
H	436	413	391	463	565	328	322	347	432	366	299
C		346	305	358	485	—	—	272	339	285	213
N			163	201	283	—	—	—	192	—	—
O				146	—	452	335	—	218	201	201
F					155	565	490	284	253	249	278
Si						222	—	293	381	310	234
P							201	—	326	—	184
S								226	255	—	—
Cl									242	216	208
Br										193	175
I											151

Multiple Bonds

N=N	418	C=C	610
N≡N	945	C≡C	835
C=N	615	C=O	745
C≡N	887	C≡O	1046
O=O (in O_2)	498		

*Sources of dissociation enthalpies: I. Klotz and R. M. Rosenberg: *Chemical Thermodynamics,* 4th Ed., p. 55, New York, John Wiley, 1994; and J. E. Huheey, E. A. Keiter, and R. L. Keiter: *Inorganic Chemistry* 4th Ed., Table E. 1, New York, Harper-Collins, 1993. See also Lange's *Handbook of Chemistry,* J. A. Dean (ed.), McGraw-Hill Inc., New York.

Bond Energy and Electronegativity Linus Pauling derived electronegativity values from a consideration of bond energies. He recognized that the energy required to break a bond between two different atoms is often greater than expected, based on an assumption that bond electrons are shared equally. He postulated that the "extra energy" arises from the fact that the atoms do not share electrons equally. One atom is slightly positive and the other slightly negative. This means there is a small coulombic force of attraction involving oppositely charged ions in addition to the force of attraction arising from the sharing of electrons. This coulombic force enhances the overall force of attraction.

In reactions between molecules, bonds in reactants are broken; new bonds are formed as products form. If the total energy released when new bonds form exceeds the energy required to break the original bonds, the overall reaction is exothermic. If the opposite is true, then the overall reaction is endothermic. Let us see how this works in practice.

Let us use bond dissociation enthalpies to estimate the enthalpy change for the hydrogenation of propene to propane:

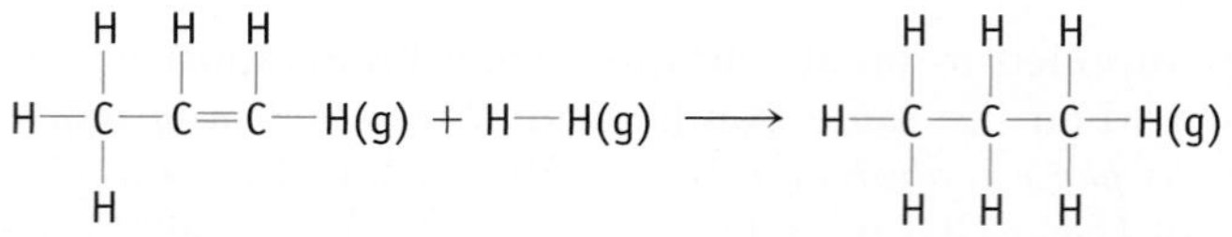

propene propane

The first step is to examine the reactants and product to see what bonds are broken and what bonds are formed. In this case, the C═C bond in propene and the H—H bond in hydrogen are broken. A C—C bond and two C—H bonds are formed.

Bonds broken: 1 mol of C═C bonds and 1 mol of H—H bonds

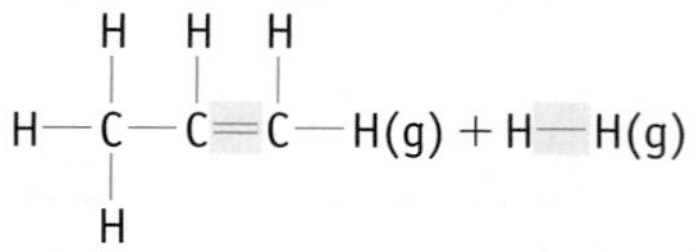

Energy required = 610 kJ for C═C bonds + 436 kJ for H—H bonds = 1046 kJ/mol-rxn

Bonds formed: 1 mol of C—C bonds and 2 mol of C—H bonds

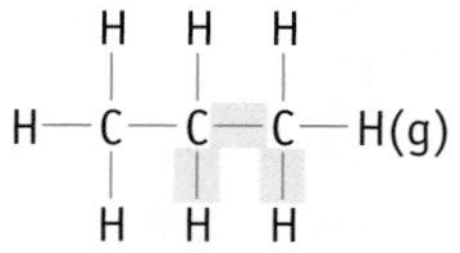

Energy evolved = 346 kJ for C—C bonds + 2 mol × 413 kJ/mol for C—H bonds = 1172 kJ/mol-rxn

Hydrogenation Reactions Adding hydrogen to a double (or triple) bond is called a hydrogenation reaction. It is commonly done to convert vegetable oils, whose molecules contain C═C double bonds, to solid fats.

By combining the energy required to break bonds and the energy evolved in making bonds, we can estimate $\Delta_r H$ for the hydrogenation of propene and see that the reaction is exothermic.

$$\Delta_r H = 1046 \text{ kJ/mol-rxn} - 1172 \text{ kJ/mol-rxn} = -126 \text{ kJ/mol-rxn}$$

The example of the propene–hydrogen reaction illustrates the fact that the enthalpy change for any reaction can be estimated using the equation

$$\Delta_r H = \Sigma\Delta H(\text{bonds broken}) - \Sigma\Delta H(\text{bonds formed}) \quad \textbf{(8.3)}$$

$\Delta_r H$ from Enthalpies of Formation Using $\Delta_f H°$ values for propane and propene, we calculate $\Delta_r H$ for the reaction of −125.1 kJ/mol-rxn. The bond dissociation enthalpy calculation is in excellent agreement with that from enthalpies of formation.

To use this equation, first identify all the bonds in the reactants that are broken, and add up their bond dissociation enthalpies. Then, identify all the new bonds formed in the products, and add up their bond dissociation enthalpies. The difference between the energy required to break bonds [= $\Sigma\Delta H$(bonds broken)] and the energy evolved when bonds are made [= $\Sigma\Delta H$(bonds formed)] gives the estimated enthalpy change for the reaction. Such calculations can give acceptable results in many cases.

Sign in at **www.thomsonedu.com/login** and go to Chapter 8 Contents to see Screen 8.18 to explore **how reactant and product bond energies influence the energy of reaction.**

EXAMPLE 8.14 Using Bond Dissociation Enthalpies

Problem Acetone, a common industrial solvent, can be converted to isopropanol, rubbing alcohol, by hydrogenation. Calculate the enthalpy change for this reaction using bond energies.

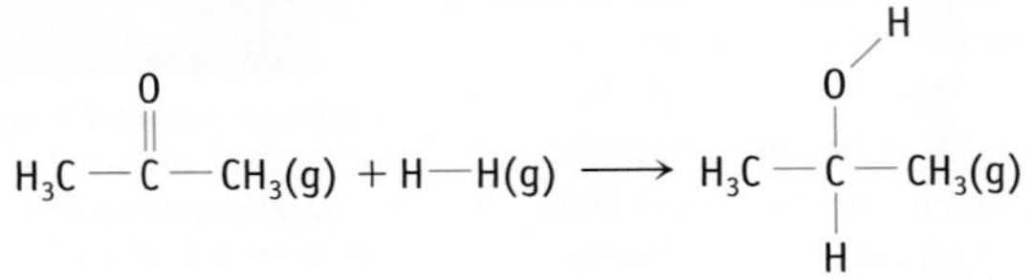

acetone isopropanol

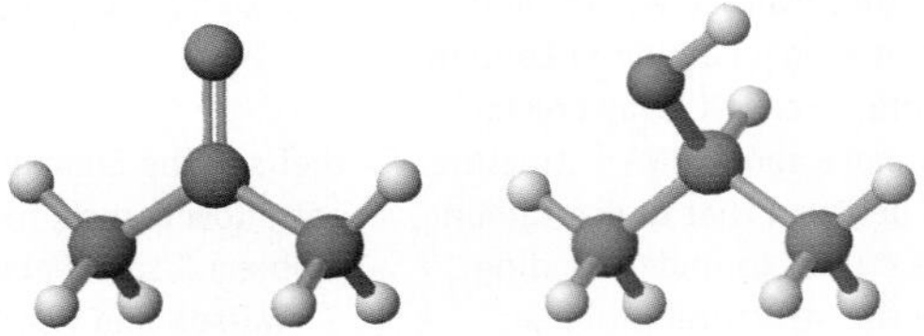

Strategy Examine the reactants and products to determine which bonds are broken and which are formed. Add up the energies required to break bonds in the reactants and the energy evolved to form bonds in the product. The difference in the sums of bond dissociation enthalpies is an estimate of the enthalpy change of the reaction (Equation 8.3).

Solution

Bonds broken: 1 mol of C═O bonds and 1 mol of H—H bonds

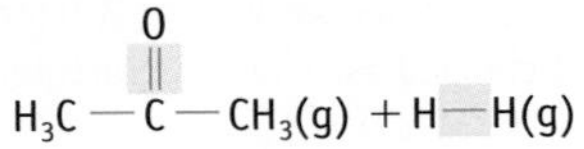

$\Sigma\Delta H$(bonds broken) = 745 kJ for C═O bonds + 436 kJ for H—H bonds = 1181 kJ/mol-rxn

Bonds formed: 1 mol of C—H bonds, 1 mol of C—O bonds, and 1 mol of O—H bonds

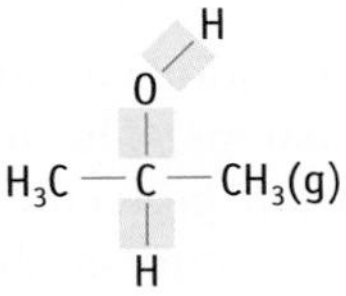

$\Sigma\Delta H$(bonds formed) = 413 kJ for C—H + 358 kJ for C—O + 463 kJ for O—H = 1234 kJ/mol-rxn

$\Delta_r H = \Sigma\Delta H$(bonds broken) $- \Sigma\Delta H$(bonds formed)

$\Delta_r H$ = 1181 kJ − 1234 kJ = −53 kJ/mol-rxn

Comment The overall reaction is predicted to be exothermic by 53 kJ per mol of product formed. This is in good agreement with the value calculated from $\Delta_f H°$ values (= −55.8 kJ/mol-rxn).

EXERCISE 8.17 Using Bond Dissociation Enthalpies

Using the bond dissociation enthalpies in Table 8.9, estimate the enthalpy of combustion of gaseous methane, CH_4. That is, estimate $\Delta_r H$ for the reaction of methane with O_2 to give water vapor and carbon dioxide gas.

Historical Perspectives — DNA—Watson, Crick, and Franklin

DNA is the substance in every plant and animal that carries the exact blueprint of that plant or animal. The structure of this molecule, the cornerstone of life, was uncovered in 1953, and James D. Watson, Francis Crick, and Maurice Wilkins shared the 1962 Nobel Prize in medicine and physiology for the work. It was one of the most important scientific discoveries of the 20th century, and the story has been told by Watson in his book *The Double Helix*.

A. Barrington Brown/Science Source/Photo Researchers, Inc.

James D. Watson and Francis Crick. In a photo taken in 1953, Watson (left) and Crick (right) stand by their model of the DNA double helix. Together with Maurice Wilkins, Watson and Crick received the Nobel Prize in medicine and physiology in 1962. (A Barrington Brown/Science/Photo Researchers, Inc.)

When Watson was a graduate student at Indiana University, he had an interest in the gene and said he hoped that its biological role might be solved "without my learning any chemistry." Later, however, he and Crick found out just how useful chemistry can be when they began to unravel the structure of DNA.

Solving important problems requires teamwork among scientists of many kinds, so Watson went to Cambridge University in England in 1951. There he met Crick, who, Watson said, talked louder and faster than anyone else. Crick shared Watson's belief in the fundamental importance of DNA, and the pair soon learned that Maurice Wilkins and Rosalind Franklin at King's College in London were using a technique called x-ray crystallography to learn more about DNA's structure. Watson and Crick believed that understanding this structure was crucial to understanding genetics. To solve the structural problem, however, they needed experimental data of the type that could come from the experiments at King's College.

The King's College group was initially reluctant to share their data; and, what is more, they did not seem to share Watson and Crick's sense of urgency. There was also an ethical dilemma: Could Watson and Crick work on a problem that others had claimed as theirs? "The English sense of fair play would not allow Francis to move in on Maurice's problem," said Watson.

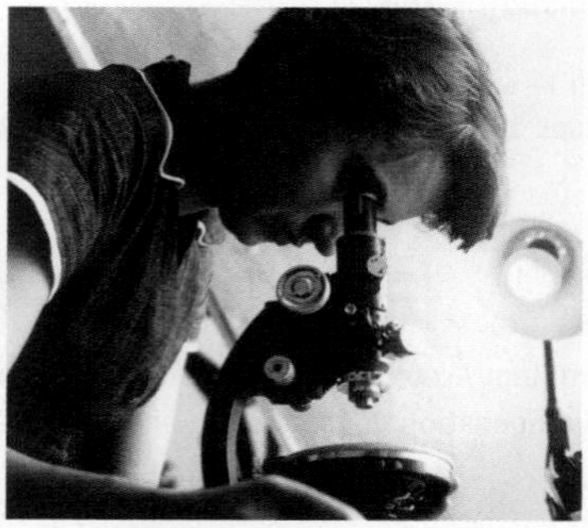
Henry Grant Collection/Museum of London

Rosalind Franklin of King's College, London. She died in 1958 at the age of 37. Because Nobel Prizes are never awarded posthumously, she did not share in this honor with Watson, Crick, and Wilkins. For more on Rosalind Franklin, read *Rosalind Franklin: The Dark Lady of DNA* by Brenda Maddox.

Watson and Crick approached the problem through a technique chemists now use frequently—model building. They built models of the pieces of the DNA chain, and they tried various chemically reasonable ways of fitting them together. Finally, they discovered that one arrangement was "too pretty not to be true." Ultimately, the experimental evidence of Wilkins and Franklin confirmed the "pretty structure" to be the real DNA structure.

8.10 DNA, Revisited

This chapter opened with some questions about the structure of DNA, one of the key molecules in all biological systems. The tools are now in place to say more about the structure of this important molecule and why it looks the way it does.

As shown in Figure 8.17, each strand of the double-stranded DNA molecule consists of three units: a phosphate, a deoxyribose molecule (a sugar molecule with a five-member ring), and a nitrogen-containing base. (The bases in DNA can be one of four molecules: adenine, guanine, cytosine, and thymine; in Figure 8.17, the base is adenine.) Two units of the backbone (without the adenine on the deoxyribose ring) are also illustrated in Figure 8.17.

The important point here is that the repeating unit in the backbone of DNA consists of the atoms O—P—O—C—C—C. Each atom has a tetrahedral electron-pair geometry. Therefore, the chain cannot be linear. In fact, the chain twists as one moves along the backbone. This twisting gives DNA its helical shape.

Why are there two strands in DNA with the O—P—O—C—C—C backbone on the outside and the nitrogen-containing bases on the inside? This structure arises from the polarity of the bonds in the base molecules attached to the backbone.

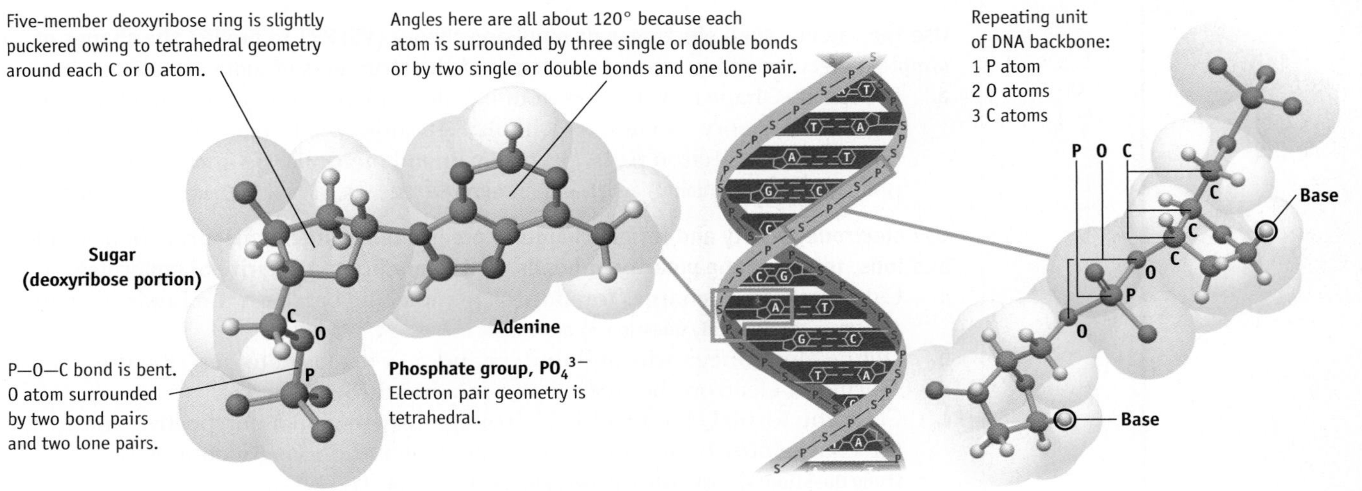

FIGURE 8.17 A portion of the DNA molecule. A repeating unit consists of a phosphate portion, a deoxyribose portion (a sugar molecule with a five-member ring), and a nitrogen-containing base (here adenine) attached to the deoxyribose ring.

For example, the N-H bonds in the adenine molecule are very polar, which leads to a special form of intermolecular forces—hydrogen bonding—to the base molecule in the neighboring chain. More about this in Chapter 12 when we explore intermolecular forces and again in *The Chemistry of Life: Biochemistry* (pages 496–512).

Chapter Goals Revisited

Now that you have studied this chapter, you should ask whether you have met the chapter goals. In particular, you should be able to:

Understand the difference between ionic and covalent bonds

a. Describe the basic forms of chemical bonding—ionic and covalent—and the differences between them, and predict from the formula whether a compound has ionic or covalent bonding, based on whether a metal is part of the formula (Section 8.1).

b. Write Lewis symbols for atoms (Section 8.2).

Draw Lewis electron dot structures for small molecules and ions

a. Draw Lewis structures for molecular compounds and ions (Section 8.2). **Study Question(s) assignable in OWL: 6, 8, 10.**

b. Understand and apply the octet rule; recognize exceptions to the octet rule (Sections 8.2–8.5). **Study Question(s) assignable in OWL: 6, 8, 10, 12, 56.**

c. Write resonance structures, understand what resonance means, and how and when to use this means of representing bonding (Section 8.4). **Study Question(s) assignable in OWL: 10.**

ChemistryNow™ Sign in at **www.thomsonedu.com/login** to:

- Assess your understanding with Study Questions in OWL keyed to each goal in the Goals and Homework menu for this chapter
- For quick review, download Go Chemistry mini-lecture flashcard modules (or purchase them at **www.ichapters.com**)
- Check your readiness for an exam by taking the Pre-Test and exploring the modules recommended in your Personalized Study plan.

Access **How Do I Solve It?** tutorials on how to approach problem solving using concepts in this chapter.

Use the valence shell electron-pair repulsion theory (VSEPR) to predict the shapes of simple molecules and ions and to understand the structures of more complex molecules.

a. Predict the shape or geometry of molecules and ions of main group elements using VSEPR theory (Section 8.6). Table 8.10 shows a summary of the relation between valence electron pairs, electron-pair and molecular geometry, and molecular polarity. **Study Question(s) assignable in OWL: 18, 20, 22, 24, 86, 88; Go Chemistry Module 12.**

Use electronegativity and formal charge to predict the charge distribution in molecules and ions, to define the polarity of bonds, and to predict the polarity of molecules.

a. Calculate formal charges for atoms in a molecule based on the Lewis structure (Section 8.3). **Study Question(s) assignable in OWL: 14, 16, 36.**

b. Define electronegativity and understand how it is used to describe the unequal sharing of electrons between atoms in a bond (Section 8.7).

c. Combine formal charge and electronegativity to gain a perspective on the charge distribution in covalent molecules and ions (Section 8.7). **Study Question(s) assignable in OWL: 28, 29, 31, 32, 34, 71.**

d. Understand why some molecules are polar whereas others are nonpolar (Section 8.8). See Table 8.7. **Study Question(s) assignable in OWL: 38.**

e. Predict the polarity of a molecule (Section 8.8). **Study Question(s) assignable in OWL: 38, 40, 78, 79, 81, 86; Go Chemistry Module 13.**

Understand the properties of covalent bonds and their influence on molecular structure

a. Define and predict trends in bond order, bond length, and bond dissociation energy (Section 8.9). **Study Question(s) assignable in OWL: 27, 42, 44, 45, 48, 58, 81.**

b. Use bond dissociation enthalpies in calculations (Section 8.9 and Example 8.14). **Study Question(s) and assignable in OWL: 50, 51, 52, 69.**

TABLE 8.10 Summary of Molecular Shapes and Molecular Polarity

Valence Electron Pairs	Electron-Pair Geometry	Number of Bond Pairs	Number of Lone Pairs	Molecular Geometry	Molecular Dipole?*	Examples
2	linear	2	0	linear	no	$BeCl_2$
3	trigonal planar	3	0	trigonal planar	no	BF_3, BCl_3
		2	1	bent	yes	$SnCl_2(g)$
4	tetrahedral	4	0	tetrahedral	no	CH_4, BF_4^-
		3	1	trigonal pyramidal	yes	NH_3, PF_3
		2	2	bent	yes	H_2O, SCl_2
5	trigonal bipyramidal	5	0	trigonal bipyramidal	no	PF_5
		4	1	seesaw	yes	SF_4
		3	2	T-shaped	yes	ClF_3
		2	3	linear	no	XeF_2, I_3^-
6	octahedral	6	0	octahedral	no	SF_6, PF_6^-
		5	1	square pyramidal	yes	ClF_5
		4	2	square planar	no	XeF_4

*For molecules of the AX_n, where the X atoms are identical.

KEY EQUATIONS

Equation 8.1 (page 359) Calculating the formal charge on an atom in a molecule

Formal charge of an atom in a molecule or ion = Group Number − [LPE + ½(BE)]

Equation 8.2 (page 387) Calculating bond order

$$\text{Bond order} = \frac{\text{number of shared pairs in all X—Y bonds}}{\text{number of X—Y links in the molecule or ion}}$$

Equation 8.3 (page 390) Estimating the enthalpy change for a reaction using bond dissociation enthalpies

$$\Delta_r H = \Sigma\Delta H(\text{bonds broken}) - \Sigma\Delta H(\text{bonds formed})$$

STUDY QUESTIONS

Online homework for this chapter may be assigned in OWL.

▲ denotes challenging questions.

■ denotes questions assignable in OWL.

Blue-numbered questions have answers in Appendix O and fully-worked solutions in the *Student Solutions Manual.*

Practicing Skills

Valence Electrons and the Octet Rule

(See Section 8.1 and ChemistryNow Screen 8.2.)

1. Give the periodic group number and number of valence electrons for each of the following atoms.
   (a) O (d) Mg
   (b) B (e) F
   (c) Na (f) S

2. Give the periodic group number and number of valence electrons for each of the following atoms.
   (a) C (d) Si
   (b) Cl (e) Se
   (c) Ne (f) Al

3. For elements in Groups 4A–7A of the periodic table, give the number of bonds an element is expected to form if it obeys the octet rule.

4. Which of the following elements are capable of forming compounds in which the indicated atom has more than four valence electron pairs?
   (a) C (d) F (g) Se
   (b) P (e) Cl (h) Sn
   (c) O (f) B

Lewis Electron Dot Structures

(See Examples 8.1, 8.2, 8.4, and 8.5, and ChemistryNow Screens 8.5–8.11.)

5. Draw a Lewis structure for each of the following molecules or ions.
   (a) NF_3 (c) HOBr
   (b) ClO_3^- (d) SO_3^{2-}

6. ■ Draw a Lewis structure for each of the following molecules or ions:
   (a) CS_2
   (b) BF_4^-
   (c) HNO_2 (where the bonding is in the order HONO)
   (d) $OSCl_2$ (where S is the central atom)

7. Draw a Lewis structure for each of the following molecules:
   (a) Chlorodifluoromethane, $CHClF_2$ (C is the central atom)
   (b) Acetic acid, CH_3CO_2H. Its basic structure is pictured.

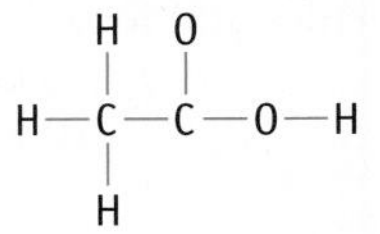

   (c) Acetonitrile, CH_3CN (the framework is $H_3C—C—N$)
   (d) Allene, H_2CCCH_2

8. ■ Draw a Lewis structure for each of the following molecules:
   (a) Methanol, CH_3OH
   (b) Vinyl chloride, $H_2C{=}CHCl$, the molecule from which PVC plastics are made.
   (c) Acrylonitrile, $H_2C{=}CHCN$, the molecule from which materials such as Orlon are made

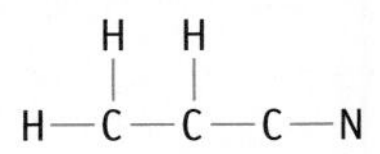

9. Show all possible resonance structures for each of the following molecules or ions:
(a) SO_2
(b) HNO_2
(c) SCN^-

10. ■ Show all possible resonance structures for each of the following molecules or ions:
(a) Nitrate ion, NO_3^-
(b) Nitric acid, HNO_3
(c) Nitrous oxide (laughing gas), N_2O (where the bonding is in the order N-N-O)

11. Draw a Lewis structure for each of the following molecules or ions:
(a) BrF_3
(b) I_3^-
(c) XeO_2F_2
(d) XeF_3^+

12. ■ Draw a Lewis structure for each of the following molecules or ions:
(a) BrF_5
(b) IF_3
(c) IBr_2^-
(d) BrF_2^+

Formal Charge

(See Example 8.3 and ChemistryNow Screen 8.11.)

13. Determine the formal charge on each atom in the following molecules or ions:
(a) N_2H_4
(b) PO_4^{3-}
(c) BH_4^-
(d) NH_2OH

14. ■ Determine the formal charge on each atom in the following molecules or ions:
(a) SCO
(b) HCO_2^- (formate ion)
(c) CO_3^{2-}
(d) HCO_2H (formic acid)

15. Determine the formal charge on each atom in the following molecules and ions:
(a) NO_2^+
(b) NO_2^-
(c) NF_3
(d) HNO_3

16. ■ Determine the formal charge on each atom in the following molecules and ions:
(a) SO_2
(b) $OSCl_2$
(c) O_2SCl_2
(d) FSO_3^-

Molecular Geometry

(See Examples 8.6 and ChemistryNow Screens 8.12–8.14. Note that many of these molecular structures are available in ChemistryNow.)

17. Draw a Lewis structure for each of the following molecules or ions. Describe the electron-pair geometry and the molecular geometry around the central atom.
(a) NH_2Cl
(b) Cl_2O (O is the central atom)
(c) SCN^-
(d) HOF

18. ■ Draw a Lewis structure for each of the following molecules or ions. Describe the electron-pair geometry and the molecular geometry around the central atom.
(a) ClF_2^+
(b) $SnCl_3^-$
(c) PO_4^{3-}
(d) CS_2

19. The following molecules or ions all have two oxygen atoms attached to a central atom. Draw a Lewis structure for each one, and then describe the electron-pair geometry and the molecular geometry around the central atom. Comment on similarities and differences in the series.
(a) CO_2
(b) NO_2^-
(c) O_3
(d) ClO_2^-

20. ■ The following molecules or ions all have three oxygen atoms attached to a central atom. Draw a Lewis structure for each one, and then describe the electron-pair geometry and the molecular geometry around the central atom. Comment on similarities and differences in the series.
(a) CO_3^{2-}
(b) NO_3^-
(c) SO_3^{2-}
(d) ClO_3^-

21. Draw a Lewis structure for each of the following molecules or ions. Describe the electron-pair geometry and the molecular geometry around the central atom.
(a) ClF_2^-
(b) ClF_3
(c) ClF_4^-
(d) ClF_5

22. ■ Draw a Lewis structure of each of the following molecules or ions. Describe the electron-pair geometry and the molecular geometry around the central atom.
(a) SiF_6^{2-}
(b) PF_5
(c) SF_4
(d) XeF_4

23. Give approximate values for the indicated bond angles.
(a) O—S—O in SO_2
(b) F—B—F angle in BF_3
(c) Cl—C—Cl angle in Cl_2CO
(d) H—C—H (angle 1) and C—C≡N (angle 2) in acetonitrile

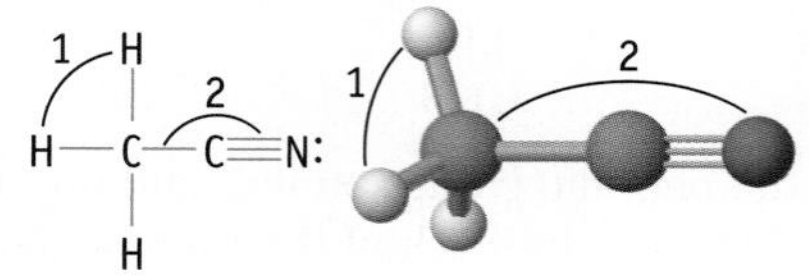

24. ■ Give approximate values for the indicated bond angles.
(a) Cl—S—Cl in SCl_2
(b) N—N—O in N_2O
(c) Bond angles 1, 2, and 3 in vinyl alcohol (a component of polymers and a molecule found in outer space).

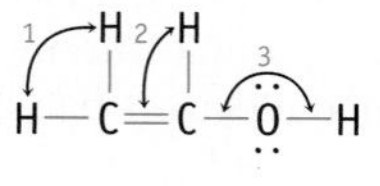

25. Phenylalanine is one of the natural amino acids and is a "breakdown" product of aspartame. Estimate the values of the indicated angles in the amino acid. Explain why the —CH_2—$CH(NH_2)$—CO_2H chain is not linear.

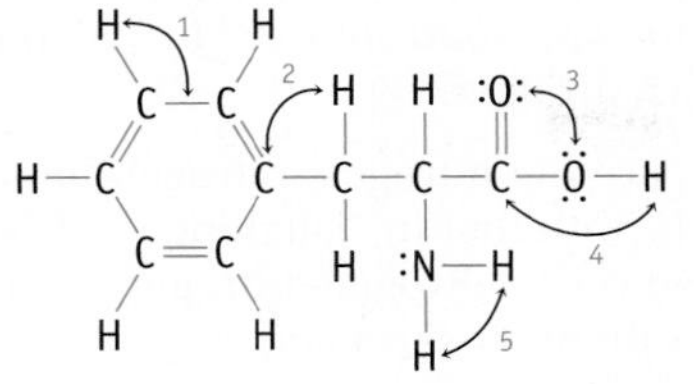

26. ■ Acetylacetone has the structure shown here. Estimate the values of the indicated angles.

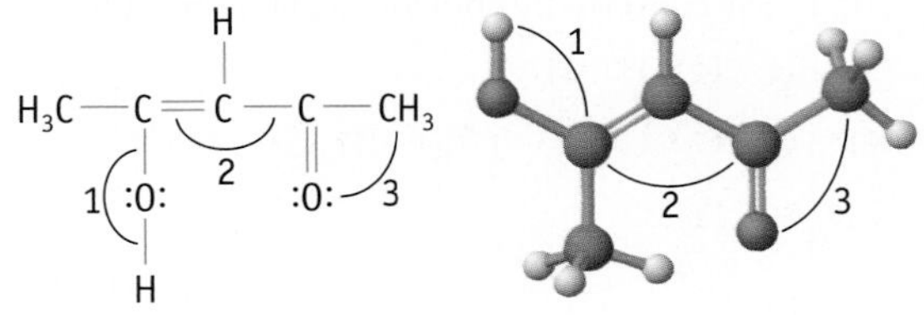

Bond Polarity, Electronegativity, and Formal Charge

(See Examples 8.10 and 8.11 and ChemistryNow Screens 8.11 and 8.15.)

27. ■ For each pair of bonds, indicate the more polar bond, and use an arrow to show the direction of polarity in each bond.
(a) C—O and C—N
(b) P—Br and P—Cl
(c) B—O and B—S
(d) B—F and B—I

28. ■ For each of the bonds listed below, tell which atom is the more negatively charged.
(a) C—N
(b) C—H
(c) C—Br
(d) S—O

29. ■ Acrolein, C_3H_4O, is the starting material for certain plastics.

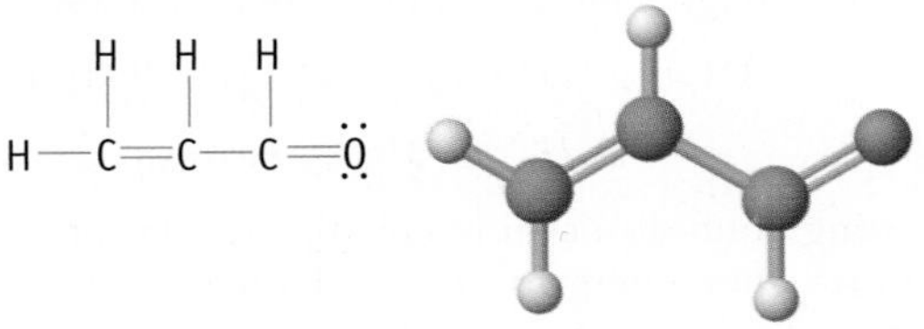

(a) Which bonds in the molecule are polar, and which are nonpolar?
(b) Which is the most polar bond in the molecule? Which is the more negative atom of this bond?

30. Urea, $(NH_2)_2CO$, is used in plastics and fertilizers. It is also the primary nitrogen-containing substance excreted by humans.
(a) Which bonds in the molecule are polar, and which are nonpolar?
(b) Which is the most polar bond in the molecule? Which atom is the negative end of the bond dipole?

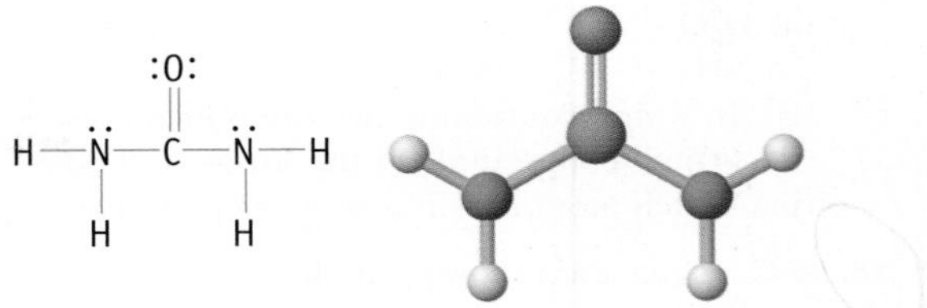

31. ■ Considering both formal charges and bond polarities, predict on which atom or atoms the negative charge resides in the following anions:
(a) OH^-
(b) BH_4^-
(c) $CH_3CO_2^-$

32. ■ Considering both formal charge and bond polarities, predict on which atom or atoms the positive charge resides in the following cations.
(a) H_3O^+
(b) NH_4^+
(c) NO_2^+
(d) NF_4^+

33. Three resonance structures are possible for dinitrogen monoxide, N_2O.
(a) Draw the three resonance structures.
(b) Calculate the formal charge on each atom in each resonance structure.
(c) Based on formal charges and electronegativity, predict which resonance structure is the most reasonable.

34. ■ Compare the electron dot structures of the carbonate (CO_3^{2-}) and borate (BO_3^{3-}) ions.
(a) Are these ions isoelectronic?
(b) How many resonance structures does each ion have?
(c) What are the formal charges of each atom in these ions?
(d) If an H^+ ion attaches to CO_3^{2-} to form the bicarbonate ion, HCO_3^-, does it attach to an O atom or to the C atom?

35. The chemistry of the nitrite ion and HNO_2:
(a) Two resonance structures are possible for NO_2^-. Draw these structures, and then find the formal charge on each atom in each resonance structure.
(b) If an H^+ ion is attached to NO_2^- (to form the acid HNO_2), it attaches to the O atom and not the N atom. Explain why you would predict this structure.
(c) Two resonance structures are possible for HNO_2. Draw these structures, and then find the formal charge on each atom in each resonance structure. Is either of these structures strongly preferred over the other?

36. ■ Draw the resonance structures for the formate ion, HCO_2^-, and find the formal charge on each atom. If an H^+ ion is attached to HCO_2^- (to form formic acid), does it attach to C or O?

Molecular Polarity

(See Examples 8.12 and 8.13 and ChemistryNow Screen 8.16.)

37. Consider the following molecules:
(a) H_2O (c) CO_2 (e) CCl_4
(b) NH_3 (d) ClF
(i) In which compound are the bonds most polar?
(ii) Which compounds in the list are *not* polar?
(iii) Which atom in ClF is more negatively charged?

38. ■ Consider the following molecules:
(a) CH_4 (c) BF_3
(b) NH_2Cl (d) CS_2
(i) Which compound has the most polar bonds?
(ii) Which compounds in the list are *not* polar?

39. Which of the following molecules is (are) polar? For each polar molecule, indicate the direction of polarity—that is, which is the negative end, and which is the positive end of the molecule.
(a) $BeCl_2$ (c) CH_3Cl
(b) HBF_2 (d) SO_3

40. ■ Which of the following molecules is (are) not polar? Which molecule has bonds with the largest polarity?
(a) CO (d) PCl_3
(b) BCl_3 (e) GeH_4
(c) CF_4

Bond Order and Bond Length

(See Exercise 8.16 and ChemistryNow Screen 8.17.)

41. Give the bond order for each bond in the following molecules or ions:
(a) CH_2O (c) NO_2^+
(b) SO_3^{2-} (d) NOCl

42. ■ Give the bond order for each bond in the following molecules or ions:
(a) CN^- (c) SO_3
(b) CH_3CN (d) $CH_3CH{=}CH_2$

43. In each pair of bonds, predict which is shorter.
(a) B—Cl or Ga—Cl (c) P—S or P—O
(b) Sn—O or C—O (d) C═O or C═N

44. ■ In each pair of bonds, predict which is shorter.
(a) Si—N or Si—O
(b) Si—O or C—O
(c) C—F or C—Br
(d) The C—N bond or the C≡N bond in $H_2NCH_2C{\equiv}N$

45. ■ Consider the nitrogen–oxygen bond lengths in NO_2^+, NO_2^-, and NO_3^-. In which ion is the bond predicted to be longest? In which is it predicted to be the shortest? Explain briefly.

46. Compare the carbon–oxygen bond lengths in the formate ion (HCO_2^-), in methanol (CH_3OH), and in the carbonate ion (CO_3^{2-}). In which species is the carbon–oxygen bond predicted to be longest? In which is it predicted to be shortest? Explain briefly.

Bond Strength and Bond Dissociation Enthalpy

(See Table 8.9, Example 8.14, and ChemistryNow Screen 8.18.)

47. ■ Consider the carbon–oxygen bond in formaldehyde (CH_2O) and carbon monoxide (CO). In which molecule is the CO bond shorter? In which molecule is the CO bond stronger?

48. ■ Compare the nitrogen–nitrogen bond in hydrazine, H_2NNH_2, with that in "laughing gas," N_2O. In which molecule is the nitrogen–nitrogen bond shorter? In which is the bond stronger?

49. Hydrogenation reactions, which involve the addition of H_2 to a molecule, are widely used in industry to transform one compound into another. For example, 1-butene (C_4H_8) is converted to butane (C_4H_{10}) by addition of H_2.

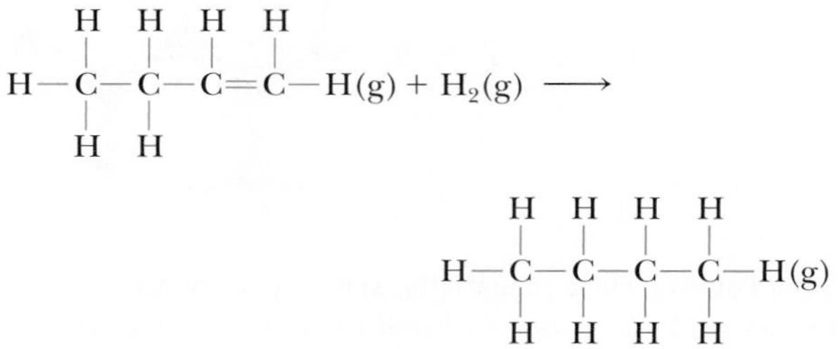

Use the bond dissociation enthalpies in Table 8.9 to estimate the enthalpy change for this hydrogenation reaction.

50. ■ Phosgene, Cl_2CO, is a highly toxic gas that was used as a weapon in World War I. Using the bond dissociation enthalpies in Table 8.9, estimate the enthalpy change for the reaction of carbon monoxide and chlorine to produce phosgene. (*Hint:* First draw the electron dot structures of the reactants and products so you know the types of bonds involved.)

$$CO(g) + Cl_2(g) \rightarrow Cl_2CO(g)$$

51. ■ The compound oxygen difluoride is quite reactive, giving oxygen and HF when treated with water:

$$OF_2(g) + H_2O(g) \rightarrow O_2(g) + 2\ HF(g)$$
$$\Delta_r H^\circ = -318\ \text{kJ/mol-rxn}$$

Using bond dissociation enthalpies, calculate the bond dissociation energy of the O—F bond in OF_2.

52. ■ Oxygen atoms can combine with ozone to form oxygen:

$$O_3(g) + O(g) \rightarrow 2\ O_2(g)$$
$$\Delta_r H^\circ = -394\ \text{kJ/mol-rxn}$$

Using $\Delta_r H^\circ$ and the bond dissociation enthalpy data in Table 8.9, estimate the bond dissociation enthalpy for the oxygen–oxygen bond in ozone, O_3. How does your estimate compare with the energies of an O—O single bond and an O═O double bond? Does the oxygen–oxygen bond dissociation enthalpy in ozone correlate with its bond order?

▲ more challenging ■ in OWL Blue-numbered questions answered in Appendix O

General Questions on Bonding and Molecular Structure

These questions are not designated as to type or location in the chapter. They may combine several concepts.

53. ■ Specify the number of valence electrons for Li, Ti, Zn, Si, and Cl.

54. In boron compounds, the B atom often is not surrounded by four valence electron pairs. Illustrate this with BCl_3. Show how the molecule can achieve an octet configuration by forming a coordinate covalent bond with ammonia (NH_3).

55. Which of the following compounds or ions do not have an octet of electrons surrounding the central atom: BF_4^-, SiF_4, SeF_4, BrF_4^-, XeF_4?

56. ■ In which of the following does the central atom obey the octet rule: NO_2, SF_4, NH_3, SO_3, ClO_2, and ClO_2^-? Are any of these species odd-electron molecules or ions?

57. Draw resonance structures for the formate ion, HCO_2^- and then determine the C—O bond order in the ion.

58. ■ Consider a series of molecules in which carbon is bonded by single bonds to atoms of second-period elements: C—O, C—F, C—N, C—C, and C—B. Place these bonds in order of increasing bond length.

59. To estimate the enthalpy change for the reaction

$$O_2(g) + 2\ H_2(g) \rightarrow 2\ H_2O(g)$$

what bond dissociation enthalpies do you need? Outline the calculation, being careful to show correct algebraic signs.

60. What is the principle of electroneutrality? Use this rule to exclude a possible resonance structure of CO_2.

61. Draw Lewis structures (and resonance structures where appropriate) for the following molecules and ions. What similarities and differences are there in this series?
(a) CO_2 (b) N_3^- (c) OCN^-

62. Draw resonance structures for the SO_2 molecule, and indicate the partial charges on the S and O atoms. Are the S—O bonds polar, and is the molecule as a whole polar? If so, what is the direction of the net dipole in SO_2? Is your prediction confirmed by the electrostatic potential surface? Explain briefly.

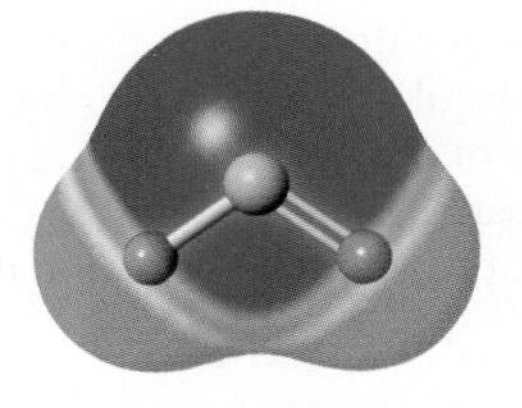

Electrostatic potential surface for sulfur dioxide.

63. What are the orders of the N—O bonds in NO_2^- and NO_2^+? The nitrogen–oxygen bond length in one of these ions is 110 pm and 124 pm in the other. Which bond length corresponds to which ion? Explain briefly.

64. Which has the greater O—N—O bond angle, NO_2^- or NO_2^+? Explain briefly.

65. Compare the F—Cl—F angles in ClF_2^+ and ClF_2^-. Using Lewis structures, determine the approximate bond angle in each ion. Decide which ion has the greater bond angle, and explain your reasoning.

66. Draw an electron dot structure for the cyanide ion, CN^-. In aqueous solution, this ion interacts with H^+ to form the acid. Should the acid formula be written as HCN or CNH?

67. Draw the electron dot structure for the sulfite ion, SO_3^{2-}. In aqueous solution, the ion interacts with H^+. Predict whether a H^+ ion will attach to the S atom or the O atom of SO_3^{2-}.

68. Dinitrogen monoxide, N_2O, can decompose to nitrogen and oxygen gas:

$$2\ N_2O(g) \rightarrow 2\ N_2(g) + O_2(g)$$

Use bond dissociation enthalpies to estimate the enthalpy change for this reaction.

69. ▲ ■ The equation for the combustion of gaseous methanol is

$$2\ CH_3OH(g) + 3\ O_2(g) \rightarrow 2\ CO_2(g) + 4\ H_2O(g)$$

(a) Using the bond dissociation enthalpies in Table 8.9, estimate the enthalpy change for this reaction. What is the enthalpy of combustion of one mole of gaseous methanol?
(b) Compare your answer in part (a) with a calculation of $\Delta_r H^\circ$ using thermochemical data and the methods of Chapter 5 (see Equation 5.6).

70. ▲ Acrylonitrile, C_3H_3N, is the building block of the synthetic fiber Orlon.

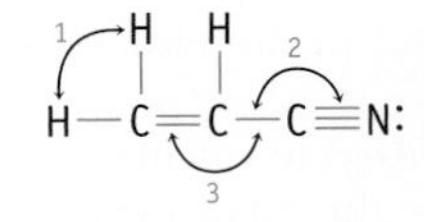

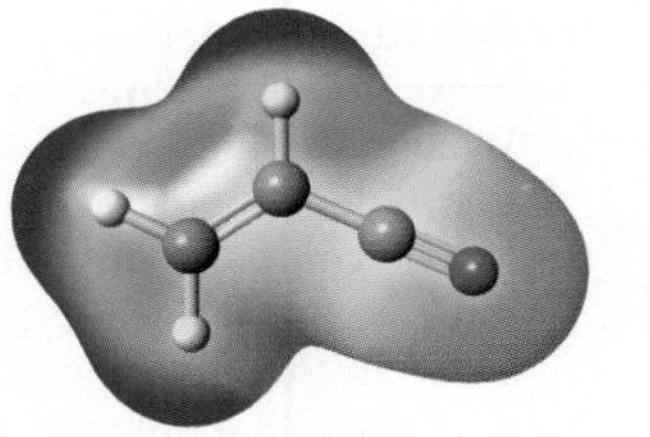

Electrostatic potential surface for acrylonitrile.

(a) Give the approximate values of angles 1, 2, and 3.
(b) Which is the shorter carbon–carbon bond?
(c) Which is the stronger carbon–carbon bond?
(d) Based on the electrostatic potential surface, where are the positive and negative charges located in the molecule?
(e) Which is the most polar bond?
(f) Is the molecule polar?

71. ▲ ■ The cyanate ion, NCO^-, has the least electronegative atom, C, in the center. The very unstable fulminate ion, CNO^-, has the same formula, but the N atom is in the center.
    (a) Draw the three possible resonance structures of CNO^-.
    (b) On the basis of formal charges, decide on the resonance structure with the most reasonable distribution of charge.
    (c) Mercury fulminate is so unstable it is used in blasting caps. Can you offer an explanation for this instability? (*Hint:* Are the formal charges in any resonance structure reasonable in view of the relative electronegativities of the atoms?)

72. Vanillin is the flavoring agent in vanilla extract and in vanilla ice cream. Its structure is shown here:

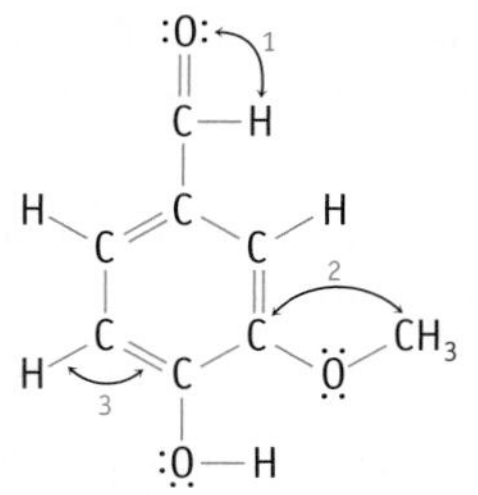

    (a) Give values for the three bond angles indicated.
    (b) Indicate the shortest carbon–oxygen bond in the molecule.
    (c) Indicate the most polar bond in the molecule.

73. ▲ Given that the spatial requirement of a lone pair is greater than that of a bond pair, explain why
    (a) XeF_2 has a linear molecular structure and not a bent one.
    (b) ClF_3 has a T-shaped structure and not a trigonal-planar one.

74. The formula for nitryl chloride is $ClNO_2$.
    (a) Draw the Lewis structure for the molecule, including all resonance structures.
    (b) What is the N—O bond order?
    (c) Describe the electron-pair and molecular geometries, and give values for all bond angles.
    (d) What is the most polar bond in the molecule? Is the molecule polar?
    (e) ▲ The computer program used to calculate electrostatic potential surfaces gave the following charges on atoms in the molecule: A = −0.03, B = −0.26, and C = +0.56. Identify the atoms A, B, and C. Are these calculated charges in accord with your predictions?

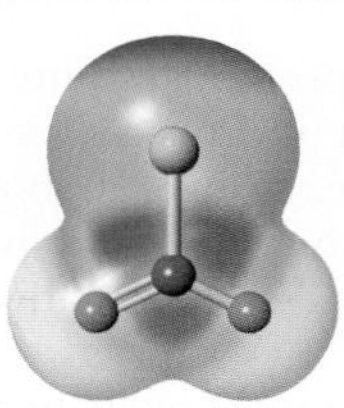

Electrostatic potential surface for $ClNO_2$.

75. Hydroxyproline is a less-common amino acid.

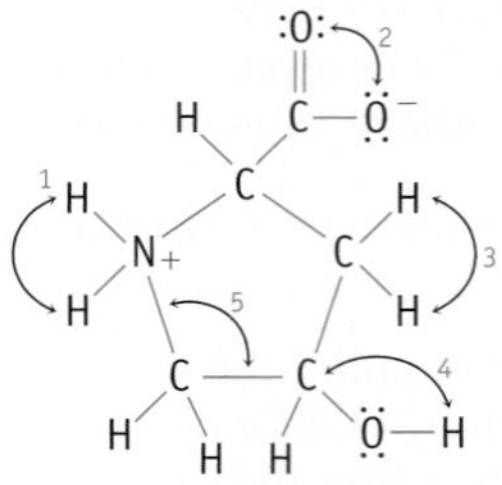

    (a) ■ Give approximate values for the indicated bond angles.
    (b) Which are the most polar bonds in the molecule?

76. Amides are an important class of organic molecules. They are usually drawn as sketched here, but another resonance structure is possible.

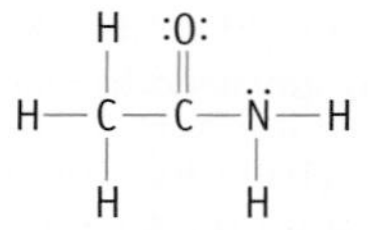

    (a) Draw that structure, and then suggest why it is usually not pictured.
    (b) Suggest a reason for the fact that the H—N—H angle is close to 120°.

77. Use the bond dissociation enthalpies in Table 8.9 to calculate the enthalpy change for the decomposition of urea (Study Question 30) to hydrazine, $H_2N—NH_2$, and carbon monoxide. (Assume all compounds are in the gas phase.)

78. The molecule shown here, 2-furylmethanethiol, is responsible for the aroma of coffee:

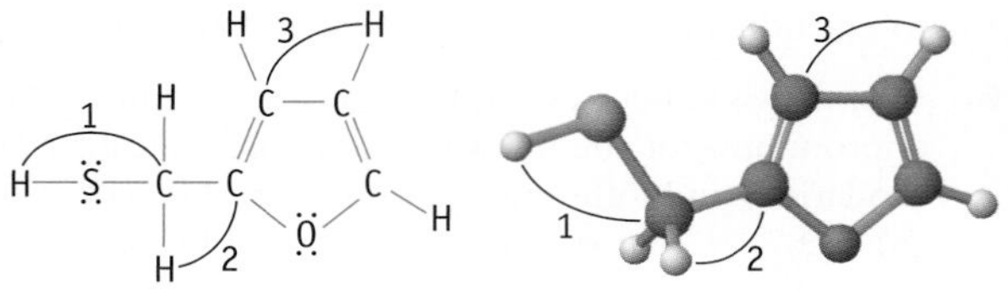

    (a) What are the formal charges on the S and O atoms?
    (b) ■ Give approximate values of angles 1, 2, and 3.
    (c) Which are the shorter carbon–carbon bonds in the molecule?
    (d) Which bond in this molecule is the most polar?
    (e) Is the molecule as a whole polar or nonpolar?
    (f) The molecular model makes it clear that the four C atoms of the ring are all in a plane. Is the O atom in that same plane (making the five-member ring planar), or is the O atom bent above or below the plane?

▲ more challenging ■ in OWL Blue-numbered questions answered in Appendix O

79. ▲ ■ Dihydroxyacetone is a component of quick-tanning lotions. (It reacts with the amino acids in the upper layer of skin and colors them brown in a reaction similar to that occurring when food is browned as it cooks.)
   (a) Supposing you can make this compound by treating acetone with oxygen, use bond dissociation enthalpies to estimate the enthalpy change for the following reaction (which is assumed to occur in the gas phase). Is the reaction exothermic or endothermic?

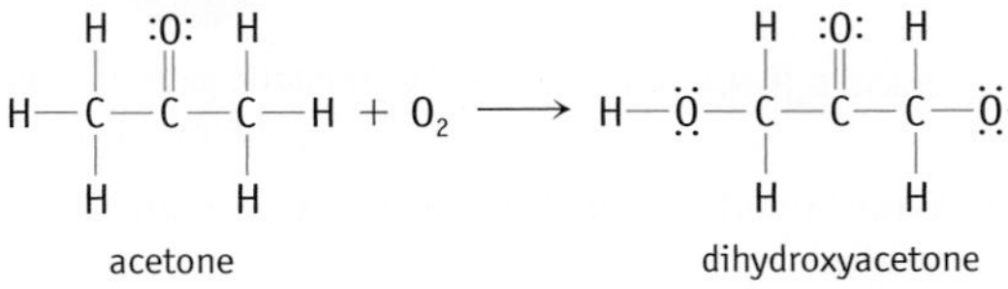

   (b) Is acetone polar?
   (c) Positive H atoms can sometimes be removed (as H^+) from molecules with strong bases (which is in part what happens in the tanning reaction). Which H atoms are the most positive in dihydroxyacetone?

80. Nitric acid, HNO_3, has three resonance structures. One of them, however, contributes much less to the resonance hybrid than the other two. Sketch the three resonance structures, and assign a formal charge to each atom. Which one of your structures is the least important?

81. ▲ ■ Acrolein is used to make plastics. Suppose this compound can be prepared by inserting a carbon monoxide molecule into the C—H bond of ethylene.

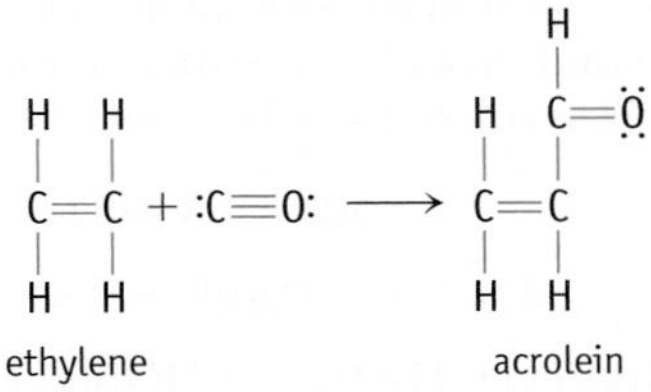

   (a) Which is the stronger carbon–carbon bond in acrolein?
   (b) Which is the longer carbon–carbon bond in acrolein?
   (c) Is ethylene or acrolein polar?
   (d) Is the reaction of CO with C_2H_4 to give acrolein endothermic or exothermic?

82. Molecules in space:
   (a) In addition to molecules such as CO, HCl, H_2O, and NH_3, glycolaldehyde has been detected in outer space. Is the molecule polar?

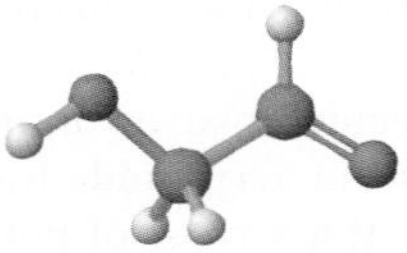

$HOCH_2CHO$, glycolaldehyde.

   (b) Where do the positive and negative charges lie in the molecule?
   (c) One molecule found in the 1995 Hale-Bopp comet is HC_3N. Suggest a structure for this molecule.

83. 1,2-Dichloroethylene can be synthesized by adding Cl_2 to the carbon–carbon triple bond of acetylene.

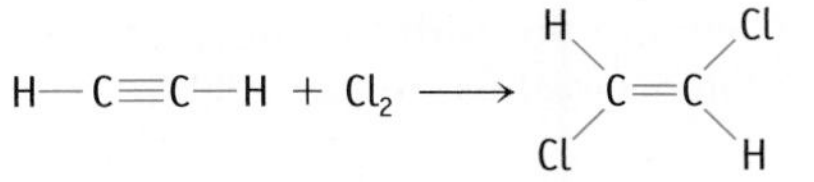

Using bond dissociation enthalpies, estimate the enthalpy change for this reaction in the gas phase.

84. The molecule pictured below is epinephrine, a compound used as a bronchodilator and antiglaucoma agent.

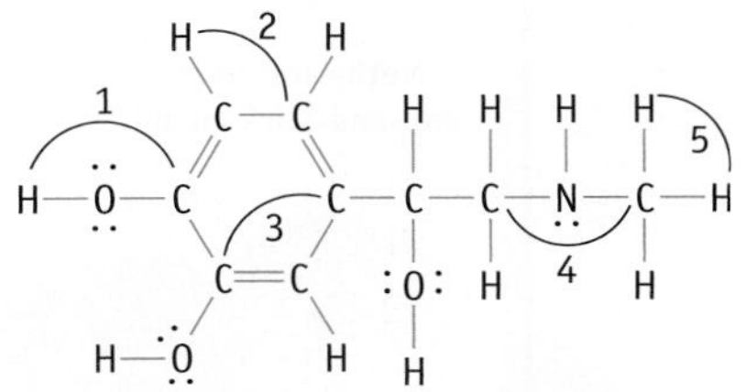

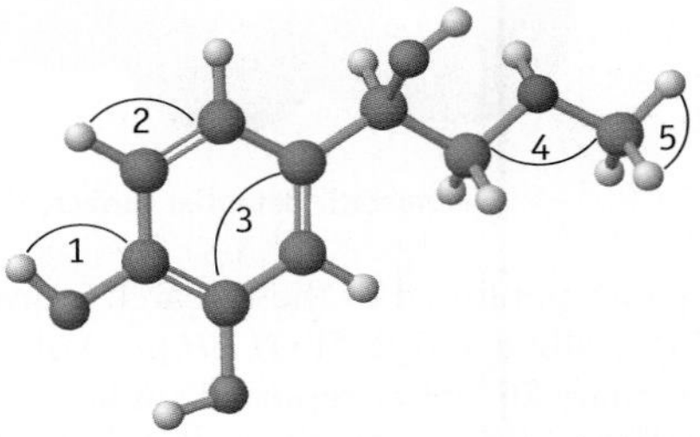

   (a) ■ Give a value for each of the indicated bond angles.
   (b) What are the most polar bonds in the molecule?

In the Laboratory

85. You are doing an experiment in the laboratory and want to prepare a solution in a polar solvent. Which solvent would you choose, methanol (CH_3OH) or toluene ($C_6H_5CH_3$)? Explain your choice.

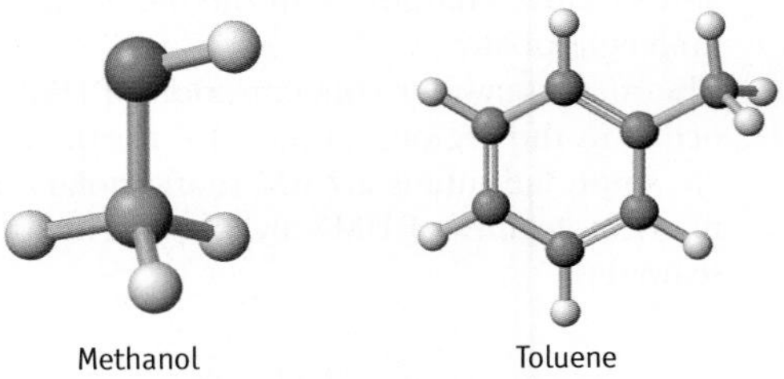

Methanol (left) and toluene (right).

86. ■ Methylacetamide, $CH_3CONHCH_3$, is a small molecule with an amide link (CO—NH), the group that binds one amino acid to another in proteins.
(a) Is this molecule polar?
(b) Where do you expect the positive and negative charges to lie in this molecule? Does the electrostatic potential surface confirm your predictions?. (Compare this with the dipeptide model in the *A Closer Look* box on page 382.)

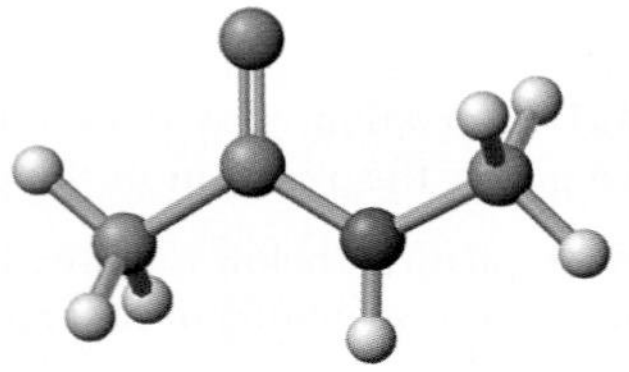

Methylacetamide Ball-and-stick model.

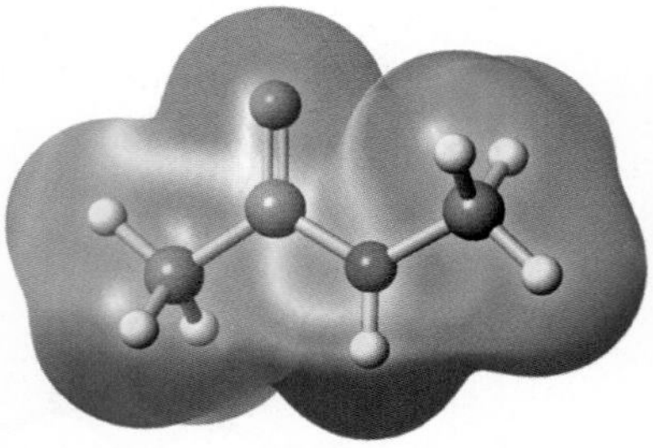

Electrostatic potential surface.

87. ▲ A paper published in the research journal *Science* in 2007 (S. Vallina and R. Simo, *Science*, Vol. 315, page 506, January 26, 2007) reported studies of dimethylsulfide (DMS), an important greenhouse gas that is released by marine phytoplankton. This gas "represents the largest natural source of atmospheric sulfur and a major precursor of hygroscopic (i.e., cloud-forming) particles in clean air over the remote oceans, thereby acting to reduce the amount of solar radiation that crosses the atmosphere and is absorbed by the ocean."
(a) Sketch the Lewis structure of dimethylsulfide, CH_3SCH_3, and give the unique bond angles in the molecule.
(b) Use electronegativities to decide where the positive and negative charges lie in the molecule. Is the molecule polar?
(c) The mean seawater concentration of DMS in the ocean in the region between 15° north latitude and 15° south latitude is 2.7 nM (nanomolar). How many molecules of DMS are present in 1.0 m^3 of seawater?

88. ■ Uracil is one of the bases in DNA.

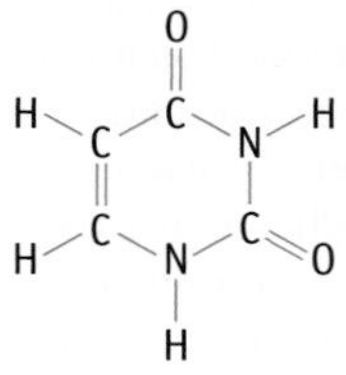

Uracil, $C_4H_4N_2O_2$.

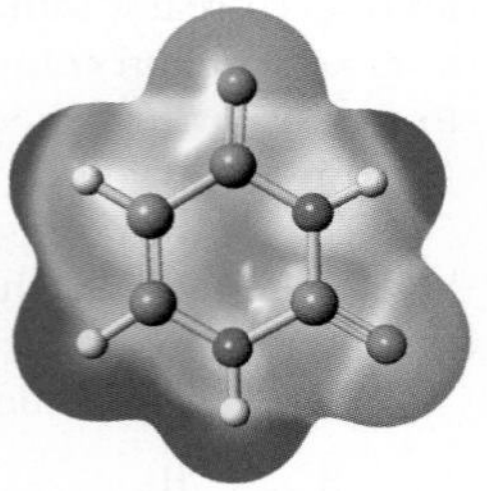

Electrostatic potential surface for uracil.

(a) What are the values of the O—C—N and C—N—H angles?
(b) There are two carbon–carbon bonds in the molecule. Which is predicted to be shorter?
(c) If a proton attacks the molecule, decide on the basis of the electrostatic potential surface to which atom or atoms it could be attached.

Summary and Conceptual Questions

The following questions may use concepts from this and previous chapters.

89. Bromine-containing species play a role in environmental chemistry. For example, they are evolved in volcanic eruptions.
(a) The following molecules are important in bromine environmental chemistry: HBr, BrO, and HOBr. Which are odd-electron molecules?
(b) Use bond dissociation enthalpies to estimate $\Delta_r H$ for three reactions of bromine:

$$Br_2(g) \longrightarrow 2\ Br(g)$$

$$2\ Br(g) + O_2(g) \longrightarrow 2\ BrO(g)$$

$$BrO(g) + H_2O(g) \longrightarrow HOBr(g) + OH(g)$$

(c) Using bond dissociation enthalpies, estimate the standard enthalpy of formation of HOBr(g) from $H_2(g)$, $O_2(g)$, and $Br_2(g)$.
(d) Are the reactions in parts (b) and (c) exothermic or endothermic?

90. Acrylamide, $H_2C{=}CHC({=}O)NH_2$, is a known neurotoxin and possible carcinogen. It was a shock to all consumers of potato chips and french fries a few years ago when it was found to occur in those products (page 90).
(a) Sketch the molecular structure of acrylamide, showing all unique bond angles.
(b) Indicate which carbon–carbon bond is the stronger of the two.
(c) Is the molecule polar or nonpolar?
(d) The amount of acrylamide found in potato chips is 1.7 mg/kg. If a serving of potato chips is 28 g, how many moles of acrylamide are you consuming?

91. See ChemistryNow Screen 8.16, Molecular Polarity. Use the Molecular Polarity tool on this screen to explore the polarity of molecules.
(a) Is BF_3 a polar molecule? Does the molecular polarity change as the F atoms of BF_3 are replaced by H atoms?
(b) Is $BeCl_2$ a polar molecule? Does the polarity change when Cl is replaced by Br?

92. Locate the molecules in the table shown here in the Molecular Models available in ChemistryNow. Measure the carbon–carbon bond length in each, and complete the table. (Note that the bond lengths are given in angstrom units, where 1 Å = 0.1 nm.)

Formula	Measured Bond Distance (Å)	Bond Order
ethane, C_2H_6	________	________
butane, C_4H_{10}	________	________
ethylene, C_2H_4	________	________
acetylene, C_2H_2	________	________
benzene, C_6H_6	________	________

What relationship between bond order and carbon–carbon bond length do you observe?

9 Bonding and Molecular Structure: Orbital Hybridization and Molecular Orbitals

©Gary J. Schrobilgen

White crystals of xenon difluoride, XeF_2, form when a mixture of Xe and F_2 gases is irradiated with UV light.

The Chemistry of the Noble Gases

It was a shock when, in 1962, we learned that the noble gases were not chemically inert as our chemistry professors had taught us. Xenon at the very least was found to form compounds! The first was an ionic compound, now known to be $XeF^+Pt_2F_{11}^-$. However, this was followed shortly thereafter with the discovery of a large number of covalently bonded compounds, including XeF_4, XeF_6, $XeOF_4$, and XeO_3.

Since 1962, the field of noble gas chemistry has expanded with the discovery of such interesting molecules as FXeOXeF and, at low temperatures, species such as HArF, HXeH, HXeCl, and even HKrF.

Initially, xenon compounds were thought to form only under the most severe conditions. Therefore, it was again a surprise when it was learned that irradiating a mixture of xenon and fluorine gases at room temperature gave crystals of XeF_2 (as seen in the photo).

Questions:

1. What is the most reasonable structure of XeF_2? Does knowing that the molecule has no dipole moment confirm your structural choice? Why or why not?
2. Describe the bonding in XeF_2 using valence bond theory.
3. Predict a structure for FXeOXeF.

Answers to these questions are in Appendix Q.

Chapter Goals

See Chapter Goals Revisited (page 433) for Study Questions keyed to these goals and assignable in OWL.

- Understand the differences between valence bond theory and molecular orbital theory.
- Identify the hybridization of an atom in a molecule or ion.
- Understand the differences between bonding and antibonding molecular orbitals and be able to write the molecular orbital configurations for simple diatomic molecules.

Chapter Outline

Just how are molecules held together? How can two molecules with distinctly different properties have the same formula? Why is oxygen paramagnetic, and how is this property connected with bonding in the molecule? These are just a few of the fundamental and interesting questions that are raised in this chapter and that require us to take a more advanced look at bonding.

Chemistry Now™

Throughout the text this icon introduces an opportunity for self-study or to explore interactive tutorials by signing in at **www.thomsonedu.com/login**.

9.1 Orbitals and Theories of Chemical Bonding

From Chapter 6, you know that the location of the valence electrons in atoms is described by an orbital model. It seems reasonable that an orbital model could also be used to describe electrons in molecules.

Two common approaches to rationalizing chemical bonding based on orbitals are **valence bond (VB) theory** and **molecular orbital (MO) theory.** The former was developed largely by Linus Pauling (page 377) and the latter by another American scientist, Robert S. Mulliken (1896–1986). The valence bond approach is closely tied to Lewis's idea of bonding electron pairs between atoms and lone pairs of electrons localized on a particular atom. In contrast, Mulliken's approach was to derive molecular orbitals that are "spread out," or *delocalized*, over the molecule. One way to do this is to combine atomic orbitals to form a set of orbitals that are the property of the molecule, and then distribute the electrons of the molecule within these orbitals.

Why are two theories used? Is one more correct than the other? Actually, both give good descriptions of the bonding in molecules and polyatomic ions, but they are used for different purposes. Valence bond theory is generally the method of choice to provide a qualitative, visual picture of molecular structure and bonding. This theory is particularly useful for molecules made up of many atoms. In contrast, molecular orbital theory is used when a more quantitative picture of bonding is needed. Furthermore, valence bond theory provides a good description of bonding for molecules in their ground, or lowest, energy state. On the other hand, MO theory is essential if we want to describe molecules in higher energy, excited states. Among other things, this is important in explaining the colors of compounds. Finally, for a few molecules such as NO and O_2, MO theory is the only theory that can describe their bonding accurately.

■ **Bonds Are a "Figment of Our Own Imagination"** C. A. Coulson, a prominent theoretical chemist at the University of Oxford, England, has said that "Sometimes it seems to me that a bond between atoms has become so real, so tangible, so friendly, that I can almost see it. Then I awake with a little shock, for a chemical bond is not a real thing. It does not exist. No one has ever seen one. No one ever can. It is a figment of our own imagination" (*Chemical and Engineering News,* January 29, 2007, page 37). Nonetheless, bonds are a useful figment, and this chapter will present some of these useful ideas.

9.2 Valence Bond Theory

The Orbital Overlap Model of Bonding

What happens if two atoms at an infinite distance apart are brought together to form a bond? This process is often illustrated with H_2 because, with two electrons and two nuclei, this is the simplest molecular compound known (Figure 9.1). Initially, when two hydrogen atoms are widely separated, they do not interact. If the atoms move closer together, however, the electron on one atom begins to experience an attraction to the positive charge of the nucleus of the other atom. Because of the attractive forces, the electron clouds on the atoms distort as the electron of one atom is drawn toward the nucleus of the second atom, and the potential energy of the system is lowered. Calculations show that when the distance between the H atoms is 74 pm, the potential energy reaches a minimum and the H_2 molecule is most stable. Significantly, 74 pm corresponds to the experimentally measured bond distance in the H_2 molecule.

■ **Bonds in Valence Bond Theory** In the language of valence bond theory, a pair of electrons of opposite spin located between a pair of atoms constitutes a bond.

Individual hydrogen atoms each have a single electron. In H_2 the two electrons pair up to form the bond. There is a net stabilization, representing the extent to which the energies of the two electrons are lowered from their value in the free atoms. The net stabilization (the extent to which the potential energy is lowered) can be calculated, and the calculated value approximates the experimentally determined bond energy. Agreement between theory and experiment on both bond distance and energy is evidence that this theoretical approach has merit.

Bond formation is depicted in Figures 9.1 and 9.2 as occurring when the electron clouds on the two atoms interpenetrate or overlap. This **orbital overlap** increases the probability of finding the bonding electrons in the region of space between the two nuclei. *The idea that bonds are formed by overlap of atomic orbitals is the basis for valence bond theory.*

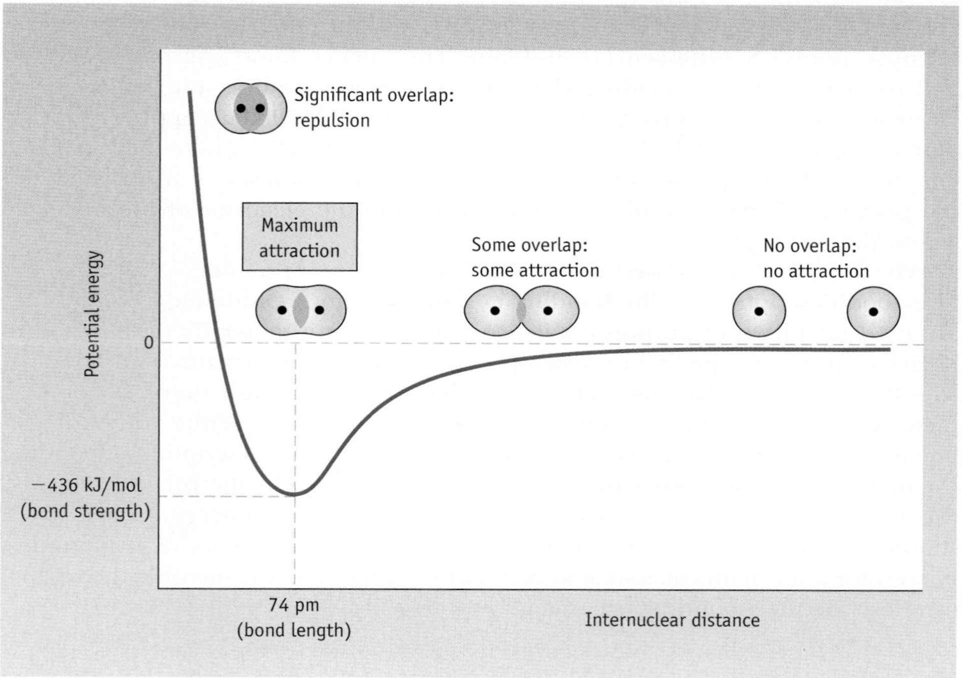

Active Figure 9.1 Potential energy change during H–H bond formation from isolated hydrogen atoms. The lowest energy is reached at an H—H separation of 74 pm, where there is overlap of 1*s* orbitals. At greater distances, the overlap is less, and the bond is weaker. At H—H distances less than 74 pm, repulsions between the nuclei and between the electrons of the two atoms increase rapidly, and the potential energy curve rises steeply.

ChemistryNow™ Sign in at www.thomsonedu.com/login and go to the Chapter Contents menu to explore an interactive version of this figure accompanied by an exercise.

When the single covalent bond is formed in H_2, the $1s$ electron cloud of each atom is distorted in a way that gives the electrons a higher probability of being in the region between the two hydrogen atoms (Figure 9.2a). This makes sense because this distortion results in the electrons being situated so they can be attracted equally to the two positively charged nuclei. Placing the electrons between the nuclei also matches the Lewis electron dot model.

The covalent bond that arises from the overlap of two s orbitals, one from each of two atoms as in H_2, is called a **sigma (σ) bond.** *The electron density of a sigma bond is greatest along the axis of the bond.*

In summary, the main points of the valence bond approach to bonding are:

- Orbitals overlap to form a bond between two atoms.
- Two electrons, *of opposite spin,* can be accommodated in the overlapping orbitals. Usually, one electron is supplied by each of the two bonded atoms.
- Because of orbital overlap, the bonding electrons have a higher probability of being found within a region of space influenced by both nuclei. Both electrons are simultaneously attracted to both nuclei.

What happens for elements beyond hydrogen? In the Lewis structure of HF, for example, a bonding electron pair is placed between H and F, and three lone pairs of electrons are depicted as localized on the F atom (Figure 9.2b). To use an orbital approach, look at the valence shell electrons and orbitals for each atom that can overlap. The hydrogen atom will use its $1s$ orbital in bond formation. The electron configuration of fluorine is $1s^22s^22p^5$, and the unpaired electron for this atom is assigned to one of the $2p$ orbitals. A sigma bond results from overlap of the hydrogen $1s$ and the fluorine $2p$ orbital.

Formation of the H—F bond is similar to formation of an H—H bond. A hydrogen atom approaches a fluorine atom along the axis containing the $2p$ orbital with

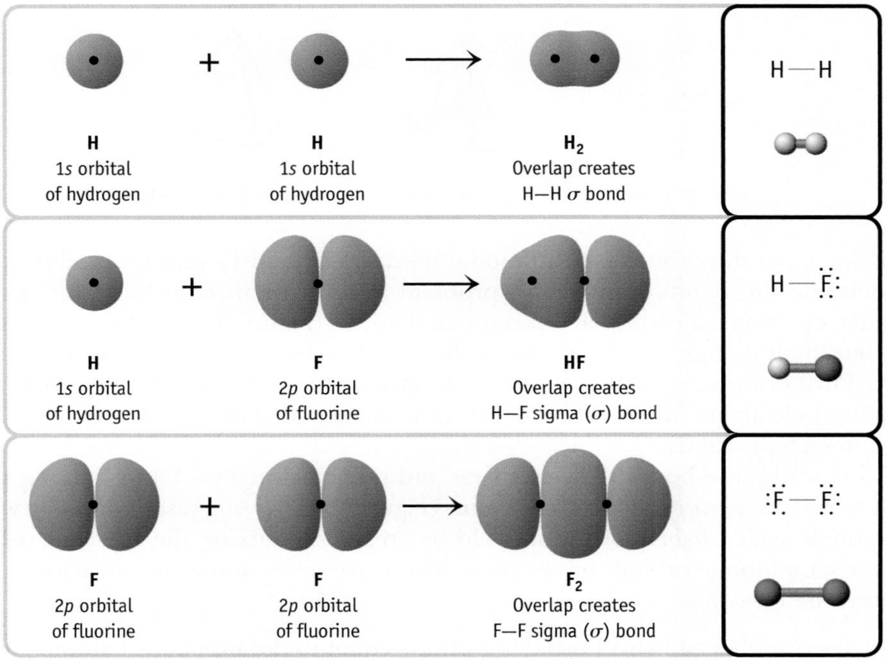

FIGURE 9.2 Covalent bond formation in H_2, HF, and F_2.

(a) Overlap of hydrogen $1s$ orbitals to form the H—H sigma (σ) bond.

(b) Overlap of hydrogen $1s$ and fluorine $2p$ orbitals to form the sigma (σ) bond in HF.

(c) Overlap of $2p$ orbitals on two fluorine atoms forming the sigma (σ) bond in F_2.

a single electron. The orbitals (1*s* on H and 2*p* on F) distort as each atomic nucleus influences the electron and orbital of the other atom. Still closer together, the 1*s* and 2*p* orbitals overlap, and the two electrons, with opposite spins, pair up to give a σ bond (Figure 9.2b). There is an optimum distance (92 pm) at which the energy is lowest, and this corresponds to the bond distance in HF. The net stabilization achieved in this process is the energy for the H—F bond.

The remaining electrons on the fluorine atom (a pair of electrons in the 2*s* orbital and two pairs of electrons in the other two 2*p* orbitals) are not involved in bonding. They are nonbonding electrons, the lone pairs associated with this element in the Lewis structure.

Extension of this model gives a description of bonding in F_2. The 2*p* orbitals on the two atoms overlap, and the single electron from each atom is paired in the resulting σ bond (Figure 9.2c). The 2*s* and the 2*p* electrons not involved in the bond are the lone pairs on each atom.

Chemistry Now™

Sign in at **www.thomsonedu.com/login** and go to Chapter 9 Contents to see Screen 9.3 for an exercise on **bond formation.**

Hybridization of Atomic Orbitals

The simple picture using orbital overlap to describe bonding in H_2, HF, and F_2 works well, but we run into difficulty when molecules with more atoms are considered. For example, a Lewis dot structure of methane, CH_4, shows four C—H covalent bonds. VSEPR theory predicts, and experiments confirm, that the electron-pair geometry of the C atom in CH_4 is tetrahedral, with an angle of 109.5° between the bond pairs. The hydrogens are identical in this structure. This means that four equivalent bonding electron pairs occur around the C atom. An orbital picture of the bonds should convey both the geometry and the fact that all C—H bonds are the same.

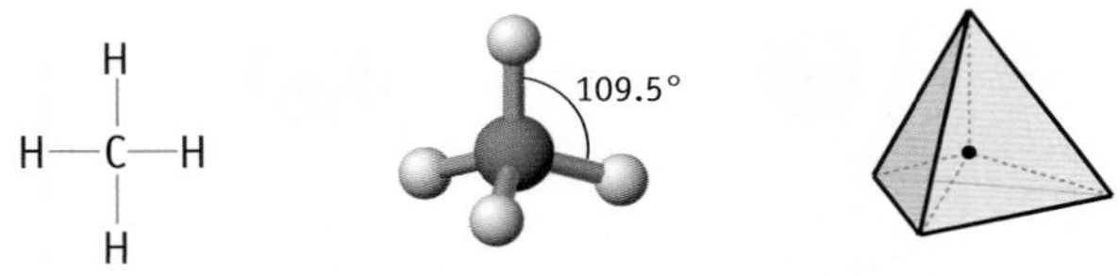

Lewis structure | Molecular model | Electron-pair geometry

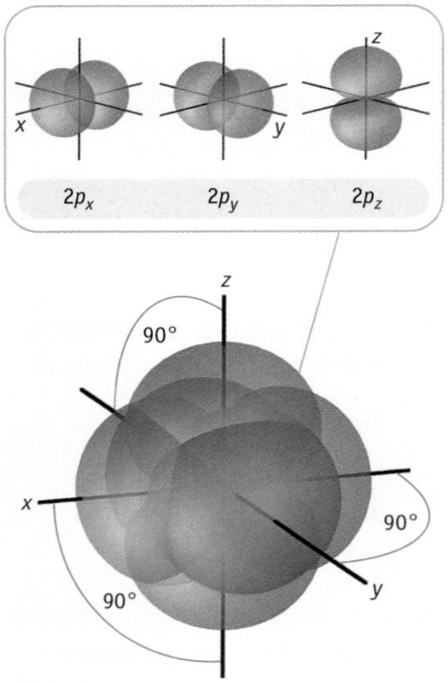

FIGURE 9.3 The 2*p* orbitals of an atom. The $2p_x$, $2p_y$, and $2p_z$ orbitals lie along the *x*-, *y*-, and *z*-axes, 90° to each other.

If we apply the orbital overlap model used for H_2 and F_2 without modification to describe the bonding in CH_4, a problem arises. The three orbitals for the 2*p* valence electrons of carbon are at right angles, 90° (Figure 9.3), and do not match the tetrahedral angle of 109.5°. The spherical 2*s* orbital could bond in any direction. Furthermore, a carbon atom in its ground state ($1s^22s^22p^2$) has only two unpaired electrons (in the 2*p* orbitals), not the four that are needed to allow formation of four bonds.

To describe the bonding in methane and other molecules, Linus Pauling proposed the theory of **orbital hybridization** (Figure 9.4). He suggested that a new set of orbitals, called **hybrid orbitals,** could be created by mixing the *s*, *p*, and (when required) *d* atomic orbitals on an atom. There are three important principles that govern the outcome.

- The number of hybrid orbitals is always equal to the number of atomic orbitals that are mixed to create the hybrid orbital set.

Charles D. Winters

FIGURE 9.4 Hybridization: an analogy. Atomic orbitals can mix, or hybridize, to form hybrid orbitals. When two atomic orbitals on an atom combine, two new orbitals are produced on that atom. The new orbitals have a different direction in space than the original orbitals. An analogy is mixing two different colors (left) to produce a third color, which is a "hybrid" of the original colors (center). After mixing, there are still two beakers (right), each containing the same volume of solution as before, but the color is a "hybrid" color.

- Hybrid orbital sets are always built by combining an *s* orbital with as many *p* orbitals (and *d* orbitals if necessary) to have enough hybrid orbitals to accommodate the bond and lone pairs on the central atom.
- The hybrid orbitals are directed toward the terminal atoms, leading to better orbital overlap and a stronger bond between the central and terminal atoms.

Chemistry Now™

Sign in at **www.thomsonedu.com/login** and go to Chapter 9 Contents to see:
- Screen 9.4 for exercises on **hybrid orbitals**
- Screen 9.6 for a tutorial on **determining hybrid orbitals**

The sets of hybrid orbitals that arise from mixing *s*, *p*, and *d* atomic orbitals are illustrated in Figure 9.5. *The hybrid orbitals required by an atom in a molecule or ion are chosen to match the electron pair geometry of the atom* because a hybrid orbital is required for each sigma bond electron pair and each lone pair. The following types of hybridization are important:

- ***sp***: If the valence shell *s* orbital on the central atom in a molecule or ion is mixed with a valence shell *p* orbital on that same atom, two *sp* hybrid orbitals are created. They are separated by 180°.
- sp^2: If an *s* orbital is combined with two *p* orbitals, all in the same valence shell, three sp^2 hybrid orbitals are created. They are in the same plane and are separated by 120°.
- sp^3: When the *s* orbital in a valence shell is combined with three *p* orbitals, the result is four hybrid orbitals, each labeled sp^3. The hybrid orbitals are separated by 109.5°, the tetrahedral angle.
- sp^3d and sp^3d^2: If one or two *d* orbitals are combined with *s* and *p* orbitals in the same valence shell, two other hybrid orbital sets are created. These are utilized by the central atom of a molecule or ion with a trigonal-bipyramidal or octahedral electron-pair geometry, respectively.

Valence Bond Theory for Methane, CH_4

In methane, four orbitals directed to the corners of a tetrahedron are needed to match the electron-pair geometry on the central carbon atom. By mixing the four valence shell orbitals, the 2*s* and all three of the 2*p* orbitals on carbon, a new set of

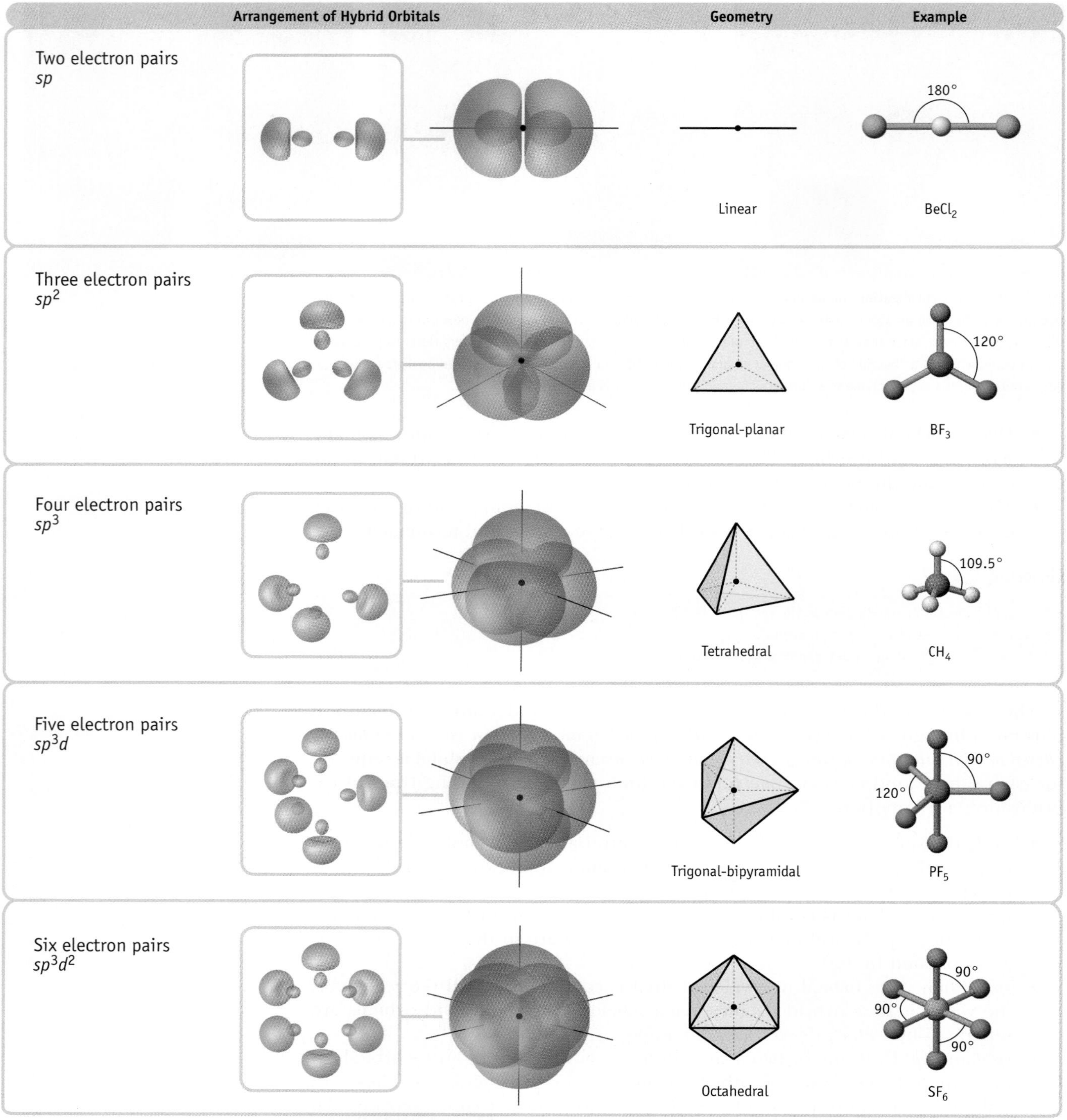

Active Figure 9.5 **Hybrid orbitals for two to six electron pairs.** The geometry of the hybrid orbital sets for two to six valence shell electron pairs is given in the right column. In forming a hybrid orbital set, the *s* orbital is always used, plus as many *p* orbitals (and *d* orbitals) as are required to give the necessary number of σ-bonding and lone-pair orbitals.

Chemistry Now™ Sign in at www.thomsonedu.com/login and go to the Chapter Contents menu to explore an interactive version of this figure accompanied by an exercise.

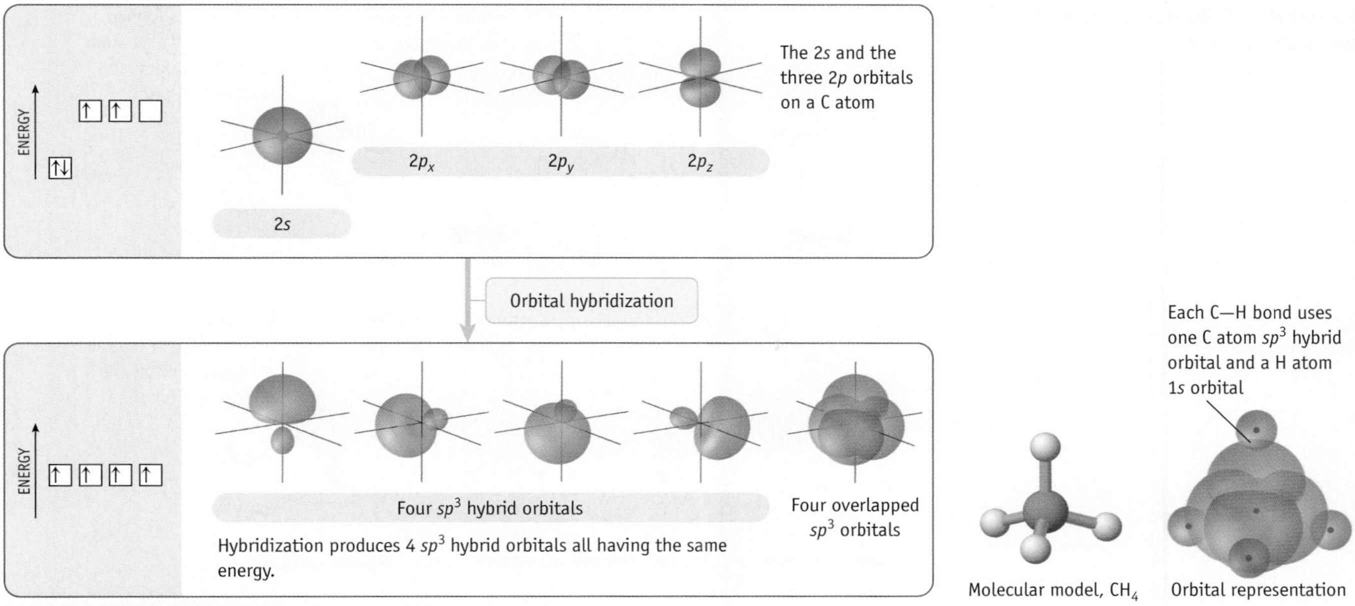

Active Figure 9.6 Bonding in the methane (CH_4) molecule.

Chemistry Now™ Sign in at www.thomsonedu.com/login and go to the Chapter Contents menu to explore an interactive version of this figure accompanied by an exercise.

four hybrid orbitals is created that has tetrahedral geometry (Figures 9.5 and 9.6). Each of the four hybrid orbitals is labeled sp^3 to indicate the atomic orbital combination (an *s* orbital and three *p* orbitals) from which they are derived. All four sp^3 orbitals have an identical shape, and the angle between them is 109.5°, the tetrahedral angle. Because the orbitals have the same energy, one electron can be assigned to each according to Hund's rule (see Section 7.3 [page 312]). Then, each C—H bond is formed by overlap of one of the carbon sp^3 hybrid orbitals with the 1*s* orbital from a hydrogen atom; one electron from the C atom is paired with an electron from an H atom.

■ **Hybrid and Atomic Orbitals** Be sure to notice that *four* atomic orbitals produce *four* hybrid orbitals. The number of hybrid orbitals produced is always the same as the number of atomic orbitals used.

Valence Bond Theory for Ammonia, NH_3

The Lewis structure for ammonia shows there are four electron pairs in the valence shell of nitrogen: three bond pairs and a lone pair (Figure 9.7). VSEPR theory predicts a tetrahedral electron-pair geometry and a trigonal-pyramidal molecular geometry. The actual structure is a close match to the predicted structure; the H—N—H bond angles are 107.5° in this molecule.

Based on the electron-pair geometry of NH_3, we predict sp^3 hybridization to accommodate the four electron pairs on the N atom. The lone pair is assigned to one of the hybrid orbitals, and each of the other three hybrid orbitals is occupied by a single electron. Overlap of each of the singly occupied, sp^3 hybrid orbitals with a 1*s* orbital from a hydrogen atom, and pairing of the electrons in these orbitals, create the N—H bonds.

Valence Bond Theory for Water, H_2O

The oxygen atom of water has two bonding pairs and two lone pairs in its valence shell, and the H—O—H angle is 104.5° (Figure 9.7). Four sp^3 hybrid orbitals are created from the 2*s* and 2*p* atomic orbitals of oxygen. Two of these sp^3 orbitals are

FIGURE 9.7 Bonding in ammonia, NH_3, and water, H_2O.

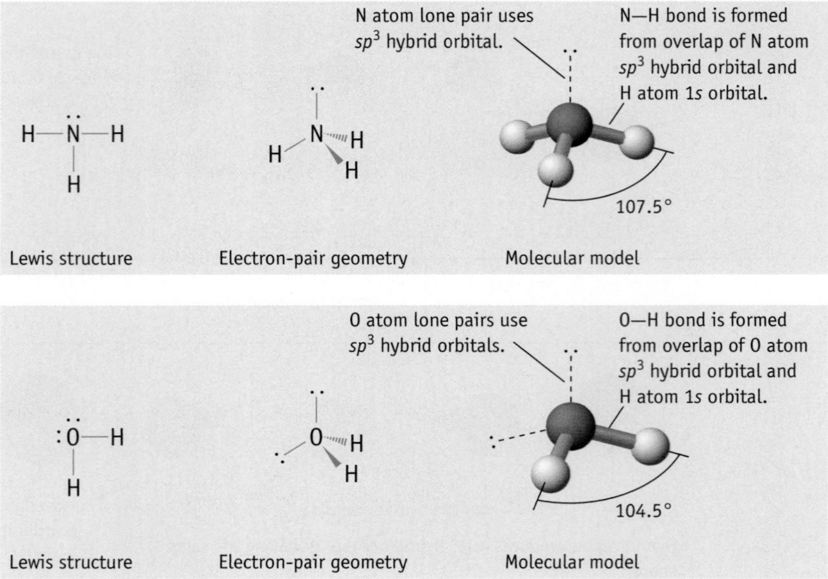

occupied by unpaired electrons and are used to form O—H bonds. Lone pairs occupy the other two hybrid orbitals.

Chemistry Now™

Sign in at **www.thomsonedu.com/login** and go to Chapter 9 Contents to see:

- Screen 9.4 for exercises on **hybrid orbitals**
- Screen 9.5 for a tutorial on **sigma bonding**
- Screen 9.6 for a tutorial on **determining hybrid orbitals**

Hybridization and Geometry Hybridization reconciles the electron-pair geometry with the orbital overlap criterion of bonding. A statement such as "the atom is tetrahedral because it is sp^3 hybridized" is backward. That the electron-pair geometry around the atom is tetrahedral is a fact. Hybridization is one way to rationalize that fact.

EXAMPLE 9.1 Valence Bond Description of Bonding in Ethane

Problem Describe the bonding in ethane, C_2H_6, using valence bond theory.

Strategy First, draw the Lewis structure, and predict the electron-pair geometry at both carbon atoms. Next, assign a hybridization to these atoms. Finally, describe covalent bonds that arise based on orbital overlap, and place electron pairs in their proper locations.

Solution Each carbon atom has an octet configuration, sharing electron pairs with three hydrogen atoms and with the other carbon atom. The electron pairs around carbon have tetrahedral geometry, so carbon is assigned sp^3 hybridization. The C—C bond is formed by overlap of sp^3 orbitals on each C atom, and each of the C—H bonds is formed by overlap of an sp^3 orbital on carbon with a hydrogen 1s orbital.

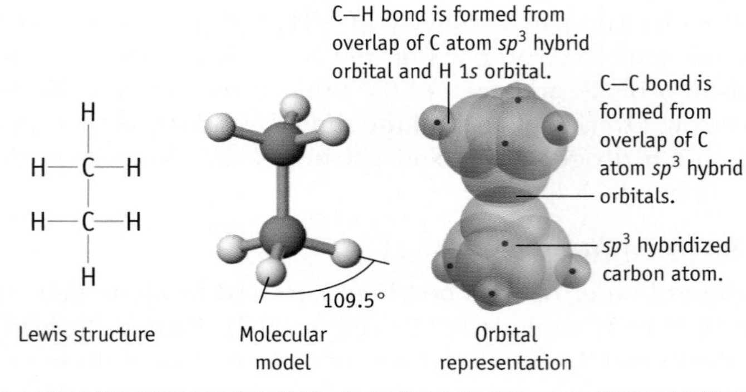

EXAMPLE 9.2 Valence Bond Description of Bonding in Methanol

Problem Describe the bonding in the methanol molecule, CH_3OH, using valence bond theory.

Strategy First, construct the Lewis structure for the molecule. The electron pair geometry around each atom determines the hybrid orbital set used by that atom.

Solution The electron-pair geometry around both the C and O atoms in CH_3OH is tetrahedral. Thus, we may assign sp^3 hybridization to each atom, and the C—O bond is formed by overlap of sp^3 orbitals on these atoms. Each C—H bond is formed by overlap of a carbon sp^3 orbital with a hydrogen 1*s* orbital, and the O—H bond is formed by overlap of an oxygen sp^3 orbital with the hydrogen 1*s* orbital. Two lone pairs on oxygen occupy the remaining sp^3 orbitals on the atom.

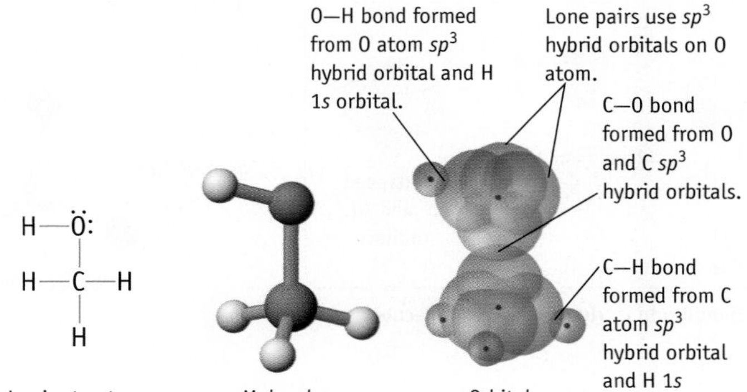

Lewis structure | Molecular model | Orbital representation

Comment Notice that one end of the CH_3OH molecule (the CH_3 or methyl group) is just like the CH_3 group in the methane molecule, and the OH group resembles the OH group in water. It is helpful to recognize pieces of molecules and their bonding descriptions.

This example also shows how to predict the structure and bonding in a complicated molecule by looking at each atom separately. This is an important principle that is essential when dealing with molecules made up of many atoms.

EXERCISE 9.1 Valence Bond Description of Bonding

Use valence bond theory to describe the bonding in the hydronium ion, H_3O^+, and methylamine, CH_3NH_2.

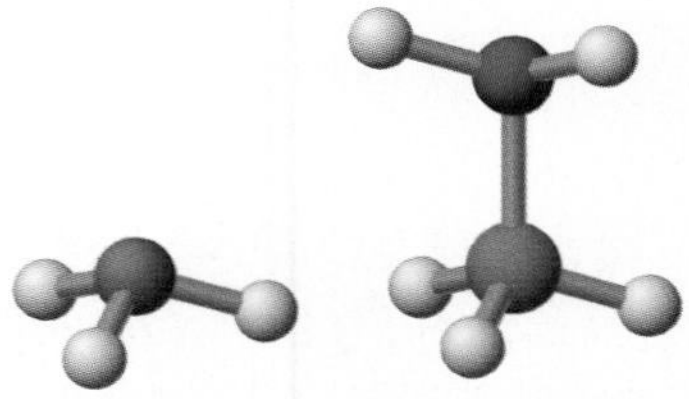

Hydronium ion, H_3O^+ Methylamine, CH_3NH_2

Hybrid Orbitals for Molecules and Ions with Trigonal-Planar Electron Pair Geometries

The central atoms in species such as BF_3, O_3, NO_3^-, and CO_3^{2-} all have a trigonal-planar electron-pair geometry, which requires a central atom with three hybrid orbitals in a plane, 120° apart. Three hybrid orbitals mean three atomic orbitals must be combined, and the combination of an *s* orbital with two *p* orbitals is appropriate (Figure 9.8). If p_x and p_y orbitals are used in hybrid orbital formation, the three hybrid sp^2 orbitals will lie in the xy-plane. The p_z orbital not used to form these hybrid orbitals is perpendicular to the plane containing the three sp^2 orbitals (Figure 9.8).

Boron trifluoride has a trigonal-planar electron-pair and molecular geometry. Each boron–fluorine bond in this compound results from overlap of an sp^2 orbital on boron with a *p* orbital on fluorine. Notice that the p_z orbital on boron, which is not used to form the sp^2 hybrid orbitals, is not occupied by electrons.

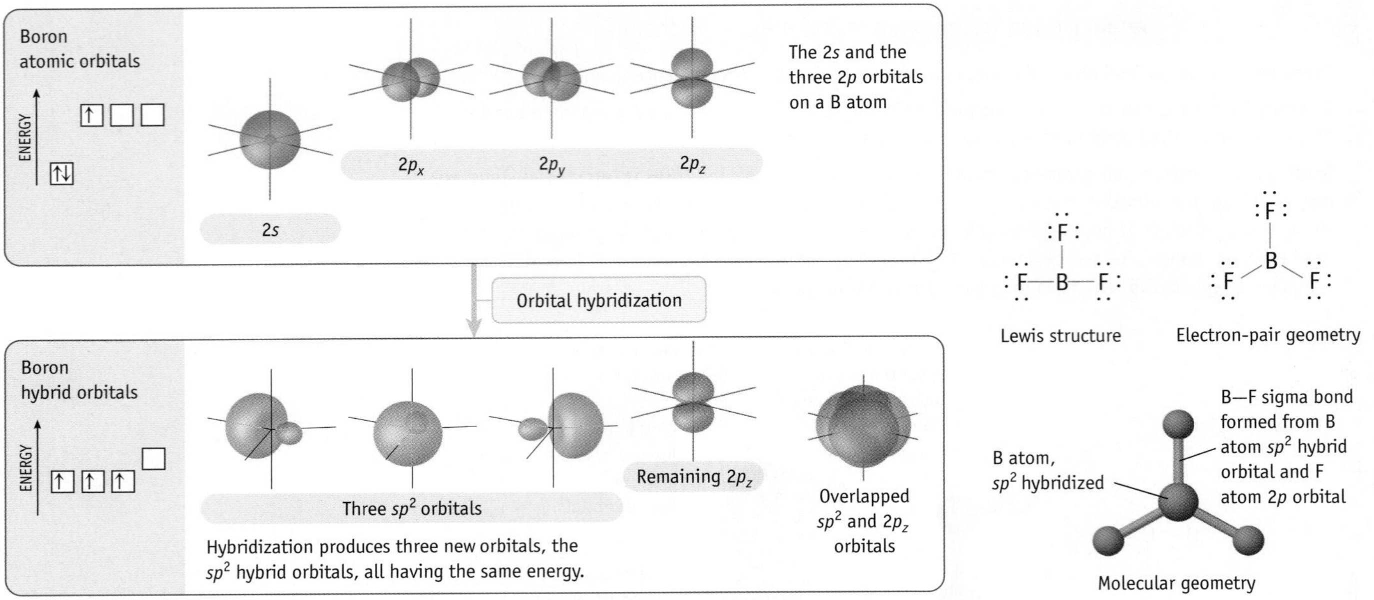

FIGURE 9.8 Bonding in a trigonal-planar molecule.

Hybrid Orbitals for Molecules and Ions with Linear Electron-Pair Geometries

For molecules in which the central atom has a linear electron-pair geometry, two hybrid orbitals, 180° apart, are required. One *s* and one *p* orbital can be hybridized to form two *sp* hybrid orbitals (Figure 9.9). If the p_z orbital is used, then the *sp* orbitals are oriented along the *z*-axis. The p_x and p_y orbitals are perpendicular to this axis.

Beryllium dichloride, $BeCl_2$, is a solid under ordinary conditions. When it is heated to over 520 °C, however, it vaporizes to give $BeCl_2$ vapor. In the gas phase, $BeCl_2$ is a linear molecule, so *sp* hybridization is appropriate for the beryllium atom in this species. Combining beryllium's 2*s* and $2p_z$ orbitals gives the two *sp* hybrid orbitals that lie along the *z*-axis. Each Be—Cl bond arises by overlap of an *sp* hybrid orbital on beryllium with a 3*p* orbital on chlorine. In this molecule, there are only two electron pairs around the beryllium atom, so the p_x and p_y orbitals are not occupied (Figure 9.9).

Hybrid Orbitals for Molecules and Ions with Trigonal-Bipyramidal or Octahedral Electron-Pair Geometries

Bonding in compounds having five or six electron pairs on a central atom (such as PF_5 or SF_6) requires the atom to have five or six hybrid orbitals, which must be created from five or six atomic orbitals. This is possible if additional atomic orbitals from the *d* subshell are used in hybrid orbital formation. The *d* orbitals are considered to be valence shell orbitals for main group elements of the third and higher periods.

To accommodate six electron pairs in the valence shell of an element, six sp^3d^2 hybrid orbitals can be created from one *s*, three *p*, and two *d* orbitals. The six sp^3d^2 hybrid orbitals are directed to the corners of an octahedron (Figure 9.5). Thus, they are oriented to accommodate the valence electron pairs for a compound that has an octahedral electron-pair geometry. Five coordination and trigonal-bipyramidal electron-pair geometry are matched to sp^3d hybridization. One *s*, three *p*, and one *d* orbital combine to produce five sp^3d hybrid orbitals.

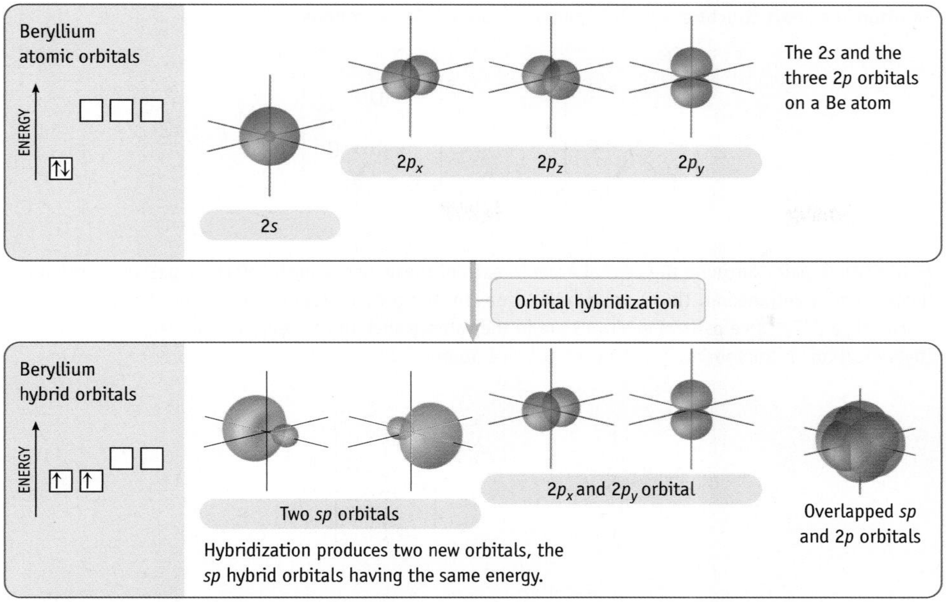

FIGURE 9.9 Bonding in a linear molecule. Because only one *p* orbital is incorporated in the hybrid orbital, two *p* orbitals remain unhybridized. These orbitals are perpendicular to each other and to the axis along which the two *sp* hybrid orbital lies.

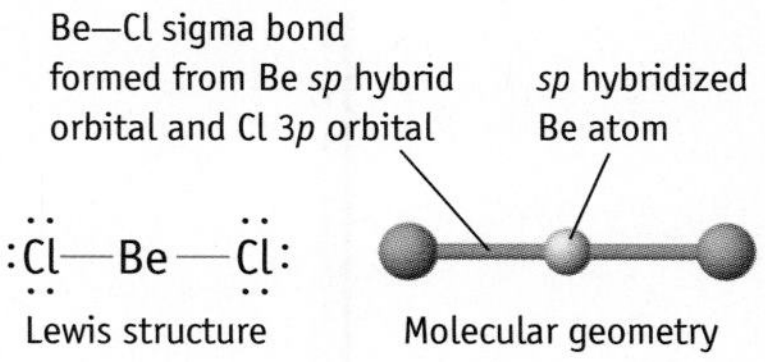

EXAMPLE 9.3 Hybridization Involving *d* Orbitals

Problem Describe the bonding in PF_5 using valence bond theory.

Strategy The first step is to establish the electron-pair and molecular geometry of PF_5. The electron-pair geometry around the P atom gives the number of hybrid orbitals required. If five hybrid orbitals are required, the combination of atomic orbitals is sp^3d.

Solution Here, the P atom is surrounded by five electron pairs, so PF_5 has a trigonal-bipyramidal electron-pair and molecular geometry. The hybridization scheme is therefore sp^3d.

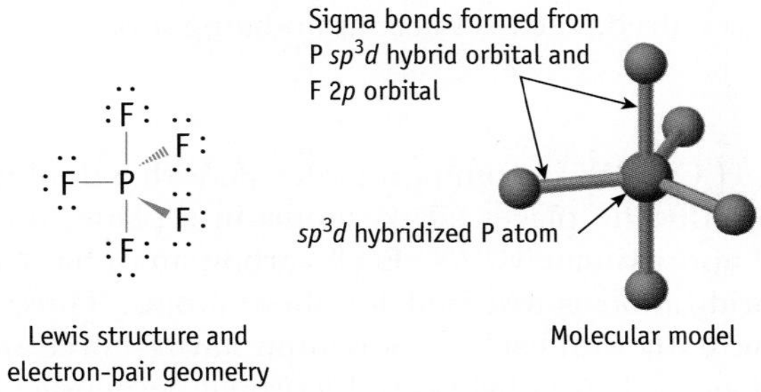

■ **Do $(n - 1)d$ orbitals participate in bonding?** Because the $(n - 1)d$ orbitals are at a relatively high energy their involvement in bonding is believed to be minimal. Using molecular orbital theory, it is possible to describe the bonding in compounds with expanded octets without using *d* orbitals to form hybrid orbital sets.

EXAMPLE 9.4 Recognizing Hybridization

Problem Identify the hybridization of the central atom in the following compounds and ions:

(a) SF_3^+ **(c)** SF_4 **(b)** SO_4^{2-} **(d)** I_3^-

Strategy The hybrid orbitals used by a central atom are determined by the electron-pair geometry (see Figure 9.5). Thus, to answer this question first write the Lewis structure, and then predict the electron-pair geometry.

Solution The Lewis structures for SF_3^+, and SO_4^{2-} are written as follows:

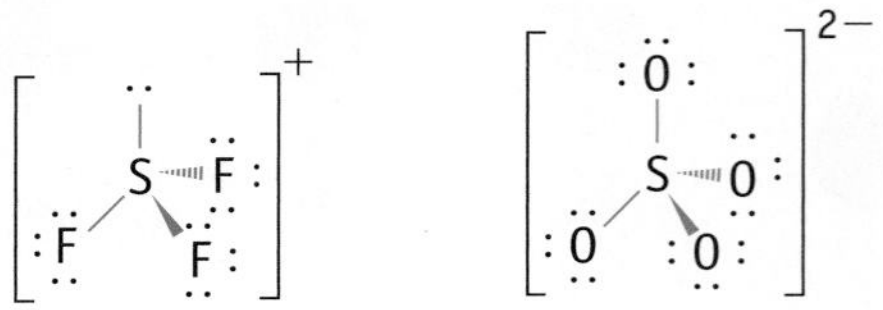

Four electron pairs surround the central atom in each of these ions, and the electron-pair geometry for these atoms is tetrahedral. Thus, sp^3 hybridization for the central atom is used to describe the bonding.

For SF_4 and I_3^-, five pairs of electrons are in the valence shell of the central atom. For these, sp^3d hybridization is appropriate for the central S or I atom.

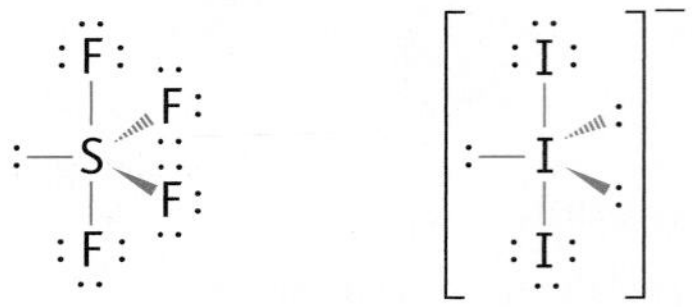

> **EXERCISE 9.2 Recognizing Hybridization**
>
> Identify the hybridization of the underlined central atom in the following compounds and ions:
>
> **(a)** $\underline{B}H_4^-$ **(d)** $\underline{Cl}F_3$
>
> **(b)** $\underline{S}F_5^-$ **(e)** $\underline{B}Cl_3$
>
> **(c)** $O\underline{S}F_4$ **(f)** $\underline{Xe}O_6^{4-}$

■ **Multiple Bonds**

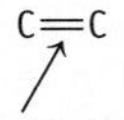

Double bond requires two sets of overlapping orbitals and two pairs of electrons.

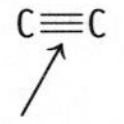

Triple bond requires three sets of overlapping orbitals and three pairs of electrons.

Multiple Bonds

According to valence bond theory, bond formation requires that two orbitals on adjacent atoms overlap. Many molecules have two or three bonds between pairs of atoms. Therefore, according to valence bond theory, a double bond requires *two* sets of overlapping orbitals and *two* electron pairs. For a triple bond, *three* sets of atomic orbitals are required, each set accommodating a pair of electrons.

Double Bonds

Consider ethylene, $H_2C{=}CH_2$, a common molecule with a double bond. The molecular structure of ethylene places all six atoms in a plane, with H—C—H and H—C—C angles of approximately 120°. Each carbon atom has trigonal-planar geometry, so sp^2 hybridization is assumed for these atoms. Thus, a description of bonding in ethylene starts with each carbon atom having three sp^2 hybrid orbitals in the molecular plane and an unhybridized p orbital perpendicular to that plane. Because each carbon atom is involved in four bonds, a single unpaired electron is placed in each of these orbitals.

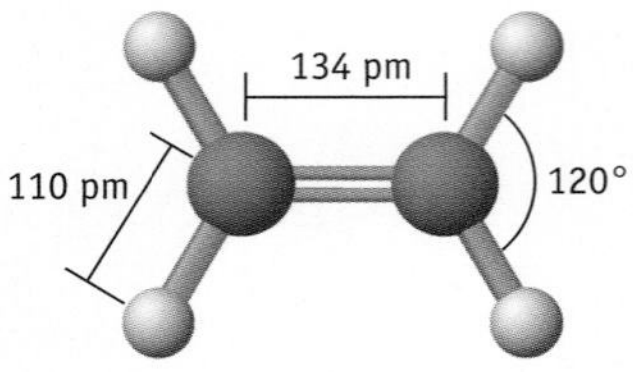

Ethylene, C_2H_4

[↑] Unhybridized p orbital. Used for π bonding in C_2H_4.

[↑][↑][↑] Three sp^2 hybrid orbitals. Used for C—H and C—C σ bonding in C_2H_4.

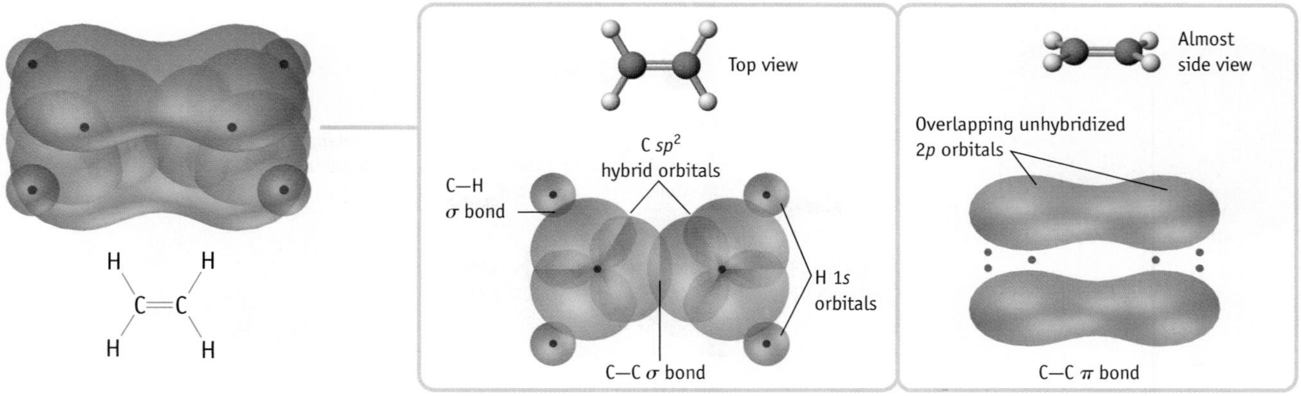

(a) Lewis structure and bonding of ethylene, C_2H_4.

(b) The C—H σ bonds are formed by overlap of C atom sp^2 hybrid orbitals with H atom 1*s* orbitals. The σ bond between C atoms arises from overlap of sp^2 orbitals.

(c) The carbon–carbon π bond is formed by overlap of an unhybridized 2*p* orbital on each atom. Note the lack of electron density along the C—C bond axis from this bond.

Active Figure 9.10 **The valence bond model of bonding in ethylene, C_2H_4.** Each C atom is assumed to be sp^2 hybridized.

Chemistry Now™ Sign in at www.thomsonedu.com/login and go to the Chapter Contents menu to explore an interactive version of this figure accompanied by an exercise.

Now we can visualize the C—H bonds, which arise from overlap of sp^2 orbitals on carbon with hydrogen 1*s* orbitals. After accounting for the C—H bonds, one sp^2 orbital on each carbon atom remains. These orbitals point toward each other and overlap to form one of the bonds linking the carbon atoms (Figure 9.10). This leaves only one other orbital unaccounted for on each carbon, an unhybridized *p* orbital, and it is these orbitals that can be used to create the second bond between carbon atoms in C_2H_4. If they are aligned correctly, the unhybridized *p* orbitals on the two carbons can overlap, allowing the electrons in these orbitals to be paired. The overlap does not occur directly along the C—C axis, however. Instead, the arrangement compels these orbitals to overlap sideways, and the electron pair occupies an orbital with electron density above and below the plane containing the six atoms.

This description results in two types of bonds in C_2H_4. One type is the C—H and C—C bonds that arise from the overlap of atomic orbitals so that the bonding electrons that lie along the bond axes form sigma (σ) bonds. The other is the bond formed by sideways overlap of *p* atomic orbitals, called a **pi (π) bond.** In a π bond, the overlap region is above and below the internuclear axis, and the electron density of the π bond is above and below the bond axis.

Be sure to notice that a π bond can form *only* if (a) there are unhybridized *p* orbitals on adjacent atoms and (b) the *p* orbitals are perpendicular to the plane of the molecule and parallel to one another. This happens only if the sp^2 orbitals of both carbon atoms are in the same plane. A consequence of this is that both atoms involved in the π bond have trigonal-planar geometry, and the six atoms in and around the π bond (the two atoms involved in the π bond and the four atoms attached to the π-bonded atoms) lie in one plane.

Double bonds between carbon and oxygen, sulfur, or nitrogen are quite common. Consider formaldehyde, CH_2O, in which a carbon–oxygen π bond occurs

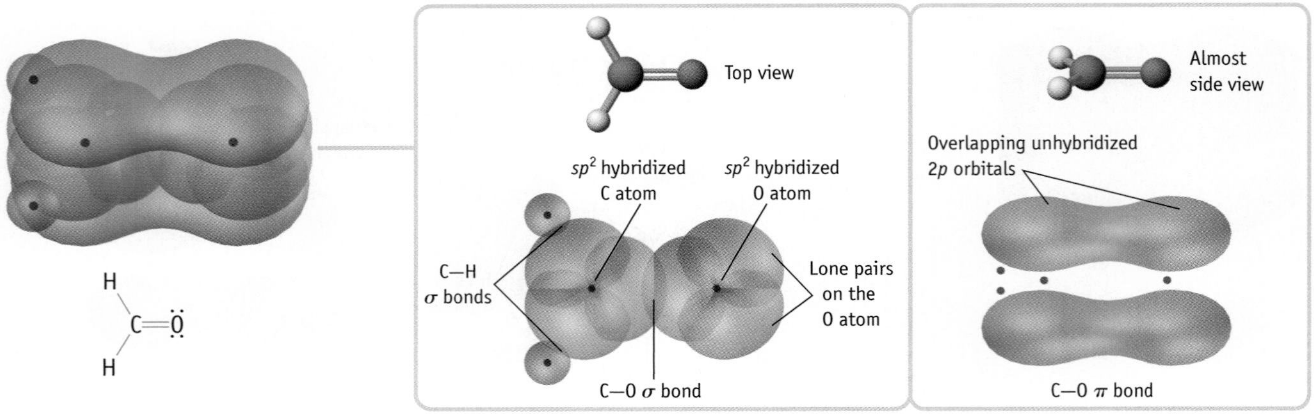

(a) Lewis structure and bonding of formaldehyde, CH_2O.

(b) The C—H σ bonds are formed by overlap of C atom sp^2 hybrid orbitals with H atom 1*s* orbitals. The σ bond between C and O atoms arises from overlap of sp^2 orbitals.

(c) The C—O π bond comes from the side-by-side overlap of *p* orbitals on the two atoms.

FIGURE 9.11 Valence bond description of bonding in formaldehyde, CH_2O.

■ **Alternative View of the C—O π Bond in CH_2O** An alternative but still satisfactory explanation of the C—O π bond is to assume the O atom is unhybridized and that the π bond is constructed from an unhybridized 2*p* oxygen orbital overlapping with a *p* orbital on the carbon atom.

(Figure 9.11). A trigonal-planar electron-pair geometry indicates sp^2 hybridization for the C atom. The σ bonds from carbon to the O atom and the two H atoms form by overlap of sp^2 hybrid orbitals with half-filled orbitals from the oxygen and two hydrogen atoms. An unhybridized *p* orbital on carbon is oriented perpendicular to the molecular plane (just as for the carbon atoms of C_2H_4). This *p* orbital is available for π bonding, this time with an oxygen orbital.

What orbitals on oxygen are used in this model? The approach in Figure 9.11 assumes sp^2 hybridization for oxygen. This uses one O atom sp^2 orbital in σ bond formation, leaving two sp^2 orbitals to accommodate lone pairs. The remaining *p* orbital on the O atom participates in the π bond.

Chemistry Now™

Sign in at **www.thomsonedu.com/login** and go to Chapter 9 Contents menu to see Screen 9.7 for exercises and a tutorial on **hybrid orbitals and σ and π bonding.**

EXAMPLE 9.5 Bonding in Acetic Acid

Problem Using valence bond theory, describe the bonding in acetic acid, CH_3CO_2H, the important ingredient in vinegar.

Strategy Write a Lewis electron dot structure, and determine the geometry around each atom using VSEPR theory. Use this geometry to decide on the hybrid orbitals used in σ bonding. If unhybridized *p* orbitals are available on adjacent C and O atoms, then C—O π bonding can occur.

Solution The carbon atom of the CH_3 group has tetrahedral electron-pair geometry, which means it is sp^3 hybridized. Three sp^3 orbitals are used to form the C—H bonds. The fourth sp^3 orbital is used to bond to the adjacent carbon atom. This carbon atom has a trigonal-planar electron-pair geometry; it must be sp^2 hybridized. The C—C bond is formed using one of these hybrid orbitals, and the other two sp^2 orbitals are used to form the σ bonds to the two oxygens. The oxygen of the O—H group has four electron pairs; it must be tetrahedral and sp^3 hybridized. Thus, this O atom uses two sp^3 orbitals to bond to the adjacent carbon and the hydrogen, and two sp^3 orbitals accommodate the two lone pairs.

Finally, the carbon–oxygen double bond can be described by assuming the C and O atoms are both sp^2 hybridized (like the C—O π bond in formaldehyde, Figure 9.11). The unhybridized *p* orbital remaining on each atom is used to form the carbon–oxygen π bond, and the lone pairs on the O atom are accommodated in sp^2 hybrid orbitals.

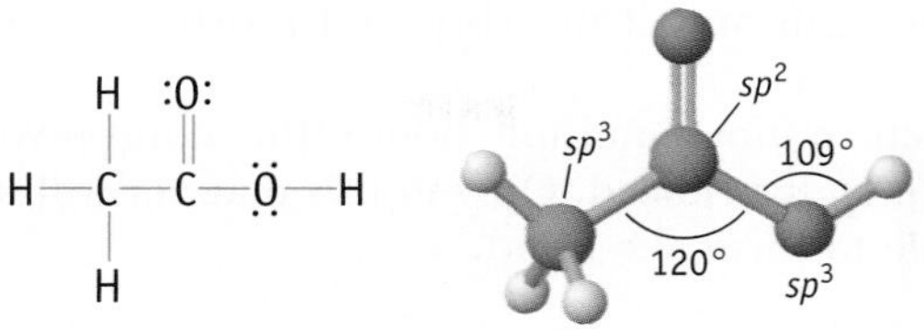

Lewis dot structure   Molecular model

EXERCISE 9.3 Bonding in Acetone

Use valence bond theory to describe the bonding in acetone, CH_3COCH_3.

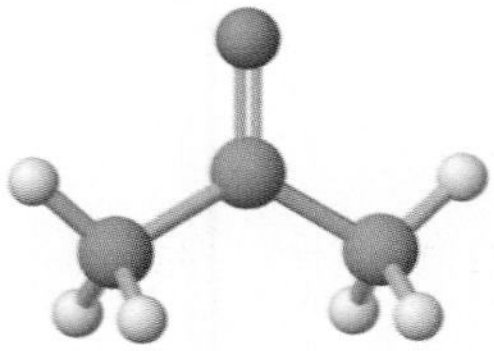

Acetone

Triple Bonds

Acetylene, H—C≡C—H, is an example of a molecule with a triple bond. VSEPR theory predicts that the four atoms lie in a straight line with H—C—C angles of 180°. This implies that the carbon atom is *sp* hybridized (Figure 9.12). For each carbon atom, there are two *sp* orbitals, one directed toward hydrogen and used to create the C—H σ bond, and the second directed toward the other carbon and used to create a σ bond between the two carbon atoms. Two unhybridized *p* orbitals remain on each carbon, and they are oriented so that it is possible to form *two* π bonds in HC≡CH.

[↑|↑] Two unhybridized *p* orbitals. Used for π bonding in C_2H_2.

[↑|↑] Two *sp* hybrid orbitals. Used for C—H and C—C σ bonding in C_2H_2.

These π bonds are perpendicular to the molecular axis and perpendicular to each other. Three electrons on each carbon atom are paired to form the triple bond consisting of a σ bond and two π bonds (Figure 9.12).

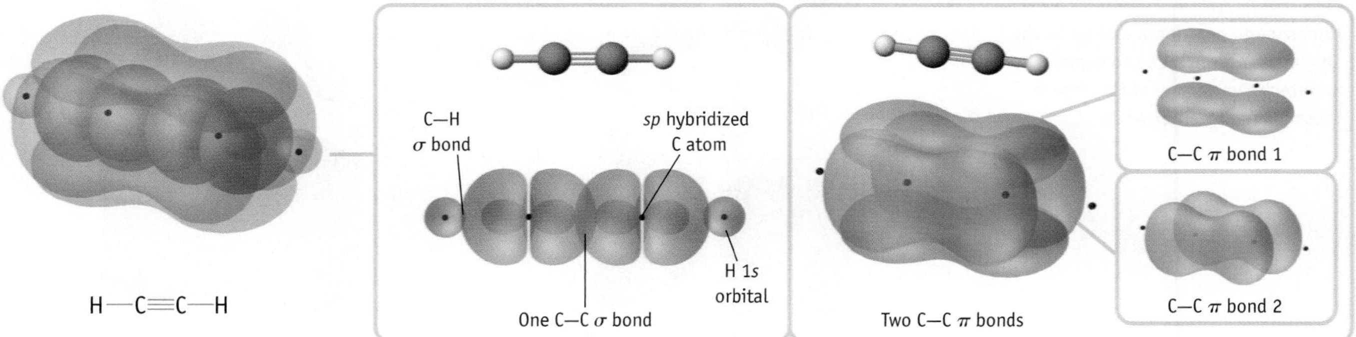

FIGURE 9.12 Bonding in acetylene.

Now that we have examined two cases of multiple bonds, let us summarize several important points:

- In valence bond theory a double bond always consists of a σ bond and a π bond. Similarly, a triple bond always consists of a σ bond and *two* π bonds.
- A π bond may form only if unhybridized p orbitals remain on the bonded atoms.
- If a Lewis structure shows multiple bonds, the atoms involved must therefore be either sp^2 or sp hybridized. Only in this manner will unhybridized p orbitals be available to form a π bond.

EXERCISE 9.4 Triple Bonds Between Atoms

Describe the bonding in a nitrogen molecule, N_2.

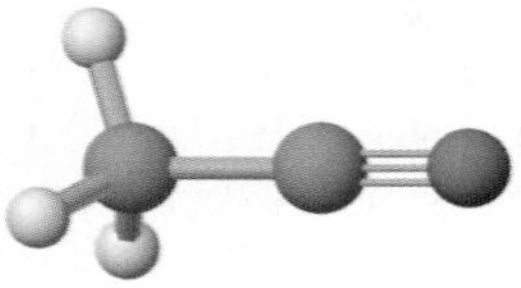

Acetonitrile, CH_3CN

EXERCISE 9.5 Bonding and Hybridization

Estimate values for the H—C—H, H—C—C, and C—C—N angles in acetonitrile, $CH_3C{\equiv}N$. Indicate the hybridization of both carbon atoms and the nitrogen atom, and analyze the bonding using valence bond theory.

Cis-Trans Isomerism: A Consequence of π Bonding

Ethylene, C_2H_4, is a planar molecule, a geometry that allows the unhybridized p orbitals on the two carbon atoms to line up and form a π bond (see Figure 9.13b). Let us speculate on what would happen if one end of the ethylene molecule were twisted relative to the other end. This action would distort the molecule away from planarity, and the p orbitals would rotate out of alignment. Rotation would diminish the extent of overlap of these orbitals, and, if a twist of 90° were achieved, the two p orbitals would no longer overlap at all; the π bond would be broken. However, so much energy is required to break this bond (about 260 kJ/mol) that rotation around a C=C bond is not expected to occur at room temperature.

Active Figure 9.13 Rotation around bonds.

Chemistry Now™ Sign in at www.thomsonedu.com/login and go to the Chapter Contents menu to explore an interactive version of this figure accompanied by an exercise.

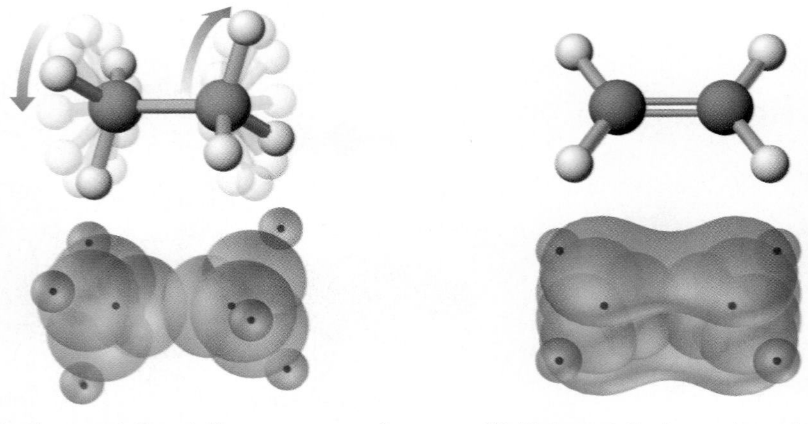

(a) In ethane nearly free rotation can occur around the axis of a single (σ) bond.

(b) Ethylene rotation is severely restricted around double bonds because doing so would break the π bond, a process generally requiring a great deal of energy.

A consequence of restricted rotation is that isomers occur for many compounds containing a C═C bond. **Isomers** are compounds that have the same formula but different structures. In this case, the two isomeric compounds differ with respect to the orientation of the groups attached to the carbons of the double bond. Two isomers of $C_2H_2Cl_2$ are *cis-* and *trans-*1,2-dichloroethylene. Their structures resemble ethylene, except that two hydrogen atoms have been replaced by chlorine atoms. Because a large amount of energy is required to break the π bond, the *cis* compound cannot rearrange to the *trans* compound under ordinary conditions. Each compound can be obtained separately, and each has its own identity. *Cis*-1,2-dichloroethylene boils at 60.3 °C, whereas *trans*-1,2-dichloroethylene boils at 47.5 °C.

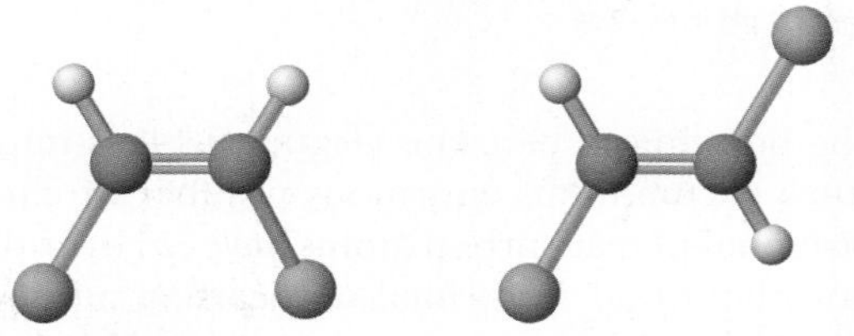

cis-1,2-dichloroethylene   *trans*-1,2-dichloroethylene

■ ***Cis* and *Trans* Isomers** Compounds having the same formula, but different structures, are isomers. *Trans* isomers have distinguishing groups on opposite sides of a double bond. *Cis* isomers have these groups on the same side of the double bond.

Although *cis* and *trans* isomers do not interconvert at ordinary temperatures, they will do so at higher temperatures. If the temperature is sufficiently high, the molecular motions can become sufficiently energetic that rotation around the C═C bond can occur. This may also occur under other special conditions, such as when the molecule absorbs light energy.

ChemistryNow™

Sign in at **www.thomsonedu.com/login** and go to Chapter 9 Contents to see Screen 9.8 for an exercise on **isomers and multiple bonds.**

Benzene: A Special Case of π Bonding

Benzene, C_6H_6, is the simplest member of a large group of substances known as *aromatic* compounds, a historical reference to their odor. It occupies a pivotal place in the history and practice of chemistry.

To 19th-century chemists, benzene was a perplexing substance with an unknown structure. Based on its chemical reactions, however, August Kekulé (1829–1896) suggested that the molecule has a planar, symmetrical ring structure. We know now he was correct. The ring is flat, and all the carbon–carbon bonds are the same length, 139 pm, a distance intermediate between the average single bond (154 pm) and double bond (134 pm) lengths. Assuming the molecule has two resonance structures with alternating double bonds, the observed structure is rationalized. The C—C bond order in C_6H_6(1.5) is the average of a single and a double bond.

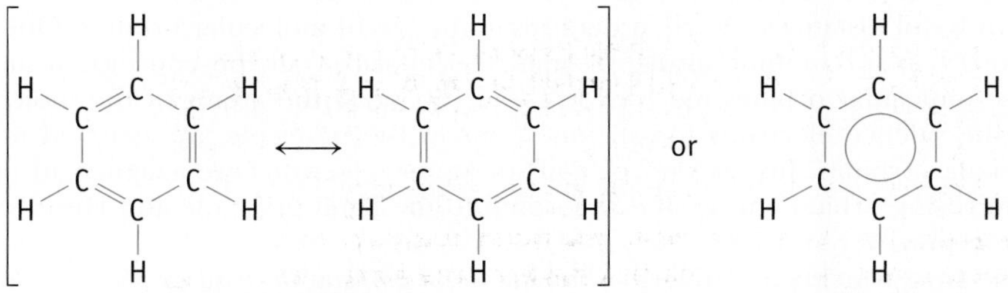

resonance structures   resonance hybrid

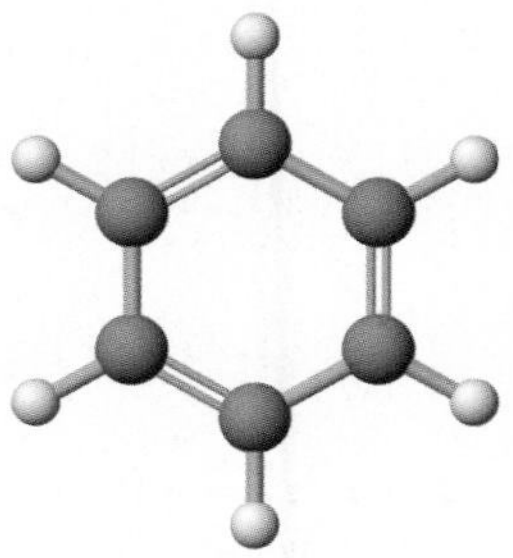

Benzene, C_6H_6

FIGURE 9.14 Bonding in benzene, C_6H_6. (left) The C atoms of the ring are bonded to each other through σ bonds using C atom sp^2 hybrid orbitals. The C—H bonds also use C atom sp^2 hybrid orbitals. The π framework of the molecule arises from overlap of C atom p orbitals not used in hybrid orbital formation. Because these orbitals are perpendicular to the ring, π electron density is above and below the plane of the ring. (right) A composite of σ and π bonding in benzene.

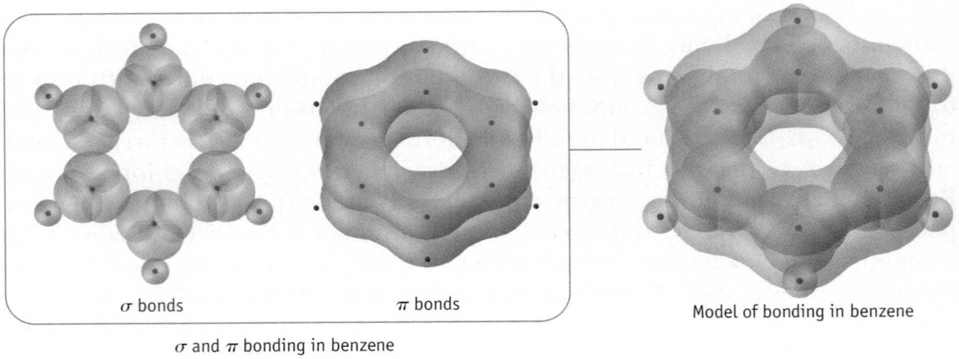

Understanding the bonding in benzene (Figure 9.14) is important because the benzene ring structure occurs in an enormous number of chemical compounds. We assume that the trigonal-planar carbon atoms have sp^2 hybridization. Each C—H bond is formed by overlap of an sp^2 orbital of a carbon atom with a 1s orbital of hydrogen, and the C—C σ bonds arise by overlap of sp^2 orbitals on adjacent carbon atoms. After accounting for the σ bonding, an unhybridized p orbital remains on each C atom, and each is occupied by a single electron. These six orbitals and six electrons form π bonds. Because all carbon–carbon bond lengths are the same, each p orbital overlaps equally well with the p orbitals of both adjacent carbons, and the π interaction is unbroken around the six-member ring.

9.3 Molecular Orbital Theory

Molecular orbital (MO) theory is an alternative way to view orbitals in molecules. In contrast to the localized bond and lone pair electrons of valence bond theory, MO theory assumes that pure s and p atomic orbitals of the atoms in the molecule combine to produce orbitals that are spread out, or delocalized, over several atoms or even over an entire molecule. These orbitals are called **molecular orbitals.**

One reason for learning about the MO concept is that it correctly predicts the electronic structures of molecules such as O_2 that do not follow the electron-pairing assumptions of the Lewis approach. The rules of Section 8.2 would guide you to draw the electron dot structure of O_2 with all the electrons paired, which fails to explain its paramagnetism (Figure 9.15). The molecular orbital approach can account for this property, but valence bond theory cannot. To see how MO theory can be used to describe the bonding in O_2 and other diatomic molecules, we shall first describe four principles used to develop the theory.

Principles of Molecular Orbital Theory

In MO theory, we begin with a given arrangement of atoms in the molecule at the known bond distances. We then determine the *sets* of molecular orbitals. One way to do this is to combine available valence orbitals on all the constituent atoms. These molecular orbitals more or less encompass all the atoms of the molecule, and the valence electrons for all the atoms in the molecule are assigned to the molecular orbitals. Just as with orbitals in atoms, electrons are assigned in order of increasing orbital energy and according to the Pauli principle and Hund's rule (see Sections 7.1 and 7.3).

The **first principle of molecular orbital theory** is that *the total number of molecular orbitals is always equal to the total number of atomic orbitals contributed by the atoms that*

(a) Making liquid O_2

(b) Liquid O_2 is a light blue color.

(c) Paramagnetic liquid O_2 clings to a magnet.

(d) Diamagnetic liquid N_2 is not attracted to a magnet.

Charles D. Winters

FIGURE 9.15 The paramagnetism of liquid oxygen. Oxygen gas condenses (a) to a pale blue liquid at −183 °C (b). Because O_2 molecules have two unpaired electrons, oxygen in the liquid state is paramagnetic and clings a relatively strong neodymium magnet (c). In contrast, liquid N_2 is diamagnetic and does not stick to the magnet (d). It just splashes on the surface when poured onto the magnet.

have combined. To illustrate this orbital conservation principle, let us consider the H_2 molecule.

Molecular Orbitals for H_2

Molecular orbital theory specifies that when the 1*s* orbitals of two hydrogen atoms overlap, *two* molecular orbitals result. One molecular orbital results from *addition* of the 1*s* atomic orbital wave functions, leading to an increased probability that electrons will reside in the bond region between the two nuclei (Figure 9.16). This is called a **bonding molecular orbital.** It is also a σ orbital because the region of electron probability lies directly along the bond axis. This molecular orbital is labeled σ_{1s}, the subscript 1*s* indicating that 1*s* atomic orbitals were used to create the molecular orbital.

■ **Orbitals and Electron Waves** Orbitals are characterized as electron waves; therefore, a way to view molecular orbital formation is to assume that two electron waves, one from each atom, interfere with each other. The interference can be constructive, giving a bonding MO, or destructive, giving an antibonding MO.

The other molecular orbital is constructed by *subtracting* one atomic orbital wave function from the other (see Figure 9.16). When this happens, the probability of finding an electron between the nuclei in the molecular orbital is reduced, and the probability of finding the electron in other regions is higher. Without significant electron density between them, the nuclei repel one another. This type of orbital is called an **antibonding molecular orbital.** Because it is also a σ orbital, it is labeled σ^*_{1s}. The asterisk signifies that it is antibonding. *Antibonding orbitals have no counterpart in valence bond theory.*

A **second principle of molecular orbital theory** is that *the bonding molecular orbital is lower in energy than the parent orbitals, and the antibonding orbital is higher in energy* (Figure 9.16). This means that the energy of a group of atoms is lower than the energy of the separated atoms when electrons are assigned to bonding molecular orbitals. Chemists say the system is "stabilized" by chemical bond formation.

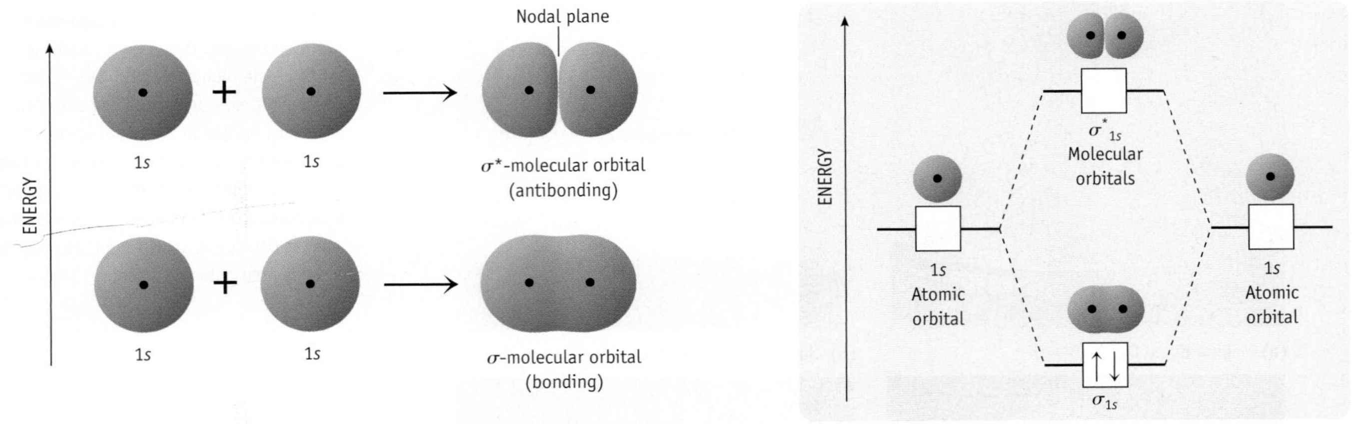

(a) (b)

FIGURE 9.16 Molecular orbitals. (a) Bonding and antibonding σ molecular orbitals are formed from two 1*s* atomic orbitals on adjacent atoms. Notice the presence of a node in the antibonding orbital. (The node is a plane on which there is zero probability of finding an electron.) (b) A molecular orbital diagram for H_2. The two electrons are placed in the σ_{1s} orbital, the molecular orbital lower in energy.

Sign in at **www.thomsonedu.com/login** and go to Chapter 9 Contents to see Screen 9.11 for an animated version of this figure.

Conversely, the system is "destabilized" when electrons are assigned to antibonding orbitals because the energy of the system is higher than that of the atoms themselves.

A **third principle of molecular orbital theory** is that the *electrons of the molecule are assigned to orbitals of successively higher energy* according to the Pauli exclusion principle and Hund's rule. This is analogous to the procedure for building up electronic structures of atoms. Thus, electrons occupy the lowest energy orbitals available, and when two electrons are assigned to an orbital, their spins must be paired. Because the energy of the electrons in the bonding orbital of H_2 is lower than that of either parent 1*s* electron (see Figure 9.16b), the H_2 molecule is more stable than two separate H atoms. We write the electron configuration of H_2 as $(\sigma_{1s})^2$.

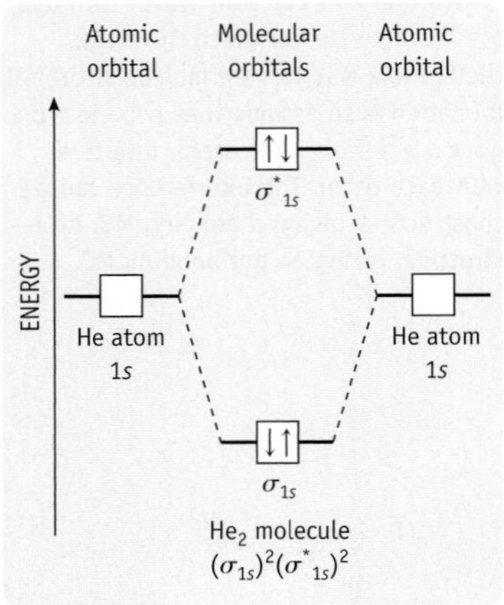

FIGURE 9.17 A molecular orbital energy level diagram for the dihelium molecule, He_2. This diagram provides a rationalization for the nonexistence of the molecule. In He_2, both the bonding (σ_{1s}) and antibonding orbitals (σ^*_{1s}) would be fully occupied.

What would happen if we tried to combine two helium atoms to form dihelium, He_2? Both He atoms have a 1*s* valence orbital that can produce the same kind of molecular orbitals as in H_2. Unlike H_2, however, four electrons need to be assigned to these orbitals (Figure 9.17). The pair of electrons in the σ_{1s} orbital stabilizes He_2. The two electrons in σ^*_{1s}, however, destabilize the He_2 molecule. The energy decrease from the electrons in the σ_{1s}-bonding molecular orbital is offset by the energy increase due to the electrons in the σ^*_{1s}-antibonding molecular orbital. Thus, molecular orbital theory predicts that He_2 has no net stability; two He atoms have no tendency to combine. This confirms what we already know, that elemental helium exists in the form of single atoms and not as a diatomic molecule.

Bond Order

Bond order was defined in Section 8.9 as the net number of bonding electron pairs linking a pair of atoms. This same concept can be applied directly to molecular orbital theory, but now bond order is defined as

$$\text{Bond order} = 1/2\ (\text{number of electrons in bonding MOs} - \text{number of electrons in antibonding MOs}) \quad (9.1)$$

In the H_2 molecule, there are two electrons in a bonding orbital and none in an antibonding orbital, so H_2 has a bond order of 1. In contrast, in He_2 the stabilizing effect of the σ_{1s} pair is canceled by the destabilizing effect of the σ^*_{1s} pair, and so the bond order is 0.

Fractional bond orders are possible. Consider the ion He_2^+. Its molecular orbital electron configuration is $(\sigma_{1s})^2(\sigma^*_{1s})^1$. In this ion, there are two electrons in a bonding molecular orbital, but only one in an antibonding orbital. MO theory predicts that He_2^+ should have a bond order of 0.5; that is, a weak bond should exist between helium atoms in such a species. Interestingly, this ion has been identified in the gas phase using special experimental techniques.

Chemistry Now™

Sign in at **www.thomsonedu.com/login** and go to Chapter 9 Contents to see:

- Screen 9.9 for exercises on **molecular orbital theory**
- Screen 9.10 for a description of **molecular electron configurations**

EXAMPLE 9.6 Molecular Orbitals and Bond Order

Problem Write the electron configuration of the H_2^- ion in molecular orbital terms. What is the bond order of the ion?

Strategy Count the number of valence electrons in the ion, and then place those electrons in the MO diagram for the H_2 molecule. Find the bond order from Equation 9.1.

Solution This ion has three electrons (one each from the H atoms plus one for the negative charge). Therefore, its electronic configuration is $(\sigma_{1s})^2(\sigma^*_{1s})^1$, identical with the configuration for He_2^+. This means H_2^- also has a net bond order of 0.5. The H_2^- ion is thus predicted to exist under special circumstances.

EXERCISE 9.6 Molecular Orbitals and Bond Order

What is the electron configuration of the H_2^+ ion? Compare the bond order of this ion with He_2^+ and H_2^-. Do you expect H_2^+ to exist?

Molecular Orbitals of Li_2 and Be_2

A **fourth principle of molecular orbital theory** is that *atomic orbitals combine to form molecular orbitals most effectively when the atomic orbitals are of similar energy.* This principle becomes important when we move past He_2 to Li_2, dilithium, and heavier molecules such as O_2 and N_2.

A lithium atom has electrons in two orbitals of the *s* type (1*s* and 2*s*), so a 1*s* ± 2*s* combination is theoretically possible. Because the 1*s* and 2*s* orbitals are quite different in energy, however, this interaction can be disregarded. Thus, the molecular orbitals come only from 1*s* ± 1*s* and 2*s* ± 2*s* combinations (Figure 9.18). This means the molecular orbital electron configuration of dilithium, Li_2, is

$$Li_2 \text{ MO Configuration:} \quad (\sigma_{1s})^2(\sigma^*_{1s})^2(\sigma_{2s})^2$$

The bonding effect of the σ_{1s} electrons is canceled by the antibonding effect of the σ^*_{1s} electrons, so these pairs make no net contribution to bonding in Li_2. Bonding in Li_2 is due to the electron pair assigned to the σ_{2s} orbital, and the bond order is 1.

The fact that the σ_{1s} and σ^*_{1s} electron pairs of Li_2 make no net contribution to bonding is exactly what you observed in drawing electron dot structures in

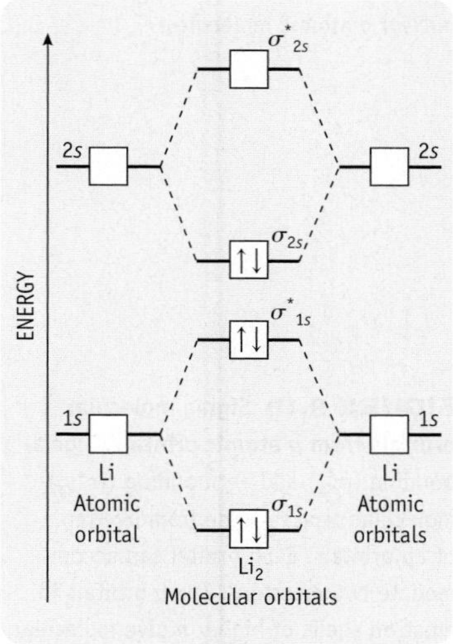

FIGURE 9.18 Energy level diagram for the combination of two Li atoms with 1s and 2s atomic orbitals. Notice that the molecular orbitals are created by combining orbitals of similar energies. The electron configuration is shown for Li_2.

Section 8.2: *core electrons are ignored*. In molecular orbital terms, core electrons are assigned to bonding and antibonding molecular orbitals that offset one another.

A diberyllium molecule, Be_2, is not expected to exist. Its electron configuration is

$$Be_2 \text{ MO Configuration:} \quad [\text{core electrons}](\sigma_{2s})^2(\sigma^*_{2s})^2$$

The effects of σ_{2s} and σ^*_{2s} electrons cancel, and there is no net bonding. The bond order is 0, so the molecule does not exist.

EXAMPLE 9.7 Molecular Orbitals in Diatomic Molecules

Problem Be_2 does not exist. But what about the Be_2^+ ion? Describe its electron configuration in molecular orbital terms, and give the net bond order. Do you expect the ion to exist?

Strategy Count the number of electrons in the ion, and place them in the MO diagram in Figure 9.18. Write the electron configuration, and calculate the bond order from Equation 9.1.

Solution The Be_2^+ ion has only seven electrons (in contrast to eight for Be_2), of which four are core electrons. (The core electrons are assigned to σ_{1s} and σ^*_{1s} molecular orbitals.) The remaining three electrons are assigned to the σ_{2s} and σ^*_{2s} molecular orbitals, so the MO electron configuration is [core electrons]$(\sigma_{2s})^2(\sigma^*_{2s})^1$. This means the net bond order is 0.5, and so Be_2^+ is predicted to exist under special circumstances.

EXERCISE 9.7 Molecular Orbitals in Diatomic Molecules

Could the anion Li_2^- exist? What is the ion's bond order?

Molecular Orbitals from Atomic *p* Orbitals

■ **Diatomic Molecules** Molecules such as H_2, Li_2, and N_2, in which two identical atoms are bonded, are examples of *homonuclear* diatomic molecules.

With the principles of molecular orbital theory in place, we are ready to account for bonding in such important homonuclear diatomic molecules as N_2, O_2, and F_2. To describe the bonding in these molecules, we will have to use both *s* and *p* valence orbitals in forming molecular orbitals.

For *p*-block elements, sigma-bonding and antibonding molecular orbitals are formed by their *s* orbitals interacting as in Figure 9.16. Similarly, it is possible for a *p* orbital on one atom to interact with a *p* orbital on the other atom to produce a pair of σ-bonding and σ*-antibonding molecular orbitals (Figure 9.19).

In addition, each *p*-block atom has *two p* orbitals in planes perpendicular to the σ bond connecting the two atoms. These *p* orbitals can interact sideways to give π-bonding and π-antibonding molecular orbitals (Figure 9.20). Combining these

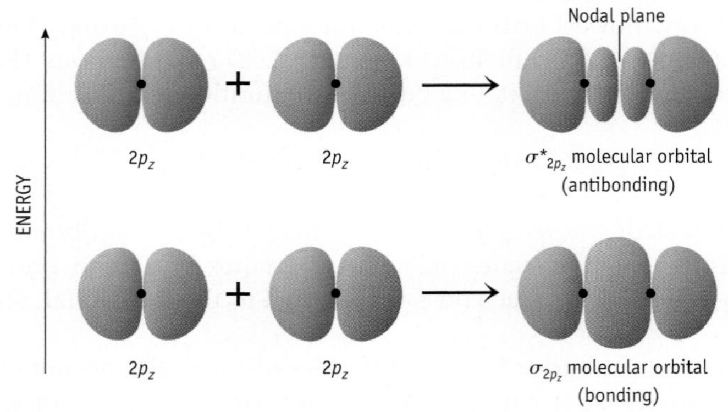

FIGURE 9.19 Sigma molecular orbitals from *p* atomic orbitals. Sigma-bonding (σ_{2p}) and antibonding (σ^*_{2p}) molecular orbitals arise from overlap of 2*p* orbitals. Each orbital can accommodate two electrons. The *p* orbitals in electron shells of higher *n* give molecular orbitals of the same basic shape.

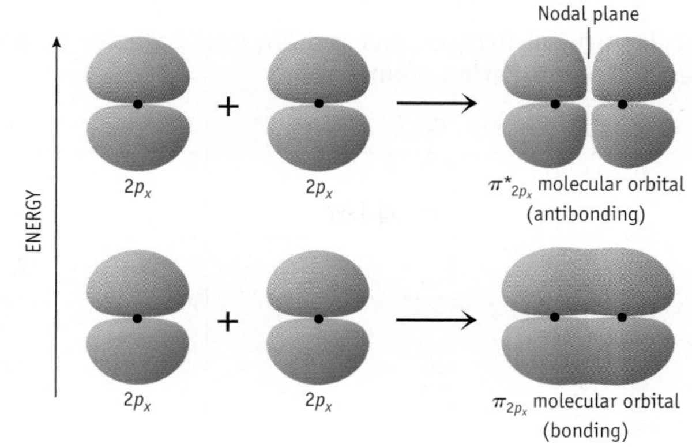

FIGURE 9.20 Formation of π molecular orbitals. Sideways overlap of atomic 2*p* orbitals that lie in the same direction in space gives rise to pi-bonding (π_{2p}) and pi-antibonding (π^*_{2p}) molecular orbitals. The *p* orbitals in shells of higher *n* give molecular orbitals of the same basic shape.

two *p* orbitals on each atom produces *two* π-bonding molecular orbitals (π_p) and *two* π-antibonding molecular orbitals (π^*_p).

Electron Configurations for Homonuclear Molecules for Boron Through Fluorine

Orbital interactions in a second-period, homonuclear, diatomic molecule lead to the energy level diagram in Figure 9.21. Electron assignments can be made using this diagram, and the results for the diatomic molecules B_2 through F_2 are assembled in Table 9.1, which has two noteworthy features.

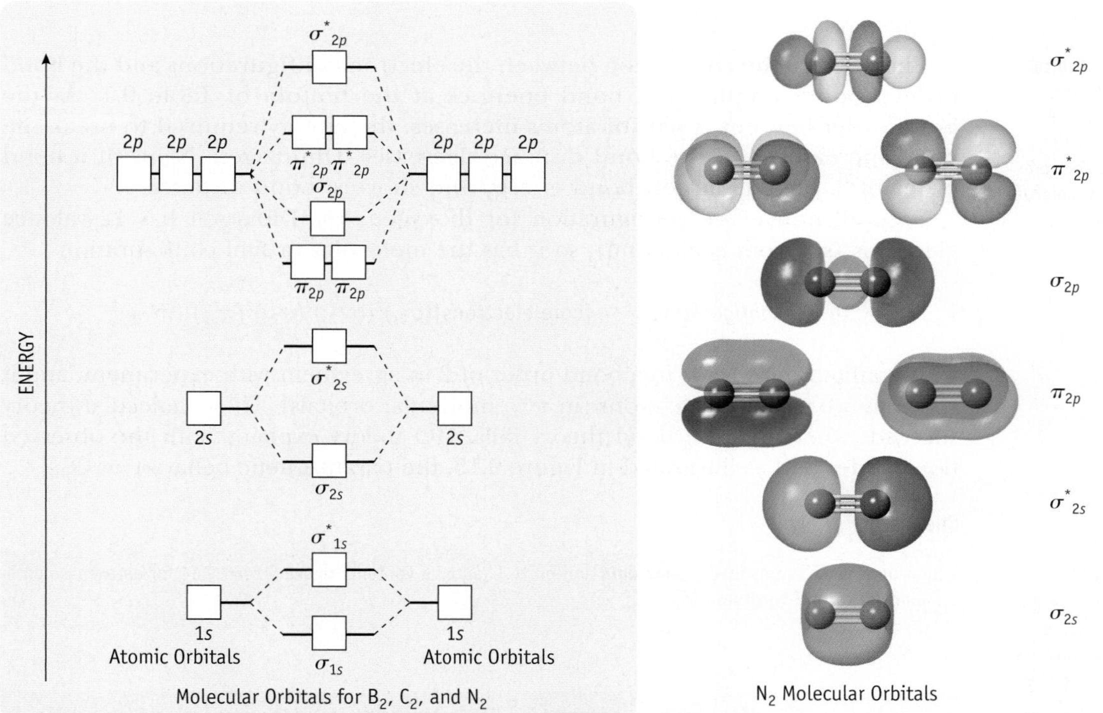

FIGURE 9.21 Molecular orbitals for homonuclear diatomic molecules of second period elements. (left) Energy level diagram. Although the diagram leads to the correct conclusions regarding bond order and magnetic behavior for O_2 and F_2, the energy ordering of the MOs in this figure is correct only for B_2, C_2, and N_2. For O_2 and F_2, the σ_{2p} MO is lower in energy than the π_{2p} MOs. See *A Closer Look,* page 429. (right) Calculated molecular orbitals for N_2. (Color scheme: occupied MOs are blue/green. Unoccupied MOs are red/yellow. The different colors reflect the different phases [positive or negative signs] of the wave functions.)

TABLE 9.1 Molecular Orbital Occupations and Physical Data for Homonuclear Diatomic Molecules of Second-Period Elements

	B_2	C_2	N_2		O_2	F_2
σ^*_{2p}	[ ]	[ ]	[ ]	σ^*_{2p}	[ ]	[ ]
π^*_{2p}	[ ][ ]	[ ][ ]	[ ][ ]	π^*_{2p}	[↑][↑]	[↑↓][↑↓]
σ_{2p}	[ ]	[ ]	[↑↓]	π_{2p}	[↑↓][↑↓]	[↑↓][↑↓]
π_{2p}	[↑][↑]	[↑↓][↑↓]	[↑↓][↑↓]	σ_{2p}	[↑↓]	[↑↓]
σ^*_{2s}	[↑↓]	[↑↓]	[↑↓]	σ^*_{2s}	[↑↓]	[↑↓]
σ_{2s}	[↑↓]	[↑↓]	[↑↓]	σ_{2s}	[↑↓]	[↑↓]
Bond order	One	Two	Three		Two	One
Bond-dissociation energy (kJ/mol)	290	620	945		498	155
Bond distance (pm)	159	131	110		121	143
Observed magnetic behavior (paramagnetic or diamagnetic)	Para	Dia	Dia		Para	Dia

■ **Highest Occupied Molecular Orbital (HOMO)** Chemists often refer to the highest energy MO that contains electrons as the HOMO. For O_2, this is the π^*_{2p} orbital. Chemists also use the term LUMO for the lowest unoccupied molecular orbital. For O_2, this would be σ^*_{2p}.

First, notice the correlation between the electron configurations and the bond orders, bond lengths, and bond energies at the bottom of Table 9.1. As the bond order between a pair of atoms increases, the energy required to break the bond increases, and the bond distance decreases. Dinitrogen, N_2, with a bond order of 3, has the largest bond energy and shortest bond distance.

Second, notice the configuration for dioxygen, O_2. Dioxygen has 12 valence electrons (six from each atom), so it has the molecular orbital configuration

O_2 MO Configuration: $$[\text{core electrons}](\sigma_{2s})^2(\sigma^*_{2s})^2(\sigma_{2p})^2(\pi_{2p})^4(\pi^*_{2p})^2$$

This configuration leads to a bond order of 2 in agreement with experiment, and it specifies two unpaired electrons (in π^*_{2p} molecular orbitals). Thus, molecular theory succeeds where valence bond theory fails. MO theory explains both the observed bond order and as illustrated in Figure 9.15, the paramagnetic behavior of O_2.

Chemistry Now™

Sign in at **www.thomsonedu.com/login** and go to Chapter 9 Contents to see Screen 9.11 for an exercise on **molecular orbital configurations.**

■ **Phases of Atomic Orbitals and Molecular Orbitals** Recall from page 289 in Chapter 6 that electron orbitals are electron waves and as such have positive and negative phases. For this reason, the atomic orbitals in Figures 9.19–9.21 are drawn with two different colors. Looking at the *p* orbitals in Figure 9.19, you see that a bonding MO is formed when *p* orbitals with the same wave function sign overlap (+ with +). An antibonding orbital arises if they overlap out of phase (+ with −).

EXAMPLE 9.8 Electron Configuration for a Homonuclear Diatomic Ion

Problem When potassium reacts with O_2, potassium superoxide, KO_2, is one of the products. This is an ionic compound, and the anion is the superoxide ion, O_2^-. Write the molecular orbital electron configuration for the ion. Predict its bond order and magnetic behavior.

Strategy Use the energy level diagram for O_2 in Table 9.1 to generate the electron configuration of this ion, and use Equation 9.1 to determine the bond order.

A Closer Look — Molecular Orbitals for Compounds Formed from *p*-Block Elements

Several features of the molecular orbital energy level diagram in Figure 9.21 should be described in more detail.

(a) The bonding and antibonding σ orbitals from 2*s* interactions are lower in energy than the σ and π MOs from 2*p* interactions. The reason is that 2*s* orbitals have a lower energy than 2*p* orbitals in the separated atoms.

(b) The energy separation of the bonding and antibonding orbitals is greater for σ_{2p} than for π_{2p}. This happens because *p* orbitals overlap to a greater extent when they are oriented head to head (to give σ_{2p} MOs) than when they are side by side (to give π_{2p} MOs). The greater the orbital overlap, the greater the stabilization of the bonding MO and the greater the destabilization of the antibonding MO.

Figure 9.21 shows an energy ordering of molecular orbitals that you might not have expected, but there are reasons for this. A more sophisticated approach takes into account the "mixing" of *s* and *p* atomic orbitals, which have similar energies. This causes the σ_{2s} and σ^*_{2s} molecular orbitals to be lower in energy than otherwise expected, and the σ_{2p} and σ^*_{2p} orbitals to be higher in energy.

The mixing of *s* and *p* orbitals is important for B_2, C_2, and N_2, so Figure 9.21 applies strictly only to these molecules. For O_2 and F_2, σ_{2p} is lower in energy than π_{2p}, and Table 9.1 takes this into account.

Solution The MO electron configuration for O_2^- is

$$O_2^-\ \textit{MO Configuration:}\quad [\text{core electrons}](\sigma_{2s})^2(\sigma^*_{2s})^2(\sigma_{2p})^2(\pi_{2p})^4(\pi^*_{2p})^3$$

The ion is predicted to be paramagnetic to the extent of one unpaired electron, a prediction confirmed by experiment. The bond order is 1.5, because there are eight bonding electrons and five antibonding electrons. The bond order for O_2^- is lower than O_2 so we predict the O—O bond length in O_2^- should be longer than the oxygen–oxygen bond length in O_2. The superoxide ion in fact has an O—O bond length of 134 pm, whereas the bond length in O_2 is 121 pm.

Comment You should quickly spot the fact that the superoxide ion (O_2^-), contains an odd number of electrons. This is another diatomic species (in addition to NO and O_2) for which it is not possible to write a Lewis structure that accurately represents the bonding.

EXERCISE 9.8 Molecular Electron Configurations

The cations O_2^+ and N_2^+ are important components of Earth's upper atmosphere. Write the electron configuration of O_2^+. Predict its bond order and magnetic behavior.

Electron Configurations for Heteronuclear Diatomic Molecules

The compounds NO, CO, and ClF, molecules containing two different elements, are examples of **heteronuclear diatomic molecules.** MO descriptions for heteronuclear diatomic molecules generally resemble those for homonuclear diatomic molecules. As a consequence, an energy level diagram like Figure 9.21 can be used to judge the bond order and magnetic behavior for heteronuclear diatomics.

Let us do this for nitrogen monoxide, NO. Nitrogen monoxide has 11 valence electrons. If these are assigned to the MOs for a homonuclear diatomic molecule, the molecular electron configuration is

$$\textit{NO MO Configuration:}\quad [\text{core electrons}](\sigma_{2s})^2(\sigma^*_{2s})^2(\pi_{2p})^4(\sigma_{2p})^2(\pi^*_{2p})^1$$

The net bond order is 2.5, in accord with bond length information. The single, unpaired electron is assigned to the π^*_{2p} molecular orbital, and the molecule is paramagnetic, as predicted for a molecule with an odd number of electrons.

Case Study — Two Chemical Bonding Mysteries

Nature is based on millions of chemical compounds, and chemists have created thousands more in the laboratory. In general, we understand their structures and bonding reasonably well, but from time to time nature gives us wonderful mysteries to solve.

Some of the mysteries are about the simplest molecules. One is B_2H_6, diborane, the simplest member of a large class of compounds. When it was discovered in the 1930s, chemists thought it must look like ethane, $H_3C{-}CH_3$. However, by the 1940s it was known that its structure had B—H—B bridges.

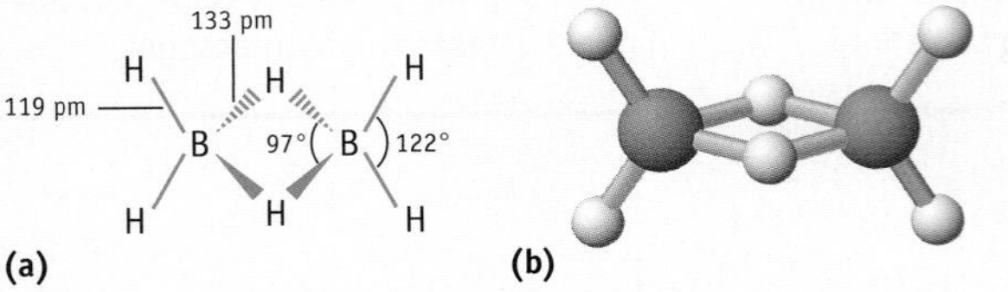

There are two mysteries. If each line in the structure represents a two-electron bond, you realize the molecule is "electron-deficient"; two B atoms and six H atoms do not contribute enough electrons for eight two-electron bonds. Also, notice that hydrogen is bonded to two different atoms, something we don't expect to happen because H has only one valence electron. One way to approach this problem is to begin by assuming each B atom is sp^3 hybridized. The four outside or terminal H atoms are then bonded to the B atoms by normal, two-electron bonds using two of the sp^3 hybrid orbitals on each B atom. The remaining two sp^3 hybrid orbitals on each B atom point into the bridging region where H atom 1*s* orbitals overlap with sp^3 orbitals from two different B atoms. Two electrons are assigned to each set of the bridging B—H—B groups, so the bridging bonds are two-electron/three-center bonds.

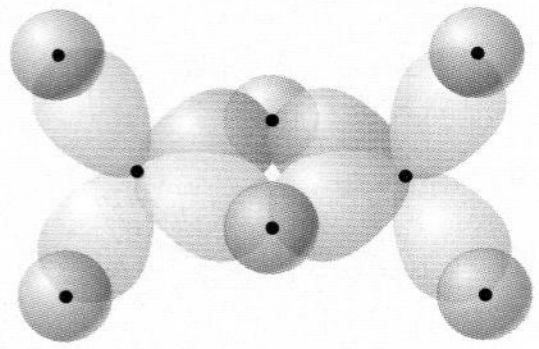

One can also apply molecular orbital theory to the molecule, and the bonding picture that emerges likewise has stable B—H—B bridges. The lowest energy molecular orbital, shown here, shows that electron density is spread over the B_2H_2 portion of the molecule.

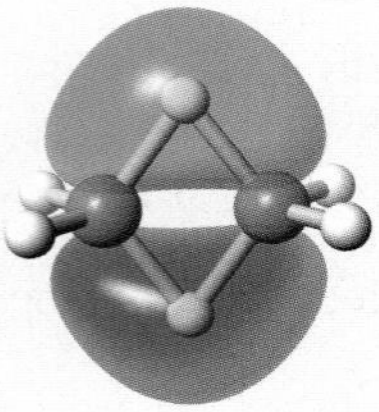

The molecular orbital that accounts for B—H—B bridge bonding.

It turns out that hydrogen bridges are also found in many other boron compounds and in other kinds of molecules. William Lipscomb received the Nobel Prize in Chemistry in 1976 for "his studies of boranes which have illuminated problems with chemical bonding."

Another bonding mystery involves Zeise's salt, a compound first discovered in the 1820s. However, it was not until the 1950s that three chemists (M. Dewar, J. Chatt, and Duncanson) devised a reasonable bonding model. The salt is based on the anion $[(C_2H_4)PtCl_3]^-$. The most interesting aspect of this ion is that the ethylene molecule, C_2H_4, is bonded sideways to the Pt^{2+} ion. How can this occur? As illustrated in the figure below, the π-bonding electrons of ethylene are donated to Pt^{2+} through overlap of the filled π orbital for C_2H_4 with an empty Pt^{2+} orbital. To account for the stability of the ion, however, there is another important aspect: a filled *d*-orbital on the Pt^{2+} ion donates electron density to the empty π-antibonding LUMO on the ethylene.

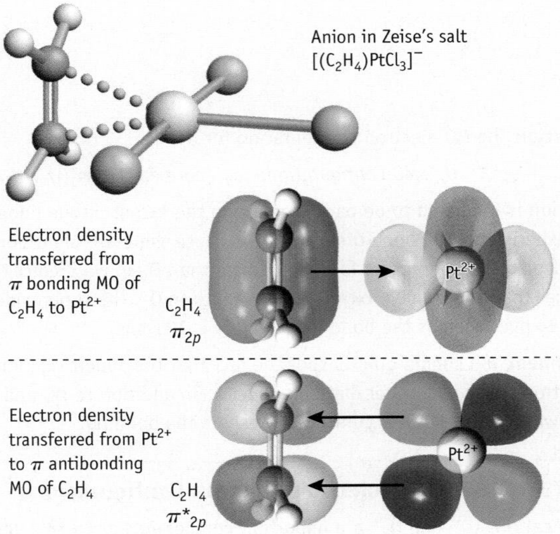

An understanding of bonding in Zeise's salt was a seminal event in chemistry because the ion is the model for the binding of other molecules like ethylene to transition metal centers.

Questions:

1. *If you treat each line in the B_2H_6 structure above as a two-electron bond, how many electrons are required for bonding in the molecule? How many electrons are actually available?*
2. *Diborane reacts with molecules such as NH_3 to give, for example, $H_3B{-}NH_3$. Draw a Lewis structure for this compound, estimate bond angles, and indicate the B and N hybridization. Is the molecule polar? What are the formal charges on the atoms?*
3. *Silver(I) ion is known to bond to ethylene, forming an ion of the formula $[Ag(C_2H_4)_x]^+$. When heated, the reaction $[Ag(C_2H_4)_x]BF_4(s) \longrightarrow AgBF_4(s) + x\,C_2H_4(g)$ occurs. If 62.1 mg of the silver(I) complex is heated, 54.3 mg of $AgBF_4$ remains. What is the value of x?*

Answers to these questions are in Appendix Q.

Resonance and MO Theory

Ozone, O_3, is a simple triatomic molecule with equal oxygen–oxygen bond lengths. Equal X—O bond lengths are also observed in other molecules and ions, such as SO_2, NO_2^-, and HCO_2^-. Valence bond theory introduced resonance to rationalize the equivalent bonding to the oxygen atoms in these structures, but MO theory provides another view of this problem.

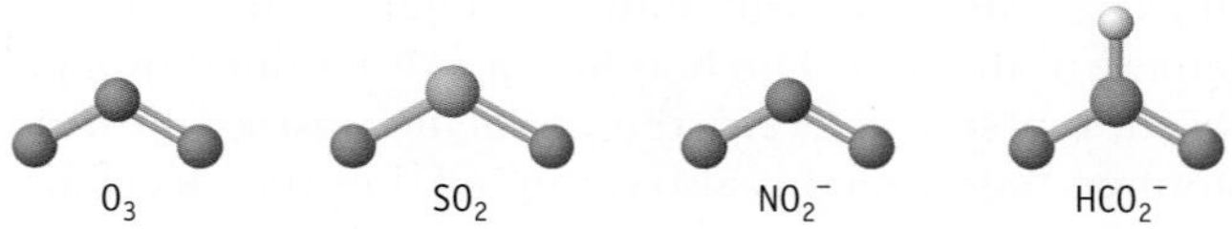

To understand the bonding in ozone, we begin by looking at the valence bond picture. Let us assume that all three O atoms are sp^2 hybridized. The central atom uses its sp^2 hybrid orbitals to form two σ bonds and to accommodate a lone pair. The terminal atoms use their sp^2 hybrid orbitals to form one σ bond and to accommodate two lone pairs. In total, the lone pairs and bonding pairs in the σ framework of O_3 account for seven of the nine valence electron pairs in O_3.

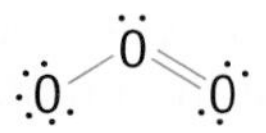

Lewis structure of O_3. All O atoms are sp^2 hybridized.

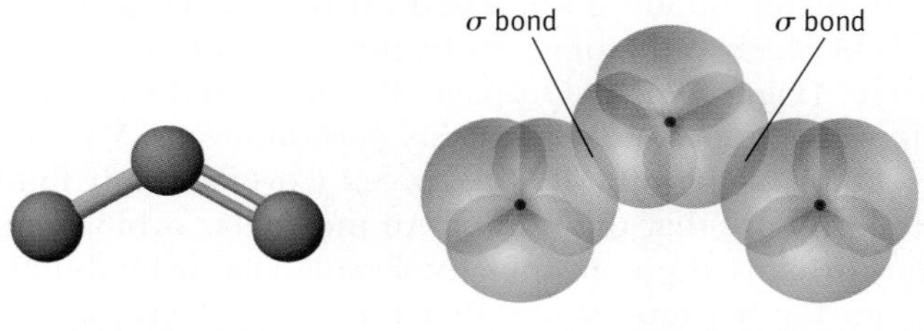

Molecular model

A representation of the sigma bonding framework of O_3 using sp^2 hybrid orbitals

The π bond in ozone arises from the two remaining pairs (Figure 9.22). Because we have assumed that each oxygen atom in O_3 is sp^2 hybridized, an unhybridized p orbital perpendicular to the O_3 plane remains on each of the three oxygen atoms. The orbitals are in the correct orientation to form π bonds.

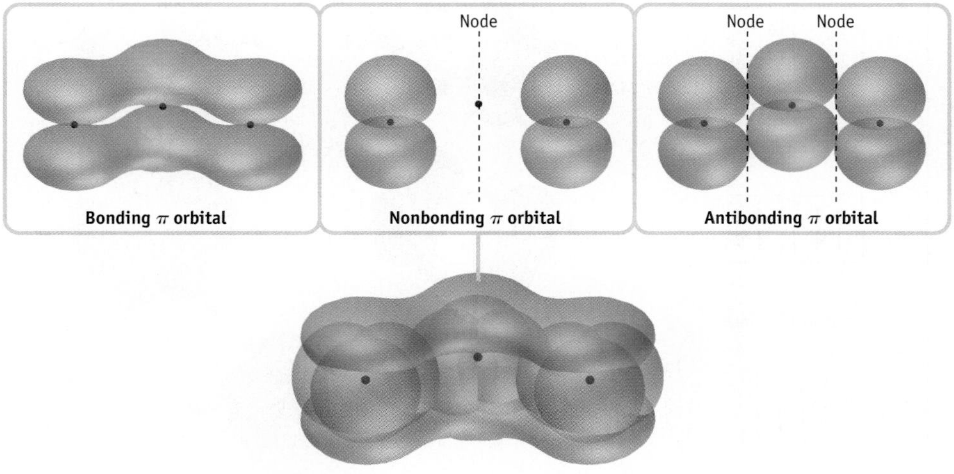

σ and π bonding in ozone

FIGURE 9.22 Pi-bonding in ozone, O_3. Each O atom in O_3 is sp^2 hybridized. The three $2p$ orbitals, one on each atom, are used to create the three π molecular orbitals. Two pairs of electrons are assigned to the orbitals: one pair in the bonding orbital and one pair in the nonbonding orbital. The π bond order is 0.5, as one bonding pair is spread across two bonds.

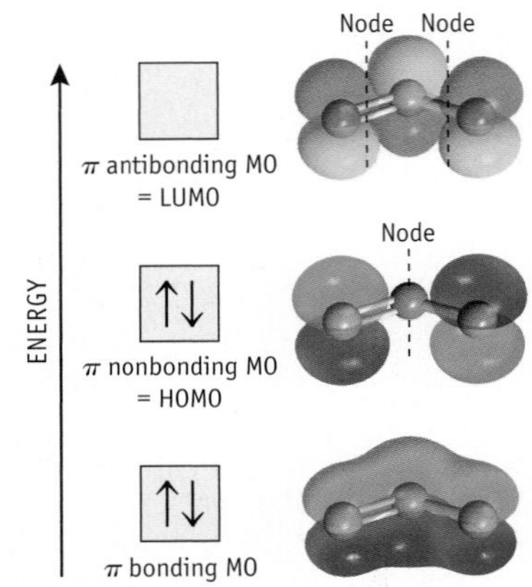

FIGURE 9.23. The π molecular diagram for ozone. Notice that, as in the other MO diagrams illustrated (especially Figure 9.21), the energy of the molecular orbitals increases as the number of nodes increases.

Now let's apply MO theory to the pi bonds. A principle of MO theory is that the number of molecular orbitals must equal the number of atomic orbitals. Thus, the three $2p$ atomic orbitals must be combined in a way that forms three molecular orbitals.

One π_p MO for ozone is a bonding orbital because the three p orbitals are "in phase" across the molecule. Another π_p MO is an antibonding orbital because the atomic orbital on the central atom that is "out of phase" with the terminal atom p orbitals. The third π_p MO is a nonbonding orbital because the middle p orbital does not participate in the MO. The bonding π_p MO is filled by a pair of electrons that is delocalized, or "spread over," the molecule, just as the resonance hybrid implies. The nonbonding orbital is also occupied, but the electrons in this orbital are concentrated near the two terminal oxygens. As the name implies, electrons in this molecular orbital neither help nor hinder the bonding in the molecule. The π bond order of O_3 is 0.5 (Figure 9.23). Because the σ bond order is 1.0 and the π bond order is 0.5, the net oxygen–oxygen bond order is 1.5—the same value given by valence bond theory.

The observation that two of the π molecular orbitals for ozone extend over three atoms illustrates an important point regarding molecular orbital theory: *Orbitals can extend beyond two atoms.* In valence bond theory, in contrast, all representations for bonding are based on being able to localize pairs of electrons in bonds between two atoms. To further illustrate the MO approach, look again at benzene (Figure 9.24). On page 421, we noted that the π electrons in this molecule were spread out over all six carbon atoms. We can now see how the same case can be made with MO theory. Six p orbitals contribute to the π system. Based on the premise that the number of molecular orbitals must equal the number of atomic orbitals, there must be six π molecular orbitals in benzene. An energy level diagram for benzene shows that the six π electrons reside in the three lowest-energy (bonding) molecular orbitals.

FIGURE 9.24 Molecular orbital energy level diagram for benzene. Because there are six unhybridized p orbitals, six π molecular orbitals can be formed—three bonding and three antibonding. The three bonding molecular orbitals accommodate the six π electrons.

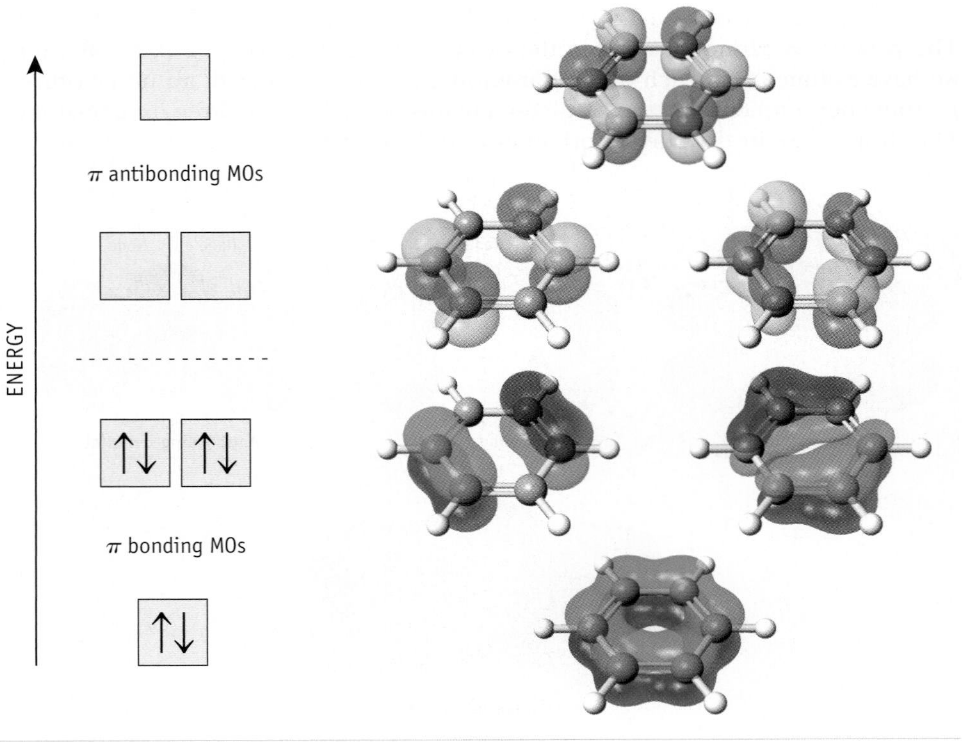

Chapter Goals Revisited

Now that you have studied this chapter, you should ask whether you have met the chapter goals. In particular, you should be able to:

Understand the differences between valence bond theory and molecular orbital theory

a. Describe the main features of valence bond theory and molecular orbital theory, the two commonly used theories for covalent bonding (Section 9.1).

b. Recognize that the premise for valence bond theory is that bonding results from the overlap of atomic orbitals. By virtue of the overlap of orbitals, electrons are concentrated (or localized) between two atoms (Section 9.2).

c. Distinguish how sigma (σ) and pi (π) bonds arise. For σ bonding, orbitals overlap in a head-to-head fashion, concentrating electrons along the bond axis. Sideways overlap of p atomic orbitals results in π bond formation, with electron density above and below the molecular plane (Section 9.2).

d. Understand how molecules having double bonds can have isomeric forms. Study Question(s) assignable in OWL: 14.

Identify the hybridization of an atom in a molecule or ion

a. Use the concept of hybridization to rationalize molecular structure (Section 9.2). Study Question(s) assignable in OWL: 2, 3, 4, 5, 6, 8, 9, 11, 12, 21, 22, 24, 27, 32, 35, 36, 38, 44, 45, 47–50, 51, 52–54; Go Chemistry Module 14.

Hybrid Orbitals	Atomic Orbitals Used	Number of Hybrid Orbitals	Electron-Pair Geometry
sp	$s + p$	2	Linear
sp^2	$s + p + p$	3	Trigonal-planar
sp^3	$s + p + p + p$	4	Tetrahedral
sp^3d	$s + p + p + p + d$	5	Trigonal bipyramidal
sp^3d^2	$s + p + p + p + d + d$	6	Octahedral

Understand the differences between bonding and antibonding molecular orbitals and be able to write the molecular orbital configurations for simple diatomic molecules.

a. Understand molecular orbital theory (Section 9.3), in which atomic orbitals are combined to form bonding orbitals, nonbonding orbitals, or antibonding orbitals that are delocalized over several atoms. In this description, the electrons of the molecule or ion are assigned to the orbitals beginning with the one at lowest energy, according to the Pauli exclusion principle and Hund's rule.

b. Use molecular orbital theory to explain the properties of O_2 and other diatomic molecules. Study Question(s) assignable in OWL: 15–20, 40, 42, 43, 57.

Chemistry.Now™ Sign in at **www.thomsonedu.com/login** to:

- Assess your understanding with Study Questions in OWL keyed to each goal in the Goals and Homework menu for this chapter
- For quick review, download Go Chemistry mini-lecture flashcard modules (or purchase them at **www.ichapters.com**)
- Check your readiness for an exam by taking the Pre-Test and exploring the modules recommended in your Personalized Study plan.

Access **How Do I Solve It?** tutorials on how to approach problem solving using concepts in this chapter.

KEY EQUATIONS

Equation 9.1 (page 424) Calculating the order of a bond from the molecular orbital electron configuration

Bond order = 1/2 (number of electrons in bonding MOs − number of electrons in antibonding MOs)

STUDY QUESTIONS

Online homework for this chapter may be assigned in OWL.

▲ denotes challenging questions.

■ denotes questions assignable in OWL.

Blue-numbered questions have answers in Appendix O and fully-worked solutions in the *Student Solutions Manual.*

Practicing Skills

Valence Bond Theory

(See Examples 9.1–9.5 and ChemistryNow Screens 9.2–9.7.)

1. Draw the Lewis structure for chloroform, $CHCl_3$. What are its electron-pair and molecular geometries? What orbitals on C, H, and Cl overlap to form bonds involving these elements?

2. ■ Draw the Lewis structure for NF_3. What are its electron-pair and molecular geometries? What is the hybridization of the nitrogen atom? What orbitals on N and F overlap to form bonds between these elements?

3. ■ Specify the electron-pair and molecular geometry for each underlined atom in the following list. Describe the hybrid orbital set used by this atom in each molecule or ion.
(a) $\underline{B}Br_3$ (b) $\underline{C}O_2$ (c) $\underline{C}H_2Cl_2$ (d) $\underline{C}O_3^{2-}$

4. ■ Specify the electron-pair and molecular geometry for each underlined atom in the following list. Describe the hybrid orbital set used by this atom in each molecule or ion.
(a) $\underline{C}Se_2$ (b) $\underline{S}O_2$ (c) $\underline{C}H_2O$ (d) $\underline{N}H_4^+$

5. ■ Describe the hybrid orbital set used by each of the indicated atoms in the molecules below:
(a) the carbon atoms and the oxygen atom in dimethyl ether, H_3COCH_3
(b) each carbon atom in propene

(c) the two carbon atoms and the nitrogen atom in the amino acid glycine

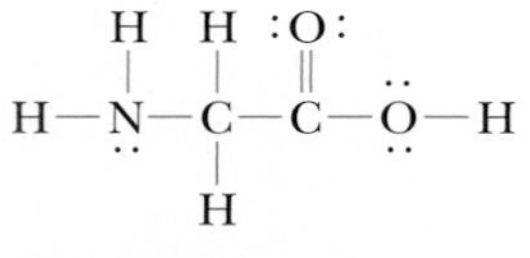

6. ■ Give the hybrid orbital set used by each of the underlined atoms in the following molecules.

(a) 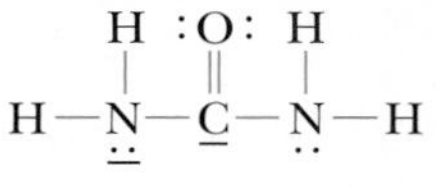

(b) 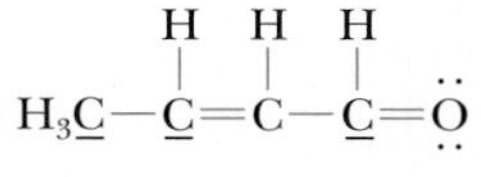

(c) 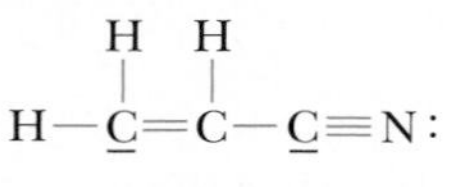

7. Draw the Lewis structure, and then specify the electron-pair and molecular geometries for each of the following molecules or ions. Identify the hybridization of the central atom.
(a) SiF_6^{2-} (b) SeF_4 (c) ICl_2^- (d) XeF_4

8. ■ Draw the Lewis structure, and then specify the electron-pair and molecular geometries for each of the following molecules or ions. Identify the hybridization of the central atom.
(a) $XeOF_4$
(b) BrF_5
(c) central S in SOF_4
(d) central Br in Br_3^-

9. ■ Draw the Lewis structures of the acid HPO_2F_2 and its anion $PO_2F_2^-$. What is the molecular geometry and hybridization for the phosphorus atom in each species? (H is bonded to an O atom in the acid.)

10. Draw the Lewis structures of HSO_3F and SO_3F^-. What is the molecular geometry and hybridization for the sulfur atom in each species? (H is bonded to an O atom in the acid.)

11. ■ What is the hybridization of the carbon atom in phosgene, Cl_2CO? Give a complete description of the σ and π bonding in this molecule.

12. ■ What is the hybridization of the sulfur atom in sulfuryl fluoride, SO_2F_2?

13. The arrangement of groups attached to the C atoms involved in a C=C double bond leads to *cis* and *trans* isomers. For each compound below, draw the other isomer.

(a) 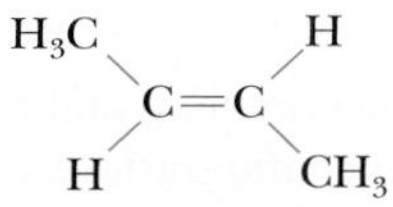

(b) 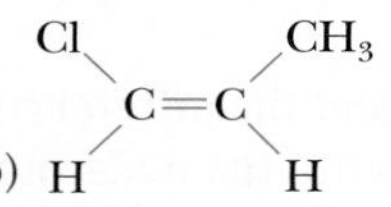

14. ■ For each compound below, decide whether *cis* and *trans* isomers are possible. If isomerism is possible, draw the other isomer.

(a) 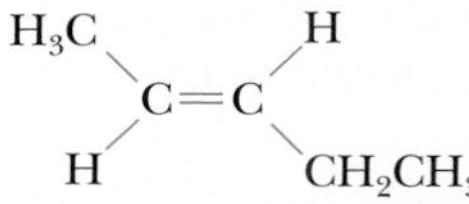

(b) 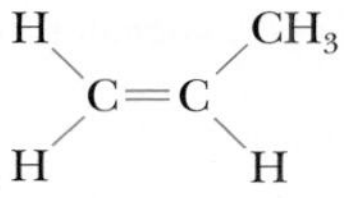

(c) 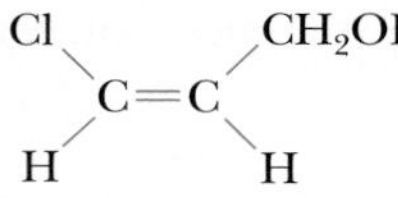

Molecular Orbital Theory

(See Examples 9.6–9.8 and ChemistryNow Screens 9.9–9.12.)

15. ■ The hydrogen molecular ion, H_2^+, can be detected spectroscopically. Write the electron configuration of the ion in molecular orbital terms. What is the bond order of the ion? Is the hydrogen–hydrogen bond stronger or weaker in H_2^+ than in H_2?

16. ■ Give the electron configurations for the ions Li_2^+ and Li_2^- in molecular orbital terms. Compare the Li—Li bond order in these ions with the bond order in Li_2.

17. ■ Calcium carbide, CaC_2, contains the acetylide ion, C_2^{2-}. Sketch the molecular orbital energy level diagram for the ion. How many net σ and π bonds does the ion have? What is the carbon–carbon bond order? How has the bond order changed on adding electrons to C_2 to obtain C_2^{2-}? Is the C_2^{2-} ion paramagnetic?

18. ■ Oxygen, O_2, can acquire one or two electrons to give O_2^- (superoxide ion) or O_2^{2-} (peroxide ion). Write the electron configuration for the ions in molecular orbital terms, and then compare them with the O_2 molecule on the following bases.
   (a) magnetic character
   (b) net number of σ and π bonds
   (c) bond order
   (d) oxygen–oxygen bond length

19. ■ Assume the energy level diagram for homonuclear diatomic molecules (Figure 9.21) can be applied to heteronuclear diatomics such as CO.
   (a) Write the electron configuration for carbon monoxide, CO.
   (b) What is the highest-energy, occupied molecular orbital?
   (c) Is the molecule diamagnetic or paramagnetic?
   (d) What is the net number of σ and π bonds? What is the CO bond order?

20. ■ The nitrosyl ion, NO^+, has an interesting chemistry.
   (a) Is NO^+ diamagnetic or paramagnetic? If paramagnetic, how many unpaired electrons does it have?
   (b) Assume the molecular orbital diagram for a homonuclear diatomic molecule (Figure 9.21) applies to NO^+. What is the highest-energy molecular orbital occupied by electrons?
   (c) What is the nitrogen–oxygen bond order?
   (d) Is the N—O bond in NO^+ stronger or weaker than the bond in NO?

General Questions on Valence Bond and Molecular Orbital Theory

These questions are not designated as to type or location in the chapter. They may combine several concepts from this and other chapters.

21. ■ Draw the Lewis structure for AlF_4^-. What are its electron-pair and molecular geometries? What orbitals on Al and F overlap to form bonds between these elements? What are the formal charges on the atoms? Is this a reasonable charge distribution?

22. ■ Draw the Lewis structure for ClF_3. What are its electron-pair and molecular geometries? What is the hybridization of the chlorine atom? What orbitals on Cl and F overlap to form bonds between these elements?

23. ■ Describe the O—S—O angle and the hybrid orbital set used by sulfur in each of the following molecules or ions:
   (a) SO_2 (b) SO_3 (c) SO_3^{2-} (d) SO_4^{2-}

   Do all have the same value for the O—S—O angle? Does the S atom in all these species use the same hybrid orbitals?

24. ■ Sketch the Lewis structures of ClF_2^+ and ClF_2^-. What are the electron-pair and molecular geometries of each ion? Do both have the same F—Cl—F angle? What hybrid orbital set is used by Cl in each ion?

25. Sketch the resonance structures for the nitrite ion, NO_2^-. Describe the electron-pair and molecular geometries of the ion. From these geometries, decide on the O—N—O bond angle, the average NO bond order, and the N atom hybridization.

26. Sketch the resonance structures for the nitrate ion, NO_3^-. Is the hybridization of the N atom the same or different in each structure? Describe the orbitals involved in bond formation by the central N atom.

27. ■ Sketch the resonance structures for the N_2O molecule. Is the hybridization of the N atoms the same or different in each structure? Describe the orbitals involved in bond formation by the central N atom.

28. Compare the structure and bonding in CO_2 and CO_3^{2-} with regard to the O—C—O bond angles, the CO bond order, and the C atom hybridization.

29. Numerous molecules are detected in deep space. Three of them are illustrated here.

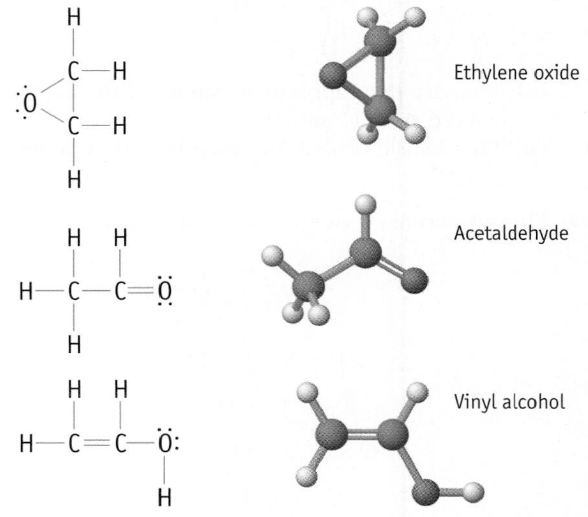

   (a) Comment on the similarities or differences in the formulas of these compounds. Are they isomers?
   (b) Indicate the hybridization of each C atom in each molecule.
   (c) Indicate the value of the H—C—H angle in each of the three molecules.
   (d) Are any of these molecules polar?
   (e) Which molecule should have the strongest carbon–carbon bond? The strongest carbon–oxygen bond?

30. Acrolein, a component of photochemical smog, has a pungent odor and irritates eyes and mucous membranes.

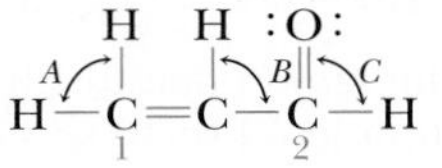

(a) What are the hybridizations of carbon atoms 1 and 2?
(b) What are the approximate values of angles *A*, *B*, and *C*?
(c) Is *cis-trans* isomerism possible here?

31. The organic compound below is a member of a class known as oximes.

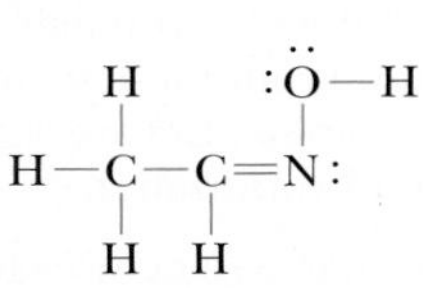

(a) What are the hybridizations of the two C atoms and of the N atom?
(b) What is the approximate C—N—O angle?

32. ■ The compound sketched below is acetylsalicylic acid, commonly known as aspirin:

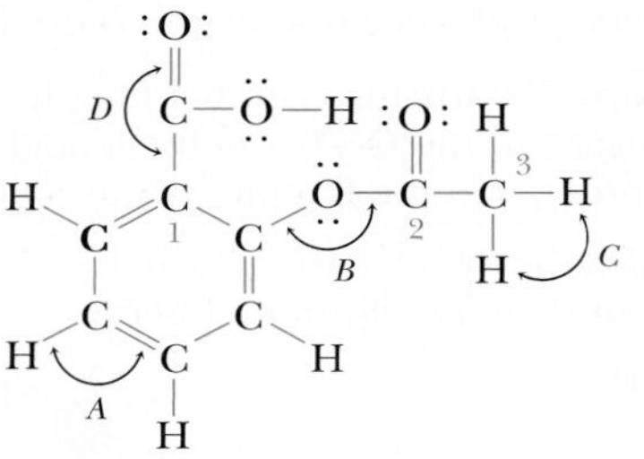

(a) What are the approximate values of the angles marked *A*, *B*, *C*, and *D*?
(b) What hybrid orbitals are used by carbon atoms 1, 2, and 3?

33. Phosphoserine is a less-common amino acid.

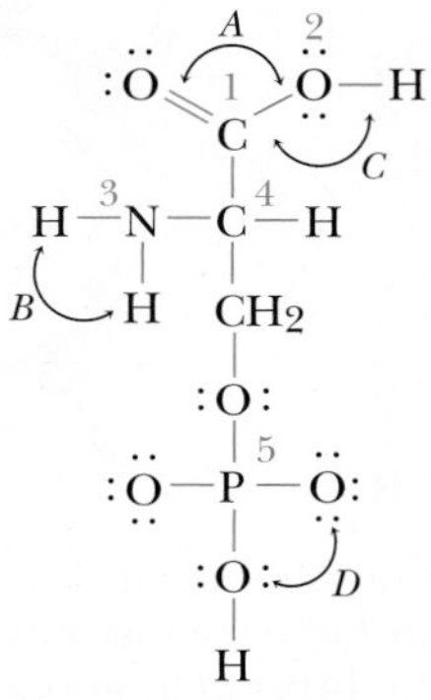

(a) Describe the hybridizations of atoms 1 through 5.
(b) What are the approximate values of the bond angles *A*, *B*, *C*, and *D*?
(c) What are the most polar bonds in the molecule?

34. ■ Lactic acid is a natural compound found in sour milk.

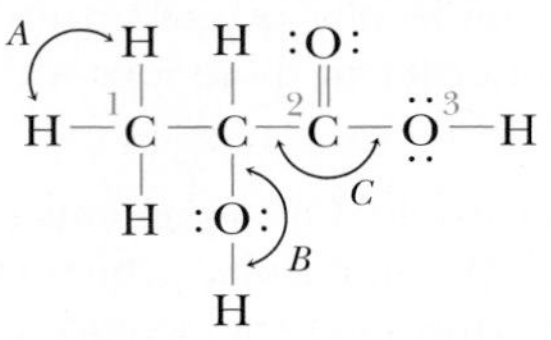

(a) How many π bonds occur in lactic acid? How many σ bonds?
(b) Describe the hybridization of atoms 1, 2, and 3.
(c) Which CO bond is the shortest in the molecule? Which CO bond is the strongest?
(d) What are the approximate values of the bond angles *A*, *B*, and *C*?

35. ■ Cinnamaldehyde occurs naturally in cinnamon oil.

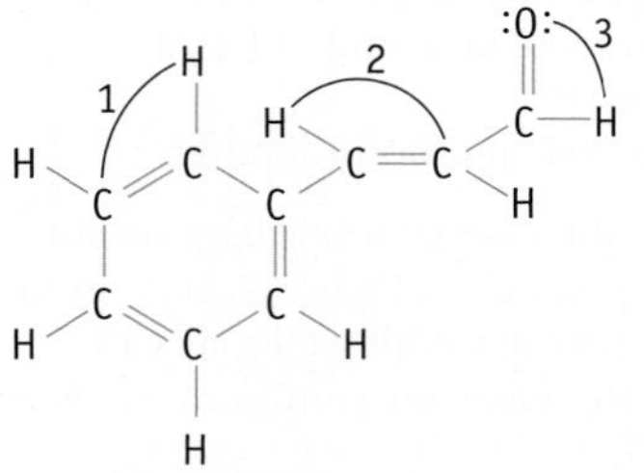

Cinnamaldehyde

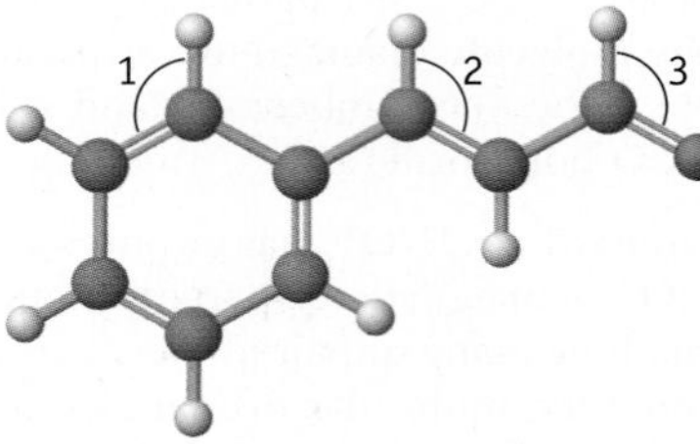

(a) What is the most polar bond in the molecule?
(b) How many sigma (σ) bonds and how many pi (π) bonds are there?
(c) Is *cis-trans* isomerism possible? If so, draw the isomers of the molecule.
(d) Give the hybridization of the C atoms in the molecule.
(e) What are the values of the bond angles 1, 2, and 3?

36. ■ Iodine and oxygen form a complex series of ions, among them IO_4^- and IO_5^{3-}. Draw the Lewis structures for these ions, and specify their electron-pair geometries and the shapes of the ions. What is the hybridization of the I atom in these ions?

37. Antimony pentafluoride reacts with HF according to the equation

$$2\ HF + SbF_5 \rightarrow [H_2F]^+[SbF_6]^-$$

(a) What is the hybridization of the Sb atom in the reactant and product?
(b) Draw a Lewis structure for H_2F^+. What is the geometry of H_2F^+? What is the hybridization of F in H_2F^+?

38. ■ Xenon forms well-characterized compounds (◀ page 404). Two xenon–oxygen compounds are XeO_3 and XeO_4. Draw the Lewis structures of these compounds, and give their electron-pair and molecular geometries. What are the hybrid orbital sets used by xenon in these two oxides?

39. The simple valence bond picture of O_2 does not agree with the molecular orbital view. Compare these two theories with regard to the peroxide ion, O_2^{2-}.
   (a) Draw an electron dot structure for O_2^{2-}. What is the bond order of the ion?
   (b) Write the molecular orbital electron configuration for O_2^{2-}. What is the bond order based on this approach?
   (c) Do the two theories of bonding lead to the same magnetic character and bond order for O_2^{2-}?

40. ■ Nitrogen, N_2, can ionize to form N_2^+ or add an electron to give N_2^-. Using molecular orbital theory, compare these species with regard to (a) their magnetic character, (b) net number of π bonds, (c) bond order, (d) bond length, and (e) bond strength.

41. Which of the homonuclear, diatomic molecules of the second-period elements (from Li_2 to Ne_2) are paramagnetic? Which have a bond order of 1? Which have a bond order of 2? Which diatomic molecule has the highest bond order?

42. ■ Which of the following molecules or molecule ions should be paramagnetic? What is the highest occupied molecular orbital (HOMO) in each one? Assume the molecular orbital diagram in Figure 9.21 applies to all of them.
   (a) NO
   (b) OF^-
   (c) O_2^{2-}
   (d) Ne_2^+
   (e) CN

43. ■ The CN molecule has been found in interstellar space. Assuming the electronic structure of the molecule can be described using the molecular orbital energy level diagram in Figure 9.21, answer the following questions.
   (a) What is the highest energy occupied molecular orbital (HOMO) to which an electron (or electrons) is (are) assigned?
   (b) What is the bond order of the molecule?
   (c) How many net σ bonds are there? How many net π bonds?
   (d) Is the molecule paramagnetic or diamagnetic?

44. ■ Amphetamine is a stimulant. Replacing one H atom on the NH_2, or amino, group with CH_3 gives methamphetamine, a particularly dangerous drug commonly known as "speed."

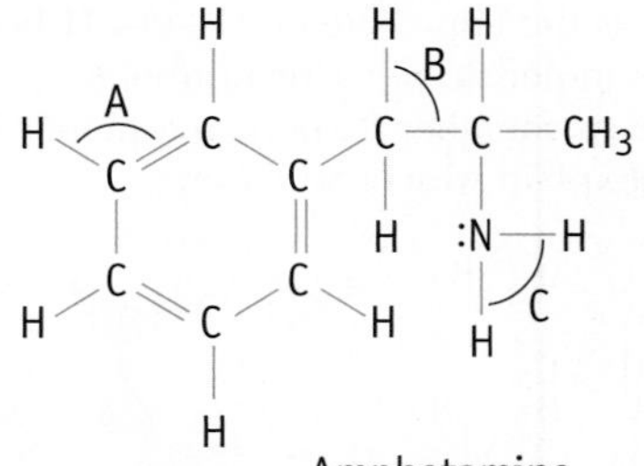

Amphetamine

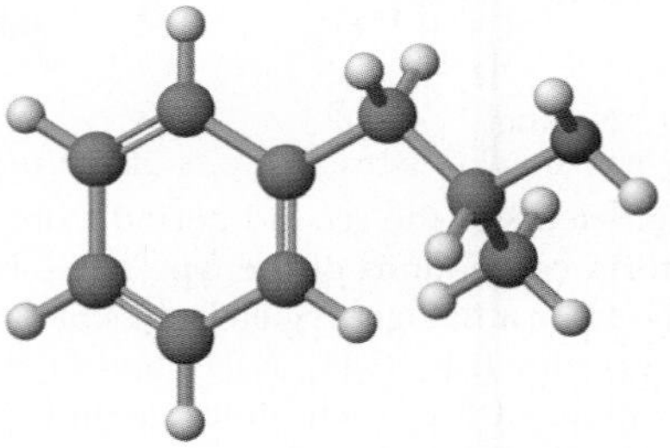

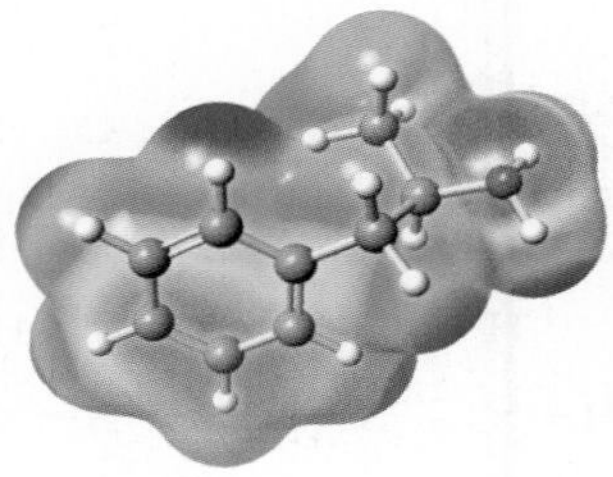

   (a) What are the hybrid orbitals used by the C atoms of the C_6 ring, by the C atoms of the side chain, and by the N atom?
   (b) Give approximate values for the bond angles *A*, *B*, and *C*.
   (c) How many σ bonds and π bonds are in the molecule?
   (d) Is the molecule polar or nonpolar?
   (e) Amphetamine reacts readily with a proton (H^+) in aqueous solution. Where does this proton attach to the molecule? Does the electrostatic potential map shown above confirm this possibility?

45. ■ Menthol is used in soaps, perfumes, and foods. It is present in the common herb mint, and it can be prepared from turpentine.
  (a) What are the hybridizations used by the C atoms in the molecule?
  (b) What is the approximate C—O—H bond angle?
  (c) Is the molecule polar or nonpolar?
  (d) Is the six-member carbon ring planar or nonplanar? Explain why or why not.

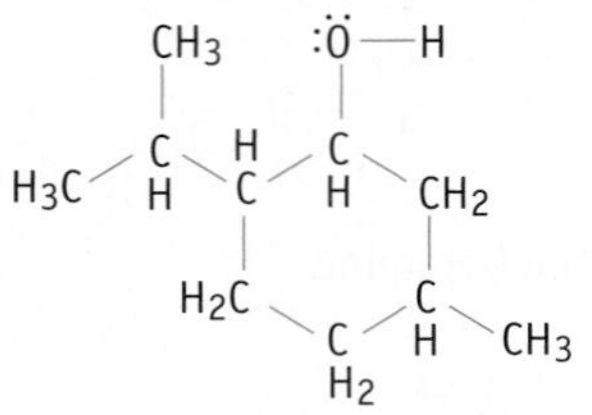
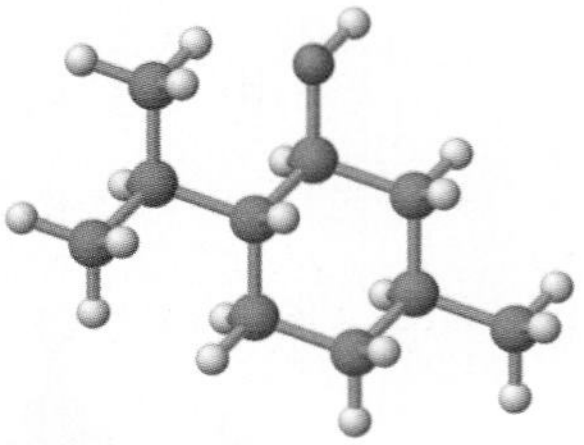

Menthol

46. The elements of the second period from boron to oxygen form compounds of the type X_nE—EX_n, where X can be H or a halogen. Sketch possible molecular structures for B_2F_4, C_2H_4, N_2H_4, and O_2H_2. Give the hybridizations of E in each molecule and specify approximate X—E—E bond angles.

In the Laboratory

47. ■ Suppose you carry out the following reaction of ammonia and boron trifluoride in the laboratory.

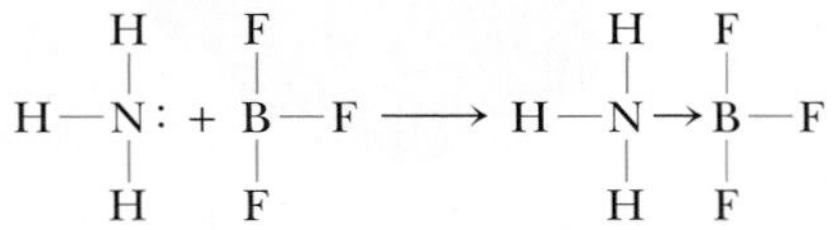

  (a) What is the geometry of the boron atom in BF_3? In $H_3N \rightarrow BF_3$?
  (b) What is the hybridization of the boron atom in the two compounds?
  (c) Does the boron atom's hybridization change on formation of the coordinate covalent bond?
  (d) Considering atom electronegativities and the bonding in NH_3 and BF_3, why do you expect the nitrogen on NH_3 to donate an electron pair to the B atom of BF_3?
  (e) BF_3 also reacts readily with water. Based on the ammonia reaction above, speculate on how water can interact with BF_3.

48. ▲ ■ Ethylene oxide is an intermediate in the manufacture of ethylene glycol (antifreeze) and polyester polymers. More than 4 million tons are produced annually in the U.S. The molecule has a three-member ring of two C atoms and an O atom.

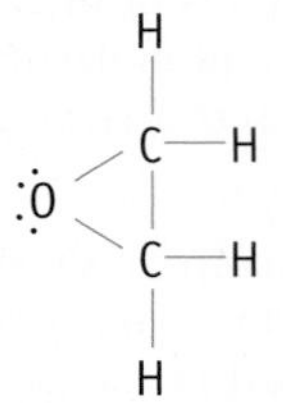
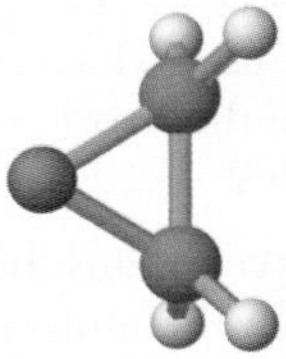

Ethylene oxide

  (a) What are the expected bond angles in the ring?
  (b) What is the hybridization of each atom in the ring?
  (c) Comment on the relation between the bond angles expected based on hybridization and the bond angles expected for a three-member ring.
  (d) Is the molecule polar? Based on the electrostatic potential map shown below, where do the negative and positive charges lie in the molecule?

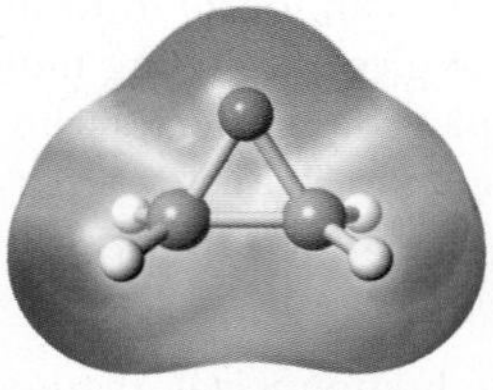

Electrostatic potential map for ethylene oxide.

49. ■ The sulfamate ion, $H_2NSO_3^-$, can be thought of as having been formed from the amide ion, NH_2^-, and sulfur trioxide, SO_3.
  (a) What are the geometries of the amide ion and of SO_3? What are the hybridizations of the N and S atoms, respectively?
  (b) Sketch a structure for the sulfamate ion, and estimate the bond angles.
  (c) What changes in hybridization do you expect for N and S in the course of the reaction

$$NH_2^- + SO_3 \rightarrow H_2N\text{—}SO_3^-?$$

  (d) Is SO_3 the donor of an electron pair or the acceptor of an electron pair in the reaction with amide ion? Does the electrostatic potential map shown below confirm your prediction?

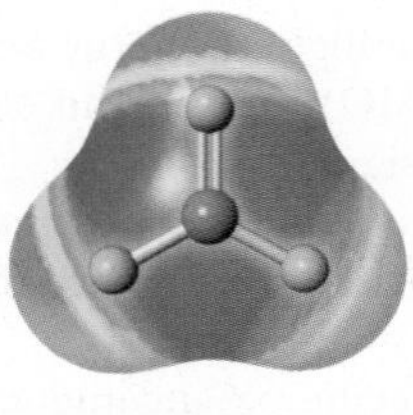

Electrostatic potential map for sulfur trioxide.

50. ▲ ■ The compound whose structure is shown here is acetylacetone. It exists in two forms: the *enol* form and the *keto* form.

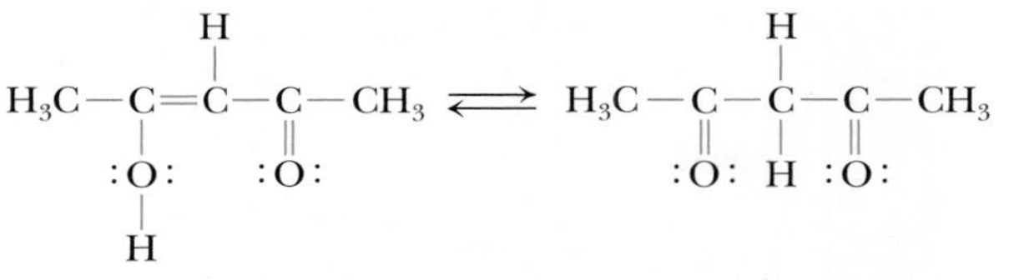

enol form *keto* form

The molecule reacts with OH^- to form an anion, $[CH_3COCHCOCH_3]^-$ (often abbreviated $acac^-$ for acetylacetonate ion). One of the most interesting aspects of this anion is that one or more of them can react with transition metal cations to give stable, highly colored compounds.

(a) Are the *keto* and *enol* forms of acetylacetone resonance forms? Explain your answer.

(b) What is the hybridization of each atom (except H) in the *enol* form? What changes in hybridization occur when it is transformed into the *keto* form?

(c) What are the electron-pair geometry and molecular geometry around each C atom in the *keto* and *enol* forms? What changes in geometry occur when the *keto* form changes to the *enol* form?

(d) Draw three possible resonance structures for the $acac^-$ ion.

(e) Is *cis-trans* isomerism possible in either the *enol* or the *keto* form?

(f) Is the enol form of acetylacetone polar? Where do the positive and negative charges lie in the molecule?

Summary and Conceptual Questions

The following questions may use concepts from this and previous chapters.

51. ■ What is the maximum number of hybrid orbitals that a carbon atom may form? What is the minimum number? Explain briefly.

52. ■ Consider the three fluorides BF_4^-, SiF_4, and SF_4.

(a) Identify a molecule that is isoelectronic with BF_4^-.

(b) Are SiF_4 and SF_4 isoelectronic?

(c) What is the hybridization of the central atom in each of these species?

53. ▲ ■ When two amino acids react with each other, they form a linkage called an amide group, or a peptide link. (If more linkages are added, a protein or polypeptide is formed.)

(a) What are the hybridizations of the C and N atoms in the peptide linkage?

(b) Is the structure illustrated the only resonance structure possible for the peptide linkage? If another resonance structure is possible, compare it with the one shown. Decide which is the more important structure.

(c) The computer-generated structure shown here, which contains a peptide linkage, shows that the linkage is flat. This is an important feature of proteins. Speculate on reasons that the CO—NH linkage is planar. What are the sites of positive and negative charge in this dipeptide?

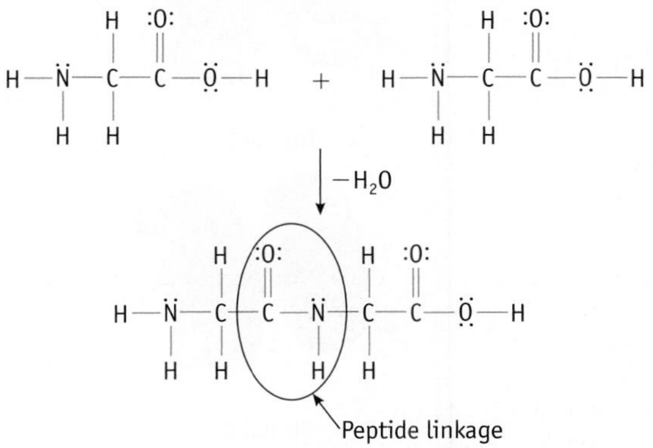

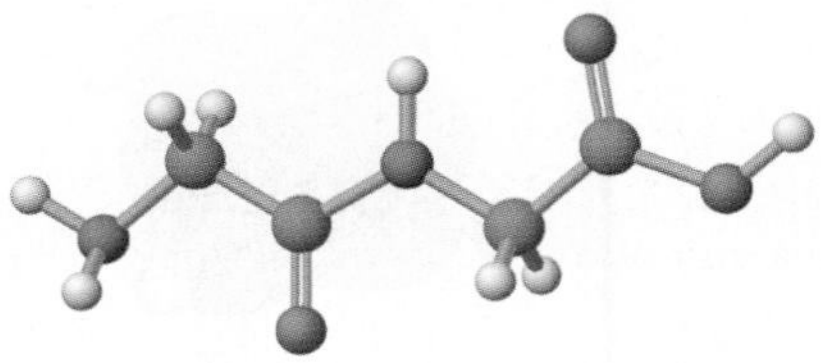

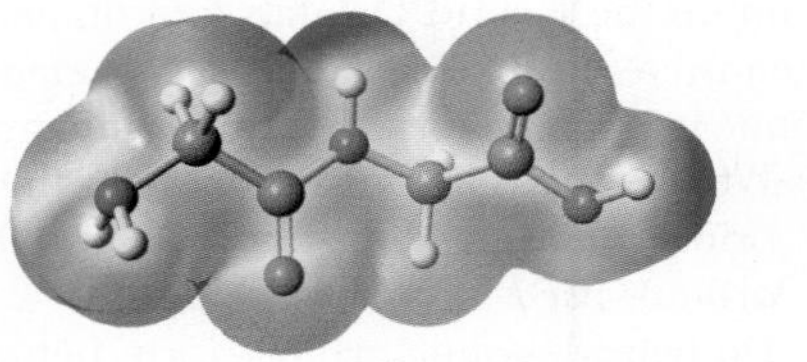

54. ■ What is the connection between bond order, bond length, and bond energy? Use ethane (C_2H_6), ethylene (C_2H_4), and acetylene (C_2H_2) as examples.

55. When is it desirable to use MO theory rather than valence bond theory?

56. How do valence bond theory and molecular orbital theory differ in their explanation of the bond order of 1.5 for ozone?

57. ■ Three of the four π molecular orbitals for cyclobutadiene are pictured here. Place them in order of increasing energy. (Remember that orbitals increase in energy in order of an increasing number of nodes. If a pair of orbitals have the same number of nodes, they have the same energy.)

Orbital A

Orbital B

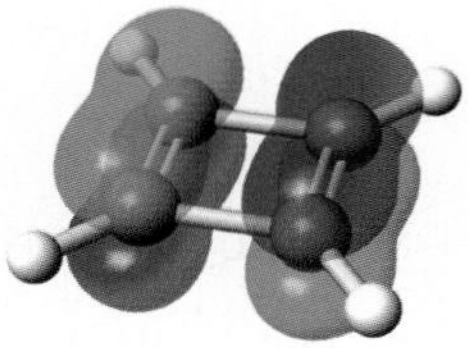

Orbital C

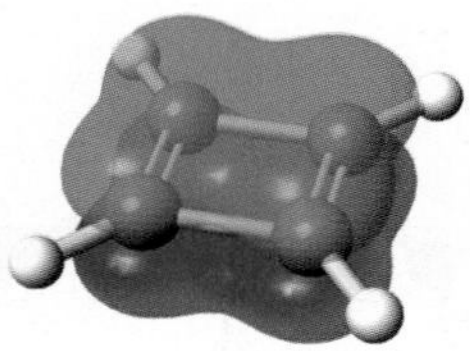

58. Examine the Hybrid Orbitals tool on Screen 9.6 of ChemistryNow. Use this tool to systematically combine atomic orbitals to form hybrid atomic orbitals.
(a) What is the relationship between the number of hybrid orbitals produced and the number of atomic orbitals used to create them?
(b) Do hybrid atomic orbitals form between different *p* orbitals without involving *s* orbitals?
(c) What is the relationship between the energy of hybrid atomic orbitals and the atomic orbitals from which they are formed?
(d) Compare the shapes of the hybrid orbitals formed from an s orbital and a p_x orbital with the hybrid atomic orbitals formed from an *s* orbital and a p_z orbital.
(e) Compare the shape of the hybrid orbitals formed from *s*, p_x, and p_y orbitals with the hybrid atomic orbitals formed from *s*, p_x, and p_z orbitals.

59. Screen 9.2 of ChemistryNow shows the change in energy as a function of the H—H distance when H_2 forms from separated H atoms.

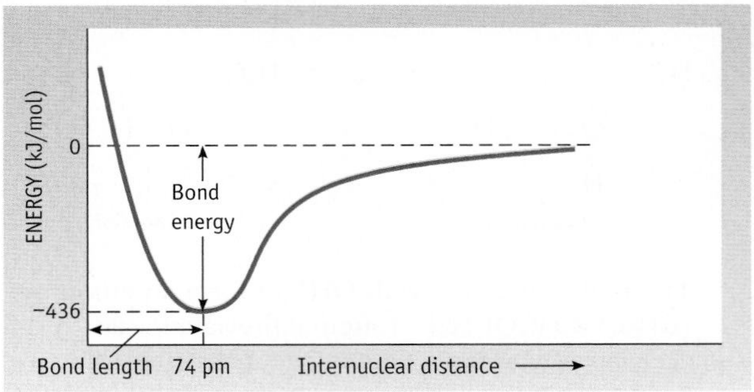

(a) Screen 9.3 describes the attractive and repulsive forces that occur when two atoms approach each other. What must be true about the relative strengths of those attractive and repulsive forces if a covalent bond is to form?
(b) When two atoms are widely separated, the energy of the system is defined as zero. As the atoms approach each other, the energy drops, reaches a minimum, and then increases as they approach still more closely. Explain these observations.
(c) For a bond to form, orbitals on adjacent atoms must overlap, and each pair of overlapping orbitals will contain two electrons. Explain why neon does not form a diatomic molecule, Ne_2, whereas fluorine forms F_2.

60. Examine the bonding in ethylene, C_2H_4, on Screen 9.7 of ChemistryNow and then go to the *A Closer Look* auxiliary screen.

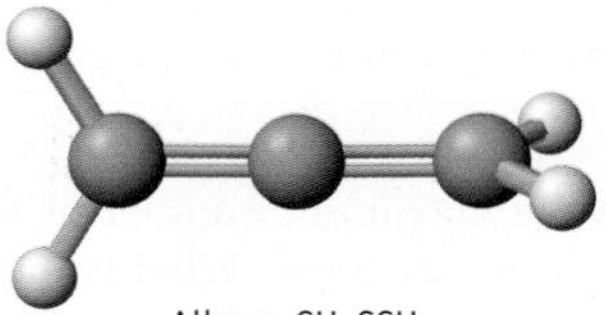

Allene, CH_2CCH_2

(a) Explain why the allene molecule is not flat. That is, explain why the CH_2 groups at opposite ends do not lie in the same plane.
(b) Based on the theory of orbital hybridization, explain why benzene is a planar, symmetrical molecule.
(c) What are the hybrid orbitals used by the three C atoms of allyl alcohol?

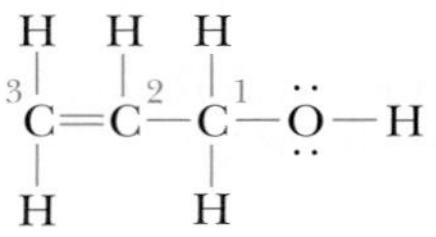

61. Screen 9.8 of ChemistryNow describes the motions of molecules.
   (a) Observe the animations of the rotations of *trans*-2-butene and butane about their carbon–carbon bonds.

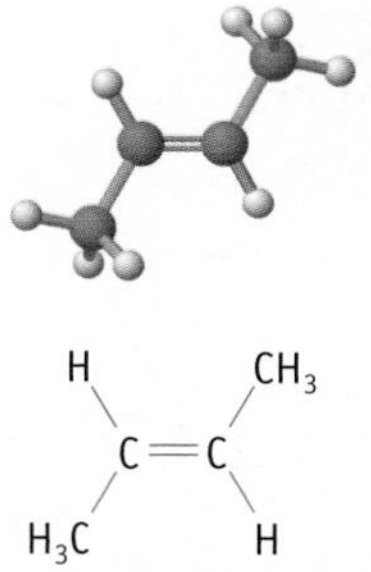

trans-2-Butene

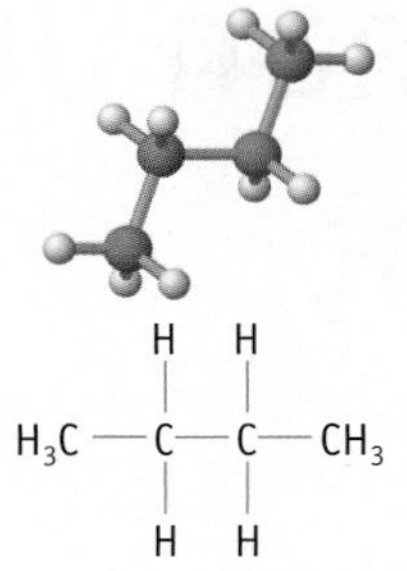

Butane

As one end of *trans*-2-butene rotates relative to the other end, the energy increases about 20-25 kJ/mol and then drops as the rotation produces *cis*-2-butene. In contrast, the rotation of the butane molecule requires much less energy (only 12.5 kJ/mol). When butane has reached the halfway point in its rotation, the energy has reached a maximum. Why does *trans*-2-butene require so much more energy to rotate about the central carbon–carbon bond than does butane?
   (b) Can the two CH_2 fragments of allene (see the sidebar of Screen 9.7) rotate with respect to each other? Briefly explain why or why not.

10 Carbon: More Than Just Another Element

Charles D. Winters

Camphor, an "Aromatic" Molecule

You might know just what this compound smells like! When you were a child and had a cold or cough, your mother might have smeared some Vick's® VapoRub on your chest. This home remedy, now more than 100 years old, contains camphor (5% by weight) as well as eucalyptus oil, turpentine oil, and menthol as active ingredients.

Camphor is said to be the first pure chemical that humans isolated and purified. Beginning in about 3000 BC, it was isolated from the camphor tree (*Cinnamomum camphora*), a native of China, Japan, and Indonesia, by chipping the wood from the tree and then steaming it. Camphor, being quite volatile, readily sublimes, and the solid is collected by cooling the resulting vapor.

An early use for camphor was as a wine additive, but the compound is toxic when taken internally, so this was clearly not a healthy practice. In the Middle Ages, it was said to be an aphrodisiac, but later it was declared an antiaphrodisiac.

Camphor was also used as a treatment during the yellow fever epidemic in 1793 that killed thousands in Philadelphia. Benjamin Rush, a Philadelphia physician, chemist (who published the first American chemistry textbook), and signer of the Declaration of Independence, recommended a mixture of vinegar and camphor to ward off the yellow fever. It did not cure the disease, but it did keep away mosquitoes, the carrier of yellow fever.

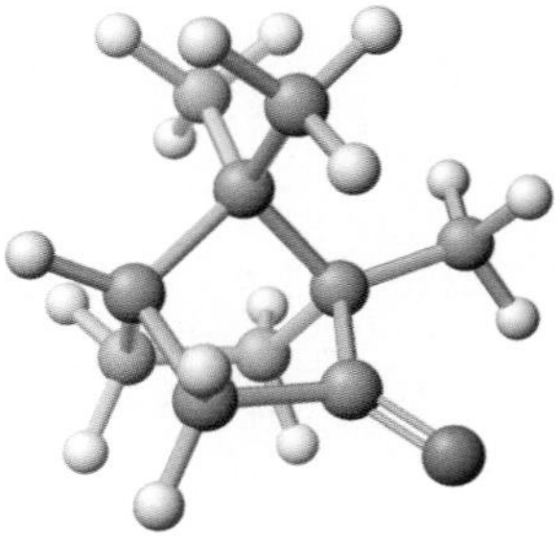

Questions:

1. The structure of camphor is interesting, although it was not known until 1893.
   - **a)** The O atom is attached to a structurally unique C atom. What is the geometry around this atom? What is its hybridization? What are the geometry and hybridization of the other C atoms in the molecule?
   - **b)** The five- and six-carbon rings in the molecule are not flat. Why is this so?
   - **c)** Is there a chiral center in this molecule? (See page 446.)
2. Organic compounds are often classified by their "functional group." To what class does this molecule belong?

Answers to these questions are in Appendix Q.

Chapter Goals

See Chapter Goals Revisited (page 488) for Study Questions keyed to these goals and assignable in OWL.

- Classify organic compounds based on formula and structure.
- Recognize and draw structures of structural isomers and stereoisomers for carbon compounds.
- Name and draw structures of common organic compounds.
- Know the common reactions of organic functional groups.
- Relate properties to molecular structure.
- Identify common polymers.

Chapter Outline

The vast majority of the millions of chemical compounds currently known are organic; that is, they are compounds built on a carbon framework. Organic compounds vary greatly in size and complexity, from the simplest hydrocarbon, methane, to molecules made up of many thousands of atoms. As you read this chapter, you will see why the range of possible materials is huge.

Chemistry.Now™

Throughout the text this icon introduces an opportunity for self-study or to explore interactive tutorials by signing in at **www.thomsonedu.com/login**.

10.1 Why Carbon?

We begin this discussion of organic chemistry with a question: What features of carbon lead to both the abundance and the complexity of organic compounds? Answers fall into two categories: structural diversity and stability.

Structural Diversity

With four electrons in its outer shell, carbon will form four bonds to reach an octet configuration. In contrast, the elements boron and nitrogen form three bonds in molecular compounds; oxygen forms two bonds; and hydrogen and the halogens form one bond. With a larger number of bonds comes the opportunity to create more complex structures. This will become increasingly evident in this brief tour of organic chemistry.

A carbon atom can reach an octet of electrons in various ways (Figure 10.1):

- *By forming four single bonds.* A carbon atom can bond to four other atoms, which can be either atoms of other elements (often H, N, or O) or other carbon atoms.
- *By forming a double bond and two single bonds.* The carbon atoms in ethylene, $H_2C{=}CH_2$, are linked to other atoms in this way.
- *By forming two double bonds,* as in carbon dioxide ($O{=}C{=}O$).
- *By forming a triple bond and a single bond,* an arrangement seen in acetylene, $HC{\equiv}CH$.

Recognize, with each of these arrangements, the various possible geometries around carbon: tetrahedral, trigonal planar, and linear. Carbon's tetrahedral

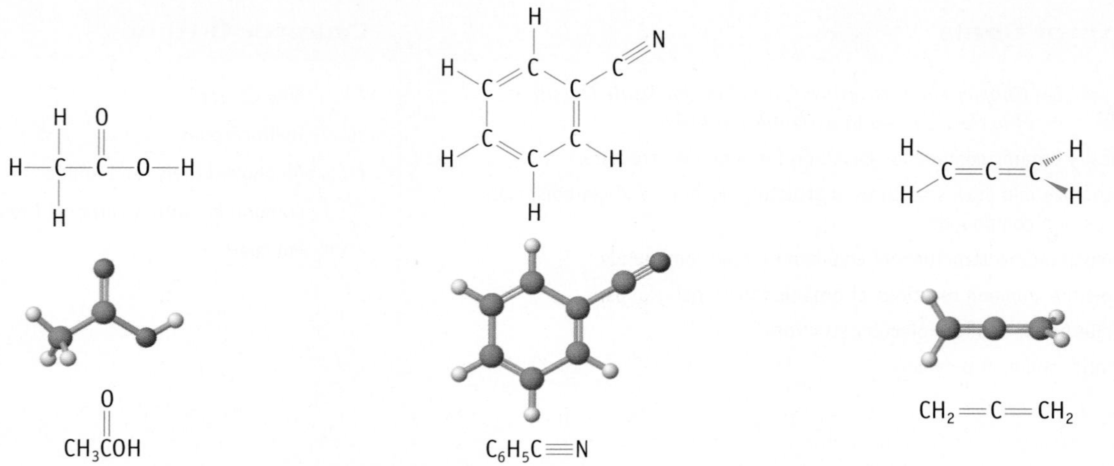

(a) Acetic acid. One carbon atom in this compound is attached to four other atoms by single bonds and has tetrahedral geometry. The second carbon atom, connected by a double bond to one oxygen and by single bonds to the other oxygen and to carbon, has trigonal-planar geometry.

(b) Benzonitrile. Six trigonal-planar carbon atoms make up the benzene ring. The seventh C atom, bonded by a single bond to carbon and a triple bond to nitrogen, has a linear geometry.

(c) Carbon is linked by double bonds to two other carbon atoms in C_3H_4, a linear molecule commonly called allene.

FIGURE 10.1 Ways that carbon atoms bond.

geometry is of special significance because it leads to three-dimensional chains and rings of carbon atoms, as in propane and cyclopentane.

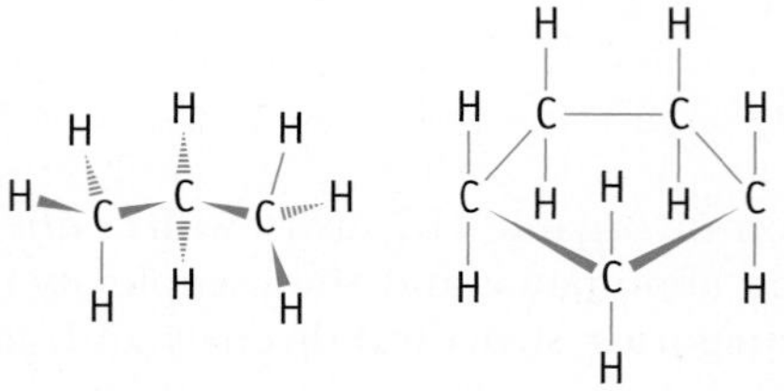

propane, C_3H_8 cyclopentane, C_5H_{10}

The ability to form multiple bonds leads to families of compounds with double and triple bonds.

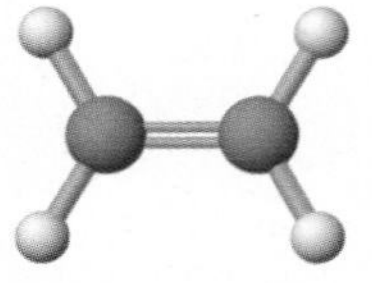

Ethylene, $H_2C{=}CH_2$

Acetylene, $HC{\equiv}CH$

Ethylene and acetylene. These two-carbon hydrocarbons can be the building blocks of more complex molecules. These are their common names, but their systematic names are ethene and ethyne.

Isomers

A hallmark of carbon chemistry is the remarkable array of isomers that can exist. **Isomers** are compounds that have identical composition but different structures. Two broad categories of isomers exist: structural isomers and stereoisomers.

Structural isomers are compounds having the same elemental composition, but the atoms are linked together in different ways. Ethanol and dimethyl ether are structural isomers, as are 1-butene and 2-methylpropene.

Ethanol	Dimethyl ether	1-Butene	2-Methylpropene
C_2H_6O	C_2H_6O	C_4H_8	C_4H_8
CH_3CH_2OH	CH_3OCH_3	$CH_3CH_2CH{=}CH_2$	$CH_3C({=}CH_2)CH_3$

A Closer Look

Writing Formulas and Drawing Structures

In Chapter 2, you learned that there are various ways of presenting structures (page 68). It is appropriate to return to this topic as we look at organic compounds. Consider methane and ethane, for example. We can represent these molecules in several ways:

1. *Molecular formula:* CH_4 or C_2H_6. This type of formula gives information on composition only.
2. *Condensed formula:* For ethane, this would be written CH_3CH_3 (or as H_3CCH_3). This method of writing the formula gives some information on the way atoms are connected.
3. *Structural formula:* You will recognize this formula as the Lewis structure. An elaboration on the condensed formula in (2), this representation defines more clearly how the atoms are connected, but it fails to describe the shapes of molecules.

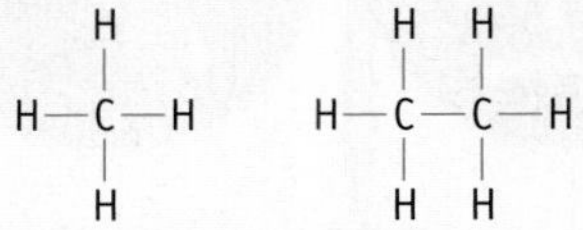

Methane, CH_4 Ethane, C_2H_6

4. *Perspective drawings:* These drawings are used to convey the three-dimensional nature of molecules. Bonds extending out of the plane of the paper are drawn with wedges, and bonds behind the plane of the paper are represented as dashed wedges (page 70). Using these guidelines, the structures of methane and ethane could be drawn as follows:

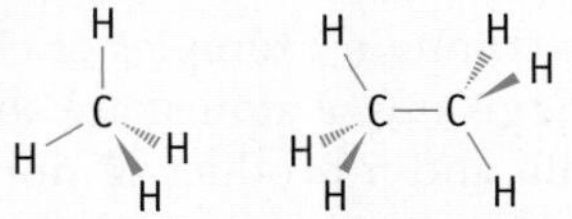

5. *Computer-drawn ball-and-stick and space-filling models.*

Ball-and-stick

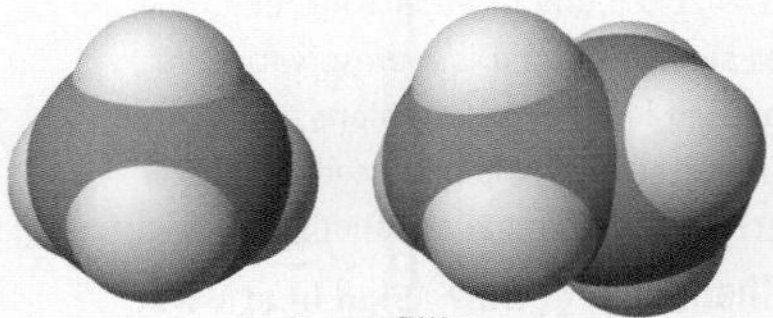

Space-filling

Stereoisomers are compounds with the same formula and in which there is a similar attachment of atoms. However, the atoms have different orientations in space. Two types of stereoisomers exist: geometric isomers and optical isomers.

Cis- and *trans*-2-butene are **geometric isomers.** Geometric isomerism in these compounds occurs as a result of the C=C double bond. Recall that the carbon atom and the attached groups cannot rotate around a double bond (page 420). Thus, the geometry around the C=C double bond is fixed in space. If identical groups occur on the adjacent carbon atoms and on the same side of the double bond, a *cis* isomer is produced. If those groups appear on opposite sides, a *trans* isomer is produced.

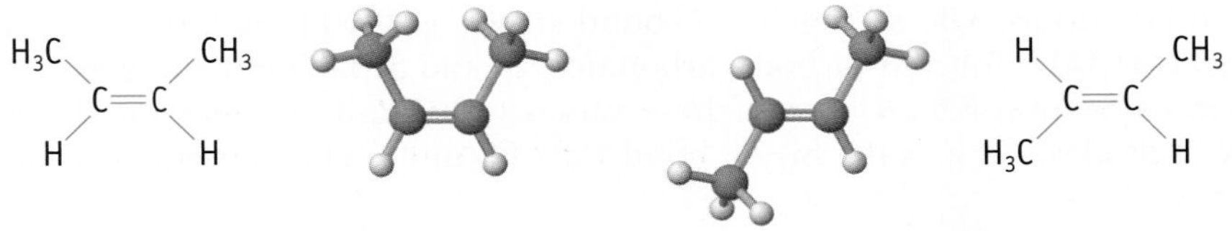

Cis-2-butene, C_4H_8 *Trans*-2-butene, C_4H_8

Optical isomerism is a second type of stereoisomerism. Optical isomers are molecules that have nonsuperimposable mirror images (Figure 10.2). Molecules (and other objects) that have nonsuperimposable mirror images are termed **chiral.** Pairs of nonsuperimposable molecules are called **enantiomers.**

Pure samples of enantiomers have the same physical properties, such as melting point, boiling point, density, and solubility in common solvents. They differ in one significant way, however: When a beam of plane-polarized light passes through a solution of a pure enantiomer, the plane of polarization rotates. The two enantiomers rotate polarized light to an equal extent, but in opposite directions (Figure 10.3). The term "optical isomerism" is used because this effect involves light.

Active Figure 10.2 Optical isomers. (a) Optical isomerism occurs if a molecule and its mirror image cannot be superimposed. The situation is seen if four different groups are attached to carbon. (b) Lactic acid is a chiral molecule because four different groups (H, OH, CH_3, and CO_2H) are attached to the central carbon atom.

Lactic acid is produced from milk when milk is fermented to make cheese. It is also found in other sour foods such as sauerkraut and is a preservative in pickled foods such as onions and olives. In our bodies, it is produced by muscle activity and normal metabolism.

Chemistry Now™ Sign in at www.thomsonedu.com/login and go to the Chapter Contents menu to explore an interactive version of this figure accompanied by an exercise.

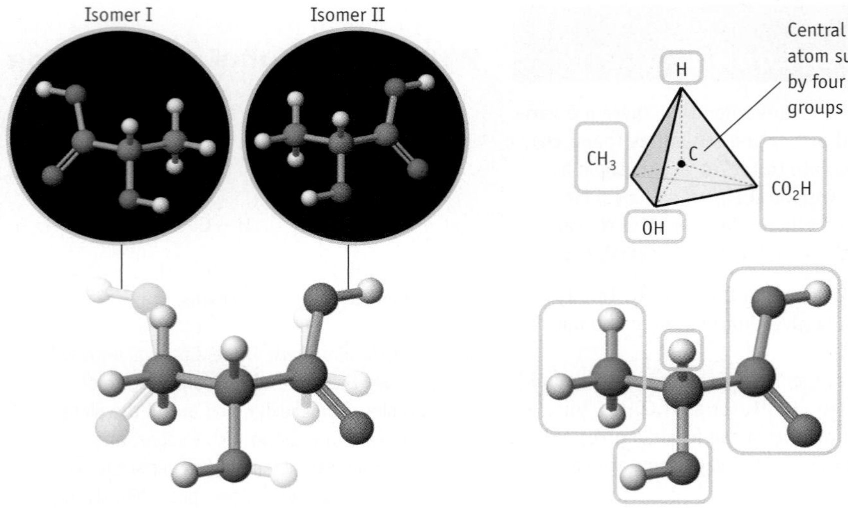

(a) Lactic acid enantiomers are nonsuperimposable

(b) Lactic acid, $CH_3CH(OH)CO_2H$

The most common examples of chiral compounds are those in which four different atoms (or groups of atoms) are attached to a tetrahedral carbon atom. Lactic acid, found in milk and a product of normal human metabolism, is an example of one such chiral compound (Figure 10.2). Optical isomerism is particularly important in the amino acids (► *The Chemistry of Life: Biochemistry*) and other biologically important molecules. Among the many interesting examples is a compound, frontalin, produced naturally by male elephants (see "Chemical Perspectives: Chirality and Elephants").

Stability of Carbon Compounds

Carbon compounds are notable for their resistance to chemical change. This resistance is a result of two things: strong bonds and slow reactions.

Strong bonds are needed for molecules to survive in their environment. Molecular collisions in gases, liquids, and solutions often provide enough energy to break some chemical bonds, and bonds can be broken if the energy associated with photons of visible and ultraviolet light exceeds the bond energy. Carbon–carbon bonds are relatively strong, however, as are the bonds between carbon and most other atoms. The average C—C bond energy is 346 kJ/mol; the C—H bond energy is 413 kJ/mol; and carbon–carbon double and triple bond energies are even higher (◄ Section 8.9). Contrast these values with bond energies for the Si—H bond (328 kJ/mol) and the Si—Si bond (222 kJ/mol). The consequence of high

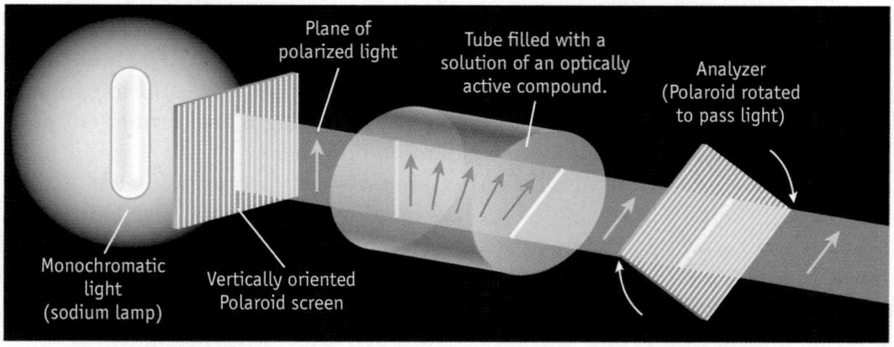

FIGURE 10.3 Rotation of plane-polarized light by an optical isomer. Monochromatic light (light of only one wavelength) is produced by a sodium lamp. After it passes through a polarizing filter, the light vibrates in only one direction—it is polarized. A solution of an optical isomer placed between the first and second polarizing filters causes rotation of the plane of polarized light. The angle of rotation can be determined by rotating the second filter until maximum light transmission occurs. The magnitude and direction of rotation are unique physical properties of the optical isomer being tested.

Chemical Perspectives — Chirality and Elephants

During a period known as musth, male elephants undergo a time of heightened sexual activity. They can become more aggressive and can work themselves into a frenzy. Aside from these physical changes, the males also produce chemical signals. A secretion containing the enantiomers of frontalin ($C_8H_{14}O_2$) is emitted from a gland between the eye and the ear. Young males produce mixtures containing more of one enantiomer than the other, whereas older elephants produce a more balanced and more concentrated mixture. When that occurs in older elephants, other males are repelled, but ovulating female elephants are more highly attracted.

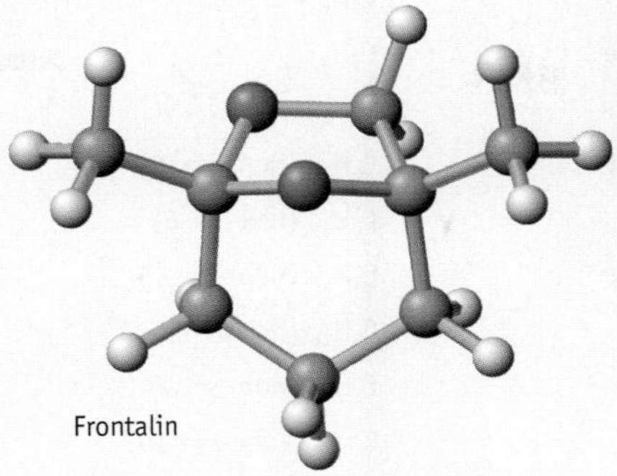

Frontalin

An African elephant in musth. Fluid containing the enantiomers of frontalin flows from a gland between the elephant's eye and ear.

bond energies for bonds to carbon is that, for the most part, organic compounds do not degrade under normal conditions.

Oxidation of most organic compounds is strongly product-favored, but most organic compounds survive contact with O_2. The reason is that these reactions occur slowly. Most organic compounds burn only if their combustion is initiated by heat or by a spark. As a consequence, oxidative degradation is not a barrier to the existence of organic compounds.

Module 15

10.2 Hydrocarbons

Hydrocarbons, compounds made of carbon and hydrogen only, are classified into several subgroups: alkanes, cycloalkanes, alkenes, alkynes, and aromatic compounds (Table 10.1). We begin our discussion by considering compounds that have carbon atoms with four single bonds, the alkanes and cycloalkanes.

Chemistry Now™

Sign in at **www.thomsonedu.com/login** and go to Chapter 10 Contents to see Screen 10.3 for a description of the **classes of hydrocarbons.**

TABLE 10.1 Some Types of Hydrocarbons

Type of Hydrocarbon	Characteristic Features	General Formula	Example
alkanes	C—C single bonds and all C atoms have four single bonds	C_nH_{2n+2}	CH_4, methane C_2H_6, ethane
cycloalkanes	C—C single bonds and all C atoms have four single bonds	C_nH_{2n}	C_6H_{12}, cyclohexane
alkenes	C=C double bond	C_nH_{2n}	$H_2C{=}CH_2$, ethylene
alkynes	C≡C triple bond	C_nH_{2n-2}	HC≡CH, acetylene
aromatics	rings with π bonding extending over several C atoms	—	benzene, C_6H_6

TABLE 10.2 Selected Hydrocarbons of the Alkane Family, C_nH_{2n+2}*

Name	Molecular Formula	State at Room Temperature
methane	CH_4	
ethane	C_2H_6	gas
propane	C_3H_8	
butane	C_4H_{10}	
pentane	C_5H_{12} (pent- = 5)	
hexane	C_6H_{14} (hex- = 6)	
heptane	C_7H_{16} (hept- = 7)	liquid
octane	C_8H_{18} (oct- = 8)	
nonane	C_9H_{20} (non- = 9)	
decane	$C_{10}H_{22}$ (dec- = 10)	
octadecane	$C_{18}H_{38}$ (octadec- = 18)	solid
eicosane	$C_{20}H_{42}$ (eicos- = 20)	

* This table lists only selected alkanes. Liquid compounds with 11 to 16 carbon atoms are also known. Many solid alkanes with more than 20 carbon atoms also exist.

Alkanes

Alkanes have the general formula C_nH_{2n+2}, with n having integer values (Table 10.2). Formulas of specific compounds can be generated from this general formula, the first four of which are CH_4 (methane), C_2H_6 (ethane), C_3H_8 (propane), and C_4H_{10} (butane) (Figure 10.4). Methane has four hydrogen atoms arranged tetrahedrally around a single carbon atom. Replacing a hydrogen atom in methane by a —CH_3 group gives ethane. If an H atom of ethane is replaced by yet another —CH_3 group, propane results. Butane is derived from propane by replacing an H atom of one of the chain-ending carbon atoms with a —CH_3 group. In all of these compounds, each C atom is attached to four other atoms, either C or H, so alkanes are often called **saturated compounds**.

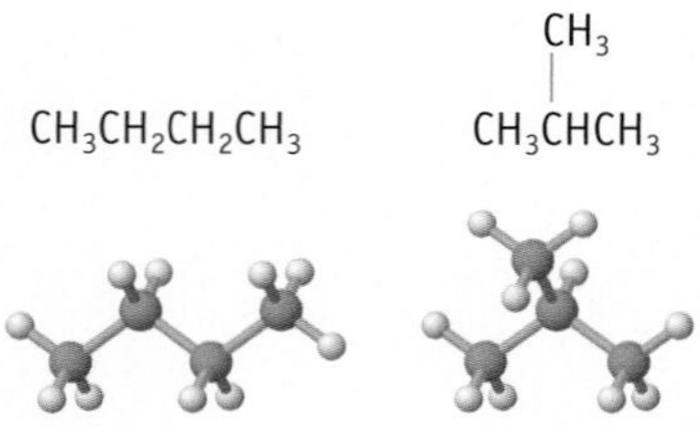

Butane 2-Methylpropane

Structural isomers of butane, C_4H_{10}.

Structural Isomers

Structural isomers are possible for all alkanes larger than propane. For example, there are two structural isomers for C_4H_{10} and three for C_5H_{12}. As the number of carbon atoms in an alkane increases, the number of possible structural isomers

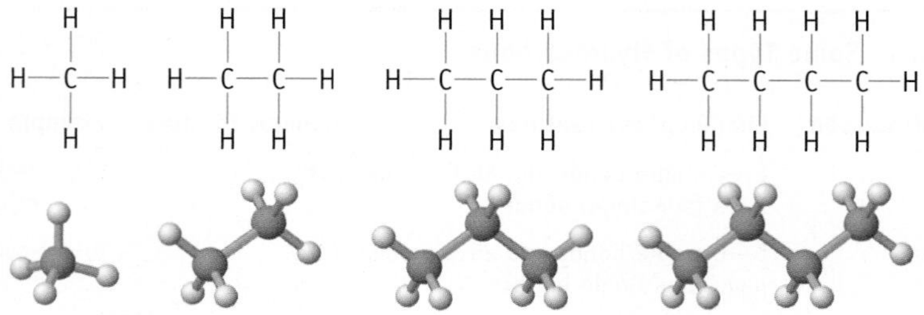

Methane Ethane Propane Butane

Active Figure 10.4 Alkanes. The lowest–molar-mass alkanes, all gases under normal conditions, are methane, ethane, propane, and butane.

Chemistry•Now™ Sign in at www.thomsonedu.com/login and go to the Chapter Contents menu to explore an interactive version of this figure accompanied by an exercise.

greatly increases; there are five isomers possible for C_6H_{14}, nine isomers for C_7H_{16}, 18 for C_8H_{18}, 75 for $C_{10}H_{22}$, and 366,319 for $C_{20}H_{42}$.

To recognize the isomers corresponding to a given formula, keep in mind the following points:

- Each alkane is built upon a framework of tetrahedral carbon atoms, and each carbon must have four single bonds.
- An effective approach is to create a framework of carbon atoms and then fill the remaining positions around carbon with H atoms so that each C atom has four bonds.
- Nearly free rotation occurs around carbon–carbon single bonds. Therefore, when atoms are assembled to form the skeleton of an alkane, the emphasis is on how carbon atoms are attached to one another and not on how they might lie relative to one another in the plane of the paper.

$CH_3CH_2CH_2CH_2CH_3$

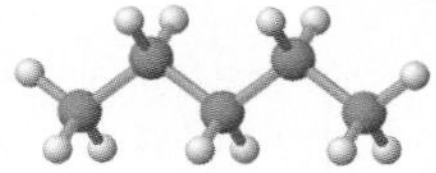

Pentane

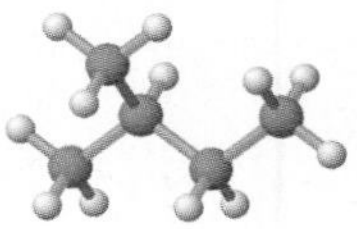

$CH_3CHCH_2CH_3$ (with CH_3 on C2)

2-Methylbutane

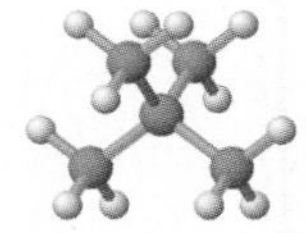

H_3CCCH_3 (with CH_3 above and below the central C)

2,2-Dimethylpropane

Structural isomers of pentane, C_5H_{12}.

EXAMPLE 10.1 Drawing Structural Isomers of Alkanes

Problem Draw structures of the five isomers of C_6H_{14}. Are any of these isomers chiral?

Strategy Focus first on the different frameworks that can be built from six carbon atoms. Having created a carbon framework, fill hydrogen atoms into the structure so that each carbon has four bonds.

Solution

Step 1. Placing six carbon atoms in a chain gives the framework for the first isomer. Now fill in hydrogen atoms: three on the carbons on the ends of the chain, two on each of the carbons in the middle. You have created the first isomer, hexane.

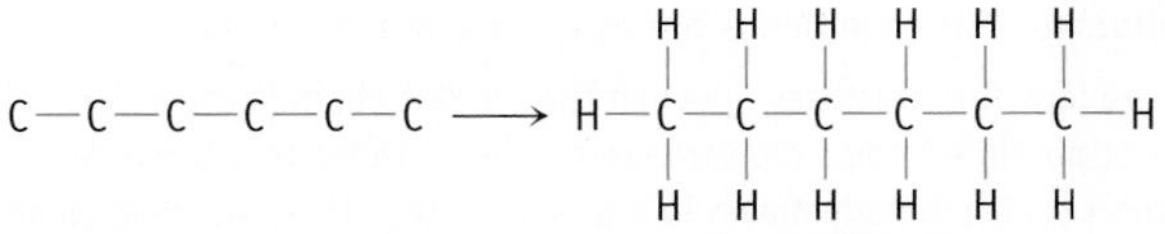

carbon framework of hexane — hexane

Step 2. Draw a chain of five carbon atoms; then add the sixth carbon atom to one of the carbons in the middle of this chain. (Adding it to a carbon at the end of the chain gives a six-carbon chain, the same framework drawn in Step 1.) Two different carbon frameworks can be built from the five-carbon chain, depending on whether the sixth carbon is linked to the 2 or 3 position. For each of these frameworks, fill in the hydrogens.

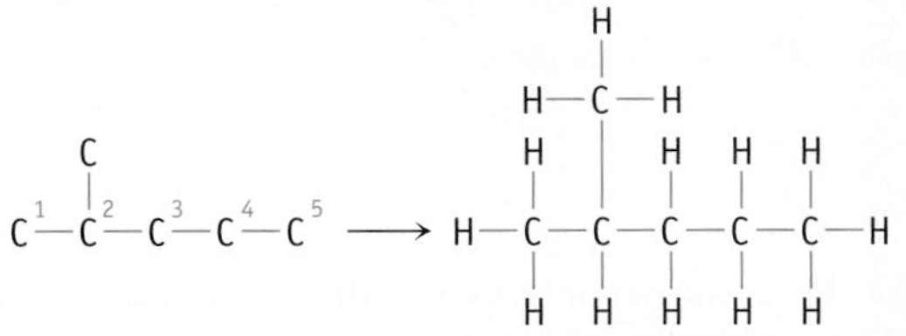

carbon framework of methylpentane isomers — 2-methylpentane

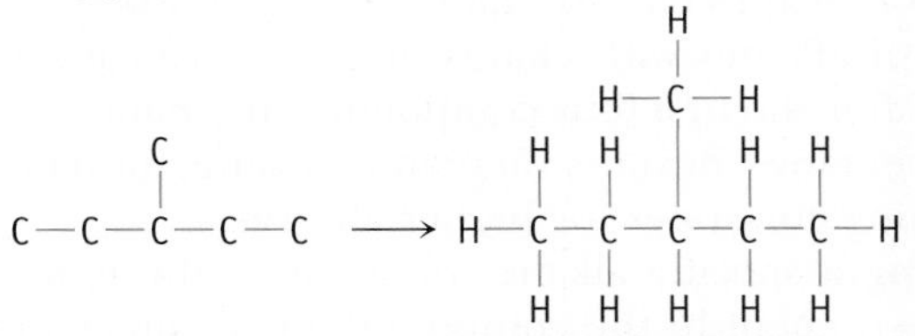

3-methylpentane

■ **Chirality in Alkanes** To be chiral, *a compound must have at least one C atom attached to four different groups.* Thus, the C_7H_{16} isomer here is chiral.

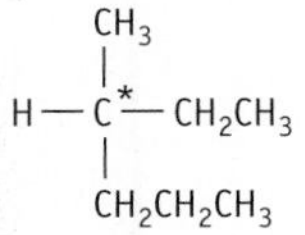

We often designate the center of chirality with an asterisk.

Step 3. Draw a chain of four carbon atoms. Add in the two remaining carbons, again being careful not to extend the chain length. Two different structures are possible: one with the remaining carbon atoms in the 2 and 3 positions, and another with both extra carbon atoms attached at the 2 position. Fill in the 14 hydrogens. You have now drawn the fourth and fifth isomers.

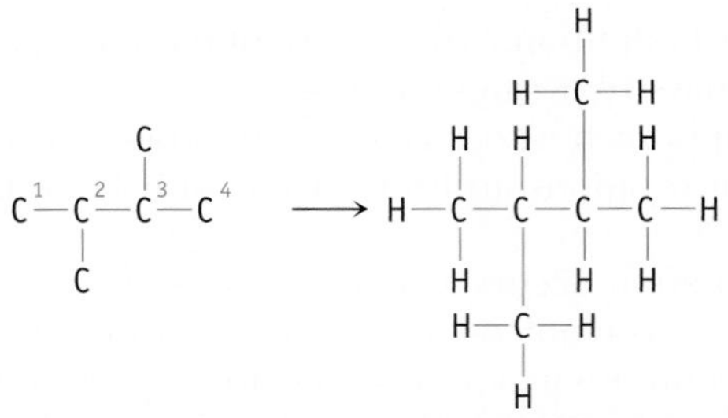

carbon atom frameworks for dimethylbutane isomers

2,3-dimethylbutane

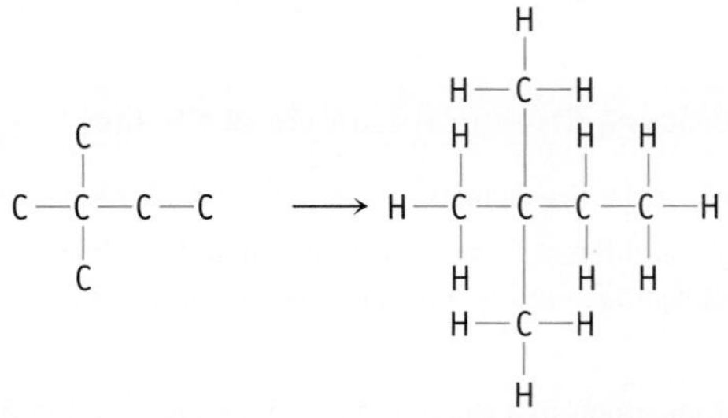

2,2-dimethylbutane

None of the isomers of C_6H_{14} is chiral. *To be chiral, a compound must have at least one C atom with four different groups attached.* This condition is not met in any of these isomers.

Comment Should we look for structures in which the longest chain is three carbon atoms? Try it, but you will see that it is not possible to add the three remaining carbons to a three-carbon chain without creating one of the carbon chains already drawn in a previous step. Thus, we have completed the analysis, with five isomers of this compound being identified.

Names have been given to each of these compounds. See the text that follows this Example, and see Appendix E for guidelines on nomenclature.

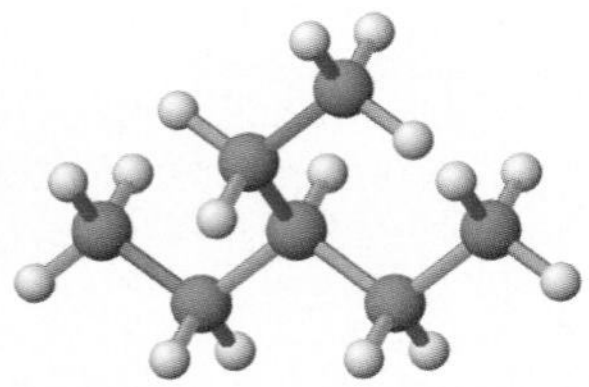

One possible isomer of an alkane with the formula C_7H_{16}.

EXERCISE 10.1 Drawing Structural Isomers of Alkanes

(a) Draw the nine isomers having the formula C_7H_{16}. (Hint: There is one structure with a seven-carbon chain, two structures with six-carbon chains, five structures in which the longest chain has five carbons [one is illustrated in the margin], and one structure with a four-carbon chain.)

(b) Identify the isomers of C_7H_{16} that are chiral.

Naming Alkanes

■ **Naming Guidelines** For more details on naming organic compounds, see Appendix E.

With so many possible isomers for a given alkane, chemists need a systematic way of naming them. The guidelines for naming alkanes and their derivatives follow:

- The names of alkanes end in "-ane."
- The names of alkanes with chains of one to 10 carbon atoms are given in Table 10.2. After the first four compounds, the names are derived from Latin numbers—pentane, hexane, heptane, octane, nonane, decane—and this regular naming continues for higher alkanes.
- When naming a specific alkane, the root of the name corresponds to the longest carbon chain in the compound. One isomer of C_5H_{12} has a three—

Problem Solving Tip 10.1 Drawing Structural Formulas

An error students sometimes make is to suggest that the three carbon skeletons drawn here are different. They are, in fact, the same. All are five-carbon chains with another C atom in the 2 position.

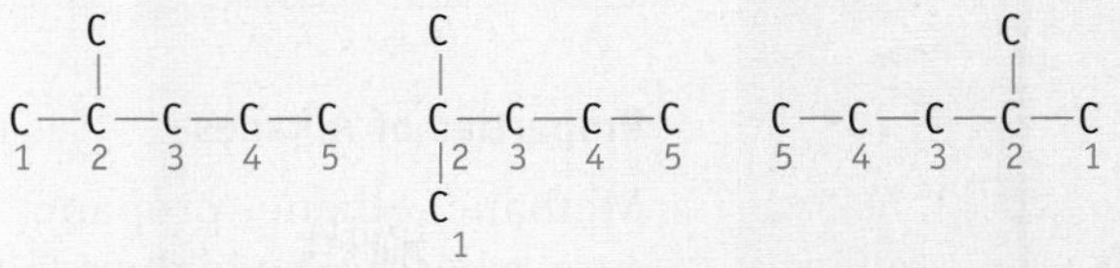

Remember that Lewis structures do not indicate the geometry of molecules.

carbon chain with two $—CH_3$ groups on the second C atom of the chain. Thus, its name is based on propane.

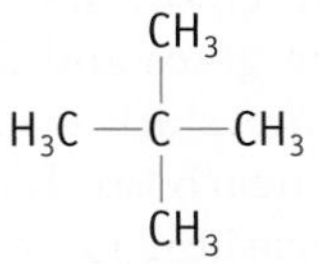

2,2-dimethylpropane

- Substituent groups on a hydrocarbon chain are identified by a name and the position of substitution in the carbon chain; this information precedes the root of the name. The position is indicated by a number that refers to the carbon atom to which the substituent is attached. (Numbering of the carbon atoms in a chain should begin at the end of the carbon chain that allows the substituent groups to have the lowest numbers.) Both $—CH_3$ groups in 2,2-dimethylpropane are located at the 2 position.
- Names of hydrocarbon substituents, called **alkyl groups,** are derived from the name of the hydrocarbon. The group $—CH_3$, derived by taking a hydrogen from methane, is called the methyl group; the C_2H_5 group is the ethyl group.
- If two or more of the same substituent groups occur, the prefixes di-, tri-, and tetra- are added. When different substituent groups are present, they are generally listed in alphabetical order.

■ **Systematic and Common Names** The IUPAC (International Union of Pure and Applied Chemistry) has formulated rules for systematic names, which are generally used in this book. (See Appendix E.) However, many organic compounds are known by common names. For example, 2,2-dimethylpropane is also called neopentane.

EXAMPLE 10.2 Naming Alkanes

Problem Give the systematic name for

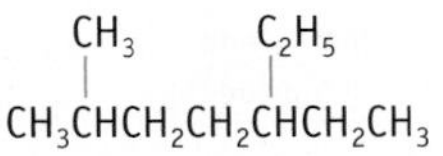

Strategy Identify the longest carbon chain and base the name of the compound on that alkane. Identify the substituent groups on the chain and their locations. When there are two or more substituents (the groups attached to the chain), number the parent chain from the end that gives the lower number to the substituent encountered first. If the substituents are different, list them in alphabetical order. (For more on naming compounds, see Appendix E.)

Solution Here, the longest chain has seven C atoms, so the root of the name is *heptane*. There is a methyl group on C-2 and an ethyl group on C-5. Giving the substituents in alphabetical order and numbering the chain from the end having the methyl group, the systematic name is 5-ethyl-2-methylheptane.

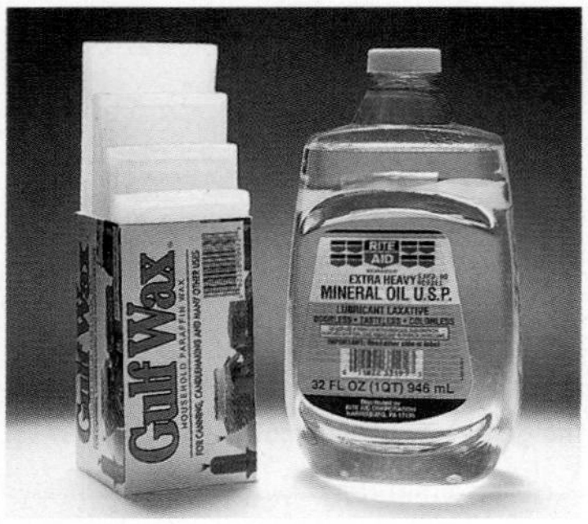

Charles D. Winters

FIGURE 10.5 Paraffin wax and mineral oil. These common consumer products are mixtures of alkanes.

EXERCISE 10.2 Naming Alkanes

Name the nine isomers of C_7H_{16} in Exercise 10.1.

Properties of Alkanes

Methane, ethane, propane, and butane are gases at room temperature and pressure, whereas the higher–molar-mass compounds are liquids or solids (Table 10.2). An increase in melting point and boiling point with molar mass is a general phenomenon that reflects the increased forces of attraction between molecules (► Section 12.2).

You already know about alkanes in a nonscientific context because several are common fuels. Natural gas, gasoline, kerosene, fuel oils, and lubricating oils are all mixtures of various alkanes. White mineral oil is also a mixture of alkanes, as is paraffin wax (Figure 10.5).

Pure alkanes are colorless. (The colors seen in gasoline and other petroleum products are due to additives.) The gases and liquids have noticeable but not unpleasant odors. All of these substances are insoluble in water, which is typical of compounds that are nonpolar or nearly so. Low polarity is expected for alkanes because the electronegativities of carbon (χ = 2.5) and hydrogen (χ = 2.2) are not greatly different (◄ Section 8.8).

All alkanes burn readily in air to give CO_2 and H_2O in very exothermic reactions. This is, of course, the reason they are widely used as fuels.

$$CH_4(g) + 2\ O_2(g) \rightarrow CO_2(g) + 2\ H_2O(\ell) \qquad \Delta_rH° = -890.3\ \text{kJ/mol-rxn}$$

Other than in combustion reactions, alkanes exhibit relatively low chemical reactivity. One reaction that does occur, however, is the replacement of the hydrogen atoms of an alkane by chlorine atoms on reaction with Cl_2. It is formally an oxidation because Cl_2, like O_2, is a strong oxidizing agent. These reactions, which can be initiated by ultraviolet radiation, are free radical reactions. Highly reactive Cl atoms are formed from Cl_2 under UV radiation. Reaction of methane with Cl_2 under these conditions proceeds in a series of steps, eventually yielding CCl_4, commonly known as carbon tetrachloride. (HCl is the other product of these reactions.)

$$CH_4 \xrightarrow[UV]{Cl_2} CH_3Cl \xrightarrow[UV]{Cl_2} CH_2Cl_2 \xrightarrow[UV]{Cl_2} CHCl_3 \xrightarrow[UV]{Cl_2} CCl_4$$

	CH_3Cl	CH_2Cl_2	$CHCl_3$	CCl_4
Systematic name	chloromethane	dichloromethane	trichloromethane	tetrachloromethane
Common name	methyl chloride	methylene chloride	chloroform	carbon tetrachloride

The last three compounds are used as solvents, albeit less frequently today because of their toxicity. Carbon tetrachloride was also once widely used as a dry cleaning fluid and, because it does not burn, in fire extinguishers.

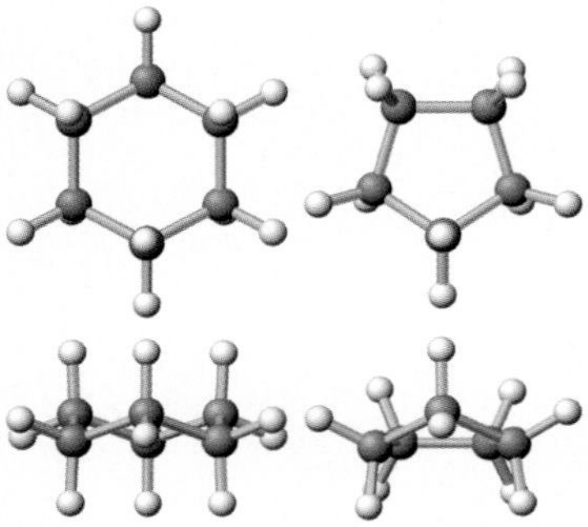

Cyclohexane, top and front views   Cyclopentane, top and front views

The structures of cyclopentane, C_5H_{10}, and cyclohexane, C_6H_{12}. The C_5 ring is nearly planar. In contrast, the tetrahedral geometry around carbon means that the C_6 ring is decidedly puckered.

Cycloalkanes, C_nH_{2n}

Cycloalkanes are constructed with tetrahedral carbon atoms joined together to form a ring. For example, cyclopentane, C_5H_{10}, consists of a ring of five carbon atoms. Each carbon atom is bonded to two adjacent carbon atoms and to two hydrogen atoms. Notice that the five carbon atoms fall very nearly in a plane because

A Closer Look — Flexible Molecules

Most organic molecules are flexible; that is, they can twist and bend in various ways. Few molecules better illustrate this behavior than cyclohexane. Two structures are possible: "chair" and "boat" forms. These forms can interconvert by partial rotation of several bonds.

The more stable structure is the chair form, which allows the hydrogen atoms to remain as far apart as possible. A side view of this form of cyclohexane reveals two sets of hydrogen atoms in this molecule. Six hydrogen atoms, called the equatorial hydrogens, lie in a plane around the carbon ring. The other six hydrogens are positioned above and below the plane and are called axial hydrogens. Flexing the ring (a rotation around the C—C single bonds) moves the hydrogen atoms between axial and equatorial environments.

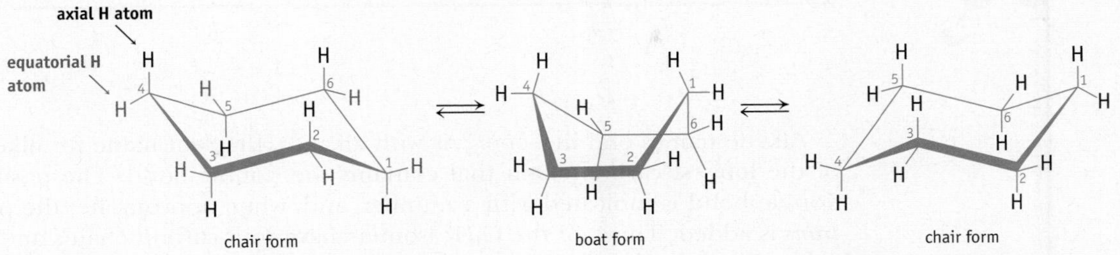

the internal angles of a pentagon, 108°, closely match the tetrahedral angle of 109.5°. The small distortion from planarity allows hydrogen atoms on adjacent carbon atoms to be a little farther apart.

Cyclohexane has a nonplanar ring with six —CH_2 groups. If the carbon atoms were in the form of a regular hexagon with all carbon atoms in one plane, the C—C—C bond angles would be 120°. To have tetrahedral bond angles of 109.5° around each C atom, the ring has to pucker. The C_6 ring is flexible, however, and exists in two interconverting forms (see *A Closer Look: Flexible Molecules*).

Interestingly, cyclobutane and cyclopropane are also known, although the bond angles in these species are much less than 109.5°. These compounds are examples of **strained hydrocarbons,** so named because an unfavorable geometry is imposed around carbon. One of the features of strained hydrocarbons is that the C—C bonds are weaker and the molecules readily undergo ring-opening reactions that relieve the bond angle strain.

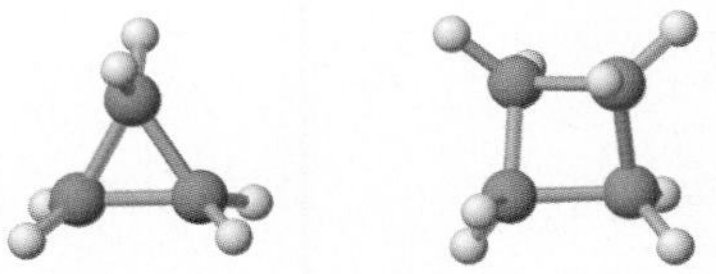

Cyclopropane, C_3H_6 Cyclobutane, C_4H_8

Cyclopropane and cyclobutane. Cyclopropane was at one time used as a general anesthetic in surgery. However, its explosive nature when mixed with oxygen soon eliminated this application. The *Columbia Encyclopedia* states that "cyclopropane allowed the transport of more oxygen to the tissues than did other common anesthetics and also produced greater skeletal muscle relaxation. It is not irritating to the respiratory tract. Because of the low solubility of cyclopropane in the blood, postoperative recovery was usually rapid but nausea and vomiting were common."

Alkenes and Alkynes

The diversity seen for alkanes is repeated with **alkenes,** hydrocarbons with one or more C=C double bonds. The presence of the double bond adds two features missing in alkanes: the possibility of geometric isomerism and increased reactivity.

The general formula for alkenes is C_nH_{2n}. The first two members of the series of alkenes are ethene, C_2H_4 (common name, ethylene), and propene, C_3H_6 (common name, propylene). Only a single structure can be drawn for these compounds. As with alkanes, the occurrence of isomers begins with species containing four carbon atoms. Four alkene isomers have the formula C_4H_8, and each has distinct chemical and physical properties (Table 10.3).

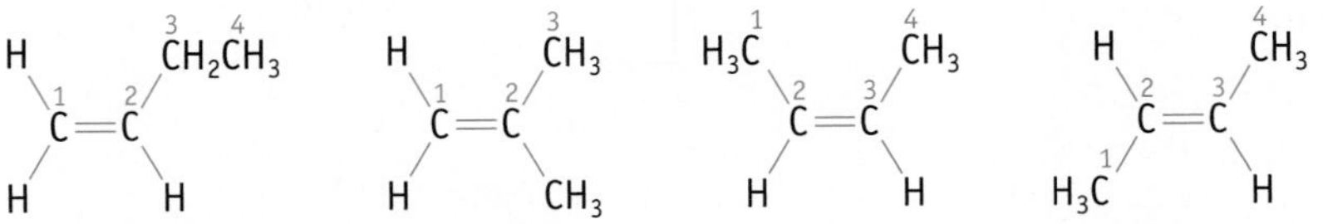

1-butene 2-methylpropene *cis*-2-butene *trans*-2-butene

C_2H_4
Systematic name:
Ethene
Common name:
Ethylene

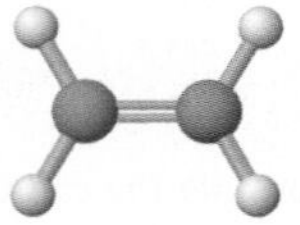

C_3H_6
Systematic name:
Propene
Common name:
Propylene

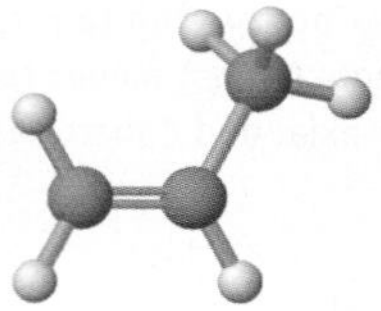

TABLE 10.3 Properties of Butene Isomers

Name	Boiling Point	Melting Point	Dipole Moment (D)	$\Delta_f H°$ (gas) (kJ/mol)
1-butene	−6.26 °C	−185.4 °C	—	−20.5
2-methylpropene	−6.95 °C	−140.4 °C	0.503	−37.5
cis-2-butene	3.71 °C	−138.9 °C	0.253	−29.7
trans-2-butene	0.88 °C	−105.5 °C	0	−33.0

Alkene names end in "-ene." As with alkanes, the root name for alkenes is that of the longest carbon chain that contains the double bond. The position of the double bond is indicated with a number, and, when appropriate, the prefix *cis* or *trans* is added. Three of the C_4H_8 isomers have four-carbon chains and so are butenes. One has a three-carbon chain and is a propene. Notice that the carbon chain is numbered from the end that gives the double bond the lowest number. In the first isomer at the left, the double bond is between C atoms 1 and 2, so the name is 1-butene and not 3-butene.

EXAMPLE 10.3 Determining Isomers of Alkenes from a Formula

Problem Draw structures for the six possible alkene isomers with the formula C_5H_{10}. Give the systematic name of each.

Strategy A procedure that involved drawing the carbon skeleton and then adding hydrogen atoms served well when drawing structures of alkanes (Example 10.1), and a similar approach can be used here. It will be necessary to put one double bond into the framework and to be alert for *cis-trans* isomerism.

Solution

1. A five-carbon chain with one double bond can be constructed in two ways. *Cis–trans* isomers are possible for 2-pentene.

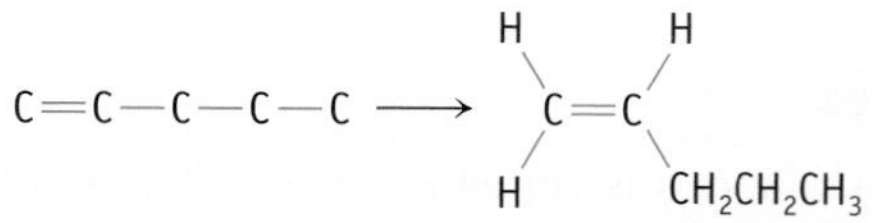

1-pentene

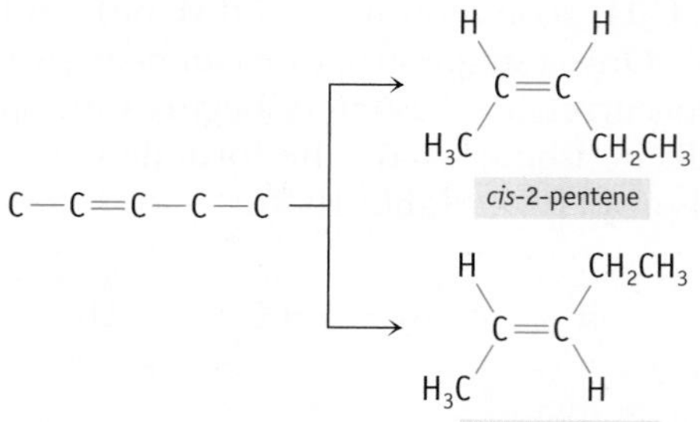

cis-2-pentene

trans-2-pentene

2. Draw the possible four-carbon chains containing a double bond. Add the fifth carbon atom to either the 2 or 3 position. When all three possible combinations are found, fill in the hydrogen atoms. This results in three more structures:

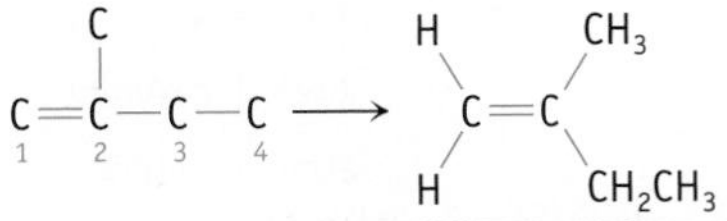

2-methyl-1-butene

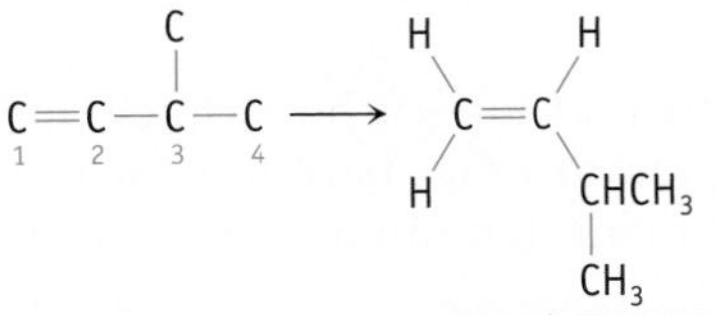

3-methyl-1-butene

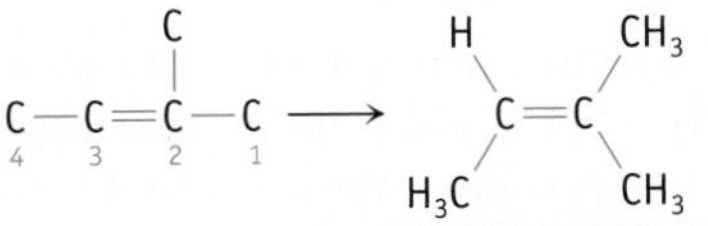

2-methyl-2-butene

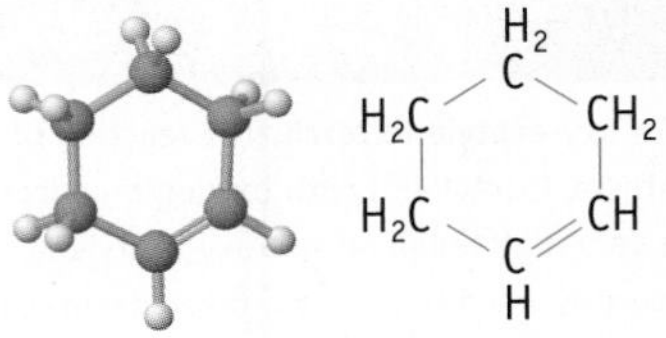

Cyclohexene, C_6H_{10}

$H_2C{=}CHCH{=}CH_2$

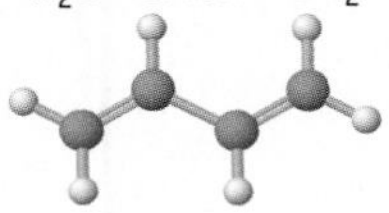

1,3-Butadiene, C_4H_6

Cycloalkenes and dienes. Cyclohexene, C_6H_{10} (*top*), and 1,3-butadiene (C_4H_6) (*bottom*).

EXERCISE 10.3 Determining Structural Isomers of Alkenes from a Formula

There are 17 possible alkene isomers with the formula C_6H_{12}. Draw structures of the five isomers in which the longest chain has six carbon atoms, and give the name of each. Which of these isomers is chiral? (There are also eight isomers in which the longest chain has five carbon atoms, and four isomers in which the longest chain has four carbon atoms. How many can you find?)

More than one double bond can be present in a hydrocarbon. Butadiene, for example, has two double bonds and is known as a *diene.* Many natural products have numerous double bonds (Figure 10.6). There are also cyclic hydrocarbons, such as cyclohexene, with double bonds.

Charles D. Winters

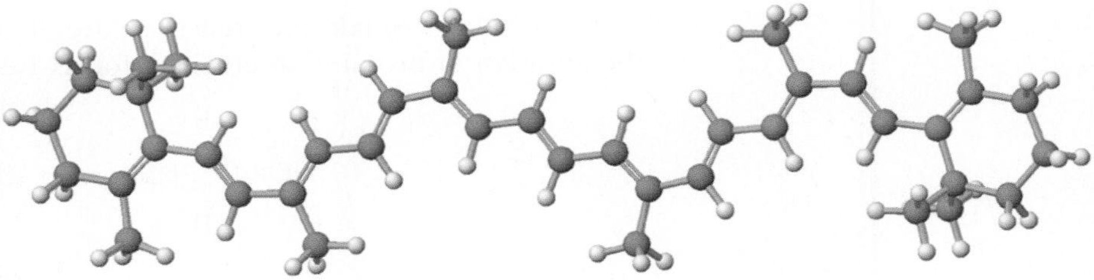

FIGURE 10.6 Carotene, a naturally occurring compound with 11 C=C bonds. The π electrons can be excited by visible light in the blue-violet region of the spectrum. As a result, carotene appears orange-yellow to the observer. Carotene or carotene-like molecules are partnered with chlorophyll in nature in the role of assisting in the harvesting of sunlight. Green leaves have a high concentration of carotene. In autumn, green chlorophyll molecules are destroyed, and the yellows and reds of carotene and related molecules are seen. The red color of tomatoes, for example, comes from a molecule very closely related to carotene. As a tomato ripens, its chlorophyll disintegrates, and the green color is replaced by the red of the carotene-like molecule.

Charles D. Winters

An oxy-acetylene torch. The reaction of ethyne (acetylene) with oxygen produces a very high temperature. Oxy-acetylene torches, used in welding, take advantage of this fact.

TABLE 10.4 Some Simple Alkynes C_nH_{2n-2}

Structure	Systematic Name	Common Name	BP (°C)
$HC{\equiv}CH$	ethyne	acetylene	−85
$CH_3C{\equiv}CH$	propyne	methylacetylene	−23
$CH_3CH_2C{\equiv}CH$	1-butyne	ethylacetylene	9
$CH_3C{\equiv}CCH_3$	2-butyne	dimethylacetylene	27

Alkynes, compounds with a carbon–carbon triple bond, have the general formula (C_nH_{2n-2}). Table 10.4 lists alkynes that have four or fewer carbon atoms. The first member of this family is ethyne (common name, acetylene), a gas used as a fuel in metal cutting torches.

Properties of Alkenes and Alkynes

Like alkanes, alkenes and alkynes are colorless. Low–molar-mass compounds are gases, whereas compounds with higher molecular weights are liquids or solids. Alkanes, alkenes, and alkynes are also oxidized by O_2 to give CO_2 and H_2O.

In contrast to alkanes, alkenes and alkynes have an elaborate chemistry. We gain an insight into their chemical behavior by noting that they are called **unsaturated compounds.** Carbon atoms are capable of bonding to a maximum of four other atoms, and they do so in alkanes and cycloalkanes. In alkenes, however, the carbon atoms linked by a double bond are bonded to only three atoms; in alkynes, they bond to two atoms. It is possible to increase the number of groups attached to carbon by **addition reactions,** in which molecules with the general formula X—Y (such as hydrogen, halogens, hydrogen halides, and water) add across the carbon–carbon double bond (Figure 10.7). The result is a compound with four groups bonded to each carbon.

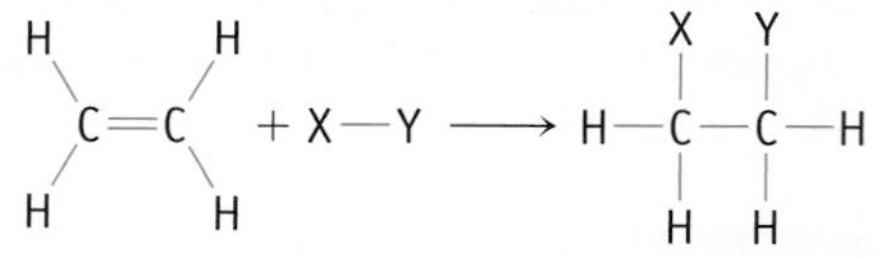

X—Y = H_2, Cl_2, Br_2; H—Cl, H—Br, H—OH, HO—Cl

The products of addition reactions are often substituted alkanes. For example, the addition of bromine to ethylene forms 1,2-dibromoethane.

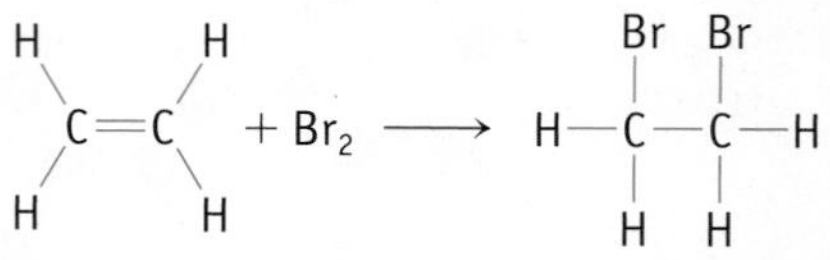

1,2-dibromoethane

The addition of 2 mol of chlorine to acetylene gives 1,1,2,2-tetrachloroethane.

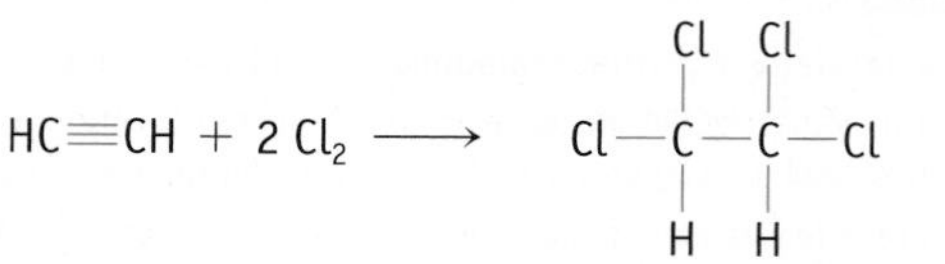

1,1,2,2-tetrachloroethane

During the 1860s, a Russian chemist, Vladimir Markovnikov, examined a large number of alkene addition reactions. In cases in which two isomeric products were

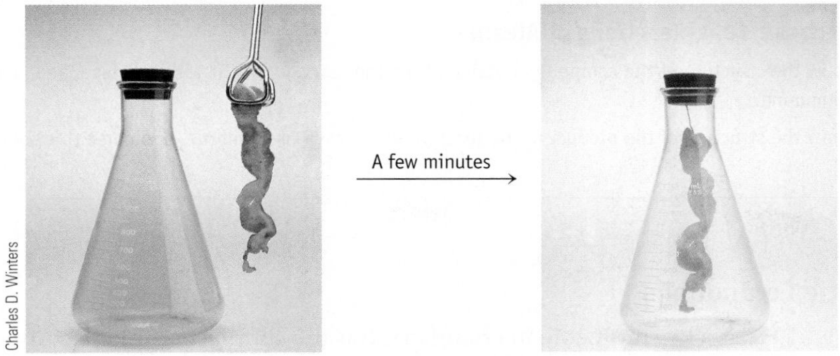

FIGURE 10.7 Bacon fat and addition reactions. The fat in bacon is partially unsaturated. Like other unsaturated compounds, bacon fat reacts with Br_2 in an addition reaction. Here, you see the color of Br_2 vapor fade when a strip of bacon is introduced.

possible, he found that one was more likely to predominate. Based on these results, Markovnikov formulated a rule (now called *Markovnikov's rule*) stating that, when a reagent HX adds to an unsymmetrical alkene, the hydrogen atom in the reagent becomes attached to the carbon that already has the largest number of hydrogens. An example of Markovnikov's rule is the reaction of 2-methylpropene with HCl that results in formation of 2-chloro-2-methylpropane rather than 1-chloro-2-methylpropane.

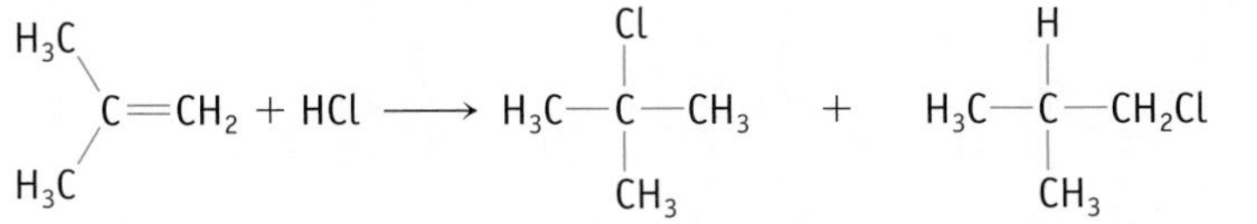

2-methylpropene — 2-chloro-2-methylpropane Sole product — 1-chloro-2-methylpropane NOT formed

■ **Nomenclature of Substituted Alkanes** The substituent groups in substituted alkanes are identified by the name and position of the substituent on the alkane chain.

If the reagent added to a double bond is hydrogen (X—Y = H_2), the reaction is called **hydrogenation.** Hydrogenation is usually a very slow reaction, but it can be speeded up in the presence of a catalyst, often a specially prepared form of a metal, such as platinum, palladium, and rhodium. You may have heard the term hydrogenation because certain foods contain "hydrogenated" or "partially hydrogenated" ingredients. One brand of crackers has a label that says, "Made with 100% pure vegetable shortening . . . (partially hydrogenated soybean oil with hydrogenated cottonseed oil)." One reason for hydrogenating an oil is to make it less susceptible to spoilage; another is to convert it from a liquid to a solid.

■ **Catalysts** A catalyst is a substance that causes a reaction to occur at a faster rate without itself being permanently changed in the reaction. We will describe catalysts in more detail in Chapter 15.

Chemistry Now™

Sign in at **www.thomsonedu.com/login** and go to Chapter 10 Contents to see Screen 10.4 for a simulation and tutorial on **alkene addition reactions.**

EXAMPLE 10.4 Reaction of an Alkene

Problem Draw the structure of the compound obtained from the reaction of Br_2 with propene, and name the compound.

Strategy Bromine adds across the C=C double bond. The name includes the name of the carbon chain and indicates the positions of the Br atoms.

Solution

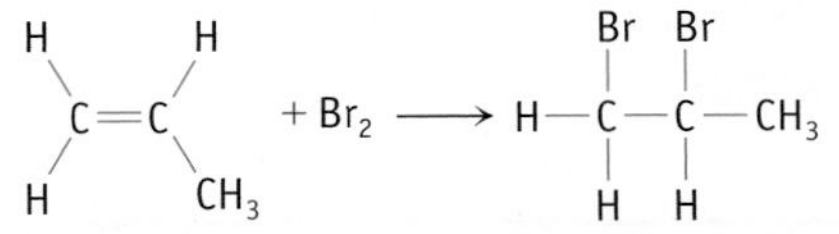

propene — 1,2-dibromopropane

EXERCISE 10.4 Reactions of Alkenes

(a) Draw the structure of the compound obtained from the reaction of HBr with ethylene, and name the compound.

(b) Draw the structure of the product of the reaction of Br_2 with *cis*-2-butene, and name this compound.

Aromatic Compounds

Benzene, C_6H_6, is a key molecule in chemistry. It is the simplest **aromatic compound,** a class of compounds so named because they have significant, and usually not unpleasant, odors. Other members of this class, which are all based on benzene, include toluene and naphthalene. A source of many aromatic compounds is coal. These compounds, and many other volatile substances, are released when coal is heated to a high temperature in the absence of air (Table 10.5).

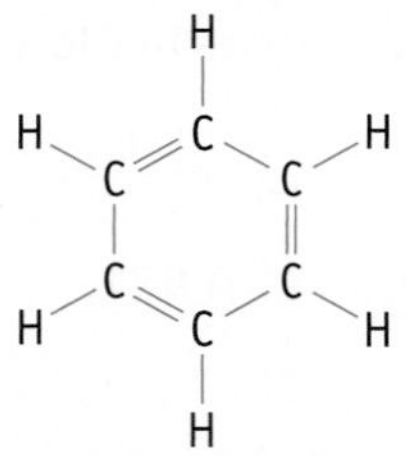
benzene

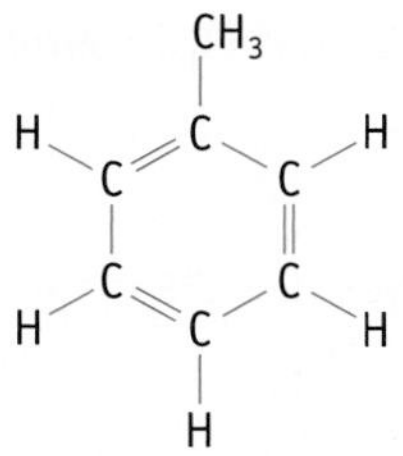
toluene

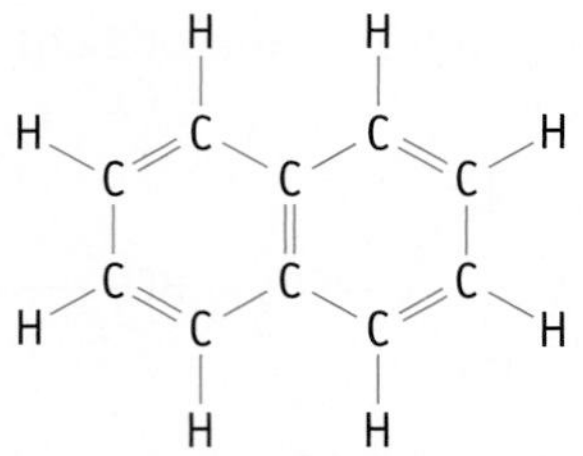
naphthalene

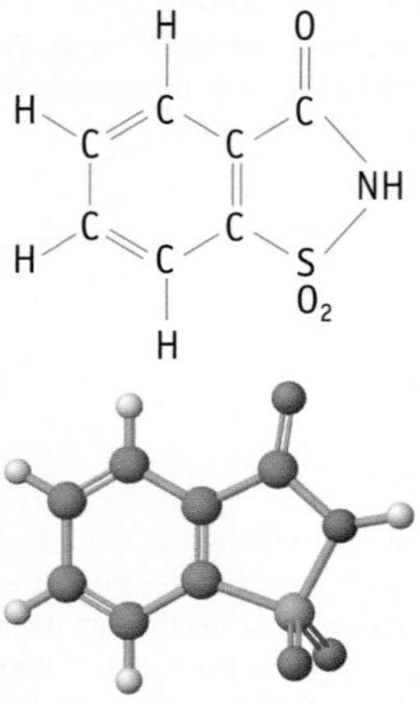
Saccharin ($C_7H_5NO_3S$). This compound, an artificial sweetener, contains an aromatic ring.

Benzene occupies a pivotal place in the history and practice of chemistry. Michael Faraday discovered this compound in 1825 as a by-product of illuminating gas, a fuel produced by heating coal. Today, benzene is an important industrial chemical, usually ranking among the top 25 chemicals in production annually in the United States. It is used as a solvent and is also the starting point for making thousands of different compounds by replacing the H atoms of the ring.

Toluene was originally obtained from tolu balsam, the pleasant-smelling gum of a South American tree, *Toluifera balsamum.* This balsam has been used in cough syrups and perfumes. Naphthalene is an ingredient in "moth balls," although 1,4-dichlorobenzene is now more commonly used. Aspartame and another artificial sweetener, saccharin, are also benzene derivatives.

TABLE 10.5 Some Aromatic Compounds from Coal Tar

Common Name	Formula	Boiling Point (°C)	Melting Point (°C)
benzene	C_6H_6	80	+6
toluene	$C_6H_5CH_3$	111	−95
o-xylene	1,2-$C_6H_4(CH_3)_2$	144	−25
m-xylene	1,3-$C_6H_4(CH_3)_2$	139	−48
p-xylene	1,4-$C_6H_4(CH_3)_2$	138	+13
naphthalene	$C_{10}H_8$	218	+80

The Structure of Benzene

The formula of benzene suggested to 19th-century chemists that this compound should be unsaturated, but, if viewed this way, its chemistry was perplexing. Whereas alkenes readily add Br_2, for example, benzene does not do so under similar conditions. The structural question was finally solved by August Kekulé (1829–1896). We now recognize that benzene's different reactivity relates to its structure and bonding, both of which are quite different from the structure and bonding in alkenes. Benzene has equivalent carbon–carbon bonds, 139 pm in length, intermediate between a C—C single bond (154 pm) and a C=C double bond (134 pm). The π bonds are formed by the continuous overlap of the p orbitals on the six carbon atoms (page 421). Using valence bond terminology, the structure is represented by a hybrid of two resonance structures.

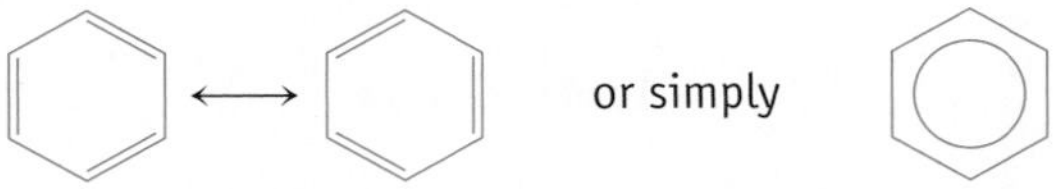

Representations of benzene, C_6H_6

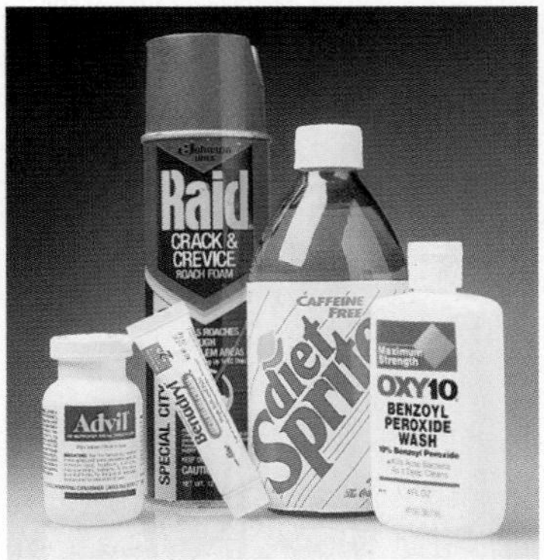

Charles D. Winters

Some products containing compounds based on benzene. Examples include sodium benzoate in soft drinks, ibuprofen in Advil, and benzoyl peroxide in Oxy-10.

Benzene Derivatives

Toluene, chlorobenzene, benzoic acid, aniline, styrene, and phenol are common examples of benzene derivatives.

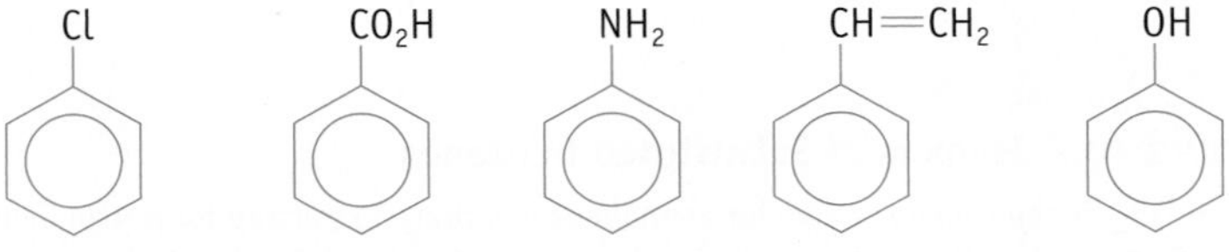

chlorobenzene | benzoic acid | aniline | styrene | phenol

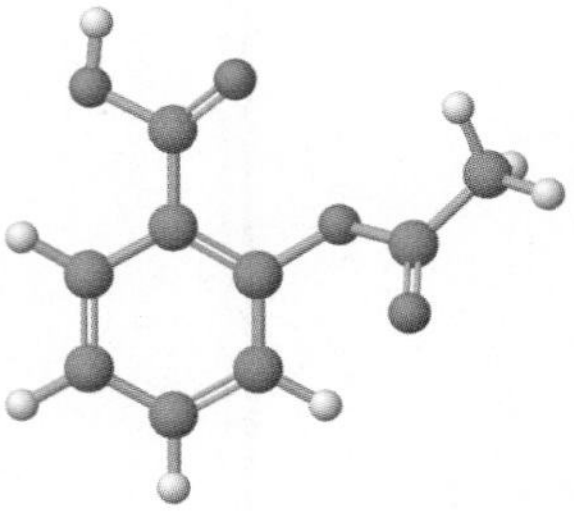

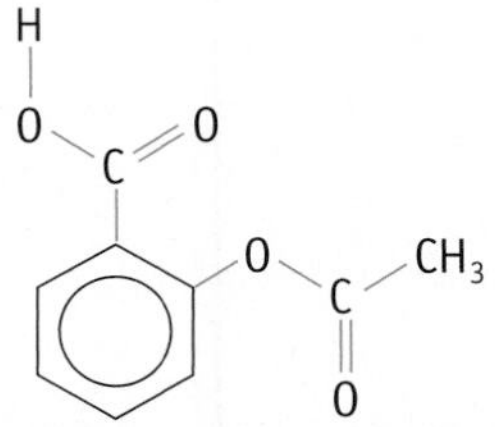

Aspirin, a commonly used analgesic. It is based on benzoic acid with an acetate group, $—O_2CCH_3$, in the *ortho* position.

If more than one H atom of benzene is replaced, isomers can arise. Thus, the systematic nomenclature for benzene derivatives involves naming substituent groups and identifying their positions on the ring by numbering the six carbon atoms (▶ Appendix E). Some common names, which are based on an older naming scheme, are also used. This scheme identified isomers of disubstituted benzenes with the prefixes *ortho* (*o*-, substituent groups on adjacent carbons in the benzene ring), *meta* (*m*-, substituents separated by one carbon atom), and *para* (*p*-, substituent groups on carbons on opposite sides of the ring).

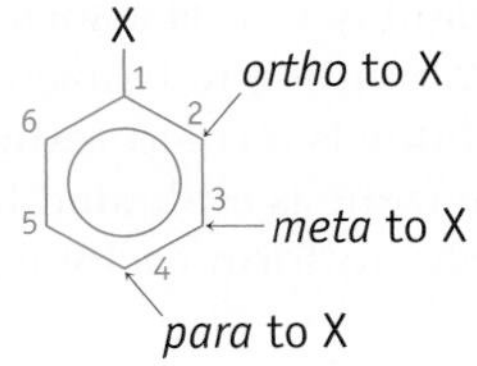

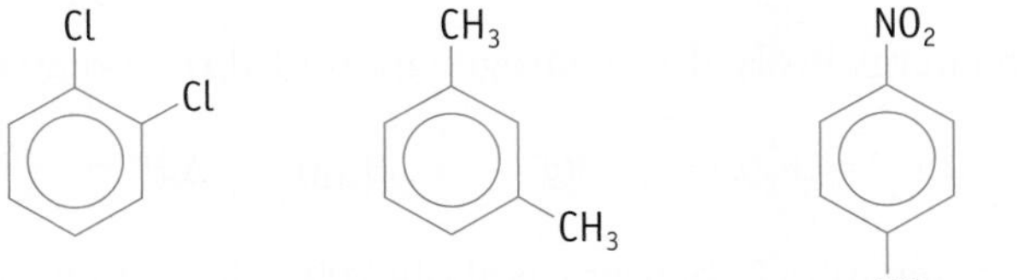

Systematic name:	1,2-dichlorobenzene	1,3-dimethylbenzene	1,4-dinitrobenzene
Common name:	*o*-dichlorobenzene	*m*-xylene	*p*-dinitrobenzene

EXAMPLE 10.5 Isomers of Substituted Benzenes

Problem Draw and name the isomers of $C_6H_3Cl_3$.

Strategy Begin by drawing the structure of C_6H_5Cl. Place a second Cl atom on the ring in the *ortho*, *meta*, and *para* positions. Add the third Cl in one of the remaining positions.

Solution The three isomers of $C_6H_3Cl_3$ are shown here. They are named as derivatives of benzene by specifying the number of substituent groups with the prefix "tri-," the name of the substituent, and the positions of the three groups around the six-member ring.

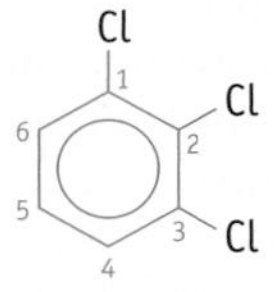

1,2,3-trichlorobenzene

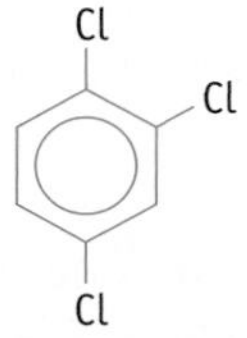

1,2,4-trichlorobenzene

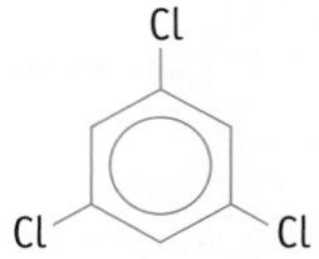

1,3,5-trichlorobenzene

EXERCISE 10.5 Isomers of Substituted Benzenes

Aniline, $C_6H_5NH_2$, is the common name for aminobenzene. Draw a structure for *p*-diaminobenzene, a compound used in dye manufacture. What is the systematic name for *p*-diaminobenzene?

Properties of Aromatic Compounds

Benzene is a colorless liquid, and simple substituted benzenes are liquids or solids under normal conditions. The properties of aromatic hydrocarbons are typical of hydrocarbons in general: They are insoluble in water, soluble in nonpolar solvents, and oxidized by O_2 to form CO_2 and H_2O.

One of the most important properties of benzene and other aromatic compounds is an unusual stability that is associated with the unique π bonding in this molecule. Because the π bonding in benzene is typically described using resonance structures, the extra stability is termed **resonance stabilization.** The extent of resonance stabilization in benzene is evaluated by comparing the energy evolved in the hydrogenation of benzene to form cyclohexane

$$C_6H_6(\ell) + 3\ H_2(g) \xrightarrow{\text{catalyst}} C_6H_{12}(\ell) \qquad \Delta_rH^\circ = -206.7\ \text{kJ/mol-rxn}$$

with the energy evolved in hydrogenation of three isolated double bonds.

$$3\ H_2C{=}CH_2(g) + 3\ H_2(g) \rightarrow 3\ C_2H_6(g) \qquad \Delta_rH^\circ = -410.8\ \text{kJ/mol-rxn}$$

The hydrogenation of benzene is about 200 kJ less exothermic than the hydrogenation of three moles of ethylene. The difference is attributable to the added stability associated with π bonding in benzene.

A Closer Look — Petroleum Chemistry

Much of the world's current technology relies on petroleum. Burning fuels derived from petroleum provides by far the largest amount of energy in the industrial world (see *The Chemistry of Fuels and Energy Sources,* pages 254–267). Petroleum and natural gas are also the chemical raw materials used in the manufacture of plastics, rubber, pharmaceuticals, and a vast array of other compounds.

The petroleum that is pumped out of the ground is a complex mixture whose composition varies greatly, depending on its source. The primary components of petroleum are always alkanes, but, to varying degrees, nitrogen- and sulfur-containing compounds are also present. Aromatic compounds are present as well, but alkenes and alkynes are not.

An early step in the petroleum refining process is distillation, in which the crude mixture is separated into a series of fractions based on boiling point: first a gaseous fraction (mostly alkanes with one to four carbon atoms; this fraction is often burned off), and then gasoline, kerosene, and fuel oils. After distillation, considerable material, in the form of a semi-solid, tar-like residue, remains.

Thomas Kitchin/Tom Stack & Associates

A modern petrochemical plant.

The petrochemical industry seeks to maximize the amounts of the higher-valued fractions of petroleum produced and to make specific compounds for which a particular need exists. This means carrying out chemical reactions involving the raw materials on a huge scale. One process to which petroleum is subjected is known as *cracking*. At very high temperatures, bond breaking or "cracking" can occur, and longer-chain hydrocarbons will fragment into smaller molecular units. These reactions are carried out in the presence of a wide array of catalysts, materials that speed up reactions and direct them toward specific products. Among the important products of cracking are ethylene and other alkenes, which serve as the raw materials for the formation of materials such as polyethylene. Cracking also produces gaseous hydrogen, a widely used raw material in the chemical industry.

Other important reactions involving petroleum are run at elevated temperatures and in the presence of specific catalysts. Such reactions include *isomerization* reactions, in which the carbon skeleton of an alkane rearranges to form a new isomeric species, and *reformation* processes, in which smaller molecules combine to form new molecules. Each process is directed toward achieving a specific goal, such as increasing the proportion of branched-chain hydrocarbons in gasoline to obtain higher octane ratings. A great amount of chemical research has gone into developing and understanding these highly specialized processes.

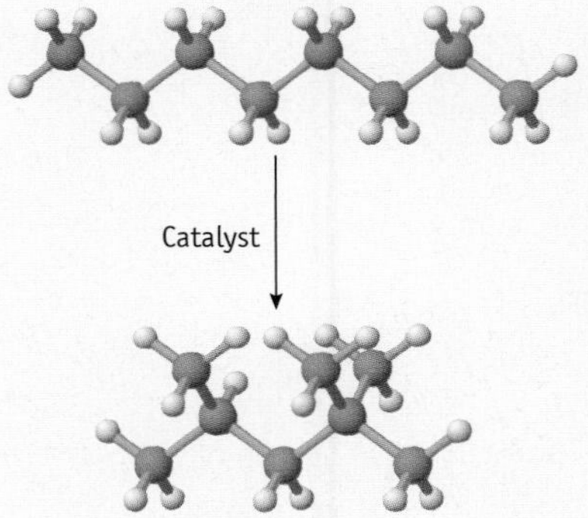

Producing gasoline. Branched hydrocarbons have a higher octane rating in gasoline. Therefore, an important process in producing gasoline is the isomerization of octane to a branched hydrocarbon such as isooctane, 2,2,4-trimethylpentane.

Although aromatic compounds are unsaturated hydrocarbons, they do not undergo the addition reactions typical of alkenes and alkynes. Instead, *substitution reactions* occur, in which one or more hydrogen atoms are replaced by other groups. Such reactions require a strong Brønsted acid such as H_2SO_4 or a Lewis acid such as $AlCl_3$ or $FeBr_3$.

$$\textit{Nitration:}\quad C_6H_6(\ell) + HNO_3(\ell) \xrightarrow{H_2SO_4} C_6H_5NO_2(\ell) + H_2O(\ell)$$

$$\textit{Alkylation:}\quad C_6H_6(\ell) + CH_3Cl(\ell) \xrightarrow{AlCl_3} C_6H_5CH_3(\ell) + HCl(g)$$

$$\textit{Halogenation:}\quad C_6H_6(\ell) + Br_2(\ell) \xrightarrow{FeBr_3} C_6H_5Br(\ell) + HBr(g)$$

10.3 Alcohols, Ethers, and Amines

Other types of organic compounds arise as elements other than carbon and hydrogen are included in the compound. Two elements in particular, oxygen and nitrogen, add a rich dimension to carbon chemistry.

TABLE 10.6 Common Functional Groups and Derivatives of Alkanes

Functional Group*	General Formula*	Class of Compound	Examples
F, Cl, Br, I	RF, RCl, RBr, RI	haloalkane	CH_3CH_2Cl, chloroethane
OH	ROH	alcohol	CH_3CH_2OH, ethanol
OR′	ROR′	ether	$(CH_3CH_2)_2O$, diethyl ether
NH_2†	RNH_2	(primary) amine	$CH_3CH_2NH_2$, ethylamine
$-\overset{O}{\overset{\Vert}{C}}H$	RCHO	aldehyde	CH_3CHO, ethanal (acetaldehyde)
$-\overset{O}{\overset{\Vert}{C}}-R'$	RCOR′	ketone	CH_3COCH_3, propanone (acetone)
$-\overset{O}{\overset{\Vert}{C}}-OH$	RCO_2H	carboxylic acid	CH_3CO_2H, ethanoic acid (acetic acid)
$-\overset{O}{\overset{\Vert}{C}}-OR'$	RCO_2R'	ester	$CH_3CO_2CH_3$, methyl acetate
$-\overset{O}{\overset{\Vert}{C}}-NH_2$	$RCONH_2$	amide	CH_3CONH_2, acetamide

* R and R′ can be the same or different hydrocarbon groups.
† Secondary amines (R_2NH) and tertiary amines (R_3N) are also possible, see discussion in the text.

Organic chemistry organizes compounds containing elements other than carbon and hydrogen as derivatives of hydrocarbons. Formulas (and structures) are represented by substituting one or more hydrogens in a hydrocarbon molecule by a **functional group.** A functional group is an atom or group of atoms attached to a carbon atom in the hydrocarbon. Formulas of hydrocarbon derivatives are then written as R—X, in which R is a hydrocarbon lacking a hydrogen atom, and X is the functional group that has replaced the hydrogen. The chemical and physical properties of the hydrocarbon derivatives are a blend of the properties associated with hydrocarbons and the group that has been substituted for hydrogen.

Table 10.6 identifies some common functional groups and the families of organic compounds resulting from their attachment to a hydrocarbon.

Chemistry Now™

Sign in at **www.thomsonedu.com/login** and go to Chapter 10 Contents to see Screen 10.5 for a description of the **types of organic functional groups** and for tutorials on **their structures, bonding, and chemistry.**

David Young/Tom Stack & Associates

Alcohol racing fuel. Methanol, CH_3OH, is used as the fuel in cars of the type that race in Indianapolis.

Alcohols and Ethers

If one of the hydrogen atoms of an alkane is replaced by a hydroxyl (—OH) group, the result is an **alcohol,** ROH. Methanol, CH_3OH, and ethanol, CH_3CH_2OH, are the most important alcohols, but others are also commercially important (Table 10.7). Notice that several have more than one OH functional group.

More than 5×10^8 kg of methanol is produced in the United States annually. Most of this production is used to make formaldehyde (CH_2O) and acetic acid

TABLE 10.7 Some Important Alcohols

Condensed Formula	BP (°C)	Systematic Name	Common Name	Use
CH_3OH	65.0	methanol	methyl alcohol	fuel, gasoline additive, making formaldehyde
CH_3CH_2OH	78.5	ethanol	ethyl alcohol	beverages, gasoline additive, solvent
$CH_3CH_2CH_2OH$	97.4	1-propanol	propyl alcohol	industrial solvent
$CH_3CH(OH)CH_3$	82.4	2-propanol	isopropyl alcohol	rubbing alcohol
$HOCH_2CH_2OH$	198	1,2-ethanediol	ethylene glycol	antifreeze
$HOCH_2CH(OH)CH_2OH$	290	1,2,3-propanetriol	glycerol (glycerin)	moisturizer in consumer products

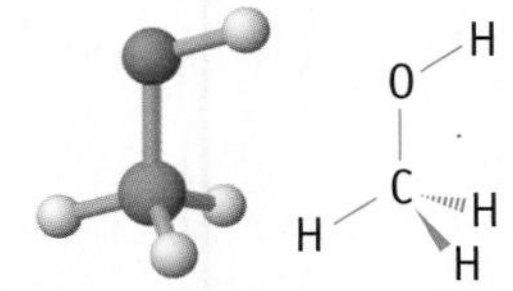

Methanol, CH_3OH, is the simplest alcohol. Methanol is often called "wood alcohol" because it was originally produced by heating wood in the absence of air.

(CH_3CO_2H), both important chemicals in their own right. Methanol is also used as a solvent, as a de-icer in gasoline, and as a fuel in high-powered racing cars. It is found in low concentration in new wine, where it contributes to the odor, or "bouquet." Like ethanol, methanol causes intoxication, but methanol differs in being more poisonous, largely because the human body converts it to formic acid (HCO_2H) and formaldehyde (CH_2O). These compounds attack the cells of the retina in the eye, leading to permanent blindness.

Ethanol is the "alcohol" of alcoholic beverages, in which it is formed by the anaerobic (without air) fermentation of sugar. For many years, industrial alcohol, which is used as a solvent and as a starting material for the synthesis of other compounds, was made by fermentation. In the last several decades, however, it has become cheaper to make ethanol from petroleum by-products—specifically, by the addition of water to ethylene.

■ **Aerobic Fermentation** Aerobic fermentation (in the presence of O_2) of sugar leads to the formation of acetic acid. This is how wine vinegar is made.

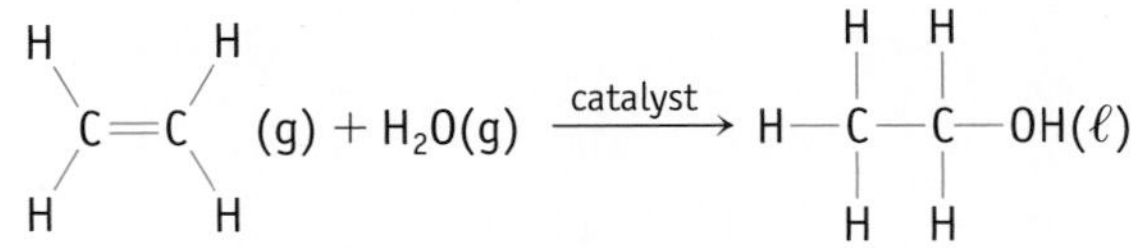

ethylene ethanol

Beginning with three-carbon alcohols, structural isomers are possible. For example, 1-propanol and 2-propanol (common name, isopropyl alcohol) are different compounds (Table 10.7).

Ethylene glycol and glycerol are common alcohols having two and three —OH groups, respectively. Ethylene glycol is used as antifreeze in automobiles. Glycerol's most common use is as a softener in soaps and lotions. It is also a raw material for the preparation of nitroglycerin (Figure 10.8).

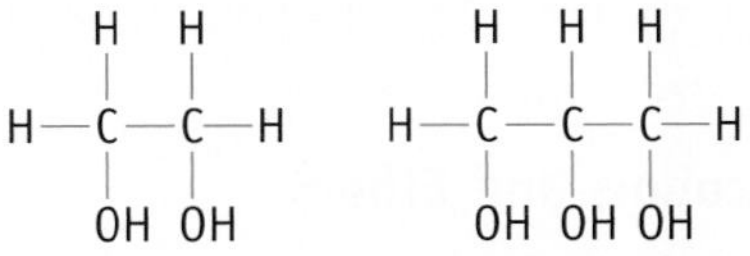

Systematic name:	1,2-ethanediol	1,2,3-propanetriol
Common name:	ethylene glycol	glycerol or glycerin

Charles D. Winters

Rubbing alcohol. Common rubbing alcohol is 2-propanol, also called isopropyl alcohol.

(a) (b) (c)

FIGURE 10.8 Nitroglycerin. (a) Concentrated nitric acid and glycerin react to form an oily, highly unstable compound called nitroglycerin, $C_3H_5(ONO_2)_3$. (b) Nitroglycerin is more stable if absorbed onto an inert solid, a combination called dynamite. (c) The fortune of Alfred Nobel (1833–1896), built on the manufacture of dynamite, now funds the Nobel Prizes.

EXAMPLE 10.6 Structural Isomers of Alcohols

Problem How many different alcohols are derivatives of pentane? Draw structures, and name each alcohol.

Strategy Pentane, C_5H_{12}, has a five-carbon chain. An —OH group can replace a hydrogen atom on one of the carbon atoms. Alcohols are named as derivatives of the alkane (pentane) by replacing the "-e" at the end with "-ol" and indicating the position of the —OH group by a numerical prefix (Appendix E).

Solution Three different alcohols are possible, depending on whether the —OH group is placed on the first, second, or third carbon atom in the chain. (The fourth and fifth positions are identical to the second and first positions in the chain, respectively.)

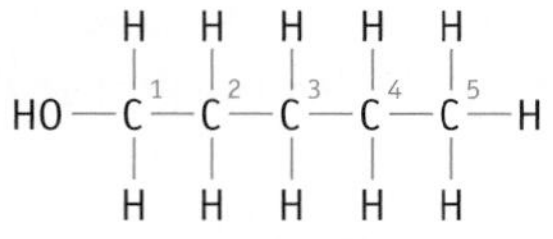

1-pentanol

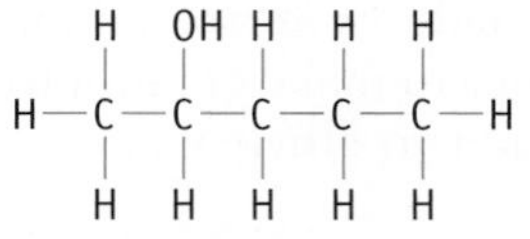

2-pentanol

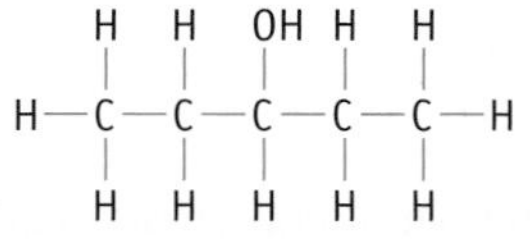

3-pentanol

Comment Additional structural isomers with the formula $C_5H_{11}OH$ are possible in which the longest carbon chain has three C atoms (one isomer) or four C atoms (four isomers).

EXERCISE 10.6 Structures of Alcohols

Draw the structure of 1-butanol and alcohols that are structural isomers of the compound.

Properties of Alcohols and Ethers

Methane, CH_4, is a gas (boiling point, −161 °C) with low solubility in water. Methanol, CH_3OH, by contrast, is a liquid that is *miscible* with water in all proportions. The boiling point of methanol, 65 °C, is 226 °C higher than the boiling point of methane. What a difference the addition of a single atom into the structure can make in the properties of simple molecules!

Alcohols are related to water, with one of the H atoms of H_2O being replaced by an organic group. If a methyl group is substituted for one of the hydrogens of water, methanol results. Ethanol has a $—C_2H_5$ (ethyl) group, and propanol has a $—C_3H_7$ (propyl) group in place of one of the hydrogens of water. Viewing alcohols as related to water also helps in understanding the properties of alcohols.

■ **Hydrogen Bonding** The intermolecular forces of attraction of compounds with hydrogen attached to a highly electronegative atom, like O, N, or F, are so exceptional that they are accorded a special name: hydrogen bonding. We will discuss hydrogen bonding in Section 12.2.

The two parts of methanol, the $—CH_3$ group and the —OH group, contribute to its properties. For example, methanol will burn, a property associated with hydrocarbons. On the other hand, its boiling point is more like that of water. The temperature at which a substance boils is related to the forces of attraction between molecules, called *intermolecular forces:* The stronger the attractive, intermolecular forces in a sample, the higher the boiling point (► Section 12.4). These forces are particularly strong in water, a result of the polarity of the —OH group in this molecule (◄ Section 8.8). Methanol is also a polar molecule, and it is the polar —OH group that leads to a high boiling point. In contrast, methane is nonpolar and its low boiling point is the result of weak intermolecular forces.

It is also possible to explain the differences in the solubility of methane and methanol in water. The solubility of methanol is conferred by the polar —OH portion of the molecule. Methane, which is nonpolar, has low water-solubility.

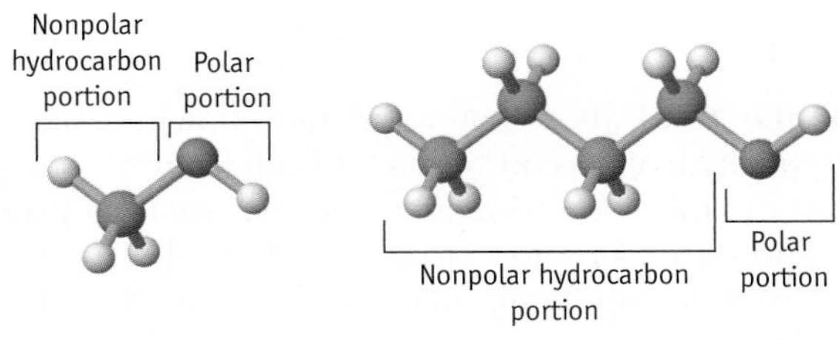

As the size of the alkyl group in an alcohol increases, the alcohol boiling point rises, a general trend seen in families of similar compounds and related to molar mass (see Table 10.7). The solubility in water in this series decreases. Methanol and ethanol are completely miscible in water, whereas 1-propanol is moderately water-soluble; 1-butanol is less soluble than 1-propanol. With an increase in the size of the hydrocarbon group, the organic group (the nonpolar part of the molecule) has become a larger fraction of the molecule, and properties associated with nonpolarity begin to dominate. Space-filling models show that in methanol, the polar and nonpolar parts of the molecule are approximately similar in size, but in 1-butanol the —OH group is less than 20% of the molecule. The molecule is less like water and more "organic."

Attaching an additional—OH group to a hydrocarbon framework has an effect on water solubility (Figure 10.9). Two —OH groups on a three-carbon framework, as found in propylene glycol, convey complete miscibility with water, in contrast to the limited solubility of 1-propanol and 2-propanol.

Ethers have the general formula ROR′. The best-known ether is diethyl ether, $CH_3CH_2OCH_2CH_3$. Lacking an —OH group, the properties of ethers are in sharp contrast to those of alcohols. Diethyl ether, for example, has a lower boiling point (34.5 °C) than ethanol, CH_3CH_2OH (78.3 °C), and is only slightly soluble in water.

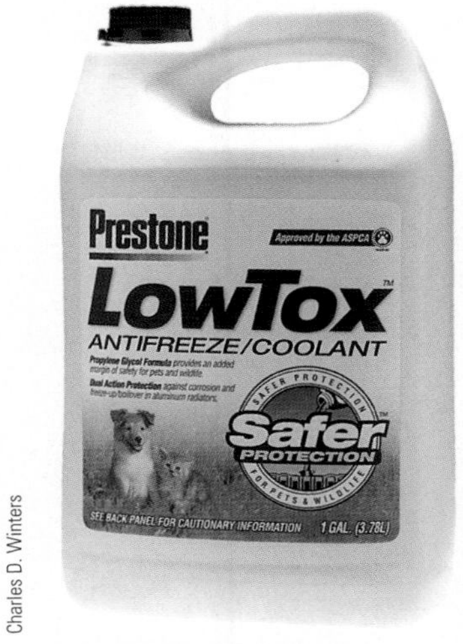

Charles D. Winters

Safe antifreeze—propylene glycol, $CH_3CHOHCH_2OH$. Most antifreeze sold today consists of about 95% ethylene glycol. Cats and dogs are attracted by the smell and taste of the compound, but it is toxic. In fact, only a few milliliters can prove fatal to a small dog or cat. In the first stage of poisoning, an animal may appear drunk, but within 12–36 hours the kidneys stop functioning, and the animal slips into a coma. To avoid accidental poisoning of domestic and wild animals, you can use propylene glycol antifreeze. This compound affords the same antifreeze protection but is much less toxic.

Chemistry Now™

Sign in at **www.thomsonedu.com/login** and go to Chapter 10 Contents to see Screen 10.6 for an exercise on **substitution and elimination reactions of alcohols.**

Methanol is often added to automobile gasoline tanks in the winter to prevent fuel lines from freezing. It is soluble in water and lowers the water's freezing point.

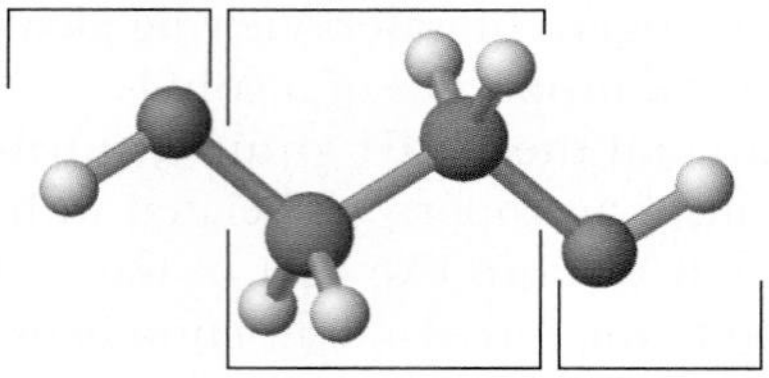

Ethylene glycol is used in automobile radiators. It is soluble in water, and lowers the freezing point and raises the boiling point of the water in the cooling system. (See Section 14.4.)

Ethylene glycol, a major component of automobile antifreeze, is completely miscible with water.

FIGURE 10.9 Properties and uses of methanol and ethylene glycol.

Amines

It is often convenient to think about water and ammonia as being similar molecules: They are the simplest hydrogen compounds of adjacent second-period elements. Both are polar and exhibit some similar chemistry, such as protonation (to give H_3O^+ and NH_4^+) and deprotonation (to give OH^- and NH_2^-).

This comparison of water and ammonia can be extended to alcohols and amines. Alcohols have formulas related to water in which one hydrogen in H_2O is replaced with an organic group (R—OH). In organic **amines,** one or more hydrogen atoms of NH_3 are replaced with an organic group. Amine structures are similar to ammonia's structure; that is, the geometry about the N atom is trigonal pyramidal.

Amines are categorized based on the number of organic substituents as primary (one organic group), secondary (two organic groups), or tertiary (three organic groups). As examples, consider the three amines with methyl groups: CH_3NH_2, $(CH_3)_2NH$, and $(CH_3)_3N$.

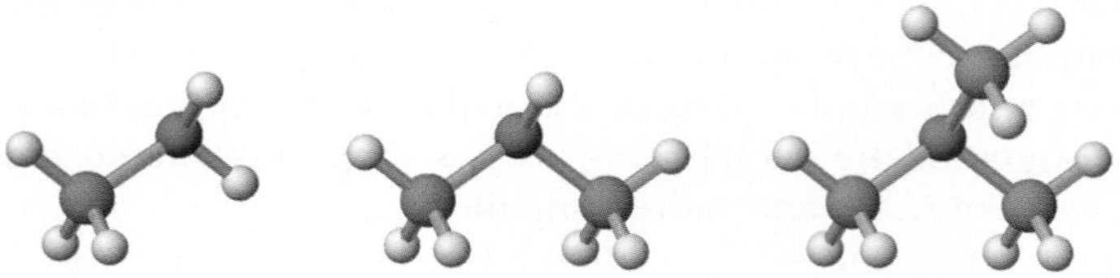

CH_3NH_2	$(CH_3)_2NH$	$(CH_3)_3N$
Primary amine	Secondary amine	Tertiary amine
Methylamine	Dimethylamine	Trimethylamine

Properties of Amines

Amines usually have offensive odors. You know what the odor is if you have ever smelled decaying fish. Two appropriately named amines, putrescine and cadaverine, add to the odor of urine, rotten meat, and bad breath.

$H_2NCH_2CH_2CH_2CH_2NH_2$	$H_2NCH_2CH_2CH_2CH_2CH_2NH_2$
putrescine	cadaverine
1,4-butanediamine	1,5-pentanediamine

The smallest amines are water-soluble, but most amines are not. All amines are bases, however, and they react with acids to give salts, many of which are water-soluble. As with ammonia, the reactions involve adding H^+ to the lone pair of electrons on the N atom. This is illustrated by the reaction of aniline (aminobenzene) with H_2SO_4 to give anilinium sulfate, a compound of some historical interest (see "Historical Perspectives: Mauvine").

$$C_6H_5NH_2(aq) + H_2SO_4(aq) \longrightarrow C_6H_5NH_3^+(aq) + HSO_4^-(aq)$$

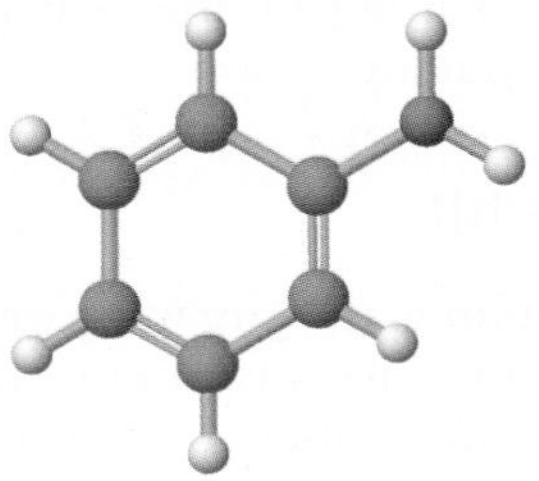

Aniline

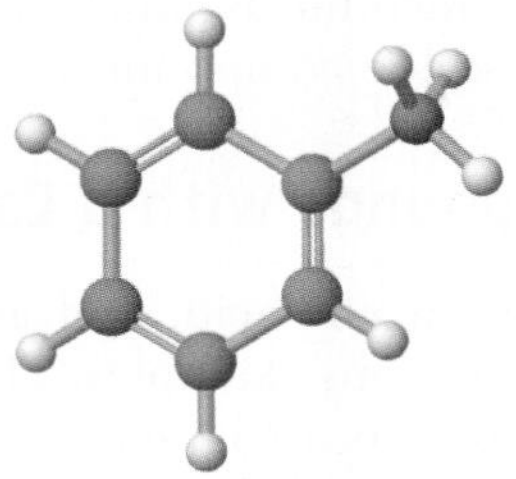

Anilinium ion

Historical Perspectives Mauveine

Among the roots of modern organic chemistry was the synthesis, in 1856, of the compound mauveine (or mauve) by William Henry Perkin (1838–1907). This discovery led to a flourishing dye industry, one of the first chemical industries.

The discovery of mauve is an interesting tale. At the age of 13, Perkin enrolled at the City of London School. His father paid an extra fee for him to attend a lunchtime chemistry course and set up a lab at home for him to do experiments. He began attending the public lectures that Michael Faraday gave on Saturdays at the Royal Institution. At 15, Perkin enrolled in the Royal College of Science in London to study chemistry under the school's Director, August Wilhelm von Hofmann. After he completed his studies at age 17, he took a position at the college as Hofmann's assistant, rather a great honor.

Science & Society Picture Library/Science Museum/London

Mauveine. The original stoppered bottle of mauveine prepared by Perkin. The structure of the mauveine cation is shown here.

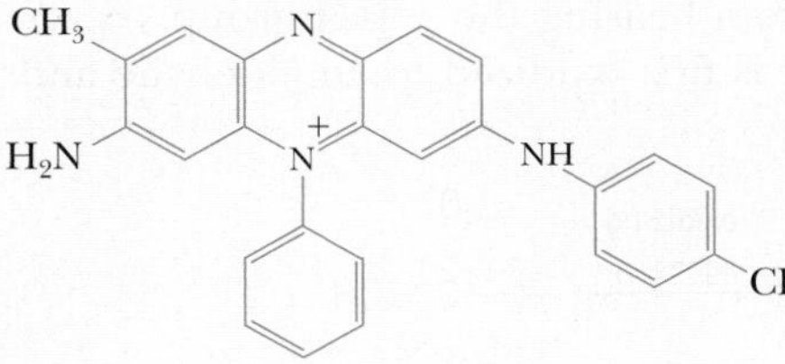

Perkin's first project was to synthesize quinine, an antimalarial drug. The route he proposed involved oxidizing anilinium sulfate. From the reaction, he obtained a black solid that dissolved in a water-ethanol mixture to give a purple solution that stained cloth a beautiful purple color. The color didn't wash out, an essential feature for a dye. Later, it was learned that the anilinium sulfate Perkin used had been impure and that the impurity was essential in the synthesis. Had Perkins used a pure sample or his starting reagent, the discovery of mauve would not have happened. A study in 1994 on samples of mauve preserved in museums determined that Perkin's mauve was actually a mixture of two very similar compounds, along with traces of several others.

At the age of 18, Perkin quit his assistantship and, with financial help from his family, set up a dye factory outside of London. By the age of 36, he was a very wealthy man. He then retired from the dye business and devoted the rest of his life to chemical research on various topics, including the synthesis of fragrances and a study of optical activity. During his lifetime, he received numerous honors for his research, but one honor came many years after his death. In 1972, when The Chemical Society (in England) renamed its journals after famous society members, it chose Perkin's name for the organic chemistry journals. (See *Mauve,* a book on Perkin's life, by S. Garfield, W. W. Norton Publishers, New York.)

Science & Society Picture Library/Science Museum/London

A silk dress dyed with Perkin's original sample of mauve in 1862, at the dawning of the synthetic dye industry. From *Mauve.*

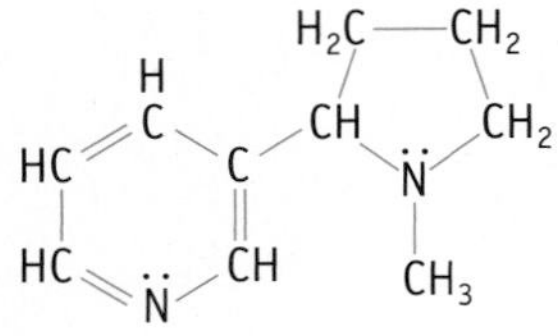

Nicotine

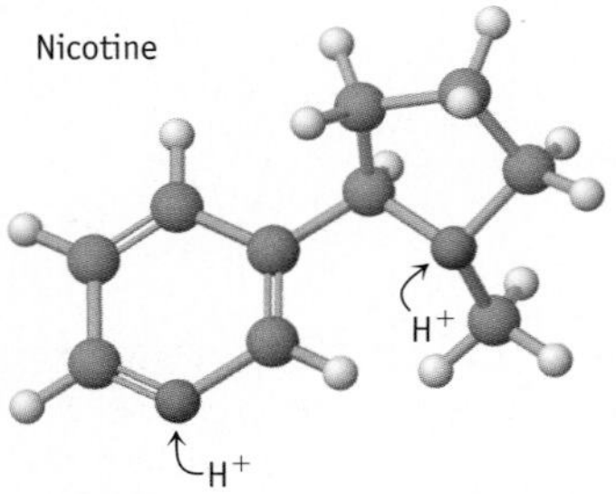

Nicotine. Two nitrogen atoms in the nicotine molecule can be protonated, which is the form in which nicotine is normally found. The protons can be removed, however, by treating it with a base. This "free-base" form is much more poisonous and addictive. See J. F. Pankow: *Environmental Science & Technology,* Vol 31, p. 2428, August 1997.

The facts that an amine can be protonated and that the proton can be removed again by treating the compound with a base have practical and physiological importance. Nicotine in cigarettes is normally found in the protonated form. (This water-soluble form is often used in insecticides.) Adding a base such as ammonia removes the H^+ ion to leave nicotine in its "free-base" form.

$$NicH_2^{2+}(aq) + 2\ NH_3(aq) \rightarrow Nic(aq) + 2\ NH_4^+(aq)$$

In this form, nicotine is much more readily absorbed by the skin and mucous membranes, so the compound is a much more potent poison.

10.4 Compounds with a Carbonyl Group

Formaldehyde, acetic acid, and acetone are among the organic compounds referred to in previous examples. These compounds have a common structural feature: Each contains a trigonal-planar carbon atom doubly bonded to an oxygen. The C=O group is called the **carbonyl group,** and all of these compounds are members of a large class of compounds called **carbonyl compounds.**

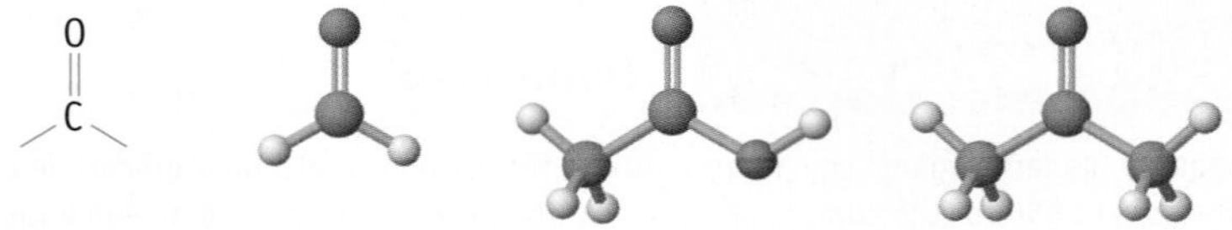

Carbonyl group	Formaldehyde	Acetic acid	Acetone
	CH_2O	CH_3CO_2H	CH_3COCH_3
	Aldehyde	*Carboxylic acid*	*Ketone*

In this section, we will examine five groups of carbonyl compounds (Table 10.6, page 462):

- *Aldehydes* (RCHO) have an organic group (—R) and an H atom attached to a carbonyl group.
- *Ketones* (RCOR′) have two —R groups attached to the carbonyl carbon; they may be the same groups, as in acetone, or different groups.
- *Carboxylic acids* (RCO_2H) have an —R group and an —OH group attached to the carbonyl carbon.
- *Esters* (RCO_2R') have —R and —OR′ groups attached to the carbonyl carbon.
- *Amides* ($RCONR_2'$, RCONHR′, and $RCONH_2$) have an —R group and an amino group (—NH_2, —NHR, —NR_2) bonded to the carbonyl carbon.

Primary alcohol: ethanol

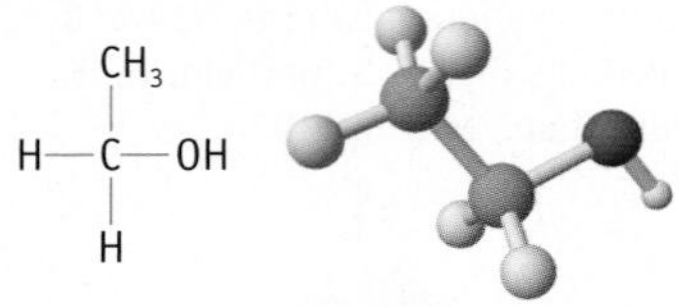

Secondary alcohol: 2-propanol

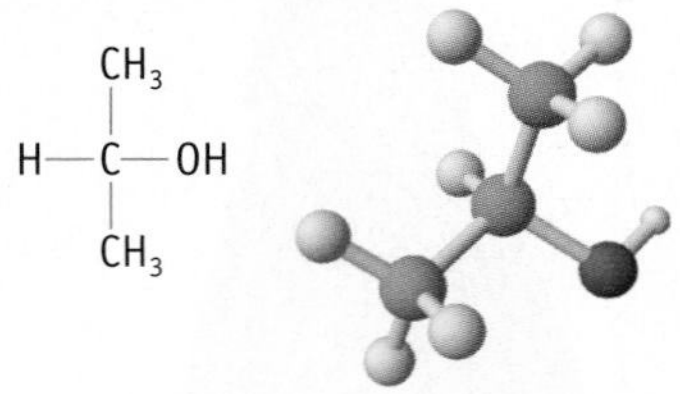

Tertiary alcohol: 2-methyl-2-propanol

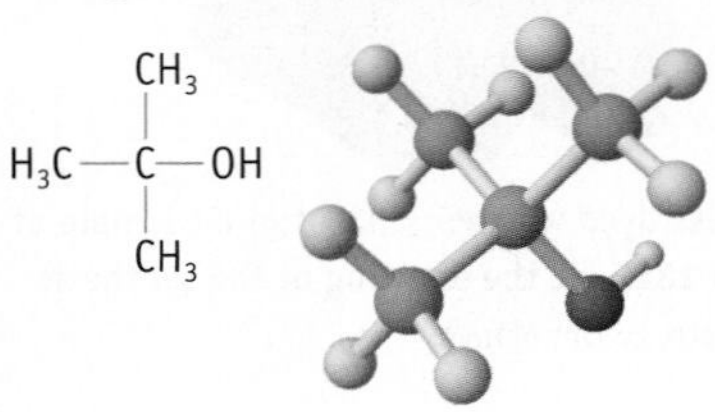

Aldehydes, ketones, and carboxylic acids are oxidation products of alcohols and, indeed, are commonly made by this route. The product obtained through oxidation of an alcohol depends on the alcohol's structure, which is classified according to the number of carbon atoms bonded to the C atom bearing the —OH group. *Primary alcohols* have one carbon and two hydrogen atoms attached, whereas *secondary alcohols* have two carbon atoms and one hydrogen atom attached. *Tertiary alcohols* have three carbon atoms attached to the C atom bearing the —OH group.

A *primary alcohol* is oxidized in two steps. It is first oxidized to an aldehyde and then in a second step to a carboxylic acid:

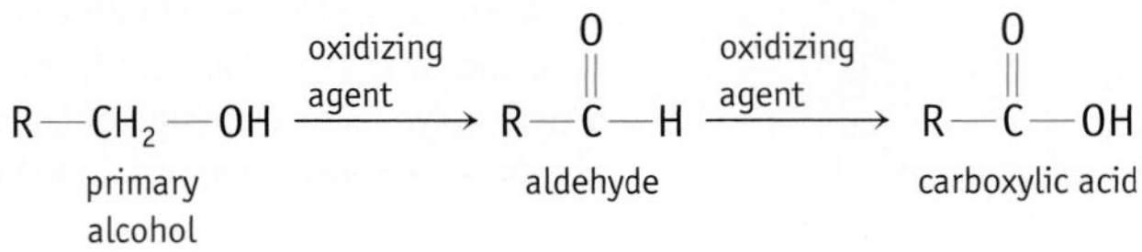

For example, the air oxidation of ethanol in wine produces wine (with excess oxygen) vinegar, the most important ingredient of which is acetic acid.

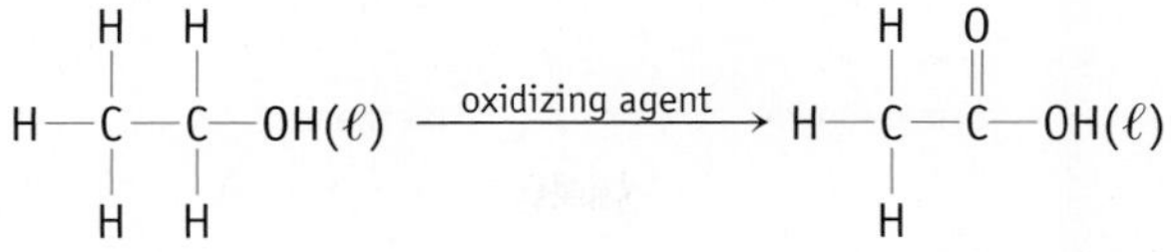

ethanol   acetic acid

Charles D. Winters

FIGURE 10.10 Alcohol tester. This device for testing a person's breath for the presence of ethanol relies on the oxidation of the alcohol. If present, ethanol is oxidized by potassium dichromate, $K_2Cr_2O_7$, to acetaldehyde, and then to acetic acid. The yellow-orange dichromate ion is reduced to green $Cr^{3+}(aq)$, the color change indicating that ethanol was present.

Acids have a sour taste. The word "vinegar" (from the French *vin aigre*) means sour wine. A device to test one's breath for alcohol relies on a similar oxidation of ethanol (Figures 3.21 and 10.10).

In contrast to primary alcohols, oxidation of a *secondary alcohol* produces a ketone:

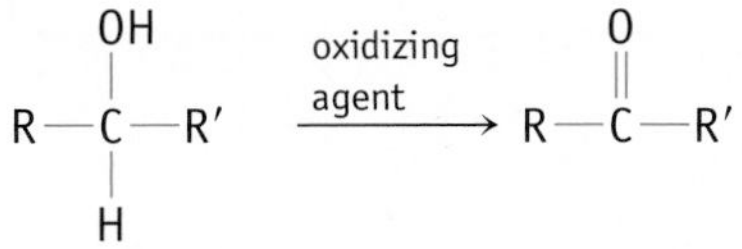

secondary alcohol   ketone

(—R and —R′ are organic groups. They may be the same or different.)

Common oxidizing agents used for these reactions are reagents such as $KMnO_4$ and $K_2Cr_2O_7$ (Table 3.4).

Finally, tertiary alcohols do *not* react with the usual oxidizing agents.

$$(CH_3)_3COH \xrightarrow{\text{oxidizing agent}} \text{no reaction}$$

Aldehydes and Ketones

Aldehydes and **ketones** have pleasant odors and are often used in fragrances. Benzaldehyde is responsible for the odor of almonds and cherries; cinnamaldehyde is found in the bark of the cinnamon tree; and the ketone 4-(*p*-hydroxyphenyl) 2-butanone is responsible for the odor of ripe raspberries (a favorite of the authors of this book). Table 10.8 lists several simple aldehydes and ketones.

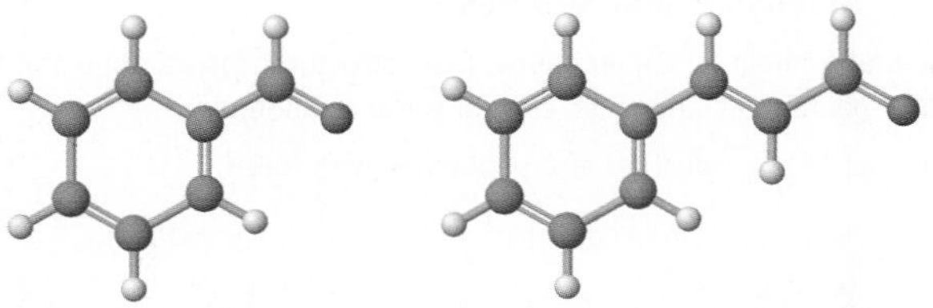

Benzaldehyde, C_6H_5CHO *trans*-Cinnamaldehyde, $C_6H_5CH{=}CHCHO$

Aldehydes and ketones are the oxidation products of primary and secondary alcohols, respectively. The reverse reactions—reduction of aldehydes to primary alcohols and reduction of ketones to secondary alcohols—are also known.

Charles D. Winters

Aldehydes and odors. The odors of almonds and cinnamon are due to aldehydes, but the odor of fresh raspberries comes from a ketone.

TABLE 10.8 Simple Aldehydes and Ketones

Structure	Common Name	Systematic Name	BP (°C)
$\mathrm{H\overset{O}{\overset{\|}{C}}H}$	formaldehyde	methanal	−19
$\mathrm{CH_3\overset{O}{\overset{\|}{C}}H}$	acetaldehyde	ethanal	20
$\mathrm{CH_3\overset{O}{\overset{\|}{C}}CH_3}$	acetone	propanone	56
$\mathrm{CH_3\overset{O}{\overset{\|}{C}}CH_2CH_3}$	methyl ethyl ketone	butanone	80
$\mathrm{CH_3CH_2\overset{O}{\overset{\|}{C}}CH_2CH_3}$	diethyl ketone	3-pentanone	102

Commonly used reagents for such reductions are $NaBH_4$ and $LiAlH_4$, although H_2 is used on an industrial scale.

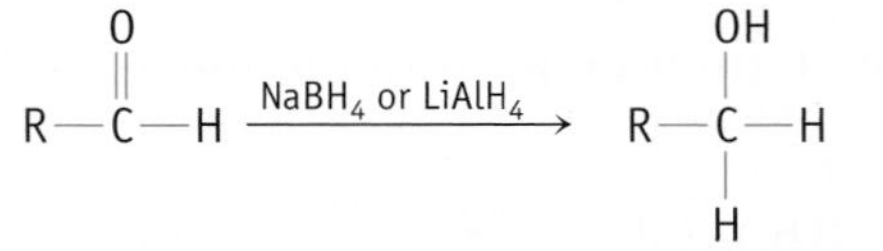

aldehyde — primary alcohol

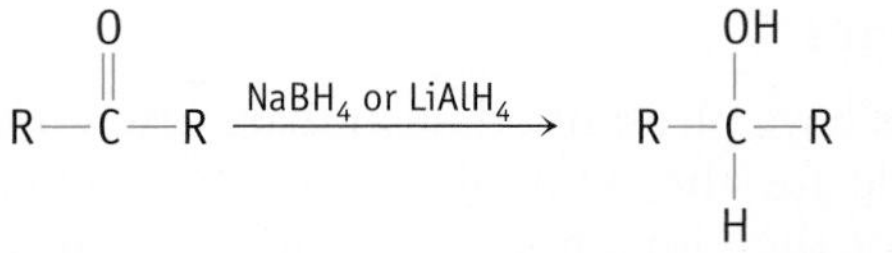

ketone — secondary alcohol

EXERCISE 10.7 Aldehydes and Ketones

(a) Draw the structural formula for 2-pentanone. Draw structures for a ketone and two aldehydes that are isomers of 2-pentanone, and name each of these compounds.

(b) What is the product of the reduction of 2-pentanone with $NaBH_4$?

EXERCISE 10.8 Aldehydes and Ketones

Draw the structures, and name the aldehyde or ketone formed upon oxidation of the following alcohols: **(a)** 1-butanol, **(b)** 2-butanol, **(c)** 2-methyl-1-propanol. Are these three alcohols structural isomers? Are the oxidation products structural isomers?

Carboxylic Acids

Acetic acid is the most common and most important **carboxylic acid.** For many years, acetic acid was made by oxidizing ethanol produced by fermentation. Now, however, acetic acid is generally made by combining carbon monoxide and methanol in the presence of a catalyst:

$$\underset{\text{methanol}}{CH_3OH(\ell)} + CO(g) \xrightarrow{\text{catalyst}} \underset{\text{acetic acid}}{CH_3CO_2H(\ell)}$$

About 1 billion kilograms of acetic acid are produced annually in the United States for use in plastics, synthetic fibers, and fungicides.

Many organic acids are found naturally (Table 10.9). Acids are recognizable by their sour taste (Figure 10.11) and are found in common foods: Citric acid in fruits, acetic acid in vinegar, and tartaric acid in grapes are just three examples.

Some carboxylic acids have common names derived from the source of the acid (Table 10.9). Because formic acid is found in ants, its name comes from the Latin word for ant (*formica*). Butyric acid gives rancid butter its unpleasant odor, and the name is related to the Latin word for butter (*butyrum*). The systematic names of acids (Table 10.10) are formed by dropping the "-e" on the name of the corresponding alkane and adding "-oic" (and the word "acid").

Because of the substantial electronegativity of oxygen, the two O atoms of the carboxylic acid group are slightly negatively charged, and the H atom of the —OH group is positively charged. This charge distribution has several important implications:

- The polar acetic acid molecule dissolves readily in water, which you already know because vinegar is an aqueous solution of acetic acid. (Acids with larger organic groups are less soluble, however.)

TABLE 10.9 Some Naturally Occurring Carboxylic Acids

Name	Structure	Natural Source
benzoic acid	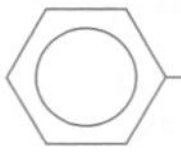	berries
citric acid	$HO_2C—CH_2—C(OH)(CO_2H)—CH_2—CO_2H$	citrus fruits
lactic acid	$H_3C—CH(OH)—CO_2H$	sour milk
malic acid	$HO_2C—CH_2—CH(OH)—CO_2H$	apples
oleic acid	$CH_3(CH_2)_7—CH{=}CH—(CH_2)_7—CO_2H$	vegetable oils
oxalic acid	$HO_2C—CO_2H$	rhubarb, spinach, cabbage, tomatoes
stearic acid	$CH_3(CH_2)_{16}—CO_2H$	animal fats
tartaric acid	$HO_2C—CH(OH)—CH(OH)—CO_2H$	grape juice, wine

FIGURE 10.11 Acetic acid in bread. Acetic acid is produced in bread when leavened with the yeast *Saccharomyces exigus*. Another group of bacteria, *Lactobacillus sanfrancisco,* contributes to the flavor of sourdough bread. These bacteria metabolize the sugar maltose, excreting acetic acid and lactic acid, $CH_3CH(OH)CO_2H$, thereby giving the bread its unique sour taste.

Formic acid, HCO_2H. This acid puts the sting in ant bites.

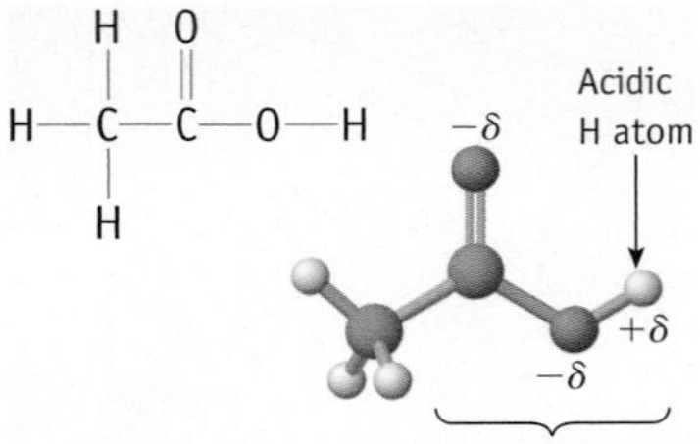

Acetic acid. The H atom of the carboxylic acid group (—CO_2H) is the acidic proton of this and other carboxylic acids.

TABLE 10.10 Some Simple Carboxylic Acids

Structure	Common Name	Systematic Name	BP (°C)
$H\overset{O}{\overset{\|}{C}}OH$	formic acid	methanoic acid	101
$CH_3\overset{O}{\overset{\|}{C}}OH$	acetic acid	ethanoic acid	118
$CH_3CH_2\overset{O}{\overset{\|}{C}}OH$	propionic acid	propanoic acid	141
$CH_3(CH_2)_2\overset{O}{\overset{\|}{C}}OH$	butyric acid	butanoic acid	163
$CH_3(CH_2)_3\overset{O}{\overset{\|}{C}}OH$	valeric acid	pentanoic acid	187

- The hydrogen of the —OH group is the acidic hydrogen. As noted in Chapter 3, acetic acid is a weak acid in water, as are most other organic acids.

Carboxylic acids undergo a number of reactions. Among these is the reduction of the acid (with reagents such as $LiAlH_4$ or $NaBH_4$) first to an aldehyde and then to an alcohol. For example, acetic acid is reduced first to acetaldehyde and then to ethanol.

$$\underset{\text{acetic acid}}{CH_3CO_2H} \xrightarrow{LiAlH_4} \underset{\text{acetaldehyde}}{CH_3CHO} \xrightarrow{LiAlH_4} \underset{\text{ethanol}}{CH_3CH_2OH}$$

Yet another important aspect of carboxylic acid chemistry is these acids' reaction with bases to give carboxylate anions. For example, acetic acid reacts with sodium hydroxide to give sodium acetate (sodium ethanoate).

$$CH_3CO_2H(aq) + OH^-(aq) \rightarrow CH_3CO_2^-(aq) + H_2O(\ell)$$

Esters

Carboxylic acids (RCO_2H) react with alcohols (R′OH) to form esters (RCO_2R') in an **esterification** reaction. (These reactions are generally run in the presence of strong acids because acids accelerate the reaction.)

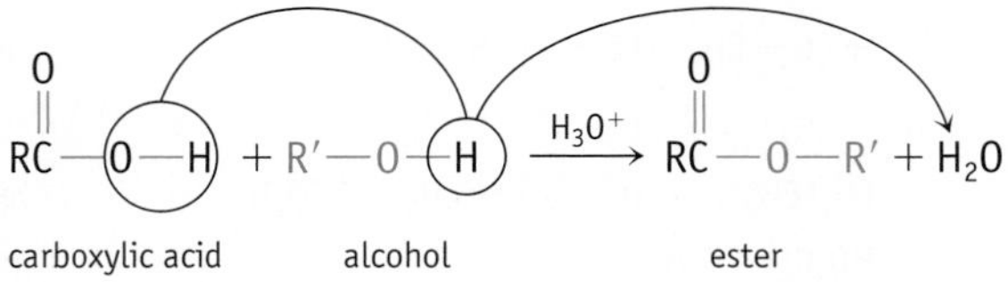

$$\underset{\text{acetic acid}}{CH_3\overset{O}{\overset{\|}{C}}OH} + \underset{\text{ethanol}}{CH_3CH_2OH} \xrightarrow{H_3O^+} \underset{\text{ethyl acetate}}{CH_3\overset{O}{\overset{\|}{C}}OCH_2CH_3} + H_2O$$

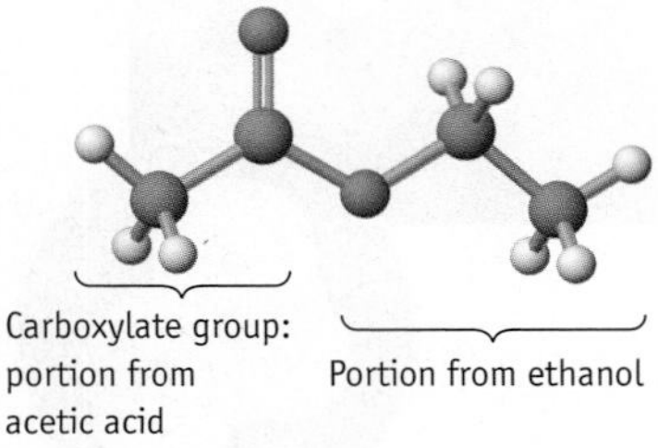

Ethyl acetate, an ester
$CH_3CO_2CH_2CH_3$

A Closer Look — Glucose and Sugars

Having described alcohols and carbonyl compounds, we now pause to look at glucose, the most common, naturally occurring carbohydrate.

As their name implies, formulas of carbohydrates can be written as though they are a combination of carbon and water, $C_x(H_2O)_y$. Thus, the formula of glucose, $C_6H_{12}O_6$, is equivalent to $C_6(H_2O)_6$. This compound is a sugar, or, more accurately, a **monosaccharide**.

Carbohydrates are polyhydroxy aldehydes or ketones. Glucose is an interesting molecule that exists in three different isomeric forms. Two of the isomers contain six-member rings; the third isomer features a chain structure. In solution, the three forms rapidly interconvert.

Notice that glucose is a chiral molecule. In the chain structure, four of the carbon atoms are bonded to four different groups.

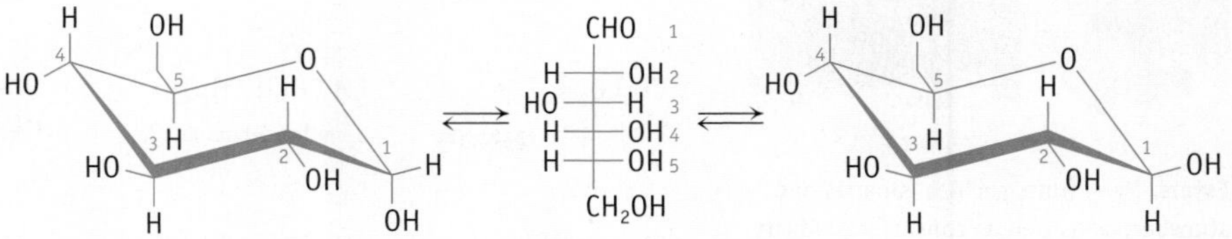

α-D-Glucose   Open-chain form   β-D-Glucose

In nature, glucose occurs in just one of its enantiomeric forms; thus, a solution of glucose rotates polarized light.

Knowing glucose's structure allows one to predict some of its properties. With five polar —OH groups in the molecule, glucose is, not surprisingly, soluble in water.

The aldehyde group is susceptible to chemical oxidation to form a carboxylic acid. Detection of glucose (in urine or blood) takes advantage of this fact; diagnostic tests for glucose involve oxidation with subsequent detection of the products.

Glucose is in a class of sugar molecules called hexoses, molecules having six carbon atoms. 2-Deoxyribose, the sugar in the backbone of the DNA molecule, is a pentose, a molecule with five carbon atoms.

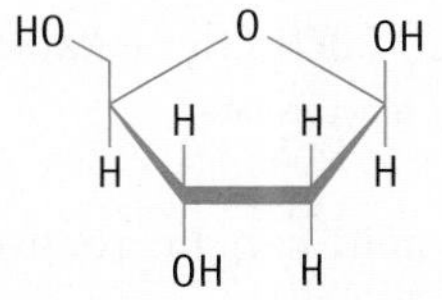

deoxyribose, a pentose, part of the DNA backbone

Glucose and other monosaccharides serve as the building blocks for larger carbohydrates. Sucrose, a disaccharide, is formed from a molecule of glucose and a molecule of fructose, another monosaccharide. Starch is a polymer composed of many monosaccharide units.

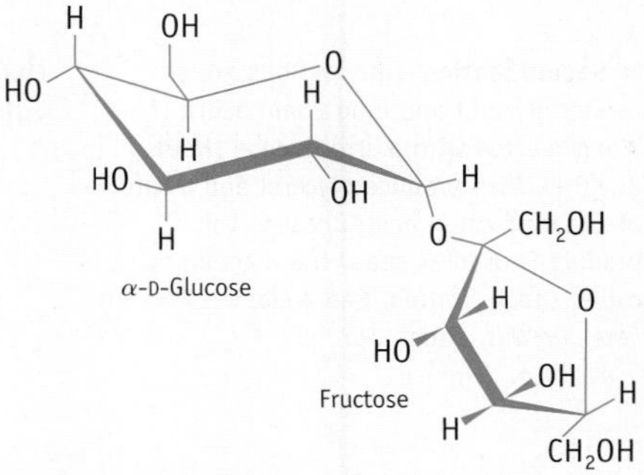

The structure of sucrose. Sucrose is formed from α-D-glucose and fructose. An ether linkage is formed by loss of H_2O from two —OH groups.

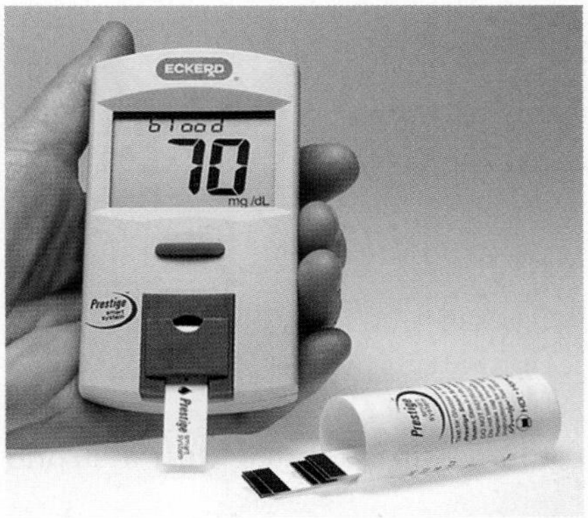

Charles D. Winters

Home test for glucose.

When a carboxylic acid and an alcohol react to form an ester, the OR group of the alcohol ends up as part of the ester (as shown above). This fact is known because of isotope labeling experiments. If the reaction is run using an alcohol in which the alcohol oxygen is ^{18}O, all of the ^{18}O ends up in the ester molecule.

Table 10.11 lists a few common esters and the acid and alcohol from which they are formed. The two-part name of an ester is given by (1) the name of the hydrocarbon group from the alcohol and (2) the name of the carboxylate group derived from the acid name by replacing "-ic" with "-ate." For example, ethanol (commonly called ethyl alcohol) and acetic acid combine to give the ester ethyl acetate.

An important reaction of esters is their **hydrolysis** (literally, reaction with water), a reaction that is the reverse of the formation of the ester. The reaction, generally

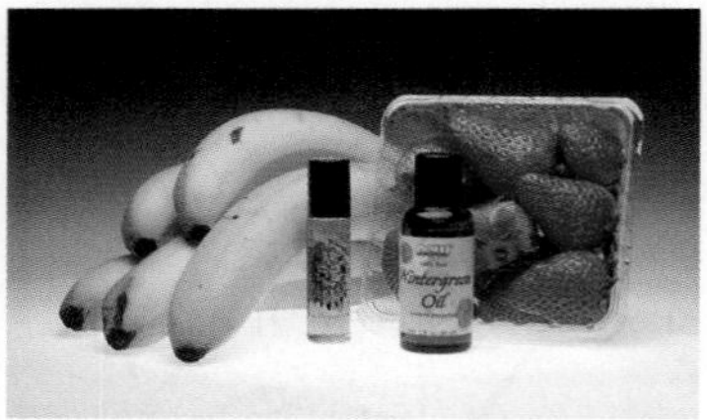

Charles D. Winters

Esters. Many fruits such as bananas and strawberries as well as consumer products (here, perfume and oil of wintergreen) contain esters.

TABLE 10.11 Some Acids, Alcohols, and Their Esters

Acid	Alcohol	Ester	Odor of Ester
CH_3CO_2H acetic acid	$CH_3\overset{CH_3}{\overset{\vert}{C}}HCH_2CH_2OH$ 3-methyl-1-butanol	$CH_3\overset{O}{\overset{\Vert}{C}}OCH_2CH_2\overset{CH_3}{\overset{\vert}{C}}HCH_3$ 3-methylbutyl acetate	banana
$CH_3CH_2CH_2CO_2H$ butanoic acid	$CH_3CH_2CH_2CH_2OH$ 1-butanol	$CH_3CH_2CH_2\overset{O}{\overset{\Vert}{C}}OCH_2CH_2CH_2CH_3$ butyl butanoate	pineapple
$CH_3CH_2CH_2CO_2H$ butanoic acid	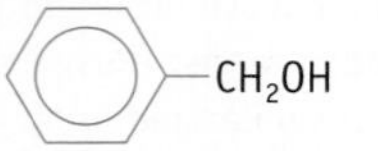 benzyl alcohol	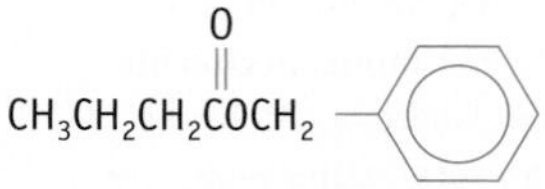 benzyl butanoate	rose

■ **Saponification** Fats and oils are esters of glycerol and long-chain acids. When reacted with a strong base (NaOH or KOH), they produce glycerol and a salt of the long-chain acid. Because this product is used as soap, the reaction is called *saponification*. See *A Closer Look: Fats and Oils,* page 476.

done in the presence of a base such as NaOH, produces the alcohol and a sodium salt of the carboxylic acid:

$$\underset{\text{ester}}{R\overset{O}{\overset{\Vert}{C}}OR'} + NaOH \xrightarrow[\text{in water}]{\text{heat}} \underset{\text{carboxylate salt}}{R\overset{O}{\overset{\Vert}{C}}O^-Na^+} + \underset{\text{alcohol}}{R'OH}$$

$$\underset{\text{ethyl acetate}}{CH_3\overset{O}{\overset{\Vert}{C}}OCH_2CH_3} + NaOH \xrightarrow[\text{in water}]{\text{heat}} \underset{\text{sodium acetate}}{CH_3\overset{O}{\overset{\Vert}{C}}O^-Na^+} + \underset{\text{ethanol}}{CH_3CH_2OH}$$

The carboxylic acid can be recovered if the sodium salt is treated with a strong acid such as HCl:

$$\underset{\text{sodium acetate}}{CH_3\overset{O}{\overset{\Vert}{C}}O^-Na^+(aq)} + HCl(aq) \longrightarrow \underset{\text{acetic acid}}{CH_3\overset{O}{\overset{\Vert}{C}}OH(aq)} + NaCl(aq)$$

Unlike the acids from which they are derived, esters often have pleasant odors (see Table 10.11). Typical examples are methyl salicylate, or "oil of wintergreen," and benzyl acetate. Methyl salicylate is derived from salicylic acid, the parent compound of aspirin.

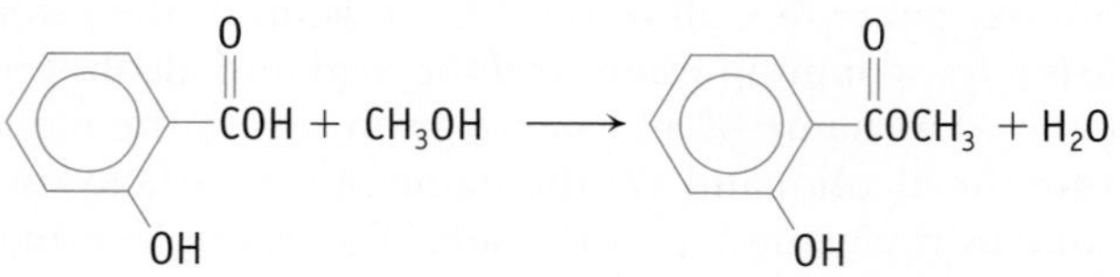

salicylic acid   methanol   methyl salicylate, oil of wintergreen

Benzyl acetate, the active component of "oil of jasmine," is formed from benzyl alcohol ($C_6H_5CH_2OH$) and acetic acid. The chemicals are inexpensive, so synthetic jasmine is a common fragrance in less-expensive perfumes and toiletries.

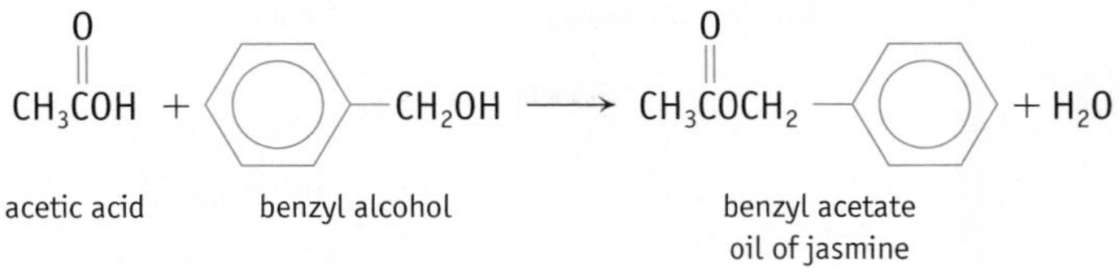

EXERCISE 10.9 Esters

Draw the structure, and name the ester formed from each of the following reactions:

(a) propanoic acid and methanol

(b) butanoic acid and 1-butanol

(c) hexanoic acid and ethanol

EXERCISE 10.10 Esters

Draw the structure, and name the acid and alcohol from which the following esters are derived:

(a) propyl acetate

(b) 3-methyl-1-pentyl benzoate

(c) ethyl salicylate

Amides

An acid and an alcohol react by loss of water to form an ester. In a similar manner, another class of organic compounds—amides—form when an acid reacts with an amine, again with loss of water.

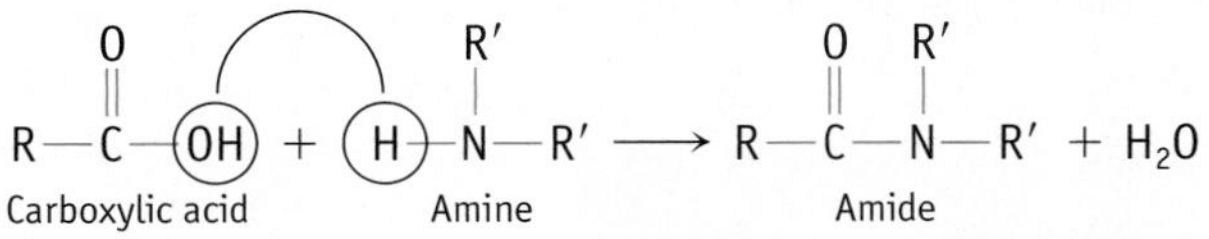

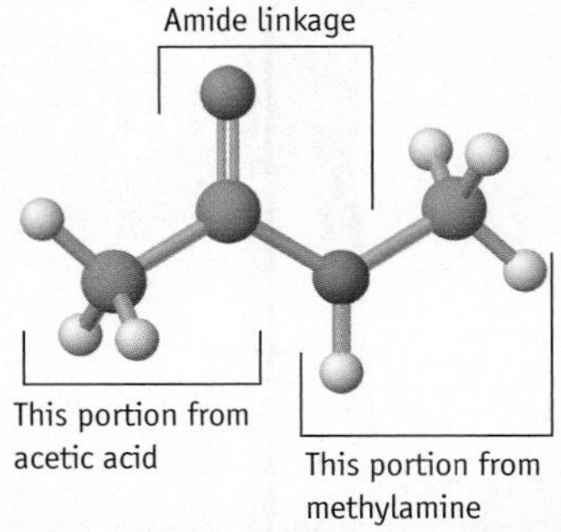

An amide, *N*-methylacetamide. The *N*-methyl portion of the name derives from the amine portion of the molecule, where the *N* indicates that the methyl group is attached to the nitrogen atom. The "-acet" portion of the name indicates the acid on which the amide is based.

Amides have an organic group and an amino group ($—NH_2$, $—NHR'$, or $—NR'R$) attached to the carbonyl group.

The C atom involved in the amide bond has three bonded groups and no lone pairs around it. We would predict it should be sp^2 hybridized with trigonal-planar geometry and bond angles of approximately 120°—and this is what is found. However, the structure of the amide group offers a surprise. The N atom is also observed to have trigonal-planar geometry with bonds to three attached atoms at 120°. Because the amide nitrogen is surrounded by four pairs of electrons, we would have predicted the N atom would have sp^3 hybridization and bond angles of about 109°.

A Closer Look | Fats and Oils

Fats and oils are among the many compounds found in plants and animal tissues. In the body, these substances serve several functions, a primary one being the storage of energy.

Fats (solids) and oils (liquids) are triesters formed from glycerol (1,2,3-propanetriol) and three carboxylic acids that can be the same or different.

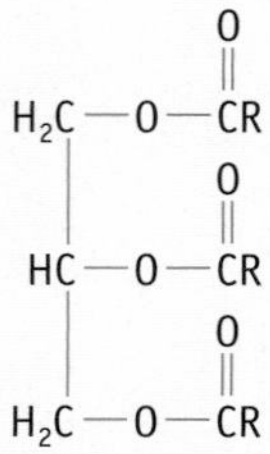

The carboxylic acids in fats and oils, known as *fatty acids*, have a lengthy carbon chain, usually containing between 12 and 18 carbon atoms. The hydrocarbon chains can be saturated or may include one or more double bonds. The latter are referred to as monounsaturated or polyunsaturated, depending on the number of double bonds. Saturated compounds are more common in animal products, while unsaturated fats and oils are more common in plants.

Charles D. Winters

About 94% of the fatty acids in olive oil are monounsaturated. The major fatty acid is oleic acid.

Common Fatty Acids

Name	Number of C Atoms	Formula
Saturated Acids		
lauric	C_{12}	$CH_3(CH_2)_{10}CO_2H$
myristic	C_{14}	$CH_3(CH_2)_{12}CO_2H$
palmitic	C_{16}	$CH_3(CH_2)_{14}CO_2H$
stearic	C_{18}	$CH_3(CH_2)_{16}CO_2H$
Unsaturated Acid		
oleic	C_{18}	$CH_3(CH_2)_7CH{=}CH(CH_2)_7CO_2H$

In general, fats containing saturated fatty acids are solids, and those containing unsaturated fatty acids are liquids at room temperature. The difference in melting point relates to the molecular structure. With only single bonds linking carbon atoms in saturated fatty acids, the hydrocarbon group is flexible, allowing the molecules to pack more closely together. The double bonds in unsaturated fats introduce kinks that make the hydrocarbon group less flexible; consequently, the molecules pack less tightly together.

Food companies often hydrogenate vegetable oils to reduce unsaturation. The chemical rationale is that double bonds are reactive and unsaturated compounds are more susceptible to oxidation, which results in unpleasant odors. There are also aesthetic reasons for this practice. Food processors often want solid fats to improve the quality and appearance of the food. If liquid vegetable oil is used in a cake icing, for example, the icing may slide off the cake.

©Randy Green/Taxi/Getty Images

Polar bear fat. Polar bears feed primarily on seal blubber and build up a huge fat reserve during winter. During summer, they maintain normal activity but eat nothing, relying entirely on body fat for sustenance. A polar bear will burn about 1 to 1.5 kg of fat per day.

The conditions under which hydrogenation occurs can also lead to the isomerization of an unsaturated fat to the *trans* configuration. Such "trans-fats" in the diet have been linked to coronary heart disease.

Like other esters, fats and oils can undergo hydrolysis. This process is catalyzed by enzymes in the body. In industry, hydrolysis is carried out using aqueous NaOH or KOH to produce a mixture of glycerol and the sodium salts of the fatty acids. This reaction is called *saponification*, a term meaning "soap making."

Glyceryl stearate, a fat
R = $—(CH_2)_{16}CH_3$

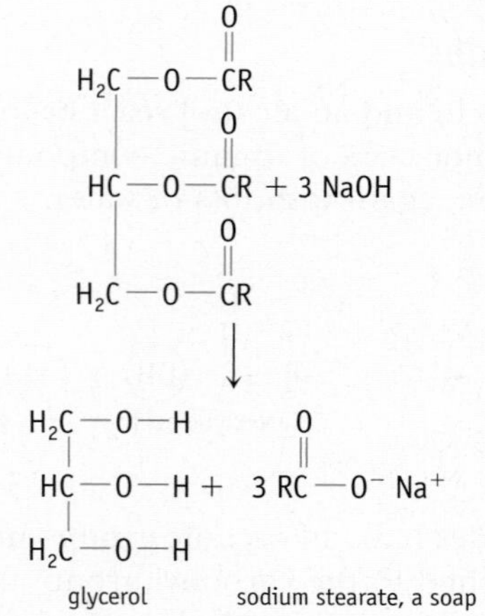

Simple soaps are sodium salts of fatty acids. The anion in these compounds has an ionic end (the carboxylate group) and a nonpolar end (the large hydrocarbon tail). The ionic end allows these molecules to interact with water, and the nonpolar end enables them to mix with oily and greasy substances to form an emulsion that can be washed away with water.

Based on the observed geometry of the amide N atom, the atom is assigned sp^2 hybridization. To explain the observed angle and to rationalize sp^2 hybridization, we can introduce a second resonance form of the amide.

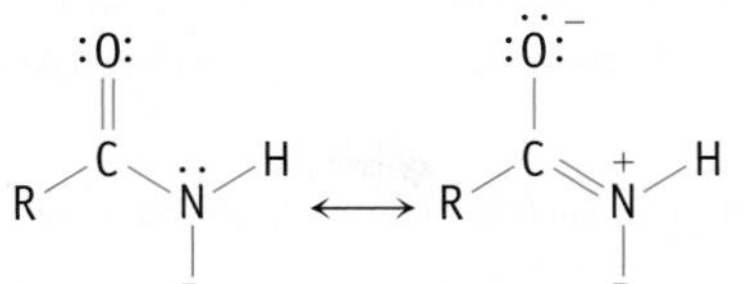

Form B contains a C=N double bond, and the O and N atoms have negative and positive charges, respectively. The N atom can be assigned sp^2 hybridization, and the π bond in B arises from overlap of *p* orbitals on C and N.

The existence of a second resonance structure for an amide link explains why the carbon–nitrogen bond is relatively short, about 132 pm, a value between that of a C—N single bond (149 pm) and a C=N double bond (127 pm). In addition, restricted rotation occurs around the C=N bond, making it possible for isomeric species to exist if the two groups bonded to N are different.

The amide grouping is particularly important in some synthetic polymers (Section 10.5) and in proteins (pages 496–513), where it is referred to as a *peptide* link. The compound *N*-acetyl-*p*-aminophenol, an analgesic known by the generic name acetaminophen and sold under the brand names Tylenol, Datril, and Momentum, among others, is another amide. Use of this compound as an analgesic was apparently discovered by accident when a common organic compound called acetanilide (like acetaminophen but without the —OH group) was mistakenly put into a prescription for a patient. Acetanilide acts as an analgesic, but it can be toxic. An —OH group *para* to the amide group makes the compound nontoxic, an interesting example of how a seemingly small structural difference affects chemical function.

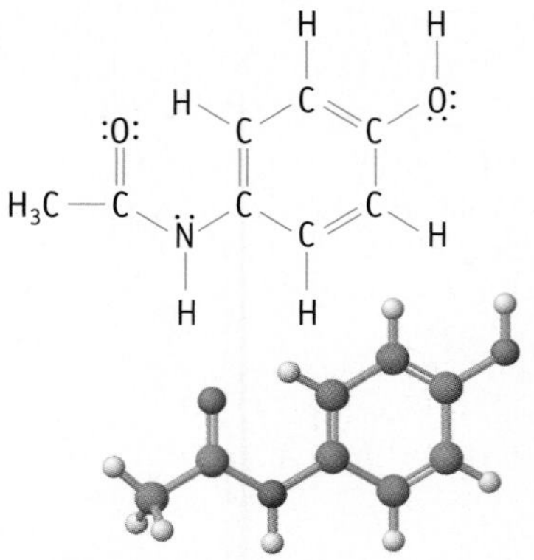

Acetaminophen, *N*-acetyl-*p*-aminophenol. This analgesic is an amide. It is used in over-the-counter painkillers such as Tylenol.

■ **Amides, Peptides, and Proteins** When amino acids combine, they form amide or peptide links. Polymers of amino acids are proteins. For more on amino acids and proteins, see *The Chemistry of Life: Biochemistry,* pages 496–513.

Chemistry Now™

Sign in at **www.thomsonedu.com/login** and go to Chapter 10 Contents to see Screen 10.5 for a description of the **types of organic functional groups** and for tutorials on **their structures, bonding, and chemistry.**

■ EXAMPLE 10.7 Functional Group Chemistry

Problem

(a) Name the product of the reaction between ethylene and HCl.

(b) Draw the structure of the product of the reaction between propanoic acid and 1-propanol. What is the systematic name of the reaction product, and what functional group does it contain?

(c) What is the result of reacting 2-butanol with an oxidizing agent? Give the name, and draw the structure of the reaction product.

Strategy Ethylene is an alkene (page 453); propanoic acid is a carboxylic acid (page 471); and 2-butanol is an alcohol (page 462). Consult the discussion regarding their chemistry.

Solution

(a) HCl will add to the double bond of ethylene to produce chloroethane.

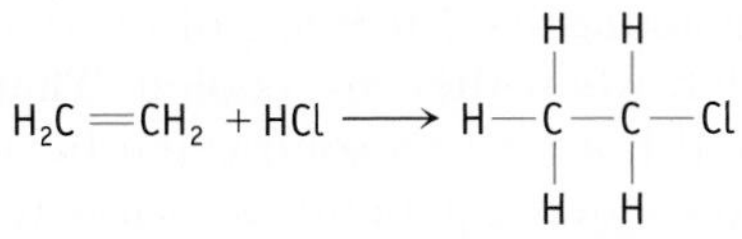

ethylene chloroethane

(b) Carboxylic acids such as propanoic acid react with alcohols to give esters.

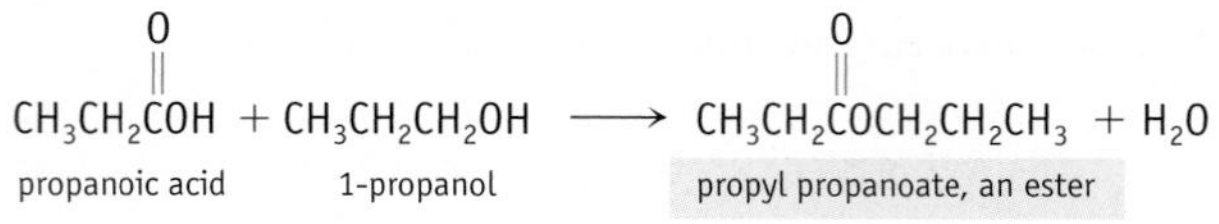

(c) 2-Butanol is a secondary alcohol. Such alcohols are oxidized to ketones.

EXERCISE 10.11 Functional Groups

(a) Name each of the following compounds and its functional group.

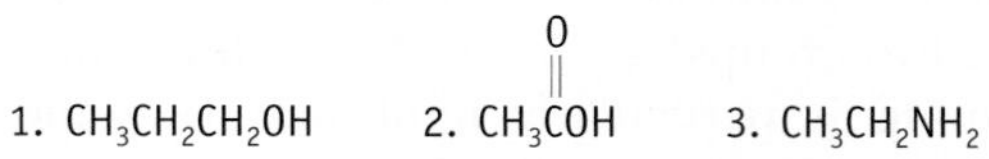

(b) Name the product from the reaction of compounds 1 and 2.

(c) What is the name and structure of the product from the oxidation of 1?

(d) What compound could result from combining compounds 2 and 3?

(e) What is the result of adding an acid (say HCl) to compound 3?

10.5 Polymers

We now turn to the very large molecules known as polymers. These can be either synthetic materials or naturally occurring substances such as proteins or nucleic acids. Although these materials have widely varying compositions, their structures and properties are understandable, based on the principles developed for small molecules.

Classifying Polymers

The word *polymer* means "many parts" (from the Greek, *poly* and *meros*). **Polymers** are giant molecules made by chemically joining many small molecules called **monomers.** Polymer molar masses range from thousands to millions.

Extensive use of synthetic polymers is a fairly recent development. A few synthetic polymers (Bakelite, rayon, and celluloid) were made early in the 20th century, but most of the products with which you are familiar originated in the last 50 years. By 1976, synthetic polymers outstripped steel as the most widely used materials in the United States. The average production of synthetic polymers in the United States is approximately 150 kg per person annually.

■ **Biochemical Polymers** Polymer chemistry extends to biochemistry, where chemists study proteins and other large molecules. See *The Chemistry of Life: Biochemistry,* pages 496–513.

The polymer industry classifies polymers in several different ways. One is their response to heating. **Thermoplastics** (such as polyethylene) soften and flow when they are heated and harden when they are cooled. **Thermosetting plastics** (such as Formica) are initially soft but set to a solid when heated and cannot be resoftened. Another classification scheme depends on the end use of the polymer—for example, plastics, fibers, elastomers, coatings, and adhesives.

Case Study — Biodiesel—A Fuel for the Future?

Biodiesel, promoted as an alternative to petroleum-based fuels used in diesel engines, is made from plant and animal oils. In the past 7 years, there has been a spectacular increase in its production and use, from under 1 million gallons in 1999 to 75 million gallons in 2005. But what is biodiesel?

© Adrian Dennis/AFP/Getty Images

Biodiesel, a mixture of long-chain esters of fatty acids.

Chemically, biodiesel is a mixture of esters of long-chain fatty acids. It is prepared from plant and animal fats and oils by *trans-esterification*. This is a reaction between an ester and an alcohol in which the —OR″ on the alcohol exchanges with the OR′ group of the ester:

$$RCO_2R' + R''OH \rightarrow RCO_2R'' + R'OH$$

Recall that fats and oils are esters (page 476), derivatives of glycerol and high-molar-mass organic acids (fatty acids). Their reaction with methanol (in the presence of a catalyst to speed up the reaction) produces a mixture of the methyl esters of long chain fatty acids and glycerol.

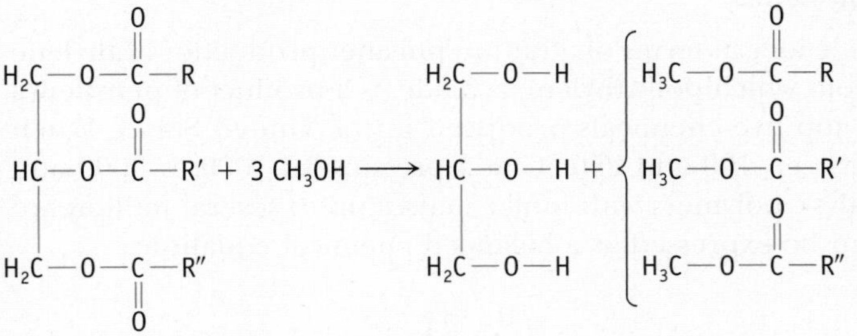

Glycerol, a by-product of the reaction, is a valuable commodity for the health care products industry, so it is separated and sold. The mixture of esters that remains can be used directly as a fuel in existing diesel engines, or it can be blended with petroleum products. In the latter case, the fuel mixture is identified by a designation such as B20 (B = biodiesel, 20 refers to 20% by volume.) The fuel has the advantage of being clean burning with fewer environmental problems associated with exhaust gases. In particular, there are no SO_2 emissions, one of the common problems associated with petroleum-based diesel fuels.

Proponents of the use of biodiesel note that biodiesel is produced from renewable resources, in contrast to petroleum. Critics point out, however, that growing crops for this purpose brings two problems. The first is that if crops are grown for biodiesel production, this has a negative effect on food supply. The case for biodiesel would be improved significantly, however, if it were possible to convert agricultural waste (corn stalks, for example) into a biofuel. Scientists are now actively trying to do this.

It is also pointed out that it would be impossible to grow enough crops to supply the raw materials to replace all petroleum-based diesel fuel—there isn't enough land available. Economics is also in the picture: Biodiesel is currently more expensive to produce; it is presently competitive only due to government subsidies.

Trans-esterification might seem like another kind of reaction, but it is actually closely related to chemistry we have already seen, the hydrolysis of an ester.

Ester hydrolysis:

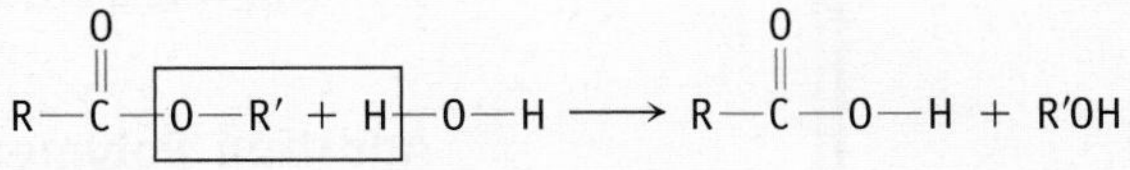

Trans-esterification:

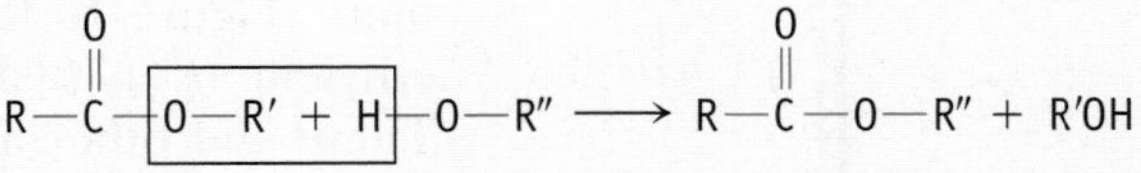

In both reactions, the OR′ group on the ester combines with hydrogen of second reagent (water or alcohol) as shown. In drawing this analogy, it is useful to recognize that there are other similarities in the chemistry of alcohols and water. For example, both can be protonated with strong acids (giving H_3O^+ and ROH_2^+) and deprotonated by strong bases (giving OH^- and OR^-).

Questions:

1. *Write a balanced chemical equation for the reaction that occurs when methyl myristate, $C_{13}H_{27}CO_2CH_3(\ell)$, is burned, forming $CO_2(g)$ and $H_2O(g)$.*
2. *Using enthalpy of formation data, calculate the standard enthalpy change per mole in the oxidation of methyl myristate ($\Delta_f H^\circ = -771.0$ kJ/mol).*
3. *Which compound, methyl myristate ($C_{15}H_{30}O_2$) or hexadecane ($C_{16}H_{34}$, one of many hydrocarbons in petroleum based diesel fuel) is predicted to provide the greater energy per mole? Per liter? ($\Delta_f H^\circ$ for $C_{16}H_{34} = -456.1$ kJ/mol) [d(methyl myristate) = 0.86 g/mL, and d($C_{16}H_{34}$) = 0.77 g/mL]*

Answers to these questions are in Appendix Q.

Charles D. Winters

(a) (b) (c)

FIGURE 10.12 Common polymer-based consumer products. (a) Packaging materials from high-density polyethylene; (b) from polystyrene; and (c) from polyvinyl chloride. Recycling information is provided on most plastics (often molded into the bottom of bottles). High-density polyethylene is designated with a "2" inside a triangular symbol and the letters "HDPE." PVC is designated with a "3" inside a triangular symbol with the letter "V" below.

A more chemically oriented approach to polymer classification is based on the method of synthesis. **Addition polymers** are made by directly adding monomer units together. **Condensation polymers** are made by combining monomer units and splitting out a small molecule, often water.

Addition Polymers

Polyethylene, polystyrene, and polyvinyl chloride (PVC) are common addition polymers (Figure 10.12). They are built by "adding together" simple alkenes such as ethylene ($CH_2{=}CH_2$), styrene ($C_6H_5CH{=}CH_2$), and vinyl chloride ($CH_2{=}CHCl$). These and other addition polymers (Table 10.12), all derived from alkenes, have widely varying properties and uses.

Chemistry Now™

Sign in at **www.thomsonedu.com/login** and go to Chapter 10 Contents to see Screen 10.9 for an animation of **addition polymerization.**

Polyethylene and Other Polyolefins

Polyethylene is by far the leader in terms of addition polymer production. Ethylene (C_2H_4), the monomer from which polyethylene is made, is a product of petroleum refining and one of the top five chemicals produced in the United States. When ethylene is heated to between 100 and 250 °C at a pressure of 1000 to 3000 atm in the presence of a catalyst, polymers with molar masses up to several million are formed. The reaction can be expressed as a balanced chemical equation:

$$\underset{\text{ethylene}}{n\,H_2C{=}CH_2} \longrightarrow \underset{\text{polyethylene}}{\left(\!-\!\!\begin{array}{cc} H & H \\ | & | \\ C\!-\! & \!C \\ | & | \\ H & H \end{array}\!\!-\!\right)_n}$$

TABLE 10.12 Ethylene Derivatives That Undergo Addition Polymerization

Formula	Monomer Common Name	Polymer Name (Trade Names)	Uses	U.S. Polymer Production (Metric tons/year)*
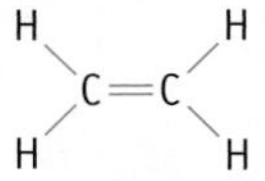	ethylene	polyethylene (polythene)	squeeze bottles, bags, films, toys and molded objects, electric insulation	7 million
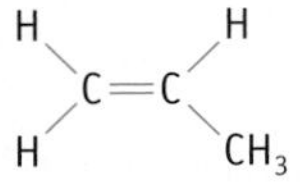	propylene	polypropylene (Vectra, Herculon)	bottles, films, indoor-outdoor carpets	1.2 million
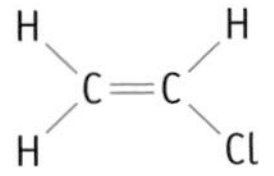	vinyl chloride	polyvinyl chloride (PVC)	floor tile, raincoats, pipe	1.6 million
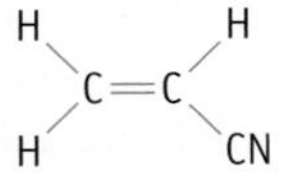	acrylonitrile	polyacrylonitrile (Orlan, Acrilan)	rugs, fabrics	0.5 million
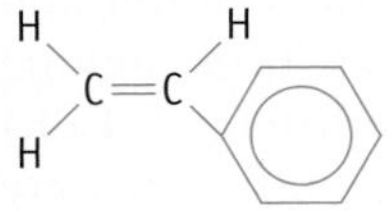	styrene	polystyrene (Styrofoam, Styron)	food and drink coolers, building material insulation	0.9 million
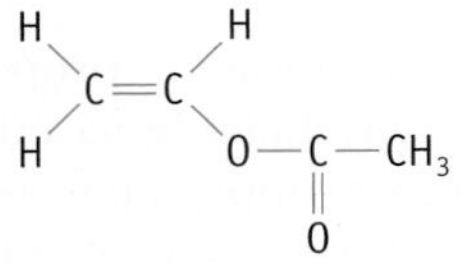	vinyl acetate	polyvinyl acetate (PVA)	latex paint, adhesives, textile coatings	200,000
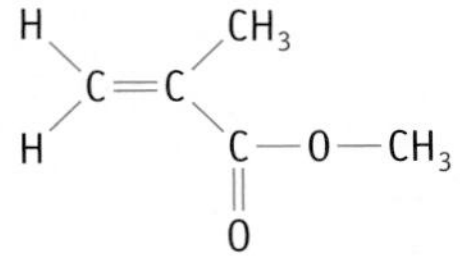	methyl methacrylate	polymethyl methacrylate (Plexiglass, Lucite)	high-quality transparent objects, latex paints, contact lenses	200,000
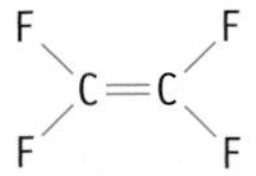	tetrafluoroethylene	polytetrafluoroethylene (Teflon)	gaskets, insulation, bearings, pan coatings	6,000

* One metric ton = 1000 kg.

The abbreviated formula of the reaction product, $\leftarrow CH_2CH_2 \rightarrow_n$, shows that polyethylene is a chain of carbon atoms, each bearing two hydrogens. The chain length for polyethylene can be very long. A polymer with a molar mass of 1 million would contain almost 36,000 ethylene molecules linked together.

Polyethylene formed under various pressures and catalytic conditions has different properties, as a result of different molecular structures. For example, when chromium oxide is used as a catalyst, the product is almost exclusively a linear chain (Figure 10.13a). If ethylene is heated to 230 °C at high pressure, however, irregular branching occurs. Still other conditions lead to cross-linked polyethylene, in which different chains are linked together (Figures 10.13b and c).

The high–molar-mass chains of linear polyethylene pack closely together and result in a material with a density of 0.97 g/cm³. This material, referred to as

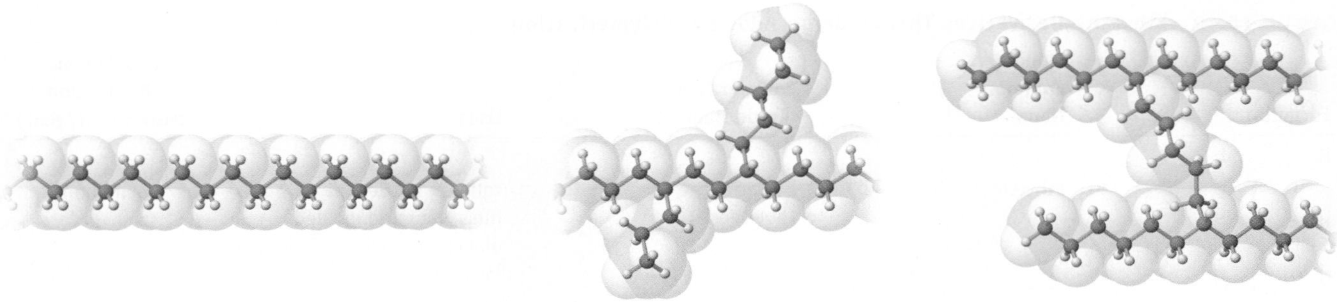

(a) (b) (c)

FIGURE 10.13 Polyethylene. (a) The linear form, high-density polyethylene (HDPE). (b) Branched chains occur in low-density polyethylene (LDPE). (c) Cross-linked polyethylene (CLPE).

high-density polyethylene (HDPE), is hard and tough, which makes it suitable for items such as milk bottles. If the polyethylene chain contains branches, however, the chains cannot pack as closely together, and a lower-density material (0.92 g/cm^3) known as low-density polyethylene (LDPE) results. This material is softer and more flexible than HDPE. It is used in plastic wrap and sandwich bags, among other things. Linking up the polymer chains in cross-linked polyethylene (CLPE) causes the material to be even more rigid and inflexible. Plastic bottle caps are often made of CLPE.

Polymers formed from substituted ethylenes ($CH_2{=}CHX$) have a range of properties and uses (see Table 10.12). Sometimes, the properties are predictable based on the molecule's structure. Polymers without polar substituent groups, such as polystyrene, often dissolve in organic solvents, a property useful for some types of fabrication (Figure 10.14).

Polymers based on substituted ethylenes, $H_2C{=}CHX$

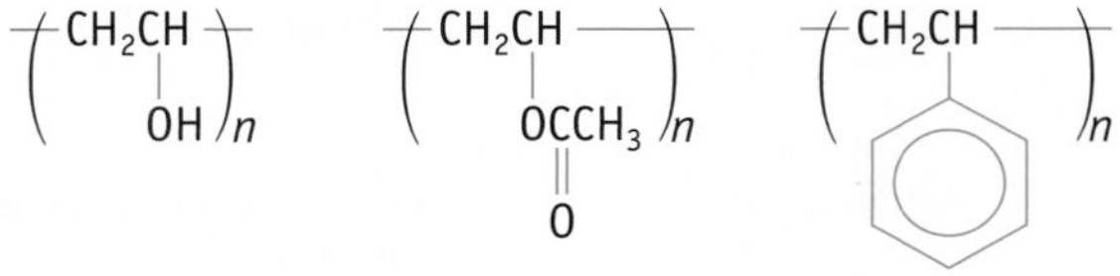

polyvinyl alcohol　　polyvinyl acetate　　polystyrene

Christopher Springmann/Corbisstockmarket.com

Polyethylene film. The polymer film is produced by extruding the molten plastic through a ring-like gap and inflating the film like a balloon.

Polyvinyl alcohol is a polymer with little affinity for nonpolar solvents but an affinity for water, which is not surprising, based on the large number of polar OH groups (Figure 10.15). Vinyl alcohol itself is not a stable compound (it isomerizes to acetaldehyde CH_3CHO), so polyvinyl alcohol cannot be made from this compound. Instead, it is made by hydrolyzing the ester groups in polyvinyl acetate.

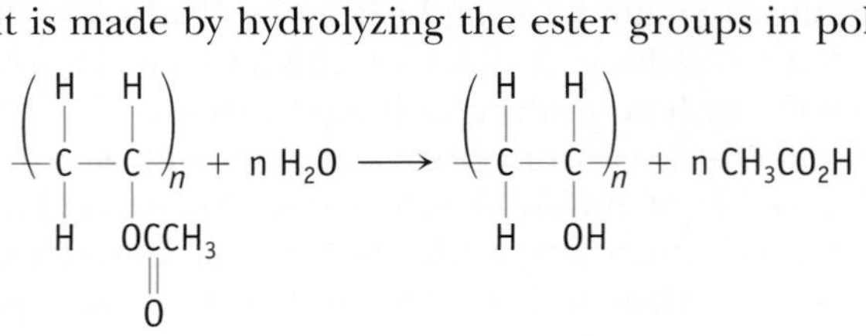

Solubility in water or organic solvents can be a liability for polymers. The many uses of polytetrafluoroethylene [Teflon, $(CF_2CF_2)_n$] stem from the fact that it does not interact with water or organic solvents.

Polystyrene, with $n = 5700$, is a clear, hard, colorless solid that can be molded easily at 250 °C. You are probably more familiar with the very light, foam-like mate-

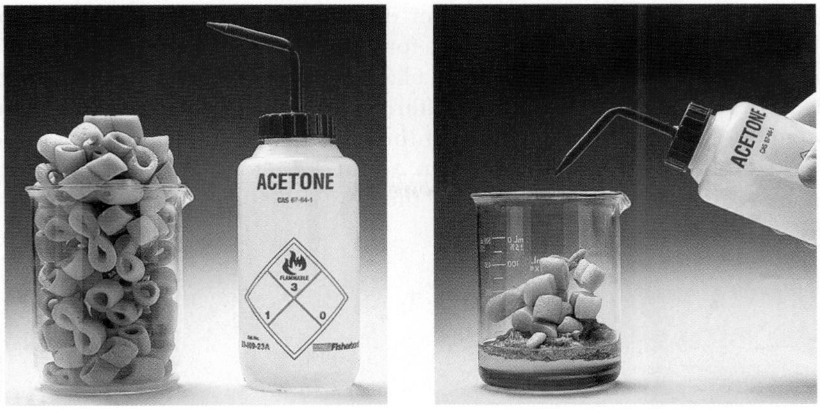

Charles D. Winters

(a) (b)

FIGURE 10.14 Polystyrene. (a) The polymer is a clear, hard, colorless solid, but it may be more familiar as a light, foam-like material called Styrofoam. (b) Styrofoam has no polar groups and thus dissolves well in organic solvents such as acetone. See also Figure 10.12b.

rial known as Styrofoam that is used widely for food and beverage containers and for home insulation (Figure 10.14). Styrofoam is produced by a process called "expansion molding." Polystyrene beads containing 4% to 7% of a low-boiling liquid like pentane are placed in a mold and heated with steam or hot air. Heat causes the solvent to vaporize, creating a foam in the molten polymer that expands to fill the shape of the mold.

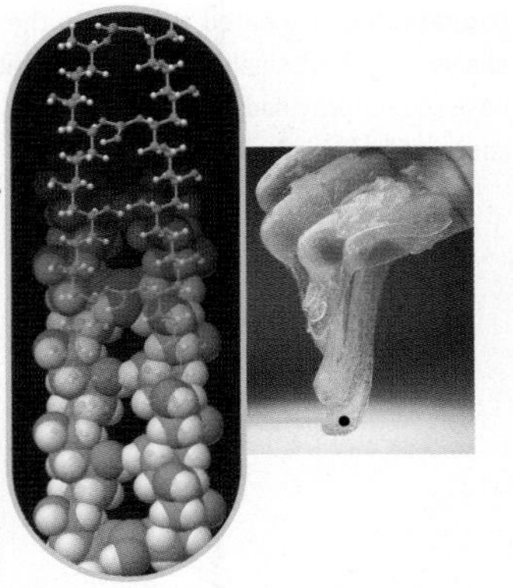

Photo, Charles D. Winters; model, S.M. Young

FIGURE 10.15 Slime. When boric acid, $B(OH)_3$, is added to an aqueous suspension of polyvinyl alcohol, $(CH_2CHOH)_n$, the mixture becomes very viscous because boric acid reacts with the —OH groups on the polymer chain, causing cross-linking to occur. (The model shows an idealized structure of a portion of the polymer.)

Natural and Synthetic Rubber

Natural rubber was first introduced in Europe in 1740, but it remained a curiosity until 1823, when Charles Macintosh invented a way of using it to waterproof cotton cloth. The mackintosh, as rain coats are still sometimes called, became popular despite major problems: Natural rubber is notably weak and is soft and tacky when warm but brittle at low temperatures. In 1839, after 5 years of research on natural rubber, the American inventor Charles Goodyear (1800–1860) discovered that heating gum rubber with sulfur produces a material that is elastic, water-repellent, resilient, and no longer sticky.

Rubber is a naturally occurring polymer, the monomers of which are molecules of 2-methyl-1,3-butadiene, commonly called *isoprene.* In natural rubber, isoprene monomers are linked together through carbon atoms 1 and 4—that is, through the end carbon atoms of the C_4 chain (Figure 10.16). This leaves a double bond between carbon atoms 2 and 3. In natural rubber, these double bonds have a *cis* configuration.

In vulcanized rubber, the material that Goodyear discovered, the polymer chains of natural rubber are cross-linked by short chains of sulfur atoms. Cross-linking helps to align the polymer chains, so the material does not undergo a permanent change when stretched and it springs back when the stress is removed. Substances that behave this way are called **elastomers**.

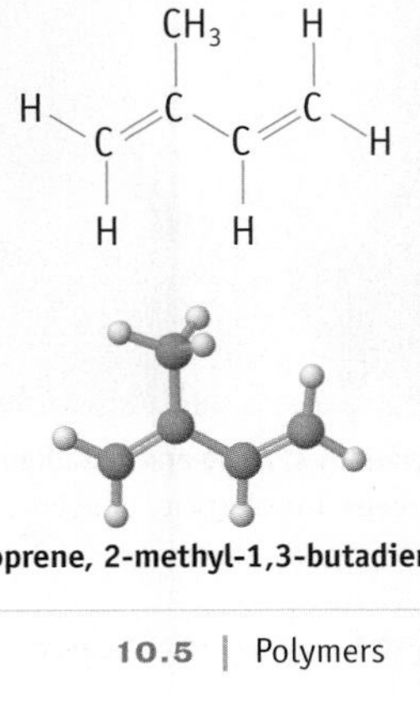

Isoprene, 2-methyl-1,3-butadiene.

With a knowledge of the composition and structure of natural rubber, chemists began searching for ways to make synthetic rubber. When they first tried to make the polymer by linking isoprene monomers together, however, what they made was sticky and useless. The problem was that synthesis procedures gave a mixture of *cis* and *trans* polyisoprene. In 1955, however, chemists at the Goodyear and Firestone companies discovered special catalysts to prepare the all-*cis* polymer. This synthetic material, which was structurally identical to natural rubber, is now manufactured cheaply. In fact, more than 8.0×10^8 kg of synthetic polyisoprene is produced annually in the United States.

© Kevin Schafer

FIGURE 10.16 Natural rubber. The sap that comes from the rubber tree is a natural polymer of isoprene. All the linkages in the carbon chain are *cis*. When natural rubber is heated strongly in the absence of air, it smells of isoprene. This observation provided a clue that rubber is composed of this building block.

Other kinds of polymers have further expanded the repertoire of elastomeric materials now available. Polybutadiene, for example, is currently used in the production of tires, hoses, and belts. Some elastomers, called **copolymers,** are formed by polymerization of two (or more) different monomers. A copolymer of styrene and butadiene, made with a 1:3 ratio of these raw materials, is the most important synthetic rubber now made; more than about 1 billion kg of styrene-butadiene rubber (SBR) is produced each year in the United States for making tires.

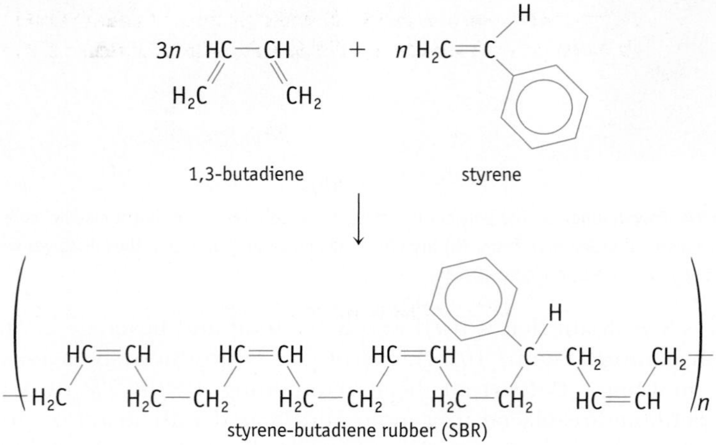

styrene-butadiene rubber (SBR)

And a little is left over each year to make bubble gum. The stretchiness of bubble gum once came from natural rubber, but SBR is now used to help you blow bubbles.

Chemistry Now™

Sign in at **www.thomsonedu.com/login** and go to Chapter 10 Contents to see Screen 10.11 for a self-study module on **the polymer used in bubble gum.**

Condensation Polymers

A chemical reaction in which two molecules react by splitting out, or eliminating, a small molecule is called a **condensation reaction.** The reaction of an alcohol with a carboxylic acid to give an ester is an example of a condensation reaction. One way to form a condensation polymer uses *two* different reactant molecules, each containing *two* functional groups. Another route uses a single molecule with two different functional groups. Commercial polyesters are made using both types of reactions.

Charles D. Winters

Copolymer of styrene and butadiene, SBR rubber. The elasticity of bubble gum comes from SBR rubber.

Chemistry Now™

Sign in at **www.thomsonedu.com/login** and go to Chapter 10 Contents to see Screen 10.10 to view an animation of **condensation polymerization** and to watch a video of **the synthesis of nylon.**

Polyesters

Terephthalic acid contains two carboxylic acid groups, and ethylene glycol contains two alcohol groups. When mixed, the acid and alcohol functional groups at both ends of these molecules can react to form ester linkages, splitting out water. The result is a polymer called polyethylene terephthalate (PET). The multiple ester linkages make this substance a **polyester.**

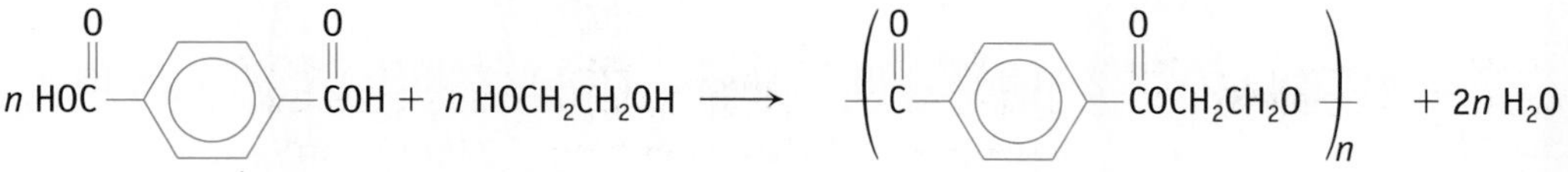

terephthalic acid — ethylene glycol — polyethylene terephthalate (PET), a polyester

Polyester textile fibers made from PET are marketed as Dacron and Terylene. The inert, nontoxic, noninflammatory, and non–blood-clotting properties of Dacron polymers make Dacron tubing an excellent substitute for human blood vessels in heart bypass operations, and Dacron sheets are sometimes used as temporary skin for burn victims. A polyester film, Mylar, has unusual strength and can be rolled into sheets one-thirtieth the thickness of a human hair. Magnetically coated Mylar films are used to make audio and video tapes (Figure 10.17).

There is considerable interest in another polyester, polylactic acid (PLA). Lactic acid contains both carboxylic acid and alcohol functional groups, so condensation between molecules of this monomer gives a polymer.

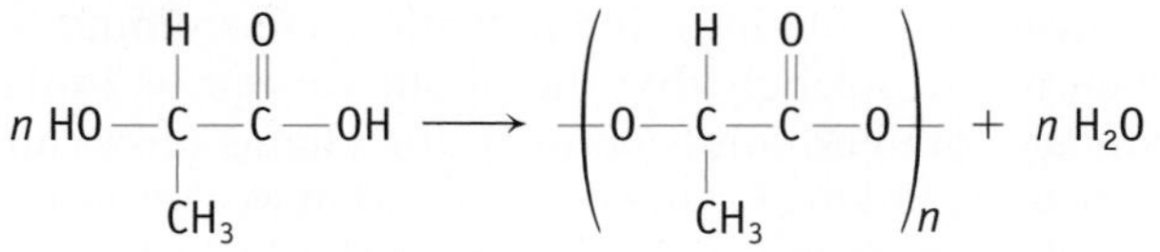

There is interest in polylactic acid for two reasons. First, the monomer used to make this polymer is obtained by biological fermentation of plant materials. (Most of the chemicals used in the manufacture of other types of polymers are derived from petroleum, and there is increased concern about the availability and cost of raw materials in the future.) Second, this polymer, which is currently being used in packaging material, is biodegradable, which has the potential to alleviate land-fill disposal problems.

Polyamides

In 1928, the DuPont Company embarked on a basic research program headed by Wallace Carothers (1896–1937). Carothers was interested in high–molar-mass compounds, such as rubbers, proteins, and resins. In 1935, his research yielded

Charles D. Winters

FIGURE 10.17 Polyesters. Polyethylene terephthalate is used to make clothing and soda bottles. The two students are wearing jackets made from recycled PET soda bottles. Mylar film, another polyester, is used to make recording tape as well as balloons. Because the film has very tiny pores, Mylar can be used for helium-filled balloons; the atoms of gaseous helium move through the pores in the film very slowly.

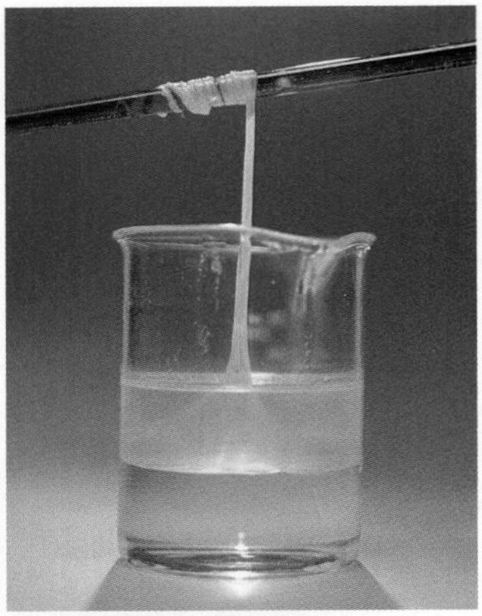

Charles D. Winters

Active Figure 10.18 Nylon-6,6. Hexamethylenediamine is dissolved in water (bottom layer), and adipoyl chloride (a derivative of adipic acid) is dissolved in hexane (top layer). The two compounds react at the interface between the layers to form nylon, which is being wound onto a stirring rod.

ChemistryNow™ Sign in at www.thomsonedu.com/login and go to the Chapter Contents menu to explore an interactive version of this figure accompanied by an exercise.

nylon-6,6 (Figure 10.18), a **polyamide** prepared from adipoyl chloride, a derivative of adipic acid, a diacid, and hexamethylenediamine, a diamine:

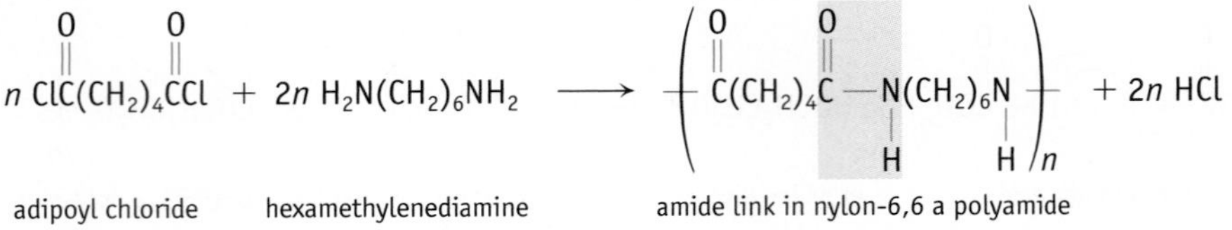

Nylon can be extruded easily into fibers that are stronger than natural fibers and chemically more inert. The discovery of nylon jolted the American textile industry at a critical time. Natural fibers were not meeting 20th-century needs. Silk was expensive and not durable; wool was scratchy; linen crushed easily; and cotton did not have a high-fashion image. Perhaps the most identifiable use for the new fiber was in nylon stockings. The first public sale of nylon hosiery took place on October 24, 1939, in Wilmington, Delaware (the site of DuPont's main office). This use of nylon in commercial products ended shortly thereafter, however, with the start of World War II. All nylon was diverted to making parachutes and other military gear. It was not until about 1952 that nylon reappeared in the consumer marketplace.

Figure 10.19 illustrates why nylon makes such a good fiber. To have good tensile strength (the ability to resist tearing), the polymer chains should be able to attract one another, albeit not so strongly that the plastic cannot be initially extended to form fibers. Ordinary covalent bonds between the chains (cross-linking) would be too strong. Instead, cross-linking occurs by a somewhat weaker intermolecular force called *hydrogen bonding* (▶ Section 12.2) between the hydrogens of N—H groups on one chain and the carbonyl oxygens on another chain. The polarities of the $N^{\delta-}$—$H^{\delta+}$ group and the $C^{\delta+}$=$O^{\delta-}$ group lead to attractive forces between the polymer chains of the desired magnitude.

EXAMPLE 10.8 Condensation Polymers

Problem What is the repeating unit of the condensation polymer obtained by combining $HO_2CCH_2CH_2CO_2H$ (succinic acid) and $H_2NCH_2CH_2NH_2$ (1,2-ethylenediamine)?

Strategy Recognize that the polymer will link the two monomer units through the amide linkage. The smallest repeating unit of the chain will contain two parts, one from the diacid and the other from the diamine.

Solution The repeating unit of this polyamide is

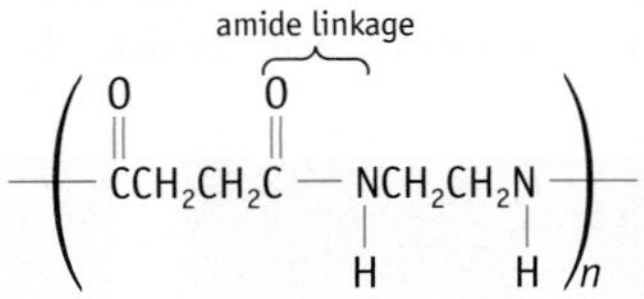

FIGURE 10.19 Hydrogen bonding between polyamide chains. Carbonyl oxygen atoms with a partial negative charge on one chain interact with an amine hydrogen with a partial positive charge on a neighboring chain. (This form of bonding is described in more detail in Section 12.3.)

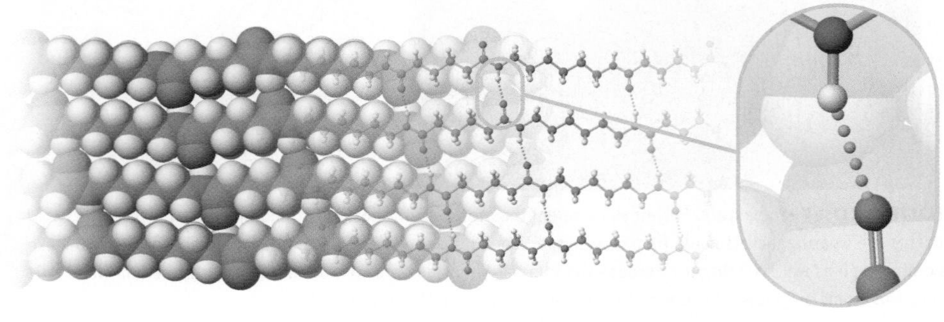

EXERCISE 10.12 Kevlar, a Condensation Polymer

A polymer that is now well known because of its use to construct sports equipment and bulletproof vests is Kevlar. This polymer has the formula shown below. Is this a condensation polymer or an addition polymer? What chemicals would be used to make this polymer? Write a balanced equation for the formation of Kevlar.

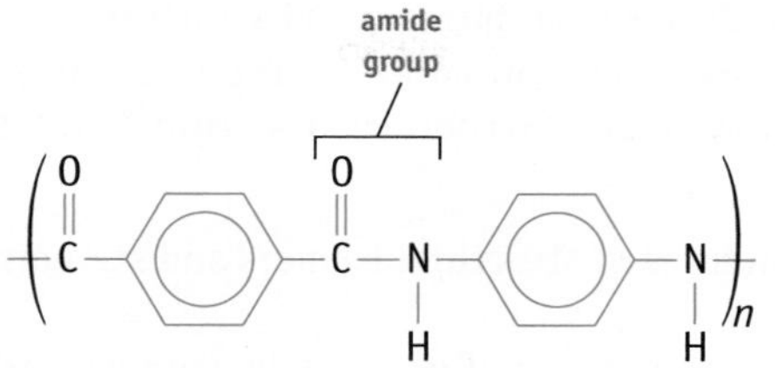

Chemical Perspectives

Super Diapers

Disposable diapers are a miracle of modern chemistry: Most of the materials used are synthetic polymers. The outer layer is mostly microporous polyethylene; it keeps the urine in but remains breathable. The inside layer is polypropylene, a material prized by winter-camping enthusiasts. It stays soft and dry while wicking moisture away from the skin. Sandwiched between these layers is powdered sodium polyacrylate combined with cellulose;

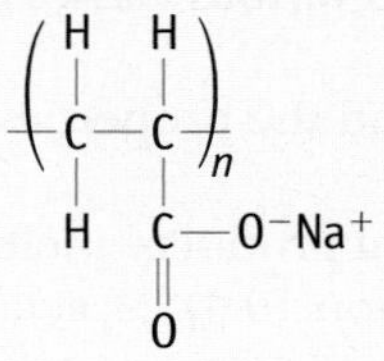

the latter is the only natural part of the materials used. The package is completed with elasticized hydrophobic polypropylene cuffs around the baby's thighs, and Velcro tabs hold the diaper on the baby.

The key ingredient in the diaper is the polyacrylate polymer filling. This substance can absorb up to 800 times its weight in water. When dry, the polymer has a carboxylate group associated with sodium ions. When placed in water, osmotic pressure causes water molecules to enter the polymer (because the ion concentration in the polymer is higher than in water; see Chapter 14). As water enters, the sodium ions dissociate from the polymer, and the polar water molecules are attracted to these positive ions and to the negative carboxylate groups of the polymer. At the same time, the negative carboxylate groups repel one another, forcing them apart and causing the polymer to unwind. Evidence for the unwinding of the polymer is seen as swelling of the diaper. In addition, because it contains so much water, the polymer becomes gel-like.

If the gelled polymer is put into a salt solution, water is attracted to the Na^+ and Cl^- ions and is drawn from the polymer. Thus, the polymer becomes solid once again. The diminished ability of sodium polyacrylate to absorb water in a salt solution is the reason that disposable diapers do not absorb urine as well as pure water.

These kinds of superabsorbent materials—sodium polyacrylate and a related material, polyacrylamide—are useful not only in diapers but also for cleaning up spills in hospitals, for protecting power and optical cables from moisture, for filtering water out of aviation gasoline, and for conditioning garden soil to retain water. You will also find them in the toy store as "gro-creatures."

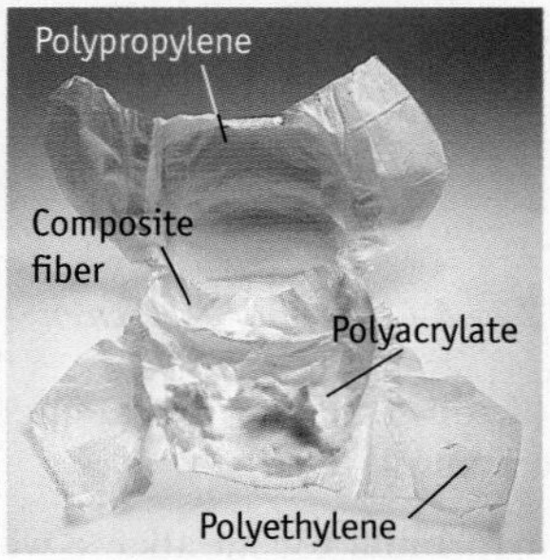

Charles D. Winters

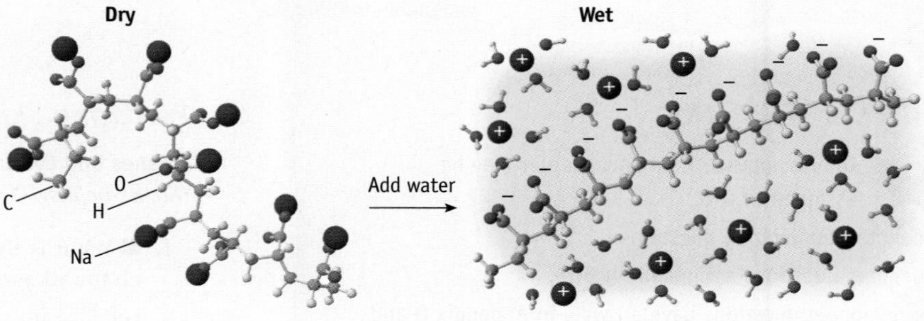

Chapter Goals Revisited

Chemistry Now™ Sign in at **www.thomsonedu.com/login** to:

- Assess your understanding with Study Questions in OWL keyed to each goal in the Goals and Homework menu for this chapter
- For quick review, download Go Chemistry mini-lecture flashcard modules (or purchase them at **www.ichapters.com**)
- Check your readiness for an exam by taking the Pre-Test and exploring the modules recommended in your Personalized Study plan.

Access **How Do I Solve It?** tutorials on how to approach problem solving using concepts in this chapter.

Now that you have studied this chapter, you should ask whether you have met the chapter goals. In particular, you should be able to:

Classify organic compounds based on formula and structure

a. Understand the factors that contribute to the large numbers of organic compounds and the wide array of structures (Section 10.1). **Study Question(s) assignable in OWL: 3.**

Recognize and draw structures of structural isomers and stereoisomers for carbon compounds

a. Recognize and draw structures of geometric isomers and optical isomers (Section 10.1). **Study Question(s) assignable in OWL: 11, 12, 15, 58.**

Name and draw structures of common organic compounds

a. Draw structural formulas, and name simple hydrocarbons, including alkanes, alkenes, alkynes, and aromatic compounds (Section 10.2). **Study Question(s) assignable in OWL: 1, 5, 7, 28, 67, 69, 70, 96; Go Chemistry Module 15.**

b. Identify possible isomers for a given formula (Section 10.2).

c. Name and draw structures of alcohols and amines (Section 10.3). **Study Question(s) assignable in OWL: 31, 32, 34.**

d. Name and draw structures of carbonyl compounds—aldehydes, ketones, acids, esters, and amides (Section 10.4). **Study Question(s) assignable in OWL: 38, 39, 40, 41, 43, 51.**

Know the common reactions of organic functional groups

a. This goal applies specifically to the reactions of alkenes, alcohols, amines, aldehydes and ketones, and carboxylic acids. **Study Question(s) assignable in OWL: 19, 21, 24, 46, 64, 76, 79, 81, 83–85, 90–92, 97.**

Relate properties to molecular structure

a. Describe the physical and chemical properties of the various classes of hydrocarbon compounds (Section 10.2).

b. Recognize the connection between the structures and the properties of alcohols (Section 10.3).

c. Know the structures and properties of several natural products, including carbohydrates (Section 10.4) and fats and oils (Section 10.4). **Study Question(s) assignable in OWL: 49, 50.**

Identify common polymers

a. Write equations for the formation of addition polymers and condensation polymers, and describe their structures (Section 10.5).

b. Relate properties of polymers to their structures (Section 10.5). **Study Question(s) assignable in OWL: 95.**

STUDY QUESTIONS

Online homework for this chapter may be assigned in OWL.

▲ denotes challenging questions.

■ denotes questions assignable in OWL.

Blue-numbered questions have answers in Appendix O and fully-worked solutions in the *Student Solutions Manual*.

Practicing Skills

Alkanes and Cycloalkanes

(See Examples 10.1 and 10.2 and ChemistryNow Screen 10.3.)

1. ■ What is the name of the straight (unbranched) chain alkane with the formula C_7H_{16}?

2. What is the molecular formula for an alkane with 12 carbon atoms?

3. ■ Which of the following is an alkane? Which could be a cycloalkane?
(a) C_2H_4 (c) $C_{14}H_{30}$
(b) C_5H_{10} (d) C_7H_8

4. Isooctane, 2,2,4-trimethylpentane, is one of the possible structural isomers with the formula C_8H_{18}. Draw the structure of this isomer, and draw and name structures of two other isomers of C_8H_{18} in which the longest carbon chain is five atoms.

5. ■ Give the systematic name for the following alkane:

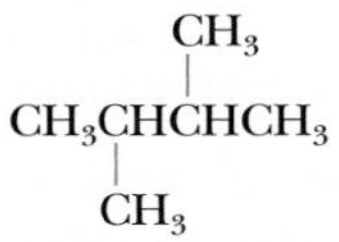

6. Give the systematic name for the following alkane. Draw a structural isomer of the compound, and give its name.

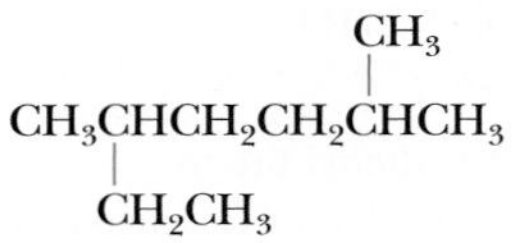

7. ■ Draw the structure of each of the following compounds:
(a) 2,3-dimethylhexane
(b) 2,3-dimethyloctane
(c) 3-ethylheptane
(d) 3-ethyl-2-methylhexane

8. Draw structures for 3-ethylpentane and 2,3-dimethylpentane.

9. Draw Lewis structures, and name all possible compounds that have a seven-carbon chain with one methyl substituent group. Which of these isomers has a chiral carbon center?

10. Draw a structure for cycloheptane. Is the seven-member ring planar? Explain your answer.

11. ■ There are two ethylheptanes (compounds with a seven-carbon chain and one ethyl substituent). Draw the structures, and name these compounds. Is either isomer chiral?

12. ■ Among the 18 structural isomers with the formula C_8H_{18} are two with a five-carbon chain having one ethyl and one methyl substituent group. Draw their structures, and name these two isomers.

13. List several typical physical properties of C_4H_{10}. Predict the following physical properties of dodecane, $C_{12}H_{26}$: color, state (s, ℓ, g), solubility in water, solubility in a nonpolar solvent.

14. Write balanced equations for the following reactions of alkanes.
(a) The reaction of methane with excess chlorine.
(b) Complete combustion of cyclohexane, C_6H_{12}, with excess oxygen.

Alkenes and Alkynes

(See Examples 10.3 and 10.4 and ChemistryNow Screens 10.3 and 10.4.)

15. ■ Draw structures for the *cis* and *trans* isomers of 4-methyl-2-hexene.

16. What structural requirement is necessary for an alkene to have *cis* and *trans* isomers? Can *cis* and *trans* isomers exist for an alkane? For an alkyne?

17. A hydrocarbon with the formula C_5H_{10} can be either an alkene or a cycloalkane.
(a) Draw a structure for each of the isomers possible for C_5H_{10}, assuming it is an alkene. Six isomers are possible. Give the systematic name of each isomer you have drawn.
(b) Draw a structure for a cycloalkane having the formula C_5H_{10}.

18. Five alkenes have the formula C_7H_{14} and a seven-carbon chain. Draw their structures, and name them.

19. ■ Draw the structure, and give the systematic name for the products of the following reactions:
(a) $CH_3CH{=}CH_2 + Br_2 \rightarrow$
(b) $CH_3CH_2CH{=}CHCH_3 + H_2 \rightarrow$

20. Draw the structure, and give the systematic name for the products of the following reactions:

(a) 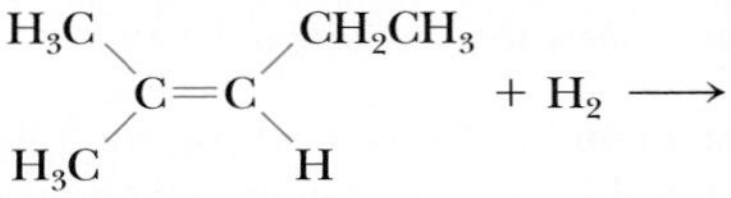

(b) $CH_3C{\equiv}CCH_2CH_3 + 2\,Br_2 \longrightarrow$

21. ■ The compound 2-bromobutane is a product of addition of HBr to an alkene. Identify the alkene and give its name.

22. The compound 2,3-dibromo-2-methylhexane is formed by addition of Br_2 to an alkene. Identify the alkene, and write an equation for this reaction.

23. Draw structures for alkenes that have the formula C_3H_5Cl, and name each compound. (These are derivatives of propene in which a chlorine atom replaces one hydrogen atom.)

24. ■ Elemental analysis of a colorless liquid has given its formula as C_5H_{10}. You recognize that this could be either a cycloalkane or an alkene. A chemical test to determine the class to which this compound belongs involves adding bromine. Explain how this would allow you to distinguish between the two classes.

Aromatic Compounds

(See Example 10.5, Exercise 10.5, and ChemistryNow Screen 10.3.)

25. Draw structural formulas for the following compounds:
(a) 1,3-dichlorobenzene (alternatively called *m*-dichlorobenzene)
(b) 1-bromo-4-methylbenzene (alternatively called *p*-bromotoluene)

26. Give the systematic name for each of the following compounds:

(a) 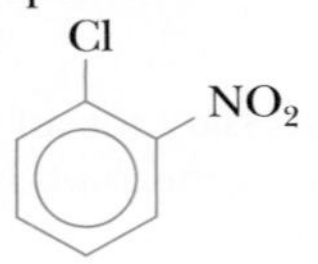

(b) 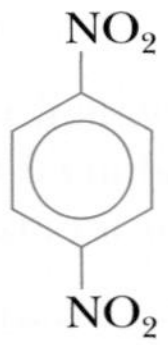

(c) 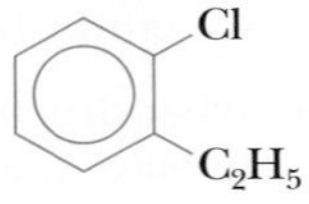

27. Write an equation for the preparation of ethylbenzene from benzene and an appropriate compound containing an ethyl group.

28. ■ Write an equation for the preparation of hexylbenzene from benzene and other appropriate reagents.

29. A single compound is formed by alkylation of 1,4-dimethylbenzene. Write the equation for the reaction of this compound with CH_3Cl and $AlCl_3$. What is the structure and name of the product?

30. Nitration of toluene gives a mixture of two products, one with the nitro group ($-NO_2$) in the *ortho* position and one with the nitro group in the *para* position. Draw structures of the two products.

Alcohols, Ethers, and Amines

(See Example 10.6 and ChemistryNow Screen 10.5.)

31. ■ Give the systematic name for each of the following alcohols, and tell if each is a primary, secondary, or tertiary alcohol:
(a) $CH_3CH_2CH_2OH$
(b) $CH_3CH_2CH_2CH_2OH$
(c) $H_3C-\underset{CH_3}{\underset{|}{\overset{CH_3}{\overset{|}{C}}}}-OH$
(d) $H_3C-\underset{OH}{\underset{|}{\overset{CH_3}{\overset{|}{C}}}}-CH_2CH_3$

32. ■ Draw structural formulas for the following alcohols, and tell if each is primary, secondary, or tertiary:
(a) 1-butanol
(b) 2-butanol
(c) 3,3-dimethyl-2-butanol
(d) 3,3-dimethyl-1-butanol

33. Write the formula, and draw the structure for each of the following amines:
(a) ethylamine
(b) dipropylamine
(c) butyldimethylamine
(d) triethylamine

34. ■ Name the following amines:
(a) $CH_3CH_2CH_2NH_2$
(b) $(CH_3)_3N$
(c) $(CH_3)(C_2H_5)NH$
(d) $C_6H_{13}NH_2$

35. Draw structural formulas for all the alcohols with the formula $C_4H_{10}O$. Give the systematic name of each.

36. Draw structural formulas for all primary amines with the formula $C_4H_9NH_2$.

37. Complete and balance the following equations:
(a) $C_6H_5NH_2(\ell) + HCl(aq) \rightarrow$
(b) $(CH_3)_3N(aq) + H_2SO_4(aq) \rightarrow$

38. ■ Aldehydes and carboxylic acids are formed by oxidation of primary alcohols, and ketones are formed when secondary alcohols are oxidized. Give the name and formula for the alcohol that, when oxidized, gives the following products:
(a) $CH_3CH_2CH_2CHO$
(b) 2-hexanone

Compounds with a Carbonyl Group

(See Exercises 10.7–10.10 and ChemistryNow Screen 10.5.)

39. ■ Draw structural formulas for (a) 2-pentanone, (b) hexanal, and (c) pentanoic acid.

40. ■ Identify the class of each of the following compounds, and give the systematic name for each:
(a) $CH_3\overset{O}{\overset{\|}{C}}CH_3$
(b) $CH_3CH_2CH_2\overset{O}{\overset{\|}{C}}H$
(c) $CH_3\overset{O}{\overset{\|}{C}}CH_2CH_2CH_3$

41. ■ Identify the class of each of the following compounds, and give the systematic name for each:
(a) $CH_3CH_2\overset{CH_3}{\overset{|}{C}}HCH_2CO_2H$
(b) $CH_3CH_2\overset{O}{\overset{\|}{C}}OCH_3$
(c) $CH_3\overset{O}{\overset{\|}{C}}OCH_2CH_2CH_2CH_3$
(d) 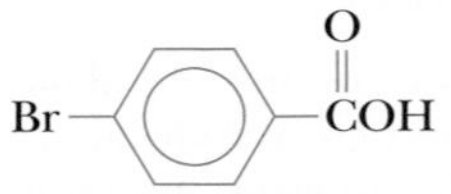

42. Draw structural formulas for the following acids and esters:
(a) 2-methylhexanoic acid
(b) pentyl butanoate (which has the odor of apricots)
(c) octyl acetate (which has the odor of oranges)

43. ■ Give the structural formula and systematic name for the product, if any, from each of the following reactions:
    (a) pentanal and $KMnO_4$
    (b) pentanal and $LiAlH_4$
    (c) 2-octanone and $LiAlH_4$
    (d) 2-octanone and $KMnO_4$

44. Describe how to prepare 2-pentanol beginning with the appropriate ketone.

45. Describe how to prepare propyl propanoate beginning with 1-propanol as the only carbon-containing reagent.

46. ■ Give the name and structure of the product of the reaction of benzoic acid and 2-propanol.

47. Draw structural formulas, and give the names for the products of the following reaction:

$$CH_3\overset{O}{\overset{\|}{C}}OCH_2CH_2CH_2CH_3 + NaOH$$

48. Draw structural formulas, and give the names for the products of the following reaction:

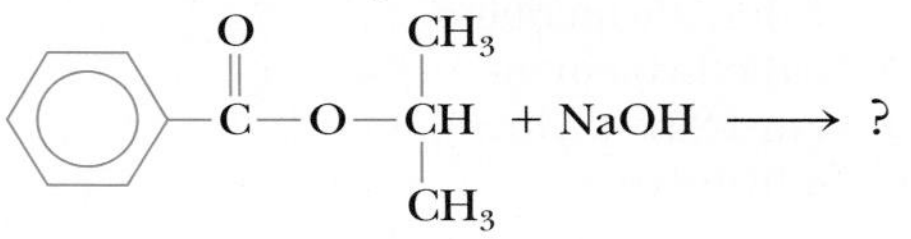

49. ■ The Lewis structure of phenylalanine, one of the 20 amino acids that make up proteins, is drawn below (without lone pairs of electrons). The carbon atoms are numbered for the purpose of this question.
    (a) What is the geometry of C_3?
    (b) What is the O—C—O bond angle?
    (c) Is this molecule chiral? If so, which carbon atom is chiral?
    (d) Which hydrogen atom in this compound is acidic?

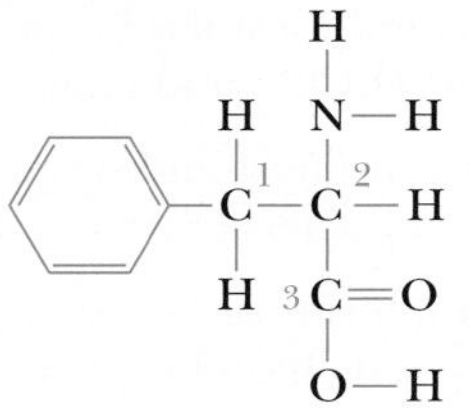

50. ■ The Lewis structure of vitamin C, whose chemical name is ascorbic acid, is drawn below (without lone pairs of electrons).

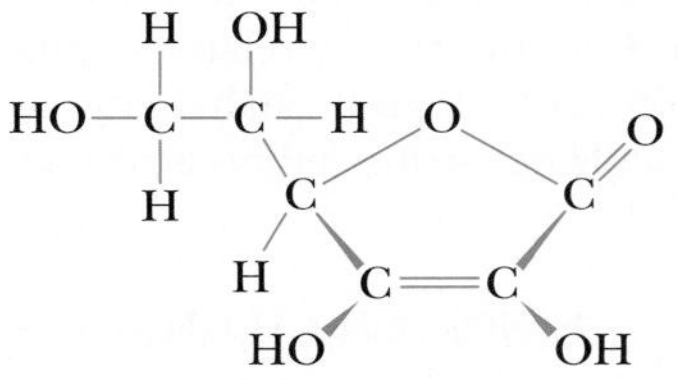

    (a) What is the approximate value for the O—C—O bond angle?
    (b) There are four OH groups in this structure. Estimate the C—O—H bond angles for these groups. Will they be the same value (more or less), or should there be significant differences in these bond angles?
    (c) Is the molecule chiral? How many chiral carbon atoms can be identified in this structure?
    (d) Identify the shortest bond in this molecule.
    (e) What are the functional groups of the molecule?

Functional Groups

(See Example 10.7 and ChemistryNow Screen 10.5.)

51. ■ Identify the functional groups in the following molecules.
    (a) $CH_3CH_2CH_2OH$
    (b) $H_3C\overset{O}{\overset{\|}{C}}NHCH_3$
    (c) $CH_3CH_2\overset{O}{\overset{\|}{C}}OH$
    (d) $CH_3CH_2\overset{O}{\overset{\|}{C}}OCH_3$

52. Consider the following molecules:
    1. $CH_3CH_2\overset{O}{\overset{\|}{C}}CH_3$
    2. $CH_3CH_2\overset{O}{\overset{\|}{C}}OH$
    3. $H_2C{=}CHCH_2OH$
    4. $CH_3CH_2\overset{OH}{\overset{|}{C}}HCH_3$

    (a) What is the result of treating compound 1 with $NaBH_4$? What is the functional group in the product? Name the product.
    (b) Draw the structure of the reaction product from compounds 2 and 4. What is the functional group in the product?
    (c) What compound results from adding H_2 to compound 3? Name the reaction product.
    (d) What compound results from adding NaOH to compound 2?

Polymers

(See Example 10.8, Exercise 10.10, and ChemistryNow Screens 10.9 and 10.10.)

53. Polyvinyl acetate is the binder in water-based paints.
    (a) Write an equation for its formation from vinyl acetate.
    (b) Show a portion of this polymer with three monomer units.
    (c) Describe how to make polyvinyl alcohol from polyvinyl acetate.

54. Neoprene (polychloroprene, a kind of rubber) is a polymer formed from the chlorinated butadiene $H_2C{=}CHCCl{=}CH_2$.
(a) Write an equation showing the formation of polychloroprene from the monomer.
(b) Show a portion of this polymer with three monomer units.

55. Saran is a copolymer of 1,1-dichloroethene and chloroethene (vinyl chloride). Draw a possible structure for this polymer.

56. The structure of methyl methacrylate is given in Table 10.12. Draw the structure of a polymethyl methacrylate (PMMA) polymer that has four monomer units. (PMMA has excellent optical properties and is used to make hard contact lenses.)

General Questions on Organic Chemistry

These questions are not designated as to type or location in the chapter. They may combine several concepts.

57. Three different compounds with the formula $C_2H_2Cl_2$ are known.
(a) Two of these compounds are geometric isomers. Draw their structures.
(b) The third compound is a structural isomer of the other two. Draw its structure.

58. ■ Draw the structure of 2-butanol. Identify the chiral carbon atom in this compound. Draw the mirror image of the structure you first drew. Are the two molecules superimposable?

59. Draw Lewis structures, and name three structural isomers with the formula C_6H_{12}. Are any of these isomers chiral?

60. Draw structures, and name the four alkenes that have the formula C_4H_8.

61. Write equations for the reactions of *cis*-2-butene with the following reagents, representing the reactants and products using structural formulas.
(a) H_2O
(b) HBr
(c) Cl_2

62. Draw the structure, and name the product formed if the following alcohols are oxidized. Assume an excess of the oxidizing agent is used. If the alcohol is not expected to react with a chemical oxidizing agent, write NR (no reaction).
(a) $CH_3CH_2CH_2CH_2OH$
(b) 2-butanol
(c) 2-methyl-2-propanol
(d) 2-methyl-1-propanol

63. Write equations for the following reactions, representing the reactants and products using structural formulas.
(a) The reaction of acetic acid and sodium hydroxide
(b) The reaction of methylamine with HCl

64. ■ Write equations for the following reactions, representing the reactants and products using structural formulas.
(a) The formation of ethyl acetate from acetic acid and ethanol
(b) The hydrolysis of glyceryl tristearate (the triester of glycerol with stearic acid, a fatty acid)

65. Write an equation for the formation of the following polymers.
(a) Polystyrene, from styrene ($C_6H_5CH{=}CH_2$)
(b) PET (polyethylene terephthalate), from ethylene glycol and terephthalic acid

66. Write equations for the following reactions, representing the reactants and products using structural formulas.
(a) The hydrolysis of the amide $C_6H_5CONHCH_3$ to form benzoic acid and methylamine
(b) The hydrolysis $\left(\!-CO(CH_2)_4CONH(CH_2)_6NH-\!\right)_n$, (nylon-6, 6, a polyamide) to give a carboxylic acid and an amine

67. ■ Draw the structure of each of the following compounds:
(a) 2,2-dimethylpentane
(b) 3,3-diethylpentane
(c) 3-ethyl-2-methylpentane
(d) 3-ethylhexane

68. ▲ Structural isomers.
(a) Draw all of the isomers possible for C_3H_8O. Give the systematic name of each, and tell into which class of compound it fits.
(b) Draw the structural formulas for an aldehyde and a ketone with the molecular formula C_4H_8O. Give the systematic name of each.

69. ▲ ■ Draw structural formulas for possible isomers of the dichlorinated propane, $C_3H_6Cl_2$. Name each compound.

70. ■ Draw structural formulas for possible isomers with the formula C_3H_6ClBr, and name each isomer.

71. Give structural formulas and systematic names for the three structural isomers of trimethylbenzene, $C_6H_3(CH_3)_3$.

72. Give structural formulas and systematic names for possible isomers of dichlorobenzene, $C_6H_4Cl_2$.

73. Voodoo lilies depend on carrion beetles for pollination. Carrion beetles are attracted to dead animals, and because dead and putrefying animals give off the horrible-smelling amine cadaverine, the lily likewise releases cadaverine (and the closely related compound putrescine) (page 466). A biological catalyst, an enzyme, converts the naturally occurring amino acid lysine to cadaverine.

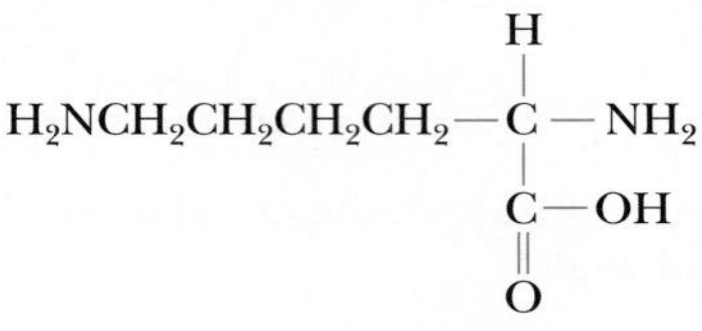

lysine

▲ more challenging ■ in OWL Blue-numbered questions answered in Appendix O

What group of atoms must be replaced in lysine to make cadaverine? (Lysine is essential to human nutrition but is not synthesized in the human body.)

74. Benzoic acid occurs in many berries. When humans eat berries, benzoic acid is converted to hippuric acid in the body by reaction with the amino acid glycine $H_2NCH_2CO_2H$. Draw the structure of hippuric acid, knowing it is an amide formed by reaction of the carboxylic acid group of benzoic acid and the amino group of glycine. Why is hippuric acid referred to as an acid?

75. ■ Consider the reaction of *cis*-2-butene with H_2 (in the presence of a catalyst).
(a) Draw the structure, and give the name of the reaction product. Is this reaction product chiral?
(b) Draw an isomer of the reaction product.

76. ■ Give the name of each compound below, and name the functional group involved.

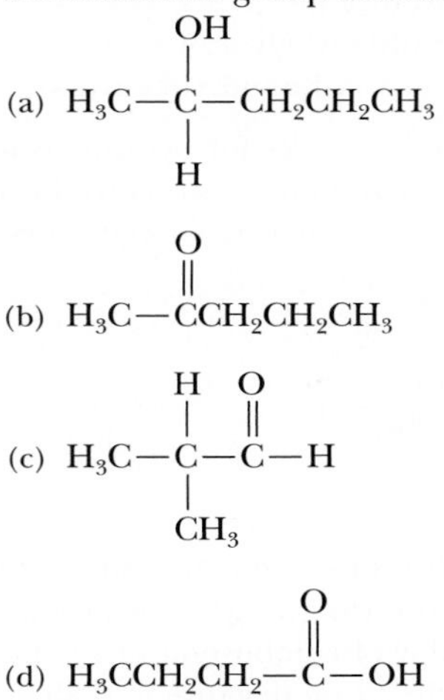

77. Draw the structure of glyceryl trilaurate. When this triester is saponified, what are the products? (See page 476.)

78. ▲ A well-known company selling outdoor clothing has recently introduced jackets made of recycled polyethylene terephthalate (PET), the principal material in many soft drink bottles. Another company makes PET fibers by treating recycled bottles with methanol to give the diester dimethylterephthalate and ethylene glycol and then repolymerizes these compounds to give new PET. Write a chemical equation to show how the reaction of PET with methanol can give dimethylterephthalate and ethylene glycol.

79. ■ Identify the reaction products, and write an equation for the following reactions of $CH_2{=}CHCH_2OH$.
(a) H_2 (hydrogenation, in the presence of a catalyst)
(b) Oxidation (excess oxidizing agent)
(c) Addition polymerization
(d) Ester formation, using acetic acid

80. Write a chemical equation describing the reaction between glycerol and stearic acid to give glyceryl tristearate.

81. ■ The product of an addition reaction of an alkene is often predicted by Markovnikov's rule.
(a) Draw the structure of the product of adding HBr to propene, and give the name of the product.
(b) Draw the structure, and give the name of the compound that results from adding H_2O to 2-methyl-1-butene.
(c) If you add H_2O to 2-methyl-2-butene, is the product the same or different than the product from the reaction in part (b)?

82. An unknown colorless liquid has the formula $C_4H_{10}O$. Draw the structures for the four alcohol compounds that have this formula.

In the Laboratory

83. ■ Which of the following compounds produces acetic acid when treated with an oxidizing agent such as $KMnO_4$?

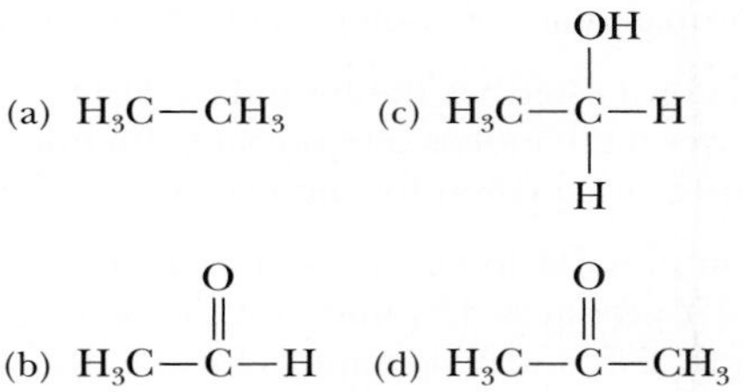

84. ■ Consider the reactions of C_3H_7OH.

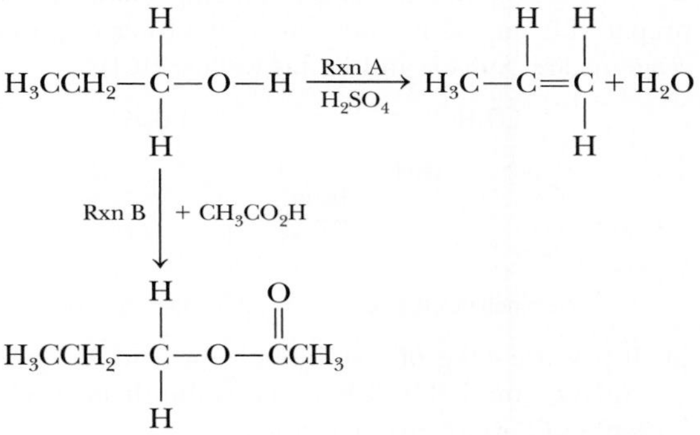

(a) Name the reactant C_3H_7OH.
(b) Draw a structural isomer of the reactant, and give its name.
(c) Name the product of reaction A.
(d) Name the product of reaction B.

85. You have a liquid that is either cyclohexene or benzene. When the liquid is exposed to dark-red bromine vapor, the vapor is immediately decolorized. What is the identity of the liquid? Write an equation for the chemical reaction that has occurred.

86. ▲ ■ Hydrolysis of an unknown ester of butanoic acid, $CH_3CH_2CH_2CO_2R$, produces an alcohol A and butanoic acid. Oxidation of this alcohol forms an acid B that is a structural isomer of butanoic acid. Give the names and structures for alcohol A and acid B.

87. ▲ You are asked to identify an unknown colorless, liquid carbonyl compound. Analysis has determined that the formula for this unknown is C_3H_6O. Only two compounds match this formula.
(a) Draw structures for the two possible compounds.
(b) To decide which of the two structures is correct, you react the compound with an oxidizing agent and isolate from that reaction a compound that is found to give an acidic solution in water. Use this result to identify the structure of the unknown.
(c) Name the acid formed by oxidation of the unknown.

88. Describe a simple chemical test to tell the difference between $CH_3CH_2CH_2CH{=}CH_2$ and its isomer cyclopentane.

89. Describe a simple chemical test to tell the difference between 2-propanol and its isomer methyl ethyl ether.

90. ▲ ■ An unknown ester has the formula $C_4H_8O_2$. Hydrolysis gives methanol as one product. Identify the ester, and write an equation for the hydrolysis reaction.

91. ▲ ■ Addition of water to alkene X gives an alcohol Y. Oxidation of Y produces 3,3-dimethyl-2-pentanone. Identify X and Y, and write equations for the two reactions.

92. ■ 2-Iodobenzoic acid, a tan, crystalline solid, can be prepared from 2-aminobenzoic acid. Other required reagents are $NaNO_2$ and KI (as well as HCl).

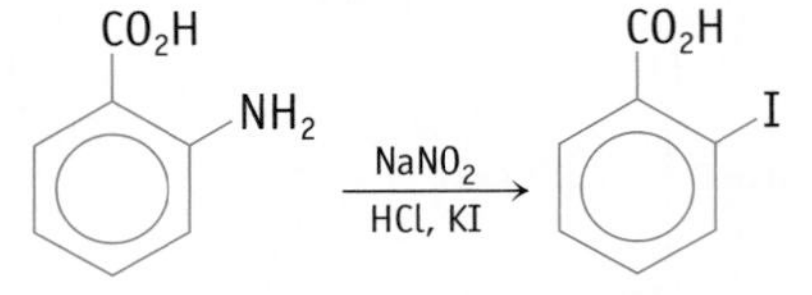

2-aminobenzoic acid 2-iodobenzoic acid

(a) If you use 4.0 g of 2-aminobenzoic acid, 2.2 g of $NaNO_2$, and 5.3 g of KI, what is the theoretical yield of 2-iodobenzoic acid?
(b) Are other isomers of 2-iodobenzoic acid possible?
(c) To verify that you have isolated 2-iodobenzoic acid, you titrate it in water/ethanol. If you use 15.62 mL of 0.101 M NaOH to titrate 0.399 g of the product, what is its molar mass? Is it in reasonable agreement with the theoretical molar mass?

Summary and Conceptual Questions

The following questions may use concepts from this and previous chapters.

93. Carbon atoms appear in organic compounds in several different ways with single, double, and triple bonds combining to give an octet configuration. Describe the various ways that carbon can bond to reach an octet, and give the name, and draw the structure of a compound that illustrates that mode of bonding.

94. There is a high barrier to rotation around a carbon–carbon double bond, whereas the barrier to rotation around a carbon–carbon single bond is considerably smaller. Use the orbital overlap model of bonding (Chapter 9) to explain why there is restricted rotation around a double bond.

95. ■ What important properties do the following characteristics impart on a polymer?
(a) Cross linking in polyethylene
(b) The OH groups in polyvinyl alcohol
(c) Hydrogen bonding in a polyamide like nylon

96. ■ One of the resonance structures for pyridine is illustrated here. Draw another resonance structure for the molecule. Comment on the similarity between this compound and benzene.

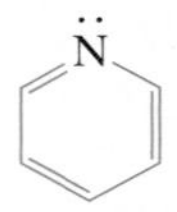

pyridine

97. ■ Write balanced equations for the combustion of ethane gas and liquid ethanol (to give gaseous products).
(a) Calculate the enthalpy of combustion of each compound. Which has the more negative enthalpy change for combustion per gram?
(b) If ethanol is assumed to be partially oxidized ethane, what effect does this have on the heat of combustion?

98. Plastics make up about 20% of the volume of landfills. There is, therefore, considerable interest in reusing or recycling these materials. To identify common plastics, a set of universal symbols is now used, five of which are illustrated here. They symbolize low- and high-density polyethylene, polyvinyl chloride, polypropylene, and polyethylene terephthalate.

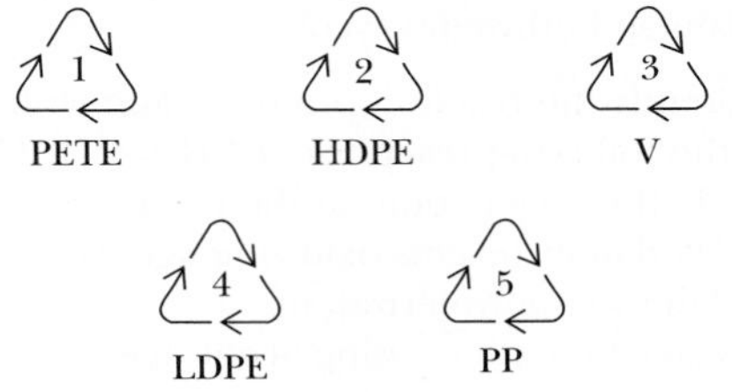

(a) Tell which symbol belongs to which type of plastic.
(b) Find an item in the grocery or drug store made from each of these plastics.
(c) Properties of several plastics are listed in the table. Based on this information, describe how to separate samples of these plastics from one another.

Plastic	**Density (g/cm^3)**	**Melting Point (°C)**
Polypropylene	0.92	170
High-density polyethylene	0.97	135
Polyethylene terephthalate	1.34–1.39	245

99. ▲ Maleic acid is prepared by the catalytic oxidation of benzene. It is a dicarboxylic acid; that is, it has two carboxylic acid groups.
(a) Combustion of 0.125 g of the acid gives 0.190 g of CO_2 and 0.0388 g of H_2O. Calculate the empirical formula of the acid.
(b) A 0.261-g sample of the acid requires 34.60 mL of 0.130 M NaOH for complete titration (so that the H ions from both carboxylic acid groups are used). What is the molecular formula of the acid?
(c) Draw a Lewis structure for the acid.
(d) Describe the hybridization used by the C atoms.
(e) What are the bond angles around each C atom?

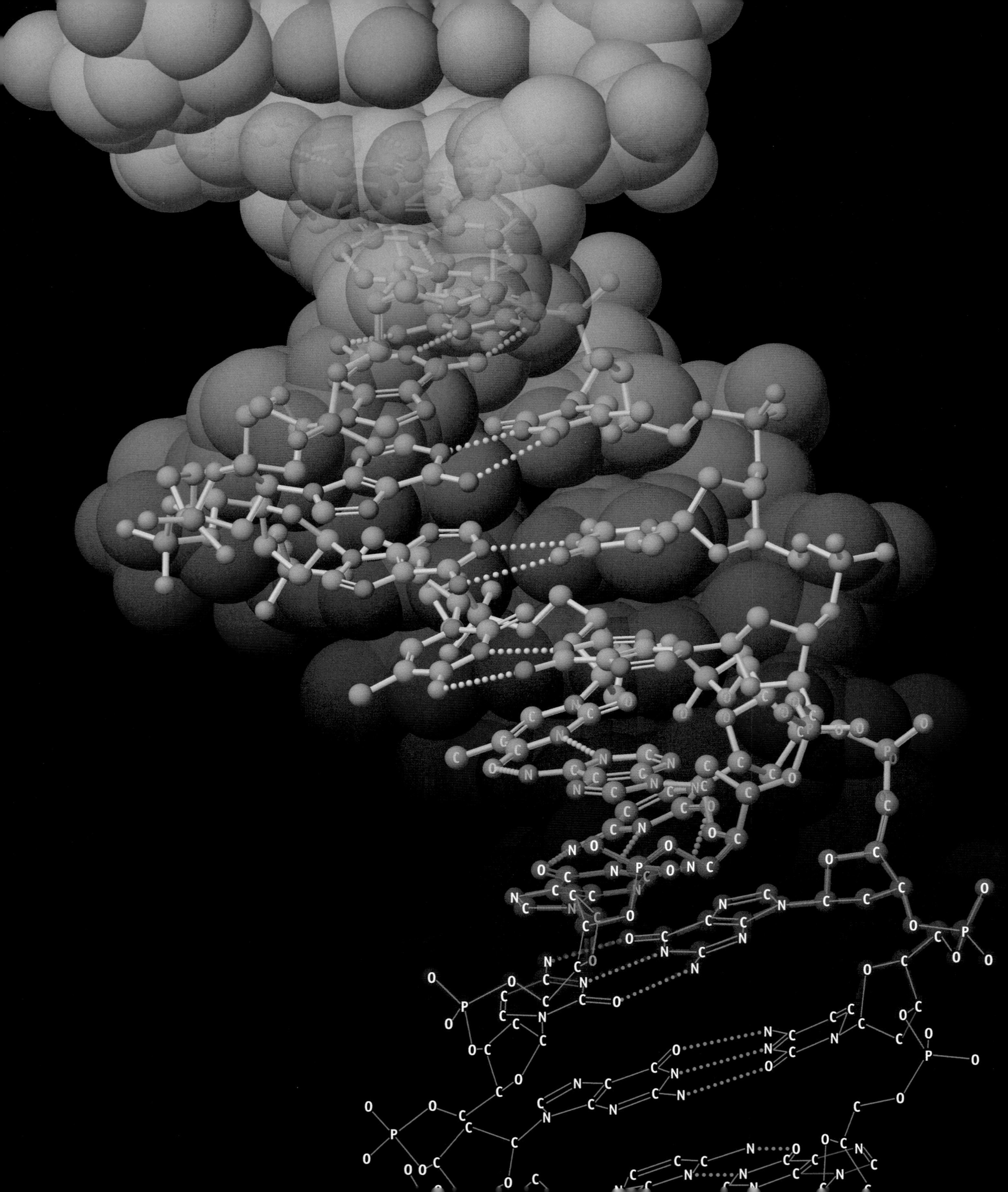

The Chemistry of Life—Biochemistry

You are a marvelously complex biological organism. So is every other living thing on Earth. What molecules are present in you, and what are their properties? How is genetic information passed from generation to generation? How does your body carry out the numerous reactions that are needed for life?

These questions and many others fall into the realm of biochemistry, one of the most rapidly expanding areas of science. As the name implies, biochemistry exists at the interface of two scientific disciplines: biology and chemistry.

What separates a biochemist's perspective of biological phenomena from a biologist's perspective? The difference is becoming less distinct, but biochemists tend to concentrate more on the specific molecules involved in biological processes and on how chemical reactions occur in an organism. They use the strategies of chemists to understand processes in living things.

The goal of this interchapter is to consider how chemistry is involved in answering important biological questions. To do so, we will examine three major classes of biological compounds: proteins, nucleic acids, and lipids. We will also discuss some chemical reactions that occur in living things, including some reactions involved in obtaining energy from food.

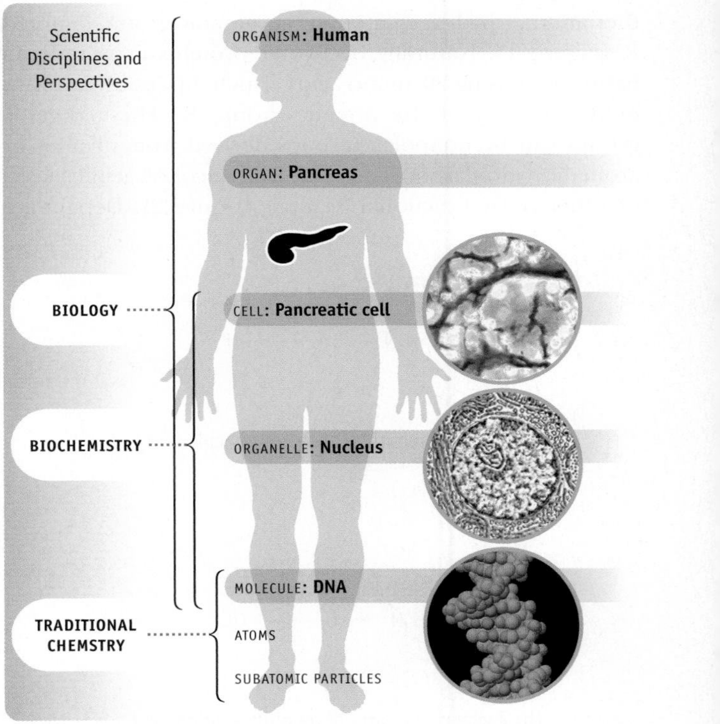

The human body with areas of interest to biologists, biochemists, and chemists.

Proteins

Your body contains thousands of different proteins, and about 50% of the dry weight of your body consists of proteins. Proteins provide structural support (muscle, collagen), help organisms move (muscle), store and transport chemicals from one area to another (hemoglobin), regulate when certain chemical reactions will occur (hormones), and catalyze a host of chemical reactions (enzymes). All of these different functions and others are accomplished using this one class of compounds.

• **The Structure of a Ribosome.** A molecule of transfer ribonucleic acid (tRNA, shown in red) binding to a molecule of messenger RNA (mRNA, shown in gold) in a ribosome.

Amino Acids Are the Building Blocks of Proteins

Proteins are condensation polymers (Section 10.5) formed from amino acids. **Amino acids** are organic compounds that contain an amino group (—NH_2) and a carboxylic acid group (—CO_2H) (Figure 1). Each of these functional groups can exist in two different states: an ionized form (—NH_3^+ and —CO_2^-) and an unionized form (—NH_2 and —CO_2H). If both groups are in their ionized forms, the resulting species contains both a positive and a negative charge and is called a **zwitterion.** In an aqueous environment at physiological pH (about 7.4), amino acids are predominantly in the zwitterionic form.

Almost all amino acids that make up proteins are α-amino acids. In an α-amino acid, the amino group is at one end of the molecule, and the acid group is at the other end. In between these two groups, a single carbon atom (the α-carbon) has attached to it a hydrogen atom and either another hydrogen atom or an organic group, denoted R (Figure 1). Naturally occurring proteins are predominantly built using 20 amino acids, which differ only in terms of the identity of the organic group, R. These organic groups can be nonpolar (groups derived from alkanes or aromatic hydrocarbons) or polar (with alcohol, acidic, basic, or other polar functional groups) (Figure 2). Depending

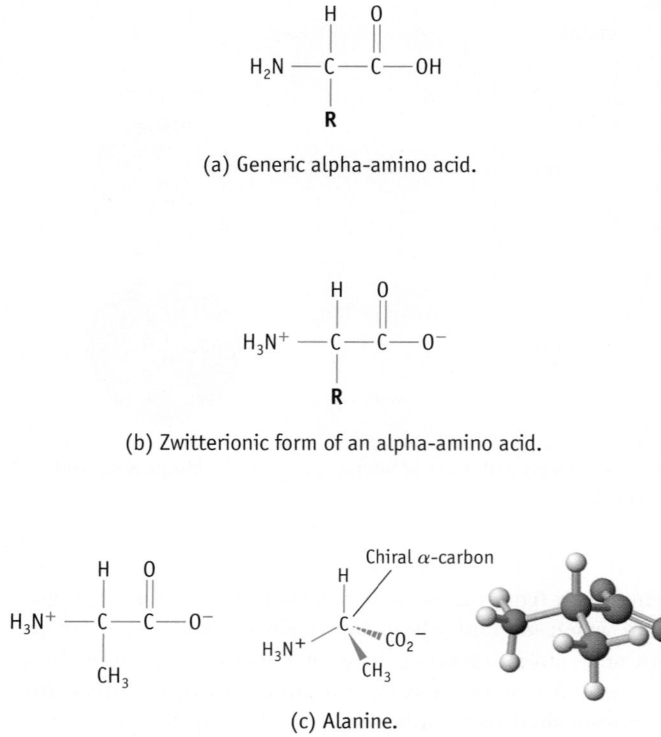

Figure 1 **α-Amino acids.** (a) α-Amino acids have a C atom to which are attached an amino group (—NH_2), a carboxylic acid group (—CO_2H), an organic group (R), and an H atom. (b) The zwitterionic form of an α-amino acid. (c) Alanine, one of the naturally occurring amino acids.

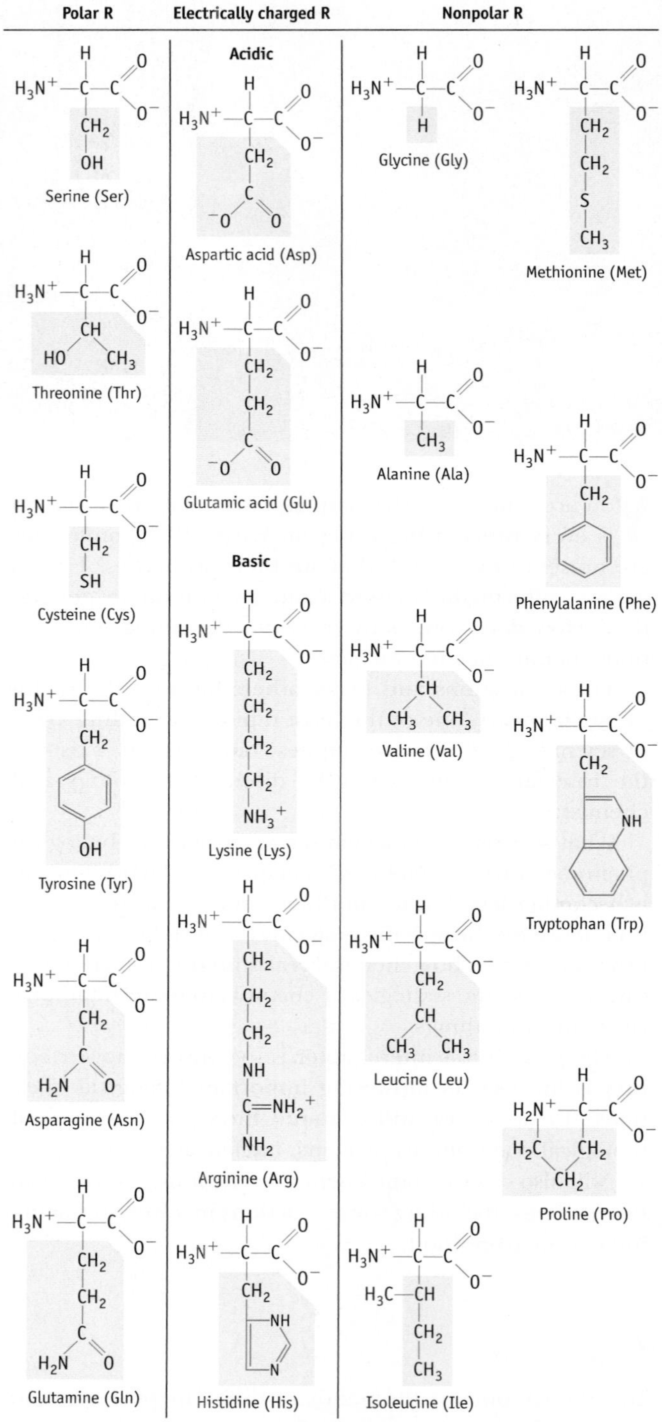

Figure 2 **The 20 most common amino acids in proteins.** All (except proline and glycine) share the characteristic that there is an NH_3^+ group, a CO_2^- group, an H atom, and an organic group attached to a chiral C atom, called the alpha (α) carbon. The organic groups may be polar, nonpolar, or electrically charged. (Histidine is shown in the electrically charged column because the unprotonated N in the organic group can easily be protonated.)

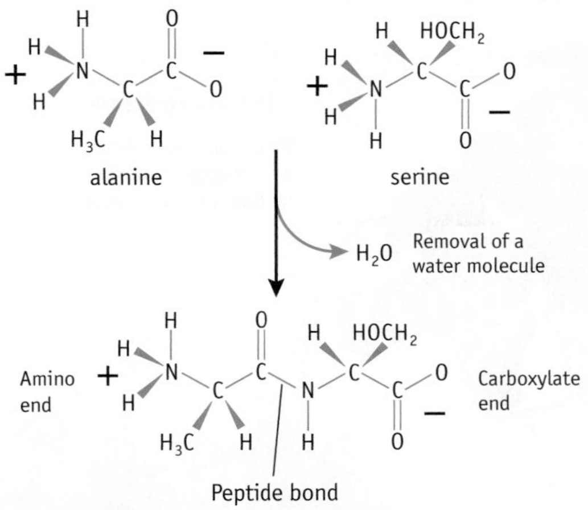

Figure 3 Formation of a peptide. Two α-amino acids condense to form an amide linkage, which is often called a peptide bond. Proteins are polypeptides, polymers consisting of many amino acid units linked through peptide bonds.

on which amino acids are present, a region in a protein may be nonpolar, very polar, or anything in between.

All α-amino acids, except glycine, have four different groups attached to the α-carbon. The α-carbon is thus a chiral center (◀ page 445), and two enantiomers exist. Interestingly, all of these amino acids occur in nature in a single enantiomeric form.

Condensation reactions between two amino acids result in the elimination of water and the formation of an amide linkage (Figure 3). The amide linkage in proteins is often referred to as a **peptide bond,** and the polymer (the protein) that results from a series of these reactions is called a **polypeptide.** The amide linkage is planar (◀ page 475), and both the carbon and the nitrogen atoms are sp^2 hybridized. There is partial double-bond character in the C—O and C—N bonds, leading to restricted rotation about the carbon–nitrogen bond. As a consequence, each peptide bond in a protein possesses a rigid, planar section, which plays a role in determining its structure.

Naturally occurring proteins consist of one or more polypeptide chains that are often hundreds of amino acids long. Their molar masses are thus often thousands of grams per mole.

Protein Structure and Hemoglobin

With this basic understanding of amino acids and peptide bonds, let us examine some larger issues related to protein structure. One of the central tenets of biochemistry is that "structure determines function." In other words, what a molecule can do is determined by which atoms or groups of atoms are present and how they are arranged in space. It is not surprising, therefore, that much effort has been devoted to determining the structures of proteins.

To simplify their discussions, biochemists describe proteins as having different structural levels. Each level of structure can be illustrated using hemoglobin.

Hemoglobin is the molecule in red blood cells that carries oxygen from the lungs to all of the body's other cells. It is a large iron-containing protein, made up of more than 10,000 atoms having a molar mass of 64,500 g/mol. Hemoglobin consists of four polypeptide segments: two identical segments called the α subunits containing 141 amino acids each and two other segments called the β subunits containing 146 amino acids each. The β subunits are identical to each other but different from the α subunits. Each subunit contains an iron(II) ion locked inside an organic ion called a **heme** unit (Figure 4). The oxygen molecules transported by hemoglobin bind to these iron(II) ions. (For more information about the heme group, see *A Closer Look* on page 1031.)

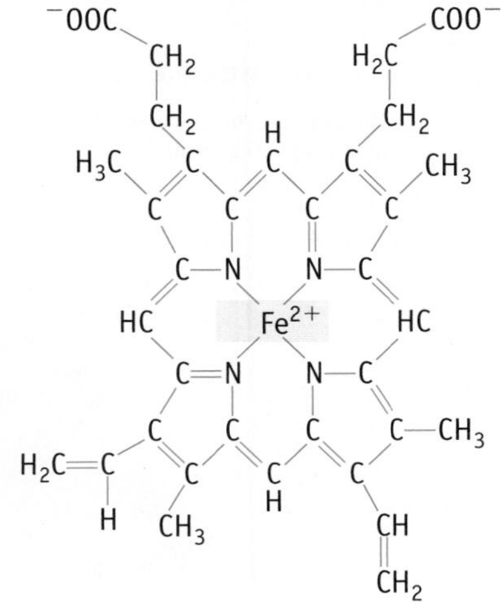

Heme
(Fe-protoporphyrin IX)

Figure 4 Heme. The heme unit in hemoglobin (and in myoglobin, a related protein) consists of an iron ion in the center of a porphyrin ring system.

Let us focus on the polypeptide part of hemoglobin (Figure 5). The first step in describing a structure is to identify how the atoms are linked together. This is called the **primary structure** of a protein, which is simply the sequence of amino acids linked together by peptide bonds. For example, a glycine unit can be followed by an alanine, followed by a valine, and so on.

The remaining levels of structure all deal with noncovalent (nonbonding) interactions between amino acids in the protein. The **secondary structure** of a protein refers to how amino acids near one another in the sequence arrange themselves in space. Some regular patterns often emerge, such as helices, sheets, and turns. In hemoglobin, it was discovered that the amino acids in large portions of the polypeptide chains arrange themselves into many helical regions, a commonly observed polypeptide secondary structure.

The **tertiary structure** of a protein refers to how the chain is folded, including how amino acids that are far apart in the sequence interact with each other. In other words, this structure deals with how the regions of the polypeptide chain fold into the overall three-dimensional structure.

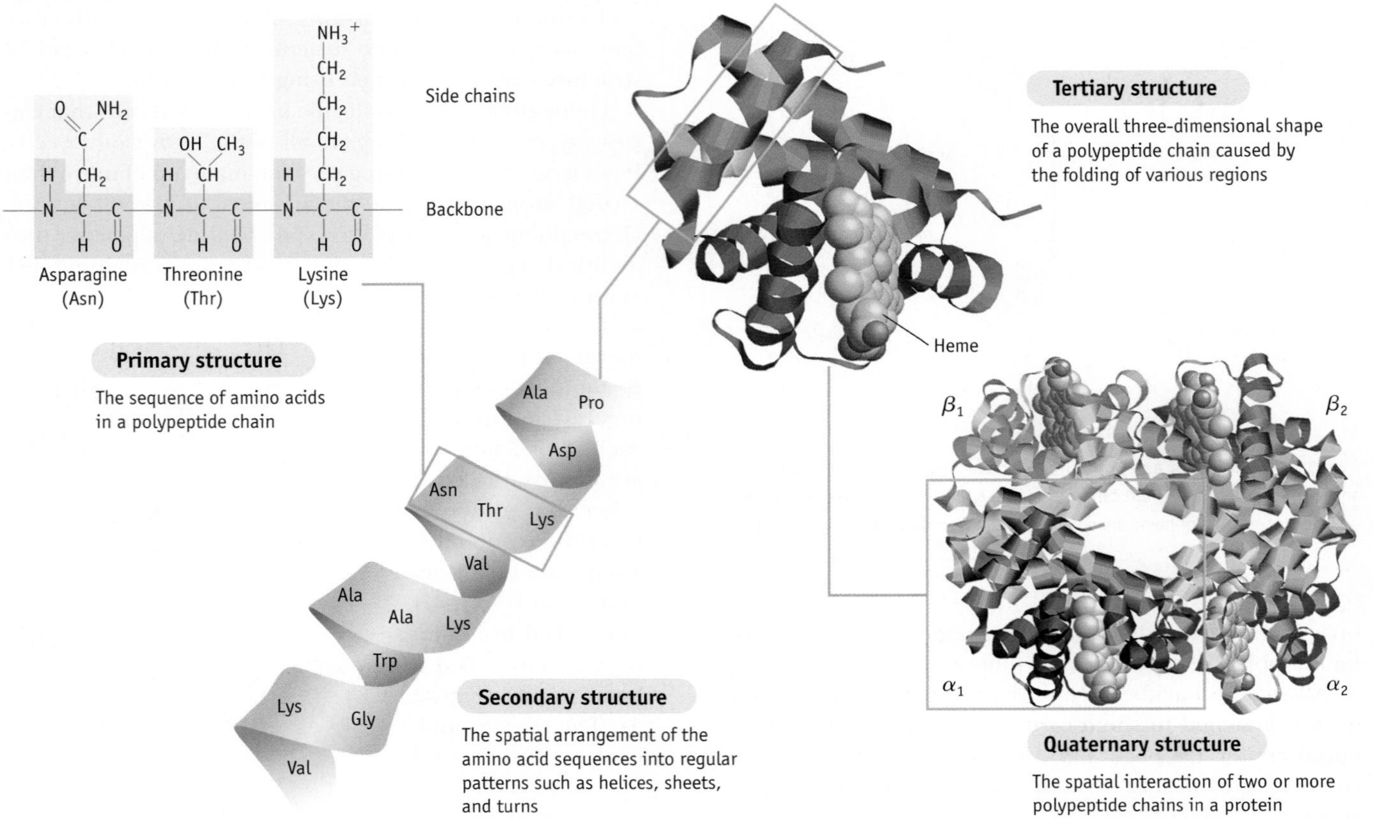

Figure 5 The primary, secondary, tertiary, and quaternary structures of hemoglobin.

For proteins consisting of only one chain, the tertiary structure is the highest level of structure present. In proteins consisting of more than one polypeptide chain, such as hemoglobin, there is a fourth level of structure, called the **quaternary structure.** It is concerned with how the different chains interact. The quaternary structure of hemoglobin shows how the four subunits are related to one another in the overall protein.

Sickle Cell Anemia

The subtleties of sequence, structure, and function are dramatically illustrated in the case of hemoglobin. Seemingly small changes in the amino acid sequence of hemoglobin and other molecules can be important in determining function, as is clearly illustrated by the disease called *sickle cell anemia.* This disease, which is sometimes fatal, affects some individuals of African descent. Persons affected by this disease are anemic; that is, they have low red blood cell counts. In addition, many of their red blood cells are elongated and curved like a sickle instead of being round disks (Figure 6a). These elongated red blood cells are more fragile than normal blood cells, leading to the anemia that is observed. They also restrict the flow of blood within the capillaries, thereby decreasing the amount of oxygen that the individual's cells receive.

The cause of sickle cell anemia has been traced to a small structural difference in hemoglobin. In the β subunits of the hemoglobin in individuals carrying the sickle cell trait, a valine has been substituted for a glutamic acid at position 6. An amino acid in this position ends up on the surface of the protein, where it is exposed to the aqueous environment of the cell. Glutamic acid and valine are quite different from each other. The side chain in glutamic acid is ionic, whereas that in valine is nonpolar. The nonpolar side chain on valine causes a nonpolar region to stick out from the molecule where one should not occur. When hemoglobin (normal or sickle cell) is in the deoxygenated state, it has a nonpolar cavity in another region. The nonpolar region around the valine on one sickle cell hemoglobin molecule fits nicely into this nonpolar cavity on another hemoglobin. The sickle cell hemoglobins thus link together, forming long chainlike structures that lead to the symptoms described (Figure 6b).

Just one amino acid substitution in each β subunit causes sickle cell anemia! While other amino acid substitutions may not lead to such severe consequences, sequence, structure, and function are intimately linked and of crucial importance throughout biochemistry.

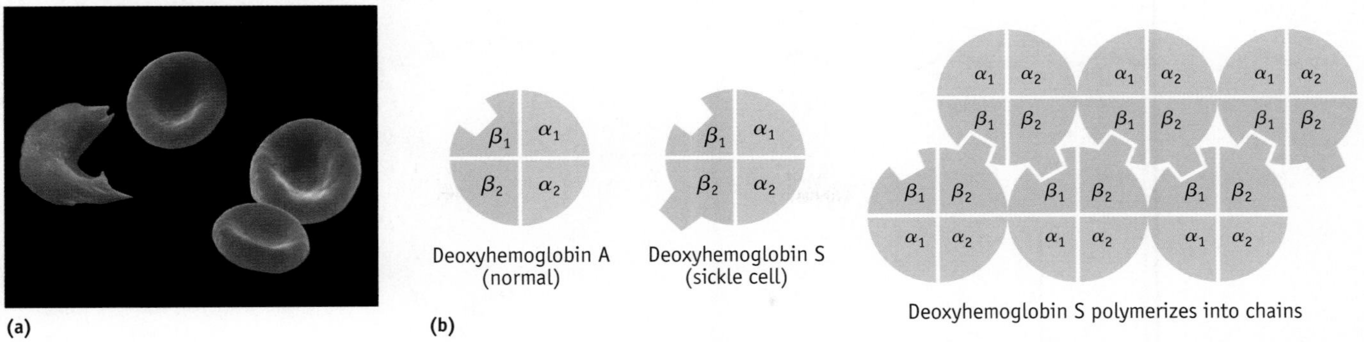

Figure 6 **Normal and sickled red blood cells.** (a) Red blood cells are normally rounded in shape, but people afflicted with sickle cell anemia have cells with a characteristic "sickle" shape. (b) Sickle cell hemoglobin has a nonpolar region that can fit into a nonpolar cavity on another hemoglobin. Sickle cell hemoglobins can link together to form long chainlike structures.

Enzymes, Active Sites, and Lysozyme

Many reactions necessary for life occur too slowly on their own, so organisms speed them up to the appropriate level using biological catalysts called **enzymes.** Almost every metabolic reaction in a living organism requires an enzyme, and most of these enzymes are proteins. Enzymes are often able to speed up reactions by tremendous amounts; catalyzed rates are typically 10^7 to 10^{14} times faster than uncatalyzed rates.

For an enzyme to catalyze a reaction, several key steps must occur:

1. A reactant (often called the **substrate**) must bind to the enzyme.
2. The chemical reaction must take place.
3. The product(s) of the reaction must leave the enzyme so that more substrate can bind and the process can be repeated.

Typically, enzymes are very specific; that is, only a limited number of compounds (often only one) serve as substrates for a given enzyme, and the enzyme catalyzes only one type of reaction. The place in the enzyme where the substrate binds and the reaction occurs is called the **active site.** The active site often consists of a cavity or cleft in the enzyme into which the substrate or part of the substrate can fit. The R groups of amino acids or the presence of metal ions in an active site, for example, are often important factors in binding a substrate and catalyzing a reaction.

Lysozyme is an enzyme that can be obtained from human mucus and tears and from other sources, such as egg whites. Alexander Fleming (1881–1955) (who later discovered penicillin) is said to have discovered its presence in mucus when he had a cold. He purposely allowed some of the mucus from his nose to drip onto a dish containing a bacteria culture and found that some of the bacteria died. The chemical in the mucus responsible for this effect was a protein. Fleming called it lysozyme because it is an enzyme that causes some bacteria to undergo lysis (rupture).

Lysozyme's antibiotic activity has been traced to its ability to catalyze a reaction that breaks down the cell walls of some bacteria. These cell walls contain a polysaccharide, a polymer of sugar molecules. This polysaccharide is composed of two alternating sugars: *N*-acetylmuramic acid (NAM) and *N*-acetylglucosamine (NAG) (Figure 7). Lysozyme speeds up the reaction that breaks the bond between C-1 of NAM and C-4 of NAG (Figure 8). Lysozyme

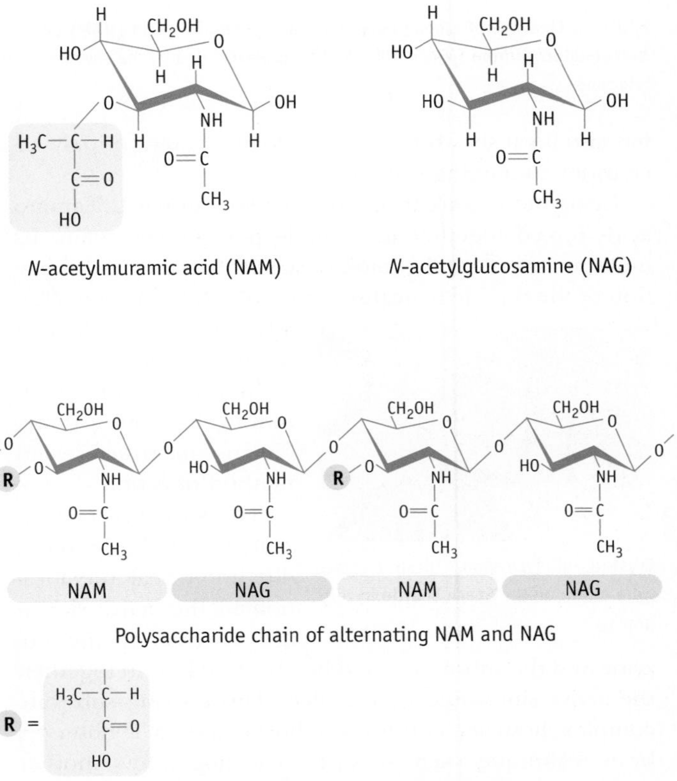

Figure 7 **The structures of *N*-acetylmuramic acid (NAM) and *N*-acetylglucosamine (NAG).** The cell walls of some bacteria contain a polysaccharide chain of alternating NAM and NAG units.

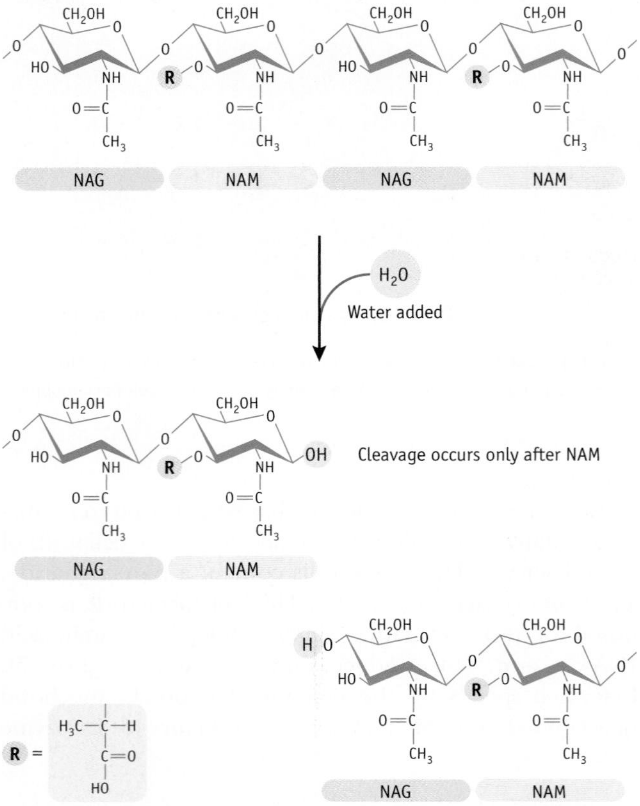

Figure 8 Cleavage of a bond between *N*-acetylmuramic acid (NAM) and *N*-acetylglucosamine (NAG). This reaction is accelerated by the enzyme lysozyme.

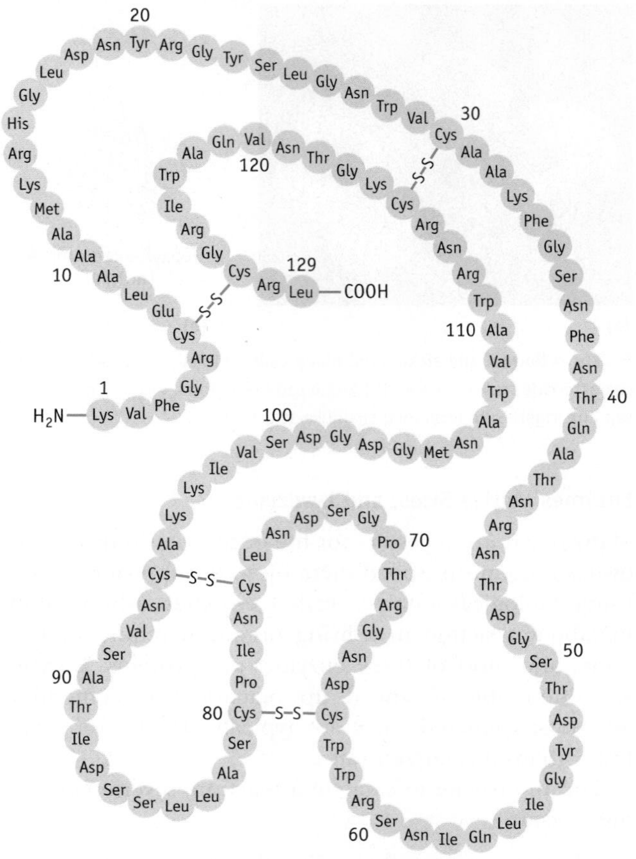

Figure 9 The primary structure of lysozyme. The cross-chain disulfide links (—S—S—) are links between cysteine amino acid residues.

has also been shown to catalyze the breakdown of polysaccharides containing only NAG.

Lysozyme (Figure 9) is a protein containing 129 amino acids linked together in a single polypeptide chain. Its molar mass is 14,000 g/mol. As was true in the determination of the double-helical structure of DNA (◀ page 392), x-ray crystallography and model building were key techniques used in determining lysozyme's three-dimensional structure and method of action.

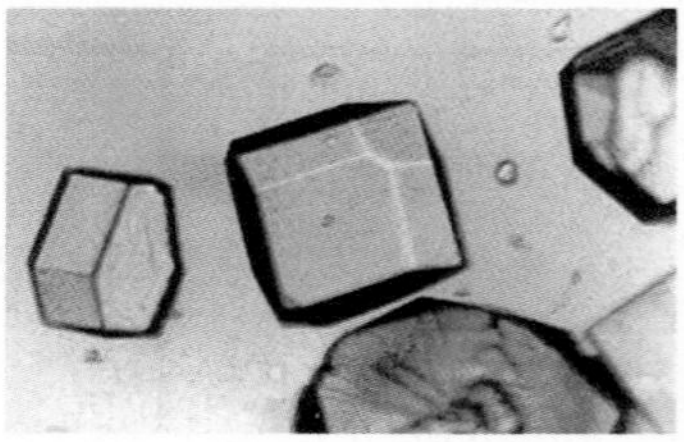

NASA

Crystals of lysozyme. These crystals were grown on the Space Shuttle in zero gravity.

The structure of lysozyme by itself, however, did not reveal the location of the active site in the enzyme. If the enzyme and the substrate could be observed bound together, the active site would be revealed. This enzyme–substrate complex, however, lasts for too short a time to be observed by a technique such as x-ray crystallography. Another method had to be used to identify the active site.

Lysozyme is not very effective in cleaving molecules consisting of only two or three NAG units [$(NAG)_3$]. In fact, these molecules act as inhibitors of the enzyme. Researchers surmised that the inhibition resulted from these small molecules binding to the active site in the enzyme. Therefore, x-ray crystallography was performed on crystals of lysozyme that had been treated with $(NAG)_3$. It revealed that $(NAG)_3$ binds to a cleft in lysozyme (Figure 10).

The cleft in lysozyme where $(NAG)_3$ binds has room for a total of six NAG units. Molecular models of the enzyme and $(NAG)_6$ showed that five of the six sugars fit nicely into the cleft but that the fourth sugar in the sequence did not fit well. To get this sugar into the active site, its structure has to be distorted in the direction that the sugar must move during the cleavage reaction (assuming the bond cleaved is the one connecting it and the next sugar). Amino acids immediately around this location could also assist in the cleavage reaction. In addition, models showed that if an alternating sequence of NAM and NAG binds to the enzyme in this cleft, NAM must bind to this location in the active site: NAM cannot fit into the sugar-binding site immediately before this one, whereas NAG can. For this reason, cleavage must occur only between C-1 of NAM and C-4 of the following NAG, not the other way around—and this is exactly what occurs.

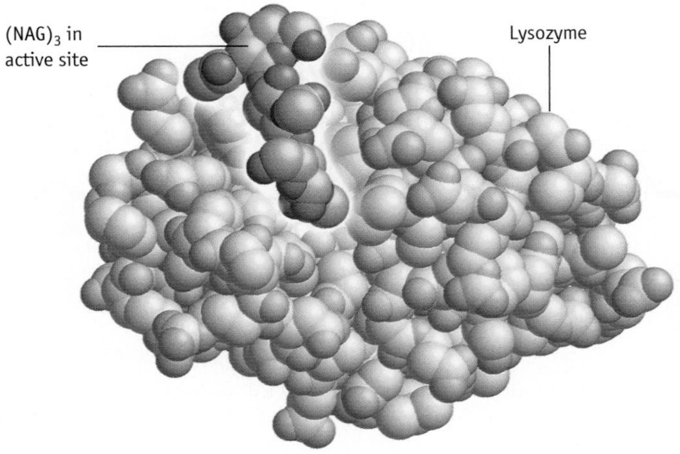

Figure 10 **Lysozyme with $(NAG)_3$.**

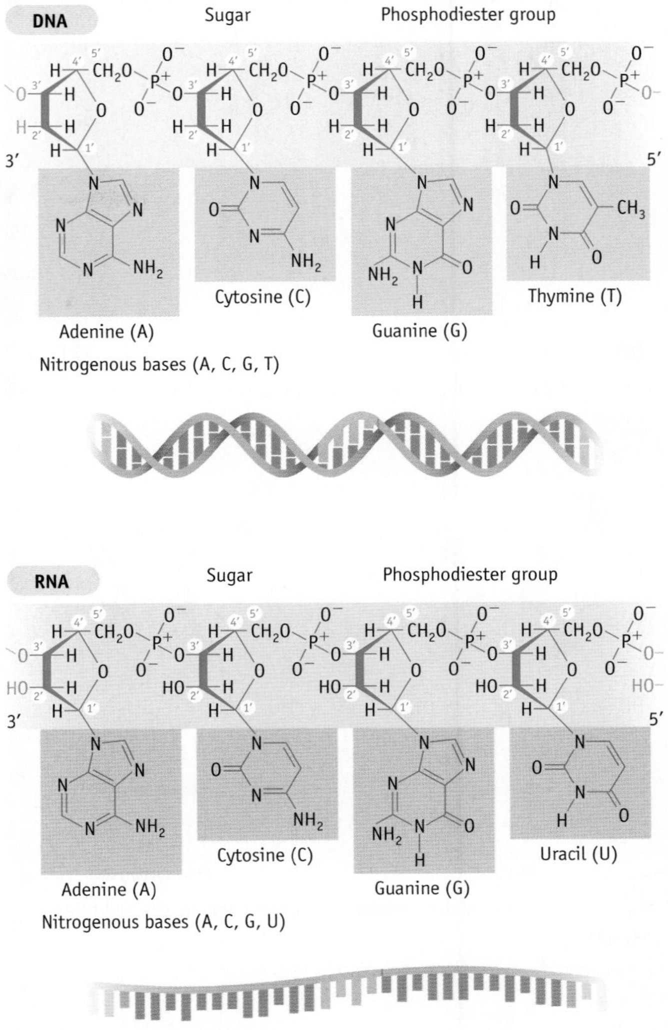

Figure 11 **DNA and RNA.**

Nucleic Acids

In the first half of the 20th century, researchers identified **deoxyribonucleic acid (DNA)** as the genetic material in cells. Also found in cells was a close relative of DNA called **ribonucleic acid (RNA).** Once it was known that DNA was the molecule involved in heredity, scientists set about determining how it accomplishes this task. Because structure determines function, to understand how a molecule works, you must first know its structure.

Nucleic Acid Structure

RNA and DNA are polymers (Figure 11). They consist of sugars having five carbons (β-D-ribose in RNA and β-D-2-deoxyribose in DNA) that are connected by phosphodiester groups. A phosphodiester group links the 3′ (pronounced "three prime") position of one sugar to the 5′ position of the next sugar. Attached at the 1′ position of each sugar is an aromatic, nitrogen-containing (nitrogenous) base. The bases in DNA are adenine (A), cytosine (C), guanine (G), and thymine (T); in RNA, the nitrogenous bases are the same as in DNA, except that uracil (U) is used rather than thymine (Figure 12). A single sugar with a nitrogenous base attached is called a **nucleoside.** If a phosphate group is also attached, then the combination is called a **nucleotide** (Figure 12).

The principal chemical difference between RNA and DNA is the identity of the sugar (Figure 13). Ribose has a hydroxyl group (—OH) at the 2 position, whereas 2-deoxyribose has only a hydrogen atom at this position. This seemingly small difference turns out to have profound effects. The polymer chain of RNA is cleaved many times faster than a corresponding chain of DNA under similar conditions due to the involvement of this hydroxyl group in the cleavage reaction. The greater stability of DNA contributes to it being a better repository for genetic information.

How does DNA store genetic information? DNA consists of a double helix; one strand of DNA is paired with another strand running in the opposite direction. The key parts of the structure of DNA for this function are the nitrogenous bases. James Watson and Francis Crick (page 392) noticed that A can form two hydrogen bonds (▶ Section 12.2) with T and that C can form three hydrogen bonds with G. The spacing in the double helix is just right for either an A–T pair or for a C–G pair to fit, but other combinations (such as A–G) do not fit properly (Figure 14). Thus, if we know the identity of a nucleotide on one strand of the double helix, then we can figure out which nucleotide must be bound to it on the other strand. The two strands are referred to as **complementary strands.**

If the two strands are separated from each other, as they are before the cell division process called *mitosis,* the cell can construct a new complementary strand for each of the original strands by placing a G wherever there is

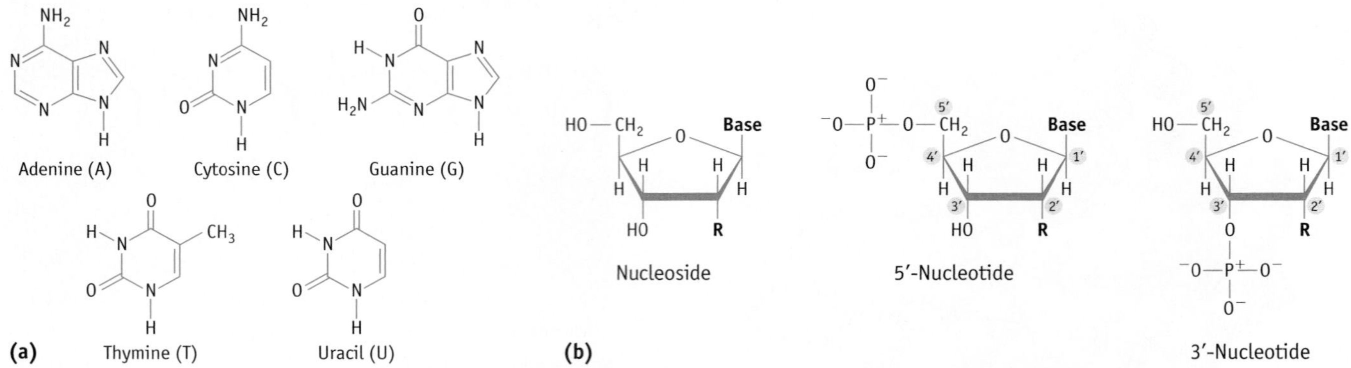

Figure 12 **Bases, nucleosides, and nucleotides.** (a) The five bases present in DNA and RNA. (b) A nucleoside, a 5′-nucleotide, and a 3′-nucleotide.

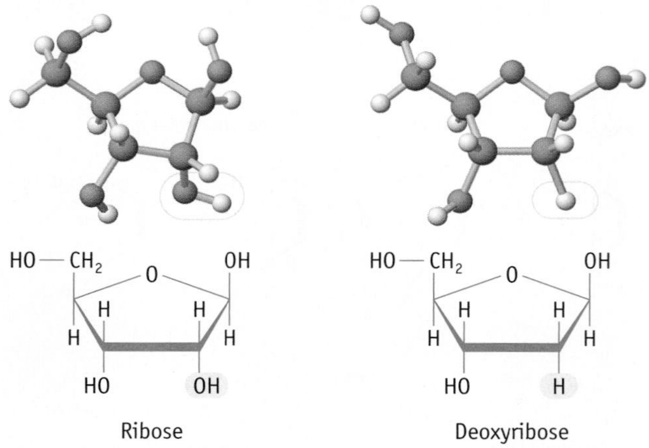

Figure 13 **Ribose and deoxyribose.** The sugars found in RNA and DNA, respectively.

a C, a T wherever there is an A, and so forth. Through this process, called **replication**, the cell ends up with two identical double-stranded DNA molecules for each molecule of DNA initially present. When the cell divides, each of the two resulting cells gets one copy of each DNA molecule (Figure 15). In this way, genetic information is passed along from one generation to the next.

Protein Synthesis

The sequence of nucleotides in a cell's DNA contains the instructions to make all of the various proteins the cell needs. DNA is the information storage molecule. To use this information, the cell first makes a complementary copy of the required portion of the DNA using RNA. This step is called **transcription.** The molecule of RNA that results is called **messenger RNA (mRNA)** because it takes this message to where protein synthesis occurs in the cell. The cell uses the less stable (more rapidly cleaved) RNA rather than DNA to carry out this function.

It makes sense to use DNA, the more stable molecule, to store the genetic information because the cell wants this information to be passed from generation to generation intact. Conversely, it makes sense to use RNA to send the message to make a particular protein. By using the less-stable RNA, the message will not be permanent but rather will be destroyed after a certain time, thus allowing the cell to turn off the synthesis of the protein.

Protein synthesis occurs in **ribosomes,** complex bodies in a cell consisting of a mixture of proteins and RNA. The new protein is made as the ribosome moves along the strand of mRNA. The sequence of nucleotides in mRNA contains informa-

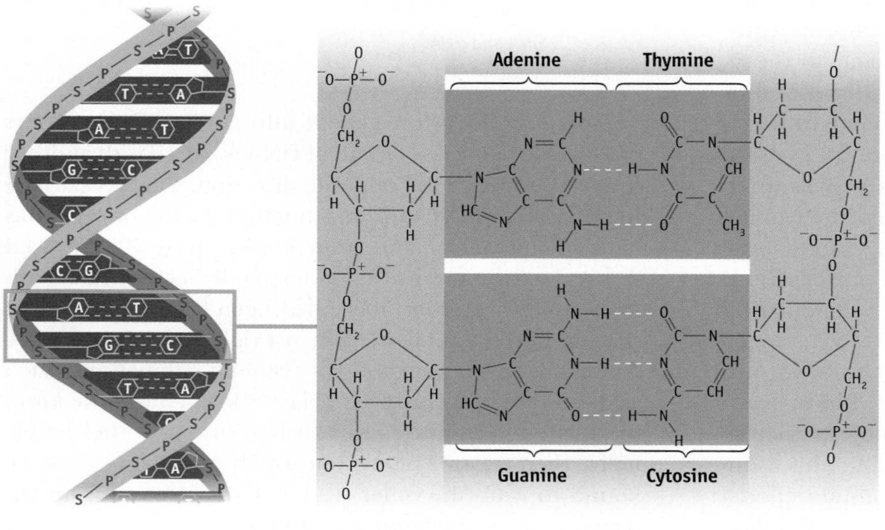

Figure 14 **Base pairs and complementary strands in DNA.** With the four bases in DNA, the usual pairings are adenine with thymine and cytosine with guanine. The pairing is promoted by hydrogen bonding, the interaction of an H atom bound to an O or N atom with an O or N atom in a neighboring molecule.

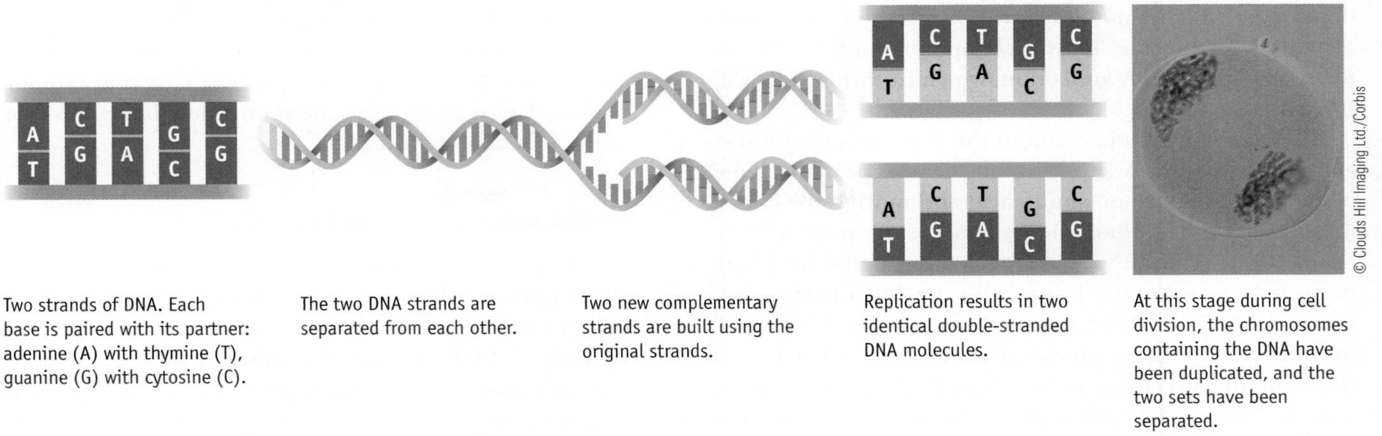

Figure 15 **The main steps in DNA replication.** The products of this replication are two identical double-helical DNA molecules. When a cell divides, each resulting cell gets one of these.

TABLE 1 Examples of the 64 Codons in the Genetic Code

Codon Base Sequence*	Amino Acid to Be Added
AAA	Lysine
AAC	Asparagine
AUG	Start
CAA	Glutamine
CAU	Histidine
GAA	Glutamic acid
GCA	Alanine
UAA	Stop
UAC	Tyrosine

* A = Adenine, C = cytosine, G = guanine, U = uracil.

tion about the order of amino acids in the desired protein. Following the signal in mRNA to start protein synthesis, every sequence of three nucleotides provides the code for an amino acid until the ribosome reaches the signal to stop (Table 1). These three-nucleotide sequences in mRNA are referred to as **codons,** and the correspondence between each codon and its message (start, a particular amino acid, or stop) is referred to as the **genetic code.**

How is the genetic code used to make a protein? In the ribosome–mRNA complex, there are two neighboring binding sites, the P site and the A site. (The ribosomes of eukaryotic cells, cells that contain nuclei, also have a third binding site, called the E site.) Each cycle that seeks to add an amino acid to a growing protein begins with that part of the protein already constructed being located in the P site. The A site is where the next amino acid is brought in. Yet another type of RNA becomes involved at this point. This **transfer RNA (tRNA)** consists of a strand of RNA to which an amino acid can be attached (Figure 16). A strand of tRNA has a particular region that contains a sequence of three nucleotides that can attempt to form base pairs to a codon in the mRNA at the ribosome's A site. This three-nucleotide sequence in the tRNA is called the **anticodon.** Only if the base pairing between the codon and anticodon is complementary (for example, A with U) will the tRNA be able to bind to the mRNA–ribosome complex. Not only does the anticodon determine to which codon a particular strand of tRNA can bind, but it also

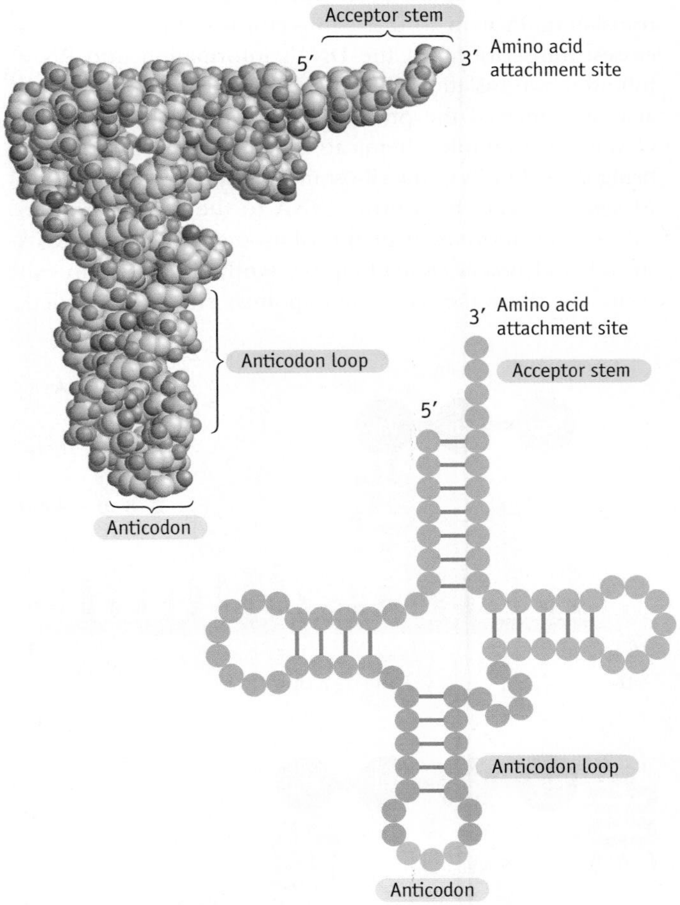

Figure 16 **tRNA structure.**

determines which amino acid will be attached to the end of the tRNA molecule. Thus, a codon in the mRNA selects for a particular tRNA anticodon, which in turn selects for the correct amino acid.

The growing protein chain in the P site reacts with the amino acid in the A site, resulting in the protein chain being elongated by one amino acid and moving the chain into the A site. The ribosome then moves down the mRNA chain, moving the tRNA with the protein strand attached from the A site into the P site and exposing a new codon in the A site. The tRNA that had been in the P site and that no longer has an amino acid attached either leaves the ribosome directly or, if there is an E site present, moves into the E site before exiting from the ribosome. The process is then repeated (Figure 17).

Converting the information from a nucleotide sequence in mRNA into an amino acid sequence in a protein is called **translation.** Protein synthesis thus consists of two main processes: transcription of the DNA's information into RNA, followed by translation of the RNA's message into the amino acid sequence of the protein. However, there is more involved. For example, chemicals called *Initiation Factors* help bring the subunits of the ribosome and the mRNA together, *Elongation Factors* help bring tRNA to the A site and also catalyze the movement of the ribosome down the mRNA strand, and *Release Factors* help the synthesized protein exit from the ribosome. At various points, energy is needed, provided by the hydrolysis of guanosine triphosphate (GTP) in a reaction similar to that discussed for ATP later in this interchapter. Nonetheless, the processes of transcription and translation as discussed here provide a basic introduction to this important topic.

The RNA World and the Origin of Life

One of the most fascinating and persistent questions scientists pursue is how life arose on earth. Plaguing those trying to answer this question is a molecular chicken-and-egg problem: Which came first, DNA or proteins? DNA is good at storing genetic information, but it is not good at catalyzing reactions. Proteins are good at catalyzing reactions, but they are not good at storing genetic information. In trying to picture an early self-replicating molecule, deciding whether it should be based on DNA or proteins seemed hopeless. Ultimately, both functions are important. These problems have caused some scientists to turn away from considering either DNA or proteins as candidates for the first molecule of life. One hypothesis that has gained support in recent years suggests that the first life on earth may have been based on RNA instead.

Like DNA, RNA is a nucleic acid and can serve as a genetic storage molecule. We have already seen how it serves as an information molecule in the process of protein synthesis. In addition, scientists have discovered that ret-

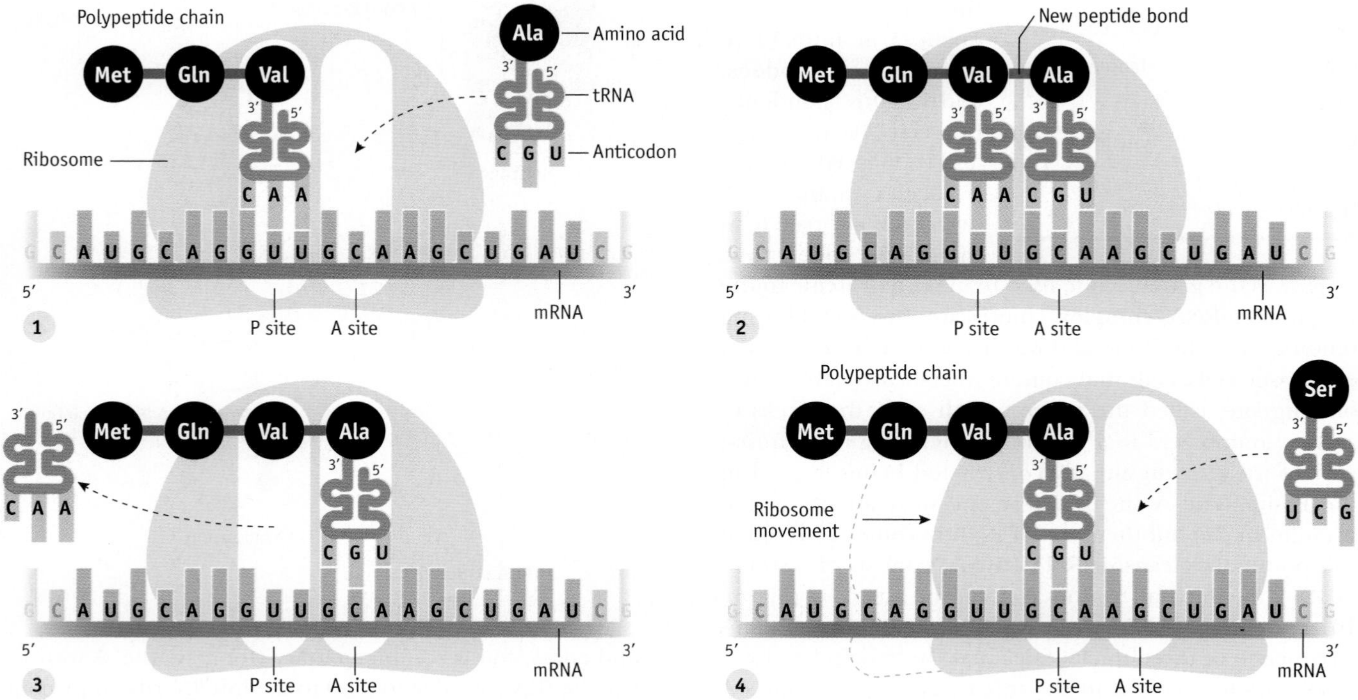

Figure 17 Protein synthesis. The tRNA with an anticodon complementary to the mRNA codon exposed in the A site of the ribosome brings the next amino acid to be added to the growing protein chain. After the new peptide bond is formed, the ribosome moves down the mRNA, exposing a new codon in the A site and transferring the previous tRNA and the protein chain to the P site.

roviruses, like the human immunodeficiency virus (HIV) that causes AIDS, use RNA as the repository of genetic information instead of DNA. Perhaps the first organisms on earth also used RNA to store genetic information.

In the 1980s, researchers discovered that particular strands of RNA catalyze some reactions involving cutting and joining together strands of RNA. Thomas Cech and Sidney Altman shared the 1989 Nobel Prize in chemistry for their independent discoveries of systems that utilize "catalytic RNA." One might imagine that an organism could use RNA both as the genetic material and as a catalyst. Information and action are thus combined in this one molecule.

According to proponents of the "RNA World" hypothesis, the first organism used RNA for both information and catalysis. At some later date, DNA evolved and had better information storage capabilities, so it took over the genetic information storage functions from RNA. Likewise, proteins eventually evolved and proved better at catalysis than RNA, so they took over this role for most reactions in a cell. RNA still plays a central role in the flow of genetic information, however. Genetic information does not go directly from DNA to proteins; it must pass through RNA along the way. Those favoring the RNA World hypothesis also point out that many enzyme cofactors, molecules that must be present for an enzyme to work, are RNA nucleotides or are based on RNA nucleotides. As we shall see, one of the most important molecules in metabolism is an RNA nucleotide, adenosine 5′-triphosphate (ATP). The importance of these nucleotides might date back to an earlier time when organisms were based on RNA alone.

The RNA World hypothesis is interesting and can answer some of the questions that arise in research on the origins of life, but it is not the only current hypothesis dealing with the origin of a self-replicating system. Much research remains to be done before we truly understand how life could have arisen on earth.

Lipids and Cell Membranes

Lipids are another important type of compound found in organisms. Among other things, they are the principal components of cell membranes and a repository of chemical energy in the form of fat. In addition, some of the chemical messengers called **hormones** are lipids.

Chemical Perspectives — AIDS and Reverse Transcriptase

One of the major health crises in modern times is the epidemic associated with the disease called acquired immune deficiency syndrome (AIDS). A person develops AIDS in the final stages of infection with the human immunodeficiency virus (HIV). At the time of this writing, an estimated 40 million people worldwide are infected with HIV. HIV is a retrovirus. Unlike all organisms and most viruses, a retrovirus does not use DNA as its genetic material, but rather, single-stranded RNA.

During the course of infection, the viral RNA is transcribed into DNA by means of an enzyme called reverse transcriptase. It is so named because the direction of information flow is in the opposite direction (RNA ⟶ DNA) than that usually found in cells. The resulting DNA is inserted into the cell's DNA. The infected cell then produces the proteins and RNA to make new virus particles.

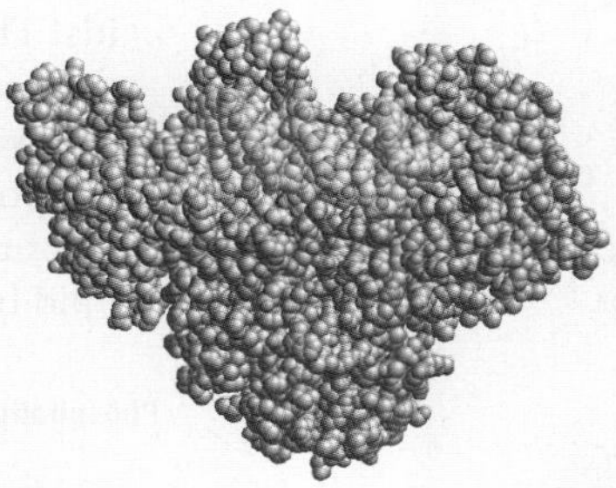

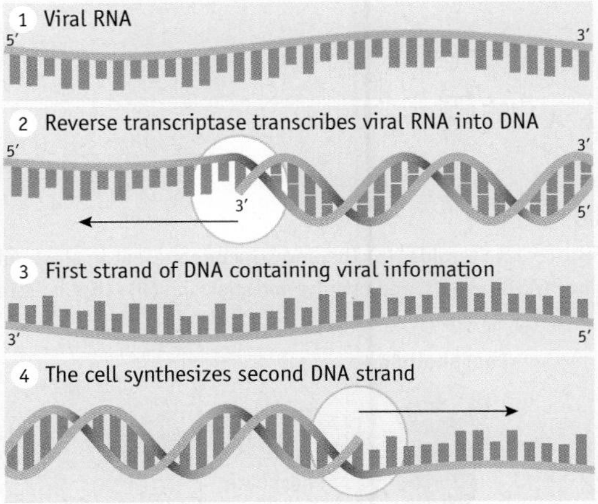

Reverse transcriptase. The reverse transcriptase enzyme consists of two subunits (shown in red and purple). Reverse transcriptase catalyzes the transcription of viral RNA into DNA. The cell then constructs a complementary strand of DNA. The resulting double-stranded DNA is inserted into the cell's DNA.

Reverse transcriptase consists of two subunits (see the accompanying figure). One subunit has a molar mass of approximately 6.6×10^4 g/mol, and the other has a molar mass of roughly 5.1×10^4 g/mol. Reverse transcriptase is not a very accurate enzyme, however. It makes an error in transcription for every 2000 to 4000 nucleotides copied. This is a much larger error rate than that for most cellular enzymes that copy DNA, which typically make one error for every 10^9 to 10^{10} nucleotides copied. The high error rate for reverse transcriptase contributes to the challenge scientists face in trying to combat HIV because these replication errors lead to frequent mutations in the virus. That is, the virus keeps changing, which means that developing a treatment that works and will continue to work is very difficult. Some treatments have been successful in significantly delaying the onset of AIDS, but none has yet proven to be a cure. More research is needed to combat this deadly disease.

Lipids include a wide range of compounds because classification of a compound as a lipid is based on its solubility rather than on a particular chemical functional group. A lipid is a compound that is at best slightly soluble in water but is soluble in organic solvents.

Polar compounds tend to be soluble in a polar solvent such as water. Nonpolar compounds tend not to be soluble in polar solvents but in nonpolar solvents instead. This tendency is sometimes referred to as "like dissolves like" (▶ Section 14.8). Compounds that are nonpolar, or at least substantally nonpolar, have limited solubility in water and are therefore lipids.

A major category of lipids consists of molecules that have one end that is polar and another end that is nonpolar. The polar end provides the slight water solubility necessary for it to be compatible with being in the aqueous environment of the cell, but the nonpolar end greatly limits the solubility. Fatty acids and triglycerides (◀ page 476), molecules important in the storage of energy in cells, are this type of lipid. The polar end in fatty acids is a carboxylic acid group, and the nonpolar end is a long hydrocarbon chain. A triglyceride is an ester formed by reaction of glycerol (1,2,3-propanetriol) with three different fatty acids.

Steroids are another category of lipids. Steroid molecules consist of four hydrocarbon rings joined together (Figure 18a). Three of the rings contain six carbon atoms, and one contains five carbon atoms. Examples of steroids are the sex hormones testosterone (◀ page 1), estradiol, and progesterone. Cholesterol (Figure 18b) is also an important steroid. You may have heard of cholesterol because of its correlation with heart disease. While some cholesterol is necessary for humans, excess cholesterol can deposit in blood vessels, thus partially blocking them and causing the heart to work harder than it should.

Lipids are very important components of cell membranes. The most prevalent molecules in most cell membranes are **phospholipids** (Figure 19a). These are similar to triglycerides in that they are based on glycerol. Two of the alcohol groups in glycerol are esterified to long-chain fatty acids. The third alcohol group, however, is bonded to a phosphate that has another hydrocarbon chain attached to it. Phosphate groups are very polar. In phospholipid molecules, the phosphate end is sometimes called the "head," and the nonpolar hydrocarbon chains comprise the "tail."

When phospholipids are placed in water, they typically arrange themselves in a **bilayer structure** (Figure 19b). This is exactly the arrangement that phospholipids have in a cell membrane. Water is present on both the inside and the outside of the bilayer, corresponding to the inside and outside of the cell. In the outside layer of the membrane, the phospholipids line up alongside each other such that their polar heads face the aqueous environment outside the cell. Moving inward, next come the tails of these lipids. The phospholipids in the second layer align themselves so that their nonpolar tails are in contact with the outer layer's nonpolar tails. Finally, the polar heads of the second layer face the aqueous environment inside the cell. The phospholipid bilayer nicely encloses the cell and provides

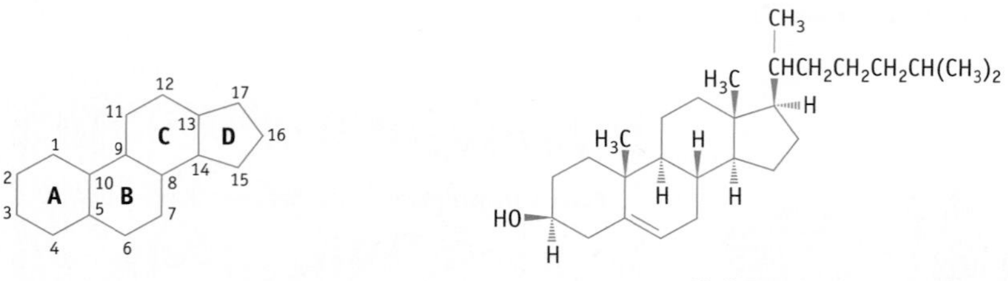

(a) (b) Cholesterol

Figure 18 **Steroids.** (a) The ring structure present in all steroids. There are three six-member rings (A, B, and C), and one five-member ring (D). (b) Cholesterol.

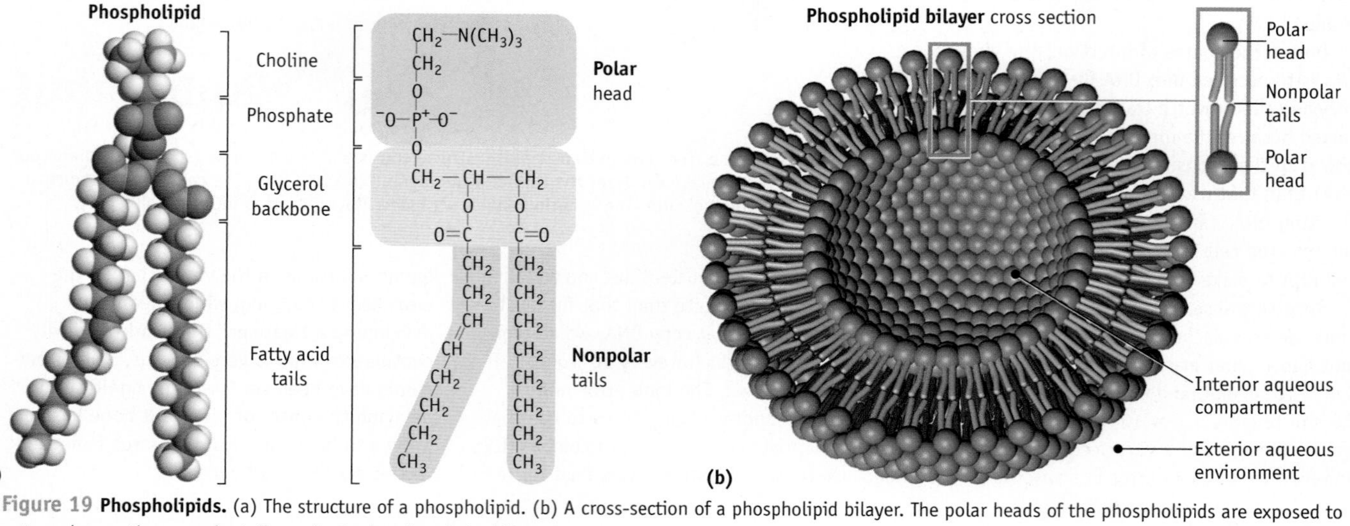

(a) (b)

Figure 19 **Phospholipids.** (a) The structure of a phospholipid. (b) A cross-section of a phospholipid bilayer. The polar heads of the phospholipids are exposed to water, whereas the nonpolar tails are in the interior of the bilayer.

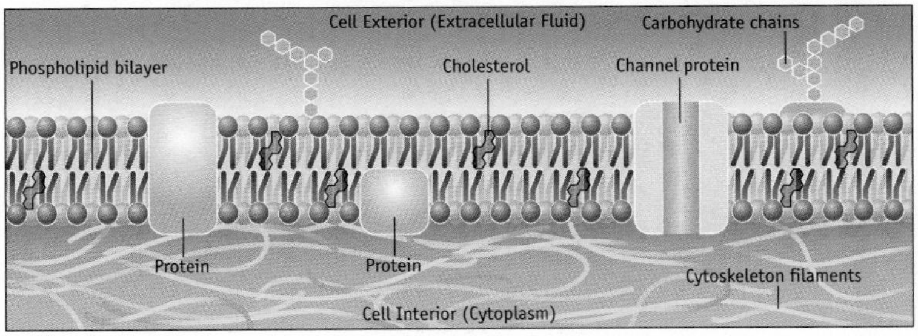

Figure 20 **The fluid-mosaic model of cell membranes.** A cell membrane is made up primarily of a phospholipid bilayer in which are embedded cholesterol, other lipids, and proteins. Movement within a layer occurs, but movement from one side of the bilayer to the other is rare.

a good barrier between the inside and the outside of the cell, due to the different solubility characteristics of the nonpolar region in the middle of the bilayer.

There are other molecules present in cell membranes, including cholesterol and proteins. Cholesterol is an important part of animal cell membranes, helping to give the membranes greater rigidity. Some proteins in the cell membrane allow select materials to cross from one side of the membrane to the other (*transport proteins*). Others accept chemical signals from other cells or respond to materials in the cell's environment (*receptor proteins*). Finally, some enzymes are also associated with the membrane.

The overall model for a cell membrane is called the **fluid-mosaic model** (Figure 20). In this model, the membrane's structure is largely that of a phospholipid bilayer described earlier. Embedded in this bilayer are molecules such as cholesterol and proteins. Movement of all of these components within each layer of the bilayer occurs readily; the membrane is thus fluid to a certain extent. On the other hand, there is little movement of components between layers. The "like dissolves like" observation provides the reason for this lack of exchange between layers. The head of a phospholipid in the outer layer, for example, would not be compatible with the very nonpolar region within the bilayer that it would need to pass through in order to traverse from one side of the bilayer to the other.

A cell membrane serves as the boundary between the cell and the rest of the universe, but an exchange of some materials between the cell and the outside world needs to occur. There are different mechanisms by which this happens (Figure 21). The simplest is *passive diffusion*. In this process, a molecule moves through the phospholipid bilayer from a region of higher concentration to a region of lower concentration, the natural direction of flow. Because the bilayer provides such a good barrier, only a few very small uncharged molecules (such as N_2, O_2, CO_2, and H_2O) can pass through the membrane this way. Many more species enter or leave the cell through a process called *facilitated diffusion*. In this process, ions or molecules still travel from a region of higher concentration to a region of lower concentration, but they do not pass directly through the bilayer. Instead, they pass through channels formed by proteins embedded in the cell membrane.

Sometimes, it is necessary for the cell to move species against the concentration gradient, from a region of lower concentration to a region of higher concentration. The cell accomplishes this by means of *active transport*. This is again mediated by transport proteins in the cell membrane. Because the species of interest must move in the opposite direction than it would normally go, the cell must expend energy in order to make this occur.

Finally, cells sometimes transport materials into themselves by means of *endocytosis*. This process is usually mediated by a receptor protein. The species of interest (called the substrate) binds to the receptor protein. A portion of the cell membrane surrounds the receptor–substrate complex. This portion of the cell membrane is then broken off, bringing the complex into the cell.

In this section, we have seen that the simple "like dissolves like" rule explains much about the formation and function of a cell membrane, the structure that defines the boundary between a cell and the rest of the universe. The phospholipid bilayer prevents many ions and molecules from entering or leaving the

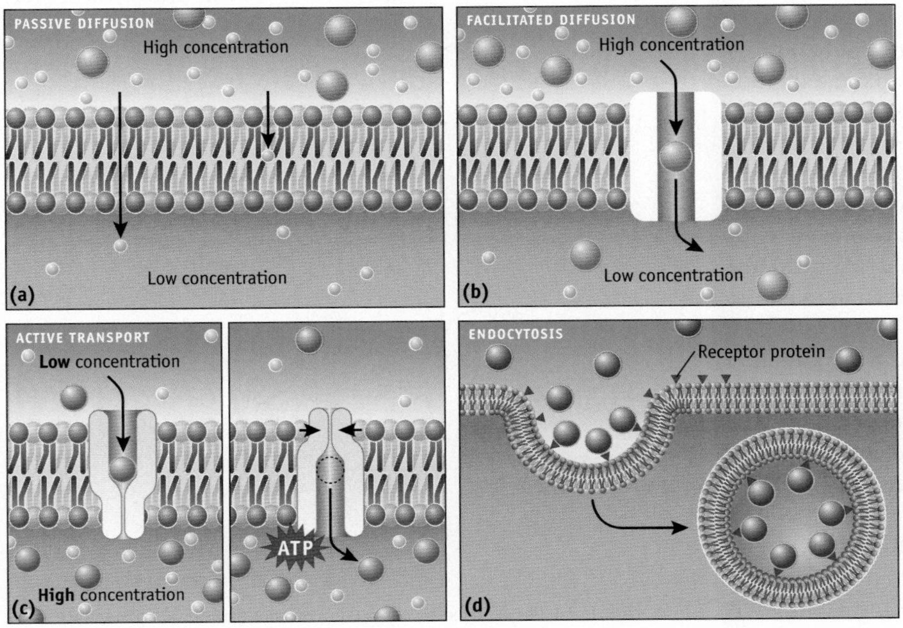

Figure 21 **Transport of materials across a cell membrane.** (a) Passive diffusion. (b) Facilitated diffusion. (c) Active transport. (d) Endocytosis.

cell directly. Finally, proteins embedded in the membrane allow the cell to exchange specific materials with the outside environment while still maintaining a barrier against others.

Metabolism

Why do we eat? Some components of our food, such as water, are used directly in our bodies. We break down other chemicals to obtain the molecular building blocks we need to make the many chemicals in our bodies. Oxidation of foods also provides the energy we need to perform the activities of life. The many different chemical reactions that foods undergo in the body to provide energy and chemical building blocks fall into the area of biochemistry called **metabolism.** We have already studied some aspects of energy changes in chemical reactions in Chapter 5 and some aspects of oxidation-reduction reactions in Chapter 3. We shall now examine some of these same considerations in biochemical reactions.

Energy and ATP

Substances in food, such as carbohydrates and fats, are oxidized in part of the metabolic process. These oxidations are energetically favorable reactions, releasing large quantities of energy. For example, the thermochemical equation for the oxidation of the sugar glucose ($C_6H_{12}O_6$) to form carbon dioxide and water is

$$C_6H_{12}O_6(s) + 6\ O_2(g) \rightarrow 6\ CO_2(g) + 6\ H_2O(\ell)$$

$$\Delta_r H° = -2803\ \text{kJ/mol-rxn}$$

Rather than carry out this reaction in one rapid and exothermic step, a cell carries out a more controlled oxidation in a series of steps so that it can obtain the energy in small increments. In addition, it would be inefficient if every part of a cell needed to have all the mechanisms necessary to carry out the oxidation of every type of molecule used for energy. Instead, it carries out the oxidation of compounds such as glucose in one location and stores the energy in a small set of compounds that can be used almost anywhere in the cell.

The principal compound used to perform this function is adenosine 5′-triphosphate (ATP). This ribonucleotide consists of a ribose molecule to which the nitrogenous base adenine is connected at the 1′ position and a triphosphate group is connected at the 5′ position (Figure 22). In aerobic respiration, the equivalent of 30–32 moles of ATP is typically produced per mole of glucose oxidized. Based on the ΔH values for the processes, a greater production of ATP might be expected, but the process is not completely efficient.

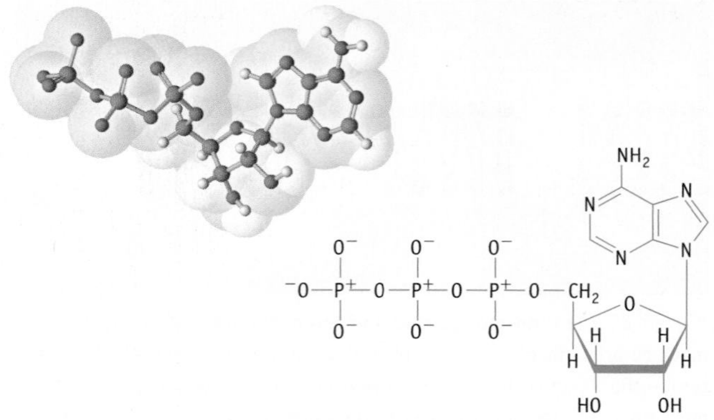

Figure 22 **Adenosine-5′-triphosphate (ATP).**

The hydrolysis of ATP to adenosine 5′-diphosphate (ADP) and inorganic phosphate (P_i) is an exothermic process (Figure 23).

$$\text{ATP} + H_2O \rightarrow \text{ADP} + P_i \qquad \Delta_r H \approx -24\ \text{kJ/mol-rxn}$$

Why is this reaction exothermic? We can assess this by evaluating bond enthalpies (page 388). In this reaction, we must break two bonds, a P—O bond in ATP and an H—O bond in water. But we also form two new bonds: a P—O bond between the phosphate group being cleaved off the ATP and the OH of the original water and an H—O bond between the hydrogen from the water and the portion of the ATP that forms ADP. In the overall process, more energy is released in forming these new bonds in the products than is required to break the necessary bonds in the reactants. Thus, the overall reaction is exothermic.

In cells, many chemical processes that would be endothermic on their own are linked with the hydrolysis of ATP.

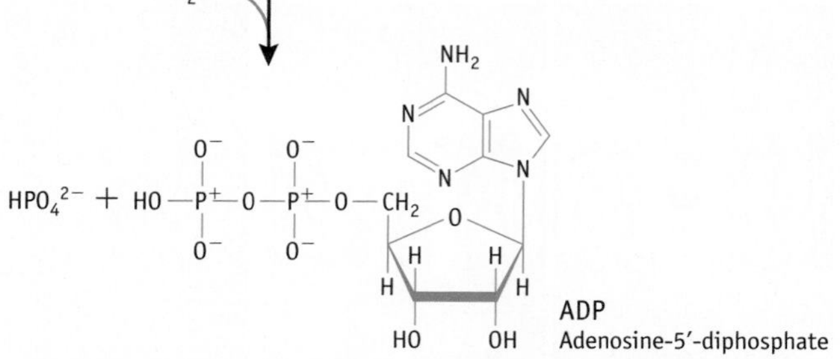

Figure 23 **The exothermic conversion of adenosine-5′-triphosphate (ATP) to adenosine-5′-diphosphate (ADP).**

The combination of an energetically unfavorable process with the energetically favorable hydrolysis of ATP can yield a process that is energetically favorable. For example, most cells have a greater concentration of potassium ions and a smaller concentration of sodium ions inside them than are present outside them. The natural tendency, therefore, is for sodium ions to flow into the cell and for potassium ions to flow out. To maintain the correct concentrations, the cell must counteract this movement and use active transport to pump sodium ions out of the cell and potassium ions into the cell. This requires energy. To accomplish this feat, the cell links this pumping process to the hydrolysis of ATP to ADP. The energy released from the hydrolysis reaction provides the energy to run a molecular pump (a transport protein) that moves the ions in the direction the cell needs.

Oxidation-Reduction and NADH

Cells also need compounds that can be used to carry out oxidation-reduction reactions. Just as ATP is a compound used in many biochemical reactions when energy is needed, so nature uses another small set of compounds to run many redox reactions. An important example is **nicotinamide adenine dinucleotide (NADH).** This compound consists of two ribonucleotides joined at their 5′ positions via a diphosphate linkage. One of the nucleotides has adenine as its nitrogenous base, whereas the other has a nicotinamide ring (Figure 24). When NADH is oxidized, changes occur in the nicotinamide ring, such that the equivalent of a hydride ion (H^-) is lost. Because this hydride ion has two electrons associated with it, the nicotinamide ring loses two electrons in the process. The resulting species, referred to as NAD^+, is shown on the right in Figure 24.

In many biochemical reactions, when a particular species needs to be reduced, it reacts with NADH. The NADH is oxidized to NAD^+, losing two electrons in the process, and the species of interest is reduced by gaining these electrons. If a species must be oxidized, the opposite process often occurs; that is, it reacts with NAD^+. The NAD^+ is reduced to NADH, and the species of interest is oxidized.

Respiration and Photosynthesis

In the process of **respiration**, a cell breaks down molecules such as glucose, oxidizing them to CO_2 and H_2O.

$$C_6H_{12}O_6(s) + 6\ O_2(g) \rightarrow 6\ CO_2(g) + 6\ H_2O(\ell)$$

The energy released in these reactions is used to generate the ATP needed by the cell. The sugars employed in this process can be traced back to green plants, where sugars are made via the process of **photosynthesis.** In photosynthesis, plants carry out the reverse of glucose oxidation—that is, the synthesis of glucose from CO_2 and H_2O.

$$6\ CO_2(g) + 6\ H_2O(\ell) \rightarrow C_6H_{12}O_6(s) + 6\ O_2(g)$$

Green plants have found a way to use light to provide the energy needed to run this endothermic reaction.

The key molecule involved in trapping the energy from light in photosynthesis is chlorophyll. Green plants contain two types of chlorophyll: chlorophyll a and chlorophyll b (Figure 25). The absorbance spectra of chlorophyll a and chlorophyll b are also shown in Figure 25. Notice that these molecules absorb best in the blue-violet and red-orange regions pf the visible spectrum. Not much light is absorbed in the green region. When white light shines on chlorophyll, red-orange and blue-violet light are absorbed by the chlorophyll; green light is not absorbed but rather is reflected. We see the reflected light, so plants appear green to us. The light energy absorbed by the chlorophyll is used to drive the process of photosynthesis.

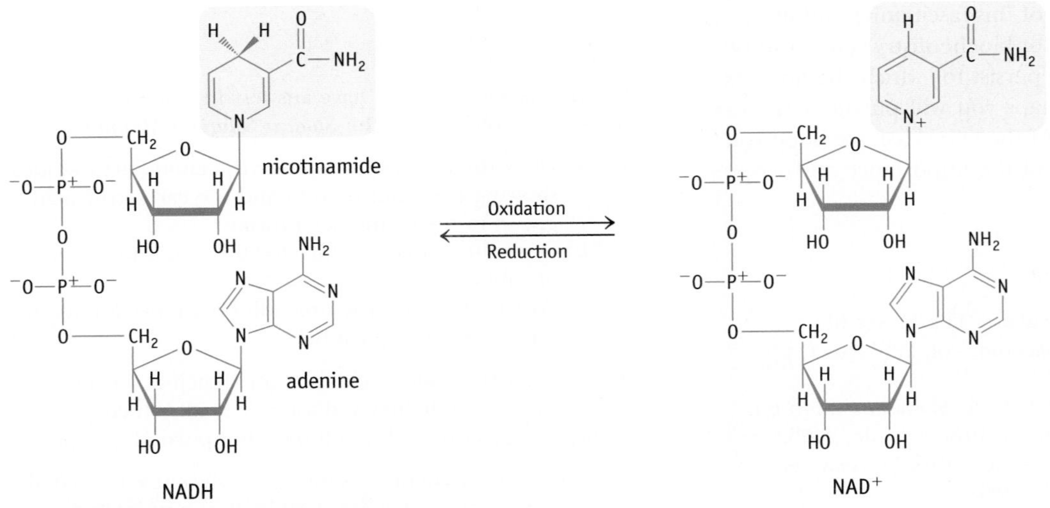

Figure 24 The structures of NADH and NAD^+.

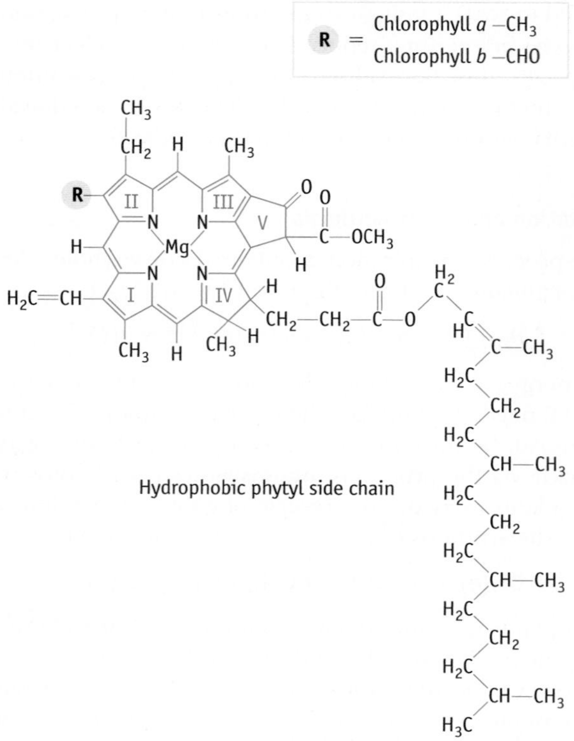

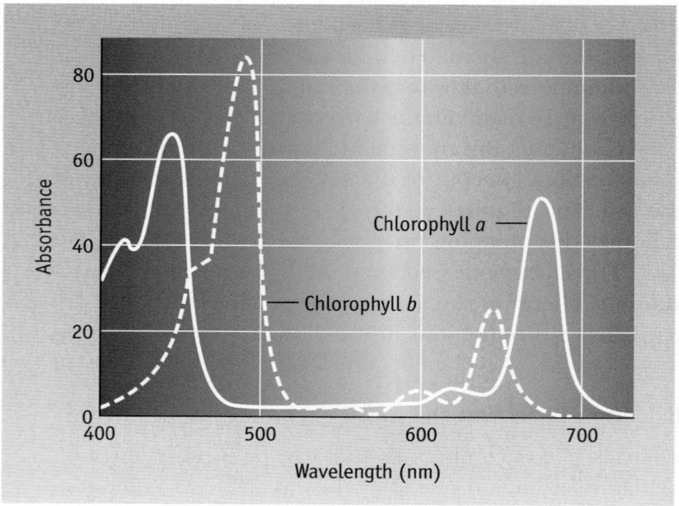

Figure 25 **The structure of chlorophyll and the visible absorbance spectra of chlorophyll a and b.**

Concluding Remarks

In this brief overview of biochemistry, we have examined proteins and their structures, nucleic acids, protein synthesis, lipids, and metabolism. As you have seen, the principles of chemistry you have been learning in general chemistry can be applied to understanding biological processes. We hope you have begun to recognize the marvelous complexity of life as well as some of the underlying patterns that exist within this complexity. This discussion has, however, merely scratched the surface of this fascinating and important field of study. Vast areas of biochemistry remain to be studied, and many questions persist for which the answers are currently unknown. Perhaps you will pursue a career doing research in this area. At the very least, we hope you have gained an appreciation of the importance and scope of this area of science.

SUGGESTED READINGS

1. J. E. Barrick and R. R. Breaker: "The Power of Riboswitches." *Scientific American*, Vol. 296, No. 1, pp. 50–57, 2007.
2. M. K. Campbell and S. O. Farrell: *Biochemistry*, 5th ed., Belmont, California: Thomson Brooks/Cole, 2006.
3. T. R. Cech: "RNA as an Enzyme." *Scientific American*, Vol. 255, No. 5, pp. 64–75, 1986.
4. W. C. Galley: "Exothermic Bond Breaking: A Persistent Misconception." *Journal of Chemical Education*, Vol. 81, pp. 523–525, 2004.
5. R. H. Garrett and C. M. Grisham: *Biochemistry*, 3rd ed., Belmont, California: Thomson Brooks/Cole, 2007.
6. D. C. Phillips: "The Three-dimensional Structure of an Enzyme Molecule." *Scientific American*, Vol. 215, No. 5, pp. 78–90, 1966.
7. J. D. Watson: *The Double Helix: A Personal Account of the Discovery of the Structure of DNA*, New York: Mentor, 1968.

STUDY QUESTIONS

Blue-numbered questions have answers in Appendix P and fully-worked solutions in the *Student Solutions Manual.*

1. (a) Draw the Lewis structure for the amino acid valine, showing the amino group and the carboxylic acid group in their unionized forms.
   (b) Draw the Lewis structure for the zwitterionic form of valine.
   (c) Which of these structures will be the predominant form at physiological pH?
2. Consider the amino acids alanine, leucine, serine, phenylalanine, lysine, and aspartic acid. Which have polar R groups, and which have nonpolar R groups?
3. Using Lewis structures, show two different ways that alanine and glycine may be combined in a peptide bond.

4. When listing the sequence of amino acids in a polypeptide or protein, the sequence always begins with the amino acid that has the free amino group and ends with the amino acid that has the free carboxylic acid group. Draw the Lewis structure for the tripeptide: serine-leucine-valine.

5. Draw two Lewis structures for the dipeptide alanine-isoleucine that show the resonance structures of the amide linkage.

6. Identify the type of structure (primary, secondary, tertiary, or quaternary) that corresponds to the following statements.
   (a) This type of structure is the amino acid sequence in the protein.
   (b) This type of structure indicates how different peptide chains in the overall protein are arranged with respect to one another.
   (c) This type of structure refers to how the polypeptide chain is folded, including how amino acids that are far apart in the sequence end up in the overall molecule.
   (d) This type of structure deals with how amino acids near one another in the sequence arrange themselves.

7. (a) Draw the Lewis structure for the sugar ribose.
   (b) Draw the Lewis structure for the nucleoside adenosine (it consists of ribose and adenine).
   (c) Draw the Lewis structure for the nucleotide adenosine 5′-monophosphate.

8. A DNA or RNA sequence is usually written from the end with a free 5′-OH to the end with a free 3′-OH. Draw the Lewis structure for the tetranucleotide AUGC.

9. Do the DNA sequences ATGC and CGTA represent the same molecule?

10. (a) What type of interaction holds DNA's double-helical strands together?
    (b) Why would it not be good for DNA's double-helical strands to be held together by covalent bonds?

11. Complementary strands of nucleic acids run in opposite directions. That is, the 5′ end of one strand will be lined up with the 3′ end of the other. Given the following nucleotide sequence in DNA:
    5′—ACGCGATTC—3′:
    (a) Determine the sequence of the complementary strand of DNA. Report this sequence by writing it from its 5′ end to its 3′ end (the usual way of reporting nucleic acid sequences).
    (b) Write the sequence (5′–3′) for the strand of mRNA that would be complementary to the original strand of DNA.
    (c) Assuming that this sequence is part of the coding sequence for a protein and that it is properly lined up so that the first codon of this sequence begins with the 5′ nucleotide of the mRNA, write the sequences for the three anticodons that would be complementary to this strand of mRNA in this region.
    (d) What sequence of amino acids is coded for by this mRNA?

12. (a) According to the genetic code in Table 1, which amino acid is coded for by the mRNA codon GAA?
    (b) What is the sequence in the original DNA that led to this codon being present in the mRNA?
    (c) If a mutation occurs in the DNA in which a G is substituted for the nucleotide at the second position of this coding region in the DNA, which amino acid will now be selected?

13. (a) Describe what occurs in the process of transcription.
    (b) Describe what occurs in the process of translation.

14. Sketch a section of a phospholipid bilayer in which you let a circle represent the polar head group and curvy lines represent the hydrocarbon tails. Label the regions of the bilayer as being polar or nonpolar.

15. What structure do all steroids have in common?

16. The section about metabolism provided a value for $\Delta_r H°$ for the oxidation of one mole of glucose. Using $\Delta_f H°$ values at 25 °C, verify that this is the correct value for the equation

$$C_6H_{12}O_6(s) + 6\ O_2(g) \rightarrow 6\ CO_2(g) + 6\ H_2O(\ell)$$

$$\Delta_f H°[C_6H_{12}O_6(s)] = -1273.3\ \text{kJ/mol}$$

17. Which of the following statements are true?
    (a) Breaking the P—O bond in ATP is an exothermic process.
    (b) Making a new bond between the phosphorus atom in the phosphate group being cleaved off ATP and the OH group of water is an exothermic process.
    (c) Breaking bonds is an endothermic process.
    (d) The energy released in the hydrolysis of ATP may be used to run endothermic reactions in a cell.

18. Consider the following reaction:
    $NADH + H^+ + \frac{1}{2}\ O_2 \rightarrow NAD^+ + H_2O$
    (a) Which species is being oxidized (NADH, H^+, or O_2)?
    (b) Which species is being reduced?
    (c) Which species is the oxidizing agent?
    (d) Which species is the reducing agent?

19. (a) Calculate the enthalpy change for the production of one mole of glucose by the process of photosynthesis at 25 °C. $\Delta_f H°$ [glucose(s)] = −1273.3 kJ/mol

$$6\ CO_2(g) + 6\ H_2O(\ell) \rightarrow C_6H_{12}O_6(s) + 6\ O_2(g)$$

    (b) What is the enthalpy change involved in producing 1 molecule of glucose by this process?
    (c) Chlorophyll molecules absorb light of various wavelengths. One wavelength absorbed is 650 nm. Calculate the energy of a photon of light having this wavelength.
    (d) Assuming that all of this energy goes toward providing the energy required for the photosynthetic reaction, can the absorption of one photon at 650 nm lead to the production of one molecule of glucose, or must multiple photons be absorbed?

11 Gases and Their Properties

©Davis Barber/PhotoEdit

The Atmosphere and Altitude Sickness

Some of you may have dreamed of climbing to the summits of the world's tallest mountains, or you may be an avid skier and visit high-mountain ski areas. In either case, "acute mountain sickness" (AMS) is a possibility. AMS is common at higher altitudes and is characterized by a headache, nausea, insomnia, dizziness, lassitude, and fatigue. It can be prevented by a slow ascent, and its symptoms can be relieved by a mild pain reliever.

AMS and more serious forms of high altitude sickness are generally due to hypoxia or oxygen deprivation. The oxygen concentration in Earth's atmosphere is 21%. As you go higher into the atmosphere, the concentration remains 21%, but the atmospheric pressure drops. When you reach 3000 m (the altitude of some ski resorts), the barometric pressure is about 70% of that at sea level. At 5000 m, barometric pressure is only 50% of sea level, and on the summit of Mt. Everest, it is only 29% of the sea level pressure. At sea level, your blood is nearly saturated with oxygen, but as the partial pressure of oxygen drops, the percent saturation drops as well. At $P(O_2)$ of 50 mm Hg, hemoglobin in the red blood cells is about 80% saturated. Other saturation levels are given in the table (for a pH of 7.4).

$P(O_2)$ (mm Hg)	Approximate Percent Saturation
90	95%
80	92%
70	90%
60	85%
50	80%
40	72%

For more on the atmosphere, see page 534.

Questions:

1. Assume a sea level pressure of 1 atm (760 mm Hg). What are the O_2 partial pressures at a 3000-m ski resort and on Mt. Everest?
2. What are the approximate blood saturation levels under these conditions?

Answers to these questions are in Appendix Q.

Chapter Goals

See Chapter Goals Revisited (page 544) for Study Questions keyed to these goals and assignable in OWL.

- Understand the basis of the gas laws and know how to use those laws (Boyle's law, Charles's law, Avogadro's hypothesis, Dalton's law).
- Use the ideal gas law.
- Apply the gas laws to stoichiometric calculations.
- Understand kinetic-molecular theory as it is applied to gases, especially the distribution of molecular speeds (energies).
- Recognize why gases do not behave like ideal gases under some conditions.

Chapter Outline

Mountain climbers, hot air balloons, SCUBA diving, and automobile air bags (Figure 11.1) depend on the properties of gases. Aside from understanding how these work, there are at least three reasons for studying gases. First, some common elements and compounds (such as oxygen, nitrogen, and methane) exist in the gaseous state under normal conditions of pressure and temperature. Furthermore, many liquids such as water can be vaporized, and the physical properties of these vapors are important. Second, our gaseous atmosphere provides one means of transferring energy and material throughout the globe, and it is the source of life-sustaining chemicals.

The third reason for studying gases is also compelling. Of the three states of matter, gases are reasonably simple when viewed at the molecular level, and, as a result, gas behavior is well understood. It is possible to describe the properties of gases *qualitatively* in terms of the behavior of the molecules that make up the gas. Even more impressive, it is possible to describe the properties of gases *quantitatively* using simple mathematical models. One objective of scientists is to develop precise mathematical and conceptual models of natural phenomena, and a study of gas behavior will introduce you to this approach. To describe gases, chemists have learned that only four quantities are needed: the pressure (P), volume (V), and temperature (T, kelvins) of the gas, and amount (n, mol).

Chemistry.Now™

Throughout the text this icon introduces an opportunity for self-study or to explore interactive tutorials by signing in at **www.thomsonedu.com/login**.

SAAB Car, USA, Inc.

FIGURE 11.1. Automobile air bags. Most automobiles are now equipped with airbags to protect the driver and passengers in the event of a head-on or side crash. Such bags are inflated with nitrogen gas, which is generated by the explosive decomposition of sodium azide:

$$2\ NaN_3(s) \longrightarrow 2\ Na(s) + 3\ N_2(g)$$

The airbag is fully inflated in about 0.050 s. This is important because the typical automobile collision lasts about 0.125 s.
See ChemistryNow Screen 11.1 for questions about automobile air bags.

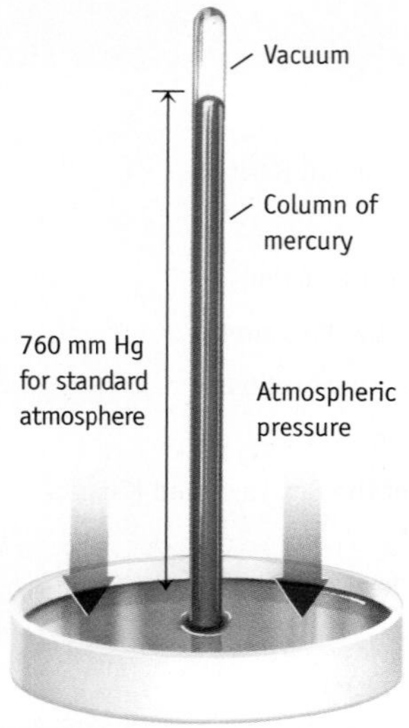

FIGURE 11.2 A barometer. The pressure of the atmosphere on the surface of the mercury in the dish is balanced by the downward pressure exerted by the column of mercury. The barometer was invented in 1643 by Evangelista Torricelli (1608–1647). A unit of pressure called the torr in his honor is equivalent to 1 mm Hg.

■ **Hectopascals** Meteorologists have long measured atmospheric pressure in millibars. However, after the SI system of units became more widespread, they began to use the unit "hectopascal," which is equivalent to the millibar.
1 hectopascal (hPa) = 100 Pa = 1 mbar
1 kilopascal (kPa) = 1000 Pa = 10 hPa

11.1 Gas Pressure

Pressure is the force exerted on an object divided by the area over which it is exerted, and a barometer depends on this to measure atmospheric pressure. A barometer can be made by filling a tube with a liquid, often mercury, and inverting the tube in a dish containing the same liquid (Figure 11.2). If the air has been removed completely from the vertical tube, the liquid in the tube assumes a level such that the pressure exerted by the mass of the column of liquid in the tube is balanced by the pressure of the atmosphere pressing down on the surface of the liquid in the dish.

Pressure is often reported in units of **millimeters of mercury (mm Hg),** the height (in mm) of the mercury column in a mercury barometer above the surface of the mercury in the dish. At sea level, this height is about 760 mm. Pressures are also reported as **standard atmospheres (atm),** a unit defined as follows:

1 standard atmosphere (1 atm) = 760 mm Hg (exactly)

The SI unit of pressure is the **pascal (Pa).**

$$1 \text{ pascal (Pa)} = 1 \text{ newton/meter}^2$$

(The newton is the SI unit of force.) Because the pascal is a very small unit compared with ordinary pressures, the unit kilopascal (kPa) is more often used. Another unit used for gas pressures is the **bar**, where 1 bar = 100,000 Pa. To summarize, the units used in science for pressure are

1 atm = 760 mm Hg (exactly) = 101.325 kilopascals (kPa) = 1.01325 bar

or

$$1 \text{ bar} = 1 \times 10^5 \text{ Pa (exactly)} = 1 \times 10^2 \text{ kPa} = 0.9872 \text{ atm}$$

EXAMPLE 11.1 Pressure Unit Conversions

Problem Convert a pressure of 635 mm Hg into its corresponding value in units of atmospheres (atm), bars, and kilopascals (kPa).

Strategy Use the relationships between millimeters of Hg, atmospheres, bars, and pascals described earlier in the text.

Solution The relationship between millimeters of mercury and atmospheres is 1 atm = 760 mm Hg.

$$635 \cancel{\text{mm Hg}} \times \frac{1 \text{ atm}}{760 \cancel{\text{mm Hg}}} = 0.836 \text{ atm}$$

The relationship between atmospheres and bars is 1 atm = 1.013 bar.

$$0.836 \cancel{\text{atm}} \times \frac{1.013 \text{ bar}}{1 \cancel{\text{atm}}} = 0.846 \text{ bar}$$

The relationship between millimeters of mercury and kilopascals is 101.325 kPa = 760 mm Hg.

$$635 \cancel{\text{mm Hg}} \times \frac{101.3 \text{ kPa}}{760 \cancel{\text{mm Hg}}} = 84.6 \text{ kPa}$$

A Closer Look — Measuring Gas Pressure

Pressure is the force exerted on an object divided by the area over which the force is exerted:

$$Pressure = force/area$$

This book, for example, weighs more than 4 lb and has an area of 82 in^2, so it exerts a pressure of about 0.05 lb/in^2 when it lies flat on a surface. (In metric units, the pressure is about 3 g/cm^2.)

Now consider the pressure that the column of mercury exerts on the mercury in the dish in the barometer shown in Figure 11.2. This pressure exactly balances the pressure of the atmosphere. Thus, the pressure of the atmosphere (or of any other gas) can be measured by relating it to the height of the column of mercury (or any other liquid) the gas can support.

Mercury is the liquid of choice for barometers because of its high density. A barometer filled with water would be over 10 m in height. [The water column is about 13.6 times as high as a column of mercury because mercury's density (13.53 g/cm^3) is 13.6 times that of water (density = 0.997 g/cm^3, at 25 °C).]

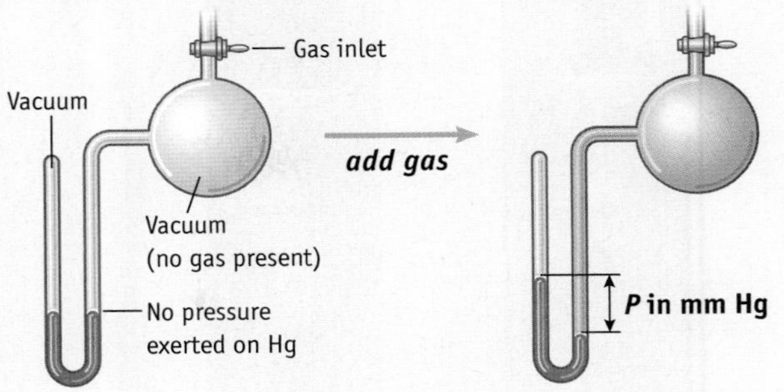

In the laboratory, we often use a U-tube manometer, which is a mercury-filled, U-shaped glass tube. The closed side of the tube has been evacuated so that no gas remains to exert pressure on the mercury on that side. The other side is open to the gas whose pressure we want to measure. When the gas presses on the mercury in the open side, the gas pressure is read directly (in mm Hg) as the difference in mercury levels on the closed and open sides.

You may have used a tire gauge to check the pressure in your car or bike tires. In the U.S., such gauges usually indicate the pressure in pounds per square inch (psi) where 1 atm = 14.7 psi. Some newer gauges give the pressure in kilopascals as well. Be sure to recognize that the reading on the scale refers to the pressure *in excess of atmospheric pressure.* (A flat tire is not a vacuum; it contains air at atmospheric pressure.) For example, if the gauge reads 35 psi (2.4 atm), the pressure in the tire is actually about 50 psi or 3.4 atm.

EXERCISE 11.1 Pressure Unit Conversions

Rank the following pressures in decreasing order of magnitude (from largest to smallest): 75 kPa, 250 mm Hg, 0.83 bar, and 0.63 atm.

11.2 Gas Laws: The Experimental Basis

Boyle's Law: The Compressibility of Gases

When you pump up the tires of your bicycle, the pump squeezes the air into a smaller volume (Figure 11.3). This property of a gas is called its **compressibility.** While studying the compressibility of gases, Robert Boyle (1627–1691) observed that the volume of a fixed amount of gas at a given temperature is inversely proportional to the pressure exerted by the gas. All gases behave in this manner, and we now refer to this relationship as **Boyle's law.**

Boyle's law can be demonstrated in many ways. In Figure 11.4, a hypodermic syringe is filled with air and sealed. When pressure is applied to the movable plunger of the syringe, the air inside is compressed. As the pressure (P) increases on the syringe, the gas volume in the syringe (V) decreases. When $1/V$ of the gas in the syringe is plotted as a function of P, a straight line results. This type of plot demonstrates that the pressure and volume of the gas are inversely proportional; that is, they change in opposite directions.

Charles D. Winters

FIGURE 11.3 A bicycle pump—Boyle's law in action. This works by compressing air into a smaller volume. You experience Boyle's law because you can feel the increasing pressure of the gas as you press down on the plunger.

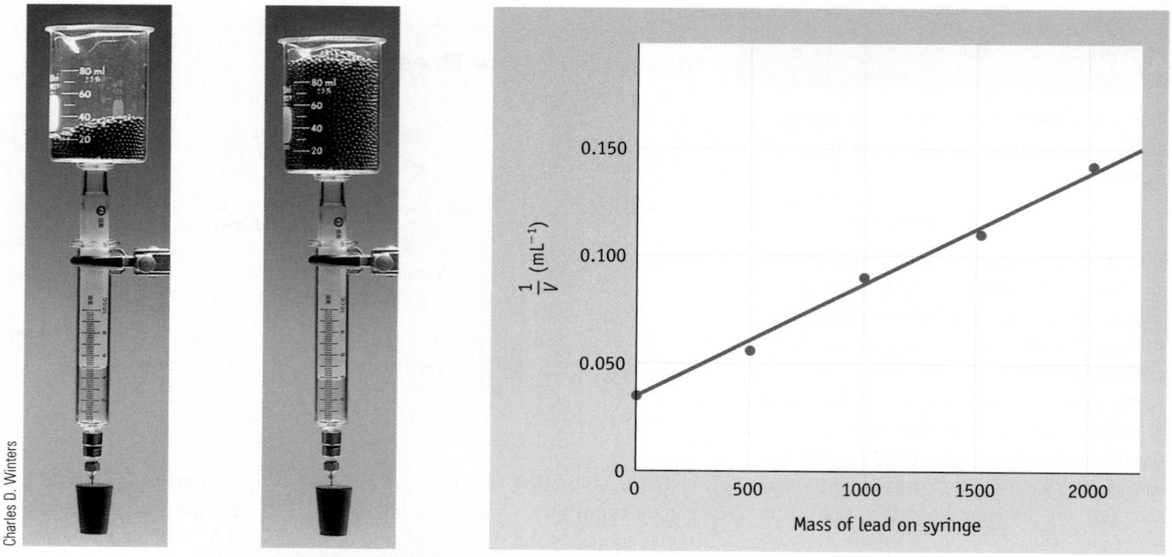

Active Figure 11.4 **An experiment to demonstrate Boyle's law.** A syringe filled with air was sealed. Pressure was applied by adding lead shot to the beaker on top of the syringe. As the mass of lead increased, the pressure on the air in the sealed syringe increased, and the gas was compressed. A plot of (1/volume of air in the syringe) versus P (as measured by the mass of lead) is a straight line.

Chemistry Now™ Sign in at www.thomsonedu.com/login and go to the Chapter Contents menu to explore an interactive version of this figure accompanied by an exercise.

Mathematically, we can write Boyle's law as:

$$P \propto \frac{1}{V} \quad \text{when } n \text{ and } T \text{ are constant}$$

where the symbol $\propto$ means "proportional to."

When two quantities are proportional to each other, they can be equated if a *proportionality constant*, here called C_B, is introduced.

$$P = C_B \times \frac{1}{V} \quad \text{or} \quad PV = C_B \quad \text{when } n \text{ and } T \text{ are constant}$$

This form of Boyle's law expresses the fact that *the product of the pressure and volume of a gas sample is a constant at a given temperature,* where the constant C_B is determined by the amount of gas (in moles) and its temperature (in kelvins). It follows from this that, if the pressure–volume product is known for a gas sample under one set of conditions (P_1 and V_1), then it is known for another set of conditions (P_2 and V_2). Under either set of conditions, the PV product is equal to C_B, so

$$P_1V_1 = P_2V_2 \text{ at constant } n \text{ and } T \tag{11.1}$$

This form of Boyle's law is useful when we want to know, for example, what happens to the volume of a given amount of gas when the pressure changes at a constant temperature.

EXAMPLE 11.2 Boyle's Law

Problem A sample of gaseous nitrogen in a 65.0-L automobile air bag has a pressure of 745 mm Hg. If this sample is transferred to a 25.0-L bag at the same temperature, what is the pressure of the gas in the 25.0-L bag?

Strategy Here, we use Boyle's law, Equation 11.1. The original pressure and volume (P_1 and V_1) and the new volume (V_2) are known.

Solution It is often useful to make a table of the information provided.

Initial Conditions	Final Conditions
$P_1 = 745$ mm Hg	$P_2 = ?$
$V_1 = 65.0$ L	$V_2 = 25.0$ L

You know that $P_1V_1 = P_2V_2$. Therefore,

$$P_2 = \frac{P_1V_1}{V_2} = \frac{(745 \text{ mm Hg})(65.0 \text{ L})}{25.0 \text{ L}} = 1940 \text{ mm Hg}$$

Comment According to Boyle's law, P and V change in opposite directions. Because the volume has decreased, the new pressure (P_2) must be greater than the original pressure (P_1). A quick way to solve these problems takes advantage of this: if the volume decreases, the pressure must increase, and the original pressure must be multiplied by a volume fraction greater than 1.

$$P_2 = P_1\left(\frac{65.0 \text{ L}}{25.0 \text{ L}}\right)$$

EXERCISE 11.2 Boyle's Law

A sample of CO_2 with a pressure of 55 mm Hg in a volume of 125 mL is compressed so that the new pressure of the gas is 78 mm Hg. What is the new volume of the gas? (Assume the temperature is constant.)

The Effect of Temperature on Gas Volume: Charles's Law

In 1787, the French scientist Jacques Charles (1746–1823) discovered that the volume of a fixed quantity of gas at constant pressure decreases with decreasing temperature (Figure 11.5).

Figure 11.6 illustrates how the volumes of two different gas samples change with temperature (at a constant pressure). When the plots of volume versus temperature

Charles D. Winters

(a) (b) (c)

FIGURE 11.5 A dramatic illustration of Charles's law. (a) Air-filled balloons are placed in liquid nitrogen (77 K). The volume of the gas in the balloons is dramatically reduced at this temperature. (b) After all of the balloons have been placed in the liquid nitrogen, (c) they are removed; as they warm to room temperature, they reinflate to their original volume.

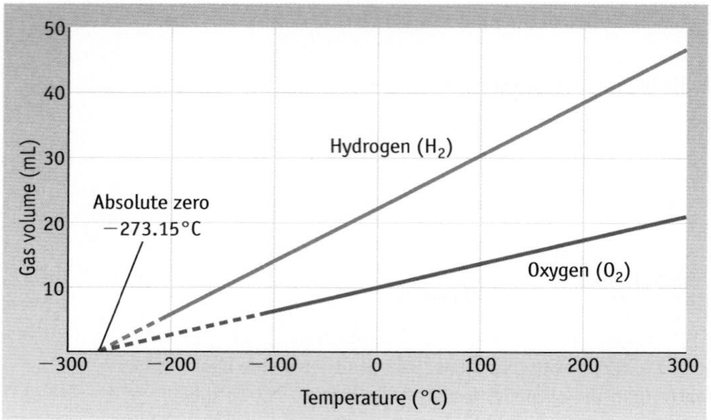

T (°C)	T (K)	Vol. H_2 (mL)	Vol. O_2 (mL)
300	573	47.0	21.1
200	473	38.8	17.5
100	373	30.6	13.8
0	273	22.4	10.1
−100	173	14.2	6.39
−200	73	6.00	—

Active Figure 11.6 **Charles's law.** The solid lines represent the volumes of the samples of hydrogen and oxygen at different temperatures. The volumes decrease as the temperature is lowered (at constant pressure). These lines, if extended, intersect the temperature axis at approximately −273 °C.

Chemistry Now™ Sign in at www.thomsonedu.com/login and go to the Chapter Contents menu to explore an interactive version of this figure accompanied by an exercise.

■ **Boyle's and Charles's Laws** Neither Boyle's law nor Charles's law depends on the identity of the gas being studied. These laws describe the behavior of any gaseous substance, regardless of its identity.

are extended to lower temperatures, they all reach zero volume at the same temperature, −273.15 °C. (Of course, gases will not actually reach zero volume; they liquefy above that temperature.) This temperature is significant, however. William Thomson (1824–1907), also known as Lord Kelvin, proposed a temperature scale—now known as the Kelvin scale—for which the zero point is −273.15 °C (◀ page 27).

When Kelvin temperatures are used with volume measurements, the volume–temperature relationship is

$$V = C_c \times T$$

where C_c is a proportionality constant (which depends on the amount of gas and its pressure). This is **Charles's law,** which states that if a given quantity of gas is held at a constant pressure, its volume is directly proportional to the Kelvin temperature.

Writing Charles's law another way, we have $V/T = C_c$; that is, the volume of a gas divided by the temperature of the gas (in kelvins) is constant for a given sample of gas at a specified pressure. Therefore, if we know the volume and temperature of a given quantity of gas (V_1 and T_1), we can find the volume, V_2, at some other temperature, T_2, using the equation

$$\frac{V_1}{T_1} = \frac{V_2}{T_2} \text{ at constant } n \text{ and } P \quad \textbf{(11.2)}$$

Calculations using Charles's law are illustrated by the following example and exercise. Be sure to notice that the temperature T *must always be expressed in kelvins.*

EXAMPLE 11.3 Charles's Law

Problem A sample of CO_2 in a gas-tight syringe (as in Figure 11.4) has a volume of 25.0 mL at room temperature (20.0 °C). What is the final volume of the gas if you hold the syringe in your hand to raise its temperature to 37 °C?

Strategy Because a given quantity of gas is heated (at a constant pressure), Charles's law applies. Because we know the original V and T, and want to calculate a new volume at a new, but known, temperature, use Equation 11.2.

Solution Organize the information in a table. Remember the temperature must be converted to kelvins.

Initial Conditions	**Final Conditions**
$V_1 = 25.0$ mL	$V_2 = ?$
$T_1 = 20.0 + 273.2 = 293.2$ K	$T_2 = 37 + 273 = 310.$ K

Substitute the known quantities into Equation 11.2, and solve for V_2:

$$V_2 = T_2\left(\frac{V_1}{T_1}\right) = 310.\ \text{K}\left(\frac{25.0\ \text{mL}}{293.2\ \text{K}}\right) = 26.5\ \text{mL}$$

Comment As expected, the volume of the gas increased with a temperature increase. The new volume (V_2) must equal the original volume (V_1) multiplied by a temperature fraction that is greater than 1 to reflect the effect of the temperature increase. That is,

$$V_2 = V_1\left(\frac{310.\ \text{K}}{293\ \text{K}}\right)$$

EXERCISE 11.3 Charles's Law

A balloon is inflated with helium to a volume of 45 L at room temperature (25 °C). If the balloon is cooled to −10 °C, what is the new volume of the balloon? Assume that the pressure does not change.

Combining Boyle's and Charles's Laws: The General Gas Law

The volume of a given amount of gas is inversely proportional to its pressure at constant temperature (Boyle's law) and directly proportional to the Kelvin temperature at constant pressure (Charles's law). But what if we need to know what happens to the gas when two of the three parameters (*P*, *V*, and *T*) change? For example, what would happen to the pressure of a sample of nitrogen in an automobile air bag if the same amount of gas were placed in a smaller bag and heated to a higher temperature? You can deal with this situation by combining the two equations that express Boyle's and Charles's laws.

$$\frac{P_1V_1}{T_1} = \frac{P_2V_2}{T_2} \text{ for a given amount of gas, } n \qquad \textbf{(11.3)}$$

This equation is sometimes called the **general gas law** or **combined gas law.** It applies specifically to situations in which the *amount of gas does not change.*

NASA/Science Source/Photo Researchers, Inc.

A weather balloon is filled with helium. As it ascends into the troposphere, does the volume increase or decrease?

EXAMPLE 11.4 General Gas Law

Problem Helium-filled balloons are used to carry scientific instruments high into the atmosphere. Suppose a balloon is launched when the temperature is 22.5 °C and the barometric pressure is 754 mm Hg. If the balloon's volume is 4.19×10^3 L (and no helium escapes from the balloon), what will the volume be at a height of 20 miles, where the pressure is 76.0 mm Hg and the temperature is −33.0 °C?

Strategy Here we know the initial volume, temperature, and pressure of the gas. We want to know the volume of the same amount of gas at a new pressure and temperature. It is most convenient to use Equation 11.3, the general gas law.

Solution Begin by setting out the information given in a table.

Initial Conditions	**Final Conditions**
$V_1 = 4.19 \times 10^3$ L	$V_2 = ?$ L
$P_1 = 754$ mm Hg	$P_2 = 76.0$ mm Hg
$T_1 = 22.5$ °C (295.7 K)	$T_2 = -33.0$ °C (240.2 K)

We can rearrange the general gas law to calculate the new volume V_2:

$$V_2 = \left(\frac{T_2}{P_2}\right) \times \left(\frac{P_1V_1}{T_1}\right) = V_1 \times \frac{P_1}{P_2} \times \frac{T_2}{T_1}$$

$$= 4.19 \times 10^3 \text{ L}\left(\frac{754 \text{ mm Hg}}{76.0 \text{ mm Hg}}\right)\left(\frac{240.2 \text{ K}}{295.7 \text{ K}}\right)$$

$$= 3.38 \times 10^4 \text{ L}$$

Comment The pressure decreased by almost a factor of 10, which should lead to about a ten-fold volume increase. This increase is partly offset by a drop in temperature that leads to a volume decrease. On balance, the volume increases because the pressure has dropped so substantially.

Notice that the solution was to multiply the original volume (V_1) by a pressure factor larger than 1 (because the volume increases with a lower pressure) and a temperature factor smaller than 1 (because volume decreases with a decrease in temperature).

EXERCISE 11.4 The General Gas Law

You have a 22.-L cylinder of helium at a pressure of 150 atm and at 31 °C. How many balloons can you fill, each with a volume of 5.0 L, on a day when the atmospheric pressure is 755 mm Hg and the temperature is 22 °C?

The general gas law leads to other, useful predictions of gas behavior. For example, if a given amount of gas is held in a closed container, the pressure of the gas will increase with increasing temperature.

$$\frac{P_1}{T_1} = \frac{P_2}{T_2} \text{ when } V_1 = V_2 \text{ and so } P_2 = P_1 \times \frac{T_2}{T_1}$$

■ **Gay-Lussac's Law** Gay-Lussac's law states that, at constant volume, the pressure of a given mass of gas is proportional to the absolute temperature. In 1779 Joseph Lambert proposed a definition of absolute zero of temperature based on this relationship.

That is, when T_2 is greater than T_1, P_2 will be greater than P_1. In fact, this is the reason tire manufacturers recommend checking tire pressures when the tires are cold. After driving for some distance, friction warms a tire and increases the internal pressure. Filling a warm tire to the recommended pressure may lead to an underinflated tire.

Avogadro's Hypothesis

Front and side air bags are now common in automobiles. In the event of an accident, a bag is rapidly inflated with nitrogen gas generated by a chemical reaction. The air bag unit has a sensor that is sensitive to sudden deceleration of the vehicle and will send an electrical signal that will trigger the reaction (Figures 11.1 and 11.7). In many types of air bags, the explosion of sodium azide generates nitrogen gas.

$$2 \text{ NaN}_3(\text{s}) \longrightarrow 2 \text{ Na(s)} + 3 \text{ N}_2(\text{g})$$

Driver-side air bags inflate to a volume of about 35–70 L, and passenger air bags inflate to about 60–160 L. The final volume of the bag will depend on the amount of nitrogen gas generated.

The relationship between volume and amount of gas was first noted by Amedeo Avogadro. In 1811, he used work on gases by the chemist (and early experimenter with hot air balloons) Joseph Gay-Lussac (1778–1850) to propose that *equal volumes of gases under the same conditions of temperature and pressure have equal numbers of particles* (either molecules or atoms, depending on the composition of the gas.) This idea came to be known as **Avogadro's hypothesis.** Stated another way, the volume

Autoliv/ASP

When a car decelerates in a collision, an electrical contact is made in the sensor unit. The propellant (green solid) detonates, releasing nitrogen gas, and the folded nylon bag explodes out of the plastic housing.

Charles D. Winters

Driver-side air bags inflate with 35–70 L of N_2 gas, whereas passenger air bags hold about 60–160 L.

Charles D. Winters

The bag deflates within 0.2 s, the gas escaping through holes in the bottom of the bag.

FIGURE 11.7 Automobile air bags. See ChemistryNow Screen 11.1 for more on air bags.

of a gas at a given temperature and pressure is directly proportional to the amount of gas in moles:

$$V \propto n \text{ at constant } T \text{ and } P$$

Chemistry Now™

Sign in at **www.thomsonedu.com/login** and go to Chapter 11 Contents to see Screen 11.3 for exercises on **the three gas laws.**

EXAMPLE 11.5 Avogadro's Hypothesis

Problem Ammonia can be made directly from the elements:

$$N_2(g) + 3\ H_2(g) \longrightarrow 2\ NH_3(g)$$

If you begin with 15.0 L of $H_2(g)$, what volume of $N_2(g)$ is required for complete reaction (both gases being at the same T and P)? What is the theoretical yield of NH_3, in liters, under the same conditions?

Strategy From Avogadro's law, we know that gas volume is proportional to the amount of gas. Therefore, we can substitute gas volumes for moles in this stoichiometry problem.

Solution Calculate the volumes of N_2 required and NH_3 produced (in liters) by multiplying the volume of H_2 available by a stoichiometric factor (also in units of liters) obtained from the chemical equation:

$$V\ (N_2 \text{ required}) = (15.0 \text{ L } H_2 \text{ available})\left(\frac{1 \text{ L } N_2 \text{ required}}{3 \text{ L } H_2 \text{ available}}\right) = 5.00 \text{ L } N_2 \text{ required}$$

$$V\ (NH_3 \text{ produced}) = (15.0 \text{ L } H_2 \text{ available})\left(\frac{2 \text{ L } NH_3 \text{ produced}}{3 \text{ L } H_2 \text{ available}}\right) = 10.0 \text{ L } NH_3 \text{ produced}$$

EXERCISE 11.5 Avogadro's Hypothesis

Methane burns in oxygen to give CO_2 and H_2O, according to the balanced equation

$$CH_4(g) + 2\ O_2(g) \longrightarrow CO_2(g) + 2\ H_2O(g)$$

If 22.4 L of gaseous CH_4 is burned, what volume of O_2 is required for complete combustion? What volumes of CO_2 and H_2O are produced? Assume all gases have the same temperature and pressure.

11.3 The Ideal Gas Law

Four interrelated quantities can be used to describe a gas: pressure, volume, temperature, and amount (moles). We know from experiments that three gas laws can be used to describe the relationship of these properties (Section 11.2).

Boyle's Law	Charles's Law	Avogadro's Hypothesis
$V \propto (1/P)$	$V \propto T$	$V \propto n$
(constant T, n)	(constant P, n)	(constant T, P)

If all three laws are combined, the result is

$$V \propto \frac{nT}{P}$$

■ **Properties of an Ideal Gas** For ideal gases, it is assumed that there are no forces of attraction between molecules and that the molecules themselves occupy no volume.

This can be made into a mathematical equation by introducing a proportionality constant, now labeled ***R***. This constant, called the **gas constant,** is a *universal constant,* a number you can use to interrelate the properties of any gas:

$$V = R\left(\frac{nT}{P}\right)$$

or **(11.4)**

$$PV = nRT$$

The equation $PV = nRT$ is called the **ideal gas law.** It describes the behavior of a so-called ideal gas. As you will learn in Section 11.9, however, there is no such thing as an "ideal" gas. Nonetheless, real gases at pressures around one atmosphere or less and temperatures around room temperature usually behave close enough to the ideal that $PV = nRT$ adequately describes their behavior.

To use the equation $PV = nRT$, we need a value for R. This is readily determined experimentally. By carefully measuring P, V, n, and T for a sample of gas, we can calculate the value of R from these values using the ideal gas law equation. For example, under conditions of **standard temperature and pressure (STP)** (a gas temperature of 0 °C or 273.15 K and a pressure of 1 atm), 1 mol of gas occupies 22.414 L, a quantity called the **standard molar volume.** Substituting these values into the ideal gas law gives a value for R:

■ **STP—What Is It?** A gas is at STP, or standard temperature and pressure, when its temperature is 0 °C or 273.15 K and its pressure is 1 atm. Under these conditions, exactly 1 mol of a gas occupies 22.414 L.

$$R = \frac{PV}{nT} = \frac{(1.0000\ \text{atm})(22.414\ \text{L})}{(1.0000\ \text{mol})(273.15)} = 0.082057\frac{\text{L}\cdot\text{atm}}{\text{K}\cdot\text{mol}}$$

With a value for R, we can now use the ideal gas law in calculations.

Chemistry Now™

Sign in at **www.thomsonedu.com/login** and go to Chapter 11 Contents to see Screen 11.4 for a simulation of the **ideal gas law.**

■ EXAMPLE 11.6 Ideal Gas Law

Problem The nitrogen gas in an automobile air bag, with a volume of 65 L, exerts a pressure of 829 mm Hg at 25 °C. What amount of N_2 gas (in moles) is in the air bag?

Strategy You are given P, V, and T and want to calculate the amount of gas (n). Use the ideal gas law, Equation 11.4.

Solution First, list the information provided.

$P = 829$ mm Hg $\quad V = 65$ L $\quad T = 25$ °C $\quad n = ?$

To use the ideal gas law with R having units of (L · atm/K · mol), the pressure must be expressed in atmospheres and the temperature in kelvins. Therefore,

$$P = 829 \text{ mm Hg} \left(\frac{1 \text{ atm}}{760 \text{ mm Hg}}\right) = 1.09 \text{ atm}$$

$$T = 25 + 273 = 298 \text{ K}$$

Now substitute the values of P, V, T, and R into the ideal gas law, and solve for the amount of gas, n:

$$n = \frac{PV}{RT} = \frac{(1.09 \text{ atm})(65 \text{ L})}{(0.082057 \text{ L} \cdot \text{atm/K} \cdot \text{mol})(298 \text{ K})} = 2.9 \text{ mol}$$

Notice that units of atmospheres, liters, and kelvins cancel to leave the answer in units of moles.

EXERCISE 11.6 Ideal Gas Law

The balloon used by Jacques Charles in his historic balloon flight in 1783 (see page 533) was filled with about 1300 mol of H_2. If the temperature of the gas was 23 °C and its pressure was 750 mm Hg, what was the volume of the balloon?

The Density of Gases

The density of a gas at a given temperature and pressure (Figure 11.8) is a useful quantity. Because the amount (n, mol) of any compound is given by its mass (m) divided by its molar mass (M), we can substitute m/M for n in the ideal gas equation.

$$PV = \left(\frac{m}{M}\right)RT$$

Density (d) is defined as mass divided by volume (m/V). We can rearrange the form of the gas law above to give the following equation, which has the term (m/V) on the left. This is the density of the gas.

$$d = \frac{m}{V} = \frac{PM}{RT} \qquad \textbf{(11.5)}$$

Charles D. Winters

(a)

Greg Gawlowski/Dembinski Associates

(b)

FIGURE 11.8 Gas density. (a) The balloons are filled with nearly equal amounts of gas at the same temperature and pressure. One yellow balloon contains helium, a low-density gas ($d = 0.179$ g/L at STP). The other balloons contain air, a higher density gas ($d = 1.2$ g/L at STP). (b) A hot-air balloon rises because the heated air has a lower density than the surrounding air.

Charles D. Winters

FIGURE 11.9 Gas density. Because carbon dioxide from fire extinguishers is denser than air, it settles on top of a fire and smothers it. (When CO_2 gas is released from the tank, it expands and cools significantly. The white cloud is condensed moisture from the air.)

Gas density is directly proportional to the pressure and molar mass and inversely proportional to the temperature. Equation 11.5 is useful because gas density can be calculated from the molar mass, or the molar mass can be found from a measurement of gas density at a given pressure and temperature.

EXAMPLE 11.7 Density and Molar Mass

Problem Calculate the density of CO_2 at STP. Is CO_2 more or less dense than air?

Strategy Use Equation 11.5, the equation relating gas density and molar mass. Here, we know the molar mass (44.0 g/mol), the pressure (P = 1.00 atm), the temperature (T = 273.15 K), and the gas constant (R). Only the density (d) is unknown.

Solution The known values are substituted into Equation 11.5, which is then solved for molar mass (M):

$$d = \frac{PM}{RT} = \frac{(1.00\text{ atm})(44.0\text{ g/mol})}{(0.082057\text{ L}\cdot\text{atm/K}\cdot\text{mol})(273\text{ K})} = 1.96\text{ g/L}$$

The density of CO_2 is considerably greater than that of dry air at STP (1.2 g/L).

EXERCISE 11.7 Gas Density and Molar Mass

The density of an unknown gas is 5.02 g/L at 15.0 °C and 745 mm Hg. Calculate its molar mass.

Gas density has practical implications. From the equation $d = PM/RT$, we recognize that the density of a gas is directly proportional to its molar mass. Dry air, which has an average molar mass of about 29 g/mol, has a density of about 1.2 g/L at 1 atm and 25 °C. Gases or vapors with molar masses greater than 29 g/mol have densities larger than 1.2 g/L under these same conditions (1 atm and 25 °C). Gases such as CO_2, SO_2, and gasoline vapor settle along the ground if released into the atmosphere (Figure 11.9). Conversely, gases such as H_2, He, CO, CH_4 (methane), and NH_3 rise if released into the atmosphere.

The significance of gas density has been revealed in several tragic events. One occurred in the African country of Cameroon in 1984 when Lake Nyos expelled a huge bubble of CO_2 into the atmosphere. Because CO_2 is denser than air, the CO_2 cloud hugged the ground, killing 1700 people nearby (page 630).

Calculating the Molar Mass of a Gas from *P*, *V*, and *T* Data

When a new compound is isolated in the laboratory, one of the first things to be done is to determine its molar mass. If the compound is in the gas phase, a classical method of determining the molar mass is to measure the pressure and volume exerted by a given mass of the gas at a given temperature.

Chemistry Now™

Sign in at **www.thomsonedu.com/login** and go to Chapter 11 Contents to see:

- Screen 11.5 for an exercise on **gas density**
- Screen 11.6 for a tutorial on **using gas laws determining molar mass**

EXAMPLE 11.8 Calculating the Molar Mass of a Gas from *P*, *V*, and *T* Data

Problem You are trying to determine, by experiment, the formula of a gaseous compound to replace chlorofluorocarbons in air conditioners. You have determined the empirical formula is CHF_2, but now you want to know the molecular formula. To do this, you need the molar mass of the compound. You therefore do another experiment and find that a 0.100-g sample of the compound exerts a pressure of 70.5 mm Hg in a 256-mL container at 22.3 °C. What is the molar mass of the compound? What is its molecular formula?

Strategy Here, you know the mass of a gas in a given volume (V), so you can calculate its density, d. Then, knowing the gas pressure and temperature, you can use Equation 11.5 to calculate the molar mass.

Solution Begin by organizing the data:

$$m = \text{mass of gas} = 0.100 \text{ g}$$
$$P = 70.5 \text{ mm Hg, or } 0.0928 \text{ atm}$$
$$V = 256 \text{ mL, or } 0.256 \text{ L}$$
$$T = 22.3 \text{ °C, or } 295.5 \text{ K}$$

The density of the gas is the mass of the gas divided by the volume:

$$d = \frac{0.100 \text{ g}}{0.256 \text{ L}} = 0.391 \text{ g/L}$$

Use this value of density along with the values of pressure and temperature in Equation 11.5 ($d = PM/RT$), and solve for the molar mass (M).

$$M = \frac{dRT}{P} = \frac{(0.391 \text{ g/L})(0.082057 \text{ L} \cdot \text{atm/K} \cdot \text{mol})(295.5 \text{ K})}{0.0928 \text{ atm}} = 102 \text{ g/mol}$$

With this result, you can compare the experimentally determined molar mass with the mass of a mole of gas having the empirical formula CHF_2.

$$\frac{\text{Experimental molar mass}}{\text{Mass of 1 mol } CHF_2} = \frac{102 \text{ g/mol}}{51.0 \text{ g/formula unit}} = 2 \text{ formula units of } CHF_2 \text{ per mol}$$

Therefore, the formula of the compound is $C_2H_2F_4$.

Comment Alternatively, you can use the ideal gas law. Here, you know the P and T of a gas in a given volume (V), so you can calculate the amount of gas (n).

$$n = \frac{PV}{RT} = \frac{(0.0928 \text{ atm})(0.256 \text{ L})}{(0.082057 \text{ L} \cdot \text{atm/K} \cdot \text{mol})(295.5 \text{ K})} = 9.80 \times 10^{-4} \text{ mol}$$

You now know that 0.100 g of gas is equivalent to 9.80×10^{-4} mol. Therefore,

$$\text{Molar mass} = \frac{0.100 \text{ g}}{9.80 \times 10^{-4} \text{ mol}} = 102 \text{ g/mol}$$

EXERCISE 11.8 Molar Mass from P, V, and T Data

A 0.105-g sample of a gaseous compound has a pressure of 561 mm Hg in a volume of 125 mL at 23.0 °C. What is its molar mass?

11.4 Gas Laws and Chemical Reactions

Many industrially important reactions involve gases. Two examples are the combination of nitrogen and hydrogen to produce ammonia,

$$N_2(g) + 3\ H_2(g) \longrightarrow 2\ NH_3(g)$$

and the electrolysis of aqueous NaCl to produce hydrogen and chlorine,

$$2\ NaCl(aq) + 2\ H_2O\ (\ell) \longrightarrow 2\ NaOH(aq) + H_2(g) + Cl_2(g)$$

If we want to understand the quantitative aspects of such reactions, we need to carry out stoichiometry calculations. The scheme in Figure 11.10 connects these calculations for gas reactions with the stoichiometry calculations in Chapter 4.

Chemistry Now™

Sign in at **www.thomsonedu.com/login** and go to Chapter 11 Contents to see Screen 11.7 for a tutorial on **gas laws and chemical reactions: stoichiometry.**

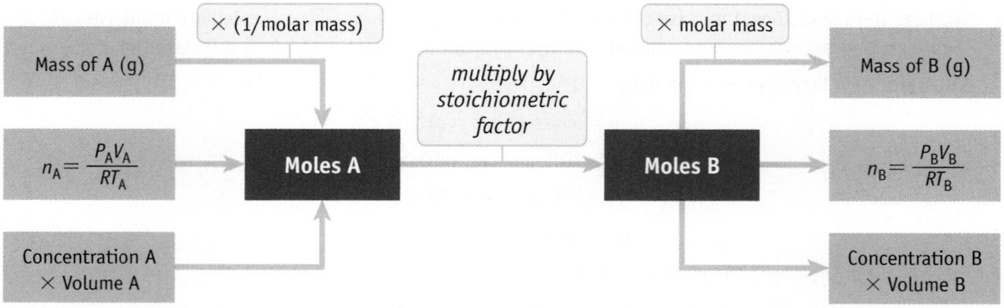

FIGURE 11.10 A scheme for stoichiometry calculations. Here, A and B may be either reactants or products. The amount of A (mol) can be calculated from its mass in grams and its molar mass, from the concentration and volume of a solution, or from *P*, *V*, and *T* data by using the ideal gas law. Once the amount of B is determined, this value can be converted to a mass or solution concentration or volume, or to a volume of gas at a given pressure and temperature.

EXAMPLE 11.9 Gas Laws and Stoichiometry

Problem You are asked to design an air bag for a car. You know that the bag should be filled with gas with a pressure higher than atmospheric pressure, say 829 mm Hg, at a temperature of 22.0 °C. The bag has a volume of 45.5 L. What quantity of sodium azide, NaN_3, should be used to generate the required quantity of gas? The gas-producing reaction is

$$2\ NaN_3(s) \longrightarrow 2\ Na(s) + 3\ N_2(g)$$

Strategy The general logic to be used here follows a pathway in Figure 11.10.

Use $PV = nRT$ with gas data ⟶ Amount of N_2 required ⟶ Use stoichiometric factor to calculate amount of NaN_3 required ⟶ Use molar mass to calculate mass of NaN_3 required

Solution The first step is to find the amount (mol) of gas required so that this can be related to the quantity of sodium azide required:

$$P = 829\ \text{mm Hg}\ (1\ \text{atm}/760\ \text{mm Hg}) = 1.09\ \text{atm}$$

$$V = 45.5\ \text{L}$$

$$T = 22.0\ °\text{C, or } 295.2\ \text{K}$$

$$n = \text{N}_2\ \text{required (mol)} = \frac{PV}{RT}$$

$$n = \frac{(1.09\ \text{atm})(45.5\ \text{L})}{(0.082057\ \text{L}\cdot\text{atm/K}\cdot\text{mol})(295.2\ \text{K})} = 2.05\ \text{mol N}_2$$

Now that the required amount of nitrogen has been calculated, we can calculate the quantity of sodium azide that will produce 2.05 mol of N_2 gas.

$$\text{Mass of NaN}_3 = 2.05\ \text{mol N}_2\left(\frac{2\ \text{mol NaN}_3}{3\ \text{mol N}_2}\right)\left(\frac{65.01\ \text{g}}{1\ \text{mol NaN}_3}\right) = 88.8\ \text{g NaN}_3$$

EXAMPLE 11.10 Gas Laws and Stoichiometry

Problem You wish to prepare some deuterium gas, D_2, for use in an experiment. One way to do this is to react heavy water, D_2O, with an active metal such as lithium.

$$2\ Li(s) + 2\ D_2O(\ell) \longrightarrow 2\ LiOD(aq) + D_2(g)$$

What amount of D_2 (in moles) can be prepared from 0.125 g of Li metal in 15.0 mL of D_2O ($d = 1.11$ g/mL). If dry D_2 gas is captured in a 1450-mL flask at 22.0 °C, what is the pressure of the gas in mm Hg? (Deuterium has an atomic weight of 2.0147 g/mol.)

Strategy You are combining two reactants with no guarantee that they are in the correct stoichiometric ratio. This example must therefore be approached as a limiting reactant problem. You have to find the amount of each substance and then see if one of them is present in a limited amount. Once the limiting reactant is known, the amount of D_2 produced and its pressure under the conditions given can be calculated.

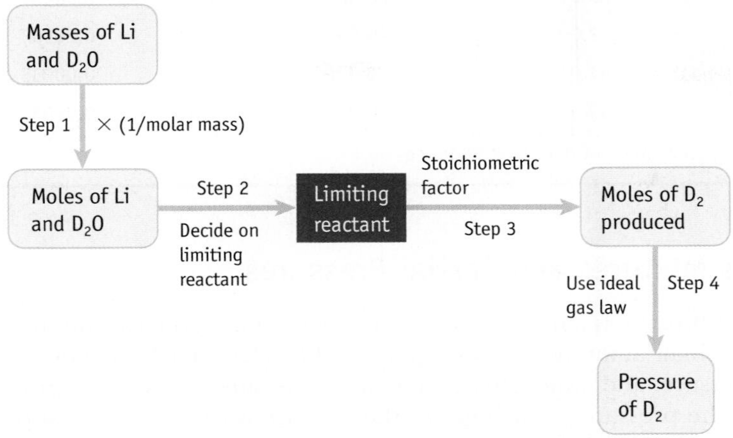

Charles D. Winters

Lithium metal (in the spoon) reacts with drops of water, H_2O, to produce LiOH and hydrogen gas, H_2. If heavy water, D_2O, is used, deuterium gas, D_2, can be produced.

Solution

Step 1. *Calculate the amount (mol) of Li and of D_2O:*

$$0.125 \text{ g Li}\left(\frac{1 \text{ mol Li}}{6.941 \text{ g Li}}\right) = 0.0180 \text{ mol Li}$$

$$15.0 \text{ mL } D_2O\left(\frac{1.11 \text{ g } D_2O}{1 \text{ mL } D_2O}\right)\left(\frac{1 \text{ mol } D_2O}{20.03 \text{ g } D_2O}\right) = 0.831 \text{ mol } D_2O$$

Step 2. *Decide which reactant is the limiting reactant:*

$$\text{Ratio of moles of reactants available} = \frac{0.831 \text{ mol } D_2O}{0.0180 \text{ mol Li}} = \frac{46.2 \text{ mol } D_2O}{1 \text{ mol Li}}$$

The balanced equation shows that the ratio should be 1 mol of D_2O to 1 mol of Li. From the calculated values, we see that D_2O is in large excess, and so Li is the limiting reactant. Therefore, further calculations are based on the amount of Li available.

Step 3. *Use the limiting reactant to calculate the quantity of D_2 produced:*

$$0.0180 \text{ mol Li}\left(\frac{1 \text{ mol } D_2 \text{ produced}}{2 \text{ mol Li}}\right) = 0.00900 \text{ mol } D_2 \text{ produced}$$

Step 4. *Calculate the pressure of D_2:*

$P = ?$ $\quad T = 22.0$ °C, or 295.2 K

$V = 1450$ mL, or 1.45 L $\quad n = 0.00900$ mol D_2

$$P = \frac{nRT}{V} = \frac{(0.00900 \text{ mol})(0.082057 \text{ L}\cdot\text{atm/K}\cdot\text{mol})(295.2 \text{ K})}{1.45 \text{ L}} = 0.150 \text{ atm}$$

EXERCISE 11.9 Gas Laws and Stoichiometry

Gaseous ammonia is synthesized by the reaction

$$N_2(g) + 3\ H_2(g) \longrightarrow 2\ NH_3(g)$$

Assume that 355 L of H_2 gas at 25.0 °C and 542 mm Hg is combined with excess N_2 gas. What amount of NH_3 gas, in moles, can be produced? If this amount of NH_3 gas is stored in a 125-L tank at 25.0 °C, what is the pressure of the gas?

TABLE 11.1 Components of Atmospheric Dry Air

Constituent	Molar Mass*	Mole Percent	Partial Pressure at STP (atm)
N_2	28.01	78.08	0.7808
O_2	32.00	20.95	0.2095
CO_2	44.01	0.0385	0.00033
Ar	39.95	0.934	0.00934

*The average molar mass of dry air = 28.960 g/mol.

11.5 Gas Mixtures and Partial Pressures

The air you breathe is a mixture of nitrogen, oxygen, argon, carbon dioxide, water vapor, and small amounts of other gases (Table 11.1). Each of these gases exerts its own pressure, and atmospheric pressure is the sum of the pressures exerted by each gas. The pressure of each gas in the mixture is called its **partial pressure.**

John Dalton (1766–1844) was the first to observe that the pressure of a mixture of ideal gases is the sum of the partial pressures of the different gases in the mixture. This observation is now known as **Dalton's law of partial pressures** (Figure 11.11). Mathematically, we can write Dalton's law of partial pressures as

$$P_{total} = P_1 + P_2 + P_3 \ldots \qquad \textbf{(11.6)}$$

where P_1, P_2, and P_3 are the pressures of the different gases in a mixture, and P_{total} is the total pressure.

In a mixture of gases, each gas behaves independently of all others in the mixture. Therefore, we can consider the behavior of each gas in a mixture separately. As an example, let us take a mixture of three ideal gases, labeled A, B, and C. There are n_A moles of A, n_B moles of B, and n_C moles of C. Assume that the mixture ($n_{total} = n_A + n_B + n_C$) is contained in a given volume (V) at a given temperature (T). We can calculate the pressure exerted by each gas from the ideal gas law equation:

$$P_AV = n_ART \qquad P_BV = n_BRT \qquad P_CV = n_CRT$$

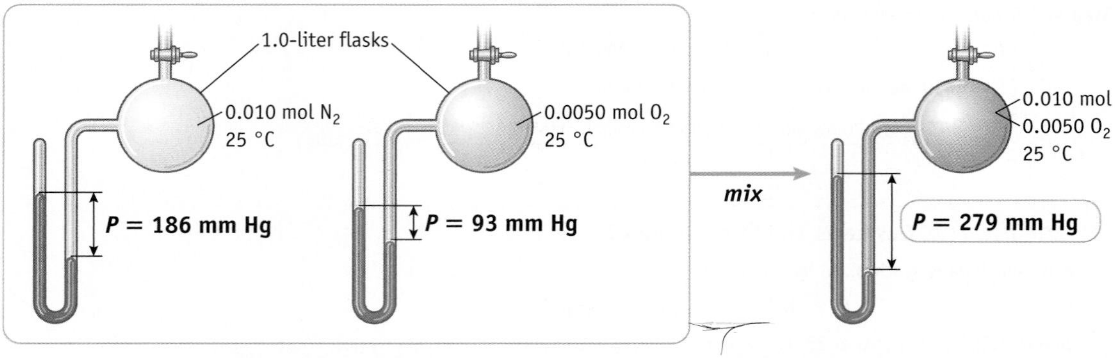

FIGURE 11.11 Dalton's law. In a 1.0-L flask at 25 °C, 0.010 mol of N_2 exerts a pressure of 186 mm Hg, and 0.0050 mol of O_2 in a 1.0-L flask at 25 °C exerts a pressure of 93 mm Hg (left and middle). The N_2 and O_2 samples are mixed in a 1.0-L flask at 25 °C (right). The total pressure, 279 mm Hg, is the sum of the pressures that each gas alone exerts in the flask.

where each gas (A, B, and C) is in the same volume V and is at the same temperature T. According to Dalton's law, the total pressure exerted by the mixture is the sum of the pressures exerted by each component:

$$P_{total} = P_A + P_B + P_C = n_A\left(\frac{RT}{V}\right) + n_B\left(\frac{RT}{V}\right) + n_C\left(\frac{RT}{V}\right)$$

$$P_{total} = (n_A + n_B + n_C)\left(\frac{RT}{V}\right)$$

$$P_{total} = (n_{total})\left(\frac{RT}{V}\right) \qquad \textbf{(11.7)}$$

For mixtures of gases, it is convenient to introduce a quantity called the **mole fraction, X,** which is defined as the number of moles of a particular substance in a mixture divided by the total number of moles of all substances present. Mathematically, the mole fraction of a substance A in a mixture with B and C is expressed as

$$X_A = \frac{n_A}{n_A + n_B + n_C} = \frac{n_A}{n_{total}}$$

Now we can combine this equation (written as $n_{total} = n_A/X_A$) with the equations for P_A and P_{total}, and derive the equation

$$P_A = X_A P_{total} \qquad \textbf{(11.8)}$$

This equation is useful because it tells us that *the pressure of a gas in a mixture of gases is the product of its mole fraction and the total pressure of the mixture.* For example, the mole fraction of N_2 in air is 0.78, so, at STP, its partial pressure is 0.78 atm or 590 mm Hg.

Chemistry Now™

Sign in at **www.thomsonedu.com/login** and go to Chapter 11 Contents to see Screen 11.8 for tutorials on **gas mixtures and partial pressures.**

EXAMPLE 11.11 Partial Pressures of Gases

Problem Halothane, $C_2HBrClF_3$, is a nonflammable, nonexplosive, and nonirritating gas that is commonly used as an inhalation anesthetic.

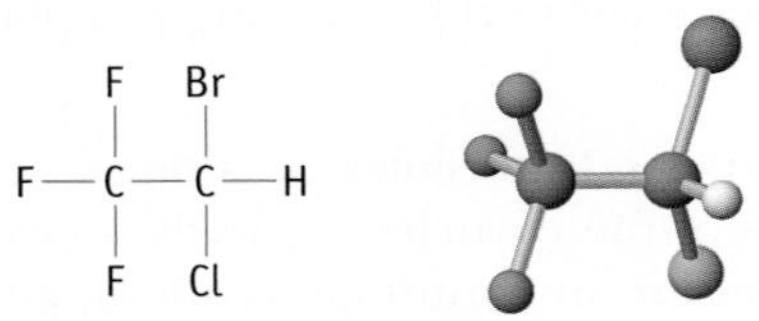

1,1,1-trifluorobromochloroethane, halothane

The total pressure of a mixture of 15.0 g of halothane vapor and 23.5 g of oxygen gas is 855 mm Hg. What is the partial pressure of each gas?

Strategy One way to solve this problem is to recognize that the partial pressure of a gas is given by the total pressure of the mixture multiplied by the mole fraction of the gas.

FIGURE 11.12 A molecular view of gases and liquids. The fact that a large volume of N_2 gas can be condensed to a small volume of liquid indicates that the distance between molecules in the gas phase is very large as compared with the distances between molecules in liquids.

Solution Let us first calculate the mole fractions of halothane and of O_2.

Step 1. *Calculate mole fractions:*

$$\text{Amount of } C_2HBrClF_3 = 15.0 \text{ g}\left(\frac{1 \text{ mol}}{197.4 \text{ g}}\right) = 0.0760 \text{ mol}$$

$$\text{Amount of } O_2 = 23.5 \text{ g}\left(\frac{1 \text{ mol}}{32.00 \text{ g}}\right) = 0.734 \text{ mol}$$

$$\text{Total amount of gas} = 0.0760 \text{ mol } C_2HBrClF_2 + 0.734 \text{ mol } O_2 = 0.810 \text{ mol}$$

$$\text{Mole fraction of } C_2HBrClF_3 = \frac{0.0760 \text{ mol } C_2HBrClF_3}{0.810 \text{ total moles}} = 0.0938$$

Because the sum of the mole fraction of halothane and of O_2 must equal 1.0000, this means that the mole fraction of oxygen is 0.906.

$$X_{\text{halothane}} + X_{\text{oxygen}} = 1.0000$$
$$0.0938 + X_{\text{oxygen}} = 1.0000$$
$$X_{\text{oxygen}} = 0.906$$

Step 2. *Calculate partial pressures:*

$$\text{Partial pressure of halothane} = P_{\text{halothane}} = X_{\text{halothane}} \cdot P_{\text{total}}$$
$$P_{\text{halothane}} = 0.0938 \cdot P_{\text{total}} = 0.0938 \text{ (855 mm Hg)}$$
$$P_{\text{halothane}} = 80.2 \text{ mm Hg}$$

The total pressure of the mixture is the sum of the partial pressures of the gases in the mixture.

$$P_{\text{halothane}} + P_{\text{oxygen}} = 855 \text{ mm Hg}$$

and so

$$P_{\text{oxygen}} = 855 \text{ mm Hg} - P_{\text{halothane}}$$
$$P_{\text{oxygen}} = 855 \text{ mm Hg} - 80.2 \text{ mm Hg} = 775 \text{ mm Hg}$$

EXERCISE 11.10 Partial Pressures

The halothane–oxygen mixture described in Example 11.11 is placed in a 5.00-L tank at 25.0 °C. What is the total pressure (in mm Hg) of the gas mixture in the tank? What are the partial pressures (in mm Hg) of the gases?

Module 16

11.6 The Kinetic-Molecular Theory of Gases

So far, we have discussed the macroscopic properties of gases, properties such as pressure and volume that result from the behavior of a system with a large number of particles. Now we turn to the kinetic-molecular theory (◄ page 7) for a description of the behavior of matter at the molecular or atomic level. Hundreds of experimental observations have led to the following postulates regarding the behavior of gases.

- Gases consist of particles (molecules or atoms) whose separation is much greater than the size of the particles themselves (see Figure 11.12).
- The particles of a gas are in continual, random, and rapid motion. As they move, they collide with one another and with the walls of their container, but they do so without loss of energy.
- The average kinetic energy of gas particles is proportional to the gas temperature. *All gases, regardless of their molecular mass, have the same average kinetic energy at the same temperature.*

Let us discuss the behavior of gases from this point of view.

Historical Perspectives — Studies on Gases: Robert Boyle and Jacques Charles

Robert Boyle (1627–1691) was born in Ireland as the 14th and last child of the first Earl of Cork. In his book *Uncle Tungsten*, Oliver Sacks tells us that "Chemistry as a true science made its first emergence with the work of Robert Boyle in the middle of the seventeenth century. Twenty years [Isaac] Newton's senior, Boyle was born at a time when the practice of alchemy still held sway, and he still maintained a variety of alchemical beliefs and practices, side by side with his scientific ones. He believed gold could be created, and that he had succeeded in creating it (Newton, also an alchemist, advised him to keep silent about this)."

Oesper Collection in the History of Chemistry, University of Cincinnati

Robert Boyle (1627–1691).

Boyle examined crystals, explored color, devised an acid-base indicator from the syrup of violets, and provided the first modern definition of an element. He was also a physiologist, and was the first to show that the healthy human body has a constant temperature. Today, Boyle is best known for his studies of gases, which were described in his book *The Sceptical Chymist*, published in 1680.

The French chemist and inventor Jacques Alexandre César Charles began his career as a clerk in the finance ministry, but his real interest was science. He developed several inventions and was best known in his lifetime for inventing the hydrogen balloon. In August 1783, Charles exploited his recent studies on hydrogen gas by inflating a balloon with this gas. Because hydrogen would escape easily from a paper bag, he made a silk bag coated with rubber. Inflating the bag took several days and required nearly 225 kg of sulfuric acid and 450 kg of iron to produce the H_2 gas. The balloon stayed aloft for almost 45 minutes and traveled about 15 miles. When it landed in a village, however, the people were so terrified they tore it to shreds. Several months later, Charles and a passenger flew a new hydrogen-filled balloon some distance across the French countryside and ascended to the then-incredible altitude of 2 miles.

Image Courtesy of Library of Congress

Jacques Alexandre César Charles (1746–1823).

Smithsonian National Air & Space Museum

Jacques Charles and A. Roberts ascended over Paris on December 1, 1783, in a hydrogen-filled balloon.

Molecular Speed and Kinetic Energy

If your friend walks into your room carrying a pizza, how do you know it? In scientific terms, we know that the odor-causing molecules of food enter the gas phase and drift through space until they reach the cells of your body that react to odors. The same thing happens in the laboratory when bottles of aqueous ammonia (NH_3) and hydrochloric acid (HCl) sit side by side (Figure 11.13). Molecules of the two compounds enter the gas phase and drift along until they encounter one another, at which time they react and form a cloud of tiny particles of solid ammonium chloride (NH_4Cl).

Charles D. Winters

FIGURE 11.13 The movement of gas molecules. Open dishes of aqueous ammonia and hydrochloric acid are placed side by side. When molecules of NH_3 and HCl escape from solution to the atmosphere and encounter one another, a cloud of solid ammonium chloride, NH_4Cl is observed.

If you change the temperature of the environment of the containers in Figure 11.13 and measure the time needed for the cloud of ammonium chloride to form, you would find the time would be longer at lower temperatures. The reason for this is that the speed at which molecules move depends on the temperature. Let us expand on this idea.

The molecules in a gas sample do not all move at the same speed. Rather, as illustrated in Figure 11.14 for O_2 molecules, there is a distribution of speeds. Figure 11.14 shows the number of particles in a gas sample that are moving at certain speeds at a given temperature, and there are two important observations we can make. First, at a given temperature some molecules have high speeds, and others have low speeds. Most of the molecules, however, have some intermediate speed, and their most probable speed corresponds to the maximum in the curve. For oxygen gas at 25 °C, for example, most molecules have speeds in the range

Chemical Perspectives The Earth's Atmosphere

Earth's atmosphere is a fascinating mixture of gases in more or less distinct layers with widely differing temperatures.

Up to the troposphere, there is a gradual decline in temperature (and pressure) with altitude. The temperature climbs again in the stratosphere due to the absorption of energy from the sun by stratospheric ozone, O_3.

Above the stratosphere, the pressure declines because there are fewer molecules present. At still higher altitudes, we observe a dramatic increase in temperature in the thermosphere. This is an illustration of the difference between *temperature* and *thermal energy*. The temperature of a gas reflects the average kinetic energy of the molecules of the gas, whereas the thermal energy present in an object is the *total* kinetic energy of the molecules. In the thermosphere, the few molecules present have a very high temperature, but the thermal energy is exceedingly small because there are so few molecules.

Gases within the troposphere are well mixed by convection. Pollutants that are evolved on Earth's surface can rise into the stratosphere, but it is said that the stratosphere acts as a "thermal lid" on the troposphere and prevents significant mixing of polluting gases into the stratosphere and beyond.

The pressure of the atmosphere declines with altitude, and so the partial pressure of O_2 declines. The figure shows why climbers have a hard time breathing on Mt. Everest, where the altitude is 29,028 ft (8848 m) and the O_2 partial pressure is only 29% of the sea level partial pressure. With proper training, a climber could reach the summit without supplemental oxygen. However, this same feat would not be possible if Everest were farther north. Earth's atmosphere thins toward the poles, and so the O_2 partial pressure would be even less if Everest's summit were in North America, for example.

(See G. N. Eby, *Environmental Geochemistry*, Thomson/Brooks/Cole, 2004.)

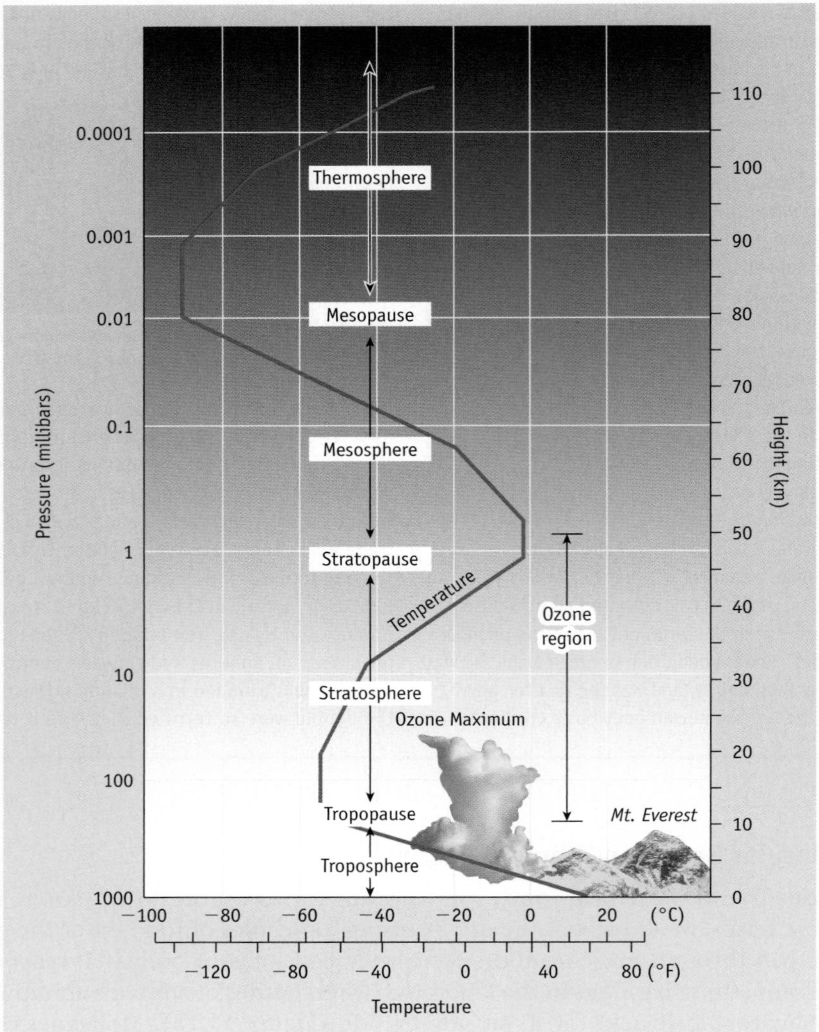

Average Composition of Earth's Atmosphere to a Height of 25 km

Gas	Volume %	Source
N_2	78.08	biologic
O_2	20.95	biologic
Ar	0.93	radioactivity
Ne	0.0018	Earth's interior
He	0.0005	radioactivity
H_2O	0 to 4	evaporation
CO_2	0.0385	biologic, industrial
CH_4	0.00017	biologic
N_2O	0.00003	biologic, industrial
O_3	0.000004	photochemical

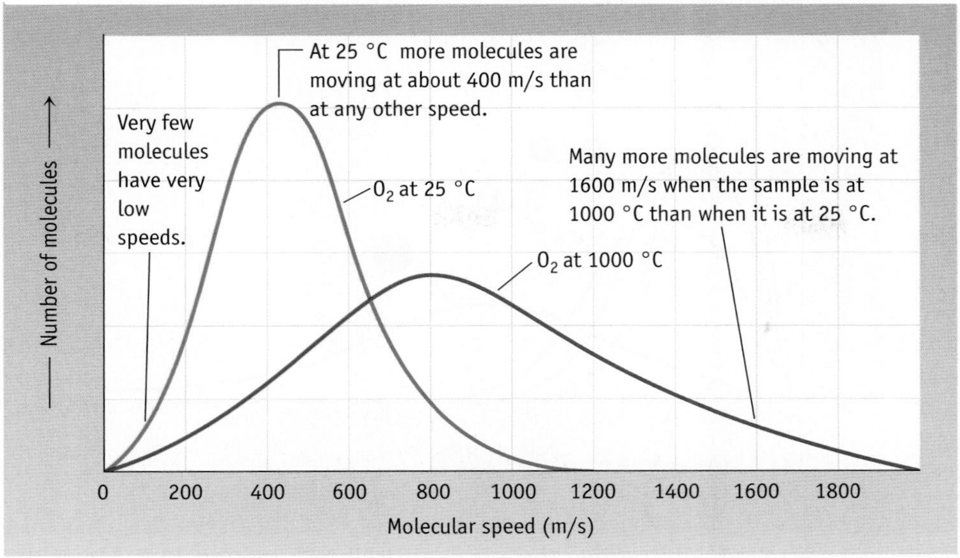

FIGURE 11.14 The distribution of molecular speeds. A graph of the number of molecules with a given speed versus that speed shows the distribution of molecular speeds. The red curve shows the effect of increased temperature. Even though the curve for the higher temperature is "flatter" and broader than the one at a lower temperature, the areas under the curves are the same because the number of molecules in the sample is fixed.

from 200 m/s to 700 m/s, and their most probable speed is about 400 m/s. (These are very high speeds, indeed. A speed of 400 m/s corresponds to about 1000 miles per hour!)

A second observation regarding the distribution of speeds is that as the temperature increases the most probable speed increases, and the number of molecules traveling at very high speeds increases greatly.

The kinetic energy of a single molecule of mass m in a gas sample is given by the equation

$$KE = \frac{1}{2}(mass)(speed)^2 = \frac{1}{2}mu^2$$

where u is the speed of that molecule. We can calculate the kinetic energy of a single gas molecule from this equation but not of a collection of molecules because not all of the molecules in a gas sample are moving at the same speed. However, we can calculate the average kinetic energy of a collection of molecules by relating it to other averaged quantities of the system. In particular, the average kinetic energy is related to the average speed:

$$\overline{KE} = \frac{1}{2}m\overline{u^2}$$

(The horizontal bar over the symbols KE and u indicate an average value.) This equation states that the average kinetic energy of the molecules in a gas sample, $\overline{KE}$, is related to $\overline{u^2}$, the average of the squares of their speeds (called the "mean square speed").

Experiments also show that the average kinetic energy, $\overline{KE}$, of a sample of gas molecules is directly proportional to temperature with a proportionality constant of $\frac{3}{2}R$,

$$\overline{KE} = \frac{3}{2}RT$$

where R is the gas constant expressed in SI units (8.314472 J/K · mol).

Now, because $\overline{KE}$ is proportional to both $1/2\ m\overline{u^2}$ and T, temperature and $1/2\ m\overline{u^2}$ must also be proportional; that is, $1/2\ m\overline{u^2} \propto T$. This relation among

FIGURE 11.15 The effect of molecular mass on the distribution of speeds. At a given temperature, molecules with higher masses have lower speeds.

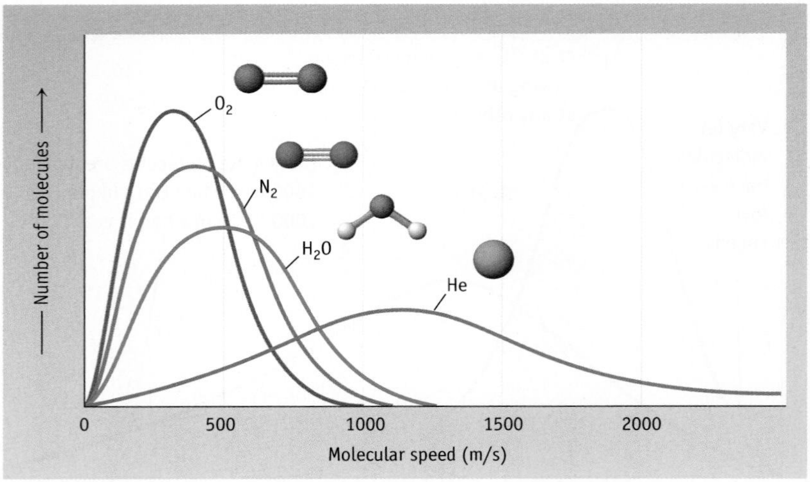

■ **Maxwell–Boltzmann Curves** Plots showing the relation between the number of molecules and their speed or energy (Figure 11.14) are often called Maxwell–Boltzmann distribution curves. They are named after James Clerk Maxwell (1831–1879) and Ludwig Boltzmann (1844–1906). The distribution of speeds (or kinetic energies) of molecules (as illustrated by Figures 11.14 and 11.15) is often used when explaining chemical phenomena.

mass, average speed, and temperature is expressed in Equation 11.9. Here, the square root of the mean square speed ($\sqrt{\overline{u^2}}$, called the **root-mean-square,** or **rms speed**), the temperature (T, in kelvins), and the molar mass (M) are related.

$$\sqrt{\overline{u^2}} = \sqrt{\frac{3RT}{M}} \tag{11.9}$$

This equation, sometimes called *Maxwell's equation* after James Clerk Maxwell (Section 6.1), shows that the speeds of gas molecules are indeed related directly to the temperature (Figure 11.14). The rms speed is a useful quantity because of its direct relationship to the average kinetic energy and because it is very close to the true average speed for a sample. (The average speed is 92% of the rms speed.)

All gases have the same average kinetic energy at the same temperature. However, if you compare a sample of one gas with another, say compare O_2 and N_2, this does not mean the molecules have the same average speed (Figure 11.15). Instead, Maxwell's equation shows that the smaller the molar mass of the gas the greater the rms speed.

Chemistry Now™

Sign in at **www.thomsonedu.com/login** and go to Chapter 11 Contents to see:

- Screen 11.9 for a self-study module on **gases at different temperatures**
- Screen 11.11 for a tutorial on **Boltzmann distribution and calculation of distribution curves**

EXAMPLE 11.12 Molecular Speed

Problem Calculate the rms speed of oxygen molecules at 25 °C.

Strategy We must use Equation 11.9 with M in units of kg/mol. The reason for this is that R is in units of J/K · mol, and 1 J = 1 kg · m^2/s^2.

Solution The molar mass of O_2 is 32.0×10^{-3} kg/mol.

$$\sqrt{\overline{u^2}} = \sqrt{\frac{3(8.3145\ \text{J/K}\cdot\text{mol})(298\ \text{K})}{32.0 \times 10^{-3}\ \text{kg/mol}}} = \sqrt{2.32 \times 10^5\ \text{J/kg}}$$

To obtain the answer in meters per second, we use the relation $1\ J = 1\ kg \cdot m^2/s^2$. This means we have

$$\sqrt{\overline{u^2}} = \sqrt{2.32 \times 10^5\ kg \cdot m^2/(kg \cdot s^2)} = \sqrt{2.32 \times 10^5\ m^2/s^2} = 482\ m/s$$

This speed is equivalent to about 1100 miles per hour!

EXERCISE 11.11 Molecular Speeds

Calculate the rms speeds of helium atoms and N_2 molecules at 25 °C.

Kinetic-Molecular Theory and the Gas Laws

The gas laws, which come from experiment, can be explained by the kinetic-molecular theory. The starting place is to describe how pressure arises from collisions of gas molecules with the walls of the container holding the gas (Figure 11.16). Remember that pressure is related to the force of the collisions (see Section 11.1).

$$\text{Gas pressure} = \frac{\text{force of collisions}}{\text{area}}$$

The force exerted by the collisions depends on the number of collisions and the average force per collision. When the temperature of a gas is increased, we know the average kinetic energy of the molecules increases. This causes the average force of the collisions with the walls to increase as well. (This is much like the difference in the force exerted by a car traveling at high speed versus one moving at only a few kilometers per hour.) Also, because the speed of gas molecules increases with temperature, more collisions occur per second. Thus, the collective force per square centimeter is greater, and the pressure increases. Mathematically, this is related to the direct proportionality between P and T when n and V are fixed, that is, $P = (nR/V)T$.

Increasing the number of molecules of a gas at a fixed temperature and volume does not change the average collision force, but it does increase the number of collisions occurring per second. Thus, the pressure increases, and we can say that P is proportional to n when V and T are constant, that is, $P = n(RT/V)$.

If the pressure is to remain constant when either the number of molecules of gas or the temperature is increased, then the volume of the container (and the area over which the collisions can take place) must increase. This is expressed by stating that V is proportional to nT when P is constant [$V = nT(R/P)$], a statement that is a *combination of Avogadro's hypothesis and Charles's law.*

Finally, if the temperature is constant, the average impact force of molecules of a given mass with the container walls must be constant. If n is kept constant while the volume of the container is made smaller, the number of collisions with the container walls per second must increase. This means the pressure increases, and so P is proportional to $1/V$ when n and T are constant, as stated by *Boyle's law,* that is, $P = (1/V)(nRT)$.

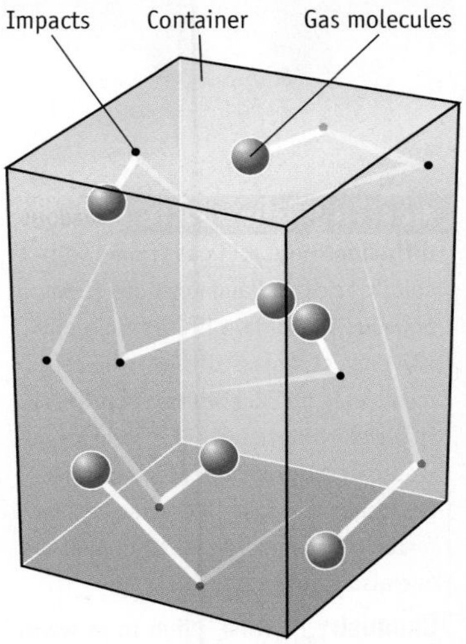

FIGURE 11.16 Gas pressure. According to the kinetic-molecular theory, gas pressure is caused by gas molecules bombarding the container walls.

Chemistry Now™

Sign in at **www.thomsonedu.com/login** and go to Chapter 11 Contents to see Screen 11.10 for simulations of **the gas laws at the molecular level.**

FIGURE 11.17 Diffusion. (a) Liquid bromine, Br_2, was placed in a small flask inside a larger container. (b) The cork was removed from the flask, and, with time, bromine vapor diffused into the larger container. Bromine vapor is now distributed evenly in the containers.

Charles D. Winters

(a) (b)

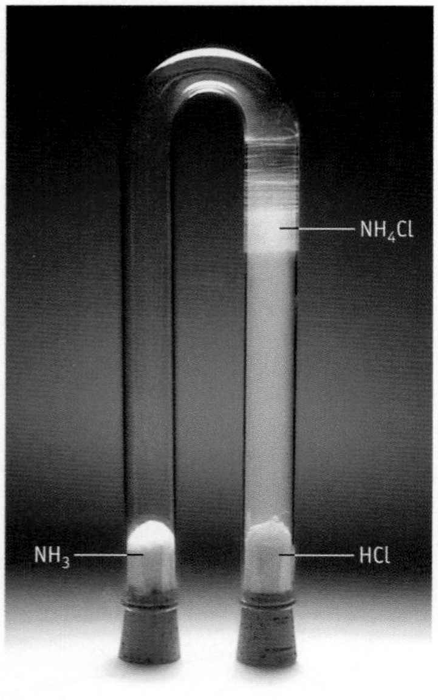

Charles D. Winters

Active Figure 11.18 Gaseous diffusion. Here, HCl gas (from hydrochloric acid) and ammonia gas (from aqueous ammonia) diffuse from opposite ends of a glass U-tube. When they meet, they produce white, solid NH_4Cl. It is clear that the NH_4Cl is formed closer to the end from which the HCl gas begins because HCl molecules move faster, on average, than NH_3 molecules. See also Figure 11.13.

Chemistry Now™ Sign in at www.thomsonedu.com/login and go to the Chapter Contents menu to explore an interactive version of this figure accompanied by an exercise.

11.7 Diffusion and Effusion

When a pizza is brought into a room, the volatile aroma-causing molecules vaporize into the atmosphere, where they mix with the oxygen, nitrogen, carbon dioxide, water vapor, and other gases present. Even if there were no movement of the air in the room caused by fans or people moving about, the odor would eventually reach everywhere in the room. This mixing of molecules of two or more gases due to their random molecular motions is the result of **diffusion.** Given time, the molecules of one component in a gas mixture will thoroughly and completely mix with all other components of the mixture (Figure 11.17).

Diffusion is also illustrated by the experiment in Figure 11.18. Here, we have placed cotton moistened with hydrochloric acid at one end of a U-tube and cotton moistened with aqueous ammonia at the other end. Molecules of HCl and NH_3 diffuse into the tube, and, when they meet, they produce white, solid NH_4Cl (just as in Figure 11.13).

$$HCl(g) + NH_3(g) \longrightarrow NH_4Cl(s)$$

We find that the gases do not meet in the middle. Rather, because the heavier HCl molecules diffuse less rapidly than the lighter NH_3 molecules, the molecules meet closer to the HCl end of the U-tube.

Closely related to diffusion is **effusion**, which is the movement of gas through a tiny opening in a container into another container where the pressure is very low (Figure 11.19). Thomas Graham (1805–1869), a Scottish chemist, studied the effusion of gases and found that the rate of effusion of a gas—the amount of gas moving from one place to another in a given amount of time—is inversely proportional to the square root of its molar mass. Based on these experimental results, the rates of effusion of two gases can be compared:

$$\frac{\text{Rate of effusion of gas 1}}{\text{Rate of effusion of gas 2}} = \sqrt{\frac{\text{molar mass of gas 2}}{\text{molar mass of gas 1}}} \qquad \textbf{(11.10)}$$

The relationship in Equation 11.10—now known as **Graham's law**—is readily derived from Maxwell's equation by recognizing that the rate of effusion depends on

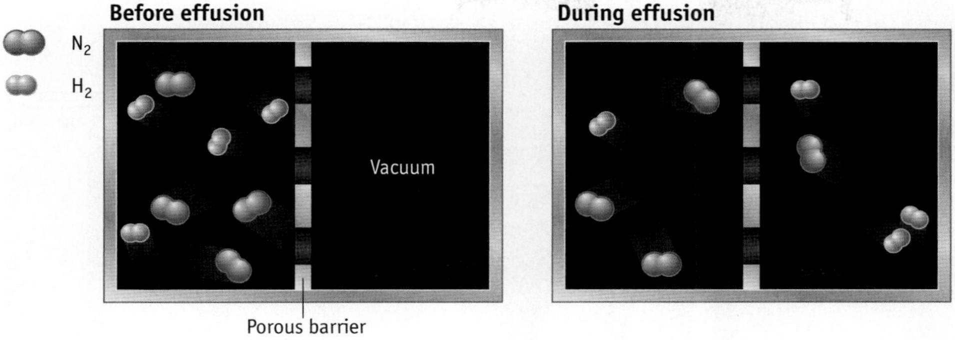

FIGURE 11.19 Effusion. H_2 and N_2 gas molecules effuse through the pores of a porous barrier. Lighter molecules (H_2) with higher average speeds strike the barrier more often and pass more often through it than heavier, slower molecules (N_2) at the same temperature. According to Graham's law, H_2 molecules effuse 3.72 times faster than N_2 molecules.

the speed of the molecules. The ratio of the rms speeds is the same as the ratio of the effusion rates:

$$\frac{\text{Rate of effusion of gas 1}}{\text{Rate of effusion of gas 2}} = \frac{\sqrt{\overline{u^2} \text{ of gas 1}}}{\sqrt{\overline{u^2} \text{ of gas 2}}} = \frac{\sqrt{3RT/(M \text{ of gas 1})}}{\sqrt{3RT/(M \text{ of gas 2})}}$$

Canceling out like terms gives the expression in Equation 11.10.

Chemistry Now™

Sign in at **www.thomsonedu.com/login** and go to Chapter 11 Contents to see screen 11.12 for an exercise and tutorial on **diffusion.**

EXAMPLE 11.13 Using Graham's Law of Effusion to Calculate a Molar Mass

Problem Tetrafluoroethylene, C_2F_4, effuses through a barrier at a rate of 4.6×10^{-6} mol/h. An unknown gas, consisting only of boron and hydrogen, effuses at the rate of 5.8×10^{-6} mol/h under the same conditions. What is the molar mass of the unknown gas?

Strategy From Graham's law, we know that a light molecule will effuse more rapidly than a heavier one. Because the unknown gas effuses more rapidly than C_2F_4 ($M = 100.0$ g/mol), the unknown must have a molar mass less than 100 g/mol. Substitute the experimental data into Graham's law equation (Equation 11.10).

Solution

$$\frac{5.8 \times 10^{-6} \text{ mol/h}}{4.6 \times 10^{-6} \text{ mol/h}} = 1.3 = \sqrt{\frac{100.0 \text{ g/mol}}{M \text{ of unknown}}}$$

To solve for the unknown molar mass, square both sides of the equation and rearrange to find M for the unknown.

$$1.6 = \frac{100.0 \text{ g/mol}}{M \text{ of unknown}}$$

$$M = 63 \text{ g/mol}$$

Comment A boron–hydrogen compound corresponding to this molar mass is B_5H_9, called pentaborane.

EXERCISE 11.12 Graham's Law

A sample of pure methane, CH_4, is found to effuse through a porous barrier in 1.50 min. Under the same conditions, an equal number of molecules of an unknown gas effuses through the barrier in 4.73 min. What is the molar mass of the unknown gas?

FIGURE 11.20 Isotope separation. Separation of uranium isotopes for use in atomic weaponry or in nuclear power plants was originally done by gas effusion. (There are still plants in use in the U.S. at Piketon, Ohio, and Paducah, Kentucky.) The more modern approach is to use a gas centrifuge, and that is what is pictured here (left). (Right) UF_6 gas is injected into the centrifuge from a tube passing down through the center of a tall, spinning cylinder. The heavier $^{238}UF_6$ molecules experience more centrifugal force and move to the outer wall of the cylinder; the lighter $^{235}UF_6$ molecules stay closer to the center. A temperature difference inside the rotor causes the $^{235}UF_6$ molecules to move to the top of the cylinder and the $^{238}UF_6$ molecules to move to the bottom. (See the *New York Times*, page F1, March 23, 2004.)

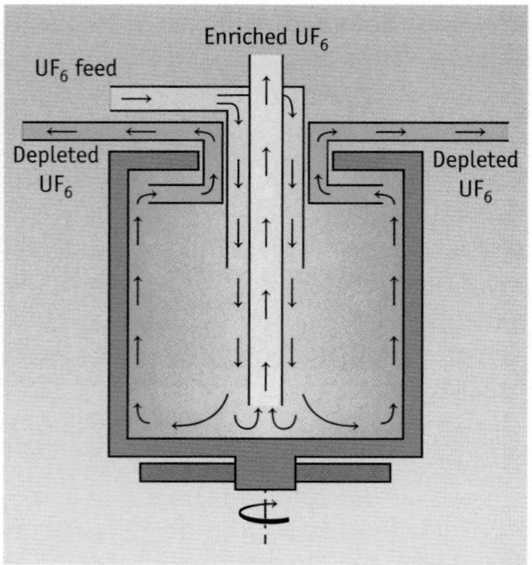

11.8 Some Applications of the Gas Laws and Kinetic-Molecular Theory

Separating Isotopes

The effusion process played a central role in the development of the atomic bomb in World War II and is still in use today to prepare fissionable uranium for nuclear power plants. Naturally occurring uranium exists primarily as two isotopes: ^{235}U (0.720% abundant) and ^{238}U (99.275% abundant). However, because only the lighter isotope, ^{235}U, is suitable as a fuel in reactors, uranium ore must be enriched in this isotope.

Gas effusion is one way to separate the ^{235}U and ^{238}U isotopes. To achieve this, a uranium oxide sample is first converted to uranium hexafluoride, UF_6. This solid fluoride sublimes readily; it has a vapor pressure of 760 mm Hg at 55.6 °C. When UF_6 vapor is placed is a chamber with porous walls, the lighter, more rapidly moving $^{235}UF_6$ molecules effuse through the walls at a greater rate than the heavier $^{238}UF_6$ molecules.

To assess the separation of uranium isotopes, let us compare the rates of effusion of $^{235}UF_6$ and $^{238}UF_6$. Using Graham's law,

$$\frac{\text{Rate of } ^{235}\text{UF}_6}{\text{Rate of } ^{238}\text{UF}_6} = \sqrt{\frac{238.051 + 6(18.998)}{235.044 + 6(18.998)}} = 1.0043$$

we find that $^{235}UF_6$ will pass through a porous barrier 1.0043 times faster than $^{238}UF_6$. In other words, if we sample the gas that passes through the barrier, the fraction of $^{235}UF_6$ molecules will be larger. If the process is carried out again on the sample now higher in $^{235}UF_6$ concentration, the fraction of $^{235}UF_6$ would again increase in the effused sample, and the separation factor is now 1.0043 × 1.0043. If the cycle is repeated over and over again, the separation factor is 1.0043^n, where n is the number of enrichment cycles. To achieve a separation of about 99%, hundreds of cycles are required!

Deep Sea Diving

Diving with a self-contained underwater breathing apparatus (SCUBA) is exciting. If you want to dive much beyond about 60 ft (18 m) or so, however, you need to take special precautions.

Case Study — You Stink!

Do those dirty old sneakers in your closet stink? Did your friends ever tell you you have halitosis, the polite term for bad breath? Did your roommates ever experience flatulence (a malodorous gaseous emission, to say it politely) after eating too many beans? Or have you ever smelled the odor from a paper-making plant or from brackish water? The bad odors in all these cases can come from several gaseous, sulfur-containing compounds. Hydrogen sulfide (H_2S) and dimethylsulfide (CH_3SCH_3) are important contributors, but methyl mercaptan (CH_3SH) is the main culprit.

Charles D. Winters

Methyl mercaptan, also called methanethiol, heads the list of things that smell bad. Sources say it smells like rotten cabbage, but you already know what it smells like even if you have not smelled rotten cabbage recently. It is a gas at room temperature, but can be condensed to a liquid in an ice bath.

Data for methyl mercaptan

Melting Point	−123 °C
Boiling point	+5.95 °C
Density (gas, 298 K, 1 atm)	1.966 g/L
$\Delta_f H°$	−22.3 kJ/mol

Current OSHA guidelines are that the compound should not exceed concentrations of 10 parts per million (ppm) in air (or about 20 mg/m^3). Concentrations over 400 ppm have been known to cause death. However, humans can detect the odor of the compound at levels of a few parts per billion and so would leave the area if possible before concentrations became dangerous.

Bad breath comes from the formation of CH_3SH and similar compounds by the action of enzymes in the mouth on sulfur-containing compounds. Two of these compounds are common amino acids, methionine and cysteine. Methyl mercaptan is also produced when you digest allicin, which is produced when garlic is chopped to put into your pizza or your salad.

How can you get rid of halitosis? One way is to use a mouthwash. This can wash away some sources of sulfur compounds and might mask the odor. A more suitable method, however, is to use a toothpaste that has anti-plaque agents such as zinc and tin salts. It is thought that these interfere with the enzymes that act on something like methionine to produce methyl mercaptan.

Methyl mercaptan is not just a source of bad odors. It is used industrially to make pesticides, to regenerate catalysts in the petroleum industry, and to make methionine, which is used as a supplement in animal feed. Finally, mercaptans are added to natural gas and tanks of cooking gas. The three hydrocarbons in natural gas and cooking gas are odorless, so if you smell the unmistakable odor of a mercaptan you know there is a gas leak.

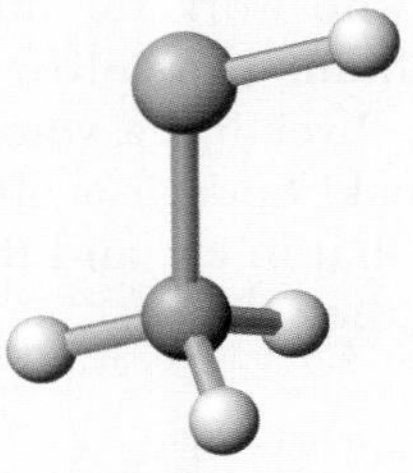

Methyl mercapatan or methanethiol.

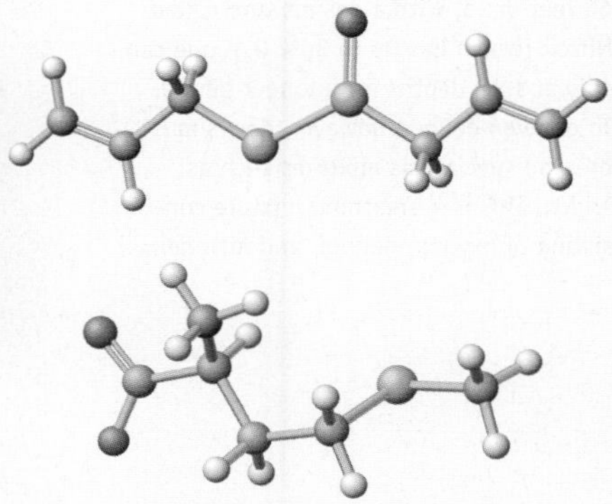

The digestion of allicin (top, from garlic) and methionine (bottom) is a source of CH_3SH in bad breath. Methionine is also made industrially using CH_3SH as one of the starting materials.

Questions:

1. *If an air sample contains CH_3SH with a concentration of 15 mg/m^3, what is its partial pressure at 25 °C? How many molecules are there per cubic meter?*
2. *What are the bond angles in CH_3SH?*
3. *Is CH_3SH polar or nonpolar?*
4. *Do you expect CH_3SH gas to behave as an ideal gas? (See Section 11.9.)*
5. *Which gas diffuses most rapidly, CH_3SH, H_2S, or CH_3SCH_3?*

Answers to these questions are in Appendix Q.

When you breathe air from a SCUBA tank (Figure 11.21), the pressure of the gas in your lungs is equal to the pressure exerted on your body. When you are at the surface, atmospheric pressure is about 1 atm, and, because air has an oxygen concentration of 21%, the partial pressure of O_2 is about 0.21 atm. If you are at a depth of about 33 ft, the water pressure is 2 atm. This means the oxygen partial pressure is double the surface partial pressure, or about 0.4 atm. Similarly, the partial pressure of N_2, which is about 0.8 atm at the surface, doubles to about 1.6 atm at a depth of 33 ft. The solubility of gases in water (and in blood) is directly proportional to pressure. Therefore, more oxygen and nitrogen dissolve in blood under these conditions, and this can lead to several problems.

FIGURE 11.21 SCUBA diving. Ordinary recreational dives can be made with compressed air to depths of about 60 feet or so. With a gas mixture called Nitrox (which has up to 36% O_2), one can stay at such depths for a longer period. To go even deeper, however, divers must breathe special gas mixtures such as Trimix. This is a breathing mixture consisting of oxygen, helium, and nitrogen.

OAR/National Undersea Research Program (NURP)

Nitrogen narcosis, also called "rapture of the deep" or the "martini effect," results from the toxic effect on nerve conduction of N_2 dissolved in blood. Its effect is comparable to drinking a martini on an empty stomach or taking laughing gas (nitrous oxide, N_2O) at the dentist; it makes you slightly giddy. In severe cases, it can impair a diver's judgment and even cause a diver to take the regulator out of his or her mouth and hand it to a fish! Some people can go as deep as 130 ft with no problem, but others experience nitrogen narcosis at 80 ft.

Another problem with breathing air at depths beyond 100 ft or so is oxygen toxicity. Our bodies are regulated for a partial pressure of O_2 of 0.21 atm. At a depth of 130 ft, the partial pressure of O_2 is comparable to breathing 100% oxygen at sea level. These higher partial pressures can harm the lungs and cause central nervous system damage. Oxygen toxicity is the reason deep dives are done not with compressed air but with gas mixtures with a much lower percentage of O_2, say about 10%.

Because of the risk of nitrogen narcosis, divers going beyond about 130 ft, such as those who work for offshore oil drilling companies, use a mixture of oxygen and helium. This solves the nitrogen narcosis problem, but it introduces another. If the diver has a voice link to the surface, the diver's speech sounds like Donald Duck! Speech is altered because the velocity of sound in helium is different from that in air, and the density of gas at several hundred feet is much higher than at the surface.

11.9 Nonideal Behavior: Real Gases

If you are working with a gas at approximately room temperature and a pressure of 1 atm or less, the ideal gas law is remarkably successful in relating the amount of gas and its pressure, volume, and temperature. At higher pressures or lower temperatures, however, deviations from the ideal gas law occur. The origin of these deviations is explained by the breakdown of the assumptions used when describing ideal gases, specifically the assumptions that the particles have no size and that there are no forces between them.

At standard temperature and pressure (STP), the volume occupied by a single molecule is *very* small relative to its share of the total gas volume. A helium atom with a radius of 31 pm has relatively about the same space to move about as a pea has inside a basketball. Now suppose the pressure is increased significantly, to 1000 atm. The volume available to each molecule is a sphere with a radius of only about 200 pm, which means the situation is now like that of a pea inside a sphere a bit larger than a Ping-Pong ball.

■ **Assumptions of the KMT—Revisited** The assumptions of the kinetic molecular theory were given on page 532.

1. Gases consist of particles (molecules or atoms) whose separation is much greater than the size of the particles themselves.
2. The particles of a gas are in continual, random, and rapid motion. As they move, they collide with one another and with the walls of their container, but they do so without loss of energy.
3. The average kinetic energy of gas particles is proportional to the gas temperature. All gases, regardless of their molecular mass, have the same average kinetic energy at the same temperature.

The kinetic-molecular theory and the ideal gas law are concerned with the volume available to the molecules to move about, not the total volume of the container. The problem is that the volume occupied by gas molecules is not negligible at higher pressures. For example, suppose you have a flask marked with a volume of 500 mL. This does not mean the space available to molecules is 500 mL. Rather, the available volume is less than 500 mL, especially at high gas pressures, because the molecules themselves occupy some of the volume.

Another assumption of the kinetic-molecular theory is that the atoms or molecules of the gas never stick to one another by some type of intermolecular force. This is clearly not true as well. All gases can be liquefied—although some gases require a very low temperature (see Figure 11.12)—and the only way this can happen is if there are forces between the molecules. When a molecule is about to strike the wall of its container, other molecules in its vicinity exert a slight pull on the molecule and pull it away from the wall. The effect of the intermolecular forces is that molecules strike the wall with less force than in the absence of intermolecular attractive forces. Thus, because collisions between molecules in a real gas and the wall are softer, the observed gas pressure is less than that predicted by the ideal gas law. This effect can be particularly pronounced when the temperature is low.

The Dutch physicist Johannes van der Waals (1837–1923) studied the breakdown of the ideal gas law equation and developed an equation to correct for the errors arising from nonideality. This equation is known as the **van der Waals equation:**

Observed pressure (↓ P) Container V (↓ V)

$$\left(P + a\left[\frac{n}{V}\right]^2\right)(V - bn) = nRT \qquad \textbf{(11.11)}$$

Correction for intermolecular forces (↑ $a[n/V]^2$) Correction for molecular volume (↑ bn)

where a and b are experimentally determined constants (Table 11.2). Although Equation 11.11 might seem complicated at first glance, the terms in parentheses are those of the ideal gas law, each corrected for the effects discussed previously. The pressure correction term, $a(n/V)^2$, accounts for intermolecular forces. Owing to intermolecular forces, the observed gas pressure is lower than the ideal pressure ($P_{observed} < P_{ideal}$ where P_{ideal} is calculated using the equation $PV = nRT$). Therefore, the term $a(n/V)^2$ is added to the observed pressure. The constant a typically has values in the range 0.01 to 10 atm · L^2/mol^2. The actual volume available to the molecules is smaller than the volume of the container because the molecules themselves take up space. Therefore, an amount is subtracted from the container volume ($= bn$) to take this into account. Here, n is the number of moles of gas, and b is an experimental quantity that corrects for the molecular volume. Typical values of b range from 0.01 to 0.1 L/mol, roughly increasing with increasing molecular size.

As an example of the importance of these corrections, consider a sample of 4.00 mol of chlorine gas, Cl_2, in a 4.00-L tank at 100.0 °C. The ideal gas law would lead you to expect a pressure of 30.6 atm. A better estimate of the pressure, obtained from the van der Waals equation, is 26.0 atm, about 4.6 atm less than the ideal pressure!

TABLE 11.2. van der Waals Constants

Gas	a Values atm · L^2/mol^2	b Values L/mol
He	0.034	0.0237
Ar	1.34	0.0322
H_2	0.244	0.0266
N_2	1.39	0.0391
O_2	1.36	0.0318
CO_2	3.59	0.0427
Cl_2	6.49	0.0562
H_2O	5.46	0.0305

EXERCISE 11.13 van der Waals's Equation

Using both the ideal gas law and van der Waals's equation, calculate the pressure expected for 10.0 mol of helium gas in a 1.00-L container at 25 °C.

Chapter Goals Revisited

Chemistry Now™ Sign in at **www.thomsonedu.com/login** to:

- Assess your understanding with Study Questions in OWL keyed to each goal in the Goals and Homework menu for this chapter
- For quick review, download Go Chemistry mini-lecture flashcard modules (or purchase them at **www.ichapters.com**)
- Check your readiness for an exam by taking the Pre-Test and exploring the modules recommended in your Personalized Study plan.

Access **How Do I Solve It?** tutorials on how to approach problem solving using concepts in this chapter.

Now that you have studied this chapter, you should ask whether you have met the chapter goals. In particular, you should be able to:

Understand the basis of the gas laws and how to use those laws.

a. Describe how pressure measurements are made and the units of pressure, especially atmospheres (atm) and millimeters of mercury (mm Hg) (Section 11.1). **Study Question(s) assignable in OWL: 1.**

b. Understand the basis of the gas laws (Boyle's Law, Charles's Law, and Avogadro's Hypothesis) and how to apply them (Section 11.2). **Study Question(s) assignable in OWL: 6, 8, 10, 12, 14.**

Use the ideal gas law.

a. Understand the origin of the ideal gas law and how to use the equation (Section 11.3). **Study Question(s) assignable in OWL: 18, 22, 24, 59, 63, 73, 81, 84, 88, 90, 96.**

b. Calculate the molar mass of a compound from a knowledge of the pressure of a known quantity of a gas in a given volume at a known temperature (Section 11.3). **Study Question(s) assignable in OWL: 26, 28, 30, 66, 85, 86, 92.**

Apply the gas laws to stoichiometric calculations.

a. Apply the gas laws to a study of the stoichiometry of reactions (Section 11.4). **Study Question(s) assignable in OWL: 32, 34, 65, 78.**

b. Use Dalton's law of partial pressures (Section 11.5). **Study Question(s) assignable in OWL: 39, 40, 70, 76, 83.**

Understand kinetic molecular theory as it is applied to gases, especially the distribution of molecular speeds (energies) (Section 11.6).

a. Apply the kinetic-molecular theory of gas behavior at the molecular level (Section 11.6). **Study Question(s) assignable in OWL: 41, 45, 101; Go Chemistry Module 16.**

b. Understand the phenomena of diffusion and effusion and how to use Graham's law (Section 11.7). **Study Question(s) assignable in OWL: 47.**

Recognize why gases do not behave like ideal gases under some conditions.

a. Appreciate the fact that gases usually do not behave as ideal gases. Deviations from ideal behavior are largest at high pressure and low temperature (Section 11.9). **Study Question(s) assignable in OWL: 51, 52.**

KEY EQUATIONS

Equation 11.1 (page 518) Boyle's law (where P is the pressure and V is the volume)

$$P_1V_1 = P_2V_2$$

Equation 11.2 (page 520) Charles's law (where T is the Kelvin temperature)

$$\frac{V_1}{T_1} = \frac{V_2}{T_2} \text{ at constant } n \text{ and } P$$

Equation 11.3 (page 521) General gas law (combined gas law)

$$\frac{P_1V_1}{T_1} = \frac{P_2V_2}{T_2} \text{ for a given amount of gas, } n$$

Equation 11.4 (page 524) Ideal gas law (where n is the amount of gas (moles) and R is the universal gas constant, 0.082057 L · atm/K · mol)

$$PV = nRT$$

Equation 11.5 (page 525) Density of gases (where d is the gas density in g/L and M is the molar mass of the gas)

$$d = \frac{m}{V} = \frac{PM}{RT}$$

Equation 11.6 (page 530) Dalton's law of partial pressures. The total pressure of a gas mixture is the sum of the partial pressures of the component gases (P_n).

$$P_{\text{total}} = P_1 + P_2 + P_3 + \ldots$$

Equation 11.7 (page 531) The total pressure of a gas mixture is equal to the total number of moles of gases multiplied by (RT/V).

$$P_{\text{total}} = (n_{\text{total}})\left(\frac{RT}{V}\right)$$

Equation 11.8 (page 531) The pressure of a gas (A) in a mixture is the product of its mole fraction (X_A) and the total pressure of the mixture.

$$P_A = X_A P_{\text{total}}$$

Equation 11.9 (page 536) Maxwell's equation relates the rms speed ($\sqrt{\overline{u^2}}$) to the molar mass of a gas (M) and its temperature (T).

$$\sqrt{\overline{u^2}} = \sqrt{\frac{3RT}{M}}$$

Equation 11.10 (page 538) Graham's law. The rate of effusion of a gas—the amount of material moving from one place to another in a given time—is inversely proportional to the square root of its molar mass.

$$\frac{\text{Rate of effusion of gas 1}}{\text{Rate of effusion of gas 2}} = \sqrt{\frac{\text{molar mass of gas 2}}{\text{molar mass of gas 1}}}$$

Equation 11.11 (page 543) The van der Waals equation: Relates pressure, volume, temperature, and amount of gas for a nonideal gas.

Observed pressure — Container V

$$\left(P + a\left[\frac{n}{V}\right]^2\right)(V - bn) = nRT$$

Correction for intermolecular forces — Correction for molecular volume

STUDY QUESTIONS

Online homework for this chapter may be assigned in OWL.

▲ denotes challenging questions.

■ denotes questions assignable in OWL.

Blue-numbered questions have answers in Appendix O and fully-worked solutions in the *Student Solutions Manual.*

PRACTICING SKILLS

Pressure

(See Example 11.1 and ChemistryNow Screen 11.2.)

1. ■ The pressure of a gas is 440 mm Hg. Express this pressure in units of (a) atmospheres, (b) bars, and (c) kilopascals.

2. The average barometric pressure at an altitude of 10 km is 210 mm Hg. Express this pressure in atmospheres, bars, and kilopascals.

3. Indicate which represents the higher pressure in each of the following pairs:
   (a) 534 mm Hg or 0.754 bar
   (b) 534 mm Hg or 650 kPa
   (c) 1.34 bar or 934 kPa

4. Put the following in order of increasing pressure: 363 mm Hg, 363 kPa, 0.256 atm, and 0.523 bar.

Boyle's Law and Charles's Law

(See Examples 11.2 and 11.3 and ChemistryNow Screen 11.3.)

5. A sample of nitrogen gas has a pressure of 67.5 mm Hg in a 500.-mL flask. What is the pressure of this gas sample when it is transferred to a 125-mL flask at the same temperature?

6. ■ A sample of CO_2 gas has a pressure of 56.5 mm Hg in a 125-mL flask. The sample is transferred to a new flask, where it has a pressure of 62.3 mm Hg at the same temperature. What is the volume of the new flask?

7. You have 3.5 L of NO at a temperature of 22.0 °C. What volume would the NO occupy at 37 °C? (Assume the pressure is constant.)

8. ■ A 5.0-mL sample of CO_2 gas is enclosed in a gas-tight syringe (see Figure 11.4) at 22 °C. If the syringe is immersed in an ice bath (0 °C), what is the new gas volume, assuming that the pressure is held constant?

The General Gas Law

(See Example 11.4.)

9. You have 3.6 L of H_2 gas at 380 mm Hg and 25 °C. What is the pressure of this gas if it is transferred to a 5.0-L flask at 0.0 °C?

10. ■ You have a sample of CO_2 in a flask A with a volume of 25.0 mL. At 20.5 °C, the pressure of the gas is 436.5 mm Hg. To find the volume of another flask, B, you move the CO_2 to that flask and find that its pressure is now 94.3 mm Hg at 24.5 °C. What is the volume of flask B?

11. You have a sample of gas in a flask with a volume of 250 mL. At 25.5 °C, the pressure of the gas is 360 mm Hg. If you decrease the temperature to −5.0 °C, what is the gas pressure at the lower temperature?

12. ■ A sample of gas occupies 135 mL at 22.5 °C; the pressure is 165 mm Hg. What is the pressure of the gas sample when it is placed in a 252-mL flask at a temperature of 0.0 °C?

13. One of the cylinders of an automobile engine has a volume of 400. cm^3. The engine takes in air at a pressure of 1.00 atm and a temperature of 15 °C and compresses the air to a volume of 50.0 cm^3 at 77 °C. What is the final pressure of the gas in the cylinder? (The ratio of before and after volumes—in this case, 400 : 50 or 8 : 1—is called the compression ratio.)

14. ■ A helium-filled balloon of the type used in long-distance flying contains 420,000 ft^3 (1.2×10^7 L) of helium. Suppose you fill the balloon with helium on the ground, where the pressure is 737 mm Hg and the temperature is 16.0 °C. When the balloon ascends to a height of 2 miles, where the pressure is only 600. mm Hg and the temperature is −33 °C, what volume is occupied by the helium gas? Assume the pressure inside the balloon matches the external pressure. Comment on the result.

Avogadro's Hypothesis

(See Example 11.5 and ChemistryNow Screen 11.3.)

15. Nitrogen monoxide reacts with oxygen to give nitrogen dioxide.

$$2\ NO(g) + O_2(g) \longrightarrow 2\ NO_2(g)$$

   (a) If you mix NO and O_2 in the correct stoichiometric ratio and NO has a volume of 150 mL, what volume of O_2 is required (at the same pressure and temperature)?
   (b) After reaction is complete between 150 mL of NO and the stoichiometric volume of O_2, what is the volume of NO_2 (at the same pressure and temperature)?

16. Ethane, C_2H_6, burns in air according to the equation

$$2\ C_2H_6(g) + 7\ O_2(g) \longrightarrow 4\ CO_2(g) + 6\ H_2O(g)$$

What volume of O_2 (L) is required for complete reaction with 5.2 L of C_2H_6? What volume of H_2O vapor (L) is produced? Assume all gases are measured at the same temperature and pressure.

Ideal Gaw Law

(See Example 11.6 and ChemistryNow Screen 11.4.)

17. A 1.25-g sample of CO_2 is contained in a 750.-mL flask at 22.5 °C. What is the pressure of the gas?

18. ■ A balloon holds 30.0 kg of helium. What is the volume of the balloon if the final pressure is 1.20 atm and the temperature is 22 °C?

19. A flask is first evacuated so that it contains no gas at all. Then, 2.2 g of CO_2 is introduced into the flask. On warming to 22 °C, the gas exerts a pressure of 318 mm Hg. What is the volume of the flask?

20. A steel cylinder holds 1.50 g of ethanol, C_2H_5OH. What is the pressure of the ethanol vapor if the cylinder has a volume of 251 cm^3 and the temperature is 250 °C? (Assume all of the ethanol is in the vapor phase at this temperature.)

21. A balloon for long-distance flying contains 1.2×10^7 L of helium. If the helium pressure is 737 mm Hg at 25 °C, what mass of helium (in grams) does the balloon contain? (See Study Question 14.)

22. ■ What mass of helium, in grams, is required to fill a 5.0-L balloon to a pressure of 1.1 atm at 25 °C?

Gas Density

(See Example 11.8 and ChemistryNow Screen 11.5.)

23. Forty miles above Earth's surface, the temperature is 250 K, and the pressure is only 0.20 mm Hg. What is the density of air (in grams per liter) at this altitude? (Assume the molar mass of air is 28.96 g/mol.)

24. ■ Diethyl ether, $(C_2H_5)_2O$, vaporizes easily at room temperature. If the vapor exerts a pressure of 233 mm Hg in a flask at 25 °C, what is the density of the vapor?

25. A gaseous organofluorine compound has a density of 0.355 g/L at 17 °C and 189 mm Hg. What is the molar mass of the compound?

26. ■ Chloroform is a common liquid used in the laboratory. It vaporizes readily. If the pressure of chloroform vapor in a flask is 195 mm Hg at 25.0 °C and the density of the vapor is 1.25 g/L, what is the molar mass of chloroform?

Ideal Gas Laws and Determining Molar Mass

(See Examples 11.7 and 11.8 and ChemistryNow Screen 11.6.)

27. A 1.007-g sample of an unknown gas exerts a pressure of 715 mm Hg in a 452-mL container at 23 °C. What is the molar mass of the gas?

28. ■ A 0.0125-g sample of a gas with an empirical formula of CHF_2 is placed in a 165-mL flask. It has a pressure of 13.7 mm Hg at 22.5 °C. What is the molecular formula of the compound?

29. A new boron hydride, B_xH_y, has been isolated. To find its molar mass, you measure the pressure of the gas in a known volume at a known temperature. The following experimental data are collected:

    Mass of gas = 12.5 mg Pressure of gas = 24.8 mm Hg

    Temperature = 25 °C Volume of flask = 125 mL

    Which formula corresponds to the calculated molar mass?

    (a) B_2H_6
    (b) B_4H_{10}
    (c) B_5H_9
    (d) B_6H_{10}
    (e) $B_{10}H_{14}$

30. ■ Acetaldehyde is a common liquid compound that vaporizes readily. Determine the molar mass of acetaldehyde from the following data:

    Sample mass = 0.107 g Volume of gas = 125 mL

    Temperature = 0.0 °C Pressure = 331 mm Hg

Gas Laws and Stoichiometry

(See Examples 11.9 and 11.10 and ChemistryNow Screen 11.7.)

31. Iron reacts with hydrochloric acid to produce iron(II) chloride and hydrogen gas:

$$Fe(s) + 2\ HCl(aq) \longrightarrow FeCl_2(aq) + H_2(g)$$

    The H_2 gas from the reaction of 2.2 g of iron with excess acid is collected in a 10.0-L flask at 25 °C. What is the pressure of the H_2 gas in this flask?

32. ■ Silane, SiH_4, reacts with O_2 to give silicon dioxide and water:

$$SiH_4(g) + 2\ O_2(g) \longrightarrow SiO_2(s) + 2\ H_2O(\ell)$$

    A 5.20-L sample of SiH_4 gas at 356 mm Hg pressure and 25 °C is allowed to react with O_2 gas. What volume of O_2 gas, in liters, is required for complete reaction if the oxygen has a pressure of 425 mm Hg at 25 °C?

33. Sodium azide, the explosive compound in automobile air bags, decomposes according to the following equation:

$$2\ NaN_3(s) \longrightarrow 2\ Na(s) + 3\ N_2(g)$$

    What mass of sodium azide is required to provide the nitrogen needed to inflate a 75.0-L bag to a pressure of 1.3 atm at 25 °C?

34. ■ The hydrocarbon octane (C_8H_{18}) burns to give CO_2 and water vapor:

$$2\ C_8H_{18}(g) + 25\ O_2(g) \longrightarrow 16\ CO_2(g) + 18\ H_2O(g)$$

    If a 0.048-g sample of octane burns completely in O_2, what will be the pressure of water vapor in a 4.75-L flask at 30.0 °C? If the O_2 gas needed for complete combustion was contained in a 4.75-L flask at 22 °C, what would its pressure be?

35. Hydrazine reacts with O_2 according to the following equation:

$$N_2H_4(g) + O_2(g) \rightarrow N_2(g) + 2\ H_2O(\ell)$$

Assume the O_2 needed for the reaction is in a 450-L tank at 23 °C. What must the oxygen pressure be in the tank to have enough oxygen to consume 1.00 kg of hydrazine completely?

36. A self-contained underwater breathing apparatus uses canisters containing potassium superoxide. The superoxide consumes the CO_2 exhaled by a person and replaces it with oxygen.

$$4\ KO_2(s) + 2\ CO_2(g) \rightarrow 2\ K_2CO_3(s) + 3\ O_2(g)$$

What mass of KO_2, in grams, is required to react with 8.90 L of CO_2 at 22.0 °C and 767 mm Hg?

Gas Mixtures and Dalton's Law

(See Example 11.11 and ChemistryNow Screen 11.8.)

37. What is the total pressure in atmospheres of a gas mixture that contains 1.0 g of H_2 and 8.0 g of Ar in a 3.0-L container at 27 °C? What are the partial pressures of the two gases?

38. A cylinder of compressed gas is labeled "Composition (mole %): 4.5% H_2S, 3.0% CO_2, balance N_2." The pressure gauge attached to the cylinder reads 46 atm. Calculate the partial pressure of each gas, in atmospheres, in the cylinder.

39. ■ A halothane–oxygen mixture ($C_2HBrClF_3$ + O_2) can be used as an anesthetic. A tank containing such a mixture has the following partial pressures: P (halothane) = 170 mm Hg and P (O_2) = 570 mm Hg.
(a) What is the ratio of the number of moles of halothane to the number of moles of O_2?
(b) If the tank contains 160 g of O_2, what mass of $C_2HBrClF_3$ is present?

40. ■ A collapsed balloon is filled with He to a volume of 12.5 L at a pressure of 1.00 atm. Oxygen, O_2, is then added so that the final volume of the balloon is 26 L with a total pressure of 1.00 atm. The temperature, which remains constant throughout, is 21.5 °C.
(a) What mass of He does the balloon contain?
(b) What is the final partial pressure of He in the balloon?
(c) What is the partial pressure of O_2 in the balloon?
(d) What is the mole fraction of each gas?

Kinetic-Molecular Theory

(See Section 11.6, Example 11.12, and ChemistryNow Screens 11.9–11.12.)

41. ■ You have two flasks of equal volume. Flask A contains H_2 at 0 °C and 1 atm pressure. Flask B contains CO_2 gas at 25 °C and 2 atm pressure. Compare these two gases with respect to each of the following:
(a) average kinetic energy per molecule
(b) average molecular velocity
(c) number of molecules
(d) mass of gas

42. Equal masses of gaseous N_2 and Ar are placed in separate flasks of equal volume at the same temperature. Tell whether each of the following statements is true or false. Briefly explain your answer in each case.
(a) There are more molecules of N_2 present than atoms of Ar.
(b) The pressure is greater in the Ar flask.
(c) The Ar atoms have a greater average speed than the N_2 molecules.
(d) The N_2 molecules collide more frequently with the walls of the flask than do the Ar atoms.

43. If the speed of an oxygen molecule is 4.28×10^4 cm/s at 25 °C, what is the speed of a CO_2 molecule at the same temperature?

44. Calculate the rms speed for CO molecules at 25 °C. What is the ratio of this speed to that of Ar atoms at the same temperature?

45. ■ Place the following gases in order of increasing average molecular speed at 25 °C: Ar, CH_4, N_2, CH_2F_2.

46. The reaction of SO_2 with Cl_2 gives dichlorine oxide, which is used to bleach wood pulp and to treat wastewater:

$$SO_2(g) + 2\ Cl_2(g) \rightarrow OSCl_2(g) + Cl_2O(g)$$

All of the compounds involved in the reaction are gases. List them in order of increasing average speed.

Diffusion and Effusion

(See Example 11.13 and ChemistryNow Screen 11.12.)

47. ■ In each pair of gases below, tell which will effuse faster:
(a) CO_2 or F_2
(b) O_2 or N_2
(c) C_2H_4 or C_2H_6
(d) two chlorofluorocarbons: $CFCl_3$ or $C_2Cl_2F_4$

48. Argon gas is 10 times denser than helium gas at the same temperature and pressure. Which gas is predicted to effuse faster? How much faster?

49. A gas whose molar mass you wish to know effuses through an opening at a rate one third as fast as that of helium gas. What is the molar mass of the unknown gas?

50. ▲ A sample of uranium fluoride is found to effuse at the rate of 17.7 mg/h. Under comparable conditions, gaseous I_2 effuses at the rate of 15.0 mg/h. What is the molar mass of the uranium fluoride? (*Hint:* Rates must be converted to units of moles per time.)

Nonideal Gases

(See Section 11.9.)

51. ■ In the text, it is stated that the pressure of 4.00 mol of Cl_2 in a 4.00-L tank at 100.0 °C should be 26.0 atm if calculated using the van der Waals equation. Verify this result, and compare it with the pressure predicted by the ideal gas law.

▲ more challenging ■ in OWL Blue-numbered questions answered in Appendix O

52. ■ You want to store 165 g of CO_2 gas in a 12.5-L tank at room temperature (25 °C). Calculate the pressure the gas would have using (a) the ideal gas law and (b) the van der Waals equation. (For CO_2, a = 3.59 atm · L^2/mol^2 and b = 0.0427 L/mol.)

General Questions

These questions are not designated as to type or location in the chapter. They may combine several concepts.

53. Complete the following table:

	atm	mm Hg	kPa	bar
Standard atmosphere	____	____	____	____
Partial pressure of N_2 in the atmosphere	____	593	____	____
Tank of compressed H_2	____	____	____	133
Atmospheric pressure at the top of Mount Everest	____	____	33.7	____

54. On combustion, 1.0 L of a gaseous compound of hydrogen, carbon, and nitrogen gives 2.0 L of CO_2, 3.5 L of H_2O vapor, and 0.50 L of N_2 at STP. What is the empirical formula of the compound?

55. ▲ You have a sample of helium gas at −33 °C, and you want to increase the average speed of helium atoms by 10.0%. To what temperature should the gas be heated to accomplish this?

56. If 12.0 g of O_2 is required to inflate a balloon to a certain size at 27 °C, what mass of O_2 is required to inflate it to the same size (and pressure) at 5.0 °C?

57. Butyl mercaptan, C_4H_9SH, has a very bad odor and is among the compounds added to natural gas to help detect a leak of otherwise odorless natural gas. In an experiment, you burn 95.0 mg of C_4H_9SH and collect the product gases (SO_2, CO_2, and H_2O) in a 5.25 L flask at 25 °C. What is the total gas pressure in the flask, and what is the partial pressure of each of the product gases?

58. A bicycle tire has an internal volume of 1.52 L and contains 0.406 mol of air. The tire will burst if its internal pressure reaches 7.25 atm. To what temperature, in degrees Celsius, does the air in the tire need to be heated to cause a blowout?

59. ■ The temperature of the atmosphere on Mars can be as high as 27 °C at the equator at noon, and the atmospheric pressure is about 8 mm Hg. If a spacecraft could collect 10. m^3 of this atmosphere, compress it to a small volume, and send it back to Earth, how many moles would the sample contain?

60. If you place 2.25 g of solid silicon in a 6.56-L flask that contains CH_3Cl with a pressure of 585 mm Hg at 25 °C, what mass of dimethyldichlorosilane, $(CH_3)_2SiCl_2(g)$, can be formed?

$$Si(s) + 2\ CH_3Cl(g) \rightarrow (CH_3)_2SiCl_2(g)$$

What pressure of $(CH_3)_2SiCl_2(g)$ would you expect in this same flask at 95 °C on completion of the reaction? (Dimethyldichlorosilane is one starting material used to make silicones, polymeric substances used as lubricants, antistick agents, and water-proofing caulk.)

61. $Ni(CO)_4$ can be made by reacting finely divided nickel with gaseous CO. If you have CO in a 1.50-L flask at a pressure of 418 mm Hg at 25.0 °C, along with 0.450 g of Ni powder, what is the theoretical yield of $Ni(CO)_4$?

62. The gas B_2H_6 burns in air to give H_2O and B_2O_3.

$$B_2H_6(g) + 3\ O_2(g) \rightarrow B_2O_3(s) + 3\ H_2O(g)$$

(a) Three gases are involved in this reaction. Place them in order of increasing rms speed. (Assume all are at the same temperature.)
(b) A 3.26-L flask contains B_2H_6 at a pressure of 256 mm Hg and a temperature of 25 °C. Suppose O_2 gas is added to the flask until B_2H_6 and O_2 are in the correct stoichiometric ratio for the combustion reaction. At this point, what is the partial pressure of O_2?

63. ■ You have four gas samples:
1. 1.0 L of H_2 at STP
2. 1.0 L of Ar at STP
3. 1.0 L of H_2 at 27 °C and 760 mm Hg
4. 1.0 L of He at 0 °C and 900 mm Hg

(a) Which sample has the largest number of gas particles (atoms or molecules)?
(b) Which sample contains the smallest number of particles?
(c) Which sample represents the largest mass?

64. Propane reacts with oxygen to give carbon dioxide and water vapor.

$$C_3H_8(g) + 5\ O_2(g) \rightarrow 3\ CO_2(g) + 4\ H_2O(g)$$

If you mix C_3H_8 and O_2 in the correct stoichiometric ratio, and if the total pressure of the mixture is 288 mm Hg, what are the partial pressures of C_3H_8 and O_2? If the temperature and volume do not change, what is the pressure of the water vapor reaction?

65. ■ Iron carbonyl can be made by the direct reaction of iron metal and carbon monoxide.

$$Fe(s) + 5\ CO(g) \rightarrow Fe(CO)_5(\ell)$$

What is the theoretical yield of $Fe(CO)_5$ if 3.52 g of iron is treated with CO gas having a pressure of 732 mm Hg in a 5.50-L flask at 23 °C?

66. ■ Analysis of a gaseous chlorofluorocarbon, CCl_xF_y, shows that it contains 11.79% C and 69.57% Cl. In another experiment, you find that 0.107 g of the compound fills a 458-mL flask at 25 °C with a pressure of 21.3 mm Hg. What is the molecular formula of the compound?

67. There are five compounds in the family of sulfur–fluorine compounds with the general formula S_xF_y. One of these compounds is 25.23% S. If you place 0.0955 g of the compound in a 89-mL flask at 45 °C, the pressure of the gas is 83.8 mm Hg. What is the molecular formula of S_xF_y?

68. A miniature volcano can be made in the laboratory with ammonium dichromate. When ignited, it decomposes in a fiery display.

$$(NH_4)_2Cr_2O_7(s) \rightarrow N_2(g) + 4\ H_2O(g) + Cr_2O_3(s)$$

If 0.95 g of ammonium dichromate is used and if the gases from this reaction are trapped in a 15.0-L flask at 23 °C, what is the total pressure of the gas in the flask? What are the partial pressures of N_2 and H_2O?

Thermal decomposition of $(NH_4)_2Cr_2O_7$.

69. The density of air 20 km above the earth's surface is 92 g/m^3. The pressure of the atmosphere is 42 mm Hg, and the temperature is −63 °C.
(a) What is the average molar mass of the atmosphere at this altitude?
(b) If the atmosphere at this altitude consists of only O_2 and N_2, what is the mole fraction of each gas?

70. ■ A 3.0-L bulb containing He at 145 mm Hg is connected by a valve to a 2.0-L bulb containing Ar at 355 mm Hg. (See the accompanying figure.) Calculate the partial pressure of each gas and the total pressure after the valve between the flasks is opened.

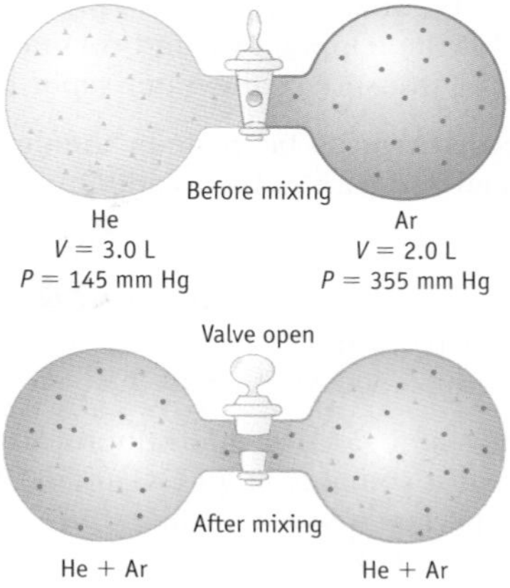

71. Chlorine dioxide, ClO_2, reacts with fluorine to give a new gas that contains Cl, O, and F. In an experiment, you find that 0.150 g of this new gas has a pressure of 17.2 mm Hg in a 1850-mL flask at 21 °C. What is the identity of the unknown gas?

72. A xenon fluoride can be prepared by heating a mixture of Xe and F_2 gases to a high temperature in a pressure-proof container. Assume that xenon gas was added to a 0.25-L container until its pressure reached 0.12 atm at 0.0 °C. Fluorine gas was then added until the total pressure reached 0.72 atm at 0.0 °C. After the reaction was complete, the xenon was consumed completely, and the pressure of the F_2 remaining in the container was 0.36 atm at 0.0 °C. What is the empirical formula of the xenon fluoride?

73. ■ A balloon at the circus is filled with helium gas to a gauge pressure of 22 mm Hg at 25 °C. The volume of the gas is 305 mL, and the barometric pressure is 755 mm Hg. What amount of helium is in the balloon? (Remember that gauge pressure = total pressure − barometric pressure. See page 517.)

74. If you have a sample of water in a closed container, some of the water will evaporate until the pressure of the water vapor, at 25 °C, is 23.8 mm Hg. How many molecules of water per cubic centimeter exist in the vapor phase?

75. You are given 1.56 g of a mixture of $KClO_3$ and KCl. When heated, the $KClO_3$ decomposes to KCl and O_2,

$$2\ KClO_3(s) \rightarrow 2\ KCl(s) + 3\ O_2(g)$$

and 327 mL of O_2 with a pressure of 735 mm Hg is collected at 19 °C. What is the weight percentage of $KClO_3$ in the sample?

76. ▲ ■ A study of climbers who reached the summit of Mount Everest without supplemental oxygen showed that the partial pressures of O_2 and CO_2 in their lungs were 35 mm Hg and 7.5 mm Hg, respectively. The barometric pressure at the summit was 253 mm Hg. Assume the lung gases are saturated with moisture at a body temperature of 37 °C [which means the partial pressure of water vapor in the lungs is $P(H_2O) = 47.1$ mm Hg]. If you assume the lung gases consist of only O_2, N_2, CO_2, and H_2O, what is the partial pressure of N_2?

77. Nitrogen monoxide reacts with oxygen to give nitrogen dioxide:

$$2\ NO(g) + O_2(g) \rightarrow 2\ NO_2(g)$$

(a) Place the three gases in order of increasing rms speed at 298 K.
(b) If you mix NO and O_2 in the correct stoichiometric ratio and NO has a partial pressure of 150 mm Hg, what is the partial pressure of O_2?
(c) After reaction between NO and O_2 is complete, what is the pressure of NO_2 if the NO originally had a pressure of 150 mm Hg and O_2 was added in the correct stoichiometric amount?

▲ more challenging ■ in OWL Blue-numbered questions answered in Appendix O

78. ▲ ■ Ammonia gas is synthesized by combining hydrogen and nitrogen:

$$3\ H_2(g) + N_2(g) \longrightarrow 2\ NH_3(g)$$

(a) If you want to produce 562 g of NH_3, what volume of H_2 gas, at 56 °C and 745 mm Hg, is required?
(b) To produce 562 g of NH_3, what volume of air (the source of N_2) is required if the air is introduced at 29 °C and 745 mm Hg? (Assume the air sample has 78.1 mole % N_2.)

79. Nitrogen trifluoride is prepared by the reaction of ammonia and fluorine.

$$4\ NH_3(g) + 3\ F_2(g) \longrightarrow 3\ NH_4F(s) + NF_3(g)$$

If you mix NH_3 with F_2 in the correct stoichiometric ratio, and if the total pressure of the mixture is 120 mm Hg, what are the partial pressures of NH_3 and F_2? When the reactants have been completely consumed, what is the total pressure in the flask? (Assume T is constant.)

80. Chlorine trifluoride, ClF_3, is a valuable reagent because it can be used to convert metal oxides to metal fluorides:

$$6\ NiO(s) + 4\ ClF_3(g) \longrightarrow 6\ NiF_2(s) + 2\ Cl_2(g) + 3\ O_2(g)$$

(a) What mass of NiO will react with ClF_3 gas if the gas has a pressure of 250 mm Hg at 20 °C in a 2.5-L flask?
(b) If the ClF_3 described in part (a) is completely consumed, what are the partial pressures of Cl_2 and of O_2 in the 2.5-L flask at 20 °C (in mm Hg)? What is the total pressure in the flask?

81. ▲ ■ Relative humidity is the ratio of the partial pressure of water in air at a given temperature to the vapor pressure of water at that temperature. Calculate the mass of water per liter of air under the following conditions:
(a) at 20 °C and 45% relative humidity
(b) at 0 °C and 95% relative humidity

Under which circumstances is the mass of H_2O per liter greater? (See Appendix G for the vapor pressure of water.)

82. ■ How much water vapor is present in a dormitory room when the relative humidity is 55% and the temperature is 23 °C? The dimensions of the room are 4.5 m^2 floor area and 3.5 m ceiling height. (See Study Question 81 for a definition of relative humidity and Appendix G for the vapor pressure of water.)

In the Laboratory

83. ▲ ■ You have a 550.-mL tank of gas with a pressure of 1.56 atm at 24 °C. You thought the gas was pure carbon monoxide gas, CO, but you later found it was contaminated by small quantities of gaseous CO_2 and O_2. Analysis shows that the tank pressure is 1.34 atm (at 24 °C) if the CO_2 is removed. Another experiment shows that 0.0870 g of O_2 can be removed chemically. What are the masses of CO and CO_2 in the tank, and what is the partial pressure of each of the three gases at 25 °C?

84. ▲ ■ Methane is burned in a laboratory Bunsen burner to give CO_2 and water vapor. Methane gas is supplied to the burner at the rate of 5.0 L/min (at a temperature of 28 °C and a pressure of 773 mm Hg). At what rate must oxygen be supplied to the burner (at a pressure of 742 mm Hg and a temperature of 26 °C)?

85. ▲ ■ Iron forms a series of compounds of the type $Fe_x(CO)_y$. In air, they are oxidized to Fe_2O_3 and CO_2 gas. After heating a 0.142-g sample of $Fe_x(CO)_y$ in air, you isolate the CO_2 in a 1.50-L flask at 25 °C. The pressure of the gas is 44.9 mm Hg. What is the empirical formula of $Fe_x(CO)_y$?

86. ▲ ■ Group 2A metal carbonates are decomposed to the metal oxide and CO_2 on heating:

$$MCO_3(s) \longrightarrow MO(s) + CO_2(g)$$

You heat 0.158 g of a white, solid carbonate of a Group 2A metal (M) and find that the evolved CO_2 has a pressure of 69.8 mm Hg in a 285-mL flask at 25 °C. Identify M.

87. One way to synthesize diborane, B_2H_6, is the reaction

$$2\ NaBH_4(s) + 2\ H_3PO_4(aq) \longrightarrow B_2H_6(g) + 2\ NaH_2PO_4(aq) + 2\ H_2(g)$$

(a) If you have 0.136 g of $NaBH_4$ and excess H_3PO_4, and you collect the B_2H_6 in a 2.75 L flask at 25 °C, what is the pressure of the B_2H_6 in the flask?
(b) A by-product of the reaction is H_2 gas. If both B_2H_6 and H_2 gas come from this reaction, what is the *total* pressure in the 2.75-L flask (after reaction of 0.136 g of $NaBH_4$ with excess H_3PO_4) at 25 °C?

88. ■ You are given a solid mixture of $NaNO_2$ and NaCl and are asked to analyze it for the amount of $NaNO_2$ present. To do so, you allow the mixture to react with sulfamic acid, HSO_3NH_2, in water according to the equation

$$NaNO_2(aq) + HSO_3NH_2(aq) \longrightarrow NaHSO_4(aq) + H_2O(\ell) + N_2(g)$$

What is the weight percentage of $NaNO_2$ in 1.232 g of the solid mixture if reaction with sulfamic acid produces 295 mL of N_2 gas with a pressure of 713 mm Hg at 21.0 °C?

89. ▲ You have 1.249 g of a mixture of $NaHCO_3$ and Na_2CO_3. You find that 12.0 mL of 1.50 M HCl is required to convert the sample completely to NaCl, H_2O, and CO_2.

$$NaHCO_3(aq) + HCl(aq) \longrightarrow NaCl(aq) + H_2O(\ell) + CO_2(g)$$

$$Na_2CO_3(aq) + 2\ HCl(aq) \longrightarrow 2\ NaCl(aq) + H_2O(\ell) + CO_2(g)$$

What volume of CO_2 is evolved at 745 mm Hg and 25 °C?

90. ▲ ■ A mixture of $NaHCO_3$ and Na_2CO_3 has a mass of 2.50 g. When treated with HCl(aq), 665 mL of CO_2 gas is liberated with a pressure of 735 mm Hg at 25 °C. What is the weight percent of $NaHCO_3$ and Na_2CO_3 in the mixture? (See Study Question 89 for the reactions that occur.)

91. ▲ Many nitrate salts can be decomposed by heating. For example, blue, anhydrous copper(II) nitrate produces nitrogen dioxide and oxygen when heated. In the laboratory, you find that a sample of this salt produced 0.195 g of mixture of NO_2 and O_2 with a pressure of 725 mm Hg at 35 °C in a 125-mL flask (and black, solid CuO was left as a residue). What is the average molar mass of the gas mixture? What are the mole fractions of NO_2 and O_2? What amount of each gas is in the mixture? Do these amounts reflect the relative amounts of NO_2 and O_2 expected based on the balanced equation? Is it possible that the fact that some NO_2 molecules combine to give N_2O_4 plays a role?

Heating copper(II) nitrate produces nitrogen dioxide and oxygen gas and leaves a residue of copper(II) oxide.

92. ▲ ■ A compound containing C, H, N, and O is burned in excess oxygen. The gases produced by burning 0.1152 g are first treated to convert the nitrogen-containing product gases into N_2, and then the resulting mixture of CO_2, H_2O, N_2, and excess O_2 is passed through a bed of $CaCl_2$ to absorb the water. The $CaCl_2$ increases in mass by 0.09912 g. The remaining gases are bubbled into water to form H_2CO_3, and this solution is titrated with 0.3283 M NaOH; 28.81 mL is required to achieve the second equivalence point. The excess O_2 gas is removed by reaction with copper metal (to give CuO). Finally, the N_2 gas is collected in a 225.0-mL flask, where it has a pressure of 65.12 mm Hg at 25 °C. In a separate experiment, the unknown compound is found to have a molar mass of 150 g/mol. What are the empirical and molecular formulas of the unknown compound?

Summary and Conceptual Questions

The following questions may use concepts from the previous chapters.

93. A 1.0-L flask contains 10.0 g each of O_2 and CO_2 at 25 °C.
(a) Which gas has the greater partial pressure, O_2 or CO_2, or are they the same?
(b) Which molecules have the greater average speed, or are they the same?
(c) Which molecules have the greater average kinetic energy, or are they the same?

94. If equal masses of O_2 and N_2 are placed in separate containers of equal volume at the same temperature, which of the following statements is true? If false, tell why it is false.
(a) The pressure in the flask containing N_2 is greater than that in the flask containing O_2.
(b) There are more molecules in the flask containing O_2 than in the flask containing N_2.

95. You have two pressure-proof steel cylinders of equal volume, one containing 1.0 kg of CO and the other containing 1.0 kg of acetylene, C_2H_2.
(a) In which cylinder is the pressure greater at 25 °C?
(b) Which cylinder contains the greater number of molecules?

96. ■ Two flasks, each with a volume of 1.00 L, contain O_2 gas with a pressure of 380 mm Hg. Flask A is at 25 °C, and flask B is at 0 °C. Which flask contains the greater number of O_2 molecules?

97. ▲ State whether each of the following samples of matter is a gas. If there is not enough information for you to decide, write "insufficient information."
(a) A material is in a steel tank at 100 atm pressure. When the tank is opened to the atmosphere, the material suddenly expands, increasing its volume by 10%.
(b) A 1.0-mL sample of material weighs 8.2 g.
(c) The material is transparent and pale green in color.
(d) One cubic meter of material contains as many molecules as 1.0 m^3 of air at the same temperature and pressure.

98. Each of the four tires of a car is filled with a different gas. Each tire has the same volume, and each is filled to the same pressure, 3.0 atm, at 25 °C. One tire contains 116 g of air, another tire has 80.7 g of neon, another tire has 16.0 g of helium, and the fourth tire has 160. g of an unknown gas.
(a) Do all four tires contain the same number of gas molecules? If not, which one has the greatest number of molecules?
(b) How many times heavier is a molecule of the unknown gas than an atom of helium?
(c) In which tire do the molecules have the largest kinetic energy? The highest average speed?

▲ more challenging ■ in OWL Blue-numbered questions answered in Appendix O

99. You have two gas-filled balloons, one containing He and the other containing H_2. The H_2 balloon is twice the size of the He balloon. The pressure of gas in the H_2 balloon is 1 atm, and that in the He balloon is 2 atm. The H_2 balloon is outside in the snow (−5 °C), and the He balloon is inside a warm building (23 °C).
(a) Which balloon contains the greater number of molecules?
(b) Which balloon contains the greater mass of gas?

100. The sodium azide required for automobile air bags is made by the reaction of sodium metal with dinitrogen oxide in liquid ammonia:

$$3\ N_2O(g) + 4\ Na(s) + NH_3(\ell) \rightarrow NaN_3(s) + 3\ NaOH(s) + 2\ N_2(g)$$

(a) You have 65.0 g of sodium and a 35.0-L flask containing N_2O gas with a pressure of 2.12 atm at 23 °C. What is the theoretical yield (in grams) of NaN_3?
(b) Draw a Lewis structure for the azide ion. Include all possible resonance structures. Which resonance structure is most likely?
(c) What is the shape of the azide ion?

101. ■ If the absolute temperature of a gas doubles, by how much does the average speed of the gaseous molecules increase? *(See ChemistryNow Screen 11.9.)*

102. ▲ Chlorine gas (Cl_2) is used as a disinfectant in municipal water supplies, although chlorine dioxide (ClO_2) and ozone are becoming more widely used. ClO_2 is a better choice than Cl_2 in this application because it leads to fewer chlorinated by-products, which are themselves pollutants.
(a) How many valence electrons are in ClO_2?
(b) The chlorite ion, ClO_2^-, is obtained by reducing ClO_2. Draw a possible electron dot structure for ClO_2^-. (Cl is the central atom.)
(c) What is the hybridization of the central Cl atom in ClO_2^-? What is the shape of the ion?
(d) Which species has the larger bond angle, O_3 or ClO_2^-? Explain briefly.
(e) Chlorine dioxide, ClO_2, a yellow-green gas, can be made by the reaction of chlorine with sodium chlorite:

$$2\ NaClO_2(s) + Cl_2(g) \rightarrow 2\ NaCl(s) + 2\ ClO_2(g)$$

Assume you react 15.6 g of $NaClO_2$ with chlorine gas, which has a pressure of 1050 mm Hg in a 1.45-L flask at 22 °C. What mass of ClO_2 can be produced?

12 Intermolecular Forces and Liquids

John Kotz

Antarctica Scene—Icebergs, Penguins, Snow, Ice, and Fog

Antarctica is a place of unique wonder and beauty. Aside from the living creatures—many species of penguins, whales, and birds—one mostly thinks of ice and icebergs. Some icebergs are as large as a small state, while others are only the size of a football field. We all know ice floats on water, but what is the reason for this?

Questions:

1. Given the density of seawater ($d = 1.026$ g/mL) and of ice ($d = 0.917$ g/cm^3), is most of the volume of an iceberg above or below the waterline?
2. What volume is above or below the waterline?

Answers to these questions are in Appendix Q.

Chapter Goals

See Chapter Goals Revisited (page 580) for Study Questions keyed to these goals and assignable in OWL.

- Describe intermolecular forces and their effects.
- Understand the importance of hydrogen bonding.
- Understand the properties of liquids.

Chapter Outline

Hundreds of common compounds can exist in the liquid and vapor states at or near room temperature and under ordinary pressures, and the one that comes to mind immediately is water. Water molecules in the atmosphere can interact at lower temperatures and come together to form clouds or fog, and still larger clusters of molecules eventually can fall as rain drops. At only slightly lower temperatures, the molecules assemble into a crystalline lattice, which you see as snowflakes and ice.

The primary objectives of this chapter are to examine the intermolecular forces that allow molecules to interact and then to look at liquids, a result of such interactions. You will find this a useful chapter because it explains, among other things, why your body is cooled when you sweat and why ice can float on liquid water, a property shared by almost no other substance in its liquid and solid states.

Chemistry.Now™

Throughout the text this icon introduces an opportunity for self-study or to explore interactive tutorials by signing in at **www.thomsonedu.com/login**.

John Kotz

Early morning fog over Lake Champlain in upstate New York.

12.1 States of Matter and Intermolecular Forces

The kinetic-molecular theory of gases (◀ Section 11.6) assumes that gas molecules or atoms are widely separated and that these particles can be considered to be independent of one another. Consequently, we can relate the properties of gases under most conditions by a simple mathematical equation, $PV = nRT$, known as the ideal gas law equation (Equation 11.4). In real gases, however, there are forces between molecules—intermolecular forces—and these require a more complex analysis of gas behavior (◀ page 543). If these intermolecular forces become strong enough, the substance can condense to a liquid and eventually to a solid. For liquids in particular, the existence of intermolecular forces makes the picture more complex, and it is not possible to create a simple "ideal liquid equation."

How different are the states of matter at the particulate level? We can get a sense of this by comparing the volumes occupied by equal numbers of molecules of a material in different states. Figure 12.1a shows a flask containing about 300 mL of liquid nitrogen. If all of the liquid were allowed to evaporate, the gaseous nitrogen, at 1 atm and room temperature, would fill a large balloon (more than 200 L). A great amount of space exists between molecules in a gas, whereas in liquids the molecules are close together.

The increase in volume when converting liquids to gases is strikingly large. In contrast, no dramatic change in volume occurs when a solid is converted to a liquid. Figure 12.1b shows the same amount of liquid and solid benzene, C_6H_6, side by side. As you see, they are not appreciably different in volume. This means that the atoms in the liquid are packed together about as tightly as the atoms in the solid phase.

■ **The Lotus Effect** The photograph on the cover of this book illustrates the importance of intermolecular forces. You can read about the lotus effect on the back of the title page and on the back cover of the book.

FIGURE 12.1 Contrasting gases, liquids, and solids. (a) When a 300-mL sample of liquid nitrogen evaporates, it will produce more than 200 L of gas at 25 °C and 1.0 atm. In the liquid phase, the molecules of N_2 are close together; in the gas phase, they are far apart. (b) The same volume of liquid benzene, C_6H_6, is placed in two test tubes, and one tube (right) is cooled, freezing the liquid. The solid and liquid states have almost the same volume, showing that the molecules are packed together almost as tightly in the liquid state as they are in the solid state.

We know that gases can be compressed easily, a process that involves forcing the gas molecules much closer together. In contrast, the molecules, ions, or atoms in liquid or solid phases strongly resist forces that would push them even closer together. Lack of compressibility is a characteristic property of liquids and solids. For example, the volume of liquid water changes only by 0.005% per atmosphere of pressure applied.

Intermolecular forces influence chemistry in many ways:

- They are directly related to properties such as melting point, boiling point, and the energy needed to convert a solid to a liquid or a liquid to a vapor.
- They are important in determining the solubility of gases, liquids, and solids in various solvents.
- They are crucial in determining the structures of biologically important molecules such as DNA and proteins.

Bonding in ionic compounds depends on the electrostatic forces of attraction between oppositely charged ions. Similarly, the intermolecular forces attracting one molecule to another are electrostatic. By comparison, the attractive forces between the ions in ionic compounds are usually in the range of 700 to 1100 kJ/mol, and most covalent bond energies are in the range of 100 to 400 kJ/mol (Table 8.9). As a rough guideline, intermolecular forces are generally less than about 15% of the values of bond energies. Nonetheless, these interactions can have a profound effect on molecular properties and are the subject of this section.

The sections that follow are organized around the polarity of the molecules involved. We shall first describe forces involving polar molecules and then those involving nonpolar molecules. In Chapter 13, we shall describe ionic and metallic solids and the bonding in those substances.

ChemistryNow™

Sign in at **www.thomsonedu.com/login** and go to Chapter 12 Contents to see:

- Screen 12.2 to view an animation of **gases, liquids, and solids at the molecular level**
- Screen 12.3 for an outline of **the important intermolecular forces**

12.2 Intermolecular Forces Involving Polar Molecules

Interactions Between Ions and Molecules with a Permanent Dipole

The distribution of bonding electrons in a molecule often results in a permanent dipole moment (◀ Section 8.7). Because polar molecules have positive and negative ends, if a polar molecule and an ionic compound are mixed, the negative end of the dipole will be attracted to a positive cation (Figure 12.2). Similarly, the positive end of the dipole will be attracted to a negative anion. Forces of attraction between a positive or negative ion and polar molecules—**ion–dipole forces**—are less than those for ion–ion attractions (which can be on the order of 500 kJ/mol), but they are greater than other types of forces between molecules, whether polar or nonpolar.

Ion–dipole attractions can be evaluated based on Coulomb's law (◀ Equation 2.3), which informs us that the force of attraction between two charged objects depends on the product of their charges divided by the square of the distance between them (◀ Section 2.7). Therefore, when a polar molecule encounters an ion, the attractive forces depend on three factors:

- The distance between the ion and the dipole. The closer the ion and dipole, the stronger the attraction.
- The charge on the ion. The higher the ion charge, the stronger the attraction.
- The magnitude of the dipole. The greater the magnitude of the dipole, the stronger the attraction.

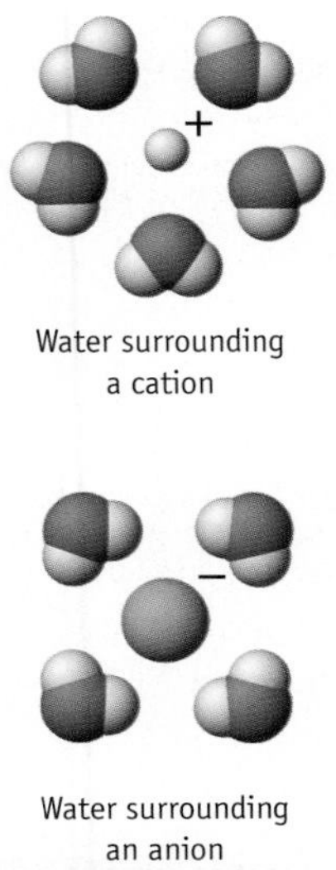

Active Figure 12.2 Ion–dipole interactions. When an ionic compound such as NaCl is placed in water, the polar water molecules surround the cations and anions.

Chemistry Now™ Sign in at www.thomsonedu.com/login and go to the Chapter Contents menu to explore an interactive version of this figure accompanied by an exercise.

The formation of hydrated ions in aqueous solution is one of the most important examples of the interaction between an ion and a polar molecule (Figure 12.3) The enthalpy change associated with the hydration of ions—which is generally called the **enthalpy of solvation** or, for ions in water, the **enthalpy of hydration**—is substantial. The solvation enthalpy for an individual ion cannot be measured directly, but values can be estimated. For example, the hydration of sodium ions is described by the following reaction:

$$Na^+(g) + x\ H_2O(\ell) \rightarrow [Na(H_2O)_x]^+(aq)\ (x \text{ probably} = 6) \qquad \Delta_r H^\circ = -405 \text{ kJ/mol}$$

The enthalpy of hydration depends on $1/d$, where d is the distance between the center of the ion and the oppositely charged "pole" of the dipole.

As the ion radius becomes larger, d increases, and the enthalpy of hydration becomes less exothermic. This trend is illustrated by the enthalpies of hydration of the alkali metal cations (Table 12.1) and by those for Mg^{2+}, Li^+, and K^+ (Figure 12.3). It is interesting to compare these values with the enthalpy of hydration of the

■ **Coulomb's Law** The **force** of attraction between oppositely charged particles depends directly on the product of their charges and inversely on the square of the distance (d) between the ions ($1/d^2$) (Equation 2.3, page 78). The **energy** of the attraction is also proportional to the charge product, but it is inversely proportional to the distance between them ($1/d$).

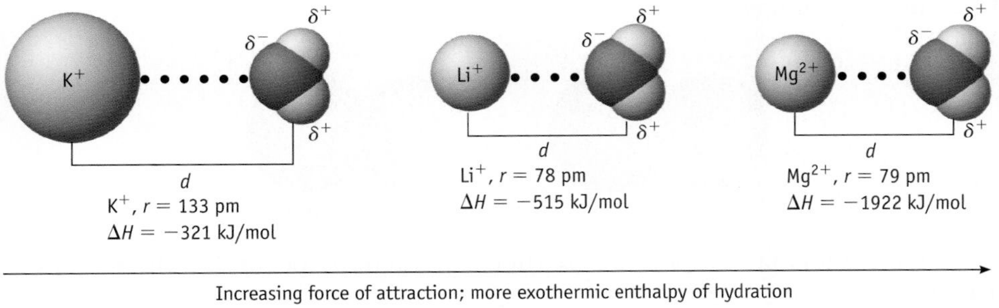

FIGURE 12.3 Enthalpy of hydration. The energy evolved when an ion is hydrated depends on the dipole moment of water, the ion charge, and the distance d between centers of the ion and the polar water molecule. The distance d increases as ion size increases.

TABLE 12.1 Radii and Enthalpies of Hydration of Alkali Metal Ions

Cation	Ion Radius (pm)	Enthalpy of Hydration (kJ/mol)
Li^+	78	−515
Na^+	98	−405
K^+	133	−321
Rb^+	149	−296
Cs^+	165	−263

H^+ ion, estimated to be −1090 kJ/mol. This extraordinarily large value is due to the tiny size of the H^+ ion.

Chemistry Now™

Sign in at **www.thomsonedu.com/login** and go to Chapter 12 Contents to see Screen 12.4 to view an animation of **ion–dipole forces.**

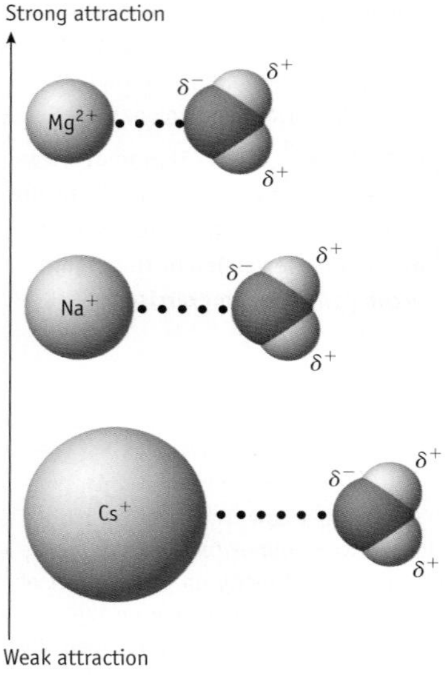

EXAMPLE 12.1 Hydration Energy

Problem Explain why the enthalpy of hydration of Na^+ (−405 kJ/mol) is somewhat more exothermic than that of Cs^+ (−263 kJ/mol), whereas that of Mg^{2+} is much more exothermic (−1922 kJ/mol) than that of either Na^+ or Cs^+.

Strategy The strength of ion–dipole attractions depends directly on the size of the ion charge and the magnitude of the dipole, and inversely on the distance between them. To judge the ion–dipole distance, we need ion sizes from Figure 7.12.

Solution The relevant ion sizes are Na^+ = 98 pm, Cs^+ = 165 pm, and Mg^{2+} = 79 pm. From these values, we can predict that the distances between the center of the positive charge on the metal ion and the negative side of the water dipole will vary in this order: $Mg^{2+} < Na^+ < Cs^+$. The hydration energy varies in the reverse order (with the hydration energy of Mg^{2+} being the most negative value). Notice also that Mg^{2+} has a 2+ charge, whereas the other ions are 1+. The greater charge on Mg^{2+} leads to a greater force of ion–dipole attraction than for the other two ions, which have only a 1+ charge. As a result, the hydration energy for Mg^{2+} is much more negative than for the other two ions.

EXERCISE 12.1 Hydration Energy

Which should have the more negative hydration energy, F^- or Cl^-? Explain briefly.

Interactions Between Molecules with Permanent Dipoles

When a polar molecule encounters another polar molecule, of the same or a different kind, the positive end of one molecule is attracted to the negative end of the other polar molecule.

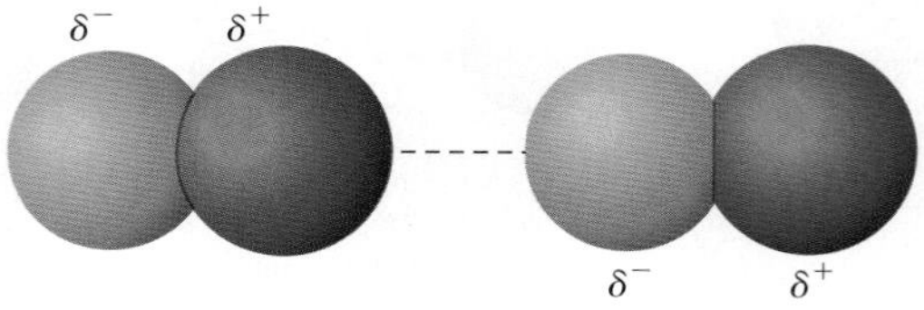

Many molecules have dipoles, and their interactions occur by **dipole–dipole attraction.**

A Closer Look — Hydrated Salts

Solid salts with waters of hydration are common. The formulas of these compounds are given by appending a specific number of water molecules to the end of the formula, as in $BaCl_2 \cdot 2\ H_2O$. Sometimes, the water molecules simply fill in empty spaces in a crystalline lattice, but often the cation in these salts is directly associated with water molecules. For example, the compound $CrCl_3 \cdot 6\ H_2O$ is better written as $[Cr(H_2O)_4Cl_2]Cl \cdot 2\ H_2O$. Four of the six water molecules are associated with the Cr^{3+} ion by ion–dipole attractive forces; the remaining two water molecules are in the lattice. Common examples of hydrated salts are listed in the table.

Compound	Common Name	Uses
$Na_2CO_3 \cdot 10\ H_2O$	Washing soda	Water softener
$Na_2S_2O_3 \cdot 5\ H_2O$	Hypo	Photography
$MgSO_4 \cdot 7\ H_2O$	Epsom salt	Cathartic, dyeing and tanning
$CaSO_4 \cdot 2\ H_2O$	Gypsum	Wallboard
$CuSO_4 \cdot 5\ H_2O$	Blue vitriol	Biocide

Photos: Charles D. Winters

Hydrated cobalt(II) chloride, $CoCl_2 \cdot 6\ H_2O$. In the solid state, the compound is best described by the formula $[Co(H_2O)_4Cl_2] \cdot 2\ H_2O$. The cobalt(II) ion is surrounded by four water molecules and two chloride ions in an octahedral arrangement. In water, the ion is completely hydrated, now being surrounded by six water molecules. Cobalt(II) ions and water molecules interact by ion–dipole forces. This is an example of a coordination compound, a class of compounds discussed in detail in Chapter 22.

For polar molecules, dipole–dipole attractions influence, among other things, the evaporation of a liquid and the condensation of a gas (Figure 12.4). An energy change occurs in both processes. Evaporation requires the input of energy, specifically the enthalpy of vaporization ($\Delta_{vap}H°$) [see Section 5.3 and Section 12.4]. The value for the enthalpy of vaporization has a positive sign, indicating that evaporation is an endothermic process. The enthalpy change for the condensation process—the reverse of evaporation—has a negative value.

The greater the forces of attraction between molecules in a liquid, the greater the energy that must be supplied to separate them. Thus, we expect polar compounds to have a higher value for their enthalpy of vaporization than nonpolar compounds with similar molar masses. For example, notice that $\Delta_{vap}H°$ for polar

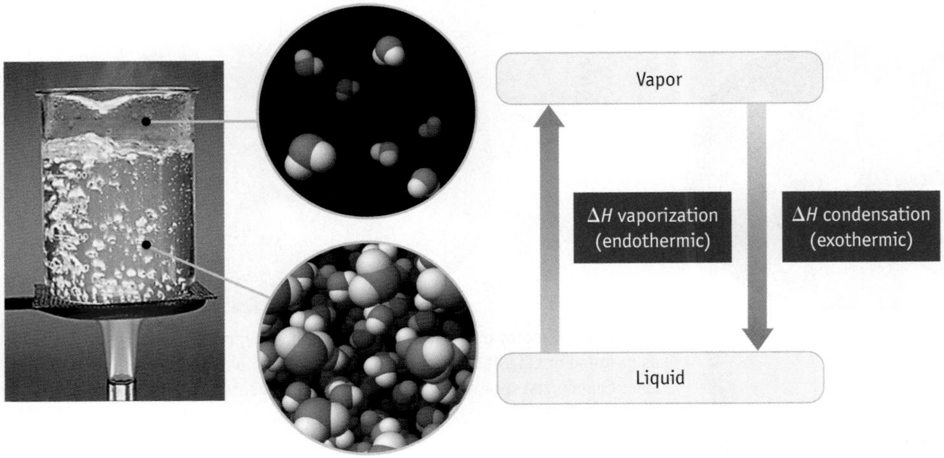

Photos: Charles D. Winters

FIGURE 12.4 Evaporation at the molecular level. Energy must be supplied to separate molecules in the liquid state against intermolecular forces of attraction.

TABLE 12.2 Molar Masses, Boiling Points, and $\Delta_{vap}H°$ of Nonpolar and Polar Substances

Nonpolar				Polar			
	M (g/mol)	BP (°C)	$\Delta_{vap}H°$ (kJ/mol)		M (g/mol)	BP (°C)	$\Delta_{vap}H°$ (kJ/mol)
N_2	28	−196	5.57	CO	28	−192	6.04
SiH_4	32	−112	12.10	PH_3	34	−88	14.06
GeH_4	77	−90	14.06	AsH_3	78	−62	16.69
Br_2	160	59	29.96	ICl	162	97	—

molecules is greater than for nonpolar molecules of approximately the same size and mass (Table 12.2).

The boiling point of a liquid also depends on intermolecular forces of attraction. As the temperature of a substance is raised, its molecules gain kinetic energy. Eventually, when the boiling point is reached, the molecules have sufficient kinetic energy to escape the forces of attraction of their neighbors. For molecules of similar molar mass, the greater the polarity, the higher the temperature required for the liquid to boil. In Table 12.2, you see that the boiling point for polar ICl is greater than that for nonpolar Br_2, for example.

Intermolecular forces also influence solubility. A qualitative observation on solubility is that "like dissolves like." In other words, polar molecules are likely to dissolve in a polar solvent, and nonpolar molecules are likely to dissolve in a nonpolar solvent (Figure 12.5) (◀ Chapter 8). The converse is also true; that is, it is unlikely that polar molecules will dissolve in nonpolar solvents or that nonpolar molecules will dissolve in polar solvents.

For example, water and ethanol (C_2H_5OH) can be mixed in any ratio to give a homogeneous mixture. In contrast, water does not dissolve in gasoline to an ap-

Photos: Charles D. Winters

(a) Ethylene glycol ($HOCH_2CH_2OH$), a polar compound used as antifreeze in automobiles, dissolves in water.

(b) Nonpolar motor oil (a hydrocarbon) dissolves in nonpolar solvents such as gasoline or CCl_4. It will not dissolve in a polar solvent such as water, however. Commercial spot removers use nonpolar solvents to dissolve oil and grease from fabrics.

FIGURE 12.5 "Like dissolves like."

preciable extent. The difference in these two situations is that ethanol and water are polar molecules, whereas the hydrocarbon molecules in gasoline (e.g., octane, C_8H_{18}) are nonpolar. The water–ethanol interactions are strong enough that the energy expended in pushing water molecules apart to make room for ethanol molecules is compensated for by the energy of attraction between the two kinds of polar molecules. In contrast, water–hydrocarbon attractions are weak. The hydrocarbon molecules cannot disrupt the stronger water–water attractions.

Chemistry Now™

Sign in at **www.thomsonedu.com/login** and go to Chapter 12 Contents to see Screen 12.4 to view an animation of **dipole–dipole forces.**

Hydrogen Bonding

Hydrogen fluoride, water, ammonia and many other compounds with O—H and N—H bonds have exceptional properties. Consider, for example, the boiling points for hydrogen compounds of elements in Groups 4A through 7A (Figure 12.6). Generally, the boiling points of related compounds increase with molar mass. This trend is seen in the boiling points of the hydrogen compounds of Group 4A elements, for example ($CH_4 < SiH_4 < GeH_4 < SnH_4$). The same effect is also operating for the heavier molecules of the hydrogen compounds of elements of Groups 5A, 6A, and 7A. The boiling points of NH_3, H_2O, and HF, however, deviate significantly from what might be expected based on molar mass alone. If we extrapolate the curve for the boiling points of H_2Te, H_2Se, and H_2S, the boiling point of water is predicted to be around −90 °C. The boiling point of water is almost 200 °C higher than this value! Similarly, the boiling points of NH_3 and HF are much higher than would be expected based on molar mass. Because the temperature at which a substance boils depends on the attractive forces between molecules, the extraordinarily high boiling points of H_2O, HF, and NH_3 indicate strong intermolecular attractions.

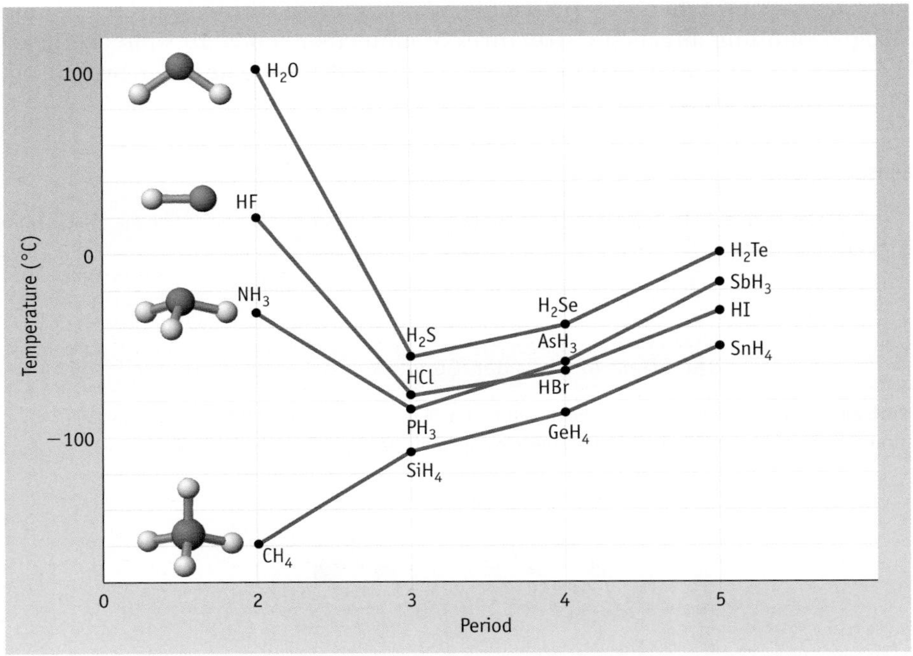

Active Figure 12.6 The boiling points of some simple hydrogen compounds. The effect of hydrogen bonding is apparent in the unusually high boiling points of H_2O, HF, and NH_3. (Also, notice that the boiling point of HCl is somewhat higher than expected based on the data for HBr and HI. It is apparent that some degree of hydrogen bonding also occurs in liquid HCl.)

Chemistry Now™ Sign in at www.thomsonedu.com/login and go to the Chapter Contents menu to explore an interactive version of this figure accompanied by an exercise.

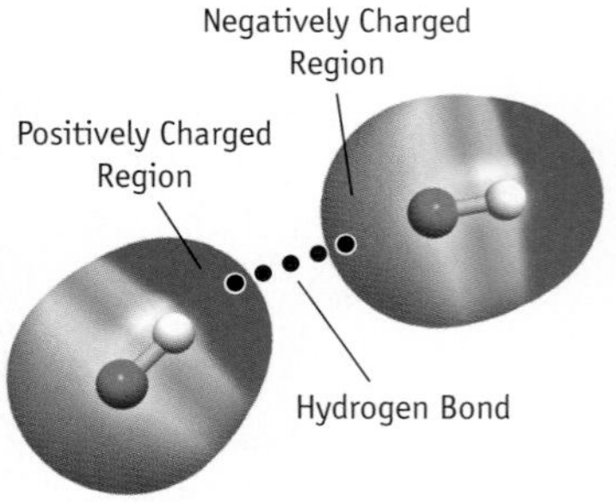

Hydrogen bonding between HF molecules. The partially negative F atom of one HF molecule interacts through hydrogen bonding with a neighboring HF molecule. (Red regions of the molecule are negatively charged, whereas blue regions are positively charged. For more on electrostatic potential surfaces, see page 382.)

The electronegativities of N (3.0), O (3.5), and F (4.0) are among the highest of all the elements, whereas the electronegativity of hydrogen is much lower (2.2). This large difference in electronegativity means that N—H, O—H, and F—H bonds are very polar. In bonds between H and N, O, or F, the more electronegative element takes on a significant negative charge (see Figure 8.11), and the hydrogen atom acquires a significant positive charge.

There is an unusually strong attraction between an electronegative atom with a lone pair of electrons (most often, an N, O, or F atom in another molecule or even in the same molecule) and the hydrogen atom of the N—H, O—H, or F—H bond. This type of interaction is known as a **hydrogen bond.** Hydrogen bonds are an extreme form of dipole–dipole interaction where one atom involved is always H and the other atom is highly electronegative, most often O, N, or F. A hydrogen bond can be represented as

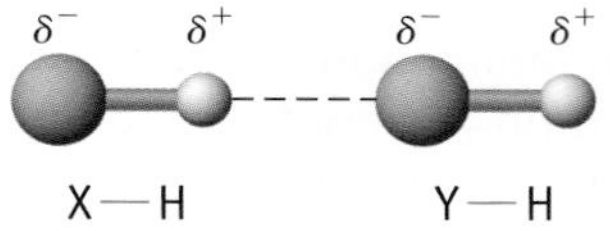

The hydrogen atom becomes a bridge between the two electronegative atoms X and Y, and the dashed line represents the hydrogen bond. The most pronounced effects of hydrogen bonding occur where both X and Y are N, O, or F. Energies associated with most hydrogen bonds involving these elements are in the range of 5 to 30 kJ/mol.

Types of Hydrogen Bonds [X—H - - - :Y]

N—H - - - :N—	O—H - - - :N—	F—H - - - :N—
N—H - - - :O—	O—H - - - :O—	F—H - - - :O—
N—H - - - :F—	O—H - - - :F—	F—H - - - :F—

Hydrogen bonding has important implications for any property of a compound that is influenced by intermolecular forces of attraction. For example, hydrogen bonding affects the structures of molecular solids. In solid acetic acid, CH_3CO_2H, for example, two molecules are joined to one another by hydrogen bonding (Figure 12.7).

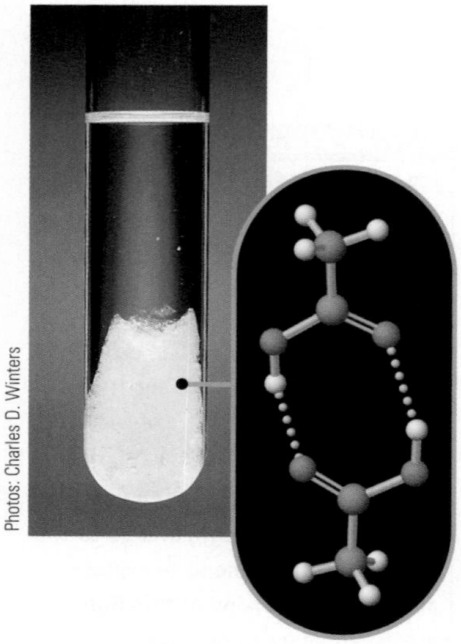

FIGURE 12.7 Hydrogen bonding. Two acetic acid molecules can interact through hydrogen bonds. This photo shows partly solid glacial acetic acid. Notice that the solid is denser than the liquid, a property shared by virtually all substances, the notable exception being water.

Chemistry•Now™

Sign in at **www.thomsonedu.com/login** and go to Chapter 12 Contents to see Screen 12.6 for a **description of hydrogen bonding.**

EXAMPLE 12.2 The Effect of Hydrogen Bonding

Problem Ethanol, CH_3CH_2OH, and dimethyl ether, CH_3OCH_3, have the same formula but a different arrangement of atoms. Predict which of these compounds has the higher boiling point.

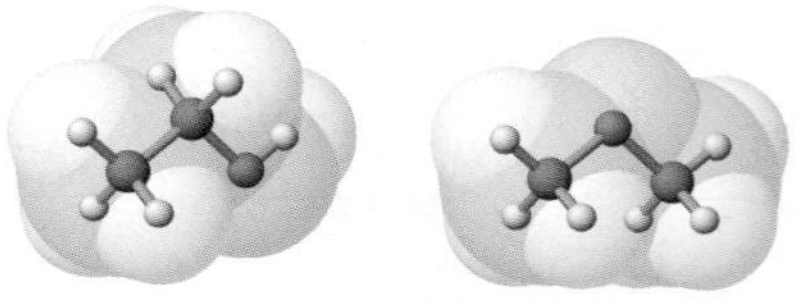

Ethanol, CH_3CH_2OH Dimethyl ether, CH_3OCH_3

Strategy Inspect the structure of each molecule to decide whether each is polar and, if polar, whether hydrogen bonding is possible.

Solution Although these two compounds have identical masses, they have different structures. Ethanol possesses an O—H group, and an electrostatic potential surface in the margin shows it to be polar with a partially negative O atom and a partially negative H atom. The result is that hydrogen bonding between ethanol molecules is possible and makes an important contribution to its intermolecular forces.

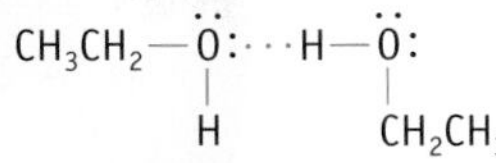

hydrogen bonding in ethanol, CH_3CH_2OH

In contrast, dimethyl ether, although a polar molecule, presents no opportunity for hydrogen bonding because there is no O—H bond. The H atoms are attached to much less electronegative C atoms. We can predict, therefore, that intermolecular forces will be larger in ethanol than in dimethyl ether and that ethanol will have the higher boiling point. Indeed, ethanol boils at 78.3 °C, whereas dimethyl ether has a boiling point of −24.8 °C, more than 100 °C lower. Under standard conditions, dimethyl ether is a gas, whereas ethanol is a liquid.

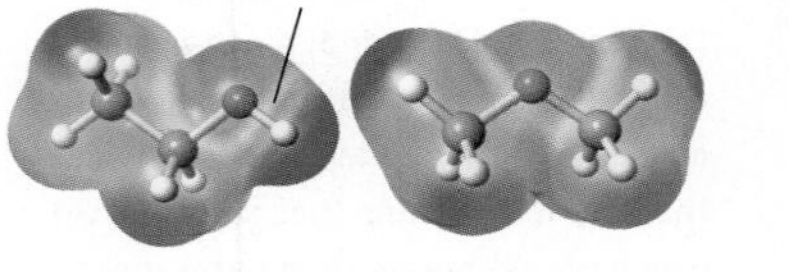

Electrostatic potential surfaces for ethanol (left) and dimethyl ether (right). The surface for ethanol clearly shows the polar O—H bond. The O atom in dimethyl ether has a partial negative charge, but there is no H atom attached. [Color coding: Red indicates a region of largest negative charge. Colors from yellow to green to turquoise indicate increasing positive charge (or decreasing negative charge). Blue indicates a region of partial positive charge.]

EXERCISE 12.2 Hydrogen Bonding

Using structural formulas, describe the hydrogen bonding between methanol (CH_3OH) molecules. What physical properties of methanol are likely to be affected by hydrogen bonding?

Hydrogen Bonding and the Unusual Properties of Water

One of the most striking differences between our planet and others in our solar system is the presence of large amounts of water on Earth. Three fourths of the planet is covered by oceans; the polar regions are vast ice fields; and even soil and rocks hold large amounts of water. Although we tend to take water for granted, almost no other substance behaves in a similar manner. Water's unique features reflect the ability of H_2O molecules to cling tenaciously to one another by hydrogen bonding.

One reason for ice's unusual structure and water's unusual properties is that each hydrogen atom of a water molecule can form a hydrogen bond to a lone pair of electrons on the oxygen atom of an adjacent water molecule. In addition, because the oxygen atom in water has two lone pairs of electrons, it can form two more hydrogen bonds with hydrogen atoms from adjacent molecules (Figure 12.8a). The result, seen particularly in ice, is a tetrahedral arrangement for the hydrogen atoms around each oxygen, involving two covalently bonded hydrogen atoms and two hydrogen-bonded hydrogen atoms.

As a consequence of the regular arrangement of water molecules linked by hydrogen bonding, ice has an open-cage structure with lots of empty space (Figure 12.8b). The result is that ice has a density about 10% less than that of liquid water, which explains why ice floats. (In contrast, virtually all other solids sink in their liquid phase.) We can also see in this structure that the oxygen atoms are arranged at the corners of puckered, hexagonal rings. Snowflakes are always based on six-sided figures (◀ page 69), a reflection of this internal molecular structure of ice.

When ice melts at 0 °C, the regular structure imposed on the solid state by hydrogen bonding breaks down, and a relatively large increase in density occurs (Figure 12.9). Another surprising thing occurs when the temperature of liquid water is raised from 0 °C to 4 °C: The density of water increases. For almost every other substance known, density decreases as the temperature is raised. Once again, hydrogen bonding is the reason for water's seemingly odd behavior. At a tempera-

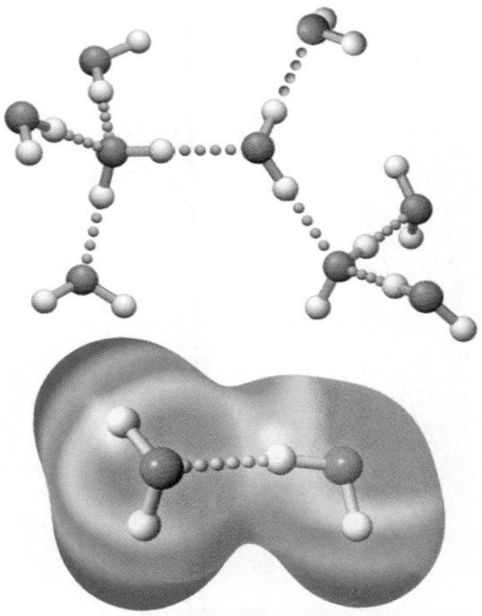

Hydrogen bonding in water. (above) Water readily forms hydrogen bonds. (below) An electrostatic potential surface for two water molecules shows the hydrogen bond involving the negatively charged O atom of one molecule and the positively charged H atom of a neighboring molecule.

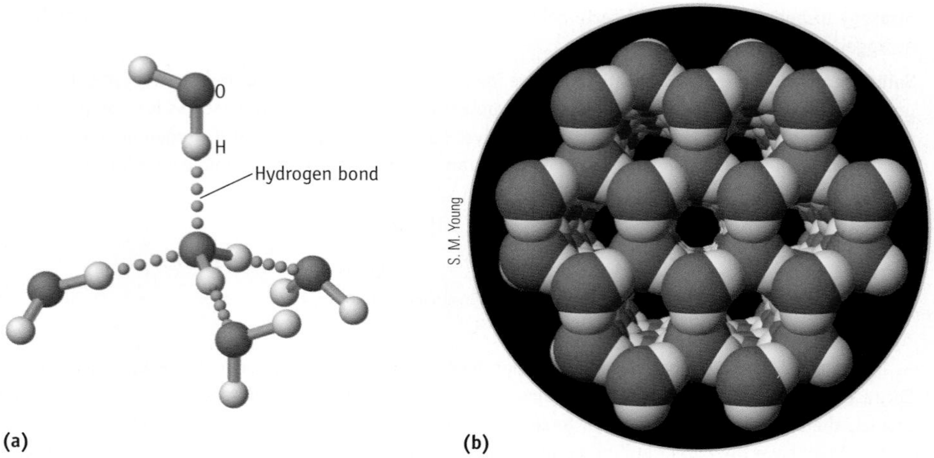

FIGURE 12.8 The structure of ice. (a) The oxygen atom of a water molecule attaches itself to two other water molecules by hydrogen bonds. Notice that the four groups that surround an oxygen atom are arranged as a distorted tetrahedron. Each oxygen atom is covalently bonded to two hydrogen atoms and hydrogen bonded to hydrogen atoms from two other molecules. The hydrogen bonds are longer than the covalent bonds. (b) In ice, the structural unit shown in part (a) is repeated in the crystalline lattice. This computer-generated structure shows a small portion of the extensive lattice. Notice the six-member, hexagonal rings. The corners of each hexagon are O atoms, and each side is composed of a normal O—H bond and a slightly longer hydrogen bond.

ture just above the melting point, some of the water molecules continue to cluster in ice-like arrangements, which require extra space. As the temperature is raised from 0 °C to 4 °C, the final vestiges of the ice structure disappear, and the volume contracts further, giving rise to the increase in density. Water's density reaches a maximum at about 4 °C. From this point, the density declines with increasing temperature in the normal fashion.

Because of the way that water's density changes as the temperature approaches the freezing point, lakes do not freeze solidly from the bottom up in the winter. When lake water cools with the approach of winter, its density increases, the cooler water sinks, and the warmer water rises. This "turn over" process continues until all the water reaches 4 °C, the maximum density. (This is the way oxygen-rich water moves to the lake bottom to restore the oxygen used during the summer and nutrients are brought to the top layers of the lake.) As the temperature decreases further, the colder water stays on the top of the lake, because water cooler than 4 °C is less dense than water at 4 °C. With further heat loss, ice can then begin to form on the surface, floating there and protecting the underlying water and aquatic life from further heat loss.

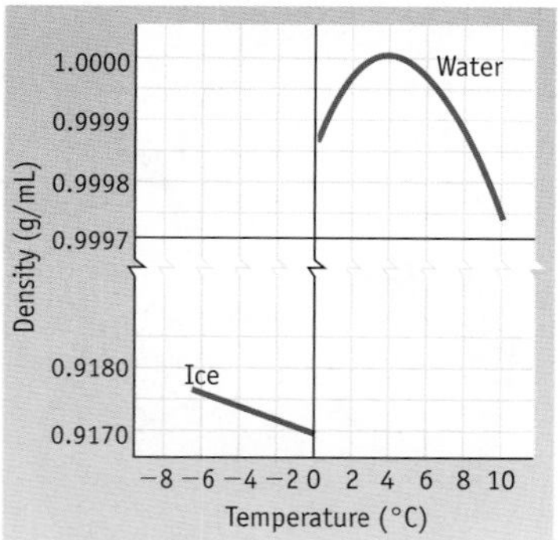

Active Figure 12.9 **The temperature dependence of the densities of ice and water.**

Chemistry Now™ Sign in at www.thomsonedu.com/login and go to the Chapter Contents menu to explore an interactive version of this figure accompanied by an exercise.

Extensive hydrogen bonding is also the origin of the extraordinarily high heat capacity of water. Although liquid water does not have the regular structure of ice, hydrogen bonding still occurs. With a rise in temperature, the extent of hydrogen bonding diminishes. Disrupting hydrogen bonds requires energy. The high heat capacity of water is, in large part, why oceans and lakes have such an enormous effect on weather. In autumn, when the temperature of the air is lower than the temperature of the ocean or lake, the water transfers energy as heat to the atmosphere, moderating the drop in air temperature. Furthermore, so much energy is available to be transferred for each degree drop in temperature that the decline in water temperature is gradual. For this reason, the temperature of the ocean or of a large lake is generally higher than the average air temperature until late in the autumn.

Chemistry Now™

Sign in at **www.thomsonedu.com/login** and go to Chapter 12 Contents to see Screen 12.7 to view an animation of **the transformation of ice to water** and for a table listing all of the **unusual properties of water.**

A Closer Look   Hydrogen Bonding in Biochemistry

It is arguable that our world is what it is because of hydrogen bonding in water and in biochemical systems. Perhaps the most important occurrence is in DNA and RNA where the organic bases adenine, cytosine, guanine, and thymine (in DNA) or uracil (in RNA) are attached to sugar-phosphate chains (**Figure A**). The chains in DNA are joined by the pairing of bases, adenine with thymine and guanine with cytosine.

Figure B illustrates the hydrogen bonding between adenine and thymine. These models show that the molecules naturally fit together to form a six-sided ring, where two of the six sides involve hydrogen bonds. One side consists of a N ··· H—N grouping, and the other side is N—H ··· O. Here, the electrostatic potential surfaces show that the N atoms of adenine and the O atoms of thymine bear partial negative charges, and the H atoms of the N—H groups bear a positive charge. These charges and the geometry of the bases lead to these very specific interactions.

The fact that base pairing through hydrogen bonding leads to the joining of the sugar-phosphate chains of DNA, and to the double helical form of DNA, was first recognized by James Watson and Francis Crick on the basis of experimental work by Rosalind Franklin and Maurice Wilkins in the 1950s. It was this development that was so important in the molecular biology revolution in the last part of the 20th century. See page 392 for more on these scientists.

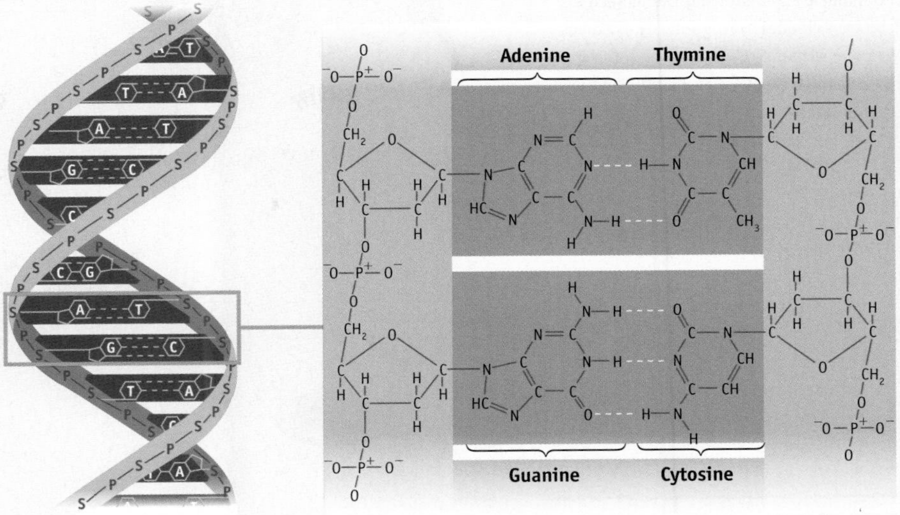

FIGURE A Hydrogen bonding in DNA. With the four bases in DNA, the usual pairings are adenine with thymine and guanine with cytosine. This pairing is promoted by hydrogen bonding.

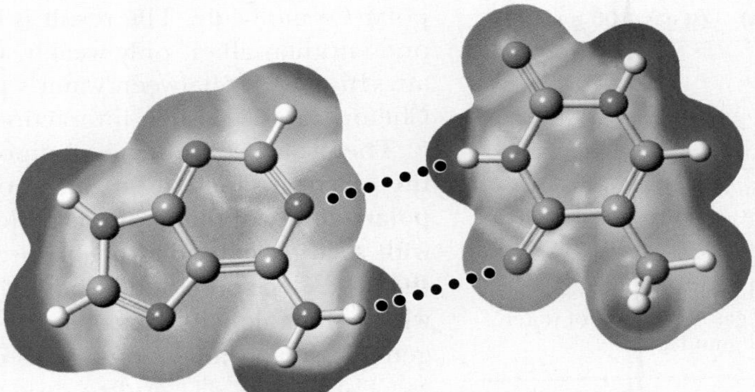

FIGURE B Hydrogen bonding between adenine and thymine. Electrostatic potential surfaces show that the polar N—H bond on one molecule can hydrogen bond to an electronegative N atom in a neighboring molecule.

12.3 Intermolecular Forces Involving Nonpolar Molecules

Module 17

Many important molecules such as O_2, N_2, and the halogens are not polar. Why, then, does O_2 dissolve in polar water? Why can the N_2 of the atmosphere be liquefied (see Figure 12.1)? Some intermolecular forces must be acting between O_2 and water and between N_2 molecules, but what is their nature?

■ **Van der Waals Forces** The name "van der Waals forces" is a general term applied to attractive intermolecular interactions. (P. W. Atkins: *Quanta: A Handbook of Concepts,* 2nd ed., p. 187, Oxford, Oxford University Press, 2000.)

Dipole/Induced Dipole Forces

Polar molecules such as water can induce, or create, a dipole in molecules that do not have a permanent dipole. To see how this situation can occur, picture a polar water molecule approaching a nonpolar molecule such as O_2 (Figure 12.10). The

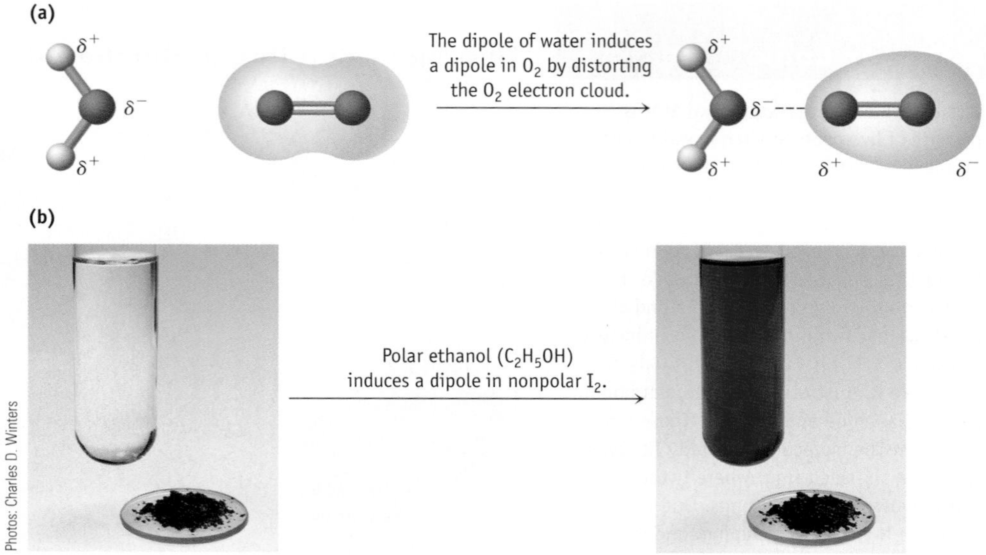

FIGURE 12.10 Dipole/induced dipole interaction. (a) A polar molecule such as water can induce a dipole in nonpolar O_2 by distorting the molecule's electron cloud. (b) Nonpolar I_2 dissolves in polar ethanol (C_2H_5OH). The intermolecular force involved is a dipole/induced dipole force.

TABLE 12.3 The Solubility of Some Gases in Water*

	Molar Mass (g/mol)	Solubility at 20 °C (g gas/100 g water)†
H_2	2.01	0.000160
N_2	28.0	0.00190
O_2	32.0	0.00434

* Data taken from J. Dean: *Lange's Handbook of Chemistry.* 14th Ed., pp. 5.3–5.8, New York, McGraw-Hill, 1992.
† Measured under conditions where pressure of gas + pressure of water vapor = 760 mm Hg.

TABLE 12.4 Enthalpies of Vaporization and Boiling Points of Some Nonpolar Substances

	$\Delta_{vap}H°$ (kJ/mol)	Element/ Compound BP (°C)
N_2	5.57	−196
O_2	6.82	−183
CH_4 (methane)	8.2	−161.5
Br_2	29.96	+58.8
C_6H_6 (benzene)	30.7	+80.1
I_2	41.95	+185

electron cloud of an isolated (gaseous) O_2 molecule is symmetrically distributed between the two oxygen atoms. As the negative end of the polar H_2O molecule approaches, however, the O_2 electron cloud becomes distorted. In this process, the O_2 molecule itself becomes polar; that is, a dipole is *induced* in the otherwise nonpolar O_2 molecule. The result is that H_2O and O_2 molecules are now attracted to one another, albeit only weakly. Oxygen can dissolve in water because a force of attraction exists between water's permanent dipole and the induced dipole in O_2. Chemists refer to such interactions as **dipole/induced dipole interactions.**

The process of inducing a dipole is called **polarization,** and the degree to which the electron cloud of an atom or a molecule can be distorted depends on the **polarizability** of that atom or molecule. The electron cloud of an atom or molecule with a large, extended electron cloud, such as I_2, can be polarized more readily than the electron cloud in a much smaller atom or molecule, such as He or H_2, in which the valence electrons are close to the nucleus and more tightly held. In general, for an analogous series of substances, say the halogens or alkanes (such as CH_4, C_2H_6, C_3H_8, and so on), *the higher the molar mass, the greater the polarizability of the molecule.*

The solubilities of common gases in water illustrate the effect of interactions between a dipole and an induced dipole. In Table 12.3, you see a trend to higher solubility with increasing mass of the nonpolar gas. As the molar mass of the gas increases, the polarizability of the electron cloud increases and the strength of the dipole/induced dipole interaction increases.

London Dispersion Forces: Induced Dipole/Induced Dipole Forces

Iodine, I_2, is a solid and not a gas around room temperature and pressure, illustrating that nonpolar molecules must also experience intermolecular forces. An estimate of these forces is provided by the enthalpy of vaporization of the substance at its boiling point. The data in Table 12.4 suggest that these forces can range from very weak (N_2, O_2, and CH_4 with low enthalpies of vaporization and very low boiling points) to more substantial (I_2 and benzene).

To understand how two nonpolar molecules can attract each other, recall that the electrons in atoms or molecules are in a state of constant motion. When two

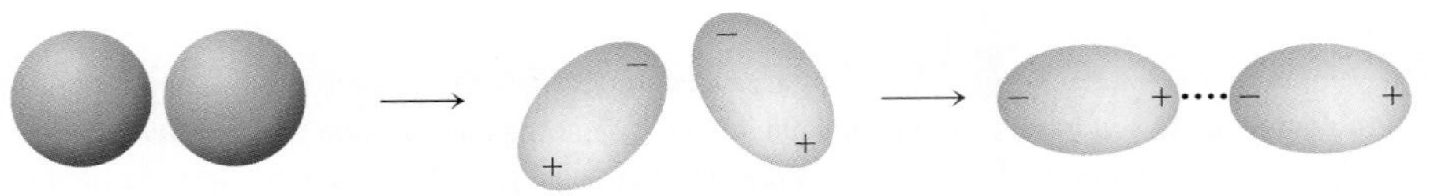

Two nonpolar atoms or molecules (depicted as having an electron cloud that has a time-averaged spherical shape).

Momentary attractions and repulsions between nuclei and electrons in neighboring molecules lead to induced dipoles.

Correlation of the electron motions between the two atoms or molecules (which are now dipolar) leads to a lower energy and stabilizes the system.

FIGURE 12.11 Induced dipole interactions. Momentary attractions and repulsions between nuclei and electrons create induced dipoles and lead to a net stabilization due to attractive forces.

atoms or nonpolar molecules approach each other, attractions or repulsions between their electrons and nuclei can lead to distortions in their electron clouds (Figure 12.11). That is, dipoles can be induced momentarily in neighboring atoms or molecules, and these induced dipoles lead to intermolecular attractions. The intermolecular force of attraction in liquids and solids composed of nonpolar molecules is an **induced dipole/induced dipole force**. Chemists often call them **London dispersion forces**. London dispersion forces actually arise between all molecules, both nonpolar and polar, but *London dispersion forces are the only intermolecular forces that allow nonpolar molecules to interact.*

A Closer Look — Methane Hydrates: An Answer to World Fuel Supplies?

Hydrogen bonds involving water are also responsible for the structure and properties of one of the strangest substances on earth (Figure). When methane (CH_4) is mixed with water at high pressures and low temperatures, solid methane hydrate forms. Although the substance has been known for years, vast deposits of methane hydrate were only recently discovered deep within sediments on the floor of Earth's oceans. How these were formed is a mystery, but what is important is their size. It is estimated that global methane hydrate deposits contain approximately 10^{13} tons of carbon, or about twice the combined amount in all known reserves of coal, oil, and natural gas. Methane hydrate is also an efficient energy storehouse; a liter of methane hydrate releases about 160 liters of methane gas.

But of course there are problems to be solved. One significant problem is how to bring commercially useful quantities to the surface from deep in the ocean. Yet another is the possibility of a large, uncontrolled release of methane. Methane is a very effective greenhouse gas, so the release of a significant quantity into the atmosphere could damage the earth's climate.

Among the many sources of information is: E. Suess, G. Bohrmann, J. Greinert, and E. Lausch, *Scientific American*, November 1999, pp. 76–83.

John Pinkston and Laura Stern/U.S. Geological Survey/*Science News*, 11-9-96

(a) Methane hydrate burns as methane gas escapes from the solid hydrate.

(b) Methane hydrate consists of a lattice of water molecules with methane molecules trapped in the cavity.

Methane hydrate. (a) When a sample is brought to the surface from the depths of the ocean, the methane oozes out of the solid, and the gas readily burns. (b) The structure of the solid methane hydrate consists of methane molecules trapped in a lattice of water molecules. Each point of the lattice shown here is an O atom of an H_2O molecule. The edges are O—H—O hydrogen bonds. Such structures are often called "clathrates." (For more on methane hydrates, see pages 259–260.)

Induced dipole/induced dipole forces. Br_2 (left) and I_2 (right) both consist of nonpolar molecules. They are a liquid and a solid, respectively, implying that there are forces between the molecules sufficient to cause them to be in a condensed phase. These forces between nonpolar substances are known as London dispersion forces or induced dipole/induced dipole forces.

EXAMPLE 12.3 Intermolecular Forces

Problem Suppose you have a mixture of solid iodine, I_2, and the liquids water and carbon tetrachloride (CCl_4). What intermolecular forces exist between each possible pair of compounds? Describe what you might see when these compounds are mixed.

Strategy First, decide whether each substance is polar or nonpolar. Second, determine the types of intermolecular forces that could exist between the different pairs. Finally, use the "like dissolves like" guideline to decide whether iodine will dissolve in water or CCl_4 and whether CCl_4 will dissolve in water.

Solution Iodine, I_2, is nonpolar. As a molecule composed of large iodine atoms, it has an extensive electron cloud. Thus, the molecule is easily polarized, and iodine could interact with water, a polar molecule, by dipole/induced dipole forces.

Carbon tetrachloride, a tetrahedral molecule, is not polar (see Figure 8.15). As a consequence, it can interact with iodine only by dispersion forces. Water and CCl_4 could interact by dipole/induced dipole forces, but the interaction is expected to be weak.

The photo here shows the result of mixing these three compounds. Iodine does dissolve to a small extent in water to give a brown solution. When this brown solution is added to a test tube containing CCl_4, the liquid layers do not mix. (Polar water does not dissolve in nonpolar CCl_4.) (Notice the more dense CCl_4 layer [$d = 1.58$ g/mL] is underneath the less dense water layer.) When the test tube is shaken, however, nonpolar I_2 dissolves preferentially in nonpolar CCl_4, as evidenced by the disappearance of the color of I_2 in the water layer (top) and the appearance of the purple I_2 color in the CCl_4 layer (bottom).

Geckos use intermolecular forces! A little gecko can climb vertically 1 m up a polished glass surface in 1 s. Geckos have millions of tiny hairs or setae on their feet, and each setae ends in 1000 or more even tinier hairs at the tip. Recent research has found that geckos are unique in that they adhere to a surface through van der Waals forces of attraction between the hairs and the surface. (K. Autumn, "How gecko toes stick." *American Scientist* Vol. 94, pages 124–132, 2006.)

EXERCISE 12.3 Intermolecular Forces

You mix water, CCl_4, and hexane ($CH_3CH_2CH_2CH_2CH_2CH_3$). What type of intermolecular forces can exist between each pair of these compounds?

Summary of Intermolecular Forces

Intermolecular forces involve molecules that are polar or those in which polarity can be induced (Table 12.5). Furthermore, several types of intermolecular forces can be at work in a single type of molecule (Figure 12.12). Also note in Figure 12.12 that while each individual induced dipole/induced dipole force is usually quite small, the sum of these forces over the entire structure of a molecule can actually be quite great, even in polar molecules.

Chemistry Now™

Sign in at **www.thomsonedu.com/login** and go to Chapter 12 Contents to see Screen 12.5 to view an animation of **induced dipole forces** and for an exercise and tutorial on **intermolecular forces.**

TABLE 12.5 Summary of Intermolecular Forces

Type of Interaction	Factors Responsible for Interaction	Approximate Energy (kJ/mol)	Example
Ion-dipole	Ion change, magnitude of dipole	40–600	$Na^+ \ldots H_2O$
Dipole-dipole	Dipole moment (depends on atom electronegativities and molecular structure)	20–30	$H_2O \ldots CH_3OH$
Hydrogen bonding, X—H . . . :Y	Very polar X—H bond (where X = F, N, O) and atom Y with lone pair of electrons	5–30	$H_2O \ldots H_2O$
Dipole/induced dipole	Dipole moment of polar molecule and polarizability of nonpolar molecule	2–10	$H_2O \ldots I_2$
Induced dipole/induced dipole (London dispersion forces)	Polarizability	0.05–40	$I_2 \ldots I_2$

EXAMPLE 12.4 Intermolecular Forces

Problem Decide which are the most important intermolecular forces involved in each of the following, and place them in order of increasing strength of interaction: (a) liquid methane, CH_4; (b) a mixture of water and methanol (CH_3OH); and (c) a solution of bromine in water.

Strategy For each molecule, we consider its structure and decide whether it is polar. If polar, we consider the possibility of hydrogen bonding.

Solution (a) Methane is a covalently bonded molecule. Based on the Lewis structure, we can conclude that it must be a tetrahedral molecule and that it cannot be polar. The only way methane molecules can interact with one another is through induced dipole/induced dipole forces.

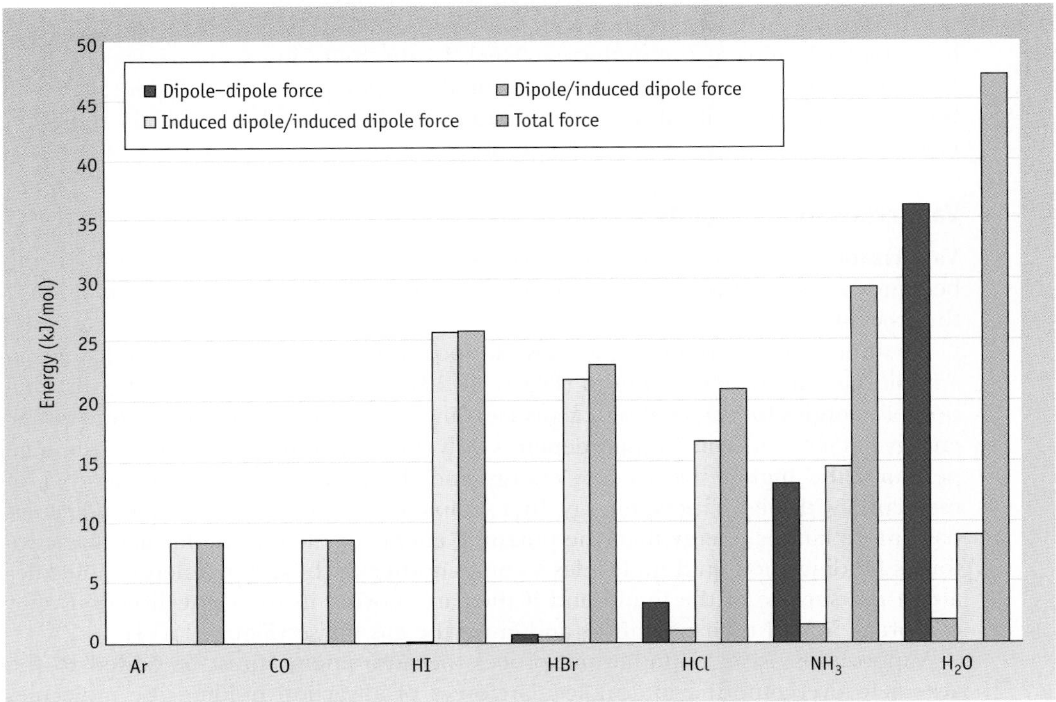

FIGURE 12.12 Energies associated with intermolecular forces.

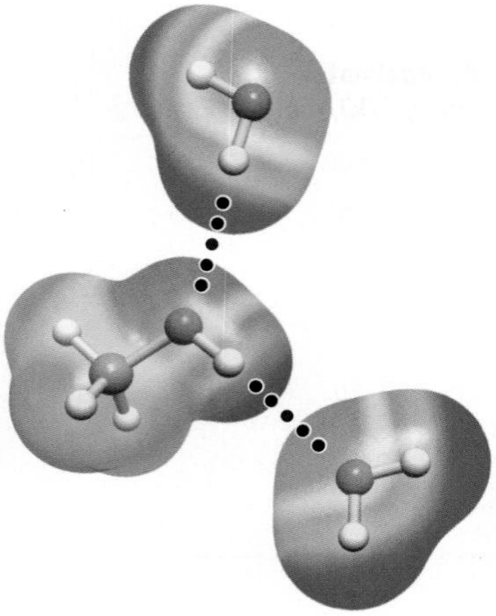

Hydrogen bonding involving methanol (CH_3OH) and water.

(b) Both water and methanol are covalently bonded molecules; both are polar; and both have an O—H bond. They therefore interact through the special dipole–dipole force called hydrogen bonding, as well as by dipole-dipole and London forces.

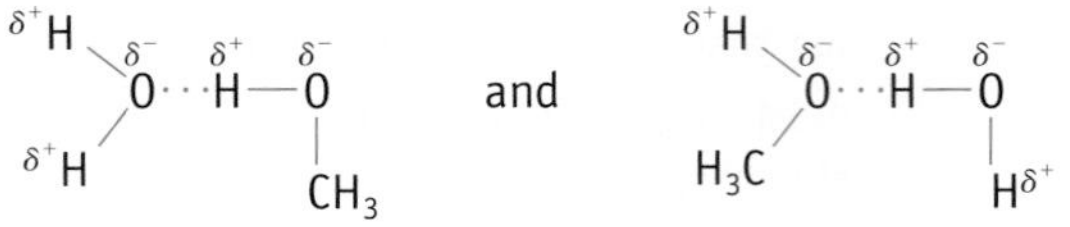

(c) Nonpolar molecules of bromine, Br_2, interact by induced dipole forces, whereas water is a polar molecule. Therefore, dipole/induced dipole forces (and London forces) are involved when Br_2 molecules interact with water. (This is similar to the I_2–ethanol interaction in Figure 12.10.)

In order of increasing strength, the likely order of interactions is

$$\text{liquid } CH_4 < H_2O \text{ and } Br_2 < H_2O \text{ and } CH_3OH$$

EXERCISE 12.4 Intermolecular Forces

Decide which type of intermolecular force is involved in **(a)** liquid O_2; **(b)** liquid CH_3OH; and **(c)** N_2 dissolved in H_2O. Place the interactions in order of increasing strength.

12.4 Properties of Liquids

Of the three states of matter, liquids are the most difficult to describe precisely. The molecules in a gas under normal conditions are far apart and may be considered more or less independent of one another. The structures of solids can be described readily because the particles that make up solids—atoms, molecules, or ions—are close together and are usually in an orderly arrangement. The particles of a liquid interact with their neighbors, like the particles in a solid, but, unlike in solids, there is little long-range order in their arrangement.

In spite of a lack of precision in describing liquids, we can still consider the behavior of liquids at the molecular level. In the following sections, we will look further at the process of vaporization, at the vapor pressure of liquids, at their boiling points and critical properties, and at the behavior that results in their surface tension, capillary action, and viscosity.

Vaporization and Condensation

Vaporization or evaporation is the process in which a substance in the liquid state becomes a gas. In this process, molecules escape from the liquid surface and enter the gaseous state.

To understand evaporation, we have to look at molecular energies. Molecules in a liquid have a range of energies (Figure 12.13) that closely resembles the distribution of energies for molecules of a gas (see Figure 11.14). As with gases, the average energy for molecules in a liquid depends only on temperature: The higher the temperature, the higher the average energy and the greater the relative number of molecules with high kinetic energy. In a sample of a liquid, at least a few molecules have more kinetic energy than the potential energy of the intermolecular attractive forces holding the liquid molecules to one another. If these high-energy molecules are at the surface of the liquid and if they are moving in the right direction, they can break free of their neighbors and enter the gas phase (Figure 12.14).

Vaporization is an endothermic process because energy must be added to the system to overcome the intermolecular forces of attraction holding the molecules together. The energy required to vaporize a sample is often given as the standard

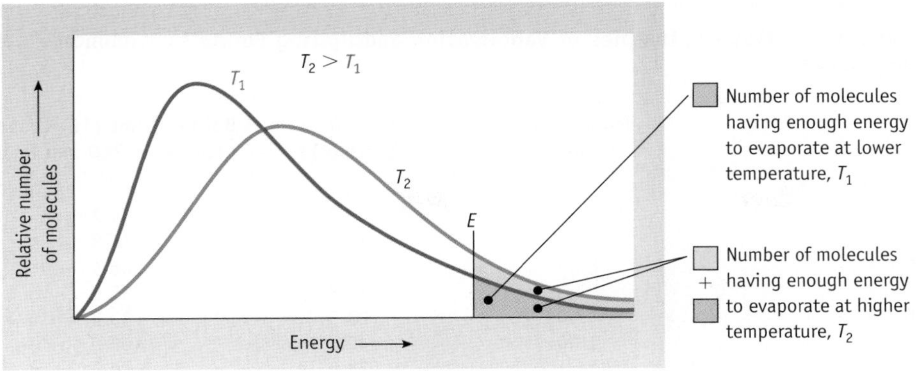

FIGURE 12.13 The distribution of energy among molecules in a liquid sample. T_2 is a higher temperature than T_1, and at the higher temperature, there are more molecules with an energy greater than E.

molar enthalpy of vaporization, $\Delta_{vap}H°$ (in units of kilojoules per mole; see Tables 12.4 and 12.6 and Figure 12.4).

$$\text{Liquid} \xrightarrow[\text{heat energy absorbed by liquid}]{\text{vaporization}} \text{Vapor} \qquad \Delta_{vap}H° = \text{molar heat of vaporization}$$

A molecule in the gas phase can transfer some of its kinetic energy by colliding with slower gaseous molecules and solid objects. If this molecule loses sufficient energy and comes in contact with the surface of the liquid, it can reenter the liquid phase in the process called **condensation**.

$$\text{Vapor} \xrightarrow[\text{heat energy released by vapor}]{\text{condensation}} \text{Liquid}$$

Condensation is the reverse of vaporization. Condensation is exothermic, so energy is transferred to the surroundings. *The enthalpy change for condensation is equal but opposite in sign to the enthalpy of vaporization.* For example, the enthalpy change for the vaporization of 1.00 mol of water at 100 °C is +40.7 kJ. On condensing 1.00 mol of water vapor to liquid water at 100 °C, the enthalpy change is −40.7 kJ.

In the discussion of intermolecular forces, we pointed out the relationship between the $\Delta_{vap}H°$ values for various substances and the temperatures at which they boil (Table 12.6). Both properties reflect the attractive forces between particles in the liquid. The boiling points of nonpolar liquids (e.g., the hydrocarbons, atmospheric gases, and the halogens) increase with increasing atomic or molecular mass,

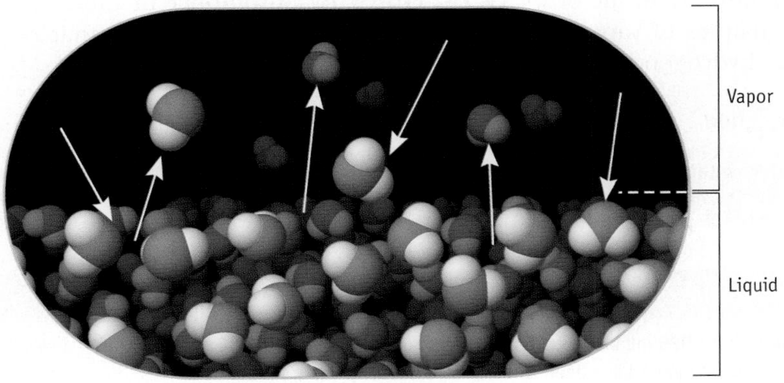

FIGURE 12.14 Evaporation. Some molecules at the surface of a liquid have enough energy to escape the attractions of their neighbors and enter the gaseous state. At the same time, some molecules in the gaseous state can reenter the liquid.

TABLE 12.6 Molar Enthalpies of Vaporization and Boiling Points for Common Substances*

Compound	Molar Mass (g/mol)	$\Delta_{vap}H°$ (kJ/mol)†	Boiling Point (°C) (Vapor pressure = 760 mm Hg)
Polar Compounds			
HF	20.0	25.2	19.7
HCl	36.5	16.2	−84.8
HBr	80.9	19.3	−66.4
HI	127.9	19.8	−35.6
NH_3	17.0	23.3	−33.3
H_2O	18.0	40.7	100.0
SO_2	64.1	24.9	−10.0
Nonpolar Compounds			
CH_4 (methane)	16.0	8.2	−161.5
C_2H_6 (ethane)	30.1	14.7	−88.6
C_3H_8 (propane)	44.1	19.0	−42.1
C_4H_{10} (butane)	58.1	22.4	−0.5
Monatomic Elements			
He	4.0	0.08	−268.9
Ne	20.2	1.7	−246.1
Ar	39.9	6.4	−185.9
Xe	131.3	12.6	−108.0
Diatomic Elements			
H_2	2.0	0.90	−252.9
N_2	28.0	5.6	−195.8
O_2	32.0	6.8	−183.0
F_2	38.0	6.6	−188.1
Cl_2	70.9	20.4	−34.0
Br_2	159.8	30.0	58.8

*Data taken from D. R. Lide: *Basic Laboratory and Industrial Chemicals,* Boca Raton, FL, CRC Press, 1993.

†$\Delta_{vap}H°$ is measured at the normal boiling point of the liquid.

a reflection of increased intermolecular dispersion forces. The alkanes (such as methane) listed in Table 12.6 show this trend clearly. Similarly, the boiling points and enthalpies of vaporization of the heavier hydrogen halides (HX, where X = Cl, Br, and I) increase with increasing molecular mass. For these molecules, hydrogen bonding is not as important as it is in HF, so dispersion forces and ordinary dipole–dipole forces account for their intermolecular attractions (see Figure 12.12). Because dispersion forces become increasingly important with increasing mass, the boiling points are in the order HCl < HBr < HI. Also notice in Table 12.6 the very high enthalpies of vaporization of water and hydrogen fluoride that result from extensive hydrogen bonding.

Chemistry Now™

Sign in at **www.thomsonedu.com/login** and go to Chapter 12 Contents to see Screen 12.8 to view an animation of the **vaporization process** and for a table of **$\Delta_{vap}H°$ values.**

EXAMPLE 12.5 Enthalpy of Vaporization

Problem You put 925 mL of water (about 4 cupsful) in a pan at 100 °C, and the water slowly evaporates. How much energy must have been transferred as heat to vaporize the water?

Strategy Three pieces of information are needed to solve this problem:

1. $\Delta_{vap}H°$ for water = +40.7 kJ/mol at 100 °C.
2. The density of water at 100 °C = 0.958 g/cm^3. (This is needed because $\Delta_{vap}H°$ has units of kilojoules per mole, so you first must find the mass of water and then the amount.)
3. Molar mass of water = 18.02 g/mol.

Solution A volume of 925 mL (or 9.25×10^2 cm^3) is equivalent to 886 g, and this mass is in turn equivalent to 49.2 mol of water.

$$925 \text{ mL}\left(\frac{0.958 \text{ g}}{1 \text{ mL}}\right)\left(\frac{1 \text{ mol}}{18.02 \text{ g}}\right) = 49.2 \text{ mol } H_2O$$

Therefore, the amount of energy required is

$$49.2 \text{ mol } H_2O\left(\frac{40.7 \text{ kJ}}{\text{mol}}\right) = 2.00 \times 10^3 \text{ kJ}$$

2000 kJ is equivalent to about one quarter of the energy in your daily food intake.

EXERCISE 12.5 Enthalpy of Vaporization

The molar enthalpy of vaporization of methanol, CH_3OH, is 35.2 kJ/mol at 64.6 °C. How much energy is required to evaporate 1.00 kg of this alcohol at 64.6 °C?

Water is exceptional among the liquids listed in Table 12.6 in that an enormous amount of heat is required to convert liquid water to water vapor. This fact is important to your own physical well-being. When you exercise vigorously, your body responds by sweating to rid itself of the excess heat. Energy from your body is transferred to sweat in the process of evaporation, and your body is cooled.

Enthalpies of vaporization and condensation of water also play a role in weather (Figure 12.15). For example, if enough water condenses from the air to fall as an inch of rain on an acre of ground, the heat released exceeds 2.0×10^8 kJ! This is equivalent to about 50 tons of exploded dynamite, the energy released by a small bomb.

The Image Bank/Getty Images

FIGURE 12.15 Rainstorms release an enormous quantity of energy. When water vapor condenses, energy is evolved to the surroundings. The enthalpy of condensation of water is large, so a large quantity of energy is released in a rainstorm.

Vapor Pressure

If you put some water in an open beaker, it will eventually evaporate completely. Air movement and gas diffusion remove the water vapor from the vicinity of the liquid surface, so many water molecules are not able to return to the liquid.

If you put water in a sealed flask (Figure 12.16), however, the water vapor cannot escape, and some will recondense to form liquid water. Eventually, the masses of liquid and of vapor in the flask remain constant. This is another example of a **dynamic equilibrium** (◀ page 119).

$$\text{Liquid} \rightleftharpoons \text{Vapor}$$

Molecules still move continuously from the liquid phase to the vapor phase and from the vapor phase back to the liquid phase. The rate at which molecules move from liquid to vapor is the same as the rate at which they move from vapor to liquid; thus, there is no net change in the masses of the two phases.

When a liquid–vapor equilibrium has been established, the equilibrium vapor pressure (often just called the vapor pressure) can be measured. The **equilibrium vapor pressure** of a substance is the pressure exerted by the vapor in equilibrium with the liquid phase. Conceptually, the vapor pressure of a liquid is a measure of the tendency of its molecules to escape from the liquid phase and enter the vapor

Active Figure 12.16 Vapor pressure. A volatile liquid is placed in an evacuated flask (left). At the beginning, no molecules of the liquid are in the vapor phase. After a short time, however, some of the liquid evaporates, and the molecules now in the vapor phase exert a pressure. The pressure of the vapor measured when the liquid and the vapor are in equilibrium is called the equilibrium vapor pressure (right).

Chemistry Now™ Sign in at www.thomsonedu.com/login and go to the Chapter Contents menu to explore an interactive version of this figure accompanied by an exercise.

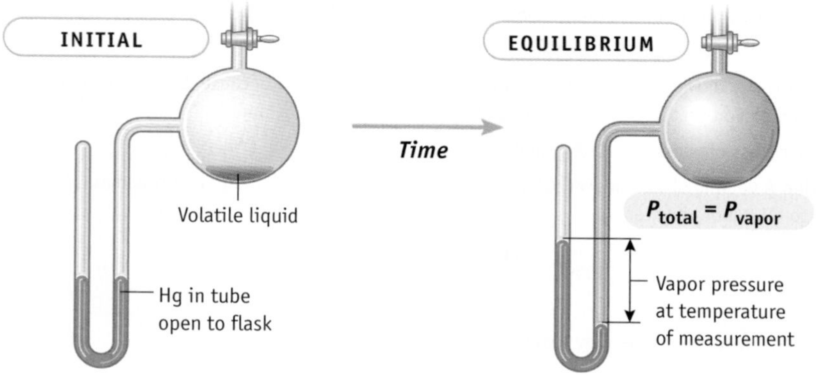

phase at a given temperature. This tendency is referred to qualitatively as the **volatility** of the compound. The higher the equilibrium vapor pressure at a given temperature, the more volatile the substance.

As described previously (see Figure 12.13), the distribution of molecular energies in the liquid phase is a function of temperature. At a higher temperature, more molecules have sufficient energy to escape the surface of the liquid. The equilibrium vapor pressure must, therefore, increase with temperature.

■ **Equilibrium Vapor Pressure** At the conditions of *T* and *P* given by any point on a curve in Figure 12.17, the pure liquid and its vapor are in dynamic equilibrium. If *T* and *P* define a point not on the curve, the system is not at equilibrium. See Appendix G for the equilibrium vapor pressures of water at various temperatures.

It is useful to represent vapor pressure as a function of temperature. Figure 12.17 shows the vapor pressure curves for several liquids as a function of temperature. *All points along the vapor pressure versus temperature curves represent conditions of pressure and temperature at which liquid and vapor are in equilibrium.* For example, at 60 °C the vapor pressure of water is 149 mm Hg (Appendix G). If water is placed in an evacuated flask that is maintained at 60 °C, liquid water will evaporate until the pressure exerted by the water vapor is 149 mm Hg (assuming enough water is in the flask so that some liquid remains when equilibrium is reached).

Chemistry Now™

Sign in at **www.thomsonedu.com/login** and go to Chapter 12 Contents to see Screen 12.9 to view an animation on **equilibrium vapor pressure** and for a simulation of **vapor pressure curves.**

Active Figure 12.17 Vapor pressure curves for diethyl ether [$(C_2H_5)_2O$], ethanol (C_2H_5OH), and water. Each curve represents conditions of *T* and *P* at which the two phases, liquid and vapor, are in equilibrium. These compounds exist as liquids for temperatures and pressures to the left of the curve and as gases under conditions to the right of the curve.

Chemistry Now™ Sign in at www.thomsonedu.com/login and go to the Chapter Contents menu to explore an interactive version of this figure accompanied by an exercise.

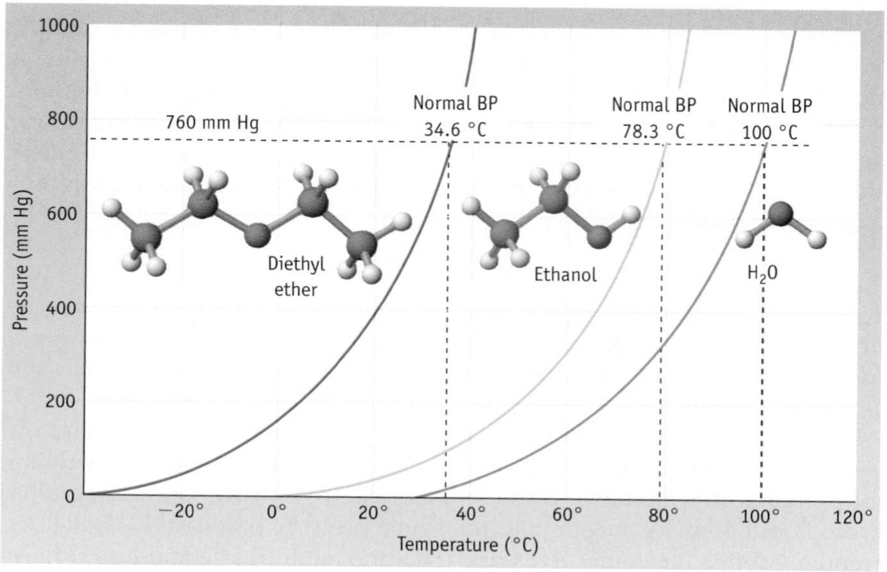

EXAMPLE 12.6 Vapor Pressure

Problem You place 2.00 L of water in an open container in your dormitory room; the room has a volume of 4.25×10^4 L. You seal the room and wait for the water to evaporate. Will all of the water evaporate at 25 °C? (At 25 °C the density of water is 0.997 g/mL, and its vapor pressure is 23.8 mm Hg.)

Strategy One approach to solving this problem is to calculate the quantity of water that must evaporate to exert a pressure of 23.8 mm Hg in a volume of 4.25×10^4 L at 25 °C. We use the ideal gas law for this calculation.

Solution Calculate the amount and then mass and volume of water that fulfills the following conditions: $P = 23.8$ mm Hg, $V = 4.25 \times 10^4$ L, $T = 25$ °C (298 K).

$$P = 23.8 \text{ mm Hg}\left(\frac{1 \text{ atm}}{760 \text{ mm Hg}}\right) = 0.0313 \text{ atm}$$

$$n = \frac{PV}{RT} = \frac{(0.0313 \text{ atm})(4.25 \times 10^4 \text{ L})}{\left(0.082057 \frac{\text{L}\cdot\text{atm}}{\text{K}\cdot\text{mol}}\right)(298 \text{ K})} = 54.4 \text{ mol}$$

$$54.4 \text{ mol } H_2O\left(\frac{18.02 \text{ g}}{1 \text{ mol } H_2O}\right) = 980. \text{ g } H_2O$$

$$980. \text{ g } H_2O\left(\frac{1 \text{ mL}}{0.997 \text{ g } H_2O}\right) = 983 \text{ mL}$$

Only about half of the available water needs to evaporate to achieve the equilibrium water vapor pressure of 23.8 mm Hg at 25 °C.

EXERCISE 12.6 Vapor Pressure Curves

Examine the vapor pressure curve for ethanol in Figure 12.17.

(a) What is the approximate vapor pressure of ethanol at 40 °C?

(b) Are liquid and vapor in equilibrium when the temperature is 60 °C and the pressure is 600 mm Hg? If not, does liquid evaporate to form more vapor, or does vapor condense to form more liquid?

EXERCISE 12.7 Vapor Pressure

If 0.50 g of pure water is sealed in an evacuated 5.0-L flask and the whole assembly is heated to 60 °C, will the pressure be equal to or less than the equilibrium vapor pressure of water at this temperature? What if you use 2.0 g of water? Under either set of conditions, is any liquid water left in the flask, or does all of the water evaporate?

Vapor Pressure, Enthalpy of Vaporization, and the Clausius–Clapeyron Equation

Plotting the vapor pressure for a liquid at a series of temperatures results in a curved line (Figure 12.17). However, the German physicist R. Clausius (1822–1888) and the Frenchman B. P. E. Clapeyron (1799–1864) showed that, for a pure liquid, a linear relationship exists between the reciprocal of the Kelvin temperature ($1/T$) and the natural logarithm of vapor pressure ($\ln P$) (Figure 12.18).

$$\ln P = -(\Delta_{vap}H°/RT) + C \qquad (12.1)$$

Here, $\Delta_{vap}H°$ is the enthalpy of vaporization of the liquid; R is the ideal gas constant (8.314472 J/K · mol); and C is a constant characteristic of the liquid in question. This equation, now called the **Clausius-Clapeyron equation,** provides a method of obtaining values for $\Delta_{vap}H°$. The equilibrium vapor pressure of a liquid can be measured at several different temperatures, and the logarithm of these pressures

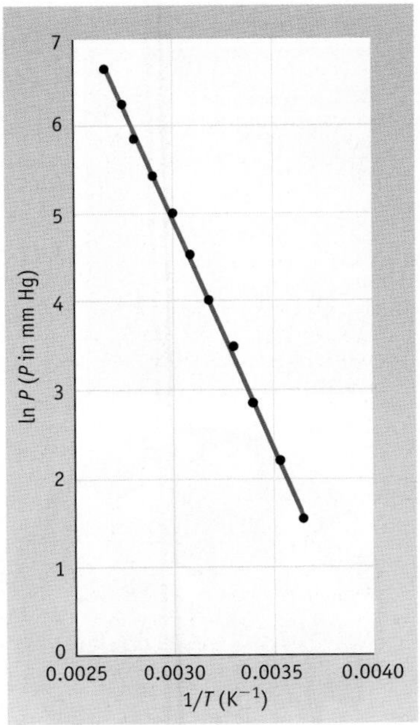

FIGURE 12.18 Clausius-Clapeyron equation. When the natural logarithm of the vapor pressure ($\ln P$) of water at various temperatures (T) is plotted against $1/T$, a straight line is obtained. The slope of the line equals $-\Delta_{vap}H°/R$. Values of T and P are from Appendix G.

is plotted versus $1/T$. The result is a straight line with a slope of $-\Delta_{vap}H°/R$. For example, plotting data for water (Figure 12.18), we find the slope of the line is -4.90×10^3, which gives $\Delta_{vap}H° = 40.7$ kJ/mol.

As an alternative to plotting ln P versus $1/T$, we can write the following equation that allows us to calculate $\Delta_{vap}H°$ knowing the vapor pressure of a liquid at two different temperatures.

$$\ln P_2 - \ln P_1 = \left[\frac{-\Delta_{vap}H°}{RT_2} + C\right] - \left[\frac{-\Delta_{vap}H°}{RT_1} + C\right]$$

This can be simplified to

$$\ln\frac{P_2}{P_1} = -\frac{\Delta_{vap}H°}{R}\left[\frac{1}{T_2} - \frac{1}{T_1}\right] \tag{12.2}$$

For example, ethylene glycol has a vapor pressure of 14.9 mm Hg (P_1) at 373 K (T_1), and a vapor pressure of 49.1 mm Hg (P_2) at 398 K (T_2).

$$\ln\left(\frac{49.1 \text{ mm Hg}}{14.9 \text{ mm Hg}}\right) = -\frac{\Delta_{vap}H°}{0.0083145 \text{ kJ/K}\cdot\text{mol}}\left[\frac{1}{398 \text{ K}} - \frac{1}{373 \text{ K}}\right]$$

$$1.192 = -\frac{\Delta_{vap}H°}{0.0083145 \text{ kJ/K}\cdot\text{mol}}\left(-\frac{0.000168}{K}\right)$$

$$\Delta_{vap}H° = 59.0 \text{ kJ/mol}$$

Chemistry Now™

Sign in at **www.thomsonedu.com/login** and go to Chapter 12 Contents to see Screen 12.9 for three tutorials on using the **Clausius–Clapeyron equation.**

EXERCISE 12.8 Clausius–Clapeyron Equation

Calculate the enthalpy of vaporization of diethyl ether, $(C_2H_5)_2O$ (see Figure 12.17). This compound has vapor pressures of 57.0 mm Hg and 534 mm Hg at −22.8 °C and 25.0 °C, respectively.

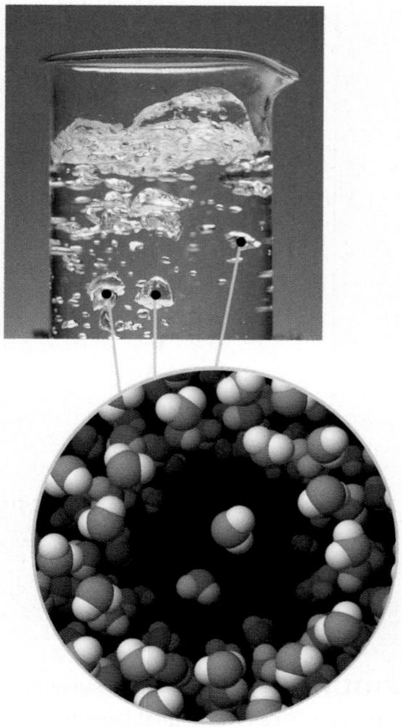

Charles D. Winters

FIGURE 12.19 Vapor pressure and boiling. When the vapor pressure of the liquid equals the atmospheric pressure, bubbles of vapor begin to form within the body of liquid, and the liquid boils.

Boiling Point

If you have a beaker of water open to the atmosphere, the atmosphere presses down on the surface. If enough energy is added, a temperature is eventually reached at which the vapor pressure of the liquid equals the atmospheric pressure. At this temperature, bubbles of the liquid's vapor will not be crushed by the atmospheric pressure. The bubbles can rise to the surface, and the liquid boils (Figure 12.19).

The boiling point of a liquid is the temperature at which its vapor pressure is equal to the external pressure. If the external pressure is 760 mm Hg, this temperature is called the **normal boiling point.** This point is highlighted on the vapor pressure curves for the substances in Figure 12.17.

The normal boiling point of water is 100 °C, and in a great many places in the United States, water boils at or near this temperature. If you live at higher altitudes, however, such as in Salt Lake City, Utah, where the barometric pressure is about 650 mm Hg, water will boil at a noticeably lower temperature. The curve in Figure 12.19 shows that a pressure of 650 mm Hg corresponds to a boiling temperature of about 95 °C. Food, therefore, has to be cooked a little longer in Salt Lake City to achieve the same result as in New York City at sea level.

Chemistry Now™

Sign in at **www.thomsonedu.com/login** and go to Chapter 12 Contents to see Screen 12.10 to watch a video of **the boiling of a liquid in a partial vacuum** and to explore **the relationship of molecular composition and structure and boiling point** in two simulations.

Critical Temperature and Pressure

On first thought, it might seem that vapor pressure–temperature curves (such as shown in Figure 12.17) should continue upward without limit, but this is not so. Instead, when a specific temperature and pressure are reached, the interface between the liquid and the vapor disappears. This point is called the **critical point.** The temperature at which this phenomenon occurs is the **critical temperature, T_c,** and the corresponding pressure is the **critical pressure, P_c** (Figure 12.20). The substance that exists under these conditions is called a **supercritical fluid.** It is like a gas under such a high pressure that its density resembles that of a liquid, while its viscosity (ability to flow) remains close to that of a gas (▶ page 580).

Consider what the substance might look like at the molecular level under these conditions. The molecules have been forced almost as close together as they are in the liquid state, but each molecule has enough kinetic energy to exceed the forces holding molecules together. As a result, the supercritical fluid has a tightly packed molecular arrangement like a liquid, but the intermolecular forces of attraction that characterize the liquid state are less than the kinetic energy of the particles.

For most substances, the critical point is at a very high temperature and pressure (Table 12.7). Water, for instance, has a critical temperature of 374 °C and a critical pressure of 217.7 atm.

Supercritical fluids can have unexpected properties, such as the ability to dissolve normally insoluble materials. Supercritical CO_2 is especially useful. Carbon dioxide is widely available, essentially nontoxic, nonflammable, and inexpensive. It is relatively easy to reach its critical temperature of 30.99 °C and critical pressure of 72.8 atm. One use of supercritical CO_2 is to extract caffeine from coffee. The coffee beans are treated with steam to bring the caffeine to the surface. The beans are then immersed in supercritical CO_2, which selectively dissolves the caffeine but leaves intact the compounds that give flavor to coffee. (Decaffeinated coffee contains less than 3% of the original caffeine.) The solution of caffeine in supercritical CO_2 is poured off, and the CO_2 is evaporated, trapped, and reused.

TABLE 12.7 Critical Temperatures and Pressures for Common Compounds*

Compound	T_c (°C)	P_c (atm)
CH_4 (methane)	−82.6	45.4
C_2H_6 (ethane)	32.3	49.1
C_3H_8 (propane)	96.7	41.9
C_4H_{10} (butane)	152.0	37.3
CCl_2F_2 (CFC-12)	111.8	40.9
NH_3	132.4	112.0
H_2O	374.0	217.7
CO_2	30.99	72.8
SO_2	157.7	77.8

*Data taken from D. R. Lide: *Basic Laboratory and Industrial Chemicals*, Boca Raton, FL, CRC Press, 1993.

■ **Green Chemistry and Supercritical CO_2** It is not surprising that other uses are being sought for supercritical CO_2. One application being investigated is its use as a dry cleaning solvent. More than 10 billion kilograms of organic and halogenated solvents are used worldwide every year in cleaning applications. These cleaning agents can have deleterious effects on the environment, so it is hoped that many can be replaced by supercritical CO_2. (For more about supercritical CO_2 see page 609.)

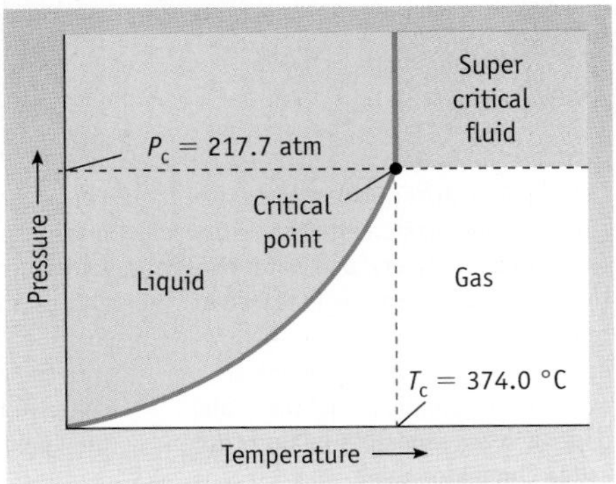

FIGURE 12.20 Critical temperature and pressure for water. The curve representing equilibrium conditions for liquid and gaseous water ends at the critical point; above that temperature and pressure, water becomes a supercritical fluid.

FIGURE 12.21 Intermolecular forces in a liquid. Forces acting on a molecule at the surface of a liquid are different than those acting on a molecule in the interior of a liquid.

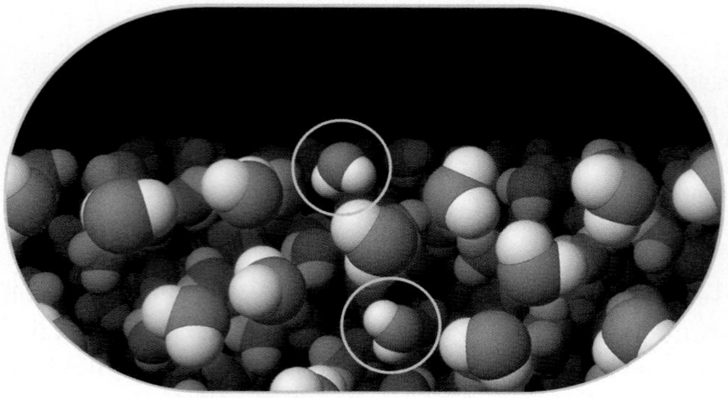

Water molecules on the surface are not completely surrounded by other water molecules.

Water molecules under the surface are completely surrounded by other water molecules.

Surface Tension, Capillary Action, and Viscosity

Molecules in the interior of a liquid interact with molecules all around them (Figure 12.21). In contrast, molecules on the surface of a liquid are affected only by those molecules located at or below the surface layer. This leads to a net inward force of attraction on the surface molecules, contracting the surface area and making the liquid behave as though it had a skin. The toughness of this skin is measured by its **surface tension**—the energy required to break through the surface or to disrupt a liquid drop and spread the material out as a film. Surface tension causes water drops to be spheres and not little cubes, for example (Figure 12.22a), because a sphere has a smaller surface area than any other shape of the same volume.

■ **Viscosity** Long chains of atoms, such as those present in oils, are floppy and become entangled with one another in the liquid; the longer the chain, the greater the tangling and the greater the viscosity.

Capillary action is closely related to surface tension. When a small-diameter glass tube is placed in water, the water rises in the tube, just as water rises in a piece of paper in water (Figure 12.22b). Because polar Si—O bonds are present on the

FIGURE 12.22 Adhesive and cohesive forces.

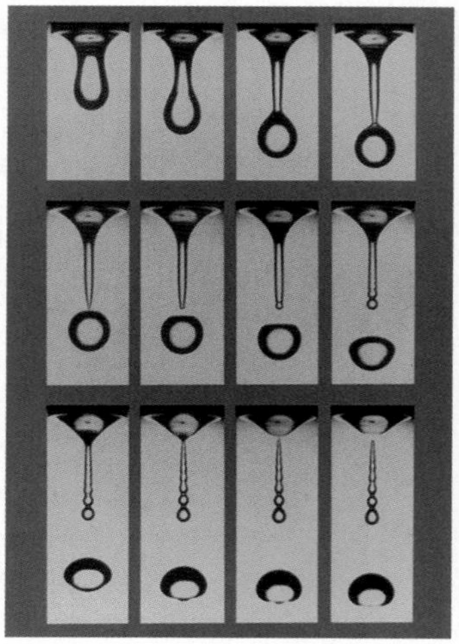

S.R. Nagel, James Frank Institute, University of Chicago

(a) A series of photographs showing the different stages when a water drop falls. The drop was illuminated by a strobe light of 5-ms duration. (The total time for this sequence was 0.05 s.) Water droplets take a spherical shape because of surface tension.

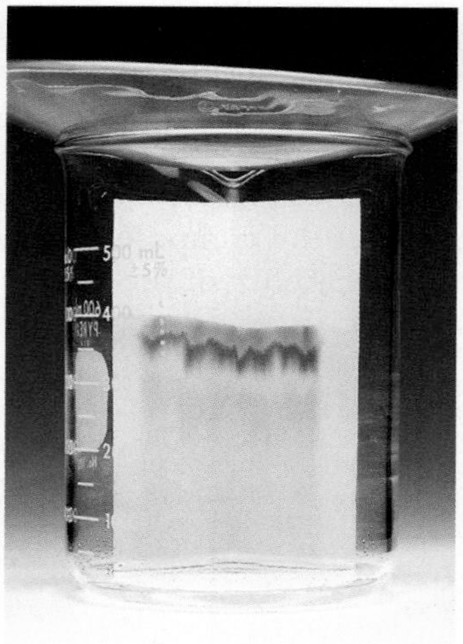

Charles D. Winters

(b) Capillary action. Polar water molecules are attracted to the —OH bonds in paper fibers, and water rises in the paper. If a line of ink is placed in the path of the rising water, the different components of the ink are attracted differently to the water and paper and are separated in a process called chromatography.

Charles D. Winters

(c) Water (top layer) forms a concave meniscus, while mercury (bottom layer) forms a convex meniscus. The different shapes are determined by the adhesive forces of the molecules of the liquid with the walls of the tube and the cohesive forces between molecules of the liquid.

Case Study — The Mystery of the Disappearing Fingerprints

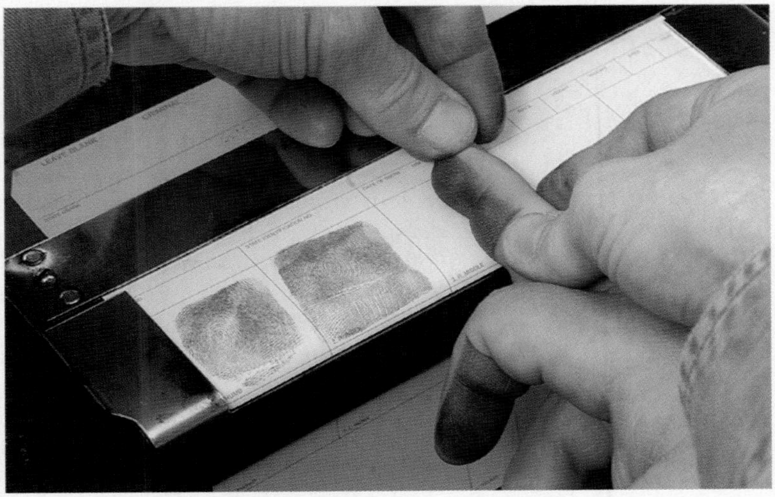

Charles D. Winters

Taking a fingerprint at the local police station.

The events of September 11, 2001, are etched in everyone's memory. The possibility of domestic terrorism, however, began almost two years before when a man was apprehended in late December 1999 at the Canadian border with bomb materials and a map of the Los Angeles International Airport. Although he claimed innocence, his fingerprints were on the bomb materials, and he was convicted of an attempt to bomb the airport.

Each of us has a unique fingerprint pattern, as first described by John Purkinji in 1823. Not long after, the English in India began using fingerprints on contracts because they believed it made the contract appear more binding. It was not until late in the 19th century, however, that fingerprinting was used as an identifier. Sir Francis Galton, a British anthropologist and cousin of Charles Darwin, established that a person's fingerprints do not change over the course of a lifetime and that no two prints are exactly the same. Fingerprinting has since become an accepted tool in forensic science.

In 1993 in Knoxville, Tennessee, detective Art Bohanan thought he could use it to solve the case of the kidnapping of a young girl. The girl had been taken from her home and driven away in a green car. The girl soon managed to escape from her attacker and was able to describe the car to the police. After four days, the police found the car and arrested its owner. But had the girl been in that car? Art Bohanan inspected the car for her fingerprints and even used the latest technique, fuming with superglue. No prints were found.

The abductor of the girl was eventually convicted on other evidence, but Bohanan wondered why he had never found her prints in the car. He decided to test the permanence of children's fingerprints compared with adults. To his amazement, he found that children's prints disappear in a few hours, whereas an adult's prints can last for days. Bohanan said, "It sounded like the compounds in children's fingerprints might simply be evaporating faster than adult's."

To answer this, you should know about the nature of fingerprints. The residue deposited by fingerprints is 99% water. The other 1% contains oils, fatty acids, esters, salts, urea [$(NH_2)_2CO$], and amino acids. An example of a fatty acid is myristic acid: $CH_3(CH_2)_{12}CO_2H$. An ester is the combination of an acid with an alcohol such as ethanol (CH_3CH_2OH), so an ester of myristic acid would be: $CH_3(CH_2)_{12}CO_2CH_2CH_3$

Scientists at Oak Ridge National Laboratory studied the fingerprints of 50 child and adult volunteers, identifying the compounds present by such techniques as mass spectrometry (◀ page 68). What they found clarified the mystery of the disappearing fingerprints.

Children's fingerprints contain more low-molecular-weight fatty acids than adult fingerprints. In contrast, adult fingerprints contain esters of long-chain fatty acids with long-chain alcohols. These are waxes, semisolid or solid organic compounds with high molecular weights. (Examples of waxes are lanolin, a component of wool, or carnauba wax used in furniture polish.)

Before puberty, children do not produce waxy compounds in their skin. However, sebaceous glands in adult skin produce sebum, a complex mixture of organic compounds (triglycerides, fatty acids, cholesterol, and waxes). There are only a few of these glands on the hands; most are on the mid-back, forehead, and chin. So, when you touch your face, this mixture of compounds is transferred to your fingers, and you can leave a fingerprint that is unique to you.

Question:

Why do children's fingerprints evaporate more readily than adult fingerprints?

Answer to this question is in Appendix Q.

surface of glass, polar water molecules are attracted by **adhesive forces** between the two different substances. These forces are strong enough that they can compete with the **cohesive forces** between the water molecules themselves. Thus, some water molecules can adhere to the walls; other water molecules are attracted to them and build a "bridge" back into the liquid. The surface tension of the water (from cohesive forces) is great enough to pull the liquid up the tube, so the water level rises in the tube. The rise will continue until the attractive forces—adhesion between water and glass, cohesion between water molecules—are balanced by the force of gravity pulling down on the water column. These forces lead to the characteristic concave, or downward-curving, meniscus seen with water in a test tube (Figure 12.22c).

In some liquids, cohesive forces (high surface tension) are much greater than adhesive forces with glass. Mercury is one example. Mercury does not climb the walls of a glass capillary. In fact, when it is in a glass tube, mercury will form a convex, or upward-curving, meniscus (Figure 12.22c).

One other important property of liquids in which intermolecular forces play a role is **viscosity,** the resistance of liquids to flow. When you turn over a glassful of water, it empties quickly. In contrast, it takes much more time to empty a glassful of olive oil or honey. Olive oil consists of molecules with long chains of carbon atoms, and it is about 70 times more viscous than ethanol, a small molecule with only two carbons and one oxygen. Longer chains have greater intermolecular forces because there are more atoms to attract one another, with each atom contributing to the total force. Honey (a concentrated aqueous solution of sugar molecules), however, is also a viscous liquid, even though the size of the molecules is fairly small. In this case, the sugar molecules have numerous —OH groups. These lead to greater forces of attraction due to hydrogen bonding.

Chemistry Now™

Sign in at **www.thomsonedu.com/login** and go to Chapter 12 Contents to see Screen 12.11 to watch videos on **surface tension, capillary action,** and **viscosity.**

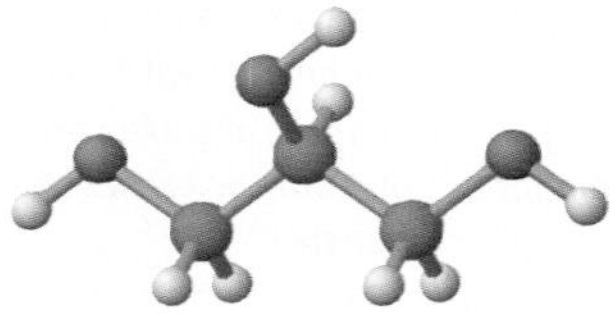
Glycerol

EXERCISE 12.9 Viscosity

Glycerol ($HOCH_2CHOHCH_2OH$) is used in cosmetics. Do you expect its viscosity to be larger or smaller than the viscosity of ethanol, CH_3CH_2OH? Explain briefly.

Chapter Goals Revisited

Chemistry Now™ Sign in at **www.thomsonedu.com/login** to:

- Assess your understanding with Study Questions in OWL keyed to each goal in the Goals and Homework menu for this chapter
- For quick review, download Go Chemistry mini-lecture flashcard modules (or purchase them at **www.ichapters.com**)
- Check your readiness for an exam by taking the Pre-Test and exploring the modules recommended in your Personalized Study plan.

Access **How Do I Solve It?** tutorials on how to approach problem solving using concepts in this chapter.

Now that you have studied this chapter, you should ask whether you have met the chapter goals. In particular, you should be able to:

Describe intermolecular forces and their effects

a. Describe the various intermolecular forces found in liquids and solids (Sections 12.2 and 12.3). **Study Question(s) assignable in OWL: 2, 4, 6, 7, 25–28, 30, 32, 39; Go Chemistry Module 17.**

b. Tell when two molecules can interact through a dipole–dipole attraction and when hydrogen bonding may occur. The latter occurs most strongly when H is attached to O, N, or F (Section 12.2). **Study Question(s) assignable in OWL: 7–10.**

c. Identify instances in which molecules interact by induced dipoles (dispersion forces) (Section 12.3). **Study Question(s) assignable in OWL: 7.**

Understand the importance of hydrogen bonding

a. Explain how hydrogen bonding affects the properties of water (Section 12.2).

Understand the properties of liquids

- **a.** Explain the processes of evaporation and condensation, and use the enthalpy of vaporization in calculations (Section 12.4). **Study Question(s) assignable in OWL: 11, 12, 18, 31, 53.**
- **b.** Define the equilibrium vapor pressure of a liquid, and explain the relationship between the vapor pressure and boiling point of a liquid (Section 12.4). **Study Question(s) assignable in OWL: 14, 15, 17, 19, 20, 29, 38, 50.**
- **c.** Describe the phenomena of the critical temperature, T_c, and critical pressure, P_c, of a substance (Section 12.4). **Study Question(s) assignable in OWL: 23.**
- **d.** Describe how intermolecular interactions affect the cohesive forces between identical liquid molecules, the energy necessary to break through the surface of a liquid (surface tension), and the resistance to flow, or viscosity, of liquids (Section 12.4). **Study Question(s) assignable in OWL: 41.**
- **e.** Use the Clausius–Clapeyron equation, which connects temperature, vapor pressure, and enthalpy of vaporization for liquids (Section 12.4). **Study Question(s) assignable in OWL: 21, 22, 34.**

KEY EQUATION

Equation 12.2 (page 576) The Clausius–Clapeyron equation relates the equilibrium vapor pressure, P, of a volatile liquid to the molar enthalpy of vaporization ($\Delta_{vap}H°$) at a given temperature, T. (R is the universal constant, 8.314472 J/K · mol.) Equation 12.2 allows you to calculate $\Delta_{vap}H°$ if you know the vapor pressures at two different temperatures. Alternatively, you may plot ln P versus $1/T$; the slope of the line is $-\Delta_{vap}H°/R$.

$$\ln\frac{P_2}{P_1} = -\frac{\Delta_{vap}H°}{R}\left[\frac{1}{T_2} - \frac{1}{T_1}\right]$$

STUDY QUESTIONS

Online homework for this chapter may be assigned in OWL.

▲ denotes challenging questions.

■ denotes questions assignable in OWL.

Blue-numbered questions have answers in Appendix O and fully-worked solutions in the *Student Solutions Manual.*

Practicing Skills

Intermolecular Forces

(See Examples 12.1–12.4 and ChemistryNow Screens 12.3–12.7.)

1. What intermolecular force(s) must be overcome to
   (a) melt ice
   (b) sublime solid I_2
   (c) convert liquid NH_3 to NH_3 vapor

2. ■ What type of forces must be overcome within solid I_2 when I_2 dissolves in methanol, CH_3OH? What type of forces must be disrupted between CH_3OH molecules when I_2 dissolves? What type of forces exist between I_2 and CH_3OH molecules in solution?

3. What type of intermolecular forces must be overcome in converting each of the following from a liquid to a gas?
   (a) liquid O_2
   (b) mercury
   (c) CH_3I (methyl iodide)
   (d) CH_3CH_2OH (ethanol)

4. ■ What type of intermolecular forces must be overcome in converting each of the following from a liquid to a gas?
   (a) CO_2
   (b) NH_3
   (c) $CHCl_3$
   (d) CCl_4

5. Rank the following atoms or molecules in order of increasing strength of intermolecular forces in the pure substance. Which exists as a gas at 25 °C and 1 atm?
   (a) Ne
   (b) CH_4
   (c) CO
   (d) CCl_4

6. ■ Rank the following in order of increasing strength of intermolecular forces in the pure substances. Which exists as a gas at 25 °C and 1 atm?
   (a) $CH_3CH_2CH_2CH_3$ (butane)
   (b) CH_3OH (methanol)
   (c) He

7. ■ Which of the following compounds would be expected to form intermolecular hydrogen bonds in the liquid state?
   (a) CH_3OCH_3 (dimethyl ether)
   (b) CH_4
   (c) HF
   (d) CH_3CO_2H (acetic acid)
   (e) Br_2
   (f) CH_3OH (methanol)

8. ■ Which of the following compounds would be expected to form intermolecular hydrogen bonds in the liquid state?
   (a) H_2Se
   (b) HCO_2H (formic acid)
   (c) HI
   (d) acetone (see structure below)

9. ■ In each pair of ionic compounds, which is more likely to have the more negative enthalpy of hydration? Briefly explain your reasoning in each case.
   (a) LiCl or CsCl
   (b) $NaNO_3$ or $Mg(NO_3)_2$
   (c) RbCl or $NiCl_2$

10. ■ When salts of Mg^{2+}, Na^+, and Cs^+ are placed in water, the positive ion is hydrated (as is the negative ion). Which of these three cations is most strongly hydrated? Which one is least strongly hydrated?

Liquids

(See Examples 12.5 and 12.6 and ChemistryNow Screens 12.8–12.11.)

11. ■ Ethanol, CH_3CH_2OH, has a vapor pressure of 59 mm Hg at 25 °C. What quantity of energy as heat is required to evaporate 125 mL of the alcohol at 25 °C? The enthalpy of vaporization of the alcohol at 25 °C is 42.32 kJ/mol. The density of the liquid is 0.7849 g/mL.

12. ■ The enthalpy of vaporization of liquid mercury is 59.11 kJ/mol. What quantity of energy as heat is required to vaporize 0.500 mL of mercury at 357 °C, its normal boiling point? The density of mercury is 13.6 g/mL.

13. Answer the following questions using Figure 12.17:
    (a) What is the approximate equilibrium vapor pressure of water at 60 °C? Compare your answer with the data in Appendix G.
    (b) At what temperature does water have an equilibrium vapor pressure of 600 mm Hg?
    (c) Compare the equilibrium vapor pressures of water and ethanol at 70 °C. Which is higher?

14. ■ Answer the following questions using Figure 12.17:
    (a) What is the equilibrium vapor pressure of diethyl ether at room temperature (approximately 20 °C)?
    (b) Place the three compounds in Figure 12.17 in order of increasing intermolecular forces.
    (c) If the pressure in a flask is 400 mm Hg and if the temperature is 40 °C, which of the three compounds (diethyl ether, ethanol, and water) are liquids, and which are gases?

15. ■ Assume you seal 1.0 g of diethyl ether (see Figure 12.17) in an evacuated 100.-mL flask. If the flask is held at 30 °C, what is the approximate gas pressure in the flask? If the flask is placed in an ice bath, does additional liquid ether evaporate, or does some ether condense to a liquid?

16. Refer to Figure 12.17 as an aid in answering these questions:
    (a) You put some water at 60 °C in a plastic milk carton and seal the top very tightly so gas cannot enter or leave the carton. What happens when the water cools?
    (b) If you put a few drops of liquid diethyl ether on your hand, does it evaporate completely or remain a liquid?

17. ■ Which member of each of the following pairs of compounds has the higher boiling point?
    (a) O_2 or N_2
    (b) SO_2 or CO_2
    (c) HF or HI
    (d) SiH_4 or GeH_4

18. ■ Place the following four compounds in order of increasing boiling point:
    (a) SCl_2
    (b) NH_3
    (c) C_2H_6
    (d) Ne

19. ■ Vapor pressure curves for CS_2 (carbon disulfide) and CH_3NO_2 (nitromethane) are drawn here.
    (a) What are the approximate vapor pressures of CS_2 and CH_3NO_2 at 40 °C?
    (b) What type of intermolecular forces exist in the liquid phase of each compound?
    (c) What is the normal boiling point of CS_2? Of CH_3NO_2?
    (d) At what temperature does CS_2 have a vapor pressure of 600 mm Hg?
    (e) At what temperature does CH_3NO_2 have a vapor pressure of 60 mm Hg?

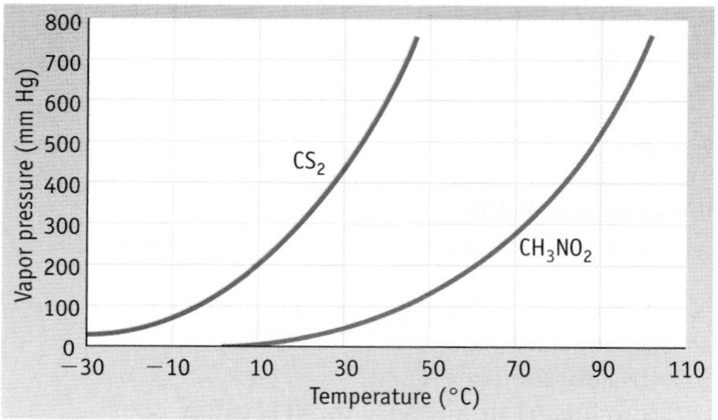

▲ more challenging ■ in OWL Blue-numbered questions answered in Appendix O

20. ■ Answer each of the following questions with *increases, decreases,* or *does not change.*
(a) If the intermolecular forces in a liquid increase, the normal boiling point of the liquid ______.
(b) If the intermolecular forces in a liquid decrease, the vapor pressure of the liquid ______.
(c) If the surface area of a liquid decreases, the vapor pressure ______.
(d) If the temperature of a liquid increases, the equilibrium vapor pressure ______.

21. ■ The following data are the equilibrium vapor pressure of benzene, C_6H_6, at various temperatures.

Temperature (°C)	Vapor Pressure (mm Hg)
7.6	40.
26.1	100.
60.6	400.
80.1	760.

(a) What is the normal boiling point of benzene?
(b) Plot these data so that you have a plot resembling the one in Figure 12.17. At what temperature does the liquid have an equilibrium vapor pressure of 250 mm Hg? At what temperature is it 650 mm Hg?
(c) Calculate the molar enthalpy of vaporization for benzene using the the Clausius–Clapeyron equation (Equation 12.2, page 576).

22. ■ Vapor pressure data are given here for octane, C_8H_{18}.

Temperature (°C)	Vapor Pressure (mm Hg)
25	13.6
50.	45.3
75	127.2
100.	310.8

Use the Clausius–Clapeyron equation (Equation 12.2, page 576) to calculate the molar enthalpy of vaporization of octane and its normal boiling point.

23. ■ Can carbon monoxide (T_c = 132.9 K; P_c = 34.5 atm) be liquefied at or above room temperature? Explain briefly.

24. Methane (CH_4) cannot be liquefied at room temperature, no matter how high the pressure. Propane (C_3H_8), another simple hydrocarbon, has a critical pressure of 42 atm and a critical temperature of 96.7 °C. Can this compound be liquefied at room temperature?

General Questions

These questions are not designated as to type or location in the chapter. They may combine several concepts.

25. ■ Rank the following substances in order of increasing strength of intermolecular forces: (a) Ar, (b) CH_3OH, and (c) CO_2.

26. ■ What types of intermolecular forces are important in the liquid phase of (a) C_2H_6 and (b) $(CH_3)_2CHOH$.

27. ■ Which of the following salts, Li_2SO_4 or Cs_2SO_4, is expected to have the more exothermic enthalpy of hydration?

28. ■ Select the substance in each of the following pairs that should have the higher boiling point:
(a) Br_2 or ICl
(b) neon or krypton
(c) CH_3CH_2OH (ethanol) or C_2H_4O (ethylene oxide, structure below)

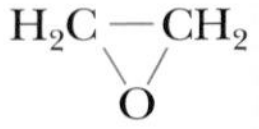

29. ■ Use the vapor pressure curves illustrated here to answer the questions that follow.

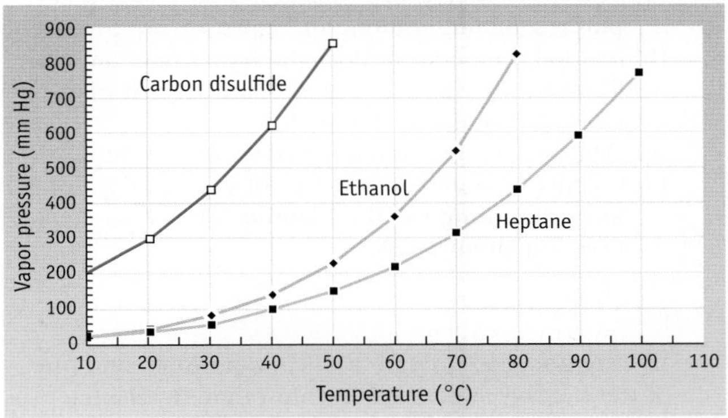

(a) What is the vapor pressure of ethanol, C_2H_5OH, at 60 °C?
(b) Considering only carbon disulfide (CS_2) and ethanol, which has the stronger intermolecular forces in the liquid state?
(c) At what temperature does heptane (C_7H_{16}) have a vapor pressure of 500 mm Hg?
(d) What are the approximate normal boiling points of each of the three substances?
(e) At a pressure of 400 mm Hg and a temperature of 70 °C, is each substance a liquid, a gas, or a mixture of liquid and gas?

30. ■ Which of the following salts will you most likely find as hydrated solids? Explain your reasoning.
(a) $Fe(NO_3)_3$ (c) NaCl
(b) $CoCl_2$ (d) $Al(NO_3)_3$

31. ■ Rank the following compounds in order of increasing molar enthalpy of vaporization: CH_3OH, C_2H_6, HCl.

32. ■ Rank the following molecules in order of increasing intermolecular forces: CH_3Cl, HCO_2H (formic acid), and CO_2.

33. Mercury and many of its compounds are dangerous poisons if breathed, swallowed, or even absorbed through the skin. The liquid metal has a vapor pressure of 0.00169 mm Hg at 24 °C. If the air in a small room is saturated with mercury vapor, how many atoms of mercury vapor occur per cubic meter?

34. ▲ ■ The following data are the equilibrium vapor pressure of limonene, $C_{10}H_{16}$, at various temperatures. (Limonene is used as a scent in commercial products.)

Temperature (°C)	Vapor Pressure (mm Hg)
14.0	1.0
53.8	10.
84.3	40.
108.3	100.
151.4	400.

(a) Plot these data as ln P versus $1/T$ so that you have plot resembling the one in Figure 12.18.
(b) At what temperature does the liquid have an equilibrium vapor pressure of 250 mm Hg? At what temperature is it 650 mm Hg?
(c) What is the normal boiling point of limonene?
(d) Calculate the molar enthalpy of vaporization for limonene using the the Clausius–Clapeyron equation (Equation 12.2).

In the Laboratory

35. You are going to prepare a silicone polymer, and one of the starting materials is dichlorodimethylsilane, $SiCl_2(CH_3)_2$. You need its normal boiling point and so measure equilibrium vapor pressures at various temperatures.

Temperature (°C)	Vapor Pressure (mm Hg)
−0.4	40.
+17.5	100.
51.9	400.
70.3	760.

(a) What is the normal boiling point of dichlorodimethylsilane?
(b) Plot these data as ln P versus $1/T$ so that you have a plot resembling the one in Figure 12.18. At what temperature does the liquid have an equilibrium vapor pressure of 250 mm Hg? At what temperature is it 650 mm Hg?
(c) Calculate the molar enthalpy of vaporization for dichlorodimethylsilane using the the Clausius–Clapeyron equation (Equation 12.2).

36. A "hand boiler" can be purchased in toy stores or at science supply companies. If you cup your hand around the bottom bulb, the volatile liquid in the boiler boils, and the liquid moves to the upper chamber. Using your knowledge of kinetic molecular theory and intermolecular forces, explain how the hand boiler works.

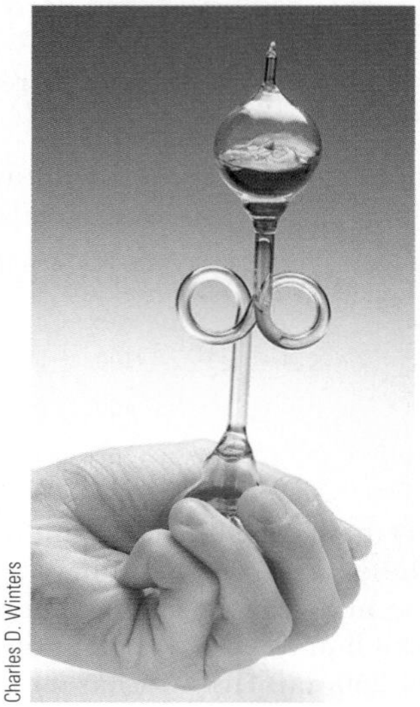

37. ▲ The photos below illustrate an experiment you can do yourself. Place 10 mL of water in an empty soda can, and heat the water to boiling. Using tongs or pliers, turn the can over in a pan of cold water, making sure the opening in the can is below the water level in the pan.
(a) Describe what happens, and explain it in terms of the subject of this chapter.

(a) (b)

(b) Prepare a molecular level sketch of the situation inside the can before heating and after heating (but prior to inverting the can).

38. ■ If you place 1.0 L of ethanol (C_2H_5OH) in a room that is 3.0 m long, 2.5 m wide, and 2.5 m high, will all the alcohol evaporate? If some liquid remains, how much will there be? The vapor pressure of ethyl alcohol at 25 °C is 59 mm Hg, and the density of the liquid at this temperature is 0.785 g/cm^3.

▲ more challenging ■ in OWL Blue-numbered questions answered in Appendix O

Summary and Conceptual Questions

The following questions may use concepts from this and previous chapters.

39. ■ Acetone, CH_3COCH_3, is a common laboratory solvent. It is usually contaminated with water, however. Why does acetone absorb water so readily? Draw molecular structures showing how water and acetone can interact. What intermolecular force(s) is (are) involved in the interaction?

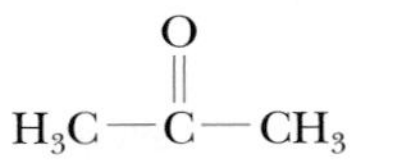

40. Cooking oil floats on top of water. From this observation, what conclusions can you draw regarding the polarity or hydrogen-bonding ability of molecules found in cooking oil?

41. ■ Liquid ethylene glycol, $HOCH_2CH_2OH$, is one of the main ingredients in commercial antifreeze. Do you predict its viscosity to be greater or less than that of ethanol, CH_3CH_2OH?

42. Liquid methanol, CH_3OH, is placed in a glass tube. Is the meniscus of the liquid concave or convex? Explain briefly.

43. Account for these facts:
(a) Although ethanol (C_2H_5OH) (bp, 80 °C) has a higher molar mass than water (bp, 100 °C), the alcohol has a lower boiling point.
(b) Mixing 50 mL of ethanol with 50 mL of water produces a solution with a volume slightly less than 100 mL.

44. Rationalize the observation that $CH_3CH_2CH_2OH$, 1-propanol, has a boiling point of 97.2 °C, whereas a compound with the same empirical formula, methyl ethyl ether ($CH_3CH_2OCH_3$), boils at 7.4 °C.

45. Cite two pieces of evidence to support the statement that water molecules in the liquid state exert considerable attractive force on one another.

46. During thunderstorms in the Midwest, very large hailstones can fall from the sky. (Some are the size of golf balls!) To preserve some of these stones, we put them in the freezer compartment of a frost-free refrigerator. Our friend, who is a chemistry student, tells us to use an older model that is not frost-free. Why?

47. Refer to Figure 12.12 to answer the following questions:
(a) Of the three hydrogen halides (HX), which has the largest total intermolecular force?
(b) Why are the dispersion forces greater for HI than for HCl?
(c) Why are the dipole–dipole forces greater for HCl than for HI?
(d) Of the seven molecules in Figure 12.12, which involves the largest dispersion forces? Explain why this is reasonable.

48. ▲ What quantity of energy is evolved (in joules) when 1.00 mol of liquid ammonia cools from −33.3 °C (its boiling point) to −43.3 °C? (The specific heat capacity of liquid NH_3 is 4.70 J/g · K.) Compare this with the quantity of heat evolved by 1.00 mol of liquid water cooling by exactly 10 °C. Which evolves more heat per mole on cooling 10 °C, liquid water or liquid ammonia? *(The underlying reason for the difference in heat evolved is scientifically illuminating and interesting. You can learn more by searching the Internet for specific heat capacity and its dependence on molecular properties.)*

49. A fluorocarbon, CF_4, has a critical temperature of −45.7 °C and a critical pressure of 37 atm. Are there any conditions under which this compound can be a liquid at room temperature? Explain briefly.

50. ▲ ■ The figure below is a plot of vapor pressure versus temperature for dichlorodifluoromethane, CCl_2F_2. The enthalpy of vaporization of the liquid is 165 kJ/g, and the specific heat capacity of the liquid is about 1.0 J/g · K.

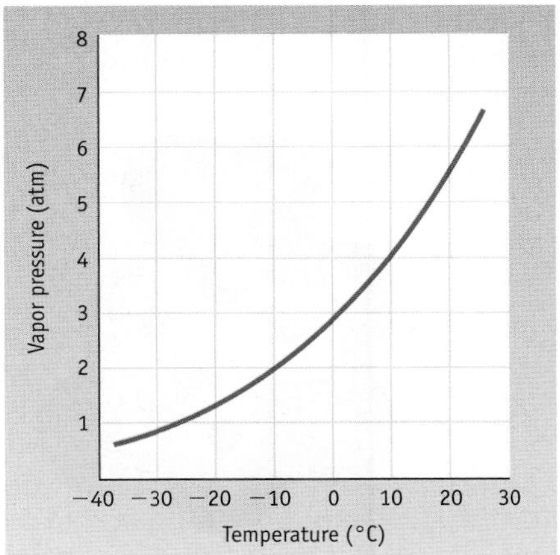

(a) What is the approximate normal boiling point of CCl_2F_2?
(b) A steel cylinder containing 25 kg of CCl_2F_2 in the form of liquid and vapor is set outdoors on a warm day (25 °C). What is the approximate pressure of the vapor in the cylinder?
(c) The cylinder valve is opened, and CCl_2F_2 vapor gushes out of the cylinder in a rapid flow. Soon, however, the flow becomes much slower, and the outside of the cylinder is coated with ice frost. When the valve is closed and the cylinder is reweighed, it is found that 20 kg of CCl_2F_2 is still in the cylinder. Why is the flow fast at first? Why does it slow down long before the cylinder is empty? Why does the outside become icy?
(d) Which of the following procedures would be effective in emptying the cylinder rapidly (and safely)? (1) Turn the cylinder upside down, and open the valve. (2) Cool the cylinder to −78 °C in dry ice, and open the valve. (3) Knock off the top of the cylinder, valve and all, with a sledge hammer.

51. Acetaminophen is used in analgesics. A model of the molecule is shown here with its electrostatic potential surface. Where are the most likely sites for hydrogen bonding?

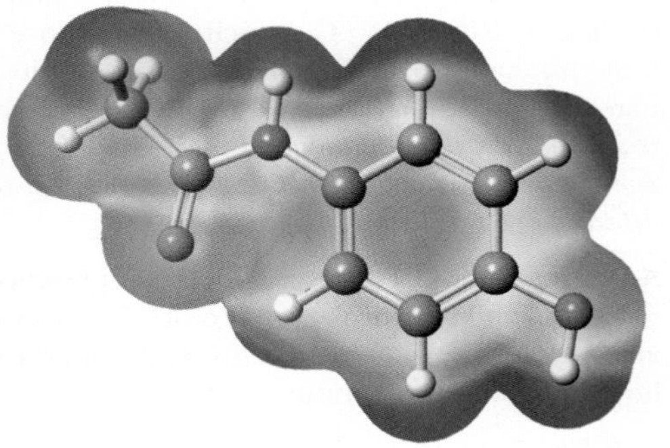

52. Shown here are models of two bases in DNA with the electrostatic potential surfaces: cytosine and guanine. What sites in these molecules are involved in hydrogen bonding with each other? Draw molecular structures showing how cytosine can hydrogen bond with guanine.

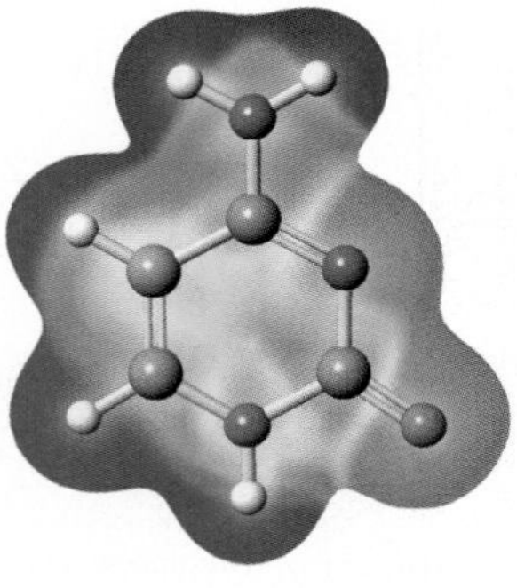

Cytosine

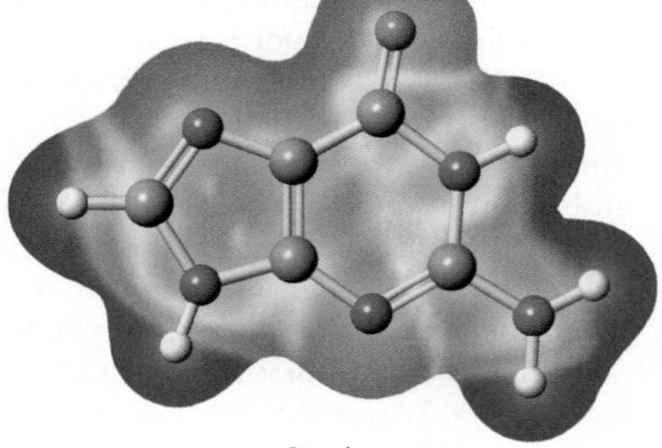

Guanine

53. List four properties of liquids that are directly determined by intermolecular forces.

54. List the following ions in order of hydration energies: Na^+, K^+, Mg^{2+}, Ca^{2+}. Explain how you determined this order.

55. Compare the boiling points of the various isomeric hydrocarbons shown in the table below. Notice the relationship between boiling point and structure; branched-chain hydrocarbons have lower boiling points than the unbranched isomer. Speculate on possible reasons for this trend. Why might the intermolecular forces be slightly different in these compounds?

Compound	Boiling point (°C)
Hexane	68.9
3-methylpentane	63.2
2-methylpentane	60.3
2,3-dimethylbutane	58.0
2,2-dimethylbutane	49.7

56. A 8.82 g sample of Br_2 is placed in an evacuated 1.00 L flask and heated to 58.8 °C, the normal boiling point of bromine. Describe the contents of the flask under these conditions.

57. Polarizability is defined as the extent to which the electron cloud surrounding an atom or molecule can be distorted by an external charge. Rank the halogens (F_2, Cl_2, Br_2, I_2) and the noble gases (He, Ne, Ar, Kr, Xe) in order of polarizability (from least polarizable to most polarizable). What properties of these substances could be used to determine this ranked order?

58. In which of the following organic molecules might we expect hydrogen bonding to occur?
(a) methyl acetate, $CH_3CO_2CH_3$
(b) acetaldehyde (ethanal), CH_3CHO
(c) acetone (2-propanone) (see Question 8)
(d) benzoic acid ($C_6H_5CO_2H$)
(e) acetamide (CH_3CONH_2 an amide formed from acetic acid and ammonia)
(f) N,N-dimethylacetamide [$CH_3CON(CH_3)_2$, an amide formed from acetic acid and dimethylamine]

59. A pressure cooker (a kitchen appliance) is a pot on which the top seals tightly, allowing pressure to build up inside. You put water in the pot and heat it to boiling. At the higher pressure, water boils at a higher temperature and this allows food to cook at a faster rate. Most pressure cookers have a setting of 15 psi, which means that the pressure in the pot is 15 psi above atmospheric pressure (1 atm = 14.70 psi). Use the Clausius-Clapeyron equation to calculate the temperature at which water boils in the pressure cooker.

60. Vapor pressures of $NH_3(\ell)$ at several temperatures are given in the table below. Use this information to calculate the enthalpy of vaporization of ammonia.

Temperature (°C)	Vapor Pressure (atm)
−68.4	0.132
−45.4	0.526
−33.6	1.000
−18.7	2.00
4.7	5.00
25.7	10.00
50.1	20.00

61. Chemists sometimes carry out reactions in liquid ammonia as a solvent. With adequate safety protection these reactions can be done above ammonia's boiling point in a sealed, thick-walled glass tube. If the reaction is being carried out at 20 °C, what is the pressure of ammonia inside the tube? (Use data from the previous question to answer this question.)

62. The data in the following table was used to create the graph shown below (vp = vapor pressure of ethanol (CH_3CH_2OH) expressed in mm Hg, T = kelvin temperature)

ln(vp)	$1/T$ (K^{-1})
2.30	0.00369
3.69	0.00342
4.61	0.00325
5.99	0.00297
6.63	0.00285

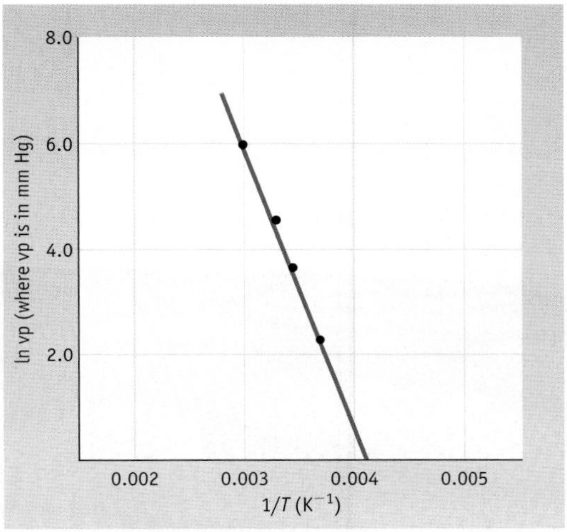

(a) Derive an equation for the straight line in this graph.
(b) Describe in words how to use the graph to determine the enthalpy of vaporization of ethanol.
(c) Calculate the vapor pressure of ethanol at 0.00 °C and at 100 °C.

13 The Chemistry of Solids

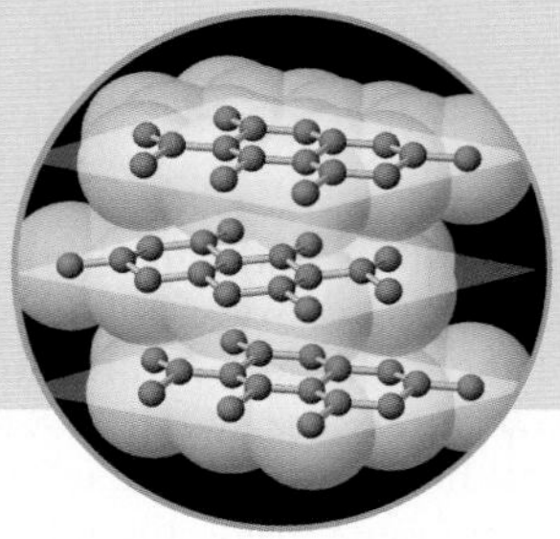

Graphite is composed of sheets of carbon atoms in six-member rings.

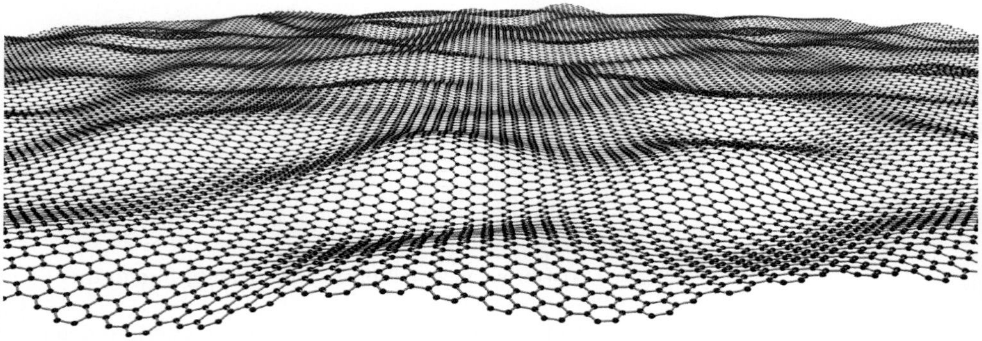

Graphene is a single sheet of six-member carbon rings. This latest material in the world of carbon chemistry has unusual electrical properties.

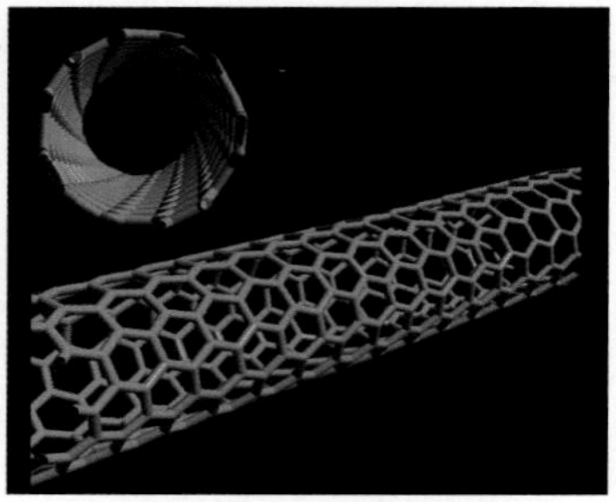

Carbon nanotubes are composed of six-member carbon rings.

Graphite to Graphene

One of the most interesting developments in chemistry in the last 20 years has been the discovery of new forms of carbon. First, there were buckyballs and then single-wall and multi-wall carbon nanotubes.

Common graphite, from which your pencil lead is made, consists of six-member rings of carbon atoms connected in sheets, and the sheets stack one on top of another like cards in a deck. But if carbon compounds are heated under the right conditions, the carbon atoms assemble into sheets, and the sheets close on themselves to form tubes. These are called **nanotubes** because the tubes are only a few nanometers in diameter. Sometimes they are single tubes, and other times there are tubes within tubes. Carbon nanotubes are at least 100 times stronger than steel but only one sixth as dense, and they conduct heat and electricity far better than copper. There has been enormous interest in their commercial applications, but there has also been difficulty in making them with consistent properties.

Now there is **graphene**, a single sheet of six-member carbon atoms. Researchers in England discovered them in a simple way: put a flake of graphite on Scotch tape, fold the tape over, and then pull it apart. The graphite layers come apart, and, if you do it enough times, only one layer—one C atom thick!—is left on the tape. This is clearly not the way to make graphene commercially, but methods have since been developed to make it in larger amounts. And now researchers are looking at ways to make graphene sheets in specific shapes, and to use them as transistors and other electronic devices.

Questions:

1. Based on a C—C distance of 139 pm, what is the side-to-side dimension of a planar, C_6 ring?
2. If a graphene sheet has a width of 1.0 micrometer, how many C_6 rings are joined across the sheet?
3. Estimate the thickness of a sheet of graphene (in pm). How did you determine this value?

Answers to these questions are in Appendix Q.

Chapter Goals

See Chapter Goals Revisited (page 610) for Study Questions keyed to these goals and assignable in OWL.

- Understand cubic unit cells.
- Relate unit cells for ionic compounds to formulas.
- Describe the properties of solids.
- Understand the nature of phase diagrams.

Chapter Outline

Many kinds of solids exist in the world around us (Figure 13.1 and Table 13.1). As the description of graphene shows, solid-state chemistry is one of the booming areas of science, especially because it relates to the development of interesting new materials. As we describe various kinds of solids, we hope to provide a glimpse of the reasons this area is exciting.

Chemistry Now™

Throughout the text this icon introduces an opportunity for self-study or to explore interactive tutorials by signing in at **www.thomsonedu.com/login**.

13.1 Crystal Lattices and Unit Cells

Module 18

In both gases and liquids, molecules move continually and randomly, and they rotate and vibrate as well. Because of this movement, an orderly arrangement of molecules in the gaseous or liquid state is not possible. In solids, however, the molecules, atoms, or ions cannot change their relative positions (although they vibrate and occasionally rotate). Thus, a regular, repeating pattern of atoms or molecules within the structure—a long-range order—is a characteristic of most solids. The beautiful, external (macroscopic) regularity of a crystal of salt (Figure 13.1) suggests it has an internal symmetry.

TABLE 13.1 Structures and Properties of Various Types of Solid Substances

Type	Examples	Structural Units	Forces Holding Units Together	Typical Properties
Ionic	$NaCl$, K_2SO_4, $CaCl_2$, $(NH_4)_3PO_4$	Positive and negative ions; no discrete molecules	Ionic; attractions among charges on positive and negative ions	Hard; brittle; high melting point; poor electric conductivity as solid, good as liquid; often water-soluble
Metallic	Iron, silver, copper, other metals and alloys	Metal atoms (positive metal ions with delocalized electrons)	Metallic; electrostatic attraction among metal ions and electrons	Malleable; ductile; good electric conductivity in solid and liquid; good heat conductivity; wide range of hardness and melting points
Molecular	H_2, O_2, I_2, H_2O, CO_2, CH_4, CH_3OH, CH_3CO_2H	Molecules	Dispersion forces, dipole–dipole forces, hydrogen bonds	Low to moderate melting points and boiling points; soft; poor electric conductivity in solid and liquid
Network	Graphite, diamond, quartz, feldspars, mica	Atoms held in an infinite two- or three-dimensional network	Covalent; directional electron-pair bonds	Wide range of hardness and melting points (three-dimensional bonding $>$ two-dimensional bonding); poor electric conductivity, with some exceptions
Amorphous	Glass, polyethylene, nylon	Covalently bonded networks with no long-range regularity	Covalent; directional electron-pair bonds	Noncrystalline; wide temperature range for melting; poor electric conductivity, with some exceptions

FIGURE 13.1
Some common solids.

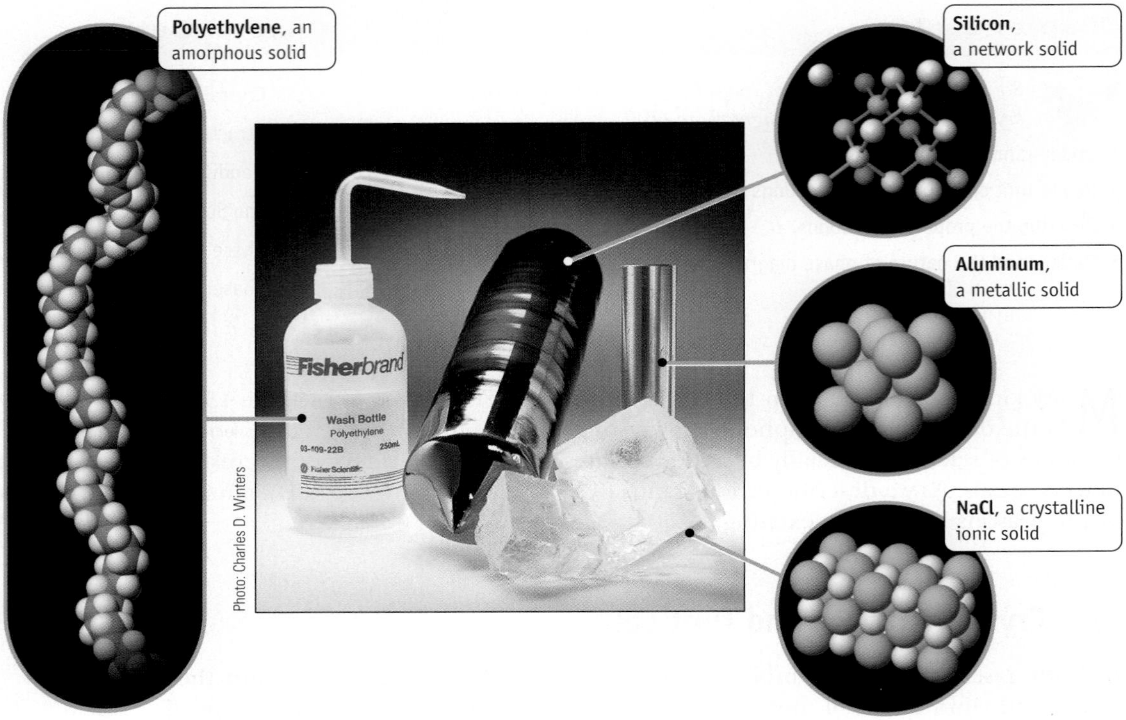

Structures of solids can be described as three-dimensional lattices of atoms, ions, or molecules. For a crystalline solid, we can identify the **unit cell**, the smallest repeating unit that has all of the symmetry characteristic of the way the atoms, ions, or molecules are arranged in the solid.

To understand unit cells, consider first a two-dimensional lattice model, the repeating pattern of circles shown in Figure 13.2. The yellow square at the left is a unit cell because the overall pattern can be created from a group of these cells by joining them edge to edge. It is also a requirement that unit cells reflect the stoichiometry of the solid. Here, the square unit cell at the left contains one smaller sphere and one fourth of each of the four larger circles, giving a total of one small and one large circle per two-dimensional unit cell.

You may recognize that it is possible to draw other unit cells for this two-dimensional lattice. One option is the square in the middle of Figure 13.2 that fully encloses a single large circle and parts of small circles that add up to one net small circle. Yet another possible unit cell is the parallelogram at the right. Other unit

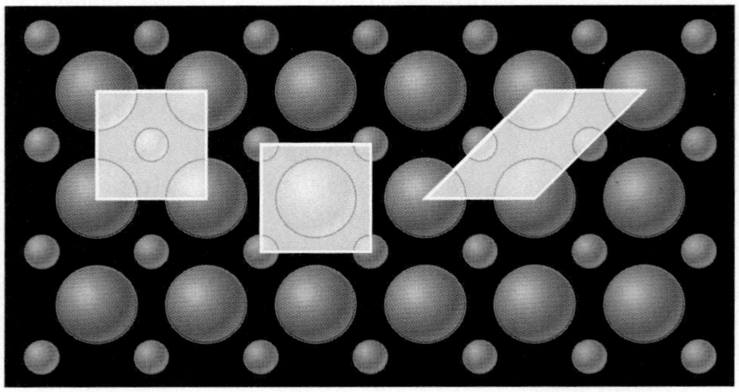

FIGURE 13.2 Unit cells for a flat, two-dimensional solid made from circular "atoms." A lattice can be represented as being built from repeating unit cells. This two-dimensional lattice can be built by translating the unit cells throughout the plane of the figure. Each cell must move by the length of one side of the unit cell. In this figure, all unit cells contain a net of one large circle and one small circle. Be sure to notice that several unit cells are possible, with two of the most obvious being squares.

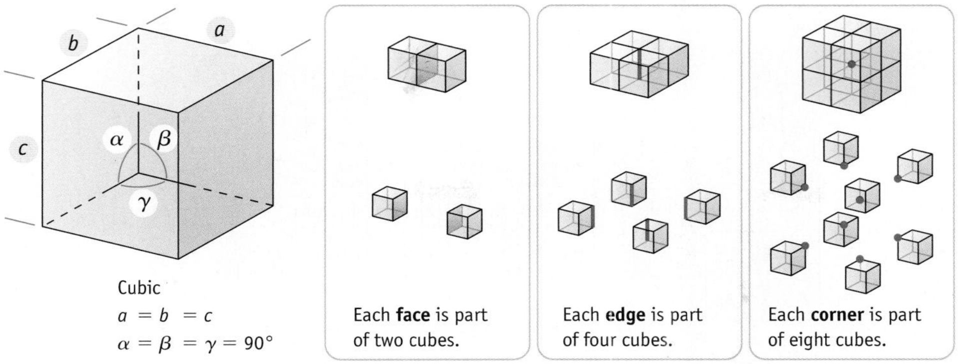

(a) **(b)**

FIGURE 13.3 Cubic unit cells. (a) The cube is one of the seven basic unit cells that describe crystal systems. In a cube, all sides are of equal length, and all angles are 90°. In noncubic unit cells, the angles are not necessarily 90°, and the sides are not equal in length. (b) Stacking cubes to build a crystal lattice. Each crystal face is part of two cubes; each edge is part of four cubes; and each corner is part of eight cubes.

cells are possible, but it is conventional to draw unit cells in which atoms or ions are placed at the **lattice points;** that is, at the corners of the cube or other geometric object that constitutes the unit cell.

The three-dimensional lattices of solids can be built by assembling three-dimensional unit cells much like building blocks (Figure 13.3). The assemblage of these three-dimensional unit cells defines the **crystal lattice.**

To construct crystal lattices, nature uses seven three-dimensional unit cells. They differ from one another in that their sides have different relative lengths and their edges meet at different angles. The simplest of the seven crystal lattices is the **cubic unit cell,** a cell with edges of equal length that meet at 90° angles. We shall look in detail at just this structure, not only because cubic unit cells are easily visualized but also because they are commonly encountered.

Within the cubic class, three cell symmetries occur: **primitive cubic (pc), body-centered cubic (bcc),** and **face-centered cubic (fcc)** (Figure 13.4). All three have

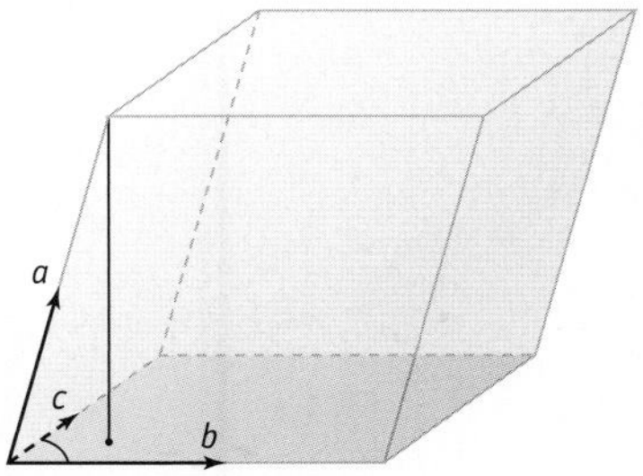

Unit cells. Of the possible unit cells, all are parallelpipeds (except for the hexagonal cell), figures in which opposite sides are parallel. In a cube, all angles (a-o-c, a-o-b, and c-o-b; where o is the origin) are 90°, and all sides are equal. In other cells, the angles and sides may be the same or different. For example, in a tetragonal cell, the angles are 90°, but $a = b \neq c$. In a triclinic cell, the sides have different lengths, the angles are different, and none equals 90°.

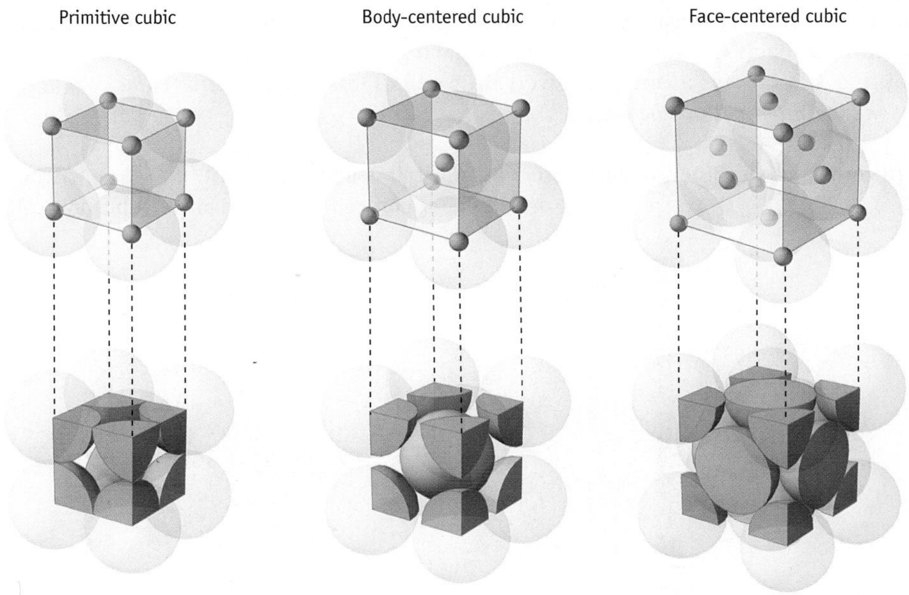

FIGURE 13.4 The three cubic unit cells. The top row shows the lattice points of the three cells, and the bottom row shows the same cells using space-filling spheres. The spheres in each figure represent identical atoms (or ions) centered on the lattice points. Because eight unit cells share a corner atom, only $\frac{1}{8}$ of each corner atom lies within a given unit cell; the remaining $\frac{7}{8}$ lies in seven other unit cells. Because each face of a fcc unit cell is shared with another unit cell, one half of each atom in the face of a face-centered cube lies in a given unit cell, and the other half lies in the adjoining cell.

FIGURE 13.5 Metals use four different unit cells. Three are based on the cube, and the fourth is the hexagonal unit cell (see page 595). (Many metals can crystallize in more than one structure.)

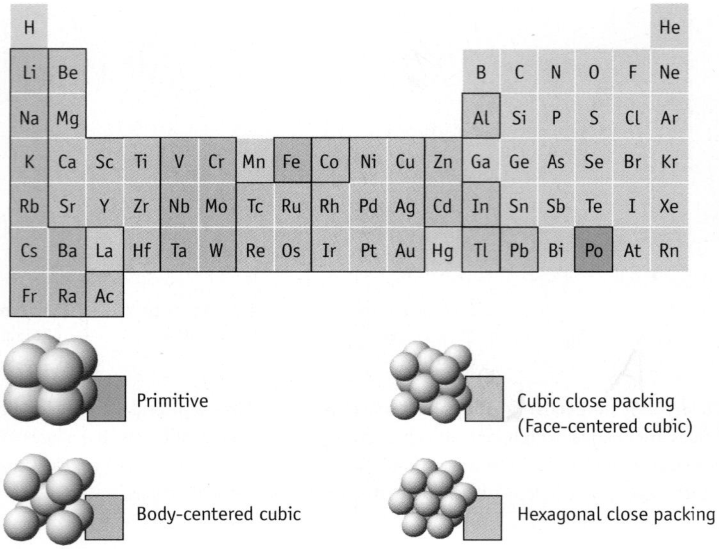

identical atoms, molecules, or ions at the corners of the cubic unit cell. The bcc and fcc arrangements, however, differ from the primitive cube in that they have additional particles at other locations. The bcc structure is called "body-centered" because it has an additional particle, of the same type as those at the corners, at the center of the cube. The fcc arrangement is called "face-centered" because it has a particle, of the same type as the corner atoms, in the center of each of the six faces of the cube. Examples of each structure are found among the crystal lattices of the metals (Figure 13.5). The alkali metals, for example, are body-centered cubic, whereas nickel, copper, and aluminum are face-centered cubic. Notice that only one metal, polonium, has a primitive cubic lattice.

When the cubes pack together to make a three-dimensional crystal of a metal, the atom at each corner is shared among eight cubes (Figures 13.3, 13.4, and 13.6a). Because of this, only one eighth of each corner atom is actually within a given unit cell. Furthermore, because a cube has eight corners, and because one eighth of the atom at each corner "belongs to" a particular unit cell, the corner atoms contribute a net of one atom to a given unit cell. Thus, *the primitive cubic arrangement has one net atom within the unit cell.*

(8 corners of a cube)($\frac{1}{8}$ of each corner atom within a unit cell) =
1 net atom per unit cell for the primitive cubic unit cell

FIGURE 13.6 Atom sharing at cube corners and faces. (a) In any cubic lattice, each corner particle is shared equally among eight cubes, so one eighth of the particle is within a particular cubic unit cell. (b) In a face-centered lattice, each particle on a cube face is shared equally between two unit cells. One half of each particle of this type is within a given unit cell.

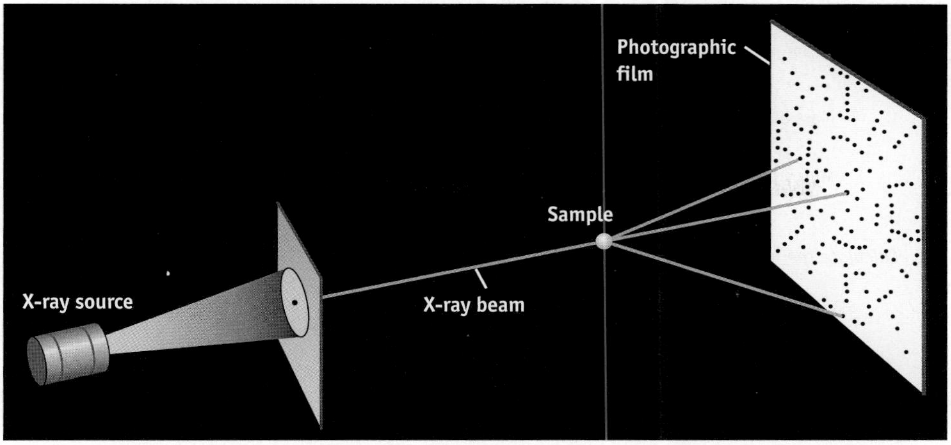

FIGURE 13.7 X-ray crystallography. In the x-ray diffraction experiment, a beam of x-rays is directed at a crystalline solid. The photons of the x-ray beam are scattered by the atoms of the solid. The scattered x-rays are detected by a photographic film or an electronic detector, and the pattern of scattered x-rays is related to the locations of the atoms or ions in the crystal.

In contrast to the primitive cubic lattice, a body-centered cube has an additional atom wholly within the unit cell at the cube's center. The center particle is present in addition to those at the cube corners, so *the body-centered cubic arrangement has a net of two atoms within the unit cell.*

In a face-centered cubic arrangement, there is an atom on each of the six faces of the cube in addition to those at the cube corners. One half of each atom on a face belongs to a given unit cell (Figure 13.6b). Three net particles are therefore contributed by the particles on the faces of the cube:

$$(6 \text{ faces of a cube})(\tfrac{1}{2} \text{ of an atom within a unit cell}) = \\ 3 \text{ net face-centered atoms within a face-centered cubic unit cell}$$

Thus, *the face-centered cubic arrangement has a net of four atoms within the unit cell,* one contributed by the corner atoms and another three contributed by the atoms centered in the six faces.

An experimental technique, x-ray crystallography, can be used to determine the structure of a crystalline substance (Figure 13.7). Once the structure is known, the information can be combined with other experimental information to calculate such useful parameters as the radius of an atom (Study Questions 13.7–13.10).

Chemistry Now™

Sign in at **www.thomsonedu.com/login** and go to Chapter 13 Contents to see Screen 13.2 for a self-study module on **crystal lattices.**

Charles D. Winters

Aluminum metal. The metal has a face-centered cubic unit cell with a net of four Al atoms in each unit cell.

EXAMPLE 13.1 Determining an Atom Radius from Lattice Dimensions

Problem Aluminum has a density of 2.699 g/cm^3, and the atoms are packed in a face-centered cubic crystal lattice. What is the radius of an aluminum atom?

Strategy Our strategy for solving this problem is as follows:

1. Find the mass of a unit cell from the knowledge that it is face-centered cubic.
2. Combine the density of aluminum with the mass of the unit cell to find the cell volume.
3. Find the length of a side of the unit cell from its volume.
4. Calculate the atom radius from the edge dimension.

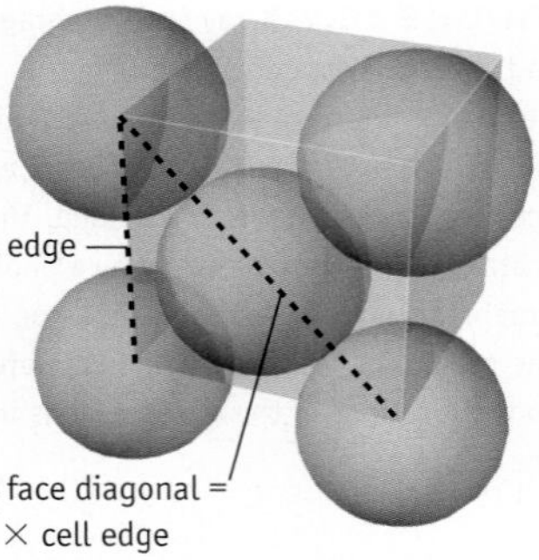

One face of a face-centered cubic unit cell. This shows the cell face diagonal, $\sqrt{2}$ × edge, is equal to four times the radius of the atoms in the lattice.

Solution

1. *Calculate the mass of the unit cell.*

$$\text{Mass of 1 Al atom} = \left(\frac{26.98\text{ g}}{1\text{ mol}}\right)\left(\frac{1\text{ mol}}{6.022 \times 10^{23}\text{ atoms}}\right) = 4.480 \times 10^{-23}\text{ g/atom}$$

$$\text{Mass of unit cell} = \left(\frac{4.480 \times 10^{-23}\text{ g}}{1\text{ Al atom}}\right)\left(\frac{4\text{ Al atoms}}{1\text{ unit cell}}\right) = 1.792 \times 10^{-22}\text{ g/unit cell}$$

2. *Calculate the volume of the unit cell.*

$$\text{Volume of unit cell} = \left(\frac{1.792 \times 10^{-22}\text{ g}}{\text{unit cell}}\right)\left(\frac{1\text{ cm}^3}{2.699\text{ g}}\right) = 6.640 \times 10^{-23}\text{ cm}^3\text{/unit cell}$$

3. *Calculate the length of a unit cell edge.* The length of the unit cell edge is the cube root of the cell volume.

$$\text{Length of unit cell edge} = \sqrt[3]{6.640 \times 10^{-23}\text{ cm}^3} = 4.049 \times 10^{-8}\text{ cm}$$

4. *Calculate the atom radius.* Notice in the model of aluminum in the margin (and in Figure 13.4) that the Al atoms at the cell corners do not touch each other. Rather, the four corner atoms touch the face-centered atom. Thus, the diagonal distance across the face of the cell is equal to four times the Al atom radius.

$$\text{Cell face diagonal} = 4 \times (\text{Al atom radius})$$

The cell diagonal is the hypotenuse of a right isosceles triangle, so, using the Pythagorean theorem,

$$(\text{Diagonal distance})^2 = 2 \times (\text{edge})^2$$

Taking the square root of both sides, we have

$$\begin{aligned}\text{Diagonal distance} &= \sqrt{2} \times (\text{cell edge}) \\ &= \sqrt{2} \times (4.049 \times 10^{-8}\text{ cm}) = 5.727 \times 10^{-8}\text{ cm}\end{aligned}$$

We divide the diagonal distance by 4 to obtain the Al atom radius in cm.

$$\text{Al atom radius} = \frac{5.727 \times 10^{-8}\text{ cm}}{4} = 1.432 \times 10^{-8}\text{ cm}$$

Atomic dimensions are often expressed in picometers, so we convert the radius to that unit.

$$1.432 \times 10^{-8}\text{ cm}\left(\frac{1\text{ m}}{100\text{ cm}}\right)\left(\frac{1\text{ pm}}{1 \times 10^{-12}\text{ m}}\right) = 143.2\text{ pm}$$

This is in excellent agree with the radius in Figure 7.8.

EXERCISE 13.1 Determining an Atom Radius from Lattice Dimensions

Gold has a face-centered unit cell, and its density is 19.32 g/cm^3. Calculate the radius of a gold atom.

EXERCISE 13.2 The Structure of Solid Iron

Iron has a density of 7.8740 g/cm^3, and the radius of an iron atom is 126 pm. Verify that solid iron has a body-centered cubic unit cell. (Be sure to note that the atoms in a body-centered cubic unit cell touch along the diagonal across the cell. They do not touch along the edges of the cell.) (Hint: the diagonal distance across the unit cell is edge × $\sqrt{3}$.)

A Closer Look   Packing Oranges

It is a "rule" that nature does things as efficiently as possible. You know this if you have ever tried to stack some oranges into a pile that doesn't fall over and that takes up as little space as possible. How did you do it? Clearly, the pyramid arrangement below on the right works, whereas the cubic one on the left does not.

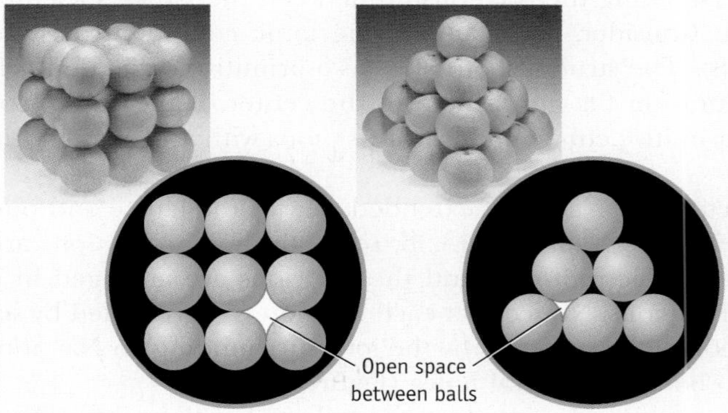

Photos: Charles D. Winters

If you could look inside the pile, you would find that less open space is left in the pyramid stacking than in the cube stacking. Only 52% of the space is filled in the cubic packing arrangement. (If you could stack oranges as a body-centered cube, that would be slightly better; 68% of the space is used.) However, the best method is the pyramid stack, which is really a face-centered cubic arrangement. Oranges, atoms, or ions packed this way occupy 74% of the available space.

To fill three-dimensional space, the most efficient way to pack oranges or atoms is to begin with a hexagonal arrangement of spheres, as in this arrangement of marbles.

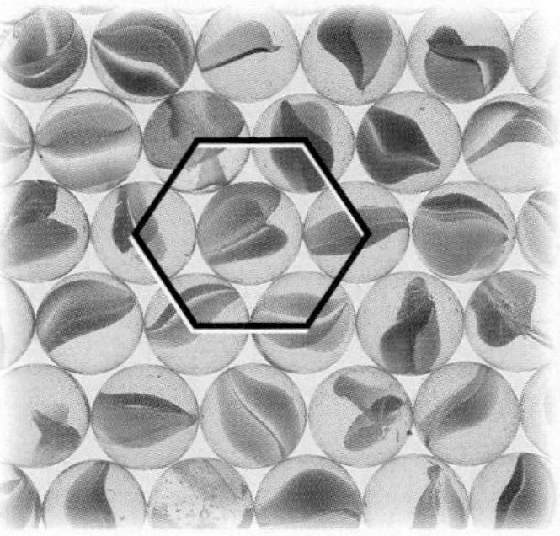

Charles D. Winters

Succeeding layers of atoms or ions are then stacked one on top of the other in two different ways. Depending on the stacking pattern (Figure 1), you will get either a **cubic close-packed (ccp)** or **hexagonal close-packed (hcp)** arrangement.

In the hcp arrangement, additional layers of particles are placed above and below a given layer, fitting into the same depressions on either side of the middle layer. In a three-dimensional crystal, the layers repeat their pattern in the manner ABABAB. . . . Atoms in each A layer are directly above the ones in another A layer; the same holds true for the B layers.

In the ccp arrangement, the atoms of the "top" layer (A) rest in depressions in the middle layer (B), and those of the "bottom" layer (C) are oriented opposite to those in the top layer. In a crystal, the pattern is repeated ABCABCABC. . . . By turning the whole crystal, you can see that the ccp arrangement is the face-centered cubic structure (Figure 2).

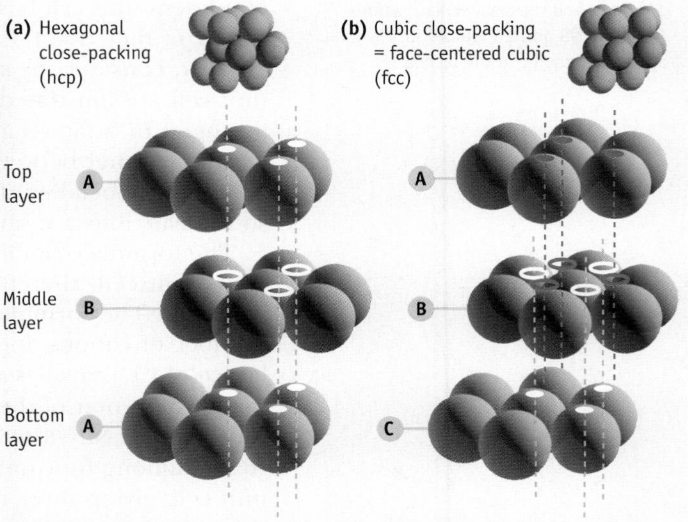

FIGURE 1 Efficient packing. The most efficient ways to pack atoms or ions in crystalline materials are hexagonal close-packing (hcp) and cubic close packing (ccp).

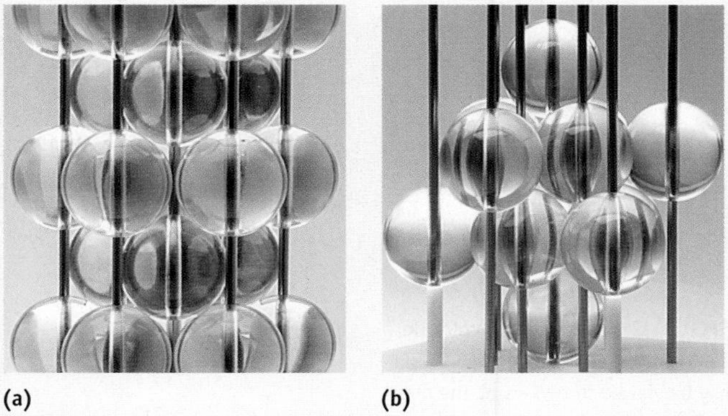

Charles D. Winters

(a) (b)

FIGURE 2 Models of close packing. (a) A model of hexagonal close-packing, where the layers repeat in the order ABABAB. . . . (b) A face-centered unit cell (cubic close-packing), where the layers repeat in the order ABCABC. . . . (A kit from which these models can be built is available from the Institute for Chemical Education at the University of Wisconsin at Madison.)

13.2 Structures and Formulas of Ionic Solids

The lattices of many ionic compounds are built by taking a primitive cubic or face-centered cubic lattice of ions of one type and placing ions of opposite charge in the holes within the lattice. This produces a three-dimensional lattice of regularly placed ions. The smallest repeating unit in these structures is, by definition, the unit cell for the ionic compound.

■ **Lattice Ions and Holes** Chemists usually think of ionic lattices as being built from the larger anions with the smaller cations located in the holes that remain. For NaCl, for example, an fcc lattice is built out of the Cl^- ions (radius = 181 pm), and the smaller Na^+ cations (radius = 98 pm) are placed in appropriate holes in the lattice.

The choice of the lattice and the number and location of the holes that are filled are the keys to understanding the relationship between the lattice structure and the formula of a salt. Consider, for example, the ionic compound cesium chloride, CsCl (Figure 13.8). The structure of CsCl has a primitive cubic unit cell of chloride ions. The cesium ion fits into a hole in the center of the cube. (An equivalent unit cell has a primitive cubic unit cell of Cs^+ ions with a Cl^- ion in the center of the cube.)

Next, consider the structure for NaCl. An extended view of the lattice and one unit cell are illustrated in Figures 13.9a and 13.9b, respectively. The Cl^- ions are arranged in a face-centered cubic unit cell, and the Na^+ ions are arranged in a regular manner between these ions. Notice that each Na^+ ion is surrounded by six Cl^- ions. An octahedral geometry is assumed by the ions surrounding an Na^+ ion, so the Na^+ ions are said to be in **octahedral holes** (Figure 13.9c).

The formula of an ionic compound must always be reflected in the composition of its unit cell; therefore, the formula can always be derived from the unit cell structure. The formula for NaCl can be related to this structure by counting the number of cations and anions contained in one unit cell. A face-centered cubic lattice of Cl^- ions has a net of four Cl^- ions within the unit cell. There is one Na^+ ion in the center of the unit cell, contained totally within the unit cell. In addition, there are 12 Na^+ ions along the edges of the unit cell. Each of these Na^+ ions is shared among four unit cells, so each contributes one fourth of an Na^+ ion to the unit cell, giving three additional Na^+ ions within the unit cell.

$$(1\ Na^+ \text{ ion in the center of the unit cell}) + (\tfrac{1}{4} \text{ of } Na^+ \text{ ion in each edge} \times 12 \text{ edges})$$
$$= \text{net of } 4\ Na^+ \text{ ions in NaCl unit cell}$$

This accounts for all of the ions contained in the unit cell: four Cl^- and four Na^+ ions. Thus, a unit cell of NaCl has a 1:1 ratio of Na^+ and Cl^- ions, as the formula requires.

Another common unit cell again has ions of one type in a face-centered cubic unit cell. Ions of the other type are located in **tetrahedral holes**, wherein each ion is surrounded by four oppositely charged ions. As illustrated in Figure 13.10, there are eight tetrahedral holes in a face-centered unit cell. In ZnS (zinc blende), the sulfide

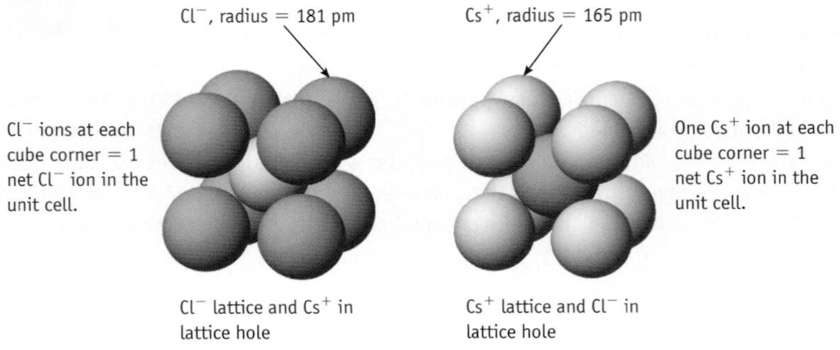

FIGURE 13.8 Cesium chloride (CsCl) unit cell. The unit cell of CsCl may be viewed in two ways. The only requirement is that the unit cell must have a net of one Cs^+ ion and one Cl^- ion. Either way, it is a simple cubic unit cell of ions of one type (Cl^- on the left or Cs^+ on the right). Generally, ionic lattices are assembled by placing the larger ions (here Cl^-) at the lattice points and placing the smaller ions (here Cs^+) in the lattice holes.

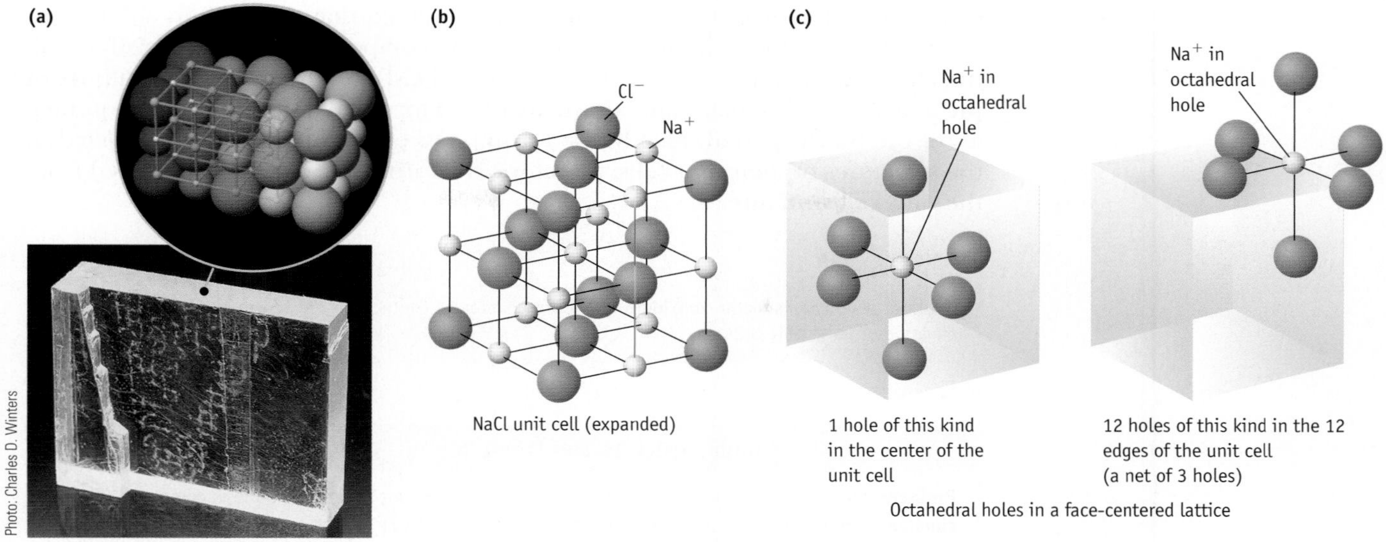

FIGURE 13.9 Sodium chloride. (a) Cubic NaCl is based on a face-centered cubic unit cell of Na^+ and Cl^- ions. (b) An expanded view of a sodium chloride lattice. (The lines represent the connections between lattice points.) The smaller Na^+ ions (silver) are packed into a face-centered cubic lattice of larger Cl^- ions (yellow). (c) A close-up view of the octahedral holes in the lattice.

ions (S^{2-}) form a face-centered cubic unit cell. The zinc ions (Zn^{2+}) then occupy one half of the tetrahedral holes, and each Zn^{2+} ion is surrounded by four S^{2-} ions. The unit cell consists of a net of four S^{2-} ions and four Zn^{2+} ions, which are contained wholly within the unit cell. This 1:1 ratio of the ions is reflected in the formula.

In summary, compounds with the formula MX commonly form one of three possible crystal structures:

1. M^{n+} ions occupying all the cubic holes of a primitive cubic X^{n-} lattice. Example, CsCl
2. M^{n+} ions in all the octahedral holes in a face-centered cubic X^{n-} lattice. Example, NaCl
3. M^{n+} ions occupying half of the tetrahedral holes in a face-centered cube lattice of X^{n-} ions. Example, ZnS

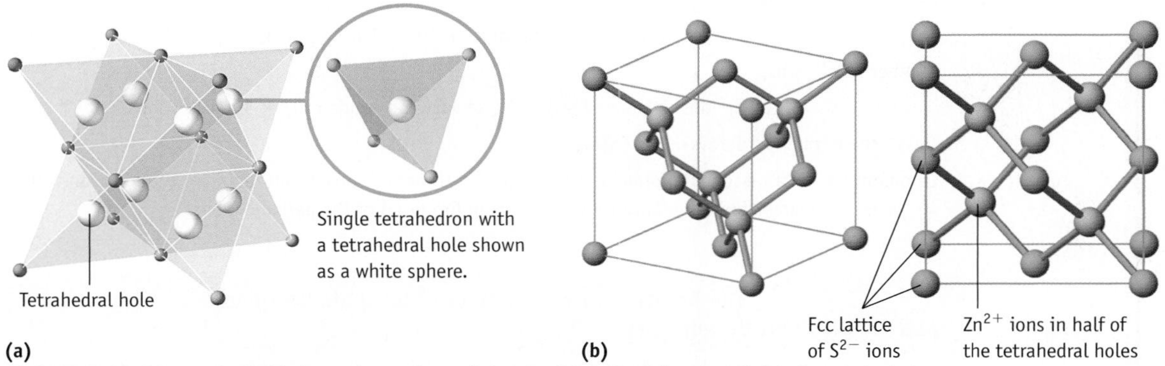

FIGURE 13.10 Tetrahedral holes and two views of the ZnS (zinc blende) unit cell. (a) The tetrahedral holes in a face-centered cubic lattice. (b) This unit cell is an example of a face-centered cubic lattice of ions of one type with ions of the opposite type in one half of the tetrahedral holes.

Chemists and geologists in particular have observed that the sodium chloride or "rock salt" structure is adopted by many ionic compounds, most especially by all the alkali metal halides (except CsCl, CsBr, and CsI), all the oxides and sulfides of the alkaline earth metals, and all the oxides of formula MO of the transition metals of the fourth period. Finally, the formulas of compounds must be reflected in the structures of their unit cells; therefore, the formula can always be derived from the unit cell structure.

Chemistry Now™

Sign in at **www.thomsonedu.com/login** and go to Chapter 13 Contents to see Screen 13.3 to view an animation of **ionic unit cells.**

EXAMPLE 13.2 Ionic Structure and Formula

Problem One unit cell of the mineral perovskite is illustrated here. This compound is composed of calcium and titanium cations and oxide anions. Based on the unit cell, what is the formula of perovskite?

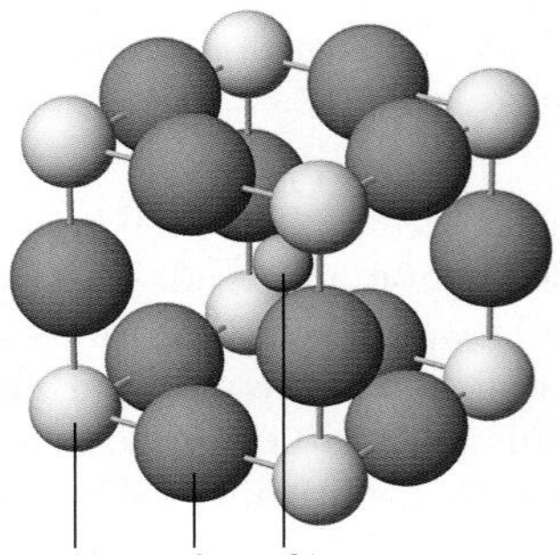

Strategy Identify the ions present in the unit cell and their locations within the unit cell. Decide on the net number of ions of each kind in the cell.

Solution The unit cell has Ti^{4+} ions at the corners of the cubic unit cell, a calcium ion in the center of the cell, and oxide ions along the edges.

Number of Ti^{4+} ions:

$$(8\ Ti^{4+} \text{ ions at cube corners}) \times (1/8 \text{ of each ion inside unit cell}) = 1 \text{ net } Ti^{4+} \text{ ion}$$

Number of Ca^{2+} ions:

$$\text{One ion is in the cube center} = 1 \text{ net } Ca^{2+} \text{ ion}$$

Number of O^{2-} ions:

$$(12\ O^{2-} \text{ ions in cube edges}) \times (1/4 \text{ of each ion inside cell}) = 3 \text{ net } O^{2-} \text{ ions}$$

Thus, the formula of perovskite is $CaTiO_3$.

Comment This is a reasonable formula. A Ca^{2+} ion and three O^{2-} ions would require a titanium ion with a 4+ charge, a reasonable value because titanium is in Group 4B of the periodic table.

EXAMPLE 13.3 The Relation of the Density of an Ionic Compound and its Unit Cell Dimensions

Problem Magnesium oxide has a face-centered cubic unit cell of oxide ions with magnesium ions in octahedral holes. If the radius of Mg^{2+} is 79 pm and the density of MgO is 3.56 g/cm^3, what is the radius of the oxide ion?

Strategy The unit cell contains 4 MgO units, so we can calculate the mass of the unit cell. Combining the unit cell mass and the density of the solid gives us the unit cell volume, from which we can find the length of one edge of the unit cell. The edge of the unit cell is twice the radius of a Mg^{2+} ion (2 times 79 pm) plus twice the radius of an O^{2-} ion (the unknown).

Solution

1. *Calculate the mass of the unit cell.* An ionic compound of formula MX and based on a face-centered cubic lattice of X^- ions with M^+ ions in the octahedral holes has 4 MX unit per unit cell.

$$\text{Unit cell mass} = \left(\frac{40.31 \text{ g}}{1 \text{ mol MgO}}\right)\left(\frac{1 \text{ mol MgO}}{6.022 \times 10^{23} \text{ units of MgO}}\right)\left(\frac{4 \text{ MgO units}}{1 \text{ unit cell}}\right)$$
$$= 2.677 \times 10^{-22} \text{ g/unit cell}$$

2. *Calculate the volume of the unit cell from the mass and density.*

$$\text{Unit cell volume} = \left(\frac{2.667 \times 10^{-22} \text{ g}}{\text{unit cell}}\right)\left(\frac{1 \text{ cm}^3}{3.56 \text{ g}}\right) = 7.49 \times 10^{-23} \text{ cm}^3/\text{unit cell}$$

3. *Calculate the edge dimension of the unit cell in pm.*

$$\text{Unit cell edge} = (7.49 \times 10^{-23} \text{ cm}^3)^{1/3} = 4.22 \times 10^{-8} \text{ cm}$$

$$\text{Unit cell edge} = 4.22 \times 10^{-8} \text{ cm}\left(\frac{1 \text{ m}}{100 \text{ cm}}\right)\left(\frac{1 \times 10^{12} \text{ pm}}{1 \text{ m}}\right) = 422 \text{ pm}$$

4. *Calculate the oxide ion radius.*

One face of the MgO unit cell is shown in the margin. The O^{2-} ions define the lattice, and the Mg^{2+} and O^{2-} ions along the cell edge just touch one another. This means that one edge of the cell is equal to one O^{2-} radius (x) plus twice the Mg^{2+} radius plus one more O^{2-} radius.

$$\text{MgO unit cell edge} = x \text{ pm} + 2(79 \text{ pm}) + x \text{ pm} = 422 \text{ pm}$$
$$x = \text{oxide ion radius} = 132 \text{ pm}$$

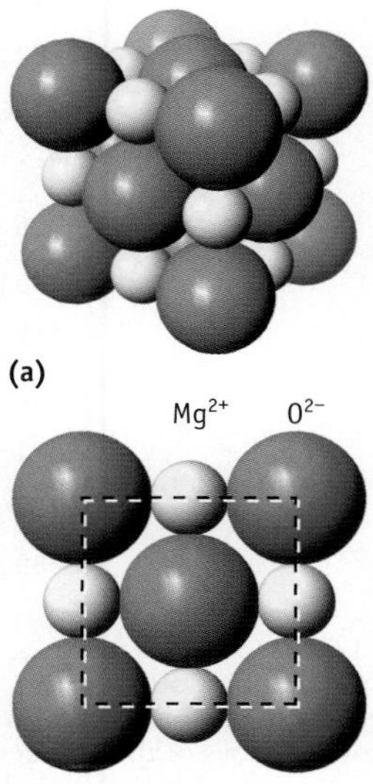

Magnesium oxide. (a) A unit cell showing oxide ions in a face-centered cubic lattice with magnesium ions in the octahedral holes. (b) One face of the cell.

EXERCISE 13.3 Structure and Formula

If an ionic solid has an fcc lattice of anions (X) and all of the tetrahedral holes are occupied by metal cations (M), is the formula of the compound MX, MX_2, or M_2X?

EXERCISE 13.4 Density from Cell Dimensions

Potassium chloride has the same unit cell as NaCl. Using the ion sizes in Figure 7.12, calculate the density of KCl.

13.3 Bonding in Ionic Compounds: Lattice Energy

Ionic compounds typically have high melting points, an indication of the strength of the bonding in the ionic crystal lattice. A measure of that is the lattice energy, the main topic of this section.

Lattice Energy

Ionic compounds exist as solids under normal conditions. Their structures contain positive and negative ions arranged in a three-dimensional lattice (Figure 13.9). In an ionic crystal lattice, there are extensive attractions between ions of opposite charge and repulsions between ions of like charge. Each of these interactions is governed by an equation related to Coulomb's law (◄ page 78). For

TABLE 13.2 Lattice Energies of Some Ionic Compounds

Compound	$\Delta_{lattice}U$ (kJ/mol)
LiF	−1037
LiCl	−852
LiBr	−815
LiI	−761
NaF	−926
NaCl	−786
NaBr	−752
NaI	−702
KF	−821
KCl	−717
KBr	−689
KI	−649

Source: D. Cubicciotti: Lattice energies of the alkali halides and electron affinities of the halogens. *Journal of Chemical Physics*, Vol. 31, p. 1646, 1959.

example, $U_{\text{ion pair}}$, the energy of attractive interactions between 1 mol of ion pairs is given by

$$U_{\text{ion pair}} = C(N_A)\left(\frac{(n^+e)(n^-e)}{d}\right)$$

The symbol C represents a constant; d is the distance between the ion centers; n^+ is the number of positive charges on the cation; n^- is the number of negative charges on the anion; and e is the charge on an electron; n^+e is assigned a positive value, and n^-e is assigned a negative value due to the respective charges of the ions. Including Avogadro's number, N_A, allows us to calculate the energy change for 1 mol of ion pairs. Be sure to notice that the energy depends directly on the charges on the ions and inversely on the distance between them.

In an extended ionic lattice, there are multiple cation–anion interactions. Let us take NaCl as an example (Figure 13.9). If we focus on an Na^+ ion in the center of the unit cell, we see it is surrounded by, and attracted to, six Cl^- ions. Just a bit farther away from this Na^+ ion, however, there are 12 other Na^+ ions, and there is a force of repulsion between the center Na^+ and these ions. (These are 12 Na^+ ions in the edges of the cube.) And if we still focus on the "center" Na^+ ion, we see there are eight more Cl^- ions, and these are attracted to the "center" Na^+ ion. If we were to take into account *all* of the interactions between the ions in a lattice, it would be possible to calculate the **lattice energy, $\Delta_{lattice}U$,** the energy of formation of one mole of a solid crystalline ionic compound when ions in the gas phase combine (see Table 13.2). For sodium chloride, this reaction would correspond to

$$Na^+(g) + Cl^-(g) \longrightarrow NaCl(s)$$

Lattice energy is a measure of the strength of ionic bonding. Often, however, chemists use **lattice enthalpy, $\Delta_{lattice}H$** rather than lattice energy because of the difficulty of estimating some energy quantities. The same trends are seen in both, though, and, because we are dealing with a condensed phase, the numerical values are nearly identical.

We shall focus here on the dependence of lattice enthalpy on ion charges and sizes. As given by Coulomb's law, the higher the ion charges, the greater the attraction between oppositely charged ions, and so $\Delta_{lattice}H$ has a larger negative value for more highly charged ions. This is illustrated by the lattice enthalpies of MgO and NaF. The value of $\Delta_{lattice}H$ for MgO (−4050 kJ/mol) is about four times more negative than the value for NaF (−926 kJ/mol) because the charges on the Mg^{2+} and O^{2-} ions [(2+) × (2−)] are twice as large as those on Na^+ and F^- ions.

Because the attraction between ions is inversely proportional to the distance between them, the effect of ion size on lattice enthalpy is also predictable: A lattice built from smaller ions generally leads to a more negative value for the lattice enthalpy (Table 13.2 and Figure 13.11). For alkali metal halides, for example, the lattice enthalpy for lithium compounds is generally more negative than that for potassium compounds because the Li^+ ion is much smaller than the K^+ cation. Similarly, fluorides are more strongly bonded than are iodides with the same cation.

■ **Born–Haber Cycles** Calculation of lattice energies by this procedure is named for Max Born (1882–1970) and Fritz Haber (1868–1934), German scientists who played prominent roles in thermodynamic research.

Calculating a Lattice Enthalpy from Thermodynamic Data

Lattice enthalpies can be calculated using a thermodynamic relationship known as a **Born–Haber cycle.** This calculation is an application of Hess's law (◀ page 233). Such a cycle is illustrated in Figure 13.12 for solid sodium chloride.

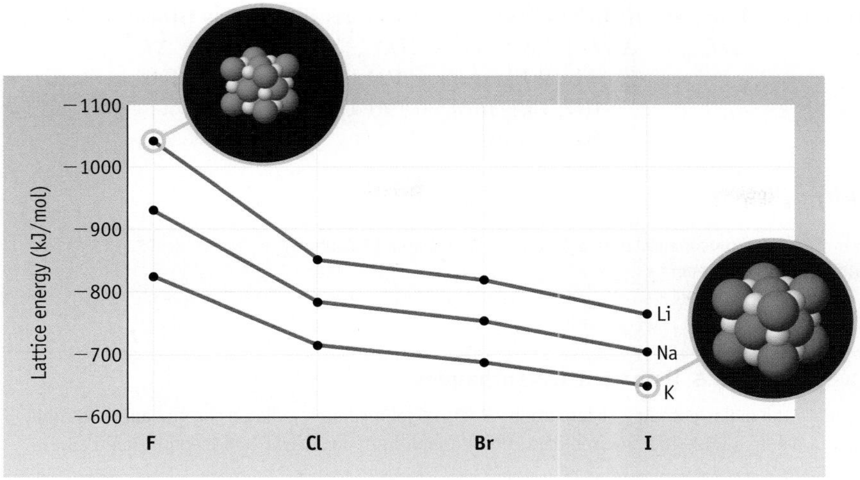

FIGURE 13.11 Lattice energy. $\Delta_{lattice}U$ is illustrated for the formation of the alkali metal halides, MX(s), from the ions $M^+(g) + X^-(g)$.

Steps 1 and 2 in Figure 13.12 involve formation of $Na^+(g)$ and $Cl^-(g)$ ions from the elements; the enthalpy change for each of these steps is known (Appendices F and L). Step 3 in Figure 13.12 gives the lattice enthalpy, $\Delta_{lattice}H$. $\Delta_f H°$ is the standard molar enthalpy of formation of NaCl(s) (Appendix L). The enthalpy values for each step are related by the following equation:

$$\Delta_f H° \text{ [NaCl(s)]} = \Delta H_{\text{Step 1a}} + \Delta H_{\text{Step 1b}} + \Delta H_{\text{Step 2a}} + \Delta H_{\text{Step 2b}} + \Delta H_{\text{Step 3}}$$

Because the values for all of these quantities are known except for $\Delta H_{\text{Step 3}}$ ($\Delta_{lattice}H$), the value for this step can be calculated.

Step 1a.	Enthalpy of formation of Cl(g)	= +121.3 kJ/mol (Appendix L)
Step 1b.	ΔH for $Cl(g) + e^- \longrightarrow Cl^-(g)$	= −349 kJ/mol (Appendix F)
Step 2a.	Enthalpy of formation of Na(g)	= +107.3 kJ/mol (Appendix L)
Step 2b.	ΔH for $Na(g) \longrightarrow Na^+(g) + e^-$	= +496 kJ/mol (Appendix F)

The standard enthalpy of formation of NaCl(s), $\Delta_f H°$, is −411.12 kJ/mol. Combining this with the known values of Steps 1 and 2, we can calculate ΔH_{step3}, which is the lattice enthalpy, $\Delta_{lattice}H$.

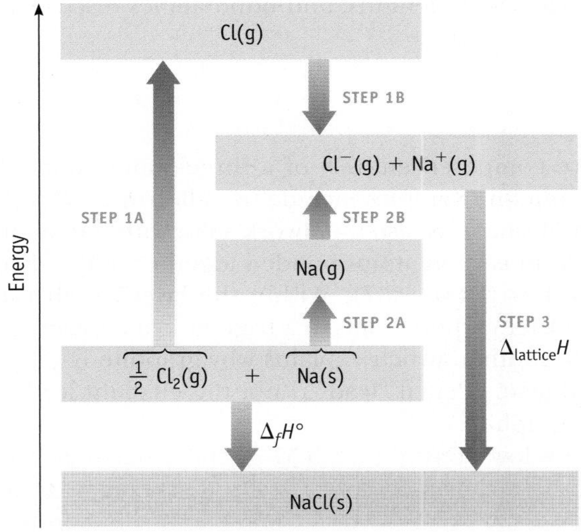

FIGURE 13.12 Born–Haber cycle for the formation of NaCl(s) from the elements. The calculation in the text uses enthalpy values, and the value obtained is the lattice enthalpy, $\Delta_{lattice}H$. The difference between $\Delta_{lattice}U$ and $\Delta_{lattice}H$ is generally not significant and can be corrected for, if desired. (Note that the energy diagram is not to scale.)

Step 3. Formation of NaCl(s) from the ions in the gas phase = ΔH_{step3}

$$\Delta H_{step3} = \Delta_f H° \,[\text{NaCl(s)}] - \Delta H_{\text{Step 1a}} - \Delta H_{\text{Step 1b}} - \Delta H_{\text{Step 2a}} - \Delta H_{\text{Step 2b}}$$
$$= -411.12 \text{ kJ/mol} - 121.3 \text{ kJ/mol} - (-349 \text{ kJ/mol}) - 107.3 \text{ kJ/mol} - 496 \text{ kJ/mol}$$
$$= -787 \text{ kJ/mol}$$

Chemistry Now™

Sign in at **www.thomsonedu.com/login** and go to Chapter 13 Contents to see Screen 13.4 for an illustration of **lattice and lattice energy.**

EXERCISE 13.5 Using Lattice Enthalpies

Calculate the molar enthalpy of formation, $\Delta_f H°$, of solid sodium iodide using the approach outlined in Figure 13.12. The required data can be found in Appendices F and L and in Table 13.2.

13.4 The Solid State: Other Kinds of Solid Materials

So far, we have described the structures of metals and simple ionic solids. Now we will look briefly at the other categories of solids: molecular solids, network solids, and amorphous solids (Table 13.1).

Molecular Solids

Compounds such as H_2O and CO_2 exist as solids under appropriate conditions. In these cases, it is molecules, rather than atoms or ions, that pack in a regular fashion in a three-dimensional lattice. You have already seen one such structure, that of ice (Figure 12.8).

The way molecules are arranged in a crystalline lattice depends on the shape of the molecules and the types of intermolecular forces. Molecules tend to pack in the most efficient manner and to align in ways that maximize intermolecular forces of attraction. Thus, the water structure was established to gain the maximum intermolecular attraction through hydrogen bonding.

It is from structural studies on molecular solids that most of the information on molecular geometries, bond lengths, and bond angles discussed in Chapter 8 was assembled.

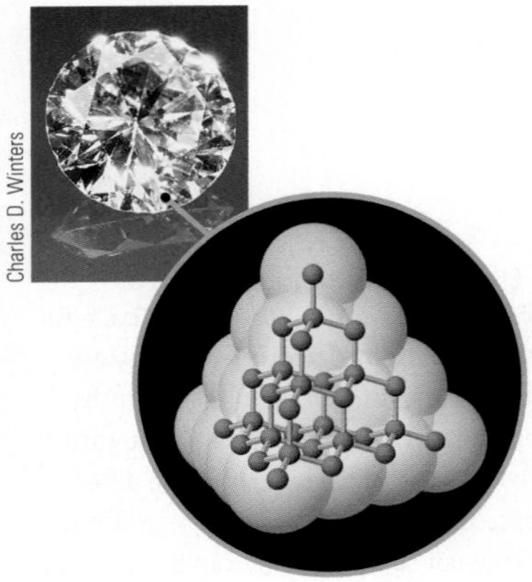

FIGURE 13.13 A diamond and the diamond lattice. The colors of diamonds may range from colorless to yellow, brown, or black. Poorer-quality diamonds are used extensively in industry, mainly for cutting or grinding tools. Industrial-quality diamonds are produced synthetically at present by heating graphite, along with a metal catalyst, to 1200–1500 °C and a pressure of 65–90 kilobars.

Network Solids

Network solids are composed entirely of a three-dimensional array of covalently bonded atoms. Common examples include two allotropes of carbon: graphite and diamond. Elemental silicon is also a network solid with a diamond-like structure.

Graphite consists of carbon atoms bonded together in flat sheets that cling only weakly to one another (Figure 2.7). Within the layers, each carbon atom is surrounded by three other carbon atoms in a trigonal planar arrangement. The layers can slip easily over another, which explains why graphite is soft, a good lubricant, and used in pencil lead. (Pencil "lead" is not the element lead, but rather a composite of clay and graphite.)

Diamonds have a low density (d = 3.51 g/cm^3), but they are also the hardest material and the best conductor of heat known. They are transparent to visible light, as well as to infrared and ultraviolet radiation. Diamonds are electrically insulating but behave as semiconductors with some advantages over silicon. In addition to their use in jewelry, many diamonds are used as abrasives and in diamond-coated cutting tools. In the structure of diamond (Figure 13.13), each carbon atom is bonded to four other carbon atoms at the corners of a tetrahedron, and this pattern extends throughout the solid.

Silicates, compounds composed of silicon and oxygen, represent an enormous class of chemical compounds. You know them in the form of sand, quartz, talc, and mica, or as a major constituent of rocks such as granite. The structure of quartz is illustrated in Figure 13.14. It consists of tetrahedral silicon atoms covalently bonded to oxygen atoms in a giant three-dimensional lattice.

Most network solids are hard and rigid and are characterized by high melting and boiling points. These characteristics reflect the fact that a great deal of energy must be provided to break the covalent bonds in the lattice. For example, silicon dioxide melts at temperatures higher than 1600 °C.

Photo: Charles D. Winters

FIGURE 13.14 Silicon dioxide. Common quartz, SiO_2, is a network solid consisting of silicon and oxygen atoms.

Amorphous Solids

A characteristic property of pure crystalline solids––whether metals, ionic solids, or molecular solids—is that they melt at a specific temperature. For example, water melts at 0 °C, aspirin at 135 °C, lead at 327.5 °C, and NaCl at 801 °C. Because they are specific and reproducible values, melting points are often used as a means of identifying chemical compounds.

Another property of crystalline solids is that they form well-defined crystals, with smooth, flat faces. When a sharp force is applied to a crystal, it will most often cleave to give smooth, flat faces. The resulting solid particles are smaller versions of the original crystal (Figure 13.15a).

Many common solids, including ones that we encounter every day, do not have these properties, however. Glass is a good example. When glass is heated, it softens over a wide temperature range, a property useful for artisans and craftsmen who can create beautiful and functional products for our enjoyment and use. Glass also possesses a property that we would rather it not have: When glass breaks, it leaves randomly shaped pieces. Other materials that behave similarly include common polymers such as polyethylene, nylon, and other plastics.

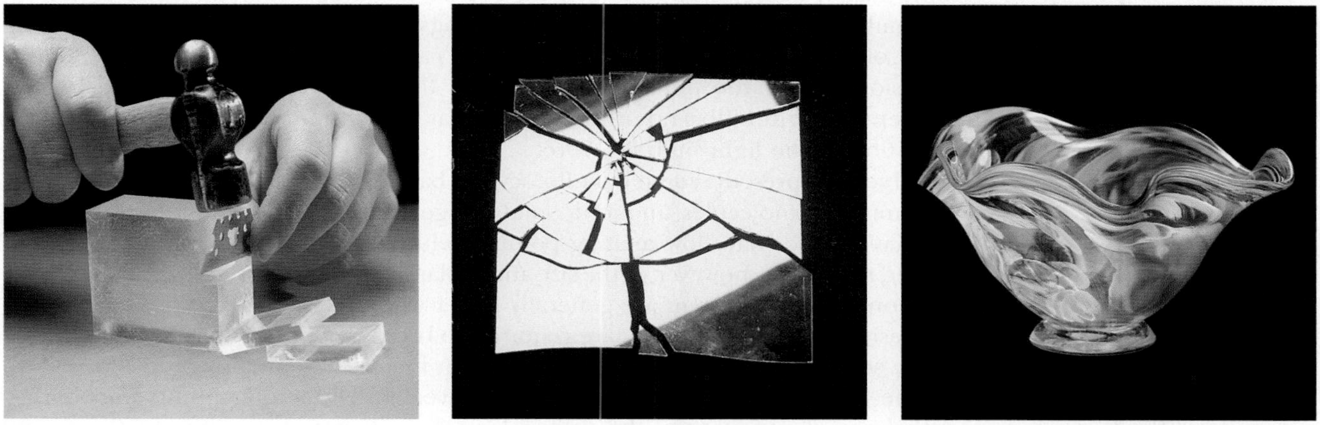

Charles D. Winters

(a) A salt crystal can be cleaved cleanly into smaller and smaller crystals that are duplicates of the larger crystal.

(b) Glass is an amorphous solid composed of silicon and oxygen atoms. It has, however, no long-range order as in crystalline quartz.

(c) Glass can be molded and shaped into beautiful forms and, by adding metal oxides, can take on wonderful colors.

FIGURE 13.15 Crystalline and amorphous solids.

The characteristics of these amorphous solids relate to their molecular structure. At the particulate level, amorphous solids do not have a regular structure. In fact, in many ways these substances look a lot like liquids. Unlike liquids, however, the forces of attraction are strong enough that movement of the molecules or ions is restricted.

Chemistry Now™

Sign in at **www.thomsonedu.com/login** and go to Chapter 13 Contents to see:

- Screen 13.5 for an exercise on **molecular solids**
- Screen 13.6 for a self-study module on **network solids**
- Screen 13.7 for a self-study module on **silicate minerals**

13.5 Phase Changes Involving Solids

The shape of a crystalline solid is a reflection of its internal structure. But what about physical properties of solids, such as the temperatures at which they melt? This and many other physical properties of solids are of interest to chemists, geologists, and engineers, among others.

Melting: Conversion of Solid into Liquid

The melting point of a solid is the temperature at which the lattice collapses and the solid is converted into a liquid. Like the liquid-to-vapor transformation, melting requires energy, called the enthalpy of fusion (given in kilojoules per mole) (◀ Chapter 5).

Energy absorbed as heat on melting = enthalpy of fusion = $\Delta_{fusion}H$ (kJ/mol)
Energy evolved as heat on freezing = enthalpy of crystallization = $-\Delta_{fusion}H$ (kJ/mol)

■ **Uncle Tungsten** *Uncle Tungsten* is the title of a book by Oliver Sacks (Alfred Knopf, New York, 2001). In it, he describes growing up with an uncle who had a light bulb factory and used tungsten. He also describes other "chemical adventures."

Enthalpies of fusion can range from just a few thousand joules per mole to many thousands of joules per mole (Table 13.3). A low melting temperature will certainly mean a low value for the enthalpy of fusion, whereas high melting points are associated with high enthalpies of fusion. Figure 13.16 shows the enthalpies of fusion for the metals of the fourth through the sixth periods. Based on this figure, we see that transition metals have high enthalpies of fusion, with many of those in the sixth period being extraordinarily high. This trend parallels the trend seen with the melting points for these elements. Tungsten, which has the highest melting point of all the known elements except for carbon, also has the highest enthalpy of fusion among the transition metals. For this reason, tungsten is used for the filaments in light bulbs; no other material has been found to work better since the invention of the light bulb in 1908.

Table 13.3 presents some data for several basic types of substances: metals, polar and nonpolar molecules, and ionic solids. In general, nonpolar substances that form molecular solids have low melting points. Melting points increase within a series of related molecules, however, as the size and molar mass increase. This happens because London dispersion forces are generally larger when the molar mass is larger. Thus, increasing amounts of energy are required to break down the intermolecular forces in the solid, a principle that is reflected in an increasing enthalpy of fusion.

The ionic compounds in Table 13.3 have higher melting points and higher enthalpies of fusion than the molecular solids. This trend is due to the strong ion–ion forces present in ionic solids, forces that are reflected in high lattice energies (page 599). Because ion–ion forces depend on ion size (as well as ion charge), there is a good correlation between lattice energy and the position of the metal or

TABLE 13.3 Melting Points and Enthalpies of Fusion of Some Elements and Compounds

Compound	Melting Point (°C)	Enthalpy of Fusion (kJ/mol)	Type of Interparticle Forces
Metals			
Hg	−39	2.29	Metal bonding; see pages 657–663.
Na	98	2.60	
Al	660	10.7	
Ti	1668	20.9	
W	3422	35.2	
Molecular Solids: Nonpolar Molecules			
O_2	−219	0.440	Dispersion forces only.
F_2	−220	0.510	
Cl_2	−102	6.41	
Br_2	−7.2	10.8	
Molecular Solids: Polar Molecules			
HCl	−114	1.99	All three HX molecules have dipole–dipole forces. Dispersion forces increase with size and molar mass.
HBr	−87	2.41	
HI	−51	2.87	
H_2O	0	6.01	Hydrogen bonding and dispersion forces
Ionic Solids			
NaF	996	33.4	All ionic solids have extended ion–ion interactions. Note the general trend is the same as for lattice energies (see Section 13.3 and Figure 13.11).
NaCl	801	28.2	
NaBr	747	26.1	
NaI	660	23.6	

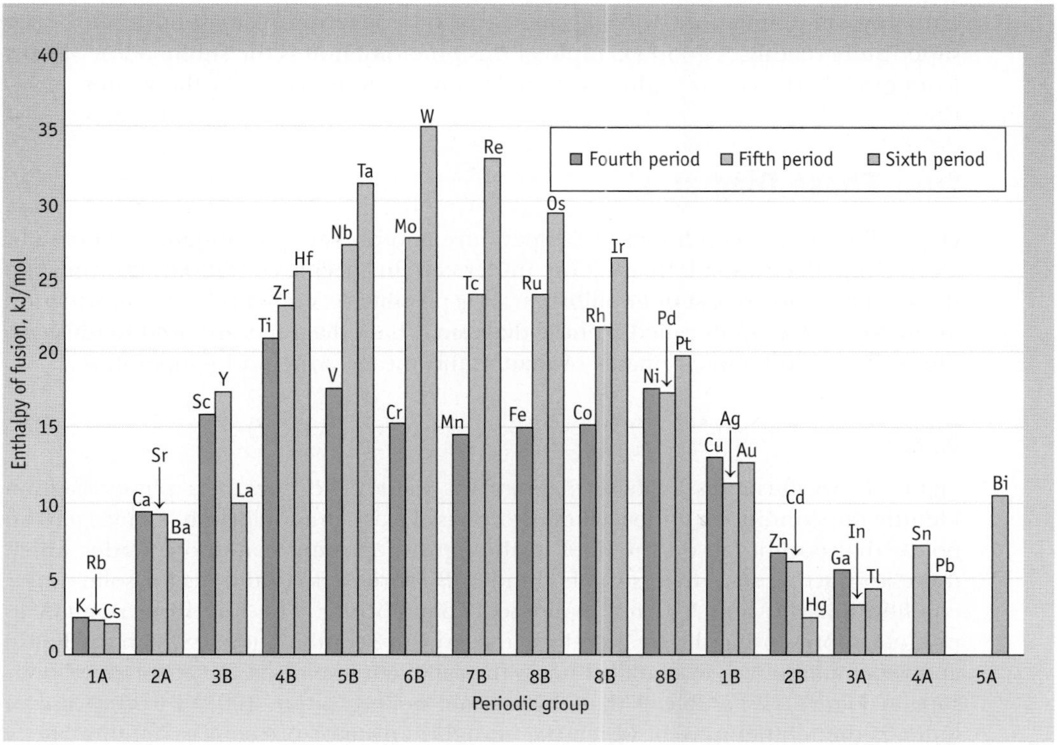

FIGURE 13.16 Enthalpy of fusion of fourth-, fifth-, and sixth-period metals. Enthalpies of fusion range from 2–5 kJ/mol for Group 1A elements to 35.2 kJ/mol for tungsten. Notice that enthalpies of fusion generally increase for group 4B–8B metals on descending the periodic table.

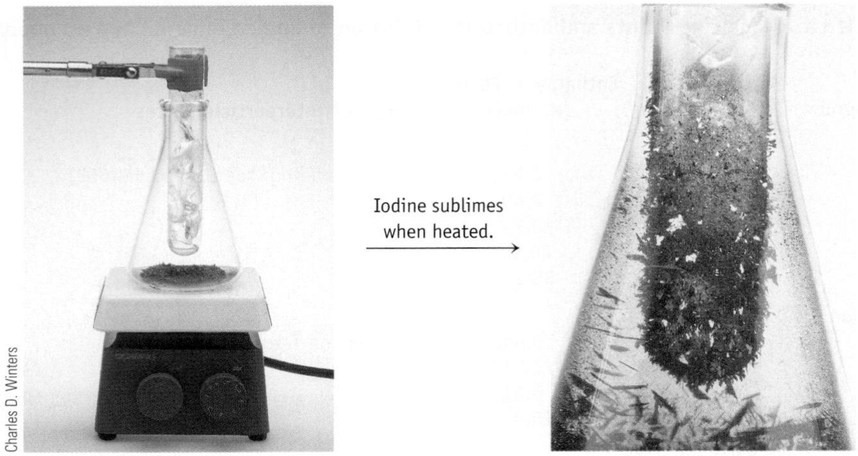

FIGURE 13.17 Sublimation. Sublimation entails the conversion of a solid directly to its vapor. Here, iodine (I_2) sublimes when warmed. If an ice-filled test tube is inserted into the flask, the vapor deposits on the cold surface.

halogen in the periodic table. For example, the data in Table 13.3 show a decrease in melting point and enthalpy of fusion for sodium salts as the halide ion increases in size. This parallels the decrease in lattice energy seen with increasing ion size.

Sublimation: Conversion of Solid into Vapor

Molecules can escape directly from the solid to the gas phase by sublimation (Figure 13.17).

$$\text{Solid} \rightarrow \text{Gas} \qquad \text{Energy required as heat} = \Delta_{\text{sublimation}}H$$

Sublimation, like fusion and evaporation, is an endothermic process. The energy required as heat is called the **enthalpy of sublimation.** Water, which has a molar enthalpy of sublimation of 51 kJ/mol, can be converted from solid ice to water vapor quite readily. A good example of this phenomenon is the sublimation of frost from grass and trees as night turns to day on a cold morning in the winter.

13.6 Phase Diagrams

Depending on the conditions of temperature and pressure, a substance can exist as a gas, a liquid, or a solid. In addition, under certain specific conditions, two (or even three) states can coexist in equilibrium. It is possible to summarize this information in the form of a graph called a **phase diagram.** Phase diagrams are used to illustrate the relationship between phases of matter and the pressure and temperature.

Water

Figure 13.18 illustrates a phase diagram for water. The lines in a phase diagram identify the conditions under which two phases exist at equilibrium. Conversely, all points that do not fall on the lines in the figure represent conditions under which there is only one state that is stable. Line A–B represents conditions for solid–vapor equilibrium, and line A–C for liquid–solid equilibrium. The line from point A to point D, representing the temperature and pressure combination at which the liquid and vapor phases are in equilibrium, is the same curve plotted for water vapor pressure in Figure 12.17. Recall that the normal boiling point, 100 °C in the case of water, is the temperature at which the equilibrium vapor pressure is 760 mm Hg.

Point A, appropriately called the **triple point**, indicates the conditions under which all three phases coexist in equilibrium. For water, the triple point is at $P = 4.6$ mm Hg and $T = 0.01$ °C.

The line A–C shows the conditions of pressure and temperature at which solid–liquid equilibrium exists. (Because no vapor pressure is involved here, the pressure referred to is the external pressure on the liquid.) For water, this line has a negative slope; the change for water is approximately −0.01 °C for each one-atmosphere increase in pressure. That is, the higher the external pressure, the lower the melting point.

The negative slope of the water solid–liquid equilibrium line can be explained from our knowledge of the structure of water and ice. When the pressure on an object increases, common sense tells us that the volume of the object will become

Case Study — The World's Lightest Solid

The *Guinness Book of Records* calls it the "world's lightest solid" and the "best thermal insulator." Even though it is 99.8% air and has a density of only about 1 mg/cm^3, it is a light blue solid that, to the touch, feels much like Styrofoam chips that are used in packaging. It is also strong structurally, able to hold over 2000 times its weight (Figure A).

"It" is a silica aerogel, a low-density substance derived from a gel in which the liquid has been replaced by air (▶ page 666). There are aerogels based silicon and carbon as well as aluminum and other metals, but the silicon-based aerogel is the most thoroughly studied. This aerogel is made by polymerizing a compound like $Si(OC_2H_5)_4$ in alcohol. The resulting long-chain molecules form a gel that is bathed in the alcohol. This substance is then placed in supercritical CO_2 (▶ page 609), which causes the alcohol in the nanopores in the gel to be replaced by CO_2. When the CO_2 is vented off as a gas, what remains is a highly porous aerogel with an incredibly low density.

Courtesy of NASA

FIGURE A Silica aerogel. A 2.5-kg brick is supported by a piece of silica aerogel weighing about 2 g. (http://stardust.jpl.nasa.gov/photo/aerogel.html)

Aerogels have been known for decades but have only recently received a lot of study. They do have amazing properties! Chief among them is their insulating ability, as illustrated in Figure B. Aerogels do not allow heat to be conducted through the lattice, and convective heat transfer is also poor because air cannot circulate throughout the lattice. One practical use for these aerogels is in insulating glass. However, before it can be truly useful for this purpose, researchers need to find a way to make completely transparent aerogel. (Silica aerogel is very light blue owing to Rayleigh scattering, the same process that makes the sky blue.) Aerogels are also biocompatible and have been studied as possible drug delivery systems.

Aerogel has been in the news in the past few years because it was used to catch comet dust in Project Stardust. A spacecraft was sent to intercept a comet in 2004 and returned to Earth in January 2006. On the spacecraft was an array holding blocks of aerogel. As the craft flew through the comet's tail, dust particles impacted the aerogel blocks and were "brought to a standstill as they tunneled through it without much heating or alteration, leaving carrot-shaped tracks." When the spacecraft was returned to Earth, scientists analyzed the particles and found that there were silicate minerals that seemed to have been formed in the inner regions of the solar system. (See *Science*, Vol. 314, 15 December 2006.)

Courtesy of NASA

FIGURE B Aerogel as an insulator. http://stardust.jpl.nasa.gov/images/gallery/aerogelmatches.jpg

Questions:

1. *Assume the repeating unit in the aerogel polymer is $OSi(OC_2H_5)_2$. If the polymer is 99.8% air, how many silicon atoms are there in 1.0 cm^3 of aerogel?*
2. *Suppose you wish to make a superinsulating window and so fill the gap between two sheets of glass with aerogel. What mass of aerogel is needed for a 180 cm × 150 cm window with a gap of 2.0 mm between the glass sheets?*

Answers to these questions are in Appendix Q.

Active Figure 13.18 Phase diagram for water. The scale is intentionally exaggerated to be able to show the triple point and the negative slope of the line representing the liquid–solid equilibrium.

Chemistry Now™ Sign in at www.thomsonedu.com/login and go to the Chapter Contents menu to explore an interactive version of this figure accompanied by an exercise.

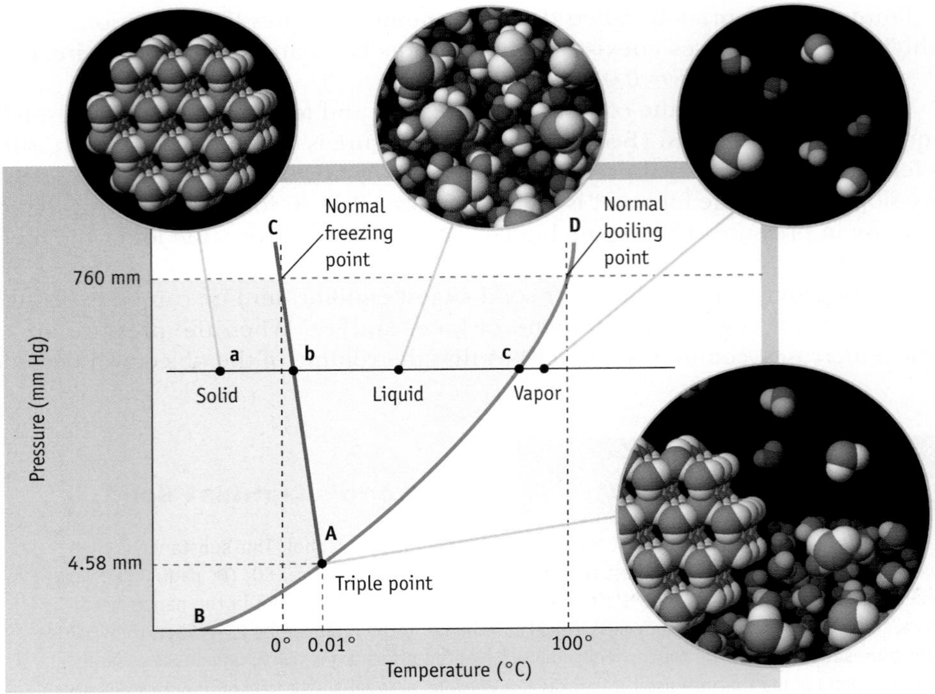

smaller, giving the substance a higher density. Because ice is less dense than liquid water (due to the open lattice structure of ice, Figure 12.8), ice and water in equilibrium respond to increased pressure (at constant *T*) by melting ice to form more water because the same mass of water requires less volume.

Phase Diagrams and Thermodynamics

Let us explore the water phase diagram further by correlating phase changes with thermodynamic data. Suppose we begin with ice at −10 °C and under a pressure of 500 mm Hg (point *a* on Figure 13.18). As ice is heated (at constant *P*), it absorbs about 2.1 J/g · K in warming from point *a* to point *b* at a temperature between 0 °C and 0.01 °C. At this point, the solid is in equilibrium with liquid water. Solid–liquid equilibrium is maintained until 333 J/g has been transferred to the sample and it has become liquid water at this temperature. If the liquid, still under a pressure of 500 mm Hg, now absorbs 4.184 J/g · K, it warms to point *c*. The temperature at point *c* is about 89 °C, and equilibrium is established between liquid water and water vapor. The equilibrium vapor pressure of the liquid water is 500 mm Hg. If 2260 J/g is transferred to the liquid–vapor sample, the equilibrium vapor pressure remains 500 mm Hg until the liquid is completely converted to vapor at 89 °C.

Carbon Dioxide

The features of the phase diagram for CO_2 (Figure 13.19) are generally the same as those for water but with some important differences.

In contrast to water, the CO_2 solid–liquid equilibrium line has a positive slope. Once again, increasing pressure on the solid in equilibrium with the liquid will shift the equilibrium to the more dense phase, but for CO_2 this will be the solid. Because solid CO_2 is denser than the liquid, the newly formed solid CO_2 sinks to the bottom in a container of liquid CO_2.

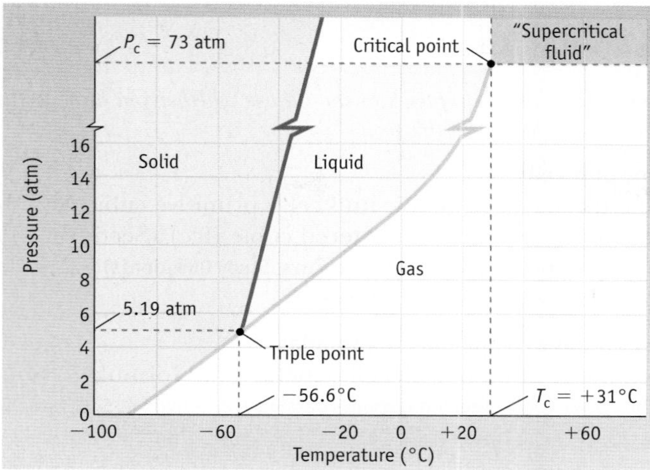

FIGURE 13.19 The phase diagram of CO_2. Notice in particular the positive slope of the solid–liquid equilibrium line. (For more on the critical point, see page 577.)

Another feature of the CO_2 phase diagram is the triple point that occurs at a pressure of 5.19 atm (3940 mm Hg) and 216.6 K (−56.6 °C). Carbon dioxide cannot be a liquid at pressures lower than this.

At pressures around normal atmospheric pressure, CO_2 will be either a solid or a gas, depending on the temperature. [At a pressure of 1 atm, solid CO_2 is in equilibrium with the gas at a temperature of 197.5 K (−78.7 °C).] As a result, as solid CO_2 warms above this temperature, it sublimes rather than melts. Carbon dioxide is called *dry ice* for this reason; it looks like water ice, but it does not melt.

From the CO_2 phase diagram, we can also learn that CO_2 gas can be converted to a liquid at room temperature (20–25 °C) by exerting a moderate pressure on the gas. In fact, CO_2 is regularly shipped in tanks as a liquid to laboratories and industrial companies.

Finally, the critical pressure and temperature for CO_2 are 73 atm and 31 °C, respectively. Because the critical temperature and pressure are easily attained in the laboratory, it is possible to observe the transformation to supercritical CO_2 (Figure 13.20).

Chemistry Now™

Sign in at **www.thomsonedu.com/login** and go to Chapter 13 Contents to see Screen 13.8 to view animations of **phase changes** and to do an exercise on **phase diagrams.**

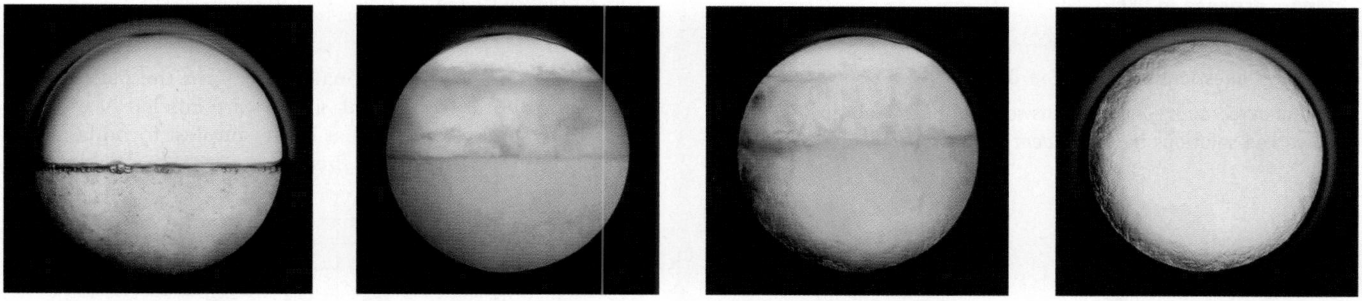

The separate phases of CO_2 are seen through the window in a high-pressure vessel.

As the sample warms and the pressure increases, the meniscus becomes less distinct.

As the temperature continues to increase, it is more difficult to distinguish the liquid and vapor phases.

Once the critical *T* and *P* are reached, distinct liquid and vapor phases are no longer in evidence. This homogeneous phase is "supercritical CO_2."

FIGURE 13.20 Transformation to supercritical CO_2.

Chapter Goals Revisited

ChemistryNow™ Sign in at **www.thomsonedu.com/login** to:

- Assess your understanding with Study Questions in OWL keyed to each goal in the Goals and Homework menu for this chapter
- For quick review, download Go Chemistry mini-lecture flashcard modules (or purchase them at **www.ichapters.com**)
- Check your readiness for an exam by taking the Pre-Test and exploring the modules recommended in your Personalized Study plan.

Access **How Do I Solve It?** tutorials on how to approach problem solving using concepts in this chapter.

Now that you have studied this chapter, you should ask whether you have met the chapter goals. In particular, you should be able to:

Understand cubic unit cells

a. Describe the three types of cubic unit cells: primitive cubic (pc), body-centered cubic (bcc), and face-centered cubic (fcc) (Section 13.1).

b. Relate atom size and unit cell dimensions. **Study Question(s) assignable in OWL: 7, 8, 10, 26, 29, 32, 34, 36, 43; Go Chemistry Module 18.**

Relate unit cells for ionic compounds to formulas

a. Understand the relation of unit cell structure and formula for ionic compounds. (Section 13.2) **Study Question(s) assignable in OWL: 4, 5, 6, 8; Go Chemistry Module 18.**

Describe the properties of solids

a. Understand lattice energy and how it is calculated (Section 13.3). **Study Question(s) assignable in OWL: 11, 13, 14, 16, 38.**

b. Characterize different types of solids: metallic (e.g., copper), ionic (e.g., NaCl and CaF_2), molecular (e.g., water and I_2), network (e.g., diamond), and amorphous (e.g., glass and many synthetic polymers) (Table 13.1). **Study Question(s) assignable in OWL: 17.**

c. Define the processes of melting, freezing, and sublimation and their enthalpies (Sections 13.4 and 13.5). **Study Question(s) assignable in OWL: 20.**

Understand the nature of phase diagrams

a. Identify the different points (triple point, normal boiling point, freezing point) and regions (solid, liquid, vapor) of a phase diagram, and use the diagram to evaluate the vapor pressure of a liquid and the relative densities of a liquid and a solid (Section 13.5). **Study Question(s) assignable in OWL: 21, 22, 23, 24.**

STUDY QUESTIONS

Online homework for this chapter may be assigned in OWL.

▲ denotes challenging questions.

■ denotes questions assignable in OWL.

Blue-numbered questions have answers in Appendix O and fully-worked solutions in the *Student Solutions Manual*.

Practicing Skills

Metallic and Ionic Solids

(See Examples 13.1–13.3 and ChemistryNow Screens 13.2 and 13.3.)

1. Outline a two-dimensional unit cell for the pattern shown here. If the black squares are labeled A and the white squares are B, what is the simplest formula for a "compound" based on this pattern?

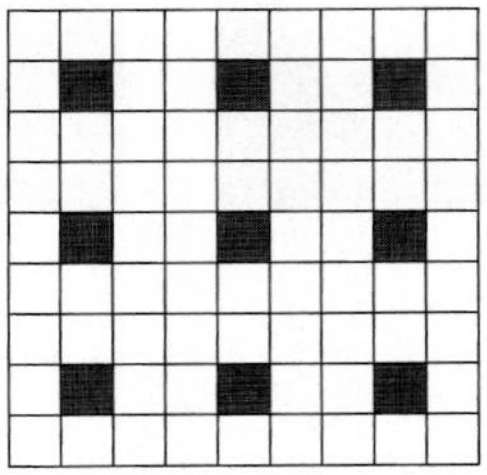

2. Outline a two-dimensional unit cell for the pattern shown here. If the black squares are labeled A and the white squares are B, what is the simplest formula for a "compound" based on this pattern?

3. One way of viewing the unit cell of perovskite was illustrated in Example 13.2. Another way is shown here. Prove that this view also leads to a formula of $CaTiO_3$.

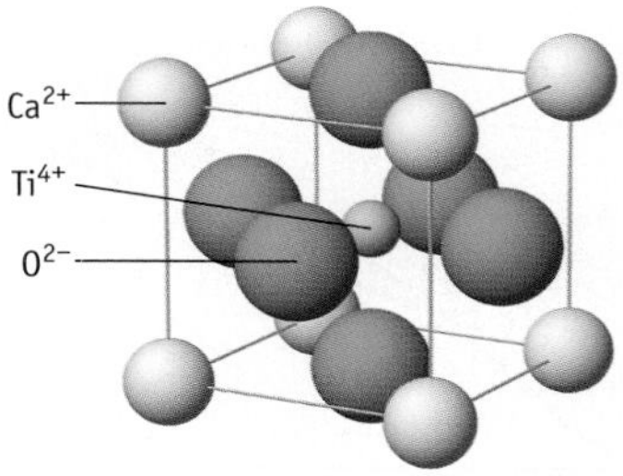

4. ■ Rutile, TiO_2, crystallizes in a structure characteristic of many other ionic compounds. How many formula units of TiO_2 are in the unit cell illustrated here? (The oxide ions marked by an *x* are wholly within the cell; the others are in the cell faces.)

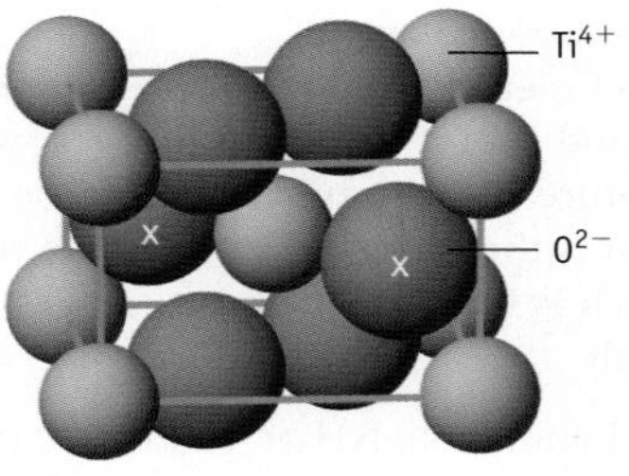

5. ■ Cuprite is a semiconductor. Oxide ions are at the cube corners and in the cube center. Copper ions are wholly within the unit cell.
   (a) What is the formula of cuprite?
   (b) What is the oxidation number of copper?

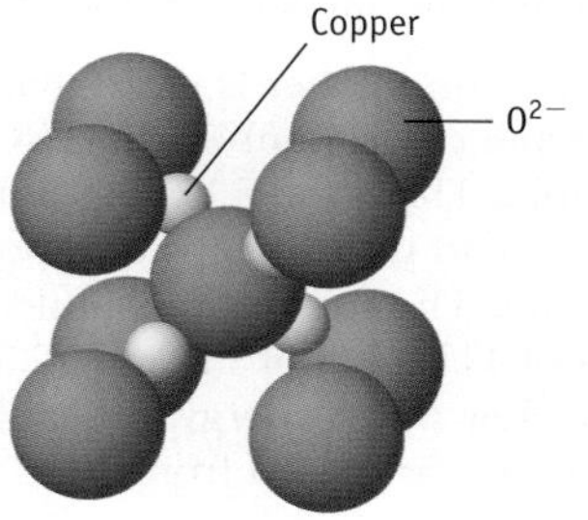

6. ■ The mineral fluorite, which is composed of calcium ions and fluoride ions, has the unit cell shown here.
   (a) What type of unit cell is described by the Ca^{2+} ions?
   (b) Where are the F^- ions located, in octahedral holes or tetrahedral holes?
   (c) Based on this unit cell, what is the formula of fluorite?

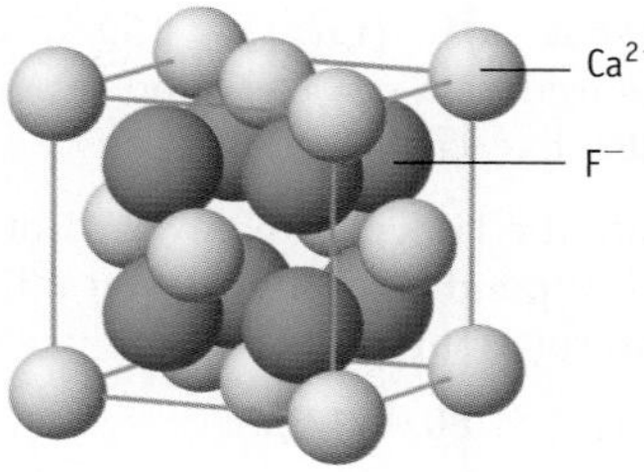

7. ■ Calcium metal crystallizes in a face-centered cubic unit cell. The density of the solid is 1.54 g/cm^3. What is the radius of a calcium atom?

8. ■ The density of copper metal is 8.95 g/cm^3. If the radius of a copper atom is 127.8 pm, is the copper unit cell primitive, body-centered cubic, or face-centered cubic?

9. Potassium iodide has a face-centered cubic unit cell of iodide ions with potassium ions in octahedral holes. The density of KI is 3.12 g/cm^3. What is the length of one side of the unit cell? (Ion sizes are found in Table 7.12.)

10. ▲ ■ A unit cell of cesium chloride is shown on page 596. The density of the solid is 3.99 g/cm^3, and the radius of the Cl^- ion is 181 pm. What is the radius of the Cs^+ ion in the center of the cell? (Assume that the Cs^+ ion touches all of the corner Cl^- ions.)

Ionic Bonding and Lattice Energy

(See ChemistryNow Screen 13.4.)

11. ■ List the following compounds in order of increasing lattice energy (from least negative to most negative): LiI, LiF, CaO, RbI.

12. Examine the trends in lattice energy in Table 13.2. The value of the lattice energy becomes somewhat more negative on going from NaI to NaBr to NaCl, and all are in the range of −700 to −800 kJ/mol. Suggest a reason for the observation that the lattice energy of NaF ($\Delta_{lattice}U = -926$ kJ/mol) is much more negative than those of the other sodium halides.

13. ■ To melt an ionic solid, energy must be supplied to disrupt the forces between ions so the regular array of ions collapses. If the distance between the anion and the cation in a crystalline solid decreases (but ion charges remain the same), should the melting point decrease or increase? Explain.

14. ■ Which compound in each of the following pairs should require the higher temperature to melt? (See Study Question 13.)
(a) NaCl or RbCl
(b) BaO or MgO
(c) NaCl or MgS

15. Calculate the molar enthalpy of formation, $\Delta_f H°$, of solid lithium fluoride using the approach outlined on pages 599-602. $\Delta_f H°$ [Li(g)] = 159.37 kJ/mol, and other required data can be found in Appendices F and L. (See also Exercise 13.5.)

16. ■ Calculate the lattice enthalpy for RbCl. In addition to data in Appendices F and L, you will need the following information:

$\Delta_f H°$ [Rb(g)] = 80.9 kJ/mol

$\Delta_f H°$ [RbCl(s)] = −435.4 kJ/mol

Other Types of Solids

(See ChemistryNow Screens 13.6 and 13.7.)

17. ■ A diamond unit cell is shown here.
(a) How many carbon atoms are in one unit cell?
(b) The unit cell can be considered as a cubic unit cell of C atoms with other C atoms in holes in the lattice. What type of unit cell is this (pc, bcc, fcc)? In what holes are other C atoms located, octahedral or tetrahedral holes?

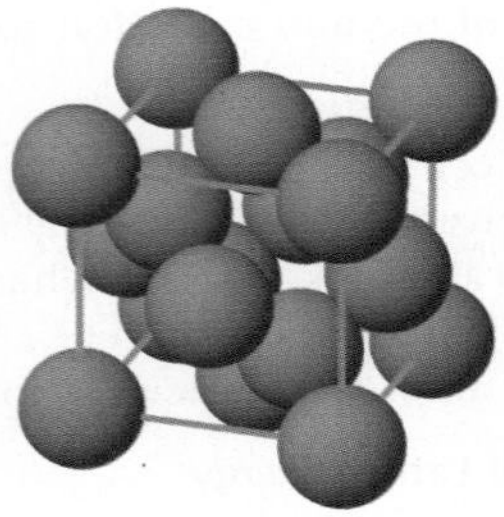

18. The structure of graphite is given in Figure 2.7.
(a) What type of intermolecular bonding forces exist between the layers of six-member carbon rings?
(b) Account for the lubricating ability of graphite. That is, why does graphite feel slippery? Why does pencil lead (which is really graphite in clay) leave black marks on paper?

Physical Properties of Solids

19. Benzene, C_6H_6, is an organic liquid that freezes at 5.5 °C (see Figure 12.1) to form beautiful, feather-like crystals. How much energy as heat is evolved when 15.5 g of benzene freezes at 5.5 °C? (The enthalpy of fusion of benzene is 9.95 kJ/mol.) If the 15.5-g sample is remelted, again at 5.5 °C, what quantity of energy as heat is required to convert it to a liquid?

20. ■ The specific heat capacity of silver is 0.235 J/g · K. Its melting point is 962 °C, and its enthalpy of fusion is 11.3 kJ/mol. What quantity of energy as heat, in joules, is required to change 5.00 g of silver from a solid at 25 °C to a liquid at 962 °C?

Phase Diagrams and Phase Changes

(See ChemistryNow Screen 13.8.)

21. ■ Consider the phase diagram of CO_2 in Figure 13.19.
(a) Is the density of liquid CO_2 greater or less than that of solid CO_2?
(b) In what phase do you find CO_2 at 5 atm and 0 °C?
(c) Can CO_2 be liquefied at 45 °C?

22. ■ Use the phase diagram given here to answer the following questions:

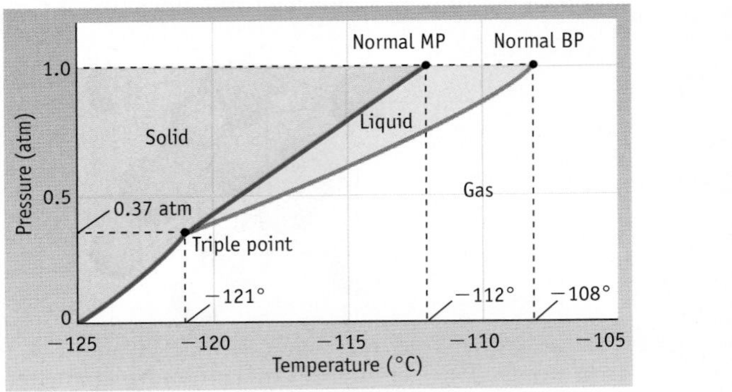

(a) In what phase is the substance found at room temperature and 1.0 atm pressure?
(b) If the pressure exerted on a sample is 0.75 atm and the temperature is −114 °C, in what phase does the substance exist?
(c) If you measure the vapor pressure of a liquid sample and find it to be 380 mm Hg, what is the temperature of the liquid phase?
(d) What is the vapor pressure of the solid at −122 °C?
(e) Which is the denser phase—solid or liquid? Explain briefly.

23. ■ Liquid ammonia, $NH_3(\ell)$, was once used in home refrigerators as the heat transfer fluid. The specific heat capacity of the liquid is 4.7 J/g · K and that of the vapor is 2.2 J/g · K. The enthalpy of vaporization is 23.33 kJ/mol at the boiling point. If you heat 12 kg of liquid ammonia from −50.0 °C to its boiling point of −33.3 °C, allow it to evaporate, and then continue warming to 0.0 °C, how much energy must you supply?

24. ■ If your air conditioner is more than several years old, it may use the chlorofluorocarbon CCl_2F_2 as the heat transfer fluid. The normal boiling point of CCl_2F_2 is −29.8 °C, and the enthalpy of vaporization is 20.11 kJ/mol. The gas and the liquid have specific heat capacities of 117.2 J/mol · K and 72.3 J/mol · K, respectively. How much energy as heat is evolved when 20.0 g of CCl_2F_2 is cooled from +40 °C to −40 °C?

▲ more challenging ■ in OWL Blue-numbered questions answered in Appendix O

General Questions

These questions are not designated as to type or location in the chapter. They may combine several concepts.

25. Construct a phase diagram for O_2 from the following information: normal boiling point, 90.18 K; normal melting point, 54.8 K; and triple point, 54.34 K at a pressure of 2 mm Hg. Very roughly estimate the vapor pressure of liquid O_2 at −196 °C, the lowest temperature easily reached in the laboratory. Is the density of liquid O_2 greater or less than that of solid O_2?

26. ▲ ■ Tungsten crystallizes in the unit cell shown here.

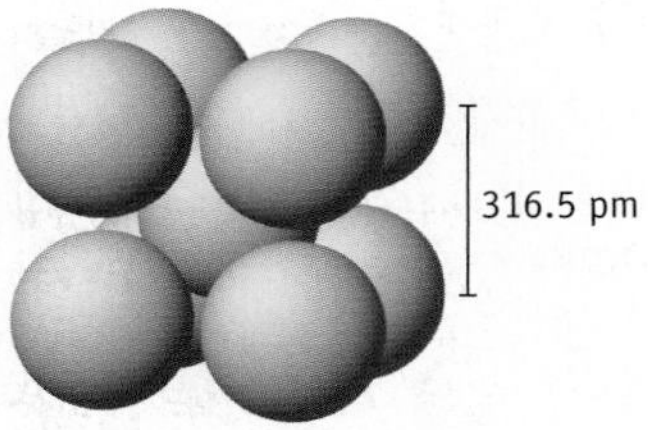

(a) What type of unit cell is this?
(b) How many tungsten atoms occur per unit cell?
(c) If the edge of the unit cell is 316.5 pm, what is the radius of a tungsten atom? (*Hint*: The W atoms touch each other along the diagonal line from one corner of the unit cell to the opposite corner of the unit cell.)

27. Silver crystallizes in a face-centered cubic unit cell. Each side of the unit cell has a length of 409 pm. What is the radius of a silver atom?

28. ▲ ■ The unit cell shown here is for calcium carbide. How many calcium atoms and how many carbon atoms are in each unit cell? What is the formula of calcium carbide? (Calcium ions are silver in color and carbon atoms are gray.)

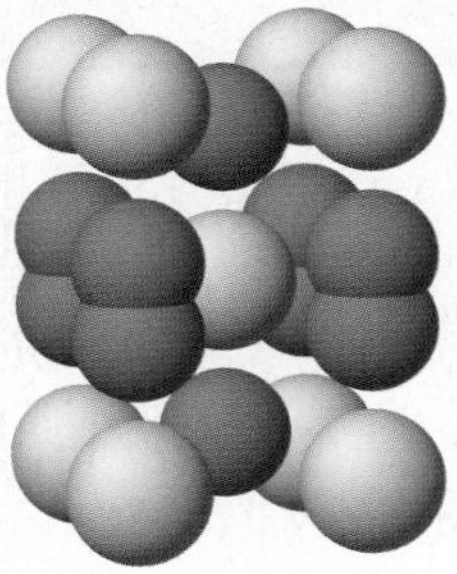

29. ■ The very dense metal iridium has a face-centered cubic unit cell and a density of 22.56 g/cm³. Use this information to calculate the radius of an atom of the element.

30. Vanadium metal has a density of 6.11 g/cm³. Assuming the vanadium atomic radius is 132 pm, is the vanadium unit cell primitive cubic, body-centered cubic, or face-centered cubic?

31. ▲ Calcium fluoride is the well-known mineral fluorite. It is known that each unit cell contains four Ca^{2+} ions and eight F^- ions and that the Ca^{2+} ions are arranged in an fcc lattice. The F^- ions fill all the tetrahedral holes in a face-centered cubic lattice of Ca^{2+} ions. The edge of the CaF_2 unit cell is 5.46295×10^{-8} cm in length. The density of the solid is 3.1805 g/cm³. Use this information to calculate Avogadro's number.

32. ▲ ■ Iron has a body-centered cubic unit cell with a cell dimension of 286.65 pm. The density of iron is 7.874 g/cm³. Use this information to calculate Avogadro's number.

33. ▲ You can get some idea of how efficiently spherical atoms or ions are packed in a three-dimensional solid by seeing how well circular atoms pack in two dimensions. Using the drawings shown here, prove that B is a more efficient way to pack circular atoms than A. A unit cell of A contains portions of four circles and one hole. In B, packing coverage can be calculated by looking at a triangle that contains portions of three circles and one hole. Show that A fills about 80% of the available space, whereas B fills closer to 90% of the available space.

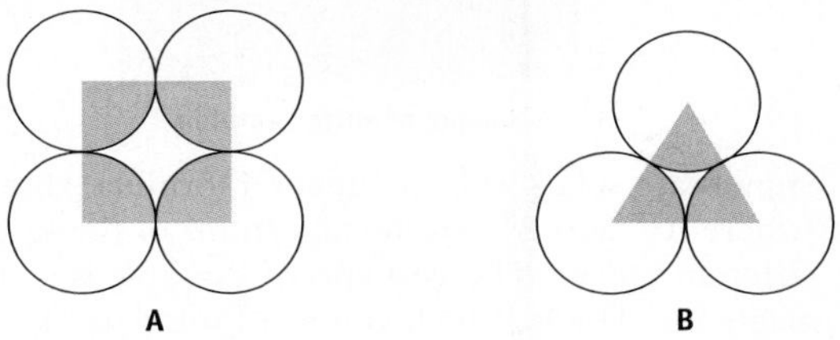

34. ▲ ■ Assuming that in a primitive cubic unit cell the spherical atoms or ions just touch along the cube's edges, calculate the percentage of empty space within the unit cell. (Recall that the volume of a sphere is $(4/3)\pi r^3$, where r is the radius of the sphere.)

35. ▲ The solid state structure of silicon is

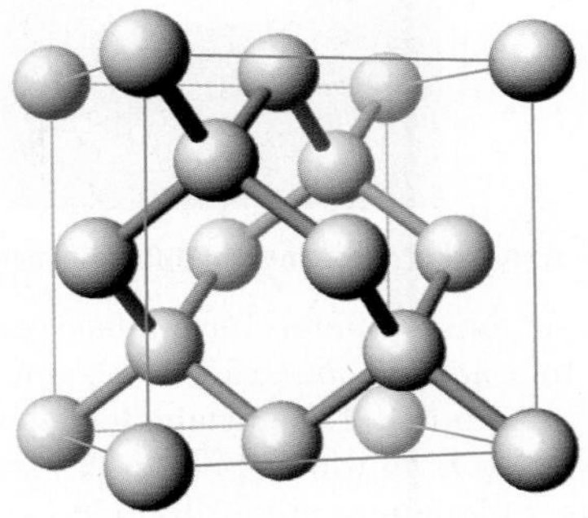

(a) Describe this crystal as pc, bcc, or fcc. What type of holes are occupied in the lattice?
(b) Calculate the density of silicon in g/cm³ (given that the cube edge has a length of 543.1 pm), and estimate the radius of the silicon atom. (Note: the Si atoms on the edges do not touch one another.)

36. ▲ ■ The solid state structure of silicon carbide, SiC, is shown below. Knowing that the Si—C bond length is 188.8 pm (and the Si—C—Si bond angle is 109.5°), calculate the density of SiC.

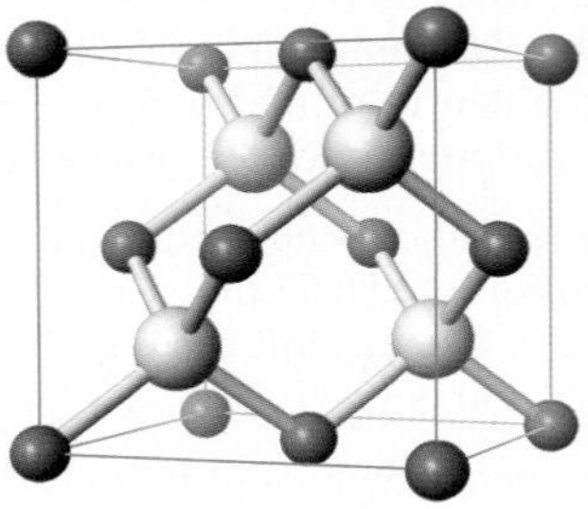

Unit cell of SiC.

Charles D. Winters

Sample of silicon carbide.

37. Spinels are solids with the general formula AB_2O_4 (where A^{2+} and B^{3+} are metal cations of the same or different metals. The best-known example is common magnetite, Fe_3O_4 [which you can formulate as $(Fe^{2+})(Fe^{3+})_2O_4$]. Another example is the mineral often referred to as spinel, $MgAl_2O_4$.

Charles D. Winters

A crystal of the spinel $MgAl_2O_4$ on a marble chip.

The oxide ions of spinels form a face-centered cubic lattice. In a *normal spinel*, cations occupy ⅛ of the tetrahedral sites and ½ of the octahedral sites.

(a) In $MgAl_2O_4$, in what type of holes are the magnesium and aluminum ions found?

(b) The mineral chromite has the formula $FeCr_2O_4$. What ions are involved, and in what type of holes are they found?

38. ■ Using the thermochemical data below, and an estimated value of −2481 kJ/mol for the lattice energy for Na_2O, calculate the value for the *second* electron affinity of oxygen [$O^-(g) + e^- \rightarrow O^{2-}(g)$].

Quantity	Numerical Value (kJ/mol)
Enthalpy of atomization of Na	107.3
Ionization Energy of Na	495.9
Enthalpy of formation of solid Na_2O	−418.0
Enthalpy of formation of O(g) from O_2	249.1
First electron affinity of O	−141.0

In the Laboratory

39. Lead sulfide, PbS (commonly called galena), has the same formula as ZnS.

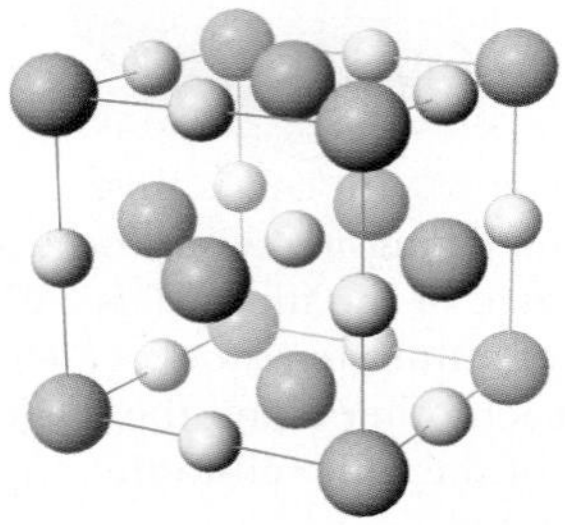

Unit cell of PbS.

Charles D. Winters

Sample of galena.

Does PbS have the same solid structure as ZnS? If different, how are they different? How is the unit cell of PbS related to its formula?

40. $CaTiO_3$, a perovskite, has the structure below.
(a) If the density of the solid is 4.10 g/cm^3, what is the length of a side of the unit cell?
(b) Calculate the radius of the Ti^{4+} ion in the center of the unit cell. How well does your calculation agree with a literature value of 75 pm?

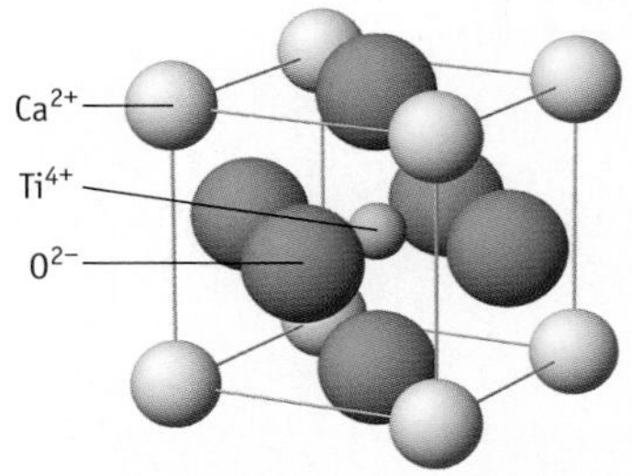

Unit cell of the perovskite $CaTiO_3$.

©DEA/C. Bevilacqua/Getty Images

A sample of perovskite $CaTiO_3$.

Summary and Conceptual Questions

The following questions may use concepts from this and previous chapters.

41.▲ Boron phosphide, BP, is a semiconductor and a hard, abrasion-resistant material. It is made by reacting boron tribromide and phosphorus tribromide in a hydrogen atmosphere at high temperature (> 750 °C).
(a) Write a balanced chemical equation for the synthesis of BP. *(Hint: Hydrogen is a reducing agent.)*
(b) Boron phosphide crystallizes in a zinc blende structure, formed from boron atoms in a face-centered cubic lattice and phosphorus atoms in tetrahedral holes. How many tetrahedral holes are filled with P atoms in each unit cell?
(c) The length of a unit cell of BP is 478 pm. What is the density of the solid in g/cm^3.
(d) Calculate the closest distance between a B and a P atom in the unit cell. (Assume the B atoms do not touch along the cell edge. The B atoms in the faces touch the B atoms at the corners of the unit cell. See page 594.)

42. ▲ Why is it not possible for a salt with the formula M_3X (Na_3PO_4, for example) to have a face-centered cubic lattice of X anions with M cations in octahedral holes?

43. ▲ ■ Two identical swimming pools are filled with uniform spheres of ice packed as closely as possible. The spheres in the first pool are the size of grains of sand; those in the second pool are the size of oranges. The ice in both pools melts. In which pool, if either, will the water level be higher? (Ignore any differences in filling space at the planes next to the walls and bottom.)

44. Spinels are described in Study Question 37. Consider two normal spinels, $CoAl_2O_4$ and $SnCo_2O_4$. What metal ions are involved in each? What are their electron configurations? Are the metal ions paramagnetic, and if so how many unpaired electrons are involved?

14 Solutions and Their Behavior

Safe Flying

You are sitting in an airport in the winter. Outside the weather is bad, with blowing snow and ice. You want to get on the plane home, but first the plane has to be de-iced. Ice and snow on the wings severely impairs the ability of wings to provide lift, so it is unsafe to try to take off unless the plane is de-iced. The fluid that is sprayed on the plane is a mixture of ethylene or propylene glycol and other substances. These are the same chemicals used in the antifreeze in your car's radiator.

Questions:

1. Why does ethylene glycol ($HOCH_2CH_2OH$) dissolve so well in water?
2. Why use a solution of ethylene glycol and water as an anti-freeze?
3. If you mix 100. g of ethylene glycol with 500. g of water, what is the freezing point of the mixture?

Answers to these questions are in Appendix Q.

Chapter Goals

See Chapter Goals Revisited (page 646) for Study Questions keyed to these goals and assignable in OWL.

- Calculate and use the solution concentration units molality, mole fractions, and weight percent.
- Understand the solution process.
- Understand and use the colligative properties of solutions.

Chapter Outline

We come into contact with solutions every day: aqueous solutions of ionic salts, gasoline with additives to improve its properties, and household cleaners such as ammonia in water. We purposely make solutions. Adding sugar, flavoring, and sometimes CO_2 to water produces a palatable soft drink. Athletes drink commercial beverages with dissolved salts to match salt concentrations in body fluids precisely, thus allowing the fluid to be taken into the body more rapidly. In medicine, saline solutions (aqueous solutions containing NaCl and other soluble salts) are infused into the body to replace lost fluids.

A **solution** is a homogeneous mixture of two or more substances in a single phase. By convention, the component present in largest amount is identified as the solvent and the other component(s) as the solute(s) (Figure 14.1). Although other types of solutions exist (such as alloys, solid solution of metals), the objective in this chapter is to develop an understanding of gases, liquids, and solids dissolved in liquid solvents.

Experience tells you that adding a solute to a pure liquid will change the properties of the liquid. Indeed, that is the reason some solutions are made. For instance, adding antifreeze to the water in your car's radiator prevents the coolant from boiling in the summer and freezing in the winter. The changes that occur in the freezing and boiling points when a substance is dissolved in a pure liquid are two observations we shall examine in detail. These properties, as well as the osmotic pressure of a solution and changes in vapor pressure, are examples of colligative properties.

Chemistry.Now™

Throughout the text this icon introduces an opportunity for self-study or to explore interactive tutorials by signing in at **www.thomsonedu.com/login**.

(a) Copper(II) chloride, the solute, is added to water, the solvent.

(b) Interactions between water molecules and Cu^{2+} and Cl^- ions allow the solid to dissolve. The ions are now sheathed with water molecules.

FIGURE 14.1 Making a solution of copper(II) chloride (the solute) in water (the solvent). When ionic compounds dissolve in water, each ion is surrounded by water molecules. The number of water molecules is usually six, but fewer are possible.

Charles D. Winters

$V_{soln} > 1.00$ L
V_{H_2O} added $= 1.00$ L
0.100 molal solution

$V_{soln} = 1.00$ L
V_{H_2O} added < 1.00 L
0.100 molar solution

FIGURE 14.2 Preparing 0.100 molal and 0.100 molar solutions. In the flask on the right, 0.100 mol (19.4 g) of K_2CrO_4 was mixed with enough water to make 1.000 L of solution. (The volumetric flask was filled to the mark on its neck, indicating that the volume is 1.000 L. Slightly less than 1.00 L of water was added.) If 1.00 kg of water was added to 0.100 mol of K_2CrO_4 in the flask on the left, the volume of solution is greater than 1.000 L. (The small pile of yellow solid in front of the flasks is 0.100 mol of K_2CrO_4.)

Colligative properties are properties of solutions that depend only on the number of solute particles per solvent molecule and not on the identity of the solute.

14.1 Units of Concentration

To analyze the colligative properties of solutions, we need ways of measuring solute concentrations that reflect the number of molecules or ions of solute per molecule of solvent.

Molarity, a concentration unit useful in stoichiometry calculations, is not useful when dealing with most colligative properties. Recall that molarity (M) is defined as the number of moles of solute per liter of solution (◀ page 174), so using molarity does not allow us to identify the exact amount of solvent used to make the solution. This fact is illustrated in Figure 14.2. The flask on the right contains a 0.100 M aqueous solution of potassium chromate. It was made by adding enough water to 0.100 mol of K_2CrO_4 to make 1.000 L of solution. There is no way to identify the amount of solvent (water) that was actually added. If 1.000 L of water had been added to 0.100 mol of K_2CrO_4, as illustrated with the flask on the left in Figure 14.2, the volume of solution would be greater than 1.000 L.

Three concentration units are described here that reflect the number of molecules or ions of solute per solvent molecule: molality, mole fraction, and weight percent.

The **molality**, *m*, of a solution is defined as the amount of solute (mol) per kilogram of solvent.

$$\text{Concentration } (c, \text{ mol/kg}) = \text{molality of solute} = \frac{\text{amount of solute (mol)}}{\text{mass of solvent (kg)}} \quad (14.1)$$

The molality of K_2CrO_4 in the flask on the left side of Figure 14.2 is 0.100 mol/kg. It was prepared from 0.100 mol (19.4 g) of K_2CrO_4 and 1.00 kg (1.000 L × 1.00 kg/L) of water.

Notice that different quantities of water were used to make the 0.100 M (0.100 molar) and 0.100 *m* (0.100 molal) solutions of K_2CrO_4. This means the *molarity and the molality of a given solution cannot be the same* (although the difference may be negligibly small when the solution is quite dilute).

The **mole fraction**, *X*, of a solution component is defined as the amount of that component (n_A) divided by the total amount of all of the components of the mixture ($n_A + n_B + n_C + ...$). Mathematically it is represented as

$$\text{Mole fraction of A } (X_A) = \frac{n_A}{n_A + n_B + n_C + \ldots} \quad (14.2)$$

Consider a solution that contains 1.00 mol (46.1 g) of ethanol, C_2H_5OH, in 9.00 mol (162 g) of water. The mole fraction of alcohol is 0.100, and that of water is 0.900.

$$X_{ethanol} = \frac{\text{1.00 mol ethanol}}{\text{1.00 mol ethanol} + \text{9.00 mol water}} = 0.100$$

$$X_{water} = \frac{\text{9.00 mol water}}{\text{1.00 mol ethanol} + \text{9.00 mol water}} = 0.900$$

Notice that the sum of the mole fractions of the components in the solution equals 1.000, a relationship that is true for all solutions.

■ **Molarity and Molality** The use of the terms "molar" (symbol M) and "molal" (symbol *m*) is common practice among chemists. Recently, however, NIST suggested that use of these terms and symbols to represent concentrations should be discontinued and replaced with the formal units of concentration (mol/L and mol/kg).

Weight percent is the mass of one component divided by the total mass of the mixture, multiplied by 100%:

$$\text{Weight \% A} = \frac{\text{mass of A}}{\text{mass of A} + \text{mass of B} + \text{mass of C} + \ldots} \times 100\% \qquad \textbf{(14.3)}$$

The alcohol–water mixture has 46.1 g of ethanol and 162 g of water, so the total mass of solution is 208 g, and the weight % of alcohol is

$$\text{Weight \% ethanol} = \frac{46.1 \text{ g ethanol}}{46.1 \text{ g ethanol} + 162 \text{ g water}} \times 100\% = 22.2\%$$

Charles D. Winters

FIGURE 14.3 Weight percent. The composition of many common products is often given in terms of weight percent. Here, the label on the cat spray indicates it contains 1.15% active ingredients.

Weight percent is a common unit in consumer products (Figure 14.3). Vinegar, for example, is an aqueous solution containing approximately 5% acetic acid and 95% water. The label on a common household bleach lists its active ingredient as 6.00% sodium hypochlorite (NaOCl) and 94.00% inert ingredients.

Naturally occurring solutions are often very dilute. Environmental chemists, biologists, geologists, oceanographers, and others frequently use **parts per million (ppm)** to express their concentrations. The unit ppm refers to relative quantities by mass; 1.0 ppm represents 1.0 g of a substance in a sample with a total mass of 1.0 million g. Because water at 25 °C has a density of 1.0 g/mL, a concentration of 1.0 mg/L is equivalent to 1.0 mg of solute in 1000 g of water or to 1.0 g of solute in 1,000,000 g of water; that is, units of ppm and mg/L are approximately equivalent.

Chemistry Now™

Sign in at **www.thomsonedu.com/login** and go to Chapter 14 Contents to see Screen 14.2 for an exercise on **calculating solution concentrations in various units.**

EXAMPLE 14.1 Calculating Mole Fractions, Molality, and Weight Percent

Problem Assume you add 1.2 kg of ethylene glycol, $HOCH_2CH_2OH$, as an antifreeze to 4.0 kg of water in the radiator of your car. What are the mole fraction, molality, and weight percent of the ethylene glycol?

Strategy Calculate the amount of ethylene glycol and water, and then use Equations 14.1–14.3.

Solution The 1.2 kg of ethylene glycol (molar mass = 62.1 g/mol) is equivalent to 19 mol, and 4.0 kg of water represents 220 mol.

Mole fraction:

$$X_{\text{glycol}} = \frac{19 \text{ mol ethylene glycol}}{19 \text{ mol ethylene glycol} + 220 \text{ mol water}} = \mathbf{0.080}$$

Molality:

$$c_{\text{glycol}} = \frac{19 \text{ mol ethylene glycol}}{4.0 \text{ kg water}} = 4.8 \text{ mol/kg} = \mathbf{4.8}\ m$$

Weight percent:

$$\text{Weight \%} = \frac{1.2 \times 10^3 \text{ g ethylene glycol}}{1.2 \times 10^3 \text{g ethylene glycol} + 4.0 \times 10^3 \text{ g water}} \times 100\% = \mathbf{23\%}$$

Charles D. Winters

Commercial antifreeze. This solution contains ethylene glycol, $HOCH_2CH_2OH$, an organic alcohol that is readily soluble in water. Regulations specify that the weight percent of ethylene glycol in ethylene glycol–based antifreeze must be at least 75%. (The remainder of the solution can be other glycols and water.)

EXERCISE 14.1 Mole Fraction, Molality, Weight Percent, and Parts per Million

(a) If you dissolve 10.0 g (about one heaping teaspoonful) of sugar (sucrose, $C_{12}H_{22}O_{11}$), in a cup of water (250. g), what are the mole fraction, molality, and weight percent of sugar?

(b) Sea water has a sodium ion concentration of 1.08×10^4 ppm. If the sodium is present in the form of dissolved sodium chloride, what mass of NaCl is in each liter of sea water? Sea water is denser than pure water because of dissolved salts. Its density is 1.05 g/mL.

14.2 The Solution Process

■ **Unsaturated** The term unsaturated is used when referring to solutions with concentrations of solute that are less than that of a saturated solution.

If solid $CuCl_2$ is added to a beaker of water, the salt will begin to dissolve (see Figure 14.1). The amount of solid diminishes, and the concentrations of $Cu^{2+}(aq)$ and $Cl^-(aq)$ in the solution increase. If we continue to add $CuCl_2$, however, we will eventually reach a point when no additional $CuCl_2$ seems to dissolve. The concentrations of $Cu^{2+}(aq)$ and $Cl^-(aq)$ will not increase further, and any additional solid $CuCl_2$ added after this point will remain as a solid at the bottom of the beaker. We say that such a solution is **saturated**.

Although no change is observed on the macroscopic level, it is a different matter on the particulate level. The process of dissolving continues, with Cu^{2+} and Cl^- ions leaving the solid state and entering solution. Concurrently, a second process is occurring: the formation of solid $CuCl_2(s)$ from $Cu^{2+}(aq)$ and $Cl^-(aq)$. The rates at which $CuCl_2$ is dissolving and reprecipitating are equal in a saturated solution, so that no net change is observed on the macroscopic level.

A Closer Look — Supersaturated Solutions

Although at first glance it may seem a contradiction, it is possible for a solution to hold more dissolved solute than the amount in a saturated solution. Such solutions are referred to as **supersaturated** solutions. Supersaturated solutions are unstable, and the excess solid eventually crystallizes from the solution until the equilibrium concentration of the solute is reached.

The solubility of substances often decreases if the temperature is lowered. Supersaturated solutions are usually made by preparing a saturated solution at a given temperature and then carefully cooling it. If the rate of crystallization is slow, the solid may not precipitate when the solubility is exceeded. Going to still lower temperatures results in a solution that has more solute than the amount defined by equilibrium conditions; it is supersaturated.

When disturbed in some manner, a supersaturated solution moves toward equilibrium by precipitating solute. This change can occur rapidly, often with the evolution of thermal energy. In fact, supersaturated solutions are used in "heat packs" to apply heat to injured muscles. When crystallization of sodium acetate ($NaCH_3CO_2$) from a supersaturated solution in a heat pack is initiated, the temperature of the heat pack rises to about 50 °C, and crystals of solid sodium acetate are detectable inside the bag.

Charles D. Winters

Supersaturated solutions. When a supersaturated solution is disturbed, the dissolved salt (here sodium acetate, $NaCH_3CO_2$) rapidly crystallizes. (See ChemistryNow Screen 14.2 to watch a video of this process.)

Charles D. Winters

Heat of crystallization. A heat pack relies on the heat evolved by the crystallization of sodium acetate. (See ChemistryNow Screen 14.6 to watch a video of a heat pack.)

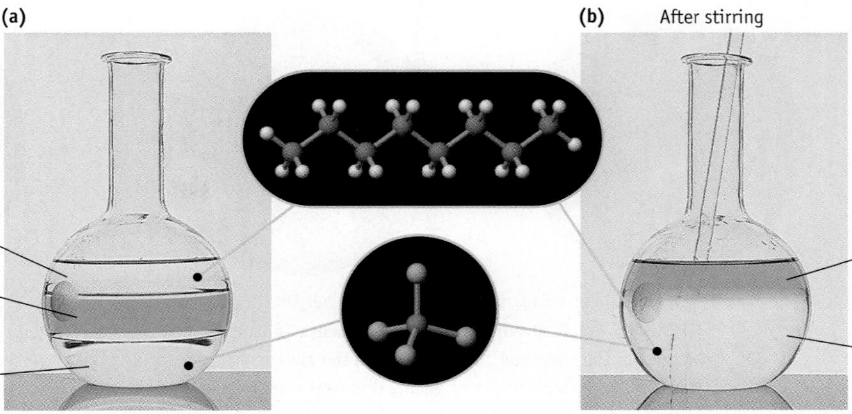

FIGURE 14.4 Miscibility. (a) The colorless, denser bottom layer is nonpolar carbon tetrachloride, CCl_4. The blue middle layer is a solution of $CuSO_4$ in water, and the colorless, less dense top layer is nonpolar octane, C_8H_{18}. This mixture was prepared by carefully layering one liquid on top of another, without mixing. (b) After stirring the mixture, the two nonpolar liquids form a homogeneous mixture. This layer of mixed liquids is under the water layer because the mixture of CCl_4 and C_8H_{18} has a greater density than water.

This process is another example of a dynamic equilibrium (◀ page 119), and we can describe the situation in terms of an equation with substances linked by a set of double arrows ($\rightleftharpoons$):

$$CuCl_2(s) \rightleftharpoons Cu^{2+}(aq) + 2\ Cl^-(aq)$$

A saturated solution gives us a way to define precisely the solubility of a solid in a liquid. **Solubility** is the concentration of solute in equilibrium with undissolved solute in a saturated solution. The solubility of $CuCl_2$, for example, is 70.6 g in 100 mL of water at 0 °C. If we add 100.0 g of $CuCl_2$ to 100 mL of water at 0 °C, we can expect 70.6 g to dissolve, and 29.4 g of solid to remain.

Liquids Dissolving in Liquids

If two liquids mix to an appreciable extent to form a solution, they are said to be **miscible**. In contrast, **immiscible** liquids do not mix to form a solution; they exist in contact with each other as separate layers (see Figures 12.5 and 14.4).

The polar compounds ethanol (C_2H_5OH) and water are miscible in all proportions as are the nonpolar liquids octane (C_8H_{18}) and carbon tetrachloride (CCl_4). On the other hand, neither C_8H_{18} nor CCl_4 is miscible with water. Observations like these have led to a familiar rule of thumb: *Like dissolves like.* That is, two or more nonpolar liquids frequently are miscible, just as are two or more polar liquids.

What is the molecular basis for the "like dissolves like" guideline? In pure water and pure ethanol, the major force between molecules is hydrogen bonding involving O—H groups. When the two liquids are mixed, hydrogen bonding between ethanol and water molecules also occurs and assists in the solution process. In contrast, molecules of pure octane or pure CCl_4, both of which are nonpolar, are held together in the liquid phase by dispersion forces (◀ Section 12.3). The energy associated with these forces of attraction is similar in value to the energy due to the forces of attraction between octane and CCl_4 molecules when these nonpolar liquids are mixed. Thus, little or no energy change occurs when octane–octane and CCl_4–CCl_4 attractive forces are replaced with octane–CCl_4 forces. The solution

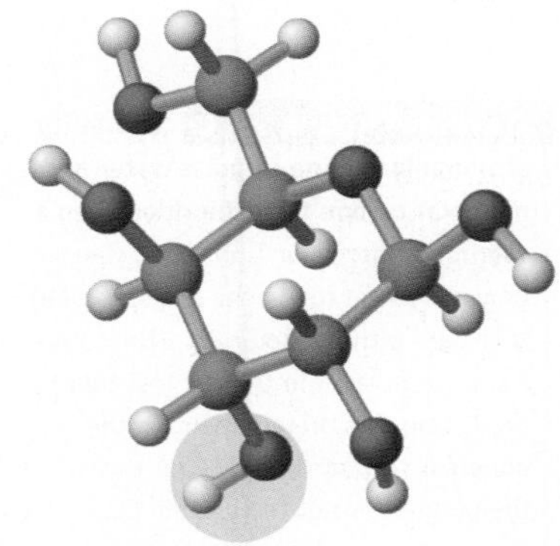

Like dissolves like. Glucose has five —OH groups on each molecule, groups that allow it to form hydrogen bonds with water molecules. As a result, glucose dissolves readily in water.

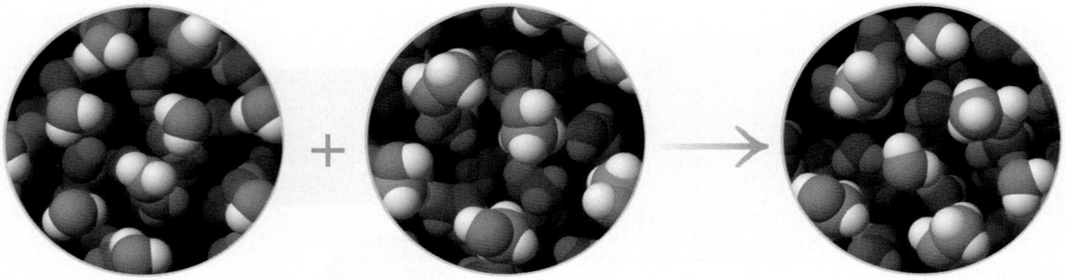

FIGURE 14.5 Driving the solution process—entropy. When two similar liquids—here water and methanol—are mixed, the molecules intermingle, and the energy of the system is more dispersed than in the two, separate pure liquids. A measure of this energy dispersal is entropy, a thermodynamic function described in more detail in Chapter 19.

process is expected to be nearly energy neutral. So, why do the liquids mix? The answer lies deeper in thermodynamics. As you shall see in Chapter 19, spontaneous changes, such as the mixing of liquids, are accompanied by an increase in entropy, a thermodynamic function that is a measure of the dispersal of the energy of the particles in the mixture relative to the pure liquids (Figure 14.5).

In contrast, polar and nonpolar liquids usually do not mix to an appreciable degree; when placed together in a container, they separate into two distinct layers (Figure 14.4). The explanation is complex and involves the interplay of the enthalpy of mixing and entropy. The enthalpy of mixing is zero or nearly so, but mixing dissimilar liquids leads to a decrease in entropy. As explained in Chapter 19, this means that mixing dissimilar liquids is not thermodynamically favorable.

■ **Entropy and the Solution Process** Although the energetics of solution formation are important, it is generally accepted that entropy is a more important contributor to the solution process. See Chapter 19 and T. P. Silverstein: "The real reason why oil and water don't mix." *Journal of Chemical Education*, Vol. 75, pp. 116–118, 1998.

Solids Dissolving in Water

The "like dissolves like" guideline also holds for molecular solids dissolving in liquids. Nonpolar solids such as naphthalene, $C_{10}H_8$, dissolve readily in nonpolar solvents such as benzene, C_6H_6, and hexane, C_6H_{14}. Iodine, I_2, a nonpolar inorganic solid, dissolves in water to some extent, but, given a choice, it dissolves to a larger extent in a nonpolar liquid such as CCl_4 (Figure 14.6). Sucrose (sugar), a polar molecular solid, is not very soluble in nonpolar solvents but is readily soluble in water, a fact that we know well because of its use to sweeten beverages. The presence of O—H groups in the structure of sugar and other substances such as glucose allows these molecules to interact with polar water molecules through hydrogen bonding.

"Like dissolves like" is a somewhat less effective but still useful guideline when considering the solubility of ionic solids. Thus, we can reasonably predict that ionic compounds, which can be considered extreme examples of polar compounds, will

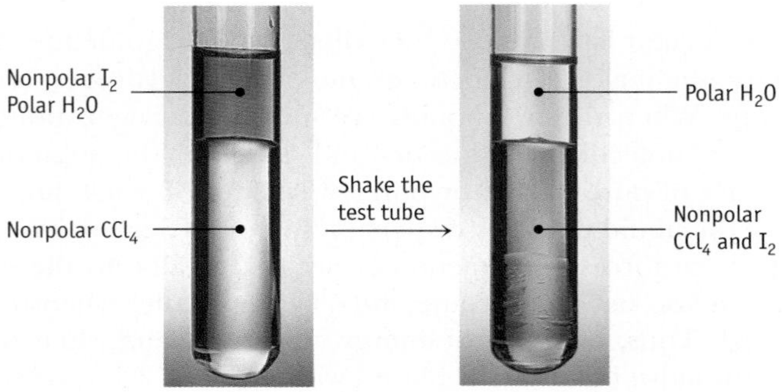

Active Figure 14.6 Solubility of nonpolar iodine in polar water and nonpolar carbon tetrachloride. When a solution of nonpolar I_2 in water (the brown layer on top in the left test tube) is shaken with nonpolar CCl_4 (the colorless bottom layer in the left test tube), the I_2 transfers preferentially to the nonpolar solvent. Evidence for this is the purple color of the bottom CCl_4 layer in the test tube on the right.

Chemistry Now™ Sign in at www.thomsonedu.com/login and go to the Chapter Contents menu to explore an interactive version of this figure accompanied by an exercise.

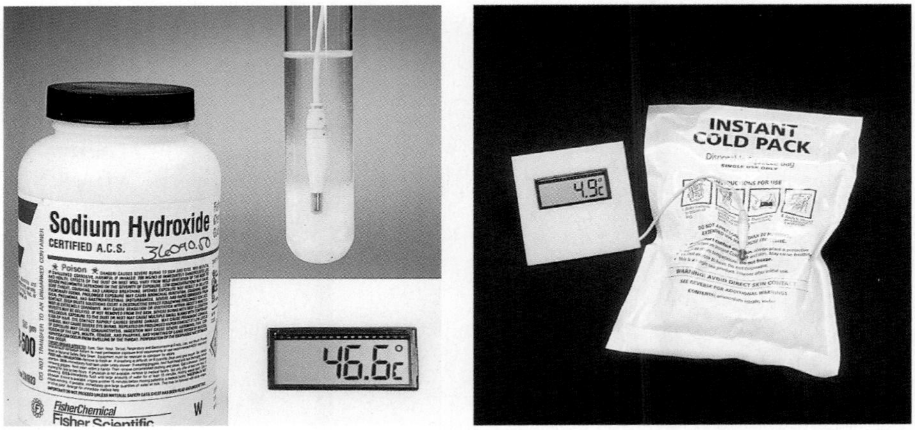

Photos: Charles D. Winters

(a) (b)

FIGURE 14.7 Dissolving ionic solids and enthalpy of solution. (a) Dissolving NaOH in water is a strongly exothermic process. (b) A "cold pack" contains solid ammonium nitrate, NH_4NO_3, and a package of water. When the water and NH_4NO_3 are mixed and the salt dissolves, the temperature of the system drops, owing to the endothermic enthalpy of solution of ammonium nitrate ($\Delta_{soln}H° = +25.7$ kJ/mol).

not dissolve in nonpolar solvents. This fact is amply borne out by observation. Sodium chloride, for example, will not dissolve in liquids such as hexane or CCl_4, but the salt does have a significant solubility in water. Many ionic compounds are soluble in water, but, according to the solubility guidelines on page 126, there are many other ionic solids that are not.

Predicting the solubility of ionic compounds in water is complicated. As mentioned earlier, two factors—enthalpy and entropy—together determine the extent to which one substance dissolves in another. For ionic compounds dissolving in water, entropy usually (but not always) favors solution. A favorable enthalpy factor (negative ΔH) generally leads to a compound being soluble. For example, when sodium hydroxide dissolves in water, the solution warms up (Figure 14.7a), and sodium hydroxide dissolves readily in water. An unfavorable enthalpy factor, however, does not guarantee that an ionic compound will not be soluble. When ammonium nitrate dissolves in water, the solution becomes colder (Figure 14.7b), but ammonium nitrate is still very soluble in water.

Network solids, including graphite, diamond, and quartz sand (SiO_2), do not dissolve in water. Indeed, where would all the beaches be if sand dissolved in water? The covalent chemical bonding in network solids is simply too strong to be broken; the lattice remains intact when in contact with water.

Enthalpy of Solution

To understand the energetics of the solution process, let us view this process at the molecular level. We will use the process of dissolving potassium fluoride, KF, in water to illustrate what occurs, and the energy-level diagram in Figure 14.8 will assist us in following the changes.

Solid potassium fluoride has an ionic crystal lattice with alternating K^+ and F^- ions held in place by attractive forces due to their opposite charges. In water, these ions are separated from each other and *hydrated*; that is, they are surrounded by water molecules (Figure 14.1). Ion–dipole forces of attraction bind water molecules strongly to each ion. The energy change to go from the reactant, KF(s), to the products, K^+(aq) and F^-(aq), can be considered to take place in two stages:

1. Energy must be supplied to separate the ions in the lattice against their attractive forces. This is the reverse of the process defining the lattice enthalpy of an ionic compound with an enthalpy equal to $-\Delta_{lattice}H$ (◀ page 600).

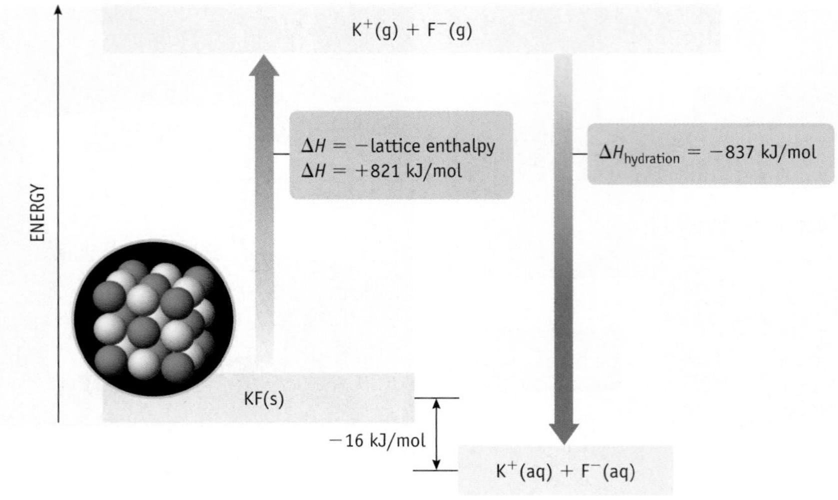

FIGURE 14.8 Model for energy changes on dissolving KF. An estimate of the magnitude of the energy change on dissolving an ionic compound in water is achieved by imagining it as occurring in two steps at the particulate level. Here, KF is first separated into cations and anions in the gas phase with an expenditure of 821 kJ per mol of KF. These ions are then hydrated, with $\Delta_{hydration}H$ estimated to be -837 kJ. Thus, the net energy change is -16 kJ, a slightly exothermic enthalpy of solution.

(See ChemistryNow Screen 14.4 for exercises on the energetics of solution formation.)

Separating the ions from one another is highly endothermic because the attractive forces between ions are strong.

2. Energy is evolved when the individual ions are transferred into water, where each ion becomes surrounded by water molecules. Again, strong forces of attraction (ion–dipole forces) are involved. This process, referred to as **hydration** when water is the solvent, is strongly exothermic.

We can therefore represent the process of dissolving KF in terms of these chemical equations:

Step 1	$KF(s) \longrightarrow K^+(g) + F^-(g)$	$-\Delta_{lattice}H$
Step 2	$K^+(g) + F^-(g) \longrightarrow K^+(aq) + F^-(aq)$	$\Delta_{hydration}H$

The overall reaction is the sum of these two steps. The enthalpy of the overall reaction, called the **enthalpy of solution** ($\Delta_{soln}H$), is the sum of the two enthalpies.

$$\text{Overall} \quad KF(s) \longrightarrow K^+(aq) + F^-(aq) \qquad \Delta_{soln}H = -\Delta_{lattice}H + \Delta_{hydration}H$$

We can use this to estimate the value of $\Delta_{hydration}H$. For example, we estimate the lattice energy for KF to be -821 kJ/mol using a Born-Haber cycle calculation (◀ page 600), and we measure the value of $\Delta_{soln}H$ in a calorimetry experiment to be -16.4 kJ/mol. From these two values, we can determine $\Delta_{hydration}H$ to be -837 kJ/mol.

As a general rule, to be soluble, a salt will have an enthalpy of solution that is exothermic or only slightly endothermic (Figure 14.9). In the latter instance, it is assumed that the enthalpy-disfavored solution process will be balanced by a favorable entropy of solution. If the enthalpy of solution is very endothermic—because of a low hydration energy, for example—then the compound is unlikely to be soluble. We can reasonably speculate that nonpolar solvents would not solvate ions strongly, and that solution formation would thus be energetically unfavorable. We therefore predict that an ionic compound, such as copper(II) sulfate, is not very soluble in nonpolar solvents such as carbon tetrachloride and octane (Figure 14.4).

It is also useful to recognize that the enthalpy of solution is the difference between two very large numbers. Small variations in either lattice energy or hydration enthalpies can determine whether a salt dissolves endothermically or exothermically.

Finally, notice that the two energy quantities, $\Delta_{lattice}H$ and $\Delta_{hydration}H$, are both affected by ion sizes and ion charges (◀ pages 557 and 600). A salt composed of smaller ions is expected to have a greater (more negative) lattice enthalpy because the ions can be closer together and experience higher attractive forces. However, the small

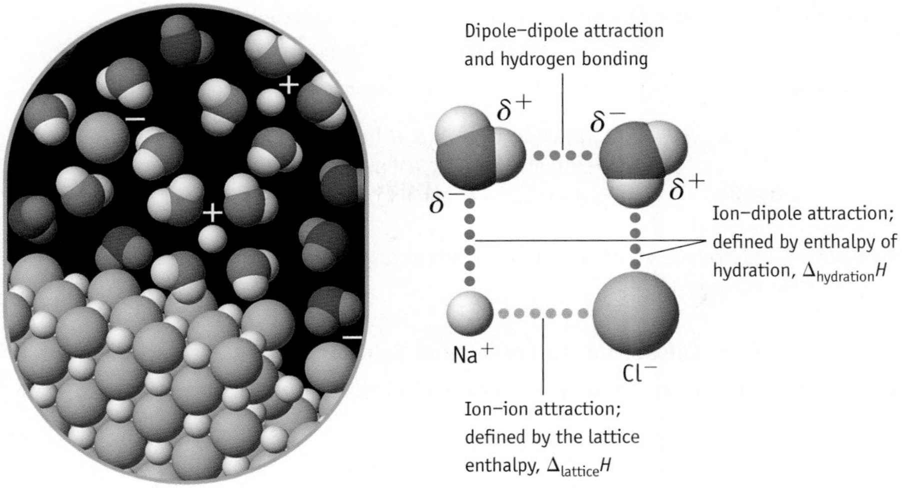

Active Figure 14.9 Dissolving an ionic solid in water. This process is a balance of forces. There are intermolecular forces between water molecules, and ion–ion forces are at work in the ionic crystal lattice. To dissolve, the ion–dipole forces between water and the ions (as measured by $\Delta_{hydration}H$) must overcome the ion–ion forces (as measured by $\Delta_{lattice}H$) and the intermolecular forces in water.

Chemistry Now™ Sign in at www.thomsonedu.com/login and go to the Chapter Contents menu to explore an interactive version of this figure accompanied by an exercise.

size will also allow a closer approach of solvent molecules and a greater solvation enthalpy. The net result is that simple correlations of solubility with structure (ionic radii) or thermodynamic parameters ($\Delta_{lattice}H$) are generally not successful.

Enthalpy of Solution: Thermodynamic Data

As mentioned earlier, the enthalpy of solution for a salt can be measured using a calorimeter. This is usually done in an open system such as the coffee-cup calorimeter described in Section 5.6. For an experiment run under standard conditions, the resulting measurement produces a value for the standard enthalpy of solution, $\Delta_{soln}H°$, where standard conditions refer to a concentration of 1 molal.

Tables of thermodynamic values often include values for the enthalpies of formation of aqueous solutions of salts. For example, a value of $\Delta_f H°$ for NaCl(aq) of −407.3 kJ/mol is listed in Table 14.1 and Appendix L. This value refers to the formation of a 1 *m* solution of NaCl from the elements. It may be considered to involve the enthalpies of two steps: (1) the formation of NaCl(s) from the elements Na(s) and Cl_2(g) in their standard states, and (2) the formation of a 1 *m* solution by dissolving solid NaCl in water:

Formation of NaCl(s):	$Na(s) + \frac{1}{2} Cl_2(g) \longrightarrow NaCl(s)$	$\Delta_f H° = -411.1$ kJ/mol
Dissolving NaCl:	$NaCl(s) \longrightarrow NaCl(aq, 1\ m)$	$\Delta_{soln}H° = +3.9$ kJ/mol
Net process:	$Na(s) + \frac{1}{2} Cl_2(g) \longrightarrow NaCl(aq, 1\ m)$	$\Delta_f H° = -407.3$ kJ/mol

TABLE 14.1 Data for Calculating Enthalpy of Solution

Compound	$\Delta_f H°$(s) (kJ/mol)	$\Delta_f H°$(aq, 1 *m*) (kJ/mol)
LiF	−616.9	−611.1
NaF	−573.6	−572.8
KF	−568.6	−585.0
RbF	−557.7	−583.8
LiCl	−408.7	−445.6
NaCl	−411.1	−407.3
KCl	−436.7	−419.5
RbCl	−435.4	−418.3
NaOH	−425.9	−469.2
NH_4NO_3	−365.6	−339.9

Chemistry Now™

Sign in at **www.thomsonedu.com/login** and go to Chapter 14 Contents to see

- Screen 14.3 for a visualization of the process and for a problem on **the solution process**
- Screen 14.4 for an analysis of the dissolution of KF and an exercise on **solution energetics**

EXAMPLE 14.2 Calculating an Enthalpy of Solution

Problem Use the data given in Table 14.1 to determine the enthalpy of solution for NH_4NO_3, the compound used in cold packs.

Strategy Use Equation 5.6 and data from Table 14.1 for reactants and products.

Solution The solution process for NH_4NO_3 is represented by the equation

$$NH_4NO_3(s) \longrightarrow NH_4NO_3(aq)$$

The enthalpy change for this process is calculated using enthalpies of formation given in Table 14.1:

$$\begin{aligned}\Delta_{soln}H^\circ &= \Sigma[\Delta_f H^\circ(\text{product})] - \Sigma[\Delta_f H^\circ(\text{reactant})] \\ &= \Delta_f H^\circ[NH_4NO_3(aq)] - \Delta_f H^\circ[NH_4NO_3(s)] \\ &= -339.9 \text{ kJ/mol} - (-365.6 \text{ kJ/mol}) = +25.7 \text{ kJ/mol}\end{aligned}$$

The process is endothermic, as indicated by the fact that $\Delta_{soln}H^\circ$ has a positive value and as verified by the experiment in Figure 14.7b.

EXERCISE 14.2 Calculating an Enthalpy of Solution

Use the data in Table 14.1 to calculate the enthalpy of solution for NaOH.

TABLE 14.2 Henry's Law Constants (25 °C)*

Gas	k_H (mol/kg · bar)
N_2	6.0×10^{-4}
O_2	1.3×10^{-3}
CO_2	0.034

*From http://webbook.nist.gov/chemistry/. Note: 1 bar = 0.9869 atm.

14.3 Factors Affecting Solubility: Pressure and Temperature

Pressure and temperature are two external factors that influence solubility. Both affect the solubility of gases in liquids, whereas only temperature is an important factor in the solubility of solids in liquids.

Dissolving Gases in Liquids: Henry's Law

The solubility of a gas in a liquid is directly proportional to the gas pressure. This is a statement of **Henry's law,**

$$S_g = k_H P_g \qquad (14.4)$$

where S_g is the gas solubility, P_g is the partial pressure of the gaseous solute, and k_H is Henry's law constant (Table 14.2), a constant characteristic of the solute and solvent.

Carbonated soft drinks illustrate how Henry's law works. These beverages are packed under pressure in a chamber filled with carbon dioxide gas, some of which dissolves in the beverage. When the can or bottle is opened, the partial pressure of CO_2 above the solution drops, which causes the solubility of CO_2 to drop. Gas bubbles out of the solution (Figure 14.10).

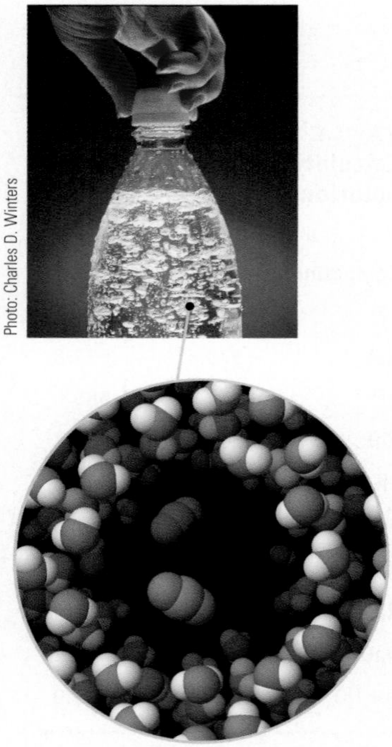

Photo: Charles D. Winters

FIGURE 14.10 Gas solubility and pressure. Carbonated beverages are bottled under CO_2 pressure. When the bottle is opened, the pressure is released, and bubbles of CO_2 form within the liquid and rise to the surface. After some time, an equilibrium between dissolved CO_2 and atmospheric CO_2 is reached. Because CO_2 provides some of the taste in the beverage, the beverage tastes flat when most of its dissolved CO_2 is lost.

Henry's law has important consequences in SCUBA diving. When you dive, the pressure of the air you breathe must be balanced against the external pressure of the water. In deeper dives, the pressure of the gases in the SCUBA gear must be several atmospheres and, as a result, more gas dissolves in the blood. This can lead to a problem. If you ascend too rapidly, you can experience a painful and potentially lethal condition referred to as "the bends," in which nitrogen gas bubbles form in the blood as the solubility of nitrogen decreases with decreasing pressure. In an effort to prevent the bends, divers may use a helium–oxygen mixture (rather than nitrogen–oxygen) because helium is not as soluble in blood as nitrogen.

We can better understand the effect of pressure on solubility by examining the system at the particulate level. The solubility of a gas is defined as the concentration of the dissolved gas in equilibrium with the substance in the gaseous state. At equilibrium, the rate at which solute gas molecules escape the solution and enter the gaseous state equals the rate at which gas molecules reenter the solution. An increase in pressure results in more molecules of gas striking the surface of the liquid and entering solution in a given time. The solution eventually reaches a new equilibrium when the concentration of gas dissolved in the solvent is high enough that the rate of gas molecules escaping the solution again equals the rate of gas molecules entering the solution.

Chemistry Now™

Sign in at **www.thomsonedu.com/login** and go to Chapter 14 Contents to see Screen 14.5 for an exercise and tutorial on **Henry's law.**

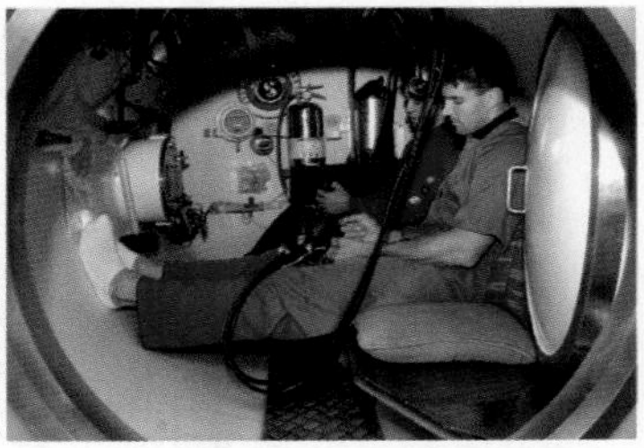

©Chris McGrath/ALLSPORT/Getty Images

Hyperbaric chamber, an illustration of Henry's law. A person is placed inside the chamber, and the partial pressure of oxygen is raised to several times the normal, atmospheric level. This raises the amount of oxygen dissolved in blood and tissues. This technique is used to treat, among other things, decompression sickness, carbon monoxide poisoning, severe anemia, and certain nonhealing wounds.

EXAMPLE 14.3 Using Henry's Law

Problem What is the concentration of O_2 in a fresh water stream in equilibrium with air at 25 °C and 1.0 bar? Express the answer in grams of O_2 per kg of solvent.

Strategy To use Henry's law to calculate the molar solubility of oxygen, the partial pressure of O_2 in air must first be calculated.

Solution The mole fraction of O_2 in air is 0.21, and, assuming the total pressure is 1.0 bar, the partial pressure of O_2 is 0.21 bar. Using this pressure for P_g in Henry's law, we have:

$$\text{Solubility of } O_2 = k_H P_g = \left(\frac{1.3 \times 10^{-3}\ \text{mol}}{\text{kg} \cdot \text{bar}}\right)(0.21\ \text{bar}) = 2.7 \times 10^{-4}\ \text{mol/l}$$

This concentration, in grams per liter, can then be calculated using the molar mass of O_2:

$$\text{Solubility of } O_2 = \left(\frac{2.7 \times 10^{-4}\ \text{mol}}{\text{kg}}\right)\left(\frac{32.0\ \text{g}}{\text{mol}}\right) = 0.0087\ \text{g/kg}$$

This concentration of O_2 (8.7 mg/kg) is quite low, but it is sufficient to provide the oxygen required by aquatic life.

Limitations of Henry's Law Henry's law holds quantitatively only for gases that do not interact chemically with the solvent. It does not accurately predict the solubility of NH_3 in water, for example, because this compound gives small concentrations of NH_4^+ and OH^- in water.

EXERCISE 14.3 Using Henry's Law

What is the concentration of CO_2 in water at 25 °C when the partial pressure is 0.33 bar? (Although CO_2 reacts with water to give traces of H^+ and HCO_3^-, the reaction occurs to such a small extent that Henry's law is obeyed at low CO_2 partial pressures.)

Temperature Effects on Solubility: Le Chatelier's Principle

The solubility of all gases in water decreases with increasing temperature. You may realize this from everyday observations such as the appearance of bubbles of air as water is heated below the boiling point.

To understand the effect of temperature on the solubility of gases, let us reexamine the enthalpy of solution. Gases that dissolve to an appreciable extent in water usually do so in an exothermic process

$$\text{Gas} + \text{liquid solvent} \xrightleftharpoons{\Delta_{soln}H < 0} \text{saturated solution} + \text{energy}$$

The reverse process, loss of dissolved gas molecules from a solution, requires energy as heat. These two processes can reach equilibrium eventually.

To understand how temperature affects solubility, we turn to **Le Chatelier's principle**, which states that a change in any of the factors determining an equilibrium causes the system to adjust by shifting in the direction that reduces or counteracts the effect of the change. If a solution of a gas in a liquid is heated, for example, the equilibrium will shift to absorb some of the added energy. That is, the reaction

$$\text{Gas} + \text{liquid solvent} \xrightleftharpoons[\text{Add energy. Equilibrium shifts left.}]{\text{Exothermic process } \Delta_{soln}H \text{ is negative.}} \text{saturated solution} + \text{energy}$$

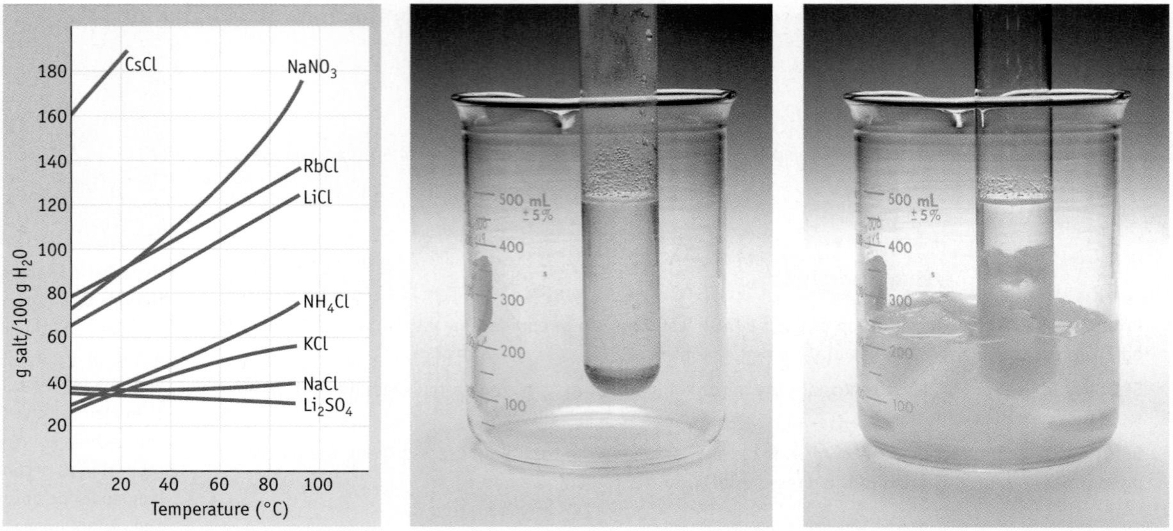

Photos: Charles D. Winters

(a) Temperature dependence of the solubility of some ionic compounds.

(b) NH_4Cl dissolved in water.

(c) NH_4Cl precipitates when the solution is cooled in ice.

FIGURE 14.11 The temperature dependence of the solubility of some ionic compounds in water. Most compounds, such as NH_4Cl, increase in solubility with increasing temperature.

Charles D. Winters

FIGURE 14.12 Giant crystals of potassium dihydrogen phosphate. The crystal being measured by this researcher at Lawrence Livermore Laboratory in California weighs 318 kg and measures 66 × 53 × 58 cm. The crystals were grown by suspending a thumbnail-sized seed crystal in a 6-foot tank of saturated KH_2PO_4. The temperature of the solution was gradually reduced from 65 °C over a period of about 50 days. The crystals are sliced into thin plates, which are used to convert light from a giant laser from infrared to ultraviolet.

shifts to the left if the temperature is raised because energy is absorbed in the process that produces free gas molecules and pure solvent. This shift corresponds to less gas dissolved and a lower solubility at higher temperature—the observed result.

The solubility of solids in water is also affected by temperature, but, unlike the situation involving solutions of gases, no general pattern of behavior is observed. In Figure 14.11, the solubilities of several salts are plotted versus temperature. The solubility of many salts increases with increasing temperature, but there are notable exceptions. Predictions based on whether the enthalpy of solution is positive or negative work most of the time, but exceptions do occur.

Chemists take advantage of the variation of solubility with temperature to purify compounds. If a compound is more soluble in a given solvent at higher temperatures, an impure sample of the compound is dissolved in the solvent at a high temperature. The solution is cooled to decrease the solubility (Figure 14.11c). When the limit of solubility is reached at the lower temperature, crystals of the pure compound form. If the process is done slowly and carefully, it is sometimes possible to obtain very large crystals (Figure 14.12).

Chemistry Now™

Sign in at **www.thomsonedu.com/login** and go to Chapter 14 Contents to see Screen 14.6 for a tutorial on **temperature and Le Chatelier's principle**.

Module 19

14.4 Colligative Properties

If you dissolve some salt in water, the vapor pressure of the water over the solution will decrease, and the water will evaporate less rapidly under the same conditions. In addition, the solution will freeze below 0 °C and boil above 100 °C. These observations refer to the colligative properties of the solution, properties that depend on the relative numbers of solute and solvent particles in a solution and not on their identity.

Changes in Vapor Pressure: Raoult's Law

The equilibrium vapor pressure at a particular temperature is the pressure of the vapor when the liquid and the vapor are in equilibrium (◄ page 573). When the vapor pressure of the solvent over a solution is measured at a given temperature, it is experimentally observed that

- The vapor pressure of the solvent over the solution is lower than the vapor pressure of the pure solvent.
- The vapor pressure of the solvent, $P_{solvent}$, is proportional to the relative number of solvent molecules in the solution; that is, the solvent vapor pressure is proportional to the solvent mole fraction, $P_{solvent} \propto X_{solvent}$.

Because solvent vapor pressure is proportional to the relative number of solvent molecules, we can write the following equation for the equilibrium vapor pressure of the solvent over a solution:

$$P_{solvent} = X_{solvent}\, P°_{solvent} \quad (14.5)$$

This equation, called **Raoult's law**, tells us that the vapor pressure of solvent over a solution ($P_{solvent}$) is some fraction of the pure solvent equilibrium vapor pressure ($P°_{solvent}$). For example, if 95% of the molecules in a solution are solvent molecules ($X_{solvent} = 0.95$), then the vapor pressure of the solvent ($P_{solvent}$) is 95% of $P°_{solvent}$.

■ **Raoult's Law** Raoult's law is named for Francois M. Raoult (1830–1901), a professor of chemistry at the University of Grenoble in France, who did the pioneering studies in this area.

Like the ideal gas law, Raoult's law describes a simplified model of a solution. We say that *an ideal solution is one that obeys Raoult's law.* No solution is ideal, however, just as no gas is truly ideal. Nevertheless, Raoult's law is a good approximation of solution behavior in many instances, especially at low solute concentration.

For Raoult's law to hold, the forces of attraction between solute and solvent molecules must be the same as those between solvent molecules in the pure solvent. This is frequently the case when molecules with similar structures are involved. Solutions of one hydrocarbon in another (hexane, C_6H_{14}, dissolved in octane, C_8H_{18}, for example) follow Raoult's law quite closely. If solvent–solute interactions are stronger than solvent–solvent interactions, the actual vapor pressure will be lower than calculated by Raoult's law. If the solvent–solute interactions are weaker than solvent–solvent interactions, the vapor pressure will be higher.

Chemistry Now™

Sign in at **www.thomsonedu.com/login** and go to Chapter 14 Contents to see Screen 14.7 for an exercise and a tutorial on **Raoult's law.**

■ EXAMPLE 14.4 Using Raoult's Law

Problem Suppose 651 g of ethylene glycol, $HOCH_2CH_2OH$, is dissolved in 1.50 kg of water. What is the vapor pressure of the water over the solution at 90 °C? Assume ideal behavior for the solution.

Strategy To use Raoult's law (Equation 14.5), we first must calculate the mole fraction of the solvent (water). We also need the vapor pressure of pure water at 90 °C (= 525.8 mm Hg, Appendix G).

Chemical Perspectives — Henry's Law and the Killer Lakes of Cameroon

It was evening on Thursday, August 21, 1986. Suddenly, people and animals around Lake Nyos in Cameroon, a small nation on the west coast of Africa, collapsed and died. By the next morning, 1700 people and hundreds of animals were dead. The calamity had no apparent cause—no fire, no earthquake, no storm. What had brought on this disaster?

Some weeks later, the mystery was solved. Lake Nyos and nearby Lake Monoun are crater lakes, which formed when cooled volcanic craters filled with water. Lake Nyos is lethal because it contains an enormous amount of dissolved carbon dioxide. The CO_2 in the lake was generated as a result of volcanic activity deep in the earth. Under the high pressures found at the bottom of the lake, a very large amount of CO_2 dissolved in the water.

On that fateful evening in 1986, something happened to disturb the lake. The CO_2-saturated water at the bottom of the lake was carried to the surface, where, under lower pressure, the gas was much less soluble. Approximately one cubic kilometer of carbon dioxide was released into the atmosphere, much like the explosive release of CO_2 from a can of carbonated beverage that has been shaken. The CO_2 shot up about 260 feet; then, because this gas is more dense than air, it hugged the ground and began to move with the prevailing breeze at about 45 miles per hour. When it reached the villages 12 miles away, vital oxygen was displaced. The result was that both people and animals were asphyxiated.

Thierry Orban/Corbis, Sygma

Lake Nyos in Cameroon (western Africa), the site of a natural disaster. In 1986, a huge bubble of CO_2 escaped from the lake and asphyxiated more than 1700 people.

In most lakes, this situation would not occur because lake water "turns over" as the seasons change. In the autumn, the top layer of water in a lake cools; its density increases; and the water sinks. This process continues, with warmer water coming to the surface and cooler water sinking. Dissolved CO_2 at the bottom of a lake would normally be expelled in this turnover process, but geologists found that the lakes in Cameroon are different. The chemocline, the boundary between deep water, rich in gas and minerals, and the upper layer, full of fresh water, stays intact. As carbon dioxide continues to enter the lake through vents in the bottom of the lake, the water becomes saturated with this gas. It is presumed that a minor disturbance—perhaps a small earthquake, a strong wind, or an underwater landslide—caused the lake water to turn over and led to the explosive and deadly release of CO_2.

Lake Nyos remains potentially deadly. Geologists estimate that the lake contains 10.6 to 14.1 billion cubic feet (300–400 million cubic meters) of carbon dioxide. This is about 16,000 times the amount found in an average lake that size.

A team of geologists from France and the United States has been working to resolve this potential threat. In early 2001, scientists lowered a pipe, about 200 meters long, into the lake. Now the pressure of escaping carbon dioxide causes a jet of water to rise as high as 165 feet in the air. Over the course of a year, about 20 million cubic meters of gas will be released. While this has been a successful first step, more gas must be removed to make the lake entirely safe, so additional vents are planned.

Solution We first calculate the amounts of water and ethylene glycol and, from these, the mole fraction of water.

$$\text{Amount of water} = 1.50 \times 10^3 \text{ g} \left(\frac{1 \text{ mol}}{18.02 \text{ g}}\right) = 83.2 \text{ mol water}$$

$$\text{Amount of ethylene glycol} = 651 \text{ g} \left(\frac{1 \text{ mol}}{62.07 \text{ g}}\right) = 10.5 \text{ mol glycol}$$

$$X_{water} = \frac{83.2 \text{ mol water}}{83.2 \text{ mol water} + 10.5 \text{ mol glycol}} = 0.888$$

Next, we apply Raoult's law, calculating the vapor pressure from the mole fraction of water and the vapor pressure of pure water:

$$P_{water} = X_{water}P^\circ_{water} = (0.888)(525.8 \text{ mm Hg}) = 467 \text{ mm Hg}$$

The dissolved solute decreases the vapor pressure by 59 mm Hg, or about 11%:

$$\Delta P_{water} = P_{water} - P^\circ_{water} = 467 \text{ mm Hg} - 525.8 \text{ mm Hg} = -59 \text{ mm Hg}$$

Comment Ethylene glycol dissolves easily in water, is noncorrosive, and is relatively inexpensive. Because of its high boiling point, it will not evaporate readily. These features make it ideal for use as antifreeze. It is, however, toxic to animals, so it is being replaced by less toxic propylene glycol for this application.

EXERCISE 14.4 Using Raoult's Law

Assume you dissolve 10.0 g of sucrose ($C_{12}H_{22}O_{11}$) in 225 mL (225 g) of water and warm the water to 60 °C. What is the vapor pressure of the water over this solution? (Appendix G lists $P^\circ(H_2O)$ at various temperatures.)

Adding a nonvolatile solute to a solvent lowers the vapor pressure of the solvent (Example 14.4). Raoult's law can be modified to calculate directly the lowering of the vapor pressure, $\Delta P_{solvent}$, as a function of the mole fraction of the solute.

$$\Delta P_{solvent} = P_{solvent} - P^\circ_{solvent}$$

Substituting Raoult's law for $P_{solvent}$, we have

$$\Delta P_{solvent} = (X_{solvent}\, P^\circ_{solvent}) - P^\circ_{solvent} = -(1 - X_{solvent})P^\circ_{solvent}$$

In a solution that has only the volatile solvent and one nonvolatile solute, the sum of the mole fraction of solvent and solute must be 1:

$$X_{solvent} + X_{solute} = 1$$

Therefore, $1 - X_{solvent} = X_{solute}$, and the equation for $\Delta P_{solvent}$ can be rewritten as

$$\Delta P_{solvent} = -X_{solute}\, P^\circ_{solvent} \qquad \textbf{(14.6)}$$

Thus, the change in the vapor pressure of the solvent is proportional to the mole fraction (the relative number of particles) of solute.

Boiling Point Elevation

Suppose you have a solution of a nonvolatile solute in the volatile solvent benzene. If the solute concentration is 0.200 mol in 100. g of benzene (C_6H_6) (= 2.00 mol/kg), this means that $X_{benzene} = 0.865$. Using $X_{benzene}$ and applying Raoult's law, we can calculate that the vapor pressure of the solvent at 60 °C will drop from 400. mm Hg for the pure solvent to 346 mm Hg for the solution:

$$P_{benzene} = X_{benzene}\,P°_{benzene} = (0.865)(400.\text{ mm Hg}) = 346\text{ mm Hg}$$

This point is marked on the vapor pressure graph in Figure 14.13. Now, what is the vapor pressure when the temperature of the solution is raised another 10 °C? The vapor pressure of pure benzene, $P°_{benzene}$, becomes larger with increasing temperature, so $P_{benzene}$ for the solution must also become larger. This new point, and additional ones calculated in the same way for other temperatures, define the vapor pressure curve for the solution (the lower curve in Figure 14.13).

An important observation we can make in Figure 14.13 is that the vapor pressure lowering caused by the nonvolatile solute leads to an increase in the boiling point. The normal boiling point of a liquid is the temperature at which its vapor pressure is equal to 1 atm or 760 mm Hg (◀ page 576). In Figure 14.13, we see that the normal boiling point of pure benzene (at 760 mm Hg) is about 80 °C. Tracing the vapor pressure curve for the solution, we also see that the vapor pressure reaches 760 mm Hg at a temperature about 5 °C higher than this value.

The vapor pressure curve and increase in the boiling point shown in Figure 14.13 refer specifically to a 2.00 *m* solution. We might wonder how the boiling point of the solution would vary with solute concentration. In fact, a simple relationship exists

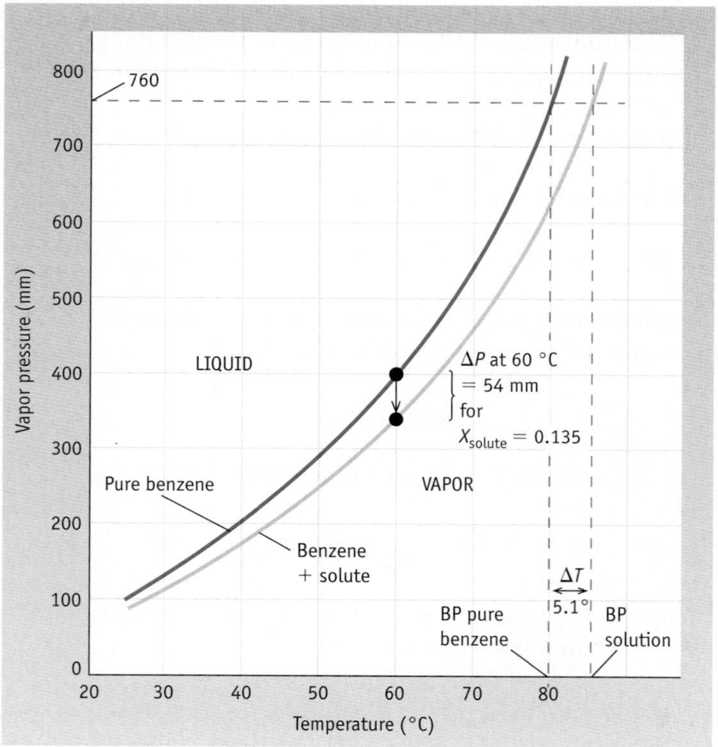

FIGURE 14.13 Lowering the vapor pressure of benzene by addition of a nonvolatile solute. The curve drawn in red represents the vapor pressure of pure benzene, and the curve in blue represents the vapor pressure of a solution containing 0.200 mol of a solute dissolved in 0.100 kg of solvent (2.00 *m*). This graph was created using a series of calculations such as those shown in the text. As an alternative, the graph could be created by measuring various vapor pressures for the solution in a laboratory experiment.

(See ChemistryNow Screen 14.8 Colligative Properties, to view an animation of this vapor pressure lowering.)

TABLE 14.3 Some Boiling Point Elevation and Freezing Point Depression Constants

Solvent	Normal Boiling Point (°C) Pure Solvent	K_{bp} (°C/m)	Normal Freezing Point (°C) Pure Solvent	K_{fp} (°C/m)
Water	100.00	+0.5121	0.0	−1.86
Benzene	80.10	+2.53	5.50	−5.12
Camphor	207.4	+5.611	179.75	−39.7
Chloroform ($CHCl_3$)	61.70	+3.63	—	—

between boiling point elevation and molal concentration: The boiling point elevation, ΔT_{bp}, is directly proportional to the molality of the solute.

$$\text{Elevation in boiling point} = \Delta T_{bp} = K_{bp} m_{solute} \qquad (14.6)$$

In this equation, K_{bp} is a proportionality constant called the **molal boiling point elevation constant**. It has the units of degrees/molal (°C/m). Values for K_{bp} are determined experimentally, and different solvents have different values (Table 14.3). Formally, the value corresponds to the elevation in boiling point for a 1 m solution.

Chemistry Now™

Sign in at **www.thomsonedu.com/login** and go to Chapter 14 Contents to see Screen 14.8 for an exercise and a tutorial on the **effect of a solute on the solution freezing point.**

EXAMPLE 14.5 Boiling Point Elevation

Problem Eugenol, the active ingredient in cloves, has the formula $C_{10}H_{12}O_2$. What is the boiling point of a solution containing 0.144 g of this compound dissolved in 10.0 g of benzene?

Strategy We can use Equation 14.6 to calculate the change in boiling point. This value is then added to the boiling point of pure benzene to provide the answer. To use Equation 14.6, you need a value of K_{bp} and the molality of the solution. The K_{bp} value for benzene is given in Table 14.3, but you need to calculate the molality, m.

Solution

$$0.144 \text{ g eugenol}\left(\frac{1 \text{ mol eugenol}}{164.2 \text{ g}}\right) = 8.77 \times 10^{-4} \text{ mol eugenol}$$

$$c_{eugenol} = \frac{8.77 \times 10^{-4} \text{ mol eugenol}}{0.0100 \text{ kg benzene}} = 8.77 \times 10^{-2}\ m$$

Use the value for the molality to calculate the boiling point elevation and then the boiling point:

$$\Delta T_{bp} = (2.53\ °\text{C}/m)(0.0877\ m) = 0.222\ °\text{C}$$

Because the boiling point rises relative to that of the pure solvent, the boiling point of the solution is

$$80.10\ °\text{C} + 0.222\ °\text{C} = 80.32\ °\text{C}$$

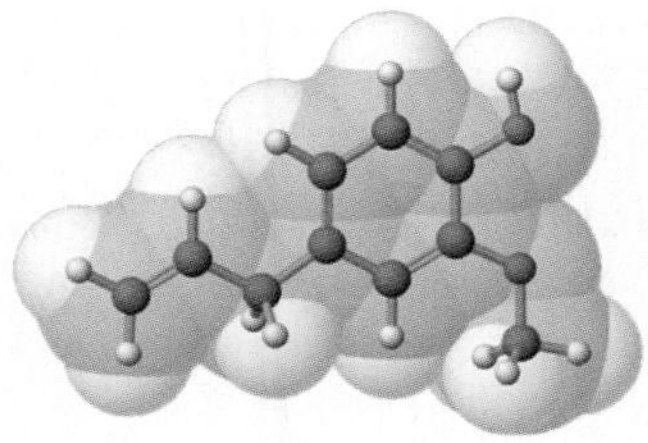

Eugenol, $C_{10}H_{12}O_2$, is an important component in oil of cloves.

EXERCISE 14.5 Boiling Point Elevation

What quantity of ethylene glycol, $HOCH_2CH_2OH$, must be added to 125 g of water to raise the boiling point by 1.0 °C? Express the answer in grams.

The elevation of the boiling point of a solvent on adding a solute has many practical consequences. One of them is the summer protection your car's engine receives from "all-season" antifreeze. The main ingredient of commercial antifreeze is ethylene glycol, $HOCH_2CH_2OH$. The car's radiator and cooling system are sealed to keep the coolant under pressure, ensuring that it will not vaporize at normal engine temperatures. When the air temperature is high in the summer, however, the radiator could "boil over" if it were not protected with "antifreeze." By adding this nonvolatile liquid, the solution in the radiator has a higher boiling point than that of pure water.

Freezing Point Depression

Another consequence of dissolving a solute in a solvent is that the freezing point of the solution is lower than that of the pure solvent (Figure 14.14). For an ideal solution, the depression of the freezing point is given by an equation similar to that for the elevation of the boiling point:

$$\text{Freezing point depression} = \Delta T_{fp} = K_{fp} m_{solute} \qquad \textbf{(14.7)}$$

where K_{fp} is the **freezing point depression constant** in degrees per molal (°C/m). Values of K_{fp} for a few common solvents are given in Table 14.3. The values are negative quantities, so the result of the calculation is a negative value for ΔT_{fp}, signifying a decrease in temperature.

The practical aspects of freezing point changes from pure solvent to solution are similar to those for boiling point elevation. The very name of the liquid you add to the radiator in your car, antifreeze, indicates its purpose (see Figure 14.14a). The label on the container of antifreeze tells you, for example, to add 6 qt (5.7 L) of antifreeze to a 12-qt (11.4-L) cooling system to lower the freezing point to −34 °C and to raise the boiling point to +109 °C.

EXAMPLE 14.6 Freezing Point Depression

Problem What mass of ethylene glycol, $HOCH_2CH_2OH$, must be added to 5.50 kg of water to lower the freezing point of the water from 0.0 °C to −10.0 °C?

Strategy To use Equation 14.7, you need K_{fp} (Table 14.3). You can then calculate the molality of the solution and, from this value, the amount and quantity of ethylene glycol required.

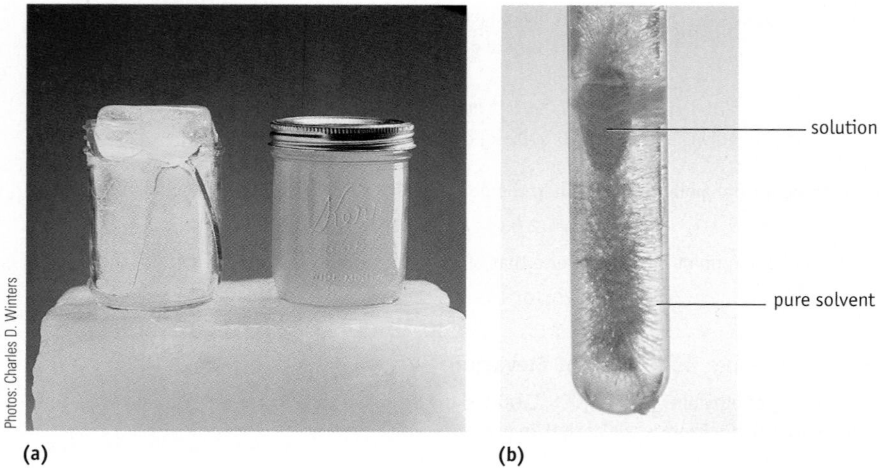

FIGURE 14.14 Freezing a solution. (a) Adding antifreeze to water prevents the water from freezing. Here, a jar of pure water (left) and a jar of water to which automobile antifreeze had been added (right) were kept overnight in the freezing compartment of a home refrigerator. (b) When a solution freezes, it is pure solvent that solidifies. To take this photo, a purple dye was dissolved in water, and the solution was frozen slowly. Pure ice formed along the walls of the tube, and the dye stayed in solution. The concentration of the solute increased as more and more solvent was frozen out, and the resulting solution had a lower and lower freezing point. Eventually, the system contains pure, colorless ice that formed along the walls of the tube and a concentrated solution of dye in the center of the tube.

Solution The solute concentration (molality) in a solution with a freezing point depression of −10.0 °C is

$$\text{Solute concentration } (m) = \frac{\Delta T_{fp}}{K_{fp}} = \frac{-10.0\ °C}{-1.86\ °C/m} = 5.38\ m$$

Because the solution contains 5.50 kg of water, we need 29.6 mol of ethylene glycol:

$$\left(\frac{5.38 \text{ mol glycol}}{1.00 \text{ kg water}}\right)(5.50 \text{ kg water}) = \boxed{29.6 \text{ mol glycol}}$$

The molar mass of ethylene glycol is 62.07 g/mol, so the mass required is 1840 g.

$$29.6 \text{ mol glycol}\left(\frac{62.07 \text{ g}}{1 \text{ mol}}\right) = \boxed{1840 \text{ g glycol}}$$

Comment The density of ethylene glycol is 1.11 kg/L, so the volume of glycol to be added is 1.84 kg (1 L/1.11 kg) = 1.66 L.

EXERCISE 14.6 Freezing Point Depression

In the northern United States, summer cottages are usually closed up for the winter. When doing so, the owners "winterize" the plumbing by putting antifreeze in the toilet tanks, for example. Will adding 525 g of $HOCH_2CH_2OH$ to 3.00 kg of water ensure that the water will not freeze at −25 °C?

Osmotic Pressure

Osmosis is the movement of solvent molecules through a semipermeable membrane from a region of lower solute concentration to a region of higher solute concentration. This movement can be demonstrated with a simple experiment. The beaker in Figure 14.15 contains pure water, and the bag and tube hold a concentrated sugar solution. The liquids are separated by a semipermeable membrane, a thin sheet of material (such as a vegetable tissue or cellophane) through

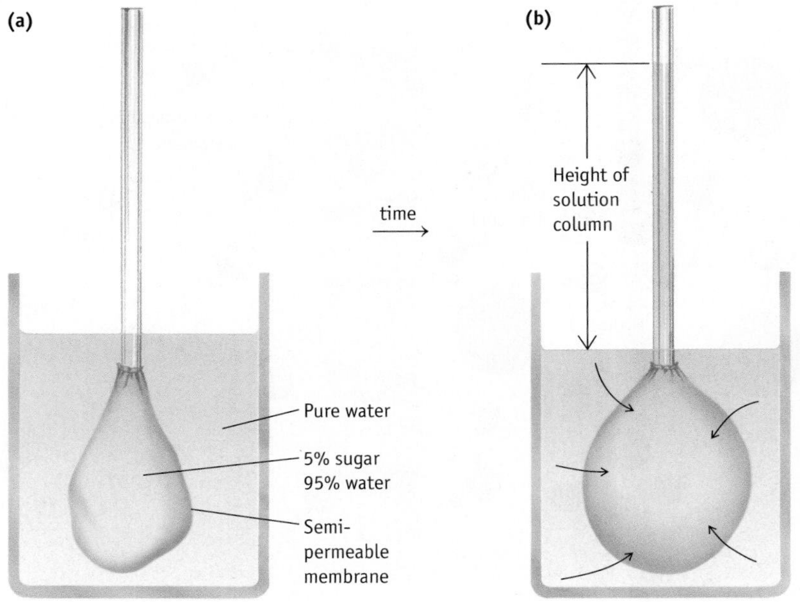

FIGURE 14.15 The process of osmosis. (a) The bag attached to the tube contains a solution that is 5% sugar and 95% water. The beaker contains pure water. The bag is made of a material that is semipermeable, meaning that it allows water, but not sugar molecules, to pass through. (b) Over time, water flows from the region of low solute concentration (pure water) to the region of higher solute concentration (the sugar solution). Flow continues until the pressure exerted by the column of solution in the tube above the water level in the beaker is great enough to result in equal rates of passage of water molecules in both directions. The height of the column of solution (b) is a measure of the osmotic pressure.

(See ChemistryNow Screen 14.9 Colligative Properties, for an animation of **osmosis.**)

which only certain types of molecules can pass. Here, water molecules can pass through the membrane, but larger sugar molecules (or hydrated ions) cannot (Figure 14.16). When the experiment is begun, the liquid levels in the beaker and the tube are the same. Over time, however, the level of the sugar solution inside the tube rises, the level of pure water in the beaker falls, and the sugar solution becomes more dilute. Eventually, no further net change occurs; equilibrium is reached.

From a molecular point of view, the semipermeable membrane does not present a barrier to the movement of water molecules, so they move through the membrane in both directions. Over time, more water molecules pass through the membrane from the pure water side to the solution side than in the opposite direction. In effect, water molecules tend to move from regions of low solute concentration to regions of high solute concentration. The same is true for any solvent, as long as the membrane allows solvent molecules but not solute molecules or ions to pass through.

Why does the system eventually reach equilibrium? Clearly, the solution in the tube in Figure 14.15 can never reach zero sugar or salt concentration, which would be required to equalize the number of water molecules moving through the membrane in each direction in a given time. The answer lies in the fact that the solution moves higher and higher in the tube as osmosis continues and water moves into the sugar solution. Eventually, the pressure exerted by this column of solution counterbalances the pressure exerted by the water moving through the membrane from the pure water side, and no further net movement of water occurs. An equilibrium of forces is achieved. The pressure created by the column of solution for the system at equilibrium is called the **osmotic pressure**, Π. A measure of this pressure is the difference between the height of the solution in the tube and the level of pure water in the beaker.

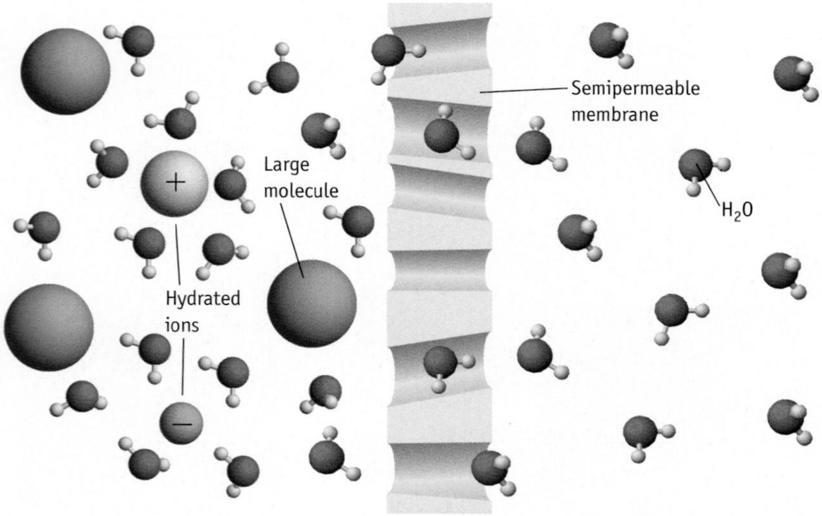

FIGURE 14.16 Osmosis at the particulate level. Osmotic flow through a membrane that is selectively permeable (semipermeable) to water. Dissolved substances such as hydrated ions or large sugar molecules cannot diffuse.

Osmotic pressure is a colligative property. From experimental measurements on dilute solutions, it is known that osmotic pressure and concentration (c) are related by the equation

$$\Pi = cRT \qquad \textbf{(14.8)}$$

In this equation, c is the molar concentration (in moles per liter); R is the gas constant; and T is the absolute temperature (in kelvins). Using a value for the gas law constant of 0.082057 L · atm/K · mol allows calculation of the osmotic pressure Π in atmospheres. This equation is analogous to the ideal gas law ($PV = nRT$), with Π taking the place of P and c being equivalent to n/V.

Because pressures on the order of 10^{-3} atm are easily measured, concentrations of very dilute solutions (as low as about 10^{-4} M) can be determined through measurements of osmotic pressure.

Other examples of osmosis are shown in Figure 14.17. In this case, the egg's membrane serves as the semipermeable membrane. Osmosis occurs in one direction if the concentration of solute is greater inside the egg than in the exterior solution and occurs in the other direction if the concentration solution is less inside the egg than it is in the exterior solution. In both cases, solvent flows from the region of low solute concentration to the region of high solute concentration.

Colligative Properties and Molar Mass Determination

Early in this book, you learned how to calculate a molecular formula from an empirical formula when given the molar mass. But how do you know the molar mass of an unknown compound? An experiment must be carried out to find this crucial piece of information, and one way to do so is to use a colligative property of a

Photos: Charles D. Winters

(a) A fresh egg is placed in dilute acetic acid. The acid reacts with the $CaCO_3$ of the shell but leaves the egg membrane intact.

(b) If the egg, with its shell removed, is placed in pure water, the egg swells.

(c) If the egg, with its shell removed, is placed in a concentrated sugar solution, the egg shrivels.

FIGURE 14.17 An experiment to observe osmosis. You can try this experiment in your kitchen. In the first step, use vinegar as a source of acetic acid.

(See ChemistryNow Screen 14.1, Puzzler, and Screen 14.9, Colligative Properties, for a video of this experiment.)

solution of the compound. The same basic logic is used for each of the colligative properties studied:

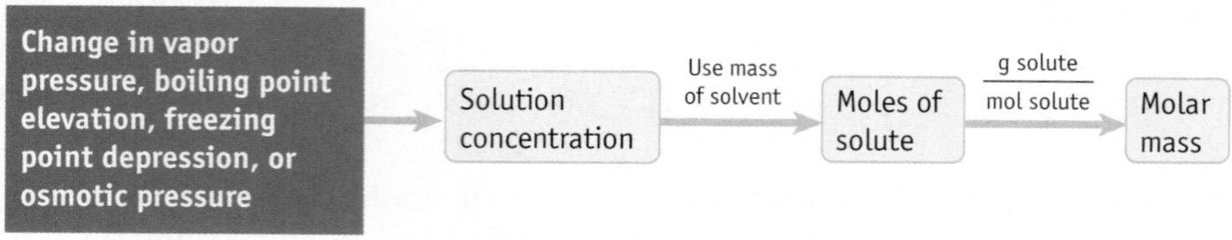

EXAMPLE 14.7 Determining Molar Mass from Boiling Point Elevation

Problem A solution prepared from 1.25 g of oil of wintergreen (methyl salicylate) in 99.0 g of benzene has a boiling point of 80.31 °C. Determine the molar mass of this compound.

Strategy Calculations using colligative properties to determine a molar mass always follow the pattern outlined in the text.

Solution We first use the boiling point elevation to calculate the solution concentration:

$$\text{Boiling point elevation } (\Delta T_{bp}) = 80.31\ °C - 80.10\ °C = 0.21\ °C$$

$$c_{solute} = \frac{\Delta T_{bp}}{K_{bp}} = \frac{0.21\ °C}{2.53\ °C/m} = 0.083\ m$$

The amount of solute in the solution is calculated from the solution concentration:

$$\text{Amount of solute} = \left(\frac{0.083\ \text{mol}}{1.00\ \text{kg}}\right)(0.099\ \text{kg solvent}) = 0.0082\ \text{mol solute}$$

Now we can combine the amount of solute with its mass to obtain its molar mass:

$$\frac{1.25\ \text{g}}{0.0082\ \text{mol}} = 150\ \text{g/mol}$$

Comment Methyl salicylate has the formula $C_8H_8O_3$ and a molar mass of 152.14 g/mol.

EXAMPLE 14.8 Osmotic Pressure and Molar Mass

Problem Beta-carotene is the most important of the A vitamins. Its molar mass can be determined by measuring the osmotic pressure generated by a given mass of β-carotene dissolved in the solvent chloroform. Calculate the molar mass of β-carotene if 10.0 mL of a solution containing 7.68 mg of β-carotene has an osmotic pressure of 26.57 mm Hg at 25.0 °C.

Strategy First, use Equation 14.8 to calculate the solution concentration from the osmotic pressure. Then, use the volume and concentration of the solution to calculate the amount of solute. Finally, find the molar mass of the solute from its mass and amount.

Solution The osmotic pressure can be used to calculate the concentration of β-carotene:

$$\text{Concentration (mol/L)} = \frac{\Pi}{RT} = \frac{(26.57\ \text{mm Hg})\left(\frac{1\ \text{atm}}{760\ \text{mm Hg}}\right)}{(0.082057\ \text{L}\cdot\text{atm/K}\cdot\text{mol})(298.2\ \text{K})}$$
$$= 1.429 \times 10^{-3}\ \text{mol/L}$$

Now the amount of β-carotene dissolved in 10.0 mL of solvent can be calculated:

$$(1.429 \times 10^{-3}\ \text{mol/L})(0.0100\ \text{L}) = 1.43 \times 10^{-5}\ \text{mol}$$

We can combine the amount of solute with its mass to calculate its molar mass:

$$\frac{7.68 \times 10^{-3}\ \text{g}}{1.43 \times 10^{-5}\ \text{mol}} = 538\ \text{g/mol}$$

Comment Beta-carotene is a hydrocarbon with the formula $C_{40}H_{56}$ (molar mass = 536.9 g/mol).

EXERCISE 14.7 Osmotic Pressure and Molar Mass

A 1.40-g sample of polyethylene, a common plastic, is dissolved in enough benzene to give exactly 100 mL of solution. The measured osmotic pressure of the solution is 1.86 mm Hg at 25 °C. Calculate the average molar mass of the polymer.

Charles D. Winters

Putting salt on ice assists in melting the ice.

Colligative Properties of Solutions Containing Ions

In the northern United States, it is common practice to scatter salt on snowy or icy roads or sidewalks. When the sun shines on the snow or patch of ice, a small amount melts, and some salt dissolves in the water. As a result of the dissolved solute, the freezing point of the solution is lower than 0 °C. The solution "eats" its way through the ice, breaking it up, and the icy patch is no longer dangerous for drivers or for people walking.

Salt (NaCl) is the most common substance used on roads because it is inexpensive and dissolves readily in water. Its relatively low molar mass means that the effect

A Closer Look — Osmosis and Medicine

Osmosis is of practical significance for people in the health professions. Patients who become dehydrated through illness often need to be given water and nutrients intravenously. Water cannot simply be dripped into a patient's vein, however. Rather, the intravenous solution must have the same overall solute concentration as the patient's blood: the solution must be isoosmotic or **isotonic** (Figures A and B, part a). If pure water was used, the inside of a blood cell would have a higher solute concentration (lower water concentration), and water would flow into the cell. This hypotonic situation would cause the red blood cells to burst (lyse) (Figure B, part c). The opposite situation, hypertonicity, occurs if the intravenous solution is more concentrated than the contents of the blood cell (Figure B, part b). In this case, the cell would lose water and shrivel up (crenate). To combat this, a dehydrated patient is rehydrated in the hospital with a sterile saline solution that is 0.16 M NaCl, a solution that is isotonic with the cells of the body.

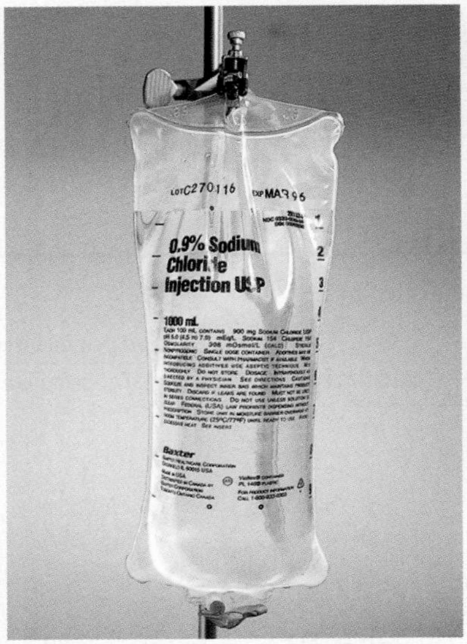

Charles D. Winters

FIGURE A An isotonic saline solution. This solution has the same molality as body fluids.

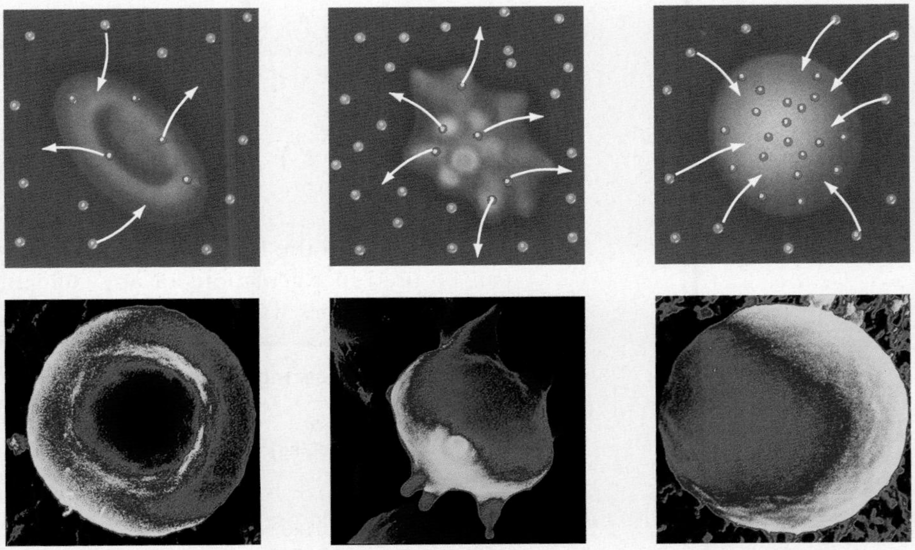

David Phillips/Science Source/Photo Researchers, Inc.

(a) Isotonic solution **(b)** Hypertonic solution **(c)** Hypotonic solution

FIGURE B Osmosis and living cells. (a) A cell placed in an isotonic solution. The net movement of water into and out of the cell is zero because the concentration of solutes inside and outside the cell is the same. (b) In a hypertonic solution, the concentration of solutes outside the cell is greater than that inside. There is a net flow of water out of the cell, causing the cell to dehydrate, shrink, and perhaps die. (c) In a hypotonic solution, the concentration of solutes outside the cell is less than that inside. There is a net flow of water into the cell, causing the cell to swell and perhaps to burst (or lyse).

per gram is large. In addition, salt is especially effective because it is an electrolyte. That is, it dissolves to give ions in solution:

$$NaCl(s) \rightarrow Na^+(aq) + Cl^-(aq)$$

Recall that colligative properties depend not on what is dissolved but *only on the number of particles of solute per solvent particle.* When 1 mol of NaCl dissolves, 2 mol of ions form, which means that the effect on the freezing point of water should be twice as large as that expected for a mole of sugar. A 0.100 *m* solution of NaCl really contains two solutes, 0.100 *m* Na^+ and 0.100 *m* Cl^-. What we should use to estimate the freezing point depression is the *total* molality of solute particles:

$$m_{total} = m(Na^+) + m(Cl^-) = (0.100 + 0.100) \text{ mol/kg} = 0.200 \text{ mol/kg}$$

$$\Delta T_{fp} = (-1.86\ °C/m)(0.200\ m) = -0.372\ °C$$

To estimate the freezing point depression for an ionic compound, first find the molality of solute from the mass and molar mass of the compound and the mass of the solvent. Then, multiply the molality by the number of ions in the formula: two for NaCl, three for Na_2SO_4, four for $LaCl_3$, five for $Al_2(SO_4)_3$, and so on.

Table 14.4 shows that as the concentration of NaCl decreases, ΔT_{fp} for NaCl approaches but does not quite reach a value that is two times larger than the value determined assuming no dissociation. Likewise, ΔT_{fp} for Na_2SO_4 approaches but does not reach a value that is three times larger. The ratio of the experimentally observed value of ΔT_{fp} to the value calculated, assuming no dissociation, is called the **van't Hoff factor** after Jacobus Henrikus van't Hoff (1852–1911), who was involved in studying this phenomenon. The van't Hoff factor is represented by *i*.

$$i = \frac{\Delta T_{fp}, \text{ measured}}{\Delta T_{fp}, \text{ calculated}} = \frac{\Delta T_{fp}, \text{ measured}}{K_{fp}\ m}$$

or

$$\Delta T_{fp} \text{ measured} = K_{fp} \times m \times i \qquad (14.9)$$

The numbers in the last column of Table 14.4 are van't Hoff factors. These values can be used in calculations of any colligative property. Vapor pressure lowering,

TABLE 14.4 Freezing Point Depressions of Some Ionic Solutions

Mass %	*m* (mol/kg)	ΔT_{fp} (measured, °C)	ΔT_{fp} (calculated, °C)	ΔT_{fp}, measured / ΔT_{fp}, calculated
NaCl				
0.00700	0.0120	−0.0433	−0.0223	1.94
0.500	0.0860	−0.299	−0.160	1.87
1.00	0.173	−0.593	−0.322	1.84
2.00	0.349	−1.186	−0.649	1.83
Na_2SO_4				
0.00700	0.00493	−0.0257	−0.00917	2.80
0.500	0.0354	−0.165	−0.0658	2.51
1.00	0.0711	−0.320	−0.132	2.42
2.00	0.144	−0.606	−0.268	2.26

Case Study — Henry's Law in a Soda Bottle

By now, everyone has seen it on YouTube or has tried it. Drop a Mentos™ into a bottle of soda (preferably diet soda), and a geyser of soda erupts from the bottle.

Charles D. Winters

A Mentos was dropped into a large bottle of Diet Coke. For more information, see J. E. Baur and M. B. Baur, *Journal of Chemical Education*, Vol. 83, pages 577–580, 2006.

It turns out this chapter, as well as the chapters on gases (Chapter 11), kinetics (Chapter 15), and equilibrium (Chapter 16) can help us explain what is happening.

Carbonated sodas are bottled under a high pressure of CO_2. Some of the gas dissolves in the soda, but some also remains in the small space above the liquid (called the "headspace"). The pressure of the CO_2 in the headspace is between 2 and 4 atm.

When the bottle cap is removed, the CO_2 in the headspace escapes rapidly. Some of the dissolved CO_2 also comes out of solution, and you observe this as bubbles of gas rising to the surface (Figure 14.10). If the bottle remains open, this continues until equilibrium is established with CO_2 in the atmosphere (where the partial pressure of CO_2 is 3.75×10^{-4} atm),

$$CO_2(\text{solution}) \rightleftharpoons CO_2(g)$$

and the soda goes "flat." If the newly opened soda bottle is undisturbed, however, the loss of CO_2 from solution is rather slow because bubble formation is not rapid, and your soda keeps its fizz.

But why is bubble formation slow? The reason for this is explained by the physics of bubble formation. For a bubble to form, nucleation sites must be available. These can be impurities in the water or the rough surface of an ice cube or bottle or drinking glass. The more nucleation sites there are available, the more rapid the bubble formation. The surface of a Mentos apparently has many such sites and promotes very rapid bubble formation.

Questions:

1. *If the headspace of a soda is 25 mL and the pressure of CO_2 in the space is 4.0 atm ($\approx$ 4.0 bar) at 25 °C, what amount of CO_2 is contained in the headspace?*
2. *If the CO_2 in the headspace escapes into the atmosphere where the partial pressure of CO_2 is 3.7×10^{-4} atm, what volume would the CO_2 occupy (at 25 °C)? By what amount did the CO_2 expand when it was released?*
3. *CO_2 obeys Henry's law to about 5 bar. What is the solubility of CO_2 in water at 25 °C when the pressure of the gas is 4.0 bar ($\approx$ 4.0 atm)? What amount of CO_2 is dissolved in 710 g of diet soda?*
4. *What is the solubility of CO_2 in water at 25 °C when the pressure of the gas is 3.7×10^{-4} bar?*

Answers to these questions are in Appendix Q.

boiling point elevation, freezing point depression, and osmotic pressure are all larger for electrolytes than for nonelectrolytes of the same molality.

The van't Hoff factor approaches a whole number (2, 3, and so on) only with very dilute solutions. In more concentrated solutions, the experimental freezing point depressions indicate that there are fewer ions in solution than expected. This behavior, which is typical of all ionic compounds, is a consequence of the strong attractions between ions. The result is as if some of the positive and negative ions are paired, decreasing the total molality of particles. Indeed, in more concentrated solutions, and especially in solvents less polar than water, ions are extensively associated in ion pairs and in even larger clusters.

EXAMPLE 14.9 Freezing Point and Ionic Solutions

Problem A 0.00200 *m* aqueous solution of an ionic compound, $Co(NH_3)_5(NO_2)Cl$, freezes at −0.00732 °C. How many moles of ions does 1.0 mol of the salt produce on being dissolved in water?

Strategy First, calculate ΔT_{fp} of the solution assuming no ions are produced. Compare this value with the actual value of ΔT_{fp}. The ratio will reflect the number of ions produced.

Solution The freezing-point depression expected for a 0.00200 m solution assuming that the salt does not dissociate into ions is

$$\Delta T_{fp} \text{ calculated} = K_{fp}m = (-1.86\ °C)(0.0200\ m) = -3.72 \times 10^{-3}\ °C$$

Now compare the calculated freezing point depression with the measured depression. This gives us the van't Hoff factor:

$$i = \frac{\Delta T_{fp}, \text{ measured}}{\Delta T_{fp}, \text{ calculated}} = \frac{-7.32 \times 10^{-3}\ °C}{-3.72 \times 10^{-3}\ °C} = 1.97 \approx 2$$

It appears that 1 mol of this compound gives 2 mol of ions. In this case, the ions are $[Co(NH_3)_5(NO_2)]^+$ and Cl^-.

EXERCISE 14.8 Freezing Point and Ionic Compounds

Calculate the freezing point of 525 g of water that contains 25.0 g of NaCl. Assume i, the van't Hoff factor, is 1.85 for NaCl.

14.5 Colloids

Charles D. Winters

Gold colloid. A water-soluble salt of $[AuCl_4]^-$ is reduced to give colloidal gold metal. The colloidal gold gives the dispersion its red color. (Similarly, colloidal gold is used to give a beautiful red color to glass.) Since the days of alchemy, some have claimed that drinking a colloidal gold solution "cleared the mind, increased intelligence and will power, and balanced the emotions."

Earlier in this chapter, we defined a solution broadly as a homogeneous mixture of two or more substances in a single phase (page 617). To this definition we should add that, in a true solution, no settling of the solute should be observed and the solute particles should be in the form of ions or relatively small molecules. Thus, NaCl and sugar form true solutions in water. You are also familiar with suspensions, which result, for example, if a handful of fine sand is added to water and shaken vigorously. Sand particles are still visible and gradually settle to the bottom of the beaker or bottle. **Colloidal dispersions**, also called **colloids**, represent a state intermediate between a solution and a suspension. Colloids include many of the foods you eat and the materials around you; among them are JELL-O®, milk, fog, and porcelain (see Table 14.5).

Around 1860, the British chemist Thomas Graham (1805–1869) found that substances such as starch, gelatin, glue, and albumin from eggs diffused only very slowly when placed in water, compared with sugar or salt. In addition, the former substances differ significantly in their ability to diffuse through a thin membrane: Sugar molecules can diffuse through many membranes, but the very large molecules that make up starch, gelatin, glue, and albumin do not. Moreover, Graham found that

TABLE 14.5 Types of Colloids

Type	Dispersing Medium	Dispersed Phase	Examples
Aerosol	Gas	Liquid	Fog, clouds, aerosol sprays
Aerosol	Gas	Solid	Smoke, airborne viruses, automobile exhaust
Foam	Liquid	Gas	Shaving cream, whipped cream
Foam	Solid	Gas	Styrofoam, marshmallow
Emulsion	Liquid	Liquid	Mayonnaise, milk, face cream
Gel	Solid	Liquid	Jelly, JELL-O®, cheese, butter
Sol	Liquid	Solid	Gold in water, milk of magnesia, mud
Solid sol	Solid	Solid	Milkglass

FIGURE 14.18 The Tyndall effect. Colloidal dispersions scatter light, a phenomenon known as the Tyndall effect. (a) Dust in the air scatters the light coming through the trees in a forest along the Oregon coast. (b) A narrow beam of light from a laser is passed through an NaCl solution (left) and then a colloidal mixture of gelatin and water (right).

he could not crystallize these substances, whereas he could crystallize sugar, salt, and other materials that form true solutions. Graham coined the word "colloid" (from the Greek, meaning "glue") to describe this class of substances that are distinctly different from true solutions and suspensions.

We now know that it is possible to crystallize some colloidal substances, albeit with difficulty, so there really is no sharp dividing line between these classes based on this property. Colloids do, however, have two distinguishing characteristics. First, colloids generally have high molar masses; this is true of proteins such as hemoglobin that have molar masses in the thousands. Second, the particles of a colloid are relatively large (say, 1000 nm in diameter). As a consequence, they exhibit the **Tyndall effect**; they scatter visible light when dispersed in a solvent, making the mixture appear cloudy (Figure 14.18). Third, even though colloidal particles are large, they are not so large that they settle out.

Graham also gave us the words **sol** for a colloidal dispersion of a solid substance in a fluid medium and **gel** for a colloidal dispersion that has a structure that prevents it from being mobile. JELL-O® is a sol when the solid is first mixed with boiling water, but it becomes a gel when cooled. Other examples of gels are the gelatinous precipitates of $Al(OH)_3$, $Fe(OH)_3$, and $Cu(OH)_2$ (Figure 14.19).

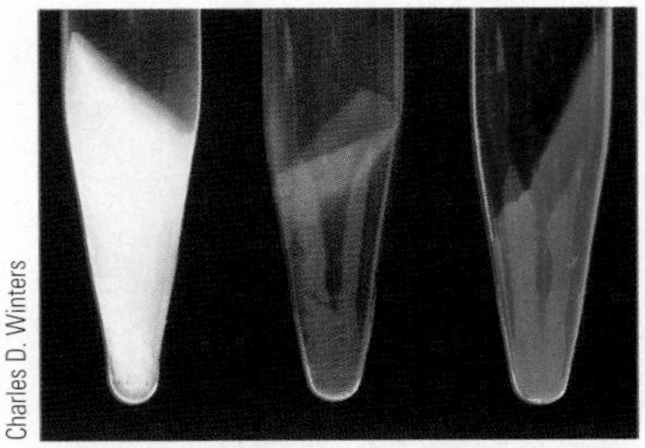

FIGURE 14.19 Gelatinous precipitates. (left) $Al(OH)_3$, (center) $Fe(OH)_3$, and (right) $Cu(OH)_2$.

Colloidal dispersions consist of finely divided particles that, as a result, have a very high surface area. For example, if you have one millionth of a mole of colloidal particles, each assumed to be a sphere with a diameter of 200 nm, the total surface area of the particles would be on the order of 200 million cm^2, or the size of several football fields. It is not surprising, therefore, that many of the properties of colloids depend on the properties of surfaces.

Types of Colloids

Colloids are classified according to the state of the dispersed phase and the dispersing medium. Table 14.5 lists several types of colloids and gives examples of each.

Colloids with water as the dispersing medium can be classified as **hydrophobic** (from the Greek, meaning "water-fearing") or **hydrophilic** ("water-loving"). A hydrophobic colloid is one in which only weak attractive forces exist between the water and the surfaces of the colloidal particles. Examples include dispersions of metals and of nearly insoluble salts in water. When compounds like AgCl precipitate, the result is often a colloidal dispersion. The precipitation reaction occurs too rapidly for ions to gather from long distances and make large crystals, so the ions aggregate to form small particles that remain suspended in the liquid.

FIGURE 14.20 Hydrophobic colloids. A hydrophobic colloid is stabilized by positive ions absorbed onto each particle and a secondary layer of negative ions. Because the particles bear similar charges, they repel one another, and precipitation is prevented.

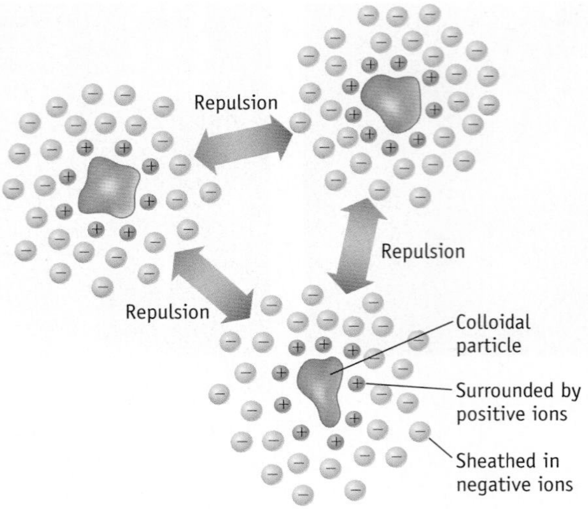

Why don't the particles come together (coagulate) and form larger particles? The answer is that the colloidal particles carry electric charges. An AgCl particle, for example, will absorb Ag^+ ions if the ions are present in substantial concentration; an attraction occurs between Ag^+ ions in solution and Cl^- ions on the surface of the particle. In this way, the colloidal particles become positively charged, allowing them to attract a secondary layer of anions. The particles, now surrounded by layers of ions, repel one another and are prevented from coming together to form a precipitate (Figure 14.20).

A stable hydrophobic colloid can be made to coagulate by introducing ions into the dispersing medium. Milk contains a colloidal suspension of protein-rich casein micelles with a hydrophobic core. When milk ferments, lactose (milk sugar) is converted to lactic acid, which forms lactate ions and hydrogen ions. The protective charges on the surfaces of the colloidal particles are overcome, and the milk coagulates; the milk solids come together in clumps called "curds."

Soil particles are often carried by water in rivers and streams as hydrophobic colloids. When river water carrying large amounts of colloidal particles meets sea water with its high concentration of salts, the particles coagulate to form the silt seen at the mouth of the river (Figure 14.21). Municipal water treatment plants often add salts such as $Al_2(SO_4)_3$ to clarify water. In aqueous solution, aluminum ions exist as $[Al(H_2O)_6]^{3+}$ cations, which neutralize the charge on the hydrophobic colloidal soil particles, causing these particles to aggregate and settle out.

Hydrophilic colloids are strongly attracted to water molecules. They often have groups such as —OH and $—NH_2$ on their surfaces. These groups form strong hydrogen bonds to water, thereby stabilizing the colloid. Proteins and starch are important examples of hydrophilic colloids, and homogenized milk is the most familiar example.

Emulsions are colloidal dispersions of one liquid in another, such as oil or fat in water. Familiar examples include salad dressing, mayonnaise, and milk. If vegetable oil and vinegar are mixed to make a salad dressing, the mixture quickly separates into two layers because the nonpolar oil molecules do not interact with the polar water and acetic acid (CH_3CO_2H) molecules. So why are milk and mayonnaise apparently homogeneous mixtures that do not separate into layers? The answer is that they contain an **emulsifying agent** such as soap or a protein. Lecithin is a phospholipid found in egg yolks, so mixing egg yolks with oil and vinegar stabilizes the colloidal dispersion known as mayonnaise. To understand this process

FIGURE 14.21 Formation of silt. Silt forms at a river delta as colloidal soil particles come in contact with salt water in the ocean. Here, the Ashley and Cooper Rivers empty into the Atlantic Ocean at Charleston, South Carolina. The high concentration of ions in sea water causes the colloidal soil particles to coagulate.

further, let us look into the functioning of soaps and detergents, substances known as surfactants.

Surfactants

Soaps and detergents are emulsifying agents. Soap is made by heating a fat with sodium or potassium hydroxide (◄ page 476), which produces the anion of a fatty acid.

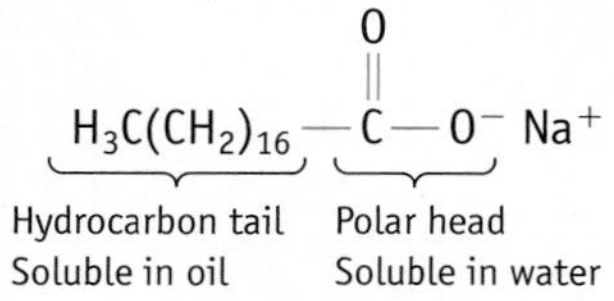

sodium stearate, a soap

The fatty acid anion has a split personality: It has a nonpolar, hydrophobic hydrocarbon tail that is soluble in other similar hydrocarbons and a polar, hydrophilic head that is soluble in water.

Oil cannot be readily washed away from dishes or clothing with water because oil is nonpolar and thus insoluble in water. Instead, we add soap to the water to clean away the oil. The nonpolar molecules of the oil interact with the nonpolar hydrocarbon tails of the soap molecules, leaving the polar heads of the soap to interact with surrounding water molecules. The oil and water then mix (Figure 14.22). If the oily material on a piece of clothing or a dish also contains some dirt particles, that dirt can now be washed away.

■ **Soaps and Surfactants** A sodium soap is a solid at room temperature, whereas potassium soaps are usually liquids. About 30 million tons of household and toilet soap, and synthetic and soap-based laundry detergents, are produced annually worldwide.

Substances such as soaps that affect the properties of surfaces, and therefore affect the interaction between two phases, are called surface-active agents, or **surfactants**, for short. A surfactant used for cleaning is called a **detergent**. One function of a surfactant is to lower the surface tension of water, which enhances the cleansing action of the detergent (Figure 14.23).

Many detergents used in the home and industry are synthetic. One example is sodium laurylbenzenesulfonate, a biodegradable compound.

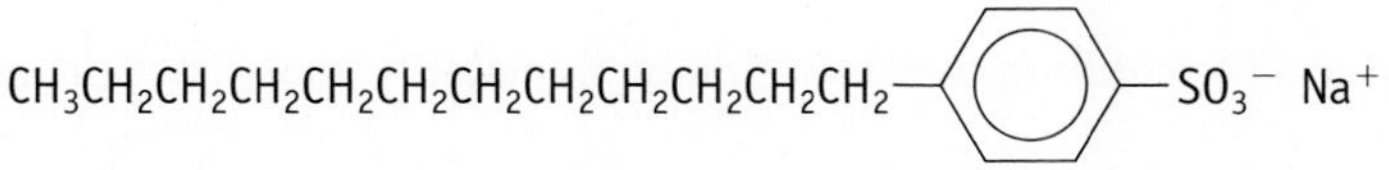

sodium laurylbenzenesulfonate

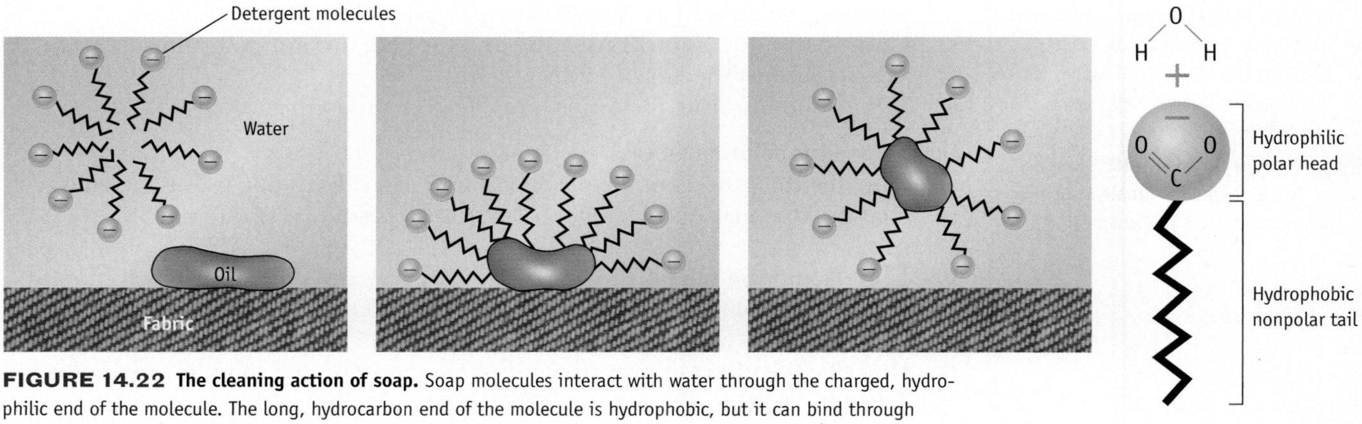

FIGURE 14.22 The cleaning action of soap. Soap molecules interact with water through the charged, hydrophilic end of the molecule. The long, hydrocarbon end of the molecule is hydrophobic, but it can bind through dispersion forces with hydrocarbons and other nonpolar substances.

FIGURE 14.23 Effect of a detergent on the surface tension of water. Sulfur (density = 2.1 g/cm³) is carefully placed on the surface of water (density, 1.0 g/cm³) (left). The surface tension of the water keeps the denser sulfur afloat. Several drops of detergent are then placed on the surface of the water (right). The surface tension of the water is reduced, and the sulfur sinks to the bottom of the beaker.

In general, synthetic detergents use the sulfonate group, $—SO_3^-$, as the polar head instead of the carboxylate group, $—CO_2^-$. The carboxylate anions form an insoluble precipitate with any Ca^{2+} or Mg^{2+} ions present in water. Because hard water is characterized by high concentrations of these ions, using soaps containing carboxylates produces bathtub rings and tell-tale gray clothing. The synthetic sulfonate detergents have the advantage that they do not form such precipitates because their calcium salts are more soluble in water.

Chapter Goals Revisited

Chemistry Now™ Sign in at **www.thomsonedu.com/login** to:

- Assess your understanding with Study Questions in OWL keyed to each goal in the Goals and Homework menu for this chapter
- For quick review, download Go Chemistry mini-lecture flashcard modules (or purchase them at **www.ichapters.com**)
- Check your readiness for an exam by taking the Pre-Test and exploring the modules recommended in your Personalized Study plan.

Access **How Do I Solve It?** tutorials on how to approach problem solving using concepts in this chapter.

Now that you have studied this chapter, you should ask whether you have met the chapter goals. In particular, you should be able to:

Calculate and use the solution concentration units molality, mole fraction, and weight percent

a. Define the terms solution, solvent, solute, and colligative properties (Section 14.1).

b. Use the following concentration units: molality, mole fraction, and weight percent. (Section 14.1). **Study Question(s) assignable in OWL: 6, 9, 10, 12, 56.**

c. Understand the distinctions between saturated, unsaturated, and supersaturated solutions (Section 14.2).

d. Define and illustrate the terms miscible and immiscible (Section 14.2).

Understand the solution process

a. Describe the process of dissolving a solute in a solvent, including the energy changes that may occur (Section 14.2). **Study Question(s) assignable in OWL: 13, 16, 88, 93.**

b. Understand the relationship of lattice enthalpy and enthalpy of hydration to the enthalpy of solution for an ionic solute (Section 14.2). **Study Question(s) assignable in OWL: 77, 88.**

c. Describe the effect of pressure and temperature on the solubility of a solute (Section 14.2).

d. Use Henry's law to calculate the solubility of a gas in a solvent (Section 14.2). **Study Question(s) assignable in OWL: 21, 22, 67.**

e. Apply Le Chatelier's principle to the change in solubility of gases with temperature changes (Section 14.2).

Understand and use the colligative properties of solutions

a. Calculate the mole fraction of a solvent ($X_{solvent}$) and the effect of a solute on solvent vapor pressure ($P_{solvent}$) using Raoult's law (Section 14.4). **Study Question(s) assignable in OWL: 24, 74, 96.**

b. Calculate the boiling point elevation or freezing point depression caused by a solute in a solvent (Section 14.4). **Study Question(s) assignable in OWL: 28, 30, 32, 45, 53, 60; Go Chemistry Module 19.**

c. Calculate the osmotic pressure (Π) for solutions (Section 14.4). **Study Question(s) assignable in OWL: 47, 78, 84, 97.**

d. Use colligative properties to determine the molar mass of a solute (Section 14.4). **Study Question(s) assignable in OWL: 35, 39, 61, 64, 84, 85.**

e. Characterize the effect of ionic solutes on colligative properties (Section 14.4). **Study Question(s) assignable in OWL: 43.**

f. Use the van't Hoff factor, *i*, in calculations involving colligative properties (Section 14.4). **Study Question(s) assignable in OWL: 41, 79, 80.**

KEY EQUATIONS

Equation 14.1 (page 618) Molality is defined as the amount of solute per kilogram of solvent.

$$\text{Concentration } (c, \text{ mol/kg}) = \text{molality of solute} = \frac{\text{amount of solute (mol)}}{\text{mass of solvent (kg)}}$$

Equation 14.2 (page 618) The mole fraction, *X*, of a solution component is defined as the number of moles of a given component of a mixture (n_A, mol) divided by the total number of moles of all of the components of the mixture.

$$\text{Mole fraction of A } (X_A) = \frac{n_A}{n_A + n_B + n_C + \ldots}$$

Equation 14.3 (page 619) Weight percent is the mass of one component divided by the total mass of the mixture (multiplied by 100%).

$$\text{Weight \% A} = \frac{\text{mass of A}}{\text{mass of A} + \text{mass of B} + \text{mass of C} + \ldots} \times 100\%$$

Equation 14.4 (page 626) Henry's law: the solubility of a gas, S_g, is equal to the product of the partial pressure of the gaseous solute (P_g) and a constant (k_H) characteristic of the solute and solvent.

$$S_g = k_H P_g$$

Equation 14.5 (page 629) Raoult's law: the equilibrium vapor pressure of a solvent over a solution at a given temperature, $P_{solvent}$, is the product of the mole fraction of the solvent ($X_{solvent}$) and the vapor pressure of the pure solvent ($P°_{solvent}$).

$$P_{solvent} = X_{solvent} P°_{solvent}$$

Equation 14.6 (page 631) The elevation in boiling point of the solvent in a solution, ΔT_{bp}, is the product of the molality of the solute, m_{solute}, and a constant characteristic of the solvent, K_{bp}.

$$\text{Elevation in boiling point} = \Delta T_{bp} = K_{bp} m_{solute}$$

Equation 14.7 (page 634) The depression of the freezing point of the solvent in a solution, ΔT_{fp}, is the product of the molality of the solute, m_{solute}, and a constant characteristic of the solvent, K_{fp}.

$$\text{Freezing point depression} = \Delta T_{fp} = K_{fp} m_{solute}$$

Equation 14.8 (page 637) The osmotic pressure, Π, is the product of the solute concentration c (in mol/L), the universal gas constant R (0.082057 L · atm/K · mol), and the temperature T (in kelvins).

$$\Pi = cRT$$

Equation 14.9 (page 639) This modified equation for freezing point depression accounts for the possible dissociation of a solute. The van't Hoff factor, i, the ratio of the measured freezing point depression and the freezing point depression calculated assuming no solute dissociation, is related to the relative number of particles produced by a dissolved solute.

$$\Delta T_{fp} \text{ measured} = K_{fp} \times m \times i$$

STUDY QUESTIONS

Online homework for this chapter may be assigned in OWL.

▲ denotes challenging questions.

■ denotes questions assignable in OWL.

Blue-numbered questions have answers in Appendix O and fully-worked solutions in the *Student Solutions Manual.*

Practicing Skills

Concentration

(See Examples 14.1 and 14.2 and ChemistryNow Screen 14.2.)

1. Suppose you dissolve 2.56 g of succinic acid, $C_2H_4(CO_2H)_2$, in 500. mL of water. Assuming that the density of water is 1.00 g/cm^3, calculate the molality, mole fraction, and weight percentage of acid in the solution.

2. ■ Assume you dissolve 45.0 g of camphor, $C_{10}H_{16}O$, in 425 mL of ethanol, C_2H_5OH. Calculate the molality, mole fraction, and weight percent of camphor in this solution. (The density of ethanol is 0.785 g/mL.)

3. Fill in the blanks in the table. Aqueous solutions are assumed.

Compound	Molality	Weight Percent	Mole Fraction
NaI	0.15	______	______
C_2H_5OH	______	5.0	______
$C_{12}H_{22}O_{11}$	0.15	______	______

4. Fill in the blanks in the table. Aqueous solutions are assumed.

Compound	Molality	Weight Percent	Mole Fraction
KNO_3	______	10.0	______
CH_3CO_2H	0.0183	______	______
$HOCH_2CH_2OH$	______	18.0	______

5. What mass of Na_2CO_3 must you add to 125 g of water to prepare 0.200 m Na_2CO_3? What is the mole fraction of Na_2CO_3 in the resulting solution?

6. ■ You want to prepare a solution that is 0.0512 m in $NaNO_3$. What mass of $NaNO_3$ must be added to 500. g of water? What is the mole fraction of $NaNO_3$ in the solution?

7. You wish to prepare an aqueous solution of glycerol, $C_3H_5(OH)_3$, in which the mole fraction of the solute is 0.093. What mass of glycerol must you add to 425 g of water to make this solution? What is the molality of the solution?

8. You want to prepare an aqueous solution of ethylene glycol, $HOCH_2CH_2OH$, in which the mole fraction of solute is 0.125. What mass of ethylene glycol, in grams, should you combine with 955 g of water? What is the molality of the solution?

9. ■ Hydrochloric acid is sold as a concentrated aqueous solution. If the molarity of commercial HCl is 12.0 and its density is 1.18 g/cm^3, calculate the following:
(a) the molality of the solution
(b) the weight percent of HCl in the solution

10. ■ Concentrated sulfuric acid has a density of 1.84 g/cm^3 and is 95.0% by weight H_2SO_4. What is the molality of this acid? What is its molarity?

11. The average lithium ion concentration in sea water is 0.18 ppm. What is the molality of Li^+ in sea water?

12. ■ Silver ion has an average concentration of 28 ppb (parts per billion) in U.S. water supplies.
(a) What is the molality of the silver ion?
(b) If you wanted 1.0×10^2 g of silver and could recover it chemically from water supplies, what volume of water in liters, would you have to treat? (Assume the density of water is 1.0 g/cm^3.)

The Solution Process

(See Example 14.3 and ChemistryNow Screens 14.3 and 14.4.)

13. ■ Which pairs of liquids will be miscible?
(a) H_2O and $CH_3CH_2CH_2CH_3$
(b) C_6H_6 (benzene) and CCl_4
(c) H_2O and CH_3CO_2H

14. Acetone, CH_3COCH_3, is quite soluble in water. Explain why this should be so.

15. Use the data of Table 14.1 to calculate the enthalpy of solution of LiCl.

16. ■ Use the following data to calculate the enthalpy of solution of sodium perchlorate, $NaClO_4$:

$\Delta_f H°(s) = -382.9$ kJ/mol and $\Delta_f H°(aq, 1\ m) = -369.5$ kJ/mol

17. You make a saturated solution of NaCl at 25 °C. No solid is present in the beaker holding the solution. What can be done to increase the amount of dissolved NaCl in this solution? (See Figure 14.11.)
(a) Add more solid NaCl.
(b) Raise the temperature of the solution.
(c) Raise the temperature of the solution, and add some NaCl.
(d) Lower the temperature of the solution, and add some NaCl.

18. Some lithium chloride, LiCl, is dissolved in 100 mL of water in one beaker, and some Li_2SO_4 is dissolved in 100 mL of water in another beaker. Both are at 10 °C, and both are saturated solutions; some solid remains undissolved in each beaker. Describe what you would observe as the temperature is raised. The following data are available to you from a handbook of chemistry:

	Solubility (g/100 mL)	
Compound	**10 °C**	**40 °C**
Li_2SO_4	35.5	33.7
LiCl	74.5	89.8

Henry's Law

(See Example 14.4 and ChemistryNow Screen 14.5.)

19. The partial pressure of O_2 in your lungs varies from 25 mm Hg to 40 mm Hg. What mass of O_2 can dissolve in 1.0 L of water at 25 °C if the partial pressure of O_2 is 40 mm Hg?

20. ■ The Henry's law constant for O_2 in water at 25 °C is given in Table 14.2. Which of the following is a reasonable constant when the temperature is 50 °C? Explain the reason for your choice.
(a) 6.7×10^{-4} mol/kg · bar (c) 1.3×10^{-3} mol/kg · bar
(b) 2.6×10^{-3} mol/kg · bar (d) 6.4×10^{-2} mol/kg · bar

21. An unopened soda can has an aqueous CO_2 concentration of 0.0506 *m* at 25 °C. What is the pressure of CO_2 gas in the can?

22. ■ Hydrogen gas has a Henry's law constant of 7.8×10^{-4} mol/kg · bar at 25 °C when dissolving in water. If the total pressure of gas (H_2 gas plus water vapor) over water is 1.0 bar, what is the concentration of H_2 in the water in grams per milliliter? (See Appendix G for the vapor pressure of water.)

Raoult's Law

(See Example 14.5 and ChemistryNow Screen 14.7.)

23. A 35.0-g sample of ethylene glycol, $HOCH_2CH_2OH$, is dissolved in 500.0 g of water. The vapor pressure of water at 32 °C is 35.7 mm Hg. What is the vapor pressure of the water–ethylene glycol solution at 32 °C? (Ethylene glycol is nonvolatile.)

24. ■ Urea, $(NH_2)_2CO$, which is widely used in fertilizers and plastics, is quite soluble in water. If you dissolve 9.00 g of urea in 10.0 mL of water, what is the vapor pressure of the solution at 24 °C? Assume the density of water is 1.00 g/mL.

25. Pure ethylene glycol, $HOCH_2CH_2OH$, is added to 2.00 kg of water in the cooling system of a car. The vapor pressure of the water in the system when the temperature is 90 °C is 457 mm Hg. What mass of glycol was added? (Assume the solution is ideal. See Appendix G for the vapor pressure of water.)

26. Pure iodine (105 g) is dissolved in 325 g of CCl_4 at 65 °C. Given that the vapor pressure of CCl_4 at this temperature is 531 mm Hg, what is the vapor pressure of the CCl_4–I_2 solution at 65 °C? (Assume that I_2 does not contribute to the vapor pressure.)

Boiling Point Elevation

(See Example 14.6 and ChemistryNow Screen 14.8.)

27. Verify that 0.200 mol of a nonvolatile solute in 125 g of benzene (C_6H_6) produces a solution whose boiling point is 84.2 °C.

28. ■ What is the boiling point of a solution composed of 15.0 g of urea, $(NH_2)_2CO$, in 0.500 kg of water?

29. What is the boiling point of a solution composed of 15.0 g of $CHCl_3$ and 0.515 g of the nonvolatile solute acenaphthene, $C_{12}H_{10}$, a component of coal tar?

30. ■ A solution of glycerol, $C_3H_5(OH)_3$, in 735 g of water has a boiling point of 104.4 °C at a pressure of 760 mm Hg. What is the mass of glycerol in the solution? What is the mole fraction of the solute?

Freezing Point Depression

(See Example 14.7 and ChemistryNow Screen 14.8.)

31. A mixture of ethanol, C_2H_5OH, and water has a freezing point of −16.0 °C.
(a) What is the molality of the alcohol?
(b) What is the weight percent of alcohol in the solution?

32. ■ Some ethylene glycol, $HOCH_2CH_2OH$, is added to your car's cooling system along with 5.0 kg of water. If the freezing point of the water–glycol solution is −15.0 °C, what mass of $HOCH_2CH_2OH$ must have been added?

33. You dissolve 15.0 g of sucrose, $C_{12}H_{22}O_{11}$, in a cup of water (225 g). What is the freezing point of the solution?

34. Assume a bottle of wine consists of an 11 weight percent solution of ethanol (C_2H_5OH) in water. If the bottle of wine is chilled to −20 °C, will the solution begin to freeze?

Colligative Properties and Molar Mass Determination

(See Example 14.8.)

35. ■ You add 0.255 g of an orange, crystalline compound whose empirical formula is $C_{10}H_8Fe$ to 11.12 g of benzene. The boiling point of the benzene rises from 80.10 °C to 80.26 °C. What are the molar mass and molecular formula of the compound?

36. Butylated hydroxyanisole (BHA) is used as an antioxidant in margarine and other fats and oils. (It prevents oxidation and prolongs the shelf-life of the food.) What is the molar mass of BHA if 0.640 g of the compound, dissolved in 25.0 g of chloroform, produces a solution whose boiling point is 62.22 °C?

37. Benzyl acetate is one of the active components of oil of jasmine. If 0.125 g of the compound is added to 25.0 g of chloroform ($CHCl_3$), the boiling point of the solution is 61.82 °C. What is the molar mass of benzyl acetate?

38. Anthracene, a hydrocarbon obtained from coal, has an empirical formula of C_7H_5. To find its molecular formula, you dissolve 0.500 g in 30.0 g of benzene. The boiling point of pure benzene is 80.10 °C, whereas the solution has a boiling point of 80.34 °C. What is the molecular formula of anthracene?

39. ■ An aqueous solution contains 0.180 g of an unknown, nonionic solute in 50.0 g of water. The solution freezes at −0.040 °C. What is the molar mass of the solute?

40. The organic compound called aluminon is used as a reagent to test for the presence of the aluminum ion in aqueous solution. A solution of 2.50 g of aluminon in 50.0 g of water freezes at −0.197 °C. What is the molar mass of aluminon?

Colligative Properties of Ionic Compounds

(See Example 14.9 and ChemistryNow Screen 14.8.)

41. ■ If 52.5 g of LiF is dissolved in 306 g of water, what is the expected freezing point of the solution? (Assume the van't Hoff factor, *i*, for LiF is 2.)

42. To make homemade ice cream, you cool the milk and cream by immersing the container in ice and a concentrated solution of rock salt (NaCl) in water. If you want to have a water–salt solution that freezes at −10. °C, what mass of NaCl must you add to 3.0 kg of water? (Assume the van't Hoff factor, *i*, for NaCl is 1.85.)

43. ■ List the following aqueous solutions in order of increasing melting point. (The last three are all assumed to dissociate completely into ions in water.)
(a) 0.1 *m* sugar
(b) 0.1 *m* NaCl
(c) 0.08 *m* $CaCl_2$
(d) 0.04 *m* Na_2SO_4

44. Arrange the following aqueous solutions in order of decreasing freezing point. (The last three are all assumed to dissociate completely into ions in water.)
(a) 0.20 *m* ethylene glycol (nonvolatile, nonelectrolyte)
(b) 0.12 *m* K_2SO_4
(c) 0.10 *m* $MgCl_2$
(d) 0.12 *m* KBr

▲ more challenging ■ in OWL Blue-numbered questions answered in Appendix O

Osmosis

(See Example 14.10 and ChemistryNow Screen 14.9.)

45. ■ An aqueous solution contains 3.00% phenylalanine ($C_9H_{11}NO_2$) by mass. Assume the phenylalanine is nonionic and nonvolatile. Find the following:
   (a) the freezing point of the solution
   (b) the boiling point of the solution
   (c) the osmotic pressure of the solution at 25 °C

   In your view, which of these values is most easily measurable in the laboratory?

46. Estimate the osmotic pressure of human blood at 37 °C. Assume blood is isotonic with a 0.154 M NaCl solution, and assume the van't Hoff factor, *i*, is 1.90 for NaCl.

47. ■ An aqueous solution containing 1.00 g of bovine insulin (a protein, not ionized) per liter has an osmotic pressure of 3.1 mm Hg at 25 °C. Calculate the molar mass of bovine insulin.

48. Calculate the osmotic pressure of a 0.0120 M solution of NaCl in water at 0 °C. Assume the van't Hoff factor, *i*, is 1.94 for this solution.

Colloids

(See Section 14.5 and ChemistryNow Screen 14.10.)

49. When solutions of $BaCl_2$ and Na_2SO_4 are mixed, the mixture becomes cloudy. After a few days, a white solid is observed on the bottom of the beaker with a clear liquid above it.
   (a) Write a balanced equation for the reaction that occurs.
   (b) Why is the solution cloudy at first?
   (c) What happens during the few days of waiting?

50. ■ The dispersed phase of a certain colloidal dispersion consists of spheres of diameter 1.0×10^2 nm.
   (a) What are the volume ($V = \frac{4}{3}\pi r^3$) and surface area ($A = 4\pi r^2$) of each sphere?
   (b) How many spheres are required to give a total volume of 1.0 cm^3? What is the total surface area of these spheres in square meters?

General Questions

These questions are not designated as to type or location in the chapter. They may combine several concepts.

51. Phenylcarbinol is used in nasal sprays as a preservative. A solution of 0.52 g of the compound in 25.0 g of water has a melting point of −0.36 °C. What is the molar mass of phenylcarbinol?

52. (a) Which aqueous solution is expected to have the higher boiling point: 0.10 *m* Na_2SO_4 or 0.15 *m* sugar?
   (b) For which aqueous solution is the vapor pressure of water higher: 0.30 *m* NH_4NO_3 or 0.15 *m* Na_2SO_4?

53. ■ Arrange the following aqueous solutions in order of (i) increasing vapor pressure of water and (ii) increasing boiling point.
   (a) 0.35 *m* $HOCH_2CH_2OH$ (a nonvolatile solute)
   (b) 0.50 *m* sugar
   (c) 0.20 *m* KBr (a strong electrolyte)
   (d) 0.20 *m* Na_2SO_4 (a strong electrolyte)

54. Making homemade ice cream is one of life's great pleasures. Fresh milk and cream, sugar, and flavorings are churned in a bucket suspended in an ice–water mixture, the freezing point of which has been lowered by adding rock salt. One manufacturer of home ice cream freezers recommends adding 2.50 lb (1130 g) of rock salt (NaCl) to 16.0 lb of ice (7250 g) in a 4-qt freezer. For the solution when this mixture melts, calculate the following:
   (a) the weight percent of NaCl
   (b) the mole fraction of NaCl
   (c) the molality of the solution

55. Dimethylglyoxime [DMG, $(CH_3CNOH)_2$] is used as a reagent to precipitate nickel ion. Assume that 53.0 g of DMG has been dissolved in 525 g of ethanol (C_2H_5OH).

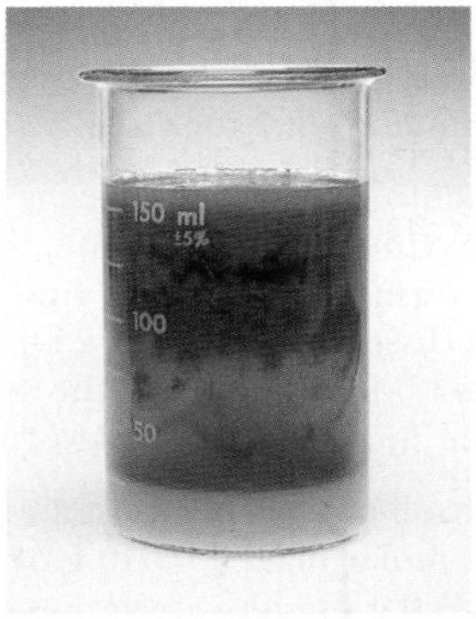

Charles D. Winters

The red, insoluble compound formed between nickel(II) ion and dimethylglyoxime (DMG) is precipitated when DMG is added to a basic solution of Ni^{2+}(aq).

   (a) What is the mole fraction of DMG?
   (b) What is the molality of the solution?
   (c) What is the vapor pressure of the ethanol over the solution at ethanol's normal boiling point of 78.4 °C?
   (d) What is the boiling point of the solution? (DMG does not produce ions in solution.) (K_{bp} for ethanol = +1.22 °C/*m*)

56. ■ A 10.7 *m* solution of NaOH has a density of 1.33 g/cm^3 at 20 °C. Calculate the following:
   (a) the mole fraction of NaOH
   (b) the weight percent of NaOH
   (c) the molarity of the solution

57. Concentrated aqueous ammonia has a molarity of 14.8 mol/L and a density of 0.90 g/cm^3. What is the molality of the solution? Calculate the mole fraction and weight percent of NH_3.

58. ■ If you dissolve 2.00 g of $Ca(NO_3)_2$ in 750 g of water, what is the molality of $Ca(NO_3)_2$? What is the total molality of ions in solution? (Assume total dissociation of the ionic solid.)

59. If you want a solution that is 0.100 *m* in ions, what mass of Na_2SO_4 must you dissolve in 125 g of water? (Assume total dissociation of the ionic solid.)

60. ■ Consider the following aqueous solutions: (i) 0.20 *m* $HOCH_2CH_2OH$ (nonvolatile, nonelectrolyte); (ii) 0.10 *m* $CaCl_2$; (iii) 0.12 *m* KBr; and (iv) 0.12 *m* Na_2SO_4.
(a) Which solution has the highest boiling point?
(b) Which solution has the lowest freezing point?
(c) Which solution has the highest water vapor pressure?

61. ■ (a) Which solution is expected to have the higher boiling point: 0.20 *m* KBr or 0.30 *m* sugar?
(b) Which aqueous solution has the lower freezing point: 0.12 *m* NH_4NO_3 or 0.10 *m* Na_2CO_3?

62. The solubility of NaCl in water at 100 °C is 39.1 g/100. g of water. Calculate the boiling point of this solution. (Assume i = 1.85 for NaCl.)

63. Instead of using NaCl to melt the ice on your sidewalk, you decide to use $CaCl_2$. If you add 35.0 g of $CaCl_2$ to 150. g of water, what is the freezing point of the solution? (Assume i = 2.7 for $CaCl_2$.)

64. ■ The smell of ripe raspberries is due to 4-(*p*-hydroxyphenyl)-2-butanone, which has the empirical formula C_5H_6O. To find its molecular formula, you dissolve 0.135 g in 25.0 g of chloroform, $CHCl_3$. The boiling point of the solution is 61.82 °C. What is the molecular formula of the solute?

65. Hexachlorophene has been used in germicidal soap. What is its molar mass if 0.640 g of the compound, dissolved in 25.0 g of chloroform, produces a solution whose boiling point is 61.93 °C?

66. The solubility of ammonium formate, NH_4CHO_2, in 100 g of water is 102 g at 0 °C and 546 g at 80 °C. A solution is prepared by dissolving NH_4CHO_2 in 200 g of water until no more will dissolve at 80 °C. The solution is then cooled to 0 °C. What mass of NH_4CHO_2 precipitates? (Assume that no water evaporates and that the solution is not supersaturated.)

67. ■ How much N_2 can dissolve in water at 25 °C if the N_2 partial pressure is 585 mm Hg?

68. Cigars are best stored in a "humidor" at 18 °C and 55% relative humidity. This means the pressure of water vapor should be 55% of the vapor pressure of pure water at the same temperature. The proper humidity can be maintained by placing a solution of glycerol $[C_3H_5(OH)_3]$ and water in the humidor. Calculate the percent by mass of glycerol that will lower the vapor pressure of water to the desired value. (The vapor pressure of glycerol is negligible.)

69. An aqueous solution containing 10.0 g of starch per liter has an osmotic pressure of 3.8 mm Hg at 25 °C.
(a) What is the average molar mass of starch? (Because not all starch molecules are identical, the result will be an average.)
(b) What is the freezing point of the solution? Would it be easy to determine the molecular weight of starch by measuring the freezing point depression? (Assume that the molarity and molality are the same for this solution.)

70. Vinegar is a 5% solution (by weight) of acetic acid in water. Determine the mole fraction and molality of acetic acid. What is the concentration of acetic acid in parts per million (ppm)? Explain why it is not possible to calculate the molarity of this solution from the information provided.

71. ■ Calculate the enthalpies of solution for Li_2SO_4 and K_2SO_4. Are the solution processes exothermic or endothermic? Compare them with LiCl and KCl. What similarities or differences do you find?

Compound	$\Delta_f H°$(s) (kJ/mol)	$\Delta_f H°$(aq, 1 *m*) (kJ/mol)
Li_2SO_4	−1436.4	−1464.4
K_2SO_4	−1437.7	−1414.0

72. ▲ Water at 25 °C has a density of 0.997 g/cm^3. Calculate the molality and molarity of pure water at this temperature.

73. ▲ If a volatile solute is added to a volatile solvent, both substances contribute to the vapor pressure over the solution. Assuming an ideal solution, the vapor pressure of each is given by Raoult's law, and the total vapor pressure is the sum of the vapor pressures for each component. A solution, assumed to be ideal, is made from 1.0 mol of toluene ($C_6H_5CH_3$) and 2.0 mol of benzene (C_6H_6). The vapor pressures of the pure solvents are 22 mm Hg and 75 mm Hg, respectively, at 20 °C. What is the total vapor pressure of the mixture? What is the mole fraction of each component in the liquid and in the vapor?

74. ■ A solution is made by adding 50.0 mL of ethanol (C_2H_5OH, d = 0.789 g/ml) to 50.0 mL of water (d = 0.998 g/mL). What is the total vapor pressure over the solution at 20 °C? (See Study Question 73.) The vapor pressure of ethanol at 20 °C is 43.6 mm Hg.

75. A 2.0% (by mass) aqueous solution of novocainium chloride ($C_{13}H_{21}ClN_2O_2$) freezes at −0.237 °C. Calculate the van't Hoff factor, i. How many moles of ions are in the solution per mole of compound?

76. A solution is 4.00% (by mass) maltose and 96.00% water. It freezes at −0.229 °C.
(a) Calculate the molar mass of maltose (which is not an ionic compound).
(b) The density of the solution is 1.014 g/mL. Calculate the osmotic pressure of the solution.

▲ more challenging ■ in OWL Blue-numbered questions answered in Appendix O

77. ▲ The following table lists the concentrations of the principal ions in sea water:

Ion	Concentration (ppm)
Cl^-	1.95×10^4
Na^+	1.08×10^4
Mg^{2+}	1.29×10^3
SO_4^{2-}	9.05×10^2
Ca^{2+}	4.12×10^2
K^+	3.80×10^2
Br^-	67

(a) Calculate the freezing point of water.
(b) Calculate the osmotic pressure of sea water at 25 °C. What is the minimum pressure needed to purify sea water by reverse osmosis?

78. ■ ▲ A tree is exactly 10 m tall.
(a) What must be the total molarity of the solutes if sap rises to the top of the tree by osmotic pressure at 20 °C? Assume the groundwater outside the tree is pure water and that the density of the sap is 1.0 g/mL. (1 mm Hg = 13.6 mm H_2O.)
(b) If the only solute in the sap is sucrose, $C_{12}H_{22}O_{11}$, what is its percent by mass?

79. ■ A 2.00% solution of H_2SO_4 in water freezes at −0.796 °C.
(a) Calculate the van't Hoff factor, *i*.
(b) Which of the following best represents sulfuric acid in a dilute aqueous solution: H_2SO_4, $H_3O^+ + HSO_4^-$, or $2\ H_3O^+ + SO_4^{2-}$?

80. ■ A compound is known to be a potassium halide, KX. If 4.00 g of the salt is dissolved in exactly 100 g of water, the solution freezes at −1.28 °C. Identify the halide ion in this formula.

In the Laboratory

81. ▲ A solution of benzoic acid in benzene has a freezing point of 3.1 °C and a boiling point of 82.6 °C. (The freezing point of pure benzene is 5.50 °C, and its boiling point is 80.1 °C.) The structure of benzoic acid is

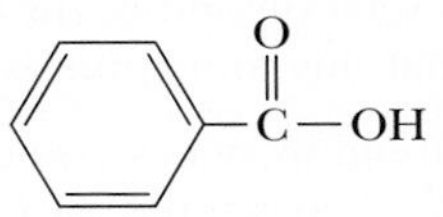

What can you conclude about the state of the benzoic acid molecules at the two different temperatures? Recall the discussion of hydrogen bonding in Section 12.2, and see Figure 12.7.

82. ▲ ■ You dissolve 5.0 mg of iodine, I_2, in 25 mL of water. You then add 10.0 mL of CCl_4 and shake the mixture. If I_2 is 85 times more soluble in CCl_4 than in H_2O (on a volume basis), what are the masses of I_2 in the water and CCl_4 layers after shaking? (See Figure 14.6.)

83. ▲ A solution of 5.00 g of acetic acid in 100. g of benzene freezes at 3.37 °C. A solution of 5.00 g of acetic acid in 100. g of water freezes at −1.49 °C. Find the molar mass of acetic acid from each of these experiments. What can you conclude about the state of the acetic acid molecules dissolved in each of these solvents? Recall the discussion of hydrogen bonding in Section 12.2 (and see Figure 12.7), and propose a structure for the species in benzene solution.

84. ▲ ■ In a police forensics lab, you examine a package that may contain heroin. However, you find the white powder is not pure heroin but a mixture of heroin ($C_{21}H_{23}O_5N$) and lactose ($C_{12}H_{22}O_{11}$). To determine the amount of heroin in the mixture, you dissolve 1.00 g of the white powdery mixture in water in a 100.0-mL volumetric flask. You find that the solution has an osmotic pressure of 539 mm Hg at 25 °C. What is the composition of the mixture?

85. ■ A newly synthesized compound containing boron and fluorine is 22.1% boron. Dissolving 0.146 g of the compound in 10.0 g of benzene gives a solution with a vapor pressure of 94.16 mm Hg at 25 °C. (The vapor pressure of pure benzene at this temperature is 95.26 mm Hg.) In a separate experiment, it is found that the compound does not have a dipole moment.
(a) What is the molecular formula for the compound?
(b) Draw a Lewis structure for the molecule, and suggest a possible molecular structure. Give the bond angles in the molecule and the hybridization of the boron atom.

86. In chemical research we often send newly synthesized compounds to commercial laboratories for analysis. These laboratories determine the weight percent of C and H by burning the compound and collecting the evolved CO_2 and H_2O. They determine the molar mass by measuring the osmotic pressure of a solution of the compound. Calculate the empirical and molecular formulas of a compound, C_xH_yCr, given the following information:
(a) The compound contains 73.94% C and 8.27% H; the remainder is chromium.
(b) At 25 °C, the osmotic pressure of a solution containing 5.00 mg of the unknown dissolved in exactly 100 mL of chloroform solution is 3.17 mm Hg.

Summary and Conceptual Questions

The following questions may use concepts from this and previous chapters.

87. In each pair of ionic compounds, which is more likely to have the more negative enthalpy of hydration? Briefly explain your reasoning in each case.
(a) LiF or RbF
(b) KNO_3 or $Ca(NO_3)_2$
(c) CsBr or $CuBr_2$

88. ■ When salts of Mg^{2+}, Ca^{2+}, and Be^{2+} are placed in water, the positive ion is hydrated (as is the negative ion). Which of these three cations is most strongly hydrated? Which one is least strongly hydrated?

89. Which salt, Li_2SO_4 or Cs_2SO_4, is expected to have the more exothermic enthalpy of hydration? Explain briefly.

90. Explain why a cucumber shrivels up when it is placed in a concentrated solution of salt.

91. If you dissolve equal molar amounts of NaCl and $CaCl_2$ in water, the $CaCl_2$ lowers the freezing point of the water almost 1.5 times as much as the NaCl. Why?

92. A 100.-gram sample of sodium chloride (NaCl) is added to 100. mL of water at 0 °C. After equilibrium is reached, about 64 g of solid remains undissolved. Describe the equilibrium that exists in this system at the particulate level.

93. ■ Which of the following substances is/are likely to dissolve in water, and which is/are likely to dissolve in benzene (C_6H_6)?
(a) $NaNO_3$
(b) diethyl ether, $CH_3CH_2OCH_2CH_3$
(c) naphthalene, $C_{10}H_8$ (see page 458 for structure)
(d) NH_4Cl

94. Account for the fact that alcohols such as methanol (CH_3OH) and ethanol (C_2H_5OH) are quite miscible with water, whereas an alcohol with a long-carbon chain, such as octanol ($C_8H_{17}OH$), is poorly soluble in water.

95. Starch contains C—C, C—H, C—O, and O—H bonds. Hydrocarbons have only C—C and C—H bonds. Both starch and hydrocarbons can form colloidal dispersions in water. Which dispersion is classified as hydrophobic? Which is hydrophilic? Explain briefly.

96. ■ Which substance would have the greater influence on the vapor pressure of water when added to 1000. g of the liquid: 10.0 g of sucrose ($C_{12}H_{22}O_{11}$) or 10.0 g of ethylene glycol [$HOCH_2CH_2OH$]?

97. ■ Suppose you have two aqueous solutions separated by a semipermeable membrane. One contains 5.85 g of NaCl dissolved in 100. mL of solution, and the other contains 8.88 g of KNO_3 dissolved in 100. mL of solution. In which direction will solvent flow: from the NaCl solution to the KNO_3 solution, or from KNO_3 to NaCl? Explain briefly.

98. A protozoan (single-celled animal) that normally lives in the ocean is placed in fresh water. Will it shrivel or burst? Explain briefly.

99. In the process of distillation, a mixture of two (or more) volatile liquids is first heated to convert the volatile materials to the vapor state. Then the vapor is condensed, reforming the liquid. The net result of this liquid→ vapor→ liquid conversion is to enrich the fraction of a more volatile component in the mixture in the condensate. We can describe how this occurs using Raoult's law. Imagine that you have a mixture of 12% (by weight) ethanol and water (as formed, for example, by fermentation of grapes.)
(a) What are the mole fractions of ethanol and water in this mixture?
(b) This mixture is heated to 78.5 °C (the normal boiling point of ethanol). What are the equilibrium vapor pressures of ethanol and water at this temperature, assuming Raoult's Law (ideal) behavior? (You will need to derive the equilibrium vapor pressure of water at 78.5 °C from data in Appendix G.)
(c) What are the mole fractions of ethanol and water in the vapor?
(d) After this vapor is condensed to a liquid, to what extent has the mole fraction of ethanol been enriched? What is the mass fraction of ethanol in the condensate?

100. Sodium chloride (NaCl) is commonly used to melt ice on roads during the winter. Calcium chloride ($CaCl_2$) is sometimes used for this purpose too. Let us compare the effectiveness of equal masses of these two compounds in lowering the freezing point of water, by calculating the freezing point lowering of solutions containing 200 g of each salt in 1.00 kg of water. (An advantage of $CaCl_2$ is that it acts more quickly because it is hygroscopic, that is, it absorbs moisture from the air to give a solution and begin the process. A disadvantage is that this compound is more costly).

101. Review the trend in values of the van't Hoff factor i as a function of concentration (Table 14.4). Use the following data to calculate the van't Hoff factor for a NaCl concentration of 5.0 mass % (for which $\Delta T = -3.05$ °C) and a Na_2SO_4 concentration of 5.0 mass % (for which $\Delta T = -1.36$ °C). Are these values in line with your expectations based on the trend in the values given in Table 14.4? Speculate on why this trend is seen.

102. The table below give experimentally determined values for freezing points of 1.00% solutions (mass %) of a series of acids.

(a) Calculate the molality of each solution, determine the calculated freezing points, and then calculate the values of the van't Hoff factor *i*. Fill these values into the table.

Acid, (1.00 mass %)	molality (mol/kg H_2O)	$T_{measured}$ (°C)	$T_{calculated}$ (°C)	*i*
HNO_3		0.56		
CH_3CO_2H		0.32		
H_2SO_4		0.42		
$H_2C_2O_4$		0.30		
HCO_2H		0.42		
CCl_3CO_2H		0.21		

(b) Analyze the results, comparing the values of *i* for the various acids. How does this data relate to acid strengths? (The discussion of strong and weak acids on pages 132 and 133 will assist you to answer this question.)

103. It is interesting how the Fahrenheit temperature scale was established. One report, given by Fahrenheit in a paper in 1724, stated that the value of 0 °F was established as the freezing temperature of saturated solutions of sea salt. From the literature we find that the freezing point of a 20% by mass solution of NaCl is −16.46 °C (This is the lowest freezing temperature reported for solutions of NaCl). Does this value lend credence to this story of the establishment of the Fahrenheit scale?

104. The osmotic pressure exerted by seawater at 25 °C is about 27 atm. Calculate the concentration of ions dissolved in seawater that is needed to give an osmotic pressure of this magnitude. (Desalinization of sea water is accomplished by reverse osmosis. In this process an applied pressure forces water through a membrane against a concentration gradient. The minimum external force needed for this process will be 27 atm. Actually, to accomplish the process at a reasonable rate, the applied pressure needs to be about twice this value.)

THE CHEMISTRY OF MODERN MATERIALS

White limestone and the minerals calcite, aragonite, and Iceland spar are all composed of calcium carbonate. So are chalk, eggshells, and sea shells. These objects have distinctly different physical characteristics (Figure 1), yet they are composed primarily of the same component particles, Ca^{2+} and CO_3^{2-} ions. What is interesting to chemists, geologists, and biologists is that the differences in the macroscopic characteristics result from small differences in composition (due to the presence of impurities) and in the arrangement of the particles.

By studying the composition and structure of synthetic and naturally occurring materials, scientists and engineers are able to gain insight into what gives each material its properties. They might then be able to use synthetic techniques to create new materials that are tailored to particular applications and that have predictable behaviors. They can also extract and purify naturally occurring substances that have properties considered desirable for specific applications. The study and synthesis of materials are the general domain of **materials science.** While chemistry serves as the foundation of materials science, understanding and working in this field may require expertise in physics, biology, and engineering.

This section explores a variety of common materials—except organic polymers, which were described in Section 10.5—and examines the connection between composition, atomic arrangements, and bulk properties. We will also look at some modern materials and their applications.

Figure 1 **Forms of calcium carbonate.** (clockwise from top) The shell of an abalone, a limestone paving block from Europe, crystalline aragonite, common blackboard chalk ($CaCO_3$ and a binder), and transparent Iceland spar.

Simon Fraser/Photo Researchers, Inc.

• **Fiberoptics.** Fibers made of glass are being used increasingly to carry information.

Metals

Bonding in Metals

Molecular orbital (MO) theory was introduced in Chapter 9 to rationalize covalent bonding. Recall the basic concepts of MO theory: Atomic orbitals from individual atoms in a molecule are combined to form molecular orbitals spanning two or more atoms, with the number of MOs being equal to the number of atomic orbitals. Electrons placed in the lower-energy, bonding molecular orbitals make the molecule more stable energetically than the individual atoms from which it is made.

MO theory can also be used to describe metallic bonding. A metal is a kind of "supermolecule," and to describe the bonding in a metal we have to look at all the atoms in a given sample.

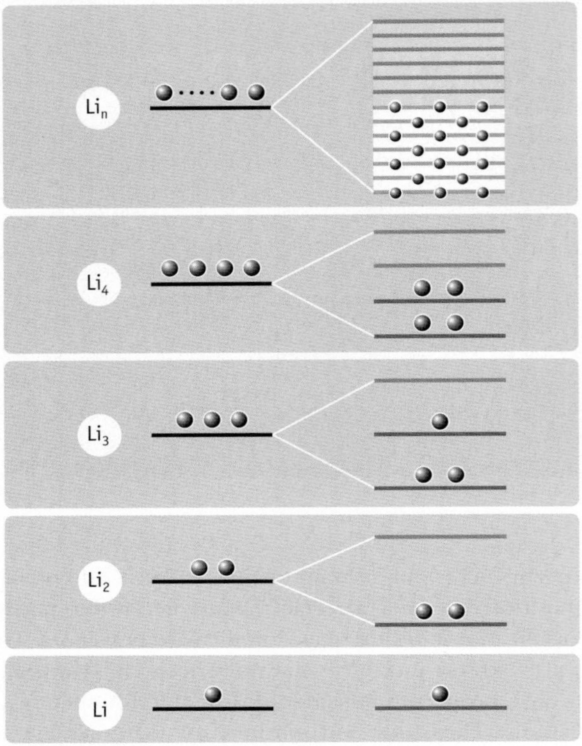

Figure 2 **Bands of molecular orbitals in a metal crystal.** Here, the 2*s* valence orbitals of Li atoms are combined to form molecular orbitals. As more and more atoms with the same valence orbitals are added, the number of molecular orbitals grows until the orbitals are so close in energy that they merge into a band of molecular orbitals. If 1 mol of Li atoms, each with its 2*s* valence orbital, is combined, 6×10^{23} molecular orbitals are formed. However, only 1 mol of electrons, or 3×10^{23} electron pairs, is available, so only half of these molecular orbitals are filled.

Even a tiny piece of metal contains a very large number of atoms and an even larger number of valence orbitals. In 1 mol of lithium atoms, for example, there are 6×10^{23} atoms). Considering only the 2*s* valence orbitals of lithium, there are 6×10^{23} atomic orbitals, from which 6×10^{23} molecular orbitals can be created. The molecular orbitals that we envision in lithium will span all the atoms in the crystalline solid. A mole of lithium has 1 mol of valence electrons, and these electrons occupy the lower-energy bonding orbitals. The bonding is described as delocalized; that is, the electrons are associated with all the atoms in the crystal and not with a specific bond between two atoms.

This theory of metallic bonding is called **band theory.** An energy-level diagram would show the bonding and antibonding molecular orbitals blending together into a band of molecular orbitals (Figure 2), with the individual MOs being so close together in energy that they are not distinguishable. The band is composed of as many molecular orbitals as there are contributing atomic orbitals, and each molecular orbital can accommodate two electrons of opposite spin.

In metals, there are not enough electrons to fill all of the molecular orbitals. In 1 mol of Li atoms, for example, 6×10^{23} electrons, or 3×10^{23} electron pairs, are sufficient to fill only half of the 6×10^{23} molecular orbitals. The lowest energy for a system occurs with all electrons in orbitals with the lowest possible energy, but this is reached only at 0 K. At 0 K, the highest filled level is called the **Fermi level** (Figure 3).

In metals at temperatures above 0 K, thermal energy will cause some electrons to occupy higher-energy orbitals. Even a small input of energy (for example, raising the temperature a few degrees above 0 K) will cause electrons to move from filled orbitals to higher-energy orbitals. For

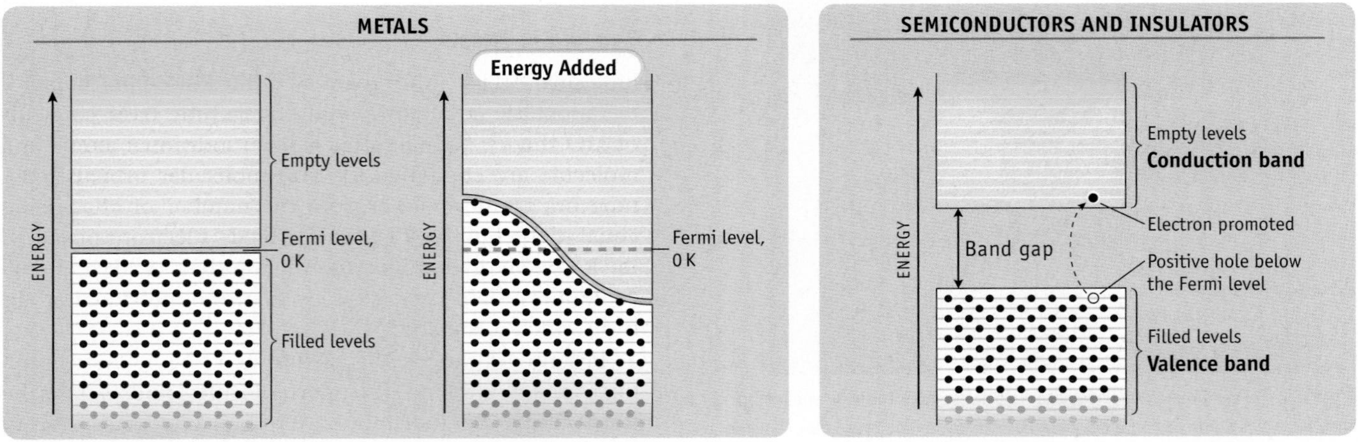

Figure 3 **Band theory applied to metals, semiconductors, and insulators.** The bonding in metals and semiconductors can be described using molecular orbital theory. Molecular orbitals are constructed from the valence orbitals on each atom and are delocalized over all the atoms. *(Metals, left and center)* The highest filled level at 0 K is referred to as the Fermi level. *(Semiconductors and insulators, right)* In contrast to metals, the band of filled levels (the valence band) is separated from the band of empty levels (the conduction band) by a band gap. In insulators, the energy of the band gap is large.

each electron promoted, two singly occupied levels result: a negative electron in an orbital above the Fermi level and a positive "hole"—from the absence of an electron—below the Fermi level.

The positive holes and negative electrons in a piece of metal account for its electrical conductivity. Electrical conductivity arises from the movement of electrons and holes in singly occupied states in the presence of an applied electric field. When an electric field is applied to the metal, negative electrons move toward the positive side, and the positive "holes" move to the negative side. (Positive holes "move" because an electron from an adjacent atom can move into the hole, thereby creating a fresh "hole.")

Because the band of unfilled energy levels in a metal is essentially continuous—that is, because the energy gaps between levels are extremely small—a metal can absorb energy of nearly any wavelength. When light is absorbed, causing an electron in a metal to move to a higher energy state, the now-excited system can immediately emit a photon of the same energy as the electron returns to the original energy level. This rapid and efficient absorption *and* reemission of light make polished metal surfaces be reflective and appear lustrous (shiny).

The molecular orbital picture for metallic bonding provides an interpretation for other physical characteristics of metals. For example, most metals are malleable and ductile, meaning they can be rolled into sheets and drawn into wires. In these processes, the metal atoms must be able to move fairly freely with respect to their nearest neighbors. This is possible because metallic bonding is delocalized—that is, nondirectional. The layers of atoms can slip past one another relatively easily, as if the delocalized electrons were ball bearings that facilitate this motion, while at the same time keeping the layers bonded through coulombic attractions between the nuclei and the electrons.

In contrast to metals, rigid network solids such as diamond, silicon, and silica (SiO_2) have localized bonding, which anchors the component atoms or ions in fixed positions. Movement of atoms in these structures relative to their neighbors requires breaking covalent bonds. As a result, such substances are typically hard and brittle. They will not deform under stress as metals do, but instead tend to cleave along crystal planes (◀ page 80).

Alloys: Mixtures of Metals

Pure metals often do not have the ideal properties needed for their typical uses. It may be possible, however, to improve their properties by adding one or more other elements to the metal to form an **alloy** (Table 1). In fact, most metallic objects we use are alloys, mixtures of a metal with one or more other metals or even with a nonmetal such as carbon (as in carbon steel). For example, sterling silver, commonly used for jewelry, is an alloy composed of 92.5% Ag and 7.5% Cu. Pure silver is soft and easily damaged, and the addition of copper makes the metal more rigid. You can confirm that an article of jewelry is sterling silver by looking for the stamp that says "925," which means 92.5% silver.

TABLE 1 Some Common Alloys

Sterling silver	92.5% Ag, 7.5% Cu
18 K "yellow" gold	75% Au, 12.5% Ag, 12.5% Cu
Pewter	91% Sn, 7.5% Sb, 1.5% Cu
Low-alloy steel	98.6% Fe, 1.0% Mn, 0.4% C
Carbon steels	Approximately 99% Fe, 0.2–1.5% C
Stainless steel	72.8% Fe, 17.0% Cr, 7.1% Ni, and approximately 1% each of Al and Mn
Alnico magnets	10% Al, 19% Ni, 12% Co, 6% Cu, remainder Fe
Brass	95–60% Cu, 5–40% Zn
Bronze	90% Cu, 10% Sn

Gold used in jewelry is rarely pure (24 Carat) gold. More often, you will find 18 K, 14 K, or 9 K stamped in a gold object, referring to alloys that are 18/24, 14/24, or 9/24 gold. The 18 K "yellow" gold is 75% gold, and the remaining 25% is copper and silver. As with sterling silver, the added metals lead to a harder and more rigid material (and one that is less costly).

Alloys fall in three general classes: solid solutions, which are homogeneous mixtures of two or more elements; heterogeneous mixtures; and intermetallic compounds.

In solid solutions, one element is usually considered the "solute" and the other the "solvent." As with solutions in liquids, the solute atoms are dispersed throughout the solvent such that the bulk structure is homogeneous. Unlike liquid solutions, however, there are limitations on the size of solvent and solute atoms. For a solid solution to form, the solute atoms must be incorporated in such a way that the original crystal structure of the solvent metal is preserved. Solid solutions can be achieved in two ways: with solute atoms as **interstitial** atoms or as **substitutional** atoms in the crystalline lattice. In interstitial alloys, the solute atoms occupy the interstices, the small "holes" between solvent atoms (Figure 4a). The solute atoms must be substan-

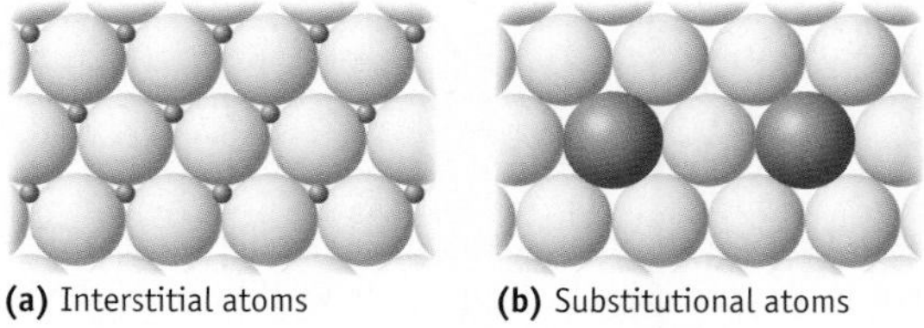

Figure 4 **Alloys.** (a) The solute atoms may be interstitial atoms, fitting into holes in the crystal lattice. (b) The solute atoms can also substitute for one of the lattice atoms.

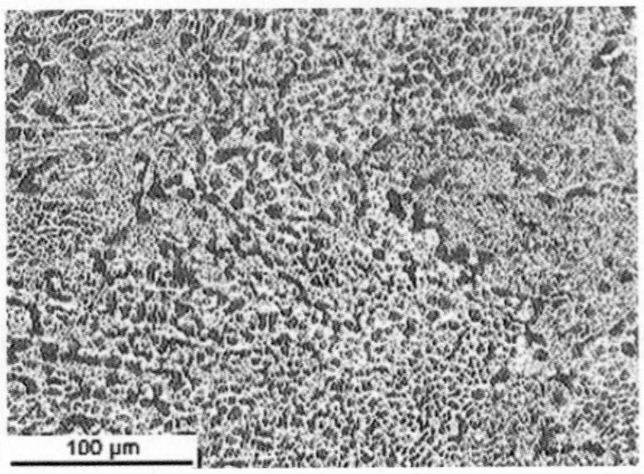

Figure 5 Photomicrograph of the surface of a heterogeneous alloy of lead and tin.

tially smaller than the metal atoms making up the lattice to fit into these positions. In substitutional alloys, the solute atoms replace one of the solvent atoms in the original crystal structure (Figure 4b). For this to occur, the solute and solvent atoms must be similar in size.

If the size constraints are not met, then the alloy will likely form a heterogeneous mixture. When viewed under a microscope, regions of different composition and crystal structure can be seen in heterogeneous alloys (Figure 5).

For a solid solution to form, the electronegativities of the alloy components must also be similar. When the two metals have different electronegativities, the possibility exists for intermetallic compounds, substances with a definite stoichiometry and formula. Examples of intermetallic compounds include $CuAl_2$, Mg_2Pb, and $AuCu_3$. In general, intermetallic compounds are likely when one element is relatively electronegative and the other is more electropositive. For Mg_2Pb, for example, χ for Pb = 2.3 and χ for Mg = 1.3 ($\Delta\chi = 1.0$).

The macroscopic properties of an alloy will vary, depending on the ratio of the elements in the mixture. For example, "stainless" steel is highly resistant to corrosion and is roughly five times stronger than carbon and low-alloy steels. Melting point, electrical resistance, thermal conductivity, ductility, and other properties can be similarly adjusted by changing the composition of the alloys.

Metals and their alloys are good examples of how changes in the atomic composition and structure of a crystalline substance can have profound effects on its macroscopic chemical and physical characteristics. The same is true in semiconductors, the next class of materials we want to explore.

Semiconductors

Semiconducting materials are at the heart of all solid-state electronic devices, including such well-known devices as computer chips and diode lasers. Semiconductors will not conduct electricity easily but can be encouraged to do so by the input of energy. This property allows devices made from semiconductors to essentially have "on" and "off" states, which form the basis of the binary logic used in computers. We can understand how semiconductors function by looking at their electronic structure, following the band theory approach used for metals.

Bonding in Semiconductors: The Band Gap

The Group 4A elements carbon (in the diamond form), silicon, and germanium have similar structures. Each atom is surrounded by four other atoms at the corners of a tetrahedron (Figure 6). Using the band model of bonding, the orbitals of each atom are combined to form molecular orbitals that are delocalized over the solid. Unlike metals, however, the result for carbon, silicon, and germanium is two bands, a lower-energy **valence band** and a higher-energy **conduction band.** In metals, there is only a small energy barrier for an electron to go from the filled molecular orbitals to empty molecular orbitals, and electricity can flow easily. In electrical insulators, such as diamond, and in semiconductors, such as silicon and germanium, the valence and conduction bands are separated from each other resulting in a **band gap,** a barrier to the promotion of electrons to higher energy levels (see Figure 3). In the Group 4A elements, the orbitals of the valence band are completely filled, but the conduction band is empty.

The band gap in diamond is 580 kJ/mol—so large that electrons are trapped in the filled valence band and cannot make the transition to the conduction band, even at elevated temperatures. Thus, it is not possible to create positive "holes," and diamond is an insulator, a nonconductor. Semiconductors, in contrast, have a smaller band gap. For common semiconducting materials, this band gap is usu-

Figure 6 The structure of diamond. The structures of silicon and germanium are similar in that each atom is bound tetrahedrally to four others.

ally in the range of 10 to 240 kJ/mol. (The band gap is 106 kJ/mol in silicon, whereas it is 68 kJ/mol in germanium.) The magnitude of the band gap in semiconductors is such that these substances are able to conduct small quantities of current under ambient conditions, but, as their name implies, they are much poorer conductors than metals.

Semiconductors can conduct a current because thermal energy is sufficient to promote a few electrons from the valence band to the conduction band (Figure 7). Conduction then occurs when the electrons in the conduction band migrate in one direction and the positive holes in the valence band migrate in the opposite direction.

Pure silicon and germanium are called **intrinsic semiconductors**, with the name referring to the fact that this is an intrinsic property of the pure material. In intrinsic semiconductors, the number of electrons in the conduction band is determined by the temperature and the magnitude of the band gap. The smaller the band gap, the smaller the energy required to promote a significant number of electrons. As the temperature increases, more electrons are promoted into the conduction band, and a higher conductivity results.

In contrast to intrinsic semiconductors are materials known as **extrinsic semiconductors.** The conductivity of these materials is controlled by adding small numbers of different atoms (typically 1 in 10^6 to 1 in 10^8) called **dopants.** That is, the characteristics of semiconductors can be changed by altering their chemical makeup, just as the properties of alloys differ from the properties of pure metals.

Suppose a few silicon atoms in the silicon lattice are replaced by aluminum atoms (or atoms of some other Group 3A element). Aluminum has only three valence electrons, whereas silicon has four. Four Si-Al bonds are created per aluminum atom in the lattice, but these bonds must be deficient in electrons. According to band theory, the Si-Al bonds form a discrete band at an energy level higher than the valence band. This level is referred to as an **acceptor level** because it can accept electrons. The gap between the valence band and the acceptor level is usually quite small, so electrons can be promoted readily to the acceptor level. The positive holes created in the valence band are able to move about under the influence of an electric potential, so current results from the hole mobility. Because positive holes are created in an aluminum-doped semiconductor, this is called a ***p*-type semiconductor** (Figure 7b, left).

Now suppose phosphorus atoms (or atoms of some other Group 5A element such as arsenic) are incorporated into the silicon lattice instead of aluminum atoms. The material is also a semiconductor, but it now has extra electrons because each phosphorus atom has one more valence electron than the silicon atom it replaces in the lattice. Semiconductors doped in this manner have a discrete, partially filled **donor level** that resides just below the conduction band. Electrons are promoted readily to the conduction band from this donor band, and electrons in the conduction band carry the charge. Such a material, consisting of negative charge carriers, is called an ***n*-type semiconductor** (Figure 7b, right).

One group of materials that have desirable semiconducting properties is the III-V semiconductors, so called because they are formed by combining elements from Group 3A (such as Ga and In) with elements from Group 5A (such as As or Sb).

GaAs is a common semiconducting material that has electrical conductivity properties that are sometimes preferable to those of pure silicon or germanium. The crystal structure of GaAs is similar to that of diamond and silicon; each Ga

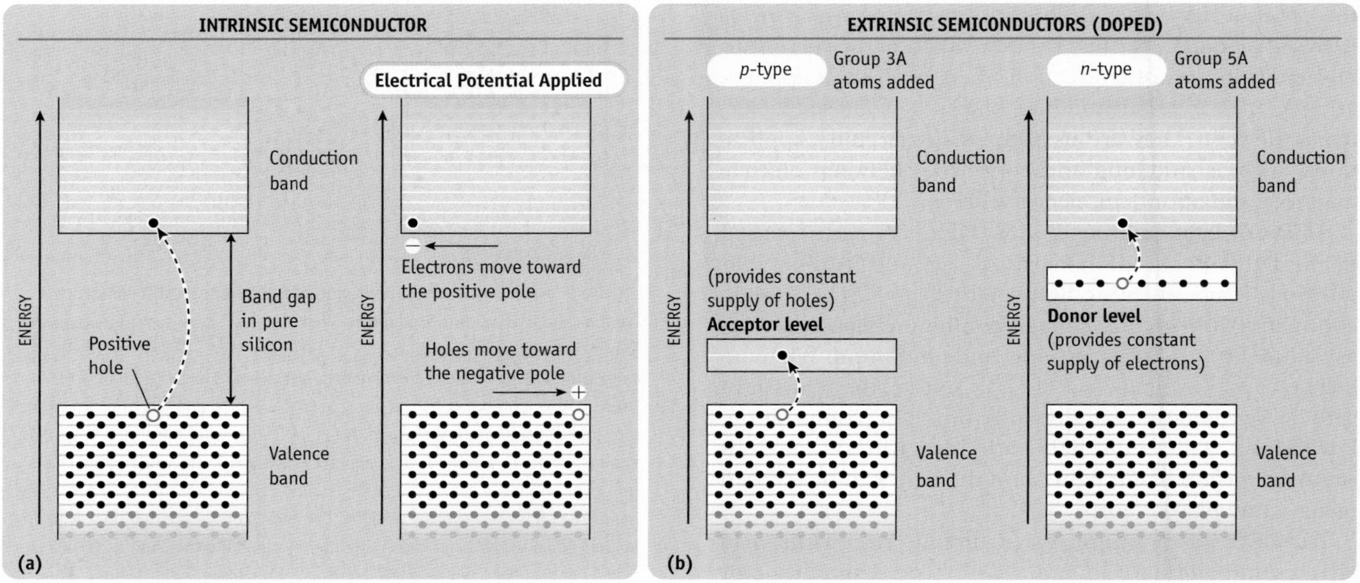

Figure 7 **Intrinsic and extrinsic semiconductors.**

atom is tetrahedrally coordinated to four As atoms, and vice versa. This structure is often referred to as the *zinc blende* structure (Figure 13.10).

It is also possible for Group 2B and 6A elements to form semiconducting compounds, such as CdS. The farther apart the elements are found in the periodic table, however, the more ionic the bonding becomes. As the ionic character of the bonding increases, the band gap will increase, and the material will become an insulator rather than a semiconductor. For example, the band gap in GaAs is 140 kJ/mol, whereas it is 232 kJ/mol in CdS.

These materials can be modified further by substituting other atoms into the structure. For example, in one widely used semiconductor, aluminum atoms are substituted for gallium atoms in GaAs, giving materials with a range of compositions ($Ga_{1-x}Al_xAs$). The importance of this modification is that the band gap depends on the relative proportions of the elements, so it is possible to control the size of the band gap by adjusting the stoichiometry. As Al atoms are substituted for Ga atoms, for example, the band gap energy increases. This consideration is important for the specific uses of these materials in devices such as LEDs.

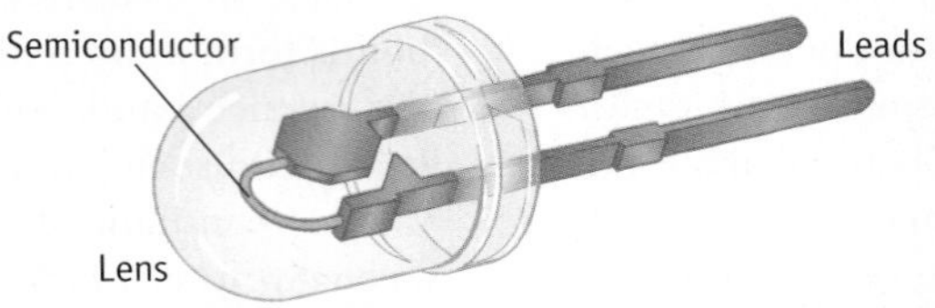

Charles D. Winters

Figure 8 Light-emitting diodes (LEDs). (left) A schematic drawing of a typical LED. (right) Traffic signs with LED light require much less energy input than incandescent lights and are now being widely used.

Applications of Semiconductors: Diodes, LEDs, and Transistors

The combination of *p*- and *n*-type semiconducting materials in a single electronic device launched the microelectronics and computer industries. When a semiconductor is created such that it is *p*-type on one half and *n*-type on the other, a marvelous device known as the *p–n rectifying junction*, or **diode**, results. Diodes, which allow current to flow easily in only one direction when a voltage is applied, are the fundamental building blocks of solid-state electronic devices. They are used for many circuitry applications, such as switching and converting between electromagnetic radiation and electric current.

LEDs, or **light-emitting diodes** (Figure 8), are now used in the lights in the dashboards of cars and in their rear warning lights, in traffic lights, and in toys. These semiconducting devices are made by combining elements such as gallium, phosphorus, arsenic, and aluminum. When attached to a low-voltage (say 6–12 V) source, they emit light with a wavelength that depends on their composition. Furthermore, they emit light with a brightness that rivals standard incandescent lights, and the light can be focused using a tiny plastic lens.

An LED has a simple construction. It consists of a *p*-type semiconductor joined to an *n*-type semiconductor (Figure 9). A voltage is applied to the material, perhaps by hooking the positive terminal of a battery to the *p*-type semiconductor and the negative terminal to the *n*-type semiconductor. Negative electrons move from the *n*-type to the *p*-type, and positive holes move from the *p*-type to the *n*-type. When electrons move across the *p–n* junction, they can drop from the conduction band into a hole in the valence band

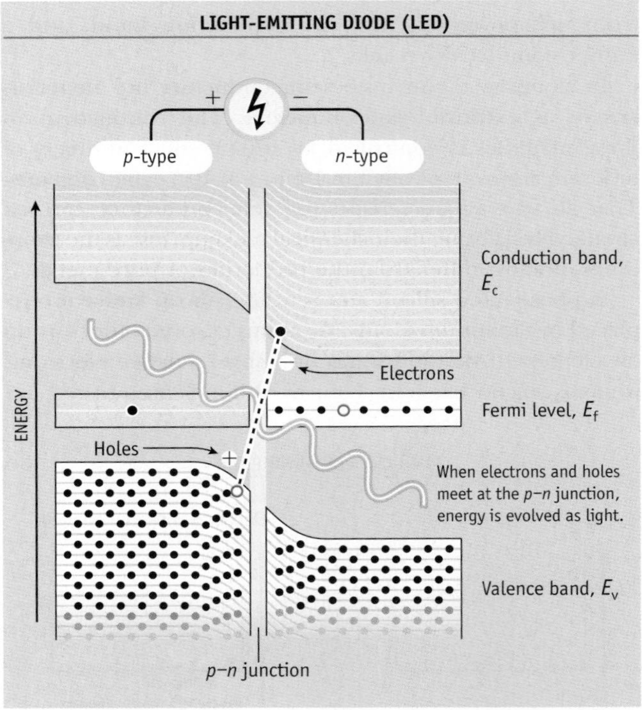

Figure 9 Mechanism for the emission of light from an LED constructed from *n*- and *p*-type semiconductors. When *p*- and *n*-type semiconductors are joined, the energy levels adjust so that the Fermi levels (E_f) are equal. This causes the energy levels of the conduction (E_c) and valence (E_v) bands to "bend." Also, holes flow from the *p* side to the *n* side, and electrons flow from *n* to *p* until equilibrium is reached. No more charge will flow until a voltage is applied. When an electric field is applied, occasionally electrons in the conduction band will move across the band gap and combine with holes in the valence band. Energy is then evolved as light. The energy of the emitted light is approximately equal to the band gap. Therefore, by adjusting the band gap, the color of the emitted light can be altered. (See S. M. Condren, et al.: *Journal of Chemical Education*, Vol. 78, pp. 1033–1040, 2001.)

Figure 10 Gallium arsenide (GaAs) solar panel. This panel was built for NASA's Deep Space 1 probe. The array uses 3600 solar cells, which convert light to electricity to power an ion propulsion system. (DS1 was launched on October 24, 1998, and sent back images of Comet Borrelly in deep space. The spacecraft was retired on December 18, 2001.)

of the *p*-type semiconductor, and energy is released as light. (The mechanism of light emission by an LED is similar to that described for excited atoms in Section 6.3.) If the band gap energy is equivalent to the energy of light in the visible region, light can be observed. Because the band gap energy can be adjusted by changing the composition of the doped semiconductor, the wavelength of the light can also be altered, giving light of different colors.

The same device that forms the LED can be run in reverse to convert light that falls on it into an electrical signal. Solar panel cells work in this manner (Figure 10). They are generally GaAs-based *p–n* junction materials that have a band gap corresponding to the energy of visible light. When sunlight falls on these devices, a current is induced. That current can be used either immediately or stored in batteries for later use. A similar technology is used in simpler devices referred to as photodiode detectors. They have an abundance of applications, ranging from the light-sensitive switches on elevator doors to sensitive detection equipment for scientific instruments.

The *p*- and *n*-type semiconductor materials can also be constructed into a sandwich structure of either *p–n–p* or *n–p–n* composition. This arrangement forms a device known as a **transistor.** A transistor amplifies an electrical signal, making it ideal for powering loudspeakers, for example. Transistors can also be used for processing and storing information, a critical function for computer chips. By combining thousands of these transistors and diodes, an integrated circuit can be made that is the basis of what we commonly refer to as computer chips, devices for controlling and storing information (Figure 11).

Figure 11 Integrated circuits. (left) A wafer on which a large number of integrated circuits has been printed. (right) A close-up of a semiconductor chip showing the complex layering of circuits that is now possible.

Ceramics

Let's go back to our original examples of various forms of $CaCO_3$, including sea shells and chalk. Chalk is so soft that it will rub off on the rough surface of a blackboard. In contrast, sea shells are inherently tough. They are designed to protect their soft and vulnerable inhabitants from the powerful jaws of sea-borne predators or rough conditions underwater. Chalk, sea shells, and the spines of sea urchins (Figure 12) are all ceramics, but they are obviously different from one another. Clearly, there is a great deal of variability in this class of materials.

You may be accustomed to thinking of "ceramics" as the objects that result from high-temperature firing, such as pottery. From a materials chemistry perspective, however, other materials such as clay, which largely consists of hydrated silicates of various compositions, are also considered ceramics. **Ceramics** are solid inorganic compounds. Their composition includes metal and nonmetal atoms, and the bonding between atoms ranges from very ionic to covalent (◄ Section 9.2). In general, ceramics are hard, relatively brittle, and inflexible, and they are usually good thermal insulators. Some ceramics can be electrically conductive, but most are electrical insulators. Some, like glass, another type of ceramic, can be optically transparent, whereas other ceramics are completely opaque.

It is also possible to have ceramics in which impurity atoms are included in the composition. As we saw with metals and semiconductors, impurity atoms can have dramatic effects on the characteristics of a material.

Glass: A Disordered Ceramic

An amorphous, or noncrystalline, solid material is generally referred to as a **glass** (see Section 13.4). Glasses are formed by melting the raw material and then cooling it from the liquid state rapidly so that the component atoms do not have time to crystallize into a regular lattice structure. A wide range of materials, including metals and organic poly-

Figure 12 **A sea urchin.** The spines of the urchin are composed primarily of $CaCO_3$, but a significant amount of $MgCO_3$ is present as well.

mers, can be coaxed into a glassy form. However, the best-known glasses are silicate glasses. These are derived from SiO_2, which is plentiful, inexpensive, and chemically unreactive. Each silicon atom is linked to four oxygen atoms in the solid structure, with a tetrahedral arrangement around each silicon atom. The SiO_2 units are linked together to form a large network of atoms (Figure 13). Over a longer distance, however, the network has no discernible order or pattern.

Glasses can be modified by the presence of alkali metal oxides (such as Na_2O and K_2O) or other metal or nonmetal oxides (such as CaO, B_2O_3, and Al_2O_3). The added impurities change the silicate network and alter the properties of the material. The oxide ions are incorporated into the silicate network structure, and the resultant negative charge is balanced by the interstitial metal cations (Figure 13c). Because the network is changed by such an addition, these network modifiers can dramatically alter the physical characteristics of the material, such as melting point, color, opacity, and strength. Soda-lime glass—made from SiO_2, Na_2O (soda), and CaO (lime)—is a common glass used in windows and for containers. The metal oxides lower the melting temperature by about a thousand degrees from that of pure silica. Pyrex glass, also called borosilicate glass, incorporates an additional component, boric oxide. The boric oxide raises the softening temperature and minimizes the coefficient of thermal expansion, enabling the glass to better withstand temperature changes. Because of its excellent thermal properties, this type of glass is used for beakers and flasks in chemistry laboratories and for ovenware for the kitchen.

An important characteristic of some glasses is their optical transparency, which allows them to be used as windows and lenses. Glasses can also be reflective. The combination of transparency and reflectivity is controlled by the material's **index of refraction.** All materials have an index of refraction that determines how much a beam of light will change its velocity when entering the material. The index of refraction is defined relative to the speed of light in a vacuum, which is defined as exactly 1. (The index of refraction = velocity of light in a vacuum/velocity in material.) On this basis, dry air has an index of refraction of 1.0003, and typical values for silicate glasses range from 1.5 to 1.9.

The change in the velocity of the electromagnetic wave once it enters the material causes the beam to bend, or change direction within the material. If light hits a surface at some incident angle relative to the line perpendicular to the surface, some of the light will be reflected at the same angle, and some will be transmitted into the material at a refracted angle (Figure 14). Both the incident angle

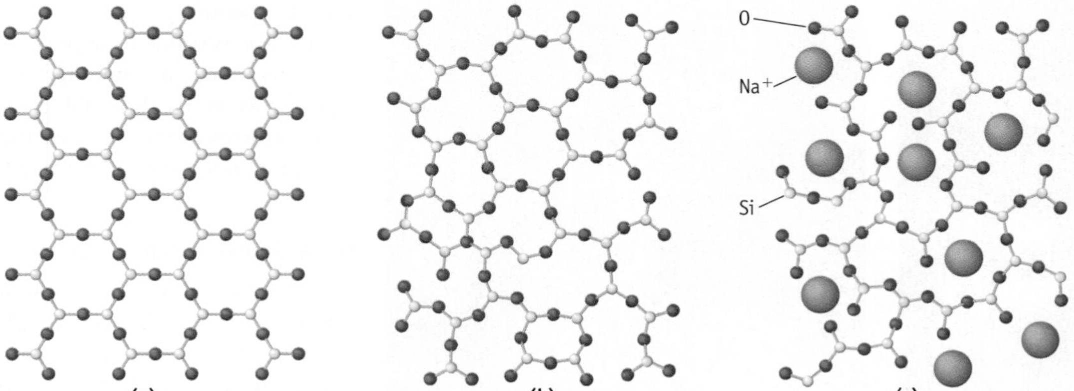

Figure 13 **Representation of glass structure.** (a) Silica glass (SiO_2) may have some order over a short distance but much less order over a larger portion of the solid (b). (c) The SiO_2 structure can be modified by adding metal oxides, which leads to a lower melting temperature and other desirable properties. (In this simple representation, the gray Si atoms are shown at the center of a planar triangle of red O atoms; in reality, each Si atom is surrounded tetrahedrally by O atoms. The structure is not planar but is three dimensional.)

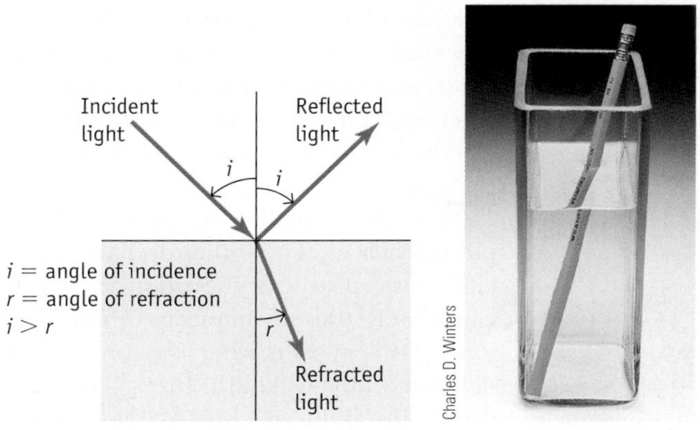

(a) (b)

Figure 14 **Refraction of light.** (a) When light enters a different medium, its velocity changes. This causes the path of a photon to change direction in the material. (b) Observing an object in a glass of water illustrates the effect of light refraction.

and the index of refraction will affect how much of the light is reflected and the angle at which it bends in the second material. You can observe this effect by putting an object in a glass of water and looking at the apparent bend that results in the object (Figure 14b).

This combination of the transmission and reflection characteristics of glass has allowed scientists and engineers to develop optical fibers (Figure 15). Optical fibers are designed to have a property called total internal reflection, whereby all the light that enters at one end of the fiber stays within the fiber through reflections with the interior surface as the light travels from one end of the fiber to the other. Total internal reflection in these fibers is achieved by controlling the ratio of the indices of refraction between the fiber's core and its outside surface. Chemically, the index of refraction is controlled by adjusting the quantity and type of cationic network modifiers that are added to the glass. The index of refraction of a glass fiber can be controlled so that it has one value at the core of the fiber but changes smoothly across the radius of the fiber to a different value at the surface. This is accomplished by an ion-exchange process during fiber production in which, for example, K^+ ions are replaced by Tl^+ ions.

Figure 15 **Optical fibers.** (left) Glass fibers transmit light along the axis of the fiber. (right) Bell Laboratory scientist Joanna Aizenberg recently discovered that a deep-sea sponge, made chiefly of silica (SiO_2), has a framework that has the characteristics of optical fibers. (For more about the structure of the sponge, see Figure 2 on page 27.)

Optical fibers are transforming the communications industry in an amazing fashion. Instead of transmitting information using electrons traveling through metallic wires, optical fibers allow communication to occur by transmitting photons through glass fiber bundles. Signal transmission by optical fibers, known as photonics, is much faster and more economical than transmission using copper wires and cables. For example, the quantity of copper required to carry the equivalent amount of information transmitted by optical fiber would weigh 300,000 times more than the optical fiber material!

Fired Ceramics for Special Purposes: Cements, Clays, and Refractories

Other classes of ceramics include cements, clays, and refractories. Unlike glasses, these ceramics are processed by shaping, drying, and then firing, without ever melting the solid.

Cements are extremely strong and are commonly used as structural materials. They can be formed into almost any shape. When mixed with water, they produce a paste that can be poured into molds and allowed to dry and harden.

Clays are generally mixtures of hydrated alumina (Al_2O_3) and silica (SiO_2), but may also contain other ingredients, such as tricalcium silicate, ($3\ CaO \cdot SiO_2$), dicalcium silicate, ($2\ CaO \cdot SiO_2$), and MgO. Their composition is irregular, and, because they are powders, their crystallinity extends for only short distances.

Clays have the useful property of becoming very plastic when water is added, a characteristic referred to as **hydroplasticity.** This plasticity, and clay's ability to hold its shape during firing, are very important for the forming processes used to create various objects.

The layered molecular structure of clays results in microscopic platelets that can slide over each other easily when wet. The layers consist of SiO_4 tetrahedra joined with AlO_6 octahedra (see Section 21.7). In addition to these basic silicon- and aluminum-based structures, different cations can be substituted into the framework to change the properties of the clay. Common substituents include Ca^{2+}, Fe^{2+}, and Mg^{2+}. Different clay materials can then be created by varying the combinations of layers and the substituent cations.

NASA JPL

Figure 16 Aerogel, a networked matrix of SiO_2. (left) NASA's Peter Tsou holds a piece of aerogel. It is 99.8% air, is 39 times more insulating than the best fiberglass insulation, and is 1000 times less dense than glass. (right) Aerogel was used on a NASA mission to collect the particles in comet dust. The particles entered the gel at a very high velocity, but were slowed gradually. Scientists studied the tracks made by the particles and later retrieved the particles and studied their composition.

Refractories constitute a class of ceramics that are capable of withstanding very high temperatures without deforming, in some cases up to 1650 °C (3000 °F), and that are thermally insulating. Because of these properties, refractory bricks are used in applications such as furnace linings and in metallurgical operations. These materials are thermally insulating largely because of the porosity of their structure; that is, holes (or pores) are dispersed evenly within the solid. However, while porosity will make a material more thermally insulating, it will also weaken it. As a consequence, refractories are not as strong as cements.

An amazing example of the use of porosity to increase the insulating capacities of a ceramic is found in a material developed at NASA called *aerogel* (Figure 16; see Case Study: The World's Lightest Solid, page 607). Aerogel is more than 99% air, with the remainder consisting of a networked matrix of SiO_2. This makes aerogel about 1000 times less dense than glass but gives the material extraordinary thermal insulating abilities. NASA used aerogel on a mission in which a spacecraft flew through the tail of the comet Wild 2 and returned to Earth with space particles embedded in the aerogel.

Modern Ceramics with Exceptional Properties

In 1880, Pierre Curie and his brother Jacques worked in a small laboratory in Paris to examine the electrical properties of certain crystalline substances. Using nothing more than tin foil, glue, wire, and magnets, they were able to confirm the presence of surface charges on samples of materials such as tourmaline, quartz, and topaz when they were subjected to mechanical stresses. This phenomenon, now called **piezoelectricity**, is the property that allows a mechanical distortion (such as a slight bending) to induce an electrical current and, conversely, an electrical current to cause a distortion in the material.

Not all crystalline ceramics exhibit piezoelectricity. Those that do have a specific unit cell structure (Section 13.2) that can loosely trap an impurity cation. The ion's position shifts when the unit cell is deformed by mechanical stress. This shift causes an induced dipole (see Section 12.3) and, therefore, a potential difference across the material that can be converted to an electrical signal.

In addition to the minerals originally tested by the Curie brothers, materials known to exhibit the piezoelectric effect include titanium compounds of barium and lead, lead zirconate ($PbZrO_3$), and ammonium dihydrogen phosphate ($NH_4H_2PO_4$).

Materials that exhibit piezoelectricity have a great many applications, ranging from home gadgets to sophisticated medical and scientific applications. One use with which you may be familiar is the automatic ignition systems on some barbecue grills and lighters (Figure 17). All digital watch beepers are based on piezoceramics, as are smoke detector alarms. A less familiar application is found in the sensing lever of some atomic force microscopes (AFMs) and scanning-tunneling microscopes (STMs), instruments that convert mechanical vibrations to electrical signals.

Scientists and engineers are always searching for materials with new and useful properties. Perhaps the most dramatic property that has been observed in newly developed ceramics is superconductivity at relatively high temperatures.

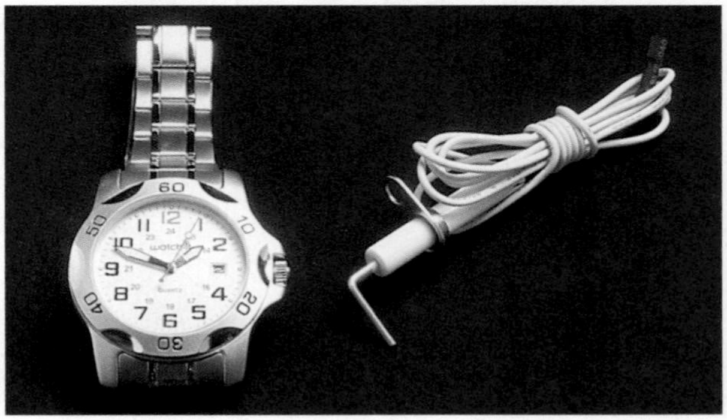

Charles D. Winters

Figure 17 Devices that depend on the piezoelectric effect. These devices work by using a mechanical stress to produce an electric current. Piezoelectric devices are widely used in ignitors and in devices that convert electric impulses to vibrations, such as in the timing circuit of a wristwatch.

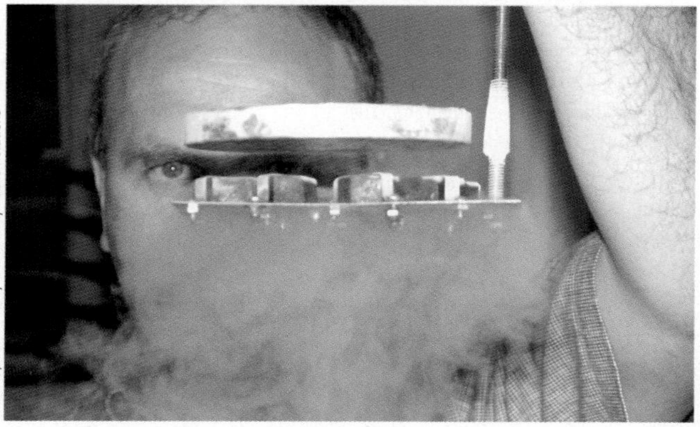

Figure 18 **Superconductivity.** When a superconducting material is cooled to a low temperature, say in liquid nitrogen (boiling point is 77 K), it generates a very strong magnetic field. In this photo, a 1-pound magnet is levitated in the field created by the cooled superconductor.

Superconductivity is a phenomenon in which the electrical resistivity of a material drops to nearly zero at a particular temperature referred to as the **critical temperature, T_c** (Figure 18). Most metals naturally have resistivities that decrease with temperature in a constant manner but still have significant resistivity even at temperatures near 0 K.

A few metals and metal alloys have been found to exhibit superconductivity. For metals, however, the critical temperatures are extremely low, between 0 and 20 K. These temperatures are costly to achieve and difficult to maintain. Recent scientific attention has, therefore, focused on a class of ceramics with superconductive critical temperatures near 100 K. These materials include $YBa_2Cu_3O_7$, with $T_c = 92$ K (Figure 19), and $HgBa_2Ca_2Cu_2O_3$, with $T_c = 153$ K.

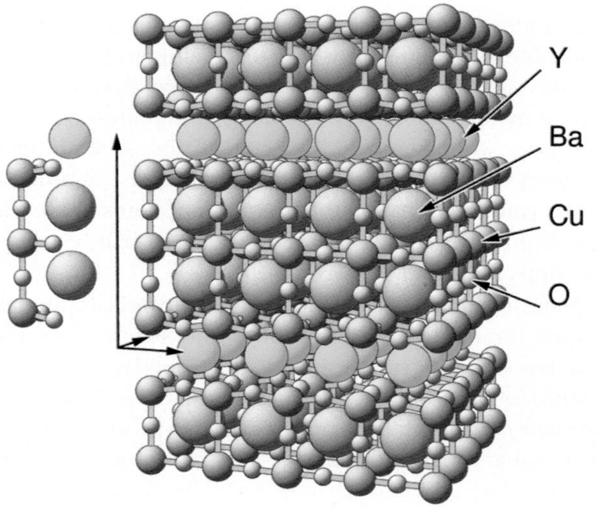

Figure 19 **The lattice of $YBa_2Cu_3O_7$, a superconductor.** Yttrium ions are yellow; barium ions are red; copper ions are green; and oxygen ions are blue. (Reprinted with permission of Dr. Klaus Hermann of the Fritz Haber Institution.)

Once again, we see that combining atoms into sometimes complex chemical compositions allows scientists to develop materials with particular properties. In ceramics, which are normally electrically insulating, this includes even the ability to conduct electricity.

Biomaterials: Learning from Nature

Most of the materials described so far in this chapter come from nonliving sources and, in many cases, are the result of laboratory syntheses. However, an important branch of materials research deals with examining, understanding, and even copying materials produced by living systems.

The study of naturally occurring materials has led to the development of synthetic materials that possess important properties. A good example is rubber (Chapter 10, page 483). The polymer we know as rubber was initially obtained from certain trees and chemically modified to convert it to a useful material. Natural rubber was found to be so useful that chemists eventually achieved the synthesis of a structurally identical material. Research on rubber, which spanned more than 200 years, has had important consequences for humans as evidenced by the myriad applications of rubber today.

Today, scientists continue to look to nature to provide new materials and to provide clues to improve the materials we already use. The sea urchin and its ceramic spines (Figure 12) and the sponge whose skeleton has the characteristics of optical fibers (Figure 15) are just two examples where biomaterials research has focused on sea life in a search for new materials. Scientists have also examined conch shells to understand their incredible fracture strength. They used scanning electron microscopy (SEM) to scrutinize the structure of the shell when it was fractured. What they discovered was a criss-crossed, layered structure that is the equivalent of a "ceramic plywood" (Figure 20). This microarchitecture prevents fractures that occur on the outside surface of the shell from

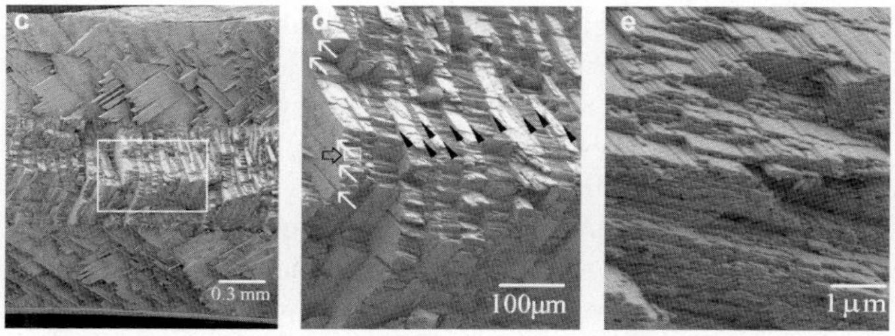

Figure 20 **A scanning electron microscope picture of the shell of the conch.** Photos from S. Kamat, X. Su, R. Ballarini, and A. H. Heuer. Structural basis for the fracture toughness of the shell of the conch *Strombus gigas*. *Nature*. Vol. 405, pp. 1036–1040, 2000.

being transferred into the inner layers. The discovery has inspired materials engineers to create materials that are significantly strengthened by incorporating a fibrous ceramic matrix, such as SiC (silicon carbide) whiskers.

In another area of research focusing on sea creatures, the connective tissues of sea cucumbers and other echinoderms (marine invertebrates with tube feet and calcite-covered, radially symmetrical bodies) have been studied in an attempt to discover how these animals can reversibly control the stiffness of their outer skin. The connective tissues of these animals include the protein collagen in a cross-linked fiber structure, similar to the dermis, an inner layer of the skin consisting of sensitive connective tissue of many mammals. At the same time, other proteins and soluble molecules in the echinoderm system allow the animals to change the characteristics of the connective tissue in response to their nervous system. As a result, creatures such as sea cucumbers can move about and, in some cases, can defend themselves by hardening their skin to an almost shell-like consistency. The ensuing laboratory research has focused on the formulation of a synthetic collagen-based polymer composite material in which the stiffness can be changed repeatedly through a series of oxidation and reduction reactions. Scientists are now developing models for synthetic skin and muscle based on their findings.

Research on adhesive materials represents another area in which sea creatures can provide some clues. Getting things to stick together is important in a multitude of applications. The loss of the space shuttle Columbia in early 2003, caused by loss of some ceramic tiles when a piece of insulating foam that fell off during launch hit them, offered a sobering lesson in adhesive failure under extreme conditions of temperature and humidity. If you look around, you will probably find something with an adhesive label, something with an attached plastic part, something with a rubber seal, or perhaps something taped together. Adhesives have also proven useful for medical applications, where specialized glues help doctors seal tissues within the human body. For every type of sticking application, different properties are needed for the adhesive material.

Nature provides numerous examples of adhesion. Geckos and flies that can walk on glass while completely inverted hold clues to the kind of biologically based adhesion that could be the basis of synthetic analogs. Marine mussels, which can stick quite well to wood, metal, and rock, also hold great interest for scientists studying adhesion (Figure 21).

Jonathan Wilker, Purdue University

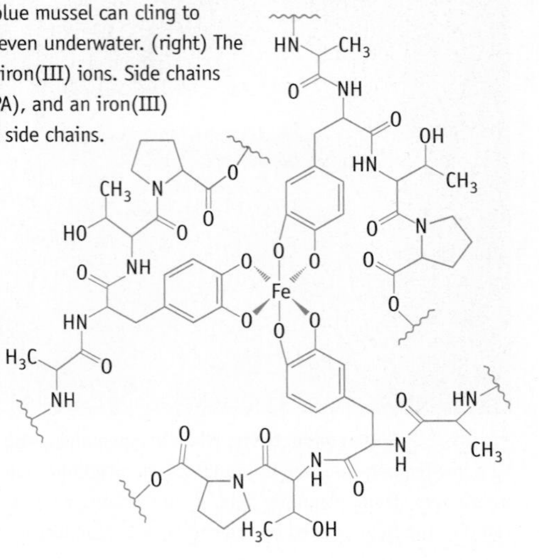

Figure 21 Strong mussels. (left) A common blue mussel can cling to almost any surface, including this Teflon sheet, even underwater. (right) The adhesive precursor is a protein interlinked with iron(III) ions. Side chains on the protein are dihydroxyphenylalanine (DOPA), and an iron(III) ion binds to the hydroxyl groups (–OH) in three side chains.

Scientists who have researched mussel adhesives have been able to determine that the amino acid 3,4-dihydroxyphenylalanine (DOPA) is the agent primarily responsible for the strength of the adhesion. But DOPA alone cannot explain the incredible strength of the mussel glues. The secret lies in the combination of an Fe^{3+} ion with DOPA to form a cross-linked matrix of the mussel's protein (Figure 21). The curing process, or hardening of the natural proteinaceous liquid produced by the mussel, is a result of the iron–protein interaction that occurs to form $Fe(DOPA)_3$ cross-links.

The Future of Materials

The modern tools and techniques of chemistry are making it possible for scientists not only to develop novel materials, but also to proceed in new and unforeseen directions. The field of **nanotechnology** is an example. In nanotechnology, structures with dimensions on the order of nanometers are used to carry out specific functions. For example, scientists can now create a tube of carbon atoms embedded in a slightly larger carbon tube to act as a ball bearing at the molecular level (Figure 22).

Nanoscience has provided profoundly important applications for medicine, computing, and energy consumption. For example, scientists have developed quantum dots (Figure 23), nanometer-scale crystals of different materials that can emit light and can even be made to function as lasers. Quantum dots have been used as biological markers by attaching them to various cells. By shining light on them, the quantum dots will fluoresce in different colors, allowing the cells to be imaged.

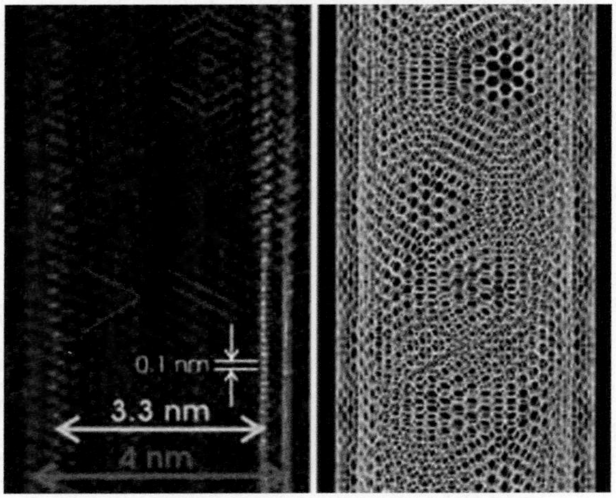

Figure 22 A molecular bearing: double-walled carbon nanotubes. (left) An electron diffraction image of a double-walled carbon nanotube. (right) A model of the material.

Other ongoing research is being carried out on nanoscale drug delivery, a technology that allows one to deliver medicinal agents directly into the cells that need them.

One way that scientists have been able to achieve such breakthroughs is by studying the structures of materials that are already known. They have been learning to manipulate atoms and molecules so that they will arrange themselves in specific ways to achieve desired shapes and functions. In a process referred to as **self-assembly,** molecules or atoms will arrange themselves based on their shapes, the intermolecular forces between them, and their interactions with their environment.

Chemistry is the key to understanding and developing materials. The atomic compositions and long-term atomic arrangements of different materials fundamentally determine their properties and characteristics. Chemists can use analytical instruments to determine these structures. They can then exploit this knowledge to manipulate or develop materials to achieve different properties for special functions. In many cases, we can look to nature to provide answers and suggestions on how to proceed.

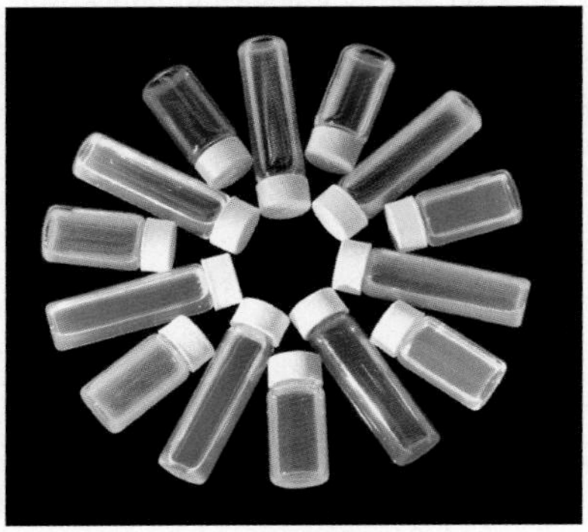

Figure 23 Quantum dots. Quantum dots are a special kind of semiconductor. They have diameters in the 2–10 nm range and have unique properties. Among these is the ability to tune the band gap so different colors result on excitation.

SUGGESTED READINGS

1. S. M. Sze: *Semiconductor Devices: Physics and Technology.* New York: John Wiley & Sons, 1985.
2. X. Su, S. Kamat, and A. H. Heuer: "The structure of sea urchin spines, large biogenic single crystals of calcite." *Journal of Materials Science,* Vol. 35, pp. 5545–5551, 2000.
3. M. J. Sever, J. T. Weisser, J. Monahan, S. Srinivasan, and J. J. Wilker: "Metal-mediated cross-linking in the generation of marine mussel adhesive." *Angewandte Chemie International Edition,* Vol. 43, pp. 447–450, 2004.

STUDY QUESTIONS

Blue-numbered questions have answers in Appendix P and fully-worked solutions in the *Student Solutions Manual.*

1. What is the maximum wavelength of light that can excite an electron transition across the band gap of GaAs? To which region of the electromagnetic spectrum does this correspond?
2. Which of the following would be good substitutional impurities for an aluminum alloy?
(a) Sn (b) P (c) K (d) Pb
3. The amount of sunlight striking the surface of the earth (when the sun is directly overhead on a clear day) is approximately 925 watts per square meter (W/m^2). The area of a typical solar cell is approximately 1.0 cm^2. If the cell is running at 25% efficiency, what is its energy output per minute?
4. Using the result of the calculation in Question 3, estimate the number of solar cells that would be needed to power a 700-W microwave oven. If the solar cells were assembled into a panel, what would be the approximate area of the panel?
5. Describe how you could calculate the density of pewter from the densities of the component elements, assuming that pewter is a substitutional alloy. Look up the densities of the constituent elements and carry out this calculation. Densities can be found on a website such as www.webelements.com, or go to ChemistryNOW and click on the periodic table tool. Click on the symbol of each of the elements in pewter. A table of atomic properties includes the element's density.
6. Calculate an approximate value for the density of aerogel using the fact that it is 99% air, by volume, and the remainder is SiO_2 (Density = 2.3 g/cm^3). What is the mass of a 1.0-cm^3 piece of aerogel?

15 Chemical Kinetics: The Rates of Chemical Reactions

Charles D. Winters

Fading of the color of phenolphthalein with time (elapsed time about 3 minutes)

Where Did the Indicator Go?

The indicator phenolphthalein is often used for the titration of a weak acid using a strong base. A change from colorless to pale pink indicates that the equivalence point in the reaction has been reached (◀ page 184). If more base is added to the solution, the color of the indicator intensifies to a bright red color.

If the solution containing phenolphthalein has a pH higher than about 12, another phenomenon is observed. Slowly, the red color fades, and the solution becomes colorless. This is due to a chemical reaction of the anion of phenolphthalein with hydroxide ion, as shown in the equation. The reaction is slow, and it is easy to measure the rate of this reaction by monitoring the intensity of color of the solution.

This chapter is about one of the fundamental areas of chemistry: the rate of reactions and how they occur. In Study Question 71, you will see some data that will allow you to discover how the rate of the phenolphthalein reaction depends on the hydroxide ion concentration, and you will derive an equation that will allow you to predict the results under other conditions.

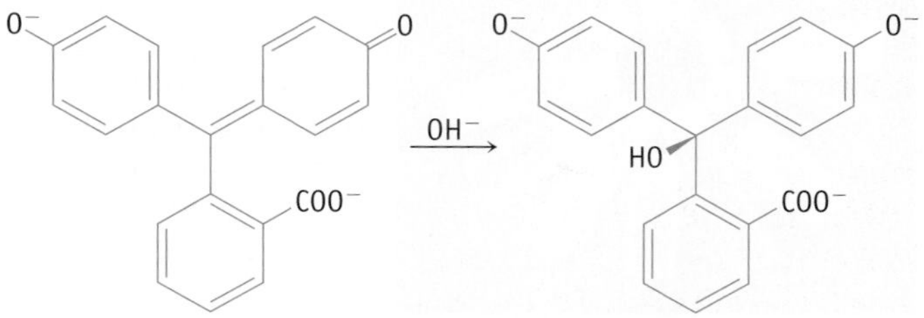

The reaction of phenolphthalein and hydroxide ion.

Chapter Goals

See Chapter Goals Revisited (page 710) for Study Questions keyed to these goals and assignable in OWL.

- Understand rates of reaction and the conditions affecting rates.
- Derive a rate equation, rate constant, and reaction order from experimental data.
- Use integrated rate laws.
- Understand the collision theory of reaction rates and the role of activation energy.
- Relate reaction mechanisms and rate laws.

Chapter Outline

When carrying out a chemical reaction, chemists are concerned with two issues: the *rate* at which the reaction proceeds and the *extent* to which the reaction is product-favored. Chapter 5 began to address the second question, and Chapters 16 and 19 will develop that topic further. In this chapter, we turn to the other part of our question, **chemical kinetics,** a study of the rates of chemical reactions.

The study of kinetics is divided into two parts. The first part is at the *macroscopic level,* which addresses rates of reactions: what reaction rate means, how to determine a reaction rate experimentally, and how factors such as temperature and the concentrations of reactants influence rates. The second part of this subject considers chemical reactions at the *particulate level.* Here, the concern is with the **reaction mechanism,** the detailed pathway taken by atoms and molecules as a reaction proceeds. The goal is to reconcile data in the macroscopic world of chemistry with an understanding of how and why chemical reactions occur at the particulate level—and then to apply this information to control important reactions.

Chemistry.Now™

Throughout the text this icon introduces an opportunity for self-study or to explore interactive tutorials by signing in at **www.thomsonedu.com/login**.

■ **Macroscopic–Particulate Connections** Recall the statement from page 10 that "Chemists carry out experiments on the macroscopic level, but they think about chemistry at the particulate level."

15.1 Rates of Chemical Reactions

The concept of rate is encountered in many nonchemical circumstances. Common examples are the speed of an automobile, given in terms of the distance traveled per unit time (for example, kilometers per hour) and the rate of flow of water from a faucet, given as volume per unit time (perhaps liters per minute). In each case, a change is measured over an interval of time. Similarly, the rate of a chemical reaction refers to the change in concentration of a reactant or product per unit of time.

$$\text{Rate of reaction} = \frac{\text{change in concentration}}{\text{change in time}}$$

Two measurements are made to determine the average speed of an automobile: distance traveled and time elapsed. Average speed is the distance traveled divided by the time elapsed, or $\Delta(\text{distance})/\Delta(\text{time})$. If an automobile travels 3.9 km in 4.5 min (0.075 h), its average speed is (3.9 km/0.075 h), or 52 kph (or 32 mph).

Average rates of chemical reactions can be determined similarly. Two quantities, concentration and time, are measured. Concentrations can be determined in a variety of ways, sometimes directly (using a pH meter for example), sometimes by measuring a property such as absorbance of light that can be related to concentration (Figure 15.1). The average rate of the reaction is the change in the concentration per unit time—that is, $\Delta(\text{concentration})/\Delta(\text{time})$.

Photos: Charles D. Winters

(a) (b) (c)

FIGURE 15.1 An experiment to measure rate of reaction. (a) A few drops of blue food dye were added to water, followed by a solution of bleach. Initially, the concentration of dye was about 3.4×10^{-5} M, and the bleach (NaOCl) concentration was about 0.034 M. (b and c) The dye faded as it reacted with the bleach. The absorbance of the solution can be measured at various times using a spectrophotometer, and these values can be used to determine the concentration of the dye.

Let's consider an example, the decomposition of N_2O_5 in a solvent. This reaction occurs according to the following equation:

$$N_2O_5 \rightarrow 2\ NO_2 + \tfrac{1}{2}\ O_2$$

Concentrations and time elapsed for a typical experiment done at 30.0 °C are presented as a graph in Figure 15.2.

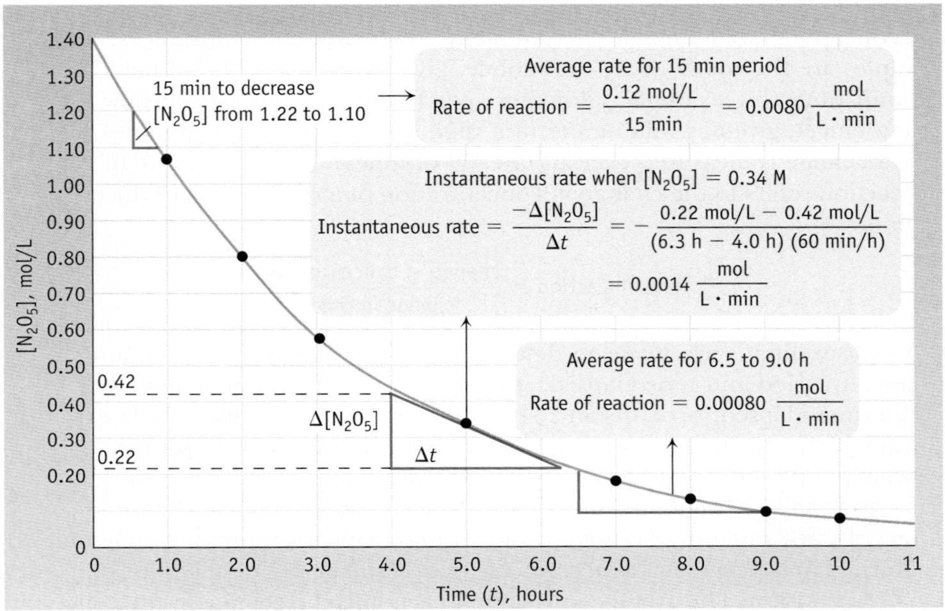

Active Figure 15.2 A plot of reactant concentration versus time for the decomposition of N_2O_5. The average rate for a 15-min interval from 45 min to 1 h is 0.0080 mol/L · min. The instantaneous rate calculated when $[N_2O_5] = 0.34$ M is 0.0014 mol/L · min.

Chemistry Now™ Sign in at www.thomsonedu.com/login and go to the Chapter Contents menu to explore an interactive version of this figure accompanied by an exercise.

The rate of this reaction for any interval of time can be expressed as the change in concentration of N_2O_5 divided by the change in time:

$$\text{Rate of reaction} = \frac{\text{change in }[N_2O_5]}{\text{change in time}} = -\frac{\Delta[N_2O_5]}{\Delta t}$$

The minus sign is required because the concentration of N_2O_5 decreases with time (that is, $\Delta[N_2O_5] = [N_2O_5](\text{final}) - [N_2O_5](\text{initial})$ is negative), and rate is always expressed as a positive quantity. Using data from Figure 15.2, the rate of disappearance of N_2O_5 between 40 min and 55 min is given by

$$\text{Rate} = -\frac{\Delta[N_2O_5]}{\Delta t} = -\frac{(1.10\text{ mol/L}) - (1.22\text{ mol/L})}{55\text{ min} - 40\text{ min}} = +\frac{0.12\text{ mol/L}}{15\text{ min}}$$

$$\text{Rate} = 0.0080\frac{\text{mol } N_2O_5 \text{ consumed}}{\text{L} \cdot \text{min}}$$

■ **Representing Concentration** Recall that square brackets around a formula indicate its concentrations in mol/L (Section 4.5).

■ **Calculating Changes** Recall that when we calculate a change in a quantity, we always do so by subtracting the initial quantity from the final quantity: $\Delta c = c_{\text{final}} - c_{\text{initial}}$.

Note the units for reaction rates; if concentration is expressed in mol/L, the units for rate will be mol/L · time.

During a chemical reaction, amounts of reactants decrease with time, and amounts of products increase. For the decomposition of N_2O_5, we could also express the rate either as $\Delta[NO_2]/\Delta t$ or as $\Delta[O_2]/\Delta t$. Rates based on changes in concentrations of products will have a positive sign because the concentration is increasing. Furthermore, the numerical values of rates defined in these ways will be different from value of $\Delta[N_2O_5]/\Delta t$. Note that the rate of decomposition of N_2O_5 is one half the rate of formation of NO_2 and twice the rate of formation of O_2. The relationship between these rate expressions is determined from the coefficients in the chemical equation.

$$\text{Rate of reaction} = -\frac{\Delta[N_2O_5]}{\Delta t} = +\frac{1}{2}\frac{\Delta[NO_2]}{\Delta t} = +2\frac{\Delta[O_2]}{\Delta t}$$

For the 15-minute interval between 40 and 55 minutes, the rates for the formation of NO_2 and O_2 are

$$\text{Rate} = \frac{\Delta[NO_2]}{\Delta t} = \frac{0.0080\text{ mol } N_2O_5\text{ consumed}}{\text{L} \cdot \text{min}} \times \frac{2\text{ mol } NO_2\text{ formed}}{1\text{ mol } N_2O_5\text{ consumed}}$$

$$= 0.016\frac{\text{mol } NO_2\text{ formed}}{\text{L} \cdot \text{min}}$$

$$\text{Rate} = \frac{\Delta[O_2]}{\Delta t} = \frac{0.0080\text{ mol } N_2O_5\text{ consumed}}{\text{L} \cdot \text{min}} \times \frac{\frac{1}{2}\text{ mol } O_2\text{ formed}}{1\text{ mol } N_2O_5\text{ consumed}}$$

$$= 0.0040\frac{\text{mol } O_2\text{ formed}}{\text{L} \cdot \text{min}}$$

The graph of $[N_2O_5]$ versus time in Figure 15.2 does not give a straight line because the rate of the reaction changes during the course of the reaction. The concentration of N_2O_5 decreases rapidly at the beginning of the reaction but more slowly near the end. We can verify this by comparing the rate of disappearance of N_2O_5 calculated previously (the concentration decreased by 0.12 mol/L in 15 min) to the rate of reaction calculated for the time interval from 6.5 h to 9.0 h (when the concentration drops by 0.12 mol/L in 150 min). The rate in this later stage of this reaction is only one tenth of the previous value.

$$-\frac{\Delta[N_2O_5]}{\Delta t} = -\frac{(0.10\ \text{mol/L}) - (0.22\ \text{mol/L})}{540\ \text{min} - 390\ \text{min}} = +\frac{0.12\ \text{mol/L}}{150\ \text{min}}$$
$$= 0.00080\frac{\text{mol}}{\text{L}\cdot\text{min}}$$

The procedure we have used to calculate the reaction rate gives the average rate over the chosen time interval.

We might also ask what the instantaneous rate is at a single point in time. In an automobile, the instantaneous rate can be read from the speedometer. For a chemical reaction, we can extract the instantaneous rate from the concentration–time graph by drawing a line tangent to the concentration–time curve at a particular time (see Figure 15.2). The instantaneous rate is obtained from the slope of this line. For example, when $[N_2O_5] = 0.34$ mol/L and $t = 5.0$ h, the rate is

■ **The Slope of a Line** The instantaneous rate in Figure 15.2 can be determined from an analysis of the slope of the line. See pages 39–41 for more on finding the slope of a line.

$$\text{Rate when } [N_2O_5] \text{ is } 0.34\ M = -\frac{\Delta[N_2O_5]}{\Delta t} = +\frac{0.20\ \text{mol/L}}{140\ \text{min}}$$
$$= 1.4 \times 10^{-3}\frac{\text{mol}}{\text{L}\cdot\text{min}}$$

At that particular moment in time, ($t = 5.0$ h), N_2O_5 is being consumed at a rate of 0.0014 mol/L · min.

Chemistry Now™

Sign in at **www.thomsonedu.com/login** and go to Chapter 15 Contents to see Screen 15.2 for a visualization of **ways to express reaction rates.**

EXAMPLE 15.1 Relative Rates and Stoichiometry

Problem Relate the rates for the disappearance of reactants and formation of products for the following reaction:

$$4\ PH_3(g) \rightarrow P_4(g) + 6\ H_2(g)$$

Strategy In this reaction, PH_3 disappears, and P_4 and H_2 are formed. Consequently, the value of $\Delta[PH_3]/\Delta t$ will be negative, whereas $\Delta[P_4]/\Delta t$ and $\Delta[H_2]/\Delta t$ will be positive. To relate the rates to each other, we divide Δ[reagent]/Δt by its stoichiometric coefficient in the balanced equation.

Solution Because four moles of PH_3 disappear for every one mole of P_4 formed, the numerical value of the rate of formation of P_4 is one fourth of the rate of disappearance of PH_3. Similarly, P_4 is formed at only one sixth of the rate that H_2 is formed.

$$-\frac{1}{4}\left(\frac{\Delta[PH_3]}{\Delta t}\right) = +\frac{\Delta[P_4]}{\Delta t} = +\frac{1}{6}\left(\frac{\Delta[H_2]}{\Delta t}\right)$$

■ **Reaction Rates and Stoichiometry** For the general reaction $a\,A + b\,B \rightarrow c\,C + d\,D$, the international convention defines the reaction rate as

$$\text{Rate} = -\frac{1}{a}\frac{\Delta[A]}{\Delta t} = -\frac{1}{b}\frac{\Delta[B]}{\Delta t} = +\frac{1}{c}\frac{\Delta[C]}{\Delta t} = +\frac{1}{d}\frac{\Delta[D]}{\Delta t}$$

EXAMPLE 15.2 Rate of Reaction

Problem Data collected on the concentration of dye as a function of time (see Figure 15.1) are given in the graph below. What is the average rate of change of the dye concentration over the first 2 min? What is the average rate of change during the fifth minute (from $t = 4$ min to $t = 5$ min)? Estimate the instantaneous rate at 4 min.

Strategy To find the average rate, calculate the difference in concentration at the beginning and end of a time period ($\Delta c = c_{\text{final}} - c_{\text{initial}}$) and divide by the elapsed time. To find the instantaneous rate at 4 minutes, draw a line tangent to the graph at the specified time. The negative of the slope of the line is the instantaneous rate.

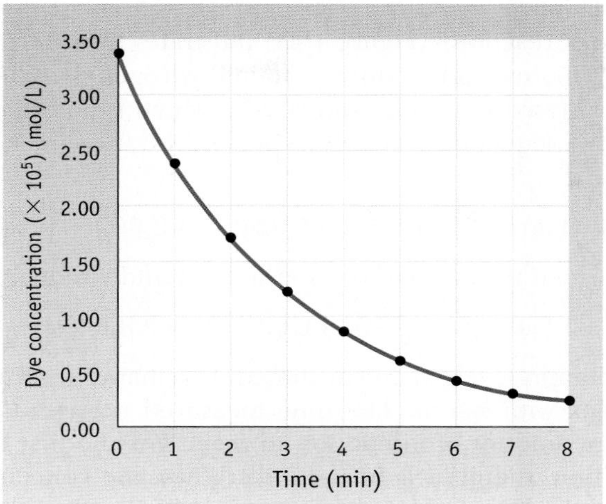

Solution The concentration of dye decreases from 3.4×10^{-5} mol/L at $t = 0$ min to 1.7×10^{-5} mol/L at $t = 2.0$ min. The average rate of the reaction in this interval of time is

$$\text{Average rate} = -\frac{\Delta[\text{Dye}]}{\Delta t} = -\frac{(1.7 \times 10^{-5}\ \text{mol/L}) - (3.4 \times 10^{-5}\ \text{mol/L})}{2.0\ \text{mol}}$$

$$\text{Average rate} = +\frac{8.5 \times 10^{-6}\ \text{mol}}{\text{L} \cdot \text{min}}$$

The concentration of dye decreases from 0.90×10^{-5} mol/L at $t = 4.0$ min to 0.60×10^{-5} mol/L at $t = 5.0$ min. The average rate of the reaction in this interval of time is

$$\text{Average rate} = -\frac{\Delta[\text{Dye}]}{\Delta t} = -\frac{(0.60 \times 10^{-5}\ \text{mol/L}) - (0.90 \times 10^{-5}\ \text{mol/L})}{1.0\ \text{mol}}$$

$$\text{Average rate} = +\frac{3.0 \times 10^{-6}\ \text{mol}}{\text{L} \cdot \text{min}}$$

From the slope of the line tangent to the curve, the instantaneous rate at 4 min is found to be $+3.5 \times 10^{-6}$ mol/L · min.

EXERCISE 15.1 Reaction Rates and Stoichiometry

What are the relative rates of appearance or disappearance of each product and reactant, respectively, in the decomposition of nitrosyl chloride, NOCl?

$$2\ \text{NOCl(g)} \rightarrow 2\ \text{NO(g)} + \text{Cl}_2\text{(g)}$$

EXERCISE 15.2 Rate of Reaction

Sucrose decomposes to fructose and glucose in acid solution. A plot of the concentration of sucrose as a function of time is given here. What is the rate of change of the sucrose concentration over the first 2 h? What is the rate of change over the last 2 h? Estimate the instantaneous rate at 4 h.

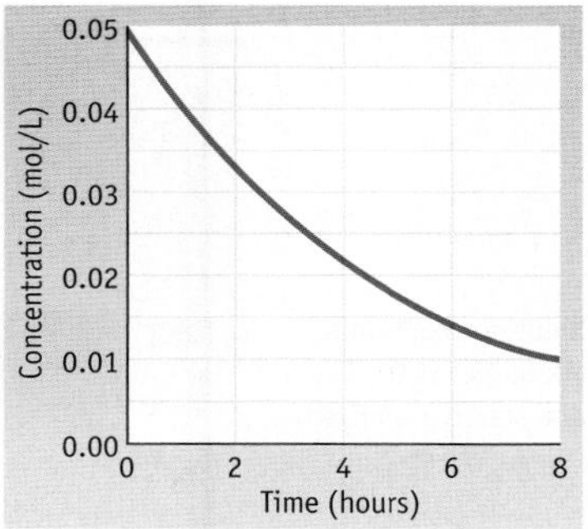

Concentration versus time for the decomposition of sucrose. (See Exercise 15.2.)

Chemistry Now™

Sign in at **www.thomsonedu.com/login** and go to Chapter 15 Contents to see Screen 15.2 for a self-study module on **rate of reaction.**

15.2 Reaction Conditions and Rate

Several factors—reactant concentrations, temperature, and presence of catalysts—affect the rate of a reaction. If the reactant is a solid, the surface area available for reaction is also a factor.

The "iodine clock reaction" (Figure 15.3) illustrates the effect of concentration and temperature. The reaction mixture contains hydrogen peroxide (H_2O_2), iodide ion (I^-), vitamin C (ascorbic acid), and starch (which is an indicator of the presence of iodine, I_2). A sequence of reactions begins with the slow oxidation of iodide ion to I_2 by H_2O_2.

$$H_2O_2(aq) + 2\ I^-(aq) + 2\ H_3O^+(aq) \rightarrow 4\ H_2O(\ell) + I_2(aq)$$

As soon as I_2 is formed in the solution, vitamin C rapidly reduces it to I^-.

$$2\ H_2O(\ell) + I_2(aq) + C_6H_8O_6(aq) \rightarrow C_6H_6O_6(aq) + 2\ H_3O^+(aq) + 2\ I^-(aq)$$

When all of the vitamin C has been consumed, I_2 remains in solution and forms a blue–black complex with starch. The time measured represents how long it has taken for the given amount of iodide ion to react. For the first experiment (A in Figure 15.3) the time required is 51 seconds. When the concentration of iodide ion is smaller (B), the time required for the vitamin C to be consumed is longer, 1 minute and 33 seconds. Finally, when the concentrations are again the same as in experiment B but the reaction mixture is heated, the reaction occurs more rapidly (56 seconds).

■ **Effect of Temperature on Reaction Rate** Cooking involves chemical reactions, and a higher temperature results in foods cooking faster. In the laboratory, reaction mixtures are often heated to make reactions occur faster.

(a) Initial Experiment. The blue color of the starch–iodine complex develops in 51 seconds.

(b) Change Concentration. The blue color of starch–iodine complex develops in 1 minute, 33 seconds when the solution is less concentrated than A.

(c) Change Temperature. The blue color of the starch–iodine complex develops in 56 seconds when the solution is the same concentration as in B but at a higher temperature.

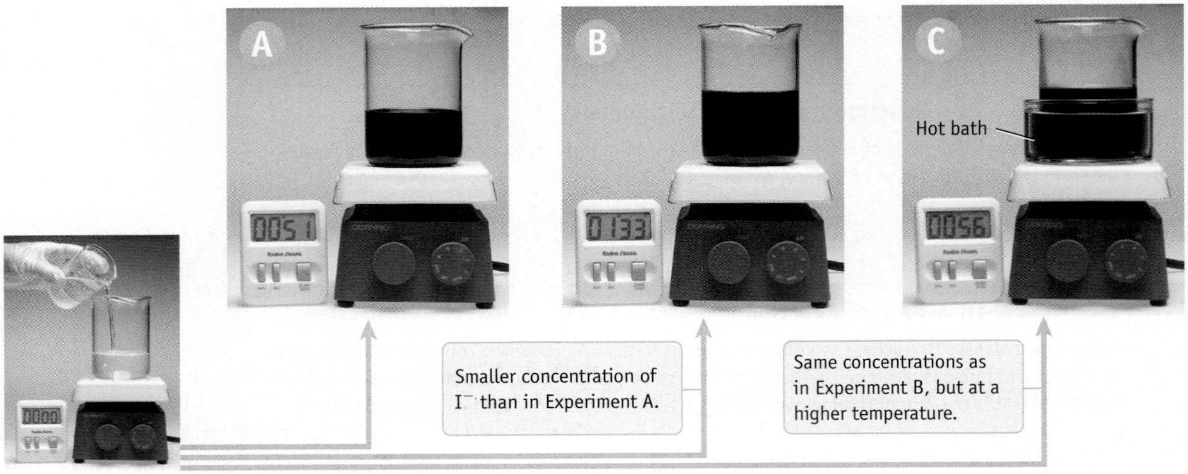

Solutions containing vitamin C, H_2O_2, I^-, and starch are mixed.

Photos: Charles D. Winters

FIGURE 15.3 The iodine clock reaction. This reaction illustrates the effects of concentration and temperature on reaction rate. (You can do these experiments yourself with reagents available in the supermarket. For details, see S. W. Wright: "The vitamin C clock reaction," *Journal of Chemical Education*, Vol. 79, p. 41, 2002.) See ChemistryNow Screen 15.11 for a video of the iodine clock reaction.

Charles D. Winters

Thomas Eisner with Daniel Aneshansley, Cornell University

Charles D. Winters

(a) (b) (c)

FIGURE 15.4 Catalyzed decomposition of H_2O_2. (a) The rate of decomposition of hydrogen peroxide is increased by the catalyst MnO_2. Here, H_2O_2 (as a 30% aqueous solution) is poured onto the black solid MnO_2 and rapidly decomposes to O_2 and H_2O. Steam forms because of the high heat of reaction. (b) A bombardier beetle uses the catalyzed decomposition of H_2O_2 as a defense mechanism. The heat of the reaction lets the insect eject hot water and other irritating chemicals with explosive force. (c) A naturally occurring catalyst, called an enzyme, decomposes hydrogen peroxide. Here, the enzyme found in a potato is used to catalyze H_2O_2 decomposition, and bubbles of O_2 gas are seen rising in the solution.

Catalysts are substances that accelerate chemical reactions but are not themselves consumed. Consider the effect of a catalyst on the decomposition of hydrogen peroxide, H_2O_2, to form water and oxygen.

$$2\ H_2O_2(aq) \rightarrow O_2(g) + 2\ H_2O(\ell)$$

This decomposition is very slow; a solution of H_2O_2 can be stored for many months with only minimal change in concentration. Adding a manganese salt, an iodide-containing salt, or a biological substance called an *enzyme* causes this reaction to occur rapidly, as shown by vigorous bubbling as gaseous oxygen escapes from the solution (Figure 15.4).

The surface area of a solid reactant can also affect the reaction rate. Only molecules at the surface of a solid can come in contact with other reactants. The smaller the particles of a solid, the more molecules are found on the solid's surface. With very small particles, the effect of surface area on rate can be quite dramatic (Figure 15.5). Farmers know that explosions of fine dust particles (suspended in the air in an enclosed silo or at a feed mill) represent a major hazard.

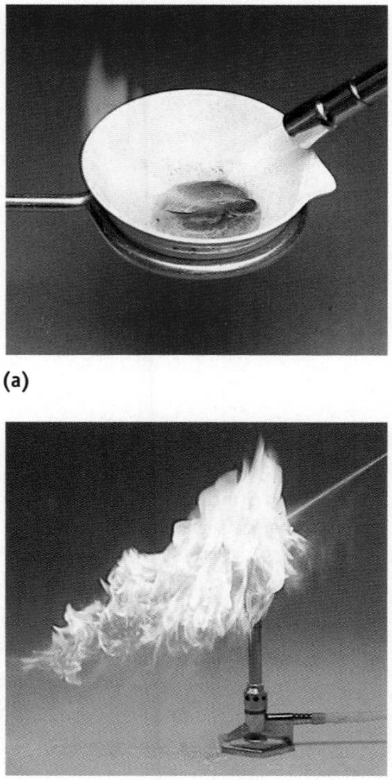

Photos: Charles D. Winters

(a)

(b)

FIGURE 15.5 The combustion of lycopodium powder. (a) The spores of this common fern burn only with difficulty when piled in a dish. (b) If the spores are ground to a fine powder and sprayed into a flame, combustion is rapid.

Chemistry Now™

Sign in at **www.thomsonedu.com/login** and go to Chapter 15 Contents to see:

- Screen 15.3 for a visualization of **the factors controlling rates**
- Screen 15.4 for a simulation of the **effect of concentration on rate**

15.3 Effect of Concentration on Reaction Rate

One important goal in studying the kinetics of a reaction is to determine its mechanism; that is, how the reaction occurs at the molecular level. The place to begin is to learn how concentrations of reactants affect the reaction rate.

The effect of concentration can be determined by evaluating the rate of a reaction using different concentrations of each reactant (with the temperature held constant).

Consider, for example, the decomposition of N_2O_5 to NO_2 and O_2. Figure 15.2 presented data on the concentration of N_2O_5 as a function of time. We previously calculated that, when $[N_2O_5] = 0.34$ mol/L, the instantaneous rate of disappearance of N_2O_5 is 0.0014 mol/L · min. An evaluation of the instantaneous rate of the reaction when $[N_2O_5] = 0.68$ mol/L reveals a rate of 0.0028 mol/L · min. That is, doubling the concentration of N_2O_5 doubles the reaction rate. A similar exercise shows that if $[N_2O_5]$ is 0.17 mol/L (half of 0.34 mol/L), the reaction rate is also halved. From these results, we know that the reaction rate must be directly proportional to the reactant concentration for this reaction:

$$N_2O_5 \rightarrow 2\ NO_2 + \tfrac{1}{2}\ O_2$$

$$\text{Rate of reaction} \propto [N_2O_5]$$

where the symbol $\propto$ means "proportional to."

Different relationships between reaction rate and reactant concentration are encountered in other reactions. For example, the reaction rate could be independent of concentration, or it may depend on the reactant concentration raised to some power (that is, $[\text{reactant}]^n$). If the reaction involves several reactants, the reaction rate may depend on the concentrations of each of them or on only one of them. Finally, if a catalyst is involved, its concentration may also affect the rate, as can the concentrations of products.

Rate Equations

The relationship between reactant concentration and reaction rate is expressed by an equation called a **rate equation**, or **rate law**. For the decomposition of N_2O_5 the rate equation is

$$N_2O_5(g) \rightarrow 2\ NO_2(g) + \tfrac{1}{2}\ O_2$$

$$\text{Rate of reaction} = k[N_2O_5]$$

where the proportionality constant, k, is called the **rate constant**. This rate equation tells us that this reaction rate is proportional to the concentration of the reactant. Based on this equation, we can determine that when $[N_2O_5]$ is doubled, the reaction rate doubles.

Generally, for a reaction such as

$$a\ A + b\ B \rightarrow x\ X$$

the rate equation has the form

$$\text{Rate of reaction} = k[A]^m[B]^n$$

■ **Exponents on Reactant Concentrations and Reaction Stoichiometry** It is important to recognize that the exponents m and n are not necessarily the stoichiometric coefficients (a and b) for the balanced chemical equation.

The rate equation expresses the fact that the rate of reaction is proportional to the reactant concentrations, each concentration being raised to some power. The exponents in this equation are often positive whole numbers, but they can also be negative numbers, fractions, or zero and they are determined by experiment.

If a homogeneous catalyst is present, its concentration might also be included in the rate equation, even though the catalytic species in not a product or reactant in the equation for the reaction. Consider, for example, the decomposition of hydrogen peroxide in the presence of a catalyst such as iodide ion.

$$H_2O_2(aq) \xrightarrow{I^-(aq)} H_2O(\ell) + \tfrac{1}{2}\ O_2(g)$$

Experiments show that this reaction has the following rate equation:

$$\text{Reaction rate} = k[H_2O_2][I^-]$$

Here, the concentration of I^- appears in the rate law, even though it is not involved in the balanced equation.

■ **The Nature of Catalysts** A catalyst does not appear as a reactant in the balanced, overall equation for the reaction, but it may appear in the rate expression. A common practice is to identify catalysts by name or symbol above the reaction arrow, as shown in the example. A homogeneous catalyst is one in the same phase as the reactants. For example, both H_2O_2 and I^- are dissolved in water.

The Order of a Reaction

The **order** of a reaction with respect to a particular reactant is the exponent of its concentration term in the rate expression, and the **overall reaction order** is the sum of the exponents on all concentration terms. Consider, for example, the reaction of NO and Cl_2:

$$2\ NO(g) + Cl_2(g) \rightarrow 2\ NOCl(g)$$

The experimentally determined rate equation for this reaction is

$$\text{Rate} = k[NO]^2[Cl_2]$$

This reaction is second order in NO, first order in Cl_2, and third order overall. How is this related to the experimental data for the rate of disappearance of NO?

■ **Overall Reaction Order** The overall reaction order is the sum of the reaction orders of the different reactants.

Experiment	[NO] mol/L	[Cl_2] mol/L	Rate mol/L · s
1	0.250	0.250	1.43×10^{-6}
	↓ × 2	↓ no change	↓ × 4
2	0.500	0.250	5.72×10^{-6}
3	0.250	0.500	2.86×10^{-6}
4	0.500	0.500	11.4×10^{-6}

- *Compare Experiments 1 and 2:* Here, [Cl_2] is held constant, and [NO] is doubled. The change in [NO] leads to a reaction rate increase by a factor of 4; that is, the rate is proportional to the *square* of the NO concentration.
- *Compare Experiments 1 and 3:* In experiments 1 and 3, [NO] is held constant, and [Cl_2] is doubled, causing the rate to double. That is, the rate is proportional to [Cl_2].
- *Compare Experiments 1 and 4:* Both [NO] and [Cl_2] are doubled from 0.250 M to 0.500 M. From previous experiments, we know that doubling [NO] should cause a four-fold increase, and doubling [Cl_2] causes a two-fold increase. Therefore, doubling both concentrations should cause an eight-fold increase, as is observed ($1.43 \times 10^{-6} \times 8 = 11.4 \times 10^{-6}$ mol/L · s).

The decomposition of ammonia on a platinum surface at 856 °C is a zero order reaction.

$$NH_3(g) \rightarrow \tfrac{1}{2}\ N_2(g) + \tfrac{3}{2}\ H_2(g)$$

This means that the reaction rate is independent of NH_3 concentration.

$$\text{Rate} = k[NH_3]^0 = k$$

Reaction order is important because it gives some insight into the most interesting question of all—how the reaction occurs. This is described further in Section 15.6.

The Rate Constant, *k*

The rate constant, *k*, is a proportionality constant that relates rate and concentration at a given temperature. It is an important quantity because it enables you to find the reaction rate for a new set of concentrations. To see how to use *k*, consider

■ **Time and Rate Constants** The time in a rate constant can be seconds, minutes, hours, days, years, or whatever time unit is appropriate. The fraction 1/time can also be written as $time^{-1}$. For example, 1/*y* is equivalent to y^{-1}, and 1/s is equivalent to s^{-1}.

the substitution of Cl^- ion by water in the cancer chemotherapy agent cisplatin, $Pt(NH_3)_2Cl_2$.

$$Pt(NH_3)_2Cl_2(aq) \quad + \quad H_2O(\ell) \quad \longrightarrow \quad [Pt(NH_3)_2(H_2O)Cl]^+(aq) \quad + \quad Cl^-(aq)$$

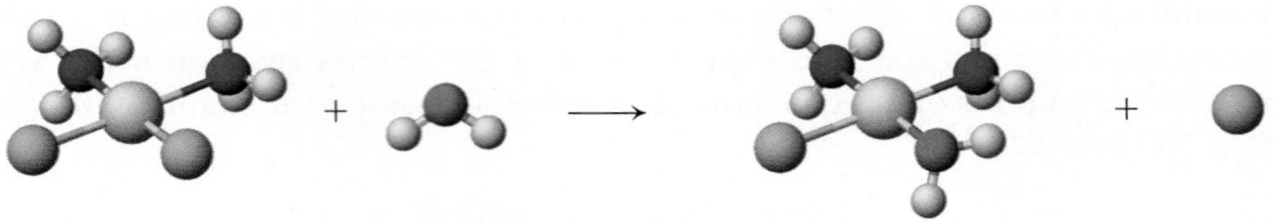

The rate law for this reaction is

$$\text{Rate} = k[Pt(NH_3)_2Cl_2]$$

and the rate constant, *k*, is 0.27/h at 25 °C. Knowing *k* allows you to calculate the rate at a particular reactant concentration—for example, when $[Pt(NH_3)_2Cl_2]$ = 0.018 mol/L:

$$\text{Rate} = (0.27/\text{h})(0.018 \text{ mol/L}) = 0.0049 \text{ mol/L} \cdot \text{h}$$

■ **Some Rate Constants**

First Order	*k*(1/s)
$2 N_2O_5(g) \rightarrow 4 NO_2(g) + O_2(g)$	3.38×10^{-5} at 25 °C
$C_2H_6(g) \rightarrow 2 CH_3(g)$	5.36×10^{-4} at 700 °C
Sucrose(aq, H_3O^+) → fructose(aq) + glucose(aq)	6.0×10^{-5} at 25 °C

Second Order	*k*(L/mol · s)
$2 NOBr(g) \rightarrow 2 NO(g) + Br_2(g)$	0.80 at 10 °C
$H_2(g) + I_2(g) \rightarrow 2 HI(g)$	0.0242 at 400 °C

As noted, earlier, reaction rates have units of mol/L · time when concentrations are given as moles per liter. Rate constants must have units consistent with the units for the other terms in the rate equation.

- First-order reactions: the units of *k* are 1/time.
- Second-order reactions: the units of *k* are L/mol · time.
- Zero-order reaction: the units of *k* are mol/L · time.

Determining a Rate Equation

One way to determine a rate equation is by using the "method of initial rates." The initial rate is the instantaneous reaction rate at the start of the reaction (the rate at $t = 0$). An approximate value of the initial rate can be obtained by mixing the reactants and determining Δ[product]/Δt or $-\Delta$[reactant]/Δt after 1% to 2% of the limiting reactant has been consumed. Measuring the rate during the initial stage of a reaction is convenient because initial concentrations are known.

As an example of the determination of a reaction rate by the method of initial rates, let us consider the reaction of sodium hydroxide with methyl acetate to produce acetate ion and methanol.

$$CH_3CO_2CH_3(aq) \quad + \quad OH^-(aq) \quad \longrightarrow \quad CH_3CO_2^-(aq) \quad + \quad CH_3OH(aq)$$

Reactant concentrations and initial rates for this reaction for several experiments at 25 °C are collected in the table below.

Experiment	Initial Concentrations (mol/L) $[CH_3CO_2CH_3]$	$[OH^-]$	Initial Reaction Rate (mol/L · s) at 25 °C
1	0.050	0.050	0.00034
	↓ no change	↓ × 2	↓ × 2
2	0.050	0.10	0.00069
	↓ × 2	↓ no change	↓ × 2
3	0.10	0.10	0.00137

As noted in the table, when the initial concentration of one reactant (either $CH_3CO_2CH_3$ or OH^-) is doubled while the concentration of the other reactant is held constant, the initial reaction rate doubles. This rate doubling shows that the reaction rate is directly proportional to the concentrations of both $CH_3CO_2CH_3$ and OH^-; thus, the reaction is first order in each of these reactants and second order overall. The rate law that reflects these experimental observations is

$$\text{Rate} = k[CH_3CO_2CH_3][OH^-]$$

Using this equation, we can predict that doubling both concentrations at the same time should cause the rate to go up by a factor of 4. What happens, however, if one concentration is doubled and the other is halved? The rate equation tells us the rate should not change!

If the rate equation is known, the value of *k*, the rate constant, can be found by substituting values for the rate and concentration into the rate equation. Using the data for the methyl acetate/hydroxide ion reaction from the first experiment, we have

$$\text{Rate} = 0.00034 \text{ mol/L} \cdot \text{s} = k(0.050 \text{ mol/L})(0.050 \text{ mol/L})$$

$$k = \frac{0.00034 \text{ mol/L} \cdot \text{s}}{(0.050 \text{ mol/L})(0.050 \text{ mol/L})} = 0.14 \text{ L/mol} \cdot \text{s}$$

Chemistry Now™

Sign in at **www.thomsonedu.com/login** and go to Chapter 15 Contents to see:

- Screen 15.4 for a self-study module on **control of reaction rates**
- Screen 15.5 for a simulation, a tutorial, and an exercise on **determining rate equations from a study of the effect of concentration on reaction rate**

EXAMPLE 15.3 Determining a Rate Equation

Problem The rate of the reaction between CO and NO_2 at 540 K

$$CO(g) + NO_2(g) \rightarrow CO_2(g) + NO(g)$$

was measured starting with various concentrations of CO and NO_2. Determine the rate equation and the value of the rate constant.

Experiment	Initial Concentrations [CO], mol/L	$[NO_2]$, mol/L	Initial Rate (mol/L · h)
1	5.10×10^{-4}	0.350×10^{-4}	3.4×10^{-8}
2	5.10×10^{-4}	0.700×10^{-4}	6.8×10^{-8}
3	5.10×10^{-4}	0.175×10^{-4}	1.7×10^{-8}
4	1.02×10^{-3}	0.350×10^{-4}	6.8×10^{-8}
5	1.53×10^{-3}	0.350×10^{-4}	10.2×10^{-8}

Strategy For a reaction involving several reactants, the general approach is to keep the concentration of one reactant constant and then decide how the rate of reaction changes as the concentration of the other reagent is varied. Because the rate is proportional to the concentration of a reactant, say A, raised to some power n (the reaction order)

$$\text{Rate} \propto [A]^n$$

we can write the general equation

$$\frac{\text{Rate in experiment 2}}{\text{Rate in experiment 1}} = \frac{[A_2]^n}{[A_1]^n} = \left(\frac{[A_2]}{[A_1]}\right)^n$$

If [A] is doubled and the rate doubles from experiment 1 to experiment 2, then $n = 1$. If [A] doubles and the rate goes up by 4, then $n = 2$.

Solution In the first three experiments, the concentration of CO is held constant. In the second experiment, the NO_2 concentration has been doubled relative to Experiment 1, leading to a twofold increase in the rate. Thus, $n = 1$ and the reaction is first order in NO_2.

$$\frac{\text{Rate in experiment 2}}{\text{Rate in experiment 1}} = \frac{6.8 \times 10^{-8}\ \text{mol/L} \cdot \text{h}}{3.4 \times 10^{-8}\ \text{mol/L} \cdot \text{h}} = \left(\frac{0.700 \times 10^{-4}}{0.350 \times 10^{-4}}\right)^n$$

$$2 = (2)^n$$

and so $n = 1$.

This finding is confirmed by experiment 3. Decreasing $[NO_2]$ to half its original value in experiment 3 causes the rate to decrease by half.

The data in experiments 1 and 4 (with constant $[NO_2]$) show that doubling [CO] doubles the rate, and the data from experiments 1 and 5 show that tripling the concentration of CO triples the rate. These results mean that the reaction is first order in [CO]. We now know the rate equation is

$$\text{Rate} = k[CO][NO_2]$$

The rate constant, k, can be found by inserting data for one of the experiments into the rate equation. Using data from experiment 1, for example,

$$\text{Rate} = 3.4 \times 10^{-8}\ \text{mol/L} \cdot \text{h} = k(5.10 \times 10^{-4}\ \text{mol/L})(0.350 \times 10^{-4}\ \text{mol/L})$$

$$k = 1.9\ \text{L/mol} \cdot \text{h}$$

EXAMPLE 15.4 Using a Rate Equation to Determine Rates

Problem Using the rate equation and rate constant determined for the reaction of CO and NO_2 at 540 K in Example 15.3, determine the initial rate of the reaction when $[CO] = 3.8 \times 10^{-4}$ mol/L and $[NO_2] = 0.650 \times 10^{-4}$ mol/L.

Strategy A rate equation consists of three parts: a rate, a rate constant (k), and the concentration terms. If two of these parts are known (here k and the concentrations), the third can be calculated.

Solution Substitute k (= 1.9 L/mol · h) and the concentration of each reactant into the rate law determined in Example 15.3.

$$\text{Rate} = k[CO][NO_2] = (1.9\ \text{L/mol} \cdot \text{h})(3.8 \times 10^{-4}\ \text{mol/L})(0.650 \times 10^{-4}\ \text{mol/L})$$

$$\text{Rate} = 4.7 \times 10^{-8}\ \text{mol/L} \cdot \text{h}$$

Comment As a check on the calculated result, it is sometimes useful to make an educated guess at the answer before carrying out the mathematical solution. We know that the reaction here is first order in both reactants. Comparing the concentration values given in this problem with the concentration values in found experiment 1 in Example 15.3, we notice that [CO] is about three fourths of the concentration value, whereas $[NO_2]$ is almost twice the value. The effects do not precisely offset each other, but we might predict that the difference in rates between this experiment and experiment 1 will be fairly small, with the rate of this experiment being just a little greater. The calculated value bears this out.

EXERCISE 15.3 Determining a Rate Equation

The initial rate of the reaction of nitrogen monoxide and oxygen

$$2\ NO(g) + O_2(g) \rightarrow 2\ NO_2(g)$$

was measured at 25 °C for various initial concentrations of NO and O_2. Determine the rate equation from these data. What is the value of the rate constant, *k*, and what are its units?

	Initial Concentrations (mol/L)		Initial Rate
Experiment	[NO]	$[O_2]$	(mol/L · s)
1	0.020	0.010	0.028
2	0.020	0.020	0.057
3	0.020	0.040	0.114
4	0.040	0.020	0.227
5	0.010	0.020	0.014

EXERCISE 15.4 Using Rate Laws

The rate constant, *k*, at 25 °C is 0.27/h for the reaction

$$Pt(NH_3)_2Cl_2(aq) + H_2O(\ell) \rightarrow [Pt(NH_3)_2(H_2O)Cl]^+(aq) + Cl^-(aq)$$

and the rate equation is

$$\text{Rate} = k[Pt(NH_3)_2Cl_2]$$

Calculate the rate of reaction when the concentration of $Pt(NH_3)_2Cl_2$ is 0.020 M.

15.4 Concentration–Time Relationships: Integrated Rate Laws

Module 20

It is often important for a chemist to know how long a reaction must proceed to reach a predetermined concentration of some reactant or product, or what the reactant and product concentrations will be after some time has elapsed. A mathematical equation that relates time and concentration—that is, an equation that describes concentration–time curves like the one shown in Figure 15.2—can be used to determine this information. With such an equation, we could calculate the concentration at any given time or the length of time needed for a given amount of reactant to react.

First-Order Reactions

Suppose the reaction "R → products" is first order. This means the reaction rate is directly proportional to the concentration of R raised to the first power, or, mathematically,

$$-\frac{\Delta[R]}{\Delta t} = k[R]$$

Using calculus, this relationship can be transformed into a very useful equation called an **integrated rate equation** (because integral calculus is used in its derivation).

$$\ln\frac{[R]_t}{[R]_0} = -kt \quad \textbf{(15.1)}$$

A Closer Look — Rate Laws, Rate Constants, and Reaction Stoichiometry

Qualitatively, the rate of a reaction is easy to understand: it represents the change in concentration of the reactants and products. When we deal with rates of reaction quantitatively, however, we need to be specific about the reaction stoichiometry.

Consider the first order decomposition of N_2O_5, a reaction that we mentioned earlier

$$2\ N_2O_5 \rightarrow 4\ NO_2 + O_2(g)$$

The rate of the reaction can be expressed (and measured in lab) as the change in concentration of either reactants or products as a function of time. The numerical values of the rates of formation of reactants and products are related, but they are different because of the reaction stoichiometry. If we equate the rate of reaction to the rate of appearance of O_2, we would write

$$\text{Rate} = \frac{\Delta[O_2]}{\Delta t} = -\frac{1}{2}\frac{\Delta[N_2O_5]}{\Delta t} = +\frac{1}{4}\frac{\Delta[NO_2]}{\Delta t}$$

This relation is written by dividing each rate by its stoichiometric coefficient (◀ page 674).

There are two different pieces of information in the equation above. First, it gives the relationship between the rates of change of concentrations of reactants and products. Based on the reaction stoichiometry, we know that the rate of appearance of O_2 is one half the rate of disappearance of N_2O_5 and one fourth the rate of NO_2 appearance.

Second, this equation specifically defines what we mean by rate; it provides a single, numerical values for this parameter; and it tells us how to calculate the value of reaction rate from experimental data. If we were to follow the disappearance of N_2O_5 as a measure of reaction rate and base our definition of rate on the stoichiometry above, we should write the following differential form of the rate law.

$$\text{Rate} = -\frac{1}{2}\frac{\Delta[N_2O_5]}{\Delta t} = k[N_2O_5]$$

From this definition, it follows that the differential rate equation is

$$\text{Rate} = -\frac{\Delta[N_2O_5]}{\Delta t} = 2k[N_2O_5]$$

the integrated rate equation is (see page 683),

$$\ln\frac{[N_2O_5]_t}{[N_2O_5]_0} = -2kt$$

and the half-life equation is

$$t_{1/2} = 0.693/2k$$

We can also write the equation for N_2O_5 decomposition as follows:

$$N_2O_5 \rightarrow 2\ NO_2 + \tfrac{1}{2}\ O_2(g)$$

Following the reasoning above, the rate laws for the reaction written this way would be:

Differential rate equation:

$$-\frac{\Delta[N_2O_5]}{\Delta t} = k'[N_2O_5]$$

Integrated rate equation:

$$\ln\frac{[N_2O_5]_t}{[N_2O_5]_0} = -k't$$

Half-life equation:

$$t_{1/2} = 0.693/k'$$

Note that the differential and integrated rate laws derived based on the two different chemical equations have the same form, but k and k' do not have the same values. In this case, $2k = k'$.

For more on this issue, see K. T. Quisenberry and J. Tellinghuisen, *Journal of Chemical Education*, Vol. 83, pp. 510–512, 2006.

■ **Initial and Final Time, t** The time $t = 0$ does not need to correspond to the actual beginning of the experiment. It can be the time when instrument readings were started, for example, even though the reaction may have already begun.

Here, $[R]_0$ and $[R]_t$ are concentrations of the reactant at time $t = 0$ and at a later time, t, respectively. The ratio of concentrations, $[R]_t/[R]_0$, is the fraction of reactant that remains after a given time has elapsed. In words, the equation says

$$\text{Natural logarithm}\left(\frac{\text{concentration of R after some time}}{\text{concentration of R at start of experiment}}\right)$$
$$= \ln(\text{fraction of R remaining at time, } t)$$
$$= -(\text{rate constant})(\text{elapsed time})$$

Notice the negative sign in the equation. The ratio $[R]_t/[R]_0$ is less than 1 because $[R]_t$ is always less than $[R]_0$; the reactant R is consumed during the reaction. This means the logarithm of $[R]_t/[R]_0$ is negative, so the other side of the equation must also bear a negative sign.

Equation 15.1 can be used to carry out the following calculations:

- If $[R]_t/[R]_0$ is measured in the laboratory after some amount of time has elapsed, then k can be calculated.
- If $[R]_0$ and k are known, then the concentration of material remaining after a given amount of time ($[R]_t$) can be calculated.
- If k is known, then the time elapsed until a specific fraction ($[R]_t/[R]_0$) remains can be calculated.

Finally, notice that k for a first-order reaction is independent of concentration; k has units of time^{-1} (y^{-1} or s^{-1}, for example). This means we can choose any convenient unit for $[R]_t$ and $[R]_0$: moles per liter, moles, grams, number of atoms, number of molecules, or gas pressure.

Chemistry•Now™

Sign in at **www.thomsonedu.com/login** and go to Chapter 15 Contents to see Screen 15.6 for a tutorial on **the use of the integrated first-order rate equation.**

EXAMPLE 15.5 The First-Order Rate Equation

Problem In the past, cyclopropane, C_3H_6, was used in a mixture with oxygen as an anesthetic. (This practice has almost ceased today, because the compound is flammable.) When heated, cyclopropane rearranges to propene in a first-order process.

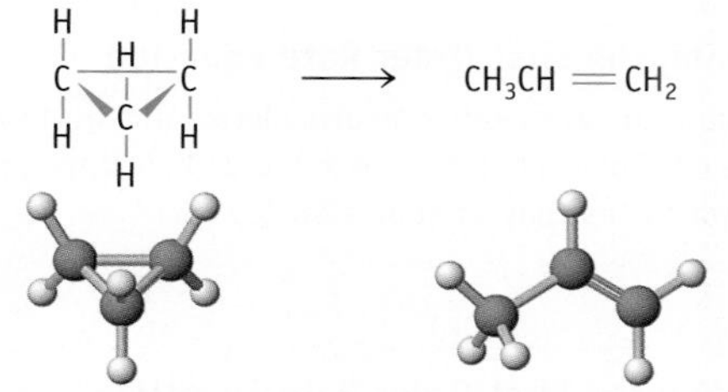

cyclopropane propene

Rate = k[cyclopropane] $k = 2.42\ h^{-1}$ at 500 °C

If the initial concentration of cyclopropane is 0.050 mol/L, how much time (in hours) must elapse for its concentration to drop to 0.010 mol/L?

Strategy The reaction is first order in cyclopropane. You know the rate constant, k, and the initial and final concentrations. Use Equation 15.1 to calculate the time (t) elapsed to reach a concentration of 0.010 mol/L.

Solution Values for $[\text{cyclopropane}]_t$, $[\text{cyclopropane}]_0$, and k are substituted into Equation 15.1; t (time) is the unknown:

$$\ln\frac{[0.010]}{[0.050]} = (2.42\ h^{-1})t$$

$$t = \frac{-\ln(0.20)}{2.42\ h^{-1}} = \frac{-(-1.61)}{2.42\ h^{-1}} = 0.665\ h$$

Comment Cycloalkanes with fewer than five carbon atoms are strained because the C—C—C bond angles cannot match the preferred 109.5°. Because of ring strain, the cyclopropane ring opens readily to form propene.

EXAMPLE 15.6 Using the First-Order Rate Equation

Problem Hydrogen peroxide decomposes in a dilute sodium hydroxide solution at 20 °C in a first-order reaction:

$$H_2O_2(aq) \rightarrow H_2O(\ell) + \tfrac{1}{2}\ O_2(g)$$

Rate = $k[H_2O_2]$ with $k = 1.06 \times 10^{-3}\ min^{-1}$

What is the fraction remaining after 100. min if the initial concentration of H_2O_2 is 0.020 mol/L? What is the concentration of the peroxide after 100. min?

Strategy Because the reaction is first order in H_2O_2, we use Equation 15.1. Here, $[H_2O_2]_0$, k, and t are known, and we are asked to find the value of $[H_2O_2]_t$ and the fraction remaining. Recall that

$$\frac{[R]_t}{[R]_0} = \text{fraction remaining}$$

Once this value is known, and knowing $[H_2O_2]_0$, we can calculate $[H_2O_2]_t$.

Solution Substitute the known values into Equation 15.1.

$$\ln \frac{[H_2O_2]_t}{[H_2O_2]_0} = -kt = -(1.06 \times 10^{-3}\ \text{min}^{-1})(100.\ \text{min})$$

$$\ln \frac{[H_2O_2]_t}{[H_2O_2]_0} = -0.106$$

Taking the antilogarithm of -0.106 [i.e., the inverse of $\ln(-0.106)$ or $e^{-0.106}$], we find the fraction remaining to be 0.90.

$$\text{Fraction remaining} = \frac{[H_2O_2]_t}{[H_2O_2]_0} = 0.90$$

Because $[H_2O_2]_0 = 0.020$ mol/L, this gives $[H_2O_2]_t = 0.018$ mol/L.

EXERCISE 15.5 Using the First-Order Rate Equation

Sucrose, a sugar, decomposes in acid solution to give glucose and fructose. The reaction is first order in sucrose, and the rate constant at 25 °C is $k = 0.21\ \text{h}^{-1}$. If the initial concentration of sucrose is 0.010 mol/L, what is its concentration after 5.0 h?

EXERCISE 15.6 Using the First-Order Rate Equation

Gaseous azomethane ($CH_3N_2CH_3$) decomposes to ethane and nitrogen when heated:

$$CH_3N_2CH_3g) \rightarrow CH_3CH_3(g) + N_2(g)$$

The disappearance of azomethane is a first-order reaction with $k = 3.6 \times 10^{-4}\ \text{s}^{-1}$ at 600 K.

(a) A sample of gaseous $CH_3N_2CH_3$ is placed in a flask and heated at 600 K for 150 s. What fraction of the initial sample remains after this time?

(b) How long must a sample be heated so that 99% of the sample has decomposed?

Chemistry Now™

Sign in at **www.thomsonedu.com/login** and go to Chapter 15 Contents to see Screen 15.6 for a self-study module on the **first-order rate equation.**

Second-Order Reactions

Suppose the reaction "R → products" is second order. The rate equation is

$$-\frac{\Delta[R]}{\Delta t} = k[R]^2$$

Using calculus, this relationship can be transformed into the following equation that relates reactant concentration and time:

$$\frac{1}{[R]_t} - \frac{1}{[R]_0} = kt \qquad \textbf{(15.2)}$$

The same symbolism used with first-order reactions applies: $[R]_0$ is the concentration of reactant at the time $t = 0$; $[R]_t$ is the concentration at a later time; and k is the second-order rate constant which has the units of L/mol · time.

EXAMPLE 15.7 Using the Second-Order Integrated Rate Equation

Problem The gas-phase decomposition of HI

$$HI(g) \rightarrow \frac{1}{2} H_2(g) + \frac{1}{2} I_2(g)$$

has the rate equation

$$-\frac{\Delta[HI]}{\Delta t} = k[HI]^2$$

where $k = 30.$ L/mol · min at 443 °C. How much time does it take for the concentration of HI to drop from 0.010 mol/L to 0.0050 mol/L at 443 °C?

Strategy Substitute the values of $[HI]_0$, $[HI]_t$, and k into Equation 15.2, and solve for the unknown, t.

Solution Here, $[HI]_0 = 0.010$ mol/L and $[HI]_t = 0.0050$ mol/L. Using Equation 15.2, we have

$$\frac{1}{0.0050 \text{ mol/L}} - \frac{1}{0.010 \text{ mol/L}} = (30.\ \text{L/mol} \cdot \text{min})t$$

$$(2.0 \times 10^2 \text{ L/mol}) - (1.0 \times 10^2 \text{ L/mol}) = (30.\ \text{L/mol} \cdot \text{min})t$$

$$t = 3.3 \text{ min}$$

EXERCISE 15.7 Using the Second-Order Integrated Rate Law Equation

Using the rate constant for HI decomposition given in Example 15.7, calculate the concentration of HI after 12 min if $[HI]_0 = 0.010$ mol/L.

Zero-Order Reactions

If a reaction (R → products) is zero order, the rate equation is

$$-\frac{\Delta[R]}{\Delta t} = k[R]^0$$

This equation leads to the integrated rate equation

$$[R]_0 - [R]_t = kt \qquad \textbf{(15.3)}$$

where the units of k are mol/L · s.

Graphical Methods for Determining Reaction Order and the Rate Constant

We can derive a convenient way to determine the order of a reaction and its rate constant using graphical methods. Equations 15.1, 15.2, and 15.3, if rearranged slightly, have the form $y = mx + b$. This is the equation for a straight line, where m is the slope of the line and b is the y-intercept. In these equations, $x = t$ in each case.

Zero order	First order	Second order
$\underset{y}{[R]_t} = \underset{mx}{-kt} + \underset{b}{[R]_0}$	$\underset{y}{\ln [R]_t} = \underset{mx}{-kt} + \underset{b}{\ln [R]_0}$	$\underset{y}{\frac{1}{[R]_t}} = \underset{mx}{+kt} + \underset{b}{\frac{1}{[R]_0}}$

■ **Finding the Slope of a Line** See pages 39–41 for a description of methods for finding the slope of a line.

Active Figure 15.6 The decomposition of azomethane, $CH_3N_2CH_3$. If data for the decomposition of azomethane,

$CH_3N_2CH_3(g) \rightarrow CH_3CH_3(g) + N_2(g)$

are plotted as the natural logarithm of the $CH_3N_2CH_3$ pressure versus time, the result is a straight line with a negative slope. This indicates a first-order reaction. The rate constant $k = -$slope.

Chemistry Now™ Sign in at www.thomsonedu.com/login and go to the Chapter Contents menu to explore an interactive version of this figure accompanied by an exercise.

t(s)	$P \times 10^2$ atm	ln P
0	8.20	−2.50
1000	5.72	−2.86
2000	3.99	−3.22
3000	2.78	−3.58
4000	1.94	−3.94

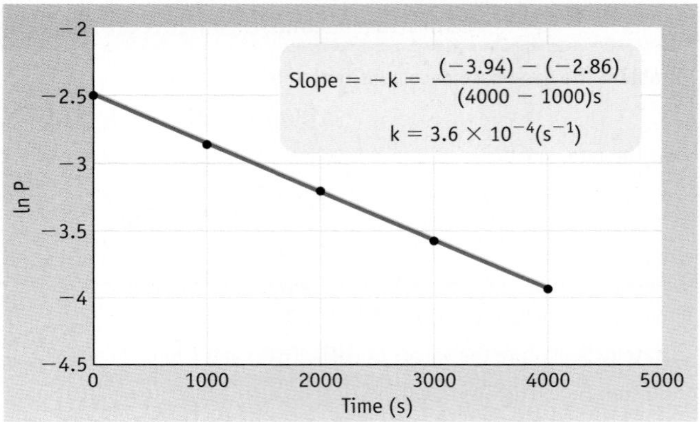

As an example of the graphical method for determining reaction order, consider the decomposition of azomethane.

$$CH_3N_2CH_3(g) \rightarrow CH_3CH_3(g) + N_2(g)$$

The decomposition of azomethane was followed at 600 K by observing the decrease in its partial pressure with time (Figure 15.6). (Recall from Chapter 11 that pressure is proportional to concentration at a given temperature and volume.) The third column lists values of ln $P(CH_3N_2CH_3)$. A plot of pressure vs. time for a first-order reaction is a curved line (see Figure 15.2). As shown in Figure 15.6, however, a graph of ln $P(CH_3N_2CH_3)$ versus time produces a straight line, showing that the reaction is first order in $CH_3N_2CH_3$. The slope of the line can be measured, and the negative of the slope equals the rate constant for the reaction, $3.6 \times 10^{-4}\ s^{-1}$.

The decomposition of NO_2 is a second-order process.

$$NO_2(g) \rightarrow NO(g) + \tfrac{1}{2}\, O_2(g)$$

$$\text{Rate} = k[NO_2]^2$$

This fact can be verified by showing that a plot of $1/[NO_2]$ versus time is a straight line (Figure 15.7). Here, the slope of the line is equal to k.

For a zero-order reaction (Figure 15.8), a plot of concentration vs. time gives a straight line with a slope equal to the negative of the rate constant.

FIGURE 15.7 A second-order reaction. A plot of $1/[NO_2]$ versus time for the decomposition of NO_2,

$NO_2(g) \rightarrow NO(g) + \tfrac{1}{2}\, O_2(g)$

results in a straight line. This confirms that this is a second-order reaction. The slope of the line equals the rate constant for this reaction.

Time (min)	$[NO_2]$ (mol/L)	$1/[NO_2]$ (L/mol)
0	0.020	50
0.50	0.015	67
1.0	0.012	83
1.5	0.010	100
2.0	0.0087	115

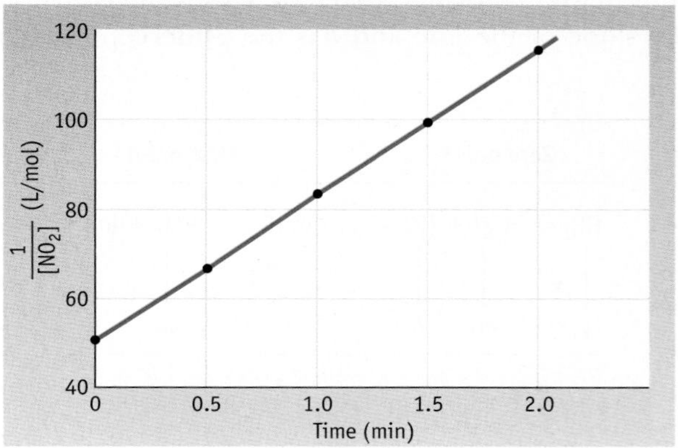

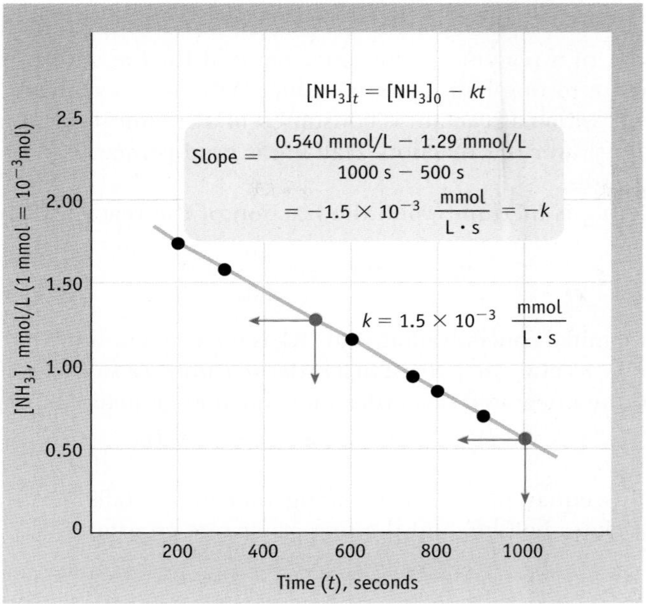

FIGURE 15.8 Plot of a zero-order reaction. A graph of the concentration of ammonia, $[NH_3]_t$, against time for the decomposition of NH_3.

$$2\ NH_3(g) \longrightarrow N_2(g) + 3\ H_2(g)$$

on a metal surface at 856 °C is a straight line, indicating that this is a zero-order reaction. The rate constant, k, for this reaction is found from the slope of the line; $k = -$slope. (The points chosen to calculate the slope are given in red.)

Table 15.1 summarizes the relationships between concentration and time for first-, second-, and zero-order processes.

Chemistry Now™

Sign in at **www.thomsonedu.com/login** and go to Chapter 15 Contents to see Screen 15.7 for a tutorial on **graphical methods.**

EXERCISE 15.8 Using Graphical Methods

Data for the decomposition of N_2O_5 in a particular solvent at 45 °C are as follows:

$[N_2O_5]$, mol/L	t, min
2.08	3.07
1.67	8.77
1.36	14.45
0.72	31.28

Plot $[N_2O_5]$, ln $[N_2O_5]$, and $1/[N_2O_5]$ versus time, t. What is the order of the reaction? What is the rate constant for the reaction?

TABLE 15.1 Characteristic Properties of Reactions of the Type "R ⟶ Products"

Order	Rate Equation	Integrated Rate Equation	Straight-Line Plot	Slope	k Units
0	$-\Delta[R]/\Delta t = k[R]^0$	$[R]_0 - [R]_t = kt$	$[R]_t$ vs. t	$-k$	mol/L · time
1	$-\Delta[R]/\Delta t = k[R]^1$	$\ln([R]_t/[R]_0) = -kt$	$\ln [R]_t$ vs. t	$-k$	time^{-1}
2	$-\Delta[R]/\Delta t = k[R]^2$	$(1/[R]_t) - (1/[R]_0) = kt$	$1/[R]_t$ vs. t	k	L/mol · time

Half-Life and First-Order Reactions

The **half-life,** $t_{1/2}$, of a reaction is the time required for the concentration of a reactant to decrease to one half its initial value. Half-life is a convenient way to describe the rate at which a reactant is consumed in a chemical reaction: The longer the half-life, the slower the reaction. Half-life is used primarily when dealing with first-order processes.

The half-life, $t_{1/2}$, is the time when the fraction of the reactant R remaining is

$$[R]_t = \tfrac{1}{2}[R]_0 \quad \text{or} \quad \frac{[R]_t}{[R]_0} = \tfrac{1}{2}$$

Here, $[R]_0$ is the initial concentration, and $[R]_t$ is the concentration after the reaction is half completed. To evaluate $t_{1/2}$ for a first-order reaction, we substitute $[R]_t/[R]_0 = \tfrac{1}{2}$ and $t = t_{1/2}$ into the integrated first-order rate equation (Equation 15.1),

$$\ln(\tfrac{1}{2}) = -kt_{1/2} \quad \text{or} \quad \ln 2 = kt_{1/2}$$

Rearranging this equation (and calculating that ln 2 = 0.693) provides a useful equation that relates half-life and the first-order rate constant:

$$t_{1/2} = \frac{0.693}{k} \qquad \textbf{(15.4)}$$

This equation identifies an important feature of first-order reactions: *$t_{1/2}$ is independent of concentration.*

To illustrate the concept of half-life, consider again the first-order decomposition of azomethane, $CH_3N_2CH_3$.

$$CH_3N_2CH_3(g) \rightarrow CH_3CH_3(g) + N_2(g)$$

$$\text{Rate} = k[CH_3N_2CH_3] \text{ with } k = 3.6 \times 10^{-4}\ s^{-1} \text{ at 600 K}$$

Given a rate constant of $3.6 \times 10^{-4}\ s^{-1}$, we calculate a half-life of 1.9×10^3 s or 32 min.

$$t_{1/2} = \frac{0.693}{3.6 \times 10^{-4}\ s^{-1}} = 1.9 \times 10^3\ s$$

The partial pressure of azomethane has been plotted as a function of time in Figure 15.9, and this graph shows that *P*(azomethane) decreases by half every 32 minutes. The initial pressure was 820 mm Hg, but it dropped to 410 mm Hg in 32 minutes, and then dropped to 205 mm Hg in another 32 minutes. That is, after two half-lives (64 min), the pressure is $(\tfrac{1}{2}) \times (\tfrac{1}{2}) = (\tfrac{1}{2})^2 = \tfrac{1}{4}$ or 25% of the initial pressure. After three half-lives, the pressure has dropped further to 102 mm Hg or 12.5% of the initial value and equal to $(\tfrac{1}{2}) \times (\tfrac{1}{2}) \times (\tfrac{1}{2}) = (\tfrac{1}{2})^3 = \tfrac{1}{8}$ of the initial value.

It can be hard to visualize whether a reaction is fast or slow from the rate constant value. Can you tell from the rate constant, $k = 3.6 \times 10^{-4}\ s^{-1}$, whether the azomethane decomposition will take seconds, hours, or days to reach completion? Probably not, but this is easily assessed from the value of the half-life for the reaction (32 min). Now we know that we would only have to wait a few hours for the reactant to be essentially consumed.

Half-Life and Radioactivity Half-life is a term often encountered when dealing with radioactive elements. Radioactive decay is a first-order process, and half-life is commonly used to describe how rapidly a radioactive element decays. See Chapter 23 and Example 15.9.

Half-Life Equations for Other Reaction Orders

For a zero-order reaction, R → products

$$t_{1/2} = \frac{[R]_0}{2k}$$

For a second-order reaction, R → products

$$t_{1/2} = \frac{1}{k[R]_0}$$

Note that in both cases the half-life depends on the initial concentration.

Chemistry Now™

Sign in at **www.thomsonedu.com/login** and go to Chapter 15 Contents to see Screen 15.8 for tutorials on **using half-life.**

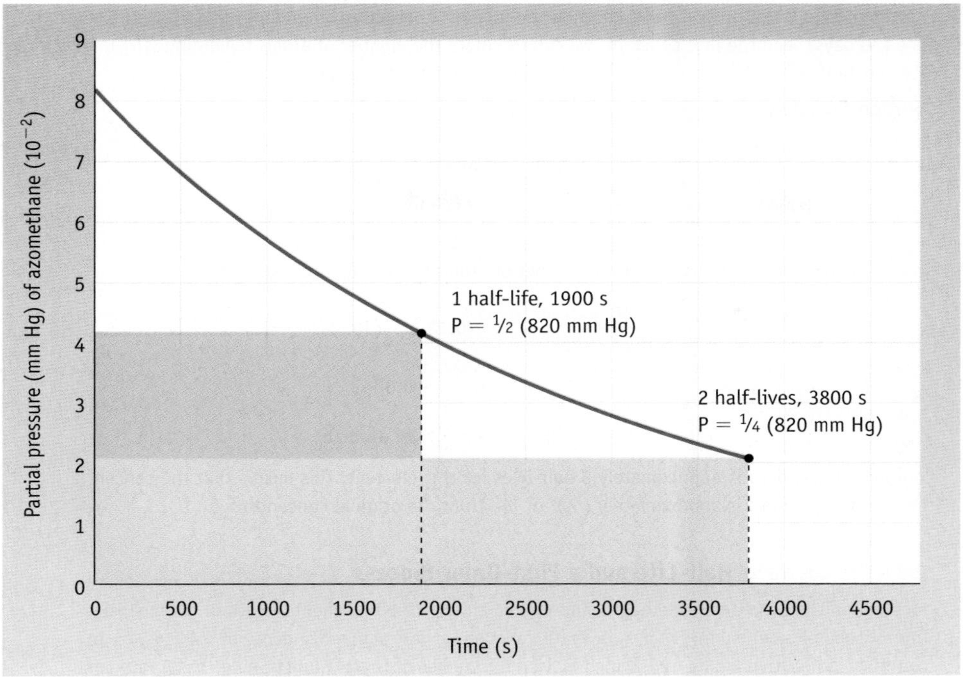

Active Figure 15.9 Half-life of a first-order reaction. The curve is a plot of the pressure of $CH_3N_2CH_3$ as a function of time. (The compound decomposes to CH_3CH_3 and N_2 with $k = 3.6 \times 10^{-4}\ s^{-1}$). The pressure of $CH_3N_2CH_3$ is halved every 1900 s (32 min). (This plot of pressure versus time is similar in shape to plots of concentration versus time for all other first-order reactions.)

Chemistry Now™ Sign in at www.thomsonedu.com/login and go to the Chapter Contents menu to explore an interactive version of this figure accompanied by an exercise.

EXAMPLE 15.8 Half-Life and a First-Order Process

Problem Sucrose, $C_{12}H_{22}O_{11}$, decomposes to fructose and glucose in acid solution with the rate law

Rate $= k[C_{12}H_{22}O_{11}]$ $\qquad k = 0.216\ h^{-1}$ at 25 °C

(a) What is the half-life of $C_{12}H_{22}O_{11}$ at this temperature?

(b) What amount of time is required for 87.5% of the initial concentration of $C_{12}H_{22}O_{11}$ to decompose?

Strategy (a) The decomposition of $C_{12}H_{22}O_{11}$ is first order in this compound, so Equation 15.4 can be used to calculate the half-life. (b) After 87.5% of the $C_{12}H_{22}O_{11}$ has decomposed, 12.5% (or one-eighth of the sample) remains. To reach this point, three half-lives are required.

Half-Life	Fraction Remaining
1	0.5
2	0.25
3	0.125

Therefore, we calculate the half-life from Equation 15.4 and then multiply by 3.

Solution

(a) The half-life for the reaction is

$$t_{1/2} = 0.693/k = 0.693/(0.216\ h^{-1}) = 3.21\ h$$

(b) Three half-lives must elapse before the fraction remaining is 0.125, so

$$\text{Time elapsed} = 3 \times 3.21\ h = 9.63\ h$$

EXAMPLE 15.9 Half-Life and First-Order Processes

Problem Radioactive radon-222 gas (^{222}Rn) from natural sources can seep into the basement of a home. The half-life of ^{222}Rn is 3.8 days. Assume that the radon gas is trapped in the basement and cannot escape. If a basement has 4.0×10^{13} atoms of ^{222}Rn per liter of air initially, how many atoms of ^{222}Rn per liter will remain after one month (30 days)?

Strategy Using Equation 15.1, and knowing the number of atoms at the beginning ($[R]_0$), the elapsed time (30 days), and the rate constant, we can calculate the number of atoms remaining ($[R]_t$). First, the rate constant, k, must be found from the half-life using Equation 15.4.

Solution The rate constant, k, is

$$k = \frac{0.693}{t_{1/2}} = \frac{0.693}{3.8 \text{ d}} = 0.18 \text{ d}^{-1}$$

Now use Equation 15.1 to calculate the number of atoms remaining after 30. days.

$$\ln \frac{[\text{Rn}]_t}{4.0 \times 10^{13} \text{ atom/L}} = -(0.18 \text{ d}^{-1})(30.\text{ d}) = -5.5$$

$$\frac{[\text{Rn}]_t}{4.0 \times 10^{13} \text{ atom/L}} = e^{-5.5} = 0.0042$$

$$[\text{Rn}]_t = 1.7 \times 10^{11} \text{ atom/L}$$

Comment Thirty days is approximately 8 half-lives for this element. This means that the concentration at the end of the month is approximately $(1/2)^8$ or $1/256$th of the original concentration.

EXERCISE 15.9 Half-Life and a First-Order Process

Americium is used in smoke detectors and in medicine for the treatment of certain malignancies. One isotope of americium, ^{241}Am, has a rate constant, k, for radioactive decay of 0.0016 y^{-1}. In contrast, radioactive iodine-125, which is used for studies of thyroid functioning, has a rate constant for decay of 0.011 d^{-1}.

(a) What are the half-lives of these isotopes?

(b) Which isotope decays faster?

(c) If you are given a dose of iodine-125 containing 1.6×10^{15} atoms, how many atoms remain after 2.0 days?

15.5 A Microscopic View of Reaction Rates

Throughout this book, we have turned to the particulate level of chemistry to understand chemical phenomena. Rates of reaction are no exception. Looking at the way reactions occur at the atomic and molecular levels can provide some insight into the various influences on rates of reactions.

Let us review the macroscopic observations we have made so far concerning reaction rates. We know that there are wide differences in rates of reactions—from very fast reactions like the explosion that occurs when hydrogen and oxygen are exposed to a spark or flame (Figure 1.18), to slow reactions like the formation of rust that occur over days, weeks, or years. For a specific reaction, factors that influence reaction rate include the concentrations of the reactants, the temperature of the reaction system, and the presence of catalysts. Let us look at each of these influences in more depth.

Concentration, Reaction Rate, and Collision Theory

Consider the gas-phase reaction of nitric oxide and ozone:

$$NO(g) + O_3(g) \rightarrow NO_2(g) + O_2(g)$$

The rate law for this product-favored reaction is first order in each reactant: Rate $= k[NO][O_3]$. How can this reaction have this rate law?

Let us consider the reaction at the particulate level and imagine a flask containing a mixture of NO and O_3 molecules in the gas phase. Both kinds of molecules are in rapid and random motion within the flask. They strike the walls of the vessel and collide with other molecules. For this or any other reaction to occur, the **collision theory of reaction rates** states that three conditions must be met:

1. The reacting molecules must collide with one another.
2. The reacting molecules must collide with sufficient energy to initiate the process of breaking and forming bonds.
3. The molecules must collide in an orientation that can lead to rearrangement of the atoms and the formation of products.

We shall discuss each of these conditions within the context of the effects of concentration and temperature on reaction rate.

To react, molecules must collide with one another. It is reasonable to propose that the rate of their reaction be related to the number of collisions, which is in turn related to their concentrations (Figure 15.10). Doubling the concentration of one reagent in the NO + O_3 reaction, say NO, will lead to twice the number of molecular collisions. Figure 15.10a shows a single molecule of one of the reactants (NO) moving randomly among sixteen O_3 molecules. In a given time period, it might collide with two O_3 molecules. The number of NO—O_3 collisions will double, however, if the concentration of NO molecules is doubled (to 2, as shown in Figure 15.10b) or if the number of O_3 molecules is doubled (to 32, as in Figure 15.10c). Thus, we can explain the dependence of reaction rate on concentration: The number of collisions between the two reactant molecules is directly proportional to the concentration of each reactant, and the rate of the reaction shows a first-order dependence on each reactant.

Chemistry Now™

Sign in at **www.thomsonedu.com/login** and go to Chapter 15 Contents to see Screen 15.9 for a visualization of **collision theory** and for tutorials on **using half-life.**

Temperature, Reaction Rate, and Activation Energy

In a laboratory or in the chemical industry, a chemical reaction is often carried out at elevated temperature because this allows the reaction to occur more rapidly. Conversely, it is sometimes desirable to lower the temperature to slow down a chemical reaction (to avoid an uncontrollable reaction or a potentially dangerous explosion). Chemists are very aware of the effect of temperature on the rate of a reaction.

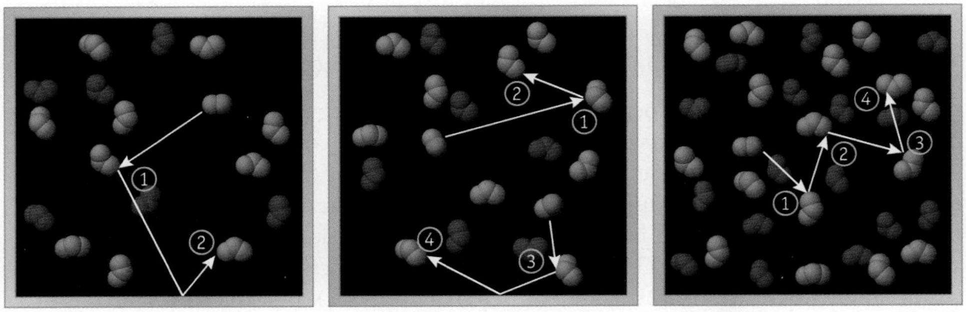

(a) 1 NO : 16 O_3 — *2 hits/second* (b) 2 NO : 16 O_3 — *4 hits/second* (c) 1 NO : 32 O_3 — *4 hits/second*

FIGURE 15.10 The effect of concentration on the frequency of molecular collisions. (a) A single NO molecule, moving among sixteen O_3 molecules, is shown colliding with two of them per second. (b) If two NO molecules move among 16 O_3 molecules, we would predict that four NO—O_3 collisions would occur per second. (c) If the number of O_3 molecules is doubled (to 32), the frequency of NO—O_3 collisions is also doubled, to four per second.

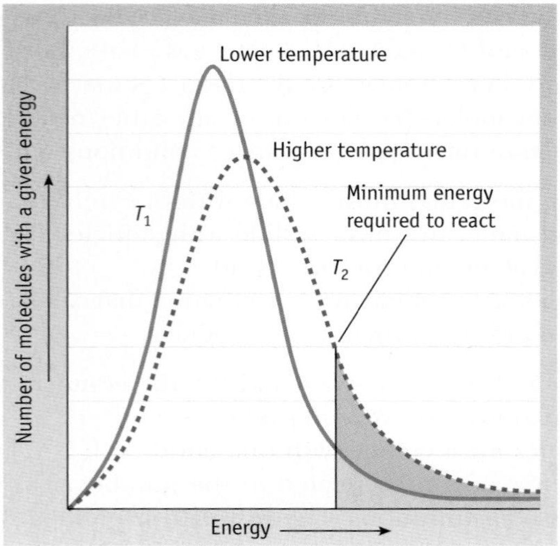

FIGURE 15.11 Energy distribution curve. The vertical axis gives the relative number of molecules possessing the energy indicated on the horizontal axis. The graph indicates the minimum energy required for an arbitrary reaction. At a higher temperature, a larger fraction of the molecules have sufficient energy to react. (Recall Figure 11.14, the Boltzmann distribution function, for a collection of gas molecules.)

A discussion of the effect of temperature on reaction rate begins with reference to distribution of energies for molecules in a sample of a gas or liquid. Recall from studying gases and liquids that the molecules in a sample have a wide range of energies, described earlier as a Boltzmann distribution of energies (◀ Figure 11.14). In any sample of a gas or liquid, some molecules have very low energies; others have very high energies; but most have some intermediate energy. As the temperature increases, the average energy of the molecules in the sample increases, as does the fraction having higher energies (Figure 15.11).

Activation Energy

Molecules require some minimum energy to react. Chemists visualize this as an energy barrier that must be surmounted by the reactants for a reaction to occur (Figure 15.12). The energy required to surmount the barrier is called the **activation energy, E_a**. If the barrier is low, the energy required is low, and a high proportion of the molecules in a sample may have sufficient energy to react. In such a case, the reaction will be fast. If the barrier is high, the activation energy is high, and only a few reactant molecules in a sample may have sufficient energy. In this case, the reaction will be slow.

Charles D. Winters

FIGURE 15.12 An analogy to chemical activation energy. For the volleyball to go over the net, the player must give it sufficient energy.

To illustrate an activation energy barrier, consider the conversion of NO_2 and CO to NO and CO_2 or the reverse reaction (Figure 15.13). At the molecular level, we imagine that the reaction involves the transfer of an O atom from an NO_2 molecule to a CO molecule (or, in the reverse reaction, the transfer of an O atom from CO_2 to NO).

$$NO_2(g) + CO(g) \rightleftharpoons NO(g) + CO_2(g)$$

We can describe this process by using an energy diagram or **reaction coordinate** diagram. The horizontal axis describes the reaction progress as the reaction proceeds, and the vertical axis represents the potential energy of the system during the reaction. When NO_2 and CO approach and O atom transfer begins, an N—O bond is being broken, and a C=O bond is forming. Energy input (the activation energy) is required for this to occur. The energy of the system reaches a maximum at the **transition state**. At the transition state, sufficient energy has been concentrated in the appropriate bonds; bonds in the reactants can now break, and new bonds can form

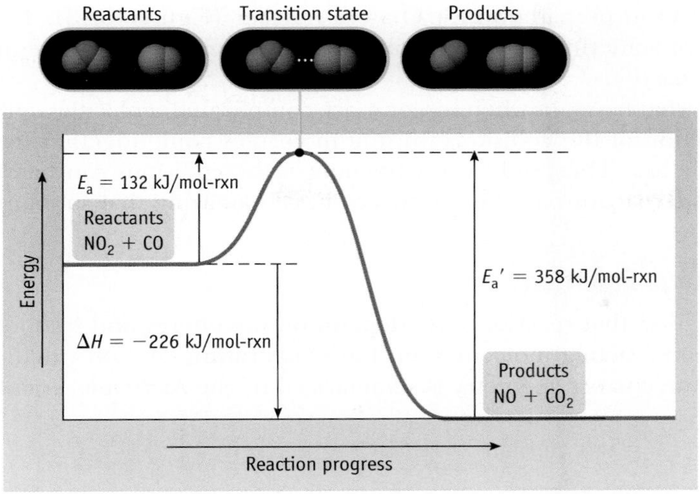

Active Figure 15.13 Activation Energy. The reaction of NO_2 and CO (to give NO and CO_2) has an activation energy barrier of 132 kJ/mol-rxn. The reverse reaction ($NO + CO_2 \rightarrow NO_2 + CO$) requires 358 kJ/mol-rxn. The net energy change for the reaction of NO_2 and CO is −226 kJ/mol-rxn.

Chemistry Now™ Sign in at www.thomsonedu.com/login and go to the Chapter Contents menu to explore an interactive version of this figure accompanied by an exercise.

to give products. The system is poised to go on to products. Alternatively, it can return to the reactants. Because the transition state is at a maximum in potential energy, it cannot be isolated. Using computer molecular modeling techniques, however, chemists can describe what the transition state must look like.

In the NO_2 + CO reaction, 132 kJ/mol is required to reach the transition state; that is, the top of the energy barrier. As the reaction continues toward the products—as the N—O bond is finally broken and a C═O bond forms—the reaction evolves energy, 358 kJ/mol-rxn. The net energy change involved in this exothermic reaction is −226 kJ/mol-rxn.

$$\Delta U = +132 \text{ kJ/mol-rxn} + (-358 \text{ kJ/mol-rxn}) = -226 \text{ kJ/mol-rxn}$$

What happens if NO and CO_2 are mixed to form NO_2 and CO? Now the reaction requires 358 kJ/mol-rxn to reach the transition state, and 132 kJ/mol-rxn is evolved on proceeding to the product, NO_2 and CO. The reaction in this direction is endothermic, requiring an input of +226 kJ/mol-rxn.

Chemistry Now™

Sign in at **www.thomsonedu.com/login** and go to Chapter 15 Contents to see Screen 15.10 for a simulation of **reaction coordinate diagrams.**

Effect of a Temperature Increase

The conversion of NO_2 and CO to products at room temperature is slow because only a small fraction of the molecules have enough energy to reach the transition state. The rate can be increased by heating the sample. Raising the temperature increases the reaction rate by increasing the fraction of molecules with enough energy to surmount the activation energy barrier (Figure 15.11).

Effect of Molecular Orientation on Reaction Rate

Having a sufficiently high energy is necessary, but it is not sufficient to ensure that reactants will form products. The reactant molecules must also come together in the correct orientation. For the reaction of NO_2 and CO, we can imagine that the transition state structure has one of the O atoms of NO_2 beginning to bind to the

C atom of CO in preparation for O atom transfer (Figure 15.13). The lower the probability of achieving the proper alignment, the smaller the value of k, and the slower the reaction.

Imagine what happens when two or more complicated molecules collide. In only a small fraction of the collisions will the molecules come together in exactly the right orientation. Thus, only a tiny fraction of the collisions can be effective. No wonder some reactions are slow. Conversely, it is amazing that so many are fast!

The Arrhenius Equation

The observation that reaction rates depend on the energy and frequency of collisions between reacting molecules, on the temperature, and on whether the collisions have the correct geometry is summarized by the **Arrhenius equation**:

$$k = \text{rate constant} = \underbrace{A}_{\text{Frequency factor}}\underbrace{e^{-E_a/RT}}_{\text{Fraction of molecules with minimum energy for reaction}} \quad (15.5)$$

■ **Interpreting the Arrhenius Equation** (a) The exponential term. This gives the fraction of molecules having sufficient energy for reaction and is a function of T.

Temperature (K)	Value of $e^{-E_a/RT}$ for E_a = 40 kJ/mol-rxn
298	9.7×10^{-8}
400	5.9×10^{-6}
600	3.3×10^{-4}

(b) Significance of A. Although a complete understanding of A goes beyond the level of this text, it can be noted that A becomes smaller as the reactants become larger. It reflects the fact that reacting molecules must come together in the appropriate geometry.

In this equation, k is the rate constant, where R is the gas constant with a value of 8.314510×10^{-3} kJ/K · mol, and T is the kelvin temperature. The parameter A is called the **frequency factor,** and it has units of L/mol · s. It is related to the number of collisions and to the fraction of collisions that have the correct geometry; A is specific to each reaction and is temperature dependent. The factor $e^{-E_a/RT}$ represents *the fraction of molecules having the minimum energy required for reaction*; its value is always less than 1. As the table in the margin shows, this fraction changes significantly with temperature.

The Arrhenius equation has significant uses.

- It can be used to calculate E_a from the temperature, dependence of the rate constant.
- It can be used to calculate the rate constant for a given temperature, if E_a and A are known.

If rate constants of a given reaction are measured at several temperatures then one can apply graphical techniques to determine the activation energy of a reaction. Taking the natural logarithm of each side of Equation 15.5, we have

$$\ln k = \ln A + \left(-\frac{E_a}{RT}\right)$$

Rearranging this expression slightly shows that ln k and $1/T$ are related linearly.

$$\ln k = -\frac{E_a}{R}\left(\frac{1}{T}\right) + \ln A \quad \leftarrow \text{Arrhenius equation} \qquad (15.6)$$

$$\downarrow \qquad \downarrow \qquad \downarrow$$

$$y = mx + b \quad \leftarrow \text{Equation for straight line}$$

This means that, if the natural logarithm of k (ln k) is plotted versus $1/T$, the result is a downward-sloping straight line with a slope of $(-E_a/R)$. The activation energy, E_a, can be obtained from the slope of this line.

Chemistry Now™

Sign in at **www.thomsonedu.com/login** and go to Chapter 15 Contents to see Screen 15.11 for a simulation and tutorials on **the temperature dependence of reaction rates** and the **Arrhenius equation.**

A Closer Look — Reaction Coordinate Diagrams

Reaction coordinate diagrams (Figure 15.13) convey a great deal of information. A reaction that would have an energy diagram like that in Figure 15.13 is the substitution of a halogen atom of CH_3Cl by an ion such as F^-. Here, the F^- ion attacks the molecule from the side opposite the Cl substituent. As F^- begins to form a bond to carbon, the C—Cl bond weakens, and the CH_3 portion of the molecule changes shape. As time progresses, the products CH_3F and Cl^- are formed.

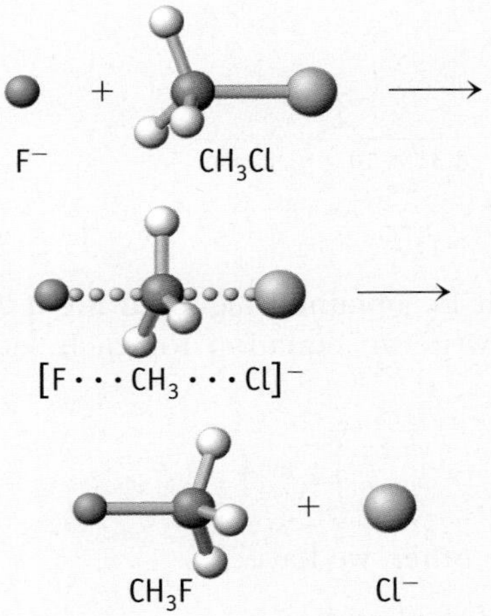

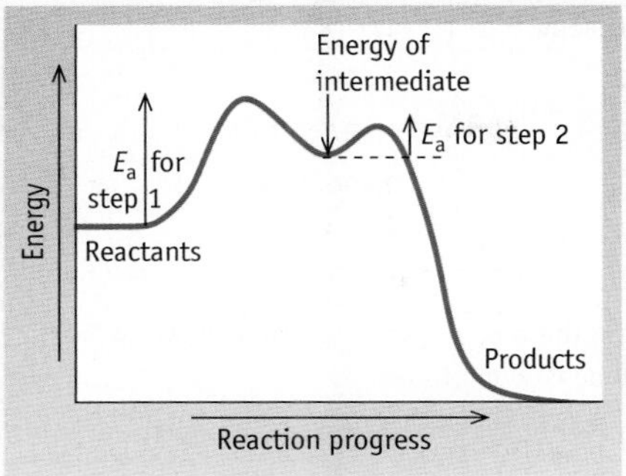

FIGURE A A reaction coordinate diagram for a two-step reaction, a process involving an intermediate.

The diagram in Figure A describes a reaction that occurs in two steps. An example of such a reaction is the substitution of the —OH group on methanol by a halide ion in the presence of acid. In the first step, an H^+ ion attaches to the O of the C—O—H group in a rapid, reversible reaction. The energy of this protonated species, $CH_3OH_2^+$, a reaction intermediate, is higher than the energies of the reactants and is represented by the dip in the curve shown in Figure A. In the second step, a halide ion, say Br^-, attacks the intermediate to produce methyl bromide, CH_3Br, and water. There is an activation energy barrier in both the first step and second step.

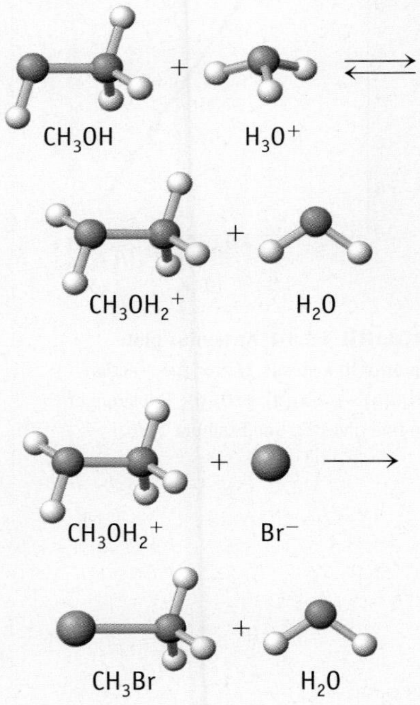

Notice in Figure A, as in Figure 15.13, that the energy of the products is lower than the energy of the reactants. The reaction is exothermic.

EXAMPLE 15.10 Determination of E_a from the Arrhenius Equation

Problem Using the experimental data shown in the table, calculate the activation energy E_a for the reaction

$$2\ N_2O(g) \rightarrow 2\ N_2(g) + O_2(g)$$

Experiment	Temperature (K)	k (L/mol · s)
1	1125	11.59
2	1053	1.67
3	1001	0.380
4	838	0.0011

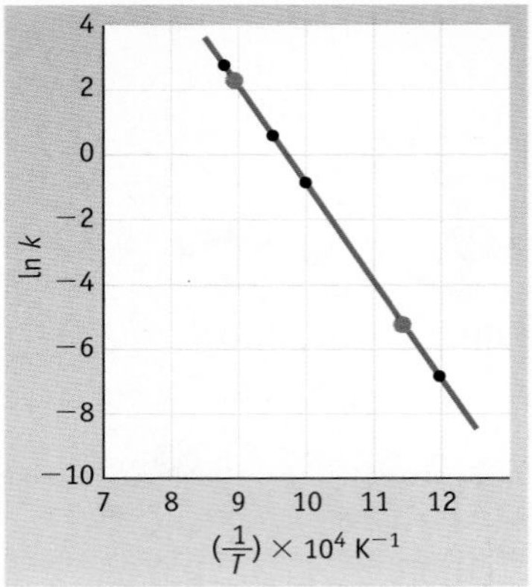

FIGURE 15.14 Arrhenius plot. A plot of ln k versus $1/T$ for the reaction $2\ N_2O(g) \rightarrow 2\ N_2(g) + O_2(g)$. The slope of the line gives E_a. See Example 15.10.

Strategy To solve this problem graphically, we first need to calculate ln k and $1/T$ for each data point. These data are then plotted, and E_a is calculated from the resulting straight line (slope $= -E_a/R$).

Solution First, calculate $1/T$ and ln k.

Experiment	$1/T$ (K^{-1})	ln k
1	8.889×10^{-4}	2.4501
2	9.497×10^{-4}	0.513
3	9.990×10^{-4}	−0.968
4	11.9×10^{-4}	−6.81

Plotting these data gives the graph shown in Figure 15.14. Choosing the large blue points on the graph, the slope is found to be

$$\text{Slope} = \frac{\Delta \ln k}{\Delta(1/T)} = \frac{2.0 - (-5.6)}{(9.0 - 11.5)(10^{-4})K^{-1}} = -3.0 \times 10^4\ \text{K}$$

The activation energy is evaluated from the slope.

$$\text{Slope} = -\frac{E_a}{R} = -3.0 \times 10^4\ \text{K} = -\frac{E_a}{8.31 \times 10^{-3}\ \text{kJ/K} \cdot \text{mol}}$$

$$E_a = 250\ \text{kJ/mol-rxn}$$

The activation energy, E_a, for a reaction can be obtained algebraically if k is known at two different temperatures. We can write an equation for each set of these conditions:

$$\ln k_1 = -\left(\frac{E_a}{RT_1}\right) + \ln A \quad \text{or} \quad \ln k_2 = -\left(\frac{E_a}{RT_2}\right) + \ln A$$

If one of these equations is subtracted from the other, we have

$$\ln k_2 - \ln k_1 = \ln \frac{k_2}{k_1} = -\frac{E_a}{R}\left[\frac{1}{T_2} - \frac{1}{T_1}\right] \qquad \textbf{(15.7)}$$

Example 15.11 demonstrates the use of this equation.

EXAMPLE 15.11 Calculating E_a Numerically

Problem Use values of k determined at two different temperatures to calculate the value of E_a for the decomposition of HI:

$$2\ HI(g) \rightarrow H_2(g) + I_2(g)$$

$$k_1 = 2.15 \times 10^{-8}\ \text{L/(mol} \cdot \text{s) at } 6.50 \times 10^2\ \text{K } (T_1)$$

$$k_2 = 2.39 \times 10^{-7}\ \text{L/(mol} \cdot \text{s) at } 7.00 \times 10^2\ \text{K } (T_2)$$

Strategy Use Equation 15.7.

Solution

$$\ln \frac{2.39 \times 10^{-7}\ \text{L/(mol} \cdot \text{s)}}{2.15 \times 10^{-8}\ \text{L/(mol} \cdot \text{s)}} = -\frac{E_a}{8.315 \times 10^{-3}\ \text{kJ/K} \cdot \text{mol}} \times \left[\frac{1}{7.00 \times 10^2\ \text{K}} - \frac{1}{6.50 \times 10^2\ \text{K}}\right]$$

Solving this equation gives $E_a = 180$ kJ/mol-rxn.

Comment Another way to write the difference in fractions in brackets is

$$\left[\frac{1}{T_2} - \frac{1}{T_1}\right] = \frac{T_1 - T_2}{T_1 T_2}$$

This expression is often easier to use.

■ **E_a, Reaction Rates, and Temperature** A often-used rule of thumb is that reaction rates double for every 10 °C rise in temperature in the vicinity of room temperature.

EXERCISE 15.10 Calculating E_a from the Temperature Dependence of k

The colorless gas N_2O_4 decomposes to the brown gas NO_2 in a first-order reaction.

$$N_2O_4(g) \rightarrow 2\ NO_2(g)$$

The rate constant $k = 4.5 \times 10^3\ s^{-1}$ at 274 K and $k = 1.00 \times 10^4\ s^{-1}$ at 283 K. What is the activation energy, E_a?

Effect of Catalysts on Reaction Rate

Catalysts are substances that speed up the rate of a chemical reaction. We have seen several examples of catalysts in earlier discussions in this chapter: MnO_2, iodide ion, an enzyme in a potato, and hydroxide ion all catalyze the decomposition of hydrogen peroxide (Figure 15.4). In biological systems, catalysts called enzymes influence the rates of most reactions (page 702).

■ **Enzymes: Biological Catalysts** Catalase is an enzyme whose function is to speed up the decomposition of hydrogen peroxide. This enzyme ensures that hydrogen peroxide, which is highly toxic, does not build up in the body.

Catalysts are not consumed in a chemical reaction. They are, however, intimately involved in the details of the reaction at the particulate level. Their function is to provide a different pathway with a lower activation energy for the reaction. To illustrate how a catalyst participates in a reaction, let us consider the isomerization of *cis*-2-butene, to the slightly more stable isomer, *trans*-2-butene.

■ **Catalysts and the Economy** "One third of [the] material gross national product in the U.S. involves a catalytic process somewhere in the production chain." (Quoted in A. Bell, *Science*, Vol. 299, page 1688, 2003.)

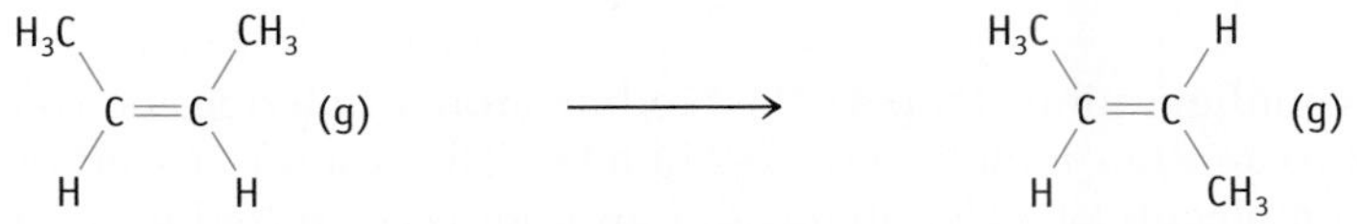

cis-2-butene   Transition state   *trans*-2-butene

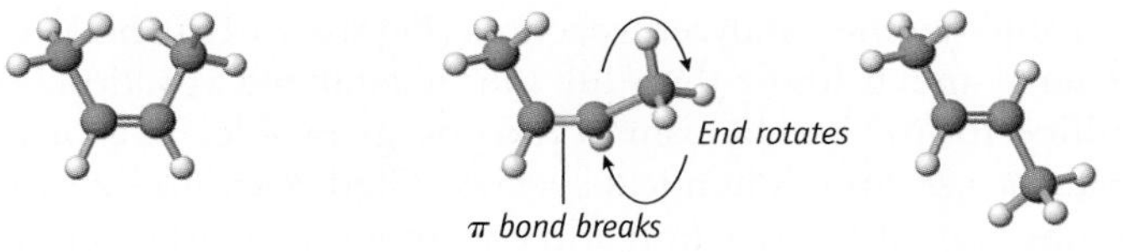

The activation energy for the uncatalyzed conversion is relatively large—264 kJ/mol-rxn—because the π bond must be broken to allow one end of the molecule to rotate into a new position. Because of the high activation energy, this is a slow reaction, and rather high temperatures are required for it to occur at a reasonable rate.

The *cis*- to *trans*-2-butene reaction is greatly accelerated by a catalyst, iodine. In the presence of iodine, this reaction can be carried out at a temperature several hundred degrees lower than for the uncatalyzed reaction. Iodine is not consumed (nor is it a product), and it does not appear in the overall balanced equation. It does appear in the reaction rate law, however; the rate of the reaction depends on the square root of the iodine concentration:

$$\text{Rate} = k[cis\text{-2-butene}][I_2]^{1/2}$$

■ **Butene Isomerization** Isomerization of *cis*-2-butene is a first order process with the rate law "Rate = k[*cis*-2-butene]." It is suggested to occur by rotation around the carbon–carbon double bond. The rate at which a molecule will isomerize is related to the fraction of molecules that have a high enough energy.
(See ChemistryNow Screen 15.8 to view an animation of the interconversion of butene isomers and the energy barrier for the process.)

The presence of I_2 changes the way the reaction occurs; that is, it changes the mechanism of the reaction (Figure 15.15). The best hypothesis is that iodine molecules first dissociate to form iodine atoms (Step 1). An iodine atom then adds to one of the C atoms of the C=C double bond (Step 2). This converts the double bond between the carbon atoms to a single bond (the π bond is broken) and allows the ends of the molecule to twist freely relative to each other (Step 3). If the iodine atom then dissociates from the intermediate, the double bond can re-form

FIGURE 15.15 The mechanism of the iodine-catalyzed isomerization of *cis*-2-butene. *Cis*-2-butene is converted to *trans*-2-butene in the presence of a catalytic amount of iodine. Catalyzed reactions are often pictured in such diagrams to emphasize what chemists refer to as a "catalytic cycle."

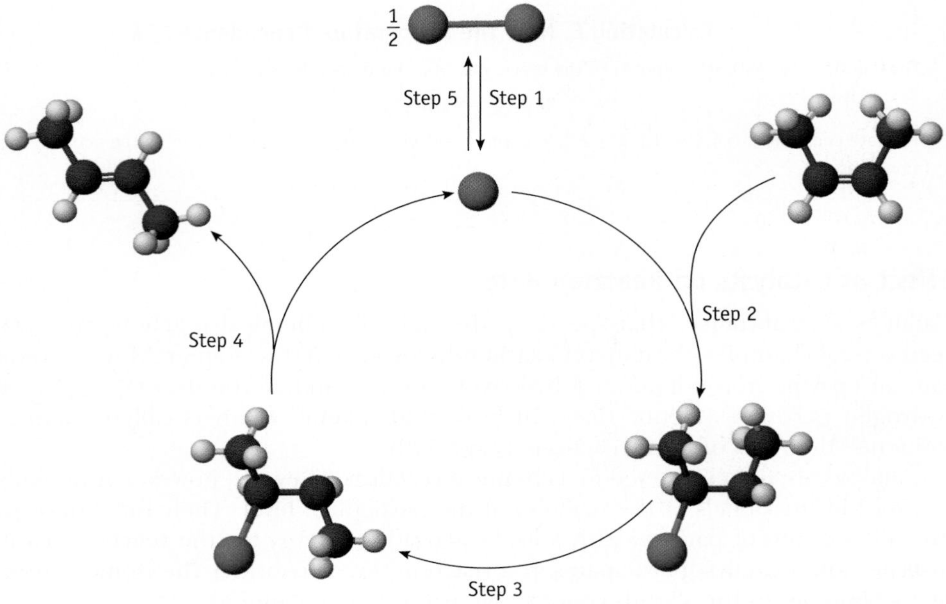

in the *trans* configuration (Step 4). The iodine atom catalyzing the rotation is now free to add to another molecule of *cis*-2-butene. The result is a kind of chain reaction, as one molecule of *cis*-2-butene after another is converted to the *trans* isomer. The chain is broken if the iodine atom recombines with another iodine atom to re-form molecular iodine.

An energy profile for the catalyzed reaction (Figure 15.16) shows that the overall energy barrier is much lower than for the uncatalyzed reaction. Five separate steps are identified for the mechanism in the energy profile. This proposed mechanism also includes a series of chemical species called **reaction intermediates,** species formed in one step of the reaction and consumed in a later step. Iodine atoms are intermediates, as are the free radical species formed when an iodine atom adds to *cis*-2-butene.

FIGURE 15.16 Energy profile for the iodine-catalyzed reaction of *cis*-2-butene. A catalyst accelerates a reaction by altering the mechanism so that the activation energy is lowered. With a smaller barrier to overcome, more reacting molecules have sufficient energy to surmount the barrier, and the reaction occurs more rapidly. The energy profile for the uncatalyzed conversion of *cis*-2-butene to *trans*-2-butene is shown by the black curve, and that for the iodine-catalyzed reaction is represented by the red curve. Notice that the shape of the barrier has changed because the mechanism has changed.

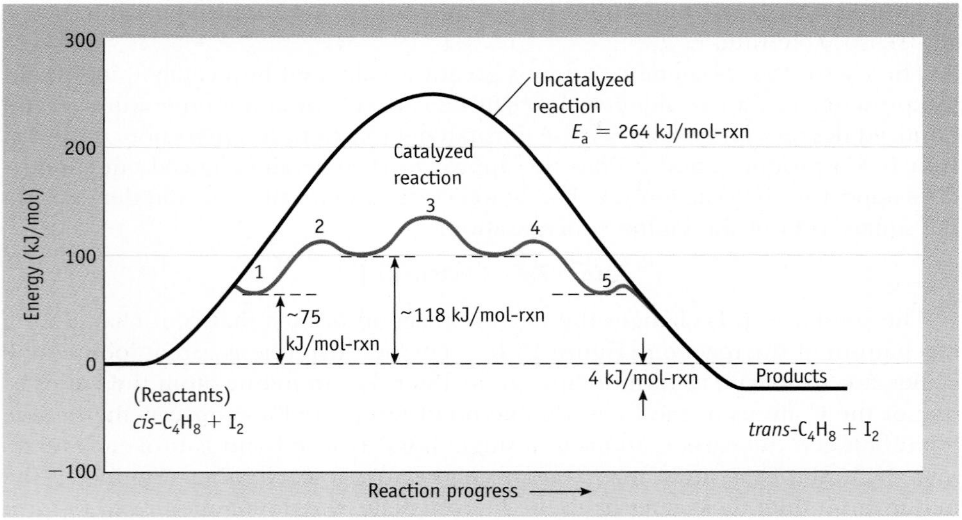

Five important points are associated with this mechanism:

- Iodine molecules, I_2, dissociate to atoms and then re-form. On the macroscopic level, the concentration of I_2 is unchanged. Iodine does not appear in the balanced, stoichiometric equation even though it appears in the rate equation. This is generally true of catalysts.
- Both the catalyst I_2 and the reactant *cis*-2-butene are in the gas phase. If a catalyst is present in the same phase as the reacting substance, it is called a *homogeneous catalyst.*
- Iodine atoms and the radical species formed by addition of an iodine atom to a 2-butene molecule are intermediates.
- The activation energy barrier to reaction is significantly lower because the mechanism changed. Dropping the activation energy from 264 kJ/mol-rxn for the uncatalyzed reaction to about 150 kJ/mol-rxn for the catalyzed process makes the catalyzed reaction 10^{15} times faster!
- The diagram of energy-versus-reaction progress has five energy barriers (five humps appear in the curve). This feature in the diagram means that the reaction occurs in a series of five steps.

What we have described here are reaction mechanisms. The uncatalyzed isomerization reaction of *cis*-2-butene is a one-step reaction mechanism, whereas the catalyzed mechanism involves a series of steps. We shall discuss reaction mechanisms in more detail in the next section.

Chemistry Now™

Sign in at **www.thomsonedu.com/login** and go to Chapter 15 Contents to see Screen 15.14 for tutorials on **using half-life various catalysts**, a visualization of **the effect of a catalyst on activation energy**, an interview of a scientist describing **catalyst use in industry**, two exercises on **reaction mechanisms and the effect of catalysts**, and a video exercise on **catalysis.**

15.6 Reaction Mechanisms

Rate laws help us understand **reaction mechanisms**, the sequence of bond-making and bond-breaking steps that occurs during the conversion of reactants to products. We want to analyze the changes that molecules undergo when they react. We then want to relate this description back to the macroscopic world, to the experimental observations of reaction rates.

Based on the rate equation for a reaction, and by applying chemical intuition, chemists can often make an educated guess about the mechanism for a reaction. In some reactions, the conversion of reactants to products in a single step is envisioned as the logical mechanism. For example, the uncatalyzed isomerization of *cis*-2-butene to *trans*-2-butene is best described as a single-step reaction (Figure 15.16).

Most chemical reactions occur in a sequence of steps, however. A multiple-step mechanism was proposed for the iodine-catalyzed 2-butene isomerization reaction. Another example of a reaction that occurs in several steps is the reaction of bromine and NO:

$$Br_2(g) + 2\ NO \rightarrow 2\ BrNO(g)$$

A single-step reaction would require that three reactant molecules collide simultaneously in just the right orientation. The probability of this occurring is small; thus, it would be reasonable to look for a mechanism that occurs in a series of steps, with each step involving only one or two molecules. In one possible mechanism, Br_2 and NO might combine in an initial step to produce an intermediate species,

■ **Rate Laws and Mechanisms** Rate laws are derived by experiment; they are macroscopic observations. Mechanisms are schemes we propose that speculate on how reactions occur at the particulate level.

Within any living organism, there are untold numbers of chemical reactions occurring, many of them extremely rapidly. In many cases, enzymes, natural catalysts, speed up reactions that would normally move at a snail's pace from reactants to products. Typically, enzymes give reaction rates 10^7 to 10^{14} times faster than the uncatalyzed rate.

Enzymes are typically large proteins, often containing metal ions such as Zn^{2+}. They are thought to function by bringing the reactants together in just the right orientation in a site where specific bonds can be broken and/or made.

Carbonic anhydrase is one of many enzymes important in biological processes (Figure A). Carbon dioxide dissolves in water to a small extent to produce carbonic acid, which ionizes to give H_3O^+ and HCO_3^- ions.

$$CO_2(g) \rightleftharpoons CO_2(aq) \quad (1)$$

$$CO_2(aq) + H_2O(\ell) \rightleftharpoons H_2CO_3(aq) \quad (2)$$

$$H_2CO_3(aq) + H_2O(\ell) \rightleftharpoons H_3O^+(aq) + HCO_3^-(aq) \quad (3)$$

Carbonic anhydrase speeds up reactions 1 and 2. Many of the H_3O^+ ions produced by ionization of H_2CO_3 (reaction 3) are picked up by hemoglobin in the blood as hemoglobin loses O_2. The resulting HCO_3^- ions are transported back to the lungs. When hemoglobin again takes on O_2, it releases H_3O^+ ions. These ions and HCO_3^- re-form H_2CO_3, from which CO_2 is liberated and exhaled.

You can do an experiment that illustrates the effect of carbonic anhydrase. First, add a small amount of NaOH to a cold, aqueous solution of CO_2. The solution becomes basic immediately because there is not enough H_2CO_3 in the solution to use up the NaOH. After some seconds, however, dissolved CO_2 slowly produces more H_2CO_3, which consumes NaOH, and the solution is again acidic.

Now try the experiment again, this time adding a few drops of blood to the solution (Figure A). Carbonic anhydrase in blood speeds up reactions 1 and 2 by a factor of about 10^7, as evidenced by the more rapid reaction under these conditions.

In 1913, Leonor Michaelis and Maud L. Menten proposed a general theory of enzyme action based on kinetic observations. They assumed that the substrate, S (the reactant), and the enzyme, E, form a complex, ES. This complex then breaks down, releasing the enzyme and the product, P.

$$E + S \rightleftharpoons ES$$

$$ES \rightarrow P$$

FIGURE A CO_2 in water. (a) A few drops of blood are added to a cold solution of CO_2 in water. (b) A few drops of a dye (bromthymol blue) are added to the solution, the yellow color indicating an acidic solution. (c and d) A less-than-stoichiometric amount of sodium hydroxide is added, converting the H_2CO_3 to HCO_3^- (and CO_3^{2-}). The blue color of the dye indicates a basic solution. (e) The blue color begins to fade after some seconds as CO_2 forms more H_2CO_3. The amount of H_2CO_3 formed is finally sufficient to consume the added NaOH, and the solution is again acidic. Blood is a source of the enzyme carbonic anhydrase, so the last steps are noticeably more rapid than the reaction in the absence of blood.

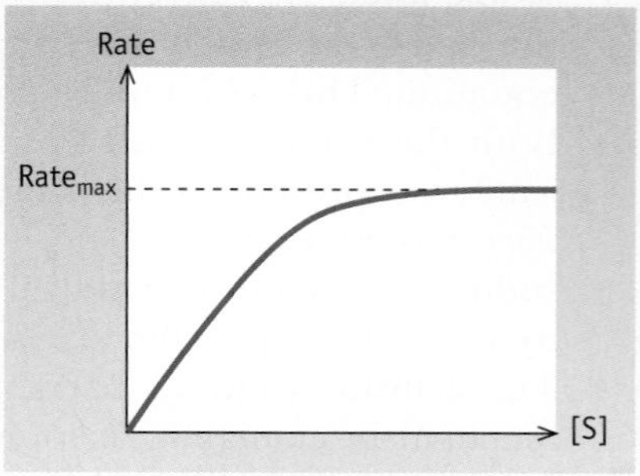

FIGURE B Rate of enzyme catalyzed reaction. This plot of substrate concentration [S] versus reaction velocity is typical of reactions catalyzed by enzymes and following the Michaelis–Menten model.

When the substrate concentration is low, the rate of the reaction is first order in S (Figure B). As [S] increases, however, the active sites in the enzyme become saturated with substrate, and the rate reaches its maximum value. Now the kinetics are zero order in substrate.

Questions:

1. *Catalase can decompose hydrogen peroxide to O_2 and water about 10^7 times faster than the uncatalyzed reaction. If the latter requires one year, how much time is required by the enzyme-catalyzed reaction?*
2. *According to the Michaelis–Menten model, if 1/Rate is plotted versus 1/[S], the intercept of the plot (when 1/[S] = 0) is $1/Rate_{max}$. Find $Rate_{max}$ for a reaction involving carbonic anhydrase.*

[S], mol/L	Rate (millimoles/min)
2.500	0.588
1.00	0.500
0.714	0.417
0.526	0.370
0.250	0.256

Answers to these questions are in Appendix Q.

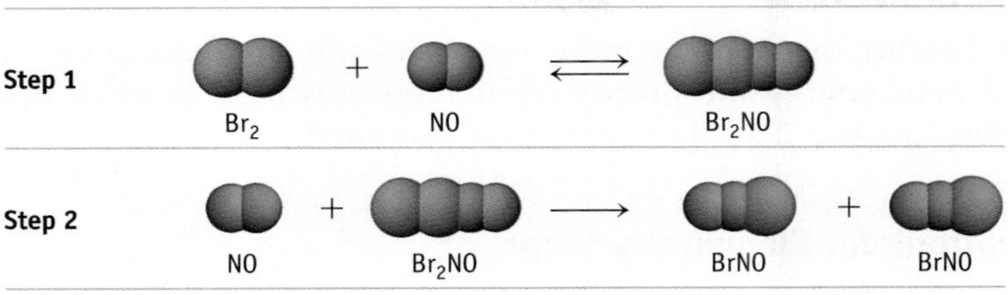

FIGURE 15.17 A reaction mechanism. A representation of the proposed two-step mechanism by which NO and Br_2 are converted to NOBr.

Br_2NO (Figure 15.17). This intermediate would then react with another NO molecule to give the reaction products. The equation for the overall reaction is obtained by adding the equations for these two steps:

Step 1.	$Br_2(g) + NO(g) \rightleftharpoons Br_2NO(g)$
Step 2.	$Br_2NO(g) + NO(g) \rightarrow 2\ BrNO(g)$
Overall Reaction:	$Br_2(g) + 2\ NO(g) \rightarrow 2\ BrNO(g)$

Each step in a multistep reaction sequence is an **elementary step**, defined by a chemical equation that describes a single molecular event such as the formation or rupture of a chemical bond resulting from a molecular collision. Each step has its own activation energy, E_a, and rate constant, k. Adding the equations for each step must give the balanced equation for the overall reaction, and the time required to complete all of the steps defines the overall reaction rate. The series of steps constitutes a possible reaction mechanism.

Mechanisms of reactions are usually postulated starting with experimental data. To see how this is done, we first describe three types of elementary steps in terms of the concept of *molecularity*.

Molecularity of Elementary Steps

Elementary steps are classified by the number of reactant molecules (or ions, atoms, or free radicals) that come together. This whole, positive number is called the **molecularity** of the elementary step. When one molecule is the only reactant in an elementary step, the reaction is a **unimolecular** process. A **bimolecular** elementary process involves two molecules, which may be identical ($A + A \rightarrow$ products) or different ($A + B \rightarrow$ products). The mechanism proposed for the decomposition of ozone in the stratosphere illustrates the use of these terms.

Step 1.	Unimolecular	$O_3(g) \rightarrow O_2(g) + O(g)$
Step 2.	Bimolecular	$O_3(g) + O(g) \rightarrow 2\ O_2(g)$
Overall Reaction:		$2\ O_3(g) \rightarrow 3\ O_2(g)$

A **termolecular** elementary step involves three molecules, which could be the same or different ($3\,A \rightarrow$products; $2\,A + B \rightarrow$ products; or $A + B + C \rightarrow$ products). Be aware, however, the simultaneous collision of three molecules has a low probability, unless one of the molecules involved is in high concentration, such as a solvent molecule. In fact, most termolecular processes involve the collision of two reactant molecules and a third, inert molecule. The function of the inert molecule is to absorb the excess energy produced when a new chemical bond is formed by the first two molecules. For example, N_2 is unchanged in a termolecular reaction between oxygen molecules and oxygen atoms that produces ozone in the upper atmosphere:

$$O(g) + O_2(g) + N_2(g) \rightarrow O_3(g) + \text{energetic } N_2(g)$$

The probability that four or more molecules will simultaneously collide with sufficient kinetic energy and proper orientation to react is so small that reaction molecularities greater than three are never proposed.

Rate Equations for Elementary Steps

The experimentally determined rate equation for a reaction cannot be predicted from its overall stoichiometry. In contrast, the rate equation for any elementary step is defined by the reaction stoichiometry. The rate equation of an elementary step is given by the product of the rate constant and the concentrations of the reactants in that step. We can therefore write the rate equation for any elementary step, as shown by examples in the following table:

Elementary Step	Molecularity	Rate Equation
$A \rightarrow$ product	unimolecular	Rate $= k[A]$
$A + B \rightarrow$ product	bimolecular	Rate $= k[A][B]$
$A + A \rightarrow$ product	bimolecular	Rate $= k[A]^2$
$2\ A + B \rightarrow$ product	termolecular	Rate $= k[A]^2[B]$

For example, the rate laws for each of the two steps in the decomposition of ozone are

$$\text{Rate for (unimolecular) Step 1} = k[O_3]$$

$$\text{Rate for (bimolecular) Step 2} = k'[O_3][O]$$

When a reaction mechanism consists of two elementary steps, the two steps will likely occur at different rates. The two rate constants (k and k' in this example) are not expected to have the same value (nor the same units, if the two steps have different molecularities).

Molecularity and Reaction Order

A unimolecular elementary step must be first order; a bimolecular elementary step must be second order; and a termolecular elementary step must be third order. Such a direct relation between molecularity and order is emphatically not true for a multistep reaction. If you learn from an experiment that a reaction is first order, you cannot conclude that it occurs in a single, unimolecular elementary step. Similarly, a second-order rate equation does not imply that the reaction occurs in a single, bimolecular elementary step. An illustration of this is the decomposition of N_2O_5:

$$2\ N_2O_5(g) \rightarrow 4\ NO_2(g) + O_2(g)$$

Here, the rate law is "Rate $= k[N_2O_5]$," but chemists are fairly certain the mechanism involves a series of unimolecular and bimolecular steps.

To see how the experimentally observed rate equation for the overall reaction is connected with a possible mechanism or sequence of elementary steps requires some chemical intuition. We will provide only a glimpse of the subject in the next section.

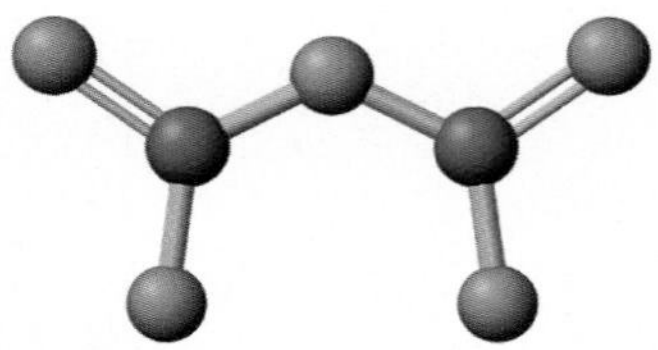

N_2O_5 Decomposition. The first step in the decomposition of N_2O_5 is thought to be the cleavage of one of the N—O bonds in the N—O—N link to give the odd-electron molecules NO_2 and NO_3. These react further to give the final products.

Chemistry Now™

Sign in at **www.thomsonedu.com/login** and go to Chapter 15 Contents to see:

- Screen 15.12 for exercises on **reaction mechanisms**
- Screen 15.13 for exercises on **reaction mechanisms**

EXAMPLE 15.12 Elementary Steps

Problem The hypochlorite ion undergoes self-oxidation–reduction to give chlorate, ClO_3^-, and chloride ions.

$$3\ ClO^-(aq) \rightarrow ClO_3^-(aq) + 2\ Cl^-(aq)$$

This reaction is thought to occur in two steps:

Step 1: $ClO^-(aq) + ClO^-(aq) \rightarrow ClO_2^-(aq) + Cl^-(aq)$

Step 2: $ClO_2^-(aq) + ClO^-(aq) \rightarrow ClO_3^-(aq) + Cl^-(aq)$

What is the molecularity of each step? Write the rate equation for each reaction step. Show that the sum of these reactions gives the equation for the net reaction.

Strategy The molecularity is the number of ions or molecules involved in a reaction step. The rate equation involves the concentration of each ion or molecule in an elementary step, raised to the power of its stoichiometric coefficient.

Solution Because two ions are involved in each elementary step, each step is bimolecular. The rate equation for any elementary step involves the product of the concentrations of the reactants. Thus, in this case, the rate equations are

Step 1: $\text{Rate} = k[ClO^-]^2$

Step 2: $\text{Rate} = k'[ClO_2^-][ClO^-]$

From the equations for the two elementary steps, we see that the ClO_2^- ion is an intermediate, a product of the first step and a reactant in the second step. It therefore cancels out, and we are left with the stoichiometric equation for the overall reaction:

Step 1: $ClO^-(aq) + ClO^-(aq) \rightarrow ClO_2^-(aq) + Cl^-(aq)$

Step 2: $ClO_2^-(aq) + ClO^-(aq) \rightarrow ClO_3^-(aq) + Cl^-(aq)$

Sum of steps: $3\ ClO^-(aq) \rightarrow ClO_3^-(aq) + 2\ Cl^-(aq)$

EXERCISE 15.11 Elementary Steps

Nitrogen monoxide is reduced by hydrogen to give nitrogen and water:

$$2\ NO(g) + 2\ H_2(g) \rightarrow N_2(g) + 2\ H_2O(g)$$

One possible mechanism for this reaction is

$$2\ NO(g) \rightarrow N_2O_2(g)$$

$$N_2O_2(g) + H_2(g) \rightarrow N_2O(g) + H_2O(g)$$

$$N_2O(g) + H_2(g) \rightarrow N_2(g) + H_2O(g)$$

What is the molecularity of each of the three steps? What is the rate equation for the third step? Identify the intermediates in this reaction; how many different intermediates are there? Show that the sum of these elementary steps gives the equation for the overall reaction.

Reaction Mechanisms and Rate Equations

The dependence of rate on concentration is an experimental fact. Mechanisms, by contrast, are constructs of our imagination, intuition, and good "chemical sense." To describe a mechanism, we need to make a guess (a good guess, we hope) about how the reaction occurs at the particulate level. Several mechanisms can always be proposed that correspond to the observed rate equation, and a postulated mechanism is often wrong. A good mechanism is a worthy goal because it allows us to understand the chemistry better. A practical consequence of a good mechanism is that it allows us to predict, for example, how to control a reaction better and how to design new experiments.

One of the important guidelines of kinetics is that *products of a reaction can never be produced at a rate faster than the rate of the slowest step.* If one step in a multistep reaction is slower than the others, then *the rate of the overall reaction is limited by the combined rates of all elementary steps up through the slowest step in the mechanism.* Often the overall reaction rate and the rate of the slow step are nearly the same. If the slow step determines the rate of the reaction, it is called the **rate-determining step,** or rate-limiting step.

Imagine that a reaction takes place with a mechanism involving two sequential steps, and assume that we know the rates of both steps. The first step is slow and the second is fast:

■ **Can You Derive a Mechanism?** At this introductory level, you cannot be expected to derive reaction mechanisms. Given a mechanism, however, you can decide whether it agrees with experimental rate laws.

Elementary Step 1 $\quad A + B \xrightarrow[\text{Slow, } E_a \text{ large}]{k_1} X + M$

Elementary Step 2 $\quad M + A \xrightarrow[\text{Fast, } E_a \text{ small}]{k_2} Y$

Overall Reaction $\quad 2A + B \longrightarrow X + Y$

In the first step, A and B come together and slowly react to form one of the products (X) plus another reactive species, M. Almost as soon as M is formed, however, it is rapidly consumed by reacting with another molecule of A to form the second product Y. The rate-determining elementary step in this example is the first step. That is, the rate of the first step is equal to the rate of the overall reaction. This step is bimolecular and so has the rate equation

$$\text{Rate} = k_1[\text{A}][\text{B}]$$

where k_1 is the rate constant for that step. The overall reaction is expected to have this same second-order rate equation.

Let us apply these ideas to the mechanism of a real reaction. Consider the reaction of nitrogen dioxide with fluorine which has a second-order rate equation:

Overall Reaction $\quad 2\ NO_2(g) + F_2(g) \rightarrow 2\ FNO_2(g)$

$$\text{Rate} = k[NO_2][F_2]$$

The rate equation immediately rules out the possibility that the reaction occurs in a single step. If the equation for the reaction represented an elementary step, the rate law would have a second-order dependence on $[NO_2]$. Because a single-step reaction is ruled out, the mechanism must include at least two steps. We can also conclude from the rate law that the rate-determining elementary step must involve NO_2 and F_2 in a 1:1 ratio. One possible mechanism proposes that molecules of NO_2 and F_2 first react to produce one molecule of the product (FNO_2) plus one F atom. In a second step, the fluorine atom produced in the first step reacts with additional NO_2 to give a second molecule of product. If the first, bimolecular step is rate determining, the rate equation would be "Rate $= k_1[NO_2][F_2]$," the same as the experimentally observed rate equation. The experimental rate constant would be the same as k_1.

Elementary Step 1 Slow $\quad NO_2(g) + F_2(g) \xrightarrow{k_1} FNO_2(g) + F(g)$

Elementary Step 2 Fast $\quad NO_2(g) + F(g) \xrightarrow{k_2} FNO_2(g)$

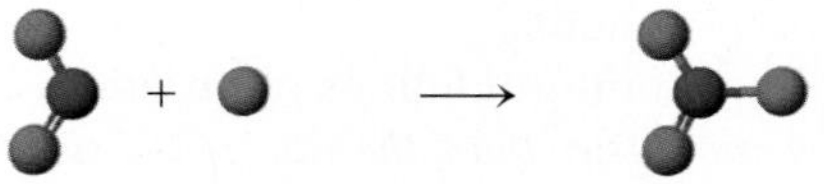

Overall Reaction $\quad 2\ NO_2(g) + F_2(g) \longrightarrow 2\ FNO_2(g)$

The fluorine atom formed in the first step of the NO_2/F_2 reaction is a reaction intermediate. It does not appear in the equation describing the overall reaction. Reaction intermediates usually have only a fleeting existence, but occasionally they have long enough lifetimes to be observed. The detection and identification of an intermediate are strong evidence for the proposed mechanism.

EXAMPLE 15.13 Elementary Steps and Reaction Mechanisms

Problem Oxygen atom transfer from NO_2 to CO produces nitrogen monoxide and carbon dioxide (Figure 15.13):

$$NO_2(g) + CO(g) \rightarrow NO(g) + CO_2(g)$$

The rate equation for this reaction at temperatures less than 500 K is:

$$\text{Rate} = k[NO_2]^2$$

Can this reaction occur in one bimolecular step?

Strategy Write the rate law based on the equation for the NO_2 + CO reaction occurring as if it were an elementary step. If this rate law corresponds to the observed rate law, then a one-step mechanism is possible.

Solution If the reaction occurs by the collision of one NO_2 molecule with one CO molecule, the rate equation would be

$$\text{Rate} = k[NO_2][CO]$$

This does not agree with experiment, so the mechanism must involve more than a single step. In one possible mechanism, the reaction occurs in two, bimolecular steps, the first one slow and the second one fast:

Elementary Step 1 Slow, rate-determining $2\ NO_2(g) \longrightarrow NO_3(g) + NO(g)$

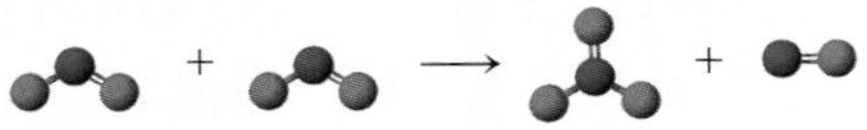

Elementary Step 2 Fast $NO_3(g) + CO(g) \longrightarrow NO_2(g) + CO_2(g)$

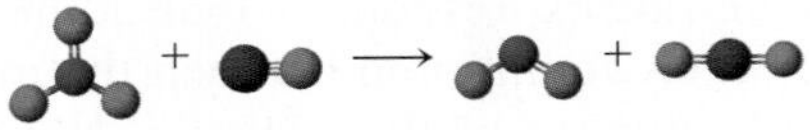

Overall Reaction $NO_2(g) + CO(g) \longrightarrow NO(g) + CO_2(g)$

The first (rate-determining) step has a rate equation that agrees with experiment, so this is a possible mechanism.

EXERCISE 15.12 Elementary Steps and Reaction Mechanisms

The Raschig reaction produces hydrazine, N_2H_4, an industrially important reducing agent, from NH_3 and OCl^- in basic, aqueous solution. A proposed mechanism is

Step 1 Fast $NH_3(aq) + OCl^-(aq) \rightarrow NH_2Cl(aq) + OH^-(aq)$

Step 2 Slow $NH_2Cl(aq) + NH_3(aq) \rightarrow N_2H_5^+(aq) + Cl^-(aq)$

Step 3 Fast $N_2H_5^+(aq) + OH^-(aq) \rightarrow N_2H_4(aq) + H_2O(\ell)$

(a) What is the overall stoichiometric equation?

(b) Which step of the three is rate determining?

(c) Write the rate equation for the rate-determining elementary step.

(d) What reaction intermediates are involved?

A common two-step reaction mechanism involves an initial fast reaction that produces an intermediate, followed by a slower second step in which the intermediate is converted to the final product. The rate of the reaction is determined by the second step, for which a rate law can be written. The rate of that step, however, depends on the concentration of the intermediate. Remember, though, that the rate law must be written with respect to the reactants only. An intermediate, whose concentration will probably not be measurable, cannot appear as a term in the overall rate equation.

The reaction of nitrogen monoxide and oxygen is an example of a two-step reaction where the first step is fast and the second step is rate determining.

$$2\ NO(g) + O_2(g) \rightarrow 2\ NO_2(g)$$

$$\text{Rate} = k[NO]^2[O_2]$$

The experimentally determined rate law shows second-order dependence on NO and first-order dependence on O_2. Although this rate law would be correct for a termolecular reaction, experimental evidence indicates that an intermediate is formed in this reaction. A possible two-step mechanism that proceeds through an intermediate is

Elementary Step 1:	Fast, Equilibrium	$NO(g) + O_2(g) \underset{k_{-1}}{\overset{k_1}{\rightleftharpoons}} OONO(g)$ intermediate
Elementary Step 2:	Slow, Rate-determining	$NO(g) + OONO(g) \xrightarrow{k_2} 2\ NO_2(g)$
Overall Reaction		$2\ NO(g) + O_2(g) \rightarrow 2\ NO_2(g)$

The second step of this reaction is the slow step, and the overall rate depends on it. We can write a rate law for the second step:

$$\text{Rate} = k_2[NO][OONO]$$

This rate law cannot be compared directly with the experimental rate law because it contains the concentration of an intermediate, OONO. To eliminate the intermediate from this rate expression, we look at the rapid first step in this reaction sequence that involves an equilibrium between the intermediate species and the reactants.

At the beginning of the reaction, NO and O_2 react rapidly and produce the intermediate OONO. The rate of formation can be defined by a rate law with a rate constant k_1:

$$\text{Rate of production of OONO} = k_1[NO][O_2]$$

Because the intermediate is consumed only very slowly in the second step, it is possible for the OONO to revert to NO and O_2 before it reacts further:

$$\text{Rate of reverse reaction } (OONO \rightarrow NO + O_2) = k_{-1}[OONO]$$

As NO and O_2 form OONO, their concentrations drop, so the rate of the forward reaction decreases. At the same time, the concentration of OONO builds up, so the rate of the reverse reaction increases. At equilibrium, the rates of the forward and reverse reactions become the same.

$$\text{Rate of forward reaction} = \text{rate of reverse reaction}$$

$$k_1[NO][O_2] = k_{-1}[OONO]$$

Rearranging this equation, we find

$$\frac{k_1}{k_{-1}} = \frac{[N_2O_2]}{[NO][O_2]} = K$$

Problem Solving Tip 15.1 Relating Rate Equations and Reaction Mechanisms

The connection between an experimental rate equation and the proposed reaction mechanism is important in chemistry.

1. Experiments must first be performed to determine the experimental rate equation.
2. A mechanism for the reaction is proposed on the basis of the experimental rate equation, the principles of stoichiometry and molecular structure and bonding, general chemical experience, and intuition.
3. The proposed reaction mechanism is used to derive a rate equation. This rate equation can contain only those species present in the overall chemical reaction. If the derived and experimental rate equations are the same, the postulated mechanism *may* be a reasonable hypothesis of the reaction sequence.
4. If more than one mechanism can be proposed, and they all predict derived rate equations in agreement with experiment, then more experiments must be done.

Both k_1 and k_{-1} are constants (they will change only if the temperature changes). We can define a new constant K equal to the ratio of these two constants and called the **equilibrium constant**, which is equal to the quotient [OONO]/[NO][O_2]. From this, we can derive an expression for the concentration of OONO:

$$[\text{OONO}] = K[\text{NO}][\text{O}_2]$$

If K[NO][O_2] is substituted for [OONO] in the rate law for the rate-determining elementary step, we have

$$\text{Rate} = k_2[\text{NO}][\text{OONO}] = k_2[\text{NO}]\{K[\text{NO}][\text{O}_2]\}$$
$$= k_2K[\text{NO}]^2[\text{O}_2]$$

Because both k_2 and K are constants, their product is another constant k', and we have

$$\text{Rate} = k'[\text{NO}]^2[\text{O}_2]$$

This is exactly the rate law derived from experiment. Thus, the sequence of reactions on which the rate law is based may be a reasonable mechanism for this reaction. It is not the only possible mechanism, however. This rate equation is also consistent with the reaction occurring in a single termolecular step. Another possible mechanism is illustrated in Example 15.14.

■ **Equilibrium Constant** The important concept of chemical equilibrium was introduced in Chapter 3 and will be described in more detail in Chapters 16–19.

■ **Mechanisms with an Initial Equilibrium** In this mechanism, the forward and reverse reactions in the first elementary step are so much faster than the second elementary step that equilibrium is established before any significant amount of OONO is consumed by NO to give NO_2. The state of equilibrium for the first step remains throughout the lifetime of the overall reaction.

EXAMPLE 15.14 Reaction Mechanism Involving an Equilibrium Step

Problem The NO + O_2 reaction described in the text could also occur by the following mechanism:

Elementary Step 1: Fast, equilibrium

$$\text{NO(g)} + \text{NO(g)} \underset{k_{-1}}{\overset{k_1}{\rightleftharpoons}} \underset{\text{intermediate}}{\text{N}_2\text{O}_2\text{(g)}}$$

Elementary Step 2: Slow, rate-determining

$$\text{N}_2\text{O}_2\text{(g)} + \text{O}_2\text{(g)} \xrightarrow{k_2} 2\ \text{NO}_2\text{(g)}$$

Overall Reaction: 2 NO(g) + O_2(g) ⟶ 2 NO_2(g)

Show that this mechanism leads to the following experimental rate law: Rate = $k[\text{NO}]^2[\text{O}_2]$.

Strategy The rate law for the rate-determining elementary step is

$$\text{Rate} = k_2[\text{N}_2\text{O}_2][\text{O}_2]$$

The intermediate N_2O_2 cannot appear in the final derived rate law. To obtain the rate law, we use the equilibrium constant expression for the first step.

Solution $[N_2O_2]$ and $[NO]$ are related by the equilibrium constant.

$$\frac{k_1}{k_{-1}} = \frac{[N_2O_2]}{[NO]^2} = K$$

Solving this equation for $[N_2O_2]$ gives $[N_2O_2] = K[NO]^2$. When this is substituted into the derived rate law

$$\text{Rate} = k_2\{K[NO_2]^2\}[O_2]$$

the resulting equation is identical with the experimental rate law where $k_2K = k$.

Comment Three mechanisms have been proposed for the $NO + O_2$ reaction. The challenge for chemists is to decide which is correct. In this case, further experimentation detected the species OONO as a short-lived intermediate, confirming the mechanism involving this intermediate.

EXERCISE 15.13 Reaction Mechanism Involving a Fast Initial Step

One possible mechanism for the decomposition of nitryl chloride, NO_2Cl, is

Elementary Step 1: Fast, Equilibrium $$NO_2Cl(g) \underset{k_{-1}}{\overset{k_1}{\rightleftharpoons}} NO_2(g) + Cl(g)$$

Elementary Step 2: Slow $$NO_2Cl(g) + Cl(g) \xrightarrow{k_2} NO_2(g) + Cl_2(g)$$

What is the overall reaction? What rate law would be derived from this mechanism? What effect does increasing the concentration of the product NO_2 have on the reaction rate?

Chapter Goals Revisited

Chemistry Now™ Sign in at **www.thomsonedu.com/login** to:

- Assess your understanding with Study Questions in OWL keyed to each goal in the Goals and Homework menu for this chapter
- For quick review, download Go Chemistry mini-lecture flashcard modules (or purchase them at **www.ichapters.com**)
- Check your readiness for an exam by taking the Pre-Test and exploring the modules recommended in your Personalized Study plan.

Access **How Do I Solve It?** tutorials on how to approach problem solving using concepts in this chapter.

Now that you have studied this chapter, you should ask whether you have met the chapter goals. In particular, you should be able to:

Understand rates of reaction and the conditions affecting rates

a. Explain the concept of reaction rate (Section 15.1).

b. Derive the average and instantaneous rates of a reaction from concentration–time data (Section 15.1). **Study Question(s) assignable in OWL: 5.**

c. Describe factors that affect reaction rate (i.e., reactant concentrations, temperature, presence of a catalyst, and the state of the reactants) (Section 15.2). **Study Question(s) assignable in OWL: 8, 10, 62, 76, 77, 81.**

Derive the rate equation, rate constant, and reaction order from experimental data

a. Define the various parts of a rate equation (the rate constant and order of reaction), and understand their significance (Section 15.3). **Study Question(s) assignable in OWL: 12, 14, 68.**

b. Derive a rate equation from experimental information (Section 15.3). **Study Question(s) assignable in OWL: 12, 14, 48, 56, 58, 71.**

Use integrated rate laws

a. Describe and use the relationships between reactant concentration and time for zero-order, first-order, and second-order reactions (Section 15.4 and Table 15.1). **Study Question(s) assignable in OWL: 16, 18, 19, 20, 24; Go Chemistry Module 20.**

b. Apply graphical methods for determining reaction order and the rate constant from experimental data (Section 15.4 and Table 15.1). **Study Question(s) assignable in OWL: 28, 30.**

c. Use the concept of half-life ($t_{1/2}$), especially for first-order reactions (Section 15.4). **Study Question(s) assignable in OWL: 22, 24, 26, 60, 71.**

Understand the collision theory of reaction rates and the role of activation energy

a. Describe the collision theory of reaction rates (Section 15.5).

b. Relate activation energy (E_a) to the rate and thermodynamics of a reaction (Section 15.5). **Study Question(s) assignable in OWL: 36, 69, 83.**

c. Use collision theory to describe the effect of reactant concentration on reaction rate (Section 15.5).

d. Understand the effect of molecular orientation on reaction rate (Section 15.5).

e. Describe the effect of temperature on reaction rate using the collision theory of reaction rates and the Arrhenius equation (Equation 15.7 and Section 15.5).

f. Use Equations 15.5, 15.6, and 15.7 to calculate the activation energy from rate constants at different temperatures (Section 15.5).

Relate reaction mechanisms and rate laws

a. Describe the functioning of a catalyst and its effect on the activation energy and mechanism of a reaction (Section 15.5).

b. Understand reaction coordinate diagrams (Section 15.5).

c. Understand the concept of a reaction mechanism (a proposed sequence of bond-making and bond-breaking steps that occurs during the conversion of reactants to products) and the relation of the mechanism to the overall, stoichiometric equation for a reaction (Section 15.6).

d. Describe the elementary steps of a mechanism, and give their molecularity (Section 15.6). **Study Question(s) assignable in OWL: 40, 42, 44, 74.**

e. Define the rate-determining step in a mechanism, and identify any reaction intermediates (Section 15.6). **Study Question(s) assignable in OWL: 44, 69, 70, 78, 80.**

KEY EQUATIONS

Equation 15.1 (page 683) Integrated rate equation for a first-order reaction (in which $-\Delta[R]/\Delta t = k[R]$).

$$\ln\frac{[R]_t}{[R]_0} = -kt$$

Here, $[R]_0$ and $[R]_t$ are concentrations of the reactant at time $t = 0$ and at a later time, t. The ratio of concentrations, $[R]_t/[R]_0$, is the fraction of reactant that remains after a given time has elapsed.

Equation 15.2 (page 686) Integrated rate equation for a second-order reaction (in which $-\Delta[R]/\Delta t = k[R]^2$).

$$\frac{1}{[R]_t} - \frac{1}{[R]_0} = kt$$

Equation 15.3 (page 687) Integrated rate equation for a zero-order reaction (in which $-\Delta[R]/\Delta t = k[R]^0$).

$$[R]_0 - [R]_t = kt$$

Equation 15.4 (page 690) The relation between the half-life ($t_{1/2}$) and the rate constant (k) for a first-order reaction.

$$t_{1/2} = \frac{0.693}{k}$$

Equation 15.5 (page 696) Arrhenius equation in exponential form

$$k = \text{rate constant} = Ae^{-E_a/RT}$$

Frequency factor (A)

Fraction of molecules with minimum energy for reaction ($e^{-E_a/RT}$)

A is the frequency factor; E_a is the activation energy; T is the temperature (in kelvins); and R is the gas constant ($= 8.314510 \times 10^{-3}$ kJ/K · mol).

Equation 15.6 (page 696) Expanded Arrhenius equation in logarithmic form.

$$\ln k = -\frac{E_a}{R}\left(\frac{1}{T}\right) + \ln A \longleftarrow \text{Arrhenius equation}$$

$$\downarrow \quad\quad \downarrow \quad\quad \downarrow$$

$$y = mx + b \longleftarrow \text{Equation for straight line}$$

Equation 15.7 (page 698) A version of the Arrhenius equation used to calculate the activation energy for a reaction when you know the values of the rate constant at two temperatures (in kelvins).

$$\ln k_2 - \ln k_1 = \ln \frac{k_2}{k_1} = -\frac{E_a}{R}\left[\frac{1}{T_2} - \frac{1}{T_1}\right]$$

STUDY QUESTIONS

Online homework for this chapter may be assigned in OWL.

▲ denotes challenging questions.

■ denotes questions assignable in OWL.

Blue-numbered questions have answers in Appendix O and fully-worked solutions in the *Student Solutions Manual.*

Practicing Skills

Reaction Rates

(See Examples 15.1–15.2, Exercises 15.1–15.2, and ChemistryNow Screen 15.2.)

1. Give the relative rates of disappearance of reactants and formation of products for each of the following reactions.
   (a) $2\ O_3(g) \longrightarrow 3\ O_2(g)$
   (b) $2\ HOF(g) \longrightarrow 2\ HF(g) + O_2(g)$

2. Give the relative rates of disappearance of reactants and formation of products for each of the following reactions.
   (a) $2\ NO(g) + Br_2(g) \longrightarrow 2\ NOBr(g)$
   (b) $N_2(g) + 3\ H_2(g) \longrightarrow 2\ NH_3(g)$

3. In the reaction $2\ O_3(g) \longrightarrow 3\ O_2(g)$, the rate of formation of O_2 is 1.5×10^{-3} mol/L · s. What is the rate of decomposition of O_3?

4. In the synthesis of ammonia, if $-\Delta[H_2]/\Delta t = 4.5 \times 10^{-4}$ mol/L·min, what is $\Delta[NH_3]/\Delta t$?

$$N_2(g) + 3\ H_2(g) \rightarrow 2\ NH_3(g)$$

5. ■ Experimental data are listed here for the reaction $A \longrightarrow 2\ B$.

Time (s)	[B] (mol/L)
0.00	0.000
10.0	0.326
20.0	0.572
30.0	0.750
40.0	0.890

   (a) Prepare a graph from these data; connect the points with a smooth line; and calculate the rate of change of [B] for each 10-s interval from 0.0 to 40.0 s. Does the rate of change decrease from one time interval to the next? Suggest a reason for this result.
   (b) How is the rate of change of [A] related to the rate of change of [B] in each time interval? Calculate the rate of change of [A] for the time interval from 10.0 to 20.0 s.
   (c) What is the instantaneous rate, $\Delta[B]/\Delta t$, when $[B] = 0.750$ mol/L?

6. Phenyl acetate, an ester, reacts with water according to the equation

$$\underset{\text{phenyl acetate}}{CH_3\overset{O}{\overset{\|}{C}}OC_6H_5} + H_2O \longrightarrow \underset{\text{acetic acid}}{CH_3\overset{O}{\overset{\|}{C}}OH} + \underset{\text{phenol}}{C_6H_5OH}$$

The data in the table were collected for this reaction at 5 °C.

Time (s)	[Phenyl acetate] (mol/L)
0	0.55
15.0	0.42
30.0	0.31
45.0	0.23
60.0	0.17
75.0	0.12
90.0	0.085

(a) Plot the phenyl acetate concentration versus time, and describe the shape of the curve observed.
(b) Calculate the rate of change of the phenyl acetate concentration during the period 15.0 s to 30.0 s and also during the period 75.0 s to 90.0 s. Why is one value smaller than the other?
(c) What is the rate of change of the phenol concentration during the time period 60.0 s to 75.0 s?
(d) What is the instantaneous rate at 15.0 s?

Concentration and Rate Equations

(See Examples 15.3–15.4, Exercises 15.3–15.4, and ChemistryNow Screens 15.4 and 15.5.)

7. Using the rate equation "Rate = $k[A]^2[B]$," define the order of the reaction with respect to A and B. What is the total order of the reaction?

8. ■ A reaction has the experimental rate equation Rate = $k[A]^2$. How will the rate change if the concentration of A is tripled? If the concentration of A is halved?

9. The reaction between ozone and nitrogen dioxide at 231 K is first order in both $[NO_2]$ and $[O_3]$.

$$2\ NO_2(g) + O_3(g) \rightarrow N_2O_5(s) + O_2(g)$$

(a) Write the rate equation for the reaction.
(b) If the concentration of NO_2 is tripled (and $[O_3]$ is not changed), what is the change in the reaction rate?
(c) What is the effect on reaction rate if the concentration of O_3 is halved (no change in $[NO_2]$)?

10. ■ Nitrosyl bromide, NOBr, is formed from NO and Br_2:

$$2\ NO(g) + Br_2(g) \rightarrow 2\ NOBr(g)$$

Experiments show that this reaction is second order in NO and first order in Br_2.
(a) Write the rate equation for the reaction.
(b) How does the initial reaction rate change if the concentration of Br_2 is changed from 0.0022 mol/L to 0.0066 mol/L?
(c) What is the change in the initial rate if the concentration of NO is changed from 0.0024 mol/L to 0.0012 mol/L?

11. The data in the table are for the reaction of NO and O_2 at 660 K.

$$2\ NO(g) + O_2(g) \rightarrow 2\ NO_2(g)$$

Reactant Concentration (mol/L) [NO]	[O₂]	Rate of Disappearance of NO (mol/L · s)
0.010	0.010	2.5×10^{-5}
0.020	0.010	1.0×10^{-4}
0.010	0.020	5.0×10^{-5}

(a) Determine the order of the reaction for each reactant.
(b) Write the rate equation for the reaction.
(c) Calculate the rate constant.
(d) Calculate the rate (in mol/L · s) at the instant when [NO] = 0.015 mol/L and $[O_2]$ = 0.0050 mol/L.
(e) At the instant when NO is reacting at the rate 1.0×10^{-4} mol/L · s, what is the rate at which O_2 is reacting and NO_2 is forming?

12. ■ The reaction

$$2\ NO(g) + 2\ H_2(g) \rightarrow N_2(g) + 2\ H_2O(g)$$

was studied at 904 °C, and the data in the table were collected.

Reactant Concentration (mol/L) [NO]	[H₂]	Rate of Appearance of N₂ (mol/L · s)
0.420	0.122	0.136
0.210	0.122	0.0339
0.210	0.244	0.0678
0.105	0.488	0.0339

(a) Determine the order of the reaction for each reactant.
(b) Write the rate equation for the reaction.
(c) Calculate the rate constant for the reaction.
(d) Find the rate of appearance of N_2 at the instant when [NO] = 0.350 mol/L and $[H_2]$ = 0.205 mol/L.

13. Data for the reaction $2\ NO(g) + O_2(g) \rightarrow 2\ NO_2(g)$ are given in the table.

Experiment	Concentration (mol/L) [NO]	[O$_2$]	Initial Rate (mol/L · h)
1	3.6×10^{-4}	5.2×10^{-3}	3.4×10^{-8}
2	3.6×10^{-4}	1.04×10^{-2}	6.8×10^{-8}
3	1.8×10^{-4}	1.04×10^{-2}	1.7×10^{-8}
4	1.8×10^{-4}	5.2×10^{-3}	?

(a) What is the rate law for this reaction?
(b) What is the rate constant for the reaction?
(c) What is the initial rate of the reaction in experiment 4?

14. ■ Data for the following reaction are given in the table below.

$$CO(g) + NO_2(g) \rightarrow CO_2(g) + NO(g)$$

Experiment	Concentration (mol/L) [CO]	[NO$_2$]	Initial Rate (mol/L · h)
1	5.0×10^{-4}	0.36×10^{-4}	3.4×10^{-8}
2	5.0×10^{-4}	0.18×10^{-4}	1.7×10^{-8}
3	1.0×10^{-3}	0.36×10^{-4}	6.8×10^{-8}
4	1.5×10^{-3}	0.72×10^{-4}	?

(a) What is the rate law for this reaction?
(b) What is the rate constant for the reaction?
(c) What is the initial rate of the reaction in experiment 4?

Concentration–Time Relationships

(See Examples 15.5–15.7, Exercises 15.5–15.7, and ChemistryNow Screen 15.6.)

15. The rate equation for the hydrolysis of sucrose to fructose and glucose

$$C_{12}H_{22}O_{11}(aq) + H_2O(\ell) \rightarrow 2\ C_6H_{12}O_6(aq)$$

is $-\Delta[\text{sucrose}]/\Delta t = k[C_{12}H_{22}O_{11}]$. After 27 min at 27 °C, the sucrose concentration decreased from 0.0146 M to 0.0132 M. Find the rate constant, k.

16. ■ The decomposition of N_2O_5 in CCl_4 is a first-order reaction. If 2.56 mg of N_2O_5 is present initially, and 2.50 mg is present after 4.26 min at 55 °C, what is the value of the rate constant, k?

17. The decomposition of SO_2Cl_2 is a first-order reaction:

$$SO_2Cl_2(g) \rightarrow SO_2(g) + Cl_2(g)$$

The rate constant for the reaction is $2.8 \times 10^{-3}\ \text{min}^{-1}$ at 600 K. If the initial concentration of SO_2Cl_2 is 1.24×10^{-3} mol/L, how long will it take for the concentration to drop to 0.31×10^{-3} mol/L?

18. ■ The conversion of cyclopropane to propene (see Example 15.5) occurs with a first-order rate constant of $2.42 \times 10^{-2}\ \text{h}^{-1}$. How long will it take for the concentration of cyclopropane to decrease from an initial concentration 0.080 mol/L to 0.020 mol/L?

19. ■ Hydrogen peroxide, $H_2O_2(aq)$, decomposes to $H_2O(\ell)$ and $O_2(g)$ in a reaction that is first order in H_2O_2 and has a rate constant $k = 1.06 \times 10^{-3}\ \text{min}^{-1}$ at a given temperature.
(a) How long will it take for 15% of a sample of H_2O_2 to decompose?
(b) How long will it take for 85% of the sample to decompose?

20. ■ The decomposition of nitrogen dioxide at a high temperature

$$NO_2(g) \rightarrow NO(g) + \tfrac{1}{2}\ O_2(g)$$

is second order in this reactant. The rate constant for this reaction is 3.40 L/mol · min. Determine the time needed for the concentration of NO_2 to decrease from 2.00 mol/L to 1.50 mol/L.

Half-Life

(See Examples 15.8 and 15.9, Exercise 15.9, and ChemistryNow Screen 15.8.)

21. The rate equation for the decomposition of N_2O_5 (giving NO_2 and O_2) is Rate $= k[N_2O_5]$. The value of k is $6.7 \times 10^{-5}\ \text{s}^{-1}$ for the reaction at a particular temperature.
(a) Calculate the half-life of N_2O_5.
(b) How long does it take for the N_2O_5 concentration to drop to one tenth of its original value?

22. ■ The decomposition of SO_2Cl_2

$$SO_2Cl_2(g) \rightarrow SO_2(g) + Cl_2(g)$$

is first order in SO_2Cl_2, and the reaction has a half-life of 245 min at 600 K. If you begin with 3.6×10^{-3} mol of SO_2Cl_2 in a 1.0-L flask, how long will it take for the amount of SO_2Cl_2 to decrease to 2.00×10^{-4} mol?

23. Gaseous azomethane, $CH_3N{=}NCH_3$, decomposes in a first-order reaction when heated:

$$CH_3N{=}NCH_3(g) \rightarrow N_2(g) + C_2H_6(g)$$

The rate constant for this reaction at 600 K is 0.0216 min^{-1}. If the initial quantity of azomethane in the flask is 2.00 g, how much remains after 0.0500 min? What quantity of N_2 is formed in this time?

24. ■ The compound $Xe(CF_3)_2$ decomposes in a first-order reaction to elemental Xe with a half-life of 30. min. If you place 7.50 mg of $Xe(CF_3)_2$ in a flask, how long must you wait until only 0.25 mg of $Xe(CF_3)_2$ remains?

25. The radioactive isotope ^{64}Cu is used in the form of copper(II) acetate to study Wilson's disease. The isotope has a half-life of 12.70 h. What fraction of radioactive copper(II) acetate remains after 64 h?

26. ■ Radioactive gold-198 is used in the diagnosis of liver problems. The half-life of this isotope is 2.7 days. If you begin with a 5.6-mg sample of the isotope, how much of this sample remains after 1.0 day?

▲ more challenging ■ in OWL Blue-numbered questions answered in Appendix O

Graphical Analysis: Rate Equations and *k*

(See Exercise 15.8 and ChemistryNow Screen 15.7.)

27. Data for the decomposition of dinitrogen oxide

$$2\ N_2O(g) \rightarrow 2\ N_2(g) + O_2(g)$$

on a gold surface at 900 °C are given below. Verify that the reaction is first order by preparing a graph of ln $[N_2O]$ versus time. Derive the rate constant from the slope of the line in this graph. Using the rate law and value of *k*, determine the decomposition rate at 900 °C when $[N_2O]$ = 0.035 mol/L.

Time (min)	$[N_2O]$ (mol/L)
15.0	0.0835
30.0	0.0680
80.0	0.0350
120.0	0.0220

28. ■ Ammonia decomposes when heated according to the equation

$$NH_3(g) \rightarrow NH_2(g) + H(g)$$

The data in the table for this reaction were collected at a high temperature.

Time (h)	$[NH_3]$ (mol/L)
0	8.00×10^{-7}
25	6.75×10^{-7}
50	5.84×10^{-7}
75	5.15×10^{-7}

Plot ln $[NH_3]$ versus time and $1/[NH_3]$ versus time. What is the order of this reaction with respect to NH_3? Find the rate constant for the reaction from the slope.

29. Gaseous NO_2 decomposes at 573 K.

$$2\ NO_2(g) \rightarrow 2\ NO(g) + O_2(g)$$

The concentration of NO_2 was measured as a function of time. A graph of $1/[NO_2]$ versus time gives a straight line with a slope of 1.1 L/mol · s. What is the rate law for this reaction? What is the rate constant?

30. ■ The decomposition of HOF occurs at 25 °C.

$$2\ HOF(g) \rightarrow 2\ HF(g) + O_2(g)$$

Using the data in the table below, determine the rate law, and then calculate the rate constant.

[HOF] (mol/L)	Time (min)
0.850	0
0.810	2.00
0.754	5.00
0.526	20.0
0.243	50.0

31. For the reaction $2\ C_2F_4 \rightarrow C_4F_8$, a graph of $1/[C_2F_4]$ versus time gives a straight line with a slope of +0.04 L/mol · s. What is the rate law for this reaction?

32. Butadiene, $C_4H_6(g)$, dimerizes when heated, forming 1,5-cyclooctadiene, C_8H_{12}. The data in the table were collected.

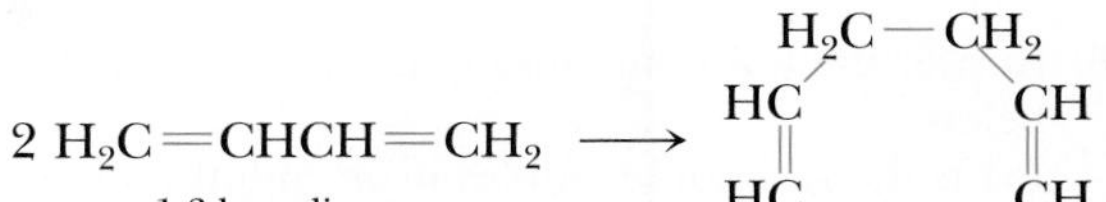

1,3-butadiene

1,5-cyclooctadiene

$[C_4H_6]$ (mol/L)	Time (s)
1.0×10^{-2}	0
8.7×10^{-3}	200.
7.7×10^{-3}	500.
6.9×10^{-3}	800.
5.8×10^{-3}	1200.

(a) Use a graphical method to verify that this is a second-order reaction.

(b) Calculate the rate constant for the reaction.

Kinetics and Energy

(See Examples 15.10 and 15.11, and ChemistryNow Screens 15.9 and 15.10.)

33. Calculate the activation energy, E_a, for the reaction

$$2\ N_2O_5(g) \rightarrow 4\ NO_2(g) + O_2(g)$$

from the observed rate constants: *k* at 25 °C = $3.46 \times 10^{-5}\ s^{-1}$ and *k* at 55 °C = $1.5 \times 10^{-3}\ s^{-1}$.

34. If the rate constant for a reaction triples when the temperature rises from 3.00×10^2 K to 3.10×10^2 K, what is the activation energy of the reaction?

35. When heated to a high temperature, cyclobutane, C_4H_8, decomposes to ethylene:

$$C_4H_8(g) \rightarrow 2\ C_2H_4(g)$$

The activation energy, E_a, for this reaction is 260 kJ/mol-rxn. At 800 K, the rate constant $k = 0.0315\ s^{-1}$. Determine the value of *k* at 850 K.

36. ■ When heated, cyclopropane is converted to propene (see Example 15.5). Rate constants for this reaction at 470 °C and 510 °C are $k = 1.10 \times 10^{-4}\ s^{-1}$ and $k = 1.02 \times 10^{-3}\ s^{-1}$, respectively. Determine the activation energy, E_a, from these data.

37. The reaction of H_2 molecules with F atoms

$$H_2(g) + F(g) \rightarrow HF(g) + H(g)$$

has an activation energy of 8 kJ/mol-rxn and an energy change of −133 kJ/mol-rxn. Draw a diagram similar to Figure 15.13 for this process. Indicate the activation energy and enthalpy change on this diagram.

38. Answer the following questions based on the diagram below.
(a) Is the reaction exothermic or endothermic?
(b) Does the reaction occur in more than one step? If so, how many?

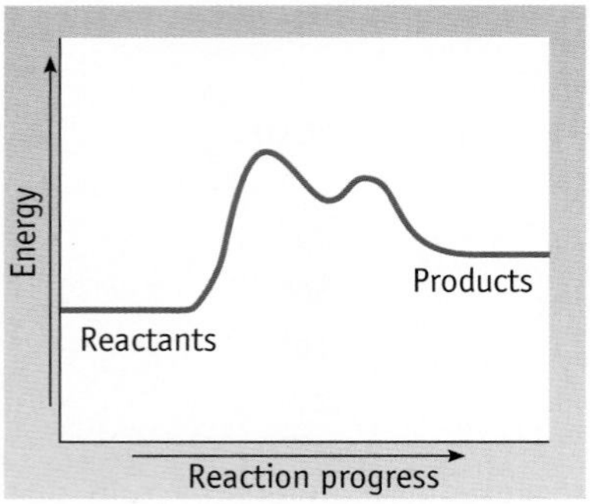

Reaction Mechanisms

(See Examples 15.12–15.14, Exercises 15.11–15.13, and ChemistryNow Screens 15.12 and 15.13.)

39. What is the rate law for each of the following elementary reactions?
(a) $NO(g) + NO_3(g) \rightarrow 2\ NO_2(g)$
(b) $Cl(g) + H_2(g) \rightarrow HCl(g) + H(g)$
(c) $(CH_3)_3CBr(aq) \rightarrow (CH_3)_3C^+(aq) + Br^-(aq)$

40. ■ What is the rate law for each of the following elementary reactions?
(a) $Cl(g) + ICl(g) \rightarrow I(g) + Cl_2(g)$
(b) $O(g) + O_3(g) \rightarrow 2\ O_2(g)$
(c) $2\ NO_2(g) \rightarrow N_2O_4(g)$

41. Ozone, O_3, in the earth's upper atmosphere decomposes according to the equation

$$2\ O_3(g) \rightarrow 3\ O_2(g)$$

The mechanism of the reaction is thought to proceed through an initial fast, reversible step followed by a slow, second step.

Step 1 Fast, reversible $O_3(g) \rightleftharpoons O_2(g) + O(g)$

Step 2 Slow $O_3(g) + O(g) \rightarrow 2\ O_2(g)$

(a) Which of the steps is rate-determining?
(b) Write the rate equation for the rate determining step.

42. ■ The reaction of $NO_2(g)$ and $CO(g)$ is thought to occur in two steps:

Step 1 Slow $NO_2(g) + NO_2(g) \rightarrow NO(g) + NO_3(g)$

Step 2 Fast $NO_3(g) + CO(g) \rightarrow NO_2(g) + CO_2(g)$

(a) Show that the elementary steps add up to give the overall, stoichiometric equation.
(b) What is the molecularity of each step?
(c) For this mechanism to be consistent with kinetic data, what must be the experimental rate equation?
(d) Identify any intermediates in this reaction.

43. A proposed mechanism for the reaction of NO_2 and CO is

Step 1 Slow, endothermic
$$2\ NO_2(g) \rightarrow NO(g) + NO_3(g)$$

Step 2 Fast, exothermic
$$NO_3(g) + CO(g) \rightarrow NO_2(g) + CO_2(g)$$

Overall Reaction Exothermic
$$NO_2(g) + CO(g) \rightarrow NO(g) + CO_2(g)$$

(a) Identify each of the following as a reactant, product, or intermediate: $NO_2(g)$, $CO(g)$, $NO_3(g)$, $CO_2(g)$, $NO(g)$.
(b) Draw a reaction coordinate diagram for this reaction. Indicate on this drawing the activation energy for each step and the overall enthalpy change.

44. ■ The mechanism for the reaction of CH_3OH and HBr is believed to involve two steps. The overall reaction is exothermic.

Step 1 Fast, endothermic
$$CH_3OH + H^+ \rightleftharpoons CH_3OH_2^+$$

Step 2 Slow
$$CH_3OH_2^+ + Br^- \rightarrow CH_3Br + H_2O$$

(a) Write an equation for the overall reaction.
(b) Draw a reaction coordinate diagram for this reaction.
(c) Show that the rate law for this reaction is Rate = $k[CH_3OH][H^+][Br^-]$.

General Questions

These questions are not designated as to type or location in the chapter. They may combine several concepts from this and other chapters.

45. A reaction has the following experimental rate equation: Rate = $k[A]^2[B]$. If the concentration of A is doubled and the concentration of B is halved, what happens to the reaction rate?

46. For a first-order reaction, what fraction of reactant remains after five half-lives have elapsed?

▲ more challenging ■ in OWL Blue-numbered questions answered in Appendix O

47. To determine the concentration dependence of the rate of the reaction

$$H_2PO_3^-(aq) + OH^-(aq) \rightarrow HPO_3^{2-}(aq) + H_2O(\ell)$$

you might measure $[OH^-]$ as a function of time using a pH meter. (To do so, you would set up conditions under which $[H_2PO_3^-]$ remains constant by using a large excess of this reactant.) How would you prove a second-order rate dependence for $[OH^-]$?

48. ■ Data for the following reaction are given in the table.

$$2\ NO(g) + Br_2(g) \rightarrow 2\ NOBr(g)$$

Experiment	[NO] (M)	[Br_2] (M)	Initial Rate (mol/L · s)
1	1.0×10^{-2}	2.0×10^{-2}	2.4×10^{-2}
2	4.0×10^{-2}	2.0×10^{-2}	0.384
3	1.0×10^{-2}	5.0×10^{-2}	6.0×10^{-2}

What is the order of the reaction with respect to [NO] and [Br_2], and what is the overall order of the reaction?

49. Formic acid decomposes at 550 °C according to the equation

$$HCO_2H(g) \rightarrow CO_2(g) + H_2(g)$$

The reaction follows first-order kinetics. In an experiment, it is determined that 75% of a sample of HCO_2H has decomposed in 72 seconds. Determine $t_{1/2}$ for this reaction.

50. Isomerization of CH_3NC occurs slowly when CH_3NC is heated.

$$CH_3NC(g) \rightarrow CH_3CN(g)$$

To study the rate of this reaction at 488 K, data on [CH_3NC] were collected at various times. Analysis led to the graph below.

(a) What is the rate law for this reaction?
(b) What is the equation for the straight line in this graph?
(c) Calculate the rate constant for this reaction.
(d) How long does it take for half of the sample to isomerize?
(e) What is the concentration of CH_3NC after 1.0×10^4 s?

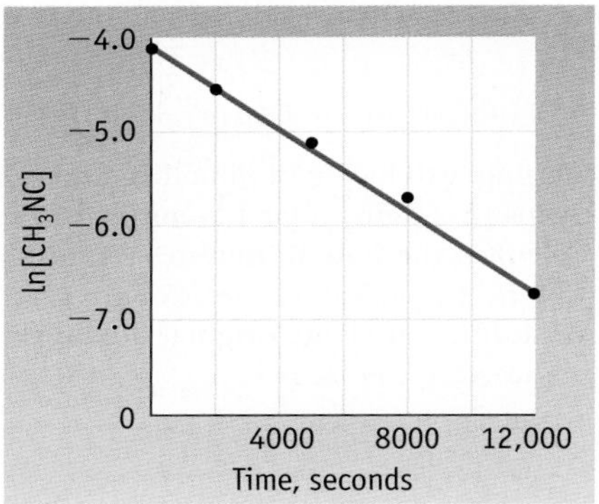

51. When heated, tetrafluoroethylene dimerizes to form octafluorocyclobutane.

$$2\ C_2F_4(g) \rightarrow C_4F_8(g)$$

To determine the rate of this reaction at 488 K, the data in the table were collected. Analysis was done graphically, as shown below:

[C_2F_4] (M)	Time (s)
0.100	0
0.080	56
0.060	150.
0.040	335
0.030	520.

(a) What is the rate law for this reaction?
(b) What is the value of the rate constant?
(c) What is the concentration of C_2F_4 after 600 s?
(d) How long will it take until the reaction is 90% complete?

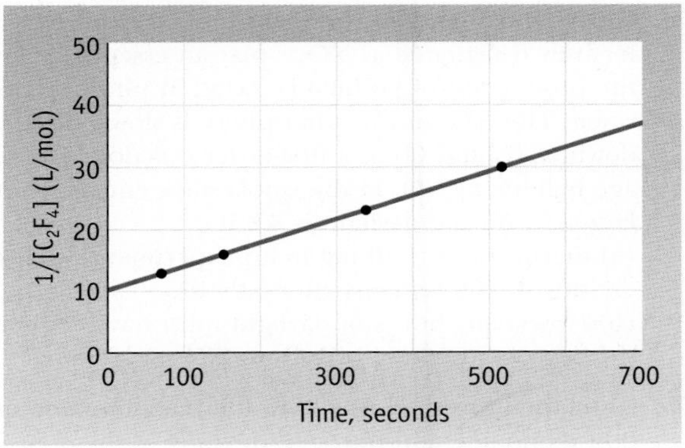

52. Data in the table were collected at 540 K for the following reaction:

$$CO(g) + NO_2(g) \rightarrow CO_2(g) + NO(g)$$

(a) Derive the rate equation.
(b) Determine the reaction order with respect to each reactant.
(c) Calculate the rate constant, giving the correct units for *k*.

Initial Concentration (mol/L)		Initial Rate
[CO]	[NO_2]	(mol/L · h)
5.1×10^{-4}	0.35×10^{-4}	3.4×10^{-8}
5.1×10^{-4}	0.70×10^{-4}	6.8×10^{-8}
5.1×10^{-4}	0.18×10^{-4}	1.7×10^{-8}
1.0×10^{-3}	0.35×10^{-4}	6.8×10^{-8}
1.5×10^{-3}	0.35×10^{-4}	10.2×10^{-8}

53. Ammonium cyanate, NH_4NCO, rearranges in water to give urea, $(NH_2)_2CO$.

$$NH_4NCO(aq) \longrightarrow (NH_2)_2CO(aq)$$

Time (min)	$[NH_4NCO]$ (mol/L)
0	0.458
4.50×10^1	0.370
1.07×10^2	0.292
2.30×10^2	0.212
6.00×10^2	0.114

Using the data in the table:
(a) Decide whether the reaction is first order or second order.
(b) Calculate k for this reaction.
(c) Calculate the half-life of ammonium cyanate under these conditions.
(d) Calculate the concentration of NH_4NCO after 12.0 h.

54. Nitrogen oxides, NO_x (a mixture of NO and NO_2 collectively designated as NO_x), play an essential role in the production of pollutants found in photochemical smog. The NO_x in the atmosphere is slowly broken down to N_2 and O_2 in a first-order reaction. The average half-life of NO_x in the smokestack emissions in a large city during daylight is 3.9 h.
(a) Starting with 1.50 mg in an experiment, what quantity of NO_x remains after 5.25 h?
(b) How many hours of daylight must have elapsed to decrease 1.50 mg of NO_x to 2.50×10^{-6} mg?

55. At temperatures below 500 K, the reaction between carbon monoxide and nitrogen dioxide

$$CO(g) + NO_2(g) \longrightarrow CO_2(g) + NO(g)$$

has the following rate equation: Rate = $k[NO_2]^2$. Which of the three mechanisms suggested here best agrees with the experimentally observed rate equation?

Mechanism 1	Single, elementary step
	$NO_2 + CO \longrightarrow CO_2 + NO$
Mechanism 2	Two steps
Slow	$NO_2 + NO_2 \longrightarrow NO_3 + NO$
Fast	$NO_3 + CO \longrightarrow NO_2 + CO_2$
Mechanism 3	Two steps
Slow	$NO_2 \longrightarrow NO + O$
Fast	$CO + O \longrightarrow CO_2$

56. ■ ▲ Nitryl fluoride can be made by treating nitrogen dioxide with fluorine:

$$2\ NO_2(g) + F_2(g) \longrightarrow 2\ NO_2F(g)$$

Use the rate data in the table to do the following:
(a) Write the rate equation for the reaction.
(b) Indicate the order of reaction with respect to each component of the reaction.
(c) Find the numerical value of the rate constant, k.

	Initial Concentrations (mol/L)			Initial Rate
Experiment	$[NO_2]$	$[F_2]$	$[NO_2F]$	(mol/L · s)
1	0.001	0.005	0.001	2.0×10^{-4}
2	0.002	0.005	0.001	4.0×10^{-4}
3	0.006	0.002	0.001	4.8×10^{-4}
4	0.006	0.004	0.001	9.6×10^{-4}
5	0.001	0.001	0.001	4.0×10^{-5}
6	0.001	0.001	0.002	4.0×10^{-5}

57. The decomposition of dinitrogen pentaoxide

$$2\ N_2O_5(g) \longrightarrow 4\ NO_2(g) + O_2(g)$$

has the following rate equation: Rate = $k[N_2O_5]$. It has been found experimentally that the decomposition is 20.5 % complete in 13.0 h at 298 K. Calculate the rate constant and the half-life at 298 K.

58. ■ The data in the table give the temperature dependence of the rate constant for the reaction $N_2O_5(g) \longrightarrow 2\ NO_2(g) + \frac{1}{2}\ O_2(g)$. Plot these data in the appropriate way to derive the activation energy for the reaction.

T(K)	k (s^{-1})
338	4.87×10^{-3}
328	1.50×10^{-3}
318	4.98×10^{-4}
308	1.35×10^{-4}
298	3.46×10^{-5}
273	7.87×10^{-7}

59. The decomposition of gaseous dimethyl ether at ordinary pressures is first order. Its half-life is 25.0 min at 500 °C:

$$CH_3OCH_3(g) \longrightarrow CH_4(g) + CO(g) + H_2(g)$$

(a) Starting with 8.00 g of dimethyl ether, what mass remains (in grams) after 125 min and after 145 min?
(b) Calculate the time in minutes required to decrease 7.60 ng (nanograms) to 2.25 ng.
(c) What fraction of the original dimethyl ether remains after 150 min?

60. ■ The decomposition of phosphine, PH_3, proceeds according to the equation

$$4\ PH_3(g) \rightarrow P_4(g) + 6\ H_2(g)$$

It is found that the reaction has the following rate equation: Rate = $k[PH_3]$. The half-life of PH_3 is 37.9 s at 120 °C.

(a) How much time is required for three-fourths of the PH_3 to decompose?

(b) What fraction of the original sample of PH_3 remains after 1.00 min?

61. The ozone in the earth's ozone layer decomposes according to the equation

$$2\ O_3(g) \rightarrow 3\ O_2(g)$$

The mechanism of the reaction is thought to proceed through an initial fast equilibrium and a slow step:

Step 1 Fast, reversible $O_3(g) \rightleftharpoons O_2(g) + O(g)$

Step 2 Slow $O_3(g) + O(g) \rightarrow 2\ O_2(g)$

Show that the mechanism agrees with this experimental rate law:

$$-\Delta[O_3]/\Delta t = k\ [O_3]^2/[O_2].$$

62. ■ Hundreds of different reactions can occur in the stratosphere, among them reactions that destroy the earth's ozone layer. The table below lists several (second-order) reactions of Cl atoms with ozone and organic compounds; each is given with its rate constant.

Reaction	Rate Constant (298 K, cm³/molecule · s)
(a) $Cl + O_3 \rightarrow ClO + O_2$	1.2×10^{-11}
(b) $Cl + CH_4 \rightarrow HCl + CH_3$	1.0×10^{-13}
(c) $Cl + C_3H_8 \rightarrow HCl + C_3H_7$	1.4×10^{-10}
(d) $Cl + CH_2FCl \rightarrow HCl + CHFCl$	3.0×10^{-18}

For equal concentrations of Cl and the other reactant, which is the slowest reaction? Which is the fastest reaction?

63. Data for the reaction

$$[Mn(CO)_5(CH_3CN)]^+ + NC_5H_5 \longrightarrow [Mn(CO)_5(NC_5H_5)]^+ + CH_3CN$$

are given in the table. Calculate E_a from a plot of ln k versus $1/T$.

T(K)	k(min⁻¹)
298	0.0409
308	0.0818
318	0.157

64. ■ The gas-phase reaction

$$2\ N_2O_5(g) \rightarrow 4\ NO_2(g) + O_2(g)$$

has an activation energy of 103 kJ/mol-rxn, and the rate constant is 0.0900 min^{-1} at 328.0 K. Find the rate constant at 318.0 K.

65. ■ ▲ Egg protein albumin is precipitated when an egg is cooked in boiling (100 °C) water. E_a for this first-order reaction is 52.0 kJ/mol. Estimate the time to prepare a 3-min egg at an altitude at which water boils at 90 °C.

66. ■ ▲ Two molecules of 1,3-butadiene (C_4H_6) form 1,5-cyclooctadiene, C_8H_{12} at higher temperatures.

$$2\ C_4H_6(g) \rightarrow C_8H_{12}(g)$$

Use the following data to determine the order of the reaction and the rate constant, k. (Note that the total pressure is the pressure of the unreacted C_4H_6 at any time and the pressure of the C_8H_{12}.)

Time (min)	Total Pressure (mm Hg)
0	436
3.5	428
11.5	413
18.3	401
25.0	391
32.0	382
41.2	371

67. ■ ▲ Hypofluorous acid, HOF, is very unstable, decomposing in a first-order reaction to give HF and O_2, with a half-life of 30. min at room temperature:

$$HOF(g) \rightarrow HF(g) + \tfrac{1}{2}\ O_2(g)$$

If the partial pressure of HOF in a 1.00-L flask is initially 1.00×10^2 mm Hg at 25 °C, what are the total pressure in the flask and the partial pressure of HOF after exactly 30 min? After 45 min?

68. ■ ▲ We know that the decomposition of SO_2Cl_2 is first order in SO_2Cl_2,

$$SO_2Cl_2(g) \rightarrow SO_2(g) + Cl_2(g)$$

with a half-life of 245 min at 600 K. If you begin with a partial pressure of SO_2Cl_2 of 25 mm Hg in a 1.0-L flask, what is the partial pressure of each reactant and product after 245 min? What is the partial pressure of each reactant after 12 h?

69. ■ ▲ Nitramide, NO_2NH_2, decomposes slowly in aqueous solution according to the following reaction:

$$NO_2NH_2(aq) \rightarrow N_2O(g) + H_2O(\ell)$$

The reaction follows the experimental rate law

$$\text{Rate} = \frac{k[NO_2NH_2]}{[H_3O^+]}$$

(a) What is the apparent order of the reaction in a buffered solution?
(b) Which of the following mechanisms is the most appropriate for the interpretation of this rate law? Explain.

Mechanism 1

$$NO_2NH_2 \xrightarrow{k_1} N_2O + H_2O$$

Mechanism 2

$$NO_2NH_2 + H_3O^+ \underset{k_2'}{\overset{k_2}{\rightleftharpoons}} NO_2NH_3^+ + H_2O \quad \textit{(rapid equilibrium)}$$

$$NO_2NH_3^+ \xrightarrow{k_3} N_2O + H_3O^+ \quad \textit{(rate limiting step)}$$

Mechanism 3

$$NO_2NH_2 + H_2O \underset{k_4'}{\overset{k_4}{\rightleftharpoons}} NO_2NH^- + H_3O^+ \quad \textit{(rapid equilibrium)}$$

$$NO_2NH^- \xrightarrow{k_5} N_2O + OH^- \quad \textit{(rate limiting step)}$$

$$H_3O^+ + OH^- \xrightarrow{k_6} 2\ H_2O \quad \textit{(very fast reaction)}$$

(c) Show the relationship between the experimentally observed rate constant, k, and the rate constants in the selected mechanism.
(d) Show that hydroxyl ions catalyze the decomposition of nitramide.

70. ■ Many biochemical reactions are catalyzed by acids. A typical mechanism consistent with the experimental results (in which HA is the acid and X is the reactant) is

Step 1 **Fast, reversible** $HA \rightleftharpoons H^+ + A^-$

Step 2 **Fast, reversible** $X + H^+ \rightleftharpoons XH^+$

Step 3 **Slow** $XH^+ \rightarrow$ products

What rate law is derived from this mechanism? What is the order of the reaction with respect to HA? How would doubling the concentration of HA affect the reaction?

In the Laboratory

71. ■ The color change accompanying the reaction of phenolphthalein with strong base is illustrated on page 670. The change in concentration of the dye can be followed by spectrophotometry (page 190), and some data collected by that approach are given below. The initial concentrations were [phenolphthalein] = 0.0050 mol/L and $[OH^-]$ = 0.61 mol/L. (Data are taken from review materials for kinetics at chemed.chem.purdue.edu.) (For more details on this reaction see L. Nicholson, *Journal of Chemical Education*, Vol. 66, page 725, 1989.)

Concentration of Phenolphthalein (mol/L)	Time (s)
0.0050	0.00
0.0045	10.5
0.0040	22.3
0.0035	35.7
0.0030	51.1
0.0025	69.3
0.0020	91.6
0.0015	120.4
0.0010	160.9
0.00050	230.3
0.00025	299.6

(a) Plot the data above as [phenolphthalein] versus time, and determine the average rate from $t = 0$ to $t = 15$ s and from $t = 100$ s to $t = 125$ s. Does the rate change? If so, why?
(b) What is the instantaneous rate at 50 s?
(c) Use a graphical method to determine the order of the reaction with respect to phenolphthalein. Write the rate law, and determine the rate constant.
(d) What is the half-life for the reaction?

72. ▲ We want to study the hydrolysis of the beautiful green, cobalt-based complex called *trans*-dichloro-bis (ethylenediamine)cobalt(III) ion,

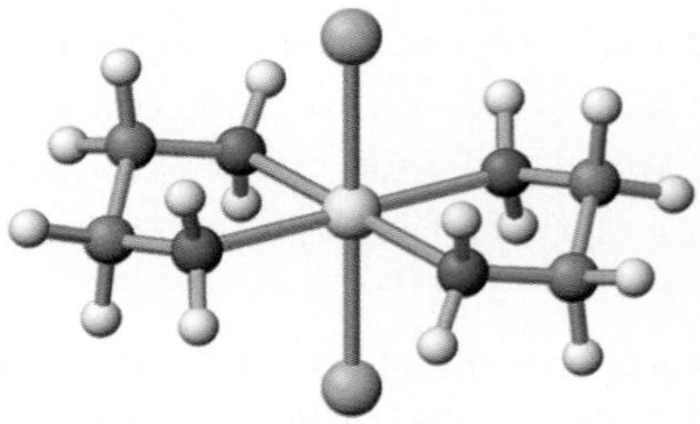

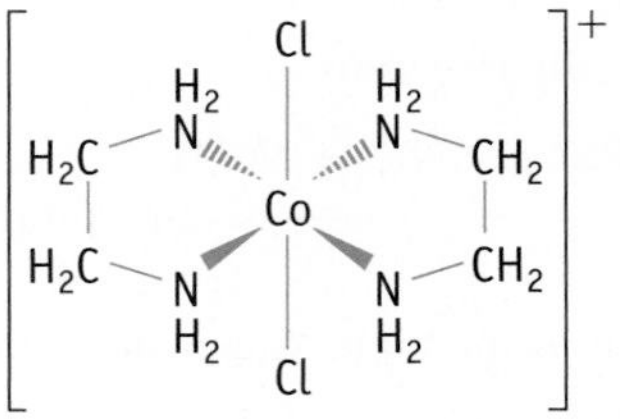

In this hydrolysis reaction, the green complex ion *trans*–$[Co(en)_2Cl_2]^+$ forms the red complex ion $[Co(en)_2(H_2O)Cl]^{2+}$ as a Cl^- ion is replaced with a water molecule on the Co^{3+} ion (en = $H_2NCH_2CH_2NH_2$).

trans–$[Co(en)_2Cl_2]^+(aq) + H_2O(\ell) \rightarrow$
green

$[Co(en)_2(H_2O)Cl]^{2+}(aq) + Cl^-(aq)$
red

The reaction progress is followed by observing the color of the solution. The original solution is green, and the final solution is red, but at some intermediate stage when both the reactant and product are present, the solution is gray.

Original solution

Intermediate solution

Final solution

Photos: Charles D. Winters

Reactions such as this have been studied extensively, and experiments suggest that the initial, slow step in the reaction is the breaking of the Co—Cl bond to give a five-coordinate intermediate. The intermediate is then attacked rapidly by water.

Slow: *trans*–$[Co(en)_2Cl_2]^+(aq) \rightarrow [Co(en)_2Cl]^{2+}(aq) + Cl^-(aq)$

Fast: $[Co(en)_2Cl]^{2+}(aq) + H_2O(aq) \rightarrow [Co(en)_2(H_2O)Cl]^{2+}(aq)$

(a) Based on the reaction mechanism, what is the predicted rate law?

(b) As the reaction proceeds, the color changes from green to red with an intermediate stage where the color is gray. The gray color is reached at the same time, no matter what the concentration of the green starting material (at the same temperature). How does this show the reaction is first order in the green form? Explain.

(c) The activation energy for a reaction can be found by plotting ln k versus $1/T$. However, here we do not need to measure k directly. Instead, because $k = -(1/t)\ln([R]/[R]_0)$, the time needed to achieve the gray color is a measure of k. Use the data below to find the activation energy.

Temperature °C	Time Needed to Achieve Gray Colors (for the same initial concentration)
56	156 s
60	114 s
65	88 s
75	47 s

73. The enzyme chymotrypsin catalyzes the hydrolysis of a peptide containing phenylalanine. Using the data below at a given temperature, calculate the maximum rate of the reaction, $Rate_{max}$. (For more information on enzyme catalysis and the Michaelis–Menten model, see page 702.)

Peptide Concentration (mol/L)	Reaction Rate (mol/L · min)
2.5×10^{-4}	2.2×10^{-6}
5.0×10^{-4}	3.8×10^{-6}
10.0×10^{-4}	5.9×10^{-6}
15.0×10^{-4}	7.1×10^{-6}

74. ■ The substitution of CO in $Ni(CO)_4$ by another molecule L [where L is an electron-pair donor such as $P(CH_3)_3$] was studied some years ago and led to an understanding of some of the general principles that govern the chemistry of compounds having metal–CO bonds. (See J. P. Day, F. Basolo, and R. G. Pearson: *Journal of the American Chemical Society*, Vol. 90, p. 6927, 1968.) A detailed study of the kinetics of the reaction led to the following mechanism:

Slow $Ni(CO)_4 \rightarrow Ni(CO)_3 + CO$

Fast $Ni(CO)_3 + L \rightarrow Ni(CO)_3L$

(a) What is the molecularity of each of the elementary reactions?

(b) Doubling the concentration of $Ni(CO)_4$ increased the reaction rate by a factor of 2. Doubling the concentration of L had no effect on the reaction rate. Based on this information, write the rate equation for the reaction. Does this agree with the mechanism described?

(c) The experimental rate constant for the reaction, when L = $P(C_6H_5)_3$, is 9.3×10^{-3} s^{-1} at 20 °C. If the initial concentration of $Ni(CO)_4$ is 0.025 M, what is the concentration of the product after 5.0 min?

Summary and Conceptual Questions

The following questions may use concepts from this and previous chapters.

75. Hydrogenation reactions, processes wherein H_2 is added to a molecule, are usually catalyzed. An excellent catalyst is a very finely divided metal suspended in the reaction solvent. Tell why finely divided rhodium, for example, is a much more efficient catalyst than a small block of the metal.

76. ■ ▲ Suppose you have 1000 blocks, each of which is 1.0 cm on a side. If all 1000 of these blocks are stacked to give a cube that is 10. cm on a side, what fraction of the 1000 blocks have at least one surface on the outside surface of the cube? Next, divide the 1000 blocks into eight equal piles of blocks and form them into eight cubes, 5.0 cm on a side. What fraction of the blocks now have at least one surface on the outside of the cubes? How does this mathematical model pertain to Study Question 75?

77. ■ The following statements relate to the reaction for the formation of HI:

$$H_2(g) + I_2(g) \longrightarrow 2\ HI(g) \qquad \text{Rate} = k[H_2][I_2]$$

Determine which of the following statements are true. If a statement is false, indicate why it is incorrect.
(a) The reaction must occur in a single step.
(b) This is a second-order reaction overall.
(c) Raising the temperature will cause the value of k to decrease.
(d) Raising the temperature lowers the activation energy for this reaction.
(e) If the concentrations of both reactants are doubled, the rate will double.
(f) Adding a catalyst in the reaction will cause the initial rate to increase.

78. ■ Chlorine atoms contribute to the destruction of the earth's ozone layer by the following sequence of reactions:

$$Cl + O_3 \longrightarrow ClO + O_2$$

$$ClO + O \longrightarrow Cl + O_2$$

where the O atoms in the second step come from the decomposition of ozone by sunlight:

$$O_3(g) \longrightarrow O(g) + O_2(g)$$

What is the net equation on summing these three equations? Why does this lead to ozone loss in the stratosphere? What is the role played by Cl in this sequence of reactions? What name is given to species such as ClO?

79. Describe each of the following statements as true or false. If false, rewrite the sentence to make it correct.
(a) The rate-determining elementary step in a reaction is the slowest step in a mechanism.
(b) It is possible to change the rate constant by changing the temperature.
(c) As a reaction proceeds at constant temperature, the rate remains constant.
(d) A reaction that is third order overall must involve more than one step.

80. ■ Identify which of the following statements are incorrect. If the statement is incorrect, rewrite it to be correct.
(a) Reactions are faster at a higher temperature because activation energies are lower.
(b) Rates increase with increasing concentration of reactants because there are more collisions between reactant molecules.
(c) At higher temperatures, a larger fraction of molecules have enough energy to get over the activation energy barrier.
(d) Catalyzed and uncatalyzed reactions have identical mechanisms.

81. ■ The reaction cyclopropane ⟶ propene occurs on a platinum metal surface at 200 °C. (The platinum is a catalyst.) The reaction is first order in cyclopropane. Indicate how the following quantities change (increase, decrease, or no change) as this reaction progresses, assuming constant temperature.
(a) [cyclopropane]
(b) [propene]
(c) [catalyst]
(d) the rate constant, k
(e) the order of the reaction
(f) the half-life of cyclopropane

82. Isotopes are often used as "tracers" to follow an atom through a chemical reaction, and the following is an example. Acetic acid reacts with methanol (Chapter 11).

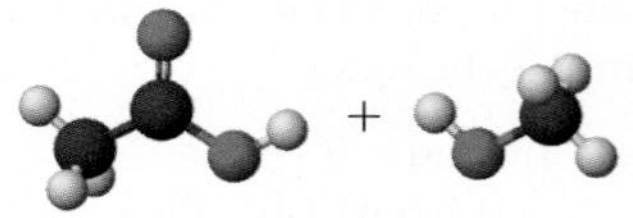

CH_3CO_2H + CH_3OH

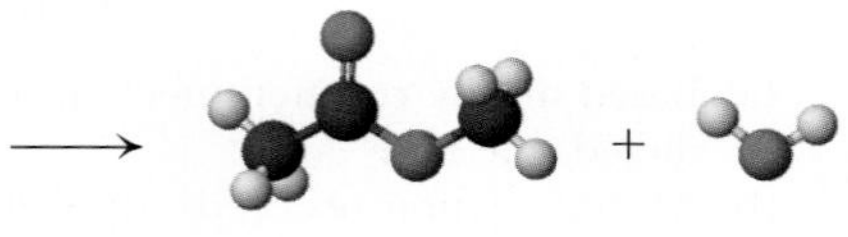

$CH_3CO_2CH_3$ + H_2O

Explain how you could use the isotope ^{18}O to show whether the oxygen atom in the water comes from the —OH of the acid or the —OH of the alcohol.

83. ■ Examine the reaction coordinate diagram given here.

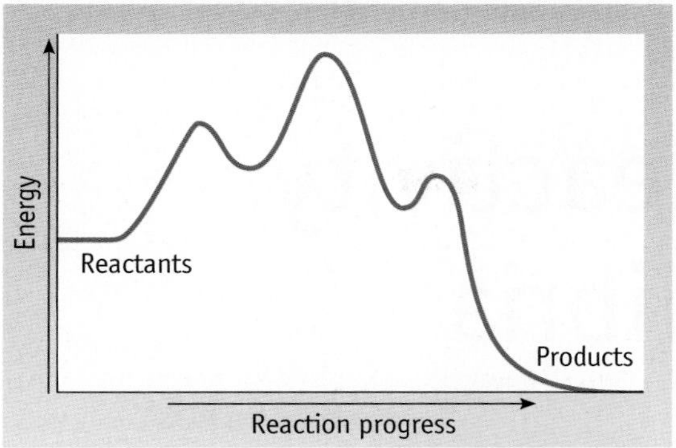

(a) How many steps are in the mechanism for the reaction described by this diagram?
(b) Is the reaction overall exothermic or endothermic?

84. Draw a reaction coordinate diagram for an exothermic reaction that occurs in a single step. Identify the activation energy and the net energy change for the reaction on this diagram. Draw a second diagram that represents the same reaction in the presence of a catalyst, assuming a single step reaction is involved here also. Identify the activation energy of this reaction and the energy change. Is the activation energy in the two drawings different? Does the energy evolved in the two reactions differ?

85. Screen 15.2 in ChemistryNow illustrates the rate at which a blue dye is bleached.
(a) What is the difference between an instantaneous rate and an average rate?
(b) Observe the graph of food dye concentration versus time on this screen. (Click the "tool" icon on this screen.) The plot shows the concentration of dye as the reaction progresses. What does the steepness of the plot at any particular time tell you about the rate of the reaction at that time?
(c) As the reaction progresses, the concentration of dye decreases as it is consumed. What happens to the reaction rate as this occurs? What is the relationship between reaction rate and dye concentration?

86. Watch the video on Screen 15.4 in ChemistryNow (Control of Reaction Rates—Concentration Dependence).
(a) How does an increase in HCl concentration affect the rate of the reaction of the acid with magnesium metal?
(b) On the second portion of this screen are data for the rate of decomposition of N_2O_5 (click "More"). The initial reaction rate is given for three separate experiments, each beginning with a different concentration of N_2O_5. How is the initial reaction rate related to $[N_2O_5]$?

87. The "Microscopic View of Reactions" is described on Screen 15.9 in ChemistryNow.
(a) According to collision theory, what three conditions must be met for two molecules to react?
(b) Examine the animations that play when numbers 1 and 2 are selected. One of these occurs at a higher temperature than the other. Which one? Explain briefly.
(c) Examine the animations that play when numbers 2 and 3 are selected. Would you expect the reaction of O_3 with N_2 to be more or less sensitive to proper orientation for reaction than the reaction displayed on this screen? Explain briefly.

88. "Reaction Mechanisms and Rate Equations" are described on Screen 15.13 in ChemistryNow.
(a) What is the relationship between the stoichiometric coefficients of the reactants in an elementary step and the rate law for that step?
(b) What is the rate law for Step 2 of mechanism 2?
(c) Examine the "Isotopic Labeling" sidebar to this screen. If the transfer of an oxygen atom from NO_2 to CO occurred in a single step, would any $N^{16}O^{18}O$ be found if the reaction is started using a mixture of $N^{16}O_2$ and $N^{18}O_2$? Why or why not?

89. The mechanism for the iodide ion–catalyzed decomposition of H_2O_2 is described on Screen 15.14 (Catalysis and Reaction Rate) in ChemistryNow.
(a) Examine the mechanism for the iodide ion–catalyzed decomposition of H_2O_2. Explain how the mechanism shows that I^- is a catalyst.
(b) How does the reaction coordinate diagram show that the catalyzed reaction is expected to be faster than the uncatalyzed reaction?

16 Principles of Reactivity: Chemical Equilibria

Photos: Charles D. Winters

Solution of cobalt(II) chloride in dilute hydrochloric acid.

Solution in an ice bath.

Solution in a boiling water bath.

Solution after adding excess hydrochloric acid.

Solution after adding excess water.

Reactions and equilibria involving cobalt (II) ions in aqueous solution.

Dynamic and Reversible!

Chemical reactions are dynamic, as proved by experiments in which reactants and products can be interconverted by making small changes in the conditions of the reaction. Dissolving cobalt(II) chloride in dilute hydrochloric acid demonstrates this very well.

$$\underset{\text{red}}{[Co(H_2O)_6]^{2+}(aq)} + 4\ Cl^-(aq) \rightleftharpoons \underset{\text{blue}}{[CoCl_4]^{2-}(aq)} + 6\ H_2O(\ell)$$

The solution in the top photo shows the system is a mixture of the red cation, $[Co(H_2O)_6]^{2+}$, and the blue anion, $[CoCl_4]^{2-}$. When the solution is placed in ice, the color changes to red, and when it is put in boiling water, the color is blue. Adding excess hydrochloric acid again changes the color as does adding water.

Questions:

1. Is the conversion of the red cation to the blue anion by changing the temperature exo- or endothermic?
2. Account for the effect of adding hydrochloric acid and excess water.
3. How do these observations prove the reaction is reversible?

Answers to these questions are in Appendix Q.

Chapter Goals

See Chapter Goals Revisited (page 750) for Study Questions keyed to these goals and assignable in OWL.

- Understand the nature and characteristics of chemical equilibria.
- Understand the significance of the equilibrium constant, *K*, and the reaction quotient, *Q*.
- Understand how to use *K* in quantitative studies of chemical equilibria.

Chapter Outline

The concept of equilibrium is fundamental in chemistry. The general concept was introduced in Chapter 3, and you have already encountered its importance in explaining such phenomena as solubility, acid–base behavior, and changes of state. These preliminary discussions of equilibrium emphasized the following concepts: that chemical reactions are reversible, that in a closed system a state of equilibrium is achieved eventually between reactants and products, and that outside forces can affect the equilibrium. A major result of our further exploration of chemical equilibria in this and the next two chapters will be an ability to describe chemical reactions in more quantitative terms.

Chemistry Now™

Throughout the text this icon introduces an opportunity for self-study or to explore interactive tutorials by signing in at **www.thomsonedu.com/login.**

16.1 Chemical Equilibrium: A Review

If you mix solutions of $CaCl_2$ and $NaHCO_3$, a chemical reaction is immediately detected: a gas (CO_2) bubbles from the mixture, and white solid (insoluble) $CaCO_3$ forms (Figure 16.1a). The reaction occurring is:

$$Ca^{2+}(aq) + 2\ HCO_3^-(aq) \rightarrow CaCO_3(s) + CO_2(g) + H_2O(\ell)$$

If you next add pieces of dry ice to the suspension of $CaCO_3$ (or if you bubble gaseous CO_2 into the mixture), you will observe that the solid $CaCO_3$ dissolves (Figure 16.1b). This happens because a reaction occurs that is the reverse of the reaction that led to precipitation of $CaCO_3$; that is:

$$CaCO_3(s) + CO_2(aq) + H_2O(\ell) \rightarrow Ca^{2+}(aq) + 2\ HCO_3^-(aq)$$

Now imagine what will happen if the solution of Ca^{2+} and HCO_3^- ions is in a *closed* container (unlike the reaction in Figure 16.1, which was done in an open container). As the reaction begins, Ca^{2+} and HCO_3^- react to give products at some rate. As the reactants are used up, the rate of this reaction slows. At the same time, however, the reaction products ($CaCO_3$, CO_2, and H_2O) begin to combine to reform Ca^{2+} and HCO_3^-, at a rate that increases as the amounts of $CaCO_3$ and CO_2 increase. Eventually, the rate of the forward reaction, the formation of $CaCO_3$, and the rate of the reverse reaction, the redissolving of $CaCO_3$, become equal. With $CaCO_3$ being formed and redissolving at the same rate, no further macroscopic change is observed. We have reached equilibrium, and no further *net* change is observed.

■ **Cave Chemistry** This same chemistry accounts for stalactites and stalagmites in caves. See page 119.

We describe an equilibrium system with an equation that connects reactants and products with double arrows. The double arrows, $\rightleftharpoons$, indicate that the reaction

(a) (b) (c)

FIGURE 16.1 Equilibria in the $CO_2/Ca^{2+}/H_2O$ system. (a) Combining solution of $NaHCO_3$ and $CaCl_2$ produces solid $CaCO_3$. (b) If excess dry ice (the white solid) is added to the $CaCO_3$ precipitated in (a), the calcium carbonate dissolves to give $Ca^{2+}(aq)$ and $HCO_3^-(aq)$ (c). (See also Figure 3.6.)

■ **Reversibility of Reactions** All chemical reactions are reversible, in theory. You may see some of these in your laboratory. A few examples include

$NH_3(g) + HCl(g) \rightleftharpoons NH_4Cl(s)$

$CuSO_4 \cdot 5H_2O(s) \rightleftharpoons CuSO_4(s) + 5\ H_2O(g)$

$(NH_4)_2CO_3(s) \rightleftharpoons 2\ NH_3(g) + H_2O(g) + CO_2(g)$

Practically speaking, some reactions cannot be reversed. Frying an egg, for example, is not a reversible process in practical terms.

is reversible and that the reaction will be studied using the concepts of chemical equilibria.

$$Ca^{2+}(aq) + 2\ HCO_3^-(aq) \rightleftharpoons CaCO_3(s) + CO_2(g) + H_2O(\ell)$$

These experiments illustrate an important feature of chemical reactions: *All chemical reactions are reversible, at least in principle.* This was a key point in our earlier discussion of equilibrium (Chapter 3).

Our next step will be to move from a qualitative to a quantitative assessment of equilibrium systems. Among other things, this will lead us to the subject of product- and reactant-favored reactions (◀ page 121). Recall that a reaction that has a greater concentration of products than reactants once it has reached equilibrium is said to be *product-favored.* Similarly, a reaction that has a greater concentration of reactants than products at equilibrium is said to be *reactant-favored.*

Chemistry Now™

Sign in at **www.thomsonedu.com/login** and go to Chapter 16 Contents to see:
- Screen 16.2 to watch a video of **a reversible reaction**
- Screen 16.3 for a simulation of **a chemical equilibrium**

16.2 The Equilibrium Constant and Reaction Quotient

Chemical equilibria can also be described in a quantitative fashion. The concentrations of reactants and products when a reaction has reached equilibrium are related. For the reaction of hydrogen and iodine to produce hydrogen iodide, for example, a very large number of experiments have shown that at equilibrium the ratio of the square of the HI concentration to the product of the H_2 and I_2 concentrations is a constant.

$$H_2(g) + I_2(g) \rightleftharpoons 2\ HI(g)$$

$$\frac{[HI]^2}{[H_2][I_2]} = \text{constant } (K) \text{ at equilibrium}$$

This constant is always the same within experimental error for all experiments done at a given temperature. Suppose, for example, the concentrations of H_2 and I_2 in a

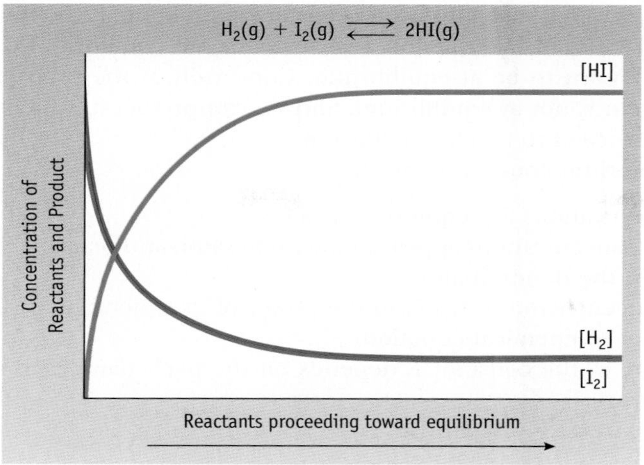

Active Figure 16.2 The reaction of H_2 and I_2 reaches equilibrium. The final concentrations of H_2, I_2, and HI depend on the initial concentrations of H_2 and I_2. If one begins with a different set of initial concentrations, the equilibrium concentrations will be different, but the quotient $[HI]^2/[H_2][I_2]$ will always be the same at a given temperature.

ChemistryNow™ Sign in at www.thomsonedu.com/login and go to the Chapter Contents menu to explore an interactive version of this figure accompanied by an exercise.

flask are each initially 0.0175 mol/L at 425 °C and that no HI is present. Over time, the concentrations of H_2 and I_2 will decrease, and the concentration of HI will increase until a state of equilibrium is reached (Figure 16.2). If the gases in the flask are then analyzed, the observed concentrations would be $[H_2] = [I_2] = 0.0037$ mol/L and $[HI] = 0.0276$ mol/L. The following table—which we call an **ICE table** for *initial*, *change*, and *equilibrium* concentrations—summarizes these results:

Equation	$H_2(g)$	+	$I_2(g)$	$\rightleftharpoons$	2 HI(g)
I = *Initial* concentration (M)	0.0175		0.0175		0
C = *Change* in concentration as reaction proceeds to equilibrium	−0.0138		−0.0138		+0.0276
E = *Equilibrium* concentration (M)	0.0037		0.0037		0.0276

■ **ICE Table: Initial, Change, and Equilibrium** Throughout our discussions of chemical equilibria, we shall express the quantitative information for reactions in an amounts table or ICE table (see Chapter 4, page 159). These tables show what the initial (I) concentrations are, how those concentrations change (C) on proceeding to equilibrium, and what the concentrations are at equilibrium (E).

The second line in the table gives the change in concentration of reactants and products on proceeding to equilibrium. Changes are always equal to the difference between the experimentally observed equilibrium and initial concentrations.

$$\text{Change in concentration} = \text{Equilibrium concentration} - \text{Initial concentration}$$

Putting the equilibrium concentration values from the ICE table into the expression for the constant (K) gives a value of 56 (to two significant figures).

$$\frac{[HI]^2}{[H_2][I_2]} = \frac{(0.0276)^2}{(0.0037)(0.0037)} = 56$$

Other experiments can be done on the H_2/I_2 reaction with different concentrations of reactants, or done using mixtures of reactants and products. Regardless of the initial amounts, when equilibrium is achieved, the ratio $[HI]^2/[H_2][I_2]$ is always the same (at the same temperature).

The observation that the product and reactant concentrations for the H_2 and I_2 reaction are always in the same ratio can be generalized to other reactions. For the general chemical reaction

$$a\,A + b\,B \rightleftharpoons c\,C + d\,D$$

we can define the equilibrium constant, K.

When reaction is at equilibrium

$$\text{Equilibrium constant} = K = \frac{[C]^c[D]^d}{[A]^a[B]^b} \qquad \textbf{(16.1)}$$

Equation 16.1 is called the **equilibrium constant expression.** If the ratio of products to reactants as defined by Equation 16.1 matches the equilibrium constant value, the system is known to be at equilibrium. Conversely, if the ratio has a different value, the system is not at equilibrium, and we can predict in which direction the reaction will proceed to reach equilibrium.

In an equilibrium constant expression,

- All concentrations are equilibrium values.
- Product concentrations appear in the numerator, and reactant concentrations appear in the denominator.
- Each concentration is raised to the power of its stoichiometric coefficient in the balanced chemical equation.
- The value of the constant K depends on the particular reaction and on the temperature.
- Units are never given with K.

Chemistry Now™

Sign in at **www.thomsonedu.com/login** and go to Chapter 16 Contents to see Screen 16.4 for a simulation of **equilibrium and determination of the constant.**

Writing Equilibrium Constant Expressions

Reactions Involving Solids

The oxidation of solid, yellow sulfur produces colorless sulfur dioxide gas in a product-favored reaction (Figure 16.3).

$$S(s) + O_2(g) \rightleftharpoons SO_2(g)$$

The general principle when writing an equilibrium constant expression is to place product concentrations in the numerator and reactant concentrations in the denominator. In reactions involving solids, however, experiments show that the equilibrium concentrations of other reactants or products—here, O_2 and SO_2—do not depend on the amount of solid present (as long as some solid is present at equilibrium). The concentration of a solid such as sulfur is determined by its density, and the density is a fixed value. Therefore, the concentration of sulfur is essentially constant and is not included in the equilibrium constant expression.

$$K = \frac{[SO_2]}{[O_2]}$$

In general, *the concentrations of any solid reactants and products are not included in the equilibrium constant expression.*

Charles D. Winters

FIGURE 16.3 Burning sulfur. Elemental sulfur burns in oxygen with a beautiful blue flame to give SO_2 gas.

Reactions in Aqueous Solutions

There are also special considerations for reactions occurring in aqueous solution in which water is either a reactant or a product. Consider ammonia, which is a weak base owing to its reaction with water.

$$NH_3(aq) + H_2O(\ell) \rightleftharpoons NH_4^+(aq) + OH^-(aq)$$

Because the water concentration is very high in a dilute ammonia solution, the concentration of water is essentially unchanged by the reaction. The general rule for such reactions in aqueous solution is that *the molar concentration of water is not included in the equilibrium constant expression.* Thus, for aqueous ammonia we write

$$K = \frac{[NH_4^+][OH^-]}{[NH_3]}$$

Reactions Involving Gases: K_c and K_p

Concentration data can be used to calculate equilibrium constants for both aqueous and gaseous systems. In these cases, the symbol K is sometimes given the subscript "c" for "concentration," as in K_c. For gases, however, equilibrium constant expressions can be written in another way—in terms of partial pressures of reactants and products. If you rearrange the ideal gas law, $[PV = nRT]$, and recognize that the "gas concentration," (n/V), is equivalent to P/RT, you see that the partial pressure of a gas is proportional to its concentration $[P = (n/V)RT]$. If reactant and product quantities are given in partial pressures (in atmospheres or, more properly, in bars), then K is given the subscript "p," as in K_p.

■ **K_c and K_p** The subscript "c" (K_c) indicates that the numerical values of concentrations in the equilibrium constant expression have units of mol/L. A subscript "p" (K_p) indicates values in units of pressure. In this chapter, we will sometimes write simply K for K_c but will always write K_p to indicate that equilibrium values are in units of pressure.

$$H_2(g) + I_2(g) \rightleftharpoons 2\ HI(g)$$

$$K_p = \frac{P_{HI}^{\ 2}}{P_{H_2}P_{I_2}}$$

Notice that the basic form of the equilibrium constant expression is the same as for K_c. In some cases, the numerical values of K_c and K_p are the same, but they are different when the numbers of moles of gaseous reactants and products are different. *A Closer Look: Equilibrium Constant Expressions for Gases—K_c and K_p* shows how K_c and K_p are related and how to convert from one to the other.

Chemistry Now™

Sign in at **www.thomsonedu.com/login** and go to Chapter 16 Contents to see Screen 16.5 for a simulation and a tutorial on **writing equilibrium expressions.**

EXAMPLE 16.1 Writing Equilibrium Constant Expressions

Problem Write the equilibrium constant expressions for the following reactions.

(a) $N_2(g) + 3\ H_2(g) \rightleftharpoons 2\ NH_3(g)$

(b) $H_2CO_3(aq) + H_2O(\ell) \rightleftharpoons HCO_3^-(aq) + H_3O^+(aq)$

Strategy Remember that product concentrations always appear in the numerator and reactant concentrations appear in the denominator. Each concentration should be raised to a power equal to the stoichiometric coefficient in the balanced equation. In reaction (b), the water concentration does not appear in the equilibrium constant expression.

Solution

(a) $K = \frac{[NH_3]^2}{[N_2][H_2]^3}$ (b) $K = \frac{[HCO_3^-][H_3O^+]}{[H_2CO_3]}$

EXERCISE 16.1 Writing Equilibrium Constant Expressions

Write the equilibrium constant expression for each of the following reactions in terms of concentrations.

(a) $CO_2(g) + C(s) \rightleftharpoons 2\ CO(g)$

(b) $[Cu(NH_3)_4]^+(aq) \rightleftharpoons Cu^{2+}(aq) + 4\ NH_3(aq)$

(c) $CH_3CO_2H(aq) + H_2O(\ell) \rightleftharpoons CH_3CO_2^-(aq) + H_3O^+(aq)$

A Closer Look — Equilibrium Constant Expressions for Gases—K_c and K_p

Many metal carbonates, such as limestone, decompose on heating to give the metal oxide and CO_2 gas.

$$CaCO_3(s) \rightleftharpoons CaO(s) + CO_2(g)$$

The equilibrium condition for this reaction can be expressed either in terms of the number of moles per liter of CO_2, $K_c = [CO_2]$ or in terms of the pressure of CO_2, $K_p = P_{CO_2}$. From the ideal gas law, you know that

$$P = (n/V)RT = (\text{concentration in mol/L}) \times RT$$

For this reaction, we can therefore say that $P_{CO_2} = [CO_2]RT = K_p$. Because $K_c = [CO_2]$, we find that $K_p = K_c(RT)$. That is, the values of K_p and K_c are not the same; for the decomposition of calcium carbonate, K_p is the product of K_c and the factor RT.

Consider the equilibrium constant for the reaction of N_2 and H_2 to produce ammonia in terms of partial pressures, K_p.

$$N_2(g) + 3\ H_2(g) \rightleftharpoons 2\ NH_3(g)$$

$$K_p = \frac{(P_{NH_3})^2}{(P_{N_2})(P_{H_2})^3} = 5.8 \times 10^5$$

Does K_c, the equilibrium constant in terms of concentrations, have the same value as or a different value than K_p? We can answer this question by substituting for each pressure in K_p the equivalent expression $[C](RT)$. That is,

$$K_p = \frac{\{[NH_3](RT)\}^2}{\{[N_2](RT)\}\{[H_2](RT)\}^3} = \frac{[NH_3]^2}{[N_2][H_2]^3} \times \frac{1}{(RT)^2} = \frac{K_c}{(RT)^2}$$

Solving for K_c we find

$$K_c = K_p(RT)^2$$

$$K_c = 5.8 \times 10^5\ [(0.08206)(298)]^2 = 3.5 \times 10^8$$

Once again, you see that K_p and K_c are not the same but are related by some function of RT.

Looking carefully at these examples, we find that

$$K_p = K_c(RT)^{\Delta n}$$

where Δn is the change in the number of moles of gas on going from reactants to products.

$$\Delta n = \text{total moles of gaseous products} - \text{total moles of gaseous reactants}$$

For the decomposition of $CaCO_3$,

$$\Delta n = 1 - 0 = 1$$

whereas the value of Δn for the ammonia synthesis is

$$\Delta n = 2 - 4 = -2$$

The Meaning of the Equilibrium Constant, K

Table 16.1 lists a few equilibrium constants for different kinds of reactants. A large value of K means that the concentration of the products is higher than the concentrations of the reactants at equilibrium. That is, the products are favored over the reactants at equilibrium.

> **$K > 1$: Reaction is product-favored at equilibrium.** The concentrations of products are greater than the concentrations of the reactants at equilibrium.

An example is the reaction of nitrogen monoxide and ozone.

$$NO(g) + O_3(g) \rightleftharpoons NO_2(g) + O_2(g)$$

$$K = \frac{[NO_2][O_2]}{[NO][O_3]} = 6 \times 10^{34} \text{ at } 25\ °C$$

The very large value of K indicates that, at equilibrium, $[NO_2][O_2] >> [NO][O_3]$. If stoichiometric amounts of NO and O_3 are mixed and allowed to come to equilibrium, virtually none of the reactants will be found (Figure 16.4a). Essentially, all will have been converted to NO_2 and O_2. A chemist would say that "the reaction has gone to completion."

Conversely, a small value of K means that very little of the products exist when equilibrium has been achieved (Figure 16.4b). That is, the reactants are favored over the products at equilibrium.

> **$K < 1$: Reaction is reactant-favored at equilibrium.** Concentrations of reactants are greater than concentrations of products at equilibrium.

This is true for the formation of ozone from oxygen.

$$3/2\ O_2(g) \rightleftharpoons O_3(g)$$

$$K = \frac{[O_3]}{[O_2]^{3/2}} = 2.5 \times 10^{-29} \text{ at } 25\ °C$$

TABLE 16.1 Selected Equilibrium Constant Values

Reaction	Equilibrium Constant, *K* (at 25 °C)	Product- or Reactant-Favored
Combination Reaction of Nonmetals		
$S(s) + O_2(g) \rightleftharpoons SO_2(g)$	4.2×10^{52}	$K > 1$; product-favored
$2\ H_2(g) + O_2(g) \rightleftharpoons 2\ H_2O(g)$	3.2×10^{81}	$K > 1$; product-favored
$N_2(g) + 3\ H_2(g) \rightleftharpoons 2\ NH_3(g)$	3.5×10^{8}	$K > 1$; product-favored
$N_2(g) + O_2(g) \rightleftharpoons 2\ NO(g)$	1.7×10^{-3} (at 2300 K)	$K < 1$; reactant-favored
Ionization of Weak Acids and Bases		
$HCO_2H(aq) + H_2O(\ell) \rightleftharpoons HCO_2^-(aq) + H_3O^+(aq)$ formic acid	1.8×10^{-4}	$K < 1$; reactant-favored
$CH_3CO_2H(aq) + H_2O(\ell) \rightleftharpoons CH_3CO_2^-(aq) + H_3O^+(aq)$ acetic acid	1.8×10^{-5}	$K < 1$; reactant-favored
$H_2CO_3(aq) + H_2O(\ell) \rightleftharpoons HCO_3^-(aq) + H_3O^+(aq)$ carbonic acid	4.2×10^{-7}	$K < 1$; reactant-favored
$NH_3(aq) + H_2O(\ell) \rightleftharpoons NH_4^+(aq) + OH^-(aq)$ ammonia	1.8×10^{-5}	$K < 1$; reactant-favored
Dissolution of "Insoluble" Solids		
$CaCO_3(s) \rightleftharpoons Ca^{2+}(aq) + CO_3^{2-}(aq)$	3.8×10^{-9}	$K < 1$; reactant-favored
$AgCl(s) \rightleftharpoons Ag^+(aq) + Cl^-(aq)$	1.8×10^{-10}	$K < 1$; reactant-favored

The very small value of *K* indicates that, at equilibrium, $[O_3] << [O_2]^{3/2}$. If O_2 is placed in a flask, very little O_2 will have been converted to O_3 when equilibrium has been achieved.

When *K* is close to 1, it may not be immediately clear whether the reactant concentrations are larger than the product concentrations, or *vice versa.* It will depend on the form of *K* and thus on the reaction stoichiometry. Calculations of the concentrations will have to be done.

Chemistry Now™

Sign in at **www.thomsonedu.com/login** and go to Chapter 16 Contents to see Screen 16.6 for a simulation of **an equilibrium.**

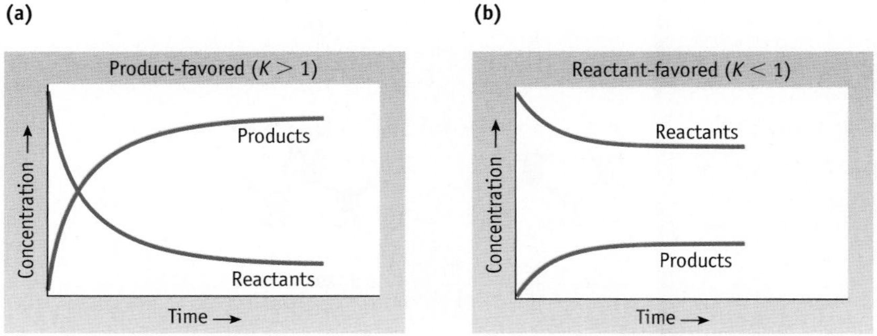

FIGURE 16.4 The difference between product- and reactant-favored reactions. In (a) the concentration of products exceeds the concentration of reactants, and the reaction is product-favored. In (b), the reactant concentrations exceed the product concentration, and the reaction is reactant-favored.

EXERCISE 16.2 The Equilibrium Constant and Extent of Reaction

Are the following reactions product- or reactant-favored at equilibrium?

(a) $[Cu(NH_3)_4]^{2+} \rightleftharpoons Cu^{2+}(aq) + 4\ NH_3(aq)$ $\quad K = 1.5 \times 10^{-13}$

(b) $[Cd(NH_3)_4]^{2+} \rightleftharpoons Cd^{2+}(aq) + 4\ NH_3(aq)$ $\quad K = 1.0 \times 10^{-7}$

If each reaction has a reactant concentration of 0.10 M, in which solution is the NH_3 concentration greater?

The Reaction Quotient, *Q*

The equilibrium constant, *K*, for a reaction has a particular numerical value when the reactants and products are at equilibrium. When the reactants and products in a reaction are not at equilibrium, however, it is convenient to calculate the reaction quotient, *Q*. For the general reaction of *A* and *B* to give *C* and *D*,

$$a\,A + b\,B \rightleftharpoons c\,C + d\,D$$

the reaction quotient is defined as

$$\text{Reaction quotient} = Q = \frac{[C]^c[D]^d}{[A]^a[B]^b} \qquad \textbf{(16.2)}$$

■ **Comparing *Q* and *K***

Relative Magnitude	Direction of Reaction
$Q < K$	Reactants → Products
$Q = K$	Reaction at equilibrium
$Q > K$	Reactants ← Products

This expression *appears* to be just like Equation 16.1, but it is not. The concentrations of reactants and products in the expression for *Q* are those that occur *at any point* as the reaction proceeds from reactants to an equilibrium mixture. *Only when the system is at equilibrium does Q = K.* For the reaction of H_2 and I_2 to give HI (Figure 16.2), any combination of reactant and product concentrations before equilibrium is achieved will give a value of *Q* different than *K*.

Determining a reaction quotient is useful for two reasons. First, it will tell you whether a system is at equilibrium (when $Q = K$) or is not at equilibrium (when $Q \neq K$). Second, by comparing *Q* and *K*, we can predict what changes will occur in reactant and product concentrations as the reaction proceeds to equilibrium.

- ***Q* < *K*** If *Q* is less than *K*, some reactants must be converted to products for the reaction to reach equilibrium. This will decrease the reactant concentrations and increase the product concentrations. (This is the case for the system in Figure 16.5a.)
- ***Q* > *K*** If *Q* is greater than *K*, some products must be converted to reactants for the reaction to reach equilibrium. This will increase the reactant concentrations and decrease the product concentrations. (See Figure 16.5c)

To illustrate these points, let us consider a reaction such as the transformation of butane to isobutane (2-methylpropane).

Butane $\rightleftharpoons$ Isobutane

$CH_3CH_2CH_2CH_3 \rightleftharpoons CH_3CH(CH_3)CH_3$

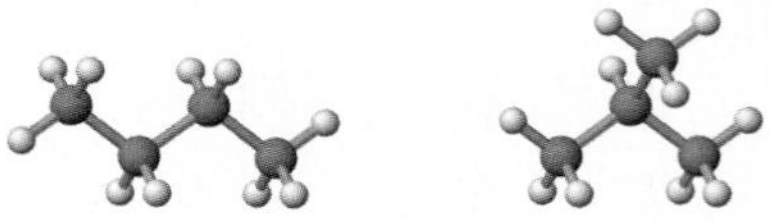

$$K_c = \frac{[\text{isobutane}]}{[\text{butane}]} = 2.50 \text{ at } 298\ K$$

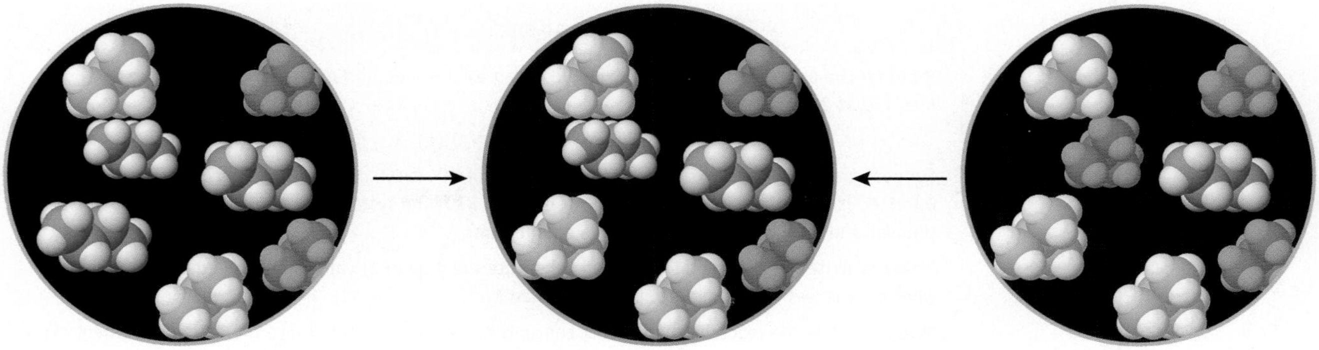

(a) Not at equilibrium. $Q < K$.

Here, four isobutane molecules and three butane molecules are present. Reaction will proceed to convert butane into isobutane to reach equilibrium.

(b) At equilibrium. $Q = K$.

Here, five isobutane molecules and two butane molecules are present. Reaction is at equilibrium.

(c) Not at equilibrium. $Q > K$.

Here, six isobutane molecules and one butane molecule are present. Reaction will proceed to convert isobutane into butane to reach equilibrium.

FIGURE 16.5 The interconversion of isobutane and butane. Only when the concentrations of isobutane and butane are in the ratio [isobutane/butane] = 2.5 is the system at equilibrium (b). With any other ratio of concentrations, there will be a net conversion of one compound into the other until equilibrium is achieved.

If the concentration of one of the compounds is known, then only one value of the other concentration will satisfy the equation for the equilibrium constant. For example, if [butane] is 1.0 mol/L, then the equilibrium concentration of isobutane, [isobutane], must be 2.5 mol/L. If [butane] is 0.63 mol/L at equilibrium, then [isobutane] is 1.6 mol/L.

$$[\text{isobutane}] = K[\text{butane}] = (2.50)(0.63 \text{ mol/L}) = 1.6 \text{ mol/L}$$

Any mixture of butane and isobutane, whether at equilibrium or not, can be represented by the reaction quotient Q (=[isobutane]/[butane]). Suppose you have a mixture composed of 3 mol/L of butane and 4 mol/L of isobutane (at 298 K) (Figure 16.5a). This means that the reaction quotient, Q, is

$$Q = \frac{[\text{isobutane}]}{[\text{butane}]} = \frac{4.0}{3.0} = 1.3$$

This set of concentrations clearly does not represent an equilibrium system because $Q < K$. To reach equilibrium, some butane molecules must be transformed into molecules of isobutane, thereby lowering [butane] and raising [isobutane]. This transformation will continue until the ratio [isobutane]/[butane] = 2.5; that is, until $Q = K$ (Figure 16.5b).

What happens when there is too much isobutane in the system relative to the amount of butane? Suppose [isobutane] = 6.0 mol/L but [butane] is only 1.0 mol/L (Figure 16.5c). Now the reaction quotient Q is greater than K ($Q > K$), and the system is again not at equilibrium. It will proceed to equilibrium by converting isobutane molecules to butane molecules.

Chemistry Now™

Sign in at **www.thomsonedu.com/login** and go to Chapter 16 Contents to see Screen 16.9 for a simulation and tutorial on ***Q*, the reaction quotient.**

EXAMPLE 16.2 The Reaction Quotient

Problem The brown gas nitrogen dioxide, NO_2, can exist in equilibrium with the colorless gas N_2O_4. $K = 170$ at 298 K for the reaction

$$2\ NO_2(g) \rightleftharpoons N_2O_4(g) \quad K = 170$$

Suppose that, at a specific time, the concentration of NO_2 is 0.015 M and the concentration of N_2O_4 is 0.025 M. Is Q larger than, smaller than, or equal to K? If the system is not at equilibrium, in which direction will the reaction proceed to achieve equilibrium?

Strategy Write the expression for Q, and substitute the numerical values into the equation. Decide whether Q is less than, equal to, or greater than K.

Solution When the reactant and product concentrations are substituted into the reaction quotient expression, we have

$$Q = \frac{[N_2O_4]}{[NO_2]^2} = \frac{(0.025)}{(0.015)^2} = 110$$

The value of Q is less than the value of K ($Q < K$), so the reaction is not at equilibrium. The system proceeds to equilibrium by converting NO_2 to N_2O_4, increasing $[N_2O_4]$, and decreasing $[NO_2]$ until $Q = K$.

Comment When calculating Q, make sure that you raise each concentration to the power of the stoichiometric coefficient.

EXERCISE 16.3 The Reaction Quotient

Answer the following questions regarding the butane $\rightleftharpoons$ isobutane equilibrium ($K = 2.50$ at 298 K).

(a) Is the system at equilibrium when [butane] = 0.97 M and [isobutane] = 2.18 M? If it is not at equilibrium, in which direction will the reaction proceed to achieve equilibrium?

(b) Is the system at equilibrium when [butane] = 0.75 M and [isobutane] = 2.60 M? If it is not at equilibrium, in which direction will the reaction proceed to achieve equilibrium?

EXERCISE 16.4 The Reaction Quotient

At 2000 K the equilibrium constant for the formation of NO(g) is 4.0×10^{-4}.

$$N_2(g) + O_2(g) \rightleftharpoons 2\ NO(g)$$

You have a flask in which, at 2000 K, the concentration of N_2 is 0.50 mol/L, that of O_2 is 0.25 mol/L, and that of NO is 4.2×10^{-3} mol/L. Is the system at equilibrium? If not, predict which way the reaction will proceed to achieve equilibrium.

16.3 Determining an Equilibrium Constant

When the experimental values of the concentrations of all of the reactants and products are known at equilibrium, an equilibrium constant can be calculated by substituting the data into the equilibrium constant expression. Consider this concept as it applies to the oxidation of sulfur dioxide.

$$2\ SO_2(g) + O_2(g) \rightleftharpoons 2\ SO_3(g)$$

In an experiment done at 852 K, the equilibrium concentrations are found to be $[SO_2] = 3.61 \times 10^{-3}$ mol/L, $[O_2] = 6.11 \times 10^{-4}$ mol/L, and $[SO_3] = 1.01 \times 10^{-2}$ mol/L. Substituting these data into the equilibrium constant expression, we can determine the value of K.

$$K = \frac{[SO_3]^2}{[SO_2]^2[O_2]} = \frac{(1.01 \times 10^{-2})^2}{(3.61 \times 10^{-3})^2(6.11 \times 10^{-4})} = 1.28 \times 10^4 \text{ at } 852\text{ K}$$

(Notice that K has a large value; at equilibrium, the oxidation of sulfur dioxide is product-favored at 852 K.)

More commonly, an experiment will provide information on the initial quantities of reactants and the concentration at equilibrium of only one of the reactants or of one of the products. The equilibrium concentrations of the rest of the reactants and products must then be inferred from the balanced chemical equation. As an example, consider again the oxidation of sulfur dioxide to sulfur trioxide. Suppose that 1.00 mol of SO_2 and 1.00 mol of O_2 are placed in a 1.00-L flask, this time at 1000 K. When equilibrium has been achieved, 0.925 mol of SO_3 has been formed. Let us use this information to calculate the equilibrium constant for the reaction. After writing the equilibrium constant expression in terms of concentrations, we set up an ICE table (page 727) showing the initial concentrations, the changes in those concentrations on proceeding to equilibrium, and the concentrations at equilibrium.

Equation	$2\ SO_2(g)$	+	$O_2(g)$	$\rightleftharpoons$	$2\ SO_3(g)$
Initial (M)	1.00		1.00		0
Change (M)	−0.925		−0.925/2		+0.925
Equilibrium (M)	1.00 − 0.925 = 0.075		1.00 − 0.925/2 = 0.54		0.925

The quantities in the ICE table result from the following analysis:

- The amount of SO_2 consumed on proceeding to equilibrium is equal to the amount of SO_3 produced (= 0.925 mol because the stoichiometric factor is [2 mol SO_2 consumed/2 mol SO_3 produced]). Because SO_2 is consumed, the change in SO_2 concentration is −0.925 M.
- The amount of O_2 consumed is half of the amount of SO_3 produced (= 0.463 mol because the stoichiometric factor is [1 mol O_2 consumed/2 mol SO_3 produced]). The amount of O_2 remaining is 0.54 M.
- The equilibrium concentration of a reactant is always the initial concentration minus the quantity consumed on proceeding to equilibrium. The equilibrium concentration of a product is always the initial concentration plus the quantity produced on proceeding to equilibrium.

With the equilibrium concentrations now known, it is possible to calculate *K*.

$$K = \frac{[SO_3]^2}{[SO_2]^2[O_2]} = \frac{(0.925)^2}{(0.075)^2(0.54)} = 2.8 \times 10^2 \text{ at } 1000\text{ K}$$

Chemistry Now™

Sign in at **www.thomsonedu.com/login** and go to Chapter 16 Contents to see:
- Screen 16.8 for a simulation and a tutorial on **determining an equilibrium constant**
- Screen 16.9 for a simulation and a tutorial on **systems at equilibrium**

EXAMPLE 16.3 Calculating an Equilibrium Constant

Problem An aqueous solution of ethanol and acetic acid, each with an initial concentration of 0.810 M, is heated to 100 °C. At equilibrium, the acetic acid concentration is 0.748 M. Calculate *K* for the reaction

$$C_2H_5OH(aq) + CH_3CO_2H(aq) \rightleftharpoons CH_3CO_2C_2H_5(aq) + H_2O(\ell)$$

ethanol — acetic acid — ethyl acetate

Strategy Always focus on defining equilibrium concentrations. The amount of acetic acid remaining is known, so the amount consumed is given by [initial concentration of reactant − concentration of reactant remaining]. Because the balanced chemical equation tells us 1 mol of ethanol reacts per mol of acetic acid, the concentration of ethanol is also known at equilibrium. Finally, the concentration of the product formed upon reaching equilibrium is calculated from the amount of reactant consumed.

■ **Ester Preparation and Equilibrium** Esters are a common and important class of organic molecules (◀ page 729). The equilibrium constant for the reaction of an acid and an alcohol is not large, so chemists know that to achieve maximum yield of the ester they remove the water formed in the reaction. As explained in Section 16.6, this means that a larger concentration of ester can exist at equilibrium.

Solution The amount of acetic acid consumed is 0.810 M − 0.748 M = 0.062 M. This is the same as the amount of ethanol consumed and the same as the amount of ethyl acetate produced. The ICE table for this reaction is therefore

Equation	C_2H_5OH	+	CH_3CO_2H	$\rightleftharpoons$	$CH_3CO_2C_2H_5$	+	H_2O
Initial (M)	0.810		0.810		0		
Change (M)	−0.062		−0.062		+0.062		
Equilibrium (M)	0.748		0.748		0.062		

The concentration of each substance at equilibrium is now known, and K can be calculated.

$$K = \frac{[CH_3CO_2C_2H_5]}{[C_2H_5OH][CH_3CO_2H]} = \frac{0.062}{(0.748)(0.748)} = 0.11$$

Comment Notice that water does not appear in the equilibrium expression.

EXAMPLE 16.4 Calculating an Equilibrium Constant (K_p) Using Partial Pressures

Problem Suppose a tank initially contains H_2S at a pressure of 10.00 atm and a temperature of 800 K. When the reaction

$$2\ H_2S(g) \rightleftharpoons 2\ H_2(g) + S_2(g)$$

has come to equilibrium, the partial pressure of S_2 vapor is 0.020 atm. Calculate K_p.

Strategy Recall from page 729 that the equilibrium constant expression can be written in terms of gas partial pressures or concentrations; here, the data are given in terms of partial pressures. In determining the value of an equilibrium constant, we again focus on defining equilibrium amounts. The partial pressure of $S_2(g)$ at equilibrium is known, so the amount produced in proceeding to equilibrium is given by [equilibrium partial pressure − initial partial pressure]. The balanced chemical equation tells us that 2 mol of $H_2(g)$ are produced per mole of $S_2(g)$ produced, so we can determine the partial pressure of $H_2(g)$ produced and then the partial pressure of $H_2(g)$ at equilibrium. Finally, the balanced equation also tells us that the partial pressure of $H_2S(g)$ consumed is twice the partial pressure of $S_2(g)$ produced. The equilibrium pressure of the $H_2S(g)$ is then given by $P_{initial} - P_{gas\ consumed}$.

Solution The equilibrium constant expression that we want to evaluate is

$$K_p = \frac{(P_{H_2})^2 P_{S_2}}{(P_{H_2S})^2}$$

We know that $P(H_2S)_{initial}$ = 10.00 atm and that $P(S_2)_{equilibrium}$ = 0.020 atm, so we can set up an ICE table that expresses the equilibrium partial pressures of each gas.

Equation	$2\ H_2S(g)$	$\rightleftharpoons$	$2\ H_2(g)$	+	$S_2(g)$
Initial (atm)	10.00		0		0
Change (atm)	−2(0.020)		+2(0.020)		+0.020
Equilibrium (atm)	9.96		0.040		0.020

Now that the partial pressures of all of the reactants and products are known, K_p can be calculated.

$$K_p = \frac{(P_{H_2})^2 P_{S_2}}{(P_{H_2S})^2} = \frac{(0.040)^2(0.020)}{(9.96)^2} = 3.2 \times 10^{-7}$$

Comment The value of K_p will be the same as the value of K_c only when the number of moles of gaseous reactants is the same as the number of moles of gaseous products. This is not true here, so $K_p \neq K_c$. See *A Closer Look* (page 730).

$C_6H_{10}I_2$

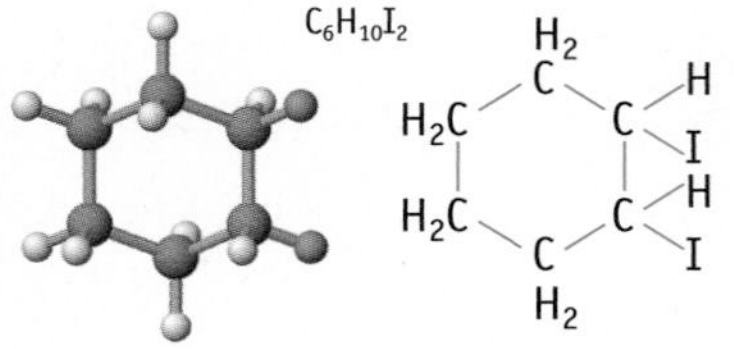

C_6H_{10}

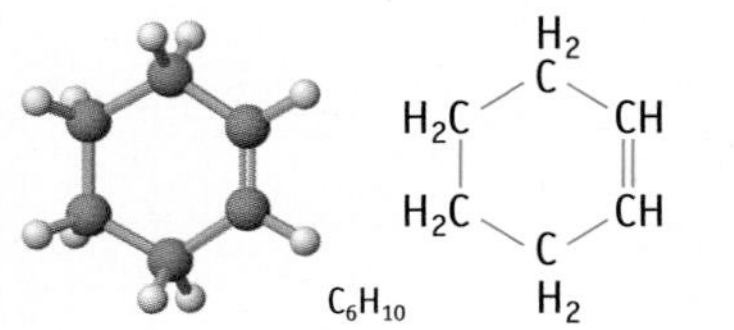

EXERCISE 16.5 Calculating an Equilibrium Constant, K

A solution is prepared by dissolving 0.050 mol of diiodocyclohexane, $C_6H_{10}I_2$, in the solvent CCl_4. The total solution volume is 1.00 L. When the reaction

$$C_6H_{10}I_2 \rightleftharpoons C_6H_{10} + I_2$$

has come to equilibrium at 35 °C, the concentration of I_2 is 0.035 mol/L.

(a) What are the concentrations of $C_6H_{10}I_2$ and C_6H_{10} at equilibrium?

(b) Calculate K_c, the equilibrium constant.

16.4 Using Equilibrium Constants in Calculations

Module 21

In many cases, the value of K and the initial amounts of reactants are known, and you want to know the amounts present at equilibrium. As we look at several examples of this situation, we will again use ICE tables that summarize initial conditions, final conditions, and changes on proceeding to equilibrium.

Chemistry Now™

Sign in at **www.thomsonedu.com/login** and go to Chapter 16 Contents to see Screen 16.10 for a tutorial on **determining equilibrium concentrations.**

EXAMPLE 16.5 Calculating Equilibrium Concentrations

Problem The equilibrium constant K_c (= 55.64) for

$$H_2(g) + I_2(g) \rightleftharpoons 2\ HI(g)$$

has been determined at 425 °C. If 1.00 mol each of H_2 and I_2 are placed in a 0.500-L flask at 425 °C, what are the concentrations of H_2, I_2, and HI when equilibrium has been achieved?

Strategy Because we know the value of K_c and the initial concentrations, we can set up the equilibrium constant expression and an ICE table. We will use the equilibrium constant expression to solve for the unknown values in the table.

Solution The first step is to write the equilibrium constant expression.

$$K = \frac{[HI]^2}{[H_2][I_2]} = 55.64$$

Next, set up an ICE table to express the concentrations of H_2, I_2, and HI before reaction and upon reaching equilibrium. Here, however, we do not know the numerical values of the changes in the H_2 and I_2 concentrations on proceeding to equilibrium. Because the change in $[H_2]$ is the same as the change in $[I_2]$ and because their coefficients in the balanced equation are each 1, we express these changes as the unknown quantity x. It follows that $2x$ is the quantity of HI produced (because the stoichiometric factor is [2 mol HI produced/1 mol H_2 consumed]).

Equation	**$H_2(g)$**	**+**	**$I_2(g)$**	**$\rightleftharpoons$**	**2 HI(g)**
Initial (M)	1.00 mol/0.500 L = 2.00 M		1.00 mol/0.500 L = 2.00 M		0 0
Change (M)	$-x$		$-x$		$+2x$
Equilibrium (M)	$2.00 - x$		$2.00 - x$		$2x$

Now the expressions for the equilibrium concentrations can be substituted into the equilibrium constant expression.

$$55.64 = \frac{(2x)^2}{(2.00 - x)(2.00 - x)} = \frac{(2x)^2}{(2.00 - x)^2}$$

In this case, the unknown quantity x can be found by taking the square root of both sides of the equation,

$$\sqrt{K} = 7.459 = \frac{2x}{2.00 - x}$$

$$7.459\ (2.00 - x) = 14.9 - 7.459x = 2x$$

$$14.9 = 9.459x$$

$$x = 1.58$$

With x known, we can solve for the equilibrium concentrations of the reactants and products.

$$[H_2] = [I_2] = 2.00 - x = 0.42\ M$$

$$[HI] = 2x = 3.16\ M$$

Comment It is always wise to verify the answer by substituting the values back into the equilibrium expression to see if your calculated K agrees with the one given in the problem. In this case, $(3.16)^2/(0.42)^2 = 57$. The slight discrepancy with the given value, $K = 55.64$, occurs because we know $[H_2]$ and $[I_2]$ to only two significant figures.

EXERCISE 16.6 Calculating Equilibrium Concentrations

At some temperature, $K_c = 33$ for the reaction

$$H_2(g) + I_2(g) \rightleftharpoons 2\ HI(g)$$

Assume the initial concentrations of both H_2 and I_2 are 6.00×10^{-3} mol/L. Find the concentration of each reactant and product at equilibrium.

Calculations Where the Solution Involves a Quadratic Expression

Suppose you are studying the decomposition of PCl_5 to form PCl_3 and Cl_2. You know that $K = 1.20$ at a given temperature.

$$PCl_5(g) \rightleftharpoons PCl_3(g) + Cl_2(g)$$

If the initial concentration of PCl_5 is 1.60 M, what will be the concentrations of reactant and products when the system reaches equilibrium? Following the procedures outlined in Example 16.5, you would set up an ICE table to define the equilibrium concentrations of reactants and products.

Reaction	$PCl_5(g)$	$\rightleftharpoons$	$PCl_3(g)$	+	$Cl_2(g)$
Initial (M)	1.60		0		0
Change (M)	$-x$		$+x$		$+x$
Equilibrium (M)	$1.60 - x$		x		x

Substituting into the equilibrium constant expression, we have

$$K = 1.20 = \frac{[PCl_3][Cl_2]}{[PCl_5]} = \frac{(x)(x)}{1.60 - x}$$

Expanding the algebraic expression results in a quadratic equation,

$$x^2 + 1.20x - 1.92 = 0$$

Using the quadratic formula (Appendix A; $a = 1$, $b = 1.20$, and $c = -1.92$), we find two roots to the equation: $x = 0.91$ and -2.11. Because a negative value of x (which represents a negative concentration) is not chemically meaningful, the answer is $x = 0.91$. Therefore, we have, at equilibrium,

$$[PCl_5] = 1.60 - 0.91 = 0.69\ M$$

$$[PCl_3] = [Cl_2] = 0.91\ M$$

Although a solution to a quadratic equation can always be obtained using the quadratic formula, in many instances an acceptable answer can be obtained by using a realistic approximation to simplify the equation. To illustrate this, let us consider another equilibrium, the dissociation of I_2 molecules to form I atoms, for which $K = 5.6 \times 10^{-12}$ at 500 K.

$$I_2(g) \rightleftharpoons 2\ I(g)$$

$$K = \frac{[I]^2}{[I_2]} = 5.6 \times 10^{-12}$$

Assuming the initial I_2 concentration is 0.45 M, and setting up the ICE table in the usual manner, we have

Reaction	$I_2(g)$	$\rightleftharpoons$	$2\ I(g)$
Initial (M)	0.45		0
Change (M)	$-x$		$+2x$
Equilibrium (M)	$0.45 - x$		$2x$

For the equilibrium constant expression, we again arrive at a quadratic equation.

$$K = 5.6 \times 10^{-12} = \frac{(2x)^2}{(0.45 - x)}$$

Although we could solve this equation using the quadratic formula, there is a simpler way to reach an answer. Notice that the value of K is very small, indicating that the amount of I_2 that will be dissociated ($= x$) is very small. In fact, K is so small that subtracting x from the original reactant concentration (0.45 mol/L) in the denominator of the equilibrium constant expression will leave the denominator essentially unchanged. That is, $(0.45 - x)$ is essentially equal to 0.45. Thus, we drop x in the denominator and have a simpler equation to solve.

$$K = 5.6 \times 10^{-12} = \frac{(2x)^2}{(0.45)}$$

The solution to this equation gives $x = 7.9 \times 10^{-7}$. From this value, we can determine that $[I_2] = 0.45 - x \approx 0.45$ mol/L and $[I] = 2x = 1.6 \times 10^{-6}$ mol/L. Notice that the answer to the I_2 dissociation problem confirms the assumption the dissociation of I_2 is so small that $[I_2]$ at equilibrium is essentially equal to the initial concentration.

When is it possible to simplify a quadratic equation? The decision depends on both the value of the initial concentration of the reactant and the value of x, which is in turn related to the value of K. Consider the general reaction

$$A \rightleftharpoons B + C$$

where $K = [B][C]/[A]$. Assume we know K and the initial concentration of A ($= [A]_0$) and wish to find the equilibrium concentrations of B and C ($= x$). The equilibrium constant expression now is

$$K = \frac{[B][C]}{[A]} = \frac{(x)(x)}{[A]_0 - x}$$

When K is very small, the value of x will be much less than $[A]_0$, so $[A]_0 - x \approx [A]_0$. Therefore, we can write the following expression.

$$K = \frac{[B][C]}{[A]} \approx \frac{(x)(x)}{[A]_0} \qquad \textbf{(16.3)}$$

If $100 \times K < [A]_0$, the approximate expression, Equation 16.3, will give acceptable values of equilibrium concentrations (to two significant figures). For more about this useful guideline, see Problem Solving Tip 16.1.

■ **Solving Quadratic Equations** Quadratic equations are usually solved using the quadratic formula (Appendix A). An alternative is the *method of successive approximations*, also outlined in Appendix A. Most equilibrium expressions can be solved quickly by this method, and you are urged to try it. This will remove the uncertainty of whether K expressions need to be solved exactly. (There are, however, rare cases in which this does not work.)

Problem Solving Tip 16.1 When Do You Need to Use the Quadratic Formula?

In most equilibrium calculations, the quantity x may be neglected in the denominator of the equation $K = x^2/([A]_0 - x)$ if x is less than 10% of the quantity of reactant initially present. The guideline presented in the text for making the approximation that $[A]_0 - x \approx [A]_0$ when $100 \times K < [A]_0$ reflects this fact.

In general, when K is about 1 or greater, the approximation cannot be made. If K is much less than 1 and $100 \times K < [A]_0$ (you will see many such cases in Chapter 17), the approximate expression ($K = x^2/[A]_0$) gives an acceptable answer. If you are not certain, then first make the assumption that the unknown (x) is small, and solve the approximate expression (Equation 16.3). Next, compare the "approximate" value of x with $[A]_0$. If x has a value equal to or less than 10% of $[A]_0$, then there is no need to solve the full equation using the quadratic formula.

EXAMPLE 16.6 Calculating Equilibrium Concentrations Using an Equilibrium Constant

Problem The reaction

$$N_2(g) + O_2(g) \rightleftharpoons 2\ NO(g)$$

contributes to air pollution whenever a fuel is burned in air at a high temperature, as in a gasoline engine. At 1500 K, $K = 1.0 \times 10^{-5}$. Suppose a sample of air has $[N_2] = 0.80$ mol/L and $[O_2] = 0.20$ mol/L before any reaction occurs. Calculate the equilibrium concentrations of reactants and products after the mixture has been heated to 1500 K.

Strategy Set up an ICE table of equilibrium concentrations, and then substitute these concentrations into the equilibrium constant expression. The result will be a quadratic equation. This expression can be solved using the methods outlined in Appendix A or by using the guideline in the text to derive an acceptable, approximate answer.

Solution We first set up an ICE table of equilibrium concentrations.

Equation	$N_2(g)$	+	$O_2(g)$	$\rightleftharpoons$	$2\ NO(g)$
Initial (M)	0.80		0.20		0
Change (M)	$-x$		$-x$		$+2x$
Equilibrium (M)	$0.80 - x$		$0.20 - x$		$2x$

Next, the equilibrium concentrations are substituted into the equilibrium constant expression.

$$K = 1.0 \times 10^{-5} = \frac{[NO]^2}{[N_2][O_2]} = \frac{[2x]^2}{(0.80 - x)(0.20 - x)}$$

We refer to the guideline (Equation 16.3) to decide whether an approximate solution is possible. Here, $100 \times K$ ($= 1.0 \times 10^{-3}$) is smaller than either of the initial reactant concentrations (0.80 and 0.20). This means we can use the approximate expression

$$K = 1.0 \times 10^{-5} = \frac{[NO]^2}{[N_2][O_2]} = \frac{(2x)^2}{(0.80)(0.20)}$$

Solving this expression, we find

$$1.6 \times 10^{-6} = 4x^2$$

$$x = 6.3 \times 10^{-4}$$

Therefore, the reactant and product concentrations at equilibrium are

$$[N_2] = 0.80 - 6.3 \times 10^{-4} \approx 0.80\ M$$

$$[O_2] = 0.20 - 6.3 \times 10^{-4} \approx 0.20\ M$$

$$[NO] = 2x = 1.26 \times 10^{-3}\ M$$

Comment The value of x obtained using the approximation is the same as that obtained from the quadratic formula. If the full equilibrium constant expression is expanded, we have

$$(1.0 \times 10^{-5})(0.80 - x)(0.20 - x) = 4x^2$$

$$(1.0 \times 10^{-5})(0.16 - 1.00x + x^2) = 4x^2$$

$$\underset{ax^2}{(4 - 1.0 \times 10^{-5})x^2} + \underset{bx}{(1.0 \times 10^{-5})x} - \underset{c}{0.16 \times 10^{-5}} = 0$$

The two roots to this equation are:

$$x = 6.3 \times 10^{-4} \text{ or } x = -6.3 \times 10^{-4}$$

The only meaningful root is identical to the approximate answer obtained above. The approximation is indeed valid in this case.

EXERCISE 16.7 Calculating an Equilibrium Concentration Using an Equilibrium Constant

Graphite and carbon dioxide are kept at constant volume at 1000 K until the reaction

$$\text{C(graphite)} + CO_2\text{(g)} \rightleftharpoons 2\text{ CO(g)}$$

has come to equilibrium. At this temperature, $K = 0.021$. The initial concentration of CO_2 is 0.012 mol/L. Calculate the equilibrium concentration of CO.

16.5 More About Balanced Equations and Equilibrium Constants

Chemical equations can be balanced using different sets of stoichiometric coefficients. For example, the equation for the oxidation of carbon to give carbon monoxide can be written

$$\text{C(s)} + \tfrac{1}{2}\,O_2\text{(g)} \rightleftharpoons \text{CO(g)}$$

In this case, the equilibrium constant expression would be

$$K_1 = \frac{[\text{CO}]}{[O_2]^{1/2}} = 4.6 \times 10^{23} \text{ at 25 °C}$$

You can write the chemical equation equally well, however, as

$$2\text{ C(s)} + O_2\text{(g)} \rightleftharpoons 2\text{ CO(g)}$$

and the equilibrium constant expression would now be

$$K_2 = \frac{[\text{CO}]^2}{[O_2]} = 2.1 \times 10^{47} \text{ at 25 °C}$$

When you compare the two equilibrium constant expressions you find that $K_2 = (K_1)^2$; that is,

$$K_2 = \frac{[\text{CO}]^2}{[O_2]} = \left\{\frac{[\text{CO}]}{[O_2]^{1/2}}\right\}^2 = K_1^{\,2}$$

When the stoichiometric coefficients of a balanced equation are multiplied by some factor, the equilibrium constant for the new equation (K_{new}) is the old equilibrium constant (K_{old}) raised to the power of the multiplication factor.

In the case of the oxidation of carbon, the second equation was obtained by multiplying the first equation by two. Therefore, K_2 is the *square* of K_1 ($K_2 = K_1^{\,2}$).

Let us consider what happens if a chemical equation is reversed. Here, we will compare the value of K for formic acid transferring an H^+ ion to water

$$HCO_2H(aq) + H_2O(\ell) \rightleftharpoons HCO_2^-(aq) + H_3O^+(aq)$$

$$K_1 = \frac{[HCO_2^-][H_3O^+]}{[HCO_2H]} = 1.8 \times 10^{-4} \text{ at } 25\ °C$$

with the opposite reaction, the gain of an H^+ ion by the formate ion, HCO_2^-.

$$HCO_2^-(aq) + H_3O^+(aq) \rightleftharpoons HCO_2H(aq) + H_2O(\ell)$$

$$K_2 = \frac{[HCO_2H]}{[HCO_2^-][H_3O^+]} = 5.6 \times 10^3 \text{ at } 25\ °C$$

Here, $K_2 = 1/K_1$.

The equilibrium constants for a reaction and its reverse are the reciprocals of one another.

It is often useful to add two equations to obtain the equation for a net process. As an example, consider the reactions that take place when silver chloride dissolves in water (to a *very* small extent) and ammonia is added to the solution. The ammonia reacts with the silver ion to form a water-soluble compound, $Ag(NH_3)_2Cl$ (Figure 16.6). Adding the equation for dissolving solid AgCl to the equation for the reaction of Ag^+ ion with ammonia gives the equation for net reaction, dissolving solid AgCl in aqueous ammonia. (All equilibrium constants are given at 25 °C.)

$AgCl(s) \rightleftharpoons Ag^+(aq) + Cl^-(aq)$ $\qquad K_1 = [Ag^+][Cl^-] = 1.8 \times 10^{-10}$

$Ag^+(aq) + 2\ NH_3(aq) \rightleftharpoons [Ag(NH_3)_2]^+(aq)$ $\qquad K_2 = \frac{[Ag(NH_3)_2^+]}{[Ag^+][NH_3]^2} = 1.6 \times 10^7$

Net reaction:

$AgCl(s) + 2\ NH_3(aq) \rightleftharpoons [Ag(NH_3)_2]^+(aq) + Cl^-(aq)$

To obtain the equilibrium constant for the net reaction, K_{net}, we *multiply* the equilibrium constants for the two reactions, K_1 by K_2.

$$K_{net} = K_1 \times K_2 = [Ag^+][Cl^-] \times \frac{[Ag(NH_3)_2^+]}{[Ag^+][NH_3]^2} = \frac{[Ag(NH_3)_2^+][Cl^-]}{[NH_3]^2}$$

$$K_{net} = K_1 \times K_2 = 2.9 \times 10^{-3}$$

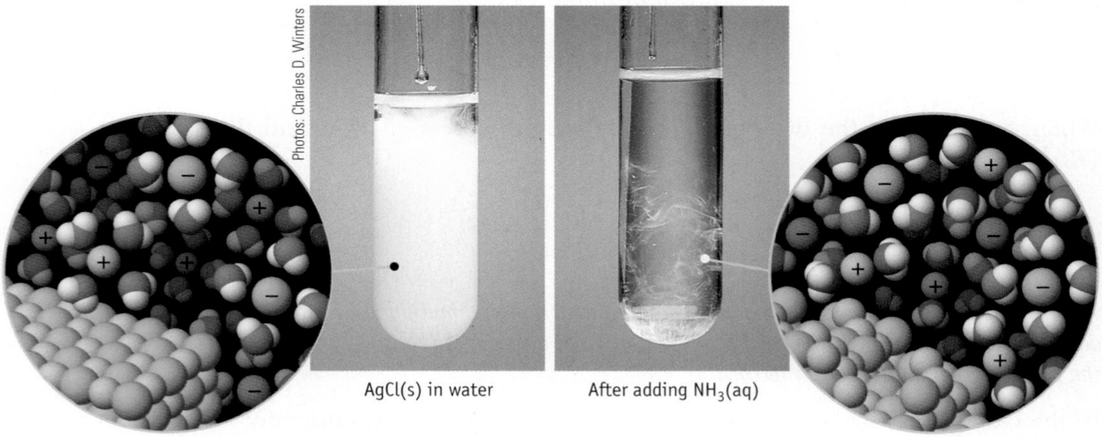

FIGURE 16.6 Dissolving silver chloride in aqueous ammonia. (left) A precipitate of AgCl(s) is suspended in water. (right) When aqueous ammonia is added, the ammonia reacts with the trace of silver ion in solution, the equilibrium shifts, and the silver chloride dissolves.

When two or more chemical equations are added to produce a net equation, the equilibrium constant for the net equation is the product of the equilibrium constants for the added equations.

■ **Complex Ions** This compound, $[Ag(NH_3)_2]Cl$, is made up of a cation (called a complex ion), $[Ag(NH_3)_2]^+$, and an anion, Cl^-. Square brackets are often used to indicate the cation is a single entity. We will discuss complex ions further in Chapters 18 and 22.

EXAMPLE 16.7 Balanced Equations and Equilibrium Constants

Problem A mixture of nitrogen, hydrogen, and ammonia is brought to equilibrium. When the equation is written using whole-number coefficients, as follows, the value of K is 3.5×10^8 at 25 °C.

Equation 1: $N_2(g) + 3\ H_2(g) \rightleftharpoons 2\ NH_3(g) \qquad K_1 = 3.5 \times 10^8$

However, the equation can also be written as given in *Equation 2*. What is the value of K_2?

Equation 2: $\frac{1}{2}\ N_2(g) + \frac{3}{2}\ H_2(g) \rightleftharpoons NH_3(g) \qquad K_2 = ?$

The decomposition of ammonia to the elements *(Equation 3)* is the reverse of its formation *(Equation 1)*. What is the value of K_3?

Equation 3: $2\ NH_3(g) \rightleftharpoons N_2(g) + 3\ H_2(g) \qquad K_3 = ?$

Strategy Review what happens to the value of K when the stoichiometric coefficients are changed or the reaction is reversed. (See Problem Solving Tip 16.2.)

Solution To see the relation between K_1 and K_2, first write the equilibrium constant expressions for these two balanced equations.

$$K_1 = \frac{[NH_3]^2}{[N_2][H_2]^3} \qquad K_2 = \frac{[NH_3]}{[N_2]^{1/2}[H_2]^{3/2}}$$

Writing these expressions makes it clear that K_2 is the square root of K_1.

$$K_2 = (K_1)^{1/2} = \sqrt{K_1} = \sqrt{3.5 \times 10^8} = \mathbf{1.9 \times 10^4}$$

Equation 3 is the reverse of *Equation* 1, and its equilibrium constant expression is

$$K_3 = \frac{[N_2][H_2]^3}{[NH_3]^2}$$

In this case, K_3 is the reciprocal of K_1. That is, $K_3 = 1/K_1$.

$$K_3 = \frac{1}{K_1} = \frac{1}{3.5 \times 10^8} = \mathbf{2.9 \times 10^{-9}}$$

Comment As a final comment, notice that the production of ammonia from the elements has a large equilibrium constant and is product-favored (see Section 16.2). As expected, the reverse reaction, the decomposition of ammonia to its elements, has a small equilibrium constant and is reactant-favored.

EXERCISE 16.8 Manipulating Equilibrium Constant Expressions

The conversion of oxygen to ozone has a very small equilibrium constant.

$3/2\ O_2(g) \rightleftharpoons O_3(g) \qquad K = 2.5 \times 10^{-29}$

(a) What is the value of K when the equation is written using whole-number coefficients?

$3\ O_2(g) \rightleftharpoons 2\ O_3(g)$

(b) What is the value of K for the conversion of ozone to oxygen?

$2\ O_3(g) \rightleftharpoons 3\ O_2(g)$

EXERCISE 16.9 Manipulating Equilibrium Constant Expressions

The following equilibrium constants are given at 500 K:

$H_2(g) + Br_2(g) \rightleftharpoons 2\ HBr(g) \qquad K_p = 7.9 \times 10^{11}$

$H_2(g) \rightleftharpoons 2\ H(g) \qquad K_p = 4.8 \times 10^{-41}$

$Br_2(g) \rightleftharpoons 2\ Br \qquad K_p = 2.2 \times 10^{-15}$

Calculate K_p for the reaction of H and Br atoms to give HBr.

$H(g) + Br(g) \rightleftharpoons HBr(g) \qquad K_p = ?$

Problem Solving Tip 16.2 **Balanced Equations and Equilibrium Constants**

You should now know

1. How to write an equilibrium constant expression from the balanced equation, recognizing that the concentrations of solids and liquids used as solvents do not appear in the expression.
2. That when the stoichiometric coefficients in a balanced equation are changed by a factor of n, $K_{new} = (K_{old})^n$.
3. That when a balanced equation is reversed, $K_{new} = 1/K_{old}$.
4. That when several balanced equations (each with its own equilibrium constant, K_1, K_2, etc.) are added to obtain a net, balanced equation, $K_{net} = K_1 \cdot K_2 \cdot K_3 \cdot \ldots$.

16.6 Disturbing a Chemical Equilibrium

The equilibrium between reactants and products may be disturbed in three ways: (1) by changing the temperature, (2) by changing the concentration of a reactant or product, or (3) by changing the volume (for systems involving gases) (Table 16.2). *A change in any of the factors that determine the equilibrium conditions of a system will cause the system to change in such a manner as to reduce or counteract the effect of the change.* This statement is often referred to as *Le Chatelier's principle* (◀ page 627). It is a shorthand way of saying how a reaction will adjust the quantities of reactants and products so that equilibrium is restored, that is, so that the reaction quotient is once again equal to the equilibrium constant.

Chemistry Now™

Sign in at **www.thomsonedu.com/login** and go to Chapter 16 Contents to see Screen 16.11 to view an animation of **Le Chatelier's principle.**

TABLE 16.2 Effects of Disturbances on Equilibrium Composition

Disturbance	Change as Mixture Returns to Equilibrium	Effect on Equilibrium	Effect on K
Reactions Involving Solids, Liquids, or Gases			
Rise in temperature	Heat energy is consumed by system	Shift in endothermic direction	Change
Drop in temperature	Heat energy is generated by system	Shift in exothermic direction	Change
Addition of reactant*	Some of added reactant is consumed	Product concentration increases	No change
Addition of product*	Some of added product is consumed	Reactant concentration increases	No change
Reactions Involving Gases			
Decrease in volume, increase in pressure	Pressure decreases	Composition changes to reduce total number of gas molecules	No change
Increase in volume, decrease in pressure	Pressure increases	Composition changes to increase total number of gas molecules	No change

* Does not apply when an insoluble solid reactant or product is added. Recall that their "concentrations" do not appear in the reaction quotient.

Effect of the Addition or Removal of a Reactant or Product

If the concentration of a reactant or product is changed from its equilibrium value *at a given temperature*, equilibrium will be reestablished eventually. The new equilibrium concentrations of reactants and products will be different, but the value of the equilibrium constant expression will still equal K (Table 16.1). To illustrate this, let us return to the butane/isobutane equilibrium (with $K = 2.5$).

$$\underset{\text{butane}}{CH_3CH_2CH_2CH_3} \rightleftharpoons \underset{\text{isobutane}}{CH_3\overset{\overset{\displaystyle CH_3}{|}}{C}HCH_3} \qquad K = 2.5$$

Suppose the equilibrium mixture consists of two molecules of butane and five molecules of isobutane (Figure 16.7). The reaction quotient, Q, is 5/2 (or 2.5/1), the value of the equilibrium constant for the reaction. Now we add seven more molecules of isobutane to the mixture to give a ratio of 12 isobutane molecules to two butane molecules. The reaction quotient is now 6/1. Q is greater than K, so the system will change to reestablish equilibrium. To do so, some molecules of isobutane must be changed into butane molecules, a process that continues until the ratio [isobutane]/[butane] is once again 2.5/1. In this particular case, if two of the 12 isobutane molecules change to butane, the ratio of isobutane to butane is again equal to K ($= 10/4 = 2.5/1$), and equilibrium is reestablished.

Chemistry Now™

Sign in at **www.thomsonedu.com/login** and go to Chapter 16 Contents to see Screen 16.13 for a simulation and a tutorial on **the effect of concentration changes on an equilibrium.**

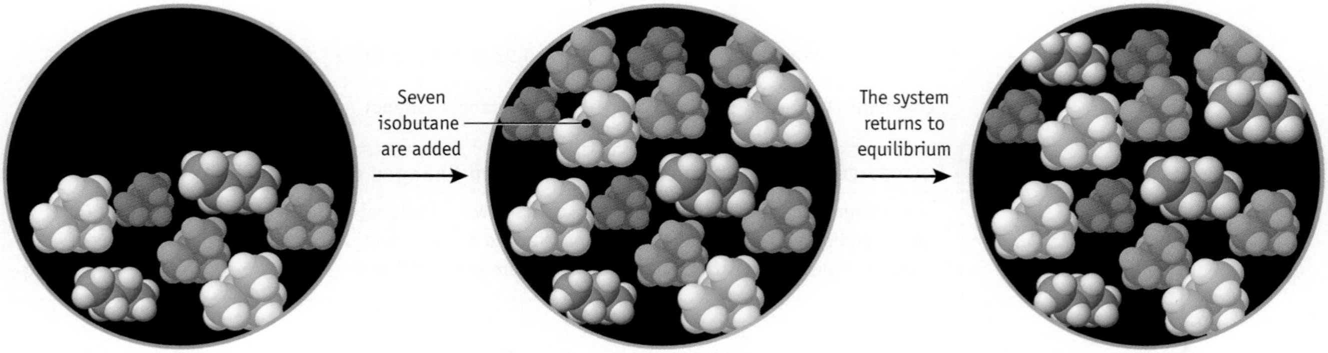

$Q = 5/2 = K$

An equilibrium mixture of five isobutane molecules and two butane molecules.

$Q = 12/2 > K$

Seven isobutane molecules are added, so the system is no longer at equilibrium.

$Q = 10/4 = K$

A net of two isobutane molecules has changed to butane molecules, to once again give an equilibrium mixture where the ratio of isobutane to butane is 5 to 2 (or 2.5/1).

Active Figure 16.7 Addition of more reactant or product to an equilibrium system.

Chemistry Now™ Sign in at www.thomsonedu.com/login and go to the Chapter Contents menu to explore an interactive version of this figure accompanied by an exercise.

EXAMPLE 16.8 Effect of Concentration Changes on Equilibrium

Problem Assume equilibrium has been established in a 1.00-L flask with [butane] = 0.500 mol/L and [isobutane] = 1.25 mol/L.

$$\text{Butane} \rightleftharpoons \text{Isobutane} \quad K = 2.50$$

Then 1.50 mol of butane is added. What are the concentrations of butane and isobutane when equilibrium is reestablished?

Strategy After adding excess butane, $Q < K$. To reestablish equilibrium, the concentration of butane must decrease, and that of isobutane must increase. Use an ICE table to track the changes. The decrease in butane concentration and the increase in isobutane concentration are both designated as x.

Solution First organize the information in a modified ICE table.

Equation	Butane	$\rightleftharpoons$	Isobutane
Initial (M)	0.500		1.25
Concentration immediately on adding butane (M)	0.500 + 1.50		1.25
Change in concentration to reestablish equilibrium (M)	$-x$		$+x$
Equilibrium (M)	$0.500 + 1.50 - x$		$1.25 + x$

The entries in this table were arrived at as follows:

(a) The concentration of butane when equilibrium is reestablished will be the original equilibrium concentration plus what was added (1.50 mol/L) minus the concentration of butane that is converted to isobutane to reestablish equilibrium. The quantity of butane converted to isobutane is unknown and so is designated as x.

(b) The concentration of isobutane when equilibrium is reestablished is the concentration that was already present (1.25 mol/L) plus the concentration formed (x mol/L) on reestablishing equilibrium.

Having defined [butane] and [isobutane] when equilibrium is reestablished and remembering that K is a constant (=2.50), we can write

$$K = 2.50 = \frac{[\text{isobutane}]}{[\text{butane}]}$$

We now calculate the new equilibrium composition:

$$2.50 = \frac{1.25 + x}{0.500 + 1.50 - x} = \frac{1.25 + x}{2.00 - x}$$

$$2.50\ (2.00 - x) = 1.25 + x$$

$$x = 1.07 \text{ mol/L}$$

[butane] = $0.500 + 1.50 - x = 0.93$ M and [isobutane] = $1.25 + x = 2.32$ M

Comment Check your answer to verify that [isobutane]/[butane] = 2.32/0.93 = 2.5.

EXERCISE 16.10 Effect of Concentration Changes on Equilibrium

Equilibrium exists between butane and isobutane when [butane] = 0.20 M and [isobutane] = 0.50 M. An additional 2.00 mol/L of isobutane is added to the mixture. What are the concentrations of butane and isobutane after equilibrium has again been attained?

Effect of Volume Changes on Gas-Phase Equilibria

For a reaction that involves gases, what happens to equilibrium concentrations or pressures if the size of the container is changed? (Such a change occurs, for example, when fuel and air are compressed in an automobile engine.) To answer this question, recall that concentrations are in moles per liter. If the volume of a gas changes, its concentration therefore must also change, and the equilibrium composition can change. As an example, consider the following equilibrium:

$$2\ NO_2(g) \rightleftharpoons N_2O_4(g)$$

brown gas colorless gas

$$K = \frac{[N_2O_4]}{[NO_2]^2} = 170 \text{ at } 298\ K$$

What happens to this equilibrium if the volume of the flask holding the gases is suddenly halved? The immediate result is that the concentrations of both gases will double. For example, assume equilibrium is established when $[N_2O_4]$ is 0.0280 mol/L and $[NO_2]$ is 0.0128 mol/L. When the volume is halved, $[N_2O_4]$ becomes 0.0560 mol/L, and $[NO_2]$ is 0.0256 mol/L. The reaction quotient, Q, under these circumstances is $(0.0560)/(0.0256)^2 = 85.5$, a value less than K. Because Q is less than K, the quantity of product must increase at the expense of the reactants to return to equilibrium, and the new equilibrium composition will have a higher concentration of N_2O_4 than immediately after the volume change.

$$2\ NO_2(g) \rightleftharpoons N_2O_4(g)$$

$$\xrightarrow[\text{new equilibrium favors product}]{\text{decrease volume of container}}$$

The concentration of NO_2 decreases twice as much as the concentration of N_2O_4 increases because one molecule of N_2O_4 is formed by consuming two molecules of NO_2. This occurs until the reaction quotient, $Q = [N_2O_4]/[NO_2]^2$, is once again equal to K. The net effect of the volume decrease is to decrease the number of molecules in the gas phase.

The conclusions for the NO_2/N_2O_4 equilibrium can be generalized:

- For reactions involving gases, the stress of a volume decrease (a pressure increase) will be counterbalanced by a change in the equilibrium composition to one having a smaller number of gas molecules.
- For a volume increase (a pressure decrease), the equilibrium composition will favor the side of the reaction with the larger number of gas molecules.
- For a reaction in which there is no change in the number of gas molecules, such as in the reaction of H_2 and I_2 to produce HI [$H_2(g) + I_2(g) \rightleftharpoons 2\ HI(g)$], a volume change will have no effect.

Chemistry Now™

Sign in at **www.thomsonedu.com/login** and go to Chapter 16 Contents to see Screen 16.14 for a simulation of **the effect of a volume change for a reaction involving gases.**

EXERCISE 16.11 Effect of Concentration and Volume Changes on Equilibria

The formation of ammonia from its elements is an important industrial process.

$$3\ H_2(g) + N_2(g) \rightleftharpoons 2\ NH_3(g)$$

(a) How does the equilibrium composition change when extra H_2 is added? When extra NH_3 is added?

(b) What is the effect on the equilibrium when the volume of the system is increased?

Effect of Temperature Changes on Equilibrium Composition

The value of the equilibrium constant for a given reaction varies with temperature. Changing the temperature of a system at equilibrium is therefore different in some ways from the other means we have studied of disturbing a chemical equilibrium because the equilibrium constant itself will be different at the new temperature from what it was at the previous temperature. Predicting the exact changes in equilibrium compositions with temperature is beyond the scope of this text, but you can make a qualitative prediction about the effect if you know whether the reaction is exothermic or endothermic. As an example, consider the endothermic reaction of N_2 with O_2 to give NO.

■ ***K* for the N_2/O_2 Reaction** We are surrounded by N_2 and O_2, but you know that they do not react appreciably at room temperature. However, if a mixture of N_2 and O_2 is heated above 700 °C, as in an automobile engine, the equilibrium mixture will contain appreciable amounts of NO.

$$N_2(g) + O_2(g) \rightleftharpoons 2\ NO(g) \qquad \Delta_rH° = +180.6\ \text{kJ/mol-rxn}$$

$$K = \frac{[NO]^2}{[N_2][O_2]}$$

Le Chatelier's principle allows us to predict how the value of *K* will vary with temperature. The formation of NO from N_2 and O_2 is endothermic; that is, energy must be provided as heat for the reaction to occur. We might imagine that heat is a "reactant." If the system is at equilibrium and the temperature then increases, the system will adjust to alleviate this "stress." The way to counteract the energy input is to use up some of the energy added as heat by consuming N_2 and O_2 and producing more NO as the system returns to equilibrium. This raises the value of the numerator ($[NO]^2$) and lowers the value of the denominator ($[N_2][O_2]$) in the reaction quotient, *Q*, resulting in a higher value of *K*.

This prediction is borne out. The following table lists the equilibrium constant for this reaction at various temperatures. As predicted, the equilibrium constant and thus the proportion of NO in the equilibrium mixture increase with temperature.

Equilibrium Constant, *K*	Temperature (K)
4.5×10^{-31}	298
6.7×10^{-10}	900
1.7×10^{-3}	2300

As another example, consider the combination of molecules of the brown gas NO_2 to form colorless N_2O_4. An equilibrium between these compounds is readily achieved in a closed system (Figure 16.8).

$$2\ NO_2(g) \rightleftharpoons N_2O_4(g) \qquad \Delta_rH° = -57.1\ \text{kJ/mol-rxn}$$

$$K = \frac{[N_2O_4]}{[NO_2]^2}$$

Equilibrium Constant, *K*	Temperature (K)
1300	273
170	298

Here, the reaction is exothermic, so we might imagine heat as being a reaction "product." By lowering the temperature of the system, as in Figure 16.8, some energy is removed as heat. The removal of energy can be counteracted if the reaction produces energy as heat by the combination of NO_2 molecules to give more N_2O_4. Thus, the equilibrium concentration of NO_2 decreases; the concentration of N_2O_4 increases; and the value of *K* is larger at lower temperatures.

Case Study

Applying Equilibrium Concepts: The Haber–Bosch Process

Nitrogen-containing substances are used around the world to stimulate the growth of field crops. Farmers from Portugal to Tibet have used animal waste for centuries as a "natural" fertilizer. In the 19th century, industrialized countries imported nitrogen-rich marine bird manure from Peru, Bolivia, and Chile, but the supply of this material was clearly limited. In 1898, William Ramsay (the discoverer of the noble gases) pointed out that the amount of "fixed nitrogen" in the world was being depleted and predicted that world food shortages would occur by the mid-20th century as a result. That Ramsay's prediction failed to materialize was due in part to the work of Fritz Haber (1868–1934). In about 1908, Haber developed a method making ammonia directly from the elements,

$$N_2(g) + 3\ H_2(g) \rightleftharpoons 2\ NH_3(g)$$

and, a few years later, Carl Bosch (1874–1940) perfected the industrial scale synthesis. Ammonia is now made for pennies per kilogram and is consistently ranked in the top five chemicals produced in the United States with 15–20 billion kilograms produced annually. Not only is ammonia used directly as a fertilizer, but it is also a starting material for making nitric acid and ammonium nitrate, among other things.

The manufacture of ammonia (Figure A) is a good example of the role that kinetics and chemical equilibria play in practical chemistry.

The $N_2 + H_2$ reaction is exothermic and product-favored ($K > 1$ at 25 °C).

At 25 °C, K (calc'd value) $= 3.5 \times 10^8$
and $\Delta_r H° = -92.2$ kJ/mol-rxn

Unfortunately, the reaction at 25 °C is slow, so it is carried out at a higher temperature to increase the reaction rate. The problem with this, however, is that the equilibrium constant declines with temperature, as predicted by Le Chatelier's Principle.

At 450 °C, K (experimental value) $= 0.16$
and $\Delta_r H° = -111.3$ kJ/mol-rxn

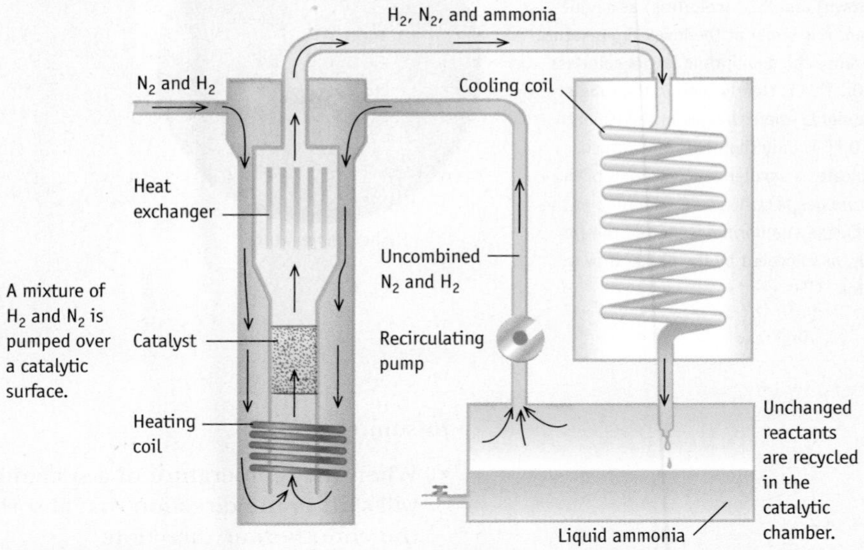

FIGURE A The Haber process for ammonia synthesis. A mixture of H_2 and N_2 is pumped over a catalytic surface. The NH_3 is collected as a liquid (at −33 °C), and unchanged reactants are recycled in the catalytic chamber.

Thus, the yield declines with increasing temperature.

There are two things that can be done. The first is to raise the pressure. This does not change the value of K, but an increase in pressure can be compensated by converting 4 mol of reactants to 2 mol of product.

In an industrial ammonia plant, it is necessary to balance reaction rate (improved at higher temperature) with product yield (K is smaller at higher temperatures). Therefore, catalysts are often used to accelerate reactions. An effective catalyst for the Haber process is Fe_3O_4 mixed with KOH, SiO_2, and Al_2O_3 (all inexpensive chemicals). Because the catalyst is not effective below 400 °C, the process is carried out at 450–500 °C and 250 atm pressure.

Questions:

1. *Anhydrous ammonia is used directly as a fertilizer, but much of it is also converted to other fertilizers, ammonium nitrate and urea.*

   (a) *How is NH_3 converted to ammonium nitrate?*

   (b) *Urea is formed in the reaction of ammonia and CO_2.*

   $$2\ NH_3(g) + CO_2(g) \rightleftharpoons (NH_2)_2CO(s) + H_2O(g)$$

   Which would favor urea production, high temperature or high pressure? ($\Delta_f H°$ for solid urea $= -333.1$ kJ/mol-rxn)

2. *One important aspect of the Haber process is the source of the hydrogen. This is made from natural gas in a process called steam reforming.*

   $$CH_4(g) + H_2O(g) \rightarrow CO(g) + 3\ H_2(g)$$

   $$CO(g) + H_2O(g) \rightarrow CO_2(g) + H_2(g)$$

   (a) *Are the two reactions above endo- or exothermic?*

   (b) *To manufacture 15 billion kilograms of NH_3, how much CH_4 is required, and what mass of CO_2 is produced as a by-product?*

Answers to these questions are in Appendix Q.

FIGURE 16.8 Effect of temperature on an equilibrium. The tubes in the photograph both contain gaseous NO_2 (brown) and N_2O_4 (colorless) at equilibrium. *K* is larger at the lower temperature because the equilibrium favors colorless N_2O_4. This is clearly seen in the tube at the right, where the gas in the ice bath at 0 °C is only slightly brown, which indicates a smaller concentration of the brown gas NO_2. At 50 °C (the tube at the left), the equilibrium is shifted toward NO_2, as indicated by the darker brown color.

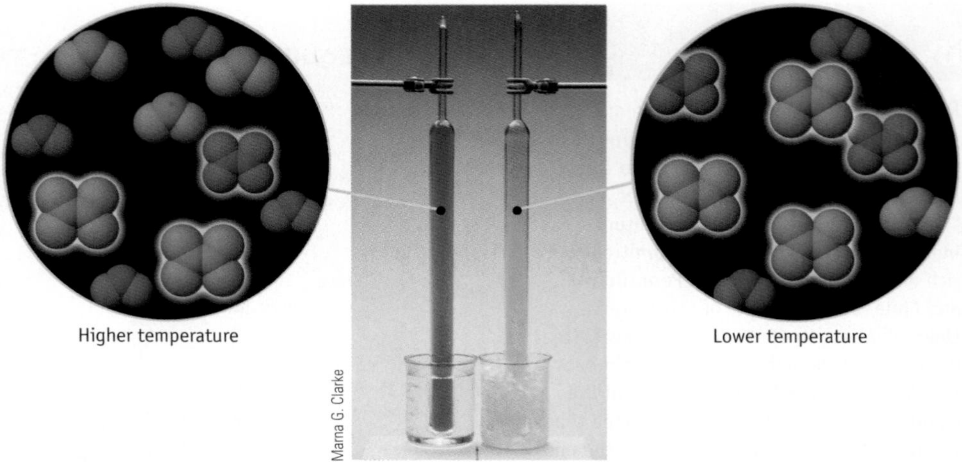

In summary,

- When the temperature of a system at equilibrium increases, the equilibrium will shift in the direction that absorbs energy as heat (Table 16.2)—that is, in the endothermic direction.
- If the temperature decreases, the equilibrium will shift in the direction that releases energy as heat—that is, in the exothermic direction.
- Changing the temperature changes the value of *K*.

Chemistry Now™

Sign in at **www.thomsonedu.com/login** and go to Chapter 16 Contents to see Screen 16.12 for a tutorial on **the effect of temperature changes.**

EXERCISE 16.12 Disturbing a Chemical Equilibrium

Does the equilibrium concentration of NOCl increase or decrease as the temperature of the system is increased?

$$2\ NOCl(g) \rightleftharpoons 2\ NO(g) + Cl_2(g) \quad \Delta_r H° = +77.1\ kJ$$

Chapter Goals Revisited

Now that you have studied this chapter, you should ask whether you have met the chapter goals. In particular, you should be able to:

Understand the nature and characteristics of chemical equilibria

a. Chemical reactions are reversible and equilibria are dynamic (Section 16.1).

Understand the significance of the equilibrium constant, *K*, and reaction quotient, *Q*.

a. Write the reaction quotient, *Q*, for a chemical reaction (Section 16.2). When the system is at equilibrium, the reaction quotient is called the equilibrium constant expression and has a constant value called the equilibrium constant, which is symbolized by *K* (Equation 16.2). **Study Question(s) assignable in OWL: 2,4.**

b. Recognize that the concentrations of solids, pure liquids, and solvents (e.g., water) are not included in the equilibrium constant expression (Equation 16.1, Section 16.2).

c. Recognize that a large value of K ($K > 1$) means the reaction is product-favored, and the product concentrations are greater than the reactant concentrations at equilibrium. A small value of K ($K < 1$) indicates a reactant-favored reaction in which the product concentrations are smaller than the reactant concentrations at equilibrium (Section 16.2). **Study Question(s) assignable in OWL: 66, 68, 70.**

d. Appreciate the fact that equilibrium concentrations may be expressed in terms of reactant and product concentrations (in moles per liter) and that K is then sometimes designated as K_c. Alternatively, concentrations of gases may be represented by partial pressures, and K for such cases is designated K_p (Section 16.2).

Understand how to use *K* in quantitative studies of chemical equilibria

a. Use the reaction quotient (Q) to decide whether a reaction is at equilibrium ($Q = K$), or if there will be a net conversion of reactants to products ($Q < K$) or products to reactants ($Q > K$) to attain equilibrium (Section 16.2).

b. Calculate an equilibrium constant given the reactant and product concentrations at equilibrium (Section 16.3). **Study Question(s) assignable in OWL: 8, 11, 29, 33, 34, 44, 61a.**

c. Use equilibrium constants to calculate the concentration (or pressure) of a reactant or a product at equilibrium (Section 16.4). **Study Question(s) assignable in OWL: 16, 17, 32, 36, 42, 46, 47, 50–54, 56, 58–62; Go Chemistry Module 21.**

d. Know how K changes as different stoichiometric coefficients are used in a balanced equation, if the equation is reversed, or if several equations are added to give a new net equation (Section 16.5). **Study Question(s) assignable in OWL: 21, 31, 37.**

e. Know how to predict, using Le Chatelier's principle, the effect of a disturbance on a chemical equilibrium—a change in temperature, a change in concentrations, or a change in volume or pressure for a reaction involving gases (Section 16.6 and Table 16.2). **Study Question(s) assignable in OWL: 25, 26, 28, 39, 41, 54, 62.**

Chemistry.♁.Now™ Sign in at **www.thomsonedu.com/login** to:

- Assess your understanding with Study Questions in OWL keyed to each goal in the Goals and Homework menu for this chapter
- For quick review, download Go Chemistry mini-lecture flashcard modules (or purchase them at **www.ichapters.com**)
- Check your readiness for an exam by taking the Pre-Test and exploring the modules recommended in your Personalized Study plan.

Access **How Do I Solve It?** tutorials on how to approach problem solving using concepts in this chapter.

KEY EQUATIONS

Equation 16.1 (page 727) The equilibrium constant expression. At equilibrium, the ratio of products to reactants has a constant value, K (at a particular temperature). For the general reaction $aA + bB \rightleftharpoons cC + dD$,

$$\text{Equilibrium constant} = K = \frac{[C]^c[D]^d}{[A]^a[B]^b}$$

Equation 16.2 (page 732) For the general reaction $aA + bB \rightleftharpoons cC + dD$, the ratio of product to reactant concentrations at any point in the reaction is the reaction quotient.

$$\text{Reaction quotient} = Q = \frac{[C]^c[D]^d}{[A]^a[B]^b}$$

STUDY QUESTIONS

Online homework for this chapter may be assigned in OWL.

▲ denotes challenging questions.

■ denotes questions assignable in OWL.

Blue-numbered questions have answers in Appendix O and fully-worked solutions in the *Student Solutions Manual*.

Practicing Skills

Writing Equilibrium Constant Expressions

(See Example 16.1 and ChemistryNow Screens 16.3, 16.4, and 16.5.)

1. Write equilibrium constant expressions for the following reactions. For gases, use either pressures or concentrations.
(a) $2\ H_2O_2(g) \rightleftharpoons 2\ H_2O(g) + O_2(g)$
(b) $CO(g) + \frac{1}{2}\ O_2(g) \rightleftharpoons CO_2(g)$
(c) $C(s) + CO_2(g) \rightleftharpoons 2\ CO(g)$
(d) $NiO(s) + CO(g) \rightleftharpoons Ni(s) + CO_2(g)$

2. ■ Write equilibrium constant expressions for the following reactions. For gases, use either pressures or concentrations.
(a) $3\ O_2(g) \rightleftharpoons 2\ O_3(g)$
(b) $Fe(s) + 5\ CO(g) \rightleftharpoons Fe(CO)_5(g)$
(c) $(NH_4)_2CO_3(s) \rightleftharpoons 2\ NH_3(g) + CO_2(g) + H_2O(g)$
(d) $Ag_2SO_4(s) \rightleftharpoons 2\ Ag^+(aq) + SO_4^{2-}(aq)$

The Equilibrium Constant and Reaction Quotient

(See Example 16.2 and ChemistryNow Screen 16.9.)

3. $K = 5.6 \times 10^{-12}$ at 500 K for the dissociation of iodine molecules to iodine atoms.

$$I_2(g) \rightleftharpoons 2\ I(g)$$

A mixture has $[I_2] = 0.020$ mol/L and $[I] = 2.0 \times 10^{-8}$ mol/L. Is the reaction at equilibrium (at 500 K)? If not, which way must the reaction proceed to reach equilibrium?

4. ■ The reaction

$$2\ NO_2(g) \rightleftharpoons N_2O_4(g)$$

has an equilibrium constant, K_c, of 170 at 25 °C. If 2.0×10^{-3} mol of NO_2 is present in a 10.-L flask along with 1.5×10^{-3} mol of N_2O_4, is the system at equilibrium? If it is not at equilibrium, does the concentration of NO_2 increase or decrease as the system proceeds to equilibrium?

5. A mixture of SO_2, O_2, and SO_3 at 1000 K contains the gases at the following concentrations: $[SO_2] = 5.0 \times 10^{-3}$ mol/L, $[O_2] = 1.9 \times 10^{-3}$ mol/L, and $[SO_3] = 6.9 \times 10^{-3}$ mol/L. Is the reaction at equilibrium? If not, which way will the reaction proceed to reach equilibrium?

$$2\ SO_2(g) + O_2(g) \rightleftharpoons 2\ SO_3(g) \quad K = 279$$

6. The equilibrium constant, K_c, for the reaction

$$2\ NOCl(g) \rightleftharpoons 2\ NO(g) + Cl_2(g)$$

is 3.9×10^{-3} at 300 °C. A mixture contains the gases at the following concentrations: $[NOCl] = 5.0 \times 10^{-3}$ mol/L, $[NO] = 2.5 \times 10^{-3}$ mol/L, and $[Cl_2] = 2.0 \times 10^{-3}$ mol/L. Is the reaction at equilibrium at 300 °C? If not, in which direction does the reaction proceed to come to equilibrium?

Calculating an Equilibrium Constant

(See Examples 16.3 and 16.4 and ChemistryNow Screens 16.4 and 16.8.)

7. The reaction

$$PCl_5(g) \rightleftharpoons PCl_3(g) + Cl_2(g)$$

was examined at 250 °C. At equilibrium, $[PCl_5] = 4.2 \times 10^{-5}$ mol/L, $[PCl_3] = 1.3 \times 10^{-2}$ mol/L, and $[Cl_2] = 3.9 \times 10^{-3}$ mol/L. Calculate K for the reaction.

8. ■ An equilibrium mixture of SO_2, O_2, and SO_3 at a high temperature contains the gases at the following concentrations: $[SO_2] = 3.77 \times 10^{-3}$ mol/L, $[O_2] = 4.30 \times 10^{-3}$ mol/L, and $[SO_3] = 4.13 \times 10^{-3}$ mol/L. Calculate the equilibrium constant, K, for the reaction.

$$2\ SO_2(g) + O_2(g) \rightleftharpoons 2\ SO_3(g)$$

9. The reaction

$$C(s) + CO_2(g) \rightleftharpoons 2\ CO(g)$$

occurs at high temperatures. At 700 °C, a 2.0-L flask contains 0.10 mol of CO, 0.20 mol of CO_2, and 0.40 mol of C at equilibrium.
(a) Calculate K for the reaction at 700 °C.
(b) Calculate K for the reaction, also at 700 °C, if the amounts at equilibrium in the 2.0-L flask are 0.10 mol of CO, 0.20 mol of CO_2, and 0.80 mol of C.
(c) Compare the results of (a) and (b). Does the quantity of carbon affect the value of K? Explain.

10. Hydrogen and carbon dioxide react at a high temperature to give water and carbon monoxide.

$$H_2(g) + CO_2(g) \rightleftharpoons H_2O(g) + CO(g)$$

(a) Laboratory measurements at 986 °C show that there are 0.11 mol each of CO and H_2O vapor and 0.087 mol each of H_2 and CO_2 at equilibrium in a 1.0-L container. Calculate the equilibrium constant for the reaction at 986 °C.
(b) Suppose 0.050 mol each of H_2 and CO_2 are placed in a 2.0-L container. When equilibrium is achieved at 986 °C, what amounts of CO(g) and H_2O(g), in moles, would be present? [Use the value of K from part (a).]

11. ■ A mixture of CO and Cl_2 is placed in a reaction flask: [CO] = 0.0102 mol/L and $[Cl_2]$ = 0.00609 mol/L. When the reaction

$$CO(g) + Cl_2(g) \rightleftharpoons COCl_2(g)$$

has come to equilibrium at 600 K, $[Cl_2]$ = 0.00301 mol/L.

(a) Calculate the concentrations of CO and $COCl_2$ at equilibrium.

(b) Calculate K_c.

12. You place 3.00 mol of pure SO_3 in an 8.00-L flask at 1150 K. At equilibrium, 0.58 mol of O_2 has been formed. Calculate K for the reaction at 1150 K.

$$2\ SO_3(g) \rightleftharpoons 2\ SO_2(g) + O_2(g)$$

Using Equilibrium Constants

(See Examples 16.5 and 16.6 and ChemistryNow Screen 16.10.)

13. The value of K_c for the interconversion of butane and isobutane is 2.5 at 25 °C.

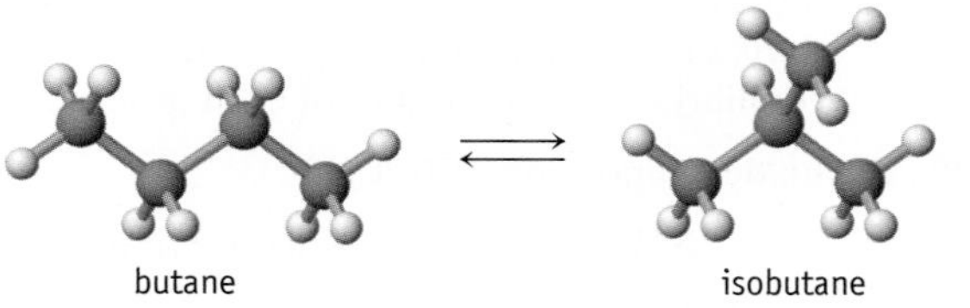

If you place 0.017 mol of butane in a 0.50-L flask at 25 °C and allow equilibrium to be established, what will be the equilibrium concentrations of the two forms of butane?

14. Cyclohexane, C_6H_{12}, a hydrocarbon, can isomerize or change into methylcyclopentane, a compound of the same formula ($C_5H_9CH_3$) but with a different molecular structure.

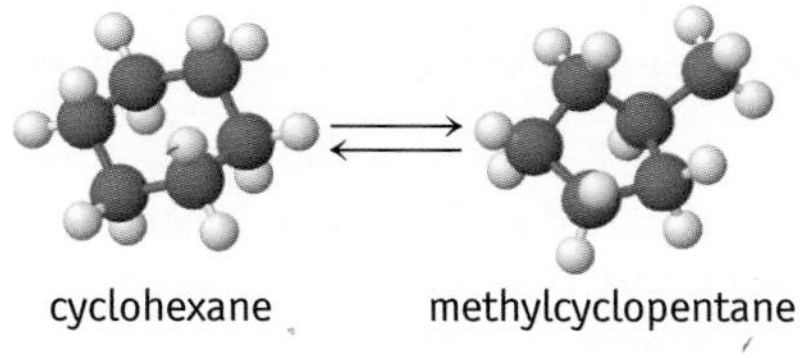

The equilibrium constant has been estimated to be 0.12 at 25 °C. If you had originally placed 0.045 mol of cyclohexane in a 2.8-L flask, what would be the concentrations of cyclohexane and methylcyclopentane when equilibrium is established?

15. The equilibrium constant for the dissociation of iodine molecules to iodine atoms

$$I_2(g) \rightleftharpoons 2\ I(g)$$

is 3.76×10^{-3} at 1000 K. Suppose 0.105 mol of I_2 is placed in a 12.3-L flask at 1000 K. What are the concentrations of I_2 and I when the system comes to equilibrium?

16. ■ The equilibrium constant for the reaction

$$N_2O_4(g) \rightleftharpoons 2\ NO_2(g)$$

at 25 °C is 5.88×10^{-3}. Suppose 15.6 g of N_2O_4 is placed in a 5.00-L flask at 25 °C. Calculate the following:

(a) the amount of NO_2 (mol) present at equilibrium;

(b) the percentage of the original N_2O_4 that is dissociated.

17. ■ Carbonyl bromide decomposes to carbon monoxide and bromine.

$$COBr_2(g) \rightleftharpoons CO(g) + Br_2(g)$$

K_c is 0.190 at 73 °C. If you place 0.500 mol of $COBr_2$ in a 2.00-L flask and heat it to 73 °C, what are the equilibrium concentrations of $COBr_2$, CO, and Br_2? What percentage of the original $COBr_2$ decomposed at this temperature?

18. Iodine dissolves in water, but its solubility in a nonpolar solvent such as CCl_4 is greater.

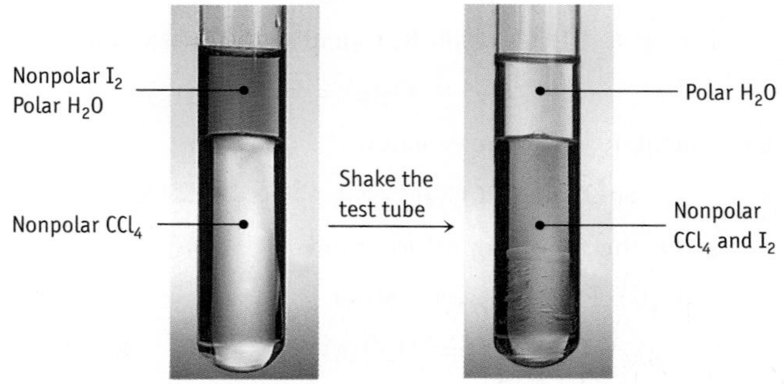

Photos: Charles D. Winters

Extracting iodine (I_2) from water with the nonpolar solvent CCl_4. I_2 is more soluble in CCl_4 and, after shaking a mixture of water and CCl_4, the I_2 has accumulated in the more dense CCl_4 layer.

The equilibrium constant is 85.0 for the reaction

$$I_2(aq) \rightleftharpoons I_2(CCl_4)$$

You place 0.0340 g of I_2 in 100.0 mL of water. After shaking it with 10.0 mL of CCl_4, how much I_2 remains in the water layer?

Manipulating Equilibrium Constant Expressions

(See Example 16.7 and ChemistryNow Screen 16.7.)

19. Which of the following correctly relates the equilibrium constants for the two reactions shown?

$$A + B \rightleftharpoons 2\ C \quad K_1$$

$$2\ A + 2\ B \rightleftharpoons 4\ C \quad K_2$$

(a) $K_2 = 2K_1$ (c) $K_2 = 1/K_1$

(b) $K_2 = K_1^2$ (d) $K_2 = 1/K_1^2$

20. Which of the following correctly relates the equilibrium constants for the two reactions shown?

$$A + B \rightleftharpoons 2\,C \quad K_1$$

$$C \rightleftharpoons \tfrac{1}{2}\,A + \tfrac{1}{2}\,B \quad K_2$$

(a) $K_2 = 1/(K_1)^{1/2}$ (c) $K_2 = K_1^2$
(b) $K_2 = 1/K_1$ (d) $K_2 = -K_1^{1/2}$

21. ■ Consider the following equilibria involving $SO_2(g)$ and their corresponding equilibrium constants.

$$SO_2(g) + \tfrac{1}{2}\,O_2(g) \rightleftharpoons SO_3(g) \quad K_1$$

$$2\,SO_3(g) \rightleftharpoons 2\,SO_2(g) + O_2(g) \quad K_2$$

Which of the following expressions relates K_1 to K_2?
(a) $K_2 = K_1^2$ (d) $K_2 = 1/K_1$
(b) $K_2^2 = K_1$ (e) $K_2 = 1/K_1^2$
(c) $K_2 = K_1$

22. The equilibrium constant K for the reaction

$$CO_2(g) \rightleftharpoons CO(g) + \tfrac{1}{2}\,O_2(g)$$

is 6.66×10^{-12} at 1000 K. Calculate K for the reaction

$$2\,CO(g) + O_2(g) \rightleftharpoons 2\,CO_2(g)$$

23. Calculate K for the reaction

$$SnO_2(s) + 2\,CO(g) \rightleftharpoons Sn(s) + 2\,CO_2(g)$$

given the following information:

$$SnO_2(s) + 2\,H_2(g) \rightleftharpoons Sn(s) + 2\,H_2O(g) \quad K = 8.12$$

$$H_2(g) + CO_2(g) \rightleftharpoons H_2O(g) + CO(g) \quad K = 0.771$$

24. Calculate K for the reaction

$$Fe(s) + H_2O(g) \rightleftharpoons FeO(s) + H_2(g)$$

given the following information:

$$H_2O(g) + CO(g) \rightleftharpoons H_2(g) + CO_2(g) \quad K = 1.6$$

$$FeO(s) + CO(g) \rightleftharpoons Fe(s) + CO_2(g) \quad K = 0.67$$

Disturbing a Chemical Equilibrium

(See Example 16.8 and ChemistryNow Screens 16.11–16.14.)

25. ■ Dinitrogen trioxide decomposes to NO and NO_2 in an endothermic process ($\Delta_r H$ = 40.5 kJ/mol-rxn).

$$N_2O_3(g) \rightleftharpoons \text{`}NO(g) + NO_2(g)$$

Predict the effect of the following changes on the position of the equilibrium; that is, state which way the equilibrium will shift (left, right, or no change) when each of the following changes is made.
(a) adding more $N_2O_3(g)$
(b) adding more $NO_2(g)$
(c) increasing the volume of the reaction flask
(d) lowering the temperature

26. ■ K_p for the following reaction is 0.16 at 25 °C:

$$2\,NOBr(g) \rightleftharpoons 2\,NO(g) + Br_2(g)$$

The enthalpy change for the reaction at standard conditions is +16.3 k/mol-rxn. Predict the effect of the following changes on the position of the equilibrium; that is, state which way the equilibrium will shift (left, right, or no change) when each of the following changes is made.
(a) adding more $Br_2(g)$
(b) removing some NOBr(g)
(c) decreasing the temperature
(d) increasing the container volume

27. Consider the isomerization of butane with an equilibrium constant of $K = 2.5$. (See Study Question 13.) The system is originally at equilibrium with [butane] = 1.0 M and [isobutane] = 2.5 M.
(a) If 0.50 mol/L of isobutane is suddenly added and the system shifts to a new equilibrium position, what is the equilibrium concentration of each gas?
(b) If 0.50 mol/L of butane is added and the system shifts to a new equilibrium position, what is the equilibrium concentration of each gas?

28. ■ The decomposition of NH_4HS

$$NH_4HS(s) \rightleftharpoons NH_3(g) + H_2S(g)$$

is an endothermic process. Using Le Chatelier's principle, explain how increasing the temperature would affect the equilibrium. If more NH_4HS is added to a flask in which this equilibrium exists, how is the equilibrium affected? What if some additional NH_3 is placed in the flask? What will happen to the pressure of NH_3 if some H_2S is removed from the flask?

General Questions

These questions are not designated as to type or location in the chapter. They may combine several concepts from this and other chapters.

29. ■ Suppose 0.086 mol of Br_2 is placed in a 1.26-L flask and heated to 1756 K, a temperature at which the halogen dissociates to atoms

$$Br_2(g) \rightleftharpoons 2\,Br(g)$$

If Br_2 is 3.7% dissociated at this temperature, calculate K_c.

30. The equilibrium constant for the reaction

$$N_2(g) + O_2(g) \rightleftharpoons 2\,NO(g)$$

is 1.7×10^{-3} at 2300 K.
(a) What is K for the reaction when written as follows?

$$\tfrac{1}{2}\,N_2(g) + \tfrac{1}{2}\,O_2(g) \rightleftharpoons NO\,(g)$$

(b) What is K for the following reaction?

$$2\,NO(g) \rightleftharpoons N_2(g) + O_2(g)$$

31. ■ K_p for the formation of phosgene, $COCl_2$, is 6.5×10^{11} at 25 °C.

$$CO(g) + Cl_2(g) \rightleftharpoons COCl_2(g)$$

What is the value of K_p for the dissociation of phosgene?

$$COCl_2(g) \rightleftharpoons CO(g) + Cl_2(g)$$

32. ■ The equilibrium constant, K_c, for the following reaction is 1.05 at 350 K.

$$2\ CH_2Cl_2(g) \rightleftharpoons CH_4(g) + CCl_4(g)$$

If an equilibrium mixture of the three gases at 350 K contains 0.0206 M $CH_2Cl_2(g)$ and 0.0163 M CH_4, what is the equilibrium concentration of CCl_4?

33. ■ Carbon tetrachloride can be produced by the following reaction:

$$CS_2(g) + 3\ Cl_2(g) \rightleftharpoons S_2Cl_2(g) + CCl_4(g)$$

Suppose 1.2 mol of CS_2 and 3.6 mol of Cl_2 are placed in a 1.00-L flask. After equilibrium has been achieved, the mixture contains 0.90 mol CCl_4. Calculate K_c.

34. ■ Equal numbers of moles of H_2 gas and I_2 vapor are mixed in a flask and heated to 700 °C. The initial concentration of each gas is 0.0088 mol/L, and 78.6% of the I_2 is consumed when equilibrium is achieved according to the equation

$$H_2(g) + I_2(g) \rightleftharpoons 2\ HI(g)$$

Calculate K_c for this reaction.

35. The equilibrium constant for the butane $\rightleftharpoons$ isobutane isomerization reaction is 2.5 at 25 °C. If 1.75 mol of butane and 1.25 mol of isobutane are mixed, is the system at equilibrium? If not, when it proceeds to equilibrium, which reagent increases in concentration? Calculate the concentrations of the two compounds when the system reaches equilibrium.

36. ■ At 2300 K the equilibrium constant for the formation of NO(g) is 1.7×10^{-3}.

$$N_2(g) + O_2(g) \rightleftharpoons 2\ NO(g)$$

(a) Analysis shows that the concentrations of N_2 and O_2 are both 0.25 M, and that of NO is 0.0042 M under certain conditions. Is the system at equilibrium?
(b) If the system is not at equilibrium, in which direction does the reaction proceed?
(c) When the system is at equilibrium, what are the equilibrium concentrations?

37. ■ Which of the following correctly relates the two equilibrium constants for the two reactions shown?

$$NOCl(g) \rightleftharpoons NO(g) + \tfrac{1}{2}\ Cl_2(g) \qquad K_1$$

$$2\ NO(g) + Cl_2(g) \rightleftharpoons 2\ NOCl(g) \qquad K_2$$

(a) $K_2 = -K_1^2$ (c) $K_2 = 1/K_1^2$
(b) $K_2 = 1/(K_1)^{1/2}$ (d) $K_2 = 2K_1$

38. Sulfur dioxide is readily oxidized to sulfur trioxide.

$$2\ SO_2(g) + O_2(g) \rightleftharpoons 2\ SO_3(g) \qquad K_c = 279$$

If we add 3.00 g of SO_2 and 5.00 g of O_2 to a 1.0-L flask, approximately what quantity of SO_3 will be in the flask once the reactants and the product reach equilibrium?
(a) 2.21 g (c) 3.61 g
(b) 4.56 g (d) 8.00 g

(Note: The full solution to this problem results in a cubic equation. Do not try to solve it exactly. Decide only which of the answers is most reasonable.)

39. ■ Heating a metal carbonate leads to decomposition.

$$BaCO_3(s) \rightleftharpoons BaO(s) + CO_2(g)$$

Predict the effect on the equilibrium of each change listed below. Answer by choosing (i) no change, (ii) shifts left, or (iii) shifts right.
(a) add $BaCO_3$ (c) add BaO
(b) add CO_2 (d) raise the temperature
(e) increase the volume of the flask containing the reaction

40. Carbonyl bromide decomposes to carbon monoxide and bromine.

$$COBr_2(g) \rightleftharpoons CO(g) + Br_2(g)$$

K_c is 0.190 at 73 °C. Suppose you place 0.500 mol of $COBr_2$ in a 2.00-L flask and heat it to 73 °C (Study Question 17). After equilibrium has been achieved, you add an additional 2.00 mol of CO.
(a) How is the equilibrium mixture affected by adding more CO?
(b) When equilibrium *is* reestablished, what are the new equilibrium concentrations of $COBr_2$, CO, and Br_2?
(c) How has the addition of CO affected the percentage of $COBr_2$ that decomposed?

41. ■ Phosphorus pentachloride decomposes at higher temperatures.

$$PCl_5(g) \rightleftharpoons PCl_3(g) + Cl_2(g)$$

An equilibrium mixture at some temperature consists of 3.120 g of PCl_5, 3.845 g of PCl_3, and 1.787 g of Cl_2 in a 1.00-L flask. If you add 1.418 g of Cl_2, how will the equilibrium be affected? What will the concentrations of PCl_5, PCl_3, and Cl_2 be when equilibrium is reestablished?

42. ■ Ammonium hydrogen sulfide decomposes on heating.

$$NH_4HS(s) \rightleftharpoons NH_3(g) + H_2S(g)$$

If K_p for this reaction is 0.11 at 25 °C (when the partial pressures are measured in atmospheres), what is the total pressure in the flask at equilibrium?

43. ■ Ammonium iodide dissociates reversibly to ammonia and hydrogen iodide if the salt is heated to a sufficiently high temperature.

$$NH_4I(s) \rightleftharpoons NH_3(g) + HI(g)$$

Some ammonium iodide is placed in a flask, which is then heated to 400 °C. If the total pressure in the flask when equilibrium has been achieved is 705 mm Hg, what is the value of K_p (when partial pressures are in atmospheres)?

44. ■ When solid ammonium carbamate sublimes, it dissociates completely into ammonia and carbon dioxide according to the following equation:

$$(NH_4)(H_2NCO_2)(s) \rightleftharpoons 2\ NH_3(g) + CO_2(g)$$

At 25 °C, experiment shows that the total pressure of the gases in equilibrium with the solid is 0.116 atm. What is the equilibrium constant, K_p?

45. The equilibrium reaction $N_2O_4(g) \rightleftharpoons 2\ NO_2(g)$ has been thoroughly studied (see Figure 16.8).
(a) If the total pressure in a flask containing NO_2 and N_2O_4 gas at 25 °C is 1.50 atm and the value of K_p at this temperature is 0.148, what fraction of the N_2O_4 has dissociated to NO_2?
(b) What happens to the fraction dissociated if the volume of the container is increased so that the total equilibrium pressure falls to 1.00 atm?

46. ■ In the gas phase, acetic acid exists as an equilibrium of monomer and dimer molecules. (The dimer consists of two molecules linked through hydrogen bonds.)

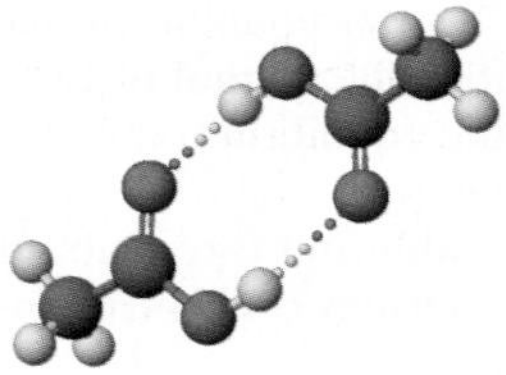

The equilibrium constant, K_c, at 25 °C for the monomer–dimer equilibrium

$$2\ CH_3CO_2H \rightleftharpoons (CH_3CO_2H)_2$$

has been determined to be 3.2×10^4. Assume that acetic acid is present initially at a concentration of 5.4×10^{-4} mol/L at 25 °C and that no dimer is present initially.
(a) What percentage of the acetic acid is converted to dimer?
(b) As the temperature increases, in which direction does the equilibrium shift? (Recall that hydrogen-bond formation is an exothermic process.)

47. ■ Assume 3.60 mol of ammonia is placed in a 2.00-L vessel and allowed to decompose to the elements.

$$2\ NH_3(g) \rightleftharpoons N_2(g) + 3\ H_2(g)$$

If the experimental value of K is 6.3 for this reaction at the temperature in the reactor, calculate the equilibrium concentration of each reagent. What is the total pressure in the flask?

48. ■ The total pressure for a mixture of N_2O_4 and NO_2 is 1.5 atm. If $K_p = 6.75$ (at 25 °C), calculate the partial pressure of each gas in the mixture.

$$2\ NO_2(g) \rightleftharpoons N_2O_4(g)$$

49. K_c for the decomposition of ammonium hydrogen sulfide is 1.8×10^{-4} at 25 °C.

$$NH_4HS(s) \rightleftharpoons NH_3(g) + H_2S(g)$$

(a) When the pure salt decomposes in a flask, what are the equilibrium concentrations of NH_3 and H_2S?
(b) If NH_4HS is placed in a flask already containing 0.020 mol/L of NH_3 and then the system is allowed to come to equilibrium, what are the equilibrium concentrations of NH_3 and H_2S?

50. ■ The equilibrium constant, K_p, is 0.15 at 25 °C for the following reaction:

$$N_2O_4(g) \rightleftharpoons 2\ NO_2(g)$$

If the total pressure of the gas mixture is 2.5 atm at equilibrium, what is the partial pressure of each gas?

51. ■ ▲ A 15-L flask at 300 K contains 64.4 g of a mixture of NO_2 and N_2O_4 in equilibrium. What is the total pressure in the flask? (K_p for NO_2 (g) $\rightleftharpoons N_2O_4(g)$ is 6.67.)

52. ■ ▲ Lanthanum oxalate decomposes when heated to lanthanum oxide, CO, and CO_2.

$$La_2(C_2O_4)_3(s) \rightleftharpoons La_2O_3(s) + 3\ CO(g) + 3\ CO_2(g)$$

(a) If, at equilibrium, the total pressure in a 10.0-L flask is 0.200 atm, what is the value of K_p?
(b) Suppose 0.100 mol of $La_2(C_2O_4)_3$ was originally placed in the 10.0-L flask. What quantity of $La_2(C_2O_4)_3$ remains unreacted at equilibrium at 373 K?

53. ■ ▲ The reaction of hydrogen and iodine to give hydrogen iodide has an equilibrium constant, K_c, of 56 at 435 °C.
(a) What is the value of K_p?
(b) Suppose you mix 0.45 mol of H_2 and 0.45 mol of I_2 in a 10.0-L flask at 425 °C. What is the total pressure of the mixture before and after equilibrium is achieved?
(c) What is the partial pressure of each gas at equilibrium?

▲ more challenging ■ in OWL Blue-numbered questions answered in Appendix O

54. ■ Sulfuryl chloride, SO_2Cl_2, is a compound with very irritating vapors; it is used as a reagent in the synthesis of organic compounds. When heated to a sufficiently high temperature, it decomposes to SO_2 and Cl_2.

$$SO_2Cl_2(g) \rightleftharpoons SO_2(g) + Cl_2(g) \qquad K_c = 0.045 \text{ at } 375\ °C$$

(a) A 1.00-L flask containing 6.70 g of SO_2Cl_2 is heated to 375 °C. What is the concentration of each of the compounds in the system when equilibrium is achieved? What fraction of SO_2Cl_2 has dissociated?
(b) What are the concentrations of SO_2Cl_2, SO_2, and Cl_2 at equilibrium in the 1.00-L flask at 375 °C if you begin with a mixture of SO_2Cl_2 (6.70 g) and Cl_2 (1.00 atm)? What fraction of SO_2Cl_2 has dissociated?
(c) Compare the fractions of SO_2Cl_2 in parts (a) and (b). Do they agree with your expectations based on Le Chatelier's principle?

55. ▲ Hemoglobin (Hb) can form a complex with both O_2 and CO. For the reaction

$$HbO_2(aq) + CO(g) \rightleftharpoons HbCO(aq) + O_2(g)$$

at body temperature, K is about 200. If the ratio [HbCO]/[HbO_2] comes close to 1, death is probable. What partial pressure of CO in the air is likely to be fatal? Assume the partial pressure of O_2 is 0.20 atm.

56. ■ ▲ Limestone decomposes at high temperatures.

$$CaCO_3(s) \rightleftharpoons CaO(s) + CO_2(g)$$

At 1000 °C, $K_p = 3.87$. If pure $CaCO_3$ is placed in a 5.00-L flask and heated to 1000 °C, what quantity of $CaCO_3$ must decompose to achieve the equilibrium pressure of CO_2?

57. At 1800 K, oxygen dissociates very slightly into its atoms.

$$O_2(g) \rightleftharpoons 2\ O(g) \qquad K_p = 1.2 \times 10^{-10}$$

If you place 1.0 mol of O_2 in a 10.-L vessel and heat it to 1800 K, how many O atoms are present in the flask?

58. ■ ▲ Nitrosyl bromide, NOBr, is prepared by the direct reaction of NO and Br_2.

$$2\ NO(g) + Br_2(g) \rightarrow 2\ NOBr(g)$$

The compound dissociates readily at room temperature, however.

$$NOBr(g) \rightleftharpoons NO(g) + \tfrac{1}{2}\ Br_2(g)$$

Some NOBr is placed in a flask at 25 °C and allowed to dissociate. The total pressure at equilibrium is 190 mm Hg and the compound is found to be 34% dissociated. What is the value of K_p?

59. ■ ▲ Boric acid and glycerin form a complex

$$B(OH)_3(aq) + \text{glycerin}(aq) \rightleftharpoons B(OH)_3 \cdot \text{glycerin}(aq)$$

with an equilibrium constant of 0.90. If the concentration of boric acid is 0.10 M, how much glycerin should be added, per liter, so that 60.% of the boric acid is in the form of the complex?

60. ■ ▲ The dissociation of calcium carbonate has an equilibrium constant of $K_p = 1.16$ at 800 °C.

$$CaCO_3(s) \rightleftharpoons CaO(s) + CO_2(g)$$

(a) What is K_c for the reaction?
(b) If you place 22.5 g of $CaCO_3$ in a 9.56-L container at 800 °C, what is the pressure of CO_2 in the container?
(c) What percentage of the original 22.5-g sample of $CaCO_3$ remains undecomposed at equilibrium?

61. ▲ A sample of N_2O_4 gas with a pressure of 1.00 atm is placed in a flask. When equilibrium is achieved, 20.0% of the N_2O_4 has been converted to NO_2 gas.
(a) ■ Calculate K_p.
(b) If the original pressure of N_2O_4 is 0.10 atm, what is the percent dissociation of the gas? Is the result in agreement with Le Chatelier's principle?

62. ■ ▲ A reaction important in smog formation is

$$O_3(g) + NO(g) \rightleftharpoons O_2(g) + NO_2(g) \qquad K = 6.0 \times 10^{34}$$

(a) If the initial concentrations are $[O_3] = 1.0 \times 10^{-6}$ M, $[NO] = 1.0 \times 10^{-5}$ M, $[NO_2] = 2.5 \times 10^{-4}$ M, and $[O_2] = 8.2 \times 10^{-3}$ M, is the system at equilibrium? If not, in which direction does the reaction proceed?
(b) If the temperature is increased, as on a very warm day, will the concentrations of the products increase or decrease? (*Hint:* You may have to calculate the enthalpy change for the reaction to find out if it is exothermic or endothermic.)

In the Laboratory

63. ▲ The ammonia complex of trimethylborane, $(NH_3)B(CH_3)_3$, dissociates at 100 °C to its components with $K_p = 4.62$ (when the pressures are in atmospheres).

$$(NH_3)B(CH_3)_3(g) \rightleftharpoons B(CH_3)_3(g) + NH_3(g)$$

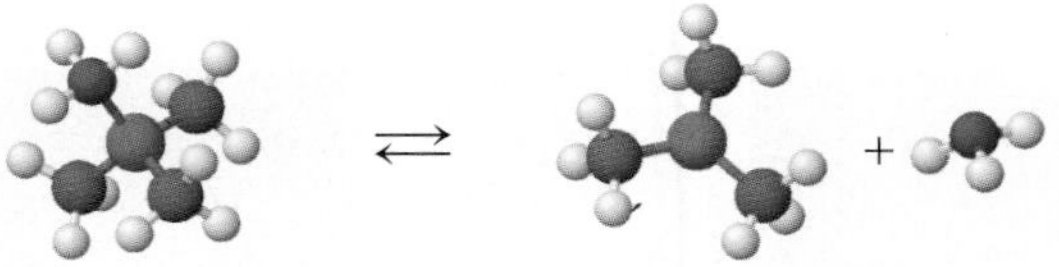

If NH_3 is changed to some other molecule, the equilibrium constant is different.

For $[(CH_3)_3P]B(CH_3)_3$ $\qquad K_p = 0.128$

For $[(CH_3)_3N]B(CH_3)_3$ $\qquad K_p = 0.472$

(a) If you begin an experiment by placing 0.010 mol of each complex in a flask, which would have the largest partial pressure of $B(CH_3)_3$ at 100 °C?
(b) If 0.73 g (0.010 mol) of $(NH_3)B(CH_3)_3$ is placed in a 100.-mL flask and heated to 100 °C, what is the partial pressure of each gas in the equilibrium mixture, and what is the total pressure? What is the percent dissociation of $(NH_3)B(CH_3)_3$?

64. The photographs below show what occurs when a solution of potassium chromate is treated with a few drops of concentrated hydrochloric acid. Some of the bright yellow chromate ion is converted to the orange dichromate ion.

$$2\ CrO_4^{2-}(aq) + 2\ H_3O^+(aq) \rightleftharpoons Cr_2O_7^{2-}(aq) + 3\ H_2O(\ell)$$

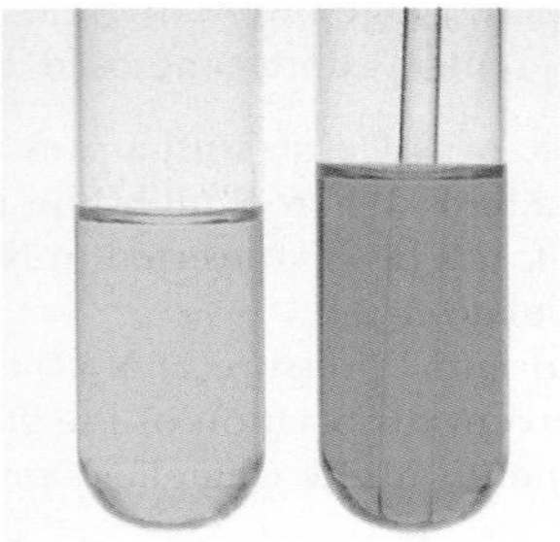

Charles D. Winters

(a) Explain this experimental observation in terms of Le Chatelier's principle.

(b) What would you observe if you treated the orange solution with sodium hydroxide? Explain your observation.

65. The photographs in (a) show what occurs when a solution of iron(III) nitrate is treated with a few drops of aqueous potassium thiocyanate. (See ChemistryNow, Screen 16.4.) The nearly colorless iron(III) ion is converted to a red $[Fe(H_2O)_5SCN]^{2+}$ ion. (This is a classic test for the presence of iron(III) ions in solution.)

$$[Fe(H_2O)_6]^{3+}(aq) + SCN^-(aq) \rightleftharpoons [Fe(H_2O)_5SCN]^{2+}(aq) + H_2O(\ell)$$

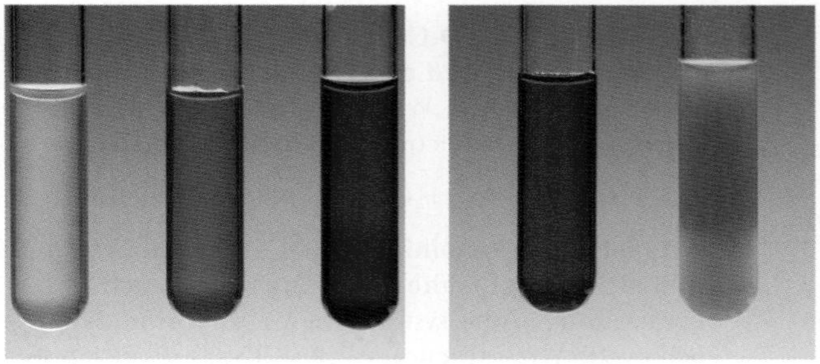

Photos: Charles D. Winters

(a) Adding KSCN **(b) Adding Ag^+**

(a) As more KSCN is added to the solution, the color becomes even more red. Explain this observation.

(b) Silver ions form a white precipitate with SCN^- ions. What would you observe on adding a few drops of aqueous silver nitrate to a red solution of $[Fe(H_2O)_5SCN]^+$ ions? Explain your observation.

66. ■ ▲ The photographs at the bottom of the page show what occurs when you add ammonia to aqueous nickel(II) nitrate and then add ethylenediamine ($NH_2CH_2CH_2NH_2$) to the intermediate blue-purple solution.

$$\underset{\text{green}}{[Ni(H_2O)_6]^{2+}(aq)} + 6\ NH_3(aq) \rightleftharpoons \underset{\text{blue-purple}}{[Ni(NH_3)_6]^{2+}(aq)} + 6\ H_2O(\ell) \quad K_1$$

$$\underset{\text{blue-purple}}{[Ni(NH_3)_6]^{2+}(aq)} + 3\ NH_2CH_2CH_2NH_2(aq) \rightleftharpoons \underset{\text{violet}}{[Ni(NH_2CH_2CH_2NH_2)_3]^{2+}(aq)} + 6\ NH_3(aq) \quad K_2$$

Which equilibrium constant is greater, K_1 or K_2? Explain.

Summary and Conceptual Questions

The following questions may use concepts from this and previous chapters.

67. Decide whether each of the following statements is true or false. If false, change the wording to make it true.
   (a) The magnitude of the equilibrium constant is always independent of temperature.
   (b) When two chemical equations are added to give a net equation, the equilibrium constant for the net equation is the product of the equilibrium constants of the summed equations.
   (c) The equilibrium constant for a reaction has the same value as K for the reverse reaction.
   (d) Only the concentration of CO_2 appears in the equilibrium constant expression for the reaction $CaCO_3(s) \rightleftharpoons CaO(s) + CO_2(g)$.
   (e) For the reaction $CaCO_3(s) \rightleftharpoons CaO(s) + CO_2(g)$, the value of K is numerically the same, whether the amount of CO_2 is expressed as moles/liter or as gas pressure.

68. ■ Neither $PbCl_2$ nor PbF_2 is appreciably soluble in water. If solid $PbCl_2$ and solid PbF_2 are placed in equal amounts of water in separate beakers, in which beaker is the concentration of Pb^{2+} greater? Equilibrium constants for these solids dissolving in water are as follows:

$$PbCl_2(s) \rightleftharpoons Pb^{2+}(aq) + 2\ Cl^-(aq) \quad K = 1.7 \times 10^{-5}$$

$$PbF_2(s) \rightleftharpoons Pb^{2+}(aq) + 2\ F^-(aq) \quad K = 3.7 \times 10^{-8}$$

69. Characterize each of the following as product- or reactant-favored.
   (a) $CO(g) + \frac{1}{2}\ O_2(g) \rightleftharpoons CO_2(g) \quad K_p = 1.2 \times 10^{45}$
   (b) $H_2O(g) \rightleftharpoons H_2(g) + \frac{1}{2}\ O_2(g) \quad K_p = 9.1 \times 10^{-41}$
   (c) $CO(g) + Cl_2(g) \rightleftharpoons COCl_2(g) \quad K_p = 6.5 \times 10^{11}$

70. ■ Consider a gas-phase reaction where a colorless compound C produces a blue compound B.

$$2\ C(g) \rightleftharpoons B(g)$$

   After reaching equilibrium, the size of the flask is halved.
   (a) What color change (if any) is observed immediately upon halving the flask size?
   (b) What color change (if any) is observed after equilibrium has been reestablished in the flask?

71. An ice cube is placed in a beaker of water at 20 °C. The ice cube partially melts, and the temperature of the water is lowered to 0 °C. At this point, both ice and water are at 0 °C, and no further change is apparent. Is the system at equilibrium? Is this a dynamic equilibrium? That is, are events still occurring at the molecular level? Suggest an experiment to test whether this is so. (*Hint:* Consider using D_2O.)

72. See the simulation on ChemistryNow, Screen 16.4.
   (a) Set the concentration of Fe^{3+} at 0.0050 M and that of SCN^- at 0.0070 M. Click the "React" button. Does the concentration of Fe^{3+} go to zero? When equilibrium is reached, what are the concentrations of the reactants and the products? What is the equilibrium constant?
   (b) Begin with $[Fe^{3+}] = [SCN^-] = 0.0$ M and $[FeSCN^{2+}] = 0.0080$ M. When equilibrium is reached, which ion has the largest concentration in solution?
   (c) Begin with $[Fe^{3+}] = 0.0010$ M, $[SCN^-] = 0.0020$ M, and $[FeSCN^{2+}] = 0.0030$ M. Describe the result of allowing this system to come to equilibrium.

17 The Chemistry of Acids and Bases

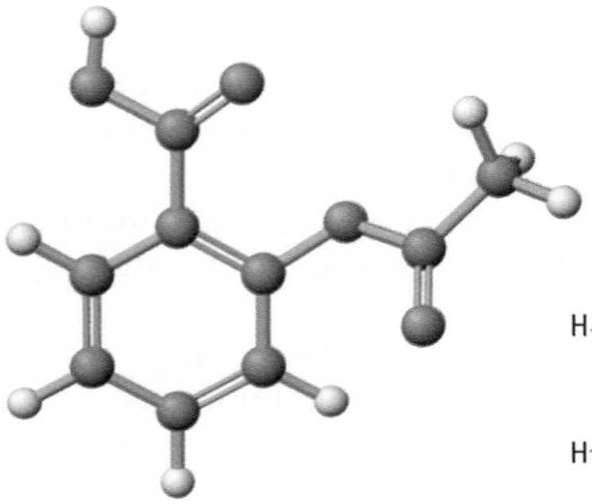

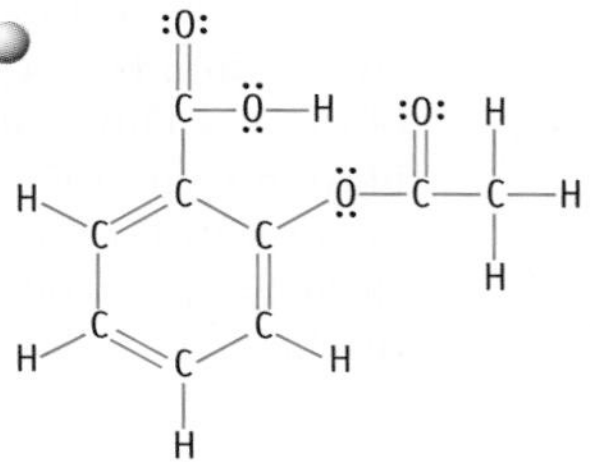

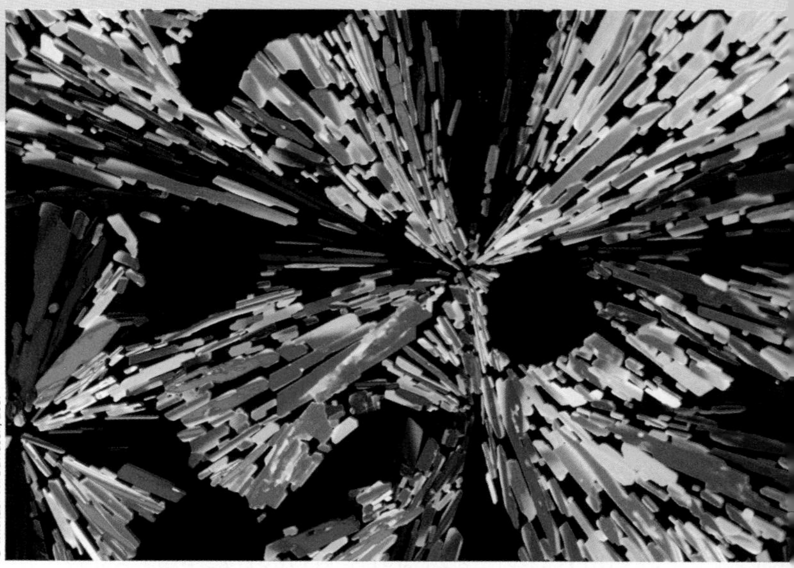

Photo of crystals of aspirin taken using a microscope and polarized light.

Aspirin Is Over 100 Years Old!

Aspirin is one of the most successful non-prescription drugs ever made. Americans swallow more than 50 million aspirin tablets a day, mostly for the pain-relieving (analgesic) effects of the drug. Aspirin also wards off heart disease and thrombosis (blood clots), and it has even been suggested as a possible treatment for certain cancers and for senile dementia.

Hippocrates (460–370 BC), the ancient Greek physician, recommended an infusion of willow bark to ease the pain of childbirth. It was not until the 19th century that an Italian chemist, Raffaele Piria, isolated salicylic acid, the active compound in the bark. Soon thereafter, it was found that the acid could be extracted from a wild flower, *Spiraea ulmaria*. It is from the name of this plant that the name "aspirin" (a + spiraea) is derived.

Hippocrates's willow bark extract, salicylic acid, is an analgesic, but it is also very irritating to the stomach lining. It was therefore an important advance when chemists at Bayer Chemicals in Germany found, in 1897, that a derivative of salicylic acid, acetylsalicylic acid, was also a useful drug and had fewer side effects. This is the compound we now call "aspirin."

Acetylsalicylic acid slowly reverts to salicylic acid, $C_6H_4(CO_2H)(OH)$, and acetic acid in the presence of moisture; therefore, if you smell the characteristic odor of acetic acid in an old bottle of aspirin tablets, they are too old and should be discarded.

Aspirin is a component of various over-the-counter medicines, such as Anacin, Ecotrin, Excedrin, and Alka-Seltzer. The latter is a combination of aspirin with citric acid and sodium bicarbonate. Sodium bicarbonate is a base and reacts with the acid to produce the sodium salt of acetylsalicylic acid, a form of aspirin that is water-soluble and quicker acting.

Questions:

1. Aspirin has a pK_a of 3.49, and that of acetic acid is 4.74. Which is the stronger acid?
2. Identify the acidic H atom in aspirin.
3. Write an equation for the ionization of aspirin.

Answers to these questions are in Appendix Q.

Chapter Goals

See Chapter Goals Revisited (page 799) for Study Questions keyed to these goals and assignable in OWL.

- Use the Brønsted–Lowry and Lewis theories of acids and bases.
- Apply the principles of chemical equilibrium to acids and bases in aqueous solution.
- Predict the outcome of reactions of acids and bases.
- Understand the influence of structure and bonding on acid–base properties.

Chapter Outline

Acids and bases are among the most common substances in nature. Amino acids are the building blocks of proteins. The pH of the lakes, rivers, and oceans is affected by dissolved acids and bases, and your bodily functions depend on acids and bases. You were introduced to the definitions of acids and bases and to some of their chemistry in Chapter 3, but this chapter and the next take up the detailed chemistry of this important class of substances.

Chemistry•Now™

Throughout the text this icon introduces an opportunity for self-study or to explore interactive tutorials by signing in at **www.thomsonedu.com/login**.

17.1 Acids and Bases: A Review

In Chapter 3, you were introduced to two definitions of acids and bases: the Arrhenius definition and the Brønsted–Lowry definition. According to the Arrhenius definition, an acid is any substance that, when dissolved in water, increases the concentration of hydrogen ions, H^+ (◀ page 132). An Arrhenius base is any substance that increases the concentration of hydroxide ions, OH^-, when dissolved in water. Based on the Arrhenius definition, hydrochloric acid was therefore classified as an acid, and sodium hydroxide was classified as a base.

$$HCl(aq) \rightarrow H^+(aq) + Cl^-(aq)$$

$$NaOH(aq) \rightarrow Na^+(aq) + OH^-(aq)$$

Using this definition, reactions between acids and bases involve the combination of H^+ and OH^- ions to form water (and a salt).

$$NaOH(aq) + HCl(aq) \rightarrow H_2O(\ell) + NaCl(aq)$$

The Brønsted–Lowry definition of acids and bases is more general and views acid–base behavior in terms of proton transfer from one substance to another. A Brønsted–Lowry acid is a proton (H^+) donor, and a Brønsted–Lowry base is a proton acceptor. The reaction between an acid and a base is viewed as involving an equilibrium in which a new acid and base (the conjugate base of the acid and the conjugate acid of the base) are formed. This definition extends the list of bases and the scope of acid–base reactions. In the following reaction, HCl acts as a

Brønsted–Lowry acid, and water acts as a Brønsted–Lowry base because HCl transfers a H^+ ion to H_2O to form the hydronium ion, H_3O^+.

$$HCl(aq) + H_2O(\ell) \rightleftharpoons H_3O^+(aq) + Cl^-(aq)$$

This equilibrium strongly favors formation of $H_3O^+(aq)$ and $Cl^-(aq)$; that is, HCl is a stronger acid than H_3O^+.

We will begin this chapter by looking at Brønsted–Lowry acid–base chemistry in more detail.

17.2 The Brønsted–Lowry Concept of Acids and Bases Extended

A wide variety of Brønsted–Lowry acids is known. These include some molecular compounds such as nitric acid,

$$\underset{\text{Acid}}{HNO_3(aq)} + H_2O(\ell) \rightleftharpoons NO_3^-(aq) + H_3O^+(aq)$$

cations such as NH_4^+,

$$\underset{\text{Acid}}{NH_4^+(aq)} + H_2O(\ell) \rightleftharpoons NH_3(aq) + H_3O^+(aq)$$

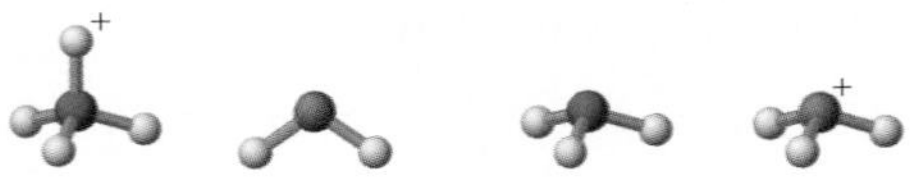

and anions.

$$H_2PO_4^-(aq) + H_2O(\ell) \rightleftharpoons HPO_4^{2-}(aq) + H_3O^+(aq)$$

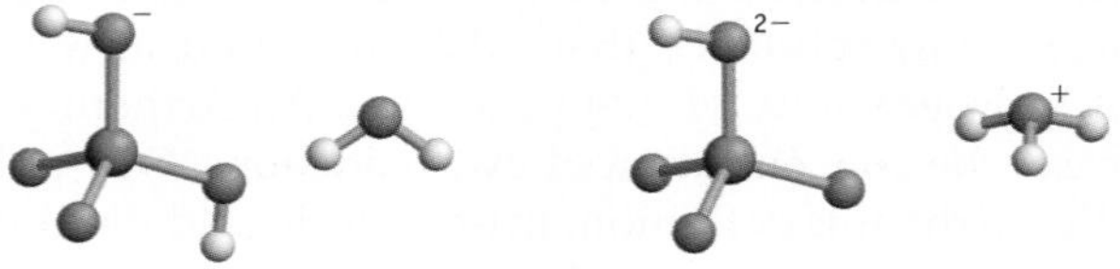

Similarly, many different types of species can act as Brønsted bases in their reactions with water. These include some molecular compounds,

$$\underset{\text{Base}}{NH_3(aq)} + H_2O(\ell) \rightleftharpoons NH_4^+(aq) + OH^-(aq)$$

anions,

$$CO_3^{2-}(aq) + H_2O(\ell) \rightleftharpoons HCO_3^-(aq) + OH^-(aq)$$

and hydrated metal cations. These cations can act as acids and bases.

$$[Fe(H_2O)_6]^{3+}(aq) + H_2O(\ell) \rightleftharpoons [Fe(H_2O)_6(OH)]^{2+}(aq) + H_3O^+(aq)$$

$$[Al(H_2O)_5(OH)]^{2+}(aq) + H_2O(\ell) \rightleftharpoons [Al(H_2O)_6]^{3+}(aq) + OH^-(aq)$$

Acids such as HF, HCl, HNO_3, and CH_3CO_2H (acetic acid) are all capable of donating one proton and so are called **monoprotic acids.** Other acids, called **polyprotic acids** (Table 17.1), are capable of donating two or more protons. A familiar example of a polyprotic acid is sulfuric acid.

$$H_2SO_4(aq) + H_2O(\ell) \rightleftharpoons HSO_4^-(aq) + H_3O^+(aq)$$

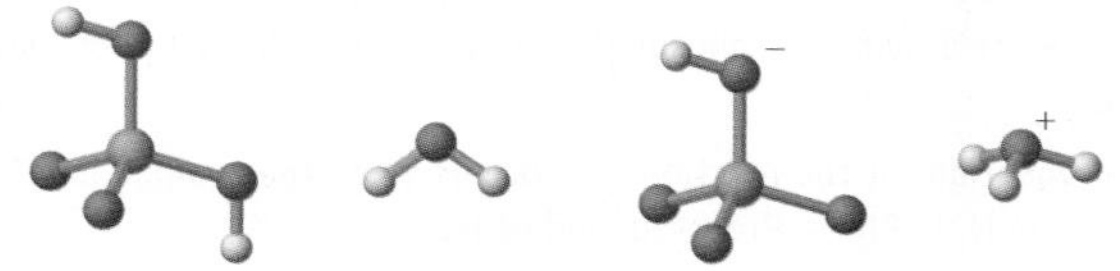

$$HSO_4^-(aq) + H_2O(\ell) \rightleftharpoons SO_4^{2-}(aq) + H_3O^+(aq)$$

Charles D. Winters

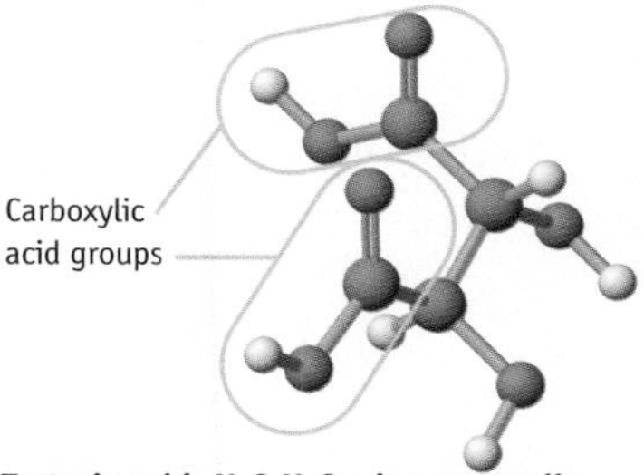

Tartaric acid, $H_2C_4H_4O_6$, is a naturally occurring diprotic acid. Tartaric acid and its potassium salt are found in many fruits. The acidic protons are the H atoms of the $-CO_2H$ or carboxylic acid groups.

Just as there are acids that can donate more than one proton, so there are **polyprotic bases** that can accept more than one proton. The fully deprotonated anions of polyprotic acids are polyprotic bases; examples include SO_4^{2-}, PO_4^{3-}, CO_3^{2-}, and $C_2O_4^{2-}$. The carbonate ion, for example, can accept two protons.

$$CO_3^{2-}(aq) + H_2O(\ell) \rightleftharpoons HCO_3^-(aq) + OH^-(aq)$$

Base

$$HCO_3^-(aq) + H_2O(\ell) \rightleftharpoons H_2CO_3(aq) + OH^-(aq)$$

Base

Some molecules (such as water) and ions can behave either as Brønsted acids or bases and are referred to as being **amphiprotic** (◀ page 136). An example of an

TABLE 17.1 Polyprotic Acids and Bases

Acid Form	Amphiprotic Form	Base Form
H_2S (hydrosulfuric acid or hydrogen sulfide)	HS^- (hydrogen sulfide ion)	S^{2-} (sulfide ion)
H_3PO_4 (phosphoric acid)	$H_2PO_4^-$ (dihydrogen phosphate ion) HPO_4^{2-} (hydrogen phosphate ion)	PO_4^{3-} (phosphate ion)
H_2CO_3 (carbonic acid)	HCO_3^- (hydrogen carbonate ion or bicarbonate ion)	CO_3^{2-} (carbonate ion)
$H_2C_2O_4$ (oxalic acid)	$HC_2O_4^-$ (hydrogen oxalate ion)	$C_2O_4^{2-}$ (oxalate ion)

amphiprotic anion that is particularly important in biochemical systems is the dihydrogen phosphate anion (Table 17.1).

$$H_2PO_4^-(aq) + H_2O(\ell) \rightleftharpoons H_3O^+(aq) + HPO_4^{2-}(aq)$$

Acid

$$H_2PO_4^{2-}(aq) + H_2O(\ell) \rightleftharpoons H_3PO_4(aq) + OH^-(aq)$$

Base

EXERCISE 17.1 Brønsted Acids and Bases

(a) Write a balanced equation for the reaction that occurs when H_3PO_4, phosphoric acid, donates a proton to water to form the dihydrogen phosphate ion. Is the dihydrogen phosphate ion an acid, a base, or amphiprotic?

(b) Write a balanced equation for the reaction that occurs when the cyanide ion, CN^-, accepts a proton from water to form HCN. Is CN^- a Brønsted acid or base?

Conjugate Acid–Base Pairs

A reaction important in the control of acidity in biological systems involves the hydrogen carbonate ion, which can act as a Brønsted base or acid in water. The equation for HCO_3^- functioning as an base exemplifies a feature of all reactions involving Brønsted acids and bases.

conjugate pair 1

conjugate pair 2

$$\underset{\text{Base}}{HCO_3^-(aq)} + \underset{\text{Acid}}{H_2O(\ell)} \rightleftharpoons \underset{\text{Base}}{OH^-(aq)} + \underset{\text{Acid}}{H_2CO_3(aq)}$$

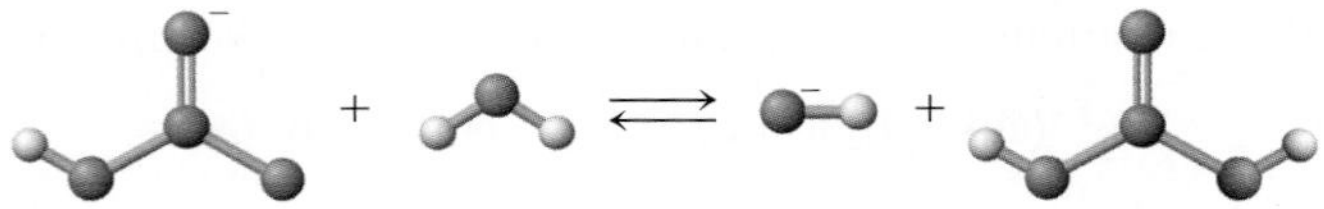

In the forward direction, HCO_3^- is the Brønsted base because it captures H^+ from the Brønsted acid, H_2O. The reverse reaction, however, is also an acid–base reaction. In this case, the H_2CO_3 is the acid, and OH^- is the base.

A **conjugate acid–base pair** consists of two species that differ from each other by the presence of one hydrogen ion. Thus, H_2CO_3 and HCO_3^- comprise a conjugate acid–base pair. In this pair, HCO_3^- is the conjugate base of the acid H_2CO_3, and H_2CO_3 is the conjugate acid of the base HCO_3^-. There is a second conjugate acid–base pair in this reaction: H_2O and H_3O^+. In fact, *every reaction between a Brønsted acid and a Brønsted base involves two conjugate acid–base pairs* (Table 17.2).

Chemistry Now™

Sign in at **www.thomsonedu.com/login** and go to Chapter 17 Contents to see Screen 17.2 for an exercise and tutorial on **acids, bases, and their conjugates.**

EXERCISE 17.2 Conjugate Acids and Bases

In the following reaction, identify the acid on the left and its conjugate base on the right. Similarly, identify the base on the left and its conjugate acid on the right.

$$HNO_3(aq) + NH_3(aq) \rightleftharpoons NH_4^+(aq) + NO_3^-(aq)$$

TABLE 17.2 Acid–Base Reactions and Conjugate Acid–Base Pairs*

Name	Acid 1		Base 2		Base 1		Acid 2
Hydrochloric acid	HCl	+	H_2O	$\rightleftharpoons$	Cl^-	+	H_3O^+
Nitric acid	HNO_3	+	H_2O	$\rightleftharpoons$	NO_3^-	+	H_3O^+
Carbonic acid	H_2CO_3	+	H_2O	$\rightleftharpoons$	HCO_3^-	+	H_3O^+
Acetic acid	CH_3CO_2H	+	H_2O	$\rightleftharpoons$	$CH_3CO_2^-$	+	H_3O^+
Hydrocyanic acid	HCN	+	H_2O	$\rightleftharpoons$	CN^-	+	H_3O^+
Hydrogen sulfide	H_2S	+	H_2O	$\rightleftharpoons$	HS^-	+	H_3O^+
Ammonia	H_2O	+	NH_3	$\rightleftharpoons$	OH^-	+	NH_4^+
Carbonate ion	H_2O	+	CO_3^{2-}	$\rightleftharpoons$	OH^-	+	HCO_3^-
Water	H_2O	+	H_2O	$\rightleftharpoons$	OH^-	+	H_3O^+

*Acid 1 and base 1 are a conjugate pair, as are base 2 and acid 2.

17.3 Water and the pH Scale

Because we generally use aqueous solutions of acids and bases, and because the acid–base reactions in your body occur in your aqueous interior, we want to consider the behavior of water in terms of chemical equilibria.

Water Autoionization and the Water Ionization Constant, K_w

An acid such as HCl does not need to be present for the hydronium ion to exist in water. In fact, two water molecules can interact with each other to produce a hydronium ion and a hydroxide ion by proton transfer from one water molecule to the other.

$$2\ H_2O(\ell) \rightleftharpoons H_3O^+(aq) + OH^-(aq)$$

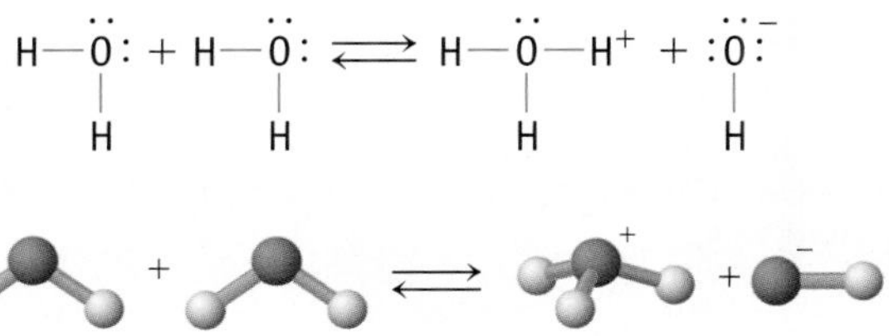

This **autoionization** reaction of water was demonstrated many years ago by Friedrich Kohlrausch (1840–1910). He found that, even after water is painstakingly purified, it still conducts electricity to a very small extent because autoionization produces very low concentrations of H_3O^+ and OH^- ions. Water autoionization is the cornerstone of our concepts of aqueous acid–base behavior.

When water autoionizes, the equilibrium lies far to the left side. In fact, in pure water at 25 °C, only about two out of a billion (10^9) water molecules are ionized at any instant. To express this idea more quantitatively, we can write the equilibrium constant expression for autoionization.

$$K_w = [H_3O^+][OH^-] = 1.0 \times 10^{-14} \text{ at } 25\ °C \quad (17.1)$$

■ **K_w and Temperature** The equation $K_w = [H_3O^+][OH^-]$ is valid for pure water and for any aqueous solution. K_w is temperature dependent. Because the autoionization reaction is endothermic, K_w increases with temperature.

T (°C)	K_w
10	0.29×10^{-14}
15	0.45×10^{-14}
20	0.68×10^{-14}
25	1.01×10^{-14}
30	1.47×10^{-14}
50	5.48×10^{-14}

There are several important aspects of this equation.

- Based on the rules we have given for writing equilibrium constants, we would not include the concentration of water.
- The equilibrium constant is given a special symbol, K_w, and is known as the **autoionization constant for water.**
- Because the autoionization of water is the only source of hydronium and hydroxide ions in pure water, we know that $[H_3O^+]$ must equal $[OH^-]$ in pure water. Electrical conductivity measurements of pure water show that $[H_3O^+] = [OH^-] = 1.0 \times 10^{-7}$ M at 25 °C, so K_w has a value of 1.0×10^{-14} at 25 °C.

In pure water, the hydronium ion and hydroxide ion concentrations are equal, and the water is said to be neutral. If some acid or base is added to pure water, however, the equilibrium

$$2\ H_2O(\ell) \rightleftharpoons H_3O^+(aq) + OH^-(aq)$$

is disturbed. Adding acid raises the concentration of the H_3O^+ ions, so the solution is acidic. To oppose this increase, Le Chatelier's principle (◀ Section 16.6) predicts that a small fraction of the H_3O^+ ions will react with OH^- ions from water autoionization to form water. This lowers $[OH^-]$ until the product of $[H_3O^+]$ and $[OH^-]$ is again equal to 1.0×10^{-14} at 25 °C. Similarly, adding a base to pure water gives a basic solution because the OH^- ion concentration has increased. Le Chatelier's principle predicts that some of the added OH^- ions will react with H_3O^+ ions present in the solution from water autoionization, thereby lowering $[H_3O^+]$ until the value of the product $[H_3O^+]$ and $[OH^-]$ equals 1.0×10^{-14} at 25 °C.

Thus, for aqueous solutions at 25 °C, we can say that

- In a neutral solution, $[H_3O^+] = [OH^-]$.
  Both are equal to 1.0×10^{-7} M.
- In an acidic solution, $[H_3O^+] > [OH^-]$.
  $[H_3O^+] > 1.0 \times 10^{-7}$ M and $[OH^-] < 1.0 \times 10^{-7}$ M.
- In a basic solution, $[H_3O^+] < [OH^-]$.
  $[H_3O^+] < 1.0 \times 10^{-7}$ M and $[OH^-] > 1.0 \times 10^{-7}$ M.

Chemistry Now™

Sign in at **www.thomsonedu.com/login** and go to Chapter 17 Contents to see Screen 17.3 for a simulation of **the effect of temperature on K_w.**

EXAMPLE 17.1 Ion Concentrations in a Solution of a Strong Base

Problem What are the hydroxide and hydronium ion concentrations in a 0.0012 M solution of NaOH at 25 °C?

Strategy NaOH, a strong base, is 100% dissociated into ions in water, so we assume that the OH^- ion concentration is the same as the NaOH concentration. The H_3O^+ ion concentration can then be calculated using Equation 17.1.

Solution The initial concentration of OH^- is 0.0012 M.

$$0.0012 \text{ mol NaOH per liter} \rightarrow 0.0012 \text{ M } Na^+(aq) + 0.0012 \text{ M } OH^-(aq)$$

Substituting the OH^- concentration into Equation 17.1, we have

$$K_w = 1.0 \times 10^{-14} = [H_3O^+][OH^-] = [H_3O^+](0.0012)$$

and so

$$[H_3O^+] = \frac{1.0 \times 10^{-14}}{0.0012} = 8.3 \times 10^{-12}\ M$$

Comment Why didn't we take into account the ions produced by water autoionization when we calculated the concentration of hydroxide ions? It should add OH^- and H_3O^+ ions to the solution. If x is equal to the concentration of OH^- ions generated by the autoionization of water, then, when equilibrium is achieved,

$$[OH^-] = (0.0012\ M + OH^- \text{ from water autoionization})$$

$$[OH^-] = (0.0012\ M + x)$$

In pure water, the concentration of OH^- ion generated is 1.0×10^{-7} M. Le Chatelier's principle (◀ Section 16.6) suggests that the concentration should be even smaller when OH^- ions are already present in solution from NaOH; that is, x should be $<< 1.0 \times 10^{-7}$ M. This means x in the term $(0.0012 + x)$ is insignificant compared with 0.0012. (Following the rules for significant figures, the sum of 0.0012 and a number even smaller than 1.0×10^{-7} is 0.0012.) Thus, the equilibrium concentration of OH^- is equivalent to the concentration of NaOH in the solution.

Lastly, what about the Na^+ ion? As described later (page 774), alkali metal ions have no effect on the acidity or basicity of a solution.

EXERCISE 17.3 Hydronium Ion Concentration in a Solution of a Strong Acid

A solution of the strong acid HCl has $[HCl] = 4.0 \times 10^{-3}$ M. What are the concentrations of H_3O^+ and OH^- in this solution at 25 °C? (Recall that because HCl is a strong acid, it is 100% ionized in water.)

The pH Scale

The **pH** of a solution is defined as the negative of the base-10 logarithm (log) of the hydronium ion concentration (◀ Section 4.6, page 179).

$$pH = -\log[H_3O^+] \qquad \textbf{(4.3 and 17.2)}$$

In a similar way, we can define the pOH of a solution as the negative of the base-10 logarithm of the hydroxide ion concentration.

$$pOH = -\log[OH^-] \qquad \textbf{(17.3)}$$

In pure water, the hydronium and hydroxide ion concentrations are both 1.0×10^{-7} M. Therefore, for pure water at 25 °C

$$pH = -\log(1.0 \times 10^{-7}) = 7.00$$

In the same way, you can show that the pOH of pure water is also 7.00 at 25 °C.

If we take the negative logarithms of both sides of the expression $K_w = [H_3O^+][OH^-]$, we obtain another useful equation.

$$\begin{aligned} K_w &= 1.0 \times 10^{-14} = [H_3O^+][OH^-] \\ -\log K_w &= -\log(1.0 \times 10^{-14}) = -\log([H_3O^+][OH^-]) \\ pK_w &= 14.00 = -\log([H_3O^+]) + (-\log[OH^-]) \\ pK_w &= 14.00 = pH + pOH \end{aligned} \qquad \textbf{(17.4)}$$

■ **The pX Scale**
In general, $-\log X = pX$, so $-\log K = pK$
$-\log[H_3O^+] = pH$
$-\log[OH^-] = pOH$

The sum of the pH and pOH of a solution must be equal to 14.00 at 25 °C.

As illustrated in Figures 4.11 and 17.1, solutions with pH less than 7.00 (at 25 °C) are acidic, whereas solutions with pH greater than 7.00 are basic. Solutions with pH = 7.00 at 25 °C are neutral.

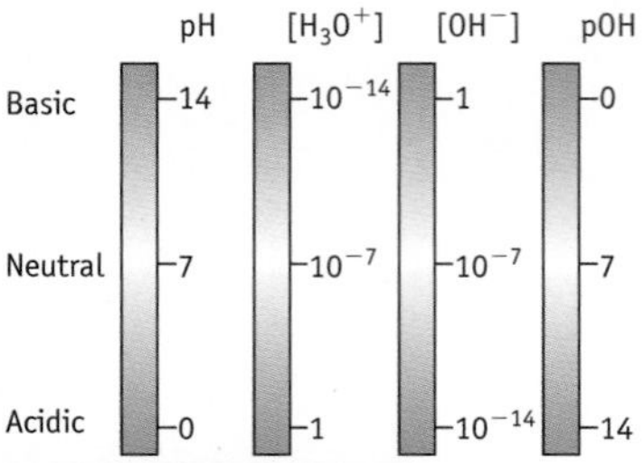

Active Figure 17.1 pH and pOH. This figure illustrates the relationship of hydronium ion and hydroxide ion concentrations and of pH and pOH.

Chemistry Now™ Sign in at www.thomsonedu.com/login and go to the Chapter Contents menu to explore an interactive version of this figure accompanied by an exercise.

Calculating pH

The calculation of pH from the hydronium ion concentration, or the concentration of hydronium ion concentration from pH, was introduced in Chapter 4 (◀ page 179). Exercise 17.4 reviews those calculations.

Chemistry Now™

Sign in at **www.thomsonedu.com/login** and go to Chapter 17 Contents to see Screen 17.4 for a simulation and tutorial on **using pH and pOH.**

EXERCISE 17.4 Reviewing pH Calculations

(a) What is the pH of a 0.0012 M NaOH solution at 25 °C?

(b) The pH of a diet soda is 4.32 at 25 °C. What are the hydronium and hydroxide ion concentrations in the soda?

(c) If the pH of a solution containing the strong base $Sr(OH)_2$ is 10.46 at 25 °C, what is the concentration of $Sr(OH)_2$?

17.4 Equilibrium Constants for Acids and Bases

In Chapter 3, it was stated that acids and bases can be divided roughly into those that are strong electrolytes (such as HCl, HNO_3, and NaOH) and those that are weak electrolytes (such as CH_3CO_2H and NH_3) (Figure 17.2) (◀ Table 3.2, Common Acids and Bases, page 132). Hydrochloric acid is a strong acid, so 100% of the acid ionizes to produce hydronium and chloride ions. In contrast, acetic acid and ammonia are weak electrolytes. They ionize to only a very small extent in water. For example, for acetic acid, the acid, its anion, and the hydronium ion are all present at equilibrium in solution, but the ions are present in very low concentration relative to the acid concentration.

One way to define the relative strengths of a series of acids would be to measure the pH of solutions of acids of equal concentration: The lower the pH, the greater the concentration of hydronium ion, the stronger the acid. Similarly, for a series

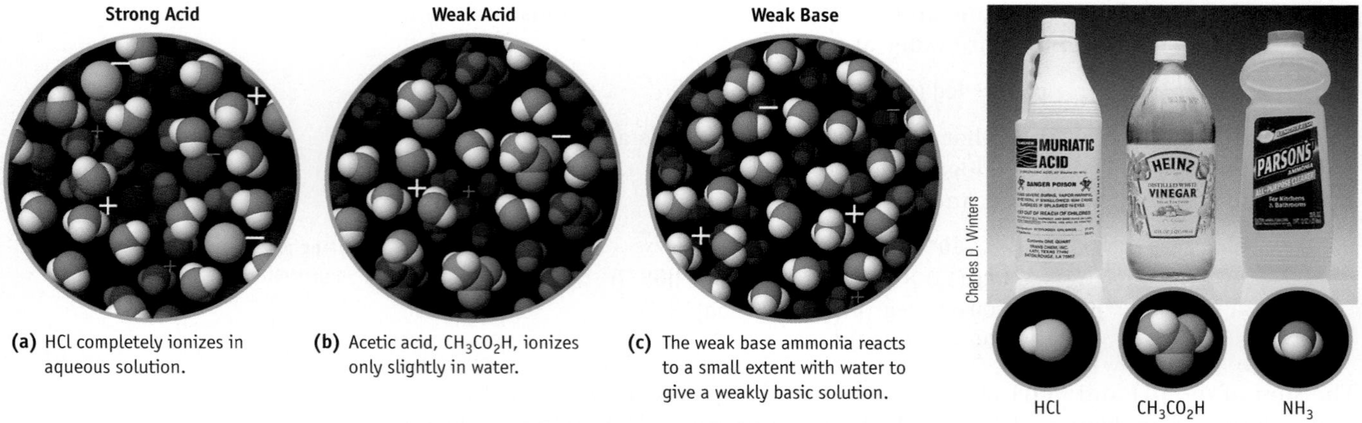

(a) HCl completely ionizes in aqueous solution.

(b) Acetic acid, CH_3CO_2H, ionizes only slightly in water.

(c) The weak base ammonia reacts to a small extent with water to give a weakly basic solution.

FIGURE 17.2 Strong and weak acids and bases. (a) Hydrochloric acid, a strong acid, is sold for household use as "muriatic acid." The acid completely ionizes in water. (b) Vinegar is a solution of acetic acid, a weak acid that ionizes only to a small extent in water. (c) Ammonia is a weak base, ionizing to a small extent in water.

of weak bases, $[OH^-]$ will increase, and the pH will increase as the bases become stronger.

- For a strong monoprotic acid, $[H_3O^+]$ in solution is equal to the original acid concentration. Similarly, for a strong monoprotic base, $[OH^-]$ will be equal to the original base concentration.
- For a weak acid, $[H_3O^+]$ will be much less than the original acid concentration. That is, $[H_3O^+]$ will be smaller than if the acid were a strong acid of the same concentration. Similarly, a weak base will give a smaller $[OH^-]$ than if the base were a strong base of the same concentration.
- For a series of weak monoprotic acids (of the type HA) of the same concentration, $[H_3O^+]$ will increase (and the pH will decline) as the acids become stronger. Similarly, for a series of weak bases, $[OH^-]$ will increase (and the pH will increase) as the bases become stronger.

The relative strength of an acid or base can also be expressed quantitatively with an equilibrium constant, often called an **ionization constant.** For the general acid HA, we can write

$$HA(aq) + H_2O(\ell) \rightleftharpoons H_3O^+(aq) + A^-(aq)$$

$$K_a = \frac{[H_3O^+][A^-]}{[HA]} \qquad (17.5)$$

where the equilibrium constant, *K*, has a subscript "a" to indicate that it is an equilibrium constant for an acid in water. For weak acids, the value of K_a is less than 1 because the product $[H_3O^+][A^-]$ is less than the equilibrium concentration of the weak acid [HA]. For a series of acids, the acid strength increases as the value of K_a increases.

Similarly, we can write the equilibrium expression for a weak base B in water. Here, we label *K* with a subscript "b." Its value is less than 1 for weak bases.

$$B(aq) + H_2O(\ell) \rightleftharpoons BH^+(aq) + OH^-(aq)$$

$$K_b = \frac{[BH^+][OH^-]}{[B]} \qquad (17.6)$$

Some acids and bases are listed in Table 17.3, each with its value of K_a or K_b. The following are important ideas concerning this table.

- Acids are listed in Table 17.3 at the left, and their conjugate bases are on the right.
- A large value of *K* indicates that ionization products are strongly favored, whereas a small value of *K* indicates that reactants are favored.
- The strongest acids are at the upper left. They have the largest K_a values. K_a values become smaller on descending the chart as the acid strength declines.
- The strongest bases are at the lower right. They have the largest K_b values. K_b values become larger on descending the chart as base strength increases.
- *The weaker the acid, the stronger its conjugate base.* That is, the smaller the value of K_a, the larger the value of K_b.
- Some acids or bases are listed as having K_a or K_b values that are large or very small. Acids that are stronger than H_3O^+ are completely ionized (HNO_3, for example), so their K_a values are "large." Their conjugate bases (such as NO_3^-) do not produce meaningful concentrations of OH^- ions, so their K_b values are "very small." Similar arguments follow for strong bases and their conjugate acids.

TABLE 17.3 Ionization Constants for Some Acids and Their Conjugate Bases at 25 °C

Acid Name	Acid	K_a	Base	K_b	Base Name
Perchloric acid	$HClO_4$	large	ClO_4^-	very small	perchlorate ion
Sulfuric acid	H_2SO_4	large	HSO_4^-	very small	hydrogen sulfate ion
Hydrochloric acid	HCl	large	Cl^-	very small	chloride ion
Nitric acid	HNO_3	large	NO_3^-	very small	nitrate ion
Hydronium ion	H_3O^+	1.0	H_2O	1.0×10^{-14}	water
Sulfurous acid	H_2SO_3	1.2×10^{-2}	HSO_3^-	8.3×10^{-13}	hydrogen sulfite ion
Hydrogen sulfate ion	HSO_4^-	1.2×10^{-2}	SO_4^{2-}	8.3×10^{-13}	sulfate ion
Phosphoric acid	H_3PO_4	7.5×10^{-3}	$H_2PO_4^-$	1.3×10^{-12}	dihydrogen phosphate ion
Hexaaquairon(III) ion	$[Fe(H_2O)_6]^{3+}$	6.3×10^{-3}	$[Fe(H_2O)_5OH]^{2+}$	1.6×10^{-12}	pentaaquahydroxoiron(III) ion
Hydrofluoric acid	HF	7.2×10^{-4}	F^-	1.4×10^{-11}	fluoride ion
Nitrous acid	HNO_2	4.5×10^{-4}	NO_2^-	2.2×10^{-11}	nitrite ion
Formic acid	HCO_2H	1.8×10^{-4}	HCO_2^-	5.6×10^{-11}	formate ion
Benzoic acid	$C_6H_5CO_2H$	6.3×10^{-5}	$C_6H_5CO_2^-$	1.6×10^{-10}	benzoate ion
Acetic acid	CH_3CO_2H	1.8×10^{-5}	$CH_3CO_2^-$	5.6×10^{-10}	acetate ion
Propanoic acid	$CH_3CH_2CO_2H$	1.3×10^{-5}	$CH_3CH_2CO_2^-$	7.7×10^{-10}	propanoate ion
Hexaaquaaluminum ion	$[Al(H_2O)_6]^{3+}$	7.9×10^{-6}	$[Al(H_2O)_5OH]^{2+}$	1.3×10^{-9}	pentaaquahydroxoaluminum ion
Carbonic acid	H_2CO_3	4.2×10^{-7}	HCO_3^-	2.4×10^{-8}	hydrogen carbonate ion
Hexaaquacopper(II) ion	$[Cu(H_2O)_6]^{2+}$	1.6×10^{-7}	$[Cu(H_2O)_5OH]^+$	6.3×10^{-8}	pentaaquahydroxocopper(II) ion
Hydrogen sulfide	H_2S	1×10^{-7}	HS^-	1×10^{-7}	hydrogen sulfide ion
Dihydrogen phosphate ion	$H_2PO_4^-$	6.2×10^{-8}	HPO_4^{2-}	1.6×10^{-7}	hydrogen phosphate ion
Hydrogen sulfite ion	HSO_3^-	6.2×10^{-8}	SO_3^{2-}	1.6×10^{-7}	sulfite ion
Hypochlorous acid	HClO	3.5×10^{-8}	ClO^-	2.9×10^{-7}	hypochlorite ion
Hexaaqualead(II) ion	$[Pb(H_2O)_6]^{2+}$	1.5×10^{-8}	$[Pb(H_2O)_5OH]^+$	6.7×10^{-7}	pentaaquahydroxolead(II) ion
Hexaaquacobalt(II) ion	$[Co(H_2O)_6]^{2+}$	1.3×10^{-9}	$[Co(H_2O)_5OH]^+$	7.7×10^{-6}	pentaaquahydroxocobalt(II) ion
Boric acid	$B(OH)_3(H_2O)$	7.3×10^{-10}	$B(OH)_4^-$	1.4×10^{-5}	tetrahydroxoborate ion
Ammonium ion	NH_4^+	5.6×10^{-10}	NH_3	1.8×10^{-5}	ammonia
Hydrocyanic acid	HCN	4.0×10^{-10}	CN^-	2.5×10^{-5}	cyanide ion
Hexaaquairon(II) ion	$[Fe(H_2O)_6]^{2+}$	3.2×10^{-10}	$[Fe(H_2O)_5OH]^+$	3.1×10^{-5}	pentaaquahydroxoiron(II) ion
Hydrogen carbonate ion	HCO_3^-	4.8×10^{-11}	CO_3^{2-}	2.1×10^{-4}	carbonate ion
Hexaaquanickel(II) ion	$[Ni(H_2O)_6]^{2+}$	2.5×10^{-11}	$[Ni(H_2O)_5OH]^+$	4.0×10^{-4}	pentaaquahydroxonickel(II) ion
Hydrogen phosphate ion	HPO_4^{2-}	3.6×10^{-13}	PO_4^{3-}	2.8×10^{-2}	phosphate ion
Water	H_2O	1.0×10^{-14}	OH^-	1.0	hydroxide ion
Hydrogen sulfide ion*	HS^-	1×10^{-19}	S^{2-}	1×10^{5}	sulfide ion
Ethanol	C_2H_5OH	very small	$C_2H_5O^-$	large	ethoxide ion
Ammonia	NH_3	very small	NH_2^-	large	amide ion
Hydrogen	H_2	very small	H^-	large	hydride ion

Increasing Acid Strength ↑ | Increasing Base Strength ↓

*The values of K_a for HS^- and K_b for S^{2-} are estimates.

Problem Solving Tip 17.1 Strong or Weak?

How can you tell whether an acid or a base is weak? The easiest way is to remember those few that are strong. All others are probably weak.

Strong acids are:

Hydrohalic acids: HCl, HBr, and HI (but not HF)

Nitric acid: HNO_3

Sulfuric acid: H_2SO_4 (for loss of first H^+ only)

Perchloric acid: $HClO_4$

Some common strong bases include the following:

All Group 1A hydroxides: LiOH, NaOH, KOH, RbOH, CsOH

Group 2A hydroxides: $Sr(OH)_2$ and $Ba(OH)_2$ [$Mg(OH)_2$ and $Ca(OH)_2$ are not considered strong bases because they do not dissolve appreciably in water.]

To illustrate some of these ideas, let us compare some common acids and bases. For example, HF is a stronger acid than HClO, which is in turn stronger than HCO_3^-,

Decreasing acid strength →

HF	HClO	HCO_3^-
$K_a = 7.2 \times 10^{-4}$	$K_a = 3.5 \times 10^{-8}$	$K_a = 4.8 \times 10^{-11}$

and their conjugate bases become stronger from F^- to ClO^- to CO_3^{2-}.

Increasing base strength →

F^-	ClO^-	CO_3^{2-}
$K_b = 1.4 \times 10^{-11}$	$K_b = 2.9 \times 10^{-7}$	$K_b = 2.1 \times 10^{-4}$

Nature abounds in acids and bases (Figure 17.3). Many naturally occurring acids are based on the carboxyl group ($-CO_2H$), and a few are illustrated here. Notice that the organic portion of the *molecule* has an effect on its relative strength (as described further in Section 17.10).

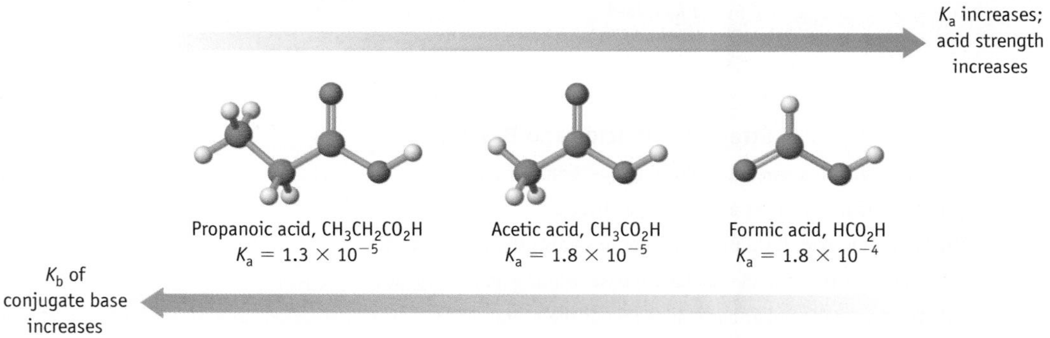

Propanoic acid, $CH_3CH_2CO_2H$ $K_a = 1.3 \times 10^{-5}$ | Acetic acid, CH_3CO_2H $K_a = 1.8 \times 10^{-5}$ | Formic acid, HCO_2H $K_a = 1.8 \times 10^{-4}$

There are many naturally occurring weak bases as well as weak acids (Figure 17.3). Ammonia and its conjugate acid, the ammonium ion, are part of the nitrogen cycle in the environment (► page 951). Biological systems reduce nitrate ion to NH_3 and NH_4^+ and incorporate nitrogen into amino acids and proteins. Many bases are derived from NH_3 by replacement of the H atoms with organic groups.

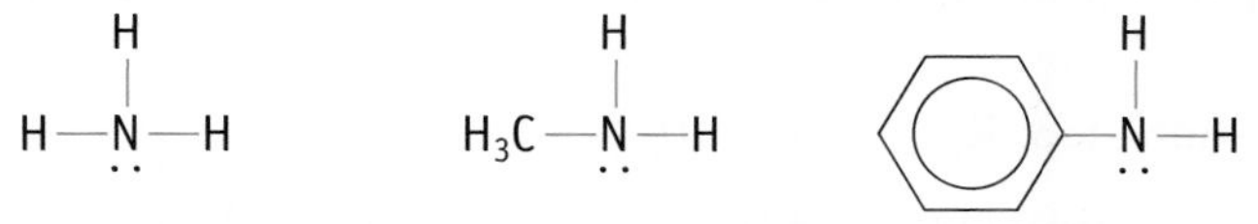

Ammonia $K_b = 1.8 \times 10^{-5}$ | Methylamine $K_b = 5.0 \times 10^{-4}$ | Aniline $K_b = 4.0 \times 10^{-10}$

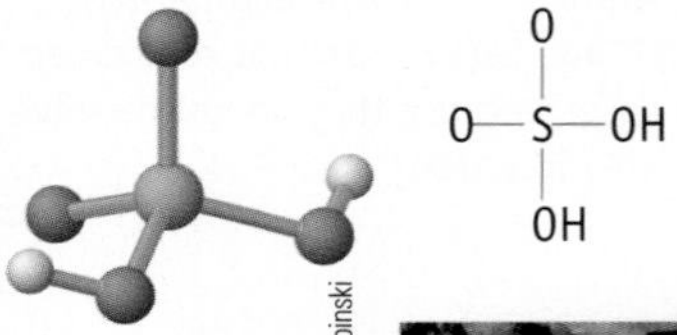

Sharksong/M. Kazmers/Dembinski Photo Associates

▲ A sea slug excretes the strong acid sulfuric acid in self-defense.

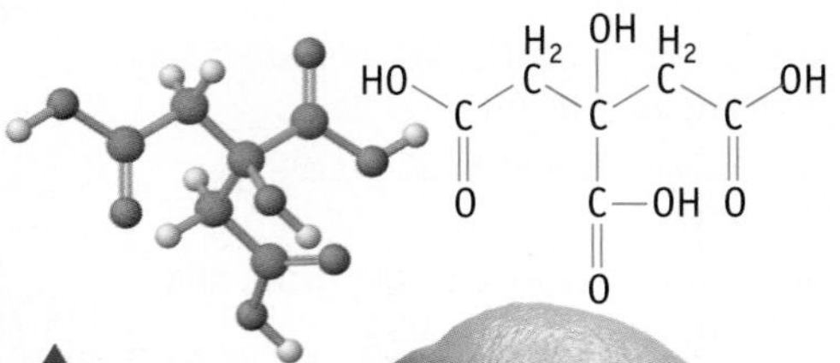

Charles D. Winters

▲ The tartness of lemons and oranges comes from the weak acid citric acid. The acid is found widely in nature and in many consumer products.

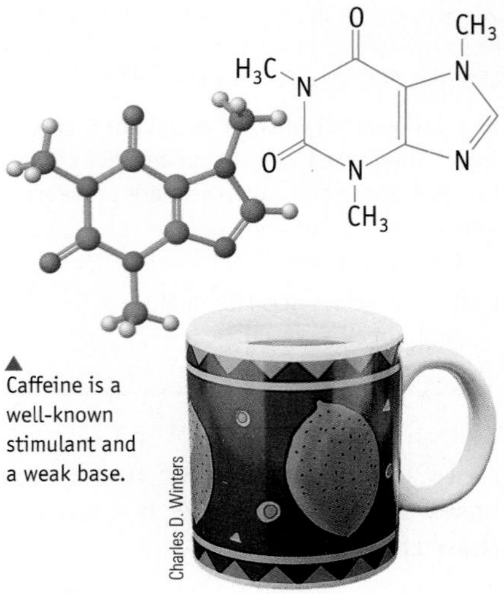

Charles D. Winters

▲ Caffeine is a well-known stimulant and a weak base.

FIGURE 17.3 Natural acids and bases. Hundreds of acids and bases occur in nature. Our foods contain a wide variety, and biochemically important molecules are often acids and bases.

Ammonia is a weaker base than methylamine (K_b for NH_3 < K_b for CH_3NH_2). This means that the conjugate acid of ammonia, NH_4^+ ($K_a = 5.6 \times 10^{-10}$) is stronger than the conjugate acid of methylamine ($CH_3NH_3^+$, $K_a = 2.0 \times 10^{-11}$).

Chemistry Now™

Sign in at **www.thomsonedu.com/login** and go to Chapter 17 Contents to see:
- Screen 17.5 for tutorials on **the pH of solutions of acids and bases**
- Screen 17.6 for a table of **K_a and K_b values**

EXERCISE 17.5 Strengths of Acids and Bases

Use Table 17.2 to answer the following questions.

(a) Which is the stronger acid, H_2SO_4 or H_2SO_3?

(b) Is benzoic acid, $C_6H_5CO_2H$, stronger or weaker than acetic acid?

(c) Which has the stronger conjugate base, acetic acid or boric acid?

(d) Which is the stronger base, ammonia or the acetate ion?

(e) Which has the stronger conjugate acid, ammonia or the acetate ion?

K_a Values for Polyprotic Acids

Like all polyprotic acids, phosphoric acid ionizes in a series of steps, three in this case.

First ionization step: $K_{a1} = 7.5 \times 10^{-3}$

$$H_3PO_4(aq) + H_2O(\ell) \rightleftharpoons H_2PO_4^-(aq) + H_3O^+(aq)$$

Second ionization step: $K_{a2} = 6.2 \times 10^{-8}$

$$H_2PO_4^-(aq) + H_2O(\ell) \rightleftharpoons HPO_4^{2-}(aq) + H_3O^+(aq)$$

Third ionization step: $K_{a3} = 3.6 \times 10^{-13}$

$$HPO_4^{2-}(aq) + H_2O(\ell) \rightleftharpoons PO_4^{3-}(aq) + H_3O^+(aq)$$

Notice that the K_a value for each successive step becomes smaller and smaller because it is more difficult to remove H^+ from a negatively charged ion, such as $H_2PO_4^-$, than from a neutral molecule, such as H_3PO_4. Similarly, the larger the negative charge of the anionic acid, the more difficult it is to remove H^+. Finally, also notice that for many inorganic polyprotic acids, K_a values become smaller by about 10^{-5} for each proton removed.

Aqueous Solutions of Salts

A number of the acids and bases listed in Table 17.2 are cations or anions. As described earlier, anions can act as Brønsted bases because they can accept a proton from an acid to form the conjugate acid of the base.

$$CO_3^{2-}(aq) + H_2O(\ell) \rightleftharpoons HCO_3^-(aq) + OH^-(aq)$$

$$K_b = 2.1 \times 10^{-4}$$

You should also notice that many metal cations in water are Brønsted acids.

$$[Al(H_2O)_6]^{3+}(aq) + H_2O(\ell) \rightleftharpoons [Al(H_2O)_5(OH)]^{2+}(aq) + H_3O^+(aq)$$

$$K_a = 7.9 \times 10^{-6}$$

Table 17.4 summarizes the acid–base properties of some of the common cations and anions. As you look over this table, notice the following points:

- Anions that are conjugate bases of strong acids (for example, Cl^- and NO_3^-) are such weak bases that they have no effect on solution pH.
- There are numerous basic anions (such as CO_3^{2-}). All are the conjugate bases of weak acids.
- The acid–base behavior of anions of polyprotic acids depends on the extent of deprotonation. For example, a fully deprotonated anion (such as CO_3^{2-}) will be basic. A partially deprotonated anion (such as HCO_3^-) is amphiprotic. Its behavior will depend on the other species in the reaction.
- Alkali metal and alkaline earth cations have no measurable effect on solution pH.
- Basic cations are conjugate bases of acidic cations such as $[Al(H_2O)_6]^{3+}$.
- Acidic cations are limited to metal cations with 2+ and 3+ charges and to ammonium ions (and their organic derivatives).
- All metal cations are hydrated in water. That is, they form ions such as $[M(H_2O)_6]^{n+}$. However, only when M is a 2+ or 3+ ion, particularly a transition metal ion, does the ion act as an acid.

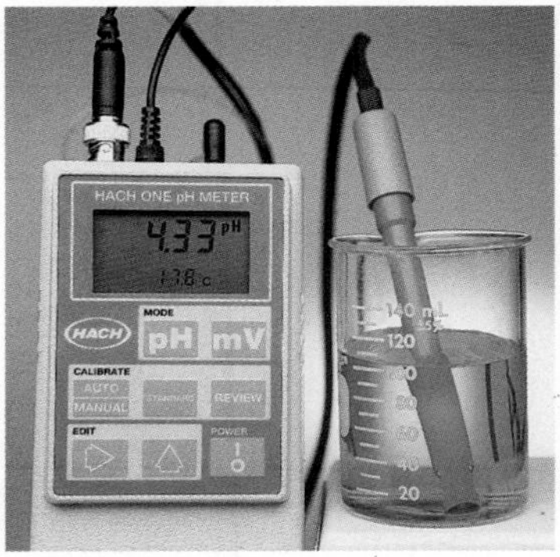

Charles D. Winters

Many aqueous metal cations are Brønsted acids. A pH measurement of a dilute solution of copper(II) sulfate shows that the solution is clearly acidic. Among the common cations, Al^{3+} and transition metal ions form acidic solutions in water.

Chemistry Now™

Sign in at **www.thomsonedu.com/login** and go to Chapter 17 Contents to see Screen 17.11 for a simulation showing **the pH of a number of cation/anion combinations.**

■ **Hydrolysis Reactions** Chemists often say that, when ions interact with water to produce acidic or basic solutions, the ions "hydrolyze" in water, or they undergo "hydrolysis." Thus, some books refer to the K_a and K_b values of ions as "hydrolysis constants," K_h.

TABLE 17.4 Acid and Base Properties of Some Ions in Aqueous Solution

	Neutral		Basic			Acidic
Anions	Cl^-	NO_3^-	$CH_3CO_2^-$	CN^-	SO_4^{2-}	HSO_4^-
	Br^-	ClO_4^-	HCO_2^-	PO_4^{3-}	HPO_4^{2-}	$H_2PO_4^-$
	I^-		CO_3^{2-}	HCO_3^-	SO_3^{2-}	HSO_3^-
			S^{2-}	HS^-	OCl^-	
			F^-	NO_2^-		
Cations	Li^+		$[Al(H_2O)_5(OH)]^{2+}$ (for example)			$[Al(H_2O)_6]^{3+}$ and hydrated transition metal cations (such as $[Fe(H_2O)_6]^{3+}$)
	Na^+	Ca^{2+}				
	K^+	Ba^{2+}				
						NH_4^+

EXAMPLE 17.2 Acid–Base Properties of Salts

Problem Decide whether each of the following will give rise to an acidic, basic, or neutral solution in water.

(a) $NaNO_3$

(b) K_3PO_4

(c) $FeCl_2$

(d) $NaHCO_3$

(e) NH_4F

Strategy First, decide on the cation and anion in each salt. Next, use Tables 17.3 and 17.4 to describe the properties of each ion.

Solution

(a) $NaNO_3$: This salt gives a neutral, aqueous solution (pH = 7). Neither the sodium ion, Na^+, nor the nitrate ion, NO_3^- (the very weak conjugate base of a strong acid), affects the solution pH.

(b) K_3PO_4: An aqueous solution of K_3PO_4 should be basic (pH > 7) because PO_4^{3-} is the conjugate base of the weak acid HPO_4^{2-}. The K^+ ion, like the Na^+ ion, does not affect the solution pH.

(c) $FeCl_2$: An aqueous solution of $FeCl_2$ should be weakly acidic (pH < 7). The Fe^{2+} ion in water, $[Fe(H_2O)_6]^{2+}$, is a Brønsted acid. In contrast, Cl^- is the very weak conjugate base of the strong acid HCl, so it does not contribute excess OH^- ions to the solution.

(d) $NaHCO_3$: Some additional information is needed concerning salts of amphiprotic anions such as HCO_3^-. Because they have an ionizable hydrogen, they can act as acids.

$$HCO_3^-(aq) + H_2O(\ell) \rightleftharpoons CO_3^{2-}(aq) + H_3O^+(aq) \qquad K_a = 4.8 \times 10^{-11}$$

They are also the conjugate bases of weak acids.

$$HCO_3^-(aq) + H_2O(\ell) \rightleftharpoons H_2CO_3(aq) + OH^-(aq) \qquad K_b = 2.4 \times 10^{-8}$$

Whether the solution is acidic or basic will depend on the relative magnitude of K_a and K_b. In the case of the hydrogen carbonate anion, K_b is larger than K_a, so $[OH^-]$ is larger than $[H_3O^+]$, and an aqueous solution of $NaHCO_3$ will be slightly basic.

(e) NH_4F: What happens if you have a salt based on an acidic cation and a basic anion? One example is ammonium fluoride. Here, the ammonium ion would decrease the pH, and the fluoride ion would increase the pH.

$$NH_4^+(aq) + H_2O(\ell) \rightleftharpoons H_3O^+(aq) + NH_3(aq) \qquad K_a(NH_4^+) = 5.6 \times 10^{-10}$$

$$F^-(aq) + H_2O(\ell) \rightleftharpoons HF(aq) + OH^-(aq) \qquad K_b(F^-) = 1.4 \times 10^{-11}$$

Because $K_a(NH_4^+) > K_b(F^-)$, the ammonium ion is a stronger acid than the fluoride ion is a base. The resulting solution should be slightly acidic.

Comment There are two important points to notice here:

- Anions that are conjugate bases of strong acids—such as Cl^- and NO_3^-—have no effect on solution pH.
- For a salt that has an acidic cation and a basic anion, the pH of the solution will be determined by the ion that is the stronger acid or base of the two.

Problem Solving Tip 17.2 Aqueous Solutions of Salts

Because aqueous solutions of salts are found in our bodies and throughout our economy and environment, it is important to know how to predict their acid and base properties. Information on the pH of an aqueous solution of a salt is summarized in Table 17.4. Consider also the following examples:

Cation	Anion	pH of the Solution
From strong base (Na^+)	From strong acid (Cl^-)	= 7 (neutral)
From strong base (K^+)	From weak acid ($CH_3CO_2^-$)	> 7 (basic)
From weak base (NH_4^+)	From strong acid (Cl^-)	< 7 (acidic)
From any weak base (BH^+)	From any weak acid (A^-)	Depends on relative strengths of BH^+ and A^-

EXERCISE 17.6 Acid–Base Properties of Salts in Aqueous Solution

For each of the following salts in water, predict whether the pH will be greater than, less than, or equal to 7.

(a) KBr **(b)** NH_4NO_3 **(c)** $AlCl_3$ **(d)** Na_2HPO_4

A Logarithmic Scale of Relative Acid Strength, pK_a

Many chemists and biochemists use a logarithmic scale to report and compare relative acid strengths.

$$pK_a = -\log K_a \quad (17.7)$$

The pK_a of an acid is the negative log of the K_a value (just as pH is the negative log of the hydronium ion concentration). For example, acetic acid has a pK_a value of 4.74.

$$pK_a = -\log(1.8 \times 10^{-5}) = 4.74$$

The pK_a value becomes smaller as the acid strength increases.

—— Acid strength increases ⟶

Propanoic acid	Acetic acid	Formic acid
$CH_3CH_2CO_2H$	CH_3CO_2H	HCO_2H
$K_a = 1.3 \times 10^{-5}$	$K_a = 1.8 \times 10^{-5}$	$K_a = 1.8 \times 10^{-4}$
$pK_a = 4.89$	$pK_a = 4.74$	$pK_a = 3.74$

⟵ pK_a increases ——

EXERCISE 17.7 A Logarithmic Scale for Acid Strength, pK_a

(a) What is the pK_a value for benzoic acid, $C_6H_5CO_2H$?

(b) Is chloroacetic acid ($ClCH_2CO_2H$), $pK_a = 2.87$, a stronger or weaker acid than benzoic acid?

(c) What is the pK_a for the conjugate acid of ammonia? Is this acid stronger or weaker than acetic acid?

Relating the Ionization Constants for an Acid and Its Conjugate Base

Let us look again at Table 17.3. From the top of the table to the bottom, the strengths of the acids decline (K_a becomes smaller), and the strengths of their conjugate bases increase (the values of K_b increase). Examining a few cases shows that the product of K_a for an acid and K_b for its conjugate base is equal to a constant, specifically K_w.

■ **A Relation Among p*K* Values** A useful relationship for an acid–conjugate base pair can be derived from Equation 17.8.

$$pK_w = pK_a + pK_b$$

$$K_a \times K_b = K_w \qquad \textbf{(17.8)}$$

Consider the specific case of the ionization of a weak acid, say HCN, and the interaction of its conjugate base, CN^-, with H_2O.

Weak acid:	$HCN(aq) + H_2O(\ell) \rightleftharpoons H_3O^+(aq) + CN^-(aq)$	$K_a = 4.0 \times 10^{-10}$
Conjugate base:	$CN^-(aq) + H_2O(\ell) \rightleftharpoons HCN(aq) + OH^-(aq)$	$K_b = 2.5 \times 10^{-5}$
	$2\ H_2O(\ell) \rightleftharpoons H_3O^+(aq) + OH^-(aq)$	$K_w = 1.0 \times 10^{-14}$

Adding the equations gives the chemical equation for the autoionization of water, and the numerical value is indeed 1.0×10^{-14}. That is,

$$K_a \times K_b = \left(\frac{[H_3O^+]\cancel{[CN^-]}}{\cancel{[HCN]}}\right)\left(\frac{\cancel{[HCN]}[OH^-]}{\cancel{[CN^-]}}\right) = [H_3O^+][OH^-] = K_w$$

Equation 17.8 is useful because K_b can be calculated from K_a. The value of K_b for the cyanide ion, for example, is

$$K_b \text{ for } CN^- = \frac{K_w}{K_a \text{ for HCN}} = \frac{1.0 \times 10^{-14}}{4.0 \times 10^{-10}} = 2.5 \times 10^{-5}$$

EXERCISE 17.8 Using the Equation $K_a \times K_b = K_w$

K_a for lactic acid, $CH_3CHOHCO_2H$, is 1.4×10^{-4}. What is K_b for the conjugate base of this acid, $CH_3CHOHCO_2^-$? Where does this base fit in Table 17.3?

17.5 Predicting the Direction of Acid–Base Reactions

According to the Brønsted–Lowry theory, all acid–base reactions can be written as equilibria involving the acid and base and their conjugates.

$$\text{Acid} + \text{Base} \rightleftharpoons \text{Conjugate base of the acid} + \text{Conjugate acid of the base}$$

In Section 17.4, we used equilibrium constants to provide quantitative information about the relative strengths of acids and bases. Now we want to show how the constants can be used to decide whether a particular acid–base reaction is product- or reactant–favored at equilibrium.

Hydrochloric acid is a strong acid. Its equilibrium constant for reaction with water is very large, with the equilibrium effectively lying completely to the right.

■ ***K* and Product- and Reactant-Favored Reactions** Reactions with an equilibrium constant greater than 1 are said to be product-favored at equilibrium. Those with $K < 1$ are reactant-favored at equilibrium.

$$HCl(aq) + H_2O(\ell) \rightleftharpoons H_3O^+(aq) + Cl^-(aq)$$

$$\text{Strong acid } (\approx 100\% \text{ ionized}),\ K >> 1$$

$$[H_3O^+] \approx \text{initial concentration of the acid}$$

Of the two acids here, HCl is stronger than H_3O^+. Of the two bases, H_2O and Cl^-, water is the stronger base and wins out in the competition for the proton. Thus, the equilibrium lies to the side of the chemical equation having the weaker acid and base.

conjugate pair 1 (HCl / Cl^-)

conjugate pair 2 (H_2O / H_3O^+)

$$HCl(aq) + H_2O(\ell) \underset{K>1}{\rightleftharpoons} H_3O^+(aq) + Cl^-(aq)$$

$HCl(aq)$	$H_2O(\ell)$	$H_3O^+(aq)$	$Cl^-(aq)$
Stronger acid than H_3O^+	Stronger base than Cl^-	Weaker acid than HCl	Weaker base than H_2O

In contrast to HCl and other strong acids, acetic acid, a *weak* acid, ionizes to only a very small extent (Table 17.3).

$$CH_3CO_2H(aq) + H_2O(\ell) \rightleftharpoons H_3O^+(aq) + CH_3CO_2^-(aq)$$

Weak acid (< 100% ionized), $K = 1.8 \times 10^{-5}$

$[H_3O^+] << \text{initial concentration of the acid}$

When equilibrium is achieved in a 0.1 M aqueous solution of CH_3CO_2H, the concentrations of $H_3O^+(aq)$ and $CH_3CO_2^-(aq)$ are each only about 0.001 M. Approximately 99% of the acetic acid is not ionized.

conjugate pair 1

conjugate pair 2

$$CH_3CO_2H(aq) + H_2O(\ell) \underset{K<1}{\rightleftharpoons} H_3O^+(aq) + CH_3CO_2^-(aq)$$

$CH_3CO_2H(aq)$	$H_2O(\ell)$	$H_3O^+(aq)$	$CH_3CO_2^-(aq)$
Weaker acid than H_3O^+	Weaker base than $CH_3CO_2^-$	Stronger acid than CH_3CO_2H	Stronger base than H_2O

Again, the equilibrium lies toward the side of the reaction having the weaker acid and base.

These two examples of the relative extent of acid–base reactions illustrate a general principle: *All proton transfer reactions proceed from the stronger acid and base to the weaker acid and base.* Using this principle and Table 17.3, you can predict which reactions are product-favored and which are reactant-favored. Consider the possible reaction of phosphoric acid and acetate ion to give acetic acid and the dihydrogen phosphate ion. Table 17.3 informs us that H_3PO_4 is a stronger acid ($K_a = 7.5 \times 10^{-3}$) than acetic acid ($K_a = 1.8 \times 10^{-5}$), and the acetate ion ($K_b = 5.6 \times 10^{-10}$) is a stronger base than the dihydrogen phosphate ion ($K_b = 1.3 \times 10^{-12}$).

Brønsted acids

Brønsted bases

$$H_3PO_4(aq) + CH_3CO_2^-(aq) \rightleftharpoons H_2PO_4^-(aq) + CH_3CO_2H(aq)$$

$H_3PO_4(aq)$	$CH_3CO_2^-(aq)$	$H_2PO_4^-(aq)$	$CH_3CO_2H(aq)$
Stronger acid than CH_3CO_2H	Stronger base than $H_2PO_4^-$	Weaker base than $CH_3CO_2^-$	Weaker acid than H_3PO_4

Thus, mixing phosphoric acid with sodium acetate would produce a significant amount of dihydrogen phosphate ion and acetic acid. That is, the equilibrium is predicted to lie to the right because the reaction proceeds from the stronger acid–base combination to the weaker acid–base combination.

Chemistry Now™

Sign in at **www.thomsonedu.com/login** and go to Chapter 17 Contents to see Screen 17.7 for a simulation on **predicting the direction of acid–base reactions.**

EXAMPLE 17.3 Reactions of Acids and Bases

Problem Write a balanced, net ionic equation for the reaction that occurs between acetic acid and sodium bicarbonate. Decide whether the equilibrium lies predominantly to the left or to the right.

Strategy First, identify the products of the acid–base reaction (which arise by H^+ transfer from the acid to the base). Next, identify the two acids (or the two bases) in the reaction. Finally, use Table 17.3 to decide which is the weaker of the two acids (or the weaker of the two bases). The reaction will proceed from the stronger acid (or base) to the weaker acid (or base).

Charles D. Winters

Reaction of vinegar and baking soda. This reaction involves the weak acid acetic acid and the weak base HCO_3^- from sodium hydrogen carbonate. Based on the values of the equilibrium constants, the reaction is predicted to proceed to the right to produce acetate ion, CO_2, and water.

Solution Acetic acid is clearly one acid involved (and its conjugate base is the acetate ion, $CH_3CO_2^-$). The other reactant, $NaHCO_3$, is a water-soluble salt that forms Na^+ and HCO_3^- ions in water. Because acetic acid can function only as an acid, the HCO_3^- ion in this case must be the Brønsted base. Thus, hydrogen ion transfer from the acid to the base (HCO_3^- ion) could lead to the following net ionic equation:

$$CH_3CO_2H(aq) + HCO_3^-(aq) \rightleftharpoons CH_3CO_2^-(aq) + H_2CO_3(aq)$$

According to Table 17.3, H_2CO_3 is a weaker acid ($K_a = 4.2 \times 10^{-7}$) than CH_3CO_2H ($K_a = 1.8 \times 10^{-5}$), and $CH_3CO_2^-$ is a weaker base ($K_b = 5.6 \times 10^{-10}$) than HCO_3^- ($K_b = 2.4 \times 10^{-8}$). The reaction favors the side having the weaker acid and base—that is, the right side.

Comment The reaction of acetic acid and $NaHCO_3$ favors the weaker acid (H_2CO_3) and base ($CH_3CO_2^-$). In the photograph in the margin, you see that the product, H_2CO_3, also dissociates into CO_2 and H_2O because the CO_2 bubbles out of the solution: the equilibrium lies far to the right.

$$H_2CO_3(aq) \rightleftharpoons CO_2(g) + H_2O(\ell)$$

See the discussion of gas-forming reactions in Chapter 3 and of Le Chatelier's principle in Section 14.3.

EXERCISE 17.9 Predicting the Direction of an Acid–Base Reaction

(a) Which is the stronger Brønsted acid, HCO_3^- or NH_4^+? Which has the stronger conjugate base?

(b) Is a reaction between HCO_3^- ions and NH_3 product- or reactant-favored?

$$HCO_3^-(aq) + NH_3(aq) \rightleftharpoons CO_3^{2-}(aq) + NH_4^+(aq)$$

(c) You mix solutions of sodium hydrogen phosphate and ammonia. The net ionic equation for a possible reaction is

$$HPO_4^{2-}(aq) + NH_3(aq) \rightleftharpoons PO_4^{3-}(aq) + NH_4^+(aq)$$

Does the equilibrium lie to the left or to the right in this reaction?

17.6 Types of Acid–Base Reactions

The reaction of hydrochloric acid and sodium hydroxide is the classic example of a strong acid–strong base reaction, whereas the reaction of citric acid and bicarbonate ion represents the reaction of a weak acid and weak base (Figure 17.4). There are four general types of acid–base reactions.

Type of Acid–Base Reaction	*Example*
Strong acid + strong base	HCl and NaOH
Strong acid + weak base	HCl and NH_3
Weak acid + strong base	CH_3CO_2H and NaOH
Weak acid + weak base	Citric acid and HCO_3^-

Because acid–base reactions are among the most important classes of chemical reactions, it is useful for you to know the outcome of the various types of these reactions (Table 17.5).

TABLE 17.5 Characteristics of Acid–Base Reactions

Type	Example	Net Ionic Equation	Species Present after Equal Molar Amounts Are Mixed; pH
Strong acid + strong base	HCl + NaOH	$H_3O^+(aq) + OH^-(aq) \rightleftharpoons 2\ H_2O(\ell)$	Cl^-, Na^+, pH = 7
Strong acid + weak base	HCl + NH_3	$H_3O^+(aq) + NH_3(aq) \rightleftharpoons NH_4^+(aq) + H_2O(\ell)$	Cl^-, NH_4^+, pH < 7
Weak acid + strong base	HCO_2H + NaOH	$HCO_2H(aq) + OH^-(aq) \rightleftharpoons HCO_2^-(aq) + H_2O(\ell)$	HCO_2^-, Na^+, pH > 7
Weak acid + weak base	HCO_2H + NH_3	$HCO_2H(aq) + NH_3(aq) \rightleftharpoons HCO_2^-(aq) + NH_4^+(aq)$	HCO_2^-, NH_4^+, pH dependent on K_a and K_b of conjugate acid and base

The Reaction of a Strong Acid with a Strong Base

Strong acids and bases are effectively 100% ionized in solution. Therefore, the total ionic equation for the reaction of HCl (strong acid) and NaOH (strong base) is

$$H_3O^+(aq) + Cl^-(aq) + Na^+(aq) + OH^-(aq) \rightleftharpoons 2\ H_2O(\ell) + Na^+(aq) + Cl^-(aq)$$

which leads to the following net ionic equation

$$H_3O^+(aq) + OH^-(aq) \rightleftharpoons 2\ H_2O(\ell) \qquad K = 1/K_w = 1.0 \times 10^{14}$$

The net ionic equation for the reaction of any strong acid with any strong base is always simply the union of hydronium ion and hydroxide ion to give water (◀ Section 3.7). Because this reaction is the reverse of the autoionization of water, it has an equilibrium constant of $1/K_w$. This very large value of K shows that, for all practical purposes, the reactants are completely consumed to form products. Thus, if equal numbers of moles of NaOH and HCl are mixed, the result is just a solution of NaCl in water. The constituents of NaCl, Na^+ and Cl^- ions, which arise from a strong base and a strong acid, respectively, produce a neutral aqueous solution. For this reason, reactions of strong acids and bases are often called "neutralizations."

Mixing equal amounts (moles) of a strong base with a strong acid produces a neutral solution (pH = 7.00 at 25 °C).

Charles D. Winters

FIGURE 17.4 Reaction of a weak acid with a weak base. The bubbles coming from the tablet are carbon dioxide. This arises from the reaction of a weak Brønsted acid (citric acid) with a weak Brønsted base (HCO_3^-). The reaction is driven to completion by gas evolution.

The Reaction of a Weak Acid with a Strong Base

Consider the reaction of the naturally occurring weak acid formic acid, HCO_2H, with sodium hydroxide. The net ionic equation is

$$HCO_2H(aq) + OH^-(aq) \rightleftharpoons H_2O(\ell) + HCO_2^-(aq)$$

In the reaction of formic acid with NaOH, OH^- is a much stronger base than HCO_2^- ($K_b = 5.6 \times 10^{-11}$), and the reaction is predicted to proceed to the right. If equal amounts of weak acid and base are mixed, the final solution will contain sodium formate ($NaHCO_2$), a salt that is 100% dissociated in water. The Na^+ ion is the cation of a strong base and so gives a neutral solution. The formate ion, however, is the conjugate base of a weak acid (Table 17.3), so the solution is basic. This example leads to a useful general conclusion:

■ **Formic Acid + NaOH** The equilibrium constant for the reaction of formic acid and sodium hydroxide is 1.8×10^{10}. Can you confirm this? (See Study Question 17.97.)

Mixing equal amounts (moles) of a strong base with a weak acid produces a salt whose anion is the conjugate base of the weak acid. The solution is basic, with the pH depending on K_b for the anion.

The Reaction of a Strong Acid with a Weak Base

The net ionic equation for the reaction of the strong acid HCl and the weak base NH_3 is

$$H_3O^+(aq) + NH_3(aq) \rightleftharpoons H_2O(\ell) + NH_4^+(aq)$$

The hydronium ion, H_3O^+, is a much stronger acid than NH_4^+ ($K_a = 5.6 \times 10^{-10}$), and NH_3 is a stronger base ($K_b = 1.8 \times 10^{-5}$) than H_2O. Therefore, reaction is predicted to proceed to the right and essentially to completion. Thus, after mixing equal amounts of HCl and NH_3, the solution contains the salt ammonium chloride, NH_4Cl. The Cl^- ion has no effect on the solution pH (Tables 17.3 and 17.4). However, the NH_4^+ ion is the conjugate acid of the weak base NH_3, so the solution at the conclusion of the reaction is acidic. In general, we can conclude that

■ **Ammonia + HCl** The equilibrium constant for the reaction of a strong acid with aqueous ammonia is 1.8×10^9. Can you confirm this? (See Study Question 17.96.)

Mixing equal amounts (moles) of a strong acid and a weak base produces a salt whose cation is the conjugate acid of the weak base. The solution is acidic, with the pH depending on K_a for the cation.

Charles D. Winters

A weak acid reacting with a weak base. Baking powder contains the weak acid calcium dihydrogen phosphate, $Ca(H_2PO_4)_2$. This can react with the basic HCO_3^- ion in baking soda to give HPO_4^{2-}, CO_2 gas, and water.

■ ***K* for Reaction of Weak Acid and Weak Base** The equilibrium constant for the reaction between a weak acid and a weak base is $K_{net} = (K_a \cdot K_b)/K_w$. Can you confirm this? (See Study Question 17.121.)

Module 22

The Reaction of a Weak Acid with a Weak Base

If acetic acid, a weak acid, is mixed with ammonia, a weak base, the following reaction occurs.

$$CH_3CO_2H(aq) + NH_3(aq) \rightleftharpoons NH_4^+(aq) + CH_3CO_2^-(aq)$$

You know that the reaction is product-favored because CH_3CO_2H is a stronger acid than NH_4^+ and NH_3 is a stronger base than $CH_3CO_2^-$ (Table 17.3). Thus, if equal amounts of the acid and base are mixed, the resulting solution contains ammonium acetate, $NH_4CH_3CO_2$. Is this solution acidic or basic? On page 775 you learned that this depends on the relative values of K_a for the conjugate acid (here, NH_4^+; $K_a = 5.6 \times 10^{-10}$) and K_b for the conjugate base (here, $CH_3CO_2^-$; $K_b = 5.6 \times 10^{-10}$). In this case, the values of K_a and K_b are the same, so the solution is predicted to be neutral.

Mixing equal amounts (moles) of a weak acid and a weak base produces a salt whose cation is the conjugate acid of the weak base and whose anion is the conjugate base of the weak acid. The solution pH depends on the relative K_a and K_b values.

EXERCISE 17.10 Acid–Base Reactions

(a) Equal amounts (moles) of HCl(aq) and NaCN(aq) are mixed. Is the resulting solution acidic, basic, or neutral?

(b) Equal amounts (moles) of acetic acid and sodium sulfite, Na_2SO_3, are mixed. Is the resulting solution acidic, basic, or neutral?

17.7 Calculations with Equilibrium Constants

Determining *K* from Initial Concentrations and Measured pH

The K_a and K_b values found in Table 17.3 and in the more extensive tables in Appendices H and I were all determined by experiment. There are several experimental methods available, but one approach, illustrated by the following example, is to determine the pH of the solution.

Chemistry Now™

Sign in at **www.thomsonedu.com/login** and go to Chapter 17 Contents to see Screen 17.8 for a tutorial on **determining K_a and K_b values.**

EXAMPLE 17.4 Calculating a K_a Value from a Measured pH

Problem A 0.10 M aqueous solution of lactic acid, $CH_3CHOHCO_2H$, has a pH of 2.43. What is the value of K_a for lactic acid?

Strategy To calculate K_a, we must know the equilibrium concentration of each species. The pH of the solution directly tells us the equilibrium concentration of H_3O^+, and we can derive the other equilibrium concentrations from this. These are used to calculate K_a.

Solution The equation for the equilibrium interaction of lactic acid with water is

$$\underset{\text{Lactic acid}}{CH_3CHOHCO_2H(aq)} + H_2O(\ell) \rightleftharpoons \underset{\text{Lactate ion}}{CH_3CHOHCO_2^-(aq)} + H_3O^+(aq)$$

and the equilibrium constant expression is

$$K_a\ (\text{lactic acid}) = \frac{[H_3O^+][CH_3CHOHCO_2^-]}{[CH_3CHOHCO_2H]}$$

We begin by converting the pH to $[H_3O^+]$.

$$[H_3O^+] = 10^{-pH} = 10^{-2.43} = 3.7 \times 10^{-3}\ M$$

Next, prepare an ICE table of the concentrations in the solution before equilibrium is established, the change that occurs as the reaction proceeds to equilibrium, and the concentrations when equilibrium has been achieved. (See Examples 16.2–16.5.)

Equilibrium	$CH_3CHOHCO_2H + H_2O$	$\rightleftharpoons CH_3CHOHCO_2^-$	$+ H_3O^+$
Initial (M)	0.10	0	0
Change (M)	$-x$	$+x$	$+x$
Equilibrium (M)	$(0.10 - x)$	x	x

The following points can be made concerning the ICE table.

- The quantity x represents the equilibrium concentrations of hydronium ion and lactate ion.

That is, at equilibrium $x = [H_3O^+] = [CH_3CHOHCO_2^-] = 3.7 \times 10^{-3}$ M.

- By stoichiometry, x is also the quantity of acid that ionized on proceeding to equilibrium.

With these points in mind, we can calculate K_a for lactic acid.

$$K_a \text{ (lactic acid)} = \frac{[H_3O^+][CH_3CHOHCO_2^-]}{[CH_3CHOHCO_2H]}$$

$$= \frac{(3.7 \times 10^{-3})(3.7 \times 10^{-3})}{0.10 - 0.0037} = 1.4 \times 10^{-4}$$

Comparing this value of K_a with others in Table 17.3, we see it is similar to formic acid in its strength.

Comment Hydronium ion, H_3O^+, is present in solution from lactic acid ionization and from water autoionization. Le Chatelier's principle informs us that the H_3O^+ added to the water by lactic acid will suppress the H_3O^+ coming from the water autoionization. However, because $[H_3O^+]$ from water must be less than 10^{-7} M, the pH is almost completely a reflection of H_3O^+ from lactic acid. (See Example 17.1.)

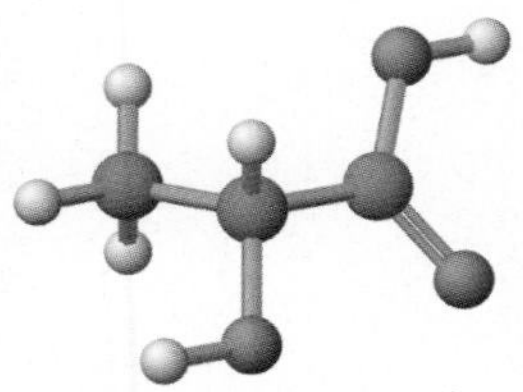

Lactic acid, $CH_3CHOHCO_2H$. Lactic acid is a weak monoprotic acid that occurs naturally in sour milk and arises from metabolism in the human body.

EXERCISE 17.11 Calculating a K_a Value from a Measured pH

A solution prepared from 0.055 mol of butanoic acid dissolved in sufficient water to give 1.0 L of solution has a pH of 2.72. Determine K_a for butanoic acid. The acid ionizes according to the balanced equation

$$CH_3CH_2CH_2CO_2H(aq) + H_2O(\ell) \rightleftharpoons H_3O^+(aq) + CH_3CH_2CH_2CO_2^-(aq)$$

There is an important point to notice in Example 17.4. The lactic acid concentration at equilibrium was given by $(0.10 - x)$ where x was found to be 3.7×10^{-3} M. By the usual rules governing significant figures $(0.10 - 0.0037)$ is equal to 0.10. The acid is weak, so very little of it ionizes (approximately 4%), and the equilibrium concentration of lactic acid is essentially equal to the initial acid concentration. Neglecting to subtract 0.0037 from 0.10 has no effect on the answer to two significant figures.

Like lactic acid, most weak acids (HA) are so weak that the equilibrium concentration of the acid, [HA], is effectively its initial concentration ($= [HA]_0$). This leads to the useful conclusion that the denominator in the equilibrium constant expression for dilute solutions of most weak acids is simply $[HA]_0$, the original or initial concentration of the weak acid.

$$HA(aq) + H_2O(\ell) \rightleftharpoons H_3O^+(aq) + A^-(aq)$$

$$K_a = \frac{[H_3O^+][A^-]}{[HA]_0 - [H_3O^+]} \approx \frac{[H_3O^+][A^-]}{[HA]_0}$$

Analysis shows that

> The approximation that $[HA]_{equilibrium}$ is effectively equal to $[HA]_0$ ($[HA]_{equilibrium} = [HA]_0 - [H_3O^+] \approx [HA]_0$) is valid whenever $[HA]_0$ is greater than or equal to $100 \cdot K_a$.

This is the same approximation we derived in Chapter 16 when deciding whether we needed to solve quadratic equations exactly (◀ Problem Solving Tip 16.1).

What Is the pH of an Aqueous Solution of a Weak Acid or Base?

Knowing values of the equilibrium constants for weak acids and bases enables us to calculate the pH of a solution of a weak acid or base.

Chemistry Now™

Sign in at **www.thomsonedu.com/login** and go to Chapter 17 Contents to see Screen 17.9 for a tutorial on **estimating the pH of weak acid solutions.**

EXAMPLE 17.5 Calculating Equilibrium Concentrations and pH from K_a

Problem Calculate the pH of a 0.020 M solution of benzoic acid ($C_6H_5CO_2H$) if $K_a = 6.3 \times 10^{-5}$ for the acid.

$$C_6H_5CO_2H(aq) + H_2O(\ell) \rightleftharpoons H_3O^+(aq) + C_6H_5CO_2^-(aq)$$

Strategy This is similar to Examples 16.5 and 16.6, where we wanted to find the concentration of a reaction product. The strategy is the same: designate the quantity of product (here $[H_3O^+]$) by x, and derive the other concentrations from that starting point.

Solution Organize the information in an ICE table

Equilibrium	$C_6H_5CO_2H + H_2O \rightleftharpoons$	$C_6H_5CO_2^-$	$+ H_3O^+$
Initial (M)	0.020	0	0
Change (M)	$-x$	$+x$	$+x$
Equilibrium (M)	$(0.020 - x)$	x	x

According to reaction stoichiometry,

$$[H_3O^+] = [C_6H_5CO_2^-] = x \text{ at equilibrium}$$

Stoichiometry also tells us that the quantity of acid ionized is x. Thus, the benzoic acid concentration at equilibrium is

$$[C_6H_5CO_2H] = \text{initial acid quantity} - \text{quantity of acid that ionized}$$

$$[C_6H_5CO_2H] = [C_6H_5CO_2H]_0 - x$$

$$[C_6H_5CO_2H] = 0.020 - x$$

Substituting these equilibrium concentrations into the K_a expression, we have

$$K_a = \frac{[H_3O^+][C_6H_5CO_2^-]}{[C_6H_5CO_2H]}$$

$$6.3 \times 10^{-5} = \frac{(x)(x)}{0.020 - x}$$

The value of x is small compared with 0.020 (because $[HA]_0 > 100 \cdot K_a$; $0.020 \text{ M} > 6.3 \times 10^{-3}$). Therefore,

$$K_a = 6.3 \times 10^{-5} = \frac{x^2}{0.020}$$

Solving for x, we have

$$x = \sqrt{K_a \times (0.020)} = 0.0011 \text{ M}$$

and we find that

$$[H_3O^+] = [C_6H_5CO_2^-] = 0.0011 \text{ M}$$

and

$$[C_6H_5CO_2H] = (0.020 - x) = 0.019 \text{ M}$$

Finally, the pH of the solution is found to be

$$\text{pH} = -\log(1.1 \times 10^{-3}) = 2.96$$

Comment Let us think again about the result. Because benzoic acid is weak, we made the approximation that $(0.020 - x) \approx 0.020$. If we do not make the approximation and instead solve the exact expression, $x = [H_3O^+] = 1.1 \times 10^{-3}$ M. This is the same answer to two significant figures that we obtained from the "approximate" expression. Finally, notice that we again ignored any H_3O^+ that arises from water ionization.

EXAMPLE 17.6 Calculating Equilibrium Concentrations and pH from K_a and Using the Method of Successive Approximations

Problem What is the pH of a 0.0010 M solution of formic acid? What is the concentration of formic acid at equilibrium? The acid is moderately weak, with $K_a = 1.8 \times 10^{-4}$.

$$HCO_2H(aq) + H_2O(\ell) \rightleftharpoons HCO_2^-(aq) + H_3O^+(aq)$$

Strategy This is similar to Example 17.5, except that an approximate solution will not be possible.

Solution The ICE table is shown here.

Equilibrium	HCO_2H + H_2O ⇌	HCO_2^- +	H_3O^+
Initial (M)	0.0010	0	0
Change (M)	$-x$	$+x$	$+x$
Equilibrium (M)	$(0.0010 - x)$	x	x

Substituting the values in the table into the K_a expression we have

$$K_a = \frac{[H_3O^+][HCO_2^-]}{[HCO_2H]} = 1.8 \times 10^{-4} = \frac{(x)(x)}{0.0010 - x}$$

Formic acid is a weak acid because it has a value of K_a much less than 1. In this example, however, $[HA]_0$ (= 0.0010 M) is *not* greater than $100 \cdot K_a$ (= 1.8×10^{-2}), so the usual approximation is not reasonable. Thus, we have to find the equilibrium concentrations by solving the "exact" expression. This can be solved with the quadratic formula (page 740) or by successive approximations (Appendix A). Let us use the successive approximation method here.

To use the successive approximations approach, begin by solving the approximate expression for x.

$$1.8 \times 10^{-4} = \frac{(x)(x)}{0.0010}$$

Solving this, we find $x = 4.2 \times 10^{-4}$. Put this value into the expression for x in the denominator of the exact expression.

$$1.8 \times 10^{-4} = \frac{(x)(x)}{0.0010 - x} = \frac{(x)(x)}{0.0010 - 4.2 \times 10^{-4}}$$

Solving this equation for x, we now find $x = 3.2 \times 10^{-4}$. Again, put this value into the denominator, and solve for x.

$$1.8 \times 10^{-4} = \frac{(x)(x)}{0.0010 - x} = \frac{(x)(x)}{0.0010 - 3.2 \times 10^{-4}}$$

Continue this procedure until the value of x does not change from one cycle to the next. In this case, two more steps give us the result that

$$x = [H_3O^+] = [HCO_2^-] = 3.4 \times 10^{-4} \text{ M}$$

Thus,

$$[HCO_2H] = 0.0010 - x \approx 0.0007 \text{ M}$$

and the pH of the formic acid solution is

$$\text{pH} = -\log(3.4 \times 10^{-4}) = 3.47$$

Comment If we would have used the approximate expression to find the H_3O^+ concentration, we would have obtained a value of $[H_3O^+] = 4.2 \times 10^{-4}$ M. A simplifying assumption led to a large error, about 24%. The approximate solution fails in this case because (a) the acid concentration is small and (b) the acid is not all that weak. These made invalid the approximation that $[HA]_{equilibrium} \approx [HA]_0$.

EXERCISE 17.12 Calculating Equilibrium Concentrations and pH from K_a

What are the equilibrium concentrations of acetic acid, the acetate ion, and H_3O^+ for a 0.10 M solution of acetic acid ($K_a = 1.8 \times 10^{-5}$)? What is the pH of the solution?

EXERCISE 17.13 Calculating Equilibrium Concentrations and pH from K_a

What are the equilibrium concentrations of HF, F^- ion, and H_3O^+ ion in a 0.015 M solution of HF? What is the pH of the solution?

Just as acids can be molecular species or ions, so too can bases be molecular or ionic (Figures 17.3–17.5). Many molecular bases are based on nitrogen, with ammonia being the simplest. Many other nitrogen-containing bases occur naturally; caffeine and nicotine are two that are well known. The anionic conjugate bases of weak acids make up another group of bases. The following example describes the calculation of the pH for a solution of sodium acetate.

Chemistry Now™

Sign in at **www.thomsonedu.com/login** and go to Chapter 17 Contents to see:

- Screen 17.7 for a simulation of **the prediction of the direction of a number of acid–base reactions**
- Screen 17.11 for a simulation **of acid–base properties of salts**

FIGURE 17.5 Examples of weak bases. Weak bases in water include molecules having one or more N atoms capable of accepting an H^+ ion. Anionic bases such as benzoate and phosphate are conjugate bases of weak acids.

EXAMPLE 17.7 The pH of a Solution of a Weakly Basic Salt, Sodium Acetate

Problem What is the pH of a 0.015 M solution of sodium acetate, $NaCH_3CO_2$?

Strategy Sodium acetate will be basic in water because the acetate ion, the conjugate base of a weak acid, acetic acid, reacts with water to form OH^- (Tables 17.3 and 17.4). (Note that the sodium ion of sodium acetate does not affect the solution pH.) We shall calculate the hydroxide ion concentration in manner parallel with that in Example 17.5.

Solution The value of K_b for the acetate ion is 5.6×10^{-10} (Table 17.3).

$$CH_3CO_2^-(aq) + H_2O(\ell) \rightleftharpoons CH_3CO_2H(aq) + OH^-(aq)$$

Set up an ICE table to summarize the initial and equilibrium concentrations of the species in solution.

Equilibrium	$CH_3CO_2^- + H_2O$	$\rightleftharpoons$	CH_3CO_2H	+	OH^-
Initial (M)	0.015		0		0
Change (M)	$-x$		$+x$		$+x$
Equilibrium (M)	$(0.015 - x)$		x		x

Next, substitute the values in the table into the K_b expression.

$$K_b = 5.6 \times 10^{-10} = \frac{[CH_3CO_2H][OH^-]}{[CH_3CO_2^-]} = \frac{x^2}{0.015 - x}$$

The acetate ion is a weak base, as reflected by the very small value of K_b. Therefore, we assume that x, the concentration of hydroxide ion generated by reaction of acetate ion with water, is very small, and we use the approximate expression to solve for x.

$$K_b = 5.6 \times 10^{-10} = \frac{x^2}{0.015}$$

$$x = [OH^-] = [CH_3CO_2H] = \sqrt{(5.6 \times 10^{-10})(0.015)} = 2.9 \times 10^{-6}\ M$$

To calculate the pH of the solution, we need the hydronium ion concentration. In aqueous solutions, it is always true that, at 25 °C,

$$K_w = 1.0 \times 10^{-14} = [H_3O^+][OH^-]$$

Therefore,

$$[H_3O^+] = \frac{K_w}{[OH^-]} = \frac{1.0 \times 10^{-14}}{2.9 \times 10^{-6}} = 3.5 \times 10^{-9}\ M$$

$$pH = -\log(3.5 \times 10^{-9}) = 8.46$$

The acetate ion gives rise to a weakly basic solution.

Comment The hydroxide ion concentration (x) is indeed quite small relative to the initial acetate ion concentration. (We would have predicted this from our "rule of thumb": that $100 \cdot K_b$ should be less than the initial base concentration if we wish to use the approximate expression.)

EXERCISE 17.14 The pH of the Solution of the Conjugate Base of a Weak Acid

Sodium hypochlorite, NaClO, is used as a disinfectant in swimming pools and water treatment plants. What are the concentrations of HClO and OH^- and the pH of a 0.015 M solution of NaClO?

Chemistry Now™

Sign in at **www.thomsonedu.com/login** and go to Chapter 17 Contents to see Screen 17.10 for a tutorial on **estimating the pH following an acid–base reaction.**

EXAMPLE 17.8 Calculating the pH After the Reaction of a Base with an Acid

Problem What is the pH of the solution that results from mixing 25 mL of 0.016 M NH_3 and 25 mL of 0.016 M HCl?

Strategy This question involves three problems in one:

(a) *Writing a Balanced Equation:* We first have to write a balanced equation for the reaction that occurs and then decide whether the reaction products are acids or bases. Here, NH_4^+ is the product of interest, and it is a weak acid.

(b) *Stoichiometry Problem:* To find the "initial" NH_4^+ concentration is a stoichiometry problem: What amount of NH_4^+ (in moles) is produced in the HCl + NH_3 reaction, and in what volume of solution is the NH_4^+ ion found?

(c) *Equilibrium Problem:* Calculating the pH involves solving an equilibrium problem. The crucial piece of information needed here is the "initial" concentration of NH_4^+ from part (b).

Solution If equal amounts (moles) of base (NH_3) and acid (HCl) are mixed, the result should be an acidic solution because the significant species remaining in solution upon completion of the reaction is NH_4^+, the conjugate acid of the weak base ammonia (see Tables 17.3 and 17.5). The chemistry can be summarized by the following net ionic equations.

(a) *Writing Balanced Equations*

Reaction of HCl (the supplier of hydronium ion) with NH_3 to give NH_4^+:

$$NH_3(aq) + H_3O^+(aq) \longrightarrow NH_4^+(aq) + H_2O(\ell)$$

The reaction of NH_4^+, the product, with water:

$$NH_4^+(aq) + H_2O(\ell) \rightleftharpoons H_3O^+(aq) + NH_3(aq)$$

(b) *Stoichiometry Problem*

Amount of HCl and NH_3 consumed:

$$(0.025 \text{ L HCl})(0.016 \text{ mol/L}) = 4.0 \times 10^{-4} \text{ mol HCl}$$

$$(0.025 \text{ L } NH_3)(0.016 \text{ mol/L}) = 4.0 \times 10^{-4} \text{ mol } NH_3$$

Amount of NH_4^+ produced upon completion of the reaction:

$$4.0 \times 10^{-4} \text{ mol } NH_3 \left(\frac{1 \text{ mol } NH_4^+}{1 \text{ mol } NH_3}\right) = 4.0 \times 10^{-4} \text{ mol } NH_4^+$$

Concentration of NH_4^+: Combining 25 mL each of HCl and NH_3 gives a total solution volume of 50. mL. Therefore, the concentration of NH_4^+ is

$$[NH_4^+] = \frac{4.0 \times 10^{-4} \text{ mol}}{0.050 \text{ L}} = 8.0 \times 10^{-3} \text{ M}$$

(c) *Acid–Base Equilibrium Problem*

With the initial concentration of ammonium ion known, set up an ICE table to find the equilibrium concentration of hydronium ion.

Equilibrium	$NH_4^+ + H_2O$	$\rightleftharpoons$	NH_3	+	H_3O^+
Initial (M)	0.0080		0		0
Change	$-x$		$+x$		$+x$
Equilibrium (M)	$(0.0080 - x)$		x		x

Next, substitute the values in the table into the K_a expression for the ammonium ion. Thus, we have

$$K_a = 5.6 \times 10^{-10} = \frac{[H_3O^+][NH_3]}{[NH_4^+]} = \frac{(x)(x)}{0.0080 - x}$$

Problem Solving Tip 17.3 What Is the pH After Mixing Equal Molar Amounts of an Acid and a Base?

Table 17.5 summarizes the outcome of mixing various types of acids and bases. But how do you calculate a numerical value for the pH, particularly in the case of mixing a weak acid with a strong base or a weak base with a strong acid? The strategy (Example 17.8) is to recognize that this involves two related calculations: a stoichiometry calculation and an equilibrium calculation. The key to this is that you need to know the concentration of the weak acid or weak base produced when the acid and base are mixed. You should ask yourself the following questions:

(a) What amounts of acid and base are used (in moles)? (This is a stoichiometry problem.)
(b) What is the total volume of the solution after mixing the acid and base solutions?
(c) What is the concentration of the weak acid or base produced on mixing the acid and base solutions?
(d) Using the concentration found in Step (c), what is the hydronium ion concentration in the solution? (This is an equilibrium problem.)
(e) Calculate the pH of the solution from $[H_3O^+]$.

The ammonium ion is a very weak acid, as reflected by the very small value of K_a. Therefore, x, the concentration of hydronium ion generated by reaction of ammonium ion with water, is assumed to be very small, and the approximate expression is used to solve for x. (Here $100 \cdot K_a$ is much less than the original acid concentration.)

$$K_a = 5.6 \times 10^{-10} \approx \frac{x^2}{0.0080}$$

$$x = \sqrt{(5.6 \times 10^{-10})(0.0080)}$$

$$= [H_3O^+] = [NH_3] = 2.1 \times 10^{-6}\ M$$

$$pH = -\log(2.1 \times 10^{-6}) = 5.67$$

Comment As predicted (Table 17.5), the solution after mixing equal amounts of a strong acid and weak base is weakly acidic.

EXERCISE 17.15 What is the pH After the Reaction of a Weak Acid and Strong Base?

Calculate the pH after mixing 15 mL of 0.12 M acetic acid with 15 mL of 0.12 M NaOH. What are the major species in solution at equilibrium (besides water), and what are their concentrations?

17.8 Polyprotic Acids and Bases

Because polyprotic acids are capable of donating more than one proton (Table 17.1), they present us with additional problems when predicting the pH of their solutions. For many inorganic polyprotic acids, such as phosphoric acid, carbonic acid, and hydrogen sulfide, each successive loss of a proton is about 10^4 to 10^6 more difficult than the previous ionization step. This means that the first ionization step of a polyprotic acid produces up to about a million times more H_3O^+ ions than the second step. For this reason, *the pH of many inorganic polyprotic acids depends primarily on the hydronium ion generated in the first ionization step; the hydronium ion produced in the second step can be neglected.* The same principle applies to the conjugate bases of polyprotic acids. This is illustrated by the calculation of the pH of a solution of carbonate ion, an important base in our environment (Example 17.9).

Charles D. Winters

A polyprotic acid. Malic acid is a diprotic acid occurring in apples. It is also classified as an alpha-hydroxy acid because it has an OH group on the C atom next to the CO_2H (in the alpha position). It is one of a larger group of natural acids such as lactic acid, citric acid, and ascorbic acid. Alpha-hydroxy acids have been touted as an ingredient in "anti-aging" skin creams. They work by accelerating the natural process by which skin replaces the outer layer of cells with new cells.

EXAMPLE 17.9 Calculating the pH of the Solution of a Polyprotic Base

Problem The carbonate ion, CO_3^{2-}, is a base in water, forming the hydrogen carbonate ion, which in turn can form carbonic acid.

$$CO_3^{2-}(aq) + H_2O(\ell) \rightleftharpoons HCO_3^-(aq) + OH^-(aq) \qquad K_{b1} = 2.1 \times 10^{-4}$$

$$HCO_3^-(aq) + H_2O(\ell) \rightleftharpoons H_2CO_3(aq) + OH^-(aq) \qquad K_{b2} = 2.4 \times 10^{-8}$$

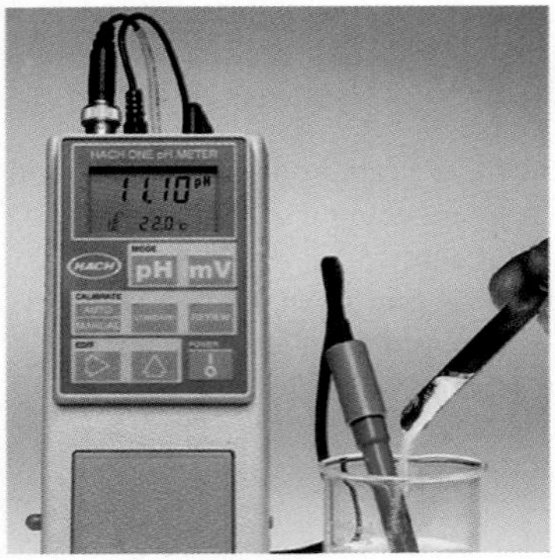

Charles D. Winters

Sodium carbonate, a polyprotic base. This common substance is a base in aqueous solution. Its primary use is in the glass industry. Although it used to be manufactured, it is now mined as the mineral trona, $Na_2CO_3 \cdot NaHCO_3 \cdot 2\,H_2O$.

What is the pH of a 0.10 M solution of Na_2CO_3?

Strategy The second ionization constant, K_{b2}, is much smaller than the first, K_{b1}, so the hydroxide ion concentration in the solution results almost entirely from the first step. Therefore, let us calculate the OH^- concentration produced in the first ionization step but test the conclusion that OH^- produced in the second step is negligible.

Solution Set up an ICE table for the reaction of the carbonate ion (Equilibrium Table 1).

Equilibrium Table 1—Reaction of CO_3^{2-} Ion

Equilibrium	CO_3^{2-}	+	H_2O	$\rightleftharpoons$	HCO_3^-	+	OH^-
Initial (M)	0.10				0		0
Change	$-x$				$+x$		$+x$
Equilibrium (M)	$(0.10 - x)$				x		x

Based on this table, the equilibrium concentration of OH^- $(= x)$ can then be calculated.

$$K_{b1} = 2.1 \times 10^{-4} = \frac{[HCO_3^-][OH^-]}{[CO_3^{2-}]} = \frac{x^2}{0.10 - x}$$

Because K_{b1} is relatively small, it is reasonable to make the approximation that $(0.10 - x) \approx 0.10$. Therefore,

$$x = [HCO_3^-] = [OH^-] = \sqrt{(2.1 \times 10^{-4})(0.10)} = 4.6 \times 10^{-3}\ M$$

Using this value of $[OH^-]$, we first calculate the pOH of the solution,

$$pOH = -\log(4.6 \times 10^{-3}) = 2.34$$

and then use the relationship pH + pOH = 14 to calculate the pH.

$$pH = 14 - pOH = 11.66$$

Finally, we see that the concentration of the carbonate ion is, to a good approximation, 0.10 M.

$$[CO_3^{2-}] = 0.10 - 0.0046 \approx 0.10\ M$$

Comment It is instructive to ask what the concentration of H_2CO_3 in the solution might be. If HCO_3^- were to react significantly with water to produce H_2CO_3, the pH of the solution would be affected. Let us set up a second ICE Table.

Equilibrium Table 2—Reaction of HCO_3^- Ion

Equilibrium	HCO_3^-	+	H_2O	$\rightleftharpoons$	H_2CO_3	+	OH^-
Initial (M)	4.6×10^{-3}				0		4.6×10^{-3}
Change	$-y$				$+y$		$+y$
Equilibrium (M)	$(4.6 \times 10^{-3} - y)$				y		$(4.6 \times 10^{-3} + y)$

Because K_{b2} is so small, the second step occurs to a much smaller extent than the first step. This means the amount of H_2CO_3 and OH^- produced in the second step $(= y)$ is much smaller than 10^{-3} M. Therefore, it is reasonable that both $[HCO_3^-]$ and $[OH^-]$ are very close to 4.6×10^{-3} M.

$$K_{b2} = 2.4 \times 10^{-8} = \frac{[H_2CO_3][OH^-]}{[HCO_3^-]} = \frac{(y)(4.6 \times 10^{-3})}{4.6 \times 10^{-3}}$$

Because $[HCO_3^-]$ and $[OH^-]$ have nearly identical values, they cancel from the expression, and we find that $[H_2CO_3]$ is simply equal to K_{b2}.

$$y = [H_2CO_3] = K_{b2} = 2.4 \times 10^{-8}\ M$$

For the carbonate ion, where K_1 and K_2 differ by about 10^4, the hydroxide ion is essentially all produced in the first equilibrium process.

EXERCISE 17.16 Calculating the pH of the Solution of a Polyprotic Acid

What is the pH of a 0.10 M solution of oxalic acid, $H_2C_2O_4$? What are the concentrations of H_3O^+, $HC_2O_4^-$, and the oxalate ion, $C_2O_4^{2-}$? (See Appendix H for K_a values.)

17.9 The Lewis Concept of Acids and Bases

The concept of acid–base behavior advanced by Brønsted and Lowry in the 1920s works well for reactions involving proton transfer. A more general acid–base concept, however, was developed by Gilbert N. Lewis in the 1930s (◀ page 352). This concept is based on the sharing of electrons pairs between an acid and a base. A **Lewis acid** is a substance that can accept a pair of electrons from another atom to form a new bond, and a **Lewis base** is a substance that can donate a pair of electrons to another atom to form a new bond. This means that an acid–base reaction in the Lewis sense occurs when a molecule (or ion) donates a pair of electrons to another molecule (or ion).

A	+	B:	$\rightarrow$	B$\rightarrow$A
Acid		Base		Adduct

The product is often called an **acid–base adduct.** In Section 8.3, this type of chemical bond was called a *coordinate covalent bond.*

Case Study — Uric Acid, Gout, and Bird Droppings

All living creatures metabolize food and dispose of the waste products, many of which contain nitrogen. Ammonia is the end product of this metabolic chain for fish and marine invertebrates, for example. And fish sometimes produce the Lewis and Brønsted base trimethylamine, $N(CH_3)_3$, a water-soluble compound and the source of the characteristic "fish odor."

Unlike fish, most terrestrial animals do not have an "infinite" supply of water. Mammals have a bladder and usually live in conditions where adequate water is available. Their mechanism of disposal for most toxins is to prepare a water-soluble compound and then excrete it through the urine. Thus, urea, NH_2CONH_2, is a major by-product of nitrogen metabolism in mammals. Reptiles and desert animals do not usually have much water available, and birds cannot afford the luxury of the weight of a bladder. These animals do not make urea; rather, they convert all of their nitrogen waste to uric acid, the concentrated white solid so familiar in bird droppings.

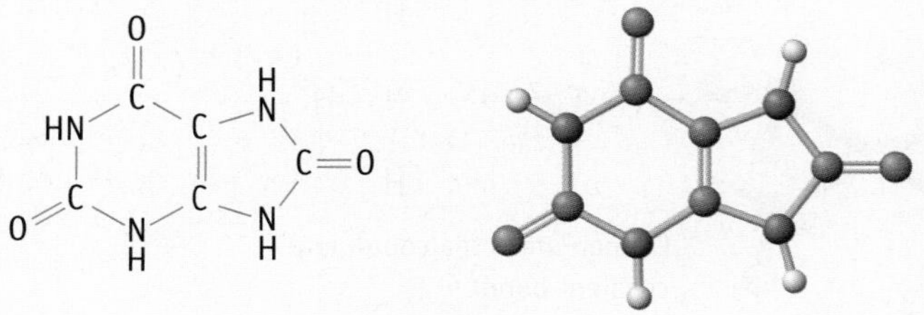

Uric acid

Uric acid can cause problems in primates because of its low water solubility. Deposits of uric acid in the joints and extremities can cause a very painful condition called *gout* or gouty arthritis. Chronic gout can also lead to lumps of uric acid around and in joints, to decreased kidney function, and to kidney stones.

You know you have gout if you experience excruciating and sudden pain with swelling and redness in a joint. Often, the first attack occurs in a big toe, but other joints such as those in the ankle, knee, wrist, fingers, and elbow can be affected as well.

John Kotz

Uric acid in bird droppings. Birds excrete uric acid, which you see as the white solid in their feces. The acidic material can cause severe environmental damage.

Although gout may be a hereditary disease, you can help avoid the problem by eating smaller amounts of foods with high levels of the purines, which are metabolized to uric acid. These include meat, fish, dry beans, mushrooms, spinach, and asparagus. On the other hand, eating fresh fruit (especially cherries and strawberries) and most fresh vegetables can help lower uric acid levels.

Finally, gout is one of the most frequently recorded medical conditions throughout history. People such as Henry VIII, Isaac Newton, Thomas Jefferson, Benjamin Franklin, John Hancock, and Karl Marx have suffered from the condition.

Questions:

1. *A diagnosis of hyperuricemia will be made when the uric acid blood level is greater than 420 μmol/L. What is this level in milligrams of uric acid per liter?*
2. *Uric acid is a polyprotic acid with one pK_a of 5.40 and a second pK_a of 5.53. Considering only the loss of the first proton, what acid in Table 17.3 has a similar acid strength?*

Answers to these questions are in Appendix Q.

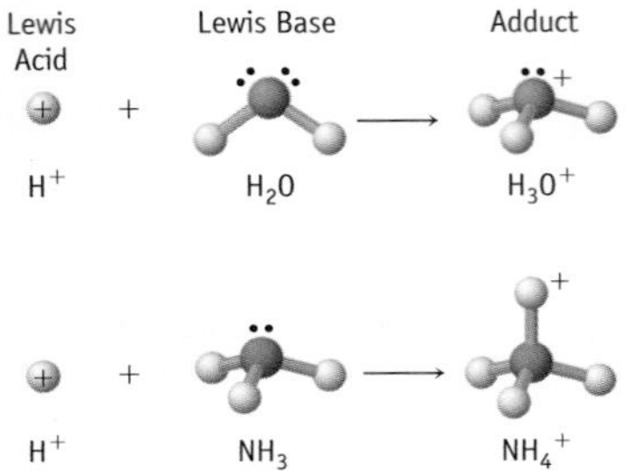

FIGURE 17.6 Protonation of water and ammonia are examples of Lewis acid–base reactions.

Formation of a hydronium ion from H^+ and water is a good example of a Lewis acid–base reaction (Figure 17.6). The H^+ ion has no electrons in its valence (1*s*) shell, and the water molecule has two unshared pairs of electrons (located in sp^3 hybrid orbitals). One of the O atom lone pairs of a water molecule can be shared with an H^+ ion, thus forming an O—H bond in an H_3O^+ ion. A similar interaction occurs between H^+ and the Lewis base ammonia to form the ammonium ion. Such reactions are very common. In general, they involve Lewis acids that are cations or neutral molecules with an available, empty valence orbital and bases that are anions or neutral molecules with a lone electron pair.

Cationic Lewis Acids

Just as H^+ and water form a Lewis acid–base adduct, metal cations interact with water molecules to form hydrated cations (Figure 17.7 and page 557). In these species, coordinate covalent bonds form between the metal cation and a lone pair of electrons on the O atom of each water. For example, an iron(II) ion, Fe^{2+}, forms six coordinate covalent bonds to water.

$$Fe^{2+}(aq) + 6\ H_2O(\ell) \rightarrow [Fe(H_2O)_6]^{2+}(aq)$$

Similar structures formed by transition metal cations are generally very colorful (Figures 17.7 and 17.8 and Section 22.3). Chemists call these **complex ions** or, because of the coordinate covalent bond, **coordination complexes.** Several are listed in Table 17.3 as acids, and their behavior is described further in Section 17.10 and Chapter 22.

Like water, ammonia is an excellent Lewis base and combines with metal cations to give adducts (complex ions), which are often very colorful. For example, copper(II) ions, light blue in aqueous solution (Figure 17.7), react with ammonia to give a deep blue adduct with four ammonia molecules surrounding each Cu^{2+} ion.

$$\underset{\textit{light blue}}{Cu^{2+}(aq)} + 4\ NH_3(aq) \longrightarrow \underset{\textit{deep blue}}{[Cu(NH_3)_4]^{2+}(aq)}$$

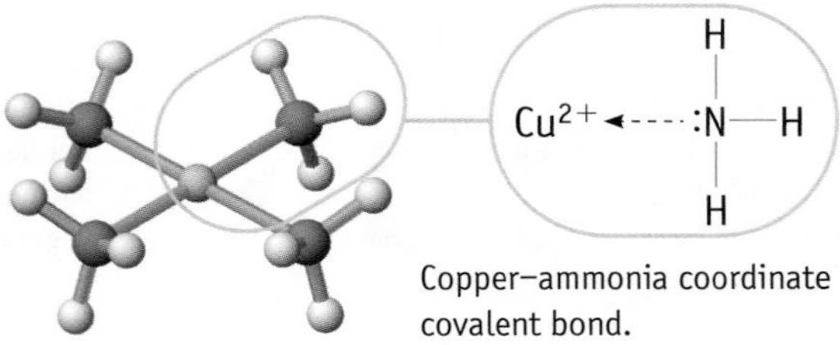

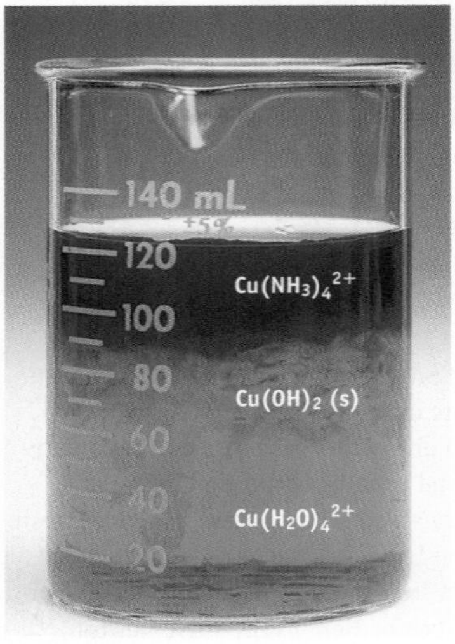

Charles D. Winters

FIGURE 17.7 The Lewis acid–base complex ion $[Cu(NH_3)_4]^{2+}$. Here, aqueous ammonia was added to aqueous $CuSO_4$ (the light blue solution at the bottom of the beaker). The small concentration of OH^- in $NH_3(aq)$ first formed insoluble blue-white $Cu(OH)_2$ (the solid in the middle of the beaker). With additional NH_3, however, the deep blue, soluble complex ion formed (the solution at the top of the beaker). The model in the text shows the copper(II)–ammonia complex ion.

Hydroxide ion, OH^-, is an excellent Lewis base and binds readily to metal cations to give metal hydroxides. An important feature of the chemistry of some metal hydroxides is that they are **amphoteric.** An amphoteric metal hydroxide can behave as an acid or a base (Table 17.6). One of the best examples of this behavior is provided by aluminum hydroxide, $Al(OH)_3$ (Figure 17.9). Adding OH^- to a precipitate of $Al(OH)_3$ produces the water-soluble $[Al(OH)_4]^-$ ion.

$$\underset{\text{Acid}}{Al(OH)_3(s)} + \underset{\text{Base}}{OH^-(aq)} \rightarrow [Al(OH)_4]^-(aq)$$

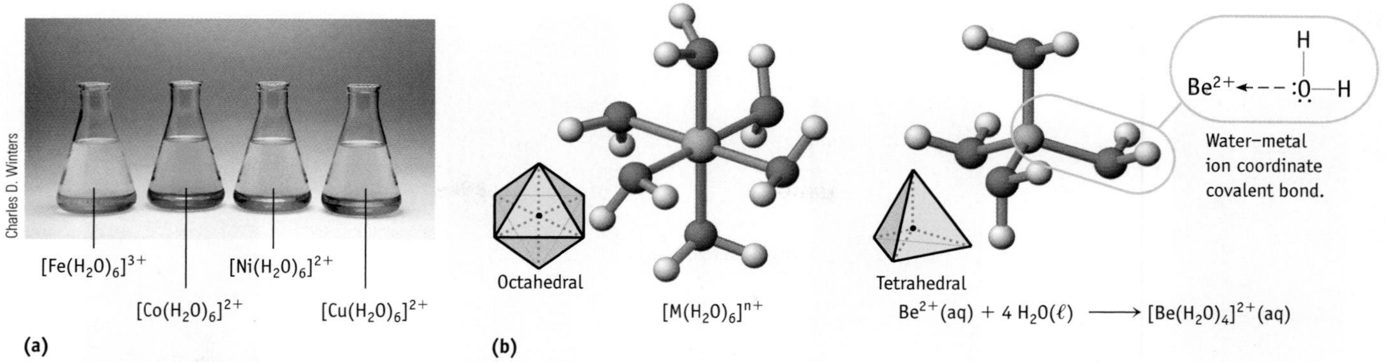

FIGURE 17.8 Metal cations in water. (a) Solutions of the nitrate salts of iron(III), cobalt(II), nickel(II), and copper(II) all have characteristic colors. (b) Models of complex ions (Lewis acid–base adducts) formed between a metal cation and water molecules. Such complexes often have six water molecules arranged octahedrally around the metal cation.

If acid is added to the $Al(OH)_3$ precipitate, it again dissolves. This time, however, aluminum hydroxide is acting as a base.

$$\underset{\text{Base}}{Al(OH)_3(s)} + \underset{\text{Acid}}{3\ H_3O^+(aq)} \rightarrow Al^{3+}(aq) + 6\ H_2O(\ell)$$

Molecular Lewis Acids

Lewis's acid–base concept also accounts for the fact that oxides of nonmetals such as CO_2 and SO_2 behave as acids (◀ Section 3.7). Because oxygen is more electronegative than C, the C—O bonding electrons in CO_2 are polarized away from carbon and toward oxygen. This causes the carbon atom to be slightly positive, and it is this atom that the negatively charged Lewis base OH^- can attack to give, ultimately, the bicarbonate ion.

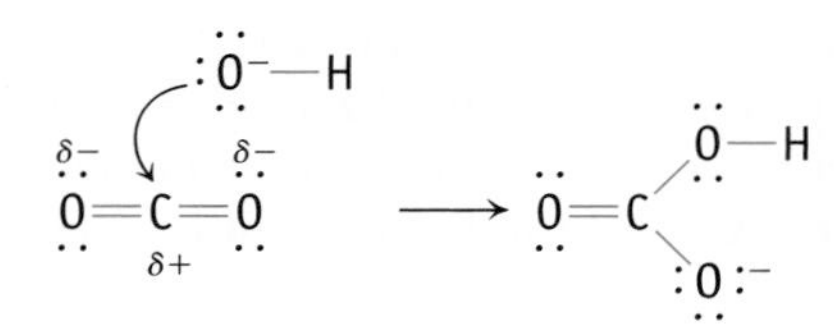

■ **CO_2 in Basic Solution** This reaction of CO_2 with OH^- is the first step in the precipitation of $CaCO_3$ when CO_2 is bubbled into a solution of $Ca(OH)_2$ (Figure 3.6, page 120).

Similarly, SO_2 reacts with aqueous OH^- to form the HSO_3^- ion.

Compounds based on the Group 3A elements boron and aluminum are among the most-studied Lewis acids. One example is a reaction in organic chemistry that is catalyzed by the Lewis acid $AlCl_3$. The mechanism of this important reaction—called the Friedel–Crafts reaction—is illustrated here. In the first step, a Lewis base,

TABLE 17.6 Some Common Amphoteric Metal Hydroxides*

Hydroxide	Reaction as a Base	Reaction as an Acid
$Al(OH)_3$	$Al(OH)_3(s) + 3\ H_3O^+(aq) \rightleftharpoons Al^{3+}(aq) + 6\ H_2O(\ell)$	$Al(OH)_3(s) + OH^-(aq) \rightleftharpoons [Al(OH)_4]^-(aq)$
$Zn(OH)_2$	$Zn(OH)_2(s) + 2\ H_3O^+(aq) \rightleftharpoons Zn^{2+}(aq) + 4\ H_2O(\ell)$	$Zn(OH)_2(s) + 2\ OH^-(aq) \rightleftharpoons [Zn(OH)_4]^{2-}(aq)$
$Sn(OH)_4$	$Sn(OH)_4(s) + 4\ H_3O^+(aq) \rightleftharpoons Sn^{4+}(aq) + 8\ H_2O(\ell)$	$Sn(OH)_4(s) + 2\ OH^-(aq) \rightleftharpoons [Sn(OH)_6]^{2-}(aq)$
$Cr(OH)_3$	$Cr(OH)_3(s) + 3\ H_3O^+(aq) \rightleftharpoons Cr^{3+}(aq) + 6\ H_2O(\ell)$	$Cr(OH)_3(s) + OH^-(aq) \rightleftharpoons [Cr(OH)_4]^-(aq)$

* The aqueous metal cations are best described as $[M(H_2O)_6]^{n+}$.

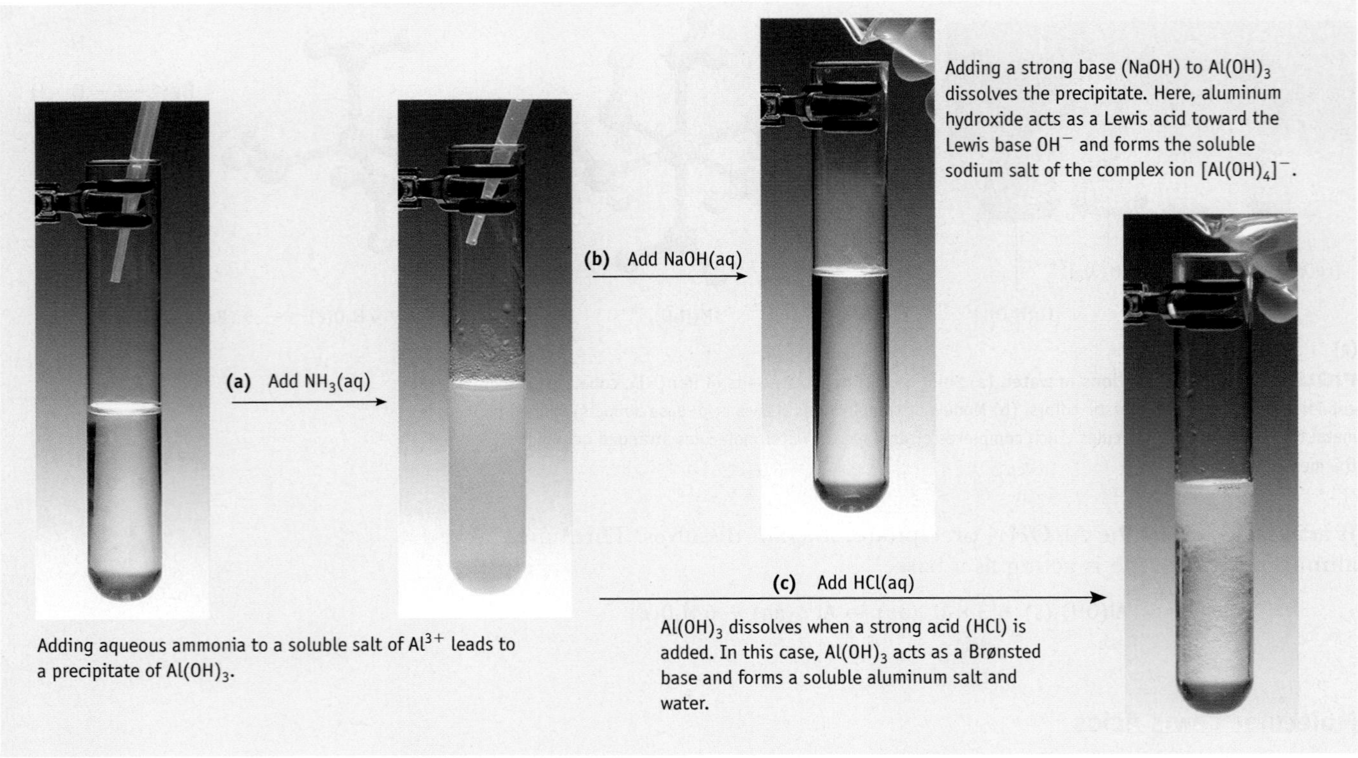

Charles D. Winters

FIGURE 17.9 The amphoteric nature of $Al(OH)_3$. Aluminum hydroxide is formed by the reaction of aqueous Al^{3+} and ammonia.

$$Al^{3+}(aq) + 3\ NH_3(aq) + 3\ H_2O(\ell) \rightleftharpoons Al(OH)_3(s) + 3\ NH_4^+(aq)$$

Reactions of solid $Al(OH)_3$ with aqueous NaOH and HCl demonstrate that aluminum hydroxide is amphoteric.

the Cl^- ion, transfers from the reactant, here CH_3COCl, to the Lewis acid to give $[AlCl_4]^-$ and an organic cation (that is stabilized by resonance). The organic cation attacks a benzene molecule to give a cationic intermediate, and this then interacts with $[AlCl_4]^-$ to produce HCl and the final organic product.

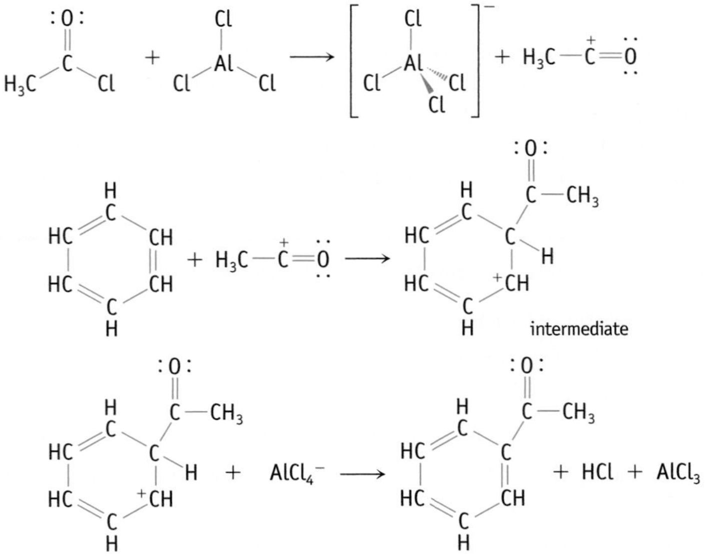

Molecular Lewis Bases

Ammonia is widely distributed in nature and is involved as a Lewis base in numerous reactions. One example where this is important is in the conversion of ammonia to urea (NH_2CONH_2) in natural systems. The process begins with the reaction of bicarbonate ion with ATP (adenosine triphosphate), and a subsequent step in the mechanism is the following:

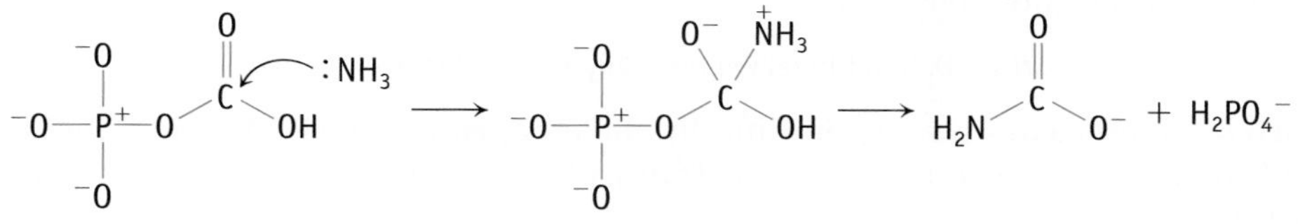

Here, the Lewis base ammonia attacks a carbon atom with a partial positive charge. The dihydrogen phosphate ion is then released, yielding the $NH_2CO_2^-$ ion, which eventually forms urea in another step in this reaction mechanism.

Chemistry•Now™

Sign in at **www.thomsonedu.com/login** and go to Chapter 17 Contents to see:
- Screen 17.12 for a tutorial on **Lewis acids and bases**
- Screen 17.13 for a description of **cationic Lewis acids**
- Screen 17.14 for a tutorial on **neutral Lewis acids**

EXERCISE 17.17 Lewis Acids and Bases

Describe each of the following as a Lewis acid or a Lewis base.

(a) PH_3 **(c)** H_2S

(b) BCl_3 **(d)** HS^-

Hint: In each case, draw the Lewis electron dot structure of the molecule or ion. Are there lone pairs of electrons on the central atom? If so, it can be a Lewis base. Does the central atom lack an electron pair? If so, it can behave as a Lewis acid.

17.10 Molecular Structure, Bonding, and Acid–Base Behavior

One of the most interesting aspects of chemistry is the correlation between a molecule's structure and bonding and its chemical properties. Because so many compounds are acids and bases, and play such a key role in chemistry, it is especially useful to see if there are some general principles governing acid–base behavior.

Chemistry•Now™

Sign in at **www.thomsonedu.com/login** and go to Chapter 17 Contents to see Screen 17.15 for a self-study module on **molecular interpretation of acid–base behavior.**

Acid Strength of the Hydrogen Halides, HX

Aqueous HF is a weak Brønsted acid in water, whereas the other hydrohalic acids—aqueous HCl, HBr, and HI—are all strong acids. Experiments show that the acid strength increases in the order HF $<<$ HCl $<$ HBr $<$ HI. A detailed analysis of

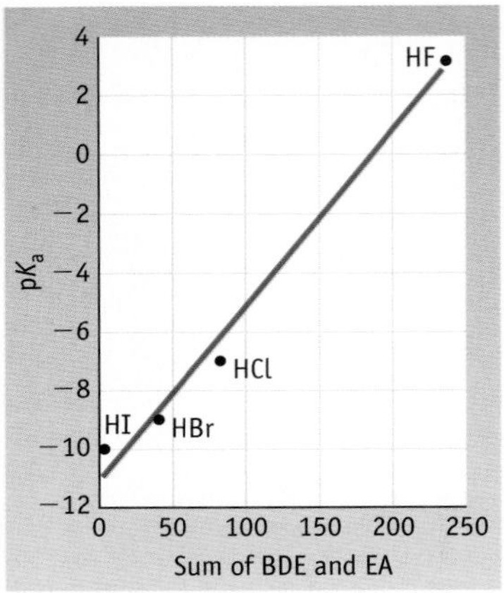

FIGURE 17.10 The effect of H—X bond energy and A electron affinity on acid strength. Stronger acids have weaker H—X bonds and more electronegative A atoms. (BDE is the bond dissociation enthalpy of the H—X bond, and EA is the electron affinity of the halogen atom.) See M. Moran, *Journal of Chemical Education*, Vol, 83, pages 800–803, 2006.

the factors that lead to these differences in acid strength in this group is complex (see *A Closer Look, Acid and Base Strength*). However, predictions about relative acid strength can be made based on the sum of two energy quantities, the energy required to break the H—X bond and the electron affinity of the halogen. That is, when a hydrohalic acid HX ionizes in water,

$$HX(aq) + H_2O(\ell) \rightarrow H_3O^+(aq) + X^-(aq)$$

the sum of the two energy terms

$$\Delta U \propto \text{HX bond breaking enthalpy} + \text{electron affinity of X}$$

correlates with acid strength. Specifically, the more easily the H—X bond is broken and the more negative the electron affinity of X, the greater the relative strength of the acid.

The bond enthalpy and electron affinity effects can work together (a weak H—X bond and a large electron affinity of the X group) to produce a strong acid, but they can also work in opposite directions. The balance of the two effects is thus important. Let us examine some data for the Group 7A binary acids, HX.

	—— Increasing acid strength ⟶			
	HF	**HCl**	**HBr**	**HI**
pK_a	+3.14	−7	−9	−10
H–X bond strength (kJ/mol)	.565	432	366	299
Electron affinity of X (kJ/mol)	−328	−349	−325	−295
Sum (kJ/mol)	237	83	41	4

In this series of acids, the bond enthalpy factor dominates, the weakest acid, HF, has the strongest H—X bond, and the strongest acid, HI, has the weakest H—X bond. However, the electron affinity of X becomes less negative from F to I. A low electron affinity should lead to a weaker acid, but it is the *sum* of the two effects that leads to the observation that HI is the strongest acid. In Figure 17.10, you see there is a good correlation between an acid's pK_a and the *sum* of the bond breaking enthalpy and electron affinity.

TABLE 17.7 Oxoacids

Acid	pK_a
Cl-Based Oxoacids	
HOCl	7.46
HOClO ($HClO_2$)	~ 2
$HOClO_2$ ($HClO_3$)	~ −3
$HOClO_3$ ($HClO_4$)	~ −8
S-Based Oxoacids	
$(HO)_2SO$ [H_2SO_3]	1.92, 7.21
$(HO)_2SO_2$ [H_2SO_4]	~ −3, 1.92

According to Linus Pauling, for oxoacids with the general formula $(HO)_nE(O)_m$, the value of pK_a is about 8–5m. When $n > 1$, the pK_a increases by about 5 for each successive loss of a proton.

Comparing Oxoacids: HNO_2 and HNO_3

Nitrous acid (HNO_2) and nitric acid (HNO_3) are representative of several series of **oxoacids.** Oxoacids contain an atom (usually a nonmetal atom) bonded to one or more oxygen atoms, some with hydrogen atoms attached. Besides those based on N, you are familiar with the sulfur- and chlorine-based oxoacids (Table 17.7). In all these series of related compounds, the acid strength increases as the number of oxygen atoms bonded to the central element increases. Thus, nitric acid (HNO_3) is a stronger acid than nitrous acid (HNO_2).

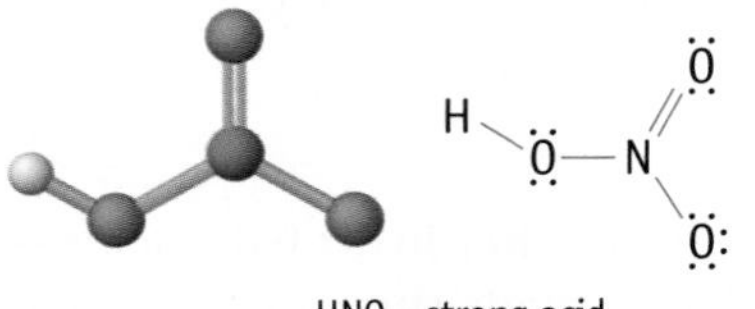

HNO_3, strong acid, pK_a = −1.4

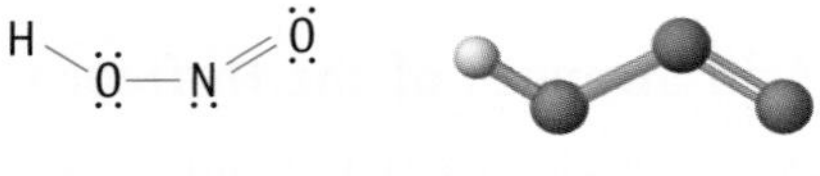

HNO_2, weak acid, pK_a = +3.35

A Closer Look

Acid Strengths and Molecular Structure

Although predictions about acid strength in aqueous solution are fairly simple to make, explanations are in fact quite complicated.

Acid strength is sometimes correlated with the strength and the polarity of the H—X bond, readily identifiable characteristics derived from the structure of the acid, the reactant in the ionization process. We need to point out, however, that when one is assessing any chemical reaction, it is necessary to consider both reactants and products. Looking only at the reactant when dealing with acid dissociation only takes you halfway.

When evaluating the strength of an acid HX(aq), we are looking at the following reaction

$$HX(aq) + H_2O(\ell) \rightleftharpoons H_3O^+(aq) + X^-(aq)$$

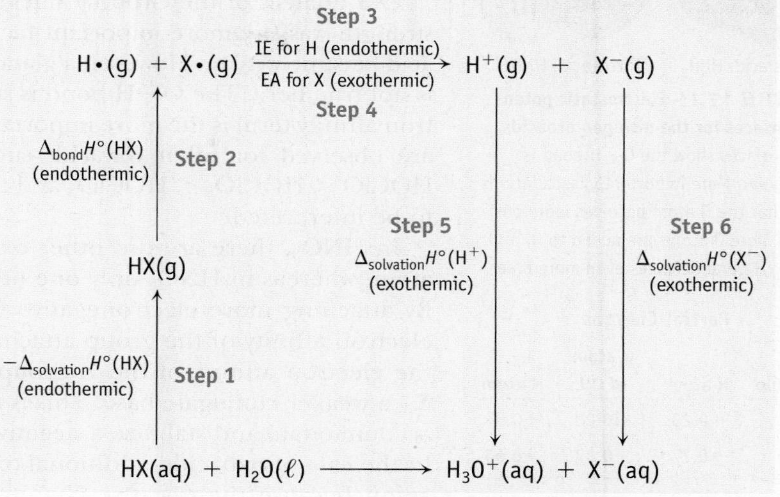

An energy diagram for the ionization of an acid HX in aqueous solution.

To fully explain the extent of ionization, we must consider characteristics of both the acid and the anion. The ability of the anion to spread out the negative charge across the ion, for example, and the solvation of the anion by the solvent are among the issues that must have some relevance in an explanation of the strength of the acid.

How enthalpy changes contribute to acid strength can be assessed using a thermochemical cycle (such as the one used to evaluate a lattice enthalpy, page 599). Consider the relative acid strengths of the hydrogen halides. The enthalpy change for the ionization of an acid in water can be related to other enthalpy changes as shown in the diagram. The solvation of H^+ (Step 5) and the ionization energy of H(g) (Step 3) are common to all of the hydrogen halides and do not contribute to the differences among hydrogen halides, but the four remaining terms are different. For the hydrogen halides, the bond dissociation enthalpy (Step 2) of HF is much larger than the dissociation enthalpies of the other hydrogen halides. However, it is compensated for significantly by the enthalpy of solvation of the anion (Step 6), which for the fluoride ion is much more exothermic than the solvation energies of the other halide ions. The electron affinity is also a contributor to the differences in overall enthalpy changes. Electron affinity values vary among the halogens, but to a smaller extent than the variation in bond energy and the enthalpy of solvation of the halide ion. Differences in solvation energy for the molecular species (Step 1) are minimal.

A complete analysis of strengths of acids in aqueous solution will include consideration of both enthalpies and entropies. Entropy has yet to be discussed in this text in detail (► Chapter 19) but we have noted earlier (◄ page 621) that entropy plays an important role in solution chemistry, specifically in determining solubilities. It is not surprising that entropy has a role in determining acid strength, too. Indeed, differences in entropy changes are significant in accounting for the differences in acid strength of the hydrogen halides.

Although all of these terms contribute to acid strength, acid strength can often be correlated with a subset of this information, as can be seen by the examples presented in this section. We point out, however, that correlations, while highly useful to a chemist because they can be used to make important predictions, are at best only partial explanations.

Let's apply the bond enthalpy/electron affinity analysis to HNO_3 and HNO_2.

—— Increasing acid strength ⟶

	HNO_2	**HNO_3**
pK_a	+3.35	−1.4
H—O bond strength (kJ/mol)	328	423
Electron affinity of X (kJ/mol)	−219	−377
Sum (kJ/mol)	109	46

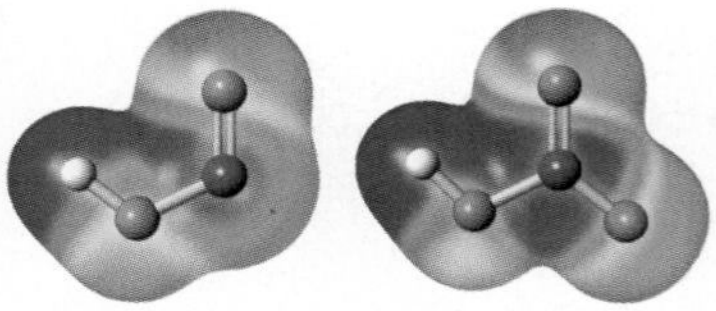

FIGURE 17.11 Electrostatic potential surfaces for the nitrogen oxoacids. Both surfaces show the O—H bond is quite polar. More importantly, calculations show that the H atom becomes more positive as more O atoms are added to N, and the O—H bond becomes even more polar.

Partial Charges

Molecule	H atom	O atom of OH	N atom
HNO_2	+0.23	−0.28	+0.34
HNO_3	+0.27	−0.37	+0.66

As in the case of the Group 7A acids, we again see that acid strength correlates with the sum of the bond breaking energy and X group (NO_2 or NO_3) electron affinity.

Our analysis of the Group 7A acids showed that, in this series, the H—X bond strength was the more important factor; as the H—X bond became stronger, the acid became weaker. However, a glance at the data for HNO_3 and HNO_2 shows this is not true here. The O—H bond is stronger in the stronger acid HNO_3. The electron affinity term is the more important term in this correlation. These same effects are observed for other oxyacids such as the chlorine-based oxoacids HOCl < HOClO < $HOClO_2$ < $HOClO_3$ and the S-based oxoacids (Table 17.7). How is this to be interpreted?

In HNO_3, there are two other oxygen atoms bonded to the central nitrogen atom, whereas in HNO_2 only one other oxygen is bonded to the nitrogen atom. By attaching more electronegative O atoms to nitrogen, we are increasing the electron affinity of the group attached to hydrogen, and anything that increases the electron affinity of the X group should also make HX a stronger acid and X^- a weaker conjugate base. This is another way of saying that if X^- has a way to accommodate and stabilize a negative charge, it will be a weaker conjugate base. In the case of oxoacids, additional oxygen atoms have the effect of stabilizing the anion because the negative charge on the anion can be dispersed over more atoms. In the nitrate ion, for example, the negative charge is shared equally over the three oxygen atoms. This is represented symbolically in the three resonance structures for this ion.

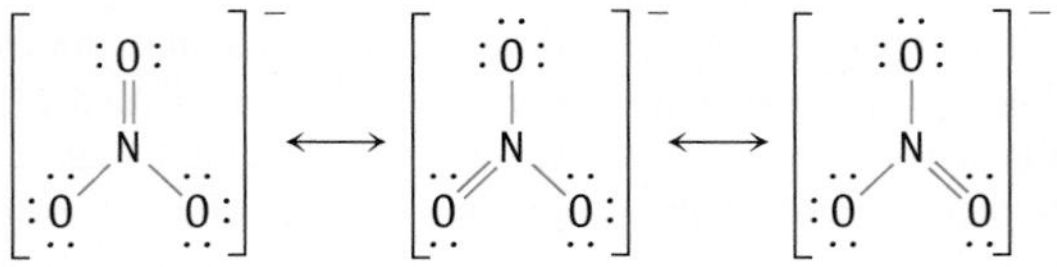

In nitrite ion, only two atoms share the negative charge. Therefore, NO_2^- is a stronger conjugate base than NO_3^-. In general, greater stabilization of the products formed by ionizing the acid contributes to increased acidity.

A glance at Table 17.7 and Figure 17.11 shows that another empirical correlation can be made between the structure of an acid and its acidity: In a series of related acids, the larger the formal charge on the central atom, the stronger the acid (Table 17.8). For example, the N atom formal charge in the weak acid HNO_2 is 0, whereas it is +1 in the strong acid HNO_3, and these are reflected by the results of theoretical calculations cited in Figure 17.11.

TABLE 17.8 Correlation of Atom Formal Charge and pK_a

	Central atom formal charge	pK_a
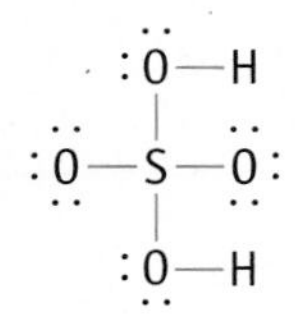	+2	−3
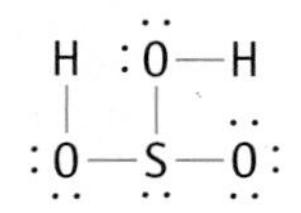	+1	1.92
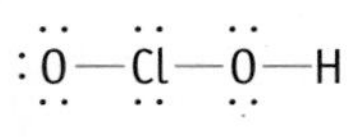	+1	2
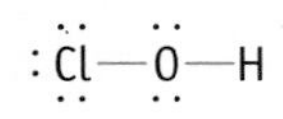	0	7.46

In summary, molecules such as the oxoacids can behave as stronger Brønsted acids when the anion created by loss of H^+ is stable and able to accommodate the negative charge. These conditions are promoted by

- the presence of electronegative atoms attached to the central atom.
- the possibility of resonance structures for the anion, which lead to delocalization of the negative charge over the anion and thus to a stable ion.

Why Are Carboxylic Acids Brønsted Acids?

There is a large class of organic acids, typified by acetic acid (CH_3CO_2H) (Figure 17.12), called carboxylic acids because all have the carboxylic acid group, $-CO_2H$. The arguments used to explain the acidity of oxoacids can also be applied to carboxylic acids. The O—H bond in these compounds is polar, a prerequisite for ionization.

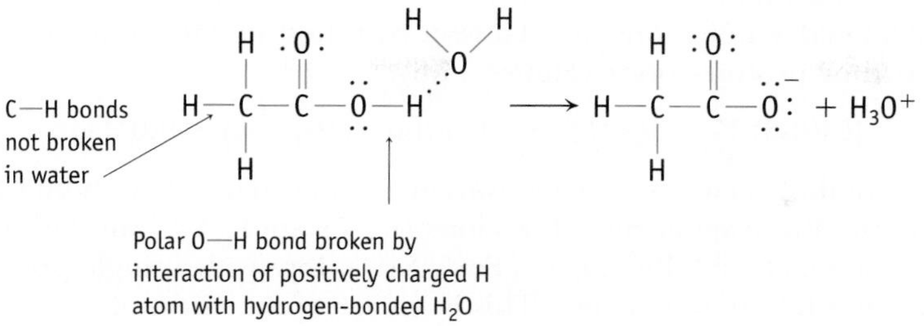

In addition, carboxylate anions are stabilized by delocalizing the negative charge over the two oxygen atoms.

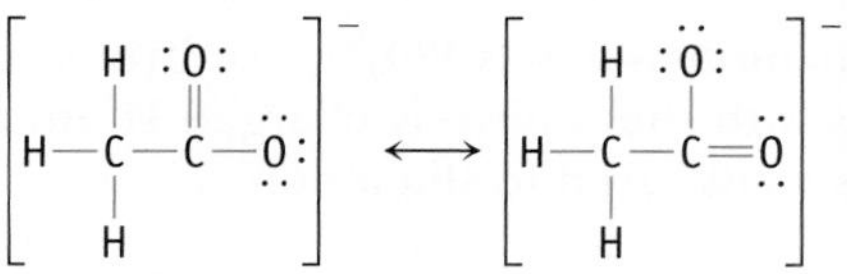

The simple carboxylic acids, RCO_2H in which R is a hydrocarbon group (◀ Section 10.4), do not differ markedly in acid strength (compare acetic acid, $pK_a = 4.74$, and propanoic acid, $pK_a = 4.89$, Table 17.3). The acidity of carboxylic acids is enhanced, however, if electronegative substituents replace the hydrogens in the alkyl group. Compare, for example, the pK_a values of a series of acetic acids in which hydrogen is replaced sequentially by the more electronegative element chlorine.

Acid		pK_a Value
CH_3CO_2H	Acetic acid	4.74
$ClCH_2CO_2H$	Chloroacetic acid	2.85
Cl_2CHCO_2H	Dichloroacetic acid	1.49
Cl_3CCO_2H	Trichloroacetic acid	0.7

increasing acid strength ↓

As in the nitrogen oxoacids, increasingly electronegative substituents lead to an increase in acid strength. Indeed, recent research has found that a key role of electronegative substituents is to stabilize the negative charge of the anion. That is, compared with $CH_3CO_2^-$, the $Cl_3CCO_2^-$ anion is stabilized by the presence of the three Cl atoms, and $Cl_3CCO_2^-$ is a weaker base than $CH_3CO_2^-$.

Finally, why are the C—H hydrogens of carboxylic acids not dissociated as H^+ instead of (or in addition to) the O—H hydrogen atom? The calculated partial positive charges listed in Figure 17.12 show that the H atoms of the CH_3 group have a much smaller positive partial charge than the O—H hydrogen atom. Furthermore, in carboxylic acids, the C atom of the CH_3 group is not sufficiently electronegative to accommodate the negative charge left if the bond breaks as $C—H \rightarrow C:^- + H^+$, and the product anion is not well stabilized.

Why Are Hydrated Metal Cations Brønsted Acids?

When a coordinate covalent bond is formed between a metal cation (a Lewis acid) and a water molecule (a Lewis base), the positive charge of the metal ion and its small size means that the electrons of the $H_2O—M^{n+}$ bond are very strongly at-

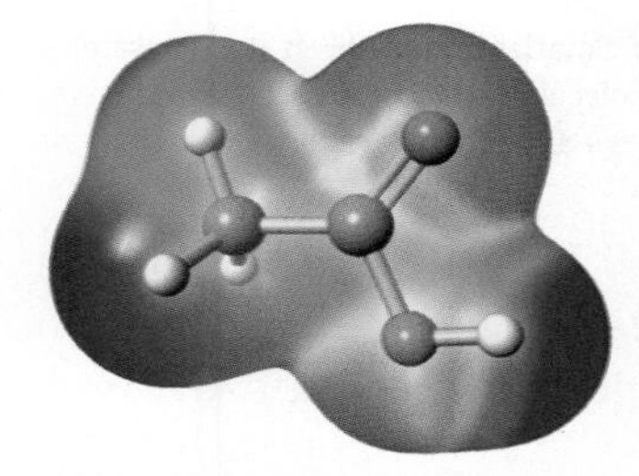

FIGURE 17.12 Electrostatic potential surface and partial charges for acetic acid. The H atoms of the molecule are all positively charged, but the H atom of the OH is much more highly charged. As expected, both of the electronegative O atoms have a partial negative charge. The table below gives the computer-calculated partial charges on the acid.

Atom or Group	Calc'd Partial Charge
H of OH	+0.24
O of OH	−0.32
H of CH_3	+0.12

Polarization of O—H Bonds Water molecules attached to a metal cation have strongly polarized O—H bonds.

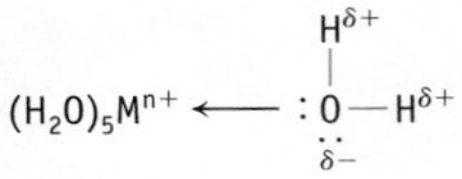

tracted to the metal. As a result, the O—H bonds of the bound water molecules are polarized, just as in oxoacids and carboxylic acids. The net effect is that a H atom of a coordinated water molecule is removed as H^+ more readily than in an uncoordinated water molecule. Thus, a hydrated metal cation functions as a Brønsted acid or proton donor (Figure 17.8).

$$[Cu(H_2O)_6]^{2+} + H_2O(\ell) \rightleftharpoons [Cu(H_2O)_5(OH)]^+(aq) + H_3O^+(aq)$$

The effect of the metal ion increases with increasing charge. Consulting Table 17.3, you see that the Brønsted acidity of +3 ions (for example Al^{3+} and Fe^{3+}) is greater than for +2 cations (Cu^{2+}, Pb^{2+}, Co^{2+}, Fe^{2+}, Ni^{2+}). Ions with a single positive charge such as Na^+ and K^+ are not acidic. (This is similar to the effect of central atom formal charge in a series of related acids. See Table 17.8.)

TABLE 17.9 Basic Oxoanions

Anion	pK_b
PO_4^{3-}	1.55
HPO_4^{2-}	6.80
$H_2PO_4^-$	11.89
CO_3^{2-}	3.68
HCO_3^-	7.62
SO_3^{2-}	6.80
HSO_3^-	12.08

Why Are Anions Brønsted Bases?

Anions, particularly oxoanions such as PO_4^{3-}, are Brønsted bases. The negatively charged anion interacts with the positively charged H atom of a polar water molecule, and an H^+ ion is transferred to the anion.

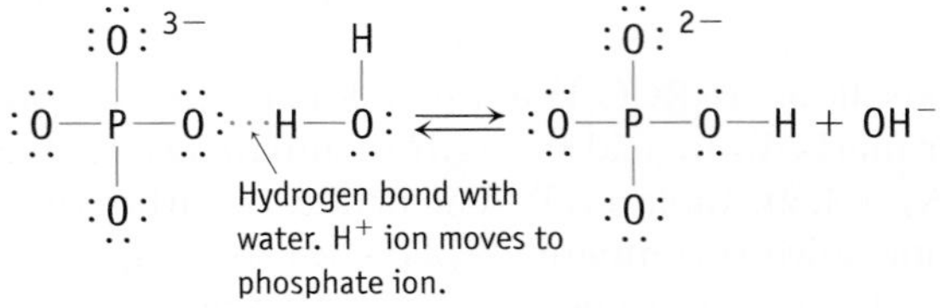

The data in Table 17.9 show that, in a series of related anions, the basicity of an anionic base increases as the negative charge of the anion increases.

Why Are Ammonia and Its Derivatives Brønsted and Lewis Bases?

Ammonia is the parent compound of an enormous number of compounds that behave as Brønsted and Lewis bases (Figure 17.13). These molecules all have an electronegative N atom with a partial negative charge surrounded by three bonds and a lone pair of electrons. Owing to this negatively charged N atom, they can extract a proton from water.

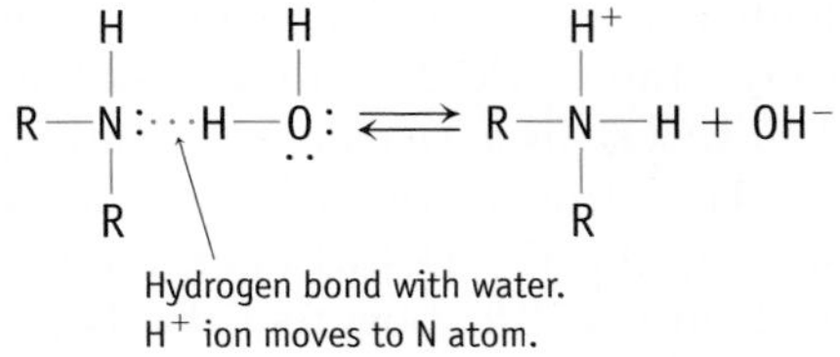

FIGURE 17.13 Nitrogen-based Lewis and Brønsted bases. All have an N atom surrounded by three bonds and a lone pair of electrons.

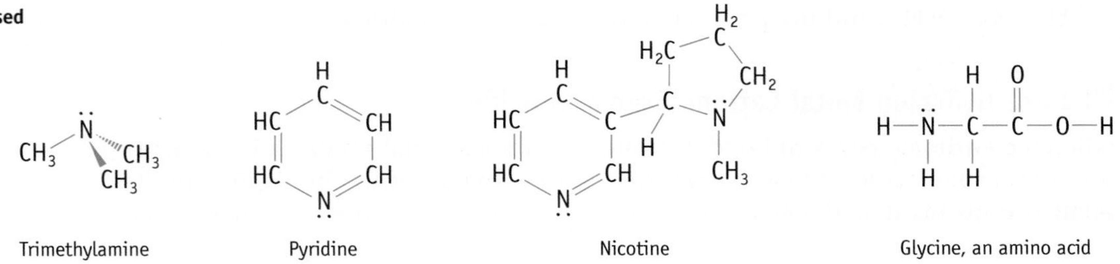

In addition, the lone pair can be used to form a coordinate covalent bond to Lewis acids such as a metal cation (◀ page 790).

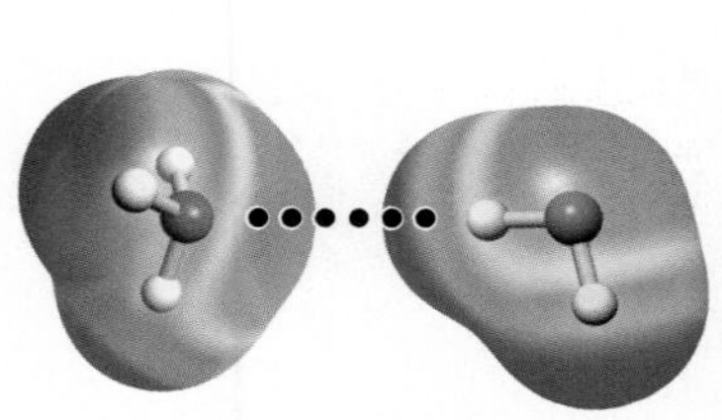

Electrostatic potential surfaces for NH_3 and H_2O. The red or negative region of these surfaces corresponds to the lone pair of electrons on N or the two pairs on O. The N atom of NH_3 has a calculated partial charge of −0.40, and the H atoms have a charge of +0.13. The N atom lone pair of NH_3 is involved in forming a hydrogen bond with the polar O—H bond of water. The NH_3 is both a Lewis and a Brønsted base and can remove the proton from water to form NH_4^+ and OH^-.

EXERCISE 17.18 Molecular Structure, Acids, and Bases

(a) Which should be the stronger acid, H_2SeO_4 or H_2SeO_3?

(b) Which should be the stronger acid, $[Fe(H_2O)_6]^{2+}$ or $[Fe(H_2O)_6]^{3+}$?

(c) Which should be the stronger acid, HOCl or HOBr?

(d) The molecule whose structure is illustrated here is amphetamine, a stimulant. Is the compound a Brønsted acid, a Lewis acid, a Brønsted base, a Lewis base, or some combination of these?

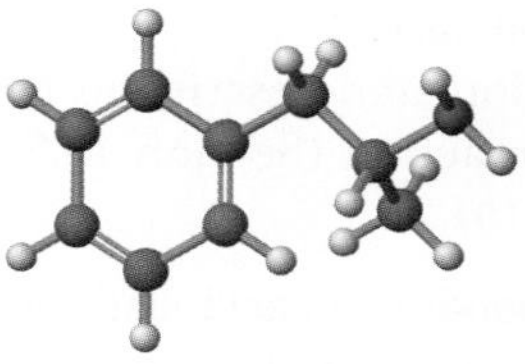

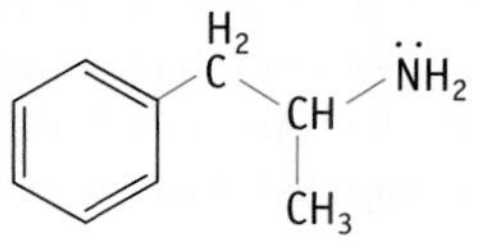

Chapter Goals Revisited

Now that you have studied this chapter, you should ask whether you have met the chapter goals. In particular, you should be able to:

Use the Brønsted–Lowry and Lewis theories of acids and bases

a. Define and use the Brønsted concept of acids and bases (Sections 17.1 and 17.2).

b. Recognize common monoprotic and polyprotic acids and bases, and write balanced equations for their ionization in water (Section 17.2).

c. Appreciate when a substance can be amphiprotic (Section 17.2).

d. Recognize the Brønsted acid and base in a reaction, and identify the conjugate partner of each (Section 17.2). Study Question(s) assignable in OWL: 2, 4, 8.

e. Understand the concept of water autoionization and its role in Brønsted acid–base chemistry. Use the water ionization constant, K_w (Section 17.3).

f. Use the pH concept (Section 17.3). Study Question(s) assignable in OWL: 10, 12.

g. Identify common strong acids and bases (Tables 3.2 and 17.3).

h. Recognize some common weak acids and understand that they can be neutral molecules (such as acetic acid), cations (NH_4^+ or hydrated metal ions such as $[Fe(H_2O)_6]^{2+}$, or anions (such as HCO_3^-) (Table 17.3).

Apply the principles of chemical equilibrium to acids and bases in aqueous solution

a. Write equilibrium constant expressions for weak acids and bases (Section 17.4).

b. Calculate pK_a from K_a (or K_a from pK_a), and understand how pK_a is correlated with acid strength (Section 17.4). Study Question(s) assignable in OWL: 26, 28, 30, 104, 106.

c. Understand the relationship between K_a for a weak acid and K_b for its conjugate base (Section 17.4). Study Question(s) assignable in OWL: 15, 18.

d. Write equations for acid–base reactions, and decide whether they are product- or reactant-favored at equilibrium (Section 17.5 and Table 17.5). Study Question(s) assignable in OWL: 36, 38.

Chemistry.Now™ Sign in at **www.thomsonedu.com/login** to:

- Assess your understanding with Study Questions in OWL keyed to each goal in the Goals and Homework menu for this chapter
- For quick review, download Go Chemistry mini-lecture flashcard modules (or purchase them at **www.ichapters.com**)
- Check your readiness for an exam by taking the Pre-Test and exploring the modules recommended in your Personalized Study plan.

Access **How Do I Solve It?** tutorials on how to approach problem solving using concepts in this chapter.

e. Calculate the equilibrium constant for a weak acid (K_a) or a weak base (K_b) from experimental information (such as pH, $[H_3O^+]$, or $[OH^-]$) (Section 17.7 and Example 17.4). Study Question(s) assignable in OWL: 42, 44, 85, 107.
f. Use the equilibrium constant and other information to calculate the pH of a solution of a weak acid or weak base (Section 17.7 and Examples 17.5 and 13.7). Study Question(s) assignable in OWL: 48, 52, 56, 65, 66, 89, 91, 99; Go Chemistry Module 22.
g. Describe the acid–base properties of salts, and calculate the pH of a solution of a salt of a weak acid or of a weak base (Section 17.7 and Example 17.7). Study Question(s) assignable in OWL: 84, 92, 93, 103, 121.

Predict the outcome of reactions of acids and bases

a. Recognize the type of acid–base reaction, and describe its result (Section 17.6).
b. Calculate the pH after an acid–base reaction (Section 17.7 and Example 17.8). Study Question(s) assignable in OWL: 62, 98, 101, 102.

Understand the influence of structure and bonding on acid–base properties

a. Characterize a compound as a Lewis base (an electron-pair donor) or a Lewis acid (an electron-pair acceptor) (Section 17.9). Study Question(s) assignable in OWL: 70, 72, 108.
b. Appreciate the connection between the structure of a compound and its acidity or basicity (Section 17.10). Study Question(s) assignable in OWL: 74.

KEY EQUATIONS

Equation 17.1 (page 765): Water ionization constant.

$$K_w = [H_3O^+][OH^-] = 1.0 \times 10^{-14} \text{ at } 25\ °C$$

Equation 17.2 (page 767): Definition of pH (see also Equation 4.3).

$$pH = -\log[H_3O^+]$$

Equation 17.3 (page 767): Definition of pOH

$$pOH = -\log[OH^-]$$

Equation 17.4 (page 767): Definition of $pK_w = pH + pOH$ (= 14.00 at 25 °C)

$$pK_w = 14.00 = pH + pOH$$

Equation 17.5 (page 769): Equilibrium expression for a general acid, HA, in water.

$$K_a = \frac{[H_3O^+][A^-]}{[HA]}$$

Equation 17.6 (page 769): Equilibrium expression for a general base, B, in water.

$$K_b = \frac{[BH^+][OH^-]}{[B]}$$

Equation 17.7 (page 775): Definition of pK_a.

$$pK_a = -\log K_a$$

Equation 17.8 (page 776): Relationship of K_a, K_b, and K_w, where K_a and K_b are for a conjugate acid–base pair.

$$K_a \times K_b = K_w$$

STUDY QUESTIONS

Online homework for this chapter may be assigned in OWL.

▲ denotes challenging questions.

■ denotes questions assignable in OWL.

Blue-numbered questions have answers in Appendix O and fully-worked solutions in the *Student Solutions Manual.*

Practicing Skills

The Brønsted Concept

(See Exercises 17.1 and 17.2 and ChemistryNow Screen 17.2.)

1. Write the formula and give the name of the conjugate base of each of the following acids.
(a) HCN (b) HSO_4^- (c) HF

2. ■ Write the formula and give the name of the conjugate acid of each of the following bases.
(a) NH_3 (b) HCO_3^- (c) Br^-

3. What are the products of each of the following acid–base reactions? Indicate the acid and its conjugate base and the base and its conjugate acid.
(a) $HNO_3 + H_2O \rightarrow$
(b) $HSO_4^- + H_2O \rightarrow$
(c) $H_3O^+ + F^- \rightarrow$

4. ■ What are the products of each of the following acid–base reactions? Indicate the acid and its conjugate base and the base and its conjugate acid.
(a) $HClO_4 + H_2O \rightarrow$
(b) $NH_4^+ + H_2O \rightarrow$
(c) $HCO_3^- + OH^- \rightarrow$

5. Write balanced equations showing how the hydrogen oxalate ion, $HC_2O_4^-$, can be both a Brønsted acid and a Brønsted base.

6. Write balanced equations showing how the HPO_4^{2-} ion of sodium hydrogen phosphate, Na_2HPO_4, can be a Brønsted acid or a Brønsted base.

7. In each of the following acid–base reactions, identify the Brønsted acid and base on the left and their conjugate partners on the right.
(a) $HCO_2H(aq) + H_2O(\ell) \rightleftharpoons HCO_2^-(aq) + H_3O^+(aq)$
(b) $NH_3(aq) + H_2S(aq) \rightleftharpoons NH_4^+(aq) + HS^-(aq)$
(c) $HSO_4^-(aq) + OH^-(aq) \rightleftharpoons SO_4^{2-}(aq) + H_2O(\ell)$

8. ■ In each of the following acid–base reactions, identify the Brønsted acid and base on the left and their conjugate partners on the right.
(a) $C_5H_5N(aq) + CH_3CO_2H(aq) \rightleftharpoons C_5H_5NH^+(aq) + CH_3CO_2^-(aq)$
(b) $N_2H_4(aq) + HSO_4^-(aq) \rightleftharpoons N_2H_5^+(aq) + SO_4^{2-}(aq)$
(c) $[Al(H_2O)_6]^{3+}(aq) + OH^-(aq) \rightleftharpoons [Al(H_2O)_5OH]^{2+}(aq) + H_2O(\ell)$

pH Calculations

(See Examples 4.7 and 17.1, Exercise 17.4, and ChemistryNow Screens 4.11 and 17.3–17.4.)

9. An aqueous solution has a pH of 3.75. What is the hydronium ion concentration of the solution? Is it acidic or basic?

10. ■ A saturated solution of milk of magnesia, $Mg(OH)_2$, has a pH of 10.52. What is the hydronium ion concentration of the solution? What is the hydroxide ion concentration? Is the solution acidic or basic?

11. What is the pH of a 0.0075 M solution of HCl? What is the hydroxide ion concentration of the solution?

12. ■ What is the pH of a 1.2×10^{-4} M solution of KOH? What is the hydronium ion concentration of the solution?

13. What is the pH of a 0.0015 M solution of $Ba(OH)_2$?

14. The pH of a solution of $Ba(OH)_2$ is 10.66 at 25 °C. What is the hydroxide ion concentration in the solution? If the solution volume is 125 mL, what mass of $Ba(OH)_2$ must have been dissolved?

Equilibrium Constants for Acids and Bases

(See Example 17.2, Exercise 17.5, and ChemistryNow Screen 17.6.)

15. ■ Several acids are listed here with their respective equilibrium constants:

$$C_6H_5OH(aq) + H_2O(\ell) \rightleftharpoons H_3O^+(aq) + C_6H_5O^-(aq) \quad K_a = 1.3 \times 10^{-10}$$

$$HCO_2H(aq) + H_2O(\ell) \rightleftharpoons H_3O^+(aq) + HCO_2^-(aq) \quad K_a = 1.8 \times 10^{-4}$$

$$HC_2O_4^-(aq) + H_2O(\ell) \rightleftharpoons H_3O^+(aq) + C_2O_4^{2-}(aq) \quad K_a = 6.4 \times 10^{-5}$$

(a) Which is the strongest acid? Which is the weakest acid?
(b) Which acid has the weakest conjugate base?
(c) Which acid has the strongest conjugate base?

16. Several acids are listed here with their respective equilibrium constants.

$$HF(aq) + H_2O(\ell) \rightleftharpoons H_3O^+(aq) + F^-(aq) \quad K_a = 7.2 \times 10^{-4}$$

$$HPO_4^-(aq) + H_2O(\ell) \rightleftharpoons H_3O^+(aq) + PO_4^{3-}(aq) \quad K_a = 3.6 \times 10^{-13}$$

$$CH_3CO_2H(aq) + H_2O(\ell) \rightleftharpoons H_3O^+(aq) + CH_3CO_2^-(aq) \quad K_a = 1.8 \times 10^{-5}$$

(a) Which is the strongest acid? Which is the weakest acid?
(b) What is the conjugate base of the acid HF?
(c) Which acid has the weakest conjugate base?
(d) Which acid has the strongest conjugate base?

17. State which of the following ions or compounds has the strongest conjugate base, and briefly explain your choice.
(a) HSO_4^- (b) CH_3CO_2H (c) HOCl

18. ■ Which of the following compounds or ions has the strongest conjugate acid? Briefly explain your choice.
(a) CN^- (b) NH_3 (c) SO_4^{2-}

19. Dissolving K_2CO_3 in water gives a basic solution. Write a balanced equation showing how this salt can produce a basic solution.

20. Dissolving ammonium bromide in water gives an acidic solution. Write a balanced equation showing how this can occur.

21. If each of the salts listed here were dissolved in water to give a 0.10 M solution, which solution would have the highest pH? Which would have the lowest pH?
(a) Na_2S
(b) Na_3PO_4
(c) NaH_2PO_4
(d) NaF
(e) $NaCH_3CO_2$
(f) $AlCl_3$

22. Which of the following common food additives would give a basic solution when dissolved in water?
(a) $NaNO_3$ (used as a meat preservative)
(b) $NaC_6H_5CO_2$ (sodium benzoate; used as a soft-drink preservative)
(c) Na_2HPO_4 (used as an emulsifier in the manufacture of pasteurized cheese)

pK_a: A Logarithmic Scale of Acid Strength

(See Exercise 17.7 and ChemistryNow Screen 17.6.)

23. A weak acid has a K_a of 6.5×10^{-5}. What is the value of pK_a for the acid?

24. If K_a for a weak acid is 2.4×10^{-11}, what is the value of pK_a?

25. Epinephrine hydrochloride has a pK_a value of 9.53. What is the value of K_a? Where does the acid fit in Table 17.3?

26. ■ An organic acid has pK_a = 8.95. What is its K_a value? Where does the acid fit in Table 17.3?

27. Which is the stronger of the following two acids?
(a) benzoic acid, $C_6H_5CO_2H$, pK_a = 4.20
(b) 2-chlorobenzoic acid, $ClC_6H_4CO_2H$, pK_a = 2.88

28. ■ Which is the stronger of the following two acids?
(a) acetic acid, CH_3CO_2H, $K_a = 1.8 \times 10^{-5}$
(b) chloroacetic acid, $ClCH_2CO_2H$, pK_a = 2.87

Ionization Constants for Weak Acids and Their Conjugate Bases

(See Exercise 17.8 and ChemistryNow Screen 17.6.)

29. Chloroacetic acid ($ClCH_2CO_2H$) has $K_a = 1.41 \times 10^{-3}$. What is the value of K_b for the chloroacetate ion ($ClCH_2CO_2^-$)?

30. ■ A weak base has $K_b = 1.5 \times 10^{-9}$. What is the value of K_a for the conjugate acid?

31. The trimethylammonium ion, $(CH_3)_3NH^+$, is the conjugate acid of the weak base trimethylamine, $(CH_3)_3N$. A chemical handbook gives 9.80 as the pK_a value for $(CH_3)_3NH^+$. What is the value of K_b for $(CH_3)_3N$?

32. The chromium(III) ion in water, $[Cr(H_2O)_6]^{3+}$, is a weak acid with pK_a = 3.95. What is the value of K_b for its conjugate base, $[Cr(H_2O)_5OH]^{2+}$?

Predicting the Direction of Acid–Base Reactions

(See Example 17.3 and ChemistryNow Screen 17.7.)

33. Acetic acid and sodium hydrogen carbonate, $NaHCO_3$, are mixed in water. Write a balanced equation for the acid–base reaction that could, in principle, occur. Using Table 17.3, decide whether the equilibrium lies predominantly to the right or to the left.

34. Ammonium chloride and sodium dihydrogen phosphate, NaH_2PO_4, are mixed in water. Write a balanced equation for the acid–base reaction that could, in principle, occur. Using Table 17.3, decide whether the equilibrium lies predominantly to the right or to the left.

35. For each of the following reactions, predict whether the equilibrium lies predominantly to the left or to the right. Explain your predictions briefly.
(a) $NH_4^+(aq) + Br^-(aq) \rightleftharpoons NH_3(aq) + HBr(aq)$
(b) $HPO_4^{2-}(aq) + CH_3CO_2^-(aq) \rightleftharpoons PO_4^{3-}(aq) + CH_3CO_2H(aq)$
(c) $[Fe(H_2O)_6]^{3+}(aq) + HCO_3^-(aq) \rightleftharpoons [Fe(H_2O)_5(OH)]^{2+}(aq) + H_2CO_3(aq)$

36. ■ For each of the following reactions, predict whether the equilibrium lies predominantly to the left or to the right. Explain your predictions briefly.
(a) $H_2S(aq) + CO_3^{2-}(aq) \rightleftharpoons HS^-(aq) + HCO_3^-(aq)$
(b) $HCN(aq) + SO_4^{2-}(aq) \rightleftharpoons CN^-(aq) + HSO_4^-(aq)$
(c) $SO_4^{2-}(aq) + CH_3CO_2H(aq) \rightleftharpoons HSO_4^-(aq) + CH_3CO_2^-(aq)$

Types of Acid–Base Reactions

(See Exercise 17.10 and ChemistryNow Screen 17.7.)

37. Equal molar quantities of sodium hydroxide and sodium hydrogen phosphate (Na_2HPO_4) are mixed.
(a) Write the balanced, net ionic equation for the acid–base reaction that can, in principle, occur.
(b) Does the equilibrium lie to the right or left?

38. ■ Equal molar quantities of hydrochloric acid and sodium hypochlorite (NaClO) are mixed.
(a) Write the balanced, net ionic equation for the acid–base reaction that can, in principle, occur.
(b) Does the equilibrium lie to the right or left?

39. Equal molar quantities of acetic acid and sodium hydrogen phosphate (Na_2HPO_4) are mixed.
(a) Write a balanced, net ionic equation for the acid–base reaction that can, in principle, occur.
(b) Does the equilibrium lie to the right or left?

40. Equal molar quantities of ammonia and sodium dihydrogen phosphate (NaH_2PO_4) are mixed.
(a) Write a balanced, net ionic equation for the acid–base reaction that can, in principle, occur.
(b) Does the equilibrium lie to the right or left?

Using pH to Calculate Ionization Constants

(See Example 17.4 and ChemistryNow Screen 17.8.)

41. A 0.015 M solution of hydrogen cyanate, HOCN, has a pH of 2.67.
(a) What is the hydronium ion concentration in the solution?
(b) What is the ionization constant, K_a, for the acid?

42. ■ A 0.10 M solution of chloroacetic acid, $ClCH_2CO_2H$, has a pH of 1.95. Calculate K_a for the acid.

43. A 0.025 M solution of hydroxylamine has a pH of 9.11. What is the value of K_b for this weak base?

$$H_2NOH(aq) + H_2O(\ell) \rightleftharpoons H_3NOH^+(aq) + OH^-(aq)$$

44. ■ Methylamine, CH_3NH_2, is a weak base.

$$CH_3NH_2(aq) + H_2O(\ell) \rightleftharpoons CH_3NH_3^+(aq) + OH^-(aq)$$

If the pH of a 0.065 M solution of the amine is 11.70, what is the value of K_b?

45. A 2.5×10^{-3} M solution of an unknown acid has a pH of 3.80 at 25 °C.
(a) What is the hydronium ion concentration of the solution?
(b) Is the acid a strong acid, a moderately weak acid (K_a of about 10^{-5}), or a very weak acid (K_a of about 10^{-10})?

46. A 0.015 M solution of a base has a pH of 10.09.
(a) What are the hydronium and hydroxide ion concentrations of this solution?
(b) Is the base a strong base, a moderately weak base (K_b of about 10^{-5}), or a very weak base (K_b of about 10^{-10})?

Using Ionization Constants

(See Examples 17.5–17.7 and ChemistryNow Screens 17.9–17.11.)

47. What are the equilibrium concentrations of hydronium ion, acetate ion, and acetic acid in a 0.20 M aqueous solution of acetic acid?

48. ■ The ionization constant of a very weak acid, HA, is 4.0×10^{-9}. Calculate the equilibrium concentrations of H_3O^+, A^-, and HA in a 0.040 M solution of the acid.

49. What are the equilibrium concentrations of H_3O^+, CN^-, and HCN in a 0.025 M solution of HCN? What is the pH of the solution?

50. Phenol (C_6H_5OH), commonly called carbolic acid, is a weak organic acid.

$$C_6H_5OH(aq) + H_2O(\ell) \rightleftharpoons C_6H_5O^-(aq) + H_3O^+(aq) \qquad K_a = 1.3 \times 10^{-10}$$

If you dissolve 0.195 g of the acid in enough water to make 125 mL of solution, what is the equilibrium hydronium ion concentration? What is the pH of the solution?

51. What are the equilibrium concentrations of NH_3, NH_4^+, and OH^- in a 0.15 M solution of ammonia? What is the pH of the solution?

52. ■ A hypothetical weak base has $K_b = 5.0 \times 10^{-4}$. Calculate the equilibrium concentrations of the base, its conjugate acid, and OH^- in a 0.15 M solution of the base.

53. The weak base methylamine, CH_3NH_2, has $K_b = 4.2 \times 10^{-4}$. It reacts with water according to the equation

$$CH_3NH_2(aq) + H_2O(\ell) \rightleftharpoons CH_3NH_3^+(aq) + OH^-(aq)$$

Calculate the equilibrium hydroxide ion concentration in a 0.25 M solution of the base. What are the pH and pOH of the solution?

54. Calculate the pH of a 0.12 M aqueous solution of the base aniline, $C_6H_5NH_2$ ($K_b = 4.0 \times 10^{-10}$).

$$C_6H_5NH_2(aq) + H_2O(\ell) \rightleftharpoons C_6H_5NH_3^+(aq) + OH^-(aq)$$

55. Calculate the pH of a 0.0010 M aqueous solution of HF.

56. ■ A solution of hydrofluoric acid, HF, has a pH of 2.30. Calculate the equilibrium concentrations of HF, F^-, and H_3O^+, and calculate the amount of HF originally dissolved per liter.

Acid–Base Properties of Salts

(See Example 17.7 and ChemistryNow Screen 17.11.)

57. Calculate the hydronium ion concentration and pH in a 0.20 M solution of ammonium chloride, NH_4Cl.

58. ▲ Calculate the hydronium ion concentration and pH for a 0.015 M solution of sodium formate, $NaHCO_2$.

59. Sodium cyanide is the salt of the weak acid HCN. Calculate the concentrations of H_3O^+, OH^-, HCN, and Na^+ in a solution prepared by dissolving 10.8 g of NaCN in enough water to make 5.00×10^2 mL of solution at 25 °C.

60. The sodium salt of propanoic acid, $NaCH_3CH_2CO_2$, is used as an antifungal agent by veterinarians. Calculate the equilibrium concentrations of H_3O^+ and OH^-, and the pH, for a solution of 0.10 M $NaCH_3CH_2CO_2$.

pH after an Acid–Base Reaction

(See Example 17.8 and ChemistryNow Screens 17.7 and 17.10.)

61. Calculate the hydronium ion concentration and pH of the solution that results when 22.0 mL of 0.15 M acetic acid, CH_3CO_2H, is mixed with 22.0 mL of 0.15 M NaOH.

62. ■ Calculate the hydronium ion concentration and the pH when 50.0 mL of 0.40 M NH_3 is mixed with 50.0 mL of 0.40 M HCl.

63. For each of the following cases, decide whether the pH is less than 7, equal to 7, or greater than 7.
(a) Equal volumes of 0.10 M acetic acid, CH_3CO_2H, and 0.10 M KOH are mixed.
(b) 25 mL of 0.015 M NH_3 is mixed with 25 mL of 0.015 M HCl.
(c) 150 mL of 0.20 M HNO_3 is mixed with 75 mL of 0.40 M NaOH.

64. For each of the following cases, decide whether the pH is less than 7, equal to 7, or greater than 7.
(a) 25 mL of 0.45 M H_2SO_4 is mixed with 25 mL of 0.90 M NaOH.
(b) 15 mL of 0.050 M formic acid, HCO_2H, is mixed with 15 mL of 0.050 M NaOH.
(c) 25 mL of 0.15 M $H_2C_2O_4$ (oxalic acid) is mixed with 25 mL of 0.30 M NaOH. (Both H^+ ions of oxalic acid are removed with NaOH.)

Polyprotic Acids and Bases

(See Example 17.9.)

65. ■ Sulfurous acid, H_2SO_3, is a weak acid capable of providing two H^+ ions.
(a) What is the pH of a 0.45 M solution of H_2SO_3?
(b) What is the equilibrium concentration of the sulfite ion, SO_3^{2-}, in the 0.45 M solution of H_2SO_3?

66. ■ Ascorbic acid (vitamin C, $C_6H_8O_6$) is a diprotic acid ($K_{a1} = 6.8 \times 10^{-5}$ and $K_{a2} = 2.7 \times 10^{-12}$). What is the pH of a solution that contains 5.0 mg of acid per milliliter of solution?

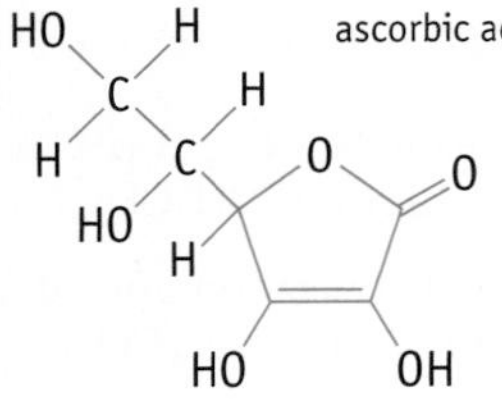

ascorbic acid

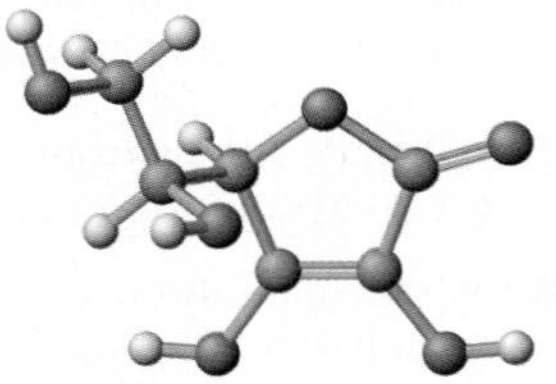

67. Hydrazine, N_2H_4, can interact with water in two steps.

$$N_2H_4(aq) + H_2O(\ell) \rightleftharpoons N_2H_5^+(aq) + OH^-(aq) \qquad K_{b1} = 8.5 \times 10^{-7}$$

$$N_2H_5^+(aq) + H_2O(\ell) \rightleftharpoons N_2H_6^{2+}(aq) + OH^-(aq) \qquad K_{b2} = 8.9 \times 10^{-16}$$

(a) What is the concentration of OH^-, $N_2H_5^+$, and $N_2H_6^{2+}$ in a 0.010 M aqueous solution of hydrazine?
(b) What is the pH of the 0.010 M solution of hydrazine?

68. Ethylenediamine, $H_2NCH_2CH_2NH_2$, can interact with water in two steps, forming OH^- in each step (see Appendix I). If you have a 0.15 M aqueous solution of the amine, calculate the concentrations of $[H_3NCH_2CH_2NH_3]^{2+}$ and OH^-.

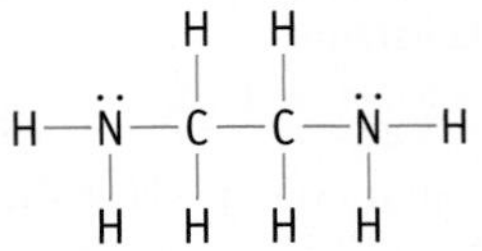

ethylenediamine

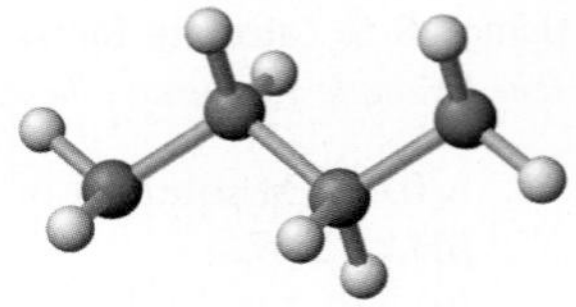

Lewis Acids and Bases

(See Exercise 17.17 and ChemistryNow Screens 17.12–17.14.)

69. Decide whether each of the following substances should be classified as a Lewis acid or a Lewis base.
(a) H_2NOH in the reaction

$$H_2NOH(aq) + HCl(aq) \longrightarrow [H_3NOH]Cl(aq)$$

(b) Fe^{2+}
(c) CH_3NH_2 (Hint: Draw the electron dot structure.)

70. ■ Decide whether each of the following substances should be classified as a Lewis acid or a Lewis base.
(a) BCl_3 (Hint: Draw the electron dot structure.)
(b) H_2NNH_2, hydrazine (Hint: Draw the electron dot structure.)
(c) the reactants in the reaction

$$Ag^+(aq) + 2\ NH_3(aq) \rightleftharpoons [Ag(NH_3)_2]^+(aq)$$

71. Carbon monoxide forms complexes with low-valent metals. For example, $Ni(CO)_4$ and $Fe(CO)_5$ are well known. CO also forms complexes with the iron(II) ion in hemoglobin, which prevents the hemoglobin from acting in its normal way. Is CO a Lewis acid or a Lewis base?

72. ■ Trimethylamine, $(CH_3)_3N$, is a common reagent. It interacts readily with diborane gas, B_2H_6. The latter dissociates to BH_3, and this forms a complex with the amine, $(CH_3)_3N{\rightarrow}BH_3$. Is the BH_3 fragment a Lewis acid or a Lewis base?

Molecular Structure, Bonding, and Acid–Base Behavior

(See Section 17.10 and Exercise 17.18.)

73. Which should be the stronger acid, HOCN or HCN? Explain briefly. (In HOCN, the H^+ ion is attached to the O atom of the OCN^- ion.)

74. ■ Which should be the stronger Brønsted acid, $[V(H_2O)_6]^{2+}$ or $[V(H_2O)_6]^{3+}$?

75. Explain why benzenesulfonic acid is a Brønsted acid.

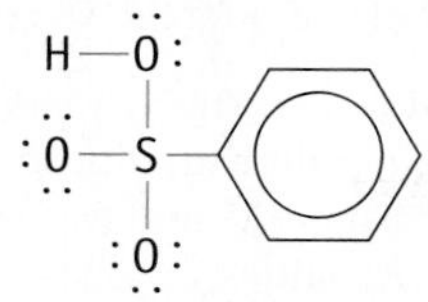

benzenesulfonic acid

76. The structure of ethylenediamine is illustrated in Study Question 68. Is this compound a Brønsted acid, a Brønsted base, a Lewis acid, or a Lewis base, or some combination of these?

General Questions on Acids and Bases

These questions are not designated as to type or location in the chapter. They may combine several concepts.

77. About this time, you may be wishing you had an aspirin. Aspirin is an organic acid (page 760) with a K_a of 3.27×10^{-4} for the reaction

$$HC_9H_7O_4(aq) + H_2O(\ell) \rightleftharpoons C_9H_7O_4^-(aq) + H_3O^+(aq)$$

If you have two tablets, each containing 0.325 g of aspirin (mixed with a neutral "binder" to hold the tablet together), and you dissolve them in a glass of water to give 225 mL of solution, what is the pH of the solution?

78. Consider the following ions: NH_4^+, CO_3^{2-}, Br^-, S^{2-}, and ClO_4^-.
   (a) Which of these ions might lead to an acidic solution, and which might lead to a basic solution?
   (b) Which of these anions will have no effect on the pH of an aqueous solution?
   (c) Which ion is the strongest base?
   (d) Write a chemical equation for the reaction of each basic anion with water.

79. A 2.50-g sample of a solid that could be $Ba(OH)_2$ or $Sr(OH)_2$ was dissolved in enough water to make 1.00 L of solution. If the pH of the solution is 12.61, what is the identity of the solid?

80. ▲ In a particular solution, acetic acid is 11% ionized at 25 °C. Calculate the pH of the solution and the mass of acetic acid dissolved to yield 1.00 L of solution.

81. Hydrogen sulfide, H_2S, and sodium acetate, $NaCH_3CO_2$, are mixed in water. Using Table 17.3, write a balanced equation for the acid–base reaction that could, in principle, occur. Does the equilibrium lie toward the products or the reactants?

82. For each of the following reactions, predict whether the equilibrium lies predominantly to the left or to the right. Explain your prediction briefly.
   (a) $HCO_3^-(aq) + SO_4^{2-}(aq) \rightleftharpoons CO_3^{2-}(aq) + HSO_4^-(aq)$
   (b) $HSO_4^-(aq) + CH_3CO_2^-(aq) \rightleftharpoons SO_4^{2-}(aq) + CH_3CO_2H(aq)$
   (c) $[Co(H_2O)_6]^{2+}(aq) + CH_3CO_2^-(aq) \rightleftharpoons [Co(H_2O)_5(OH)]^+(aq) + CH_3CO_2H(aq)$

83. A monoprotic acid HX has $K_a = 1.3 \times 10^{-3}$. Calculate the equilibrium concentrations of HX and H_3O^+ and the pH for a 0.010 M solution of the acid.

84. ■ Arrange the following 0.10 M solutions in order of increasing pH.
   (a) NaCl
   (b) NH_4Cl
   (c) HCl
   (d) $NaCH_3CO_2$
   (e) KOH

85. ■ *m*-Nitrophenol, a weak acid, can be used as a pH indicator because it is yellow at a pH above 8.6 and colorless at a pH below 6.8. If the pH of a 0.010 M solution of the compound is 3.44, calculate its pK_a.

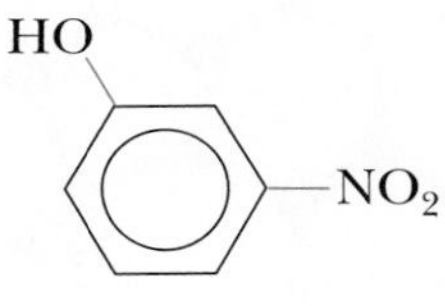

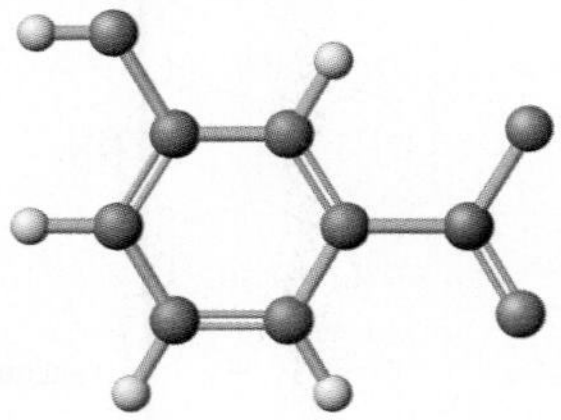

m-nitrophenol

86. The butylammonium ion, $C_4H_9NH_3^+$, has a K_a of 2.3×10^{-11}.

$$C_4H_9NH_3^+(aq) + H_2O(\ell) \rightleftharpoons H_3O^+(aq) + C_4H_9NH_2(aq)$$

   (a) Calculate K_b for the conjugate base, $C_4H_9NH_2$ (butylamine).
   (b) Place the butylammonium ion and its conjugate base in Table 17.3. Name an acid weaker than $C_4H_9NH_3^+$ and a base stronger than $C_4H_9NH_2$.
   (c) What is the pH of a 0.015 M solution of the butylammonium chloride?

87. The local anesthetic novocaine is the hydrogen chloride salt of an organic base, procaine.

$$\underset{\text{procaine}}{C_{13}H_{20}N_2O_2(aq)} + HCl(aq) \longrightarrow \underset{\text{novocaine}}{[HC_{13}H_{20}N_2O_2]^+Cl^-(aq)}$$

The pK_a for novocaine is 8.85. What is the pH of a 0.0015 M solution of novocaine?

88. Pyridine is a weak organic base and readily forms a salt with hydrochloric acid.

$$\underset{\text{pyridine}}{C_5H_5N(aq)} + HCl(aq) \longrightarrow \underset{\text{pyridinium ion}}{C_5H_5NH^+(aq)} + Cl^-(aq)$$

What is the pH of a 0.025 M solution of pyridinium hydrochloride, $[C_5H_5NH^+]Cl^-$?

89. ■ The base ethylamine ($CH_3CH_2NH_2$) has a K_b of 4.3×10^{-4}. A closely related base, ethanolamine ($HOCH_2CH_2NH_2$), has a K_b of 3.2×10^{-5}.
(a) Which of the two bases is stronger?
(b) Calculate the pH of a 0.10 M solution of the stronger base.

90. Chloroacetic acid, $ClCH_2CO_2H$, is a moderately weak acid ($K_a = 1.40 \times 10^{-3}$). If you dissolve 94.5 mg of the acid in water to give 125 mL of solution, what is the pH of the solution?

91. ■ Saccharin ($HC_7H_4NO_3S$) is a weak acid with $pK_a = 2.32$ at 25 °C. It is used in the form of sodium saccharide, $NaC_7H_4NO_3S$. What is the pH of a 0.10 M solution of sodium saccharide at 25 °C?

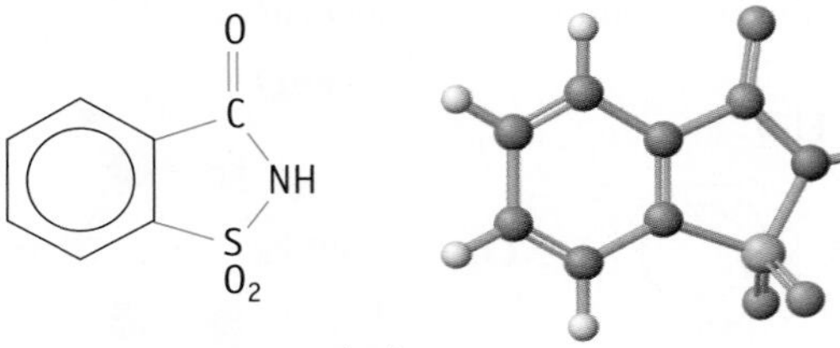

saccharin

92. ■ Given the following solutions:
(a) 0.1 M NH_3
(b) 0.1 M Na_2CO_3
(c) 0.1 M NaCl
(d) 0.1 M CH_3CO_2H
(e) 0.1 M NH_4Cl
(f) 0.1 M $NaCH_3CO_2$
(g) 0.1 M $NH_4CH_3CO_2$
(i) Which of the solutions are acidic?
(ii) Which of the solutions are basic?
(iii) Which of the solutions is most acidic?

93. ■ For each of the following salts, predict whether a 0.10 M solution has a pH less than, equal to, or greater than 7.
(a) $NaHSO_4$
(b) NH_4Br
(c) $KClO_4$
(d) Na_2CO_3
(e) $(NH_4)_2S$
(f) $NaNO_3$
(g) Na_2HPO_4
(h) LiBr
(i) $FeCl_3$
Which solution has the highest pH? The lowest pH?

94. Nicotine, $C_{10}H_{14}N_2$, has two basic nitrogen atoms (page 798), and both can react with water.

$$Nic(aq) + H_2O(\ell) \rightleftharpoons NicH^+(aq) + OH^-(aq)$$

$$NicH^+(aq) + H_2O(\ell) \rightleftharpoons NicH_2^{2+}(aq) + OH^-(aq)$$

K_{b1} is 7.0×10^{-7} and K_{b2} is 1.1×10^{-10}. Calculate the approximate pH of a 0.020 M solution.

95. ■ Oxalic acid is a relatively weak diprotic acid. Calculate the equilibrium constant for the reaction shown below from K_{a1} and K_{a2}. (See Appendix H for the required K_a values.)

$$H_2C_2O_4(aq) + 2\,H_2O(\ell) \rightleftharpoons C_2O_4^{2-}(aq) + 2\,H_3O^+(aq)$$

96. ▲ The equilibrium constant for the reaction of hydrochloric acid and ammonia is 1.8×10^9 (page 779). Confirm this value.

97. ▲ The equilibrium constant for the reaction of formic acid and sodium hydroxide is 1.8×10^{10} (page 779). Confirm this value.

98. ■▲ Calculate the pH of the solution that results from mixing 25.0 mL of 0.14 M formic acid and 50.0 mL of 0.070 M sodium hydroxide.

99. ■▲ To what volume should 1.00×10^2 mL of any weak acid, HA, with a concentration 0.20 M be diluted to double the percentage ionization?

100. ▲ The hydrogen phthalate ion, $C_8H_5O_4^-$, is a weak acid with $K_a = 3.91 \times 10^{-6}$.

$$C_8H_5O_4^-(aq) + H_2O(\ell) \rightleftharpoons C_8H_4O_4^{2-}(aq) + H_3O^+(aq)$$

What is the pH of a 0.050 M solution of potassium hydrogen phthalate, $KC_8H_5O_4$? *Note:* To find the pH for a solution of the anion, we must take into account that the ion is amphiprotic. It can be shown that, for most cases of amphiprotic ions, the H_3O^+ concentration is

$$[H_3O^+] = \sqrt{K_1 \times K_2}$$

For phthalic acid, $C_8H_6O_4$, K_1 is 1.12×10^{-3}, and K_2 is 3.91×10^{-6}.

101. ■▲ You prepare a 0.10 M solution of oxalic acid, $H_2C_2O_4$. What molecules and ions exist in this solution? List them in order of decreasing concentration.

102. ■▲ You mix 30.0 mL of 0.15 M NaOH with 30.0 mL of 0.15 M acetic acid. What molecules and ions exist in this solution? List them in order of decreasing concentration.

In the Laboratory

103. ■ Describe an experiment that will allow you to place the following three bases in order of increasing base strength: NaCN, CH_3NH_2, Na_2CO_3.

▲ more challenging ■ in OWL Blue-numbered questions answered in Appendix O

104. ■ The data below compare the strength of acetic acid with a related series of acids, where the H atoms of the CH_3 group in acetic acid are successively replaced by Br.

Acid	pK_a
CH_3CO_2H	4.74
$BrCH_2CO_2H$	2.90
Br_2CHCO_2H	1.39
Br_3CCO_2H	−0.147

(a) What trend in acid strength do you observe as H is successively replaced by Br? Can you suggest a reason for this trend?
(b) Suppose each of the acids above were present as a 0.10 M aqueous solution. Which would have the highest pH? The lowest pH?

105. ▲ You have three solutions labeled A, B, and C. You know only that each contains a different cation—Na^+, NH_4^+, or H^+. Each has an anion that does not contribute to the solution pH (e.g., Cl^-). You also have two other solutions, Y and Z, each containing a different anion, Cl^- or OH^-, with a cation that does not influence solution pH (e.g., K^+). If equal amounts of B and Y are mixed, the result is an acidic solution. Mixing A and Z gives a neutral solution, whereas B and Z give a basic solution. Identify the five unknown solutions. (Adapted from D. H. Barouch: *Voyages in Conceptual Chemistry*, Boston, Jones and Bartlett, 1997.)

	Y	Z
A		neutral
B	acidic	basic
C		

106. ■ A hydrogen atom in the organic base pyridine, C_5H_5N, can be substituted by various atoms or groups to give XC_5H_4N, where X is an atom such as Cl or a group such as CH_3. The following table gives K_a values for the conjugate acids of a variety of substituted pyridines.

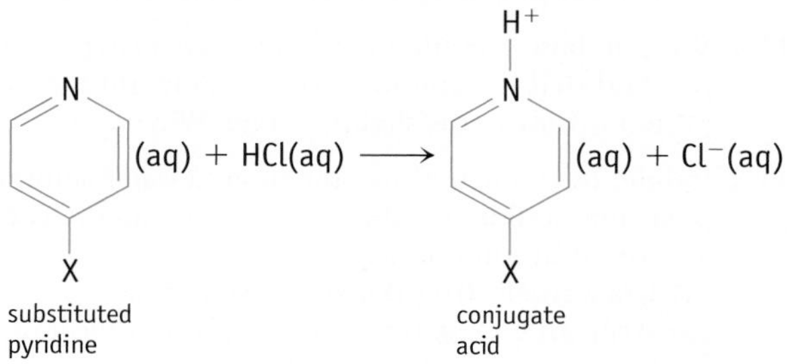

Atom or Group X	K_a of Conjugate Acid
NO_2	5.9×10^{-2}
Cl	1.5×10^{-4}
H	6.8×10^{-6}
CH_3	1.0×10^{-6}

(a) Suppose each conjugate acid is dissolved in sufficient water to give a 0.050 M solution. Which solution would have the highest pH? The lowest pH?
(b) Which of the substituted pyridines is the strongest Brønsted base? Which is the weakest Brønsted base?

107. ■ Nicotinic acid, $C_6H_5NO_2$, is found in minute amounts in all living cells, but appreciable amounts occur in liver, yeast, milk, adrenal glands, white meat, and corn. Whole-wheat flour contains about 60. μg per 1g of flour. One gram (1.00 g) of the acid dissolves in water to give 60. mL of solution having a pH of 2.70. What is the approximate value of K_a for the acid?

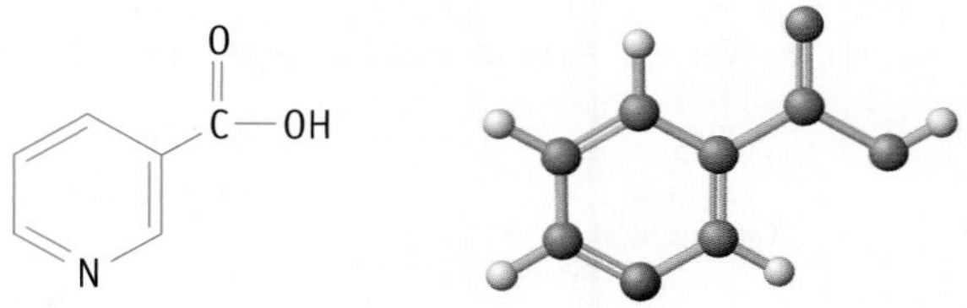

nicotinic acid

108. ■ ▲ Equilibrium constants can be measured for the dissociation of Lewis acid–base complexes such as the dimethyl ether complex of BF_3, $(CH_3)_2O{\rightarrow}BF_3$. The value of K (here K_p) for the reaction is 0.17.

$$(CH_3)_2O{\rightarrow}BF_3(g) \rightleftharpoons BF_3(g) + (CH_3)_2O(g)$$

(a) Describe each product as a Lewis acid or a Lewis base.
(b) If you place 1.00 g of the complex in a 565-mL flask at 25 °C, what is the total pressure in the flask? What are the partial pressures of the Lewis acid, the Lewis base, and the complex?

109. ▲ Sulfanilic acid, which is used in making dyes, is made by reacting aniline with sulfuric acid.

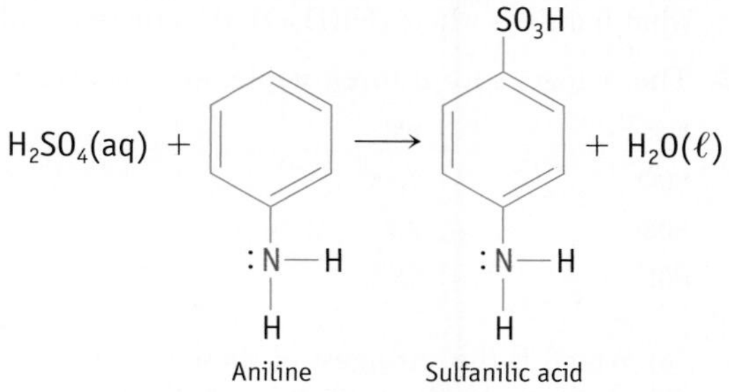

(a) Is aniline a Brønsted base, a Lewis base, or both? Explain, using its possible reactions with HCl, BF_3, or other acid.

(b) Sulfanilic acid has a pK_a value of 3.23. The sodium salt of the acid, $Na(H_2NC_6H_4SO_3)$, is quite soluble in water. If you dissolve 1.25 g of the salt in water to give 125 mL of solution, what is the pH of the solution?

110. Amino acids are an important group of compounds (see page 498). At low pH, both the carboxylic acid group ($—CO_2H$) and the amine group (—NHR) are protonated. However, as the pH of the solution increases (say by adding base), the carboxylic acid proton is removed, usually at a pH between 2 and 3. In a middle range of pHs, the amine group is protonated, but the carboxulic acid group has lost the proton. (This is called a *zwitterion.*) At more basic pH values, the amine proton is dissociated.

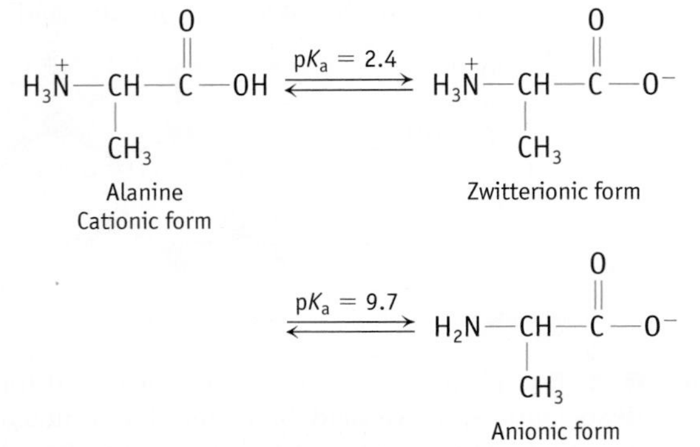

What is the pH of a 0.20 M solution of alanine hydrochloride, $[NH_3CHCH_3CO_2H]Cl$?

Summary and Conceptual Questions

The following questions may use concepts from this and previous chapters.

111. How can water be both a Brønsted base and a Lewis base? Can water be a Brønsted acid? A Lewis acid?

112. The nickel(II) ion exists as $[Ni(H_2O)_6]^{2+}$ in aqueous solution. Why is such a solution acidic? As part of your answer, include a balanced equation depicting what happens when $[Ni(H_2O)_6]^{2+}$ interacts with water.

113. The halogens form three stable, weak acids HOX.

Acid	pK_a
HOCl	7.46
HOBr	8.7
HOI	10.6

(a) Which is the strongest of these acids?

(b) Explain why the acid strength changes as the halogen atom is changed.

114. The acidity of the oxoacids was described on page 794, and a larger number of acids are listed in the table below.

$E(OH)_m$	pK_a	$EO(OH)_m$	pK_a	$EO_2(OH)_m$	pK_a	$EO_3(OH)_m$	pK_a
Very weak		**Weak**		**Strong**		**Very strong**	
Cl(OH)	7.5	ClO(OH)	2	$ClO_2(OH)$	−3	$ClO_3(OH)$	−10
Br(OH)	8.7	NO(OH)	3.4	$NO_2(OH)$	−1.4		
I(OH)	10.6	IO(OH)	1.6	$IO_2(OH)$	0.8		
$Si(OH)_4$	9.7	$SO(OH)_2$	1.8	$SO_2(OH)_2$	−3		
$Sb(OH)_4$	11.0	$SeO(OH)_2$	2.5	$SeO_2(OH)_2$	−3		
$As(OH)_3$	9.2	$AsO(OH)_3$	2.3				
		$PO(OH)_3$	2.1				
		$HPO(OH)_2$	1.8				
		$H_2PO(OH)$	2.0				

(a) What general trends do you see in these data?

(b) What has a greater effect on acidity, the number of O atoms bonded directly to the central atom E or the number of OH groups?

(c) Look at the acids based on Cl, N, and S. Is there a correlation of acidity with the formal charge on the central atom, E?

(d) The acid H_3PO_3 has a pK_a of 1.8, and this led to some insight into its structure. If the structure of the acid were $P(OH)_3$, what would be its predicted pK_a value? Given that this is a diprotic acid, which H atoms are lost as H^+ ions?

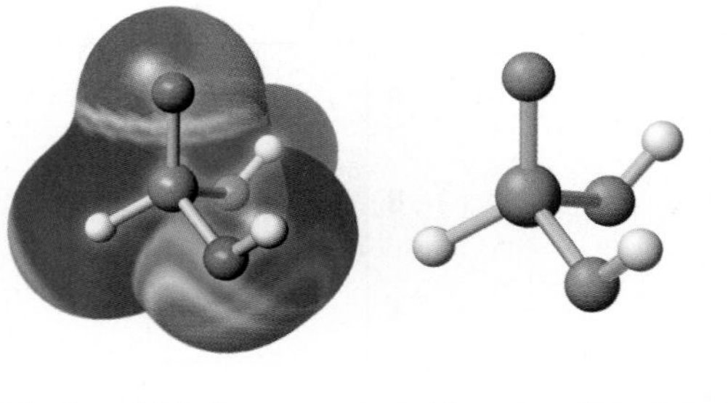

115. Perchloric acid behaves as an acid, even when it is dissolved in sulfuric acid.

(a) Write a balanced equation showing how perchloric acid can transfer a proton to sulfuric acid.

(b) Draw a Lewis electron dot structure for sulfuric acid. How can sulfuric acid function as a base?

116. You purchase a bottle of water. On checking its pH, you find that it is not neutral, as you might have expected. Instead, it is slightly acidic. Why?

117. Iodine, I_2, is much more soluble in a water solution of potassium iodide, KI, than it is in pure water. The anion found in solution is I_3^-.

(a) Draw an electron dot structure for I_3^-.

(b) Write an equation for this reaction, indicating the Lewis acid and the Lewis base.

118. ▲ Uracil is a base found in RNA (see page 504). Indicate sites in the molecule where hydrogen bonding is possible or that are sites of Lewis basicity. The electrostatic potential surface shows that one of the four C atoms in uracil has a partial negative charge. Designate that carbon atom.

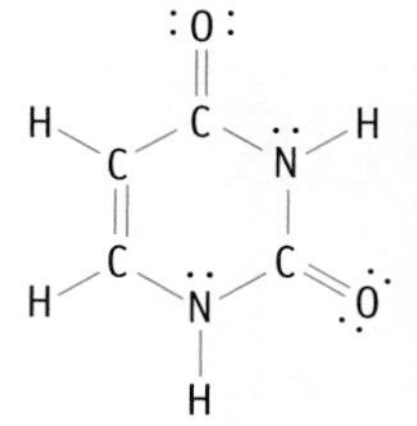

Uracil

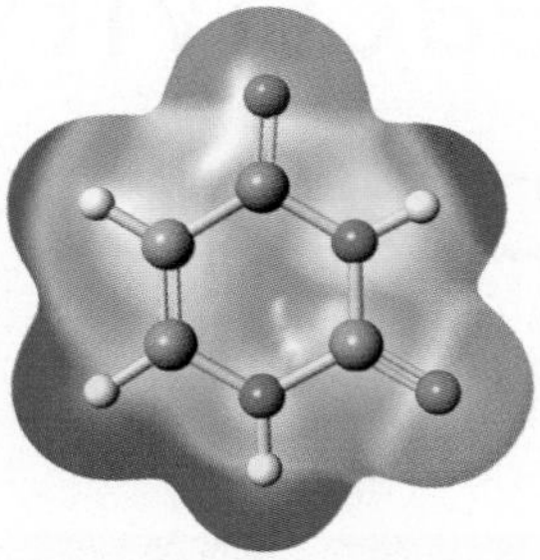

119. Chemists often refer to the *degree of ionization* of a weak acid or base and give it the symbol α. The equilibrium constant, in terms of α and C_o, the initial acid or base concentration, is given by the useful equation

$$K = \frac{\alpha^2 C_o}{(1 - \alpha)}$$

As an example, the degree of ionization for 0.010 M acetic acid is 0.013.

(a) Show how we can arrive at the general equation given above.

(b) Calculate the degree of ionization for the ammonium ion in 0.10 M NH_4Cl.

120. ▲ Exploring the degree of ionization equation in Study Question 119.

(a) Calculate the degree of ionization, α, for formic acid at the following concentrations: 0.0100 M, 0.0200 M, 0.0400 M, 0.100 M, 0.200 M, 0.400 M, 1.00 M, 2.00 M, and 4.00 M.

(b) Plot the results of the calculation as α versus formic acid concentration. Is there a linear relationship? If not, try plotting the logarithm of C_o versus α.

(c) What can you conclude about the relationship of the degree of ionization and the initial acid or base concentration?

121. ■ ▲ Consider a salt of a weak base and a weak acid such as ammonium cyanide. Both the NH_4^+ and CN^- ions interact with water in aqueous solution, but the net reaction can be considered as a proton transfer from NH_4^+ to CN^-.

$$NH_4^+(aq) + CN^-(aq) \rightleftharpoons NH_3(aq) + HCN(aq)$$

(a) Show that the equilibrium constant for this reaction, K_{net}, is

$$K_{net} = \frac{K_w}{K_a K_b}$$

where K_a is the ionization constant for the weak acid HCN and K_b is the constant for the weak base NH_3.

(b) Prove that the hydronium ion concentration in this solution must be given by

$$[H_3O^+] = \sqrt{\frac{K_w K_a}{K_b}}$$

(c) What is the pH of a 0.15 M solution of ammonium cyanide?

(d) What is the pH of a 0.10 M solution of ammonium acetate?

(e) Do you need to do a calculation to decide if solutions of salts are acidic or basic when one ion is a base and the other is an acid? Can you predict the relative pH of the solution? Explain.

18 Principles of Reactivity: Other Aspects of Aqueous Equilibria

Charles D. Winters

Mineral samples (clockwise from the top center): red rhodochrosite, yellow orpiment, golden iron pyrite, green-blue turquoise, black stibnite, purple fluorite, and blue azurite. Formulas are in the text.

Minerals and Gems—The Importance of Solubility

Minerals and gems are among nature's most beautiful creations. Many, such as rubies, are metal oxides, and the various types of quartz are based on silicon dioxide. Another large class of gemstones consists largely of metal silicates. These include emerald, topaz, aquamarine, and tourmaline.

Yet another large class of minerals and of a few gemstones is carbonates. Rhodochrosite, one of the most beautiful red stones, is manganese(II) carbonate. And one of the most abundant minerals on earth is limestone, calcium carbonate, which also largely composes sea shells and corals.

Hydroxides are represented by azurite, which is a mixed carbonate/hydroxide with the formula $Cu_3(OH)_2(CO_3)_2$. Turquoise is a mixed hydroxide/phosphate based on copper(II), the source of the blue color of turquoise.

Among the most common minerals are sulfides such as golden iron pyrite (FeS_2), black stibnite (Sb_2S_3), red cinnabar (HgS), and yellow orpiment (As_2S_3).

Other smaller classes of minerals exist; one of the smallest is the class based on the halides, and the best example is fluorite. Fluorite, CaF_2, exhibits a wide range of colors from purple to green to yellow.

What do all of these minerals and gems have in common? They are all insoluble or poorly soluble in water. If they were more soluble, they would be dissolved in the world's lakes and oceans.

Questions:

1. Which is more soluble in water, $CaCO_3$ or $MnCO_3$?
2. Which is more soluble in water, HgS or PbS?
3. What is the calculated solubility of fluorite (in g/L)?

Answers to these questions are in Appendix Q.

Chapter Goals

See Chapter Goals Revisited (page 849) for Study Questions keyed to these goals and assignable in OWL.

- Understand the common ion effect.
- Understand the control of pH in aqueous solution with buffers.
- Evaluate the pH in the course of acid–base titrations.
- Apply chemical equilibrium concepts to the solubility of ionic compounds.

Chapter Outline

In Chapter 3, we described four fundamental types of chemical reactions: acid–base reactions, precipitation reactions, gas-forming reactions, and oxidation-reduction reactions. In the present chapter, we want to apply the principles of chemical equilibria to an understanding of the first two of these kinds of reactions.

With regard to acid–base reactions, we are looking for answers to the following questions:

- How can we control the pH in a solution?
- What happens when an acid and base are mixed in any amount?

Precipitation reactions can also be understood in terms of chemical equilibria. The following questions are discussed in this chapter:

- If aqueous solutions of two ionic compounds are mixed, will precipitation occur?
- To what extent does an insoluble substance actually dissolve?
- What chemical reactions can be used to redissolve a precipitate?

Chemistry Now™

Throughout the text this icon introduces an opportunity for self-study or to explore interactive tutorials by signing in at **www.thomsonedu.com/login**.

18.1 The Common Ion Effect

In the previous chapter, we examined the behavior of weak acids and bases in aqueous solution. There are many cases, however, where the weak acid solution also contains a significant concentration of its conjugate base or where a weak base solution has a significant concentration of conjugate acid. The pH of such solutions will be different than those of solutions of a weak acid or base with very small amounts of conjugate bases or acids produced by ionization. The effect of a significant concentration of conjugate base on the pH of a weak acid solution, for example, is called the **common ion effect**, and it is particularly important in buffer solutions, as we shall see in Section 18.2.

Let us see how the common ion effect works. If 1.0 L of a 0.25 M acetic acid solution has a pH of 2.67, what is the pH after adding 0.10 mol of sodium acetate? Sodium acetate, $NaCH_3CO_2$, is 100% dissociated into its ions, Na^+ and $CH_3CO_2^-$, in water. Sodium ion has no effect on the pH of a solution (◀ Table 17.4 and Example 17.2). Thus, the important components of the solution are a weak acid (CH_3CO_2H) and its conjugate base ($CH_3CO_2^-$); the added acetate ion is "common" to the ionization equilibrium reaction of acetic acid.

$$CH_3CO_2H(aq) + H_2O(\ell) \rightleftharpoons H_3O^+(aq) + CH_3CO_2^-(aq)$$

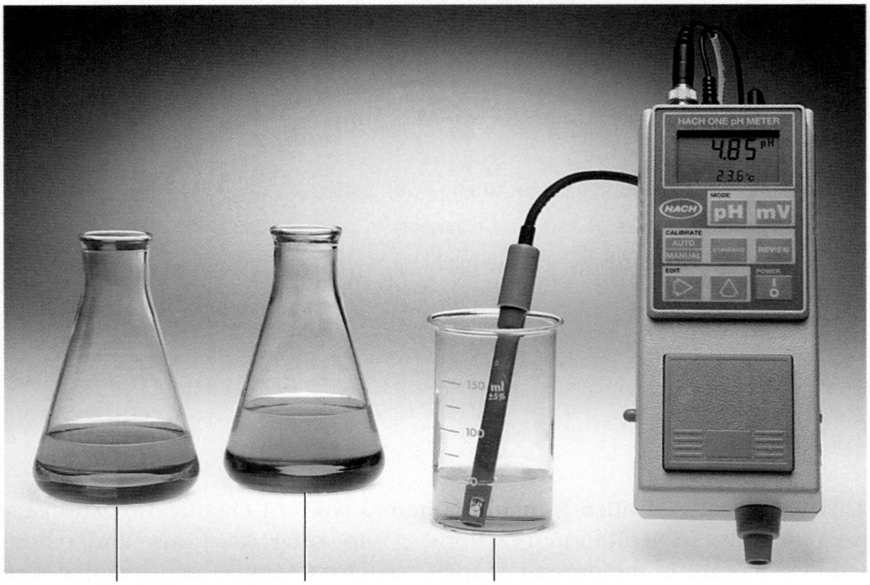

Aqueous acetic acid pH 2.7

Aqueous sodium acetate pH 9

Mixture of acetic acid and sodium acetate

Charles D. Winters

FIGURE 18.1 The common ion effect. Approximately equal amounts of acetic acid (left flask, pH about 2.7) and sodium acetate, a base (right flask, pH about 9) were mixed in the beaker. The pH meter shows that the resulting solution in the beaker has a lower hydronium ion concentration (a pH of about 5) than the acetic acid solution, owing to the presence of acetate ion, the conjugate base of the acid and an ion common to the ionization reaction of the acid. (The acid and base solutions each had about the same concentration. Each solution contains universal indicator. This dye is red in low pH, yellow in slightly acidic media, and green in neutral to weakly basic media.)

Assume the acid ionizes to give H_3O^+ and $CH_3CO_2^-$, both in the amount *x*. This means that, relative to their initial concentrations, CH_3CO_2H decreases in concentration slightly (by an amount *x*) and $CH_3CO_2^-$ increases slightly (by an amount *x*).

■ **The Common Ion Effect** In this ICE table, the first row (Initial) reflects the assumption that no ionization (or hydrolysis of the conjugate base) has yet occurred. Ionization of the acid in the presence of the conjugate base then produces *x* mol/L of hydronium ion and *x* mol/L more of the conjugate base.

Equation	CH_3CO_2H	+	$H_2O \rightleftharpoons$	H_3O^+	+	$CH_3CO_2^-$
Initial (M)	0.25			0		0.10
Change (M)	$-x$			$+x$		$+x$
Equilibrium (M)	$(0.25 - x)$			x		$0.10 + x$

Because we have been able to define the equilibrium concentrations of acid and conjugate base and we know K_a, the hydronium ion concentration ($= x$) can be calculated from the usual equilibrium constant expression.

$$K_a = 1.8 \times 10^{-5} = \frac{[H_3O^+][CH_3CO_2^-]}{[CH_3CO_2H]} = \frac{(x)(0.10 + x)}{0.25 - x}$$

Now, because acetic acid is a weak acid and because it is ionizing in the presence of a significant concentration of its conjugate base, let us assume *x* is quite small. That is, it is reasonable to assume that $(0.10 + x)\text{M} \approx 0.10$ M and that $(0.25 - x)\text{M} \approx 0.25$ M. This leads to the "approximate" expression.

$$K_a = 1.8 \times 10^{-5} = \frac{[H_3O^+][CH_3CO_2^-]}{[CH_3CO_2H]} = \frac{(x)(0.10)}{0.25}$$

■ **Equilibrium Constants and Temperature** Unless specified otherwise, all equilibrium constants and all calculations in this chapter are at 25 °C.

Solving this, we find that $x = [H_3O^+] = 4.5 \times 10^{-5}$ M and the pH is 4.35.

Without added $NaCH_3CO_2$, which provides the "common ion" $CH_3CO_2^-$, ionization of 0.25 M acetic acid will produce H_3O^+ and $CH_3CO_2^-$ ions in a concentration of 0.0021 M (to give a pH of 2.68). Le Chatelier's principle, however, predicts that the added common ion causes the reaction to proceed less far to the right. Hence, as we have found, $x = [H_3O^+]$ is less than 0.0021 M in the presence of added acetate ion.

Chemistry Now™

Sign in at **www.thomsonedu.com/login** and go to Chapter 18 Contents to see Screen 18.2 for a simulation of the **common ion effect.**

EXAMPLE 18.1 Reaction of Lactic Acid with a Deficiency of Sodium Hydroxide

Problem What is the pH of the solution that results from adding 25.0 mL of 0.0500 M NaOH to 25.0 mL of 0.100 M lactic acid? (K_a for lactic acid = 1.4×10^{-4})

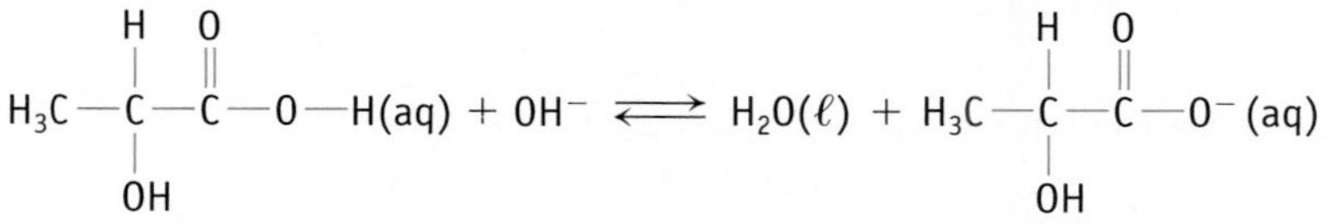

lactic acid ($HC_3H_5O_3$)
$K_a = 1.4 \times 10^{-4}$

lactate ion ($C_3H_5O_3^-$)

Strategy There are two parts to this problem: a stoichiometry problem followed by an equilibrium problem. We first calculate the concentrations of lactic acid and lactate ion that are present following the reaction of lactic acid with NaOH. Then, with the acid and conjugate base concentrations known, we follow the strategy in the text above to determine the pH.

Solution

Part 1: *Stoichiometry Problem*

First, consider what species remain in solution after the acid–base reaction and what the concentrations of those species are.

(a) Amounts of NaOH and lactic acid used in the reaction

$$(0.0250 \text{ L NaOH})(0.0500 \text{ mol/L}) = 1.25 \times 10^{-3} \text{ mol NaOH}$$

$$(0.0250 \text{ L lactic acid})(0.100 \text{ mol/L}) = 2.50 \times 10^{-3} \text{ mol lactic acid}$$

(b) Amount of lactate ion produced by the acid-base reaction

Recognizing that NaOH is the limiting reactant, we have

$$(1.25 \times 10^{-3} \text{ mol NaOH})\left(\frac{1 \text{ mol lactate ion}}{1 \text{ mol NaOH}}\right) = 1.25 \times 10^{-3} \text{ mol lactate ion produced}$$

(c) Amount of lactic acid consumed

$$(1.25 \times 10^{-3} \text{ mol NaOH})\left(\frac{1 \text{ mol lactic acid}}{1 \text{ mol NaOH}}\right) = 1.25 \times 10^{-3} \text{ mol lactic acid consumed}$$

(d) Amount of lactic acid remaining when reaction is complete.

2.50×10^{-3} mol lactic acid available $-$ 1.25×10^{-3} mol lactic acid consumed

$= 1.25 \times 10^{-3}$ mol lactic acid remaining

(e) Concentrations of lactic acid and lactate ion after reaction. Note that the total solution volume after reaction is 50.0 mL or 0.050 L.

$$[\text{lactic acid}] = \frac{1.25 \times 10^{-3} \text{ mol lactic acid}}{0.0500 \text{ L}} = 2.50 \times 10^{-2} \text{ M}$$

Because the amount of lactic acid remaining is the same as the amount of lactate ion produced, we have

$$[\text{lactic acid}] = [\text{lactate ion}] = 2.50 \times 10^{-2} \text{ M}$$

Part 2: *Equilibrium Calculation*

With the "initial" concentrations known, construct a table summarizing the equilibrium concentrations.

Equilibrium	$HC_3H_5O_3 + H_2O$	$\rightleftharpoons$	H_3O^+	+	$C_3H_5O_3^-$
Initial (M)	0.0250		0		0.0250
Change (M)	$-x$		$+x$		$+x$
Equilibrium (M)	$(0.0250 - x)$		x		$(0.0250 + x)$

Substituting the concentrations into the equilibrium expression, we have

$$K_a \text{ (lactic acid)} = 1.4 \times 10^{-4} = \frac{[H_3O^+][C_3H_5O_2^-]}{[HC_3H_5O_2]} = \frac{(x)(0.0250 + x)}{0.0250 - x}$$

Making the assumption that x is small with respect to 0.0250 M, we see that

$$x = [H_3O^+] = K_a = 1.4 \times 10^{-4} \text{ M}$$

which gives a pH of 3.85.

Comment There are two final points to be made:

- Our assumption that $x << 0.0250$ is valid.
- The pH of a solution containing only 0.100 M lactic acid solution is 2.43. Adding a base (lactate ion) increases the pH.

EXERCISE 18.1 Common Ion Effect

Assume you have a 0.30 M solution of formic acid (HCO_2H) and have added enough sodium formate ($NaHCO_2$) to make the solution 0.10 M in the salt. Calculate the pH of the formic acid solution before and after adding solid sodium formate.

EXERCISE 18.2 Mixing an Acid and a Base

What is the pH of the solution that results from adding 30.0 mL of 0.100 M NaOH to 45.0 mL of 0.100 M acetic acid?

Module 23

18.2 Controlling pH: Buffer Solutions

The normal pH of human blood is 7.4. However, the addition of a small quantity of strong acid or base, say 0.010 mol, to a liter of blood leads to a change in pH of only about 0.1 pH units. In comparison, if you add 0.010 mol of HCl to 1.0 L of pure water, the pH drops from 7 to 2. Addition of 0.010 mol of NaOH to pure water increases the pH from 7 to 12. Blood, and many other body fluids, are said to be buffered. A **buffer** causes solutions to be resistant to a change in pH when a strong acid or base is added (Figure 18.2).

There are two requirements for a buffer:

- Two substances are needed: an acid capable of reacting with added OH^- ions and a base that can consume added H_3O^+ ions.
- The acid and base must not react with each another.

These requirements mean a buffer is usually prepared from a conjugate acid–base pair: (1) a weak acid and its conjugate base (acetic and acetate ion, for example), or (2) a weak base and its conjugate acid (ammonia and ammonium ion, for example). Some buffers commonly used in the laboratory are given in Table 18.1.

Before | After adding 0.10 M HCl

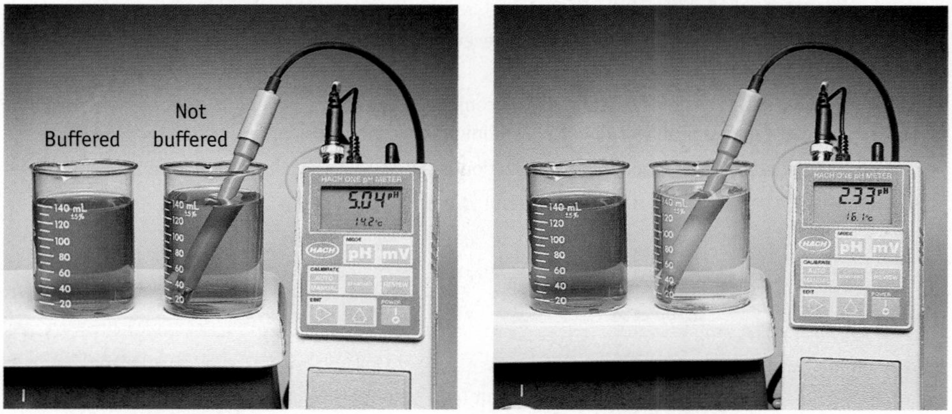

Charles D. Winters

(a) The pH electrode is indicating the pH of water that contains a trace of acid (and bromphenol blue acid–base indicator). The solution at the left is a buffer solution with a pH of about 7. (It also contains bromphenol blue dye.)

(b) When 5 mL of 0.10 M HCl is added to each solution, the pH of the water drops several units, whereas the pH of the buffer stays constant, as implied by the fact that the indicator color did not change.

Active Figure 18.2
Buffer solutions.
ChemistryNow™ Sign in at www.thomsonedu.com/login and go to the Chapter Contents menu to explore an interactive version of this figure accompanied by an exercise.

To see how a buffer works, let us consider an acetic acid/acetate ion buffer. Acetic acid, a weak acid, is needed to consume any added hydroxide ions.

$$CH_3CO_2H(aq) + OH^-(aq) \rightleftharpoons CH_3CO_2^-(aq) + H_2O(\ell) \qquad K = 1.8 \times 10^9$$

The equilibrium constant for the reaction is very large because OH^- is a much stronger base than acetate ion, $CH_3CO_2^-$ (see Section 17.6 and Table 17.3). This means that any OH^- entering the solution from an outside source is consumed completely. In a similar way, any hydronium ion added to the solution reacts with the acetate ion present in the buffer.

$$H_3O^+(aq) + CH_3CO_2^-(aq) \rightleftharpoons H_2O(\ell) + CH_3CO_2H(aq) \qquad K = 5.6 \times 10^4$$

The equilibrium constant for this reaction is also quite large because H_3O^+ is a much stronger acid than CH_3CO_2H.

The next several examples illustrate how to calculate the pH of a buffer solution, how to prepare a buffer, and how a buffer can control the pH of a solution.

■ **Buffers and the Common Ion Effect** The common ion effect is observed for an acid (or base) ionizing in the presence of its conjugate base (or acid). A buffer is a solution of an acid, for example, and its conjugate base.

ChemistryNow™

Sign in at **www.thomsonedu.com/login** and go to Chapter 18 Contents to see:
- Screen 18.3 for a simulation and tutorials on **buffer solutions**
- Screen 18.4 for a simulation and tutorials on **pH of buffer solutions**

TABLE 18.1 Some Commonly Used Buffer Systems

Weak Acid	Conjugate Base	Acid K_a (pK_a)	Useful pH Range
Phthalic acid, $C_6H_4(CO_2H)_2$	Hydrogen phthalate ion, $C_6H_4(CO_2H)(CO_2)^-$	1.3×10^{-3} (2.89)	1.9–3.9
Acetic acid, CH_3CO_2H	Acetate ion, $CH_3CO_2^-$	1.8×10^{-5} (4.74)	3.7–5.8
Dihydrogen phosphate ion, $H_2PO_4^-$	Hydrogen phosphate ion, HPO_4^{2-}	6.2×10^{-8} (7.21)	6.2–8.2
Hydrogen phosphate ion, HPO_4^{2-}	Phosphate ion, PO_4^{3-}	3.6×10^{-13} (12.44)	11.3–13.3

EXAMPLE 18.2 pH of a Buffer Solution

Problem What is the pH of an acetic acid/sodium acetate buffer with $[CH_3CO_2H] = 0.700$ M and $[CH_3CO_2^-] = 0.600$ M?

Strategy The concentrations of the weak acid, its conjugate base, and K_a are all known, so we can use the usual equilibrium expression to calculate the hydronium ion concentration.

Solution Write a balanced equation for the ionization of acetic acid and set up an ICE table.

Equilibrium	$CH_3CO_2H + H_2O$	$\rightleftharpoons$	H_3O^+	+	$CH_3CO_2^-$
Initial (M)	0.700		0		0.600
Change (M)	$-x$		$+x$		$+x$
Equilibrium (M)	$0.700 - x$		x		$0.600 + x$

The appropriate equilibrium constant expression is

$$K_a = 1.8 \times 10^{-5} = \frac{[H_3O^+][CH_3CO_2^-]}{[CH_3CO_2H]} = \frac{(x)(0.600 + x)}{0.700 - x}$$

As explained on page 814, the value of x will be very small with respect to 0.700 or 0.600, so we can use the "approximate expression" to find x, the hydronium ion concentration.

$$K_a = 1.8 \times 10^{-5} = \frac{[H_3O^+][CH_3CO_2^-]}{[CH_3CO_2H]} = \frac{(x)(0.600)}{0.700}$$

$$x = 2.1 \times 10^{-5} \text{ M}$$

$$\text{pH} = -\log (2.1 \times 10^{-5}) = 4.68$$

Comment The pH of the buffer has a value between the pH of 0.700 M acetic acid (2.45) and 0.600 M sodium acetate (9.26).

EXERCISE 18.3 pH of a Buffer Solution

What is the pH of a buffer solution composed of 0.50 M formic acid (HCO_2H) and 0.70 M sodium formate ($NaHCO_2$)?

General Expressions for Buffer Solutions

In Example 18.2, we found the hydronium ion concentration of the acetic acid/acetate ion buffer solution by solving for x in the equation

$$K_a = 1.8 \times 10^{-5} = \frac{[H_3O^+][CH_3CO_2^-]}{[CH_3CO_2H]} = \frac{(x)(0.600)}{0.700}$$

If this equation is rearranged, we have a very useful equation that can help you better understand how a buffer works.

$$[H_3O^+] = \frac{[CH_3CO_2H]}{[CH_3CO_2^-]} \times K_a$$

That is, the hydrogen ion concentration in the acetic acid/acetate ion buffer is given by the ratio of the acid and conjugate base concentrations times the acid ionization constant. Indeed, this is true for all solutions of *a weak acid and its conjugate base.*

$$[H_3O^+] = \frac{[\text{acid}]}{[\text{conjugate base}]} \times K_a \quad \textbf{(18.1)}$$

It is often convenient to use Equation 18.1 in a different form. If we take the negative logarithm of each side of the equation, we have

$$-\log[H_3O^+] = \left\{-\log\frac{[\text{acid}]}{[\text{conjugate base}]}\right\} + (-\log K_a)$$

■ **Buffer Solutions** You will find it generally useful to consider all buffer solutions as composed of a weak acid and its conjugate base. Suppose, for example, a buffer is composed of the weak base ammonia and its conjugate acid ammonium ion. The hydronium ion concentration can be found from Equation 18.1 by assuming the buffer is composed of the weak acid NH_4^+ and its conjugate base, NH_3.

You know that $-\log[H_3O^+]$ is defined as pH, and $-\log K_a$ is defined as pK_a (◀ Sections 17.3 and 17.4). Furthermore, because

$$-\log\frac{[\text{acid}]}{[\text{conjugate base}]} = +\log\frac{[\text{conjugate base}]}{[\text{acid}]}$$

the preceding equation can be rewritten as

$$\text{pH} = \text{p}K_a + \log\frac{[\text{conjugate base}]}{[\text{acid}]} \qquad \textbf{(18.2)}$$

This equation is known as the **Henderson–Hasselbalch equation.**

Both Equations 18.1 and 18.2 show that the pH of a buffer solution is controlled by two factors: the strength of the acid (as expressed by K_a or pK_a) and the relative amounts of acid and conjugate base. The solution pH is established primarily by the value of K_a or pK_a, and the pH is fine-tuned by adjusting the acid-to-conjugate base ratio.

When the concentrations of conjugate base and acid are the same in a solution, the ratio [conjugate base]/[acid] is 1. The log of 1 is zero, so pH = pK_a under these circumstances. If there is more of the conjugate base in the solution than acid, for example, then pH > pK_a. Conversely, if there is more acid than conjugate base in solution, then pH < pK_a.

■ **The Henderson–Hasselbalch Equation** Many handbooks of chemistry list acid ionization constants in terms of pK_a values, so the approximate pH values of possible buffer solutions are readily apparent.

Chemistry Now™

Sign in at **www.thomsonedu.com/login** and go to Chapter 18 Contents to see Screen 18.4 for a simulation and tutorial on **the Henderson–Hasselbalch equation.**

EXAMPLE 18.3 Using the Henderson–Hasselbalch Equation

Problem Benzoic acid ($C_6H_5CO_2H$, 2.00 g) and sodium benzoate ($NaC_6H_5CO_2$, 2.00 g) are dissolved in enough water to make 1.00 L of solution. Calculate the pH of the solution using the Henderson–Hasselbalch equation.

Strategy The Henderson–Hasselbalch equation requires the pK_a of the acid, and this is obtained from the K_a for the acid (see Table 17.3 or Appendix H). You will also need the acid and conjugate base concentrations.

Solution K_a for benzoic acid as 6.3×10^{-5}. Therefore,

$$\text{p}K_a = -\log(6.3 \times 10^{-5}) = 4.20$$

Next, we need the concentrations of the acid (benzoic acid) and its conjugate base (benzoate ion).

$$2.00 \text{ g benzoic acid}\left(\frac{1 \text{ mol}}{122.1 \text{ g}}\right) = 0.0164 \text{ mol benzoic acid}$$

$$2.00 \text{ g sodium benzoate}\left(\frac{1 \text{ mol}}{144.1 \text{ g}}\right) = 0.0139 \text{ mol sodium benzoate}$$

Because the solution volume is 1.00 L, the concentrations are [benzoic acid] = 0.0164 M and [sodium benzoate] = 0.0139 M. Therefore, using Equation 18.2, we have

$$\text{pH} = 4.20 + \log\frac{0.0139}{0.0164} = 4.20 + \log(0.848) = 4.13$$

Comment Notice that the pH is less than the pK_a because the concentration of acid is greater than the concentration of the conjugate base (the ratio of conjugate base to acid concentration is less than 1).

EXERCISE 18.4 Using the Henderson–Hasselbalch Equation

Use the Henderson–Hasselbalch equation to calculate the pH of 1.00 L of a buffer solution containing 15.0 g of $NaHCO_3$ and 18.0 g of Na_2CO_3. (Consider this buffer as a solution of the weak acid HCO_3^- with CO_3^{2-} as its conjugate base.)

Preparing Buffer Solutions

To be useful, a buffer solution must have two characteristics:

- *pH Control:* It should control the pH at the desired value. The Henderson–Hasselbalch equation shows us how this can be done.

$$pH = pK_a + \log\frac{[\text{conjugate base}]}{[\text{acid}]}$$

  First, an acid is chosen whose pK_a (or K_a) is near the intended value of pH (or $[H_3O^+]$). Second, the exact value of pH (or $[H_3O^+]$) is then achieved by adjusting the acid-to-conjugate base ratio. (Example 18.4 illustrates this approach.)
- *Buffer capacity:* The buffer should have the ability to keep the pH approximately constant after the addition of reasonable amounts of acid and base. For example, the concentration of acetic acid in an acetic acid/acetate ion buffer must be sufficient to consume all the hydroxide ion that may be added and still control the pH (see Example 18.4). Buffers are usually prepared as 0.10 M to 1.0 M solutions of reagents. However, any buffer will lose its capacity if too much strong acid or base is added.

Chemistry.Now™

Sign in at **www.thomsonedu.com/login** and go to Chapter 18 Contents to see Screen 18.5 for a simulation and tutorials on **preparing buffer solutions.**

EXAMPLE 18.4 Preparing a Buffer Solution

Problem You wish to prepare 1.0 L of a buffer solution with a pH of 4.30. A list of possible acids (and their conjugate bases) follows:

Acid	Conjugate Base	K_a	pK_a
HSO_4^-	SO_4^{2-}	1.2×10^{-2}	1.92
CH_3CO_2H	$CH_3CO_2^-$	1.8×10^{-5}	4.75
HCO_3^-	CO_3^{2-}	4.8×10^{-11}	10.32

Which combination should be selected, and what should the ratio of acid to conjugate base be?

Strategy Use either the general equation for a buffer (Equation 18.1) or the Henderson–Hasselbalch equation (Equation 18.2). Equation 18.1 informs you that $[H_3O^+]$ should be close to the acid K_a value, and Equation 18.2 tells you that pH should be close to the acid pK_a value. This will establish which acid you will use.

Having decided which acid to use, convert pH to $[H_3O^+]$ to use Equation 18.1. If you use Equation 18.2, use the pK_a value in the table. Finally, calculate the ratio of acid to conjugate base.

Solution The hydronium ion concentration for the buffer is found from the targeted pH.

$$pH = 4.30, \text{ so } [H_3O^+] = 10^{-pH} = 10^{-4.30} = 5.0 \times 10^{-5} \text{ M}$$

Of the acids given, only acetic acid (CH_3CO_2H) has a K_a value close to that of the desired $[H_3O^+]$ (or a pK_a close to a pH of 4.30). Now you need only to adjust the ratio $[CH_3CO_2H]/[CH_3CO_2^-]$ to achieve the desired hydronium ion concentration.

$$[H_3O^+] = 5.0 \times 10^{-5} \text{ M} = \frac{[CH_3CO_2H]}{[CH_3CO_2^-]}(1.8 \times 10^{-5})$$

Rearrange this equation to find the ratio $[CH_3CO_2H]/[CH_3CO_2^-]$.

$$\frac{[CH_3CO_2H]}{[CH_3CO_2^-]} = \frac{K_a}{[H_3O^+]} = \frac{5.0 \times 10^{-5}}{1.8 \times 10^{-5}} = \frac{2.8 \text{ mol/L}}{1.0 \text{ mol/L}}$$

Therefore, if you add 0.28 mol of acetic acid and 0.10 mol of sodium acetate (or any other pair of molar quantities in the ratio 2.8/1) to enough water to make 1.0 L of solution, the buffer solution will have a pH of 4.30.

Comment If you prefer to use the Henderson–Hasselbalch equation, you would have

$$pH = 4.30 = 4.74 + \log\frac{[CH_3CO_2^-]}{[CH_3CO_2H]}$$

$$\log\frac{[CH_3CO_2^-]}{[CH_3CO_2H]} = 4.30 - 4.74 = -0.44$$

$$\frac{[CH_3CO_2^-]}{[CH_3CO_2H]} = 10^{-0.44} = 0.36$$

The ratio of conjugate base to acid, $[CH_3CO_2^-]/[CH_3CO_2H]$, is 0.36. The reciprocal of this ratio $\{= [CH_3CO_2H]/[CH_3CO_2^-] = 1/0.36)\}$ is 2.8/1. This is the same result obtained previously using Equation 18.1.

EXERCISE 18.5 Preparing a Buffer Solution

Using an acetic acid/sodium acetate buffer solution, what ratio of acid to conjugate base will you need to maintain the pH at 5.00? Describe how you would make up such a solution.

Example 18.4 illustrates several important points concerning buffer solutions. The hydronium ion concentration depends not only on the K_a value of the acid but also on the ratio of acid and conjugate base concentrations. However, even though we write these ratios in terms of reagent concentrations, it is *the relative number of moles of acid and conjugate base that is important in determining the pH of a buffer solution.* Because both reagents are dissolved in the same solution, their concentrations depend on the same solution volume. In Example 18.4, the ratio 2.8/1 for acetic acid and sodium acetate implies that 2.8 times as many moles of acid were dissolved per liter as moles of sodium acetate.

$$\frac{[CH_3CO_2H]}{[CH_3CO_2^-]} = \frac{2.8 \text{ mol } CH_3CO_2H/L}{1.0 \text{ mol } CH_3CO_2^-/L} = \frac{2.8 \text{ mol } CH_3CO_2H}{1.0 \text{ mol } CH_3CO_2^-}$$

Notice that on dividing one concentration by the other, the volumes "cancel." This means that we only need to ensure that the ratio of moles of acid to moles of conjugate base is 2.8 to 1 in this example. The acid and its conjugate base could have been dissolved in any reasonable amount of water. This also means that *diluting a buffer solution will not change its pH.* Commercially available buffer solutions are often sold as premixed, dry ingredients. To use them, you only need to mix the ingredients in some volume of pure water (Figure 18.3).

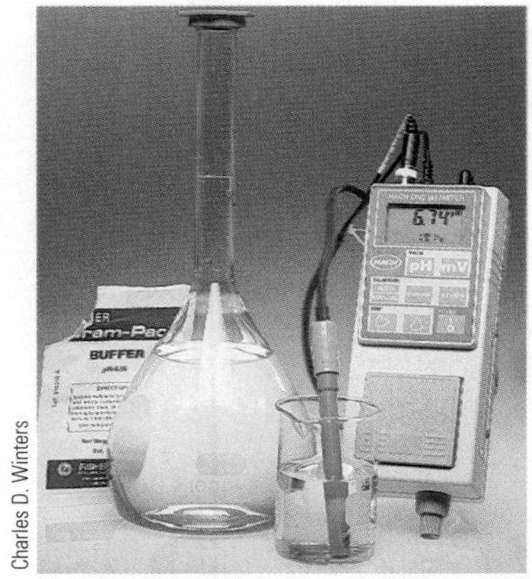

Charles D. Winters

FIGURE 18.3 A commercial buffer solution. The solid acid and conjugate base in the packet are mixed with water to give a solution with the indicated pH. The quantity of water used does not matter because the ratio [acid]/[conjugate base] does not depend on the solution volume. (However, if too much water is added, the acid and conjugate base concentrations will be too low, and the buffer capacity could be exceeded. Again, buffer solutions usually have solute concentrations around 0.1 M to 1.0 M.)

Problem Solving Tip 18.1 Buffer Solutions

The following is a summary of important aspects of buffer solutions.

- A buffer resists changes in pH on adding small quantities of strong acid or base.
- A buffer contains a weak acid and its conjugate base.
- The hydronium ion concentration of a buffer solution can be calculated from Equation 18.1,

$$[H_3O^+] = \frac{[\text{acid}]}{[\text{conjugate base}]} \times K_a$$

or the pH can be calculated from the Henderson–Hasselbalch equation (Equation 18.2).

$$pH = pK_a + \log\frac{[\text{conjugate base}]}{[\text{acid}]}$$

- The pH depends primarily on the K_a of the weak acid and secondarily on the relative amounts of acid and conjugate base.
- The function of the weak acid of a buffer is to consume added base, and the conjugate base consumes added acid. Such reactions affect the relative quantities of weak acid and its conjugate base. Because this ratio of acid to its conjugate base has only a secondary effect on the pH, the pH can be maintained relatively constant.
- The buffer must have sufficient capacity to react with reasonable quantities of added acid or base.

How Does a Buffer Maintain pH?

Now let us explore quantitatively how a given buffer solution can maintain the pH of a solution on adding a small amount of strong acid.

Chemistry Now™

Sign in at **www.thomsonedu.com/login** and go to Chapter 18 Contents to see Screen 18.6 for a tutorial on **adding reagents to a buffer solution.**

EXAMPLE 18.5 How Does a Buffer Maintain a Constant pH?

Problem What is the change in pH when 1.00 mL of 1.00 M HCl is added to (1) 1.000 L of pure water and to (2) 1.000 L of acetic acid/sodium acetate buffer with $[CH_3CO_2H] = 0.700$ M and $[CH_3CO_2^-] = 0.600$ M? (See Example 18.2, where the pH of this acetic acid/acetate ion buffer was found to be 4.68.)

Strategy HCl is a strong acid and therefore ionizes completely to supply H_3O^+ ions. Part 1 involves two steps: (a) Find the H_3O^+ concentration when diluting 1.00 mL of acid to 1.00 L. (b) Convert the value of $[H_3O^+]$ for the dilute solution to pH.

Part 2 involves three steps: (a) A stoichiometry calculation to find how the concentrations of acid and conjugate base change on adding H_3O^+. (b) An equilibrium calculation to find $[H_3O^+]$ for a buffer solution where the concentrations of CH_3CO_2H and $CH_3CO_2^-$ are slightly altered owing to the reaction of $CH_3CO_2^-$ with added H_3O^+. (c) Conversion of $[H_3O^+]$ to pH.

Solution

Part 1: *Adding Acid to Pure Water*

1.00 mL of 1.00 M HCl represents 0.00100 mol of acid. If this is added to 1.000 L of pure water, the H_3O^+ concentration of the water changes from 10^{-7} to 10^{-3},

$$c_1 \times V_1 = c_2 \times V_2$$

$$(1.00\text{ M})(0.00100\text{ L}) = c_2 \times (1.001\text{ L})$$

$$c_2 = [H_3O^+] \text{ in diluted solution} = 1.00 \times 10^{-3}\text{ M}$$

and so the pH falls from 7.00 to 3.00.

Part 2: *Adding Acid to an Acetic Acid/Acetate Buffer Solution*

HCl is a strong acid that is 100% ionized in water and supplies H_3O^+, which reacts completely with the base (acetate ion) in the buffer solution according to the following equation:

$$H_3O^+(aq) + CH_3CO_2^-(aq) \rightleftharpoons H_2O(\ell) + CH_3CO_2H(aq)$$

	$[H_3O^+]$ from Added HCl	$[CH_3CO_2^-]$ from Buffer	$[CH_3CO_2H]$ from Buffer
Initial amount of acid or base (mol = $c \times V$)	0.00100	0.600	0.700
Change (mol)	−0.00100	−0.00100	+0.00100
Equilibrium (mol)	0	0.599	0.700
Concentrations after reaction (c = mol/V)	0	0.598	0.700

Because the added HCl reacts completely with acetate ion to produce acetic acid, the solution after this reaction (with $V = 1.001$ L) is once again a buffer containing only the weak acid and its salt. Now we only need to use Equation 18.1 (or the Henderson–Hasselbalch equation) to find $[H_3O^+]$ and the pH in the buffer solution as in Examples 18.2 and 18.3.

Equilibrium	CH_3CO_2H + H_2O	$\rightleftharpoons$	H_3O^+	+	$CH_3CO_2^-$
Initial (M)	0.700		0		0.598
Change (M)	$-x$		$+x$		$+x$
Equilibrium (M)	$0.700 - x$		x		$0.598 + x$

As usual, we make the approximation that x, the amount of H_3O^+ formed by ionizing acetic acid in the presence of acetate ion, is very small compared with 0.700 M or 0.598 M. Using Equation 18.1, we calculate a pH of 4.68.

$$[H_3O^+] = x = \frac{[CH_3CO_2H]}{[CH_3CO_2^-]} \times K_a = \left(\frac{0.700 \text{ mol}}{0.598 \text{ mol}}\right)(1.8 \times 10^{-5}) = 2.1 \times 10^{-5} \text{ M}$$

$$\text{pH} = 4.68$$

Comment Within the number of significant figures allowed, the pH of the buffer solution does not change after adding HCl. The buffer solution contains the conjugate base of the weak acid, and the base consumed the added HCl. In contrast, pH changed by 4 units when 1 mL of 1.0 M HCl was added to 1.0 L of pure water.

EXERCISE 18.6 Buffer Solutions

Calculate the pH of 0.500 L of a buffer solution composed of 0.50 M formic acid (HCO_2H) and 0.70 M sodium formate ($NaHCO_2$) before and after adding 10.0 mL of 1.0 M HCl.

18.3 Acid–Base Titrations

A titration is one of the most important ways of determining accurately the quantity of an acid, a base, or some other substance in a mixture or of ascertaining the purity of a substance. You learned how to perform the stoichiometry calculations involved in titrations in Chapter 4 (◀ Section 4.7). In Chapter 17, we described the following important points regarding acid–base reactions (◀ Section 17.6):

- The pH at the equivalence point of a strong acid–strong base titration is 7. The solution at the equivalence point is truly "neutral" *only* when a strong acid is titrated with a strong base and vice versa.
- If the substance being titrated is a weak acid or base, then the pH at the equivalence point is not 7 (see Table 17.5).
  (a) A weak acid titrated with strong base leads to pH > 7 at the equivalence point due to the conjugate base of the weak acid.
  (b) A weak base titrated with strong acid leads to pH < 7 at the equivalence point due to the conjugate acid of the weak base.

A knowledge of buffer solutions and how they work will now allow us to more fully understand how the pH changes in the course of an acid–base reaction.

■ **Equivalence Point** The equivalence point for a reaction is the point at which one reactant has been completely consumed by addition of another reactant. See page 185.

■ **Weak Acid–Weak Base Titrations** Titrations combining a weak acid and weak base are generally not done because the equivalence point often cannot be accurately judged.

Case Study — Take a Deep Breath!

Maintenance of pH is vital to the cells of all living organisms because enzyme activity is influenced by pH. The primary protection against harmful pH changes in cells is provided by buffer systems. The intracellular pH of most cells is maintained in a range between 6.9 and 7.4. Two important biological buffer systems control pH in this range: the bicarbonate/carbonic acid system (HCO_3^-/H_2CO_3) and the phosphate system ($HPO_4^{2-}/H_2PO_4^-$).

The bicarbonate/carbonic acid buffer is important in blood plasma. Three equilibria are important here.

$$CO_2(g) \rightleftharpoons CO_2(\text{dissolved})$$

$$CO_2(\text{dissolved}) + H_2O(\ell) \rightleftharpoons H_2CO_3(aq)$$

$$H_2CO_3(aq) + H_2O(\ell) \rightleftharpoons H_3O^+(aq) + HCO_3^-(aq)$$

The overall equilibrium constant for the second and third steps is $pK_{overall} = 6.3$ at 37 °C, the temperature of the human body. Thus,

$$7.4 = 6.3 + \log\frac{[HCO_3^-]}{[CO_2(\text{dissolved})]}$$

Although the value of $pK_{overall}$ is about 1 pH unit away from the blood pH, the natural partial pressure of CO_2 in the alveoli of the lungs (about 40 mm Hg) is sufficient to keep $[CO_2(\text{dissolved})]$ at about 1.2×10^{-3} M and $[HCO_3^-]$ at about 1.5×10^{-2} M.

If blood pH rises above about 7.45, you can suffer from a condition called *alkalosis*. *Respiratory alkalosis* can arise from hyperventilation when a person breathes quickly to expel CO_2 from the lungs. This has the effect of lowering the CO_2 concentration, which in turn leads to a lower H_3O^+ concentration and a higher pH. This same condition can also arise from severe anxiety or from an oxygen deficiency at high altitude. It can ultimately lead to overexcitability of the central nervous system, muscle spasms, convulsions, and death. One way to treat acute respiratory alkalosis is to breathe into a paper bag. The CO_2 you exhale is recycled. This raises the blood CO_2 level and causes the equilibria above to shift to the right, thus raising the hydronium ion concentration and lowering the pH.

Charles D. Winters

Alkalosis. If blood pH is too high, alkalosis results. Respiratory alkalosis can be reversed by breathing into a bag, an action that recycles exhaled CO_2. This affects the carbonic acid buffer system in the body, raising the blood hydronium ion concentration. The blood pH drops back to a more normal level of 7.4.

Metabolic alkalosis can occur if you take large amounts of sodium bicarbonate to treat stomach acid (which is mostly HCl at a pH of about 4). It also commonly occurs when a person vomits profusely. This depletes the body of hydrogen ions, which leads to an increase in bicarbonate ion concentration.

Athletes can use the H_2CO_3/HCO_3^- equilibrium to enhance their performance. Strenuous activity produces high levels of lactic acid, and this can lower blood pH and cause muscle cramps. To counteract this, athletes will prepare before the race by hyperventilating for some seconds to raise blood pH, thereby helping to neutralize the acidity from the lactic acid.

Acidosis is the opposite of alkalosis. A toddler who came to the hospital with viral gastroenteritis had *metabolic acidosis*. He had severe diarrhea, was dehydrated, and had a high rate of respiration. One function of the bicarbonate ion is to neutralize stomach acid in the intestines. However, because of his diarrhea, the toddler was losing bicarbonate ions in his stool, and his blood pH was too low. To compensate, the toddler was breathing rapidly and blowing off CO_2 through the lungs (the effect of which is to lower $[H_3O^+]$ and raise the pH).

Respiratory acidosis results from a buildup of CO_2 in the body. This can be caused by pulmonary problems, by head injuries, or by drugs such as anesthetics and sedatives. It can be reversed by breathing rapidly and deeply. Doubling the breathing rate increases the blood pH by about 0.23 units.

Questions:

Phosphate ions are abundant in cells, both as the ions themselves and as important substituents on organic molecules. Most importantly, the pK_a for the $H_2PO_4^-$ ion is 7.20, which is very close to the high end of the normal pH range in the body.

$$H_2PO_4^-(aq) + H_2O(\ell) \rightleftharpoons H_3O^+(aq) + HPO_4^{2-}(aq)$$

1. *What should the ratio $[HPO_4^{2-}]/[H_2PO_4^-]$ be to control the pH at 7.4?*
2. *A typical total phosphate concentration in a cell, $[HPO_4^{2-}] + [H_2PO_4^-]$, is 2.0×10^{-2} M. What are the concentrations of HPO_4^{2-} and $H_2PO_4^-$?*

Answers to these questions are in Appendix Q.

Titration of a Strong Acid with a Strong Base

Figure 18.4 illustrates what happens to the pH as 0.100 M NaOH is slowly added to 50.0 mL of 0.100 M HCl.

$$HCl(aq) + NaOH(aq) \rightarrow NaCl(aq) + H_2O(\ell)$$

$$\text{Net ionic equation: } H_3O^+(aq) + OH^-(aq) \rightarrow 2\,H_2O(\ell)$$

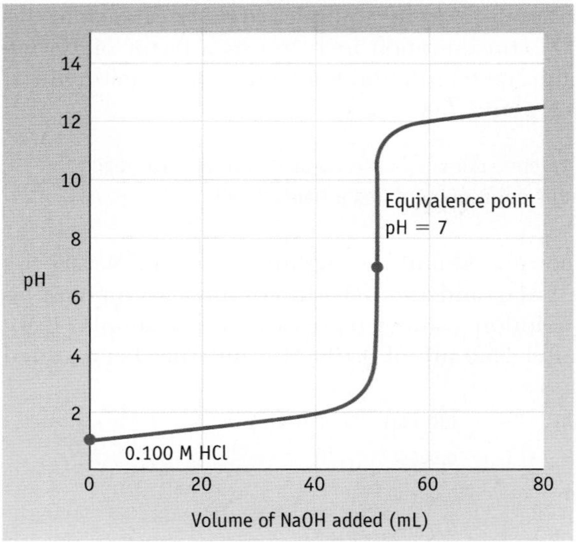

50.0 mL of 0.100 M HCl titrated with 0.100 M NaOH

Volume of base added	pH
0.0	1.00
10.0	1.18
20.0	1.37
40.0	1.95
45.0	2.28
48.0	2.69
49.0	3.00
50.0	7.00
51.0	11.00
55.0	11.68
60.0	11.96
80.0	12.36
100.0	12.52
very large amount	13.00 (maximum)

FIGURE 18.4 The change in pH as a strong acid is titrated with a strong base. In this case, 50.0 mL of 0.100 M HCl is titrated with 0.100 M NaOH. The pH at the equivalence point is 7.0 for the reaction of a strong acid with a strong base.

Let us focus on four regions on this plot.

- pH of the initial solution
- pH as NaOH is added to the HCl solution before the equivalence point
- pH at the equivalence point
- pH after the equivalence point

Before beginning the titration, the 0.100 M solution of HCl has a pH of 1.00. As NaOH is added to the acid solution, the amount of HCl declines, and the acid remaining is dissolved in an ever-increasing volume of solution. Thus, $[H_3O^+]$ decreases, and the pH slowly increases. As an example, let us find the pH of the solution after 10.0 mL of 0.100 M NaOH has been added to 50.0 mL of 0.100 M HCl. Here, we set up a table to list the amounts of acid and base before reaction, the changes in those amounts, and the amounts remaining after reaction. Be sure to notice that the volume of the solution after reaction is the sum of the combined volumes of NaOH and HCl (60.0 mL or 0.0600 L in this case).

	$H_3O^+(aq)$	+	$OH^-(aq)$	$\rightarrow$ $2\ H_2O(\ell)$
Initial amount (mol = $c \times V$)	0.00500		0.00100	
Change (mol)	−0.00100		−0.00100	
After reaction (mol)	0.00400		0	
After reaction (c = mol/V)	0.00400 mol/0.0600 L = 0.0667 M		0	

After addition of 10.0 mL of NaOH, the final solution has a hydronium ion concentration of 0.0667 M, and so the pH is

$$pH = -\log[H_3O^+] = -\log(0.0667) = 1.176$$

After 49.5 mL of base has been added—that is, just before the equivalence point—we can use the same approach to show that the pH is 3.3. The solution being titrated is still quite acidic, even very close to the equivalence point.

The pH of the equivalence point in an acid–base titration is taken as the midpoint in the vertical portion of the pH versus volume of titrant curve. (The **titrant** is the substance being added during the titration.) In the HCl/NaOH titration illustrated in Figure 18.4, you see that the pH increases very rapidly near the equivalence point. In fact, in this case the pH rises 7 units (the H_3O^+ concentration increases by a factor of 10 million!) when only a drop or two of the NaOH solution is added, and the midpoint of the vertical portion of the curve is a pH of 7.00.

The pH of the solution at the equivalence point in a strong acid–strong base reaction is always 7.00 (at 25 °C) because the solution contains a neutral salt.

After all of the HCl has been consumed and the slightest excess of NaOH has been added, the solution will be basic, and the pH will continue to increase as more NaOH is added (and the solution volume increases). For example, if we calculate the pH of the solution after 55.0 mL of 0.100 M NaOH has been added to 50.0 mL of 0.100 M HCl, we find

	$H_3O^+(aq)$	+	$OH^-(aq)$	$\rightarrow$	$2\ H_2O(\ell)$
Initial amount (mol = $c \times V$)	0.00500		0.00550		
Change (mol)	−0.00500		−0.00500		
After reaction (mol)	0		0.00050		
After reaction (c = mol/V)	0		0.00050 mol/0.1050 L = 0.0048 M		

At this point, the solution has a hydroxide ion concentration of 0.0048 M. Calculate the pOH from this value, then use this to calculate pH.

$$pOH = -\log[OH^-] = -\log(0.0048) = 2.32$$

$$pH = 14.00 - pOH = 11.68$$

EXERCISE 18.7 Titration of a Strong Acid with a Strong Base

What is the pH after 25.0 mL of 0.100 M NaOH has been added to 50.0 mL of 0.100 M HCl? What is the pH after 50.50 mL of NaOH has been added?

Titration of a Weak Acid with a Strong Base

The titration of a weak acid with a strong base is somewhat different from the strong acid–strong base titration. Look carefully at the curve for the titration of 100.0 mL of 0.100 M acetic acid with 0.100 M NaOH (Figure 18.5).

$$CH_3CO_2H(aq) + NaOH(aq) \rightarrow NaCH_3CO_2(aq) + H_2O(\ell)$$

Let us focus on three important points on this curve:

- *The pH before titration begins.* The pH before any base is added can be calculated from the weak acid K_a value and the acid concentration (◀ Example 17.5).
- *The pH at the equivalence point.* At the equivalence point, the solution contains only sodium acetate, the CH_3CO_2H and NaOH having been consumed. The pH is controlled by the acetate ion, the conjugate base of acetic acid (◀ Table 17.5, page 728).
- *The pH at the halfway point (half-equivalence point) of the titration.* Here, the pH is equal to the pK_a of the weak acid, a conclusion that is discussed in more detail below.

As NaOH is added to acetic acid the base is consumed and sodium acetate is produced. Thus, at every point between the beginning of the titration (when only acetic acid is present) and the equivalence point (when only sodium acetate is present), the solution contains both acetic acid and its salt, sodium acetate. These

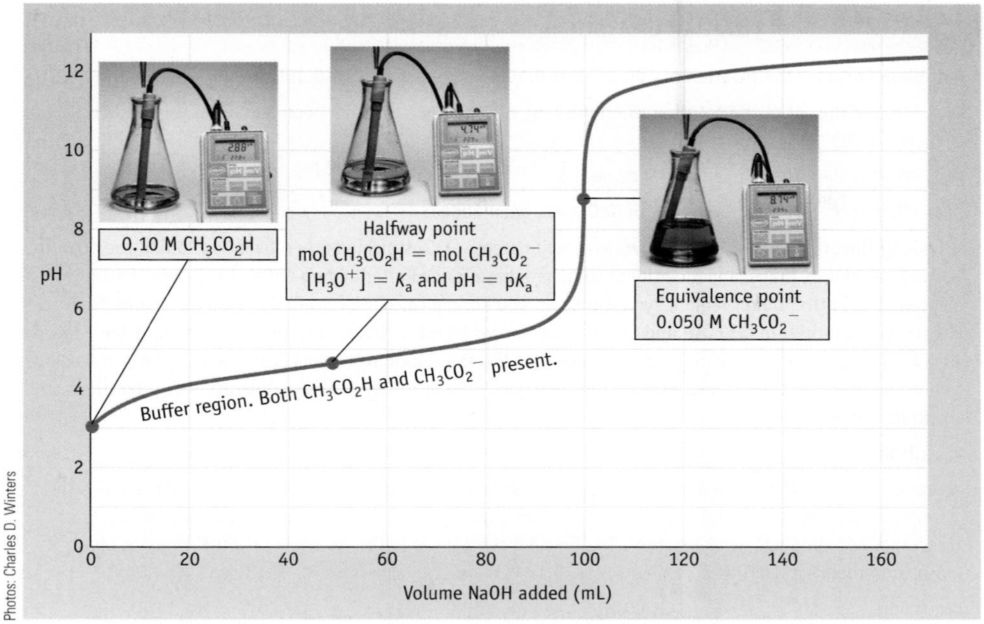

Active Figure 18.5 The change in pH during the titration of a weak acid with a strong base. Here, 100.0 mL of 0.100 M acetic acid is titrated with 0.100 M NaOH. Note especially the following: (a) Acetic acid is a weak acid, so the pH of the original solution is 2.87. (b) The pH at the point at which half the acid has reacted with base is equal to the pK_a for the acid (pH = pK_a = 4.74). (c) At the equivalence point, the solution contains the acetate ion, a weak base. Therefore, the solution is basic, with a pH of 8.72.

Chemistry Now™ Sign in at www.thomsonedu.com/login and go to the Chapter Contents menu to explore an interactive version of this figure accompanied by an exercise.

are the components of a *buffer solution,* and the hydronium ion concentration can be calculated from Equation 18.3 or 18.4.

$$[H_3O^+] = \frac{[\text{weak acid remaining}]}{[\text{conjugate base produced}]} \times K_a \quad (18.3)$$

or

$$pH = pK_a + \log\frac{[\text{conjugate base produced}]}{[\text{weak acid remaining}]} \quad (18.4)$$

The fact that a buffer is present at any point between the beginning of the titration and the equivalence point is the reason that the pH of the solution only rises slowly after a few milliliters of titrant has been added.

What happens when *exactly* half of the acid has been consumed by base? Half of the acid (CH_3CO_2H) has been converted to the conjugate base ($CH_3CO_2^-$), and half remains. Therefore, the concentration of weak acid remaining is equal to the concentration of conjugate base produced ($[CH_3CO_2H] = [CH_3CO_2^-]$). Using Equations 18.3 or 18.4, we see that

$$[H_3O^+] = (1) \times K_a \quad \text{or} \quad pH = pK_a + \log(1)$$

Because log(1) = 0, we come to the following general conclusion:

At the halfway point in the titration of a weak acid with a strong base

$$[H_3O^+] = K_a \text{ and } pH = pK_a \quad (18.5)$$

In the particular case of the titration of acetic acid with a strong base, $[H_3O^+] = 1.8 \times 10^{-5}$ M at the halfway point, and so the pH is 4.74. This is equal to the pK_a of acetic acid.

Chemistry Now™

Sign in at **www.thomsonedu.com/login** and go to Chapter 18 Contents to see Screen 18.7 for a simulation and tutorial on **titration curves.**

EXAMPLE 18.6 Titration of Acetic Acid with Sodium Hydroxide

Problem Consider the titration of 100.0 mL of 0.100 M acetic acid with 0.100 M NaOH (see Figure 18.5).

(a) What is the pH of the solution when 90.0 mL of 0.100 M NaOH has been added to 100.0 mL of 0.100 M acetic acid?

(b) What is the pH at the equivalence point?

(c) What is the pH after 110.0 mL of NaOH has been added?

Strategy The problem generally involves two major steps: (1) A stoichiometry calculation to find the quantity of acid remaining, if any, and quantity of conjugate base formed after adding NaOH. (2) Before the equivalence point, in the "buffer region," you will do an equilibrium calculation to find $[H_3O^+]$ for a buffer solution where the quantities of CH_3CO_2H and $CH_3CO_2^-$ are known from the first part of the calculation. At the equivalence point, only the conjugate base will remain, so the calculation resembles Example 17.7. After the equivalence point, the solution contains both the conjugate base of the weak acid and excess NaOH, but the latter controls the pH.

Solution

Part a: *pH before the equivalence point.* Let us first calculate the amounts of reactants before reaction (= concentration × volume) and then use the principles of stoichiometry to calculate the amounts of reactants and products after reaction. The limiting reactant is NaOH, so some CH_3CO_2H remains along with the product, $CH_3CO_2^-$.

Equation	CH_3CO_2H	+	OH^-	$\rightleftharpoons$	$CH_3CO_2^-$	+	H_2O
Initial (mol)	0.0100		0.00900		0		
Change (mol)	−0.00900		−0.00900		+0.00900		
After reaction (mol)	0.0010		0		0.00900		

The ratio of amounts (moles) of acid to conjugate base is the same as the ratio of their concentrations. Therefore, we can use the amounts of weak acid remaining and conjugate base formed to find the pH from Equation 18.3 (where we use amounts and not concentrations).

$$[H_3O^+] = \frac{\text{mol } CH_3CO_2H}{\text{mol } CH_3CO_2^-} \times K_a = \left(\frac{0.0010 \text{ mol}}{0.0090 \text{ mol}}\right)(1.8 \times 10^{-5}) = 2.0 \times 10^{-6} \text{ M}$$

$$pH = -\log(2.0 \times 10^{-6}) = 5.70$$

The pH is 5.70, in agreement with Figure 18.5. Notice that this pH is appropriate for a point after the halfway point (4.74) but before the equivalence point (8.72).

Part b: *pH at the equivalence point.* To reach the equivalence point, 0.0100 mol of NaOH has been added to 0.0100 mol of CH_3CO_2H and 0.0100 mol of $CH_3CO_2^-$ has been formed.

Equation	CH_3CO_2H	+	OH^-	$\rightleftharpoons$	$CH_3CO_2^-$	+	H_2O
Initial (mol)	0.0100		0.0100		0		
Change (mol)	−0.0100		−0.0100		+0.0100		
After reaction (mol)	0		0		0.0100		

Because two solutions, each with a volume of 100.0 mL, have been combined, the concentration of $CH_3CO_2^-$ at the equivalence point is (0.0100 mol/0.200 L) = 0.0050 M. Next, we set up an ICE table for the hydrolysis of this weak base,

Equation	$CH_3CO_2^-$	+	H_2O	$\rightleftharpoons$	CH_3CO_2H	+	OH^-
Initial (M)	0.00500				0		0
Change (M)	$-x$				$+x$		$+x$
After reaction (M)	$0.00500 - x$				x		x

and calculate the concentration of OH^- ion using K_b for the weak base.

$$K_b \text{ for } CH_3CO_2^- = 5.6 \times 10^{-10} = \frac{[CH_3CO_2H][OH^-]}{[CH_3CO_2^-]} = \frac{(x)(x)}{0.00500 - x}$$

Making the usual assumption that x is small with respect to 0.00500 M,

$$x = [OH^-] = 1.7 \times 10^{-6} \text{ M (pOH} = 5.78)$$

$$pH = 14.00 - 5.78 = 8.22$$

Part c: *pH after the equivalence point.* Now the limiting reactant is the CH_3CO_2H, and the solution contains excess OH^- ion from the unused NaOH as well as from the hydrolysis of $CH_3CO_2^-$.

Equation	CH_3CO_2H	+	OH^-	$\rightleftharpoons$	$CH_3CO_2^-$	+	H_2O
Initial (mol)	0.0100		0.0110		0		
Change (mol)	−0.0100		−0.0100		+0.0100		
After reaction (mol)	0		0.0010		0.0100		

The amount of OH^- produced by $CH_3CO_2^-$ hydrolysis is very small (see part b), so the pH of the solution after the equivalence point is determined by the excess NaOH (in 210 mL of solution).

$$[OH^-] = 0.0010 \text{ mol}/0.210 \text{ L} = 4.8 \times 10^{-3} \text{ M (pOH} = 2.32)$$

$$pH = 14.00 - 2.32 = 11.68$$

EXERCISE 18.8 Titration of a Weak Acid with a Strong Base

The titration of 0.100 acetic acid with 0.100 M NaOH is described in the text. What is the pH of the solution when 35.0 mL of the base has been added to 100.0 mL of 0.100 M acetic acid?

Titration of Weak Polyprotic Acids

The titrations illustrated thus far have been for the reaction of a monoprotic acid (HA) with a base such as NaOH. It is possible to extend the discussion of titrations to polyprotic acids such as oxalic acid, $H_2C_2O_4$.

$$H_2C_2O_4(aq) + H_2O(\ell) \rightleftharpoons HC_2O_4^-(aq) + H_3O^+(aq) \qquad K_{a1} = 5.9 \times 10^{-2}$$

$$HC_2O_4^-(aq) + H_2O(\ell) \rightleftharpoons C_2O_4^-(aq) + H_3O^+(aq) \qquad K_{a2} = 6.4 \times 10^{-5}$$

Figure 18.6 illustrates the curve for the titration of 100 mL of 0.100 M oxalic acid with 0.100 M NaOH. The first significant rise in pH is experienced after 100 mL of base has been added, indicating that the first proton of the acid has been titrated.

$$H_2C_2O_4(aq) + OH^-(aq) \rightleftharpoons HC_2O_4^-(aq) + H_2O(aq)$$

When the second proton of oxalic acid is titrated, the pH again rises significantly.

$$HC_2O_4^-(aq) + OH^-(aq) \rightleftharpoons C_2O_4^{2-}(aq) + H_2O(\ell)$$

The pH at this second equivalence point is controlled by the oxalate ion, $C_2O_4^{2-}$.

$$C_2O_4^{2-}(aq) + H_2O(\ell) \rightleftharpoons HC_2O_4^-(aq) + OH^-(aq)$$

$$K_b = K_w/K_{a2} = 1.6 \times 10^{-10}$$

Calculation of the pH at the equivalence point indicates that it should be about 8.4, as observed.

Chemistry Now™

Sign in at **www.thomsonedu.com/login** and go to Chapter 18 Contents to see Screen 18.8 for a simulation on **titration of a weak polyprotic acid.**

FIGURE 18.6 Titration curve for a diprotic acid. The curve for the titration of 100.0 mL of 0.100 M oxalic acid ($H_2C_2O_4$, a weak diprotic acid) with 0.100 M NaOH. The first equivalence point (at 100 mL) occurs when the first hydrogen ion of $H_2C_2O_4$ is titrated, and the second (at 200 mL) occurs at the completion of the reaction. The curve for pH versus volume of NaOH added shows an initial rise at the first equivalence point and then another rise at the second equivalence point.

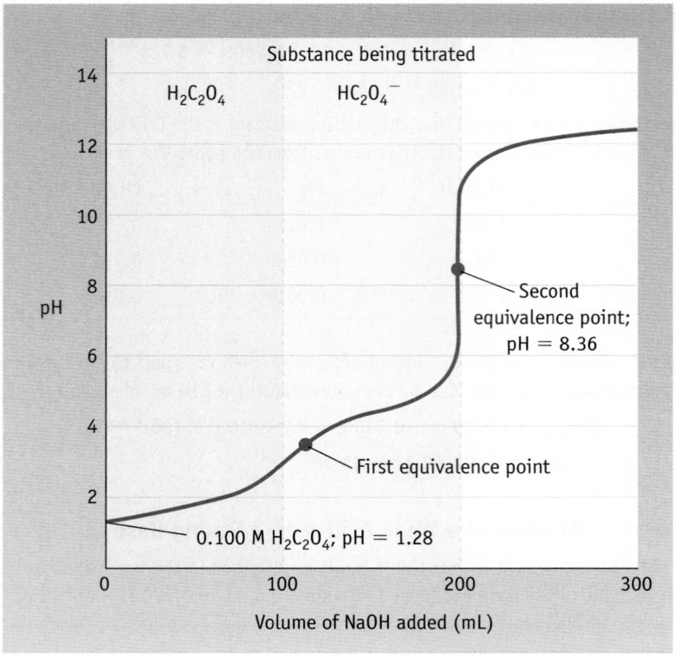

Titration of a Weak Base with a Strong Acid

Finally, it is useful to consider the titration of a weak base with a strong acid. Figure 18.7 illustrates the pH curve for the titration of 100.0 mL of 0.100 M NH_3 with 0.100 M HCl.

$$NH_3(aq) + H_3O^+(aq) \rightleftharpoons NH_4^+(aq) + H_2O(\ell)$$

The initial pH for a 0.100 M NH_3 solution is 11.12. As the titration progresses, the important species in solution are the weak acid NH_4^+ and its conjugate base, NH_3.

$$NH_4^+(aq) + H_2O(\ell) \rightleftharpoons NH_3(aq) + H_3O^+(aq) \qquad K_a = 5.6 \times 10^{-10}$$

FIGURE 18.7 Titration of a weak base with a strong acid. The change in pH during the titration of a weak base (100.0 mL of 0.100 M NH_3) with a strong acid (0.100 M HCl). The pH at the half-neutralization point is equal to the pK_a for the conjugate acid (NH_4^+) of the weak base (NH_3) (pH = pK_a = 9.26). At the equivalence point, the solution contains the NH_4^+ ion, a weak acid, so the pH is about 5.

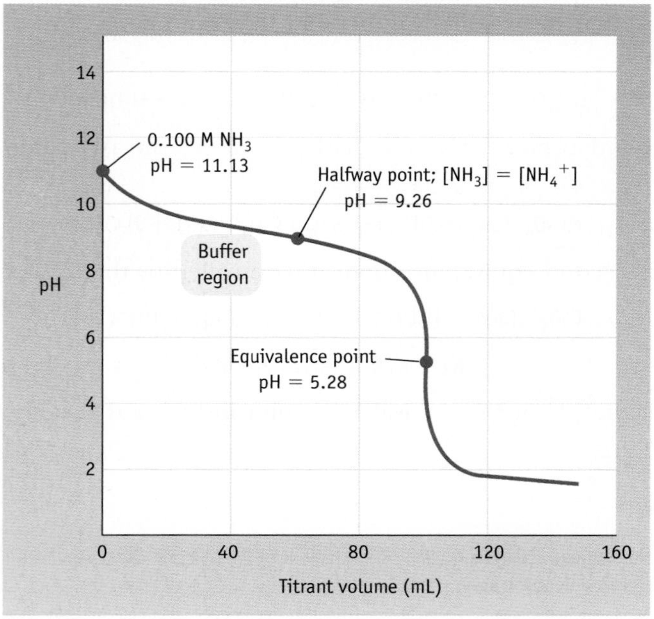

Problem Solving Tip 18.2 Calculating the pH at Various Stages of an Acid–Base Reaction

Finding the pH at or before the equivalence point for an acid–base reaction always involves several calculation steps. There are no shortcuts. Consider the *titration of a weak base, B, with a strong acid* as in Example 18.7. (The same principles apply to other acid–base reactions.)

$$H_3O^+(aq) + B(aq) \rightleftharpoons BH^+(aq) + H_2O(\ell)$$

Step 1. *Solve the stoichiometry problem.* Up to the equivalence point, acid is consumed completely to leave a solution containing some base (B) and its conjugate acid (BH^+). Use the principles of stoichiometry to calculate (a) the amount of acid added, (b) the amount of base consumed, and (c) the amount of conjugate base (BH^+) formed.

Step 2. *Calculate the concentrations of base, [B], and conjugate acid, [BH^+].* Recognize that the volume of the solution at any point is the sum of the original volume of base solution plus the volume of acid solution added.

Step 3. *Calculate the pH before the equivalence point.* At any point before the equivalence point, the solution is a buffer solution because both the base and its conjugate acid are present. Calculate [H_3O^+] using the concentrations of Step 2 and the value of K_a for the conjugate acid of the weak base.

Step 4. *Calculate the pH at the equivalence point.* Calculate the concentration of the conjugate acid using the procedure of Steps 1 and 2. Use the value of K_a for the conjugate acid of the weak base and the procedure outlined in Example 18.7 (or in Example 18.6 for a weak acid). (For a titration of a weak acid with a strong base, use the value of K_b for the conjugate base of the acid and follow the procedure outlined in Example 18.6.)

At the halfway point, the concentrations of NH_4^+ and NH_3 are the same, so

$$[H_3O^+] = \frac{[NH_4^+]}{[NH_3]} \times K_a = 5.6 \times 10^{-10}$$

$$[H_3O^+] = K_a$$

$$pH = pK_a = -\log(5.6 \times 10^{-10}) = 9.25$$

As the addition of HCl to NH_3 continues, the pH declines slowly because of the buffering action of the NH_3/NH_4^+ combination. Near the equivalence point, however, the pH drops rapidly. At the equivalence point, the solution contains only ammonium chloride, a weak Brønsted acid, and the solution is weakly acidic.

Chemistry Now™

Sign in at **www.thomsonedu.com/login** and go to Chapter 18 Contents to see Screen 18.9 for a simulation on **titration of a weak base with a strong acid.**

EXAMPLE 18.7 Titration of Ammonia with HCl

Problem What is the pH of the solution at the equivalence point in the titration of 100.0 mL of 0.100 M ammonia with 0.100 M HCl (see Figure 18.7)?

Strategy This problem has two steps: (a) A stoichiometry calculation to find the concentration of NH_4^+ at the equivalence point. (b) An equilibrium calculation to find [H_3O^+] for a solution of the weak acid NH_4^+.

Solution

Part 1: *Stoichiometry Problem*

Here, we are titrating 0.0100 mol of NH_3 ($= c \times V$), so 0.0100 mol of HCl is required. Thus, 100.0 mL of 0.100 M HCl (= 0.0100 mol HCl) must be used in the titration.

Equation	NH_3	+	H_3O^+	$\rightleftharpoons$	NH_4^+	+	H_2O
Initial (mol $= c \times V$)	0.0100		0.0100		0		
Change on reaction (mol)	−0.0100		−0.0100		+0.0100		
After reaction (mol)	0		0		0.0100		
Concentration (M)	0		0		0.0100 mol (in 0.200L) = 0.0500 M		

Part 2: *Equilibrium Problem*

When the equivalence point is reached, the solution consists of 0.0500 M NH_4^+. The pH is determined by the hydrolysis of the acid.

Equation	NH_4^+ + H_2O	$\rightleftharpoons$	NH_3	+ H_3O^+
Initial (M)	0.0500		0	0
Change (M)	$-x$		$+x$	$+x$
Equilibrium (M)	$0.0500 - x$		x	x

Using K_a for the weak acid NH_4^+, we have

$$K_a = 5.6 \times 10^{-10} = \frac{[NH_3][H_3O^+]}{[NH_4^+]} = \frac{x}{0.0500 - x}$$

$$\text{Simplifying, } x = [H_3O^+] = \sqrt{(5.6 \times 10^{-10})(0.0500)} = 5.3 \times 10^{-6} \text{ M}$$

$$\text{pH} = 5.28$$

The pH at the equivalence point in this weak base–strong acid titration is indeed slightly acidic, as expected.

EXERCISE 18.9 Titration of a Weak Base with a Strong Acid

Calculate the pH after 75.0 mL of 0.100 M HCl has been added to 100.0 mL of 0.100 M NH_3. See Figure 18.7.

pH Indicators

Many organic compounds, both natural and synthetic, have a color that changes with pH (Figure 18.8). Not only does this add beauty and variety to our world, but it is also a useful property in chemistry.

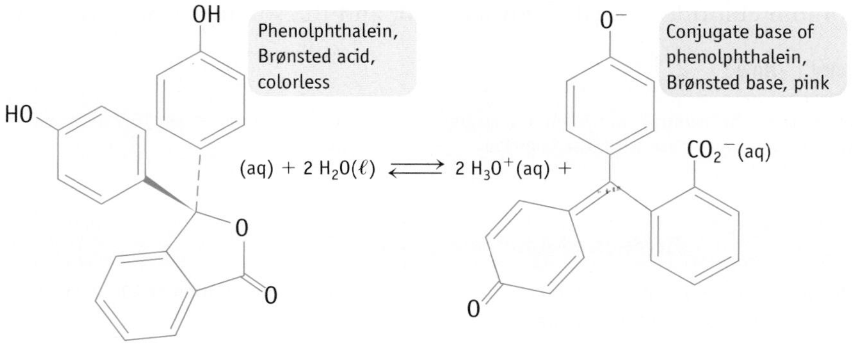

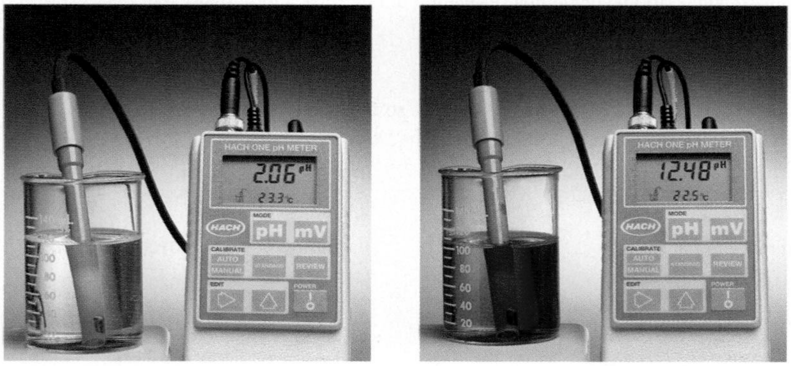

Photos: Charles D. Winters

FIGURE 18.8 Phenolphthalein, a common acid–base indicator. Phenolphthalein, a weak acid, is colorless. As the pH increases, the pink conjugate base form predominates, and the color of the solution changes. The change in color is most noticeable around pH 9. The dye is commonly used for strong acid + strong base or weak acid + strong base titrations because the pH changes from 3–4 to 10–12 in these cases. For other suitable indicator dyes, see Figure 18.10.

You have likely carried out an acid–base titration in the laboratory, and, before starting the titration, you would have added an **indicator.** The acid–base indicator is usually an organic compound that is itself a weak acid or weak base (similar to the compounds that give the color to flowers). In aqueous solution, the acid form is in equilibrium with its conjugate base. Abbreviating the indicator's acid formula as HInd and the formula of its conjugate base as Ind^-, we can write the equilibrium equation

$$\text{HInd(aq)} + \text{H}_2\text{O}(\ell) \rightleftharpoons \text{H}_3\text{O}^+\text{(aq)} + \text{Ind}^-\text{(aq)}$$

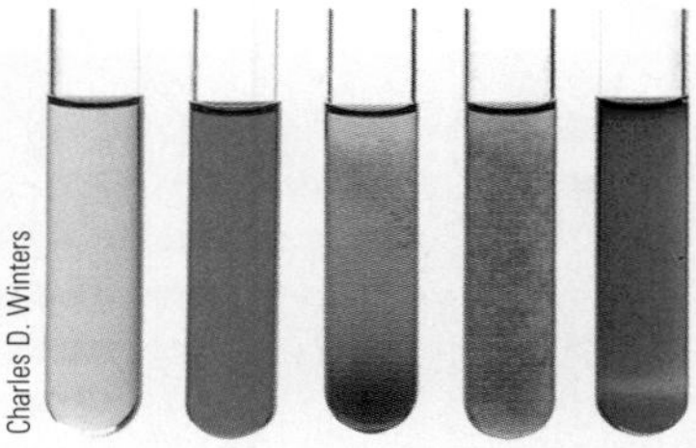

Extract of red rose at various pH's. (a) The pigment in red rose petals was extracted with ethanol; the extract was a faint red. (b) After adding one drop of 6 M HCl, the color changed to a vivid red. (c) Adding two drops of 6 M NH_3 produced a green color, and (d) adding 1 drop each of HCl and NH_3 (to give a buffer solution) gave a blue solution. (e) Finally, adding a few milligrams of $Al(NO_3)_3$ turned the solution deep purple. (The deep purple color with aluminum ions was so intense that the solution had to be diluted significantly to take the photo.)

The important characteristic of acid–base indicators is that the acid form of the compound (HInd) has one color and the conjugate base (Ind^-) has another. To see how such compounds can be used as equivalence point indicators, let us write the usual equilibrium constant expressions for the dependence of hydronium ion concentration or pH on the indicator's ionization constant (K_a) and on the relative quantities of the acid and conjugate base.

$$[\text{H}_3\text{O}^+] = \frac{[\text{HInd}]}{[\text{Ind}^-]} \times K_a \quad \text{or} \quad \text{pH} = \text{p}K_a + \log\frac{[\text{Ind}^-]}{[\text{HInd}]} \qquad \textbf{(18.6)}$$

These equations inform us that

- when the hydronium ion concentration is equivalent to the value of K_a (or when pH = pK_a), then [HInd] = [Ind^-]
- when [H_3O^+] > K_a (or pH < pK_a), then [HInd] > [Ind^-]
- when [H_3O^+] < K_a (or pH > pK_a), then [HInd] < [Ind^-]

Now let us apply these conclusions to, for example, the titration of an acid with a base using an indicator whose pK_a value is nearly the same as the pH at the equivalence point (Figure 18.9). At the beginning of the titration, the pH is low and [H_3O^+] is high; the acid form of the indicator (HInd) predominates. Its color is the one observed. As the titration progresses and the pH increases ([H_3O^+] decreases), less of the acid HInd and more of its conjugate base exist in solution. Finally, just after we reach the equivalence point, [Ind^-] is much larger than [HInd], and the color of [Ind^-] is observed.

Several obvious questions remain to be answered. If you are trying to analyze for an acid and add an indicator that is a weak acid, won't this affect the analysis? Recall that you use only a tiny amount of an indicator in a titration. Although the acidic indicator molecules also react with the base as the titration progresses, so little indicator is present that the analysis is not in error.

Another question is whether you could accurately determine the pH by observing the color change of an indicator. In practice, your eyes are not quite that good. Usually, you see the color of HInd when [HInd]/[Ind^-] is about 10/1, and the color of Ind^- when [HInd]/[Ind^-] is about 1/10. This means the color change is observed over a hydronium ion concentration interval of about 2 pH units. However, as you can see in Figures 18.4–18.7, on passing through the equivalence point of these titrations, the pH changes by as many as 7 units.

As Figure 18.10 shows, a variety of indicators are available, each changing color in a different pH range. If you are analyzing a weak acid or base by titration, you must choose an indicator that changes color in a range that includes the pH to be observed at the equivalence point. This means that an indicator that changes color in the pH range 7 ± 2 should be used for a strong acid–strong base titration. On the other hand, the pH at the equivalence point in the titration of a weak acid with a strong base is greater than 7, and you should choose an indicator that changes color at a pH near the anticipated equivalence point.

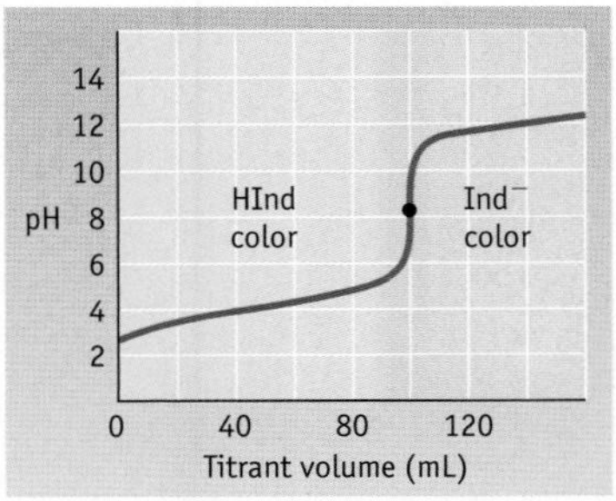

FIGURE 18.9 Indicator color changes in the course of a titration when the pK_a of the indicator HInd is about 8.

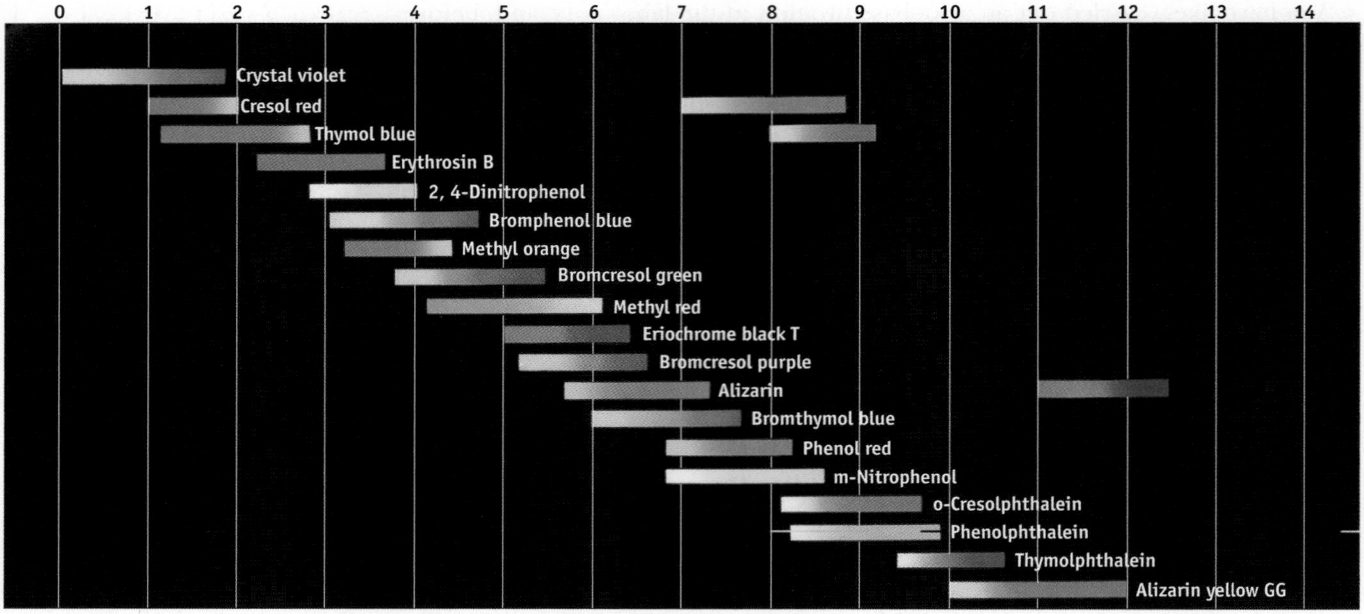

Hach Company

FIGURE 18.10 Common acid–base indicators. The color changes occur over a range of pH values. Notice that a few indicators have color changes over two different pH ranges.

Chemistry Now™

Sign in at **www.thomsonedu.com/login** and go to Chapter 18 Contents to see Screen 18.10 for a simulation on **acid–base indicators.**

EXERCISE 18.10 Indicators

Use Figure 18.10 to decide which indicator is best to use in the titration of NH_3 with HCl shown in Figure 18.7.

18.4 Solubility of Salts

Precipitation reactions (◀ Section 3.6) are exchange reactions in which one of the products is a water-insoluble compound such as $CaCO_3$,

$$CaCl_2(aq) + Na_2CO_3(aq) \rightarrow CaCO_3(s) + 2\ NaCl(aq)$$

that is, a compound having a water solubility of less than about 0.01 mole of dissolved material per liter of solution (Figure 18.11).

How do you know when to predict an insoluble compound as the product of a reaction? In Chapter 3, we listed some guidelines for predicting solubility (Figure 3.10) and mentioned a few important minerals that are insoluble in water. Now we want to make our estimates of solubility more quantitative and to explore conditions under which some compounds precipitate and others do not.

The Solubility Product Constant, K_{sp}

Silver bromide, AgBr, is used in photographic film (Figure 18.11c). If some AgBr is placed in pure water, a tiny amount of the compound dissolves, and an equilibrium is established.

$$AgBr(s) \rightleftharpoons Ag^+(aq,\ 7.35 \times 10^{-7}\ M) + Br^-(aq,\ 7.35 \times 10^{-7}\ M)$$

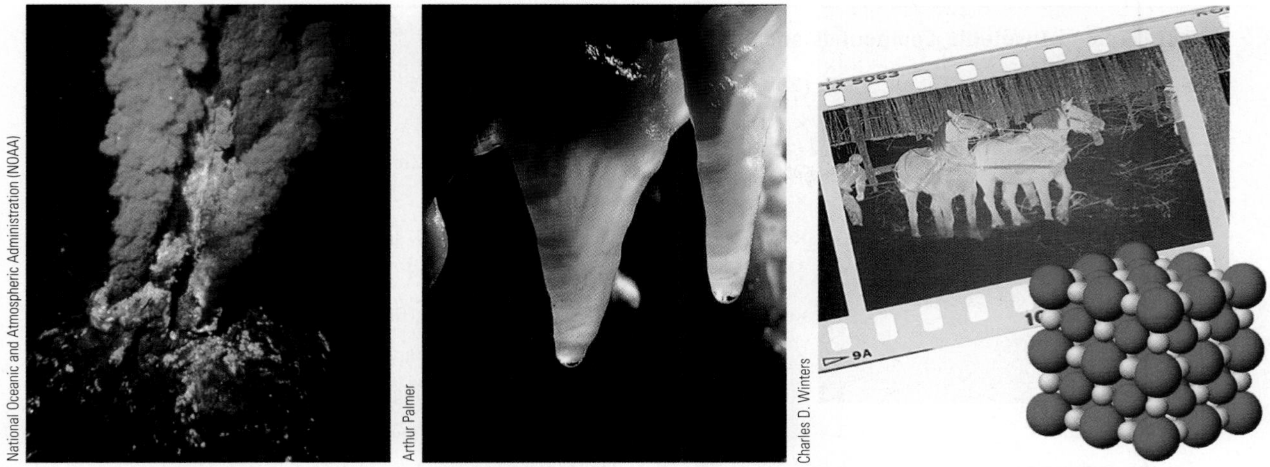

(a) Metal sulfides (and hydroxides) in a black smoker (◀ page 140).

(b) $CaCO_3$ stalactites.

(c) Black-and-white film is coated with water-insoluble silver bromide. The image is formed by metallic silver particles.

FIGURE 18.11 Some insoluble substances.

When sufficient AgBr has dissolved and equilibrium is attained, the solution is said to be **saturated** (◀ Section 14.2), and experiments show that the concentrations of the silver and bromide ions in the solution are each about 7.35×10^{-7} M at 25 °C. The extent to which an insoluble salt dissolves can be expressed in terms of the equilibrium constant for the dissolving process. In this case, the appropriate expression is

$$K_{sp} = [Ag^+][Br^-]$$

■ **Writing Equilibrium Constant Expressions** Solids are not included in these equations. See page 728.

The value of the equilibrium constant that reflects the solubility of a compound is often referred to as its **solubility product constant.** Chemists often use the notation $\boldsymbol{K_{sp}}$ for such constants, the subscript "sp" denoting a "solubility product."

The water solubility of a compound, and thus its K_{sp} value, can be estimated by determining the concentration of the cation or anion when the compound dissolves. For example, if we find that AgBr dissolves to give a silver ion concentration of 7.35×10^{-7} mol/L, we know that 7.35×10^{-7} mol of AgBr must have dissolved per liter of solution (and that the bromide ion concentration also equals 7.35×10^{-7} M). Therefore, the calculated value of the equilibrium constant for the dissolving process is

$$K_{sp} = [Ag^+][Br^-] = (7.35 \times 10^{-7})(7.35 \times 10^{-7}) = 5.40 \times 10^{-13} \text{ (at 25 °C)}$$

Equilibrium constants for the dissolving of other insoluble salts can be calculated in the same manner.

The solubility product constant, K_{sp}, for any salt always has the form

$$A_xB_y(s) \rightleftharpoons x\,A^{y+}(aq) + y\,B^{x-}(aq) \qquad K_{sp} = [A^{y+}]^x[B^{x-}]^y$$

For example,

$$CaF_2(s) \rightleftharpoons Ca^{2+}(aq) + 2\,F^-(aq) \qquad K_{sp} = [Ca^{2+}][F^-]^2 = 5.3 \times 10^{-11}$$

$$Ag_2SO_4(s) \rightleftharpoons 2\,Ag^+(aq) + SO_4^{2-}(aq) \qquad K_{sp} = [Ag^+]^2[SO_4^{2-}] = 1.2 \times 10^{-5}$$

The numerical values of K_{sp} for a few salts are given in Table 18.2, and more values are collected in Appendix J.

TABLE 18.2 Some Common Insoluble Compounds and Their K_{sp} Values*

Formula	Name	K_{sp} (25 °C)	Common Names/Uses
$CaCO_3$	Calcium carbonate	3.4×10^{-9}	Calcite, iceland spar
$MnCO_3$	Manganese(II) carbonate	2.3×10^{-11}	Rhodochrosite (forms rose-colored crystals)
$FeCO_3$	Iron(II) carbonate	3.1×10^{-11}	Siderite
CaF_2	Calcium fluoride	5.3×10^{-11}	Fluorite (source of HF and other inorganic fluorides)
AgCl	Silver chloride	1.8×10^{-10}	Chlorargyrite
AgBr	Silver bromide	5.4×10^{-13}	Used in photographic film
$CaSO_4$	Calcium sulfate	4.9×10^{-5}	The hydrated form is commonly called gypsum
$BaSO_4$	Barium sulfate	1.1×10^{-10}	Barite (used in "drilling mud" and as a component of paints)
$SrSO_4$	Strontium sulfate	3.4×10^{-7}	Celestite
$Ca(OH)_2$	Calcium hydroxide	5.5×10^{-5}	Slaked lime

* The values in this table were taken from *Lange's Handbook of Chemistry,* 15th edition, McGraw-Hill Publishers, New York, NY (1999). Additional K_{sp} values are given in Appendix J.

Do not confuse the *solubility* of a compound with its *solubility product constant.* The *solubility* of a salt is the quantity present in some volume of a saturated solution, expressed in moles per liter, grams per 100 mL, or other units. The *solubility product constant* is an equilibrium constant. Nonetheless, there is a connection between them: If one is known, the other can be calculated.

Chemistry Now™

Sign in at **www.thomsonedu.com/login** and go to Chapter 18 Contents to see:
- Screen 18.11 for a review of **precipitation reactions**
- Screen 18.12 for a simulation on **solubility product constant**

EXERCISE 18.11 Writing K_{sp} Expressions

Write K_{sp} expressions for the following insoluble salts and look up numerical values for the constant in Appendix J.

(a) AgI **(b)** BaF_2 **(c)** Ag_2CO_3

Charles D. Winters

Fluorite. The mineral fluorite is water-insoluble calcium fluoride. The mineral can vary widely in color from purple to green and to colorless. The colors are likely due to impurities.

Relating Solubility and K_{sp}

Solubility product constants are determined by careful laboratory measurements of the concentrations of ions in solution.

Chemistry Now™

Sign in at **www.thomsonedu.com/login** and go to Chapter 18 Contents to see Screen 18.13 for a tutorial on **determining K_{sp} experimentally.**

EXAMPLE 18.8 K_{sp} from Solubility Measurements

Problem Calcium fluoride, the main component of the mineral fluorite, dissolves to a slight extent in water.

$$CaF_2(s) \rightleftharpoons Ca^{2+}(aq) + 2\,F^-(aq) \qquad K_{sp} = [Ca^{2+}]\,[F^-]^2$$

Calculate the K_{sp} value for CaF_2 if the calcium ion concentration has been found to be 2.3×10^{-4} mol/L.

Strategy We first write the K_{sp} expression for CaF_2 and then substitute the numerical values for the equilibrium concentrations of the ions.

Solution When CaF_2 dissolves to a small extent in water, the balanced equation shows that the concentration of F^- ion must be twice the Ca^{2+} ion concentration.

$$\text{If } [Ca^{2+}] = 2.3 \times 10^{-4} \text{ M, then } [F^-] = 2 \times [Ca^{2+}] = 4.6 \times 10^{-4} \text{ M}$$

This means the solubility product constant is

$$K_{sp} = [Ca^{2+}]\,[F^-]^2 = (2.3 \times 10^{-4})(4.6 \times 10^{-4})^2 = 4.9 \times 10^{-11}$$

EXERCISE 18.12 K_{sp} from Solubility Measurements

The barium ion concentration, $[Ba^{2+}]$, in a saturated solution of barium fluoride is 7.5×10^{-3} M. Calculate the value of the K_{sp} for BaF_2.

$$BaF_2(s) \rightleftharpoons Ba^{2+}(aq) + 2\,F^-(aq)$$

K_{sp} values for insoluble salts can be used to estimate the solubility of a solid salt or to determine whether a solid will precipitate when solutions of its anion and cation are mixed. Let us first look at an example of the estimation of the solubility of a salt from its K_{sp} value.

Chemistry Now™

Sign in at **www.thomsonedu.com/login** and go to Chapter 18 Contents to see Screen 18.14 for a tutorial on **estimating salt solubility using K_{sp}.**

EXAMPLE 18.9 Solubility from K_{sp}

Problem The K_{sp} for $BaSO_4$ (as the mineral barite, Figure 18.12) is 1.1×10^{-10} at 25 °C. Calculate the solubility of barium sulfate in pure water in (a) moles per liter and (b) grams per liter.

Strategy When 1 mol of $BaSO_4$ dissolves, 1 mol of Ba^{2+} ions and 1 mol of SO_4^{2-} ions are produced. Thus, the solubility of $BaSO_4$ can be estimated by calculating the concentration of either Ba^{2+} or SO_4^{2-} from the solubility product constant.

Solution The equation for the solubility of $BaSO_4$ is

$$BaSO_4(s) \rightleftharpoons Ba^{2+}(aq) + SO_4^{2-}(aq) \qquad K_{sp} = [Ba^{2+}][SO_4^{2-}] = 1.1 \times 10^{-10}$$

Let us denote the solubility of $BaSO_4$ (in mol/L) by x; that is, x moles of $BaSO_4$ dissolve per liter. Therefore, both $[Ba^{2+}]$ and $[SO_4^{2-}]$ must also equal x at equilibrium.

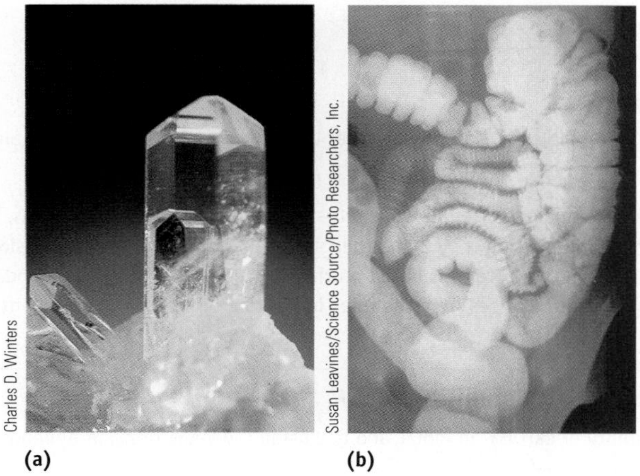

FIGURE 18.12 Barium sulfate. Barium sulfate, a white solid, is quite insoluble in water ($K_{sp} = 1.1 \times 10^{-10}$) (see Example 18.9). (a) A sample of the mineral barite, which is mostly barium sulfate. (b) Barium sulfate is opaque to x-rays, so it is used by physicians to examine the digestive tract. A patient drinks a "cocktail" containing $BaSO_4$, and the progress of the $BaSO_4$ through the digestive organs can be followed by x-ray analysis. This photo is an x-ray of a gastrointestinal tract after a person ingested barium sulfate. It is fortunate that $BaSO_4$ is so insoluble, because water- and acid-soluble barium salts are toxic.

Equation	$BaSO_4(s)$	$\rightleftharpoons$	$Ba^{2+}(aq)$	+	$SO_4^{2-}(aq)$
Initial (M)			0		0
Change (M)			$+x$		$+x$
Equilibrium (M)			x		x

Because K_{sp} is the product of the barium and sulfate ion concentrations, K_{sp} is the square of the solubility, x,

$$K_{sp} = [Ba^{2+}][SO_4^{2-}] = 1.1 \times 10^{-10} = (x)(x) = x^2$$

and so the value of x is

$$x = [Ba^{2+}] = [SO_4^{2-}] = \sqrt{1.1 \times 10^{-10}} = 1.0 \times 10^{-5}\text{ M}$$

The solubility of $BaSO_4$ in pure water is 1.0×10^{-5} mol/L. To find its solubility in g/L, we need only multiply by the molar mass of $BaSO_4$.

$$\text{Solubility in g/L} = (1.0 \times 10^{-5}\text{ mol/L})(233\text{ g/mol}) = 0.0024\text{ g/L}$$

EXAMPLE 18.10 Solubility from K_{sp}

Problem Knowing that the K_{sp} value for MgF_2 is 5.2×10^{-11}, calculate the solubility of the salt in (a) moles per liter and (b) grams per liter.

Strategy The problem is to define the salt solubility in terms that will allow us to solve the K_{sp} expression for this value. We know that, if 1 mol of MgF_2 dissolves, 1 mol of Mg^{2+} and 2 mol of F^- appear in the solution. This means the MgF_2 solubility (in moles dissolved per liter) is equivalent to the concentration of Mg^{2+} ions in the solution. Thus, if the solubility of MgF_2 is x mol/L, then $[Mg^{2+}] = x$ and $[F^-] = 2x$.

Solution We begin by writing the equilibrium equation and the K_{sp} expression,

$$MgF_2(s) \rightleftharpoons Mg^{2+}(aq) + 2\,F^-(aq) \qquad K_{sp} = [Mg^{2+}][F^-]^2 = 5.2 \times 10^{-11}$$

and then set up an ICE table.

Equation	$MgF_2(s)$	$\rightleftharpoons$	$Mg^{2+}(aq)$	+	$2\,F^-(aq)$
Initial (M)			0		0
Change (M)			$+x$		$+2x$
Equilibrium (M)			x		$2x$

Substituting these values into the K_{sp} expression, we find

$$K_{sp} = [Mg^{2+}][F^-]^2 = (x)(2x)^2 = 4x^3$$

Solving the equation for x,

$$x = \sqrt[3]{\frac{K_{sp}}{4}} = \sqrt[3]{\frac{5.2 \times 10^{-11}}{4}} = 2.4 \times 10^{-4}$$

we find that 2.4×10^{-4} moles of MgF_2 dissolve per liter. The solubility of MgF_2 in grams per liter is

$$(2.4 \times 10^{-4}\text{ mol/L})(62.3\text{ g/mol}) = 0.015\text{ g } MgF_2/L$$

Comment Problems like this one often provoke our students to ask such questions as, "Aren't you counting things twice when you multiply x by 2 and then square it as well?" in the expression $K_{sp} = (x)(2x)^2$. The answer is no. The 2 in the $2x$ term is based on the stoichiometry of the compound. The exponent of 2 on the F^- ion concentration arises from the rules for writing equilibrium expressions.

EXERCISE 18.13 Salt Solubility from K_{sp}

Calculate the solubility of $Ca(OH)_2$ in mol/L and g/L using the value of K_{sp} in Appendix J.

A Closer Look — Solubility Calculations

The K_{sp} value reported for lead(II) chloride, $PbCl_2$, is 1.7×10^{-5}. If we assume the appropriate equilibrium in solution is

$$PbCl_2(s) \rightleftharpoons Pb^{2+}(aq) + 2\ Cl^-(aq)$$

the calculated solubility of $PbCl_2$ is 0.016 M. The experimental value for the solubility of the salt, however, is 0.036 M, more than twice the calculated value! What is the problem? There are several, as summarized by the diagram below.

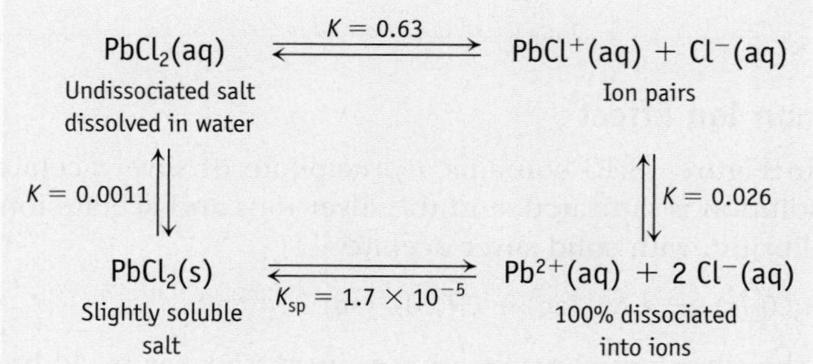

The main problem in the lead(II) chloride case, and in many others, is that the compound dissolves but is not 100% dissociated into its constituent ions. Instead, it dissolves as the undissociated salt or forms ion pairs.

Other problems that lead to discrepancies between calculated and experimental solubilities are the reactions of ions (particularly anions) with water and complex ion formation. An example of the former effect is the reaction of sulfide ion with water, that is, hydrolysis.

$$S^{2-}(aq) + H_2O(\ell) \rightleftharpoons HS^-(aq) + OH^-(aq)$$

This means that the solubility of a metal sulfide is better described by a chemical equation such as

$$NiS(s) + H_2O(\ell) \rightleftharpoons Ni^{2+}(aq) + HS^-(aq) + OH^-(aq)$$

Complex ion formation is illustrated by the fact that lead chloride is more soluble in the presence of excess chloride ion, owing to the formation of the complex ion $PbCl_4^{2-}$.

$$PbCl_2(s) + 2\ Cl^-(aq) \rightleftharpoons PbCl_4^{2-}(aq)$$

Hydrolysis and complex ion formation are discussed further on pages 840–841 and 847–848, respectively.

For further information on these issues see:

(a) L. Meites, J. S. F. Pode, and H. C. Thomas: *Journal of Chemical Education*, Vol. 43, pp. 667–672, 1966.
(b) S. J. Hawkes: *Journal of Chemical Education*, Vol. 75, pp. 1179–1181, 1998.
(c) R. W. Clark and J. M. Bonicamp: *Journal of Chemical Education*, Vol. 75, pp. 1182–1185, 1998.
(d) R. J. Myers: *Journal of Chemical Education*, Vol. 63, pp. 687–690, 1986.

The *relative* solubilities of salts can often be deduced by comparing values of solubility product constants, but you must be careful! For example, the K_{sp} for silver chloride is

$$AgCl(s) \rightleftharpoons Ag^+(aq) + Cl^-(aq) \qquad K_{sp} = 1.8 \times 10^{-10}$$

whereas that for silver chromate is

$$Ag_2CrO_4(s) \rightleftharpoons 2\ Ag^+(aq) + CrO_4^{2-}(aq) \qquad K_{sp} = 9.0 \times 10^{-12}$$

In spite of the fact that Ag_2CrO_4 has a numerically smaller K_{sp} value than AgCl, the chromate salt is about 10 times *more* soluble than the chloride salt. If you determine solubilities from K_{sp} values as in the examples above, you would find the solubility of AgCl is 1.3×10^{-5} mol/L, whereas that of Ag_2CrO_4 is 1.3×10^{-4} mol/L. From this example and countless others, we conclude that

> Direct comparisons of the solubility of two salts on the basis of their K_{sp} values can be made only for salts having the same cation-to-anion ratio.

This means, for example, that you can directly compare solubilities of 1:1 salts such as the silver halides by comparing their K_{sp} values.

$$AgI\ (K_{sp} = 8.5 \times 10^{-17}) < AgBr\ (K_{sp} = 5.4 \times 10^{-13}) < AgCl\ (K_{sp} = 1.8 \times 10^{-10})$$

increasing K_{sp} and increasing solubility →

■ **Comparing Solubilities** When two salts have different cation-anion ratios, you will have to do a calculation to find their solubilities. Simply comparing their K_{sp} values will not necessarily give their relative solubilities.

■ **Complex Ions** The solubility of a salt will increase if addition of the common ion leads to formation of a complex ion. (See Section 18.6, page 846)

Similarly, you could compare 1:2 salts such as the lead halides.

$$PbI_2\ (K_{sp} = 9.8 \times 10^{-9}) < PbBr_2\ (K_{sp} = 6.6 \times 10^{-6}) < PbCl_2\ (K_{sp} = 1.7 \times 10^{-5})$$

——— increasing K_{sp} and increasing solubility ———→

but you cannot directly compare a 1:1 salt (AgCl) with a 2:1 salt (Ag_2CrO_4).

EXERCISE 18.14 Comparing Solubilities

Using K_{sp} values, predict which salt in each pair is more soluble in water.

(a) AgCl or AgCN

(b) $Mg(OH)_2$ or $Ca(OH)_2$

(c) $Ca(OH)_2$ or $CaSO_4$

Solubility and the Common Ion Effect

The test tube on the left in Figure 18.13 contains a precipitate of silver acetate, $AgCH_3CO_2$, in water. The solution is saturated, and the silver ions and acetate ions in the solution are in equilibrium with solid silver acetate.

$$AgCH_3CO_2(s) \rightleftharpoons Ag^+(aq) + CH_3CO_2^-(aq)$$

But what would happen if the silver ion concentration is increased, say by adding silver nitrate? Le Chatelier's principle (◄ Section 16.6) suggests—and we observe—that more silver acetate precipitate should form because a product ion has been added, causing the equilibrium to shift to form more silver acetate.

The ionization of weak acids and bases is affected by the presence of an ion common to the equilibrium process (Section 18.1), and the effect of adding silver ions to a saturated silver acetate solution is another example of the common ion effect. Adding a common ion to a saturated solution of a salt will lower the salt solubility.

Chemistry Now™

Sign in at **www.thomsonedu.com/login** and go to Chapter 18 Contents to see Screen 18.15 for a simulation and tutorial on the **common ion effect.**

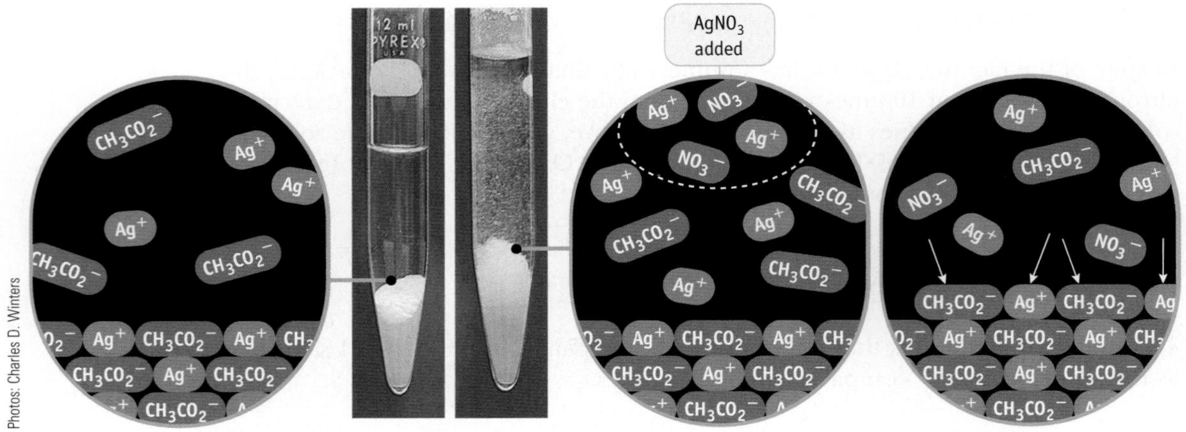

FIGURE 18.13 The common ion effect. The tube at the left contains a saturated solution of silver acetate, $AgCH_3CO_2$. When 1.0 M $AgNO_3$ is added to the tube (right), more solid silver acetate forms.

EXAMPLE 18.11 The Common Ion Effect and Salt Solubility

Problem If solid AgCl is placed in 1.00 L of 0.55 M NaCl, what mass of AgCl will dissolve?

Strategy The presence of an ion common to the equilibrium suppresses the solubility of a salt. To determine the solubility of the salt under these circumstances, calculate the concentration of the ion (Ag^+ ion in this case) other than the common ion (here Cl^-).

Solution In pure water, the solubility of AgCl is equal to either $[Ag^+]$ or $[Cl^-]$.

$$AgCl(s) \rightleftharpoons Ag^+(aq) + Cl^-(aq)$$

$$\text{Solubility of AgCl in pure water} = [Ag^+] \text{ or } [Cl^-] = \sqrt{K_{sp}} = 1.3 \times 10^{-5} \text{ mol/L or } 0.0019 \text{ g/L}$$

However, in water already containing a common ion, here the Cl^- ion, Le Chatelier's principle predicts that the solubility is less than 1.3×10^{-5} mol/L. In this case, the solubility of AgCl is equivalent to the concentration of Ag^+ ion in solution, so we set up an ICE table to show the concentrations of Ag^+ and Cl^- when equilibrium is attained.

Equation	AgCl(s)	$\rightleftharpoons$	$Ag^+(aq)$	+	$Cl^-(aq)$
Initial (M)			0		0.55
Change (M)			$+x$		$+x$
Equilibrium (M)			x		$0.55 + x$

Some AgCl dissolves in the presence of chloride ion and produces Ag^+ and Cl^- ion concentrations of x mol/L. Because some chloride ion was already present, the total chloride ion concentration is what was already there (0.55 M) plus the amount supplied by AgCl dissociation ($= x$).

The equilibrium concentrations from the table are substituted into the K_{sp} expression,

$$K_{sp} = 1.8 \times 10^{-10} = [Ag^+][Cl^-] = (x)(0.55 + x)$$

and rearranged to

$$x^2 + 0.55x - K_{sp} = 0$$

This is a quadratic equation and can be solved by the methods in Appendix A. An easier approach, however, is to make the approximation that x is very small with respect to 0.55 [and so $(0.55 + x) \approx 0.55$]. This is a reasonable assumption because we know that the solubility equals 1.3×10^{-5} M without the common ion Cl^- and that it will be even smaller in the presence of added Cl^-. Therefore,

$$K_{sp} = 1.8 \times 10^{-10} = (x)(0.55)$$

$$x = [Ag^+] = 3.3 \times 10^{-10} \text{ M}$$

The solubility in grams per liter is then

$$(3.3 \times 10^{-10} \text{ mol/L})(143 \text{ g/mol}) = 4.7 \times 10^{-8} \text{ g/L}$$

As predicted by Le Chatelier's principle, the solubility of AgCl in the presence of added Cl^- is less (3.3×10^{-10} M) than in pure water (1.3×10^{-5} M).

Comment As a final step, check the approximation by substituting the calculated value of x into the exact expression $K_{sp} = (x)(0.55 + x)$. If the product $(x)(0.55 + x)$ is the same as the given value of K_{sp}, the approximation is valid.

$$K_{sp} = (x)(0.55 + x) = (3.3 \times 10^{-10})(0.55 + 3.3 \times 10^{-10}) = 1.8 \times 10^{-10}$$

The approximation we made here is similar to the approximations we make in acid–base equilibrium problems.

EXAMPLE 18.12 The Common Ion Effect and Salt Solubility

Problem Calculate the solubility of silver chromate, Ag_2CrO_4, at 25 °C in the presence of 0.0050 M K_2CrO_4 solution.

$$Ag_2CrO_4(s) \rightleftharpoons 2\, Ag^+(aq) + CrO_4^{2-}(aq)$$

$$K_{sp} = [Ag^+]^2[CrO_4^{2-}] = 9.0 \times 10^{-12}$$

For comparison, the solubility of Ag_2CrO_4 in pure water is 1.3×10^{-4} mol/L.

Strategy In the presence of chromate ion from the water-soluble salt K_2CrO_4, the concentration of Ag^+ ions produced by Ag_2CrO_4 will be less than in pure water. Assume the solubility of Ag_2CrO_4 is x mol/L. This means the concentration of Ag^+ ions will be $2x$ mol/L, whereas the concentration of CrO_4^{2-} ions will be x mol/L plus the amount of CrO_4^{2-} already in the solution.

Solution

Equation	$Ag_2CrO_4(s) \rightleftharpoons$ 2 $Ag^+(aq)$	+	$CrO_4^{2-}(aq)$
Initial (M)	0		0.0050
Change	$+2x$		$+x$
Equilibrium (M)	$2x$		$0.0050 + x$

Substituting the equilibrium amounts into the K_{sp} expression, we have

$$K_{sp} = 9.0 \times 10^{-12} = [Ag^+]^2[CrO_4^{2-}]$$

$$K_{sp} = (2x)^2(0.0050 + x)$$

As in Example 18.11, you can make the approximation that x is very small with respect to 0.0050, and so $(0.0050 + x) \approx 0.0050$. (This is reasonable because $[CrO_4^{2-}]$ is 0.00013 M without added chromate ion, and it is certain that x is even smaller in the presence of extra chromate ion.) Therefore, the approximate expression is

$$K_{sp} = 9.0 \times 10^{-12} = [Ag^+]^2[CrO_4^{2-}] = (2x)^2(0.0050)$$

Solving, we find x, the solubility of silver chromate in the presence of excess chromate ion, is

$$x = \text{Solubility of } Ag_2CrO_4 = 2.1 \times 10^{-5}\text{ M}$$

Comment The silver ion concentration in the presence of the common ion is

$$[Ag^+] = 2x = 4.2 \times 10^{-5}\text{ M}$$

This silver ion concentration is indeed less than its value in pure water (2.6×10^{-4} M), owing to the presence of an ion "common" to the equilibrium.

EXERCISE 18.15 The Common Ion Effect and Salt Solubility

Calculate the solubility of $BaSO_4$ (a) in pure water and (b) in the presence of 0.010 M $Ba(NO_3)_2$. K_{sp} for $BaSO_4$ is 1.1×10^{-10}.

EXERCISE 18.16 The Common Ion Effect and Salt Solubility

Calculate the solubility of $Zn(CN)_2$ at 25 °C (a) in pure water and (b) in the presence of 0.10 M $Zn(NO_3)_2$. (K_{sp} for $Zn(CN)_2$ is 8.0×10^{-12}.)

Examples 18.11 and 18.12 allow us to propose two important general ideas:

- The solubility of a salt will be reduced by the presence of a common ion, in accordance with Le Chatelier's principle.
- We made the approximation that the amount of common ion added to the solution was very large in comparison with the amount of that ion coming from the insoluble salt, and this allowed us to simplify our calculations. This is almost always the case, but you should check to be sure.

The Effect of Basic Anions on Salt Solubility

The next time you are tempted to wash a supposedly insoluble salt down the kitchen or laboratory drain, stop and consider the consequences. Many metal ions such as lead, chromium, and mercury are toxic in the environment. Even if a so-called insoluble salt of one of these cations does not appear to dissolve, its solubility in water may be greater than you think, in part owing to the possibility that the anion of the salt is a weak base or the cation is a weak acid.

Lead sulfide, PbS, which is found in nature as the mineral galena (Figure 18.14), provides an example of the effect of the acid–base properties of an ion on salt solubility. When placed in water, a trace amount dissolves,

$$PbS(s) \rightleftharpoons Pb^{2+}(aq) + S^{2-}(aq)$$

and one product of the reaction is the sulfide ion. This anion is a strong base,

$$S^{2-}(aq) + H_2O(\ell) \rightleftharpoons HS^-(aq) + OH^-(aq) \qquad K_b = 1 \times 10^5$$

and it undergoes extensive hydrolysis (reaction with water) (◀ Table 17.3). The equilibrium process for dissolving PbS thus shifts to the right, and the lead ion concentration in solution is greater than expected from the simple ionization of the salt.

The lead sulfide example leads to the following general observation:

Any salt containing an anion that is the conjugate base of a weak acid will dissolve in water to a greater extent than given by K_{sp}.

This means that salts of phosphate, acetate, carbonate, and cyanide, as well as sulfide, can be affected, because all of these anions undergo the general hydrolysis reaction

$$X^-(aq) + H_2O(\ell) \rightleftharpoons HX(aq) + OH^-(aq)$$

The observation that ions from insoluble salts can undergo hydrolysis is related to another useful, general conclusion:

Insoluble salts in which the anion is the conjugate base of a weak acid dissolve in strong acids.

Insoluble salts containing such anions as acetate, carbonate, hydroxide, phosphate, and sulfide dissolve in strong acids. For example, you know that if a strong acid is added to a water-insoluble metal carbonate such as $CaCO_3$, the salt dissolves (◀ Section 3.4).

$$CaCO_3(s) + 2\ H_3O^+(aq) \rightarrow Ca^{2+}(aq) + 3\ H_2O(\ell) + CO_2(g)$$

You can think of this as the result of a series of reactions.

$$CaCO_3(s) \rightleftharpoons Ca^{2+}(aq) + CO_3^{2-}(aq) \qquad K_{sp} = 3.4 \times 10^{-9}$$

$$CO_3^{2-}(aq) + H_3O^+(aq) \rightleftharpoons HCO_3^-(aq) + H_2O(\ell) \qquad 1/K_{a2} = 1/4.8 \times 10^{-11} = 2.1 \times 10^{10}$$

$$HCO_3^-(aq) + H_3O^+(aq) \rightleftharpoons H_2CO_3(aq) + H_2O(\ell) \qquad 1/K_{a1} = 1/4.2 \times 10^{-7} = 2.4 \times 10^6$$

Overall: $CaCO_3(s) + 2\ H_3O^+(aq) \rightleftharpoons Ca^{2+}(aq) + 2\ H_2O(\ell) + H_2CO_3(aq)$

$$K_{net} = (K_{sp})(1/K_{a2})(1/K_{a1}) = 1.7 \times 10^8$$

Carbonic acid, a product of this reaction, is not stable,

$$H_2CO_3(aq) \rightleftharpoons CO_2(g) + H_2O(\ell) \qquad K \approx 10^5$$

and you see CO_2 bubbling out of the solution, a process that moves the $CaCO_3 + H_3O^+$ equilibrium even further to the right. Calcium carbonate dissolves completely in strong acid!

FIGURE 18.14 Lead sulfide (galena). This and other metal sulfides dissolve in water to a greater extent than expected because the sulfide ion reacts with water to form the very stable species HS^- and OH^-.

$$PbS(s) + H_2O(\ell) \rightleftharpoons Pb^{2+}(aq) + HS^-(aq) + OH^-(aq)$$

The model of PbS shows that the unit cell is cubic, a feature reflected by the cubic crystals of the mineral galena.

■ **Metal Sulfide Solubility** The true solubility of a metal sulfide is better represented by a modified solubility product constant, K_{spa}, which is defined as follows:

$$MS(s) \rightleftharpoons M^{2+}(aq) + S^{2-}(aq) \qquad K_{sp} = [M^{2+}][S^{2-}]$$

$$S^{2-}(aq) + H_2O(\ell) \rightleftharpoons HS^-(aq) + OH^-(aq) \qquad K_b = [HS^-][OH^-]/[S^{2-}]$$

Net reaction:

$$MS(s) + H_2O(\ell) \rightleftharpoons HS^-(aq) + M^{2+}(aq) + OH^-(aq)$$

$$K_{spa} = [M^{2+}][HS^-][OH^-] = K_{sp} \times K_b$$

Values for K_{spa} for several metal sulfides are included in Appendix J.

FIGURE 18.15 The effect of the anion on salt solubility in acid. (left) A precipitate of AgCl (white) and Ag_3PO_4 (yellow). (right) Adding a strong acid (HNO_3) dissolves Ag_3PO_4 (and leaves insoluble AgCl). The basic anion PO_4^{3-} reacts with acid to give H_3PO_4, whereas Cl^- is too weakly basic to form HCl.

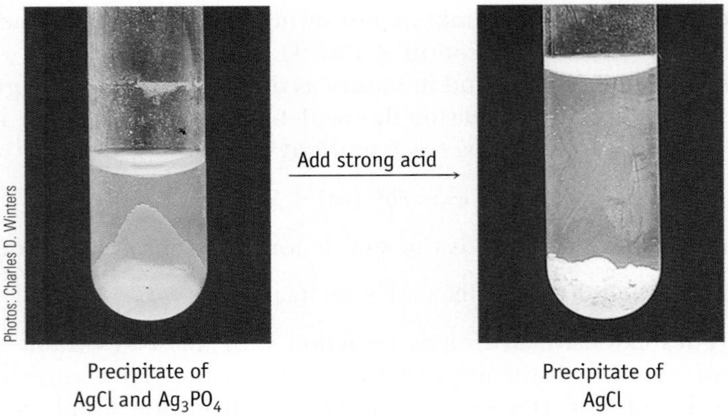

Many metal sulfides are also soluble in strong acids

$$FeS(s) + 2\ H_3O^+(aq) \rightleftharpoons Fe^{2+}(aq) + H_2S(aq) + 2\ H_2O(\ell)$$

as are metal phosphates (Figure 18.15),

$$Ag_3PO_4(s) + 3\ H_3O^+(aq) \rightleftharpoons 3\ Ag^+(aq) + H_3PO_4(aq) + 3\ H_2O(\ell)$$

and metal hydroxides.

$$Mg(OH)_2(s) + 2\ H_3O^+(aq) \rightleftharpoons Mg^{2+}(aq) + 4\ H_2O(\ell)$$

In general, the solubility of a salt containing the conjugate base of a weak acid is increased by addition of a stronger acid to the solution. In contrast, salts are not soluble in strong acid if the anion is the conjugate base of a strong acid. For example, AgCl is not soluble in strong acid

$$AgCl(s) \rightleftharpoons Ag^+(aq) + Cl^-(aq) \qquad K_{sp} = 1.8 \times 10^{-10}$$

$$H_3O^+(aq) + Cl^-(aq) \rightleftharpoons HCl(aq) + H_2O(\ell) \qquad K << 1$$

because Cl^- is a very weak base (◀ Table 17.3), and so its concentration is not lowered by a reaction with the strong acid H_3O^+ (Figure 18.15). This same conclusion would also apply to insoluble salts of Br^- and I^-.

Chemistry.Now™

Sign in at **www.thomsonedu.com/login** and go to Chapter 18 Contents to see Screen 18.16 for a self-study module on **solubility and pH.**

FIGURE 18.16 Minerals. Minerals are insoluble salts. The minerals shown here are light purple fluorite (calcium fluoride), black hematite [iron(III) oxide], and rust brown goethite, a mixture of iron(III) oxide and iron(III) hydroxide.

18.5 Precipitation Reactions

Metal-bearing ores contain the metal in the form of an insoluble salt (Figure 18.16), and, to complicate matters, ores often contain several such metal salts. Many industrial methods for separating metals from their ores involve dissolving metal salts to obtain the metal ion or ions in solution. The solution is then concentrated in some manner, and a precipitating agent is added to precipitate selectively only one type of metal ion as an insoluble salt. In the case of nickel, for example, the Ni^{2+} ion can be precipitated as insoluble nickel(II) sulfide or nickel(II) carbonate.

$$Ni^{2+}(aq) + HS^-(aq) + H_2O(\ell) \rightleftharpoons NiS(s) + H_3O^+(aq) \qquad K = 1.7 \times 10^{18}$$

$$Ni^{2+}(aq) + CO_3^{2-}(aq) \rightleftharpoons NiCO_3(s) \qquad K = 3.2 \times 10^{10}$$

The final step in obtaining the metal itself is to reduce the metal cation to the metal either chemically or electrochemically (► Chapter 20).

Our immediate goal is to work out methods to determine whether a precipitate will form under a given set of conditions. For example, if Ag^+ and Cl^- are present at some given concentrations, will AgCl precipitate from the solution?

K_{sp} and the Reaction Quotient, Q

Silver chloride, like silver bromide, is used in photographic films. It dissolves to a very small extent in water and has a correspondingly small value of K_{sp}.

$$AgCl(s) \rightleftharpoons Ag^+(aq) + Cl^-(aq) \qquad K_{sp} = [Ag^+][Cl^-] = 1.8 \times 10^{-10}$$

But let us look at the problem from the other direction: If a solution contains Ag^+ and Cl^- ions at some concentration, will AgCl precipitate from solution? This is the same question we asked in Section 16.3 when we wanted to know if a given mixture of reactants and products was an equilibrium mixture, if the reactants continued to form products, or if products would revert to reactants. The procedure there was to calculate the reaction quotient, Q.

For silver chloride, the expression for the reaction quotient, Q, is

$$Q = [Ag^+][Cl^-]$$

Recall that *the difference between Q and K is that the concentrations in the reaction quotient expression may or may not be those at equilibrium.* For the case of a slightly soluble salt such as AgCl, we can reach the following conclusions (◄ Section 16.3).

1. If $Q = K_{sp}$, the solution is saturated.

When the product of the ion concentrations is equal to K_{sp}, the ion concentrations have reached their maximum value.

2. If $Q < K_{sp}$, the solution is not saturated.

This can mean two things: (i) If solid AgCl is present, more will dissolve until equilibrium is achieved (when $Q = K_{sp}$). (ii) If solid AgCl is not already present, more $Ag^+(aq)$ or more $Cl^-(aq)$ (or both) could be added to the solution until precipitation of solid AgCl begins (when $Q > K_{sp}$).

3. If $Q > K_{sp}$, the system is not at equilibrium; precipitation will occur.

The concentrations of Ag^+ and Cl^- in solution are too high, and AgCl will precipitate until $Q = K_{sp}$.

Chemistry Now™

Sign in at **www.thomsonedu.com/login** and go to Chapter 18 Contents to see Screen 18.17 for a simulation and tutorial on **when a precipitation reaction can occur.**

EXAMPLE 18.13 Solubility and the Reaction Quotient

Problem Solid AgCl has been placed in a beaker of water. After some time, the concentrations of Ag^+ and Cl^- are each 1.2×10^{-5} mol/L. Has the system reached equilibrium? If not, will more AgCl dissolve?

Strategy Use the experimental ion concentrations to calculate the reaction quotient Q. Compare Q and K_{sp} to decide if the system is at equilibrium (if $Q = K_{sp}$).

Solution For this AgCl case,

$$Q = [Ag^+][Cl^-] = (1.2 \times 10^{-5})(1.2 \times 10^{-5}) = 1.4 \times 10^{-10}$$

Here, Q is less than K_{sp} (1.8×10^{-10}). The solution is not yet saturated, and AgCl will continue to dissolve until $Q = K_{sp}$, at which point $[Ag^+] = [Cl^-] = 1.3 \times 10^{-5}$ M. That is, an additional 0.1×10^{-5} mol of AgCl (about 1.9 mg) will dissolve per liter.

EXERCISE 18.17 Solubility and the Reaction Quotient

Solid PbI_2 ($K_{sp} = 9.8 \times 10^{-9}$) is placed in a beaker of water. After a period of time, the lead(II) concentration is measured and found to be 1.1×10^{-3} M. Has the system yet reached equilibrium? That is, is the solution saturated? If not, will more PbI_2 dissolve?

K_{sp}, the Reaction Quotient, and Precipitation Reactions

With some knowledge of the reaction quotient, we can decide (1) if a precipitate will form when the ion concentrations are known or (2) what concentrations of ions are required to begin the precipitation of an insoluble salt.

Suppose the concentration of aqueous magnesium ion in a solution is 1.5×10^{-6} M. If enough NaOH is added to make the solution 1.0×10^{-4} M in hydroxide ion, OH^-, will precipitation of $Mg(OH)_2$ occur ($K_{sp} = 5.6 \times 10^{-12}$)? If not, will it occur if the concentration of OH^- is increased to 1.0×10^{-2} M?

Our strategy will be as in Example 18.13. That is, use the ion concentrations to calculate the value of Q and then compare Q with K_{sp} to decide if the system is at equilibrium. Let us begin with the equation for the dissolution of insoluble $Mg(OH)_2$.

$$Mg(OH)_2(s) \rightleftharpoons Mg^{2+}(aq) + 2\ OH^-(aq)$$

When the concentrations of magnesium and hydroxide ions are those stated above, we find that Q is less than K_{sp}.

$$Q = [Mg^{2+}][OH^-]^2 = (1.5 \times 10^{-6})(1.0 \times 10^{-4})^2 = 1.5 \times 10^{-14}$$

$$Q\ (1.5 \times 10^{-14}) < K_{sp}\ (5.6 \times 10^{-12})$$

This means the solution is not yet saturated, and precipitation does not occur.

When $[OH^-]$ is increased to 1.0×10^{-2} M, the reaction quotient is 1.5×10^{-10},

$$Q = (1.5 \times 10^{-6})(1.0 \times 10^{-2})^2$$

$$Q = 1.5 \times 10^{-10} > K_{sp}\ (5.6 \times 10^{-12})$$

The reaction quotient is now *larger* than K_{sp}. Precipitation of $Mg(OH)_2$ occurs and will continue until the Mg^{2+} and OH^- ion concentrations have declined to the point where their product is equal to K_{sp}.

EXERCISE 18.18 Deciding Whether a Precipitate Will Form

Will $SrSO_4$ precipitate from a solution containing 2.5×10^{-4} M strontium ion, Sr^{2+}, if enough of the soluble salt Na_2SO_4 is added to make the solution 2.5×10^{-4} M in SO_4^{2-}? K_{sp} for $SrSO_4$ is 3.4×10^{-7}.

Now that we know how to decide if a precipitate will form when the concentration of each ion is known, let us turn to the problem of deciding how much of the precipitating agent is required to begin the precipitation of an ion at a given concentration level.

EXAMPLE 18.14 Ion Concentrations Required to Begin Precipitation

Problem The concentration of barium ion, Ba^{2+}, in a solution is 0.010 M.

(a) What concentration of sulfate ion, SO_4^{2-}, is required once $BaSO_4$ has started to precipitate?

(b) When the concentration of sulfate ion in the solution reaches 0.015 M, what concentration of barium ion will remain in solution?

Strategy There are three variables in the K_{sp} expression: K_{sp} and the anion and cation concentrations. Here, we know K_{sp} (1.1×10^{-10}) and one of the ion concentrations. We can then calculate the other ion concentration.

Solution Let us begin by writing the balanced equation for the equilibrium that will exist when $BaSO_4$ has been precipitated.

$$BaSO_4(s) \rightleftharpoons Ba^{2+}(aq) + SO_4^{2-}(aq) \qquad K_{sp} = [Ba^{2+}][SO_4^{2-}] = 1.1 \times 10^{-10}$$

(a) When the product of the ion concentrations exceeds the K_{sp} ($= 1.1 \times 10^{-10}$)—that is, when $Q > K_{sp}$—precipitation will occur. The Ba^{2+} ion concentration is known (0.010 M), so the SO_4^{2-} ion concentration necessary for precipitation can be calculated.

$$[SO_4^{2-}] = \frac{K_{sp}}{[Ba^{2+}]} = \frac{1.1 \times 10^{-10}}{0.010} = 1.1 \times 10^{-8}\ M$$

The result tells us that if the sulfate ion is just slightly greater than 1.1×10^{-8} M, $BaSO_4$ will begin to precipitate; $Q = [Ba^{2+}][SO_4^{2-}]$ would then be greater than K_{sp}.

(b) If the sulfate ion concentration is increased to 0.015 M, the maximum concentration of Ba^{2+} ion that can exist in solution (in equilibrium with $BaSO_4$) is

$$[Ba^{2+}] = \frac{K_{sp}}{[SO_4^{2-}]} = \frac{1.1 \times 10^{-10}}{0.015} = 7.3 \times 10^{-9}\ M$$

Comment The fact that the barium ion concentration is so small under these circumstances means that the Ba^{2+} ion has been essentially completely removed from solution. (It began at 0.010 M and has dropped by a factor of about 10 million.) The Ba^{2+} ion precipitation is, for all practical purposes, complete.

EXAMPLE 18.15 K_{sp} and Precipitations

Problem Suppose you mix 100.0 mL of 0.0200 M $BaCl_2$ with 50.0 mL of 0.0300 M Na_2SO_4. Will $BaSO_4$ ($K_{sp} = 1.1 \times 10^{-10}$) precipitate?

Strategy Here, we mix two solutions, one containing Ba^{2+} ions and the other SO_4^{2-} ions. First, find the concentration of each of these ions after mixing. Then, knowing the ion concentrations in the diluted solution, calculate Q and compare it with the K_{sp} value for $BaSO_4$.

Solution First, use the equation $c_1V_1 = c_2V_2$ (◀ Section 4.5) to calculate c_2, the concentration of the Ba^{2+} or SO_4^{2-} ions after mixing, to give a new solution with a volume of 150.0 mL ($= V_2$).

$$[Ba^{2+}] \text{ after mixing} = \frac{(0.0200\ mol/L)(0.1000\ L)}{0.1500\ L} = 0.0133\ M$$

$$[SO_4^{2-}] \text{ after mixing} = \frac{(0.0300\ mol/L)(0.0500\ L)}{0.1500\ L} = 0.0100\ M$$

Now the reaction quotient can be calculated.

$$Q = [Ba^{2+}][SO_4^{2-}] = (0.0133)(0.0100) = 1.33 \times 10^{-4}$$

Q is much larger than K_{sp}, so $BaSO_4$ precipitates.

EXERCISE 18.19 Ion Concentrations Required to Begin Precipitation

What is the minimum concentration of I^- that can cause precipitation of PbI_2 from a 0.050 M solution of $Pb(NO_3)_2$? K_{sp} for PbI_2 is 8.7×10^{-9}. What concentration of Pb^{2+} ions remains in solution when the concentration of I^- is 0.0015 M?

Dimethylglyoximate complex of Ni^{2+} ion

Charles D. Winters

FIGURE 18.17 Complex ions. The green solution contains soluble $Ni(H_2O)_6{}^{2+}$ ions in which water molecules are bound to Ni^{2+} ions by ion–dipole forces. This complex ion gives the solution its green color. The Ni^{2+}-ammonia complex ion is purple. The red, insoluble solid is the dimethylglyoximate complex of the Ni^{2+} ion $[Ni(C_4H_7O_2N_2)_2]$ (model at top). Formation of this beautiful red insoluble compound is the classical test for the presence of the aqueous Ni^{2+} ion.

■ **Complex Ions** Complex ions are prevalent in chemistry and are the basis of such biologically important substances as hemoglobin and vitamin B_{12}. They are described in more detail in Chapter 22.

EXERCISE 18.20 K_{sp} and Precipitation

You have 100.0 mL of 0.0010 M silver nitrate. Will AgCl precipitate if you add 5.0 mL of 0.025 M HCl?

18.6 Equilibria Involving Complex Ions

Metal ions exist in aqueous solution as complex ions (◀ Section 17.10) (Figure 18.17). Complex ions consist of the metal ion and other molecules or ions bound into a single entity. In water, metal ions are always surrounded by water molecules, with the negative end of the polar water molecule, the oxygen atom, attracted to the positive metal ion. In the case of Ni^{2+}, the ion exists as $[Ni(H_2O)_6]^{2+}$ in water. On adding ammonia, water molecules are displaced successively, and in the presence of a high enough concentration of ammonia, the complex ion $[Ni(NH_3)_6]^{2+}$ exists. Many organic molecules also form complex ions with metal ions, one example being the complex with the dimethylglyoximate ion in Figure 18.17.

The molecules or ions that bind to metal ions are called **ligands** (▶ Chapter 22). In aqueous solution, metal ions and ligands exist in equilibrium, and the equilibrium constants for these reactions are referred to as **formation constants, K_f** (Appendix K). For example,

$$Cu^{2+}(aq) + NH_3(aq) \rightleftharpoons [Cu(NH_3)]^{2+}(aq) \qquad K_{f1} = 2.0 \times 10^4$$

$$[Cu(NH_3)]^{2+}(aq) + NH_3(aq) \rightleftharpoons [Cu(NH_3)_2]^{2+}(aq) \qquad K_{f2} = 4.7 \times 10^3$$

$$[Cu(NH_3)_2]^{2+}(aq) + NH_3(aq) \rightleftharpoons [Cu(NH_3)_3]^{2+}(aq) \qquad K_{f3} = 1.1 \times 10^3$$

$$[Cu(NH_3)_3]^{2+}(aq) + NH_3(aq) \rightleftharpoons [Cu(NH_3)_4]^{2+}(aq) \qquad K_{f4} = 2.0 \times 10^2$$

(In these reactions, Cu^{2+} begins as $[Cu(H_2O)_4]^{2+}$. Ammonia successively displaces the water molecules.) Overall, the formation of the tetraammine copper(II) complex ion has an equilibrium constant of 2.1×10^{13} ($= K_{f1} \times K_{f2} \times K_{f3} \times K_{f4}$).

$$Cu^{2+}(aq) + 4\ NH_3(aq) \rightleftharpoons [Cu(NH_3)_4]^{2+}(aq) \qquad K_f = 2.1 \times 10^{13}$$

■ **EXAMPLE 8.16 Complex Ion Equilibria**

Problem What is the concentration of Cd^{2+} ions in a solution prepared by adding 0.00100 mol of $Cd(NO_3)_2$ to 1.00 L of 1.50 M NH_3? K_f for complex ion $[Cd(NH_3)_4]^{2+}$ is 1.3×10^7.

Strategy The formation constant for the complex ion is very large, so we can assume nearly all of the Cd^{2+} ions are in that form. That is, the initial concentration of the complex ion, $[Cd(NH_3)_4]^{2+}$, is 0.00100 M. This then dissociates to produce Cd^{2+} ions in solution.

Solution Let us set up an ICE table for the dissociation of $[Cd(NH_3)_4]^{2+}$.

Equation	$[Cd(NH_3)_4]^{2+}(aq)$	$\rightleftharpoons$	$Cd^{2+}(aq)$	+	$4\ NH_3(aq)$
Initial (M)	0.00100		0		1.50 − 0.00400 M
Change	$-x$		$+x$		$+4x$
Equilibrium (M)	$0.00100 - x \approx 0.00100$		x		$1.50 - 0.00400 + 4x \approx 1.50$

$$K = \frac{1}{K_f} = \frac{1}{1.3 \times 10^7} = \frac{[Cd^{2+}][NH_3]^4}{\{[Cd(NH_3)_4]^{2+}\}} = \frac{(x)(1.50)}{0.000100}$$

$$x = [Cd^{2+}] = 5.1 \times 10^{-11}\ M$$

EXERCISE 18.21 Complex Ion Equilibria

Silver nitrate (0.0050 mol) is added to 1.00 L of 1.00 M NH_3. What is the concentration of Ag^+ ions at equilibrium?

$$Ag^+(aq) + 2\ NH_3(aq) \rightleftharpoons [Ag(NH_3)_2]^+(aq) \qquad K_f = 1.6 \times 10^7$$

18.7 Solubility and Complex Ions

Silver chloride dissolves neither in water nor in strong acid, but it does dissolve in ammonia because it forms a water-soluble complex ion, $[Ag(NH_3)_2]^+$ (Figure 18.18).

$$AgCl(s) + 2\ NH_3(aq) \rightleftharpoons [Ag(NH_3)_2]^+(aq) + Cl^-(aq)$$

We can view dissolving AgCl(s) in this way as a two-step process. First, AgCl dissolves minimally in water, giving $Ag^+(aq)$ and $Cl^-(aq)$ ion. Then, the $Ag^+(aq)$ ion combines with NH_3 to give the ammonia complex. Lowering the $Ag^+(aq)$ concentration through complexation with NH_3 shifts the solubility equilibrium to the right, and more solid AgCl dissolves.

$$AgCl(s) \rightleftharpoons Ag^+(aq) + Cl^-(aq) \qquad K_{sp} = 1.8 \times 10^{-10}$$

$$Ag^+(aq) + 2\ NH_3(aq) \rightleftharpoons [Ag(NH_3)_2]^+(aq) \qquad K_f = 1.6 \times 10^7$$

This is an example of combining or "coupling" two (or more) equilibria where one is a product-favored reaction and the other is reactant-favored.

The large value of the formation constant for $[Ag(NH_3)_2]^+$ means that the equilibrium lies well to the right, and AgCl dissolves in the presence of NH_3. If we combine K_f with K_{sp}, we obtain the net equilibrium constant for the interaction of AgCl and aqueous ammonia.

$$K_{net} = K_{sp} \times K_f = (1.8 \times 10^{-10})(1.6 \times 10^7) = 2.9 \times 10^{-3}$$

$$K_{net} = 2.9 \times 10^{-3} = \frac{\{[Ag(NH_3)_2]^+\}[Cl^-]}{[NH_3]^2}$$

Even though the value of K_{net} seems small, if we use a large concentration of NH_3, the concentration of $[Ag(NH_3)_2]^+$ in solution can be high. Silver chloride is thus much more soluble in the presence of ammonia than in pure water.

The stabilities of various complex ions involving silver(I) can be compared by comparing values of their formation constants.

Formation Equilibrium	K_f
$Ag^+(aq) + 2\ Cl^-(aq) \rightleftharpoons [AgCl_2]^-(aq)$	2.5×10^5
$Ag^+(aq) + 2\ S_2O_3^{2-}(aq) \rightleftharpoons [Ag(S_2O_3)_2]^{3-}(aq)$	2.0×10^{13}
$Ag^+(aq) + 2\ CN^-(aq) \rightleftharpoons [Ag(CN)_2]^-(aq)$	5.6×10^{18}

The formation of all three silver complexes is strongly product-favored. The cyanide complex ion $[Ag(CN)_2]^-$ is the most stable of the three.

Figure 18.18 shows what happens as complex ions are formed. Beginning with a precipitate of AgCl, adding aqueous ammonia dissolves the precipitate to give the soluble complex ion $[Ag(NH_3)_2]^+$. Silver bromide is even more stable than $[Ag(NH_3)_2]^+$, so AgBr ($K_{sp} = 5.4 \times 10^{-13}$) forms in preference to the complex ion on adding bromide ion. If thiosulfate ion, $S_2O_3^{2-}$, is then added, AgBr dissolves due to the formation of $[Ag(S_2O_3)_2]^{3-}$, a complex ion with a large formation constant (2.0×10^{13}).

Photos: Charles D. Winters

AgCl(s), $K_{sp} = 1.8 \times 10^{-10}$

(a) AgCl precipitates on adding NaCl(aq) to $AgNO_3(aq)$ (see Figure 3.7).

$[Ag(NH_3)_2]^+(aq)$

(b) The precipitate of AgCl dissolves on adding aqueous NH_3 to give water-soluble $[Ag(NH_3)_2]^+$.

AgBr(s), $K_{sp} = 5.4 \times 10^{-13}$

(c) The silver-ammonia complex ion is changed to insoluble AgBr on adding NaBr(aq).

$[Ag(S_2O_3)_2]^{3-}(aq)$

(d) Solid AgBr is dissolved on adding $Na_2S_2O_3(aq)$. The product is the water-soluble complex ion $[Ag(S_2O_3)_2]^{3-}$.

FIGURE 18.18 Forming and dissolving precipitates. Insoluble compounds often dissolve upon addition of a complexing agent.

Chemistry Now™

Sign in at **www.thomsonedu.com/login** and go to Chapter 18 Contents to see:

- Screen 18.18 for more information on **combining equilibria**
- Screen 18.19 for a tutorial on **complex ion formation and solubility**

EXAMPLE 18.17 Complex Ions and Solubility

Problem What is the value of the equilibrium constant, K_{net}, for dissolving AgBr in a solution containing the thiosulfate ion, $S_2O_3^{2-}$ (Figure 18.18)? Explain why AgBr dissolves readily on adding aqueous sodium thiosulfate to the solid.

Strategy Summing several equilibrium processes gives the net chemical equation. K_{net} is the product of the values of K of the summed chemical equations. (See the preceding text and Section 16.10.)

Solution The overall reaction for dissolving AgBr in the presence of the thiosulfate anion is the sum of two equilibrium processes.

$$AgBr(s) \rightleftharpoons Ag^+(aq) + Br^-(aq) \qquad K_{sp} = 5.0 \times 10^{-13}$$

$$Ag^+(aq) + 2\ S_2O_3^{2-}(aq) \rightleftharpoons [Ag(S_2O_3)_2]^{3-}(aq) \qquad K_f = 2.0 \times 10^{13}$$

Net chemical equation:

$$AgBr(aq) + 2\ S_2O_3^{2-}(aq) \rightleftharpoons [Ag(S_2O_3)_2]^{3-}(aq) + Br^-(aq) \qquad K_{net} = K_{sp} \times K_f = 1.0 \times 10^1$$

The value of K_{net} is greater than 1, indicating a product-favored reaction. AgBr is predicted to dissolve readily in aqueous $Na_2S_2O_3$, as observed (Figure 18.18).

EXERCISE 18.22 Complex Ions and Solubility

Calculate the value of the equilibrium constant, K_{net}, for dissolving $Cu(OH)_2$ in aqueous ammonia (to form the complex ion $[Cu(NH_3)_4]^{2+}$) (see Figure 17.7).

Chapter Goals Revisited

Now that you have studied this chapter, you should ask whether you have met the chapter goals. In particular, you should be able to:

Understand the common ion effect

(a) Predict the effect of the addition of a "common ion" on the pH of the solution of a weak acid or base (Section 18.1). **Study Question(s) assignable in OWL: 2, 4, 109.**

Understand the control of pH in aqueous solutions with buffers

(a) Describe the functioning of buffer solutions. (Section 18.2) **Go Chemistry Module 23.**

(b) Use the Henderson–Hasselbalch equation (Equation 18.2) to calculate the pH of a buffer solution of given composition. **Study Question(s) assignable in OWL: 6, 7, 14.**

(c) Describe how a buffer solution of a given pH can be prepared. **Study Question(s) assignable in OWL: 9, 16, 83, 90, 101, 102, 109.**

(d) Calculate the pH of a buffer solution before and after adding acid or base. **Study Question(s) assignable in OWL: 20, 22, 74, 76, 82.**

Evaluate the pH in the course of acid–base titrations

(a) Predict the pH of an acid–base reaction at its equivalence point (Section 18.3; see also Sections 17.5 and 17.6).

Acid	Base	pH at Equivalence Point
Strong	Strong	= 7 (neutral)
Strong	Weak	< 7 (acidic)
Weak	Strong	> 7 (basic)

Study Question(s) assignable in OWL: 24, 74, 76, 98, 108.

(b) Understand the differences between the titration curves for a strong acid–strong base titration and titrations in which one of the substances is weak.

(c) Describe how an indicator functions in an acid–base titration. **Study Question(s) assignable in OWL: 32, 72, 80, 96, 105.**

Apply chemical equilibrium concepts to the solubility of ionic compounds

(a) Write the equilibrium constant expression—relating concentrations of ions in solutions to K_{sp}—for any insoluble salt (Section 18.4).

(b) Calculate K_{sp} values from experimental data (Section 18.4). **Study Question(s) assignable in OWL: 40, 42.**

(c) Estimate the solubility of a salt from the value of K_{sp} (Section 18.4). **Study Question(s) assignable in OWL: 46, 48, 72, 80, 96, 105.**

(d) Calculate the solubility of a salt in the presence of a common ion (Section 18.4). **Study Question(s) assignable in OWL: 54.**

(e) Understand how hydrolysis of basic anions affects the solubility of a salt (Section 18.4). **Study Question(s) assignable in OWL: 58.**

(f) Decide if a precipitate will form when the ion concentrations are known Section 18.5). **Study Question(s) assignable in OWL: 36, 60, 64, 70, 88, 89.**

(g) Calculate the ion concentrations that are required to begin the precipitation of an insoluble salt (Section 18.5).

(h) Understand that the formation of a complex ion can increase the solubility of an insoluble salt (Section 18.6). **Study Question(s) assignable in OWL: 66, 70, 92.**

Chemistry•Now™ Sign in at **www.thomsonedu.com/login** to:

- Assess your understanding with Study Questions in OWL keyed to each goal in the Goals and Homework menu for this chapter
- For quick review, download Go Chemistry mini-lecture flashcard modules (or purchase them at **www.ichapters.com**)
- Check your readiness for an exam by taking the Pre-Test and exploring the modules recommended in your Personalized Study plan.

Access **How Do I Solve It?** tutorials on how to approach problem solving using concepts in this chapter.

KEY EQUATIONS

Equation 18.1 (page 816) Hydronium ion concentration in a buffer solution composed of a weak acid and its conjugate base.

$$[H_3O^+] = \frac{[\text{acid}]}{[\text{conjugate base}]} \times K_a$$

Equation 18.2 (page 817) Henderson–Hasselbalch equation. To calculate the pH of a buffer solution composed of a weak acid and its conjugate base.

$$pH = pK_a + \log\frac{[\text{conjugate base}]}{[\text{acid}]}$$

Equation 18.3 (page 825) Equation to calculate the hydronium ion concentration before the equivalence point in the titration of a weak acid with a strong base. See also Equation 18.4 for the version of the equation based on the Henderson–Hasselbalch equation.

$$[H_3O^+] = \frac{[\text{weak acid remaining}]}{[\text{conjugate base produced}]} \times K_a$$

Equation 18.5 (page 825) The relationship between the pH of the solution and the pK_a of the weak acid (or $[H_3O^+]$ and K_a) at the halfway or half-neutralization point in the titration of a weak acid with a strong base (or of a weak base with a strong acid).

$$[H_3O^+] = K_a \text{ and } pH = pK_a$$

STUDY QUESTIONS

Online homework for this chapter may be assigned in OWL.

▲ denotes challenging questions.

■ denotes questions assignable in OWL.

Blue-numbered questions have answers in Appendix O and fully-worked solutions in the *Student Solutions Manual*.

Practicing Skills

The Common Ion Effect and Buffer Solutions

(See Examples 18.1 and 18.2 and ChemistryNow Screens 18.2–18.4.)

1. Does the pH of the solution increase, decrease, or stay the same when you
   (a) Add solid ammonium chloride to a dilute aqueous solution of NH_3?
   (b) Add solid sodium acetate to a dilute aqueous solution of acetic acid?
   (c) Add solid NaCl to a dilute aqueous solution of NaOH?

2. ■ Does the pH of the solution increase, decrease, or stay the same when you
   (a) Add solid sodium oxalate, $Na_2C_2O_4$, to 50.0 mL of 0.015 M oxalic acid, $H_2C_2O_4$?
   (b) Add solid ammonium chloride to 75 mL of 0.016 M HCl?
   (c) Add 20.0 g of NaCl to 1.0 L of 0.10 M sodium acetate, $NaCH_3CO_2$?

3. What is the pH of a solution that consists of 0.20 M ammonia, NH_3, and 0.20 M ammonium chloride, NH_4Cl?

4. ■ What is the pH of 0.15 M acetic acid to which 1.56 g of sodium acetate, $NaCH_3CO_2$, has been added?

5. What is the pH of the solution that results from adding 30.0 mL of 0.015 M KOH to 50.0 mL of 0.015 M benzoic acid?

6. ■ What is the pH of the solution that results from adding 25.0 mL of 0.12 M HCl to 25.0 mL of 0.43 M NH_3?

7. ■ What is the pH of the buffer solution that contains 2.2 g of NH_4Cl in 250 mL of 0.12 M NH_3? Is the final pH lower or higher than the pH of the 0.12 M ammonia solution?

8. Lactic acid ($CH_3CHOHCO_2H$) is found in sour milk, in sauerkraut, and in muscles after activity. (K_a for lactic acid = 1.4×10^{-4}.)
   (a) If 2.75 g of $NaCH_3CHOHCO_2$, sodium lactate, is added to 5.00×10^2 mL of 0.100 M lactic acid, what is the pH of the resulting buffer solution?
   (b) Is the pH of the buffered solution lower or higher than the pH of the lactic acid solution?

9. ■ What mass of sodium acetate, $NaCH_3CO_2$, must be added to 1.00 L of 0.10 M acetic acid to give a solution with a pH of 4.50?

10. What mass of ammonium chloride, NH_4Cl, must be added to exactly 5.00×10^2 mL of 0.10 M NH_3 solution to give a solution with a pH of 9.00?

Using the Henderson–Hasselbalch Equation

(See Example 18.3 and ChemistryNow Screen 18.4.)

11. Calculate the pH of a solution that has an acetic acid concentration of 0.050 M and a sodium acetate concentration of 0.075 M.

12. Calculate the pH of a solution that has an ammonium chloride concentration of 0.050 M and an ammonia concentration of 0.045 M.

13. A buffer is composed of formic acid and its conjugate base, the formate ion.
    (a) What is the pH of a solution that has a formic acid concentration of 0.050 M and a sodium formate concentration of 0.035 M?
    (b) What must the ratio of acid to conjugate base be to increase the pH by 0.5 unit from the value calculated in part (a)?

14. ■ A buffer solution is composed of 1.360 g of KH_2PO_4 and 5.677 g of Na_2HPO_4.
    (a) What is the pH of the buffer solution?
    (b) What mass of KH_2PO_4 must be added to decrease the buffer solution pH by 0.5 unit from the value calculated in part (a)?

Preparing a Buffer Solution

(See Example 18.4 and ChemistryNow Screen 18.5.)

15. Which of the following combinations would be the best to buffer the pH of a solution at approximately 9?
    (a) HCl and NaCl
    (b) NH_3 and NH_4Cl
    (c) CH_3CO_2H and $NaCH_3CO_2$

16. ■ Which of the following combinations would be the best choice to buffer the pH of a solution at approximately 7?
    (a) H_3PO_4 and NaH_2PO_4
    (b) NaH_2PO_4 and Na_2HPO_4
    (c) Na_2HPO_4 and Na_3PO_4

17. Describe how to prepare a buffer solution from NaH_2PO_4 and Na_2HPO_4 to have a pH of 7.5.

18. Describe how to prepare a buffer solution from NH_3 and NH_4Cl to have a pH of 9.5.

Adding an Acid or a Base to a Buffer Solution

(See Example 18.5 and ChemistryNow Screen 18.6.)

19. A buffer solution was prepared by adding 4.95 g of sodium acetate, $NaCH_3CO_2$, to 2.50×10^2 mL of 0.150 M acetic acid, CH_3CO_2H.
    (a) What is the pH of the buffer?
    (b) What is the pH of 1.00×10^2 mL of the buffer solution if you add 82 mg of NaOH to the solution?

20. ■ You dissolve 0.425 g of NaOH in 2.00 L of a buffer solution that has $[H_2PO_4^-] = [HPO_4^{2-}] = 0.132$ M. What is the pH of the solution before adding NaOH? After adding NaOH?

21. A buffer solution is prepared by adding 0.125 mol of ammonium chloride to 5.00×10^2 mL of 0.500 M solution of ammonia.
    (a) What is the pH of the buffer?
    (b) If 0.0100 mol of HCl gas is bubbled into 5.00×10^2 mL of the buffer, what is the new pH of the solution?

22. ■ What will be the pH change when 20.0 mL of 0.100 M NaOH is added to 80.0 mL of a buffer solution consisting of 0.169 M NH_3 and 0.183 M NH_4Cl?

More about Acid–Base Reactions: Titrations

(See Examples 18.6 and 18.7 and ChemistryNow Screen 18.7.)

23. Phenol, C_6H_5OH, is a weak organic acid. Suppose 0.515 g of the compound is dissolved in exactly 125 mL of water. The resulting solution is titrated with 0.123 M NaOH.

$$C_6H_5OH(aq) + OH^-(aq) \rightleftharpoons C_6H_5O^-(aq) + H_2O(\ell)$$

    (a) What is the pH of the original solution of phenol?
    (b) What are the concentrations of all of the following ions at the equivalence point: Na^+, H_3O^+, OH^-, and $C_6H_5O^-$?
    (c) What is the pH of the solution at the equivalence point?

24. ■ Assume you dissolve 0.235 g of the weak acid benzoic acid, $C_6H_5CO_2H$, in enough water to make 1.00×10^2 mL of solution and then titrate the solution with 0.108 M NaOH.

$$C_6H_5CO_2H(aq) + OH^-(aq) \rightleftharpoons C_6H_5CO_2^-(aq) + H_2O(\ell)$$

    (a) What was the pH of the original benzoic acid solution?
    (b) What are the concentrations of all of the following ions at the equivalence point: Na^+, H_3O^+, OH^-, and $C_6H_5CO_2^-$?
    (c) What is the pH of the solution at the equivalence point?

25. You require 36.78 mL of 0.0105 M HCl to reach the equivalence point in the titration of 25.0 mL of aqueous ammonia.
(a) What was the concentration of NH_3 in the original ammonia solution?
(b) What are the concentrations of H_3O^+, OH^-, and NH_4^+ at the equivalence point?
(c) What is the pH of the solution at the equivalence point?

26. A solution of the weak base aniline, $C_6H_5NH_2$, in 25.0 mL of water requires 25.67 mL of 0.175 M HCl to reach the equivalence point.

$$C_6H_5NH_2(aq) + H_3O^+(aq) \rightleftharpoons C_6H_5NH_3^+(aq) + H_2O(\ell)$$

(a) What was the concentration of aniline in the original solution?
(b) What are the concentrations of H_3O^+, OH^-, and $C_6H_5NH_3^+$ at the equivalence point?
(c) What is the pH of the solution at the equivalence point?

Titration Curves and Indicators

(See Figures 18.4–18.10 and ChemistryNow Screen 18.7.)

27. Without doing detailed calculations, sketch the curve for the titration of 30.0 mL of 0.10 M NaOH with 0.10 M HCl. Indicate the approximate pH at the beginning of the titration and at the equivalence point. What is the total solution volume at the equivalence point?

28. Without doing detailed calculations, sketch the curve for the titration of 50 mL of 0.050 M pyridine, C_5H_5N (a weak base), with 0.10 M HCl. Indicate the approximate pH at the beginning of the titration and at the equivalence point. What is the total solution volume at the equivalence point?

29. You titrate 25.0 mL of 0.10 M NH_3 with 0.10 M HCl.
(a) What is the pH of the NH_3 solution before the titration begins?
(b) What is the pH at the equivalence point?
(c) What is the pH at the halfway point of the titration?
(d) What indicator in Figure 18.10 could be used to detect the equivalence point?
(e) Calculate the pH of the solution after adding 5.00, 15.0, 20.0, 22.0, and 30.0 mL of the acid. Combine this information with that in parts (a)–(c) and plot the titration curve.

30. Construct a rough plot of pH versus volume of base for the titration of 25.0 mL of 0.050 M HCN with 0.075 M NaOH.
(a) What is the pH before any NaOH is added?
(b) What is the pH at the halfway point of the titration?
(c) What is the pH when 95% of the required NaOH has been added?
(d) What volume of base, in milliliters, is required to reach the equivalence point?
(e) What is the pH at the equivalence point?
(f) What indicator would be most suitable for this titration? (See Figure 18.10.)
(g) What is the pH when 105% of the required base has been added?

31. Using Figure 18.10, suggest an indicator to use in each of the following titrations:
(a) The weak base pyridine is titrated with HCl.
(b) Formic acid is titrated with NaOH.
(c) Ethylenediamine, a weak diprotic base, is titrated with HCl.

32. ■ Using Figure 18.10, suggest an indicator to use in each of the following titrations.
(a) $NaHCO_3$ is titrated to CO_3^{2-} with NaOH.
(b) Hypochlorous acid is titrated with NaOH.
(c) Trimethylamine is titrated with HCl.

Solubility Guidelines

(Review Section 3.5, Figure 3.10, and Example 3.2; also see ChemistryNow.)

33. Name two insoluble salts of each of the following ions.
(a) Cl^-
(b) Zn^{2+}
(c) Fe^{2+}

34. Name two insoluble salts of each of the following ions.
(a) SO_4^{2-}
(b) Ni^{2+}
(c) Br^-

35. Using the solubility guidelines (Figure 3.10), predict whether each of the following is insoluble or soluble in water.
(a) $(NH_4)_2CO_3$
(b) $ZnSO_4$
(c) NiS
(d) $BaSO_4$

36. ■ Predict whether each of the following is insoluble or soluble in water.
(a) $Pb(NO_3)_2$
(b) $Fe(OH)_3$
(c) $ZnCl_2$
(d) CuS

Writing Solubility Product Constant Expressions

(See Exercise 18.11 and ChemistryNow Screen 18.9.)

37. For each of the following insoluble salts, (1) write a balanced equation showing the equilibrium occurring when the salt is added to water, and (2) write the K_{sp} expression.
(a) AgCN
(b) $NiCO_3$
(c) $AuBr_3$

38. For each of the following insoluble salts, (1) write a balanced equation showing the equilibrium occurring when the salt is added to water, and (2) write the K_{sp} expression.
(a) $PbSO_4$
(b) BaF_2
(c) Ag_3PO_4

Calculating K_{sp}

(See Example 18.8 and ChemistryNow Screen 18.10.)

39. When 1.55 g of solid thallium(I) bromide is added to 1.00 L of water, the salt dissolves to a small extent.

$$TlBr(s) \rightleftharpoons Tl^+(aq) + Br^-(aq)$$

The thallium(I) and bromide ions in equilibrium with TlBr each have a concentration of 1.9×10^{-3} M. What is the value of K_{sp} for TlBr?

40. ■ At 20 °C, a saturated aqueous solution of silver acetate, $AgCH_3CO_2$, contains 1.0 g of the silver compound dissolved in 100.0 mL of solution. Calculate K_{sp} for silver acetate.

$$AgCH_3CO_2(s) \rightleftharpoons Ag^+(aq) + CH_3CO_2^-(aq)$$

41. When 250 mg of SrF_2, strontium fluoride, is added to 1.00 L of water, the salt dissolves to a very small extent.

$$SrF_2(s) \rightleftharpoons Sr^{2+}(aq) + 2\ F^-(aq)$$

At equilibrium, the concentration of Sr^{2+} is found to be 1.03×10^{-3} M. What is the value of K_{sp} for SrF_2?

42. ■ Calcium hydroxide, $Ca(OH)_2$, dissolves in water to the extent of 1.3 g per liter. What is the value of K_{sp} for $Ca(OH)_2$?

$$Ca(OH)_2(s) \rightleftharpoons Ca^{2+}(aq) + 2\ OH^-(aq)$$

43. You add 0.979 g of $Pb(OH)_2$ to 1.00 L of pure water at 25 °C. The pH is 9.15. Estimate the value of K_{sp} for $Pb(OH)_2$.

44. You place 1.234 g of solid $Ca(OH)_2$ in 1.00 L of pure water at 25 °C. The pH of the solution is found to be 12.68. Estimate the value of K_{sp} for $Ca(OH)_2$.

Estimating Salt Solubility from K_{sp}

(See Examples 18.9 and 18.10, Exercise 8.14, and ChemistryNow Screen 18.11.)

45. Estimate the solubility of silver iodide in pure water at 25 °C (a) in moles per liter and (b) in grams per liter.

$$AgI(s) \rightleftharpoons Ag^+(aq) + I^-(aq)$$

46. ■ What is the molar concentration of $Au^+(aq)$ in a saturated solution of AuCl in pure water at 25 °C?

$$AuCl(s) \rightleftharpoons Au^+(aq) + Cl^-(aq)$$

47. Estimate the solubility of calcium fluoride, CaF_2, (a) in moles per liter and (b) in grams per liter of pure water.

$$CaF_2(s) \rightleftharpoons Ca^{2+}(aq) + 2\ F^-(aq)$$

48. ■ Estimate the solubility of lead(II) bromide (a) in moles per liter and (b) in grams per liter of pure water.

49. The K_{sp} value for radium sulfate, $RaSO_4$, is 4.2×10^{-11}. If 25 mg of radium sulfate is placed in 1.00×10^2 mL of water, does all of it dissolve? If not, how much dissolves?

50. If 55 mg of lead(II) sulfate is placed in 250 mL of pure water, does all of it dissolve? If not, how much dissolves?

51. Use K_{sp} values to decide which compound in each of the following pairs is the more soluble.
(a) $PbCl_2$ or $PbBr_2$
(b) HgS or FeS
(c) $Fe(OH)_2$ or $Zn(OH)_2$

52. Use K_{sp} values to decide which compound in each of the following pairs is the more soluble.
(a) AgBr or AgSCN
(b) $SrCO_3$ or $SrSO_4$
(c) AgI or PbI_2
(d) MgF_2 or CaF_2

The Common Ion Effect and Salt Solubility

(See Examples 18.11 and 18.12 and ChemistryNow Screen 18.12.)

53. Calculate the molar solubility of silver thiocyanate, AgSCN, in pure water and in water containing 0.010 M NaSCN.

54. ■ Calculate the solubility of silver bromide, AgBr, in moles per liter, in pure water. Compare this value with the molar solubility of AgBr in 225 mL of water to which 0.15 g of NaBr has been added.

55. Compare the solubility, in milligrams per milliliter, of silver iodide, AgI, (a) in pure water and (b) in water that is 0.020 M in $AgNO_3$.

56. What is the solubility, in milligrams per milliliter, of BaF_2, (a) in pure water and (b) in water containing 5.0 mg/mL KF?

The Effect of Basic Anions on Salt Solubility

(See pages 882–883 and ChemistryNow Screen 18.13.)

57. Which insoluble compound in each pair should be more soluble in nitric acid than in pure water?
(a) $PbCl_2$ or PbS
(b) Ag_2CO_3 or AgI
(c) $Al(OH)_3$ or AgCl

58. ■ Which compound in each pair is more soluble in water than is predicted by a calculation from K_{sp}?
(a) AgI or Ag_2CO_3
(b) $PbCO_3$ or $PbCl_2$
(c) AgCl or AgCN

Precipitation Reactions

(See Examples 18.13–18.15 and ChemistryNow Screen 18.14.)

59. You have a solution that has a lead(II) ion concentration of 0.0012 M.

$$PbCl_2(s) \rightleftharpoons Pb^{2+}(aq) + 2\ Cl^-(aq)$$

If enough soluble chloride-containing salt is added so that the Cl^- concentration is 0.010 M, will $PbCl_2$ precipitate?

60. ■ Sodium carbonate is added to a solution in which the concentration of Ni^{2+} ion is 0.0024 M.

$$NiCO_3(s) \rightleftharpoons Ni^{2+}(aq) + CO_3^{2-}(aq)$$

Will precipitation of $NiCO_3$ occur (a) when the concentration of the carbonate ion is 1.0×10^{-6} M or (b) when it is 100 times greater (or 1.0×10^{-4} M)?

61. If the concentration of Zn^{2+} in 10.0 mL of water is 1.63×10^{-4} M, will zinc hydroxide, $Zn(OH)_2$, precipitate when 4.0 mg of NaOH is added?

62. You have 95 mL of a solution that has a lead(II) concentration of 0.0012 M. Will $PbCl_2$ precipitate when 1.20 g of solid NaCl is added?

63. If the concentration of Mg^{2+} ion in seawater is 1350 mg per liter, what OH^- concentration is required to precipitate $Mg(OH)_2$?

64. ■ Will a precipitate of $Mg(OH)_2$ form when 25.0 mL of 0.010 M NaOH is combined with 75.0 mL of a 0.10 M solution of magnesium chloride?

Equilibria Involving Complex Ions

(See Examples 18.16 and 18.17 and ChemistryNow Screen 18.16.)

65. Zinc hydroxide is amphoteric (page 791). Use equilibrium constants to show that, given sufficient OH^-, $Zn(OH)_2$ can dissolve in NaOH.

66. ■ Solid silver iodide, AgI, can be dissolved by adding aqueous sodium cyanide to it. Calculate K_{net} for the following reaction.

$$AgI(s) + 2\ CN^-(aq) \rightleftharpoons [Ag(CN)_2]^-(aq) + I^-(aq)$$

67. ▲ What amount of ammonia (moles) must be added to dissolve 0.050 mol of AgCl suspended in 1.0 L of water?

68. Can you dissolve 15.0 mg of AuCl in 100.0 mL of water if you add 15.0 mL of 6.00 M NaCN?

69. What is the solubility of AgCl (a) in pure water and (b) in 1.0 M NH_3?

70. ■ ▲ Suppose you mix 50.0 mL of 0.200 M NaCN with 10.0 mL of 0.100 M $AgNO_3$. Will the compound $Ag[Ag(CN)_2]$ precipitate? The equilibria involved are:

$$Ag[Ag(CN)_2](s) \rightleftharpoons Ag^+(aq) + [Ag(CN)_2]^-(aq) \quad K_{sp} = 5.0 \times 10^{-12}$$

$$Ag^+(aq) + 2\ CN^-(aq) \rightleftharpoons [Ag(CN)_2]^-(aq) \quad K_f = 5.0 \times 10^{21}$$

General Questions

These questions are not designated as to type or location in the chapter. They may combine several concepts.

71. In each of the following cases, decide whether a precipitate will form when mixing the indicated reagents, and write a balanced equation for the reaction.
(a) $NaBr(aq) + AgNO_3(aq)$
(b) $KCl(aq) + Pb(NO_3)_2(aq)$

72. ■ In each of the following cases, decide whether a precipitate will form when mixing the indicated reagents, and write a balanced equation for the reaction.
(a) $Na_2SO_4(aq) + Mg(NO_3)_2(aq)$
(b) $K_3PO_4(aq) + FeCl_3(aq)$

73. If you mix 48 mL of 0.0012 M $BaCl_2$ with 24 mL of 1.0×10^{-6} M Na_2SO_4, will a precipitate of $BaSO_4$ form?

74. ■ Calculate the hydronium ion concentration and the pH of the solution that results when 20.0 mL of 0.15 M acetic acid, CH_3CO_2H, is mixed with 5.0 mL of 0.17 M NaOH.

75. Calculate the hydronium ion concentration and the pH of the solution that results when 50.0 mL of 0.40 M NH_3 is mixed with 25.0 mL of 0.20 M HCl.

76. ■ For each of the following cases, decide whether the pH is less than 7, equal to 7, or greater than 7.
(a) Equal volumes of 0.10 M acetic acid, CH_3CO_2H, and 0.10 M KOH are mixed.
(b) 25 mL of 0.015 M NH_3 is mixed with 12 mL of 0.015 M HCl.
(c) 150 mL of 0.20 M HNO_3 is mixed with 75 mL of 0.40 M NaOH.
(d) 25 mL of 0.45 M H_2SO_4 is mixed with 25 mL of 0.90 M NaOH.

77. Rank the following compounds in order of increasing solubility in water: Na_2CO_3, $BaCO_3$, Ag_2CO_3.

78. A sample of hard water contains about 2.0×10^{-3} M Ca^{2+}. A soluble fluoride-containing salt such as NaF is added to "fluoridate" the water (to aid in the prevention of dental caries). What is the maximum concentration of F^- that can be present without precipitating CaF_2?

Dietary sources of fluoride ion. Adding fluoride ion to drinking water (or toothpaste) prevents the formation of dental caries.

79. What is the pH of a buffer solution prepared from 5.15 g of NH_4NO_3 and 0.10 L of 0.15 M NH_3? What is the new pH if the solution is diluted with pure water to a volume of 5.00×10^2 mL?

80. ■ If you place 5.0 mg of $SrSO_4$ in 1.0 L of pure water, will all of the salt dissolve before equilibrium is established, or will some salt remain undissolved?

Charles D. Winters

Celestite, $SrSO_4$
Strontium sulfate

81. Describe the effect on the pH of the following actions:
(a) Adding sodium acetate, $NaCH_3CO_2$, to 0.100 M CH_3CO_2H
(b) Adding $NaNO_3$ to 0.100 M HNO_3
(c) Explain why there is or is not an effect in each case.

82. ■ What volume of 0.120 M NaOH must be added to 100. mL of 0.100 M $NaHC_2O_4$ to reach a pH of 4.70?

83. ■ ▲ A buffer solution is prepared by dissolving 1.50 g each of benzoic acid, $C_6H_5CO_2H$, and sodium benzoate, $NaC_6H_5CO_2$, in 150.0 mL of solution.
(a) What is the pH of this buffer solution?
(b) Which buffer component must be added, and what quantity is needed to change the pH to 4.00?
(c) What quantity of 2.0 M NaOH or 2.0 M HCl must be added to the buffer to change the pH to 4.00?

84. What volume of 0.200 M HCl must be added to 500.0 mL of 0.250 M NH_3 to have a buffer with a pH of 9.00?

85. What is the equilibrium constant for the following reaction?

$$AgCl(s) + I^-(aq) \rightleftharpoons AgI(s) + Cl^-(aq)$$

Does the equilibrium lie predominantly to the left or to the right? Will AgI form if iodide ion, I^-, is added to a saturated solution of AgCl?

86. Calculate the equilibrium constant for the following reaction.

$$Zn(OH)_2(s) + 2\ CN^-(aq) \rightleftharpoons Zn(CN)_2(s) + 2\ OH^-(aq)$$

Does the equilibrium lie predominantly to the left or to the right? Can zinc hydroxide be transformed into zinc cyanide by adding a soluble salt of the cyanide ion?

87. ▲ In principle, the ions Ba^{2+} and Ca^{2+} can be separated by the difference in solubility of their fluorides, BaF_2 and CaF_2. If you have a solution that is 0.10 M in both Ba^{2+} and Ca^{2+}, CaF_2 will begin to precipitate first as fluoride ion is added slowly to the solution.
(a) What concentration of fluoride ion will precipitate the maximum amount of Ca^{2+} ion without precipitating BaF_2?
(b) What concentration of Ca^{2+} remains in solution when BaF_2 just begins to precipitate?

88. ■ ▲ A solution contains 0.10 M iodide ion, I^-, and 0.10 M carbonate ion, CO_3^{2-}.
(a) If solid $Pb(NO_3)_2$ is slowly added to the solution, which salt will precipitate first, PbI_2 or $PbCO_3$?
(b) What will be the concentration of the first ion that precipitates (CO_3^{2-} or I^-) when the second, more soluble salt begins to precipitate?

Charles D. Winters

Lead iodide ($K_{sp} = 9.8 \times 10^{-9}$) is a bright yellow solid.

89. ■ ▲ A solution contains Ca^{2+} and Pb^{2+} ions, both at a concentration of 0.010 M. You wish to separate the two ions from each other as completely as possible by precipitating one but not the other using aqueous Na_2SO_4 as the precipitating agent.
(a) Which will precipitate first as sodium sulfate is added, $CaSO_4$ or $PbSO_4$?
(b) What will be the concentration of the first ion that precipitates (Ca^{2+} or Pb^{2+}) when the second, more soluble salt begins to precipitate?

90. ■ Buffer capacity is defined as the number of moles of a strong acid or strong base that are required to change the pH of one liter of the buffer solution by one unit. What is the buffer capacity of a solution that is 0.10 M in acetic acid and 0.10 M in sodium acetate?

91. The Ca^{2+} ion in hard water can be precipitated as $CaCO_3$ by adding soda ash, Na_2CO_3. If the calcium ion concentration in hard water is 0.010 M and if the Na_2CO_3 is added until the carbonate ion concentration is 0.050 M, what percentage of the calcium ions have been removed from the water? (You may neglect carbonate ion hydrolysis.)

Charles D. Winters

This sample of calcium carbonate ($K_{sp} = 3.4 \times 10^{-9}$) was deposited in a cave formation.

92. ■ Some photographic film is coated with crystals of AgBr suspended in gelatin. Some of the silver ions are reduced to silver metal on exposure to light. Unexposed AgBr is then dissolved with sodium thiosulfate in the "fixing" step.

$$AgBr(s) + 2\ S_2O_3^{2-}(aq) \rightleftharpoons [Ag(S_2O_3)_2]^{3-}(aq) + Br^-(aq)$$

(a) What is the equilibrium constant for the reaction above?
(b) What mass of $Na_2S_2O_3$ must be added to dissolve 1.00 g of AgBr suspended in 1.00 L of water?

In the Laboratory

93. Each pair of ions below is found together in aqueous solution. Using the table of solubility product constants in Appendix J, devise a way to separate these ions by precipitating one of them as an insoluble salt and leaving the other in solution.
(a) Ba^{2+} and Na^+
(b) Ni^{2+} and Pb^{2+}

94. ■ Each pair of ions below is found together in aqueous solution. Using the table of solubility product constants in Appendix J, devise a way to separate these ions by adding one reagent to precipitate one of them as an insoluble salt and leave the other in solution.
(a) Cu^{2+} and Ag^+
(b) Al^{3+} and Fe^{3+}

95. ▲ The cations Ba^{2+} and Sr^{2+} can be precipitated as very insoluble sulfates.
(a) If you add sodium sulfate to a solution containing these metal cations, each with a concentration of 0.10 M, which is precipitated first, $BaSO_4$ or $SrSO_4$?
(b) What will be the concentration of the first ion that precipitates (Ba^{2+} or Sr^{2+}) when the second, more soluble salt begins to precipitate?

96. ■ ▲ You will often work with salts of Fe^{3+}, Pb^{2+}, and Al^{3+} in the laboratory. (All are found in nature, and all are important economically.) If you have a solution containing these three ions, each at a concentration of 0.10 M, what is the order in which their hydroxides precipitate as aqueous NaOH is slowly added to the solution?

97. Aniline hydrochloride, $(C_6H_5NH_3)Cl$, is a weak acid. (Its conjugate base is the weak base aniline, $C_6H_5NH_2$.) The acid can be titrated with a strong base such as NaOH.

$$C_6H_5NH_3^+(aq) + OH^-(aq) \rightleftharpoons C_6H_5NH_2(aq) + H_2O(\ell)$$

Assume 50.0 mL of 0.100 M aniline hydrochloride is titrated with 0.185 M NaOH. (K_a for aniline hydrochloride is 2.4×10^{-5}.)
(a) What is the pH of the $(C_6H_5NH_3)Cl$ solution before the titration begins?
(b) What is the pH at the equivalence point?
(c) What is the pH at the halfway point of the titration?
(d) Which indicator in Figure 18.10 could be used to detect the equivalence point?
(e) Calculate the pH of the solution after adding 10.0, 20.0, and 30.0 mL of base.
(f) Combine the information in parts (a), (b), and (e), and plot an approximate titration curve.

98. ■ The weak base ethanolamine, $HOCH_2CH_2NH_2$, can be titrated with HCl.

$$HOCH_2CH_2NH_2(aq) + H_3O^+(aq) \rightleftharpoons HOCH_2CH_2NH_3^+(aq) + H_2O(\ell)$$

Assume you have 25.0 mL of a 0.010 M solution of ethanolamine and titrate it with 0.0095 M HCl. (K_b for ethanolamine is 3.2×10^{-5}.)
(a) What is the pH of the ethanolamine solution before the titration begins?
(b) What is the pH at the equivalence point?
(c) What is the pH at the halfway point of the titration?
(d) Which indicator in Figure 18.10 would be the best choice to detect the equivalence point?
(e) Calculate the pH of the solution after adding 5.00, 10.0, 20.0, and 30.0 mL of the acid.
(f) Combine the information in parts (a), (b), and (e), and plot an approximate titration curve.

99. For the titration of 50.0 mL of 0.150 M ethylamine, $C_2H_5NH_2$, with 0.100 M HCl, find the pH at each of the following points, and then use that information to sketch the titration curve and decide on an appropriate indicator.
(a) At the beginning, before HCl is added
(b) At the halfway point in the titration
(c) When 75% of the required acid has been added
(d) At the equivalence point
(e) When 10.0 mL more HCl has been added than is required
(f) Sketch the titration curve.
(g) Suggest an appropriate indicator for this titration.

100. A buffer solution with a pH of 12.00 consists of Na_3PO_4 and Na_2HPO_4. The volume of solution is 200.0 mL.
(a) Which component of the buffer is present in a larger amount?
(b) If the concentration of Na_3PO_4 is 0.400 M, what mass of Na_2HPO_4 is present?
(c) Which component of the buffer must be added to change the pH to 12.25? What mass of that component is required?

101. ■ To have a buffer with a pH of 2.50, what volume of 0.150 M NaOH must be added to 100. mL of 0.230 M H_3PO_4?

102. ■ ▲ What mass of Na_3PO_4 must be added to 80.0 mL of 0.200 M HCl to obtain a buffer with a pH of 7.75?

103. You have a solution that contains $AgNO_3$, $Pb(NO_3)_2$, and $Cu(NO_3)_2$. Devise a separation method that results in having Ag^+ in one test tube, Pb^{2+} ions in another, and Cu^{2+} in a third test tube. Use solubility guidelines and K_{sp} and K_f values.

104. Once you have separated the three salts in Study Question 103 into three test tubes, you now need to confirm their presence.
(a) For Pb^{2+} ion, one way to do this is to treat a precipitate of $PbCl_2$ with K_2CrO_4 to produce $PbCrO_4$. Using K_{sp} values, confirm that the chloride salt should be converted to the chromate salt.

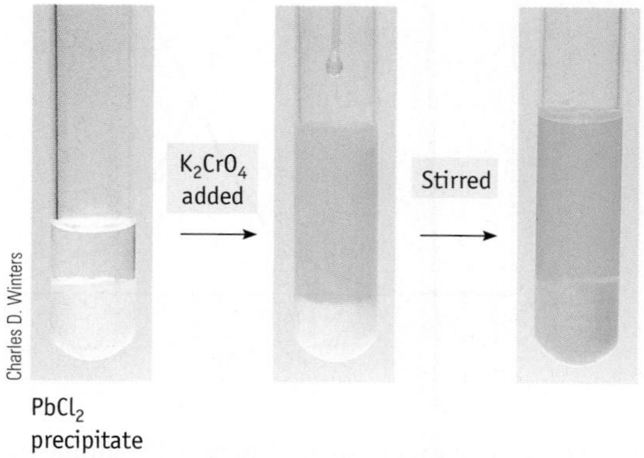

White $PbCl_2$ is converted to yellow $PbCrO_4$ on adding K_2CrO_4.

(b) Suggest a method of confirming the presence of Ag^+ and Cu^{2+} ions using complex ions.

Summary and Conceptual Questions

The following questions may use concepts from this and previous chapters.

105. ■ Suggest a method for separating a precipitate consisting of a mixture of solid CuS and solid $Cu(OH)_2$.

106. Which of the following barium salts should dissolve in a strong acid such as HCl: $Ba(OH)_2$, $BaSO_4$, or $BaCO_3$?

107. Explain why the solubility of Ag_3PO_4 can be greater in water than is calculated from the K_{sp} value of the salt.

108. ■ Two acids, each approximately 0.01 M in concentration, are titrated separately with a strong base. The acids show the following pH values at the equivalence point: HA, pH = 9.5, and HB, pH = 8.5.
(a) Which is the stronger acid, HA or HB?
(b) Which of the conjugate bases, A^- or B^-, is the stronger base?

109. ■ Composition diagrams, commonly known as "alpha plots," are often used to visualize the species in a solution of an acid or base as the pH is varied. The diagram for 0.100 M acetic acid is shown here.

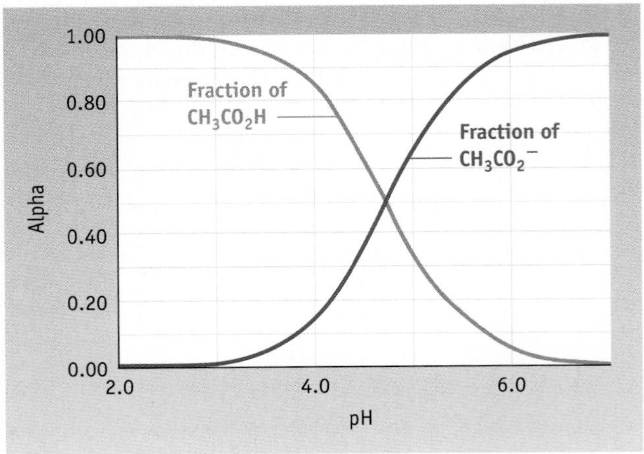

The plot shows how the fraction [= alpha (α)] of acetic acid in solution,

$$\alpha = \frac{[CH_3CO_2H]}{[CH_3CO_2H] + [CH_3CO_2^-]}$$

changes as the pH increases (blue curve). (The red curve shows how the fraction of acetate ion, $CH_3CO_2^-$, changes as the pH increases.) Alpha plots are another way of viewing the relative concentrations of acetic acid and acetate ion as a strong base is added to a solution of acetic acid in the course of a titration.
(a) Explain why the fraction of acetic acid declines and that of acetate ion increases as the pH increases.
(b) Which species predominates at a pH of 4, acetic acid or acetate ion? What is the situation at a pH of 6?
(c) Consider the point where the two lines cross. The fraction of acetic acid in the solution is 0.5, and so is that of acetate ion. That is, the solution is half acid and half conjugate base; their concentrations are equal. At this point, the graph shows the pH is 4.74. Explain why the pH at this point is 4.74.

110. The composition diagram, or alpha plot, for the important acid–base system of carbonic acid, H_2CO_3, is illustrated below. (See Study Question 109 for more information on such diagrams.)

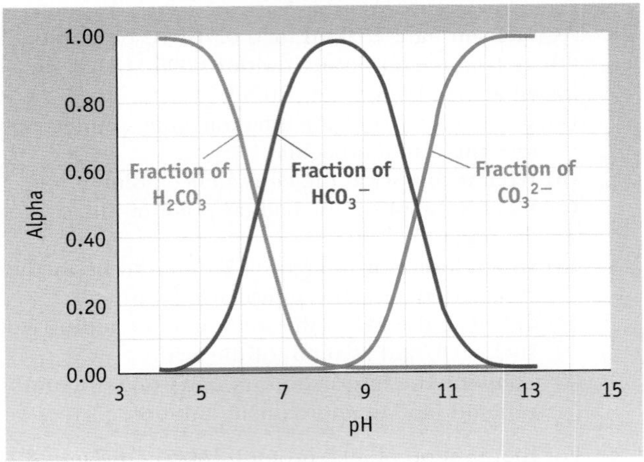

(a) Explain why the fraction of bicarbonate ion, HCO_3^-, rises and then falls as the pH increases.
(b) What is the composition of the solution when the pH is 6.0? When the pH is 10.0?
(c) If you wanted to buffer a solution at a pH of 11.0, what should be the ratio of HCO_3^- to CO_3^{2-}?

111. The chemical name for aspirin is acetylsalicylic acid. It is believed that the analgesic and other desirable properties of aspirin are due not to the aspirin itself but rather to the simpler compound salicylic acid, $C_6H_4(OH)CO_2H$, that results from the breakdown of aspirin in the stomach.

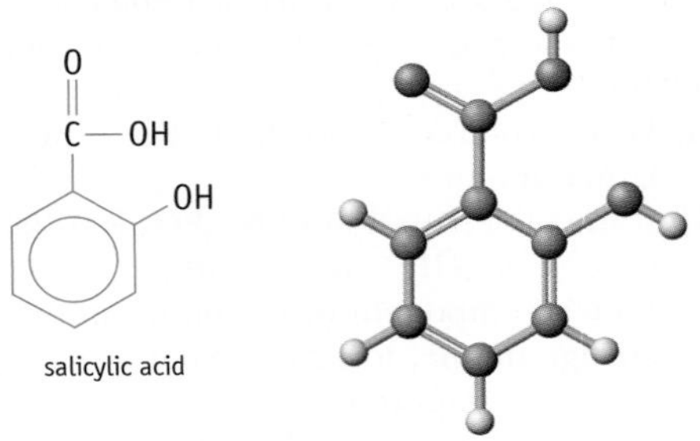

(a) Give approximate values for the following bond angles in the acid: (i) C—C—C in the ring; (ii) O—C=O; (iii) either of the C—O—H angles; and (iv) C—C—H.
(b) What is the hybridization of the C atoms of the ring? Of the C atom in the $—CO_2H$ group?
(c) Experiment shows that 1.00 g of the acid will dissolve in 460 mL of water. If the pH of this solution is 2.4, what is K_a for the acid?

(d) If you have salicylic acid in your stomach and if the pH of gastric juice is 2.0, calculate the percentage of salicylic acid that will be present in the stomach in the form of the salicylate ion, $C_6H_4(OH)CO_2^-$.

(e) Assume you have 25.0 mL of a 0.014 M solution of salicylic acid and titrate it with 0.010 M NaOH. What is the pH at the halfway point of the titration? What is the pH at the equivalence point?

112. Aluminum hydroxide reacts with phosphoric acid to give $AlPO_4$. The substance is used industrially as adhesives, binders, and cements.

(a) Write the balanced equation for the preparation of $AlPO_4$ from aluminum chloride and phosphoric acid.

(b) If you begin with 152 g of aluminum chloride and 3.00 L of 0.750 M phosphoric acid, what is the theoretical yield of $AlPO_4$?

(c) If you place 25.0 g of $AlPO_4$ in 1.00 L of water, what are the concentrations of Al^{3+} and PO_4^{3-} at equilibrium? (Neglect hydrolysis of aqueous Al^{3+} and PO_4^{3-} ion.) K_{sp} for $AlPO_4$ is 1.3×10^{-20}.

(d) Does the solubility of $AlPO_4$ increase or decrease on adding HCl? Explain.

Charles D. Winters

This is a sample of hydrated aluminum phosphate, a mineral known as augelite.

19 Principles of Reactivity: Entropy and Free Energy

Charles D. Winters

Thus far, the ethanol used as a fuel in the United States is largely derived from corn.

Can Ethanol Contribute to Energy and Environmental Goals?

About 3.4 billion gallons of ethanol were blended into gasoline in 2004. According to A. E. Farrell, et al. (*Science*, Vol. 311, pp. 506–508, 2006), this amounted to about 2% of all gasoline sold by volume and 1.3% of its energy content. This is a significant contribution to the energy used in the United States, and it promises to be larger in the future. Therefore, the use of ethanol as a substitute for gasoline has been widely discussed in the popular press and in the scientific literature.

Let us focus on two points at the beginning of this chapter: the energetics of ethanol combustion and the production of the greenhouse gas, CO_2. Let us compare the energy and amount of CO_2 released by burning 1.00 kg of liquid ethanol (C_2H_5OH) and, to represent gasoline, 1.00 kg of liquid octane (C_8H_{18}).

Questions:

1. Calculate $\Delta_rH°$ and $\Delta_rG°$ for the combustion of 1.00 kg each of liquid ethanol and liquid octane. Which fuel releases more energy per kilogram? [Values for $\Delta_fH°$ and $S°$ are located in Appendix L. In addition, for liquid octane at 298 K, $\Delta_fH° = -250.0$ kJ/mol and $S° = 361.2$ J/K · mol.]
2. Compare the two fuels on the basis of the release of CO_2, a common greenhouse gas. Which fuel produces more CO_2 per kilogram?
3. On the basis of this simple comparison and neglecting the energy costs involved in producing 1.00 kg each of ethanol and octane, which is the better fuel in terms of energy production and greenhouse gases?

Answers to these questions are in Appendix Q.

Chapter Goals

See Chapter Goals Revisited (page 886) for Study Questions keyed to these goals and assignable in OWL.

- Understand the concept of entropy and its relationship to spontaneity.
- Calculate the change in entropy for system, surroundings, and the universe to determine whether a process is spontaneous under standard conditions.
- Understand and use the Gibbs free energy to calculate free energy changes from standard free energies of formation and relate free energy changes to the equilibrium constant to decide if a reaction is product- or reactant-favored at equilibrium.

Chapter Outline

Change is central to chemistry, so it is important to understand the factors that determine whether a change will occur. In chemistry, we encounter many examples of chemical change (chemical reactions) and physical change (the formation of mixtures, expansion of gases, and changes of state, to name a few). Chemists use the term **spontaneous** to represent a change that occurs without outside intervention. *Spontaneous changes occur only in the direction that leads to equilibrium.* Whether or not the process is spontaneous does not tell us anything about the rate of the change or the extent to which a process will occur before equilibrium is reached. It says only that the change will occur in a specific direction (toward equilibrium) and will occur naturally and unaided.

If a piece of hot metal is placed in a beaker of cold water, energy is transferred spontaneously as heat from the hot metal to the cooler water (Figure 19.1), and energy transfer will continue until the two objects are at the same temperature and thermal equilibrium is attained. Similarly, chemical reactions proceed spontaneously until equilibrium is reached, regardless of whether the position of the equilibrium favors products or reactants. We readily recognize that, starting with pure reactants, all product-favored reactions must be spontaneous, such as the formation of water from gaseous hydrogen and oxygen and the neutralization of H_3O^+(aq) and OH^-(aq) (◀ Chapter 3). Notice, however, that reactant-favored reactions are also spontaneous until equilibrium is achieved. Even though the dissolution of $CaCO_3$ is reactant-favored at equilibrium, if you place a handful of $CaCO_3$ in water, the process of dissolving will proceed spontaneously until equilibrium is reached.

All physical and chemical changes occur in a direction toward achieving equilibrium. Systems never change spontaneously in a direction that takes them farther from equilibrium. Given two objects at the same temperature, in contact but thermally isolated from their surroundings, it will never happen that one will heat up while the other becomes colder. Gas molecules will never spontaneously congregate at one end of a flask. Similarly, once equilibrium is established, the small amount of dissolved $CaCO_3$ in equilibrium with solid $CaCO_3$ will not spontaneously precipitate from solution.

The factors that determine the directionality and extent of change are among the topics of this chapter.

Chemistry Now™

Throughout the text this icon introduces an opportunity for self-study or to explore interactive tutorials by signing in at **www.thomsonedu.com/login.**

Charles D. Winters

FIGURE 19.1 A spontaneous process. The heated metal cylinder is placed in water. Energy transfers as heat spontaneously from the metal to water, that is, from the hotter object to the cooler object.

Problem Solving Tip 19.1 A Review of Concepts of Thermodynamics

To understand the thermodynamic concepts introduced in this chapter, be sure to review the ideas of Chapter 5.

System: The part of the universe under study.

Surroundings: The rest of the universe exclusive of the system, capable of exchanging energy and/or matter with the system.

Exothermic: Energy transfers as heat from the system to the surroundings.

Endothermic: Energy transfers as heat from the surroundings to the system.

First law of thermodynamics: The law of conservation of energy, $\Delta U = q + w$. The change in internal energy of a system is the sum of energy transferred as heat to or from the system and the work done on or by the system. The first law states that energy cannot be created or destroyed.

Enthalpy change: The energy transferred as heat at constant pressure.

State function: A quantity whose value depends only on the state of the system; changes in a state function can be calculated by taking into account the initial and final states of a system.

Standard conditions: Pressure of 1 bar (1 bar = 0.98692 atm) and solution concentration of 1 *m*.

Standard enthalpy of formation, $\Delta_f H°$: The enthalpy change occurring when a compound is formed from its elements in their standard states.

19.1 Spontaneity and Energy Transfer as Heat

■ **Spontaneous Processes** A spontaneous physical or chemical change proceeds to equilibrium without outside intervention. Such a process may or may not be product-favored at equilibrium.

We can readily recognize many chemical reactions that are spontaneous, such as hydrogen and oxygen combining to form water, methane burning to give CO_2 and H_2O, Na and Cl_2 reacting to form NaCl, HCl(aq), and NaOH(aq), reacting to form H_2O and NaCl(aq). A common feature of these reactions is that they are exothermic, so it would be tempting to conclude that evolution of energy as heat is the criterion that determines whether a reaction or process is spontaneous. Further inspection, however, reveals significant flaws in this reasoning. This is especially evident with the inclusion of some common spontaneous changes that are endothermic or energy neutral:

- *Dissolving NH_4NO_3.* The ionic compound NH_4NO_3 dissolves spontaneously in water. The process is endothermic ($\Delta_r H° = +25.7$ kJ/mol).
- *Expansion of a gas into a vacuum.* A system is set up with two flasks connected by a valve (Figure 19.2). One flask is filled with a gas, and the other is evacuated. When the valve is opened, the gas will flow spontaneously from one flask to the other until the pressure is the same throughout. The expansion of an ideal gas is energy neutral (although expansion of most real gases is endothermic).
- *Phase changes.* Melting of ice is an endothermic process. Above 0 °C, the melting of ice is spontaneous. Below 0 °C, melting of ice is not spontaneous. At 0 °C, no net change will occur; liquid water and ice coexist at equilibrium. This example illustrates that temperature can have a role in determining spontaneity and that equilibrium is somehow an important aspect of the problem.
- *Energy transfer as heat.* The temperature of a cold soft drink sitting in a warm environment will rise until the beverage reaches the ambient temperature. The energy required for this process comes from the surroundings. Energy transfer as heat from a hotter object (the surroundings) to a cooler object (the soft drink) is spontaneous.
- The reaction of H_2 and I_2 to form HI is endothermic, and the reverse reaction, the decomposition of HI to form H_2 and I_2, is exothermic. If $H_2(g)$ and $I_2(g)$ are mixed, a reaction forming HI will occur [$H_2(g) + I_2(g) \rightleftharpoons 2\ HI(g)$] until equilibrium is reached. Furthermore, if HI(g) is placed in a container, there will also be a reaction, but in the reverse direction, until equilibrium is achieved. Notice that approach to equilibrium occurs spontaneously from either direction.

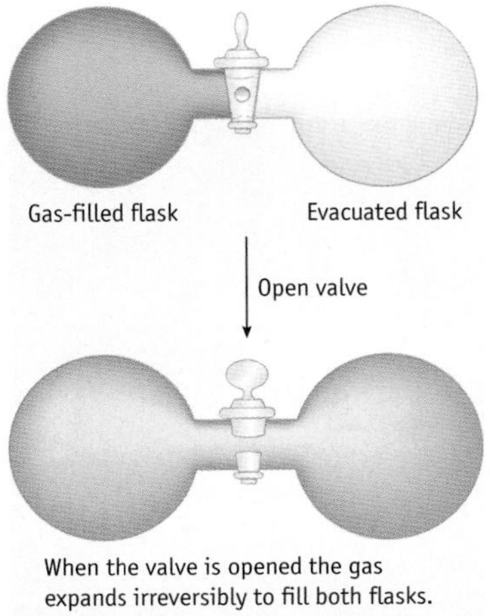

FIGURE 19.2 Spontaneous expansion of a gas. (See ChemistryNow Screen 19.3 to view an animation of this figure.)

On further reflection, it is logical to conclude that evolution of heat cannot be a sufficient criterion in determining spontaneity. The first law of thermodynamics tells us that in any process energy must be conserved. If energy is transferred out of a system, then the same amount of energy must be transferred to the surroundings. Exothermicity of the system is always accompanied by an endothermic change in the surroundings. If energy evolution were the only factor determining whether a change is spontaneous, then for every spontaneous process there would be a corresponding nonspontaneous change in the surroundings. We must search further than the first law of thermodynamics to determine whether a change is spontaneous.

19.2 Dispersal of Energy: Entropy

We have shown that we cannot use energy itself as an indicator of spontaneity because energy is conserved in any process; we always end up with the same amount of energy as we had at the beginning. Imagine dropping this book on the floor. (But don't actually do it!) It would fall spontaneously. The initial potential energy it has from being a certain distance above the floor is converted to kinetic energy. When the book hits the floor, the kinetic energy of the book is converted into other forms of kinetic energy: acoustic energy and thermal energy of the book, floor, and air, since these are all heated up somewhat. Through all of these processes, the total energy is conserved. There is a directionality to this process, however. The book will spontaneously fall to the ground, but we will not observe a book on its own jump spontaneously from the floor up to a desk. Is there a way to predict this directionality?

■ **Entropy** For a more complete discussion of entropy, see the papers by F. L. Lambert, such as "Entropy Is Simple, Qualitatively," *Journal of Chemical Education*, Vol. 79, pp. 1241–1246, 2002, and references therein. See also Lambert's site: **www.entropysite.com.** Finally, see A. H. Jungermann, "Entropy and the Shelf Model," *Journal of Chemical Education*, Vol. 83, pp. 1686–1694, 2006.

Let us consider the initial and final states for this process. Initially, the energy is concentrated in the book—the book has a certain potential energy as it is held above the floor. At the end of the process, the energy has been dispersed to the air, floor, and book. The energy has gone from being concentrated to being more dispersed. This is the indicator for which we have been searching. *In a spontaneous process, energy goes from being more concentrated to being more dispersed.*

There is a state function (◀ Section 5.4) called **entropy (S)** that allows us to quantify this. The **second law of thermodynamics** states that *a spontaneous process is one that results in an increase of the entropy of the universe.* In a spontaneous process, therefore, ΔS(universe) is greater than zero; this corresponds to energy being dispersed in the process.

Because thermal energy is caused by the random motion of particles, potential energy is dispersed when it is converted to thermal energy. This conversion occurs when energy is transferred as heat, q. It is therefore not surprising that q is a part of the mathematical definition of ΔS. This is not the whole picture, however, in part because q is not a state function. In addition, the effect of a given quantity of energy transferred as heat on energy dispersal is different at different temperatures. It turns out that a given q has a greater effect at a lower temperature than at a higher temperature; that is, the extent of energy dispersal is inversely proportional to the temperature.

Our proposed definition for ΔS is thus q/T, but this is still not quite correct. We must be a little more specific about q. The value of q used in the calculation of an entropy change must be the energy transferred as heat under what are called *reversible conditions*, q_{rev} (see *A Closer Look: Reversible and Irreversible Processes*). Adding energy by heating an object slowly (adding energy in very small increments) approximates a reversible process. Our mathematical definition of ΔS is therefore q_{rev} divided by the absolute (Kelvin) temperature:

$$\Delta S = \frac{q_{rev}}{T} \qquad \textbf{(19.1)}$$

As this equation predicts, the units for entropy are J/K.

A Closer Look

Reversible and Irreversible Processes

To determine entropy changes experimentally, the energy transferred by heating and cooling must be measured for a reversible process. But what is a reversible process?

The test for reversibility is that after carrying out a change along a given path (for example, energy added as heat), it must be possible to return to the starting point by the same path (energy taken away as heat) without altering the surroundings. Melting of ice and freezing of water at 0 °C are examples of reversible processes. Given a mixture of ice and water at equilibrium, adding energy as heat in small increments will convert ice to water; removing energy as heat in small increments will convert water back to ice.

Reversibility is closely associated with equilibrium. Assume that we have a system at equilibrium. Reversible changes can then be made by very slightly perturbing the equilibrium and letting the system readjust.

Spontaneous processes are often not reversible. Suppose a gas is allowed to expand into a vacuum. No work is done in this process because there is no force to resist expansion. To return the system to its original state, it is necessary to compress the gas. Doing so means doing work on the system, however, because the system will not return to its original state on its own. In this process, the energy content of the surroundings decreases by the amount of work expended by the surroundings. The system can be restored to its original state, but the surroundings will be altered in the process.

In summary, there are two important points concerning reversibility:

- At every step along a reversible pathway between two states, the system remains at equilibrium.
- Spontaneous processes often follow irreversible pathways and involve nonequilibrium conditions.

To determine the entropy change for a process, it is necessary to identify a reversible pathway. Only then can an entropy change for the process be calculated from q_{rev} and the Kelvin temperature.

19.3 Entropy: A Microscopic Understanding

We have stated that entropy is a measure of the extent of energy dispersal. In all spontaneous physical and chemical changes, energy changes from being localized or concentrated in a system to becoming dispersed or spread out in the system and its surroundings. In a spontaneous process, the change in entropy, ΔS, of the universe indicates the extent to which energy is dispersed in a process carried out at a temperature T. So far, however, we have not explained why dispersal of energy occurs, nor have we given an equation for entropy itself. In order to do this, we will need to consider energy in its quantized form and matter on the atomic level.

Dispersal of Energy

We can explore the dispersal of energy using a simple example: energy being transferred as heat between hot and cold gaseous atoms. Consider an experiment involving two containers, one holding hot atoms and the other with cold atoms (Figure 19.3). Because they have translational energy, the atoms move randomly in each container and collide with the walls. When the containers are in contact, energy is transferred through the container walls. Eventually, both containers will be at the same temperature; the energy originally localized in the hotter atoms is distributed over a greater number of atoms; and the atoms in each container will have the same distribution of energies.

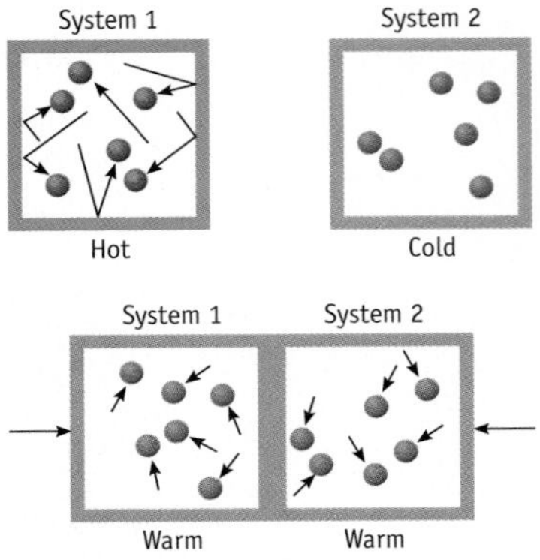

FIGURE 19.3 Energy transfer between molecules in the gas phase.

For further insight, let us look at a statistical explanation to show why energy is dispersed in a system. With statistical arguments, systems must include large numbers of particles for the arguments to be accurate. It will be easiest, however, if we look first at simple examples to understand the underlying concept and then extrapolate our conclusions to larger systems.

Consider a simple system in which, initially, there is one atom (1) with two discrete packets, or quanta, of energy and three other atoms (2, 3, and 4) with no energy (Figure 19.4). When these four atoms are brought together, the total energy in the system is 2 quanta. Collisions among the atoms allow energy to be transferred so that,

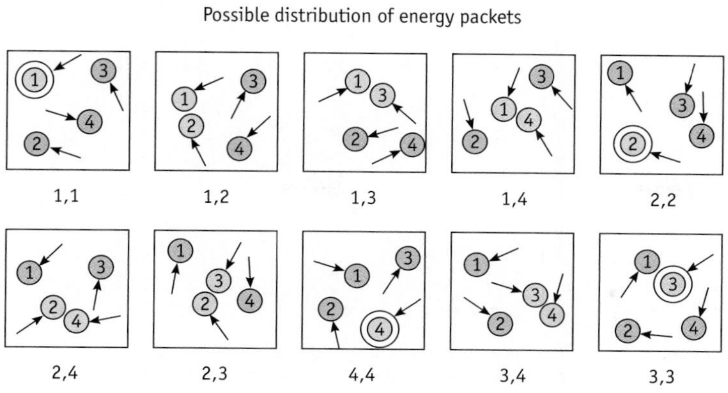

FIGURE 19.4 Energy dispersal. Possible ways of distributing two packets of energy among four atoms. To keep our analysis simple, we assume that initially there is one atom with two quanta of energy and three atoms (2, 3 and 4) with no energy. The figure shows that there are 10 different ways to distribute the two quanta of energy over four atoms.

over time, all distributions of the two packets of energy over the four atoms are seen. There are 10 different ways to distribute these 2 quanta of energy over the four atoms. Each of these 10 different ways to distribute the energy is called a **microstate.** In only one of the microstates do the 2 quanta remain on atom 1. In fact, only in 4 of the 10 microstates [1, 1; 2, 2; 3, 3; and 4, 4] is the energy concentrated on a single atom. In the majority of cases, 6 out of 10, the energy is distributed to two different atoms. Even in a small sample (four atoms) with only two packets of energy, it is more likely that at any given time the energy will be distributed to two atoms rather than concentrated on a single atom. There is a distinct preference that the energy will be dispersed over a greater number of atoms.

Let us now add more atoms to our system. We again consider one atom (1) with 2 quanta of energy present but now have five other atoms (2, 3, 4, 5, and 6) with no energy initially. Collisions let the energy be transferred between the atoms, and we now find there are 21 possible microstates (Figure 19.5). There are six microstates in which the energy is concentrated on one atom, including one in which the energy is still on atom 1, but there are now 15 microstates in which the energy is present on two atoms. As the number of particles increases, the number of microstates available increases dramatically, and the fraction of microstates in which the energy is concentrated rather than dispersed goes down dramatically. It is much more likely that the energy will be dispersed than concentrated.

Now let us return to an example using a total of four atoms, but in which we have increased the quantity of energy from 2 quanta to 6 quanta. Assume that we start with two atoms having 3 quanta of energy each. The other two atoms initially have zero energy (Figure 19.6). Through collisions, energy can be transferred to achieve different distributions of energy among the four atoms. In all, there are 84 microstates, falling into nine basic patterns. For example, one possible arrangement has one atom with 3 quanta of energy, and three atoms with 1 quantum each. There are four microstates in which this is true (Figure 19.6c). Increasing the number of quanta from 2 to 6 with the same number of atoms increased the num-

Number of Microstates	Distribution of 2 Quanta of Energy Among Six Atoms					
6	1:1	2:2	3:3	4:4	5:5	6:6
5	1:2	2:3	3:4	4:5	5:6	
4	1:3	2:4	3:5	4:6		
3	1:4	2:5	3:6			
2	1:5	2:6				
1	1:6					

FIGURE 19.5 Distributing 2 quanta of energy among six atoms. The result is a total of 21 ways—21 microstates—of distributing 2 quanta of energy among six atoms.

FIGURE 19.6 Energy dispersal. Possible ways of distributing 6 quanta of energy among four atoms. A total of 84 microstates is possible.

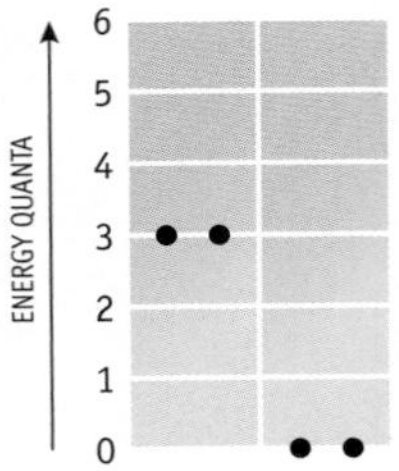

(a) Initially, four particles are separated from each other. Two particles each have 3 quanta of energy, and the other two have none. A total of 6 quanta of energy will be distributed once the particles interact.

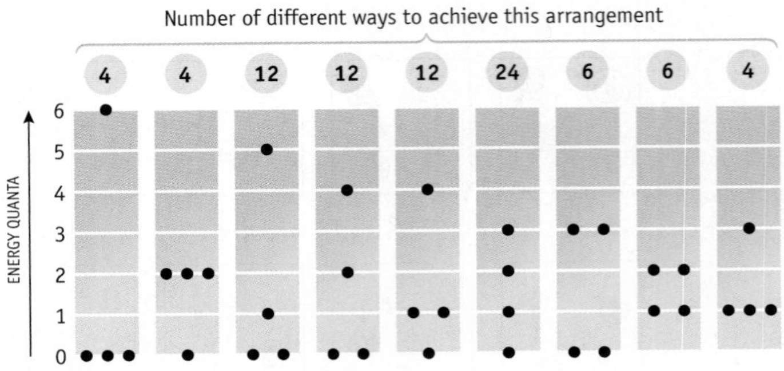

(b) Once the particles begin to interact, there are nine ways to distribute the 6 available quanta. Each of these arrangements will have multiple ways of distributing the energy among the four atoms. Part (c) shows how the arrangement on the right can be achieved four ways.

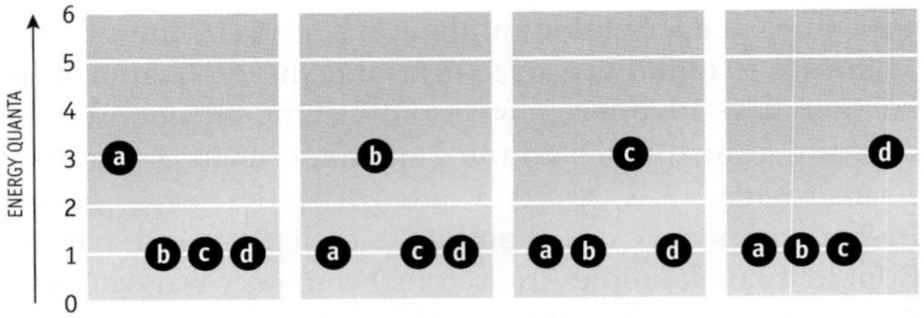

(c) There are four different ways to have four particles (a, b, c, and d) such that one particle has 3 quanta of energy and the other three each have 1 quantum of energy.

Ludwig Boltzmann (1844–1906). Engraved on his tombstone in Vienna, Austria, is his equation defining entropy. The constant k is now known as Boltzmann's constant.

Oesper Collection in the History of Chemistry, University of Cincinnati

■ **How Many Microstates?** To give you a sense of the number of microstates available to a substance, consider a mole of ice at 273 K, where $S° = 41.3$ J/K · mol. Using Boltzmann's equation, we find that $W = 10^{1,299,000,000,000,000,000,000,000}$. That is, there are many, many more microstates for 1 mol of ice than there are atoms in the universe (about 10^{80}).

ber of possible microstates from 10 to 84. In this case, the amount of energy that is dispersed resulted in an increase in the number of microstates.

A statistical analysis for larger aggregates of atoms and energy quanta is increasingly complex, but the conclusions are even more compelling. As the number of particles and/or quanta increases, the number of energy microstates grows rapidly. These allow the extent over which the energy is dispersed and/or the amount of the energy dispersed to increase. Ludwig Boltzmann proposed that the entropy of a system (the dispersal of energy at a given temperature) results from the number of microstates available. *As the number of microstates increases, so does the entropy of the system.* He enunciated this idea over 100 years ago in the equation

$$S = k \ln W \tag{19.2}$$

which states that the entropy of a system, S, is proportional to the natural logarithm of the number of accessible microstates, W, that belong to a given energy of a system or substance. (The proportionality constant, k, is now known as the **Boltzmann's constant.**) Within these microstates, it turns out that those states that disperse energy over the largest number of atoms are vastly more probable than the others.

Dispersal of Matter: Dispersal of Energy Revisited

In many processes, it appears that the dispersal of matter also contributes to spontaneity. We shall see, however, that these effects can also be explained in terms of energy dispersal. Let us examine a specific case. Matter dispersal was illustrated in

Figure 19.2 by the expansion of a gas into a vacuum. How is this spontaneous expansion of a gas related to energy dispersal and entropy?

We begin with the premise that all energy is quantized and that this applies to any system, including gas molecules in a room or in a reaction flask. You know from the previous discussion of kinetic molecular theory that the molecules in a gas sample have a distribution of energies (◀ Figure 11.14) (often referred to as a Boltzmann distribution). The molecules are assigned to (or "occupy") quantized microstates, most of them in states near the average energy of the system, but fewer of them in states of high or low energy. (For a gas in a laboratory-sized container, the energy levels are so closely spaced that, for most purposes, there is a continuum of energy states.)

When the gas expands to fill a larger container, the average energy of the sample and the energy for the particles in a given energy range are constant. However, quantum mechanics shows (for now, you will have to take our word for it) that as a consequence of having a larger volume in which the molecules can move in the expanded state, there is an increase in the number of microstates and that those microstates are even more closely spaced than before (Figure 19.7). The result of this greater density of microstates is that the number of microstates available to the gas particles increases when the gas expands. Gas expansion, a dispersal of matter, leads to the dispersal of energy over a larger number of more closely spaced microstates and thus to an increase in entropy.

The logic applied to the expansion of a gas into a vacuum can be used to rationalize the mixing of two gases, the mixing of two liquids, or the dissolution of a solid in a liquid (Figure 19.8). For example, if flasks containing O_2 and N_2 are connected (in an experimental setup like that in Figure 19.2), the two gases diffuse together, eventually leading to a mixture in which O_2 and N_2 molecules are evenly distributed throughout the total volume. A mixture of O_2 and N_2 will never separate into samples of each component of its own accord. The gases spontaneously move toward a situation in which each gas and its energy are maximally dispersed. The energy of the system is dispersed over a larger number of microstates, and the entropy of the system increases. Indeed, this is a large part of the explanation for the fact that similar liquids (such as oil and gasoline or water and ethanol) will readily form homogeneous solutions. Recall the rule of thumb that "like dissolves like" (◀ Chapter 14).

■ **Statistical Thermodynamics** The arguments presented here come from a branch of chemistry called statistical thermodynamics. For an accessible treatment see H. Jungermann, *Journal of Chemical Education*, Vol. 83, pp. 1686–1694, 2006.

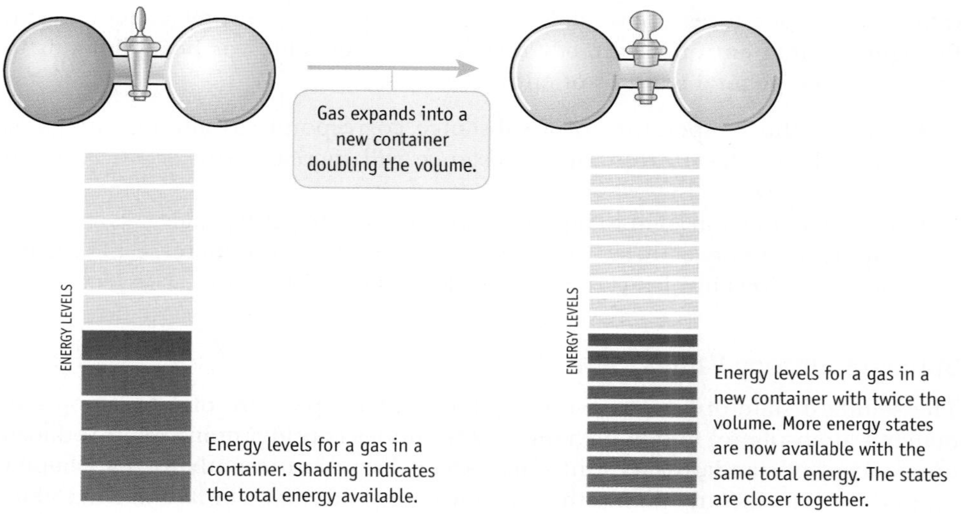

FIGURE 19.7 Energy (and matter) dispersal. As the size of the container for the chemical or physical change increases, the number of microstates accessible to the atoms or molecules of the system increases, and the density of states increases. A consequence of the distribution of molecules over a greater number of microstates is an increase in entropy.

Note that for a gas in a container of the size likely to be found in a laboratory, the energy levels are so closely spaced that we do not need to think in terms of quantization of energy levels. For most purposes, the system can be regarded as having a continuum of energy levels.

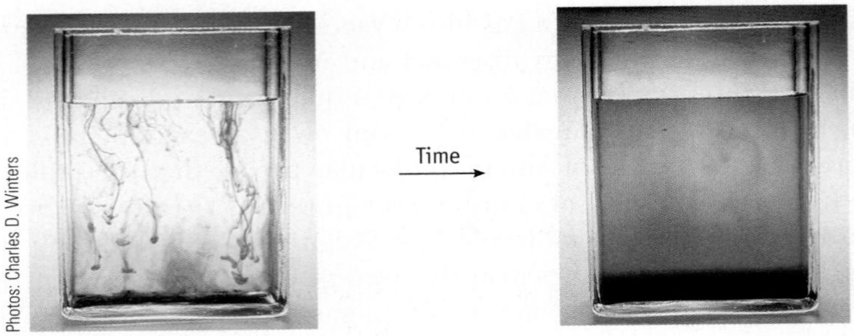

FIGURE 19.8 Dissolving $KMnO_4$ in water. A small quantity of solid, purple $KMnO_4$ is added to water (left). With time, the solid dissolves, and the highly colored MnO_4^- ions (and the K^+ ions) become dispersed throughout the solution. Entropy makes a large contribution to the mixing of liquids and solutions.

A Summary: Entropy, Entropy Change, and Energy Dispersal

According to Boltzmann's equation for entropy, entropy is proportional to the number of ways that energy can be dispersed in a substance, that is, to the number of microstates available to the system (W). Thus, there will be a change in entropy, ΔS, if there is a change in the number of microstates over which energy can be dispersed.

$$\Delta S = S_{final} - S_{initial} = k\ (\ln W_{final} - \ln W_{initial}) = k \ln(W_{final}/W_{initial})$$

■ **Entropy Change on Gas Expansion** The entropy change for a gas expansion can be calculated from

$$\Delta S = nR\ln(V_{final}/V_{initial})$$

At a given temperature, V is proportional to the number of microstates, so the equation is related to $k \ln(W_{final}/W_{initial})$.

Our focus as chemists is on ΔS, and we shall be mainly concerned with the dispersion of energy in systems and surroundings during a physical or chemical change.

19.4 Entropy Measurement and Values

For any substance under a given set of conditions, a numerical value for entropy can be determined. The greater the dispersal of energy, the greater the entropy and the larger the value of S. The point of reference for entropy values is established by the **third law of thermodynamics.** Defined by Ludwig Boltzmann, the third law states that *a perfect crystal at 0 K has zero entropy; that is, $S = 0$.* The entropy of an element or compound under any other set of conditions is the entropy gained by converting the substance from 0 K to those conditions.

To determine the value of S, it is necessary to measure the energy transferred as heat under reversible conditions for the conversion from 0 K to the defined conditions and then to use Equation 19.1 ($\Delta S = q_{rev}/T$). Because it is necessary to add energy as heat to raise the temperature, all substances have positive entropy values at temperatures above 0 K. Negative values of entropy cannot occur. Recognizing that entropy is directly related to energy added as heat allows us to predict several general features of entropy values:

■ **Negative Entropy Values** A glance at thermodynamic tables indicates that ions in aqueous solution can and do have negative entropy values listed. However, these are not absolute entropies. For ions, the entropy of $H^+(aq)$ is arbitrarily assigned a standard entropy of zero.

- Raising the temperature of a substance corresponds to adding energy as heat. Thus, the entropy of a substance will increase with an increase in temperature.
- Conversions from solid to liquid and from liquid to gas typically require large inputs of energy as heat. Consequently, there is a large increase in entropy in conversions involving changes of state (Figure 19.9).

Standard Entropy Values, $S°$

The standard state of a substance is its state under a pressure of 1 bar (approximately 1 atmosphere) and at the temperature under consideration. We introduced the concept of standard states into the earlier discussion of enthalpy (◄ Chapter 5), and we can similarly define the entropy of any substance in its standard state.

Photos: Charles D. Winters

FIGURE 19.9 Entropy and states of matter. (a) The entropy of liquid bromine, $Br_2(\ell)$, is 152.2 J/K · mol, and that for bromine vapor is 245.47 J/K · mol. (b) The entropy of ice, which has a highly ordered molecular arrangement, is smaller than the entropy of liquid water. (Sign in at **www.thomsonedu.com/login** and go to Chapter 19 Contents to see Screen 19.4 for a self-study module on entropy.)

The standard entropy, $S°$, of a substance is the entropy gained by converting it from a perfect crystal at 0 K to standard state conditions (1 bar, 1 molal for a solution). The units for standard entropy values are J/K · mol. Generally, values of $S°$ found in tables of data refer to a temperature of 298 K. Standard entropies at 298 K for a few substances are given in Table 19.1, and many more are available in Appendix L. More extensive lists of $S°$ values can be found in standard reference sources such as the NIST tables (**webbook.nist.gov**).

Scanning a list of standard entropies (such as those in Appendix L) will show that *large molecules generally have larger entropies than small molecules.* This seems reasonable; with a larger molecule, there are more ways for the molecule to rotate and vibrate, which provides a larger number of energy microstates over which energy can be distributed. As an example, consider the standard entropies for methane (CH_4), ethane (C_2H_6), and propane (C_3H_8), whose values are 186.3, 229.2, and 270.3 J/K · mol, respectively. Also, *molecules with more complex structures have larger entropies than molecules with simpler structures.* The effect of molecular structure can also be seen when comparing atoms or molecules of similar molar mass: Gaseous argon, CO_2, and C_3H_8 have entropies of 154.9, 213.7, and 270.3 J/K · mol, respectively.

Tables of entropy values also confirm the hypothesis that *entropies of gases are larger than those for liquids, and entropies of liquids are larger than those for solids.* In a solid, the particles have fixed positions in the solid lattice. When a solid melts, these particles have more freedom to assume different positions, resulting in an increase in the number of microstates available and an increase in entropy. When a liquid evaporates, constraints due to forces between the particles nearly disappear; the volume increases greatly; and a large entropy increase occurs. For example, the standard entropies of $I_2(s)$, $Br_2(\ell)$, and $Cl_2(g)$ are 116.1, 152.2, and 223.1 J/K · mol, respectively.

Finally, as illustrated in Figure 19.9, *for a given substance, a large increase in entropy accompanies changes of state,* reflecting the relatively large energy transfer as heat required to carry out these processes (as well as the dispersion of energy over a larger number of available microstates). For example, the entropies of liquid and gaseous water are 65.95 and 188.84 J/K · mol.

TABLE 19.1 Standard Molar Entropy Values at 298 K

Substance	Entropy, $S°$ (J/K · mol)
C(graphite)	5.6
C(diamond)	2.377
C(vapor)	158.1
$H_2(g)$	130.7
$O_2(g)$	205.1
$H_2O(g)$	188.84
$H_2O(\ell)$	69.95

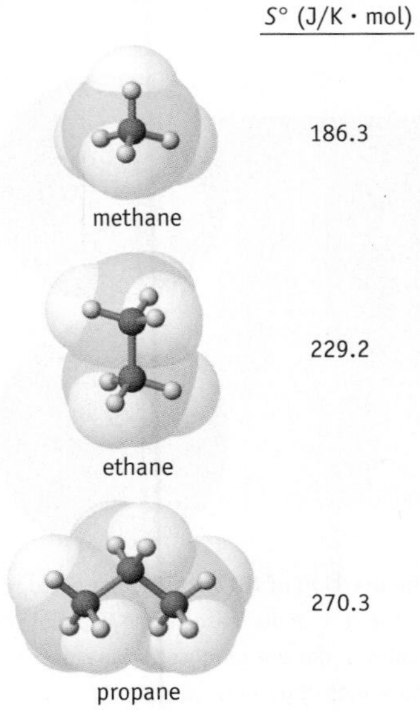

EXAMPLE 19.1 Entropy Comparisons

Problem Which substance has the higher entropy under standard conditions? Explain your reasoning. Check your answer against data in Appendix L.

(a) $NO_2(g)$ or $N_2O_4(g)$

(b) $I_2(g)$ or $I_2(s)$

Strategy Entropy decreases in the order gas > liquid > solid, and larger molecules of related substances have greater entropy than smaller molecules.

Solution

(a) Both NO_2 and N_2O_4 are gases. N_2O_4 is a larger molecule than NO_2 and so is expected to have the higher standard entropy. $S°$ values in Appendix L confirm this prediction: $S°$ for $NO_2(g)$ is 240.04 J/K · mol, and $S°$ for $N_2O_4(g)$ is 304.38 J/K · mol.

(b) For a given substance, gases have higher entropies than solids. $S°$ for $I_2(g)$ is 260.69 J/K · mol; $S°$ for $I_2(s)$ is 116.135 J/K · mol.

EXERCISE 19.1 Entropy Comparisons

Predict which substance has the higher entropy and explain your reasoning.

(a) $O_2(g)$ or $O_3(g)$

(b) $SnCl_4(\ell)$ or $SnCl_4(g)$

Determining Entropy Changes in Physical and Chemical Processes

It is also possible to use standard entropy values quantitatively to calculate the change in entropy that occurs in various processes under standard conditions. The standard entropy change is the sum of the standard entropies of the products minus the sum of the standard entropies of the reactants.

$$\Delta_r S° = \Sigma S°(\text{products}) - \Sigma S°(\text{reactants}) \qquad (19.3)$$

This equation allows us to calculate entropy changes for a *system* in which reactants are completely converted to products, under standard conditions. To illustrate, let us calculate $\Delta_r S°$ for the oxidation of NO with O_2.

$$2\ NO(g) + O_2(g) \rightarrow 2\ NO_2(g)$$

Here, we subtract the entropies of the reactants (2 mol NO and 1 mol O_2) from the entropy of the products (2 mol NO_2).

$$\begin{aligned}\Delta_r S° &= (2\ \text{mol}\ NO_2/\text{mol-rxn})(240.0\ \text{J/K}\cdot\text{mol}) - \\ &\quad [(2\ \text{mol}\ NO(g)/\text{mol-rxn})(210.8\ \text{J/K}\cdot\text{mol}) + (1\ \text{mol}\ O_2/\text{mol-rxn})(205.1\ \text{J/K}\cdot\text{mol})] \\ &= -146.7\ \text{J/K}\cdot\text{mol-rxn}\end{aligned}$$

The entropy of the system decreases, as is generally observed when some number of gaseous reactants has been converted to fewer molecules of gaseous products.

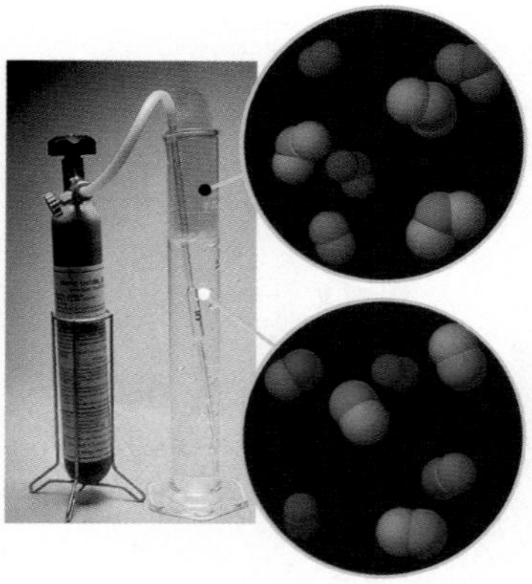

Photos: Charles D. Winters

The reaction of NO with O_2. The entropy of the system decreases when two molecules of gas are produced from three molecules of gaseous reactants.

EXAMPLE 19.2 Predicting and Calculating $\Delta_r S°$ for a Reaction

Problem Calculate the standard entropy changes for the following processes. Do the calculations match predictions?

(a) Evaporation of 1.00 mol of liquid ethanol to ethanol vapor

$$C_2H_5OH(\ell) \rightarrow C_2H_5OH(g)$$

(b) Formation of ammonia from hydrogen and nitrogen.

$$N_2(g) + 3\ H_2(g) \rightarrow 2\ NH_3(g)$$

Problem Solving Tip 19.2 A List of Common Entropy-Favored Processes

The discussion to this point, along with the examples and exercises, allows the listing of several general principles involving entropy changes:

- The entropy of a substance will increase in going from a solid to a liquid to a gas.
- The entropy of any substance increases as the temperature is raised. Energy must be added to a system to increase its temperature (that is, $q > 0$), so q_{rev}/T is necessarily positive.
- The entropy of a gas increases with an increase in volume. A larger volume provides a larger number of energy states over which to disperse energy.
- Reactions that increase the number of moles of gases in a system are accompanied by an increase in entropy.

Strategy Entropy changes for each system can be calculated from values of standard entropies (Appendix L) using Equation 19.3. Predictions are made using the guidelines given in the text: An entropy increase is predicted going from solid to liquid to gas and if there is an increase in the number of moles of gas.

Solution

(a) Evaporation of ethanol

$$\Delta_r S° = \Sigma S°(\text{products}) - \Sigma S°(\text{reactants})$$

$$= (1 \text{ mol } C_2H_5OH(g)/\text{mol-rxn})\{S°[C_2H_5OH(g)]\} - (1 \text{ mol } C_2H_5OH(\ell)/\text{mol-rxn})\{S°[C_2H_5OH(\ell)]\}$$

$$= (1 \text{ mol } C_2H_5OH(g)/\text{mol-rxn})(282.70 \text{ J/K} \cdot \text{mol}) - (1 \text{ mol } C_2H_5OH(\ell)/\text{ mol-rxn})(160.7 \text{ J/K} \cdot \text{mol})$$

$$= +122.0 \text{ J/K} \cdot \text{mol-rxn}$$

A large positive value for the entropy change is expected because the process converts ethanol from a liquid to a vapor.

(b) Formation of ammonia

$$\Delta_r S° = \Sigma S°(\text{products}) - \Sigma S°(\text{reactants})$$

$$= (2 \text{ mol } NH_3(g)/\text{mol-rxn})\{S°[NH_3(g)]\} - [(1 \text{ mol } N_2(g)/\text{mol-rxn})\{S°[N_2(g)]\} + (3 \text{ mol } H_2(g)/\text{mol-rxn})\{S°[H_2(g)]\}]$$

$$= (2 \text{ mol } NH_3(g)/\text{mol-rxn})(192.77 \text{ J/K} \cdot \text{mol}) - [(1 \text{ mol } N_2(g)/\text{mol-rxn})(191.56 \text{ J/K} \cdot \text{mol}) + (3 \text{ mol } H_2(g)/\text{mol-rxn})(130.7 \text{ J/K} \cdot \text{mol})]$$

$$= -198.12 \text{ J/K} \cdot \text{mol-rxn}$$

A decrease in entropy is predicted for this reaction because the number of moles of gases decreases from four to two.

Comment Values of entropies in tables are based on 1 mol of the compound. In part (b), the number of moles of reactants and products per mole of reaction is defined by the stoichiometric coefficients in the balanced chemical equation.

■ **Amount of Substance and Thermodynamic Calculations** In the calculation here and in all others in this chapter, when we write, for example,

282.70 J/K · mol

for the standard entropy of ethanol at 298 K, we mean

282.70 J/K · mol $C_2H_5OH(\ell)$

The identifying formula has been left off for the sake of simplicity.

EXERCISE 19.2 Calculating the Entropy Change for a Reaction, $\Delta_r S°$

Calculate the standard entropy changes for the following processes using the entropy values in Appendix L. Is the sign of the calculated values of $\Delta_r S°$ in accord with predictions?

(a) Dissolving 1 mol of $NH_4Cl(s)$ in water: $NH_4Cl(s) \rightarrow NH_4Cl(aq)$

(b) Oxidation of ethanol: $C_2H_5OH(g) + 3\ O_2(g) \rightarrow 2\ CO_2(g) + 3\ H_2O(g)$

19.5 Entropy Changes and Spontaneity

As illustrated by Example 19.2, the standard entropy change *for the system* in a physical or chemical change can be either positive (evaporation of ethanol) or negative (synthesis of ammonia from nitrogen and hydrogen). How does this information contribute to determining the spontaneity of the process?

Spontaneity and the Second Law Spontaneous change is accompanied by an increase in entropy in the universe. This is in contrast to enthalpy and internal energy. According to the first law, the energy contained in the universe is constant.

As discussed previously (page 863), spontaneity is determined by the *second law of thermodynamics,* which states that *a spontaneous process is one that results in an increase of entropy in the universe.* The universe has two parts: the system and its surroundings (◀ Section 5.1), and it makes sense that the entropy change for the universe is the sum of the entropy changes for the system and the surroundings. Under standard conditions, the entropy change for the universe, $\Delta S°$(universe) is

$$\Delta S°(\text{universe}) = \Delta S°(\text{system}) + \Delta S°(\text{surroundings}) \quad \textbf{(19.4)}$$

The calculation in Example 19.2 gave us the entropy change under standard conditions for a system, only half of the information needed. Now we will have to determine how the change being studied affects the entropy of the surroundings so that we can then find the entropy change for the universe.

The value of $\Delta S°$(universe) calculated from Equation 19.4 is the entropy change when reactants are converted *completely* to products, with all species at standard conditions. *A process is spontaneous under standard conditions if $\Delta S°$(universe) is greater than zero.* As an example of the determination of reaction spontaneity, let us calculate $\Delta S°$(universe) for the reaction currently used to manufacture methanol, CH_3OH.

$$CO(g) + 2\ H_2(g) \rightarrow CH_3OH(\ell)$$

If $\Delta S°$(universe) is positive, the conversion of 1 mol of CO(g) and 2 mol of H_2(g) to 1 mol of $CH_3OH(\ell)$ will be spontaneous under standard conditions.

Using $\Delta S°$(universe) For a process that is spontaneous under standard conditions:

$$\Delta S°(\text{universe}) > 0$$

For a process at equilibrium under standard conditions:

$$\Delta S°(\text{universe}) = 0$$

For a process that is not spontaneous under standard conditions:

$$\Delta S°(\text{universe}) < 0$$

Note that these conclusions refer to the complete conversion of reactants to products.

Calculating $\Delta S°$(system) To calculate $\Delta S°$(system), we start by defining the system to include the reactants and products. This means that $\Delta S°$(system) corresponds to the entropy change for the reaction, $\Delta_r S°$. Calculation of this entropy change follows the procedure given in Example 19.2.

$$\begin{aligned}
\Delta S°(\text{system}) &= \Delta_r S° = \Sigma\Delta S°(\text{products}) - \Sigma\Delta S°(\text{reactants}) \\
&= (1\text{ mol CH}_3\text{OH}(\ell)/\text{mol-rxn})\{S°[\text{CH}_3\text{OH}(\ell)]\} - \\
&\qquad [(1\text{ mol CO(g)}/\text{mol-rxn})\{S°[\text{CO(g)}]\} + (2\text{ mol H}_2\text{(g)}/\text{mol-rxn})\{S°[\text{H}_2\text{(g)}]\}] \\
&= (1\text{ mol CH}_3\text{OH}(\ell)/\text{mol-rxn})(127.2\text{ J/K}\cdot\text{mol}) - \\
&\qquad [(1\text{ mol CO(g)}/\text{mol-rxn})(197.7\text{ J/K}\cdot\text{mol}) + (2\text{ mol H}_2\text{(g)}/\text{mol-rxn})(130.7\text{ J/K}\cdot\text{mol})] \\
&= -331.9\text{ J/K}\cdot\text{mol-rxn}
\end{aligned}$$

A decrease in entropy for the system is expected because three moles of gaseous reactants are converted to one mole of a liquid product.

Calculating $\Delta S°$(surroundings) We now need to calculate the entropy change for the surroundings. Recall from Equation 19.1 that for a reversible change, ΔS is equal to q_{rev}/T. Under constant pressure conditions and assuming a reversible process, the entropy change in the surroundings results from the fact that the enthalpy change for the reaction ($q_p = \Delta H$) affects the surroundings. That is, the energy associated with an exothermic chemical reaction is dispersed into the surroundings. Recognizing that $\Delta H°$(surroundings) $= -\Delta_r H°$(system), the entropy change for the surroundings can be calculated by the equation

$$\Delta S°(\text{surroundings}) = \Delta H°(\text{surroundings})/T = -\Delta_r H°(\text{system})/T$$

For the synthesis of methanol by the reaction given, the enthalpy change is −127.9 kJ/mol-rxn, calculated from enthalpy of formation data.

$$\Delta H°(\text{system}) = \Sigma\Delta_f H°(\text{products}) - \Sigma\Delta_f H°(\text{reactants})$$
$$= (1 \text{ mol } CH_3OH(\ell)/\text{mol-rxn})\{\Delta_f H°[CH_3OH(\ell)]\} - [(1 \text{ mol } CO(g)/\text{mol-rxn})\{\Delta_f H°[CO(g)]\} + (2 \text{ mol } H_2(g)/\text{mol-rxn})\{\Delta_f H°[H_2(g)]\}]$$
$$= (1 \text{ mol } CH_3OH(\ell)/\text{mol-rxn})(-238.4 \text{ kJ/mol}) - [(1 \text{ mol } CO(g)/\text{mol-rxn})(-110.5 \text{ kJ/mol}) + (2 \text{ mol } H_2(g)/\text{mol-rxn})(0 \text{ kJ/ mol})]$$
$$= -127.9 \text{ kJ/mol-rxn}$$

Assuming that the process is reversible and occurs at a constant temperature and pressure, the entropy change for the surroundings in the methanol synthesis is +429.2 J/K · mol-rxn, calculated as follows.

$$\Delta S°(\text{surroundings}) = -\Delta_r H°(\text{system})/T$$
$$= -[(-127.9 \text{ kJ/mol-rxn})/298 \text{ K})](1000 \text{ J/kJ})$$
$$= +429.2 \text{ J/K} \cdot \text{mol-rxn}$$

Calculating ΔS°(universe), the Entropy Change for the System and Surroundings
The pieces are now in place to calculate the entropy change in the universe. For the formation of $CH_3OH(\ell)$ from CO(g) and $H_2(g)$, ΔS°(universe), is

$$\Delta S°(\text{universe}) = \Delta S°(\text{system}) + \Delta S°(\text{surroundings})$$
$$= -331.9 \text{ J/K} \cdot \text{mol-rxn} + 429.2 \text{ J/K} \cdot \text{mol-rxn}$$
$$= +97.3 \text{ J/K} \cdot \text{mol-rxn}$$

The positive value indicates an increase in the entropy of the universe. It follows from the second law of thermodynamics that this reaction is spontaneous.

Chemistry Now™

Sign in at **www.thomsonedu.com/login** and go to Chapter 19 Contents to see:
- Screen 19.5 for a simulation and tutorial on **calculating ΔS for a reaction**
- Screen 19.6 for a simulation and tutorial on **the second law of thermodynamics**

EXAMPLE 19.3 Determining Whether a Process Is Spontaneous

Problem Show that ΔS°(universe) is positive for the process of dissolving NaCl in water at 298 K.

Strategy The process occurring is NaCl(s) → NaCl(aq). The entropy change for the system, ΔS°(system), can be calculated from values of S° for the two species using Equation 19.3. ΔS°(surroundings) is determined by dividing $-\Delta_r H°$ for the process by the Kelvin temperature. The sum of these two entropy changes is ΔS°(universe). Values of $\Delta_f H°$ and S° for NaCl(s) and NaCl(aq) are obtained from Appendix L.

Solution

Calculate ΔS°(system)

$$\Delta S°(\text{system}) = \Sigma S°(\text{products}) - \Sigma S°(\text{reactants})$$
$$= (1 \text{ mol NaCl(aq)/mol-rxn})\{S°[\text{NaCl(aq)}]\} - (1 \text{ mol NaCl(s)/mol-rxn})\{S°[\text{NaCl(s)}]\}$$
$$= (1 \text{ mol NaCl(aq)/mol-rxn})(115.5 \text{ J/K} \cdot \text{mol}) - (1 \text{ mol NaCl(s)/mol-rxn})(72.11 \text{ J/K} \cdot \text{mol})$$
$$= +43.4 \text{ J/K} \cdot \text{mol-rxn}$$

■ **$\Delta S°$(universe), Spontaneity, and Standard Conditions** It is important to reiterate that when we calculate $\Delta H°$ or $\Delta S°$ for a reaction, this is the value for the complete conversion of reactants to products under standard conditions. If $\Delta S°$(universe) is > 0, the reaction as written is spontaneous *under standard conditions*. However, one can calculate values for ΔS(universe) (without the superscript zero) for *nonstandard* conditions. If ΔS(universe) > 0, the reaction is spontaneous under those conditions. However, it is possible that this same reaction under standard conditions is not spontaneous [$\Delta S°$(universe) < 0]. We will return to this point in Sections 19.6 and 19.7.

Calculate $\Delta S°$(surroundings)

$$\begin{aligned}\Delta_r H°(\text{system}) &= \Sigma\Delta_f H°(\text{products}) - \Sigma\Delta_f H°(\text{reactants})\\ &= (1\text{ mol NaCl(aq)/mol-rxn})\{\Delta_f H°[\text{NaCl(aq)}]\} - (1\text{ mol NaCl(s)/mol-rxn})\{\Delta_f H°[\text{NaCl(s)}]\}\\ &= (1\text{ mol NaCl(aq)/mol-rxn})(-407.27\text{ kJ/mol}) - (1\text{ mol NaCl(s)/mol-rxn})(-411.12\text{ kJ/mol})\\ &= +3.85\text{ kJ/mol-rxn}\end{aligned}$$

The entropy change of the surroundings is determined by dividing $-\Delta_r H°$(system) by the Kelvin temperature.

$$\begin{aligned}\Delta S°(\text{surroundings}) &= -\Delta_r H°(\text{system})/T\\ &= [(-3.85\text{ kJ/mol-rxn}/298\text{ K})](1000\text{ J/1 kJ})\\ &= -12.9\text{ J/K}\cdot\text{mol-rxn}\end{aligned}$$

Calculate $\Delta S°$(universe)

The overall entropy change—the change of entropy in the universe—is the sum of the values for the system and the surroundings.

$$\begin{aligned}\Delta S°(\text{universe}) &= \Delta S°(\text{system}) + \Delta S°(\text{surroundings})\\ &= +43.4\text{ J/K}\cdot\text{mol-rxn} - 12.9\text{ J/K}\cdot\text{mol-rxn}\\ &= +30.5\text{ J/K}\cdot\text{mol-rxn}\end{aligned}$$

Comment The sum of the two entropy quantities is positive, indicating that the entropy in the universe increases; thus, the process is spontaneous under standard conditions. Notice, however, that the spontaneity of the process results from $\Delta S°$(system) and not from $\Delta S°$(surroundings).

In Summary: Spontaneous or Not?

In the preceding examples, predictions about the spontaneity of a process under standard conditions were made using values of $\Delta S°$(system) and $\Delta H°$(system) calculated from tables of thermodynamic data. It will be useful to look at all possibilities that result from the interplay of these two quantities. There are four possible outcomes when these two quantities are paired (Table 19.2). In two, $\Delta H°$(system) and $\Delta S°$(system) work in concert (Types 1 and 4 in Table 19.2). In the other two, the two quantities are opposed (Types 2 and 3).

Processes in which both the standard enthalpy and entropy changes favor energy dispersal (Type 1) are always spontaneous under standard conditions. Processes disfavored by both their standard enthalpy and entropy changes in the system (Type 4) can *never* be spontaneous under standard conditions. Let us consider examples that illustrate each situation.

Combustion reactions are always exothermic and often produce a larger number of product molecules from a few reactant molecules. They are Type 1 reactions. The equation for the combustion of butane is an example.

$$2\ C_4H_{10}(g) + 13\ O_2(g) \rightarrow 8\ CO_2(g) + 10\ H_2O(g)$$

TABLE 19.2 Predicting Whether a Reaction Will Be Spontaneous Under Standard Conditions

Reaction Type	$\Delta H°$(system)	$\Delta S°$(system)	Spontaneous Process (Standard Conditions)
1	Exothermic, < 0	Positive, > 0	Spontaneous at all temperatures. $\Delta S°$(universe) > 0.
2	Exothermic, < 0	Negative, < 0	Depends on relative magnitudes of $\Delta H°$ and $\Delta S°$. Spontaneous at lower temperatures.
3	Endothermic, > 0	Positive, > 0	Depends on relative magnitudes of $\Delta H°$ and $\Delta S°$. Spontaneous at higher temperatures.
4	Endothermic, > 0	Negative, < 0	Not spontaneous at any temperature. $\Delta S°$(universe) < 0.

For this reaction, $\Delta_r H° = -5315.1$ kJ, and $\Delta_r S° = 312.4$ J/K. Both contribute to this reaction's being spontaneous under standard conditions.

Hydrazine, N_2H_4, is used as a high-energy rocket fuel. Synthesis of N_2H_4 from gaseous N_2 and H_2 would be attractive because these reactants are inexpensive.

$$N_2(g) + 2\ H_2(g) \rightarrow N_2H_4(\ell)$$

However, this reaction fits into the Type 4 category. The reaction is endothermic ($\Delta_r H° = +50.63$ kJ/mol-rxn), and the entropy change is negative ($\Delta_r S° = -331.4$ J/K·mol-rxn) (1 mol of liquid is produced from 3 mol of gases), so the reaction is not spontaneous under standard conditions, and complete conversion of reactants to products will not occur without outside intervention.

In the two other possible outcomes, entropy and enthalpy changes oppose each other. A process could be favored by the enthalpy change but disfavored by the entropy change (Type 2), or vice versa (Type 3). In either instance, whether a process is spontaneous depends on which factor is more important.

Temperature also influences the value of $\Delta S°$(universe). Because the enthalpy change for the surroundings is divided by the temperature to obtain $\Delta S°$(surroundings), the numerical value of $\Delta S°$(surroundings) will be smaller (either less positive or less negative) at higher temperatures. In contrast, $\Delta S°$(system) and $\Delta H°$(system) do not vary much with temperature. Thus, the effect of $\Delta S°$(surroundings) relative to $\Delta S°$(system) is diminished at higher temperature. Stated another way, at higher temperature, the enthalpy change becomes a less important factor in determining the overall entropy change. Consider the two cases where $\Delta H°$(system) and $\Delta S°$(system) are in opposition (Table 19.2):

- Type 2: Exothermic processes with $\Delta S°$(system) < 0. Such processes become less favorable with an increase in temperature.
- Type 3: Endothermic processes with $\Delta S°$(system) > 0. These processes become more favorable as the temperature increases.

The effect of temperature is illustrated by two examples. The first is the reaction of N_2 and H_2 to form NH_3. The reaction is exothermic, and thus it is favored by energy dispersal to the surroundings. The entropy change for the system is unfavorable, however, because the reaction, $N_2(g) + 3\ H_2(g) \rightarrow 2\ NH_3(g)$, converts four moles of gaseous reactants to two moles of gaseous products. The favorable enthalpy effect [$\Delta_r S°(\text{surroundings}) = -\Delta_r H°(\text{system})/T$] becomes less important at higher temperatures. Furthermore, it is reasonable to expect that the reaction will not be spontaneous if the temperature is high enough.

The second example considers the thermal decomposition of NH_4Cl (Figure 19.10). At room temperature, NH_4Cl is a stable, white, crystalline salt. When heated strongly, it decomposes to $NH_3(g)$ and HCl(g). The reaction is endothermic (enthalpy-disfavored) but entropy-favored because of the formation of two moles of gas from one mole of a solid reactant. The reaction is increasingly favored at higher temperatures.

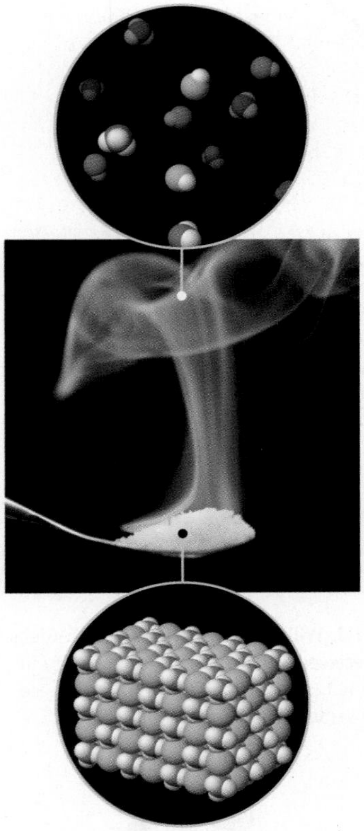

Charles D. Winters

FIGURE 19.10 Thermal decomposition of $NH_4Cl(s)$. White, solid ammonium chloride, $NH_4Cl(s)$, is heated in a spoon. At high temperatures, decomposition to form $NH_3(g)$ and HCl(g) is spontaneous. At lower temperatures, the reverse reaction, forming $NH_4Cl(s)$, is spontaneous. As gaseous HCl(g) and $NH_3(g)$ cool, they recombine to form solid NH_4Cl, the white "smoke" seen in this photo.

EXERCISE 19.3 Is a Reaction Spontaneous?

Classify the following reactions as one of the four types of reactions summarized in Table 19.2.

Reaction	$\Delta_r H°$ (at 298 K) (kJ/mol-rxn)	$\Delta_r S°$ (at 298 K) (J/K · mol-rxn)
(a) $CH_4(g) + 2\ O_2(g) \rightarrow 2\ H_2O(\ell) + CO_2(g)$	−890.6	−242.8
(b) $2\ Fe_2O_3(s) + 3\ C(\text{graphite}) \rightarrow 4\ Fe(s) + 3\ CO_2(g)$	+467.9	+560.7
(c) $C(\text{graphite}) + O_2(g) \rightarrow CO_2(g)$	−393.5	+3.1
(d) $N_2(g) + 3\ F_2(g) \rightarrow 2\ NF_3(g)$	−264.2	−277.8

EXERCISE 19.4 Is a Reaction Spontaneous?

Is the reaction of hydrogen and chlorine to give hydrogen chloride gas predicted to be spontaneous under standard conditions?

$$H_2(g) + Cl_2(g) \rightarrow 2\ HCl(g)$$

Predict the spontaneity of the reaction based on $\Delta_r H°$ and $\Delta_r S°$, and then calculate $\Delta S°$(universe) to verify your prediction.

EXERCISE 19.5 Effect of Temperature on Spontaneity

Iron is produced in a blast furnace by reducing iron oxide using carbon. For this reaction,

$$2\ Fe_2O_3(s) + 3\ C(\text{graphite}) \rightarrow 4\ Fe(s) + 3\ CO_2(g),$$

the following parameters are determined: $\Delta_r H° = +467.9$ kJ/mol-rxn, and $\Delta_r S° = +560.7$ J/K · mol-rxn. Show that it is necessary that this reaction be carried out at a high temperature.

Module 24

19.6 Gibbs Free Energy

■ **J. Willard Gibbs (1839–1903)** Gibbs received a Ph.D. from Yale University in 1863. His was the first Ph.D. in science awarded from an American university.

Burndy Library/Courtesy Emilio Segre Visual Archives

The method used so far to determine whether a process is spontaneous required evaluation of two quantities, $\Delta S°$(system) and $\Delta S°$(surroundings). Wouldn't it be convenient to have a single thermodynamic function that serves the same purpose? A function associated with the system only—one that does not require assessment of the surroundings—would be even better. Such a function exists. It is called the **Gibbs free energy**, with the name honoring J. Willard Gibbs (1839–1903). Gibbs free energy, G, often referred to simply as "free energy," is defined mathematically as

$$G = H - TS$$

where H is enthalpy, T is the Kelvin temperature, and S is entropy. In this equation, G, H, and S all refer to the system. Because enthalpy and entropy are state functions (◀ Section 5.4), free energy is also a state function.

Every substance possesses free energy, but the actual quantity is seldom known. Instead, just as with enthalpy (H) and internal energy (U), we are concerned with *changes* in free energy, ΔG, that occur in chemical and physical processes.

Let us first see how to use free energy as a way to determine whether a reaction is spontaneous. We can then ask further questions about the meaning of the term "free energy" and its use in deciding whether a reaction is product- or reactant-favored.

The Change in the Gibbs Free Energy, ΔG

Recall the equation defining the entropy change for the universe:

$$\Delta S(\text{universe}) = \Delta S(\text{surroundings}) + \Delta S(\text{system})$$

The entropy change of the surroundings equals the negative of the change in enthalpy of the system divided by T. Thus,

$$\Delta S(\text{universe}) = -\Delta H(\text{system})/T + \Delta S(\text{system})$$

Multiplying through this equation by $-T$, gives the equation

$$-T\Delta S(\text{universe}) = \Delta H(\text{system}) - T\Delta S(\text{system})$$

Gibbs defined the free energy function so that $\Delta G(\text{system}) = -T\Delta S(\text{universe})$. Combining terms and simplifying give the general expression relating changes in free energy to the enthalpy and entropy changes in the system.

$$\Delta G = \Delta H - T\Delta S$$

Under standard conditions, we can rewrite the Gibbs free energy equation as

$$\Delta G^\circ = \Delta H^\circ - T\Delta S^\circ \qquad \text{(19.5)}$$

Gibbs Free Energy, Spontaneity, and Chemical Equilibrium

Because $\Delta_r G^\circ$ is related directly to ΔS°(universe), the Gibbs free energy can be used as a criterion of spontaneity for physical and chemical changes. As shown earlier, the signs of $\Delta_r G^\circ$ and ΔS°(universe) will be opposites [$\Delta_r G^\circ$(system) = $-T\Delta S^\circ$(universe)]. Therefore, we find the following relationships:

$\Delta_r G^\circ < 0$	The process is spontaneous in the direction written under standard conditions.
$\Delta_r G^\circ = 0$	The process is at equilibrium under standard conditions.
$\Delta_r G^\circ > 0$	The process is not spontaneous in the direction written under standard conditions.

To better understand the Gibbs function, let us examine the diagrams in Figure 19.11. The free energy of pure reactants is plotted on the left, and the free energy of the pure products on the right. The extent of reaction, plotted on the *x*-axis, goes from zero (pure reactants) to one (pure products). In both cases in Figure 19.11, the free energy initially declines as reactants begin to form products; it reaches a minimum at equilibrium and then increases again as we move from the equilibrium position to pure products. *The free energy at equilibrium, where there is a mixture of reactants and products, is always lower than the free energy of the pure reactants and of the pure products. A reaction proceeds spontaneously toward the minimum in free energy, which corresponds to equilibrium.*

$\Delta_r G^\circ$ is the change in free energy accompanying the chemical reaction in which the reactants are converted completely to the products under standard

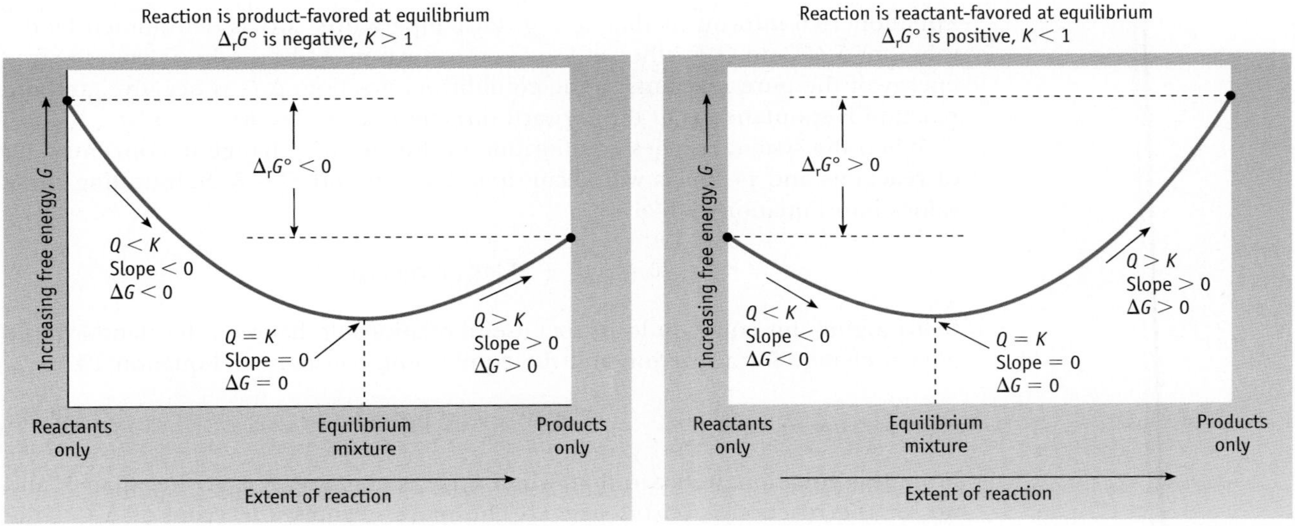

(a) (b)

FIGURE 19.11 Free energy changes in the course of a reaction. The difference in free energy between the pure reactants in their standard states and the pure products in their standard states is $\Delta_r G^\circ$. Here, Q is the reaction quotient, and K is the equilibrium constant.

conditions. Mathematically, it is the difference in free energy between the products and the reactants under standard conditions. If the free energy of the products is less than that of the reactants, then $\Delta_r G^\circ < 0$, and the reaction is spontaneous under standard conditions (Figure 19.11a). Conversely, if the free energy of the products is greater than that of the reactants, then $\Delta_r G^\circ$ is positive ($\Delta_r G^\circ > 0$), and the reaction is not spontaneous under standard conditions (Figure 19.11b).

Finally, notice that in Figure 19.11a, the equilibrium position occurs closer to the product side than to the reactant side. This is a product-favored reaction at equilibrium. In Figure 19.11b, we find the opposite. The reaction is reactant-favored at equilibrium. It is no accident that the reaction with a negative $\Delta_r G^\circ$ is product-favored, whereas the one with the positive $\Delta_r G^\circ$ is reactant-favored at equilibrium. It turns out that this is always true as the discussion below will show.

Now let us now consider what happens to the instantaneous slope of the curve in Figure 19.11 as the reaction proceeds. Initially, this slope is negative, corresponding to a negative ΔG in moving from point to point. Eventually, however, the free energy reaches a minimum. At this point, the instantaneous slope of the graph is zero ($\Delta G = 0$) and the reaction has reached equilibrium. If we move past the equilibrium point, the instantaneous slope is positive ($\Delta G > 0$); proceeding further to products is not spontaneous. In fact, the reverse reaction will occur spontaneously; the reaction will once again proceed toward equilibrium.

The relationship of $\Delta_r G^\circ$ (the value of $\Delta_r G$ under standard conditions) and the value of $\Delta_r G$ under nonstandard conditions is given by Equation 19.6

$$\Delta_r G = \Delta_r G^\circ + RT \ln Q \tag{19.6}$$

where R is the universal gas constant, T is the temperature in kelvins, and Q is the reaction quotient (◀ Section 16.2).

$$Q = \frac{[C]^c[D]^d}{[A]^a[B]^b} \text{ for } aA + bB \rightarrow cC + dD$$

Equation 19.6 informs us that, at a given temperature, $\Delta_r G$ is determined by the values of $\Delta_r G^\circ$ and Q. Further, as long as the reaction is "descending" from the free energy of the pure reactants to the equilibrium position, $\Delta_r G$ is negative, and the reaction is spontaneous in the forward direction (and $Q < K$).

When the system reaches equilibrium, no further net change in concentration of reactants and products will occur, and $\Delta_r G = 0$ and $Q = K$. Substituting these values into Equation 19.6 gives

$$0 = \Delta_r G^\circ + RT \ln K \text{ (at equilibrium)}$$

Rearranging this equation leads to a useful relationship between the standard free energy change for a reaction and the equilibrium constant, K, Equation 19.7:

$$\Delta_r G^\circ = -RT \ln K \tag{19.7}$$

From this equation, we learn that, when $\Delta_r G^\circ$ is negative, K is greater than 1, and we say the reaction is *product-favored.* The more negative the value of $\Delta_r G^\circ$, the larger the equilibrium constant. This makes sense because, as described in Chapter 16, large equilibrium constants are associated with product-favored reactions. The converse is also true: For *reactant-favored* reactions, $\Delta_r G^\circ$ is positive, and K is less than 1. Finally, if $K = 1$ (a special set of conditions), then $\Delta_r G^\circ = 0$.

A Summary: Gibbs Free Energy (Δ_rG and $\Delta_rG°$), the Reaction Quotient (Q) and Equilibrium Constant (K), and Reaction Favorability

Let us now summarize the relationships among $\Delta_rG°$, Δ_rG, Q, and K.

- In Figure 19.11, you see that free energy decreases to a minimum as a system approaches equilibrium. The free energy of the mixture of reactants and products at equilibrium is always lower than the free energy of the pure reactants or of the pure products.
- When $\Delta_rG < 0$, the reaction is proceeding spontaneously toward equilibrium and $Q < K$.
- When $\Delta_rG > 0$, the reaction is beyond the equilibrium point and is not spontaneous in the forward direction. It will be spontaneous in the reverse direction; $Q > K$.
- When $\Delta_rG = 0$, the reaction is at equilibrium; $Q = K$.
- When $\Delta_rG° < 0$, the reaction is *spontaneous under standard conditions.* The system will proceed to an equilibrium position at which point the products will dominate in the reaction mixture because $K > 1$. That is, the reaction is product-favored at equilibrium.
- When $\Delta_rG° > 0$, the reaction is *not spontaneous under standard conditions.* The system will proceed to the equilibrium position at which the reactants will dominate in the equilibrium mixture because $K < 1$. That is, the reaction is *reactant-favored.*
- For the special condition where $\Delta_rG° = 0$, the reaction is at equilibrium at standard conditions, with $K = 1$.

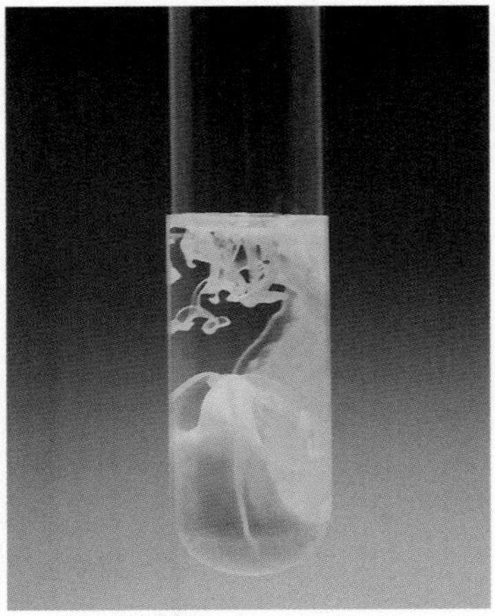

A reactant-favored process. If a sample of yellow lead iodide is placed in pure water, a small amount of the compound will dissolve spontaneously ($\Delta_rG < 0$ and $Q < K$) until equilibrium is reached. Because PbI_2 is quite insoluble ($K_{sp} = 9.8 \times 10^{-9}$), however, the process of dissolving the compound is reactant-favored. We may conclude, therefore, that the value of $\Delta_rG°$ is positive.

What Is "Free" Energy?

The term "free energy" was not arbitrarily chosen. In any given process, the free energy represents the maximum energy available to do useful work (mathematically, $\Delta G = w_{max}$). In this context, the word "free" means "available."

To illustrate the reasoning behind this relationship, consider a reaction carried out under standard conditions and in which energy is evolved as heat ($\Delta_rH° < 0$) and entropy decreases ($\Delta_rS° < 0$).

$$2\ H_2(g) + O_2(g) \rightarrow 2\ H_2O(g)$$

$$\Delta_rH° = -483.6 \text{ kJ/mol-rxn and } \Delta_rS° = -88.8 \text{ J/K} \cdot \text{mol-rxn}$$

$$\Delta_rG° = -483.6 \text{ kJ/mol-rxn} - (298 \text{ K})(-0.0888 \text{ kJ/mol-rxn}) = -457.2 \text{ kJ/mol-rxn}$$

At first glance, it might seem reasonable that all the energy released as heat (−483.6 kJ/mol-rxn) would be available. This energy could be transferred to the surroundings and would thus be available to do work. This is not the case, however. A negative entropy change in this reaction means that energy is less dispersed in the products than in the reactants. A portion of the energy released as heat from the reaction must be used to reverse energy dispersal in the system; that is, to concentrate energy in the product. The energy left over is "free," or available to perform work. Here, the free energy change amounts to −457.2 kJ/mol-rxn.

19.7 Calculating and Using Free Energy

Standard Free Energy of Formation

The standard free energy of formation of a compound, $\Delta_fG°$, is the free energy change that occurs to form one mole of the compound from the component elements, with products and reactants in their standard states. By defining $\Delta_fG°$ in this way, *the free energy of formation of an element in its standard state is zero.*

Just as the standard enthalpy or entropy change for a reaction can be calculated using values of $\Delta_f H°$ (Equation 5.6) or $S°$ (Equation 19.3), the standard free energy change for a reaction can be calculated from values of $\Delta_f G°$:

$$\Delta_r G° = \Sigma \Delta_f G°(\text{products}) - \Sigma \Delta_f G°(\text{reactants}) \qquad (19.8)$$

Calculating $\Delta_r G°$, the Free Energy Change for a Reaction Under Standard Conditions

The free energy change for a reaction under standard conditions can be calculated from thermodynamic data in two ways, either from standard enthalpy and entropy changes using values of $\Delta_f H°$ and $S°$ (as we did for the formation of H_2O just above) or directly from values of $\Delta_f G°$ found in tables. These calculations are illustrated in the following two examples.

Chemistry Now™

Sign in at **www.thomsonedu.com/login** and go to Chapter 19 Contents to see Screen 19.7 for an exercise and tutorials on **Gibbs free energy.**

EXAMPLE 19.4 Calculating $\Delta_r G°$ from $\Delta_r H°$ and $\Delta_r S°$

Problem Calculate the standard free energy change, $\Delta_r G°$, for the formation of methane from carbon and hydrogen at 298 K, using tabulated values of $\Delta_f H°$ and $\Delta S°$. Is the reaction spontaneous under standard conditions? Is it product-favored or reactant-favored at equilibrium?

$$\text{C(graphite)} + 2\ \text{H}_2\text{(g)} \rightarrow \text{CH}_4\text{(g)}$$

Strategy The values of $\Delta_f H°$ and $S°$ needed in this problem are found in Appendix L. These are first combined to find $\Delta_r H°$ and $\Delta_r S°$. With these values known, $\Delta_r G°$ can be calculated using Equation 19.5. When doing the calculation, remember that $S°$ values are given in units of J/K · mol, whereas $\Delta_f H°$ values are given in units of kJ/mol.

Solution

	C(graphite)	H_2(g)	CH_4(g)
$\Delta_f H°$ (kJ/mol)	0	0	−74.9
$S°$ (J/K·mol)	+5.6	+130.7	+186.3

From these values, we can find both $\Delta_r H°$ and $\Delta_r S°$ for the reaction:

$$\Delta_r H° = \Sigma \Delta_f H°(\text{products}) - \Sigma \Delta_f H°(\text{reactants})$$

$$= (1\ \text{mol CH}_4\text{(g)/mol-rxn})\{\Delta_f H°[\text{CH}_4\text{(g)}]\} - [(1\ \text{mol C(graphite)/mol-rxn})\{\Delta_f H°[\text{C(graphite)}]\} + (2\ \text{mol H}_2\text{(g)/mol-rxn})\{\Delta_f H°[\text{H}_2\text{(g)}]\}]$$

$$= (1\ \text{mol CH}_4\text{(g)/mol-rxn})(-74.9\ \text{kJ/mol}) - [(1\ \text{mol C(graphite)/mol-rxn})(0\ \text{kJ/mol}) + (2\ \text{mol H}_2\text{(g)/mol-rxn})(0\ \text{kJ/mol})]$$

$$= -74.9\ \text{kJ/mol-rxn}$$

$$\Delta_r S° = \Sigma \Delta S°(\text{products}) - \Sigma \Delta S°(\text{reactants})$$

$$= (1\ \text{mol CH}_4\text{(g)/mol-rxn})\{S°[\text{CH}_4\text{(g)}]\} - [(1\ \text{mol C(graphite)/mol-rxn})\{S°[\text{C(graphite)}]\} + (2\ \text{mol H}_2\text{(g)/mol-rxn})\{S°[\text{H}_2\text{(g)}]\}]$$

$$= (1\ \text{mol CH}_4\text{(g)/mol-rxn})(186.3\ \text{J/K}\cdot\text{mol}) - [1\ \text{mol C(graphite)/mol-rxn}](5.6\ \text{J/K}\cdot\text{mol}) + (2\ \text{mol H}_2\text{(g)/mol-rxn})(130.7\ \text{J/K}\cdot\text{mol})]$$

$$= -80.7\ \text{J/K}\cdot\text{mol-rxn}$$

Combining the values of $\Delta_r H°$ and $\Delta_r S°$ using Equation 19.5 gives $\Delta_r G°$.

$$\Delta_r G° = \Delta_r H° - T\Delta_r S°$$
$$= -74.9 \text{ kJ/mol-rxn} - [(298 \text{ K})(-80.7 \text{ J/K} \cdot \text{mol-rxn})](1 \text{ kJ/1000 J})$$
$$= -50.9 \text{ kJ/mol-rxn}$$

$\Delta_r G°$ is negative at 298 K, so the reaction is predicted to be spontaneous under standard conditions at this temperature. It is also predicted to be product-favored at equilibrium.

■ **Enthalpy- and Entropy-Driven Reactions** In Example 19.4, the product $T\Delta_r S°$ is negative (−24.0 J/mol-rxn) and disfavors the reaction. However, the entropy change is relatively small, and $\Delta_r H° = -74.9$ kJ/mol-rxn) is the dominant term. Chemists call this an "enthalpy-driven reaction."

EXAMPLE 19.5 Calculating $\Delta_r G°$ Using Free Energies of Formation

Problem Calculate the standard free energy change for the combustion of one mole of methane using values for standard free energies of formation of the products and reactants. Is the reaction spontaneous under standard conditions? Is it product-favored or reactant-favored at equilibrium?

Strategy Write a balanced equation for the reaction. Then, use Equation 19.8 with values of $\Delta_f G°$ obtained from Appendix L.

Solution The balanced equation and values of $\Delta_f G°$ for each reactant and product are:

	$CH_4(g)$	+ 2 $O_2(g)$	→ 2 $H_2O(g)$	+ $CO_2(g)$
$\Delta_f G°$(kJ/mol)	−50.8	0	−228.6	−394.4

Because $\Delta_f G°$ values are given for 1 mol of each substance (the units are kJ/mol), each value of $\Delta_f G°$ must be multiplied by the number of moles defined by the stoichiometric coefficient in the balanced chemical equation.

$$\Delta_r G°(\text{system}) = \Sigma\Delta_f G°(\text{products}) - \Sigma\Delta_f G°(\text{reactants})$$
$$= [(2 \text{ mol } H_2O(g)/\text{mol-rxn})\{\Delta_f G°[H_2O(g)]\} + (1 \text{ mol } CO_2(g)/\text{mol-rxn})\{\Delta_f G°[CO_2(g)]\}]$$
$$- [(1 \text{ mol } CH_4(g)/\text{mol-rxn})\{\Delta_f G°[CH_4(g)]\} + (2 \text{ mol } O_2(g)/\text{mol-rxn})\{\Delta_f G°[O_2(g)]\}]$$
$$= [(2 \text{ mol } H_2O(g)/\text{mol-rxn})(-228.6 \text{ kJ/mol}) + (1 \text{ mol } CO_2(g)/\text{mol-rxn})(-394.4 \text{ kJ/mol})\}$$
$$- [(1 \text{ mol } CH_4(g)/\text{mol-rxn})(-50.8 \text{ kJ/mol}) - (2 \text{ mol } O_2(g)/\text{mol-rxn})(0 \text{ kJ/mol})]$$
$$= -800.8 \text{ kJ/mol-rxn}$$

The large negative value of $\Delta_r G°$ indicates that the reaction is spontaneous under standard conditions and that it is product-favored at equilibrium.

Comment Common errors made by students in this calculation are (1) overlooking the stoichiometric coefficients in the equation and (2) confusing the signs for the terms when using Equation 19.8.

EXERCISE 19.6 Calculating $\Delta_r G°$ from $\Delta_r H°$ and $\Delta_r S°$

Using values of $\Delta_f H°$ and $S°$ to find $\Delta_r H°$ and $\Delta_r S°$, calculate the free energy change, $\Delta_r G°$, for the formation of 2 mol of $NH_3(g)$ from the elements at standard conditions and 25 °C. $N_2(g) + 3\ H_2(g) \rightarrow 2\ NH_3(g)$.

EXERCISE 19.7 Calculating $\Delta_r G°$ from $\Delta_f G°$

Calculate the standard free energy change for the oxidation of 1.00 mol of $SO_2(g)$ to form $SO_3(g)$ using values of $\Delta_f G°$.

Free Energy and Temperature

The definition for free energy, $G = H - TS$, informs us that free energy is a function of temperature, so $\Delta_r G°$ will change as the temperature changes (Figure 19.12). A consequence of this dependence on temperature is that, in certain instances, reactions can be product-favored at equilibrium at one temperature

■ **Entropy- or Enthalpy-Favored** Table 19.2 describes the balance of $\Delta H°$ and $\Delta S°$ and the effect of temperature on reaction spontaneity.

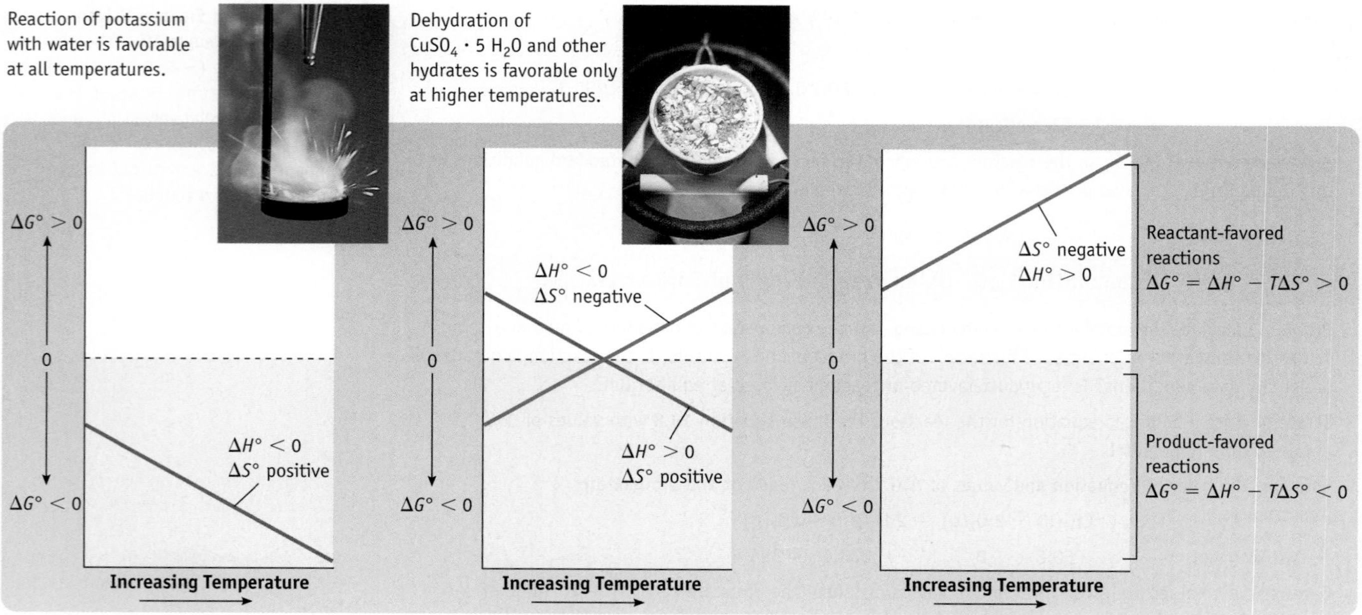

$\Delta H° < 0$ and $\Delta S° > 0$. Product-favored at all temperatures.

Blue line: $\Delta H° < 0$ and $\Delta S° < 0$. Favored at low T. Red line: $\Delta H° > 0$ and $\Delta S° > 0$. Favored at high T.

$\Delta H° > 0$ and $\Delta S° < 0$. Reactant-favored at all temperatures.

Photos: Charles D. Winters

Active Figure 19.12 **The variation in $\Delta_r G°$ with temperature.**

Chemistry Now™ Sign in at www.thomsonedu.com/login and go to the Chapter Contents menu to explore an interactive version of this figure accompanied by an exercise.

and reactant-favored at another. Those instances arise when the $\Delta_r H°$ and $T\Delta_r S°$ terms work in opposite directions:

- Processes that are entropy-favored ($\Delta_r S° > 0$) and enthalpy-disfavored ($\Delta_r H° > 0$)
- Processes that are enthalpy-favored ($\Delta_r H° < 0$) and entropy-disfavored ($\Delta_r S° < 0$)

Let us explore the relationship of $\Delta G°$ and T further and illustrate how it can be used to advantage.

Calcium carbonate is the primary component of limestone, marble, and seashells. Heating $CaCO_3$ produces lime, CaO, an important chemical, along with gaseous CO_2. The data below from Appendix L are at 298 K (25 °C).

	$CaCO_3(s)$	→	CaO(s)	+	$CO_2(g)$
$\Delta_f G°$ (kJ/mol)	−1129.16		−603.42		−394.36
$\Delta_f H°$ (kJ/mol)	−1207.6		−635.09		−393.51
$S°$ (J/K · mol)	91.7		38.2		213.74

For the conversion of 1 mol of $CaCO_3(s)$ to 1 mol of CaO(s) under standard conditions, $\Delta_r G° = +131.38$ kJ, $\Delta_r H° = +179.0$ kJ, and $\Delta_r S° = +160.2$ J/K. Although the reaction is entropy-favored, the large positive and unfavorable enthalpy change dominates at 298 K. Thus, the standard free energy change is positive at 298 K and 1 bar, indicating that the reaction is reactant-favored at equilibrium under the given conditions.

The temperature dependence of $\Delta_r G°$ provides a means to turn the $CaCO_3$ decomposition into a product-favored reaction. Notice that the entropy change in the reaction is positive as a result of the formation of CO_2 gas in the reaction. Thus, raising the temperature results in the value of $T\Delta_r S°$ becoming increasingly large. At a high enough temperature, $T\Delta_r S°$ will outweigh the enthalpy effect, and the process will become product-favored at equilibrium.

How high must the temperature be for this reaction to become product-favored? An estimate of the temperature can be obtained using Equation 19.5, by calculating the temperature at which $\Delta_r G° = 0$. Above that temperature, $\Delta_r G°$ will have a negative value.

$$\Delta_r G° = \Delta_r H° - T\Delta_r S°$$

$$0 = (179.0 \text{ kJ/mol-rxn})(1000 \text{ J/kJ}) - T(160.2 \text{ J/K} \cdot \text{mol-rxn})$$

$$T = 1117 \text{ K (or 844 °C)}$$

How accurate is this result? As noted earlier, we can obtain only an approximate answer from this calculation. One source of error is the assumption that $\Delta_r H°$ and $\Delta S°$ do not vary with temperature, which is not strictly true. There is always a small variation in these values when the temperature changes—not large enough to be important if the temperature range is narrow, but potentially a problem over wider temperature ranges such as seen in this example. As an estimate, however, a temperature in the range of 850 °C for this reaction is reasonable.

■ **$CaCO_3$ Decomposition** Experiments show that the pressure of CO_2 in an equilibrium system [$CaCO_3(s) \rightleftharpoons CaO(s) + CO_2(g)$; $\Delta G° = 0$] is 1 bar at about 900 °C, close to our estimated temperature.

Chemistry Now™

Sign in at **www.thomsonedu.com/login** and go to Chapter 19 Contents to see Screen 19.8 for a simulation and tutorial on **the relationship of $\Delta H°$, $\Delta S°$, and T.**

EXAMPLE 19.6 Effect of Temperature on $\Delta_r G°$

Problem The decomposition of liquid $Ni(CO)_4$ to produce nickel metal and carbon monoxide has a $\Delta_r G°$ value of 40 kJ/mol-rxn at 25 °C.

$$Ni(CO)_4(\ell) \rightarrow Ni(s) + 4\,CO(g)$$

Use values of $\Delta_f H°$ and $S°$ for the reactant and products to estimate the temperature at which the reaction becomes product-favored at equilibrium.

Strategy The reaction is reactant-favored at equilibrium at 298 K. However, if the entropy change is positive for the reaction and the reaction is endothermic (with a positive value of $\Delta_r H°$), then a higher temperature may allow the reaction to become product-favored at equilibrium. Therefore, we first find $\Delta_r H°$ and $\Delta_r S°$ to see if their values meet our criteria for spontaneity at a higher temperature, and then we calculate the temperature at which the following condition is met: $0 = \Delta_r H° - T\Delta_r S°$.

Solution Values for $\Delta_f H°$ and $S°$ are obtained from the chemical literature for the substances involved.

	$Ni(CO)_4(\ell)$	→	$Ni(s)$	+	$4\,CO(g)$
$\Delta_f H°$(kJ/mol)	−632.0		0		−110.525
$S°$(J/K · mol)	320.1		29.87		197.67

For a process in which 1 mol of liquid $Ni(CO)_4$ is converted to 1 mol of Ni(s) and 4 mol of CO(g), we find

$$\Delta_r H° = +189.9 \text{ kJ/mol-rxn}$$

$$\Delta_r S° = +500.5 \text{ J/K mol-rxn}$$

At 298 K, the reaction is reactant-favored at equilibrium largely because it is quite endothermic. However, a positive entropy change should allow the reaction to be product-favored at equilibrium at a higher temperature. Therefore, we use the values of $\Delta_r H°$ and $\Delta_r S°$ to find the temperature at which $\Delta_r G° = 0$.

$$\Delta_r G° = \Delta_r H° - T\Delta_r S°$$

$$0 = (189.9 \text{ kJ/mol-rxn})(1000 \text{ J/kJ}) - T(500.5 \text{ J/K} \cdot \text{mol-rxn})$$

$$T = 379.4 \text{ K (or 106.2 °C)}$$

Case Study — Thermodynamics and Living Things

The laws of thermodynamics apply to all chemical reactions. It should come as no surprise, therefore, that issues of spontaneity and ΔG arise in studies of biochemical reactions. For biochemical processes, however, a different standard state is often used. Most of the usual definition is retained: 1 bar pressure for gases and 1 *m* concentration for aqueous solutes with the exception of one very important solute. Rather than using a standard state of 1 molal for hydronium ions (corresponding to a pH of about 0), biochemists use a hydronium concentration of 1×10^{-7} M, corresponding to a pH of 7. This pH is much more useful for biochemical reactions. When biochemists use this as the standard state, they write the symbol ′ next to the thermodynamic function. For example, they would write $\Delta G°'$ (pronounced *delta G zero prime*).

Living things require energy to perform their many functions. One of the main reactions involved in providing this energy is the reaction of adenosine triphosphate (ATP) with water, a reaction for which $\Delta_r G°' = -30.5$ kJ/mol–rxn (◀ *The Chemistry of Life: Biochemistry*).

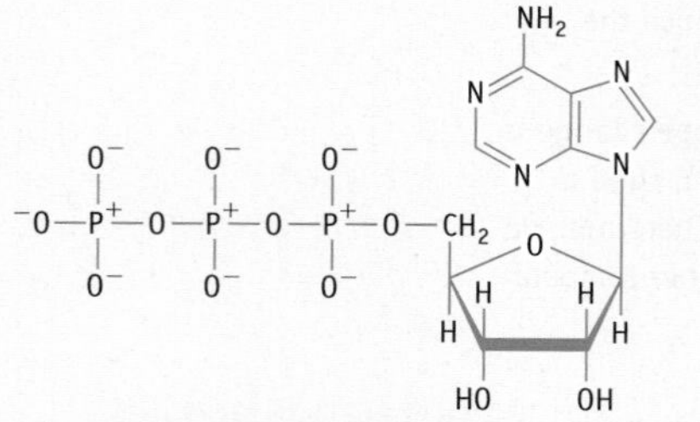

ATP, adenosine triphosphate

One of the key functions of the process of respiration is to produce molecules of ATP for our bodies to use. ATP is produced in the reaction of adenosine diphosphate (ADP) with hydrogen phosphate ($HP_i = HPO_4^{2-}$),

$$\text{ADP} + HP_i + H^+ \rightarrow \text{ATP} + H_2O \qquad \Delta_r G°' = +30.5 \text{ kJ/mol}$$

a reaction that is reactant-favored at equilibrium. How then do our bodies get this reaction to occur? The answer is to couple the production of ATP with another reaction that is even more product-favored than ATP production is reactant-favored. For example, organisms carry out the oxidation of carbohydrates in a multistep process, producing energy. One of the compounds produced in the process called *glycolysis* is phosphoenolpyruvate (PEP).

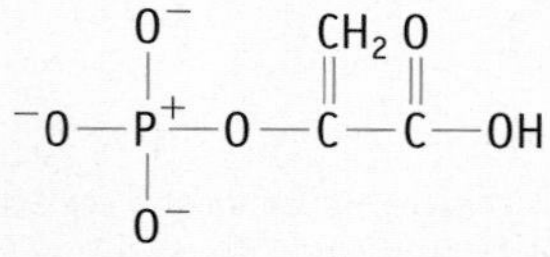

PEP, phosphoenolpyruvate

Its reaction with water is product-favored at equilibrium

$$\text{PEP} + H_2O \rightarrow \text{Pyruvate} + HP_i \qquad \Delta_r G°' = -61.9 \text{ kJ/mol}$$

This reaction and the ATP formation are linked through the HP_i that is produced in the PEP reaction. If both reactions are carried out, we obtain the following

$$\text{PEP} + H_2O \rightarrow \text{Pyruvate} + HP_i \qquad \Delta_r G°' = -61.9 \text{ kJ/mol}$$

$$\text{ADP} + HP_i + H^+ \rightarrow \text{ATP} + H_2O \qquad \Delta_r G°' = +30.5 \text{ kJ/mol}$$

$$\text{PEP} + \text{ADP} + H^+ \rightarrow \text{Pyruvate} + \text{ATP} \qquad \Delta_r G°' = -31.4 \text{ kJ/mol}$$

The overall reaction has a negative value for $\Delta_r G°'$ and thus is product-favored at equilibrium. ATP is formed in this process.

The coupling of reactions to produce a system that is product-favored is used in a multitude of reactions that occur in our bodies.

John Kotz

Living things use ATP to produce energy.

Questions:

1. *Consider the hydrolysis reactions of creatine phosphate and adenosine-5′-monophosphate.*

   Creatine Phosphate $+ H_2O \rightarrow$ *Creatine* $+ HP_i$

   $\Delta_r G°' = -43.3$ *kJ/mol*

   Adenosine-5′-Monophosphate $+ H_2O \rightarrow$ *adenosine* $+ HP_i + H^+$

   $\Delta_r G°' = -9.2$ *kJ/mol*

   In which direction does a reaction that is product-favored at equilibrium result: for creatine phosphate to transfer phosphate to adenosine or for adenosine-5′-monophosphate to transfer phosphate to creatine?
2. *Assume the reaction* $A(aq) + B(aq) \rightarrow C(aq) + H_3O^+(aq)$ *produces one hydronium ion. What is the mathematical relationship between* $\Delta G°'$ *and* $\Delta G°$ *at 25° C? (Hint: Use the equation* $\Delta G = \Delta G° + RT \ln Q$ *and substitute* $\Delta G°'$ *for* ΔG.*)*

Answers to these questions are in Appendix Q.

EXERCISE 19.8 Effect of Temperature on $\Delta_r G°$

Oxygen was first prepared by Joseph Priestley (1733–1804) by heating HgO. Use data in Appendix L to estimate the temperature required to decompose HgO(s) into Hg(ℓ) and O_2(g).

Using the Relationship Between $\Delta_r G°$ and K

Equation 19.7 provides a direct route to determine the standard free energy change from experimentally determined equilibrium constants. Alternatively, it allows calculation of an equilibrium constant from thermochemical data contained in tables or obtained from an experiment.

Chemistry Now™

Sign in at **www.thomsonedu.com/login** and go to Chapter 19 Contents to see Screen 19.9 for a simulation and tutorial on **$\Delta G°$ and K.**

EXAMPLE 19.7 Calculating K_p from $\Delta_r G°$

Problem Determine the standard free energy change, $\Delta_r G°$, for the formation of 1.00 mol of NH_3(g) from nitrogen and hydrogen, and use this value to calculate the equilibrium constant for this reaction at 25 °C.

Strategy The free energy of formation of ammonia represents the free energy change to form 1.00 mol of NH_3(g) from the elements. The equilibrium constant for this reaction is calculated from $\Delta_r G°$ using Equation 19.7. Because the reactants and products are gases, the calculated value will be K_p.

Solution Begin by specifying a balanced equation for the chemical reaction under investigation.

$$\tfrac{1}{2}\, N_2(g) + \tfrac{3}{2}\, H_2(g) \rightleftharpoons NH_3(g)$$

The free energy change for this reaction is −16.37 kJ/mol-rxn ($\Delta_r G° = \Delta_f G°$ for NH_3(g); Appendix L). In a calculation of K_p using Equation 19.7, we will need consistent units. The gas constant, R, is 8.3145 J/K · mol, so the value of $\Delta_r G°$ must be in J/mol-rxn (and not kJ/mol-rxn). The temperature is 298 K.

$$\Delta_r G° = -RT \ln K$$

$$-16{,}370 \text{ J/K} \cdot \text{mol-rxn} = (-8.3145 \text{ J/K} \cdot \text{mol-rxn})(298.15 \text{ K}) \ln K_p$$

$$\ln K_p = 6.604$$

$$K_p = 7.38 \times 10^2$$

Comment This example illustrates how to calculate equilibrium constants from thermodynamic data. In fact, many equilibrium constants you find in the chemical literature are not experimentally determined but are instead calculated from thermodynamic data in this way.

EXAMPLE 19.8 Calculating $\Delta_r G°$ from K_{sp} for an Insoluble Solid

Problem The value of K_{sp} for AgCl(s) at 25 °C is 1.8×10^{-10}. Use this value in Equation 19.7 to determine $\Delta_r G°$ for the process Ag^+(aq) + Cl^-(aq) ⇄ AgCl(s) at 25 °C.

Strategy The chemical equation given is the opposite of the equation used to define K_{sp}; therefore, the equilibrium constant for this reaction is $1/K_{sp}$. This value is used to calculate $\Delta_r G°$.

Solution For Ag^+(aq) + Cl^-(aq) ⇄ AgCl(s),

$$K = 1/K_{sp} = 1/\ 1.8 \times 10^{-10} = 5.6 \times 10^9$$

$$\Delta_r G° = -RT \ln K = -(8.3145 \text{ J/K} \cdot \text{mol-rxn})(298.15 \text{ K}) \ln(5.6 \times 10^9)$$

$$= -56 \text{ kJ/mol-rxn}$$

Comment The negative value of $\Delta_r G°$ indicates that the precipitation of AgCl from Ag^+(aq) and Cl^-(aq) is product-favored at equilibrium.

EXERCISE 19.9 Calculating K_p from $\Delta_r G°$

Determine the value of $\Delta_r G°$ for the reaction $C(s) + CO_2(g) \rightleftharpoons 2\ CO(g)$ from data in Appendix L. Use this result to calculate the equilibrium constant.

Chapter Goals Revisited

Chemistry Now™ Sign in at **www.thomsonedu.com/login** to:

- Assess your understanding with Study Questions in OWL keyed to each goal in the Goals and Homework menu for this chapter
- For quick review, download Go Chemistry mini-lecture flashcard modules (or purchase them at **www.ichapters.com**)
- Check your readiness for an exam by taking the Pre-Test and exploring the modules recommended in your Personalized Study plan.

Access **How Do I Solve It?** tutorials on how to approach problem solving using concepts in this chapter.

Now that you have studied this chapter, you should ask whether you have met the chapter goals. In particular, you should be able to:

Understand the concept of entropy and its relationship to reaction spontaneity

a. Understand that entropy is a measure of energy dispersal (Section 19.2). **Study Question(s) assignable in OWL: 2.**

b. Recognize that an entropy change can be determined experimentally as the energy transferred as heat for a reversible process divided by the Kelvin temperature. (*A Closer Look,* Section 19.3.) **Study Question(s) assignable in OWL: 40.**

c. Identify common processes that are entropy favored (Section 19.4).

Calculate the change in entropy for system, surroundings, and the universe to determine whether a process is spontaneous

a. Calculate entropy changes from tables of standard entropy values for compounds (Section 19.4). **Study Question(s) assignable in OWL: 4, 6, 11.**

b. Use standard entropy and enthalpy changes to predict whether a reaction will be spontaneous under standard conditions (Section 19.5 and Table 19.2). **Study Question(s) assignable in OWL: 10, 12, 33, 35, 47, 55, 67, 72.**

c. Recognize how temperature influences whether a reaction is spontaneous (Section 19.5). **Study Question(s) assignable in OWL: 14, 25, 26.**

Understand and use the Gibbs free energy

a. Understand the connection between enthalpy and entropy changes and the Gibbs free energy change for a process (Section 19.6).

b. Understand the relationship of $\Delta_r G$, $\Delta_r G°$, Q, K, reaction spontaneity, and product- or reactant-favorability (Section 19.6).

Q	ΔG	Spontaneous?
$Q < K$	$\Delta G < 0$	Spontaneous to the right as the equation is written
$Q = K$	$\Delta G = 0$	Reaction is at equilibrium
$Q > K$	$\Delta G > 0$	Not spontaneous to the right; spontaneous to the left

K	$\Delta G°$	Reactant-Favored or Product-Favored at Equilibrium?	Spontaneous Under Standard Conditions?
$K >> 1$	$\Delta G° < 0$	Product-favored	Spontaneous under standard conditions
$K = 1$	$\Delta G° = 0$	$[C]^c[D]^d = [A]^a[B]^b$ at equilibrium	At equilibrium under standard conditions
$K << 1$	$\Delta G° > 0$	Reactant-favored	Not spontaneous under standard conditions

c. Describe and use the relationship between the free energy change under standard conditions and equilibrium constants, and calculate K from $\Delta_r G°$ (Sections 19.6 and 19.7). **Study Question(s) assignable in OWL: 28, 30, 44, 48, 50, 61; Go Chemistry Module 24.**

d. Calculate the change in free energy at standard conditions for a reaction from the enthalpy and entropy changes under standard conditions or from the standard free energy of formation of reactants and products ($\Delta_f G°$) (Section 19.7). **Study Question(s) assignable in OWL: 16, 18, 20, 22, 46, 56, 75, 79.**

e. Know how free energy changes with temperature (Section 19.7). **Study Question(s) assignable in OWL: 24, 57, 59, 60, 63, 65, 70.**

KEY EQUATIONS

Equation 19.1 (page 863): Calculate the entropy change from the energy transferred as heat for a reversible process and the temperature at which it occurs.

$$\Delta S = \frac{q_{rev}}{T}$$

Equation 19.2 (page 866) The Boltzmann equation: The entropy of a system, S, is proportional to the number of accessible microstates, W, belonging to a given energy of a system or substance.

$$S = k \ln W$$

Equation 19.3 (page 870): Calculate the standard entropy change under standard conditions for a process from the tabulated entropies of the products and reactants.

$$\Delta_r S° = \Sigma S°(\text{products}) - \Sigma S°(\text{reactants})$$

Equation 19.4 (page 872): Calculate the total entropy change for a system and its surroundings, to determine whether a process is spontaneous under standard conditions.

$$\Delta S°(\text{universe}) = \Delta S°(\text{system}) + \Delta S°(\text{surroundings})$$

Equation 19.5 (page 877): Calculate the free energy change for a process from enthalpy and entropy changes.

$$\Delta_r G° = \Delta_r H° - T\Delta_r S°$$

Equation 19.6 (page 878): Relates the free energy change under nonstandard conditions ($\Delta_r G$) to the standard free energy change ($\Delta_r G°$) and the reaction quotient Q.

$$\Delta_r G = \Delta_r G° + RT \ln Q$$

Equation 19.7 (page 878): Relates the standard free energy change for a reaction and its equilibrium constant.

$$\Delta_r G° = -RT \ln K$$

Equation 19.8 (page 880): Calculate the standard free energy change for a reaction using tabulated values of $\Delta_f G°$.

$$\Delta_r G° = \Sigma\Delta_f G°(\text{products}) - \Sigma\Delta_f G°(\text{reactants})$$

STUDY QUESTIONS

Online homework for this chapter may be assigned in OWL.

▲ denotes challenging questions.

■ denotes questions assignable in OWL.

Blue-numbered questions have answers in Appendix O and fully-worked solutions in the *Student Solutions Manual*.

Practicing Skills

Entropy

(See Examples 19.1 and 19.2 and ChemistryNow Screens 19.4 and 19.5.)

1. Which substance has the higher entropy?
   (a) dry ice (solid CO_2) at −78 °C or CO_2(g) at 0 °C
   (b) liquid water at 25 °C or liquid water at 50 °C
   (c) pure alumina, Al_2O_3(s), or ruby (Ruby is Al_2O_3 in which some Al^{3+} ions in the crystalline lattice are replaced with Cr^{3+} ions.)
   (d) one mole of N_2(g) at 1 bar pressure or one mole of N_2(g) at 10 bar pressure (both at 298 K)

2. ■ Which substance has the higher entropy?
(a) a sample of pure silicon (to be used in a computer chip) or a piece of silicon containing a trace of another element such as boron or phosphorus
(b) $O_2(g)$ at 0 °C or $O_2(g)$ at −50 °C
(c) $I_2(s)$ or $I_2(g)$, both at room temperature
(d) one mole of $O_2(g)$ at 1 bar pressure or one mole of $O_2(g)$ at 0.01 bar pressure (both at 298 K)

3. Use $S°$ values to calculate the standard entropy change, $\Delta_r S°$, for each of the following processes and comment on the sign of the change.
(a) $KOH(s) \longrightarrow KOH(aq)$
(b) $Na(g) \longrightarrow Na(s)$
(c) $Br_2(\ell) \longrightarrow Br_2(g)$
(d) $HCl(g) \longrightarrow HCl(aq)$

4. ■ Use $S°$ values to calculate the standard entropy change, $\Delta_r S°$, for each of the following changes, and comment on the sign of the change.
(a) $NH_4Cl(s) \longrightarrow NH_4Cl(aq)$
(b) $CH_3OH(\ell) \longrightarrow CH_3OH(g)$
(c) $CCl_4(g) \longrightarrow CCl_4(\ell)$
(d) $NaCl(s) \longrightarrow NaCl(g)$

5. Calculate the standard entropy change for the formation of 1.0 mol of the following compounds from the elements at 25 °C.
(a) $HCl(g)$ (b) $Ca(OH)_2(s)$

6. ■ Calculate the standard entropy change for the formation of 1.0 mol of the following compounds from the elements at 25 °C.
(a) $H_2S(g)$ (b) $MgCO_3(s)$

7. ■ Calculate the standard entropy change for each of the following reactions at 25 °C. Comment on the sign of $\Delta_r S°$.
(a) $2\ Al(s) + 3\ Cl_2(g) \longrightarrow 2\ AlCl_3(s)$
(b) $2\ CH_3OH(\ell) + 3\ O_2(g) \longrightarrow 2\ CO_2(g) + 4\ H_2O(g)$

8. Calculate the standard entropy change for each of the following reactions at 25 °C. Comment on the sign of $\Delta_r S°$.
(a) $2\ Na(s) + 2\ H_2O(\ell) \longrightarrow 2\ NaOH(aq) + H_2(g)$
(b) $Na_2CO_3(s) + 2\ HCl(aq) \longrightarrow 2\ NaCl(aq) + H_2O(\ell) + CO_2(g)$

$\Delta_r S°$(universe) and Spontaneity

(See Example 19.3 and ChemistryNow Screen 19.6.)

9. Is the reaction $Si(s) + 2\ Cl_2(g) \longrightarrow SiCl_4(g)$ spontaneous under standard conditions at 298 K? Answer this question by calculating $\Delta S°$(system), $\Delta S°$(surroundings), and $\Delta S°$(universe). (Define reactants and products as the system.)

10. ■ Is the reaction $Si(s) + 2\ H_2(g) \longrightarrow SiH_4(g)$ spontaneous under standard conditions at 298 K? Answer this question by calculating calculating $\Delta S°$(system), $\Delta S°$(surroundings), and $\Delta S°$(universe). (Define reactants and products as the system.)

11. Calculate $\Delta S°$(universe) for the decomposition of 1 mol of liquid water to form gaseous hydrogen and oxygen. Is this reaction spontaneous under these conditions at 25 °C? Explain your answer briefly.

12. ■ Calculate $\Delta S°$(universe) for the formation of 1 mol HCl(g) from gaseous hydrogen and chlorine. Is this reaction spontaneous under these conditions at 25 °C? Explain your answer briefly.

13. Classify each of the reactions according to one of the four reaction types summarized in Table 19.2.
(a) $Fe_2O_3(s) + 2\ Al(s) \longrightarrow 2\ Fe(s) + Al_2O_3(s)$

$\Delta_r H° = -851.5$ kJ/mol-rxn
$\Delta_r S° = -375.2$ J/K · mol-rxn

(b) $N_2(g) + 2\ O_2(g) \longrightarrow 2\ NO_2(g)$

$\Delta_r H° = 66.2$ kJ/mol-rxn
$\Delta_r S° = -121.6$ J/K · mol-rxn

14. ■ Classify each of the reactions according to one of the four reaction types summarized in Table 19.2.
(a) $C_6H_{12}O_6(s) + 6\ O_2(g) \longrightarrow 6\ CO_2(g) + 6\ H_2O(\ell)$

$\Delta_r H° = -673$ kJ/mol-rxn
$\Delta_r S° = 60.4$ J/K · mol-rxn

(b) $MgO(s) + C(graphite) \longrightarrow Mg(s) + CO(g)$

$\Delta_r H° = 490.7$ kJ/mol-rxn
$\Delta_r S° = 197.9$ J/K · mol-rxn

Gibbs Free Energy

(See Example 19.4; see ChemistryNow Screen 19.7.)

15. Using values of $\Delta_f H°$ and $S°$, calculate $\Delta_r G°$ for each of the following reactions at 25 °C.
(a) $2\ Pb(s) + O_2(g) \longrightarrow 2\ PbO(s)$
(b) $NH_3(g) + HNO_3(aq) \longrightarrow NH_4NO_3(aq)$

Which of these reactions is (are) predicted to be product-favored at equilibrium? Are the reactions enthalpy- or entropy-driven?

16. ■ Using values of $\Delta_f H°$ and $S°$, calculate $\Delta_r G°$ for each of the following reactions at 25 °C.
(a) $2\ Na(s) + 2\ H_2O(\ell) \longrightarrow 2\ NaOH(aq) + H_2(g)$
(b) $6\ C(graphite) + 3\ H_2(g) \longrightarrow C_6H_6(\ell)$

Which of these reactions is (are) predicted to be product-favored at equilibrium? Are the reactions enthalpy- or entropy-driven?

17. Using values of $\Delta_f H°$ and $S°$, calculate the standard molar free energy of formation, $\Delta_f G°$, for each of the following compounds:
(a) $CS_2(g)$
(b) $NaOH(s)$
(c) $ICl(g)$

Compare your calculated values of $\Delta_f G°$ with those listed in Appendix L. Which of these formation reactions are predicted to be spontaneous under standard conditions at 25 °C?

18. ■ Using values of $\Delta_f H°$ and $S°$, calculate the standard molar free energy of formation, $\Delta_f G°$, for each of the following:
(a) $Ca(OH)_2(s)$
(b) $Cl(g)$
(c) $Na_2CO_3(s)$

Compare your calculated values of $\Delta_f G°$ with those listed in Appendix L. Which of these formation reactions are predicted to be spontaneous under standard conditions at 25 °C?

Free Energy of Formation

(See Example 19.5; see ChemistryNow Screen 19.7.)

19. Using values of $\Delta_f G°$, calculate $\Delta_r G°$ for each of the following reactions at 25 °C. Which are product-favored at equilibrium?
(a) $2\ K(s) + Cl_2(g) \longrightarrow 2\ KCl(s)$
(b) $2\ CuO(s) \longrightarrow 2\ Cu(s) + O_2(g)$
(c) $4\ NH_3(g) + 7\ O_2(g) \longrightarrow 4\ NO_2(g) + 6\ H_2O(g)$

20. ■ Using values of $\Delta_f G°$, calculate $\Delta_r G°$ for each of the following reactions at 25 °C. Which are product-favored at equilibrium?
(a) $HgS(s) + O_2(g) \longrightarrow Hg(\ell) + SO_2(g)$
(b) $2\ H_2S(g) + 3\ O_2(g) \longrightarrow 2\ H_2O(g) + 2\ SO_2(g)$
(c) $SiCl_4(g) + 2\ Mg(s) \longrightarrow 2\ MgCl_2(s) + Si(s)$

21. For the reaction $BaCO_3(s) \longrightarrow BaO(s) + CO_2(g)$, $\Delta_r G° = +219.7$ kJ. Using this value and other data available in Appendix L, calculate the value of $\Delta_f G°$ for $BaCO_3(s)$.

22. ■ For the reaction $TiCl_2(s) + Cl_2(g) \longrightarrow TiCl_4(\ell)$, $\Delta_r G° = -272.8$ kJ. Using this value and other data available in Appendix L, calculate the value of $\Delta_f G°$ for $TiCl_2(s)$.

Effect of Temperature on ΔG

(See Example 19.6 and ChemistryNow Screen 19.8.)

23. Determine whether the reactions listed below are entropy-favored or disfavored under standard conditions. Predict how an increase in temperature will affect the value of $\Delta_r G°$.
(a) $N_2(g) + 2\ O_2(g) \longrightarrow 2\ NO_2(g)$
(b) $2\ C(s) + O_2(g) \longrightarrow 2\ CO(g)$
(c) $CaO(s) + CO_2(g) \longrightarrow CaCO_3(s)$
(d) $2\ NaCl(s) \longrightarrow 2\ Na(s) + Cl_2(g)$

24. ■ Determine whether the reactions listed below are entropy-favored or disfavored under standard conditions. Predict how an increase in temperature will affect the value of $\Delta_r G°$.
(a) $I_2(g) \longrightarrow 2\ I(g)$
(b) $2\ SO_2(g) + O_2(g) \longrightarrow 2\ SO_3(g)$
(c) $SiCl_4(g) + 2\ H_2O(\ell) \longrightarrow SiO_2(s) + 4\ HCl(g)$
(d) $P_4(s, \text{white}) + 6\ H_2(g) \longrightarrow 4\ PH_3(g)$

25. ■ Heating some metal carbonates, among them magnesium carbonate, leads to their decomposition.

$$MgCO_3(s) \longrightarrow MgO(s) + CO_2(g)$$

(a) Calculate $\Delta_r H°$ and $\Delta_r S°$ for the reaction.
(b) Is the reaction spontaneous under standard conditions at 298 K?
(c) Is the reaction predicted to be spontaneous at higher temperatures?

26. ■ Calculate $\Delta_r H°$ and $\Delta_r S°$ for the reaction of tin(IV) oxide with carbon.

$$SnO_2(s) + C(s) \longrightarrow Sn(s) + CO_2(g)$$

(a) Is the reaction spontaneous under standard conditions at 298 K?
(b) Is the reaction predicted to be spontaneous at higher temperatures?

Free Energy and Equilibrium Constants

(See Example 19.7; use $\Delta G° = -RT \ln K$; see ChemistryNow Screen 19.9.)

27. The standard free energy change, $\Delta_r G°$, for the formation of NO(g) from its elements is +86.58 kJ/mol at 25 °C. Calculate K_p at this temperature for the equilibrium

$$\tfrac{1}{2}\ N_2(g) + \tfrac{1}{2}\ O_2(g) \rightleftharpoons NO(g)$$

Comment on the sign of $\Delta G°$ and the magnitude of K_p.

28. ■ The standard free energy change, $\Delta_r G°$, for the formation of $O_3(g)$ from $O_2(g)$ is +163.2 kJ/mol at 25 °C. Calculate K_p at this temperature for the equilibrium

$$3\ O_2(g) \rightleftharpoons 2\ O_3(g)$$

Comment on the sign of $\Delta G°$ and the magnitude of K_p.

29. Calculate $\Delta_r G°$ at 25 °C for the formation of one mol of $C_2H_6(g)$ from $C_2H_4(g)$ and $H_2(g)$. Use this value to calculate K_p for the equilibrium.

$$C_2H_4(g) + H_2(g) \rightleftharpoons C_2H_6(g)$$

Comment on the sign of $\Delta_r G°$ and the magnitude of K_p.

30. ■ Calculate $\Delta_r G°$ at 25 °C for the formation of 1 mol of $C_2H_5OH(g)$ from $C_2H_4(g)$ and $H_2O(g)$. Use this value to calculate K_p for the equilibrium.

$$C_2H_4(g) + H_2O(g) \rightleftharpoons C_2H_5OH(g)$$

Comment on the sign of $\Delta_r G°$ and the magnitude of K_p.

General Questions

These questions are not designated as to type or location in the chapter. They may combine several concepts.

31. Compare the formulas in each set of compounds, and decide which is expected to have the higher entropy. Assume all are at the same temperature. Check your answers using data in Appendix L.
(a) HF(g), HCl(g), or HBr(g)
(b) $NH_4Cl(s)$ or $NH_4Cl(aq)$
(c) $C_2H_4(g)$ or $N_2(g)$ (two substances with the same molar mass)
(a) NaCl(s) or NaCl(g)

32. Using standard entropy values, calculate $\Delta_r S°$ for the formation of 1.0 mol of $NH_3(g)$ from $N_2(g)$ and $H_2(g)$ at 25 °C.

33. ■ About 5 billion kilograms of benzene, C_6H_6, are made each year. Benzene is used as a starting material for many other compounds and as a solvent (although it is also a carcinogen, and its use is restricted). One compound that can be made from benzene is cyclohexane, C_6H_{12}.

$$C_6H_6(\ell) + 3\ H_2(g) \rightarrow C_6H_{12}(\ell)$$

$$\Delta_r H° = -206.7\ \text{kJ};\ \Delta_r S° = -361.5\ \text{J/K}$$

Is this reaction predicted to be product-favored at equilibrium at 25 °C? Is the reaction enthalpy- or entropy-driven?

34. Hydrogenation, the addition of hydrogen to an organic compound, is an industrially important reaction. Calculate $\Delta_r H°$, $\Delta_r S°$, and $\Delta_r G°$ for the hydrogenation of octene, C_8H_{16}, to give octane, C_8H_{18}, at 25 °C. Is the reaction product- or reactant-favored at equilibrium?

$$C_8H_{16}(g) + H_2(g) \rightarrow C_8H_{18}(g)$$

Along with data in Appendix L, the following information is needed for this calculation.

Compound	$\Delta_f H°$ (kJ/mol)	$S°$ (J/K · mol)
Octene	−82.93	462.8
Octane	−208.45	463.639

35. ■ Is the combustion of ethane, C_2H_6, product-favored at equilibrium at 25 °C?

$$C_2H_6(g) + \tfrac{7}{2}\,O_2(g) \rightarrow 2CO_2(g) + 3\ H_2O(g)$$

Answer the question by calculating the value of $\Delta S°$ (universe) at 298 K, using values of $\Delta_f H°$ and $S°$ in Appendix L. Does the answer agree with your preconceived idea of this reaction?

36. Write a balanced equation that depicts the formation of 1 mol of $Fe_2O_3(s)$ from its elements. What is the standard free energy of formation of 1.00 mol of $Fe_2O_3(s)$? What is the value of $\Delta G°$ when 454 g (1 lb) of $Fe_2O_3(s)$ is formed from the elements?

37. When vapors from hydrochloric acid and aqueous ammonia come in contact, they react, producing a white "cloud" of solid NH_4Cl (Figure 19.10).

$$HCl(g) + NH_3(g) \rightleftharpoons NH_4Cl(s)$$

Defining the reactants and products as the system under study:
(a) Predict whether $\Delta S°$(system), $\Delta S°$(surroundings), $\Delta S°$(universe), $\Delta_r H°$, and $\Delta_r G°$ (at 298 K) are greater than zero, equal to zero, or less than zero; and explain your prediction. Verify your predictions by calculating values for each of these quantities.
(b) Calculate the value of K_p for this reaction at 298 K.

38. Calculate $\Delta S°$(system), $\Delta S°$(surroundings), $\Delta S°$(universe) for each of the following processes at 298 K, and comment on how these systems differ.
(a) $HNO_3(g) \rightarrow HNO_3(aq)$
(b) $NaOH(s) \rightarrow NaOH(aq)$

39. Methanol is now widely used as a fuel in race cars. Consider the following reaction as a possible synthetic route to methanol.

$$C(\text{graphite}) + \tfrac{1}{2}\,O_2(g) + 2\ H_2(g) \rightleftharpoons CH_3OH(\ell)$$

Calculate K_p for the formation of methanol at 298 K using this reaction. Would a different temperature be better suited to this reaction?

40. ■ The enthalpy of vaporization of liquid diethyl ether, $(C_2H_5)_2O$, is 26.0 kJ/mol at the boiling point of 35.0 °C. Calculate $\Delta S°$ for a vapor-to-liquid transformation at 35.0 °C.

41. Calculate the entropy change, $\Delta S°$, for the vaporization of ethanol, C_2H_5OH, at its normal boiling point, 78.0 °C. The enthalpy of vaporization of ethanol is 39.3 kJ/mol.

42. Using thermodynamic data, estimate the normal boiling point of ethanol. (Recall that liquid and vapor are in equilibrium at 1.0 atm pressure at the normal boiling point.) The actual normal boiling point is 78 °C. How well does your calculated result agree with the actual value?

43. The following reaction is reactant-favored at equilibrium at room temperature.

$$COCl_2(g) \rightarrow CO(g) + Cl_2(g)$$

Will raising or lowering the temperature make it product-favored?

44. ■ When calcium carbonate is heated strongly, CO_2 gas is evolved. The equilibrium pressure of CO_2 is 1.00 bar at 897 °C, and $\Delta_r H°$ at 298 K is 179.0 kJ.

$$CaCO_3(s) \rightarrow CaO(s) + CO_2(g)$$

Estimate the value of $\Delta_r S°$ at 897 °C for the reaction.

45. Sodium reacts violently with water according to the equation

$$Na(s) + H_2O(\ell) \rightarrow NaOH(aq) + \tfrac{1}{2}\,H_2(g)$$

Without doing calculations, predict the signs of $\Delta_r H°$ and $\Delta_r S°$ for the reaction. Verify your prediction with a calculation.

46. ■ Yeast can produce ethanol by the fermentation of glucose ($C_6H_{12}O_6$), which is the basis for the production of most alcoholic beverages.

$$C_6H_{12}O_6(aq) \rightarrow 2\ C_2H_5OH(\ell) + 2\ CO_2(g)$$

Calculate $\Delta_r H°$, $\Delta_r S°$, and $\Delta_r G°$ for the reaction at 25 °C. Is the reaction product- or reactant-favored? In addition to the thermodynamic values in Appendix L, you will need the following data for $C_6H_{12}O_6(aq)$:

$\Delta_f H° = -1260.0$ kJ/mol; $S° = 289$ J/K · mol; and $\Delta_f G° = -918.8$ kJ/mol.

47. ■ Elemental boron, in the form of thin fibers, can be made by reducing a boron halide with H_2.

$$BCl_3(g) + \tfrac{3}{2}\,H_2(g) \rightarrow B(s) + 3HCl(g)$$

Calculate $\Delta_r H°$, $\Delta_r S°$, and $\Delta_r G°$ at 25 °C for this reaction. Is the reaction predicted to be product-favored at equilibrium at 25 °C? If so, is it enthalpy- or entropy-driven? [$S°$ for B(s) is 5.86 J/K · mol.]

48. ■ ▲ Estimate the vapor pressure of ethanol at 37 °C using thermodynamic data. Express the result in mm of mercury.

49. The equilibrium constant, K_p, for $N_2O_4(g) \rightleftharpoons 2\ NO_2(g)$ is 0.14 at 25 °C. Calculate $\Delta_r G°$ for the conversion of $N_2O_4(g)$ to $NO_2(g)$ from this constant, and compare this value with that determined from the $\Delta_f G°$ values in Appendix L.

50. ■ ▲ Estimate the boiling point of water in Denver, Colorado (where the altitude is 1.60 km and the atmospheric pressure is 630 mm 0.840 bar).

51. The equilibrium constant for the butane ⇌ isobutane equilibrium at 25 °C is 2.50. Calculate $\Delta_r G°$ at this temperature in units of kJ/mol.

butane ⇌ isobutane

$CH_3CH_2CH_2CH_3 \rightleftharpoons CH_3CH(CH_3)CH_3$

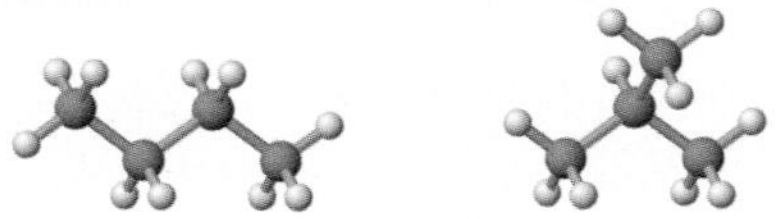

$$K_c = \frac{[\text{isobutane}]}{[\text{butane}]} = 2.50 \text{ at } 298\ K$$

52. ■ A crucial reaction for the production of synthetic fuels is the conversion of coal to H_2 with steam. The chemical reaction is

$$C(s) + H_2O(g) \rightarrow CO(g) + H_2(g)$$

(a) Calculate $\Delta_r G°$ for this reaction at 25 °C, assuming C(s) is graphite.
(b) Calculate K_p for the reaction at 25 °C.
(c) Is the reaction predicted to be product-favored at equilibrium at 25 °C? If not, at what temperature will it become so?

53. Calculate $\Delta_r G°$ for the decomposition of sulfur trioxide to sulfur dioxide and oxygen.

$$2\ SO_3(g) \rightleftharpoons 2\ SO_2(g) + O_2(g)$$

(a) Is the reaction product-favored at equilibrium at 25 °C?
(b) If the reaction is not product-favored at 25 °C, is there a temperature at which it will become so? Estimate this temperature.
(c) What is the equilibrium constant for the reaction at 1500 °C?

54. Methanol can be made by partial oxidation of methane by $O_2(g)$.

$$CH_4(g) + \tfrac{1}{2}\,O_2(g) \rightleftharpoons CH_3OH(\ell)$$

(a) Determine $\Delta S°$(system), $\Delta S°$(surroundings), and $\Delta S°$(universe) for this process.
(b) Is this reaction product-favored at equilibrium at 25 °C?

55. ■ A cave in Mexico was recently discovered to have some interesting chemistry. Hydrogen sulfide, H_2S, reacts with oxygen in the cave to give sulfuric acid, which drips from the ceiling in droplets with a pH less than 1. The reaction occurring is

$$H_2S(g) + 2\ O_2(g) \rightarrow H_2SO_4(\ell)$$

Calculate $\Delta_r H°$, $\Delta_r S°$, and $\Delta_r G°$. Is the reaction product-favored at equilibrium at 25 °C? Is it enthalpy- or entropy-driven?

56. ■ Wet limestone is used to scrub SO_2 gas from the exhaust gases of power plants. One possible reaction gives hydrated calcium sulfite:

$$CaCO_3(s) + SO_2(g) + \tfrac{1}{2}\,H_2O(\ell) \rightleftharpoons CaSO_3 \cdot \tfrac{1}{2}\,H_2O(s) + CO_2(g)$$

Another reaction gives hydrated calcium sulfate:

$$CaCO_3(s) + SO_2(g) + \tfrac{1}{2}\,H_2O(\ell) + \tfrac{1}{2}\,O_2(g) \rightleftharpoons CaSO_4 \cdot \tfrac{1}{2}\,H_2O(s) + CO_2(g)$$

(a) Which is the more product-favored reaction? Use the data in the table on the next page and any other information needed in Appendix L to calculate $\Delta_r G°$ for each reaction at 25 °C.

	$CaSO_3 \cdot \frac{1}{2} H_2O(s)$	$CaSO_4 \cdot \frac{1}{2} H_2O(s)$
$\Delta_f H°$ (kJ/mol)	−1311.7	−1574.65
$S°$ (J/K · mol)	121.3	134.8

(b) Calculate $\Delta_r G°$ for the reaction

$$CaSO_3 \cdot \tfrac{1}{2} H_2O(s) + \tfrac{1}{2} O_2(g) \rightleftharpoons CaSO_4 \cdot \tfrac{1}{2} H_2O(s)$$

Is this reaction product- or reactant-favored at equilibrium?

57. ■ Sulfur undergoes a phase transition between 80 and 100 °C.

$$S_8(\text{rhombic}) \longrightarrow S_8(\text{monoclinic})$$

$$\Delta_r H° = 3.213 \text{ kJ/mol-rxn} \quad \Delta_r S° = 8.7 \text{ J/K} \cdot \text{mol-rxn}$$

(a) Estimate $\Delta_r G°$ for the transition at 80.0 °C and 110.0 °C. What do these results tell you about the stability of the two forms of sulfur at each of these temperatures?
(b) Calculate the temperature at which $\Delta_r G° = 0$. What is the significance of this temperature?

58. Calculate the entropy change for dissolving HCl gas in water at 25 °C. Is the sign of $\Delta S°$ what you expected? Why or why not?

In the Laboratory

59. ■ Some metal oxides can be decomposed to the metal and oxygen under reasonable conditions. Is the decomposition of silver(I) oxide product-favored at 25 °C?

$$2\, Ag_2O(s) \longrightarrow 4\, Ag(s) + O_2(g)$$

If not, can it become so if the temperature is raised? At what temperature does the reaction become product-favored?

60. ■ Copper(II) oxide, CuO, can be reduced to copper metal with hydrogen at higher temperatures.

$$CuO(s) + H_2(g) \longrightarrow Cu(s) + H_2O(g)$$

Is this reaction product- or reactant-favored at equilibrium at 298 K?

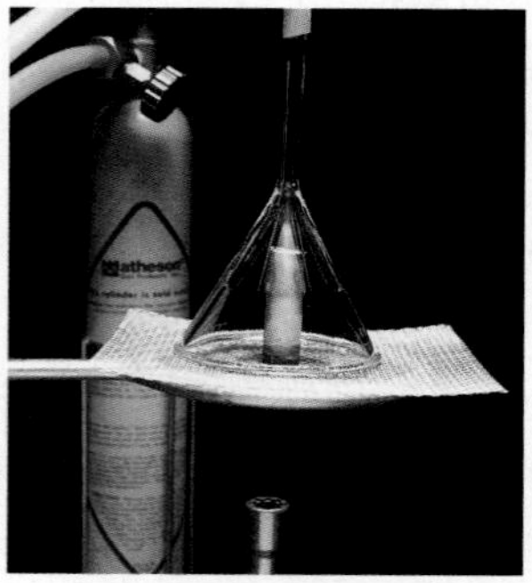

Charles D. Winters

If copper metal is heated in air, a black film of CuO forms on the surface. In this photo, the heated bar, covered with a black CuO film, has been bathed in hydrogen gas. Black, solid CuO is reduced rapidly to copper at higher temperatures.

61. ■ Calculate $\Delta_f G°$ for HI(g) at 350 °C, given the following equilibrium partial pressures: $P(H_2) = 0.132$ bar, $P(I_2) = 0.295$ bar, and $P(HI) = 1.61$ bar. At 350 °C and 1 bar, I_2 is a gas.

$$\tfrac{1}{2} H_2(g) + \tfrac{1}{2} I_2(g) \rightleftharpoons HI(g)$$

62. ■ Calculate the equilibrium constant for the formation of NiO at 1627 °C. Can the reaction proceed in the forward direction if the initial pressure of O_2 is below 1.00 mm Hg? {$\Delta_f G°$ [NiO(s)] = −72.1 kJ/mol at 1627 °C}

$$Ni(s) + \tfrac{1}{2} O_2(g) \rightleftharpoons NiO(s)$$

63. ■ Titanium(IV) oxide is converted to titanium carbide with carbon at a high temperature.

$$TiO_2(s) + 3\, C(s) \longrightarrow 2\, CO(g) + TiC(s)$$

Compound	Free Energies of Formation at 727 °C, kJ/mol
$TiO_2(s)$	−757.8
TiC(s)	−162.6
CO(g)	−200.2

(a) Calculate $\Delta_r G°$ and K at 727 °C
(b) Is the reaction product-favored at equilibrium at this temperature?
(c) How can the reactant or product concentrations be adjusted for the reaction to be product-favored at 727 °C?

64. Cisplatin [*cis*-diamminedichloroplatinum(II)] is a potent treatment for certain types of cancers, but the *trans* isomer is not effective. (They are called *isomers* because the two compounds have the same formula but a different arrangement of atoms.) What is the equilibrium constant at 298 K for the transformation of the *cis* to the *trans* isomer? Which is more thermodynamically stable, the *cis* or the *trans* isomer?

Compound	$\Delta_f H°$ (kJ/mol, 298 K)	$\Delta_f G°$ (kJ/mol, 298 K)
Cis-$Pt(NH_3)_2Cl_2$	−467.4	−228.7
Trans-$Pt(NH_3)_2Cl_2$	−480.3	−222.8

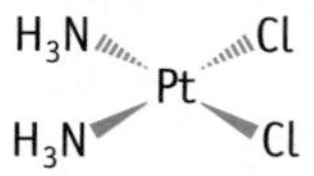

cis isomer

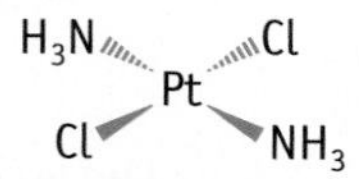

trans isomer

Summary and Conceptual Questions

The following questions may use concepts from this and previous chapters.

65. ■ ▲ Mercury vapor is dangerous because it can be breathed into the lungs. We wish to estimate the vapor pressure of mercury at two different temperatures from the following data:

	$\Delta_f H°$ (kJ/mol)	$S°$ (J/K · mol)	$\Delta_f G°$ (kJ/mol)
Hg(ℓ)	0	76.02	0
Hg(g)	61.38	174.97	31.88

Estimate the temperature at which K_p for the process Hg(ℓ) $\rightleftharpoons$ Hg(g) is equal to (a) 1.00 and (b) 1/760. What is the vapor pressure at each of these temperatures? (Experimental vapor pressures are 1.00 mm Hg at 126.2 °C and 1.00 bar at 356.6 °C.) (Note: The temperature at which $P = 1.00$ bar can be calculated from thermodynamic data. To find the other temperature, you will need to use the temperature for $P = 1.00$ bar and the Clausius–Clapeyron equation on page 576.)

66. Explain why each of the following statements is incorrect.
    (a) Entropy increases in all spontaneous reactions.
    (b) Reactions with a negative free energy change ($\Delta_r G° < 0$) are product-favored and occur with rapid transformation of reactants to products.
    (c) All spontaneous processes are exothermic.
    (d) Endothermic processes are never spontaneous.

67. ■ Decide whether each of the following statements is true or false. If false, rewrite it to make it true.
    (a) The entropy of a substance increases on going from the liquid to the vapor state at any temperature.
    (b) An exothermic reaction will always be spontaneous.
    (c) Reactions with a positive $\Delta_r H°$ and a positive $\Delta_r S°$ can never be product-favored.
    (d) If $\Delta_r G°$ for a reaction is negative, the reaction will have an equilibrium constant greater than 1.

68. Under what conditions is the entropy of a pure substance 0 J/K · mol? Could a substance at standard conditions have a value of 0 J/K · mol? A negative entropy value? Are there any conditions under which a substance will have negative entropy? Explain your answer.

69. In Chapter 14, you learned that entropy, as well as enthalpy, plays a role in the solution process. If $\Delta H°$ for the solution process is zero, explain how the process can be driven by entropy.

70. ■ ▲ Consider the formation of NO(g) from its elements.

$$N_2(g) + O_2(g) \rightleftharpoons 2\ NO(g)$$

    (a) Calculate K_p at 25 °C. Is the reaction product-favored at this temperature?
    (b) Assuming $\Delta_r H°$ and $\Delta_r S°$ are nearly constant with temperature, calculate $\Delta_r G°$ at 700 °C. Estimate K_p from the new value of $\Delta_r G°$ at 700 °C. Is the reaction product-favored at 700 °C?
    (c) Using K_p at 700 °C, calculate the equilibrium partial pressures of the three gases if you mix 1.00 bar each of N_2 and O_2.

71. Write a chemical equation for the oxidation of $C_2H_6(g)$ by $O_2(g)$ to form $CO_2(g)$ and $H_2O(g)$. Defining this as the system:
    (a) Predict whether the signs of $\Delta S°$ (system), $\Delta S°$ (surroundings), and $\Delta S°$ (universe) will be greater than zero, equal to zero, or less than zero. Explain your prediction.
    (b) Predict the signs of $\Delta_r H°$ and $\Delta_r G°$. Explain how you made this prediction.
    (c) Will the value of K_p be very large, very small, or near 1? Will the equilibrium constant, K_p, for this system be larger or smaller at temperatures greater than 298 K? Explain how you made this prediction.

72. ■ The normal melting point of benzene, C_6H_6, is 5.5 °C. For the process of melting, what is the sign of each of the following?
    (a) $\Delta H°$ (b) $\Delta S°$ (c) $\Delta G°$ at 5.5 °C (d) $\Delta G°$ at 0.0 °C (e) $\Delta G°$ at 25.0 °C

73. Calculate the standard molar entropy change, $\Delta_r S°$, for each of the following reactions at 25 °C:
    1. $C(s) + 2\ H_2(g) \rightarrow CH_4(g)$
    2. $CH_4(g) + \frac{1}{2}\ O_2(g) \rightarrow CH_3OH(\ell)$
    3. $C(s) + 2\ H_2(g) + \frac{1}{2}\ O_2(g) \rightarrow CH_3OH(\ell)$

    Verify that these values are related by the equation $\Delta_r S°_1 + \Delta_r S°_2 = \Delta_r S°_3$. What general principle is illustrated here?

74. For each of the following processes, predict the algebraic sign of $\Delta_r H°$, $\Delta_r S°$, and $\Delta_r G°$. No calculations are necessary; use your common sense.
    (a) The decomposition of liquid water to give gaseous oxygen and hydrogen, a process that requires a considerable amount of energy.
    (b) Dynamite is a mixture of nitroglycerin, $C_3H_5N_3O_9$, and diatomaceous earth. The explosive decomposition of nitroglycerin gives gaseous products such as water, CO_2, and others; much heat is evolved.
    (c) The combustion of gasoline in the engine of your car, as exemplified by the combustion of octane.

$$2\ C_8H_{18}(g) + 25\ O_2(g) \rightarrow 16\ CO_2(g) + 18\ H_2O(g)$$

75. ■ "Heater Meals" are food packages that contain their own heat source. Just pour water into the heater unit, wait a few minutes, and voilà! You have a hot meal.

$$Mg(s) + 2\ H_2O(\ell) \longrightarrow Mg(OH)_2(s) + H_2(g)$$

Charles D. Winters

The heat for the heater unit is produced by the reaction of magnesium with water.

(a) Confirm that this is a spontaneous reaction under standard conditions.
(b) What mass of magnesium is required to produce sufficient energy to heat 225 mL of water (density = 0.995 g/mL) from 25 °C to the boiling point?

76. *Abba's Refrigerator* was described on page 222 is an example of the validity of the second law of thermodynamics. Explain how the second law applies to this simple but useful device.

77. Oxygen dissolved in water can cause corrosion in hot-water heating systems. To remove oxygen, hydrazine (N_2H_4) is often added. Hydrazine reacts with dissolved O_2 to form water and N_2.
(a) Write a balanced chemical equation for the reaction of hydrazine and oxygen. Identify the oxidizing and reducing agents in this redox reaction.
(b) Calculate $\Delta_rH°$, $\Delta_rS°$, and $\Delta_rG°$ for this reaction involving 1 mol of N_2H_4 at 25 °C.
(c) Because this is an exothermic reaction, energy is evolved as heat. What temperature change is expected in a heating system containing 5.5×10^4 L of water? (Assume no energy is lost to the surroundings.)
(d) The mass of a hot-water heating system is 5.5×10^4 kg. What amount of O_2 (in moles) would be present in this system if it is filled with water saturated with O_2? (The solubility of O_2 in water at 25 °C is 0.000434 g per 100 g of water.)
(e) Assume hydrazine is available as a 5.0% solution in water. What mass of this solution should be added to totally consume the dissolved O_2 [described in part (d)]?
(f) Assuming the N_2 escapes as a gas, calculate the volume of $N_2(g)$ (measured at 273 K and 1.00 bar) that will be produced.

78. The formation of diamond from graphite is a process of considerable importance.

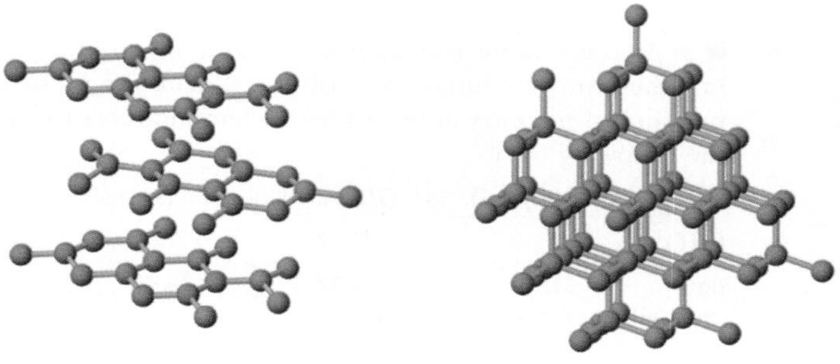

graphite   diamond

(a) Using data in Appendix L, calculate $\Delta_rS°$, $\Delta_rH°$, and $\Delta_rG°$ for this process at 25 °C.
(b) The calculations will suggest that this process is not possible at any temperature. However, the synthesis of diamonds by this reaction is a commercial process. How can this contradiction be rationalized? (Note: In the industrial synthesis, high pressure and high temperatures are used.)

79. ■ Iodine, I_2, dissolves readily in carbon tetrachloride. For this process, $\Delta H° = 0$ kJ/mol.

$$I_2(s) \longrightarrow I_2\ (\text{in } CCl_4 \text{ solution})$$

What is the sign of $\Delta_rG°$? Is the dissolving process entropy driven or enthalpy driven? Explain briefly.

80. Use the simulation on Screen 19.8 of ChemistryNow to answer the following questions.
(a) Consider the reaction of Fe_2O_3 and C. How does $\Delta_rG°$ vary with temperature? Is there a temperature at which the reaction is spontaneous?
(b) Consider the reaction of HCl and Na_2CO_3. Is there a temperature at which the reaction is no longer spontaneous?
(c) Is the spontaneity of a reaction dependent on or independent of temperature?

81. Use the simulation on Screen 19.6 of ChemistryNow to answer the following questions.
(a) Does the spontaneity of the decomposition of CH_3OH to the elements change as the temperature increases?
(b) Is there a temperature between 400 K and 1000 K at which the decomposition is spontaneous?

82. Use the simulation on Screen 19.6 of ChemistryNow to answer the following questions regarding the reaction of NO and Cl_2 to produce NOCl.
(a) What is $\Delta S°$ (system) at 400 K for this reaction?
(b) Does $\Delta S°$ (system) change with temperature?
(c) Does $\Delta S°$ (surroundings) change with temperature?

(d) Does $\Delta S°$(universe) always change with an increase in temperature?
(e) Do exothermic reactions always lead to positive values of $\Delta S°$(universe)?
(f) Is the NO + Cl_2 reaction spontaneous at 400 K? At 700 K?

83. ■ ▲ The Haber–Bosch process for the production of ammonia is one of the key processes in industrialized countries.

$$N_2(g) + 3\ H_2(g) \rightleftharpoons 2\ NH_3(g)$$

(a) Calculate $\Delta_r G°$ for the reaction at 298 K, 800 K, and 1300 K. Data at 298 K are in Appendix L. Data for the other temperatures are as follows:

Temperature	$\Delta_r H_o$ (kJ/mol)	$\Delta_r S_o$ (J/K · mol)
800 K	−107.4	−225.4
1300 K	−112.4	−228.0

How does the free energy change for the reaction change with temperature?
(b) Calculate the equilibrium constant for the reaction at 298 K, 800 K, and 1300 K.
(c) Calculate the mole fraction of ammonia in the equilibrium mixture at each of the temperatures. At what temperature is the mole fraction of NH_3 the largest?

84. ▲ Muscle cells need energy to contract. One biochemical pathway for energy transfer is the breakdown of glucose to pyruvate in a process called glycolysis. In the presence of sufficient oxygen in the cell, pyruvate is oxidized to CO_2 and H_2O to make further energy available. However, under extreme conditions not enough oxygen can be supplied to the cells, so muscle cells produce lactate ion according to the reaction

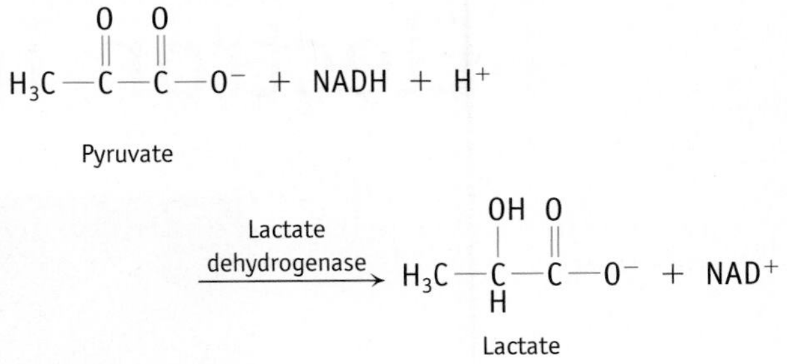

where $\Delta_r G°' = -25.1$ kJ/mol. In living cells, the pH value is about 7. The hydronium ion concentration is constant and is included in $\Delta G°$, which is then called $\Delta_r G°'$ (as explained on page 884). *(This problem is taken from the problems for the 36th International Chemistry Olympiad for high school students held in Kiel, Germany in 2004.)*
(a) Calculate $\Delta_r G°$ for the reaction at 25 °C.
(b) Calculate the equilibrium constant K'. (The hydronium ion concentration is included in the constant. That is, $K' = K \cdot [H_3O^+]$ for the reaction at 25 °C and pH = 7.0.)
(c) $\Delta_r G°'$ is the free energy change under standard conditions; that is, the concentrations of all reactants (except H_3O^+) are 1.00 mol/L. Calculate $\Delta_r G'$ at 25 °C, assuming the following concentrations in the cell: pyruvate, 380 μmol/L; NADH, 50 μmol/L; lactate ion, 3700 μmol/L; and NAD^+ ion, 540 μmol/L.

20 Principles of Reactivity: Electron Transfer Reactions

Charles D. Winters

New euro coins that use different metal alloys.

Don't Hold onto That Money!

Nickel Allergy from Coins

Shortly after the new euro coins were introduced in European countries in 2002, people showed up in hospitals with strange allergies. These allergic reactions were soon identified as arising from nickel, presumably from the coins.

The surface of the 1- and 2-euro coins consists of a white Cu-Ni alloy (75% Cu and 25% Ni) and a yellow nickel-brass (75% Cu, 20% Zn, and 5% Ni). Skin tests with these coins showed that patients did get a rash from the coins, so there was initially quite a controversy about the coins. More extensive testing, however, showed that the euro coins did not release any more nickel than other coins with a similar composition.

Nickel allergy, caused by nickel(II) ions, is one of the most common causes of contact dermatitis, a skin inflammation that results in swollen, reddened, and itchy skin. Up to 15% of women are allergic to nickel, probably from nickel in less expensive costume jewelry. Unfortunately, there is no cure. One can only prevent it by staying away from nickel-containing objects, which is not easy to do because so many articles (such as watchbands, hairpins, eyeglass frames, paper clips, keys, and tools) contain nickel.

Questions:

Experiments show that metal ions, Ni^{2+} and Cu^{2+}, are present on the surface of coins, presumably from corrosion. In one experiment, the bimetallic euro coins were put in salty "artificial sweat," and the coins showed signs of significant corrosion. Under these conditions, the yellow outer ring of the 1-euro coin had a tiny negative charge, and the white inner portion had a slight positive charge.

1. What electrochemical reactions are possible that could involve Cu^{2+} and Ni or Ni^{2+} and Cu?
2. Could either reaction produce an electric current?
3. If such a reaction occurs under standard conditions, what is the value of the electric potential, $E°_{cell}$?

Answers to these questions are in Appendix Q.

Chapter Goals

See Chapter Goals Revisited (page 939) for Study Questions keyed to these goals and assignable in OWL.

- Balance equations for oxidation–reduction reactions in acidic or basic solutions using the half-reaction approach.
- Understand the principles underlying voltaic cells.
- Understand how to use electrochemical potentials.
- Explore electrolysis, the use of electrical energy to produce chemical change.

Chapter Outline

Let us introduce you to electrochemistry and electron transfer reactions with a simple experiment. Place a piece of copper in an aqueous solution of silver nitrate. After a short time, metallic silver deposits on the copper, and the solution takes on the blue color typical of aqueous Cu^{2+} ions (Figure 20.1). The following oxidation–reduction (redox) reaction has occurred:

$$Cu(s) + 2\ Ag^{+}(aq) \rightarrow Cu^{2+}(aq) + 2\ Ag(s)$$

At the particulate level, Ag^{+} ions in solution come into direct contact with the copper surface where the transfer of electrons occurs. Two electrons are transferred

Chemistry Now™

Throughout the text this icon introduces an opportunity for self-study or to explore interactive tutorials by signing in at **www.thomsonedu.com/login**.

FIGURE 20.1 The oxidation of copper by silver ions. Note that water molecules are not shown for clarity. (See ChemistryNow for photographs of this reaction sequence.)

from a Cu atom to two Ag^+ ions. Copper ions, Cu^{2+}, enter the solution, and silver atoms are deposited on the copper surface. This product-favored reaction continues until one or both of the reactants is consumed.

The reaction between copper metal and silver ions can be used to generate an electric current if the experiment is carried out in a different way. If the reactants, Cu(s) and Ag^+(aq), are in direct contact, electrons will be transferred directly from copper atoms to silver ions, and an increase in temperature (heating) rather than electrical work will result. Instead, the reaction has to be done in an apparatus that allows electrons to be transferred from one reactant to the other through an electrical circuit. The movement of electrons through the circuit constitutes an electric current that can be used to light a light bulb or to run a motor.

Devices that use chemical reactions to produce an electric current are called **voltaic cells** or **galvanic cells**, names honoring Count Alessandro Volta (1745–1827) and Luigi Galvani (1737–1798). All voltaic cells work in the same general way. They use product-favored redox reactions composed of an oxidation and a reduction. The cell is constructed so that electrons produced by the reducing agent are transferred through an electric circuit to the oxidizing agent.

A voltaic cell converts chemical energy to electrical energy. The opposite process, the use of electric energy to effect a chemical change, occurs in **electrolysis**. An example is the electrolysis of water (Figure 1.12), in which electrical energy is used to split water into its component elements, hydrogen and oxygen. Electrolysis is also used to electroplate one metal onto another, to obtain aluminum from its common ore (bauxite, mostly Al_2O_3), and to prepare important chemicals such as chlorine.

■ **Two Types of Electrochemical Processes**

- Chemical change can produce an electric current in a voltaic cell.
- Electric energy can cause chemical change in the process of electrolysis.

Electrochemistry is the field of chemistry that considers chemical reactions that produce or are caused by electrical energy. Because all electrochemical reactions are oxidation–reduction (redox) reactions, we begin our exploration of this subject by first describing electron transfer reactions in more detail.

Module 25

20.1 Oxidation–Reduction Reactions

In an oxidation–reduction reaction, electron transfer occurs between a reducing agent and an oxidizing agent [see Section 3.9]. The essential features of all electron transfer reactions are as follows:

- One reactant is oxidized, and one is reduced.
- The extent of oxidation and reduction must balance.
- The oxidizing agent (the chemical species causing oxidation) is reduced.
- The reducing agent (the chemical species causing reduction) is oxidized.

These aspects of oxidation–reduction or redox reactions are illustrated for the reaction of copper metal and silver ion (Figure 20.1).

■ **Oxidation Numbers** Oxidation numbers (page 144) can be used to determine whether a substance is oxidized or reduced. An element is oxidized if its oxidation number increases. The oxidation number decreases in a reduction.

Cu oxidized, oxidation number increases;
Cu is the reducing agent.

$$Cu(s) + 2\,Ag^+(aq) \longrightarrow Cu^{2+}(aq) + 2\,Ag(s)$$

Ag^+ reduced, oxidation number decreases;
Ag^+ is the oxidizing agent.

Balancing Oxidation–Reduction Equations

All equations for oxidation–reduction reactions must be balanced for both mass and charge. The same number of atoms appear in the products and reactants of an equation, and the sum of electric charges of all the species on each side of the equation arrow must be the same. Charge balance guarantees that the number of electrons produced in oxidation equals the number of electrons consumed in reduction.

Balancing some redox equations can be complicated, but fortunately some systematic procedures can be used in these cases. Here, we describe the **half-reaction method**, a procedure that involves writing separate, balanced equations for the oxidation and reduction processes called half-reactions. One half-reaction describes the oxidation part of the reaction, and a second half-reaction describes the reduction part.

When a reaction has been determined to involve oxidation and reduction (by noting, for example, that oxidation state changes have occurred), the equation is separated into two half-reactions, which are then balanced for mass and charge. The equation for the overall reaction is then the sum of these two balanced half-reactions, after adjustments have been made (if necessary) in one or both half-reaction equations so that the numbers of electrons transferred from reducing agent to oxidizing agent balance. For example, the half-reactions for the reaction of copper metal with silver ions are

Reduction half-reaction:	$Ag^+(aq) + e^- \rightarrow Ag(s)$
Oxidation half-reaction:	$Cu(s) \rightarrow Cu^{2+}(aq) + 2\ e^-$

Notice that the equations for the half-reactions are themselves balanced for mass and charge. In the copper half-reaction, there is one Cu atom on each side of the equation (mass balance). The electric charge on the right side of the equation is 0 (the sum of +2 for the ion and −2 for two electrons), as it is on the left side (charge balance).

To produce the net chemical equation, we will add the two half-reactions. First, however, we must multiply the silver half-reaction by 2.

$$2\ Ag^+(aq) + 2\ e^- \rightarrow 2\ Ag(s)$$

Each mole of copper atoms produces two moles of electrons, and two moles of Ag^+ ions are required to consume those electrons.

Finally, adding the two half-reactions and canceling electrons from both sides leads to the net ionic equation for the reaction.

Reduction half-reaction:	$2\ [Ag^+(aq) + e^- \rightarrow Ag(s)]$
Oxidation half-reaction:	$Cu(s) \rightarrow Cu^{2+}(aq) + 2\ e^-$
Net ionic equation	$Cu(s) + 2\ Ag^+(aq) \rightarrow Cu^{2+}(aq) + 2\ Ag(s)$

The resulting net ionic equation is balanced for mass and charge.

Chemistry Now™

Sign in at **www.thomsonedu.com/login** and go to Chapter 20 Contents to see:
- Screen 20.2 for a self-study module on various types of **oxidation–reduction reactions**
- Screen 20.3 for a self-study module on **methods of balancing redox reactions in acidic solution**

FIGURE 20.2 Reduction of Cu^{2+} by Al. (a) A ball of aluminum foil is added to a solution of $Cu(NO_3)_2$ and NaCl. (b) A coating of copper is soon seen on the surface of the aluminum, and the reaction generates a significant amount of heat. (Aluminum always has a thin coating of Al_2O_3 on the surface, which protects the metal from further reaction. However, in the presence of Cl^- ion, the coating is breached, and reaction occurs.) See Example 20.1.

Photos: Charles D. Winters

(a) (b)

EXAMPLE 20.1 Balancing Oxidation–Reduction Equations

Problem Balance the following net ionic equation

$$Al(s) + Cu^{2+}(aq) \rightarrow Al^{3+}(aq) + Cu(s)$$

Identify the oxidizing agent, the reducing agent, the substance oxidized, and the substance reduced. Write balanced half-reactions and the balanced net ionic equation. See Figure 20.2 for photographs of this reaction.

Strategy First, make sure the reaction is an oxidation–reduction reaction by checking each element to see whether the oxidation numbers change. Next, separate the equation into half-reactions, identifying what has been reduced (oxidizing agent) and what has been oxidized (reducing agent). Then balance the half-reactions, first for mass and then for charge. Finally, add the two half-reactions, after ensuring that the reducing agent half-reaction involves the same number of moles of electrons as the oxidizing agent half-reaction.

Solution

Step 1. Recognize the reaction as an oxidation–reduction reaction.

Here, the oxidation number for aluminum changes from 0 to +3, and the oxidation number of copper changes from +2 to 0. Aluminum is oxidized and serves as the reducing agent. Copper(II) ions are reduced, and Cu^{2+} is the oxidizing agent.

Step 2. Separate the process into half-reactions.

Reduction: $Cu^{2+}(aq) \rightarrow Cu(s)$
(Oxidation number of Cu decreases.)

Oxidation: $Al(s) \rightarrow Al^{3+}(aq)$
(Oxidation number of Al increases.)

Step 3. Balance each half-reaction for mass.

Both half-reactions are already balanced for mass.

Step 4. Balance each half-reaction for charge.

To balance the equations for charge, add electrons to the more positive side of each half-reaction.

Reduction: $2\,e^- + Cu^{2+}(aq) \rightarrow Cu(s)$
Each Cu^{2+} ion requires two electrons.

Oxidation: $Al(s) \rightarrow Al^{3+}(aq) + 3\,e^-$
Each Al atom releases three electrons.

Step 5. Multiply each half-reaction by an appropriate factor.

The reducing agent must donate as many electrons as the oxidizing agent acquires. Three Cu^{2+} ions are required to take on the six electrons produced by two Al atoms. Thus, we multiply the Cu^{2+}/Cu half-reaction by 3 and the Al/Al^{3+} half-reaction by 2.

Reduction: $3[2\,e^- + Cu^{2+}(aq) \rightarrow Cu(s)]$

Oxidation: $2[Al(s) \rightarrow Al^{3+}(aq) + 3\,e^-]$

Step 6. Add the half-reactions to produce the overall balanced equation.

Reduction:	$6\,e^- + 3\,Cu^{2+}(aq) \rightarrow 3\,Cu(s)$
Oxidation:	$2\,Al(s) \rightarrow 2\,Al^{3+}(aq) + 6\,e^-$
Net ionic equation:	$3\,Cu^{2+}(aq) + 2\,Al(s) \rightarrow 3\,Cu(s) + 2\,Al^{3+}(aq)$

Step 7. Simplify by eliminating reactants and products that appear on both sides.

This step is not required here.

Comment You should always check the overall equation to ensure there is mass and charge balance. In this case, three Cu atoms and two Al atoms appear on each side. The net electric charge on each side is +6. The equation is balanced.

EXERCISE 20.1 Balancing Oxidation–Reduction Equations

Aluminum reacts with nonoxidizing acids to give $Al^{3+}(aq)$ and $H_2(g)$. The (unbalanced) equation is

$$Al(s) + H^+(aq) \rightarrow Al^{3+}(aq) + H_2(g)$$

Write balanced equations for the half-reactions and the balanced net ionic equation. Identify the oxidizing agent, the reducing agent, the substance oxidized, and the substance reduced.

■ **Balancing Equations in Acid Solution** To simplify equations, we shall use H^+ instead of H_3O^+ when balancing equations in acid solution.

Balancing Equations in Acid Solution

When balancing equations for redox reactions in aqueous solution, it is sometimes necessary to add water molecules (H_2O) and either $H^+(aq)$ in acidic solution or $OH^-(aq)$ in basic solution to the equation. Equations that include oxoanions such as SO_4^{2-}, NO_3^-, ClO^-, CrO_4^{2-}, and MnO_4^- and organic compounds fall into this category. The process is outlined in Example 20.2 for the reduction of an oxocation in acid solution and in Example 20.3 for a reaction in basic solution.

EXAMPLE 20.2 Balancing Equations for Oxidation–Reduction Reactions in Acid Solution

Problem Balance the net ionic equation for the reaction of the dioxovanadium(V) ion, VO_2^+, with zinc in acid solution to form VO^{2+} (see Figure 20.3).

$$VO_2^+(aq) + Zn(s) \rightarrow VO^{2+}(aq) + Zn^{2+}(aq)$$

Strategy Follow the strategy outlined in the text and Example 20.1. Note that water and H^+ ions will appear as product and reactant, respectively, in the half-reaction for the reduction of VO_2^+ ion. (But H_2O and H^+ will never appear on the same side of the balanced half-reaction.)

Solution

Step 1. Recognize the reaction as an oxidation–reduction reaction.

The oxidation number of V changes from +5 in VO_2^+ to +4 in VO^{2+}. The oxidation number of Zn changes from 0 in the metal to +2 in Zn^{2+}.

Step 2. Separate the process into half-reactions.

Oxidation: $Zn(s) \rightarrow Zn^{2+}(aq)$

Zn(s) is oxidized and is the reducing agent.

Reduction: $VO_2^+(aq) \rightarrow VO^{2+}(aq)$

VO_2^+(aq) is reduced and is the oxidizing agent.

Step 3. Balance the half-reactions for mass.

Begin by balancing all atoms except H and O. (These atoms are always the last to be balanced because they often appear in more than one reactant or product.)

Zinc half-reaction: $Zn(s) \rightarrow Zn^{2+}(aq)$

This half-reaction is already balanced for mass.

Vanadium half-reaction: $VO_2^+(aq) \rightarrow VO^{2+}(aq)$

FIGURE 20.3 Reduction of vanadium(V) with zinc. See Example 20.2 for the balanced equation for the first stage of the reaction.

The V atoms in this half-reaction are already balanced. An oxygen-containing species must be added to the right side of the equation to achieve an O atom balance, however.

$$VO_2^+(aq) \rightarrow VO^{2+}(aq) + \text{(need 1 O atom)}$$

In acid solution, add H_2O to the side requiring O atoms, one H_2O molecule for each O atom required.

$$VO_2^+(aq) \rightarrow VO^{2+}(aq) + H_2O(\ell)$$

There are now two unbalanced H atoms on the right. Because the reaction occurs in an acidic solution, H^+ ions are present. Therefore, a mass balance for H can be achieved by adding H^+ to the side of the equation deficient in H atoms. Here, two H^+ ions are added to the left side of the equation.

$$2\ H^+(aq) + VO_2^+(aq) \rightarrow VO^{2+}(aq) + H_2O(\ell)$$

Step 4. Balance the half-reactions for charge by adding electrons to the more positive side to make the charges equal on both sides.

Two electrons are added to the right side of the zinc half-reaction to bring its charge down to the same value as is present on the left side (in this case, zero).

Zinc half-reaction: $Zn(s) \rightarrow Zn^{2+}(aq) + 2\ e^-$

The mass-balanced VO_2^+ equation has a net charge of 3+ on the left side and 2+ on the right. Therefore, 1 e^- is added to the more positive left side.

Vanadium half-reaction: $e^- + 2\ H^+(aq) + VO_2^+(aq) \rightarrow VO^{2+}(aq) + H_2O(\ell)$

As a check on your work, notice that the vanadium atom changes in oxidation number from +5 to +4 and so needs to acquire one electron.

Step 5. Multiply the half-reactions by appropriate factors so that the reducing agent donates as many electrons as the oxidizing agent consumes.

Here, the oxidation half-reaction supplies 2 mol of electrons per mol of Zn, and the reduction half-reaction consumes 1 mol of electrons per mol of VO_2^+. Therefore, the reduction half-reaction must be multiplied by 2. Now 2 mol of the oxidizing agent (VO_2^+) consumes the 2 mol of electrons provided per mole of the reducing agent (Zn).

$$Zn(s) \rightarrow Zn^{2+}(aq) + 2\ e^-$$

$$2[e^- + 2\ H^+(aq) + VO_2^+(aq) \rightarrow VO^{2+}(aq) + H_2O(\ell)]$$

Step 6. Add the half-reactions to give the balanced, overall equation.

Oxidation half-reaction: $Zn(s) \rightarrow Zn^{2+}(aq) + 2\ e^-$

Reduction half-reaction: $2\ e^- + 4\ H^+(aq) + 2\ VO_2^+(aq) \rightarrow 2\ VO^{2+}(aq) + 2\ H_2O(\ell)$

Net ionic equation: $Zn(s) + 4\ H^+(aq) + 2\ VO_2^+(aq) \rightarrow Zn^{2+}(aq) + 2\ VO^{2+}(aq) + 2\ H_2O(\ell)$

Step 7. Simplify by eliminating reactants and products that appear on both sides.

This step is not required here.

Comment Check the overall equation to ensure that there is a mass and charge balance.

Mass balance: 1 Zn, 2 V, 4 H, and 4 O

Charge balance: Each side has a net charge of 6+.

Charles D. Winters

FIGURE 20.4 The reaction of purple permanganate ion (MnO_4^-) with iron(II) ions in acid solution. The products are the nearly colorless Mn^{2+} and Fe^{3+} ions.

EXERCISE 20.2 Balancing Oxidation–Reduction Equations

The yellow dioxovanadium(V) ion, $VO_2^+(aq)$, is reduced by zinc metal in three steps. The first step reduces it to blue $VO^{2+}(aq)$ (Example 20.2). This ion is further reduced to green $V^{3+}(aq)$ in the second step, and V^{3+} can be reduced to violet $V^{2+}(aq)$ in a third step. In each step, zinc is oxidized to $Zn^{2+}(aq)$. Write balanced net ionic equations for Steps 2 and 3. (This reduction sequence is shown in Figure 20.3.)

EXERCISE 20.3 Balancing Equations for Oxidation–Reduction Reactions in Acid Solution

A common laboratory analysis for iron is to titrate aqueous iron(II) ion with a solution of potassium permanganate of precisely known concentration. Use the half-reaction method to write the balanced net ionic equation for the reaction in acid solution.

$$MnO_4^-(aq) + Fe^{2+}(aq) \rightarrow Mn^{2+}(aq) + Fe^{3+}(aq)$$

Identify the oxidizing agent, the reducing agent, the substance oxidized, and the substance reduced. See Figure 20.4.

Balancing Equations in Basic Solution

Example 20.2 and Exercises 20.2 and 20.3 illustrate the technique of balancing equations for redox reactions involving oxocations and oxoanions that occur in acid solution. Under these conditions, H^+ ion or the H^+/H_2O pair can be used to achieve a balanced equation if required. Conversely, in basic solution, only OH^- ion or the OH^-/H_2O pair can be used.

Problem Solving Tip 20.1 Balancing Oxidation–Reduction Equations: A Summary

- Hydrogen balance can be achieved only with H^+/H_2O (in acid) or OH^-/H_2O (in base). Never add H or H_2 to balance hydrogen.
- Use H_2O or OH^- as appropriate to balance oxygen. Never add O atoms, O^{2-} ions, or O_2 for O balance.
- Never include $H^+(aq)$ and $OH^-(aq)$ in the same equation. A solution can be either acidic or basic, never both.
- The number of electrons in a half-reaction reflects the change in oxidation number of the element being oxidized or reduced.
- Electrons are always a component of half-reactions but should never appear in the overall equation.
- Include charges in the formulas for ions. Omitting the charge, or writing the charge incorrectly, is one of the most common errors seen on student papers.
- The best way to become competent in balancing redox equations is to practice, practice, practice.

EXAMPLE 20.3 Balancing Equations for Oxidation–Reduction Reactions in Basic Solution

Problem Aluminum metal is oxidized in aqueous base, with water serving as the oxidizing agent. The products of the reaction are $[Al(OH)_4]^-(aq)$ and $H_2(g)$. Write a balanced net ionic equation for this reaction.

Strategy First, identify the oxidation and reduction half-reactions, and then balance them for mass and charge. Finally, add the balanced half-reactions to obtain the balanced net ionic equation for the reaction.

Solution

Step 1. Recognize the reaction as an oxidation–reduction reaction.

The unbalanced equation is

$$Al(s) + H_2O(\ell) \rightarrow [Al(OH)_4]^-(aq) + H_2(g)$$

Here, aluminum is oxidized, with its oxidation number changing from 0 to +3. Hydrogen is reduced, with its oxidation number decreasing from +1 to zero.

Step 2. Separate the process into half-reactions.

Oxidation half-reaction: $Al(s) \rightarrow [Al(OH)_4]^-(aq)$

(Al oxidation number increased from 0 to +3.)

Reduction half-reaction: $H_2O(\ell) \rightarrow H_2(g)$

(H oxidation number decreased from +1 to 0.)

Step 3. Balance the half-reactions for mass.

Addition of OH^- and/or H_2O is required for mass balance in both half-reactions. In the case of the aluminum half-reaction, we simply add OH^- ions to the left side.

Oxidation half-reaction: $Al(s) + 4\,OH^-(aq) \rightarrow [Al(OH)_4]^-(aq)$

To balance the half-reaction for water reduction, notice that an oxygen-containing species must appear on the right side of the equation. Because H_2O is a reactant, we use OH^-, which is present in this basic solution, as the other product.

Reduction half-reaction: $2\,H_2O(\ell) \rightarrow H_2(g) + 2\,OH^-(aq)$

Step 4. Balance the half-reactions for charge.

Electrons are added to balance charge.

Oxidation half-reaction: $Al(s) + 4\,OH^-(aq) \rightarrow [Al(OH)_4]^-(aq) + 3\,e^-$

Reduction half-reaction: $2\,H_2O(\ell) + 2\,e^- \rightarrow H_2(g) + 2\,OH^-(aq)$

Step 5. Multiply the half-reactions by appropriate factors so that the reducing agent donates as many electrons as the oxidizing agent consumes.

Here, electron balance is achieved by using 2 mol of Al to provide 6 mol of e^-, which are then acquired by 6 mol of H_2O.

Oxidation half-reaction: $2[Al(s) + 4\,OH^-(aq) \rightarrow [Al(OH)_4]^-(aq) + 3\,e^-]$

Reduction half-reaction: $3[2\,H_2O(\ell) + 2\,e^- \rightarrow H_2(g) + 2\,OH^-(aq)]$

Step 6. Add the half-reactions.

$$2\,Al(s) + 8\,OH^-(aq) \rightarrow 2\,[Al(OH)_4]^-(aq) + 6\,e^-$$

$$6\,H_2O(\ell) + 6\,e^- \rightarrow 3\,H_2(g) + 6\,OH^-(aq)]$$

Net equation: $2\,Al(s) + 8\,OH^-(aq) + 6\,H_2O(\ell) \rightarrow 2\,[Al(OH)_4]^-(aq) + 3\,H_2(g) + 6\,OH^-(aq)$

Step 7. Simplify by eliminating reactants and products that appear on both sides.

Six OH^- ions can be canceled from the two sides of the equation:

$$2\,Al(s) + 2\,OH^-(aq) + 6\,H_2O(\ell) \rightarrow 2\,[Al(OH)_4]^-(aq) + 3\,H_2(g)$$

Comment The final equation is balanced for mass and charge.

Mass balance: 2 Al, 14 H, and 8 O

Charge balance: There is a net −2 charge on each side.

Problem Solving Tip 20.2 An Alternative Method of Balancing Equations in Basic Solution

Balancing redox equations in basic solution, which may require you to use OH^- and H_2O, can sometimes be more challenging than doing so in acidic solution. One of the ways to balance such equations for reactions in basic solution is to first balance it as if it were in acidic solution and then add enough OH^- ions to both sides of the equation so that the H^+ ions are converted to water. Taking the half-reaction for the reduction of ClO^- ion to Cl_2, we have:

(a) Balance in acid

$$4\ H^+(aq) + 2\ ClO^-(aq) + 2\ e^- \rightarrow Cl_2(g) + 2\ H_2O(\ell)$$

(b) Add four OH^- ions to both sides

$$4\ OH^-(aq) + 4\ H^+(aq) + 2\ ClO^-(aq) + 2\ e^- \rightarrow Cl_2(g) + 2\ H_2O(\ell) + 4\ OH^-(aq)$$

(c) Combine OH^- and H^+ to form water where appropriate

$$4\ H_2O(\ell) + 2\ ClO^-(aq) + 2\ e^- \rightarrow Cl_2(g) + 2\ H_2O(\ell) + 4\ OH^-(aq)$$

(d) Simplify

$$2\ H_2O(\ell) + 2\ ClO^-(aq) + 2\ e^- \rightarrow Cl_2(g) + 4\ OH^-(aq)$$

EXERCISE 20.4 Balancing Equations for Oxidation–Reduction Reactions in Basic Solution

Voltaic cells based on the reduction of sulfur are under development. One such cell involves the reaction of sulfur with aluminum under basic conditions.

$$Al(s) + S(s) \rightarrow Al(OH)_3(s) + HS^-(aq)$$

(a) Balance this equation, showing each balanced half-reaction.

(b) Identify the oxidizing and reducing agents, the substance oxidized, and the substance reduced.

20.2 Simple Voltaic Cells

Let us use the reaction of copper metal and silver ions (Figure 20.1) as the basis of a voltaic cell. To do so, we place the components of the two half-reactions in separate compartments (Figure 20.5). This prevents the copper metal from transferring electrons directly to silver ions. Instead, electrons will be transferred through an external circuit, and useful work can potentially be done.

The copper half-cell (on the left in Figure 20.5) holds copper metal that serves as one electrode and a solution containing copper(II) ions. The half-cell on the right uses a silver electrode and a solution containing silver(I) ions. Important features of this simple cell are as follows:

- *The two half-cells are connected with a* ***salt bridge*** *that allows cations and anions to move between the two half-cells.* The electrolyte chosen for the salt bridge should contain ions that will not react with chemical reagents in both half-cells. In the example in Figure 20.5, $NaNO_3$ is used.
- *In all electrochemical cells, the* ***anode*** *is the electrode at which oxidation occurs. The electrode at which reduction occurs is always the* ***cathode***. (In Figure 20.5, the copper electrode is the anode, and the silver electrode is the cathode.)
- *A negative sign can be assigned to the anode in a voltaic cell, and the cathode is marked with a positive sign.* The chemical oxidation occurring at the anode, which produces electrons, gives it a negative charge. Electric current in the external circuit of a voltaic cell consists of electrons moving from the negative to the positive electrode.
- *In all electrochemical cells, electrons flow in the external circuit from the anode to the cathode.*

■ **Salt Bridges** A simple salt bridge can be made by adding gelatin to a solution of an electrolyte. Gelatin makes the contents semi-rigid so that the salt bridge is easier to handle. Porous glass disks and permeable membranes are alternatives to a salt bridge. These devices allow ions to traverse from one half-cell to the other while keeping the two solutions from mixing.

FIGURE 20.5 A voltaic cell using $Cu(s)$ | $Cu^{2+}(aq)$ and $Ag(s)$ | $Ag^{+}(aq)$ half cells. Electrons flow through the external circuit from the anode (the copper electrode) to the cathode (silver electrode). In the salt bridge, which contains aqueous $NaNO_3$, negative $NO_3^-(aq)$ ions migrate toward the copper half-cell, and positive $Na^+(aq)$ ions migrate toward the silver half-cell. Using 1.0 M $Cu^{2+}(aq)$ and 1.0 M $Ag^+(aq)$ solutions, this cell will generate 0.46 volts.

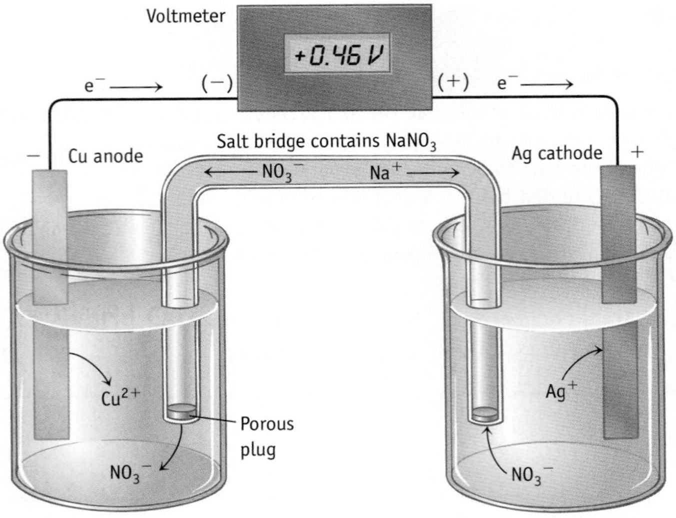

Net reaction: $Cu(s) + 2\ Ag^+(aq) \longrightarrow Cu^{2+}(aq) + 2\ Ag(s)$

The chemistry occurring in the cell pictured in Figure 20.5 is summarized by the following half-reactions and net ionic equation:

Cathode (reduction):	$2\ Ag^+(aq) + 2\ e^- \rightarrow 2\ Ag(s)$
Anode (oxidation):	$Cu(s) \rightarrow Cu^{2+}(aq) + 2\ e^-$
Net ionic equation:	$Cu(s) + 2\ Ag^+(aq) \rightarrow Cu^{2+}(aq) + 2\ Ag(s)$

The salt bridge is required in a voltaic cell for the reaction to proceed. In the Cu/Ag^+ voltaic cell, anions move in the salt bridge toward the copper half-cell, and cations move toward the silver half-cell (Figure 20.5). As $Cu^{2+}(aq)$ ions are formed in the copper half-cell by oxidation of copper metal, negative ions enter that cell from the salt bridge (and positive ions leave the cell), so that the numbers of positive and negative charges in the half-cell compartment remain in balance. Likewise, in the silver half-cell, negative ions move out of the half-cell into the salt bridge, and positive ions move into the cell as Ag^+ (aq) ions are reduced to silver metal. A complete circuit is required for current to flow. If the salt bridge is removed, reactions at the electrodes will cease.

In Figure 20.5, the electrodes are connected by wires to a voltmeter. In an alternative set-up, the connections might be to a light bulb or other device that uses electricity. Electrons are produced by oxidation of copper, and $Cu^{2+}(aq)$ ions enter the solution. The electrons traverse the external circuit to the silver electrode, where they reduce $Ag^+(aq)$ ions to silver metal. To balance the extent of oxidation and reduction, two $Ag^+(aq)$ ions are reduced for every $Cu^{2+}(aq)$ ion formed. The main features of this and of all other voltaic cells are summarized in Figure 20.6.

■ **Electron and Ion Flow** It is helpful to notice that in an electrochemical cell the negative electrons and negatively charged anions make a "circle." That is, electrons move from anode to cathode in the external circuit, and negative anions move from the cathode compartment, through the salt bridge, to the anode compartment.

Chemistry Now™

Sign in at **www.thomsonedu.com/login** and go to Chapter 20 Contents to see Screen 20.4 for an animation of **a cell based on zinc and copper.**

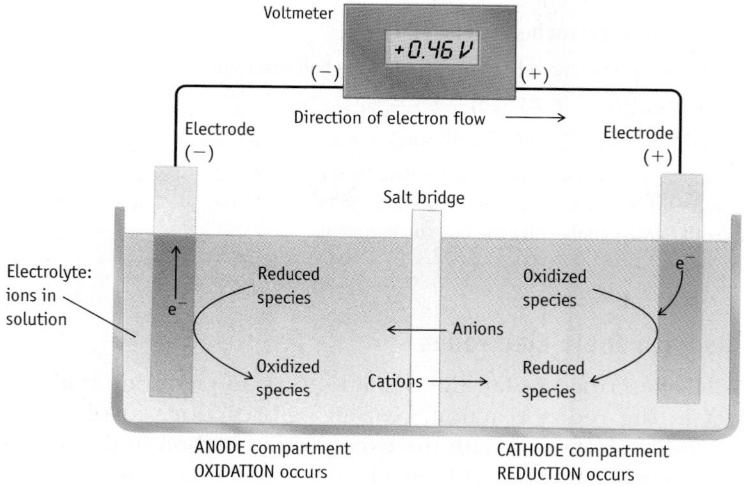

FIGURE 20.6 Summary of terms used in a voltaic cell. Electrons move from the anode, the site of oxidation, through the external circuit to the cathode, the site of reduction. Charge balance in each half-cell is achieved by migration of ions through the salt bridge. Negative ions move from the reduction half-cell to the oxidation half-cell and positive ions move in the opposite direction.

EXAMPLE 20.4 Electrochemical Cells

Problem Describe how to set up a voltaic cell to generate an electric current using the reaction

$$Fe(s) + Cu^{2+}(aq) \rightarrow Cu(s) + Fe^{2+}(aq)$$

Which electrode is the anode, and which is the cathode? In which direction do electrons flow in the external circuit? In which direction do the positive and negative ions flow in the salt bridge? Write equations for the half-reactions that occur at each electrode.

Strategy First, identify the two different half-cells that make up the cell. Next, decide in which half-cell oxidation occurs and in which reduction occurs.

Solution This voltaic cell is similar to the one diagrammed in Figure 20.5. One half-cell contains an iron electrode and a solution of an iron(II) salt such as $Fe(NO_3)_2$. The other half-cell contains a copper electrode and a soluble copper(II) salt such as $Cu(NO_3)_2$. The two half-cells are linked with a salt bridge containing an electrolyte such as KNO_3. Iron is oxidized, so the iron electrode is the anode:

Oxidation, anode: $Fe(s) \rightarrow Fe^{2+}(aq) + 2\,e^-$

Because copper(II) ions are reduced, the copper electrode is the cathode. The cathodic half-reaction is

Reduction, cathode: $Cu^{2+}(aq) + 2\,e^- \rightarrow Cu(s)$

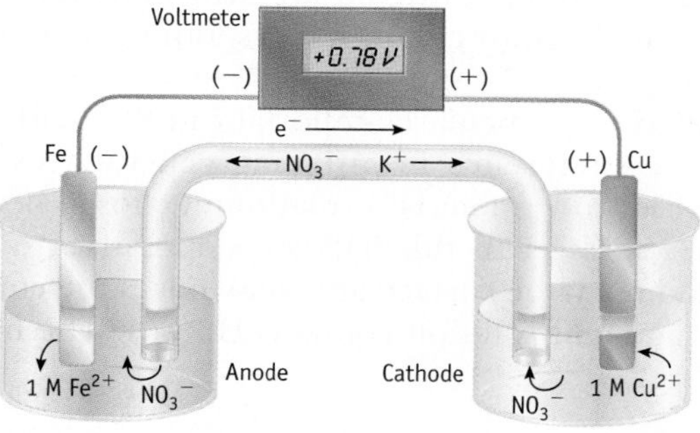

In the external circuit, electrons flow from the iron electrode (anode) to the copper electrode (cathode). In the salt bridge, negative ions flow toward the Fe | Fe^{2+}(aq) half-cell, and positive ions flow in the opposite direction.

EXERCISE 20.5 Electrochemical Cells

Describe how to set up a voltaic cell using the following half-reactions:

Reduction half-reaction: $Ag^+(aq) + e^- \rightarrow Ag(s)$

Oxidation half-reaction: $Ni(s) \rightarrow Ni^{2+}(aq) + 2\ e^-$

Which is the anode, and which is the cathode? What is the overall cell reaction? What is the direction of electron flow in an external wire connecting the two electrodes? Describe the ion flow in a salt bridge (with $NaNO_3$) connecting the cell compartments.

Voltaic Cells with Inert Electrodes

In the half-cells described so far, the metal used as an electrode is also a reactant or a product in the redox reaction. Not all half-reactions involve a metal as a reactant or product, however. With the exception of carbon in the form of graphite, most nonmetals are unsuitable as electrode materials because they do not conduct electricity. It is not possible to make an electrode from a gas, a liquid (except mercury), or a solution. Ionic solids do not make satisfactory electrodes because the ions are locked tightly in a crystal lattice, and these materials do not conduct electricity.

In situations where reactants and products cannot serve as the electrode material, an inert or chemically unreactive electrode must be used. Such electrodes are made of materials that conduct an electric current but that are neither oxidized nor reduced in the cell.

Consider constructing a voltaic cell to accommodate the following product-favored reaction:

$$2\ Fe^{3+}(aq) + H_2(g) \rightarrow 2\ Fe^{2+}(aq) + 2\ H^+(aq)$$

Reduction half reaction: $Fe^{3+}(aq) + e^- \rightarrow Fe^{2+}(aq)$

Oxidation half reaction: $H_2(g) \rightarrow 2\ H^+(aq) + 2\ e^-$

Neither the reactants nor the products can be used as an electrode material. Therefore, the two half-cells are set up so that the reactants and products come in contact with an electrode such as graphite where they can accept or give up electrons. Graphite is a commonly used electrode material: It is a conductor of electricity, and it is inexpensive (essential in commercial cells) and not readily oxidized under the conditions encountered in most cells. Mercury is used in certain types of cells. Platinum and gold are also commonly used because both are chemically inert under most circumstances, but they are generally too costly for commercial cells.

The *hydrogen electrode* is particularly important in the field of electrochemistry because it is used as a reference in assigning cell voltages (see Section 20.4) (Figure 20.7). The electrode material is platinum, chosen because hydrogen adsorbs on the metal's surface. In this half-cell's operation, hydrogen is bubbled over the electrode, and a large surface area maximizes the contact of the gas and the electrode. The aqueous solution contains $H^+(aq)$. The half-reactions involving $H^+(aq)$ and $H_2(g)$

$$2\ H^+(aq) + 2\ e^- \rightarrow H_2(g) \quad \text{or} \quad H_2(g) \rightarrow 2\ H^+(aq) + 2\ e^-$$

take place at the electrode surface, and the electrons involved in the reaction are conducted to or from the reaction site by the metal electrode.

A half-cell using the reduction of $Fe^{3+}(aq)$ to $Fe^{2+}(aq)$ can also be set up with a platinum electrode. In this case, the solution surrounding the electrode contains

FIGURE 20.7 Hydrogen electrode. Hydrogen gas is bubbled over a platinum electrode in a solution containing H^+ ions. Such electrodes function best if they have a large surface area. Often, platinum wires are woven into a gauze, or the metal surface is roughened either by abrasion or by chemical treatment to increase the surface area.

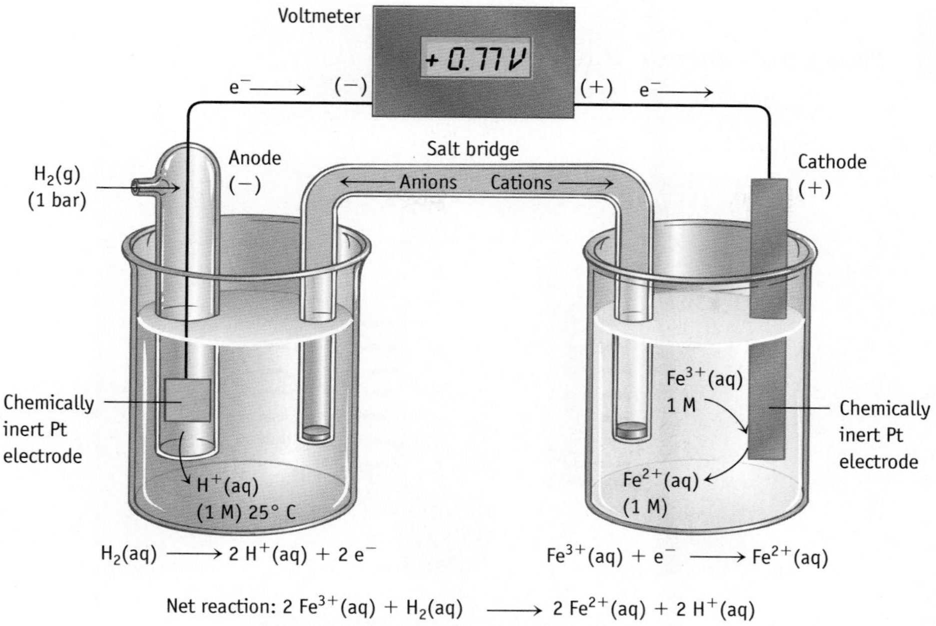

FIGURE 20.8 A voltaic cell with a hydrogen electrode. This cell has Fe^{2+}(aq, 1.0 M) and Fe^{3+}(aq, 1.0 M) in the cathode compartment and H_2(g) and H^+(aq, 1.0 M) in the anode compartment. At 25 °C, the cell generates 0.77 V.

iron ions in two different oxidation states. Transfer of electrons to or from the reactant occurs at the electrode surface.

A voltaic cell involving the reduction of Fe^{3+}(aq, 1.0 M) to Fe^{2+}(aq, 1.0 M) with H_2 gas is illustrated in Figure 20.8. In this cell, the hydrogen electrode is the anode (H_2 is oxidized to H^+), and the iron-containing compartment is the cathode (Fe^{3+} is reduced to Fe^{2+}). The cell produces 0.77 V.

Electrochemical Cell Notations

Chemists often use a shorthand notation to simplify cell descriptions. For example, the cell involving the reduction of silver ion with copper metal is written as

$$Cu(s)\,|\,Cu^{2+}(aq,\ 1.0\ M)\,||\,Ag^+(aq,\ 1.0\ M)\,|\,Ag(s)$$

The cell using H_2 gas to reduce Fe^{3+} ions is written as

$$\underbrace{Pt\,|\,H_2(P = 1\ bar)\,|\,H^+(aq,\ 1.0\ M)}_{\text{Anode information}}\,||\,\underbrace{Fe^{3+}(aq,\ 1.0\ M),\ Fe^{2+}(aq,\ 1.0\ M)\,|\,Pt}_{\text{Cathode information}}$$

By convention, on the left we write the anode and information with respect to the solution with which it is in contact. A single vertical line (|) indicates a phase boundary, and double vertical lines (||) indicate a salt bridge.

20.3 Commercial Voltaic Cells

The cells described so far are unlikely to have practical use. They are neither compact nor robust, high priorities for most applications. In most situations, it is also important that the cell produce a constant voltage, but a problem with the cells described so far is that the voltage produced varies as the concentrations of ions in solution change (see Section 20.5). Also, the current production is low.

Historical Perspectives — Frogs and Voltaic Piles

Voltaic cells are also called galvanic cells after an Italian physician Luigi Galvani (1737–1798), who carried out early studies of what he called "animal electricity," studies that brought new words into our language—among them "galvanic" and "galvanize."

Around 1780, Galvani observed that the electric current from a static electricity generator caused the contraction of the muscles in a frog's leg. Investigating further, he found he could induce contractions when the muscle was in contact with two different metals. Because no external source of electricity was applied to the muscles, Galvani concluded the frog's muscles were themselves generating electricity. This was evidence, he believed, of a kind of "vital energy" or "animal electricity," which was related to but different from "natural electricity" generated by machines or lightning.

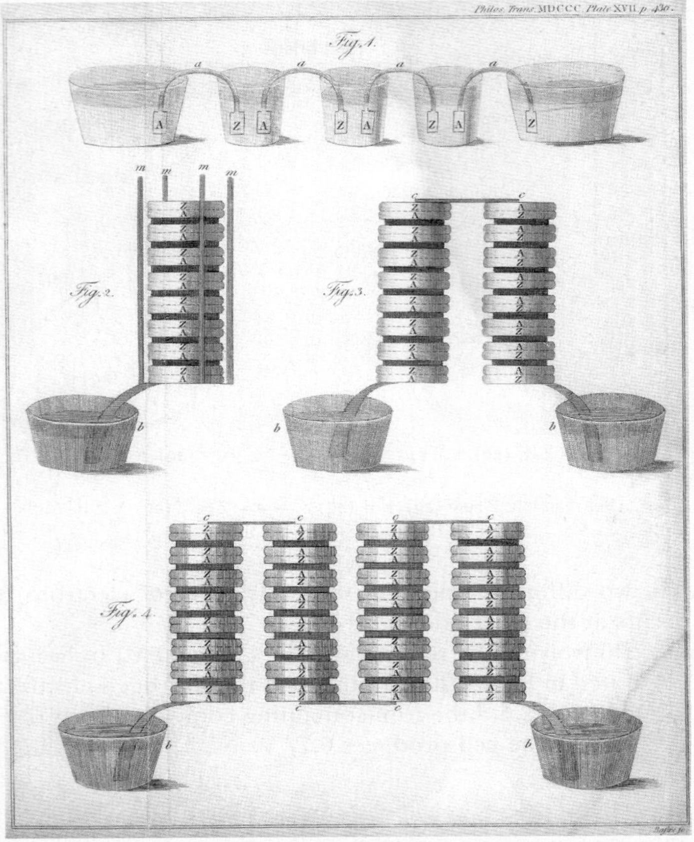

© Smithsonian Institution

Volta's "voltaic pile." These drawings done by Volta show the arrangement of silver and zinc disks used to generate an electric current.

©Bettman/Corbis

Alessandro Volta, 1745–1827.

Alessandro Volta repeated Galvani's experiments, with the same results, but he came to different conclusions. Volta proposed that an electric current was generated by the contact between two different metals—an explanation we now know to be correct—and that the frog muscle was simply detecting the small current generated.

To prove his hypothesis, Volta built the first "electric pile" in 1800. This was a series of metal disks of two kinds (silver and zinc), separated by paper disks soaked in acid or salt solutions. Soon after Volta announced his discovery, Carlisle and Nicholson in England used the electricity from a "pile" to decompose water into hydrogen and oxygen. Within a few years, the great English chemist Humphry Davy used a more powerful voltaic pile to isolate potassium and sodium metals by electrolysis.

Attempting to draw a large current results in a drop in voltage because the current depends on how fast ions in solution migrate to the electrode. Ion concentrations near the electrode become depleted if current is drawn rapidly, resulting in a decline in voltage.

■ **Cell Potentials and Reactant and Product Concentrations** Concentrations of species in a cell affect the potential, as discussed in Section 20.5.

The electrical work that can be drawn from a voltaic cell depends on the quantity of reagents consumed. A voltaic cell must have a large mass of reactants to produce current over a prolonged period. In addition, a voltaic cell that can be recharged is

FIGURE 20.9 Some commercial voltaic cells. Commercial voltaic cells provide energy for a wide range of devices, come in a myriad of sizes and shapes, and produce different voltages. Some are rechargeable; others are discarded after use. One might think that there is nothing further to learn about batteries, but this is not true. Research on these devices is actively pursued in the chemical community.

attractive. Recharging a cell means returning the reagents to their original sites in the cell. In the cells described so far, the movement of ions in the cell mixes the reagents, and they cannot be "unmixed" after the cell has been running.

Batteries can be classified as primary and secondary. **Primary batteries** cannot be returned to their original state by recharging, so when the reactants are consumed, the battery is "dead" and must be discarded. **Secondary batteries** are often called **storage batteries** or **rechargeable batteries**. The reactions in these batteries can be reversed; thus, the batteries can be recharged.

Years of development have led to many different commercial voltaic cells to meet specific needs (Figure 20.9), and several common ones are described below. All adhere to the principles that have been developed in earlier discussions.

Primary Batteries: Dry Cells and Alkaline Batteries

If you buy an inexpensive flashlight battery or dry cell battery, it might be a modern version of a voltaic cell invented by George LeClanché in 1866 (Figure 20.10). Zinc serves as the anode, and the cathode is a graphite rod placed down the center of the device. These cells are often called "dry cells" because there is no visible liquid phase. However, water is present, so the cell contains a moist paste of NH_4Cl, $ZnCl_2$, and MnO_2. The moisture is necessary because the ions present must be in a medium in which they can migrate from one electrode to the other. The cell generates a potential of 1.5 V using the following half-reactions:

Cathode, reduction: $2\ NH_4^+(aq) + 2\ e^- \rightarrow 2\ NH_3(g) + H_2(g)$

Anode, oxidation: $Zn(s) \rightarrow Zn^{2+}(aq) + 2\ e^-$

The two gases formed at the cathode will build up pressure and could cause the cell to rupture. This problem is avoided, however, by two other reactions that take place in the cell. Ammonia molecules bind to Zn^{2+} ions, and hydrogen gas is oxidized by MnO_2 to water.

$$Zn^{2+}(aq) + 2\ NH_3(g) + 2\ Cl^-(aq) \rightarrow Zn(NH_3)_2Cl_2(s)$$

$$2\ MnO_2(s) + H_2(g) \rightarrow Mn_2O_3(s) + H_2O(\ell)$$

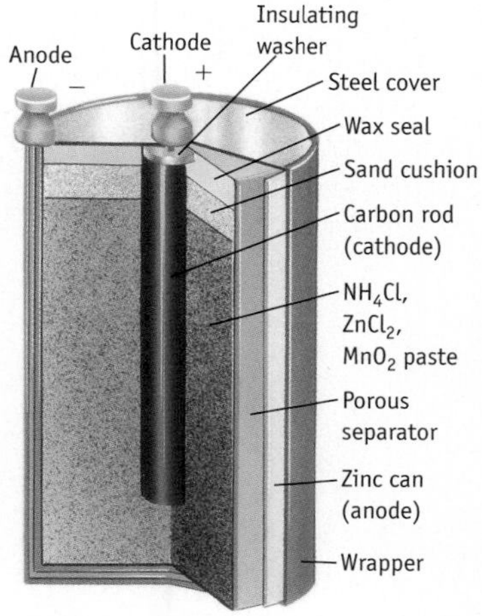

FIGURE 20.10 The common LeClanché dry cell battery.

■ **Batteries** The word battery has become part of our common language, designating any self-contained device that generates an electric current. The term battery has a more precise scientific meaning, however. It refers to a collection of two or more voltaic cells. For example, the 12-volt battery used in automobiles is made up of six voltaic cells. Each voltaic cell develops a voltage of 2 volts. Six cells connected in series produce 12 volts.

LeClanché cells were widely used because of their low cost, but they have several disadvantages. If current is drawn from the battery rapidly, the gaseous products cannot be consumed rapidly enough, so the cell resistance rises, and the voltage drops. In addition, the zinc electrode and ammonium ions are in contact in the cell, and these chemicals react slowly. Recall that zinc reacts with acid to form hydrogen. The ammonium ion, $NH_4^+(aq)$, is a weak Brønsted acid and reacts slowly with zinc. Because of this reaction, these voltaic cells cannot be stored indefinitely, a fact you may have learned from experience. When the zinc outer shell deteriorates, the battery can leak acid and perhaps damage the appliance in which it is contained.

At the present time, you are more likely to use **alkaline batteries** in your camera or flashlight. They generate current up to 50% longer than a dry cell of the same size. The chemistry of alkaline cells is quite similar to that in a LeClanché cell, except that the material inside the cell is basic (alkaline). Alkaline cells use the oxidation of zinc and the reduction of MnO_2 to generate a current, but NaOH or KOH is used in the cell instead of the acidic salt NH_4Cl.

Cathode, reduction: $2\ MnO_2(s) + H_2O(\ell) + 2\ e^- \rightarrow Mn_2O_3(s) + 2\ OH^-(aq)$

Anode, oxidation: $Zn(s) + 2\ OH^-(aq) \rightarrow ZnO(s) + H_2O(\ell) + 2\ e^-$

Alkaline cells, which produce 1.54 V (approximately the same voltage as the LeClanché cell), have the further advantage that the cell potential does not decline under high current loads because no gases are formed.

Prior to 2000, mercury-containing batteries were widely used in calculators, cameras, watches, heart pacemakers, and other devices. However, these small batteries were banned in the United States in the 1990s because of environmental problems. They have been replaced by several other types of batteries, such as silver oxide batteries and zinc-oxygen batteries. Both operate under alkaline conditions, and both have zinc anodes. In the silver oxide battery, which produces a voltage of about 1.5 V, the cell reactions are

Cathode, reduction: $Ag_2O(s) + H_2O(\ell) + 2\ e^- \rightarrow 2\ Ag(s) + 2\ OH^-(aq)$

Anode, oxidation: $Zn(s) + 2\ OH^-(aq) \rightarrow ZnO(s) + H_2O(\ell) + 2\ e^-$

The zinc-oxygen battery, which produces about 1.15–1.35 V, is unique in that atmospheric oxygen and not a metal oxide is the oxidizing agent.

Cathode, reduction: $O_2(g) + 2\ H_2O(\ell) + 4\ e^- \rightarrow 4\ OH^-(aq)$

Anode, oxidation: $Zn(s) + 2\ OH^-(aq) \rightarrow ZnO(s) + H_2O(\ell) + 2\ e^-$

These batteries have found use in hearing aids, pagers, and medical devices.

Secondary or Rechargeable Batteries

When a LeClanché cell or an alkaline cell ceases to produce a usable electric current, it is discarded. In contrast, some types of cells can be recharged, often hundreds of times. Recharging requires applying an electric current from an external source to restore the cell to its original state.

An automobile battery—the **lead storage battery**—is probably the best-known rechargeable battery (Figure 20.11). The 12-V version of this battery contains six voltaic cells, each generating about 2 V. The lead storage battery can produce a large initial current, an essential feature when starting an automobile engine.

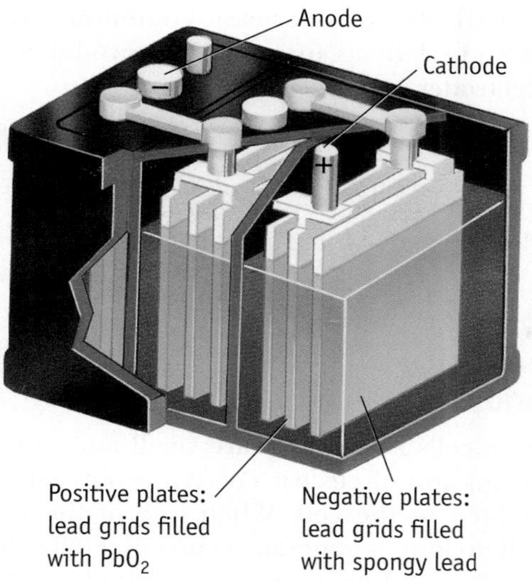

FIGURE 20.11 Lead storage battery, a secondary or rechargeable battery. The negative plates (anode) are lead grids filled with spongy lead. The positive plates (cathode) are lead grids filled with lead(IV) oxide, PbO_2. Each cell of the battery generates 2 V.

The anode of a lead storage battery is metallic lead. The cathode is also made of lead, but it is covered with a layer of compressed, insoluble lead(IV) oxide, PbO_2. The electrodes, arranged alternately in a stack and separated by thin fiberglass sheets, are immersed in aqueous sulfuric acid. When the cell supplies electrical energy, the lead anode is oxidized to lead(II) sulfate, an insoluble substance that adheres to the electrode surface. The two electrons produced per lead atom move through the external circuit to the cathode, where PbO_2 is reduced to Pb^{2+} ions that, in the presence of H_2SO_4, also form lead(II) sulfate.

Cathode, reduction: $PbO_2(s) + 4\ H^+(aq) + SO_4^{2-}(aq) + 2\ e^- \rightarrow PbSO_4(s) + 2\ H_2O(\ell)$

Anode, oxidation: $Pb(s) + SO_4^{2-}(aq) \rightarrow PbSO_4(s) + 2\ e^-$

Net ionic equation: $Pb(s) + PbO_2(s) + 2\ H_2SO_4(aq) \rightarrow 2\ PbSO_4(s) + 2\ H_2O(\ell)$

When current is generated, sulfuric acid is consumed and water is formed. Because water is less dense than sulfuric acid, the density of the solution decreases during this process. Therefore, one way to determine whether a lead storage battery needs to be recharged is to measure the density of the solution.

A lead storage battery is recharged by supplying electrical energy. The $PbSO_4$ coating the surfaces of the electrodes is converted back to metallic lead and PbO_2, and sulfuric acid is regenerated. Recharging this battery is possible because the reactants and products remain attached to the electrode surface. The lifetime of a lead storage battery is limited, however, because, with time, the coatings of PbO_2 and $PbSO_4$ flake off of the surface and fall to the bottom of the battery case.

Scientists and engineers would like to find an alternative to lead storage batteries, especially for use in cars. Lead storage batteries have the disadvantage of being large and heavy. In addition, lead and its compounds are toxic and their disposal adds a further complication. Nevertheless, at this time, the advantages of lead storage batteries outweigh their disadvantages.

Nickel-cadmium ("Ni-cad") batteries, used in a variety of cordless appliances such as telephones, video camcorders, and cordless power tools, are lightweight and rechargeable. The chemistry of the cell utilizes the oxidation of cadmium and the reduction of nickel(III) oxide under basic conditions. As with the lead storage battery, the reactants and products formed when producing a current are solids that adhere to the electrodes.

Cathode, reduction: $NiO(OH)(s) + H_2O(\ell) + e^- \rightarrow Ni(OH)_2(s) + OH^-(aq)$

Anode, oxidation: $Cd(s) + 2\ OH^-(aq) \rightarrow Cd(OH)_2(s) + 2\ e^-$

Ni-cad batteries produce a nearly constant voltage. However, their cost is relatively high, and there are restrictions on their disposal because cadmium compounds are toxic and present an environmental hazard.

Fuel Cells and Hybrid Cars

An advantage of voltaic cells is that they are small and portable, but their size is also a limitation. The amount of electric current produced is limited by the quantity of reagents contained in the cell. When one of the reactants is completely consumed, the cell will no longer generate a current. Fuel cells avoid this limitation because the reactants (fuel and oxidant) can be supplied continuously to the cell from an external reservoir.

Although the first fuel cells were constructed more than 150 years ago, little was done to develop this technology until the space program rekindled interest in these devices. Hydrogen-oxygen fuel cells have been used in NASA's Gemini, Apollo, and Space Shuttle programs. Not only are they lightweight and efficient, but they also have the added benefit that they generate drinking water for the ship's crew. The fuel cells on board the Space Shuttle deliver the same power as batteries weighing 10 times as much.

In a hydrogen-oxygen fuel cell (Figure 20.12), hydrogen is pumped onto the anode of the cell, and O_2 (or air) is directed to the cathode where the following reactions occur:

Cathode, reduction: $O_2(g) + 2\ H_2O(\ell) + 4\ e^- \rightarrow 4\ OH^-(aq)$

Anode, oxidation: $H_2(g) \rightarrow 2\ H^+(aq) + 2\ e^-$

■ **Energy for Automobiles** Energy available from systems that can be used to power an automobile.

Chemical System	W · h/kg* (1 W · h = 3600 J)
Lead-acid battery	18–56
Nickel-cadmium battery	33–70
Sodium-sulfur battery	80–140
Lithium polymer battery	150
Gasoline-air combustion engine	12,200

* watt-hour/kilogram

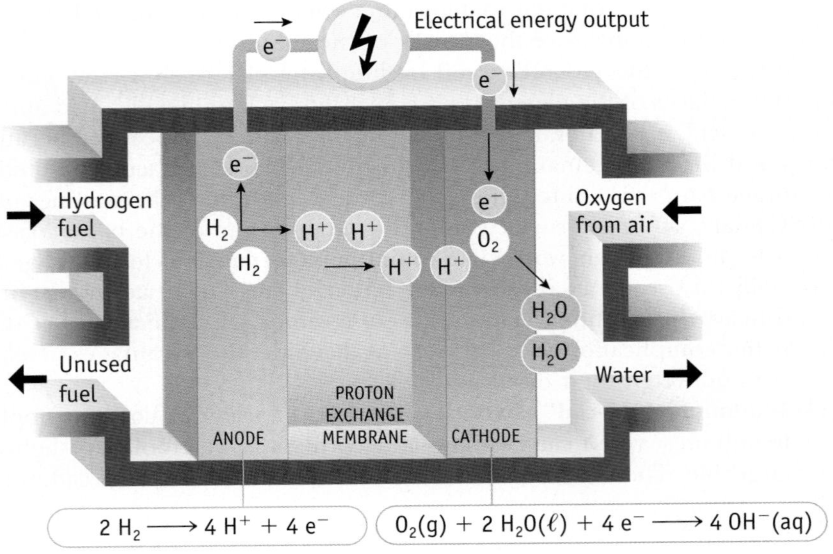

FIGURE 20.12 Fuel cell design. Hydrogen gas is oxidized to $H^+(aq)$ at the anode surface. On the other side of the proton exchange membrane (PEM), oxygen gas is reduced to $OH^-(aq)$. The $H^+(aq)$ ions travel through the PEM and combine with $OH^-(aq)$, forming water.

The two halves of the cell are separated by a special material called a proton exchange membrane (PEM). Protons, $H^+(aq)$, formed at the anode traverse the PEM and react with the hydroxide ions produced at the cathode, forming water. The net reaction in the cell is thus the formation of water from H_2 and O_2. Cells currently in use run at temperatures of 70–140 °C and produce about 0.9 V.

Hydrogen-oxygen fuel cells operate at 40–60% efficiency and meet most of the requirements for use in automobiles: they operate at room temperature or slightly above, start rapidly, and develop a high current density. Cost is a serious problem, however, and it appears that a substantial shift away from the internal combustion engine remains a long way off. For this reason, several major car manufacturers have designed electric cars that use various types of batteries to provide the power to drive the car. The most commonly employed type for automotive use is the lead storage battery, but these devices are problematic, owing to their mass. To produce one mole of electrons requires 321 g of reactants in lead storage batteries. As a result, these batteries rank very low among various options in power per kilogram of battery weight. In fact, the power available from any type of battery is much less than that available from an equivalent mass of gasoline.

Hybrid cars appear to offer an interim solution. These vehicles combine a small gasoline-fueled engine with an electric motor and batteries for storage of electric energy. Currently, hybrid cars use rechargeable nickel-metal hydride batteries. Electrons are generated when H atoms interact with OH^- ions at the metal alloy anode.

$$\text{Alloy(H)} + OH^- \rightarrow \text{Alloy} + H_2O + e^-$$

The reaction at the cathode is the same as in Ni-cad batteries.

$$NiOOH + H_2O + e^- \rightarrow Ni(OH)_2 + OH^-$$

Chemistry Now™

Sign in at **www.thomsonedu.com/login** and go to Chapter 20 Contents to see Screen 20.5 to view animations of **various types of batteries.**

©AP Photo/Shizuo Kambayashi, File

Hybrid car. This car combines a gasoline-fueled engine with an electric motor and rechargable batteries. Its fuel efficiency is about double that of the current generation of cars using only gasoline engines.

20.4 Standard Electrochemical Potentials

Different electrochemical cells produce different voltages: 1.5 V for the LeClanché and alkaline cells, about 1.25 V for a Ni-Cd battery, and about 2.0 V for the individual cells in a lead storage battery. In this section, we want to identify the various factors affecting cell voltages and develop procedures to calculate the voltage of a cell based on the chemistry in the cell and the conditions used.

Electromotive Force

Electrons generated at the anode of an electrochemical cell move through the external circuit toward the cathode, and the force needed to move the electrons arises from a difference in the potential energy of electrons at the two electrodes. This difference in potential energy per electrical charge is called the **electromotive force** or **emf**, for which the literal meaning is "force causing electrons to move." Emf has units of volts (V); 1 volt is the potential difference needed to impart one joule of energy to an electric charge of one coulomb (1 J = 1 V × 1 C). *One coulomb is the quantity of charge that passes a point in an electric circuit when a current of 1 ampere flows for 1 second (1 C = 1 A × 1 s).*

Electrochemical Units

- The coulomb (abbreviated C) is the standard (SI) unit of electrical charge (Appendix C.3).
- 1 joule = 1 volt × 1 coulomb.
- 1 coulomb = 1 ampere × 1 second.

Measuring Standard Potentials

Imagine you planned to study cell voltages in a laboratory with two objectives: (1) to understand the factors that affect these values and (2) to be able to predict the potential of a voltaic cell. You might construct a number of different half-cells, link them together in various combinations to form voltaic cells (as in Figure 20.13), and measure the cell potentials. After a few experiments, it would become apparent that cell voltages depend on a number of factors: the half-cells used (i.e., the reaction in each half-cell and the overall or net reaction in the cell), the concentrations of reactants and products in solution, the pressure of gaseous reactants, and the temperature.

So that we can later compare the potential of one half-cell with another, let us measure all cell voltages under **standard conditions**:

- Reactants and products are present in their standard states.
- Solutes in aqueous solution have a concentration of 1.0 M.
- Gaseous reactants or products have a pressure of 1.0 bar.

A cell potential measured under these conditions is called the **standard potential** and is denoted by $\boldsymbol{E°_{cell}}$. Unless otherwise specified, all values of $E°_{cell}$ refer to measurements at 298 K (25 °C).

Suppose you set up a number of standard half-cells and connect each in turn to a **standard hydrogen electrode (SHE)**. Your apparatus would look like the voltaic cell in Figure 20.13. For now, we will concentrate on three aspects of this cell:

1. *The reaction that occurs.* The reaction occurring in the cell pictured in Figure 20.13 could be *either* the reduction of Zn^{2+} ions with H_2 gas

$$Zn^{2+}(aq) + H_2(g) \rightarrow Zn(s) + 2\ H^+(aq)$$

$Zn^{2+}(aq)$ is the oxidizing agent, and H_2 is the reducing agent.

Standard hydrogen electrode would be the anode (negative electrode).

or the reduction of $H^+(aq)$ ions by Zn(s).

$$Zn(s) + 2\ H^+(aq) \rightarrow Zn^{2+}(aq) + H_2(g)$$

Zn is the reducing agent, and $H^+(aq)$ is the oxidizing agent.

Standard hydrogen electrode would be the cathode (positive electrode).

Charles D. Winters

(a)

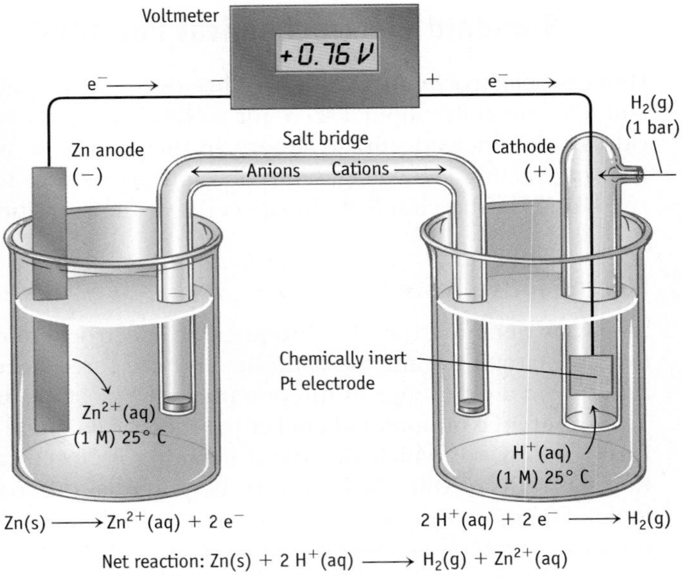

(b)

Active Figure 20.13 A voltaic cell using $Zn|Zn^{2+}$(aq, 1.0 M) and $H_2|H^+$(aq, 1.0 M) half cells. (a) Zinc metal reacts readily with aqueous HCl. (b) When zinc and acid are combined in an electrochemical cell, the cell generates a potential of 0.76 V under standard conditions. The electrode in the $H_2|H^+$(aq, 1.0 M) half-cell is the cathode, and the Zn electrode is the anode. Electrons flow in the external circuit to the hydrogen half-cell from the zinc half-cell. The positive sign of the measured voltage indicates that the hydrogen electrode is the cathode or positive electrode.

Chemistry Now™ Sign in at www.thomsonedu.com/login and go to the Chapter Contents menu to explore an interactive version of this figure accompanied by an exercise.

All the substances named in these equations are present in the cell. The reaction that actually occurs is the one that is product-favored. That is, the reaction occurring is the one in which the reactants are the stronger reducing and oxidizing agents.

2. *Direction of electron flow in the external circuit.* In a voltaic cell, electrons always flow from the anode (negative electrode) to the cathode (positive electrode). That is, *electrons move from the electrode of higher potential energy to the one of lower potential energy.* We can tell the direction of electron movement by placing a voltmeter in the circuit. A positive potential is observed if the voltmeter terminal with a plus sign (+) is connected to the positive electrode [and the terminal with the minus sign (−) is connected to the negative electrode]. Connected in the opposite way (plus to minus and minus to plus), the voltmeter will show a negative value on the digital readout.
3. *Cell potential.* In Figure 20.13, the voltmeter is hooked up with its positive terminal connected to the hydrogen half-cell, and a reading of +0.76 V is observed. The hydrogen electrode is thus the positive electrode or cathode, and the reactions occurring in this cell must be

Reduction, cathode:	$2\ H^+(aq) + 2\ e^- \rightarrow H_2(g)$
Oxidation, anode:	$Zn(s) \rightarrow Zn^{2+}(aq) + 2\ e^-$
Net cell reaction:	$Zn(s) + 2\ H^+(aq) \rightarrow Zn^{2+}(aq) + H_2(g)$

This result confirms that, of the two oxidizing agents present in the cell, $H^+(aq)$ is better than $Zn^{2+}(aq)$ and that Zn metal is a better reducing agent than H_2 gas.

A potential of +0.76 V is measured for the oxidation of zinc with hydrogen ion. This value reflects the difference in potential energy of an electron at each electrode. From the direction of flow of electrons in the external circuit (Zn electrode → H_2 electrode), we conclude that the potential energy of an electron at the zinc electrode is higher than the potential energy of the electron at the hydrogen electrode.

Hundreds of electrochemical cells like that shown in Figure 20.13 can be set up, allowing us to determine the relative oxidizing or reducing ability of various chemical species and to determine the electrical potential generated by the reaction under standard conditions. A few results are given in Figure 20.14, where half-reactions are listed as reductions. That is, the chemical species on the left are oxidizing agents and are listed in descending oxidizing ability.

Standard Reduction Potentials

By doing experiments such as that illustrated by Figure 20.13, we not only have a notion of the relative oxidizing and reducing abilities of various chemical species, but we can also rank them quantitatively.

If E°_{cell} is a measure of the standard potential for the cell, then $E^\circ_{cathode}$ and E°_{anode} can be taken as measures of electrode potential. Because E°_{cell} reflects the difference in electrode potentials, E°_{cell} must be the difference between $E^\circ_{cathode}$ and E°_{anode}.

$$E^\circ_{cell} = E^\circ_{cathode} - E^\circ_{anode} \qquad (20.1)$$

Here, $E^\circ_{cathode}$ and E°_{anode} are the standard *reduction* potentials for the half-cell reactions that occur at the cathode and anode, respectively. Equation 20.1 is important for three reasons:

- If we have values for $E^\circ_{cathode}$ and E°_{anode}, we can calculate the standard potential, E°_{cell}, for a voltaic cell.

■ **Equation 20.1** Equation 20.1 is another example of calculating a change from $X_{final} - X_{initial}$. Electrons move to the cathode (the "final" state) from the anode (the "initial" state). Thus, Equation 20.1 resembles equations you have seen previously in this book (such as Equations 5.6 and 6.5).

A Closer Look — EMF, Cell Potential, and Voltage

Emf and cell potential (E_{cell}) are often used synonymously, but the two are subtly different. E_{cell} is a measured quantity, so its value is affected by how the measurement is made. To understand this point, consider as an analogy water in a pipe under pressure. Water pressure can be viewed as analogous to emf; it represents a force that will cause water in the pipe to move. If we open a faucet, water will flow. Opening the faucet will, however, decrease the pressure in the system.

Emf is the potential difference when no current flows. To measure E_{cell}, a voltmeter is placed in the external circuit. Although voltmeters have high internal resistance to minimize current flow, a small current flows nonetheless. As a result, the value of E_{cell} will be slightly different than the emf.

Finally, there is a difference between a potential and a voltage. The voltage of a cell has a magnitude but no sign. In contrast, the potential of a half-reaction or a cell has a sign (+ or −) and a magnitude.

- *When the calculated value of $E°_{cell}$ is positive, the reaction as written is predicted to be product-favored at equilibrium.* Conversely, if the calculated value of $E°_{cell}$ is negative, the reaction is predicted to be reactant-favored at equilibrium. Such a reaction will be product-favored at equilibrium in a direction opposite to the way it is written.
- If we measure $E°_{cell}$ and know either $E°_{cathode}$ or $E°_{anode}$, we can calculate the other value. This value would tell us how one half-cell reaction compares with others in terms of relative oxidizing or reducing ability.

But here is a dilemma. One cannot measure individual half-cell potentials. Just as values for $\Delta_f H°$ and $\Delta_f G°$ were established by choosing a reference point (the elements in their standard states), scientists have selected a reference point for half-reactions. We assign a potential of exactly 0 V to the half-reaction that occurs at a standard hydrogen electrode (SHE).

$$2\ H^+(aq,\ 1\ M) + 2\ e^- \rightarrow H_2(g,\ 1\ bar)\quad E° = 0.00\ V$$

With this standard, we can now determine $E°$ values for half-cells by measuring $E°_{cell}$ in experiments such as those described in Figures 20.8 and 20.13, where one of the electrodes is the standard hydrogen electrode. We can then quantify the information with reduction potential tables such as Figure 20.14 and use these values to make predictions about $E°_{cell}$ for new voltaic cells.

Tables of Standard Reduction Potentials

The experimental approach just described leads to lists of $E°$ values such as seen in Figure 20.14, Table 20.1, and Appendix M. Let us list some important points concerning these tables and then illustrate them in the discussion and examples that follow.

1. Reactions are written as "oxidized form + electrons → reduced form." The species on the left side of the reaction arrow is an oxidizing agent, and the species on the right side of the reaction arrow is a reducing agent. Therefore, *all potentials are for reduction reactions.*
2. The more positive the value of $E°$ for the reactions in Figure 20.14, Table 20.1, and similar tables, the better the oxidizing ability of the ion or compound on the left side of the reaction. This means *$F_2(g)$ is the best oxidizing agent in the table.* Lithium ion at the lower-left corner of Table 20.1 is the poorest oxidizing agent because its $E°$ value is the most negative.

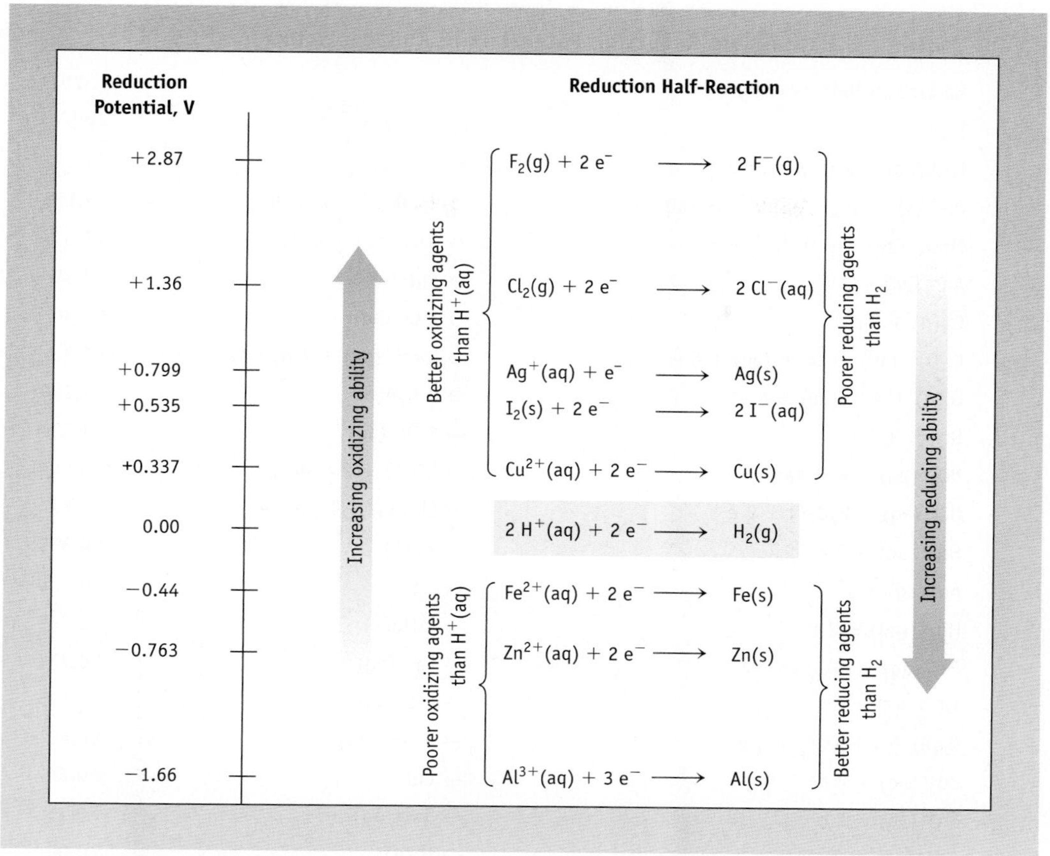

FIGURE 20.14 A potential ladder for reduction half-reactions. The relative position of a half-reaction on this potential ladder reflects the relative ability of the species at the left to act as an oxidizing agent. The higher the compound or ion is in list, the better it is as an oxidizing agent. Conversely, the atoms or ions on the right are reducing agents. The lower they are in the list, the better they are as a reducing agent. The potential for each half-reaction is given with its reduction potential, $E°_{cathode}$. (For more information see J. R. Runo and D. G. Peters, *Journal of Chemical Education*, Volume 70, page 708, 1993.)

3. The more negative the value of the reduction potential, $E°$, the less likely the half-reaction will occur as a reduction, and the more likely the reverse half-reaction will occur (as an oxidation). Thus, Li(s) is the strongest reducing agent in the table, and F^- is the weakest reducing agent. The reducing agents in the table (the ions, elements, and compounds at the right) increase in strength from the top to the bottom.
4. When a reaction is reversed (to give "reduced form → oxidized form + electrons"), the sign of $E°$ is reversed, but the magnitude of $E°$ is unaffected.

$$Fe^{3+}(aq, 1\ M) + e^- \rightarrow Fe^{2+}(aq, 1\ M) \qquad E° = +0.771\ V$$

$$Fe^{2+}(aq, 1\ M) \rightarrow Fe^{3+}(aq, 1\ M) + e^- \qquad E° = -0.771\ V$$

5. The reaction between any substance on the left in this table (an oxidizing agent) with any substance *lower* than it on the right (a reducing agent) is product-favored at equilibrium. This has been called the *northwest–southeast*

■ **$E°$ Values** An extensive listing of $E°$ values is found in Appendix M, and still larger tables of data can be found in chemistry reference books. A common convention, used in Appendix M, lists standard reduction potentials in two groups, one for acid and neutral solutions and the other for basic solutions.

TABLE 20.1 Standard Reduction Potentials in Aqueous Solution at 25 °C*

Reduction Half-Reaction		$E°$ (V)
$F_2(g) + 2\,e^-$	$\rightarrow 2\,F^-(aq)$	+2.87
$H_2O_2(aq) + 2\,H^+(aq) + 2\,e^-$	$\rightarrow 2\,H_2O(\ell)$	+1.77
$PbO_2(s) + SO_4^{2-}(aq) + 4\,H^+(aq) + 2\,e^-$	$\rightarrow PbSO_4(s) + 2\,H_2O(\ell)$	+1.685
$MnO_4^-(aq) + 8\,H^+(aq) + 5\,e^-$	$\rightarrow Mn^{2+}(aq) + 4\,H_2O(\ell)$	+1.51
$Au^{3+}(aq) + 3\,e^-$	$\rightarrow Au(s)$	+1.50
$Cl_2(g) + 2\,e^-$	$\rightarrow 2\,Cl^-(aq)$	+1.36
$Cr_2O_7^{2-}(aq) + 14\,H^+(aq) + 6\,e^-$	$\rightarrow 2\,Cr^{3+}(aq) + 7\,H_2O(\ell)$	+1.33
$O_2(g) + 4\,H^+(aq) + 4\,e^-$	$\rightarrow 2\,H_2O(\ell)$	+1.229
$Br_2(\ell) + 2\,e^-$	$\rightarrow 2\,Br^-(aq)$	+1.08
$NO_3^-(aq) + 4\,H^+(aq) + 3\,e^-$	$\rightarrow NO(g) + 2\,H_2O(\ell)$	+0.96
$OCl^-(aq) + H_2O(\ell) + 2\,e^-$	$\rightarrow Cl^-(aq) + 2\,OH^-(aq)$	+0.89
$Hg^{2+}(aq) + 2\,e^-$	$\rightarrow Hg(\ell)$	+0.855
$Ag^+(aq) + e^-$	$\rightarrow Ag(s)$	+0.799
$Hg_2^{2+}(aq) + 2\,e^-$	$\rightarrow 2\,Hg(\ell)$	+0.789
$Fe^{3+}(aq) + e^-$	$\rightarrow Fe^{2+}(aq)$	+0.771
$I_2(s) + 2\,e^-$	$\rightarrow 2\,I^-(aq)$	+0.535
$O_2(g) + 2\,H_2O(\ell) + 4\,e^-$	$\rightarrow 4\,OH^-(aq)$	+0.40
$Cu^{2+}(aq) + 2\,e^-$	$\rightarrow Cu(s)$	+0.337
$Sn^{4+}(aq) + 2\,e^-$	$\rightarrow Sn^{2+}(aq)$	+0.15
$2\,H^+(aq) + 2\,e^-$	$\rightarrow H_2(g)$	0.00
$Sn^{2+}(aq) + 2\,e^-$	$\rightarrow Sn(s)$	−0.14
$Ni^{2+}(aq) + 2\,e^-$	$\rightarrow Ni(s)$	−0.25
$V^{3+}(aq) + e^-$	$\rightarrow V^{2+}(aq)$	−0.255
$PbSO_4(s) + 2\,e^-$	$\rightarrow Pb(s) + SO_4^{2-}(aq)$	−0.356
$Cd^{2+}(aq) + 2\,e^-$	$\rightarrow Cd(s)$	−0.40
$Fe^{2+}(aq) + 2\,e^-$	$\rightarrow Fe(s)$	−0.44
$Zn^{2+}(aq) + 2\,e^-$	$\rightarrow Zn(s)$	−0.763
$2\,H_2O(\ell) + 2\,e^-$	$\rightarrow H_2(g) + 2\,OH^-(aq)$	−0.8277
$Al^{3+}(aq) + 3\,e^-$	$\rightarrow Al(s)$	−1.66
$Mg^{2+}(aq) + 2\,e^-$	$\rightarrow Mg(s)$	−2.37
$Na^+(aq) + e^-$	$\rightarrow Na(s)$	−2.714
$K^+(aq) + e^-$	$\rightarrow K(s)$	−2.925
$Li^+(aq) + e^-$	$\rightarrow Li(s)$	−3.045

↑ Increasing strength of oxidizing agents

↓ Increasing strength of reducing agents

* In volts (V) versus the standard hydrogen electrode.

rule: Product-favored reactions will always involve a reducing agent that is "southeast" of the proposed oxidizing agent.

Reduction Half-Reaction

$$I_2(s) + 2\,e^- \longrightarrow 2\,I^-(aq)$$
$$Cu^{2+}(aq) + 2\,e^- \longrightarrow Cu(s)$$
$$2\,H^+(aq) + 2\,e^- \longrightarrow H_2(g)$$
$$Fe^{2+}(aq) + 2\,e^- \longrightarrow Fe(s)$$
$$Zn^{2+}(aq) + 2\,e^- \longrightarrow Zn(s)$$

The northwest–southeast rule: The reducing agent always lies to the southeast of the oxidizing agent in a product-favored reaction.

For example, Zn can reduce Fe^{2+}, H^+, Cu^{2+}, and I_2, but, of the species on this list, Cu can reduce only I_2.

6. The algebraic sign of the half-reaction reduction potential is the sign of the electrode when it is attached to the H_2/H^+ standard cell (see Figures 20.8 and 20.13).
7. Electrochemical potentials depend on the nature of the reactants and products and their concentrations, not on the quantities of material used. Therefore, changing the stoichiometric coefficients for a half-reaction does not change the value of $E°$. For example, the reduction of Fe^{3+} has an $E°$ of +0.771 V, whether the reaction is written as

$$Fe^{3+}(aq,\ 1\ M) + e^- \rightarrow Fe^{2+}(aq,\ 1\ M) \qquad E° = +0.771\ V$$

or as

$$2\,Fe^{3+}(aq,\ 1\ M) + 2\,e^- \rightarrow 2\,Fe^{2+}(aq,\ 1\ M) \qquad E° = +0.771\ V$$

■ **Northwest–Southeast Rule** This guideline is a reflection of the idea of moving down a potential energy "ladder" in a product-favored reaction.

■ **Changing Stoichiometric Coefficients** The volt is defined as "energy/charge" (V = J/C). Multiplying a reaction by some number causes both the energy and the charge to be multiplied by that number. Thus, the ratio "energy/charge = volt" does not change.

Using Tables of Standard Reduction Potentials

Tables or "ladders" of standard reduction potentials are immensely useful. They allow you to predict the potential of a new voltaic cell, provide information that can be used to balance redox equations, and help predict which redox reactions are product-favored.

Calculating Standard Cell Potentials, $E°_{cell}$

The standard reduction potentials for half-reactions were obtained by measuring cell potentials. It makes sense, therefore, that these values can be combined to give the potential of some new cell.

The net reaction occurring in a voltaic cell using silver and copper half-cells is

$$2\,Ag^+(aq) + Cu(s) \rightarrow 2\,Ag(s) + Cu^{2+}(aq)$$

The silver electrode is the cathode, and the copper electrode is the anode. We know this because silver ion is reduced (to silver metal) and copper metal is oxidized (to Cu^{2+} ions). (Recall that oxidations always occur at the anode and reductions at the cathode.) Also notice that the $Cu^{2+}|Cu$ half-reaction is "southeast" of the $Ag^+|Ag$ half-reaction in the potential ladder (Figure 20.14 and Table 20.1).

$$E°_{cathode} = +0.799\ V \qquad Ag^+(aq) + e^- \longrightarrow Ag(s)$$

"Distance" from $E°_{cathode}$ to $E°_{anode}$ is 0.799 V − 0.337 V = 0.462 V.

Cu is "southeast" of Ag^+

$$E°_{anode} = +0.337\ V \qquad Cu^{2+}(aq) + 2e^- \longrightarrow Cu(s)$$

The potential for the voltaic cell is the difference between the standard reduction potentials for the two half-reactions.

$$E^\circ_{cell} = E^\circ_{cathode} - E^\circ_{anode}$$

$$E^\circ_{cell} = (+0.799\text{ V}) - (+0.337\text{ V})$$

$$E^\circ_{cell} = +0.462\text{ V}$$

Notice that the value of E°_{cell} is related to the "distance" between the cathode and anode reactions on the potential ladder. The products have a lower potential energy than the reactants, and the cell potential, E°_{cell}, has a positive value.

A positive potential calculated for the $Ag^+|Ag$ and $Cu^{2+}|Cu$ cell ($E^\circ_{cell} = +0.462$ V) confirms that the reduction of silver ions in water with copper metal is product-favored at equilibrium (Figure 20.1). We might ask, however, about the value of E°_{cell} if a reactant-favored equation had been selected. For example, what is E°_{cell} for the reduction of copper ions with silver metal?

Cathode, reduction:	$Cu^{2+}(aq) + 2\,e^- \rightarrow Cu(s)$
Anode, oxidation:	$2\,Ag(s) \rightarrow 2\,Ag^+(aq) + 2\,e^-$
Net ionic equation:	$2\,Ag\,(s) + Cu^{2+}(aq) \rightarrow 2\,Ag^+(aq) + Cu(s)$

Cell Voltage Calculation

$$E^\circ_{cathode} = +\,0.337\text{ V and } E^\circ_{anode} = +\,0.799\text{ V}$$

$$E^\circ_{cell} = E^\circ_{cathode} - E^\circ_{anode} = (+0.337\text{ V}) - (0.799\text{ V})$$

$$E^\circ_{cell} = -\,0.462\text{ V}$$

The negative sign for E°_{cell} indicates that the reaction as written is reactant-favored at equilibrium. The products of the reaction (Ag^+ and Cu) have a higher potential energy than the reactants (Ag and Cu^{2+}). For the indicated reaction to occur, a potential of at least 0.462 V would have to be imposed on the system by an external source of electricity (see Section 20.7).

Chemistry Now™

Sign in at **www.thomsonedu.com/login** and go to Chapter 20 Contents to see:

- Screen 20.5 for demonstration of **the potentials of various cells**
- Screen 20.6 for a simulation and tutorial on **standard potentials**

EXERCISE 20.6 Calculating Standard Cell Potentials

The net reaction that occurs in a voltaic cell is

$$Zn(s) + 2\,Ag^+(aq) \rightarrow Zn^{2+}(aq) + 2\,Ag(s)$$

Identify the half-reactions that occur at the anode and the cathode, and calculate a potential for the cell assuming standard conditions.

Relative Strengths of Oxidizing and Reducing Agents

Five half-reactions, selected from Appendix M, are arranged from the half-reaction with the highest (most positive) $E°$ value to the one with the lowest (most negative) value.

$E°$, V	Reduction Half-Reaction
+1.36	$Cl_2(g) + 2\,e^- \longrightarrow 2\,Cl^-(aq)$
+0.80	$Ag^+(aq) + e^- \longrightarrow Ag(s)$
+0.00	$2\,H^+(aq) + 2\,e^- \longrightarrow H_2(g)$
−0.25	$Ni^{2+}(aq) + 2\,e^- \longrightarrow Ni(s)$
−0.76	$Zn^{2+}(aq) + 2\,e^- \longrightarrow Zn(s)$

Increasing strength as oxidizing agents (↑)

- The list on the left is headed by Cl_2, an element that is a strong oxidizing agent and thus is easily reduced. At the bottom of the list is $Zn^{2+}(aq)$, an ion not easily reduced and thus a poor oxidizing agent.
- On the right, the list is headed by $Cl^-(aq)$, an ion that can be oxidized to Cl_2 only with difficulty. It is a very poor reducing agent. At the bottom of the list is zinc metal, which is quite easy to oxidize and a good reducing agent.

By arranging these half-reactions based on $E°$ values, we have also arranged the chemical species on the two sides of the equation in order of their strengths as oxidizing or reducing agents. In this list, from strongest to weakest, the order is

$$\textit{Oxidizing agents: } Cl_2 > Ag^+ > H^+ > Ni^{2+} > Zn^{2+}$$

strong ⟶ *weak*

$$\textit{Reducing agents: } Zn > Ni > H_2 > Ag > Cl^-$$

strong ⟶ *weak*

Finally, notice that the value of $E°_{cell}$ is greater the farther apart the oxidizing and reducing agents are on the potential ladder. For example,

$$Zn(s) + Cl_2(g) \rightarrow Zn^{2+}(aq) + 2\,Cl^-(aq) \qquad E° = +2.12\ V$$

is more strongly product-favored than the reduction of hydrogen ions with nickel metal

$$Ni(s) + 2\,H^+(aq) \rightarrow Ni^{2+}(aq) + H_2(g) \qquad E° = +0.25\ V$$

EXAMPLE 20.5 Ranking Oxidizing and Reducing Agents

Problem Use the table of standard reduction potentials (Table 20.1) to do the following:

(a) Rank the halogens in order of their strength as oxidizing agents.

(b) Decide whether hydrogen peroxide (H_2O_2) in acid solution is a stronger oxidizing agent than Cl_2.

(c) Decide which of the halogens is capable of oxidizing gold metal to $Au^{3+}(aq)$.

Strategy The ability of a species on the left side of a reduction potential table to function as an oxidizing agent declines on descending the list (see points 2 and 3, pages 918-919).

Solution

(a) Ranking halogens according to oxidizing ability. The halogens (F_2, Cl_2, Br_2, and I_2) appear in the upper-left portion of the table, with F_2 being highest, followed in order by the other three species. Their strengths as oxidizing agents are $F_2 > Cl_2 > Br_2 > I_2$. (The ability of bromine to oxidize iodide ions to molecular iodine is illustrated in Figure 20.15.)

FIGURE 20.15 The reaction of bromine and iodide ion. This experiment proves that Br_2 is a better oxidizing agent than I_2. The presence of I_2 in the bottom layer in the photo on the right indicates that the added Br_2 was able to oxidize the iodide ions originally present to molecular iodine (I_2).

(b) Comparing hydrogen peroxide and chlorine. H_2O_2 lies just below F_2 but well above Cl_2 in the potential ladder (Table 20.1). Thus, H_2O_2 is a weaker oxidizing agent than F_2 but a stronger one than Cl_2. (Note that the $E°$ value for H_2O_2 refers to an acidic solution and standard conditions.)

(c) Which halogen will oxidize gold metal to gold(III) ions? The $Au^{3+}|Au$ half-reaction is listed below the $F_2|F^-$ half-reaction and just above the $Cl_2|Cl^-$ half-reaction. This tells us that, among the halogens, only F_2 is capable of oxidizing Au to Au^{3+} under standard conditions. That is, for the reaction of Au and F_2,

Oxidation, anode:	$Au(s) \rightarrow Au^{3+}(aq) + 3\,e^-$
Reduction, cathode:	$F_2(g) + 2\,e^- \rightarrow 2\,F^-(aq)$
Net ionic equation:	$3\,F_2(g) + 2\,Au(s) \rightarrow 6\,F^-(aq) + 2\,Au^{3+}(aq)$

$$E°_{cell} = E°_{cathode} - E°_{anode} = +2.87\text{ V} - (+1.50\text{ V}) = +1.37\text{ V}$$

F_2 is a stronger oxidizing agent than Au^{3+}, so the reaction proceeds from left to right as written. (This is confirmed by a positive value of $E°_{cell}$.) For the reaction of Cl_2 and Au, Table 20.1 shows us that Cl_2 is a weaker oxidizing agent than Au^{3+}, so the reaction would be expected to proceed in the opposite direction.

Oxidation, anode:	$Au(s) \rightarrow Au^{3+}(aq) + 3\,e^-$
Reduction, cathode:	$Cl_2(aq) + 2\,e^- \rightarrow 2\,Cl^-(aq)$
Net ionic equation:	$3\,Cl_2(aq) + 2\,Au(s) \rightarrow 6\,Cl^-(aq) + 2\,Au^{3+}(aq)$

$$E°_{cell} = E°_{cathode} - E°_{anode} = +1.36\text{ V} - (+1.50\text{ V}) = -0.14\text{ V}$$

This is confirmed by the negative value for $E°_{cell}$.

Comment In part (c), we calculated $E°_{cell}$ for two reactions. To achieve a balanced net ionic equation, we added the half-reactions, but only after multiplying the gold half-reaction by 2 and the halogen, half-reaction by 3. (This means 6 mol of electrons was transferred from 2 mol Au to 3 mol Cl_2.) Notice that this multiplication does not change the value of $E°$ for the half-reactions because cell potentials do not depend on the quantity of material.

EXERCISE 20.7 Relative Oxidizing and Reducing Ability

Which metal in the following list is easiest to oxidize: Fe, Ag, Zn, Mg, Au? Which metal is the most difficult to oxidize?

EXERCISE 20.8 Using a Table of Standard Reduction Potentials to Predict Chemical Reactions

Determine whether the following redox equations are product-favored at equilibrium.

(a) $Ni^{2+}(aq) + H_2(g) \rightarrow Ni(s) + 2\,H^+(aq)$

(b) $2\,Fe^{3+}(aq) + 2\,I^-(aq) \rightarrow 2\,Fe^{2+}(aq) + I_2(s)$

(c) $Br_2(\ell) + 2\,Cl^-(aq) \rightarrow 2\,Br^-(aq) + Cl_2(g)$

(d) $Cr_2O_7^{2-}(aq) + 6\,Fe^{2+}(aq) + 14\,H^+(aq) \rightarrow 2\,Cr^{3+}(aq) + 6\,Fe^{3+}(aq) + 7\,H_2O(\ell)$

Chemical Perspectives — An Electrochemical Toothache!

It was recently reported that a 66-year-old woman had intense pain that was traced to her dental work (*New England Journal of Medicine,* Vol. 342, p. 2000, 2003). A root canal job had moved a mercury amalgam filling on one tooth slightly closer to a gold alloy crown on an adjacent tooth. Eating acidic foods caused her intense pain. When dental amalgams of dissimilar metals come in contact with saliva, a voltaic cell is formed that generates potentials up to several hundred millivolts—and you feel it! You can do it, too, if you chew an aluminum foil gum wrapper with teeth that have been filled with a dental amalgam. Ouch!

20.5 Electrochemical Cells Under Nonstandard Conditions

Electrochemical cells seldom operate under standard conditions in the real world. Even if the cell is constructed with all dissolved species at 1 M, reactant concentrations decrease and product concentrations increase in the course of the reaction. Changing concentrations of reactants and products will affect the cell voltage. Thus, we need to ask what happens to cell potentials under nonstandard conditions.

The Nernst Equation

Based on both theory and experimental results, it has been determined that cell potentials are related to concentrations of reactants and products and to temperature, as follows:

$$E = E° - (RT/nF) \ln Q \qquad \textbf{(20.2)}$$

In this equation, which is known as the **Nernst equation**, R is the gas constant (8.314472 J/K · mol); T is the temperature (K); and n is the number of moles of electrons transferred between oxidizing and reducing agents (as determined by the balanced equation for the reaction). The symbol F represents the **Faraday constant** (9.6485338×10^4 C/mol). *One Faraday is the quantity of electric charge carried by one mole of electrons.* The term Q is the reaction quotient, an expression relating the concentrations of the products and reactants raised to an appropriate power as defined by the stoichiometric coefficients in the balanced, net equation [see Equation 16.2, Section 16.2]. Substituting values for the constants in Equation 20.2, and using 298 K as the temperature, gives

■ **Walther Nernst (1864–1941)** Nernst was a German physicist and chemist known for his work relating to the third law of thermodynamics.

$$E = E° - \frac{0.0257}{n}\ln Q \quad \text{at 25 °C} \qquad \textbf{(20.3)}$$

or, in a commonly used form using base-10 logarithms,

$$E = E° - \frac{0.0591}{n}\log Q$$

In essence, the term $(RT/nF)\ln Q$ "corrects" the standard potential $E°$ for nonstandard conditions or concentrations.

■ **Units of R and F** The gas constant R has units of J/K · mol, and F has units of coulombs per mol (C/mol). Because 1 J = 1 C · V, the factor RT/nF has units of volts.

Chemistry Now™

Sign in at **www.thomsonedu.com/login** and go to Chapter 20 Contents to see Screen 20.8 for a tutorial on **the Nernst equation.**

EXAMPLE 20.6 Using the Nernst Equation

Problem A voltaic cell is set up at 25 °C with the half-cells: Al^{3+}(0.0010 M) | Al and Ni^{2+}(0.50 M) | Ni. Write an equation for the reaction that occurs when the cell generates an electric current, and determine the cell potential.

Strategy The first step is to determine which substance is oxidized (Al or Ni) by looking at the appropriate half-reactions in Table 20.1 and deciding which is the better reducing agent (Example 20.5). Next, add the half-reactions to determine the net ionic equation, and calculate $E°$. Finally, use the Nernst equation to calculate E, the nonstandard potential.

Solution Aluminum metal is a stronger reducing agent than Ni metal. (Conversely, Ni^{2+} is a better oxidizing agent than Al^{3+}.) Therefore, Al is oxidized, and the Al^{3+} | Al compartment is the anode.

Cathode, reduction:	$Ni^{2+}(aq) + 2\,e^- \rightarrow Ni(s)$
Anode, oxidation:	$Al(s) \rightarrow Al^{3+}(aq) + 3\,e^-$
Net ionic equation:	$2\,Al(s) + 3\,Ni^{2+}(aq) \rightarrow 2\,Al^{3+}(aq) + 3\,Ni(s)$

$E°_{cell} = E°_{cathode} - E°_{anode}$

$E°_{cell} = (-0.25\text{ V}) - (-1.66\text{ V}) = 1.41\text{ V}$

The expression for Q is written based on the cell reaction. In the net reaction, Al^{3+}(aq) has a coefficient of 2, so this concentration is squared. Similarly, $[Ni^{2+}(aq)]$ is cubed. Solids are not included in the expression for Q (◀ Section 16.1).

$$Q = \frac{[Al^{3+}]^2}{[Ni^{2+}]^3}$$

The net equation requires transfer of six electrons from two Al atoms to three Ni^{2+} ions, so $n = 6$. Substituting for $E°$, n and Q in the Nernst equation gives

$$E_{cell} = E°_{cell} - \frac{0.0257}{n}\ln\frac{[Al^{3+}]^2}{[Ni^{2+}]^3}$$

$$= +1.41\text{ V} - \frac{0.0257}{6}\ln\frac{[0.0010]^2}{[0.50]^3}$$

$$= +1.41\text{ V} - 0.00428\ln(8.0 \times 10^{-6})$$

$$= +1.41\text{ V} - 0.00428(-11.7)$$

$$= 1.46\text{ V}$$

Comment Notice that E_{cell} is larger than $E°_{cell}$ because the product concentration, $[Al^{3+}]$, is much smaller than 1.0 M. Generally, when product concentrations are smaller initially than the reactant concentrations in a product-favored reaction, the cell potential is more positive than $E°$.

EXERCISE 20.9 Variation of E_{cell} with Concentration

A voltaic cell is set up with an aluminum electrode in a 0.025 M $Al(NO_3)_3$(aq) solution and an iron electrode in a 0.50 M $Fe(NO_3)_2$(aq) solution. Determine the cell potential, E_{cell}, at 298 K.

Example 20.6 demonstrates the calculation of a cell potential if concentrations are known. It is also useful to apply the Nernst equation in the opposite sense, using a measured cell potential to determine an unknown concentration. A device that does just this is the pH meter (Figure 20.16). In an electrochemical cell in which H^+(aq) is a reactant or product, the cell voltage will vary predictably with the hydrogen ion concentration. The cell voltage is measured and the value used to calculate pH. Example 20.7 illustrates how E_{cell} varies with the hydrogen ion concentration in a simple cell.

EXAMPLE 20.7 Variation of E_{cell} with Concentration

Problem A voltaic cell is set up with copper and hydrogen half-cells. Standard conditions are employed in the copper half-cell, Cu^{2+}(aq, 1.00 M) | Cu(s). The hydrogen gas pressure is 1.00 bar, and [H^+(aq)] in the hydrogen half-cell is the unknown. A value of 0.490 V is recorded for E_{cell} at 298 K. Determine the pH of the solution.

Strategy We first decide which is the better oxidizing and reducing agent so as to decide what net reaction is occurring in the cell. With this known, $E°_{cell}$ can be calculated. The only unknown quantity in the Nernst equation is the concentration of hydrogen ion, from which we can calculate the solution pH.

Solution Hydrogen is a better reducing agent than copper metal, so Cu(s) | Cu^{2+}(aq, 1.00 M) is the cathode, and H_2(g, 1.00 bar) | H^+(aq, ? M) is the anode.

Cathode, reduction:	$Cu^{2+}(aq) + 2\ e^- \rightarrow Cu(s)$
Anode, oxidation:	$H_2(g) \rightarrow 2\ H^+(aq) + 2\ e^-$
Net ionic equation:	$H_2(g) + Cu^{2+}(aq) \rightarrow Cu(s) + 2\ H^+(aq)$

$$E°_{cell} = E°_{cathode} - E°_{anode}$$

$$E°_{cell} = (+0.337\ V) - (0.00\ V) = +0.337\ V$$

The reaction quotient, Q, is derived from the balanced net ionic equation.

$$Q = \frac{[H^+]^2}{[Cu^{2+}]P_{H_2}}$$

The net equation requires the transfer of two electrons, so $n = 2$. The value of [Cu^{2+}] is 1.00 M, but [H^+] is unknown. Substitute this information into the Nernst equation (and don't overlook the fact that [H^+] is squared in the expression for Q).

$$E = E° - \frac{0.0257}{n}\ln\frac{[H^+]^2}{[Cu^{2+}]P_{H_2}}$$

$$0.490\ V = 0.337\ V - \frac{0.0257}{2}\ln\frac{[H^+]^2}{(1.00)(1.00)}$$

$$-11.9 = \ln\ [H^+]^2$$

$$[H^+] = 2.6 \times 10^{-3}\ M$$

$$pH = 2.59$$

EXERCISE 20.10 Using the Nernst Equation

The half-cells Fe^{2+}(aq, 0.024 M) | Fe(s) and H^+(aq, 0.056 M) | H_2(1.0 bar) are linked by a salt bridge to create a voltaic cell. Determine the cell potential, E_{cell}, at 298 K.

In the real world, using a hydrogen electrode in a pH meter is not practical. The apparatus is clumsy; it is anything but robust; and platinum (for the electrode) is costly. Common pH meters use a glass electrode, so called because it contains a thin glass membrane separating the cell from the solution whose pH is to be measured (Figure 20.16). Inside the glass electrode is a silver wire coated with AgCl and a solution of HCl; outside is the solution of unknown pH to be evaluated. A Ag/AgCl or calomel electrode—a common reference electrode using a mercury(I)–mercury redox couple (Hg_2Cl_2|Hg)—serves as the second electrode of the cell. The potential across the glass membrane depends on [H^+]. Common pH meters give a direct readout of pH.

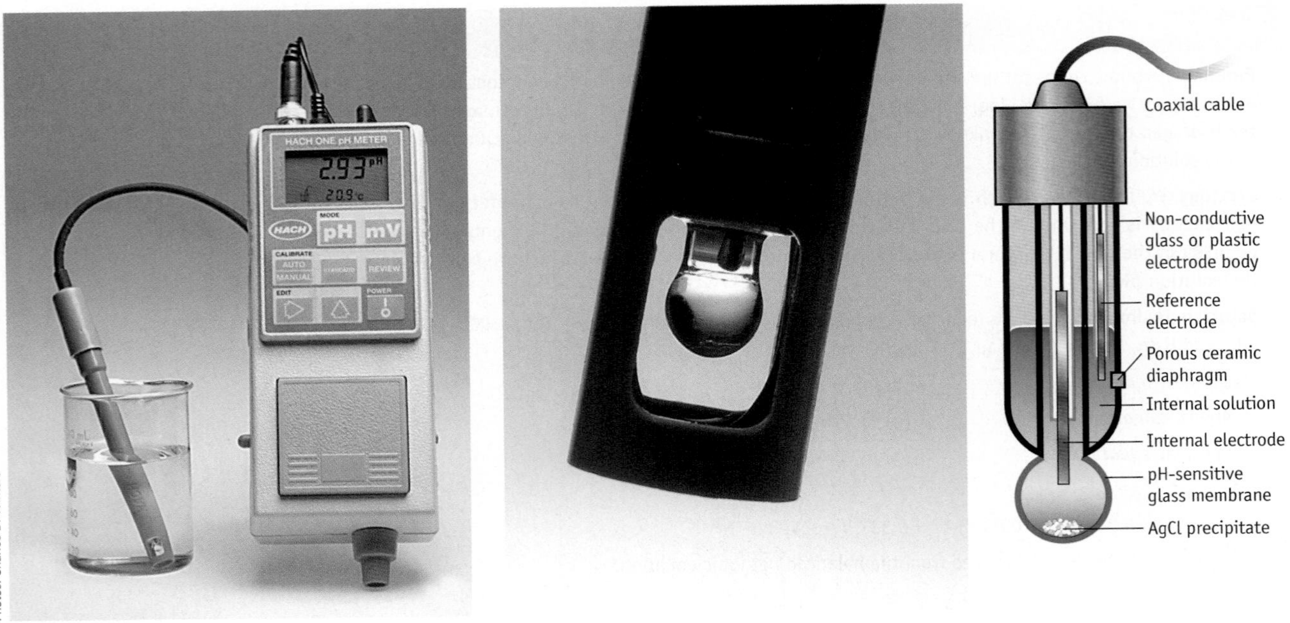

Photos: Charles D. Winters

(a) (b) (c)

FIGURE 20.16 Measuring pH. (a) A pH portable meter that can be used in the field. (b) The tip of a glass electrode for measuring pH. (c) A schematic diagram of a glass electrode. (See ChemistryNow Screen 17.4 for an animation of the operation of a glass electrode for pH measurement.)

20.6 Electrochemistry and Thermodynamics

Work and Free Energy

The first law of thermodynamics [Section 5.4] states that the internal energy change in a system (ΔU) is related to two quantities, heat (q) and work (w): $\Delta U = q + w$. This equation also applies to chemical changes that occur in a voltaic cell. As current flows, energy is transferred from the system (the voltaic cell) to the surroundings.

In a voltaic cell, the decrease in internal energy in the system will manifest itself ideally as electrical work done on the surroundings by the system. In practice, however, some energy is usually evolved as heat by the voltaic cell. The maximum work done by an electrochemical system (ideally, assuming no heat is generated) is proportional to the potential difference (volts) and the quantity of charge (coulombs):

$$w_{max} = nFE \qquad \textbf{(20.4)}$$

In this equation, E is the cell voltage, and nF is the quantity of electric charge transferred from anode to cathode.

The free energy change for a process is, by definition, the maximum amount of work that can be obtained [Section 19.6]. Because the maximum work and the cell potential are related, $E°$ and $\Delta_r G°$ can be related mathematically (taking care to assign signs correctly). The maximum work done on the surroundings when electricity is produced by a voltaic cell is $+nFE$, with the positive sign denoting an increase in energy in the surroundings. The energy content of the cell decreases by this amount. Thus, $\Delta_r G$ for the voltaic cell has the opposite sign.

$$\Delta_r G = -nFE \qquad \textbf{(20.5)}$$

Under standard conditions, the appropriate equation is

$$\Delta_r G° = -nFE° \quad (20.6)$$

This expression shows that, the more positive the value of $E°$, the larger and more negative the value of $\Delta_r G°$ for the reaction. Also, because of the relationship between $\Delta_r G°$ and K, the farther apart the half-reactions on the potential ladder, the more strongly product-favored the reaction is at equilibrium.

■ **Units in Equation 20.6** n has units of mol e^-, and F has units of (C/mol e^-). Therefore, nF has units of coulombs (C). Because 1 J = 1 C · V, the product nFE will have units of energy (J).

EXAMPLE 20.8 The Relation Between $E°$ and $\Delta_r G°$

Problem The standard cell potential, $E°_{cell}$ for the reduction of silver ions with copper metal (Figure 20.5) is +0.462 V at 25 °C. Calculate $\Delta_r G°$ for this reaction.

Strategy We use Equation 20.6, where F is a constant and $E°_{cell}$ is given. The only problem here is to determine the value of n, the number of moles of electrons transferred between copper metal and silver ions in the balanced equation.

Solution In this cell, copper is the anode, and silver is the cathode. The overall cell reaction is

$$Cu(s) + 2\ Ag^+(aq) \rightarrow Cu^{2+}(aq) + 2\ Ag(s)$$

which means that each mole of copper transfers two moles of electrons to two moles of Ag^+ ions. That is, $n = 2$. Now use Equation 20.6.

$$\Delta_r G° = -nFE° = -(2 \text{ mol } e^-)(96{,}485 \text{ C/mol } e^-)(0.462 \text{ V}) = -89{,}200 \text{ C} \cdot \text{V}$$

Because 1 C · V = 1 J, we have

$$\Delta_r G° = -89{,}200 \text{ J or } -89.2 \text{ kJ}$$

Comment This example demonstrates a very effective method of obtaining thermodynamic values from relatively simple electrochemical experiments.

EXERCISE 20.11 The Relationship Between $E°$ and $\Delta_r G°$

The following reaction has an $E°$ value of −0.76 V:

$$H_2(g) + Zn^{2+}(aq) \rightarrow Zn(s) + 2\ H^+(aq)$$

Calculate $\Delta_r G°$ for this reaction. Is the reaction product- or reactant-favored at equilibrium?

$E°$ and the Equilibrium Constant

When a voltaic cell produces an electric current, the reactant concentrations decrease, and the product concentrations increase. The cell voltage also changes; as reactants are converted to products, the value of E_{cell} decreases. Eventually, the cell potential reaches zero; no further net reaction occurs; and equilibrium is achieved.

This situation can be analyzed using the Nernst equation. When $E_{cell} = 0$, the reactants and products are at equilibrium, and the reaction quotient, Q, is equal to the equilibrium constant, K. Substituting the appropriate symbols and values into the Nernst equation,

$$E = 0 = E° - \frac{0.0257}{n} \ln K$$

and collecting terms gives an equation that relates the cell potential and equilibrium constant:

$$\ln K = \frac{nE°}{0.0257} \quad \text{at 25 °C} \quad (20.7)$$

■ **K and $E°$** The farther apart half-reactions for a product-favored reaction are on the potential ladder, the larger the value of K.

Equation 20.7 can be used to determine values for equilibrium constants, as illustrated in Example 20.9 and Exercise 20.12.

EXAMPLE 20.9 $E°$ and Equilibrium Constants

Problem Calculate the equilibrium constant for the reaction

$$Fe(s) + Cd^{2+}(aq) \rightleftharpoons Fe^{2+}(aq) + Cd(s)$$

Strategy First, determine $E°_{cell}$ from $E°$ values for the two half-reactions (see Example 20.5) and from those the value of n, the other parameter required in Equation 20.7.

Solution The half-reactions and $E°$ values are

Cathode, reduction:	$Cd^{2+}(aq) + 2\,e^- \rightarrow Cd(s)$
Anode, oxidation:	$Fe(s) \rightarrow Fe^{2+}(aq) + 2\,e^-$
Net ionic equation:	$Fe(s) + Cd^{2+}(aq) \rightleftharpoons Fe^{2+}(aq) + Cd(s)$

$$E°_{cell} = E°_{cathode} - E°_{anode}$$

$$E°_{cell} = (-0.40\text{ V}) - (-0.44\text{ V}) = +0.04\text{ V}$$

Now substitute $n = 2$ and $E°_{cell}$ into Equation 20.7.

$$\ln K = \frac{nE°}{0.0257} = \frac{(2)(0.04\text{ V})}{0.0257} = 3.1$$

$$K = 20$$

Comment The relatively small positive voltage (0.04 V) for the cell indicates that the cell reaction is only mildly product-favored. A value of 20 for the equilibrium constant is in accord with this observation.

EXERCISE 20.12 $E°$ and Equilibrium Constants

Calculate the equilibrium constant at 25 °C for the reaction

$$2\,Ag^+(aq) + Hg(\ell) \rightleftharpoons 2\,Ag(s) + Hg^{2+}(aq)$$

The relationships between $E°$, K, and $\Delta_rG°$, which is summarized in Table 20.2, can be used to obtain equilibrium constants for many different chemical systems. For example, let us construct an electrode in which an insoluble ionic compound (such as AgCl) is a component of the half-cell. For this purpose, a silver electrode with a surface layer of AgCl can be prepared. The reaction occurring at this electrode is then

$$AgCl(s) + e^- \rightarrow Ag(s) + Cl^-(aq)$$

The standard reduction potential for this half-cell (Appendix M) is +0.222 V. When this half-reaction is paired with a standard silver electrode in an electrochemical cell, the cell reactions are

Cathode, reduction:	$AgCl(s) + e^- \rightarrow Ag(s) + Cl^-(aq)$
Anode, oxidation:	$Ag(s) \rightarrow Ag^+(aq) + e^-$
Net ionic equation:	$AgCl(s) \rightarrow Ag^+(aq) + Cl^-(aq)$

$$E°_{cell} = E°_{cathode} - E°_{anode} = (+0.222\text{ V}) - (+0.799\text{ V}) = -0.577\text{ V}$$

TABLE 20.2 Summary of the Relationship of K, $\Delta_rG°$, and $E°$

K	$\Delta_rG°$	$E°$	Reactant-Favored or Product-Favored at Equilibrium?	Spontaneous under Standard Conditions?
$K >> 1$	$\Delta_rG° < 0$	$E° > 0$	Product-favored	Spontaneous under standard conditions
$K = 1$	$\Delta_rG° = 0$	$E° = 0$	$[C]^c[D]^d = [A]^a[B]^b$ at equilibrium	At equilibrium under standard conditions
$K << 1$	$\Delta_rG° > 0$	$E° < 0$	Reactant-favored	Not spontaneous under standard conditions.

The equation for the net reaction represents the equilibrium of solid AgCl and its ions. The cell potential is negative, indicating a reactant-favored process, as would be expected based on the low solubility of AgCl. Using Equation 20.7, the value of the equilibrium constant [K_{sp}, Section 18.4] can be obtained from $E°_{cell}$.

$$\ln K = \frac{nE°}{0.0257} = \frac{(1)(-0.577 \text{ V})}{0.0257} = -22.5$$

$$K_{sp} = e^{-22.5} = 2 \times 10^{-10}$$

EXERCISE 20.13 Determining an Equilibrium Constant

In Appendix M, the following standard reduction potential is reported:

$$[Zn(CN)_4]^{2-}(aq) + 2\,e^- \rightarrow Zn(s) + 4\,CN^-(aq) \qquad E° = -1.26 \text{ V}$$

Use this information, along with the data on the $Zn^{2+}(aq)|Zn$ half-cell, to calculate the equilibrium constant for the reaction

$$Zn^{2+}(aq) + 4\,CN^-(aq) \rightarrow [Zn(CN)_4]^{2-}(aq)$$

The value calculated is the formation constant for this complex ion at 25 °C.

20.7 Electrolysis: Chemical Change Using Electrical Energy

Thus far, we have described electrochemical cells that use product-favored redox reactions to generate an electric current. Equally important, however, is the opposite process, **electrolysis**, the use of electrical energy to bring about chemical change.

Electrolysis of water is a classic chemistry experiment, and the electroplating of metals is another example of electrolysis (Figure 20.17). In electroplating, an electric current is passed through a solution containing a salt of the metal to be plated. The object to be plated is the cathode. When metal ions in solution are reduced, the metal deposits on the object's surface.

Electrolysis is an important procedure because it is widely used in the refining of metals such as aluminum and in the production of chemicals such as chlorine.

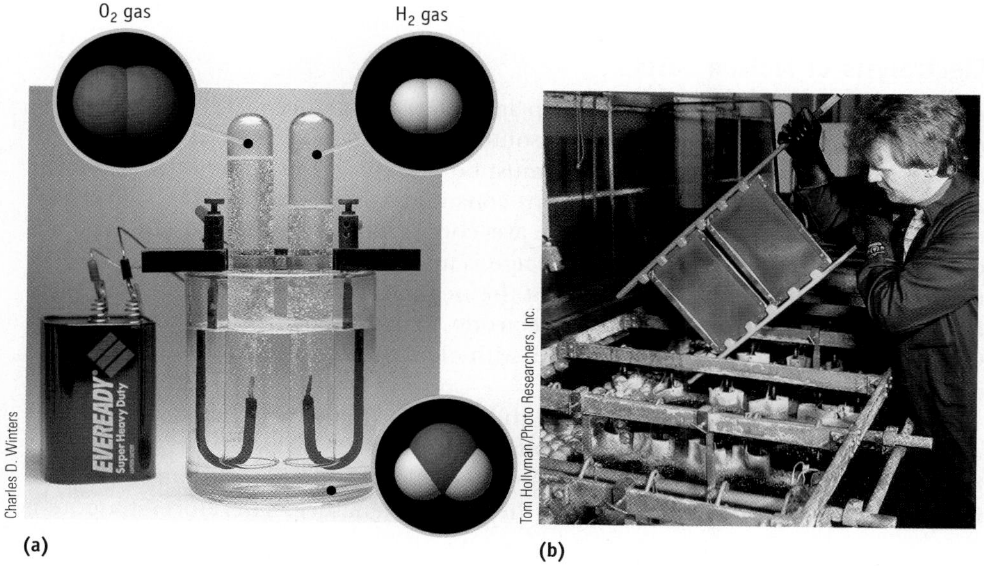

FIGURE 20.17 Electrolysis. (a) Electrolysis of water produces hydrogen and oxygen gas. (b) Electroplating adds a layer of metal to the surface of an object, either to protect the object from corrosion or to improve its physical appearance. The procedure uses an electrolysis cell, set up with the object to be plated as the cathode and a solution containing a salt of the metal to be plated.

Case Study — Manganese in the Oceans

Manganese is a key component of some oxidation–reduction cycles in the oceans. According to an article in the journal *Science*, it "can perform this role because it exists in multiple oxidation states and is recycled rapidly between these states by bacterial processes."

Figure A shows how this cycle was thought to work. Manganese(II) ions in subsurface water are oxidized to form manganese(IV) oxide, MnO_2. Particles of this insoluble solid sink toward the ocean floor. However, some encounter hydrogen sulfide, which is produced in the ocean depths, rising toward the surface. Another redox reaction occurs, producing sulfur and manganese(II) ions. The newly formed Mn^{2+} ions diffuse upward, where they are again oxidized.

The manganese cycle had been thought to involve only the +2 and +4 oxidation states of manganese, and analyses of water samples assumed the dissolved manganese existed only as Mn^{2+} ions. One reason for this is that the intermediate oxidation state, Mn^{3+}, is not predicted to be stable in water. It should disproportionate to the +2 and +4 states.

$$2\ Mn^{3+}(aq) + 2\ H_2O(\ell) \rightarrow Mn^{2+}(aq) + MnO_2(s) + 4\ H^+(aq)$$

It is known, however, that Mn^{3+} can exist when complexed with species such as pyrophosphate ions, $P_2O_7^{4-}$.

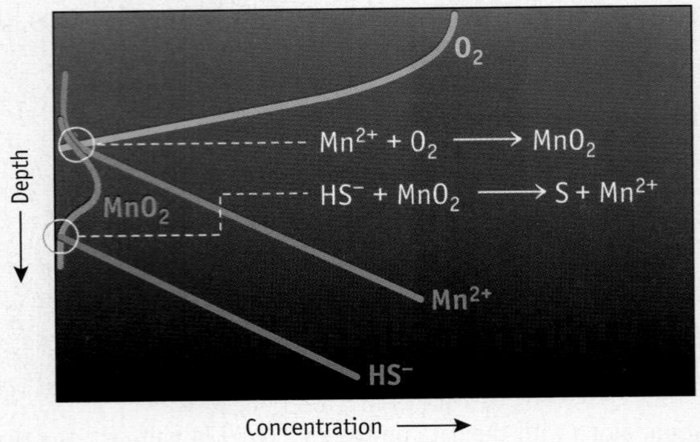

Figure A Manganese chemistry in the oceans. Relative concentrations of important species as a function of depth in the oceans. See K. S. Johnson, *Science*, Vol. 313, p. 1896, 2006 and R. E. Trouwborst, et al., *Science*, Vol. 313, pp. 1955–1957, 2006.

Several years ago, geochemists suggested that Mn^{3+} ions could exist in natural water. They could be produced by bacterial action and stabilized by phosphate from ATP or ADP. They speculated that the Mn^{3+} ion could play an important part of the natural manganese cycle.

Now, other researchers have indeed discovered that, in oxygen-poor waters, the manganese(III) ion, Mn^{3+}, can persist. These ions were found in anoxic zones (zones without dissolved oxygen) below 100 m in the Black Sea and below about 15 m in the Chesapeake Bay. It is now clear that Mn^{3+} ions, which had previously been known only in the laboratory, can exist in natural waters under the right circumstances and that the manganese cycle may have to be revised.

Questions:

1. *Given the following reduction potentials, show that Mn^{3+} should disproportionate to Mn^{2+} and MnO_2 at standard conditions.*

   $$4\ H^+(aq) + MnO_2(s) + e^- \rightarrow Mn^{3+}(aq) + 2\ H_2O(\ell)$$
   $E° = 0.95$ V

   $$Mn^{3+}(aq) + e^- \rightarrow Mn^{2+}(aq)$$
   $E° = 1.50$ V

2. *Balance the following equations in acid solution.*

   (a) Reduction of MnO_2 with HS^- to Mn^{2+} and S

   (b) Oxidation of Mn^{2+} with O_2 to MnO_2

3. *Calculate $E°$ for the oxidation of Mn^{2+} with O_2 to MnO_2.*

Answers to these questions are in Appendix Q.

Electrolysis of Molten Salts

All electrolysis experiments are set up in a similar manner. The material to be electrolyzed, either a molten salt or a solution, is contained in an electrolysis cell. As was the case with voltaic cells, ions must be present in the liquid or solution for a current to flow. The movement of ions constitutes the electric current within the cell. The cell has two electrodes that are connected to a source of DC (direct-current) voltage. If a high enough voltage is applied, chemical reactions occur at the two electrodes. Reduction occurs at the negatively charged cathode, with electrons being transferred from that electrode to a chemical species in the cell. Oxidation occurs at the positive anode, with electrons from a chemical species being transferred to that electrode.

Let us first focus our attention on the chemical reactions that occur at each electrode in the electrolysis of a molten salt. Sodium chloride melts at about 800 °C, and in the molten state sodium ions (Na^+) and chloride ions (Cl^-) are freed from their rigid arrangement in the crystalline lattice. Therefore, if a poten-

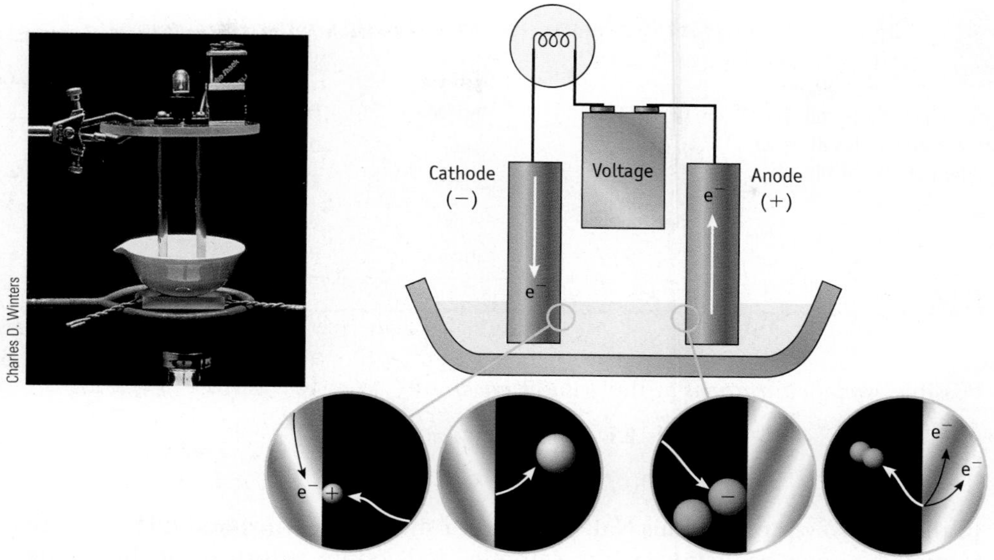

FIGURE 20.18 The preparation of sodium and chlorine by the electrolysis of molten NaCl. In the molten state, sodium ions migrate to the negative cathode, where they are reduced to sodium metal. Chloride ions migrate to the positive anode, where they are oxidized to elemental chlorine.

tial is applied to the electrodes, sodium ions are attracted to the negative electrode, and chloride ions are attracted to the positive electrode (Figure 20.18). If the potential is high enough, chemical reactions occur at each electrode. At the negative cathode, Na^+ ions accept electrons and are reduced to sodium metal (a liquid at this temperature). Simultaneously, at the positive anode, chloride ions give up electrons and form elemental chlorine.

Cathode (−), reduction:	$2\ Na^+ + 2\ e^- \rightarrow 2\ Na(\ell)$
Anode (+), oxidation:	$2\ Cl^- \rightarrow Cl_2(g) + 2\ e^-$
Net ionic equation:	$2\ Na^+ + 2\ Cl^- \rightarrow 2\ Na(\ell) + Cl_2(g)$

Electrons move through the external circuit under the force exerted by the applied potential, and the movement of positive and negative ions in the molten salt constitutes the current within the cell. Finally, it is important to recognize that the reaction is not spontaneous. The energy required for this reaction to occur has been provided by the electric current.

Electrolysis of Aqueous Solutions

Sodium ions (Na^+) and chloride ions (Cl^-) are the primary species present in molten NaCl. Only chloride ions can be oxidized, and only sodium ions can be reduced. Electrolyses of aqueous solutions are more complicated than electrolyses of molten salts, however, because water is now present. Water is an *electroactive* substance; that is, it can be oxidized or reduced in an electrochemical process.

Consider the electrolysis of aqueous sodium iodide (Figure 20.19). In this experiment, the electrolysis cell contains $Na^+(aq)$, $I^-(aq)$, and H_2O molecules. Possible *reduction reactions* at the *negative cathode* include

$$Na^+(aq) + e^- \rightarrow Na(s)$$

$$2\ H_2O(\ell) + 2\ e^- \rightarrow H_2(g) + 2\ OH^-(aq)$$

Problem Solving Tip 20.3 — Electrochemical Conventions: Voltaic Cells and Electrolysis Cells

Whether you are describing a voltaic cell or an electrolysis cell, the terms "anode" and "cathode" always refer to the electrodes at which oxidation and reduction occur, respectively. The polarity of the electrodes is reversed, however.

Type of Cell	Electrode	Function	Polarity
Voltaic	Anode	Oxidation	−
	Cathode	Reduction	+
Electrolysis	Anode	Oxidation	+
	Cathode	Reduction	−

Possible *oxidation reactions* at the *positive anode* are

$$2\ I^-(aq) \rightarrow I_2(aq) + 2\ e^-$$

$$2\ H_2O(\ell) \rightarrow O_2(g) + 4\ H^+(aq) + 4\ e^-$$

In the electrolysis of aqueous NaI, experiment shows that $H_2(g)$ and $OH^-(aq)$ are formed by water reduction at the cathode, and iodine is formed at the anode (Figure 20.19). Thus, the overall cell process can be summarized by the following equations:

Cathode (−), reduction:	$2\ H_2O(\ell) + 2\ e^- \rightarrow H_2(g) + 2\ OH^-(aq)$
Anode (+), oxidation:	$2\ I^-(aq) \rightarrow I_2(aq) + 2\ e^-$
Net ionic equation:	$2\ H_2O(\ell) + 2\ I^-(aq) \rightarrow H_2(g) + 2\ OH^-(aq) + I_2(aq)$

where $E°_{cell}$ has a negative value.

$$E°_{cell} = E°_{cathode} - E°_{anode} = (-0.8277\ V) - (+0.621\ V) = -1.449\ V$$

This process is not spontaneous under standard conditions, and a potential of *at least* 1.45 V must be *applied* to the cell for these reactions to occur. If the process had involved the oxidation of water instead of iodide ion at the anode, the required potential would be −2.057 V [$E°_{cathode} - E°_{anode} = (-0.8277\ V) - (+1.229\ V)$], and if the reaction involving the reduction of Na^+ and the oxidation of I^- had occurred, the

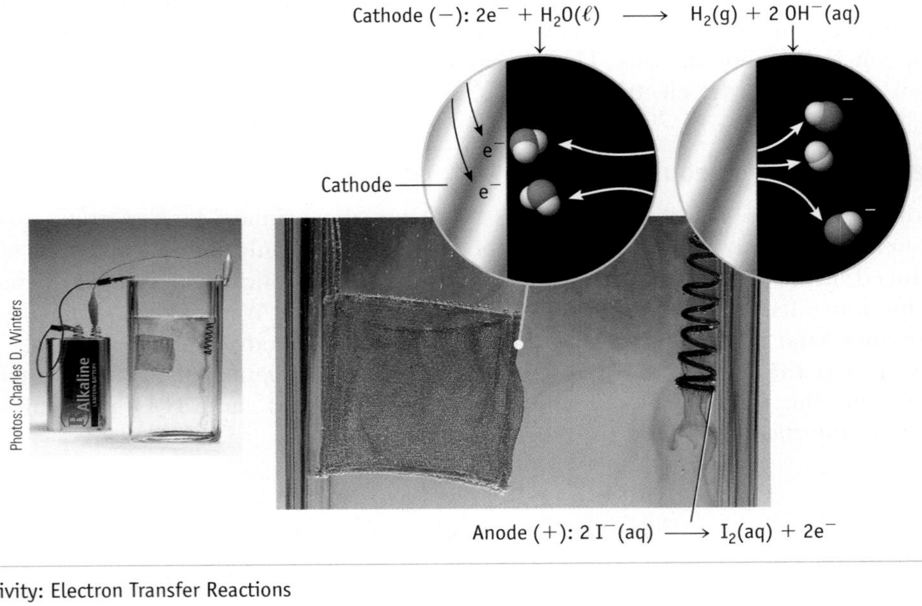

FIGURE 20.19 Electrolysis of aqueous NaI. A solution of NaI(aq) is electrolyzed, a potential applied using an external source of electricity. A drop of phenolphthalein has been added to the solution in this experiment so that the formation of $OH^-(aq)$ can be detected (by the red color of the indicator in basic solution). Iodine forms at the anode, and H_2 and OH^- form at the cathode.

required potential would be -3.335 V [$E°_{cathode} - E°_{anode} = (-2.714 \text{ V}) - (+0.621 \text{ V})$]. The reaction occurring is the one requiring the smaller applied potential, so the net cell reaction in the electrolysis of NaI(aq) is the oxidation of iodide and reduction of water.

What happens if an aqueous solution of some other metal halide such as $SnCl_2$ is electrolyzed? As before, consult Appendix M, and consider all possible half-reactions. In this case, aqueous Sn^{2+} ion is much more easily reduced ($E° = -0.14$ V) than water ($E° = -0.83$ V) at the cathode, so tin metal is produced. At the anode, two oxidations are possible: Cl^-(aq) to Cl_2(g) or H_2O to O_2(g). Experiments show that chloride ion is generally oxidized in preference to water, so the reactions occurring on electrolysis of aqueous tin(II) chloride are (Figure 20.20)

■ **Overvoltage** Voltages higher than the minimum are typically used to speed up reactions that would otherwise be slow. The term overvoltage is often used when describing experiments; this refers to the voltage needed to make a reaction occur at a reasonable rate.

Cathode (−), reduction: $Sn^{2+}(aq) + 2\,e^- \rightarrow Sn(s)$

Anode (+), oxidation: $2\,Cl^-(aq) \rightarrow Cl_2(g) + 2\,e^-$

Net ionic equation: $Sn^{2+}(aq) + 2\,Cl^-(aq) \rightarrow Sn(s) + Cl_2(g)$

$$E°_{cell} = E°_{cathode} - E°_{anode} = (-0.14 \text{ V}) - (+1.36 \text{ V}) = -1.50 \text{ V}$$

Formation of Cl_2 at the anode in the electrolysis of $SnCl_2$(aq) is contrary to a prediction based on $E°$ values. If the electrode reactions were

Cathode (−), reduction: $Sn^{2+}(aq) + 2\,e^- \rightarrow Sn(s)$

Anode (+), oxidation: $2\,H_2O(\ell) \rightarrow O_2(g) + 4\,H^+(aq) + 4\,e^-$

$$E°_{cell} = (-0.14 \text{ V}) - (+1.23 \text{ V}) = -1.37 \text{ V}$$

a smaller applied potential would seemingly be required. To explain the formation of chlorine instead of oxygen, we must take into account rates of reaction. This problem occurs in the commercially important electrolysis of aqueous NaCl, where a voltage high enough to oxidize both Cl^- and H_2O is used. However, because chloride ion is oxidized much faster than H_2O, the result is that Cl_2 is the major product in this electrolysis.

Another instance in which rates are important concerns electrode materials. Graphite, commonly used to make inert electrodes, can be oxidized. For the half-reaction $CO_2(g) + 4\,H^+(aq) + 4\,e^- \rightarrow C(s) + 2\,H_2O(\ell)$, $E°$ is $+0.20$ V, indicating that carbon is slightly easier to oxidize than copper ($E° = +0.34$ V). Based on this value, oxidation of a graphite electrode might reasonably be expected to occur during an electrolysis. And indeed it does, albeit slowly; graphite electrodes used in electrolysis cells slowly deteriorate and eventually have to be replaced.

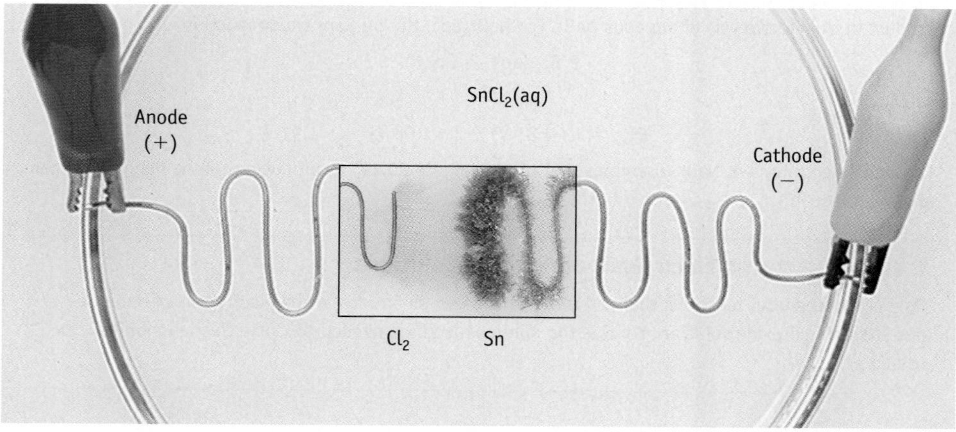

FIGURE 20.20 Electrolysis of aqueous tin(II) chloride. Tin metal collects at the negative cathode. Chlorine gas is formed at the positive anode. Elemental chlorine is formed in the cell, in spite of the fact that the potential for the oxidation of Cl^- is more negative than that for oxidation of water. (That is, chlorine should be less readily oxidized than water.) This is the result of chemical kinetics and illustrates the complexity of some aqueous electrochemistry.

One other factor—the concentration of electroactive species in solution—must be taken into account when discussing electrolyses. As shown in Section 20.6, the potential at which a species in solution is oxidized or reduced depends on concentration. Unless standard conditions are used, predictions based on $E°$ values are merely qualitative. In addition, the rate of a half-reaction depends on the concentration of the electroactive substance at the electrode surface. At a very low concentration, the rate of the redox reaction may depend on the rate at which an ion diffuses from the solution to the electrode surface.

Chemistry Now™

Sign in at **www.thomsonedu.com/login** and go to Chapter 20 Contents to see Screen 20.11 for an illustration of **water electrolysis.**

EXAMPLE 20.10 Electrolysis of Aqueous Solutions

Problem Predict how products of electrolyses of aqueous solutions of NaF, NaCl, NaBr, and NaI are likely to be different. (The electrolysis of NaI is illustrated in Figure 20.19.)

Strategy The main criterion used to predict the chemistry in an electrolytic cell should be the ease of oxidation and reduction, an assessment based on $E°$ values.

Solution The cathode reaction in all four examples presents no problem—water is reduced to hydroxide ion and H_2 gas in preference to reduction of $Na^+(aq)$ (as in the electrolysis of aqueous NaI). Thus, the primary cathode reaction in all cases is

$$2\ H_2O(\ell) + 2\ e^- \rightarrow H_2(g) + 2\ OH^-(aq)$$

$$E°_{cathode} = -0.83\ V$$

At the anode, we need to assess the ease of oxidation of the halide ions relative to water. Based on $E°$ values, the ease of oxidation of halide ions is $I^-(aq) > Br^-(aq) > Cl^-(aq) >> F^-(aq)$. Fluoride ion is much more difficult to oxidize than water, and electrolysis of an aqueous solution containing this ion results exclusively in O_2 formation. The primary anode reaction for NaF(aq) is

$$2\ H_2O(\ell) \rightarrow O_2(g) + 4\ H^+(aq) + 4\ e^-$$

Therefore, in this case,

$$E°_{cell} = (-0.83\ V) - (+1.23\ V) = -2.06\ V$$

Recall that chlorine is the primary product at the anode in the electrolysis of aqueous solutions of chloride salts. Therefore, the anode reaction is

$$2\ Cl^-(aq) \rightarrow Cl_2(g) + 2\ e^-$$

$$E°_{cell} = (-0.83\ V) - (+1.36\ V) = -2.19\ V$$

Bromide ions are considerably easier to oxidize than chloride ions. Br_2 may be expected as the primary product in the electrolysis of aqueous NaBr. For NaBr(aq), the primary anode reaction is

$$2\ Br^-(aq) \rightarrow Br_2(\ell) + 2\ e^-$$

so $E°_{cell}$ is

$$E°_{cell} = (-0.83\ V) - (+1.08\ V) = -1.91\ V$$

Thus, the electrolysis of NaBr resembles that of NaI (Figure 20.19) in producing the halogen, hydrogen, and hydroxide ion.

EXERCISE 20.14 Electrolysis of Aqueous Solutions

Predict the chemical reactions that will occur at the two electrodes in the electrolysis of an aqueous sodium hydroxide solution. What is the minimum voltage needed to cause this reaction to occur?

20.8 Counting Electrons

Metallic silver is produced at the cathode in the electrolysis of aqueous $AgNO_3$ in which one mole of electrons is required to produce one mole of silver. In contrast, two moles of electrons are required to produce one mole of tin (Figure 20.20):

$$Sn^{2+}(aq) + 2\ e^- \rightarrow Sn(s)$$

It follows that if the number of moles of electrons flowing through the electrolysis cell could be measured, the number of moles of silver or tin produced could be calculated. Conversely, if the amount of silver or tin produced is known, then the number of moles of electrons moving through the circuit could be calculated.

The number of moles of electrons consumed or produced in an electron transfer reaction is obtained by measuring the current flowing in the external electric circuit in a given time. The **current** flowing in an electrical circuit is the amount of charge (in units of coulombs, C) per unit time, and the usual unit for current is the ampere (A). (One ampere equals the passage of one coulomb of charge per second.)

$$\text{Current (amperes, A)} = \frac{\text{electric charge (coulombs, C)}}{\text{time, } t \text{ (seconds, s)}} \qquad \textbf{(20.8)}$$

The current passing through an electrochemical cell and the time for which the current flows are easily measured quantities. Therefore, the charge (in coulombs) that passes through a cell can be obtained by multiplying the current (in amperes) by the time (in seconds). Knowing the charge and using the Faraday constant as a conversion factor, we can calculate the number of moles of electrons that passed through an electrochemical cell. In turn, we can use this quantity to calculate the quantities of reactants and products. The following example illustrates this type of calculation.

Historical Perspectives — Electrochemistry and Michael Faraday

The terms anion, cation, electrode, and electrolyte originated with Michael Faraday (1791–1867), one of the most influential people in the history of chemistry. Faraday was apprenticed to a bookbinder in London when he was 13. This situation suited him perfectly, as he enjoyed reading the books sent to the shop for binding. By chance, one of these volumes was a small book on chemistry, which whetted his appetite for science, and he began performing experiments on electricity. In 1812, a patron of the shop invited Faraday to accompany him to the Royal Institute to attend a lecture by one of the most famous chemists of the day, Sir Humphry Davy. Faraday was so intrigued by Davy's lecture that he wrote to ask Davy for a position as an assistant. He was accepted and began work in 1813. Faraday was so talented that his work proved extraordinarily fruitful, and he was made the director of the laboratory of the Royal Institute about 12 years later.

It has been said that Faraday's contributions were so enormous that, had there been Nobel Prizes when he was alive, he would have received at least six. These prizes could have been awarded for discoveries such as the following:

- Electromagnetic induction, which led to the first transformer and electric motor
- The laws of electrolysis (the effect of electric current on chemicals)
- The magnetic properties of matter
- Benzene and other organic chemicals (which led to important chemical industries)
- The "Faraday effect" (the rotation of the plane of polarized light by a magnetic field)
- The introduction of the concept of electric and magnetic fields

Oesper Collection in the History of Chemistry, University of Cincinnati

Michael Faraday (1791–1867)

In addition to making discoveries that had profound effects on science, Faraday was an educator. He wrote and spoke about his work in memorable ways, especially in lectures to the general public that helped to popularize science.

EXAMPLE 20.11 Using the Faraday Constant

Problem A current of 2.40 A is passed through a solution containing Cu^{2+}(aq) for 30.0 minutes, with copper metal being deposited at the cathode. What mass of copper, in grams, is deposited?

Strategy A roadmap for this calculation is as follows:

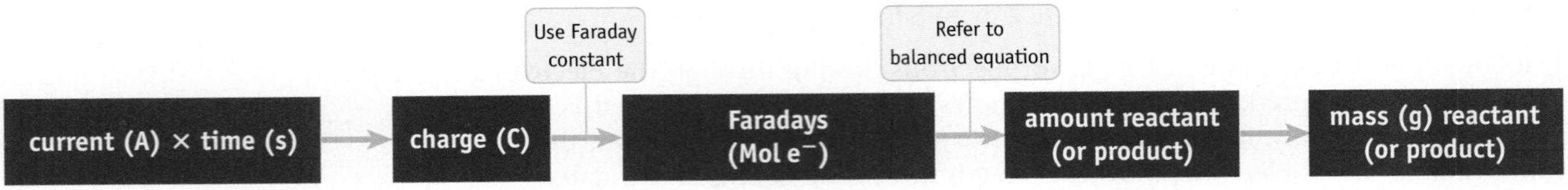

Solution

1. Calculate the charge (number of coulombs) passing through the cell in 30.0 min.

$$\begin{aligned}\text{Charge (C)} &= \text{current (A)} \times \text{time (s)} \\ &= (2.40\ \text{A})(30.0\ \text{min})(60.0\ \text{s/min}) \\ &= 4.32 \times 10^3\ \text{C}\end{aligned}$$

2. Calculate the number of moles of electrons (i.e., the number of Faradays of electricity).

$$(4.32 \times 10^3\ \text{C})\left(\frac{1\ \text{mol}\ e^-}{96{,}485\ \text{C}}\right) = 4.48 \times 10^{-2}\ \text{mol}\ e^-$$

3. Calculate the amount of copper and, from this, the mass of copper.

$$\text{mass of copper} = (4.48 \times 10^{-2}\ \text{mol}\ e^-)\left(\frac{1\ \text{mol Cu}}{2\ \text{mol}\ e^-}\right)\left(\frac{63.55\ \text{g Cu}}{1\ \text{mol Cu}}\right) = \mathbf{1.42\ g}$$

Comment The key relation in this calculation is current = charge/time. Most situations will involve knowing two of these three quantities from experiment and calculating the third.

■ **Faraday Constant** The Faraday constant is the charge carried by 1 mol of electrons:

$$9.6485338 \times 10^4\ \text{C/mol}\ e^-$$

EXERCISE 20.15 Using the Faraday Constant

Calculate the mass of O_2 produced in the electrolysis of water, using a current of 0.445 A for a period of 45 minutes.

EXERCISE 20.16 Using the Faraday Constant

In the commercial production of sodium by electrolysis, the cell operates at 7.0 V and a current of 25×10^3 A. What mass of sodium can be produced in one hour?

Chapter Goals Revisited

Now that you have studied this chapter, you should ask whether you have met the chapter goals. In particular, you should be able to:

Balance equations for oxidation–reduction reactions in acidic or basic solutions using the half-reaction approach Study Question(s) assignable in OWL: 2, 4-6, 16; Go Chemistry Module 25.

Understand the principles underlying voltaic cells

a. In a voltaic cell, identify the half-reactions occurring at the anode and the cathode, the polarity of the electrodes, the direction of electron flow in the external circuit, and the direction of ion flow in the salt bridge (Section 20.2). Study Question(s) assignable in OWL: 8, 10, 16, 51.

b. Appreciate the chemistry and advantages and disadvantages of dry cells, alkaline batteries, lead storage batteries, and Ni-cad batteries (Section 20.3).

c. Understand how fuel cells work, and recognize the difference between batteries and fuel cells (Section 20.3).

Understand how to use electrochemical potentials

a. Understand the process by which standard reduction potentials are determined, and identify standard conditions as applied to electrochemistry (Section 20.4).

b. Describe the standard hydrogen electrode ($E° = 0.00$ V), and explain how it is used as the standard to determine the standard potentials of half-reactions (Section 20.4).

c. Know how to use standard reduction potentials to determine cell voltages for cells under standard conditions (Equation 20.1). Study Question(s) assignable in OWL: 14, 16, 18, 56, 57.

d. Know how to use a table of standard reduction potentials (Table 20.1 and Appendix M) to rank the strengths of oxidizing and reducing agents, to predict which substances can reduce or oxidize another species, and to predict whether redox reactions will be product-favored or reactant-favored (Sections 20.4 and 20.5). Study Question(s) assignable in OWL: 20, 22, 24, 55, 56, 82.

e. Use the Nernst equation (Equations 20.2 and 20.3) to calculate the cell potential under nonstandard conditions (Section 20.5). Study Question(s) assignable in OWL: 26, 28, 30, 36.

f. Explain how cell voltage relates to ion concentration, and explain how this allows the determination of pH (Section 20.5) and other ion concentrations.

g. Use the relationships between cell voltage ($E°_{cell}$) and free energy ($\Delta_r G°$) (Equations 20.5 and 20.6) and between $E°_{cell}$ and an equilibrium constant for the cell reaction (Equation 20.7) (Section 20.6 and Table 20.2). Study Question(s) assignable in OWL: 32, 34, 60, 61, 63, 78, 86, 88.

Explore electrolysis, the use of electrical energy to produce chemical change

a. Describe the chemical processes occurring in an electrolysis. Recognize the factors that determine which substances are oxidized and reduced at the electrodes (Section 20.7). Study Question(s) assignable in OWL: 40, 41.

b. Relate the amount of a substance oxidized or reduced to the amount of current and the time the current flows (Section 20.8). Study Question(s) assignable in OWL: 44, 46, 48, 64-66, 73, 81.

Chemistry Now™ Sign in at **www.thomsonedu.com/login** to:

- Assess your understanding with Study Questions in OWL keyed to each goal in the Goals and Homework menu for this chapter
- For quick review, download Go Chemistry mini-lecture flashcard modules (or purchase them at **www.ichapters.com**)
- Check your readiness for an exam by taking the Pre-Test and exploring the modules recommended in your Personalized Study plan.

Access **How Do I Solve It?** tutorials on how to approach problem solving using concepts in this chapter.

KEY EQUATIONS

Equation 20.1 (page 917): Calculating a standard cell potential, $E°_{cell}$, from standard half-cell potentials.

$$E°_{cell} = E°_{cathode} - E°_{anode}$$

Equation 20.3 (page 925): Nernst equation (at 298 K).

$$E = E° - \frac{0.0257}{n} \ln Q$$

E is the cell potential under nonstandard conditions; n is the number of electrons transferred from the reducing agent to the oxidizing agent (according to the balanced equation); and Q is the reaction quotient.

Equation 20.6 (page 929): Relationship between standard free energy change and the standard cell potential.

$$\Delta G° = -nFE°$$

F is the Faraday constant, 96,485 C/mol e^-.

Equation 20.7 (page 929): Relationship between the equilibrium constant and the standard cell potential for a reaction (at 298 K).

$$\ln K = \frac{nE°}{0.0257}$$

Equation 20.8 (page 937): Relationship between current, electric charge, and time.

$$\text{Current (amperes, A)} = \frac{\text{electric charge (coulombs, C)}}{\text{time, } t \text{ (seconds, s)}}$$

STUDY QUESTIONS

Online homework for this chapter may be assigned in OWL.

▲ denotes challenging questions.

■ denotes questions assignable in OWL.

Blue-numbered questions have answers in Appendix O and fully-worked solutions in the *Student Solutions Manual.*

Practicing Skills

Balancing Equations for Oxidation–Reduction Reactions

(See Examples 20.1–20.3 and ChemistryNow Screen 20.3.)

When balancing the following redox equations, it may be necessary to add H^+ (aq) or H^+ (aq) plus H_2O for reactions in acid, and OH^- (aq) or OH^- (aq) plus H_2O for reactions in base.

1. Write balanced equations for the following half-reactions. Specify whether each is an oxidation or reduction.
   (a) $Cr(s) \rightarrow Cr^{3+}(aq)$ (in acid)
   (b) $AsH_3(g) \rightarrow As(s)$ (in acid)
   (c) $VO_3^-(aq) \rightarrow V^{2+}(aq)$ (in acid)
   (d) $Ag(s) \rightarrow Ag_2O(s)$ (in base)

2. ■ Write balanced equations for the following half-reactions. Specify whether each is an oxidation or reduction.
   (a) $H_2O_2(aq) \rightarrow O_2(g)$ (in acid)
   (b) $H_2C_2O_4(aq) \rightarrow CO_2(g)$ (in acid)
   (c) $NO_3^-(aq) \rightarrow NO(g)$ (in acid)
   (d) $MnO_4^-(aq) \rightarrow MnO_2(s)$ (in base)

3. Balance the following redox equations. All occur in acid solution.
   (a) $Ag(s) + NO_3^-(aq) \rightarrow NO_2(g) + Ag^+(aq)$
   (b) $MnO_4^-(aq) + HSO_3^-(aq) \rightarrow Mn^{2+}(aq) + SO_4^{2-}(aq)$
   (c) $Zn(s) + NO_3^-(aq) \rightarrow Zn^{2+}(aq) + N_2O(g)$
   (d) $Cr(s) + NO_3^-(aq) \rightarrow Cr^{3+}(aq) + NO(g)$

4. ■ Balance the following redox equations. All occur in acid solution.
   (a) $Sn(s) + H^+(aq) \rightarrow Sn^{2+}(aq) + H_2(g)$
   (b) $Cr_2O_7^{2-}(aq) + Fe^{2+}(aq) \rightarrow Cr^{3+}(aq) + Fe^{3+}(aq)$
   (c) $MnO_2(s) + Cl^-(aq) \rightarrow Mn^{2+}(aq) + Cl_2(g)$
   (d) $CH_2O(aq) + Ag^+(aq) \rightarrow HCO_2H(aq) + Ag(s)$

5. ■ Balance the following redox equations. All occur in basic solution.
   (a) $Al(s) + H_2O(\ell) \rightarrow Al(OH)_4^-(aq) + H_2(g)$
   (b) $CrO_4^{2-}(aq) + SO_3^{2-}(aq) \rightarrow Cr(OH)_3(s) + SO_4^{2-}(aq)$
   (c) $Zn(s) + Cu(OH)_2(s) \rightarrow [Zn(OH)_4]^{2-}(aq) + Cu(s)$
   (d) $HS^-(aq) + ClO_3^-(aq) \rightarrow S(s) + Cl^-(aq)$

6. ■ Balance the following redox equations. All occur in basic solution.
(a) $Fe(OH)_3(s) + Cr(s) \longrightarrow Cr(OH)_3(s) + Fe(OH)_2(s)$
(b) $NiO_2(s) + Zn(s) \longrightarrow Ni(OH)_2(s) + Zn(OH)_2(s)$
(c) $Fe(OH)_2(s) + CrO_4^{2-}(aq) \longrightarrow Fe(OH)_3(s) + [Cr(OH)_4]^-(aq)$
(d) $N_2H_4(aq) + Ag_2O(s) \longrightarrow N_2(g) + Ag(s)$

Constructing Voltaic Cells

(See Example 20.4 and ChemistryNow Screen 20.4.)

7. A voltaic cell is constructed using the reaction of chromium metal and iron(II) ion.

$$2\ Cr(s) + 3\ Fe^{2+}(aq) \longrightarrow 2\ Cr^{3+}(aq) + 3\ Fe(s)$$

Complete the following sentences: Electrons in the external circuit flow from the ___ electrode to the ___ electrode. Negative ions move in the salt bridge from the ___ half-cell to the ___ half-cell. The half-reaction at the anode is ___, and that at the cathode is ___.

8. ■ A voltaic cell is constructed using the reaction

$$Mg(s) + 2\ H^+(aq) \longrightarrow Mg^{2+}(aq) + H_2(g)$$

(a) Write equations for the oxidation and reduction half-reactions.
(b) Which half-reaction occurs in the anode compartment, and which occurs in the cathode compartment?
(c) Complete the following sentences: Electrons in the external circuit flow from the ___ electrode to the ___ electrode. Negative ions move in the salt bridge from the ___ half-cell to the ___ half-cell. The half-reaction at the anode is ___, and that at the cathode is ___.

9. The half-cells $Fe^{2+}(aq) \mid Fe(s)$ and $O_2(g) \mid H_2O$ (in acid solution) are linked to create a voltaic cell.
(a) Write equations for the oxidation and reduction half-reactions and for the overall (cell) reaction.
(b) Which half-reaction occurs in the anode compartment, and which occurs in the cathode compartment?
(c) Complete the following sentences: Electrons in the external circuit flow from the ___ electrode to the ___ electrode. Negative ions move in the salt bridge from the ___ half-cell to the ___ half-cell.

10. ■ The half-cells $Sn^{2}(aq) \mid Sn(s)$ and $Cl_2(g) \mid Cl^-(aq)$ are linked to create a voltaic cell.
(a) Write equations for the oxidation and reduction half-reactions and for the overall (cell) reaction.
(b) Which half-reaction occurs in the anode compartment, and which occurs in the cathode compartment?
(c) Complete the following sentences: Electrons in the external circuit flow from the ___ electrode to the ___ electrode. Negative ions move in the salt bridge from the ___ half-cell to the ___ half-cell.

Commercial Cells

11. What are the similarities and differences between dry cells, alkaline batteries, and Ni-cad batteries?

12. What reactions occur when a lead storage battery is recharged?

Standard Electrochemical Potentials

(See Example 20.5 and ChemistryNow Screens 20.6 and 20.7.)

13. Calculate the value of $E°$ for each of the following reactions. Decide whether each is product-favored in the direction written.
(a) $2\ I^-(aq) + Zn^{2+}(aq) \longrightarrow I_2(s) + Zn(s)$
(b) $Zn^{2+}(aq) + Ni(s) \longrightarrow Zn(s) + Ni^{2+}(aq)$
(c) $2\ Cl^-(aq) + Cu^{2+}(aq) \longrightarrow Cu(s) + Cl_2(g)$
(d) $Fe^{2+}(aq) + Ag^+(aq) \longrightarrow Fe^{3+}(aq) + Ag(s)$

14. ■ Calculate the value of $E°$ for each of the following reactions. Decide whether each is product-favored in the direction written. [Reaction (d) occurs in basic solution.]
(a) $Br_2(\ell) + Mg(s) \longrightarrow Mg^{2+}(aq) + 2\ Br^-(aq)$
(b) $Zn^{2+}(aq) + Mg(s) \longrightarrow Zn(s) + Mg^{2+}(aq)$
(c) $Sn^{2+}(aq) + 2\ Ag^+(aq) \longrightarrow Sn^{4+}(aq) + 2\ Ag(s)$
(d) $2\ Zn(s) + O_2(g) + 2\ H_2O(\ell) + 4\ OH^-(aq) \longrightarrow 2\ [Zn(OH)_4]^{2-}(aq)$

15. Balance each of the following unbalanced equations; then calculate the standard potential, $E°$, and decide whether each is product-favored as written. (All reactions occur in acid solution.)
(a) $Sn^{2+}(aq) + Ag(s) \longrightarrow Sn(s) + Ag^+(aq)$
(b) $Al(s) + Sn^{4+}(aq) \longrightarrow Sn^{2+}(aq) + Al^{3+}(aq)$
(c) $ClO_3^-(aq) + Ce^{3+}(aq) \longrightarrow Cl_2(g) + Ce^{4+}(aq)$
(d) $Cu(s) + NO_3^-(aq) \longrightarrow Cu^{2+}(aq) + NO(g)$

16. ■ Balance each of the following unbalanced equations; then calculate the standard potential, $E°$, and decide whether each is product-favored as written. (All reactions occur in acid solution.)
(a) $I_2(s) + Br^-(aq) \longrightarrow I^-(aq) + Br_2(\ell)$
(b) $Fe^{2+}(aq) + Cu^{2+}(aq) \longrightarrow Cu(s) + Fe^{3+}(aq)$
(c) $Fe^{2+}(aq) + Cr_2O_7^{2-}(aq) \longrightarrow Fe^{3+}(aq) + Cr^{3+}(aq)$
(d) $MnO_4^-(aq) + HNO_2(aq) \longrightarrow Mn^{2+}(aq) + NO_3^-(aq)$

17. Consider the following half-reactions:

Half-Reaction	$E°$ (V)
$Cu^{2+}(aq) + 2\ e^- \longrightarrow Cu(s)$	+0.34
$Sn^{2+}(aq) + 2\ e^- \longrightarrow Sn(s)$	−0.14
$Fe^{2+}(aq) + 2\ e^- \longrightarrow Fe(s)$	−0.44
$Zn^{2+}(aq) + 2\ e^- \longrightarrow Zn(s)$	−0.76
$Al^{3+}(aq) + 3\ e^- \longrightarrow Al(s)$	−1.66

(a) Based on $E°$ values, which metal is the most easily oxidized?
(b) Which metals on this list are capable of reducing $Fe^{2+}(aq)$ to $Fe(s)$?
(c) Write a balanced chemical equation for the reaction of $Fe^{2+}(aq)$ with $Sn(s)$. Is this reaction product-favored or reactant-favored?
(d) Write a balanced chemical equation for the reaction of $Zn^{2+}(aq)$ with $Sn(s)$. Is this reaction product-favored or reactant-favored?

18. ■ Consider the following half-reactions:

Half-Reaction	$E°$ (V)
$MnO_4^-(aq) + 8\ H^+(aq) + 5\ e^- \longrightarrow Mn^{2+}(aq) + 4\ H_2O(\ell)$	+1.51
$BrO_3^-(aq) + 6\ H^+(aq) + 6\ e^- \longrightarrow Br^-(aq) + 3\ H_2O(\ell)$	+1.47
$Cr_2O_7^{2-}(aq) + 14\ H^+(aq) + 6\ e^- \longrightarrow 2\ Cr^{3+}(aq) + 7\ H_2O(\ell)$	+1.33
$NO_3^-(aq) + 4\ H^+(aq) + 3\ e^- \longrightarrow NO(g) + 2\ H_2O(\ell)$	+0.96
$SO_4^{2-}(aq) + 4\ H^+(aq) + 2\ e^- \longrightarrow SO_2(g) + 2\ H_2O(\ell)$	+0.20

(a) Choosing from among the reactants in these half-reactions, identify the strongest and weakest oxidizing agents.
(b) Which of the oxidizing agents listed is (are) capable of oxidizing $Br^-(aq)$ to $BrO_3^-(aq)$ (in acid solution)?
(c) Write a balanced chemical equation for the reaction of $Cr_2O_7^{2-}(aq)$ with $SO_2(g)$ in acid solution. Is this reaction product-favored or reactant-favored?
(d) Write a balanced chemical equation for the reaction of $Cr_2O_7^{2-}(aq)$ with $Mn^{2+}(aq)$. Is this reaction product-favored or reactant-favored?

Ranking Oxidizing and Reducing Agents

(See Example 20.5 and ChemistryNow Screen 20.7.) Use a table of standard reduction potentials (Table 20.1 or Appendix M) to answer Study Questions 19–24.

19. Which of the following elements is the best reducing agent under standard conditions?
(a) Cu (d) Ag
(b) Zn (e) Cr
(c) Fe

20. ■ From the following list, identify those elements that are easier to oxidize than $H_2(g)$.
(a) Cu (d) Ag
(b) Zn (e) Cr
(c) Fe

21. Which of the following ions is most easily reduced?
(a) $Cu^{2+}(aq)$ (d) $Ag^+(aq)$
(b) $Zn^{2+}(aq)$ (e) $Al^{3+}(aq)$
(c) $Fe^{2+}(aq)$

22. ■ From the following list, identify the ions that are more easily reduced than $H^+(aq)$.
(a) $Cu^{2+}(aq)$ (d) $Ag^+(aq)$
(b) $Zn^{2+}(aq)$ (e) $Al^{3+}(aq)$
(c) $Fe^{2+}(aq)$

23. (a) Which halogen is most easily reduced: F_2, Cl_2, Br_2, or I_2 in acidic solution?
(b) Identify the halogens that are better oxidizing agents than $MnO_2(s)$ in acidic solution.

24. ■ (a) Which ion is most easily oxidized to the elemental halogen: F^-, Cl^-, Br^-, or I^- in acidic solution?
(b) Identify the halide ions that are more easily oxidized than $H_2O(\ell)$ in acidic solution.

Electrochemical Cells Under Nonstandard Conditions

(See Examples 20.6 and 20.7 and ChemistryNow Screen 20.8.)

25. Calculate the potential delivered by a voltaic cell using the following reaction if all dissolved species are 2.5×10^{-2} M and the pressure of H_2 is 1.0 bar.

$$Zn(s) + 2\ H_2O(\ell) + 2\ OH^-(aq) \longrightarrow [Zn(OH)_4]^{2-}(aq) + H_2(g)$$

26. ■ Calculate the potential developed by a voltaic cell using the following reaction if all dissolved species are 0.015 M.

$$2\ Fe^{2+}(aq) + H_2O_2(aq) + 2\ H^+(aq) \longrightarrow 2\ Fe^{3+}(aq) + 2\ H_2O(\ell)$$

27. One half-cell in a voltaic cell is constructed from a silver wire dipped into a 0.25 M solution of $AgNO_3$. The other half-cell consists of a zinc electrode in a 0.010 M solution of $Zn(NO_3)_2$. Calculate the cell potential.

28. ■ One half-cell in a voltaic cell is constructed from a copper wire dipped into a 4.8×10^{-3} M solution of $Cu(NO_3)_2$. The other half-cell consists of a zinc electrode in a 0.40 M solution of $Zn(NO_3)_2$. Calculate the cell potential.

29. One half-cell in a voltaic cell is constructed from a silver wire dipped into a $AgNO_3$ solution of unknown concentration. The other half-cell consists of a zinc electrode in a 1.0 M solution of $Zn(NO_3)_2$. A potential of 1.48 V is measured for this cell. Use this information to calculate the concentration of $Ag^+(aq)$.

30. ■ One half-cell in a voltaic cell is constructed from an iron wire dipped into an $Fe(NO_3)_2$ solution of unknown concentration. The other half-cell is a standard hydrogen electrode. A potential of 0.49 V is measured for this cell. Use this information to calculate the concentration of $Fe^{2+}(aq)$.

Electrochemistry, Thermodynamics, and Equilibrium

(See Examples 20.8 and 20.9 and ChemistryNow Screen 20.9.)

31. Calculate $\Delta_r G°$ and the equilibrium constant for the following reactions.
    (a) $2\ Fe^{3+}(aq) + 2\ I^-(aq) \rightleftharpoons 2\ Fe^{2+}(aq) + I_2(aq)$
    (b) $I_2(aq) + 2\ Br^-(aq) \rightleftharpoons 2\ I^-(aq) + Br_2(aq)$

32. ■ Calculate $\Delta_r G°$ and the equilibrium constant for the following reactions.
    (a) $Zn^{2+}(aq) + Ni(s) \rightleftharpoons Zn(s) + Ni^{2+}(aq)$
    (b) $Cu(s) + 2\ Ag^+(aq) \rightleftharpoons Cu^{2+}(aq) + 2\ Ag(s)$

33. Use standard reduction potentials (Appendix M) for the half-reactions $AgBr(s) + e^- \longrightarrow Ag(s) + Br^-(aq)$ and $Ag^+(aq) + e^- \longrightarrow Ag(s)$ to calculate the value of K_{sp} for AgBr.

34. ■ Use the standard reduction potentials (Appendix M) for the half-reactions $Hg_2Cl_2(s) + 2\ e^- \longrightarrow 2\ Hg(\ell) + 2\ Cl^-(aq)$ and $Hg_2^{\ 2+}(aq) + 2\ e^- \longrightarrow 2\ Hg(\ell)$ to calculate the value of K_{sp} for Hg_2Cl_2.

35. Use the standard reduction potentials (Appendix M) for the half-reactions $[AuCl_4]^-(aq) + 3\ e^- \longrightarrow Au(s) + 4\ Cl^-(aq)$ and $Au^{3+}(aq) + 3\ e^- \longrightarrow Au(s)$ to calculate the value of $K_{formation}$ for the complex ion $[AuCl_4]^-(aq)$.

36. ■ Use the standard reduction potentials (Appendix M) for the half-reactions $[Zn(OH)_4]^{2-}(aq) + 2\ e^- \longrightarrow Zn\ (s) + 4\ OH^-(aq)$ and $Zn^{2+}(aq) + 2\ e^- \longrightarrow Zn(s)$ to calculate the value of $K_{formation}$ for the complex ion $[Zn(OH)_4]^{2-}$.

Electrolysis

(See Section 20.7, Example 20.10, and ChemistryNow Screen 20.10.)

37. Diagram the apparatus used to electrolyze molten NaCl. Identify the anode and the cathode. Trace the movement of electrons through the external circuit and the movement of ions in the electrolysis cell.

38. Diagram the apparatus used to electrolyze aqueous $CuCl_2$. Identify the reaction products, the anode, and the cathode. Trace the movement of electrons through the external circuit and the movement of ions in the electrolysis cell.

39. Which product, O_2 or F_2, is more likely to form at the anode in the electrolysis of an aqueous solution of KF? Explain your reasoning.

40. ■ Which product, Ca or H_2, is more likely to form at the cathode in the electrolysis of $CaCl_2$? Explain your reasoning.

41. ■ An aqueous solution of KBr is placed in a beaker with two inert platinum electrodes. When the cell is attached to an external source of electrical energy, electrolysis occurs.
    (a) Hydrogen gas and hydroxide ion form at the cathode. Write an equation for the half-reaction that occurs at this electrode.
    (b) Bromine is the primary product at the anode. Write an equation for its formation.

42. An aqueous solution of Na_2S is placed in a beaker with two inert platinum electrodes. When the cell is attached to an external battery, electrolysis occurs.
    (a) Hydrogen gas and hydroxide ion form at the cathode. Write an equation for the half-reaction that occurs at this electrode.
    (b) Sulfur is the primary product at the anode. Write an equation for its formation.

Counting Electrons

(See Example 20.11 and ChemistryNow Screen 20.12.)

43. In the electrolysis of a solution containing $Ni^{2+}(aq)$, metallic Ni(s) deposits on the cathode. Using a current of 0.150 A for 12.2 min, what mass of nickel will form?

44. ■ In the electrolysis of a solution containing $Ag^+(aq)$, metallic Ag(s) deposits on the cathode. Using a current of 1.12 A for 2.40 h, what mass of silver forms?

45. Electrolysis of a solution of $CuSO_4(aq)$ to give copper metal is carried out using a current of 0.66 A. How long should electrolysis continue to produce 0.50 g of copper?

46. ■ Electrolysis of a solution of $Zn(NO_3)_2(aq)$ to give zinc metal is carried out using a current of 2.12 A. How long should electrolysis continue in order to prepare 2.5 g of zinc?

47. A voltaic cell can be built using the reaction between Al metal and O_2 from the air. If the Al anode of this cell consists of 84 g of aluminum, how many hours can the cell produce 1.0 A of electricity, assuming an unlimited supply of O_2?

48. ■ Assume the specifications of a Ni-Cd voltaic cell include delivery of 0.25 A of current for 1.00 h. What is the minimum mass of the cadmium that must be used to make the anode in this cell?

General Questions

These questions are not designated as to type or location in the chapter. They may combine several concepts.

49. Write balanced equations for the following half-reactions.
    (a) $UO_2^{\ +}(aq) \longrightarrow U^{4+}(aq)$ (acid solution)
    (b) $ClO_3^{\ -}(aq) \longrightarrow Cl^-(aq)$ (acid solution)
    (c) $N_2H_4(aq) \longrightarrow N_2(g)$ (basic solution)
    (d) $ClO^-(aq) \longrightarrow Cl^-(aq)$ (basic solution)

50. Balance the following equations.
(a) $Zn(s) + VO^{2+}(aq) \rightarrow Zn^{2+}(aq) + V^{3+}(aq)$ (acid solution)
(b) $Zn(s) + VO_3^-(aq) \rightarrow V^{2+}(aq) + Zn^{2+}(aq)$ (acid solution)
(c) $Zn(s) + ClO^-(aq) \rightarrow Zn(OH)_2(s) + Cl^-(aq)$ (basic solution)
(d) $ClO^-(aq) + [Cr(OH)_4]^-(aq) \rightarrow Cl^-(aq) + CrO_4^{2-}(aq)$ (basic solution)

51. ■ Magnesium metal is oxidized, and silver ions are reduced in a voltaic cell using Mg^{2+}(aq, 1 M) | Mg and Ag^+(aq, 1 M) | Ag half-cells.

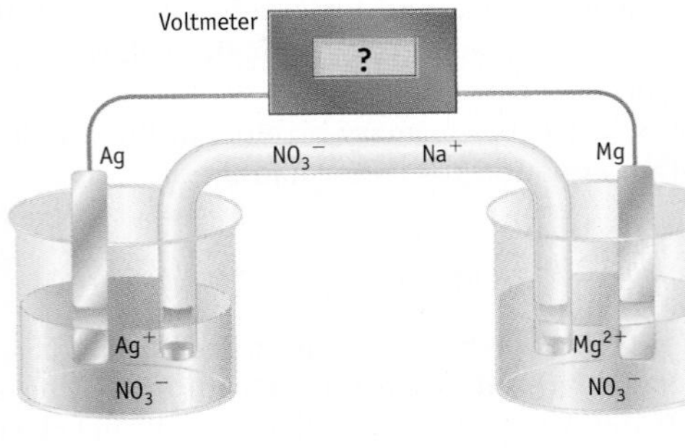

(a) Label each part of the cell.
(b) Write equations for the half-reactions occurring at the anode and the cathode, and write an equation for the net reaction in the cell.
(c) Trace the movement of electrons in the external circuit. Assuming the salt bridge contains $NaNO_3$, trace the movement of the Na^+ and NO_3^- ions in the salt bridge that occurs when a voltaic cell produces current. Why is a salt bridge required in a cell?

52. You want to set up a series of voltaic cells with specific cell voltages. A Zn^{2+}(aq, 1.0 M) | Zn(s) half-cell is in one compartment. Identify several half-cells that you could use so that the cell potential will be close to (a) 1.1 V and (b) 0.50 V. Consider cells in which zinc can be either the cathode or the anode.

53. You want to set up a series of voltaic cells with specific cell potentials. The Ag^+(aq, 1.0 M) | Ag(s) half-cell is one of the compartments. Identify several half-cells that you could use so that the cell potential will be close to (a) 1.7 V and (b) 0.50 V. Consider cells in which silver can be either the cathode or the anode.

54. Which of the following reactions is (are) product-favored?
(a) $Zn(s) + I_2(s) \rightarrow Zn^{2+}(aq) + 2\,I^-(aq)$
(b) $2\,Cl^-(aq) + I_2(s) \rightarrow Cl_2(g) + 2\,I^-(aq)$
(c) $2\,Na^+(aq) + 2\,Cl^-(aq) \rightarrow 2\,Na(s) + Cl_2(g)$
(d) $2\,K(s) + H_2O(\ell) \rightarrow 2\,K^+(aq) + H_2(g) + 2\,OH^-(aq)$

55. ■ In the table of standard reduction potentials, locate the half-reactions for the reductions of the following metal ions to the metal: $Sn^{2+}(aq)$, $Au^+(aq)$, $Zn^{2+}(aq)$, $Co^{2+}(aq)$, $Ag^+(aq)$, $Cu^{2+}(aq)$. Among the metal ions and metals that make up these half-reactions:
(a) Which metal ion is the weakest oxidizing agent?
(b) Which metal ion is the strongest oxidizing agent?
(c) Which metal is the strongest reducing agent?
(d) Which metal is the weakest reducing agent?
(e) Will Sn(s) reduce $Cu^{2+}(aq)$ to Cu(s)?
(f) Will Ag(s) reduce $Co^{2+}(aq)$ to Co(s)?
(g) Which metal ions on the list can be reduced by Sn(s)?
(h) What metals can be oxidized by $Ag^+(aq)$?

56. ■ ▲ In the table of standard reduction potentials, locate the half-reactions for the reductions of the following nonmetals: F_2, Cl_2, Br_2, I_2 (reduction to halide ions), and O_2, S, Se (reduction to H_2X in aqueous acid). Among the elements, ions, and compounds that make up these half-reactions:
(a) Which element is the weakest oxidizing agent?
(b) Which element is the weakest reducing agent?
(c) Which of the elements listed is (are) capable of oxidizing H_2O to O_2?
(d) Which of these elements listed is (are) capable of oxidizing H_2S to S?
(e) Is O_2 capable of oxidizing I^- to I_2, in acid solution?
(f) Is S capable of oxidizing I^- to I_2?
(g) Is the reaction $H_2S(aq) + Se(s) \rightarrow H_2Se(aq) + S(s)$ product-favored?
(h) Is the reaction $H_2S(aq) + I_2(s) \rightarrow 2\,H^+(aq) + 2\,I^-(aq) + S(s)$ product-favored?

57. ■ Four voltaic cells are set up. In each, one half-cell contains a standard hydrogen electrode. The second half-cell is one of the following: Cr^{3+}(aq, 1.0 M) | Cr(s), Fe^{2+}(aq, 1.0 M) | Fe(s), Cu^{2+}(aq, 1.0 M) | Cu(s), or Mg^{2+}(aq, 1.0 M) | Mg(s).
(a) In which of the voltaic cells does the hydrogen electrode serve as the cathode?
(b) Which voltaic cell produces the highest voltage? Which produces the lowest voltage?

58. The following half-cells are available: Ag^+(aq, 1.0 M) | Ag(s), Zn^{2+}(aq, 1.0 M) | Zn(s), Cu^{2+}(aq, 1.0 M) | Cu(s), and Co^{2+}(aq, 1.0 M) | Co(s). Linking any two half-cells makes a voltaic cell. Given four different half-cells, six voltaic cells are possible. These are labeled, for simplicity, Ag-Zn, Ag-Cu, Ag-Co, Zn-Cu, Zn-Co, and Cu-Co.
(a) In which of the voltaic cells does the copper electrode serve as the cathode? In which of the voltaic cells does the cobalt electrode serve as the anode?
(b) Which combination of half-cells generates the highest potential? Which combination generates the lowest potential?

▲ more challenging ■ in OWL Blue-numbered questions answered in Appendix O

59. The reaction occurring in the cell in which Al_2O_3 and aluminum salts are electrolyzed is $Al^{3+}(aq) + 3\ e^- \rightarrow Al(s)$. If the electrolysis cell operates at 5.0 V and 1.0×10^5 A, what mass of aluminum metal can be produced in a 24-h day?

60. ■ ▲ A cell is constructed using the following half-reactions:

$$Ag^+(aq) + e^- \rightarrow Ag(s)$$

$$Ag_2SO_4(s) + 2\ e^- \rightarrow 2\ Ag(s) + SO_4^{2-}(aq) \quad E° = 0.653\ V$$

(a) What reactions should be observed at the anode and cathode?
(b) Calculate the solubility product constant, K_{sp}, for Ag_2SO_4.

61. ■ ▲ A potential of 0.142 V is recorded (under standard conditions) for a voltaic cell constructed using the following half reactions:

Cathode: $Pb^{2+}(aq) + 2\ e^- \rightarrow Pb(s)$

Anode: $PbCl_2(s) + 2\ e^- \rightarrow Pb(s) + 2\ Cl^-(aq)$

Net: $Pb^{2+}(aq) + 2\ Cl^-(aq) \rightarrow PbCl_2(s)$

(a) What is the standard reduction potential for the anode reaction?
(b) Estimate the solubility product, K_{sp}, for $PbCl_2$.

62. What is the value of $E°$ for the following half-reaction?

$$Ag_2CrO_4(s) + 2\ e^- \rightarrow 2\ Ag(s) + CrO_4^{2-}(aq)$$

63. ■ The standard voltage, $E°$, for the reaction of Zn(s) and Cl_2(g) is +2.12 V. What is the standard free energy change, $\Delta_r G°$, for the reaction?

64. ■ ▲ An electrolysis cell for aluminum production operates at 5.0 V and a current of 1.0×10^5 A. Calculate the number of kilowatt-hours of energy required to produce 1 metric ton (1.0×10^3 kg) of aluminum. (1 kWh = 3.6×10^6 J and 1 J = 1 C · V.)

65. ■ ▲ Electrolysis of molten NaCl is done in cells operating at 7.0 V and 4.0×10^4 A. What mass of Na(s) and Cl_2(g) can be produced in one day in such a cell? What is the energy consumption in kilowatt-hours? (1 kWh = 3.6×10^6 J and 1 J = 1 C · V.)

66. ■ ▲ A current of 0.0100 A is passed through a solution of rhodium sulfate, causing reduction of the metal ion to the metal. After 3.00 h, 0.038 g of Rh has been deposited. What is the charge on the rhodium ion, Rh^{n+}? What is the formula for rhodium sulfate?

67. ▲ A current of 0.44 A is passed through a solution of ruthenium nitrate causing reduction of the metal ion to the metal. After 25.0 min, 0.345 g of Ru has been deposited. What is the charge on the ruthenium ion, Ru^{n+}? What is the formula for ruthenium nitrate?

68. The total charge that can be delivered by a large dry cell battery before its voltage drops too low is usually about 35 amp-hours. (One amp-hour is the charge that passes through a circuit when 1 A flows for 1 h.) What mass of Zn is consumed when 35 amp-hours are drawn from the cell?

69. Chlorine gas is obtained commercially by electrolysis of brine (a concentrated aqueous solution of NaCl). If the electrolysis cells operate at 4.6 V and 3.0×10^5 A, what mass of chlorine can be produced in a 24-h day?

70. ▲ Write balanced equations for the following reduction half-reactions involving organic compounds.
(a) $HCO_2H \rightarrow CH_2O$ (acid solution)
(b) $C_6H_5CO_2H \rightarrow C_6H_5CH_3$ (acid solution)
(c) $CH_3CH_2CHO \rightarrow CH_3CH_2CH_2OH$ (acid solution)
(d) $CH_3OH \rightarrow CH_4$ (acid solution)

71. ▲ Balance the following equations involving organic compounds.
(a) $Ag^+(aq) + C_6H_5CHO(aq) \rightarrow Ag(s) + C_6H_5CO_2H(aq)$ (acid solution)
(b) $CH_3CH_2OH + Cr_2O_7^{2-}(aq) \rightarrow CH_3CO_2H(aq) + Cr^{3+}(aq)$ (acid solution)

72. A voltaic cell is constructed in which one half-cell consists of a silver wire in an aqueous solution of $AgNO_3$. The other half-cell consists of an inert platinum wire in an aqueous solution containing $Fe^{2+}(aq)$ and $Fe^{3+}(aq)$.
(a) Calculate the cell potential, assuming standard conditions.
(b) Write the net ionic equation for the reaction occurring in the cell.
(c) In this voltaic cell, which electrode is the anode, and which is the cathode?
(d) If $[Ag^+]$ is 0.10 M, and $[Fe^{2+}]$ and $[Fe^{3+}]$ are both 1.0 M, what is the cell potential? Is the net cell reaction still that used in part (a)? If not, what is the net reaction under the new conditions?

73. ■ An expensive but lighter alternative to the lead storage battery is the silver-zinc battery.

$$Ag_2O(s) + Zn(s) + H_2O(\ell) \rightarrow Zn(OH)_2(s) + 2\ Ag(s)$$

The electrolyte is 40% KOH, and silver–silver oxide electrodes are separated from zinc–zinc hydroxide electrodes by a plastic sheet that is permeable to hydroxide ions. Under normal operating conditions, the battery has a potential of 1.59 V.
(a) How much energy can be produced per gram of reactants in the silver-zinc battery? Assume the battery produces a current of 0.10 A.
(b) How much energy can be produced per gram of reactants in the standard lead storage battery? Assume the battery produces a current of 0.10 A at 2.0 V.
(c) Which battery (silver-zinc or lead storage) produces the greater energy per gram of reactants?

74. The specifications for a lead storage battery include delivery of a steady 1.5 A of current for 15 h.
(a) What is the minimum mass of lead that will be used in the anode?
(b) What mass of PbO_2 must be used in the cathode?
(c) Assume that the volume of the battery is 0.50 L. What is the minimum molarity of H_2SO_4 necessary?

75. Manganese may play an important role in chemical cycles in the oceans (page 932). Two reactions involving manganese are the reduction of nitrate ions (to NO) with Mn^{2+} ions and the oxidation of ammonia (to N_2) with MnO_2.
(a) Write balanced chemical equations for these reactions (in acid solution).
(b) Calculate $E°_{cell}$ for the reactions. (One half-reaction potential you need is for reduction of N_2 to NH_4^+, $E° = -0.272$ V.)

76. ▲ You want to use electrolysis to plate a cylindrical object (radius = 2.50 and length = 20.00 cm) with a coating of nickel metal, 4.0 mm thick. You place the object in a bath containing a salt (Na_2SO_4). One electrode is impure nickel, and the other is the object to be plated. The electrolyzing potential is 2.50 V.
(a) Which is the anode, and which is the cathode in the experiment? What half-reaction occurs at each electrode?
(b) Calculate the number of kilowatt-hours (kWh) of energy required to carry out the electrolysis. (1 kWh = 3.6×10^6 J and 1 J = 1 C × 1 V).

77. ▲ Iron(II) ion undergoes a disproportionation reaction to give Fe(s) and the iron(III) ion. That is, iron(II) ion is both oxidized and reduced within the same reaction.

$$3\ Fe^{2+}(aq) \rightarrow Fe(s) + 2\ Fe^{3+}(aq)$$

(a) What two half-reactions make up the disproportionation reaction?
(b) Use the values of the standard reduction potentials for the two half-reactions in part (a) to determine whether this disproportionation reaction is product-favored.
(c) What is the equilibrium constant for this reaction?

78. ■ ▲ Copper(I) ion disproportionates to copper metal and copper(II) ion. (See Study Question 77.)

$$2\ Cu^{+}(aq) \rightarrow Cu(s) + Cu^{2+}(aq)$$

(a) What two half-reactions make up the disproportionation reaction?
(b) Use values of the standard reduction potentials for the two half-reactions in part (a) to determine whether this disproportionation reaction is spontaneous.
(c) What is the equilibrium constant for this reaction? If you have a solution that initially contains 0.10 mol of Cu^+ in 1.0 L of water, what are the concentrations of Cu^+ and Cu^{2+} at equilibrium?

In the Laboratory

79. Consider an electrochemical cell based on the half-reactions $Ni^{2+}(aq) + 2\ e^- \rightarrow Ni(s)$ and $Cd^{2+}(aq) + 2\ e^- \rightarrow Cd(s)$.
(a) Diagram the cell, and label each of the components (including the anode, cathode, and salt bridge).
(b) Use the equations for the half-reactions to write a balanced, net ionic equation for the overall cell reaction.
(c) What is the polarity of each electrode?
(d) What is the value of $E°_{cell}$?
(e) In which direction do electrons flow in the external circuit?
(f) Assume that a salt bridge containing $NaNO_3$ connects the two half-cells. In which direction do the $Na^+(aq)$ ions move? In which direction do the $NO_3^-(aq)$ ions move?
(g) Calculate the equilibrium constant for the reaction.
(h) If the concentration of Cd^{2+} is reduced to 0.010 M and $[Ni^{2+}]$ = 1.0 M, what is the value of E_{cell}? Is the net reaction still the reaction given in part (b)?
(i) If 0.050 A is drawn from the battery, how long can it last if you begin with 1.0 L of each of the solutions and each was initially 1.0 M in dissolved species? Each electrode weighs 50.0 g in the beginning.

80. An old method of measuring the current flowing in a circuit was to use a "silver coulometer." The current passed first through a solution of $Ag^+(aq)$ and then into another solution containing an electroactive species. The amount of silver metal deposited at the cathode was weighed. From the mass of silver, the number of atoms of silver was calculated. Since the reduction of a silver ion requires one electron, this value equalled the number of electrons passing through the circuit. If the time was noted, the average current could be calculated. If, in such an experiment, 0.052 g of Ag is deposited during 450 s, what was the current flowing in the circuit?

81. ■ A "silver coulometer" (Study Question 80) was used in the past to measure the current flowing in an electrochemical cell. Suppose you found that the current flowing through an electrolysis cell deposited 0.089 g of Ag metal at the cathode after exactly 10 min. If this same current then passed through a cell containing gold(III) ion in the form of $[AuCl_4]^-$, how much gold was deposited at the cathode in that electrolysis cell?

82. ■ ▲ Four metals, A, B, C, and D, exhibit the following properties:
(a) Only A and C react with 1.0 M hydrochloric acid to give $H_2(g)$.
(b) When C is added to solutions of the ions of the other metals, metallic B, D, and A are formed.
(c) Metal D reduces B^{n+} to give metallic B and D^{n+}.

Based on this information, arrange the four metals in order of increasing ability to act as reducing agents.

83. ▲ A solution of KI is added dropwise to a pale blue solution of $Cu(NO_3)_2$. The solution changes to a brown color, and a precipitate of CuI forms. In contrast, no change is observed if solutions of KCl and KBr are added to aqueous $Cu(NO_3)_2$. Consult the table of standard reduction potentials to explain the dissimilar results seen with the different halides. Write an equation for the reaction that occurs when solutions of KI and $Cu(NO_3)_2$ are mixed.

84. ▲ The amount of oxygen, O_2, dissolved in a water sample at 25 °C can be determined by titration. The first step is to add solutions of $MnSO_4$ and NaOH to the water to convert the dissolved oxygen to MnO_2. A solution of H_2SO_4 and KI is then added to convert the MnO_2 to Mn^{2+}, and the iodide ion is converted to I_2. The I_2 is then titrated with standardized $Na_2S_2O_3$.
(a) Balance the equation for the reaction of Mn^{2+} ions with O_2 in basic solution.
(b) Balance the equation for the reaction of MnO_2 with I^- in acid solution.
(c) Balance the equation for the reaction of $S_2O_3^{2-}$ with I_2.
(d) Calculate the amount of O_2 in 25.0 mL of water if the titration requires 2.45 mL of 0.0112 M $Na_2S_2O_3$ solution.

Summary and Conceptual Questions

The following questions may use concepts from this and previous chapters.

85. Fluorinated organic compounds are important commercially, since they are used as herbicides, flame retardants, and fire-extinguishing agents, among other things. A reaction such as

$$CH_3SO_2F + 3\ HF \longrightarrow CF_3SO_2F + 3\ H_2$$

is carried out electrochemically in liquid HF as the solvent.
(a) If you electrolyze 150 g of CH_3SO_2F, what mass of HF is required, and what mass of each product can be isolated?
(b) Is H_2 produced at the anode or the cathode of the electrolysis cell?
(c) A typical electrolysis cell operates at 8.0 V and 250 A. How many kilowatt-hours of energy does one such cell consume in 24 h?

86. ■ ▲ The free energy change for a reaction, $\Delta_r G°$, is the maximum energy that can be extracted from the process, whereas $\Delta_r H°$ is the total chemical potential energy change. The efficiency of a fuel cell is the ratio of these two quantities.

$$\text{Efficiency} = \frac{\Delta_r G°}{\Delta_r H°} \times 100\%$$

Consider the hydrogen-oxygen fuel cell, where the net reaction is

$$H_2(g) + \tfrac{1}{2}\ O_2(g) \longrightarrow H_2O(\ell)$$

(a) Calculate the efficiency of the fuel cell under standard conditions.
(b) Calculate the efficiency of the fuel cell if the product is water vapor instead of liquid water.
(c) Does the efficiency depend on the state of the reaction product? Why or why not?

87. A hydrogen-oxygen fuel cell operates on the simple reaction

$$H_2(g) + \tfrac{1}{2}\ O_2(g) \longrightarrow H_2O(\ell)$$

If the cell is designed to produce 1.5 A of current and if the hydrogen is contained in a 1.0-L tank at 200 atm pressure at 25 °C, how long can the fuel cell operate before the hydrogen runs out? (Assume there is an unlimited supply of O_2.)

88. ■ ▲ (a) Is it easier to reduce water in acid or base? To evaluate this, consider the half-reaction

$$2\ H_2O(\ell) + 2\ e^- \longrightarrow 2\ OH^-(aq) + H_2(g) \qquad E° = -0.83\ V$$

(b) What is the reduction potential for water for solutions at pH = 7 (neutral) and pH = 1 (acid)? Comment on the value of $E°$ at pH = 1.

89. ▲ Living organisms derive energy from the oxidation of food, typified by glucose.

$$C_6H_{12}O_6(aq) + 6\ O_2(g) \longrightarrow 6\ CO_2(g) + 6\ H_2O(\ell)$$

Electrons in this redox process are transferred from glucose to oxygen in a series of at least 25 steps. It is instructive to calculate the total daily current flow in a typical organism and the rate of energy expenditure (power). (See T. P. Chirpich: *Journal of Chemical Education*, Vol. 52, p. 99, 1975.)
(a) The molar enthalpy of combustion of glucose is −2800 kJ/mol-rxn. If you are on a typical daily diet of 2400 Cal (kilocalories), what amount of glucose (in moles) must be consumed in a day if glucose is the only source of energy? What amount of O_2 must be consumed in the oxidation process?
(b) How many moles of electrons must be supplied to reduce the amount of O_2 calculated in part (a)?
(c) Based on the answer in part (b), calculate the current flowing, per second, in your body from the combustion of glucose.
(d) If the average standard potential in the electron transport chain is 1.0 V, what is the rate of energy expenditure in watts?

THE CHEMISTRY OF THE ENVIRONMENT

John Emsley University of Cambridge

Planet Earth can be discussed from many points of view: geological, political, historical, and environmental. The last of these has now become a main topic of concern. If we are to understand why this is so, then we need to understand the chemical composition of our environment and the way this is changing.

Land, sea, air—continents, oceans, atmosphere—solid, liquid, gas. These are the three main components of the environment, and each is governed by the laws of chemistry that we associate with these physical states. Look a little closer at the continents, and we can see more and more space being devoted to human occupation and farming. We may think the water on the planet is either seawater or fresh water, but how do we get from one to the other? We know the atmosphere is a mixture of gases and vapors and is restlessly in motion, sometimes violently so, but what will happen if its composition changes? Understanding the environment is crucial, but without a chemical perspective we are floundering in the dark.

In this interchapter, we will look particularly at those parts of the environment that are most susceptible to change: the atmosphere and the aqueous sphere (water). What humans are doing to the atmosphere appears to be leading to climate change and in a direction that could destroy much of what we have created. The water of the planet is also of concern: we often pollute it, and we need to purify it. Nature can supply water in abundance as rain, rivers, and lakes but not necessarily where it is most needed. While we in the developed world have a seemingly abundant supply of pure, fresh water, others obtain water only with difficulty—and even then it may not be fit to drink. As the world's population increases, many more people will find themselves having to rely on seawater for their supply of freshwater. Seawater, of course, is not drinkable, but chemistry can make it so.

If we are to understand atmospheric pollution and its effects on climate, and to understand water supplies and the need to make them drinkable, then we need to understand the chemistry that is involved.

The Atmosphere

The atmosphere may be the least weighty part of the environment, but this does not mean it has the least effect. In fact, it has a disproportionately large influence on our standard of living. We know it is changing because of what we are adding to it, and we fear the consequences of that change. Earth has changed dramatically over the billions of years the planet has existed, and so has the atmosphere. The first atmosphere is thought to have been mainly ammonia, water vapor, and carbon dioxide. Today, it is mainly nitrogen, oxygen, and argon. At some stage, most of the water condensed onto the surface of the planet, and seas were formed. It was in the seas that life probably emerged, and life had to adapt to conditions as they changed. Sometimes the changes were so great that many species could not adapt and were wiped out. But, despite these mass extinctions, there were always a few survivors.

Table 1 gives the composition of the atmosphere in terms of the gases it contains. The values in the table are for dry air at sea level. Normally, air also contains water vapor, and this can vary considerably from place to place and from day to day and can be as high as 40,000 ppm (parts per million), but is generally about half this.

NASA

• Earth as seen from space.

TABLE 1. The gases of the atmosphere*

Gases	Proportion (ppm)
Nitrogen (N_2)	780,840
Oxygen (O_2)	209,460
Argon (Ar)	9,340
Carbon dioxide (CO_2)	385
Neon (Ne)	18
Helium (He)	5.2
Methane (CH_4)	1.7
Krypton (Kr)	1.1
Hydrogen (H_2)	0.5
Dinitrogen oxide (nitrous oxide) (N_2O)	0.55
Ozone (O_3)	0.4
Carbon monoxide (CO)	0.25
Xenon (Xe)	0.086
Radon (Rn)	traces

*Data on gas concentrations refer to relative numbers of particles (and hence are related to mole fractions).

In fact, there are many more components of the atmosphere than those listed in Table 1. There are trace amounts of chemicals emitted by plants, animals, and microbes, as well as those released by human activity. Our additions to the atmosphere come from burning fuels to generate electricity, from vehicle exhausts, from industry, and from our day-to-day activities. The atmosphere is changing, and we need to understand what this means for the future.

We might imagine that air is a homogeneous mixture of gases and particles because it is constantly being stirred by the motion of the planet and the heat of the sun. In fact, this mixing is far from complete. There is relatively little exchange of air between the north and south regions of the planet, and there are distinct layers in the atmosphere (◀ page 535). As far as humans are concerned, the most important layer is the lower layer in which we live—the troposphere, which goes up to around 7 km at the poles and 17 km at the equator. Above the troposphere is the stratosphere; this extends to about 30 km, getting thinner and thinner as we rise to higher altitudes. There are even higher regions, but there the air is very thin, although not so thin it cannot interact with incoming space debris or vehicles like the Space Shuttle and cause them to glow red hot. In this interchapter, we will concern ourselves only with the troposphere and the stratosphere.

The total mean mass of the atmosphere is estimated to be around 5.15×10^{18} kg (5.15×10^{15} metric tons, or 5.15×10^{6} gigatons). (Gigatons are often used in environmental studies because the masses we are discussing are so large. A *metric ton* is 1000 kg, but a gigaton is a billion metric tons, that is, 10^9 metric tons.) Three quarters of the mass of the atmosphere is in the troposphere, and there its average temperature is 14 °C (but that is an average of a wide range of temperatures). It can be as low as −89 °C, the lowest temperature ever recorded, in Antarctica in 1983, or as high as 58 °C, the highest temperature ever recorded, in North Africa in 1922. After temperature, the next most obvious property of the atmosphere is the pressure it exerts, which at sea level is around 101 kPa (1 atmosphere).

Some of the gases of the atmosphere are unreactive and never change chemically. These are the noble gases: helium, neon, argon, krypton, and xenon. Two of these, namely helium and argon, are being added continuously to the atmosphere because they are products of radioactive decay—helium comes from alpha particles emitted by elements such as uranium, and argon is produced by decay of ^{40}K, a long-lived radioactive isotope of potassium. Helium is such a light gas that it is continually being lost into space.

The other gases of the atmosphere, namely oxygen, nitrogen, and carbon dioxide, are chemically reactive and can be changed naturally into other molecules by reactions driven by lightning, ultraviolet rays from the sun, or the influence of living things, and especially the actions of humans.

Helium. Ultraprecise electrical measurements require extremely stable temperature, humidity, and vibration control. In this photo, NIST physicist Rand Elmquist fills a cryogenic chamber with liquid helium in preparation for measuring the international standard for electrical resistance. U.S. laboratories and manufacturers use NIST electrical standards and calibrations for all kinds of measurements from home electricity usage to electrocardiograms to gene-mapping.

Nitrogen and Nitrogen Oxides

The nitrogen of the atmosphere arose from the out-gassing of Earth when it was simply a molten mass. Even today, some nitrogen escapes when volcanoes erupt. Nitrogen gas is relatively unreactive, but every living thing on the planet needs to incorporate nitrogen into its cells as, among other compounds, amino acids, the building blocks of proteins. The process by which nitrogen gas is converted to useful nitrogen-containing compounds, such as ammonia (NH_3) and nitrate ions (NO_3^-) is known as **nitrogen fixation** (Figure 1). Bacteria in the soil and algae in the oceans are the primary "fixers" of nitrogen. On a global scale, the conversion of nitrogen to various compounds is accomplished to the extent of around 50% by biological processes, 10% by fertilizer manufacture, 10% by lightning, and 30% from the burning of fossil fuels. Once nitrogen is "fixed" it is incorporated into living things, passes through many other life forms, and ends up in the soil, which contains significant amounts of nitrogen-containing chemicals such as ammonium salts, nitrates, and nitrites. The two anions are converted back to nitrogen gas or to N_2O and returned to the atmosphere.

Dinitrogen monoxide (nitrous oxide), N_2O, is only one of several nitrogen oxides important in our environment. Others are nitrogen monoxide (nitric oxide), NO, and nitrogen dioxide, NO_2. Collectively, they are referred to as NO_x compounds. There are a number of natural sources of NO_x compounds.

- NO produced from the combination of N_2 and O_2 when air is heated or sparked by a bolt of lighting
- NO and NO_2 produced by the photochemical oxidation of N_2O in the atmosphere
- Oxidation of NH_3 in the atmosphere
- Production of NO and N_2O in soil by microbial processes

In addition, fossil fuel burning and biomass combustion by humans produce about 65% of the NO_x in the atmosphere.

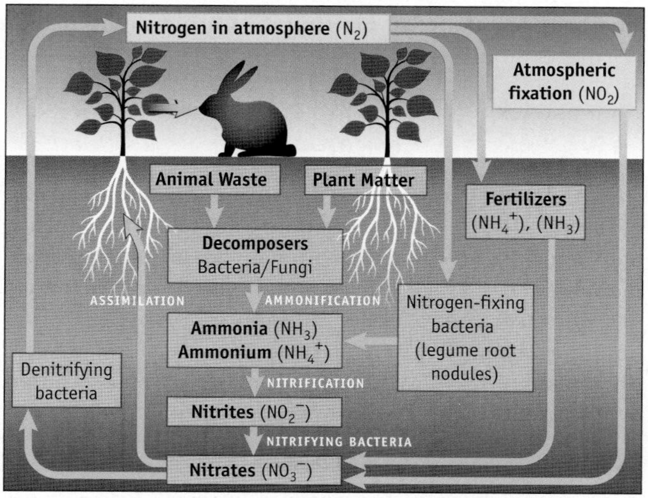

Figure 1 **The nitrogen cycle.** The nitrogen cycle involves nitrogen-fixation by soil bacteria or, in aquatic environments, by cyanobacteria. The NH_4^+ ions produced are converted to nitrate ions, the main form of nitrogen absorbed by plants. Nitrogen is returned to the atmosphere by denitrifying bacteria, which convert nitrate ions to N_2.

The most abundant nitrogen oxide, because it is the least reactive, is dinitrogen monoxide. Most of this oxide comes from microbes in the soil. Ten million tons of N_2O are released each year, some from human activity.

The other two nitrogen oxides make up only 0.00005 ppm of the atmosphere. While 100 million tons of NO and NO_2 come each year from the exhausts of vehicles, they are quickly washed out of the atmosphere by rain. When they do linger, and in particular over cities in sunny climates, they can react with hydrocarbons in the atmosphere, such as traces of unburnt fuel, to produce irritating photochemical smog. Among the most polluting components of smog are PANs, peroxyacyl nitrates, represented here by peroxyacetyl nitrate.

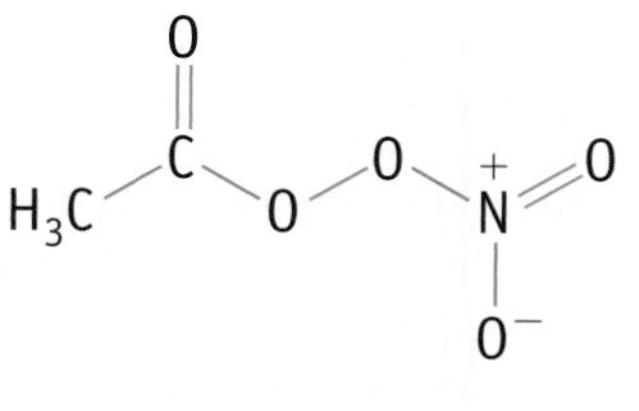

One route to its formation is as follows:

a) Organic compounds called aldehydes (such as CH_3CHO) are produced by complex reactions in smog. These molecules react with hydroxyl radicals ($OH\cdot$), which are also the result of photochemical reactions in the atmosphere. Their interaction produces another radical species $CH_3CO\cdot$.

$$CH_3CHO + OH\cdot \rightarrow CH_3CO\cdot + H_2O$$

b) The newly produced radical reacts with oxygen to produce the peroxyacetyl radical.

$$CH_3CO\cdot + O_2 \rightarrow CH_3COOO\cdot$$

c) The peroxyacetyl radical reacts with NO_2 in polluted air to give peroxyacetyl nitrate.

$$CH_3COOO\cdot + NO_2 \rightarrow CH_3COOONO_2$$

PANs are toxic and irritating. At low concentrations, they irritate the eyes, but at higher concentrations they can cause more serious damage to both animals and vegetation. They are relatively stable, dissociating only slowly into radicals and NO_2. Thus, it is possible that they can travel some distance from where they are formed before dissociating into harmful radicals.

Oxygen

Plant life had a dramatic effect on the early atmosphere, changing it from a reducing to an oxidizing one, in other words, from there being no oxygen present to one where this reactive gas is the second most abundant species present. This change meant life had to change. The percent of oxygen in the air is now midway between two extremes that would make life on Earth impossible for humans: below 17%

we would suffocate, and above 25% all organic material burns very easily, although there is evidence that oxygen in Earth's atmosphere did at one time exceed this level.

The total mass of oxygen in the atmosphere is a million gigatons. Even though the burning of 7 gigatons of fossil fuel carbon a year uses up 18 gigatons of oxygen, this makes almost no discernable difference to the amount of oxygen in the atmosphere. Oxygen is also moderately soluble in water (◀ page 626), which makes life in rivers, lakes and oceans possible.

Oxygen is a by-product of plant **photosynthesis.** Carbon dioxide is the source of the carbon plants need and which they capture from the air and turn into carbohydrates such as glucose ($C_6H_{12}O_6$) by photosynthesis. The overall chemical reaction is:

$$6\ CO_2 + 6\ H_2O \rightarrow C_6H_{12}O_6 + 6\ O_2$$

and the net result is the release of an oxygen molecule back into the atmosphere for each CO_2 absorbed. Oxygen molecules released by photosynthesis remain in the atmosphere, on average, for around 3000 years before being reabsorbed.

Blue-green algae or cyanobacteria first began producing oxygen as long ago as 3.5 billion years. But, mysteriously, hundreds of millions of years elapsed before there was a significant amount of oxygen in the atmosphere. Astrobiologists are not at all certain why this occurred, but it does seem plausible that the oxygen first produced did not remain in the atmosphere because it reacted with metals, especially converting iron(II) to iron(III). This process continued for about a billion years, but then the level in the atmosphere began to rise 2 billion years ago, and about 500 million years ago it rose relatively rapidly to around 20% when the first land plants started to appear.

There are three naturally occurring isotopes of oxygen: oxygen-16 accounts for 99.76% of O atoms, oxygen-17 for a mere 0.04%, and oxygen-18 for 0.2%. Because oxygen-18 is 12% heavier than oxygen-16, it can influence the behavior of water in the environment. The ratio of oxygen-18 to oxygen-16 in the world's oceans has varied slightly over geological time, and this has left an imprint on parts of the environment, providing evidence of past climates. When the world is in a cooler period, water molecules with oxygen-16 evaporate more easily from the oceans than their heavier oxygen-18 counterpart. Thus, snow that falls is very slightly richer in oxygen-16, and the water that remains in the oceans is very slightly richer in oxygen-18. Marine creatures therefore lay down shells that have more oxygen-18 than expected, and these are preserved in sediments. Analyzing the ratio of the two isotopes in such deposits reveals the cycle of global cooling and warming that has characterized the past half million years with its five ice ages.

Ozone

Ozone (O_3) has a key role to play for life on this planet, but it is also a threat. In the troposphere, it is a pollutant, while in the stratosphere it acts as a shield, protecting the planet from damaging ultraviolet rays from the sun. This shield is known as the **ozone layer.**

There is a natural low level of ozone in the air we breathe, about 0.02 ppm (parts per million), but in summer this can increase to 0.1 ppm or more as a result of sunlight acting on the nitrogen dioxide emitted by vehicles. In the presence of sunlight, NO_2 dissociates to NO and O atoms, and the O atoms can react with O_2 (in the presence of a third molecule acting as an energy "sink") to produce ozone.

$$NO_2 + \text{energy} \rightarrow NO + O\ (\lambda < 240\ \text{nm})$$

$$O + O_2 \rightarrow O_3$$

Because ozone damages the lungs, the legal limit for exposure to ozone in the work place is 0.1 ppm. Some growing plants are also susceptible to the gas, and even though they do not show visible signs of stress, their growth is reduced in proportion to the level of ozone in the air.

In contrast with ozone in the troposphere, O_3 in the stratosphere is vital to the planet because the molecules absorb ultraviolet radiation before it reaches Earth's surface. Radiation with wavelengths shorter than 240 nm interacts with O_2 molecules and cleaves them into two O atoms. Each O atom combines with another O_2 molecule to produce an ozone molecule.

$$O_2 + \text{energy} \rightarrow O + O$$

$$O + O_2 \rightarrow O_3$$

Ozone in turn absorbs ultraviolet radiation with wavelengths of less than 320 nm, and the O_3 is decomposed to O_2 and O atoms.

$$O_3 + \text{energy} \rightarrow O_2 + O\ (\lambda < 320\ \text{nm})$$

$$O + O_3 \rightarrow 2\ O_2$$

Without the ozone layer, dangerous radiation capable of harming living cells would penetrate to Earth's surface and could cause numerous human health and environmental effects. These effects include increased incidence of skin cancer and cataracts and suppression of the human immune response system. Damage to crops and marine phytoplankton and weathering of plastics are also caused by increased levels of ultraviolet radiation.

Chlorofluorocarbons (CFCs) and Ozone

Refrigerators from the late 1800s until 1929 used ammonia (NH_3), chloromethane (CH_3Cl), and sulfur dioxide (SO_2) as refrigerants. However, leakage from refrigerators caused several fatal accidents in the 1920s, so three American corporations, Frigidaire, General Motors, and DuPont, collaborated on the search for a less dangerous fluid. In 1928, Thomas Midgley, Jr., and his coworkers invented a "miracle compound" as a substitute. This compound, composed of carbon, chlorine, and fluorine, is a member of a large family of compounds called chlorofluorocarbons, or CFCs. The two compounds shown here are CFC-114 ($C_2Cl_2F_4$) and CFC-12 (CCl_2F_2).

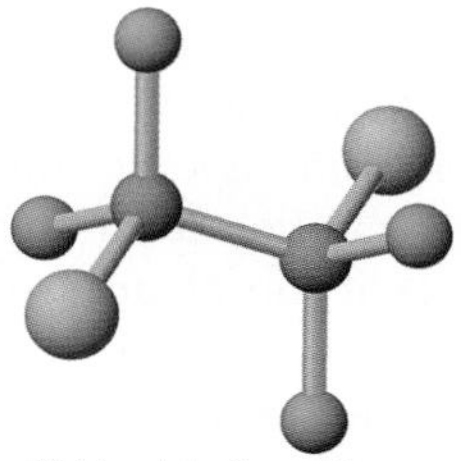

Dichlorotetrafluoroethane

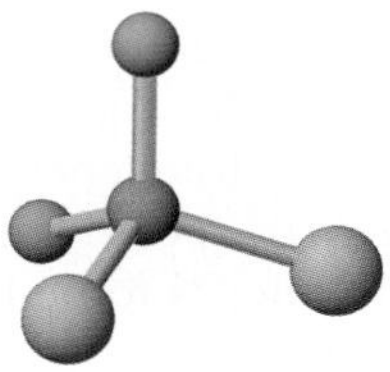

Dichlorodifluoromethane

These compounds had exactly the physical and chemical properties needed for a refrigerant: appropriate critical temperatures and pressures, no toxicity, and apparent chemical inertness.

The uses of CFCs grew dramatically, not only for air conditioning and refrigeration equipment, but also in applications such as propellants for aerosol cans, foaming agents in the production of expanded plastic foams, and inhalers for asthma sufferers.

Unfortunately, as the world has often learned about other "miracle compounds," the properties that made CFCs so useful also led to environmental problems. CFCs are unreactive in Earth's troposphere, which allows them to remain in the troposphere for hundreds of years. Over time, however, they slowly diffuse into the stratosphere where, as proposed by Sherwood Rowland and Mario Molina in 1974, solar radiation could lead to their decomposition. They also proposed that the Cl atoms and ClO radicals released in the decomposition of CFCs and other chlorine-containing compounds could destroy large numbers of ozone molecules (Figure 2). In 1986, Susan Solomon and a team from the National Center for Atmospheric Research went to Antarctica to investigate the "hole" in the ozone layer over that continent (Figure 3) and found that there were indeed higher levels of ClO than expected in the stratosphere.

Because of the apparent damage caused to the stratospheric ozone layer by CFCs and related compounds, the United States banned the use of CFCs as aerosol propellants in 1978, and 68 nations followed suit in 1987 by signing the Montreal Protocol. This Protocol called for an immediate reduction in nonessential uses of CFCs. In 1990, 100 nations met in London and decided to ban the production of CFCs. The United States and 140 other countries agreed to a complete halt in CFC manufacture as of December 31, 1995, and Rowland and Molina received the Nobel Prize for their work in that same year.

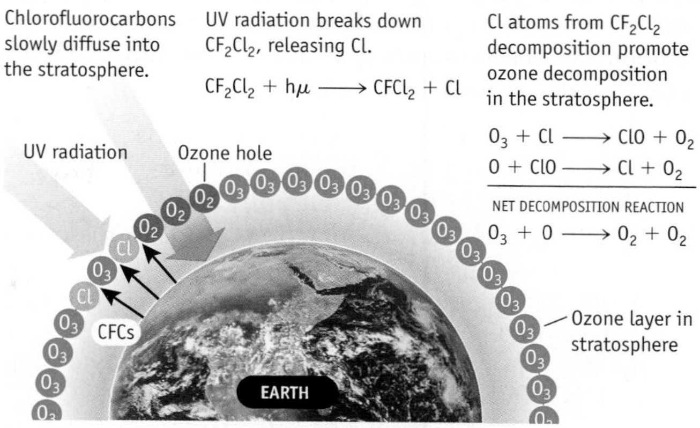

Figure 2 The interaction of CFCs, chlorine atoms, and ozone in the stratosphere. The O atoms come from the decomposition of O_3 by solar radiation.

Carbon Dioxide

No element is more essential to life than carbon, and the reason is that only carbon has the ability to form long chains and rings of atoms that are stable. This is the basis of the structures for many compounds that comprise the living cell, of which the most important is DNA (◀ page 503). The food we eat—carbohydrates, oils, proteins, and fiber—is made up of compounds of carbon, and this carbon eventually returns to the atmosphere as CO_2 as part of the natural cycle. This cycle moves carbon between various compartments of the terrestrial ecosphere and so rules the tempo of life on Earth. The carbon cycle

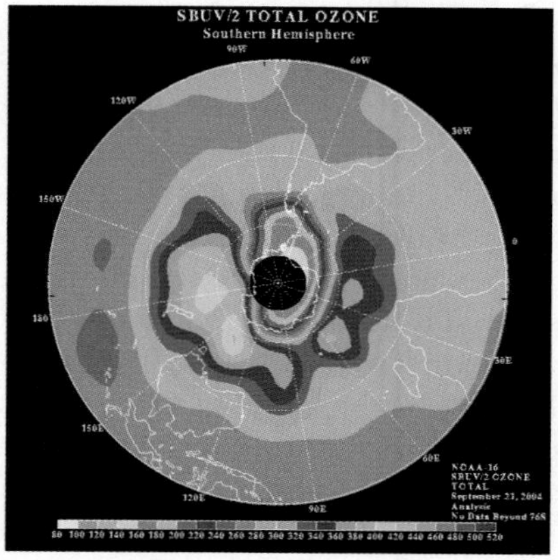

NOAA

Figure 3 The "ozone hole" over the Antarctic continent. During the Antarctic winter, when there are 24 hours of darkness, aerosols of HCl and $ClONO_2$ freeze and accumulate in the polar stratospheric clouds. During the Antarctic spring, these crystals melt, and Cl and ClO radicals are rapidly formed and lead to a depletion of stratospheric ozone over the continent.

turns over 200 gigatons of carbon every year, and the amounts of carbon residing in the various compartments are, in order of increasing amounts:

40 gigatons in living things in the oceans
725 gigatons as CO_2 in the atmosphere
2000 gigatons in living things on land
39,000 gigatons as carbonate ions (CO_3^{2-}) dissolved in the oceans
20,000,000 gigatons as reduced carbon in fossil fuel deposits
100,000,000+ gigatons as carbonate minerals such as limestone

Reduced carbon in Earth's crust comes in various forms such as natural gas, which is CH_4; oil, which approximates to CH_2; and coal, a complex material with a carbon–hydrogen ratio corresponding to CH. Known reserves of these fossil fuels amount to a vast 1 trillion tons of coal, 160 billion tons of oil, and 180 trillion cubic meters of natural gas. A much larger amount of carbon in Earth's crust is also present, but is too widely dispersed and inaccessible to be regarded as a potential energy resource. Human use of fossil fuels each year releases around 25 gigatons of CO_2 into the atmosphere.

Climate Change

The atmosphere impacts the environment of the whole planet and determines the overall temperature and weather we experience. If we pollute the air, then we may be punished by a change in the climate that we do not want, and that change is **global warming.**

The *Intergovernmental Panel on Climate Change* (IPCC) was set up in 1988 by members of the United Nations Environment Program and the World Meteorological Organization. To date, it has issued four Assessment Reports warning of the dangers of climate change and assessing whether human activities are the cause of these.

Earth began to emerge from a cold period around 150 years ago. Would the natural cycle of temperature change create a warm period like those at the time of the Roman Empire and during the Middle Ages? Or has the increase in human population and its burning of fossil fuels, which increases the level of CO_2 in the atmosphere, made an additional contribution that will result in a much warmer period than those earlier ages experienced?

The IPCC issued its first, second, and third Assessment Reports in 1990, 1995, and 2001 and warned that human activity might be partly responsible for raising the global temperature. While some heeded the IPCC warnings, little was done to check the use of fossil fuels, although some effort is now being directed into generating electricity by sustainable means such as wind power and solar panels. In 2007, the IPCC issued its *Fourth Assessment Report.* This report went further by saying that there is now little doubt the planet is warming and that additions of greenhouse gases to the atmosphere by humans are having a noticeable effect.

Greenhouse Gases

The greenhouse effect is the process by which radiation from the sun warms Earth's surface (◀ page 265). This is vital to life on Earth because the surface temperature is about 35 °C warmer than it would be without the effect.

The greenhouse effect depends on the gases in the troposphere. These gases are small molecules whose bonds vibrate with frequencies in the infrared region of the electromagnetic spectrum. Such a gas can trap infrared radiation that needs to be lost to space if the planet is not to get too hot.

Which gases can act as greenhouse gases depends solely on their chemistry. Gases that are single atoms, like argon, have no chemical bonds and so cannot interact with infrared light. Homonuclear diatomic molecules such as nitrogen (N_2) and oxygen (O_2) also do not absorb infrared radiation. Since these three gases make up 99% of the atmosphere, it is clearly beneficial for us that they are not greenhouse gases.

On the other hand, molecules with bonds between different atoms, such as water (H_2O) and carbon dioxide (CO_2) are greenhouse gases and allow Earth to warm by an average of 35 °C. (Water vapor is the main greenhouse gas and accounts for most of the observed effect.) In this respect, these gases are vital. If Earth did not retain some of the sun's heat, then Earth's average temperature would be like that of the moon, too hot when feeling the full glare of the sun and too cold when in shadow. Life would be impossible.

Other gases in the atmosphere can also act as greenhouse gases. Methane (CH_4) is a greenhouse gas much more powerful than CO_2, and nitrous oxide (N_2O) is also very powerful at retaining heat energy. These come partly from natural sources but in amounts too small to pose a threat. Methane is released by anaerobic bacteria (that is, those that do not use oxygen). Some anaerobic bacteria live at the bottom of lakes and swamps, some in termite mounds, and some in the guts of animals such as cows and humans. Non-natural activities can also release large amounts of these gases. For example, methane is the natural gas we use in vast amounts, and some of this escapes from leaking pipelines. It also leaks from landfill sites and coal mines, as well as from agricultural sites such as rice fields. Nitrous oxide release is also stimulated by our abundant use of ammonium nitrate fertilizers.

The methane concentration in the atmosphere is now around 1.7 ppm compared to the "natural" level, which is less than half this amount, and the concentration of N_2O is around 0.55 ppm instead of 0.27 ppm. More significantly, the carbon dioxide concentration has increased from

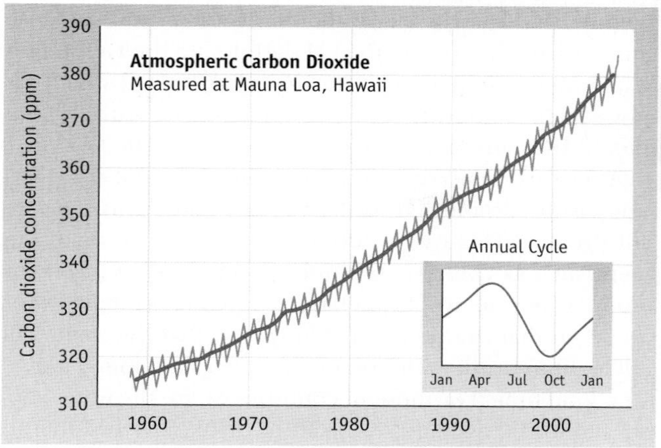

Figure 4 Mean monthly concentrations of CO_2 at Mauna Loa, Hawaii.

around 300 to 385 ppm (since about 1900) (Figure 4). The IPCC sees these continuing to rise but says we must strive to stabilize these concentrations if irreversible damage to the planet is to be avoided.

Industry has added new greenhouse gases to the atmosphere such as the many CFCs and various chemicals known as *volatile organic compounds* (VOCs). Although VOCs, especially isoprene and terpenes, are also given off by trees, VOCs tend to be higher indoors than outdoors; volatile organic compounds are given off indoors by paints, adhesives, plastics, air fresheners, and cosmetics. The family of CFCs has been replaced by a new generation of fluorocarbon compounds (HFCs such as $F_3C–CH_2F$ or HFC-134a, a compound now commonly used in air conditioners and refrigerators). Although HFCs behave as greenhouse gases, they are much less persistent in the atmosphere than the long-lived CFCs.

The IPCC estimates that in the past 100 years the atmosphere has warmed by 0.74 °C, and this might well have been higher had it not been offset by the release of soot and other chemicals into the atmosphere. These cause a certain amount of "global dimming."

So how much hotter will the world become if we continue to rely on fossil fuels? The IPCC has considered several computer predictions and says there are two possible "scenarios," one of which predicts a relatively low level of warming, between 1.8 and 2.9 °C. The other scenario predicts a range of between 4 and 6.4 °C and is much more worrisome. The lower scenario would lead to sea level rises of 18–38 cm; the higher range would cause a rise of 26–59 cm. While advanced nations could easily protect vulnerable areas against such rises—the Dutch have been living below sea level for hundreds of years thanks to building dikes—less-developed nations and especially those that occupy large river deltas in Asia and small island states would literally be swamped by the sea. We still have time to decide what action to take, but the longer we continue with our heavy reliance on natural gas, oil, and coal, the worse the risk becomes.

Some people say that we can offset the amount of carbon dioxide we release into the atmosphere when we drive our car or travel by air, by paying an extra levy so that trees can be planted to reabsorb the CO_2. Therefore, our journeys would be "carbon-neutral." In fact, there have been very large areas planted with trees around the world. Unfortunately, it now appears that trees planted in temperate regions and especially those at higher latitudes might actually contribute to climate warming by absorbing more heat from the sun than the land they occupy would otherwise absorb. However, trees planted in the tropics do have the desired effect and absorb carbon dioxide while not contributing to this heating effect.

The Aqua Sphere (Water)

Most of the planet is covered by water, and for centuries humans have treated it as an inexhaustible source of food and the easiest place in which to dump our waste. That is now changing. The aqua sphere is a major asset but will remain so only if we treat it with respect.

The Oceans

The quantity of minerals dissolved in the oceans is vast (Table 2), and even when the concentration is low, the total can still be quite surprising. For example, though something may be present at only 5 ppm in seawater, its total mass is almost 7 trillion tons (7000 gigatons.) The lure of vast amounts of dissolved metals has tantalized people down the ages, and the amount of gold in the sea particularly so. At a concentration of only 10 parts per trillion (ppt; or 1 g in 1 trillion g of water), even this amounts to more than 13 million tons. Although some have tried, it cannot be profitably exploited.

Water has fascinated scientists through the ages, in part because water has some unusual features that are of critical importance to the environment. For example, when it freezes it expands in volume unlike most other materials

TABLE 2. Concentrations of Some Cations and Anions in Seawater

Element	Dissolved Species	Amounts (mmol/L)
Chlorine	Cl^-	550
Sodium	Na^+	460
Magnesium	Mg^{2+}	52
Calcium	Ca^{2+}	10
Potassium	K^+	10
Carbon	HCO_3^-, CO_3^{2-}	30
Phosphorus	HPO_4^{2-}	<1

that shrink as they get colder (◀ page 563). This is why ice floats instead of sinking. In fact, this was essential for the emergence of early forms of life. Because water freezes from the top down, it actually protects the creatures that live underneath; otherwise, only a few microbes that can exist in ice would have survived.

The fact that ice is less dense than water also means that when it melts the volume of water it creates is smaller than the volume of ice. The result of this is that if the North Polar ice melts, it will not raise sea levels, whereas the ice stacked on Greenland will. Thankfully, the ice at the South Pole is not melting (yet), and computer models show that this part of the planet is not likely to be affected by global warming to the same extent as the northern hemisphere.

Drinking Water

Water is as vital to life as food; indeed, we can survive for weeks without anything to eat, but we would not last a week if we had no water to drink. Those in developed countries have clean, safe water supplied to their homes by pipe (although curiously many prefer to buy it in bottles). Sadly, a large number of our fellow humans, and generally those in less-developed countries, have to make do with whatever water they can find, taking it from a well or even from a river, and then they are at risk of contracting a water-borne disease such as cholera, typhoid, gastroenteritis, or meningitis. More than 5000 people a day are thought to die from such diseases due to the water they drink.

Water Purification

Water purification began with the ancient Egyptians 3500 years ago. They discovered that when alum—potassium aluminum sulfate, $KAl(SO_4)_2$—was dissolved in it, all visible impurities were removed. This salt forms a voluminous precipitate of insoluble aluminum hydroxide, $Al(OH)_3$, which carries down any cloudiness as it settles. This is still the way that modern treatment plants produce crystal-clear water (Figure 5). Adding a small amount of *slaked lime* (CaO, calcium hydroxide) to raise the pH slightly and then a solution of aluminum sulfate creates a *floc* of $Al(OH)_3$, which removes many impurities as it precipitates. The water is then filtered through sand to remove any trace of the floc.

After treatment with alum, the water is crystal clear and ready to be disinfected to destroy any pathogens that are still present. This can be done with a powerful oxidizing agent such as chlorine, sodium hypochlorite (NaOCl), or ozone. Chlorine and hypochlorite have an advantage over ozone, in that they persist for longer so that the water will still be fit to drink even when it has been standing in a pipe for a long time. For those in countries where the water supply cannot be assumed to be fit to drink, it is possible to treat local supplies with sanitizing tablets that react with the water to form a solution of hypochlorite ions (ClO^-) strong enough to make the water safe to drink. Boiling water also makes it fit to drink, but this might not be an option.

Ozone (O_3) is the most commonly used water supply disinfectant in Europe. It is also currently used in Los Angeles and some smaller communities in the United States. Its advantages are that it does not produce trihalomethanes, by-products of chlorine treatment; is more effective than chlorine at killing *Cryptosporidium* bacteria and viruses; and does not leave a residual "chlorine" taste. However, due to its reactivity, ozone must be generated on site and is therefore more expensive. Also, it does not provide residual disinfection in the distribution system.

Ultraviolet (UV) light has been used to disinfect food products, such as milk, for some time. Wavelength ranges from 200 to 295 nm are used with maximum disinfection occurring at 253.7 nm. However, UV radiation is not frequently used to disinfect drinking water. Installation of radiation units is physically complicated and expensive, and it provides no residual disinfection in the water distribution system.

In the U.S., tap water from a municipality must meet standard at least as strict as those for bottled water. All

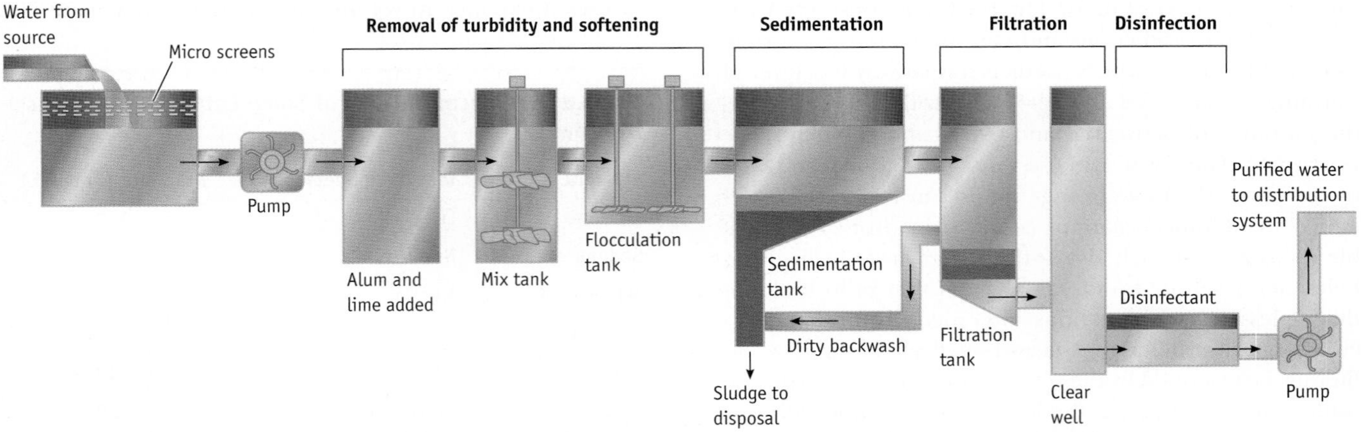

Figure 5 A municipal water treatment plant. The general steps in this process are (1) the removal of turbidity, (2) softening, (3) sedimentation and filtration, and (4) disinfection. The removal of turbidity (by addition of alum) and the disinfection of water are discussed in the text.

municipalities publish annual reports of the water quality, many of them on the internet. Check the quality of the tap water in your municipality.

Reverse Osmosis for Water Purification

Pure water can also be obtained from undrinkable water in other ways. The one most used in areas where there is access to seawater is to desalinate it by **reverse osmosis.** Osmosis is the process whereby water in one solution will move across a membrane in such a way as to dilute a solution with a higher concentration (◀ page 635). The osmotic pressure can be measured by seeing to what extent water will continue to pass across a membrane, even when it has to push against a growing pressure on the other side. Eventually, the pressure is such that the process of osmosis stops and can be reversed if more pressure is applied. This is what reverse osmosis is all about, and it is the way that some cities near the sea get their water supplies (Figure 6).

Chemistry is an important part of the desalination apparatus because of the membranes that allow the passage of water molecules but not of dissolved salts. The membranes are made of polymers such as polyvinylidene fluoride and polypropylene with ultra fine pores. They have to withstand high pumping pressures, pressures that are needed to force water against the osmotic pressure gradient. The cost is not excessive, and a cubic meter of water (1000 L) produced this way need cost only \$1. Desalination systems are now a billion dollar a year industry and growing at the rate of 7% per year. And it is not only seawater that can be purified this way; any waste water from a suspect source can be treated.

Diminishing Water Supplies

People also want water for washing themselves, their clothing, and their bed linen, or for watering the garden, filling the pool, and flushing the lavatory. In developed countries with adequate rainfall, water can be collected and stored in vast reservoirs above or below ground. But such is the increase in global population that even those regions may one day have to conserve supplies; there are times when water use is restricted, and such restrictions are likely to increase as the effects of global warming come into play this century. Already 500 million people worldwide are affected by water shortages. Even in the Great Plains of the U.S. the underground sources of water, known as aquifers, are becoming seriously depleted, and lakes such as Lake Mead, which was created by the Hoover Dam, that were once full to overflowing, are now around 20 meters (60 feet) below the level of water they held in the 1980s.

Water Pollution

Water has been polluted by many things, mainly from activities such as farming, mining, papermaking, inadequate or nonexistent sewage treatment facilities, and the chemical industry. Protecting natural bodies of water is now high on the agenda of most developed societies, and this has been achieved mainly by banning the use of dangerous chemicals and by installing better sewage treatment facilities.

Phosphates

In the past, phosphate pollution on a massive scale has been an environmental problem, such as that which afflicted one of the Great Lakes in the 1960s when it appeared it was no longer capable of supporting life owing to a lack of dissolved oxygen. The lake had become eutrophic due to unrestricted growth of algae. This was due in turn to an excess of nutrients supplied by the rivers that drained into the lake. The rivers were carrying run-off from farmland and water from sewage works, and, while the latter was removing most undesirables, they were not removing phosphates that had become a major component of detergents. Phosphate fertilizers were also being used by farmers. Similar problems were being experienced in other parts of the world, notably in Europe.

Campaigns to control the use of phosphates were successful. These chemicals were taken out of household products, and phosphate was extracted at sewage works before the waste water was released to rivers. The result was to discourage the growth of algae, and slowly the lake began to regenerate. In fact, phosphate was not entirely to blame for the pollution. Lack of zooplankton, which feed on algae, also contributed. These creatures were being killed by traces of pesticides and heavy metals. Without the zooplankton, the algae could breed unchecked.

Arsenic in Drinking Water

Even when drinking water appears to be safe, it may harbor unsuspected dangers acquired from the environment. One recent example is the occurrence of arsenic-containing compounds in water supplies. This is the case in the Lagunera region of

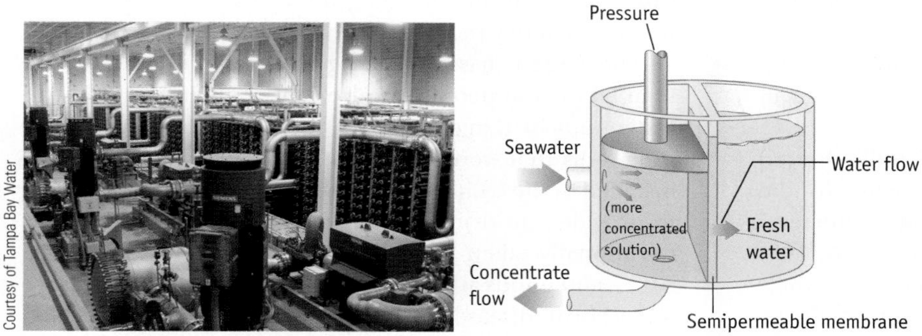

Figure 6 Water purification by reverse osmosis. (left) This plant in Tampa Bay, Florida, will eventually produce about 95 million liters of potable water per day from seawater. (right) Reverse osmosis. The osmotic pressure of seawater is approximately 27 atm. To obtain fresh water at a reasonable rate requires a pressure of over 50 atm. For comparison, bicycle tires are usually pumped up to a pressure of 2–3 atm.

Chemical Perspectives — Chlorination of Water Supplies

Chlorination is the oldest and most commonly used water disinfection method in the United States. Originally, chlorine gas was dissolved in water to yield hypochlorous acid (HClO):

$$Cl_2(g) + H_2O(\ell) \longrightarrow HClO(aq) + HCl(aq)$$

which dissociates to yield the hypochlorite ion (ClO^-),

$$HClO(aq) + H_2O(\ell) \rightleftharpoons H_3O^+(aq) + ClO^-(aq)$$

$$K_a = 3.5 \times 10^{-8}$$

Two of the chemical species formed by chlorine in water—hypochlorous acid and the hypochlorite ion—are known to environmental engineers as **free available chlorine,** meaning they are available for disinfection. Because chlorine gas is quite toxic and can easily escape from damaged containers, most water treatment facilities are now switching to solid hypochlorite salts such as sodium or calcium hypochlorite [NaClO and $Ca(ClO)_2$, respectively].

Hypochlorous acid and hypochlorite ion are effective in killing bacteria. However, they will also oxidize reduced inorganic ions such as iron(II), manganese(II), and nitrite ion, as well as organic impurities. These compounds are collectively known as the **chlorine demand.** To be available to kill bacteria, sufficient chlorine must be added to exceed the chlorine demand.

The advantages of chlorination include its relatively low cost and simple application as well as its ability to maintain residual disinfection throughout the distribution system. However, there are also disadvantages. Chlorine can react with dissolved, naturally occurring organic compounds in the source water to form carcinogenic compounds such as trihalomethanes (of which chloroform, $CHCl_3$, is one example). Chlorination has also been blamed for causing an unpleasant taste and odor in some drinking water.

In 1998, the Disinfection Byproducts Rule (DBR), which restricts the presence of chlorinated organic compounds in water, was implemented in the United States. One way to comply with the DBR is to add ammonia to the water, and about 30% of major U.S. water companies currently use this technique. Ammonia and hypochlorous acid react to form chloramines such as NH_2Cl:

$$HClO(aq) + NH_3(aq) \longrightarrow NH_2Cl(aq) + H_2O(\ell)$$

Chloramines are called **combined available chlorine.** Like free available chlorine, combined available chlorine is retained as a disinfectant throughout the water distribution system, but it is a weaker disinfectant overall. Combined available chlorine, however, has the advantage that it is less likely to lead to chlorinated organic compounds.

Recently, however, chemists in Washington, D.C., which uses the ammonia/chlorine method of water treatment, found increasing lead levels in the drinking water in area homes. [Several hundred homes had lead levels of 300 parts per billion (ppb), whereas the allowed level is 15 ppb.] Water treated with only chlorine is highly oxidizing, and lead in the water is oxidized to lead(IV) oxide. This insoluble oxide is trapped in the mineral scale that inevitably forms inside water pipes, and

Charles D. Winters

Chlorine tanks in the water treatment plant of a small city.

lead is thus removed from the water system. However, switching to the less-oxidizing ammonia/chlorine method apparently means lead ions are not removed by oxidation, so lead(II) ions remain in the system.

The odd-electron molecule chlorine dioxide (ClO_2) is also an effective disinfectant. The primary advantage of ClO_2 is that it completely oxidizes organic compounds (forming CO_2 and H_2O) and therefore does not produce trihalomethanes. However, ClO_2 provides no residual disinfection in the distribution system. Also, ClO_2 must be generated on site because it is explosive when exposed to air.

Mexico, the Antofagasta region of northern Chile, the Codoban region of Argentina, and especially in the West Bengal region of India and the neighboring country of Bangladesh. This had gone unnoticed for many years, but the environmental effects of arsenic in drinking water can be very serious.

In Taiwan, an epidemiological study in the 1960s showed that the prevalence of skin cancer was related to the amount of arsenic in water. In one region where the level of arsenic was 500 micrograms per liter (500 ppb), 10% of people aged 60 or over had cancers, mainly on their skin. Thankfully, these types of cancers are not usually fatal and can be removed. Internal cancers, however, are not as easy to diagnose or treat, and in one part of Taiwan where the level of arsenic was 800 ppb, there was a high incidence of bladder cancer. Unfortunately, even when measures are taken to purify the water, the incidence of cancer can continue to be high for many years because of the long latency period before the disease appears.

The largest mass poisoning by arsenic involved more than 30 million people (Figure 7). People in West Bengal, India, and in Bangladesh got their drinking water from tube wells that were drilled in the 1970s. The wells were installed by the United Nation's Children's Fund (UNICEF) to provide safe drinking water for a population that had traditionally taken its water from contaminated streams, rivers, and ponds and which had suffered accordingly from water-borne diseases such as gastroenteritis, typhoid, and cholera. Eventually, 5 million such wells were drilled, and they indeed served the purpose of bringing cholera under control. In its place, however, they brought low-level arsenic poisoning to large areas of Bangladesh.

Tube wells are 5 cm in diameter and tap the abundant ground water about 200 meters below the surface. A few had been installed in Bangladesh in the 1930s when the region was part of the British Empire. Those villages that had installed them showed a decrease in illness, especially among children. UNICEF continued to support the program after independence, and by 1997 it could report that it had already surpassed its goal of providing 80% of the population in those regions with "safe" water. Such had been the success of the program that many villagers installed private wells. Then doctors began to notice a worrying increase in skin cancer and suspected that arsenic was the cause. Analysis of the water from the wells showed that more than 20 million people were drinking water containing arsenic at the 50 ppb level, and as many as 5 million were drinking water containing 300 ppb arsenic. It does appear, though, that 80% of the Bangladeshi population has access to safe drinking water.

© Sucheta Das/Reuters/Corbis

Figure 7 A Bangladeshi villager shows his hands affected by arsenic contamination in the water supply.

In 1997, the Bangladeshi government instituted a Rapid Action Program that focused on a group of villages where contamination was severe. Almost two thirds of 30,000 tube wells in these villages were delivering water with arsenic in excess of 100 ppb. A team from the Dhaka Community Hospital visited 18 affected areas and examined 2000 adults and children; they concluded that more than half had skin lesions due to arsenic.

So what can the people of this region do? One answer is to dig wells that go deeper than 200 meters or shallower wells that reach down only 20 meters. In this way, the contaminated layer of ground water can be avoided. Once people have access to uncontaminated water, the arsenic in their bodies is quickly lost.

Remediation steps have to be taken when drinking water is contaminated, and yet it is a problem that can be costly to solve. When half a million people in New Mexico were exposed to high arsenic levels in their drinking water, action was taken but at a cost of $100 million. Taking similar action in Bangladesh would be beyond the ability of that country to pay. Even replacing a shallow well with one that draws water from the lower, safer, water table costs $1000. (Shallow tube wells cost only $100 to drill, which is why they have proved so popular.) The alternative is to remove the arsenic from the water. Various simple and cheap devices have been designed by chemists for doing this, such as passing the water through layers of sand, pieces of iron, and charcoal. The inorganic arsenic, which is present as arsenite ions, AsO_3^{3-}, sticks to the pieces of iron while the organic forms of arsenic are absorbed into the charcoal.

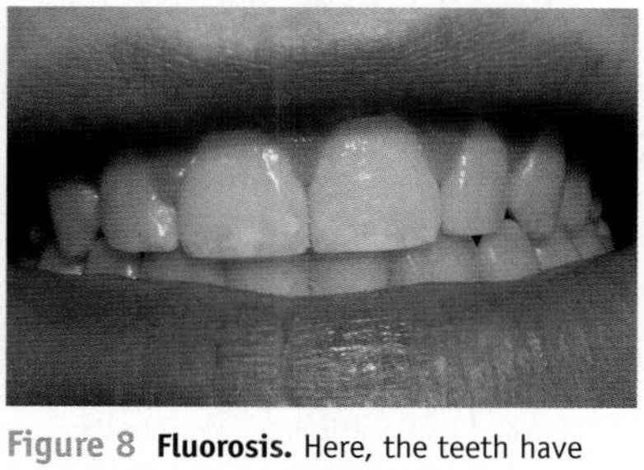

© Hardy Limeback. Used by permission of Kevin and Hardy Limeback.

Figure 8 Fluorosis. Here, the teeth have whitish patches, indicating mild fluorosis. Teeth and bones are hydroxyapatite, $Ca_5(OH)(PO_4)_3$. If one drinks water with a high F^- ion concentration, the OH^- ion in apatite can be replaced by the F^- ion. This can have the beneficial effect of hardening the surface of teeth, but in excessive cases it can lead to mottling. And in very high concentrations, skeletal deformities can result.

Fluoride in Drinking Water

It is generally acknowledged that a little fluoride in drinking water strengthens tooth enamel and prevents decay, and to this end many public water supplies are fluoridated to a level of 1 ppm. Nevertheless, there are dangers in drinking water that has much higher levels, and in some localities the level of fluoride is naturally high. Humans, as well as animals, can be affected and suffer from *fluorosis* (Figure 8), which is a hardening of the bones leading to a deformed skeleton. In certain parts of India, such as the Punjab, the condition is endemic, especially where villagers drink water from wells with high levels of fluoride, up to 15 ppm. About 25 million Indians suffer a mild form of fluorosis, with many thousands showing skeletal deformities.

Green Chemistry

Our environment is a complex chemical system but not one that is so complex that we can never understand it. It is also a robust system that may be capable of being healed when it is damaged. (One example is the slow reversal of the Antarctic ozone hole when CFC production was stopped.) Our stewardship of Earth demands a good knowledge of chemistry and sensible behavior. Nothing less will suffice.

One way chemists have responded to environmental problems is the so-called green chemistry movement. In 1996, the U.S. Environmental Protection Agency (EPA) initiated its Green Chemistry Program, which includes research, education, and outreach efforts as well as the Presidential Green Chemistry Challenge Awards, an annual program recognizing innovations in "cleaner, cheaper, smarter chemistry." The American Chemical Society actively promotes green chemistry (Figure 9), and the Royal Society of Chemistry in England publishes the research journal *Green Chemistry*. Some universities now offer degrees in green chemistry.

If we assume the word "green" means environmentally sound or friendly, how can chemistry be green? People have

Chemical Perspectives

Particulates and Air Pollution

The words fog, haze, mist, and smoke all describe atmospheric **particulates** that can range in size from about 0.1 millimeter to less than 1 μm (1 μm = 1 micrometer).

Large particulates, more than 1 μm in diameter, generally originate from the disintegration or dispersion of even larger particles from volcanoes, wind blown dust, or sea spray. Many of these originate as soil or rock, and their composition is similar to the Earth's crust; that is, they have high concentrations of Al, Ca, Si, and O.

Fine particulates are mainly formed by chemical reactions and by the coalescing of small molecules. These particulates often contain carbon that originates in power plants burning fossil fuels, incinerators, home furnaces, fireplaces, combustion engines, and forest fires. Black soot particles, mainly crystallites of carbon, are present in the exhaust from diesel trucks and coal-fired power plants.

Most other fine particulates in the atmosphere consist of inorganic sulfur and nitrogen compounds. Sulfur compounds originate as sulfur dioxide, SO_2, a gas produced naturally by volcanoes as well as by coal-fired power plants, metal smelters, and, in some parts of the world, by cars. Sulfur dioxide is oxidized in air to yield gaseous sulfur trioxide, SO_3, in a process catalyzed by particulates. The trioxide then reacts with water to form sulfuric acid, H_2SO_4, which travels in the air as an aerosol of fine droplets.

Nitric acid can be formed by the oxidation of nitrogen-containing compounds such as NO (page 991) and ammonia. Automobile exhaust gases provide a major source of NO, which forms in a reaction between N_2 and O_2 at the high temperatures reached in an internal combustion engine. Sulfuric and nitric acids react with ammonia in the atmosphere to produce particles of ammonium sulfate or ammonium nitrate, respectively.

A common measurement of atmospheric particulates is the **particulate matter** or **PM** index, which is defined as the mass of particulate matter in a given volume of air, usually micrograms per cubic meter (μg/m^3). Smaller particulates have a greater detrimental effect on human health than do larger ones. Therefore, only those particulates having a diameter smaller than a given value are measured. This cut-off diameter is reported as a subscript to the PM symbol. In the past, the U.S. has monitored particulates having diameters less than 10 μm (PM_{10}). Monitoring of even smaller particulates, such as those with $PM_{2.5}$, has begun more recently.

Particulates can have several effects, the most obvious of which is a decrease in visibility. Particles with a diameter near the wavelength of visible light, 0.3 to 0.8 μm (300–800 nm), can interfere with the transmission of light through air or scatter the light, which reduces the amount of light reaching the ground. Particulates also provide active surfaces for chemical reactions, such as the oxidation of SO_2. Finally, particulates provide a surface for the condensation of water vapor, thereby exerting a significant influence on weather.

Particulate air pollution has caused health problems for centuries. Many studies have linked breathing particulate matter to such health problems as aggravated asthma, increases in respiratory symptoms such as coughing, difficult or painful breathing, chronic bronchitis, and decreased lung function.

Particles smaller than 10 μm in diameter are considered *inhalable particulates* because they may enter our respiratory system when we breathe. However, particles larger than 2.5 μm in diameter are efficiently filtered by the cilia and generally do not enter the lungs. These larger particles can still pose a problem, however, because SO_2, produced along with particulates from burning coal, paralyzes the cilia in the respiratory tract and thus increases the damage caused by particulates.

Respirable particulates (those smaller than 2.5 μm in diameter, $PM_{2.5}$) seem to be responsible for most health problems. Indeed, $PM_{2.5}$ correlates most strongly with increases in the rate of disease or mortality in most regions.

Particulate air pollution levels were recently related to mortality across 151 metropolitan areas in the U.S. for the period 1982-1985. This study of 500,000 people found mortality rates (deaths/yr/1,000,000 people) to be correlated to both sulfate particulate concentrations and to $PM_{2.5}$.

© Sheldan Collins/Corbis

Haze. A high particulate concentration produces haze. This view of the Taj Mahal in India is partly obscured by haze from natural and human sources.

Courtesy of American Chemical Society

Figure 9 Green chemistry. The American Chemical Society has been actively encouraging programs in green chemistry. See www.acs.org.

many different opinions concerning chemistry and chemicals. Some use the word "chemical" as a synonym for a toxic substance. Others believe that all our problems can be solved through research in chemistry and other sciences. Chemists realize that all *matter*, including our food, water, and bodies as well as toxic compounds, is composed of chemicals. The study *and* practice of chemistry, like most aspects of life, have created problems at times, but chemistry has also solved many, many more.

SUGGESTED READINGS

1. J. Emsley, *Nature's Building Blocks.* England: Oxford University Press, 2001.
2. S. Manahan, editor: *Environmental Chemistry*, 8th ed., Boca Raton: CRC Press, Florida, 2004.
3. Safe Drinking Water Act, including a list of current maximum contaminant levels: www.epa.gov/safewater/sdwa/sdwa.html
4. Green Chemistry Program: www.epa.gov/greenchemistry
5. To check on the quality of air in your community: airnow.gov; see also www.arl.noaa.gov/ready/aq.html
6. Intergovernmental Panel on Climate Change: www.ipcc.ch/

STUDY QUESTIONS

Blue-numbered questions have answers in Appendix P and fully-worked solutions in the *Student Solutions Manual.*

1. Use the data in Table 2 (page 955) to calculate the molar concentrations of Na^+ and Cl^- in seawater. Which amount is larger?

2. Using the data in Table 2, estimate the minimum mass of solid that would remain after evaporation of 1.0 L. of seawater.

3. What mass of NaCl could be obtained by evaporating 1.0 L of seawater? (Note: the amount of Na^+ ions in seawater limits the amount of NaCl that can be obtained; see Question 1.)

4. In any solution, there must be a balance between the positive and negative charges of the ions that are present. Use the data in Table 2 to determine how close to a balance of charge is achieved for the ions in seawater. (The table lists only the major ions in seawater and the balance between positive and negative charge will only be approximate.)

5. In the discussion on the composition of air, mention is made of the fact that water vapor may have a concentration as high as 40,000 ppm. Calculate the pressure exerted by water vapor at this concentration. Assume that this represents a situation with 100% humidity. What temperature would be needed to achieve this value? (See Appendix G.)

6. Although there are a number of magnesium containing minerals, the commercial source of this metal is seawater. Treatment of seawater with $Ca(OH)_2$ gives solid insoluble $Mg(OH)_2$. This reacts with HCl to produce $MgCl_2$, which is dried, and Mg is obtained by electrolysis of the molten salt. Write balanced net ionic equations for the three reactions described here.

7. Calculate the mass of Mg that could be obtained by the process described in the previous question from 1.0 L of seawater. To prepare 100. kg of Mg, what volume of sea water would be needed? (The density of seawater is 1.025 g/cm^3.)

8. The quoted CO_2 concentration in the atmosphere, 385 ppm, is an average value. The actual concentration of CO_2 at different sites will vary. Speculate on whether the concentration of CO_2 would be expected to be higher, lower, or the same as this average value in a typical large city.

9. Imagine the following experiment. You have a large graduated cylinder containing 100. mL of liquid water at 0 °C. You drop an ice cube with a volume of 25 cm^3 into the cylinder. Ice has a density of 0.92 g/cm^3, less than the density of liquid water, so it floats with 92% being under water.
   a) To what level will the water in the graduated cylinder rise after adding the ice?
   b) Allow the ice to melt. What volume will now be occupied by the liquid water? (One consequence of global warming will be a rise in sea level as ice in the northern and southern regions of the planets melts. However, the effect relates only to melting of ice on land.) Melting of floating ice will have no effect on sea levels.

21 The Chemistry of the Main Group Elements

Charles D. Winters

Quartz (SiO_2), dry ice (solid CO_2), and pure silicon.

Carbon and Silicon

Carbon and silicon are both highly significant elements in the earth's crust. Mendeleev placed them in the same periodic group based on the similarity in stoichiometry of their simple compounds such as their oxides (CO_2 and SiO_2). We know now that carbon is the backbone of millions of organic compounds. As we will see in this chapter, silicon, the second most abundant element in the earth's crust, is vitally important in the structures of a large number of minerals.

Probe a bit further into silicon chemistry, however, and what emerges are the surprising differences between these elements. We get a hint of this when we compare the two oxides, CO_2, a gas above −78 °C and 1 atm pressure, and SiO_2, a rock hard solid with a very high melting point. Another interesting contrast can be made between the reactions of the two simple hydrogen compounds, CH_4 and SiH_4, with water. The reaction of methane with water is reactant-favored, whereas silane will explode on contact with water.

In this chapter, we will evaluate and summarize the inorganic chemistry of the main group elements and explore the similarities and differences between elements in each periodic group.

Questions:

1. Write balanced chemical equations for the reactions of water with CH_4 (forming CO_2 and H_2O) and SiH_4 (forming SiO_2 and H_2).
2. Using enthalpy of formation data, calculate the standard enthalpy change for the reactions in Question 1.
3. Look up the electronegativities of carbon, silicon, and hydrogen. What conclusion can you draw concerning the polarity of C—H and Si—H bonds?
4. Carbon and silicon compounds with the empirical formulas $(CH_3)_2CO$ (acetone) and $(CH_3)_2SiO$ (a silicone polymer) also have quite different structures (Chapter 10). Draw Lewis structures for these species. This difference, along with the difference between structures of CO_2 and SiO_2, suggests a general observation about silicon compounds. Based on that observation, do you expect that a silicon compound with a structure similar to ethene (C_2H_4) exists?

Answers to these questions are in Appendix Q.

Chapter Goals

See Chapter Goals Revisited (page 1010) for Study Questions keyed to these goals and assignable in OWL.

- Relate the formulas and properties of compounds to the periodic table.
- Describe the chemistry of the main group or A-Group elements, particularly H; Na and K; Mg and Ca; B and Al; Si; N and P; O and S; and F and Cl.
- Apply the principles of stoichiometry, thermodynamics, and electrochemistry to the chemistry of the main group elements.

Chapter Outline

The main group or A-Group elements occupy an important place in the world of chemistry. Eight of the 10 most abundant elements on the earth are in these groups. Likewise, the top 10 chemicals produced by the U.S. chemical industry are all main group elements or their compounds.

Because main group elements and their compounds are economically important—and because they have interesting chemistries—we devote this chapter to a brief survey of these elements.

Chemistry Now™

Throughout the text this icon introduces an opportunity for self-study or to explore interactive tutorials by signing in at **www.thomsonedu.com/login**.

21.1 Element Abundances

The abundance of the first 18 elements in the solar system is plotted against their atomic numbers in Figure 21.1. As you can see, hydrogen and helium are the most abundant by a wide margin because most of the mass of the solar system resides in the sun, and these elements are the sun's primary components. Lithium, beryllium, and boron are low in abundance, but carbon's abundance is very high. From this point on, with the exception of iron and nickel, elemental abundances gradually decline as the atomic number increases.

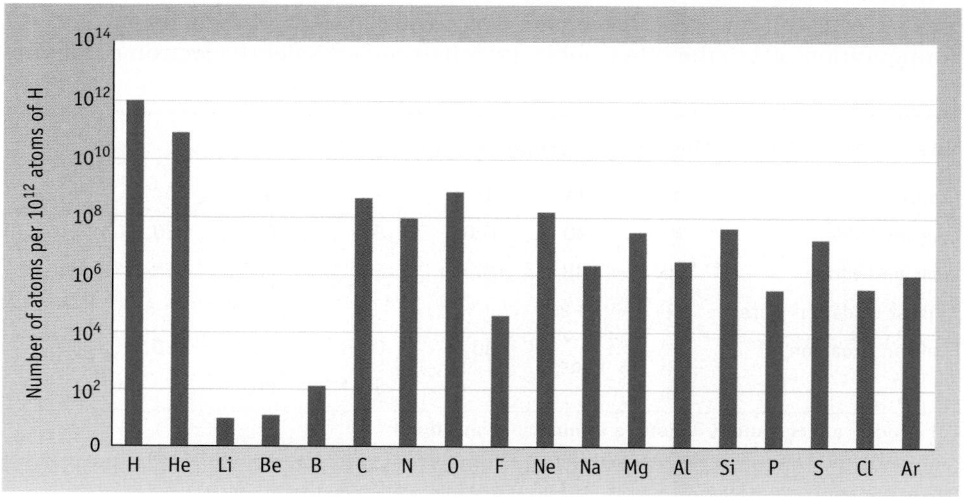

FIGURE 21.1 Abundance of elements 1–18. Li, Be, and B have relatively low abundances because they are circumvented when elements are made in stars. The common elements such as C, O, and Ne are made in stars by the accretion of alpha particles (helium nuclei). Helium has an atomic number of 2. If three He atoms combine, they produce an atom with atomic number 6 (carbon). Adding yet another He atom gives an atom with atomic number 8 (oxygen), and so on. (Notice that the vertical axis uses a logarithmic scale. This means, for example, there are 10^{12} H atoms for every 100 B atoms.)

TABLE 21.1 The 10 Most Abundant Elements in Earth's Crust

Rank	Element	Abundance (ppm)*
1	Oxygen	474,000
2	Silicon	277,000
3	Aluminum	82,000
4	Iron	56,300
5	Calcium	41,000
6	Sodium	23,600
7	Magnesium	23,300
8	Potassium	21,000
9	Titanium	5,600
10	Hydrogen	1,520

*ppm = g per 1000 kg. Most abundance data taken from J. Emsley: *The Elements*, New York, Oxford University Press, 3rd edition, 1998.

Ten elements account for 99% of the aggregate mass of our planet (Table 21.1). Oxygen, silicon, and aluminum represent more than 80% of this mass. Oxygen and nitrogen are the primary components of the atmosphere, and oxygen-containing water is highly abundant on the surface, underground, and as a vapor in the atmosphere. Many common minerals also contain these elements, including limestone ($CaCO_3$) and quartz or sand (SiO_2, Figure 2.8). Aluminum and silicon occur together in many minerals; among the more common ones are feldspar, granite, and clay.

21.2 The Periodic Table: A Guide to the Elements

The similarities in the properties of certain elements guided Mendeleev when he created the first periodic table (◀ page 59). He placed elements in groups based partly on the composition of their common compounds with oxygen and hydrogen (see Table 21.2). We now understand that the elements are grouped according to the arrangements of their valence electrons.

Recall that the metallic character of the elements declines on moving from left to right in the periodic table. Elements in Group 1A, the alkali metals, are the most metallic elements in the periodic table. Elements on the far right are nonmetals, and in between are the metalloids. Metallic character also increases from the top of a group to the bottom. This is especially well illustrated by Group 4A. Carbon, at the top of the group, is a nonmetal; silicon and germanium are metalloids; and tin and lead are metals (Figure 21.2). The significance of metallic character in a discussion of the chemistry of the elements is readily apparent; metals typically form ionic compounds, whereas compounds composed only of nonmetals are covalent. Typically, ionic compounds are crystalline solids that have high melting points and conduct electricity in the molten state. Covalent compounds, on the other hand, can be gases, liquids, or solids and have low melting and boiling points.

Valence Electrons

The *ns* and *np* electrons are the valence electrons for main group elements (where *n* is the period in which the element is found) (◀ Section 7.4). The chemical behavior of an element is determined by the valence electrons.

When considering electronic structure, a useful reference point is the noble gases (Group 8A), elements having filled electron subshells. Helium has an electron configuration of $1s^2$; the other noble gases have ns^2np^6 valence electron configura-

Charles D. Winters

FIGURE 21.2 Group 4A elements. A nonmetal, carbon (graphite crucible); a metalloid, silicon (round, lustrous bar); and metals tin (chips of metal) and lead (a bullet, a toy, and a sphere).

TABLE 21.2 Similarities within Periodic Groups*

Group	1A	2A	3A	4A	5A	6A	7A
Common oxide	M_2O	MO	M_2O_3	EO_2	E_4O_{10}	EO_3	E_2O_7
Common hydride	MH	MH_2	MH_3	EH_4	EH_3	EH_2	EH
Highest oxidation state	+1	+2	+3	+4	+5	+6	+7
Common oxoanion			BO_3^{3-}	CO_3^{2-}	NO_3^-	SO_4^{2-}	ClO_4^-
				SiO_4^{4-}	PO_4^{3-}		

*M denotes a metal and E denotes a nonmetal or metalloid.

tions. The dominant characteristic of the noble gases is their lack of reactivity. Indeed, the first two elements in the group do not form any compounds that can be isolated. The other four elements are now known to have limited chemistry, however, and the discovery of xenon compounds in the 1960s ranks as one of the most interesting developments in modern chemistry.

Ionic Compounds of Main Group Elements

Ions of main group elements having filled *s* and *p* subshells are very common—justifying the often-seen statement that elements react in ways that achieve a "noble gas configuration." The elements in Groups 1A and 2A form 1+ and 2+ ions with electron configurations that are the same as those for the previous noble gases. All common compounds of these elements (e.g., NaCl, $CaCO_3$) are ionic. The metallic elements in Group 3A (aluminum, gallium, indium, and thallium, but not the metalloid boron) form compounds containing 3+ ions.

Elements of Groups 6A and 7A can also achieve a noble gas configuration by adding electrons. The Group 7A elements (halogens) form anions with a 1− charge (the halide ions, F^-, Cl^-, Br^-, I^-), and the Group 6A elements form anions with a 2− charge (O^{2-}, S^{2-}, Se^{2-}, Te^{2-}). In Group 5A chemistry, 3− ions with a noble gas configuration (such as the nitride ion, N^{3-}) are also known. The energy required to form highly charged anions is large, however, which means that other types of chemical behavior will usually take precedence.

EXAMPLE 21.1 Reactions of Group 1A–3A Elements

Problem Give the formula and name for the product in each of the following reactions. Write a balanced chemical equation for the reaction.

(a) Ca(s) + S_8(s)

(b) Rb(s) + I_2(s)

(c) lithium and chlorine

(d) aluminum and oxygen

Strategy Predictions are based on the assumption that ions are formed with the electron configuration of the nearest noble gas. Group 1A elements form 1+ ions; Group 2A elements form 2+ ions; and metals in Group 3A form 3+ ions. In their reactions with metals, halogen atoms typically add a single electron to give anions with a 1− charge; Group 6A elements add two electrons to form anions with a 2− charge. For names of products, refer to the nomenclature discussion on page 77.

Solution

	Balanced Equation	Product Name
(a)	8 Ca(s) + S_8(s) → 8 CaS(s)	Calcium sulfide
(b)	2 Rb(s) + I_2(s) → 2 RbI(s)	Rubidium iodide
(c)	2 Li(s) + Cl_2(g) → 2 LiCl(s)	Lithium chloride
(d)	4 Al(s) + 3 O_2(g) → 2 Al_2O_3(s)	Aluminum oxide

EXERCISE 21.1 Main Group Element Chemistry

Write a balanced chemical equation for a reaction forming the following compounds from the elements.

(a) NaBr

(b) CaSe

(c) PbO

(d) $AlCl_3$

Charles D. Winters

FIGURE 21.3 Boron halides. Liquid BBr_3 (left) and solid BI_3 (right). Formed from a metalloid and a nonmetal, both are molecular compounds. Both are sealed in glass ampules to prevent these boron compounds from reacting with H_2O in the air.

Molecular Compounds of Main Group Elements

Many avenues of reactivity are open to nonmetallic main group elements. Reactions with metals generally result in formation of ionic compounds, whereas compounds containing only metalloids and nonmetallic elements are generally molecular in nature.

Molecular compounds are encountered with the Group 3A element boron (Figure 21.3), and the chemistry of carbon in Group 4A is dominated by molecular compounds with covalent bonds (◀ Chapters 8 and 10). Similarly, nitrogen chemistry is dominated by molecular compounds. Consider ammonia, NH_3; the various nitrogen oxides; and nitric acid, HNO_3. In each of these species, nitrogen bonds covalently to another nonmetallic element. Also in Group 5A, phosphorus reacts with chlorine to produce the molecular compounds PCl_3 and PCl_5 (◀ page 113).

The valence electron configurations of an element determine the composition of its molecular compounds. Involving all the valence electrons in the formation of a compound is a frequent occurrence in main group element chemistry. We should not be surprised to discover compounds in which the central element has the highest possible oxidation number (such as P in PF_5). The highest oxidation number is readily apparent: it equals the group number. Thus, the highest (and only) oxidation number of sodium in its compounds is +1; the highest oxidation number of C is +4; and the highest oxidation number of phosphorus is +5 (Tables 21.2 and 21.3).

TABLE 21.3 Fluorine Compounds Formed by Main Group Elements

Group	Compound	Bonding
1A	NaF	Ionic
2A	MgF_2	Ionic
3A	AlF_3	Ionic
4A	SiF_4	Covalent
5A	PF_5	Covalent
6A	SF_6	Covalent
7A	IF_7	Covalent
8A	XeF_4	Covalent

EXAMPLE 21.2 Predicting Formulas for Compounds of Main Group Elements

Problem Predict the formula for each of the following:

(a) the product of the reaction between germanium and excess oxygen

(b) the product of the reaction of arsenic and fluorine

(c) a compound formed from phosphorus and excess chlorine

(d) an anion of selenic acid

Strategy We will predict that in each reaction the element other than the halogen or oxygen in each product achieves its most positive oxidation number, a value equal to the number of its periodic group.

Solution

(a) The Group 4A element germanium should have a maximum oxidation number of +4. Thus, its oxide has the formula GeO_2.

(b) Arsenic, in Group 5A, reacts vigorously with fluorine to form AsF_5, in which arsenic has an oxidation number of +5.

(c) PCl_5 is formed when the Group 5A element phosphorus reacts with excess chlorine.

(d) The chemistries of S and Se are similar. Sulfur, in Group 6A, has a maximum oxidation number of +6, so it forms SO_3 and sulfuric acid, H_2SO_4. Selenium, also in Group 6A, has analogous chemistry, forming SeO_3 and selenic acid, H_2SeO_4. The anion of this acid is the selenate ion, SeO_4^{2-}.

EXERCISE 21.2 Predicting Formulas for Main Group Compounds

Write the formula for each of the following:

(a) hydrogen telluride (c) selenium hexachloride

(b) sodium arsenate (d) perbromic acid

There are many similarities among elements in the same periodic group. This means you can use compounds of more common elements as examples when you encounter compounds of elements with which you are not familiar. For example, water, H_2O, is the simplest hydrogen compound of oxygen. You can reasonably expect the hydrogen compounds of other Group 6A elements to be H_2S, H_2Se, and H_2Te; all are well known.

EXAMPLE 21.3 Predicting Formulas

Problem Predict the formula for each of the following:

(a) a compound of hydrogen and phosphorus

(b) the hypobromite ion

(c) germane (the simplest hydrogen compound of germanium)

(d) two oxides of tellurium

Strategy Recall as examples some of the compounds of lighter elements in a group, and then assume other elements in that group will form analogous compounds.

Solution

(a) Phosphine, PH_3, has a composition analogous to ammonia, NH_3.

(b) Hypobromite ion, BrO^-, is similar to the hypochlorite ion, ClO^-, the anion of hypochlorous acid (HClO).

(c) GeH_4 is analogous to CH_4 and SiH_4, other Group 4A hydrogen compounds.

(d) Te and S are in Group 6A. TeO_2 and TeO_3 are analogs of the oxides of sulfur, SO_2 and SO_3.

EXAMPLE 21.4 Recognizing Incorrect Formulas

Problem One formula is incorrect in each of the following groups. Pick out the incorrect formula, and indicate why it is incorrect.

(a) $CsSO_4$, KCl, $NaNO_3$, Li_2O

(b) MgO, CaI_2, $BaPO_4$, $CaCO_3$

(c) CO, CO_2, CO_3

(d) PF_5, PF_4^+, PF_2, PF_6^-

Strategy Look for errors such as incorrect charges on ions or an oxidation number exceeding the maximum possible for the periodic group.

Solution

(a) $CsSO_4$. Sulfate ion has a 2− charge, so this formula would require a Cs^{2+} ion. Cesium, in Group 1A, forms only 1+ ions. The formula of cesium sulfate is Cs_2SO_4.

(b) $BaPO_4$. This formula implies a Ba^{3+} ion (because the phosphate anion is PO_4^{3-}). The cation charge does not equal the group number. The formula of barium phosphate is $Ba_3(PO_4)_2$.

(c) CO_3. Given that O has an oxidation number of −2, carbon would have an oxidation number of +6. Carbon is in Group 4A, however, and can have a maximum oxidation number of +4.

(d) PF_2. This species has an odd number of electrons. Very few odd electron molecules are commonly seen. Examples include NO, NO_2, and ClO_2.

Comment To chemists, this exercise is second nature. Incorrect formulas stand out. You will find that your ability to write and recognize correct formulas will grow as you learn more chemistry.

EXERCISE 21.3 Predicting Formulas

Identify a compound or ion based on a second-period element that has a formula and Lewis structure analogous to each of the following:

(a) PH_4^+

(b) S_2^{2-}

(c) P_2H_4

(d) PF_3

A Closer Look — Hydrogen, Helium, and Balloons

In 1783, Jacques Charles first used hydrogen to fill a balloon large enough to float above the French countryside (◀ page 516), In World War I, hydrogen-filled observation balloons were used. The Graf Zeppelin, a passenger-carrying dirigible built in Germany in 1928, was also filled with hydrogen. It carried more than 13,000 people between Germany and the United States until 1937, when it was replaced by the Hindenburg. The Hindenburg was designed to be filled with helium. At that time, World War II was approaching, and the United States, which has much of the world's supply of helium, would not sell the gas to Germany. As a consequence, the Hindenburg had to use hydrogen. The Hindenburg exploded and burned when landing in Lakehurst, New Jersey, in May 1937. Of the 62 people on board, only about half escaped uninjured. As a result of this disaster, hydrogen has acquired a reputation as being a very dangerous substance. Actually, it is as safe to handle as any other fuel, as evidenced by the large quantities used in rockets today.

Mary Evans Picture Library/Photo Researchers, Inc.

The Hindenburg. This hydrogen-filled dirigible crashed in Lakehurst, New Jersey, in May 1937. Some have speculated that the aluminum paint coating the skin of the dirigible was involved in sparking the fire.

EXERCISE 21.4 Recognizing Incorrect Formulas

Explain why compounds with the following formulas would not be expected to exist: ClO, Na_2Cl, $CaCH_3CO_2$, C_3H_7.

21.3 Hydrogen

Chemical and Physical Properties of Hydrogen

Hydrogen has three isotopes, two of them stable (protium and deuterium) and one radioactive (tritium).

Isotopes of Hydrogen

Isotope Mass (u)	Symbol	Name
1.0078	^1H (H)	Hydrogen (protium)
2.0141	^2H (D)	Deuterium
3.0160	^3H (T)	Tritium

Of the three isotopes, only H and D are found in nature in significant quantities. Tritium, which is produced by cosmic ray bombardment of nitrogen in the atmosphere, is found to the extent of 1 atom per 10^{18} atoms of ordinary hydrogen. Tritium has a half-life of 12.26 years.

Under standard conditions, hydrogen is a colorless gas. Its very low boiling point, 20.7 K, reflects its nonpolar character and low molar mass. As the least dense gas known, it is ideal for filling lighter-than-air craft.

Deuterium compounds have been the subject of much research. One important observation is that, because D has twice the mass of H, reactions involving D atom transfer are slightly slower than those involving H atoms. This knowledge led to a way to produce D_2O or "heavy water." Hydrogen can be produced, albeit expensively, by electrolysis of water (Figure 21.4).

$$2\ H_2O(\ell) + \text{electrical energy} \rightarrow 2\ H_2(g) + O_2(g)$$

Any sample of natural water always contains a tiny concentration of D_2O. When electrolyzed, H_2O is electrolyzed more rapidly than D_2O. Thus, as the electrolysis proceeds, the liquid remaining is enriched in D_2O. Repeating this process many times will eventually give pure D_2O, often called "heavy water." Large amounts of D_2O are now produced because this compound is used as a moderator in some nuclear reactors that are used for power generation.

Hydrogen combines chemically with virtually every other element, except the noble gases. There are three different types of binary hydrogen-containing compounds.

Ionic metal hydrides are formed in the reaction of H_2 with a Group 1A or 2A metal.

$$2\ Na(s) + H_2(g) \rightarrow 2\ NaH(s)$$

$$Ca(s) + H_2(g) \rightarrow CaH_2(s)$$

These compounds contain the hydride ion, H^-, in which hydrogen has a -1 oxidation number.

Molecular compounds (such as H_2O, HF, and NH_3) are generally formed by direct combination of hydrogen with nonmetallic elements (Figure 21.5). The oxidation number of the hydrogen atom in these compounds is $+1$, but covalent bonds to hydrogen are the rule.

$$N_2(g) + 3\ H_2(g) \rightarrow 2\ NH_3(g)$$

$$F_2(g) + H_2(g) \rightarrow 2\ HF(g)$$

Hydrogen is absorbed by many metals to form *interstitial hydrides*, the third general class of hydrogen compounds. This name refers to the structures of these species, in which the hydrogen atoms reside in the spaces between the metal atoms (called *interstices*) in the crystal lattice. Palladium metal, for example, can soak up 1000 times its volume of hydrogen (at STP). Most interstitial hydrides are non-stoichiometric; that is, the ratio of metal and hydrogen is not a whole number. When interstitial metal hydrides are heated, H_2 is driven out. This phenomenon allows these materials to store H_2, just as a sponge can store water. It suggests one way to store hydrogen for use as a fuel in automobiles (◀ page 264).

Preparation of Hydrogen

About 300 billion liters (STP) of hydrogen gas are produced annually worldwide, and virtually all is used immediately in the manufacture of ammonia (▶ Section 21.8), methanol (◀ Section 10.3), or other chemicals.

Some hydrogen is made from coal and steam, a reaction that has been used for more than 100 years.

$$C(s) + H_2O(g) \longrightarrow \underbrace{H_2(g) + CO(g)}_{\text{water gas or synthesis gas}}$$

$$\Delta_r H° = +131\ \text{kJ/mol-rxn}$$

The reaction is carried out by injecting water into a bed of red-hot coke. The mixture of gases produced, called *water gas* or *synthesis gas*, was used until about 1950 as a fuel for cooking, heating, and lighting. However, it has serious drawbacks. It

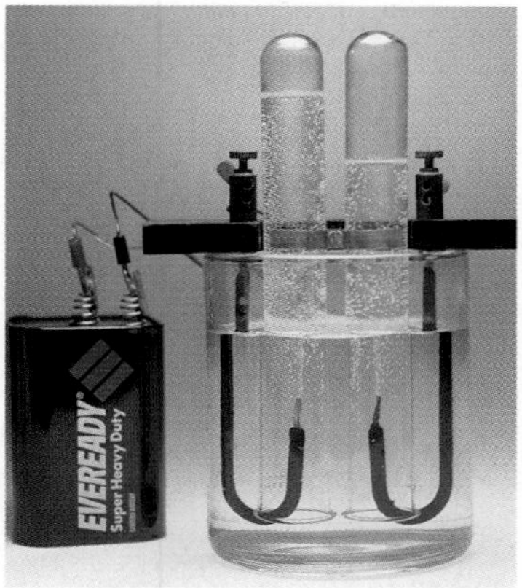

Charles D. Winters

FIGURE 21.4 Electrolysis of water. Electrolysis of water (containing dilute H_2SO_4 as an electrolyte) gives O_2 (left) and H_2 (right).

Charles D. Winters

FIGURE 21.5 The reaction of H_2 and Br_2. Hydrogen gas burns in an atmosphere of bromine vapor to give hydrogen bromide.

FIGURE 21.6 Production of water gas. Water gas, also called synthesis gas, is a mixture of CO and H_2. It is produced by treating coal, coke, or a hydrocarbon like methane with steam at high temperatures in plants such as that pictured here. Methane has the advantage that it gives more total H_2 per gram than other hydrocarbons, and the ratio of the by-product CO_2 to H_2 is lower.

produces only about half as much heat as an equal amount of methane does, and the flame is nearly invisible. Moreover, because it contains carbon monoxide, water gas is toxic.

The largest quantity of hydrogen is now produced by the *catalytic steam reformation of hydrocarbons* such as methane in natural gas (Figure 21.6). Methane reacts with steam at high temperature to give H_2 and CO.

$$CH_4(g) + H_2O(g) \rightarrow 3\ H_2(g) + CO(g) \qquad \Delta_rH° = +206\ \text{kJ/mol-rxn}$$

The reaction is rapid at 900–1000 °C and goes nearly to completion. More hydrogen can be obtained in a second step in which the CO formed in the first step reacts with more water. This so-called *water gas shift reaction* is run at 400–500 °C and is slightly exothermic.

$$H_2O(g) + CO(g) \rightarrow H_2(g) + CO_2(g) \qquad \Delta_rH° = -41\ \text{kJ/mol-rxn}$$

The CO_2 formed in the process is removed by reaction with CaO (to give solid $CaCO_3$), leaving fairly pure hydrogen.

Perhaps the cleanest way to make hydrogen on a relatively large scale is the electrolysis of water (Figure 21.4). This approach provides not only hydrogen gas but also high-purity O_2. Because electricity is quite expensive, however, this method is not generally used commercially.

Table 21.4 and Figure 21.7 give examples of reactions used to produce H_2 gas in the laboratory. The most often used method is the reaction of a metal with an acid. Alternatively, the reaction of aluminum with aqueous NaOH (Figure 21.7b) also generates hydrogen. During World War II, this reaction was used to obtain hydrogen to inflate small balloons for weather observation and to raise radio antennas. Metallic aluminum was plentiful at the time because it came from damaged aircraft.

The combination of a metal hydride and water (Figure 21.7c) is an efficient but expensive way to synthesize H_2 in the laboratory. The reaction is more commonly used in laboratories to dry organic solvents because the metal hydride reacts with traces of water present in the solvent.

> **EXERCISE 21.5 Hydrogen Chemistry**
>
> Use bond enthalpies (page 389) to estimate the enthalpy change for the reaction of methane and water to give hydrogen and carbon monoxide (with all compounds in the gas phase).

FIGURE 21.7 Producing hydrogen gas.

Photos: Charles D. Winters

(a) The reaction of magnesium and acid. The products are hydrogen gas and a magnesium salt.

(b) The reaction of aluminum and NaOH. The products of this reaction are hydrogen gas and a solution of $Na[Al(OH)_4]$.

(c) The reaction of CaH_2 and water. The products are hydrogen gas and $Ca(OH)_2$.

TABLE 21.4 Methods for Preparing H_2 in the Laboratory

1. Metal + Acid → metal salt + H_2

   $Mg(s) + 2\ HCl(aq) \rightarrow MgCl_2(aq) + H_2(g)$

2. Metal + H_2O → metal hydroxide or oxide + H_2

   $2\ Na(s) + 2\ H_2O(\ell) \rightarrow 2\ NaOH(aq) + H_2(g)$

   $2\ Fe(s) + 3\ H_2O(\ell) \rightarrow Fe_2O_3(s) + 3\ H_2(g)$

   $2\ Al(s) + 2\ KOH(aq) + 6\ H_2O(\ell) \rightarrow 2\ K[Al(OH)_4](aq) + 3\ H_2(g)$

3. Metal hydride + H_2O → metal hydroxide + H_2

   $CaH_2(s) + 2\ H_2O(\ell) \rightarrow Ca(OH)_2(s) + 2\ H_2(g)$

Charles D. Winters

FIGURE 21.8 The importance of salt. All animals, including humans, need a certain amount of salt in their diet. Sodium ions are important in maintaining electrolyte balance and in regulating osmotic pressure. For an interesting account of the importance of salt in society, culture, history, and economy, see *Salt, A World History*, by M. Kurlansky, New York, Penguin Books, 2002.

21.4 The Alkali Metals, Group 1A

Sodium and potassium are, respectively, the sixth and eighth most abundant elements in the earth's crust by mass. In contrast, lithium is relatively rare, as are rubidium and cesium. Only traces of radioactive francium occur in nature. Its longest-lived isotope (^{223}Fr) has a half-life of only 22 minutes.

The Group 1A elements are metals, and all are highly reactive with oxygen, water, and other oxidizing agents (see Figure 7.13, page 329). In all cases, compounds of the Group 1A metals contain the element as a 1+ ion. The free metal is never found in nature.

Most sodium and potassium compounds are water-soluble (◀ solubility guidelines, Figure 3.10), so it is not surprising that sodium and potassium compounds are found either in the oceans or in underground deposits that are the residue of ancient seas. To a much smaller extent, these elements are also found in minerals, such as Chilean saltpeter ($NaNO_3$).

Despite the fact that sodium is only slightly more abundant than potassium on the earth, sea water contains significantly more sodium than potassium (2.8% NaCl versus 0.8% KCl). Why the great difference? Most compounds of both elements are water-soluble, so why didn't rain dissolve Na- and K-containing minerals over the centuries and carry them down to the sea, so that they appear in the same proportions in the oceans as on land? The answer lies in the fact that potassium is an important factor in plant growth. Most plants contain four to six times as much combined potassium as sodium. Thus, most of the potassium ions in groundwater from dissolved minerals are taken up preferentially by plants, whereas sodium salts continue on to the oceans. (Because plants require potassium, commercial fertilizers usually contain a significant amount of potassium salts.)

Some NaCl is essential in the diet of humans and other animals because many biological functions are controlled by the concentrations of Na^+ and Cl^- ions (Figure 21.8). The fact that salt has long been recognized as important is evident in surprising ways. For example, we are paid a "salary" for work done. This word is derived from the Latin *salarium*, which meant "salt money" because Roman soldiers were paid in salt.

Group 1A
Alkali metals

Element	Atomic number	Symbol	Abundance
Lithium	3	Li	20 ppm
Sodium	11	Na	23,600 ppm
Potassium	19	K	21,000 ppm
Rubidium	37	Rb	90 ppm
Cesium	55	Cs	0.0003 ppm
Francium	87	Fr	trace

Element abundances are in parts per million in the earth's crust.

Preparation of Sodium and Potassium

Sodium is produced by reducing sodium ions in sodium salts. However, common chemical reducing agents are not powerful enough to convert sodium ions to sodium metal. Because of this, the metals are usually prepared by electrolysis.

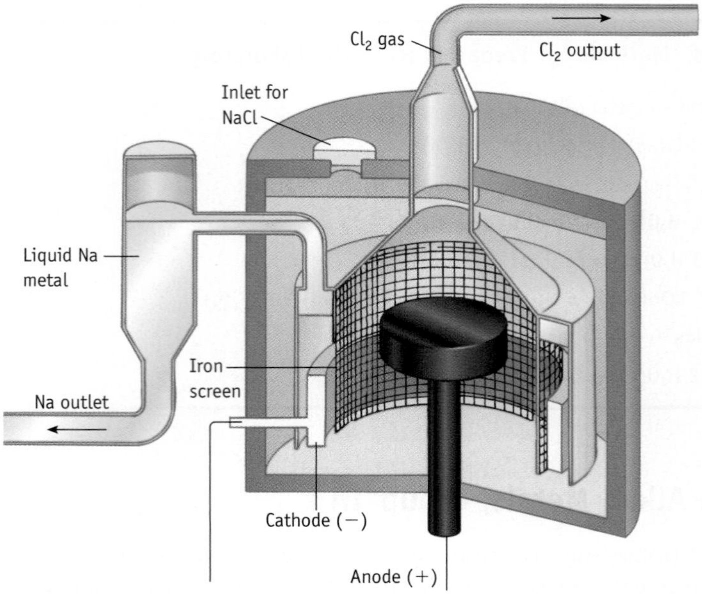

FIGURE 21.9. A Downs cell for preparing sodium. A circular iron cathode is separated from the graphite anode by an iron screen. At the temperature of the electrolysis, about 600 °C, sodium is a liquid. It floats to the top and is drawn off periodically. Chlorine gas is produced at the anode and collected inside the inverted cone in the center of the cell.

The English chemist Sir Humphry Davy first isolated sodium in 1807 by the electrolysis of molten sodium carbonate. However, the element remained a laboratory curiosity until 1824, when it was found sodium could be used to reduce aluminum chloride to aluminum metal. At that time, metallic aluminum was rare and very valuable, so this discovery inspired considerable interest in manufacturing sodium. By 1886, a practical method of sodium production had been devised (the reduction of NaOH with carbon). Unfortunately for sodium producers, in this same year Charles Hall and Paul Heroult invented the electrolytic method for aluminum production (► page 980), thereby eliminating this market for sodium.

Sodium is currently produced by the electrolysis of molten NaCl (◄ Section 20.7). The Downs cell for the electrolysis of molten NaCl operates at 7 to 8 V with currents of 25,000 to 40,000 amps (Figure 21.9). The cell is filled with a mixture of dry NaCl, $CaCl_2$, and $BaCl_2$. Adding other salts to NaCl lowers the melting point from that of pure NaCl (800.7 °C) to about 600 °C. [Recall that solutions have lower melting points than pure solvents (Chapter 14).] Sodium is produced at a copper or iron cathode that surrounds a circular graphite anode. Directly over the cathode is an inverted trough in which the low-density, molten sodium (melting point, 97.8 °C) collects. Chlorine, a valuable by-product, collects at the anode.

Potassium can also be made by electrolysis. Molten potassium is soluble in molten KCl, however, making separation of the metal difficult. The preferred method for preparation of potassium uses the reaction of sodium vapor with molten KCl, with potassium being continually removed from the equilibrium mixture.

$$Na(g) + KCl(\ell) \rightleftharpoons K(g) + NaCl(\ell)$$

■ **Alkali Metals React with Water—Thermodynamics and Kinetics** The alkali metals all react vigorously with water, and it is easily observed that the violence of the reaction increases with atomic number. This seems counter to the argument presented in *A Closer Look* that lithium is the best reducing agent. However, the reducing ability of a metal is a thermodynamic property, whereas the violence of the reaction is mainly a consequence of reaction rate.

Properties of Sodium and Potassium

Sodium and potassium are silvery metals that are soft and easily cut with a knife (see Figure 2.6). They are just a bit less dense than water. Their melting points are quite low, 97.8 °C for sodium and 63.7 °C for potassium.

All of the alkali metals are highly reactive. When exposed to moist air, the metal surface quickly becomes coated with a film of oxide or hydroxide. Consequently, the metals must be stored in a way that avoids contact with air, typically by placing them in kerosene or mineral oil.

A Closer Look

The Reducing Ability of the Alkali Metals

The uses of the Group 1A metals depend on their reducing ability. The values of $E°$ reveal that Li is the best reducing agent in the group, whereas Na is the poorest; the remainder of these metals have roughly comparable reducing ability.

Reduction Potential

Element	E° (V)
$Li^+ + e^- \rightarrow Li$	−3.045
$Na^+ + e^- \rightarrow Na$	−2.714
$K^+ + e^- \rightarrow K$	−2.925
$Rb^+ + e^- \rightarrow Rb$	−2.925
$Cs^+ + e^- \rightarrow Cs$	−2.92

Analysis of $E°$ is a thermodynamic problem, and to understand it better we can break the process of metal oxidation, $M(s) \rightarrow M^+(aq) + e^-$, into a series of steps. Here, we imagine that the metal sublimes to vapor, an electron is removed to form the gaseous cation, and the cation is hydrated. The first two steps require energy, but the last is exothermic.

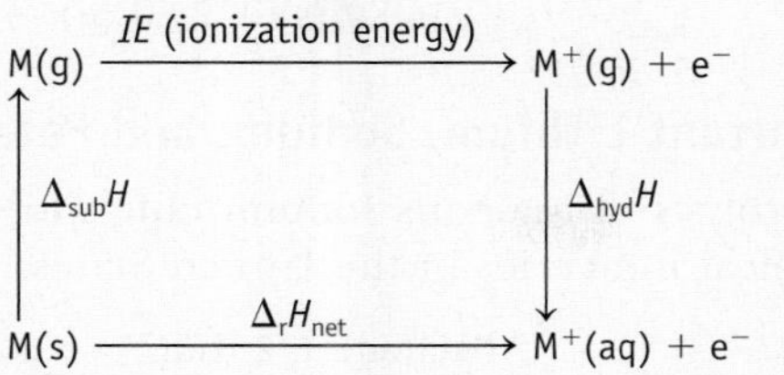

From Hess's law (page 233), we know that the overall energy change should be

$$\Delta_r H_{net} = \Delta_{sub} H + IE + \Delta_{hyd} H$$

The element that is the best reducing agent should have the most negative (or least positive) value of ΔH. That is, the best reducing agent should be the metal that has the most exothermic value for its hydration energy because this can offset the energy of the endothermic steps ($\Delta_{sub}H$ and IE). For the alkali metals, enthalpies of hydration range from −506 kJ/mol for Li^+ to −180 kJ/mol for Cs^+. The fact that $\Delta_{hyd}H$ is so much greater for Li^+ than for Cs^+ largely accounts for the difference in reducing ability.

While this analysis of the problem gives us a reasonable explanation for the great reducing ability of lithium, recall that $E°$ is directly related to $\Delta_r G°$ and not to $\Delta H°$. However, $\Delta_r G°$ is largely determined by $\Delta_r H°$, so it is possible to relate variations in $E°$ to variations in $\Delta_r H°$.

Charles D. Winters

Potassium is a very good reducing agent and reacts vigorously with water.

The high reactivity of Group 1A metals is exemplified by their reaction with water, which generates an aqueous solution of the metal hydroxide and hydrogen gas (Figure 7.13, page 329),

$$2\ Na(s) + 2\ H_2O(\ell) \rightarrow 2\ Na^+(aq) + 2\ OH^-(aq) + H_2(g)$$

and their reaction with any of the halogens to yield a metal halide (Figure 1.4),

$$2\ Na(s) + Cl_2(g) \rightarrow 2\ NaCl(s)$$

$$2\ K(s) + Br_2(\ell) \rightarrow 2\ KBr(s)$$

Chemistry often produces surprises. Group 1A metal oxides, M_2O, are known, but they are not the principal products of reactions between the Group 1A elements and oxygen. Instead, the primary product of the reaction of sodium and oxygen is sodium *peroxide*, Na_2O_2, whereas the principal product from the reaction of potassium and oxygen is KO_2, potassium *superoxide*.

$$2\ Na(s) + O_2(g) \rightarrow Na_2O_2(s)$$

$$K(s) + O_2(g) \rightarrow KO_2(s)$$

Both Na_2O_2 and KO_2 are ionic compounds. The Group 1A cation is paired with either the peroxide ion (O_2^{2-}) or the superoxide ion (O_2^-). These compounds are not merely laboratory curiosities. They are used in oxygen generation devices in places where people are confined, such as submarines, aircraft, and spacecraft, or when an emergency supply is needed. When a person breathes, 0.82 L of CO_2 is exhaled for every 1 L of O_2 inhaled. Thus, a requirement of an O_2 generation system is that it should produce a larger volume of O_2 than the volume of CO_2

Courtesy of Mine Safety Appliances Company

FIGURE 21.10. A closed-circuit breathing apparatus that generates its own oxygen. One source of oxygen is potassium superoxide (KO_2). Both carbon dioxide and moisture exhaled by the wearer into the breathing tube react with the KO_2 to generate oxygen. Because the rate of the chemical reaction is determined by the quantity of moisture and carbon dioxide exhaled, the production of oxygen is regulated automatically. With each exhalation, more oxygen is produced by volume than is required by the user.

taken in. This requirement is met with superoxides (Figure 21.10). With KO_2 the reaction is

$$4\ KO_2(s) + 2\ CO_2(g) \rightarrow 2\ K_2CO_3(s) + 3\ O_2(g)$$

Important Lithium, Sodium, and Potassium Compounds

Electrolysis of aqueous sodium chloride (*brine*) is the basis of one of the largest chemical industries in the United States.

$$2\ NaCl(aq) + 2\ H_2O(\ell) \rightarrow Cl_2(g) + 2\ NaOH(aq) + H_2(g)$$

Two of the products from this process—chlorine and sodium hydroxide—give the industry its name: the *chlor-alkali industry*. More than 10 billion kilograms of Cl_2 and NaOH is produced annually in the United States.

Sodium carbonate, Na_2CO_3, is another commercially important compound of sodium. It is also known by two common names, *soda ash* and *washing soda*. In the past, it was largely manufactured by combining NaCl, ammonia, and CO_2 in the *Solvay process* (which remains the method of choice in many countries). In the United States, however, sodium carbonate is obtained from naturally occurring deposits of the mineral *trona*, $Na_2CO_3 \cdot NaHCO_3 \cdot 2\ H_2O$ (Figure 21.11).

Owing to the environmental problems associated with the chlor-alkali process, considerable interest has arisen in the possibility of manufacturing sodium hydroxide by other methods. This has led to a revival of the old "soda-lime process," which produces NaOH from inexpensive lime (CaO) and soda ash (Na_2CO_3).

$$Na_2CO_3(aq) + CaO(s) + H_2O(\ell) \rightarrow 2\ NaOH(aq) + CaCO_3(s)$$

The insoluble calcium carbonate by-product is filtered off, then heated (calcining) to convert it to lime, which is recycled into the reaction system.

$$CaCO_3(s) \rightarrow CaO(s) + CO_2(g)$$

Sodium bicarbonate, $NaHCO_3$, also known as *baking soda*, is another common compound of sodium. Not only is $NaHCO_3$ used in cooking, but it is also added in small amounts to table salt. NaCl is often contaminated with small amounts of $MgCl_2$. The magnesium salt is hygroscopic; that is, it picks water up from the air

FIGURE 21.11 Producing soda ash. Trona mined in Wyoming and California is processed into soda ash (Na_2CO_3) and other sodium-based chemicals. Soda ash is the ninth most widely used chemical in the United States. Domestically, about half of all soda ash production is used in making glass. The remainder goes to make chemicals such as sodium silicate, sodium phosphate, and sodium cyanide. Some is also used to make detergents, in the pulp and paper industry, and in water treatment.

Jack Dermid/Photo Researchers, Inc.

(a) (Above) A mine in California. The mineral trona is taken from a mine 1600 feet deep.

Courtesy of FMC Wyoming Corp.

(b) (Right) Blocks of trona are cut from the face of the mine.

and, in doing so, causes the NaCl to clump. Adding $NaHCO_3$ converts $MgCl_2$ to magnesium carbonate, a nonhygroscopic salt.

$$MgCl_2(s) + 2\ NaHCO_3(s) \rightarrow MgCO_3(s) + 2\ NaCl(s) + H_2O(\ell) + CO_2(g)$$

Large deposits of sodium nitrate, $NaNO_3$, are found in Chile, which explains its common name of "Chile saltpeter." These deposits are thought to have formed by bacterial action on organisms in shallow seas. The initial product was ammonia, which was subsequently oxidized to nitrate ion; combination with sea salt led to sodium nitrate. Because nitrates in general, and alkali metal nitrates in particular, are highly water-soluble, deposits of $NaNO_3$ are found only in areas with very little rainfall.

Sodium nitrate is important because it can be converted to potassium nitrate by an exchange reaction.

$$NaNO_3(aq) + KCl(aq) \rightleftharpoons KNO_3(aq) + NaCl(s)$$

Equilibrium favors the products here because, of the four salts involved in this reaction, NaCl is least soluble in hot water. Sodium chloride precipitates, and the KNO_3 that remains in solution can be recovered by evaporating the water.

Potassium nitrate has been used for centuries as the oxidizing agent in gunpowder. A mixture of KNO_3, charcoal, and sulfur will spontaneously react when ignited.

$$2\ KNO_3(s) + 4\ C(s) \rightarrow K_2CO_3(s) + 3\ CO(g) + N_2(g)$$

$$2\ KNO_3(s) + 2\ S(s) \rightarrow K_2SO_4(s) + SO_2(g) + N_2(g)$$

Notice that both reactions (which are doubtless more complex than those written here) produce gases. These gases propel the bullet from a gun or cause a firecracker to explode.

Lithium carbonate, Li_2CO_3, has been used for more than 40 years as a treatment for bipolar disorder, an illness that involves alternating periods of depression or overexcitement that can extend over a few weeks to a year or more. Although the alkali metal salt is efficient in controlling the symptoms of bipolar disorder, its mechanism of action is not understood.

EXERCISE 21.6 Brine Electrolysis

What current must be used in a Downs cell operating at 7.0 V to produce 1.00 metric ton (exactly 1000 kg) of sodium per day? Assume 100% efficiency.

21.5 The Alkaline Earth Elements, Group 2A

The "earth" part of the name alkaline earth dates back to the days of medieval alchemy. To alchemists, any solid that did not melt and was not changed by fire into another substance was called an "earth." Compounds of the Group 2A elements, such CaO, were alkaline according to experimental tests conducted by the alchemists: they had a bitter taste and neutralized acids. With very high melting points, these compounds were unaffected by fire.

Calcium and magnesium rank fifth and eighth, respectively, in abundance on the earth. Both elements form many commercially important compounds, and we shall focus our attention on these species.

Like the Group 1A elements, the Group 2A elements are very reactive, so they are found in nature as compounds. Unlike most of the compounds of the Group 1A metals, however, many compounds of the Group 2A elements have low water solubility, which explains their occurrence in various minerals (Figure 21.12).

Group 2A
Alkaline earths

Name	Atomic number	Symbol	Abundance
Beryllium	4	Be	2.6 ppm
Magnesium	12	Mg	23,300 ppm
Calcium	20	Ca	41,000 ppm
Strontium	38	Sr	370 ppm
Barium	56	Ba	500 ppm
Radium	88	Ra	6×10^{-7} ppm

Element abundances are in parts per million in the earth's crust.

Charles D. Winters

Common minerals of Group 2A elements.

Charles D. Winters

Icelandic spar. This mineral, one of a number of crystalline forms of $CaCO_3$, displays birefringence, a property in which a double image is formed when light passes through the crystal.

James Cowlin/Images Enterprises, Phoenix, AZ

The walls of the Grand Canyon in Arizona are largely limestone or dolomite.

FIGURE 21.12 Various minerals containing calcium and magnesium.

Calcium minerals include limestone ($CaCO_3$), gypsum ($CaSO_4 \cdot 2\ H_2O$), and fluorite (CaF_2). Magnesite ($MgCO_3$), talc or soapstone ($3\ MgO \cdot 4\ SiO_2 \cdot H_2O$), and asbestos ($3\ MgO \cdot 4\ SiO_2 \cdot 2\ H_2O$) are common magnesium-containing minerals. The mineral dolomite, $MgCa(CO_3)_2$, contains both magnesium and calcium.

Limestone, a sedimentary rock, is found widely on the earth's surface. Many of these deposits contain the fossilized remains of marine life. Other forms of calcium carbonate include marble and Icelandic spar, the latter occurring as large, clear crystals with the interesting optical property of birefringence.

Properties of Calcium and Magnesium

Calcium and magnesium are fairly high-melting, silvery metals. The chemical properties of these elements present few surprises. They are oxidized by a wide range of oxidizing agents to form ionic compounds that contain the M^{2+} ion. For example, these elements combine with halogens to form MX_2, with oxygen or sulfur to form MO or MS, and with water to form hydrogen and the metal hydroxide, $M(OH)_2$ (Figure 21.13). With acids, hydrogen is evolved (see Figure 21.6), and a salt of the metal cation and the anion of the acid results.

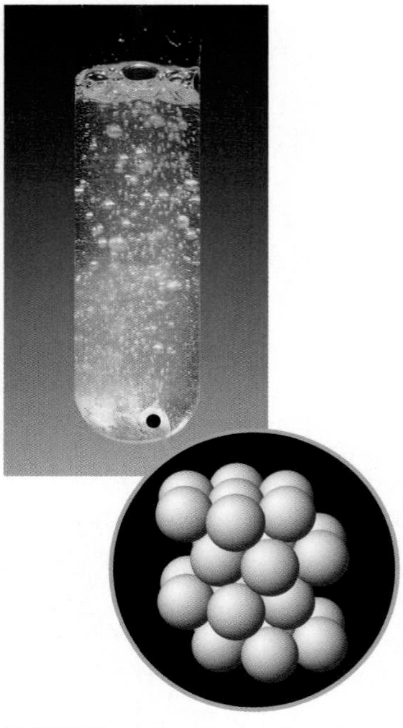

Charles D. Winters

FIGURE 21.13 The reaction of calcium and warm water. Hydrogen bubbles are seen rising from the metal surface. The other reaction product is $Ca(OH)_2$. The inset is a model of hexagonal close-packed calcium metal (see page 595).

Metallurgy of Magnesium

Several hundred thousand tons of magnesium are produced annually, largely for use in lightweight alloys. (Magnesium has a very low density, 1.74 g/cm^3.) Most aluminum used today contains about 5% magnesium to improve its mechanical properties and to make it more resistant to corrosion. Other alloys having more magnesium than aluminum are used when a high strength-to-weight ratio is needed and when corrosion resistance is important, such as in aircraft and automotive parts and in lightweight tools.

Interestingly, magnesium-containing minerals are not the source of this element. Most magnesium is obtained from sea water, in which Mg^{2+} ion is present in a concentration of about 0.05 M. To obtain magnesium metal, magnesium ions in sea water are first precipitated (Figure 21.14) as the relatively insoluble hydroxide [K_{sp} for $Mg(OH)_2 = 5.6 \times 10^{-12}$]. Calcium hydroxide, the source of OH^- in this reaction, is prepared in a sequence of reactions beginning with $CaCO_3$, which may be in the form of seashells. Heating $CaCO_3$ gives CO_2 and CaO, and addition of

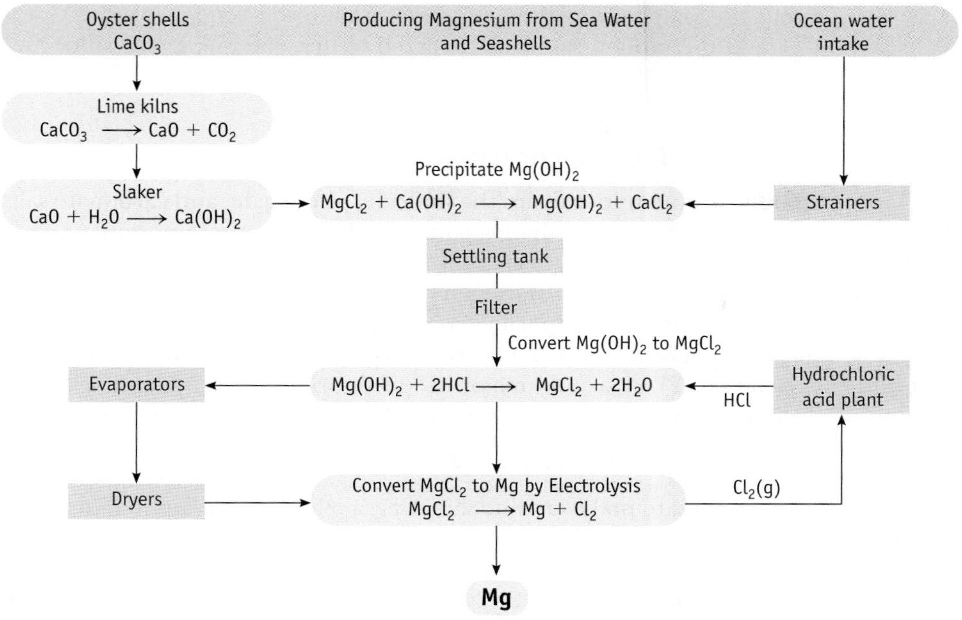

FIGURE 21.14 The process used to produce magnesium metal from the magnesium in sea water.

water to CaO gives calcium hydroxide. When $Ca(OH)_2$ is added to sea water, $Mg(OH)_2$ precipitates:

$$Mg^{2+}(aq) + Ca(OH)_2(s) \rightleftharpoons Mg(OH)_2(s) + Ca^{2+}(aq)$$

Magnesium hydroxide is isolated by filtration and then converted to magnesium chloride by reaction with hydrochloric acid.

$$Mg(OH)_2(s) + 2\ HCl(aq) \rightarrow MgCl_2(aq) + 2\ H_2O(\ell)$$

Chemical Perspectives — Alkaline Earth Metals and Biology

Plants and animals derive energy from the oxidation of a sugar, glucose, with oxygen. Plants are unique, however, in being able to synthesize glucose from CO_2 and H_2O by using sunlight as an energy source. This process is initiated by chlorophyll, a very large, magnesium-based molecule.

In your body, the metal ions Na^+, K^+, Mg^{2+}, and Ca^{2+} serve regulatory functions. Although the two alkaline earth metal ions are required by living systems, the other Group 2A elements are toxic. Beryllium compounds are carcinogenic, and soluble barium salts are poisons. You may be concerned if your physician asks you to drink a "barium cocktail" to check the condition of your digestive tract. Don't be afraid, because the "cocktail" contains very insoluble $BaSO_4$ ($K_{sp} = 1.1 \times 10^{-10}$), so it passes through your digestive tract without a significant amount being absorbed. Barium sulfate is opaque to x-rays, so its path through your organs appears on the developed x-ray.

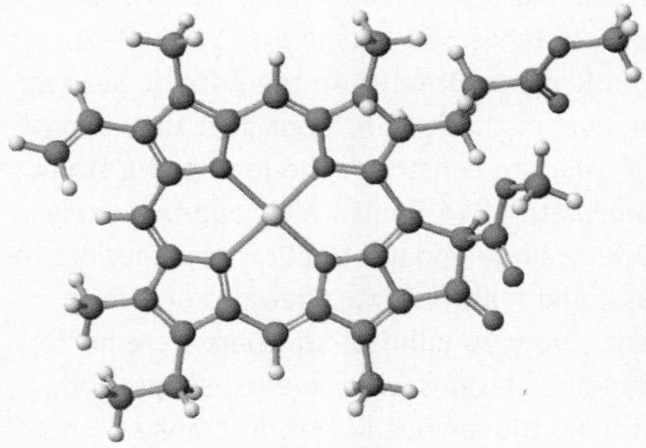

A molecule of chlorophyll. Magnesium is its central element.

The calcium-containing compound hydroxyapatite is the main component of tooth enamel. Cavities in your teeth form when acids (such as soft drinks) decompose the weakly basic hydroxyapatite coating.

$$Ca_5(OH)(PO_4)_3(s) + 4\ H_3O^+(aq) \rightarrow 5\ Ca^{2+}(aq) + 3\ HPO_4^{2-}(aq) + 5\ H_2O(\ell)$$

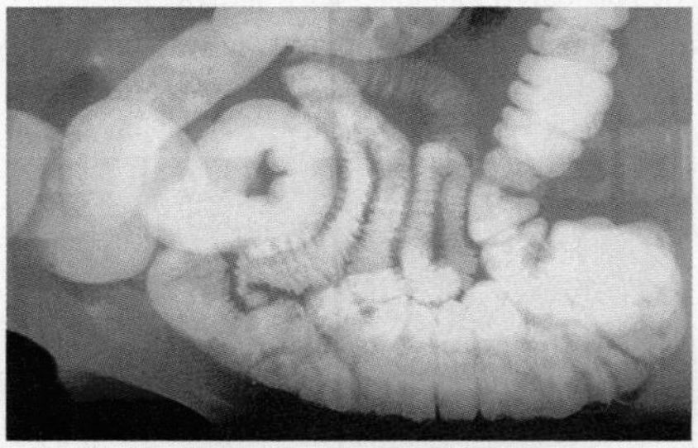

Susan Leavines/Science Source/Photo Researchers, Inc.

X-ray of a gastrointestinal tract using $BaSO_4$ to make the organs visible.

This reaction can be prevented by converting hydroxyapatite to the much more acid-resistant coating of fluoroapatite.

$$Ca_5(OH)(PO_4)_3(s) + F^-(aq) \rightarrow Ca_5F(PO_4)_3(s) + OH^-(aq)$$

The source of the fluoride ion can be sodium fluoride or sodium monofluorophosphate (Na_2FPO_3, commonly known as MFP) in your toothpaste.

After evaporating the water, anhydrous magnesium chloride remains. Solid $MgCl_2$ melts at 714 °C, and the molten salt is electrolyzed to give the metal and chlorine.

$$MgCl_2(\ell) \rightarrow Mg(s) + Cl_2(g)$$

Calcium Minerals and Their Applications

The most common calcium minerals are the fluoride, phosphate, and carbonate salts of the element. Fluorite, CaF_2, and fluoroapatite, $Ca_5F(PO_4)_3$, are important as commercial sources of fluorine. Almost half of the CaF_2 mined is used in the steel industry, where it is added to the mixture of materials that is melted to make crude iron. The CaF_2 acts to remove some impurities and improves the separation of molten metal from silicates and other by-products resulting from the reduction of iron ore to the metal (Chapter 22). A second major use of fluorite is in the manufacture of hydrofluoric acid by a reaction of the mineral with concentrated sulfuric acid.

$$CaF_2(s) + H_2SO_4(\ell) \rightarrow 2\ HF(g) + CaSO_4(s)$$

Hydrofluoric acid is used to make cryolite, Na_3AlF_6, a material needed in aluminum production (▶ Section 21.6) and in the manufacture of fluorocarbons such as tetrafluoroethylene, the precursor to Teflon (Table 10.12).

Apatites have the general formula $Ca_5X(PO_4)_3$ (X = F, Cl, OH). More than 100 million tons of apatite is mined annually, with Florida alone accounting for about one third of the world's output. Most of this material is converted to phosphoric acid by reaction with sulfuric acid. Phosphoric acid is needed in the manufacture of a multitude of products, including fertilizers and detergents, baking powder, and various food products (▶ Section 21.8.).

Chemical Perspectives — Of Romans, Limestone, and Champagne

The stones of the Appian Way in Italy, a road conceived by the Roman senate in about 310 B.C., are cemented with mortar made from limestone. The Appian Way was intended to serve as a military road linking Rome to seaports from which soldiers could embark to Greece and other Mediterranean ports. The road stretches 560 kilometers (350 miles) from Rome to Brindisi on the Adriatic Sea (at the heel of the Italian "boot"). It took almost 200 years to construct. The road had a standard width of 14 Roman feet, approximately 20 feet, large enough to allow two chariots to pass, and featured two sidewalks of 4 feet each. Every 10 miles or so, there were horse-changing stations with taverns, shops, and *latrinae*, the famous Roman restrooms.

All over the Roman Empire, buildings, temples, and aqueducts were constructed of blocks of limestone and marble, and the mortar to cement the blocks was made by heating the chips from stone cutting (to give CaO). In central France, the Romans dug chalk (also $CaCO_3$) from the ground for cementing sandstone blocks. This activity created huge caves that remain to this day and are used for aging and storing champagne.

Louis Goldman/Photo Researchers, Inc.

The Appian Way in Italy.

Roy/Explorer/Photo Researchers, Inc.

Champagne in a limestone cave in France.

Calcium carbonate and calcium oxide (*lime*) are of special interest. The thermal decomposition of $CaCO_3$ to give lime (and CO_2) is one of the oldest chemical reactions known. Lime is one of the top 10 industrial chemicals produced today, with about 20 billion kilograms produced annually.

Limestone, which consists mostly of calcium carbonate, has been used in agriculture for centuries. It is spread on fields to neutralize acidic compounds in the soil and to supply Ca^{2+}, an essential nutrient. Because magnesium carbonate is often present in limestone, "liming" a field also supplies Mg^{2+}, another important nutrient for plants.

For several thousand years, lime has been used in *mortar* (a lime, sand, and water paste) to secure stones to one another in building houses, walls, and roads. The Chinese used it to set stones in the Great Wall. The Romans perfected its use, and the fact that many of their constructions still stand today is testament both to their skill and to the usefulness of lime. The famous Appian Way used lime mortar between several layers of its stones.

The utility of mortar depends on some simple chemistry. Mortar consists of one part lime to three parts sand, with water added to make a thick paste. The first reaction, referred to as *slaking*, occurs after the solids are mixed with water. This produces a slurry containing calcium hydroxide, which is known as *slaked lime*.

$$CaO(s) + H_2O(\ell) \rightleftharpoons Ca(OH)_2(s)$$

When the wet mortar mix is placed between bricks or stone blocks, it slowly reacts with CO_2 from the air, and the slaked lime is converted to calcium carbonate.

$$Ca(OH)_2(s) + CO_2(g) \rightleftharpoons CaCO_3(s) + H_2O(\ell)$$

The sand grains are bound together by the particles of calcium carbonate.

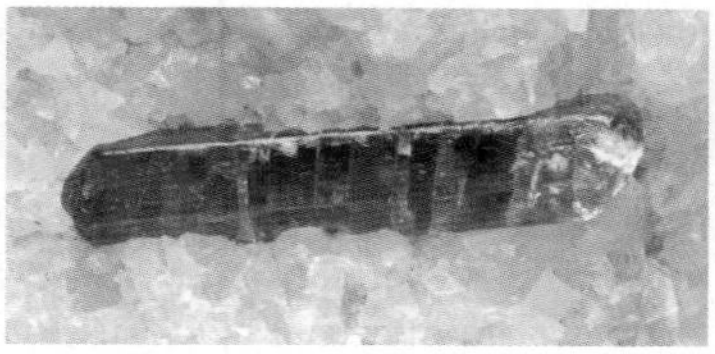
Charles D. Winters

Apatite. The mineral has the general formula of $Ca_5X(PO_4)_3$ (X = F, Cl, OH). (The apatite is the elongated crystal in the center of a matrix of other rock.)

■ **Dissolving Limestone** Figure 3.6 illustrates the equilibrium involving $CaCO_3$, CO_2, H_2O, Ca^{2+}, and HCO_3^-.

EXERCISE 21.7 Beryllium Chemistry

Beryllium, the lightest element in Group 2A, has some important industrial applications, but exposure (by breathing) to some of its compounds can cause berylliosis. Search the World Wide Web for the uses of the element and the causes and symptoms of berylliosis.

21.6 Boron, Aluminum, and the Group 3A Elements

With Group 3A, we see the first evidence of a change from metallic behavior of the elements at the left side of the periodic table to nonmetal behavior on the right side of the table. Boron is a metalloid, whereas all the other elements of Group 3A are metals.

The elements of Group 3A vary widely in their relative abundances on earth. Aluminum is the third most abundant element in the earth's crust (82,000 ppm), whereas the other elements of the group are relatively rare, and, except for boron, their compounds have limited commercial uses.

Group 3A
Boron 5 B 10 ppm
Aluminum 13 Al 82,000 ppm
Gallium 31 Ga 18 ppm
Indium 49 In 0.05 ppm
Thallium 81 Tl 0.6 ppm

Element abundances are in parts per million in the earth's crust.

Chemistry of the Group 3A Elements

It is generally recognized that a chemical similarity exists between some elements diagonally situated in the periodic table. This diagonal relationship means that lithium and magnesium share some chemical properties, as do Be and Al, and B and Si. For example:

- Boric oxide, B_2O_3, and boric acid, $B(OH)_3$, are weakly acidic, as are SiO_2 and its acid, orthosilic acid (H_4SiO_4). Boron–oxygen compounds, borates, are often chemically similar to silicon–oxygen compounds, silicates.
- Both $Be(OH)_2$ and $Al(OH)_3$ are amphoteric, dissolving in a strong base such as aqueous NaOH (◀ page 791).

Case Study — Hard Water

No, hard water doesn't refer to ice. It is the name given to water containing high concentrations of $Ca^{2+}(aq)$ and, in some instances, $Mg^{2+}(aq)$ and other divalent metal cations. Accompanying these cations will be various anions, including in particular the hydrogen carbonate anion, $HCO_3^-(aq)$.

Water obtained from reservoirs that store rainwater does not usually contain high concentrations of these ions, and so is not classified as being "hard water." In many parts of the country, however, the municipal water supply is obtained from aquifers deep underground. If rainwater, containing some dissolved CO_2, has to percolate down through layers of limestone ($CaCO_3$) to get into the aquifer, a small amount of the solid will dissolve because of the following equilibrium:

$$CaCO_3(s) + CO_2(aq) + H_2O(\ell) \rightleftharpoons Ca^{2+}(aq) + 2\ HCO_3^-(aq)$$

If hard water containing $Ca^{2+}(aq)$ and $HCO_3^-(aq)$ is heated, or even if it is left to stand in an open vessel, CO_2 will be expelled, and the equilibrium will shift, precipitating $CaCO_3(s)$ (◀ page 120). This can be a small problem in a teakettle but a bigger problem in an industrial setting, where the solid $CaCO_3$ can clog pipes. Another consequence of hard water is the formation of a scum when soaps are added to water. Recall that soap is made by hydrolysis of fats (saponification, page 476); soap scum is a precipitate of the calcium salt of a long chain carboxylic acid.

To avoid the problems associated with hard water, chemists and engineers have devised ways to soften water, that is, to decrease the concentration of the offending cations. In a water treatment plant for a municipality or a large industrial facility, most of the water hardness will be removed chemically, mainly by treatment with calcium oxide (lime, CaO). If HCO_3^- ions are present along with Ca^{2+} and/or Mg^{2+}, the following reactions occur:

$$Ca^{2+}(aq) + 2\ HCO_3^-(aq) + CaO(s) \rightarrow 2\ CaCO_3(s) + H_2O(\ell)$$

$$Mg^{2+}(aq) + 2\ HCO_3^-(aq) + CaO(s) \rightarrow CaCO_3(s) + MgCO_3(s) + H_2O(\ell)$$

Although it seems odd to add CaO to remove calcium ions, notice that adding one mole of CaO leads to the precipitation of two moles of Ca^{2+} as $CaCO_3$.

On a smaller scale, most home water purification systems use ion exchange to soften water (Figure). This process involves the replacement of an ion adsorbed onto a solid ion-exchange resin by an ion in solution. Zeolites (▶ page 989), naturally occurring aluminosilicate minerals, were at one time used as ion exchange materials. Now, synthetic organic polymers with negatively charged functional groups (such as carboxylate groups, $—CO_2^-$) are the most commonly used resins for this purpose. Sodium ions, Na^+, are present in the resin to balance the negative charges of the carboxylate groups. The affinity of a surface for multicharged cations is greater than for monovalent cations. Therefore, when a solution of hard water is passed over the surface of the ion exchange resin, Ca^{2+} and Mg^{2+} (and other divalent ions if present) readily replace Na^+ ions in an ion exchange column. This process can be illustrated in a general way by the equilibrium

$$2\ NaX + Ca^{2+}(aq) \rightleftharpoons CaX_2 + 2\ Na^+(aq)$$

where X represents an adsorption site on the ion exchange resin. The equilibrium favors adsorption of Ca^{2+} and release of Na^+. However, the equilibrium is reversed if $Na^+(aq)$ ions are present in high concentration, and this allows regeneration of the ion-exchange resin. A solution containing a high concentration of Na^+ (usually from salt, NaCl) is passed through the resin to convert the resin back to its initial form.

Questions:

1. *Assume that a sample of hard water contains 50 mg/L of Mg^{2+} and 150 mg/L of Ca^{2+}, with HCO_3^- as the accompanying anion. What mass of CaO should be added to 1.0 L of this aqueous solution to cause precipitation of $CaCO_3$ and $MgCO_3$? What is the total mass of the two solids formed?*
2. *One way to remove the calcium carbonate residue in a teakettle is to add vinegar (a dilute solution of acetic acid). Write a chemical equation to explain this process. What kind of a reaction is this?*

Answers to these questions are in Appendix Q.

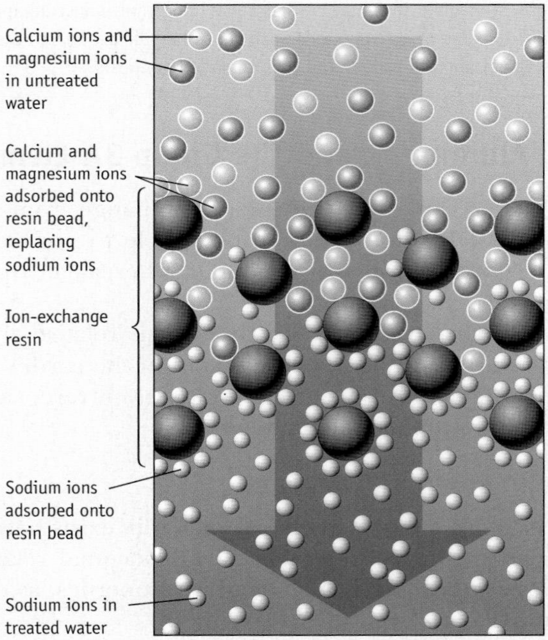

(a) The general operation of an ion-exchange resin. The ion-exchange material is usually a polymeric material formed into small beads.

Water softening by ion exchange in the home. The ion exchange resin is usually a polymeric material formed into small beads.

- Chlorides, bromides, and iodides of boron and silicon (such as BCl_3 and $SiCl_4$) react vigorously with water.
- The hydrides of boron and silicon are simple, molecular species; are volatile and flammable; and react readily with water.
- Beryllium hydride and aluminum hydride are colorless, nonvolatile solids that are extensively polymerized through Be—H—Be and Al—H—Al three-centered, two-electron bonds (see page 431).

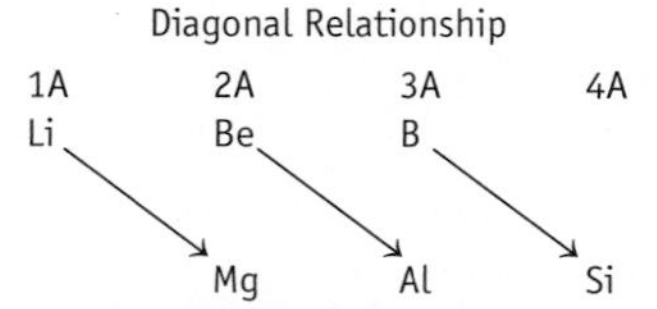

■ **Diagonal Relationship** The chemistries of elements diagonally situated in the periodic table are often quite similar.

Finally, the Group 3A elements are characterized by electron configurations of the type ns^2np^1. This means that each may lose three electrons to have a +3 oxidation number, although the heavier elements, especially thallium, also form compounds with an oxidation number of +1.

Boron Minerals and Production of the Element

Although boron has a low abundance on earth, its minerals are found in concentrated deposits. Large deposits of borax, $Na_2B_4O_7 \cdot 10\ H_2O$, are currently mined in the Mojave Desert near the town of Boron, California (Figure 21.15).

Isolation of pure, elemental boron from boron-containing minerals is extremely difficult and is done in small quantities. Like most metals and metalloids, boron can be obtained by chemically or electrolytically reducing an oxide or halide. Magnesium has often been used for chemical reductions, but the product of this reaction is a noncrystalline boron of low purity.

$$B_2O_3(s) + 3\ Mg(s) \rightarrow 2\ B(s) + 3\ MgO(s)$$

Boron has several allotropes, all characterized by having an icosahedron of boron atoms as one structural element (Figure 21.15c). Partly as a result of extended covalent bonding, elemental boron is a very hard and refractory (resistant to heat) semiconductor. In this regard, it differs from the other Group 3A elements; Al, Ga, In, and Tl are all relatively low-melting, rather soft metals with high electrical conductivity.

Charles D. Winters

Gallium. Gallium, with a melting point of 29.8 °C, is one of the few metals that can be a liquid at or near room temperature. (Others are Hg and Cs.)

Metallic Aluminum and Its Production

The low cost of aluminum and the excellent characteristics of its alloys with other metals (low density, strength, ease of handling in fabrication, and inertness toward corrosion, among others), have led to its widespread use. You know it best in the form of aluminum foil, aluminum cans, and parts of aircraft.

© George Gerster/Photo Researchers, Inc.

(a) (b) (c)

FIGURE 21.15 Boron. (a) A borax mine near the town of Boron, California. (b) Crystalline borax, $Na_2B_4O_7 \cdot 10\ H_2O$. (c) All allotropes of elemental boron have an icosahedron (a 20-sided polyhedron) of 12 covalently linked boron atoms as a structural element.

Pure aluminum is soft and weak; moreover, it loses strength rapidly at temperatures higher than 300 °C. What we call "aluminum" is actually aluminum alloyed with small amounts of other elements to strengthen the metal and improve its properties. A typical alloy may contain about 4% copper with smaller amounts of silicon, magnesium, and manganese. Softer, more corrosion-resistant alloys for window frames, furniture, highway signs, and cooking utensils may include only manganese.

The standard reduction potential of aluminum [$Al^{3+}(aq) + 3\ e^- \rightarrow Al(s)$; $E° = -1.66$ V] tells you that aluminum is easily oxidized. From this, we might expect aluminum to be highly susceptible to corrosion but, in fact, it is quite resistant. Aluminum's corrosion resistance is due to the formation of a thin, tough, and transparent skin of Al_2O_3 that adheres to the metal surface. An important feature of the protective oxide layer is that it rapidly self-repairs. If you penetrate the surface coating by scratching it or using some chemical agent, the exposed metal surface immediately reacts with oxygen (or another oxidizing agent) to form a new layer of oxide over the damaged area (Figure 21.16).

■ **Charles Martin Hall (1863–1914)** Hall was only 22 years old when he worked out the electrolytic process for extracting aluminum from Al_2O_3 in a woodshed behind the family home in Oberlin, Ohio. He went on to found a company that eventually became ALCOA, the Aluminum Corporation of America.

Oesper Collection in the History of Chemistry/ University of Cincinnati

Aluminum was first prepared by reducing $AlCl_3$ using sodium or potassium. This was a costly process, and, in the 19th century, aluminum was a precious metal. At the 1855 Paris Exposition, in fact, a sample of aluminum was exhibited along with the crown jewels of France. In an interesting coincidence, in 1886 Frenchman Paul Heroult (1863–1914) and American Charles Hall (1863–1914) simultaneously and independently conceived of the electrochemical method used today. The Hall–Heroult method bears the names of the two discoverers.

Aluminum is found in nature as aluminosilicates, minerals such as clay that are based on aluminum, silicon, and oxygen. As these minerals weather, they break down to various forms of hydrated aluminum oxide, $Al_2O_3 \cdot n\ H_2O$, called *bauxite.* Mined in huge quantities, bauxite is the raw material from which aluminum is obtained. The first step is to purify the ore, separating Al_2O_3 from iron and silicon oxides. This is done by the *Bayer process,* which relies on the amphoteric, basic, or acidic nature of the various oxides. Silica, SiO_2, is an acidic oxide; Al_2O_3 is amphoteric; and Fe_2O_3 is a basic oxide. Silica and Al_2O_3 dissolve in a hot concentrated solution of caustic soda (NaOH), leaving insoluble Fe_2O_3 to be filtered out.

$$Al_2O_3(s) + 2\ NaOH(aq) + 3\ H_2O(\ell) \rightarrow 2\ Na[Al(OH)_4](aq)$$

$$SiO_2(s) + 2\ NaOH(aq) + 2\ H_2O(\ell) \rightarrow Na_2[Si(OH)_6](aq)$$

If a solution containing aluminate and silicate anions is treated with CO_2, Al_2O_3 precipitates, and the silicate ion remains in solution. Recall that CO_2 is an acidic

Photos: Charles D. Winters

(a) (b)

FIGURE 21.16 Corrosion of aluminum. (a) A ball of aluminum foil is added to a solution of copper(II) nitrate and sodium chloride. Normally, the coating of chemically inert Al_2O_3 on the surface of aluminum protects the metal from further oxidation. (b) In the presence of the Cl^- ion, the coating of Al_2O_3 is breached, and aluminum reduces copper(II) ions to copper metal. The reaction is rapid and so exothermic that the water can boil on the surface of the foil. [The blue color of aqueous copper(II) ions will fade as these ions are consumed in the reaction.]

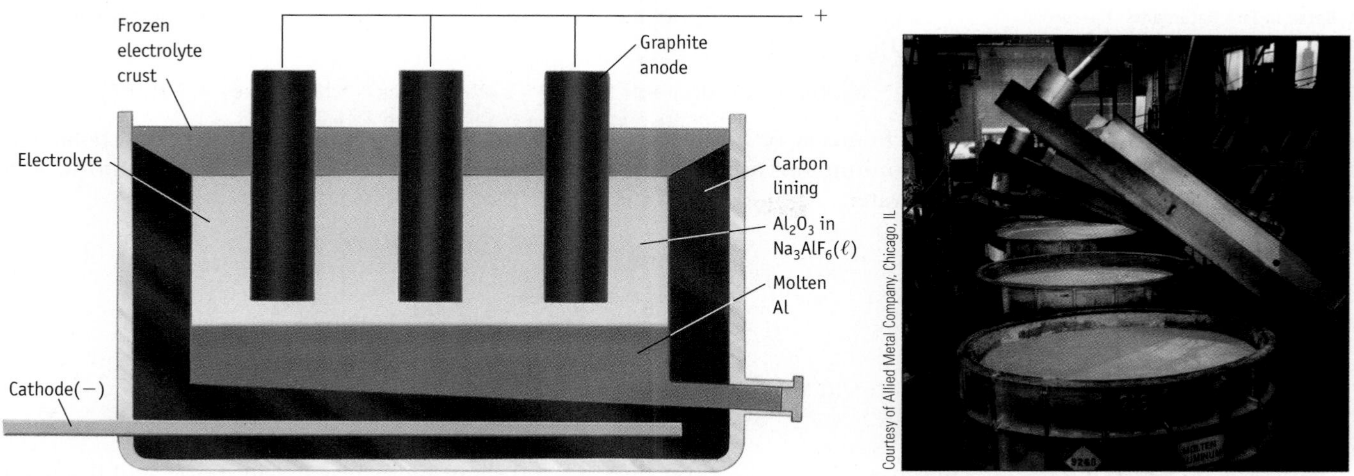

(a) Electrolysis of aluminum oxide to produce aluminum metal.

(b) Molten aluminum from recycled metal.

Active Figure 21.17 **Industrial production of aluminum.** (a) Purified aluminum-containing ore (bauxite), essentially Al_2O_3, is mixed with cryolite (Na_3AlF_6) to give a mixture that melts at a lower temperature than Al_2O_3 alone. The aluminum-containing substances are reduced at the steel cathode to give molten aluminum. Oxygen is produced at the graphite anode, and the gas reacts slowly with the carbon to give CO_2, leading to eventual destruction of the electrode. (b) Molten aluminum alloy, produced from recycled metal, at 760 °C, in 1.6×10^4-kg capacity crucibles.

Chemistry Now™ Sign in at www.thomsonedu.com/login and go to the Chapter Contents menu to explore an interactive version of this figure accompanied by an exercise.

oxide that forms the weak acid H_2CO_3 in water, so the Al_2O_3 precipitation in this step is an acid–base reaction.

$$H_2CO_3(aq) + 2\ Na[Al(OH)_4](aq) \rightarrow Na_2CO_3(aq) + Al_2O_3(s) + 5\ H_2O(\ell)$$

Metallic aluminum is obtained from purified bauxite by electrolysis (Figure 21.17). Bauxite is first mixed with cryolite, Na_3AlF_6, to give a lower-melting mixture (melting temperature = 980 °C) that is electrolyzed in a cell with graphite electrodes. The cell operates at a relatively low voltage (4.0–5.5 V) but with an extremely high current (50,000–150,000 A). Aluminum is produced at the cathode and oxygen at the anode. To produce 1 kg of aluminum requires 13 to 16 kilowatt-hours of energy plus the energy required to maintain the high temperature.

Boron Compounds

Borax, $Na_2B_4O_7 \cdot 10\ H_2O$, is the most important boron–oxygen compound and is the form of the element most often found in nature. It has been used for centuries in metallurgy because of the ability of molten borax to dissolve other metal oxides. That is, borax is used as a *flux* that cleans the surfaces of metals to be joined and permits a good metal-to-metal contact.

The formula for borax gives little information about its structure. The anion is better described by the formula $[B_4O_5(OH)_4]^{2-}$, the structure of which illustrates two commonly observed structural features in inorganic chemistry. First, many minerals consist of MO_n groups that share O atoms. Second, the sharing of O atoms between two metals or metalloids often leads to MO rings.

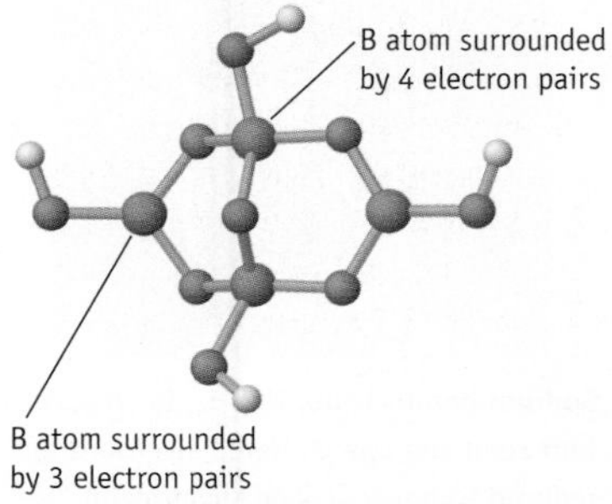

The borate ion of borax, $[B_4O_5(OH)_4]^{2-}$.

■ **Borax in Fire Retardants** The second largest use for boric acid and borates is as a flame retardant for cellulose home insulation. Such insulation is often made of scrap paper, which is inexpensive but flammable. To control the flammability, 5–10% of the weight of the insulation is boric acid.

After refinement, borax can be treated with sulfuric acid to produce boric acid, $B(OH)_3$.

$$Na_2B_4O_7 \cdot 10\ H_2O(s) + H_2SO_4(aq) \rightarrow 4\ B(OH)_3(aq) + Na_2SO_4(aq) + 5\ H_2O(\ell)$$

The chemistry of boric acid incorporates both Lewis and Brønsted acid behavior. Hydronium ions are produced by a Lewis–acid base interaction between boric acid and water.

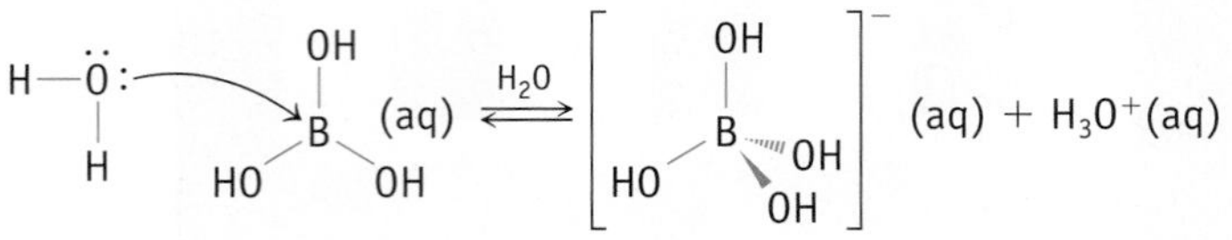

$$K_a = 7.3 \times 10^{-10}$$

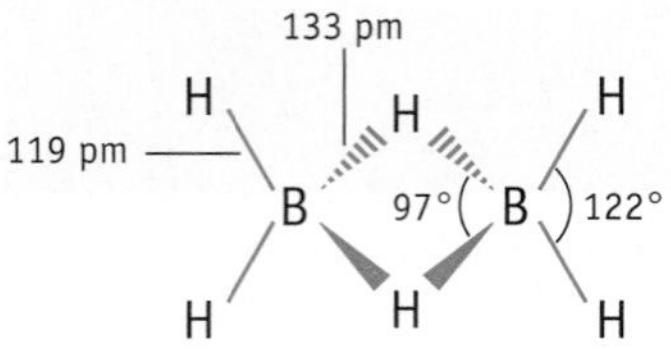

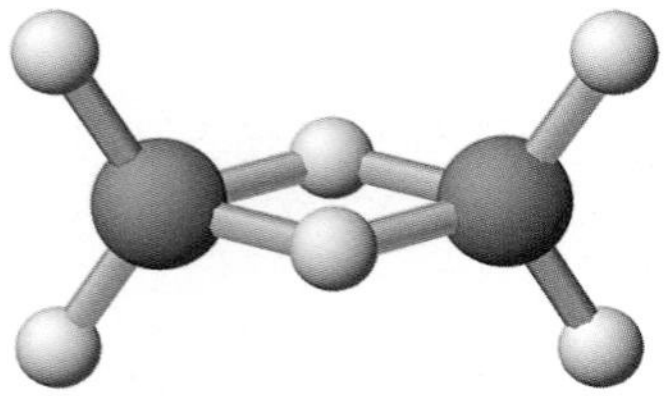

The structure of diborane, B_2H_6, the simplest member of a family of boron hydrides. See also page 431.

Because of its weak acid properties and slight biological activity, boric acid has been used for many years as an antiseptic. Furthermore, because boric acid is a weak acid, salts of borate ions, such as the $[B_4O_5(OH)_4]^{2-}$ ion in borax, are weak bases.

Boric acid is dehydrated to boric oxide when strongly heated.

$$2\ B(OH)_3(s) \rightarrow B_2O_3(s) + 3\ H_2O(\ell)$$

By far the largest use for the oxide is in the manufacture of borosilicate glass. This type of glass is composed of 76% SiO_2, 13% B_2O_3, and much smaller amounts of Al_2O_3 and Na_2O. The presence of boric oxide gives the glass a higher softening temperature, imparts a better resistance to attack by acids, and makes the glass expand less on heating.

Like its metalloid neighbor silicon, boron forms a series of molecular compounds with hydrogen. Because boron is slightly less electronegative than hydrogen, these compounds are described as hydrides, in which the H atoms bear a partial negative charge. More than 20 neutral boron hydrides, or boranes, with the general formula B_xH_y are known. The simplest of these is diborane, B_2H_6, where x is 2 and y is 6. This colorless, gaseous compound has a boiling point of −92.6 °C. This molecule is described as electron deficient since there are apparently not enough electrons to attach all of the atoms using two-electron bonds. Instead, the description of bonding uses 3-center 2-electron bonds in the B-H-B bridges (◄ page 431).

Diborane has an endothermic enthalpy of formation ($\Delta_f H° = +41.0$ kJ/mol), which contributes to the high exothermicity of oxidation of the compound. Diborane burns in air to give boric oxide and water vapor in an extremely exothermic reaction. It is not surprising that diborane and other boron hydrides were once considered as possible rocket fuels.

$$B_2H_6(g) + 3\ O_2(g) \rightarrow B_2O_3(s) + 3\ H_2O(g) \quad \Delta_r H° = -2038\ \text{kJ/mol-rxn}$$

Diborane can be synthesized from sodium borohydride, $NaBH_4$, the only B—H compound produced in ton quantities.

$$2\ NaBH_4(s) + I_2(s) \rightarrow B_2H_6(g) + 2\ NaI(s) + H_2(g)$$

Sodium borohydride, $NaBH_4$, a white, crystalline, water-soluble solid, is made from NaH and borate esters such as $B(OCH_3)_3$.

$$4\ NaH(s) + B(OCH_3)_3(g) \rightarrow NaBH_4(s) + 3\ NaOCH_3(s)$$

Charles D. Winters

Sodium borohydride, $NaBH_4$, is an excellent reducing agent. Here, silver ions are reduced to finely divided silver metal.

The main use of $NaBH_4$ is as a reducing agent in organic synthesis. We previously encountered its use to reduce aldehydes, carboxylic acids, and ketones in Chapter 10.

Chemistry Now™

Sign in at **www.thomsonedu.com/login** and go to Chapter 21 Contents to see Screen 21.4 for an exercise on **the structures of the family of boron-hydrogen compounds.**

Aluminum Compounds

Aluminum is an excellent reducing agent, so it reacts readily with hydrochloric acid. In contrast, it does not react with nitric acid, a stronger oxidizing agent than hydrochloric acid. It turns out that nitric acid does rapidly oxidize the surface of aluminum, but the resulting film of Al_2O_3 that is produced protects the metal from further attack. This protection allows nitric acid to be shipped in aluminum tanks.

Charles D. Winters

Aluminum does not react with nitric acid. Nitric acid, a strongly oxidizing acid, reacts vigorously with copper (left), but aluminum (right) is untouched.

Various salts of aluminum dissolve in water, giving the hydrated Al^{3+}(aq) ion. These solutions are acidic (◀ Table 17.3, page 770) because the hydrated ion is a weak Brønsted acid.

$$[Al(H_2O)_6]^{3+}(aq) + H_2O(\ell) \rightleftharpoons [Al(H_2O)_5(OH)]^{2+}(aq) + H_3O^+(aq)$$

Adding acid shifts the equilibrium to the left, whereas adding base causes the equilibrium to shift to the right. Addition of sufficient hydroxide ion results ultimately in precipitation of the hydrated oxide $Al_2O_3 \cdot 3\ H_2O$.

Aluminum oxide, Al_2O_3, formed by dehydrating the hydrated oxide, is quite insoluble in water and generally resistant to chemical attack. In the crystalline form, aluminum oxide is known as *corundum.* This material is extraordinarily hard, a property that leads to its use as an abrasive in grinding wheels, "sandpaper," and toothpaste.

Some gems are impure aluminum oxide. Rubies, beautiful red crystals prized for jewelry and used in some lasers, are composed of Al_2O_3 contaminated with a small amount of Cr^{3+} (Figure 21.18). The Cr^{3+} ions replacing some of the Al^{3+} ions in the crystal lattice is the source of the red color. Synthetic rubies were first made in 1902, and the worldwide capacity is now about 200,000 kg/year; much of this production is used for jewel bearings in watches and instruments. Blue sapphires consist of Al_2O_3 with Fe^{2+} and Ti^{4+} impurities in place of Al^{3+} ions.

Boron forms halides such as gaseous BF_3 and BCl_3 that have the expected planar, trigonal molecular geometry of halogen atoms surrounding an sp^2 hybridized boron atom. In contrast, the aluminum halides are all solids and have more interesting structures. Aluminum bromide, which is made by the very exothermic reaction of aluminum metal and bromine (Figure 2.12, page 67),

$$2\ Al(s) + 3\ Br_2(\ell) \rightarrow Al_2Br_6(s)$$

is composed of two units of $AlBr_3$. That is, Al_2Br_6 is a dimer of $AlBr_3$ units. The structure resembles that of diborane in that bridging atoms appear between the two Al atoms. However, Al_2Br_6 is not electron deficient; the bridge is formed when a Br atom on one $AlBr_3$ uses a lone pair to form a coordinate covalent bond to a neighboring tetrahedral, sp^3-hybridized aluminum atom.

Chip Clark, Smithsonian Museum of Natural History

FIGURE 21.18 Forms of corundum. Corundum is a crystalline form of aluminum oxide. Both rubies and sapphires are a form of corundum in which a few Al^{3+} ions have been replaced by ions such as Cr^{3+}, Fe^{2+} or Ti^{4+}. (*top*) The Star of Asia sapphire. (*middle*) Various sapphires. (*bottom*) Uncut corundum.

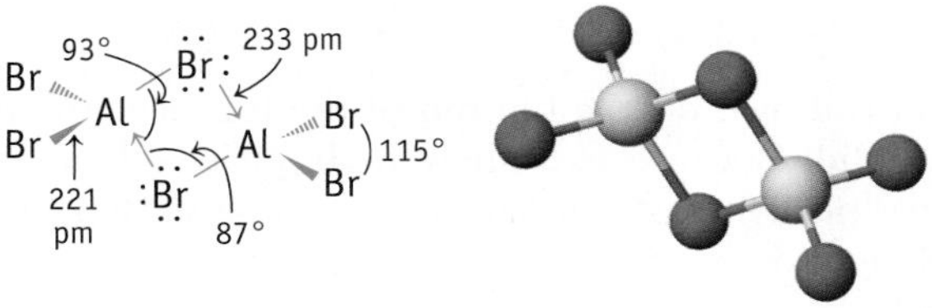

The structure of Al_2Br_6. The bonding in Al_2Br_6 is not unique to the aluminum halides. Metal–halogen–metal bridges are found in many other metal–halogen compounds.

Both aluminum bromide and aluminum iodide have this structure. Aluminum chloride has a different solid state structure, but it exists as dimeric molecules in the vapor state.

Aluminum chloride can react with a chloride ion to form the anion $[AlCl_4]^-$. Aluminum fluoride, in contrast, can accommodate three additional F^- ions to form an octahedral $[AlF_6]^{3-}$ ion. This anion is found in cryolite, Na_3AlF_6, the compound added to aluminum oxide in the electrolytic production of aluminum metal. Apparently, the Al^{3+} ion can bind to six of the smaller F^- ions, whereas only four of the larger Cl^-, Br^-, or I^- ions can surround an Al^{3+} ion.

Chemistry Now™

Sign in at **www.thomsonedu.com/login** and go to Chapter 21 Contents to see Screen 21.5 for an exercise on **the chemistry of aluminum compounds.**

EXERCISE 21.8 Gallium Chemistry

(a) Gallium hydroxide, like aluminum hydroxide, is amphoteric. Write a balanced equation to show how this hydroxide can dissolve in both HCl(aq) and NaOH(aq).

(b) Gallium ion in water, $Ga^{3+}(aq)$, has a K_a value of 1.2×10^{-3}. Is this ion a stronger or a weaker acid than $Al^{3+}(aq)$?

21.7 Silicon and the Group 4A Elements

Carbon is a nonmetal; silicon and germanium are classified as metalloids; tin and lead are metals. As a result, the elements of Group 4A have a broad range of chemical behavior.

The Group 4A elements are characterized by half-filled valence shells with two electrons in the *ns* orbital and two electrons in *np* orbitals. The bonding in carbon and silicon compounds is largely covalent and involves sharing four electron pairs with neighboring atoms. In germanium compounds, the +4 oxidation state is common (GeO_2 and $GeCl_4$), but some +2 oxidation state compounds exist (GeI_2). Oxidation numbers of both +2, and +4 are common in compounds of tin and lead (such as $SnCl_2$, $SnCl_4$, PbO, and PbO_2). Oxidation numbers two units less than the group number are often encountered for heavier elements in Groups 3A–7A.

Group 4A

Element	Atomic number	Symbol	Abundance
Carbon	6	C	480 ppm
Silicon	14	Si	277,000 ppm
Germanium	32	Ge	1.8 ppm
Tin	50	Sn	2.2 ppm
Lead	82	Pb	14 ppm

Element abundances are in parts per million in the earth's crust.

Silicon

Silicon is second after oxygen in abundance in the earth's crust, so it is not surprising that we are surrounded by silicon-containing materials: bricks, pottery, porcelain, lubricants, sealants, computer chips, and solar cells. The computer revolution is based on the semiconducting properties of silicon.

Reasonably pure silicon can be made in large quantities by heating pure silica sand with purified coke to approximately 3000 °C in an electric furnace.

$$SiO_2(s) + 2\ C(s) \rightarrow Si(\ell) + 2\ CO(g)$$

The molten silicon is drawn off the bottom of the furnace and allowed to cool to a shiny blue-gray solid. Because extremely high-purity silicon is needed for the electronics industry, purifying raw silicon requires several steps. First, the silicon in the impure sample is allowed to react with chlorine to convert the silicon to liquid silicon tetrachloride.

$$Si(s) + 2\ Cl_2(g) \rightarrow SiCl_4(\ell)$$

Silicon tetrachloride (boiling point of 57.6 °C) is carefully purified by distillation and then reduced to silicon using magnesium.

$$SiCl_4(g) + 2\ Mg(s) \rightarrow 2\ MgCl_2(s) + Si(s)$$

The magnesium chloride is washed out with water, and the silicon is remelted and cast into bars. A final purification is carried out by zone refining, a process in which a special heating device is used to melt a narrow segment of the silicon rod. The heater is moved slowly down the rod. Impurities contained in the silicon tend to remain in the liquid phase because the melting point of a mixture is lower than that of the pure element (Chapter 14). The silicon that crystallizes above the heated zone is therefore of a higher purity (Figure 21.19).

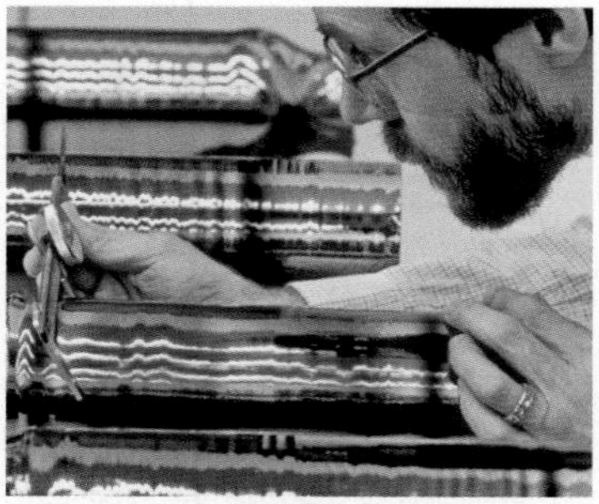
©Science Vu/Visuals Unlimited

FIGURE 21.19 Pure silicon. The manufacture of very pure silicon begins with producing the volatile liquid silanes $SiCl_4$ or $SiHCl_3$. After carefully purifying these by distillation, they are reduced to elemental silicon with extremely pure Mg or Zn. The resulting spongy silicon is purified by zone refining. The end result is a cylindrical rod of ultrapure silicon such as those seen in this photograph. Thin wafers of silicon are cut from the bars and are the basis for the semiconducting chips in computers and other devices.

Silicon Dioxide

The simplest oxide of silicon is SiO_2, commonly called *silica*, a constituent of many rocks such as granite and sandstone. Quartz is a pure crystalline form of silica, but impurities in quartz produce gemstones such as amethyst (Figure 21.20).

Silica and CO_2 are oxides of two elements in the same chemical group, so similarities between them might be expected. In fact, SiO_2 is a high-melting solid (quartz melts at 1610 °C), whereas CO_2 is a gas at room temperature and 1 bar. This great disparity arises from the different structures of the two oxides. Carbon dioxide is a molecular compound, with the carbon atom linked to each oxygen atom by a double bond. In contrast, SiO_2 is a network solid, which is the preferred structure because the bond energy of two Si═O double bonds is much less than the bond energy of four Si—O single bonds. The contrast between SiO_2 and CO_2 exemplifies a more general phenomenon. Multiple bonds, often encountered between second-period elements, are rare among elements in the third and higher periods. There are many compounds with multiple bonds to carbon but very few compounds featuring multiple bonds to silicon.

Quartz crystals are used to control the frequency of radio and television transmissions. Because these and related applications use so much quartz, there is not enough natural quartz to fulfill demand, and quartz is therefore synthesized. Noncrystalline, or vitreous, quartz, made by melting pure silica sand, is placed in a steel "bomb," and dilute aqueous NaOH is added. A "seed" crystal is placed in the mixture, just as you might use a seed crystal in a hot sugar solution to grow

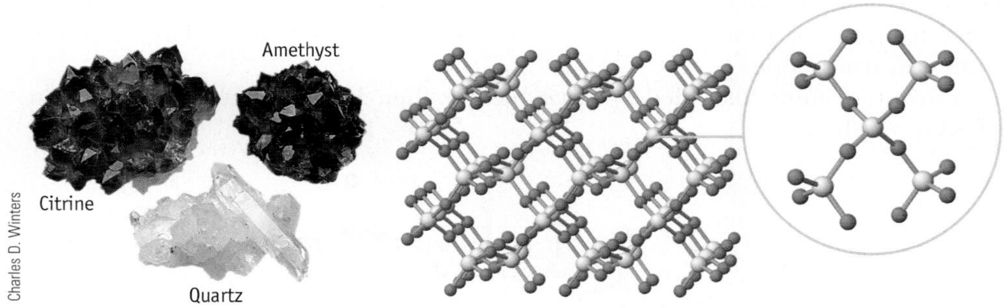

(a) Pure quartz is colorless, but the presence of small amounts of impurities adds color. Purple amethyst and brown citrine crystals are quartz with iron impurities.

(b) Quartz is a network solid in which each Si atom is bound tetrahedrally to four O atoms, each O atom linked to another Si atom. The basic structure consists of a lattice of Si and O atoms.

FIGURE 21.20 Various forms of quartz.

Charles D. Winters

Synthetic quartz. These crystals were grown from silica in sodium hydroxide. The colors come from added Co^{2+} ions (blue) or Fe^{2+} ions (brown).

rock candy. When the mixture is heated above the critical temperature of water (above 400 °C and 1700 atm) over a period of days, pure quartz crystallizes.

Silicon dioxide is resistant to attack by all acids except HF, with which it reacts to give SiF_4 and H_2O.

$$SiO_2(s) + 4\ HF(\ell) \rightarrow SiF_4(g) + 2\ H_2O(\ell)$$

Silicon dioxide also dissolves slowly in hot, molten NaOH or Na_2CO_3 to give Na_4SiO_4, sodium silicate.

$$SiO_2(s) + 2\ Na_2CO_3(\ell) \rightarrow Na_4SiO_4(s) + 2\ CO_2(g)$$

After the molten mixture has cooled, hot water under pressure is added. This partially dissolves the material to give a solution of sodium silicate. After filtering off insoluble sand or glass, the solvent is evaporated to leave sodium silicate, called *water glass.* The biggest single use of this material is in household and industrial detergents, in which it is included because a sodium silicate solution maintains pH by its buffering ability. Additionally, sodium silicate is used in various adhesives and binders, especially for gluing corrugated cardboard boxes.

If sodium silicate is treated with acid, a gelatinous precipitate of SiO_2 called *silica gel* is obtained. Washed and dried, silica gel is a highly porous material with dozens of uses. It is a drying agent, readily absorbing up to 40% of its own weight of water. Small packets of silica gel are often placed in packing boxes of merchandise during storage. The material is frequently stained with $(NH_4)_2CoCl_4$, a humidity detector that is pink when hydrated and blue when dry.

Silicate Minerals with Chain and Ribbon Structures

The structure and chemistry of silicate minerals is an enormous topic in geology and chemistry. Although all silicates are built from tetrahedral SiO_4 units, they have different properties and a wide variety of structures because of the way these tetrahedral SiO_4 units link together.

The simplest silicates, *orthosilicates,* contain SiO_4^{4-} anions. The 4− charge of the anion is balanced by four M^+ ions, two M^{2+} ions, or a combination of ions. Olivine, an important mineral in the earth's mantle, contains Mg^{2+} and Fe^{2+}, with the Fe^{2+} ion giving the mineral its characteristic olive color, and gem-like zircons are $ZrSiO_4$. Calcium orthosilicate, Ca_2SiO_4, is a component of Portland cement, the most common type of cement used in many parts of the world. (It consists mostly of a mixture of CaO and SiO_2 with the remainder largely aluminum and iron oxides.)

A group of minerals called *pyroxenes* have as their basic structural unit a chain of SiO_4 tetrahedra.

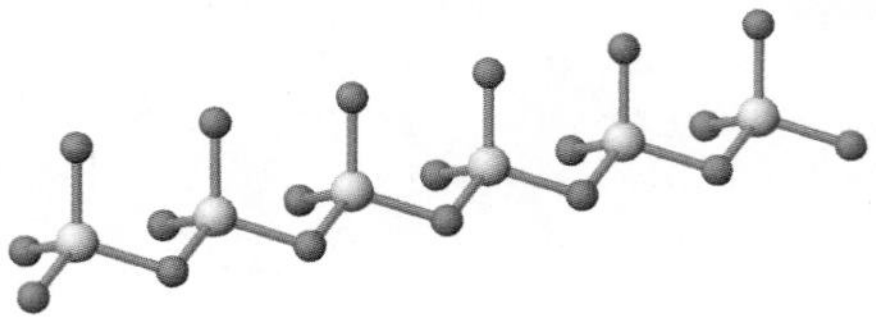

If two such chains are linked together by sharing oxygen atoms, the result is an *amphibole,* of which the asbestos minerals are one example. The molecular chain results in asbestos being a fibrous material.

Charles D. Winters

Silica gel. Silica gel is solid, noncrystalline SiO_2. Packages of the material are often used to keep electronic equipment dry when stored. Silica gel is also used to clarify beer; passing beer through a bed of silica gel removes minute particles that would otherwise make the brew cloudy. Yet another use is in kitty litter.

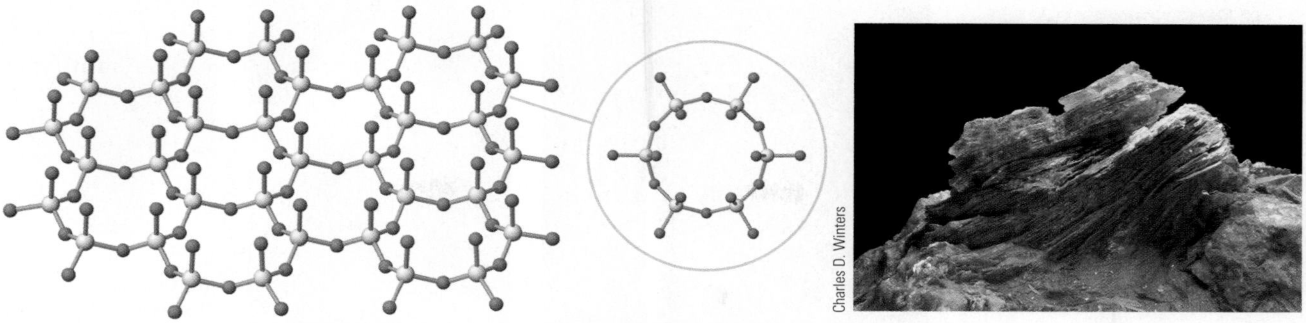

FIGURE 21.21 Mica, a sheet silicate. The sheet-like molecular structure of mica explains its physical appearance. As in the pyroxenes, each silicon is bonded to four oxygen atoms, but the Si and O atoms form a sheet of six-member rings of Si atoms with O atoms in each edge. The ratio of Si to O in this structure is 1 to 2.5. A formula of $SiO_{2.5}$ requires a positive ion, such as Na^+, to counterbalance the charge. Thus, mica and other sheet silicates, and aluminosilicates such as talc and many clays, have positive ions between the sheets. The sheet structure leads to the characteristic feature of mica, which is often found as "books" of thin, silicate sheets. Mica is used in furnace windows and as insulation, and flecks of mica give the glitter to "metallic" paints.

Silicates with Sheet Structures and Aluminosilicates

Linking many silicate chains together produces a sheet of SiO_4 tetrahedra (Figure 21.21). This sheet is the basic structural feature of some of the earth's most important minerals, particularly the clay minerals (such as china clay), mica, talc, and the chrysotile form of asbestos. However, these minerals do not contain just silicon and oxygen. Rather, they are often referred to as *aluminosilicates* because they frequently have Al^{3+} ions in place of Si^{4+} (which means that other positive ions such as Na^+, K^+, and Mg^{2+} must also be present in the lattice to balance the net negative and positive charges). In kaolinite clay, for example, the sheet of SiO_4 tetrahedra is bonded to a sheet of AlO_6 octahedra. In addition, some Si^{4+} ions can be replaced by Al^{3+} atoms. Another example is muscovite, a form of mica. Aluminum ions have replaced some Si^{4+} ions, and there are charge-balancing K^+ ions, so it is best represented by the formula $KAl_2(OH)_2(Si_3AlO_{10})$.

There are some interesting uses of clays, one being in medicine (Figure 21.22). In certain cultures, clay is eaten for medicinal purposes. Several remedies for the relief of upset stomach contain highly purified clays that absorb excess stomach acid as well as potentially harmful bacteria and their toxins by exchanging the intersheet cations in the clays for the toxins, which are often organic cations.

Other aluminosilicates include the feldspars, common minerals that make up about 60% of the earth's crust, and zeolites (Figure 21.22). Both materials are composed of SiO_4 tetrahedra in which some of the Si atoms have been replaced by Al atoms, along with alkali and alkaline earth metal ions for charge balance. The main feature of zeolite structures is their regularly shaped tunnels and cavities. Hole diameters are between 300 and 1000 pm, and small molecules such as water can fit into the cavities of the zeolite structure. As a result, zeolites can be used as drying agents to selectively absorb water from air or a solvent. Small amounts of zeolites are often sealed into multipane windows to keep the air dry between the panes.

Zeolites are also used as catalysts. ExxonMobil, for example, has patented a process in which methanol, CH_3OH, is converted to gasoline in the presence of specially tailored zeolites. In addition, zeolites are added to detergents, where they function as water-softening agents because the sodium ions of the zeolite can be exchanged for Ca^{2+} ions in hard water, effectively removing Ca^{2+} ions from the water.

Kaolinite Clay. The basic structural feature of many clays, and kaolinite in particular, is a sheet of SiO_4 tetrahedra (black and red spheres) bonded to a sheet of AlO_6 octahedra (gray and green spheres).

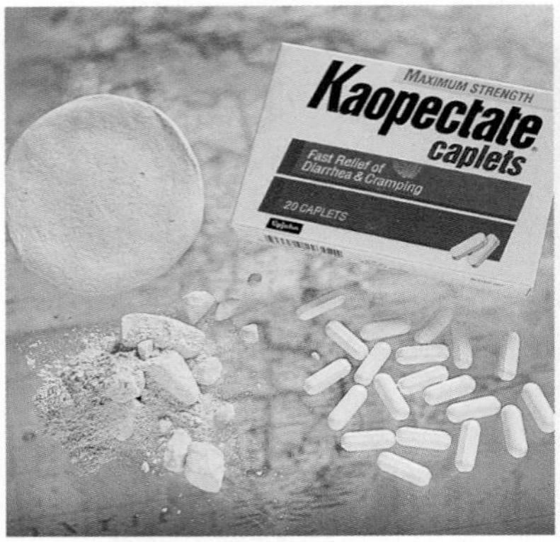

Charles D. Winters

(a) **Remedies for stomach upset.** One of the ingredients in Kaopectate is kaolin, one form of clay. The off-white objects are pieces of clay purchased in a market in Ghana, West Africa. This clay was made to be eaten as a remedy for stomach ailments. Eating clay is widespread among the world's different cultures.

Alfred Pasieka/Science Photo Library/Photo Researchers, Inc.

(b) **The stucture of a zeolite.** Zeolites, which have Si, Al, and O linked in a polyhedral framework, are often portrayed in drawings like this. Each edge consists of a Si—O—Si, Al—O—Si, or Al—O—Al bond. The channels in the framework can selectively capture small molecules or ions or act as catalytic sites.

Charles D. Winters

(c) **Apophyllite**, a crystalline zeolite.

Charles D. Winters

(d) **Consumer products** that remove odor-causing molecules from the air often contain zeolites.

FIGURE 21.22. Aluminosilicates.

Chemistry Now™

Sign in at **www.thomsonedu.com/login** and go to Chapter 21 Contents to see Screen 21.6 for an exercise on **the structural chemistry of silicon–oxygen compounds.**

Silicone Polymers

Silicon and chloromethane (CH_3Cl) react at 300 °C in the presence of a catalyst, Cu powder. The primary product of this reaction is $(CH_3)_2SiCl_2$.

$$Si(s) + 2\ CH_3Cl(g) \rightarrow (CH_3)_2SiCl_2(\ell)$$

Halides of Group 4A elements other than carbon hydrolyze readily. Thus, the reaction of $(CH_3)_2SiCl_2$ with water initially produces $(CH_3)_2Si(OH)_2$. On standing, these molecules combine to form a condensation polymer by eliminating water. The polymer is called polydimethylsiloxane, a member of the *silicone* family of polymers.

$$(CH_3)_2SiCl_2 + 2\ H_2O \rightarrow (CH_3)_2Si(OH)_2 + 2\ HCl$$

$$n\ (CH_3)_2Si(OH)_2 \rightarrow [\text{—}(CH_3)_2SiO\text{—}]_n + n\ H_2O$$

Silicone polymers are nontoxic and have good stability to heat, light, and oxygen; they are chemically inert and have valuable antistick and antifoam properties. They can take the form of oils, greases, and resins. Some have rubber-like properties ("Silly Putty," for example, is a silicone polymer). More than 1 million tons of silicone polymers are made worldwide annually. These materials are used in a wide variety of products: lubricants, peel-off labels, lipstick, suntan lotion, car polish, and building caulk.

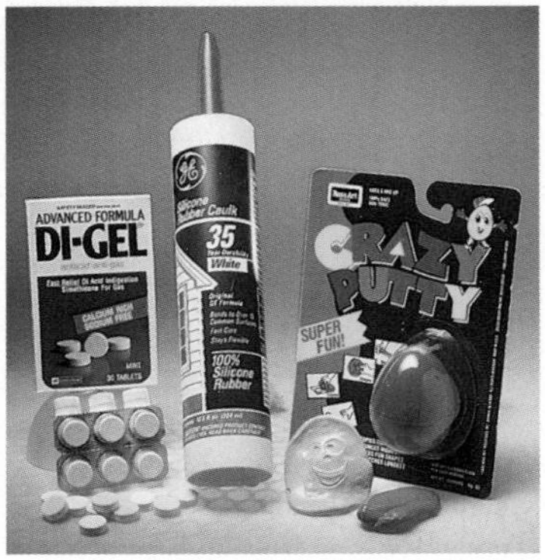

Charles D. Winters

Silicone. Some examples of products containing silicones, polymers with repeating —R_2Si—O— units.

EXERCISE 21.9 Silicon Chemistry

Silicon-oxygen rings are a common structural feature in silicate chemistry. Draw the structure for the anion $Si_3O_9^{6-}$, which is found in minerals such as benitoite. The ring has three Si atoms and three O atoms, and there are two other O atoms on each Si atom.

Case Study — Lead, Beethoven, and a Mystery Solved

Lead anchors the bottom of Group 4A. One of a handful of elements known since ancient times, it has a variety of modern uses. It ranks fifth among metals in usage behind iron, copper, aluminum, and zinc. The major uses of the metal and its compounds are in storage batteries (page 913), pigments, ammunition, solders, plumbing, and bearings.

Unfortunately, lead and its compounds are cumulative poisons, particularly in children. At a blood level as low as 50 ppb (parts per billion), blood pressure is elevated; intelligence is affected at 100 ppb; and blood levels higher than 800 ppb can lead to coma and possible death. Health experts believe that more than 200,000 children become ill from lead poisoning annually, a problem caused chiefly by children eating paint containing lead-based pigments. Older homes often contain lead-based paint because white lead [$2\ PbCO_3 \cdot Pb(OH)_2$] was the pigment used in white paint until about 40 years ago, when it was replaced by TiO_2. Lead salts have a sweet taste, which may contribute to the tendency of children to chew on painted objects.

The symptoms of lead poisoning are, among others, nausea, abdominal pain, irritability, headaches, and excess lethargy or hyperactivity. Indeed, these describe some of the symptoms of the illness that affected Ludwig van Beethoven. As a child, he was recognized as a musical prodigy and was thought to be the greatest pianist in Europe by the time he was 19. But then he fell ill, and, by the time he was 29, he wrote to his brother to say he was considering suicide. By the time he died in 1827 at the age of 56, his belly, arms, and legs were swollen, and he complained constantly of pain in his joints and in his big toe. It is said he wandered the streets of Vienna with long, uncombed hair, dressed in a top hat and long coat, and scribbling in a notebook.

Erich Lessing/Art Resource, NY

Ludwig van Beethoven (1770–1827).

An autopsy at the time showed he died of kidney failure. Kidney stones had destroyed his kidneys, stones that presumably came from gout, the buildup of uric acid in his body. (Gout leads to joint pain, among other things. See the Case Study, page 789.) But why did he have gout?

It was well known in the time of the Roman Empire that lead and its salts are toxic. The Romans drank wine sweetened with a very concentrated grape juice syrup that was prepared by boiling the juice in a lead kettle. The resulting syrup, called Sapa, had a very high concentration of lead, and many Romans contracted gout. So, if Beethoven enjoyed drinking wine, which was often kept in lead-glass decanters, he could have contracted gout and lead poisoning. One scientist has also noted that he may have been one of a small number of people who have a "metal metabolism disorder," a condition that prevents the excretion of toxic metals like lead.

In 2005, scientists at Argonne National Laboratory examined fragments of Beethoven's hair and skull and found both were extremely high in lead. The hair sample, for example, had 60 ppm lead, about 100 times higher than normal.

The mystery of what caused Beethoven's death has been solved. But what remains a mystery is how he contracted lead poisoning.

Questions:

1. *If blood contains 50 ppb lead, how many atoms of lead are there in 1.0 L of blood?*
2. *Research has found that port wine stored for a year in lead-glass decanters contains 2000 ppm lead. If the decanter contains 750 mL of wine ($d = 1.0$ g/mL), what mass of lead has been extracted into the wine?*

Answers to these question are in Appendix Q.

21.8 Nitrogen, Phosphorus, and the Group 5A Elements

Group 5A elements are characterized by the ns^2np^3 configuration with its half-filled np subshell. In compounds of the Group 5A elements, the primary oxidation numbers are +3 and +5, although common nitrogen compounds display a range of oxidation numbers from −3 to +3 and +5. Once again, as in Groups 3A and 4A, the most positive oxidation number is less common for the heavier elements. In many arsenic, antimony, and bismuth compounds, the element has an oxidation number of +3 state. Not surprisingly, compounds of these elements with oxidation numbers of +5 are powerful oxidizing agents.

This part of our tour of the main group elements will concentrate on the chemistries of nitrogen and phosphorus. Nitrogen is found primarily as N_2 in the atmosphere, where it constitutes 78.1% by volume (75.5% by weight). In contrast, phosphorus occurs in the earth's crust in solids. More than 200 different phosphorus-containing minerals are known; all contain the tetrahedral phosphate ion, PO_4^{3-}, or a derivative

Group 5A

Nitrogen 7 N 25 ppm
Phosphorus 15 P 1000 ppm
Arsenic 33 As 1.5 ppm
Antimony 51 Sb 0.2 ppm
Bismuth 83 Bi 0.048 ppm

Element abundances are in parts per million in the earth's crust.

of this ion. By far, the most abundant phosphorus-containing minerals are apatites (◄ page 978).

Nitrogen and its compounds play a key role in our economy, with ammonia making a particularly notable contribution. Phosphoric acid is an important commodity chemical, and it finds its greatest use in producing fertilizers.

Both phosphorus and nitrogen are part of every living organism. Phosphorus is contained in nucleic acids and phospholipids, and nitrogen occurs in proteins and nucleic acids (see *The Chemistry of Life: Biochemistry*, page 497).

Properties of Nitrogen and Phosphorus

Nitrogen (N_2) is a colorless gas that liquifies at 77 K (−196 °C) (Figure 12.1, page 556). Its most notable feature is its reluctance to react with other elements or compounds because the N≡N triple bond has a large bond enthalpy (945 kJ/mol) and because the molecule is nonpolar. Nitrogen does, however, react with hydrogen to give ammonia in the presence of a catalyst (◄ Case Study, page 749) and with a few metals (notably lithium and magnesium) to give metal nitrides, compounds containing the N^{3-} ion.

$$3\ Mg(s) + N_2(g) \longrightarrow \underset{\text{magnesium nitride}}{Mg_3N_2(s)}$$

Elemental nitrogen is a very useful material. Because of its lack of reactivity, it is used to provide a nonoxidizing atmosphere for packaged foods and wine and to pressurize electric cables and telephone wires. Liquid nitrogen is valuable as a coolant in freezing biological samples such as blood and semen, in freeze-drying food, and for other applications that require extremely low temperatures.

Elemental phosphorus was first derived from human waste (see *A Closer Look: Making Phosphorus*, page 997), but it is now produced by the reduction of phosphate minerals in an electric furnace.

$$2\ Ca_3(PO_4)_2(s) + 10\ C(s) + 6\ SiO_2(s) \rightarrow P_4(g) + 6\ CaSiO_3(s) + 10\ CO(g)$$

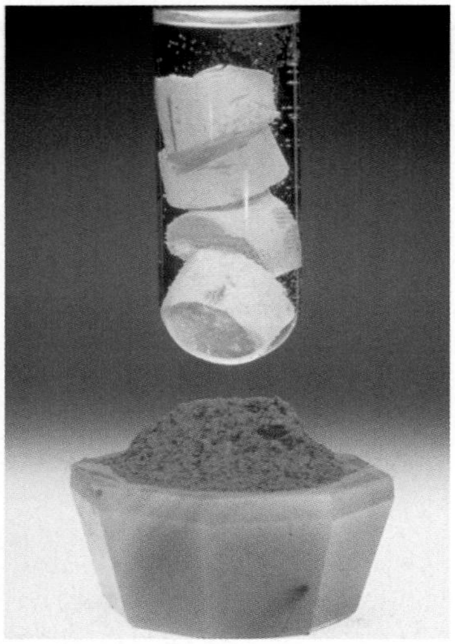

Charles D. Winters

The red and white allotropes of phosphorus.

White phosphorus is the most stable allotrope of phosphorus. Rather than occurring as a diatomic molecule with a triple bond, like its second-period relative nitrogen (N_2), phosphorus is made up of tetrahedral P_4 molecules in which each P atom is joined to three others via single bonds. Red phosphorus is a polymer of P_4 units.

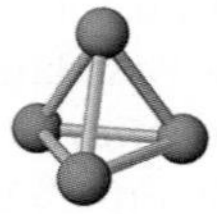

White phosphorus, P_4

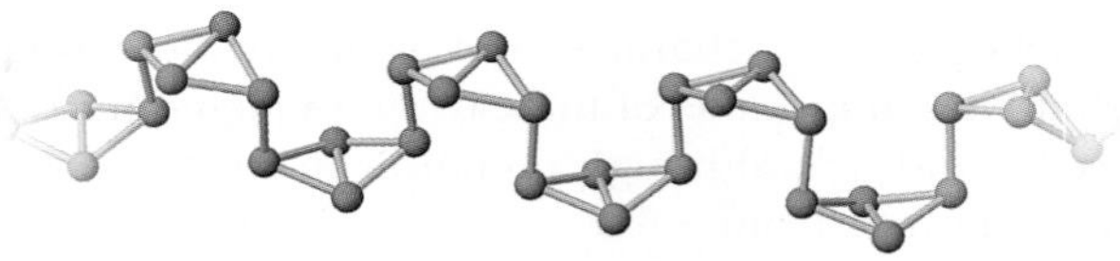

Polymeric red phosphorus

Nitrogen Compounds

A notable feature of the chemistry of nitrogen is the wide diversity of its compounds. Compounds with nitrogen in all oxidation numbers between −3 and +5 are known (Figure 21.23).

Hydrogen Compounds of Nitrogen: Ammonia and Hydrazine

Ammonia is a gas at room temperature and pressure. It has a very penetrating odor and condenses to a liquid at −33 °C under 1 bar of pressure. Solutions in water, often referred to as ammonium hydroxide, are basic due to the reaction of ammonia with water (◀ Section 17.5 and Figure 3.13).

$$NH_3(aq) + H_2O(\ell) \rightleftharpoons NH_4^+(aq) + OH^-(aq) \quad K_b = 1.8 \times 10^{-5} \text{ at } 25\ °C$$

Ammonia is a major industrial chemical and is prepared by the Haber process (◀ page 749), largely for use as a fertilizer.

Hydrazine, N_2H_4, is a colorless, fuming liquid with an ammonia-like odor (mp, 2.0 °C; bp, 113.5 °C). Almost 1 million kilograms of hydrazine is produced annually by the Raschig process—the oxidation of ammonia with alkaline sodium hypochlorite in the presence of gelatin (which is added to suppress metal-catalyzed side reactions that lower the yield of hydrazine).

$$2\ NH_3(aq) + NaClO(aq) \rightleftharpoons N_2H_4(aq) + NaCl(aq) + H_2O(\ell)$$

Hydrazine, like ammonia, is also a base,

$$N_2H_4(aq) + H_2O(\ell) \rightleftharpoons N_2H_5^+(aq) + OH^-(aq) \quad K_b = 8.5 \times 10^{-7}$$

and it is a strong reducing agent, as reflected in the reduction potential for the following reaction of N_2 in basic solution:

$$N_2(g) + 4\ H_2O(\ell) + 4\ e^- \rightarrow N_2H_4(aq) + 4\ OH^-(aq) \quad E° = -1.15\ V$$

Hydrazine's reducing ability is exploited in its use in wastewater treatment for chemical plants. It removes oxidizing ions such as CrO_4^{2-} by reducing them, thus preventing them from entering the environment. A related use is the treatment of water boilers in large electric-generating plants. Oxygen dissolved in the water presents a serious problem in these plants because the dissolved gas can oxidize (corrode) the metal of the boiler and pipes. Hydrazine reduces the amount of dissolved oxygen in water.

$$N_2H_4(aq) + O_2(g) \rightarrow N_2(g) + 2\ H_2O(\ell)$$

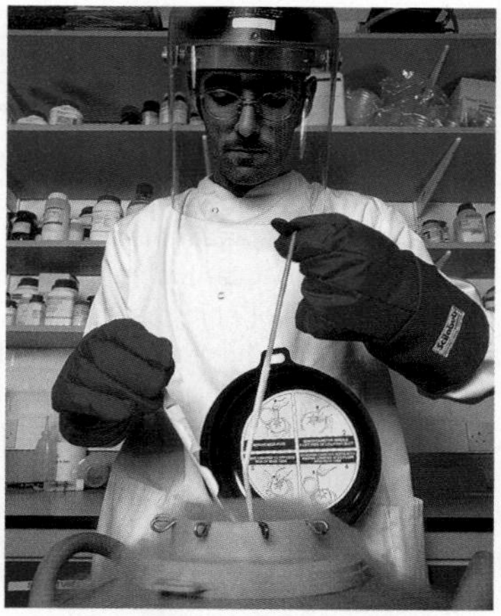

Simon Fraser/MRC Unit, Newcastle General Hospital/Science Photo Library/Photo Researchers, Inc.

Liquid nitrogen. Biological samples—such as embryos or semen from animals or humans—can be stored in liquid nitrogen (at −196 °C) for long periods of time.

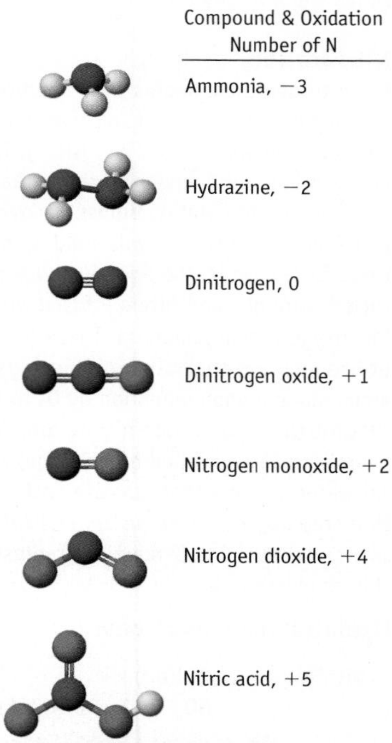

Active Figure 21.23

Compounds and oxidation numbers for nitrogen. In its compounds, the N atom can have oxidation states ranging from −3 to +5.

Chemistry Now™ Sign in at www.thomsonedu.com/login and go to the Chapter 21 Contents menu to explore an interactive version of this figure accompanied by an exercise.

Oxides and Oxoacids of Nitrogen

Nitrogen is unique among all elements in the number of binary oxides it forms (Table 21.5). All are thermodynamically unstable with respect to decomposition to N_2 and O_2; that is, all have positive $\Delta_f G°$ values. Most are slow to decompose, however, and so are described as kinetically stable.

Dinitrogen monoxide, N_2O, commonly called *nitrous oxide,* is a nontoxic, odorless, and tasteless gas in which nitrogen has the lowest oxidation number (+1) among nitrogen oxides. It can be made by the careful decomposition of ammonium nitrate at 250 °C.

$$NH_4NO_3(s) \rightarrow N_2O(g) + 2\ H_2O(g)$$

It is used as an anesthetic in minor surgery and has been called "laughing gas" because of its euphoriant effects. Because it is soluble in vegetable fats, the largest commercial use of N_2O is as a propellant and aerating agent in cans of whipped cream.

Nitrogen monoxide, NO, is an odd-electron molecule. It has 11 valence electrons, giving it one unpaired electron and making it a free radical. The compound has recently been the subject of intense research because it has been found to be important in a number of biochemical processes (◀ page 367).

Case Study — A Healthy Saltwater Aquarium and the Nitrogen Cycle

by Jeffrey Keaffaber, University of Florida

Large saltwater aquariums like Sea World in Florida, the Shedd Aquarium in Chicago, and the new Georgia Aquarium in Atlanta are a continual source of enjoyment. So are smaller aquariums in your home. Maintaining these facilities is not trivial, however; a healthy environment for its marine inhabitants is essential. For this, chemistry plays an important role.

A key part of aquarium maintenance involves control of the concentrations of various dissolved nitrogen-containing species, including ammonia, nitrite ion, and nitrate ion, all of which are stressful to fish at low concentrations and toxic in higher concentrations. The chemistry that relates to maintaining proper balance among these species is called the *nitrogen cycle*.

Nitrification

The nitrogen cycle begins with the production of ammonia (and, in acid solution, its conjugate acid the ammonium ion, NH_4^+), fundamental waste products of protein metabolism in an aquarium habitat. Unless removed, the ammonia concentration will build up over time. To remove it, the aquarium water is cycled through sand filters infused with aerobic, oxygen-loving bacteria. These bacteria utilize enzymes that catalyze the oxidation of ammonia and ammonium ion by O_2 to form first nitrite ion and then nitrate ion. The overall process is called *nitrification*, and the saltwater bacteria that mediate each oxidation step are *Nitrosococcus sp.* and *Nitrococcus sp.*, respectively. Half-reactions representing this chemistry are:

Oxidation half-reactions:

$$NH_4^+(aq) + 8\ OH^-(aq) \rightarrow NO_2^-(aq) + 6\ H_2O(\ell) + 6\ e^-$$

$$NO_2^-(aq) + 2\ OH^-(aq) \rightarrow NO_3^-(aq) + H_2O(\ell) + 2\ e^-$$

Reduction half-reaction:

$$O_2(aq) + 2\ H_2O(\ell) + 4\ e^- \rightarrow 4\ OH^-(aq)$$

When setting up an aquarium at home, it is appropriate to monitor the concentrations of the various nitrogen species. Initially the NH_3/NH_4^+ concentration rises, but then it begins to fall as oxidation occurs. With time, the nitrite ion concentration builds up, peaks, and then decreases, with an accompanying increase in nitrate ion concentration. To reach a stable situation, as much as six weeks may be required.

Denitrification

Nitrate is much less toxic than ammonia and the nitrite ion, but its buildup must also be limited. In a small aquarium, nitrate ion concentration can be controlled by partial exchange of water. However, because of environmental restrictions this is not possible for large aquariums; they must use a closed water treatment process.

To remedy the build up the nitrate ion, another biologically catalyzed process is used that reduces the nitrate ion to nitrogen gas, N_2. A reducing agent is required, and early designs of denitrifying filters utilized methanol, CH_3OH, as the reducing agent. Among naturally occurring saltwater bacteria capable of nitrate reduction under low oxygen (anoxic) conditions, *Pseudomonas sp.*, is commonly used. The bacteria, utilizing enzymes to catalyze reaction of nitrate and methanol to form N_2 and CO_2, are introduced to sand filters where methanol is added.

A stable pH is also important to the health of aquarium fish. Therefore, the marine environment in a saltwater aquarium is maintained at a relatively constant pH of 8.0-8.2. To aid in this, the CO_2 produced by methanol oxidation remains dissolved in solution, and increases the buffer capacity of the seawater.

Questions:

1. *Write a balanced net ionic equation for the oxidation of NH_4^+ by O_2 to produce H_2O and NO_2^-.*
2. *Write half-reactions for the reduction of NO_3^- to N_2 and for the oxidation of CH_3OH to CO_2 in basic solution. Then, combine these half-reactions to obtain the balanced equation for the reduction of NO_3^- by CH_3OH.*
3. *Consider the carbon-containing species CO_2, H_2CO_3, HCO_3^-, and CO_3^{2-}. Which one is present in largest concentration at the pH conditions of the aquarium? Give a short explanation. K_a of $H_2CO_3 = 4.2 \times 10^{-7}$ and K_a of $HCO_3^- = 4.8 \times 10^{-11}$. (See Study Question 18-110.)*
4. *A large, 2.2×10^7 L, aquarium contains 1.7×10^4 kg of dissolved NO_3^-. Calculate nitrate concentrations in ppm (mg/L) N, ppm NO_3^-, and the molar concentration of NO_3^-.*

Answers to these questions are in Appendix Q.

TABLE 21.5 Some Oxides of Nitrogen

Formula	Name	Structure	Nitrogen Oxidation Number	Description
N_2O	Dinitrogen monoxide (nitrous oxide)	:N≡N—O: linear	+1	Colorless gas (laughing gas)
NO	Nitrogen monoxide (nitric oxide)	*	+2	Colorless gas; odd-electron molecule (paramagnetic)
N_2O_3	Dinitrogen trioxide	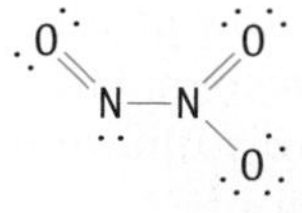 planar	+3	Blue solid (mp, −100.7 °C); reversibly dissociates to NO and NO_2
NO_2	Nitrogen dioxide	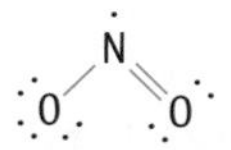	+4	Brown, paramagnetic gas; odd-electron molecule
N_2O_4	Dinitrogen tetraoxide	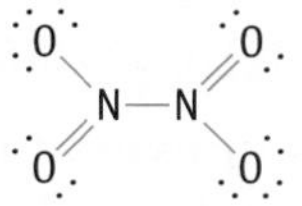 planar	+4	Colorless liquid/gas; dissociates to NO_2 (see Figure 16.8)
N_2O_5	Dinitrogen pentaoxide	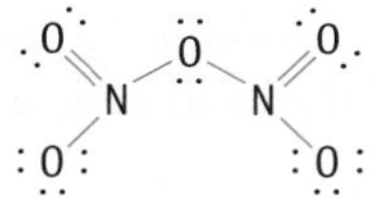	+5	Colorless solid

*It is not possible to draw a Lewis structure that accurately represents the electronic structure of NO. See Chapter 8. Also note that only one resonance structure is shown for each structure.

Nitrogen dioxide, NO_2, is the brown gas you see when a bottle of nitric acid is allowed to stand in the sunlight.

$$2\ HNO_3(aq) \rightarrow 2\ NO_2(g) + H_2O(\ell) + \frac{1}{2}\ O_2(g)$$

Nitrogen dioxide is also a culprit in air pollution (◀ page 951). Nitrogen monoxide forms when atmospheric nitrogen and oxygen are heated in internal combustion engines. Released into the atmosphere, NO rapidly reacts with O_2 to form NO_2.

$$2\ NO(g) + O_2(g) \rightarrow 2\ NO_2(g)$$

Nitrogen dioxide has 17 valence electrons, so it is also an odd-electron molecule. Because the odd electron largely resides on the N atom, two NO_2 molecules can combine, forming an N—N bond and producing N_2O_4, *dinitrogen tetraoxide.*

$$\underset{\text{deep brown gas}}{2\ NO_2(g)} \longrightarrow \underset{\text{colorless (mp, −11.2 °C)}}{N_2O_4(g)}$$

Solid N_2O_4 is colorless and consists entirely of N_2O_4 molecules. However, as the solid melts and the temperature increases to the boiling point, the color darkens as N_2O_4 dissociates to form brown NO_2. At the normal boiling point (21.5 °C), the distinctly brown gas consists of 15.9% NO_2 and 84.1% N_2O_4.

Charles D. Winters

Nitrous oxide, N_2O. This oxide readily dissolves in fats, so the gas is added, under pressure, to cans of cream. When the valve is opened, the gas expands, whipping the cream. N_2O is also an anesthetic and is considered safe for medical uses. However, significant dangers arise from using it as a recreational drug. Long-term use can induce nerve damage and cause such problems as weakness and loss of feeling.

When NO_2 is bubbled into water, nitric acid and nitrous acid form.

$$2\ NO_2(g) + H_2O(\ell) \rightarrow \underset{\text{nitric acid}}{HNO_3(aq)} + \underset{\text{nitrous acid}}{HNO_2(aq)}$$

Nitric acid has been known for centuries and has become an important compound in our modern economy. The oldest way to make the acid is to treat $NaNO_3$ with sulfuric acid (Figure 21.24).

$$2\ NaNO_3(s) + H_2SO_4(\ell) \rightarrow 2\ HNO_3(\ell) + Na_2SO_4(s)$$

Enormous quantities of nitric acid are now produced industrially by the oxidation of ammonia in the multistep *Ostwald process.* The acid has many applications, but by far the greatest amount is turned into ammonium nitrate (for use as a fertilizer) by the reaction of nitric acid and ammonia.

Nitric acid is a powerful oxidizing agent, as the large, positive $E°$ values for the following half-reactions illustrate:

$$NO_3^-(aq) + 4\ H_3O^+(aq) + 3\ e^- \rightarrow NO(g) + 6\ H_2O(\ell) \qquad E° = +0.96\ V$$

$$NO_3^-(aq) + 2\ H_3O^+(aq) + e^- \rightarrow NO_2(g) + 3\ H_2O(\ell) \qquad E° = +0.80\ V$$

Concentrated nitric acid attacks and oxidizes most metals. (Aluminum is an exception; see page 985.) In this process, the nitrate ion is reduced to one of the nitrogen oxides. Which oxide is formed depends on the metal and on reaction conditions. In the case of copper, for example, either NO or NO_2 is produced, depending on the concentration of the acid (Figure 21.24b).

In dilute acid:

$$3\ Cu(s) + 8\ H_3O^+(aq) + 2\ NO_3^-(aq) \rightarrow 3\ Cu^{2+}(aq) + 12\ H_2O(\ell) + 2\ NO(g)$$

In concentrated acid:

$$Cu(s) + 4\ H_3O^+(aq) + 2\ NO_3^-(aq) \rightarrow Cu^{2+}(aq) + 6\ H_2O(\ell) + 2\ NO_2(g)$$

Four metals (Au, Pt, Rh, and Ir) that are not attacked by nitric acid are often described as the "noble metals." The alchemists of the 14th century, however, knew that if they mixed HNO_3 with HCl in a ratio of about 1 : 3, this aqua regia, or "kingly water," would attack even gold, the noblest of metals.

$$10\ Au(s) + 6\ NO_3^-(aq) + 40\ Cl^-(aq) + 36\ H_3O^+(aq) \rightarrow 10\ [AuCl_4]^-(aq) + 3\ N_2(g) + 54\ H_2O(\ell)$$

FIGURE 21.24 The preparation and properties of nitric acid. (a) Nitric acid is prepared by the reaction of sulfuric acid and sodium nitrate. Pure HNO_3 is colorless, but samples of the acid are often brown because of NO_2 formed by decomposition of the acid. This gas fills the apparatus and colors the liquid in the distillation flask. (b) When concentrated nitric acid reacts with copper, the metal is oxidized to copper(II) ions, and NO_2 gas is a reaction product.

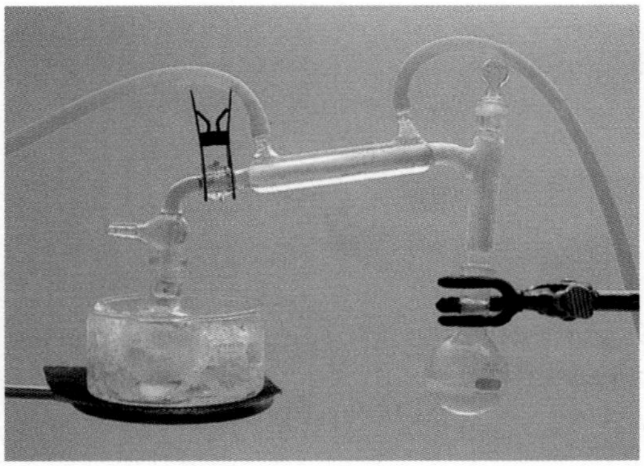

(a) Preparation of nitric acid.

(b) Reaction of HNO_3 with copper.

Photos: Charles D. Winters

A Closer Look — Making Phosphorus

He stoked his small furnace with more charcoal and pumped the bellows until his retort glowed red hot. Suddenly something strange began to happen. Glowing fumes filled the vessel and from the end of the retort dripped a shining liquid that burst into flames.

J. Emsley: *The 13th Element,* p. 5. New York, John Wiley, 2000.

John Emsley begins his story of phosphorus, its discovery, and its uses, by imagining what the German alchemist Hennig Brandt must have seen in his laboratory that day in 1669. (See page 338 for an artist's conception of the discovery of phosphorus by Brandt.) He was in search of the philosopher's stone, the magic elixir that would turn the crudest substance into gold.

Brandt was experimenting with urine, which had served as the source of useful chemicals since Roman times. It is not surprising that phosphorus could be extracted from this source. Humans consume much more phosphorus, in the form of phosphate, than they require, and the excess phosphorus (about 1.4 g per day) is excreted in the urine. It is nonetheless extraordinary that Brandt was able to isolate the element. According to an 18th-century chemistry book, about 30 g of phosphorus could be obtained from 60 gallons of urine. And the process was not simple. Another 18th-century recipe states that "50 or 60 pails full" of urine was to be used. "Let it lie steeping . . . till it putrefy and breed worms." The chemist was then to reduce the whole to a paste and finally to heat the paste very strongly in a retort. After some days, phosphorus distilled from the mixture and was collected in water. (We know now that carbon from the organic compounds in the urine would have reduced the phosphate to phosphorus.) Phosphorus was made in this manner for more than 100 years.

Charles D. Winters

The glow of phosphorus burning in air.

EXERCISE 21.10 Nitrogen Oxide Chemistry

Dinitrogen monoxide can be made by the decomposition of NH_4NO_3.

(a) A Lewis electron dot structure of N_2O is given in Table 21.5. Is it the only possible structure? If other structures are possible, is the one in Table 21.5 the most important?

(b) Is the decomposition of $NH_4NO_3(s)$ to give $N_2O(g)$ and $H_2O(g)$ endothermic or exothermic?

Hydrogen Compounds of Phosphorus and Other Group 5A Elements

The phosphorus analog of ammonia, phosphine (PH_3), is a poisonous, highly reactive gas with a faint garlic-like odor. Industrially, it is made by the reaction of white phosphorus and aqueous NaOH.

$$P_4(s) + 3\ KOH(aq) + 3\ H_2O(\ell) \rightarrow PH_3(g) + 3\ KH_2PO_2(aq)$$

The other hydrides of the heavier Group 5A elements are also toxic and become more unstable as the atomic number of the element increases. Nonetheless, arsine (AsH_3) is used in the semiconductor industry as a starting material in the preparation of gallium arsenide (GaAs) semiconductors.

Phosphorus Oxides and Sulfides

The most important compounds of phosphorus are those with oxygen, and there are at least six simple binary compounds containing just phosphorus and oxygen. All of them can be thought of as being derived structurally from the P_4 tetrahedron of white phosphorus. For example, if P_4 is carefully oxidized, P_4O_6 is formed; an O atom has been placed into each P—P bond in the tetrahedron (Figure 21.25).

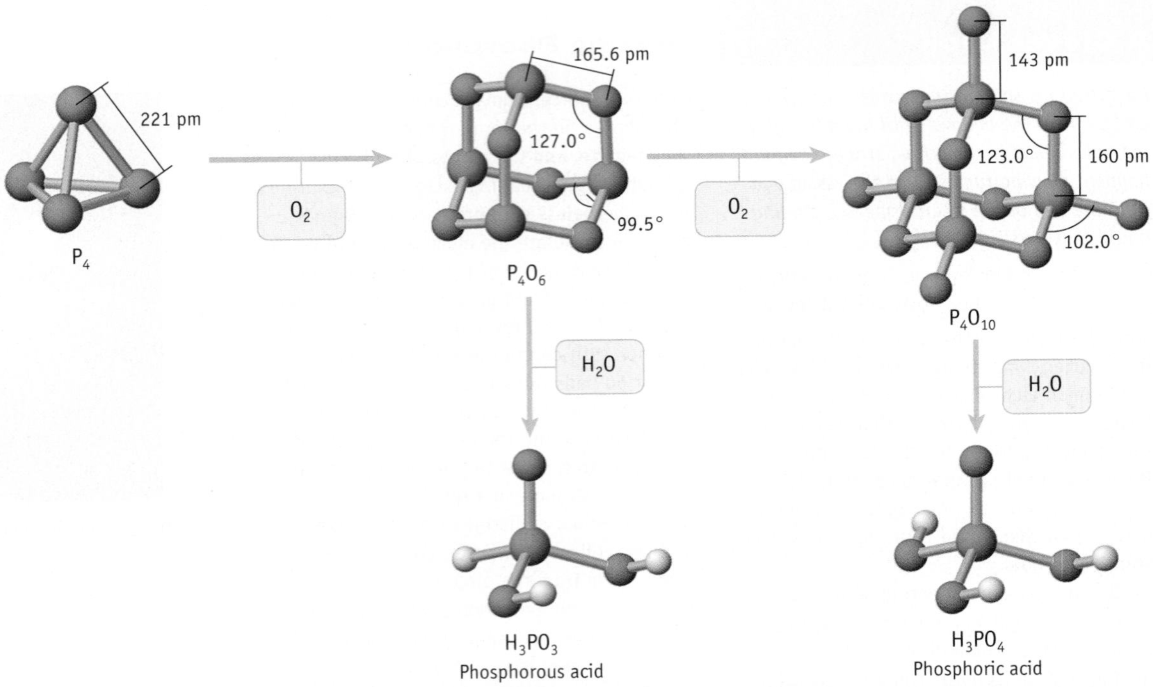

FIGURE 21.25 Phosphorus oxides. Other binary P—O compounds have formulas between P_4O_6 and P_4O_{10}. They are formed by starting with P_4O_6 and adding O atoms successively to the P atom vertices.

The most common and important phosphorus oxide is P_4O_{10}, a fine white powder commonly called "phosphorus pentaoxide" because its empirical formula is P_2O_5. In P_4O_{10}, each phosphorus atom is surrounded tetrahedrally by O atoms.

Phosphorus also forms a series of compounds with sulfur. Of these, the most important is P_4S_3.

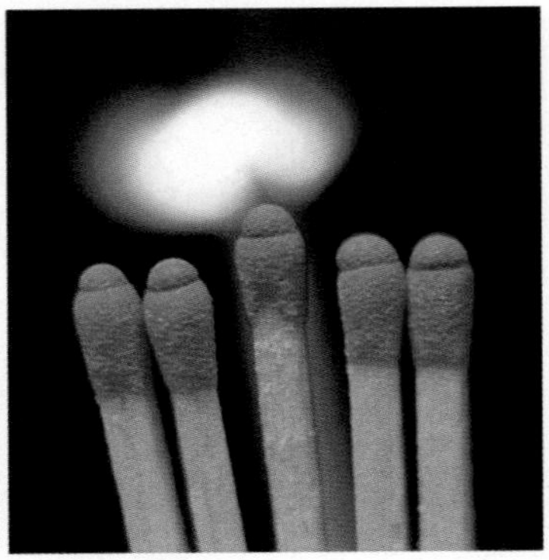

Charles D. Winters

Matches. The head of a "strike anywhere" match contains P_4S_3 and the oxidizing agent $KClO_3$. (Other components are ground glass, Fe_2O_3, ZnO, and glue.) Safety matches have sulfur (3–5%) and $KClO_3$ (45–55%) in the match head and red phosphorus in the striking strip.

$$4\ P(s,\text{red allotrope}) + \tfrac{3}{8}\ S_8(s) \longrightarrow$$

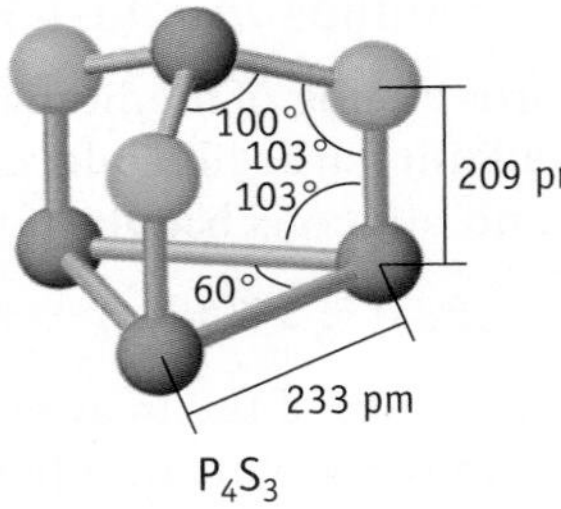

In this phosphorus sulfide, S atoms are placed into only three of the P—P bonds. The principal use of P_4S_3 is in "strike anywhere" matches, the kind that light when you rub the head against a rough object. The active ingredients are P_4S_3 and the powerful oxidizing agent potassium chlorate, $KClO_3$. The "safety match" is now more common than the "strike anywhere" match. In safety matches, the head is predominantly $KClO_3$, and the material on the match book is red phosphorus (about 50%), Sb_2S_3, Fe_2O_3, and glue.

Phosphorus Oxoacids and Their Salts

A few of the many known phosphorus oxoacids are illustrated in Table 21.6. Indeed, there are so many acids and their salts in this category that structural principles have been developed to organize and understand them.

(a) All P atoms in the oxoacids and their anions (conjugate bases) are four-coordinate and tetrahedral.

(b) All the P atoms in the acids have at least one P—OH group (and this occurs often in the anions as well). In every case, the H atom is ionizable as H^+.

(c) Some oxoacids have one or more P—H bonds. This H atom is not ionizable as H^+.

(d) Polymerization can occur by P—O—P bond formation to give both linear and cyclic species. Two P atoms are never joined by more than one P—O—P bridge.

(e) When a P atom is surrounded only by O atoms (as in H_3PO_4), its oxidation number is +5. For each P—OH that is replaced by P—H, the oxidation number drops by 2 (because P is considered more electronegative than H). For example, the oxidation number of P in H_3PO_2 is +1.

Charles D. Winters

FIGURE 21.26 Reaction of P_4O_{10} and water. The white solid oxide reacts vigorously with water to give orthophosphoric acid, H_3PO_4. (The heat generated vaporizes the water, so steam is visible.)

Orthophosphoric acid, H_3PO_4, and its salts are far more important commercially than other P—O acids. Millions of tons of phosphoric acid are made annually, some using white phosphorus as the starting material. The element is burned in oxygen to give P_4O_{10}, and the oxide reacts with water to produce the acid (Figure 21.26).

$$P_4O_{10}(s) + 6\ H_2O(\ell) \rightarrow 4\ H_3PO_4(aq)$$

TABLE 21.6 Phosphorus Oxoacids

Formula	Name	Structure
H_3PO_4	Orthophosphoric acid	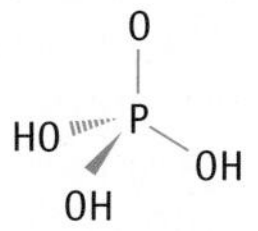
$H_4P_2O_7$	Pyrophosphoric acid (diphosphoric acid)	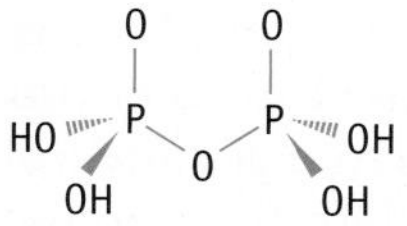
$(HPO_3)_3$	Metaphosphoric acid	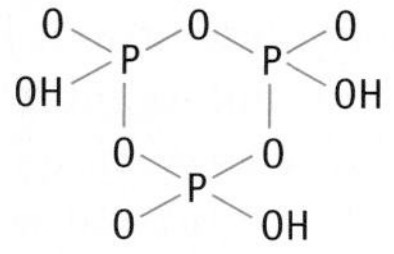
H_3PO_3	Phosphorous acid (phosphonic acid)	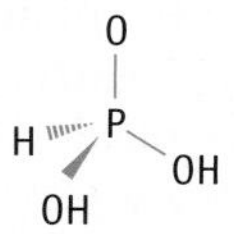
H_3PO_2	Hypophosphorous acid (phosphinic acid)	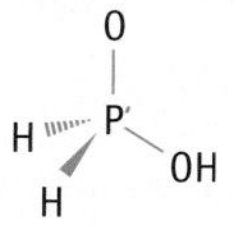

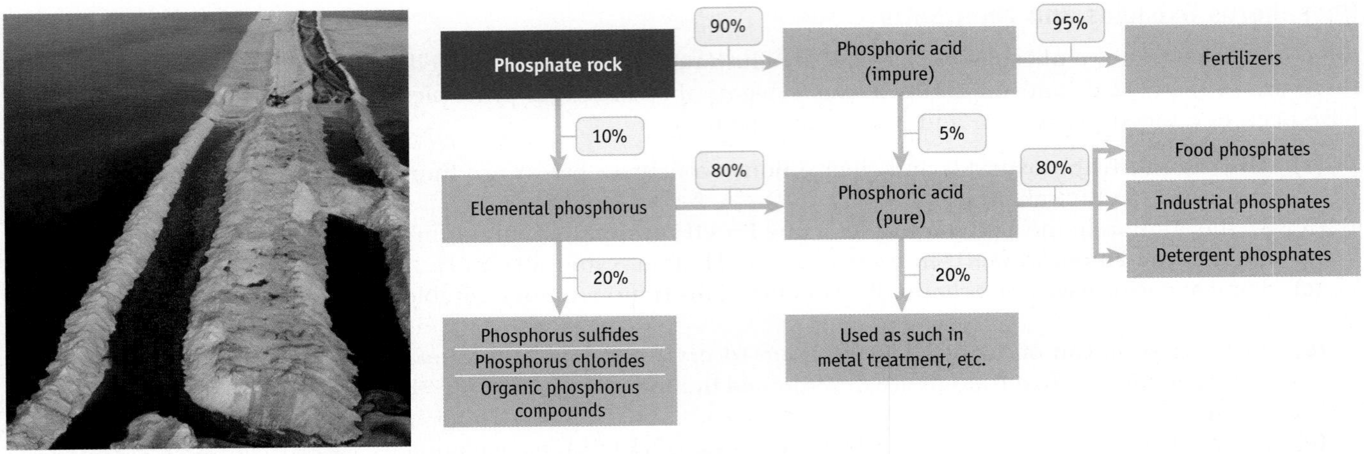

(a) Mining phosphate rock. Phosphate rock is primarily $Ca_3(PO_4)_2$, and most mined in the United States comes from Florida.

(b) Uses of phosphorus and phosphoric acid.

FIGURE 21.27 Uses of phosphate rock, phosphorus, and phosphoric acid.

This approach gives a pure product, so it is employed to make phosphoric acid for use in food products in particular. The acid is nontoxic, and it gives the tart or sour taste to carbonated "soft drinks," such as various colas (about 0.05% H_3PO_4) or root beer (about 0.01% H_3PO_4).

A major use for phosphoric acid is to impart corrosion resistance to metal objects such as nuts and bolts, tools, and car-engine parts by plunging the object into a hot acid bath. Car bodies are similarly treated with phosphoric acid containing metal ions such as Zn^{2+}, and aluminum trim is "polished" by treating it with the acid.

The reaction of H_3PO_4 with strong bases produces salts such as NaH_2PO_4, Na_2HPO_4, and Na_3PO_4. In industry, the monosodium and disodium salts are produced using Na_2CO_3 as the base, but an excess of the stronger (and more expensive) base NaOH is required to remove the third proton to give Na_3PO_4.

Sodium phosphate (Na_3PO_4) is used in scouring powders and paint strippers because the anion PO_4^{3-} is a relatively strong base in water ($K_b = 2.8 \times 10^{-2}$). Sodium monohydrogen phosphate, Na_2HPO_4, which has a less basic anion than PO_4^{3-}, is widely used in food products. Kraft has patented a process using the salt in the manufacture of pasteurized cheese, for example. Thousands of tons of Na_2HPO_4 are still used for this purpose, even though the function of the salt in this process is not completely understood. In addition, a small amount of Na_2HPO_4 in pudding mixes enables the mix to gel in cold water, and the basic anion raises the pH of cereals to provide "quick-cooking" breakfast cereal. (The OH^- ion from HPO_4^{2-} hydrolysis accelerates the breakdown of the cellulose material in the cereal.)

Calcium phosphates are used in a broad spectrum of products. For example, the weak acid $Ca(H_2PO_4)_2 \cdot H_2O$ is used as the acid leavening agent in baking powder. A typical baking powder contains (along with inert ingredients) 28% $NaHCO_3$, 10.7% $Ca(H_2PO_4)_2 \cdot H_2O$, and 21.4% $NaAl(SO_4)_2$ (also a weak acid). The weak acids react with sodium bicarbonate to produce CO_2 gas. For example,

$$Ca(H_2PO_4)_2 \cdot H_2O(s) + 2\ NaHCO_3(aq) \rightarrow 2\ CO_2(g) + 3\ H_2O(\ell) + Na_2HPO_4(aq) + CaHPO_4(aq)$$

Finally, calcium monohydrogen phosphate, $CaHPO_4$, is used as an abrasive and polishing agent in toothpaste.

21.9 Oxygen, Sulfur, and the Group 6A Elements

Oxygen is by far the most abundant element in the earth's crust, representing slightly less than 50% of it by weight. It is present as elemental oxygen in the atmosphere and is combined with other elements in water and in many minerals. Scientists believe that elemental oxygen did not appear on this planet until about 3.5 billion years ago, when it was formed on the planet by plants through the process of photosynthesis.

Sulfur, seventeenth in abundance in the earth's crust, is also found in its elemental form in nature, but only in certain concentrated deposits. Sulfur-containing compounds occur in natural gas and oil. In minerals, sulfur occurs as the sulfide ion (for example, in cinnabar, HgS, and galena, PbS), as the disulfide ion (in iron pyrite, FeS_2, or "fool's gold"), and as sulfate ion (e.g., in gypsum, $CaSO_4 \cdot 2\ H_2O$). Sulfur oxides (SO_2 and SO_3) also occur in nature, primarily as products of volcanic activity (Figure 21.28).

In the United States, most sulfur—about 10 million tons per year—is obtained from deposits of the element found along the Gulf of Mexico. These deposits occur typically at a depth of 150 to 750 m below the surface in layers about 30 m thick. They are thought to have been formed by anaerobic ("without elemental oxygen") bacteria acting on sedimentary sulfate deposits such as gypsum.

Group 6A

Group 6A
Oxygen 8 O 474,000 ppm
Sulfur 16 S 260 ppm
Selenium 34 Se 0.5 ppm
Tellurium 52 Te 0.005 ppm
Polonium 84 Po trace

Element abundances are in parts per million in the earth's crust.

Preparation and Properties of the Elements

Pure oxygen is obtained by the fractional distillation of air and is among the top five industrial chemicals produced in the United States. Oxygen can be made in the laboratory by electrolysis of water (Figure 21.4) and by the catalyzed decomposition of metal chlorates such as $KClO_3$.

$$2\ KClO_3(s) \xrightarrow{\text{catalyst}} 2\ KCl(s) + 3\ O_2(g)$$

At room temperature and pressure, oxygen is a colorless gas, but it is pale blue when condensed to the liquid at −183 °C (◀ Figure 9.15). As described in Section 9.3, diatomic oxygen is paramagnetic because it has two unpaired electrons.

An allotrope of oxygen, ozone (O_3), is a blue, diamagnetic gas with an odor so strong that it can be detected in concentrations as low as 0.05 ppm. Ozone is synthesized by passing O_2 through an electric discharge or by irradiating O_2 with ultraviolet light. It is often in the news because of the realization that the earth's protective layer of ozone in the stratosphere is being disrupted by chlorofluorocarbons and other chemicals (◀ page 953).

Sulfur has numerous allotropes. The most common and most stable allotrope is the yellow, orthorhombic form, which consists of S_8 molecules with the sulfur atoms arranged in a crown-shaped ring (Figure 21.29a). Less stable allotropes are known that have rings of 6 to 20 sulfur atoms. Another form of sulfur, called plastic sulfur, has a molecular structure with chains of sulfur atoms (Figure 21.29b).

Sulfur is obtained from underground deposits by a process developed by Herman Frasch (1851–1914) about 1900. Superheated water (at 165 °C) and then air are forced into the deposit. The sulfur melts (mp, 113 °C) and is forced to the surface as a frothy, yellow stream, from which it solidifies.

Selenium and tellurium are comparatively rare on earth, having abundances about the same as those of silver and gold, respectively. Because their chemistry is similar to that of sulfur, they are often found in minerals associated with the sulfides of copper, silver, iron, and arsenic, and they are recovered as by-products of the industries devoted to those metals.

FIGURE 21.28 Sulfur spewing from a volcano in Indonesia.

FIGURE 21.29 **Sulfur allotropes.** (a) At room temperature, sulfur exists as a bright yellow solid composed of S_8 rings. (b) When heated, the rings break open, and eventually form chains of S atoms in a material described as "plastic sulfur."

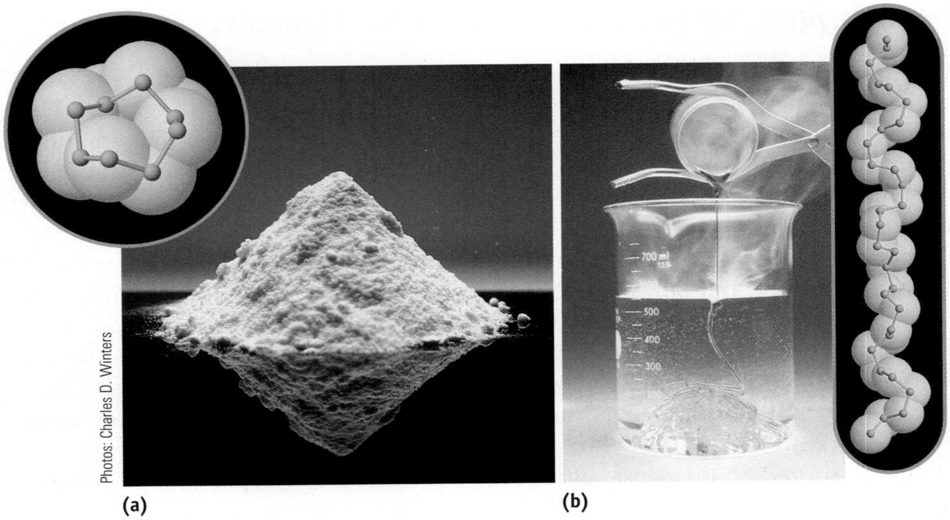

Selenium has a range of uses, including in glass making. A cadmium sulfide/selenide mixture is added to glass to give it a brilliant red color (Figure 21.30a). The most familiar use of selenium is in xerography, a word meaning "dry printing" and a process at the heart of the modern copy machine. Most photocopy machines use an aluminum plate or roller coated with selenium. Light coming from the imaging lens selectively discharges a static electric charge on the selenium surface, and the black toner sticks only on the areas that remain charged. A copy is made when the toner is transferred to a sheet of plain paper.

The heaviest element of Group 6A, polonium, is radioactive and found only in trace amounts on earth. It was discovered in Paris, France, in 1898 by Marie Sklodowska Curie (1867–1934) and her husband Pierre Curie (1859-1906). The Curies painstakingly separated this element from a large quantity of pitchblende, a uranium-containing ore.

Chemistry Now™

Sign in at **www.thomsonedu.com/login** and go to Chapter 21 Contents to see Screen 21.8 for an exercise on **sulfur chemistry.**

FIGURE 21.30 **Uses of selenium.** (a) Glass takes on a brilliant red color when a mixture of cadmium sulfide/selenide (CdS, CdSe) is added to it. (b) These sample bottles hold suspensions of quantum dots, nanometer-sized crystals of CdSe dispersed in a polymer matrix. The crystals emit light in the visible range when excited by ultraviolet light. Light emission at different wavelengths is achieved by changing the particle size. Crystals of PbS and PbSe can be made that emit light in the infrared range.

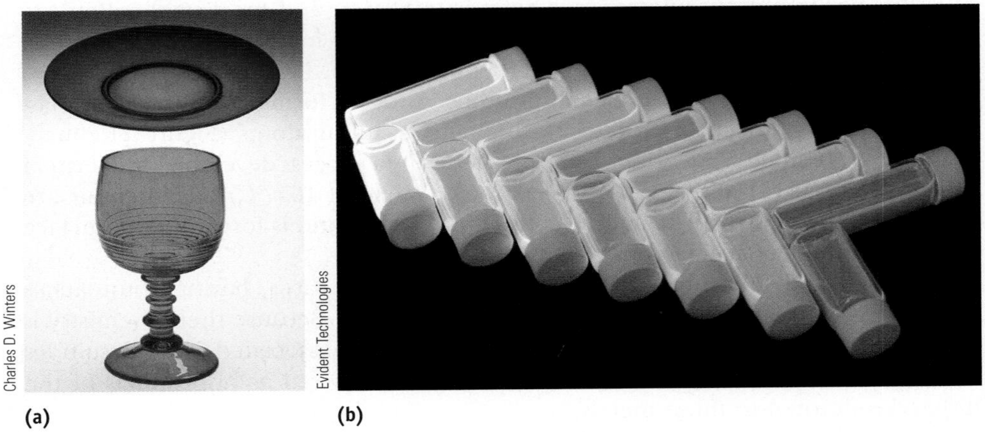

Sulfur Compounds

Hydrogen sulfide, H_2S, has a bent molecular geometry, like water. Unlike water, however, H_2S is a gas under standard conditions (mp, −85.6 °C; bp, −60.3 °C) because its intermolecular forces are weak compared with the strong hydrogen bonding in water (see Figure 12.6). Hydrogen sulfide is poisonous, comparable in toxicity to hydrogen cyanide, but fortunately it has a terrible odor and is detectable in concentrations as low as 0.02 ppm. You must be careful with H_2S, though. Because it has an anesthetic effect, your nose rapidly loses its ability to detect it. Death occurs at H_2S concentrations of 100 ppm.

■ **Bad Breath** Halitosis or "bad breath" is due to three sulfur-containing compounds: H_2S, CH_3SH (methyl mercaptan), and $(CH_3)_2S$ (dimethyl sulfide). All three can be detected in very tiny concentrations. For example, your nose knows if as little as 0.2 microgram of CH_3SH is present per liter of air. The compounds result from bacteria's attack on the sulfur-containing amino acids cysteine and methionine in food particles in the mouth. A general rule: if something smells bad, it probably contains sulfur! (See Case Study, page 524.)

Sulfur is often found as the sulfide ion in conjunction with metals because all metal sulfides (except those based on Group 1A metals) are insoluble. The recovery of metals from their sulfide ores usually begins by heating the ore in air.

$$2\ PbS(s) + 3\ O_2(g) \rightarrow 2\ PbO(s) + 2\ SO_2(g)$$

Here, lead sulfide is converted to lead(II) oxide, and this is further reduced to lead using carbon or carbon monoxide in a blast furnace.

$$PbO(s) + CO(g) \rightarrow Pb(\ell) + CO_2(g)$$

Alternatively, the oxide can be reduced to elemental lead by combining it with fresh lead sulfide.

$$2\ PbO(s) + PbS(s) \rightarrow 3\ Pb(s) + SO_2(g)$$

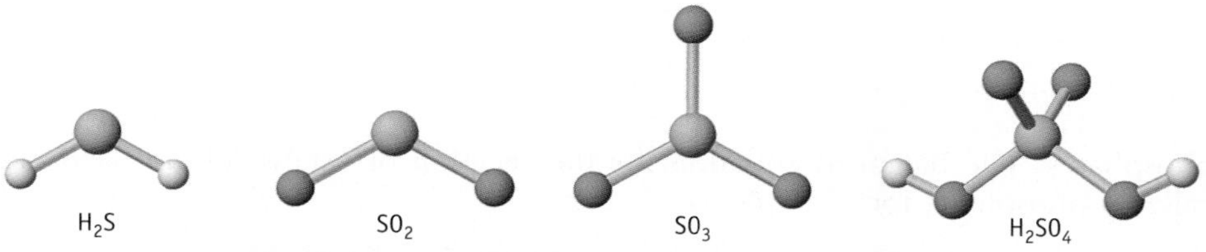

Models of some common sulfur-containing molecules: H_2S, SO_2, SO_3, and H_2SO_4.

Sulfur dioxide (SO_2), a colorless, toxic gas with a sharp odor, is produced on an enormous scale by the combustion of sulfur and by roasting sulfide ores in air. The combustion of sulfur in sulfur-containing coal and fuel oil creates particularly large environmental problems. It has been estimated that about 2.0×10^8 tons of sulfur oxides (primarily SO_2) are released into the atmosphere each year by human activities; this is more than half of the total emitted by all other natural sources of sulfur in the environment.

Sulfur dioxide readily dissolves in water. The most important reaction of this gas is its oxidation to SO_3.

$$SO_2(g) + \tfrac{1}{2}\ O_2(g) \rightarrow SO_3(g) \qquad \Delta_r H° = -98.9\ \text{kJ/mol-rxn}$$

Sulfur trioxide is almost never isolated but is converted directly to sulfuric acid by reaction with water in the "contact process."

The largest use of sulfur is the production of sulfuric acid, H_2SO_4, the compound produced in largest quantity by the chemical industry (◄ page 135). In the United States, roughly 70% of the acid is used to manufacture superphosphate fertilizer from phosphate rock. Plants need a soluble form of phosphorus for growth, but calcium phosphate and apatite [$Ca_5X(PO_4)_3$, X = F, OH, Cl] are insoluble. Treating phosphate-containing minerals with sulfuric acid produces a mixture of soluble

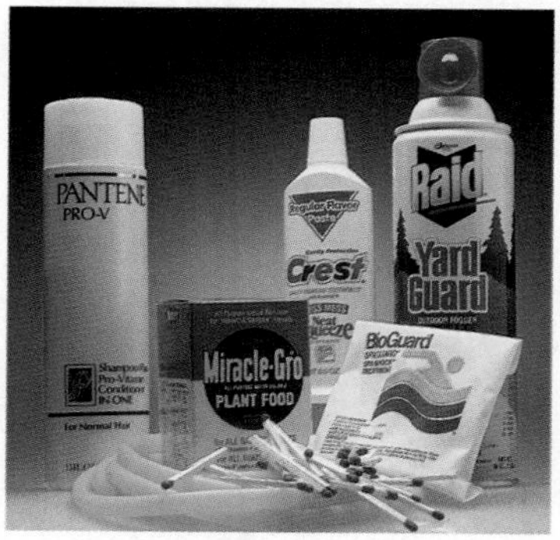

Charles D. Winters

Common household products containing sulfur or sulfur-based compounds.

A Closer Look — Snot-tites and Sulfur Chemistry

Sulfur chemistry can be important in cave formation, as a spectacular example in the jungles of southern Mexico amply demonstrates. Toxic hydrogen sulfide gas spews from the Cueva de Villa Luz along with water that is milky white with suspended sulfur particles. The cave can be followed downward to a large underground stream and a maze of actively enlarging cave passages. Water rises into the cave from underlying sulfur-bearing strata, releasing hydrogen sulfide at concentrations up to 150 ppm. Yellow sulfur crystallizes on the cave walls around the inlets. The sulfur and sulfuric acid are produced by the following reactions:

$$2\ H_2S(g) + O_2(g) \rightarrow 2\ S(s) + 2\ H_2O(\ell)$$

$$2\ S(s) + 2\ H_2O(\ell) + 3\ O_2(g) \rightarrow 2\ H_2SO_4(aq)$$

The cave atmosphere is poisonous to humans, so gas masks are essential for would-be explorers. But surprisingly, the cave is teeming with life. Several species of bacteria thrive on sulfur compounds in acidic environments. The chemical energy released in their metabolism is used to obtain carbon for their bodies from calcium carbonate and carbon dioxide, both of which are abundant in the cave. One result is that bacterial filaments hang from the walls and ceilings in bundles. Because the filaments look like something coming from a runny nose, cave explorers refer to them as "snot-tites." Other microbes feed on the bacteria, and so on up the food chain—which includes spiders, gnats, and pygmy snails—all the way to sardine-like fish that swim in the cave stream. This entire ecosystem is supported by reactions involving sulfur within the cave.

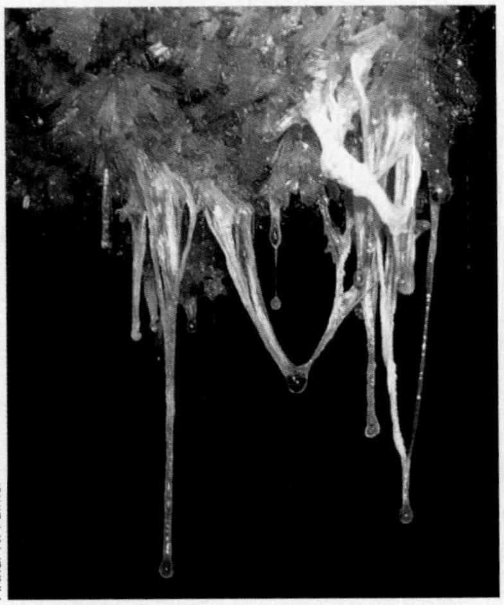

Snot-tites. Filaments of sulfur-oxidizing bacteria (dubbed "snot-tites") hang from the ceiling of a Mexican cave containing an atmosphere rich in hydrogen sulfide. The bacteria thrive on the energy released by oxidation of the hydrogen sulfide, forming the base of a complex food chain. Droplets of sulfuric acid on the filaments have an average pH of 1.4, with some as low as zero! Drops that landed on explorers in the cave burned their skin and disintegrated their clothing.

Arthur N. Palmer

phosphates. The balanced equation for the reaction of excess sulfuric acid and calcium phosphate, for example, is

$$Ca_3(PO_4)_2(s) + 3\ H_2SO_4(\ell) \rightarrow 2\ H_3PO_4(\ell) + 3\ CaSO_4(s)$$

but it does not tell the whole story. Concentrated superphosphate fertilizer is actually mostly $CaHPO_4$ or $Ca(H_2PO_4)_2$ plus some H_3PO_4 and $CaSO_4$. (Notice that the chemical principle behind this reaction is that sulfuric acid is a stronger acid than H_3PO_4 (Table 17.3), so the PO_4^{3-} ion is protonated by sulfuric acid.)

Smaller amounts of sulfuric acid are used in the conversion of ilmenite, a titanium-bearing ore, to TiO_2, which is then used as a white pigment in paint, plastics, and paper. The acid is also used to manufacture iron and steel as well as petroleum products, synthetic polymers, and paper.

Chemistry Now™

Sign in at **www.thomsonedu.com/login** and go to Chapter 21 Contents to see Screen 21.9 for an exercise on **the structural chemistry of sulfur compounds.**

EXERCISE 21.11 Sulfur Chemistry

Metal sulfides roasted in air produce metal oxides.

$$2\ ZnS(s) + 3\ O_2(g) \rightarrow 2\ ZnO(s) + 2\ SO_2(g)$$

Use thermodynamics to decide if the reaction is product- or reactant-favored at equilibrium at 298 K. Will the reaction be more or less product-favored at a high temperature?

21.10 The Halogens, Group 7A

Fluorine and chlorine are the most abundant halogens in the earth's crust, with fluorine somewhat more abundant than chlorine. If their abundance in sea water is measured, however, the situation is quite different. Chlorine has an abundance in sea water of 18,000 ppm, whereas the abundance of fluorine in the same source is only 1.3 ppm. This variation is a result of the differences in the solubility of their salts and plays a role in the methods used to recover the elements themselves.

Group 7A Halogens

Group 7A Halogens
Fluorine 9 F 950 ppm
Chlorine 17 Cl 130 ppm
Bromine 35 Br 0.37 ppm
Iodine 53 I 0.14 ppm
Astatine 85 At trace

Element abundances are in parts per million in the earth's crust.

Preparation of the Elements

Fluorine

The water-insoluble mineral fluorspar (calcium fluoride, CaF_2) is one of the many sources of fluorine. Because the mineral was originally used as a flux in metalworking, its name comes from the Latin word meaning "to flow." In the 17th century, it was discovered that solid CaF_2 would emit light when heated, and the phenomenon was called *fluorescence.* In the early 1800s, when it was recognized that a new element was contained in fluorspar, A. M. Ampère (1775–1836) suggested that the element be called fluorine.

Although fluorine was recognized as an element by 1812, it was not until 1886 that it was isolated by the French chemist Henri Moisson (1852–1907) in elemental form as a very pale yellow gas by the electrolysis of KF dissolved in anhydrous HF. Indeed, because F_2 is such a powerful oxidizing agent, chemical oxidation of F^- to F_2 is not feasible, and electrolysis is the only practical way to obtain gaseous F_2 (Figure 21.31).

The preparation of F_2 is difficult because F_2 is so reactive. It oxidizes (corrodes) the equipment and reacts violently with traces of grease or other contaminants. Furthermore, the products of electrolysis, F_2 and H_2, can recombine explosively, so they must not be allowed to come into contact with each other. (Compare with the reaction of H_2 and Br_2 in Figure 21.5.) Current U.S. production of fluorine is approximately 5000 metric tons per year.

Chlorine

Chlorine is a strong oxidizing agent, and to prepare this element from chloride ion by a chemical reaction requires a stronger oxidizing agent. Permanganate or dichromate ion in acid solution will serve this purpose (Figure 21.32). Elemental chlorine was first made by the Swedish chemist Karl Wilhelm Scheele (1742–1786) in 1774, who combined sodium chloride with an oxidizing agent in an acidic solution.

Industrially, chlorine is made by electrolysis of brine (concentrated aqueous NaCl). The other product of the electrolysis, NaOH, is also a valuable industrial chemical. About 80% of the chlorine produced is made using an electrochemical cell similar to the one depicted in Figure 21.33. Oxidation of chloride ion to Cl_2 gas occurs at the anode and reduction of water occurs at the cathode.

Anode reaction (oxidation) $\quad 2\ Cl^-(aq) \rightarrow Cl_2(g) + 2\ e^-$

Cathode reaction (reduction) $\quad 2\ H_2O(\ell) + 2\ e^- \rightarrow H_2(g) + 2\ OH^-(aq)$

Activated titanium is used for the anode, and stainless steel or nickel is preferred for the cathode in the electrolytic cell. The anode and cathode compartments are separated by a membrane that is not permeable to water but allows Na^+ ions to pass to maintain the charge balance. Thus, the membrane functions as a "salt" bridge between the anode and cathode compartments. The energy consumption of these cells is in the range of 2000–2500 kWh per ton of NaOH produced.

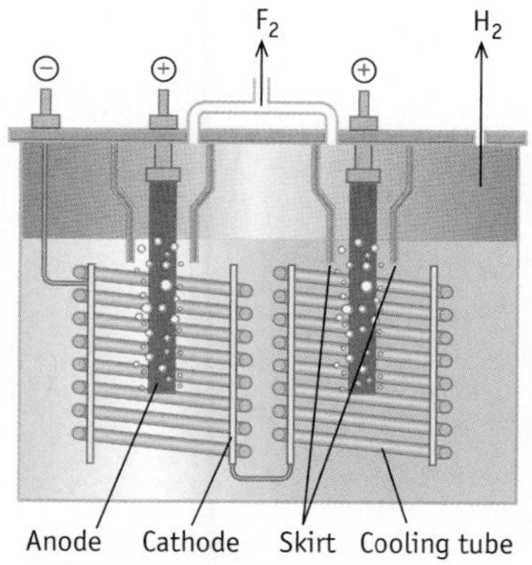

FIGURE 21.31 Schematic of an electrolysis cell for producing fluorine.

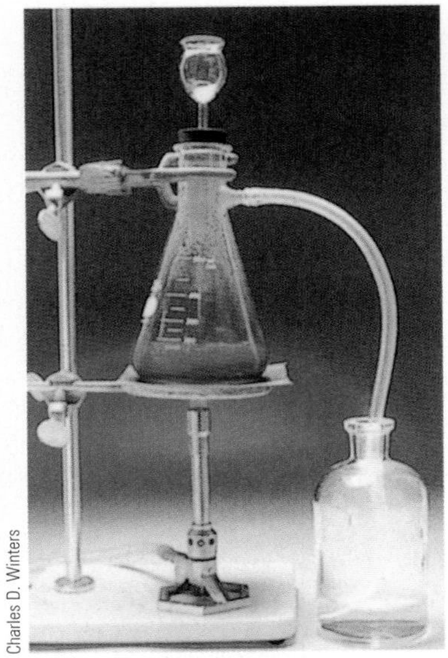

FIGURE 21.32 Chlorine preparation. Chlorine is prepared by oxidation of chloride ion using a strong oxidizing agent. Here, oxidation of NaCl is accomplished using $K_2Cr_2O_7$ in H_2SO_4. (The Cl_2 gas is bubbled into water in a receiving flask.)

Bromine

The standard reduction potentials of the halogens indicate that their strength as oxidizing agents decreases going from F_2 to I_2.

Half-Reaction	Reduction Potential ($E°$, V)
$F_2(g) + 2\,e^- \rightarrow 2\,F^-(aq)$	2.87
$Cl_2(g) + 2\,e^- \rightarrow 2\,Cl^-(aq)$	1.36
$Br_2(\ell) + 2\,e^- \rightarrow 2\,Br^-(aq)$	1.08
$I_2(s) + 2\,e^- \rightarrow 2\,I^-(aq)$	0.535

This means that Cl_2 will oxidize Br^- ions to Br_2 in aqueous solution, for example.

$$Cl_2(aq) + 2\,Br^-(aq) \rightarrow 2\,Cl^-(aq) + Br_2(aq)$$

$$E°_{net} = E°_{cathode} - E°_{anode} = 1.36\text{ V} - (1.08\text{ V}) = +0.28\text{ V}$$

In fact, this is the commercial method of preparing bromine when NaBr is obtained from natural brine wells in Arkansas and Michigan.

Iodine

Iodine is a lustrous, purple-black solid, easily sublimed at room temperature and atmospheric pressure (Figure 13.17). The element was first isolated in 1811 from seaweed and kelp, extracts of which had long been used for treatment of goiter, the enlargement of the thyroid gland. It is now known that the thyroid gland produces a growth-regulating hormone (thyroxine) that contains iodine. Consequently, most table salt in the United States has 0.01% NaI added to provide the necessary iodine in the diet.

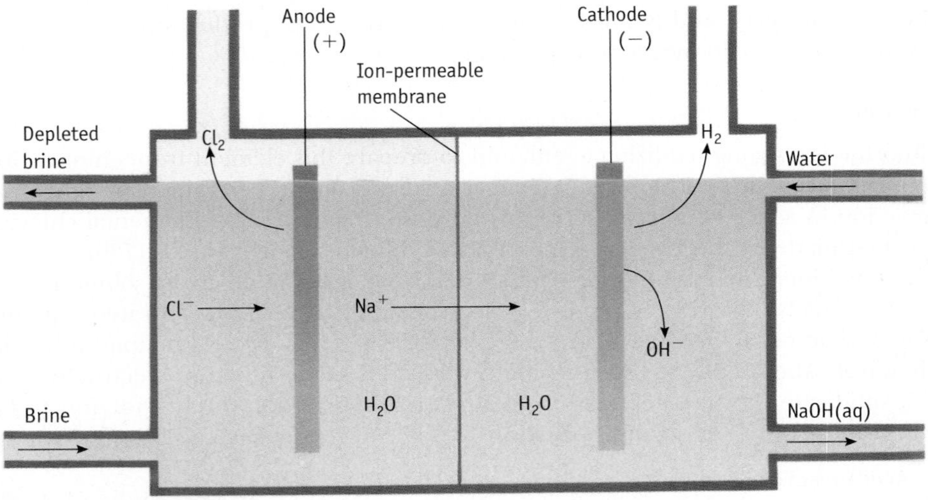

Active Figure 21.33 A membrane cell for the production of NaOH and Cl_2 gas from a saturated, aqueous solution of NaCl (brine). Here, the anode and cathode compartments are separated by a water-impermeable but ion-conducting membrane. A widely used membrane is made of Nafion, a fluorine-containing polymer that is a relative of polytetrafluoroethylene (Teflon). Brine is fed into the anode compartment and dilute sodium hydroxide or water into the cathode compartment. Overflow pipes carry the evolved gases and NaOH away from the chambers of the electrolysis cell.

Chemistry Now™ Sign in at www.thomsonedu.com/login and go to the Chapter 21 Contents menu to explore an interactive version of this figure accompanied by an exercise.

FIGURE 21.34 The preparation of iodine. A mixture of sodium iodide and manganese(IV) oxide was placed in the flask (left). On adding concentrated sulfuric acid (right), brown iodine vapor is evolved.

$$2\ NaI(s) + 2\ H_2SO_4(aq) + MnO_2(s) \rightarrow Na_2SO_4(aq) + MnSO_4(aq) + 2\ H_2O(\ell) + I_2(g)$$

A laboratory method for preparing I_2 is illustrated in Figure 21.34. The commercial preparation depends on the source of I^- and its concentration. One method is interesting because it involves some chemistry described earlier in this book. Iodide ions are first precipitated with silver ions to give insoluble AgI.

$$I^-(aq) + Ag^+(aq) \rightarrow AgI(s)$$

This is reduced by clean scrap iron to give iron(II) iodide and metallic silver.

$$2\ AgI(s) + Fe(s) \rightarrow FeI_2(aq) + 2\ Ag(s)$$

The silver is recycled by oxidizing it with nitric acid (forming silver nitrate) which is then reused. Finally, iodide ion from water-soluble FeI_2 is oxidized to iodine with chlorine [with iron(III) chloride as a by-product].

$$2\ FeI_2(aq) + 3\ Cl_2(aq) \rightarrow 2\ I_2(s) + 2\ FeCl_3(aq)$$

Fluorine Compounds

Fluorine is the most reactive of all of the elements, forming compounds with every element except He and Ne. In most cases, the elements combine directly, and some reactions can be so vigorous as to be explosive. This reactivity can be explained by at least two features of fluorine chemistry: the relatively weak F—F bond compared with chlorine and bromine, and, in particular, the relatively strong bonds formed by fluorine to other elements. This is illustrated by the table of bond dissociation enthalpies in the margin.

In addition to its oxidizing ability, another notable characteristic of fluorine is its small size. These properties lead to the formation of compounds where a number of F atoms can be bonded to a central element in a high oxidation state. Examples include PtF_6, UF_6, IF_7, and XeF_4.

Hydrogen fluoride is an important industrial chemical. More than 1 million tons of hydrogen fluoride is produced annually worldwide, almost all by the action of concentrated sulfuric acid on fluorspar.

$$CaF_2(s) + H_2SO_4(\ell) \rightarrow CaSO_4(s) + 2\ HF(g)$$

The U.S. capacity for HF production is approximately 210,000 metric tons, but currently demand is exceeding supply for this chemical. Anhydrous HF is used in

Bond Dissociation Enthalpies of Some Halogen Compounds (kJ/mol)

X	X—X	H—X	C—X (in CX_4)
F	155	565	485
Cl	242	432	339
Br	193	366	285
I	151	299	213

a broad range of industries: in the production of refrigerants, herbicides, pharmaceuticals, high-octane gasoline, aluminum, plastics, electrical components, and fluorescent lightbulbs.

The fluorspar used to produce HF must be very pure and free of SiO_2 because HF reacts readily with silicon dioxide.

$$SiO_2(s) + 4\ HF(aq) \rightarrow SiF_4(g) + 2\ H_2O(\ell)$$

$$SiF_4(g) + 2\ HF(aq) \rightarrow H_2SiF_6(aq)$$

This series of reactions explains why HF can be used to etch or frost glass (such as the inside of fluorescent light bulbs). It also explains why HF is not shipped in glass containers (unlike HCl, for example).

The aluminum industry consumes about 10–40 kg of cryolite, Na_3AlF_6, per metric ton of aluminum produced. The reason is that cryolite is added to aluminum oxide to produce a lower-melting mixture that can be electrolyzed. Cryolite is found in only small quantities in nature, so it is made in various ways, among them the following reaction:

$$6\ HF(aq) + Al(OH)_3(s) + 3\ NaOH(aq) \rightarrow Na_3AlF_6(s) + 6\ H_2O(\ell)$$

About 3% of the hydrofluoric acid produced is used in uranium fuel production. To separate uranium isotopes in a gas centrifuge (◀ page 523), the uranium must be in the form of a volatile compound. Naturally occurring uranium is processed to give UO_2. This oxide is treated with hydrogen fluoride to give UF_4, which is then reacted with F_2 to produce the volatile solid UF_6.

$$UO_2(s) + 4\ HF(aq) \rightarrow UF_4(s) + 2\ H_2O(\ell)$$

$$UF_4(s) + F_2(g) \rightarrow UF_6(s)$$

This last step consumes 70–80% of fluorine produced annually.

Chlorine Compounds

Hydrogen Chloride

Hydrochloric acid, an aqueous solution of hydrogen chloride, is a valuable industrial chemical. Hydrogen chloride gas can be prepared by the reaction of hydrogen and chlorine, but the rapid, exothermic reaction is difficult to control. The classical method of making HCl in the laboratory uses the reaction of NaCl and sulfuric acid, a procedure that takes advantage of the facts that HCl is a gas and that H_2SO_4 will not oxidize the chloride ion.

$$2\ NaCl(s) + H_2SO_4(\ell) \rightarrow Na_2SO_4(s) + 2\ HCl(g)$$

Hydrogen chloride gas has a sharp, irritating odor. Both gaseous and aqueous HCl react with metals and metal oxides to give metal chlorides and, depending on the reactant, hydrogen or water.

$$Mg(s) + 2\ HCl(aq) \rightarrow MgCl_2(aq) + H_2(g)$$

$$ZnO(s) + 2\ HCl(aq) \rightarrow ZnCl_2(aq) + H_2O(\ell)$$

Oxoacids of Chlorine

Oxoacids of chlorine range from HClO, in which chlorine has an oxidation number of +1, to $HClO_4$, in which the oxidation number is equal to the group number, +7. All are strong oxidizing agents.

Oxoacids of Chlorine

Acid	Name	Anion	Name
$HClO$	Hypochlorous	ClO^-	Hypochlorite
$HClO_2$	Chlorous	ClO_2^-	Chlorite
$HClO_3$	Chloric	ClO_3^-	Chlorate
$HClO_4$	Perchloric	ClO_4^-	Perchlorate

Hypochlorous acid, HClO, forms when chlorine dissolves in water. In this reaction, half of the chlorine is oxidized to hypochlorite ion and half is reduced to chloride ion in a **disproportionation reaction**.

■ **Disproportionation** A reaction in which an element or compound is simultaneously oxidized and reduced is called a disproportionation reaction. Here, Cl_2 is oxidized to ClO^- and reduced to Cl^-.

$$Cl_2(g) + 2\ H_2O(\ell) \rightleftharpoons H_3O^+(aq) + HClO(aq) + Cl^-(aq)$$

If Cl_2 is dissolved in cold aqueous NaOH instead of in pure water, hypochlorite ion and chloride ion form.

$$Cl_2(g) + 2\ OH^-(aq) \rightleftharpoons ClO^-(aq) + Cl^-(aq) + H_2O(\ell)$$

Under basic conditions, the equilibrium lies far to the right. The resulting alkaline solution is the "liquid bleach" used in home laundries. The bleaching action of this solution is a result of the oxidizing ability of ClO^-. Most dyes are colored organic compounds, and hypochlorite ion oxidizes dyes to colorless products.

When calcium hydroxide is combined with Cl_2, solid $Ca(ClO)_2$ is the product. This compound is easily handled and is the "chlorine" that is sold for swimming pool disinfection.

When a basic solution of hypochlorite ion is heated, another disproportionation occurs, forming chlorate ion and chloride ion:

$$3\ ClO^-(aq) \rightarrow ClO_3^-(aq) + 2\ Cl^-(aq)$$

Sodium and potassium chlorates are made in large quantities this way. The sodium salt can be reduced to ClO_2, a compound used for bleaching paper pulp. Some $NaClO_3$ is also converted to potassium chlorate, $KClO_3$, the preferred oxidizing agent in fireworks and a component of safety matches.

Perchlorates, salts containing ClO_4^-, are powerful oxidants. Pure perchloric acid, $HClO_4$, is a colorless liquid that explodes if shocked. It explosively oxidizes organic materials and rapidly oxidizes silver and gold. Dilute aqueous solutions of the acid are safe to handle, however.

Perchlorate salts of most metals are usually relatively stable, albeit unpredictable. Great care should be used when handling any perchlorate salt. Ammonium perchlorate, for example, bursts into flame if heated above 200 °C.

$$2\ NH_4ClO_4(s) \rightarrow N_2(g) + Cl_2(g) + 2\ O_2(g) + 4\ H_2O(g)$$

The strong oxidizing ability of the ammonium salt accounts for its use as the oxidizer in the solid booster rockets for the Space Shuttle. The solid propellant in these rockets is largely NH_4ClO_4, the remainder being the reducing agent, powdered aluminum. Each launch requires about 750 tons of ammonium perchlorate, and more than half of the sodium perchlorate currently manufactured is converted to the ammonium salt. The process for making this conversion is an exchange reaction that takes advantage of the fact that ammonium perchlorate is less soluble in water than sodium perchlorate:

$$NaClO_4(aq) + NH_4Cl(aq) \rightleftharpoons NaCl(aq) + NH_4ClO_4(s)$$

NASA

Use of a perchlorate salt. The solid-fuel booster rockets of the Space Shuttle utilize a mixture of NH_4ClO_4 (oxidizing agent) and Al powder (reducing agent).

EXERCISE 21.12 Reactions of the Halogen Acids

Metals generally react with hydrogen halides to give the metal halide and hydrogen. Determine whether this is true for silver by calculating $\Delta_rG°$ for the reaction with each of the hydrogen halides.

$$Ag(s) + HX(g) \rightarrow AgX(s) + \frac{1}{2} H_2(g)$$

The reaction is spontaneous under standard conditions (and product-favored at equilibrium) if $\Delta_rG°$ is negative. The required free energies of formation are (in kJ/mol)

HX, $\Delta_fG°$ (kJ/mol)	AgX, $\Delta_fG°$ (kJ/mol)
HF, −273.2	AgF, −193.8
HCl, −95.09	AgCl, −109.76
HBr, −53.45	AgBr, −96.90
HI, +1.56	AgI, −66.19

Chapter Goals Revisited

Chemistry Now™ Sign in at **www.thomsonedu.com/login** to:

- Assess your understanding with Study Questions in OWL keyed to each goal in the Goals and Homework menu for this chapter
- For quick review, download Go Chemistry mini-lecture flashcard modules (or purchase them at **www.ichapters.com**)
- Check your readiness for an exam by taking the Pre-Test and exploring the modules recommended in your Personalized Study plan.

Access **How Do I Solve It?** tutorials on how to approach problem solving using concepts in this chapter.

Now that you have studied this chapter, you should ask whether you have met the chapter goals. In particular, you should be able to:

Relate the formulas and properties of compounds to the periodic table

a. Predict several chemical reactions of the Group A elements (Section 21.2). Study Question(s) assignable in OWL: 12.

b. Predict similarities and differences among the elements in a given group, based on the periodic properties (Section 21.2). Study Question(s) assignable in OWL: 22, 32, 34, 35.

c. Know which reactions produce ionic compounds, and predict formulas for common ions and common ionic compounds based on electron configurations (Section 21.2). Study Question(s) assignable in OWL: 14, 16, 26.

d. Recognize when a formula is incorrectly written, based on general principles governing electron configurations (Section 21.2).

Describe the chemistry of the main group or A-Group elements, particularly H; Na and K; Mg and Ca; B and Al; Si; N and P; O and S; and F and Cl

a. Identify the most abundant elements, know how they are obtained, and list some of their common chemical and physical properties.

b. Be able to summarize briefly a series of facts about the most common compounds of main group elements (ionic or covalent bonding, color, solubility, simple reaction chemistry) (Sections 21.3–21.10). Study Question(s) assignable in OWL: 86, 88.

c. Identify uses of common elements and compounds, and understand the chemistry that relates to their usage (Sections 21.3–21.10).

Apply the principles of stoichiometry, thermodynamics, and electrochemistry to the chemistry of the main group elements Study Question(s) assignable in OWL: 18, 24, 29, 32, 34, 43, 48, 50, 54, 58, 61, 75, 86, 88.

STUDY QUESTIONS

Online homework for this chapter may be assigned in OWL.

▲ denotes challenging questions.

■ denotes questions assignable in OWL.

Blue-numbered questions have answers in Appendix O and fully-worked solutions in the *Student Solutions Manual.*

Practicing Skills

Properties of the Elements

1. Give examples of two basic oxides. Write equations illustrating the formation of each oxide from its component elements. Write another chemical equation that illustrates the basic character of each oxide.

2. Give examples of two acidic oxides. Write equations illustrating the formation of each oxide from its component elements. Write another chemical equation that illustrates the acidic character of each oxide.

3. Give the name and symbol of each element having the valence configuration [noble gas] ns^2np^1.

4. Give symbols and names for four monatomic ions that have the same electron configuration as argon.

5. Select one of the alkali metals, and write a balanced chemical equation for its reaction with chlorine. Is the reaction likely to be exothermic or endothermic? Is the product ionic or molecular?

6. Select one of the alkaline earth metals and write a balanced chemical equation for its reaction with oxygen. Is the reaction likely to be exothermic or endothermic? Is the product ionic or molecular?

7. For the product of the reaction you selected in Study Question 5, predict the following physical properties: color, state of matter (s, ℓ, or g), solubility in water.

8. For the product of the reaction you selected in Study Question 6, predict the following physical properties: color, state of matter (s, ℓ, or g), solubility in water.

9. Would you expect to find calcium occurring naturally in the earth's crust as a free element? Why or why not?

10. Which of the first 10 elements in the periodic table are found as free elements in the earth's crust? Which elements in this group occur in the earth's crust only as part of a chemical compound?

11. Place the following oxides in order of increasing basicity: CO_2, SiO_2, SnO_2.

12. ■ Place the following oxides in order of increasing basicity: Na_2O, Al_2O_3, SiO_2, SO_3.

13. Complete and balance the equations for the following reactions. [Assume an excess of oxygen for (d).]
    (a) $Na(s) + Br_2(\ell) \rightarrow$
    (b) $Mg(s) + O_2(g) \rightarrow$
    (c) $Al(s) + F_2(g) \rightarrow$
    (d) $C(s) + O_2(g) \rightarrow$

14. ■ Complete and balance the equations for the following reactions:
    (a) $K(s) + I_2(g) \rightarrow$
    (b) $Ba(s) + O_2(g) \rightarrow$
    (c) $Al(s) + S_8(s) \rightarrow$
    (d) $Si(s) + Cl_2(g) \rightarrow$

Hydrogen

15. Write balanced chemical equations for the reaction of hydrogen gas with oxygen, chlorine, and nitrogen.

16. ■ Write an equation for the reaction of potassium and hydrogen. Name the product. Is it ionic or covalent? Predict one physical property and one chemical property of this compound.

17. Write a balanced chemical equation for the preparation of H_2 (and CO) by the reaction of CH_4 and water. Using data in Appendix L, calculate $\Delta_r H°$, $\Delta_r G°$, and $\Delta_r S°$ for this reaction.

18. ■ Using data in Appendix L, calculate $\Delta_r H°$, $\Delta_r G°$, and $\Delta_r S°$ for the reaction of carbon and water to give CO and H_2.

19. A method recently suggested for the preparation of hydrogen (and oxygen) from water proceeds as follows:
    (a) Sulfuric acid and hydrogen iodide are formed from sulfur dioxide, water, and iodine.
    (b) The sulfuric acid from the first step is decomposed by heat to water, sulfur dioxide, and oxygen.
    (c) The hydrogen iodide from the first step is decomposed with heat to hydrogen and iodine.

    Write a balanced equation for each of these steps, and show that their sum is the decomposition of water to form hydrogen and oxygen.

20. Compare the mass of H_2 expected from the reaction of steam (H_2O) per mole of methane, petroleum, and coal. (Assume complete reaction in each case. Use CH_2 and CH as representative formulas for petroleum and coal, respectively.)

Alkali Metals

21. Write equations for the reaction of sodium with each of the halogens. Predict at least two physical properties that are common to all of the alkali metal halides.

22. ■ Write balanced equations for the reaction of lithium, sodium, and potassium with O_2. Specify which metal forms an oxide, which forms a peroxide, and which forms a superoxide.

23. The electrolysis of aqueous NaCl gives NaOH, Cl_2, and H_2.
(a) Write a balanced equation for the process.
(b) In the United States, 1.19×10^{10} kg of NaOH and 1.14×10^{10} kg of Cl_2 were produced in a recent year. Does the ratio of masses of NaOH and Cl_2 produced agree with the ratio of masses expected from the balanced equation? If not, what does this tell you about the way in which NaOH and Cl_2 are actually produced? Is the electrolysis of aqueous NaCl the only source of these chemicals?

24. ■ (a) Write equations for the half-reactions that occur at the cathode and the anode when an aqueous solution of KCl is electrolyzed. Which chemical species is oxidized, and which chemical species is reduced in this reaction?
(b) Predict the products formed when an aqueous solution of CsI is electrolyzed.

Alkaline Earth Elements

25. When magnesium burns in air, it forms both an oxide and a nitride. Write balanced equations for the formation of both compounds.

26. ■ Calcium reacts with hydrogen gas at 300—400 °C to form a hydride. This compound reacts readily with water, so it is an excellent drying agent for organic solvents.
(a) Write a balanced equation showing the formation of calcium hydride from Ca and H_2.
(b) Write a balanced equation for the reaction of calcium hydride with water (Figure 21.7).

27. Name three uses of limestone. Write a balanced equation for the reaction of limestone with CO_2 in water.

28. Explain what is meant by "hard water." What causes hard water, and what problems are associated with it?

29. ■ Calcium oxide, CaO, is used to remove SO_2 from power plant exhaust. These two compounds react to give solid $CaSO_3$. What mass of SO_2 can be removed using 1.2×10^3 kg of CaO?

30. $Ca(OH)_2$ has a K_{sp} of 5.5×10^{-5}, whereas K_{sp} for $Mg(OH)_2$ is 5.6×10^{-12}. Calculate the equilibrium constant for the reaction

$$Ca(OH)_2(s) + Mg^{2+}(aq) \rightleftharpoons Ca^{2+}(aq) + Mg(OH)_2(s)$$

Explain why this reaction can be used in the commercial isolation of magnesium from sea water.

Boron and Aluminum

31. Draw a possible structure for the cyclic anion in the salt $K_3B_3O_6$ and the anion in $Ca_2B_2O_5$.

32. ■ The boron trihalides (except BF_3) hydrolyze completely to boric acid and the acid HX.
(a) Write a balanced equation for the reaction of BCl_3 with water.
(b) Calculate $\Delta_r H°$ for the hydrolysis of BCl_3 using data in Appendix L and the following information: $\Delta_f H°$ [BCl_3(g)] = −403 kJ/mol; $\Delta_f H°$ [$B(OH)_3$(s)] = −1094 kJ/mol.

33. When boron hydrides burn in air, the reaction is very exothermic.
(a) Write a balanced equation for the combustion of B_5H_9(g) in air to give B_2O_3(s) and H_2O(ℓ).
(b) Calculate the enthalpy of combustion for B_5H_9(g) ($\Delta_f H°$ = 73.2 kJ/mol), and compare it with the enthalpy of combustion of B_2H_6 on page 984. (The enthalpy of formation of B_2O_3(s) is −1271.9 kJ/mol.)
(c) Compare the enthalpy of combustion of C_2H_6(g) with that of B_2H_6(g). Which transfers more energy as heat per gram?

34. ■ Diborane can be prepared by the reaction of $NaBH_4$ and I_2. Which substance is oxidized, and which is reduced?

35. ■ Write balanced equations for the reactions of aluminum with HCl(aq), Cl_2, and O_2.

36. (a) Write a balanced equation for the reaction of Al and H_2O(ℓ) to produce H_2 and Al_2O_3.
(b) Using thermodynamic data in Appendix L, calculate $\Delta_r H°$, $\Delta_r S°$, and $\Delta_r G°$ for this reaction. Do these data indicate that the reaction should favor the products?
(c) Why is aluminum metal unaffected by water?

37. Aluminum dissolves readily in hot aqueous NaOH to give the aluminate ion, $[Al(OH)_4]^-$, and H_2. Write a balanced equation for this reaction. If you begin with 13.2 g of Al, what volume (in milliliters) of H_2 gas is produced when the gas is measured at 22.5 °C and a pressure of 735 mm Hg?

38. Alumina, Al_2O_3, is amphoteric. Among examples of its amphoteric character are the reactions that occur when Al_2O_3 is heated strongly or "fused" with acidic oxides and basic oxides.
(a) Write a balanced equation for the reaction of alumina with silica, an acidic oxide, to give aluminum metasilicate, $Al_2(SiO_3)_3$.
(b) Write a balanced equation for the reaction of alumina with the basic oxide CaO to give calcium aluminate, $Ca(AlO_2)_2$.

39. Aluminum sulfate is the most commercially important aluminum compound, after aluminum oxide and aluminum hydroxide. It is produced from the reaction of aluminum oxide and sulfuric acid. What mass (in kilograms) of aluminum oxide and sulfuric acid must be used to manufacture 1.00 kg of aluminum sulfate?

40. "Aerated" concrete bricks are widely used building materials. They are obtained by mixing gas-forming additives with a moist mixture of lime, cement, and possibly sand. Industrially, the following reaction is important:

$$2\ Al(s) + 3\ Ca(OH)_2(s) + 6\ H_2O(\ell) \rightarrow$$
$$3\ CaO \cdot Al_2O_3 \cdot 6\ H_2O(s) + 3\ H_2(g)$$

Assume that the mixture of reactants contains 0.56 g of Al (as well as excess calcium hydroxide and water) for each brick. What volume of hydrogen gas do you expect at 26 °C and a pressure of 745 mm Hg)?

Silicon

41. Describe the structure of pyroxenes (see page 988). What is the ratio of silicon to oxygen in this type of silicate?

42. Describe how ultrapure silicon can be produced from sand.

43. ■ Silicate structures: Draw a structure, and give the charge for a silicate anion with the formula $[Si_6O_{18}]^{n-}$.

44. Silicates often have chain, ribbon, or sheet structures. One of the simpler ribbon structures is $[Si_2O_5{}^{2-}]_n$. Draw a structure for this anionic material.

Nitrogen and Phosphorus

45. Consult the data in Appendix L. Are any of the nitrogen oxides listed there stable with respect to decomposition to N_2 and O_2?

46. Use data in Appendix L to calculate the enthalpy and free energy change for the reaction

$$2\ NO_2(g) \rightarrow N_2O_4(g)$$

Is this reaction exothermic or endothermic? Is the reaction product- or reactant-favored?

47. Use data in Appendix L to calculate the enthalpy and free energy change for the reaction

$$2\ NO(g) + O_2(g) \rightarrow 2\ NO_2(g)$$

Is this reaction exothermic or endothermic? Is the reaction product- or reactant-favored?

48. ■ The overall reaction involved in the industrial synthesis of nitric acid is

$$NH_3(g) + 2\ O_2(g) \rightarrow HNO_3(aq) + H_2O(\ell)$$

Calculate $\Delta_r G°$ for this reaction and its equilibrium constant at 25 °C.

49. A major use of hydrazine, N_2H_4, is in steam boilers in power plants.
(a) The reaction of hydrazine with O_2 dissolved in water gives N_2 and water. Write a balanced equation for this reaction.
(b) O_2 dissolves in water to the extent of 0.0044 g in 100. mL of water at 20 °C. To consume all of the dissolved O_2 in 3.00×10^4 L of water (enough to fill a small swimming pool), what mass of N_2H_4 is needed?

50. ■ Before hydrazine came into use to remove dissolved oxygen in the water in steam boilers, Na_2SO_3 was commonly used for this purpose:

$$2\ Na_2SO_3(aq) + O_2(aq) \rightarrow 2\ Na_2SO_4(aq)$$

What mass of Na_2SO_3 is required to remove O_2 from 3.00×10^4 L of water as outlined in Study Question 49?

51. Review the structure of phosphorous acid in Table 21.6.
(a) What is the oxidation number of the phosphorus atom in this acid?
(b) Draw the structure of diphosphorous acid, $H_4P_2O_5$. What is the maximum number of protons this acid can dissociate in water?

52. Unlike carbon, which can form extended chains of atoms, nitrogen can form chains of very limited length. Draw the Lewis electron dot structure of the azide ion, $N_3{}^-$. Is the ion linear or bent?

Oxygen and Sulfur

53. In the "contact process" for making sulfuric acid, sulfur is first burned to SO_2. Environmental restrictions allow no more than 0.30% of this SO_2 to be vented to the atmosphere.
(a) If enough sulfur is burned in a plant to produce 1.80×10^6 kg of pure, anhydrous H_2SO_4 per day, what is the maximum amount of SO_2 that is allowed to be exhausted to the atmosphere?
(b) One way to prevent any SO_2 from reaching the atmosphere is to "scrub" the exhaust gases with slaked lime, $Ca(OH)_2$:

$$Ca(OH)_2(s) + SO_2(g) \rightarrow CaSO_3(s) + H_2O(\ell)$$

$$2\ CaSO_3(s) + O_2(g) \rightarrow 2\ CaSO_4(s)$$

What mass of $Ca(OH)_2$ (in kilograms) is needed to remove the SO_2 calculated in part (a)?

54. ■ A sulfuric acid plant produces an enormous amount of heat. To keep costs as low as possible, much of this heat is used to make steam to generate electricity. Some of the electricity is used to run the plant, and the excess is sold to the local electrical utility. Three reactions are important in sulfuric acid production: (1) burning S to SO_2; (2) oxidation of SO_2 to SO_3; and (3) reaction of SO_3 with H_2O:

$$SO_3(g) + H_2O\ (\text{in 98\% } H_2SO_4) \rightarrow H_2SO_4(\ell)$$

The enthalpy change of the third reaction is −130 kJ/mol. Estimate the enthalpy change when 1.00 mol of S is used to produce 1.00 mol of H_2SO_4. How much energy is produced per metric ton of H_2SO_4?

55. Sulfur forms anionic chains of S atoms called polysulfides. Draw a Lewis electron dot structure for the S_2^{2-} ion. The S_2^{2-} ion is the disulfide ion, an analogue of the peroxide ion. It occurs in iron pyrites, FeS_2.

56. Sulfur forms a range of compounds with fluorine. Draw Lewis electron dot structures for S_2F_2 (connectivity is FSSF), SF_2, SF_4, SF_6, and S_2F_{10}. What is the oxidation number of sulfur in each of these compounds?

Fluorine and Chlorine

57. The halogen oxides and oxoanions are good oxidizing agents. For example, the reduction of bromate ion has an $E°$ value of 1.44 V in acid solution:

$$2\ BrO_3^-(aq) + 12\ H^+(aq) + 10\ e^- \rightarrow Br_2(aq) + 6\ H_2O(\ell)$$

Is it possible to oxidize aqueous 1.0 M Mn^{2+} to aqueous MnO_4^- with 1.0 M bromate ion?

58. ■ The hypohalite ions, XO^-, are the anions of weak acids. Calculate the pH of a 0.10 M solution of NaClO. What is the concentration of HClO in this solution?

59. Bromine is obtained from brine wells. The process involves treating water containing bromide ion with Cl_2 and extracting the Br_2 from the solution using an organic solvent. Write a balanced equation for the reaction of Cl_2 and Br^-. What are the oxidizing and reducing agents in this reaction? Using the table of standard reduction potentials (Appendix M), verify that this is a product-favored reaction.

60. To prepare chlorine from chloride ion a strong oxidizing agent is required. The dichromate ion, $Cr_2O_7^{2-}$, is one example (see Figure 21.32). Consult the table of standard reduction potentials (Appendix M), and identify several other oxidizing agents that may be suitable. Write balanced equations for the reactions of these substances with chloride ion.

61. ■ If an electrolytic cell for producing F_2 (Figure 21.31) operates at 5.00×10^3 A (at 10.0 V), what mass of F_2 can be produced per 24-hour day? Assume the conversion of F^- to F_2 is 100%.

62. Halogens combine with one another to produce *interhalogens* such as BrF_3. Sketch a possible molecular structure for this molecule, and decide if the F—Br—F bond angles will be less than or greater than ideal.

General Questions

The questions are not designated as to type or location in the chapter. They may combine several concepts.

63. For each of the third-period elements (Na through Ar), identify the following:
(a) whether the element is a metal, nonmetal, or metalloid
(b) the color and appearance of the element
(c) the state of the element (s, ℓ, or g) under standard conditions

For help in this question, consult Figure 2.4 or use the periodic table "tool" in ChemistryNow. The latter provides a picture of each element and a listing of its properties.

64. Consider the chemistries of C, Si, Ge, and Sn.
(a) Write a balanced chemical equation to depict the reaction of each element with elemental chlorine.
(b) Describe the bonding in each of the products of the reactions with chlorine as ionic or covalent.
(c) Compare the reactions, if any, of some Group 4A chlorides —CCl_4, $SiCl_4$, and $SnCl_4$— with water.

65. Consider the chemistries of the elements potassium, calcium, gallium, germanium, and arsenic.
(a) Write a balanced chemical equation depicting the reaction of each element with elemental chlorine.
(b) Describe the bonding in each of the products of the reactions with chlorine as ionic or covalent.
(c) Draw Lewis electron dot structures for the products of the reactions of gallium and arsenic with chlorine. What are their electron-pair and molecular geometries?

66. When BCl_3 gas is passed through an electric discharge, small amounts of the reactive molecule B_2Cl_4 are produced. (The molecule has a B—B covalent bond.)
(a) Draw a Lewis electron dot structure for B_2Cl_4.
(b) Describe the hybridization of the B atoms in the molecule and the geometry around each B atom.

67. Complete and balance the following equations.
(a) $KClO_3$ + heat →
(b) $H_2S(g) + O_2(g) \rightarrow$
(c) $Na(s) + O_2(g) \rightarrow$
(d) $P_4(s) + KOH(aq) + H_2O(\ell) \rightarrow$
(e) $NH_4NO_3(s)$ + heat →
(f) $In(s) + Br_2(\ell) \rightarrow$
(g) $SnCl_4(\ell) + H_2O(\ell) \rightarrow$

68. (a) Heating barium oxide in pure oxygen gives barium peroxide. Write a balanced equation for this reaction.
(b) Barium peroxide is an excellent oxidizing agent. Write a balanced equation for the reaction of iron with barium peroxide to give iron(III) oxide and barium oxide.

69. Worldwide production of silicon carbide, SiC, is several hundred thousand tons annually. If you want to produce 1.0×10^5 metric tons of SiC, what mass (metric tons) of silicon sand (SiO_2) will you use if 70% of the sand is converted to SiC?

70. To store 2.88 kg of gasoline with an energy equivalence of 1.43×10^8 J requires a volume of 4.1 L. In comparison, 1.0 kg of H_2 has the same energy equivalence. What volume is required if this quantity of H_2 is to be stored at 25 °C and 1.0 atm of pressure?

71. Using data in Appendix L, calculate $\Delta_r G°$ values for the decomposition of MCO_3 to MO and CO_2 where M = Mg, Ca, Ba. What is the relative tendency of these carbonates to decompose?

72. Ammonium perchlorate is used as the oxidizer in the solid-fuel booster rockets of the Space Shuttle. Assume that one launch requires 700 tons (6.35×10^5 kg) of the salt, and the salt decomposes according to the equation on page 1009.
(a) What mass of water is produced? What mass of O_2 is produced?
(b) If the O_2 produced is assumed to react with the powdered aluminum present in the rocket engine, what mass of aluminum is required to use up all of the O_2?
(c) What mass of Al_2O_3 is produced?

73. ▲ Metals react with hydrogen halides (such as HCl) to give the metal halide and hydrogen:

$$M(s) + n\,HX(g) \longrightarrow MX_n(s) + \tfrac{1}{2}n\,H_2(g)$$

The free energy change for the reaction is $\Delta_r G° = \Delta_f G°(MX_n) - n\,\Delta_f G°[HX(g)]$
(a) $\Delta_f G°$ for HCl(g) is −95.1 kJ/mol. What must be the value for $\Delta_f G°$ for MX_n for the reaction to be product-favored?
(b) Which of the following metals is (are) predicted to have product-favored reactions with HCl(g): Ba, Pb, Hg, Ti?

74. Halogens form polyhalide ions. Sketch Lewis electron dot structures and molecular structures for the following ions:
(a) I_3^-
(b) $BrCl_2^-$
(c) ClF_2^+
(d) An iodide ion and two iodine molecules form the I_5^- ion. Here, the ion has five I atoms in a row, but the ion is not linear. Draw the Lewis dot structure for the ion, and propose a structure for the ion.

75. ■ The standard enthalpy of formation of OF_2 gas is +24.5 kJ/mol. Calculate the average O—F bond enthalpy.

76. Calcium fluoride can be used in the fluoridation of municipal water supplies. If you want to achieve a fluoride ion concentration of 2.0×10^{-5} M, what mass of CaF_2 must you use for 1.0×10^6 L of water?

77. The steering rockets in the Space Shuttle use N_2O_4 and a derivative of hydrazine, 1,1-dimethylhydrazine (page 249). This mixture is called a *hypergolic fuel* because it ignites when the reactants come into contact:

$$H_2NN(CH_3)_2(\ell) + 2\,N_2O_4(\ell) \longrightarrow 3\,N_2(g) + 4\,H_2O(g) + 2\,CO_2(g)$$

(a) Identify the oxidizing agent and the reducing agent in this reaction.
(b) The same propulsion system was used by the Lunar Lander on moon missions in the 1970s. If the Lander used 4100 kg of $H_2NN(CH_3)_2$, what mass (in kilograms) of N_2O_4 was required to react with it? What mass (in kilograms) of each of the reaction products was generated?

78. ▲ Liquid HCN is dangerously unstable with respect to trimer formation—that is, formation of $(HCN)_3$ with a cyclic structure.
(a) Propose a structure for this cyclic trimer.
(b) Estimate the energy of the trimerization reaction using bond dissociation enthalpies (Table 8.10).

79. Use $\Delta_f H°$ data in Appendix L to calculate the enthalpy change of the reaction

$$2\,N_2(g) + 5\,O_2(g) + 2\,H_2O(\ell) \longrightarrow 4\,HNO_3(aq)$$

Speculate on whether such a reaction could be used to "fix" nitrogen. Would research to find ways to accomplish this reaction be a useful endeavor?

80. ▲ Phosphorus forms an extensive series of oxoanions.
(a) Draw a structure, and give the charge for an oxophosphate anion with the formula $[P_4O_{13}]^{n-}$. How many ionizable H atoms should the completely protonated acid have?
(b) Draw a structure, and give the charge for an oxophosphate anion with the formula $[P_4O_{12}]^{n-}$. How many ionizable H atoms should the completely protonated acid have?

81. Boron and hydrogen form an extensive family of compounds, and the diagram below shows how they are related by reaction.

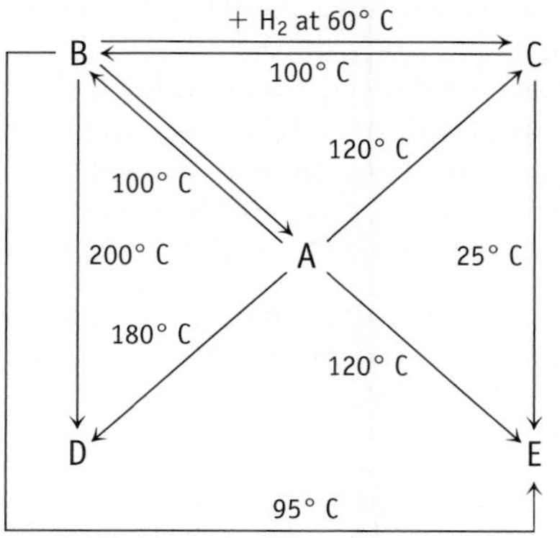

The following table gives the weight percent of boron in each of the compounds. Derive the empirical and molecular formulas of compounds A–D.

Substance	State (at STP)	Mass Percent B	Molar mass (g/mol)
A	Gas	78.3	27.7
B	Gas	81.2	53.3
C	Liquid	83.1	65.1
D	Liquid	85.7	63.1
E	Solid	88.5	122.2

82. ▲ In 1774, C. Scheele obtained a gas by reacting pyrolusite (MnO_2) with sulfuric acid. The gas, which had been obtained that same year by Joseph Priestley by a different method, was an element, **A**.
(a) What is the element isolated by Scheele and Priestley?
(b) Element **A** combines with almost all other elements. For example, with cesium it gives a compound in which the mass percent of **A** is 19.39%. The element combines with hydrogen to give a compound with a mass percent of element **A** of 94.12%. Determine the formulas of the cesium and hydrogen compounds.
(c) The compounds of cesium and hydrogen with element **A** react with one another. Write a balanced equation for the reaction.

In the Laboratory

83. One material needed to make silicones is dichlorodimethylsilane, $(CH_3)_2SiCl_2$. It is made by treating silicon powder at about 300 °C with CH_3Cl in the presence of a copper-containing catalyst.
(a) Write a balanced equation for the reaction.
(b) Assume you carry out the reaction on a small scale with 2.65 g of silicon. To measure the CH_3Cl gas, you fill a 5.60-L flask at 24.5 °C. What pressure of CH_3Cl gas must you have in the flask to have the stoichiometrically correct amount of the compound?
(c) What mass of $(CH_3)_2SiCl_2$ can be produced from 2.65 g of Si and excess CH_3Cl?

84. Sodium borohydride, $NaBH_4$, reduces many metal ions to the metal.
(a) Write a balanced equation for the reaction of $NaBH_4$ with $AgNO_3$ in water to give silver metal, H_2 gas, boric acid, and sodium nitrate (page 984).
(b) What mass of silver can be produced from 575 mL of 0.011 M $AgNO_3$ and 13.0 g of $NaBH_4$?

85. A common analytical method for hydrazine involves its oxidation with iodate ion, IO_3^-, in acid solution. In the process, hydrazine acts as a four-electron reducing agent.

$$N_2(g) + 5\ H_3O^+(aq) + 4\ e^- \longrightarrow N_2H_5^+(aq) + 5\ H_2O(\ell) \qquad E° = -0.23\ V$$

Write the balanced equation for the reaction of hydrazine in acid solution ($N_2H_5^+$) with $IO_3^-(aq)$ to give N_2 and I_2. Calculate $E°$ for this reaction.

86. ■ When 1.00 g of a white solid **A** is strongly heated, you obtain another white solid, **B,** and a gas. An experiment is carried out on the gas, showing that it exerts a pressure of 209 mm Hg in a 450-mL flask at 25 °C. Bubbling the gas into a solution of $Ca(OH)_2$ gives another white solid, **C.** If the white solid **B** is added to water, the resulting solution turns red litmus paper blue. Addition of aqueous HCl to the solution of **B** and evaporation of the resulting solution to dryness yield 1.055 g of a white solid **D.** When **D** is placed in a Bunsen burner flame, it colors the flame green. Finally, if the aqueous solution of **B** is treated with sulfuric acid, a white precipitate, **E,** forms. Identify the lettered compounds in the reaction scheme.

Charles D. Winters

The salts $CaCl_2$, $SrCl_2$, and $BaCl_2$ were suspended in methanol. When the methanol is set ablaze, the heat of combustion causes the salts to emit light of characteristic wavelengths: calcium salts are yellow; strontium salts are red; and barium salts are green-yellow.

87. ▲ In 1937, R. Schwartz and M. Schmiesser prepared a yellow-orange bromine oxide (BrO_2) by treating Br_2 with ozone in a fluorocarbon solvent. Many years later, J. Pascal found that, on heating, this oxide decomposed to two other oxides, a less volatile golden yellow oxide (**A**) and a more volatile deep-brown oxide (**B**). Oxide **B** was later identified as Br_2O. To determine the formula for oxide **A,** a sample was treated with iodide ion. The reaction liberated iodine, which was titrated to an equivalence point with 17.7 mL of 0.065 M sodium thiosulfate.

$$I_2(aq) + 2\ S_2O_3^{2-}(aq) \longrightarrow 2\ I^-(aq) + S_4O_6^{2-}(aq)$$

Compound **A** was also treated with $AgNO_3$, and 14.4 mL of 0.020 M $AgNO_3$ was required to completely precipitate the bromine from the sample.
(a) What is the formula of the unknown bromine oxide A?
(b) Draw Lewis structures for **A** and Br_2O. Speculate on their molecular geometry.

88. ■ A mixture of PCl_5 (12.41 g) and excess NH_4Cl was heated at 145 °C for six hours. The two reacted in equimolar amounts and evolved 5.14 L of HCl (at STP). Three substances (**A, B,** and **C**) were isolated from the reaction mixture. The three substances had the same elemental composition but differed in their molar mass. Substance **A,** had a molar mass of 347.7 g/mol and **B** had a molar mass of 463.5 g/mol. Give the empirical and molecular formulas for **A** and **B** and draw a reasonable Lewis structure for **A.**

Summary and Conceptual Questions

The following questions may use concepts from this and previous chapters.

89. Dinitrogen trioxide, N_2O_3, has the structure shown here.

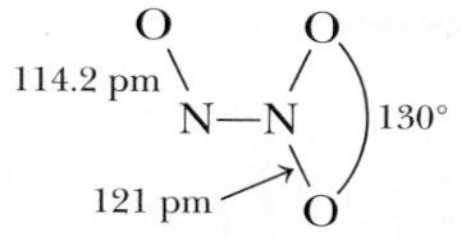

The oxide is unstable, decomposing to NO and NO_2 in the gas phase at 25 °C.

$$N_2O_3(g) \longrightarrow NO(g) + NO_2(g)$$

(a) Explain why one N—O bond distance in N_2O_3 is 114.2 pm, whereas the other two bonds are longer (121 pm) and nearly equal to each other.
(b) For the decomposition reaction, $\Delta_r H° = +40.5$ kJ/mol and $\Delta_r G° = -1.59$ kJ/mol. Calculate $\Delta S°$ and K for the reaction at 298 K.
(c) Calculate $\Delta_f H°$ for $N_2O_3(g)$.

90. ▲ The density of lead is 11.350 g/cm³, and the metal crystallizes in a face-centered cubic unit cell. Estimate the radius of the lead atom.

91. You have a 1.0-L flask that contains a mixture of argon and hydrogen. The pressure inside the flask is 745 mm Hg, and the temperature is 22 °C. Describe an experiment that you could use to determine the percentage of hydrogen in this mixture.

92. The boron atom in boric acid, $B(OH)_3$, is bonded to three —OH groups. In the solid state, the —OH groups are in turn hydrogen-bonded to —OH groups in neighboring molecules.
(a) Draw the Lewis structure for boric acid.
(b) What is the hybridization of the boron atom in the acid?
(c) Sketch a picture showing how hydrogen bonding can occur between neighboring molecules.

93. How would you extinguish a sodium fire in the laboratory? What is the worst thing you could do?

94. Tin(IV) oxide, cassiterite, is the main ore of tin. It crystallizes in a rutile-like unit cell (Question 4, page 611).
(a) How many tin(IV) ions and oxide ions are there per unit cell of this oxide?
(b) Is it thermodynamically feasible to transform solid SnO_2 into liquid $SnCl_4$ by reaction of the oxide with gaseous HCl? What is the equilibrium constant for this reaction at 25 °C?

95. You are given a stoppered flask that contains hydrogen, nitrogen, or oxygen. Suggest an experiment to identify the gas.

96. The structure of nitric acid is illustrated on page 993.
(a) Why are the N—O bonds the same length, and why are both shorter than the N—OH bond length?
(b) Rationalize the bond angles in the molecule.
(c) What is the hybridization of the central N atom? Which orbitals overlap to form the N—O π bond?

97. Assume an electrolysis cell that produces chlorine from aqueous sodium chloride operates at 4.6 V (with a current of 3.0×10^5 A). Calculate the number of kilowatt-hours of energy required to produce 1.00 kg of chlorine (1 kWh = 1 kilowatt-hour = 3.6×10^6 J).

98. Sodium metal is produced by electrolysis of molten sodium chloride. The cell operates at 7.0 V with a current of 25×10^3 A.
(a) What mass of sodium can be produced in 1 h?
(b) How many kilowatt-hours of electricity are used to produce 1.00 kg of sodium metal (1 kWh = 3.6×10^6 J)?

99. The reduction potentials for the Group 3A metals, $E°$, are given below. What trend or trends do you observe in these data? What can you learn about the chemistry of the Group 3A elements from these data?

Half-Reaction	**Reduction Potential ($E°$, V)**
$Al^{3+}(aq) + 3\,e^- \longrightarrow Al(s)$	−1.66
$Ga^{3+}(aq) + 3\,e^- \longrightarrow Ga(s)$	−0.53
$In^{3+}(aq) + 3\,e^- \longrightarrow In(s)$	−0.338
$Tl^{3+}(aq) + 3\,e^- \longrightarrow Tl(s)$	+0.72

100. (a) Magnesium is obtained from sea water. If the concentration of Mg^{2+} in sea water is 0.050 M, what volume of sea water (in liters) must be treated to obtain 1.00 kg of magnesium metal? What mass of lime (CaO; in kilograms) must be used to precipitate the magnesium in this volume of sea water?
(b) When 1.2×10^3 kg of molten $MgCl_2$ is electrolyzed to produce magnesium, what mass (in kilograms) of metal is produced at the cathode? What is produced at the anode? What is the mass of this product? What is the total number of faradays of electricity used in the process?
(c) One industrial process has an energy consumption of 18.5 kWh/kg of Mg. How many joules are required per mole (1 kWh = 1 kilowatt-hour = 3.6×10^6 J)? How does this energy compare with the energy of the following process?

$$MgCl_2(s) \longrightarrow Mg(s) + Cl_2(g)?$$

22 The Chemistry of the Transition Elements

NASA/Science Source/Photo Researchers, Inc.

Nitinol frames for glasses. Can you do this with your eye glasses? If you can, it is likely they are made of nitinol, a nickel–titanium alloy. They snap back to the proper fit even after being twisted like a pretzel.

Memory Metal

In the 1960s, William J. Buehler, an engineer at the Naval Ordnance Laboratory in Maryland, was looking for a material for use in the nose cone of a Navy missile. It was important that the material be impact- and heat-resistant and not lose these properties when bent and shaped. He prepared long, thin strips of an alloy of nickel and titanium to demonstrate that it could be folded and unfolded many times without breaking. At a meeting to discuss this material, one of his associates held a cigarette lighter to a folded-up piece of metal and was amazed to observe that the metal strip immediately unfolded and assumed its original shape. Thus, memory metal was discovered. This unusual alloy is now called *nitinol*, a name constructed out of "nickel," "titanium," and "Naval Ordnance Laboratory."

Memory metal is an alloy with roughly the same number of Ni and Ti atoms. When the atoms are arranged in the highly symmetrical austenite phase, the alloy is relatively rigid. In this phase a specific shape is established that will be "remembered." If the alloy is cooled below its "phase transition temperature," it enters a less symmetrical but flexible phase (martensite). Below this transition temperature, the metal is fairly soft and may be bent and twisted out of shape. When warmed above the phase transition temperature, nitinol returns to its original shape. The temperature at which the change in shape occurs varies with small differences in the nickel-to-titanium ratio.

Besides eye glasses frames, nitinol is now used in stents to reinforce blood vessels and in orthodontics.

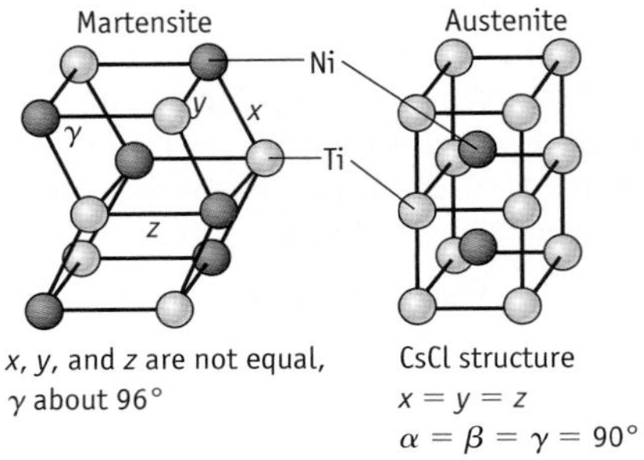

Two phases of nitinol. The austenite form has a structure like CsCl (page 622).

Questions:

1. What are the dimensions of the austenite unit cell? Assume the Ti and Ni atoms are just touching along the unit cell diagonal. (Atom radii: Ti = 145 pm; Ni = 125 pm.)
2. Calculate the density of nitinol based on the austenite unit cell parameters. Does the calculated density of the austenite unit cell agree with the reported density of 6.5 g/cm^3?
3. Are Ti and Ni atoms paramagnetic or diamagnetic?

Answers to these questions are in Appendix Q.

Chapter Goals

See Chapter Goals Revisited (page 1054) for Study Questions keyed to these goals and assignable in OWL.

- Identify and explain the chemical and physical properties of the transition elements.
- Understand the composition, structure, and bonding in coordination compounds.
- Relate ligand field theory to the magnetic and spectroscopic properties of the complexes.
- Apply the effective atomic number (EAN) rule to simple organometallic complexes of the transition metals.

Chapter Outline

The transition elements are the large block of elements in the central portion of the periodic table. All are metals and bridge the *s*-block elements at the left and the *p*-block elements on the right (Figure 22.1). The transition elements are often divided into two groups, depending on the valence electrons involved in their chemistry. The first group are the ***d*-block elements,** because their occurrence in the periodic table coincides with the filling of the *d* orbitals. The second group are the ***f*-block elements,** characterized by filling of the *f* orbitals. Contained within this group of elements are two subgroups: the *lanthanides,* elements that occur between La and Hf, and the *actinides,* elements that occur between Ac and Rf.

This chapter focuses primarily on the *d*-block elements, and within this group we concentrate mainly on the elements in the fourth period, that is, the elements of the first transition series, scandium to zinc.

Chemistry Now™

Throughout the text this icon introduces an opportunity for self-study or to explore interactive tutorials by signing in at **www.thomsonedu.com/login**.

22.1 Properties of the Transition Elements

The *d*-block metals include elements with a wide range of properties. They encompass the most common metal used in construction and manufacturing (iron), metals that are valued for their beauty (gold, silver, and platinum), and metals used

Fourth-period transition metals: left to right, Ti, V, Cr, Mn, Fe, Co, Ni, Cu

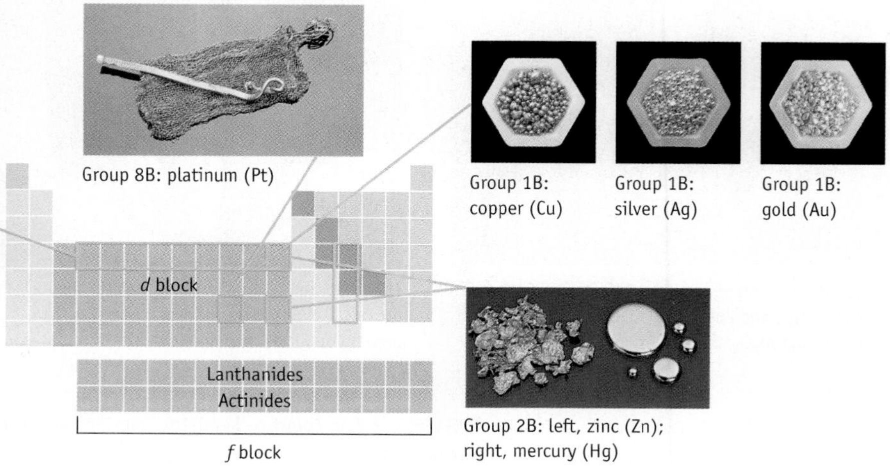

FIGURE 22.1 The transition metals. The *d*-block elements (transition elements) and *f*-block elements are highlighted in a darker shade of purple.

in coins (nickel, copper, and zinc). There are metals used in modern technology (titanium) and metals known and used in early civilizations (copper, silver, gold, and iron). The *d*-block contains the densest elements (osmium, $d = 22.49\ g/cm^3$, and iridium, $d = 22.41\ g/cm^3$), the metals with the highest and lowest melting points (tungsten, mp = 3410 °C, and mercury, mp = −38.9 °C), and one of two radioactive elements with atomic numbers less than 83 [technetium (Tc), atomic number 43; the other is promethium (Pm), atomic number 61, in the *f*-block].

With the exception of mercury, the transition elements are solids, often with high melting and boiling points. They have a metallic sheen and conduct electricity and heat. They react with various oxidizing agents to give ionic compounds. There is considerable variation in such reactions among the elements, however. Because silver, gold, and platinum resist oxidation, for example, they are used for jewelry and decorative items.

Certain *d*-block elements are particularly important in living organisms. Cobalt is the crucial element in vitamin B_{12}, which is part of a catalyst essential for several biochemical reactions. Hemoglobin and myoglobin, oxygen-carrying and storage proteins, contain iron (see page 1033). Molybdenum and iron, together with sulfur, form the reactive portion of nitrogenase, a biological catalyst used by nitrogen-fixing organisms to convert atmospheric nitrogen into ammonia.

Many transition metal compounds are highly colored, which makes them useful as pigments in paints and dyes (Figure 22.2). Prussian blue, $Fe_4[Fe(CN)_6]_3 \cdot 14\ H_2O$ is a "bluing agent" used in engineering blueprints and in the laundry to brighten yellowed white cloth. A common pigment (artist's cadmium yellow) contains cadmium sulfide (CdS), and the white in most white paints is titanium(IV) oxide, TiO_2.

The presence of transition metal ions in crystalline silicates or alumina transforms these common materials into gemstones. Iron(II) ions cause the yellow color in citrine, and chromium(III) ions produce the red color of a ruby. Transition metal complexes in small quantities add color to glass. Blue glass contains a small amount of a cobalt(III) oxide, and addition of chromium(III) oxide to glass gives a green color. Old window panes sometimes take on a purple color over time as a

Charles D. Winters

(a) Paint pigments: yellow, CdS; green, Cr_2O_3; white, TiO_2 and ZnO; purple, $Mn_3(PO_4)_2$; blue, Co_2O_3 and Al_2O_3; ochre, Fe_2O_3.

(b) Small amounts of transition metal compounds are used to color glass: blue, Co_2O_3; green, copper or chromium oxides; purple, nickel or cobalt oxides; red, copper oxide; iridescent green, uranium oxide.

(c) Traces of transition metal ions are responsible for the colors of green jade (iron), red corundum (chromium), blue azurite, blue-green turquoise (copper), and purple amethyst (iron).

FIGURE 22.2 Colorful chemistry. Transition metal compounds are often colored, a property that leads to specific uses.

consequence of oxidation of traces of manganese(II) ion present in the glass to permanganate ion (MnO_4^-).

In the next few pages, we will examine the properties of the transition elements, concentrating on the underlying principles that govern these properties.

TABLE 22.1 Electron Configurations of the Fourth-Period Transition Elements

	spdf Configu-ration	Box Notation 3d	4s
Sc	$[Ar]3d^14s^2$	[↑][ ][ ][ ][ ]	[↑↓]
Ti	$[Ar]3d^24s^2$	[↑][↑][ ][ ][ ]	[↑↓]
V	$[Ar]3d^34s^2$	[↑][↑][↑][ ][ ]	[↑↓]
Cr	$[Ar]3d^54s^1$	[↑][↑][↑][↑][↑]	[↑]
Mn	$[Ar]3d^54s^2$	[↑][↑][↑][↑][↑]	[↑↓]
Fe	$[Ar]3d^64s^2$	[↑↓][↑][↑][↑][↑]	[↑↓]
Co	$[Ar]3d^74s^2$	[↑↓][↑↓][↑][↑][↑]	[↑↓]
Ni	$[Ar]3d^84s^2$	[↑↓][↑↓][↑↓][↑][↑]	[↑↓]
Cu	$[Ar]3d^{10}4s^1$	[↑↓][↑↓][↑↓][↑↓][↑↓]	[↑]
Zn	$[Ar]3d^{10}4s^2$	[↑↓][↑↓][↑↓][↑↓][↑↓]	[↑↓]

Electron Configurations

Because chemical behavior is related to electron structure, it is important to know the electron configurations of the *d*-block elements (Table 22.1) and their common ions (◀ Section 7.4). Recall that the configuration of these metals has the general form [noble gas core]$(n - 1)d^b\, ns^a$; that is, valence electrons for the transition elements reside in the *ns* and $(n - 1)d$ subshells.

Oxidation and Reduction

A characteristic chemical property of all metals is that they undergo oxidation by a wide range of oxidizing agents such as oxygen, the halogens, and aqueous acids. Standard reduction potentials for the elements of the first transition series can be used to predict which elements will be oxidized by a given oxidizing agent. For example, all of these metals except vanadium and copper are oxidized by aqueous HCl (Table 22.2). This feature, which dominates the chemistry of these elements, is sometimes highly undesirable (see *Chemical Perspectives: Corrosion of Iron,* page 1023.)

When a transition metal is oxidized, the outermost *s* electrons are removed, followed by one or more *d* electrons. With a few exceptions, transition metal ions have the electron configuration [noble gas core]$(n - 1)d^x$. In contrast to ions formed by main group elements, transition metal cations often possess unpaired electrons, resulting in paramagnetism (◀ page 292). They are frequently colored as well, due to the absorption of light in the visible region of the electromagnetic spectrum. Color and magnetism figure prominently in a discussion of the properties and bonding of these elements, as you shall see shortly.

In the first transition series, the most commonly encountered metal ions have oxidation numbers of +2 and +3 (Table 22.2). With iron, for example, oxidation

TABLE 22.2 Products from Reactions of the Elements in the First Transition Series with O_2, Cl_2, or Aqueous HCl

Element	Reaction with O_2*	Reaction with Cl_2	Reaction with Aqueous HCl
Scandium	Sc_2O_3	$ScCl_3$	$Sc^{3+}(aq)$
Titanium	TiO_2	$TiCl_4$	$Ti^{3+}(aq)$
Vanadium	V_2O_5	VCl_4	NR†
Chromium	Cr_2O_3	$CrCl_3$	$Cr^{2+}(aq)$
Manganese	MnO_2	$MnCl_2$	$Mn^{2+}(aq)$
Iron	Fe_2O_3	$FeCl_3$	$Fe^{2+}(aq)$
Cobalt	Co_2O_3	$CoCl_2$	$Co^{2+}(aq)$
Nickel	NiO	$NiCl_2$	$Ni^{2+}(aq)$
Copper	CuO	$CuCl_2$	NR†
Zinc	ZnO	$ZnCl_2$	$Zn^{2+}(aq)$

* Product obtained with excess oxygen.
† NR = no reaction.

Charles D. Winters

Prussian blue. When Fe^{3+} ions are added to $[Fe(CN)_6]^{4-}$ ions in water (or Fe^{2+} ions are added to $[Fe(CN)_6]^{3-}$ ions), a deep blue compound called Prussian blue forms. The formula of the compound is $Fe_4[Fe(CN)_6]_3 \cdot 14\ H_2O$. The color arises from electron transfer between the Fe(II) and Fe(III) ions in the compound.

(a) (b) (c)

FIGURE 22.3 Typical reactions of transition metals. These metals react with oxygen, with halogens, and with acids under appropriate conditions. (a) Steel wool reacts with O_2; (b) steel wool reacts with chlorine gas, Cl_2; and (c) iron chips react with aqueous HCl.

converts Fe([Ar]$3d^64s^2$) to either Fe^{2+}([Ar]$3d^6$) or Fe^{3+}([Ar]$3d^5$). Iron reacts with chlorine to give $FeCl_3$, and it reacts with aqueous acids to produce Fe^{2+}(aq) and H_2 (Figure 22.3). Despite the preponderance of 2+ and 3+ ions in compounds of the first transition metal series, the range of possible oxidation states for these compounds is broad (Figure 22.4). Earlier in this text, we encountered chromium with a +6 oxidation number (CrO_4^{2-}, $Cr_2O_7^{2-}$), manganese with an oxidation number of +7 (MnO_4^-), silver and copper as 1+ ions, and vanadium oxidation numbers that can range from +5 to +2 (Figure 20.3).

Higher oxidation numbers are more common in compounds of the elements in the second and third transition series. For example, the naturally occurring

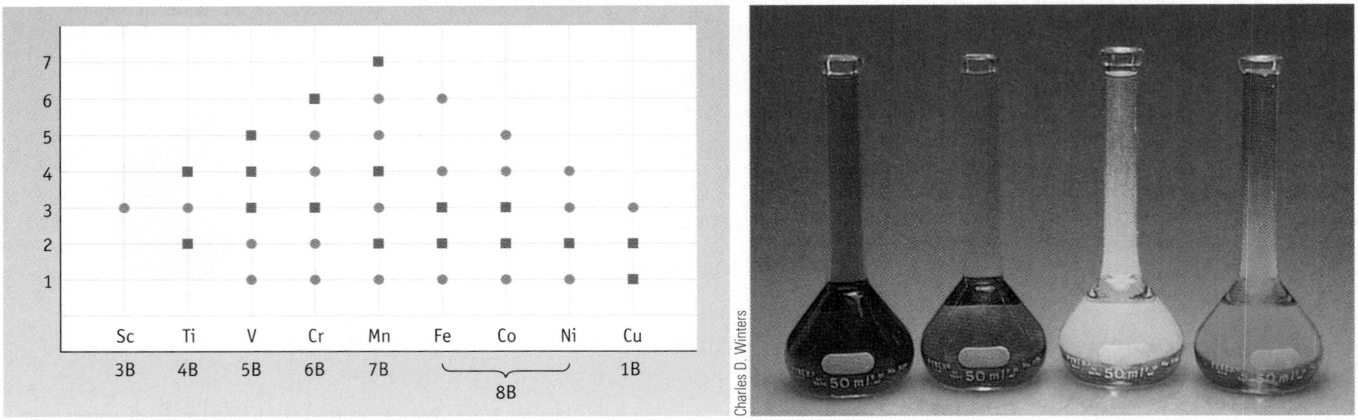

(a) (b)

FIGURE 22.4 Oxidation states of the transition elements in the first transition series. (a) The most common oxidation states are indicated with red squares; less common oxidation states are indicated with blue dots. (b) Aqueous solutions of chromium compounds with two different oxidation numbers: +3 in $Cr(NO_3)_3$ (violet) and $CrCl_3$ (green), and +6 in K_2CrO_4 (yellow) and $K_2Cr_2O_7$ (orange).

Chemical Perspectives — Corrosion of Iron

It is hard not to be aware of corrosion. Those of us who live in the northern part of the United States are well aware of the problems of rust on our automobiles. It is estimated that 20% of iron production each year goes solely to replace iron that has rusted away.

Qualitatively, we describe corrosion as the deterioration of metals by a product-favored oxidation reaction. The corrosion of iron, for example, converts iron metal to red-brown rust, which is hydrated iron(III) oxide, $Fe_2O_3 \cdot H_2O$. This process requires both air and water, and it is enhanced if the water contains dissolved ions and if the metal is stressed (e.g., if it has dents, cuts, and scrapes on the surface.)

The corrosion process occurs in what is essentially a small electrochemical cell. There is an anode and a cathode, an electrical connection between the two (the metal itself), and an electrolyte in contact with both anode and cathode. When a metal corrodes, the metal is oxidized on anodic areas of the metal surface.

Anode, oxidation $M(s) \longrightarrow M^{n+} + n\,e^-$

The electrons are consumed by several possible half-reactions in cathodic areas.

Cathode, reduction

$$2\,H_3O^+(aq) + 2\,e^- \longrightarrow H_2(g) + 2\,H_2O(\ell)$$
$$2\,H_2O(\ell) + 2\,e^- \longrightarrow H_2(g) + 2\,OH^-(aq)$$
$$O_2(g) + 2\,H_2O(\ell) + 4\,e^- \longrightarrow 4\,OH^-(aq)$$

The rate of iron corrosion is controlled by the rate of the cathodic process. Of the three possible cathodic reactions, the one that is fastest is determined by acidity and the amount of oxygen present. If little or no oxygen is present—as when a piece of iron is buried in soil such as moist clay—hydrogen ion or water is reduced, and $H_2(g)$ and hydroxide ions are the products. Iron(II) hydroxide is relatively insoluble and will precipitate on the metal surface, inhibiting the further formation of Fe^{2+}.

Anode	$Fe(s) \longrightarrow Fe^{2+}(aq) + 2\,e^-$
Cathode	$2\,H_2O(\ell) + 2\,e^- \longrightarrow H_2(g) + 2\,OH^-(aq)$
Precipitation	$Fe^{2+}(aq) + 2\,OH^-(aq) \longrightarrow Fe(OH)_2(s)$
Net reaction	$Fe(s) + 2\,H_2O(\ell) \longrightarrow H_2(g) + Fe(OH)_2(s)$

If both water and O_2 are present, the chemistry of iron corrosion is somewhat different, and the corrosion reaction is about 100 times faster than without oxygen.

Anode	$2\,Fe(s) \longrightarrow 2\,Fe^{2+}(aq) + 4\,e^-$
Cathode	$O_2(g) + 2\,H_2O(\ell) + 4\,e^- \longrightarrow 4\,OH^-(aq)$
Precipitation	$2\,Fe^{2+}(aq) + 4\,OH^-(aq) \longrightarrow 2\,Fe(OH)_2(s)$
Net reaction	$2\,Fe(s) + 2\,H_2O(\ell) + O_2(g) \longrightarrow 2\,Fe(OH)_2(s)$

If oxygen is present but not in excess, further oxidation of the iron(II) hydroxide leads to the formation of magnetic iron oxide Fe_3O_4, (which can be thought of as a mixed oxide of Fe_2O_3 and FeO).

$$6\,Fe(OH)_2(s) + O_2(g) \longrightarrow \underbrace{2\,Fe_3O_4 \cdot H_2O(s)}_{\text{green hydrated magnetite}} + 4\,H_2O(\ell)$$

$$Fe_3O_4 \cdot H_2O(s) \longrightarrow H_2O(\ell) + \underbrace{Fe_3O_4(s)}_{\text{black magnetite}}$$

It is the black magnetite that you find coating an iron object that has corroded by resting in moist soil.

If the iron object has free access to oxygen and water, as in the open or in flowing water, red-brown iron(III) oxide will form.

$$4\,Fe(OH)_2(s) + O_2(g) \longrightarrow \underbrace{2\,Fe_2O_3 \cdot H_2O(s)}_{\text{red-brown}} + 2\,H_2O(\ell)$$

This is the familiar rust you see on cars and buildings, and the substance that colors the water red in some mountain streams or in your home.

Charles D. Winters

The corrosion or rusting of iron results in major economic loss.

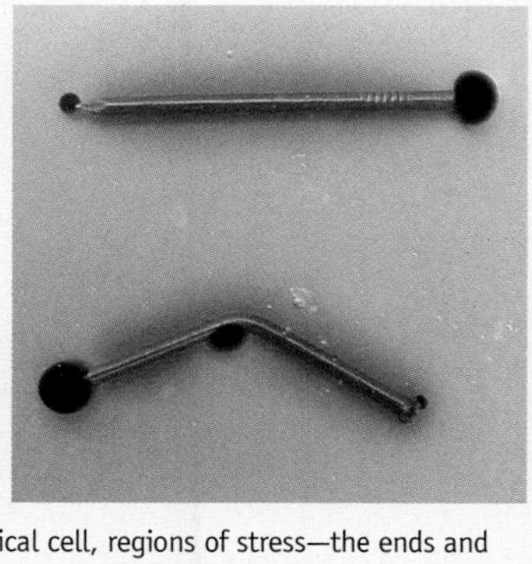

Charles D. Winters

Anode and cathode reactions in iron corrosion. Two iron nails were placed in an agar gel that contains phenolphthalein and $K_3[Fe(CN)_6]$. Iron(II) ion, formed at the tip and where the nail is bent, reacts with $[Fe(CN)_6]^{3-}$ to form blue-green $Fe_4[Fe(CN_6)]_3 \cdot 14\,H_2O$ (Prussian blue). Hydrogen and $OH^-(aq)$ are formed at the other parts of the surface of the nail, the latter being detected by the red color of the acid–dase indicator. In this electrochemical cell, regions of stress—the ends and the bent region of the nail—act as anodes, and the remainder of the surface serves as the cathode.

sources of molybdenum and tungsten are the ores molybdenite (MoS_2) and wolframite (WO_3). This general trend is carried over in the *f*-block. The lanthanides form primarily 3+ ions. In contrast, actinide elements usually have higher oxidation numbers in their compounds; +4 and even +6 are typical. For example, UO_3 is a common oxide of uranium, and UF_6 is a compound important in processing uranium fuel for nuclear reactors [▶ Section 23.6].

Chemistry Now™

Sign in at **www.thomsonedu.com/login** and go to Chapter 22 Contents to see Screen 22.2 for an exercise on **transition metal compounds.**

Periodic Trends in the *d*-Block: Size, Density, Melting Point

The periodic table is the most useful single reference source for a chemist. Not only does it provide data that have everyday use, but it also organizes the elements with respect to their chemical and physical properties. Let us look at three physical properties of the transition elements that vary periodically: atomic radii, density and melting point.

Metal Atom Radii

The variation in atomic radii for the transition elements in the fourth, fifth, and sixth periods is illustrated in Figure 7.9. The radii of the transition elements vary over a fairly narrow range, with a small decrease to a minimum being observed around the middle of this group of elements. This similarity of radii can be understood based on electron configurations. Atom size is determined by the electrons in the outermost orbital, which for these elements is the *ns* orbital ($n = 4$, 5, or 6). Progressing from left to right in the periodic table, the size decline expected from increasing the number of protons in the nucleus is mostly canceled out by an opposing effect, repulsion from additional electrons in the $(n - 1)d$ orbitals.

The radii of the *d*-block elements in the fifth and sixth periods in each group are almost identical. The reason is that the lanthanide elements immediately precede the third series of *d*-block elements. The filling of 4*f* orbitals is accompanied by a steady contraction in size, consistent with the general trend of decreasing size from left to right in the periodic table. At the point where the 5*d* orbitals begin to fill again, the radii have decreased to a size similar to that of elements in the previous period. The decrease in size that results from the filling of the 4*f* orbitals is given a specific name, the **lanthanide contraction.**

The similar sizes of the second- and third-period *d*-block elements have significant consequences for their chemistry. For example, the "platinum group metals" (Ru, Os, Rh, Ir, Pd, and Pt) form similar compounds. Thus, it is not surprising that minerals containing these metals are found in the same geological zones on earth. Nor is it surprising that it is difficult to separate these elements from one another.

Density

The variation in metal radii causes the densities of the transition elements to first increase and then decrease across a period (Figure 22.5a). Although the overall change in radii among these elements is small, the effect is magnified because the volume is actually changing with the cube of the radius [$V = (4/3)\pi r^3$].

The lanthanide contraction explains why elements in the sixth period have the highest density. The relatively small radii of sixth-period transition metals, combined with the fact that their atomic masses are considerably larger than

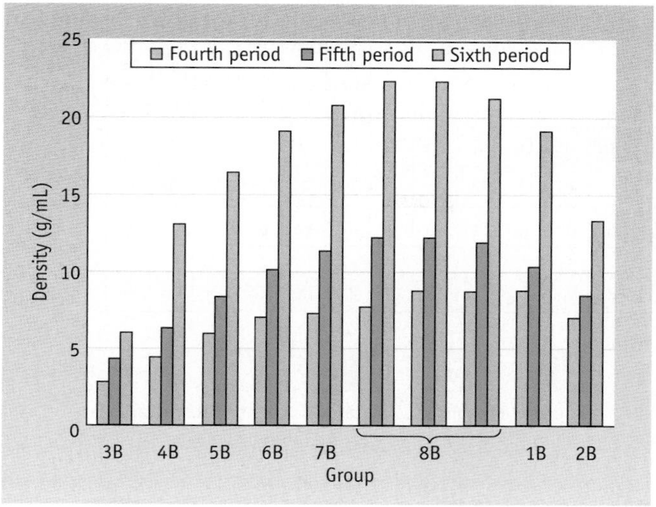

(a)

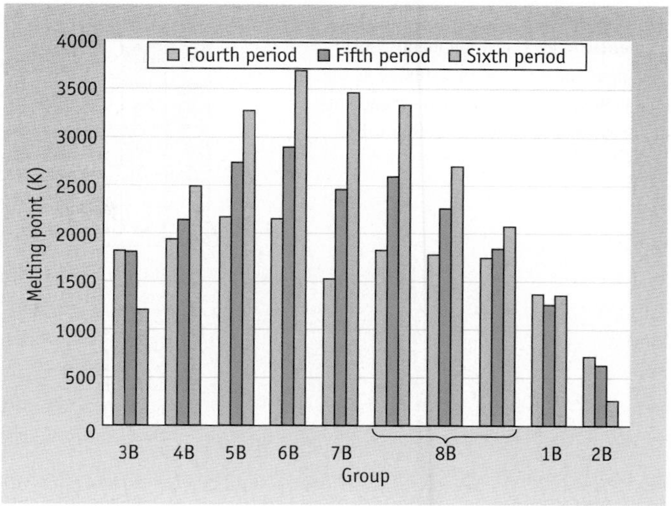

(b)

FIGURE 22.5 Periodic properties in the transition series. Density (a) and melting point (b) of the *d*-block elements.

their counterparts in the fifth period, causes sixth-period metal densities to be very large.

Melting Point

The melting point of any substance reflects the forces of attraction between the atoms, molecules, or ions that compose the solid. With transition elements, the melting points rise to a maximum around the middle of the series (Figure 22.5b), then descend. Again, these elements' electron configurations provide us with an explanation. The variation in melting point indicates that the strongest metallic bonds occur when the *d* subshell is about half filled. This is also the point at which the largest number of electrons occupy the bonding molecular orbitals in the metal. (See the discussion of bonding in metals on page 657.)

Chemistry Now™

Sign in at **www.thomsonedu.com/login** and go to Chapter 22 Contents to see Screen 22.3 for more on **transition metal chemistry.**

22.2 Metallurgy

A few metals occur in nature as the free elements. This group includes copper (Figure 22.6), silver, and gold. Most metals, however, are found as oxides, sulfides, halides, carbonates, or other ionic compounds (Figure 22.7). Some metal-containing mineral deposits have little economic value, either because the concentration of the metal is too low or because the metal is difficult to separate from impurities. The relatively few minerals from which elements can be obtained profitably are called *ores* (Figure 22.7). **Metallurgy** is the general name given to the process of obtaining metals from their ores.

Very few ores are chemically pure substances. Instead, the desired mineral is usually mixed with large quantities of impurities such as sand and clay, called **gangue** (pronounced "gang"). Generally, the first step in a metallurgical process is to sep-

Charles D. Winters

FIGURE 22.6 Naturally occurring copper. Copper occurs as the metal (native copper) and as minerals such as blue azurite [$2\ CuCO_3 \cdot Cu(OH)_2$] and malachite [$CuCO_3 \cdot Cu(OH)_2$].

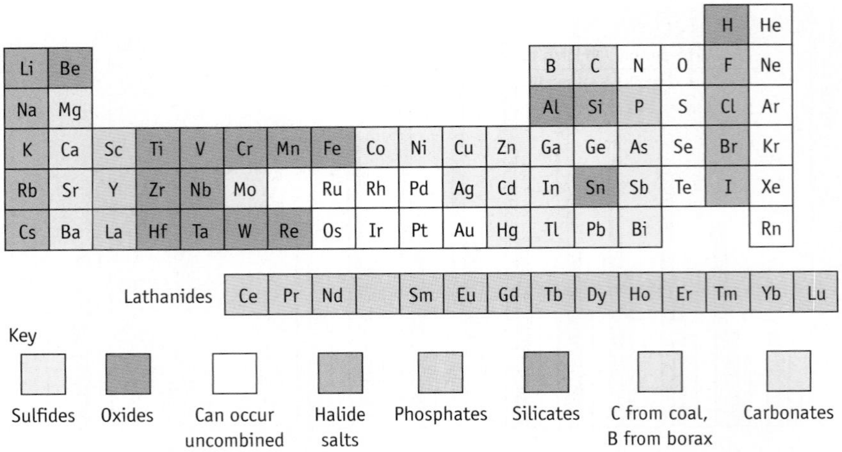

FIGURE 22.7 Major sources of the elements. A few transition metals, such as copper and gold, occur naturally as the metal. Most other elements are found naturally as oxides, sulfides, or other salts.

arate the mineral from the gangue. Then the ore is converted to the metal, a reduction process. Pyrometallurgy and hydrometallurgy are two methods of recovering metals from their ores. As the names imply, **pyrometallurgy** involves high temperatures and **hydrometallurgy** uses aqueous solutions (and thus is limited to the relatively low temperatures at which water is a liquid). Iron and copper metallurgy illustrate these two methods of metal production.

Pyrometallurgy: Iron Production

The production of iron from its ores is carried out in a blast furnace (Figure 22.8). The furnace is charged with a mixture of ore (usually hematite, Fe_2O_3), coke (which is primarily carbon), and limestone ($CaCO_3$). A blast of hot air forced in at the bottom of the furnace causes the coke to burn with such an intense heat that the temperature at the bottom is almost 1500 °C. The quantity of air input is controlled so that carbon monoxide is the primary product. Both carbon and carbon monoxide participate in the reduction of iron(III) oxide to give impure metal.

$$Fe_2O_3(s) + 3\,C(s) \longrightarrow 2\,Fe(\ell) + 3\,CO(g)$$

$$Fe_2O_3(s) + 3\,CO(g) \longrightarrow 2\,Fe(\ell) + 3\,CO_2(g)$$

■ **Coke: A Reducing Agent** Coke is made by heating coal in a tall, narrow oven that is sealed to keep out oxygen. Heating drives off volatile chemicals, including benzene and ammonia. What remains is nearly pure carbon.

Much of the carbon dioxide formed in the reduction process (and from heating the limestone) is reduced on contact with unburned coke and produces more reducing agent.

$$CO_2(g) + C(s) \longrightarrow 2\,CO(g)$$

The molten iron flows down through the furnace and collects at the bottom, where it is tapped off through an opening in the side. This impure iron is called *cast iron* or *pig iron.* Usually, the impure metal is either brittle or soft (undesirable properties for most uses) due to the presence of impurities such as elemental carbon, phosphorus, and sulfur.

Iron ores generally contain silicate minerals and silicon dioxide. Lime (CaO), formed when limestone is heated, reacts with these materials to give calcium silicate.

$$SiO_2(s) + CaO(s) \longrightarrow CaSiO_3(\ell)$$

This is an acid–base reaction because CaO is a basic oxide and SiO_2 is an acidic oxide. The calcium silicate, molten at the temperature of the blast furnace and less

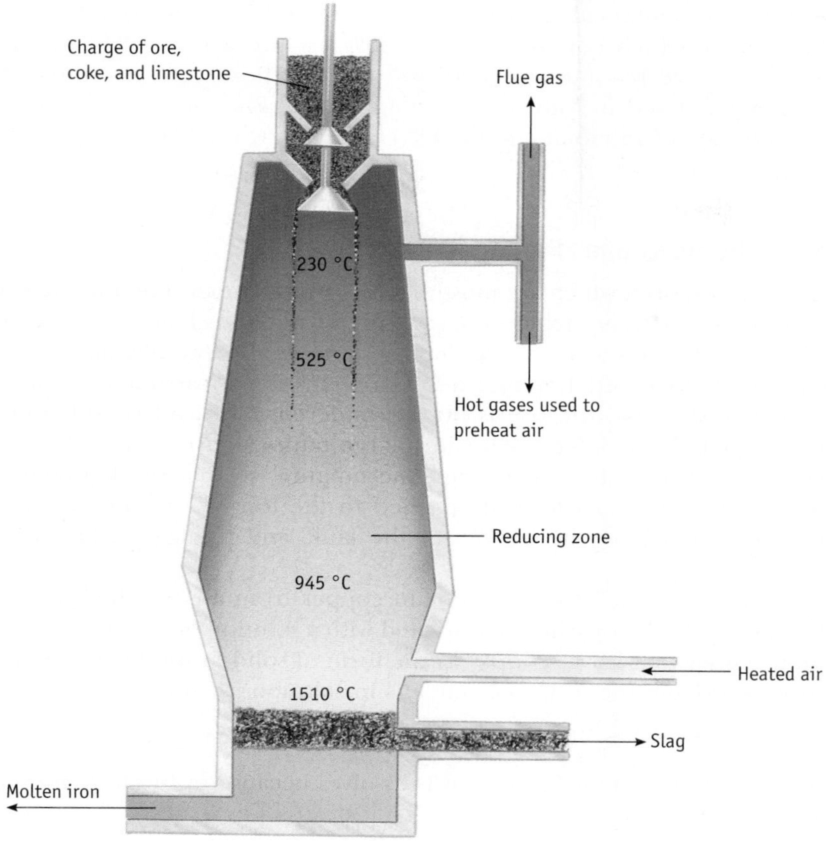

Active Figure 22.8 A blast furnace. The largest modern furnaces have hearths 14 meters in diameter. They can produce as much as 10,000 tons of iron per day.

Chemistry Now™ Sign in at www.thomsonedu.com/login and go to the Chapter Contents menu to explore an interactive version of this figure accompanied by an exercise.

dense than molten iron, floats on the iron. Other nonmetal oxides dissolve in this layer and the mixture, called *slag*, is easily removed.

Pig iron from the blast furnace may contain as much as 4.5% carbon, 0.3% phosphorus, 0.04% sulfur, 1.5% silicon, and some other elements as well. The impure iron must be purified to remove these nonmetal components. Several processes are available to accomplish this task, but the most important uses the *basic oxygen furnace* (Figure 22.9). The process in the furnace removes much of the carbon and all of the phosphorus, sulfur, and silicon. Pure oxygen is blown into the molten pig iron and oxidizes phosphorus to P_4O_{10}, sulfur to SO_2, and carbon to CO_2. These nonmetal oxides either escape as gases or react with basic oxides such as CaO that are added or are used to line the furnace. For example,

$$P_4O_{10}(g) + 6\,CaO(s) \longrightarrow 2\,Ca_3(PO_4)_2(\ell)$$

The result is ordinary *carbon steel.* Almost any degree of flexibility, hardness, strength, and malleability can be achieved in carbon steel by reheating and cooling in a process called *tempering.* The resulting material can then be used in a wide variety of applications. The major disadvantages of carbon steel are that it corrodes easily and that it loses its properties when heated strongly.

Other transition metals, such as chromium, manganese, and nickel, can be added during the steel-making process, giving *alloys* (solid solutions of two or more metals; see *The Chemistry of Modern Materials,* page 656) that have

Bethlehem Steel Corp.

FIGURE 22.9 Molten iron being poured from a basic oxygen furnace.

FIGURE 22.10 Enriching copper ore by the flotation process. The less dense particles of Cu_2S are trapped in the soap bubbles and float. The denser gangue settles to the bottom.

specific physical, chemical, and mechanical properties. One well-known alloy is stainless steel, which contains 18% to 20% Cr and 8% to 12% Ni. Stainless steel is much more resistant to corrosion than carbon steel. Another alloy of iron is Alnico V. Used in loudspeaker magnets because of its permanent magnetism, it contains five elements: Al (8%), Ni (14%), Co (24%), Cu (3%), and Fe (51%).

Hydrometallurgy: Copper Production

In contrast to iron ores, which are mostly oxides, most copper minerals are sulfides. Copper-bearing minerals include chalcopyrite ($CuFeS_2$), chalcocite (Cu_2S), and covellite (CuS). Because ores containing these minerals generally have a very low percentage of copper, enrichment is necessary. This step is carried out by a process known as *flotation.* First, the ore is finely powdered. Next, oil is added and the mixture is agitated with soapy water in a large tank (Figure 22.10). At the same time, compressed air is forced through the mixture, so that the lightweight, oil-covered copper sulfide particles are carried to the top as a frothy mixture. The heavier gangue settles to the bottom of the tank, and the copper-laden froth is skimmed off.

Hydrometallurgy can be used to obtain copper from an enriched ore. In one method, enriched chalcopyrite ore is treated with a solution of copper(II) chloride. A reaction ensues that leaves copper in the form of solid, insoluble CuCl, which is easily separated from the iron that remains in solution as aqueous $FeCl_2$.

$$CuFeS_2(s) + 3\ CuCl_2(aq) \longrightarrow 4\ CuCl(s) + FeCl_2(aq) + 2\ S(s)$$

Aqueous NaCl is then added, and CuCl dissolves because of the formation of the complex ion $[CuCl_2]^-$.

$$CuCl(s) + Cl^-(aq) \longrightarrow [CuCl_2]^-(aq)$$

Copper(I) compounds in solution are unstable with respect to Cu(0) and Cu(II). Thus, $[CuCl_2]^-$ disproportionates to the metal and $CuCl_2$, and the latter is used to treat further ore.

$$2\ [CuCl_2]^-(aq) \longrightarrow Cu(s) + Cu^{2+}(aq) + 4\ Cl^-(aq)$$

Approximately 10% of the copper produced in the United States is obtained with the aid of bacteria. Acidified water is sprayed onto copper-mining wastes that contain low levels of copper. As the water trickles down through the crushed rock, the bacterium *Thiobacillus ferrooxidans* breaks down the iron sulfides in the rock and converts iron(II) to iron(III). Iron(III) ions oxidize the sulfide ion of copper sulfide to sulfate ions, leaving copper(II) ions in solution. Then the copper(II) ion is reduced to metallic copper by reaction with iron.

$$Cu^{2+}(aq) + Fe(s) \longrightarrow Cu(s) + Fe^{2+}(aq)$$

The purity of the copper obtained via these metallurgical processes is about 99%, but this is not acceptable because even traces of impurities greatly diminish the electrical conductivity of the metal. Consequently, a further purification step is needed—one involving electrolysis (Figure 22.11). Thin sheets of pure copper metal and slabs of impure copper are immersed in a solution containing $CuSO_4$ and H_2SO_4. The pure copper sheets serve as the cathode of an electrolysis cell, and the impure slabs are the anode. Copper in the impure sample is oxidized to copper(II) ions at the anode, and copper(II) ions in solution are reduced to pure copper at the cathode.

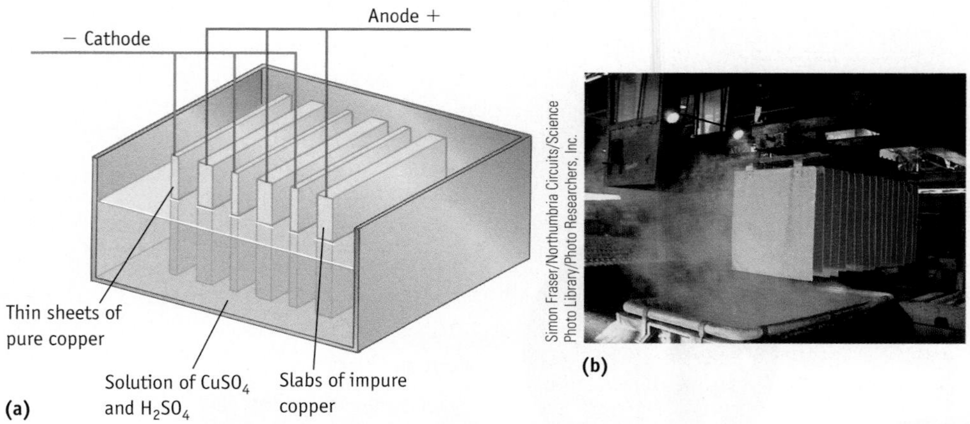

FIGURE 22.11 Electrolytic refining of copper. (a) Slabs of impure copper, called "blister copper," form the anode, and pure copper is deposited at the cathode. (b) The electrolysis cells at a copper refinery.

22.3 Coordination Compounds

When metal salts dissolve, water molecules cluster around the ions (page 617). The negative end of each polar water molecule is attracted to the positively charged metal ion, and the positive end of the water molecule is attracted to the anion. As noted earlier (◀ Section 12.2), the energy of the ion–solvent interaction (solvation energy) is an important aspect of the solution process. But there is much more to this story.

Complexes and Ligands

A green solution formed by dissolving nickel(II) chloride in water contains Ni^{2+} (aq) and Cl^-(aq) ions (Figure 22.12). If the solvent is removed, a green crystalline solid is obtained. The formula of this solid is often written as $NiCl_2 \cdot 6\,H_2O$, and the compound is called nickel(II) chloride hexahydrate. Addition of ammonia to the aqueous nickel(II) chloride solution gives a lilac-colored solution from which another compound, $NiCl_2 \cdot 6\,NH_3$, can be isolated. This formula looks very similar to the formula for the hydrate, with ammonia substituted for water.

What are these two nickel species? The formulas identify the compositions of the compounds but fail to give information about their structures. Because properties of compounds derive from their structures, we need to evaluate the structures in more detail. Typically, metal compounds are ionic, and solid ionic compounds have structures with cations and anions arranged in a regular array. The structure of hydrated nickel chloride contains cations with the formula $[Ni(H_2O)_6]^{2+}$ and chloride anions. The structure of the ammonia-containing compound is similar to the hydrate; it is made up of $[Ni(NH_3)_6]^{2+}$ cations and chloride anions.

Ions such as $[Ni(H_2O)_6]^{2+}$ and $[Ni(NH_3)_6]^{2+}$, in which a metal ion and either water or ammonia molecules compose a single structural unit, are examples of **coordination complexes,** also known as **complex ions** (Figure 22.13). Compounds containing a coordination complex as part of the structure are called **coordination compounds,** and their chemistry is known as **coordination chemistry.** Although the older "hydrate" formulas are still used, the preferred method of writing the formula for coordination compounds places the metal atom or ion and the molecules or anions directly bonded to it within brackets to show that it is a single structural entity. Thus, the formula for the nickel(II)–ammonia compound is better written as $[Ni(NH_3)_6]Cl_2$.

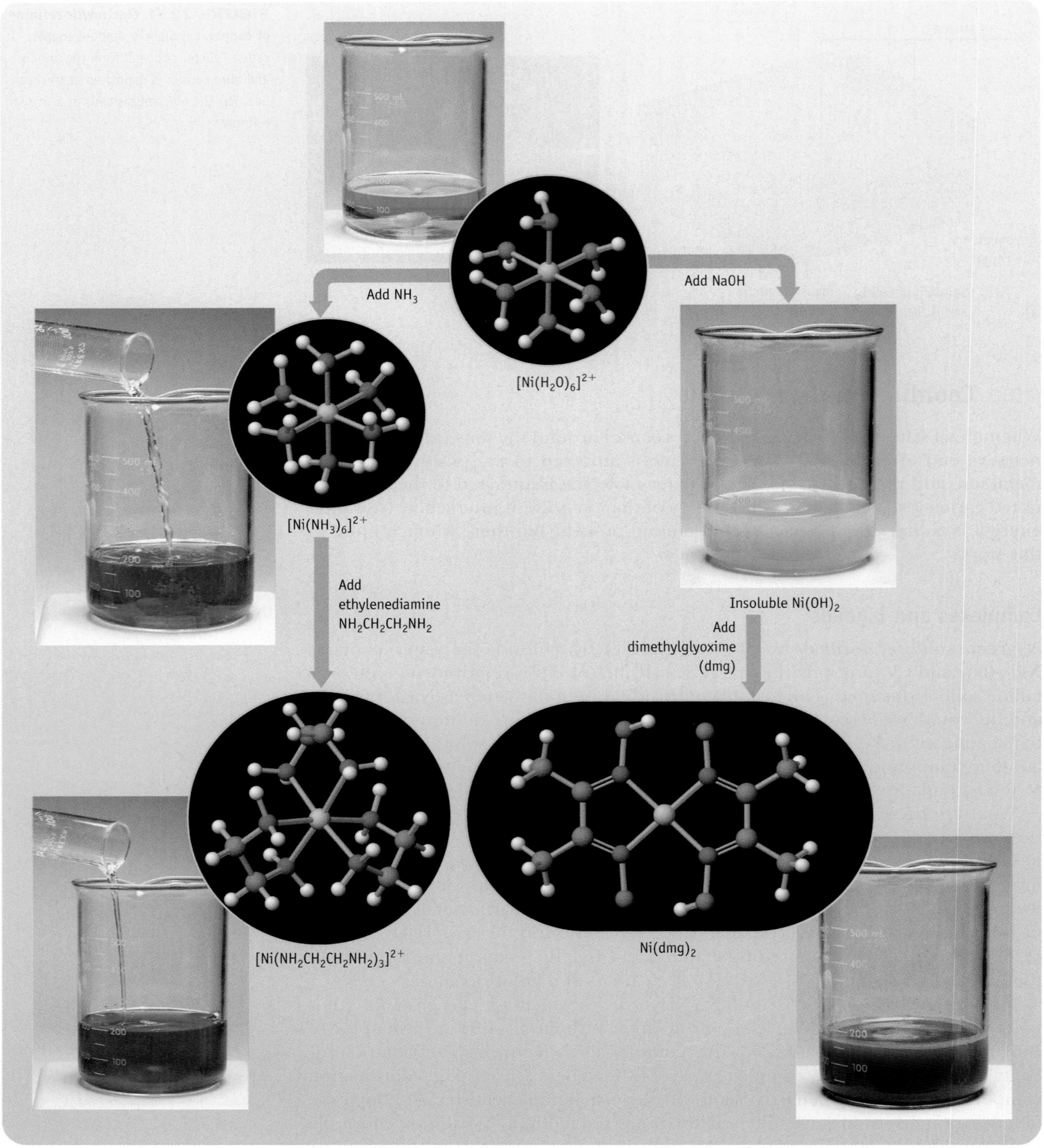

FIGURE 22.12 Coordination compounds of Ni^{2+} ion. The transition metals and their ions form a wide range of compounds, often with beautiful colors and interesting structures. One purpose of this chapter is to explore some commonly observed structures and explain how these compounds can be so colorful.

All coordination complexes contain a metal atom or ion as the central part of the structure. Bonded to the metal are molecules or ions called **ligands** (from the Latin verb *ligare*, meaning "to bind"). In the preceding examples, water and ammonia are the ligands. The number of ligand atoms attached to the metal defines the **coordination number** of the metal. The geometry described by the attached ligands is called the **coordination geometry.** In the nickel complex ion $[Ni(NH_3)_6]^{2+}$ (Figure 22.12), the six ligands are arranged in a regular octahedral geometry around the central metal ion.

Ligands can be either neutral molecules or anions (or, in rare instances, cations). The characteristic feature of a ligand is that it contains a lone pair of electrons. In the classic description of bonding in a coordination complex, the lone pair of electrons on a ligand is shared with the metal ion. The attachment is a coordinate covalent bond (◀ Section 8.5), because the electron pair being shared was originally on the ligand. The name "coordination complex" derives from the name given to this kind of bonding.

The net charge on a coordination complex is the sum of the charges on the metal and its attached groups. Complexes can be cations (as in the two nickel complexes used as examples here), anions, or uncharged.

Ligands such as H_2O and NH_3, which coordinate to the metal via a single Lewis base atom, are termed **monodentate.** The word "dentate" comes from the Latin *dentis*, meaning "tooth," so NH_3 is a "one-toothed" ligand. Some ligands attach to the metal with more than one donor atom. These ligands are called **polydentate.** Ethylenediamine (1,2-diaminoethane), $H_2NCH_2CH_2NH_2$, often abbreviated as en; oxalate ion, $C_2O_4^{2-}$ (ox^{2-}); and phenanthroline, $C_{12}H_8N_2$ (phen), are examples of the wide variety of bidentate ligands (Figure 22.14). Structures and examples of some complex ions with **bidentate** ligands are shown in Figure 22.15.

Polydentate ligands are also called **chelating ligands,** or just chelates (pronounced "key-lates"). The name derives from the Greek *chele*, meaning "claw." Because two or more bonds are broken to separate the ligand from the metal, complexes with chelated ligands have greater stability than those with monodentate ligands. Chelated complexes are important in everyday life. One way to clean the rust out of water-cooled automobile engines and steam boilers is to add a solution of oxalic acid. Iron oxide reacts with oxalic acid to give a water-soluble iron oxalate complex ion:

$$3\ H_2O(\ell) + Fe_2O_3(s) + 6\ H_2C_2O_4(aq) \longrightarrow 2\ [Fe(C_2O_4)_3]^{3-}(aq) + 6\ H_3O^+(aq)$$

Ethylenediaminetetraacetate ion ($EDTA^{4-}$), a hexadentate ligand, is an excellent chelating ligand (Figure 22.16). It can wrap around a metal ion, encapsulating it. Salts of this anion are often added to commercial salad dressings to remove traces of free metal ions from solution; otherwise, these metal ions can act as catalysts for the oxidation of the oils in the dressing. Without $EDTA^{4-}$, the dressing would

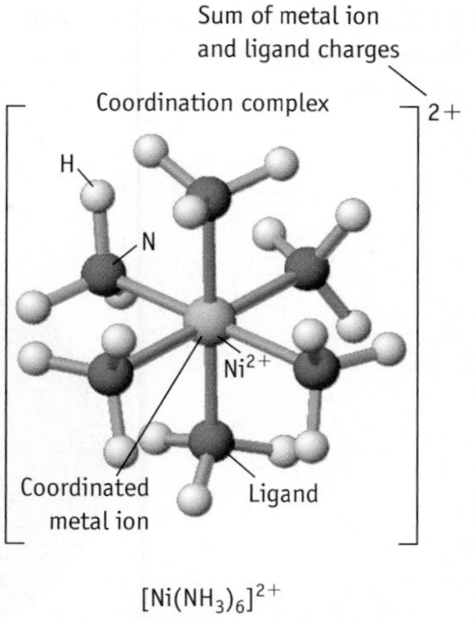

FIGURE 22.13 A coordination complex. In the $[Ni(NH_3)_6]^{2+}$ ion, the ligands are NH_3 molecules. Because the metal has a 2+ charge and the ligands have no charge, the charge on the complex ion is 2+.

■ **Ligands Are Lewis Bases** Ligands are Lewis bases because they furnish the electron pair; the metal ion is a Lewis acid because it accepts electron pairs (see Section 17.9). Thus, the coordinate covalent bond between ligand and metal can be viewed as a Lewis acid–Lewis base interaction.

■ **Bidentate Ligands** All common bidentate ligands bind to *adjacent* sites on the metal.

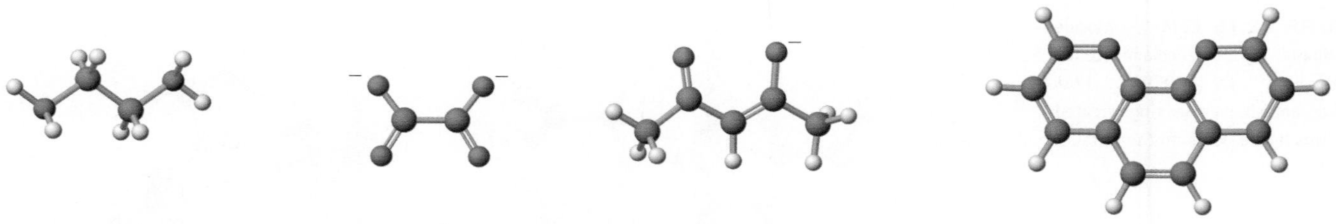

(a) $H_2NCH_2CH_2NH_2$, en **(b)** $C_2O_4^{2-}$, ox **(c)** $CH_3COCHCOCH_3^-$, acac$^-$ **(d)** $C_{12}H_8N_2$, phen

FIGURE 22.14 Common bidentate ligands. (a) Ethylenediamine, $H_2NCH_2CH_2NH_2$; (b) oxalate ion, $C_2O_4^{2-}$; (c) acetylacetonate ion, $CH_3COCHCOCH_3^-$; (d) phenanthroline, $C_{12}H_8N_2$. Coordination of these bidentate ligands to a transition metal ion results in five- or six-member metal-containing rings and no ring strain.

FIGURE 22.15 Complex ions with bidentate ligands. See Figure 22.14 for abbreviations.

$[Fe(C_2O_4)_3]^{3-}$ $[Co(en)_3]^{3+}$ $Cr(acac)_3$

quickly become rancid. Another use is in bathroom cleansers. The $EDTA^{4-}$ ion removes deposits of $CaCO_3$ and $MgCO_3$ left by hard water by coordinating to Ca^{2+} or Mg^{2+} to create soluble complex ions.

Complexes with polydentate ligands play particularly important roles in biochemistry, one example of which is described in *A Closer Look: Hemoglobin.*

Formulas of Coordination Compounds

It is useful to be able to predict the formula of a coordination complex, given the metal ion and ligands, and to derive the oxidation number of the coordinated metal ion, given the formula in a coordination compound. The following examples explore these issues.

EXAMPLE 22.1 Formulas of Coordination Complexes

Problem Give the formulas of the following coordination complexes:

(a) A Ni^{2+} ion is bound to two water molecules and two bidentate oxalate ions.

(b) A Co^{3+} ion is bound to one Cl^- ion, one ammonia molecule, and two bidentate ethylenediamine (en) molecules.

Strategy The problem requires determining the net charge, which equals the sum of the charges of the various component parts of the complex ion. With that information, the metal and ligands can be assembled in the formula, which is placed in brackets, and the net charge indicated.

FIGURE 22.16 $EDTA^{4-}$, a hexadentate ligand. (a) Ethylenediaminetetraacetate, $EDTA^{4-}$. (b) $[Co(EDTA)]^-$. Notice the five- and six-member rings created when this ligand bonds to the metal.

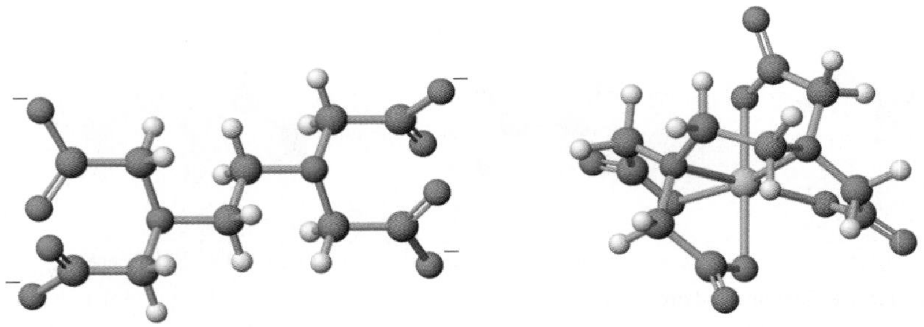

(a) Ethylenediaminetetraacetate, $EDTA^{4-}$ **(b)** $[Co(EDTA)]^-$

A Closer Look — Hemoglobin

Metal-containing coordination compounds figure prominently in many biochemical reactions. Perhaps the best-known example is hemoglobin, the chemical in the blood responsible for O_2 transport. It is also one of the most thoroughly studied bioinorganic compounds.

As described in *The Chemistry of Life: Biochemistry* (page 496), hemoglobin (Hb) is a large iron-containing protein. It includes four polypeptide segments, each containing an iron(II) ion locked inside a porphyrin ring system and coordinated to a nitrogen atom from another part of the protein. A sixth site is available to attach to oxygen.

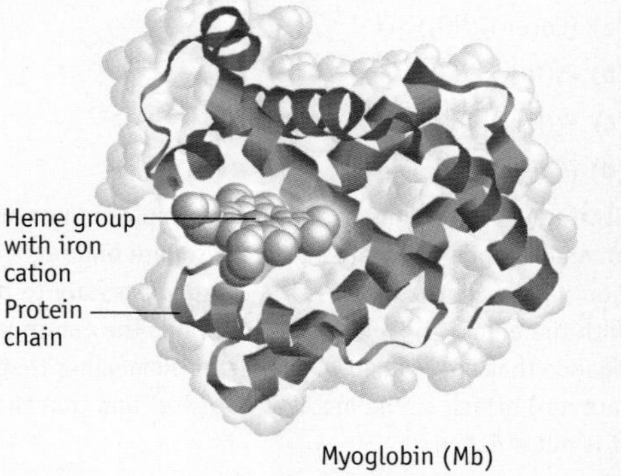

The heme group in myoglobin. This protein is a close relative of hemoglobin. The heme group with its iron ion is shown.

One segment of the hemoglobin molecule resembles the myoglobin structure shown above. (Myoglobin, the oxygen-storage protein in muscle, has only one polypeptide chain with an enclosed heme group. It is the oxygen storage protein in muscle.) In this case, the iron-containing heme group is enclosed with a polypeptide chain. The iron ion in the porphyrin ring is shown. The first and sixth coordination positions are taken up by nitrogen atoms from amino acids of the polypeptide chain.

Hemoglobin functions by reversibly adding oxygen to the sixth coordination position of each iron, giving a complex called oxyhemoglobin.

Because hemoglobin features four iron centers, a maximum of four molecules of oxygen can bind to the molecule. The binding to oxygen is cooperative; that is, binding one molecule enhances the tendency to bind the second, third, and fourth molecules.

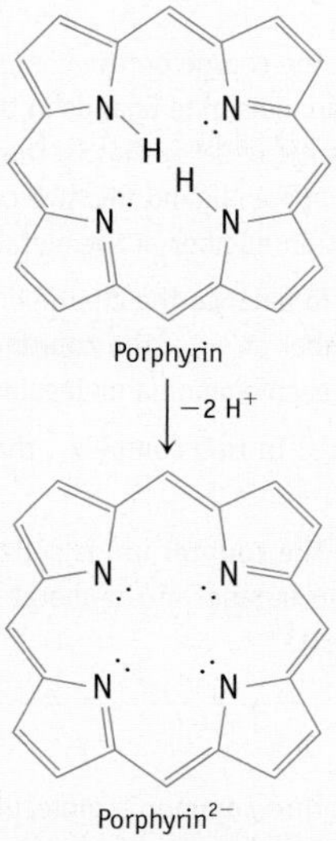

Porphyrin ring of the heme group. The tetradentate ligand surrounding the iron(II) ion in hemoglobin is a dianion of a substituted molecule called a porphyrin. Because of the double bonds in this structure, all of the carbon and nitrogen atoms in the dianion of the porphyrin lie in a plane. In addition, the nitrogen lone pairs are directed toward the center of the molecule, and the molecular dimensions are such that a metal ion may fit nicely into the cavity.

Formation of the oxygenated complex is favored, but not too highly, because oxygen must also be released by the molecule to body tissues. Interestingly, an increase in acidity leads to a decrease in the stability of the oxygenated complex. This phenomenon is known as the *Bohr effect*, named for Christian Bohr, Niels Bohr's father. Release of oxygen in tissues is facilitated by an increase in acidity that results from the presence of CO_2 formed by metabolism.

Among the notable properties of hemoglobin is its ability to form a complex with carbon monoxide. This complex is very stable, with the equilibrium constant for the following reaction being about 200 (where Hb is hemoglobin):

$$HbO_2(aq) + CO(g) \rightleftharpoons HbCO(aq) + O_2(g)$$

When CO complexes with iron, the oxygen-carrying capacity of hemoglobin is lost. Consequently, CO is highly toxic to humans. Exposure to even small amounts greatly reduces the capacity of the blood to transport oxygen.

Hemoglobin abnormalities are well known. One of the most common abnormalities causes sickle cell anemia and was described on page 500.

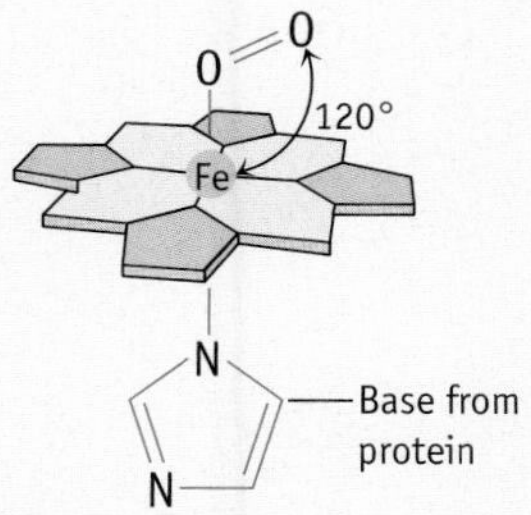

Oxygen binding. Oxygen binds to the iron of the heme group in oxyhemoglobin (and in myoglobin). Interestingly, the Fe—O—O angle is bent.

Solution

(a) This complex ion is constructed from two neutral H_2O molecules, two $C_2O_4^{2-}$ ions, and one Ni^{2+} ion, so the net charge on the complex is 2−. The formula for the complex ion is

$$[Ni(C_2O_4)_2(H_2O)_2]^{2-}$$

(b) This cobalt(III) complex combines two en molecules and one NH_3 molecule, both having no charge, as well as one Cl^- ion and a Co^{3+} ion. The net charge is 2+. The formula for this complex (writing out the entire formula for ethylenediamine) is

$$[Co(H_2NCH_2CH_2NH_2)_2(NH_3)Cl]^{2+}$$

EXAMPLE 22.2 Coordination Compounds

Problem In each of the following complexes, determine the metal's oxidation number and coordination number.

(a) $[Co(en)_2(NO_2)_2]Cl$

(b) $Pt(NH_3)_2(C_2O_4)$

(c) $Pt(NH_3)_2Cl_4$

(d) $[Co(NH_3)_5Cl]SO_4$

Strategy Each formula consists of a complex ion or molecule made up of the metal ion, neutral and/or anionic ligands (the part inside the square brackets), and a counterion (outside the brackets). The oxidation number of the metal is the charge necessary to balance the sum of the negative charges associated with the anionic ligands and counterion. The coordination number is the number of donor atoms in the ligands that are bonded to the metal. Remember that the bidentate ligands in these examples (en, oxalate ion) attach to the metal at two sites and that the counterion is not part of the complex ion—that is, it is not a ligand.

Solution

(a) The chloride ion with a 1− charge, outside the brackets, shows that the charge on the complex ion must be 1+. There are two nitrite ions (NO_2^-) and two neutral bidentate ethylenediamine ligands in the complex. To give a 1+ charge on the complex ion, the cobalt ion must have a charge of 3+; that is, the sum of 2− (two nitrites), 0 (two en ligands), and 3+ (the cobalt ion) equals 1+. Each en ligand fills two coordination positions, and the two nitrite ions fill two more positions. The coordination number of the metal is 6.

(b) There is an oxalate ion ($C_2O_4^{2-}$) and two neutral ammonia ligands. To balance the charge on the oxalate ion, platinum must have a 2+ charge; that is, it has an oxidation number of +2. The coordination number is 4, with an oxalate ligand filling two coordination positions and each ammonia molecule filling one.

(c) There are four chloride ions (Cl^-) and two neutral ammonia ligands. In this complex, the oxidation number of the metal is +4, and the coordination number is 6.

(d) There is one chloride ion (Cl^-) and five neutral ammonia ligands. The counter ion is sulfate with a 2− charge, so the overall charge on the complex is 2+. The oxidation number of the metal is +3 and the coordination number is 6 (sulfate is not coordinated to the metal).

EXERCISE 22.1 Formulas of Coordination Compounds

(a) What is the formula of a complex ion composed of one Co^{3+} ion, three ammonia molecules, and three Cl^- ions?

(b) Determine the metal's oxidation number and coordination number in (i) $K_3[Co(NO_2)_6]$ and in (ii) $Mn(NH_3)_4Cl_2$.

Naming Coordination Compounds

Just as rules govern naming of inorganic and organic compounds, coordination compounds are named according to an established system. The three compounds below are named according to the rules that follow.

Compound	Systematic Name
$[Ni(H_2O)_6]SO_4$	Hexaaquanickel(II) sulfate
$[Cr(en)_2(CN)_2]Cl$	Dicyanobis(ethylenediamine)chromium(III) chloride
$K[Pt(NH_3)Cl_3]$	Potassium amminetrichloroplatinate(II)

1. In naming a coordination compound that is a salt, name the cation first and then the anion. (This is how all salts are commonly named.)
2. When giving the name of the complex ion or molecule, name the ligands first, in alphabetical order, followed by the name of the metal. (When determining alphabetical order, the prefix is not considered part of the name.)

3. Ligands and their names:
   (a) If a ligand is an anion whose name ends in *-ite* or *-ate*, the final *e* is changed to *o* (sulfate ⟶ sulfato or nitrite ⟶ nitrito).
   (b) If the ligand is an anion whose name ends in *-ide*, the ending is changed to *o* (chloride ⟶ chloro, cyanide ⟶ cyano).
   (c) If the ligand is a neutral molecule, its common name is usually used with several important exceptions: Water as a ligand is referred to as *aqua;* ammonia is called *ammine;* and CO is called *carbonyl.*
   (d) When there is more than one of a particular monodentate ligand with a simple name, the number of ligands is designated by the appropriate prefix: *di, tri, tetra, penta,* or *hexa.* If the ligand name is complicated, the prefix changes to *bis, tris, tetrakis, pentakis,* or *hexakis,* followed by the ligand name in parentheses.
4. If the coordination complex is an anion, the suffix *-ate* is added to the metal name.
5. Following the name of the metal, the oxidation number of the metal is given in Roman numerals.

EXAMPLE 22.3 Naming Coordination Compounds

Problem Name the following compounds:

(a) $[Cu(NH_3)_4]SO_4$ (c) $Co(phen)_2Cl_2$

(b) $K_2[CoCl_4]$ (d) $[Co(en)_2(H_2O)Cl]Cl_2$

Strategy Apply the rules for nomenclature given above.

Solution

(a) The complex ion (in square brackets) is composed of four NH_3 molecules (named *ammine* in a complex) and the copper ion. To balance the 2− charge on the sulfate counterion, copper must have a 2+ charge. The compound's name is

tetraamminecopper(II) sulfate

(b) The complex ion $[CoCl_4]^{2-}$ has a 2− charge. With four Cl^- ligands, the cobalt ion must have a +2 charge, so the sum of charges is 2−. The name of the compound is

potassium tetrachlorocobaltate(II)

(c) This is a neutral coordination compound. The ligands include two Cl^- ions and two neutral bidentate *phen* (phenanthroline) ligands. The metal ion must have a 2+ charge (Co^{2+}). The name, listing ligands in alphabetical order, is

dichlorobis(phenanthroline)cobalt(II)

(d) The complex ion has a 2− charge because it is paired with two uncoordinated Cl^- ions. The cobalt ion is Co^{3+} because it is bonded to two neutral en ligands, one neutral water, and one Cl^-. The name is

aquachlorobis(ethylenediamine)cobalt(III) chloride

EXERCISE 22.2 Naming Coordination Compounds

Name the following coordination compounds.

(a) $[Ni(H_2O)_6]SO_4$ (c) $K[Pt(NH_3)Cl_3]$

(b) $[Cr(en)_2(CN)_2]Cl$ (d) $K[CuCl_2]$

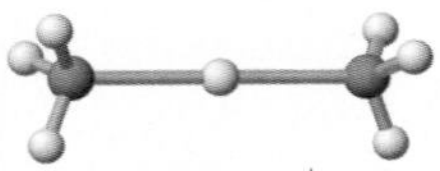

$[Ag(NH_3)_2]^+$

Linear

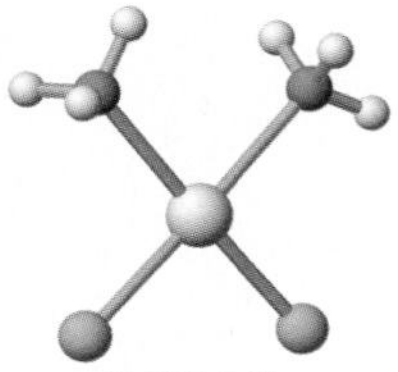

$Pt(NH_3)_2Cl_2$

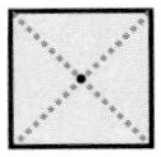

Square planar

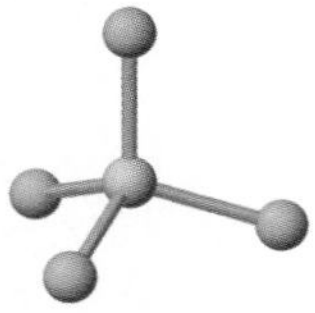

$[NiCl_4]^{2-}$

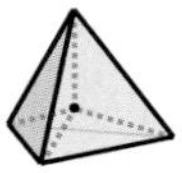

Tetrahedral

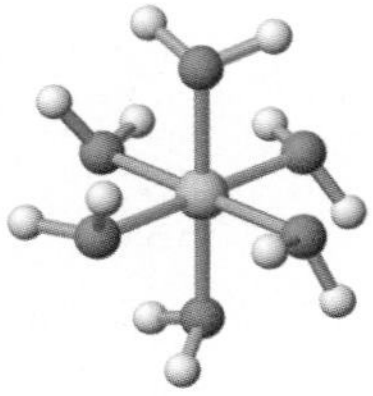

$[Ni(H_2O)_6]^{2+}$

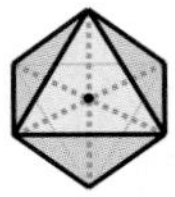

Octahedral

22.4 Structures of Coordination Compounds

Common Coordination Geometries

The geometry of a coordination complex is defined by the arrangement of donor atoms of the ligands around the central metal ion. Metal ions in coordination compounds can have coordination numbers ranging from 2 to 12. Only complexes with coordination numbers of 2, 4, and 6 are common, however, so we will concentrate on species such as $[ML_2]^{n\pm}$, $[ML_4]^{n\pm}$, and $[ML_6]^{n\pm}$, where M is the metal ion and L is a monodentate ligand. Within these stoichiometries, the following geometries are encountered:

- All $[ML_2]^{n\pm}$ complexes are linear. The two ligands are on opposite sides of the metal, and the L—M—L bond angle is 180°. Common examples include $[Ag(NH_3)_2]^+$ and $[CuCl_2]^-$.
- Tetrahedral geometry occurs in many $[ML_4]^{n\pm}$ complexes. Examples include $TiCl_4$, $[CoCl_4]^{2-}$, $[NiCl_4]^{2-}$, and $[Zn(NH_3)_4]^{2+}$.
- Some $[ML_4]^{n\pm}$ complexes can have square planar geometry. This geometry is most often seen with metal ions that have eight *d* electrons. Examples include $Pt(NH_3)_2Cl_2$, $[Ni(CN)_4]^{2-}$, and the nickel complex with the dimethylglyoximate (dmg^-) ligand in Figure 22.12.
- Octahedral geometry is found in complexes with the stoichiometry $[ML_6]^{n\pm}$ (Figure 22.12).

Chemistry Now™

Sign in at **www.thomsonedu.com/login** and go to Chapter 22 Contents to see Screen 22.5 for an exercise on **compound geometries.**

Isomerism

Isomerism is one of the most interesting aspects of molecular structure. Recall that the chemistry of organic compounds is greatly enlivened by the multitude of isomeric compounds that are known.

Isomers are classified as follows:

- *Structural isomers* have the same molecular formula but different bonding arrangements of atoms.
- *Stereoisomers* have the same atom-to-atom bonding sequence, but the atoms differ in their arrangement in space. There are two types of stereoisomers: geometric isomers (such as *cis* and *trans* alkenes, page 421) and optical isomers (non-superimposable mirror images that have the unique property that they rotate planar polarized light, page 445).

All three types of isomerism, structural, geometric, and optical, are encountered in coordination chemistry.

Structural Isomerism

The two most important types of structural isomerism in coordination chemistry are coordination isomerism and linkage isomerism. **Coordination isomerism** occurs when it is possible to exchange a coordinated ligand and the uncoordinated counterion. For example, dark violet $[Co(NH_3)_5Br]SO_4$ and red $[Co(NH_3)_5SO_4]Br$ are coordination isomers. In the first compound, bromide ion is a ligand and sulfate is a counterion; in the second, sulfate is a ligand and bromide is the counterion. A diagnostic test for this kind of isomer is often made based on chemical reactions.

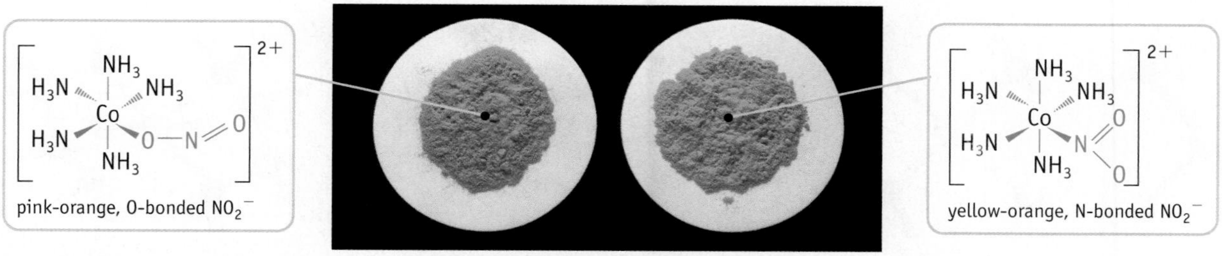

Photo: Charles D. Winters

FIGURE 22.17 Linkage isomers, $[Co(NH_3)_5ONO]^{2+}$ and $[Co(NH_3)_5NO_2]^{2+}$. These complexes, whose systematic names are pentaamminenitritocobalt(III) and pentaamminenitrocobalt(III), were the first known examples of this type of isomerism.

For example, these two compounds can be distinguished by precipitation reactions. Addition of $Ba^{2+}(aq)$ to a solution of $[Co(NH_3)_5Br]SO_4$ gives a precipitate of $BaSO_4$, indicating the presence of sulfate ion in solution. In contrast, no reaction occurs if $Ba^{2+}(aq)$ is added to a solution of $[Co(NH_3)_5SO_4]Br$. In this complex, sulfate ion is attached to Co^{3+} and is not a free ion in solution.

$$[Co(NH_3)_5Br]SO_4 + Ba^{2+}(aq) \longrightarrow BaSO_4(s) + [CO(NH_3)_5Br]^{2+}(aq)$$

$$[Co(NH_3)_5SO_4]Br + Ba^{2+}(aq) \longrightarrow \text{no reaction}$$

Linkage isomerism occurs when it is possible to attach a ligand to the metal through different atoms. The two most common ligands with which linkage isomerism arises are thiocyanate, SCN^-, and nitrite, NO_2^-. The Lewis structure of the thiocyanate ion shows that there are lone pairs of electrons on sulfur and nitrogen. The ligand can attach to a metal either through sulfur (called S-bonded thiocyanate) or through nitrogen (called N-bonded thiocyanate). Nitrite ion can attach either at oxygen or at nitrogen. The former are called nitrito complexes; the latter are nitro complexes (Figure 22.17).

Ligands forming linkage isomers

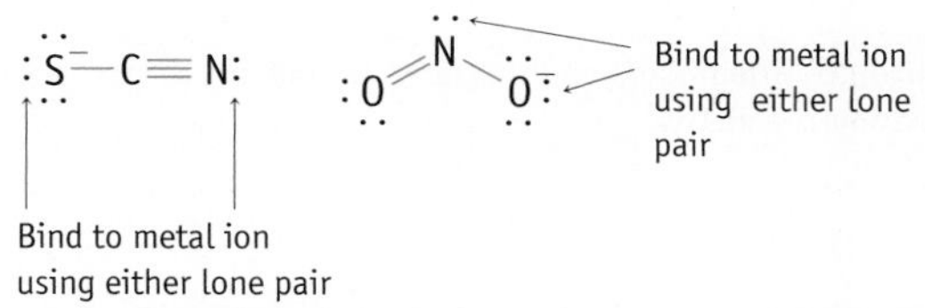

Geometric Isomerism

Geometric isomerism results when the atoms bonded directly to the metal have a different spatial arrangement. The simplest example of geometric isomerism in coordination chemistry is *cis-trans* isomerism, which occurs in both square-planar and octahedral complexes. An example of *cis-trans* isomerism is seen in the square-planar complex $Pt(NH_3)_2Cl_2$ (Figure 22.18a). In this complex, the two Cl^- ions can be either adjacent to each other (*cis*) or on opposite sides of the metal (*trans*). The *cis* isomer is effective in the treatment of testicular, ovarian, bladder, and osteogenic sarcoma cancers, but the *trans* isomer has no effect on these diseases.

Cis-trans isomerism in an octahedral complex is illustrated by a complex ion with two bidentate ethylenediamine ligands and two chloride ligands, $[Co(H_2NCH_2CH_2NH_2)_2Cl_2]^+$. In this complex, the two Cl^- ions occupy positions that are either adjacent (the purple *cis* isomer) or opposite (the green *trans* isomer) (Figure 22.18b).

■ ***Cis-Trans* Isomerism** *Cis-trans* isomerism is not possible for tetrahedral complexes. All L—M—L angles in tetrahedral geometry are 109.5°, and all positions are equivalent in this three-dimensional structure.

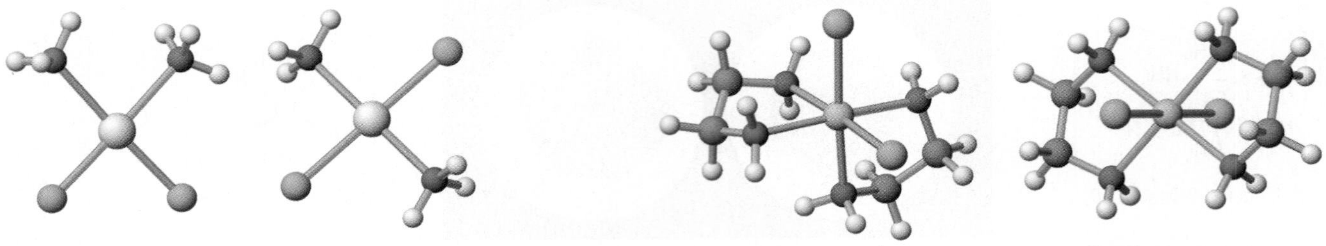

FIGURE 22.18 ***Cis-trans* isomers.** (a) The square planar complex $Pt(NH_3)_2Cl_2$ can exist in two geometries, *cis* and *trans*. (b) Similarly, *cis* and *trans* octahedral isomers are possible for $[Co(en)_2Cl_2]^+$.

Another common type of geometric isomerism occurs for octahedral complexes with the general formula MX_3Y_3. A *fac* isomer has three identical ligands lying at the corners of a triangular face of an octahedron defined by the ligands (*fac* = facial), whereas the ligands follow a meridian in the *mer* isomer (*mer* = meridional). *Fac* and *mer* isomers of $Cr(NH_3)_3Cl_3$ are shown in Figure 22.19.

Optical Isomerism

Optical isomerism (chirality) occurs for octahedral complexes when the metal ion coordinates to three bidentate ligands or when the metal ion coordinates to two bidentate ligands and two monodentate ligands in a *cis* position. The complexes $[Co(en)_3]^{3+}$ and *cis*-$[Co(en)_2Cl_2]^+$, illustrated in Figure 22.20, are examples of chiral complexes. The diagnostic test for chirality is met with both species: Mirror images of these molecules are not superimposable (page 446). Solutions of the optical isomers rotate plane-polarized light in opposite directions.

■ **Optical Isomerism** Square-planar complexes are incapable of optical isomerism based at the metal center; mirror images are always superimposable. Chiral tetrahedral complexes are possible, but examples of complexes with a metal bonded tetrahedrally to four different monodentate ligands are rare.

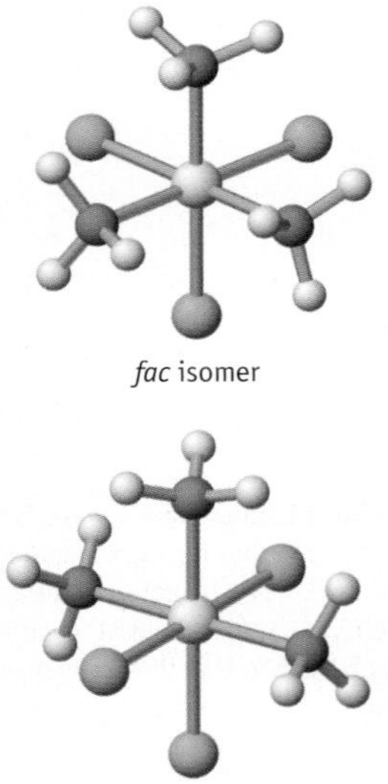

FIGURE 22.19 ***Fac* and *mer* isomers of $Cr(NH_3)_3Cl_3$.** In the *fac* isomer, the three chloride ligands (and the three ammonia ligands) are arranged at the corners of a triangular face. In the *mer* isomer, the three similar ligands follow a meridian.

Chemistry Now™

Sign in at **www.thomsonedu.com/login** and go to Chapter 22 Contents to see Screen 22.6 for an exercise on **isomerism in coordination chemistry.**

EXAMPLE 22.4 Isomerism in Coordination Chemistry

Problem For which of the following complexes do isomers exist? If isomers are possible, identify the type of isomerism (structural or geometric). Determine whether the coordination complex is capable of exhibiting optical isomerism.

(a) $[Co(NH_3)_4Cl_2]^+$

(b) $Pt(NH_3)_2(CN)_2$ (square planar)

(c) $Co(NH_3)_3Cl_3$

(e) $K_3[Fe(C_2O_4)_3]$

(d) $Zn(NH_3)_2Cl_2$ (tetrahedral)

(f) $[Co(NH_3)_5SCN]^{2+}$

Strategy Determine the number of ligands attached to the metal, and decide whether the ligands are monodentate or bidentate. Knowing how many donor atoms are coordinated to the metal (the coordination number) will allow you to establish the metal geometry. At that point, it is necessary to recall the possible types of isomers for each geometry. The only isomerism possible for square–planar complexes is geometric (*cis* and *trans*). Tetrahedral complexes do not have isomers. Six-coordinate metals of the formula MA_4B_2 can be either *cis* or *trans*. *Mer* and *fac* isomers are possible with a stoichiometry of MA_3B_3. Optical activity arises for metal complexes of the formula *cis*-M(bidentate)$_2$X$_2$ and M(bidentate)$_3$ (among others). Drawing pictures of the molecules will help you visualize the isomers.

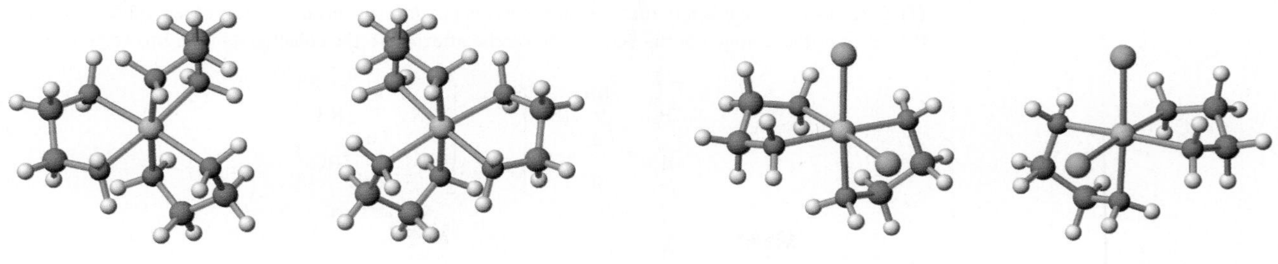

$[Co(en)_3]^{3+}$ $[Co(en)_3]^{3+}$ mirror image *cis*-$[Co(en)_2Cl_2]^+$ *cis*-$[Co(en)_2Cl_2]^+$ mirror image

FIGURE 22.20 Chiral metal complexes. Both $[Co(en)_3]^{3+}$ and *cis*-$[Co(en)_2Cl_2]^+$ are chiral. Notice that the mirror images of the two compounds are not superimposable.

Solution

(a) Two geometric isomers can be drawn for octahedral complexes with a formula of MA_4B_2, such as this one. One isomer has two Cl^- ions in *cis* positions (adjacent positions, at a 90° angle), and the other isomer has the Cl^- ligands in *trans* positions (with a 180° angle between the ligands). Optical isomers are not possible.

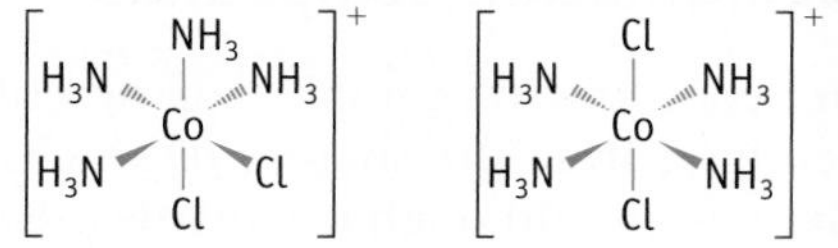

cis isomer *trans* isomer

(b) In this square-planar complex, the two NH_3 ligands (and the two CN^- ligands) can be either *cis* or *trans*. These are geometric isomers. Optical isomers are not possible.

cis isomer *trans* isomer

(c) Two geometric isomers of this octahedral complex, with chloride ligands either *fac* or *mer*, are possible. In the *fac* isomer, the three Cl^- ligands are all at 90° to each other; in the *mer* isomer, two Cl^- ligands are at 180°, and the third is 90° from the other two. Optical isomers are not possible.

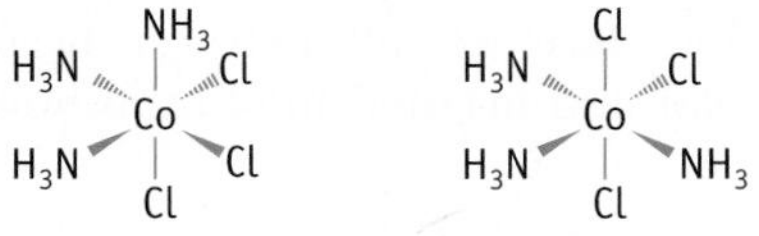

fac isomer *mer* isomer

(d) Only a single structure is possible for tetrahedral complexes such as $Zn(NH_3)_2Cl_2$.

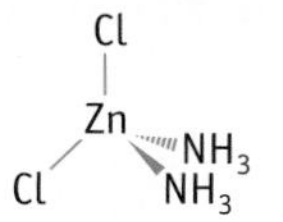

(e) Ignore the counterions, K^+. The anion is an octahedral complex—remember that the bidentate oxalate ion occupies two coordination sites of the metal, and that three oxalate ligands means that the metal has a coordination number of 6. Mirror images of complexes of the stoichiometry $M(bidentate)_3$ are not superimposable; therefore, two optical isomers are possible. (Here the ligands, $C_2O_4^{2-}$, are drawn abbreviated as O—O.)

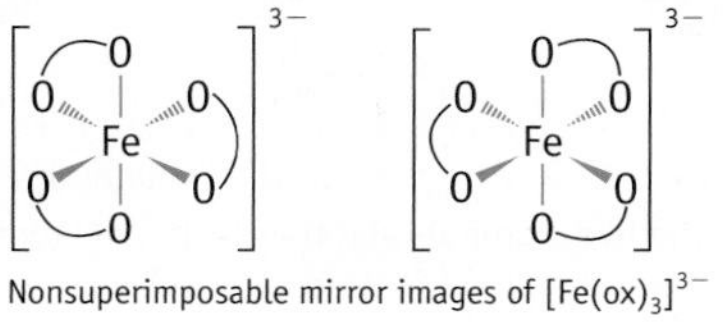

Nonsuperimposable mirror images of $[Fe(ox)_3]^{3-}$

(f) Only linkage isomerism (structural isomerism) is possible for this octahedral cobalt complex. Either the sulfur or the nitrogen of the SCN^- anion can be attached to the cobalt(III) ion in this complex.

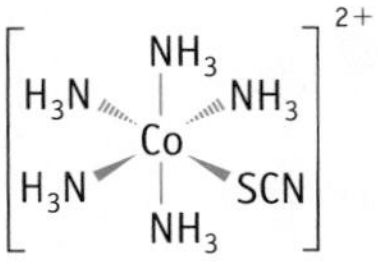

S-bonded SCN^-

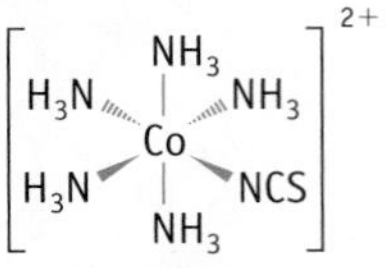

N-bonded SCN^-

EXERCISE 22.3 Isomers in Coordination Complexes

What types of isomers are possible for the following complexes?

(a) $K[Co(NH_3)_2Cl_4]$

(b) $Pt(en)Cl_2$ (square planar)

(c) $[Co(NH_3)_5Cl]^{2+}$

(d) $[Ru(phen)_3]Cl_3$

(e) $Na_2[MnCl_4]$ (tetrahedral)

(f) $[Co(NH_3)_5NO_2]^{2+}$

22.5 Bonding in Coordination Compounds

Metal–ligand bonding in a coordination complex was described earlier in this chapter as being covalent, resulting from the sharing of an electron pair between the metal and the ligand donor atom. Although frequently used, this description is not capable of explaining the color and magnetic behavior of complexes. As a consequence, the covalent bonding picture has now largely been superseded by two other bonding models: molecular orbital theory and ligand field theory.

The bonding model based on molecular orbital theory assumes that the metal and the ligand bond through the molecular orbitals formed by atomic orbital overlap between metal and ligand. The **ligand field model,** in contrast, focuses on repulsion (and destabilization) of electrons in the metal coordination sphere. The ligand field model also assumes that the positive metal ion and the negative ligand lone pair are attracted electrostatically; that is, the bond arises when a positively charged metal ion attracts a negative ion or the negative end of a polar molecule. For the most part, the molecular orbital and ligand field models predict similar, qualitative results regarding color and magnetic behavior. Here, we will focus on the ligand field approach and illustrate how it explains color and magnetism of transition metal complexes.

The *d* Orbitals: Ligand Field Theory

To understand ligand field theory, it is necessary to look at the *d* orbitals, particularly with regard to their orientation relative to the positions of ligands in a metal complex.

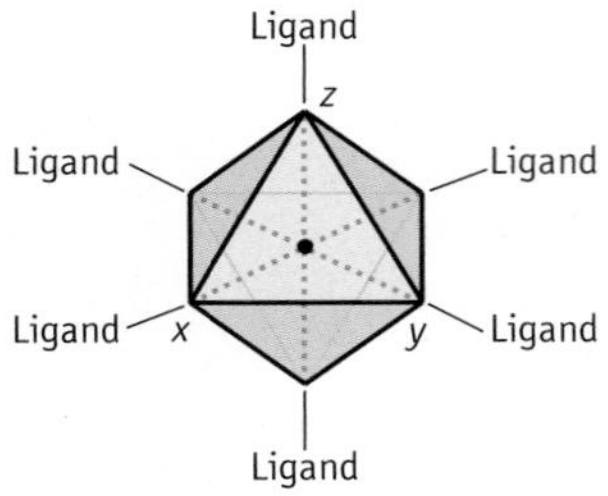

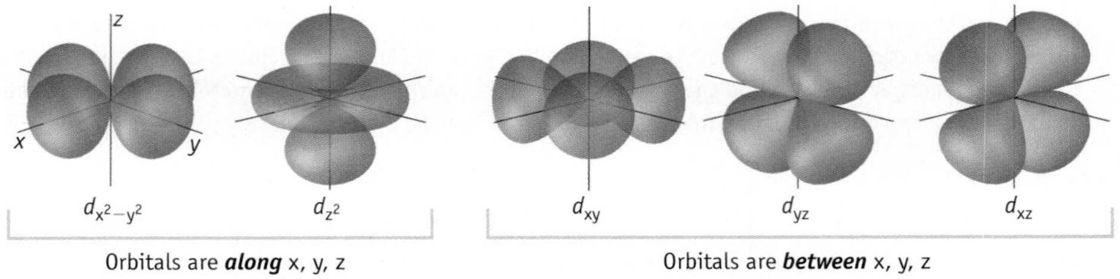

FIGURE 22.21 The *d* orbitals. The five *d* orbitals and their spatial relation to the ligands on the *x*-, *y*-, and *z*-axes.

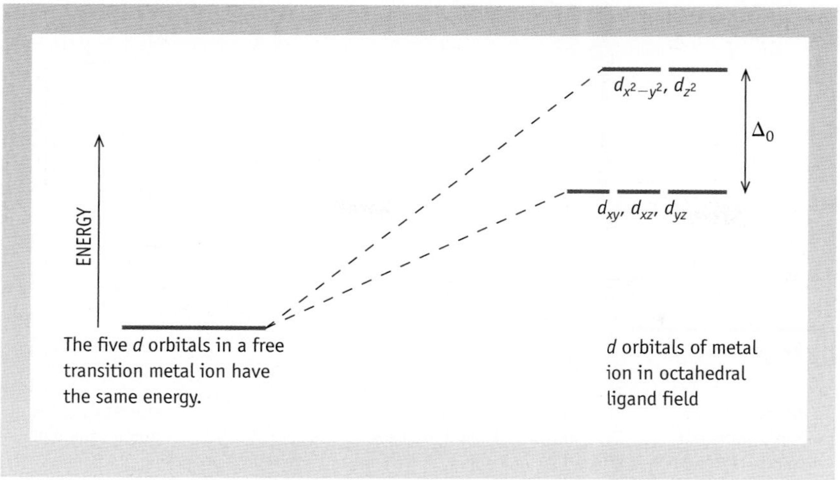

FIGURE 22.22 Ligand field splitting for an octahedral complex. The *d*-orbital energies increase as the ligands approach the metal along the *x*-, *y*-, and *z*-axes. The d_{xy}, d_{xz}, and d_{yz} orbitals, not pointed toward the ligands, are less destabilized than the $d_{x^2-y^2}$ and d_{z^2} orbitals. Thus, the d_{xy}, d_{xz}, and d_{yz} orbitals are at lower energy. (Δ_0 stands for the splitting in a octahedral ligand field.)

We look first at octahedral complexes. Assume the ligands in an octahedral complex lie along the *x*-, *y*-, and *z*-axes. This results in the five *d* orbitals (Figure 22.21) being subdivided into two sets: the $d_{x^2-y^2}$ and d_{z^2} orbitals in one set and the d_{xy}, d_{xz}, and d_{yz} orbitals in the second. The $d_{x^2-y^2}$ and d_{z^2} orbitals are directed along the *x*-, *y*-, and *z*-axes, whereas the orbitals of the second group are aligned between these axes.

In an isolated atom or ion, the five *d* orbitals have the same energy. For a metal atom or ion in a coordination complex, however, the *d* orbitals have different energies. According to the ligand field model, repulsion between *d* electrons on the metal and electron pairs of the ligands destabilizes electrons that reside in the *d* orbitals; that is, it causes their energy to increase. Electrons in the various *d* orbitals are not affected equally, however, because of their different orientations in space relative to the position of the ligand lone pairs (Figure 22.22). Electrons in the $d_{x^2-y^2}$ and d_{z^2} orbitals experience a larger repulsion because these orbitals point directly at the incoming ligand electron pairs. A smaller repulsive effect is experienced by electrons in the d_{xy}, d_{xz}, and d_{yz} orbitals. The difference in degree of repulsion means that an energy difference exists between the two sets of orbitals. This difference, called the **ligand field splitting** and denoted by the symbol Δ_0, is a function of the metal and the ligands and varies predictably from one complex to another.

A different splitting pattern is encountered with square-planar complexes (Figure 22.23). Assume that the four ligands are along the *x*- and *y*-axes. The $d_{x^2-y^2}$ orbital also points along these axes, so it has the highest energy. The d_{xy} orbital (which also lies in the *xy*-plane, but does not point at the ligands) is next highest in energy, followed by the d_{z^2} orbital. The d_{xz} and d_{yz} orbitals, both of which partially point in the *z*-direction, have the lowest energy.

The *d*-orbital splitting pattern for a tetrahedral complex is the reverse of the pattern observed for octahedral complexes. Three orbitals (d_{xz}, d_{xy}, d_{yz}) are higher in energy, whereas the $d_{x^2-y^2}$ and d_{z^2} orbitals are below them in energy (Figure 22.23).

Electron Configurations and Magnetic Properties

The *d*-orbital splitting in coordination complexes provides the means to explain both the magnetic behavior and the color of these complexes. To understand this explanation, however, we must first understand how to assign electrons to the various orbitals in each geometry.

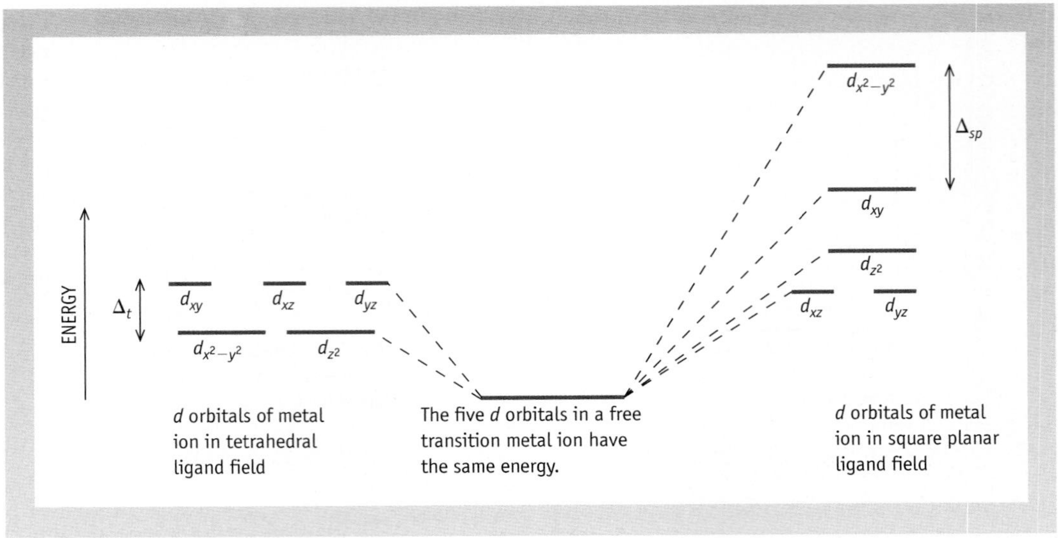

FIGURE 22.23 Splitting of the *d* orbitals in (*left*) tetrahedral and (*right*) square planar geometries. (Δ_t and Δ_{sp} are, respectively, the splitting in tetrahedral and square planar ligand fields.)

A gaseous Cr^{2+} ion has the electron configuration [Ar]3d^4. The term "gaseous" in this context is used to denote a single, isolated atom or ion with all other particles located an infinite distance away. In this situation, the five 3*d* orbitals have the same energy. The four electrons reside singly in different *d* orbitals, according to Hund's rule, and the Cr^{2+} ion has four unpaired electrons.

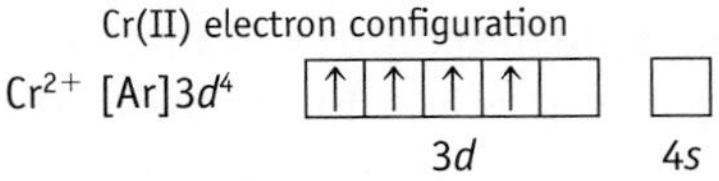

When the Cr^{2+} ion is part of an octahedral complex, the five *d* orbitals do not have identical energies. As illustrated in Figure 22.22, these orbitals divide into two sets, with the d_{xy}, d_{xz}, and d_{yz} orbitals having a lower energy than the $d_{x^2-y^2}$ and d_{z^2} orbitals. Having two sets of orbitals means that two different electron configurations are possible (Figure 22.24). Three of the four *d* electrons in Cr^{2+} are assigned to the lower-energy d_{xy}, d_{xz}, and d_{yz} orbitals. The fourth electron either can be assigned to an orbital in the higher-energy $d_{x^2-y^2}$ and d_{z^2} set or can pair up with an electron

FIGURE 22.24 High- and low-spin cases for an octahedral chromium(II) complex. (*left, high spin*) If the ligand field splitting (Δ_o) is smaller than the pairing energy (P), the electrons are placed in different orbitals, and the complex has four unpaired electrons. (*right, low spin*) If the splitting is larger than the pairing energy, all four electrons will be in the lower-energy orbital set. This requires pairing two electrons in one of the orbitals, so the complex will have two unpaired electrons.

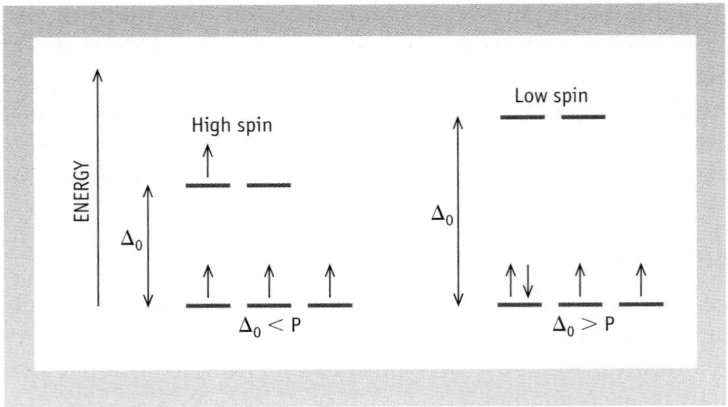

already in the lower-energy set. The first arrangement is called **high spin,** because it has the maximum number of unpaired electrons, four in the case of Cr^{2+}. The second arrangement is called **low spin,** because it has the minimum number of unpaired electrons possible.

At first glance, a high-spin configuration appears to contradict conventional thinking. It seems logical that the most stable situation would occur when electrons occupy the lowest-energy orbitals. A second factor intervenes, however. Because electrons are negatively charged, repulsion increases when they are assigned to the same orbital. This destabilizing effect bears the name **pairing energy.** The preference for an electron to be in the lowest-energy orbital and the pairing energy have opposing effects (Figure 22.24).

Low-spin complexes arise when the splitting of the *d* orbitals by the ligand field is large—that is, when Δ_0 has a large value. The energy gained by putting all of the electrons in the lowest-energy level is the dominant effect. In contrast, high-spin complexes occur if the value of Δ_0 is small.

For octahedral complexes, high- and low-spin complexes can occur only for configurations d^4 through d^7 (Figure 22.25). Complexes of the d^6 metal ion, Fe^{2+}, for example, can have either high spin or low spin. The complex formed when the Fe^{2+} ion is placed in water, $[Fe(H_2O)_6]^{2+}$, is high spin, whereas the $[Fe(CN)_6]^{4-}$ complex ion is low spin.

Electron configuration for Fe^{2+} in an octahedral complex

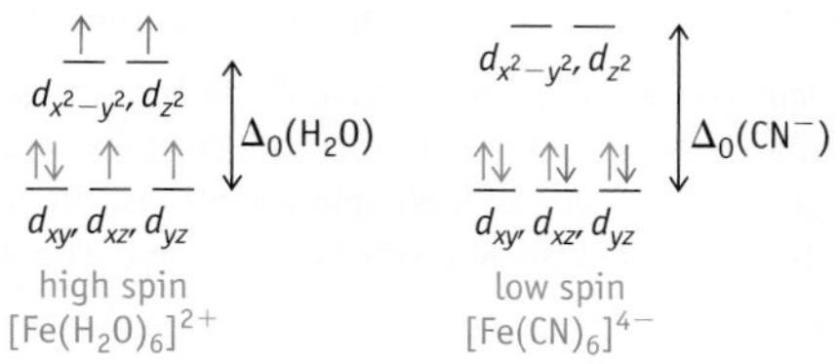

It is possible to tell whether a complex is high or low spin by examining its magnetic behavior. The high-spin complex $[Fe(H_2O)_6]^{2+}$ has four unpaired electrons and is *paramagnetic* (attracted by a magnet), whereas the low-spin $[Fe(CN)_6]^{4-}$ complex has no unpaired electrons and is *diamagnetic* (repelled by a magnet) (◄ page 293).

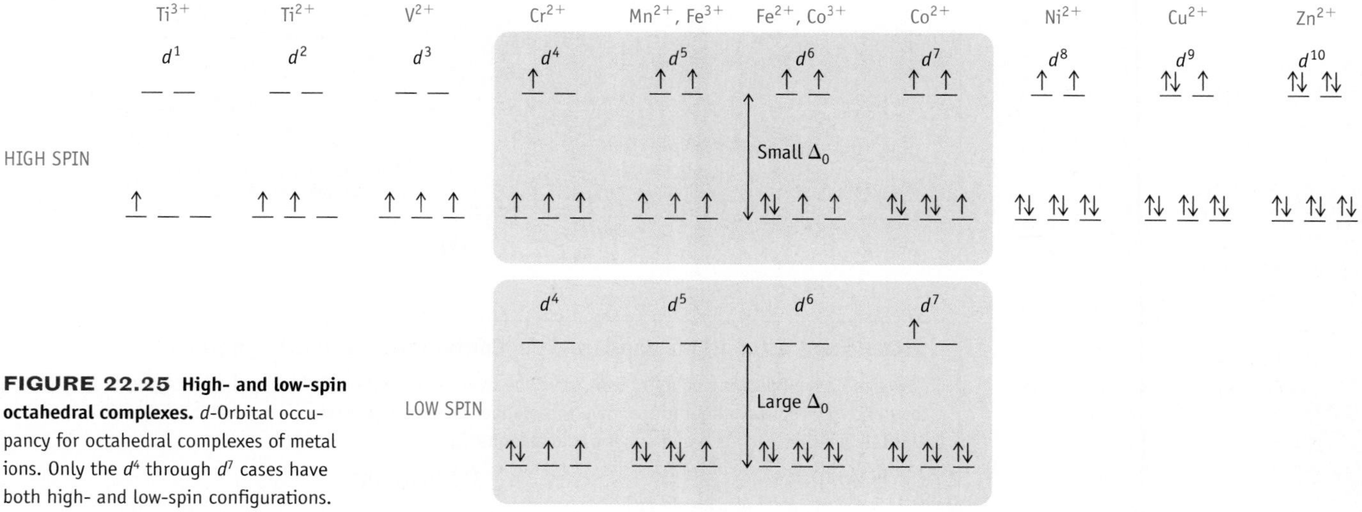

FIGURE 22.25 High- and low-spin octahedral complexes. *d*-Orbital occupancy for octahedral complexes of metal ions. Only the d^4 through d^7 cases have both high- and low-spin configurations.

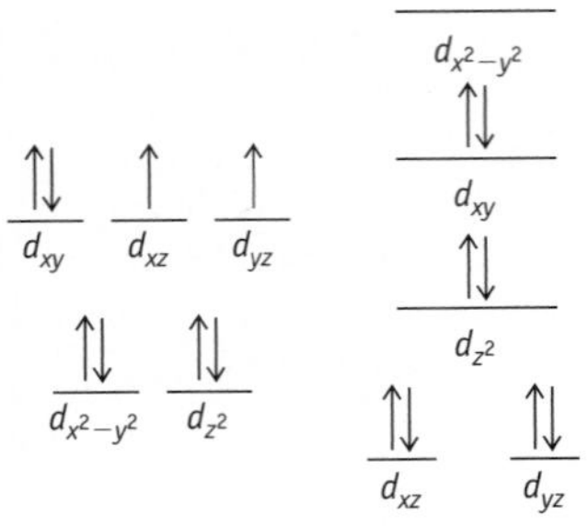

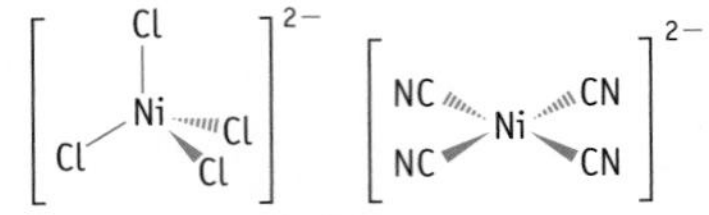

Nickel(II) complexes and magnetism. The anion $[NiCl_4]^{2-}$ is a paramagnetic tetrahedral complex. In contrast, $[Ni(CN)_4]^{2-}$ is a diamagnetic square-planar complex.

Most complexes of Pd^{2+} and Pt^{2+} ions are square planar, the electron configuration of these metals being [noble gas] $(n-1)d^8$. In a square-planar complex, there are four sets of orbitals (Figure 22.22). All except the highest-energy orbital are filled, and all electrons are paired, resulting in diamagnetic (low-spin) complexes.

Nickel, which is found above palladium in the periodic table, forms both square-planar and tetrahedral complexes. For example, the complex ion $[Ni(CN)_4]^{2-}$ is square planar, whereas the $[NiCl_4]^{2-}$ ion is tetrahedral. Magnetism allows us to differentiate between these two geometries. Based on the ligand field splitting pattern, the cyanide complex is expected to be diamagnetic, whereas the chloride complex is paramagnetic with two unpaired electrons.

Chemistry Now™

Sign in at **www.thomsonedu.com/login** and go to Chapter 22 Contents to see simulation of **bonding and electronic structure in transition metal complexes.**

EXAMPLE 22.5 High- and Low-Spin Complexes and Magnetism

Problem Give the electron configuration for each of the following complexes. How many unpaired electrons are present in each complex? Are the complexes paramagnetic or diamagnetic?

(a) low-spin $[Co(NH_3)_6]^{3+}$ **(b)** high-spin $[CoF_6]^{3-}$

Strategy These ions are complexes of Co^{3+}, which has a d^6 valence electron configuration. Set up an energy-level diagram for an octahedral complex. In low-spin complexes, the electrons are added preferentially to the lower-energy set of orbitals. In high-spin complexes, the first five electrons are added singly to each of the five orbitals, then additional electrons are paired with electrons in orbitals in the lower-energy set.

Solution

(a) The six electrons of the Co^{3+} ion fill the lower-energy set of orbitals entirely. This d^6 complex ion has no unpaired electrons and is diamagnetic.

(b) To obtain the electron configuration in high-spin $[CoF_6]^{3-}$, place one electron in each of the five *d* orbitals, and then place the sixth electron in one of the lower-energy orbitals. The complex has four unpaired electrons and is paramagnetic.

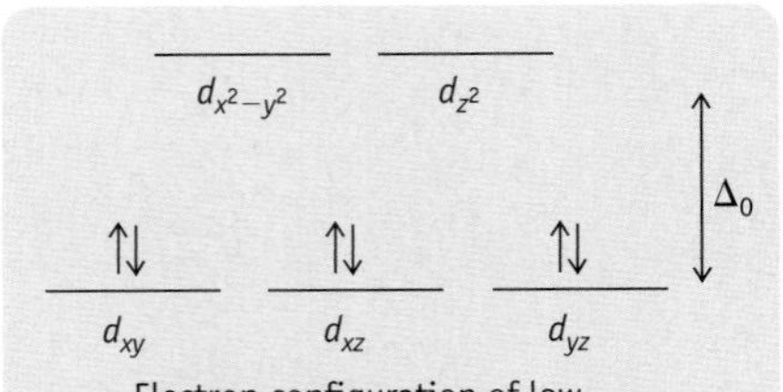

Electron configuration of low-spin, octahedral $[Co(NH_3)_6]^{3+}$

(a)

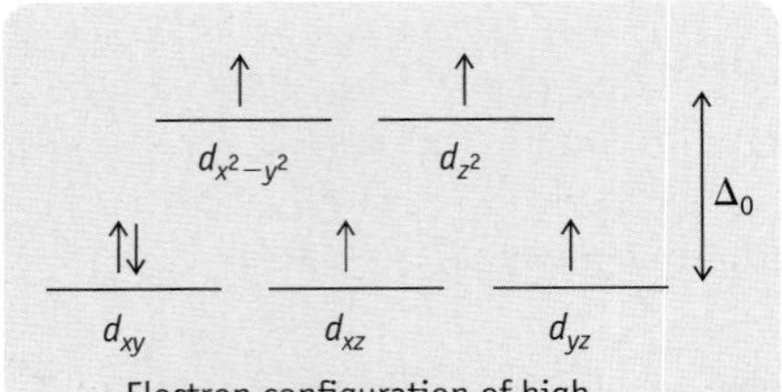

Electron configuration of high-spin, octahedral $[CoF_6]^{3-}$

(b)

EXERCISE 22.4 High- and Low-Spin Configurations and Magnetism

For each of the following complex ions, give the oxidation number of the metal, depict possible low- and high-spin configurations, give the number of unpaired electrons in each configuration, and tell whether each is paramagnetic or diamagnetic.

(a) $[Ru(H_2O)_6]^{2+}$ **(b)** $[Ni(NH_3)_6]^{2+}$

Charles D. Winters

FIGURE 22.26 Aqueous solutions of some transition metal ions. Compounds of transition metal elements are often colored, whereas those of main group metals are usually colorless. Pictured here, from left to right, are solutions of the nitrate salts of Fe^{3+}, Co^{2+}, Ni^{2+}, Cu^{2+}, and Zn^{2+}.

22.6 Colors of Coordination Compounds

One of the most interesting features of compounds of the transition elements is their colors (Figure 22.26). In contrast, compounds of main group metals are usually colorless. The color of transition metal compounds results from *d*-orbital splitting. Before discussing how *d*-orbital splitting is involved, we need to look more closely at what we mean by color.

Color

Visible light, consisting of radiation with wavelengths from 400 to 700 nm (◄ Section 6.1), represents a very small portion of the electromagnetic spectrum. Within this region are all the colors you see when white light passes through a prism: red, orange, yellow, green, blue, indigo, and violet (ROY G BIV). Each color is identified with a portion of the wavelength range.

Isaac Newton did experiments with light and established that the mind's perception of color requires only three colors! When we see white light, we are seeing a mixture of all of the colors—in other words, the superposition of red, green, and blue. If one or more of these colors is absent, the light of the other colors that reaches your eyes is interpreted by your mind as color.

Figure 22.27 will help you in analyzing perceived colors. This color wheel shows the three primary colors—red, green, and blue—as overlapping disks arranged in a triangle. The secondary colors—cyan, magenta, and yellow—appear where two disks overlap. The overlap of all three disks in the center produces white light.

The colors we perceive are determined as follows:

- Light of a single primary color is perceived as that color: Red light is perceived as red, green light as green, blue light as blue.
- Light made up of two primary colors is perceived as the color shown where the disks in Figure 22.27 overlap: Red and green light together appear yellow, green and blue light together are perceived as cyan; and red and blue light are perceived as magenta.
- Light made up of the three primary colors is white (colorless).

In discussing the color of a substance such as a coordination complex *in solution,* these guidelines are turned around.

- Red color is the result of the absence of green and blue light from white light.
- Green color results if red and blue light are absent from white light.
- Blue color results if red and green light are absent.

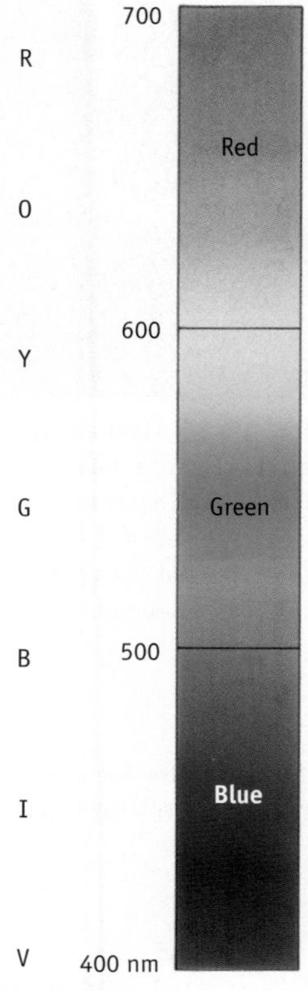

The ROY G BIV spectrum of colors of visible light. The colors used in printing this book are cyan, magenta, yellow, and black. The blue in ROY G BIV is actually cyan, according to color industry standards. Magenta doesn't have its own wavelength region. Rather, it is a mixture of blue and red.

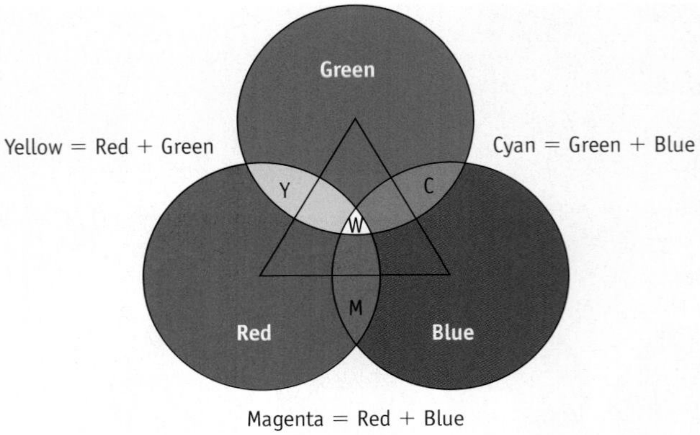

FIGURE 22.27 Using color disks to analyze colors. The three primary colors are red, green, and blue. Adding light of two primary colors gives the secondary colors yellow (= red + green), cyan (= green + blue), and magenta (= red + blue). Adding all three primary colors results in white light.

Charles D. Winters

FIGURE 22.28 Light absorption and color. The color of a solution is due to the color of the light *not* absorbed by the solution. Here, a solution of Ni^{2+} ion in water absorbs red and blue light and so appears green. See also Figures 4.15–4.19.

The secondary colors are rationalized similarly. Absorption of blue light gives yellow (the color across from it in Figure 22.27); absorption of red light results in cyan; and absorption of green light results in magenta.

Now we can apply these ideas to explain colors in transition metal complexes. Focus on what kind of light is *absorbed.* A solution of $[Ni(H_2O)_6]^{2+}$ is green. Green light is the result of removing red and blue light from white light. As white light passes through an aqueous solution of Ni^{2+}, red and blue light are absorbed, and green light is allowed to pass (Figure 22.28). Similarly, the $[Co(NH_3)_6]^{3+}$ ion is yellow because blue light has been absorbed and red and green light pass through.

The Spectrochemical Series

Recall that atomic spectra are obtained when electrons are excited from one energy level to another (◀ Section 6.3). The energy of the light absorbed or emitted is related to the energy levels of the atom or ion under study. The concept that light is absorbed when electrons move from lower to higher energy levels applies to all substances, not just atoms. It is the basic premise for the absorption of light for transition metal coordination complexes.

■ **Spectrophotometry** See Section 4.8 for a description of spectrophotometry.

In coordination complexes, the splitting between *d* orbitals often corresponds to the energy of visible light, so light in the visible region of the spectrum is absorbed when electrons move from a lower-energy *d* orbital to a higher-energy *d* orbital. This change, as an electron moves between two orbitals having different energies in a complex, is called a ***d*-to-*d* transition.** Qualitatively, such a transition for $[Co(NH_3)_6]^{3+}$ might be represented using an energy-level diagram such as that shown here.

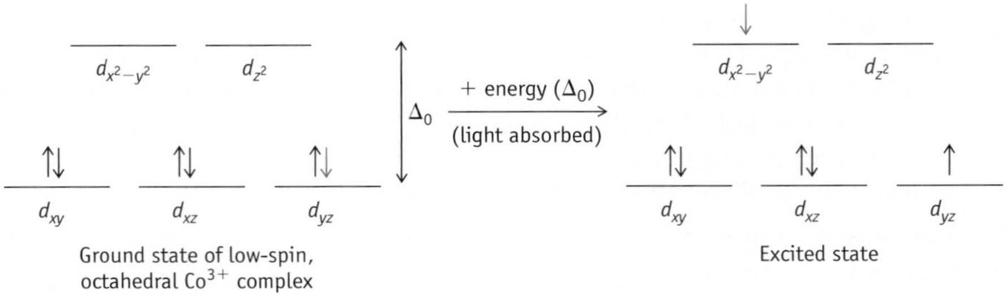

Experiments with coordination complexes reveal that, for a given metal ion, some ligands cause a small energy separation of the *d* orbitals, whereas others cause a large separation. In other words, some ligands create a small ligand field, and others create a large one. An example is seen in the spectroscopic data for several cobalt(III) complexes presented in Table 22.3.

- Both $[Co(NH_3)_6]^{3+}$ and $[Co(en)_3]^{3+}$ are yellow-orange, because they absorb light in the blue portion of the visible spectrum. These compounds have very similar spectra, to be expected because both have six amine-type donor atoms (H—NH_2 or R—NH_2).
- Although $[Co(CN)_6]^{3-}$ does not have an absorption band in the visible region, it is pale yellow. Light absorption occurs in the ultraviolet region, but the absorption is broad and extends minimally into the visible (blue) region.
- $[Co(C_2O_4)_3]^{3-}$ and $[Co(H_2O)_6]^{3+}$ have similar absorptions, in the yellow and violet regions. Their colors are shades of green with a small difference due to the relative amount of light of each color being absorbed.

The absorption maxima among the listed complexes range from 700 nm for $[CoF_6]^{3-}$ to 310 nm for $[Co(CN)_6]^{3-}$. The ligands change from member to member of this series, and we can conclude that the energy of the light absorbed by the complex is related to the different ligand field splittings, Δ_o, caused by the different ligands. Fluoride ion causes the smallest splitting of the *d* orbitals among the complexes listed in Table 22.3, whereas cyanide causes the largest splitting.

Spectra of complexes of other metals provide similar results. Based on this information, ligands can be listed in order of their ability to split the *d* orbitals. This list is called the **spectrochemical series** because it was determined by spectroscopy. A short list, with some of the more common ligands, follows:

$$F^-, Cl^-, Br^-, I^- < C_2O_4^{2-} < H_2O < NH_3 = en < phen < CN^-$$

small orbital splitting
small Δ_o

large orbital splitting
large Δ_o

The spectrochemical series is applicable to a wide range of metal complexes. Indeed, the ability of ligand field theory to explain the differences in the colors of the transition metal complexes is one of the strengths of this theory.

Based on the relative position of a ligand in the series, predictions can be made about a compound's magnetic behavior. Recall that d^4, d^5, d^6, and d^7 complexes can be high or low spin, depending on the ligand field splitting, Δ_o. Complexes

TABLE 22.3 The Colors of Some Co^{3+} Complexes*

Complex Ion	Wavelength of Light Absorbed (nm)	Color of Light Absorbed	Color of Complex
$[CoF_6]^{3-}$	700	Red	Green
$[Co(C_2O_4)_3]^{3-}$	600, 420	Yellow, violet	Dark green
$[Co(H_2O)_6]^{3+}$	600, 400	Yellow, violet	Blue-green
$[Co(NH_3)_6]^{3+}$	475, 340	Blue, ultraviolet	Yellow-orange
$[Co(en)_3]^{3+}$	470, 340	Blue, ultraviolet	Yellow-orange
$[Co(CN)_6]^{3-}$	310	Ultraviolet	Pale yellow

*The complex with fluoride ion, $[CoF_6]^{3-}$, is high spin and has one absorption band. The other complexes are low spin and have two absorption bands. In all but one case, one of these absorptions occurs in the visible region of the spectrum. The wavelengths are measured at the top of that absorption band.

Problem Solving Tip 22.1 Ligand Field Theory

This summary of the concepts of ligand field theory may help you to keep the broader picture in mind.

- Ligand–metal bonding results from the electrostatic attraction of a metal cation and an anion or polar molecule.
- The ligands define a coordination geometry. Common geometries are linear (coordination number = 2), tetrahedral and square planar (coordination number = 4), and octahedral (coordination number = 6).
- The placement of the ligands around the metal causes the *d* orbitals on the metal to have different energies. In an octahedral complex, for example, the *d* orbitals divide into two groups: a higher-energy group ($d_{x^2-y^2}$ and d_{z^2}) and a lower-energy group (d_{xy}, d_{xz}, and d_{yz}).
- Electrons are placed in the metal *d* orbitals in a manner that leads to the lowest total energy. Two competing features determine the placement: the relative energy of the sets of orbitals and the electron pairing energy.
- For the electron configurations d^4, d^5, d^6, and d^7, two electron configurations are possible: high spin, which occurs when the orbital splitting is small, and low spin, which occurs with a large orbital splitting. To determine whether a complex is high spin or low spin, measure its magnetism to determine the number of unpaired electrons.
- The *d*-orbital splitting (the energy difference between the metal *d*-orbital energies) often corresponds to the energy associated with visible light. As a consequence, many metal complexes absorb visible light and thus are colored.

formed with ligands near the left end of the spectrochemical series are expected to have small Δ_0 values and, therefore, are likely to be high spin. In contrast, complexes with ligands near the right end are expected to have large Δ_0 values and low-spin configurations. The complex $[CoF_6]^{3-}$ is high spin, whereas $[Co(NH_3)_6]^{3+}$ and the other complexes in Table 22.3 are low spin.

Chemistry Now™

Sign in at **www.thomsonedu.com/login** and go to Chapter 22 Contents to see Screen 22.8 for a simulation on **the absorption and transmission of light by transition metal complexes.**

EXAMPLE 22.6 Spectrochemical Series

Problem An aqueous solution of $[Fe(H_2O)_6]^{2+}$ is light blue-green. Do you expect the d^6 Fe^{2+} ion in this complex to have a high- or low-spin configuration? How would you make this determination by conducting an experiment?

Strategy Use the color wheel in Figure 22.27. The color of the complex, blue-green, tells us what kind of light is transmitted (blue and green), from which we learn what kind of light has been absorbed (red). Red light is at the low-energy end of the visible spectrum. From this fact, we can predict that the *d*-orbital splitting must be small. Our answer to the question derives from that conclusion.

Solution The low energy of the light absorbed suggests that $[Fe(H_2O)_6]^{2+}$ is likely to be a high-spin complex.

If the complex is high spin, it will have four unpaired electrons and be paramagnetic; if it is low spin, it will have no unpaired electrons and be diamagnetic. Identifying the presence of four unpaired electrons by measuring the compound's magnetism can be used to verify the high-spin configuration experimentally.

22.7 Organometallic Chemistry: The Chemistry of Low-Valent Metal–Organic Complexes

One of the largest and most active areas of chemistry over the past half-century has been the field of organometallic chemistry, the study of molecules having metal–carbon bonds. Thousands of such compounds have been made and characterized, and much of the activity has involved transition metals. Many have unique bonding modes and structures. In recent years, organometallic compounds have also found

Case Study — Accidental Discovery of a Chemotherapy Agent

There are many naturally occurring metal-based molecules such as heme, vitamin B_{12}, and the enzyme involved in fixing nitrogen (nitrogenase). Chemists have also synthesized various metal-based compounds for medical purposes. One of these, *cisplatin* $[PtCl_2(NH_3)_2]$, was known for many years, but it was discovered serendipitously to be effective in treatment of certain kinds of cancers.

In 1965, Barnett Rosenberg, a biophysicist at Michigan State University, set out to study the effect of electric fields on living cells, but the results of his experiments were very different from his expectations. He and his students had placed an aqueous suspension of live *Escherichia coli* bacteria in an electric field between supposedly inert platinum electrodes. Much to their surprise, they found that cell growth was significantly affected. After careful experimentation, the effect on cell division was found to be due to a trace of the complex $PtCl_4(NH_3)_2$ formed by an electrolytic process involving the platinum electrode in the presence of ammonia in the growth medium.

To follow up on this interesting discovery, Rosenberg and his students tested the effect of *cis*- and *trans*-$PtCl_2(NH_3)_2$ on cell growth and found that only the *cis* isomer was effective. This led Rosenberg and others to study the effect of so-called *cisplatin* on cancer cell growth, and the result is that cisplatin and similar compounds are now used to treat genitourinary tumors. In fact, testicular cancer in particular is now considered largely curable because of cisplatin chemotherapy.

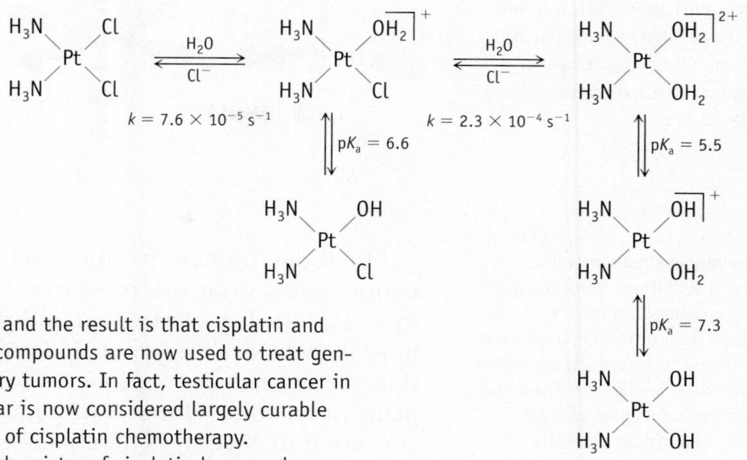

The chemistry of cisplatin has now been thoroughly studied and illustrates many of the principles of transition metal coordination chemistry. It has been found that cisplatin has a half-life of 2.5 h for the replacement of a Cl^- ligand by water at 310 K (in a first order reaction) and that the replacement of a second Cl^- ligand by water is slightly faster.

The aqua species are acidic and damaging to the kidneys, so cisplatin is generally used in a saline solution to prevent the hydrolysis reactions. It has been found that, in blood plasma at pH 7.4 and with a Cl^- ion concentration of about 1.04×10^{-5} M, $PtCl_2(NH_3)_2$ and $PtCl(OH)(NH_3)_2$ are the dominant species. In the cell nucleus, however, the Cl^- ion concentration is lower, and the aqua species are present in higher concentration.

Questions:

1. *If a patient is given 10.0 mg of cisplatin, what quantity remains as cisplatin at 24 hours?*
2. *At a pH of 7.4, what is the ratio of concentrations of $[PtCl(NH_3)_2(H_2O)]^+$ and $PtCl(NH_3)_2(OH)$?*

Answers to these questions are in Appendix Q.

particularly widespread use as reagents in organic synthesis and as catalysts for economically important chemical reactions.

Carbon Monoxide Complexes of Metals

In the earlier discussion of hemoglobin (page 1033), you learned that the iron in this biologically important complex binds not only to O_2 but also to CO. Our understanding of metal–CO complexes, however, emerged from the solution to a problem in industrial chemistry at the end of the 19th century.

The synthesis of sodium carbonate from NaCl, CO_2, NH_3, and H_2O by the Solvay process was an important chemical industry in the late 19th century (and remains so today in some parts of the world; page 974). In the late 1800s, a Solvay plant in England had a problem: the valves used to conduct the gaseous reactants and products rapidly corroded. A German chemist, Ludwig Mond (1839–1909), traced this problem to a small quantity of CO in the gas stream. Carbon monoxide gas reacted with nickel metal in the valves to form $Ni(CO)_4$ (Figure 22.29), tetracarbonyl nickel, a volatile liquid with a low boiling point (bp 47 °C). The deterioration of the valves occurred when the gaseous $Ni(CO)_4$ that formed was carried away in the effluent gas stream.

$$Ni(s) + 4\ CO(g) \rightleftharpoons Ni(CO)_4(g)$$

FIGURE 22.29 Metals carbonyls. $Ni(CO)_4$ is a tetrahedral molecule. Pentacarbonyliron(0) [$Fe(CO)_5$] is trigonal bipyramidal, whereas $Mo(CO)_6$ is octahedral, and the geometry around the manganese atom in $Mn_2(CO)_{10}$ is likewise octahedral (with the sixth position being an Mn—Mn bond).

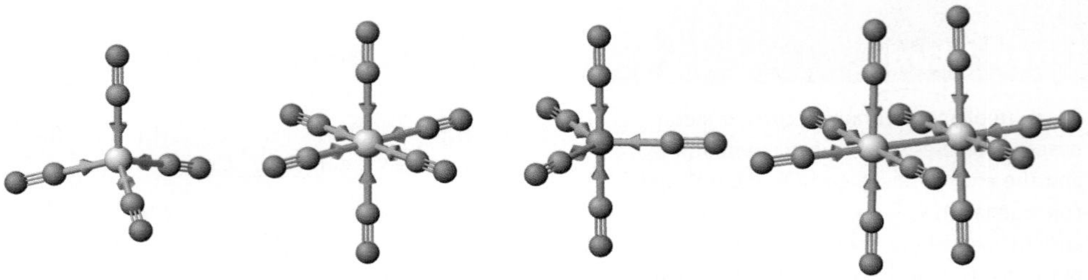

■ **Terminology in Organometallic Chemistry** The standard terminology of coordination chemistry applies to metal carbonyls. In tetracarbonylnickel(0), for example, the metal is formally zerovalent, and CO molecules are ligands. The metal has a coordination number of 4 and a tetrahedral coordination geometry.

Mond determined that the reaction forming $Ni(CO)_4$ is reversible; at moderate temperatures or at low pressures, CO is liberated, and nickel metal reforms. He then exploited the formation and decomposition reactions in a process to obtain pure nickel. Nickel and cobalt are generally found together in nature. Because of this, the two metals are obtained together when ores are refined, and they are difficult to separate. However, if a mixture of the metals is exposed to CO, nickel is converted to $Ni(CO)_4$ whereas cobalt is unchanged. Volatile $Ni(CO)_4$ is easily separated from solid cobalt metal. The $Ni(CO)_4$ can then be decomposed to give pure nickel. The Mond process, as it is now known, was the preferred procedure to obtain pure nickel for the first half of the 20th century.

Nickel is unique among metals in its facile reaction with CO, but many other transition metal carbonyl compounds are now known and can be synthesized by a variety of procedures. One of the most common is reductive carbonylation, in which a metal salt is reduced in the presence of CO, usually under high pressure. In effect, reduction of the metal salt gives metal atoms. Before these very reactive atoms aggregate to form the unreactive bulk metal, however, they react with CO. Metal carbonyls of most of the transition metals can be made by this route. Simple examples include hexacarbonyls of the Group 6B metals [$Cr(CO)_6$, $Mo(CO)_6$, and $WCO)_{6}$,] as well as $Fe(CO)_5$ and $Mn_2(CO)_{10}$ (Figure 22.29).

The Effective Atomic Number Rule and Bonding in Organometallic Compounds

An important observation has guided researchers studying these metal–CO complexes and other organometallic compounds. The **effective atomic number (EAN) rule,** now often referred to as the **18-electron rule,** states that compounds in which the sum of the metal valence electrons plus the electrons donated by the ligand groups totals 18 are likely to be stable. Thus, the 18-electron rule predicts the stoichiometry of a number of compounds. Iron(0), for example, which has eight valence electrons, would be expected to coordinate to five CO ligands. Each CO donates two electrons for a total of 10; adding these to the eight valence electrons on the metal gives 18 electrons around the Fe atom (Figure 22.30). We now recognize that the 18-electron rule is similar to the octet rule in main group chemistry, in that it defines the likely number of bonds to an element.

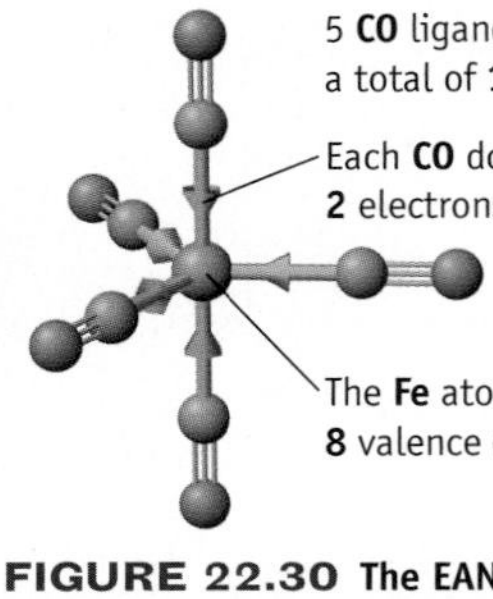

FIGURE 22.30 The EAN rule and $Fe(CO)_5$. The EAN or 18-electron rule states that stable organometallic compounds frequently have 18 valence electrons around the central metal. (There are also many 16-electron molecules, particularly of the heavier transition metals. See Study Question 22.34.) Here, the zerovalent Fe atom has the configuration $[Ar]3d^64s^2$, so eight valence electrons are available for bonding.

The bonding in traditional coordination compounds is described as being due to attractive forces between a positively charged metal ion and a polar molecule or an anion, and the properties of these species are in accord with substantial ionic character to the metal-ligand bond. In metal carbonyls, the zerovalent metal lacks a charge, and CO is only slightly polar. Both features argue against ionic bonding. The best model for these species instead describes the bonding between

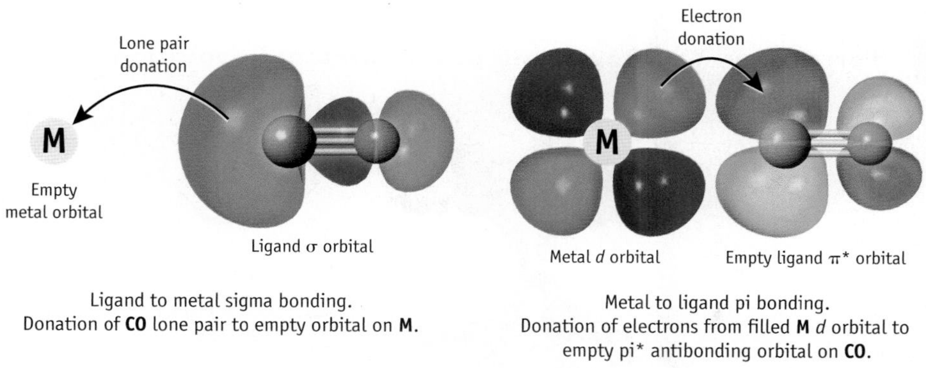

FIGURE 22.31 Bonding in metal carbonyls. The current understanding is that the CO ligand donates a lone pair of electrons to the low-valent metal to form a sigma bond. The electron-rich metal then donates electrons from a *d* orbital to the antibonding π^* orbital of the CO. There is a "synergistic" effect; the sigma and pi bonds complement each other.

the metal and CO as covalent. Each CO donates a C-atom lone pair to the metal atom to form a sigma bond (Figure 22.31). However, carbon monoxide is a very poor donor, and this alone would not lead to these species being stable. Rather, in conjunction with the sigma bond, the electron-rich metal donates a pair of electrons to form a π bond formed by overlap of a *d* orbital of the metal and a π^* antibonding orbital of CO. The latter interaction is described as $d\pi$–$p\pi$ bonding. Based on this model of bonding, carbon monoxide in these compounds is best described as a σ donor and π acceptor ligand.

Ligands in Organometallic Compounds

The bonding model for metal–CO complexes leads to important conclusions: only metals of low charge and with filled or partially filled *d* orbitals can form stable bonds to CO, and only ligands capable of forming π bonds with the metal are capable of forming low-valent metal compounds. Thus, this area is dominated by low-valent metals and special types of ligands that are capable of π bonding.

Some of the most common ligands in coordination chemistry cannot engage in π bonding and thus cannot form low-valent compounds with transition metals. For example, low-valent metal complexes with NH_3 or amines (such as $N(CH_3)_3$) do not exist, but phosphine complexes of zerovalent metals are well known. Phosphines, such as $P(CH_3)_3$, are the phosphorus analogs of amines and so have a lone pair of electrons on the P atom. Thus, phosphines can donate this lone pair to a metal to form a sigma bond. In addition, phosphorus atoms have empty 3*d* orbitals, and these can form a $d\pi$–$p\pi$ bond with the filled *d*-orbitals of a low-valent metal. Ammonia, with no empty valence orbitals to overlap with a filled metal *d* orbitals, cannot form such π bonds.

Yet another class of molecules that can serve as ligands in organometallic compounds are organic species such as ethylene (C_2H_4) and benzene (C_6H_6). For example, in the anion of Zeise's salt, ethylene binds to a Pt^{2+} ion through donation of the two π electrons of the double bond (Figure 22.32). As in metal carbonyls, there is also a metal-to-ligand π bond formed by overlap of filled metal *d* orbitals with the antibonding π orbitals from the ligand. This combination of bonding modes strengthens the metal-ligand bond.

Benzene can be thought of as a tridentate ligand capable of donating three π electron pairs to a metal atom (which then donates *d* electrons back to the ligand in a π-type interaction). Such molecules also obey the 18-electron rule. For example, in dibenzenechromium (Figure 22.33), the Cr(0) atom with six valence electrons is bound to two ligands, each donating six electrons.

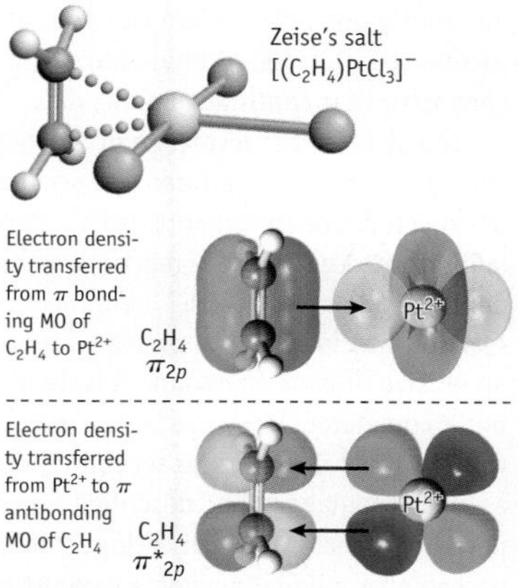

FIGURE 22.32 Bonding in the anion of Zeise's salt.

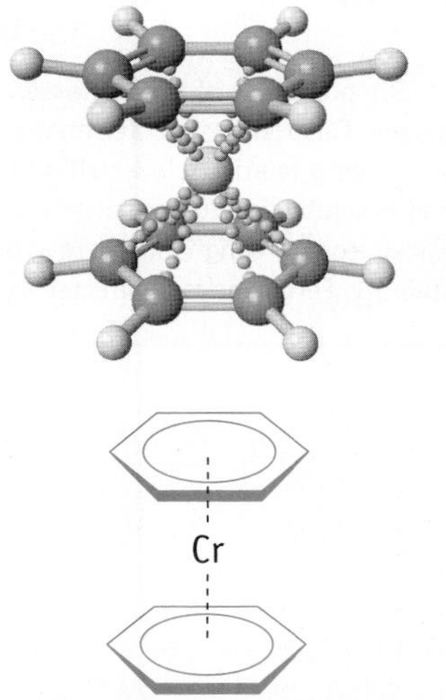

FIGURE 22.33 Dibenzenechromium. Two views of the molecule. Top: a computer-generated model. Bottom: a line drawing typically used by chemists.

Case Study — Ferrocene—The Beginning of a Chemical Revolution

Unexpected discoveries open up new areas of science. One example was the synthesis of compounds of Xe (page 405), a result that destroyed the myth that the noble gases were unreactive and led to a rich chemistry of these elements. At approximately the same time, another discovery, the synthesis of ferrocene, $Fe(C_5H_5)_2$, set the stage for the rapid and exciting growth of organometallic chemistry, the chemistry of compounds containing metal–carbon bonds. This destroyed another common myth of the time, that metal–carbon bonds are inherently unstable. Xenon compounds and ferrocene spearheaded a renaissance in inorganic chemistry that continues to this day.

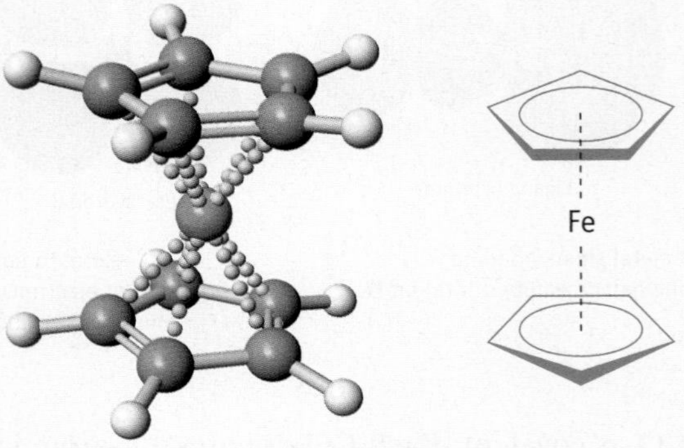

The structure of ferrocene, $Fe(\eta^5\text{-}C_5H_5)_2$

The synthesis of ferrocene in two quite dissimilar research laboratories was accidental and unexpected. The first report, from an academic laboratory, described the reaction of the cyclopentadienide anion, $[C_5H_5]^-$ with $FeCl_2$. This reaction was intended to provide a precursor to an elusive organic compound fulvalene, $(C_5H_4)_2$ but a completely new substance—ferrocene—was obtained instead. The second report, from an industrial laboratory, described a high temperature process in which cyclopentadiene was passed over a liquid ammonia catalyst that contained, among other things, iron(II) oxide.

$$2\ C_5H_5MgCl + FeCl_2 \rightarrow Fe(\eta^5\text{-}C_5H_5)_2 + MgCl_2$$
(in diethyl ether)

$$2\ C_5H_6(g) + FeO(s) \rightarrow Fe(\eta^5\text{-}C_5H_5)_2(s) + H_2O(g)$$
(at high temperature)

The properties of ferrocene were unexpected. Ferrocene is a diamagnetic orange solid, has a relatively low melting (mp 173 °C), and is soluble in organic solvents but not in water. Most striking is its thermal and oxidative stability. Ferrocene is unaffected by oxygen, water, aqueous bases and nonoxidizing acids under ambient conditions, and can be heated to over 450 °C. These observations contradicted conventional wisdom that held that metal–carbon bonds were weak and unstable, hard to make, and reactive.

The structure of ferrocene was quickly established by x-ray crystallography. Like dibenzenechromium (Figure 22.23), ferrocene is a molecular "sandwich" compound, containing an iron atom sandwiched between two planar hydrocarbon rings (Figure). Iron(II), a d^6 metal ion, is in its low spin configuration. The cyclopentadienyl anions can be thought of as tridentate ligands with the six π electrons (three pairs) of each organic ring being donated to the metal ion.

Once the structure of ferrocene was known, the race was on to make other complexes with unsaturated hydrocarbon ligands. Synthesis of cyclopentadienyl compounds of other metals soon followed, and now hundreds are known in combination with CO, benzene, ethylene, and many other carbon-containing compounds. The discovery of ferrocene was one of the keys that unlocked the field of organometallic chemistry.

Questions:

1. *Rationalize ferrocene's diamagnetism.*
2. *Dibenzenechromium is also a "sandwich" compound. What is the oxidation state of chromium in this compound? Is it diamagnetic or paramagnetic?*
3. *Do ferrocene and dibenzenechromium obey the 18-electron rule (page 1048)?*
4. *Ferrocene can be oxidized to form the ferrocenium cation, $[Fe(\eta\text{-}C_5H_5)_2]^+$. The standard reduction potential for the ferrocenium–ferrocene half-reaction is 0.400 V. Identify several oxidizing agents in Appendix M that are capable of oxidizing ferrocene. Is elemental chlorine, Cl_2, a sufficiently strong oxidizing agent to carry out this oxidation?*
5. *Write an equation for one way to synthesize nickelocene, $Ni(\eta^5\text{-}C_5H_5)_2$. Nickelocene is paramagnetic. Predict the number of unpaired electrons in this compound.*

Answers to these questions are in Appendix Q.

Dibenzenechromium is one of many organometallic compounds often referred to as "sandwich" compounds. (See *Case Study: Ferrocene.*) These are molecular compounds in which a low-valent metal atom is "sandwiched" between two organic ligands. In addition, they are often referred to as π complexes because the π electrons of the ligand are involved in bonding. Modern terminology uses the Greek letter eta (η) to indicate this type of attachment, so dibenzenechromium is properly symbolized as $Cr(\eta^6\text{-}C_6H_6)_2$ (where the superscript 6 indicates the number of carbon atoms involved in bonding on each ligand).

EXAMPLE 22.7 The Effective Atomic Number Rule

Problem Show that each of the following molecules or ions satisfies the EAN rule.

(a) $[Fe(CO)_2](\eta^5\text{-}C_5H_5)]^-$

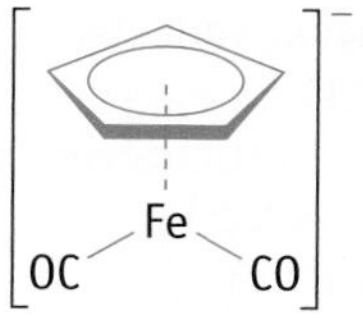

(b) $[Mn(CO)_5]^-$

(c) $Co(C_2H_4)_2(\eta^5\text{-}C_5H_5)$

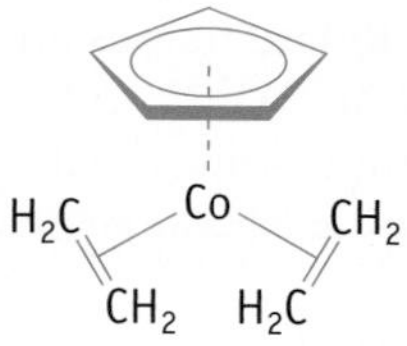

Strategy A complex obeys the EAN rule if the total number of electrons around the metal (valence electrons from the metal itself + electrons donated by the ligands) equals 18. Recognize that CO and C_2H_4 are both two-electron donors and that $C_5H_5^-$ is a six-electron donor. For the metal center, take its total number of valence electrons, and add or subtract electrons as necessary to adjust for negative or positive charges.

Solution

(a) The overall charge is 1−, which is equal to the charge on the $C_5H_5^-$ group. Thus, the Fe atom must have no charge.

6 electrons for $C_5H_5^-$ +8 electrons for Fe + 4 electrons for 2 CO groups = 18 electrons

(b) Because CO is neutral the Mn center must have a negative charge. This means the manganese center has effectively eight valence electrons. Together with the five CO groups, each donating two electrons, the total is 18 electrons.

(c) This is a neutral molecule, so the negative charge on the $C_5H_5^-$ group must be balanced by a positive charge on Co. The Co^+ ion has eight valence electrons. Each C_2H_4 molecule donates two electrons.

6 electrons for $C_5H_5^-$ +8 electrons for Co^+ +4 electrons for 2 C_2H_4 molecules = 18 electrons

EXERCISE 22.5 The Effective Atomic Number Rule

Show that the two molecules below satisfy the EAN rule.

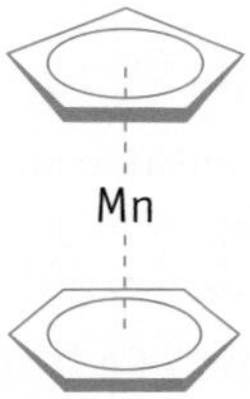

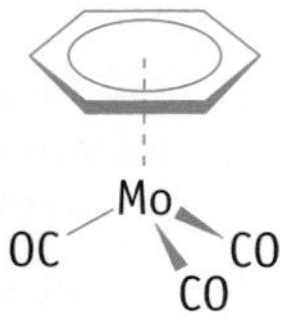

Chapter Goals Revisited

Chemistry•Now™ Sign in at **www.thomsonedu.com/login** to:

- Assess your understanding with Study Questions in OWL keyed to each goal in the Goals and Homework menu for this chapter
- For quick review, download Go Chemistry mini-lecture flashcard modules (or purchase them at **www.ichapters.com**)
- Check your readiness for an exam by taking the Pre-Test and exploring the modules recommended in your Personalized Study plan.

Access **How Do I Solve It?** tutorials on how to approach problem solving using concepts in this chapter.

Now that you have studied this chapter, you should ask whether you have met the chapter goals. In particular, you should be able to:

Identify and explain the chemical and physical properties of the transition elements

a. Identify the general classes of transition elements (Section 22.1).

b. Identify the transition metals from their symbols and positions in the periodic table, and recall some physical and chemical properties (Section 22.1).

c. Understand the electrochemical nature of corrosion (Section 22.1).

d. Describe the metallurgy of iron and copper (Section 22.2).

Understand the composition, structure, and bonding in coordination compounds

a. Given the formula for a coordination complex, identify the metal and its oxidation state, the ligands, the coordination number and coordination geometry, and the overall charge on the complex (Section 22.3). Relate names and formulas of complexes. **Study Question(s) assignable in OWL: 18, 20, 48, 57.**

b. Given the formula for a complex, be able to recognize whether isomers will exist, and draw their structures (Section 22.4). **Study Question(s) assignable in OWL: 28, 40, 48, 50.**

c. Describe the bonding in coordination complexes (Section 22.5).

d. Apply the principles of stoichiometry, thermodynamics, and equilibrium to transition metal compounds. **Study Question(s) assignable in OWL: 42, 55, 59, 64, 67, 68, 70.**

Relate ligand field theory to the magnetic and spectroscopic properties of complexes

a. Understand why substances are colored (Section 22.6). **Study Question(s) assignable in OWL: 32.**

b. Understand the relationship between the ligand field splitting, magnetism, and color of complexes (Section 22.6). **Study Question(s) assignable in OWL: 26, 28, 59, 60, 63.**

Apply the effective atomic number rule to simple organometallic compounds of the transition metals.

a. Apply the EAN rule to molecules containing a low-valent metal and ligands such as C_6H_6, C_2H_4, and CO.

STUDY QUESTIONS

Online homework for this chapter may be assigned in OWL.

▲ denotes challenging questions.

■ denotes questions assignable in OWL.

Blue-numbered questions have answers in Appendix O and fully-worked solutions in the *Student Solutions Manual.*

Practicing Skills

Properties of Transition Elements

(See Section 22.1 and Example 7.3.)

1. Give the electron configuration for each of the following ions, and tell whether each is paramagnetic or diamagnetic.
   (a) Cr^{3+} (b) V^{2+} (c) Ni^{2+} (d) Cu^+

2. Identify two transition metal cations with the following electron configurations.
   (a) $[Ar]3d^6$ (b) $[Ar]3d^{10}$ (c) $[Ar]3d^5$ (d) $[Ar]3d^8$

3. Identify a cation of a first series transition metal that is isoelectronic with each of the following.
   (a) Fe^{3+} (b) Zn^{2+} (c) Fe^{2+} (d) Cr^{3+}

4. Match up the isoelectronic ions on the following list.

   Cu^+ Mn^{2+} Fe^{2+} Co^{3+} Fe^{3+} Zn^{2+} Ti^{2+} V^{3+}

5. The following equations represent various ways of obtaining transition metals from their compounds. Balance each equation.
(a) $Cr_2O_3(s) + Al(s) \longrightarrow Al_2O_3(s) + Cr(s)$
(b) $TiCl_4(\ell) + Mg(s) \longrightarrow Ti(s) + MgCl_2(s)$
(c) $[Ag(CN)_2]^-(aq) + Zn(s) \longrightarrow Ag(s) + [Zn(CN)_4]^{2-}(aq)$
(d) $Mn_3O_4(s) + Al(s) \longrightarrow Mn(s) + Al_2O_3(s)$

6. Identify the products of each reaction, and balance the equation.
(a) $CuSO_4(aq) + Zn(s) \longrightarrow$
(b) $Zn(s) + HCl(aq) \longrightarrow$
(c) $Fe(s) + Cl_2(g) \longrightarrow$
(d) $V(s) + O_2(g) \longrightarrow$

Formulas of Coordination Compounds

(See Examples 22.1 and 22.2.)

7. Which of the following ligands is expected to be monodentate, and which might be polydentate?
(a) CH_3NH_2
(b) CH_3CN
(c) N_3^-
(d) en
(e) Br^-
(f) phen

8. One of the following nitrogen compounds or ions is not capable of serving as a ligand: NH_4^+, NH_3, NH_2^-. Identify this species, and explain your answer.

9. Give the oxidation number of the metal ion in each of the following compounds.
(a) $[Mn(NH_3)_6]SO_4$
(b) $K_3[Co(CN)_6]$
(c) $[Co(NH_3)_4Cl_2]Cl$
(d) $Cr(en)_2Cl_2$

10. Give the oxidation number of the metal ion in each of the following complexes.
(a) $[Fe(NH_3)_6]^{2+}$
(b) $[Zn(CN)_4]^{2-}$
(c) $[Co(NH_3)_5(NO_2)]^+$
(d) $[Cu(en)_2]^{2+}$

11. Give the formula of a complex constructed from one Ni^{2+} ion, one ethylenediamine ligand, three ammonia molecules, and one water molecule. Is the complex neutral or is it charged? If charged, give the charge.

12. Give the formula of a complex constructed from one Cr^{3+} ion, two ethylenediamine ligands, and two ammonia molecules. Is the complex neutral or is it charged? If charged, give the charge.

Naming Coordination Compounds

(See Example 22.3.)

13. Write formulas for the following ions or compounds.
(a) dichlorobis(ethylenediamine)nickel(II)
(b) potassium tetrachloroplatinate(II)
(c) potassium dicyanocuprate(I)
(d) tetraamminediaquairon(II)

14. Write formulas for the following ions or compounds.
(a) diamminetriaquahydroxochromium(II) nitrate
(b) hexaammineiron(III) nitrate
(c) pentacarbonyliron(0) (where the ligand is CO)
(d) ammonium tetrachlorocuprate(II)

15. Name the following ions or compounds.
(a) $[Ni(C_2O_4)_2(H_2O)_2]^{2-}$
(b) $[Co(en)_2Br_2]^+$
(c) $[Co(en)_2(NH_3)Cl]^{2+}$
(d) $Pt(NH_3)_2(C_2O_4)$

16. Name the following ions or compounds.
(a) $[Co(H_2O)_4Cl_2]^+$
(b) $Co(H_2O)_3F_3$
(c) $[Pt(NH_3)Br_3]^-$
(d) $[Co(en)(NH_3)_3Cl]^{2+}$

17. Give the name or formula for each ion or compound, as appropriate.
(a) pentaaquahydroxoiron(III) ion
(b) $K_2[Ni(CN)_4]$
(c) $K[Cr(C_2O_4)_2(H_2O)_2]$
(d) ammonium tetrachloroplatinate(II)

18. ■ Give the name or formula for each ion or compound, as appropriate.
(a) tetraaquadichlorochromium(III) chloride
(b) $[Cr(NH_3)_5SO_4]Cl$
(c) sodium tetrachlorocobaltate(II)
(d) $[Fe(C_2O_4)_3]^{3-}$

Isomerism

(See Example 22.4.)

19. Draw all possible geometric isomers of the following.
(a) $Fe(NH_3)_4Cl_2$
(b) $Pt(NH_3)_2(SCN)(Br)$ (SCN^- is bonded to Pt^{2+} through S)
(c) $Co(NH_3)_3(NO_2)_3$ (NO_2^- is bonded to Co^{3+} through N)
(d) $[Co(en)Cl_4]^-$

20. ■ In which of the following complexes are geometric isomers possible? If isomers are possible, draw their structures and label them as *cis* or *trans*, or as *fac* or *mer*.
(a) $[Co(H_2O)_4Cl_2]^+$
(b) $Co(NH_3)_3F_3$
(c) $[Pt(NH_3)Br_3]^-$
(d) $[Co(en)_2(NH_3)Cl]^{2+}$

21. Determine whether the following complexes have a chiral metal center.
(a) $[Fe(en)_3]^{2+}$
(b) trans-$[Co(en)_2Br_2]^+$
(c) *fac*-$[Co(en)(H_2O)Cl_3]$
(d) square-planar $Pt(NH_3)(H_2O)(Cl)(NO_2)$

22. Four geometric isomers are possible for $[Co(en)(NH_3)_2(H_2O)Cl]^+$. Draw the structures of all four. (Two of the isomers are chiral, meaning that each has a non-superimposable mirror image.)

Magnetic Properties of Complexes

(See Example 22.5.)

23. The following are low-spin complexes. Use the ligand field model to find the electron configuration of each ion. Determine which are diamagnetic. Give the number of unpaired electrons for the paramagnetic complexes.
(a) $[Mn(CN)_6]^{4-}$
(b) $[Co(NH_3)_6]Cl_3$
(c) $[Fe(H_2O)_6]^{3+}$
(d) $[Cr(en)_3]SO_4$

24. The following are high-spin complexes. Use the ligand field model to find the electron configuration of each ion, and determine the number of unpaired electrons in each.
(a) $K_4[FeF_6]$ (c) $[Cr(H_2O)_6]^{2+}$
(b) $[MnF_6]^{4-}$ (d) $(NH_4)_3[FeF_6]$

25. Determine the number of unpaired electrons in the following tetrahedral complexes. All tetrahedral complexes are high spin.
(a) $[FeCl_4]^{2-}$ (c) $[MnCl_4]^{2-}$
(b) $Na_2[CoCl_4]$ (d) $(NH_4)_2[ZnCl_4]$

26. ■ Determine the number of unpaired electrons in the following tetrahedral complexes. All tetrahedral complexes are high spin.
(a) $[Zn(H_2O)_4]^{2+}$ (c) $Mn(NH_3)_2Cl_2$
(b) $VOCl_3$ (d) $[Cu(en)_2]^{2+}$

27. For the high-spin complex $[Fe(H_2O)_6]SO_4$, identify the following:
(a) the coordination number of iron
(b) the coordination geometry for iron
(c) the oxidation number of iron
(d) the number of unpaired electrons
(e) whether the complex is diamagnetic or paramagnetic

28. ■ For the low-spin complex $[Co(en)(NH_3)_2Cl_2]ClO_4$, identify the following:
(a) the coordination number of cobalt
(b) the coordination geometry for cobalt
(c) the oxidation number of cobalt
(d) the number of unpaired electrons
(e) whether the complex is diamagnetic or paramagnetic
(f) Draw any geometric isomers.

29. The anion $[NiCl_4]^{2-}$ is paramagnetic, but when CN^- ions are added, the product, $[Ni(CN)_4]^{2-}$, is diamagnetic. Explain this observation.

$$\underset{\text{paramagnetic}}{[NiCl_4]^{2-}(aq)} + 4\,CN^-(aq) \longrightarrow \underset{\text{diamagnetic}}{[Ni(CN)_4]^{2-}(aq)} + 4\,Cl^-(aq)$$

30. An aqueous solution of iron(II) sulfate is paramagnetic. If NH_3 is added, the solution becomes diamagnetic. Why does the magnetism change?

Spectroscopy of Complexes

(See Example 22.6.)

31. In water, the titanium(III) ion, $[Ti(H_2O)_6]^{3+}$, has a broad absorption band at about 500 nm. What color light is absorbed by the ion?

32. ■ In water, the chromium(II) ion, $[Cr(H_2O)_6]^{2+}$, absorbs light with a wavelength of about 700 nm. What color is the solution?

Organometallic Compounds

(See Example 22.7.)

33. Show that the molecules and ions below satisfy the EAN rule.
(a) $[Mn(CO)_6]^+$ (b)

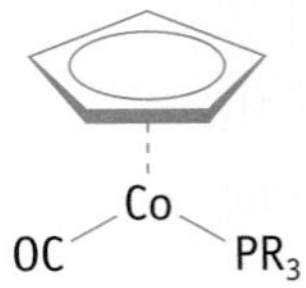

(c)

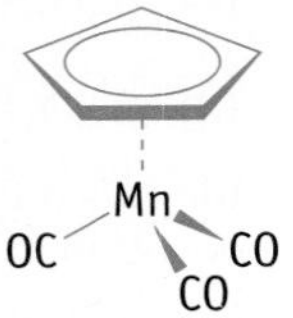

(In (b) PR_3 is a phosphine such as $P(C_6H_5)_3$, a two-electron donor ligand.)

34. Many organometallic compounds of the transition metals satisfy the 18-electron rule. However, there also are many other stable molecules that have only 16-valence electrons. These often have the capability of adding more ligands and function as catalysts in chemical reactions (see Study Question 71). (In the molecules below, PR_3 is a phosphine such as $P(C_6H_5)_3$, a two-electron donor ligand. For electron counting purposes, the CH_3 or methyl group is considered an anion, $(CH_3)^-$, and also a two-electron donor ligand.) Which molecules below have 18-valence electrons and which have 16?

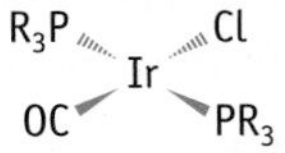

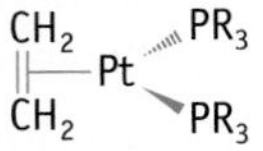

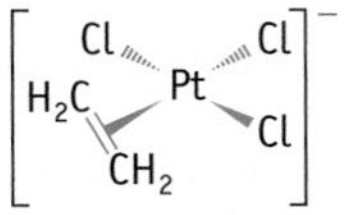

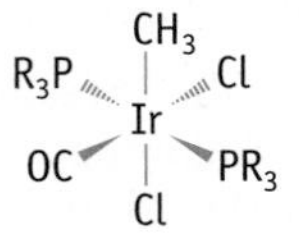

General Questions

These questions are not designated as to type or location in the chapter. They may contain several concepts.

35. Describe an experiment that would determine whether nickel in $K_2[NiCl_4]$ is square planar or tetrahedral.

36. Which of the following low-spin complexes has the greatest number of unpaired electrons?
(a) $[Cr(H_2O)_6]^{3+}$ (c) $[Fe(H_2O)_6]^{2+}$
(b) $[Mn(H_2O)_6]^{2+}$ (d) $[Ni(H_2O)_6]^{2+}$

37. How many unpaired electrons are expected for high-spin and low-spin complexes of Fe^{2+}?

38. Excess silver nitrate is added to a solution containing 1.0 mol of $[Co(NH_3)_4Cl_2]Cl$. What amount of AgCl (in moles) will precipitate?

39. Which of the following complexes is (are) square planar?
(a) $[Ti(CN)_4]^{2-}$ (c) $[Zn(CN)_4]^{2-}$
(b) $[Ni(CN)_4]^{2-}$ (d) $[Pt(CN)_4]^{2-}$

▲ more challenging ■ in OWL Blue-numbered questions answered in Appendix O

40. ■ Which of the following complexes containing the oxalate ion is (are) chiral?
(a) $[Fe(C_2O_4)Cl_4]^{2-}$
(b) *cis*-$[Fe(C_2O_4)_2Cl_2]^{2-}$
(c) *trans*-$[Fe(C_2O_4)_2Cl_2]^{2-}$

41. How many geometric isomers are possible for the square-planar complex $[Pt(NH_3)(CN)Cl_2]^-$?

42. ■ For a tetrahedral complex of a metal in the first transition series, which of the following statements concerning energies of the 3*d* orbitals is correct?
(a) The five *d* orbitals have the same energy.
(b) The $d_{x^2-y^2}$ and d_{z^2} orbitals are higher in energy than the d_{xz}, d_{yz}, and d_{xy} orbitals.
(c) The d_{xz}, d_{yz}, and d_{xy} orbitals are higher in energy than the $d_{x^2-y^2}$ and d_{z^2} orbitals.
(d) The *d* orbitals all have different energies.

43. A transition metal complex absorbs 425-nm light. What is its color?
(a) red
(b) green
(c) yellow
(d) blue

44. ■ For the low-spin complex $[Fe(en)_2Cl_2]Cl$, identify the following.
(a) the oxidation number of iron
(b) the coordination number for iron
(c) the coordination geometry for iron
(d) the number of unpaired electrons per metal atom
(e) whether the complex is diamagnetic or paramagnetic
(f) the number of geometric isomers

45. For the high-spin complex $Mn(NH_3)_4Cl_2$, identify the following.
(a) the oxidation number of manganese
(b) the coordination number for manganese
(c) the coordination geometry for manganese
(d) the number of unpaired electrons per metal atom
(e) whether the complex is diamagnetic or paramagnetic
(f) the number of geometric isomers

46. ■ A platinum-containing compound, known as Magnus's green salt, has the formula $[Pt(NH_3)_4][PtCl_4]$ (in which both platinum ions are Pt^{2+}). Name the cation and the anion.

47. Early in the 20th century, complexes sometimes were given names based on their colors. Two compounds with the formula $CoCl_3 \cdot 4\ NH_3$ were named praseocobalt chloride (*praseo* = green) and violio-cobalt chloride (violet color). We now know that these compounds are octahedral cobalt complexes and that they are *cis* and *trans* isomers. Draw the structures of these two compounds, and name them using systematic nomenclature.

48. ■ Give the formula and name of a square-planar complex of Pt^{2+} with one nitrite ion (NO_2^-, which binds to Pt^{2+} through N), one chloride ion, and two ammonia molecules as ligands. Are isomers possible? If so, draw the structure of each isomer, and tell what type of isomerism is observed.

49. Give the formula of the complex formed from one Co^{3+} ion, two ethylenediamine molecules, one water molecule, and one chloride ion. Is the complex neutral or charged? If charged, give the net charge on the ion.

50. ■ ▲ How many geometric isomers of the complex $[Cr(dmen)_3]^{3+}$ can exist? (dmen is the bidentate ligand 1,1-dimethylethylenediamine.)

$$(CH_3)_2\ddot{N}CH_2CH_2\ddot{N}H_2$$

1,1-Dimethylethylenediamine, dmen

51. ▲ Diethylenetriamine (dien) is capable of serving as a tridentate ligand.

$$H_2\ddot{N}CH_2CH_2\text{—}\underset{\displaystyle H}{\underset{|}{\ddot{N}}}\text{—}CH_2CH_2\ddot{N}H_2$$

Diethylenetriamine, dien

(a) Draw the structures of *fac*-$Cr(dien)Cl_3$ and *mer*-$Cr(dien)Cl_3$.
(b) Two different geometric isomers of *mer*-$Cr(dien)Cl_2Br$ are possible. Draw the structure for each.
(c) Three different geometric isomers are possible for $[Cr(dien)_2]^{3+}$. Two have the dien ligand in a *fac* configuration, and one has the ligand in a *mer* orientation. Draw the structure of each isomer.

52. From experiment, we know that $[CoF_6]^{3-}$ is paramagnetic and $[Co(NH_3)_6]^{3+}$ is diamagnetic. Using the ligand field model, depict the electron configuration for each ion, and use this model to explain the magnetic property. What can you conclude about the effect of the ligand on the magnitude of Δ_0?

53. Three geometric isomers are possible for $[Co(en)(NH_3)_2(H_2O)_2]^{3+}$. One of the three is chiral; that is, it has a non-superimposable mirror image. Draw the structures of the three isomers. Which one is chiral?

54. The square-planar complex $Pt(en)Cl_2$ has chloride ligands in a *cis* configuration. No *trans* isomer is known. Based on the bond lengths and bond angles of carbon and nitrogen in the ethylenediamine ligand, explain why the *trans* compound is not possible.

55. ■ The complex $[Mn(H_2O)_6]^{2+}$ has five unpaired electrons, whereas $[Mn(CN)_6]^{4-}$ has only one. Using the ligand field model, depict the electron configuration for each ion. What can you conclude about the effects of the different ligands on the magnitude of Δ_0?

56. Experiments show that $K_4[Cr(CN)_6]$ is paramagnetic and has two unpaired electrons. The related complex $K_4[Cr(SCN)_6]$ is paramagnetic and has four unpaired electrons. Account for the magnetism of each compound using the ligand field model. Predict where the SCN^- ion occurs in the spectrochemical series relative to CN^-.

57. ■ Give a systematic name or the formula for the following:
(a) $(NH_4)_2[CuCl_4]$
(b) $Mo(CO)_6$
(c) tetraaquadichlorochromium(III) chloride
(d) aquabis(ethylenediamine)thiocyanatocobalt(III) nitrate

58. When $CrCl_3$ dissolves in water, three different species can be obtained.
(a) $[Cr(H_2O)_6]Cl_3$, violet
(b) $[Cr(H_2O)_5Cl]Cl_2$, pale green
(c) $[Cr(H_2O)_4Cl_2]Cl$, dark green

If diethyl ether is added, a fourth complex can be obtained: $Cr(H_2O)_3Cl_3$ (brown). Describe an experiment that will allow you to differentiate these complexes.

59. ■ ▲ The complex ion $[Co(CO_3)_3]^{3-}$, an octahedral complex with bidentate carbonate ions as ligands, has one absorption in the visible region of the spectrum at 640 nm. From this information:
(a) Predict the color of this complex, and explain your reasoning.
(b) Is the carbonate ion as weak- or strong-field ligand?
(c) Predict whether $[Co(CO_3)_3]^{3-}$ will be paramagnetic or diamagnetic.

60. ■ The glycinate ion, $H_2NCH_2CO_2^-$, formed by deprotonation of the amino acid glycine, can function as a bidentate ligand, coordinating to a metal through the nitrogen of the amino group and one of the oxygen atoms.

Glycinate ion, a bidentate ligand

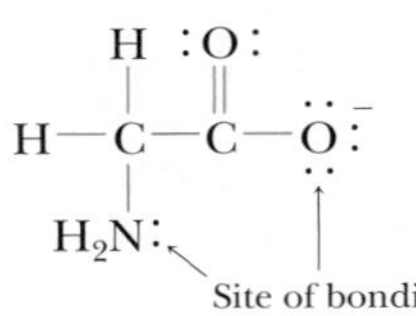

A copper complex of this ligand has the formula $Cu(H_2NCH_2CO_2)_2(H_2O)_2$. For this complex, determine the following.
(a) the oxidation state of copper
(b) the coordination number of copper
(c) the number of unpaired electrons
(d) whether the complex is diamagnetic or paramagnetic

61. ▲ Draw structures for the five possible geometric isomers of $Cu(H_2NCH_2CO_2)_2(H_2O)_2$. Are any of these species chiral? (See the structure of the ligand in Study Question 60.)

62. Chromium forms two anionic carbonyls, $[Cr(CO)_5]^{x-}$ and $[Cr(CO)_4]^{y-}$. What are the values of x and y?

63. ■ Nickel and palladium both form complexes of the general formula $M(PR_3)_2Cl_2$ (where PR_3 is a phosphine such as $P(C_6H_5)_3$, triphenylphosphine). The nickel(II) compound is paramagnetic whereas the palladium(II) compound is diamagnetic.
(a) Explain the magnetic properties of these compounds.
(b) How many isomers of each compound are expected?

64. ■ ▲ The transition metals form a class of compounds called metal carbonyls, an example of which is the tetrahedral complex $Ni(CO)_4$. Given the following thermodynamic data (at 298 K):

	$\Delta_f H°$ (kJ/mol)	$S°$ (J/K · mol)
Ni(s)	0	29.87
CO(g)	−110.525	+197.674
$Ni(CO)_4(g)$	−602.9	+410.6

(a) Calculate the equilibrium constant for the formation of $Ni(CO)_4(g)$ from nickel metal and CO gas.
(b) Is the reaction of Ni(s) and CO(g) product- or reactant-favored?
(c) Is the reaction more or less product-favored at higher temperatures? How could this reaction be used in the purification of nickel metal?

In the Laboratory

65. Two different coordination compounds containing one cobalt(III) ion, five ammonia molecules, one bromide ion, and one sulfate ion exist. The dark violet form (A) gives a precipitate upon addition of aqueous $BaCl_2$. No reaction is seen upon addition of aqueous $BaCl_2$ to the violet-red form (B). Suggest structures for these two compounds, and write a chemical equation for the reaction of (A) with aqueous $BaCl_2$.

66. Three different compounds of chromium(III) with water and chloride ion have the same composition: 19.51% Cr, 39.92% Cl, and 40.57% H_2O. One of the compounds is violet and dissolves in water to give a complex ion with a 3+ charge and three chloride ions. All three chloride ions precipitate immediately as AgCl on adding $AgNO_3$. Draw the structure of the complex ion, and name the compound. Write a net ionic equation for the reaction of this compound with silver nitrate.

67. ■ ▲ A 0.213-g sample of uranyl(VI) nitrate, $UO_2(NO_3)_2$, is dissolved in 20.0 mL of 1.0 M H_2SO_4 and shaken with Zn. The zinc reduces the uranyl ion, UO_2^{2+}, to a uranium ion, U^{n+}. To determine the value of n, this solution is titrated with $KMnO_4$. Permanganate is reduced to Mn^{2+} and U^{n+} is oxidized back to UO_2^{2+}.
    (a) In the titration, 12.47 mL of 0.0173 M $KMnO_4$ was required to reach the equivalence point. Use this information to determine the charge on the ion U^{n+}.
    (b) With the identity of U^{n+} now established, write a balanced net ionic equation for the reduction of UO_2^{2+} by zinc (assume acidic conditions).
    (c) Write a balanced net ionic equation for the oxidation of U^{n+} to UO_2^{2+} by MnO_4^- in acid.

68. ■ ▲ You have isolated a solid organometallic compound containing manganese, some number of CO ligands, and one or more CH_3 ligands. To find the molecular formula of the compound, you burn 0.225 g of the solid in oxygen and isolate 0.283 g of CO_2 and 0.0290 g of H_2O. The molar mass of the compound is 210 g/mol. Suggest a plausible formula and structure for the molecule. (Make sure it satisfies the EAN rule. The CH_3 group can be thought of as a CH_3^- ion and is a two-electron donor ligand.)

Summary and Conceptual Questions

The following questions may use concepts from this and previous chapters.

69. The stability of analogous complexes $[ML_6]^{n+}$ (relative to ligand dissociation) is in the general order Mn^{2+}, Fe^{2+}, Co^{2+}, Ni^{2+}, Cu^{2+}, Zn^{2+}. This order of ions is called the Irving–Williams series. Look up the values of the formation constants for the ammonia complexes of Co^{2+}, Ni^{2+}, Cu^{2+}, and Zn^{2+} in Appendix K, and verify this statement.

70. ■ ▲ In this question, we explore the differences between metal coordination by monodentate and bidentate ligands. Formation constants, K_f, for $[Ni(NH_3)_6]^{2+}(aq)$ and $[Ni(en)_3]^{2+}(aq)$ are as follows:

$$Ni^{2+}(aq) + 6\,NH_3(aq) \longrightarrow [Ni(NH_3)_6]^{2+}(aq) \qquad K_f = 10^8$$

$$Ni^{2+}(aq) + 3\,en(aq) \longrightarrow [Ni(en)_3]^{2+}(aq) \qquad K_f = 10^{18}$$

The difference in K_f between these complexes indicates a higher thermodynamic stability for the chelated complex, caused by the *chelate effect*. Recall that K is related to the standard free energy of the reaction by $\Delta_r G° = -RT \ln K$ and $\Delta_r G° = \Delta_r H° - T\Delta S°$. We know from experiment that $\Delta_r H°$ for the NH_3 reaction is −109 kJ/mol-rxn, and $\Delta_r H°$ for the ethylenediamine reaction is −117 kJ/mol-rxn. Is the difference in $\Delta_r H°$ sufficient to account for the 10^{10} difference in K_f? Comment on the role of entropy in the second reaction.

71. As mentioned on page 1047, transition metal organometallic compounds have found use as catalysts. One example is Wilkinson's catalyst, a rhodium compound $[RhCl(PR_3)_3]$ used in the hydrogenation of alkenes. The steps involved in the catalytic process are outlined below.

Indicate whether the rhodium compounds in each step have 18- or 16-valence electrons. (See Study Question 34.)

Step 1—Addition of H_2 to the rhodium center of Wilkinson's catalyst. (For electron counting purposes H is considered a hydride ion, H^-, a two-electron donor.)

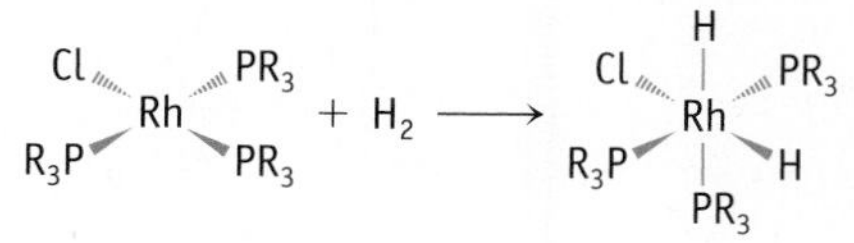

Step 2—Loss of a PR_3 ligand (a two-electron donor) to open a coordination site. (PR_3 is a phosphine such as $P(C_6H_5)_3$, triphenylphosphine.)

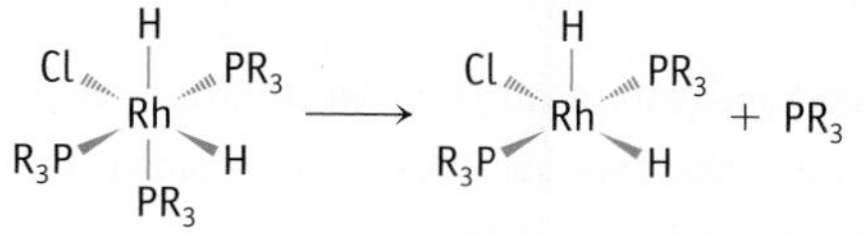

Step 3—Addition of the alkene to the open site.

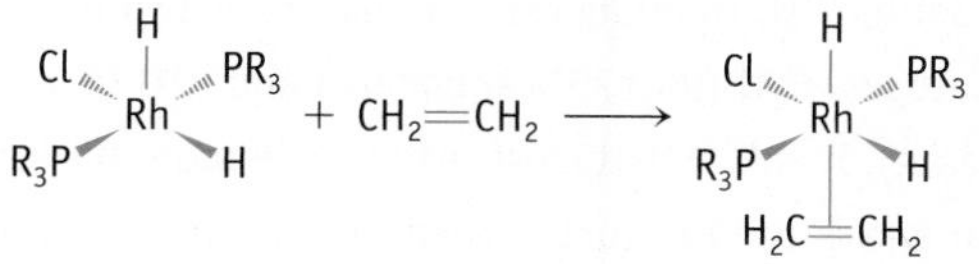

Step 4—Rearrangement to add H to the double bond. (Here the $—CH_2CH_3$ group is a two-electron donor and can be thought of as a $[CH_2CH_3]^-$ anion for electron counting purposes.)

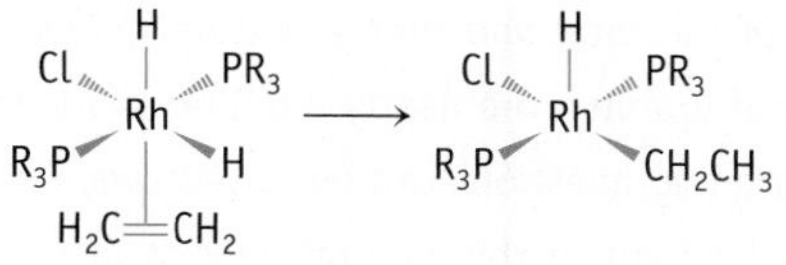

Step 5—Loss of the alkane

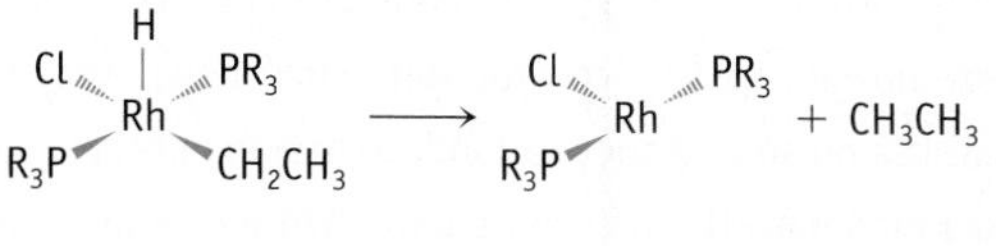

Step 6—Regeneration of the catalyst.

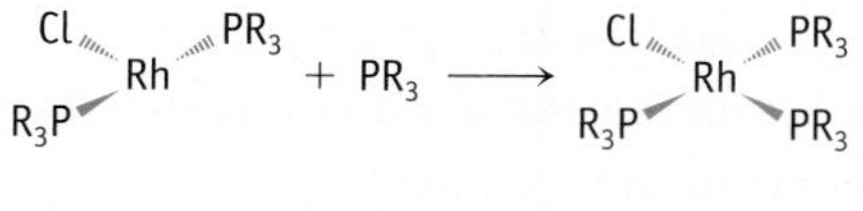

Net reaction: $CH_2{=}CH_2 + H_2 \longrightarrow CH_3CH_3$

23 Nuclear Chemistry

Courtesy of Francois Gauthier-Lafaye.

The natural nuclear reactor in Oklo, Gabon (West Africa). Nearly two billion years ago, a natural formation containing uranium oxide (the yellow material) underwent fission that started and stopped over a period of a million years.

A Primordial Nuclear Reactor

The natural nuclear reactors in West Africa have been called "one of the greatest natural phenomena that ever occurred." In 1972, a French scientist noticed that the uranium taken from a mine in Oklo, Gabon was strangely deficient in ^{235}U. Uranium exists in nature as two principal isotopes, ^{238}U (99.275% abundant) and ^{235}U (0.72% abundant). It is ^{235}U that most readily undergoes nuclear fission and is used to fuel nuclear power plants around the world. But the uranium found in the Oklo mines had an isotope ratio like that of the spent fuel that comes from modern reactors. On this and other evidence, scientists concluded that ^{235}U at the Oklo mine was once about 3% of the total and that a "natural" fission process occurred in the bed of uranium ore nearly 2 billion years ago.

But intriguing questions can be raised: why fission could occur to a significant extent in this natural deposit of uranium and why the "reactor" didn't explode. Apparently, there must have been a moderator of neutron energy and a regulation mechanism. In a modern nuclear power reactor, control rods slow down the neutrons from nuclear fission so that they can induce fission in other ^{235}U nuclei; that is, they moderate the neutron energy. Without a moderator, the neutrons just fly off. In the Oklo reactors, water seeping into the bed of uranium ore could have acted as a moderator.

The reason the Oklo reactor did not explode is that water could have also been the regulator. As the fission process heated the water, it boiled off as steam. This caused the fission to stop, but it began again when more water seeped in. Scientists now believe this natural reactor would turn on for about 30 minutes and then shut down for several hours before turning on again. There is evidence these natural reactors functioned intermittently for about 1 million years, until the concentration of uranium isotopes was too low to keep the reaction going.

Questions:

1. How many protons and neutrons are there in the ^{235}U and ^{238}U nuclei?
2. Although plutonium does not occur naturally on earth now, it is thought to have been produced in the Oklo reactor (and then decayed). Write balanced nuclear equations for:
   (a) the reaction of ^{238}U and a neutron to give ^{239}U
   (b) the decay of ^{239}Pu to ^{239}Np and then to ^{239}Pu by beta emission
   (c) the decay of ^{239}Pu to ^{235}U by alpha emission

Answers to these questions are in Appendix Q.

Chapter Goals

See Chapter Goals Revisited (page 1090) for Study Questions keyed to these goals and assignable in OWL.

- Identify radioactive elements, and describe natural and artificial nuclear reactions.
- Calculate the binding energy and binding energy per nucleon for a particular isotope.
- Understand rates of radioactive decay.
- Understand artificial nuclear reactions.
- Understand issues of health and safety with respect to radioactivity.
- Be aware of some uses of radioactive isotopes in science and medicine.

Chapter Outline

History of science scholars cite three pillars of modern chemistry: technology, medicine, and alchemy. The third of these pillars, alchemy, was pursued in many cultures on three continents for well over 1000 years. Simply stated, the goal of the ancient alchemists was to turn less valuable materials into gold. We now recognize the futility of these efforts, because this goal is not reachable by chemical processes. We also know that it *is* possible to transmute one element into another. This happens naturally in the decomposition of uranium and other radioactive elements, and scientists can intentionally carry out such reactions in the laboratory. The goal is no longer to make gold, however. Far more important and valuable products of nuclear reactions are possible.

Nuclear chemistry encompasses a wide range of topics that share one thing in common: they involve changes in the nucleus of an atom. While "chemistry" is a major focus in this chapter, the subject cuts across many areas of science. Radioactive isotopes are used in medicine. Nuclear power provides a sizable fraction of energy for modern society. And, then there are nuclear weapons. . . .

ChemistryNow™

Throughout the text this icon introduces an opportunity for self-study or to explore interactive tutorials by signing in at **www.thomsonedu.com/login.**

23.1 Natural Radioactivity

In the late 19th century, while studying radiation emanating from uranium and thorium, Ernest Rutherford (1871–1937) stated, "There are present at least two distinct types of radiation—one that is readily absorbed, which will be termed for convenience **α (alpha) radiation,** and the other of a more penetrative character, which will be termed **β (beta) radiation.**" Subsequently, charge-to-mass ratio measurements showed that α radiation is composed of helium nuclei (He^{2+}) and β radiation is composed of electrons (e^-) (Table 23.1).

Rutherford hedged his bet when he said at least two types of radiation existed. A third type was later discovered by the French scientist Paul Villard (1860–1934);

■ **Discovery of Radioactivity** The discovery of radioactivity by Henri Becquerel and the isolation of radium and polonium from pitchblende, a uranium ore, by Marie Curie were described in *Milestones in the Development of Chemistry,* page 338.

TABLE 23.1 Characteristics of α, β, and γ Radiation

Name	Symbols	Charge	Mass (g/particle)
Alpha	$^{4}_{2}He$, $^{4}_{2}\alpha$	2+	6.65×10^{-24}
Beta	$^{0}_{-1}e$, $^{0}_{-1}\beta$	1−	9.11×10^{-28}
Gamma	γ	0	0

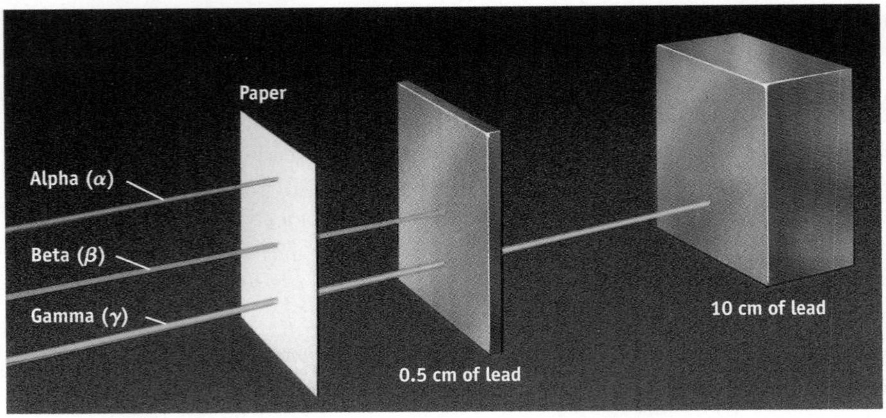

FIGURE 23.1 The relative penetrating ability of α, β, and γ radiation. Highly charged α particles interact strongly with matter and are stopped by a piece of paper. Beta particles, with less mass and a lower charge, interact to a lesser extent with matter and thus can penetrate farther. Gamma radiation is the most penetrating.

■ **Common Symbols: α and β** Symbols used to represent alpha and beta particles do not include a superscript to show that they have a charge.

he named it **γ (gamma) radiation,** using the third letter in the Greek alphabet in keeping with Rutherford's scheme. Unlike α and β radiation, γ radiation is not affected by electric and magnetic fields. Rather, it is a form of electromagnetic radiation like x-rays but even more energetic.

Early studies measured the penetrating power of the three types of radiation (Figure 23.1). Alpha radiation is the least penetrating; it can be stopped by several sheets of ordinary paper or clothing. Aluminum that is at least 0.5 cm thick is needed to stop β particles; they can penetrate several millimeters of living bone or tissue. Gamma radiation is the most penetrating. Thick layers of lead or concrete are required to shield the body from this radiation, and γ-rays can pass completely through the human body.

Alpha and β particles typically possess high kinetic energies. The energy of γ radiation is similarly very high. The energy associated with this radiation is transferred to any material used to stop the particle or absorb the radiation. This fact is important because the damage caused by radiation is related to the energy absorbed (see Section 23.8).

23.2 Nuclear Reactions and Radioactive Decay

Equations for Nuclear Reactions

In 1903, Rutherford and Frederick Soddy (1877–1956) proposed that radioactivity is the result of a natural change of an isotope of one element into an isotope of a different element. Such processes are called **nuclear reactions.**

Consider a reaction in which radium-226 (the isotope of radium with mass number 226) emits an α particle to form radon-222. The equation for this reaction is

$$^{226}_{88}\text{Ra} \rightarrow {}^{4}_{2}\alpha + {}^{222}_{86}\text{Rn}$$

■ **Symbols Used in Nuclear Equations** The mass number is included as a superscript, and the atomic number is included as a subscript preceding the symbols for reactants and products. This is done to facilitate balancing these equations.

In a nuclear reaction, the sum of the mass numbers of reacting particles must equal the sum of the mass numbers of products. Furthermore, to maintain nuclear charge balance, the sum of the atomic numbers of the products must equal the sum of the atomic numbers of the reactants. These principles are illustrated using the preceding nuclear equation:

	$^{226}_{88}\text{Ra}$	$\rightarrow$	$^{4}_{2}\alpha$	+	$^{222}_{86}\text{Rn}$
	radium-226	$\rightarrow$	α particle	+	radon-222
Mass number: (protons + neutrons)	226	=	4	+	222
Atomic number: (protons)	88	=	2	+	86

Alpha particle emission causes a decrease of two units in atomic number and four units in the mass number.

Similarly, nuclear mass and nuclear charge balance accompany β particle emission, as illustrated by the decomposition of uranium-239:

	$^{239}_{92}U$	$\rightarrow$	$^{0}_{-1}\beta$	+	$^{239}_{93}Np$
	uranium-239	$\rightarrow$	β particle	+	neptunium-239
Mass number: (protons + neutrons)	239	=	0	+	239
Atomic number: (protons)	92	=	−1	+	93

The β particle has a charge of 1−. Charge balance requires that the atomic number of the product be one unit greater than the atomic number of the reacting nucleus. The mass number does not change in this process.

How does a nucleus, composed of protons and neutrons, eject an electron? It is a complex process, but the net result is the conversion within the nucleus of a neutron to a proton and an electron.

$$^{1}_{0}n \longrightarrow {}^{0}_{-1}e + {}^{1}_{1}p$$

neutron electron proton

Notice that the mass and charge numbers balance in this equation.

What is the origin of the gamma radiation that accompanies most nuclear reactions? Recall that a photon of visible light is emitted when an atom undergoes a transition from an excited electronic state to a lower energy state (◀ Section 6.3). Gamma radiation originates from transitions between nuclear energy levels. Nuclear reactions often result in the formation of a product nucleus in an excited nuclear state. One option is to return to the ground state by emitting the excess energy as a photon. The high energy of γ radiation is a measure of the large energy difference between the energy levels in the nucleus.

Chemistry Now™

Sign in at **www.thomsonedu.com/login** and go to Chapter 23 Contents to see:
- Screen 23.2 for a tutorial on **balancing equations for nuclear reactions**
- Screen 23.3 for a tutorial on **modes of radioactive decay**

EXERCISE 23.1 Mass and Charge Balance in Nuclear Reactions

Write equations for the following nuclear reactions, and confirm that they are balanced with respect to nuclear mass and nuclear charge.

(a) the emission of an α particle by radon-222 to form polonium-218

(b) the emission of a β particle by polonium-218 to form astatine-218

EXERCISE 23.2 Gamma Ray Energies

Calculate the energy per photon, and the energy per mole of photons, for γ radiation with a wavelength of 2.0×10^{-12} m. (*Hint:* Review similar calculations on the energy of photons of visible light, Section 6.2.)

■ **Energy Units** Gamma ray energies are often reported with the unit *MeV*, which stands for 1 million electron volts. One electron volt (1 eV) is the energy of an electron that has been accelerated by a potential of one volt. The conversion factor between electron volts and joules is $1 \text{ eV} = 1.60218 \times 10^{-19}$ J.

Radioactive Decay Series

Several naturally occurring radioactive isotopes are found to decay to form a product that is also radioactive. When this happens, the initial nuclear reaction is followed by a second nuclear reaction; if the situation is repeated, a third and a fourth nuclear reaction occur; and so on. Eventually, a nonradioactive isotope is formed to end the series. Such a sequence of nuclear reactions is called a **radioactive decay series.** In each step of this nuclear reaction sequence, the reactant nucleus is called the *parent,* and the product called the *daughter.*

Uranium-238, the most abundant of three naturally occurring uranium isotopes, heads one of four radioactive decay series. This series begins with the loss of an α particle from $^{238}_{92}U$ to form radioactive $^{234}_{90}Th$. Thorium-234 then decomposes by β emission to $^{234}_{91}Pa$, which emits a β particle to give $^{234}_{92}U$. Uranium-234 is an α emitter, forming $^{230}_{90}Th$. Further α and β emissions follow, until the series ends with formation of the stable, nonradioactive isotope, $^{206}_{82}Pb$. In all, this radioactive decay series converting $^{238}_{92}U$ to $^{206}_{82}Pb$ is made up of 14 reactions, with eight α and six β particles being emitted. The series is portrayed graphically by plotting atomic number versus mass number (Figure 23.2). An equation can be written for each step in the sequence. Equations for the first four steps in the uranium-238 radioactive decay series are:

Step 1. $^{238}_{92}U \rightarrow ^{234}_{90}Th + ^{4}_{2}\alpha$

Step 2. $^{234}_{90}Th \rightarrow ^{234}_{91}Pa + ^{0}_{-1}\beta$

Step 3. $^{234}_{91}Pa \rightarrow ^{234}_{92}U + ^{0}_{-1}\beta$

Step 4. $^{234}_{92}U \rightarrow ^{230}_{90}Th + ^{4}_{2}\alpha$

Uranium ore contains trace quantities of the radioactive elements formed in the radioactive decay series. A significant development in nuclear chemistry was Marie Curie's discovery in 1898 of radium and polonium as trace components of pitchblende, a uranium ore. The amount of each of these elements is small because the isotopes of these elements have short half-lives. Marie Curie isolated only a single gram of radium from 7 tons of ore. It is a credit to her skills as a chemist that she

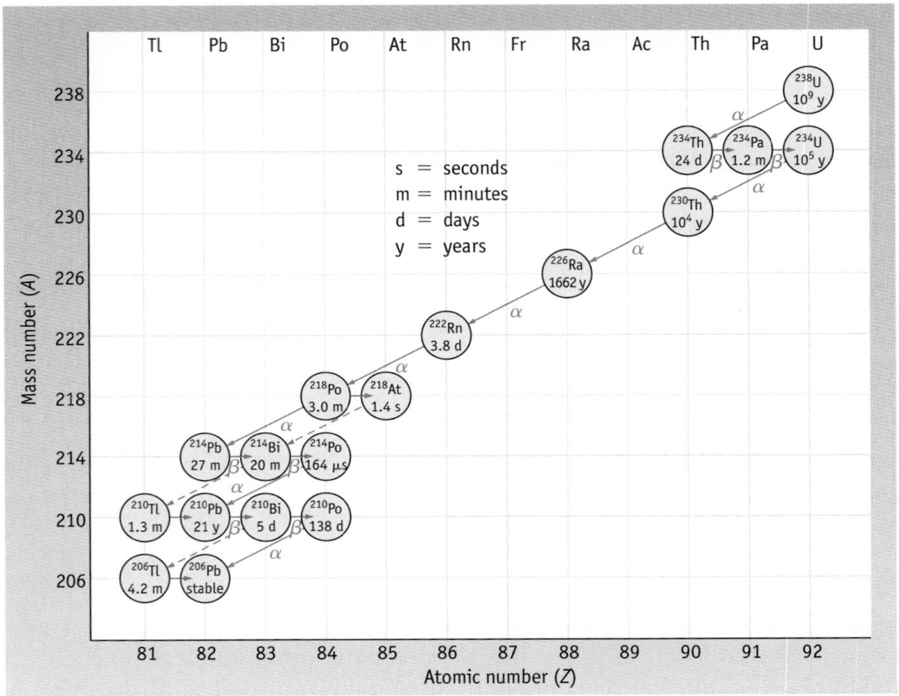

FIGURE 23.2 The uranium-238 radioactive decay series. The steps in this radioactive decay series are shown graphically in this plot of mass number versus atomic number. Each α decay step lowers the atomic number by two units and the mass number by four units. Beta particle emission does not change the mass but raises the atomic number by one unit. Half-lives of the isotopes are included on the chart.

extracted sufficient amounts of radium and polonium from uranium ore to identify these elements.

The uranium-238 radioactive decay series is also the source of the environmental hazard radon. Trace quantities of uranium are often present naturally in the soil and rocks, and radon-222 is being continuously formed. Because radon is chemically inert, it is not trapped by chemical processes in soil or water and is free to seep into mines or into homes through pores in cement block walls, through cracks in the basement floor or walls, or around pipes. Because it is more dense than air, radon tends to collect in low spots, and its concentration can build up in a basement if steps are not taken to remove it.

The major health hazard from radon, when it is inhaled by humans, arises not from radon itself but from its decomposition product, polonium.

$$^{222}_{86}\text{Rn} \longrightarrow {}^{218}_{84}\text{Po} + {}^{4}_{2}\alpha \qquad t_{1/2} = 3.82 \text{ days}$$

$$^{218}_{84}\text{Po} \longrightarrow {}^{214}_{82}\text{Pb} + {}^{4}_{2}\alpha \qquad t_{1/2} = 3.04 \text{ minutes}$$

Radon does not undergo chemical reactions or form compounds that can be taken up in the body. Polonium, however, is not chemically inert. Polonium-218 can lodge in body tissues, where it undergoes α decay to give lead-214, another radioactive isotope. The range of an α particle in body tissue is quite small, perhaps 0.7 mm. This is approximately the thickness of the epithelial cells of the lungs, however, so α particle radiation can cause serious damage to lung tissues.

Virtually every home in the United States has some level of radon, and kits can be purchased to test for the presence of this gas. If radon gas is detected in your home, you should take corrective actions such as sealing cracks around the foundation and in the basement. It may be reassuring to know that the health risks associated with radon are low. The likelihood of getting lung cancer from exposure to radon is about the same as the likelihood of dying in an accident in your home.

Charles D. Winters

Radon detector. This kit is intended for use in the home to detect radon gas. The small device is placed in the home's basement for a given time period and is then sent to a laboratory to measure the amount of radon that might be present.

EXAMPLE 23.1 Radioactive Decay Series

Problem A second radioactive decay series begins with $^{235}_{92}\text{U}$ and ends with $^{207}_{82}\text{Pb}$.

(a) How many α and β particles are emitted in this series?

(b) The first three steps of this series are (in order) α, β, and α emission. Write an equation for each of these steps.

Strategy First, find the total change in atomic number and mass number. A combination of α and β particles is required that will decrease the total nuclear mass by 28 (235 − 207) and at the same time decrease the atomic number by 10 (92 − 82).

Each equation must give symbols for the parent and daughter nuclei and the emitted particle. In the equations, the sums of the atomic numbers and mass numbers for reactants and products must be equal.

Solution

(a) Mass declines by 28 mass units (235 − 207). Because a decrease of 4 mass units occurs with each α emission, 7 α particles must be emitted. Also, for each α emission, the atomic number decreases by 2. Emission of 7 α particles would cause the atomic number to decrease by 14, but the actual decrease in atomic number is 10 (92 − 82). This means that 4 β particles must also have been emitted because each β emission *increases* the atomic number of the product by one unit. Thus, the radioactive decay sequence involves emission of 7 α and 4 β particles.

(b) Step 1. $^{235}_{92}\text{U} \longrightarrow {}^{231}_{90}\text{Th} + {}^{4}_{2}\alpha$

Step 2. $^{231}_{90}\text{Th} \longrightarrow {}^{231}_{91}\text{Pa} + {}^{0}_{-1}\beta$

Step 3. $^{231}_{91}\text{Pa} \longrightarrow {}^{227}_{89}\text{Ac} + {}^{4}_{2}\alpha$

Comment Notice in Figure 23.2 that all daughter nuclei for the series beginning with $^{238}_{92}U$ have mass numbers differing by four units: 238, 234, 230, . . . , 206. This series is sometimes called the *4n + 2 series* because each mass number (*M*) fits the equation $4n + 2 = M$, where *n* is an integer (*n* is 59 for the first member of this series). For the series headed by $^{235}_{92}U$, the mass numbers are 235, 231, 227, . . . , 207; this is the *4n + 3 series.*

Two other decay series are possible. One, called the 4*n* series and beginning with ^{232}Th, is found in nature; the other, the $4n + 1$ series, is not. No member of this series has a very long half-life. During the 5 billion years since this planet was formed, all members of this series have completely decayed.

EXERCISE 23.3 Radioactive Decay Series

(a) Six α and four β particles are emitted in the thorium-232 radioactive decay series before a stable isotope is reached. What is the final product in this series?

(b) The first three steps in the thorium-232 decay series (in order) are α, β, and β emission. Write an equation for each step.

Other Types of Radioactive Decay

■ **Positrons** Positrons were discovered by Carl Anderson (1905–1991) in 1932. The positron is one of a group of particles that are known as *antimatter*. If matter and antimatter particles collide, mutual annihilation occurs, with energy being emitted.

■ **Neutrinos and Antineutrinos** Beta particles having a wide range of energies are emitted. To balance the energy associated with β decay, it is necessary to postulate the concurrent emission of another particle, the *antineutrino*. Similarly, neutrino emission accompanies positron emission. Much study has gone into detecting neutrinos and antineutrinos. These massless, chargeless particles are not included when writing nuclear equations.

Most naturally occurring radioactive elements decay by emission of α, β, and γ radiation. Other nuclear decay processes became known, however, when new radioactive elements were synthesized by artificial means. These include **positron ($^{0}_{+1}\beta$) emission** and **electron capture.**

Positrons ($^{0}_{+1}\beta$) and electrons have the same mass but opposite charge. The positron is the antimatter analogue to an electron. Positron emission by polonium-207, for example, results in the formation of bismuth-207.

	$^{207}_{84}Po$	→	$^{0}_{+1}\beta$	+	$^{207}_{83}Bi$
	polonium-207	→	positron	+	bismuth-207
Mass number: (protons + neutrons)	207	=	0	+	207
Atomic number: (protons)	84	=	1	+	83

To retain charge balance, positron decay results in a decrease in the atomic number.

In *electron capture,* an extranuclear electron is captured by the nucleus. The mass number is unchanged, and the atomic number is reduced by 1. (In an old nomenclature, the innermost electron shell was called the K shell, and electron capture was called *K capture.*)

	$^{7}_{4}Be$	+	$^{0}_{-1}e$	→	$^{7}_{3}Li$
	beryllium-7	+	electron	→	lithium-7
Mass number: (protons + neutrons)	7	+	0	=	7
Atomic number: (protons)	4	+	−1	=	3

In summary, most unstable nuclei decay by one of four paths: α or β decay, positron emission, or electron capture. Gamma radiation often accompanies these processes. Section 23.6 introduces a fifth way that nuclei decompose, *fission.*

EXAMPLE 23.2 Nuclear Reactions

Problem Complete the following equations. Give the symbol, mass number, and atomic number of the product species.

(a) $^{37}_{18}Ar + ^{0}_{-1}e \rightarrow ?$

(b) $^{11}_{6}C \rightarrow ^{11}_{5}B + ?$

(c) $^{35}_{16}S \rightarrow ^{35}_{17}Cl + ?$

(d) $^{30}_{15}P \rightarrow ^{0}_{+1}\beta + ?$

Strategy The missing product in each reaction can be determined by recognizing that the sums of mass numbers and atomic numbers for products and reactants must be equal. When you know the nuclear mass and nuclear charge of the product, you can identify it with the appropriate symbol.

Solution

(a) This is an electron capture reaction. The product has a mass number of 37 + 0 = 37 and an atomic number of 18 − 1 = 17. Therefore, the symbol for the product is $^{37}_{17}Cl$.

(b) This missing particle has a mass of zero and a charge of 1+; these are the characteristics of a positron, $^{0}_{+1}\beta$. If this particle is included in the equation, the sums of the atomic numbers (6 = 5 + 1) and the mass numbers (11) on either side of the equation are equal.

(c) A beta particle, $^{0}_{-1}\beta$, is required to balance the mass numbers (35) and atomic numbers (16 = 17 − 1) in the equation.

(d) The product nucleus has mass number 30 and atomic number 14. This identifies the unknown as $^{30}_{14}Si$.

EXERCISE 23.4 Nuclear Reactions

Indicate the symbol, the mass number, and the atomic number of the missing product in each of the following nuclear reactions.

(a) $^{13}_{7}N \rightarrow ^{13}_{6}C + ?$

(b) $^{41}_{20}Ca + ^{0}_{-1}e \rightarrow ?$

(c) $^{90}_{38}Sr \rightarrow ^{90}_{39}Y + ?$

(d) $^{22}_{11}Na \rightarrow ? + ^{0}_{-1}\beta$

23.3 Stability of Atomic Nuclei

We can learn something about nuclear stability from Figure 23.3. In this plot, the horizontal axis represents the number of protons, and the vertical axis gives the number of neutrons for known isotopes. Each circle represents an isotope identified by the number of neutrons and protons contained in its nucleus. The black circles represent stable (nonradioactive) isotopes, about 300 in number, and the red circles represent some of the known radioactive isotopes. For example, the three isotopes of hydrogen are $^{1}_{1}H$ and $^{2}_{1}H$ (neither is radioactive) and $^{3}_{1}H$ (tritium, radioactive). For lithium, the third element, isotopes with mass numbers 4, 5, 6, and 7 are known. The isotopes with masses of 6 and 7 (shown in black) are stable, whereas the other two isotopes (in red) are radioactive.

Figure 23.3 contains the following information about nuclear stability:

- Stable isotopes fall in a very narrow range called the **band of stability.** It is remarkable how few isotopes are stable.
- Only two stable isotopes ($^{1}_{1}H$ and $^{3}_{2}He$) have more protons than neutrons.
- Up to calcium ($Z = 20$), stable isotopes often have equal numbers of protons and neutrons or only one or two more neutrons than protons.
- Beyond calcium, the neutron–proton ratio is always greater than 1. As the mass increases, the band of stable isotopes deviates more and more from a line in which $N = Z$.
- Beyond bismuth (83 protons and 126 neutrons), all isotopes are unstable and radioactive. There is apparently no nuclear "superglue" strong enough to hold heavy nuclei together.
- The lifetimes of unstable nuclei are shorter for the heaviest nuclei. For example, half of a sample of $^{238}_{92}U$ disintegrates in 4.5 billion years, whereas half of a sample of $^{257}_{103}Lr$ is gone in only 0.65 second. Isotopes that fall farther from the band of stability tend to have shorter half-lives than do unstable isotopes nearer to the band of stability.
- Elements of even atomic number have more stable isotopes than do those of odd atomic number. More stable isotopes have an even number of neutrons

FIGURE 23.3 Stable and unstable isotopes. A graph of the number of neutrons (N) versus the number of protons (Z) for stable (black circles) and radioactive (red circles) isotopes from hydrogen to bismuth. This graph is used to assess criteria for nuclear stability and to predict modes of decay for unstable nuclei.

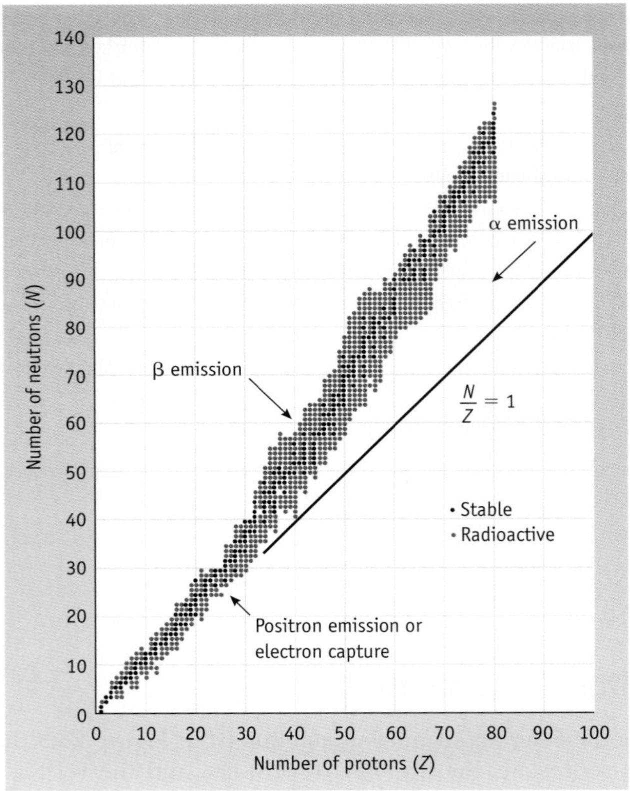

than have an odd number. Roughly 200 isotopes have an even number of neutrons and an even number of protons, whereas only about 120 isotopes have an odd number of either protons or neutrons. Only five stable isotopes ($^{1}_{1}H$, $^{6}_{3}Li$, $^{10}_{5}B$, $^{14}_{7}N$, and $^{180}_{73}Ta$) have odd numbers of both protons and neutrons.

The Band of Stability and Radioactive Decay

Besides being a criterion for stability, the neutron–proton ratio can assist in predicting what type of radioactive decay will be observed. Unstable nuclei decay in a manner that brings them toward a stable neutron–proton ratio—that is, toward the band of stability.

- All elements beyond bismuth ($Z = 83$) are unstable. To reach the band of stability starting with these elements, a process that decreases the atomic number is needed. Alpha emission is an effective way to lower Z, the atomic number, because each emission decreases the atomic number by 2. For example, americium, the radioactive element used in smoke detectors, decays by α emission:

$$^{243}_{95}Am \rightarrow {}^{4}_{2}\alpha + {}^{239}_{93}Np$$

- Beta emission occurs for isotopes that have a high neutron–proton ratio—that is, isotopes above the band of stability. With β decay, the atomic number increases by 1, and the mass number remains constant, resulting in a lower n/p ratio.

$$^{60}_{27}Co \rightarrow {}^{0}_{-1}\beta + {}^{60}_{28}Ni$$

- Isotopes with a low neutron–proton ratio, below the band of stability, decay by positron emission or by electron capture. Both processes lead to product nuclei with a lower atomic number and the same mass number:

$$^{13}_{7}\text{N} \rightarrow {}^{0}_{+1}\beta + {}^{13}_{6}\text{C}$$

$$^{41}_{20}\text{Ca} + {}^{0}_{-1}\text{e} \rightarrow {}^{41}_{19}\text{K}$$

Chemistry Now™

Sign in at **www.thomsonedu.com/login** and go to Chapter 23 Contents to see Screen 23.3 for a tutorial on **predicting modes of radioactive decay.**

EXAMPLE 23.3 Predicting Modes of Radioactive Decay

Problem Identify probable mode(s) of decay for each isotope and write an equation for the decay process.

(a) oxygen-15, $^{15}_{8}\text{O}$ **(b)** uranium-234, $^{234}_{92}\text{U}$ **(c)** fluorine-20, $^{20}_{9}\text{F}$ **(d)** manganese-56, $^{56}_{25}\text{Mn}$

Strategy In parts (a), (c), and (d), compare the mass number with the atomic mass. If the mass number of the isotope is higher than the atomic weight, then there are too many neutrons, and β emission is likely. If the mass number is lower than the atomic weight, then there are too few neutrons, and positron emission or electron capture is the more likely process. It is not possible to choose between the latter two modes of decay without further information. For part (b), note that isotopes with atomic number greater than 83 are likely to be α emitters.

Solution

(a) Oxygen-15 has 7 neutrons and 8 protons, so the n/p ratio is less than 1—too low for ^{15}O to be stable. Nuclei with too few neutrons are expected to decay by either positron emission or electron capture. In this instance, the process is ${}^{0}_{+1}\beta$ emission, and the equation is $^{15}_{8}\text{O} \rightarrow {}^{0}_{+1}\beta + {}^{15}_{7}\text{N}$.

(b) Alpha emission is a common mode of decay for isotopes of elements with atomic numbers higher than 83. The decay of uranium-234 is one example:

$$^{234}_{92}\text{U} \rightarrow {}^{230}_{90}\text{Th} + {}^{4}_{2}\alpha$$

(c) Fluorine-20 has 11 neutrons and 9 protons, a high n/p ratio. The ratio is lowered by β emission:

$$^{20}_{9}\text{F} \rightarrow {}^{0}_{-1}\beta + {}^{20}_{10}\text{Ne}$$

(d) The atomic weight of manganese is 54.85. The higher mass number, 56, suggests that this radioactive isotope has an excess of neutrons, in which case it would be expected to decay by β emission:

$$^{56}_{25}\text{Mn} \rightarrow {}^{0}_{-1}\beta + {}^{56}_{26}\text{Fe}$$

Comment Be aware that predictions made in this manner will be right much of the time, but exceptions will sometimes occur.

EXERCISE 23.5 Predicting Modes of Radioactive Decay

Write an equation for the probable mode of decay for each of the following unstable isotopes, and write an equation for that nuclear reaction.

(a) silicon-32, $^{32}_{14}\text{Si}$ **(c)** plutonium-239, $^{239}_{94}\text{Pu}$

(b) titanium-45, $^{45}_{22}\text{Ti}$ **(d)** potassium-42, $^{42}_{19}\text{K}$

Nuclear Binding Energy

An atomic nucleus can contain as many as 83 protons and still be stable. For stability, nuclear binding (attractive) forces must be greater than the electrostatic repulsive forces between the closely packed protons in the nucleus. **Nuclear binding energy,** E_b, is defined as the energy required to separate the nucleus of an atom into protons and neutrons. For example, the nuclear binding energy for deuterium is the energy required to convert one mole of deuterium ($^{2}_{1}H$) nuclei into one mole of protons and one mole of neutrons.

$$^{2}_{1}H \rightarrow ^{1}_{1}p + ^{1}_{0}n \qquad E_b = 2.15 \times 10^8 \text{ kJ/mol}$$

The positive sign for E_b indicates that energy is required for this process. A deuterium nucleus is more stable than an isolated proton and an isolated neutron, just as the H_2 molecule is more stable than two isolated H atoms. Recall, however, that the H—H bond energy is only 436 kJ/mol. The energy holding a proton and a neutron together in a deuterium nucleus, 2.15×10^8 kJ/mol, is about 500,000 times larger than the typical covalent bond energies.

To further understand nuclear binding energy, we turn to an experimental observation and a theory. The experimental observation is that the mass of a nucleus is always less than the sum of the masses of its constituent protons and neutrons. The theory is that the "missing mass," called the **mass defect,** is equated with energy that holds the nuclear particles together.

The mass defect for deuterium is the difference between the mass of a deuterium nucleus and the sum of the masses of a proton and a neutron. Mass spectrometric measurements (◄ Section 2.3) give the accurate masses of these particles to a high level of precision, providing the numbers needed to carry out calculations of mass defects.

Masses of atomic nuclei are not generally listed in reference tables, but masses of atoms are. Calculation of the mass defect can be carried out using masses of atoms instead of masses of nuclei. By using atomic masses, we are including in this calculation the masses of extranuclear electrons in the reactants and the products. Because the same number of extranuclear electrons appears in products and reactants, this does not affect the result. Thus, for one mole of deuterium nuclei, the mass defect is found as follows:

$$\begin{array}{ccccc} ^{2}_{1}H & \rightarrow & ^{1}_{1}H & + & ^{1}_{0}n \\ 2.01410 \text{ g/mol} & & 1.007825 \text{ g/mol} & & 1.008665 \text{ g/mol} \end{array}$$

$$\begin{aligned} \text{Mass defect} = \Delta m &= \text{mass of products} - \text{mass of reactants} \\ &= [1.007825 \text{ g/mol} + 1.008665 \text{ g/mol}] - 2.01410 \text{ g/mol} \\ &= 0.00239 \text{ g/mol} \end{aligned}$$

The relationship between mass and energy is contained in Albert Einstein's 1905 theory of special relativity, which holds that mass and energy are different manifestations of the same quantity. Einstein defined the energy–mass relationship: energy is equivalent to mass times the square of the speed of light; that is, $E = mc^2$. In the case of atomic nuclei, it is assumed that the missing mass (the mass defect, Δm) is equated with the binding energy holding the nucleus together.

$$E_b = (\Delta m)c^2 \qquad \textbf{(23.1)}$$

If Δm is given in kilograms and the speed of light is given in meters per second, E_b will have units of joules (because $1 \text{ J} = 1 \text{ kg} \cdot \text{m}^2/\text{s}^2$). For the decomposition of one mole of deuterium nuclei to one mole of protons and one mole of neutrons, we have

$$\begin{aligned} E_b &= (2.39 \times 10^{-6} \text{ kg/mol})(2.998 \times 10^8 \text{ m/s})^2 \\ &= 2.15 \times 10^{11} \text{ J/mol of } ^{2}_{1}\text{H nuclei } (= 2.15 \times 10^8 \text{ kJ/mol of } ^{2}_{1}\text{H nuclei}) \end{aligned}$$

The nuclear stabilities of different elements are compared using the **binding energy per mole of nucleons.** (**Nucleon** is the general name given to nuclear particles—that is, protons and neutrons.) A deuterium nucleus contains two nucle-

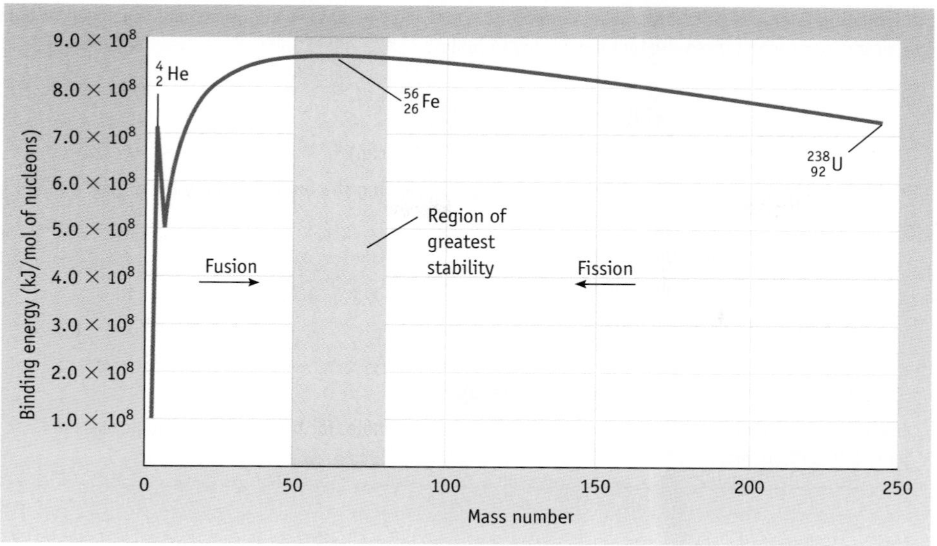

FIGURE 23.4 Relative stability of nuclei. Binding energy per nucleon for the most stable isotope of elements between hydrogen and uranium is plotted as a function of mass number. (Fission and fusion are discussed on pages 1080–1082.)

ons, so the binding energy per mole of nucleons, E_b/n, is 2.15×10^8 kJ/mol divided by 2, or 1.08×10^8 kJ/mol nucleon.

$$E_b/n = \left(\frac{2.15 \times 10^8 \text{ kJ}}{\text{mol } {}^2_1\text{H nuclei}}\right)\left(\frac{1 \text{ mol } {}^2_1\text{H nuclei}}{2 \text{ mol nucleons}}\right)$$

$$E_b/n = 1.08 \times 10^8 \text{ kJ/mol nucleons}$$

The binding energy per nucleon can be calculated for any atom whose mass is known. Then, to compare nuclear stabilities, binding energies per nucleon are plotted as a function of mass number (Figure 23.4). The greater the binding energy per nucleon, the greater the stability of the nucleus. From the graph in Figure 23.4, the point of maximum nuclear stability occurs at a mass of 56 (i.e., at iron in the periodic table).

Chemistry Now™

Sign in at **www.thomsonedu.com/login** and go to Chapter 23 Contents to see:

- Screen 23.4 for a simulation on **isotope stability**
- Screen 23.5 for a tutorial on **calculating binding energy**

EXAMPLE 23.4 Nuclear Binding Energy

Problem Calculate the binding energy, E_b (in kJ/mol), and the binding energy per nucleon, E_b/n (in kJ/mol nucleon), for carbon-12.

Strategy First, determine the mass defect, then use Equation 23.1 to determine the binding energy. There are 12 nuclear particles in carbon-12, so dividing the nuclear binding energy by 12 will give the binding energy per nucleon.

Solution The mass of ${}^1_1\text{H}$ is 1.007825 g/mol, and the mass of ${}^1_0\text{n}$ is 1.008665 g/mol. Carbon-12, ${}^{12}_6\text{C}$, is the standard for the atomic masses in the periodic table, and its mass is defined as exactly 12 g/mol

$$\Delta m = [(6 \times \text{mass } {}^1_1\text{H}) + (6 \times \text{mass } {}^1_0\text{n})] - \text{mass } {}^{12}_6\text{C}$$

$$= [(6 \times 1.007825 \text{ g/mol}) + (6 \times 1.008665 \text{ g/mol})] - 12.000000 \text{ g/mol}$$

$$= 9.8940 \times 10^{-2} \text{ g/mol nuclei}$$

The binding energy is calculated using Equation 23.1. Using the mass in kilograms and the speed of light in meters per second gives the binding energy in joules:

$$E_b = (\Delta m)c^2$$
$$= (9.8940 \times 10^{-5} \text{ kg/mol})(2.99792 \times 10^8 \text{ m/s})^2$$
$$= 8.89 \times 10^{12} \text{ J/mol nuclei } (= 8.89 \times 10^9 \text{ kJ/mol nuclei})$$

The binding energy per nucleon, E_b/n, is determined by dividing the binding energy by 12 (the number of nucleons)

$$\frac{E_b}{n} = \frac{8.89 \times 10^9 \text{ kJ/mol nuclei}}{12 \text{ mol nucleons/mol nuclei}}$$
$$= 7.41 \times 10^8 \text{ kJ/mol nucleons}$$

EXERCISE 23.6 Nuclear Binding Energy

Calculate the binding energy per nucleon, in kilojoules per mole, for lithium-6. The molar mass of ^6_3Li is 6.015125 g/mol.

23.4 Rates of Nuclear Decay

Half-Life

When a new radioactive isotope is identified, its *half-life* is usually measured. Half-life ($t_{1/2}$) is used in nuclear chemistry in the same way it is used when discussing the kinetics of first-order chemical reactions (◀ Section 15.4): It is the time required for half of a sample to decay to products (Figure 23.5). Recall that for first-order kinetics the half-life is independent of the amount of sample.

Half-lives for radioactive isotopes cover a wide range of values. Uranium-238 has one of the longer half-lives, 4.47×10^9 years, a length of time close to the age of the earth (estimated at $4.5–4.6 \times 10^9$ years). Roughly half of the uranium-238 present when the planet was formed is still around. At the other end of the range of half-lives are isotopes such as element 112, whose 277 isotope has a half-life of 240 microseconds ($1\ \mu s = 1 \times 10^{-6}$ s).

Half-life provides an easy way to estimate the time required before a radioactive element is no longer a health hazard. Strontium-90, for example, is a β emitter

FIGURE 23.5 Decay of 20.0 mg of oxygen-15. After each half-life period of 2.0 min, the mass of oxygen-15 decreases by one half.

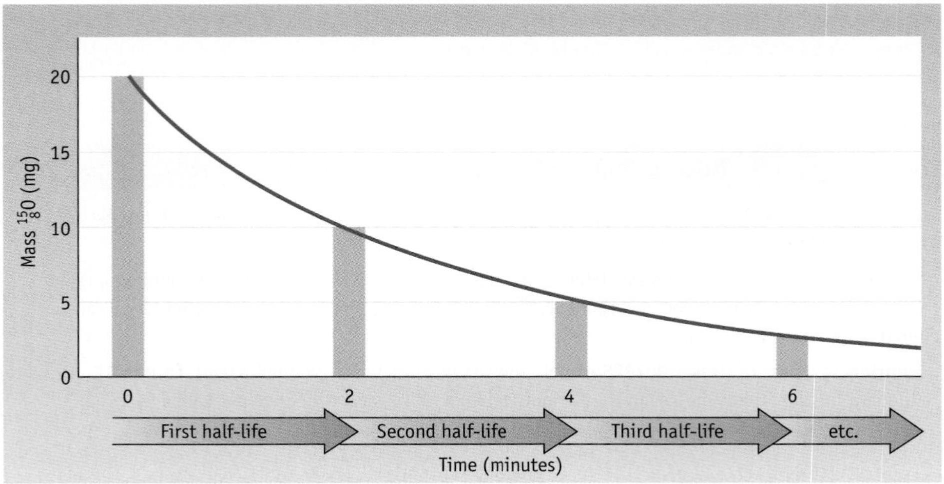

with a half-life of 29.1 years. Significant quantities of strontium-90 were dispersed into the environment in atmospheric nuclear bomb tests in the 1960s and 1970s, and from the half-life, we know that a little less than half is still around. The health problems associated with strontium-90 arise because calcium and strontium have similar chemical properties. Strontium-90 is taken into the body and deposited in bone, taking the place of calcium. Radiation damage by strontium-90 in bone has been directly linked to bone-related cancers.

■ **Half-Life and Temperature** Unlike what is observed in chemical kinetics, temperature does not affect the rate of nuclear decay.

EXAMPLE 23.5 Using Half-Life

Problem Radioactive iodine-131, used to treat hyperthyroidism, has a half-life of 8.04 days.

(a) If you have 8.8 μg (micrograms) of this isotope, what mass remains after 32.2 days?

(b) How long will it take for a sample of iodine-131 to decay to one eighth of its activity?

(c) Estimate the length of time necessary for the sample to decay to 10% of its original activity.

Strategy This problem asks you to use half-life to qualitatively assess the rate of decay. After one half-life, half of the sample remains. After another half-life, the amount of sample is again decreased by half to one fourth of its original value. (This situation is illustrated in Figure 23.5.) To answer these questions, assess the number of half-lives that have elapsed and use this information to determine the amount of sample remaining.

Solution

(a) The time elapsed, 32.2 days, is 4 half-lives (32.2/8.04 = 4). The amount of iodine-131 has decreased to 1/16 of the original amount [$1/2 \times 1/2 \times 1/2 \times 1/2 = (1/2)^4 = 1/16$]. The amount of iodine remaining is 8.8 μg $\times$ $(1/2)^4$ or 0.55 μg .

(b) After 3 half-lives , the amount of iodine-131 remaining is 1/8 ($= 1/2)^3$ of the original amount. The amount remaining is 8.8 μg $\times$ $(1/2)^3 = 1.1$ μg.

(c) After 3 half-lives, 1/8 (12.5%) of the sample remains; after 4 half-lives, 1/16 (6.25%) remains. It will take between 3 and 4 half-lives, between 24.15 and 32.2 days, to decrease the amount of sample to 10% of its original value.

Comment You will find it useful to make approximations as we have done in (c). An exact time can be calculated from the first-order rate law (pages 683 and 1074).

EXERCISE 23.7 Using Half-Life

Tritium ($^3_1$H), a radioactive isotope of hydrogen has a half-life of 12.3 years.

(a) Starting with 1.5 mg of this isotope, how many milligrams remain after 49.2 years?

(b) How long will it take for a sample of tritium to decay to one eighth of its activity?

(c) Estimate the length of time necessary for the sample to decay to 1% of its original activity.

Kinetics of Nuclear Decay

The rate of nuclear decay is determined from measurements of the **activity** (A) of a sample. Activity refers to the number of disintegrations observed per unit time, a quantity that can be measured readily with devices such as a Geiger–Müller counter (Figure 23.6). *Activity is proportional to the number of radioactive atoms present (N).*

$$A \propto N \qquad \textbf{(23.2)}$$

If the number of radioactive nuclei N is reduced by half, the activity of the sample will be half as large. Doubling N will double the activity. This evidence indicates that the rate of decomposition is first order with respect to N. Consequently, the equations describing rates of radioactive decay are the same as those used to de-

■ **Ways of Expressing Activity** Common units for activity are dps (disintegrations per second) and dpm (disintegrations per minute).

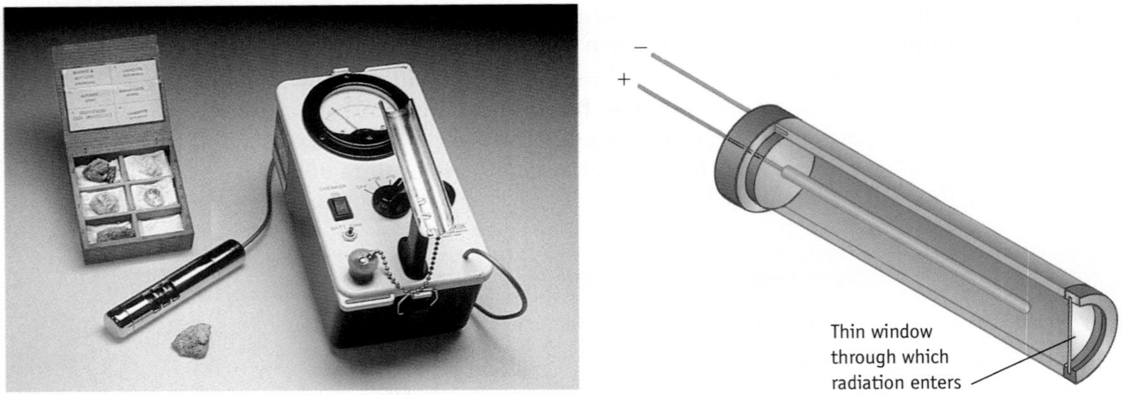

Charles D. Winters

FIGURE 23.6 A Geiger–Müller counter. A charged particle (an α or β particle) enters the gas-filled tube (diagram at the right) and ionizes the gas. The gaseous ions migrate to electrically charged plates and are recorded as a pulse of electric current. The current is amplified and used to operate a counter. A sample of carnotite, a mineral containing uranium oxide, is also shown in the photograph.

scribe first-order chemical reactions; the change in the number of radioactive atoms N per unit of time is proportional to N:

$$\frac{\Delta N}{\Delta t} = -kN \tag{23.3}$$

The integrated rate equation can be written in two ways depending on the data used:

$$\ln\left(\frac{N}{N_0}\right) = -kt \tag{23.4}$$

or

$$\ln\left(\frac{A}{A_0}\right) = -kt \tag{23.5}$$

Here, N_0 and A_0 are the number of atoms and the activity of the sample initially, respectively, and N and A are the number of atoms and the activity of the sample after time t, respectively. Thus, N/N_0 is the fraction of atoms remaining after a given time (t), and A/A_0 is the fraction of the activity remaining after the same period. In these equations, k is the rate constant (decay constant) for the isotope in question. The relationship between half-life and the first-order rate constant is the same as seen with chemical kinetics (Equation 15.4, page 690):

$$t_{1/2} = \frac{0.693}{k} \tag{23.6}$$

Equations 23.3–23.6 are useful in several ways:

- If the activity (A) or the number of radioactive nuclei (N) is measured in the laboratory over some period t, then k can be calculated. The decay constant k can then be used to determine the half-life of the sample.
- If k is known, the fraction of a radioactive sample (N/N_0) still present after some time t has elapsed can be calculated.
- If k is known, the time required for that isotope to decay to a fraction of the original activity (A/A_0) can be calculated.

EXAMPLE 23.6 Determination of Half-Life

Problem A sample of radon-222 has an initial α particle activity (A_0) of 7.0×10^4 dps (disintegrations per second). After 6.6 days, its activity (A) is 2.1×10^4 dps. What is the half-life of radon-222?

Strategy Values for A, A_0, and t are given. The problem can be solved using Equation 23.5 with k as the unknown. Once k is found, the half-life can be calculated using Equation 23.6.

Solution

$$\ln (2.1 \times 10^4 \text{ dps}/7.0 \times 10^4 \text{ dps}) = -k\,(6.6 \text{ day})$$
$$\ln (0.30) = -k(6.6 \text{ day})$$
$$k = 0.18 \text{ days}^{-1}$$

From k we obtain $t_{1/2}$:

$$t_{1/2} = 0.693/0.18 \text{ days}^{-1} = 3.8 \text{ days}$$

Comments Notice that the activity decreased to between one half and one fourth of its original value. The 6.6 days of elapsed time represents one full half-life and part of another half-life.

EXAMPLE 23.7 Time Required for a Radioactive Sample to Partially Decay

Problem Gallium citrate, containing the radioactive isotope gallium-67, is used medically as a tumor-seeking agent. It has a half-life of 78.2 h. How long will it take for a sample of gallium citrate to decay to 10.0% of its original activity?

Strategy Use Equation 23.5 to solve this problem. In this case, the unknown is the time t. The rate constant k is calculated from the half-life using Equation 23.6. Although we do not have specific values of activity, the value of A/A_0 is known. Because A is 10.0% of A_0, the value of A/A_0 is 0.100.

Solution First, determine k:

$$k = 0.693/t_{1/2} = 0.693/78.2 \text{ h}$$
$$k = 8.86 \times 10^{-3} \text{ h}^{-1}$$

Then substitute the given values of A/A_0 and k into Equation 23.5:

$$\ln (A/A_0) = -kt$$
$$\ln (0.100) = -(8.86 \times 10^{-3} \text{ h}^{-1})t$$
$$t = 2.60 \times 10^2 \text{ h}$$

Comments The time required is between three half-lives (3×78.2 h = 235 h) and four half-lives (4×78.2 h = 313 h).

EXERCISE 23.8 Determination of Half-Life

A sample of $Ca_3(PO_4)_2$ containing phosphorus-32 has an activity of 3.35×10^3 dpm. Two days later, the activity is 3.18×10^3 dpm. Calculate the half-life of phosphorus-32.

EXERCISE 23.9 Time Required for a Radioactive Sample to Partially Decay

A highly radioactive sample of nuclear waste products with a half-life of 200. years is stored in an underground tank. How long will it take for the activity to diminish from an initial activity of 6.50×10^{12} dpm to a fairly harmless activity of 3.00×10^3 dpm?

Radiocarbon Dating

In certain situations, the age of a material can be determined based on the rate of decay of a radioactive isotope. The best-known example of this procedure is the use of carbon-14 to date historical artifacts.

Carbon is primarily carbon-12 and carbon-13 with isotopic abundances of 98.9% and 1.1%, respectively. In addition, traces of a third isotope, carbon-14, are present to the extent of about 1 in 10^{12} atoms in atmospheric CO_2 and in living materials.

FIGURE 23.7 Variation of atmospheric carbon-14 activity. The amount of carbon-14 has varied with variation in cosmic ray activity. To obtain the data for the pre-1990 part of the curve shown in this graph, scientists carried out carbon-14 dating of artifacts for which the age was accurately known (often through written records). Similar results can be obtained using carbon-14 dating of tree rings.

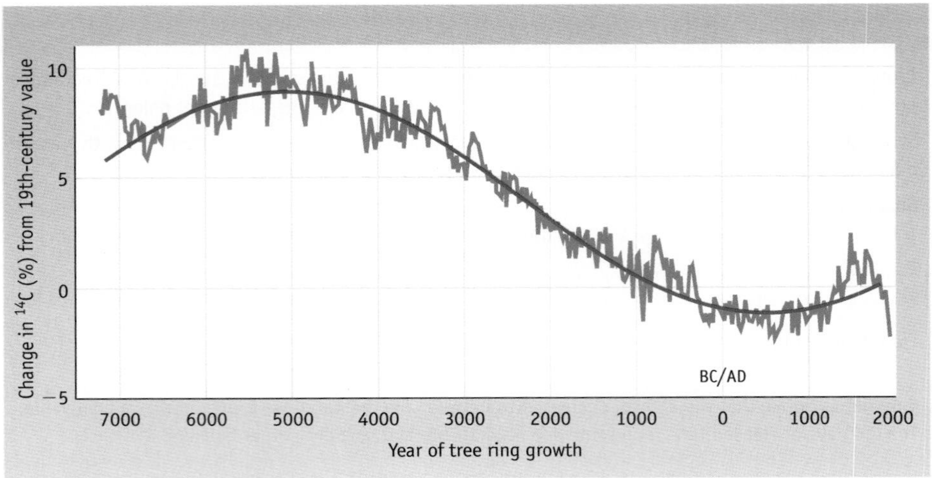

Source: Hans E. Suess, La Jolla Radiocarbon Laboratory

Oesper Collection in the History of Chemistry, University of Cincinnati

■ **Willard Libby (1908–1980)** Libby received the 1960 Nobel Prize in chemistry for developing carbon-14 dating techniques. Carbon-14 dating is widely used in fields such as anthropology.

Carbon-14 is a β emitter with a half-life of 5730 years. A 1-gram sample of carbon from living material will show about 14 disintegrations per minute, not a lot of radioactivity but nevertheless detectable by modern methods.

Carbon-14 is formed in the upper atmosphere by nuclear reactions initiated by neutrons in cosmic radiation:

$$^{14}_{7}\text{N} + ^{1}_{0}\text{n} \rightarrow ^{14}_{6}\text{C} + ^{1}_{1}\text{H}$$

Once formed, carbon-14 is oxidized to $^{14}CO_2$. This product enters the carbon cycle, circulating through the atmosphere, oceans, and biosphere.

The usefulness of carbon-14 for dating comes about in the following way. Plants absorb CO_2 and convert it to organic compounds, thereby incorporating carbon-14 into living tissue. As long as a plant remains alive, this process will continue, and the percentage of carbon that is carbon-14 in the plant will equal the percentage in the atmosphere. When the plant dies, carbon-14 will no longer be taken up. Radioactive decay continues, however, with the carbon-14 activity decreasing over time. After 5730 years, the activity will be 7 dpm/g; after 11,460 years, it will be 3.5 dpm/g; and so on. By measuring the activity of a sample, and knowing the half-life of carbon-14, it is possible to calculate when a plant (or an animal that was eating plants) died.

As with all experimental procedures, carbon-14 dating has limitations. The procedure assumes that the amount of carbon-14 in the atmosphere hundreds or thousands of years ago is the same as it is now. We know that this isn't exactly true; the percentage has varied by as much as 10% (Figure 23.7). Furthermore, it is not possible to use carbon-14 to date an object that is less than about 100 years old; the radiation level from carbon-14 will not change enough in this short time period to permit accurate detection of a difference from the initial value. In most instances, the accuracy of the measurement is, in fact, only about ±100 years. Finally, it is not possible to determine ages of objects much older than about 40,000 years. By then, after nearly seven half-lives, the radioactivity will have decreased virtually to zero. But for the span of time between 100 and 40,000 years, this technique has provided important information (Figure 23.8).

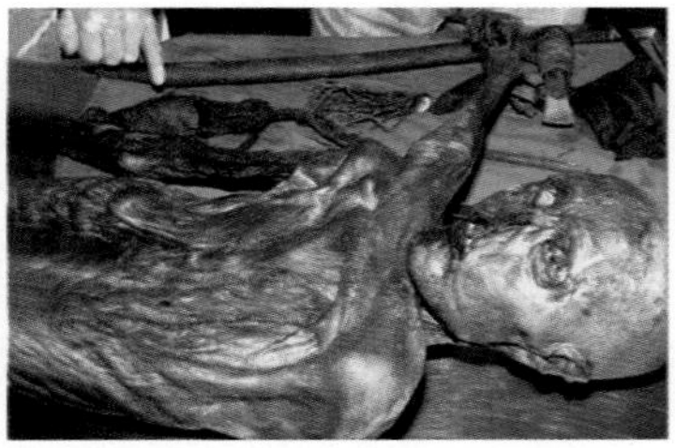

Reuters NewMedia Inc./©Corbis

FIGURE 23.8 The Ice Man. The world's oldest preserved human remains were discovered in the ice of a glacier high in the Alps. Carbon-14 dating techniques allowed scientists to determine that he lived about 5300 years ago.

Chemistry Now™

Sign in at **www.thomsonedu.com/login** and go to Chapter 23 Contents to see:

- Screen 23.3 for an exercise on **the Geiger counter**
- Screen 23.6 for a tutorial on **half-life and radiochemical dating**

EXAMPLE 23.8 Radiochemical Dating

Problem To test the concept of carbon-14 dating, J. R. Arnold and W. F. Libby applied this technique to analyze samples of acacia and cyprus wood whose ages were already known. (The acacia wood, which was supplied by the Metropolitan Museum of Art in New York, came from the tomb of Zoser, the first Egyptian pharaoh to be entombed in a pyramid. The cyprus wood was from the tomb of Sneferu.) The average activity based on five determinations was 7.04 dpm per gram of carbon. Assume (as Arnold and Libby did) that the activity of carbon-14, A_0, is 12.6 dpm per gram of carbon. Calculate the approximate age of the sample.

Strategy First, determine the rate constant for the decay of carbon-14 from its half-life ($t_{1/2}$ for ^{14}C is 5.73×10^3 years). Then, use Equation 23.5.

Solution

$$k = 0.693/t_{1/2} = 0.693/5730 \text{ yr}$$
$$= 1.21 \times 10^{-4} \text{ yr}^{-1}$$
$$\ln (A/A_0) = -kt$$
$$\ln\left(\frac{7.04 \text{ dpm/g}}{12.6 \text{ dpm/g}}\right) = (-1.21 \times 10^{-4} \text{ yr}^{-1})t$$
$$t = 4.8 \times 10^3 \text{ yr}$$

The wood is about 4800 years old.

Comment This problem uses real data from an early research paper in which the carbon-14 dating method was being tested. The age of the wood was known to be 4750 ± 250 years. (See J. R. Arnold and W. F. Libby: *Science,* Vol. 110, p. 678, 1949.)

EXERCISE 23.10 Radiochemical Dating

A sample of the inner part of a redwood tree felled in 1874 was shown to have ^{14}C activity of 9.32 dpm/g. Calculate the approximate age of the tree when it was cut down. Compare this age with that obtained from tree ring data, which estimated that the tree began to grow in 979 ± 52 BC. Use 13.4 dpm/g for the value of A_0.

23.5 Artificial Nuclear Reactions

How many different isotopes are found on earth? All of the stable isotopes occur naturally. A few unstable (radioactive) isotopes that have long half-lives are found in nature; the best-known examples are uranium-235, uranium-238, and thorium-232. Trace quantities of other radioactive isotopes with short half-lives are present because they are being formed continuously by nuclear reactions. They include isotopes of radium, polonium, and radon, along with other elements produced in various radioactive decay series, and carbon-14, formed in a nuclear reaction initiated by cosmic radiation.

Naturally occurring isotopes account for only a very small fraction of the currently known radioactive isotopes, however. The rest—several thousand—have been synthesized via artificial nuclear reactions, sometimes referred to as transmutation.

The first artificial nuclear reaction was identified by Rutherford about 80 years ago. Recall the classic experiment that led to the nuclear model of the atom (See *Milestones in the Development of Chemistry,* page 338) in which gold foil was bombarded with α particles. In the years following that experiment, Rutherford and his coworkers bombarded many other elements with α particles. In 1919, one of these experiments led to an unexpected result: when nitrogen atoms were bombarded with α particles, protons were detected among the products. Rutherford correctly concluded that a nuclear reaction had occurred. Nitrogen had undergone a *transmutation* to oxygen:

$$^{4}_{2}He + ^{14}_{7}N \rightarrow ^{17}_{8}O + ^{1}_{1}H$$

During the next decade, other nuclear reactions were discovered by bombarding other elements with α particles. Progress was slow, however, because in most cases α particles are simply scattered by target nuclei. The bombarding particles cannot get close enough to the nucleus to react because of the strong repulsive forces between the positively charged α particle and the positively charged atomic nucleus.

Two advances were made in 1932 that greatly extended nuclear reaction chemistry. The first involved the use of particle accelerators to create high-energy particles as projectiles. The second was the use of neutrons as the bombarding particles.

The α particles used in the early studies on nuclear reactions came from naturally radioactive materials such as uranium and had relatively low energies, at least by today's standards. Particles with higher energy were needed, so J. D. Cockcroft (1897–1967) and E. T. S. Walton (1903–1995), working in Rutherford's laboratory in Cambridge, England, turned to protons. Protons are formed when hydrogen atoms ionize in a cathode-ray tube, and it was known that they could be accelerated to higher energy by applying a high voltage. Cockcroft and Walton found that when energetic protons struck a lithium target, the following reaction occurs:

$$^{7}_{3}\text{Li} + ^{1}_{1}\text{p} \rightarrow 2\ ^{4}_{2}\text{He}$$

■ **Discovery of Neutrons** Neutrons had been predicted to exist for more than a decade before they were identified in 1932 by James Chadwick (1891–1974). Chadwick produced neutrons in a nuclear reaction between α particles and beryllium: $^{4}_{2}\alpha + ^{9}_{4}\text{Be} \rightarrow ^{12}_{6}\text{C} + ^{1}_{0}\text{n}$.

This was the first example of a reaction initiated by a particle that had been artificially accelerated to high energy. Since this experiment was conducted, the technique has been developed much further, and the use of particle accelerators in nuclear chemistry is now commonplace. Particle accelerators operate on the principle that a charged particle placed between charged plates will be accelerated to a high speed and high energy. Modern examples of this process are seen in the synthesis of the transuranium elements, several of which are described in more detail in *A Closer Look: The Search for New Elements.*

Experiments using neutrons as bombarding particles were first carried out in both the United States and Great Britain in 1932. Nitrogen, oxygen, fluorine, and neon were bombarded with energetic neutrons, and α particles were detected among the products. Using neutrons made sense: because neutrons have no charge, it was reasoned that these particles would not be repelled by the positively charged nucleus particles. Thus, neutrons did not need high energies to react.

Lawrence Berkeley Laboratory

■ **Glenn T. Seaborg (1912–1999)** Seaborg figured out that thorium and the elements that followed it fit under the lanthanides in the periodic table. For this insight, he and Edwin McMillan shared the 1951 Nobel Prize in chemistry. Over a 21-year period, Seaborg and his colleagues synthesized 10 new transuranium elements (Pu through Lr). To honor Seaborg's scientific contributions, the name "seaborgium" was assigned to element 106. It marked the first time an element was named for a living person.

In 1934, Enrico Fermi (1901–1954) and his coworkers showed that nuclear reactions using neutrons are more favorable if the neutrons have low energy. A low energy neutron is simply captured by the nucleus, giving a product in which the mass number is increased by one unit. Because of the low energy of the bombarding particle, the product nucleus does not have sufficient energy to fragment in these reactions. The new nucleus is produced in an excited state, however; when the nucleus returns to the ground state, a γ-ray is emitted. Reactions in which a neutron is captured and a γ-ray is emitted are called **(n, γ) reactions.**

The (n, γ) reactions are the source of many of the radioisotopes used in medicine and chemistry. An example is radioactive phosphorus, $^{32}_{15}\text{P}$, which is used in chemical studies such as tracing the uptake of phosphorus in the body.

$$^{31}_{15}\text{P} + ^{1}_{0}\text{n} \rightarrow ^{32}_{15}\text{P} + \gamma$$

■ **Transuranium Elements in Nature** Neptunium, plutonium, and americium were unknown prior to their preparation via these nuclear reactions. Later, these elements were found to be present in trace quantities in uranium ores.

Transuranium elements, elements with an atomic number greater than 92, were first made in a nuclear reaction sequence beginning with an (n, γ) reaction. Scientists at the University of California at Berkeley bombarded uranium-238 with neutrons. Among the products identified were neptunium-239 and plutonium-239. These new elements were formed when ^{239}U decayed by β radiation.

$$^{238}_{92}\text{U} + ^{1}_{0}\text{n} \rightarrow ^{239}_{92}\text{U}$$
$$^{239}_{92}\text{U} \rightarrow ^{239}_{93}\text{Np} + ^{0}_{-1}\beta$$
$$^{239}_{93}\text{Np} \rightarrow ^{239}_{94}\text{Pu} + ^{0}_{-1}\beta$$

A Closer Look — The Search for New Elements

By 1936, guided first by Mendeleev's predictions and later by atomic theory, chemists had identified all but two of the elements with atomic numbers between 1 and 92. From this point onward, all new elements to be discovered came from artificial nuclear reactions. Two gaps in the periodic table were filled when radioactive technetium and promethium, the last two elements with atomic numbers less than 92, were identified in 1937 and 1942, respectively. The first success in the search for elements with atomic numbers higher than 92 came with the 1940 discovery of neptunium and plutonium.

Since 1950, laboratories in the United States (Lawrence Berkeley National Laboratory), Russia (Joint Institute for Nuclear Research at Dubna, near Moscow), and Europe (Institute for Heavy Ion Research at Darmstadt, Germany) have competed to make new elements. Syntheses of new transuranium elements use a standard methodology. An element of fairly high atomic number is bombarded with a beam of high-energy particles. Initially, neutrons were used; later, helium nuclei and then larger nuclei such as ^{11}B and ^{12}C were employed; and, more recently, highly charged ions of elements such as calcium, chromium, cobalt, and zinc have been chosen. The bombarding particle fuses with the nucleus of the target atom, forming a new nucleus that lasts for a short time before decomposing. New elements are detected by their decomposition products, a signature of particles with specific masses and energies.

Fermilab Visual Media Services, Batavia, IL

Fermilab. The tunnel housing the four-mile-long particle accelerator at Batavia, Illinois

By using bigger particles and higher energies, the list of known elements reached 106 by the end of the 1970s. To further extend the search, Russian scientists employed a new idea, matching precisely the energy of the bombarding particle with the energy required to fuse the nuclei. This technique enabled the synthesis of elements 107, 108, and 109 in Darmstadt in the early 1980s, and the synthesis of elements 110, 111, and 112 in the following decade. Lifetimes of these elements were in the millisecond range; the 277 isotope of element 112, for example, mass has a half-life of 240 μs.

Yet another breakthrough was needed to extend the list further. Scientists have long known that isotopes with specific *magic numbers* of neutrons and protons are more stable. Elements with 2, 8, 20, 50, and 82 protons are members of this category, as are elements with 126 neutrons. The magic numbers correspond to filled shells in the nucleus. Their significance is analogous to the significance of filled shells for electronic structure. Theory had predicted that the next magic numbers would be 114 protons and 184 neutrons. Using this information, researchers discovered element 114 in early 1999. The Dubna group reporting this discovery found that the mass 289 isotope had an exceptionally long half-life, about 20 s.

At the time this book was written, 117 elements were known. Will research yield further new elements? It would be hard to say no, given past successes in this area of research, but the quest becomes ever more difficult as scientists venture to the very limits of nuclear stability.

Four years later, a similar reaction sequence was used to make americium-241. Plutonium-239 was found to add two neutrons to form plutonium-241, which decays by β emission to give americium-241.

EXAMPLE 23.9 Nuclear Reactions

Problem Write equations for the nuclear reactions described below.

(a) Fluorine-19 undergoes an (n, γ) reaction to give a radioactive product that decays by $_{-1}^{0}\beta$ emission. (Write equations for both nuclear reactions.)

(b) A common neutron source is a plutonium–beryllium alloy. Plutonium-239 is an α emitter. When beryllium-9 (the only stable isotope of beryllium) reacts with α particles emitted by plutonium, neutrons are ejected. (Write equations for both reactions.)

Strategy The equations are written so that both mass and charge are balanced.

Solution

(a) $^{19}_{9}F + ^{1}_{0}n \rightarrow ^{20}_{9}F + \gamma$

$^{20}_{9}F \rightarrow ^{20}_{10}Ne + ^{0}_{-1}\beta$

(b) $^{239}_{94}Pu \rightarrow ^{235}_{92}U + ^{4}_{2}\alpha$

$^{4}_{2}\alpha + ^{9}_{4}Be \rightarrow ^{12}_{6}C + ^{1}_{0}n$

23.6 Nuclear Fission

■ **Fission Reactions** In the fission of uranium-236, a large number of different fission products (elements) are formed. Barium was the element first identified, and its identification provided the key that led to recognition that fission had occurred.

In 1938, two chemists, Otto Hahn (1879–1968) and Fritz Strassman (1902–1980), isolated and identified barium in a sample of uranium that had been bombarded with neutrons. How was barium formed? The answer to that question explained one of the most significant scientific discoveries of the 20th century. The uranium nucleus had split into smaller pieces in the process we now call **nuclear fission.**

The details of nuclear fission were unraveled through the work of a number of scientists. They determined that a uranium-235 nucleus initially captured a neutron to form uranium-236. This isotope underwent nuclear fission to produce two new nuclei, one with a mass number around 140 and the other with a mass around 90, along with several neutrons (Figure 23.9). The nuclear reactions that led to formation of barium when a sample of ^{235}U was bombarded with neutrons are

$$^{235}_{92}U + ^{1}_{0}n \rightarrow ^{236}_{92}U$$
$$^{236}_{92}U \rightarrow ^{141}_{56}Ba + ^{92}_{36}Kr + 3\,^{1}_{0}n$$

■ **Lise Meitner (1878–1968)** Meitner's greatest contribution to 20th-century science was her explanation of the process of nuclear fission. She and her nephew, Otto Frisch, also a physicist, published a paper in 1939 that was the first to use the term "nuclear fission." Element number 109 is named meitnerium to honor Meitner's contributions. The leader of the team that discovered this element said that "She should be honored as the most significant woman scientist of [the 20th] century."

AIP-Emilio Segré Visual Archives, Herzfeld Collection

An important aspect of fission reactions is that they produce more neutrons than are used to initiate the process. Under the right circumstances, these neutrons then serve to continue the reaction. If one or more of these neutrons are captured by another ^{235}U nucleus, then a further reaction can occur, releasing still more neutrons. This sequence repeats over and over. Such a mechanism, in which each step generates a reactant to continue the reaction, is called a **chain reaction.**

A nuclear fission chain reaction has three general steps:

1. *Initiation.* The reaction of a single atom is needed to start the chain. Fission of ^{235}U is initiated by the absorption of a neutron.
2. *Propagation.* This part of the process repeats itself over and over, with each step yielding more product. The fission of ^{236}U releases neutrons that initiate the fission of other uranium atoms.
3. *Termination.* Eventually, the chain will end. Termination could occur if the reactant (^{235}U) is used up, or if the neutrons that continue the chain escape from the sample without being captured by ^{235}U.

■ **The Atomic Bomb** In an atomic bomb, each nuclear fission step produces 3 neutrons, which leads to about 3 more fissions and 9 more neutrons, which leads to 9 more fission steps and 27 more neutrons, and so on. The rate depends on the number of neutrons, so the nuclear reaction occurs faster and faster as more and more neutrons are formed, leading to an enormous output of energy in a short time span.

To harness the energy produced in a nuclear reaction, it is necessary to control the rate at which a fission reaction occurs. This is managed by balancing the propagation and termination steps by limiting the number of neutrons available. In a nuclear reactor, this balance is accomplished by using cadmium rods to absorb neutrons. By withdrawing or inserting the rods, the number of neutrons available to propagate the chain can be changed, and the rate of the fission reaction (and the rate of energy production) can be increased or decreased.

Uranium-235 and plutonium-239 are the fissionable isotopes most commonly used in power reactors. Natural uranium contains only 0.72% of uranium-235; more than 99% of the natural element is uranium-238. The percentage of uranium-235 in natural uranium is too small to sustain a chain reaction, however, so the uranium used for nuclear fuel must be enriched in this isotope. One way to do so is by

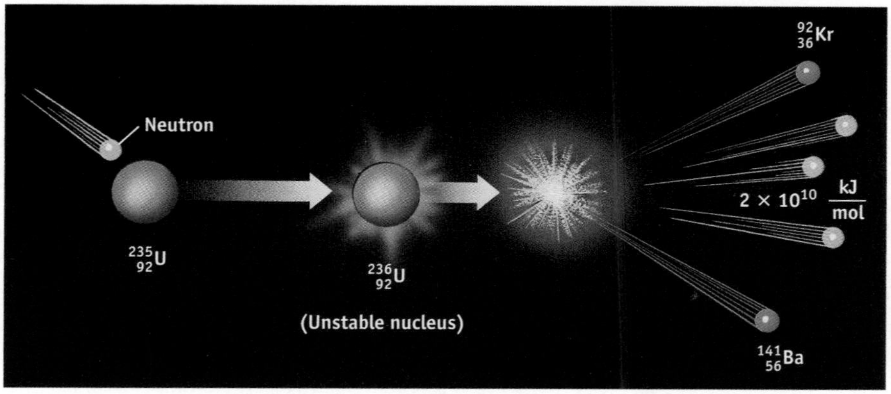

FIGURE 23.9 Nuclear fission. Neutron capture by $^{235}_{92}U$ produces $^{236}_{92}U$. This isotope undergoes fission, which yields several fragments along with several neutrons. These neutrons initiate further nuclear reactions by adding to other $^{235}_{92}U$ nuclei. The process is highly exothermic, producing about 2×10^{10} kJ/mol.

gaseous diffusion (◀ Section 11.7). Plutonium, which occurs naturally in only trace quantities, must be made via a nuclear reaction. The raw material for this nuclear synthesis is the more abundant uranium isotope, ^{238}U. Addition of a neutron to ^{238}U gives ^{239}U, which, as noted earlier, undergoes two β emissions to form ^{239}Pu.

Currently, there are over 100 operating nuclear power plants in the United States and more than 400 worldwide. About 20% of this country's electricity (and 17% of the world's energy) comes from nuclear power (Table 23.2). Although one might imagine that nuclear energy would be called upon to meet the ever-increasing needs of society, no new nuclear power plants are under construction in the United States. Among other things, the disasters at Chernobyl (in the former Soviet Union) in 1986 and Three Mile Island (in Pennsylvania) in 1979 have sensitized the public to the issue of safety. The cost to construct a nuclear power plant (measured in terms of dollars per kilowatt-hour of power) is considerably more than the cost for a natural gas–powered facility, and the regulatory restrictions for nuclear power are burdensome. Disposal of highly radioactive nuclear waste is another thorny problem, with 20 metric tons of waste being generated per year at each reactor.

In addition to technical problems, nuclear energy production brings with it significant geopolitical security concerns. The process for enriching uranium for use in a reactor is the same process used for generating weapons-grade uranium. Also, some nuclear reactors are designed so that one by-product of their operation is the isotope plutonium-239, which can be removed and used in a nuclear weapon. Despite these problems, nuclear fission is an important part of the energy profile in a number of countries. For example, three fourths of power production in France and one third in Japan is nuclear generated.

TABLE 23.2 Percentage of Electricity Produced Using Nuclear Power Plants

Country (rank)	Total Power from Nuclear Energy (%)
1. France	75.0
2. Lithuania	73.1
3. Belgium	57.7
4. Bulgaria	47.1
5. Slovak Republic	47.0
6. Sweden	46.8
. . .	
19. United States	19.9
20. Russia	14.4
21. Canada	12.7

Source: Chemical and Engineering News, p. 42, Oct. 2, 2000.

23.7 Nuclear Fusion

In a **nuclear fusion** reaction, several small nuclei react to form a larger nucleus. Tremendous amounts of energy can be generated by such reactions. An example is the fusion of deuterium and tritium nuclei to form $^{4}_{2}He$ and a neutron:

$$^{2}_{1}H + ^{3}_{1}H \rightarrow ^{4}_{2}He + ^{1}_{0}n \qquad \Delta E = -1.7 \times 10^{9} \text{ kJ/mol}$$

Fusion reactions provide the energy of our sun and other stars. Scientists have long dreamed of being able to harness fusion to provide power. To do so, a temperature of 10^6 to 10^7 K, like that in the interior of the sun, would be required to bring the positively charged nuclei together with enough energy to overcome nuclear repulsions. At the very high temperatures needed for a fusion reaction, mat-

■ **Radon in Your Home** You should test your home for the presence of radon gas (◀ page 1065). The activity level should be less than 4 picocuries per liter of air.

ter does not exist as atoms or molecules; instead, matter is in the form of a *plasma* made up of unbound nuclei and electrons.

Three critical requirements must be met before nuclear fusion could represent a viable energy source. First, the temperature must be high enough for fusion to occur. The fusion of deuterium and tritium, for example, requires a temperature of 10^7 K or more. Second, the plasma must be confined long enough to release a net output of energy. Third, the energy must be recovered in some usable form.

Harnessing a nuclear fusion reaction for a peaceful use has not yet been achieved. Nevertheless, many attractive features encourage continuing research in this field. The hydrogen used as "fuel" is cheap and available in almost unlimited amounts. As a further benefit, most radioisotopes produced by fusion have short half-lives, so they remain a radiation hazard for only a short time.

23.8 Radiation Health and Safety

Units for Measuring Radiation

Several units of measurement are used to describe levels and doses of radioactivity. As is the case in everyday life, the units used in the United States are not the same as the SI units of measurement.

In the United States, the degree of radioactivity is often measured in **curies** (Ci). Less commonly used in the United States is the SI unit, the **becquerel** (Bq). Both units measure the number of disintegrations per second; 1 Ci is 3.7×10^{10} dps (disintegrations per second), while 1 Bq represents 1 dps. The curie and the becquerel are used to report the amount of radioactivity when multiple kinds of unstable nuclei are decaying and to report amounts necessary for medical purposes.

By itself, the degree of radioactivity does not provide a good measure of the amount of energy in the radiation or the amount of damage that the radiation can cause to living tissue. Two additional kinds of information are necessary. The first is the amount of energy absorbed; the second is the effectiveness of the particular kind of radiation in causing tissue damage. The amount of energy absorbed by living tissue is measured in **rads.** *Rad* is an acronym for "radiation absorbed dose." One rad represents 0.01 J of energy absorbed per kilogram of tissue. Its SI equivalent is the **gray** (Gy); 1 Gy denotes the absorption of 1 J per kilogram of tissue.

Different forms of radiation cause different amounts of biological damage. The amount of damage depends on how strongly a form of radiation interacts with matter. Alpha particles cannot penetrate the body any farther than the outer layer of skin. If α particles are emitted within the body, however, they will do between 10 and 20 times the amount of damage done by γ-rays, which can go entirely through a human body without being stopped. In determining the amount of biological damage to living tissue, differences in damaging power are accounted for using a "quality factor." This quality factor has been set at 1 for β and γ radiation, 5 for low-energy protons and neutrons, and 20 for α particles or high-energy protons and neutrons.

Biological damage is quantified in a unit called the **rem** (an acronym for "roentgen equivalent man"). A dose of radiation in rem is determined by multiplying the energy absorbed in rads by the quality factor for that kind of radiation. The rad and the rem are very large in comparison to normal exposures to radiation, so it is more common to express exposures in millirems (mrem). The SI equivalent of the rem is the **sievert** (Sv), determined by multiplying the dose in grays by the quality factor.

■ **The Roentgen** The roentgen (R) is an older unit of radiation exposure. It is defined as the amount of x-rays or γ radiation that will produce 2.08×10^9 ions in 1 cm^3 of dry air. The roentgen and the rad are similar in size. Wilhelm Roentgen (1845–1923) first produced and detected x-radiation. Element 111 has been named roentgenium in his honor.

Radiation: Doses and Effects

Exposure to a small amount of radiation is unavoidable. Earth is constantly being bombarded with radioactive particles from outer space. There is also some exposure to radioactive elements that occur naturally on earth, including ^{14}C, ^{40}K (a radioactive isotope that occurs naturally in 0.0117% abundance), ^{238}U, and ^{232}Th. Radioactive elements in the environment that were created artificially (in the fallout from nuclear bomb tests, for example) also contribute to this exposure. For some people, medical procedures using radioisotopes are a major contributor.

The average dose of background radioactivity to which a person in the U.S. is exposed is about 200 mrem per year (Table 23.3). Well over half of that amount comes from natural sources over which we have no control. Of the 60–70 mrem per year exposure that comes from artificial sources, nearly 90% is delivered in medical procedures such as x-ray examinations and radiation therapy. Considering the controversy surrounding nuclear power, it is interesting to note that less than 0.5% of the total annual background dose of radiation that the average person receives can be attributed to the nuclear power industry.

Describing the biological effects of a dose of radiation precisely is not a simple matter. The amount of damage done depends not only on the kind of radiation and the amount of energy absorbed, but also on the particular tissues exposed and the rate at which the dose builds up. A great deal has been learned about the effects of radiation on the human body by studying the survivors of the bombs dropped over Japan in World War II and the workers exposed to radiation from

TABLE 23.3 Radiation Exposure of an Individual for One Year from Natural and Artificial Sources

	Millirem/Year	Percentage
Natural Sources		
Cosmic radiation	50.0	25.8
The earth	47.0	24.2
Building materials	3.0	1.5
Inhaled from the air	5.0	2.6
Elements found naturally in human tissue	21.0	10.8
Subtotal	**126.0**	**64.9**
Medical Sources		
Diagnostic x-rays	50.0	25.8
Radiotherapy	10.0	5.2
Internal diagnosis	1.0	0.5
Subtotal	**61.0**	**31.5**
Other Artificial Sources		
Nuclear power industry	0.85	0.4
Luminous watch dials, TV tubes	2.0	1.0
Fallout from nuclear tests	4.0	2.1
Subtotal	**6.9**	**3.5**
Total	**193.9**	**99.9**

A Closer Look — What Is a Safe Exposure?

Is the exposure to natural background radiation totally without effect? Can you equate the effect of a single dose and the effect of cumulative, smaller doses that are spread out over a long period of time? The assumption generally made is that no "safe maximum dose," or level below which absolutely no damage will occur, exists. However, the accuracy of this assumption has come into question. These issues are not testable with human subjects, and tests based on animal studies are not completely reliable because of the uncertainty of species-to-species variations.

The model used by government regulators to set exposure limits assumes that the relationship between exposure to radiation and incidence of radiation-induced problems, such as cancer, anemia, and immune system problems, is linear. Under this assumption, if a dose of 2*x* rem causes damage in 20% of the population, then a dose of *x* rem will cause damage in 10% of the population. But is this true? Cells do possess mechanisms for repairing damage. Many scientists believe that this self-repair mechanism renders the human body less susceptible to damage from smaller doses of radiation, because the damage will be repaired as part of the normal course of events. They argue that, at extremely low doses of radiation, the self-repair response results in less damage.

The bottom line is that much still remains to be learned in this area. And the stakes are significant.

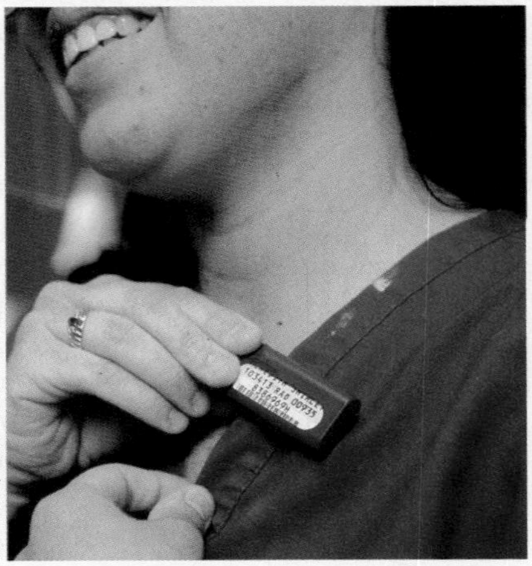
Cliff Moore/Photo Researchers, Inc.

The film badge. These badges, worn by scientists using radioactive materials, are used to monitor cumulative exposure to radiation.

the reactor disaster at Chernobyl. From studies of the health of these survivors, we have learned that the effects of radiation are not generally observable below a single dose of 25 rem. At the other extreme, a single dose of >200 rem will be fatal to about half the population (Table 23.4).

Our information is more accurate when dealing with single, large doses than it is for the effects of chronic, smaller doses of radiation. One current issue of debate in the scientific community is how to judge the effects of multiple smaller doses or long-term exposure (see *A Closer Look: What Is A Safe Exposure?*).

23.9 Applications of Nuclear Chemistry

We tend to think about nuclear chemistry in terms of power plants and bombs. In truth, radioactive elements are now used in all areas of science and medicine, and they are of ever-increasing importance to our lives. Because describing all of their uses would take several books, we have selected just a few examples to illustrate the diversity of applications of radioactivity.

TABLE 23.4 Effects of a Single Dose of Radiation

Dose (rem)	Effect
0–25	No effect observed
26–50	Small decrease in white blood cell count
51–100	Significant decrease in white blood cell count, lesions
101–200	Loss of hair, nausea
201–500	Hemorrhaging, ulcers, death in 50% of population
500	Death

Nuclear Medicine: Medical Imaging

Diagnostic procedures using nuclear chemistry are essential in medical imaging, which entails the creation of images of specific parts of the body. There are three principal components to constructing a radioisotope-based image:

- A radioactive isotope, administered as the element or incorporated into a compound, that concentrates the radioactive isotope in the tissue to be imaged
- A method of detecting the type of radiation involved
- A computer to assemble the information from the detector into a meaningful image

The choice of a radioisotope and the manner in which it is administered are determined by the tissue in question. A compound containing the isotope must be absorbed more by the target tissue than by the rest of the body. Table 23.5 lists radioisotopes that are commonly used in nuclear imaging processes, their half-lives, and the tissues they are used to image. All of the isotopes in Table 23.5 are γ emitters; γ radiation is preferred for imaging because it is less damaging to the body in small doses than either α or β radiation.

Technetium-99m is used in more than 85% of the diagnostic scans done in hospitals each year (see *A Closer Look: Technetium-99m*). The "m" stands for *metastable*, a term used to identify an unstable state that exists for a finite period of time. Recall that atoms in excited electronic states emit visible, infrared, and ultraviolet radiation (◀ Chapter 6). Similarly, a nucleus in an excited state gives up its excess energy, but in this case a much higher energy is involved, and the emission occurs as γ radiation. The γ-rays given off by ^{99m}Tc are detected to produce the image (Figure 23.10).

Another medical imaging technique based on nuclear chemistry is positron emission tomography (PET). In PET, an isotope that decays by positron emission is incorporated into a carrier compound and given to the patient. When emitted, the positron travels no more than a few millimeters before undergoing matter–antimatter annihilation.

$$ {}^{0}_{+1}\beta + {}^{0}_{-1}e \rightarrow 2\gamma $$

The two emitted γ-rays travel in opposite directions. By determining where high numbers of γ-rays are being emitted, one can construct a map showing where the positron emitter is located in the body.

An isotope often used in PET is ^{15}O. A patient is given gaseous O_2 that contains ^{15}O. This isotope travels throughout the body in the bloodstream, allowing images of the brain and bloodstream (Figure 23.11) to be obtained. Because positron emitters are typically very short-lived, PET facilities must be located near a cyclotron where the radioactive nuclei are prepared and then immediately incorporated into a carrier compound.

TABLE 23.5 Radioisotopes Used in Medical Diagnostic Procedures

Radioisotope	Half-Life (h)	Imaging
^{99m}Tc	6.0	Thyroid, brain, kidneys
^{201}Tl	73.0	Heart
^{123}I	13.2	Thyroid
^{67}Ga	78.2	Various tumors and abscesses
^{18}F	1.8	Brain, sites of metabolic activity

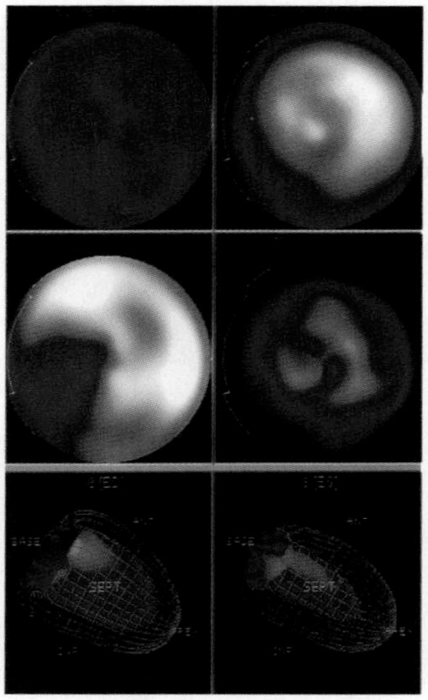

David Parker/Photo Researchers, Inc.

FIGURE 23.10 Heart imaging with technetium-99m. The radioactive element technetium-99m, a gamma emitter, is injected into a patient's vein in the form of the pertechnetate ion (TcO_4^-) or as a complex ion with an organic ligand. A series of scans of the gamma emissions of the isotope are made while the patient is resting and then again after strenuous exercise. Bright areas in the scans indicate that the isotope is binding to the tissue in that area. The scans in this figure show a normal heart function.

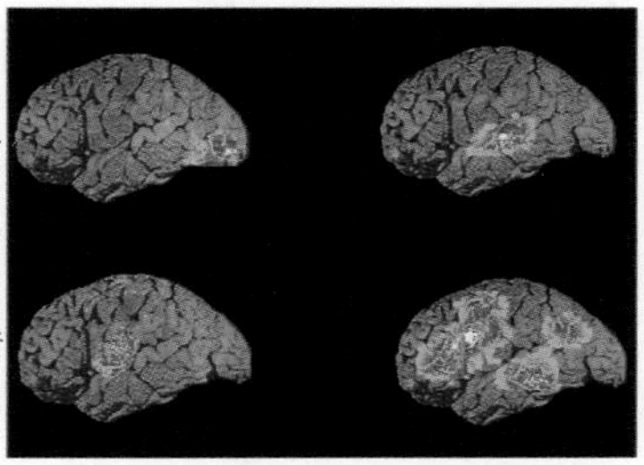

Wellcome Department of Neurology/Science Photo Library/Photo Researchers, Inc.

FIGURE 23.11 PET scans of the brain. These scans show the left side of the brain; red indicates an area of highest activity. (*upper left*) *Sight* activates the visual area in the occipital cortex at the back of the brain. (*upper right*) *Hearing* activates the auditory area in the superior temporal cortex of the brain. (*lower left*) *Speaking* activates the speech centers in the insula and motor cortex. (*lower right*) *Thinking* about verbs, and speaking them, generates high activity, including in the hearing, speaking, temporal, and parietal areas.

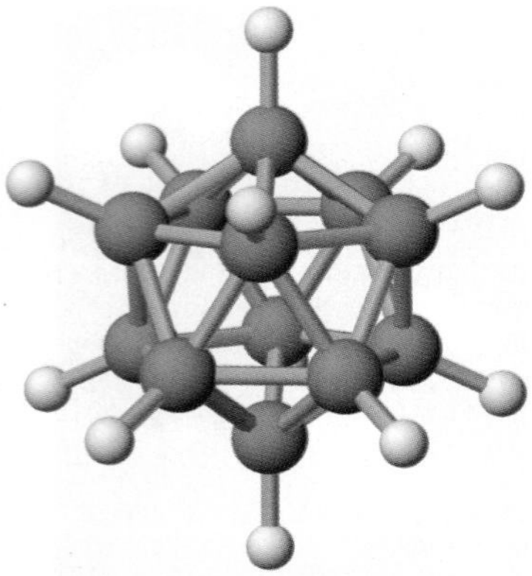

One of the compounds used in BNCT is $Na_2[B_{12}H_{12}]$. The structure of the $B_{12}H_{12}^{2-}$ anion is a regular polyhedron with 20 sides, called an icosahedron.

Nuclear Medicine: Radiation Therapy

To treat most cancers, it is necessary to use radiation that can penetrate the body to the location of the tumor. Gamma radiation from a cobalt-60 source is commonly used. Unfortunately, the penetrating ability of γ-rays makes it virtually impossible to destroy diseased tissue without also damaging healthy tissue in the process. Nevertheless, this technique is a regularly sanctioned procedure, and its successes are well known.

To avoid the side effects associated with more traditional forms of radiation therapy, a new form of treatment has been explored in the last 10 to 15 years, called *boron neutron capture therapy* (BNCT). BNCT is unusual in that boron-10, the isotope of boron used as part of the treatment, is not radioactive. Boron-10 is highly effective in capturing neutrons, however—2500 times better than boron-11, and eight times better than uranium-235. When the nucleus of a boron-10 atom captures a neutron, the resulting boron-11 nucleus has so much energy that it fragments to form an α particle and a lithium-7 atom. Although the α particles do a great deal of damage, because their penetrating power is so low, the damage remains confined to an area not much larger than one or two cells in diameter.

In a typical BNCT treatment, a solution of a boron compound is injected into the tumor. After a few hours, the tumor is bombarded with neutrons. The α particles are produced only at the site of the tumor, and the production stops when the neutron bombardment ends.

Analytical Methods: The Use of Radioactive Isotopes as Tracers

Radioactive isotopes can be used to help determine the fate of compounds in the body or in the environment. These studies begin with a compound that contains a radioactive isotope of one of its component elements. In biology, for example, scientists can use radioactive isotopes to measure the uptake of nutrients. Plants take up phosphorus-containing compounds from the soil through their roots. By adding a small amount of radioactive ^{32}P, a β emitter with a half-life of 14.3 days, to fertilizer and then measuring the rate at which the radioactivity appears in the leaves, plant biologists can determine the rate at which phosphorus is taken up. The outcome can assist scientists in identifying hybrid strains of plants that can absorb phosphorus quickly, resulting in faster-maturing crops, better yields per acre, and more food or fiber at less expense.

To measure pesticide levels, a pesticide can be tagged with a radioisotope and then applied to a test field. By counting the disintegrations of the radioactive tracer, information can be obtained about how much pesticide accumulates in the soil, is taken up by the plant, and is carried off in runoff surface water. After these tests are completed, the radioactive isotope decays to harmless levels in a few days or a few weeks because of the short half-lives of the isotopes used.

Analytical Methods: Isotope Dilution

Imagine, for the moment, that you wanted to estimate the volume of blood in an animal subject. How might you do this? Obviously, draining the blood and measuring its volume in volumetric glassware is not a desirable option.

One technique uses a method called isotope dilution. In this process, a small amount of radioactive isotope is injected into the bloodstream. After a period of time to allow the isotope to become distributed throughout the body, a blood sample is taken and its radioactivity measured. The calculation used to determine the total blood volume is illustrated in the next example.

A Closer Look — Technetium-99m

Technetium was the first new element to be made artificially. One might think that this element would be a chemical rarity, but this is not so: its importance in medical imaging has brought a great deal of attention to it. Although all the technetium in the world must be synthesized by nuclear reactions, the element is readily available and even inexpensive; its price in 1999 was \$60 per gram, only about four times the price of gold.

Technetium-99m is formed when molybdenum-99 decays by β emission ($t_{1/2} = 6.01h$). Technetium-99m then decays to its ground state with a half-life of 6.01 h, giving off a 140-KeV γ-ray in the process. (Technetium-99 is itself unstable, decaying to stable ^{99}Ru with a half-life of 2.1×10^5 years.)

Technetium-99m is produced in hospitals using a molybdenum–technetium generator. Sheathed in lead shielding, the generator contains the artificially synthesized isotope ^{99}Mo in the form of molybdate ion, MoO_4^{2-}, adsorbed on a column of alumina, Al_2O_3. The MoO_4^{2-} is continually being converted into the pertechnate ion, $^{99m}TcO_4^-$ by β emission. When it is needed, the $^{99m}TcO_4^-$ is washed from the column using a saline solution. Technetium-99m may be used directly as the pertechnate ion or converted into other compounds. The pertechnate ion or radiopharmaceuticals made from it are administered intravenously to the patient. Such small quantities are needed that 1 μg (microgram) of technetium-99m is sufficient for the average hospital's daily imaging needs.

One use of ^{99m}Tc is for imaging the thyroid gland. Because $I^-(aq)$ and $TcO_4^-(aq)$ ions have very similar sizes, the thyroid will (mistakenly) take up $TcO_4^-(aq)$ along with iodide ion. This uptake concentrates ^{99m}Tc in the thyroid and allows a physician to obtain images such as the one shown in the accompanying figure.

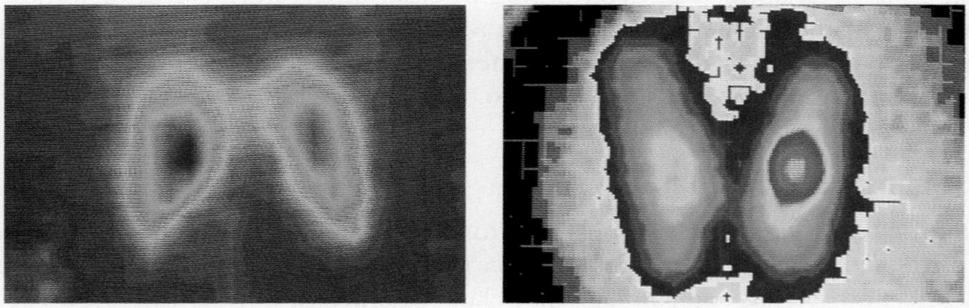

CNRI/Science Photo Library/Photo Researchers, Inc.

(a) Healthy human thyroid gland. **(b)** Thyroid gland showing effect of hyperthyroidism.

Thyroid imaging. Technetium-99m concentrates in sites of high activity. Images of this gland, which is located at the base of the neck, were obtained by recording γ-ray emission after the patient was given radioactive technetium-99m. Current technology creates a computer color-enhanced scan.

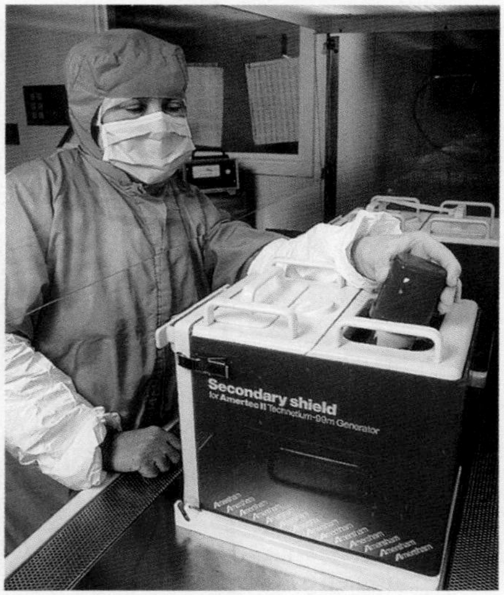

David Parker/Science Photo Library/Photo Researchers, Inc.

A technetium-99m generator. A technician is loading a sample containing the MoO_4^{2-} ion into the device that will convert molybdate ion to $^{99m}TcO_4^-$.

EXAMPLE 23.10 Analysis Using Isotope Dilution

Problem A 1.00-mL solution containing 0.240 μCi of tritium is injected into a dog's bloodstream. After a period of time to allow the isotope to be dispersed, a 1.00-mL sample of blood is drawn. The radioactivity of this sample is found to be 4.3×10^{-4} μCi/mL. What is the total volume of blood in the dog?

Strategy For any solution, concentration equals the amount of solute divided by volume of solution. In this problem, we relate the activity of the sample (in Ci) to concentration. The total amount of solute is 0.240 μCi, and the concentration (measured on the small sample of blood) is 4.3×10^{-4} μCi/mL. The unknown is the volume.

Solution The blood contains a total of 0.240 μCi of the radioactive material. We can represent its concentration as 0.240 μCi/x, where x is the total blood volume. After dilution in the bloodstream, 1.00 mL of blood is found to have an activity of 4.3×10^{-4} μCi/mL.

$$0.240\ \mu\text{Ci}/x\ \text{mL} = 4.3 \times 10^{-4}\ \mu\text{Ci}/1.00\ \text{mL}$$

$$x = 560\ \text{mL}$$

EXERCISE 23.11 Analysis by Isotope Dilution

Suppose you hydrolyze a 10.00-g sample of a protein. Next, you add to it a 3.00-mg sample of ^{14}C-labeled threonine, an amino acid with a specific activity of 3000 dpm. After mixing, part of the threonine (60.0 mg) is separated and isolated from the mixture. The activity of the isolated sample is 1200 dpm. How much threonine was present in the original sample?

TABLE 23.6 Rare Earth Analysis of Moon Rock Sample 10022 (a fine-grain igneous rock)

Element	Concentration (ppm)
La	26.4
Ce	68
Nd	66
Sm	21.2
Eu	2.04
Gd	25
Tb	4.7
Dy	31.2
Ho	5.5
Er	16
Yb	17.7
Lu	2.55

Source: L. A. Haskin, P. A, Helmke, and R. O. Allen: *Science,* Vol. 167, p. 487, 1970. The concentrations of rare earths in moon rocks were quite similar to the values in terrestrial rocks except that the europium concentration is much depleted. The January 30, 1970, issue of *Science* was devoted to analysis of moon rocks.

Space Science: Neutron Activation Analysis and the Moon Rocks

The first manned space mission to the moon brought back a number of samples of soil and rock—a treasure trove for scientists. One of their first tasks was to analyze these samples to determine their identity and composition. Most analytical methods require chemical reactions using at least a small amount of material; however, this was not a desirable option, considering that the moon rocks were at the time the most valuable rocks in the world.

A few scientists got a chance to work on this unique project, and one of the analytical tools they used was **neutron activation analysis.** In this nondestructive process, a sample is irradiated with neutrons. Most isotopes add a neutron to form a new isotope that is one mass unit higher in an excited nuclear state. When the nucleus decays to its ground state, it emits a γ-ray. The energy of the γ-ray identifies the element, and the number of γ-rays can be counted to determine the amount of the element in the sample. Using neutron activation analysis, it is possible to analyze for a number of elements in a single experiment (Table 23.6).

Neutron activation analysis has many other uses. This analytical procedure yields a kind of fingerprint that can be used to identify a substance. For example, this technique has been applied in determining whether an art work is real or fraudulent. Analysis of the pigments in paints on a painting can be carried out without damaging the painting to determine whether the composition resembles modern paints or paints used hundreds of years ago.

Food Science: Food Irradiation

Refrigeration, canning, and chemical additives provide significant protection in terms of food preservation, but in some parts of the world these procedures are unavailable, and stored-food spoilage may claim as much as 50% of the food crop. Irradiation with γ-rays from sources such as ^{60}Co and ^{137}Cs is an option for prolonging the shelf life of foods. Relatively low levels of radiation may retard the growth of organisms, such as bacteria, molds, and yeasts, that can cause food spoilage. After irradiation, milk in a sealed container has a minimum shelf life of 3 months without refrigeration. Chicken normally has a 3-day refrigerated shelf life; after irradiation, it may have a 3-week refrigerated shelf life.

Higher levels of radiation, in the 1- to 5-Mrad (1 Mrad $= 1 \times 10^6$ rad) range, will kill every living organism. Foods irradiated at these levels will keep indefinitely when sealed in plastic or aluminum-foil packages. Ham, beef, turkey, and corned beef sterilized by radiation have been used on many Space Shuttle flights, for example. An astronaut said, "The beautiful thing was that it didn't disturb the taste, which made the meals much better than the freeze-dried and other types of foods we had."

These procedures are not without their opponents, and the public has not fully embraced irradiation of foods yet. An interesting argument favoring this technique is that radiation is less harmful than other methodologies for food preservation. This type of sterilization offers greater safety to food workers because it lessens chances of exposure to harmful chemicals, and it protects the environment by avoiding contamination of water supplies with toxic chemicals.

Food irradiation is commonly used in European countries, Canada, and Mexico. Its use in the United States is currently regulated by the U.S. Food and Drug Administration (FDA) and Department of Agriculture (USDA). In 1997, the FDA approved the irradiation of refrigerated and frozen uncooked meat to control pathogens and extend shelf-life, and in 2000 the USDA approved the irradiation of eggs to control *Salmonella* infection (Table 23.7).

Case Study — Nuclear Medicine and Hyperthyroidism

The primary function of the thyroid gland, located in your neck, is the production of thyroxine (3,5,3′,5′-tetraiodothyronine) and 3,5,3′-triiodothyronine. These chemical compounds are hormones that help to regulate the rate of metabolism (page 510), a term that refers to all of the chemical reactions that take place in the body. In particular, the thyroid hormones play an important role in the processes that release energy from food.

Abnormally low levels of thyroxine result in a condition known as hypothyroidism. Its symptoms include lethargy and feeling cold much of the time. The remedy for this condition is medication, consisting of thyroxine pills. The opposite condition, hyperthyroidism, also occurs in some people. In this condition, the body produces too much of the hormone. Hyperthyroidism is diagnosed by symptoms such as nervousness, heat intolerance, increased appetite, and muscle weakness and fatigue when blood sugar is too rapidly depleted. The standard remedy for hyperthyroidism is to destroy part of the thyroid gland, and one way to do so is to use a compound containing radioactive iodine-123 or -131.

To understand this procedure, you need to know something about iodine in the body. Iodine is an essential element. Some diets provide iodine naturally (seaweed, for example, is a good source of iodine), but in the Western world most iodine taken up by the body comes from iodized salt, NaCl containing a few percent of NaI. An adult man or woman of average size should take in about 150 μg (micrograms) of iodine (1 $\mu g = 10^{-6}$ g) in the daily diet. In the body, iodide ion is transported to the thyroid, where it serves as one of the raw materials in making thyroxine.

The fact that iodine concentrates in the thyroid tissue is essential to the procedure for using radioiodine therapy as a treatment for hyperthyroidism. Typically, an aqueous NaI solution in which a small fraction of iodide consists of the radioactive isotope iodine-131 or iodine-123, and the rest is nonradioactive iodine-127. The radioactivity destroys thyroid tissue, resulting in a decrease in thyroid activity.

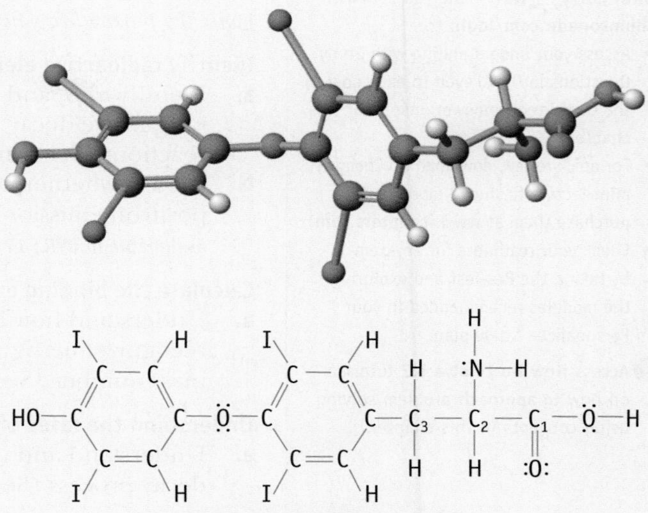

3,5,3′,5′-tetraiodothyronine (thyroxine)

Thyroxine. The hormone 3,5,3′,5′-tetraiodothyronine (thyroxine) exerts a stimulating effect on metabolism.

Questions:

1. *^{131}I decays by beta emission. What is the product of this decay? Write a balanced nuclear equation for this process.*
2. *The half-life of ^{131}I is 8.04 d, and that of ^{123}I is 13.3 h. If you begin with an equal number of moles of each isotope, at the end of 7 days what are their relative amounts?*

Answers to these questions are in Appendix Q.

TABLE 23.7 Approvals of Food Irradiation

1963	FDA approves irradiation to control insects in wheat and flour.
1964	FDA approves irradiation to inhibit sprouting of white potatoes.
1985	FDA approves irradiation at specific doses to control *Trichinella spiralis* infection of pork.
1986	FDA approves irradiation at specific doses to delay maturation, inhibit growth, and disinfect foods including vegetables and spices.
1992	USDA approves irradiation of poultry to control infection by *Salmonella* and bacteria.
1997	FDA permits use of ionizing radiation to treat refrigerated and frozen meat and other meat products.
2000	USDA approves irradiation of eggs to control *Salmonella* infection.

Source: **www.foodsafety.gov/~fsg/irradiat.html.**

Chapter Goals Revisited

Chemistry Now™ Sign in at **www.thomsonedu.com/login** to:

- Assess your understanding with Study Questions in OWL keyed to each goal in the Goals and Homework menu for this chapter
- For quick review, download Go Chemistry mini-lecture flashcard modules (or purchase them at **www.ichapters.com**)
- Check your readiness for an exam by taking the Pre-Test and exploring the modules recommended in your Personalized Study plan.

Access **How Do I Solve It?** tutorials on how to approach problem solving using concepts in this chapter.

Now that you have studied this chapter, you should ask whether you have met the chapter goals. In particular, you should be able to:

Identify radioactive elements, and describe natural and artificial nuclear reactions

a. Identify α, β, and γ radiation, the three major types of radiation in natural radioactive decay (Section 23.1) and write balanced equations for nuclear reactions (Section 23.2). **Study question(s) assignable in OWL: 12, 16, 32, 41.**

b. Predict whether a radioactive isotope will decay by α or β emission, or by positron emission or electron capture (Sections 23.2 and 23.3). **Study question(s) assignable in OWL: 17, 18, 20.**

Calculate the binding energy and binding energy per nucleon for a particular isotope

a. Understand how binding energy per nucleon is defined (Section 23.3) and recognize the significance of a graph of binding energy per nucleon versus mass number (Section 23.3). **Study question(s) assignable in OWL: 23, 52, 61, 62.**

Understand the rates of radioactive decay

a. Understand and use mathematical equations that characterize the radioactive decay process (Section 23.4). **Study question(s) assignable in OWL: 30, 32, 34, 36, 47, 50, 54, 58.**

Understand artificial nuclear reactions

a. Describe nuclear chain reactions, nuclear fission, and nuclear fusion (Sections 23.6 and 23.7).

Understand issues of health and safety with respect to radioactivity

a. Describe the units used to measure intensity, and understand how they pertain to health issues (Section 23.8 and 23.9).

Be aware of some uses of radioactive isotopes in science and medicine

KEY EQUATIONS

Equation 23.1 (page 1070): The equation relating interconversion of mass (m) and energy (E). This equation is applied in the calculation of binding energy (E_b) for nuclei.

$$E_b = (\Delta m)c^2$$

Equation 23.2 (page 1073): The activity of a radioactive sample (A) is proportional to the number of radioactive atoms (N).

$$A \propto N$$

Equation 23.4 (page 1074): The rate law for nuclear decay based on number of radioactive atoms initially present (N_0) and the number N after time t.

$$\ln(N/N_0) = -kt$$

Equation 23.5 (page 1074): The rate law for nuclear decay based on the measured activity of a sample (A).

$$\ln(A/A_0) = -kt$$

Equation 23.6 (page 1074): The relationship between the half-life and the rate constant for a nuclear decay process.

$$t_{1/2} = 0.693/k$$

STUDY QUESTIONS

Online homework for this chapter may be assigned in OWL.

▲ denotes challenging questions.

■ denotes questions assignable in OWL.

Blue-numbered questions have answers in Appendix O and fully-worked solutions in the *Student Solutions Manual.*

Practicing Skills

Important Concepts

1. Some important discoveries in scientific history that contributed to the development of nuclear chemistry are listed below. Briefly, describe each discovery, identify prominent scientists who contributed to it, and comment on the significance of the discovery to the development of this field.
(a) 1896, the discovery of radioactivity
(b) 1898, the identification of radium and polonium
(c) 1918, the first artificial nuclear reaction
(d) 1932, (n, γ) reactions
(e) 1939, fission reactions

2. In Chapter 3, the law of conservation of mass was introduced as an important principle in chemistry. The discovery of nuclear reactions forced scientists to modify this law. Explain why, and give an example illustrating that mass is not conserved in a nuclear reaction.

3. A graph of binding energy per nucleon is shown in Figure 23.4. Explain how the data used to construct this graph were obtained.

4. How is Figure 23.3 used to predict the type of decomposition for unstable (radioactive) isotopes?

5. Outline how nuclear reactions are carried out in the laboratory. Describe the artificial nuclear reactions used to make an element with an atomic number greater than 92.

6. What mathematical equations define the rates of decay for radioactive elements?

7. Explain how carbon-14 is used to estimate the ages of archeological artifacts. What are the limitations for use of this technique?

8. Describe how the concept of half-life for nuclear decay is used.

9. What is a radioactive decay series? Explain why radium and polonium are found in uranium ores.

10. The interaction of radiation with matter has both positive and negative consequences. Discuss briefly the hazards of radiation and the way that radiation can be used in medicine.

Nuclear Reactions

(See Examples 23.1 and 23.2.)

11. Complete the following nuclear equations. Write the mass number and atomic number for the remaining particle, as well as its symbol.
(a) ${}^{54}_{26}\text{Fe} + {}^{4}_{2}\text{He} \longrightarrow 2\ {}^{1}_{0}\text{n} + ?$
(b) ${}^{27}_{13}\text{Al} + {}^{4}_{2}\text{He} \longrightarrow {}^{30}_{15}\text{P} + ?$
(c) ${}^{32}_{16}\text{S} + {}^{1}_{0}\text{n} \longrightarrow {}^{1}_{1}\text{H} + ?$
(d) ${}^{96}_{42}\text{Mo} + {}^{2}_{1}\text{H} \longrightarrow {}^{1}_{0}\text{n} + ?$
(e) ${}^{98}_{42}\text{Mo} + {}^{1}_{0}\text{n} \longrightarrow {}^{99}_{43}\text{Tc} + ?$
(f) ${}^{18}_{9}\text{F} \longrightarrow {}^{18}_{8}\text{O} + ?$

12. ■ Complete the following nuclear equations. Write the mass number, atomic number, and symbol for the remaining particle.
(a) ${}^{9}_{4}\text{Be} + ? \longrightarrow {}^{6}_{3}\text{Li} + {}^{4}_{2}\text{He}$
(b) $? + {}^{1}_{0}\text{n} \longrightarrow {}^{24}_{11}\text{Na} + {}^{4}_{2}\text{He}$
(c) ${}^{40}_{20}\text{Ca} + ? \longrightarrow {}^{40}_{19}\text{K} + {}^{1}_{1}\text{H}$
(d) ${}^{241}_{95}\text{Am} + {}^{4}_{2}\text{He} \longrightarrow {}^{243}_{97}\text{Bk} + ?$
(e) ${}^{246}_{96}\text{Cm} + {}^{12}_{6}\text{C} \longrightarrow 4\ {}^{1}_{0}\text{n} + ?$
(f) ${}^{238}_{92}\text{U} + ? \longrightarrow {}^{249}_{100}\text{Fm} + 5\ {}^{1}_{0}\text{n}$

13. Complete the following nuclear equations. Write the mass number, atomic number, and symbol for the remaining particle.
(a) ${}^{111}_{47}\text{Ag} \longrightarrow {}^{111}_{48}\text{Cd} + ?$
(b) ${}^{87}_{36}\text{Kr} \longrightarrow {}^{0}_{-1}\beta + ?$
(c) ${}^{231}_{91}\text{Pa} \longrightarrow {}^{227}_{89}\text{Ac} + ?$
(d) ${}^{230}_{90}\text{Th} \longrightarrow {}^{4}_{2}\text{He} + ?$
(e) ${}^{82}_{35}\text{Br} \longrightarrow {}^{82}_{36}\text{Kr} + ?$
(f) $? \longrightarrow {}^{24}_{12}\text{Mg} + {}^{0}_{-1}\beta$

14. Complete the following nuclear equations. Write the mass number, atomic number, and symbol for the remaining particle.
(a) ${}^{19}_{10}\text{Ne} \longrightarrow {}^{0}_{+1}\beta + ?$
(b) ${}^{59}_{26}\text{Fe} \longrightarrow {}^{0}_{-1}\beta + ?$
(c) ${}^{40}_{19}\text{K} \longrightarrow {}^{0}_{-1}\beta + ?$
(d) ${}^{37}_{18}\text{Ar} + {}^{0}_{-1}\text{e}$ (electron capture) $\longrightarrow ?$
(e) ${}^{55}_{26}\text{Fe} + {}^{0}_{-1}\text{e}$ (electron capture) $\longrightarrow ?$
(f) ${}^{26}_{13}\text{Al} \longrightarrow {}^{25}_{12}\text{Mg} + ?$

15. The uranium-235 radioactive decay series, beginning with ${}^{235}_{92}\text{U}$ and ending with ${}^{207}_{82}\text{Pb}$, occurs in the following sequence: $\alpha, \beta, \alpha, \beta, \alpha, \alpha, \alpha, \alpha, \beta, \beta, \alpha$. Write an equation for each step in this series.

16. ■ The thorium-232 radioactive decay series, beginning with ${}^{232}_{90}\text{Th}$ and ending with ${}^{208}_{82}\text{Pb}$, occurs in the following sequence: $\alpha, \beta, \beta, \alpha, \alpha, \alpha, \alpha, \beta, \beta, \alpha$. Write an equation for each step in this series.

Nuclear Stability and Nuclear Decay

(See Examples 23.3 and 23.4.)

17. ■ What particle is emitted in the following nuclear reactions? Write an equation for each reaction.
   (a) Gold-198 decays to mercury-198.
   (b) Radon-222 decays to polonium-218.
   (c) Cesium-137 decays to barium-137.
   (d) Indium-110 decays to cadmium-110.

18. ■ What is the product of the following nuclear decay processes? Write an equation for each process.
   (a) Gallium-67 decays by electron capture.
   (b) Potassium-38 decays with positron emission.
   (c) Technetium-99m decays with γ emission.
   (d) Manganese-56 decays by β emission.

19. Predict the probable mode of decay for each of the following radioactive isotopes, and write an equation to show the products of decay.
   (a) Bromine-80m
   (b) Californium-240
   (c) Cobalt-61
   (d) Carbon-11

20. ■ Predict the probable mode of decay for each of the following radioactive isotopes, and write an equation to show the products of decay.
   (a) Manganese-54
   (b) Americium-241
   (c) Silver-110
   (d) Mercury-197m

21. (a) Which of the following nuclei decay by ${}_{-1}^{0}\beta$ decay?

   $^{3}H \quad ^{16}O \quad ^{20}F \quad ^{13}N$

   (b) Which of the following nuclei decays by ${}_{+1}^{0}\beta$ decay?

   $^{238}U \quad ^{19}F \quad ^{22}Na \quad ^{24}Na$

22. (a) Which of the following nuclei decay by ${}_{-1}^{0}\beta$ decay?

   $^{1}H \quad ^{23}Mg \quad ^{32}P \quad ^{20}Ne$

   (b) Which of the following nuclei decay by ${}_{+1}^{0}\beta$ decay?

   $^{235}U \quad ^{35}Cl \quad ^{38}K \quad ^{24}Na$

23. ■ Boron has two stable isotopes, ^{10}B and ^{11}B. Calculate the binding energies per nucleon of these two nuclei. The required masses (in g/mol) are ${}_{1}^{1}H$ = 1.00783, ${}_{0}^{1}n$ = 1.00867, ${}_{5}^{10}B$ = 10.01294, and ${}_{5}^{11}B$ = 11.00931.

24. Calculate the binding energy in kilojoules per mole of nucleons of P for the formation of ^{30}P and ^{31}P. The required masses (in g/mol) are ${}_{1}^{1}H$ = 1.00783, ${}_{0}^{1}n$ = 1.00867, ${}_{15}^{30}P$ = 29.97832, and ${}_{15}^{31}P$ = 30.97376.

25. Calculate the binding energy per nucleon for calcium-40, and compare your result with the value in Figure 23.4. Masses needed for this calculation are (in g/mol) ${}_{1}^{1}H$ = 1.00783, ${}_{0}^{1}n$ = 1.00867, and ${}_{20}^{40}Ca$ = 39.96259.

26. Calculate the binding energy per nucleon for iron-56. Masses needed for this calculation (in g/mol) are ${}_{1}^{1}H$ = 1.00783, ${}_{0}^{1}n$ = 1.00867, and ${}_{26}^{56}Fe$ = 55.9349. Compare the result of your calculation to the value for iron-56 in the graph in Figure 23.4.

27. Calculate the binding energy per mole of nucleons for ${}_{8}^{16}O$. Masses needed for this calculations are ${}_{1}^{1}H$ = 1.00783, ${}_{0}^{1}n$ = 1.00867, and ${}_{8}^{16}O$ = 15.99492.

28. Calculate the binding energy per nucleon for nitrogen-14. The mass of nitrogen-14 is 14.003074.

Rates of Radioactive Decay

(See Examples 23.5 and 23.6.)

29. Copper acetate containing ^{64}Cu is used to study brain tumors. This isotope has a half-life of 12.7 h. If you begin with 25.0 μg of ^{64}Cu, what mass in micrograms remains after 63.5 h?

30. ■ Gold-198 is used in the diagnosis of liver problems. The half-life of ^{198}Au is 2.69 days. If you begin with 2.8 μg of this gold isotope, what mass remains after 10.8 days?

31. Iodine-131 is used to treat thyroid cancer.
   (a) The isotope decays by β particle emission. Write a balanced equation for this process.
   (b) Iodine-131 has a half-life of 8.04 days. If you begin with 2.4 μg of radioactive ^{131}I, what mass remains after 40.2 days?

32. ■ Phosphorus-32 is used in the form of Na_2HPO_4 in the treatment of chronic myeloid leukemia, among other things.
   (a) The isotope decays by β particle emission. Write a balanced equation for this process.
   (b) The half-life of ^{32}P is 14.3 days. If you begin with 4.8 μg of radioactive ^{32}P in the form of Na_2HPO_4, what mass remains after 28.6 days (about one month)?

33. Gallium-67 ($t_{1/2}$ = 78.25 h) is used in the medical diagnosis of certain kinds of tumors. If you ingest a compound containing 0.015 mg of this isotope, what mass (in milligrams) remains in your body after 13 days? (Assume none is excreted.)

34. ■ Iodine-131 ($t_{1/2}$ = 8.04 days), a β emitter, is used to treat thyroid cancer.
   (a) Write an equation for the decomposition of ^{131}I.
   (b) If you ingest a sample of NaI containing ^{131}I, how much time is required for the activity to decrease to 35.0% of its original value?

35. Radon has been the focus of much attention recently because it is often found in homes. Radon-222 emits α particles and has a half-life of 3.82 d.
   (a) Write a balanced equation to show this process.
   (b) How long does it take for a sample of ^{222}Rn to decrease to 20.0% of its original activity?

36. ■ Strontium-90 is a hazardous radioactive isotope that resulted from atmospheric testing of nuclear weapons. A sample of strontium carbonate containing ^{90}Sr is found to have an activity of 1.0×10^3 dpm. One year later, the activity of this sample is 975 dpm.

(a) Calculate the half-life of strontium-90 from this information.
(b) How long will it take for the activity of this sample to drop to 1.0% of the initial value?

37. Radioactive cobalt-60 is used extensively in nuclear medicine as a γ-ray source. It is made by a neutron capture reaction from cobalt-59 and is a β emitter; β emission is accompanied by strong γ radiation. The half-life of cobalt-60 is 5.27 years.
(a) How long will it take for a cobalt-60 source to decrease to one eighth of its original activity?
(b) What fraction of the activity of a cobalt-60 source remains after 1.0 year?

38. Scandium occurs in nature as a single isotope, scandium-45. Neutron irradiation produces scandium-46, a β emitter with a half-life of 83.8 days. If the initial activity is 7.0×10^4 dpm, draw a graph showing disintegrations per minute as a function of time during a period of one year.

Nuclear Reactions

(See Example 23.9.)

39. Americium-240 is made by bombarding plutonium-239 with α particles. In addition to ^{240}Am, the products are a proton and two neutrons. Write a balanced equation for this process.

40. There are two isotopes of americium, both with half-lives sufficiently long to allow the handling of large quantities. Americium-241, with a half-life of 432 years, is an α emitter. It is used in smoke detectors. The isotope is formed from ^{239}Pu by absorption of two neutrons followed by emission of a β particle. Write a balanced equation for this process.

41. ■ The superheavy element ^{287}Uuq was made by firing a beam of ^{48}Ca ions at ^{242}Pu. Three neutrons were ejected in the reaction. Write a balanced nuclear equation for the synthesis of ^{287}Uuq.

42. To synthesize the heavier transuranium elements, a nucleus must be bombarded with a relatively large particle. If you know the products are californium-246 and four neutrons, with what particle would you bombard uranium-238 atoms?

43. Deuterium nuclei (^2_1H) are particularly effective as bombarding particles to carry out nuclear reactions. Complete the following equations:
(a) $^{114}_{48}Cd + ^2_1H \longrightarrow ? + ^1_1H$
(b) $^6_3Li + ^2_1H \longrightarrow ? + ^1_0n$
(c) $^{40}_{20}Ca + ^2_1H \longrightarrow ^{38}_{19}K + ?$
(d) $? + ^2_1H \longrightarrow ^{65}_{30}Zn + \gamma$

44. Element $^{287}114$ decayed by α emission with a half-life of about 5 s. Write an equation for this process.

45. Boron is an effective absorber of neutrons. When boron-10 adds a neutron, an α particle is emitted. Write an equation for this nuclear reaction.

46. Some of the reactions explored by Rutherford and others are listed below. Identify the unknown species in each reaction.
(a) $^{14}_7N + ^4_2He \longrightarrow ^{17}_8O + ?$
(b) $^9_4Be + ^4_2He \longrightarrow ? + ^1_0n$
(c) $? + ^4_2He \longrightarrow ^{30}_{15}P + ^1_0n$
(d) $^{239}_{94}Pu + ^4_2He \longrightarrow ? + ^1_0n$

General Questions

These questions are not designated as to type or location in the chapter. They may combine several concepts.

47. ■ ▲ A technique to date geological samples uses rubidium-87, a long-lived radioactive isotope of rubidium ($t_{1/2} = 4.8 \times 10^{10}$ years). Rubidium-87 decays by β emission to strontium-87. If the rubidium-87 is part of a rock or mineral, then strontium-87 will remain trapped within the crystalline structure of the rock. The age of the rock dates back to the time when the rock solidified. Chemical analysis of the rock gives the amounts of ^{87}Rb and ^{87}Sr. From these data, the fraction of ^{87}Rb that remains can be calculated.

Analysis of a stony meteorite determined that 1.8 mmol of ^{87}Rb and 1.6 mmol of ^{87}Sr were present. Estimate the age of the meteorite. (*Hint:* The amount of ^{87}Rb at t_0 is moles ^{87}Rb + moles ^{87}Sr.)

48. Tritium, ^3_1H, is one of the nuclei used in fusion reactions. This isotope is radioactive, with a half-life of 12.3 years. Like carbon-14, tritium is formed in the upper atmosphere from cosmic radiation, and it is found in trace amounts on earth. To obtain the amounts required for a fusion reaction, however, it must be made via a nuclear reaction. The reaction of ^6_3Li with a neutron produces tritium and an α particle. Write an equation for this nuclear reaction.

49. Phosphorus occurs in nature as a single isotope, phosphorus-31. Neutron irradiation of phosphorus-31 produces phosphorus-32, a β emitter with a half-life of 14.28 days. Assume you have a sample containing phosphorus-32 that has a rate of decay of 3.2×10^6 dpm. Draw a graph showing disintegrations per minute as a function of time during a period of one year.

50. ■ In June 1972, natural fission reactors, which operated billions of years ago, were discovered in Oklo, Gabon (page 1060). At present, natural uranium contains 0.72% ^{235}U. How many years ago did natural uranium contain 3.0% ^{235}U, the amount needed to sustain a natural reactor? ($t_{1/2}$ for ^{235}U is 7.04×10^8 years.)

51. If a shortage in worldwide supplies of fissionable uranium arose, it would be possible to use other fissionable nuclei. Plutonium, one such fuel, can be made in "breeder" reactors that manufacture more fuel than they consume. The sequence of reactions by which plutonium is made is as follows:
(a) A ^{238}U nucleus undergoes an (n, γ) to produce ^{239}U.
(b) ^{239}U decays by β emission ($t_{1/2}$ = 23.5 min) to give an isotope of neptunium.
(c) This neptunium isotope decays by β emission to give a plutonium isotope.
(d) The plutonium isotope is fissionable. On collision of one of these plutonium isotopes with a neutron, fission occurs, with at least two neutrons and two other nuclei as products.

Write an equation for each of the nuclear reactions.

52. ■ When a neutron is captured by an atomic nucleus, energy is released as γ radiation. This energy can be calculated based on the change in mass in converting reactants to products.

For the nuclear reaction $^{6}_{3}Li + ^{1}_{0}n \longrightarrow ^{7}_{3}Li + \gamma$:

(a) Calculate the energy evolved in this reaction (per atom). Masses needed (in g/mol) are $^{6}_{3}Li$ = 6.01512, $^{1}_{0}n$ = 1.00867, and $^{7}_{3}Li$ = 7.01600.
(b) Use the answer in part (a) to calculate the wavelength of the γ-rays emitted in the reaction.

In the Laboratory

53. A piece of charred bone found in the ruins of a Native American village has a $^{14}C : ^{12}C$ ratio that is 72% of the ratio found in living organisms. Calculate the age of the bone fragment ($t_{1/2}$ for ^{14}C is 5.73 × 103 y).

54. ■ A sample of wood from a Thracian chariot found in an excavation in Bulgaria has a ^{14}C activity of 11.2 dpm/g. Estimate the age of the chariot and the year it was made ($t_{1/2}$ for ^{14}C is 5.73 × 10^3 years, and the activity of ^{14}C in living material is 14.0 dpm/g).

55. The isotope of polonium that was most likely isolated by Marie Curie in her pioneering studies is polonium-210. A sample of this element was prepared in a nuclear reaction. Initially, its activity (α emission) was 7840 dpm. Measuring radioactivity over time produced the data below. Determine the half-life of polonium-210.

Activity (dpm)	Time (days)
7840	0
7570	7
7300	14
5920	56
5470	72

56. Sodium-23 (in a sample of NaCl) is subjected to neutron bombardment in a nuclear reactor to produce ^{24}Na. When removed from the reactor, the sample is radioactive, with β activity of 2.54 × 10^4 dpm. The decrease in radioactivity over time was studied, producing the following data:

Activity (dpm)	Time (h)
2.54 × 10^4	0
2.42 × 10^4	1
2.31 × 10^4	2
2.00 × 10^4	5
1.60 × 10^4	10
1.01 × 10^4	20

(a) Write equations for the neutron capture reaction and for the reaction in which the product of this reaction decays by β emission.
(b) Determine the half-life of sodium-24.

57. The age of minerals can sometimes be determined by measuring the amounts of ^{206}Pb and ^{238}U in a sample. This determination assumes that all of the ^{206}Pb in the sample comes from the decay of ^{238}U. The date obtained identifies when the rock solidified. Assume that the ratio of ^{206}Pb to ^{238}U in an igneous rock sample is 0.33. Calculate the age of the rock. ($t_{1/2}$ for ^{238}U is 4.5 × 10^9 y.)

58. ■ The oldest-known fossil found in South Africa has been dated based on the decay of Rb-87.

$$^{87}Rb \longrightarrow ^{87}Sr + ^{0}_{-1}\beta \qquad t_{1/2} = 4.8 \times 10^{10}\ y$$

If the ratio of the present quantity of ^{87}Rb to the original quantity is 0.951, calculate the age of the fossil.

59. To measure the volume of the blood system of an animal, the following experiment was done. A 1.0-mL sample of an aqueous solution containing tritium, with an activity of 2.0 × 10^6 dps, was injected into the animal's bloodstream. After time was allowed for complete circulatory mixing, a 1.0-mL blood sample was withdrawn and found to have an activity of 1.5 × 10^4 dps. What was the volume of the circulatory system? (The half-life of tritium is 12.3 y, so this experiment assumes that only a negligible amount of tritium has decayed in the time of the experiment.)

60. Suppose that you hydrolyze 4.644 g of a protein to form a mixture of different amino acids. To this is added a 2.80-mg sample of ^{14}C-labeled threonine (one of the amino acids present). The activity of this small sample is 1950 dpm. A chromatographic separation of the amino acids is carried out, and a small sample of pure threonine is separated. This sample has an activity of 550 dpm. What fraction of the threonine present was separated? What is the total amount of threonine in the sample?

Summary and Conceptual Questions

The following questions may use concepts from this and previous chapters.

61. ■ The average energy output of a good grade of coal is 2.6×10^7 kJ/ton. Fission of 1 mol of ^{235}U releases 2.1×10^{10} kJ. Find the number of tons of coal needed to produce the same energy as 1 lb of ^{235}U. (See Appendix C for conversion factors.)

62. ■ Collision of an electron and a positron results in formation of two γ-rays. In the process, their masses are converted completely into energy.
(a) Calculate the energy evolved from the annihilation of an electron and a positron, in kilojoules per mole.
(b) Using Planck's equation (Equation 6.2), determine the frequency of the γ-rays emitted in this process.

63. The principle underlying the isotope dilution method can be applied to many kinds of problems. Suppose that you, a marine biologist, want to estimate the number of fish in a lake. You release 1000 tagged fish, and after allowing an adequate amount of time for the fish to disperse evenly in the lake, you catch 5250 fish and find that 27 of them have tags. How many fish are in the lake?

64. ▲ Radioactive isotopes are often used as "tracers" to follow an atom through a chemical reaction. The following is an example of this process: acetic acid reacts with methanol, CH_3OH, by eliminating a molecule of H_2O to form methyl acetate, $CH_3CO_2CH_3$. Explain how you would use the radioactive isotope ^{15}O to show whether the oxygen atom in the water product comes from the —OH of the acid or the —OH of the alcohol.

65. ▲ Radioactive decay series begin with a very long-lived isotope. For example, the half-life of ^{238}U is 4.5×10^9 y. Each series is identified by the name of the long-lived parent isotope of highest mass.
(a) The uranium-238 radioactive decay series is sometimes referred to as the $4n + 2$ series because the masses of all 13 members of this series can be expressed by the equation $m = 4n + 2$, where m is the mass number and n is an integer. Explain why the masses are correlated in this way. (See the Comment in Example 23.1.)
(b) Two other radioactive decay series identified in minerals in the earth's crust are the thorium-232 series and the uranium-235 series. Do the masses of the isotopes in these series conform to a simple mathematical equation? If so, identify the equation.
(c) Identify the radioactive decay series to which each of the following isotopes belongs: $^{226}_{88}Ra$, $^{215}_{86}At$, $^{228}_{90}Th$, $^{210}_{83}Bi$.
(d) Evaluation reveals that one series of elements, the $4n + 1$ series, is not present in the earth's crust. Speculate why.

66. ▲ The thorium decay series includes the isotope $^{228}_{90}Th$. Determine the sequence of nuclei on going from $^{232}_{90}Th$ to $^{228}_{90}Th$.

67. ▲ The last unknown element between bismuth and uranium was discovered by Lise Meitner (1878–1968) and Otto Hahn (1879–1968) in 1918. They obtained ^{231}Pa by chemical extraction of pitchblende, in which its concentration is about 1 ppm. This isotope, an α emitter, has a half-life of 3.27×10^4 y.
(a) Which radioactive decay series (the uranium-235, uranium-238, or thorium-232 series) contains ^{231}Pa as a member? (See the Comment in Example 23.1.)
(b) Suggest a possible sequence of nuclear reactions starting with the long-lived isotope that eventually forms this isotope.
(c) What quantity of ore would be required to isolate 1.0 g of ^{231}Pa, assuming 100% yield?
(d) Write an equation for the radioactive decay process for ^{231}Pa.

68. ▲ You might wonder how it is possible to determine the half-life of long-lived radioactive isotopes such as ^{238}U. With a half-life of more than 10^9 y, the radioactivity of a sample of uranium will not measurably change in your lifetime. In fact, you can calculate the half-life using the mathematics governing first-order reactions.

It can be shown that a 1.0-mg sample of ^{238}U decays at the rate of 12 α emissions per second. Set up a mathematical equation for the rate of decay, $\Delta N/\Delta t = -kN$, where N is the number of nuclei in the 1.0-mg sample and $\Delta N/\Delta t$ is 12 dps. Solve this equation for the rate constant for this process, and then relate the rate constant to the half-life of the reaction. Carry out this calculation, and compare your result with the literature value, 4.5×10^9 years.

69. ▲ Marie and Pierre Curie isolated radium and polonium from uranium ore (pitchblende, which contains ^{238}U and ^{235}U). Which of the following isotopes of radium and polonium can be found in the uranium ore? (Hint: consider both the isotope half lives and the decay series starting with ^{238}U and ^{235}U.)

Isotope	Half-life
^{226}Ra	1620 y
^{225}Ra	14.8 d
^{228}Ra	6.7 y
^{216}Po	0.15 s
^{210}Po	138.4 d

List of Appendices

A Using Logarithms and the Quadratic Equation

An introductory chemistry course requires basic algebra plus a knowledge of (1) exponential (or scientific) notation, (2) logarithms, and (3) quadratic equations. The use of exponential notation was reviewed on pages 32–35, and this appendix reviews the last two topics.

A.1 Logarithms

Two types of logarithms are used in this text: (1) common logarithms (abbreviated log) whose base is 10 and (2) natural logarithms (abbreviated ln) whose base is *e* (= 2.71828):

$$\log x = n, \text{ where } x = 10^n$$
$$\ln x = m, \text{ where } x = e^m$$

Most equations in chemistry and physics were developed in natural, or base *e*, logarithms, and we follow this practice in this text. The relation between log and ln is

$$\ln x = 2.303 \log x$$

Despite the different bases of the two logarithms, they are used in the same manner. What follows is largely a description of the use of common logarithms.

A common logarithm is the power to which you must raise 10 to obtain the number. For example, the log of 100 is 2, since you must raise 10 to the second power to obtain 100. Other examples are

$$\log 1000 = \log (10^3) = 3$$
$$\log 10 = \log (10^1) = 1$$
$$\log 1 = \log (10^0) = 0$$
$$\log 0.1 = \log (10^{-1}) = -1$$
$$\log 0.0001 = \log (10^{-4}) = -4$$

To obtain the common logarithm of a number other than a simple power of 10, you must resort to a log table or an electronic calculator. For example,

$$\log 2.10 = 0.3222, \text{ which means that } 10^{0.3222} = 2.10$$
$$\log 5.16 = 0.7126, \text{ which means that } 10^{0.7126} = 5.16$$
$$\log 3.125 = 0.49485, \text{ which means that } 10^{0.49485} = 3.125$$

To check this on your calculator, enter the number, and then press the "log" key.

To obtain the natural logarithm ln of the numbers shown here, use a calculator having this function. Enter each number, and press "ln:"

$$\ln 2.10 = 0.7419, \text{ which means that } e^{0.7419} = 2.10$$
$$\ln 5.16 = 1.6409, \text{ which means that } e^{1.6409} = 5.16$$

To find the common logarithm of a number greater than 10 or less than 1 with a log table, first express the number in scientific notation. Then find the log of each part of the number and add the logs. For example,

$$\begin{aligned} \log 241 &= \log (2.41 \times 10^2) = \log 2.41 + \log 10^2 \\ &= 0.382 + 2 = 2.382 \\ \log 0.00573 &= \log (5.73 \times 10^{-3}) = \log 5.73 + \log 10^{-3} \\ &= 0.758 + (-3) = -2.242 \end{aligned}$$

Significant Figures and Logarithms

Notice that the mantissa has as many significant figures as the number whose log was found. (So that you could more clearly see the result obtained with a calculator or a table, this rule was not strictly followed until the last two examples.)

■ **Logarithms and Nomenclature**
The number to the left of the decimal in a logarithm is called the **characteristic,** and the number to the right of the decimal is the **mantissa.**

Obtaining Antilogarithms

If you are given the logarithm of a number, and find the number from it, you have obtained the "antilogarithm," or "antilog," of the number. Two common procedures used by electronic calculators to do this are:

Procedure A	Procedure B
1. Enter the log or ln.	1. Enter the log or ln.
2. Press 2ndF.	2. Press INV.
3. Press 10^x or e^x.	3. Press log or ln x.

Test one or the other of these procedures with the following examples:

1. Find the number whose log is 5.234:

Recall that $\log x = n$, where $x = 10^n$. In this case, $n = 5.234$. Enter that number in your calculator, and find the value of 10^n, the antilog. In this case,

$$10^{5.234} = 10^{0.234} \times 10^5 = 1.71 \times 10^5$$

Notice that the characteristic (5) sets the decimal point; it is the power of 10 in the exponential form. The mantissa (0.234) gives the value of the number x.

2. Find the number whose log is −3.456:

$$10^{-3.456} = 10^{0.544} \times 10^{-4} = 3.50 \times 10^{-4}$$

Notice here that −3.456 must be expressed as the sum of −4 and +0.544.

Mathematical Operations Using Logarithms

Because logarithms are exponents, operations involving them follow the same rules used for exponents. Thus, multiplying two numbers can be done by adding logarithms:

$$\log xy = \log x + \log y$$

For example, we multiply 563 by 125 by adding their logarithms and finding the antilogarithm of the result:

$$\begin{aligned}
\log 563 &= 2.751 \\
\log 125 &= \underline{2.097} \\
\log xy &= 4.848 \\
xy &= 10^{4.848} = 10^4 \times 10^{0.848} = 7.05 \times 10^4
\end{aligned}$$

One number (x) can be divided by another (y) by subtraction of their logarithms:

$$\log \frac{x}{y} = \log x - \log y$$

For example, to divide 125 by 742,

$$\begin{aligned}
\log 125 &= 2.097 \\
-\log 742 &= \underline{2.870} \\
\log \frac{x}{y} &= -0.773 \\
\frac{x}{y} &= 10^{-0.773} = 10^{0.227} \times 10^{-1} = 1.68 \times 10^{-1}
\end{aligned}$$

Similarly, powers and roots of numbers can be found using logarithms.

$$\log x^y = y(\log x)$$

$$\log \sqrt[y]{x} = \log x^{1/y} = \frac{1}{y} \log x$$

As an example, find the fourth power of 5.23. We first find the log of 5.23 and then multiply it by 4. The result, 2.874, is the log of the answer. Therefore, we find the antilog of 2.874:

$$\begin{aligned}
(5.23)^4 &= ? \\
\log (5.23)^4 &= 4 \log 5.23 = 4(0.719) = 2.874 \\
(5.23)^4 &= 10^{2.874} = 748
\end{aligned}$$

As another example, find the fifth root of 1.89×10^{-9}:

$$\sqrt[5]{1.89 \times 10^{-9}} = (1.89 \times 10^{-9})^{1/5} = ?$$

$$\log(1.89 \times 10^{-9})^{1/5} = \frac{1}{5}\log(1.89 \times 10^{-9}) = \frac{1}{5}(-8.724) = -1.745$$

The answer is the antilog of -1.745:

$$(1.89 \times 10^{-9})^{1/5} = 10^{-1.745} = 1.8 \times 10^{-2}$$

A.2 Quadratic Equations

Algebraic equations of the form $ax^2 + bx + c = 0$ are called **quadratic equations.** The coefficients a, b, and c may be either positive or negative. The two roots of the equation may be found using the *quadratic formula:*

$$x = \frac{-b \pm \sqrt{b^2 - 4ac}}{2a}$$

As an example, solve the equation $5x^2 - 3x - 2 = 0$. Here $a = 5$, $b = -3$, and $c = -2$. Therefore,

$$x = \frac{3 \pm \sqrt{(-3)^2 - 4(5)(-2)}}{2(5)}$$

$$= \frac{3 \pm [2(5) / \sqrt{9 - (-40)}]}{10} = \frac{3 \pm \sqrt{49}}{10} = \frac{3 \pm 7}{10}$$

$$= 1 \text{ and } -0.4$$

How do you know which of the two roots is the correct answer? You have to decide in each case which root has physical significance. It is *usually* true in this course, however, that negative values are not significant.

When you have solved a quadratic expression, you should always check your values by substitution into the original equation. In the previous example, we find that $5(1)^2 - 3(1) - 2 = 0$ and that $5(-0.4)^2 - 3(-0.4) - 2 = 0$.

The most likely place you will encounter quadratic equations is in the chapters on chemical equilibria, particularly in Chapters 16 through 18. Here, you will often be faced with solving an equation such as

$$1.8 \times 10^{-4} = \frac{x^2}{0.0010 - x}$$

This equation can certainly be solved using the quadratic equation (to give $x = 3.4 \times 10^{-4}$). You may find the **method of successive approximations** to be especially convenient, however. Here we begin by making a reasonable approximation of x. This approximate value is substituted into the original equation, which is then solved to give what is hoped to be a more correct value of x. This process is repeated until the answer converges on a particular value of x—that is, until the value of x derived from two successive approximations is the same.

Step 1: First, assume that x is so small that $(0.0010 - x) \approx 0.0010$. This means that

$$x^2 = 1.8 \times 10^{-4}\ (0.0010)$$

$$x = 4.2 \times 10^{-4} \text{ (to 2 significant figures)}$$

Step 2: Substitute the value of x from Step 1 into the denominator of the original equation, and again solve for x:

$$x^2 = 1.8 \times 10^{-4}(0.0010 - 0.00042)$$

$$x = 3.2 \times 10^{-4}$$

Step 3: Repeat Step 2 using the value of x found in that step:

$$x = \sqrt{1.8 \times 10^{-4}(0.0010 - 0.00032)} = 3.5 \times 10^{-4}$$

Step 4: Continue repeating the calculation, using the value of x found in the previous step:

$$x = \sqrt{1.8 \times 10^{-4}(0.0010 - 0.00035)} = 3.4 \times 10^{-4}$$

Step 5: $$x = \sqrt{1.8 \times 10^{-4}(0.0010 - 0.00034)} = 3.4 \times 10^{-4}$$

Here, we find that iterations after the fourth step give the same value for x, indicating that we have arrived at a valid answer (and the same one obtained from the quadratic formula).

Here are several final thoughts on using the method of successive approximations. First, in some cases the method does not work. Successive steps may give answers that are random or that diverge from the correct value. In Chapters 16 through 18, you confront quadratic equations of the form $K = x^2/(C - x)$. The method of approximations works as long as $K < 4C$ (assuming one begins with $x = 0$ as the first guess, that is, $K \approx x^2/C$). This is always going to be true for weak acids and bases (the topic of Chapters 17 and 18), but it may *not* be the case for problems involving gas phase equilibria (Chapter 16), where K can be quite large.

Second, values of K in the equation $K = x^2/(C - x)$ are usually known only to two significant figures. We are therefore justified in carrying out successive steps until two answers are the same to two significant figures.

Finally, we highly recommend this method of solving quadratic equations, especially those in Chapters 17 and 18. If your calculator has a memory function, successive approximations can be carried out easily and rapidly.

B* Some Important Physical Concepts

B.1 Matter

The tendency to maintain a constant velocity is called inertia. Thus, unless acted on by an unbalanced force, a body at rest remains at rest, and a body in motion remains in motion with uniform velocity. Matter is anything that exhibits inertia; the quantity of matter is its mass.

B.2 Motion

Motion is the change of position or location in space. Objects can have the following classes of motion:

- Translation occurs when the center of mass of an object changes its location. Example: a car moving on the highway.
- Rotation occurs when each point of a moving object moves in a circle about an axis through the center of mass. Examples: a spinning top, a rotating molecule.
- Vibration is a periodic distortion of and then recovery of original shape. Examples: a struck tuning fork, a vibrating molecule.

B.3 Force and Weight

Force is that which changes the velocity of a body; it is defined as

$$\text{Force} = \text{mass} \times \text{acceleration}$$

The SI unit of force is the **newton,** N, whose dimensions are kilograms times meter per second squared ($kg \cdot m/s^2$). A newton is therefore the force needed to change the velocity of a mass of 1 kilogram by 1 meter per second in a time of 1 second.

*Adapted from F. Brescia, J. Arents, H. Meislich, et al.: *General Chemistry,* 5th ed. Philadelphia, Harcourt Brace, 1988.

Because the earth's gravity is not the same everywhere, the weight corresponding to a given mass is not a constant. At any given spot on earth, gravity is constant, however, and therefore weight is proportional to mass. When a balance tells us that a given sample (the "unknown") has the same weight as another sample (the "weights," as given by a scale reading or by a total of counterweights), it also tells us that the two masses are equal. The balance is therefore a valid instrument for measuring the mass of an object independently of slight variations in the force of gravity.

B.4 Pressure*

Pressure is force per unit area. The SI unit, called the pascal, Pa, is

$$1\text{ pascal} = \frac{1\text{ newton}}{\text{m}^2} = \frac{1\text{ kg}\cdot\text{m/s}^2}{\text{m}^2} = \frac{1\text{ kg}}{\text{m}\cdot\text{s}^2}$$

The International System of Units also recognizes the bar, which is 10^5 Pa and which is close to standard atmospheric pressure (Table 1).

TABLE 1 Pressure Conversions

From	To	Multiply By
atmosphere	mm Hg	760 mm Hg/atm (exactly)
atmosphere	lb/in^2	$14.6960\text{ lb/(in}^2\cdot\text{atm)}$
atmosphere	kPa	101.325 kPa/atm
bar	Pa	10^5 Pa/bar (exactly)
bar	lb/in^2	$14.5038\text{ lb/(in}^2\cdot\text{bar)}$
mm Hg	torr	1 torr/mm Hg (exactly)

Chemists also express pressure in terms of the heights of liquid columns, especially water and mercury. This usage is not completely satisfactory, because the pressure exerted by a given column of a given liquid is not a constant but depends on the temperature (which influences the density of the liquid) and the location (which influences gravity). Such units are therefore not part of the SI, and their use is now discouraged. The older units are still used in books and journals, however, and chemists must be familiar with them.

The pressure of a liquid or a gas depends only on the depth (or height) and is exerted equally in all directions. At sea level, the pressure exerted by the earth's atmosphere supports a column of mercury about 0.76 m (76 cm, or 760 mm) high.

One **standard atmosphere** (atm) is the pressure exerted by exactly 76 cm of mercury at 0 °C (density, 13.5951 g/cm^3) and at standard gravity, 9.80665 m/s^2. The **bar** is equivalent to 0.9869 atm. One **torr** is the pressure exerted by exactly 1 mm of mercury at 0 °C and standard gravity.

B.5 Energy and Power

The SI unit of energy is the product of the units of force and distance, or kilograms times meter per second squared ($\text{kg}\cdot\text{m/s}^2$) times meters ($\times$ m), which is $\text{kg}\cdot\text{m}^2/\text{s}^2$; this unit is called the **joule,** J. The joule is thus the work done when a force of 1 newton acts through a distance of 1 meter.

*See Section 11.1.

Work may also be done by moving an electric charge in an electric field. When the charge being moved is 1 coulomb (C), and the potential difference between its initial and final positions is 1 volt (V), the work is 1 joule. Thus,

$$1 \text{ joule} = 1 \text{ coulomb volt (CV)}$$

Another unit of electric work that is not part of the International System of Units but is still in use is the **electron volt,** eV, which is the work required to move an electron against a potential difference of 1 volt. (It is also the kinetic energy acquired by an electron when it is accelerated by a potential difference of 1 volt.) Because the charge on an electron is 1.602×10^{-19} C, we have

$$1 \text{ eV} = 1.602 \times 10^{-19} \text{ CV} \times \frac{1 \text{ J}}{1 \text{ CV}} = 1.602 \times 10^{-19} \text{ J}$$

If this value is multiplied by Avogadro's number, we obtain the energy involved in moving 1 mole of electron charges (1 faraday) in a field produced by a potential difference of 1 volt:

$$1 \frac{\text{eV}}{\text{particle}} = \frac{1.602 \times 10^{-19} \text{ J}}{\text{particle}} \times \frac{6.022 \times 10^{23} \text{particles}}{\text{mol}} \cdot \frac{1 \text{ kJ}}{1000 \text{ J}} = 96.49 \text{ kJ/mol}$$

Power is the amount of energy delivered per unit time. The SI unit is the watt, W, which is 1 joule per second. One kilowatt, kW, is 1000 W. Watt hours and kilowatt hours are therefore units of energy (Table 2). For example, 1000 watts, or 1 kilowatt, is

$$1.0 \times 10^{3} \text{W} \times \frac{1 \text{ J}}{1 \text{ W} \cdot \text{s}} \cdot \frac{3.6 \times 10^{3} \text{ s}}{1 \text{ h}} = 3.6 \times 10^{6} \text{ J}$$

TABLE 2 Energy Conversions

From	To	Multiply By
calorie (cal)	joule	4.184 J/cal (exactly)
kilocalorie (kcal)	cal	10^3 cal/kcal (exactly)
kilocalorie	joule	4.184×10^3 J/kcal (exactly)
liter atmosphere (L · atm)	joule	101.325 J/L · atm
electron volt (eV)	joule	1.60218×10^{-19} J/eV
electron volt per particle	kilojoules per mole	96.485 kJ · particle/eV · mol
coulomb volt (CV)	joule	1 CV/J (exactly)
kilowatt hour (kWh)	kcal	860.4 kcal/kWh
kilowatt hour	joule	3.6×10^6 J/kWh (exactly)
British thermal unit (Btu)	calorie	252 cal/Btu

APPENDIX

C Abbreviations and Useful Conversion Factors

TABLE 3 Some Common Abbreviations and Standard Symbols

Term	Abbreviation	Term	Abbreviation
Activation energy	E_a	Entropy	S
Ampere	A	Standard entropy	$S°$
Aqueous Solution	aq	Entropy change for reaction	$\Delta_r S°$
Atmosphere, unit of pressure	atm	Equilibrium constant	K
Atomic mass unit	u	Concentration basis	K_c
Avogadro's constant	N_A	Pressure basis	K_p
Bar, unit of pressure	bar	Ionization weak acid	K_a
Body-centered cubic	bcc	Ionization weak base	K_b
Bohr radius	a_0	Solubility product	K_{sp}
Boiling point	bp	Formation constant	K_{form}
Celsius temperature, °C	T	Ethylenediamine	en
Charge number of an ion	z	Face-centered cubic	fcc
Coulomb, electric charge	C	Faraday constant	F
Curie, radioactivity	Ci	Gas constant	R
Cycles per second, hertz	Hz	Gibbs free energy	G
Debye, unit of electric dipole	D	Standard free energy	$G°$
Electron	e^-	Standard free energy of formation	$\Delta_f G°$
Electron volt	eV	Free energy change for reaction	$\Delta_r G°$
Electronegativity	χ	Half-life	$t_{1/2}$
Energy	E	Heat	q
Enthalpy	H	Hertz	Hz
Standard enthalpy	$H°$	Hour	h
Standard enthalpy of formation	$\Delta_f H°$	Joule	J
Standard enthalpy of reaction	$\Delta_r H°$	Kelvin	K

TABLE 3 Some Common Abbreviations and Standard Symbols (continued)

Term	Abbreviation	Term	Abbreviation
Kilocalorie	kcal	Pressure	
Liquid	ℓ	Pascal, unit of pressure	Pa
Logarithm, base 10	log	In atmospheres	atm
Logarithm, base *e*	ln	In millimeters of mercury	mm Hg
Minute	min	Proton number	*Z*
Molar	M	Rate constant	*k*
Molar mass	*M*	Primitive cubic (unit cell)	pc
Mole	mol	Standard temperature and pressure	STP
Osmotic pressure	Π	Volt	V
Planck's constant	*h*	Watt	W
Pound	lb	Wavelength	λ

C.1 Fundamental Units of the SI System

The metric system was begun by the French National Assembly in 1790 and has undergone many modifications. The International System of Units or *Système International* (SI), which represents an extension of the metric system, was adopted by the 11th General Conference of Weights and Measures in 1960. It is constructed from seven base units, each of which represents a particular physical quantity (Table 4).

TABLE 4 SI Fundamental Units

Physical Quantity	Name of Unit	Symbol
Length	meter	m
Mass	kilogram	kg
Time	second	s
Temperature	kelvin	K
Amount of substance	mole	mol
Electric current	ampere	A
Luminous intensity	candela	cd

The first five units listed in Table 4 are particularly useful in general chemistry and are defined as follows:

1. The *meter* was redefined in 1960 to be equal to 1,650,763.73 wavelengths of a certain line in the emission spectrum of krypton-86.
2. The *kilogram* represents the mass of a platinum–iridium block kept at the International Bureau of Weights and Measures at Sèvres, France.
3. The *second* was redefined in 1967 as the duration of 9,192,631,770 periods of a certain line in the microwave spectrum of cesium-133.

4. The *kelvin* is 1/273.15 of the temperature interval between absolute zero and the triple point of water.
5. The *mole* is the amount of substance that contains as many entities as there are atoms in exactly 0.012 kg of carbon-12 (12 g of ^{12}C atoms).

C.2 Prefixes Used with Traditional Metric Units and SI Units

Decimal fractions and multiples of metric and SI units are designated by using the prefixes listed in Table 5. Those most commonly used in general chemistry appear in italics.

C.3 Derived SI Units

In the International System of Units, all physical quantities are represented by appropriate combinations of the base units listed in Table 4. A list of the derived units frequently used in general chemistry is given in Table 6.

TABLE 5 Traditional Metric and SI Prefixes

Factor	Prefix	Symbol	Factor	Prefix	Symbol
10^{12}	tera	T	10^{-1}	*deci*	d
10^{9}	giga	G	10^{-2}	*centi*	c
10^{6}	mega	M	10^{-3}	*milli*	m
10^{3}	*kilo*	k	10^{-6}	micro	μ
10^{2}	hecto	h	10^{-9}	*nano*	n
10^{1}	deka	da	10^{-12}	*pico*	p
			10^{-15}	femto	f
			10^{-18}	atto	a

TABLE 6 Derived SI Units

Physical Quantity	Name of Unit	Symbol	Definition
Area	square meter	m^2	
Volume	cubic meter	m^3	
Density	kilogram per cubic meter	kg/m^3	
Force	newton	N	$kg \cdot m/s^2$
Pressure	pascal	Pa	N/m^2
Energy	joule	J	$kg \cdot m^2/s^2$
Electric charge	coulomb	C	$A \cdot s$
Electric potential difference	volt	V	$J/(A \cdot s)$

TABLE 7 Common Units of Mass and Weight

1 Pound = 453.39 Grams
1 kilogram = 1000 grams = 2.205 pounds
1 gram = 1000 milligrams
1 gram = 6.022×10^{23} atomic mass units
1 atomic mass unit = 1.6605×10^{-24} gram
1 short ton = 2000 pounds = 907.2 kilograms
1 long ton = 2240 pounds
1 metric tonne = 1000 kilograms = 2205 pounds

TABLE 8 Common Units of Length

1 inch = 2.54 centimeters (Exactly)
1 mile = 5280 feet = 1.609 kilometers
1 yard = 36 inches = 0.9144 meter
1 meter = 100 centimeters = 39.37 inches = 3.281 feet = 1.094 yards
1 kilometer = 1000 meters = 1094 yards = 0.6215 mile
1 Ångstrom = 1.0×10^{-8} centimeter = 0.10 nanometer = 100 picometers
= 1.0×10^{-10} meter = 3.937×10^{-9} inch

TABLE 9 Common Units of Volume

1 quart = 0.9463 liter 1 liter = 1.0567 quarts
1 liter = 1 cubic decimeter = 1000 cubic centimeters = 0.001 cubic meter
1 milliliter = 1 cubic centimeter = 0.001 liter = 1.056×10^{-3} quart
1 cubic foot = 28.316 liters = 29.924 quarts = 7.481 gallons

D Physical Constants

TABLE 10

Quantity	Symbol	Traditional Units	SI Units
Acceleration of gravity	g	980.6 cm/s	9.806 m/s
Atomic mass unit (1/12 the mass of ^{12}C atom)	u	1.6605×10^{-24} g	1.6605×10^{-27} kg
Avogadro's number	N	$6.02214179 \times 10^{23}$ particles/mol	$6.02214179 \times 10^{23}$ particles/mol
Bohr radius	a_0	0.052918 nm	5.2918×10^{-11} m
		5.2918×10^{-9} cm	
Boltzmann constant	k	1.3807×10^{-16} erg/K	1.3807×10^{-23} J/K
Charge-to-mass ratio of electron	e/m	1.7588×10^{8} C/g	1.7588×10^{11} C/kg
Electronic charge	e	1.6022×10^{-19} C	1.6022×10^{-19} C
		4.8033×10^{-10} esu	
Electron rest mass	m_e	9.1094×10^{-28} g	9.1094×10^{-31} kg
		0.00054858 amu	
Faraday constant	F	96,485 C/mol e^-	96,485 C/mol e^-
		23.06 kcal/V · mol e^-	96,485 J/V · mol e^-
Gas constant	R	$0.082057 \dfrac{\text{L} \cdot \text{atm}}{\text{mol} \cdot \text{K}}$	$8.3145 \dfrac{\text{Pa} \cdot \text{dm}^3}{\text{mol} \cdot \text{K}}$
		$1.987 \dfrac{\text{cal}}{\text{mol} \cdot \text{K}}$	8.3145 J/mol · K
Molar volume (STP)	V_m	22.414 L/mol	22.414×10^{-3} m^3/mol
			22.414 dm^3/mol
Neutron rest mass	m_n	1.67493×10^{-24} g	1.67493×10^{-27} kg
		1.008665 amu	
Planck's constant	h	6.6261×10^{-27} erg · s	$6.6260693 \times 10^{-34}$ J · s

TABLE 10 (continued)

Quantity	Symbol	Traditional Units	SI Units
Proton rest mass	m_p	1.6726×10^{-24} g	1.6726×10^{-27} kg
		1.007276 amu	
Rydberg constant	R_a	3.289×10^{15} cycles/s	$1.0974 \times 10^{7}\ m^{-1}$
	Rhc		2.1799×10^{-18} J
Velocity of light (in a vacuum)	c	2.9979×10^{10} cm/s (186,282 miles/s)	2.9979×10^{8} m/s

$\pi = 3.1416$

$e = 2.7183$

$\ln X = 2.303 \log X$

TABLE 11 Specific Heats and Heat Capacities for Some Common Substances at 25 °C

Substance	Specific Heat (J/g · K)	Molar Heat Capacity (J/mol · K)
Al(s)	0.897	24.2
Ca(s)	0.646	25.9
Cu(s)	0.385	24.5
Fe(s)	0.449	25.1
$Hg(\ell)$	0.140	28.0
$H_2O(s)$, ice	2.06	37.1
$H_2O(\ell)$, water	4.184	75.4
$H_2O(g)$, steam	1.86	33.6
$C_6H_6(\ell)$, benzene	1.74	136
$C_6H_6(g)$, benzene	1.06	82.4
$C_2H_5OH(\ell)$, ethanol	2.44	112.3
$C_2H_5OH(g)$, ethanol	1.41	65.4
$(C_2H_5)_2O(\ell)$, diethyl ether	2.33	172.6
$(C_2H_5)_2O(g)$, diethyl ether	1.61	119.5

TABLE 12 Heats of Transformation and Transformation Temperatures of Several Substances

		Heat of Fusion			Heat of Vaporization	
Substance	**MP (°C)**	*J/g*	*kJ/mol*	**BP (°C)**	*J/g*	*kJ/mol*
*Elements**						
Al	660	395	10.7	2518	12083	294
Ca	842	212	8.5	1484	3767	155
Cu	1085	209	13.3	2567	4720	300
Fe	1535	267	13.8	2861	6088	340
Hg	−38.8	11	2.29	357	295	59.1
Compounds						
H_2O	0.00	333	6.01	100.0	2260	40.7
CH_4	−182.5	58.6	0.94	−161.5	511	8.2
C_2H_5OH	−114	109	5.02	78.3	838	38.6
C_6H_6	5.48	127.4	9.95	80.0	393	30.7
$(C_2H_5)_2O$	−116.3	98.1	7.27	34.6	357	26.5

*Data for the elements are taken from J. A. Dean: *Lange's Handbook of Chemistry*, 15th Edition. New York, McGraw-Hill Publishers, 1999.

E A Brief Guide to Naming Organic Compounds

It seems a daunting task—to devise a systematic procedure that gives each organic compound a unique name—but that is what has been done. A set of rules was developed to name organic compounds by the International Union of Pure and Applied Chemistry (IUPAC). The IUPAC nomenclature allows chemists to write a name for any compound based on its structure or to identify the formula and structure for a compound from its name. In this book, we have generally used the IUPAC nomenclature scheme when naming compounds.

In addition to the systematic names, many compounds have common names. The common names came into existence before the nomenclature rules were developed, and they have continued in use. For some compounds, these names are so well entrenched that they are used most of the time. One such compound is acetic acid, which is almost always referred to by that name and not by its systematic name, ethanoic acid.

The general procedure for systematic naming of organic compounds begins with the nomenclature for hydrocarbons. Other organic compounds are then named as derivatives of hydrocarbons. Nomenclature rules for simple organic compounds are given in the following section.

E.1 Hydrocarbons

Alkanes

The names of alkanes end in "-ane." When naming a specific alkane, the root of the name identifies the longest carbon chain in a compound. Specific substituent groups attached to this carbon chain are identified by name and position.

Alkanes with chains of from one to ten carbon atoms are given in Table 10.2. After the first four compounds, the names derive from Latin numbers—pentane, hexane, heptane, octane, nonane, decane—and this regular naming continues for higher alkanes. For substituted alkanes, the substituent groups on a hydrocarbon chain must be identified both by a name and by the position of substitution; this information precedes the root of the name. The position is indicated by a number that refers to the carbon atom to which the substituent is attached. (Numbering of the carbon atoms in a chain should begin at the end of the carbon chain that allows the substituent groups to have the lowest numbers.)

Names of hydrocarbon substituents are derived from the name of the hydrocarbon. The group —CH_3, derived by taking a hydrogen from methane, is called the methyl group; the C_2H_5 group is the ethyl group. The nomenclature scheme is easily extended to derivatives of hydrocarbons with other substituent groups such as —Cl (chloro), —NO_2 (nitro), —CN (cyano), —D (deuterio), and so on (Table 13). If two or more of the same substituent groups occur, the prefixes "di-," "tri-," and "tetra-" are added. When different substituent groups are present, they are generally listed in alphabetical order.

TABLE 13 Names of Common Substituent Groups

Formula	Name	Formula	Name
—CH_3	methyl	—D	deuterio
—C_2H_5	ethyl	—Cl	chloro
—$CH_2CH_2CH_3$	1-propyl (*n*-propyl)	—Br	bromo
—$CH(CH_3)_2$	2-propyl (isopropyl)	—F	fluoro
—$CH{=}CH_2$	ethenyl (vinyl)	—CN	cyano
—C_6H_5	phenyl	—NO_2	nitro
—OH	hydroxo		
—NH_2	amino		

Example:

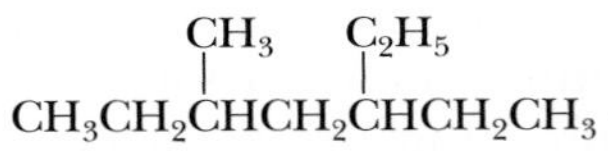

Step	Information to include	Contribution to name
1.	An alkane	name will end in "-ane"
2.	Longest chain is 7 carbons	name as a *heptane*
3.	—CH_3 group at carbon 3	3-*methyl*
4.	—C_2H_5 group at carbon 5	5-*ethyl*
Name:	5-ethyl-3-methylheptane	

Cycloalkanes are named based on the ring size and by adding the prefix "cyclo"; for example, the cycloalkane with a six-member ring of carbons is called cyclohexane.

Alkenes

Alkenes have names ending in "-ene." The name of an alkene must specify the length of the carbon chain and the position of the double bond (and when appropriate, the configuration, either *cis* or *trans*). As with alkanes, both identity and position of substituent groups must be given. The carbon chain is numbered from the end that gives the double bond the lowest number.

Compounds with two double bonds are called dienes, and they are named similarly—specifying the positions of the double bonds and the name and position of any substituent groups.

For example, the compound $H_2C{=}C(CH_3)CH(CH_3)CH_2CH_3$ has a five-carbon chain with a double bond between carbon atoms 1 and 2 and methyl groups on carbon atoms 2 and 3. Its name using IUPAC nomenclature is **2,3-dimethyl-1-pentene.** The compound $CH_3CH{=}CHCCl_3$ with a *cis* configuration around the double bond is named **1,1,1-trichloro-*cis*-2-butene.** The compound $H_2C{=}C(Cl)CH{=}CH_2$ is **2-chloro-1,3-butadiene.**

Alkynes

The naming of alkynes is similar to the naming of alkenes, except that *cis–trans* isomerism isn't a factor. The ending "-yne" on a name identifies a compound as an alkyne.

Benzene Derivatives

The carbon atoms in the six-member ring are numbered 1 through 6, and the name and position of substituent groups are given. The two examples shown here are **1-ethyl-3-methylbenzene** and **1,4-diaminobenzene.**

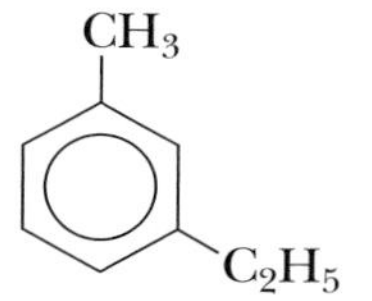

1-ethyl-3-methylbenzene

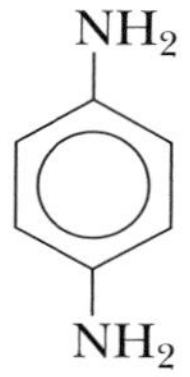

1,4-diaminobenzene

E.2 Derivatives of Hydrocarbons

The names for alcohols, aldehydes, ketones, and acids are based on the name of the hydrocarbon with an appropriate suffix to denote the class of compound, as follows:

- **Alcohols:** Substitute "-ol" for the final "-e" in the name of the hydrocarbon, and designate the position of the —OH group by the number of the carbon atom. For example, $CH_3CH_2CHOHCH_3$ is named as a derivative of the 4-carbon hydrocarbon butane. The —OH group is attached to the second carbon, so the name is 2-butanol.
- **Aldehydes:** Substitute "-al" for the final "-e" in the name of the hydrocarbon. The carbon atom of an aldehyde is, by definition, carbon-1 in the hydrocarbon chain. For example, the compound $CH_3CH(CH_3)CH_2CH_2CHO$ contains a 5-carbon chain with the aldehyde functional group being carbon-1 and the $—CH_3$ group at position 4; thus, the name is **4-methylpentanal.**
- **Ketones:** Substitute "-one" for the final "-e" in the name of the hydrocarbon. The position of the ketone functional group (the carbonyl group) is indicated by the number of the carbon atom. For example, the compound $CH_3COCH_2CH(C_2H_5)CH_2CH_3$ has the carbonyl group at the 2 position and an ethyl group at the 4 position of a 6-carbon chain; its name is **4-ethyl-2-hexanone.**
- **Carboxylic acids (organic acids):** Substitute "-oic" for the final "-e" in the name of the hydrocarbon. The carbon atoms in the longest chain are counted beginning with the carboxylic carbon atom. For example, *trans-*

$CH_3CH{=}CHCH_2CO_2H$ is named as a derivative of *trans*-3-pentene—that is, ***trans*-3-pentenoic acid.**

An **ester** is named as a derivative of the alcohol and acid from which it is made. The name of an ester is obtained by splitting the formula RCO_2R' into two parts, the RCO_2— portion and the —R' portion. The —R' portion comes from the alcohol and is identified by the hydrocarbon group name; derivatives of ethanol, for example, are called *ethyl* esters. The acid part of the compound is named by dropping the "-oic" ending for the acid and replacing it by "-oate." The compound $CH_3CH_2CO_2CH_3$ is named **methyl propanoate.**

Notice that an anion derived from a carboxylic acid by loss of the acidic proton is named the same way. Thus, $CH_3CH_2CO_2^-$ is the **propanoate anion,** and the sodium salt of this anion, $Na(CH_3CH_2CO_2)$, is **sodium propanoate.**

F Values for the Ionization Energies and Electron Affinities of the Elements

1A (1)	2A (2)	3B (3)	4B (4)	5B (5)	6B (6)	7B (7)	8B (8,9,10)			1B (11)	2B (12)	3A (13)	4A (14)	5A (15)	6A (16)	7A (17)	8 (18)
H 1312																	He 2371
Li 520	Be 899											B 801	C 1086	N 1402	O 1314	F 1681	Ne 2081
Na 496	Mg 738											Al 578	Si 786	P 1012	S 1000	Cl 1251	Ar 1521
K 419	Ca 599	Sc 631	Ti 658	V 650	Cr 652	Mn 717	Fe 759	Co 758	Ni 757	Cu 745	Zn 906	Ga 579	Ge 762	As 947	Se 941	Br 1140	Kr 1351
Rb 403	Sr 550	Y 617	Zr 661	Nb 664	Mo 685	Tc 702	Ru 711	Rh 720	Pd 804	Ag 731	Cd 868	In 558	Sn 709	Sb 834	Te 869	I 1008	Xe 1170
Cs 377	Ba 503	La 538	Hf 681	Ta 761	W 770	Re 760	Os 840	Ir 880	Pt 870	Au 890	Hg 1007	Tl 589	Pb 715	Bi 703	Po 812	At 890	Rn 1037

TABLE 14 Electron Affinity Values for Some Elements (kJ/mol)*

H						
−72.77						
Li	Be	B	C	N	O	F
−59.63	0†	−26.7	−121.85	0	−140.98	−328.0
Na	Mg	Al	Si	P	S	Cl
−52.87	0	−42.6	−133.6	−72.07	−200.41	−349.0
K	Ca	Ga	Ge	As	Se	Br
−48.39	0	−30	−120	−78	−194.97	−324.7
Rb	Sr	In	Sn	Sb	Te	I
−46.89	0	−30	−120	−103	−190.16	−295.16
Cs	Ba	Tl	Pb	Bi	Po	At
−45.51	0	−20	−35.1	−91.3	−180	−270

*Data taken from H. Hotop and W. C. Lineberger: *Journal of Physical Chemistry, Reference Data,* Vol. 14, p. 731, 1985. (This paper also includes data for the transition metals.) Some values are known to more than two decimal places.

†Elements with an electron affinity of zero indicate that a stable anion A^- of the element does not exist in the gas phase.

G Vapor Pressure of Water at Various Temperatures

TABLE 15 Vapor Pressure of Water at Various Temperatures

Temperature (°C)	Vapor Pressure (torr)	Temperature (°C)	Vapor Pressure (torr)	Temperature (°C)	Vapor Pressure (torr)	Temperature (°C)	Vapor Pressure (torr)
−10	2.1	21	18.7	51	97.2	81	369.7
−9	2.3	22	19.8	52	102.1	82	384.9
−8	2.5	23	21.1	53	107.2	83	400.6
−7	2.7	24	22.4	54	112.5	84	416.8
−6	2.9	25	23.8	55	118.0	85	433.6
−5	3.2	26	25.2	56	123.8	86	450.9
−4	3.4	27	26.7	57	129.8	87	468.7
−3	3.7	28	28.3	58	136.1	88	487.1
−2	4.0	29	30.0	59	142.6	89	506.1
−1	4.3	30	31.8	60	149.4	90	525.8
0	4.6	31	33.7	61	156.4	91	546.1
1	4.9	32	35.7	62	163.8	92	567.0
2	5.3	33	37.7	63	171.4	93	588.6
3	5.7	34	39.9	64	179.3	94	610.9
4	6.1	35	42.2	65	187.5	95	633.9
5	6.5	36	44.6	66	196.1	96	657.6
6	7.0	37	47.1	67	205.0	97	682.1
7	7.5	38	49.7	68	214.2	98	707.3
8	8.0	39	52.4	69	223.7	99	733.2
9	8.6	40	55.3	70	233.7	100	760.0
10	9.2	41	58.3	71	243.9	101	787.6
11	9.8	42	61.5	72	254.6	102	815.9
12	10.5	43	64.8	73	265.7	103	845.1
13	11.2	44	68.3	74	277.2	104	875.1
14	12.0	45	71.9	75	289.1	105	906.1
15	12.8	46	75.7	76	301.4	106	937.9
16	13.6	47	79.6	77	314.1	107	970.6
17	14.5	48	83.7	78	327.3	108	1004.4
18	15.5	49	88.0	79	341.0	109	1038.9
19	16.5	50	92.5	80	355.1	110	1074.6
20	17.5						

H Ionization Constants for Weak Acids at 25 °C

TABLE 16 Ionization Constants for Weak Acids at 25 °C

Acid	Formula and Ionization Equation	K_a
Ascetic	$CH_3CO_2H \rightleftharpoons H^+ + CH_3CO_2^-$	1.8×10^{-5}
Arsenic	$H_3AsO_4 \rightleftharpoons H^+ + H_2AsO_4^-$	$K_1 = 2.5 \times 10^{-4}$
	$H_2AsO_4^- \rightleftharpoons H^+ + HAsO_4^{2-}$	$K_2 = 5.6 \times 10^{-3}$
	$HAsO_4^{2-} \rightleftharpoons H^+ + AsO_4^{3-}$	$K_3 = 3.0 \times 10^{-13}$
Arsenous	$H_3AsO_3 \rightleftharpoons H^+ + H_2AsO_3^-$	$K_1 = 6.0 \times 10^{-10}$
	$H_2AsO_3^- \rightleftharpoons H^+ + HAsO_3^{2-}$	$K_2 = 3.0 \times 10^{-14}$
Benzoic	$C_6H_5CO_2H \rightleftharpoons H^+ + C_6H_5CO_2^-$	6.3×10^{-5}
Boric	$H_3BO_3 \rightleftharpoons H^+ + H_2BO_3^-$	$K_1 = 7.3 \times 10^{-10}$
	$H_2BO_3 \rightleftharpoons H^+ + HBO_3^{2-}$	$K_2 = 1.8 \times 10^{-13}$
	$HBO_3^{2-} \rightleftharpoons H^+ + BO_3^{3-}$	$K_3 = 1.6 \times 10^{-14}$
Carbonic	$H_2CO_3 \rightleftharpoons H^+ + HCO_3^-$	$K_1 = 4.2 \times 10^{-7}$
	$HCO_3^- \rightleftharpoons H^+ + CO_3^{2-}$	$K_2 = 4.8 \times 10^{-11}$
Citric	$H_3C_6H_5O_7 \rightleftharpoons H^+ + H_2C_6H_5O_7^-$	$K_1 = 7.4 \times 10^{-3}$
	$H_2C_6H_5O_7^- \rightleftharpoons H^+ + HC_6H_5O_7^{2-}$	$K_2 = 1.7 \times 10^{-5}$
	$HC_6H_5O_7^{2-} \rightleftharpoons H^+ + C_6H_5O_7^{3-}$	$K_3 = 4.0 \times 10^{-7}$
Cyanic	$HOCN \rightleftharpoons H^+ + OCN^-$	3.5×10^{-4}
Formic	$HCO_2H \rightleftharpoons H^+ + HCO_2^-$	1.8×10^{-4}
Hydrazoic	$HN_3 \rightleftharpoons H^+ + N_3^-$	1.9×10^{-5}
Hydrocyanic	$HCN \rightleftharpoons H^+ + CN^-$	4.0×10^{-10}
Hydrofluoric	$HF \rightleftharpoons H^+ + F^-$	7.2×10^{-4}
Hydrogen peroxide	$H_2O_2 \rightleftharpoons H^+ + HO_2^-$	2.4×10^{-12}
Hydrosulfuric	$H_2S \rightleftharpoons H^+ + HS^-$	$K_1 = 1 \times 10^{-7}$
	$HS^- \rightleftharpoons H^+ + S^{2-}$	$K_2 = 1 \times 10^{-19}$
Hypobromous	$HOBr \rightleftharpoons H^+ + OBr^-$	2.5×10^{-9}

(continued)

TABLE 16 Ionization Constants for Weak Acids at 25 °C *(continued)*

Acid	Formula and Ionization Equation	K_a
Hypochlorous	$HOCl \rightleftharpoons H^+ + OCl^-$	3.5×10^{-8}
Nitrous	$HNO_2 \rightleftharpoons H^+ + NO_2^-$	4.5×10^{-4}
Oxalic	$H_2C_2O_4 \rightleftharpoons H^+ + HC_2O_4^-$	$K_1 = 5.9 \times 10^{-2}$
	$HC_2O_4^- \rightleftharpoons H^+ + C_2O_4^{2-}$	$K_2 = 6.4 \times 10^{-5}$
Phenol	$C_6H_5OH \rightleftharpoons H^+ + C_6H_5O^-$	1.3×10^{-10}
Phosphoric	$H_3PO_4 \rightleftharpoons H^+ + H_2PO_4^-$	$K_1 = 7.5 \times 10^{-3}$
	$H_2PO_4^- \rightleftharpoons H^+ + HPO_4^{2-}$	$K_2 = 6.2 \times 10^{-8}$
	$HPO_4^{2-} \rightleftharpoons H^+ + PO_4^{3-}$	$K_3 = 3.6 \times 10^{-13}$
Phosphorous	$H_3PO_3 \rightleftharpoons H^+ + H_2PO_3^-$	$K_1 = 1.6 \times 10^{-2}$
	$H_2PO_3 \rightleftharpoons H^+ + HPO_3^{2-}$	$K_2 = 7.0 \times 10^{-7}$
Selenic	$H_2SeO_4 \rightleftharpoons H^+ + HSeO_4^-$	K_1 = very large
	$HSeO_4^- \rightleftharpoons H^+ + SeO_4^{2-}$	$K_2 = 1.2 \times 10^{-2}$
Selenous	$HSeO_3 \rightleftharpoons H^+ + HSeO_3^-$	$K_1 = 2.7 \times 10^{-3}$
	$HSeO_3^- \rightleftharpoons H^+ + SeO_3^{2-}$	$K_2 = 2.5 \times 10^{-7}$
Sulfuric	$H_2SO_4 \rightleftharpoons H^+ + HSO_4^-$	K_1 = very large
	$HSO_4^- \rightleftharpoons H^+ + SO_4^{2-}$	$K_2 = 1.2 \times 10^{-2}$
Sulfurous	$H_2SO_3 \rightleftharpoons H^+ + HSO_3^-$	$K_1 = 1.2 \times 10^{-2}$
	$HSO_3^- \rightleftharpoons H^+ + SO_3^{2-}$	$K_2 = 6.2 \times 10^{-8}$
Tellurous	$H_2TeO_3 \rightleftharpoons H^+ + HTeO_3^-$	$K_1 = 2 \times 10^{-3}$
	$HTeO_3^- \rightleftharpoons H^+ + TeO_3^{2-}$	$K_2 = 1 \times 10^{-8}$

I Ionization Constants for Weak Bases at 25 °C

TABLE 17 Ionization Constants for Weak Bases at 25 °C

Base	Formula and Ionization Equation	K_b
Ammonia	$NH_3 + H_2O \rightleftharpoons NH_4^+ + OH^-$	1.8×10^{-5}
Aniline	$C_6H_5NH_2 + H_2O \rightleftharpoons C_6H_5NH_3^+ + OH^-$	4.0×10^{-10}
Dimethylamine	$(CH_3)_2NH + H_2O \rightleftharpoons (CH_3)_2NH_2^+ + OH^-$	7.4×10^{-4}
Ethylenediamine	$H_2NCH_2CH_2NH_2 + H_2O \rightleftharpoons H_2NCH_2CH_2NH_3^+ OH^-$	$K_1 = 8.5 \times 10^{-5}$
	$H_2NCH_2CH_2NH_3^+ + H_2O \rightleftharpoons H_3NCH_2CH_2NH_3^{2+} OH^-$	$K_2 = 2.7 \times 10^{-8}$
Hydrazine	$N_2H_4 + H_2O \rightleftharpoons N_2H_5^+ + OH^-$	$K_1 = 8.5 \times 10^{-7}$
	$N_2H_5^+ + H_2O \rightleftharpoons N_2H_6^{2+} + OH^-$	$K_2 = 8.9 \times 10^{-16}$
Hydroxylamine	$NH_2OH + H_2O \rightleftharpoons NH_3OH^+ + OH^-$	6.6×10^{-9}
Methylamine	$CH_3NH_2 + H_2O \rightleftharpoons CH_3NH_3^+ + OH^-$	5.0×10^{-4}
Pyridine	$C_5H_5N + H_2O \rightleftharpoons C_5H_5NH^+ + OH^-$	1.5×10^{-9}
Trimethylamine	$(CH_3)_3N + H_2O \rightleftharpoons (CH_3)_3NH^+ + OH^-$	7.4×10^{-5}
Ethylamine	$C_2H_5NH_2 + H_2O \rightleftharpoons C_2H_5NH_3^+ + OH^-$	4.3×10^{-4}

J Solubility Product Constants for Some Inorganic Compounds at 25 °C

TABLE 18A Solubility Produce Constants (25 °C)

Cation	Compound	K_{sp}	Cation	Compound	K_{sp}
Ba^{2+}	*$BaCrO_4$	1.2×10^{-10}	Mg^{2+}	$MgCO_3$	6.8×10^{-6}
	$BaCO_3$	2.6×10^{-9}		MgF_2	5.2×10^{-11}
	BaF_2	1.8×10^{-7}		$Mg(OH)_2$	5.6×10^{-12}
	*$BaSO_4$	1.1×10^{-10}	Mn^{2+}	$MnCO_3$	2.3×10^{-11}
Ca^{2+}	$CaCO_3$ (calcite)	3.4×10^{-9}		*$Mn(OH)_2$	1.9×10^{-13}
	*CaF_2	5.3×10^{-11}	$Hg_2{}^{2+}$	*Hg_2Br_2	6.4×10^{-23}
	*$Ca(OH)_2$	5.5×10^{-5}		Hg_2Cl_2	1.4×10^{-18}
	$CaSO_4$	4.9×10^{-5}		*Hg_2I_2	2.9×10^{-29}
$Cu^{+,2+}$	CuBr	6.3×10^{-9}		Hg_2SO_4	6.5×10^{-7}
	CuI	1.3×10^{-12}	Ni^{2+}	$NiCO_3$	1.4×10^{-7}
	$Cu(OH)_2$	2.2×10^{-20}		$Ni(OH)_2$	5.5×10^{-16}
	CuSCN	1.8×10^{-13}	Ag^{+}	*AgBr	5.4×10^{-13}
Au^{+}	AuCl	2.0×10^{-13}		*$AgBrO_3$	5.4×10^{-5}
Fe^{2+}	$FeCO_3$	3.1×10^{-11}		$AgCH_3CO_2$	1.9×10^{-3}
	$Fe(OH)_2$	4.9×10^{-17}		AgCN	6.0×10^{-17}
Pb^{2+}	$PbBr_2$	6.6×10^{-6}		Ag_2CO_3	8.5×10^{-12}
	$PbCO_3$	7.4×10^{-14}		*$Ag_2C_2O_4$	5.4×10^{-12}
	$PbCl_2$	1.7×10^{-5}		*AgCl	1.8×10^{-10}
	$PbCrO_4$	2.8×10^{-13}		Ag_2CrO_4	1.1×10^{-12}
	PbF_2	3.3×10^{-8}		*AgI	8.5×10^{-17}
	PbI_2	9.8×10^{-9}		AgSCN	1.0×10^{-12}
	$Pb(OH)_2$	1.4×10^{-15}		*Ag_2SO_4	1.2×10^{-5}
	$PbSO_4$	2.5×10^{-8}			

(continued)

TABLE 18A Solubility Produce Constants (25 °C) *(continued)*

Cation	Compound	K_{sp}	Cation	Compound	K_{sp}
Sr^{2+}	$SrCO_3$	5.6×10^{-10}	Zn^{2+}	$Zn(OH)_2$	3×10^{-17}
	SrF_2	4.3×10^{-9}		$Zn(CN)_2$	8.0×10^{-12}
	$SrSO_4$	3.4×10^{-7}			
Tl^{+}	TlBr	3.7×10^{-6}			
	TlCl	1.9×10^{-4}			
	TlI	5.5×10^{-8}			

The values reported in this table were taken from J. A. Dean: *Lange's Handbook of Chemistry,* 15th Edition. New York, McGraw-Hill Publishers, 1999. Values have been rounded off to two significant figures.
*Calculated solubility from these K_{sp} values will match experimental solubility for this compound within a factor of 2. Experimental values for solubilities are given in R. W. Clark and J. M. Bonicamp: *Journal of Chemical Education,* Vol. 75, p. 1182, 1998.

TABLE 18B K_{spa} Values* for Some Metal Sulfides (25 °C)

Substance	K_{spa}
HgS (red)	4×10^{-54}
HgS (black)	2×10^{-53}
Ag_2S	6×10^{-51}
CuS	6×10^{-37}
PbS	3×10^{-28}
CdS	8×10^{-28}
SnS	1×10^{-26}
FeS	6×10^{-19}

*The equilibrium constant value K_{spa} for metal sulfides refers to the equilibrium $MS(s) + H_2O(\ell) \rightleftarrows M^{2+}(aq) + OH^-(aq) + HS^-(aq)$; see R. J. Myers, *Journal of Chemical Education,* Vol. 63, p. 687, 1986.

K Formation Constants for Some Complex Ions in Aqueous Solution

TABLE 19 Formation Constants for Some Complex Ions in Aqueous Solution*

Formation Equilibrium	K
$Ag^+ + 2\ Br^- \rightleftarrows [AgBr_2]^-$	2.1×10^7
$Ag^+ + 2\ Cl^- \rightleftarrows [AgCl_2]^-$	1.1×10^5
$Ag^+ + 2\ CN^- \rightleftarrows [Ag(CN)_2]^-$	1.3×10^{21}
$Ag^+ + 2\ S_2O_3^{2-} \rightleftarrows [Ag(S_2O_3)_2]^{3-}$	2.9×10^{13}
$Ag^+ + 2\ NH_3 \rightleftarrows [Ag(NH_3)_2]^+$	1.1×10^7
$Al^{3+} + 6\ F^- \rightleftarrows [AlF_6]^{3-}$	6.9×10^{19}
$Al^{3+} + 4\ OH^- \rightleftarrows [Al(OH)_4]^-$	1.1×10^{33}
$Au^+ + 2\ CN^- \rightleftarrows [Au(CN)_2]^-$	2.0×10^{38}
$Cd^{2+} + 4\ CN^- \rightleftarrows [Cd(CN)_4]^{2-}$	6.0×10^{18}
$Cd^{2+} + 4\ NH_3 \rightleftarrows [Cd(NH_3)_4]^{2+}$	1.3×10^7
$Co^{2+} + 6\ NH_3 \rightleftarrows [Co(NH_3)_6]^{2+}$	1.3×10^5
$Cu^+ + 2\ CN^- \rightleftarrows [Cu(CN)_2]^-$	1.0×10^{24}
$Cu^+ + 2\ Cl^- \rightleftarrows [Cu(Cl)_2]^-$	3.2×10^5
$Cu^{2+} + 4\ NH_3 \rightleftarrows [Cu(NH_3)_4]^{2+}$	2.1×10^{13}
$Fe^{2+} + 6\ CN^- \rightleftarrows [Fe(CN)_6]^{4-}$	1.0×10^{35}
$Hg^{2+} + 4\ Cl^- \rightleftarrows [HgCl_4]^{2-}$	1.2×10^{15}
$Ni^{2+} + 4\ CN^- \rightleftarrows [Ni(CN)_4]^{2-}$	2.0×10^{31}
$Ni^{2+} + 6\ NH_3 \rightleftarrows [Ni(NH_3)_6]^{2+}$	5.5×10^8
$Zn^{2+} + 6\ NH_3 \rightleftarrows [Ni(NH_3)_6]^{2+}$	4.6×10^{17}
$Zn^{2+} + 4\ NH_3 \rightleftarrows [Zn(NH_3)_4]^{2+}$	2.9×10^9

*Data reported in this table are taken from J. A. Dean: *Lange's Handbook of Chemistry,* 15th Edition. New York, McGraw-Hill Publishers, 1999.

L Selected Thermodynamic Values

TABLE 20 Selected Thermodynamic Values*

Species	ΔH_f° (298.15 K) (kJ/mol)	S° (298.15 K) (J/K · mol)	ΔG_f° (298.15 K) (kJ/mol)
Aluminum			
$Al(s)$	0	28.3	0
$AlCl_3(s)$	−705.63	109.29	−630.0
$Al_2O_3(s)$	−1675.7	50.92	−1582.3
Barium			
$BaCl_2(s)$	−858.6	123.68	−810.4
$BaCO_3(s)$	−1213	112.1	−1134.41
$BaO(s)$	−548.1	72.05	−520.38
$BaSO_4(s)$	−1473.2	132.2	−1362.2
Beryllium			
$Be(s)$	0	9.5	0
$Be(OH)_2(s)$	−902.5	51.9	−815.0
Boron			
$BCl_3(g)$	−402.96	290.17	−387.95
Bromine			
$Br(g)$	111.884	175.022	82.396
$Br_2(\ell)$	0	152.2	0
$Br_2(g)$	30.91	245.47	3.12
$BrF_3(g)$	−255.60	292.53	−229.43
$HBr(g)$	−36.29	198.70	−53.45

(continued)

*Most thermodynamic data are taken from the NIST Webbook at **http://webbook.nist.gov**.

TABLE 20 Selected Thermodynamic Values* *(continued)*

Species	ΔH_f° (298.15 K) (kJ/mol)	S° (298.15 K) (J/K · mol)	ΔG_f° (298.15 K) (kJ/mol)
Calcium			
Ca(s)	0	41.59	0
Ca(g)	178.2	158.884	144.3
Ca^{2+}(g)	1925.90	—	—
CaC_2(s)	−59.8	70.	−64.93
$CaCO_3$(s, calcite)	−1207.6	91.7	−1129.16
$CaCl_2$(s)	−795.8	104.6	−748.1
CaF_2(s)	−1219.6	68.87	−1167.3
CaH_2(s)	−186.2	42	−147.2
CaO(s)	−635.09	38.2	−603.42
CaS(s)	−482.4	56.5	−477.4
$Ca(OH)_2$(s)	−986.09	83.39	−898.43
$Ca(OH)_2$(aq)	−1002.82		−868.07
$CaSO_4$(s)	−1434.52	106.5	−1322.02
Carbon			
C(s, graphite)	0	5.6	0
C(s, diamond)	1.8	2.377	2.900
C(g)	716.67	158.1	671.2
CCl_4(ℓ)	−128.4	214.39	−57.63
CCl_4(g)	−95.98	309.65	−53.61
$CHCl_3$(ℓ)	−134.47	201.7	−73.66
$CHCl_3$(g)	−103.18	295.61	−70.4
CH_4(g, methane)	−74.87	186.26	−50.8
C_2H_2(g, ethyne)	226.73	200.94	209.20
C_2H_4(g, ethene)	52.47	219.36	68.35
C_2H_6(g, ethane)	−83.85	229.2	−31.89
C_3H_8(g, propane)	−104.7	270.3	−24.4
C_6H_6(ℓ, benzene)	48.95	173.26	124.21
CH_3OH(ℓ, methanol)	−238.4	127.19	−166.14
CH_3OH(g, methanol)	−201.0	239.7	−162.5
C_2H_5OH(ℓ, ethanol)	−277.0	160.7	−174.7
C_2H_5OH(g, ethanol)	−235.3	282.70	−168.49
CO(g)	−110.525	197.674	−137.168
CO_2(g)	−393.509	213.74	−394.359
CS_2(ℓ)	89.41	151	65.2
CS_2(g)	116.7	237.8	66.61
$COCl_2$(g)	−218.8	283.53	−204.6

(continued)

TABLE 20 Selected Thermodynamic Values* *(continued)*

Species	ΔH_f° (298.15 K) (kJ/mol)	S° (298.15 K) (J/K · mol)	ΔG_f° (298.15 K) (kJ/mol)
Cesium			
Cs(s)	0	85.23	0
Cs^+(g)	457.964	—	—
CsCl(s)	−443.04	101.17	−414.53
Chlorine			
Cl(g)	121.3	165.19	105.3
Cl^-(g)	−233.13	—	—
Cl_2(g)	0	223.08	0
HCl(g)	−92.31	186.2	−95.09
HCl(aq)	−167.159	56.5	−131.26
Chromium			
Cr(s)	0	23.62	0
Cr_2O_3(s)	−1134.7	80.65	−1052.95
$CrCl_3$(s)	−556.5	123.0	−486.1
Copper			
Cu(s)	0	33.17	0
CuO(s)	−156.06	42.59	−128.3
$CuCl_2$(s)	−220.1	108.07	−175.7
$CuSO_4$(s)	−769.98	109.05	−660.75
Fluorine			
F_2(g)	0	202.8	0
F(g)	78.99	158.754	61.91
F^-(g)	−255.39	—	—
F^-(aq)	−332.63		−278.79
HF(g)	−273.3	173.779	−273.2
HF(aq)	−332.63	88.7	−278.79
Hydrogen			
H_2(g)	0	130.7	0
H(g)	217.965	114.713	203.247
H^+(g)	1536.202	—	—
$H_2O(\ell)$	−285.83	69.95	−237.15
H_2O(g)	−241.83	188.84	−228.59
$H_2O_2(\ell)$	−187.78	109.6	−120.35
Iodine			
I_2(s)	0	116.135	0
I_2(g)	62.438	260.69	19.327
I(g)	106.838	180.791	70.250

(continued)

TABLE 20 Selected Thermodynamic Values* *(continued)*

Species	ΔH_f° (298.15 K) (kJ/mol)	S° (298.15 K) (J/K · mol)	ΔG_f° (298.15 K) (kJ/mol)
I^-(g)	−197	—	—
ICl(g)	17.51	247.56	−5.73
Iron			
Fe(s)	0	27.78	0
FeO(s)	−272	—	—
Fe_2O_3(s, hematite)	−825.5	87.40	−742.2
Fe_3O_4(s, magnetite)	−1118.4	146.4	−1015.4
$FeCl_2$(s)	−341.79	117.95	−302.30
$FeCl_3$(s)	−399.49	142.3	−344.00
FeS_2(s, pyrite)	−178.2	52.93	−166.9
$Fe(CO)_5(\ell)$	−774.0	338.1	−705.3
Lead			
Pb(s)	0	64.81	0
$PbCl_2$(s)	−359.41	136.0	−314.10
PbO(s, yellow)	−219	66.5	−196
PbO_2(s)	−277.4	68.6	−217.39
PbS(s)	−100.4	91.2	−98.7
Lithium			
Li(s)	0	29.12	0
Li^+(g)	685.783	—	—
LiOH(s)	−484.93	42.81	−438.96
LiOH(aq)	−508.48	2.80	−450.58
LiCl(s)	−408.701	59.33	−384.37
Magnesium			
Mg(s)	0	32.67	0
$MgCl_2$(s)	−641.62	89.62	−592.09
$MgCO_3$(s)	−1111.69	65.84	−1028.2
MgO(s)	−601.24	26.85	−568.93
$Mg(OH)_2$(s)	−924.54	63.18	−833.51
MgS(s)	−346.0	50.33	−341.8
Mercury			
$Hg(\ell)$	0	76.02	0
$HgCl_2$(s)	−224.3	146.0	−178.6
HgO(s, red)	−90.83	70.29	−58.539
HgS(s, red)	−58.2	82.4	−50.6

(continued)

TABLE 20 Selected Thermodynamic Values* *(continued)*

Species	ΔH_f° (298.15 K) (kJ/mol)	S° (298.15 K) (J/K · mol)	ΔG_f° (298.15 K) (kJ/mol)
Nickel			
Ni(s)	0	29.87	0
NiO(s)	−239.7	37.99	−211.7
$NiCl_2$(s)	−305.332	97.65	−259.032
Nitrogen			
N_2(g)	0	191.56	0
N(g)	472.704	153.298	455.563
NH_3(g)	−45.90	192.77	−16.37
N_2H_4(ℓ)	50.63	121.52	149.45
NH_4Cl(s)	−314.55	94.85	−203.08
NH_4Cl(aq)	−299.66	169.9	−210.57
NH_4NO_3(s)	−365.56	151.08	−183.84
NH_4NO_3(aq)	−339.87	259.8	−190.57
NO(g)	90.29	210.76	86.58
NO_2(g)	33.1	240.04	51.23
N_2O(g)	82.05	219.85	104.20
N_2O_4(g)	9.08	304.38	97.73
NOCl(g)	51.71	261.8	66.08
HNO_3(ℓ)	−174.10	155.60	−80.71
HNO_3(g)	−135.06	266.38	−74.72
HNO_3(aq)	−207.36	146.4	−111.25
Oxygen			
O_2(g)	0	205.07	0
O(g)	249.170	161.055	231.731
O_3(g)	142.67	238.92	163.2
Phosphorus			
P_4(s, white)	0	41.1	0
P_4(s, red)	−17.6	22.80	−12.1
P(g)	314.64	163.193	278.25
PH_3(g)	22.89	210.24	30.91
PCl_3(g)	−287.0	311.78	−267.8
P_4O_{10}(s)	−2984.0	228.86	−2697.7
H_3PO_4(ℓ)	−1279.0	110.5	−1119.1
Potassium			
K(s)	0	64.63	0
KCl(s)	−436.68	82.56	−408.77
$KClO_3$(s)	−397.73	143.1	−296.25
KI(s)	−327.90	106.32	−324.892

(continued)

TABLE 20 Selected Thermodynamic Values* *(continued)*

Species	ΔH_f° (298.15 K) (kJ/mol)	S° (298.15 K) (J/K · mol)	ΔG_f° (298.15 K) (kJ/mol)
KOH(s)	−424.72	78.9	−378.92
KOH(aq)	−482.37	91.6	−440.50
Silicon			
Si(s)	0	18.82	0
$SiBr_4(\ell)$	−457.3	277.8	−443.9
SiC(s)	−65.3	16.61	−62.8
$SiCl_4(g)$	−662.75	330.86	−622.76
$SiH_4(g)$	34.31	204.65	56.84
$SiF_4(g)$	−1614.94	282.49	−1572.65
SiO_2(s, quartz)	−910.86	41.46	−856.97
Silver			
Ag(s)	0	42.55	0
$Ag_2O(s)$	−31.1	121.3	−11.32
AgCl(s)	−127.01	96.25	−109.76
$AgNO_3(s)$	−124.39	140.92	−33.41
Sodium			
Na(s)	0	51.21	0
Na(g)	107.3	153.765	76.83
$Na^+(g)$	609.358	—	—
NaBr(s)	−361.02	86.82	−348.983
NaCl(s)	−411.12	72.11	−384.04
NaCl(g)	−181.42	229.79	−201.33
NaCl(aq)	−407.27	115.5	−393.133
NaOH(s)	−425.93	64.46	−379.75
NaOH(aq)	−469.15	48.1	−418.09
$Na_2CO_3(s)$	−1130.77	134.79	−1048.08
Sulfur			
S(s, rhombic)	0	32.1	0
S(g)	278.98	167.83	236.51
$S_2Cl_2(g)$	−18.4	331.5	−31.8
$SF_6(g)$	−1209	291.82	−1105.3
$H_2S(g)$	−20.63	205.79	−33.56
$SO_2(g)$	−296.84	248.21	−300.13
$SO_3(g)$	−395.77	256.77	−371.04
$SOCl_2(g)$	−212.5	309.77	−198.3
$H_2SO_4(\ell)$	−814	156.9	−689.96
$H_2SO_4(aq)$	−909.27	20.1	−744.53

(continued)

TABLE 20 Selected Thermodynamic Values* *(continued)*

Species	ΔH_f° (298.15 K) (kJ/mol)	S° (298.15 K) (J/K · mol)	ΔG_f° (298.15 K) (kJ/mol)
Tin			
Sn(s, white)	0	51.08	0
Sn(s, gray)	−2.09	44.14	0.13
$SnCl_4(\ell)$	−511.3	258.6	−440.15
$SnCl_4(g)$	−471.5	365.8	−432.31
$SnO_2(s)$	−577.63	49.04	−515.88
Titanium			
Ti(s)	0	30.72	0
$TiCl_4(\ell)$	−804.2	252.34	−737.2
$TiCl_4(g)$	−763.16	354.84	−726.7
$TiO_2(s)$	−939.7	49.92	−884.5
Zinc			
Zn(s)	0	41.63	0
$ZnCl_2(s)$	−415.05	111.46	−369.398
ZnO(s)	−348.28	43.64	−318.30
ZnS(s, sphalerite)	−205.98	57.7	−201.29

M Standard Reduction Potentials in Aqueous Solution at 25 °C

TABLE 21 Standard Reduction Potentials in Aqueous Solution at 25 °C

Acidic Solution	Standard Reduction Potential $E°$ (volts)
$F_2(g) + 2\ e^- \longrightarrow 2\ F^-(aq)$	2.87
$Co^{3+}(aq) + e^- \longrightarrow Co^{2+}(aq)$	1.82
$Pb^{4+}(aq) + 2\ e^- \longrightarrow Pb^{2+}(aq)$	1.8
$H_2O_2(aq) + 2\ H^+(aq) + 2\ e^- \longrightarrow 2\ H_2O$	1.77
$NiO_2(s) + 4\ H^+(aq) + 2\ e^- \longrightarrow Ni^{2+}(aq) + 2\ H_2O$	1.7
$PbO_2(s) + SO_4^{2-}(aq) + 4\ H^+(aq) + 2\ e^- \longrightarrow PbSO_4(s) + 2\ H_2O$	1.685
$Au^+(aq) + e^- \longrightarrow Au(s)$	1.68
$2\ HClO(aq) + 2\ H^+(aq) + 2\ e^- \longrightarrow Cl_2(g) + 2\ H_2O$	1.63
$Ce^{4+}(aq) + e^- \longrightarrow Ce^{3+}(aq)$	1.61
$NaBiO_3(s) + 6\ H^+(aq) + 2\ e^- \longrightarrow Bi^{3+}(aq) + Na^+(aq) + 3\ H_2O$	≈1.6
$MnO_4^-(aq) + 8\ H^+(aq) + 5\ e^- \longrightarrow Mn^{2+}(aq) + 4\ H_2O$	1.51
$Au^{3+}(aq) + 3\ e^- \longrightarrow Au(s)$	1.50
$ClO_3^-(aq) + 6\ H^+(aq) + 5\ e^- \longrightarrow \frac{1}{2}\ Cl_2(g) + 3\ H_2O$	1.47
$BrO_3^-(aq) + 6\ H^+(aq) + 6\ e^- \longrightarrow Br^-(aq) + 3\ H_2O$	1.44
$Cl_2(g) + 2\ e^- \longrightarrow 2\ Cl^-(aq)$	1.36
$Cr_2O_7^{2-}(aq) + 14\ H^+(aq) + 6\ e^- \longrightarrow 2\ Cr^{3+}(aq) + 7\ H_2O$	1.33
$N_2H_5^+(aq) + 3\ H^+(aq) + 2\ e^- \longrightarrow 2\ NH_4^+(aq)$	1.24
$MnO_2(s) + 4\ H^+(aq) + 2\ e^- \longrightarrow Mn^{2+}(aq) + 2\ H_2O$	1.23
$O_2(g) + 4\ H^+(aq) + 4\ e^- \longrightarrow 2\ H_2O$	1.229
$Pt^{2+}(aq) + 2\ e^- \longrightarrow Pt(s)$	1.2
$IO_3^-(aq) + 6\ H^+(aq) + 5\ e^- \longrightarrow \frac{1}{2}\ I_2(aq) + 3\ H_2O$	1.195

(continued)

TABLE 21 Standard Reduction Potentials in Aqueous Solution at 25 °C *(continued)*

Acidic Solution	Standard Reduction Potential $E°$ (volts)
$ClO_4^-(aq) + 2\ H^+(aq) + 2\ e^- \longrightarrow ClO_3^-(aq) + H_2O$	1.19
$Br_2(\ell) + 2\ e^- \longrightarrow 2\ Br^-(aq)$	1.08
$AuCl_4^-(aq) + 3\ e^- \longrightarrow Au(s) + 4\ Cl^-(aq)$	1.00
$Pd^{2+}(aq) + 2\ e^- \longrightarrow Pd(s)$	0.987
$NO_3^-(aq) + 4\ H^+(aq) + 3\ e^- \longrightarrow NO(g) + 2\ H_2O$	0.96
$NO_3^-(aq) + 3\ H^+(aq) + 2\ e^- \longrightarrow HNO_2(aq) + H_2O$	0.94
$2\ Hg^+(aq) + 2\ e^- \longrightarrow Hg_2^{2+}(aq)$	0.920
$Hg^{2+}(aq) + 2\ e^- \longrightarrow Hg(\ell)$	0.855
$Ag^+(aq) + e^- \longrightarrow Ag(s)$	0.7994
$Hg_2^{2+}(aq) + 2\ e^- \longrightarrow 2\ Hg(\ell)$	0.789
$Fe^{3+}(aq) + e^- \longrightarrow Fe^{2+}(aq)$	0.771
$SbCl_6^-(aq) + 2\ e^- \longrightarrow SbCl_4^-(aq) + 2\ Cl^-(aq)$	0.75
$[PtCl_4]^{2+}(aq) + 2\ e^- \longrightarrow Pt(s) + 4\ Cl^-(aq)$	0.73
$O_2(g) + 2\ H^+(aq) + 2\ e^- \longrightarrow H_2O_2(aq)$	0.682
$[PtCl_6]^{2-}(aq) + 2\ e^- \longrightarrow [PtCl_4]^{2-}(aq) + 2\ Cl^-(aq)$	0.68
$I_2(aq) + 2\ e^- \longrightarrow 2\ I^-(aq)$	0.621
$H_3AsO_4(aq) + 2\ H^+(aq) + 2\ e^- \longrightarrow H_3AsO_3(aq) + H_2O$	0.58
$I_2(s) + 2\ e^- \longrightarrow 2\ I^-(aq)$	0.535
$TeO_2(s) + 4\ H^+(aq) + 4\ e^- \longrightarrow Te(s) + 2\ H_2O$	0.529
$Cu^+(aq) + e^- \longrightarrow Cu(s)$	0.521
$[RhCl_6]^{3-}(aq) + 3\ e^- \longrightarrow Rh(s) + 6\ Cl^-(aq)$	0.44
$Cu^{2+}(aq) + 2\ e^- \longrightarrow Cu(s)$	0.337
$Hg_2Cl_2(s) + 2\ e^- \longrightarrow 2\ Hg(\ell) + 2\ Cl^-(aq)$	0.27
$AgCl(s) + e^- \longrightarrow Ag(s) + Cl^-(aq)$	0.222
$SO_4^{2-}(aq) + 4\ H^+(aq) + 2\ e^- \longrightarrow SO_2(g) + 2\ H_2O$	0.20
$SO_4^{2-}(aq) + 4\ H^+(aq) + 2\ e^- \longrightarrow H_2SO_3(aq) + H_2O$	0.17
$Cu^{2+}(aq) + e^- \longrightarrow Cu^+(aq)$	0.153
$Sn^{4+}(aq) + 2\ e^- \longrightarrow Sn^{2+}(aq)$	0.15
$S(s) + 2\ H^+ + 2\ e^- \longrightarrow H_2S(aq)$	0.14
$AgBr(s) + e^- \longrightarrow Ag(s) + Br^-(aq)$	0.0713
$2\ H^+(aq) + 2\ e^- \longrightarrow H_2(g)$(reference electrode)	0.0000
$N_2O(g) + 6\ H^+(aq) + H_2O + 4\ e^- \longrightarrow 2\ NH_3OH^+(aq)$	−0.05
$Pb^{2+}(aq) + 2\ e^- \longrightarrow Pb(s)$	−0.126
$Sn^{2+}(aq) + 2\ e^- \longrightarrow Sn(s)$	−0.14
$AgI(s) + e^- \longrightarrow Ag(s) + I^-(aq)$	−0.15
$[SnF_6]^{2-}(aq) + 4\ e^- \longrightarrow Sn(s) + 6\ F^-(aq)$	−0.25
$Ni^{2+}(aq) + 2\ e^- \longrightarrow Ni(s)$	−0.25
$Co^{2+}(aq) + 2\ e^- \longrightarrow Co(s)$	−0.28

(continued)

TABLE 21 Standard Reduction Potentials in Aqueous Solution at 25 °C *(continued)*

Acidic Solution	Standard Reduction Potential $E°$ (volts)
$Tl^{+}(aq) + e^{-} \longrightarrow Tl(s)$	−0.34
$PbSO_4(s) + 2\ e^{-} \longrightarrow Pb(s) + SO_4^{2-}(aq)$	−0.356
$Se(s) + 2\ H^{+}(aq) + 2\ e^{-} \longrightarrow H_2Se(aq)$	−0.40
$Cd^{2+}(aq) + 2\ e^{-} \longrightarrow Cd(s)$	−0.403
$Cr^{3+}(aq) + e^{-} \longrightarrow Cr^{2+}(aq)$	−0.41
$Fe^{2+}(aq) + 2\ e^{-} \longrightarrow Fe(s)$	−0.44
$2\ CO_2(g) + 2\ H^{+}(aq) + 2\ e^{-} \longrightarrow H_2C_2O_4(aq)$	−0.49
$Ga^{3+}(aq) + 3\ e^{-} \longrightarrow Ga(s)$	−0.53
$HgS(s) + 2\ H^{+}(aq) + 2\ e^{-} \longrightarrow Hg(\ell) + H_2S(g)$	−0.72
$Cr^{3+}(aq) + 3\ e^{-} \longrightarrow Cr(s)$	−0.74
$Zn^{2+}(aq) + 2\ e^{-} \longrightarrow Zn(s)$	−0.763
$Cr^{2+}(aq) + 2\ e^{-} \longrightarrow Cr(s)$	−0.91
$FeS(s) + 2\ e^{-} \longrightarrow Fe(s) + S^{2-}(aq)$	−1.01
$Mn^{2+}(aq) + 2\ e^{-} \longrightarrow Mn(s)$	−1.18
$V^{2+}(aq) + 2\ e^{-} \longrightarrow V(s)$	−1.18
$CdS(s) + 2\ e^{-} \longrightarrow Cd(s) + S^{2-}(aq)$	−1.21
$ZnS(s) + 2\ e^{-} \longrightarrow Zn(s) + S^{2-}(aq)$	−1.44
$Zr^{4+}(aq) + 4\ e^{-} \longrightarrow Zr(s)$	−1.53
$Al^{3+}(aq) + 3\ e^{-} \longrightarrow Al(s)$	−1.66
$Mg^{2+}(aq) + 2\ e^{-} \longrightarrow Mg(s)$	−2.37
$Na^{+}(aq) + e^{-} \longrightarrow Na(s)$	−2.714
$Ca^{2+}(aq) + 2\ e^{-} \longrightarrow Ca(s)$	−2.87
$Sr^{2+}(aq) + 2\ e^{-} \longrightarrow Sr(s)$	−2.89
$Ba^{2+}(aq) + 2\ e^{-} \longrightarrow Ba(s)$	−2.90
$Rb^{+}(aq) + e^{-} \longrightarrow Rb(s)$	−2.925
$K^{+}(aq) + e^{-} \longrightarrow K(s)$	−2.925
$Li^{+}(aq) + e^{-} \longrightarrow Li(s)$	−3.045
Basic Solution	
$ClO^{-}(aq) + H_2O + 2\ e^{-} \longrightarrow Cl^{-}(aq) + 2\ OH^{-}(aq)$	0.89
$OOH^{-}(aq) + H_2O + 2\ e^{-} \longrightarrow 3\ OH^{-}(aq)$	0.88
$2\ NH_2OH(aq) + 2\ e^{-} \longrightarrow N_2H_4(aq) + 2\ OH^{-}(aq)$	0.74
$ClO_3^{-}(aq) + 3\ H_2O + 6\ e^{-} \longrightarrow Cl^{-}(aq) + 6\ OH^{-}(aq)$	0.62
$MnO_4^{-}(aq) + 2\ H_2O + 3\ e^{-} \longrightarrow MnO_2(s) + 4\ OH^{-}(aq)$	0.588
$MnO_4^{-}(aq) + e^{-} \longrightarrow MnO_4^{2-}(aq)$	0.564
$NiO_2(s) + 2\ H_2O + 2\ e^{-} \longrightarrow Ni(OH)_2(s) + 2\ OH^{-}(aq)$	0.49
$Ag_2CrO_4(s) + 2\ e^{-} \longrightarrow 2\ Ag(s) + CrO_4^{2-}(aq)$	0.446
$O_2(g) + 2\ H_2O + 4\ e^{-} \longrightarrow 4\ OH^{-}(aq)$	0.40

(continued)

TABLE 21 Standard Reduction Potentials in Aqueous Solution at 25 °C *(continued)*

Acidic Solution	Standard Reduction Potential $E°$ (volts)
$ClO_4^-(aq) + H_2O + 2\ e^- \longrightarrow ClO_3^-(aq) + 2\ OH^-(aq)$	0.36
$Ag_2O(s) + H_2O + 2\ e^- \longrightarrow 2\ Ag(s) + 2\ OH^-(aq)$	0.34
$2\ NO_2^-(aq) + 3\ H_2O + 4\ e^- \longrightarrow N_2O(g) + 6\ OH^-(aq)$	0.15
$N_2H_4(aq) + 2\ H_2O + 2\ e^- \longrightarrow 2\ NH_3(aq) + 2\ OH^-(aq)$	0.10
$[Co(NH_3)_6]^{3+}(aq) + e^- \longrightarrow [Co(NH_3)_6]^{2+}(aq)$	0.10
$HgO(s) + H_2O + 2\ e^- \longrightarrow Hg(\ell) + 2\ OH^-(aq)$	0.0984
$O_2(g) + H_2O + 2\ e^- \longrightarrow OOH^-(aq) + OH^-(aq)$	0.076
$NO_3^-(aq) + H_2O + 2\ e^- \longrightarrow NO_2^-(aq) + 2\ OH^-(aq)$	0.01
$MnO_2(s) + 2\ H_2O + 2\ e^- \longrightarrow Mn(OH)_2(s) + 2\ OH^-(aq)$	−0.05
$CrO_4^{2-}(aq) + 4\ H_2O + 3\ e^- \longrightarrow Cr(OH)_3(s) + 5\ OH^-(aq)$	−0.12
$Cu(OH)_2(s) + 2\ e^- \longrightarrow Cu(s) + 2\ OH^-(aq)$	−0.36
$S(s) + 2\ e^- \longrightarrow S^{2-}(aq)$	−0.48
$Fe(OH)_3(s) + e^- \longrightarrow Fe(OH)_2(s) + OH^-(aq)$	−0.56
$2\ H_2O + 2\ e^- \longrightarrow H_2(g) + 2\ OH^-(aq)$	−0.8277
$2\ NO_3^-(aq) + 2\ H_2O + 2\ e^- \longrightarrow N_2O_4(g) + 4\ OH^-(aq)$	−0.85
$Fe(OH)_2(s) + 2\ e^- \longrightarrow Fe(s) + 2\ OH^-(aq)$	−0.877
$SO_4^{2-}(aq) + H_2O + 2\ e^- \longrightarrow SO_3^{2-}(aq) + 2\ OH^-(aq)$	−0.93
$N_2(g) + 4\ H_2O + 4\ e^- \longrightarrow N_2H_4(aq) + 4\ OH^-(aq)$	−1.15
$[Zn(OH)_4]^{2-}(aq) + 2\ e^- \longrightarrow Zn(s) + 4\ OH^-(aq)$	−1.22
$Zn(OH)_2(s) + 2\ e^- \longrightarrow Zn(s) + 2\ OH^-(aq)$	−1.245
$[Zn(CN)_4]^{2-}(aq) + 2\ e^- \longrightarrow Zn(s) + 4\ CN^-(aq)$	−1.26
$Cr(OH)_3(s) + 3\ e^- \longrightarrow Cr(s) + 3\ OH^-(aq)$	−1.30
$SiO_3^{2-}(aq) + 3\ H_2O + 4\ e^- \longrightarrow Si(s) + 6\ OH^-(aq)$	−1.70

N Answers to Exercises

Chapter 1

1.1 (a) Na = sodium; Cl = chlorine; Cr = chromium
(b) Zinc = Zn; nickel = Ni; potassium = K

1.2 (a) Iron: lustrous solid, metallic, good conductor of heat and electricity, malleable, ductile, attracted to a magnet
(b) Water: colorless liquid (at room temperature); melting point is 0 °C, and boiling point is 100 °C, density ~ 1 g/cm^2
(c) Table salt: solid, white crystals, soluble in water
(d) Oxygen: colorless gas (at room temperature), low solubility in water

1.3 Chemical changes: the fuel in the campfire burns in air (combustion). Physical changes: water boils. Energy evolved in combustion is transferred to the water, to the water container, and to the surrounding air.

Let's Review

LR 1 [77 K − 273.15 K] (1 °C/K) = −196 °C

LR 2 Convert thickness to cm: 0.25 mm(1 cm/10 mm) = 0.025 cm

Volume = length × width × thickness

V = (2.50 cm)(2.50 cm)(0.025 cm) = 0.16 cm^3
(answer has 2 significant figures.)

LR 3 (a) (750 mL)(1 L/1000 mL) = 0.75 L
(0.75 L)(10 dL/L) = 7.5 dL
(b) 2.0 qt = 0.50 gal
(0.50 gal)(3.786 L/gal) = 1.9 L
(1.9 L)(1 dm^3/1 L) = 1.9 dm^3

LR 4 (a) Mass in kilograms = (5.59 g)(1 kg/1000 g) = 0.00559 kg
Mass in milligrams = 5.59 g (10^3 mg/g) = 5.59×10^3 mg
(b) (0.02 μg/L)(1 g/10^6 μg) = 2×10^{-8} g/L

LR 5 Student A: average = −0.1 °C; average deviation = 0.2 °C; error = −0.1 °C. Student B: average = +0.01 °C; average deviation = 0.02 °C; error = +0.01 °C. Student B's values are more accurate and less precise.

LR 6 (a) 2.33×10^7 has three significant figures; 50.5 has three significant figures; 200 has one significant figure. (200. or 2.00×10^2 would express this number with three significant figures.)
(b) The product of 10.26 and 0.063 is 0.65, a number with two significant figures. (10.26 has four significant figures, whereas 0.063 has two.)
The sum of 10.26 and 0.063 is 10.32. The number 10.26 has only two numbers to the right of the decimal, so the sum must also have two numbers after the decimal.
(c) $x = 3.9 \times 10^6$. The difference between 110.7 and 64 is 47. Dividing 47 by 0.056 and 0.00216 gives an answer with two significant figures.

LR 7 (a) (198 cm)(1 m/100 cm) = 1.98 m;
(198 cm)(1 ft/30.48 cm) = 6.50 ft
(b) (2.33×10^7 m^2)(1 km^2/10^6 m^2) = 23.3 km^2
(c) (19,320 kg/m^3)(10^3 g/1 kg)(1 m^3/10^6 cm^3) = 19.32 g/cm^3
(d) (9.0×10^3 pc)(206,265 AU/1 pc)(1.496×10^8 km/1 AU) = 2.8×10^{17} km

LR 8 Read from the graph, the mass of 50 beans is about 123 g.

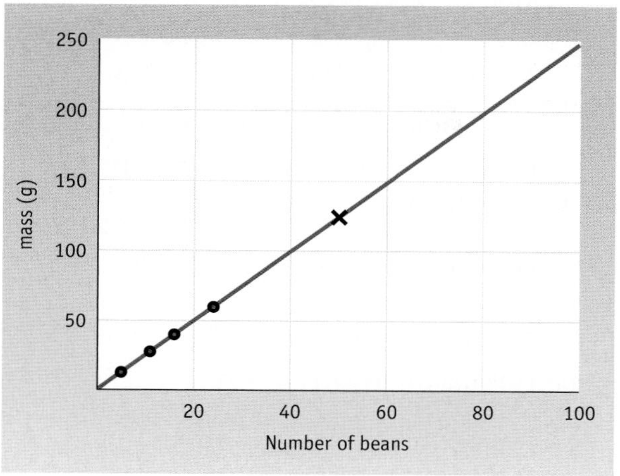

LR 9 Change all dimensions to centimeters: 7.6 m = 760 cm; 2.74 m = 274 cm; 0.13 mm = 0.013 cm.

Volume of paint = (760 cm)(274 cm)(0.013 cm) = 2.7×10^3 cm^3

Volume (L) = $(2.7 \times 10^3 \text{ cm}^3)(1 \text{ L}/10^3 \text{ cm}^3) = 2.7$ L

Mass = $(2.7 \times 10^3 \text{ cm}^3)(0.914 \text{ g/cm}^3) = 2.5 \times 10^3$ g

Chapter 2

2.1 (a) Mass number with 26 protons and 30 neutrons is 56
(b) $(59.930788 \text{ u})(1.661 \times 10^{-24} \text{ g/u}) = 9.955 \times 10^{-23}$ g
(c) ^{64}Zn has 30 protons, 30 electrons, and (64 − 30) = 34 neutrons.
(d) The mass of a ^{64}Zn atom is 63.929/12.0... . or 5.3274 times the mass of a ^{12}C atom. (Note that mass of ^{12}C is defined as an exact value.)

2.2 The mass number of the second silver isotope is 109 (62 + 47). Symbol: ^{109}Ag, abundance = 48.161%.

2.3 Use Equation 2.2 for the calculation.
Atomic mass = (34.96885)(75.77/100) + (36.96590)(24.23/100) = 35.45. (Accuracy is limited by the value of the percent abundance to 4 significant figures.)

2.4 There are eight elements in the third period. Sodium (Na), magnesium (Mg), and aluminum (Al) are metals. Silicon (Si) is a metalloid. Phosphorus (P), sulfur (S), chlorine (Cl), and argon (Ar) are nonmetals.

2.5 The molecular formula is $C_3H_7NO_2S$. You will often see its formula written as $HSCH_2CH(N^+H_3)CO_2^-$ to better identify the molecule's structure.

2.6 (a) K^+ is formed if K loses one electron. K^+ has the same number of electrons as Ar.
(b) Se^{2-} is formed by adding two electrons to an atom of Se. It has the same number of electrons as Kr.
(c) Ba^{2+} is formed if Ba loses two electrons; Ba^{2+} has the same number of electrons as Xe.
(d) Cs^+ is formed if Cs loses one electron. It has the same number of electrons as Xe.

2.7 (a) (1) NaF: 1 Na^+ and 1 F^- ion. (2) $Cu(NO_3)_2$: 1 Cu^{2+} and 2 NO_3^- ions. (3) $NaCH_3CO_2$: 1 Na^+ and 1 $CH_3CO_2^-$ ion.
(b) $FeCl_2$, $FeCl_3$
(c) Na_2S, Na_3PO_4, BaS, $Ba_3(PO_4)_2$

2.8 (1) (a) NH_4NO_3; (b) $CoSO_4$; (c) $Ni(CN)_2$; (d) V_2O_3; (e) $Ba(CH_3CO_2)_2$; (f) $Ca(ClO)_2$
(2) (a) magnesium bromide; (b) lithium carbonate; (c) potassium hydrogen sulfite; (d) potassium permanganate; (e) ammonium sulfide; (f) copper(I) chloride and copper(II) chloride

2.9 The force of attraction between ions is proportional to the product of the ion charges (Coulomb's law). The force of attraction between Mg^{2+} and O^{2-} ions in MgO is approximately four times greater than the force of attraction between Na^+ and Cl^- ions in NaCl, so a much higher temperature is required to disrupt the orderly array of ions in crystalline MgO.

2.10 (1) (a) CO_2; (b) PI_3; (c) SCl_2; (d) BF_3; (e) O_2F_2; (f) XeO_3
(2) (a) dinitrogen tetrafluoride; (b) hydrogen bromide; (c) sulfur tetrafluoride; (d) boron trichloride; (e) tetraphosphorus decaoxide; (f) chlorine trifluoride

2.11 (a) (1.5 mol Si)(28.1 g/mol) = 42 g Si
(b) (454 g S)(1.00 mol S/32.07 g) = 14.2 mol S
$(14.2 \text{ mol S})(6.022 \times 10^{23} \text{ atoms/mol}) = 8.53 \times 10^{24}$ atoms S

2.12 $(2.6 \times 10^{24} \text{ atoms})(1.000 \text{ mol}/6.022 \times 10^{23} \text{ atoms})(197.0 \text{ g Au}/1.000 \text{ mol}) = 850$ g Au

Volume = (850 g Au)(1.00 cm^3/19.32 g) = 44 cm^3

Volume = 44 cm^3 = (thickness)(area) = (0.10 cm)(area)

Area = 440 cm^2

Length = width = $\sqrt{440 \text{ cm}^2}$ = 21 cm

2.13 (a) Citric acid: 192.1 g/mol; magnesium carbonate: 84.3 g/mol
(b) 454 g citric acid (1.000 mol/192.1 g) = 2.36 mol citric acid
(c) 0.125 mol $MgCO_3$ (84.3 g/mol) = 10.5 g $MgCO_3$

2.14 (a) 1.00 mol $(NH_4)_2CO_3$ (molar mass 96.09 g/mol) has 28.0 g of N (29.2%), 8.06 g of H (8.39%), 12.0 g of C (12.5%), and 48.0 g of O (50.0%)
(b) 454 g C_8H_{18} (1 mol C_8H_{18}/114.2 g)(8 mol C/1 mol C_8H_{18})(12.01 g C/1 mol C) = 382 g C

2.15 (a) C_5H_4 (b) $C_2H_4O_2$

2.16 (88.17 g C)(1 mol C/12.011 g C) = 7.341 mol C

$(11.83 \text{ g H})(1 \text{ mol H}/1.008 \text{ g H}) = 11.74 \text{ mol H}$

$11.74 \text{ mol H}/7.341 \text{ mol C} = 1.6 \text{ mol H}/1 \text{ mol C} = (8/5)$; $(\text{mol H}/1 \text{ mol C}) = 8 \text{ mol H}/5 \text{ mol C}$

The empirical formula is C_5H_8. The molar mass, 68.11 g/mol, closely matches this formula, so C_5H_8 is also the molecular formula.

2.17 $(78.90 \text{ g C})(1 \text{ mol C}/12.011 \text{ g C}) = 6.569 \text{ mol C}$
$(10.59 \text{ g H})(1 \text{ mol H}/1.008 \text{ g H}) = 10.51 \text{ mol H}$
$(10.51 \text{ g O})(1 \text{ mol O}/16.00 \text{ g O}) = 0.6569 \text{ mol O}$

$10.51 \text{ mol H}/0.6569 \text{ mol O} = 16 \text{ mol H}/1 \text{ mol O}$

$6.569 \text{ mol C}/0.6569 \text{ mol O} = 10 \text{ mol C}/1 \text{ mol O}$

The empirical formula is $C_{10}H_{16}O$.

2.18 $(0.586 \text{ g K})(1 \text{ mol K}/39.10 \text{ g K}) = 0.0150 \text{ mol K}$
$(0.480 \text{ g O})(1 \text{ mol O}/16.00 \text{ g O}) = 0.0300 \text{ mol O}$

The ratio of moles K to moles O atoms is 1 to 2; the empirical formula is KO_2.

2.19 Mass of water lost on heating is 0.235 g − 0.128 g = 0.107 g; 0.128 g $NiCl_2$ remain

$(0.107 \text{ g } H_2O)(1 \text{ mol } H_2O/18.016 \text{ g } H_2O) = 0.00594 \text{ mol } H_2O$

$(0.128 \text{ g } NiCl_2)(1 \text{ mol } NiCl_2/129.6 \text{ g } NiCl_2) = 0.000988 \text{ mol } NiCl_2$

Mole ratio = $0.00594 \text{ mol } H_2O/0.000988 \text{ mol } NiCl_2 = 6.01$: Therefore $x = 6$

The formula for the hydrate is $NiCl_2 \cdot 6\,H_2O$.

Chapter 3

3.1 (a) Stoichiometric coefficients: 2 for Al, 3 for Br_2, and 1 for Al_2Br_6
(b) 8000 atoms of Al requires (3/2)8000 = 12,000 molecules of Br_2

3.2 (a) $2\,C_4H_{10}(g) + 13\,O_2(g) \longrightarrow 8\,CO_2(g) + 10\,H_2O(\ell)$
(b) $2\,Pb(C_2H_5)_4(\ell) + 27\,O_2(g) \longrightarrow 2\,PbO(s) + 16\,CO_2(g) + 20\,H_2O(\ell)$

3.3 Epsom salt is an electrolyte, and methanol is a nonelectrolyte.

3.4 (a) $LiNO_3$ is soluble and gives $Li^+(aq)$ and $NO_3^-(aq)$ ions.
(b) $CaCl_2$ is soluble and gives $Ca^{2+}(aq)$ and $Cl^-(aq)$ ions.
(c) CuO is not water-soluble.
(d) $NaCH_3CO_2$ is soluble and gives $Na^+(aq)$ and $CH_3CO_2^-(aq)$ ions.

3.5 (a) $Na_2CO_3(aq) + CuCl_2(aq) \longrightarrow 2\,NaCl(aq) + CuCO_3(s)$
(b) No reaction; no insoluble compound is produced.
(c) $NiCl_2(aq) + 2\,KOH(aq) \longrightarrow Ni(OH)_2(s) + 2\,KCl(aq)$

3.6 (a) $AlCl_3(aq) + Na_3PO_4(aq) \longrightarrow AlPO_4(s) + 3\,NaCl(aq)$

$Al^{3+}(aq) + PO_4^{3-}(aq) \longrightarrow AlPO_4(s)$
(b) $FeCl_3(aq) + 3\,KOH(aq) \longrightarrow Fe(OH)_3(s) + 3\,KCl(aq)$

$Fe^{3+}(aq) + 3\,OH^-(aq) \longrightarrow Fe(OH)_3(s)$
(c) $Pb(NO_3)_2(aq) + 2\,KCl(aq) \longrightarrow PbCl_2(s) + 2\,KNO_3(aq)$

$Pb^{2+}(aq) + 2\,Cl^-(aq) \rightleftharpoons PbCl_2(s)$

3.7 (a) $H_3O(aq)$ and $NO_3^-(aq)$
(b) $Ba^{2+}(aq)$ and $2\,OH^-(aq)$

3.8 (a) $H_3PO_4(aq) + H_2O(\ell) \rightleftharpoons H_3O^+(aq) + H_2PO_4^-(aq)$
(b) Acting as a base:
$H_2PO_4^-(aq) + H_2O(\ell) \rightleftharpoons H_3PO_4(aq) + OH^-(aq)$

Acting as an acid:
$H_2PO_4^-(aq) + H_2O(\ell) \rightleftharpoons HPO_4^{2-}(aq) + H_3O^+(\ell)$

Because $H_2PO_4^-(aq)$ can react as a Brønsted acid and as a base, it is said to be amphiprotic.
(c) $CN^-(aq) + H_2O(\ell) \rightleftharpoons HCN(aq) + OH^-(aq)$; cyanide ion is a Brønsted base.

3.9 $Mg(OH)_2(s) + 2\,HCl(aq) \longrightarrow MgCl_2(aq) + 2\,H_2O(\ell)$

Net ionic equation: $Mg(OH)_2(s) + 2\,H^+(aq) \longrightarrow Mg^{2+}(aq) + 2\,H_2O(\ell)$

3.10 Metals form basic oxides; nonmetals form acidic oxides.
(a) SeO_2 is an acidic oxide; (b) MgO is a basic oxide; and (c) P_4O_{10} is an acidic oxide.

3.11 (a) $BaCO_3(s) + 2\,HNO_3(aq) \longrightarrow Ba(NO_3)_2(aq) + CO_2(g) + H_2O(\ell)$

Barium carbonate and nitric acid produce barium nitrate, carbon dioxide, and water.
(b) $(NH_4)_2SO_4(aq) + 2\,NaOH(aq) \longrightarrow 2\,NH_3(g) + Na_2SO_4(aq) + 2\,H_2O(\ell)$

3.12 (a) Fe in Fe_2O_3, +3; (b) S in H_2SO_4, +6; (c) C in CO_3^{2-}, +4; (d) N in NO_2^+, +5

3.13 Dichromate ion is the oxidizing agent and is reduced. (Cr with a +6 oxidation number is reduced to Cr^{3+} with a +3 oxidation number.) Ethanol is the reducing agent and is oxidized. (The C atoms in ethanol have an oxidation number of −2. The oxidation number is 0 in acetic acid.)

3.14 (b) Cu is the reducing agent and Cl_2 is the oxidizing agent.
(d) $S_2O_3^{2-}$ is the reducing agent and I_2 is the oxidizing agent.

3.15 (a) Gas-forming reaction:

$CuCO_3(s) + H_2SO_4(aq) \longrightarrow CuSO_4(aq) + H_2O(\ell) + CO_2(g)$

Net ionic equation:

$CuCO_3(s) + 2\ H_3O(aq) \longrightarrow Cu^{2+}(aq) + 3\ H_2O(\ell) + CO_2(g)$

(b) Oxidation-reduction: $Ga(s) + O_2(g) \longrightarrow Ga_2O_3(s)$

(c) Acid–base reaction:

$Ba(OH)_2(s) + 2\ HNO_3(aq) \longrightarrow Ba(NO_3)_2(aq) + 2\ H_2O(\ell)$

Net ionic equation:

$Ba(OH)_2(s) + 2\ H_3O(aq) \longrightarrow Ba^{2+}(aq) + 4\ H_2O(\ell)$

(d) Precipitation reaction:

$CuCl_2(aq) + (NH_4)_2S(aq) \longrightarrow CuS(s) + 2\ NH_4Cl(aq)$

Net ionic equation:

$Cu^{2+}(aq) + S^{2-}(aq) \longrightarrow CuS(s)$

Chapter 4

4.1 $(454\ g\ C_3H_8)(1\ mol\ C_3H_8/44.10\ g\ C_3H_8) = 10.3\ mol\ C_3H_8$

$10.3\ mol\ C_3H_8\ (5\ mol\ O_2/1\ mol\ C_3H_8)(32.00\ g\ O_2/1\ mol\ O_2) = 1650\ g\ O_2$

$(10.3\ mol\ C_3H_8)(3\ mol\ CO_2/1\ mol\ C_3H_8)(44.01\ g\ CO_2/1\ mol\ CO_2) = 1360\ g\ CO_2$

$(10.3\ mol\ C_3H_8)(4\ mol\ H_2O/1\ mol\ C_3H_8)(18.02\ g\ H_2O/1\ mol\ H_2O) = 742\ g\ H_2O$

4.2 (a) Amount Al = $(50.0\ g\ Al)(1\ mol\ Al/26.98\ g\ Al) = 1.85\ mol\ Al$

Amount Fe_2O_3 = $(50.0\ g\ Fe_2O_3)(1\ mol\ Fe_2O_3/159.7\ g\ Fe_2O_3) = 0.313\ mol\ Fe_2O_3$

Mol Al/mol Fe_2O_3 = 1.853/0.3131 = 5.92

This is more than the 2:1 ratio required, so the limiting reactant is Fe_2O_3.

(b) Mass Fe = $(0.313\ mol\ Fe_2O_3)(2\ mol\ Fe/1\ mol\ Fe_2O_3)(55.85\ g\ Fe/1\ mol\ Fe) = 35.0\ g\ Fe$

4.3 Theoretical yield = $125\ g\ Al_4C_3(1\ mol\ Al_4C_3/143.95\ g\ Al_4C_3)(3\ mol\ CH_4/1\ mol\ Al_4C_3)(16.04\ g\ CH_4/1\ mol\ CH_4) = 41.8\ g\ CH_4$

Percent yield = (13.6 g/41.8 g)(100%) = 33.0%

4.4 $(0.143\ g\ O_2)(1\ mol\ O_2/32.00\ g\ O_2)(3\ mol\ TiO_2/3\ mol\ O_2)(79.88\ g\ TiO_2/1\ mol\ TiO_2) = 0.357\ g\ TiO_2$

Percent TiO_2 in sample = (0.357 g/2.367 g)(100%) = 15.1%

4.5 $(1.612\ g\ CO_2)(1\ mol\ CO_2/44.01\ g\ CO_2)(1\ mol\ C/1\ mol\ CO_2) = 0.03663\ mol\ C$

$(0.7425\ g\ H_2O)(1\ mol\ H_2O/18.01\ g\ H_2O)(2\ mol\ H/1\ mol\ H_2O) = 0.08243\ mol\ H$

0.08243 mol H/0.03663 mol = 2.250 H/1 C = 9 H/4 C

The empirical formula is C_4H_9, which has a molar mass of 57 g/mol. This is one half of the measured value of molar mass, so the molecular formula is C_8H_{18}.

4.6 $(0.240\ g\ CO_2)(1\ mol\ CO_2/44.01\ g\ CO_2)(1\ mol\ C/1\ mol\ CO_2)(12.01\ g\ C/1\ mol\ C) = 0.06549\ g\ C$

$(0.0982\ g\ H_2O)(1\ mol\ H_2O/18.02\ g\ H_2O)(2\ mol\ H/1\ mol\ H_2O)(1.008\ g\ H/1\ mol\ H) = 0.01099\ g\ H$

Mass O (by difference) = 0.1342 g − 0.06549 g − 0.01099 g = 0.05772 g

Amount C = 0.06549 g(1 mol C/12.01 g C) = 0.00545 mol C

Amount H = 0.01099 g H(1 mol H/1.008 g H) = 0.01090 mol H

Amount O = 0.05772 g O(1 mol O/16.00 g O) = 0.00361 mol O

To find a whole-number ratio, divide each value by 0.00361; this gives 1.51 mol C : 3.02 mol H : 1 mol O. Multiply each value by 2, and round off to 3 mol C : 6 mol H : 2 mol O. The empirical formula is $C_3H_6O_2$; given the molar mass of 74.1, this is also the molecular formula.

4.7 $(26.3\ g)(1\ mol\ NaHCO_3/84.01\ g\ NaHCO_3) = 0.313\ mol\ NaHCO_3$

$0.313\ mol\ NaHCO_3/0.200\ L = 1.57\ M$

Ion concentrations: $[Na^+] = [HCO_3^-] = 1.57\ M$

4.8 First, determine the mass of $AgNO_3$ required.

Amount of $AgNO_3$ required = $(0.0200\ M)(0.250\ L) = 5.00 \times 10^{-3}\ mol$

Mass of $AgNO_3$ = $(5.00 \times 10^{-3}\ mol)(169.9\ g/mol) = 0.850\ g\ AgNO_3$

Weigh out 0.850 g $AgNO_3$. Then, dissolve it in a small amount of water in the volumetric flask. After the solid is dissolved, fill the flask to the mark.

4.9 $(2.00\ M)(V_{conc}) = (1.00\ M)(0.250\ L)$; $V_{conc} = 0.125\ L$

To prepare the solution, measure accurately 125 mL of 2.00 M NaOH into a 250-mL volumetric flask, and add water to give a total volume of 250 mL.

4.10 (a) $pH = -\log(2.6 \times 10^{-2}) = 1.59$

(b) $-\log[H^+] = 3.80$; $[H^+] = 1.5 \times 10^{-4}\ M$

4.11 HCl is the limiting reagent.

$(0.350\ mol\ HCl/1\ L)(0.0750\ L)(1\ mol\ CO_2/2\ mol\ HCl)(44.01\ g\ CO_2/1\ mol\ CO_2) = 0.578\ g\ CO_2$

4.12 $(0.953\ mol\ NaOH/1\ L)(0.02833\ L\ NaOH) = 0.0270\ mol\ NaOH$

$(0.0270\ mol\ NaOH)(1\ mol\ CH_3CO_2H/1\ mol\ NaOH) = 0.0270\ mol\ CH_3CO_2H$

$(0.0270\ mol\ CH_3CO_2H)(60.05\ g/mol) = 1.62\ g\ CH_3CO_2H$

$0.0270\ mol\ CH_3CO_2H/0.0250\ L = 1.08\ M$

4.13 (0.100 mol HCl/1 L)(0.02967 L) = 0.00297 mol HCl

(0.00297 mol HCl)(1 mol NaOH/1 mol HCl) = 0.00297 mol NaOH

0.00297 mol NaOH/0.0250 L = 0.119 M NaOH

4.14 Mol acid = mol base = (0.323 mol/L)(0.03008 L) = 9.716×10^{-3} mol

Molar mass = 0.856 g acid/9.716×10^{-3} mol acid = 88.1 g/mol

4.15 (0.196 mol $Na_2S_2O_3$/1 L)(0.02030 L) = 0.00398 mol $Na_2S_2O_3$

(0.00398 mol $Na_2S_2O_3$)(1 mol I_2/2 mol $Na_2S_2O_3$) = 0.00199 mol I_2

0.00199 mol I_2 is in excess, and was not used in the reaction with ascorbic acid.

I_2 originally added = (0.0520 mol I_2/1 L)(0.05000 L) = 0.00260 mol I_2

I_2 used in reaction with ascorbic acid = 0.00260 mol − 0.00199 mol = 6.1×10^{-4} mol I_2

(6.1×10^{-4} mol I_2)(1 mol $C_6H_8O_6$/1 mol I_2)(176.1 g/1 mol) = 0.11 g $C_6H_8O_6$

Chapter 5

5.1 (a) (3800 calories)(4.184 J/calorie) = 1.6×10^4 J
(b) (250 calories)(1000 calories/calorie)(4.184 J/calorie)(1 kJ/1000 J) = 1.0×10^3 kJ

5.2 $C = 59.8\ \text{J}/[(25.0\ \text{g})(1.00\ \text{K})] = 2.39\ \text{J/g} \cdot \text{K}$

5.3 (15.5 g)(C_{metal})(18.9 °C − 100.0 °C) + (55.5 g)(4.184 J/g · K)(18.9 °C − 16.5 °C) = 0

C_{metal} = 0.44 J/g · K

5.4 Energy transferred as heat from tea + energy as heat expended to melt ice = 0

(250 g)(4.2 J/g · K)(273.2 K − 291.4 K) + x g (333 J/g) = 0

x = 57 g

57 g of ice melts with energy as heat supplied by cooling 250 g of tea from 18.2 °C (291.4 K) to 0 °C (273.2 K)

Mass of ice remaining = mass of ice initially − mass of ice melted

Mass of ice remaining = 75 g − 57 g = 18 g

5.5 (15.0 g C_2H_6)(1 mol C_2H_6/30.07 g C_2H_6) = 0.4988 mol C_2H_6

$\Delta_r H$ = 0.4988 mol C_2H_6(1 mol-rxn/2 mol C_2H_6)(−2857.3 kJ/mol-rxn)

= −713 kJ

5.6 Mass of final solution = 400. g

ΔT = 27.78 °C − 25.10 °C = 2.68 °C = 2.68 K

Amount of HCl used = amount of NaOH used = $C \times V$ = (0.400 mol/L) × 0.200 L = 0.0800 mol

Energy transferred as heat by acid–base reaction + energy gained as heat to warm solution = 0

q_{rxn} + (4.20 J/g · K)(400. g)(2.68 K) = 0

$q_{rxn} = -4.50 \times 10^3$ J

This represents the energy transferred as heat in the reaction of 0.0800 mol HCl.

Energy transferred as heat per mole = $\Delta_r H$ = −4.50 kJ/0.0800 mol HCl = −56.3 kJ/mol HCl

5.7 (a) Energy evolved as heat in reaction + energy as heat absorbed by H_2O + energy as heat absorbed by bomb = 0

q_{rxn} + (1.50×10^3 g)(4.20 J/g · K)(27.32 °C − 25.00 °C) + (837 J/K)(27.32 K − 25.00 K) = 0

q_{rxn} = −16,600 J (energy as heat evolved in burning 1.0 g sucrose)

(b) Energy evolved as heat per mole = (−16.6 kJ/g sucrose)(342.2 g sucrose/1 mol sucrose) = −5650 kJ/mol sucrose

5.8 $C(s) + O_2(g) \longrightarrow CO_2(g)$ $\Delta_r H^\circ_1 = -393.5$ kJ

$2\,[S(s) + O_2(g) \longrightarrow SO_2(g)]$

$\Delta_r H^\circ_2 = 2(-296.8) = -593.6$ kJ

$CO_2(g) + 2\,SO_2(g) \longrightarrow CS_2(g) + 3\,O_2(g)$

$\Delta_r H^\circ_3 = +1103.9$ kJ

Net: $C(s) + 2\,S(s) \longrightarrow CS_2(g)$

$\Delta_r H^\circ_{net} = \Delta_r H^\circ_1 + \Delta_r H^\circ_2 + \Delta_r H^\circ_3 = +116.8$ kJ

5.9 $Fe(s) + \frac{3}{2}\,Cl_2(g) \longrightarrow FeCl_3(s)$

$12\,C(\text{s, graphite}) + 11\,H_2(g) + \frac{11}{2}\,O_2(g) \longrightarrow C_{12}H_{22}O_{11}(s)$

5.10 $\Delta_r H^\circ$ = (6 mol/mol-rxn)$\Delta_f H^\circ$ [$CO_2(g)$] + (3 mol/mol-rxn)$\Delta_f H^\circ$[$H_2O(\ell)$] − {(1 mol/1 mol-rxn)$\Delta_f H^\circ$ [$C_6H_6(\ell)$] + ($\frac{15}{2}$ mol/mol-rxn) $\Delta_f H^\circ$ [$O_2(g)$]}= (6 mol/mol-rxn)(−393.5 kJ/mol) + (3 mol/mol-rxn)(−285.8 J/mol) − (1 mol/mol-rxn)(+49.0 kJ/mol) − 0

= −3267.4 kJ/mol-rxn

Chapter 6

6.1 (a) Highest frequency, violet; lowest frequency, red
(b) The FM radio frequency, 91.7 MHz, is lower than the frequency of a microwave oven, 2.45 GHz.
(c) The wavelength of x-rays is shorter than the wavelength of ultraviolet light.

6.2 Orange light: 6.25×10^2 nm $= 6.25 \times 10^{-7}$ m

$$\nu = (2.998 \times 10^8 \text{ m/s})/6.25 \times 10^{-7} \text{ m} = 4.80 \times 10^{14} \text{ s}^{-1}$$

$$E = (6.626 \times 10^{-34} \text{ J} \cdot \text{s/photon})(4.80 \times 10^{14} \text{ s}^{-1})(6.022 \times 10^{23} \text{ photons/mol})$$
$$= 1.92 \times 10^5 \text{ J/mol}$$

Microwave: $E = (6.626 \times 10^{-34} \text{ J} \cdot \text{s/photon})(2.45 \times 10^9 \text{ s}^{-1})(6.022 \times 10^{23} \text{ photons/mol})$
$$= 0.978 \text{ J/mol}$$

Orange (625-nm) light is about 200,000 times more energetic than 2.45-GHz microwaves.

6.3 (a) E (per atom) $= -Rhc/n^2$
$$= (-2.179 \times 10^{-18})/(3^2) \text{ J/atom}$$
$$= -2.421 \times 10^{-19} \text{ J/atom}$$
(b) E (per mol) $= (-2.421 \times 10^{-19} \text{ J/atom})(6.022 \times 10^{23} \text{ atoms/mol})(1 \text{ kJ}/10^3 \text{ J})$
$$= -145.8 \text{ kJ/mol}$$

6.4 The least energetic line is from the electron transition from $n = 2$ to $n = 1$.

$$\Delta E = -Rhc[1/1^2 - 1/2^2]$$
$$= -(2.179 \times 10^{-18} \text{ J/atom})(3/4)$$
$$= -1.634 \times 10^{-18} \text{ J/atom}$$

$$\nu = \Delta E/h$$
$$= (-1.634 \times 10^{-18} \text{ J/atom})/(6.626 \times 10^{-34} \text{ J} \cdot \text{s})$$
$$= 2.466 \times 10^{15} \text{ s}^{-1}$$

$$\lambda = c/\nu = (2.998 \times 10^8 \text{ m/s}^{-1})/(2.466 \times 10^{15} \text{ s}^{-1})$$
$$= 1.216 \times 10^{-7} \text{ m (or 121.6 nm)}$$

6.5 Energy per atom $= \Delta E = -Rhc[1/\infty^2 - 1/1^2]$
$= 2.179 \times 10^{-18}$ J/atom

Energy per mole $= (2.179 \times 10^{-18} \text{ J/atom})(6.022 \times 10^{23} \text{ atoms/mol})$
$= 1.312 \times 10^6$ J/mol ($= 1312$ kJ/mol)

6.6 First, calculate the velocity of the neutron:

$$v = [2E/m]^{1/2} = [2(6.21 \times 10^{-21} \text{ kg} \cdot \text{m}^2 \text{ s}^{-2})/(1.675 \times 10^{-27} \text{ kg})]^{1/2}$$
$$= 2720 \text{ m} \cdot \text{s}^{-1}$$

Use this value in the de Broglie equation:

$$\lambda = h/mv = (6.626 \times 10^{-34} \text{ kg} \cdot \text{m}^2 \text{ s}^{-2})/(1.675 \times 10^{-31} \text{ kg})(2720 \text{ m s}^{-1})$$
$$= 1.45 \times 10^{-6} \text{ m}$$

6.7 (a) $\ell = 0$ or 1; (b) $m_\ell = -1$, 0, or +1, p subshell; (c) d subshell; (d) $\ell = 0$ and $m_\ell = 0$; (e) 3 orbitals in the p subshell; (f) 7 values of m_ℓ and 7 orbitals

6.8 (a)

Orbital	n	ℓ
6*s*	6	0
4*p*	4	1
5*d*	5	2
4*f*	4	3

(b) A 4*p* orbital has one nodal plane; a 6*d* orbital has two nodal planes.

Chapter 7

7.1 (a) 4*s* ($n + \ell = 4$) filled before 4*p* ($n + \ell = 5$)
(b) 6*s* ($n + \ell = 6$) filled before 5*d* ($n + \ell = 7$)
(c) 5*s* ($n + \ell = 5$) filled before 4*f* ($n + \ell = 7$)

7.2 (a) chlorine (Cl)
(b) $1s^2 2s^2 2p^6 3s^2 3p^3$

[Ne] 3*s*: [↑↓] 3*p*: [↑][↑][↑]

(c) Calcium has two valence electrons in the 4s subshell. Quantum numbers for these two electrons are $n = 4$, $\ell = 0$, $m_\ell = 0$, and $m_s = \pm 1/2$

7.3 Obtain the answers from Table 7.3.

7.4

V^{2+} [Ar] 3*d*: [↑][↑][↑][ ][ ] 4*s*: [ ]

V^{3+} [Ar] 3*d*: [↑][↑][ ][ ][ ] 4*s*: [ ]

Co^{3+} [Ar] 3*d*: [↑↓][↑][↑][↑][↑] 4*s*: [ ]

All three ions are paramagnetic with three, two, and four unpaired electrons, respectively.

7.5 Increasing atomic radius: C < Si < Al

7.6 (a) Increasing atomic radius: C < B < Al
(b) Increasing ionization energy: Al < B < C
(c) Carbon is predicted to have the most negative electron affinity.

7.7 Trend in ionic radii: $S^{2-} > Cl^- > K^+$. These ions are isoelectronic (they all have the Ar configuration). The size decreases with increased nuclear charge, the higher nuclear charge resulting in a greater force of attraction of the electrons by the nucleus.

7.8 $MgCl_3$, if it existed, would presumably contain one Mg^{3+} ion (and three Cl^- ions). The formation of Mg^{3+} is energetically unfavorable, with a huge input of energy being required to remove the third electron (a core electron).

Chapter 8

8.1 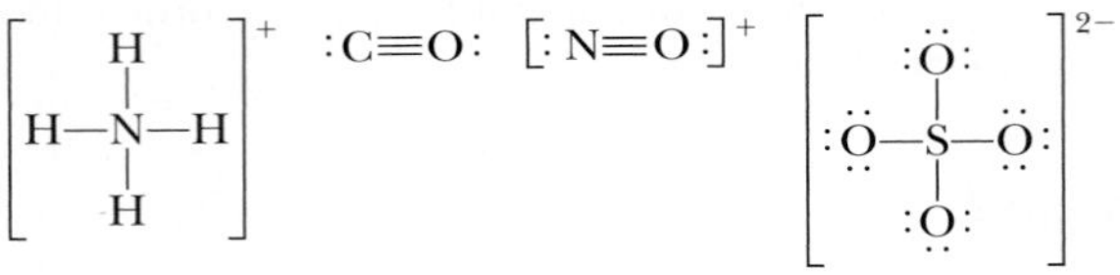

8.2 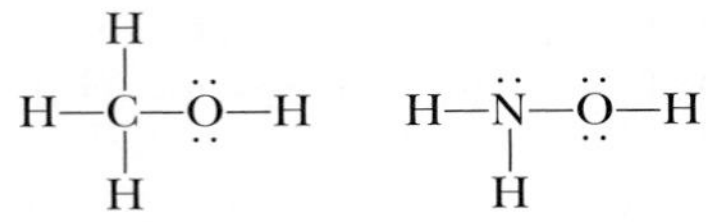

methanol hydroxylamine

8.3 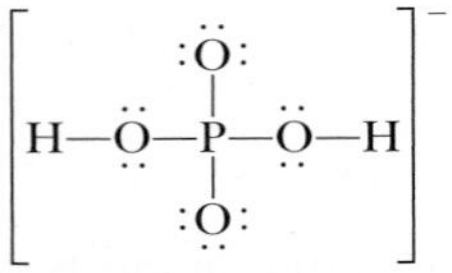

8.4 (a) The acetylide ion, C_2^{2-}, and the N_2 molecule have the same number of valence electrons (10) and identical electronic structures; that is, they are isoelectronic.
(b) Ozone, O_3, is isoelectronic with NO_2^-; hydroxide ion, OH^-, is isoelectronic with HF.

8.5 (a) CN^- : formal charge on C is −1; formal charge on N is 0.
(b) SO_3^{2-}: formal charge on S is +2; formal charge on each O is −1.

8.6 Resonance structures for the HCO_3^- ion:

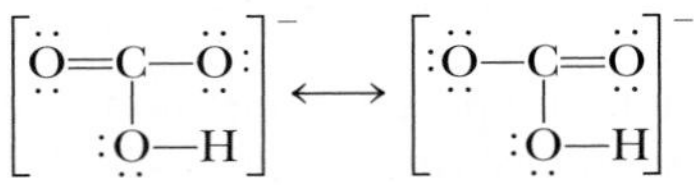

(a) No. Three resonance structures are needed in the description of CO_3^{2-}; only two are needed to describe HCO_3^-.
(b) In each resonance structure, Carbon's formal charge is 0. The oxygen of the −OH group and the double-bonded oxygen have a formal charge of zero; the singly bonded oxygen has a formal charge of −1. The average formal charge on the latter two oxygen atoms is −½. In the carbonate ion, the three oxygen atoms have an average formal charge of −⅔.
(c) H^+ would be expected to add to one of the oxygens with a negative formal charge; that is, one of the oxygens with formal charge of −½ in this structure.

8.7 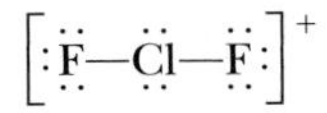 ClF_2^+, 2 bond pairs and 2 lone pairs.

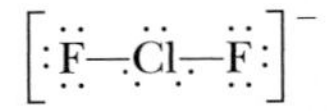 ClF_2^-, 2 bond pairs and 3 lone pairs.

8.8 Tetrahedral geometry around carbon. The Cl—C—Cl bond angle will be close to 109.5°.

8.9 For each species, the electron-pair geometry and the molecular shape are the same. BF_3: trigonal planar; BF_4^-: tetrahedral. Adding F^- to BF_3 adds an electron pair to the central atom and changes the shape.

8.10 The electron-pair geometry around I is trigonal bipyramidal. The molecular geometry of the ion is linear.

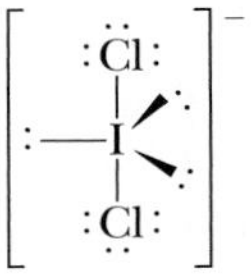

8.11 (a) In PO_4^{3-}, there is tetrahedral electron-pair geometry. The molecular geometry is also tetrahedral.

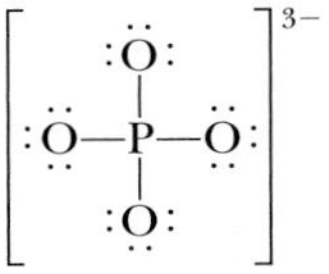

(b) In SO_3^{2-}, there is tetrahedral electron-pair geometry. The molecular geometry is trigonal pyramidal.

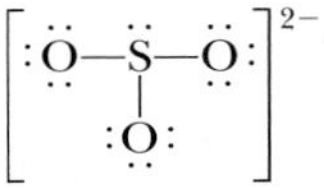

(c) In IF_5, there is octahedral electron-pair geometry. The molecular geometry is square pyramidal.

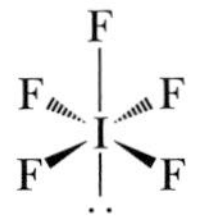

8.12 (a) The H atom is positive in each case. H—F ($\Delta\chi = 1.8$) is more polar than H—I ($\Delta\chi = 0.5$).
(b) B—F ($\Delta\chi = 2.0$) is more polar than B—C ($\Delta\chi = 0.5$). In B—F, F is the negative pole, and B is the positive pole. In B—C, C is the negative pole, and B is the positive pole.
(c) C—Si ($\Delta\chi = 0.6$) is more polar than C—S ($\Delta\chi = 0.1$). In C—Si, C is the negative pole, and Si is the positive pole. In C—S, S is the negative pole, and C the positive pole.

8.13 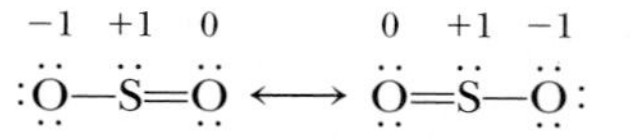

The S—O bonds are polar, with the negative end being the O atom. (The O atom is more electronegative than the S atom.) Formal charges show that these bonds are, in fact, polar, with the O atom being the more negative atom.

8.14 (a) $BFCl_2$, polar, negative side is the F atom because F is the most electronegative atom in the molecule.

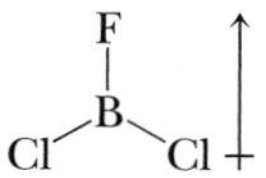

(b) NH_2Cl, polar, negative side is the Cl atom.

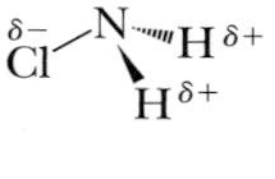

(c) SCl_2, polar, Cl atoms are on the negative side.

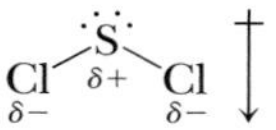

8.15 (a)

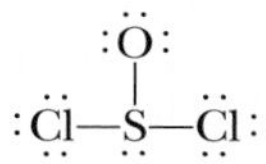

Formal charges: S = +1, 0 = −1, Cl = 0

(Lewis structure of $SOCl_2$ with formal charges indicated)

(b) Geometry: trigonal pyramidal

(c) The molecule is polar. The positive charge is on sulfur, the negative charge on oxygen.

8.16 (a) C—N: bond order 1; C=N: bond order 2; C≡N: bond order 3. Bond length: C—N > C=N > C≡N

(b) 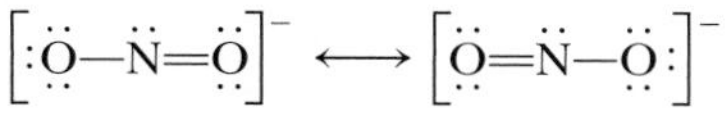

The N-O bond order in NO_2^- is 1.5. Therefore, the NO bond length (124 pm) should be between the length of a N—O single bond (136 pm) and a N=O double bond (115 pm).

8.17 $CH_4(g) + 2\ O_2(g) \longrightarrow CO_2(g) + 2\ H_2O(g)$

Break 4 C—H bonds and 2 O=O bonds:
(4 mol)(413 kJ/mol) + (2 mol)(498 kJ/mol) = 2648 kJ

Make 2 C=O bonds and 4 H—O bonds:
(2 mol)(745 kJ/mol) + (4 mol)(463 kJ/mol) = 3342 kJ

$\Delta_r H° = 2648\ kJ - 3342\ kJ = -694$ kJ/mol-rxn
(value calculated using enthalpies of formation = −797 kJ/mol-rxn)

Chapter 9

9.1 The oxygen atom in H_3O^+ is sp^3 hybridized. The three O—H bonds are formed by overlap of oxygen sp^3 and hydrogen 1*s* orbitals. The fourth sp^3 orbital contains a lone pair of electrons.

The carbon and nitrogen atoms in CH_3NH_2 are sp^3 hybridized. The C—H bonds arise from overlap of carbon sp^3 orbitals and hydrogen 1*s* orbitals. The bond between C and N is formed by overlap of sp^3 orbitals from these atoms. Overlap of nitrogen sp^3 and hydrogen 1*s* orbitals gives the two N—H bonds, and there is a lone pair in the remaining sp^3 orbital on nitrogen.

9.2 (a) BH_4^-, tetrahedral electron-pair geometry, sp^3
(b) SF_5^-, octahedral electron-pair geometry, sp^3d^2
(c) SOF_4, trigonal-bipyramidal electron-pair geometry, sp^3d
(d) ClF_3, trigonal-bipyramidal electron-pair geometry, sp^3d
(e) BCl_3, trigonal-planar electron-pair geometry, sp^2
(f) XeO_6^{4-}, octahedral electron-pair geometry, sp^3d^2

9.3 The two CH_3 carbon atoms are sp^3 hybridized, and the center carbon atom is sp^2 hybridized. For each of the carbon atoms in the methyl groups, the sp^3 orbitals overlap with hydrogen 1*s* orbitals to form the three C—H bonds, and the fourth sp^3 orbital overlaps with an sp^2 orbital on the central carbon atom, forming a carbon–carbon sigma bond. Overlap of an sp^2 orbital on the central carbon and an oxygen sp^2 orbital gives the sigma bond between these elements. The pi bond between carbon and oxygen arises by overlap of a *p* orbital from each element.

9.4 A triple bond links the two nitrogen atoms, each of which also has one lone pair. Each nitrogen is *sp* hybridized. One *sp* orbital contains the lone pair; the other is used to form the sigma bond between the two atoms. Two pi bonds arise by overlap of *p* orbitals on the two atoms, perpendicular to the molecular axis.

9.5 Bond angles: H—C—H = 109.5°, H—C—C = 109.5°, C—C—N = 180°. Carbon in the CH_3 group is sp^3 hybridized; the central C and the N are *sp* hybridized. The three C—H bonds form by overlap of an H 1*s* orbital with one of the sp^3 orbitals of the CH_3 group; the fourth sp^3 orbital overlaps with an *sp* orbital on the central C to form a sigma bond. The triple bond between C and N is a combination of a sigma bond (the *sp* orbital on C overlaps with the *sp* orbital on N) and two pi bonds (overlap of two sets of *p* orbitals on these elements). The remaining *sp* orbital on N contains a lone pair.

9.6 H_2^+: $(\sigma_{1s})^1$ The ion has a bond order of ½ and is expected to exist. A bond order of ½ is predicted for He_2^+ and H_2^-, both of which are predicted to have electron configurations $(\sigma_{1s})^2\ (\sigma^*_{1s})^1$.

9.7 Li_2^- is predicted to have an electron configuration $(\sigma_{1s})^2\ (\sigma^*_{1s})^2\ (\sigma_{2s})^2\ (\sigma^*_{2s})^1$ and a bond order of ½, the positive value implying that the ion might exist.

9.8 O_2^+: [core electrons] $(\sigma_{2s})^2\ (\sigma^*_{2s})^2\ (\pi_{2p})^4\ (\sigma_{2p})^2\ (\pi^*_{2p})^1$. The bond order is 2.5. The ion is paramagnetic with one unpaired electron.

Chapter 10

10.1 (a) Isomers of C_7H_{16}

$CH_3CH_2CH_2CH_2CH_2CH_2CH_3$ heptane

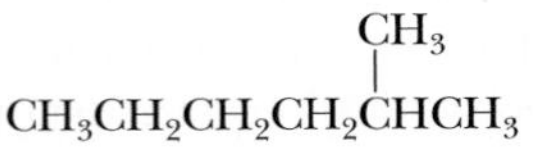
2-methylhexane

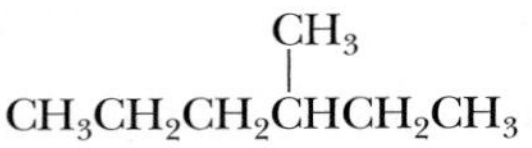
3-methylhexane

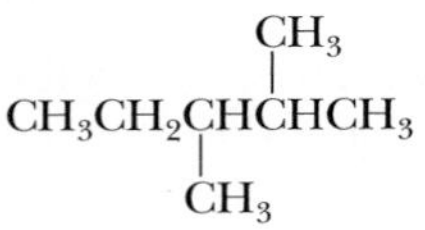
2,3-dimethylpentane

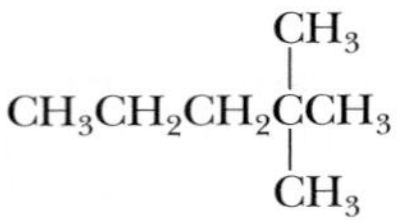
2,2-dimethylpentane

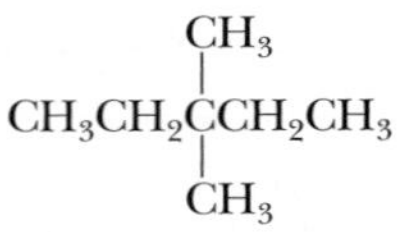
3,3-dimethylpentane

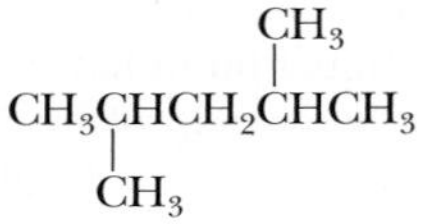
2,4-dimethylpentane

2-Ethylpentane is pictured on page 450.

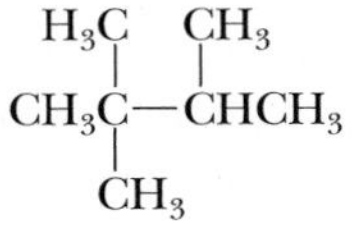
2,2,3-trimethylbutane

(b) Two isomers, 3-methylhexane, and 2,3-dimethylpentane, are chiral.

10.2 The names accompany the structures in the answer to Exercise 10.1.

10.3 Isomers of C_6H_{12} in which the longest chain has six C atoms:

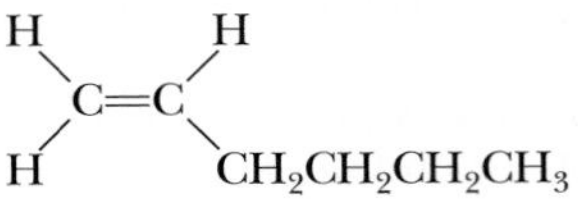

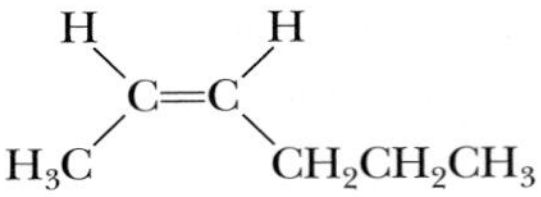

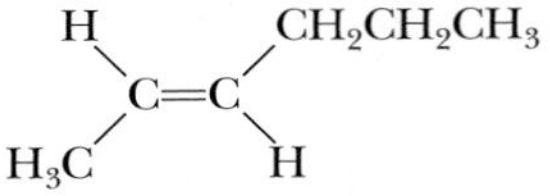

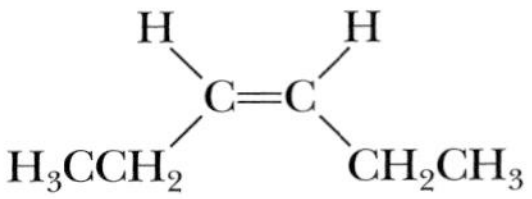

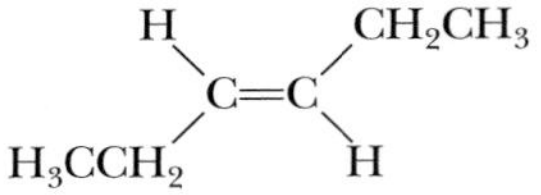

Names (in order, top to bottom): 1-hexene, *cis*-2-hexene, *trans*-2-hexene, *cis*-3-hexene, *trans*-3-hexene. None of these isomers is chiral.

10.4 (a)

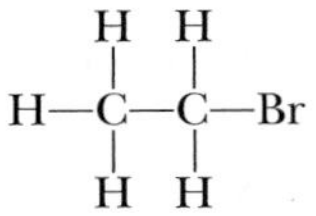
bromoethane

(b)

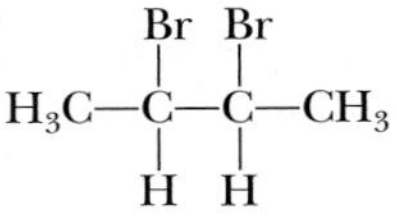
2,3-dibromobutane

10.5 1,4-diaminobenzene

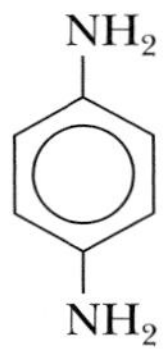

10.6 $CH_3CH_2CH_2CH_2OH$ 1-butanol

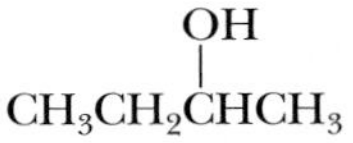
2-butanol

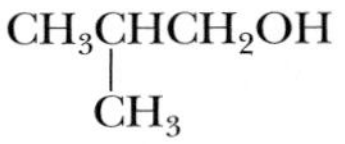
2-methyl-1-propanol

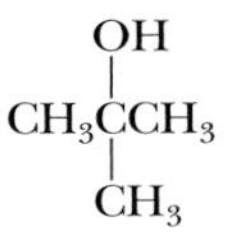
2-methyl-2-propanol

10.7 (a)

$CH_3CH_2CH_2\overset{O}{\overset{\|}{C}}CH_3$ 2-pentanone

$CH_3CH_2\overset{O}{\overset{\|}{C}}CH_2CH_3$ 3-pentanone

$CH_3CH_2CH_2CH_2\overset{O}{\overset{\|}{C}}H$ pentanal

$CH_3\underset{CH_3}{\underset{|}{C}}HCH_2\overset{O}{\overset{\|}{C}}H$ 3-methylbutanal

(b)

$CH_3\overset{OH}{\overset{|}{C}}HCH_2CH_2CH_3$, 2-pentanol

10.8 (a)

1-butanol gives butanal $CH_3CH_2CH_2\overset{O}{\overset{\|}{C}}H$

(b)

2-butanol gives butanone $CH_3CH_2\overset{O}{\overset{\|}{C}}CH_3$

(c) 2-methyl-1-propanol gives 2-methylpropanal

$$CH_3\underset{CH_3}{\underset{|}{\overset{H}{\overset{|}{C}}}}—\overset{O}{\overset{\|}{C}}H$$

The oxidation products from these three reactions are structural isomers.

10.9 (a)

$CH_3CH_2\overset{O}{\overset{\|}{C}}OCH_3$ methyl propanoate

(b)

$CH_3CH_2CH_2\overset{O}{\overset{\|}{C}}OCH_2CH_2CH_2CH_3$

butyl butanoate

(c)

$CH_3CH_2CH_2CH_2CH_2\overset{O}{\overset{\|}{C}}OCH_2CH_3$

ethyl hexanoate

10.10 (a) Propyl acetate is formed from acetic acid and propanol:

$CH_3\overset{O}{\overset{\|}{C}}OH + CH_3CH_2CH_2OH$

(b) 3-Methylpentyl benzoate is formed from benzoic acid and 3-methylpentanol:

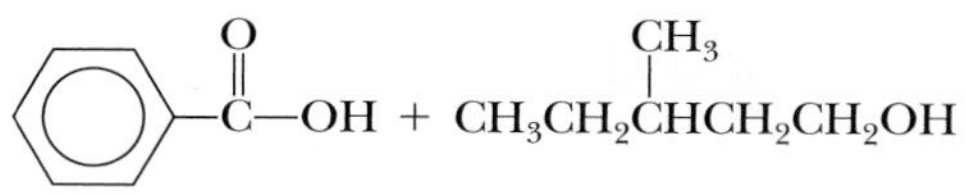

(c) Ethyl salicylate is formed from salicylic acid and ethanol:

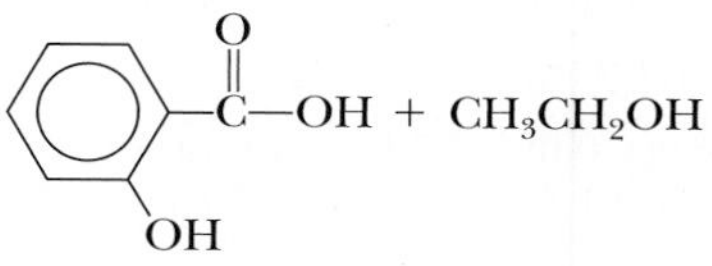

10.11 (a) $CH_3CH_2CH_2OH$: 1-propanol, has an alcohol (—OH) group

CH_3CO_2H: ethanoic acid (acetic acid), has a carboxylic acid (—CO_2H) group

$CH_3CH_2NH_2$: ethylamine, has an amino (—NH_2) group

(b) 1-propyl ethanoate (propyl acetate)

(c) Oxidation of this primary alcohol first gives propanal, CH_3CH_2CHO. Further oxidation gives propanoic acid, $CH_3CH_2CO_2H$.

(d) *N*-ethylacetamide, $CH_3CONHCH_2CH_3$

(e) The amine is protonated by hydrochloric acid, forming ethylammonium chloride, $[CH_3CH_2NH_3]Cl$.

10.12 Kevlar is a polyamide polymer, prepared by the reaction of terephthalic acid and 1,4-diaminobenzene.

$$n\ H_2NC_6H_4NH_2 + n\ HO_2CC_6H_4CO_2H \longrightarrow \text{-(-}HNC_6H_4NHCOC_6H_4CO\text{-)}_n\text{-} + 2n\ H_2O$$

Chapter 11

11.1 0.83 bar (0.82 atm) > 75 kPa (0.74 atm) > 0.63 atm > 250 mm Hg (0.33 atm)

11.2 $P_1 = 55$ mm Hg and $V_1 = 125$ mL; $P_2 = 78$ mm Hg and $V_2 = ?$

$V_2 = V_1(P_1/P_2) = (125\text{ mL})(55\text{ mm Hg}/78\text{ mm Hg}) = 88\text{ mL}$

11.3 $V_1 = 45$ L and $T_1 = 298$ K; $V_2 = ?$ and $T_2 = 263$ K

$V_2 = V_1(T_2/T_1) = (45\text{ L})(263\text{ K}/298\text{ K}) = 40.\text{ L}$

11.4 $V_2 = V_1(P_1/P_2)(T_2/T_1)$

$= (22\text{ L})(150\text{ atm}/0.993\text{ atm})(295\text{ K}/304\text{ K})$

$= 3200\text{ L}$

At 5.0 L per balloon, there is sufficient He to fill 640 balloons.

11.5 44.8 L of O_2 is required; 44.8 L of $H_2O(g)$ and 22.4 L $CO_2(g)$ are produced.

11.6 $PV = nRT$

$(750/760\text{ atm})(V) =$

$(1300\text{ mol})(0.08206\text{ L} \cdot \text{atm/mol} \cdot \text{K})(296\text{ K})$

$V = 3.2 \times 10^4$ L

11.7 $d = PM/RT$; $M = dRT/P$

$M = (5.02\ g/L)(0.082057\ L \cdot atm/mol \cdot K)$
$(288.2\ K)/(745/760\ atm) = 121\ g/mol$

11.8 $PV = (m/M)RT$; $M = mRT/PV$

$M = (0.105\ g)(0.082057\ L \cdot atm/mol \cdot K)$
$(296.2\ K)/[(561/760)\ atm\ 0.125\ L)] = 27.7\ g/mol$

11.9 $n(H_2) = PV/RT$

$= (542/760\ atm)(355\ L)/(0.08206\ L \cdot atm/mol \cdot K)$
$(298.2\ K)$

$n(H_2) = 10.3\ mol$

$n(NH_3) = (10.3\ mol\ H_2)(2\ mol\ NH_3/3\ mol\ H_2)$

$= 6.87\ mol\ NH_3$

$P\ (125\ L) = (6.87\ mol)(0.082057\ L \cdot atm/mol \cdot K)$
$(298.2\ K)$

$P(NH_3) = 1.35\ atm$

11.10 $P_{halothane}\ (5.00\ L)$
$= (0.0760\ mol)(0.08206\ L \cdot atm/mol \cdot K)\ (298.2\ K)$

$P_{halothane} = 0.372\ atm$ (or 283 mm Hg)

$P_{oxygen}\ (5.00\ L)$
$= (0.734\ mol)(0.08206\ L \cdot atm/mol \cdot K)(298.2\ K)$

$P_{oxygen} = 3.59\ atm$ (or 2730 mm Hg)

$P_{total} = P_{halothane} + P_{oxygen}$

$= 283\ mm\ Hg + 2730\ mm\ Hg = 3010\ mm\ Hg$

11.11 For He: Use Equation 11.9, with $M = 4.00 \times 10^{-3}$ kg/mol, $T = 298$ K, and $R = 8.314$ J/mol · K to calculate the rms speed of 1360 m/s. A similar calculation for N_2, with $M = 28.01 \times 10^{-3}$ kg/mol, gives an rms speed of 515 m/s.

11.12 The molar mass of CH_4 is 16.0 g/mol.

$$\frac{\text{Rate for } CH_4}{\text{Rate for unknown}} = \frac{n \text{ molecules}/1.50 \text{ min}}{n \text{ molecules}/4.73 \text{ min}} = \sqrt{\frac{M_{unknown}}{16.0}}$$

$M_{unknown} = 159\ g/mol$

11.13 $P(1.00\ L)$
$= (10.0\ mol)(0.082057\ L \cdot atm/mol \cdot K)\ (298\ K)$

$P = 245$ atm (calculated by PV = nRT)

$P = 320$ atm (calculated by van der Waals equation)

Chapter 12

12.1 Because F^- is the smaller ion, water molecules can approach most closely and interact more strongly. Thus, F^- should have the more negative enthalpy of hydration.

12.2

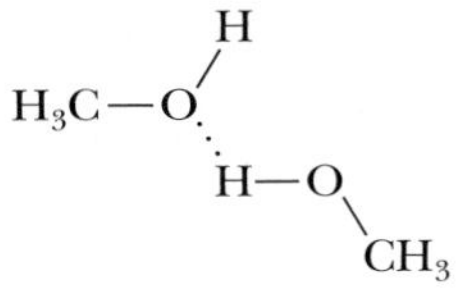

Hydrogen bonding in methanol entails the attraction of the hydrogen atom bearing a partial positive charge (δ^+) on one molecule to the oxygen atom bearing a partial negative charge (δ^-) on a second molecule. The strong attractive force of hydrogen bonding will cause the boiling point and the enthalpy of vaporization of methanol to be quite high.

12.3 Water is a polar solvent, while hexane and CCl_4 are nonpolar. London dispersion forces are the primary forces of attraction between all pairs of dissimilar solvents. For mixtures of water with the other solvents, dipole–induced dipole forces will also be important.

12.4 (a) O_2: induced dipole–induced dipole forces only.
(b) CH_3OH: strong hydrogen bonding (dipole–dipole forces) as well as induced dipole–induced dipole forces.
(c) Forces between water molecules: strong hydrogen bonding and induced dipole–induced dipole forces. Between N_2 and H_2O: dipole–induced dipole forces and induced dipole–induced dipole forces.

Relative strengths: a < forces between N_2 and H_2O in c < b < forces between water molecules in c.

12.5 $(1.00 \times 10^3\ g)(1\ mol/32.04\ g)(35.2\ kJ/mol)$
$= 1.10 \times 10^3\ kJ$

12.6 (a) At 40 °C, the vapor pressure of ethanol is about 120 mm Hg.
(b) The equilibrium vapor pressure of ethanol at 60 °C is about 320 mm Hg. At 60 °C and 600 mm Hg, ethanol is a liquid. If vapor is present, it will condense to a liquid.

12.7 $PV = nRT$

$P = 0.50\ g\ (1\ mol/18.02\ g)(0.0821\ L \cdot atm/mol \cdot K)$
$(333\ K)/5.0\ L$

$P = 0.15$ atm.

Convert to mm Hg: $P = (0.15\ atm)(760\ mm\ Hg/1\ atm) = 120$ mm Hg. The vapor pressure of water at 60 °C is 149.4 mm Hg (Appendix G). The calculated pressure is lower than this, so all the water (0.50 g) evaporates. If 2.0 g of water is used, the calculated pressure, 460 mm Hg, exceeds the vapor pressure. In this case, only part of the water will evaporate.

12.8 Use the Clausius–Clapeyron equation, with $P_1 = 57.0$ mm Hg, $T_1 = 250.4$ K, $P_2 = 534$ mm Hg, and $T_2 = 298.2$ K.

$\ln [P_2/P_1] = \Delta_{vap}H/R\ [1/T_1 - 1/T_2]$
$= [\Delta_{vap}H/R][(T_2 - T_1)/T_1T_2]$

$\ln [534/57.0] = \Delta_{vap}H/(0.0083145\ kJ/K \cdot mol)$
$[47.8/(250.4)(298.2)]$

$\Delta_{vap}H = 29.1\ kJ/mol$

12.9 Glycerol is predicted to have a higher viscosity than ethanol. It is a larger molecule than ethanol, and there are higher forces of attraction between molecules because each molecule has three OH groups that hydrogen-bond to other molecules.

Chapter 13

13.1 The strategy to solve this problem is given in Example 13.1.

Step 1. Mass of the unit cell

$= (197.0 \text{ g/mol})(1 \text{ mol}/6.022 \times 10^{23} \text{ atom/mol})(4 \text{ atoms/unit cell})$

$= 1.309 \times 10^{-21}$ g/unit cell

Step 2. Volume of unit cell

$= (1.309 \times 10^{-21} \text{ g/unit cell})(1 \text{ cm}^3/19.32 \text{ g})$

$= 6.773 \times 10^{-23} \text{ cm}^3$/unit cell

Step 3. Length of side of unit cell

$= [6.773 \times 10^{-23} \text{ cm}^3/\text{unit cell}]^{1/3} = 4.076 \times 10^{-8}$ cm

Step 4. Calculate the radius from the edge dimension.

Diagonal distance $= 4.076 \times 10^{-8} \text{ cm } (2^{½}) = 4 \ (r_{Au})$

$r_{Au} = 1.441 \times 10^{-8}$ cm (= 144.1 pm)

13.2 To verify a body centered cubic structure, calculate the mass contained in the unit cell. If the structure is bcc, then the mass will be the mass of 2 Fe atoms. (Other possibilities: fcc – mass of 4 Fe; primitive cubic – mass of 1 Fe atom). This calculation uses the four steps from the previous exercise in reverse order.

Step 1. Use radius of Fe to calculate cell dimensions. In a body-centered cube, atoms touch across the diagonal of the cube.

Diagonal distance = side dimension ($\sqrt{3}$) = 4 r_{Fe}

Side dimension of cube $= 4 \ (1.26 \times 10^{-8} \text{ cm})/(\sqrt{3}) = 2.910 \times 10^{-8}$ cm

Step 2. Calculate unit cell volume

Unit cell volume $= (2.910 \times 10^{-8} \text{ cm})^3 = 2.464 \times 10^{-23} \text{ cm}^3$

Step 3. Combine unit cell volume and density to find the mass of the unit cell.

Mass of unit cell $= 2.464 \times 10^{-23} \text{ cm}^3 \ (7.8740 \text{ g/cm}^3) = 1.940 \times 10^{-22}$ g

Step 4. Calculate the mass of 2 Fe atoms, and compare this to the answer from step 3.

Mass of 2 Fe atoms

$= 55.85 \text{ g/mol } (1 \text{ mol}/6.022 \times 10^{23} \text{ atoms})(2 \text{ atoms})$

$= 1.85 \times 10^{-22}$ g).

This is a fairly good match, and clearly much better than the two other possibilities, primitive and fcc.

13.3 M_2X; In a face-centered cubic unit cell, there are four anions and eight tetrahedral holes in which to place metal ions. All of the tetrahedral holes are inside the unit cell, so the ratio of atoms in the unit cell is 2 : 1.

13.4 We need to calculate the mass and volume of the unit cell from the information given. The density of KCl will then be mass/volume. Select units so the density is calculated as g/cm^3

Step 1. Mass: the unit cell contains 4 K^+ ions and 4 Cl^- ions

Unit cell mass $= (39.10 \text{ g/mol})(1 \text{ mol}/6.022 \times 10^{23} \text{ K}^+ \text{ ions})(4 \text{ K}^+ \text{ ions}) + (35.45 \text{ g/mol})(1 \text{ mol}/6.022 \times 10^{23} \text{ Cl}^- \text{ ions})(4 \text{ Cl}^- \text{ ions})$

$= 2.355 \times 10^{-22} \text{ g} + 2.597 \times 10^{-22} \text{ g} = 4.952 \times 10^{-22}$ g

Step 2. Volume: assuming K^+ and Cl^- ions touch along one edge of the cube, the side dimension $= 2 \ r_{K^+} + 2 \ r_{Cl^-}$. The volume of the cube is the cube of this value. (Convert the ionic radius from pm to cm.)

$V = [2(1.33 \times 10^{-8} \text{ cm}) + 2(1.81 \times 10^{-8} \text{ cm})]^3 = 2.477 \times 10^{-22} \text{ cm}^3$

Step 3: density = mass/volume $= 4.952 \times 10^{-22} \text{ g}/2.477 \times 10^{-22} \text{ cm}^3) = 2.00 \text{ g/cm}^3$

13.5 Use the Born–Haber cycle equation shown on pages 600–602. The unknown in this problem is the enthalpy of formation of NaI(s).

$\Delta_f H°$ [NaI(s)] $= \Delta H_{\text{Step 1a}} + \Delta H_{\text{Step 1b}} + \Delta H_{\text{Step 2a}} + \Delta H_{\text{Step 2b}} + \Delta_{\text{lattice}}H$

Step 1a. Enthalpy of formation of I(g) = +106.8 kJ/mol (Appendix L)

Step 1b. ΔH for $I(g) + e^- \rightarrow I^-(g) = -295$ kJ/mol (Appendix F)

Step 2a. Enthalpy of formation of Na(g) = +107.3 kJ/mol (Appendix L)

Step 2b. ΔH for $Na(g) \rightarrow Na^+(g) + e^-$ = +496 kJ/mol (Appendix F)

Step 3 = $\Delta_{\text{lattice}}H = -702$ kJ/mol (Table 13.2)

$\Delta_f H°$ [NaI(s)] = −287 kJ/mol

Chapter 14

14.1 (a) 10.0 g sucrose = 0.0292 mol; 250 g H_2O = 13.9 mol

$X_{\text{sucrose}} = (0.0292 \text{ mol})/(0.0292 \text{ mol} + 13.9 \text{ mol}) = 0.00210$

$c_{\text{sucrose}} = (0.0292 \text{ mol sucrose})/(0.250 \text{ kg solvent}) = 0.117$ m

Weight % sucrose = (10.0 g sucrose/260 g soln) (100%) = 3.85%

(b) 1.08×10^4 ppm = 1.08×10^4 mg NaCl per 1000 g soln

= $(1.08 \times 10^4$ mg Na/1000 g soln) (1050 g soln/1 L)

= 1.13×10^4 mg Na/L

= 11.3 g Na/L

(11.3 g Na/L) (58.44 g NaCl/23.0 g Na) = 28.7 g NaCl/L

14.2 $\Delta_{soln}H° = \Delta_f H°$ [NaOH(aq)] − $\Delta_f H°$ [NaOH(s)] = −469.2 kJ/mol − (−425.9 kJ/mol) = −43.3 kJ/mol

14.3 Solubility of $CO_2 = k_H P_g$ = 0.034 mol/kg · bar × 0.33 bar = 1.1×10^{-2} M

14.4 The solution contains sucrose [(10.0 g) (1 mol/342.3 g) = 0.0292 mol] in water [(225 g) (1 mol/18.02 g) = 12.5 mol].

X_{water} = (12.5 mol H_2O)/(12.5 mol + 0.0292 mol) = 0.998

$P_{water} = x_{water}P°_{water}$ = 0.998(149.4 mm Hg) = 149 mm Hg

14.5 $c_{glycol} = \Delta T_{bp}/K_{bp}$ = 1.0 °C/(0.512 °C/m) = 1.95 m = 1.95 mol/kg

$mass_{glycol}$ = (1.95 mol/kg) (0.125 kg) (62.02 g/mol) = 15 g

14.6 c_{glycol} = (525 g) (1 mol/62.07 g)/(3.00 kg) = 2.82 m

$\Delta T_{fp} = K_{fp} \times m$ = (−1.86 °C/m) (2.82 m) = −5.24 °C

You will be protected only to about −5 °C and not to −25 °C.

14.7 c (mol/L) = Π/RT = [(1.86 mm Hg) (1 atm/760 mm Hg)]/[(0.08206 L · atm/mol · K) (298 K)]

= 1.00×10^{-4} M

$(1.00 \times 10^{-4}$ mol/L) (0.100 L) = 1.0×10^{-5} mol

Molar mass = 1.40 g/1.00×10^{-5} mol = 1.4×10^5 g/mol

(Assuming the polymer is composed of CH_2 units, the polymer is about 10,000 units long.)

14.8 c_{NaCl} = (25.0 g NaCl) (1 mol/58.44 g)/(0.525 kg) = 0.815 m

$\Delta T_{fp} = K_{fp} \times m \times i$ = (−1.86 °C/m) (0.815 m) (1.85) = −2.80 °C

Chapter 15

15.1 −½ (Δ[NOCl]/Δt) = ½(Δ[NO]/Δt) = Δ[Cl_2]/Δt

15.2 For the first two hours:

−Δ[sucrose]/Δt = [(0.033 − 0.050) mol/L]/(2.0 h) = 0.0080 mol/L · h

For the last two hours:

Δ[sucrose]/Δt = −[(0.010 − 0.015) mol/L]/(2.0 h) = 0.0025 mol/L · h

Instantaneous rate at 4 h = 0.0045 mol/L · h. (Calculated from the slope of a line tangent to the curve at the defined concentration.)

15.3 Compare experiments 1 and 2: Doubling [O_2] causes the rate to double, so the rate is first order in [O_2]. Compare experiments 2 and 4: Doubling [NO] causes the rate to increase by a factor of 4, so the rate is second order in [NO]. Thus, the rate law is

Rate = k[NO]^2[O_2]

Using the data in experiment 1 to determine k:

0.028 mol/L · s = k[0.020 mol/L]^2[0.010 mol/L]

$k = 7.0 \times 10^3$ L^2/mol^2 · s

15.4 Rate = k[Pt(NH$_3$)$_2$Cl$_2$] = (0.27 h^{-1}) (0.020 mol/L) = 0.0054 mol/L · h

15.5 ln ([sucrose]/[sucrose]$_o$) = −kt

ln ([sucrose]/[0.010]) = −(0.21 h^{-1}) (5.0 h)

[sucrose] = 0.0035 mol/L

15.6 (a) The fraction remaining is [$CH_3N_2CH_3$]/[$CH_3N_2CH_3$]$_o$.

ln ([$CH_3N_2CH_3$]/[$CH_3N_2CH_3$]$_o$) = −$(3.6 \times 10^{-4}$ s^{-1}) (150 s)

[$CH_3N_2CH_3$]/[$CH_3N_2CH_3$]$_o$= 0.95

(b) After the reaction is 99% complete [$CH_3N_2CH_3$]/[$CH_3N_2CH_3$]$_o$ = 0.010.

ln (0.010) = −$(3.6 \times 10^{-4}$ s^{-1}) (t)

$t = 1.3 \times 10^4$ s (220 min)

15.7 1/[HI] − 1/[HI]$_o$ = kt

1/[HI] − 1/[0.010 M] = (30. L/mol · min) (12 min)

[HI] = 0.0022 M

15.8

Concentration versus time

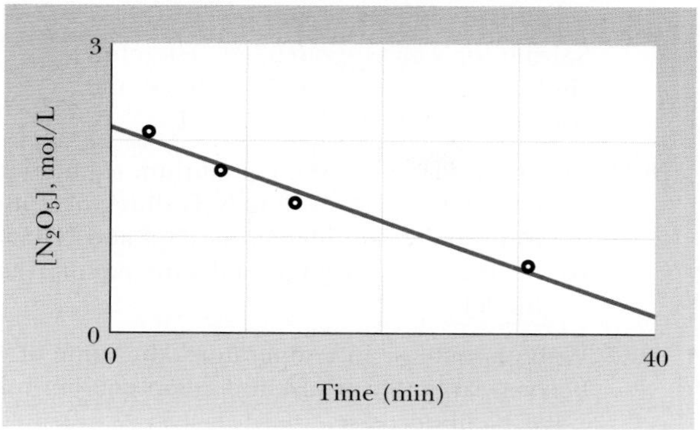

ln $[N_2O_5]$ versus time

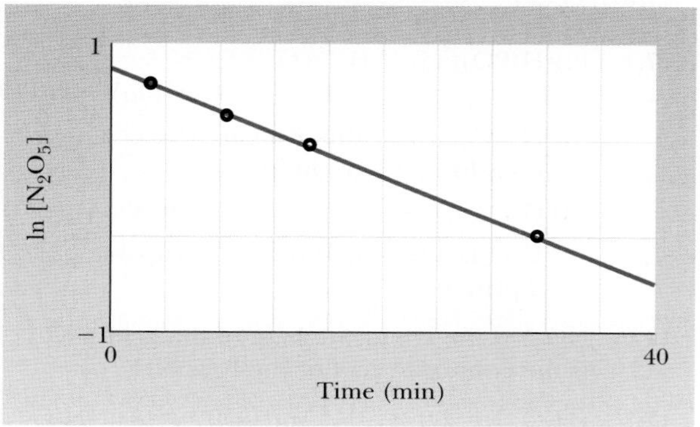

$1/[N_2O_5]$ versus time

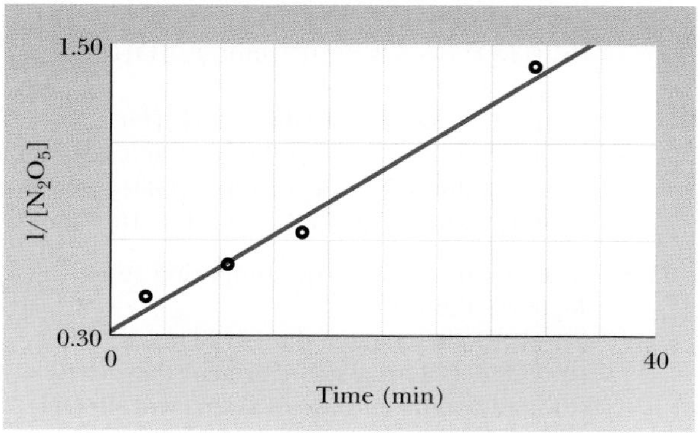

The plot of ln $[N_2O_5]$ versus time has the best linear fit, indicating that this is a first-order reaction. The rate constant is determined from the slope: $k = -\text{slope} = 0.038\ \text{min}^{-1}$.

15.9 (a) For ^{241}Am, $t_{1/2} = 0.693/k = 0.693/(0.0016\ \text{y}^{-1}) = 430$ y

For ^{125}I, $t_{1/2} = 0.693/(0.011\ \text{d}^{-1}) = 63$ d

(b) ^{125}I decays much faster.

(c) $\ln[(n)/(1.6 \times 10^{15}\ \text{atoms})] = -(0.011\ \text{d}^{-1})(2.0\ \text{d})$

$n/1.6 \times 10^{15}$ atoms $= 0.978$; $n = 1.57 \times 10^{15}$ atoms

Since the answer should have two significant figures, we should round this off to 1.6×10^{15} atoms. The approximately 2% that has decayed is not discernable within the limits of accuracy of the data presented.

15.10 $\ln(k_2/k_1) = (-E_a/R)(1/T_2 - 1/T_1)$

$\ln[(1.00 \times 10^4)/(4.5 \times 10^3)] = -(E_a/8.315 \times 10^{-3}\ \text{kJ/mol} \cdot \text{K})(1/283\ \text{K} - 1/274\ \text{K})$

$E_a = 57$ kJ/mol

15.11 All three steps are bimolecular.

For step 3: Rate $= k[N_2O][H_2]$.

There are two intermediates, $N_2O_2(g)$ and $N_2O(g)$.

When the three equations are added, N_2O_2 (a product in the first step and a reactant in the second step) and N_2O (a product in the second step and a reactant in the third step) cancel, leaving the net equation: $2\ NO(g) + 2\ H_2(g) \longrightarrow N_2(g) + 2\ H_2O(g)$.

15.12 (a) $2\ NH_3(aq) + OCl^-(aq) \longrightarrow N_2H_4(aq) + Cl^-(aq) + H_2O(\ell)$

(b) The second step is the rate-determining step.

(c) Rate $= k[NH_2Cl][NH_3]$

(d) NH_2Cl, $N_2H_5^+$, and OH^- are intermediates.

15.13 Overall reaction: $2\ NO_2Cl(g) \longrightarrow 2\ NO_2(g) + Cl_2(g)$

Rate $= k'[NO_2Cl]^2/[NO_2]$ (where $k' = k_1k_2/k_1$)

Increasing $[NO_2]$ causes the reaction rate to decrease.

Chapter 16

16.1 (a) $K = [CO]^2/[CO_2]$

(b) $K = [Cu^{2+}][NH_3]^4/[Cu(NH_3)_4{}^{2+}]$

(c) $K = [H_3O^+][CH_3CO_2{}^-]/[CH_3CO_2H]$

16.2 (a) Both reactions are reactant-favored ($K < 1$).

(b) $[NH_3]$ in the second solution is greater. K for this reaction is larger, so the reactant, $Cd(NH_3)_4{}^{2+}$, dissociates to a greater extent.

16.3 (a) $Q = [2.18]/[0.97] = 2.3$. The system is not at equilibrium; $Q < K$. To reach equilibrium, [isobutane] will increase and [butane] will decrease.

(b) $Q = [2.60]/[0.75] = 3.5$. The system is not at equilibrium; $Q > K$. To reach equilibrium, [butane] will increase and [isobutane] will decrease.

16.4 $Q = [NO]^2/[N_2][O_2] = [4.2 \times 10^{-3}]^2/[0.50][0.25] = 1.4 \times 10^{-4}$

$Q < K$, so the reaction is not at equilibrium. To reach equilibrium, [NO] will increase and $[N_2]$ and $[O_2]$ will decrease.

16.5 (a)

Equation	$C_6H_{10}I_2$ ⇌	C_6H_{10} +	I_2
Initial (M)	0.050	0	0
Change (M)	−0.035	+0.035	+0.035
Equilibrium (M)	0.015	0.035	0.035

(b) $K = (0.035)\ (0.035)/(0.015) = 0.082$

16.6

Equation	H_2 +	I_2 ⇌	2 HI
Initial (M)	6.00×10^{-3}	6.00×10^{-3}	0
Change (M)	$-x$	$-x$	$+2x$
Equilibrium (M)	$0.00600 - x$	$0.00600 - x$	$+2x$

$$K_c = 33 = \frac{(2x)^2}{(0.00600 - x)^2}$$

$x = 0.0045$ M, so $[H_2] = [I_2] = 0.0015$ M and [HI] $= 0.0090$ M.

16.7

Equation	C(s) + CO_2 (g) ⇌	2 CO(g)
Initial (M)	0.012	0
Change (M)	$-x$	$+2x$
Equilibrium (M)	$0.012 - x$	$2x$

$$K_c = 0.021 = \frac{(2x)^2}{(0.012 - x)}$$

$x = [CO_2] = 0.0057$ M and $2x = [CO] = 0.011$ M

16.8 (a) $K' = K^2 = (2.5 \times 10^{-29})^2 = 6.3 \times 10^{-58}$
(b) $K'' = 1/K^2 = 1/(6.3 \times 10^{-58}) = 1.6 \times 10^{57}$

16.9 Manipulate the equations and equilibrium constants as follows:

$\tfrac{1}{2} H_2(g) + \tfrac{1}{2} Br_2(g) \rightleftharpoons HBr(g)$
$K'_1 = (K_1)^{1/2} = 8.9 \times 10^5$

$H(g) \rightleftharpoons \tfrac{1}{2} H_2(g)$ $\quad K'_2 = 1/(K_2)^{1/2} = 1.4 \times 10^{20}$

$Br(g) \rightleftharpoons \tfrac{1}{2} Br_2(g)$ $\quad K'_3 = 1/(K_3)^{1/2} = 2.1 \times 10^7$

Net: $H(g) + Br(g) \rightleftharpoons HBr(g$

$K_{net} = K'_1 K'_2 K'_3 = 2.6 \times 10^{33}$

16.10

Equation	butane ⇌	isobutane
Initial (M)	0.20	0.50
After adding 2.0 M more isobutene	0.20	2.0 + 0.50
Change (M)	$+x$	$-x$
Equilibrium (M)	$0.20 + x$	$2.50 - x$

$$K = \frac{[\text{isobutane}]}{[\text{butane}]} = \frac{(2.50 - x)}{(0.20 + x)} = 2.50$$

Solving for x gives $x = 0.57$ M. Therefore, [isobutene] $= 2.50 - 0.55 = 1.93$ M and [butane] $= 0.20 + 0.57 = 0.77$ M.

16.11 (a) Adding H_2 shifts the equilibrium to the right, increasing $[NH_3]$. Adding NH_3 shifts the equilibrium to the left, increasing $[N_2]$ and $[H_2]$.
(b) An increase in volume shifts the equilibrium to the left.

16.12 With an increase in temperature, the value of K will become larger. To adjust and attain equilibrium, [NOCl]will decrease.

Chapter 17

17.1 (a) $H_3PO_4(aq) + H_2O(\ell) \rightleftharpoons H_3O^+(aq) + H_2PO_4^-(aq)$

$H_2PO_4^-$ can either donate or accept a proton; therefore, it is amphiprotic.

(b) $CN^-(aq) + H_2O(\ell) \rightleftharpoons HCN(aq) + OH^-(aq)$

CN^- is a Brønsted base; it is capable of accepting a proton.

17.2 NO_3^- is the conjugate base of the acid HNO_3; NH_4^+ is the conjugate acid of the base NH_3.

17.3 $[H_3O^+] = 4.0 \times 10^{-3}$ M; $[OH^-] = K_w/[H_3O^+] = 2.5 \times 10^{-12}$ M

17.4 (a) pOH $= -\log[0.0012] = 2.92$; pH $= 14.00 -$ pOH $= 11.08$
(b) $[H_3O^+] = 4.8 \times 10^{-5}$ mol/L; $[OH^-] = 2.1 \times 10^{-10}$ mol/L
(c) pOH $= 14.00 - 10.46 = 3.54$; $[OH^-] = 2.9 \times 10^{-4}$ mol/L. The solubility of $Sr(OH)_2$ is half of this value (because 1 mol $Sr(OH)_2$ gives two mol OH^- when dissolved), or 1.4×10^{-4} mol/L.

17.5 Answer this question by comparing values of K_a and K_b from Table 17.3.
(a) H_2SO_4 is stronger than H_2SO_3.
(b) $C_6H_5CO_2H$ is a stronger acid than CH_3CO_2H.
(c) The conjugate base of boric acid, $B(OH)_4^-$, is a stronger base than the conjugate base of acetic acid, $CH_3CO_2^-$.
(d) Ammonia is a stronger base than acetate ion.
(e) The conjugate acid of acetate ion, CH_3CO_2H, is a stronger acid than the conjugate acid of ammonia, NH_4^+.

17.6 (a) pH = 7
(b) pH < 7 (NH_4^+ is an acid)
(c) pH < 7 $[Al(H_2O)_6]^{3+}$ is an acid
(d) pH > 7 (HPO_4^{2-} is a stronger base than it is an acid)

17.7 (a) $pK_a = -\log [6.3 \times 10^{-5}] = 4.20$
(b) $ClCH_2CO_2H$ is stronger (the pK_a of 2.87 is less than a pK_a of 4.20)
(c) pK_a for NH_4^+, the conjugate acid of NH_3, is $-\log [5.6 \times 10^{-10}] = 9.26$. It is a weaker acid than acetic acid, for which $K_a = 1.8 \times 10^{-5}$.

17.8 K_b for the lactate ion $= K_w/K_a = 7.1 \times 10^{-11}$. It is a slightly stronger base than the formate, nitrite, and fluoride ions, and a weaker base than the benzoate ion.

17.9 (a) NH_4^+ is a stronger acid than HCO_3^-. CO_3^{2-}, the conjugate base of HCO_3^-, is a stronger base than NH_3, the conjugate base of NH_4^+.
(b) Reactant-favored; the reactants are the weaker acid and base.
(c) Reactant-favored; the reactants are the weaker acid and base.

17.10 (a) The two compounds react and form a solution containing HCN and NaCl. The solution is acidic (HCN is an acid).
(b) $CH_3CO_2H(aq) + SO_3^{2-}(aq) \rightleftharpoons HSO_3^-(aq) + CH_3CO_2^-(aq)$

The solution is acidic, because HSO_3^- is a stronger acid than $CH_3CO_2^-$ is a base.

17.11 From the pH, we can calculate $[H_3O^+] = 1.9 \times 10^{-3}$ M. Also, [butanoate$^-$] $= [H_3O^+] = 1.9 \times 10^{-3}$ M. Use these values along with [butanoic acid] to calculate K_a.

$K_a = [1.9 \times 10^{-3}]\,[1.9 \times 10^{-3}]/(0.055 - 1.9 \times 10^{-3}) = 6.8 \times 10^{-5}$

17.12 $K_a = 1.8 \times 10^{-5} = [x][x]/(0.10 - x)$

$x = [H_3O^+] = [CH_3CO_2^-] = 1.3 \times 10^{-3}$ M; $[CH_3CO_2H] = 0.099$ M; pH = 2.89

17.13 $K_a = 7.2 \times 10^{-4} = [x][x]/(0.015 - x)$

The x in the denominator cannot be dropped. This equation must be solved with the quadratic formula or by successive approximations.

$x = [H_3O^+] = [F^-] = 2.9 \times 10^{-3}$ M

$[HF] = 0.015 - 2.9 \times 10^{-3} = 0.012$ M

pH = 2.54

17.14 $OCl^-(aq) + H_2O(\ell) \rightleftharpoons HOCl(aq) + OH^-(aq)$

$K_b = 2.9 \times 10^{-7} = [x][x]/(0.015 - x)$

$x = [OH^-] = [HOCl] = 6.6 \times 10^{-5}$ M

pOH = 4.18; pH = 9.82

17.15 Equivalent amounts of acid and base react to form water, $CH_3CO_2^-$ and Na^+. Acetate ion hydrolyzes to a small extent, giving CH_3CO_2H and OH^-. We need to determine $[CH_3CO_2^-]$ and then solve a weak base equilibrium problem to determine $[OH^-]$.

Amount $CH_3CO_2^-$ = mol base = 0.12 mol/L × 0.015 L $= 1.8 \times 10^{-3}$ mol

Total volume = 0.030 L, so $[CH_3CO_2^-] = (1.8 \times 10^{-3}$ mol$)/0.030$ L $= 0.060$ M

$CH_3CO_2^-(aq) + H_2O(\ell) \rightleftharpoons CH_3CO_2H(aq) + OH^-(aq)$

$K_b = 5.6 \times 10^{-10} = [x][x]/(0.060 - x)$

$x = [OH^-] = [CH_3CO_2H] = 5.8 \times 10^{-6}$ M

pOH = 5.24; pH = 8.76

17.16 $H_2C_2O_4(aq) + H_2O(\ell) \rightleftharpoons H_3O^+(aq) + HC_2O_4^-(aq)$

$K_{a1} = 5.9 \times 10^{-2} = [x][x]/(0.10 - x)$

The x in the denominator cannot be dropped. This equation must be solved with the quadratic formula or by successive approximations.

$x = [H_3O^+] = [HC_2O_4^-] = 5.3 \times 10^{-2}$ M

pH = 1.28

$K_{b2} = [H_3O^+][C_2O_4^{2-}]/[HC_2O_4^-]$; because $[H_3O^+] = [HC_2O_4^-]$

$[C_2O_4^{2-}] = K_{a2} = 6.4 \times 10^{-5}$ M

17.17 (a) Lewis base (electron-pair donor)
(b) Lewis acid (electron-pair acceptor)
(c) Lewis base (electron-pair donor)
(d) Lewis base (electron-pair donor)

17.18 (a) H_2SeO_4
(b) $Fe(H_2O)_6^{3+}$
(c) HOCl
(d) Amphetamine is a primary amine and a (weak) base. It is both a Brønsted base and a Lewis base.

Chapter 18

18.1 pH of 0.30 M HCO_2H:

$K_a = [H_3O^+][HCO_2^-]/[HCO_2H]$

$1.8 \times 10^{-4} = [x][x]/[0.30 - x]$

$x = 7.3 \times 10^{-3}$ M; pH = 2.14

pH of 0.30 M formic acid + 0.10 M $NaHCO_2$

$K_a = [H_3O^+][HCO_2^-]/[HCO_2H]$

$1.8 \times 10^{-4} = [x][0.10 + x]/(0.30 - x)$

$x = 5.4 \times 10^{-4}$ M; pH = 3.27

18.2 NaOH: (0.100 mol/L)(0.0300 L) $= 3.00 \times 10^{-3}$ mol

CH_3CO_2H: (0.100 mol/L)(0.0450 L) $= 4.50 \times 10^{-3}$ mol

3.00×10^{-3} mol NaOH reacts with 3.00×10^{-3} mol CH_3CO_2H, forming 3.00×10^{-3} mol $CH_3CO_2^-$; 1.50×10^{-3} mol unreacted CH_3CO_2H remains in solution. The total volume is 75.0 mL. Use these values to calculate $[CH_3CO_2H]$ and $[CH_3CO_2^-]$, and use these concentrations in a weak acid equilibrium calculation to obtain $[H_3O^+]$ and pH.

$[CH_3CO_2H] = 1.5 \times 10^{-3}$ mol/0.075 L = 0.0200 M

$[CH_3CO_2^-] = 3.0 \times 10^{-3}$ mol/0.075 L = 0.0400 M

$K_a = [H_3O^+][CH_3CO_2^-]/[CH_3CO_2H]$

$1.8 \times 10^{-5} = [x][0.0400 + x]/(0.0200 - x)$

$x = [H_3O^+] = 9.0 \times 10^{-6}$ M; pH = 5.05

18.3 $pH = pK_a + \log\{[base]/[acid]\}$

$pH = -\log(1.8 \times 10^{-4}) + \log\{[0.70]/[0.50]\}$

pH = 3.74 + 0.15 = 3.89

18.4 (15.0 g $NaHCO_3$)(1 mol/84.01 g) = 0.179 mol $NaHCO_3$, and (18.0 g Na_2CO_3)(1 mol/106.0 g) = 0.170 mol Na_2CO_3

$pH = pK_a + \log\{[base]/[acid]\}$

$pH = -\log(4.8 \times 10^{-11}) + \log\{[0.170]/[0.179]\}$

pH = 10.32 − 0.02 = 10.30

18.5 $pH = pK_a + \log\{[base]/[acid]\}$

$5.00 = -\log(1.8 \times 10^{-5}) + \log\{[base]/[acid]\}$

$5.00 = 4.74 + \log\{[base]/[acid]\}$

[base]/[acid] = 1.8

To prepare this buffer solution, the ratio [base]/[acid] must equal 1.8. For example, you can dissolve 1.8 mol (148 g) of $NaCH_3CO_2$ and 1.0 mol (60.05 g) of CH_3CO_2H in enough water to make 1.0 L of solution.

18.6 Initial pH (before adding acid):

$pH = pK_a + \log\{[base]/[acid]\}$

$= -\log(1.8 \times 10^{-4}) + \log\{[0.70]/[0.50]\}$

= 3.74 + 0.15 = 3.89

After adding acid, the added HCl will react with the weak base (formate ion) and form more formic acid. The net effect is to change the ratio of [base]/[acid] in the buffer solution.

Initial amount HCO_2H = 0.50 mol/L × 0.500 L = 0.250 mol

Initial amount HCO_2^- = 0.70 mol/L × 0.50 L = 0.350 mol

Amount HCl added = 1.0 mol/L × 0.010 L = 0.010 mol

Amount HCO_2H after HCl addition = 0.250 mol + 0.010 mol = 0.26 mol

Initial amount HCO_2^- after HCl addition = 0.350 mol − 0.010 mol = 0.34 mol

$pH = pK_a + \log\{[base]/[acid]\}$

$pH = -\log(1.8 \times 10^{-4}) + \log\{[0.340]/[0.260]\}$

pH = 3.74 + 0.12 = 3.86

18.7 After addition of 25.0 mL base, half of the acid has been neutralized.

Initial amount HCl = 0.100 mol/L × 0.0500 L = 0.00500 mol

Amount NaOH added = 0.100 mol/L × 0.0250 L = 0.00250 mol

Amount HCl after reaction: 0.00500 − 0.00250 = 0.00250 mol HCl

[HCl] after reaction = 0.00250 mol/0.0750 L = 0.0333 M

This is a strong acid and completely ionized, so $[H_3O^+]$ = 0.0333 M and pH = 1.48.

After 50.50 mL base is added, a small excess of base is present in the 100.5 mL (0.1005 L) of solution. (Volume of excess base added is 0.50 mL = 5.0×10^{-4} L.)

Amount excess base = 0.100 mol/L × 5.0×10^{-4} L = 5.0×10^{-5} mol

$[OH^-] = 5.0 \times 10^{-5}$ mol/0.1005 L = 4.9×10^{-4} M

$pOH = -\log(4.9 \times 10^{-4}) = 3.31$; pH = 14.00 − pOH = 10.69

18.8 35.0 mL base will partially neutralize the acid.

Initial amount CH_3CO_2H = (0.100 mol/L)(0.1000 L) = 0.0100 mol

Amount NaOH added = (0.10 mol/L)(0.035 L) = 0.0035 mol

Amount CH_3CO_2H after reaction = 0.0100 − 0.0035 = 0.0065 mol

Amount $CH_3CO_2^-$ after reaction = 0.0035 mol

$[CH_3CO_2H]$ after reaction = 0.0065 mol/0.135 L = 0.0481 M

$[CH_3CO_2^-]$ after reaction = 0.00350 mol/0.135 L = 0.0259 M

$K_a = [H_3O^+][CH_3CO_2^-]/[CH_3CO_2H]$

$1.8 \times 10^{-5} = [x][0.0259 + x]/[0.0481 - x]$

$x = [H_3O^+] = 3.34 \times 10^{-5}$ M; pH = 4.48

18.9 75.0 mL acid will partially neutralize the base.

Initial amount NH_3 = (0.100 mol/L)(0.1000 L) = 0.0100 mol

Amount HCl added = (0.100 mol/L)(0.0750 L) = 0.00750 mol

Amount NH_3 after reaction = 0.0100 − 0.00750 = 0.0025 mol

Amount NH_4^+ after reaction = 0.00750 mol

Solve using the Henderson–Hasselbach equation; use K_a for the weak acid NH_4^+:

$pH = pK_a + \log\{[base]/[acid]\}$

$pH = -\log(5.6 \times 10^{-10}) + \log\{[0.0025]/[0.00750]\}$

pH = 9.25 − 0.48 = 8.77

18.10 An indicator that changes color near the pH at the equivalence point is required. Possible indicators include methyl red, bromcresol green, and Eriochrome black T; all change color in the pH range of 5–6.

18.11 (a) $AgI(s) \rightleftharpoons Ag^+(aq) + I^-(aq)$

$K_{sp} = [Ag^+][I^-]$; $K_{sp} = 8.5 \times 10^{-17}$

(b) $BaF_2(s) \rightleftharpoons Ba^{2+}(aq) + 2\ F^-(aq)$

$K_{sp} = [Ba^{2+}][F^-]^2$; $K_{sp} = 1.8 \times 10^{-7}$

(c) $Ag_2CO_3(s) \rightleftharpoons 2\ Ag^+(aq) + CO_3^{2-}(aq)$

$K_{sp} = [Ag^+]^2[CO_3^{2-}]$; $K_{sp} = 8.5 \times 10^{-12}$

18.12 $[Ba^{2+}] = 3.6 \times 10^{-3}$ M; $[F^-] = 7.2 \times 10^{-3}$ M

$K_{sp} = [Ba^{2+}][F^-]^2$

$K_{sp} = [3.6 \times 10^{-3}][7.2 \times 10^{-3}]^2 = 1.9 \times 10^{-7}$

18.13 $Ca(OH)_2(s) \rightleftharpoons Ca^{2+}(aq) + 2\ OH^-(aq)$

$K_{sp} = [Ca^{2+}][OH^-]^2$; $K_{sp} = 5.5 \times 10^{-5}$

$5.5 \times 10^{-5} = [x][2x]^2$ (where x = solubility in mol/L)

$x = 2.4 \times 10^{-2}$ mol/L

Solubility in g/L = $(2.4 \times 10^{-2}$ mol/L)(74.1 g/mol) = 1.8 g/L

18.14 (a) AgCl
(b) $Ca(OH)_2$
(c) Because these compounds have different stoichiometries, the most soluble cannot be identified without doing a calculation. The solubility of $Ca(OH)_2$ is 2.4×10^{-2} M (from Exercise 18.13); $Ca(OH)_2$ is more soluble than $CaSO_4$, whose solubility is 7.0×10^{-3} M $\{K_{sp} = [Ca^{2+}][SO_4^{2-}]$; $4.9 \times 10^{-5} = [x][x]$; $x = 7.0 \times 10^{-3}$ M$\}$.

18.15 (a) In pure water:

$K_{sp} = [Ba^{2+}][SO_4^{2-}]$; $1.1 \times 10^{-10} = [x][x]$;

$x = 1.0 \times 10^{-5}$ mol/L

(b) In 0.010 M $Ba(NO_3)_2$, which furnishes 0.010 M Ba^{2+} in solution:

$K_{sp} = [Ba^{2+}][SO_4^{2-}]$; $1.1 \times 10^{-10} = [0.010 + x][x]$; $x = 1.1 \times 10^{-8}$ mol/L

18.16 (a) In pure water:

$K_{sp} = [Zn^{2+}][CN^-]^2$; $8.0 \times 10^{-12} = [x][2x]^2 = 4x^3$

Solubility = $x = 1.3 \times 10^{-4}$ mol/L

(b) In 0.10 M $Zn(NO_3)_2$, which furnishes 0.10 M Zn^{2+} in solution:

$K_{sp} = [Zn^{2+}][CN^-]^2$; $8.0 \times 10^{-12} = [0.10 + x][2x]^2$

Solubility = $x = 4.5 \times 10^{-6}$ mol/L

18.17 When $[Pb^{2+}] = 1.1 \times 10^{-3}$ M, $[I^-] = 2.2 \times 10^{-3}$ M.

$Q = [Pb^{2+}][I^-]^2 = [1.1 \times 10^{-3}][2.2 \times 10^{-3}]^2 = 5.3 \times 10^{-9}$

This value is less than K_{sp}, which means that the system has not yet reached equilibrium and more PbI_2 will dissolve.

18.18 $Q = [Sr^{2+}][SO_4^{2-}] = [2.5 \times 10^{-4}][2.5 \times 10^{-4}] = 6.3 \times 10^{-8}$

This value is less than K_{sp}, which means that the system has not yet reached equilibrium. Precipitation will not occur.

18.19 $K_{sp} = [Pb^{2+}][I^-]^2$. Let x be the concentration of I^- required at equilibrium.

$9.8 \times 10^{-9} = [0.050][x]^2$

$x = [I^-] = 4.4 \times 10^{-5}$ mol/L. A concentration greater than this value will result in precipitation of PbI_2.

Let x be the concentration of Pb^{2+} in solution, in equilibrium with 0.0015 M I^-.

$9.8 \times 10^{-9} = [x][1.5 \times 10^{-3}]^2$

$x = [Pb^{2+}] = 4.4 \times 10^{-3}$ M

18.20 First, determine the concentrations of Ag^+ and Cl^-; then calculate Q, and see whether it is greater than or less than K_{sp}. Concentrations are calculated using the final volume, 105 mL, in the equation $C_{dil} \times V_{dil} = C_{conc} \times V_{conc}$.

$[Ag^+](0.105\ L) = (0.0010\ mol/L)(0.100\ L)$

$[Ag^+] = 9.5 \times 10^{-4}$ M

$[Cl^-](0.105\ L) = (0.025\ M)(0.005\ L)$

$[Cl^-] = 1.2 \times 10^{-3}$ M

$Q = [Ag^+][Cl^-] = [9.5 \times 10^{-4}][1.2 \times 10^{-3}] = 1.1 \times 10^{-6}$

Because $Q > K_{sp}$, precipitation occurs.

18.21 The logic for the solution of this exercise is outlined in Example 8.16.

Equation	$Ag(NH_3)_2^+ \rightleftharpoons$	Ag^+ +	$2\ NH_3$
Initial (M)	0.005		1.0 − 2(0.005)
Change	−x	+x	+2x
Equilibrium (M)	0.005 − x	x	0.99

$K = 1/K_f = 1/1.1 \times 10^7 = [x][0.99]^2/0.005$

$x = [Ag^+] = 4.6 \times 10^{-10}$ mol/L

18.22 $Cu(OH)_2(s) \rightleftharpoons Cu^{2+}(aq) + 2\ OH^-(aq)$

$K_{sp} = [Cu^{2+}][OH^-]^2$

$Cu^{2+}(aq) + 4\ NH_3(aq) \rightleftharpoons Cu(NH_3)_4^{2+}(aq)$

$K_{form} = [Cu(NH_3)_4^{2+}]/[Cu^{2+}][NH_3]^4$

Net: $Cu(OH)_2(s) + 4\ NH_3(aq) \rightleftharpoons Cu(NH_3)_4^{2+}(aq) + 2\ OH^-(aq)$

$K_{net} = K_{sp} \times K_{form} = (2.2 \times 10^{-20})(6.8 \times 10^{12}) = 1.5 \times 10^{-7}$

Chapter 19

19.1 (a) O_3; larger molecules generally have higher entropies than smaller molecules.
(b) $SnCl_4(g)$; gases have higher entropies than liquids.

19.2 (a) $\Delta_r S^\circ = \Sigma S^\circ(\text{products}) - \Sigma S^\circ(\text{reactants})$

$\Delta_r S^\circ = S^\circ\ [NH_4Cl(aq)]) - S^\circ\ [NH_4Cl(s)]$

$\Delta_r S° = (1\ \text{mol/mol-rxn})(169.9\ \text{J/mol} \cdot \text{K}) - (1\ \text{mol/mol-rxn})(94.85\ \text{J/mol} \cdot \text{K})$

$= 75.1\ \text{J/K} \cdot \text{mol-rxn}$

A gain in entropy for the formation of a mixture (solution) is expected.

(b) $\Delta_r S° = 2\ S°(CO_2) + 3\ S°(H_2O) - [S°(C_2H_5OH) + 3\ S°(O_2)]$

$\Delta_r S° = (2\ \text{mol/mol-rxn})(213.74\ \text{J/mol} \cdot \text{K}) + (3\ \text{mol/mol-rxn})(188.84\ \text{J/mol} \cdot \text{K}) - [(1\ \text{mol/mol-rxn})(282.70\ \text{J/mol} \cdot \text{K}) + (3\ \text{mol/mol-rxn})(205.07\ \text{J/mol} \cdot \text{K})]$

$\Delta_r S° = +96.09\ \text{J/K} \cdot \text{mol-rxn}$

An increase in entropy is expected because there is a increase in the number of moles of gases.

19.3 (a) Type 2
(b) Type 3
(c) Type 1
(d) Type 2

19.4 $\Delta S°(\text{system}) = 2\ S°(HCl) - [S°(H_2) + S°(Cl_2)]$

$\Delta S°(\text{system}) = (2\ \text{mol/mol-rxn})(186.2\ \text{J/mol} \cdot \text{K}) - [(1\ \text{mol/mol-rxn})(130.7\ \text{J/mol} \cdot \text{K}) + (1\ \text{mol/mol-rxn})(223.08\ \text{J/mol} \cdot \text{K})]$

$= 18.6\ \text{J/K} \cdot \text{mol-rxn}$

$\Delta S°(\text{surroundings}) = -\Delta H°(\text{system})\ /T = -(-184{,}620\ \text{J/mol-rxn}/298\ \text{K}) = 619.5\ \text{J/K} \cdot \text{mol-rxn}$

$\Delta S°(\text{universe}) = \Delta S°(\text{system}) + \Delta S°(\text{surroundings}) = 18.6\ \text{J/K} \cdot \text{mol-rxn} + 619.5\ \text{J/K} \cdot \text{mol-rxn} = 638.1\ \text{J/K} \cdot \text{mol-rxn}$

19.5 $\Delta S°(\text{system}) = \Delta_r S° = 560.7\ \text{J/K} \cdot \text{mol-rxn}$

At 298 K, $\Delta S°(\text{surroundings}) = -(467{,}900\ \text{J/mol-rxn})/298\ \text{K} = -1570\ \text{J/K} \cdot \text{mol-rxn}$

$\Delta S°_{univ} = \Delta S°(\text{system}) + \Delta S°(\text{surroundings})$

$= 560.7\ \text{J/K} \cdot \text{mol-rxn} - 1570\ \text{J/K} \cdot \text{mol-rxn}$

$= -1010\ \text{J/K} \cdot \text{mol-rxn}$

The negative sign indicates that the process is not spontaneous. At higher temperature, the value of $-\Delta H°(\text{system})/T$ will be less negative. At a high enough temperature, $\Delta S°(\text{surroundings})$ will outweigh $\Delta S°(\text{system})$ and the reaction will be spontaneous.

19.6 For the reaction $N_2(g) + 3\ H_2(g) \longrightarrow 2\ NH_3(g)$:

$\Delta_r H° = 2\ \Delta_f H°$ for $NH_3(g) = (2\ \text{mol/mol-rxn})(-45.90\ \text{kJ/mol}) = -91.80\ \text{kJ/mol-rxn}$

$\Delta_r S° = 2\ S°(NH_3) - [S°(N_2) + 3\ S°(H_2)]$

$\Delta_r S° = (2\ \text{mol/mol-rxn})(192.77\ \text{J/ mol} \cdot \text{K}) - [(1\ \text{mol/mol-rxn})(191.56\ \text{J/mol} \cdot \text{K}) + (3\ \text{mol/mol-rxn})(130.7\ \text{J/mol} \cdot \text{K})]$

$\Delta_r S° = -198.1\ \text{J/K} \cdot \text{mol-rxn}\ (= 0.198\ \text{kJ/K} \cdot \text{mol-rxn})$

$\Delta_r G° = \Delta_r H° - T\Delta_r S° = -91.80\ \text{kJ/mol-rxn} - (298\ \text{K})(-0.198\ \text{kJ/K} \cdot \text{mol-rxn})$

$\Delta_r G° = -32.8\ \text{kJ/mol-rxn}$

19.7 $SO_2(g) + \frac{1}{2}\ O_2(g) \longrightarrow SO_3(g)$

$\Delta_r G° = \Sigma\ \Delta_f G°(\text{products}) - \Sigma\ \Delta_f G°(\text{reactants})$

$\Delta_r G° = (1\ \text{mol/mol-rxn})\Delta G°[SO_3(g)] - \{(1\ \text{mol/mol-rxn})\Delta G°[SO_2(g)] + 0.5\ \text{mol/mol-rxn})\Delta G°[O_2(g)]\}$

$\Delta_r G° = -371.04\ \text{kJ/mol-rxn} - (-300.13\ \text{kJ/mol-rxn} + 0)$

$= -70.91\ \text{kJ/mol-rxn}$

19.8 $HgO(s) \longrightarrow Hg(\ell) + \frac{1}{2}\ O_2(g)$; determine the temperature at which $\Delta_r G° = \Delta_r H° - T\Delta_r S° = 0$. T is the unknown in this problem.

$\Delta_r H° = [-\Delta_f H°$ for $HgO(s)] = 90.83\ \text{kJ/mol-rxn}$

$\Delta_r S° = S°[Hg(\ell)] + \frac{1}{2}\ S°(O_2) - S°[HgO(s)]$

$\Delta_r S° = (1\ \text{mol/mol-rxn})(76.02\ \text{J/mol} \cdot \text{K}) + [(0.5\ \text{mol/mol-rxn})(205.07\ \text{J/mol} \cdot \text{K}) - (1\ \text{mol/mol-rxn})(70.29\ \text{J/mol} \cdot \text{K})]$

$= 108.26\ \text{J/K} \cdot \text{mol-rxn}$

$\Delta_r H° - T(\Delta_r S°) = 90{,}830\ \text{J/mol-rxn} - T(108.27\ \text{J/mol-rxn})/\text{K}) = 0$

$T = 839\ \text{K}\ (566\ °\text{C})$

19.9 $C(s) + CO_2(g) \rightleftharpoons 2\ CO(g)$

$\Delta_r G° = \Sigma\ \Delta_f G°(\text{products}) - \Sigma\ \Delta_f G°(\text{reactants})$

$\Delta_r G° = 2\ \Delta_f G°(CO) - \Delta_f G°(CO_2)$

$\Delta_r G° = (2\ \text{mol/mol-rxn})(-137.17\ \text{kJ/mol}) - (1\ \text{mol/mol-rxn})(-394.36\ \text{kJ/mol})$

$\Delta_r G° = 120.02\ \text{kJ/mol-rxn}$

$\Delta_r G° = -RT \ln K$

$120{,}020\ \text{J/mol-rxn} = -(8.3145\ \text{J/mol-rxn} \cdot \text{K})(298\ \text{K})(\ln K)$

$K = 8.94 \times 10^{-22}$

Chapter 20

20.1 Oxidation half-reaction: $Al(s) \longrightarrow Al^{3+}(aq) + 3\ e^-$

Reduction half-reaction: $2\ H^+(aq) + 2\ e^- \longrightarrow H_2(g)$

Overall reaction: $2\ Al(s) + 6\ H^+(aq) \longrightarrow 2\ Al^{3+}(aq) + 3\ H_2(g)$

Al is the reducing agent and is oxidized; $H^+(aq)$ is the oxidizing agent and is reduced.

20.2 $2\ VO^{2+}(aq) + Zn(s) + 4\ H^+(aq) \longrightarrow Zn^{2+}(aq) + 2\ V^{3+}(aq) + 2\ H_2O(\ell)$

$2\ V^{3+}(aq) + Zn(s) \longrightarrow 2\ V^{2+}(aq) + Zn^{2+}(aq)$

20.3 Oxidation (Fe^{2+}, the reducing agent, is oxidized): $Fe^{2+}(aq) \longrightarrow Fe^{3+}(aq) + e^-$

Reduction (MnO_4^-, the oxidizing agent, is reduced)

$MnO_4^-(aq) + 8\ H^+(aq) + 5\ e^- \longrightarrow Mn^{2+}(aq) + 4\ H_2O(\ell)$

Overall reaction:

$MnO_4^-(aq) + 8\ H^+(aq) + 5\ Fe^{2+}(aq) \longrightarrow Mn^{2+}(aq) + 5\ Fe^{3+}(aq) + 4\ H_2O(\ell)$

20.4 (a) Oxidation half-reaction:

$Al(s) + 3\ OH^-(aq) \longrightarrow Al(OH)_3(s) + 3\ e^-$

Reduction half-reaction:

$S(s) + H_2O(\ell) + 2\ e^- \longrightarrow HS^-(aq) + OH^-(aq)$

Overall reaction:

$2\ Al(s) + 3\ S(s) + 3\ H_2O(\ell) + 3\ OH^-(aq) \longrightarrow 2\ Al(OH)_3(s) + 3\ HS^-(aq)$

(b) Aluminum is the reducing agent and is oxidized; sulfur is the oxidizing agent and is reduced.

20.5 Construct two half-cells, the first with a silver electrode and a solution containing $Ag^+(aq)$, and the second with a nickel electrode and a solution containing $Ni^{2+}(aq)$. Connect the two half-cells with a salt bridge. When the electrodes are connected through an external circuit, electrons will flow from the anode (the nickel electrode) to the cathode (the silver electrode). The overall cell reaction is $Ni(s) + 2\ Ag^+(aq) \longrightarrow Ni^{2+}(aq) + 2\ Ag(s)$. To maintain electrical neutrality in the two half-cells, negative ions will flow from the Ag | Ag^+ half-cell to the Ni | Ni^{2+} half-cell, and positive ions will flow in the opposite direction.

20.6 Anode reaction: $Zn(s) \longrightarrow Zn^{2+}(aq) + 2\ e^-$

Cathode reaction: $2\ Ag^+(aq) + 2\ e^- \longrightarrow 2\ Ag(s)$

$E°_{cell} = E°_{cathode} - E°_{anode} = 0.80\ V - (-0.76\ V) = 1.56\ V$

20.7 Mg is easiest to oxidize, and Au is the most difficult. (See Table 20.1.)

20.8 Use the "northwest–southeast rule" or calculate the cell voltage to determine whether a reaction is product-favored. Reactions (a) and (c) are reactant-favored; reactions (b) and (d) are product-favored.

20.9 Overall reaction: $2\ Al(s) + 3\ Fe^{2+}(aq) \longrightarrow 2\ Al^{3+}(aq) + 3\ Fe(s)$

($E°_{cell} = 1.22$ V, $n = 6$)

$E_{cell} = E°_{cell} - (0.0257/n)\ \ln\ \{[Al^{3+}]^2/[Fe^{2+}]^3\}$

$= 1.22 - (0.0257/6)\ \ln\ \{[0.025]^2/[0.50]^3\}$

$= 1.22\ V - (-0.023)\ V = 1.24\ V$

20.10 Overall reaction: $Fe(s) + 2\ H^+(aq) \longrightarrow Fe^{2+}(aq) + H_2(g)$

($E°_{cell} = 0.44$ V, $n = 2$)

$E_{cell} = E°_{cell} - (0.0257/n)\ \ln\ \{[Fe^{2+}]P_{H_2}/[H^+]^2\}$

$= 0.44 - (0.0257/2)\ \ln\ \{[0.024]1.0/[0.056]^2\}$

$= 0.44\ V - 0.026\ V = 0.41\ V$

20.11 $\Delta_r G° = -nFE° = -(2\ mol\ e^-)(96{,}500\ C/mol\ e^-)(-0.76\ V)(1\ J/1\ C \cdot V)$

$= 146{,}680\ J\ (= 150\ kJ)$

The negative value of $E°$ and the positive value of $\Delta G°$ both indicate a reactant-favored reaction.

20.12 $E°_{cell} = E°_{cathode} - E°_{anode} = 0.80\ V - 0.855\ V = -0.055\ V$; $n = 2$

$E° = (0.0257/n)\ \ln K$

$-0.055 = (0.0257/2)\ \ln K$

$K = 0.014$

20.13 Cathode: $Zn^{2+}(aq) + 2\ e^- \longrightarrow Zn(s)\quad E°_{cathode} = -0.76\ V$

Anode: $Zn(s) + 4\ CN^-(aq) \longrightarrow [Zn(CN)_4^{2-}] + 2\ e^-\quad E°_{anode} = -1.26\ V$

Overall: $Zn^{2+}(aq) + 4\ CN^-(aq) \longrightarrow [Zn(CN)_4^{2-}]\quad E°_{cell} = 0.50\ V$

$E° = (0.0257/n)\ \ln K$

$0.50 = (0.0257/2)\ \ln K$

$K = 7.9 \times 10^{16}$

20.14 Cathode: $2\ H_2O(\ell) + 2\ e^- \longrightarrow 2\ OH^-(aq) + H_2(g)$

$E°_{cathode} = -0.83\ V$

Anode: $4\ OH^-(aq) \longrightarrow O_2(g) + 2\ H_2O(\ell) + 4\ e^-$

$E°_{anode} = 0.40\ V$

Overall: $2\ H_2O(\ell) \longrightarrow 2\ H_2(g) + O_2(g)$

$E°_{cell} = E°_{cathode} - E°_{anode} = -0.83\ V - 0.40\ V = -1.23\ V$

This is the minimum voltage needed to cause this reaction to occur.

20.15 O_2 is formed at the anode, by the reaction $2\ H_2O(\ell) \longrightarrow 4\ H^+(aq) + O_2(g) + 4\ e^-$.

(0.445 A)(45 min)(60 s/min)(1 C/1 A · s)(1 mol e^-/96,500 C)(1 mol O_2/4 mol e^-)(32 g O_2/1 mol O_2) = 0.10 g O_2

20.16 The cathode reaction (electrolysis of molten NaCl) is $Na^+(melt) + e^- \longrightarrow Na(\ell)$.

(25×10^3 A)(60 min)(60 s/min)(1 C/1 A · s)(1 mol e^-/96,500 C)(1 mol Na/mol e^-)(23 g Na/1 mol Na)

= 21,450 g Na = 21 kg

Chapter 21

21.1 (a) $2\ Na(s) + Br_2(\ell) \longrightarrow 2\ NaBr(s)$
(b) $Ca(s) + Se(s) \longrightarrow CaSe(s)$
(c) $2\ Pb(s) + O_2(g) \longrightarrow 2\ PbO(s)$

Lead(II) oxide, a red compound commonly called litharge, is the most widely used inorganic lead compound. Maroon-colored lead(IV) oxide is the product of lead oxidation in lead-acid storage batteries (Chapter 21). Other oxides such as Pb_3O_4 also exist.

(d) $2\ Al(s) + 3\ Cl_2(g) \longrightarrow 2\ AlCl_3(s)$

21.2 (a) H_2Te
(b) Na_3AsO_4
(c) $SeCl_6$
(d) $HBrO_4$

21.3 (a) NH_4^+ (ammonium ion)
(b) O_2^{2-} (peroxide ion)
(c) N_2H_4 (hydrazine)
(d) NF_3 (nitrogen trifluoride)

21.4 (a) ClO is an odd-electron molecule, and Cl has the unlikely oxidation number of +2.
(b) In Na_2Cl, chlorine would have the unlikely charge of 2− (to balance the two positive charges of the two Na^+ ions).
(c) This compound would require either the calcium ion to have the formula Ca^+ or the acetate ion to have the formula $CH_3CO_2^{2-}$. In all of its compounds, calcium occurs as the Ca^{2+} ion. The acetate ion, formed from acetic acid by loss of H^+, has a 1− charge.
(d) No octet structure for C_3H_7 can be drawn. This species has an odd number of electrons.

21.5 $CH_4(g) + H_2O(g) \longrightarrow 3\ H_2(g) + CO(g)$

Bonds broken: 4 C—H and 2 O—H (sum = 2578 kJ)

$4\ \Delta H(\text{C—H}) = 4(413\ \text{kJ}) = 1652\ \text{kJ}$

$2\ \Delta H(\text{O—H}) = 4(463\ \text{kJ}) = 926\ \text{kJ}$

Bonds formed: 3 H—H and 1 C≡O (sum = 2354 kJ)

$3\ \Delta H(\text{H—H}) = 3(436\ \text{kJ}) = 1308\ \text{kJ}$

$\Delta H(\text{CO}) = 1046\ \text{kJ}$

Estimated energy of reaction = 2578 kJ − 2354 kJ = +224 kJ

21.6 Cathode reaction: $Na^+ + e^- \longrightarrow Na(\ell)$; 1 F, or 96,500 C, is required to form 1 mol of Na. There are (24 h) (60 min/h)(60 s/min) = 86,400 s in 1 day.

1000 kg = 1.000×10^6 g, so

(1.000×10^6 g Na)(1 mol Na/23.00 g Na)(96,500 C/mol Na)(1 A · s/1 C)(1/86,400 s) = 4.855×10^4 A

21.7 Some interesting topics: gemstones of the mineral beryl; uses of Be in the aerospace industry and in nuclear reactors; beryllium–copper alloys; severe health hazards when beryllium or its compounds get into the lungs.

21.8 (a) $Ga(OH)_3(s) + 3\ H^+(aq) \longrightarrow Ga^{3+}(aq) + 3\ H_2O(\ell)$

$Ga(OH)_3(s) + OH^-(aq) \longrightarrow Ga(OH)_4^-(aq)$

(b) $Ga^{3+}(aq)$ ($K_a = 1.2 \times 10^{-3}$) is stronger acid than $Al^{3+}(aq)$ ($K_a = 7.9 \times 10^{-6}$)

21.9

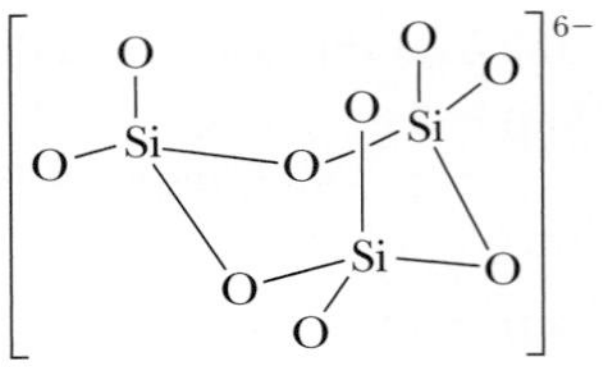

21.10 (a) $:\!N{\equiv}N{-}\ddot{\underset{..}{O}}\!: \longleftrightarrow \ddot{N}{=}N{=}\ddot{O} \longleftrightarrow :\!\ddot{\underset{..}{N}}{-}N{\equiv}O\!:$

A B C

Resonance structure A has formal charges of N = 0, N = +1, O = −1 (from left to right) and is the most favorable structure. For B, N = −1, N = +1, O = 0, and for C, N = −2, N = +1, O = +1. Structure C is clearly unfavorable. B is not as favorable as A because the more electronegative atom (O) has a 0 charge whereas N is −1.

(b) $NH_4NO_3(s) \longrightarrow N_2O(g) + 2\ H_2O(g)$

$\Delta_rH° = \Sigma\Delta_fH°(\text{products}) - \Sigma\Delta_fH°(\text{reactants})$

$\Delta_r\text{H}° = \Delta_fH°(N_2O) + 2\ \Delta_fH°(H_2O) - \Delta_fH°(NH_4NO_3)$

$= 82.05\ \text{kJ} + 2(-241.83\ \text{kJ}) - (-365.56\ \text{kJ}) = -36.05\ \text{kJ}$

The reaction is exothermic.

21.11 First, calculate $\Delta_rG°$, $\Delta_rH°$, and $\Delta_rS°$ for this reaction, using data from Appendix L.

$\Delta_rG° = \Sigma\Delta_fG°(\text{products}) - \Sigma\Delta_fG°(\text{reactants})$

$\Delta_rG° = 2\ \Delta_fG°(ZnO) + 2\ \Delta_fG°(SO_2) - 2\ \Delta_fG°(ZnS) - 3\ \Delta_fG°(O_2)$

$= 2(-318.30\ \text{kJ}) + 2(-300.13\ \text{kJ}) - 2(-201.29\ \text{kJ}) - 0$

$= -834.28$ kJ. The reaction is product-favored at 298 K.

$\Delta_rH° = 2\ \Delta_fH°(ZnO) + 2\ \Delta_fH°(SO_2) - 2\ \Delta_fH°(ZnS) - 3\ \Delta_fH°(O_2)$

$= 2(-348.28\ \text{kJ}) + 2(-296.84\ \text{kJ}) - [2(-205.98\ \text{kJ}) + 0] = -878.28\ \text{kJ}$

$\Delta_rS° = 2\ S°(ZnO) + 2\ S°(SO_2) - 2\ S°(ZnS) - 3\ S°(O_2)$

$= 2(43.64\ \text{J/K}) + 2(248.21\ \text{J/K}) - [2(57.7\ \text{J/K}) + 3(205.07\ \text{J/K})]$

$= -146.9\ \text{J/K}$

This reaction is enthalpy favored and entropy disfavored. The reaction will become less favored at higher temperatures. See Table 19.2.

21.12 For the reaction $HX + Ag \longrightarrow AgX + \frac{1}{2}\ H_2$:

$\Delta_rG° = \Sigma\Delta_fG°(\text{products}) - \Sigma\Delta_fG°(\text{reactants})$

$\Delta_rG° = \Delta_fG°(AgX) - \Delta_fG°(HX)$

For HF: $\Delta_rG° = +79.4$ kJ; reactant favored

For HCl: $\Delta_rG° = -14.67$ kJ; product favored

For HBr: $\Delta G° = -43.45$ kJ; product favored

For HI: $\Delta_rG° = -67.75$ kJ; product favored

Chapter 22

22.1 (a) $Co(NH_3)_3Cl_3$
(b) (i) $K_3[Co(NO_2)_6]$: a complex of cobalt(III) with a coordination number of 6
(ii) $Mn(NH_3)_4Cl_2$: a complex of manganese(II) with a coordination number of 6

22.2 (a) hexaaquanickel(II) sulfate
(b) dicyanobis(ethylenediamine)chromium(III) chloride
(c) potassium amminetrichloroplatinate(II)
(d) potassium dichlorocuprate(I)

22.3 (a) Geometric isomers are possible (with the NH_3 ligands in *cis* and *trans* positions).
(b) Only a single structure is possible.
(c) Only a single structure is possible.
(d) This compound is chiral; there are two optical isomers.
(e) Only a single structure is possible.
(f) Two structural isomers are possible based on coordination of the NO_2^- ligand through oxygen or nitrogen.

22.4 (a) $[Ru(H_2O)_6]^{2+}$: An octahedral complex of ruthenium(II) (d^6). A low-spin complex has no unpaired electrons and is diamagnetic. A high-spin complex has four unpaired electrons and is paramagnetic.

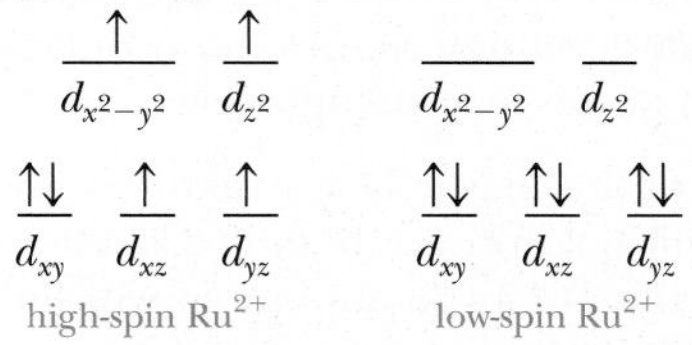

(b) $[Ni(NH_3)_6]^{2+}$: An octahedral complex of nickel(II) (d^8). Only one electron configuration is possible; it has two unpaired electrons and is paramagnetic.

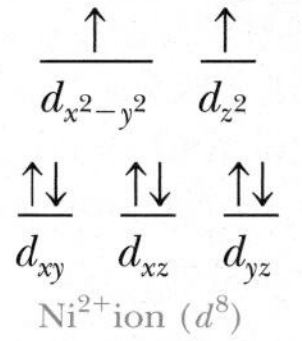

22.5 (a) The $C_5H_5^-$ ligand is an anion (6 electrons), C_6H_6 is a neutral ligand (6 electrons), so Mn must be +1 (6 valence electrons). There is a total of 18 valence electrons.
(b) The ligands in this complex are all neutral, so the Mo atom must have no charge. The C_6H_6 ligand contributes six electrons, each Co contributes two electrons for a total of six, and Mo has six valence electrons. The total is 18 electrons.

Chapter 23

23.1 (a) $^{222}_{86}Rn \longrightarrow \, ^{218}_{84}Po + \, ^{4}_{2}\alpha$
(b) $^{218}_{84}Po \longrightarrow \, ^{218}_{85}At + \, ^{0}_{-1}\beta$

23.2 (a) E (per photon) $= h\nu = hc/\lambda$
$E = [(6.626 \times 10^{-34}\ J \cdot s/photon)(3.00 \times 10^8\ m/s)]/(2.0 \times 10^{-12})$
$E = 9.94 \times 10^{-14}\ J/photon$
E (per mole) $= (9.94 \times 10^{-14}\ J/photon)(6.022 \times 10^{23}\ photons/mol)$
E (per mole) $= 5.99 \times 10^{10}\ J/mol$

23.3 (a) Emission of six α particles leads to a decrease of 24 in the mass number and a decrease of 12 in the atomic number. Emission of four β particles increases the atomic number by 4, but doesn't affect the mass. The final product of this process has a mass number of $232 - 24 = 208$ and an atomic number of $90 - 12 + 4 = 82$, identifying it as $^{208}_{82}Pb$.
(b) Step 1: $^{232}_{90}Th \longrightarrow \, ^{228}_{88}Ra + \, ^{4}_{2}\alpha$
Step 2: $^{228}_{88}Ra \longrightarrow \, ^{228}_{89}Ac + \, ^{0}_{-1}\beta$
Step 3: $^{228}_{89}Ac \longrightarrow \, ^{228}_{90}Th + \, ^{0}_{-1}\beta$

23.4 (a) $^{0}_{+1}\beta$ (b) $^{41}_{19}K$ (c) $^{0}_{-1}\beta$ (d) $^{22}_{12}Mg$

23.5 (a) $^{32}_{14}Si \longrightarrow \, ^{32}_{15}P + \, ^{0}_{-1}\beta$
(b) $^{45}_{22}Ti \longrightarrow \, ^{45}_{21}Sc + \, ^{0}_{+1}\beta$ or $^{45}_{22}Ti + \, ^{0}_{-1}e \longrightarrow \, ^{45}_{21}Sc$
(c) $^{239}_{94}Pu \longrightarrow \alpha + \, ^{45}_{22}U$
(d) $^{42}_{19}K \longrightarrow \, ^{42}_{20}Ca + \, ^{0}_{-1}\beta$

23.6 $\Delta m = 0.03438\ g/mol$
$\Delta E = (3.438 \times 10^{-5}\ kg/mol)(2.998 \times 10^8)^2$
$= 3.090 \times 10^{12}\ J/mol$ ($= 3.090 \times 10^9\ kJ/mol$)
$E_b = 5.150 \times 10^8\ kJ/mol$ nucleons

23.7 (a) 49.2 years is exactly 4 half-lives; quantity remaining $= 1.5\ mg(1/2)^4 = 0.094\ mg$
(b) 3 half-lives, 36.9 years
(c) 1% is between 6 half-lives, 73.8 years (1/64 remains), and 7 half-lives, 86.1 years (1/128 remains)

23.8 $\ln([A]/[A_o]) = -kt$
$\ln([3.18 \times 10^3]/[3.35 \times 10^3]) = -k(2.00\ d)$
$k = 0.0260\ d^{-1}$
$t_{1/2} = 0.693/k = 0.693/(0.0260\ d^{-1}) = 26.7\ d$

23.9 $k = 0.693/t_{1/2} = 0.693/200\ y = 3.47 \times 10^{-3}\ y^{-1}$
$\ln([A]/[A_o]) = -kt$
$\ln([3.00 \times 10^3]/[6.50 \times 10^{12}]) = -(3.47 \times 10^{-3}\ y^{-1})t$
$\ln(4.62 \times 10^{-10}) = -(3.47 \times 10^{-3}\ y^{-1})t$
$t = 6190\ y$

23.10 $\ln([A]/[A_o]) = -kt$
$\ln([9.32]/[13.4]) = -(1.21 \times 10^{-4}\ y^{-1})t$
$t = 3.00 \times 10^3\ y$

23.11 $3000\ dpm/x = 1200\ dpm/\ 60.0\ mg$
$x = 150\ mg$

O Answers to Selected Study Questions

CHAPTER 1

1.1 (a) C, carbon
(b) K, potassium
(c) Cl, chlorine
(d) P, phosphorus
(e) Mg, magnesium
(f) Ni, nickel

1.3 (a) Ba, barium
(b) Ti, titanium
(c) Cr, chromium
(d) Pb, lead
(e) As, arsenic
(f) Zn, zinc

1.5 (a) Na (element) and NaCl (compound)
(b) Sugar (compound) and carbon (element)
(c) Gold (element) and gold chloride (compound)

1.7 (a) Physical property
(b) Chemical property
(c) Chemical property
(d) Physical property
(e) Physical property
(f) Physical property

1.9 (a) Physical (colorless liquid) and chemical (burns in air)
(b) Physical (shiny metal, orange liquid) and chemical (reacts with bromine)

1.11 (a) Qualitative: blue-green color, solid physical state
Quantitative: density = 2.65 g/cm^3 and mass = 2.5 g
(b) Density, physical state, and color are intensive properties, whereas mass is an extensive property.
(c) Volume = 0.94 cm^3

1.13 Observations c, e, and f are chemical properties

1.15 calcium, Ca; fluorine, F

The crystals are cubic in shape because the atoms are arranged in cubic structures.

1.17 The macroscopic view is the photograph of NaCl, and the particulate view is the drawing of the ions in a cubic arrangement. The structure of the compound at the particulate level determines the properties that are observed at the macroscopic level.

1.19 The density of the plastic is less than that of CCl_4, so the plastic will float on the liquid CCl_4. Aluminum is more dense than CCl_4, so aluminum will sink when placed in CCl_4.

1.21 The three liquids will form three separate layers with hexane on the top, water in the middle, and perfluorohexane on the bottom. The HDPE will float at the interface of the hexane and water layers. The PVC will float at the interface of the water and perfluorohexane layers. The Teflon will sink to the bottom of the cylinder.

1.23 HDPE will float in ethylene glycol, water, acetic acid, and glycerol.

1.25

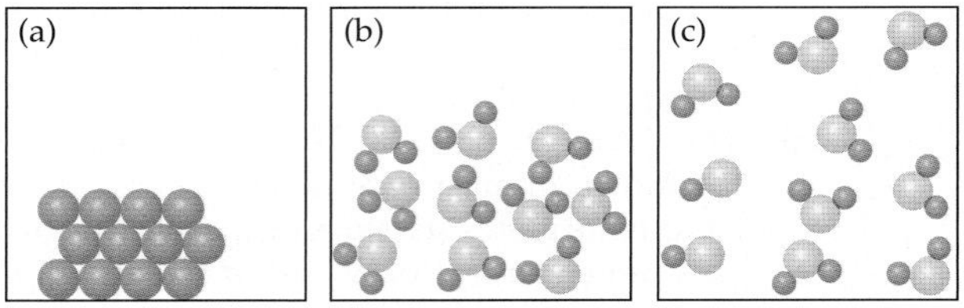

1.27 The sample's density and melting point could be compared to those of pure silver.

1.29 If too much sugar is excreted, the density of the urine would be higher than normal. If too much water is excreted, the density would be lower than normal.

1.31 (a) Solid potassium metal reacts with liquid water to produce gaseous hydrogen and a homogeneous mixture (solution) of potassium hydroxide in liquid water.
(b) The reaction is a chemical change.
(c) The reactants are potassium and water. The products are hydrogen gas and a water (aqueous) solution of potassium hydroxide. Heat and light are also evolved.
(d) Among the qualitative observations are (i) the reaction is violent, and (ii) heat and light (a purple flame) are produced.

1.33 (a) The water could be evaporated by heating the solution, leaving the salt behind.
(b) Use a magnet to attract the iron away from lead, which is not magnetic.
(c) Mixing the solids with water will dissolve only the sugar. Filtration would separate the solid sulfur from the sugar solution. Finally, the sugar could be separated from the water by evaporating the water.

1.35 Separate the iron from a weighed sample of cereal by passing a magnet through a mixture of the cereal and water after the flakes have become a gooey paste. Remove the iron flakes from the magnet and weigh them to determine the mass of iron in this mass of cereal.

1.37 Physical change

LET'S REVIEW: THE TOOLS OF QUANTITATIVE CHEMISTRY

1 298 K

3 (a) 289 K
(b) 97 °C
(c) 310 K (3.1×10^2 K)

5 42,195 m; 26.219 miles

7 5.3 cm^2; 5.3×10^{-4} m^2

9 250. cm^3; 0.250 L, 2.50×10^{-4} m^3; 0.250 dm^3

11 2.52×10^3 g

13 555 g

15 Choice (c), zinc

17 (a) Method A with all data included:
average = 2.4 g/cm^3

Method B with all data included:
average = 3.480 g/cm^3

For B, the 5.811 g/cm^3 data point can be excluded because it is more than twice as large as all other points for case Method B. Using only the first three points, average = 2.703 g/cm^3

(b) Method A: error = 0.3 g/cm^3 or about 10%

Method B: error = 0.001 g/cm^3 or about 0.04%

(c) Method A: standard deviation = 0.2 g/cm^3

Method B (including all data points):
st. dev. = 1.554 g/cm^3

Method B (excluding the 5.811 g/cm^3 data point):
st. dev. = 0.002 g/cm^3

(d) Method B's average value is both more precise and more accurate so long as the 5.811 g/cm^3 data point is excluded.

19 (a) 5.4×10^{-2} g, two significant figures
(b) 5.462×10^3 g, four significant figures
(c) 7.92×10^{-4} g, three significant figures
(d) 1.6×10^3 mL, two significant figures

21 (a) 9.44×10^{-3}
(b) 5694
(c) 11.9
(d) 0.122

23

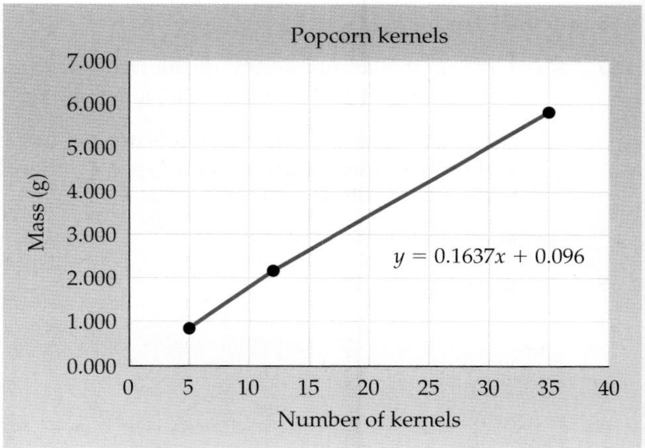

Slope: 0.1637 g/kernel

The slope represents the average mass of a popcorn kernel.

Mass of 20 popcorn kernels = 3.370 g

There are 127 kernels in a sample with a mass of 20.88 g.

25 (a) $y = -4.00x + 20.00$
(b) $y = -4.00$

27 $C = 0.0823$

29 $T = 295$

31 0.197 nm; 197 pm

33 (a) 7.5×10^{-6} m; (b) 7.5×10^3 nm; (c) 7.5×10^6 pm

35 50. mg procaine hydrochloride

37 The volume of the marbles is 99 mL − 61 mL = 38 mL. This yields a density of 2.5 g/cm^3.

39 (a) 0.178 nm^3; 1.78×10^{-22} cm^3
(b) 3.86×10^{-22} g
(c) 9.68×10^{-23} g

41 Your normal body temperature (about 98.6 °F) is 37 °C. As this is higher than gallium's melting point, the metal will melt in your hand.

43 (a) 15%
(b) 3.63×10^3 kernels

45 8.0×10^4 kg of sodium fluoride per year

47 245 g sulfuric acid

49 (a) 272 mL ice
(b) The ice cannot be contained in the can.

51 7.99 g/cm^3

53 (a) 8.7 g/cm^3
(b) The metal is probably cadmium, but the calculated density is close to that of cobalt, nickel, and copper. Further testing should be done on the metal.

55 0.0927 cm

57 (a) 1.143×10^{21} atoms; 54.9% of the lattic is filled with atoms; 24% of the lattice is open space.

Atoms are spheres. When spheres are packed together, they touch only at certain points, therefore leaving spaces in the structure.

(b) Four atoms

59 Al, aluminum

61

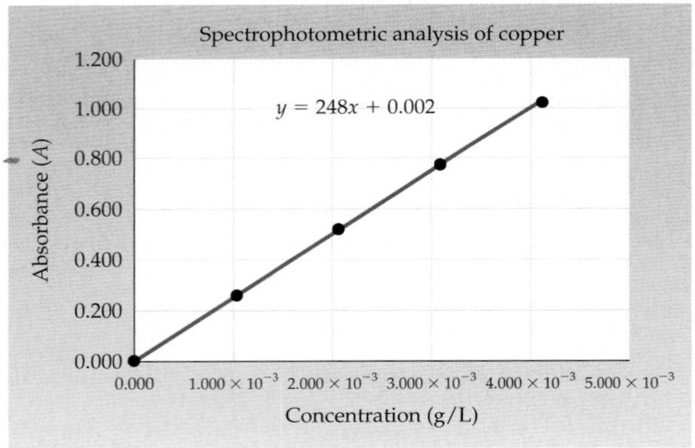

When absorbance = 0.635, concentration = 2.55×10^{-3} g/L = 2.55×10^{-3} mg/mL

63

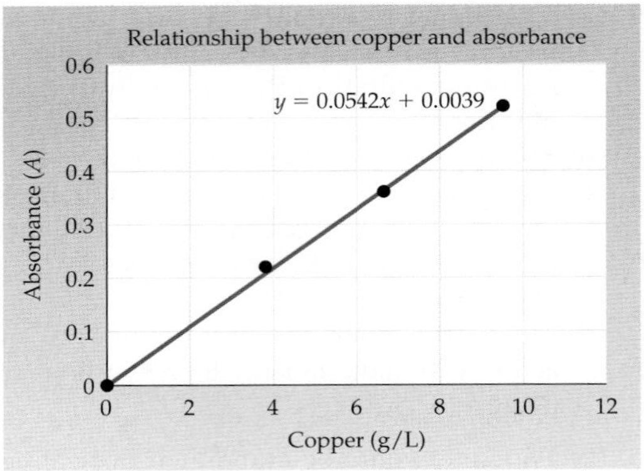

Slope = 0.054; y-intercept = 0.004; the absorbance for 5.00 g/L of copper is 0.27

CHAPTER 2

2.1 Atoms contain the following fundamental particles: protons (+1 charge), neutrons (zero charge), and electrons (−1 charge). Protons and neutrons are in the nucleus of an atom. Electrons are the least massive of the three particles.

2.3 (a) $^{27}_{12}Mg$
(b) $^{48}_{22}Ti$
(c) $^{62}_{30}Zn$

2.5

Element	Electrons	Protons	Neutrons
^{24}Mg	12	12	12
^{119}Sn	50	50	69
^{232}Th	90	90	142
^{13}C	6	6	7
^{63}Cu	29	29	34
^{205}Bi	83	83	122

2.7 $^{57}_{27}Co$, $^{58}_{27}Co$, $^{60}_{27}Co$

2.9 ^{205}Tl is more abundant than ^{203}Tl. The atomic mass of thallium is closer to 205 than to 203.

2.11 (0.0750)(6.015121) + (0.9250)(7.016003) = 6.94

2.13 (c), About 50%. Actual percent ^{107}Ag = 51.839%

2.15 ^{69}Ga, 60.12%; ^{71}Ga, 39.88%

2.17

	Symbol	Atomic No.	Atomic Mass	Group	Period	
Titanium	Ti	22	47.867	4B(IUPAC 4)	4	Metal
Thallium	Tl	81	204.3833	3A(IUPAC 13)	6	Metal

2.19 Eight elements: periods 2 and 3. 18 elements: periods 4 and 5. 32 elements: period 6.

2.21 (a) Nonmetals: C, Cl
(b) Main group elements: C, Ca, Cl, Cs
(c) Lanthanides: Ce
(d) Transition elements: Cr, Co, Cd, Ce, Cm, Cu, Cf
(e) Actinides: Cm, Cf
(f) Gases: Cl

2.23 Metals: Na, Ni, Np
Metalloids: None in this list
Nonmetals: N, Ne

2.25 Molecular Formula: H_2SO_4.
Structural Formula:

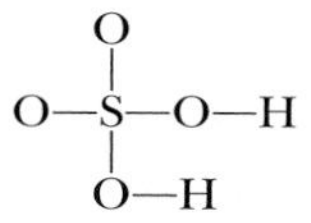

The structure is not flat. The O atoms are arranged around the sulfur at the corners of a tetrahedron. The hydrogen atoms are connected to two of the oxygen atoms.

2.27 (a) Mg^{2+}
(b) Zn^{2+}
(c) Ni^{2+}
(d) Ga^{3+}

2.29 (a) Ba^{2+}
(b) Ti^{4+}
(c) PO_4^{3-}
(d) HCO_3^-
(e) S^{2-}
(f) ClO_4^-
(g) Co^{2+}
(h) SO_4^{2-}

2.31 K loses one electron per atom to form a K^+ ion. It has the same number of electrons as an Ar atom.

2.33 Ba^{2+} and Br^- ions. The compound's formula is $BaBr_2$.

2.35 (a) Two K^+ ions and one S^{2-} ion
(b) One Co^{2+} ion and one SO_4^{2-} ion
(c) One K^+ ion and one MnO_4^- ion
(d) Three NH_4^+ ions and one PO_4^{3-} ion
(e) One Ca^{2+} ion and two ClO^- ions
(f) One Na^+ ion and one $CH_3CO_2^-$ ion

2.37 Co^{2+} gives CoO and Co^{3+} gives Co_2O_3

2.39 (a) $AlCl_2$ should be $AlCl_3$ (based on an Al^{3+} ion and three Cl^- ions).
(b) KF_2 should be KF (based on a K^+ ion and an F^- ion).
(c) Ga_2O_3 is correct.
(d) MgS is correct.

2.41 (a) potassium sulfide
(b) cobalt(II) sulfate
(c) ammonium phosphate
(d) calcium hypochlorite

2.43 (a) $(NH_4)_2CO_3$
(b) CaI_2
(c) $CuBr_2$
(d) $AlPO_4$
(e) $AgCH_3CO_2$

2.45 Compounds with Na^+: Na_2CO_3 (sodium carbonate) and NaI (sodium iodide). Compounds with Ba^{2+}: $BaCO_3$ (barium carbonate) and BaI_2 (barium iodide).

2.47 The force of attraction is stronger in NaF than in NaI because the distance between ion centers is smaller in NaF (235 pm) than in NaI (322 pm).

2.49 (a) nitrogen trifluoride
(b) hydrogen iodide
(c) boron triiodide
(d) phosphorus pentafluoride

2.51 (a) SCl_2
(b) N_2O_5
(c) $SiCl_4$
(d) B_2O_3

2.53 (a) 67 g Al
(b) 0.0698 g Fe
(c) 0.60 g Ca
(d) 1.32×10^4 g Ne

2.55 (a) 1.9998 mol Cu
(b) 0.0017 mol Li
(c) 2.1×10^{-5} mol Am
(d) 0.250 mol Al

2.57 Of these elements, He has the smallest molar mass, and Fe has the largest molar mass. Therefore, 1.0 g of He has the largest number of atoms in these samples, and 1.0 g of Fe has the smallest number of atoms.

2.59 (a) 159.7 g/mol
(b) 117.2 g/mol
(c) 176.1 g/mol

2.61 (a) 290.8 g/mol
(b) 249.7 g/mol

2.63 (a) 1.53 g
(b) 4.60 g
(c) 4.60 g
(d) 1.48 g

2.65 Amount of SO_3 = 12.5 mol
Number of molecules = 7.52×10^{24} molecules
Number of S atoms = 7.52×10^{24} atoms
Number of O atoms = 2.26×10^{25} atoms

2.67 (a) 86.60% Pb and 13.40% S
(b) 81.71% C and 18.29% H
(c) 79.96% C, 9.394% H, and 10.65% O

2.69 66.46% copper in CuS. 15.0 g of CuS is needed to obtain 10.0 g of Cu.

2.71 $C_4H_6O_4$

2.73 (a) CH, 26.0 g/mol; C_2H_2
(b) CHO, 116.1 g/mol; $C_4H_4O_4$
(c) CH_2, 112.2 g/mol, C_8H_{16}

2.75 Empirical formula, CH; molecular formula, C_2H_2

2.77 Empirical formula, C_3H_4; molecular formula, C_9H_{12}

2.79 Empirical and molecular formulas are both $C_8H_8O_3$

2.81 XeF_2

2.83 ZnI_2

2.85

Symbol	^{58}Ni	^{33}S	^{20}Ne	^{55}Mn
Protons	28	16	10	25
Neutrons	30	17	10	30
Electrons	28	16	10	25
Name	nickel	sulfur	neon	manganese

2.87

S	N
B	I

2.89 (a) 1.0552×10^{-22} g for 1 Cu atom
(b) 6.286×10^{-22} dollars for 1 Cu atom

2.91 (a) strontium
(b) zirconium
(c) carbon
(d) arsenic
(e) iodine
(f) magnesium
(g) krypton
(h) sulfur
(i) germanium or arsenic

2.93 (a) 0.25 mol U
(b) 0.50 mol Na
(c) 10 atoms of Fe

2.95 40.2 g H_2 (b) < 103 g C (c) < 182 g Al
(f) < 210 g Si (d) < 212 g Na (e) < 351 g Fe
(a) < 650 g $Cl_2(g)$

2.97 (a) Atomic mass of O = 15.873 u; Avogadro's number = 5.9802×10^{23} particles per mole
(b) Atomic mass of H = 1.00798 u; Avogadro's number = 6.0279×10^{23} particles per mole

2.99 $(NH_4)_2CO_3$, $(NH_4)_2SO_4$, $NiCO_3$, $NiSO_4$

2.101 All of these compounds have one atom of some element plus three Cl atoms. The highest mass percent of chlorine will occur in the compound having the lightest central element. Here, that element is B, so BCl_3 should have the highest mass percent of Cl (90.77%).

2.103 The molar mass of adenine ($C_5H_5N_5$) is 135.13 g/mol. 3.0×10^{23} molecules represents 67 g. Thus, 3.0×10^{23} molecules of adenine has a larger mass than 40.0 g of the compound.

2.105 1.7×10^{21} molecules of water

2.107 245.75 g/mol. Mass percent: 25.86% Cu, 22.80% N, 5.742% H, 13.05% S, and 32.55% O. In 10.5 g of compound there are 2.72 g Cu and 0.770 g H_2O.

2.109 Empirical formula of malic acid: $C_4H_6O_5$

2.111 $Fe_2(CO)_9$

2.113 (a) $C_7H_5NO_3S$

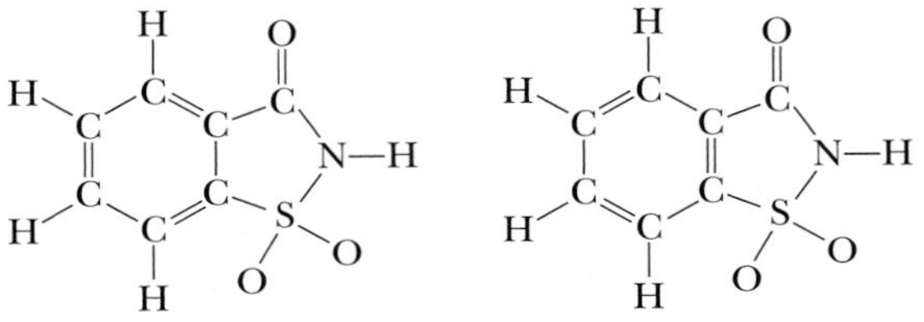

(b) 6.82×10^{-4} mol saccarin
(c) 21.9 mg S

2.115 (a) NaClO, ionic
(b) BI_3
(c) $Al(ClO_4)_3$, ionic
(d) $Ca(CH_3CO_2)_2$, ionic
(e) $KMnO_4$, ionic
(f) $(NH_4)_2SO_3$, ionic
(g) KH_2PO_4, ionic
(h) S_2Cl_2
(i) ClF_3
(j) PF_3

2.117 (a) Empirical formula = molecular formula = CF_2O_2
(b) Empirical formula = C_5H_4; molecular formula = $C_{10}H_8$

2.119 Empirical formula and molecular formula = $C_5H_{14}N_2$

2.121 $C_9H_7MnO_3$

2.123 68.42% Cr; 1.2×10^3 kg Cr_2O_3

2.125 Empirical formula = ICl_3; molecular formula = I_2Cl_6

2.127 7.35 kg of iron

2.129 (d) Na_2MoO_4

2.131 5.52×10^{-4} mol $C_{21}H_{15}Bi_3O_{12}$; 0.346 g Bi

2.133 The molar mass of the compound is 154 g/mol. The unknown element is carbon.

2.135 $n = 19$.

2.137 (a) 2.3×10^{14} g/cm^3
(b) 3.34×10^{-3} g/cm^3
(c) The nucleus is much more dense than the space occupied by the electrons.

2.139 (a) 0.0130 mol Ni
(b) NiF_2
(c) nickel(II) fluoride

2.141 Formula is $MgSO_4 \cdot 7\ H_2O$

2.143 Volume = 3.0 cm^3; length of side = 1.4 cm

2.145 1.0028×10^{23} atoms C. If the accuracy is ±0.0001 g, the maximum mass could be 2.0001 g, which also represents 1.0028×10^{23} atoms C.

2.147 Choice c. The calculated mole ratio is 0.78 mol H_2O per mol $CaCl_2$. The student should heat the crucible again and then reweigh it. More water might be driven off.

2.149 Required data: density of iron, molar mass of iron, Avogadro's number.

$$1.00\ cm^3\left(\frac{7.87\ g}{1\ cm^3}\right)\left(\frac{1\ mol}{55.85\ g}\right)\left(\frac{6.02 \times 10^{23}\ atoms}{1\ mol}\right) = 8.49 \times 10^{22}\ atoms\ Fe$$

2.151 Barium would be more reactive than calcium, so a more vigorous evolution of hydrogen should occur. Reactivity increases on descending the periodic table, at least for Groups 1A and 2A.

2.153 When words are written with the pink, hydrated compound, the words are not visible. However, when heated, the hydrated salt loses water to form anhydrous $CoCl_2$, which is deep blue. The words are then visible.

CHAPTER 3

3.1 $C_5H_{12}(\ell) + 8\ O_2(g) \rightarrow 5\ CO_2(g) + 6\ H_2O(g)$

3.3 (a) $4\ Cr(s) + 3\ O_2(g) \rightarrow 2\ Cr_2O_3(s)$
(b) $Cu_2S(s) + O_2(g) \rightarrow 2\ Cu(s) + SO_2(g)$
(c) $C_6H_5CH_3(\ell) + 9\ O_2(g) \rightarrow 4\ H_2O(\ell) + 7\ CO_2(g)$

3.5 (a) $Fe_2O_3(s) + 3\ Mg(s) \rightarrow 3\ MgO(s) + 2\ Fe(s)$
Reactants = iron(III) oxide, magnesium
Products = magnesium oxide, iron
(b) $AlCl_3(s) + 3\ NaOH(aq) \rightarrow Al(OH)_3(s) + 3\ NaCl(aq)$
Reactants = aluminum chloride, sodium hydroxide
Products = aluminum hydroxide, sodium chloride
(c) $2\ NaNO_3(s) + H_2SO_4(\ell) \rightarrow Na_2SO_4(s) + 2\ HNO_3(\ell)$
Reactants = sodium nitrate, sulfuric acid
Products = sodium sulfate, nitric acid
(d) $NiCO_3(s) + 2\ HNO_3(aq) \rightarrow Ni(NO_3)_2(aq) + CO_2(g) + H_2O(\ell)$
Reactants = nickel(II) carbonate, nitric acid
Products = nickel(II) nitrate, carbon dioxide, water

3.7 The reaction involving HCl is more product-favored at equilibrium.

3.9 Electrolytes are compounds whose aqueous solutions conduct electricity. Given an aqueous solution containing a strong electrolyte and another aqueous solution containing a weak electrolyte at the same concentration, the solution containing the strong electrolyte (such as NaCl) will conduct electricity much better than will be the one containing the weak electrolyte (such as acetic acid).

3.11 (a) $CuCl_2$
(b) $AgNO_3$
(c) All are water-soluble

3.13 (a) K^+ and OH^- ions
(b) K^+ and SO_4^{2-} ions
(c) Li^+ and NO_3^- ions
(d) NH_4^+ and SO_4^{2-} ions

3.15 (a) Soluble, Na^+ and CO_3^{2-} ions
(b) Soluble, Cu^{2+} and SO_4^{2-} ions
(c) Insoluble
(d) Soluble, Ba^{2+} and Br^- ions

3.17 $CdCl_2(aq) + 2\ NaOH(aq) \rightarrow Cd(OH)_2(s) + 2\ NaCl(aq)$

$Cd^{2+}(aq) + 2\ OH^-(aq) \rightarrow Cd(OH)_2(s)$

3.19 (a) $NiCl_2(aq) + (NH_4)_2S(aq) \rightarrow NiS(s) + 2\ NH_4Cl(aq)$

$Ni^{2+}(aq) + S^{2-}(aq) \rightarrow NiS(s)$

(b) $3\ Mn(NO_3)_2(aq) + 2\ Na_3PO_4(aq) \rightarrow Mn_3(PO_4)_2(s) + 6\ NaNO_3(aq)$

$3\ Mn^{2+}(aq) + 2\ PO_4^{3-}(aq) \rightarrow Mn_3(PO_4)_2(s)$

3.21 $HNO_3(aq) + H_2O(\ell) \rightarrow H_3O^+(aq) + NO_3^-(aq)$

3.23 $H_2C_2O_4(aq) + H_2O(\ell) \rightarrow H_3O^+(aq) + HC_2O_4^-(aq)$

$HC_2O_4^-(aq) + H_2O(\ell) \rightarrow H_3O^+(aq) + C_2O_4^{2-}(aq)$

3.25 $MgO(s) + H_2O(\ell) \rightarrow Mg(OH)_2(s)$

3.27 (a) Acetic acid reacts with magnesium hydroxide to give magnesium acetate and water.

$2\ CH_3CO_2H(aq) + Mg(OH)_2(s) \rightarrow Mg(CH_3CO_2)_2(aq) + 2\ H_2O(\ell)$

Brønsted acid: acetic acid; Brønsted base: magnesium hydroxide
(b) Perchloric acid reacts with ammonia to give ammonium perchlorate

$HClO_4(aq) + NH_3(aq) \rightarrow NH_4ClO_4(aq)$

Brønsted acid: perchloric acid; Brønsted base: ammonia

3.29 $Ba(OH)_2(aq) + 2\ HNO_3(aq) \rightarrow Ba(NO_3)_2(aq) + 2\ H_2O(\ell)$

3.31 Strong Brønsted acid examples: hydrochloric acid, nitric acid

Strong Brønsted base example: sodium hydroxide

3.33 (a) $(NH_4)_2CO_3(aq) + Cu(NO_3)_2(aq) \rightarrow CuCO_3(s) + 2\ NH_4NO_3(aq)$

$CO_3^{2-}(aq) + Cu^{2+}(aq) \rightarrow CuCO_3(s)$

(b) $Pb(OH)_2(s) + 2\ HCl(aq) \rightarrow PbCl_2(s) + 2\ H_2O(\ell)$

$Pb(OH)_2(s) + 2\ H_3O^+(aq) + 2\ Cl^-(aq) \rightarrow PbCl_2(s) + 4\ H_2O(\ell)$

(c) $BaCO_3(s) + 2\ HCl(aq) \rightarrow BaCl_2(aq) + H_2O(\ell) + CO_2(g)$

$BaCO_3(s) + 2\ H_3O^+(aq) \rightarrow Ba^{2+}(aq) + 3\ H_2O(\ell) + CO_2(g)$

(d) $2\ CH_3CO_2H(aq) + Ni(OH)_2(s) \rightarrow Ni(CH_3CO_2)_2(aq) + 2\ H_2O(\ell)$

$2\ CH_3CO_2H(aq) + Ni(OH)_2(s) \rightarrow Ni^{2+}(aq) + 2\ CH_3CO_2^-(aq) + 2\ H_2O(\ell)$

3.35 (a) $AgNO_3(aq) + KI(aq) \rightarrow AgI(s) + KNO_3(aq)$

$Ag^+(aq) + I^-(aq) \rightarrow AgI(s)$

(b) $Ba(OH)_2(aq) + 2\ HNO_3(aq) \rightarrow Ba(NO_3)_2(aq) + 2\ H_2O(\ell)$

$OH^-(aq) + H_3O^+(aq) \rightarrow 2\ H_2O(\ell)$

(c) $2\ Na_3PO_4(aq) + 3\ Ni(NO_3)_2(aq) \rightarrow Ni_3(PO_4)_2(s) + 6\ NaNO_3(aq)$

$2\ PO_4^{3-}(aq) + 3\ Ni^{2+}(aq) \rightarrow Ni_3(PO_4)_2(s)$

3.37 $FeCO_3(s) + 2\ HNO_3(aq) \rightarrow Fe(NO_3)_2(aq) + CO_2(g) + H_2O(\ell)$

Iron(II) carbonate reacts with nitric acid to give iron(II) nitrate, carbon dioxide, and water.

3.39 $(NH_4)_2S(aq) + 2\ HBr(aq) \rightarrow 2\ NH_4Br(aq) + H_2S(g)$

Ammonium sulfide reacts with hydrobromic acid to give ammonium bromide and hydrogen sulfide.

3.41 (a) Br = +5 and O = −2
(b) C = +3 each and O = −2
(c) F = −1
(d) Ca = +2 and H = −1
(e) H = +1, Si = +4, and O = −2
(f) H = +1, S = +6, and O = −2

3.43 (a) Oxidation–reduction
Zn is oxidized from 0 to +2, and N in NO_3^- is reduced from +5 to +4 in NO_2.
(b) Acid–base reaction
(c) Oxidation–reduction

Calcium is oxidized from 0 to +2 in $Ca(OH)_2$, and H is reduced from +1 in H_2O to 0 in H_2.

3.45 (a) O_2 is the oxidizing agent (as it always is) and so C_2H_4 is the reducing agent. In this process, C_2H_4 is oxidized, and O_2 is reduced.

(b) Si is oxidized from 0 in Si to +4 in $SiCl_4$. Cl_2 is reduced from 0 in Cl_2 to −1 in Cl^-. Si is the reducing agent, and Cl_2 is the oxidizing agent.

3.47 (a) Acid–base

$Ba(OH)_2(aq) + 2\ HCl(aq) \rightarrow BaCl_2(aq) + 2\ H_2O(\ell)$

(b) Gas-forming

$2\ HNO_3(aq) + CoCO_3(s) \rightarrow Co(NO_3)_2(aq) + H_2O(\ell) + CO_2(g)$

(c) Precipitation

$2\ Na_3PO_4(aq) + 3\ Cu(NO_3)_2(aq) \rightarrow Cu_3(PO_4)_2(s) + 6\ NaNO_3(aq)$

3.49 a) Precipitation

$MnCl_2(aq) + Na_2S(aq) \rightarrow MnS(s) + 2\ NaCl(aq)$

$Mn^{2+}(aq) + S^{2-}(aq) \rightarrow MnS(s)$

(b) Precipitation

$K_2CO_3(aq) + ZnCl_2(aq) \rightarrow ZnCO_3(s) + 2\ KCl(aq)$

$CO_3^{2-}(aq) + Zn^{2+}(aq) \rightarrow ZnCO_3(s)$

3.51 (a) $CuCl_2(aq) + H_2S(aq) \rightarrow CuS(s) + 2\ HCl(aq)$
precipitation

(b) $H_3PO_4(aq) + 3\ KOH(aq) \rightarrow 3\ H_2O(\ell) + K_3PO_4(aq)$

acid–base

(c) $Ca(s) + 2\ HBr(aq) \rightarrow H_2(g) + CaBr_2(aq)$

oxidation–reduction and gas-forming

(d) $MgCl_2(aq) + 2\ H_2O(\ell) \rightarrow Mg(OH)_2(s) + 2\ HCl(aq)$

precipitation

3.53 (a) $CO_2(g) + 2\ NH_3(g) \rightarrow NH_2CONH_2(s) + H_2O(\ell)$
(b) $UO_2(s) + 4\ HF(aq) \rightarrow UF_4(s) + 2\ H_2O(\ell)$

$UF_4(s) + F_2(g) \rightarrow UF_6(s)$

(c) $TiO_2(s) + 2\ Cl_2(g) + 2\ C(s) \rightarrow TiCl_4(\ell) + 2\ CO(g)$

$TiCl_4(\ell) + 2\ Mg(s) \rightarrow Ti(s) + 2\ MgCl_2(s)$

3.55 (a) NaBr, KBr, or other alkali metal bromides; Group 2A bromides; other metal bromides except AgBr, Hg_2Br_2, and $PbBr_2$
(b) $Al(OH)_3$ and transition metal hydroxides
(c) Alkaline earth carbonates ($CaCO_3$) or transition metal carbonates ($NiCO_3$)
(d) Metal nitrates are generally water-soluble [e.g., $NaNO_3$, $Ni(NO_3)_2$].
(e) CH_3CO_2H, other acids containing the $-CO_2H$ group

3.57 Water soluble: $Cu(NO_3)_2$, $CuCl_2$. Water-insoluble: $CuCO_3$, $Cu_3(PO_4)_2$

3.59 Spectator ion, NO_3^-. Acid–base reaction.

$2\ H_3O^+(aq) + Mg(OH)_2(s) \longrightarrow 4\ H_2O(\ell) + Mg^{2+}(aq)$

3.61 (a) Cl_2 is reduced (to Cl^-) and Br^- is oxidized (to Br_2).
(b) Cl_2 is the oxidizing agent and Br^- is the reducing agent.

3.63 (a) $MgCO_3(s) + 2\ H_3O^+(aq) \longrightarrow CO_2(g) + Mg^{2+}(aq) + 3\ H_2O(\ell)$

Chloride ion (Cl^-) is the spectator ion.

(b) Gas-forming reaction

3.65 (a) H_2O, NH_3, NH_4^+, and OH^- (and a trace of H_3O^+)

weak Brønsted base

(b) H_2O, CH_3CO_2H, $CH_3CO_2^-$, and H_3O^+ (and a trace of OH^-)

weak Brønsted acid

(c) H_2O, Na^+, and OH^- (and a trace of H_3O^+)

strong Brønsted base

(d) H_2O, H_3O^+, and Br^- (and a trace of OH^-)

strong Brønsted acid

3.67 (a) $K_2CO_3(aq) + 2\ HClO_4(aq) \longrightarrow 2\ KClO_4(aq) + CO_2(g) + H_2O(\ell)$

gas-forming

Potassium carbonate and perchloric acid react to form potassium perchlorate, carbon dioxide, and water

$CO_3^{2-}(aq) + 2\ H_3O^+(aq) \longrightarrow CO_2(g) + 3\ H_2O(\ell)$

(b) $FeCl_2(aq) + (NH_4)_2S(aq) \longrightarrow FeS(s) + 2\ NH_4Cl(aq)$

precipitation

Iron(II) chloride and ammonium sulfide react to form iron(II) sulfide and ammonium chloride

$Fe^{2+}(aq) + S^{2-}(aq) \longrightarrow FeS(s)$

(c) $Fe(NO_3)_2(aq) + Na_2CO_3(aq) \longrightarrow FeCO_3(s) + 2\ NaNO_3(aq)$

precipitation

Iron(II) nitrate and sodium carbonate react to form iron(II) carbonate and sodium nitrate

$Fe^{2+}(aq) + CO_3^{2-}(aq) \longrightarrow FeCO_3(s)$

(d) $3\ NaOH(aq) + FeCl_3(aq) \longrightarrow 3\ NaCl(aq) + Fe(OH)_3(s)$

precipitation

Sodium hydroxide and iron(III) chloride react to form sodium chloride and iron(III) hydroxide

$3\ OH^-(aq) + Fe^{3+}(aq) \longrightarrow Fe(OH)_3(s)$

3.69 (a) Reactants: Na(+1), I(−1), H(+1), S(+6), O(−2), Mn(+4)

Products: Na(+1), S(+6), O(−2), Mn(+2), I(0), H(+1)

(b) The oxidizing agent is MnO_2, and NaI is oxidized. The reducing agent is NaI, and MnO_2 is reduced.
(c) Based on the picture, the reaction is product-favored.
(d) Sodium iodide, sulfuric acid, and manganese(IV) oxide react to form sodium sulfate, manganese(II) sulfate, and water.

3.71 Among the reactions that could be used are the following:

$MgCO_3(s) + 2\ HCl(aq) \longrightarrow MgCl_2(aq) + CO_2(g) + H_2O(\ell)$

$MgS(s) + 2\ HCl(aq) \longrightarrow MgCl_2(aq) + H_2S(g)$

$MgSO_3(s) + 2\ HCl(aq) \longrightarrow MgCl_2(aq) + SO_2(g) + H_2O(\ell)$

In each case, the resulting solution could be evaporated to obtain the desired magnesium chloride.

3.73 The Ag^+ was reduced (to silver metal), and the glucose was oxidized (to $C_6H_{12}O_7$). The Ag^+ is the oxidizing agent, and the glucose is the reducing agent.

3.75 Weak electrolyte test: Compare the conductivity of a solution of lactic acid and that of an equal concentration of a strong acid. The conductivity of the lactic acid solution should be significantly less.

Reversible reaction: The fact that lactic acid is an electrolyte indicates that the reaction proceeds in the forward direction. To test whether the ionization is reversible, one could prepare a solution containing as much lactic acid as it will hold and then add a strong acid (to provide H_3O^+). If the reaction proceeds in the reverse direction, this will cause some lactic acid to precipitate.

3.77 (a) Several precipitation reactions are possible:

i. $BaCl_2(aq) + H_2SO_4(aq) \longrightarrow BaSO_4(s) + 2\ HCl(aq)$

ii. $BaCl_2(aq) + Na_2SO_4(aq) \longrightarrow BaSO_4(s) + 2\ NaCl(aq)$

iii. $Ba(OH)_2(aq) + H_2SO_4(aq) \longrightarrow BaSO_4(s) + 2\ H_2O(\ell)$

(b) Gas-forming reaction:

$BaCO_3(s) + H_2SO_4(aq) \longrightarrow BaSO_4(s) + CO_2(g) + H_2O(\ell)$

CHAPTER 4

4.1 4.5 mol O_2; 310 g Al_2O_3

4.3 22.7 g Br_2; 25.3 g Al_2Br_6

4.5 (a) CO_2, carbon dioxide, and H_2O, water
(b) $CH_4(g) + 2\ O_2(g) \rightarrow CO_2(g) + 2\ H_2O(\ell)$
(c) 102 g O_2
(d) 128 g products

4.7

Equation	2 PbS(s) +	3 O_2(g) →	2 PbO(s) +	2 SO_2(g)
Initial (mol)	2.5	3.8	0	0
Change (mol)	−2.5	−3/2(2.5) = −3.8	+2/2(2.5) = +2.5	+2/2(2.5) = +2.5
Final (mol)	0	0	2.5	2.5

The amounts table shows that 2.5 mol of PbS requires 3/2(2.5) = 3.8 mol of O_2 and produces 2.5 mol of PbO and 2.5 mol of SO_2.

4.9 (a) Balanced equation: $4\ Cr(s) + 3\ O_2(g) \rightarrow 2\ Cr_2O_3(s)$
(b) 0.175 g of Cr is equivalent to 0.00337 mol

Equation	4 Cr(s) +	3 O_2(g) →	2 Cr_2O_3(s)
Initial (mol)	0.00337	0.00252 mol	0
Change (mol)	−0.00337	−3/4(0.00337) = −0.00252	2/4(0.00337) = +0.00168
Final (mol)	0	0	0.00168

The 0.00168 mol Cr_2O_3 produced corresponds to 0.256 g Cr_2O_3.
(c) 0.081 g O_2

4.11 0.11 mol of Na_2SO_4 and 0.62 mol of C are mixed. Sodium sulfate is the limiting reactant. Therefore, 0.11 mol of Na_2S is formed, or 8.2 g.

4.13 F_2 is the limiting reactant.

4.15 (a) CH_4 is the limiting reactant.
(b) 375 g H_2
(c) Excess H_2O = 1390 g

4.17 (a) $2\ C_6H_{14}(\ell) + 19\ O_2(g) \rightarrow 12\ CO_2(g) + 14\ H_2O(g)$
(b) O_2 is the limiting reactant. Products are 187 g of CO_2 and 89.2 g of H_2O.
(c) 154 g of hexane remains
(d)

Equation	2 $C_6H_{14}(\ell)$ +	19 O_2(g) →	12 CO_2(g) +	14 H_2O(g)
Initial (mol)	2.49	6.72	0	0
Change (mol)	−0.707	−6.72	+4.24	+4.95
Final (mol)	1.78	0	4.24	4.95

4.19 (332 g/407 g)100% = 81.6%

4.21 (a) 14.3 g $Cu(NH_3)_4SO_4$
(b) 88.3% yield

4.23 91.9% hydrate

4.25 84.3% $CaCO_3$

4.27 1.467% Tl_2SO_4

4.29 Empirical formula = CH

4.31 Empirical formula = CH_2; molecular formula = C_5H_{10}

4.33 Empirical formula = CH_3O; molecular formula = $C_2H_6O_2$

4.35 $Ni(CO)_4$

4.37 $[Na_2CO_3]$ = 0.254 M; $[Na^+]$ = 0.508 M; $[CO_3^{2-}]$ = 0.254 M

4.39 0.494 g $KMnO_4$

4.41 5.08×10^3 mL

4.43 (a) 0.50 M NH_4^+ and 0.25 M SO_4^{2-}
(b) 0.246 M Na^+ and 0.123 M CO_3^{2-}
(c) 0.056 M H^+ and 0.056 M NO_3^-

4.45 A mass of 1.06 g of Na_2CO_3 is required. After weighing out this quantity of Na_2CO_3, transfer it to a 500.-mL volumetric flask. Rinse any solid from the neck of the flask while filling the flask with distilled water. Dissolve the solute in water. Add water until the bottom of the meniscus of the water is at the top of the scribed mark on the neck of the flask. Thoroughly mix the solution.

4.47 0.0750 M

4.49 Method (a) is correct. Method (b) gives an acid concentration of 0.15 M.

4.51 0.00340 M

4.53 $[H_3O^+] = 10^{-pH} = 4.0 \times 10^{-4}$ M; the solution is acidic.

4.55 HNO_3 is a strong acid, so $[H_3O^+]$ = 0.0013 M. pH = 2.89.

4.57

pH	$[H_3O^+]$	Acidic/Basic
(a) 1.00	0.10 M	Acidic
(b) 10.50	3.2×10^{-11} M	Basic
(c) 4.89	1.3×10^{-5} M	Acidic
(d) 7.64	2.3×10^{-8} M	Basic

4.59 268 mL

4.61 210 g NaOH and 190 g Cl_2

4.63 174 mL of $Na_2S_2O_3$

4.65 1.50×10^3 mL of $Pb(NO_3)_2$

4.67 44.6 mL

4.69 1.052 M HCl

4.71 104 g/mol

4.73 12.8% Fe

4.75

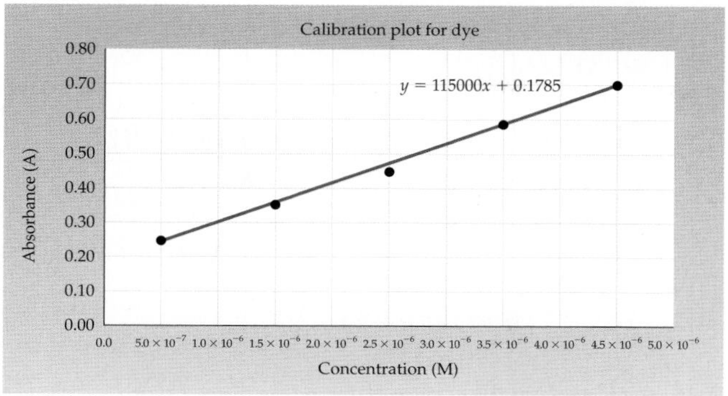

(a) slope = 1.2×10^5 M^{-1}; *y*-intercept = 0.18 M
(b) 3.0×10^{-6} M

4.77 (a) Products = $CO_2(g)$ and $H_2O(g)$
(b) $2\ C_6H_6(\ell) + 15\ O_2(g) \longrightarrow 12\ CO_2(g) + 6\ H_2O(g)$
(c) 49.28 g O_2
(d) 65.32 g products (= sum of C_6H_6 mass and O_2 mass)

4.79 0.28 g arginine, 0.21 g ornithine

4.81 (a) titanium(IV) chloride, water, titanium(IV) oxide, hydrogen chloride
(b) 4.60 g H_2O
(c) 10.2 TiO_2, 18.6 g HCl

4.83 8.33 g NaN_3

4.85 Mass percent saccharin = 75.92%

4.87 SiH_4

4.89 C_3H_2O

4.91 1.85 kg H_2SO_4

4.93 The calculated molar mass of the metal is 1.2×10^2 g/mol. The metal is probably tin (118.67 g/mol).

4.95 479 kg Cl_2

4.97 66.5 kg CaO

4.99 1.29 g C_4H_8 (45.1%) and 1.57 g C_4H_{10} (54.9%)

4.101 62.2% Cu_2S and 26.8% CuS

4.103 (a) $MgCO_3(s) + 2\ H_3O^+(aq) \longrightarrow CO_2(g) + Mg^{2+}(aq) + 2\ H_2O(\ell)$
(b) Gas-forming reaction
(c) 0.15 g

4.105 15.0 g of $NaHCO_3$ require 1190 mL of 0.15 M acetic acid. Therefore, acetic acid is the limiting reactant. (Conversely, 125 mL of 0.15 M acetic acid requires only 1.58 g of $NaHCO_3$.) 1.54 g of $NaCH_3CO_2$ produced.

4.107 3.13 g $Na_2S_2O_3$, 96.8%

4.109 (a) pH = 0.979
(b) $[H_3O^+]$ = 0.0028 M; the solution is acidic.
(c) $[H_3O^+] = 2.1 \times 10^{-10}$ M; the solution is basic.
(d) The new solution's concentration is 0.102 M HCl; the pH = 0.990

4.111 The concentration of hydrochloric acid is 2.92 M; the pH is −0.465

4.113 1.56 g of $CaCO_3$ required; 1.00 g $CaCO_3$ remain; 1.73 g $CaCl_2$ produced.

4.115 Volume of water in the pool = 7.6×10^4 L

4.117 (a) Au, gold, has been oxidized and is the reducing agent.

O_2, oxygen, has been reduced and is the oxidizing agent.

(b) 26 L NaCN solution

4.119 The concentration of Na_2CO_3 in the first solution prepared is 0.0275 M, in the second solution prepared the concentration of Na_2CO_3 is 0.00110 M.

4.121 (a) First reaction: oxidizing agent = Cu^{2+} and reducing agent = I^-

Second reaction: oxidizing agent = I_3^- and reducing agent = $S_2O_3^{2-}$

(b) 67.3% copper

4.123 x = 6; $Co(NH_3)_6Cl_3$.

4.125 11.48% 2,4-D

4.127 3.3 mol H_2O/mol $CaCl_2$

4.129 (a) Slope = 2.06×10^5; 0.024
(b) 1.20×10^{-4} g/L
(c) 0.413 mg PO_4^{3-}

4.131 The total mass of the beakers and products after reaction is equal to the total mass before the reaction (161.170 g) because no gases were produced in the reaction and there is conservation of mass in chemical reactions.

4.133 The balanced chemical equation indicates that the stoichiometric ratio of HCl to Zn is 2 mol HCl/1 mol Zn. In each reaction, there is 0.100 mol of HCl present. In reaction 1, there is 0.107 mol of Zn present. This gives a 0.93 mole HCl/mol Zn ratio, indicating that HCl is the limiting reactant. In reaction 2, there is 0.050 mol of Zn, giving a 2.0 mol HCl/mol Zn ratio. This indicates that the two reactants are present in exactly the correct stoichiometric ratio. In reaction 3, there is 0.020 mol of Zn, giving a 5.0 mol HCl/mol Zn ratio. This indicates that the HCl is present in excess and that the zinc is the limiting reactant.

4.135 If both students base their calculations on the amount of HCl solution pipeted into the flask (20 mL), then the second student's result will be (e), the same as the first student's. However, if the HCl concentration is calculated using the diluted solution volume, student 1 will use a volume of 40 mL, and student 2 will use a volume of 80 mL in the calculation. The second student's result will be (c), half that of the first student's.

4.137 150 mg/dL. Person is intoxicated.

CHAPTER 5

5.1 Mechanical energy is used to move the lever, which in turn moves gears. The device produces electrical energy and radiant energy.

5.3 5.0×10^6 J

5.5 170 kcal is equivalent to 710 kJ, considerably greater than 280 kJ.

5.7 0.140 J/g · K

5.9 2.44 kJ

5.11 32.8 °C

5.13 20.7 °C

5.15 47.8 °C

5.17 0.40 J/g · K

5.19 330 kJ

5.21 49.3 kJ

5.23 273 J

5.25 9.97×10^5 J

5.27 Reaction is exothermic because $\Delta_r H°$ is negative. The heat evolved is 2.38 kJ.

5.29 3.3×10^4 kJ

5.31 $\Delta H = -56$ kJ/mol CsOH

5.33 0.52 J/g · K

5.35 $\Delta_r H = +23$ kJ/mol-rxn

5.37 297 kJ/mol SO_2

5.39 3.09×10^3 kJ/mol $C_6H_5CO_2H$

5.41 0.236 J/g · K

5.43 (a) $\Delta_r H° = -126$ kJ/mol-rxn
(b)

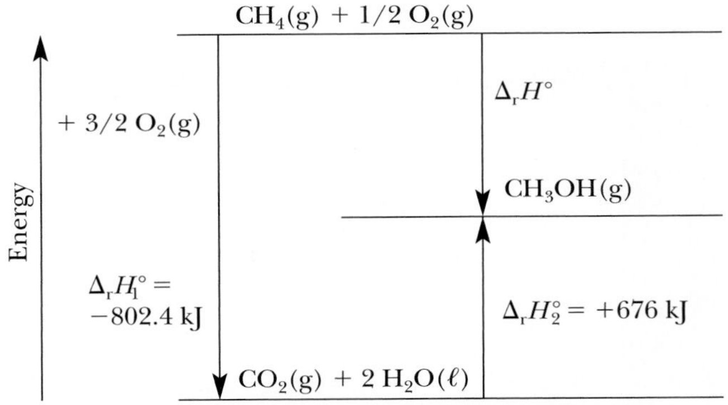

5.45 $\Delta_r H° = +90.3$ kJ/mol-rxn

5.47 $C(s) + 2\ H_2(g) + 1/2\ O_2(g) \rightarrow CH_3OH(\ell)$

$\Delta_f H° = -238.4$ kJ/mol

5.49 (a) $2\ Cr(s) + 3/2\ O_2(g) \rightarrow Cr_2O_3(s)$

$\Delta_f H° = -1134.7$ kJ/mol

(b) 2.4 g is equivalent to 0.046 mol of Cr. This will produce 26 kJ of energy transferred as heat.

5.51 (a) $\Delta H° = -24$ kJ for 1.0 g of phosphorus
(b) $\Delta H° = -18$ kJ for 0.2 mol NO
(c) $\Delta H° = -16.9$ kJ for the formation of 2.40 g of NaCl(s)
(d) $\Delta H° = -1.8 \times 10^3$ kJ for the oxidation of 250 g of iron

5.53 (a) $\Delta_r H° = -906.2$ kJ
(b) The heat evolved is 133 kJ for the oxidation of 10.0 g of NH_3

5.55 (a) $\Delta_r H° = +161.6$ kJ/mol-rxn; the reaction is endothermic.
(b)

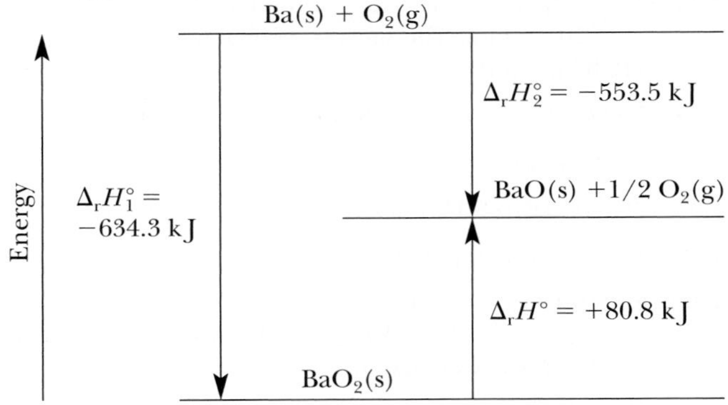

5.57 $\Delta_f H° = +77.7$ kJ/mol for naphthalene

5.59 (a) Exothermic: a process in which energy is transferred as heat from a system to its surroundings. (The combustion of methane is exothermic.)

Endothermic: a process in which energy is transferred as heat from the surroundings to the system. (Ice melting is endothermic.)

(b) System: the object or collection of objects being studied. (A chemical reaction—the system—taking place inside a calorimeter—the surroundings.)

Surroundings: everything outside the system that can exchange mass or energy with the system. (The calorimeter and everything outside the calorimeter comprise the surroundings.)

(c) Specific heat capacity: the quantity of energy that must be transferred as heat to raise the temperature of 1 gram of a substance 1 kelvin. (The specific heat capacity of water is 4.184 J/g · K).

(d) State function: a quantity that is characterized by changes that do not depend on the path chosen to go from the initial state to the final state. (Enthalpy and internal energy are state functions.)

(e) Standard state: the most stable form of a substance in the physical state that exists at a pressure of 1 bar and at a specified temperature. (The standard state of carbon at 25 °C is graphite.)

(f) Enthalpy change, ΔH: the energy transferred as heat at constant pressure. (The enthalpy change for melting ice at 0 °C is 6.00 kJ/mol.)

(g) Standard enthalpy of formation: the enthalpy change for the formation of 1 mol of a compound in its standard state directly from the component elements in their standard states. ($\Delta_f H°$ for liquid water is −285.83 kJ/mol)

5.61 (a) System: reaction between methane and oxygen

Surroundings: the furnace and the rest of the universe. Energy is transferred as heat from the system to the surroundings.

(b) System: water drops

Surroundings: skin and the rest of the universe

Energy is transferred as heat from the surroundings to the system

(c) System: water

Surroundings: freezer and the rest of the universe

Energy is transferred as heat from the system to the surroundings

(d) System: reaction of aluminum and iron(III) oxide

Surroundings: flask, laboratory bench, and rest of the universe

Energy is transferred as heat from the system to the surroundings.

5.63 Standard state of oxygen is gas, $O_2(g)$.

$O_2(g) \rightarrow 2\ O(g)$, $\Delta_r H° = +498.34$ kJ, endothermic

$3/2\ O_2(g) \rightarrow O_3(g)$, $\Delta_r H° = +142.67$ kJ

5.65

$SnBr_2(s) + TiCl_2(s) \rightarrow SnCl_2(s) + TiBr_2(s)$	$\Delta_r H° = -4.2$ kJ
$SnCl_2(s) + Cl_2(g) \rightarrow SnCl_4(\ell)$	$\Delta_r H° = -195$ kJ
$TiCl_4(\ell) \rightarrow TiCl_2(s) + Cl_2(g)$	$\Delta_r H° = +273$ kJ
$SnBr_2(s) + TiCl_4(\ell) \rightarrow SnCl_4(\ell) + TiBr_2(s)$	$\Delta_r H° = +74$ kJ

5.67 $C_{Ag} = 0.24$ J/g · K

5.69 Mass of ice melted = 75.4 g

5.71 Final temperature = 278 K (4.8 °C)

5.73 (a) When summed, the following equations give the balanced equation for the formation of $B_2H_6(g)$ from the elements.

$2\ B(s) + 3/2\ O_2(g) \rightarrow B_2O_3(s)$	$\Delta_r H° = -1271.9$ kJ
$3\ H_2(g) + 3/2\ O_2(g) \rightarrow 3\ H_2O(g)$	$\Delta_r H° = -725.4$ kJ
$B_2O_3(s) + 3\ H_2O(g) \rightarrow B_2H_6(g) + 3\ O_2(g)$	$\Delta_r H° = +2032.9$ kJ
$2\ B(s) + 3\ H_2(g) \rightarrow B_2H_6(g)$	$\Delta_r H° = +35.6$ kJ

(b) The enthalpy of formation of $B_2H_6(g)$ is +35.6 kJ/mol

(c)

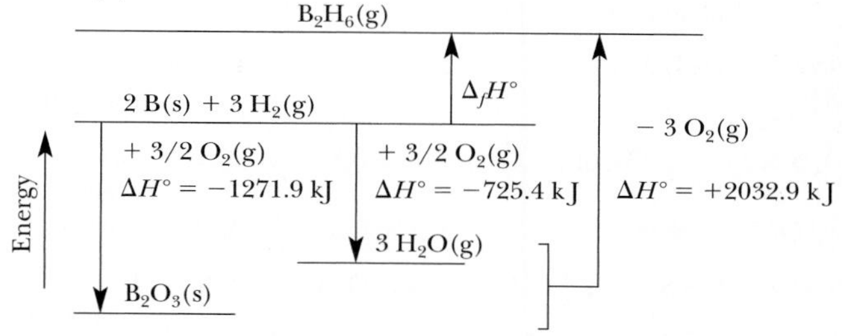

(d) The formation of $B_2H_6(g)$ is reactant-favored.

5.75 (a) $\Delta_r H° = +131.31$ kJ
(b) Reactant-favored
(c) 1.0932×10^7 kJ

5.77 Assuming $CO_2(g)$ and $H_2O(\ell)$ are the products of combustion:

$\Delta_r H°$ for isooctane is −5461.3 kJ/mol or −47.81 kJ per gram

$\Delta_r H°$ for liquid methanol is −726.77 kJ/mol or −22.682 kJ per gram

5.79 (a) Adding the equations as they are given in the question results in the desired equation for the formation of $SrCO_3(s)$. The calculated $\Delta_r H° = -1220.$ kJ/mol.

(b)

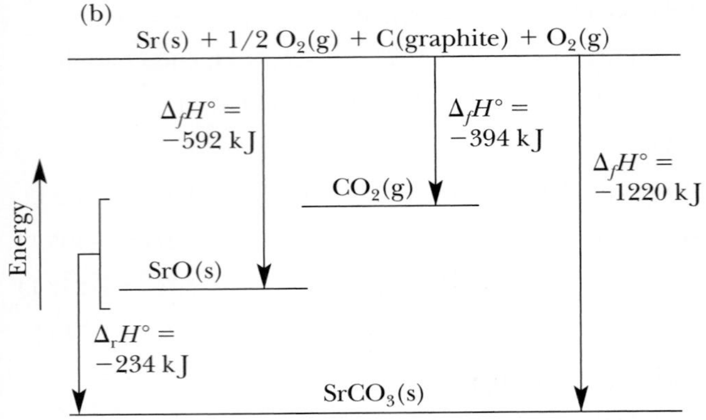

5.81 $\Delta_r H° = -305.3$ kJ

5.83 $C_{Pb} = 0.121$ J/g · K

5.85 $\Delta_r H = -69$ kJ/mol AgCl

5.87 36.0 kJ evolved per mol of NH_4NO_3

5.89 The standard enthalpy change, $\Delta_r H°$, is −352.88 kJ. The quantity of magnesium needed is 0.43 g.

5.91 (a) product-favored
(b) reactant-favored

5.93 The enthalpy change for each of the three reactions below is known or can be measured by calorimetry. The three equations sum to give the enthalpy of formation of $CaSO_4$(s).

Ca(s) + 1/2 O_2(g)	→	CaO(s)	$\Delta_r H° = \Delta_f H° = -635.09$ kJ
1/8 S_8(s) + 3/2 O_2(g)	→	SO_3(g)	$\Delta_r H° = \Delta_f H° = -395.77$ kJ
CaO(s) + SO_3(g)	→	$CaSO_4$(s)	$\Delta_r H° = -402.7$ kJ
Ca(s) + 1/8 S_8(s) + 2 O_2(g)	→	$CaSO_4$(s)	$\Delta_r H° = \Delta_f H° = -1433.6$ kJ

5.95

Metal	Molar Heat Capacity (J/mol · K)
Al	24.2
Fe	25.1
Cu	24.5
Au	25.4

All the metals have a molar heat capacity of 24.8 J/mol · K plus or minus 0.6 J/mol · K. Therefore, assuming the molar heat capacity of Ag is 24.8 J/mol · K, its specific heat capacity is 0.230 J/g · K. This is very close to the experimental value of 0.236 J/g · K.

5.97 120 g of CH_4 required (assuming H_2O(g) as product)

5.99 1.6×10^{11} kJ released to the surroundings. This is equivalent to 3.8×10^4 tons of dynamite.

5.101 (a)

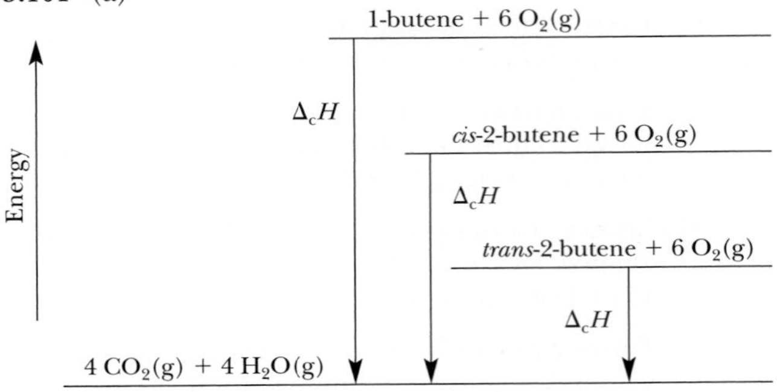

(b) *cis*-2-butene: $\Delta_f H° = 146.1$ kJ/mol
trans-2-butene: $\Delta_f H° = 142.8$ kJ/mol
1-butene: $\Delta_f H° = 155.3$ kJ/mol

(c)

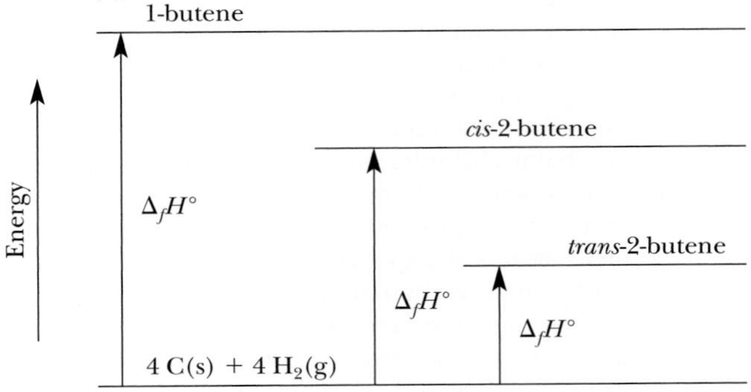

(d) −3.3 kJ/mol-rxn

5.103 (a) −726 kJ/mol Mg
(b) 25.0 °C

5.105 (a) Methane
(b) Methane
(c) −279 kJ
(d) $CH_4(g) + 2\ O_2(g) \rightarrow CH_3OH(\ell)$

5.107 (a) Metal Heated = 100.0 g of Al; Metal Cooled = 50.0 g of Au; Final Temperature = 26 °C
(b) Metal Heated = 50.0 g of Zn; Metal Cooled = 50.0 g of Al; Final Temperature = 21 °C

CHAPTER 6

6.1 (a) microwaves
(b) red light
(c) infrared

6.3 (a) Green light has a higher frequency than amber light
(b) 5.04×10^{14} s^{-1}

6.5 Frequency = 6.0×10^{14} s^{-1}; energy per photon = 4.0×10^{-19} J; energy per mol of photons = 2.4×10^5 J

6.7 Frequency = $7.5676 \times 10^{14}\ s^{-1}$; energy per photon = 5.0144×10^{-19} J; 302 kJ/mol of photons

6.9 In order of increasing energy: FM station < microwaves < yellow light < x-rays

6.11 Light with a wavelength as long as 600 nm would be sufficient. This is in the visible region.

6.13 (a) The light of shortest wavelength has a wavelength of 253.652 nm.
(b) Frequency = $1.18190 \times 10^{15}\ s^{-1}$. Energy per photon = 7.83139×10^{-19} J/photon.
(c) The lines at 404 (violet) and 436 nm (blue) are in the visible region of the spectrum.

6.15 The color is violet. $n_{initial} = 6$ and $n_{final} = 2$

6.17 (a) 10 lines possible
(b) Highest frequency (highest energy), $n = 5$ to $n = 1$
(c) Longest wavelength (lowest energy), $n = 5$ to $n = 4$

6.19 (a) $n = 3$ to $n = 2$
(b) $n = 4$ to $n = 1$; The energy levels are progressively closer at higher levels, so the energy difference from $n = 4$ to $n = 1$ is greater than from $n = 5$ to $n = 2$.

6.21 Wavelength = 102.6 nm and frequency = $2.923 \times 10^{15}\ s^{-1}$. Light with these properties is in the ultraviolet region.

6.23 Wavelength = 0.29 nm

6.25 The wavelength is 2.2×10^{-25} nm. (Calculated from $\lambda = h/m \cdot v$, where m is the ball's mass in kg and v is the velocity.) To have a wavelength of 5.6×10^{-3} nm, the ball would have to travel at 1.2×10^{-21} m/s.

6.27 (a) $n = 4$, $\ell = 0, 1, 2, 3$
(b) When $\ell = 2$, $m_\ell = -2, -1, 0, 1, 2$
(c) For a 4s orbital, $n = 4$, $\ell = 0$, and $m_\ell = 0$
(d) For a 4f orbital, $n = 4$, $\ell = 3$, and $m_\ell = -3, -2, -1, 0, 1, 2, 3$

6.29 Set 1: $n = 4$, $\ell = 1$, and $m_\ell = -1$

Set 2: $n = 4$, $\ell = 1$, and $m_\ell = 0$

Set 3: $n = 4$, $\ell = 1$, and $m_\ell = +1$

6.31 Four subshells. (The number of subshells in a shell is always equal to n.)

6.33 (a) ℓ must have a value no greater than $n - 1$.
(b) When $\ell = 0$, m_ℓ can only equal 0.
(c) When $\ell = 0$, m_ℓ can only equal 0.

6.35 (a) None. The quantum number set is not possible. When $\ell = 0$, m_ℓ can only equal 0.
(b) 3 orbitals
(c) 11 orbitals
(d) 1 orbital

6.37 (a) $m_s = 0$ is not possible. m_s may only have values of $\pm 1/2$.

One possible set of quantum numbers: $n = 4$, $\ell = 2$, $m_\ell = 0$, $m_s = +1/2$

(b) m_ℓ cannot equal -3 in this case. If $\ell = 1$, m_ℓ can only be -1, 0, or 1.

One possible set of quantum numbers: $n = 3$, $\ell = 1$, $m_\ell = -1$, $m_s = -1/2$

(c) $\ell = 3$ is not possible in this case. The maximum value of ℓ is $n - 1$.

One possible set of quantum numbers: $n = 3$, $\ell = 2$, $m_\ell = -1$, $m_s = +1/2$

6.39 2d and 3f orbitals cannot exist. The $n = 2$ shell consists only of s and p subshells. The $n = 3$ shell consists only of s, p, and d subshells.

6.41 (a) For 2p: $n = 2$, $\ell = 1$, and $m_\ell = -1, 0$, or $+1$
(b) For 3d: $n = 3$, $\ell = 2$, and $m_\ell = -2, -1, 0, +1$, or $+2$
(c) For 4f: $n = 4$, $\ell = 3$, and $m_\ell = -3, -2, -1, 0, +1, +2$, or $+3$

6.43 4d

6.45 (a) 2s has 0 nodal surfaces that pass through the nucleus ($\ell = 0$).
(b) 5d has 2 nodal surfaces that pass through the nucleus ($\ell = 2$).
(c) 5f has three nodal surfaces that pass through the nucleus ($\ell = 3$).

6.47 (a) Correct
(b) Incorrect. The intensity of a light beam is independent of frequency and is related to the number of photons of light with a certain energy.
(c) Correct

6.49 Considering only angular nodes (nodal surfaces that pass through the nucleus):

s orbital	0 nodal surfaces
p orbitals	1 nodal surface or plane passing through the nucleus
d orbitals	2 nodal surfaces or planes passing through the nucleus
f orbitals	3 nodal surfaces or planes passing through the nucleus

6.51

ℓ value	Orbital Type
3	f
0	s
1	p
2	d

6.53 Considering only angular nodes (nodal surfaces that pass through the nucleus):

Orbital Type	Number of Orbitals in a Given Subshell	Number of Nodal Surfaces
s	1	0
p	3	1
d	**5**	**2**
f	7	3

6.55 (a) Green light
(b) Red light has a wavelength of 680 nm, and green light has a wavelength of 500 nm.
(c) Green light has a higher frequency than red light.

6.57 (a) Wavelength = 0.35 m
(b) Energy = 0.34 J/mol
(c) Blue light (with λ = 420 nm) has an energy of 280 kJ/mol of photons.
(d) Blue light has an energy (per mol of photons) that is 840,000 times greater than a mole of photons from a cell phone.

6.59 The ionization energy for He^+ is 5248 kJ/mol. This is four times the ionization energy for the H atom.

6.61 $1s < 2s = 2p < 3s = 3p = 3d < 4s$

In the H atom orbitals in the same shell (e.g., 2*s* and 2*p*) have the same energy.

6.63 Frequency = $2.836 \times 10^{20}\ s^{-1}$ and wavelength = 1.057×10^{-12} m

6.65 260 *s* or 4.3 min

6.67 (a) size and energy
(b) ℓ
(c) more
(d) 7 (when ℓ = 3 these are *f* orbitals)
(e) one orbital
(f) (left to right) *d*, *s*, and *p*
(g) ℓ = 0, 1, 2, 3, 4
(h) 16 orbitals (1*s*, 3*p*, 5*d*, and 7*f*) (= n^2)
(i) paramagnetic

6.69 (a) Drawing (a) is a ferromagnetic solid, (b) is a diamagnetic solid, and (c) is a paramagnetic solid.
(b) Substance (a) would be most strongly attracted to a magnet, whereas (b) would be least strongly attracted.

6.71 The pickle glows because it was made by soaking a cucumber in brine, a concentrated solution of NaCl. The sodium atoms in the pickle are excited by the electric current and release energy as yellow light as they return to the ground state. Excited sodium atoms are the source of the yellow light you see in fireworks and in certain kinds of street lighting.

6.73 (a) λ = 0.0005 cm = 5 μm
(b) The left side is the higher energy side, and the right side is the lower energy side.
(c) The interaction with O—H requires more energy.

6.75 (c)

6.77 An experiment can be done that shows that the electron can behave as a particle, and another experiment can be done to show that it has wave properties. (However, no single experiment shows both properties of the electron.) The modern view of atomic structure is based on the wave properties of the electron.

6.79 (a) and (b)

6.81 Radiation with a wavelength of 93.8 nm is sufficient to raise the electron to the *n* = 6 quantum level (see Figure 6.10). There should be 15 emission lines involving transitions from *n* = 6 to lower energy levels. (There are five lines for transitions from *n* = 6 to lower levels, four lines for *n* = 5 to lower levels, three for *n* = 4 to lower levels, two lines for *n* = 3 to lower levels, and one line for *n* = 2 to *n* = 1.) Wavelengths for many of the lines are given in Figure 6.10. For example, there will be an emission involving an electron moving from *n* = 6 to *n* = 2 with a wavelength of 410.2 nm.

6.83 (a) Group 7B (IUPAC Group 7); Period 5
(b) $n = 5,\ \ell = 0,\ m_\ell = 0,\ m_s = +1/2$
(c) $\lambda = 8.79 \times 10^{-12}$ m; $\nu = 3.41 \times 10^{19}\ s^{-1}$
(d) (i) $HTcO_4(aq) + NaOH(aq) \rightarrow H_2O(\ell) + NaTcO_4(aq)$
(ii) 8.5×10^{-3} g $NaTcO_4$ produced; 1.8×10^{-3} g NaOH needed
(e) 0.28 mg $NaTcO_4$; 0.00015 M

6.85 Six emission lines are observed. More than one line is observed because the following changes in energy levels are possible: from *n* = 4 to *n* = 3, *n* = 2, and *n* = 1 (three lines), from *n* = 3 to *n* = 2 and *n* = 1 (2 lines), and from *n* = 2 to *n* = 1 (one line).

CHAPTER 7

7.1 (a) Phosphorus: $1s^22s^22p^63s^23p^3$

[↑↓] [↑↓] [↑↓|↑↓|↑↓] [↑↓] [↑|↑|↑]
1*s* 2*s* 2*p* 3*s* 3*p*

The element is in the third period in Group 5A. Therefore, it has five electrons in the third shell.

(b) Chlorine: $1s^22s^22p^63s^23p^5$

[↑↓] [↑↓] [↑↓|↑↓|↑↓] [↑↓] [↑↓|↑↓|↑]
1*s* 2*s* 2*p* 3*s* 3*p*

The element is in the third period and in Group 7A. Therefore, it has seven electrons in the third shell.

7.3 (a) Chromium: $1s^22s^22p^63s^23p^63d^54s^1$
(b) Iron: $1s^22s^22p^63s^23p^63d^64s^2$

7.5 (a) Arsenic: $1s^22s^22p^63s^23p^63d^{10}4s^24p^3$; $[Ar]3d^{10}4s^24p^3$
(b) Krypton: $1s^22s^22p^63s^23p^63d^{10}4s^24p^6$; $[Ar]3d^{10}4s^24p^6 = [Kr]$

7.7 (a) Tantalum: This is the third element in the transition series in the sixth period. Therefore, it has a core equivalent to Xe plus two 6*s* electrons, 14 4*f* electrons, and three electrons in 5*d*: $[Xe]4f^{14}5d^36s^2$
(b) Platinum: This is the eighth element in the transition series in the sixth period. Therefore, it is predicted to have a core equivalent to Xe plus two 6*s* electrons, 14 4*f* electrons, and eight electrons in 5*d*: $[Xe]4f^{14}5d^86s^2$. In reality, its actual configuration (Table 7.3) is $[Xe]4f^{14}5d^96s^1$.

7.9 Americium: $[Rn]5f^77s^2$ (see Table 7.3)

7.11 (a) 2
(b) 1
(c) none (because ℓ cannot equal n)

7.13 Magnesium: $1s^22s^22p^63s^2$

[Ne] [↑↓]
3*s*

Quantum numbers for the two electrons in the 3*s* orbital:

$n = 3$, $\ell = 0$, $m_\ell = 0$, and $m_s = +1/2$

$n = 3$, $\ell = 0$, $m_\ell = 0$, and $m_s = -1/2$

7.15 Gallium: $1s^22s^22p^63s^23p^63d^{10}4s^24p^1$

[Ar] [↑↓|↑↓|↑↓|↑↓|↑↓] [↑↓] [↑| |]
3*d* 4*s* 4*p*

Quantum numbers for the 4p electron:

$n = 4$, $\ell = 1$, $m_\ell = -1, 0,$ or $+1$, and $m_s = +\frac{1}{2}$ or $-\frac{1}{2}$

7.17 (a) Mg^{2+} ion

[↑↓] [↑↓] [↑↓|↑↓|↑↓]
1*s* 2*s* 2*p*

(b) K^+ ion

[↑↓] [↑↓] [↑↓|↑↓|↑↓] [↑↓] [↑↓|↑↓|↑↓]
1*s* 2*s* 2*p* 3*s* 3*p*

(c) Cl^- ion (Note that both Cl^- and K^+ have the same configuration; both are equivalent to Ar.)

[↑↓] [↑↓] [↑↓|↑↓|↑↓] [↑↓] [↑↓|↑↓|↑↓]
1*s* 2*s* 2*p* 3*s* 3*p*

(d) O^{2-} ion

[↑↓] [↑↓] [↑↓|↑↓|↑↓]
1*s* 2*s* 2*p*

7.19 (a) V (paramagnetic; three unpaired electrons)

[Ar] [↑|↑|↑| |] [↑↓]
3*d* 4*s*

(b) V^{2+} ion (paramagnetic, three unpaired electrons)

[Ar] [↑|↑|↑| |] [ ]
3*d* 4*s*

(c) V^{5+} ion. This ion has an electron configuration equivalent to argon, [Ar]. It is diamagnetic with no unpaired electrons.

7.21 (a) Manganese

[Ar] [↑|↑|↑|↑|↑] [↑↓]
3*d* 4*s*

(b) Mn^{4+}

[Ar] [↑|↑|↑| |] [ ]
3*d* 4*s*

(c) The 4^+ ion is paramagnetic to the extent of three unpaired electrons.
(d) 3

7.23 Increasing size: C < B < Al < Na < K

7.25 (a) Cl^-
(b) Al
(c) In

7.27 (c)

7.29 (a) Largest radius, Na
(b) Most negative electron affinity: O
(c) Ionization energy: Na < Mg < P < O

7.31 (a) Increasing ionization energy: S < O < F. S is less than O because the IE decreases down a group. F is greater than O because IE generally increases across a period.
(b) Largest IE: O. IE decreases down a group.
(c) Most negative electron affinity: Cl. Electron affinity becomes more negative across the periodic table and on ascending a group.
(d) Largest Size: O^{2-}. Negative ions are larger than their corresponding neutral atoms. F^- is thus larger than F. O^{2-} and F^- are isoelectronic, but the O^{2-} ion has only eight protons in its nucleus to attract the 10 electrons, whereas the F^- has nine protons, making the O^{2-} ion larger.

7.33 Uranium configuration: [Rn]$5f^36d^17s^2$

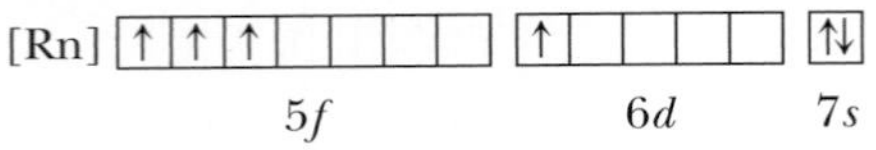

Uranium(IV) ion, U^{4+}: [Rn]$5f^2$

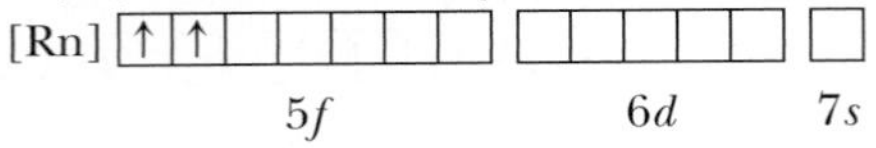

Both U and U^{4+} are paramagnetic.

7.35 (a) Atomic number = 20
(b) Total number of *s* electrons = 8
(c) Total number of *p* electrons = 12
(d) Total number of *d* electrons = 0
(e) The element is Ca, calcium, a metal.

7.37 (a) Valid. Possible elements are Li and Be.
(b) Not valid. The maximum value of ℓ is $(n - 1)$.
(c) Valid. Possible elements are B through Ne.
(d) Valid. Possible elements are Y through Cd.

7.39 (a) Neodymium, Nd: [Xe]$4f^46s^2$ (Table 7.3)

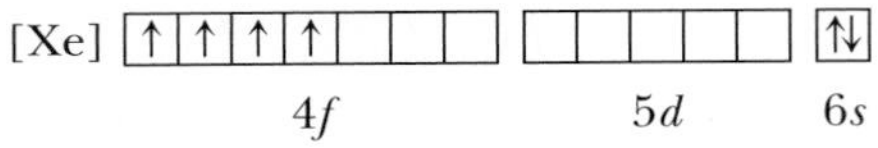

Iron, Fe: [Ar]$3d^64s^2$

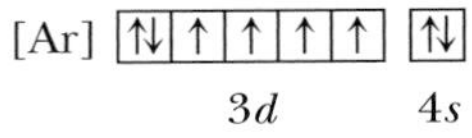

Boron, B: [He]$2s^22p^1$

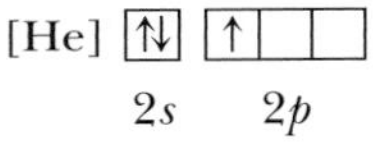

(b) All three elements have unpaired electrons and so should be paramagnetic.
(c) Neodymium(III) ion, Nd^{3+}: [Xe]$4f^3$

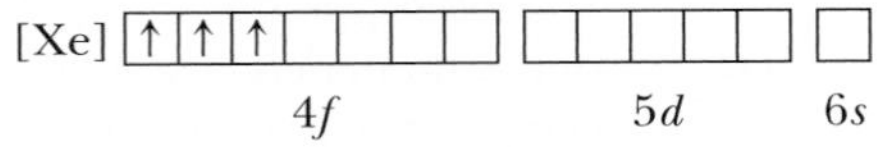

Iron(III) ion, Fe^{3+}: [Ar]$3d^5$

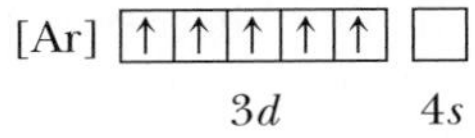

Both neodymium(III) and iron(III) have unpaired electrons and are paramagnetic.

7.41 K < Ca < Si < P

7.43 (a) metal
(b) B
(c) A
(d) A
(e) Rb_2Se

7.45 In^{4+}: Indium has three outer shell electrons and so is unlikely to form a 4^+ ion.

Fe^{6+}: Although iron has eight electrons in its 3*d* and 4*s* orbitals, ions with a 6^+ charge are highly unlikely. The ionization energy is too large.

Sn^{5+}: Tin has four outer shell electrons and so is unlikely to form a 5^+ ion.

7.47 (a) Se
(b) Br^-
(c) Na
(d) N
(e) N^{3-}

7.49 (a) Na
(b) C
(c) Na < Al < B < C

7.51 (a) Cobalt
(b) Paramagnetic
(c) Four unpaired electrons

7.53 (a) 0.421 g
(b) paramagnetic; 2 unpaired electrons
(c) 99.8 mg; the nickel powder will stick to a magnet.

7.55 Li has three electrons ($1s^22s^1$) and Li^+ has only two electrons ($1s^2$). The ion is smaller than the atom because there are only two electrons to be held by three protons in the ion. Also, an electron in a larger orbital has been removed. Fluorine atoms have nine electrons and nine protons ($1s^22s^22p^5$). The anion, F^-, has one additional electron, which means that 10 electrons must be held by only nine protons, and the ion is larger than the atom.

7.57 Element 1 comes from Group 4A (IUPAC Group 14). The first two IEs correspond to removing electrons from a *p* subshell. With the third IE, there is a fairly large jump in IE corresponding to removing an electron from an *s* subshell. The fourth electron removed comes from the same *s* subshell and therefore does not increase the IE by as much. None of the IEs are large enough to correspond to removing an electron from a lower energy level.

Element 2 comes from Group 3A (IUPAC Group 13). There is a large change in IE between the third and fourth IEs. The first three IEs correspond to removing electrons from the same energy level. The large jump at the fourth IE corresponds to having to remove the electron from a lower energy level.

7.59 Most stable: (d) The two electrons are in separate orbitals, following Hund's rule, and are of the same spin.

Least stable: (a) In this case the electrons violate both Hund's rule and the Pauli exclusion principle.

7.61 K ($1s^22s^22p^63s^23p^64s^1$) $\longrightarrow$ K^+($1s^22s^22p^63s^23p^6$)

K^+ ($1s^22s^22p^63s^23p^6$) $\longrightarrow$ K^{2+}($1s^22s^22p^63s^23p^5$)

The first ionization is for the removal of an electron from the valence shell of electrons. The second electron, however, is removed from the $3p$ subshell. This subshell is significantly lower in energy than the $4s$ subshell, and considerably more energy is required to remove this second electron.

7.63 (a) In going from one element to the next across the period, the effective nuclear charge increases slightly and the attraction between the nucleus and the electrons increases.
(b) The size of fourth period transition elements, for example, is a reflection of the size of the $4s$ orbital. As d electrons are added across the series, protons are added to the nucleus. Adding protons should lead to a decreased atom size, but the effect of the protons is balanced by repulsions of the $3d$ electrons and $4s$ electrons, and the atom size is changed little.

7.65 Among the arguments for a compound composed of Mg^{2+} and O^{2-} are:
(a) Chemical experience suggests that all Group 2A elements form 2^+ cations, and that oxygen is typically the O^{2-} ion in its compounds.
(b) Other alkaline earth elements form oxides such as BeO, CaO, and BaO.

A possible experiment is to measure the melting point of the compound. An ionic compound such as NaF (with ions having 1^+ and 1^- charges) melts at 990 °C, whereas a compound analogous to MgO, CaO, melts at a much higher temperature (2580 °C).

7.67 (a) The effective nuclear charge increases, causing the valence orbital energies to become more negative on moving across the period.
(b) As the valence orbital energies become more negative, it is increasingly difficult to remove an electron from the atom, and the IE increases. Toward the end of the period, the orbital energies have become so negative that removing an electron requires significant energy. Instead, the effective nuclear charge has reached the point that it is energetically more favorable for the atom to gain an electron, corresponding to a more negative electron affinity.
(c) The valence orbital energies are in the order:

Li (−520.7 kJ) < Be (−899.3 kJ) > B (−800.8 kJ) < C (−1029 kJ)

This means it is more difficult to remove an electron from Be than from either Li or B. The energy is more negative for C than for B, so it is more difficult to remove an electron from C than from B.

7.69 The size declines across this series of elements while their mass increases. Thus, the mass per volume, the density, increases.

7.71 (a) Element 113: [Rn]$5f^{14}6d^{10}7s^27p^1$
Element 115: [Rn]$5f^{14}6d^{10}7s^27p^3$
(b) Element 113 is in Group 3A (with elements such as boron and aluminum), and element 115 is in Group 5A (with elements such as nitrogen and phosphorus).
(c) Americium (Z = 95) + argon (Z = 18) = element 113

7.73 (a) Sulfur electron configuration

[↑↓] [↑↓] [↑↓|↑↓|↑↓] [↑↓] [↑↓|↑|↑]
$1s$ $2s$ $2p$ $3s$ $3p$

(b) n = 3, ℓ = 1, m_ℓ = 1, and m_s = +1/2
(c) S has the smallest ionization energy and O has the smallest radius.
(d) S is smaller than S^{2-} ion
(e) 584 g SCl_2
(f) 10.0 g of SCl_2 is the limiting reactant, and 11.6 g of $SOCl_2$ can be produced.
(g) $\Delta_f H°[SCl_2(g)] = -17.6$ kJ/mol

7.75 (a) Z* for F is 5.2; Z* for Ne is 5.85. The effective nuclear charge increases from O to F to Ne. As the effective nuclear charge increases, the atomic radius decreases, and the first ionization energy increases.
(b) Z* for a $3d$ electron in Mn is 13.7; for a $4s$ electron it is only 3.1. The effective nuclear charge experienced by a $4s$ electron is much smaller than that experienced by a $3d$ electron. A $4s$ electron in Mn is thus more easily removed.

CHAPTER 8

8.1. (a) Group 6A, six valence electrons
(b) Group 3A, three valence electrons
(c) Group 1A, one valence electron
(d) Group 2A, two valence electrons
(e) Group 7A, seven valence electrons
(f) Group 6A, six valence electrons

8.3 Group 3A, three bonds
Group 4A, four bonds
Group 5A, three bonds (for a neutral compound)
Group 6A, two bonds (for a neutral compound)
Group 7A, one (for a neutral compound)

8.5 (a) NF_3, 26 valence electrons

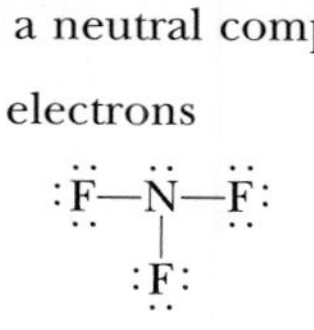

(b) ClO_3^-, 26 valence electrons

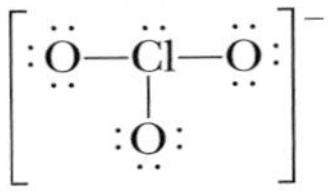

(c) HOBr, 14 valence electrons

(d) SO_3^{2-}, 26 valence electrons

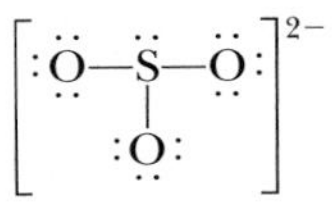

8.7 (a) $CHClF_2$, 26 valence electrons

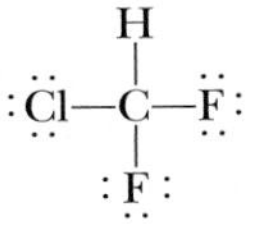

(b) CH_3CO_2H, 24 valence electrons

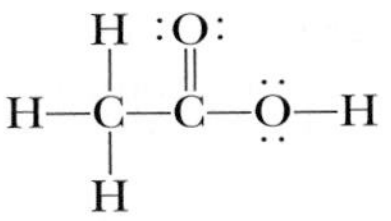

(c) CH_3CN, 16 valence electrons

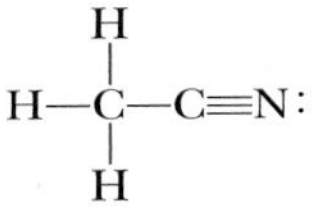

(d) H_2CCCH_2, 16 valence electrons

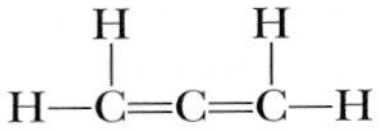

8.9 (a) SO_2, 18 valence electrons

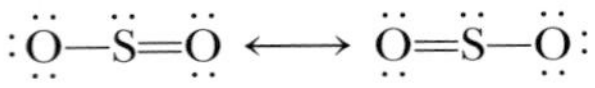

(b) HNO_2, 18 valence electrons

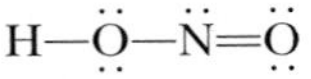

(c) SCN^-, 16 valence electrons

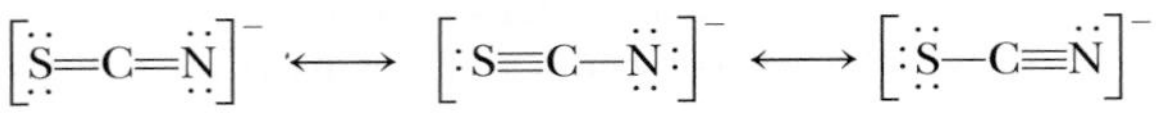

8.11 (a) BrF_3, 28 valence electrons

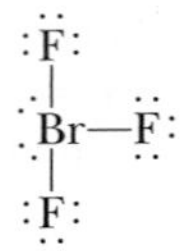

(b) I_3^-, 22 valence electrons

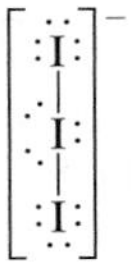

(c) XeO_2F_2, 34 valence electrons

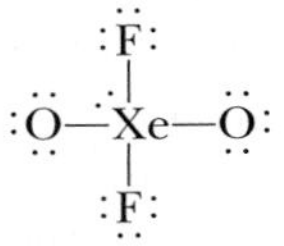

(d) XeF_3^+, 28 valence electrons

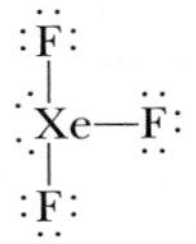

8.13 (a) N = 0; H = 0
(b) P = +1; O = −1
(c) B = −1; H = 0
(d) All are zero.

8.15 (a) N = +1; O = 0
(b) The central N is 0. The singly bonded O atom is −1, and the doubly bonded O atom is 0.

(c) N and F are both 0.
(d) The central N atom is +1, one of the O atoms is −1, and the other two O atoms are both 0.

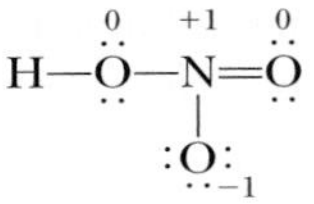

8.17 (a) Electron-pair geometry around N is tetrahedral. Molecular geometry is trigonal pyramidal.

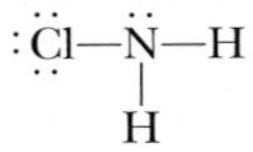

(b) Electron-pair geometry around O is tetrahedral. Molecular geometry is bent.

(c) Electron-pair geometry around C is linear. Molecular geometry is linear.

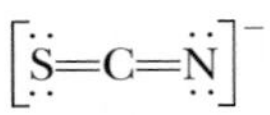

(d) Electron-pair geometry around O is tetrahedral. The molecular geometry is bent.

8.19 (a) Electron-pair geometry around C is linear. Molecular geometry is linear.

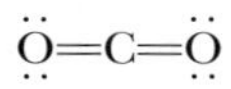

(b) Electron-pair geometry around N is trigonal planar. Molecular geometry is bent.

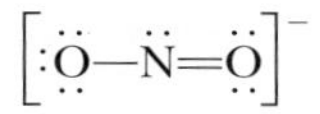

(c) Electron-pair geometry around O is trigonal planar. Molecular geometry is bent.

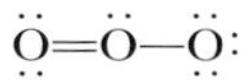

(d) Electron-pair geometry around Cl atom is tetrahedral. Molecular geometry is bent.

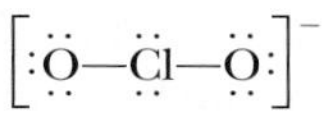

All have two atoms attached to the central atom. As the bond and lone pairs vary, the electron-pair geometries vary from linear to tetrahedral, and the molecular geometries vary from linear to bent.

8.21 (a) Electron-pair geometry around Cl is trigonal bipyramidal. Molecular geometry is linear.

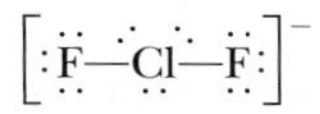

(b) Electron-pair geometry around Cl is trigonal bipyramidal. Molecular geometry is T-shaped.

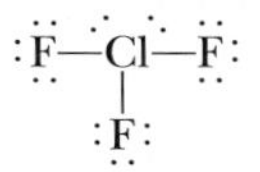

(c) Electron-pair geometry around Cl is octahedral. Molecular geometry is square planar.

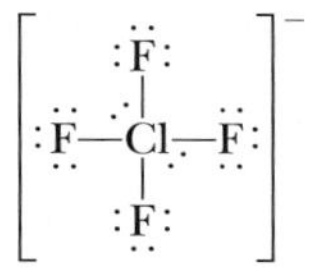

(d) Electron-pair geometry around Cl is octahedral. Molecular geometry is a square pyramid.

8.23 (a) Ideal O—S—O angle = 120°
(b) 120°
(c) 120°
(d) H—C—H = 109° and C—C—N angle = 180°

8.25 1 = 120°; 2 = 109°; 3 = 120°; 4 = 109°; 5 = 109°

The chain cannot be linear because the first two carbon atoms in the chain have bond angles of 109° and the final one has a bond angle of 120°. These bond angles do not lead to a linear chain.

8.27

$\overrightarrow{\text{C—O}}$ ($+\delta$ $-\delta$) $\quad$ $\overrightarrow{\text{C—N}}$ ($+\delta$ $-\delta$)

CO is more polar

$\overrightarrow{\text{P—Cl}}$ ($+\delta$ $-\delta$) $\quad$ $\overrightarrow{\text{P—Br}}$ ($+\delta$ $-\delta$)

PCl is more polar

$\overrightarrow{\text{B—O}}$ ($+\delta$ $-\delta$) $\quad$ $\overrightarrow{\text{B—S}}$ ($+\delta$ $-\delta$)

BO is more polar

$\overrightarrow{\text{B—F}}$ ($+\delta$ $-\delta$) $\quad$ $\overrightarrow{\text{B—I}}$ ($+\delta$ $-\delta$)

BF is more polar

8.29 (a) CH and CO bonds are polar.
(b) The CO bond is most polar, and O is the mos[…] negative atom.

8.31 (a) OH^-: The formal charge on O is −1 and […] is 0.
(b) BH_4^-: Even though the formal charge […] and on H is 0, H is slightly more el[…] than B. The four H atoms are the[…] likely to bear the −1 charge of t[…] bonds are polar with the H at[…] end.
(c) The CH and CO bonds are […] C—C bond is n[…] CO bonds l[…]

8.33 Structure […] small […] th[…]

8.37 (i) The most polar bonds are in H_2O (because O and H have the largest difference in electronegativity).
(ii) Not polar: CO_2 and CCl_4
(iii) The F atom is more negatively charged.

8.39 (a) $BeCl_2$, nonpolar linear molecule
(b) HBF_2, polar trigonal planar molecule with F atoms the negative end of the dipole and the H atom the positive end.
(c) CH_3Cl, polar tetrahedral molecule. The Cl atom is the negative end of the dipole and the three H atoms are on the positive side of the molecule.
(d) SO_3, a nonpolar trigonal planar molecule

8.41 (a) Two C—H bonds, bond order is 1; 1 C=O bond, bond order is 2.
(b) Three S—O single bonds, bond order is 1.
(c) Two nitrogen–oxygen double bonds, bond order is 2.
(d) One N=O double bond, bond order is 2; one N—Cl bond, bond order is 1.

8.43 (a) B—Cl
(b) C—O
(c) P—O
(d) C=O

45 NO bond orders: 2 in NO_2^+, 1.5 in NO_2^-; 1.33 in NO_3^-. The NO bond is longest in NO_3^- and shortest in NO_2^+.

Th CO bond in carbon monoxide is a triple bond, so it is both shorter and stronger than the CO double on in H_2CO.

H −126 kJ

nd dissociation energy = 192 kJ/mol

	Valence Electrons
	1
	4
	2
	4
	7

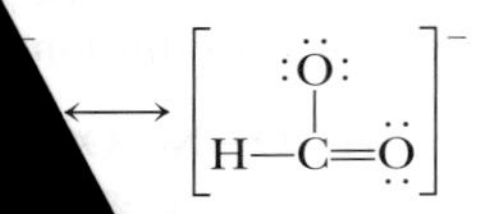

e, we need energies H—H, and H—O.

for O=O) + 2 ×

Energy evolved when bonds are made = 4 × 463 kJ (for O—H) = −1852 kJ

Total energy = −482 kJ

8.61 All the species in the series have 16 valence electrons and all are linear.

(a) 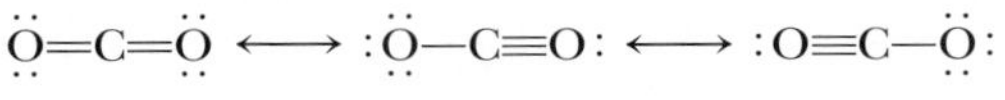

(b)

(c)

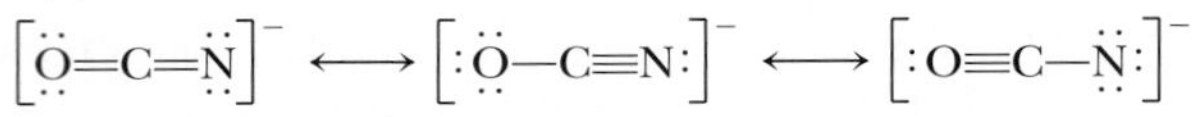

8.63 The N—O bonds in NO_2^- have a bond order of 1.5, whereas in NO_2^+ the bond order is 2. The shorter bonds (110 pm) are the NO bonds with the higher bond order (in NO_2^+), whereas the longer bonds (124 pm) in NO_2^- have a lower bond order.

8.65 The F—Cl—F bond angle in ClF_2^+, which has a tetrahedral electron-pair geometry, is approximately 109°.

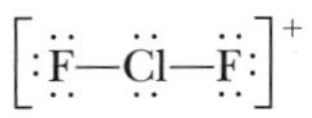

The ClF_2^- ion has a trigonal-bipyramidal electron-pair geometry with F atoms in the axial positions and the lone pairs in the equatorial positions. Therefore, the F—C—F angle is 180°.

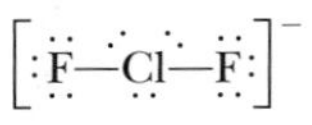

8.67 An H^+ ion will attach to an O atom of SO_3^{2-} and not to the S atom. The O atoms each have a formal charge of −1, whereas the S atom formal charge is +1.

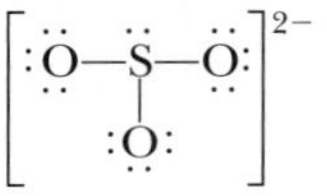

8.69 (a) Calculation from bond energies: $\Delta_r H° =$ −1070 kJ/mol-rxn; $\Delta H° = -535$ kJ/mol CH_3OH
(b) Calculation from thermochemical data: $\Delta_r H° =$ −1352.3 kJ/mol-rxn; $\Delta H° = -676$ kJ/mol CH_3OH

8.71 (a)

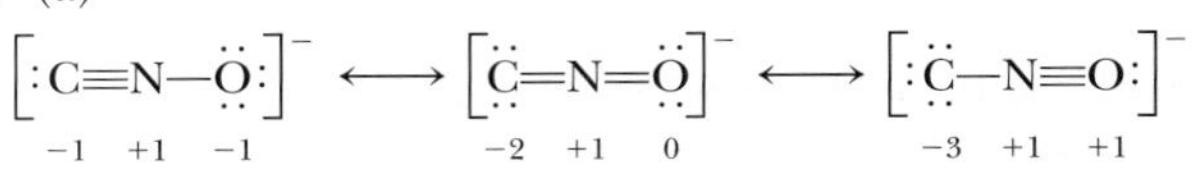

(b) The first resonance structure is the most reasonable because oxygen, the most electronegative atom, has a negative formal charge, and the unfavorable negative charge on the least electronegative atom, carbon, is smallest.

(c) This species is so unstable because carbon, the least electronegative element in the ion, has a negative formal charge. In addition, all three resonance structures have an unfavorable charge distribution.

8.73 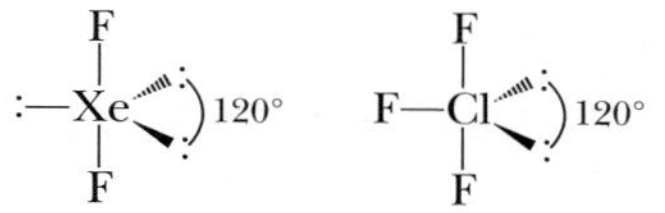

(a) XeF_2 has three lone pairs around the Xe atom. The electron-pair geometry is trigonal bipyramidal. Because lone pairs require more space than bond pairs, it is better to place the lone pairs in the equator of the bipyramid where the angles between them are 120°.
(b) Like XeF_2, ClF_3 has a trigonal bipyramidal electron-pair geometry, but with only two lone pairs around the Cl. These are again placed in the equatorial plane where the angle between them is 120°.

8.75 (a) Angle 1 = 109°; angle 2 = 120°; angle 3 = 109°; angle 4 = 109°; and angle 5 = 109°.
(b) The O—H bond is the most polar bond.

8.77 $\Delta_r H = +146\ kJ = 2\ (\Delta H_{C-N}) + \Delta H_{C=O} - [\Delta H_{N-N} + \Delta H_{C\equiv O}]$

8.79 (a) Two C—H bonds and one O=O are broken and two O—C bonds and two H—O bonds are made in the reaction. $\Delta_r H = -318$ kJ. The reaction is exothermic.
(b) Acetone is polar.
(c) The O—H hydrogen atoms are the most positive in dihydroxyacetone.

8.81 (a) The C=C bond is stronger than the C—C bond.
(b) The C—C single bond is longer than the C=C double bond.
(c) Ethylene is nonpolar, whereas acrolein is polar.
(d) The reaction is exothermic ($\Delta_r H = -45$ kJ).

8.83 $\Delta_r H = -211$ kJ

8.85 Methanol is a polar solvent. Methanol contains two bonds of significant polarity, the C—O bond and the O—H bond. The C—O—H atoms are in a bent configuration, leading to a polar molecule. Toluene contains only carbon and hydrogen atoms, which have similar electronegativites and which are arranged in tetrahedral or trigonal planar geometries, leading to a molecule that is largely nonpolar.

8.87 (a) 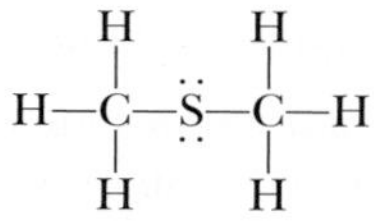

The bond angles are all approximately 109°.

(b) The sulfur atom should have a slight partial negative charge, and the carbons should have slight partial positive charges. The molecule has a bent shape and is polar.
(c) 1.6×10^{18} molecules

8.89 (a) Odd electron molecules: BrO (13 electrons)
(b) $Br_2(g) \rightarrow 2\ Br(g)$ $\quad \Delta_r H = +193$ kJ
$2\ Br(g) + O_2(g) \rightarrow 2\ BrO(g)$ $\quad \Delta_r H = +96$ kJ
$BrO(g) + H_2O(g) \rightarrow HOBr(g) + OH(g)$ $\quad \Delta_r H = 0$ kJ
(c) ΔH of formation [HOBr(g)] = −101 kJ/mol
(d) The reactions in part (b) are endothermic (or thermal-neutral for the third reaction), and the enthalpy of formation in part (c) is exothermic.

8.91 (a) BF_3 is a nonpolar molecule, but replacing one or two F atoms with an H atom (HBF_2 and H_2BF) gives polar molecules.
(b) $BeCl_2$ is not polar, whereas replacing a Cl atom with a Br atom gives a polar molecule (BeClBr).

CHAPTER 9

9.1 The electron-pair and molecular geometry of $CHCl_3$ are both tetrahedral. Each C—Cl bond is formed by the overlap of an sp^3 hybrid orbital on the C atom with a 3p orbital on a Cl atom to form a sigma bond. A C—H sigma bond is formed by the overlap of an sp^3 hybrid orbital on the C atom with an H atom 1s orbital.

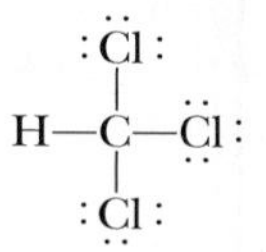

9.3

	Electron-Pair Geometry	Molecular Geometry	Hybrid Orbital Set
(a)	trigonal planar	trigonal planar	sp^2
(b)	linear	linear	sp
(c)	tetrahedral	tetrahedral	sp^3
(d)	trigonal planar	trigonal planar	sp^2

9.5 (a) C, sp^3; O, sp^3
(b) CH_3, sp^3; middle C, sp^2; CH_2, sp^2
(c) CH_2, sp^3; CO_2H, sp^2; N, sp^3

9.7 (a) Electron-pair geometry is octahedral. Molecular geometry is octahedral. S: sp^3d^2

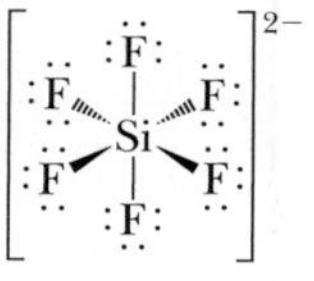

(b) Electron-pair geometry is trigonal-bipyramidal. Molecular geometry is seesaw. Se: sp^3d

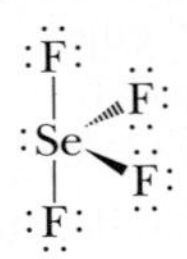

(c) Electron-pair geometry is trigonal-bipyramidal. Molecular geometry is linear. I: sp^3d

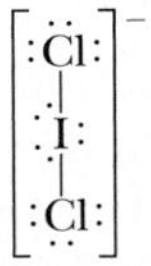

(d) Electron-pair geometry is octahedral. Molecular geometry is square-planar. Xe: sp^3d^2

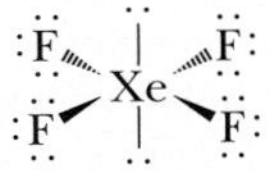

9.9 There are 32 valence electrons in both HPO_2F_2 and its anion. Both have a tetrahedral molecular geometry, and so the P atom in both is sp^3 hybridized.

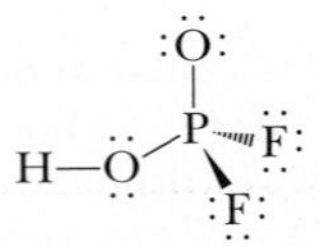

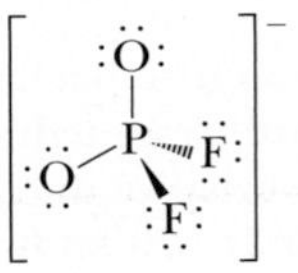

9.11 The C atom is sp^2-hybridized. Two of the sp^2 hybrid orbitals are used to form C—Cl sigma bonds, and the third is used to form the C—O sigma bond. The *p* orbital not used in the C atom hybrid orbitals is used to form the CO pi bond.

9.13

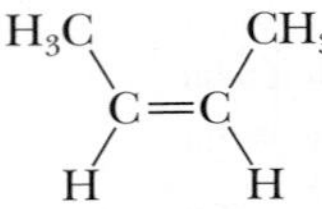

cis isomer

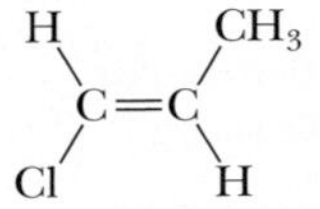

trans isomer

9.15 H_2^+ ion: $(\sigma_{1s})^1$. Bond order is 0.5. The bond in H_2^+ is weaker than in H_2 (bond order = 1).

9.17 MO diagram for C_2^{2-} ion

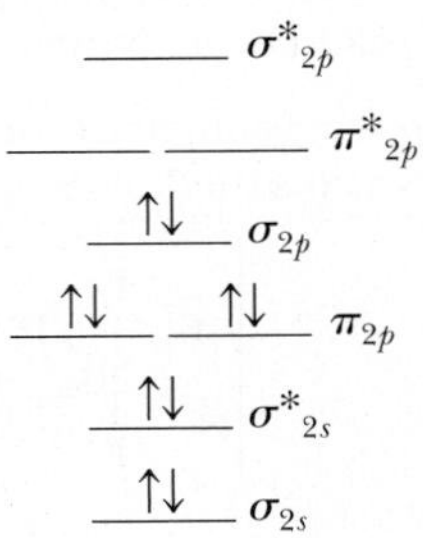

The ion has 10 valence electrons (isoelectronic with N_2). There are one net sigma bond and two net pi bonds, for a bond order of 3. The bond order increases by 1 on going from C_2 to C_2^{2-}. The ion is not paramagnetic.

9.19 (a) CO has 10 valence electrons

$$[\text{core}](\sigma_{2s})^2(\sigma^*_{2s})^2(\pi_{2p})^4(\sigma_{2p})^2$$

(b) σ_{2p}

(c) Diamagnetic

(d) There are net 1 σ bond and 2 π bonds; bond order is 3.

9.21

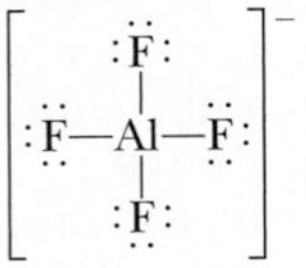

The electron pair and molecular geometries are both tetrahedral. The Al atom is sp^3 hybridized, and so the Al—F bonds are formed by overlap of an Al sp^3 orbital with a *p* orbital on each F atom. The formal charge on each of the fluorines is zero, and that on the Al is −1. This is not a reasonable charge distribution because the less electronegative atom, aluminum, has the negative charge.

9.23

Molecule/Ion	O—S—O Angle	Hybrid Orbitals
SO_2	120°	sp^2
SO_3	120°	sp^2
SO_3^{2-}	109°	sp^3
SO_4^{2-}	109°	sp^3

9.25

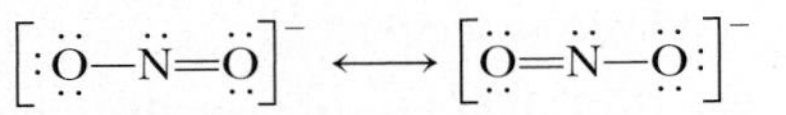

The electron-pair geometry is trigonal planar. The molecular geometry is bent (or angular). The O—N—O angle will be about 120°, the average N—O bond order is 3/2, and the N atom is sp^2 hybridized.

9.27 The resonance structures of N_2O, with formal charges, are shown here.

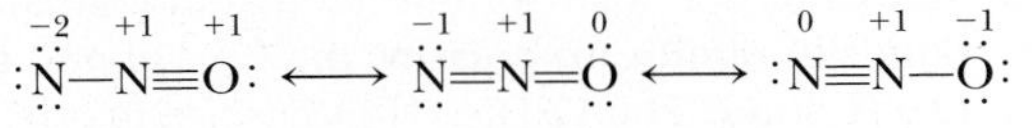

A B C

The central N atom is *sp* hybridized in all structures. The two *sp* hybrid orbitals on the central N atom are used to form N—N and N—O σ bonds. The two *p* orbitals not used in the N atom hybridization are used to form the required π bonds.

9.29 (a) All three have the formula C_2H_4O. They are usually referred to as structural isomers.

(b) *Ethylene oxide:* Both C atoms are sp^3 hybridized. *Acetaldehyde:* The CH_3 carbon atom has sp^3 hybridization, and the other C atom is sp^2 hybridized. *Vinyl alcohol:* Both C atoms are sp^2 hybridized.

(c) *Ethylene oxide:* 109°.
Acetaldehyde: 109°
Vinyl alcohol: 120°.
(d) All are polar.
(e) Acetaldehyde has the strongest CO bond, and vinyl alcohol has the strongest C—C bond.

9.31 (a) CH_3 carbon atom: sp^3
C=N carbon atom: sp^2
N atom: sp^2
(b) C—N—O bond angle = 120°

9.33 (a) C(1) = sp^2; O(2) = sp^3; N(3) = sp^3; C(4) = sp^3; P(5) = sp^3
(b) Angle A = 120°; angle B = 109°; angle C = 109°; angle D = 109°
(c) The P—O and O—H bonds are most polar ($\Delta\chi$ = 1.3).

9.35 (a) C=O bond is most polar.
(b) 18 sigma bonds and five pi bonds
(c)

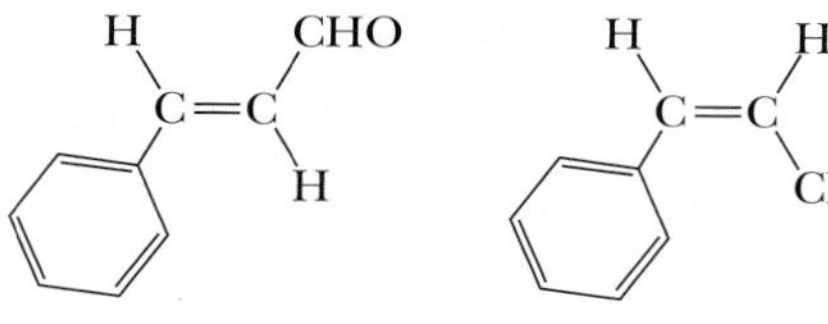

trans isomer   *cis* isomer

(d) All C atoms are sp^2 hybridized.
(e) All bond angles are 120°.

9.37 (a) The Sb in SbF_5 is sp^3d hybridized; whereas it is sp^3d^2 hybridized in SbF_6^-.
(b) The molecular geometry of the H_2F^+ ion is bent or angular, and the F atom is sp^3 hybridized.

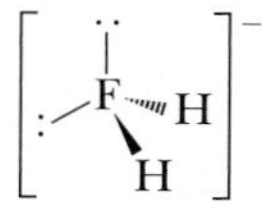

9.39 (a) The peroxide ion has a bond order of 1.

(b) [core electrons] $(\sigma_{2s})^2(\sigma^*_{2s})^2(\sigma_{2p})^2(\pi_{2p})^4(\pi^*_{2p})^4$

This configuration also leads to a bond order of 1.

(c) Both theories lead to a diamagnetic ion with a bond order of 1.

9.41 Paramagnetic diatomic molecules: B_2 and O_2

Bond order of 1: Li_2, B_2, F_2; Bond order of 2: C_2 and O_2; Highest bond order: N_2

9.43 CN has nine valence electrons
[core electrons] $(\sigma_{2s})^2(\sigma^*_{2s})^2(\pi_{2p})^4(\sigma_{2p})^1$
(a) HOMO, σ_{2p}
(b, c) Bond order = 2.5 (0.5 σ bond and 2 π bonds)
(d) Paramagnetic

9.45 (a) All C atoms are sp^3 hybridized
(b) About 109°
(c) Polar
(d) The six-membered ring cannot be planar, owing to the tetrahedral C atoms of the ring. The bond angles are all 109°.

9.47 (a) The geometry about the boron atom is trigonal planar in BF_3, but tetrahedral in $H_3N—BF_3$.
(b) Boron is sp^2 hybridized in BF_3 but sp^3 hybridized in $H_3N—BF_3$.
(c) Yes
(d) The ammonia molecule is polar with the N atom partially negative. While the BF_3 molecule is nonpolar overall, each of the B—F bonds is polarized such that the B has a partial positive charge. The partially negative N in NH_3 is attracted to the partially positive B in BF_3.
(e) One of the lone pairs on the oxygen of H_2O can form a coordinate covalent bond with the B in BF_3. The resulting compound would be (the lone pairs on the F's not shown):

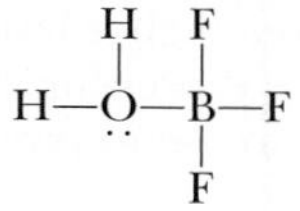

9.49 (a) NH_2^-: electron-pair geometry = tetrahedral, molecular geometry = bent, hybridization of N = sp^3

SO_3: electron-pair geometry = molecular geometry = trigonal planar, hybridization of S = sp^2

(b)

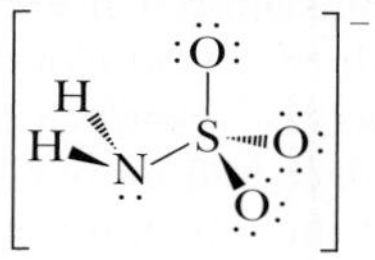

The bond angles around the N and the S are all approximately 109°.
(c) The N does not undergo any change in its hybridization; the S changes from sp^2 to sp^3.
(d) The SO_3 is the acceptor of an electron pair in this reaction. The electrostatic potential map confirms this to be reasonable because the sulfur has a partial positive charge.

9.51 A C atom may form, at most, four hybrid orbitals (sp^3). The minimum number is two, for example, the *sp* hybrid orbitals used by carbon in CO. Carbon has only four valence orbitals, so it cannot form more than four hybrid orbitals.

9.53 (a) C, sp^2; N, sp^3

(b) The amide or peptide link has two resonance structures (shown here with formal charges on the O and N atoms). Structure B is less favorable, owing to the separation of charge.

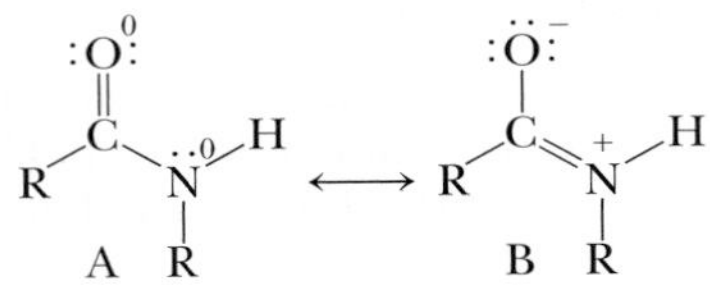

(c) The fact that the amide link is planar indicates that structure B has some importance.

The principal sites of positive charge are the nitrogen in the amide linkage, and the hydrogen of the —O—H group. The principal regions of negative charge are oxygen atoms and the nitrogen of the free $-NH_2$ group.

9.55 MO theory is better to use when explaining or understanding the effect of adding energy to molecules. A molecule can absorb energy and an electron can thus be promoted to a higher level. Using MO theory, one can see how this can occur. Additionally, MO theory is a better model to use to predict whether a molecule is paramagnetic.

9.57 Lowest Energy = Orbital C < Orbital B < Orbital A = Highest Energy

9.59 (a) The attractive forces must be greater than the repulsive forces if a covalent bond is to form.

(b) As the atoms approach each other, the energy drops as the electron clouds overlap and electron density increases between the two nuclei. If the atoms approach still more closely, electrostatic repulsion of the nuclei for each other and of the electrons for each other increases dramatically.

(c) In neon, all of the orbitals in the 2*s* and 2*p* sublevels are filled with paired electrons; there is no orbital available that can overlap with another orbital on another atom. In the case of fluorine, there is an orbital on each atom that is not completely filled that can overlap with another orbital to form a bond.

9.61 (a) The molecule with the double bond requires a great deal more energy because the π bond must be broken in order for the ends of the molecules to rotate relative to each other.

(b) No. The carbon–carbon double bonds in the molecule prevent the CH_2 fragments from rotating.

CHAPTER 10

10.1 Heptane

10.3 $C_{14}H_{30}$ is an alkane and C_5H_{10} could be a cycloalkane.

10.5 2,3-dimethylbutane

10.7 (a) 2,3-Dimethylhexane

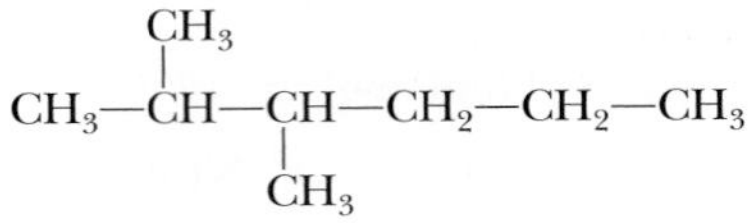

(b) 2,3-Dimethyloctane

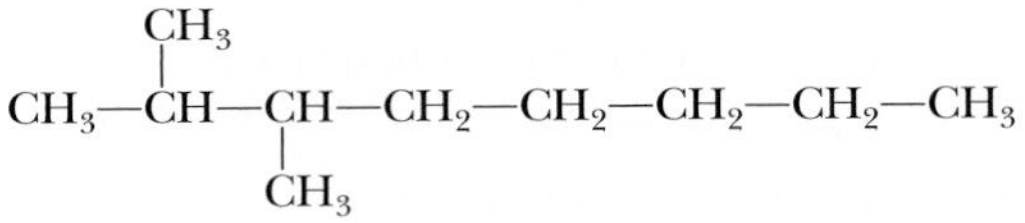

(c) 3-Ethylheptane

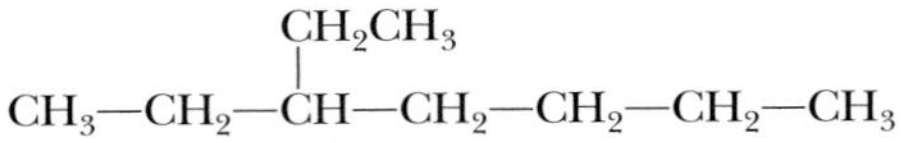

(d) 3-Ethyl-2-methylhexane

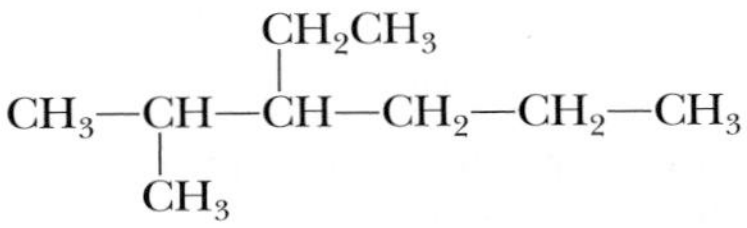

10.9

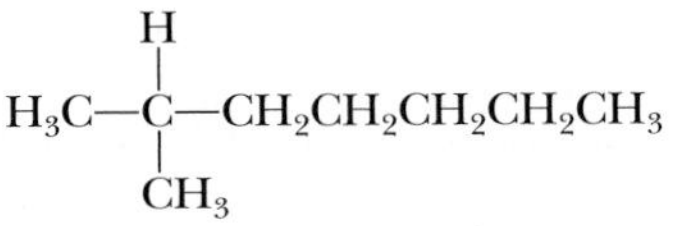

2-methylheptane

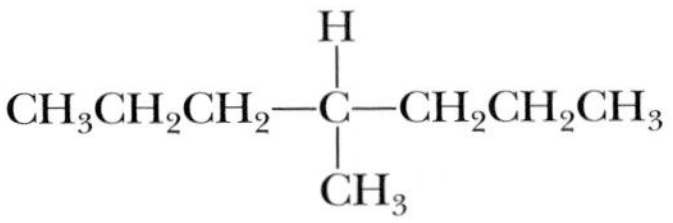

4-methylheptane

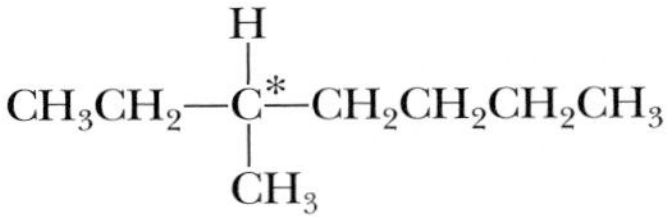

3-methylheptane. The C atom with an asterisk is chiral.

10.11

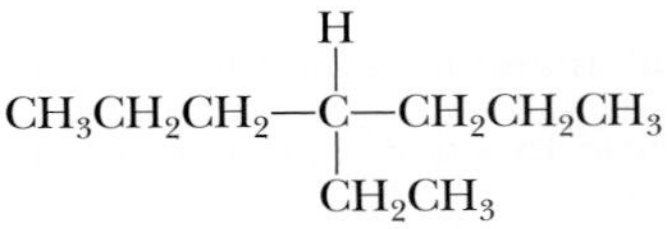

4-ethylheptane. The compound is not chiral.

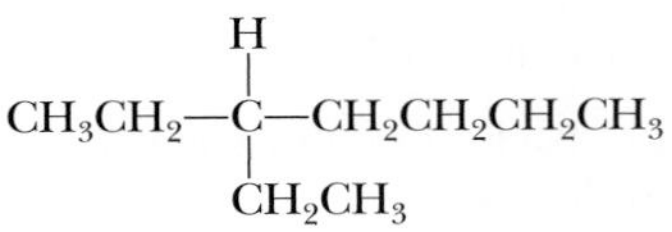

3-ethylheptane. Not chiral.

10.13 C_4H_{10}, butane: a low-molecular–weight fuel gas at room temperature and pressure. Slightly soluble in water.

$C_{12}H_{26}$, dodecane: a colorless liquid at room temperature. Expected to be insoluble in water but quite soluble in nonpolar solvents.

10.15

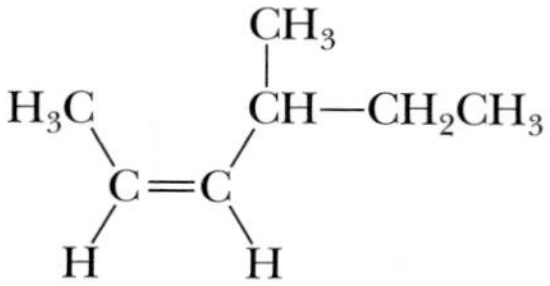

cis-4-methyl-2-hexene

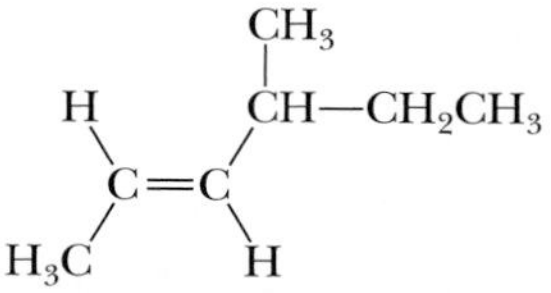

trans-4-methyl-2-hexene

10.17 (a)

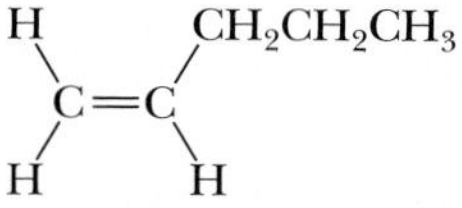

1-pentene

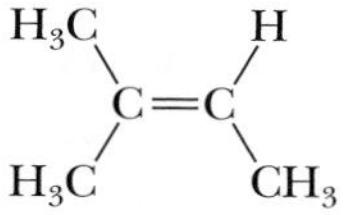

2-methyl-2-butene

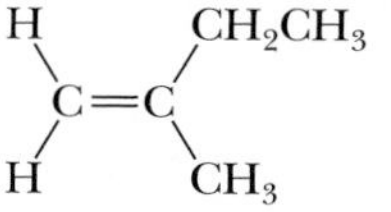

2-methyl-1-butene

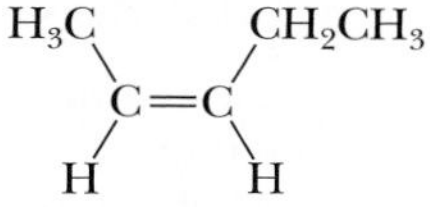

cis-2-pentene

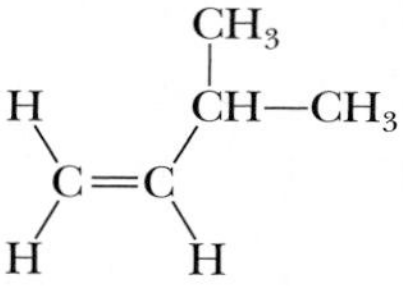

3-methyl-1-butene

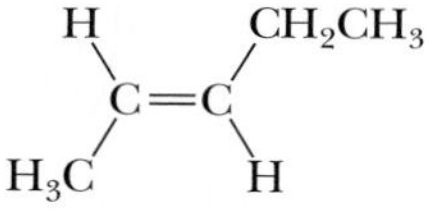

trans-2-pentene

(b)

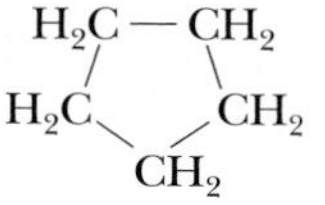

cyclopentane

10.19 (a) 1,2-Dibromopropane, $CH_3CHBrCH_2Br$
(b) Pentane, C_5H_{12}

10.21 1-Butene, $CH_3CH_2CH{=}CH_2$, or 1-butene

10.23 Four isomers are possible.

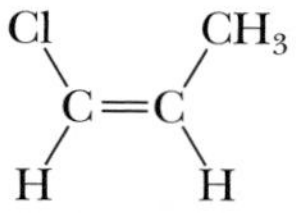

cis-1-chloropropene

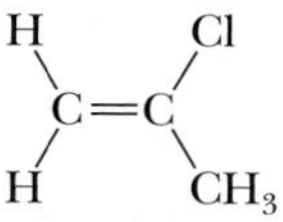

2-chloropropene

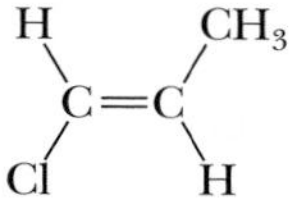

trans-1-chloropropene

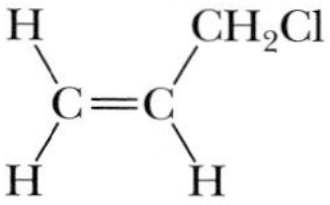

3-chloro-1-propene

10.25

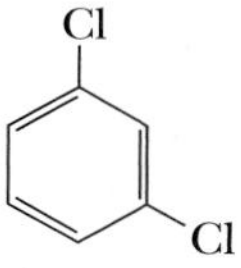

m-dichlorobenzene

o-bromotoluene

10.27

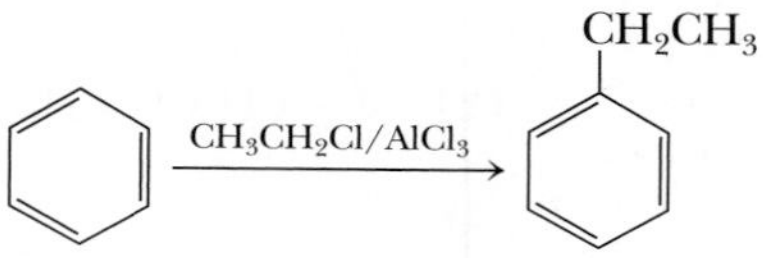

ethylbenzene

10.29

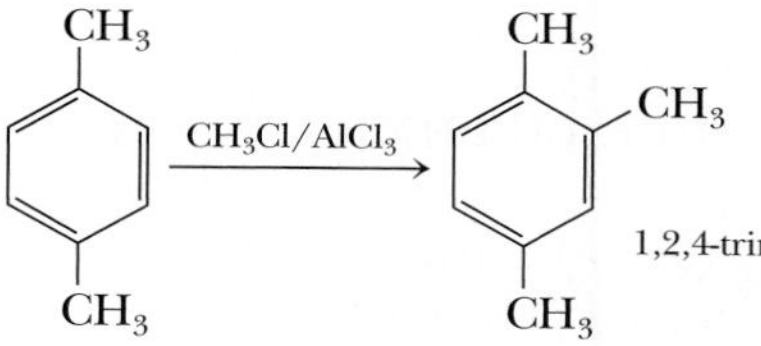

1,2,4-trimethylbenzene

10.31 (a) 1-Propanol, primary
(b) 1-Butanol, primary
(c) 2-Methyl-2-propanol, tertiary
(d) 2-Methyl-2-butanol, tertiary

10.33 (a) Ethylamine, $CH_3CH_2NH_2$
(b) Dipropylamine, $(CH_3CH_2CH_2)_2NH$

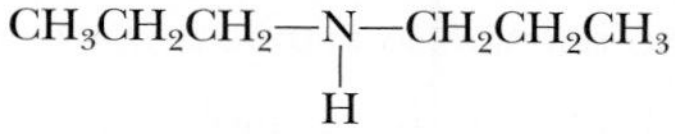

(c) Butyldimethylamine

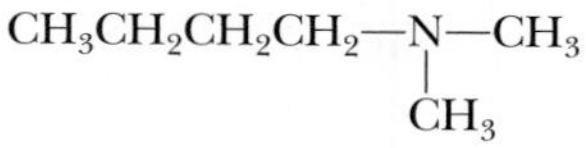

(d) triethylamine

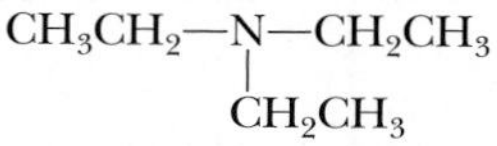

10.33 (a) 1-butanol, $CH_3CH_2CH_2CH_2OH$
(b) 2-butanol

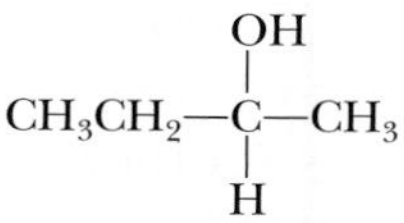

(c) 2-methyl-1-propanol

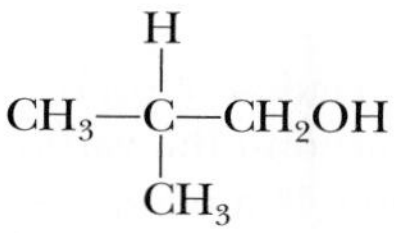

(d) 2-methyl-2-propanol

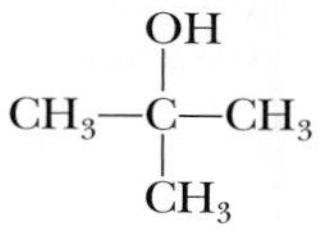

10.37 (a) $C_6H_5NH_2(aq) + HCl(aq) \rightarrow (C_6H_5NH_3)Cl(aq)$
(b) $(CH_3)_3N(aq) + H_2SO_4(aq) \rightarrow [(CH_3)_3NH]HSO_4(aq)$

10.39

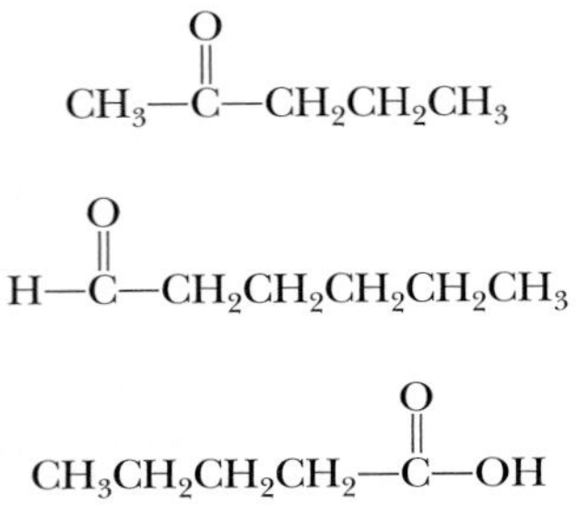

10.41 (a) Acid, 3-methylpentanoic acid
(b) Ester, methyl propanoate
(c) Ester, butyl acetate (or butyl ethanoate)
(d) Acid, *p*-bromobenzoic acid

10.43 (a) Pentanoic acid (see Question 39c)
(b) 1-Pentanol, $CH_3CH_2CH_2CH_2CH_2OH$
(c)

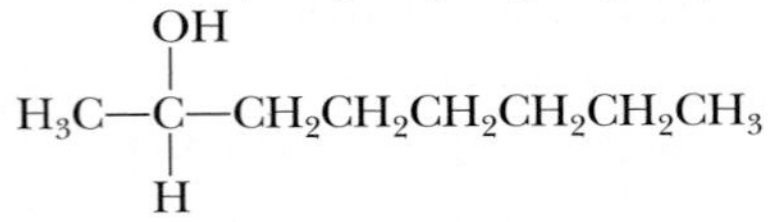

(d) No reaction. A ketone is not oxidized by $KMnO_4$.

10.45 Step 1: Oxidize 1-propanol to propanoic acid.

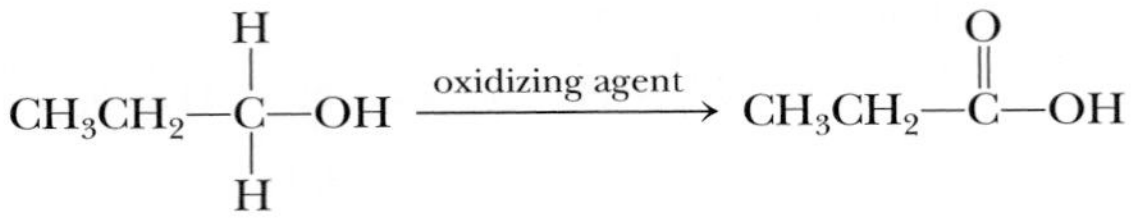

Step 2: Combine propanoic acid and 1-propanol.

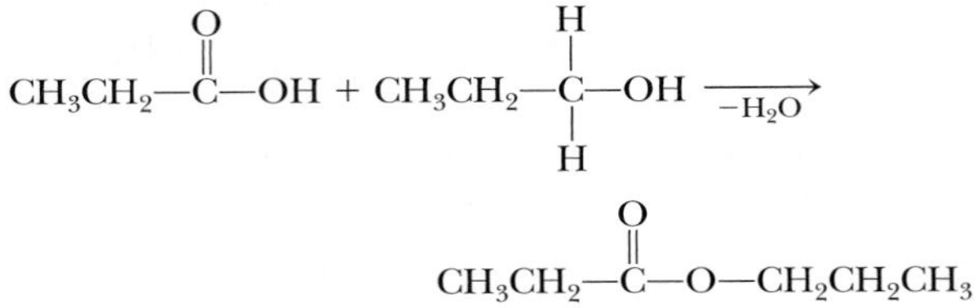

10.47 Sodium acetate, $NaCH_3CO_2$, and 1-butanol, $CH_3CH_2CH_2CH_2OH$

10.49 (a) Trigonal planar
(b) 120°
(c) The molecule is chiral. There are four different groups around the carbon atom marked 2.
(d) The acidic H atom is the H attached to the CO_2H (carboxyl) group.

10.51 (a) Alcohol (c) Acid
(b) Amide (d) Ester

10.53 (a) Prepare polyvinyl acetate (PVA) from vinylacetate.

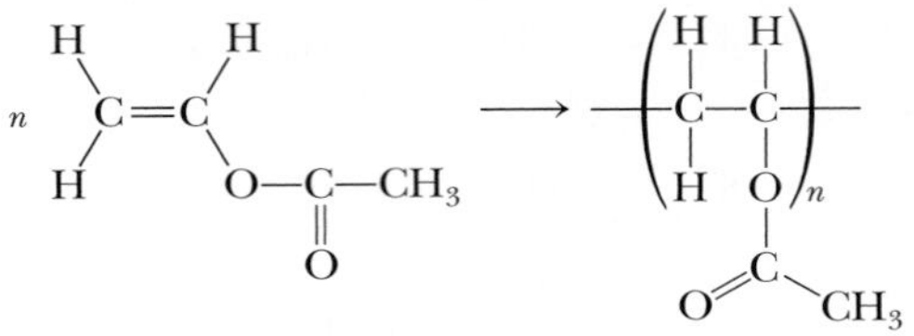

(b) The three units of PVA:

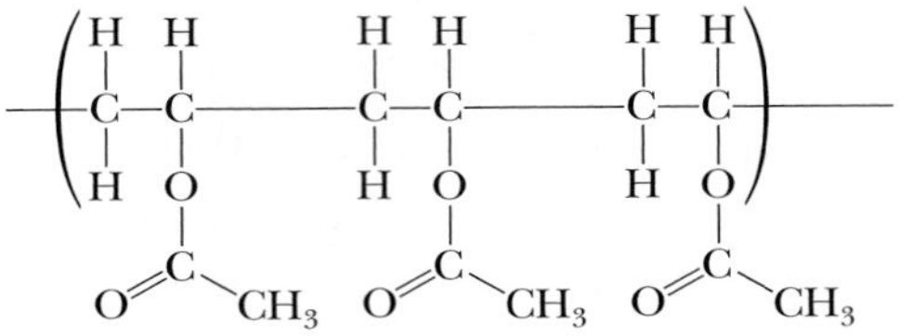

(c) Hydrolysis of polyvinyl alcohol

10.55 Illustrated here is a segment of a copolymer composed of two units of 1,1–dichloroethylene and two units of chloroethylene.

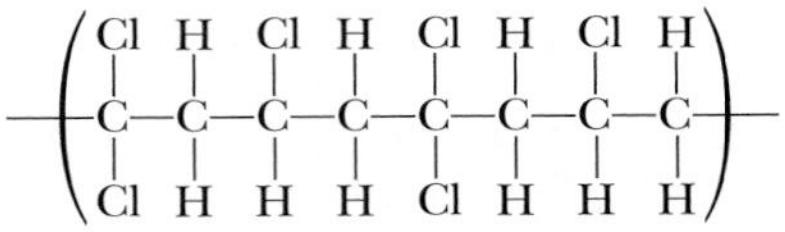

10.57 (a)

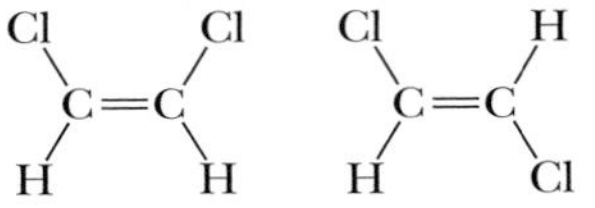

cis isomer *trans* isomer

(b)

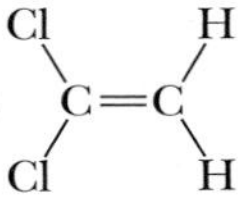

10.59

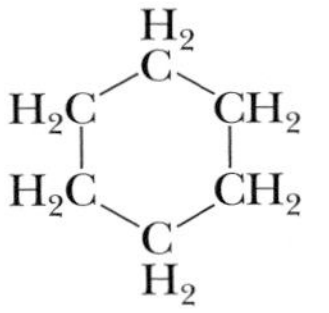

cyclohexane

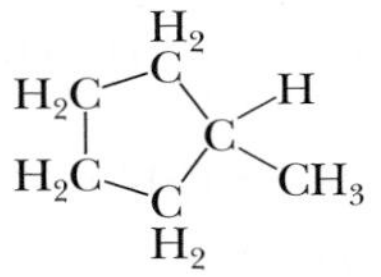

methylcyclopentane

$CH_3CH{=}CHCH_2CH_2CH_3$

2-hexene
Other isomers are possible by moving the double bond and with a branched chain.

10.61

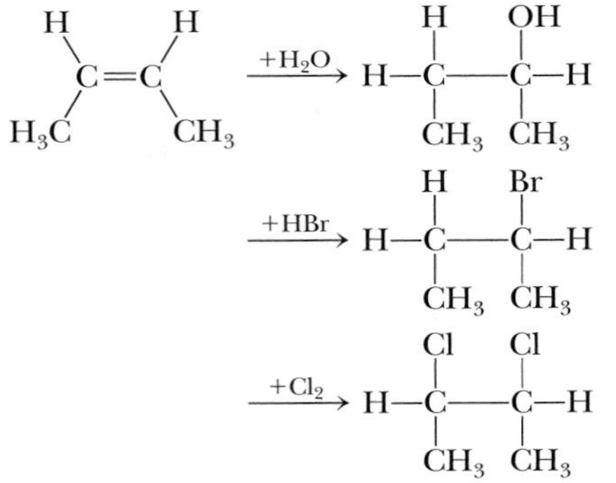

10.63 (a)

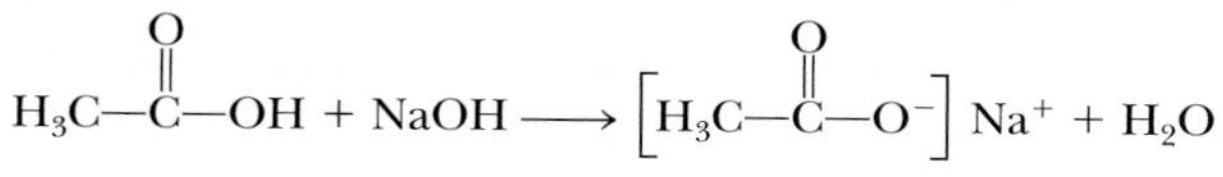

(b)

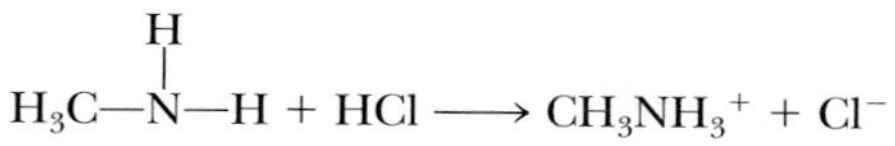

10.65

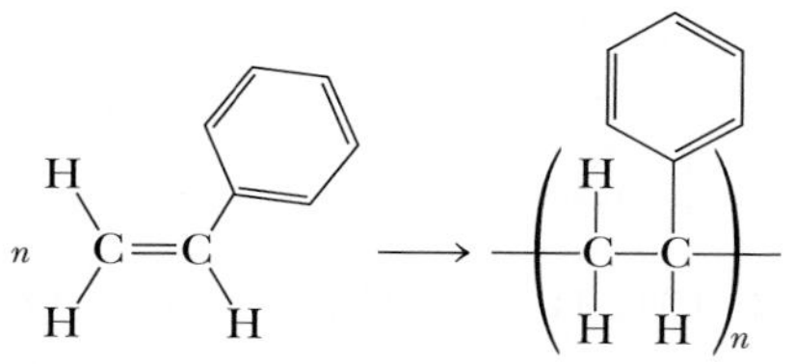

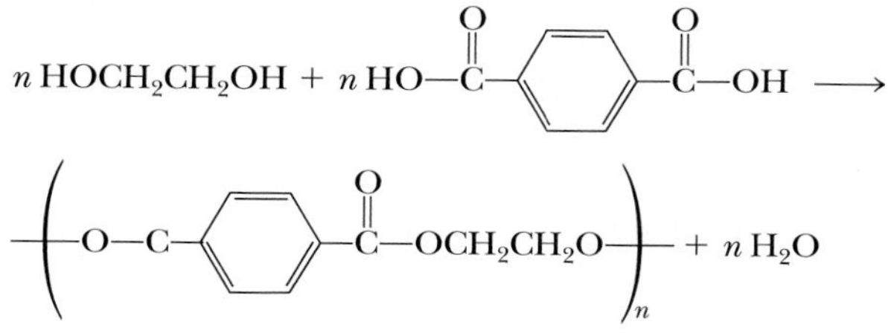

10.67 (a) 2, 3-Dimethylpentane

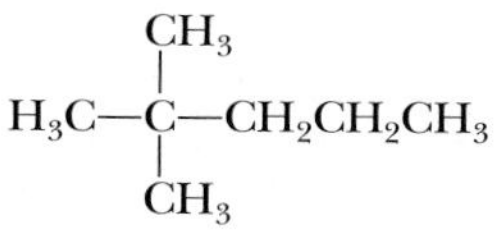

(b) 3, 3-Dimethylpentane

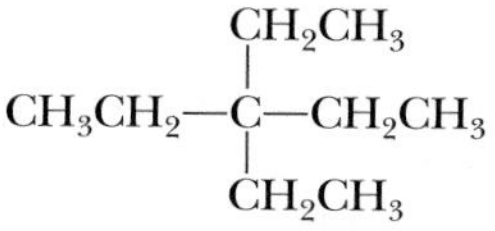

(c) 3-Ethyl-2-methylpentane

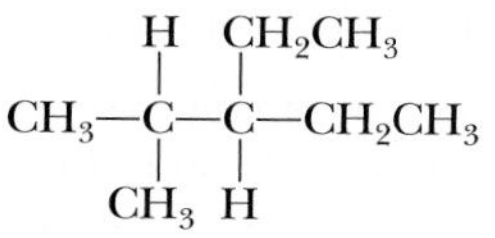

(d) 3-Ethylhexane

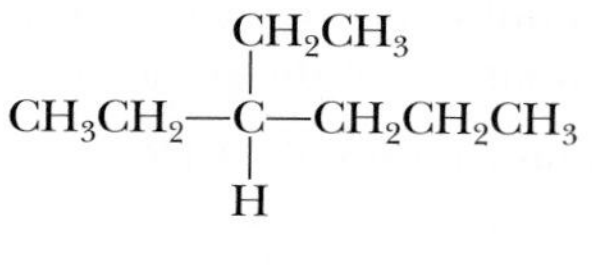

10.69

1,1-Dichloropropane

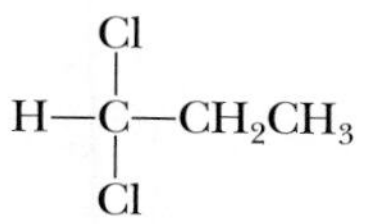

1,2-Dichloropropane

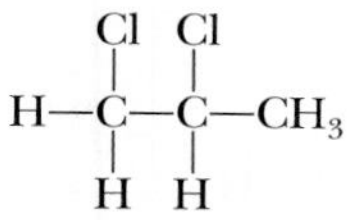

1,3-Dichloropropane

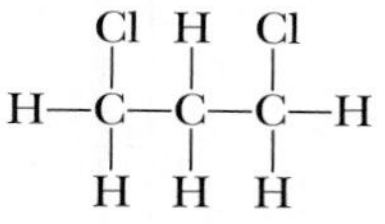

2,2-Dichloropropane

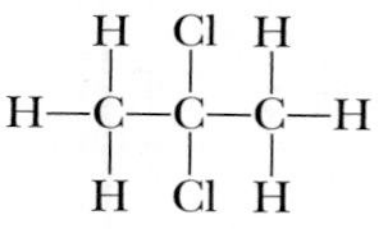

10.71

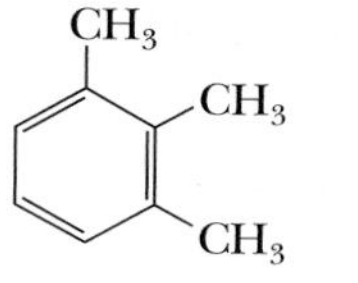

1,2,3-trimethylbenzene

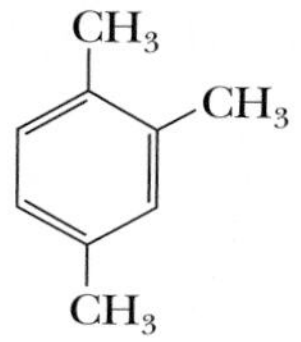

1,2,4-trimethylbenzene

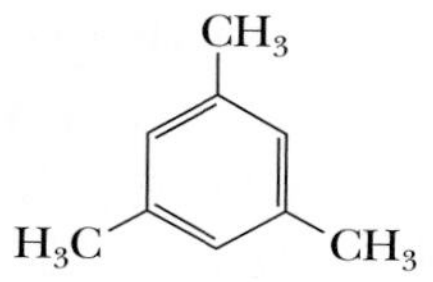

1,3,5-trimethylbenzene

10.73 Replace the carboxylic acid group with an H atom.

10.75

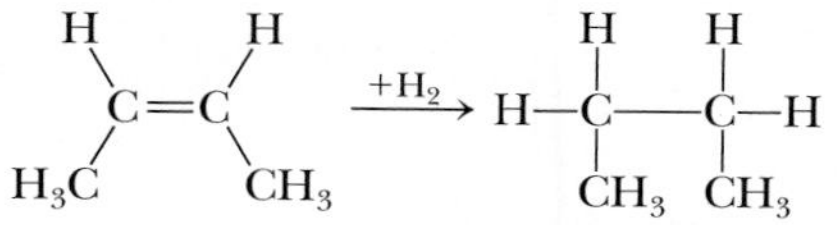

butane (not chiral)

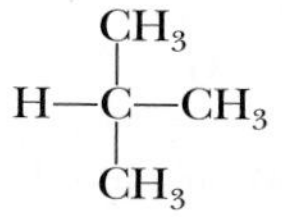

10.77

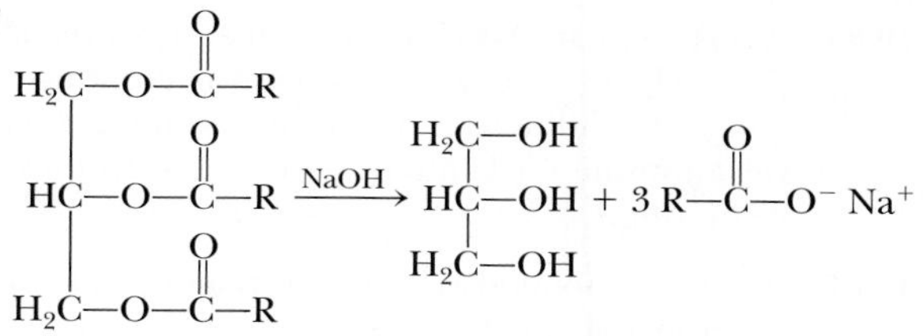

glyceryl trilaurate

glycerol

sodium laurate

10.79

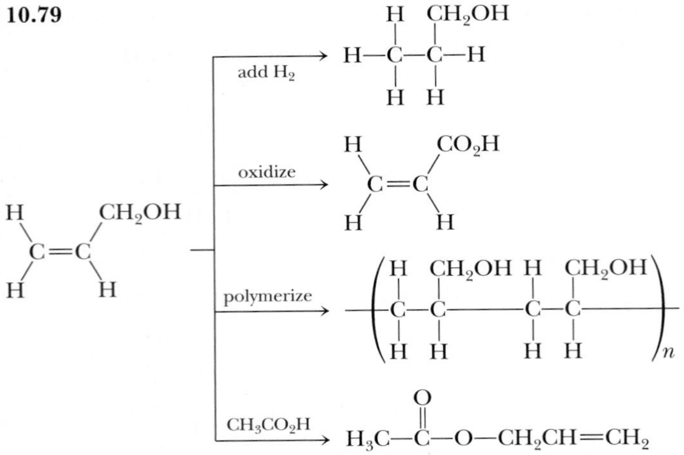

10.81 (a)

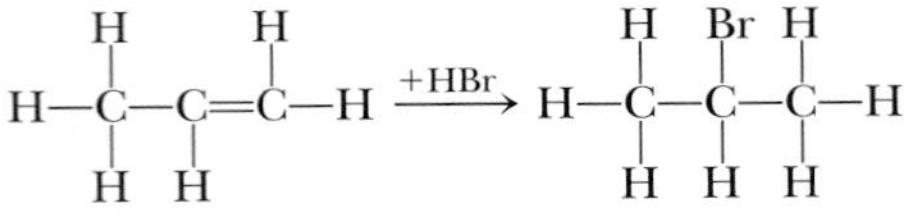

2-bromopropane

(b)

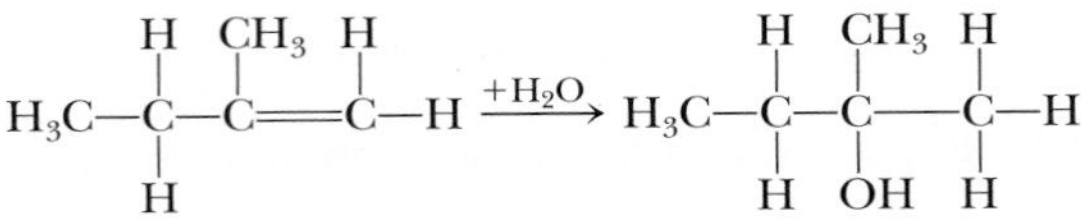

2-methyl-2-butanol

(c)

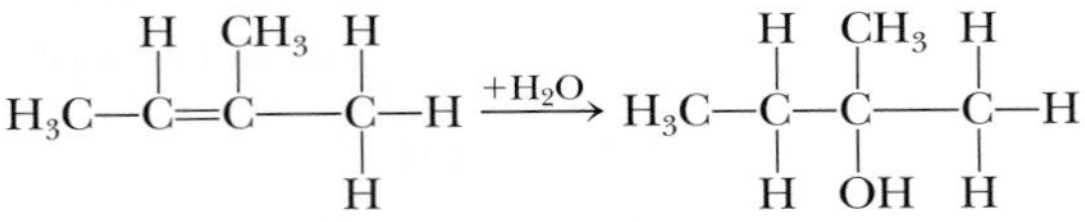

10.83 Compound (b), acetaldehyde, and (c), ethanol, produce acetic acid when oxidized.

10.85 Cyclohexene, a cyclic alkene, will add Br_2 readily (to give $C_6H_{12}Br_2$). Benzene, however, needs much more stringent conditions to react with bromine; then Br_2 will substitute for H atoms on benzene and not add to the ring.

10.87 (a) The compound is either propanone, a ketone, or propanal, an aldehyde.

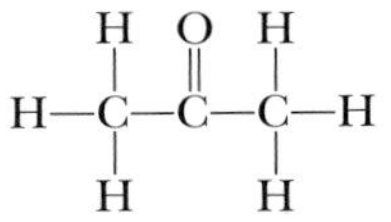

propanone
(a ketone)

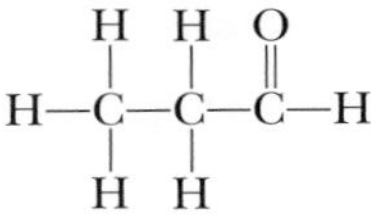

propanal
(an aldehyde)

(b) The ketone will not undergo oxidation, but the aldehyde will be oxidized to the acid, $CH_3CH_2CO_2H$. Thus, the unknown is likely propanal.

(c) Propanoic acid

10.89 2-Propanol will react with an oxidizing agent such as $KMnO_4$ (to give the ketone), whereas methyl ethyl ether ($CH_3OC_2H_5$) will not react. In addition, the alcohol should be more soluble in water than the ether.

10.91

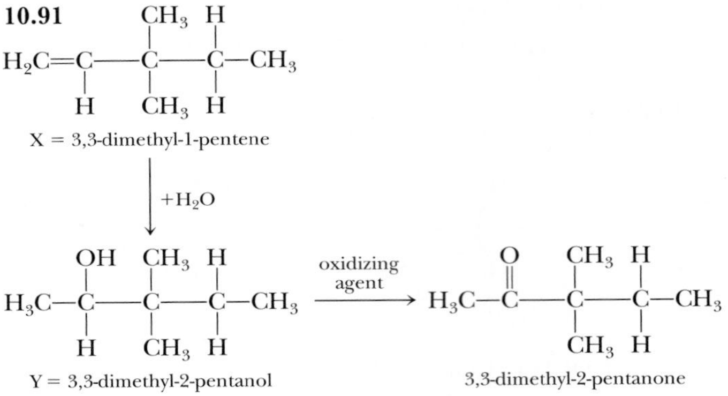

10.93

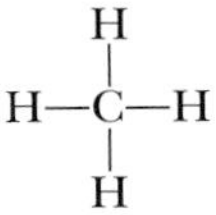	methane	four single bonds
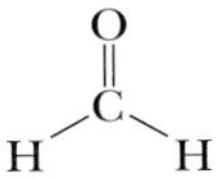	formaldehyde	one double bond and two single bonds
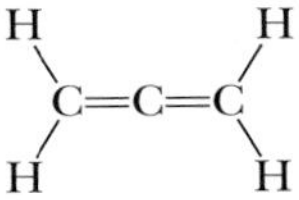	allene	two double bonds
H—C≡C—H	acetylene	one single bond and one triple bond

10.95 (a) Cross-linking makes the material very rigid and inflexible.

(b) The OH groups give the polymer a high affinity for water.

(c) Hydrogen bonding allows the chains to form coils and sheets with high tensile strength.

10.97 (a) Ethane heat of combustion = −47.51 kJ/g
Ethanol heat of combustion = −26.82 kJ/g

(b) The heat obtained from the combustion of ethanol is less negative than for ethane, so partially oxidizing ethane to form ethanol decreases the amount of energy per mole available from the combustion of the substance.

10.99 (a) Empirical formula, CHO

(b) Molecular formula, $C_4H_4O_4$

(c)

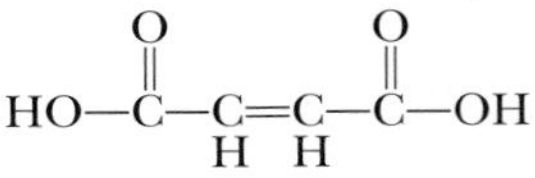

(d) All four C atoms are sp^2 hybridized.

(e) 120°

CHAPTER 11

11.1 (a) 0.58 atm
(b) 0.59 bar
(c) 59 kPa

11.3 (a) 0.754 bar
(b) 650 kPa
(c) 934 kPa

11.5 2.70×10^2 mm Hg

11.7 3.7 L

11.9 250 mm Hg

11.11 3.2×10^2 mm Hg

11.13 9.72 atm

11.15 (a) 75 mL O_2
(b) 150 mL NO_2

11.17 0.919 atm

11.19 $V = 2.9$ L

11.21 1.9×10^6 g He

11.23 3.7×10^{-4} g/L

11.25 34.0 g/mol

11.27 57.5 g/mol

11.29 Molar mass = 74.9 g/mol; B_6H_{10}

11.31 0.039 mol H_2; 0.096 atm; 73 mm Hg

11.33 170 g NaN_3

11.35 1.7 atm O_2

11.37 4.1 atm H_2; 1.6 atm Ar; total pressure = 5.7 atm

11.39 (a) 0.30 mol halothane/1 mol O_2
(b) 3.0×10^2 g halothane

11.41 (a) CO_2 has the higher kinetic energy.
(b) The average speed of the H_2 molecules is greater than the average speed of the CO_2 molecules.
(c) The number of CO_2 molecules is greater than the number of H_2 molecules [$n(CO_2) = 1.8n(H_2)$].
(d) The mass of CO_2 is greater than the mass of H_2.

11.43 Average speed of CO_2 molecule = 3.65×10^4 cm/s

11.45 Average speed increases (and molar mass decreases) in the order $CH_2F_2 < Ar < N_2 < CH_4$.

11.47 (a) F_2 (38 g/mol) effuses faster than CO_2 (44 g/mol).
(b) N_2 (28 g/mol) effuses faster than O_2 (32 g/mol).
(c) C_2H_4 (28.1 g/mol) effuses faster than C_2H_6 (30.1 g/mol).
(d) $CFCl_3$ (137 g/mol) effuses faster than $C_2Cl_2F_4$ (171 g/mol).

11.49 36 g/mol

11.51 P from the van der Waals equation = 26.0 atm

P from the ideal gas law = 30.6 atm

11.53 (a) Standard atmosphere: 1 atm; 760 mm Hg; 101.325 kPa; 1.013 bar.
(b) N_2 partial pressure: 0.780 atm; 593 mm Hg; 79.1 kPa; 0.791 bar
(c) H_2 pressure: 131 atm; 9.98×10^4 mm Hg; 1.33×10^4 kPa; 133 bar
(d) Air: 0.333 atm; 253 mm Hg; 33.7 kPa; 0.337 bar

11.55 $T = 290.$ K or 17 °C

11.57 $2\ C_4H_9SH(g) + 15\ O_2(g) \longrightarrow 8\ CO_2(g) + 10\ H_2O(g) + 2\ SO_2(g)$

Total pressure = 37.3 mm Hg. Partial pressures: CO_2 = 14.9 mm Hg, H_2O = 18.6 mm Hg, and SO_2 = 3.73 mm Hg.

11.59 4 mol

11.61 Ni is the limiting reactant; 1.31 g $Ni(CO)_4$

11.63 (a, b) Sample 4 (He) has the largest number of molecules and sample 3 (H_2 at 27 °C and 760 mm Hg) has the fewest number of molecules.
(c) Sample 2 (Ar)

11.65 8.54 g $Fe(CO)_5$

11.67 S_2F_{10}

11.69 (a) 28.7 g/mol ≃ 29 g/mol
(b) X of O_2 = 0.17 and X of N_2 = 0.83

11.71 Molar mass = 86.4 g/mol. The gas is probably ClO_2F.

11.73 n(He) = 0.0128 mol

11.75 Weight percent $KClO_3$ = 69.1%

11.77 (a) $NO_2 < O_2 < NO$
(b) $P(O_2)$ = 75 mm Hg
(c) $P(NO_2)$ = 150 mm Hg

11.79 $P(NH_3)$ = 69 mm Hg and $P(F_2)$ = 51 mm Hg

Pressure after reaction = 17 mm Hg

11.81 At 20 °C, there is 7.8×10^{-3} g H_2O/L. At 0 °C, there is 4.6×10^{-3} g H_2O/L.

11.83 The mixture contains 0.22 g CO_2 and 0.77 g CO.

$P(CO_2)$ = 0.22 atm; $P(O_2)$ = 0.12 atm; $P(CO)$ = 1.22 atm

11.85 The formula of the iron compound is $Fe(CO)_5$.

11.87 (a) $P(B_2H_6)$ = 0.0160 atm
(b) $P(H_2)$ = 0.0320 atm, so P_{total} = 0.0480 atm

11.89 Amount of Na_2CO_3 = 0.00424 mol
Amount of $NaHCO_3$ = 0.00951 mol
Amount of CO_2 produced = 0.0138 mol
Volume of CO_2 produced = 0.343 L

11.91 Decomposition of 1 mol of $Cu(NO_3)_2$ should give 2 mol NO_2 and ½ mol of O_2. Total actual amount = 4.72×10^{-3} mol of gas.
(a) Average molar mass = 41.3 g/mol.
(b) Mole fractions: $X(NO_2)$ = 0.666 and $X(O_2)$ = 0.334
(c) Amount of each gas: 3.13×10^{-3} mol NO_2 and 1.57×10^{-3} mol O_2
(d) If some NO_2 molecules combine to form N_2O_4, the apparent mole fraction of NO_2 would be smaller than expected (= 0.8). As this is the case, it is apparent that some N_2O_4 has been formed (as is observed in the experiment).

11.93 (a) 10.0 g of O_2 represents more molecules than 10.0 g of CO_2. Therefore, O_2 has the greater partial pressure.
(b) The average speed of the O_2 molecules is greater than the average speed of the CO_2 molecules.
(c) The gases are at the same temperature and so have the same average kinetic energy.

11.95 (a) $P(C_2H_2) > P(CO)$
(b) There are more molecules in the C_2H_2 container than in the CO container.

11.97 (a) Not a gas. A gas would expand to an infinite volume.
(b) Not a gas. A density of 8.2 g/mL is typical of a solid.
(c) Insufficient information
(d) Gas

11.99 (a) There are more molecules of H_2 than atoms of He.
(b) The mass of He is greater than the mass of H_2.

11.101 The speed of gas molecules is related to the square root of the absolute temperature, so a doubling of the temperature will lead to an increase of about $(2)^{1/2}$ or 1.4.

CHAPTER 12

12.1 (a) Dipole–dipole interactions (and hydrogen bonds)
(b) Induced dipole–induced dipole forces
(c) Dipole–dipole interactions (and hydrogen bonds)

12.3 (a) Induced dipole–induced dipole forces
(b) Induced dipole–induced dipole forces
(c) Dipole–dipole forces
(d) Dipole–dipole forces (and hydrogen bonding)

12.5 The predicted order of increasing strength is $Ne < CH_4 < CO < CCl_4$. In this case, prediction does not quite agree with reality. The boiling points are Ne (−246 °C) < CO (−192 °C) < CH_4 (−162 °C) < CCl_4 (77 °C).

12.7 (c) HF; (d) acetic acid; (f) CH_3OH

12.9 (a) LiCl. The Li^+ ion is smaller than Cs^+ (Figure 7.12), which makes the ion–ion forces of attraction stronger in LiCl.
(b) $Mg(NO_3)_2$. The Mg^{2+} ion is smaller than the Na^+ ion (Figure 7.12), and the magnesium ion has a 2+ charge (as opposed to 1+ for sodium). Both of these effects lead to stronger ion–ion forces of attraction in magnesium nitrate.
(c) $NiCl_2$. The nickel(II) ion has a larger charge than Rb^+ and is considerably smaller. Both effects mean that there are stronger ion–ion forces of attraction in nickel(II) chloride.

12.11 q = +90.1 kJ

12.13 (a) Water vapor pressure is about 150 mm Hg at 60 °C. (Appendix G gives a value of 149.4 mm Hg at 60 °C.)
(b) 600 mm Hg at about 93 °C
(c) At 70 °C, ethanol has a vapor pressure of about 520 mm Hg, whereas that of water is about 225 mm Hg.

12.15 At 30 °C, the vapor pressure of ether is about 590 mm Hg. (This pressure requires 0.23 g of ether in the vapor phase at the given conditions, so there is sufficient ether in the flask.) At 0 °C, the vapor pressure is about 160 mm Hg, so some ether condenses when the temperature declines.

12.17 (a) O_2 (−183 °C) (bp of N_2 = −196 °C)
(b) SO_2 (−10 °C) (CO_2 sublimes at −78 °C)
(c) HF (+19.7 °C) (HI, −35.6 °C)
(d) GeH_4 (−90.0 °C) (SiH_4, −111.8 °C)

12.19 (a) CS_2, about 620 mm Hg; CH_3NO_2, about 80 mm Hg
(b) CS_2, induced dipole–induced dipole forces; CH_3NO_2, dipole–dipole forces
(c) CS_2, about 46 °C; CH_3NO_2, about 100 °C
(d) About 39 °C
(e) About 34 °C

12.21 (a) 80.1 °C
(b) At about 48 °C, the liquid has a vapor pressure of 250 mm Hg.

The vapor pressure is 650 mm Hg at 75 °C.

(c) 33.5 kJ/mol (from slope of plot)

12.23 No, CO cannot be liquefied at room temperature because the critical temperature is lower than room temperature.

12.25 $Ar < CO_2 < CH_3OH$

12.27 Li^+ ions are smaller than Cs^+ ions (78 pm and 165 pm, respectively; see Figure 7.12). Thus, there will be a stronger attractive force between Li^+ ion and water molecules than between Cs^+ ions and water molecules.

12.29 (a) 350 mm Hg
(b) Ethanol (lower vapor pressure at every temperature)
(c) 84 °C
(d) CS_2, 46 °C; C_2H_5OH, 78 °C; C_7H_{16}, 99 °C
(e) CS_2, gas; C_2H_5OH, gas; C_7H_{16}, liquid

12.31 Molar enthalpy of vaporization increases with increasing intermolecular forces: C_2H_6 (14.69 kJ/mol; induced dipole) < HCl (16.15 kJ/mol; dipole) < CH_3OH (35.21 kJ/mol, hydrogen bonds). (The molar enthalpies of vaporization here are given at the boiling point of the liquid.)

12.33 5.49×10^{19} atoms/m^3

12.35 (a) 70.3 °C
(b)

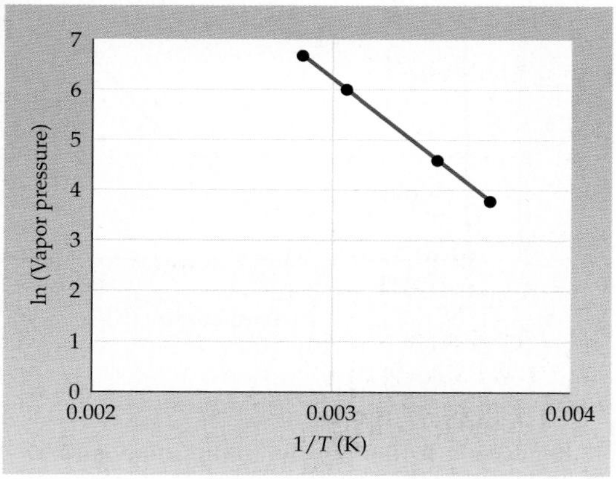

Using the equation for the straight line in the plot

$$\ln P = -3885\ (1/T) + 17.949$$

we calculate that $T = 312.6$ K (39.5 °C) when $P = 250$ mm Hg. When $P = 650$ mm Hg, $T = 338.7$ K (65.5 °C).
(c) Calculated $\Delta_{vap}H = 32.3$ kJ/mol

12.37 When the can is inverted in cold water, the water vapor pressure in the can, which was approximately 760 mm Hg, drops rapidly—say, to 9 mm Hg at 10 °C. This creates a partial vacuum in the can, and the can is crushed because of the difference in pressure inside the can and the pressure of the atmosphere pressing down on the outside of the can.

12.39 Acetone and water can interact by hydrogen bonding.

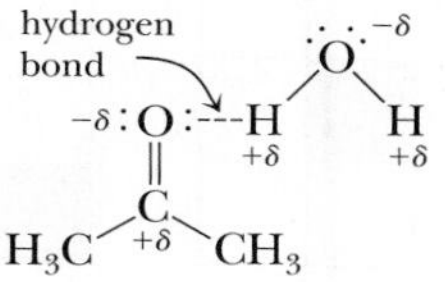

12.41 Glycol's viscosity will be greater than ethanol's, owing to the greater hydrogen-bonding capacity of glycol.

12.43 (a) Water has two OH bonds and two lone pairs, whereas the O atom of ethanol has only one OH bond (and two lone pairs). More extensive hydrogen bonding is likely for water.
(b) Water and ethanol interact extensively through hydrogen bonding, so the volume is expected to be slightly smaller than the sum of the two volumes.

12.45 Two pieces of evidence for $H_2O(\ell)$ having considerable intermolecular attractive forces:
(a) Based on the boiling points of the Group 6A hydrides (Figure 12.6), the boiling point of water should be approximately −80 °C. The actual boiling point of 100 °C reflects the significant hydrogen bonding that occurs.
(b) Liquid water has a specific heat capacity that is higher than almost any other liquid. This reflects the fact that a relatively larger amount of energy is necessary to overcome intermolecular forces and raise the temperature of the liquid.

12.47 (a) HI, hydrogen iodide
(b) The large iodine atom in HI leads to a significant polarizability for the molecule and thus to a large dispersion force.
(c) The dipole moment of HCl (1.07 D, Table 9.8) is larger than for HI (0.38 D).
(d) HI. See part (b).

12.49 A gas can be liquefied at or below its critical temperature. The critical temperature for CF_4 (−45.7 °C) is below room temperature (25 °C), so it cannot be liquefied at room temperature.

12.51 Hydrogen bonding is most likely at the O—H group at the "right" end of the molecule, and at the C=O and N—H groups in the amide group (—NH—CO—).

CHAPTER 13

13.1 Two possible unit cells are illustrated here. The simplest formula is AB_8.

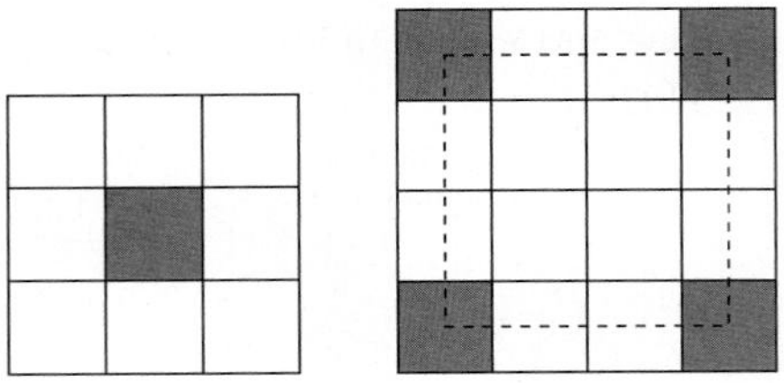

13.3 Ca^{2+} ions at eight corners = 1 net Ca^{2+} ion

O^{2-} ions in six faces = 3 net O^{2-} ions

Ti^{4+} ion in center of unit cell = 1 net Ti^{4+} ion

Formula = $CaTiO_3$

13.5 (a) There are eight O^{2-} ions at the corners and one in the center for a net of two O^{2-} ions per unit cell. There are four Cu ions in the interior in tetrahedral holes. The ratio of ions is Cu_2O.
(b) The oxidation number of copper must be +1.

13.7 Calcium atom radius = 197 pm

13.9 There are three ways the edge dimensions can be calculated:
(a) Calculate mass of unit cell (= 1.103×10^{-21} g/uc)

Calculate volume of unit cell from mass (= 3.53×10^{-22} cm^3/uc)

Calculate edge length from volume (= 707 pm)

(b) Assume I^- ions touch along the cell diagonal (see Exercise 13.2) and use I^- radius to find the edge length. Radius I^- = 220 pm

Edge = $4(220 \text{ pm})/2^{1/2}$ = 622 pm

(c) Assume the I^- and K^+ ions touch along the cell edge (page 599)

Edge = 2 × I^- radius + 2 × K^+ radius = 706 pm

Methods (a) and (c) agree. It is apparent that the sizes of the ions are such that the I^- ions cannot touch along the cell diagonal.

13.11 Increasing lattice energy: RbI < LiI < LiF < CaO

13.13 As the ion–ion distance decreases, the force of attraction between ions increases. This should make the lattice more stable, and more energy should be required to melt the compound.

13.15 $\Delta_f H^\circ = -607$ kJ/mol

13.17 (a) Eight C atoms per unit cell. There are eight corners (= 1 net C atom), six faces (= 3 net C atoms), and four internal C atoms.
(b) Face-centered cubic (fcc) with C atoms in the tetrahedral holes.

13.19 q (for fusion) = −1.97 kJ; q (for melting) = +1.97 kJ

13.21 (a) The density of liquid CO_2 is less than that of solid CO_2.
(b) CO_2 is a gas at 5 atm and 0 °C.
(c) Critical temperature = 31 °C, so CO_2 cannot be liquefied at 45 °C.

13.23 q (to heat the liquid) = 9.4×10^2 kJ

q (to vaporize NH_3) = 1.6×10^4 kJ

q (to heat the vapor) = 8.8×10^2 kJ

$q_{total} = 1.83 \times 10^4$ kJ

13.25 O_2 phase diagram. (i) Note the slight positive slope of the solid–liquid equilibrium line. It indicates that the density of solid O_2 is greater than that of liquid O_2. (ii) Using the diagram here, the vapor pressure of O_2 at 77 K is between 150 mm Hg and 200 mm Hg.

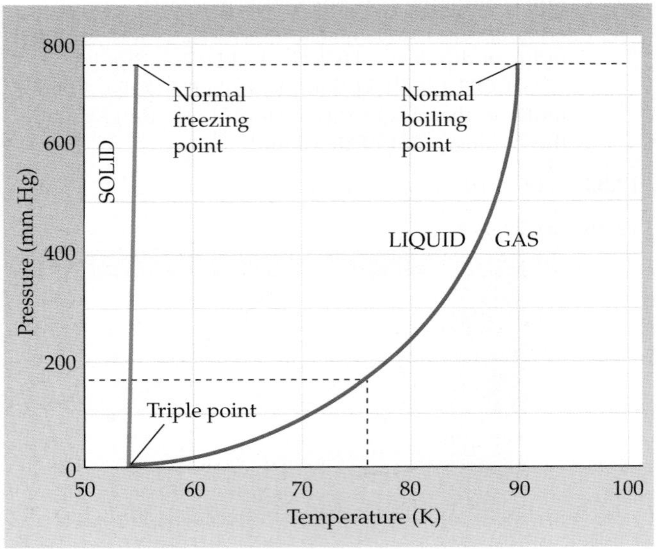

13.27 Radius of silver = 145 pm

13.29 1.356×10^{-8} cm (literature value is 1.357×10^{-8} cm)

13.31 Mass of 1 CaF_2 unit calculated from crystal data = 1.2963×10^{-22} g. Divide molar mass of CaF_2 (78.077 g/mol) by mass of 1 CaF_2 to obtain Avogadro's number. Calculated value = 6.0230×10^{23} CaF_2/mol.

13.33 Diagram A leads to a surface coverage of 78.5%. Diagram B leads to 90.7% coverage.

13.35 (a) The lattice can be described as an fcc lattice of Si atoms with Si atoms in one half of the tetrahedral holes.

(b) There are eight Si atoms in the unit cell.

Mass of unit cell = 3.731×10^{-22} g

Volume of unit cell = 1.602×10^{-22} cm^3

Density = 2.329 g/cm^3 (which is the same as the literature value)

In the Si unit cell we cannot assume the atoms touch along the edge or along the face diagonal. Instead, we know that the Si atoms in the tetahedral holes are bonded to the Si atoms at the corner.

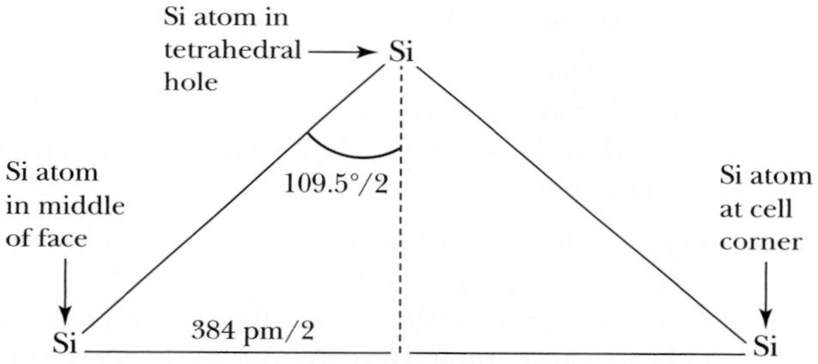

Distance = 1/2 (cell diagonal) = 384 pm

Distance across cell face diagonal = 768 pm

Sin (109.5°/2) = 0.817 = (768 pm/2)/(Si-Si distance)

Distance from Si in tetrahedral hole to face or corner Si = 235 pm

Si radius = 118 pm

Table 7.8 gives Si radius as 117 pm

13.37 (a) Mg^{2+} ions are in ⅛ of the eight possible tetrahedral holes, and Al^{3+} ions are in ½ of the four available octahedral holes.

(b) Fe^{2+} ions are in ⅛ of the eight possible tetrahedral holes, and Cr^{3+} ions are in ½ of the four available octahedral holes.

13.39 Lead sulfide has the same structure as sodium chloride, not the same structure as ZnS. There are four Pb^{2+} ions and four S^{2-} ions per unit cell, a 1:1 ratio that matches the compound formula.

13.41 (a) $BBr_3(g) + PBr_3(g) + 3\ H_2(g) \longrightarrow BP(s) + 6\ HBr(g)$

(b) If B atoms are in an fcc lattice, then the P atoms must be in ½ of the tetrahedral holes. (In this way it resembles Si in Question 13.35.)

(c) Unit cell volume = 1.092×10^{-22} cm^3
Unit cell mass = 2.775×10^{-22} g
Density = 2.54 g/cm^3

(d) The solution to this problem is identical to Question 13.35. In the BP lattice, the cell face diagonal is 676 pm. Therefore, the calculated BP distance is 207 pm.

13.43 Assuming the spheres are packed in an identical way, the water levels are the same. A face-centered cubic lattice, for example, uses 74% of the available space, regardless of the sphere size.

CHAPTER 14

14.1 (a) Concentration (m) = 0.0434 m
(b) Mole fraction of acid = 0.000781
(c) Weight percent of acid = 0.509%

14.3 NaI: 0.15 m; 2.2%; $X = 2.7 \times 10^{-3}$

CH_3CH_2OH: 1.1 m; 5.0%; $X = 0.020$

$C_{12}H_{22}O_{11}$: 0.15 m; 4.9%; $X = 2.7 \times 10^{-3}$

14.5 2.65 g Na_2CO_3; $X(Na_2CO_3) = 3.59 \times 10^{-3}$

14.7 220 g glycerol; 5.7 m

14.9 16.2 m; 37.1%

14.11 Molality = 2.6×10^{-5} m (assuming that 1 kg of seawater is equivalent to 1 kg of solvent)

14.13 (b) and (c)

14.15 $\Delta_{soln}H°$ for LiCl = −36.9 kJ/mol. This is an exothermic enthalpy of solution, as compared with the very slightly endothermic value for NaCl.

14.17 Above about 40 °C the solubility increases with temperature; therefore, add more NaCl and raise the temperature.

14.19 2×10^{-3} g O_2

14.21 1130 mm Hg or 1.49 bar

14.23 35.0 mm Hg

14.25 $X(H_2O) = 0.869$; 16.7 mol glycol; 1040 g glycol

14.27 Calculated boiling point = 84.2 °C

14.29 ΔT_{bp} = 0.808 °C; solution boiling point = 62.51 °C

14.31 Molality = 8.60 m; 28.4%

14.33 Molality = 0.195 m; ΔT_{fp} = −0.362 °C

14.35 Molar mass = 360 g/mol; $C_{20}H_{16}Fe_2$

14.37 Molar mass = 150 g/mol

14.39 Molar mass = 170 g/mol

14.41 Freezing point = −24.6 °C

14.43 0.080 *m* $CaCl_2$ < 0.10 *m* NaCl < 0.040 *m* Na_2SO_4 < 0.10 sugar

14.45 (a) $\Delta T_{fp} = -0.348$ °C; fp = −0.348 °C
(b) $\Delta T_{bp} = +0.0959$ °C; bp = 100.0959 °C
(c) $\Pi = 4.58$ atm

The osmotic pressure is large and can be measured with a small experimental error.

14.47 Molar mass = 6.0×10^3 g/mol

14.49 (a) $BaCl_2(aq) + Na_2SO_4(aq) \longrightarrow BaSO_4(s) + 2\ NaCl(aq)$
(b) Initially, the $BaSO_4$ particles form a colloidal suspension.
(c) Over time, the particles of $BaSO_4(s)$ grow and precipitate.

14.51 Molar mass = 110 g/mol

14.53 (a) Increase in vapor pressure of water

0.20 *m* Na_2SO_4 < 0.50 *m* sugar < 0.20 *m* KBr < 0.35 *m* ethylene glycol

(b) Increase in boiling point

0.35 *m* ethylene glycol < 0.20 *m* KBr < 0.50 *m* sugar < 0.20 *m* Na_2SO_4

14.55 (a) 0.456 mol DMG and 11.4 mol ethanol; $X(\text{DMG}) = 0.0385$
(b) 0.869 *m*
(c) VP ethanol over the solution at 78.4 °C = 730.7 mm Hg
(d) bp = 79.5 °C

14.57 For ammonia: 23 *m*; 28%; $X(NH_3) = 0.29$

14.59 0.592 g Na_2SO_4

14.61 (a) 0.20 *m* KBr; (b) 0.10 *m* Na_2CO_3

14.63 Freezing point = −11 °C

14.65 4.0×10^2 g/mol

14.67 4.7×10^{-4} mol/kg

14.69 (a) Molar mass = 4.9×10^4 g/mol
(b) $\Delta T_{fp} = -3.8 \times 10^{-4}$ °C

14.71 $\Delta_{soln}H°\ [Li_2SO_4] = -28.0$ kJ/mol

$\Delta_{soln}H°\ [LiCl] = -36.9$ kJ/mol

$\Delta_{soln}H°\ [K_2SO_4] = +23.7$ kJ/mol

$\Delta_{soln}H°\ [KCl] = +17.2$ kJ/mol

Both lithium compounds have exothermic enthalpies of solution, whereas both potassium compounds have endothermic values. Consistent with this is the fact that lithium salts (LiCl) are often more water-soluble than potassium salts (KCl) (see Figure 14.11).

14.73 X(benzene in solution) = 0.67 and X(toluene in solution) = 0.33

$$P_{total} = P_{toluene} + P_{benzene} = 7.3\text{ mm Hg} + 50.\text{ mm Hg} = 57\text{ mm Hg}$$

$$X(\text{toluene in vapor}) = \frac{7.3\text{ mm Hg}}{57\text{ mm Hg}} = 0.13$$

$$X(\text{benzene in vapor}) = \frac{50.\text{ mm Hg}}{57\text{ mm Hg}} = 0.87$$

14.75 $i = 1.7$. That is, there is 1.7 mol of ions in solution per mole of compound.

14.77 (a) Calculate the number of moles of ions in 10^6 g H_2O: 550. mol Cl^-; 470. mol Na^+; 53.1 mol Mg^{2+}; 9.42 mol SO_4^{2-}; 10.3 mol Ca^{2+}; 9.72 mol K^+; 0.84 mol Br^-. Total moles of ions = 1.103×10^3 per 10^6 g water. This gives ΔT_{fp} of −2.05 °C.
(b) $\Pi = 27.0$ atm. This means that a minimum pressure of 27 atm would have to be used in a reverse osmosis device.

14.79 (a) $i = 2.06$
(b) There are approximately two particles in solution, so $H^+ + HSO_4^-$ best represents H_2SO_4 in aqueous solution.

14.81 The calculated molality at the freezing point of benzene is 0.47 *m*, whereas it is 0.99 *m* at the boiling point. A higher molality at the higher temperature indicates more molecules are dissolved. Therefore, assuming benzoic acid forms dimers like acetic acid (Figure 12.7), dimer formation is more prevalent at the lower temperature. In this process two molecules become one entity, lowering the number of separate species in solution and lowering the molality.

14.83 Molar mass in benzene = 1.20×10^2 g/mol; molar mass in water = 62.4 g/mol. The actual molar mass of acetic acid is 60.1 g/mol. In benzene, the molecules of acetic acid form "dimers." That is, two molecules form a single unit through hydrogen bonding. See Figure 12.7 on page 562.

14.85 (a) Molar mass = 97.6 g/mol; empirical formula, BF_2, and molecular formula, B_2F_4
(b)

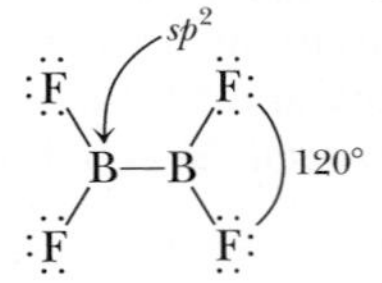

14.87 See the discussion and data on page 558.
(a) The enthalpy of hydration of LiF is more negative than that for RbF because the Li^+ ion is much smaller than the Rb^+ ion.

(b) The enthalpy of hydration for $Ca(NO_3)_2$ is larger than that for KNO_3 owing to the +2 charge on the Ca^{2+} ion (and its smaller size).
(c) The enthalpy of hydration is greater for $CuBr_2$ than for CsBr because Cu^{2+} has a larger charge than Cs^+, and the Cu^{2+} ion is smaller than the Cs^+ ion.

14.89 Li_2SO_4 should have a more negative enthalpy of hydration than Cs_2SO_4 because the Li^+ ion is smaller than the Cs^+ ion.

14.91 Colligative properties depend on the number of ions or molecules in solution. Each mole of $CaCl_2$ provides 1.5 times as many ions as each mole of NaCl.

14.93 Benzene is a nonpolar solvent. Thus, ionic substances such as $NaNO_3$ and NH_4Cl will certainly not dissolve. However, naphthalene is also nonpolar and resembles benzene in its structure; it should dissolve very well. (A chemical handbook gives a solubility of 33 g naphthalene per 100 g benzene.) Diethyl ether is weakly polar and will also be miscible to some extent with benzene.

14.95 The C—C and C—H bonds in hydrocarbons are nonpolar or weakly polar and tend to make such dispersions hydrophobic (water-hating). The C—O and O—H bonds in starch present opportunities for hydrogen bonding with water. Hence, starch is expected to be more hydrophilic.

14.97 [NaCl] = 1.0 M and [KNO_3] = 0.88 M. The KNO_3 solution has a higher solvent concentration, so solvent will flow from the KNO_3 solution to the NaCl solution.

CHAPTER 15

15.1 (a) $-\frac{1}{2}\frac{\Delta[O_3]}{\Delta t} = \frac{1}{3}\frac{\Delta[O_2]}{\Delta t}$

(b) $-\frac{1}{2}\frac{\Delta[HOF]}{\Delta t} = \frac{1}{2}\frac{\Delta[HF]}{\Delta t} = \frac{\Delta[O_2]}{\Delta t}$

15.3 $\frac{1}{3}\frac{\Delta[O_2]}{\Delta t} = -\frac{1}{2}\frac{\Delta[O_3]}{\Delta t}$ or $\frac{\Delta[O_2]}{\Delta t} = -\frac{2}{3}\frac{\Delta[O_2]}{\Delta t}$

so $\Delta[O_3]/\Delta t = -1.0 \times 10^{-3}$ mol/L · s.

15.5 (a) The graph of [B] (product concentration) versus time shows [B] increasing from zero. The line is curved, indicating the rate changes with time; thus the rate depends on concentration. Rates for the four 10–s intervals are as follows: 0–10 s, 0.0326 mol/L · s; from 10–20 s, 0.0246 mol/L · s; 20–30 s, 0.0178 mol/L · s; 30–40 s, 0.0140 mol/L · s.

(b) $-\frac{\Delta[A]}{\Delta t} = \frac{1}{2}\frac{\Delta[B]}{\Delta t}$ throughout the reaction

In the interval 10–20 s, $\frac{\Delta[A]}{\Delta t} = -0.0123 \frac{mol}{L \cdot s}$

(c) Instantaneous rate when [B] = 0.750 mol/L

$= \frac{\Delta[B]}{\Delta t} = 0.0163 \frac{mol}{L \cdot s}$

15.7 The reaction is second order in A, first order in B, and third order overall.

15.9 (a) Rate = $k[NO_2][O_3]$
(b) If [NO_2] is tripled, the rate triples.
(c) If [O_3] is halved, the rate is halved.

15.11 (a) The reaction is second order in [NO] and first order in [O_2].

(b) $\frac{-\Delta[NO]}{\Delta t} = k[NO]^2[O_2]$

(c) $k = 13\ L^2/mol^2 \cdot s$

(d) $\frac{-\Delta[NO]}{\Delta t} = -1.4 \times 10^{-5}$ mol/L · s

(e) When $-\Delta[NO]/\Delta t = 1.0 \times 10^{-4}$ mol/L · s, $\Delta[O_2]/\Delta t = 5.0 \times 10^{-5}$ mol/L · s and $\Delta[NO_2]/\Delta t = 1.0 \times 10^{-4}$ mol/L · s.

15.13 (a) Rate = $k[NO]^2[O_2]$
(b) $k = 50.\ L^2/mol^2 \cdot h$
(c) Rate = 8.5×10^{-9} mol/L · h

15.15 $k = 3.73 \times 10^{-3}\ min^{-1}$

15.17 5.0×10^2 min

15.19 (a) 153 min
(b) 1790 min

15.21 (a) $t_{1/2}$ = 10,000 s (b) 34,000 s

15.23 0.180 g azomethane remains; 0.0965 g N_2 formed

15.25 Fraction of ^{64}Cu remaining = 0.030

15.27 The straight line obtained in a graph of ln[N_2O] versus time indicates a first-order reaction.

k = (–slope) = 0.0127 min^{-1}

The rate when [N_2O] = 0.035 mol/L is 4.4×10^{-4} mol/L · min.

15.29 The graph of 1/[NO_2] versus time gives a straight line, indicating the reaction is second order with respect to [NO_2] (see Table 15.1 on page 689). The slope of the line is k, so k = 1.1 L/mol · s.

15.31 $-\Delta[C_2F_4]/\Delta t = k[C_2F_4]^2 = (0.04\ L/mol \cdot s)[C_2F_4]^2$

15.33 Activation energy = 102 kJ/mol

15.35 $k = 0.3\ s^{-1}$

15.37

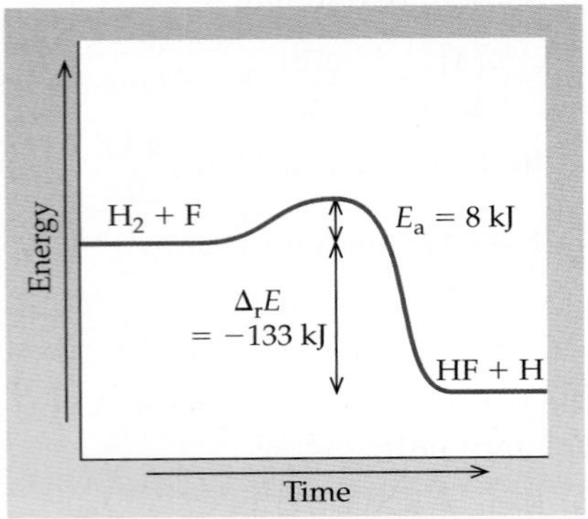

15.39 (a) Rate $= k[NO_3][NO]$
(b) Rate $= k[Cl][H_2]$
(c) Rate $= k[(CH_3)_3CBr]$

15.41 (a) The Second step (b) Rate $= k[O_3][O]$

15.43 (a) NO_2 is a reactant in the first step and a product in the second step. CO is a reactant in the second step. NO_3 is an intermediate, and CO_2 is a product. NO is a product.
(b) Reaction coordinate diagram

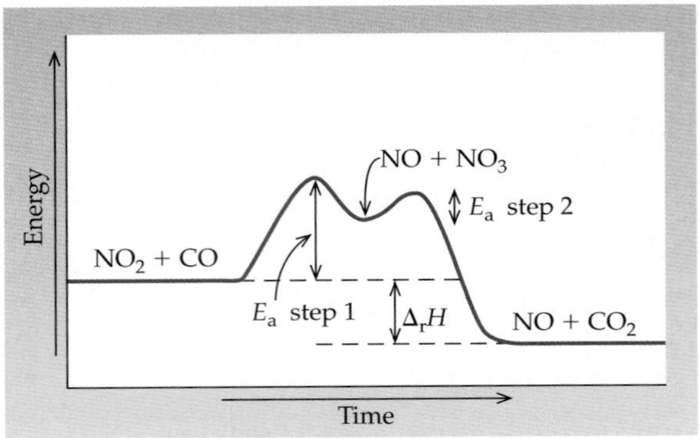

15.45 Doubling the concentration of A will increase the rate by a factor of 4 because the concentration of A appears in the rate law as $[A]^2$. Halving the concentration of B will halve the rate The net result is that the rate of the reaction will double.

15.47 After measuring pH as a function of time, one could then calculate pOH and then $[OH^-]$. Finally, a plot of $1/[OH^-]$ versus time would give a straight line with a slope equal to k.

15.49 72 s represents two half-lives, so $t_{1/2} = 36$ s.

15.51 (a) A plot of $1/[C_2F_4]$ versus time indicates the reaction is second order with respect to $[C_2F_4]$. The rate law is Rate $= k[C_2F_4]^2$.
(b) The rate constant (= slope of the line) is about 0.045 L/mol · s. (The graph does not allow a very accurate calculation.)
(c) Using $k = 0.045$ L/mol · s, the concentration after 600 s is 0.03 M (to 1 significant figure).
(d) Time = 2000 s (using k from part a).

15.53 (a) A plot of $1/[NH_4NCO]$ versus time is linear, so the reaction is second order with respect to NH_4NCO.
(b) Slope $= k = 0.0109$ L/mol · min.
(c) $t_{1/2} = 200.$ min
(d) $[NH_4NCO] = 0.0997$ mol/L

15.55 Mechanism 2

15.57 $k = 0.018\ h^{-1}$ and $t_{1/2} = 39$ h

15.59 (a) After 125 min, 0.251 g remains. After 145, 0.144 g remains.
(b) Time = 43.9 min
(c) Fraction remaining = 0.016

15.61 The rate equation for the slow step is Rate = $k[O_3][O]$. The equilibrium constant, K, for step 1 is $K = [O_2][O]/[O_3]$. Solving this for [O], we have $[O] = K[O_3]/[O_2]$. Substituting the expression for [O] into the rate equation we find

Rate $= k[O_3]\{K[O_3]/[O_2]\} = kK[O_3]^2/[O_2]$

15.63 The slope of the ln k versus $1/T$ plot is -6370. From slope $= -E_a/R$, we derive $E_a = 53.0$ kJ/mol.

15.65 Estimated time at 90 °C = 4.76 min

15.67 After 30 min (one half-life), $P_{HOF} = 50.0$ mm Hg and $P_{total} = 125.0$ mm Hg. After 45 min, $P_{HOF} =$ 35.4 mm Hg and $P_{total} = 132$ mm Hg.

15.69 (a) Reaction is first-order in NO_2NH_2 and -1 for H_3O^+.

(b, c) Mechanism 3.

In step 1, $K = k_4/k_4'$
$= [NO_2NH^-][H_3O^+]/[NO_2NH_2]$

Rearrange this and substitute into the rate law for the slow step.

Rate $= k_5[NO_2NH^-] = k_5K[NO_2NH_2]/[H_3O^+]$

This is the same as the experimental rate law, where the overall rate constant $k = k_5K$.

(d) Addition of OH^- ions will shift the equilibrium in step 1 to produce a larger concentration of NO_2NH^-, the reactant in the rate-determining step.

15.71 (a) Average rate for $t = 0$ to $t = 15$ is about 4.7×10^{-5} M/s. For $t = 100$ s to 125 s, the average rate is about 1.6×10^{-5} M/s. The rate slows because the rate of the reaction is dependent on the concentration of reactant and this concentration is declining with time.
(b) Instantaneous rate at 50 s is about 2.7×10^{-5} M/s.

(c) A plot of ln (concentration) versus time is a straight line with an equation of $y = 0.010\,x - 5.2984$. The slope, which is equal to $-k$, is -0.010, so $k = 0.010\ s^{-1}$.
(d) From the data the half-life is 69.3 s, and the same value comes from the relation $t_{1/2} = \ln 2/k$.

15.73 A plot of 1/[S] versus 1/Rate gives the equation

$1/\text{Rate} = 94\ (1/[S]) + 7.5 \times 10^4$

and so $\text{Rate}_{max} = 1/(7.5 \times 10^4) = 1.3 \times 10^{-5}$ M min^{-1}.

15.75 The finely divided rhodium metal will have a significantly greater surface area than the small block of metal. This leads to a large increase in the number of reaction sites and vastly increases the reaction rate.

15.77 (a) False. The reaction may occur in a single step but this does not have to be true.
(b) True
(c) False. Raising the temperature increases the value of k.
(d) False. Temperature has no effect on the value of E_a.
(e) False. If the concentrations of both reactants are doubled, the rate will increase by a factor of 4.
(f) True

15.79 (a) True
(b) True
(c) False. As a reaction proceeds, the reactant concentration decreases and the rate decreases.
(d) False. It is possible to have a one-step mechanism for a third-order reaction if the slow, rate-determining step is termolecular.

15.81 (a) Decrease
(b) Increase
(c) No change
(d) No change
(e) No change
(f) No change

15.83 (a) There are three mechanistic steps.
(b) The overall reaction is exothermic.

15.85 (a) The average rate is calculated over a period of time, whereas the instantaneous rate is the rate of reaction at some instant in time.
(b) The reaction rate decreases with time as the dye concentration decreases.
(c) See part (b).

15.87 (a) Molecules must collide with enough energy to overcome the activation energy, and they must be in the correct orientation.
(b) In animation 2 the molecules are moving faster, so they are at a higher temperature.
(c) Less sensitive. The O_3 must collide with NO in the correct orientation for a reaction to occur. The O_3 and N_2 collisions do not depend to the same extent on orientation because N_2 is a symmetrical, diatomic molecule.

15.89 (a) I^- is regenerated during the second step in the mechanism.
(b) The activation energy is smaller for the catalyzed reaction.

CHAPTER 16

16.1 (a) $K = \dfrac{[H_2O]^2[O_2]}{[H_2O_2]^2}$

(b) $K = \dfrac{[CO_2]}{[CO][O_2]^{1/2}}$

(c) $K = \dfrac{[CO]}{[CO_2]}$

(d) $K = \dfrac{[CO_2]}{[CO]}$

16.3 $Q = (2.0 \times 10^{-8})^2/(0.020) = 2.0 \times 10^{-14}$
$Q < K$ so the reaction proceeds to the right.

16.5 $Q = 1.0 \times 10^3$, so $Q > K$ and the reaction is not at equilibrium. It proceeds to the left to convert products to reactants.

16.7 $K = 1.2$

16.9 (a) $K = 0.025$
(b) $K = 0.025$
(c) The amount of solid does not affect the equilibrium.

16.11 (a) $[COCl_2] = 0.00308$ M; $[CO] = 0.00712$ M
(b) $K = 144$

16.13 [isobutane] = 0.024 M; [butane] = 0.010 M

16.15 $[I_2] = 6.14 \times 10^{-3}$ M; $[I] = 4.79 \times 10^{-3}$ M

16.17 $[COBr_2] = 0.107$ M; $[CO] = [Br_2] = 0.143$ M
57.1% of the $COBr_2$ has decomposed.

16.19 (b)

16.21 (e) $K_2 = 1/(K_1)^2$

16.23 $K = 13.7$

16.25 (a) Equilibrium shifts to the right
(b) Equilibrium shifts to the left
(c) Equilibrium shifts to the right
(d) Equilibrium shifts to the left

16.27 Equilibrium concentrations are the same under both circumstances: [butane] = 1.1 M and [isobutane] = 2.9 M.

16.29 $K = 3.9 \times 10^{-4}$

16.31 For decomposition of $COCl_2$, $K = 1/(K \text{ for } COCl_2 \text{ formation}) = 1/(6.5 \times 10^{11}) = 1.5 \times 10^{-12}$

16.33 $K = 4$

16.35 Q is less than K, so the system shifts to form more isobutane.

At equilibrium, [butane] = 0.86 M and [isobutane] = 2.14 M.

16.37 The second equation has been reversed and multiplied by 2.
(c) $K_2 = 1/K_1^2$

16.39 (a) No change (d) Shifts right
(b) Shifts left (e) Shifts right
(c) No change

16.41 (a) The equilibrium will shift to the left on adding more Cl_2.
(b) K is calculated (from the quantities of reactants and products at equilibrium) to be 0.0470. After Cl_2 is added, the concentrations are: $[PCl_5]$ = 0.0199 M, $[PCl_3]$ = 0.0231 M, and $[Cl_2]$ = 0.0403 M.

16.43 $K_p = 0.215$

16.45 (a) Fraction dissociated = 0.15
(b) Fraction dissociated = 0.189. If the pressure decreases, the equilibrium shifts to the right, increasing the fraction of N_2O_4 dissociated.

16.47 $[NH_3]$ = 0.67 M; $[N_2]$ = 0.57 M; $[H_2]$ = 1.7 M; P_{total} = 180 atm

16.49 (a) $[NH_3] = [H_2S]$ = 0.013 M
(b) $[NH_3]$ = 0.027 M and $[H_2S]$ = 0.0067 M

16.51 $P(NO_2)$ = 0.379 atm and $P(N_2O_4)$ = 0.960 atm; P(total) = 1.339 atm

16.53 (a) $K_p = K_c = 56$. Because 2 mol of reactants gives 2 mol of product, Δn does not change and $K_p = K_c$ (see page 730).

(b, c) Initial $P(H_2) = P(I_2)$ = 2.6 atm and P_{total} = 5.2 atm

At equilibrium, $P(H_2) = P(I_2)$ = 0.54 atm and $P(HI)$ = 4.1 atm. Therefore, P_{total} = 5.2 atm. The initial total pressure and the equilibrium total pressure are the same owing to the reaction stoichiometry. Percent dissociation = 69%

16.55 $P(CO)$ = 0.0010 atm

16.57 1.7×10^{18} O atoms

16.59 Glycerin concentration should be 1.7 M

16.61 (a) $K_p = 0.20$
(b) When initial $[N_2O_4]$ = 1.00 atm, the equilibrium pressures are $[N_2O_4]$ = 0.80 atm and $[NO_2]$ = 0.40 atm. When initial $[N_2O_4]$ = 0.10 atm, the equilibrium pressures are $[N_2O_4]$ = 0.050 atm and $[NO_2]$ = 0.10 atm. The percent dissociation is now 50.%. This is in accord with Le Chatelier's principle: If the initial pressure of the reactant decreases, the equilibrium shifts to the right, increasing the fraction of the reactant dissociated. See also Question 16.45.

16.63 (a) The flask containing $(H_3N)B(CH_3)_3$ will have the largest partial pressure of $B(CH_3)_3$.
(b) $P[B(CH_3)_3] = P(NH_3)$ = 2.1 and $P[(H_3N)B(CH_3)_3]$ = 1.0 atm

P_{total} = 5.2 atm

Percent dissociation = 69%

16.65 (a) As more KSCN is added, Le Chatelier's principle predicts more of the red complex ion $[Fe(H_2O)_5(SCN]^+$ will form.
(b) Adding Ag^+ ions leads to a precipitate of AgSCN, thus removing SCN^- ions from solution. The equilibrium shifts left, dropping the concentration of the red complex ion.

16.67 (a) False. The magnitude of K is always dependent on temperature.
(b) True
(c) False. The equilibrium constant for a reaction is the reciprocal of the value of K for its reverse.
(d) True
(e) False. $\Delta n = 1$ so $K_p = K_c(RT)$

16.69 (a) Product-favored, $K >> 1$
(b) Reactant-favored, $K << 1$
(c) Product-favored, $K >> 1$

16.71 The system is not at equilibrium because it continues to gain energy from the surroundings. The temperature of the water/ice mixture will remain at 0 °C until all the ice is melted, then the temperature will rise as more energy is gained. Only if the beaker of water/ice were moved to a perfectly insulated compartment, also at 0 °C, would it attain equilibrium at 0 °C. In this case, it would be a dynamic equilibrium with water molecules moving from the solid to the liquid phase and from the liquid to the solid phase. The quantity of ice would not change. If a D_2O ice cube was added to some $H_2O(\ell)$, an equilibrium would be obtained. The amount of D_2O in the liquid phase would increase due to the continuing molecular exchange. The water could then be sampled for the presence of D_2O.

CHAPTER 17

17.1 (a) CN^-, cyanide ion
(b) SO_4^{2-}, sulfate ion
(c) F^-, fluoride ion

17.3 (a) $H_3O^+(aq) + NO_3^-(aq)$; $H_3O^+(aq)$ is the conjugate acid of H_2O, and $NO_3^-(aq)$ is the conjugate base of HNO_3.
(b) $H_3O^+(aq) + SO_4^{2-}(aq)$; $H_3O^+(aq)$ is the conjugate acid of H_2O, and $SO_4^{2-}(aq)$ is the conjugate base of HSO_4^-.
(c) $H_2O + HF$; H_2O is the conjugate base of H_3O^+, and HF is the conjugate acid of F^-.

17.5 Brønsted acid: $HC_2O_4^-(aq) + H_2O(\ell) \rightleftharpoons H_3O^+(aq) + C_2O_4^{2-}(aq)$

Brønsted base: $HC_2O_4^-(aq) + H_2O(\ell) \rightleftharpoons H_2C_2O_4(aq) + OH^-(aq)$

17.7

	Acid (A)	Base (B)	Conjugate Base of A	Conjugate Acid of B
(a)	HCO_2H	H_2O	HCO_2^-	H_3O^+
(b)	H_2S	NH_3	HS^-	NH_4^+
(c)	HSO_4^-	OH^-	SO_4^{2-}	H_2O

17.9 $[H_3O^+] = 1.8 \times 10^{-4}$ M; acidic

17.11 HCl is a strong acid, so $[H_3O^+]$ = concentration of the acid. $[H_3O^+] = 0.0075$ M and $[OH^-] = 1.3 \times 10^{-12}$ M. pH = 2.12.

17.13 $Ba(OH)_2$ is a strong base, so $[OH^-] = 2 \times$ concentration of the base.

$[OH^-] = 3.0 \times 10^{-3}$ M; pOH = 2.52; and pH = 11.48

17.15 (a) The strongest acid is HCO_2H (largest K_a) and the weakest acid is C_6H_5OH (smallest K_a).
(b) The strongest acid (HCO_2H) has the weakest conjugate base.
(c) The weakest acid (C_6H_5OH) has the strongest conjugate base.

17.17 (c) HClO, the weakest acid in this list (Table 17.3), has the strongest conjugate base.

17.19 $CO_3^{2-}(aq) + H_2O(\ell) \rightarrow HCO_3^-(aq) + OH^-(aq)$

17.21 Highest pH, Na_2S; lowest pH, $AlCl_3$ (which gives the weak acid $[Al(H_2O)_6]^{3+}$ in solution)

17.23 $pK_a = 4.19$

17.25 $K_a = 3.0 \times 10^{-10}$

17.27 2-Chlorobenzoic acid has the smaller pK_a value.

17.29 $K_b = 7.09 \times 10^{-12}$

17.31 $K_b = 6.3 \times 10^{-5}$

17.33 $CH_3CO_2H(aq) + HCO_3^-(aq) \rightleftharpoons CH_3CO_2^-(aq) + H_2CO_3(aq)$

Equilibrium lies predominantly to the right because CH_3CO_2H is a stronger acid than H_2CO_3.

17.35 (a) Left; NH_3 and HBr are the stronger base and acid, respectively.
(b) Left; PO_4^{3-} and CH_3CO_2H are the stronger base and acid, respectively.
(c) Right; $[Fe(H_2O)_6]^{3+}$ and HCO_3^- are the stronger acid and base, respectively.

17.37 (a) $OH^-(aq) + HPO_4^{2-}(aq) \rightleftharpoons H_2O(\ell) + PO_4^{3-}(aq)$
(b) OH^- is a stronger base than PO_4^{3-}, so the equilibrium will lie to the right. (The predominant species in solution is PO_4^{3-}, so the solution is likely to be basic because PO_4^{3-} is the conjugate base of a weak acid.)

17.39 (a) $CH_3CO_2H(aq) + HPO_4^{2-}(aq) \rightleftharpoons CH_3CO_2^-(aq) + H_2PO_4^-(aq)$
(b) CH_3CO_2H is a stronger acid than $H_2PO_4^-$, so the equilibrium will lie to the right.

17.41 (a) 2.1×10^{-3} M; (b) $K_a = 3.5 \times 10^{-4}$

17.43 $K_b = 6.6 \times 10^{-9}$

17.45 (a) $[H_3O^+] = 1.6 \times 10^{-4}$ M
(b) Moderately weak; $K_a = 1.1 \times 10^{-5}$

17.47 $[CH_3CO_2^-] = [H_3O^+] = 1.9 \times 10^{-3}$ M and $[CH_3CO_2H] = 0.20$ M

17.49 $[H_3O^+] = [CN^-] = 3.2 \times 10^{-6}$ M; $[HCN] = 0.025$ M; pH = 5.50

17.51 $[NH_4^+] = [OH^-] = 1.64 \times 10^{-3}$ M; $[NH_3] = 0.15$ M; pH = 11.22

17.53 $[OH^-] = 0.0102$ M; pH = 12.01; pOH = 1.99

17.55 pH = 3.25

17.57 $[H_3O^+] = 1.1 \times 10^{-5}$ M; pH = 4.98

17.59 $[HCN] = [OH^-] = 3.3 \times 10^{-3}$ M; $[H_3O^+] = 3.0 \times 10^{-12}$ M; $[Na^+] = 0.441$ M

17.61 $[H_3O^+] = 1.5 \times 10^{-9}$ M; pH = 8.81

17.63 (a) The reaction produces acetate ion, the conjugate base of acetic acid. The solution is weakly basic. pH is greater than 7.
(b) The reaction produces NH_4^+, the conjugate acid of NH_3. The solution is weakly acidic. pH is less than 7.
(c) The reaction mixes equal molar amounts of strong base and strong acid. The solution will be neutral. pH will be 7.

17.65 (a) pH = 1.17; (b) $[SO_3^{2-}] = 6.2 \times 10^{-8}$ M

17.67 (a) $[OH^-] = [N_2H_5^+] = 9.2 \times 10^{-5}$ M; $[N_2H_6^{2+}] = 8.9 \times 10^{-16}$ M
(b) pH = 9.96

17.69 (a) Lewis base
(b) Lewis acid

(c) Lewis base (owing to lone pair of electrons on the N atom)

17.71 CO is a Lewis base in its reactions with transition metal atoms. It donates a lone pair of electrons on the C atom.

17.73 HOCN should be a stronger acid than HCN because the H atom in HOCN is attached to a highly electronegative O atom. This induces a positive charge on the H atom, making it more readily removed by an interaction with water.

17.75 The S atom is surrounded by four highly electronegative O atoms. The inductive effect of these atoms induces a positive charge on the H atom, making it susceptible to removal by water.

17.77 pH = 2.671

17.79 Both $Ba(OH)_2$ and $Sr(OH)_2$ dissolve completely in water to provide M^{2+} and OH^- ions. 2.50 g $Sr(OH)_2$ in 1.00 L of water gives $[Sr^{2+}] = 0.021$ M and $[OH^-] = 0.041$ M. The concentration of OH^- is reflected in a pH of 12.61.

17.81 $H_2S(aq) + CH_3CO_2^-(aq) \rightleftarrows CH_3CO_2H(aq) + HS^-(aq)$

The equilibrium lies to the left and favors the reactants.

17.83 $[X^-] = [H_3O^+] = 3.0 \times 10^{-3}$ M; $[HX] = 0.007$ M; pH = 2.52

17.85 $K_a = 1.4 \times 10^{-5}$; $pK_a = 4.86$

17.87 pH = 5.84

17.89 (a) Ethylamine is a stronger base than ethanolamine. (b) For ethylamine, the pH of the solution is 11.82.

17.91 pH = 7.66

17.93 Acidic: $NaHSO_4$, NH_4Br, $FeCl_3$
Neutral: $KClO_4$, $NaNO_3$, LiBr
Basic: Na_2CO_3, $(NH_4)_2S$, Na_2HPO_4

17.95 $K_{net} = K_{a1} \times K_{a2} = 3.8 \times 10^{-6}$

17.97 For the reaction $HCO_2H(aq) + OH^-(aq) \rightarrow H_2O(\ell) + HCO_2^-(aq)$, $K_{net} = K_a$ (for HCO_2H) $\times [1/K_w] = 1.8 \times 10^{10}$

17.99 To double the percent ionization, you must dilute 100 mL of solution to 400 mL.

17.101 $H_2O > H_2C_2O_4 > HC_2O_4^- = H_3O^+ > C_2O_4^{2-} > OH^-$

17.103 Measure the pH of the 0.1 M solutions of the three bases. The solution containing the strongest base will have the highest pH. The solution having the weakest base will have the lowest pH.

17.105 The possible cation–anion combinations are NaCl (neutral), NaOH (basic), NH_4Cl (acidic), NH_4OH (basic), HCl (acidic), and H_2O (neutral).

A = H^+ solution; B = NH_4^+ solution; C = Na^+ solution; Y = Cl^- solution; Z = OH^- solution

17.107 $K_a = 3.0 \times 10^{-5}$

17.109 (a) Aniline is both a Brønsted and a Lewis base. As a proton acceptor it gives $C_6H_5NH_3^+$. The N atom can also donate an electron pair to give a Lewis acid–base adduct, $F_3B \leftarrow NH_2C_6H_5$.
(b) pH = 7.97

17.111 Water can both accept a proton (a Brønsted base) and donate a lone pair (a Lewis base). Water can also donate a proton (Brønsted acid), but it cannot accept a pair of electrons (and act as a Lewis acid).

17.113 (a) HOCl is the strongest acid (smallest pK_a and largest K_a), and HOI is the weakest acid.
(b) Cl is more electronegative than I, so the OCl^- anion is more stable than the OI^- anion.

17.115 (a) $HClO_4 + H_2SO_4 \rightleftarrows ClO_4^- + H_3SO_4^+$
(b) The O atoms on sulfuric acid have lone pairs of electrons that can be used to bind to an H^+ ion.

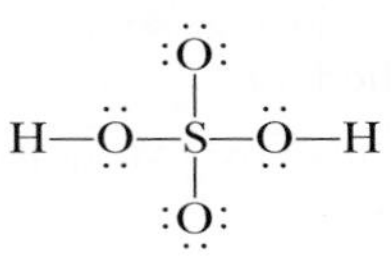

17.117 (a) 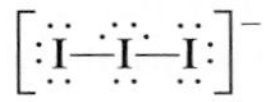
(b) $I^-(aq)$ [Lewis base] + $I_2(aq)$ [Lewis acid] $\rightarrow I_3^-(aq)$

17.119 (a) For the weak acid HA, the concentrations at equilibrium are $[HA] = C_0 - \alpha C_0$, $[H_3O^+] = [A^-] = \alpha C_0$. Putting these into the usual expression for K_a we have $K_a = \alpha^2 C_0/(1-\alpha)$.
(b) For 0.10 M NH_4^+, $\alpha = 7.5 \times 10^{-5}$ (reflecting the fact that NH_4^+ is a much weaker acid than acetic acid).

17.121 (a) Add the three equations.

$NH_4^+(aq) + H_2O(\ell) \rightleftarrows NH_3(aq) + H_3O^+(aq)$

$K_1 = K_w/K_b$

$CN^-(aq) + H_2O(\ell) \rightleftarrows HCN(aq) + OH^-(aq)$

$K_2 = K_w/K_a$

$H_3O^+(aq) + OH^-(aq) \rightleftarrows 2\ H_2O(\ell)$ $K_3 = 1/K_w$

$NH_4^+(aq) + CN^-(aq) \rightleftarrows NH_3(aq) + HCN(aq)$

$K_{net} = K_1K_2K_3 = K_w/K_aK_b$

(b) Substitute expressions for K_w, K_a, and K_b into the equation.

$$[H_3O^+] = \sqrt{\frac{K_w K_a}{K_b}}$$

$$\sqrt{\frac{K_w K_a}{K_b}} = \sqrt{\frac{[H_3O^+][OH^-]\left(\frac{[H_3O^+][NH_3]}{[NH_4^+]}\right)}{\frac{[OH^-][HCN]}{[CN^-]}}}$$

In a solution of NH_4CN, we have $[NH_4^+] = [CN^-]$ and $[NH_3] = [HCN]$. When these and $[OH^-]$, are canceled from the expression, we see it is equal to $[H_3O^+]$.

(c) pH = 9.33

(d) K_a for NH_4^+ and K_b for acetate ion are identical. Therefore, $[H_3O^+] = (K_w)^{1/2}$ or 1.0×10^{-7} M. The pH is 7.0.

(e) The pH of a solution will depend on the relative strengths of the anionic base and the cationic acid. In part (d) the anion and cation were equal in strength, so the solution was neutral. For NH_4CN, the CN^- ion is a stronger base ($K_b = 2.5 \times 10^{-5}$) than NH_4^+ is an acid ($K_a = 5.6 \times 10^{-10}$), so the solution is predicted to be basic.

CHAPTER 18

18.1 (a) Decrease pH; (b) increase pH; (c) no change in pH

18.3 pH = 9.25

18.5 pH = 4.38

18.7 pH = 9.12; pH of buffer is lower than the pH of the original solution of NH_3(pH = 11.17).

18.9 4.7 g

18.11 pH = 4.92

18.13 (a) pH = 3.59; (b) $[HCO_2H]/[HCO_2^-] = 0.45$

18.15 (b) $NH_3 + NH_4Cl$

18.17 The buffer must have a ratio of 0.51 mol NaH_2PO_4 to 1 mol Na_2HPO_4. For example, dissolve 0.51 mol NaH_2PO_4 (61 g) and 1.0 mol Na_2HPO_4 (140 g) in some amount of water.

18.19 (a) pH = 4.95; (b) pH = 5.05

18.21 (a) pH = 9.55; (b) pH = 9.50

18.23 (a) Original pH = 5.62

(b) $[Na^+] = 0.0323$ M, $[OH^-] = 1.5 \times 10^{-3}$ M, $[H_3O^+] = 6.5 \times 10^{-12}$ M, and $[C_6H_5O^-] = 0.0308$ M

(c) pH = 11.19

18.25 (a) Original NH_3 concentration = 0.0154 M

(b) At the equivalence point $[H_3O^+] = 1.9 \times 10^{-6}$ M, $[OH^-] = 5.3 \times 10^{-9}$ M, $[NH_4^+] = 6.25 \times 10^{-3}$ M.

(c) pH at equivalence point = 5.73

18.27 The titration curve begins at pH = 13.00 and drops slowly as HCl is added. Just before the equivalence point (when 30.0 mL of acid has been added), the curve falls steeply. The pH at the equivalence point is exactly 7. Just after the equivalence point, the curve flattens again and begins to approach the final pH of just over 1.0. The total volume at the equivalence point is 60.0 mL.

18.29 (a) Starting pH = 11.12

(b) pH at equivalence point = 5.28

(c) pH at midpoint (half-neutralization point) = 9.25

(d) Methyl red, bromcresol green

(e)

Acid (mL)	Added pH
5.00	9.85
15.0	9.08
20.0	8.65
22.0	8.39
30.0	2.04

18.31 See Figure 18.10 on page 832.

(a) Thymol blue or bromphenol blue

(b) Phenolphthalein

(c) Methyl red; thymol blue

18.33 (a) Silver chloride, AgCl; lead chloride, $PbCl_2$

(b) Zinc carbonate, $ZnCO_3$; zinc sulfide, ZnS

(c) Iron(II) carbonate, $FeCO_3$; iron(II) oxalate, FeC_2O_4

18.35 (a) and (b) are soluble, (c) and (d) are insoluble.

18.37 (a) $AgCN(s) \longrightarrow Ag^+(aq) + CN^-(aq)$, $K_{sp} = [Ag^+][CN^-]$

(b) $NiCO_3(s) \longrightarrow Ni^{2+}(aq) + CO_3^{2-}(aq)$, $K_{sp} = [Ni^{2+}][CO_3^{2-}]$

(c) $AuBr_3(s) \longrightarrow Au^{3+}(aq) + 3\ Br^-(aq)$, $K_{sp} = [Au^{3+}][Br^-]^3$

18.39 $K_{sp} = (1.9 \times 10^{-3})^2 = 3.6 \times 10^{-6}$

18.41 $K_{sp} = 4.37 \times 10^{-9}$

18.43 $K_{sp} = 1.4 \times 10^{-15}$

18.45 (a) 9.2×10^{-9} M; (b) 2.2×10^{-6} g/L

18.47 (a) 2.4×10^{-4} M; (b) 0.018 g/L

18.49 Only 2.1×10^{-4} g dissolves.

18.51 (a) $PbCl_2$; (b) FeS; (c) $Fe(OH)_2$

18.53 Solubility in pure water = 1.0×10^{-6} mol/L; solubility in 0.010 M SCN^- = 1.0×10^{-10} mol/L

18.55 (a) Solubility in pure water = 2.2×10^{-6} mg/mL
(b) Solubility in 0.020 M $AgNO_3$ = 1.0×10^{-12} mg/mL

18.57 (a) PbS
(b) Ag_2CO_3
(c) $Al(OH)_3$

18.59 $Q < K_{sp}$, so no precipitate forms.

18.61 $Q > K_{sp}$; $Zn(OH)_2$ will precipitate.

18.63 $[OH^-]$ must exceed 1.0×10^{-5} M.

18.65 Using K_{sp} for $Zn(OH)_2$ and K_{form} for $Zn(OH)_4{}^{2-}$, K_{net} for

$$Zn(OH)_2(s) + 2\ OH^-(aq) \rightleftarrows Zn(OH)_4{}^{2-}(aq)$$

is 13.8. This indicates that the reaction is definitely product-favored.

18.67 K_{net} for $AgCl(s) + 2\ NH_3(aq) \rightleftarrows Ag(NH_3)_2{}^+(aq) + Cl^-(aq)$ is 2.0×10^{-3}. When all the AgCl dissolves, $[Ag(NH_3)_2{}^+] = [Cl^-] = 0.050$ M. To achieve these concentrations, $[NH_3]$ must be 1.25 M. Therefore, the amount of NH_3 added must be 2×0.050 mol/L (to react with the AgCl) plus 1.25 mol/L (to achieve the proper equilibrium concentration). The total is 1.35 mol/L NH_3.

18.69 (a) Solubility in pure water = 1.3×10^{-5} mol/L or 0.0019 g/L.
(b) K_{net} for $AgCl(s) + 2\ NH_3(aq) \rightleftarrows Ag(NH_3)_2{}^+(aq) + Cl^-(aq)$ is 2.0×10^{-3}. When using 1.0 M NH_3, the concentrations of species in solution are $[Ag(NH_3)_2{}^+] = [Cl^-] = 0.041$ M and so $[NH_3] = 1.0 - 2(0.041)$ M or about 0.9 M. The amount of AgCl dissolved is 0.041 mol/L or 5.88 g/L.

18.71 (a) $NaBr(aq) + AgNO_3(aq) \rightarrow NaNO_3(aq) + AgBr(s)$
(b) $2\ KCl(aq) + Pb(NO_3)_2(aq) \rightarrow 2\ KNO_3(aq) + PbCl_2(s)$

18.73 $Q > K_{sp}$, so $BaSO_4$ precipitates.

18.75 $[H_3O^+] = 1.9 \times 10^{-10}$ M; pH = 9.73

18.77 $BaCO_3 < Ag_2CO_3 < Na_2CO_3$

18.79 Original pH = 8.62; dilution will not affect the pH.

18.81 (a) 0.100 M acetic acid has a pH of 2.87. Adding sodium acetate slowly raises the pH.
(b) Adding $NaNO_3$ to 0.100 M HNO_3 has no effect on the pH.
(c) In part (a), adding the conjugate base of a weak acid creates a buffer solution. In part (b), HNO_3 is a strong acid, and its conjugate base ($NO_3{}^-$) is so weak that the base has no effect on the complete ionization of the acid.

18.83 (a) pH = 4.13
(b) 0.6 g of $C_6H_5CO_2H$
(c) 8.2 mL of 2.0 M HCl should be added

18.85 $K = 2.1 \times 10^6$; yes, AgI forms

18.87 (a) $[F^-] = 1.3 \times 10^{-3}$ M; (b) $[Ca^{2+}] = 2.9 \times 10^{-5}$ M

18.89 (a) $PbSO_4$ will precipitate first.
(b) $[Pb^{2+}] = 5.1 \times 10^{-6}$ M

18.91 When $[CO_3{}^{2-}] = 0.050$ M, $[Ca^{2+}] = 6.8 \times 10^{-8}$ M. This means only 6.8×10^{-4}% of the ions remain, or that essentially all of the calcium ions have been removed.

18.93 (a) Add H_2SO_4, precipitating $BaSO_4$ and leaving $Na^+(aq)$ in solution.
(b) Add HCl or another source of chloride ion. $PbCl_2$ will precipitate, but $NiCl_2$ is water-soluble.

18.95 (a) $BaSO_4$ will precipitate first.
(b) $[Ba^{2+}] = 1.8 \times 10^{-7}$ M

18.97 (a) pH = 2.81
(b) pH at equivalence point = 8.72
(c) pH at the midpoint = pK_a = 4.62
(d) Phenolphthalein
(e) After 10.0 mL, pH = 4.39.
After 20.0 mL, pH = 5.07.
After 30.0 mL, pH = 11.84.
(f) A plot of pH versus volume of NaOH added would begin at a pH of 2.81, rise slightly to the midpoint at pH = 4.62, and then begin to rise more steeply as the equivalence point is approached (when the volume of NaOH added is 27.0 mL). The pH rises vertically through the equivalence point, and then begins to level off above a pH of about 11.0.

18.99 The K_b value for ethylamine (4.27×10^{-4}) is found in Appendix I.
(a) pH = 11.89
(b) Midpoint pH = 10.63
(c) pH = 10.15
(d) pH = 5.93 at the equivalence point
(e) pH = 2.13

(f) Titration curve

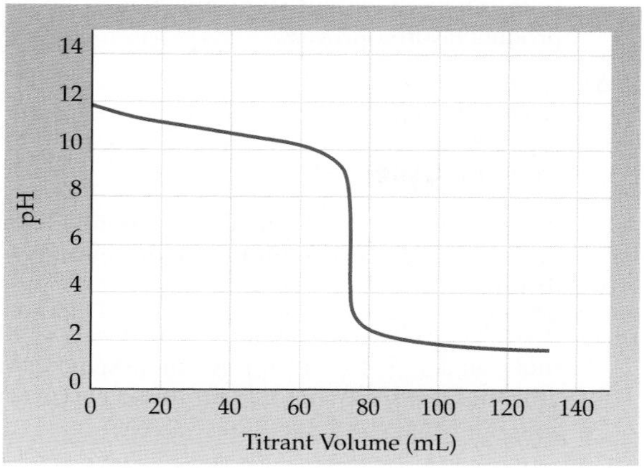

(g) Alizarin or bromcresol purple (see Figure 18.10)

18.101 110 mL NaOH

18.103 Add dilute HCl, say 1 M HCl, to a solution of the salts. Both AgCl and $PbCl_2$ will precipitate, but Cu^{2+} ions will stay in solution (as $CuCl_2$ is water-soluble). Decant off the copper-containing solution to leave a precipitate of white AgCl and $PbCl_2$. Lead(II) chloride ($K_{sp} = 1.7 \times 10^{-5}$) is much more soluble than AgCl ($K_{sp} = 1.8 \times 10^{-10}$). Warming the precipitates in water will dissolve the $PbCl_2$ and leave the AgCl as a white solid.

18.105 $Cu(OH)_2$ will dissolve in a nonoxidizing acid such as HCl, whereas CuS will not.

18.107 When Ag_3PO_4 dissolves slightly, it produces a small concentration of the phosphate ion, PO_4^{3-}. This ion is a strong base and hydrolyzes to HPO_4^{2-}. As this reaction removes the PO_4^{3-} ion from equilibrium with Ag_3PO_4, the equilibrium shifts to the right, producing more PO_4^{3-} and Ag^+ ions. Thus, Ag_3PO_4 dissolves to a greater extent than might be calculated from a K_{sp} value (unless the K_{sp} value was actually determined experimentally).

18.109 (a) Base is added to increase the pH. The added base reacts with acetic acid to form more acetate ions in the mixture. Thus, the fraction of acid declines and the fraction of conjugate base rises (i.e., the ratio $[CH_3CO_2H]/[CH_3CO_2^-]$ decreases) as the pH rises.
(b) At pH = 4, acid predominates (85% acid and 15% acetate ions). At pH = 6, acetate ions predominate (95% acetate ions and 5% acid).
(c) At the point the lines cross, $[CH_3CO_2H] = [CH_3CO_2^-]$. At this point pH = pK_a, so pK_a for acetic acid is 4.75.

18.111 (a) C—C—C angle, 120°; O—C=O, 120°; C—O—H, 109°; C—C—H, 120°
(b) Both the ring C atoms and the C in CO_2H are sp^2 hybridized.
(c) $K_a = 1 \times 10^{-3}$
(d) 10%
(e) pH at half-way point = pK_a = 3.0; pH at equivalence point = 7.3

CHAPTER 19

19.1 (a) For a given substance at a given temperature, a gas always has a greater entropy than the liquid. Matter and energy are more dispersed.
(b) Liquid water at 50 °C
(c) Ruby
(d) One mole of N_2 at 1 bar

19.3 (a) $\Delta_r S° = +12.7$ J/K · mol-rxn. Entropy increases.
(b) $\Delta_r S° = -102.55$ J/K · mol-rxn. Significant decrease in entropy.
(c) $\Delta_r S° = +93.2$ J/K · mol-rxn. Entropy increases.
(d) $\Delta_r S° = -129.7$ J/K · mol-rxn. The solution has a smaller entropy (with H^+ forming H_3O^+ and hydrogen bonding occurring) than HCl in the gaseous state.

19.5 (a) $\Delta_r S° = +9.3$ J/K · mol-rxn; (b) $\Delta_r S° = -293.97$ J/K · mol-rxn

19.7 (a) $\Delta_r S° = -507.3$ J/K · mol-rxn; entropy declines as a gaseous reactant is incorporated in a solid compound.
(b) $\Delta_r S° = +313.25$ J/K · mol-rxn; entropy increases as five molecules (two of them in the gas phase) form six molecules of products (all gases).

19.9 $\Delta_{sys}S° = -134.18$ J/K · mol-rxn; $\Delta_{sys}H° = -662.75$ kJ/mol-rxn; $\Delta_{surr}S° = +2222.9$ J/K · mol-rxn; $\Delta_{univ}S° = +2088.7$ J/K · mol-rxn

19.11 $\Delta_{sys}S° = +163.3$ J/K · mol-rxn; $\Delta_{sys}H° = +285.83$ kJ/mol-rxn; $\Delta_{surr}S° = -958.68$ J/K · mol-rxn; $\Delta_{univ}S° = -795.4$ J/K

The reaction is not spontaneous, because the overall entropy change in the universe is negative. The reaction is disfavored by energy dispersal.

19.13 (a) Type 2. The reaction is enthalpy-favored but entropy-disfavored. It is more favorable at low temperatures.
(b) Type 4. This endothermic reaction is not favored by the enthalpy change nor is it favored by the entropy change. It is not spontaneous under any conditions.

19.15 (a) $\Delta_r H° = -438$ kJ/mol-rxn; $\Delta_r S° = -201.7$ J/K · mol-rxn; $\Delta_r G° = -378$ kJ/mol-rxn. The reaction is product-favored and is enthalpy-driven.
(b) $\Delta_r H° = -86.61$ kJ/mol-rxn; $\Delta_r S° = -79.4$ J/K · mol-rxn; $\Delta_r G° = -62.9$ kJ/mol-rxn

The reaction is product-favored. The enthalpy change favors the reaction.

19.17 (a) $\Delta_r H° = +116.7$ kJ/mol-rxn; $\Delta_r S° = +168.0$ J/K mol-rxn; $\Delta_f G° = +66.6$ kJ/mol
(b) $\Delta_r H° = -425.93$ kJ/mol-rxn; $\Delta_r S° = -154.6$ J/K · mol-rxn; $\Delta_f G° = -379.82$ kJ/mol
(c) $\Delta_r H° = +17.51$ kJ/mol-rxn; $\Delta_r S° = +77.95$ J/K mol-rxn; $\Delta_f G° = -5.73$ kJ/mol

19.19 (a) $\Delta_r G° = -817.54$ kJ/mol-rxn; spontaneous
(b) $\Delta_r G° = +256.6$ kJ/mol-rxn; not spontaneous
(c) $\Delta_r G° = -1101.14$ kJ/mol-rxn; spontaneous

19.21 $\Delta_f G°$ [$BaCO_3$(s)] = −1134.4 kJ/mol

19.23 (a) $\Delta_r H° = +66.2$ kJ/mol-rxn; $\Delta_r S° = -121.62$ J/K · mol-rxn; $\Delta_r G° = +102.5$ kJ/mol-rxn

Both the enthalpy and the entropy changes indicate the reaction is not spontaneous. There is no temperature to which it will become spontaneous. This is a case like that in the right panel in Figure 19.12 and is a Type 4 reaction (Table 19.2).

(b) $\Delta_r H° = -221.05$ kJ/mol-rxn; $\Delta_r S° = +179.1$ J/K · mol-rxn; $\Delta_r G° = -283.99$ kJ/mol-rxn

The reaction is favored by both enthalpy and entropy and is product-favored at all temperatures. This is a case like that in the left panel in Figure 19.12 and is a Type 1 reaction.

(c) $\Delta_r H° = -179.0$ kJ/mol-rxn; $\Delta_r S° = -160.2$ J/K · mol-rxn; $\Delta_r G° = -131.4$ kJ/mol-rxn

The reaction is favored by the enthalpy change but disfavored by the entropy change. The reaction becomes less product-favored as the temperature increases; it is a case like the upper line in the middle panel of Figure 19.12.

(d) $\Delta_r H° = +822.2$ kJ/mol-rxn; $\Delta_r S° = +181.28$ J/K · mol-rxn; $\Delta_r G° = +768.08$ kJ/mol-rxn

The reaction is not favored by the enthalpy change but favored by the entropy change. The reaction becomes more product-favored as the temperature increases; it is a case like the lower line in the middle panel of Figure 19.12.

19.25 (a) $\Delta_r S° = +174.75$ J/K · mol-rxn; $\Delta_r H° = +116.94$ kJ/mol-rxn
(b) $\Delta_r G° = +64.87$ kJ/mol-rxn. The reaction is not spontaneous at 298 K.
(c) As the temperature increases, $\Delta_r S°$ becomes more important, so $\Delta_r G°$ can become negative at a sufficiently high temperature.

19.27 $K = 6.8 \times 10^{-16}$. Note that K is very small and that $\Delta G°$ is positive. Both indicate a reactant-favored process.

19.29 $\Delta_r G° = -100.24$ kJ/mol-rxn and $K_p = 3.64 \times 10^{17}$. Both the free energy change and K indicate a product-favored process.

19.31 (a) HBr
(b) NH_4Cl(aq)
(c) C_2H_4(g)
(d) NaCl(g)

19.33 $\Delta_r G° = -98.9$ kJ/mol-rxn. The reaction is spontaneous under standard conditions and is enthalpy-driven.

19.35 $\Delta_r H° = -1428.66$ kJ/mol-rxn; $\Delta_r S° = +47.1$ J/K · mol-rxn; $\Delta_{univ} S° = +4840$ J/K · mol-rxn. Combustion reactions are spontaneous, and this is confirmed by the sign of $\Delta_{univ} S°$.

19.37 (a) The reaction occurs spontaneously and is product-favored. Therefore, $\Delta_{univ} S°$ is positive and $\Delta_r G°$ is negative. The reaction is likely to be exothermic, so $\Delta_r H°$ is negative, and $\Delta_{surr} S°$ is positive. $\Delta_{sys} S°$ is expected to be negative because two moles of gas form one mole of solid. The calculated values are as follows:

$\Delta_{sys} S° = -284.2$ J/K · mol-rxn

$\Delta_r H° = -176.34$ kJ/mol-rxn

$\Delta_{surr} S° = +591.45$ J/K · mol-rxn

$\Delta_{univ} S° = +307.3$ J/K · mol-rxn

$\Delta_r G° = -91.64$ kJ/mol-rxn

(b) $K_p = 1.13 \times 10^{16}$

19.39 $K_p = 1.3 \times 10^{29}$ at 298 K ($\Delta G° = -166.1$ kJ/mol-rxn). The reaction is already extremely product-favored at 298 K. A higher temperature, however, would make the reaction less product-favored because $\Delta_r S°$ has a negative value (−242.3 J/K · mol-rxn).

19.41 At the boiling point, $\Delta G° = 0 = \Delta H° - T\Delta S°$.

Here $\Delta S° = \Delta H°/T = 112$ J/K · mol-rxn at 351.15 K.

19.43 $\Delta_r S°$ is +137.2 J/K · mol-rxn. A positive entropy change means that raising the temperature will increase the product favorability of the reaction (because $T\Delta S°$ will become more negative).

19.45 The reaction is exothermic, so $\Delta_r H°$ should be negative. Also, a gas and an aqueous solution are formed, so $\Delta_r S°$ should be positive. The calculated values are $\Delta_r H° = -183.32$ kJ/mol-rxn (with a negative sign as expected) and $\Delta_r S° = -7.7$ J/K · mol-rxn

The entropy change is slightly negative, not positive as predicted. The reason for this is the negative entropy change upon dissolving NaOH. Apparently the OH^- ions in water hydrogen-bond with water molecules, an effect that also leads to a small, negative entropy change.

19.47 $\Delta_r H° = +126.03$ kJ/mol-rxn; $\Delta_r S° = +78.2$ J/K · mol-rxn; and $\Delta_r G° = +103$ kJ/mol-rxn. The reaction is not predicted to be spontaneous under standard conditions.

19.49 $\Delta_r G°$ from K value = 4.87 kJ/mol-rxn

$\Delta_r G°$ from free energies of formation = 4.73 kJ/mol-rxn

19.51 $\Delta_r G° = -2.27$ kJ/mol-rxn

19.53 (a) $\Delta_r G° = +141.82$ kJ/mol-rxn, so the reaction is not spontaneous.

(b) $\Delta_r H° = +197.86$ kJ/mol-rxn; $\Delta_r S° = +187.95$ J/K · mol-rxn

$T = \Delta_r H° / \Delta_r S° = 1052.7$ K or 779.6 °C

(c) $\Delta_r G°$ at 1500 °C (1773 K) = −135.4 kJ/mol-rxn

K_p at 1500 °C = 1×10^4

19.55 $\Delta_r S° = -459.0$ J/K · mol-rxn; $\Delta_r H° = -793$ kJ/mol-rxn;

$\Delta_r G° = -657$ kJ/mol-rxn

The reaction is spontaneous and enthalpy-driven.

19.57 (a) $\Delta_r G°$ at 80.0 °C = +0.14 kJ/mol-rxn

$\Delta_r G°$ at 110.0 °C = −0.12 kJ/mol-rxn

Rhombic sulfur is more stable than monoclinic sulfur at 80 °C, but the reverse is true at 110 °C.

(b) $T = 370$ K or about 96 °C

19.59 $\Delta_r G°$ at 298 K = 22.64 kJ/mol; reaction is not product-favored.

It does become product-favored above 469 K (196 °C).

19.61 $\Delta_f G°$ [HI(g)] = −10.9 kJ/mol

19.63 (a) $\Delta_r G° = +194.8$ kJ/mol-rxn and $K = 6.68 \times 10^{-11}$
(b) The reaction is not spontaneous at 727 °C.
(c) Keep the pressure of CO as low as possible (by removing it during the course of the reciton).

19.65 $K_p = P_{Hg(g)}$ at any temperature.

$K_p = 1$ at 620.3 K or 347.2 °C when $P_{Hg(g)}$ = 1.000 bar.

T when $P_{Hg(g)}$ = (1/760) bar is 393.3 K or 125.2 °C.

19.67 (a) True
(b) False. Whether an exothermic system is spontaneous also depends on the entropy change for the system.
(c) False. Reactions with + $\Delta_r H°$ and + $\Delta_r S°$ are spontaneous at higher temperatures.
(d) True

19.69 Dissolving a solid such as NaCl in water is a spontaneous process. Thus, $\Delta G° < 0$. If $\Delta H° = 0$, then the only way the free energy change can be negative is if $\Delta S°$ is positive. Generally the entropy change is the important factor in forming a solution.

19.71 $2\ C_2H_6(g) + 7\ O_2(g) \rightarrow 4\ CO_2(g) + 6\ H_2O(g)$
(a) Not only is this an exothermic combustion reaction, but there is also an increase in the number of molecules from reactants to products. Therefore, we would predict an increase in $\Delta S°$ for both the system and the surroundings and thus for the universe as well.
(b) The exothermic reaction has $\Delta_r H° < 0$. Combined with a positive $\Delta_{sys} S°$, the value of $\Delta_r G°$ is negative.
(c) The value of K_p is likely to be much greater than 1. Further, because $\Delta_{sys} S°$ is positive, the value of K_p will be even larger at a higher temperature. (See the left panel of Figure 19.12.)

19.73 Reaction 1: $\Delta_r S°_1 = -80.7$ J/K · mol-rxn

Reaction 2: $\Delta_r S°_2 = -161.60$ J/K · mol-rxn

Reaction 3: $\Delta_r S°_3 = -242.3$ J/K · mol-rxn

$\Delta_r S°_1 + \Delta_r S°_2 = \Delta_r S°_3$

19.75 (a) $\Delta_r H° = -352.88$ kJ/mol-rxn and $\Delta_r S° = +21.31$ J/K · mol-rxn. Therefore, at 298 K, $\Delta_r G° = -359.23$ kJ/mol-rxn.
(b) 4.84 g of Mg is required.

19.77 (a) $N_2H_4(\ell) + O_2(g) \rightarrow 2\ H_2O(\ell) + N_2(g)$

O_2 is the oxidizing agent and N_2H_4 is the reducing agent.

(b) $\Delta_r H° = -622.29$ kJ/mol-rxn and $\Delta_r S° = +4.87$ J/K · mol-rxn. Therefore, at 298 K, $\Delta_r G° = -623.77$ kJ/mol-rxn.
(c) 0.0027 K
(d) 7.5 mol O_2
(e) 4.8×10^3 g solution
(f) 7.5 mol $N_2(g)$ occupies 170 L at 273 K and 1.0 atm of pressure.

19.79 Iodine dissolves readily, so the process is spontaneous and $\Delta G°$ must be less than zero. Because $\Delta H° = 0$, the process is entropy-driven.

19.81 (a) The spontaneity decreases as temperature increases.
(b) There is no temperature between 400 K and 1000 K at which the decomposition is spontaneous.

19.83 (a, b)

Temperature (K)	$\Delta_r G°$ (kJ)	K
298 K	−32.74 *K*	5.48×100^5
800 K	+72.92 *K*	1.73×10^{-5}
1300 K	+184.0 *K*	4.05×10^{-8}

(c) The largest mole fraction of NH_3 in an equilibrium mixture will be at 298 K.

CHAPTER 20

20.1 (a) $Cr(s) \rightarrow Cr^{3+}(aq) + 3\ e^-$

Cr is a reducing agent; this is an oxidation reaction.

(b) $AsH_3(g) \rightarrow As(s) + 3\ H^+(aq) + 3\ e^-$

AsH_3 is a reducing agent; this is an oxidation reaction.

(c) $VO_3^-(aq) + 6\ H^+(aq) + 3\ e^- \rightarrow V^{2+}(aq) + 3\ H_2O(\ell)$

$VO_3^-(aq)$ is an oxidizing agent; this is a reduction reaction.

(d) $2\ Ag(s) + 2\ OH^-(aq) \rightarrow Ag_2O(s) + H_2O(\ell) + 2e^-$

Silver is a reducing agent; this is an oxidation reaction.

20.3 (a) $Ag(s) \rightarrow Ag^+(aq) + e^-$

$e^- + NO_3^-(aq) + 2\ H^+(aq) \rightarrow NO_2(g) + H_2O(\ell)$

$Ag(s) + NO_3^-(aq) + 2\ H^+(aq) \rightarrow Ag^+(aq) + NO_2(g) + H_2O(\ell)$

(b) $2[MnO_4^-(aq) + 8\ H^+(aq) + 5\ e^- \rightarrow Mn^{2+}(aq) + 4\ H_2O(\ell)]$

$5[HSO_3^-(aq) + H_2O(\ell) \rightarrow SO_4^{2-}(aq) + 3\ H^+(aq) + 2\ e^-]$

$2\ MnO_4^-(aq) + H^+(aq) + 5\ HSO_3^-(aq) \rightarrow 2\ Mn^{2+}(aq) + 3\ H_2O(\ell) + 5\ SO_4^{2-}(aq)$

(c) $4[Zn(s) \rightarrow Zn^{2+}(aq) + 2\ e^-]$

$2\ NO_3^-(aq) + 10\ H^+(aq) + 8\ e^- \rightarrow N_2O(g) + 5\ H_2O(\ell)$

$4\ Zn(s) + 2\ NO_3^-(aq) + 10\ H^+(aq) \rightarrow 4\ Zn^{2+}(aq) + N_2O(g) + 5\ H_2O(\ell)$

(d) $Cr(s) \rightarrow Cr^{3+}(aq) + 3\ e^-$

$3\ e^- + NO_3^-(aq) + 4\ H^+(aq) \rightarrow NO(g) + 2\ H_2O(\ell)$

$Cr(s) + NO_3^-(aq) + 4\ H^+(aq) \rightarrow Cr^{3+}(aq) + NO(g) + 2\ H_2O(\ell)$

20.5 (a) $2[Al(s) + 4\ OH^-(aq) \rightarrow Al(OH)_4^-(aq) + 3\ e^-]$

$3[2\ H_2O(\ell) + 2\ e^- \rightarrow H_2(g) + 2\ OH^-(aq)]$

$2\ Al(s) + 2\ OH^-(aq) + 6\ H_2O(\ell) \rightarrow 2\ Al(OH)_4^-(aq) + 3\ H_2(g)$

(b) $2[CrO_4^{2-}(aq) + 4\ H_2O(\ell) + 3\ e^- \rightarrow Cr(OH)_3(s) + 5\ OH^-(aq)]$

$3[SO_3^{2-}(aq) + 2\ OH^-(aq) \rightarrow SO_4^{2-}(aq) + H_2O(\ell) + 2\ e^-]$

$2\ CrO_4^{2-}(aq) + 3\ SO_3^{2-}(aq) + 5\ H_2O(\ell) \rightarrow 2\ Cr(OH)_3(s) + 3\ SO_4^{2-}(aq) + 4\ OH^-(aq)$

(c) $Zn(s) + 4\ OH^-(aq) \rightarrow Zn(OH)_4^{2-}(aq) + 2\ e^-$

$Cu(OH)_2(s) + 2\ e^- \rightarrow Cu(s) + 2\ OH^-(aq)$

$Zn(s) + 2\ OH^-(aq) + Cu(OH)_2(s) \rightarrow Zn(OH)_4^{2-}(aq) + Cu(s)$

(d) $3[HS^-(aq) + OH^-(aq) \rightarrow S(s) + H_2O(\ell) + 2\ e^-]$

$ClO_3^-(aq) + 3\ H_2O(\ell) + 6\ e^- \rightarrow Cl^-(aq) + 6\ OH^-(aq)$

$3\ HS^-(aq) + ClO_3^-(aq) \rightarrow 3\ S(s) + Cl^-(aq) + 3\ OH^-(aq)$

20.7 Electrons flow from the Cr electrode to the Fe electrode. Negative ions move via the salt bridge from the Fe/Fe^{2+} half-cell to the Cr/Cr^{3+} half-cell (and positive ions move in the opposite direction).

Anode (oxidation): $Cr(s) \rightarrow Cr^{3+}(aq) + 3\ e^-$

Cathode (reduction): $Fe^{2+}(aq) + 2\ e^- \rightarrow Fe(s)$

20.9 (a) Oxidation: $Fe(s) \rightarrow Fe^{2+}(aq) + 2\ e^-$
Reduction: $O_2(g) + 4\ H^+(aq) + 4\ e^- \rightarrow 2\ H_2O(\ell)$
Overall: $2\ Fe(s) + O_2(g) + 4\ H^+(aq) \rightarrow 2\ Fe^{2+}(aq) + 2\ H_2O(\ell)$

(b) Anode, oxidation: $Fe(s) \rightarrow Fe^{2+}(aq) + 2\ e^-$
Cathode, reduction: $O_2(g) + 4\ H^+(aq) + 4\ e^- \rightarrow 2\ H_2O(\ell)$

(c) Electrons flow from the negative anode (Fe) to the positive cathode (site of the O_2 half-reaction). Negative ions move through the salt bridge from the cathode compartment in which the O_2 reduction occurs to the anode compartment in which Fe oxidation occurs (and positive ions move in the opposite direction).

20.11 (a) All are primary batteries, not rechargeable.
(b) Dry cells and alkaline batteries have Zn anodes. Ni-Cd batteries have a cadmium anode.
(c) Dry cells have an acidic environment, whereas the environment is alkaline for alkaline and Ni-Cd cells.

20.13 (a) $E°_{cell} = -1.298$ V; not product-favored
(b) $E°_{cell} = -0.51$ V; not product-favored
(c) $E°_{cell} = -1.023$ V; not product-favored
(d) $E°_{cell} = +0.029$ V; product-favored

20.15 (a) $Sn^{2+}(aq) + 2\ Ag(s) \longrightarrow Sn(s) + 2\ Ag^+(aq)$

$E°_{cell} = -0.94$ V; not product-favored

(b) $3\ Sn^{4+}(aq) + 2\ Al(s) \longrightarrow 3\ Sn^{2+}(aq) + 2\ Al^{3+}(aq)$

$E°_{cell} = +1.81$ V; product-favored

(c) $2\ ClO_3^-(aq) + 10\ Ce^{3+}(aq) + 12\ H^+(aq) \longrightarrow Cl_2(aq) + 10\ Ce^{4+}(aq) + 6\ H_2O(\ell)$

$E°_{cell} = -0.14$ V; not product-favored

(d) $3\ Cu(s) + 2\ NO_3^-(aq) + 8\ H^+(aq) \longrightarrow 3\ Cu^{2+}(aq) + 2\ NO(g) + 4\ H_2O(\ell)$

$E°_{cell} = +0.62$ V; product-favored

20.17 (a) Al
(b) Zn and Al
(c) $Fe^{2+}(aq) + Sn(s) \longrightarrow Fe(s) + Sn^{2+}(aq)$; reactant-favored
(d) $Zn^{2+}(aq) + Sn(s) \longrightarrow Zn(s) + Sn^{2+}(aq)$; reactant-favored

20.19 Best reducing agent, Cr(s). (Use Appendix M)

20.21 Ag^+

20.23 See Example 20.5
(a) F_2, most readily reduced
(b) F_2 and Cl_2

20.25 $E°_{cell} = +0.3923$ V. When $[Zn(OH)_4^{2-}] = [OH^-] = 0.025$ M and $P(H_2) = 1.0$ bar, $E_{cell} = 0.345$ V.

20.27 $E°_{cell} = +1.563$ V and $E_{cell} = +1.58$ V.

20.29 When $E°_{cell} = +1.563$ V, $E_{cell} = 1.48$ V, $n = 2$, and $[Zn^{2+}] = 1.0$ M, the concentration of $Ag^+ = 0.040$ M.

20.31 (a) $\Delta_r G° = -29.0$ kJ; $K = 1 \times 10^5$
(b) $\Delta_r G° = +88.6$ kJ; $K = 3 \times 10^{-16}$

20.33 $E°_{cell}$ for $AgBr(s) \longrightarrow Ag^+(aq) + Br^-(aq)$ is -0.7281.

$K_{sp} = 4.9 \times 10^{-13}$

20.35 $K_{formation} = 2 \times 10^{25}$

20.37 See Figure 20.18. Electrons from the battery or other source enter the cathode where they are transferred to Na^+ ions, reducing the ions to Na metal. Chloride ions move toward the positively charged anode where an electron is transferred from each Cl^- ion, and Cl_2 gas is formed.

20.39 O_2 from the oxidation of water is more likely than F_2. See Example 20.10.

20.41 See Example 20.10.
(a) Cathode: $2\ H_2O(\ell) + 2\ e^- \longrightarrow H_2(g) + 2\ OH^-(aq)$
(b) Anode: $2\ Br^-(aq) \longrightarrow Br_2(\ell) + 2\ e^-$

20.43 Mass of Ni = 0.0334 g

20.45 Time = 2300 s or 38 min

20.47 Time = 250 h

20.49 (a) $UO_2^+(aq) + 4\ H^+(aq) + e^- \longrightarrow U^{4+}(aq) + 2\ H_2O(\ell)$
(b) $ClO_3^-(aq) + 6\ H^+(aq) + 6\ e^- \longrightarrow Cl^-(aq) + 3\ H_2O(\ell)$
(c) $N_2H_4(aq) + 4\ OH^-(aq) \longrightarrow N_2(g) + 4\ H_2O(\ell) + 4\ e^-$
(d) $ClO^-(aq) + H_2O(\ell) + 2\ e^- \longrightarrow Cl^-(aq) + 2\ OH^-(aq)$

20.51 (a,c) The electrode at the right is a magnesium anode. (Magnesium metal supplies electrons and is oxidized to Mg^{2+} ions.) Electrons pass through the wire to the silver cathode, where Ag^+ ions are reduced to silver metal. Nitrate ions move via the salt bridge from the $AgNO_3$ solution to the $Mg(NO_3)_2$ solution (and Na^+ ions move in the opposite direction).
(b) Anode: $Mg(s) \longrightarrow Mg^{2+}(aq) + 2\ e^-$

Cathode: $Ag^+(aq) + e^- \longrightarrow Ag(s)$

Net reaction: $Mg(s) + 2\ Ag^+(aq) \longrightarrow Mg^{2+}(aq) + 2\ Ag(s)$

20.53 (a) For 1.7 V:

Use chromium as the anode to reduce $Ag^+(aq)$ to Ag(s) at the cathode. The cell potential is +1.71 V.

(b) For 0.5 V:
(i) Use copper as the anode to reduce silver ions to silver metal at the cathode. The cell potential is +0.46 V.

(ii) Use silver as the anode to reduce chlorine to chloride ions. The cell potential would be +0.56 V. (In practice, this setup is not likely to work well because the product would be insoluble silver chloride.)

20.55 (a) $Zn^{2+}(aq)$ (c) Zn(s)
(b) $Au^+(aq)$ (d) Au(s)
(e) Yes, Sn(s) will reduce Cu^{2+} (as well as Ag^+ and Au^+).
(f) No, Ag(s) can only reduce $Au^+(aq)$.
(g) See part (e).
(h) $Ag^+(aq)$ can oxidize Cu, Sn, Co, and Zn.

20.57 (a) The cathode is the site of reduction, so the half-reaction must be $2\ H^+(aq) + 2\ e^- \longrightarrow H_2(g)$. This is the case with the following half-reactions: $Cr^{3+}(aq)|Cr(s)$, $Fe^{2+}(aq)|Fe(s)$, and $Mg^{2+}(aq)|Mg(s)$.
(b) Choosing from the half-cells in part (a), the reaction of Mg(s) and $H^+(aq)$ would produce the most positive potential (2.37 V), and the reaction of H_2 with Cu^{2+} would produce the least positive potential (+0.337 V).

20.59 8.1×10^5 g Al

20.61 (a) $E°_{anode} = -0.268$ V
(b) $K_{sp} = 2 \times 10^{-5}$

20.63 $\Delta_r G° = -409$ kJ

20.65 6700 kWh; 820 kg Na; 1300 kg Cl_2

20.67 Ru^{2+}, $Ru(NO_3)_2$

20.69 9.5×10^6 g Cl_2 per day

20.71 (a) $2[Ag^+(aq) + e^+ \longrightarrow Ag(s)]$

$$C_6H_5CHO(aq) + H_2O(\ell) \longrightarrow C_6H_5CO_2H(aq) + 2\ H^+(aq) + 2\ e^-$$

$$2Ag^+(aq) + C_6H_5CHO(aq) + H_2O(\ell) \longrightarrow C_6H_5CO_2H(aq) + 2\ H^+(aq) + 2\ Ag(s)$$

(b) $3[CH_3CH_2OH(aq) + H_2O(\ell) \longrightarrow CH_3CO_2H(aq) + 4\ H^+(aq) + 4\ e^-]$

$$2[Cr_2O_7^{2-}(aq) + 14\ H^+(aq) + 6\ e^- \longrightarrow 2\ Cr^{3+}(aq) + 7\ H_2O(\ell)]$$

$$3\ CH_3CH_2OH(aq) + 2\ Cr_2O_7^{2-}(aq) + 16\ H^+(aq) \longrightarrow 3\ CH_3CO_2H(aq) + 4\ Cr^{3+}(aq) + 11\ H_2O(\ell)$$

20.73 (a) 0.974 kJ/g
(b) 0.60 kJ/g
(c) The silver-zinc battery produces more energy per gram of reactants.

20.75 (a) $2\ NO_3^-(aq) + 3\ Mn^{2+}(aq) + 2\ H_2O(\ell) \longrightarrow 2\ NO(g) + 3\ MnO_2(s) + 4\ H^+(aq)$

$$3\ MnO_2(s) + 4\ H^+(aq) + 2\ NH_4^+(aq) \longrightarrow N_2(g) + 3\ Mn^{2+}(aq) + 6\ H_2O(\ell)$$

(b) $E°$ for the reduction of NO_3^- to NO is -0.27 V. $E°$ for the oxidation of NH_4^+ to N_2 is $+0.96$ V.

20.77 (a) $Fe^{2+}(aq) + 2\ e^- \longrightarrow Fe(s)$
$2[Fe^{2+}(aq) \longrightarrow Fe^{3+}(aq) + e^-]$
$3\ Fe^{2+}(aq) \longrightarrow Fe(s) + 2\ Fe^{3+}(aq)$
(b) $E°_{cell} = -1.21$ V; not product-favored
(c) $K = 1 \times 10^{-41}$

20.79 (a)

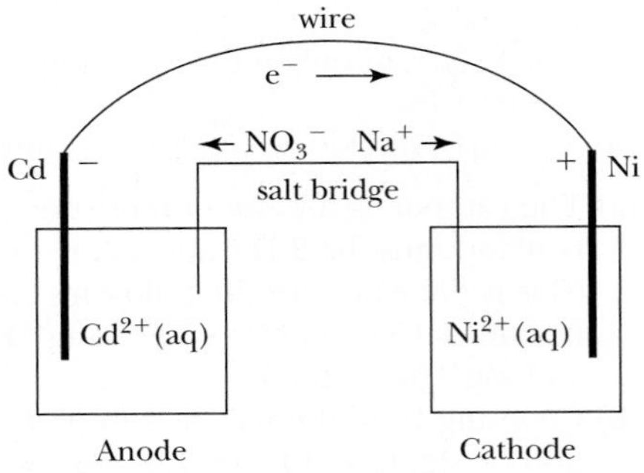

(b) Anode: $Cd(s) \longrightarrow Cd^{2+}(aq) + 2\ e^-$
Cathode: $Ni^{2+}(aq) + 2\ e^- \longrightarrow Ni(s)$
Net: $Cd(s) + Ni^{2+}(aq) \longrightarrow Cd^{2+}(aq) + Ni(s)$
(c) The anode is negative and the cathode is positive.
(d) $E°_{cell} = E°_{cathode} - E°_{anode} = (-0.25\ V) - (-0.40\ V) = +0.15$ V
(e) Electrons flow from anode (Cd) to cathode (Ni).
(f) Na^+ ions move from the anode compartment to the cathode compartment. Anions move in the opposite direction.
(g) $K = 1 \times 10^5$
(h) $E_{cell} = 0.21$ V
(i) 480 h

20.81 0.054 g Au

20.83 I^- is the strongest reducing agent of the three halide ions. Iodide ion reduces Cu^{2+} to Cu^+, forming insoluble CuI(s).

$$2\ Cu^{2+}(aq) + 4\ I^-(aq) \longrightarrow 2\ CuI(s) + I_2(aq)$$

20.85 (a) 92 g HF required; 230 g CF_3SO_2F and 9.3 g H_2 isolated
(b) H_2 is produced at the cathode.
(c) 48 kWh

20.87 290 h

20.89 (a) 3.6 mol glucose and 22 mol O_2
(b) 86 mole electrons
(c) 96 amps
(d) 96 watts

CHAPTER 21

21.1 $4\ Li(s) + O_2(g) \longrightarrow 2\ Li_2O(s)$

$Li_2O(s) + H_2O(\ell) \longrightarrow 2\ LiOH(aq)$

$2\ Ca(s) + O_2(g) \longrightarrow 2\ CaO(s)$

$CaO(s) + H_2O(\ell) \longrightarrow Ca(OH)_2(s)$

21.3 These are the elements of Group 3A: boron, B; aluminum, Al; gallium, Ga; indium, In; and thallium, Tl.

21.5 $2\ Na(s) + Cl_2(g) \longrightarrow 2\ NaCl(s)$

The reaction is exothermic and the product is ionic. See Figure 1.4.

21.7 The product, NaCl, is a colorless solid and is soluble in water. Other alkali metal chlorides have similar properties.

21.9 Calcium will not exist in the earth's crust because the metal reacts with water.

21.11 Increasing basicity: $CO_2 < SiO_2 < SnO_2$

21.13 (a) $2\ Na(s) + Br_2(\ell) \longrightarrow 2\ NaBr(s)$
(b) $2\ Mg(s) + O_2(g) \longrightarrow 2\ MgO(s)$
(c) $2\ Al(s) + 3\ F_2(g) \longrightarrow 2\ AlF_3(s)$
(d) $C(s) + O_2(g) \longrightarrow CO_2(g)$

21.15 $2\ H_2(g) + O_2(g) \longrightarrow 2\ H_2O(g)$

$H_2(g) + Cl_2(g) \longrightarrow 2\ HCl(g)$

$3\ H_2(g) + N_2(g) \longrightarrow 2\ NH_3(g)$

21.17 $CH_4(g) + H_2O(g) \longrightarrow CO(g) + 3\ H_2(g)$

$\Delta_r H° = +206.2$ kJ; $\Delta_r S° = +214.7$ J/K; $\Delta_r G° = +142.2$ kJ (at 298.15 K).

21.19 Step 1: $2\ SO_2(g) + 4\ H_2O(\ell) + 2\ I_2(s) \longrightarrow 2\ H_2SO_4(\ell) + 4\ HI(g)$

Step 2: $2\ H_2SO_4(\ell) \longrightarrow 2\ H_2O(\ell) + 2\ SO_2(g) + O_2(g)$

Step 3: $4\ HI(g) \longrightarrow 2\ H_2(g) + 2\ I_2(g)$

Net: $2\ H_2O(\ell) \longrightarrow 2\ H_2(g) + O_2(g)$

21.21 $2\ Na(s) + F_2(g) \longrightarrow 2\ NaF(s)$

$2\ Na(s) + Cl_2(g) \longrightarrow 2\ NaCl(s)$

$2\ Na(s) + Br_2(\ell) \longrightarrow 2\ NaBr(s)$

$2\ Na(s) + I_2(s) \longrightarrow 2\ NaI(s)$

The alkali metal halides are white, crystalline solids. They have high melting and boiling points, and are soluble in water.

21.23 (a) $2\ Cl^-(aq) + 2\ H_2O(\ell) \longrightarrow Cl_2(g) + H_2(g) + 2\ OH^-(aq)$

(b) If this were the only process used to produce chlorine, the mass of Cl_2 reported for industrial production would be 0.88 times the mass of NaOH produced (2 mol NaCl, 117 g, would yield 2 mol NaOH, 80 g, and 1 mol Cl_2, 70 g). The amounts quoted indicate a Cl_2-to-NaOH mass ratio 0.96. Chlorine is presumably also prepared by other routes than this one.

21.25 $2\ Mg(s) + O_2(g) \longrightarrow 2\ MgO(s)$

$3\ Mg(s) + N_2(g) \longrightarrow Mg_3N_2(s)$

21.27 $CaCO_3$ is used in agriculture to neutralize acidic soil, to prepare CaO for use in mortar, and in steel production.

$CaCO_3(s) + H_2O(\ell) + CO_2(g) \longrightarrow Ca^{2+}(aq) + 2\ HCO_3^-(aq)$

21.29 1.4×10^6 g SO_2

21.31

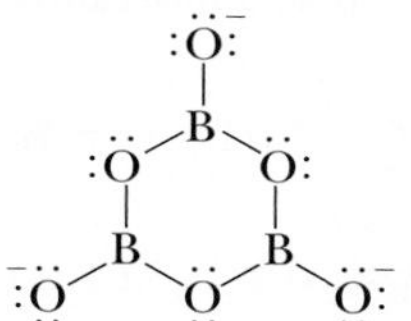

$B_3O_6^{3-}$

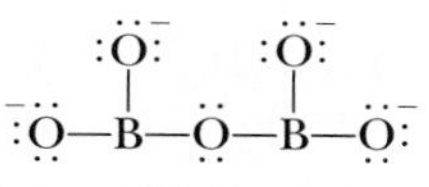

$B_2O_5^{4-}$

21.33 (a) $2\ B_5H_9(g) + 12\ O_2(g) \longrightarrow 5\ B_2O_3(s) + 9\ H_2O(g)$

(b) Enthalpy of combustion of $B_5H_9 = -4341.2$ kJ/mol. This is more than double the enthalpy of combustion of B_2H_6.

(c) Enthalpy of combustion of $C_2H_6(g)$ [to give $CO_2(g)$ and $H_2O(g)$] = −1428.7 kJ/mol. C_2H_6 produces 47.5 kJ/g, whereas diborane produces much more (73.7 kJ/g).

21.35 $2\ Al(s) + 6\ HCl(aq) \longrightarrow 2\ Al^{3+}(aq) + 6\ Cl^-(aq) + 3\ H_2(g)$

$2\ Al(s) + 3\ Cl_2(g) \longrightarrow 2\ AlCl_3(s)$

$4\ Al(s) + 3\ O_2(g) \longrightarrow 2\ Al_2O_3(s)$

21.37 $2\ Al(s) + 2\ OH^-(aq) + 6\ H_2O(\ell) \longrightarrow 2\ Al(OH)_4^-(aq) + 3\ H_2(g)$

Volume of H_2 obtained from 13.2 g Al = 18.4 L

21.39 $Al_2O_3(s) + 3\ H_2SO_4(aq) \longrightarrow Al_2(SO_4)_3(s) + 3\ H_2O(\ell)$

Mass of H_2SO_4 required = 860 g and mass of Al_2O_3 required = 298 g.

21.41 Pyroxenes have as their basic structural unit an extended chain of linked SiO_4 tetrahedra. The ratio of Si to O is 1 to 3.

21.43 This structure has a six-member ring of Si atoms with O atom bridges. Each Si also has two O atoms attached. The basic unit is SiO_3^{2-}, and the overall charge is −12 in $[(SiO_3)_6]^{-12}$. (Electron lone pairs are omitted in the following structure.)

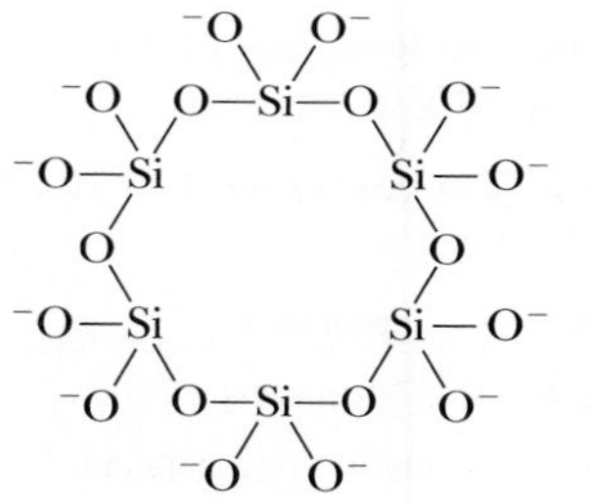

21.45 Consider the general decomposition reaction:

$N_xO_y \longrightarrow {}^x/_2\ N_2 + {}^y/_2\ O_2$

The value of $\Delta G°$ can be obtained for all N_xO_y molecules because $\Delta_r G° = -\Delta_f G°$. These data show that the decomposition reaction is spontaneous for all of the nitrogen oxides. All are unstable with respect to decomposition to the elements.

Compound	$-\Delta_f G°$ (kJ/mol)
NO(g)	−86.58
NO_2	−51.23
N_2O	−104.20
N_2O_4	−97.73

21.47 $\Delta_r H° = -114.4$ kJ; exothermic $\Delta_r G° = -70.7$ kJ, product-favored at equilibrium

21.49 (a) $N_2H_4(aq) + O_2(g) \rightarrow N_2(g) + 2\ H_2O(\ell)$
(b) 1.32×10^3 g

21.51 (a) Oxidation number = +3
(b) Diphosphonic acid ($H_4P_2O_5$) should be a diprotic acid (losing the two H atoms attached to O atoms).

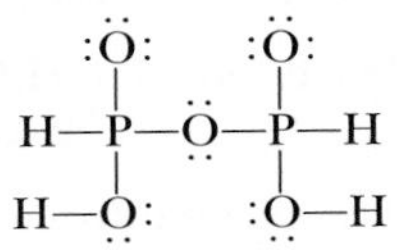

21.53 (a) 3.5×10^3 kg SO_2
(b) 4.1×10^3 kg $Ca(OH)_2$

21.55

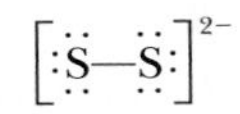

disulfide ion

21.57 $E°_{cell} = E°_{cathode} - E°_{anode} = +1.44\ V - (+1.51\ V) = -0.07\ V$

The reaction is not product-favored under standard conditions.

21.59 $Cl_2(aq) + 2\ Br^-(aq) \rightarrow 2\ Cl^-(aq) + Br_2(aq)$

Cl_2 is the oxidizing agent, Br^- is the reducing agent; $E°_{cell} = 0.28$ V.

21.61 The reaction consumes 4.32×10^8 C to produce 8.51×10^4 g F_2.

21.63

Element	Appearance	State
Na, Mg, Al	Silvery metal	Solids
Si	black, shiny metalloid	Solid
P	White, red, and black allotropes; nonmetal	Solid
S	Yellow nonmetal	Solid
Cl	Pale green nonmetal	Gas
Ar	Colorless nonmetal	Gas

21.65 (a) $2\ K(s) + Cl_2(g) \rightarrow 2\ KCl(s)$

$Ca(s) + Cl_2(g) \rightarrow CaCl_2(s)$

$2\ Ga(s) + 3\ Cl_2(g) \rightarrow 2\ GaCl_3(s)$

$Ge(s) + 2\ Cl_2(g) \rightarrow GeCl_4(\ell)$

$2\ As(s) + 3\ Cl_2(g) \rightarrow 2\ AsCl_3(\ell)$

($AsCl_5$ has been prepared but is not stable.)

(b) KCl and $CaCl_2$ are ionic; the other products are covalent.

(c) $GaCl_3$ is planar trigonal; $AsCl_3$ is pyramidal.

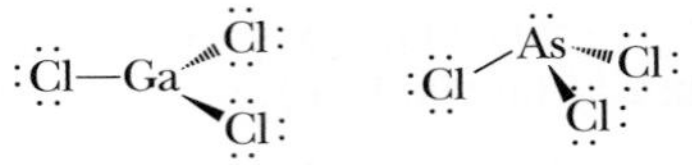

21.67 (a) $2\ KClO_3(s) \rightarrow 2\ KCl(s) + 3\ O_2(g)$
(b) $2\ H_2S(g) + 3\ O_2(g) \rightarrow 2\ H_2O(g) + 2\ SO_2(g)$
(c) $2\ Na(s) + O_2(g) \rightarrow Na_2O_2(s)$
(d) $P_4(s) + 3\ KOH(aq) + 3\ H_2O(\ell) \rightarrow PH_3(g) + 3\ KH_2PO_4(aq)$
(e) $NH_4NO_3(s) \rightarrow N_2O(g) + 2\ H_2O(g)$
(f) $2\ In(s) + 3\ Br_2(\ell) \rightarrow 2\ InBr_3(s)$
(g) $SnCl_4(\ell) + 2\ H_2O(\ell) \rightarrow SnO_2(s) + 4\ HCl(aq)$

21.69 1.4×10^5 metric tons

21.71 Mg: $\Delta_r G° = +64.9$ kJ

Ca: $\Delta_r G° = +131.40$ kJ

Ba: $\Delta_r G° = +219.4$ kJ

Relative tendency to decompose:
$MgCO_3 > CaCO_3 > BaCO_3$

21.73 (a) $\Delta_f G°$ should be more negative than $(-95.1\ kJ) \times n$.
(b) Ba, Pb, Ti

21.75 O—F bond energy = 190 kJ/mol

21.77 (a) N_2O_4 is the oxidizing agent (N is reduced from +4 to 0 in N_2), and $H_2NN(CH_3)_2$ is the reducing agent.
(b) 1.3×10^4 kg N_2O_4 is required. Product masses: 5.7×10^3 kg N_2; 4.9×10^3 kg H_2O; 6.0×10^3 kg CO_2.

21.79 $\Delta_r H° = -257.78$ kJ. This reaction is entropy-disfavored, however, with $\Delta_r S° = -963$ J/K because of the decrease in the number of moles of gases. Combining these values gives $\Delta_r G° = +29.19$ kJ, indicating that under standard conditions at 298 K the reaction is not spontaneous. (The reaction has a favorable $\Delta_r G°$ at temperatures less than 268 K, indicating that further research on this system might be worthwhile. Note that at that temperature water is a solid.)

21.81 A = B_2H_6; B = B_4H_{10}; C = B_5H_{11}; D = B_5H_9; E = $B_{10}H_{14}$

21.83 (a) $2\ CH_3Cl(g) + Si(s) \rightarrow (CH_3)_2SiCl_2(\ell)$
(b) 0.823 atm
(c) 12.2 g

21.85 $5\ N_2H_5^+(aq) + 4\ IO_3^-(aq) \rightarrow 5\ N_2(g) + 2\ I_2(aq) + H^+(aq) + 12\ H_2O(\ell)$

$E°_{net} = 1.43$ V

21.87 (a) Br_2O_3
(b) The structure of Br_2O is reasonably well known. Several possible structures for Br_2O_3 can be

imagined, but experiment confirms the structure below.

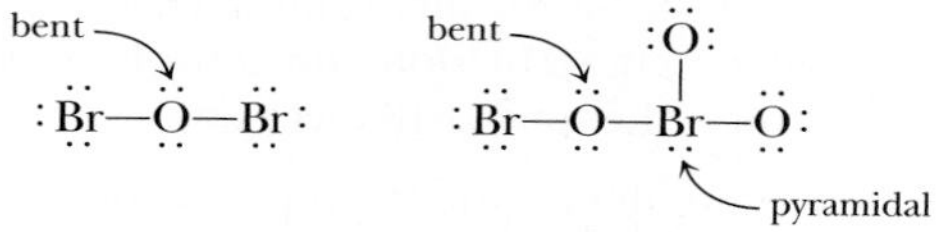

21.89 (a) The NO bond with a length of 114.2 pm is a double bond. The other two NO bonds (with a length of 121 pm) have a bond order of 1.5 (as there are two resonance structures involving these bonds).

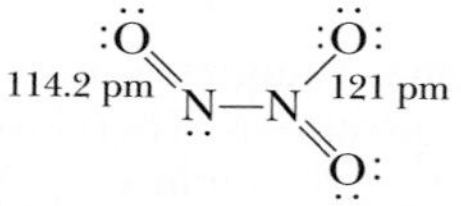

(b) $K = 1.90$; $\Delta_r S° = 141$ J/K · mol-rxn
(c) $\Delta_f H° = 82.9$ kJ/mol

21.91 The flask contains a fixed number of moles of gas at the given pressure and temperature. One could burn the mixture because only the H_2 will combust; the argon is untouched. Cooling the gases from combustion would remove water (the combustion product of H_2) and leave only Ar in the gas phase. Measuring its pressure in a calibrated volume at a known temperature would allow one to calculate the amount of Ar that was in the original mixture.

21.93 Generally, a sodium fire can be extinguished by smothering it with sand. The worst choice is to use water (which reacts violently with sodium to give H_2 gas and NaOH).

21.95 Nitrogen is a relatively unreactive gas, so it will not participate in any reaction typical of hydrogen or oxygen. The most obvious property of H_2 is that it burns, so attempting to burn a small sample of the gas would immediately confirm or deny the presence of H_2. If O_2 is present, it can be detected by allowing it to react as an oxidizing agent. There are many reactions known with low-valent metals, especially transition metal ions in solution, that can be detected by color changes.

21.97 3.5 kWh

21.99 The reducing ability of the Group 3A metals declines considerably on descending the group, with the largest drop occurring on going from Al to Ga. The reducing ability of gallium and indium are similar, but another large change is observed on going to thallium. In fact, thallium is most stable in the +1 oxidation state. This same tendency for elements to be more stable with lower oxidation numbers is seen in Groups 4A (Ge and Pb) and 5A (Bi).

CHAPTER 22

22.1 (a) Cr^{3+}: $[Ar]3d^3$, paramagnetic
(b) V^{2+}: $[Ar]3d^3$, paramagnetic
(c) Ni^{2+}: $[Ar]3d^8$, paramagnetic
(d) Cu^+: $[Ar]3d^{10}$, diamagnetic

22.3 (a) Fe^{3+}: $[Ar]3d^5$, isoelectronic with Mn^{2+}
(b) Zn^{2+}: $[Ar]3d^{10}$, isoelectronic with Cu^+
(c) Fe^{2+}: $[Ar]3d^6$, isoelectronic with Co^{3+}
(d) Cr^{3+}: $[Ar]3d^3$, isoelectronic with V^{2+}

22.5 (a) $Cr_2O_3(s) + 2\ Al(s) \rightarrow Al_2O_3(s) + 2\ Cr(s)$
(b) $TiCl_4(\ell) + 2\ Mg(s) \rightarrow Ti(s) + 2\ MgCl_2(s)$
(c) $2\ [Ag(CN)_2]^-(aq) + Zn(s) \rightarrow 2\ Ag(s) + [Zn(CN)_4]^{2-}(aq)$
(d) $3\ Mn_3O_4(s) + 8\ Al(s) \rightarrow 9\ Mn(s) + 4\ Al_2O_3(s)$

22.7 Monodentate: CH_3NH_2, CH_3CN, N_3^-, Br^-

Bidentate: en, phen (see Figure 22.14)

22.9 (a) Mn^{2+} (b) Co^{3+} (c) Co^{3+} (d) Cr^{2+}

22.11 $[Ni(en)(NH_3)_3(H_2O)]^{2+}$

22.13 (a) $Ni(en)_2Cl_2$ (en = $H_2NCH_2CH_2NH_2$)
(b) $K_2[PtCl_4]$
(c) $K[Cu(CN)_2]$
(d) $[Fe(NH_3)_4(H_2O)_2]^{2+}$

22.15 (a) Diaquabis(oxalato)nickelate(II) ion
(b) Dibromobis(ethylenediamine)cobalt(III) ion
(c) Amminechlorobis(ethylenediamine)cobalt(III) ion
(d) Diammineoxalatoplatinum(II)

22.17 (a) $[Fe(H_2O)_5OH]^{2+}$
(b) Potassium tetracyanonickelate(II)
(c) Potassium diaquabis(oxalato)chromate(III)
(d) $(NH_4)_2[PtCl_4]$

22.19

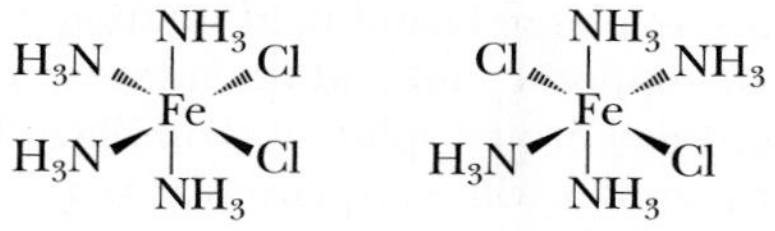

cis *trans*

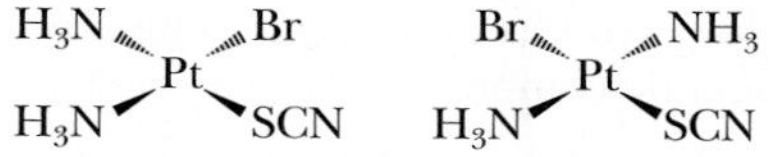

cis *trans*

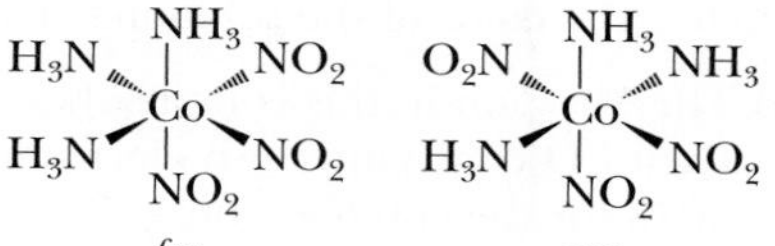

fac *mer*

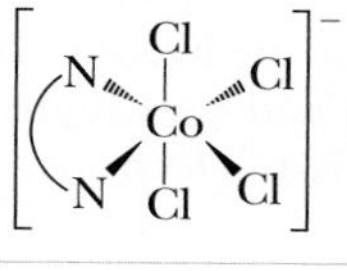

Only one structure possible. (N—N is the bidentate ethylenediamine ligand.)

22.21 For a discussion of chirality, see Chapter 10, page 446).
(a) Fe^{2+} is a chiral center.
(b) Co^{3+} is not a chiral center.
(c) Neither of the two possible isomers is chiral.
(d) No. Square-planar complexes are never chiral.

22.23 (a) $[Mn(CN)_6]^{4-}$: d^5, low-spin Mn^{2+} complex is paramagnetic.

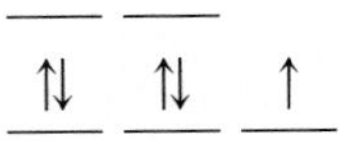

(b) $[Co(NH_3)_6]^{3+}$: d^6, low-spin Co^{3+} complex is diamagnetic.

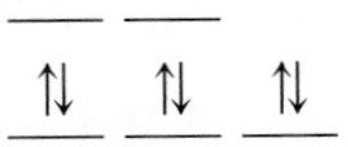

(c) $[Fe(H_2O)_6]^{3+}$: d^5, low-spin Fe^{3+} complex is paramagnetic (1 unpaired electron; same as part a).
(d) $[Cr(en)_3]^{2+}$: d^4, Cr^{3+} complex is paramagnetic (2 unpaired electrons).

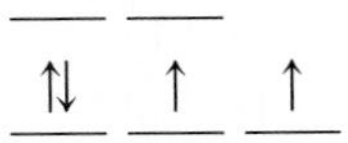

22.25 (a) Fe^{2+}, d^6, paramagnetic, four unpaired electrons
(b) Co^{2+}, d^7, paramagnetic, three unpaired electrons
(c) Mn^{2+}, d^5, paramagnetic, five unpaired electrons
(d) Zn^{2+}, d^{10}, diamagnetic, zero unpaired electrons

22.27 (a) 6
(b) octahedral
(c) +2
(d) four unpaired electrons (high spin)
(e) paramagnetic

22.29 With four ligands, complexes of the d^8 Ni^{2+} ion can be either tetrahedral or square planar. The CN^- ligand is at one end of the spectrochemical series and leads to a large ligand field splitting, whereas Cl^- is at the opposite end and often leads to complexes with small orbital splitting. With ligands such as CN^- the complex will be square planar (and for a d^8 ion it will be diamagnetic). With a weak field ligand (Cl^-) the complex will be tetrahedral and, for the d^8 ion, two electrons will be unpaired, giving a paramagnetic complex.

22.31 The light absorbed is in the blue region of the spectrum (page 271). Therefore, the light transmitted—which is the color of the solution—is yellow.

22.33 (a) The Mn^+ ion in this complex has six d electrons. Each CO contributes two electrons, giving a total of 18 for the complex ion.
(b) The $C_5H_5^-$ ligand contributes six electrons, CO and PR_3 each contribute two electrons, for a total of 10 electrons for the ligands. The cobalt is effectively a Co^+ ion and contributes eight d electrons. The total is 18 electrons.
(c) The $C_5H_5^-$ ligand contributes six electrons, each CO contributes two electrons, for a total of 12 electrons for the ligands. The manganese is effectively a Mn^+ ion and contributes six d electrons. The total is 18 electrons.

22.35 Determine the magnetic properties of the complex. Square-planar Ni^{2+} (d^8) complexes are diamagnetic, whereas tetrahedral complexes are paramagnetic.

22.37 Fe^{2+} has a d^6 configuration. Low-spin octahedral complexes are diamagnetic, whereas high-spin octahedral complexes of this ion have four unpaired electrons and are paramagnetic.

22.39 Square-planar complexes most often arise from d^8 transition metal ions. Therefore, it is likely that $[Ni(CN)_4]^{2-}$ (Ni^{2+}) and $[Pt(CN)_4]^{2-}$ (Pt^{2+}) are square planar. (See also Study Questions 22.29 and 22.65.)

22.41 Two geometric isomers are possible.

22.43 Absorbing at 425 nm means the complex is absorbing light in the blue-violet end of the spectrum. Therefore, red and green light are transmitted, and the complex appears yellow (see Figure 22.27).

22.45 (a) Mn^{2+}; (b) 6; (c) octahedral; (d) 5; (e) paramagnetic; (f) *cis* and *trans* isomers exist.

22.47 Name: tetraamminedichlorocobalt(III) ion

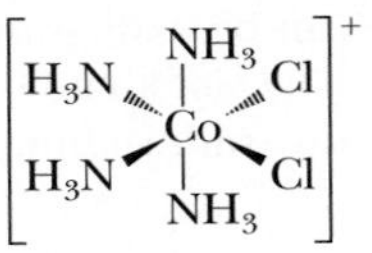
cis

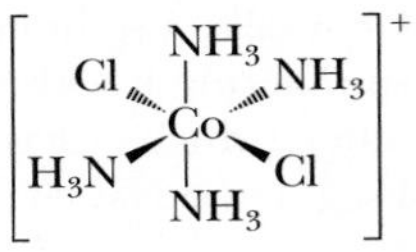
trans

22.49 $[Co(en)_2(H_2O)Cl]^{2+}$

22.51

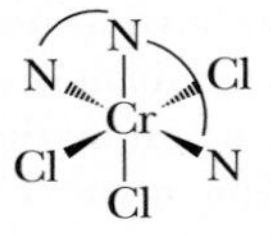
mer

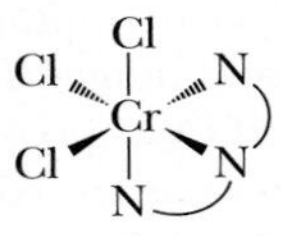
fac

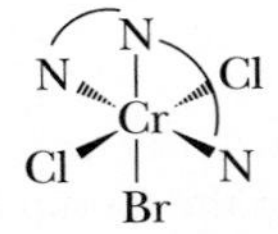
trans chlorides

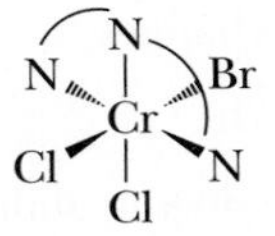
cis chlorides

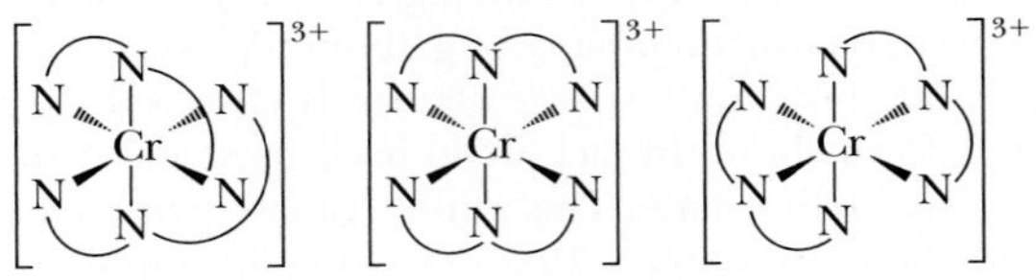

22.53

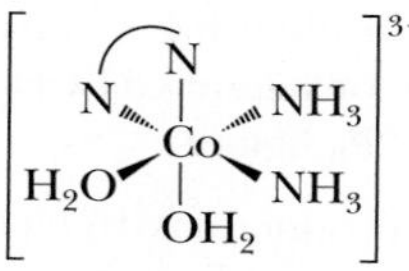

H_2O and NH_3 *cis*, chiral

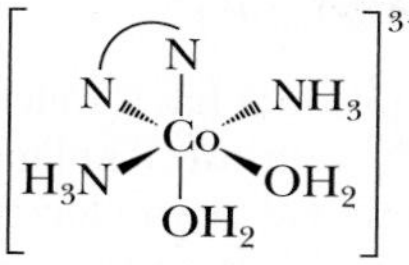

H_2O *cis* and NH_3 *trans*, not chiral

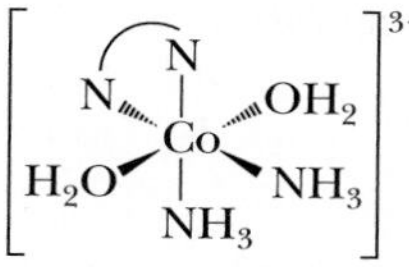

H_2O *trans* and NH_3 *cis*, not chiral

22.55 In $[Mn(H_2O)_6]^{2+}$ and $[Mn(CN)_6]^{4-}$, Mn has an oxidation number of +2 (Mn is a d^5 ion).

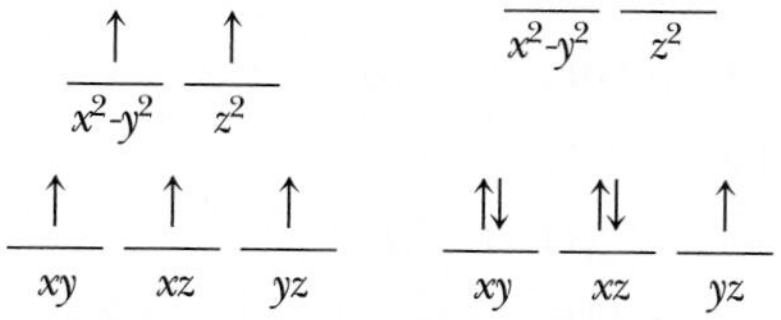

$[Mn(H_2O)_6]^{2+}$ paramagnetic, 5 unpaired e^-

$[Mn(CN)_6]^{4-}$ paramagnetic, 1 unpaired e^-

This shows that Δ_o for CN^- is greater than for H_2O.

22.57 (a) ammonium tetrachlorocuprate(II)
(b) hexacarbonylmolybdenum(0)
(c) $[Cr(H_2O)_4Cl_2]Cl$
(d) $[Co(H_2O)(NH_2CH_2CH_2NH_2)_2(SCN)](NO_3)_2$

22.59 (a) The light absorbed is in the orange region of the spectrum (page 1043). Therefore, the light transmitted (the color of the solution) is blue or cyan.
(b) Using the cobalt(III) complexes in Table 22.3 as a guide, we might place CO_3^{2-} between F^- and the oxalato ion, $C_2O_4^{2-}$.
(c) Δ_o is small, so the complex should be high spin and paramagnetic.

22.61

$\widehat{N\ O} = H_2N{-}CH_2{-}CO_2^-$

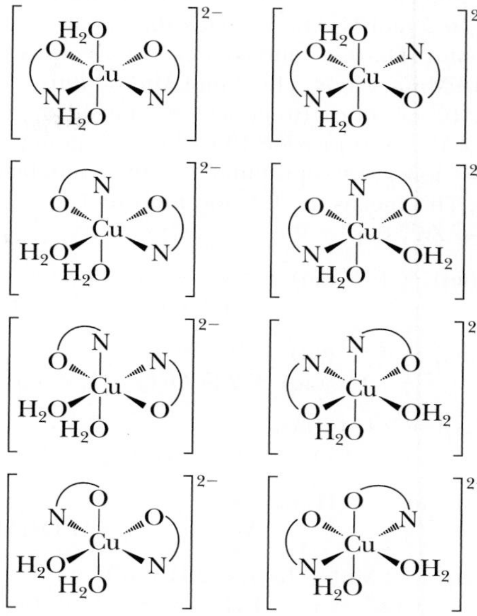

22.63 (a) In complexes such as $M(PR_3)Cl_2$ the metal is Ni^{2+} or Pd^{2+}, both of which are d^8 metal ions. If an Ni^{2+} complex is paramagnetic it must be tetrahedral, whereas the Pd^{2+} must be square planar. (A d^8 metal complex cannot be diamagnetic if it has a tetrahedral structure.)

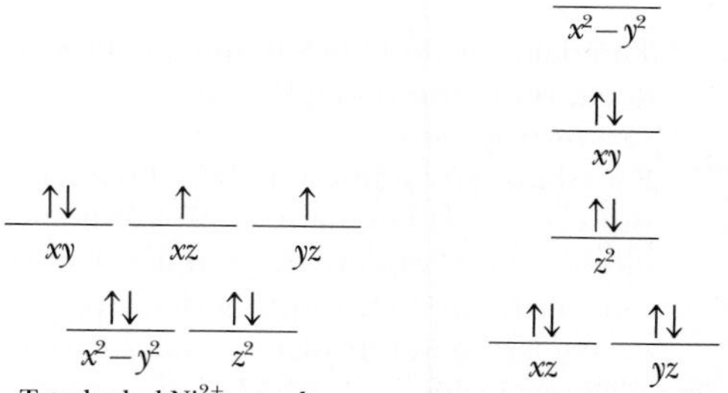

Tetrahedral Ni^{2+} complex, paramagnetic

Square planar Pd^{2+} complex, diamagnetic

(b) A tetrahedral Ni^{2+} complex cannot have isomers, whereas a square planar complex of the type $M(PR_3)_2Cl_2$ can have *cis* and *trans* isomers. See page 1036.

22.65 A, dark violet isomer: $[Co(NH_3)_5Br]SO_4$

B, violet-red isomer: $[Co(NH_3)_5(SO_4)]Br$

$$[Co(NH_3)_5Br]SO_4(aq) + BaCl_2(aq) \longrightarrow [Co(NH_3)_5Br]Cl_2(aq) + BaSO_4(s)$$

22.67 (a) There is 5.41×10^{-4} mol of $UO_2(NO_3)_2$, and this provides 5.41×10^{-4} mol of U^{n+} ions on reduction by Zn. The 5.41×10^{-4} mol U^{n+} requires 2.16×10^{-4} mol MnO_4^- to reach the equivalence point. This is a ratio of 5 mol of U^{n+} ions to 2 mol MnO_4^- ions. The 2 mol MnO_4^- ions require 10 mol of e^- (to go to Mn^{2+} ions), so 5 mol of U^{n+} ions provide 10 mol e^- (on going to UO_2^{2+} ions, with a uranium oxidation number of +6). This means the U^{n+} ion must be U^{4+}.

(b) $Zn(s) \longrightarrow Zn^{2+}(aq) + 2\,e^-$

$$UO_2^{2+}(aq) + 4\,H^+(aq) + 2\,e^- \longrightarrow U^{4+}(aq) + 2\,H_2O(\ell)$$

$$UO_2^{2+}(aq) + 4\,H^+(aq) + Zn(s) \longrightarrow U^{4+}(aq) + 2\,H_2O(\ell) + Zn^{2+}(aq)$$

(c) $$5[U^{4+}(aq) + 2\,H_2O(\ell) \longrightarrow UO_2^{2+}(aq) + 4\,H^+(aq) + 2\,e^-]$$

$$2[MnO_4^-(aq) + 8\,H^+(aq) + 5\,e^- \longrightarrow Mn^{2+}(aq) + 4\,H_2O(\ell)]$$

$$5\,U^{4+}(aq) + 2\,MnO_4^-(aq) + 2\,H_2O(\ell) \longrightarrow 5\,UO_2^{2+}(aq) + 4\,H^+(aq) + 2\,Mn^{2+}(aq)$$

22.69

Ion	$K_{formation}$ (ammine complexes)
Co^{2+}	1.3×10^5
Ni^{2+}	5.5×10^8
Cu^{2+}	2.1×10^{13}
Zn^{2+}	2.9×10^9

The data for these hexammine complexes do indeed, verify the Irving-Williams series. In the book *Chemistry of the Elements* (N. N. Greenwood and A. Earnshaw: 2nd edition, p. 908, Oxford, England, Butterworth-Heinemann, 1997), it is stated: "the stabilities of corresponding complexes of the bivalent ions of the first transition series, irrespective of the particular ligand involved, usually vary in the Irving-Williams order, . . . , which is the reverse of the order for the cation radii. These observations are consistent with the view that, at least for metals in oxidation states +2 and +3, the coordinate bond is largely electrostatic. This was a major factor in the acceptance of the crystal field theory."

22.71 Wilkinson's catalyst and the EAN rule.

Step 1: Rhodium-containing reactant: Each PR_3 ligand donates two electrons as does the Cl^- ligand. The Rh^+ ion has 8 *d* electrons. Total electrons = 16.

Step 1: Rhodium-containing product. Assuming that the H ligand is H^-, each donates two electrons, as do Cl^- and the two PR_3 ligands. The metal center is Rh^{3+}, a d^6 metal ion. Total electrons = 18.

Step 2: The product has 16 electrons because one PR_3 ligand has been dissociated.

Step 3: The product has 18 electrons. The two-electron donor ligand CH_2CH_2 has replaced the dissociated PR_3 ligand.

Step 4: The product has 16 electrons. There are three anionic, two-electron donor ligands (Cl^-, H^-, and $CH_3CH_2^-$), two two-electron donor PR_3 ligands, and an Rh^{3+} ion (d^6).

Step 5: The product has 14 electrons. It is an Rh^+ complex (d^8) with two PR_3 ligands (two electrons each) and one Cl^- (two electrons). (It is likely a solvent molecule fills the vacant site here to give a transient, 16-electron complex.)

Step 6: The product is once again the active, 16-electron catalyst.

CHAPTER 23

23.11 (a) $^{56}_{28}Ni$; (b) $^{1}_{0}n$; (c) $^{32}_{15}P$; (d) $^{97}_{43}Tc$; (e) $^{\ 0}_{-1}\beta$; (f) $^{0}_{1}e$ (positron)

23.13 (a) $^{\ 0}_{-1}\beta$; (b) $^{87}_{37}Rb$; (c) $^{4}_{2}\alpha$; (d) $^{226}_{88}Ra$; (e) $^{\ 0}_{-1}\beta$; (f) $^{24}_{11}Na$

23.15 $^{235}_{92}U \longrightarrow {}^{231}_{90}Th + {}^{4}_{2}\alpha$

$^{231}_{90}Th \longrightarrow {}^{231}_{91}Pa + {}^{\ 0}_{-1}\beta$

$^{231}_{91}Pa \longrightarrow {}^{227}_{89}Ac + {}^{4}_{2}\alpha$

$^{227}_{89}Ac \longrightarrow {}^{227}_{90}Th + {}^{\ 0}_{-1}\beta$

$^{227}_{90}Th \longrightarrow {}^{223}_{88}Ra + {}^{4}_{2}\alpha$

$^{223}_{88}Ra \longrightarrow {}^{219}_{86}Rn + {}^{4}_{2}\alpha$

$^{219}_{86}Rn \longrightarrow {}^{215}_{84}Po + {}^{4}_{2}\alpha$

$^{215}_{84}Po \longrightarrow {}^{211}_{82}Pb + {}^{4}_{2}\alpha$

$^{211}_{82}Pb \longrightarrow {}^{211}_{83}Bi + {}^{\ 0}_{-1}\beta$

$^{211}_{83}Bi \longrightarrow {}^{211}_{84}Po + {}^{\ 0}_{-1}\beta$

$^{211}_{84}Po \longrightarrow {}^{207}_{82}Pb + {}^{4}_{2}\alpha$

23.17 (a) $^{198}_{79}Au \longrightarrow {}^{198}_{80}Hg + {}^{\ 0}_{-1}\beta$
(b) $^{222}_{86}Rn \longrightarrow {}^{218}_{84}Po + {}^{4}_{2}\alpha$
(c) $^{137}_{55}Cs \longrightarrow {}^{137}_{56}Ba + {}^{\ 0}_{-1}\beta$
(d) $^{110}_{49}In \longrightarrow {}^{110}_{48}Cd + {}^{0}_{1}e$

23.19 (a) $^{80}_{35}Br$ has a high neutron/proton ratio of 45/35. Beta decay will allow the ratio to decrease: $^{80}_{35}Br \longrightarrow {}^{80}_{36}Kr + {}^{\ 0}_{-1}\beta$. Some ^{80m}Br decays by gamma emission.
(b) Alpha decay is likely: $^{240}_{98}Cf \longrightarrow {}^{236}_{96}Cm + {}^{4}_{2}\alpha$
(c) Cobalt-61 has a high *n/p* ratio so beta decay is likely:

$^{61}_{27}Co \longrightarrow {}^{61}_{28}Kr + {}^{\ 0}_{-1}\beta$

(d) Carbon-11 has only 5 neutrons so K-capture or positron emission may occur:

$^{11}_{6}C + {}^{0}_{-1}e \rightarrow {}^{11}_{5}B$

$^{11}_{6}C \rightarrow {}^{11}_{5}B + {}^{0}_{1}e$

23.21 Generally beta decay will occur when the n/p ratio is high, whereas positron emission will occur when the n/p ratio is low.
(a) Beta decay: $^{20}_{9}F \rightarrow {}^{20}_{10}Ne + {}^{0}_{-1}\beta$

$^{3}_{1}H \rightarrow {}^{3}_{2}He + {}^{0}_{-1}\beta$

(b) Positron emission

$^{22}_{11}Na \rightarrow {}^{22}_{10}Ne + {}^{0}_{1}\beta$

23.23 Binding energy per nucleon for $^{11}B = 6.70 \times 10^8$ kJ

Binding energy per nucleon for $^{10}B = 6.26 \times 10^8$ kJ

23.25 8.256×10^8 kJ/nucleon

23.27 7.700×10^8 kJ/nucleon

23.29 0.781 micrograms

23.31 (a) $^{131}_{53}I \rightarrow {}^{131}_{54}Xe + {}^{0}_{-1}\beta$
(b) 0.075 micrograms

23.33 9.5×10^{-4} mg

23.35 (a) $^{222}_{86}Rn \rightarrow {}^{218}_{84}Po + {}^{4}_{2}\alpha$
(b) Time = 8.87 d

23.37 (a) 15.8 y; (b) 88%

23.39 $^{239}_{94}Pu + + {}^{4}_{2}\alpha \rightarrow {}^{240}_{95}Am + {}^{1}_{1}H + 2\,{}^{1}_{0}n$

23.41 $^{48}_{20}Ca + {}^{242}_{94}Pu \rightarrow {}^{287}_{114}Uuq + 3\,{}^{1}_{0}n$

23.43 (a) $^{115}_{48}Cd$ (b) $^{7}_{4}Be$ (c) $^{4}_{2}\alpha$ (d) $^{63}_{29}Cu$

23.45 $^{10}_{5}B + {}^{1}_{0}n \rightarrow {}^{7}_{3}Li + {}^{4}_{2}\alpha$

23.47 Time = 4.4×10^{10} y

23.49 If $t_{1/2} = 14.28$ d, then $k = 4.854 \times 10^{-2}$ d^{-1}. If the original disintegration rate is 3.2×10^6 dpm, then (from the integrated first order rate equation), the rate after 365 d is 0.065 dpm. The plot will resemble Figure 23.5.

23.51 (a) $^{238}_{92}U + {}^{1}_{0}n \rightarrow {}^{239}_{92}U$
(b) $^{239}_{92}U \rightarrow {}^{239}_{93}Np + {}^{0}_{-1}\beta$
(c) $^{239}_{93}Np \rightarrow {}^{239}_{94}Pu + {}^{0}_{-1}\beta$
(d) $^{239}_{94}Pu + {}^{1}_{0}n \rightarrow 2\,{}^{1}_{0}n$ + energy + other nuclei

23.53 About 2700 years old

23.55 Plot ln(activity) versus time. The slope of the plot is $-k$, the rate constant for decay. Here, $k = 0.0050$ d^{-1}, so $t_{1/2} = 140$ d.

23.57 Time = 1.9×10^9 y

23.59 130 mL

23.61 Energy obtained from 1.000 lb (452.6 g) of $^{235}U = 4.05 \times 10^{-10}$ kJ

Mass of coal required = 1.6×10^3 ton (or about 3 million pounds of coal)

23.63 27 fish tagged fish out of 5250 fish caught represents 0.51% of the fish in the lake. Therefore, 1000 fish put into the lake represent 0.51% of the fish in the lake, or 0.51% of 190,000 fish.

23.65 (a) The mass decreases by 4 units (with an $^{4}_{2}a$ emission) or is unchanged (with a $^{0}_{-1}\beta$ emission) so the only masses possible are 4 units apart.
(b) ^{232}Th series, $m = 4n$; ^{235}U series $m = 4n + 3$
(c) ^{226}Ra and ^{210}Bi, $4n + 2$ series; ^{215}At, $4n + 3$ series; ^{228}Th, $4n$ series)
(d) Each series is headed by a long-lived isotope (in the order of 10^9 years, the age of the Earth.) The $4n + 1$ series is missing because there is no long-lived isotope in this series. Over geologic time, all the members of this series have decayed completely.

23.67 (a) ^{231}Pa isotope belongs to the ^{235}U decay series (see Question 23.65b).
(b) $^{235}_{92}U \rightarrow {}^{231}_{90}Th + {}^{4}_{2}\alpha$

$^{231}_{90}Th \rightarrow {}^{231}_{91}Pa + {}^{0}_{-1}\beta$

(c) Pa-231 is present to the extent of 1 part per million. Therefore, 1 million grams of pitchblende need to be used to obtain 1 g of Pa-231.
(d) $^{231}_{91}Pa \rightarrow {}^{227}_{89}Ac + {}^{4}_{2}\alpha$

23.69 Pitchblende contains $^{238}_{92}U$ and $^{235}_{92}U$. Thus, both radium and polonium isotopes must belong to either the $4n + 2$ or $4n + 3$ decay series. Furthermore, the isotopes must have sufficiently long half-lives in order to survive the separation and isolation process. These criteria are satisfied by ^{226}Ra and ^{210}Po.

P Answers to Selected Interchapter Study Questions

THE CHEMISTRY OF FUELS AND ENERGY SOURCES

1. (a) From methane: $H_2O(g) + CH_4(g) \rightarrow 3\ H_2(g) + CO(g)$

100. g CH_4(1 mol CH_4/16.043 g) (3 mol H_2/mol CH_4) (2.016 g H_2/1 mol H_2) = 37.7 g of H_2 produced

(b) From petroleum: $H_2O(g) + CH_2(\ell) \rightarrow 2\ H_2(g) + CO\ (g)$

100. g CH_2(1 mol CH_2/14.026 g CH_2) (2 mol H_2/1 mol CH_2) (2.016 g H_2/ 1 mol H_2) = 28.7 g H_2 produced

(c) From coal: $H_2O(g) + C(s) \rightarrow H_2(g) + CO(g)$

100. g C(1 mol C/12.011 g C) (1 mol H_2/mole C) (2.016 g H_2/ 1 mol H_2) = 16.8 g H_2 produced.

3. 70. lb(453.6 g/lb) (33 kJ/g) = 1.0×10^6 kJ

5. Assume burning oil produces 43 kJ/g (the value for crude petroleum in Table 2)

7.0 gal(3.785 L/gal) (1000 cm^3/L) (0.8 g/cm^3) (43 kJ/g) = 0.9×10^6 kJ. Uncertainty in the numbers is one significant figure. This value is close to the value for the energy obtained by burning from 70 kg of coal (calculated in Q.3.)

7. Per gram: (5.45×10^3 kJ/mol) (1 mol/114.26 g) = 47.7 kJ/g

Per liter: (47.7 kJ/g) (688 g/L) = 3.28×10^4 kJ/L

9. The factor for converting kW-h to kJ is 1 kW-h = 3600 kJ

(940 kW-h/yr) (3600 kJ/kW-h) = 3.4×10^6 kJ/yr

11. First, calculate $\Delta_r H°$ for the reaction $CH_3OH(\ell) + 1.5\ O_2(g) \rightarrow CO_2(g) + 2\ H_2O(\ell)$, using enthalpies of formation ($\Delta_r H° = 726.7$ kJ/mol-rxn). Use molar mass and density to calculate energy per L [−726.7 kJ/mol-rxn (1 mol-rxn/32.04 g) (787 g/ L) = 17.9×10^3 kJ/L]. Then use the kW-h to kJ conversion factor from Q. 9 to obtain the answer [(17.9×10^3 kJ/L) (1 kW-h/3600 kJ) = 4.96 kW-h/L].

13. Area of parking lot = 325 m × 50.0 m = 1.63×10^4 m^2

2.6×10^7 J/m^2(1.63×10^4 m^2) = 4.3×10^{11} J

15. Amount of Pd = 1.0 cm^3(12.0 g/cm^3) (1 mol/106.4 g) = 0.113 mol

amount of H = 0.084 g(1 mol/1.008 g) = 0.0833 mol

mol H per mol Pd = 0.083/0.113 = 0.74: Simplest formula for this compound is $PdH_{0.74}$ (Because the compound is nonstoichiometric, we will not write a formula with a whole number ratio. For these compounds, it is common practice to set the amount of metal [Pd] to be an integer and H as a non-integer.)

17. Energy per gal. of gas = (48.0 kJ/g)(0.737 g/cm^3)(1000 cm^3/L)(3.785 L/gal) = 1.34 × 10^5 kJ/gal

Energy to travel 1 mile = (1.00 mile/55.0 gal/mile)(1.34 × 10^5 kJ/gal) = 2440 kJ

MILESTONES IN THE DEVELOPMENT OF CHEMISTRY AND THE MODERN VIEW OF ATOMS AND MOLECULES

1. Atoms are not solid, hard, or impenetrable. They have mass (an important aspect of Dalton's hypothesis), and we now know that atoms are in rapid motion at all temperatures above absolute zero (the kinetic-molecular theory).

3. mass *e*/mass *p* = 9.109383×10^{-28} g/1.672622 $\times 10^{-24}$ g = 5.446170×10^{-4}. (Mass of *p* and *e* obtained from Table 2.1, page 52.) The proton is 1,834 times more massive than an electron. Dalton's estimate was off by a factor of about 2.

THE CHEMISTRY OF LIFE: BIOCHEMISTRY

1. (a)

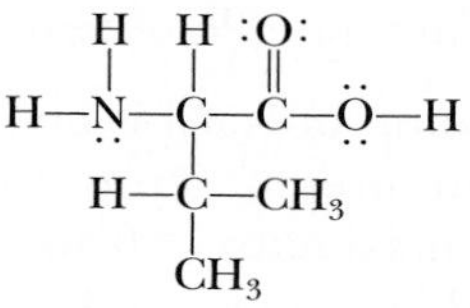

(b)

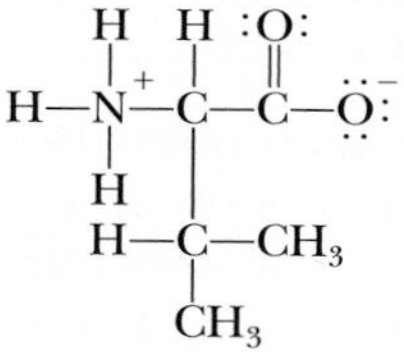

(c) The zwitterionic form is the predominant form at physiological pH.

3.

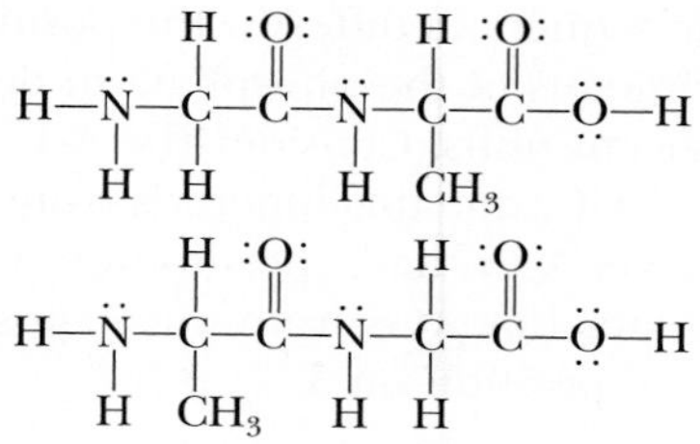

5.

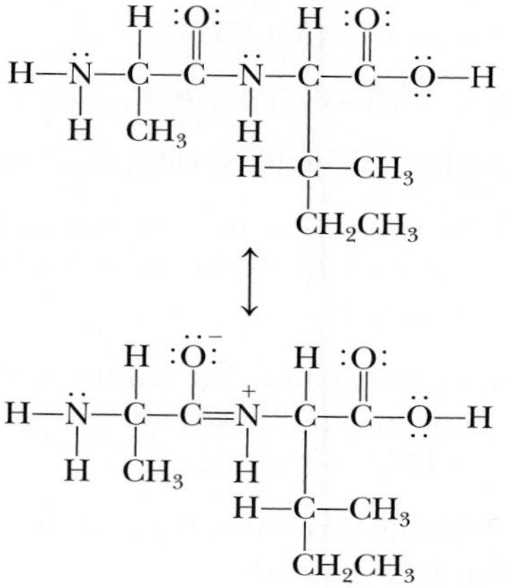

7. (a) The structure of ribose is given in Figure 13.

(b) Adenosine

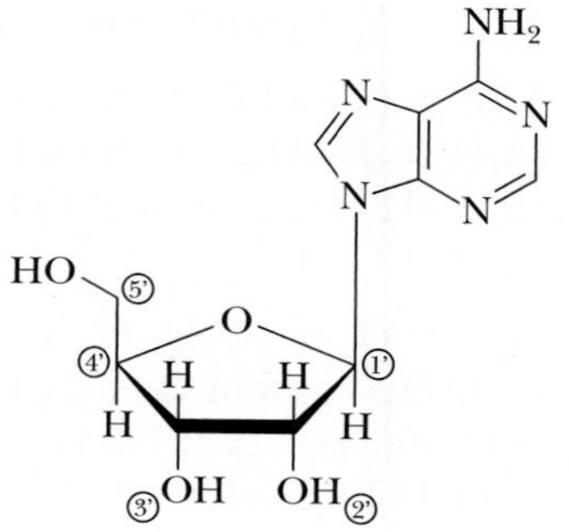

(c) Adenosine-5′-monophosphate

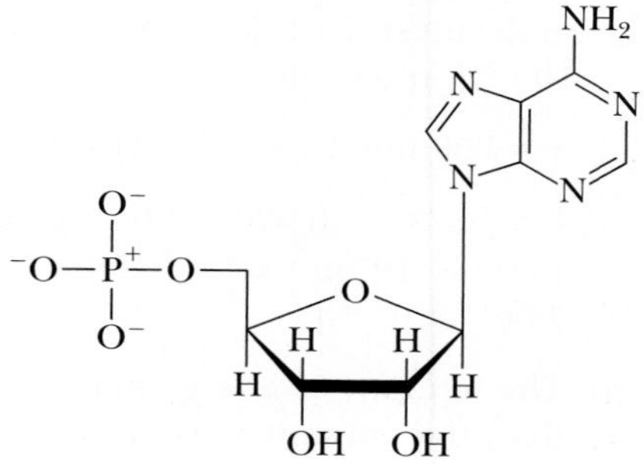

9. The sequences differ in the positions of attachments of the phosphate to deoxyribose on adjacent units. Consider the A-T attachments. In ATGC, the attachment is from the 5′ position on A to the 3′ position on T. In CGTA, the attachment is from the 3′ position on A to the 5′ position on T.

11. (a) 5′-GAATCGCGT-3′

    (b) 5′-GAAUCGCGU-3′

    (c) 5′-UUC-3′, 5′-CGA-3′, and 5′-ACG-3′

    (d) glutamic acid, serine, and arginine

13. (a) In transcription, a strand of RNA complementary to the segment of DNA is constructed.

    (b) In translation, an amino acid sequence is constructed based on the information in a mRNA sequence.

15. The 4-ring structure present in all steroids is given in Figure 18a.

17. (a) False (b) True (c) True (d) True

19. (a) $6\ CO_2(g) + 6\ H_2O(\ell) \rightarrow C_6H_{12}O_6(s) + 6\ O_2(g)$

    $\Delta_r H° = \Delta_f H°(\text{products}) - \Delta_f H°(\text{reactants})$

    $\Delta_r H° = (1\ \text{mol}\ C_6H_{12}O_6/\text{mol-rxn})[\Delta_f H°(C_6H_{12}O_6)] - (6\ \text{mol}\ H_2O/\text{mol-rxn})[\Delta_f H°(H_2O)] - (6\ \text{mol}\ CO_2/\text{mol-rxn})[\Delta_f H°(CO_2)]$

    $\Delta_r H° = (1\ \text{mol}\ C_6H_{12}O_6/\text{mol-rxn})(-1273.3\ \text{kJ/mol}\ C_6H_{12}O_6) - (6\ \text{mol}\ H_2O/\text{mol-rxn})(-285.8\ \text{kJ/mol}\ H_2O) - (6\ \text{mol}\ CO_2/\text{mol-rxn})(-393.5\ \text{kJ/mol}\ CO_2)$

    $\Delta_r H° = +2{,}803\ \text{kJ/mol-rxn}$

    (b) $(2803\ \text{kJ/mol})(1\ \text{mol}/6.022 \times 10^{23}\ \text{molecules})(1000\ \text{J}/1\ \text{kJ}) = 4.655 \times 10^{-18}\ \text{J/molecule}$

    (c) $\lambda = 650\ \text{nm}(1\ \text{m}/10^9\ \text{nm}) = 6.50 \times 10^{-7}\ \text{m}$

    $E = hc/\lambda = (6.626 \times 10^{-34}\ \text{J} \cdot \text{s})(3.00 \times 10^8\ \text{m} \cdot \text{s}^{-1})/(6.50 \times 10^{-7}\ \text{m}) = 3.06 \times 10^{-19}\ \text{J}$

    (d) The amount of energy per photon is less than the amount of required per molecule of glucose, therefore multiple photons must be absorbed.

THE CHEMISTRY OF MODERN MATERIALS

1. The GaAs band gap is 140 kJ/mol. Use the equations $E = h\nu$ and $\lambda \times \nu = c$ to calculate a wavelength of 854 nm corresponding to this energy. Radiation of this wavelength is in the infrared portion of the spectrum.

3. The amount of light falling on a single solar cell = 925 W/m² [(1 m²/10⁴ cm²)(1.0 cm²/cell) = 0.0925 W/cell.] Using the conversion factor 1 W = 1 J/s, the energy incident on the cell is (0.0925 W/cell)(1J/W · s)(60 sec/min) = 5.55 J/(min · cell). At 25% efficiency, the energy absorbed for each cell is 1.39 J/min.

5. The density of dry air at 25 °C and 1.0 atm. is 1.2 g/L (see page 526), so the mass of air in aerogel is $0.99(1.2 \times 10^{-3}\ \text{g}) = 1.2 \times 10^{-3}\ \text{g}$. Add to this 0.023 g, the mass of 0.010 cm³ of SiO_2 (density of SiO_2, from web, is 2.3 g/cm³, mass of 0.010 cm³ is 0.023 g). Thus, the total mass is 0.0012 g + 0.023 g = 0.024 g, and the density of aerogel is 0.024 g/cm³.

ENVIRONMENTAL CHEMISTRY

1. $[Na^+] = 0.460$ mol/L, $[Cl^-] = 0.550$ mol/L; a larger amount of chloride than sodium ion is present in a sample of seawater.

3. The amount of NaCl is limited by the amount of sodium present. From 1.0 L sample of seawater, a maximum of 0.460 mol NaCl could be obtained. The mass of this amount of NaCl is 26.9 g [(0.460 mol/L)(1.00 L)(58.43 g NaCl/1 mol NaCl) = 26.9 g].

5. For gases, ppm refers to numbers of particles, and hence to mole fractions (see footnote to Table 1). Gas pressure exerted is directly proportional to mole fraction. Thus, 40,000 ppm water vapor would exert a pressure of 40,000/1,000,000th of one atmosphere, or 30.4 mm Hg (0.040 × 760 mm Hg). This would be the case at a little over 29 °C, at 100% humidity.

7. The concentration of Mg^{2+} in seawater is 52 mmol/L (Table 2). Assuming that all this is converted to Mg metal, one would expect to obtain 1.26 g from 1.0 L of seawater [0.052 mol(24.31 g/mol) = 1.26 g]. To obtain 100 kg of Mg, 79,000 L of seawater [100. kg(1000 g/kg) (1 L/1.26 g) = 7.9×10^4 L] would be needed.

9. (a) The volume occupied by 25 g of ice is 33 cm^3 [25 g(1 cm^3/0.92 g) = 33 cm^3]. However, only 92% of the ice is submerged and the water displaced by ice (the volume of ice under the surface of water) is 25 cm^3 (0.92 $\times$ 33 cm^3 = 25 cm^3). Thus, the liquid level in the graduated cylinder will be 125 mL.

   (b) Melting 25 g of ice will produce 25 mL of liquid water. The water level will be 125 mL (the same as in (a), that is the water level won't rise as the ice melts).

Q Answers to Chapter Opening Puzzler and Case Study Questions

CHAPTER 1

Puzzler:

1. Sports drinks: colored, liquid, homogeneous, slightly more dense than pure water. (Dissolved salts raise the density of a solution: e.g., seawater is more dense than pure water.)

2. These drinks are often sold in 500-mL bottles. This is equivalent to 0.50 L or 5.0 dL.

Case Study: Ancient and Modern Hair Coloring

1. Lead (Pb); calcium (Ca)

2. $d = 11.35\ \text{g/cm}^3$

3. S

4. Calcium hydroxide, known as slaked lime, is made by adding water (slaking) to lime, CaO

5. Sulfide ions (S^{2-}) are on the corners and faces of a cube; lead ions Pb^{2+} lie along each edge.

6. The overall structures are identical. The yellow spheres (S^{2-} in PbS, and Cl^- in NaCl) are at the corners and on the faces of a cube; the spheres representing Pb^{2+} and Na^+ lie along the cube's edges. The small difference in appearance is due to the relative sizes of the spheres.

LET'S REVIEW

Case Study: Out of Gas!

1. Fuel density in kg/L: (1.77 lb/L) (0.4536 kg/lb) = 0.803 kg/L

2. Mass of fuel already in tank: 7682 L (0.803 kg/L) = 6170 kg

   Mass of fuel needed: 22,300 kg − 6,170 kg = 16,100 kg (Answer has three significant figures.)

   Volume of fuel needed: 16,130 kg (1 L/0.803 kg) = 20,100 L

CHAPTER 2

Puzzler:

1. Eka-silicon is germanium. Its atomic weight is 72.61 (predicted 72), and its density is $5.32\ \text{g/cm}^3$ (predicted value $5.5\ \text{g/cm}^3$).

2. Other elements missing from Mendeleev's periodic table include Sc, Ga, the noble gases (He, Ne, Ar, Kr, Xe), and all of the radioactive elements except Th and U.

Case Study: Catching Cheaters with Isotopes

1. 7 neutrons

2. 8 neutrons

3. ^{14}C is formed in the upper atmosphere by a nuclear reaction initiated by cosmic radiation. The equation for its formation is $^{14}_{7}N + ^{1}_{0}n \longrightarrow ^{14}_{6}C + ^{1}_{1}H$. (See discussion in Chapter 23 on equations for nuclear reactions.)

Case Study: What's in Those French Fries?

1. Acrylamide: C_3H_5NO, molar mass = 71.08; % N = (14.00/71.07)(100%) = 19.70 %.

   Asparagine, $C_4H_8O_3N_2$, molar mass = 132.12; % N = (28.00/132.12)(100%) = 21.20 %. Asparagine has the higher percent nitrogen.

2. Body mass in kg = 150 lb(0.4536 kg/1 lb) = 68.0 kg

   Total mass ingested = (0.0002 mg/kg body wt) (68.0 kg body wt) = 1.4×10^{-2} mg

   Number of molecules = 1.4×10^{-2} mg (1 g/1000 mg)(1 mol/71.08 g)(6.022×10^{23} molecules/mol) = 1×10^{17} molecules (1 significant figure)

CHAPTER 3

Puzzler:

$Fe^{2+}(aq) + H_2S(aq) \longrightarrow FeS(s) + 2\ H^+(aq)$
$2\ Bi^{3+}(aq) + 3\ H_2S(aq) \longrightarrow Bi_2S_3(s) + 6\ H^+(aq)$
$Ca^{2+}(aq) + SO_4^{2-}(aq) \longrightarrow CaSO_4(s)$

Case Study: Killing Bacteria with Silver

1. 100×10^{15} Ag^+ ions(1 mol/6.022 x 10^{23} ions) = 2×10^{-7} mol Ag^+

2. 2×10^{-7} mol Ag^+(107.9 g Ag^+/1 mol Ag^+) = 2×10^{-5} g Ag^+ ions

CHAPTER 4

Puzzler:

1. Oxidation-reduction reactions.

2. Oxidation of Fe gives Fe_2O_3; oxidation of Al gives Al_2O_3.

3. The mass of Al_2O_3 formed by oxidation of 1.0 g Al = 1.0 g Al (1 mol Al/26.98 g Al) (1 mol Al_2O_3/2 mol Al)(102.0 g Al_2O_3/1 mol Al_2O_3) = 1.9 g Al_2O_3.

Case Study: How Much Salt Is There in Seawater?

1. Step 1: Calculate the amount of Cl^- in the diluted solution from titration data.

   Mol Cl^- in 50 mL of dilute solution = mol Ag^+ = (0.100 mol/L)(0.02625 L) = 2.63×10^{-3} mol Cl^-

   Step 2: Calculate the concentration of Cl^- in the dilute solution.

   Concentration of Cl^- in dilute solution = 2.63×10^{-3} mol/0.0500 L = 5.26×10^{-2} M

   Step 3: Calculate the concentration of Cl^- in seawater.

   Seawater was initially diluted to one hundredth its original concentration. Thus, the concentration of Cl^- in seawater (undiluted) = 5.25 M

Case Study: Forensic Chemistry: Titrations and Food Tampering

1. Step 1: Calculate the amount of I_2 in solution from titration data:

   Amount I_2 = (0.0425 mol $S_2O_3^{2-}$/L)(0.0253 L) (1 mol I_2/2 mol $S_2O_3^{2-}$) = 5.38×10^{-4} mol I_2

   Step 2: Calculate the amount of NaClO present based on the amount of I_2 formed, and from that value calculate the mass of NaClO.

   Mass NaClO = 5.38×10^{-4} mol I_2(1 mol HClO/1 mol I_2)(1 mol NaClO/1 mol HClO) (74.44 g NaClO/1 mol NaClO) = 0.0400 g NaClO

CHAPTER 5

Puzzler:

Step 1: Calculate mass of air in the balloon

Mass air = 1100 m^3(1,200 g/m^3) = 1.3×10^6 g

Step 2: Calculate energy as heat needed to raise the temperature of air in the balloon.

Energy as heat = $C \times m \times \Delta T$ = (1.01 J/g · K) (1.3 × 10^6 g) (383 K − 295 K) = 1.2 × 10^8 J (= 1.2 × 10^5 kJ)

Step 3: Calculate enthalpy change for the oxidation of 1.00 g propane from enthalpy of formation data. Assume formation of water vapor, $H_2O(g)$, in this reaction.

$C_3H_8(g) + 5\ O_2(g) \rightarrow 3\ CO_2(g) + 4\ H_2O(g)$

$\Delta_r H° = \Delta_f H°$(products) − $\Delta_f H°$(reactants) = (3 mol CO_2/mol-rxn) [$\Delta_f H°(CO_2)$] + (4 mol H_2O/mol-rxn) [$\Delta_f H°(H_2O)$] − (1 mol C_3H_8/mol-rxn) [$\Delta_f H°(C_3H_8)$]

$\Delta_r H°$ = (3 mol CO_2/mol-rxn) [−393.5 kJ/mol CO_2] + (4 mol H_2O/mol-rxn) [− 241.8 kJ/mol H_2O] − (1 mol C_3H_8/mol-rxn) [−104.7 kJ/mol C_3H_8)] = −2043 kJ/mol-rxn

q = −2043 kJ/mol-rxn (1 mol C_3H_8/mol-rxn) (1 mol C_3H_8/44.09 g C_3H_8) = −46.33 kJ/g C_3H_8

Step 4: Use answers from Steps 2 and 3 to calculate mass of propane needed to produce energy as heat needed.)

mass C_3H_8 = 1.2 × 10^5 kJ (1 g C_3H_8/46.33 kJ) = 2.5 × 10^3 g C_3H_8

(Answer has two significant figures.)

Case Study: Abba's Refrigerator

1. To evaporate 95 g H_2O: q = 44.0 kJ/mol (1 mol/18.02 g) (95 g) = 232 kJ (= 232,000 J)

   Temperature change if 750 g H_2O gives up 232 kJ of energy as heat:

   $Q = C \times m \times \Delta T$

   −232,000 J = (4.184 J/ g · K) (750 g) (ΔT); ΔT = −74 K

Case Study: The Fuel Controversy: Alcohol and Gasoline

In the following, we assume water vapor, $H_2O(g)$, is formed upon oxidation.

1. Burning ethanol: $C_2H_5OH(\ell) + 3\ O_2(g) \rightarrow 2\ CO_2(g) + 3\ H_2O(g)$

   $\Delta_r H°$ = (2 mol CO_2/mol-rxn) [$\Delta_f H°(CO_2)$] + (3 mol H_2O/mol-rxn) [$\Delta_f H°(H_2O)$] − (1 mol C_2H_5OH/mol-rxn) [$\Delta_f H°(C_2H_5OH)$]

   $\Delta_r H°$ = (2 mol CO_2/mol-rxn) [−393.5 kJ/mol CO_2] + (3 mol H_2O/mol-rxn) [−241.8 kJ/mol H_2O] − (1 mol C_2H_5OH/mol-rxn) [−277.0 kJ/mol C_2H_5OH)] = −1235.4 kJ/mol-rxn

   1 mol ethanol per 1 mol-rxn; therefore, q per mol is −1235.4 kJ/mol

   q per gram: −1235.4 kJ/mol (1mol C_2H_5OH/46.07 g C_2H_5OH) = −26.80 kJ/g C_2H_5OH

   Burning octane: $C_8H_{18}(\ell) + 12.5\ O_2(g) \rightarrow 8\ CO_2(g) + 9\ H_2O(g)$

   $\Delta_r H°$ = (8 mol CO_2/mol-rxn) [$\Delta_f H°(CO_2)$] + (9 mol H_2O/mol-rxn) [$\Delta_f H°(H_2O)$] − (1 mol C_8H_{18}/mol-rxn) [$\Delta_f H°(C_8H_{18})$]

   $\Delta_r H°$ = (8 mol CO_2/mol-rxn) [−393.5 kJ/mol CO_2] + (9 mol H_2O/mol-rxn) [−241.8 kJ/mol H_2O] − (1 mol C_8H_{18}/mol-rxn) [−250.1 kJ/mol C_8H_{18})] = −5070.1 kJ/mol-rxn

   1 mol octane per mol-rxn; therefore, q per mol is −5070.1 kJ/mol

   q per gram: −5070.1 kJ/1 mol C_8H_{18} (1 mol C_8H_{18}/114.2 g C_8H_{18}) = −44.40 kJ/g C_8H_{18}

2. For ethanol, per liter: q = −26.80 kJ/g (785 g/L) = −2.10 × 10^4 kJ/L

   For octane, per liter: q = −44.40 kJ/g (699 g/L) = 3.10 × 10^4 kJ/L

   Octane produces almost 50% more energy per liter of fuel.

3. Mass of CO_2 per liter of ethanol: = 1.000 L (785 g C_2H_5OH/L) (1 mol C_2H_5OH /46.07 g C_2H_5OH) (2 mol CO_2/1 mol C_2H_5OH) (44.01 g CO_2/1 mol CO_2) = 1.50 × 10^3 g CO_2

   Mass of CO_2 per liter of octane: = 1.000 L (699 g C_8H_{18}/L) (1 mol C_8H_{18} /114.2 g C_8H_{18}) (8 mol CO_2/1 mol C_8H_{18}) (44.01 g CO_2/1 mol CO_2) = 2.15 × 10^3 g CO_2

4. Volume of ethanol needed to obtain 3.10 × 10^4 kJ of energy from oxidation: 2.10 × 10^4 kJ/L C_2H_5OH) (x) = 3.10 × 10^4 kJ (x is volume of ethanol)

   Volume of ethanol = x = 1.48 L

   Mass of CO_2 produced by burning 1.48 L of ethanol = (1.50 × 10^3 g CO_2/L C_2H_5OH) (1.48 L C_2H_5OH) = 2.22 × 10^3 g CO_2

   To obtain the same amount of energy, slightly more CO_2 is produced by burning ethanol than for octane.

5. Your car will travel about 50% farther on a liter of octane, and it will produce slightly less CO_2 emissions, than if you burned 1.0 L of ethanol.

CHAPTER 6

Puzzler:

1. Red light has the longer wavelength.
2. Green light has the higher energy.
3. The energy of light emitted by atoms is determined by the energy levels of the electrons in an atom. See discussion in the text, page 275.

Case Study: What Makes the Colors in Fireworks?

1. Yellow light is from 589 and 590 nm emissions.
2. Primary emission for Sr is red: this has a longer wavelength than yellow light.
3. $4\ Mg(s) + KClO_4(s) \longrightarrow KCl(s) + 4\ MgO(s)$

CHAPTER 7

Puzzler:

1. Cr: $[Ar]3d^5 4s^1$; Cr^{3+}: $[Ar]3d^3$; CrO_4^{2-} (chromium(VI)): [Ar]
2. Cr^{3+} is paramagnetic with three unpaired electrons.
3. Pb in $PbCrO_4$ is present as Pb^{2+}: $[Xe]4f^{14}5d^{10}6s^2$

Case Study: Metals in Biochemistry and Medicine

1. Fe^{2+}: $[Ar]3d^6$; Fe^{3+}: $[Ar]3d^5$
2. Both iron ions are paramagnetic.
3. Cu^+: $[Ar]3d^{10}$; Cu^{2+}: $[Ar]3d^9$; Cu^{2+} is paramagnetic; Cu^+ is diamagnetic.
4. The slightly larger size of Cu compared to Fe is related to greater electron–electron repulsions.
5. Fe^{2+} is larger than Fe^{3+} and will fit less well into the structure. As a result, some distortion of the ring structure from planarity occurs.

CHAPTER 8

Puzzler:

1. Carbon and phosphorus (in phosphate) achieve the noble gas configuration by forming four bonds.
2. In each instance, there are four bonds to the element; VSEPR predicts that these atoms will have tetrahedral geometry with 109.5° angles.
3. Bond angles in these rings are 120°. To achieve this preferred bond angle, the rings must be planar.
4. Thymine and cytosine are polar molecules.

Case Study: The Importance of an Odd Electron Molecule, NO

1. 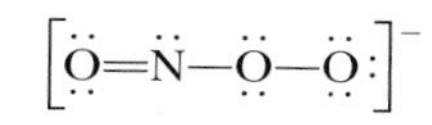
2. There is a double bond between N and the terminal O.
3. Resonance structures are not needed to describe the bonding in this ion.

CHAPTER 9

Puzzler:

1. XeF_2 is linear. The electron pair geometry is trigonal bipyramidal. Three lone pairs are located in the equatorial plane, and the two F atoms are located in the axial positions. This symmetrical structure will not have a dipole.
2. The Xe atom is sp^3d hybridized. Xe-F bonds: overlap of Xe sp^3d orbitals with F 2p orbital. 3 lone pairs in Xe sp^3d orbitals.
3. This 36-electron molecule has a bent molecular structure.

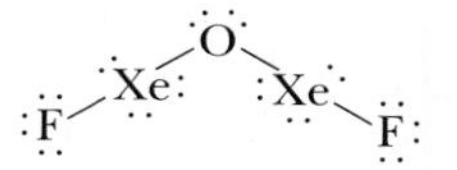

Case Study: Two Chemical Bonding Mysteries

1. Eight two-electron bonds, which would require 16 electrons, four more than the 12 available.

2. The compound has sp^3 hybridized B and N atoms and is polar. The B atom has a −1 formal charge and N has a +1 form charge. All bond angles are about 109°.

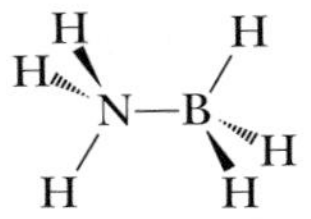

3. $(54.3 \times 10^{-3}\ g\ AgBF_4)(1\ mol/194.7\ g) = 2.79 \times 10^{-4}\ mol$

The amount of $Ag(C_2H_4)_xBF_4$ that must have decomposed is 2.79×10^{-4} mol.

Molar mass of unknown $= (62.1 \times 10^{-3}\ g)/(2.79 \times 10^{-4}\ mol) = 223\ g/mol$

The compound $Ag(C_2H_4)BF_4$ where $x = 1$ has a molar mass of 223 g/mol.

CHAPTER 10

Puzzler:

1. (a) Trigonal planar, sp^2 hybridized. The other carbon atoms in this molecule have tetrahedral geometry and sp^3 hybridization.

   (b) The non-planarity allows all of the atoms in the molecule to assume an unstrained geometry.

   (c) Actually, there are two chiral centers, labeled with an asterisk (*) in the drawing below.

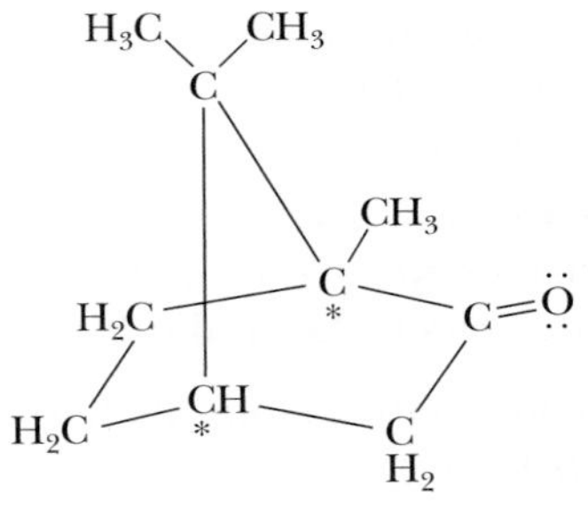

2. Camphor is a ketone.

Case Study: Biodiesel, a Fuel for the Future

1. $C_{12}H_{25}CO_2CH_3(\ell) + 20\ O_2(g) \longrightarrow 14\ CO_2(g) + 14\ H_2O(g)$

2. $\Delta_rH° = (14\ mol\ CO_2/mol\text{-}rxn)[\Delta_fH°(CO_2)] + (14\ mol\ H_2O/mol\text{-}rxn)[\Delta_fH°(H_2O)] - (1\ mol\ C_{12}H_{25}CO_2CH_3/mol\text{-}rxn)[\Delta_fH°(C_{12}H_{25}CO_2CH_3)]$

   $\Delta_rH° = (14\ mol\ CO_2/mol\text{-}rxn)[-393.5\ kJ/mol\ CO_2] + (14\ mol\ H_2O/mol\text{-}rxn)[-241.8\ kJ/mol\ H_2O] - (1\ mol\ C_{12}H_{25}CO_2CH_3/mol\text{-}rxn)[-771.0\ kJ/mol\ C_{12}H_{25}CO_2CH_3)] = -8123.2\ kJ/mol\text{-}rxn$

   1 mol methyl myristate per mol-rxn, so q per mol = −8123.2 kJ/mol

3. Burning hexadecane: $C_{16}H_{34}(\ell) + 24.5\ O_2(g) \rightarrow 16\ CO_2(g) + 17\ H_2O(g)$

   $\Delta_rH° = (16\ mol\ CO_2/mol\text{-}rxn)[\Delta_fH°(CO_2)] + (17\ mol\ H_2O/mol\text{-}rxn)[\Delta_fH°(H_2O)] - (1\ mol\ C_{16}H_{34}/mol\text{-}rxn)[\Delta_fH°(C_{16}H_{34})]$

   $\Delta_rH° = (16\ mol\ CO_2/mol\text{-}rxn)[-393.5\ kJ/mol\ CO_2] + (17\ mol\ H_2O/mol\text{-}rxn)[-241.8\ kJ/mol\ H_2O] - (1\ mol\ C_{16}H_{34}/mol\text{-}rxn)[-456.1\ kJ/mol\ C_{16}H_{34})] = -9950.5\ kJ/mol\text{-}rxn$

   1 mole hexadecane per mol-rxn, so q per mol: −9950.5 kJ/mol

   For methyl myristate, q per liter $= (-8123.2\ kJ/mol)(1\ mol/228.4\ g)(0.86\ g/L) = -30.6\ kJ/L$

   For hexadecane, q per liter $= (-9950.5\ kJ/mol)(1\ mol/226.43\ g)(0.77\ g/1\ L) = -33.8\ kJ/L$

CHAPTER 11

Puzzler:

1. $P(O_2)$ at 3000 m is 70% of 0.21 atm, the value $P(O_2)$ at sea level, thus $P(O_2)$ at 3000 m = 0.21 atm × 0.70 = 0.15 atm (110 mm Hg). At the top of Everest, $P(O_2)$ = 0.21 atm × 0.29 = 0.061 atm (46 mm Hg).

2. Blood saturation levels (estimated from table): at 3000 m, >95%; at top of Everest, 75%.

Case Study: You Stink

1. To calculate $P(CH_3SH)$, use the ideal gas law: $V = 1.00\ m^3(10^6\ cm^3/m^3)(1\ L/10^3\ cm^3) = 1.00 \times 10^3\ L$;

   $n = 1.5 \times 10^{-3}\ g(1\ mol/48.11\ g) = 3.1 \times 10^{-5}\ mol$

   $P = nRT/V = [3.1 \times 10^{-5}\ mol(0.08205\ L\ atm/mol \cdot K)(298\ K)]/1.0 \times 10^3\ L = 7.6 \times 10^{-7}$ atm. (5.8×10^{-4} mm Hg)

   Molecules per $m^3 = 3.1 \times 10^{-5}$ mol $(6.022 \times 10^{23}$ molecules/mol$) = 1.9 \times 10^{19}$ molecules

2. Bond angles: H—C—H and H—C—S, 109.5°; C—S—H somewhat less than 109°.
3. Polar
4. It should behave as an ideal gas at moderate pressures and temperatures well above its boiling point.
5. H_2S (with the lowest molar mass) will diffuse fastest.

CHAPTER 12

Puzzler:

1. In ice, water molecules are not packed as closely as they are in liquid water. This structure results so that hydrogen bonding interactions between water molecules are maximized.

   A piece of ice floats at a level where it will displace its weight of seawater. Most of the volume of an iceberg is below the water line.

2. A 1000 cm^3 piece of ice (mass = 917 g) would float with $[0.917/1.026](100\%) = 89.4\%$ below the surface or 10.6% above the surface.

Case Study: The Mystery of the Disappearing Fingerprints

1. The chemical compounds in a child's fingerprints are more volatile because they have lower molecular weights than compounds in adults' fingerprints.

CHAPTER 13

Puzzler:

1. This geometry problem is solved using the numbers shown on the drawing.

   $x^2 + (69.5)^2 = (139)^2$

   Solving, $x = 120$ pm.

   The side-to-side distance is twice this value or 240. pm

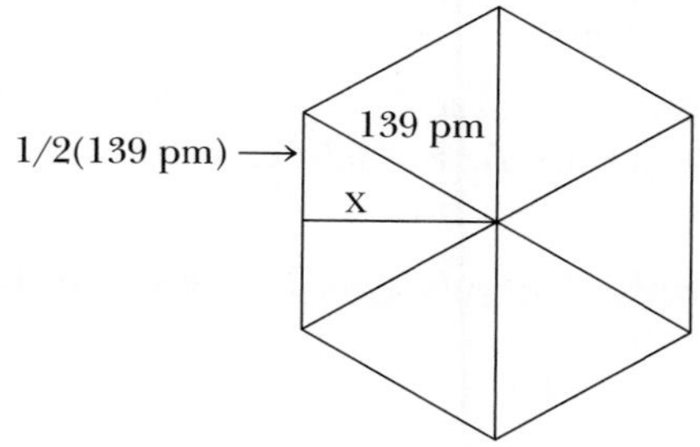

2. $1.00\ \mu m = 1.00 \times 10^{-4}$ cm and 240 pm = 2.4×10^{-8} cm.

   The number of C_6 rings spanning 1 μm is $1.00 \times 10^{-4}\ cm/2.40 \times 10^{-8}\ cm = 4.17 \times 10^3$ rings

3. Graphene is described as being one carbon atom thick so the thickness is twice the radius of a carbon atom or 154 pm.

Case Study: The World's Lightest Solid

1. The mass of 1.00 cm^3 of aerogel is 1.00 mg (1.00×10^{-3} g) and 0.2% of this, 2.00×10^{-6} g, is the mass of the polymer. The number of silicon atoms in 1.00 $cm^3 = 2.00 \times 10^{-6}$ g (1 mol $(C_2H_5O)_2SiO$/134.2 g)(1 mol Si/1 mol $(C_2H_5O)_2SiO$)(6.022×10^{23} atoms Si/mol Si) = 9.0×10^{15} atoms Si.

2. Volume between glass panes = 150 cm × 180 cm × 0.2 cm = 5,400 cm^3

   Density of aerogel = $1.00 \times 10^{-3}\ g/cm^3$, so the mass of aerogel needed = $5400\ cm^3(1.00 \times 10^{-3}\ g/cm^3) = 5.4$ g.

CHAPTER 14

Puzzler:

1. "Like dissolves like." Both liquids are polar, and both are capable of strong hydrogen bonding.
2. Ethylene glycol, a nonvolatile solute, lowers the freezing point. It is not corrosive. The liquids are miscible in all proportions.
3. c_{glycol} = 100. g(1 mol/62.07 g)/0.500 kg = 3.22 m

   $\Delta T_{fp} = -k_{fp} m_{solute} = -1.86\ °C/m \times 3.22\ m = -6.0\ °C$; $T_{fp} = -6.0\ °C$

Case Study: Henry's Law in a Soda Bottle

1. $PV = nRT$

   4.0 atm(0.025 L) = n(0.08205 L · atm/mol · K) 298 K; $n = 4.1 \times 10^{-3}$ mol
2. $P_1V_1 = P_2V_2$

   4.0 atm(0.025 L) = 3.7×10^{-4} atm (V_2); V_2 = 270 L
3. Solubility of $CO_2 = k_H P_g$ = 0.034 mol/kg · bar (4.0 bar) = 0.14 mol/kg

   Amount dissolved in 710 mL of diet cola (assume density is 1.0 g/cm^3) = 0.14 × 0.71 kg = 0.099 mol
4. Solubility of $CO_2 = k_H P_g$ = 0.034 mol/kg · bar (3.7×10^{-4} bar) = 1.3×10^{-5} mol/kg

CHAPTER 15

Puzzler:

See the answer to Study Question 71 for this chapter in Appendix O.

Case Study: Enzymes: Nature's Catalysts

1. To decompose an equivalent amount of H_2O_2 catalytically would take 1.0×10^{-7} years; this is equivalent to 3.2 s.
2. 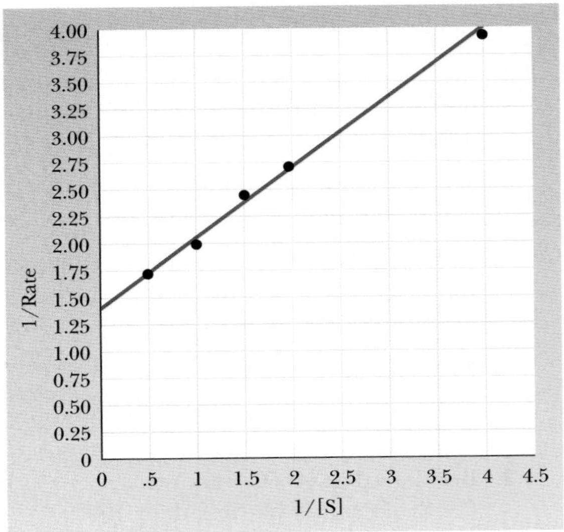

[S]	1/[S]	Rate	1/Rate
2.50	0.400	0.588	1.70
1.00	1.00	0.500	2.00
0.714	1.40	0.417	2.40
0.526	1.90	0.370	2.70
0.250	4.00	0.256	3.91

From the graph, we obtain a value of 1/Rate = 1.47 when 1/[S] = 0. From this, R_{max} = 0.68 mmol/min.

CHAPTER 16

Puzzler:

1. Endothermic. Raising the temperature (adding energy as heat) leads to conversion of reactants to products.
2. Apply LeChatelier's Principle: the system adjusts to addition of Cl^- by forming more $[CoCl_4]^{2-}$ and to addition of water by forming more $[Co(H_2O)_6]^{2+}$.
3. The various stresses applied have caused the system to adjust in either direction. (Better evidence: show that when heating and cooling the system is *repeated several times* the system cycles back and forth between the two colors.)

Case Study: Applying Equilibrium Concepts—The Haber-Bosch Process

1. (a) Oxidize part of the NH_3 to HNO_3, then react NH_3 and HNO_3 (an acid–base reaction) to form NH_4NO_3.

$$4\ NH_3 + 5O_2 \longrightarrow 4\ NO_2 + 6\ H_2O$$

$$2\ NO_2 + H_2O \longrightarrow HNO_3 + HNO_2$$

$$HNO_3 + NH_3 \longrightarrow NH_4NO_3$$

(b) $\Delta_r H° =$ (1 mol $[(NH_2)_2CO]$ /mol-rxn) $[\Delta_f H°\{(NH_2)_2CO\}]$ + (1 mol H_2O/mol-rxn) $[\Delta_f H°(H_2O)]$ − (2 mol NH_3/mol-rxn) $[\Delta_f H°(NH_3)]$ − (1 mol CO_2/mol-rxn) $[\Delta_f H°(CO_2)]$

$\Delta_r H° =$ (1 mol $[(NH_2)_2CO]$ /mol-rxn) (−333.1 kJ/mol) + (1 mol H_2O/mol-rxn) (− 241.8 kJ/mol) − (2 mol NH_3/mol-rxn) (−45.90 kJ/mol) − (1 mol CO_2/mol-rxn) (−393.5 kJ/mol)

$\Delta_r H° = -89.6$ kJ/mol-rxn.

The reaction as written is exothermic so the equilibrium will be more favorable for product formation at a low temperature. The reaction converts three moles of gaseous reactants to one mole of gaseous products; thus, high pressure will be more favorable to product formation.

2. (a) For $CH_4(g) + H_2O(g) \longrightarrow CO(g) + 3\ H_2(g)$

$\Delta_r H° =$ (1 mol CO/mol-rxn) $[\Delta_f H°(CO)]$ − (1 mol CH_4/mol-rxn) $[\Delta_f H°(CH_4)]$ − (1 mol H_2O/mol-rxn) $[\Delta_f H°(H_2O)]$

$\Delta_r H° =$ (1 mol CO/mol-rxn)(−110.5 kJ/mol) − (1 mol CH_4/mol-rxn)(−74.87 kJ/mol)] − (1 mol H_2O/mol-rxn)(−241.8 kJ/mol)] = 206.2 kJ/mol-rxn (endothermic)

For $CO(g) + H_2O(g) \longrightarrow CO_2(g) + H_2(g)$

$\Delta_r H° =$ (1 mol CO_2/mol-rxn) $[\Delta_f H°(CO_2)]$ − (1 mol CO/mol-rxn) $[\Delta_f H°(CO)]$ − (1 mol H_2O/mol-rxn) $[\Delta_f H°(H_2O)]$

$\Delta_r H° =$ (1 mol CO_2/mol-rxn)(−393.5 kJ/mol) − (1 mol CO/mol-rxn)(−110.5 kJ/mol)] − (1 mol H_2O/mol-rxn)(−241.8 kJ/mol)] = −41.2 kJ/mol-rxn (exothermic)

(b) (15 billion kg = 1.5×10^{13} g)

Add the two equations:
$CH_4(g) + 2\ H_2O(g) \longrightarrow CO_2(g) + 4\ H_2(g)$

CH_4 required = (1.5×10^{13} g NH_3)
(1 mole NH_3/17.03 g NH_3)
(3 moles H_2/2 moles NH_3)
(1 mole CH_4/4 moles H_2)
(16.04 g CH_4/1 mole CH_4)
= 5.3×10^{12} g CH_4

CO_2 formed = (1.5×10^{13} g NH_3)
(1 mole NH_3/17.03 g NH_3)
(3 moles H_2/2 moles NH_3)
(1 mole CO_2/4 moles H_2)
(44.01 g CO_2/1 mole CO_2)
= 1.5×10^{13} g CO_2

CHAPTER 17

Puzzler:

1. Aspirin, with a larger pK_a, is a stronger acid than acetic acid.

2. The acid hydrogen is the H on the $-CO_2H$ (carboxylic acid) functional group.

3. $C_6H_4(OCOCH_3)CO_2H + H_2O \longrightarrow C_6H_4(OH)CO_2H + CH_3CO_2H$

Case Study: Uric Acid, Gout, and Bird Droppings

1. (420 μmol/L)(10^{-6} mol/μmol)(168.12 g/mol)(10^3 mg/g) = 71 mg/L

2. The closest match to this pK_a is for $[Al(H_2O)_6]^{3+}$, $pK_a = 5.10$

CHAPTER 18

Puzzler:

1. $CaCO_3$ (K_{sp} for $CaCO_3 = 3.4 \times 10^{-9} > K_{sp}$ for $MnCO_3 = 2.3 \times 10^{-11}$)

2. PbS

3. $CaF_2(s) \rightleftharpoons Ca^{2+}(aq) + 2\ F^-(aq)$ ($K_{sp} = 5.3 \times 10^{-11}$)

Define solubility (mol/L) as *x;* then $[Ca^{2+}] = x$ and $[F^-] = 2x$

$5.3 \times 10^{-11} = [Ca^{2+}][F^-]^2 = [x][2x]^2$

$x = 2.4 \times 10^{-4}$; solubility = 2.4×10^{-4} mol/L

Case Study: Take a Deep Breath!

1. $pH = pK_a + \log[HPO_4^{2-}]/[H_2PO_4^-]$

   $7.4 = 7.20 + \log[HPO_4^{2-}]/[H_2PO_4^-]$

   $[HPO_4^{2-}]/[H_2PO_4^-] = 1.6$

2. Assign $x = [HPO_4^{2-}]$, then $[H_2PO_4^-] = (0.020 - x)$

   $1.6 = x/(0.020 - x)$; $x = 0.012$

   $[HPO_4^{2-}] = x = 0.012$ mol/L

   $[H_2PO_4^-] = 0.20 - x = 0.0080$ mol/L

CHAPTER 19

Puzzler:

1. Ethanol oxidation: $C_2H_5OH(\ell) + 3\ O_2(g) \rightarrow 2\ CO_2(g) + 3\ H_2O(g)$

   The enthalpy change per gram, −26.80 kJ/g, was calculated for the Case Study in Chapter 5 (see page A-124). From this, the enthalpy change per kg is -2.680×10^4 kJ/kg.

   C_8H_{18} oxidation: $C_8H_{18}(\lambda) + 12.5\ O_2(g) \rightarrow 8\ CO_2(g) + 9\ H_2O(g)$

   The enthalpy change per gram, −44.4 kJ/g, was calculated for the Case Study in Chapter 5 (see page A-124). From this, the enthalpy change per kg is 4.44×10^4 kJ/kg

2. Ethanol oxidation, free energy change:

   $\Delta_rG° = (2\text{ mol } CO_2/\text{mol-rxn})[\Delta_fG°(CO_2)] + (3\text{ mol } H_2O/\text{mol-rxn})[\Delta_fG°(H_2O)] - (1\text{ mol } C_2H_5OH/\text{mol-rxn})[\Delta_fG°(C_2H_5OH)]$

   $\Delta_rG° = (2\text{ mol } CO_2/\text{mol-rxn})[-394.4\text{ kJ/mol } CO_2] + (3\text{ mol } H_2O/\text{mol-rxn})[-228.6\text{ kJ/mol } H_2O] - (1\text{ mol } C_2H_5OH/\text{mol-rxn})[-174.7\text{ kJ/mol } C_2H_5OH)] = -1300.\text{ kJ/mol-rxn}$

   1 mol ethanol per 1 mol-rxn; therefore $\Delta_rG°$ per mol: −1300. kJ/mol

   $\Delta_rG°$ per kg: $-1300.\text{ kJ/mol}(1\text{ mol } C_2H_5OH/46.07\text{ g } C_2H_5OH)(1000\text{ g/kg}) = -2.822 \times 10^4\text{ kJ/kg } C_2H_5OH$

   C_8H_{18} oxidation, free energy change: $\Delta_rG° = \Delta_rH° - T\,\Delta_rS°$. ($\Delta_rH°$ from above. We need to calculate $\Delta_rS°$ for this reaction.)

   $\Delta_rS° = (8\text{ mol } CO_2/\text{mol-rxn})\,S°(CO_2)] + (9\text{ mol } H_2O/\text{mol})[S°(H_2O)] - (1\text{ mol } C_8H_{18}/\text{mol})[S°(C_8H_{18})] - (12.5\text{ mol } O_2)[S°(O_2)]$

   $\Delta_rS° = (8\text{ mol } CO_2/\text{mol-rxn})[213.74\text{ J/K} \cdot \text{mol } CO_2] + (9\text{ mol } H_2O/\text{mol-rxn})[188.84\text{ J/K} \cdot \text{mol } H_2O] - (1\text{ mol } C_8H_{18}/\text{mol-rxn})[361.2\text{ kJ/K} \cdot \text{mol } C_8H_{18})] - (12.5\text{ mol } O_2/\text{mol-rxn})[205.07\text{ J/K} \cdot \text{mol } O_2] = 587.5\text{ J/K} \cdot \text{mol-rxn} (= 0.5875\text{ kJ/K} \cdot \text{mol-rxn})$

   $\Delta_rG° = \Delta_rH° - T\,\Delta_rS° = -5.070\text{ kJ/mol} - 298.2\text{ K}(0.5875\text{ kJ/K} \cdot \text{mol-rxn}) = -5{,}245\text{ kJ/mol-rxn}$

   $\Delta_rG°$ per kg $= (-5{,}275\text{ kJ/mol-rxn})(1\text{ mol-rxn/mol } C_8H_{18})(1\text{ mol}/114.3\text{ g})(1\text{ kg}/1000\text{ g}) = -4.59 \times 10^4\text{ kJ/kg}$

3. For the oxidation of ethanol, entropy changes increase the energy available to do useful work. For oxidation of hydrogen, the opposite is true.

4. The difference in values of $\Delta_rH°$ and $\Delta_rG°$ result because energy is expended or acquired to achieve a higher dispersion of energy in the universe (system and surroundings).

Case Study: Thermodynamics and Living Things

1. Creatine phosphate + $H_2O \rightarrow$ Creatine + HP_i + H^+ $\quad \Delta G° = -43.3$ kJ/mol

   Adenosine + $HP_i \rightarrow$ Adenosine monophosphate + H_2O $\quad \Delta G° = +9.2$ kJ/mol

   Net reaction (sum of the two reactions):

   Creatine phosphate + Adenosine $\rightarrow$ Creatine + Adenosine monophosphate

   For this $\Delta G° = -43.3$ kJ/mol + 9.2 kJ/mol = −34.1 kJ/mol; the negative value indicates that the transfer of phosphate from creatine phosphate to adenosine is product-favored.

2. $\Delta G°' = \Delta G° + RT\ln[C][H_3O^+]/[A][B] = \Delta G° + (8.31 \times 10^{-3})(298)\ln[1][1 \times 10^{-7}]/[1][1]$

   $\Delta G°' = \Delta G° - 34.2$ kJ/mol

CHAPTER 20

Puzzler:

Possible reactions are:

1) $Cu^{2+}(aq) + Ni(s) \longrightarrow Cu(s) + Ni^{2+}(aq)$

2) $Cu(s) + Ni^{2+}(aq) \longrightarrow Cu^{2+}(aq) + Ni(s)$.

A reaction given by equation (1) would produce an electric current of 0.59 volts. [$E°_{cell} = E°(\text{cathode}) - E°(\text{anode}) = 0.34\text{ V} - (-0.25\text{ V})$]

Case Study: Manganese in the Oceans

1. Cathode reaction: $Mn^{3+} + e^- \longrightarrow Mn^{2+}$

Anode reaction: $Mn^{3+} + 2\ H_2O \longrightarrow MnO_2 + 4\ H^+ + e^-$

Net reaction: $2\ Mn^{3+} + 2\ H_2O \longrightarrow MnO_2 + Mn^{2+} + 4\ H^+$

$E°_{cell} = E°(\text{cathode}) - E°(\text{anode}) = 1.50\text{ V} - 0.95\text{ V}] = 0.55\text{ V}$

The positive value associated with disproportionation (the net reaction) is positive, indicating a product-favored reaction.

2. (a) $MnO_2 + H_2S + 2\ H^+ \longrightarrow Mn^{2+} + S + 2\ H_2O$

(b) $2\ Mn^{2+} + O_2 + 2\ H_2O \longrightarrow 2\ MnO_2 + 4\ H^+$

3. Cathode reaction: $O_2 + 4\ H^+ + 4\ e^- \longrightarrow 2\ H_2O$ ($E° = 1.229$ V)

Anode reaction: $Mn^{2+} + 2\ H_2O \longrightarrow MnO_2 + 4\ H^+ + 2\ e^-$ ($E° = 1.23$ V, from Appendix M)

$E°_{cell} = E°(\text{cathode}) - E°(\text{anode}) = 1.229\text{ V} - 1.23\text{ V} = 0\text{ V}$.

CHAPTER 21

Puzzler:

1. $CH_4(g) + 2\ H_2O(\ell) \longrightarrow CO_2(g) + 4\ H_2(g)$

$SiH_4(g) + 2\ H_2O(\ell) \longrightarrow SiO_2(s) + 4\ H_2(g)$

2. For CH_4: assumes $H_2O(\ell)$ is the reactant

$\Delta_r H° = (1\text{ mol }CO_2/\text{mol-rxn})[\Delta_f H°(CO_2)] - (1\text{ mol }CH_4/\text{mol-rxn})[\Delta_f H°(CH_4)] - (2\text{ mol }H_2O/\text{mol-rxn})[\Delta_f H°(H_2O)]$

$\Delta_r H° = (1\text{ mol }CO_2/\text{mol-rxn})(-393.5\text{ kJ/mol}) - (1\text{ mol }CH_4/\text{mol-rxn})(-74.87\text{ kJ/mol})] - (2\text{ mol }H_2O/\text{mol-rxn})(-285.8\text{ kJ/mol})] = 252.4\text{ kJ/mol-rxn}$

For SiH_4: assumes $H_2O(\ell)$ is the reactant

$\Delta_r H° = (1\text{ mol }SiO_2/\text{mol-rxn})[\Delta_f H°(SiO_2)] - (1\text{ mol }SiH_4/\text{mol-rxn})[\Delta_f H°(SiH_4)] - (2\text{ mol }H_2O/\text{mol-rxn})[\Delta_f H°(H_2O)]$

$\Delta_r H° = (1\text{ mol }SiO_2)/\text{mol-rxn})(-910.86\text{ kJ/mol}) - (1\text{ mol }SiH_4/\text{mol-rxn})(34.31\text{ kJ/mol})] - (2\text{ mol }H_2O/\text{mol-rxn})(-285.8\text{ kJ/mol})] = -373.6\text{ kJ/mol-rxn}$

3. Electronegativities: C 2.5, Si, 1.9, H 2.2. From this we conclude that polarities of C—H and Si—H bonds are in the opposite directions: in SiH_4, the H has a slight negative charge (it is hydriodic) and in CH_4 the H has a slight positive charge.

4. General observation from these examples: Carbon often bonds to other atoms via double bonds, whereas Si does not. We would not expect $H_2Si{=}SiH_2$ to exist as a molecular species; instead a polymeric structure $[-SiH_2SiH_2-]_x$ is predicted.

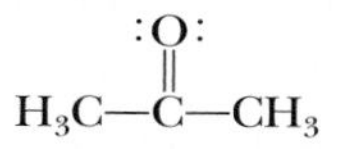

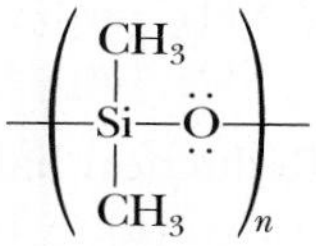

Case Study: Hard Water

1. For Mg^{2+}: (50 mg)(1 mmol Mg^{2+}/24.31 mg)(1 mmol CaO/mmol Mg^{2+})(56.07 mg CaO/1 mmol CaO) = 115.3 mg CaO

For Ca^{2+}: (50 mg)(1 mmol Ca^{2+}/40.08 mg)(1 mmol CaO/mmol Ca^{2+})(56.07 mg CaO/1 mmol CaO) = 209.8 mg CaO

Total CaO = 115.3 mg + 209.9 mg = 330 mg (2 significant figures)

We get 2 mol $CaCO_3$ per mole Ca^{2+} and 1 mol each of $CaCO_3$ and $MgCO_3$ per mol Mg^{2+}

$CaCO_3$ from Ca^{2+} reaction: (0.150 g Ca^{2+})(1 mol/40.08 g Ca^{2+})(2 mol $CaCO_3$/1 mol Ca^{2+})(100.1 g $CaCO_3$/1 mol $CaCO_3$) = 0.749 g

$CaCO_3$ from Mg^{2+} reaction: (0.050 g Mg^{2+}) (1 mol/24.31 g Mg^{2+}) (1 mol $CaCO_3$/1 mol Mg^{2+}) (100.1 g $CaCO_3$/1 mol $CaCO_3$) = 0.0.206 g

$MgCO_3$ from Mg^{2+} reaction: (0.050 g Mg^{2+}) (1 mol/24.31 g Mg^{2+}) (1 mol $MgCO_3$/1 mol Mg^{2+}) (84.31 g $CaCO_3$/1 mol $CaCO_3$) = 0.173 g

Total mass of solids = 0.747 g + 0.206 g + 0.173 g = 1.1 g (2 significant figures)

Case Study: Lead, a Mystery Solved

1. 50 ppb is 50 g in 1×10^9 g of blood. Assume the density of blood is 1.0 g/mL. In 1.0×10^3 mL (i.e., 1.0 L) of blood, there will be 50×10^{-6} g of Pb. From this:

   50×10^{-6} g (1 mol Pb/207.2 g Pb) (6.022×10^{23} atoms Pb/mol Pb) = 1.5×10^{17} atoms Pb

Case Study: A Healthy Aquarium and the Nitrogen Cycle

1. $2\ NH_4^+(aq) + 4\ OH^-(aq) + 3\ O_2(aq) \rightarrow 2\ NO_2^-(aq) + 6\ H_2O(\ell)$
2. Reduction half-reaction: $2\ NO_3^-(aq) + 6\ H_2O(\ell) + 10\ e^- \rightarrow N_2(g) + 12\ OH^-(aq)$

   Oxidation half-reaction: $CH_3OH(aq) + 6\ OH^-(aq) \rightarrow CO_2(aq) + 5\ H_2O(\ell) + 6\ e^-$

   Net: $6\ NO_3^-(aq) + 5\ CH_3OH \rightarrow 3\ N_2(g) + 5\ CO_2(aq) + 6\ OH^-(aq) + 7\ H_2O(\ell)$
3. HCO_3^- is the predominant species. Recall that when acid and base concentrations are equal, pH = pK_a. If H_2CO_3 and HCO_3^- are present in equal concentrations, the pH would be about 6.4. If HCO_3^- and CO_3^{2-} are present in equal concentrations, the pH would be 10.2. For the pH to be about 8 (in a salt water aquarium), [HCO_3^-] would have to be higher than either of the other carbonate species.
4. Conc. of N in ppm (mg N/L) = [(1.7×10^4 kg NO_3^-) (10^6 mg NO_3^-/kg NO_3^-) (14.0 mg N/ 62.0 mg NO_3^-)]/(2.2×10^7 L) = 1.7×10^2 mg/L = 1.7×10^2 ppm

   Conc. of NO_3^- in ppm (mg/L) = (1.7×10^4 kg) (10^6 mg/kg)/(2.2×10^7 L) = 770 mg/L

   Conc. of NO_3^- in mol/L = [(1.7×10^4 kg) (10^3 g/kg) (1 mol/62.0 g)]/(2.2×10^7 L) = 0.012 mol/L

CHAPTER 22

Puzzler:

1. Define length of the side of the cube as x, then the length of the diagonal across the cube is $x\sqrt{3}$. This is set equal to $2\ r_{Ti} + 2\ r_{Ni}$, i.e., $x\sqrt{3} = 2\ r_{Ti} + 2\ r_{Ni} = 540$ pm; $x = 312$ pm ($a = b = c = 3.12 \times 10^{-8}$ cm)
2. Calculated density:

   Mass of one unit cell is the mass of one Ti and one Ni atom = (47.87 g/mol) (1 mol/ 6.022×10^{23} atoms Ti) + (58.69 g/mol) (1 mol/ 6.022×10^{23} atoms Ti) = 1.77×10^{-22} g

   Volume of the unit cell is x^3 = (3.12×10^{-8} cm)^3 = 3.04×10^{-23} cm^3

   Calculated density = 1.77×10^{-22} g/ 3.04×10^{-23} cm^3 = 5.82 g/cm^3

   The agreement is not very good, probably because atoms don't pack together as tightly as is assumed.
3. As free atoms, both Ti and Ni are paramagnetic.

Case Study: Accidental Discoveries

1. First order kinetics: $\ln[x/x_o] = -\ kt$

   $\ln[x/10\ \text{mg}] = -\ 7.6 \times 10^{-5}\ s^{-1}$ [24 h $\times$ 3600 s/h]

   x/10 mg = 1.4×10^{-3}; x = 1.4 x 10^{-2} mg remain
2. Use Henderson-Hasselbalch equation for this acid dissociation equilibrium.

   pH = pK_a + log[base/acid]

   7.4 = 6.6 + log{[$PtCl(NH_3)_2OH$]/ [$PtCl(NH_3)_2(H_2O)$]}

   [$PtCl(NH_3)_2OH$]/[$PtCl(NH_3)_2(H_2O)$] = 6.3

Case Study: Ferrocene

1. Fe^{2+} in ferrocene has an electron configuration [Ar] $3d^6$ and is present in the low spin state.

2. Cr (0). Cr(0) in this compound is assumed to have an electron configuration [Ar] $3d^6$ and is present in the low spin state.

3. Both are in accord the 18 electron rule.

4. Select oxidizing agents from Table 20.1 (page 920) based on the northeast-southwest rule (above $E° = 0.400$ v). Common oxidizing agents include the halogens, H_2O_2 and MnO_4^-.

5. $NiCl_2 + 2\ Na[C_5H_5] \longrightarrow Ni(\eta\text{-}C_5H_5)_2 + 2\ NaCl$. Nickelocene is predicted to have 2 unpaired electrons (Ni^{2+}, with a d^8 configuration, in an octahedral environment.)

CHAPTER 23

Puzzler:

1. ^{235}U: 92 protons, 235 − 92 = 143 neutrons

   ^{238}U: 92 protons, 146 neutrons

2. (a) $^{238}U + {}^{1}_{0}n \longrightarrow {}^{239}U$

   (b) $^{239}U \longrightarrow {}^{239}Np + {}^{0}_{-1}\beta$

   $^{239}Np \longrightarrow {}^{239}Pu + {}^{0}_{-1}\beta$

   (c) $^{239}Pu \longrightarrow {}^{235}U + {}^{4}_{2}\alpha$

Case Study: Nuclear Medicine and Hypothyroidism:

1. $^{131}I \longrightarrow {}^{131}Xe + {}^{0}_{-1}\beta$

2. Calculate the fraction (= f) of each remaining after 7 days

   For ^{123}I: $k = 0.693/t_{1/2} = 0.693/13.3\ h = 0.0521\ h^{-1}$

   $\ln(f) = -\ 0.0521\ h^{-1}[7\ d(24\ h/d)]$

   $f = 1.6 \times 10^{-4}$

   For ^{131}I: $k = 0.693/t_{1/2} = 0.693/8.04\ d = 0.0862\ d^{-1}$

   $\ln(f) = -\ 0.0862\ d^{-1}(7\ d)$

   $f = 0.55$

   Ratio of amounts remaining, $[^{131}I]/[^{123}I] = 0.55/(1.6 \times 10^{-4}) = 3400$

   The amount of the ^{131}I isotope is 3400 times greater than the amount of ^{123}I.

Index/Glossary

Italicized page numbers indicate pages containing illustrations, and those followed by "t" indicate tables. Glossary terms, printed in boldface, are defined here as well as in the text.